The World Guide 1997/98

A view from the South

INSTITUTO DEL TERCER MUNDO

The World Guide 1997/98

Third World Guide (English language edition) 1986, 1988, 1990, 1992.
The World Guide (English language edition) 1995.

Copyright © 1997 by Instituto del Tercer Mundo

Paperback ISBN 1-869847 - 43 - 1
Hardback ISBN 1-869847 - 42 - 3

A catalogue record for this book is available from the British Library.

New Internationalist Publications Ltd.,
Registered Office: 55 Rectory Road,
Oxford OX4 1BW, United Kingdom.
Fax: 01865 793152
E-mail: newint@gn.apc.org

Printed by C&C Offset Printing Co. Ltd., Hong Kong.

Contents

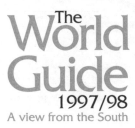

World Guide 1997/98 - General Index

1997/98
A view from the South

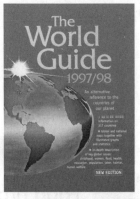

The World Guide (then known as Third World Guide) was published for the first time in English language in Rio de Janeiro in 1984.

GLOBAL PROBLEMS

19 DEMOGRAPHY

Economies in transition: striking changes

The transition from a centrally-planned economy to a market economy is at the centre of a process involving and affecting *inter alia* demographic features such as fertility, mortality, migration and population growth.

20 Population density
TABLE

Indicator: inhabitants per sq km, 1993.
Source: Statistical Yearbook 40th Issue, United Nations, 1995.

21 Fertility in figures
TABLE

Indicators: percentage of married women of childbearing age using contraceptives; crude death rate per 1,000 people; crude birth rate per 1,000 people; children per woman.
Source: The State of the World's Children 1996, UNICEF, 1996.

22 CHILDHOOD

From poverty to sexual exploitation

Children are not only the human beings most affected by starvation, work, war and poverty, in recent years they have also increasingly become victims of torture, slaughter, trafficking and sexual exploitation.

22 Breast is best
TABLE

Indicator: percentage of mothers breast-feeding at 6 months, 1980-92.
Source: Human Development Report 1995, UNDP, 1995.

23 Children's health
TABLE

Indicators: infant mortality rate (per 1,000 live births), 1994; prevalence of malnutrition (under 5 years old), 1989-95; maternal mortality rate (per 100,000 live births), 1989-95. Source: World Development Indicators 1996, World Bank, 1996.

24 Birth
TABLE

Indicators: percentage of low birth-weight babies, 1990; percentage of births attended by health personnel, 1993-94; life expectancy at birth, 1993.
Sources: The State of the World's Children 1996, UNICEF, 1996; Human Development Report 1996, UNDP, 1996.

Contents

Contents

Contents

68 GLOBAL WARMING

Climate change/global warming: not quite the same thing

Global warming might be a sign of profound changes experienced by the earth's climate as a result of human activity. Rising sea levels - which in the medium term are likely to result in flooding some populated areas - should be counted among global warming's many possible consequences.

Sea level rises of one meter or more could result in several small island states, many of which are only 1.5-2 meters above sea level, being wholly submerged within the next century.

70 WATER

Use more, save more

Over the next years water scarcity may become a source of international conflicts and major shifts in national societies, affecting prospects for peace - in unstable regions like the Middle East - as well as food security, people's ways of life, the present growth of cities and the location of industries.

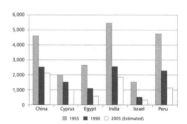

72 INTERNATIONAL ORGANIZATIONS

The United Nations' crisis

The approval of the United Nations' reform project, anticipated along with the organization's 50th anniversary in 1995, remains to be concluded. The UN's material situation further complicates matters since important contributors such as the US are large debtors to the organization, leaving it on the brink of bankruptcy.

Directory

State and/or governmental organizations; also economic, financial, military, environmental, labor, and non-governmental organizations; political parties and other organizations.

80 INDIGENOUS PEOPLES

Pending indigenous rights

The United Nations' Indigenous Rights document has remained in draft form since 1982. Many states questioned the draft's recognition of self-determination for indigenous peoples, as well as in the concept of group or community rights.

The Inupiat call themselves the "People of the Whales", since they have hunted the arctic stocks for at least 2,500 years.

Countries

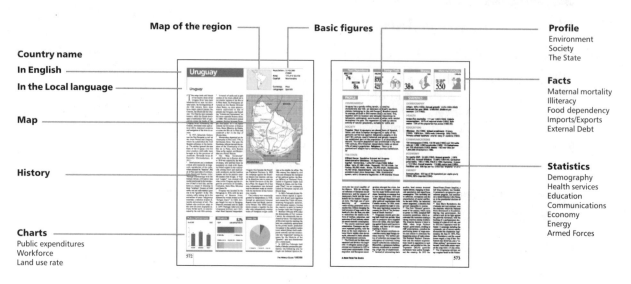

Map of the region — Basic figures — Profile
Environment
Society
The State

Country name

In English — Facts
Maternal mortality
In the Local language — Illiteracy
Food dependency
Imports/Exports
Map — External Debt

History — Statistics
Demography
Health services
Education
Communications
Charts — Economy
Public expenditures — Energy
Workforce — Armed Forces
Land use rate

PROFILE

ENVIRONMENT

Estimates of area are based on UN official estimates according to internationally recognized border; these include inland waters but not territorial ocean waters. Except where otherwise specified, territories claimed by certain countries but not under their effective jurisdiction are not taken into account, though this implies no judgment as to the validity of the claims. Geographical descriptions and data on environmental problems are based on various authoritative sources listed in the general bibliography at the end of this Guide.

SOCIETY

Peoples, Languages and Religions: Very few countries keep official ethnic and religious data, and UN-related institutions definitely do not do so. The ethnic and religious makeup of a society is a historic-cultural factor which is in constant change. Certain forces promote tribal or ethnic divisions to favor their own plans of domination, while others favor whatever makes for integration and national unity. In many countries the political and social problems cannot be grasped without reference to these factors. When including them here, we have sought to consult the most reliable sources (mentioned in the general bibliography) and to avoid the racist connotations often associated with such information.

Political parties and Social movements: It is practically impossible to make a complete listing of all parties and other movements of any country, since they are permanently changing and, in most cases, it would add up to several hundred names. Thus only major organizations are mentioned, even though 'major' is a subjective idea when votes, parliamentary representation or number of affiliates cannot be produced as solid criteria, where parties are outlawed or the right to associate is restricted.

THE STATE

Official Name: Complete name of the state in the official language.

Administrative divisions, Capital and other cities: Population figures are per last available year. The legal limits of a city frequently do not match its real borders. Thus the information may refer only to the core area, which is a minor part of the whole city.

Government: Names of the major authorities and institutional bodies as of October 1996.

National Holiday: When more than one holiday is commemorated, Independence Day (if it is a holiday) is indicated.

Armed Forces and Paramilitaries: The total number of personnel for the year indicated. Paramilitaries are considered trained and equipped beyond the level of a Police Force (though some fulfill this role) and whose constitution and control means they can be used as regular troops. The source used is the most reliable on military statistics, but the data is based on official government information which is not always clear on total number of troops and military expenditure.

CHARTS

PUBLIC EXPENDITURES

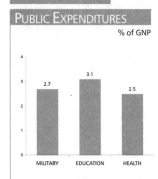

PUBLIC EXPENDITURES
% of GNP

MILITARY 2.7 · EDUCATION 3.1 · HEALTH 2.5

PUBLIC EXPENDITURES

Expressed as a percentage of the GNP of each country.
Military: generally military expenditure on internal security (police, etc.); weapons production and the civil service linked to defense administration are not taken into account in this figure.

Sources: c, d.

WORKERS

WORKERS
1994   4.2%
              UNEMPLOYMENT

■ MEN 68%   ■ WOMEN 32%

■ AGRICULTURE 5%   ■ INDUSTRY 22%   ■ SERVICES 73%

WORKERS

The unemployed and the economically active population (EAP), differentiated between men and women, but excluding housewives and other unpaid workers, and by sector of activity

(agriculture, industry and services).

LAND USE

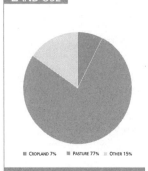

LAND USE

■ CROPLAND 7%   ■ PASTURE 77%   ■ OTHER 15%

LAND USE RATE:

Use of land in 1989 as a percentage of the country's total territory, based on agricultural production. Both lands unsuitable for agriculture and space occupied by industrial plants, roads, railways, inland waterways, urban areas, etc., are considered 'non-productive'. The category 'pastures and meadows/prairies' does not necessarily mean that they are used for cattle-raising. Similarly, 'woods and forests' does not imply commercial forestry activities.

Source: d

HISTORY

[1] The most reliable sources available have been used to update the texts on each country. In many cases this meant checking with local sources, like the searchers and correspondents of Third World Network, documentation centers linked to electronic mail networks and grass-roots organizations throughout the world. This input went into our database in Montevideo, where the final editing was done.

[2] Efforts were made to avoid the most frequent biases of Western reference books, such as to appearing to make history start with the arrival of the Europeans (particularly in the case of African and Latin American countries) or ignoring the role of women.

[3] The last overall updating of the database before printing was done in October 1996, but in several cases events happening as late as November were included.

INDICATORS

DEMOGRAPHY

Population: Total inhabitants, according to latest UN estimates (middle of the year shown in parentheses). In some cases, figures have been corrected through consultation with local sources.

Sources: d, h.

Urban population: Shown as a percentage of the total population. Figures are to be treated with care since definitions of "urban zones" differ in every country.

Sources: d.

Demographic growth: Average annual rate of population growth during the period shown in parentheses.

Sources: g.

Estimates for the year 2000: The population projections for the year 2000 are calculations based on current estimated population, rates of growth, and trends of these rates to increase or decrease.

Source: d.

Children per woman: Total fertility rate - the number of children that would be born per woman if she were to live to the end of her child-bearing years and bear children at each age in accordance with prevailing age-specific fertility rates.

Source: c.

HEALTH

Physicians per number of inhabitants: Traditional medicine and community health care by health personnel that are not officially recognized are not included in these statistics, although they may be the only health service available for the majority of the population in many countries.

Source: f.

Under-5 child mortality: Probability of dying between birth and exactly five years of age expressed per 1,000 live births.

Source: f.

Calorie consumption: Daily per capita consumption as a percentage of the minimum requirement. Shown as a national average, though a country's income distribution may create a wide gap between the average, the highest and the lowest. Minimum calorie requirements vary in different countries, depending on climate and nature of the main activities.

Sources: a, f.

Safe water: The United Nations includes treated surface waters and untreated but uncontaminated water from springs, wells and protected boreholes in the 'reasonably safe water' category.

Source: f.

EDUCATION

Illiteracy: Indicates the estimated percentage of people over the age of fifteen who cannot read and write.

Source: d.

School enrolment: Percentage of age group enrolled in education. The gross enrolment ratios may exceed 100 per cent because some pupils are older or younger than the country's standard primary, secondary or tertiary age. The level of female enrolment compared with male can be taken as an indication of the status of women within the country.

Source: d.

Primary school teachers: Ratio of teachers to students in primary school.

Source: d.

CONTINUES ON NEXT PAGE

Contents

COMMUNICATIONS

Estimation of the print run of daily **newspapers** and the number of working radio receivers and **TV sets** per 1,000 households. The latter figure may rely on the number of licences granted or the number of declared receivers. Number of **telephones** per 100 inhabitants.

Sources: a, h.

Books: Number of new titles published in the last year per 1,000,000 inhabitants.

Source: a.

ECONOMY

Per capita GNP: *GDP* is the value of the total production of goods and services of a country's economy within the national territory. *GNP* is GDP plus the income received from abroad by residents in the country (such as remittances from migrant workers and income from investments abroad), minus income obtained in the internal economy which go into the hands of persons abroad (such as profit remittances of foreign companies). These data are calculated in local currencies. Converting them into US dollars (or into any other currency) may lead to distortions. For example, if a country devalued its currency, conversion into US dollars would seem to cut its GDP or GNP while actual production was not reduced. The World Bank corrects such distortions by calculating per capita GNP on the basis of the purchasing power of the local currency and converting it into US dollar purchasing power (which is the figure we have used in this Guide).

Source: d.

Annual growth: Average annual growth in per capita GNP during the cited period.

Source: d .

Annual inflation: Average annual inflation rate calculated, in the national currency, for each year of the period indicated.

Source: d.

Consumer price index: Variation in relation to base year. Sources: **h, b. Currency:** Name and exchange rate with US dollar on given date.

Source: h.

Cereal imports: The cereals are wheat, flour, rice, unprocessed grains and the cereal components of combined foods. This figure includes cereals donated by other countries, and those distributed by international agencies.

Source: d.

Food import dependency: Imported food in relation to the food available for internal distribution (the total of food production, plus food imports, minus food exports), for 1970-92.

Source: g.

Fertilizer use: Refers to purchases of nitrate, potassium and phosphate based fertilizers used on arable land, according to the FAO's 1987-88 definition.

Source: d.

Imports & Exports: Annual value in US dollars f.o.b. (free on board) for exports and c.i.f. (costs, insurance and freight) for imports.

Source: d.

CONTINUES ON NEXT PAGE

Index to countries

STATISTICS

External debt (total and per capita): Public and state guaranteed private foreign debt accumulated by the indicated date.

Source: d.

Debt servicing: The service of a foreign debt is the sum of interest payments and repayment of principal (capital loaned, regardless of yield). The relation between debt service and exports of goods and services is a practical measurement commonly used to evaluate capacity to pay the debt or obtain new credits. These coefficients do not include private foreign debts without state guarantees -a considerable amount in some countries.

Sources: d, e.

Development aid received: Official development assistance (ODA) consists of money flows from official governmental or international institutions for the purpose of promoting economic development or social welfare in developing countries. These funds are supplied in the form of grants or 'soft' loans, i.e., long-maturity loans at interest rates lower than those prevailing on the international market.

Source: d.

ENERGY

These statistics refer only to commercial energy, and do not include, for example, that which rural people in the Third World produce by their own means (mainly coal and wood). Energy consumption of the country is measured in kilograms of 'oil-equivalent' per capita. Energy imports are given as a percentage of energy consumption. Figures are negative in those countries that are net exporters of energy products.

Source: d.

SOURCES

a) Quality of Life, from a common people's point of view, 1995 by PapyRossa Verlags GmbH & Co. KG, Köln & World Data Research Center, Ernst Fidel Fürntratt-Kloep, Box 112, S-830 23 Hackås (Sweden).

b) Yearbook of Labor Statistics 1995, 54th issue, International Labour Office, Geneva.

c) Fact Book 1996, World Bank, Washington, DC, 1996.

d) World Development Report 1996, World Bank, Washington, DC, 1996, World Development Report 1995, World Bank, Washington, DC, 1995.

e) World Debt Tables 1996, World Bank, Washington, DC, 1996.

f) The State of the World's Children 1996, UNICEF, New York, 1996.

g) Human Development Report 1995, UNDP, New York, 1995 and Human Development Report 1996, UNDP, New York, 1996.

h) Statistical Yearbook 1993, 40th issue, United Nations, New York, 1995.

Contents

Index to special boxes

The Globe 1997/98 (a full colour world map is enclosed)

PREFACE

One day Nasreddin Hodja, the 13th century Turkish philosopher, got on his donkey the wrong way, facing towards the back.
" Hodja," the people said, "you are sitting on your donkey backwards!"
"No," he replied. "It's not that I am sitting on the donkey backwards, the donkey's facing the wrong way."

Since it was first published in 1979, the "World Guide" (then known as the "Third World Guide") has been looking at the planet in an unconventional way.

Nowadays, when the world seems to be headed toward uniformity, our reports on over two hundred countries and territories emphasize the specific and unique traits of each. At a time when reference books compete to gather more and more information, the editorial staff of this edition has striven to omit overwhelming details and achieve an improved synthesis.

While most perspectives that reach the public usually praise "globalization" as the great stride toward the new millennium, whose main actors are a handful of billionaires and corporations, this Guide shows the daily difficulties faced by billions of people to obtain basic medical care or a bowl of soup.

Reference books like to be regarded as "objective". We do not conceal our sympathy for women in their quest for equity, native peoples who cling to their identities, peasants who defend their seeds, workers who demand the rights conquered by their parents and grandparents.

However, like Hodja, we recognize the donkey looks the other way. And as we are riding on it, we can not avoid moving at the pace of the times.

The greatest effort and merit of the first editions of the Guide was to gather and spread unknown or inaccessible information to readers. Nowadays, the amount of information available when one looks for a name or subject on the Internet is huge.

However, this does not mean it is any easier to know what has happened or is happening. The duty of the informant is to pick out and select from the extensive mass of data at hand. Information today resembles a vast jungle in which we have sources of water and fruit, if we know how to find them and discriminate between those which can poison us.

To help in that search - believing that research is not only justified by the usefulness of the discovery but also by the pleasure of exploring - this edition has modified the graphic presentation of data and the system of indexes and references.

All things considered, we may look one way, the donkey may head another way, but what this book offers is information about the terrain. It is up to the readers to choose their own directions.

Roberto Bissio
Editor-in-chief

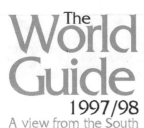

The World Guide 1997/98

A view from the South

By Instituto del Tercer Mundo

Editor-in-chief:
Roberto Bissio.

Managing editor:
Victor Bacchetta.

Translation Editor:
Amir Hamed.

Co-editor:
Roberto Elissalde.

Editing staff:
Daniel Martínez, Javier Lyonnet,
María Isabel Sans.

Contributors:
Alina González, Alexandru R. Savulescu
and Gustavo Alzugaray.

Translators:
Alvaro Queiruga, Sarah Mason, Roberto
Echavarren, Beatriz Sosa Martínez
and Raquel Núñez.

Graphic design & layout:
Carlos M. Velázquez.

Maps designer:
Mario Burgueño.

Data entry:
Leticia Kelbauskas and Carmen Orguet.

Computers and communications:
Jaime Vázquez, Gabriel Seré, Juan Carlos
Alonso, Mónica Soliño and Mauricio León.

Administrator:
Bernardo Dabezies.

Administration staff:
Ana Filippini, Rosario Lema and Walter Sena.

**By New Internationalist
Publications Ltd**
(English language edition)

Editor:
Troth Wells.

Production:
Fran Harvey, Andrew Kokotka and Ian Nixon.

Marketing & Distribution:
Dexter Tiranti.

This book is the result of documentation, research, writing, editing and design work done by the Instituto del Tercer Mundo (Third World Institute), a non-profit making institution, devoted to information, communication and education, based in Montevideo, Uruguay.

This reference book was published for the first time in Mexico in 1979, on the initiative of Neiva Moreira, to complement the dissemination work of the Third World magazine he had begun publishing. Robert Cornford encouraged us to go on with the effort of editing a new English version, and provided valuable ideas to make it viable. The initiative was made possible thanks to an agreement with New Internationalist Publications Ltd in England. Dexter Tiranti committed his enthusiasm and skills to put dreams into reality and the New Internationalist collective shared the vision and worked hard to meet tight deadlines.

The Third World Guide is a cumulative work, building on the efforts of former staff members. We would like to mention the contributions of: Mohiuddin Ahmad, Claude Alvares, Iván Alves, Juan José Argeriz, Maria Teresa Armas, Marcos Arruda, Iqbal Asaria, Gonzalo Abella, Susana de Avila, Edouard Balby, Artur Baptista, Roberto Bardini, Luis Barrios, Alicia Bidegaray, Beatriz Bissio, Ricardo de Bittencourt, Samuel Blixen, Gerardo Bocco, José Bottaro, José Cabral, Luis Caldera, Juan Cammá, Altair Campos, Paulo Cannabrava Filho, Carmen Canoura, Cristina Canoura, Gerónimo Cardozo, Diana Cariboni, Gustavo Carrier, Virgilio Caturra, Macário Costa, Carlos María Domínguez, Wáshington Estellano, Marta Etcheverrigaray, Carlos Fabião, Marcelo Falca, Helena Falcão, Wilson Fernández, Alejandro Flores, Lidia Freitas, Sonia Freitas, Marc Fried, Héctor García, José Carlos Gondim, Mario Handler, David Hathaway, Ann Heidenreich, Bill Hinchberger, Etevaldo Hipólito, Heraclio Labandera, Peter Lenny, Cecilia Lombardo, Linda Llosa, Geoffrey Lloyd Gilbert, Fernando López, Hakan Lundgren, Pablo Mazzini, Daniel Mazzone, Carol Milk, João Murteira, Abdul Naffey, Claudia Neiva, Ruben Olivera, Hattie Ortega, Pablo Piacentini, Ana Pérez, Christopher Peterson, Virginia Piera, Carlos Pinto Santos, Sieni C. Platino, Artur José Poerner, Graciela Pujol, Roberto Raposo, Felisberto Reigado, Maria da Gloria Rodrigues, Julio Rosiello, Ash Narain Roy, Ana Sadetzky, John Sayer, Dieter Schonebohm, Irene Selser, Eunice H. Senna, Baptista da Silva, Yesse Jane V. de Souza, José Steinleger, Firiel Suijker, Carolina Trujillo, Pedro Velozo, Horacio Verbitsky, Amelia Villaverde, Anatoli Voronov and Asa Zatz.

We have received the help, encouragement and support of Carlos Abín, Carlos Afonso, Cedric Belfrage, Max van den Berg, Dirk-Jan Broertjes, Alberto Brusa, Anne-Pieter van Dijk, Darryl D´Monte, Rodrigo Egaña, Cecilia Ferraría, Goran Hammer, Mohammed Idris, Sytse Kujik, Arne Lindquist, Carlos Mañosa, Fernando Molina, Hans Poelgrom, Malva Rodríguez, John Schlanger, Gregorio Selser, Herbet de Souza, Sjef Theunis and Germán Wettstein. Leo van Grunsven gave us editorial suggestions, together with his support for the project.

The editorial staff acknowledges the valuable contributions made by the libraries of the following institutions: UNDP, UNESCO, CLAEH, FAO, the UN Information Center in Montevideo, Iepala in Madrid and UNCTAD in Geneva.

The Third World Network, Penang, Malaysia provided ideas, documents and links with many contributors.

To all of them we offer our thanks. Thanks are also due to the hundreds of readers who have given us their comments, suggestions and detailed information.

The judgements and values contained in this book are the exclusive responsibility of the editors and do not represent the opinion of any of the people or institutions listed, except for the Third World Institute.

Instituto del Tercer Mundo
Juan D. Jackson 1136, Montevideo
11200, Uruguay
Tel: (598) 2 496192
Fax: (598) 2 419222.
E-mail: item@chasque.apc.org

New Internationalist Publications Ltd
55 Rectory Road
Oxford OX4 1BW, United Kingdom
Tel: (44) 01865 728181
Fax: (44) 01865 793152
E-mail: newint@gn.apc.org

INSTITUTO DEL TERCER MUNDO

Global Problems

THE TRANSITION from a centrally-planned economy to a market economy is at the centre of a process involving and affecting *inter alia* demographic features such as fertility, mortality, migration and population growth.

Economies in transition: striking changes

Many of the countries with economies in transition (CEIT) have exhibited particularly striking demographic changes. There is, however, no common pattern of demographic transition for CEIT-countries. Most of the European CEIT-countries are characterized by accelerated fertility declines to levels well below replacement, stagnating or even declining life expectancy, and increased net out-migration. As a result, many of them are currently experiencing a net loss of population. In contrast, Asian CEIT-countries, especially those in south-central Asia, exhibit higher fertility above replacement level and moderate gains in life expectancy, but also suffer net out-migration. These countries still show positive population growth, which translates into much younger populations than the CEIT-countries in Europe.

European successor states of the former USSR

[2] Six of the 7 European successor states of the former Soviet Union (Belarus, Estonia, Latvia, Lithuania, Moldova, Russia and Ukraine) have exhibited population declines since 1990; only the Republic of Moldova is estimated to have a positive population growth rate. Fertility declines have accelerated considerably during this period of economic transition. Between 1985-1995, these states experienced fertility declines between 12 per cent (Lithuania) and 27 per cent (Russian Federation). All of the European successor states, except Moldova, now exhibit fertility levels well below the replacement level of 2.1 children per woman; and Moldova's fertility is currently at replacement level (2.13 children per woman). The steepest fertility decline is observed for the Russian Fed-

eration, which shows an average fertility level for 1990-1995 of 1.5 children per woman.

[3] The ongoing transformation process has had particularly serious implications for the health status of the populations of this region, as illustrated by declines in average life expectancy at birth. For example, male life expectancy at birth in the Russian Federation is estimated to have decreased from 64.3 years in 1985-1990 to 61.7 for the current 1990-1995 period. Female life expectancy has been less effected by the transition process; in the Russian Federation it decreased from 74.3 to 73.6 between 1985-1995. Other countries exhibit similar trends in life expectancy. Belarus and the three Baltic states, Estonia, Latvia and Lithuania, all experienced a decline in male life expectancy between 1985-1995 by more than two years: in Latvia, for instance, life expectancy decreased from 65.7 years in 1985-1990 to 63.3 years for 1990-1995.

Asian states of the former USSR

[4] The demographic picture is different for the Asian successor states of the former Soviet Union. These countries showed relatively high fertility prior to the breakup of the Soviet Union, and, although fertility has declined since then, population growth rates remain strongly positive. All but Georgia and Kazakhstan grew by more than 5 per cent between 1990 and 1994. Population increase between 1990 and 1994 was highest in Tajikistan (12 per cent), Turkmenistan (10 per cent) and Uzbekistan (9 per cent). This indicator shows relatively large variation among the eight Asian successor states of the former Soviet Union: male life expectancy currently (1990-1995) ranges from 61.5 years

(Turkmenistan) to 69.5 years (Armenia), while it is between 68.5 years (Turkmenistan) to 76.7 years (Georgia) for females.

[5] All of the eight Asian successor States have exhibited a net outflow of population. The 1994 Revision estimates out-migration at more than 3 million people for the ten years between 1985 and 1995. Kazakhstan and Uzbekistan have experienced net out-migration flows of more than one million persons and 700,000 persons, respectively for the ten year period since 1985.

Other Eastern European countries

[6] Between 1990 and 1994, the total population of Bulgaria, Czech Republic, Hungary, Poland, Romania and Slovakia decreased by about 375 thousand people (0.4 per cent). However, the trend does not show uniformity: the populations of Hungary and Bulgaria each declined by 2 per cent, while Slovakia and Poland exhibit population growth.

[7] Mortality patterns in these countries have been similar to those of the European successor states of the former Soviet Union: male life expectancy has stagnated or even decreased, while female survivorship improved only slightly. Between 1985-1995, male life expectancy declined for Hungary by one year (from 65.5 to 64.5 years) and by 0.6 years for Bulgaria (from 68.4 to 67.8 years) and Slovakia (from 67.1 to 66.5 years). Female life expectancy in these countries for 1990-1995 ranges between 73.3 years (Romania) and 75.4 years (Slovakia). Between 1985-1995, only small increases between 0.2 and 0.7 years have been achieved by these countries, with the exception of the Czech Republic, where female life expectancy even declined slightly.

[8] These countries also exhibit sharp fertility declines since 1990. While in the second half of the 1980s three countries still exhibited fertility levels above or at replacement (Poland with 2.15 children per woman, Romania with 2.27 children per woman and Slovakia with 2.15 children per woman), all countries are now below replacement level fertility. In Romania, for instance, fertility fell by one third, to 1.5 for 1990-1995.

[9] A substantial net out-migration from these six countries of almost 700 thousand persons, mostly to Western Europe, was observed for the period 1985-1990. The 1994 Revision estimates an increase of net out-migration to more that 850 thousand people for the period 1990-1995.

Albania

[10] Albania's population growth rates are strongly positive; between 1990 and 1994 its population grew by 3.8 per cent. This is mainly due to Albania's high fertility, which is, with 2.85 children per woman for the period 1990-1995, highest in Europe. Life expectancy at birth in Albania is estimated to have stagnated since 1985-1990 at 69.2 years for males and 75 years for females.

Countries of the former socialist Yugoslavia

[11] The demographic situation of Bosnia and Herzegovina, Croatia, Slovenia, the former Yugoslav Republic of Macedonia and Yugoslavia is especially complicated and unstable. Because of civil war, ethnic strife and territorial disputes, casualties and involuntary population movements on a large scale have occurred. It is extremely difficult to give a proper account of demographic trends in this war-ridden area.

[12] The current events show very different impacts on the population dynamics in that region. While Bosnia and Herzegovina has experienced net population loss between 1990 and 1994 of 18 per cent, population size has remained nearly unchanged in Croatia. Population has continued to increase in Slovenia (by 1 per cent), TFYR Macedonia (5 per cent) and Yugoslavia (6 per cent)

[13] Movement of population to other countries accounts for much of the population loss for Bosnia and Herzegovina. Net out-migration from Bosnia and Herzegovina for the 1990-1995 period is figured at 975,000 persons. That is, approximately one-of-every-five of the 4.3 million persons who lived in Bosnia and Herzegovina in 1990 has left for another country.

Source: United Nations
Population Division

Population density
(Inhabitants per square kilometer) 1993

Country	Density	Country	Density	Country	Density
Afghanistan	27	Germany	228	Oman	9
Albania	122	Ghana	69	Pakistan	154
Algeria	11	Greece	78	Palau	35
Andorra	135	Grenada	267	Panama	34
Angola	8	Guadeloupe	242	Papua New Guinea	8
Anguilla	96	Guam	262	Paraguay	11
Antigua	147	Guatemala	92	Peru	17
Aotearoa/NZ	13	Guinea	26	Philippines	219
Aotearoa-Cook Is.	81	Guinea-Bissau	28	Poland	119
Aotearoa-Niue	8	Guyana	4	Portugal	107
Argentina	12	Haiti	249	Puerto Rico	407
Armenia	125	Honduras	50	Qatar	51
Aruba	358	Hungary	111	Reunion	252
Australia	2	Iceland	3	Romania	95
Austria	95	India	274	Russia	9
Azerbaijan	85	Indonesia	99	Rwanda	287
Bahamas	19	Iran	39	St Helena	53
Bahrain	777	Iraq	44	St Kitts	161
Bangladesh	800	Ireland	51	St Lucia	223
Barbados	613	Israel	250	St Vincent	284
Belarus	49	Italy	189	Samoa	59
Belgium	328	Jamaica	219	Samoa-American	256
Belize	9	Japan	327	San Marino	393
Benin	46	Jordan	51	São Tomé Principe	127
Bermuda	1,189	Kanaky/New Caledonia	10	Saudi Arabia	8
Bhutan	34	Kazakhstan	6	Senegal	40
Bolivia	6	Kenya	48	Seychelles	159
Bosnia and Herzegovina	73	Kiribati	105	Sierra Leone	60
Botswana	2	Korea Dem. Rep.	191	Singapore	4,650
Brazil	18	Korea Rep.	444	Slovakia	108
Brunei	48	Kuwait	80	Slovenia	98
Bulgaria	76	Kyrgyzstan	23	Solomon Is.	12
Burkina Faso	35	Laos	19	Somalia	14
Burundi	214	Latvia	40	South Africa	32
Cambodia	51	Lebanon	270	Spain	77
Cameroon	26	Lesotho	64	Sri Lanka	269
Canada	3	Liberia	24	Sudan	11
Cape Verde	92	Libya	3	Suriname	3
Cayman Is.	110	Liechtenstein	188	Swaziland	47
Central African Rep.	5	Lithuania	57	Sweden	19
Chad	5	Luxembourg	147	Switzerland	168
Chile	18	Macau	21,560	Syria	72
China	125	Macedonia, TFYR of	82	Tajikistan	40
China-Hong Kong	5,506	Madagascar	24	Tanzania	32
China-Macau	21,560	Malawi	77	Thailand	114
Colombia	30	Malaysia	58	Togo	68
Comoros	272	Maldives	800	Tokelau	167
Congo	7	Mali	8	Tonga	131
Costa Rica	63	Malta	1,142	Trinidad and Tobago	246
Côte d'Ivoire	41	Marshall Is.	287	Tunisia	52
Croatia	80	Martinique	337	Turkey	78
Cuba	98	Mauritania	2	Turkmenistan	8
Cyprus	78	Mauritius	535	Turks and Caicos	30
Czech Republic	131	Mauritius-Diego García	589	Tuvalu	346
Denmark	120	Mexico	47	Uganda	83
Djibouti	24	Micronesia	168	Ukraine	86
Dominica	95	Moldova	129	United Arab Emirates	14
Dominican Republic	156	Monaco	31,000	United Kingdom	238
East Timor	53	Mongolia	1	United States	28
Ecuador	39	Montserrat	108	Uruguay	18
Egypt	56	Morocco	58	Uzbekistan	49
El Salvador	262	Mozambique	19	Vanuatu	13
Equatorial Guinea	14	Myanmar/Burma	66	Venezuela	23
Estonia	34	Namibia	2	Vietnam	215
Fiji	41	Nauru	476	Virgin Is. (Am)	300
Finland	15	Nepal	148	Virgin Is. (Br)	118
France	104	Netherlands	375	Western Sahara	1
Fr. St Pierre and Miquelon	25	Netherlands Antilles	244	Yemen, Rep.	25
French Guyana	2	Nicaragua	33	Yugoslavia Fed. Rep.	103
French Polynesia	53	Niger	7	Zaire	18
Gabon	5	Nigeria	114	Zambia	12
Gambia	92	Northern Mariana	101	Zimbabwe	27
Georgia	78	Norway	13		

Source: Statistical Yearbook 40th Issue, UN, 1995.

	% of married women of childbearing age using contraceptives 1980–94	Crude death rate per thousand population 1994	Crude birth rate per thousand population 1994	Children per woman 1994		% of married women of childbearing age using contraceptives 1980–94	Crude death rate per thousand population 1994	Crude birth rate per thousand population 1994	Children per woman 1994
Afghanistan	2	22	50	6.7	Laos	..	15	45	6.5
Albania	..	6	24	2.8	Latvia	..	13	12	1.6
Algeria	51	7	30	3.7	Lebanon	55	7	27	3
Angola	1	19	51	7	Lesotho	23	10	37	5.1
Aotearoa/NZ	70	8	17	2.1	Liberia	6	14	47	6.7
Argentina	74	8	21	2.7	Libya	..	8	42	6.2
Armenia	..	7	21	2.5	Lithuania	..	11	14	1.8
Australia	76	7	15	1.9	Macedonia	..	7	16	2
Austria	71	11	12	1.6	Madagascar	17	12	44	6
Azerbaijan	..	6	23	2.4	Malawi	13	20	51	7
Bangladesh	45	12	36	4.2	Malaysia	48	5	29	3.5
Belarus	..	12	12	1.7	Mali	5	19	51	7
Belgium	79	11	12	1.7	Mauritania	4	15	40	5.3
Benin	9	18	49	7	Mauritius	75	7	21	2.3
Bhutan	2	15	40	5.7	Mexico	53	5	28	3.1
Bolivia	45	10	36	4.7	Moldova	..	10	17	2.1
Bosnia and Herzegovina	..	7	14	1.6	Mongolia	..	8	28	3.5
Botswana	33	7	37	4.7	Morocco	42	8	29	3.6
Brazil	66	8	25	2.8	Mozambique	4	19	45	6.4
Bulgaria	76	13	11	1.5	Myanmar/Burma	13	11	33	4.1
Burkina Faso	8	18	47	6.4	Namibia	29	11	37	5.1
Burundi	9	16	46	6.6	Nepal	23	13	39	5.3
Cambodia	..	15	44	5.1	Nicaragua	49	7	41	4.9
Cameroon	16	12	41	5.6	Niger	4	19	53	7.3
Canada	73	8	15	1.9	Nigeria	6	16	45	6.3
Central African Rep.	15	17	41	5.6	Norway	76	11	14	2
Chad	1	18	44	5.8	Oman	9	5	44	7
Chile	43	6	22	2.5	Pakistan	12	9	41	6
China	83	7	19	2	Panama	58	5	25	2.8
China-Hong Kong	81	6	11	1.2	Papua New Guinea	4	11	34	4.9
Colombia	66	6	24	2.6	Paraguay	48	6	33	
Congo	..	15	45	6.2	Peru	59	7	28	3.3
Costa Rica	75	4	27	3.1	Philippines	40	6	31	3.8
Côte d'Ivoire	..	15	50	7.3	Poland	75	10	13	1.9
Croatia	..	12	11	1.7	Portugal	66	10	12	1.6
Cuba	70	7	17	1.8	Romania	57	11	12	1.5
Czech Republic	69	13	13	1.8	Russia	..	12	11	1.5
Denmark	78	12	12	1.7	Rwanda	21	17	44	6.4
Dominican Republic	56	6	27	3	Saudi Arabia	..	5	35	6.2
Ecuador	53	6	29	3.4	Senegal	7	16	43	5.9
Egypt	47	8	30	3.7	Sierra Leone	4	25	49	6.4
El Salvador	53	7	34	3.9	Singapore	74	6	16	1.7
Eritrea	..	15	43	5.7	Slovakia	74	11	15	1.9
Estonia	..	13	11	1.6	Slovenia	..	11	11	1.5
Ethiopia	2	18	49	6.9	Somalia	1	19	50	6.9
Finland	80	10	13	1.9	South Africa	50	9	31	4
France	80	10	13	1.7	Spain	59	9	10	1.2
Gabon	..	16	37	5.4	Sri Lanka	62	6	21	2.4
Gambia	..	19	44	5.5	Sudan	9	13	40	5.6
Georgia	..	9	16	2.1	Sweden	78	11	14	2.1
Germany	75	12	10	1.3	Switzerland	71	9	13	1.6
Ghana	20	12	42	5.8	Syria	52	6	41	5.7
Greece	..	10	10	1.4	Tajikistan	..	6	37	4.8
Guatemala	23	8	39	5.2	Tanzania	18	14	43	5.8
Guinea	1	20	51	6.9	Thailand	66	6	20	2.1
Guinea-Bissau	1	21	43	5.7	Togo	12	13	45	6.4
Haiti	18	12	35	4.7	Trinidad and Tobago	53	6	21	2.4
Honduras	47	6	37	4.7	Tunisia	50	6	26	3
Hungary	73	15	12	1.7	Turkey	63	8	28	3.3
India	43	10	29	3.7	Turkmenistan	..	8	32	3.9
Indonesia	55	9	25	2.8	Uganda	5	19	52	7.1
Iran	49	7	36	4.9	Ukraine	..	13	12	1.6
Iraq	18	7	38	5.6	United Arab Emirates	..	3	24	4.1
Ireland	..	9	15	2.1	United Kingdom	72	11	14	1.8
Israel	..	7	21	2.8	United States	74	9	16	2.1
Italy	78	10	10	1.3	Uruguay	..	10	17	2.3
Jamaica	66	6	22	2.3	Uzbekistan	..	6	32	3.8
Japan	64	7	10	1.5	Venezuela	49	5	28	3.2
Jordan	35	6	39	5.4	Vietnam	53	8	31	3.8
Kazakhstan	..	8	20	2.5	Yemen	7	16	49	7.5
Kenya	33	12	45	6.1	Yugoslavia Fed. Rep.	..	10	14	2
Kyrgyzstan	..	7	29	3.6	Zaire	1	15	48	6.6
Korea Dem. Rep.	..	5	24	2.3	Zambia	15	15	45	5.8
Korea	79	6	16	1.8	Zimbabwe	43	12	39	4.9
Kuwait	35	2	25	3					

Source: The State of the World's Children, 1996, UNICEF, 1996.

CHILDHOOD

CHILDREN ARE NOT ONLY the human beings most affected by starvation, work, war and poverty, in recent years they have also increasingly become victims of torture, slaughter, trafficking and sexual exploitation.

From poverty to sexual exploitation

The first World Congress against the Commercial Sexual Exploitation of Children, held in Stockholm, Sweden, in August 1996, reported how the sexual exploitation of girls and boys is now an issue of global concern.

[2] It is estimated that more than one million children under the age of sixteen are sold each year for sexual favours to attend more than 12 million 'customers', according to the International Coordinator of Philippine organisation ECPAT (End Child Prostitution in Asian Tourism).

[3] According to the ECPAT, many children are coerced, kidnapped, sold, deceived or otherwise trafficked into enforced sexual encounters. Some may be pushed into prostitution by circumstances, as a way of surviving on the streets, helping to support their families, or to pay for clothes and goods.

[4] Throughout Asia, Africa, Latin America and Europe there has been a growth in the number of children involved in the sex trade. This includes not only prostitution, but child trafficking and the use of children in pornography. Poverty is a critical factor in the growth of the child sex industry. When employment opportunities are limited, women and girls in particular often have to invent informal jobs. In some cases they have no other recourse but to fall back on prostitution. The only income some poor families have comes from prostituting their children. The spread of HIV/Aids has also led to an increase in exploitation, as men are increasingly seeking to have sex with younger girls who they believe are more likely to be free from the virus.

[5] The commercial sexual exploitation of children is not a problem unique to the developing world. For example, the collapse of communism in eastern Europe and the former Soviet Union has led to a steep rise in female unemployment and a cutback in state services, which has created an environment where sex work is flourishing.

[6] The United States it is estimated that between 100,000 and 300,000 children are sexually exploited through prostitution and pornography. Most of these children are runaways who have suffered incest, rape, or abuse at home.

[7] In 1988, Canada strengthened the criminal prohibition against the sexual abuse of children, and made it easier for child victims to give evidence. Despite these new laws, child prostitution and child pornography still exist in Canada. Canada was one of the few countries to publicly admit to the ongoing sexual abuse of children in their country.

[8] In a study released by ECPAT on sex tourism to Costa Rica, an estimated 3,000 women and girls are formally employed as prostitutes in San Jose, the capital city. According to the recent study, entitled "Child Prostitution and Sex Tourism in Costa Rica", many of the male tourists who are ostensibly drawn to Costa Rica for reasons of traditional tourism, as well as some of the 30,000 US citizens who have retired to Costa Rica, also sexually exploit local women and children during the course of their stay.

[9] According to the South African Police Service Child Protection Unit (CPU) statistics, during the period 1993 to 1994, there was an increase of 60 per cent in reported child rape cases. In 1995 alone, the CPU had to deal with 16,083 cases of child sexual offences, including rape, sodomy, and incest.

[10] Latin America Programme at Casa Alianza is dedicated to working for Mexican, Honduran and Guatemalan street children, offering basic medical attention and counselling, coupled with compassion and respect. In Guatemala City alone, the organisation serves an estimated 5,000 children living on the streets. Casa Alianza now has more than 200 cases pending against the military police, security forces and private security firms.

[11] Over the last few years, Amnesty International has documented an escalating number of violations against these children, including beatings, torture, "disappearances" and killings, reportedly carried out by security forces agents. Often abandoned by parents too poor to feed themselves, or forced to flee political oppression or sexual and physical abuse, the street children face a future of crime, prostitution and a premature - often violent - death.

Breast is best

% of mothers breast-feeding at 6 months

	1980-92		1980-92
Guinea-Bissau	100	Jamaica	82
Nigeria	99	Mauritania	82
Papua N. Guinea	99	Côte d'Ivoire	81
Zaire	99	Sri Lanka	81
Zambia	99	Thailand	80
Burkina Faso	98	Guatemala	79
Congo	98	Somalia	78
Laos	98	El Salvador	77
Bangladesh	97	India	75
Rwanda	97	Liberia	75
Cameroon	95	Ecuador	73
Indonesia	95	Jordan	72
Madagascar	95	Guinea	70
Mali	95	Paraguay	69
Cambodia	93	Colombia	65
Mozambique	93	China	60
Burundi	92	Saudi Arabia	57
Ghana	92	Mauritius	55
Kenya	92	Seychelles	55
Zimbabwe	92	Panama	53
Senegal	91	Mexico	50
Turkey	91	Tr. & Tobago	49
Botswana	90	Dominican Rep.	45
Sudan	90	Iraq	45
Tanzania	90	Brazil	43
Benin	89	Lebanon	40
Pakistan	88	Costa Rica	38
Uganda	88	Argentina	36
Vietnam	88	Cuba	33
Peru	87	Uruguay	33
Swaziland	87	Honduras	28
Togo	87	Nicaragua	25
Namibia	86	Chile	18
Bolivia	84	Barbados	17
Egypt	83	St Kitts	3

Source: Human Development Report, UNDP, 1995.

Children's health

	Infant mortality rate (per 1,000 live births) 1994	% Prevalence malnutrition (under 5) 1989-95	Maternal mortality ratio (per 100,000 live births) 1989-95		Infant mortality rate (per 1,000 live births) 1994	% Prevalence malnutrition (under 5) 1989-95	Maternal mortality ratio (per 100,000 live births) 1989-95
Albania	31	..	..	Latvia	16	..	..
Algeria	35	9	140	Lesotho	44	21	598
Aotearoa/NZ	7	..	..	Lithuania	14	..	29
Argentina	23	..	140	Macedonia, FYR	24	..	..
Armenia	15	..	35	Madagascar	90	32	660
Australia	6	..	..	Malawi	134	27	620
Austria	6	..	..	Malaysia	12	23	34
Azerbaijan	25	..	29	Mali	125	..	1,249
Bangladesh	81	84	887	Mauritania	98	..	800
Belarus	13	..	25	Mauritius	17	..	112
Belgium	8	..	..	Mexico	35	..	..
Benin	96	36	2,500	Moldova	23	..	34
Bolivia	71	13	373	Mongolia	53	10	240
Botswana	34	..	220	Morocco	56	9	..
Brazil	56	18	200	Mozambique	146	..	1,512
Bulgaria	15	..	..	Myanmar/Burma	80	31	518
Burkina Faso	128	..	939	Namibia	57	..	..
Burundi	99	..	1,327	Nepal	95	70	..
Cameroon	57	14	511	Netherlands	6	..	..
Canada	6	..	..	Nicaragua	51	12	..
Central Afr. Rep.	100	..	649	Niger	120	..	593
Chad	119	..	1,594	Nigeria	81	43	1,027
Chile	12	1	..	Norway	5	..	..
China	30	17	115	Oman	18	..	184
China-Hong Kong	5			Pakistan	92	40	..
Colombia	20	10	107	Panama	20	7	..
Congo	112	..	887	Papua N. Guinea	65	..	700
Costa Rica	13	2	..	Paraguay	34	4	180
Côte d'Ivoire	90	..	822	Peru	48	16	..
Croatia	11	..	..	Philippines	40	30	208
Czech Republic	8	..	..	Poland	15	..	..
Denmark	6	..	..	Portugal	8	..	..
Dominican Rep.	38	10	..	Romania	24	..	..
Ecuador	37	45	..	Russian Fed.	19	..	52
Egypt	52	9	..	Rwanda	..	28	..
El Salvador	42	22	..	Saudi Arabia	26	..	108
Estonia	15	..	41	Senegal	64	20	510
Ethiopia	120	47	1,528	Sierra Leone	163	23	1,800
Finland	5	..	..	Singapore	5	14	..
France	6	..	..	Slovakia	11	..	..
Gabon	89	..	438	Slovenia	7	..	..
Gambia	128	..	1,050	South Africa	50	..	404
Georgia	18	..	55	Spain	7	..	..
Germany	6	..	..	Sri Lanka	16	38	30
Ghana	74	27	742	Sweden	4	..	..
Greece	8	..	..	Switzerland	6	..	..
Guatemala	44	..	464	Tajikistan	41	..	39
Guinea	131	18	880	Tanzania	84	28	748
Guinea-Bissau	138	..	..	Thailand	36	13	155
Haiti	86	27	600	Togo	81	..	626
Honduras	47	19	221	Trinidad and Tobago	14	..	..
Hungary	12	..	..	Tunisia	40	..	139
India	70	63	437	Turkey	62	..	183
Indonesia	53	39	..	Turkmenistan	46	..	55
Iran	47	16	..	Uganda	122	23	550
Ireland	6	..	..	Ukraine	14	..	33
Israel	8	..	..	United Arab Emirates	16	..	20
Italy	7	..	..	United Kingdom	6	..	..
Jamaica	13	10	..	United States	8	..	..
Japan	4	3	..	Uruguay	19	..	36
Jordan	32	17	132	Uzbekistan	28	..	43
Kazakhstan	27	..	53	Venezuela	32	6	200
Kenya	59	23	646	Vietnam	42	45	105
Korea, Rep.	12	..	30	Yemen, Rep.	102	30	1,471
Kuwait	11	..	18	Zambia	108	27	229
Kyrgyzstan	29	..	43	Zimbabwe	54	16	80
Laos	92	40	660				

Source: World Development Indicators 1996, World Bank, 1996.

Street children in Guatemala

A survey conducted by the children's welfare organization Covenant House Latin America - known locally as Casa Alianza - reports that nearly 40 per cent of street kids interviewed in Guatemala City had their first sexual experience with someone they didn't know. All children interviewed have sold their bodies to survive; all suffer sexually transmitted diseases. The investigation tried to establish the prevalence of sexually transmitted diseases (STDs) in the street children in Guatemala City, a group exposed to "high risk" factors which include: drug consumption and high levels of sexual promiscuity.

[2] The socio-economic situation of the country, the political violence and delinquency have provoked a genuine epidemic of orphaned and abandoned children. The children who live in the street are obliged on many occasions to practice conducts which are dangerous for their health and life, like commercial sex, drug addiction, hunger and delinquency. The study included 143 street children between the ages of 7 and 18.

[3] Age range: 7.7% were between the ages of 7 and 10; 32.9% were between the ages of 11 and 14; 59.4% were between the ages of 15 and 18. Gender: 72.73% were males; 27.27% were females. Ethnicity: 95.8 mestizo 1.2 indigenous. Literacy/Education: 18.1% were illiterate 81.7% had an incomplete primary school education.

[4] Partners per day: About 70% had one to two partners per day; 4.2% had 3 to 4 partners per day; and 25.1% (92.31% girls), reported more than 4 partners per day. Contraceptive use: None used contraceptives.

[5] Prevalence of sexually transmitted disease: 93% had previously contracted sexually transmitted diseases (STD), including: genital herpes, 78.3%; scabies, 69.9%; gonorrhea, 46.65%; papilomatosis, 27.3%; vaginal trichomoniasis, 13.29%; and chancroids, 11.7%.

[6] Drug use: 100% inhale solvents; 96.5% use drugs daily; 3.5%, weekly. 97.9% use sedatives; 97% use hallucinogens; 92% use barbiturates; 83% use cocaine, 77% have used crack.

[7] Sexual abuse history: 100% of children interviewed had been sexually abused, of whom: 53.2% by family members; 5.9% by friends; 2.7% by neighbors; 38.46% by people they did not know.

Country	% of low birth weight babies 1990	% of births attended by health personnel 1983-94	Life expectancy at birth (years) 1993
Afghanistan	20	9	43.7
Albania	7	99	72.0
Algeria	9	15	67.3
Angola	19	15	46.8
Antigua			74.0
Aotearoa/NZ	6	99	75.6
Argentina	6	87	72.2
Armenia			72.8
Australia	6	99	77.8
Austria	6		76.3
Azerbaijan			70.7
Bahamas			73.2
Bahrain			71.7
Bangladesh	50	10	55.9
Barbados			75.7
Belarus			69.7
Belgium	6	100	76.5
Belize			73.7
Benin		45	47.8
Bhutan		7	51.0
Bolivia	12	47	59.7
Botswana	8	78	65.2
Brazil	11	95	66.5
Brunei			74.3
Bulgaria	6	100	71.2
Burkina Faso	21	42	47.5
Burundi		19	50.3
Cambodia		47	51.9
Cameroon	13	64	56.3
Canada	6	99	77.5
Cape Verde			64.9
Cent. Afric. Rep.	15	46	49.5
Chad		15	47.7
Chile	7	98	73.9
China	9	94	68.6
China-Hong Kong	8	100	78.7
Colombia	10	81	69.4
Comoros			56.2
Congo	16		51.2
Costa Rica	6	93	76.4
Côte d'Ivoire	14	45	50.9
Cuba	9	90	75.4
Cyprus			77.1
Czech Republic			71.3

Country	% of low birth weight babies 1990	% of births attended by health personnel 1983-94	Life expectancy at birth (years) 1993
Denmark	6	100	75.3
Djibouti			48.4
Dominica			72.0
Dominican Rep.	16	92	69.7
Ecuador	11	84	69.0
Egypt	10	41	63.9
El Salvador	11	66	66.8
Eq. Guinea			48.2
Estonia			69.2
Ethiopia	16	14	47.8
Fiji			71.6
Finland	4	100	75.8
France	5	94	77.0
Gabon		80	53.7
Gambia		80	45.2
Georgia			72.9
Germany		99	76.1
Ghana	17	59	56.2
Greece	6	97	77.7
Grenada			71.0
Guatemala	14	51	65.1
Guinea	21	36	44.7
Guinea-Bissau	20	27	43.7
Guyana			65.4
Haiti	15	20	56.8
Honduras	9	81	67.9
Hungary	9	99	69.0
Iceland			78.2
India	33	33	60.7
Indonesia	14	36	63.0
Iran	9	70	67.7
Iraq	15	50	66.1
Ireland	4		75.4
Israel	7	99	76.6
Italy	5		77.6
Jamaica	11	82	73.7
Japan	6	100	79.6
Jordan	7	87	68.1
Kazakhstan			69.7
Kenya	16	54	55.5
Kyrgyzstan			69.2
Korea Dem. Rep.		100	71.2
Korea Rep.	9	89	71.3
Kuwait	7	99	75.0

Country	% of low birth weight babies 1990	% of births attended by health personnel 1983-94	Life expectancy at birth (years) 1993
Laos	18		51.3
Latvia			69.0
Lebanon	10	45	68.7
Lesotho	11	40	60.8
Liberia		58	55.6
Libya		76	63.4
Lithuania			70.3
Luxembourg			75.8
Madagascar	17	56	56.8
Malawi	20	55	45.5
Malaysia	10	87	70.9
Maldives			62.4
Mali	17	32	46.2
Malta			76.2
Mauritania	11	40	51.7
Mauritius	9	85	70.4
Mexico	12	77	71.0
Moldova			67.6
Mongolia	10	99	63.9
Morocco	9	31	63.6
Mozambique	20	25	46.4
Myanmar/Burma	16	57	57.9
Namibia	16	68	59.1
Nepal		6	53.8
Netherlands		100	77.5
Nicaragua	15	73	67.1
Niger	15	15	46.7
Nigeria	16	37	50.6
Norway	4		77.0
Oman	10	60	69.8
Pakistan	25	35	61.8
Panama	10	96	72.9
Papua New Guinea	23	20	56.0
Paraguay	8	66	70.1
Peru	11	52	66.3
Philippines	15	53	66.5
Poland		100	71.1
Portugal	5	90	74.7
Qatar			70.6
Romania	7	100	69.9
Russia			67.4
Rwanda	17	26	47.2
St Kitts			70.0
St Lucia			72.0

Country	% of low birth weight babies 1990	% of births attended by health personnel 1983-94	Life expectancy at birth (years) 1993
St Vincent			71.0
Samoa			67.8
Sao T. & Principe			67.0
Saudi Arabia	7	90	69.9
Senegal	11	46	49.5
Seychelles			71.0
Sierra Leone	17	25	39.2
Singapore	7	100	74.9
Slovakia			70.9
Solomon Is.			70.5
Somalia	16	2	47.2
South Africa			63.2
Spain	4	96	77.7
Sri Lanka	25	94	72.0
Sudan	15	69	53.2
Suriname			70.5
Swaziland			57.8
Sweden	5	100	78.3
Switzerland	5	99	78.1
Syria	11	61	67.3
Tajikistan			70.4
Tanzania	14	53	52.1
Thailand	13	71	69.2
Togo	20	54	55.2
Trin. and Tobago	10	98	71.7
Tunisia	8	69	68.0
Turkey	8	76	66.7
Turkmenistan			65.1
Uganda		38	44.7
Ukraine			69.3
United Arab Emir	6	99	73.9
United Kingdom	7	100	76.3
United States	7	99	76.1
Uruguay	8	96	72.6
Uzbekistan			69.4
Vanuatu			65.4
Venezuela	9	69	71.8
Vietnam	17	95	65.5
Yemen	19	16	50.4
Zaire	15		52.0
Zambia	13	51	48.6
Zimbabwe	14	70	53.4

Sources: The State of the World's Children 1996, UNICEF 1996; Human Development Report 1996, UNDP, 1996.

Two cases

Victoria was 15 years old, anaemic, exhausted and eight months pregnant when social workers found her sleeping in a rubbish heap behind the market in Accra, Ghana. Like many of the 10,000 children - including 4,000 girls - living on the streets of Ghana's capital, Victoria had come to the city from a poor village hoping to earn some money and eventually to return home.

2.2 When she arrived in Accra, she found work selling bags of ice water in the marketplace by day. But at night she needed protection. So, like other girls in her situation, she acquired a male minder. In exchange for sex, the girl got some security and a little extra food. But before long, she was pregnant, and her minder had moved on.

2.3 On the morning of 6 June 1990 in Guatemala City, a group of street children were sitting chatting in a park when a black van pulled up. Three armed men got out, dragged four boys and a girl into the back of the van and drove off. The men sprayed a chemical in their faces and placed a bag over each of their heads, causing the children to inhale and pass out. They were taken to a cemetery where the others came round to see one boy hung by his hands from a tree and being beaten.

2.4 The girl was the only one not handcuffed and so another boy told her: "Run, because only you and God can save us now". She ran through the graveyard, chased by one man in the truck, and escaped. She was later found and taken to hospital. The four boys were found 11 days later. Two of them were without ears; their eyes had been gouged out and their tongues cut off. They had been shot in the head. The police blame the kids for the soaring crime rate and so they take justice into their own hands. They call it teaching them a lesson.

FOOD

UNTIL THE LATE 1960s, the world sales of farming produce showed that countries in the South were supplying those in the North, but now that trend has been completely reversed.

Food production and democratic control

The exports of developed countries towards the Third World have increased. The Third World was a net exporter of grain in 1939, but is presently a net importer, in global terms. Its balance of trade in the food sector became negative as its external debt increased.

2 World grain use in 1995 was 305 kg per person. The amount was a drop of 2.6 per cent from the previous year and a decline of 8 per cent from the peak in 1986. Before 1986 grain use per person had increased, although somewhat

unsteadily, for many years. The decline in per capita consumption in 1995 was partly a result of the temporary drop in grain supplies and the dynamic changes that were taking place in world agriculture, however. Though it might first have seemed that there was some cause for concern, these changes did not necessarily imply that the world was becoming less capable of feeding its people. Rather, two forces explained most of the decline: more efficient meat production, and the restructuring of the economies.

3 About 37 per cent of the world's grain crop was fed to livestock, including cattle, hogs, poultry, sheep, horses and goats. Altough the quantity of grain fed to the world's livestock had not increased since 1986, world meat production had increased 22 per cent. This increase was explained by an increase in the efficiency of converting grain to meat.

4 The gap between world food-aid needs and food-aid deliveries from donor nations widened in 1995, and the gap was expected to grow in 1996.

Food Trade

After seven years of negotiations, known as the Uruguay Round, member nations in 1994 agreed to significant modifications of the General Agreement on Tariffs and Trade (GATT), the set of rules governing international trade. One component of the agreement was the creation of the World Trade Organization, effective in January 1995, to oversee the implementation of the trade rules.

1.2 The new rules would have major long-term implications for agricultural trade and world food security. The countries' representatives who took part in the meeting stated that a reliable trading system was essential for moving food efficiently from food-surplus to food-deficit countries. Most countries had erected barriers to trade in agricultural products to protect their farmers. The net effect of each country's actions was an inefficient global system of agriculture, in which some overproduced, others underproduced, and trade was more difficult than it needed to be. Past trade agreements greatly reduced barriers to trade in manufactured products, and as a result trade flourished. Little progress was made in agriculture, however. The Uruguay Round agreement, for the first time, provided a framework for halting the escalation of agricultural trade barriers set by the richest countries and for gradually bringing them

down. The long-term effect should be an improved global food system.

1.3 The basic principles of the trade rules were: (1) trading should take place between countries without discrimination; (2) there should be predictable and growing access to each country's markets; (3) fair trade should be promoted; and (4) industrial countries were encouraged to assist the trade of Least Developed Countries (LDCs).

1.4 The main components of the GATT agreement on agriculture were the following principles. All nontariff barriers to trade were to be converted to equivalent tariffs, with all tariffs reduced an average of at least 36% over six years. Countries must allow duty-free imports of at least 3% to 5% of the domestic consumption of agricultural products. Export subsidies were to be reduced at least 36% and the volume of subsidized exports reduced at least 21% over six years. Subsidies to domestic producers of traded products would be reduced at least 20% over six years. Sanitary regulations (human health standards and plant and animal safety standards) were to be based on science rather than on arbitrary rules that tended to discriminate against imports.

Global food-aid needs increased, while aid shipments from donor nations declined. Aid needs existed in Africa, Asia, the former Soviet Republics, Bosnia and Herzegovina, and Latin America. Chronic food shortages and emergencies were caused by a combination of natural and man-made disasters. The decline in food-aid shipments was caused by smaller aid budgets, mainly in the United States, and higher grain prices.

[5] It was estimated that global grain stocks declined to record low levels in 1995, and they were expected to decline further by the end of the 1995/96 marketing year.

NGOs: market is incapable of solving food security

[6] Consensus has been reached by NGOs in all regions - says a report of the United Nations Non Governmental Liaison Service - that the market alone is incapable of solving food security problems, and can even, at times, be directly harmful. NGOs have reiterated their concern about the ability of foreign trade to complement rather than supplant national production. In this context, NGOs in Asia recommended a freeze on the implementation of further agricultural liberalization until after a thorough study of the impact of the GATT Agricultural Accord had been undertaken and the Accord renegotiated.

[7] In Asia, NGOs insisted that democratic control of the food system is the ultimate test of democracy. All across the globe, organizations have reiterated that the vital resources that make food security possible should stay under the democratic control of food producers and local communities, with the effective participation of civil society in agricultural policy-making. On this basis, the dissolution of small farming households, as a result of indiscriminate liberalization policies enabling the entry and dominance of extremely powerful multinational agribusiness, should be prevented and reversed.

[8] Structural adjustment programmes (SAPs) have been widely criticized as undermining food security, and international financial institutions and governments have been urged to prevent SAPs from endangering access to water, sanitation, food and nutrition. In Africa, NGOs recommended that international financial institutions adopt food security as a major indicator for the success of adjustment programmes. In Europe and North America, the inadequate commitment of governments to ensure social safety nets in the process of economic adjustment, as endorsed by governments participating in the 1995 World Summit for Social Development, was deplored. NGOs in Latin America have proposed creating debt swaps for food security programmes.

[9] Calls have been made to alter a pattern of overconsumption by industrialized nations and local elites which may perpetuate and increase the disparity between tne North and South. The importance of promoting more efficient water use and sustainable, environmentally sound food production was stressed. In Asia, NGOs emphasized the urgency of reintegrating agricultural production into the local ecology and abandoning techno-fixes like the green revolution. In Africa, NGOs highlighted the necessity to engage in organic agriculture whenever appropriate and improve local technologies through harnessing indigenous knowledge.

Global food security and organic agriculture

Over 1,000 delegates from 92 countries participated in the 11th International Scientific Conference on Organic Agriculture held in August 1996 in Copenhagen, Denmark. IFOAM is a 530 member organization in almost 100 countries. Its first conference was held in Switzerland in 1977.

[2.2] The problems associated to the intensive agriculture, derived from the Green Revolution, were discussed in the conference. Participants highlighted the need to put an end to unbridled economic growth with measures to reduce environmental impact. In any case, this is not the way the World Trade Organization sees the point, the market being the main donor of privileges and solutions.

[2.3] There are more than 800 million people in the world today who do not have adequate food to eat. Ironically, a vast majority of of them are located in developing countries which are predominantly agricultural.

[2.4] The international organic conference concluded that FAO documents emphasize wrong issues in analyzing the source of the crisis in world food production and availability, and therefore that its commitments are not likely to produce effective results.

[2.5] "In particular, protection of the environment, quality and safety of food, and local sef-reliance are not recognized as essential elements of food security", stated the conference's final declaration.

[2.6] "We see the danger that the failure of the Green Revolution will be repeated by promoting the even more destructive gene-revolution. In addition, key issues of access to resources, equitable land tenure, education, and the rights of women have not been given the place they deserve. As a consequence, people will continue to suffer".

[2.7] "The practical development of organic agricultures throughout the world -said the delegates- is living proof that it is a realistic and achievable approach. Organic agriculture can ensure the long-term security of human and material resources, producing sufficient food of high quality while protecting the environment and conserving biodiversity".

[2.8] The declaration advocates that the FAO place organic agriculture on its agenda. "Organic agriculture provides the means to achieve food security for all. Local, regional and national self-sufficiency in food is our shared goal. We urge the FAO to recognize the part organic production can play in achieving this goal".

[2.9] In 1993 the world consumption of insecticides, herbicides, fungicides and plant growth regulators rose to 1,893,300 metric tons. Nearly half of it were consumed in Europe, while the US with 248,000 metric tons and China with 240,000 remained in the first and second place.

Food in figures

A = Food production per capita index (1979-81=100) 1993
B = Cereal imports (1,000 metric tons) 1993
C = Food import dependency (%) 1992

	A	B	C		A	B	C		A	B	C
Afghanistan	59	322	..	Georgia		500		Norway		302	
Albania		647		Germany		3,533		Oman		369	19
Algeria	119	5,821	26	Ghana	115	396	10	Pakistan	118	2,893	15
Angola	72			Greece		708		Panama	87	159	10
Aotearoa/NZ		282		Grenada	78	..		Papua N. Guinea	103	227	17
Argentina	94	8	6	Guatemala	94	486	12	Paraguay	109	82	13
Armenia		350		Guinea	98	335	..	Peru		1,920	20
Australia		32		Guinea-Bissau	110	70	35	Philippines	88	2,036	8
Austria		184		Guyana	94	53	..	Poland		3,142	
Azerbaijan		480		Haiti	67	236	..	Portugal		2,147	
Bangladesh	97	1,175	16	Honduras	89	197	11	Romania		2,649	
Barbados	64			Hungary		137		Russia		11,238	
Belarus		1,250		India	123	694	5	Rwanda	70	115	..
Belgium		2,291		Indonesia	145	3,105	6	Saudi Arabia	340	5,186	16
Belize	95	..		Iran	126	4,840	12	Senegal	111	579	29
Benin	119	134	25	Iraq	87	2,834	..	Sierra Leone	86	136	21
Bhutan	37	..		Ireland		409		Singapore	47	798	6
Bolivia	107	298	11	Israel		2,293		Slovakia	50		
Botswana	69	133	..	Italy		6,249		Slovenia	549		
Brazil	114	7,848	9	Jamaica	111	429	19	Solomon	88	..	
Brunei	100			Japan		28,035		Somalia	53	296	20
Bulgaria		241		Jordan	121	1,596	21	South Africa	74	2,275	5
Burkina Faso	132	121	25	Kazakhstan		100		Spain		4,955	
Burundi	92	22	18	Kenya	83	569	6	Sri Lanka	81	1,149	16
Cambodia	141	20	..	Korea Dem. Rep.	76	..	..	Sudan	76	654	19
Cameroon	79	281	15	Korea Rep.	94	11,271	6	Suriname	81	..	
Canada		1,095		Kuwait	120	251		Swaziland	82	7	
Cen. Afr. Rep.	94	32	19	Laos		8	33	Sweden		202	
Chad	99	59	18	Latvia		11		Switzerland		455	
Chile	118	983	6	Lebanon	186	356	..	Syria	89	1,440	17
China	145	7,332	5	Lesotho	70	131	..	Tajikistan		450	
China-Hong Kong	87	640	7	Liberia	58	70	..	Tanzania	76	215	6
Colombia	114	1,702	9	Libya	81	2,290		Thailand	102	638	6
Comoros	83	25	..	Lithuania		415		Togo	106	63	22
Congo	79	148	..	Macedonia		117		Trin. and Tobago	85	232	17
Costa Rica	104	535	8	Madagascar	86	111	11	Tunisia	123	1,044	8
Côte d'Ivoire	89	590	19	Malawi	70	514	8	Turkey	102	2,107	6
Cuba	65	..		Malaysia	203	3,288	7	Turkmenistan		940	
Cyprus	94	498	..	Maldives	84	..		Uganda	109	76	8
Czech Republic		519		Mali	91	83	20	Ukraine		1,500	
Denmark		579		Mauritania	81	286	23	Utd. A. Emirates		583	17
Djibouti	44	..		Mauritius	99	240	13	Utd. Kingdom		3,534	
Dominican Rep.	104	961	6	Mexico	94	6,223	11	United States		4,684	
Ecuador	110	428	5	Moldova		200		Uruguay	113	110	9
Egypt	114	7,206	29	Mongolia	63	182	..	Uzbekistan		4,151	
El Salvador	95	286	16	Morocco	106	3,653	14	Vanuatu	80	..	
Equatorial Guinea	11			Mozambique	77	507	..	Venezuela	101	2,314	10
Estonia		46		Myanmar/Burma	107	21	8	Vietnam	133	289	..
Ethiopia	86	1,045	15	Namibia	72	141	..	Yemen	75	1,843	..
Fiji	97	..		Nepal	114	27	9	Zaire	100	336	..
Finland		108		Netherlands		4,431		Zambia	99	353	8
France		1,188		Nicaragua	64	125	23	Zimbabwe	78	538	3
Gabon	78	77	17	Niger	77	136	17				
Gambia	76	87	..	Nigeria	129	1,584	18				

Sources: Human Development Report 1996, UNDP 1996; World Development Report, World Bank, 1995.

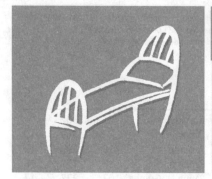

HEALTH

THE PRINCIPLES of primary health care to provide health services for all (irrespective of whether the patient can pay) have guided global and national health policies, but they are under threat due to a dramatic change in health strategies.

The concept of primary health care was adopted by virtually all governments at the landmark 1978 WHO-UNICEF global health conference that endorsed the Alma Ata Declaration (so named after the town where the Conference was held).

[2] The Declaration's centerpiece was primary health care, a comprehensive strategy that includes an equitable, consumer-centered approach to health services and also addresses underlying social factors that influence health.

[3] Today it is painfully evident that the goal of Health For All is growing more distant not just for the poor but for humanity. Three factors have undermined primary health care:

[4] Firstly, instead of adopting a comprehensive health and social program as envisaged, UNICEF compromised (due to a funds shortage) and opted for selective primary health care through selected technological interventions, particularly oral rehydration therapy (ORT) and immunization.

[5] These programs were helpful but were of limited success and are proving difficult to sustain, especially in view of economic recession. ORT usage and immunization coverage have recently declined significantly in many countries. The second setback to primary health care was the introduction in the 1980s of structural adjustment programs (SAPs) imposed on indebted Third World countries by the World Bank and IMF as a condition for debt rescheduling.

[6] Budgets for so-called non-productive government activities such as health, education and food subsidies were ruthlessly slashed. Public hospitals and health centers were sold to the private sector, thus pricing their services out of the reach of the poor. Falling real wages, food scarcity and growing unemployment pushed low-income families into worsening poverty.

[7] A third factor has put the nail in the coffin of the Alma Ata Declaration: the increasing role of the World Bank in global health policy planning, based on the same philosophy as structural adjustment.

[8] The Bank's sectoral health policy is spelled out in its World Development Report 1993. Under the guise of promoting an equitable, cost-effective, decentralized and country-appropriate health system, the World Bank's key recommendations spring from the same sort of structural adjustment paradigm that has worsened poverty and further jeopardized the health of the world's neediest people.

[9] Thus the strategy of providing public health on the basis of need is being replaced by a new strategy of providing health on the basis of cost recovery. In other words, medical care would be available only to those who can afford to pay. And because of structural adjustment, there are more people who won't be able to pay.

Source: Third World Resurgence, April 1996.

Infectious diseases and cancer

Sexually transmitted human papilloma viruses are responsible for most of the 529,000 cases of cervical cancer a year - 65 per cent of the cases in industrialized countries, and 87 per cent of those in developing countries.

About 434,000 cases a year of liver cancer - 82 per cent of the world total - are due to hepatitis B or hepatitis C viruses. Hepatitis B causes 316,000 and hepatitis C causes 118,000 of the cases. The viruses are transmitted in several ways, including through contaminated blood and sexual relations.

Some 550,000 new cases a year of stomach cancer are attributed to the bacterium helicobacter pylori, transmitted in foods. The figure equals about 55 per cent of all cases of this cancer worldwide.

Life expectancy

Global average life expectancy at birth in 1995 was more than 65 years, an increase of about 3 years since 1985. It was over 75 years in developed countries, 64 years in developing countries, and 52 years in least developed countries.

The world's lowest life expectancy at birth, just 40 years, is in Sierra Leone - barely half of the world's highest, in Japan, where it is 79.7 years.

At least 18 countries in Africa have a life expectancy at birth of 50 years or less.

The number of countries with a life expectancy at birth of over 60 years has increased from at least 98 (with a total population of 2.7 billion) in 1980 to at least 120 (with a total population of 4.9 billion) in 1995.

On average, women today can expect to live over 4 years longer than men - 67.2 years versus 63 years. The female advantage is greatest in Europe - almost 8 more years - and smallest in South-East Asia, where it is just one year.

Source: Fifty Facts from the World Health Report 1996, World

Some facts and figures

- About 52 million people died in 1995. The number is almost the same as it was 35 years ago, but the global population has almost doubled in that time.
- More than 17 million of the 52 million deaths in 1995 were due to infectious diseases.
- Of more than 11 million deaths among children under 5 in the developing world, about 9 million were attributed to infectious diseases, 25 per cent of them preventable through vaccination.
- Of the 52 million deaths, almost 34 million occurred at the extremes of life - over 11 million children died before the age of 5; over 22 million people survived until at least 65.
- Cancer killed about 6.6 million people in 1995.
- In Africa, more than 40 per cent of all deaths were among children under 5.
- About 8.4 million infants died last year before their first birthday. Developed market economies had only 6.9 infant deaths per 1,000 live births, compared to 106.2 infant deaths per 1,000 live births in the least developed countries.

Emerging and re-emerging infectious diseases

- At least 30 new diseases have been scientifically recognized around the world in the last 20 years.
- A completely new strain of cholera, called Vibrio cholerae 0139, appeared in south-eastern India in 1992 and has since spread to other areas of India and parts of South-East Asia.
- The Ebola virus was unknown 20 years ago. The Ebola-haemorrhagic fever outbreak in Zaire in 1995 was fatal in about 80 per cent of cases. The natural host of the virus remains a mystery.
- Since hantavirus infections were first recognized in the United States in 1993, they have been detected in more than 20 American states. They can cause a pulmonary syndrome with a fatality rate of over 50 per cent. Cases have also occurred in Canada and in Argentina, Brazil and Paraguay.
- Recently-recognized organisms such as cryptosporidium or new strains of bacteria such as Escherichia coli cause epidemics of foodborne and waterborne diseases in both industrialized and developing countries.
- Tuberculosis, once regarded as virtually under control, is making a deadly comeback, killing about 3.1 million people a year. Drug-resistant tuberculosis is spreading in many countries.
- Cholera, absent in South America for decades, struck Peru in 1991 and has since spread throughout the continent. Worldwide it is endemic in at least 80 countries and causes 120,000 deaths a year.
- Diphtheria epidemics that began in the Russian Federation in 1990 have since struck in 15 eastern European countries. WHO estimates that there are 100,000 diphtheria cases and up to 8,000 deaths a year worldwide.
- The biggest epidemic of yellow fever in the Americas since 1950 struck Peru in 1995.
- More than 2 billion people worldwide are at risk of dengue fever.
- There were at least 333 million new cases of sexually transmitted diseases worldwide in 1995, excluding HIV infections.

Source: Fifty Facts from the World Health Report 1996, World Health Organization (WHO), New York, 1996.

Health services

1990-95	With access to health services (%)	With access to safe water (%)		With access to health services (%)	With access to safe water (%)
Afghanistan	29	12	Lesotho	00	52
Algeria	98	79	Liberia	39	46
Angola	30	32	Libya	..	97
Aotearoa/NZ	..	97	Madagascar	65	29
Argentina	71	71	Malawi	80	47
Bangladesh	45	97	Mali	30	37
Benin	18	50	Mauritania	63	66
Bhutan	65	..	Mauritius	100	99
Bolivia	67	55	Mexico	78	83
Botswana	89		Mongolia	95	80
Brazil	..	87	Morocco	70	55
Burkina Faso	90	78	Mozambique	39	33
Burundi	80	70	Myanmar/Burma	60	38
Cambodia	53	36	Namibia	62	57
Cameroon	70	50	Nepal	..	46
Central African			Nicaragua	83	58
Republic	45	18	Niger	32	54
Chad	30	24	Nigeria	66	40
Chile	97	85	Oman	96	93
China	92	67	Pakistan	55	79
China-Hong			Panama	80	83
Kong	99	100	Papua New		
Colombia	60	87	Guinea	96	28
Congo	83	38	Paraguay	63	35
Costa Rica	80	92	Peru	75	71
Côte d'Ivoire	30	72	Philippines	76	85
Cuba	98	93	Rwanda	80	66
Dominican			Saudi Arabia	97	95
Republic	80	76	Senegal	40	52
Ecuador	88	71	Sierra Leone	38	34
Egypt	99		Singapore	100	100
El Salvador	40	55	Somalia	27	37
Ethiopia	46	25	South Africa	..	70
Gabon	90	68	Sri Lanka	93	53
Gambia	93	48	Sudan	70	60
Ghana	60	56	Syria	90	85
Guatemala	36	62	Tanzania	80	50
Guinea	80	55	Thailand	90	86
Guinea-Bissau	40	53	Togo	61	63
Haiti	50	28	Trinidad and		
Honduras	64		Tobago	100	97
India	85	81	Tunisia	90	99
Indonesia	80	62	Turkey	..	80
Iran	80		Uganda	49	34
Iraq	93	44	United Arab		
Jamaica	90	86	Emirates	99	95
Japan	..	97	Uruguay	82	75
Jordan	97	89	Venezuela	..	79
Kenya	77	53	Vietnam	90	3
Korea Rep.	100	93	Yemen	38	55
Kuwait	100	..	Zaire	26	27
Laos	67	45	Zambia	75	50
Lebanon	95	94	Zimbabwe	85	77

Source: The State of the World's Children 1996, UNICEF, New York, 1996.

1988-91	Population			Population	
	per doctor	per nurse		per doctor	per nurse
Afghanistan	7,692	11,111	Laos	4,545	
Albania	730		Lebanon	413	2,174
Algeria	1,064		Lesotho	25,000	2,000
Angola	25,000		Libya	962	328
Aotearoa/NZ	521		Madagascar	8,333	3,846
Argentina	329	1,786	Malawi	50,000	33,333
Bahrain	775		Malaysia	2,564	
Bangladesh	12,500	20,000	Mali	20,000	5,882
Belarus	282		Mauritania	16,667	2,273
Belgium	298		Mauritius	1,176	398
Benin	14,286	3,226	Mexico	621	
Bhutan	11,111	6,667	Mongolia	389	209
Bolivia	2,564	7,692	Mozambique	33,333	5,000
Botswana	4,762	469	Myanmar/Burma	12,500	
Brazil	847	3,448	Namibia	4,545	339
Bulgaria	315		Nepal	16,667	33,333
Burkina Faso	33,333	10,000	Netherlands	398	
Burundi	16,667		Nicaragua	2,000	3,125
Cameroon	12,500	1,852	Niger	50,000	3,846
Canada	446		Nigeria	5,882	1,639
Central African			Norway	309	
Republic	25,000	11,111	Pakistan	2,000	3,448
Chad	33,333	50,000	Panama	562	1,064
Chile	943	3,846	Papua New Guinea		1,587
Colombia	1,064	2,632	Paraguay	1,587	7,143
Comoros	10,000	3,448	Peru	1,031	
Congo	3,571	1,370	Philippines	8,333	
Costa Rica	1,136	2,222	Poland	467	
Côte d'Ivoire	11,111	3,226	Portugal	352	
Cuba	332	180	Romania	552	
Cyprus	585		Rwanda	25,000	8,333
Denmark	360		Saudi Arabia	704	310
Dominican Republic	935	9,091	Senegal	16,667	12,500
Ecuador	671	1,818	Singapore	725	
El Salvador	1,563	3,333	Spain	262	
Ethiopia	33,333	14,286	Sri Lanka	7,143	1,754
Finland	405		Swaziland	9,091	595
France	333		Sweden	395	
Gabon	2,500	1,471	Switzerland	585	
Ghana	25,000	3,704	Syria	1,220	1,031
Greece	313		Thailand	4,762	1,064
Guatemala	4,000	7,143	Togo	11,111	3,030
Guinea	7,692		Trinidad and		
Haiti	7,143	9,091	Tobago	1,370	
Honduras	1,266	4,545	Tunisia	1,852	407
Hungary	312		Turkey	1,176	
India	2,439	3,333	Uganda	25,000	7,143
Indonesia	7,143	2,857	Ukraine	259	
Iraq	1,667	1,370	United Arab		
Ireland	633		Emirates	1,042	568
Italy	211		Vietnam	247	1,149
Jamaica	7,143		Yemen	4,348	1,818
Jordan	649	641	Zaire	14,286	1,351
Kenya	20,000	9,091	Zambia	11,111	5,000
Korea Rep.	1,205	1,538	Zimbabwe	7,692	1,639

Sources: Human Development Report 1996, UNDP, New York, 1996.

"**WE HAVE TO** confront and thus be better able to overcome the main tensions that, although not new, will be central to the problems of the 21st century", (from the Report of the Commission created by UNESCO on Education for the 21st Century).

The dream of the 'unfinished' being

The private enterprises' sponsorship of literacy limits learning to a neoliberal point of view", asserted 75-year-old Brazilian educator Paulo Freire, at a meeting with professors from the Axé Project developed in Salvador, Bahia, in October 1996.

[2] This meeting reunited the famous Brazilian educator and a well known Brazilian education project. During its 6 year activity, the Axé Project has already worked with almost 7,000 street kids, being internationally regarded as a model pedagogical initiative for low-income boys and girls.

[3] The course that education has presently taken in Brazil is paradigmatic. Freire was regarded during the 1950s and 1960s as an enemy both of law and family, even facing imprisonment while working in favour of his country's literacy. But now, in the 1990s, the industrial elite itself is teaching literacy to its employees.

[4] However, this teaching "is confined within the neoliberal perspective, the hegemonic ideology and the way of organizing the economy in most of the world's countries", Freire said. "Basic literacy is purely mechanics. It was discovered as a necessity for the productive process".

[5] "A cook needs a technique to succeed in the profession. But s/he also needs political empowerment to exercise her or his citizenship, to participate in society", he added.

[6] These objections notwithstanding, Freire defended teaching in private enterprises: "I want more than that, but I

Education imbalances

A = Primary pupil to teacher ratio (1992)
B = Percentage of primary school children reaching grade 5 (1986-93)

Country	A	B	Country	A	B	Country	A	B	Country	A	B
Afghanistan		43	Djibouti	43		Madagascar	38	28	Suriname	23	
Albania	19	98	Ecuador		67	Malawi	68	46	Sweden	10	98
Algeria	27	93	Egypt	26	98	Malaysia	20	98	Switzerland		100
Angola	32	34	El Salvador	44	58	Mali	47	76	Syria	24	92
Aotearoa/NZ	16	94	Estonia	25	93	Mauritania	51	72	Tajikistan	21	
Argentina	16		Ethiopia	27	22	Mauritius	21	100	Tanzania	36	83
Australia	17	99	Fiji	31		Mexico	30	84	Thailand	17	88
Austria	11	97	Finland	14	100	Mongolia	28		Togo	59	70
Bahamas	21		France	12	94	Morocco	28	80	Trinidad and Tobago	26	95
Bahrain	18		Gabon	44	50	Mozambique	53	35	Tunisia	26	90
Bangladesh	63	47	Gambia	30	87	Myanmar/Burma	36		Turkey	29	92
Barbados	17		Germany	16		Namibia	32	64	Uganda	35	55
Belgium	10		Ghana	29	80	Nepal	39	52	United Arab Emirates	17	99
Benin	40	55	Greece	19	99	Nicaragua	37	55	Uruguay	21	95
Bhutan	31		Guatemala	34		Niger	38	82	Venezuela	23	78
Bolivia	25	66	Guinea	49	80	Nigeria	39	87	Vietnam	35	
Botswana	29	84	Guinea-Bissau		20	Norway	6	100	Yemen	29	
Brazil	23	72	Haiti	29	47	Oman	27	96	Zaire	42	64
Brunei	15		Honduras	38		Pakistan	41	48	Zambia	44	
Bulgaria	14	88	Hungary	12	97	Panama	23	82	Zimbabwe	38	76
Burkina Faso	60	70	India	63	62	Papua New Guinea	31	69			
Burundi	63	74	Indonesia	23	86	Paraguay	23	74			
Cambodia	93		Iran	32	89	Peru	28				
Cameroon	51	66	Iraq	22	72	Philippines	36	75			
Canada	17	96	Ireland	25	100	Poland	17	98			
Cape Verde	97		Israel	16	100	Portugal	14				
Central African Rep.	90	65	Italy	12	100	Qatar	10				
Chad	64	49	Jamaica	38	96	Romania	21	93			
Chile	25	95	Japan	20	100	Rwanda	58	59			
China	22	88	Jordan	22	98	Saudi Arabia	14	96			
Colombia	28	59	Kenya	31	77	Senegal	59	88			
Congo	66	72	Korea Rep.	33	100	Sierra Leone	34				
Costa Rica	32	86	Kuwait	16	11	Singapore	26	100			
Côte d'Ivoire	37	73	Laos	29	53	Slovakia	22	97			
Cuba	12	95	Lebanon	21		Slovenia	18	100			
Cyprus	18		Lesotho	51	60	Spain	21	96			
Czech Republic	18	95	Libya	12		Sri Lanka	29	92			
Denmark	11	100	Macedonia	20	95	Sudan	34	94			

Source: World Development Report 1995, World Bank, 1995; The State of the World's Children 1996, UNICEF, 1996.

would not put it aside. If employees read a little that just enables them not to damage the machines, later on they could themselves improve on that education."

[7] The meeting was on "Education, wishes and dreams", words that inspire both Freire and the Axé Project. The latter has termed its teaching methodology as "Pedagogy of Desire".

[8] With respect to the "dreams and wishes", Freire stated that touching this question implies the recognition "by all of us, men and women, as 'unfinished' historical beings and as beings aware of this 'unfinishedness'".

[9] For him, the contact of this awareness with curiosity provokes a "presence in the world". But this presence "cannot go on if it lacks a relationship with a certain tomorrow. The question is how we relate with tomorrow and what do we want from tomorrow". Hence the necessity of dreams and desires.

[10] "Neoliberalism fights against Dream and Utopia", said Freire, "being shaped as it is over pragmatic ideas and dominent interests".

Source: Folha de São Paulo,
October 1996.

Educational tension

We have to confront, and thus be better able to overcome, the main tensions that, although they are not new, will be central to the problems of the twenty-first century" says the October 1995 UNESCO Commission on Education report.

[1,2] According to the Commission, in the next century the tension between the global and the local will be increased due to the people's need to become world citizens, without losing their roots, while continuing to play an active part in the life of their nations and their local communities. "We cannot disregard the promises of globalization nor its risks, including the tendency to forget the unique character of each human being. Yet, we are summoned to choose our future and achieve our full potential within the carefully tended wealth of our traditions and our own cultures which, unless we are careful, can be endangered by contemporary developments."

[1,3] Three main actors contribute to the success of educational reforms: first of all, the local community, including parents, school heads and teachers; but also the public authorities; and the international community. Much past failure has been due to insufficient involvement of one or more of these partners. Attempts to impose educational reforms from the top down, or from outside, have obviously failed. Countries where the process has been relatively successful are those that obtained a determined commitment from local communities, parents and teachers, backed up by continuing dialogue and various forms of financial, technical and/or vocational assistance.

Illiteracy

(Adult illiteracy %)

	Total 1995	Female 1995	Male 1995		Total 1995	Female 1995	Male 1995		Total 1995	Female 1995	Male 1995		Total 1995	Female 1995	Male 1995
Afghanistan	69			Cuba	4			Laos	43	56	31	Senegal	67	77	57
Algeria	38	51	26	Djibouti	54			Lebanon	8			Seychelles	21		
Argentina	4	4	4	Dominican Rep.	18	18	18	Lesotho	29	38	19	Sierra Leone	69	82	55
Bahamas	2			Ecuador	10	12	8	Malawi	44	58	28	Singapore	9	14	4
Bahrain	15			Egypt	49	61	36	Malaysia	17	22	11	South Africa	18	18	18
Bangladesh	62	74	51	El Salvador	29	30	27	Maldives	7			Sri Lanka	10	13	7
Barbados	3			Ethiopia	65	75	55	Mali	69	77	61	Sudan	54		
Benin	63	74	51	Fiji	8			Mauritania	62	74	50	Surinam	7		
Bhutan	58			Gabon	37	47	26	Mauritius	17	21	13	Swaziland	23		
Bolivia	17	24	10	Gambia	61	75	47	Mexico	10	13	8	Tanzania	32	43	21
Botswana	30	40	20	Ghana	36	47	24	Morocco	56	69	43	Thailand	6	8	4
Brazil	17	17	17	Guatemala	44	51	38	Mozambique	60	77	42	Togo	48	63	33
Brunei	12			Guinea	64	78	50	Myanmar/Burma	17	22	11	Trin. and Tobago	2	3	1
Burkina Faso	81	91	71	Guinea-Bissau	45	58	32	Nepal	73	86	59	Tunisia	33	45	21
Burundi	65	78	51	Guyana	2			Nicaragua	34	33	35	Turkey	18	28	8
Cameroon	37	48	25	Haiti	55	58	52	Niger	86	93	79	Uganda	38	50	26
Cape Verde	28			Honduras	27	27	27	Nigeria	43	53	33	Unit. Arab Emir.	21	20	21
Cen. Afr. Rep.	40	48	32	Hong Kong	8	12	4	Pakistan	62	76	50	Uruguay	3	2	3
Chad	52	65	38	India	48	62	35	Panama	9	10	9	Venezuela	9	10	8
Chile	5	5	5	Indonesia	16	22	10	Papua N. Guinea	28	37	19	Vietnam	6	9	4
China	19	27	10	Iran	28	34	22	Paraguay	8	9	7	Zaire	33		
Colombia	9	9	9	Iraq	42			Peru	11	17	6	Zambia	22	29	14
Comoros	43			Jamaica	15	11	19	Philippines	5	6	5	Zimbabwe	15	20	10
Congo	25	33	17	Jordan	13	21	7	Qatar	21						
Costa Rica	5	5	5	Kenya	22	30	14	Rwanda	40	48	30				
Côte d'Ivoire	60	70	50	Kuwait	21	25	18	Saudi Arabia	37	50	29				

Source: World Development Indicators 1996, World Bank, 1996.

Education enrolment in figures

(Percentage of age group enrolled in education. The gross enrolment ratios may exceed 100 per cent because some pupils are younger or older than the country's standard primary, secondary or tertiary age.)

	Primary Female 1993	Primary Male 1993	Secondary Female 1993	Secondary Male 1993	Tertiary 1993		Primary Female 1993	Primary Male 1993	Secondary Female 1993	Secondary Male 1993	Tertiary 1993
Albania	97	95	..	..	10	Laos	92	123	19	31	2
Algeria	96	111	55	66	11	Latvia	82	83	90	84	39
Aotearoa/NZ	101	102	104	103	58	Lesotho	105	90	31	21	2
Argentina	107	108	75	70	41	Lithuania	90	95	79	76	39
Armenia	93	87	90	80	49	Macedonia, FYR	87	88	55	53	16
Australia	107	108	86	83	42	Madagascar	72	75	14	14	4
Austria	103	103	104	109	43	Malawi	77	84	3	6	1
Azerbaijan	87	91	88	89	26	Malaysia	93	93	61	56	..
Bangladesh	105	128	12	26	..	Mali	24	38	6	12	..
Belarus	95	96	96	89	44	Mauritania	62	76	11	19	4
Belgium	100	99	104	103	..	Mauritius	106	107	60	58	4
Benin	44	88	7	17	..	Mexico	110	114	58	57	14
Bolivia	..	..	..	..	23	Moldova	77	78	72	67	35
Botswana	120	113	55	49	3	Morocco	60	85	29	40	10
Brazil	..	..	..	..	12	Mozambique	51	69	6	9	0
Bulgaria	84	87	70	66	32	Namibia	138	134	61	49	3
Burkina Faso	30	47	6	11	..	Nepal	85	129	23	46	3
Burundi	63	76	5	9	1	Netherlands	99	96	120	126	45
Cameroon	..	..	..	..	2	Nicaragua	105	101	44	39	9
Canada	104	106	103	104	103	Niger	21	35	4	9	1
Central African Republic	51	92	..	..	2	Nigeria	82	105	27	32	..
Chad	38	80	..	..	1	Norway	99	99	114	118	54
Chile	98	99	70	65	27	Oman	82	87	57	64	5
China	116	120	51	60	4	Pakistan	49	80	..	..	..
China-Hong Kong	..	..	..	..	21	Panama	..	..	..	..	23
Colombia	120	118	68	57	16	Papua New Guinea	67	80	10	15	..
Costa Rica	105	106	49	45	30	Paraguay	110	114	38	36	10
Côte d'Ivoire	58	80	17	33	..	Peru	..	..	..	..	40
Croatia	87	87	86	80	27	Philippines	..	..	..	..	26
Czech Republic	100	99	88	85	16	Poland	97	98	87	82	26
Denmark	98	97	115	112	41	Portugal	118	122	..	..	23
Dominican Republic	99	95	43	30	..	Romania	86	87	82	83	12
Ecuador	122	124	56	54	..	Russian Federation	107	107	91	84	45
Egypt	89	105	69	81	17	Rwanda	50	50	9	11	..
El Salvador	80	79	30	27	15	Saudi Arabia	73	78	43	54	14
Estonia	83	84	96	87	38	Senegal	50	67	11	21	3
Ethiopia	19	27	11	12	1	Slovakia	101	101	90	87	17
Finland	100	100	130	110	63	Slovenia	97	97	90	88	28
France	105	107	107	104	50	South Africa	110	111	84	71	13
Gabon	136	132	..	..	3	Spain	105	104	120	107	41
Gambia	61	84	13	25	..	Sri Lanka	105	106	78	71	6
Germany	98	97	100	101	36	Sweden	100	100	100	99	38
Ghana	70	83	28	44	..	Switzerland	102	100	89	93	31
Guatemala	78	89	23	25	..	Tajikistan	88	91	101	98	25
Guinea	30	61	6	17	..	Tanzania	69	71	5	6	..
Guinea-Bissau	..	..	2	10	..	Thailand	97	98	37	38	19
Honduras	112	111	37	29	9	Togo	81	122	12	34	3
Hungary	94	94	82	79	17	Trinidad and Tobago	94	94	78	74	8
India	91	113	38	59	..	Tunisia	113	123	49	55	11
Indonesia	112	116	39	48	10	Turkey	98	107	48	74	16
Iran	101	109	58	74	15	Uganda	83	99	10	17	1
Ireland	103	103	110	101	34	Ukraine	87	87	95	65	46
Israel	96	95	91	84	35	United Arab Emirates	108	112	94	84	11
Italy	99	98	82	81	37	United Kingdom	113	112	94	91	37
Jamaica	108	109	70	62	6	United States	106	107	97	98	81
Japan	102	102	97	95	30	Uruguay	108	109	..	..	30
Jordan	95	94	54	52	19	Uzbekistan	79	80	92	96	33
Kazakhstan	86	86	91	89	42	Venezuela	97	95	41	29	29
Kenya	91	92	23	28	..	Vietnam	..	..	..	..	2
Korea, Rep.	102	100	92	93	48	Zambia	99	109	..	..	..
Kuwait	65	65	60	60	16	Zimbabwe	114	123	40	51	6
Kyrgyzstan	..	..	..	..	21						

Source: World Development Indicators 1996, World Bank, 1996.

WOMEN

THE 4TH WORLD CONFERENCE on Women, held in Beijing in 1995, helped to generate many government ideas to improve women's life conditions - though few funds are available for achieving this goal, according to a survey carried out in 51 countries.

Improving conditions for women

The 4th World Conference on Women was the largest meeting ever sponsored by the United Nations, with 17,000 participants, including 5,000 delegates from 189 countries and the European Union, 4,000 NGO representatives and over 3,200 media staff.

[2] Argentina and Colombia are among Latin American countries which have retreated in promoting policies for women, cutting the budget of their governments' women's agencies. Peru, with the creation of a specialized agency, is among those countries which have progressed.

[3] The survey, carried out by the non-governmental Women, Environment and Development Organization (WEDO), showed that most countries were actively involved in the Conference but few steps forward have been taken. After the meeting, many countries formed special agencies to deal with women's issues.

[4] Peru has created a women's commission in the Congress and South Africa has opened an office at the Presidency seeking to improve women's situation. Several Asian countries, including Indonesia, Pakistan and South Korea, are forming committees to involve women's NGOs in the decision-making process.

[5] However, "there is very little money on the table" said Bella Abzug, WEDO president and Susan Davis, director of the organization, at the report's presentation. Political decisions have not been followed by funds to aid organizations in their work.

[6] "Most governments try to obtain funds from existing resources, but a mere redistribution of a shrinking pie will not satisfy the hunger of women for substantial change", they added.

[7] Among the new programs is that of Puerto Rico, a free state associated to the United States, which promised to allot $1,800 million for housing, legal and psychological services to tackle violence against women.

[8] South Korea added $5.1 million to its Ministry of Political Affairs to set up a women's information agency. More often, however, the governments surveyed pledged to implement Beijing's proposals according to existing funds, or they did not answer at all.

[9] WEDO considers that an essential factor which delays women's progress throughout the world is the World Bank's program of economic structural adjustment.

> The survey, carried out by the non-governmental Women, Environment and Development Organization (WEDO), showed that most countries were actively involved in the Conference but few steps forward have been taken.

[10] WEDO's report warns that "in spite of a few encouraging measures aimed at showing greater sensibility and response to gender issues, evidence of concrete changes in the Bank's policy are yet to be revealed".

[11] The United States has tried to estimate the value of non-paid work with a joint investigation carried out by its Departments of Labour and Commerce. These first steps, according to the WEDO report, indicate that the Beijing conference might have been more than a simple meeting for discussions, turning into the beginning of concrete measures to help women's development.

[12] "This is a contract with the women of the world. It might not be legally bonding, but it is politically", said Abzug.

[13] Apart from the steps taken, compiled by the WEDO report, at the March Preparatory Committee meeting at the United Nations in New York, Australia proposed the writing of a Chart of intentions, in which each country would include the measures it would pledge to implement regarding the situation of women.

[14] Over 80 countries voted in favour of the motion and many presented their own programs. Ivory Coast announced the creation of a bank for women, to facilitate their access to loans. The government of Fiji pledged to work for a 50 per cent of government posts held by women and to encourage the gender's participation in the private sector. Lebanon's representative revealed her government's goal for the year 2000: 30 per cent of decision making positions held by women.

[15] Also for the year 2000, Tanzania pledged to achieve a 100 per cent of girls' enrolment in school. Only 18 per cent went to school in 1990.

[16] India assured it would invest 6 per cent of its GDP in education and would appoint a commissioner for women's rights while Nepal will reform its legislation to guarantee equal rights for women with regard to private property.

[17] Among developed countries, the objectives were focused on assistance at home and sexual violence victims, the anticipation of these acts and the improvement of conditions at the workplace.

[18] The Holy See informed it would instruct 300,000 Catholic institutions worldwide devoted to social welfare to dedicate their efforts to improve women's literacy rates, health, education and nutrition.

Sources: IPS, WEDO

Mothers who die

Third World countries have been the scene of 99 per cent of the cases of maternal mortality in recent years. A recent joint study by the World Health Organization (WHO) and the United Nations Children's Fund (UNICEF) indicates that the number of maternal deaths per year has reached 585,000.

1.2 These figures provide evidence of a health problem that underscores the social vulnerability of women. Maternal mortality was underestimated in the past, according to recent studies that found that there are some 80,000 additional pregnancy-related deaths each year worldwide.

1.3 Maternal mortality is an "especially sensitive indicator of inequity" according to Susan Holck, a doctor who works in the WHO Reproductive Health Program. This index, she said, reflects the condition of women, their access to health care and the availability of the health system to attend to their needs.

1.4 In Asia, the continent which accounts for 61 per cent of births worldwide, occur 55 per cent of deaths of women during pregnancy or childbirth. Nevertheless, Africa, with only 20 per cent of births, accounts for 40 per cent of the deaths. In contrast, the developed countries, with 11 per cent of births, account for only 1 per cent of deaths.

1.5 The maternal mortality rate in the developing nations, ranges from 200 per 100,000 live births in Latin America and the Caribbean, to 870 per 100,000 in Africa. Maternal mortality takes into account deaths during pregnancy or childbirth.

1.6 The regions having the highest indices, in excess of 1,000 per 100,000 live births, are in West and East Africa. The risk of pregnancy-related death is 100 times greater in Sub-Saharan Africa than in Europe.

1.7 The risk of maternal death has increased significantly in the developing world, according to the recent WHO/UNICEF study. Only five subregions - North Africa, Southern Africa, East Asia, Central America and South America - show new estimates slightly lower than those found in earlier research.

1.8 The situation in East Africa, Central Africa and West Africa, where earlier assessments had underestimated maternal mortality by approximately one-third, has turned out to be much worse. Sierra Leone, with 1,800 deaths per 100,000 live births, and Afghanistan, with 1,700 are the two countries at the top of the list. According to these figures, one out of every seven women in Sierra Leone dies due to complications arising during pregnancy or childbirth.

1.9 At the other end of the spectrum, the most favorable indices are in Norway, Sweden and Switzerland.

1.10 The WHO/UNICEF study indicates that it is extremely difficult to establish infant mortality rates. Even in developed countries, with advanced civil statistics systems, there is a tendency to under-report such deaths. In countries where there is no accounting of births and deaths it is very difficult to arrive at maternal mortality estimates.

1.11 Not many countries keep an account of births and deaths, and fewer still record the cause of death. Even fewer systematically indicate pregnancy on death records. Establishment of maternal mortality indices requires data on deaths of women in the reproductive age group (15 to 49 years of age), on the causes of death, and also an indication as to whether the women was pregnant at the time of death.

1.12 Only 78 of the 190 WHO member countries systematically report the cause of deaths. This covers barely 35 per cent of the world population.

Source: Social Watch,
Instituto del Tercer Mundo, Uruguay.

Women in high level jobs (1995)

A = % administration and managerial posts occupied
B = % professional and technical women
C = Salaries of women as a percentage of men's incomes

	A	B	C		A	B	C		A	B	C
Algeria	5.9	27.6	16	Fiji	9.6	44.7	18	Netherlands	15.0	44.2	33
Aotearoa/NZ	32.3	47.8	38	Finland	26.4	62.3	41	Niger	8.3	0.1	37
Australia	43.3	25.0	39	France	9.4	41.4	38	Nigeria	5.5	26.0	30
Austria	19.2	48.6	34	Gambia	15.5	23.7	37	Norway	30.9	57.5	41
Bahamas	26.3	56.9	28	Germany	19.2	43.0	35	Pakistan	3.4	20.1	19
Bangladesh	5.1	23.1	23	Ghana	8.8	35.7	32	Panama	27.6	49.2	26
Barbados	37.0	25.1	39	Greece	12.1	44.2	30	Papua New Guinea	11.6	29.5	35
Belgium	18.8	50.5	33	Guatemala	32.4	45.2	19	Peru	20.0	41.1	22
Belize	36.6	38.8	17	Guyana	12.8	47.5	25	Philippines	33.7	62.7	30
Bolivia	16.8	41.9	17	Haiti	32.6	39.3	36	Poland	15.6	60.4	38
Botswana	36.1	61.4	29	Honduras	30.6	49.8	23	Portugal	36.6	52.4	28
Brazil	17.3	57.2	29	Hungary	58.2	49.0	39	Rwanda	8.2	32.1	41
Bulgaria	28.9	57.0	41	India	2.3	20.5	25	Singapore	34.3	16.1	31
Burkina Faso	13.5	25.8	40	Indonesia	6.6	40.8	32	Solomon	2.6	27.4	30
Burundi	13.4	30.4	42	Iran	3.5	32.6	16	South Africa	17.4	46.7	30
Cameroon	10.1	24.4	30	Iraq	12.7	43.9	17	Spain	12.0	48.1	28
Canada	42.2	56.1	37	Ireland	17.3	48.0	25	Sri Lanka	16.9	24.5	33
Cape Verde	23.3	48.4	32	Israel	18.7	54.1	31	Sudan	2.4	28.8	21
Central African Rep.	9.0	18.9	39	Italy	37.6	46.3	30	Surinam	21.5	69.9	24
Chile	17.4	34.0	21	Japan	8.5	41.2	33	Sweden	38.9	64.4	45
China	11.6	45.0	38	Korea Rep.	4.2	45.0	27	Switzerland	27.8	23.8	30
Colombia	27.2	41.8	32	Kuwait	5.2	36.8	18	Thailand	21.8	52.4	37
Comoros	0.1	22.3	35	Lesotho	33.4	56.6	30	Togo	7.9	21.2	32
Congo	6.1	28.5	36	Luxembourg	8.6	37.7	29	Trinidad and Tobago	23.3	53.3	28
Costa Rica	21.1	44.9	26	Malawi	4.8	34.7	33	Tunisia	7.3	17.6	23
Cuba	18.5	47.8	30	Malaysia	11.9	44.5	29	Turkey	6.6	29.3	32
Cyprus	10.2	40.8	27	Maldives	14.0	34.6	17	United Arab Emirates	1.6	25.1	9
Denmark	20.0	62.8	42	Mali	19.7	19.0	12	United Kingdom	33.0	43.7	34
Dominican Republic	21.2	49.5	22	Mauritania	7.7	20.7	18	United States	42.0	52.7	40
Ecuador	31.5	48.0	17	Mauritius	14.3	41.4	25	Uruguay	25.3	62.6	32
Egypt	16.0	28.7	23	Mexico	20.0	43.6	24	Venezuela	17.6	55.2	26
El Salvador	25.3	44.5	26	Morocco	25.6	31.3	27	Zaire	9.0	16.6	29
Equatorial Guinea	1.6	26.8	28	Mozambique	11.3	20.4	42	Zambia	6.1	31.9	25
Ethiopia	11.2	23.9	33	Namibia	20.8	40.9	19	Zimbabwe	15.4	40.0	27

Source: Human Development Report 1996, UNDP, 1996.

Women in political life

(1995)

A = % parliamentary seats occupied by women
B = % governmental posts occupied by women, ministerial level
C = % governmental posts occupied by women, total

Country	A	B	C	Country	A	B	C	Country	A	B	C
Algeria	6.7	0.0	1.6	Germany	25.5	16.0	6.8	Pakistan	1.6	3.7	1.6
Angola		7.4	6.2	Ghana	8.0	10.7	10.5	Panama	8.3	11.1	10.7
Antigua		0.0	30.0	Greece	6.0	0.0	6.3	Papua New Guinea	0.0	0.0	1.6
Aotearoa/NZ	21.2	7.4	16.8	Grenada		10.0	19.4	Paraguay	5.6	0.0	3.3
Argentina		0.0	3.2	Guatemala	7.5	18.8	18.2	Peru	10.0	5.6	9.7
Australia	13.5	13.3	23.7	Guinea		14.8	4.8	Philippines	9.5	8.3	23.9
Austria	23.2	21.1	6.8	Guinea-Bissau		8.3	11.6	Poland	13.0	6.3	8.0
Bahamas	10.8	20.0	33.9	Guyana	20.0	11.1	16.2	Portugal	8.7	9.1	17.5
Bahrain		0.0	0.0	Haiti	3.0	17.4	13.8	Qatar		0.0	1.7
Bangladesh	10.6	4.5	3.4	Honduras	7.0	10.5	17.0	Rwanda	4.3	7.7	10.2
Barbados	18.4	33.3	22.9	Hungary	11.4	5.3	7.7	St Kitts		10.0	21.4
Belgium	15.4	10.5	8.3	India	8.0	4.2	6.1	St Lucia		7.7	4.5
Belize	10.3	0.0	9.8	Indonesia	12.2	3.6	1.8	St Vincent		10.0	25.0
Benin		15.0	10.3	Iran	3.4	0.0	0.4	Samoa		6.7	7.1
Bhutan		12.5	5.0	Iraq	10.8	0.0	0.0	São Tomé & Principe		0.0	4.3
Bolivia	9.6	0.0	9.4	Ireland	12.8	18.2	11.1	Saudi Arabia		0.0	0.0
Botswana	10.0	0.0	10.9	Israel	9.2			Senegal		3.6	2.3
Brazil	7.1	3.6	13.1	Italy	13.0	3.4	9.6	Seychelles		30.8	21.3
Brunei		0.0	2.3	Jamaica	5.6	13.4		Sierra Leone		3.8	4.9
Bulgaria	13.3	9.1	8.5	Japan	6.7	6.7	8.3	Singapore	3.7	0.0	5.1
Burkina Faso	3.7	11.1	9.6	Jordan		3.2	1.6	Solomon Is.	2.1	0.0	0.0
Burundi	12.3	7.7	4.3	Kenya		0.0	4.7	Somalia		0.0	0.0
Cambodia		0.0	5.1	Korea Dem. Rep.		1.2	0.8	South Africa	23.7	9.4	7.0
Cameroon	12.2	2.7	5.4	Korea Rep.	2.0	3.4	1.5	Spain	14.6	15.0	9.7
Canada	18.0	19.2	19.1	Kuwait	0.0	0.0	6.0	Sri Lanka	5.3	12.5	8.7
Cape Verde	7.6	12.5	11.5	Laos		0.0	2.7	Sudan	8.2	0.0	0.8
Central African Rep.	3.5	5.3	4.9	Lebanon		0.0	0.0	Suriname	5.9	0.0	13.6
Chad		5.0	2.5	Lesotho	11.2	6.7	13.8	Swaziland	8.4	0.0	7.1
Chile	7.2	15.8	12.2	Liberia		9.5	8.8	Sweden	40.4	47.8	33.3
China	21.0	6.4	4.0	Libya		0.0	0.0	Switzerland		16.7	7.0
Colombia	9.3	10.5	24.7	Luxembourg	20.0	16.7	7.7	Syria		6.8	3.7
Comoros	2.4	6.7	2.5	Madagascar		0.0	0.0	Tanzania	10.6	15.6	9.1
Congo	1.6	6.3	4.3	Malawi	5.6	4.5	6.1	Thailand	4.8	3.8	4.4
Costa Rica	14.0	14.8	20.8	Malaysia	11.1	7.7	5.8	Togo	1.2	4.2	2.7
Côte d'Ivoire		8.0	2.9	Maldives	6.3	5.3	10.1	Trinidad and Tobago	20.6	20.0	13.6
Cuba	22.8	3.6	8.4	Mali	2.3	9.5	6.9	Tunisia	6.7	3.4	5.3
Cyprus	3.6	7.7	4.5	Mauritania	0.0	3.6	4.7	Turkey	1.8	2.9	5.2
Denmark		33.0	19.0	Mauritius	2.9	4.0	7.4	Uganda		12.5	9.8
Djibouti		0.0	1.4	Mexico	13.9	14.3	6.7	United Arab Emirates	0.0	0.0	0.0
Dominica		8.3	31.4	Mongolia		0.0	4.7	United Kingdom	7.8	9.1	8.4
Dominican Republic	10.0	3.4	11.5	Morocco	0.6	0.0	1.2	United States	10.4	21.1	30.1
Ecuador	4.5	6.7	9.8	Mozambique	25.2	3.6	13.2	Uruguay	7.0	0.0	2.9
Egypt	2.2	3.2	2.2	Myanmar/Burma		0.0	0.0	Vanuatu		0.0	0.0
El Salvador	10.7	5.9	18.4	Namibia	18.1	9.5	6.6	Venezuela	6.3	3.6	6.0
Equatorial Guinea	7.5	3.8	2.5	Nepal		0.0	0.0	Vietnam		6.5	3.9
Ethiopia	5.0	11.5	10.5	Netherlands	28.4	26.3	19.7	Yemen		0.0	0.0
Fiji	5.8	8.7	9.8	Nicaragua		10.5	10.5	Zaire	5.0	3.4	1.7
Finland	33.5	35.0	16.3	Niger	3.6	9.5	9.1	Zambia	6.7	7.4	8.5
France	5.9	6.5	8.8	Nigeria	2.0	3.7	4.1	Zimbabwe	14.7	3.0	10.8
Gabon		3.2	6.0	Norway	39.4	40.9	44.1				
Gambia	7.8	22.2	6.7	Oman		0.0	3.7				

Source: Human Development Report 1996, UNDP, 1996.

Women, citizens

The Programme for Action emerging from the 4th World Conference on Women includes a proviso on women in power and decision-making, and another on the advancement of women. Both refer, from different angles and emphasizing different aspects, to women's practice of full citizenship.

2.2 Because of their ethnic make-up and their different forms and styles of government and exercise of authority, the position of Latin American women presents an extraordinary diversity. However, large sectors are excluded from active practice of their citizenship. The acuteness of this problem merited the hosting of a panel by the Regional NGO Coordination for Latin America and the Caribbean at the NGO Forum at Beijing.

2.3 The State is the public space par excellence. State policies concerning the practice of citizenship, particularly those on social, economic and cultural rights, affect women directly. Early socialization, mainly carried out in the home, is already strongly discriminatory towards women; the same patterns are later reinforced throughout individual lives. Public-political participation of women remains closely tied to the type of relationship and power distribution existing within the home.

2.4 The liberal individualistic tradition which has characterized State organization in Latin American countries leaves aside other forms of conceiving the subject of rights. For a good number of indigenous groups the minimal units - that is the meaning of the word individual - are the village, the lineage, the family, hence the indivisibility of certain indigenous "private properties".

Source: Social Watch,
Instituto del Tercer Mundo, Uruguay.

AMIDST A GENERAL POLICY of casualization and de-regulation of the labour market, many forms of protection for workers which were traditionally provided by governments in statutory ways are being questioned.

De-regulated labour

According to the International Labour Organization (ILO), "in economically troubled times, when there are persistently high levels of unemployment, and when increasing globalization of markets compels each country to improve its competitiveness, state participation is likely to be much more vigorous".

[2] "Governments are increasingly prevailed on to intervene; either by employers seeking more flexible labour conditions, lower social costs and fewer taxes; or by workers demanding wage increases or compensation as a result of structural adjustment".

[3] The limitation of working hours, the curtailment of security and health standards, minimum wages, protection against arbitrary discharge, maternity and paternity rights, social security, vacations and contract clauses are reasons for worldwide concern.

[4] As a direct consequence of rising unemployment and falling wages, the labour force among Third World countries in the last two decades has undergone a process of "feminization" as more women seek paid work.

[5] Unemployment rates reflect only a lack of jobs but do not register less visible situations such as underemployment, low incomes, partial use of working capacity or low productivity. According to the ILO, these issues deserve as much attention as full unemployment. In its 1995 report, the Organization claims unemployment indicators are not accurate enough but the "tip of the iceberg".

[6] Difficulties in creating new jobs for a growing flow of people in the "economically active population" category have led countries to seek new employment alternatives.

[7] In a summit meeting of European heads of State carried out in December 1993, the development of at-home jobs was discussed as a priority issue. The

> **THE SYSTEMATIC encouragement of at-home jobs, with a more "flexible" regulative context, is another way to reduce employers' costs.**

group warned that "the first countries to enter the information society will reap the greatest rewards...countries which temporize, or favor half-hearted solutions, could, in less than a decade, face disastrous declines in investment and a squeeze on jobs".

[8] The systematic encouragement of at-home jobs, with a more "flexible" regulative context, is another way to reduce employers' costs.

[9] Large transnational corporations have made their productive structures much more flexible, achieving greater integration between their different production units, and within each unit. In response to increasingly specialized demands, from markets that are increasingly seg-

> **UNEMPLOYMENT RATES reflect only a lack of jobs but do not register less visible situations such as underemployment, low incomes, partial use of working capacity or low productivity.**

mented, production units have also had to subdivide into smaller teams and even personal units.

[10] Governments have the responsibility to find the ways to adapt regulations which will guarantee the protection of at-home workers. The European Union countries are inclined to establish a series of basic principles, without regulating the implementation of details - although they have prepared "guidelines".

[11] These conventions need to be structured according to each country's specific labour relations.

[12] The promotion of at-home work in Third World countries entails that the necessary elements for its implementation should be widely used and available. Workers must have a basic communication infrastructure: telephone, fax, computers, modem, which implies that integrated systems must be functioning in the country.

[13] Those countries which have been relegated in the development of telephone and electric networks, integrated communication systems and access to information, will find more obstacles to adapt to new methods of labour. At the same time, the necessary levels of training for these tasks imply that the demand for less specialized jobs will fall.

[14] In regions such as Latin America, the rapid increase of privatizations has been concentrated precisely in the fields of telecommunications and power generation.

[15] The ILO believes that industrial transformation is already in full swing. Within the last three decades, those employed in management-level positions increased from 10 to 20 per cent; technicians and engineers from 6 to 40 per cent; clerical workers from 11 to 15 per cent, while the figures for blue collar workers dropped during the same period from 73 to 25 per cent.

Unions in Africa

Trade unionism has grown in Africa in a parallel way to the continent's decolonization. Throughout the last years, trade unions from many countries, and the supranational organizations that reunite them, have actively participated in the democratization processes, thus recuperating the political dimension of their social role.

1.2 At the same time, governmental violationss against freedom of work and association have also registered a growth. The implementation of estructural adjustment policies together with the CFA Franc - official currency in 14 African countries, mainly the ex French colonies, the French Franc being its point of reference - in the beginning of 1994 hampered the unions' efforts to make the most of the political pluralism they were achieving.

1.3 According to two of Africa's leading lights in the trade union movement, Hassan Sunmonu, Secretary General of the Organization of African Trade Unity, and Andrew Kailembo, Secretary General of the Africa regional section of the International Confederation of Free Trade Unions, the 50 per cent devalution of the CFA Franc had profound consequences for the non-salaried, debilitating the unions' credibility.

1.4 There were initial protest strikes and demonstrations in Benin, Burkina, Gabon and Niger. Other unions held meetings and demands for wage increases ranging from 35 per cent in Chad and 50 per cent in Burkina, Mali, Senegal and Togo, to 100 per cent in Gabon. Trade unionism was an important element in precipitating an end to one-party regimes notably in Mali, Zambia and Malawi, and trade union pluralism has grown with the dismantling of one-party systems.

1.5 Nairobi-based ICFTU-AFRO has 44 affiliated organizations in 40 African countries and a membership of 27 million.

1.6 "Trade unions are saying that they played an important role during the pre-independence period, fighting against the colonial past. Now they want to see that proper democracy is established. When you have democracy, you also have free and independent democratic unions", said Sunmonu in the ICFTU 1995 annual meeting in Brussels.

1.7 Their fears that the 'social partners' are being ignored in the design of IMF and World Bank programmes are also underlined in a recent study by the Brussels-based ICFTU - which has 127 million members in 124 countries - of 13 French-speaking African countries. This study asserts that structural adjustment has set back trade unionism's ambitions of pluralism, as well as progress on labour rights.The offer of a 'docile workforce' facilitates external investment adds the study.

1.8 It also claims that the 1960s scenario is being repeated. This was when the politicians' quest for centralization to meet the goals of the initial five-year plans first destroyed pluralistic stuctures in French-speaking Africa. Trade unions, says the ICFTU, were swallowed up and private sector employers marginalized. The state assumed almost complete control of industrial relations.

1.9 The ICFTU organized several structural adjustment workshops in Ghana, Chad, Senegal, Gabon and Uganda during 1996. Participants included Ministries of Finance and Economics, representatives of civil society and officials of the World Bank and IMF. "Structural adjustment means sacrifice but how can we share the fruits of that sacrifice equitably. At present, less than 2 per cent of the people - who have not made any sacrifice - are reaping 98 per cent of the fruits" says Sunmonu.

1.10 According to the trade unions "there should be a safety net for those moving into the informal sector, so that they have something to live on. If you retrench overnight, you create chaos. In the African continent, where you have no social security, if you take away one person's livelihood, you are effectively retrenching 10-20 people".

1.11 African union leaders feel Europe can cooperate in supporting labour rights in Africa by linking it with trade. The idea is that if any government disrespects labour standards, sanctions may follow. Kailembo indicated that "Africa is on the threshold of a new political, economic and social era where trade unions can once again play the positive role they once enjoyed in the struggle for independence".

Women's Working Time
(Hours per week)

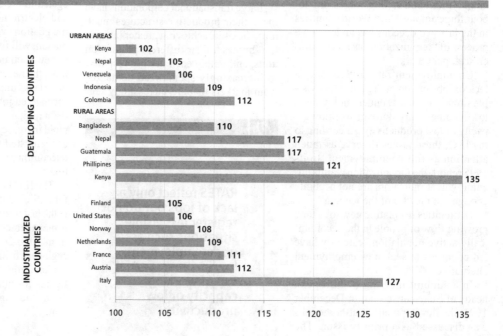

	Labor force (% of total population) 1990-93	Women in labor force (% of total) 1994	Percent of workers per economic sector			% Unemp. 1992-94		Labor force (% of total population) 1990-93	Women in labor force (% of total) 1994	Percent of workers per economic sector			% Unemp. 1992-94
			Agriculture 1990-92	Industry 1990-92	Services 1990-92					Agriculture 1990-92	Industry 1990-92	Services 1990-92	
Afghanistan	..	9	61	14	25		Laos	..	45	76	7	17	
Algeria	24	10	18	33	49	23.8	Lebanon	..	27	14	27	59	
Angola	..	38	73	10	17		Lesotho	..	43	23	33	44	
Antigua	45	..	..	..	..		Liberia	..	29	75	9	16	
Argentina	38	29	13	34	53	10.1	Libya	..	10	20	30	50	
Bahamas	45	34	5	4	91	13.3	Madagascar	..	38	81	6	13	
Bahrain	45	12	3	14	83		Malawi	43	40	87	5	8	
Bangladesh	47	41	59	13	28		Malaysia	38	36	26	28	46	
Barbados	41	46	7	11	82	21.9	Maldives	27	22	25	32	43	
Belize	.	26	.	.	.	11.1	Mali	..	15	85	2	13	
Benin	.	47	70	7	23		Mauritania	..	23	69	9	22	
Bhutan	.	32	92	3	5		Mauritius	41	26	16	30	54	
Bolivia	39	25	47	19	34	19	Mexico	39	28	23	29	48	3.8
Botswana	33	35	28	11	61		Mongolia	..	45	40	21	39	
Brazil	44	28	25	25	47	3.7	Morocco	..	21	46	25	29	
Brunei	..	33	.	.	.		Mozambique	..	47	85	7	8	
Burkina Faso	51	45	87	4	9		Myanmar/Burma	..	36	70	9	21	
Burundi	53	47	92	2	6		Namibia	..	24	43	22	35	
Cambodia	43	41	74	7	19		Nepal	..	32	93	1	6	
Cameroon	.	32	79	7	14		Nicaragua	35	30	46	16	38	14
Cape Verde	35	32	31	6	63		Niger	..	47	85	3	12	
Central African Rep.	48	45	81	3	16		Nigeria	31	35	48	7	45	
Chad	..	21	83	5	12		Oman	..	9	49	22	29	
Chile	39	29	19	26	55	5.9	Pakistan	28	13	47	20	33	4.7
China	..	43	73	14	13	2.8	Panama	48	28	27	14	59	13.8
China-Hong Kong	50	37	1	35	64	1.9	Papua New Guinea	..	38	76	10	14	
Colombia	45	23	10	24	66		Paraguay	45	20	48	21	31	4.4
Comoros	..	38	83	6	11		Peru	40	24	35	12	53	8.9
Congo	..	40	62	12	26		Philippines	..	31	45	16	39	8.4
Costa Rica	38	22	25	27	48	4.2	Qatar	..	7	3	28	69	
Côte d'Ivoire	39	34	65	8	27		Rwanda	46	46	90	2	8	
Cuba	44	33	24	29	47		St Vincent	39	..	..	..	..	
Cyprus	46	36	15	21	64	2.7	Samoa	..	37	..	..	..	
Djibouti	..	40	..	..	..		Saudi Arabia	..	7	48	14	37	
Dominica	38	42	31	13	56		Senegal	34	38	81	6	13	
Dominican Republic	..	15	46	15	39		Seychelles	44	42	..	..	..	
Ecuador	35	19	33	19	48	7.1	Sierra Leone	..	32	70	14	16	
Egypt	29	10	42	21	37	9	Singapore	49	36		35	65	2.6
El Salvador	41	28	11	23	66	7.9	Solomon	..	36	.	..	..	..
Equatorial Guinea	..	40	77	2	21		Somalia	..	38	76	8	16	
Ethiopia	41	36	88	2	10		South Africa	38	36	13	25	62	
Fiji	35	21	44	20	36	6	Sri Lanka	41	27	49	21	30	13.6
Gabon	..	37	75	11	14		Sudan	..	23	72	5	23	
Gambia	..	39	84	7	9		Suriname	48	30	20	20	60	16.3
Ghana	..	39	59	11	30		Swaziland	24	41	74	9	17	
Grenada	40	49	..	..	..		Syria	..	..	..	..	..	6.8
Guatemala	34	18	50	18	32		Tanzania	..	47	85	5	10	
Guinea	..	38	78	1	21		Thailand	56	44	67	11	22	2.7
Guinea-Bissau	..	39	82	4	14		Togo	..	35	65	6	29	
Guyana	..	26	27	26	47		Trinidad and Tobago	41	30	10	33	57	19.8
Haiti	41	41	68	9	23		Tunisia	30	24	26	34	40	
Honduras	35	21	38	15	47		Turkey	35	34	47	20	33	7.9
India	38	24	62	11	27		Uganda	..	40	86	4	10	
Indonesia	..	31	56	14	30		United Arab Emirates	..	9	5	38	57	
Iran	26	19	30	26	44		Uruguay	45	32	5	22	73	8.3
Iraq	24	22	14	19	67		Vanuatu	47	38	68	8	24	
Jamaica	45	46	26	24	50	15.9	Venezuela	36	28	13	25	62	6.4
Jordan	24	11	10	26	64		Vietnam	..	47	67	12	21	
Kenya	..	39	81	7	12		Yemen, Rep.	..	12	63	11	26	
Korea Dem. Rep.	..	46	43	30	27		Zaire	..	35	71	13	16	
Korea Rep.	..	34	17	36	47	2.4	Zambia	..	30	38	8	54	
Kuwait	39	23	..	26	73		Zimbabwe	..	33	71	8	21	

Sources: Yearbook of Labour Statistics 1995, ILO, 1995.

HABITAT

THE WORLD'S URBAN POPULATION is growing swiftly. For the first time in human history half of the earth's 6,500 million people will be living in cities by the year 2005.

Growing problems

In 1950, urban population amounted to just 29 per cent of world population. That year, 83 per cent of the population in developing countries lived in rural areas. The number of inhabitants of Third World countries grows at a much more sustained rate than that of the industrialized countries. In 1990, 77 per cent of the world's 5,300 million residents lived in developing countries.

2   Among the "rich" countries, population growth between 1970 and 1990 has been relatively moderate, with a 15 per cent increase. In those decades, the "poor" countries' population increased almost 55 per cent, from 2,650 million to 4,100 million. According to estimates by the United Nations Population Division, that gap will be enlarged and by the year 2025, approximately 84 per cent of human beings will live in the Third World.

3   Prior to World War II, population growth was low. Swift progress in health and sewage services has led to a reduction of death rates while birth rates have remained high. In southern countries, the proportion of young people is higher than in northern countries. This is directly reflected by a larger number of births in the former.

4   Even though urban areas as a whole have lower infant and children's death rates than rural areas, health levels among the cities' sub-populations vary widely. The poorest urban inhabitants, whom often live in illegal settlements, are exposed to crowding, housing deficiencies, contaminated water supplies, non-existent or scarce waste and sewage services, as well as industrial waste.

5   Large cities tend to have the biggest concentrations of water, sewage and health public services although 30 per cent to 60 per cent of the poorest urban inhabitants have no access to them. Children living in illegal settlements are 50 times as likely to die before the age of five as those living in developed countries.

6   Urban population increase in the Third World's 30,000 urban centers was uneven during the eighties. In general, the highest growth rates was evidenced in developing countries. 199 of the 281 cities with more than one million inhabitants are located in the 25 countries with the largest markets. It is estimated that 200 million Chinese will emigrate from rural areas to the cities in the next decade.

7   In an unprecedented development, some cities of the underdeveloped world have reached one million inhabitants: their population has increased tenfold in the last 4 decades, such as Amman, Curitiba, Dar es Salaam, Dhaka, Khartoum, Lagos and Nairobi.

8   In 1940, only New York and London had over five million people. In the 1990s, 22 cities surpassed the eight million figure, ten of which are in Asia. In 1995, 15 per cent of the urban population lived in cities with over 5 million inhabitants while over 60 per cent lived in cities with one million or less. Development and access to communications and transport are factors which either lead to the growth of mega-cities or their dispersion into urban regional groups.

9   World concern on the housing problem has led to a second United Nations Conference on Human Settlements (Habitat II), held in Istanbul in June 1996. Approximately 11,000 government delegates, NGO members, technicians and private agents met for 12 days in the last summit to be sponsored by the United Nations this century.

10   The magnitude of the housing problem led the conference documents to be bogged down in generalities, with a lack of contingency action plans and specific goals to eliminate the growing deterioration of living conditions among urban settlements. The Habitat Program points at two objectives: adequate housing for all and sustainable human settlements in an increasingly urbanized world.

11   Wally N'Dow, the secretary general of Habitat II, has stated: "This conference has recognized that life trends are changing worldwide and admitted that solutions should be sought locally. National governments and international organizations cannot solve nor take charge of the solutions of large-scale urban problems".

12   Around 500 million people lack housing or live in inadequate conditions. 20 per cent of the world population use 80 per cent of world resources due to problems arising from unsustainable production and consumption modes - particularly among industrialized countries.

13   "If every nation reduces its military budget by 5 per cent, the world could resolve its housing needs", said N'Dow.

Urban Population			(millions)		
Region	1950	1970	1990	2000	2025
The World	2,516	3,698	5,292	6,261	8,504
Industrialized countries	832	1,049	1,207	1,264	1,354
Developing countries	1,684	2,649	4,086	4,997	7,150
Africa	222	362	642	867	1,597
North America	166	226	276	295	332
Latin America	166	286	448	538	757
Asia	1,377	2,102	3,113	3,713	4,912
Europe	393	460	498	510	515
Oceania	13	19	26	30	38
Former USSR	180	243	289	308	352

Source: Population Division of the United Nations: World Population Prospects 1990. UN, New York, 1991.

HABITAT II

Everyday needs

In Habitat's Agenda, one of the documents signed at the Istanbul Conference, a global evaluation is made of development problems among cities and rural settlements, stressing unemployment, poverty and violence. Other problems include increasing migration, the situation of women, children, refugees and indigenous groups.

1.2 The conclusions emphasized the right to safe and healthy housing. Reference was also made to sustainable settlements, from the perspective of the ecosystem's capacity for equilibrium.

1.3 In order for housing projects to be feasible, the need for decentralization, citizen participation and government transparency was outlined. The document states that international cooperation is directly linked to the strengthening of access to financing and the creation of new alternatives.

1.4 Other peripheral issues that were addressed in the conference: lack of political participation - weakening democracy - urban violence, illegal evictions, barriers to land access, gender equality and youth access to housing.

1.5 In spite of good intentions, the general opinion that "large scale urbanization is the only way in which the world could survive large scale demographic growth", constitutes a way around the real issues.

Townbound

Some estimates indicate that migrations lead to 40-60 per cent of urban population's annual growth. Emigration is clearly seen in countries going through swift industrialization, like Southeast Asia. In other regions in which the population shift has already taken place - Latin America, Europe, North America - the incidence is less.

2.2 Cross-border migration is encouraged by the breakdown of economic barriers. Migration toward the north, from Central America to Mexico and from Mexico to the United States, grows each year. One fifth of El Salvador's population left the country during the last decade's civil war.

2.3 Though many migrants work in the informal sector, with low pay and little security, surveys suggest that the move to the city does improve their situation. One survey in New Delhi found that poor migrants from the countryside found their income was more than twice what they could earn in the village.

2.4 At the same time, the relative concentration of the poor in towns and cities is increasing. By the year 2000, United Nations estimates suggest that half the world's poorest people, or 420 million, will be living in urban settlements.

Miscalculations

Most UN estimates of the growth of cities during the 1970s have not reflected actual population behaviour. It was projected that Mexico City would have 31 million inhabitants by the year 2000. 1994 estimates indicate 16.4 million. Estimates for other cities have also been adjusted: Rio de Janeiro from 19.4 to 10.6 million, Calcutta from 19.7 to 12.7 million, Cairo from 16.4 to 10.7 million, and Seoul from 18.7 to 12.3 million. In Bombay's case, the estimates that 18 million people would live in the city by the turn of the century, have turned out to be correct.

Projected growth in the number of cities with more than 1 million inhabitants, 1950–2025

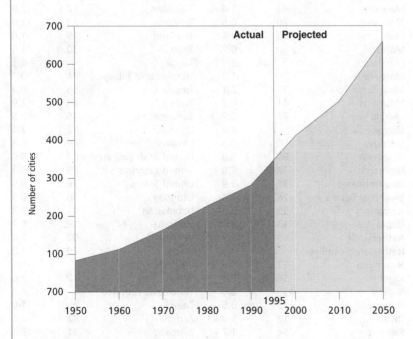

Source: World Bank Data

	% of total population 1993	% annual growth 2000		% of total population 1993	% annual growth 2000		% of total population 1993	% annual growth 2000
Afghanistan	19	1.6	Finland	62	0.4	Panama	53	0.7
Albania	37	1.8	France	73	0.6	Papua New Guinea	16	1.6
Algeria	54	1.4	French Polynesia		2.7	Paraguay	51	1.4
Andorra		4.6	Gabon	48	1.5	Peru	71	0.6
Angola	31	2.3	Haiti	30	1.9	Philippines	52	1.8
Antigua		0.3	Honduras	43	1.5	Poland	64	0.4
Aotearoa/NZ	86	0.8	Hungary	64	-0.5	Portugal	35	-0.6
Argentina	87	0.3	Iceland	91	1.2	Puerto Rico	73	0.8
Armenia	68	1.4	India	26	1.1	Qatar	91	0.4
Aruba		0.4	Indonesia	33	2.7	Reunion		1.5
Australia	85	1.5	Iran	58	1.0	Romania	55	0.0
Austria	55	0.7	Iraq	74	0.7	Russia	75	0.4
Azerbaijan	55	1.4	Ireland	57	0.0	Rwanda	6	1.9
Bahamas	86	0.6	Israel	90	2.9	St Kitts	41	1.4
Bahrain	89	0.5	Italy	67	0.2	St Lucia	47	1.0
Bangladesh	17	3.0	Jamaica	53	0.9	St Vincent	45	2.3
Barbados	46	1.2	Japan	77	0.4	Samoa	21	0.5
Belarus	69	0.4	Jordan	70	0.7	São Tomé & Principe	43	1.7
Belgium	97	0.3	Kazakhstan	59	1.0	Saudi Arabia	79	0.5
Belize	47	-0.1	Kenya	26	2.8	Senegal	41	1.2
Benin	30	1.5	Kiribati		2.0	Seychelles	53	1.7
Bermuda		1.3	Korea Dem. Rep.		0.5	Sierra Leone	35	2.1
Bhutan	6	3.8	Korea Rep.	78	1.4	Singapore	100	
Bolivia	59	1.5	Kuwait	97	0.4	Slovakia	58	0.4
Bosnia and Herzeg.		0.1	Kyrgyzstan	39	1.6	Slovenia	62	0.6
Botswana	26	3.5	Laos	20	2.9	Solomon Is.	16	3.0
Brazil	71	0.8	Latvia	72	-0.1	Somalia	25	1.4
Brunei	58	0.3	Lebanon	86	0.4	South Africa	50	0.8
Bulgaria	70	-0.8	Lesotho	22	3.2	Spain	76	0.2
Burkina Faso	23	6.9	Liberia	44	1.3	Sri Lanka	22	1.3
Burundi	7	3.5	Libya		0.6	Sudan	24	1.9
Cambodia	19	3.1	Lithuania	71	0.7	Suriname	47	1.3
Cameroon	43	1.9	Luxembourg	88	1.0	Swaziland		3.0
Canada	77	1.3	Macedonia	59	1.0	Sweden	83	0.6
Cape Verde	51	3.1	Madagascar	26	2.5	Switzerland	60	1.0
Central African Rep.	39	1.0	Malawi	13	2.9	Syria		0.8
Chad	21	1.1	Malaysia	52	1.4	Tajikistan	32	2.8
Chile	84	0.2	Maldives	26	0.9	Tanzania	23	2.9
China	29	2.7	Mali	26	2.4	Thailand	19	1.6
China-Hong Kong	95	0.2	Malta		0.7	Togo	30	1.6
China-Macau		3.2	Marshall Is.		3.9	Tonga		-0.3
Colombia	72	0.7	Martinique		1.0	Trinidad and Tobago	71	0.7
Comoros	30	1.9	Mauritania	51	2.1	Tunisia	56	0.9
Congo	57	1.6	Mauritius	41	0.3	Turkey	66	1.9
Costa Rica	49	1.1	Mexico	74	0.7	Turkmenistan	45	2.5
Côte d'Ivoire	42	1.4	Micronesia		2.4	Uganda	12	2.4
Croatia		0.4	Moldova	50	0.5	Ukraine	69	0.3
Cuba		0.6	Mongolia	60	1.0	United Arab Emirates	83	0.6
Cyprus		1.0	Morocco	47	1.0	United Kingdom	89	0.3
Czech Republic	65	0.0	Mozambique	31	3.9	United States	76	0.9
Denmark	85	0.2	Myanmar/Burma	26	1.4	Uruguay	90	0.3
Djibouti	82	0.5	Namibia	35	2.8	Uzbekistan	41	2.4
Dominica			Nepal	13	4.1	Vanuatu	19	1.1
Dominican Republic	63	1.2	Netherlands	89	0.7	Venezuela	92	0.4
Ecuador	57	1.2	Netherlands Antilles		0.9	Vietnam	20	1.2
Egypt	44	0.6	Nicaragua	62	0.9	Virgin Is. (Am)		-1.1
El Salvador	45	0.6	Niger	16	2.3	Yemen, Rep.	32	2.6
Equatorial Guinea	40	2.8	Nigeria	38	2.0	Yugoslavia Fed. Rep.		0.8
Estonia	73	0.1	Norway	73	0.5	Zaire	29	1.0
Ethiopia	13	1.9	Oman	12	3.4	Zambia	42	0.6
Fiji	40	0.8	Pakistan	34	1.7	Zimbabwe	31	2.3

Sources: World Development Report 1995, World Bank, 1995; Human Development Report 1996, UNDP, 1996.

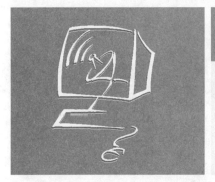

COMMUNICATIONS

THE ASSOCIATION for Progressive Communications is a global computer communications and information network dedicated to serving non-governmental organizations (NGOs) and citizens working for social justice, environmental sustainability and related issues.

APC: a global computer network for change

The Association for Progressive Communications (APC) is composed of a consortium of 21 international member networks. APC provides effective and efficient communications and information-sharing tools to NGOs and individuals. Through this global association, APC offers links of communication to over 40,000 NGOs, activists, educators, policy-makers, and community leaders in 133 countries.
2   APC member networks share a common mission: to develop and maintain the informational system that allows for geographically dispersed groups who are working for social and environmental change to coordinate activities on-line at a much cheaper rate than can be done by fax, telephone, or for-profit computer networks. APC is committed to making these tools available to people from all regions in the world.

Association for Progressive Comunication (APC)

Map of networks and connected systems. *(August, 1994)*

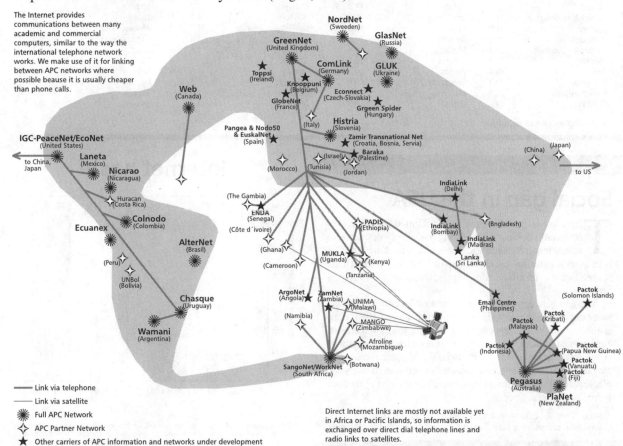

Communications

Telephones per 100 inhabitants (1993)

Bermuda	67.9	Antigua	28.9	Kazakhstan	9.1	Pakistan	1.3
Sweden	67.8	Latvia	26.8	Saudi Arabia	9.1	Philippines	1.3
Liechtenstein	62.4	Bulgaria	26.3	South Africa	9.0	Tuvalu	1.3
San Marino	61.3	Slovenia	25.9	Mexico	8.8	Yemen, Rep.	1.2
Switzerland	61.1	Netherlands Antilles	25.5	Azerbaijan	8.7	Zimbabwe	1.2
Virgin Is. (Am)	59.3	Aotearoa-Cook	25.1	Oman	8.6	Papua New Guinea	1.0
Canada	59.2	Turks and Caicos	24.9	Bolivia	7.5	India	0.9
Denmark	58.9	Kuwait	24.5	Brazil	7.5	Indonesia	0.9
United States	57.4	Virgin Is. (Br)	23.5	Kyrgyzstan	7.5	Zambia	0.9
France-St Pierre & M.	57.0	Estonia	23.2	Dominican Republic	7.4	Comoros	0.8
Finland	54.4	Bahrain	22.9	Fiji	7.1	Congo	0.8
Iceland	54.4	Lithuania	22.9	Jordan	7.0	Guinea-Bissau	0.8
Norway	54.2	Grenada	22.1	Turkmenistan	6.7	Kenya	0.8
Luxembourg	54.1	St Helena	21.9	Uzbekistan	6.6	Senegal	0.8
France	53.6	Croatia	21.5	Tonga	6.4	Côte d'Ivoire	0.7
China-Hong Kong	51.0	Qatar	21.4	France-Wallis and Futuna	6.3	Haiti	0.7
Netherlands	49.9	Brunei	19.7	Iran	5.9	Eritrea	0.6
Denmark-Faeroe	49.5	Dominica	19.1	Micronesia	5.6	Lesotho	0.6
Cyprus	49.4	Czech Republic	19.0	Ecuador	5.3	Angola	0.5
United Kingdom	49.4	Turkey	18.4	Guyana	5.1	Cameroon	0.5
Spain-Gibraltar	48.7	Yugoslavia Fed. Rep.	18.0	Tunisia	4.9	Benin	0.4
Australia	48.2	Belarus	17.6	Korea Dem. Rep.	4.8	Malawi	0.4
Japan	46.8	Uruguay	16.8	Libya	4.8	Mauritania	0.4
Guam	46.4	Slovakia	16.7	Tajikistan	4.6	Mozambique	0.4
Aotearoa/NZ	46.0	Seychelles	16.2	Namibia	4.5	Nepal	0.4
Germany	45.7	Russia	15.8	Marshall Is.	4.4	Togo	0.4
Greece	45.7	Armenia	15.6	Egypt	4.3	Vietnam	0.4
Austria	45.1	St Lucia	15.4	Maldives	4.2	Burundi	0.3
Cayman Is.	44.9	Samoa-American	15.4	Syria	4.1	Equatorial Guinea	0.3
Belgium	43.7	Trinidad and Tobago	15.0	Algeria	4.0	Ghana	0.3
Singapore	43.4	Ukraine	15.0	Cape Verde	3.8	Madagascar	0.3
Malta	43.0	Macedonia	14.8	Thailand	3.7	Nigeria	0.3
Andorra	42.2	St Vincent	14.8	Iraq	3.4	Tanzania	0.3
Italy	41.8	Hungary	14.6	Cuba	3.2	Bangladesh	0.2
Martinique	40.5	Belize	14.0	El Salvador	3.2	Bhutan	0.2
Montserrat	39.4	Bosnia and Herzeg.	13.7	Botswana	3.1	Burkina Faso	0.2
Anguilla	38.0	Malaysia	12.6	Morocco	3.1	Central African Rep.	0.2
Korea Rep.	37.8	Argentina	12.3	Paraguay	3.1	Ethiopia	0.2
Israel	37.1	Moldava	12.0	Peru	2.9	Guinea	0.2
Guadeloupe	36.7	Nauru	12.0	Mongolia	2.8	Laos	0.2
Spain	36.4	Suriname	11.6	Vanuatu	2.5	Liberia	0.2
China-Macau	34.4	Poland	11.5	Gabon	2.4	Rwanda	0.2
Puerto Rico	33.5	Romania	11.5	Guatemala	2.3	Sierra Leone	0.2
Ireland	32.8	Colombia	11.3	Kiribati	2.3	Somalia	0.2
United Arab Emirates	32.1	Costa Rica	11.1	Honduras	2.1	Sudan	0.2
Reunion	32.0	Chile	11.0	São Tomé & Principe	1.9	Afghanistan	0.1
Barbados	31.8	Jamaica	10.6	Swaziland	1.8	Cambodia	0.1
Denmark-Greenland	31.7	Monaco	10.6	Nicaragua	1.7	Chad	0.1
Portugal	31.1	Georgia	10.5	Gambia	1.6	Mali	0.1
Aruba	30.4	Panama	10.2	China	1.5	Niger	0.1
Bahamas	30.3	Venezuela	9.9	Solomon	1.5	Uganda	0.1
French Guyana	29.8	Mauritius	9.6	Albania	1.4	Zaire	0.1
Saint Kitts	29.6	Lebanon	9.3	Djibouti	1.3	Sri Lanka	0.0

Source: UN Statistical Yearbook, 40th issue, 1995.

Social gap in the USA

There are large differences in the United States in both household computer access and use of network services across income categories. These differences are due partly to other socioeconomic characteristics, but they remain highly significant even after controlling for those other characteristics.

[1,2] The gap between high-income and low-income individuals is not only large, it also widened between 1989 and 1993; higher-income individuals appear to be adopting the new technologies at a faster pace. Interestingly, the net differences are smaller for network usage than for household computer access. This may be due to broader access to network services in the workplace, where no investments in hardware on the part of the individual user are required.

Source: Tora K. Bikson and Constantjin W. A. Panis in "Universal access to e-mail Feasibility and Societal Implications"

Internet hosts

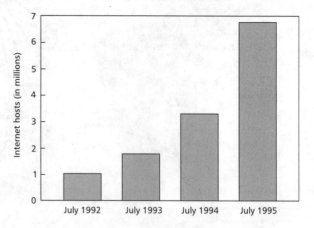

NOTE: Figures for July 1995 are estimates. Each Internet host represents from one to hundreds of users; the total number of individuals it connects is not known.
Source: The Internet Society, file ftp://ftp.isoc.org/charts/90s~host.txt

REFUGEES

LARGE-SCALE MOVEMENTS of refugees and other forced migrants have become a defining characteristic of the contemporary world. Few times in recent history have such large numbers of people in so many parts of the globe been forced to leave their own countries and communities to seek safety elsewhere.

Moving peoples

The number of people of concern to the United Nations High Commisioner for Refugees (UNHCR) has risen from 17 million in 1991 to more than 27 million at the beginning of 1995. Other estimates indicate that there are over 100 million people "on the move", one in every 60 human beings. Refugee populations in excess of 10,000 can be found in 70 countries around the world.

2 The term refugees has a specific definition in international law: "a person unable or unwilling to return to his|her homeland for fear of persecution based on reasons of race, religion, ethnicity, membership of a particular social group, or political opinion". Other circumstances also compel migration, such as denial of access to land, and the consequences of development models and economic systems which have failed to provide for people's most basic survival needs.

3 It is these unconventional "environmental" refugees that have shown most increase in recent years, until they may now number more than 25 million (10 million recognized, 15 million unrecognized) according to the NGO publication "People & the Planet".

4 The surge in refugee numbers is far outpacing the ability of the world community to respond. The annual budget of the UNCHR was increased to $1 billion in 1994, yet government costs to assist asylum seekers (a small part of the problem overall) has long surpassed $5 billion per year. UNCHR is quite unable to meet the demand for food and shelter, much less to invest in repatriation or rehabilitation of refugees.

5 Refugees often come from a home environment different from that in which they seek sanctuary. They bring alien customs, dietary preferences and religious practices. They tend to congregate in destitute sections of such megacities as Dhaka, Calcutta, Karachi, Lagos, São Paulo and Mexico City. In these localities, they depend on city services that are already over-loaded. Or they are obliged to crowd into refugee camps with inadequate subsistence conditions. Resettlement of refugees is never easy and assimilation is rare.

6 For many years, the refugee problem was considered as essentially African, Southeast Asian or Latin American. Nevertheless, the movements of refugees have increased in the east and center of Europe, the Caribbean, the Caucasus and the south of Asia. Within Africa, the most important migrations have moved from the north-east and the south into the regions west and center of the continent.

7 In the beginning of 1996, about 1.7 million Rwandans and more than 200,000 Burundians were confined in refugee camps in Zaire and Tanzania. Another 230,000 Rwandans remained in Burundi. Since the 1994 genocide in Rwanda when about 1 million people died, another 2 million Rwandans sought refuge in Zaire, Tanzania, Uganda and Burundi. Many people only feel safe in refugee camps. The absence of security and justice, plus the manipulation of news about the actual situation in Rwanda and Burundi, are the main obstacles preventing most refugees from returning voluntarily.

8 War in the former Yugoslavia caused the movement of 3,700,000 people. In Serbia, hundreds of thousands - Yugoslavs from Bosnia and Croatia, largely of Serb descent, but some Croat and Muslim - remained in Refugee Centres. The scarce job opportunities and the reduction in the rural areas of places suitable for settling have generated frictions that make the resettlement of the refugees difficult.

9 Thousands of other emigrés found sanctuary in different European countries - Germany alone has received more than 350,000 refugees from the former Yugoslavia - but in the majority of the cases only temporary permits have been extended. In recent years, many European countries have tightened their immigration laws. This trend coincides with the resurgence of xenophobic movements, some of whose members have carried out violent attacks against foreigners, some of whom were refugees.

10 The world's refugee burden is carried overwhelmingly by the poorest sectors of the global community. The 20 countries with the highest ratios of refugees have an annual per-capita income average of $700. Mozambique's neighbours have scant means to support people fleeing from drought and agricultural failure, both problems exacerbated by civil war. In Malawi - which has given sanctuary to 800,000 Mozambicans, and is among the dozen poorest countries with a per-capita GNP of $170 - one person in nine is a refugee.

11 Environmental refugees could become one of the foremost human crises of our

Minorities

1.1 In June,1996, more than 100,000 Nepalese who had formerly lived in Bhutan remained in Nepalese refugee camps. The "Citizenship Act" issued by the Bhutan government in 1985 and the subsequent harrassment and expulsion of many of the ethnic Nepalese is surely a unique case of preemptive ethnic cleansing. Many of those expelled from the country were legal foreigners of Nepalese origin.

1.2 Bhutan's monarchic government fears that this Nepalese minority - already over 40 per cent of the population - could become the motor for change, led to the decision to expel them. The harrassment, torture and murder of many pro-democracy activists forced thousands of Nepalese - who were forced to sign "Voluntary Migration Forms" - to abandon the country.

times. The phenomenon is an outward manifestation of profound change, often marked by extreme deprivation, fear and despair. While it derives from environmental problems, it is equally a crisis of social, political and economic dimensions.

[12] Countries such as Jordan, Egypt and Pakistan can find themselves suddenly suffering acute shortages of water. Countries such as the Philippines, Kenya and Costa Rica can, in the wake of ultra-rapid population growth, switch from land abundance to land shortage. Whole regions can find that their protective ozone layer is critically depleted within a single generation. The entire earth seems set to experience the rigours of global warming in what is, comparatively speaking, super-short order. Any of these environmental catastrophes can create refugees in exceptionally large numbers.

Immigration

Illegal immigration is a phenomenon that may share its roots with those of the refugees. The "rich" countries receive each year thousands of migrants seeking to integrate into societies capable of offering them better life conditions than the ones they can obtain in their countries of origin.

[2.2] The requirements for entering the United States and the majority of European countries get stricter day by day. The European Union has pressed some East European countries - such as Hungary and the Czech Republic - to tighten immigration controls, thus reducing the possibilites for these countries to become transit camps for migrants trying to flee from poverty. In April 1996 it was estimated that half a million future illegal immigrants were waiting in Eastern Europe for their chance to enter the US or any European Union country.

[2.3] According to the International Labour Office estimates, there are 2.6 million illegal migrants in Western Europe. The immigrant traffic has grown in the present decade. Hungarian police arrested, in 1995, 285 people involved in this business, doubling the arrests in the previous year.

The world's major refugee situations

UNHCR is providing protection and assistance to 27.4 million people around the world, of whom 14.5 million are refugees.

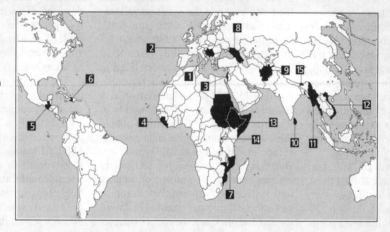

1
War in former Yugoslavia
Some 3.7 million people who have been displaced or affected by the war are receiving humanitarian assistance from the United Nations, 2.7 million of them in Bosnia and Herzegovina alone.

2
Asylum in Europe
Since the early 1980s, around five million applications for refugee status have been submitted in Western Europe. UNHCR tries to ensure that any measures taken to control this phenomenon are consistent with the principles of refugee protection.

3
The Palestinian question
Around 2.8 million people are registered with UNRWA, the agency responsible for Palestinian refugees, Their future remains one of the most complex issues which must be addressed in the Middle East peace process.

4
West African refugees
The conflicts in Liberia and Sierra Leone have forced almost a million people into exile in Guinea and Côte d'Ivoire. Large numbers are also displaced within their own countries, beyond the reach of international assistance.

5
Guatemalan repatriation
Some 20,000 Guatemalans have returned to their homeland over the past 10 years. Up to a quarter of the 45,000 who remain in Mexico are expected to repatriate in 1995 with assistance from UNHCR.

6
Haitian asylum seekers
UNHCR is assisting with procedures designed to determine the status of asylum seekers from Haiti and to monitor the situation of those who return.

7
Reintegration in Mozambique
More than 1.6 million refugees returned to Mozambique from six neighbouring states between late 1992 and early 1995. They must now begin to support themselves and to reintegrate within their own communities.

8
Conflicts in the Caucasus
Recent years have witnessed a succession of population displacements within and between Armenia, Azerbaijan, Georgia and the Russian Federation, involving around 1.5 million people. Many of this number are unable or unwilling to return to their former place of residence.

9
Reconstruction in Afghanistan
Half of the Afghan refugees have been repatriated since 1992, leaving nearly three million in Iran and Pakistan. Additional reconstruction efforts are needed within Afghanistan to enable their return.

10
Displaced Sri Lankans
More than 30,000 Sri Lankan refugees have returned from India since 1992, leaving nearly 75,000 in their country of asylum. UNHCR provides assistance to the returnees and to other people who are threatened or displaced by the war.

11
Repatriation to Myanmar/Burma
By mid-1995, only 50,000 of the 250,000 people who fled from Myanmar/Burma in 1991 and 1992 remained in Bangladesh. The homeward movement, organized by UNHCR, is scheduled for completion by the end of the year.

12
Vietnamese boat people
Although the departure of boat people has effectively come to a halt, just over 40,000 Vietnamese asylum seekers remain in camps throughout South-East Asia. More than 70,000 have gone back to their own country, where their situation is monitored by UNHCR.

13
The Horn of Africa: exile and repatriation
UNHCR continues to assist around 1.6 million people from the Horn of Africa and the Sudan, traditionally one of the most important refugee-producing regions. The repatriation to Eritrea from Sudan is finally under way, more than 30 years after the first refugees left that country.

14
The Rwanda/Burundi emergency
More than a million Rwandans poured into Zaire in mid-1994, one of the largest and fastest refugee movements ever seen. UNHCR is now providing protection and assistance to some 2.2 million displaced people in Burundi, Rwanda, Tanzania, Uganda and Zaire.

15
Ethnic cleansing in Bhutan
In 1985 the government began to expel the Nepalese, many of them with Bhutanese citizenship. More than 100.000 live now in refugee camps in Nepal.

SOCIAL DEVELOPMENT

THE WORLD SUMMIT for Social Development (WSSD) in Copenhagen made a solemn and public commitment to eradicate poverty and to achieve equality between women and men. The heads of state and governments stated that they were committed to these goals as an "ethical, social, political and economic imperative of humankind".

Some promises of help

Social development, full employment and the well-being of humanity are objectives included in the UN Charter since the creation of the organization half a century ago. Placing "people at the centre of our concerns" was the major achievement of WSSD. The World Summit for Social Development (WSSD) recognized in 1995 that the market by itself does not solve social problems and that economic growth will not by itself provide full employment, education, health care and other social services.

[2] Therefore, among other things, the policies of the IMF (International Monetary Fund) and the World Bank will have to be adjusted. The loans granted to poor countries under Structural Adjustment Programs (SAPs) have put a heavy burden on social expenditure in many countries of the South. Poor countries need to be supported in striving for social development.

[3] But the Social Summit, while acknowledging these principles, did not yield concrete measures to increase aid or reduce debts. The meeting did encourage countries to spend 20 per cent of Official Development Assistance (ODA) and 20 per cent of the budget of the recipient country on social provisions. However, this plan means that capital needed for social development can be found by making changes in the existing budgets of both the donor and the recipient country.

Social Development

[4] According to UN secretary-general Boutros Boutros-Ghali, "Social Development should be understood in a broad sense, implying progress towards higher living standards, greater equality of opportunity and securing of certain basic human rights... enhancing the abilities of individuals to control their own lives through economic, social and political actions..." In development thinking, social, economic, political and cultural aspects are often described as separate fields. Economic development problems are often referred to as "hard" issues, whereas social subjects are designated as "soft" issues.

[5] In reality, however, the division between policy fields is not so obvious. They are strongly interrelated. The most recent World Development Report of the World Bank attributes the success of rising employment, decreasing poverty and decreasing income inequality in East Asia to a combination of export-oriented economic growth and high investments in education, health care and nutrition of the population.

[6] Social development is primarily the responsibility of national governments

Basic measures

The following measures are usually considered as the basis of a social development policy:
• Providing basic health care (including family planning), primary education, food security, clean drinking water and sanitation;
• Income-generating and income-supporting activities for the poor (especially women): small-scale credit facilities, work-guarantee programs, agricultural extension programs and support to small-scale agricultural production.
• Strengthening social organizations; eg farmers' associations, women's organisations, cooperatives, trade unions, human rights organizations.

and, in economic terms, it is ultimately a problem of distribution. One of the most simple income distribution indicators is obtained by comparing the share of national wealth perceived by the top fifth of the population with the lower fifth. Brazil heads the list of countries with most inequitable distribution according to this ratio (among the countries for which data are available). Thus, even when the national income average of Brazil places the country among the world middle-class, the number of Brazilians living in poverty is several dozens of millions. At the other end of the scale, very poor countries like India or Bangladesh distribute better their national income.

[7] The basic data to monitor progress towards social development targets are available for many countries. But sometimes they are not. Very frequently this lack of data is in itself an indicator. The follow-up of the UN Conference on Environment and Development had to invent from scratch indicators for the new concept of sustainable development. And actual numbers and indexes still have to be produced.

[8] The first conditionality imposed by the Bretton Woods Institutions on the Countries in Transition when they requested their assistance was to produce reliable economic indicators. Yet, the World Bank lacks the very basic demographic and social indicators of many of the countries it assists. Is structural adjustment helping or hurting the poor? Data are not available to support a definitive conclusion. Is this anti-poverty strategy efficient? The situation at the start is not known, neither that at the end. The disbursement of loans might be stopped if a country fails to produce or distorts the figures about the economy, but nothing happens if budget cuts force countries to postpone a census one or two years therefore making it difficult to assess the situation of the people.

Source: Social Watch, "The starting point", Instituto del Tercer Mundo, Montevideo, 1996.

		GNP per capita US$ 1994			GNP per capita US$ 1994			GNP per capita US$ 1994
1	Luxembourg	39,600	53	Botswana	2,800	105	Armenia	680
2	Switzerland	37,930	54	Dominica	2,800	106	Cameroon	680
3	Japan	34,630	55	Venezuela	2,760	107	Sri Lanka	640
4	Denmark	27,970	56	Russia	2,650	108	Kyrgyzstan	630
5	Norway	26,390	57	Grenada	2,630	109	Congo	620
6	United States	25,880	58	Panama	2,580	110	Côte d'Ivoire	610
7	Germany	25,580	59	Croatia	2,560	111	Honduras	600
8	Austria	24,630	60	Belize	2,530	112	Senegal	600
9	Iceland	24,630	61	Turkey	2,500	113	China	530
10	Sweden	23,530	62	Poland	2,410	114	Guyana	530
11	France	23,420	63	Thailand	2,410	115	Guinea	520
12	Belgium	22,870	64	Costa Rica	2,400	116	Comoros	510
13	Singapore	22,500	65	Latvia	2,320	117	Azerbaijan	500
14	Netherlands	22,010	66	Fiji	2,250	118	Zimbabwe	500
15	China-Hong Kong	21,650	67	Slovakia	2,250	119	Mauritania	480
16	Canada	19,510	68	Belarus	2,160	120	Equatorial Guinea	430
17	Kuwait	19,420	69	Peru	2,110	121	Pakistan	430
18	Italy	19,300	70	Namibia	1,970	122	Ghana	410
19	Finland	18,850	71	Ukraine	1,910	123	Bhutan	400
20	United Kingdom	18,340	72	Tunisia	1,790	124	Albania	380
21	Australia	18,000	73	Colombia	1,670	125	Benin	370
22	Israel	14,530	74	Algeria	1,650	126	Central African Rep.	370
23	Brunei	14,240	75	Tonga	1,590	127	Tajikistan	360
24	Ireland	13,530	76	Paraguay	1,580	128	Zambia	350
25	Spain	13,440	77	Jamaica	1,540	129	Nicaragua	340
26	Aotearoa/NZ	13,350	78	Jordan	1,440	130	Gambia	330
27	Qatar	12,820	79	El Salvador	1,360	131	India	320
28	Bahamas	11,800	80	Lithuania	1,350	132	Laos	320
29	Cyprus	10,260	81	Dominican Republic	1,330	133	Togo	320
30	Portugal	9,320	82	Ecuador	1,280	134	Burkina Faso	300
31	Korea Rep.	8,260	83	Romania	1,270	135	Mongolia	300
32	Argentina	8,110	84	Bulgaria	1,250	136	Nigeria	280
33	Greece	7,700	85	Papua New Guinea	1,240	137	Yemen, Rep.	280
34	Bahrain	7,460	86	Guatemala	1,200	138	Kenya	250
35	Saudi Arabia	7,050	87	Kazakhstan	1,160	139	Mali	250
36	Slovenia	7,040	88	Vanuatu	1,150	140	São Tome & Principe	250
37	Antigua	6,770	89	Morocco	1,140	141	Guinea Bissau	240
38	Seychelles	6,680	90	Swaziland	1,100	142	Haiti	230
39	Barbados	6,560	91	Samoa	1,000	143	Niger	230
40	Oman	5,140	92	Uzbekistan	960	144	Bangladesh	220
41	Uruguay	4,660	93	Maldives	950	145	Madagascar	200
42	Mexico	4,180	94	Philippines	950	146	Nepal	200
43	Gabon	3,880	95	Cape Verde	930	147	Vietnam	200
44	Hungary	3,840	96	Indonesia	880	148	Uganda	190
45	Trinidad and Tobago	3,740	97	Moldova	870	149	Chad	180
46	Chile	3,520	98	Suriname	860	150	Malawi	170
47	Malaysia	3,480	99	Macedonia	820	151	Burundi	160
48	Czech Republic	3,200	100	Solomon	810	152	Sierra Leone	160
49	Mauritius	3,150	101	Bolivia	770	153	Tanzania	140
50	South Africa	3,040	102	Kiribati	740	154	Ethiopia	100
51	Brazil	2,970	103	Egypt	720	155	Mozambique	90
52	Estonia	2,820	104	Lesotho	720	156	Rwanda	80

Source: World Development Indicators 1996, World Bank, 1996

The international calendar

A checklist of targets, mainly in the field of basic health, primary education and sanitation was also agreed in Copenhagen and Beijing.
For example:

- By the year 2000, life expectancy of not less than 60 years in any country;

- By the year 2000, reduction of mortality rates for infants and children under five years of age by one third of the 1990 level, or 50 to 70 per 1,000 live births, whichever is less; by the year 2015, achievement of an infant mortality rate below 35 per 1,000 live births and an under five mortality rate below 45 per 1,000;

- By the year 2000, a reduction in maternal mortality by one half of the 1990 level; by the year 2015 a further reduction by one half.

Some social development goals for the Year 2000

The Northern countries agreed to devote at least 0.7 per cent of their GNP to international cooperation. However, official development aid from OECD countries dropped close to 2 per cent in real terms from 1992 to 1993. This was the result of major cutbacks in aid budgets by numerous member countries, particularly the two most important contributors, the US and Japan. (See aid chapter).

2.2 At the country level, all OECD members have reduced the percentage of GNP allocated to financing both to developing countries and to international organizations, as compared with the beginning of the 1980s. The drop was particularly significant in the cases of Switzerland, France, Denmark and the United Kingdom.

2.3 The only OECD countries granting 0.7 per cent or more of their GNP to such financing are the Netherlands, Sweden, Switzerland, Denmark and Norway. The countries that allocate most of such resources to developing countries are Norway, Denmark, Sweden and the Netherlands.

2.4 Only a very small part of the Official Development Assistance (ODA) is spent on social development. According to data provided by UNICEF, the proportion for basic social priorities only amounts to 10 per cent of the total aid flow.

2.5 The proportion of the aid spent on primary education in Sub-Saharan Africa even declined during the 1980s. ODA is often used for higher vocational education rather than for primary education, for expensive hospitals rather than basic health care and for expensive urban facilities rather than cheap rural provisions. Even when the policy aims at unconditional spending of the aid one often finds that a large part of the bilateral aid is spent in the donor country. The result is that between 60 and 90 per cent of the bilateral aid is currently spent on goods and services of the donor country.

2.6 Some development aid payments and loans are even counter-productive to social development. The enormous resettlement programs needed to execute the building of big dams is the most obvious example.

2.7 Meanwhile, private funds have quadrupled over the last five years, although most of them are geared to certain sectors and regions (China received 38 per cent of such funding in 1993) and their relevance to social development still has to be proven.

20/20 Compact

2.8 UNICEF and UNDP launched a plan for the Social Summit to spend a greater proportion of the aid on basic social provisions. The so-called 20/20 compact means that on average one fifth of the funds provided by the donor community is spent on basic social priorities: education, health care, family planning, clean water and sanitation, and that developing countries, too, spend one fifth of their national budget on these priorities.

2.9 In its annual report for 1994, UNDP gave figures for basic social priorities, which show that overall bilateral development aid only allocates 7 per cent of the budget to these priorities. Developing countries do better, namely 16 per cent. Here it must be observed that it is not only low expenditure in the social sector which causes problems, but in particular the lopsided spending in that sector. For instance of the $7 billion which governments in developing countries spend each year on education, less than 20 per cent goes towards primary education.

Human development index

1993 Countries ranked by social development (includes life expectancy, literacy etc.)

1	Canada	0.951	88	Samoa	0.700
2	United States	0.940	89	Sri Lanka	0.698
3	Japan	0.938	90	Turkmenistan	0.695
4	Netherlands	0.938	91	Peru	0.694
5	Norway	0.937	92	Syria	0.690
6	Finland	0.935	93	Armenia	0.680
7	France	0.935	94	Uzbekistan	0.679
8	Iceland	0.934	95	Philippines	0.665
9	Sweden	0.933	96	Azerbaijan	0.665
10	Spain	0.933	97	Lebanon	0.664
11	Australia	0.929	98	Moldavia	0.663
12	Belgium	0.929	99	Kyrgyzstan	0.663
13	Austria	0.928	100	South Africa	0.649
14	Aotearoa/NZ	0.927	101	Georgia	0.645
15	Switzerland	0.926	102	Indonesia	0.641
16	United Kingdom	0.924	103	Guyana	0.633
17	Denmark	0.924	104	Albania	0.633
18	Germany	0.920	105	Tajikistan	0.616
19	Ireland	0.919	106	Egypt	0.611
20	Italy	0.914	107	Maldives	0.610
21	Greece	0.909	108	China	0.609
22	China-Hong Kong	0.909	109	Iraq	0.599
23	Cyprus	0.909	110	Swaziland	0.586
24	Israel	0.908	111	Bolivia	0.584
25	Barbados	0.906	112	Guatemala	0.580
26	Bahamas	0.895	113	Mongolia	0.578
27	Luxembourg	0.895	114	Honduras	0.576
28	Malta	0.886	115	El Salvador	0.576
29	Korea Rep.	0.886	116	Namibia	0.573
30	Argentina	0.885	117	Nicaragua	0.568
31	Costa Rica	0.884	118	Solomon Is.	0.563
32	Uruguay	0.883	119	Vanuatu	0.562
33	Chile	0.882	120	Gabon	0.557
34	Singapore	0.881	121	Vietnam	0.523
35	Portugal	0.878	122	Cape Verde	0.539
36	Brunei	0.872	123	Morocco	0.534
37	Czech Republic	0.872	124	Zimbabwe	0.534
38	Trinidad and Tobago	0.872	125	Congo	0.517
39	Bahrain	0.866	126	Papua New Guinea	0.504
40	Antigua	0.866	127	Cameroon	0.481
41	Slovakia	0.864	128	Kenya	0.473
42	United Arab Emirates	0.864	129	Ghana	0.467
43	Panama	0.859	130	Lesotho	0.464
44	Venezuela	0.859	131	Equatorial Guinea	0.461
45	St Kitts	0.858	132	São Tome & Principe	0.458
46	Hungary	0.855	133	Myanmar/Burma	0.451
47	Fiji	0.853	134	Pakistan	0.442
48	Mexico	0.845	135	India	0.436
49	Colombia	0.840	136	Zambia	0.411
50	Qatar	0.839	137	Nigeria	0.400
51	Kuwait	0.836	138	Laos	0.400
52	Thailand	0.832	139	Comoros	0.399
53	Malaysia	0.826	140	Togo	0.385
54	Mauritius	0.825	141	Zaire	0.371
55	Latvia	0.820	142	Yemen, Rep.	0.366
56	Poland	0.819	143	Bangladesh	0.365
57	Russia	0.804	144	Tanzania	0.364
58	Brazil	0.796	145	Haiti	0.359
59	Libya	0.792	146	Sudan	0.359
60	Seychelles	0.792	147	Côte d'Ivoire	0.357
61	Belarus	0.787	148	Central African Rep.	0.355
62	Bulgaria	0.773	149	Mauritania	0.353
63	Saudi Arabia	0.771	150	Madagascar	0.349
64	Ecuador	0.764	151	Nepal	0.332
65	Dominica	0.764	152	Rwanda	0.332
66	Iran	0.754	153	Senegal	0.331
67	Belize	0.754	154	Benin	0.327
68	Estonia	0.749	155	Uganda	0.326
69	Algeria	0.746	156	Cambodia	0.325
70	Jordan	0.741	157	Malawi	0.321
71	Botswana	0.741	158	Liberia	0.311
72	Kazakhstan	0.740	159	Bhutan	0.307
73	St Vincent	0.738	160	Guinea	0.306
74	Romania	0.738	161	Guinea-Bissau	0.297
75	Suriname	0.737	162	Gambia	0.292
76	St Lucia	0.733	163	Chad	0.291
77	Grenada	0.729	164	Djibouti	0.287
78	Tunisia	0.727	165	Angola	0.283
79	Cuba	0.726	166	Burundi	0.282
80	Ukraine	0.719	167	Mozambique	0.261
81	Lithuania	0.719	168	Ethiopia	0.237
82	Oman	0.716	169	Afghanistan	0.229
83	Korea Dem. Rep.	0.714	170	Burkina Faso	0.225
84	Turkey	0.711	171	Mali	0.223
85	Paraguay	0.704	172	Somalia	0.221
86	Jamaica	0.702	173	Sierra Leone	0.219
87	Dominican Republic	0.701	174	Niger	0.204

Source: Human Development Report 1996, UNDP, 1996.

Measuring health, sanitation and education in sub-Saharan Africa

The mortality rate among children under 5 years of age is used by UNICEF as the chief indicator for measuring the level of and the changes in children's well-being. It is taken to be the result of the status of nutritional health and basic health knowledge of the mother; a certain level of immunization coverage and use of oral rehydration therapy; access to maternal-infant care (including prenatal care); certain availability of income and food per family; access to safe drinking water and effective sanitation; and a degree of security in the child's environment.

[2] Sub-Saharan Africa has the highest rates of under-five mortality. The situation is worse in Niger, Angola, Sierra Leone and Mozambique. Even in countries like Madagascar, Zambia and Zimbabwe that risk has increased in recent years. Life expectancy in the region continues to be the shortest in the world. One third of the child population suffers from rickets, and malnutrition does not let up. The vast majority of the population does not have access to health in Benin, Zaire and Somalia.

[3] Slightly under half the region's population has access to safe drinking water. In countries like the Central African Republic, Chad, Ethiopia, Zaire, Madagascar and Angola less than one third of the inhabitants have access to drinking water. Access to sanitation is practically nonexistent in Madagascar, Sierra Leone, Niger, Angola, Burkina Faso, Somalia and Ethiopia.

[4] The school enrolment rate for children 6 to 11 years of age has dropped in recent years in almost half the countries of the region. Although the gap in the primary school enrolment rate by sex is closing, it continues to be significant. The literacy rate among adults is extremely low in Niger, Burkina Faso, Somalia and Mali, while gross school enrolment of women does not reach one-fifth in Somalia, Ethiopia and Mali. In such circumstances, compliance with the specific goals of WSSD and WCW seems unlikely.

Under-five mortality (UNICEF) 1994

	Under-five mortality rate (per 1,000 live births)			Under-five mortality rate (per 1,000 live births)			Under-five mortality rate (per 1,000 live births)			Under-five mortality rate (per 1,000 live births)
1	Niger	320	39	India	119	77	Vietnam	46	115 Poland	16
2	Angola	292	40	Nepal	118	78	Dominican Republic	45	116 Chile	15
3	Sierra Leone	284	41	Bangladesh	117	79	China	43	117 Malaysia	15
4	Mozambique	277	42	Senegal	115	80	Albania	41	118 Slovakia	15
5	Afghanistan	257	43	Yemen	112	81	Lebanon	40	119 Croatia	14
6	Guinea-Bissau	231	44	Indonesia	111	82	Syria	38	120 Hungary	14
7	Guinea	223	45	Bolivia	110	83	Moldova	36	121 Kuwait	14
8	Malawi	221	46	Cameroon	109	84	Saudi Arabia	36	122 Jamaica	13
9	Liberia	217	47	Congo	109	85	Paraguay	34	123 Portugal	11
10	Mali	214	48	Myanmar/Burma	109	86	Tunisia	34	124 Belgium	10
11	Gambia	213	49	Libya	95	87	Armenia	32	125 Cuba	10
12	Somalia	211	50	Papua New Guinea	95	88	Macedonia	32	126 Czech Republic	10
13	Zambia	203	51	Kenya	90	89	Mexico	32	127 Greece	10
14	Chad	202	52	Turkmenistan	87	90	Thailand	32	128 United States	10
15	Eritrea	200	53	Tajikistan	81	91	Korea Dem. Rep.	31	129 Aotearoa/NZ	9
16	Ethiopia	200	54	Zimbabwe	81	92	Russia	31	130 France	9
17	Mauritania	199	55	Namibia	78	93	Romania	29	131 Israel	9
18	Bhutan	193	56	Mongolia	76	94	Argentina	27	132 Korea Rep.	9
19	Nigeria	191	57	Iraq	71	95	Georgia	27	133 Spain	9
20	Zaire	186	58	Guatemala	70	96	Oman	27	134 Australia	8
21	Uganda	185	59	Nicaragua	68	97	Latvia	26	135 Canada	8
22	Cambodia	177	60	South Africa	68	98	Jordan	25	136 Italy	8
23	Burundi	176	61	Algeria	65	99	Ukraine	25	137 Netherland	8
24	Central Afr. Rep.	175	62	Uzbekistan	64	100	Venezuela	24	138 Norway	8
25	Burkina Faso	169	63	Brazil	61	101	Estonia	23	139 Slovenia	8
26	Madagascar	164	64	Peru	58	102	Mauritius	23	140 Austria	7
27	Tanzania	159	65	Ecuador	57	103	Yugo. Fed. Rep.	23	141 Denmark	7
28	Lesotho	156	66	Philippines	57	104	Belarus	21	142 Germany	7
29	Gabon	151	67	El Salvador	56	105	Uruguay	21	143 Ireland	7
30	Côte d'Ivoire	150	68	Kyrgyzstan	56	106	Lithuania	20	144 Switzerland	7
31	Benin	142	69	Morocco	56	107	Panama	20	145 United Kingdom	7
32	Rwanda	139	70	Turkey	55	108	Trinidad and Tbgo.	20	146 China-Hong Kong	6
33	Laos	138	71	Botswana	54	109	Unit. Arab Emirates	20	147 Japan	6
34	Pakistan	137	72	Honduras	54	110	Bulgaria	19	148 Singapore	6
35	Togo	132	73	Egypt	52	111	Colombia	19	149 Finland	5
36	Ghana	131	74	Azerbaijan	51	112	Sri Lanka	19	150 Sweden	5
37	Haiti	127	75	Iran	51	113	Bosnia and Herz.	17		
38	Sudan	122	76	Kazakhstan	48	114	Costa Rica	16		

Source: The State of the World's Children 1996, UNICEF, 1996.

The World according to...

... The World Bank Ranked by income (GNP per capita). See page 48.

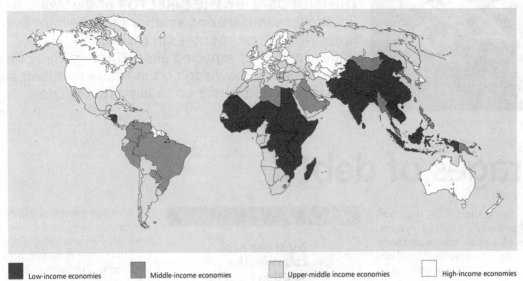

Low-income economies   Middle-income economies   Upper-middle income economies   High-income economies

... UNDP (UN development program) Ranked by Human Development Index. See page 49.

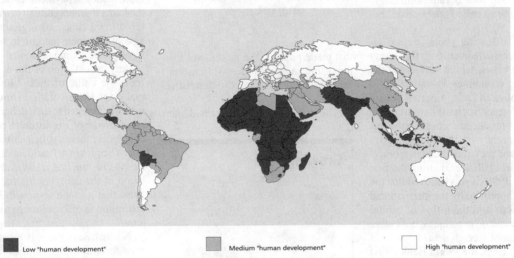

Low "human development"   Medium "human development"   High "human development"

... UNICEF (UN children's agency) Ranked by under-5 mortality rate. See page 50.

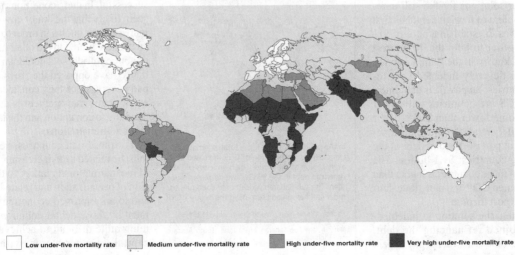

Low under-five mortality rate   Medium under-five mortality rate   High under-five mortality rate   Very high under-five mortality rate

DEBT

THE PROPOSAL BY THE DIRECTOR of the World Bank, Jim Wolfensohn, seeking solutions for the debt burdens of the 40 poorest countries in the world, did not achieve any of its expected aims. The very idea of a new plan implied acknowledging the failure of a long series of loans granted earlier on to alleviate the crisis.

Mirages of debt

During the 1970s, the international banking system, which had large capital seeking investments, stimulated the Southern countries to borrow in order to carry out ambitious infrastructural works. Governments incurred heavy debt and, by the middle of the following decade, the situation became unmanageable, and the so called "debt crisis" began.

[2] In the following years several attempts were made in order to find a way out of the crisis, which suffocated the developing economies. Privatization - a recipe applied in Latin America as well as in the former Socialist countries and in Africa - produced a considerable inflow of currency for the Southern governments and they succeeded in modifying their debt balances. During the first half of the 1990s it was believed that the soft loans approved by the international organizations and the influx of capital from privatization programs would solve the problem of foreign debt.

[3] In July, 1995, the World Bank (WB) shuffled its management. Jim Wolfensohn, its new Director, concentrated on looking for solutions for the still unresolved problem of foreign debt. His first initiative was to establish a pool of $11 billion in order to help the 40 poorest countries - known in the Bank jargon as SILICS, or Severely Indebted Low Income Countries - to pay their debt interests. SILICS are countries with a per capita income lower than $695. Their combined debt totals $160,000 million - a substantial part of the total debt of the developing countries: $1.1 billion. The total debt of those countries is higher than their combined GDP and more than double their export income.

[4] Until then, the solution to indebtedness combined refinancing the debt, granting new loans in order to pay unpaid quotas, and a series of economic reforms

> **For those rich countries the solution must be related to an increasing openness of the international market...**

that made possible the increase of their importing capacities. But Wolfensohn's proposals did not have the desired results.

[5] The rich countries that granted loans

Growing debt

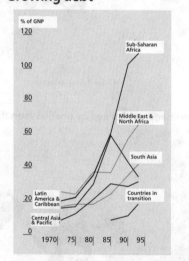

In the period 1971-1993, the total external debt expressed as a proportion of GNP, increased in every region of the world. The high levels of Latin American debt in the 1980s have decreased since then. The sub-Saharan African debt continues to grow and has already surpassed the region's GNP.

Source: World Bank, World Debt Tables 1994-95, vol. 2, Washington DC, 1994.
Note: In the Sub-Saharan Africa data, South Africa is not included.

for development had already decided to reduce their help. On the other hand, the Bank itself is the creditor of at least a fifth (or $30,000 million) of the poorest countries' debt. (The International Monetary Fund is the creditor of another fifth of that debt, and, in the particular case of the African countries, the institution receives more money in payment for the debt than the loans it grants annually). And the international financial organizations do also obtain their income in the market through the selling of bonds.

[6] Any cancellation of the debt that affects the value of these bonds would make for losses, which should themselves be charged to their debtors through an increase of the interest rates, thus countering their explicit initial motivation. The previous President of the Bank, Lewis Preston, had resisted any cancellation of the debt, but his successor affirms that it is the responsibility of the institution to offer solutions to the problem.

[7] Nevertheless, the only mention of the need for a new plan implies the admission that a long series of loans granted earlier on to alleviate the crisis were unsuccessful. In these conditions it does not seem likely that the donor countries will increase their quotas in order to strengthen the new fund. On the other hand some donor countries, such as Japan and Germany, have opposed the "rewarding" of bad payers, since they consider that bad econonomic management is due fundamentally to corruption and the inefficiency of administrations.

[8] For those rich countries the solution must be related to an increasing openness of the international market, to the extension of privatization and related reforms, and to an improved economic management by those debtor countries - something quite difficult to achieve without capital, in other words, without new loans.

	Total external debt (million $) 1994	Debt service as % of exports 1994	External debt $ per capita 1994		Total external debt (million $) 1994	Debt service as % of exports 1994	External debt $ per capita 1994
Albania	925	2.5	289	Lesotho	600	16.9	309
Algeria	29,898	56.0	1,090	Lithuania	438	2.8	118
Argentina	77,388	35.1	2,263	Macedonia, FYR	924	12.7	440
Armenia	214	1.7	57	Madagascar	4,134	9.5	316
Azerbaijan	113	0.0	15	Malawi	2,015	17.4	211
Bangladesh	16,569	15.8	140	Malaysia	24,767	7.9	1,259
Belarus	1,272	4.3	123	Mali	2,781	27.5	292
Benin	1,619	10.1	304	Mauritania	2,326	23.3	1,050
Bolivia	4,749	28.2	656	Mauritius	1,355	7.3	1,215
Botswana	691	4.3	479	Mexico	128,302	35.4	1,449
Brazil	151,104	35.8	950	Moldova	492	2.2	113
Bulgaria	10,468	14.0	1,241	Mongolia	443	9.6	187
Burkina Faso	1,125	..	111	Morocco	22,512	33.3	854
Burundi	1,125	41.7	182	Mozambique	5,491	23.0	355
Cameroon	7,275	16.7	560	Myanmar/Burma	6,502	15.4	143
Central African Rep.	891	12.9	276	Nepal	2,320	7.9	111
Chad	816	8.1	130	Nicaragua	11,019	38.0	2,651
Chile	22,939	20.3	1,639	Niger	1,569	26.1	180
China	100,536	9.3	84	Nigeria	33,485	18.5	310
Colombia	19,416	30.3	534	Oman	3,084	..	1,470
Congo	5,275	51.5	2,047	Pakistan	29,579	35.1	234
Costa Rica	3,843	15.0	1,163	Panama	7,107	..	2,721
Côte d'Ivoire	18,452	40.1	1,333	Papua New Guinea	2,878	30.0	686
Croatia	2,304	4.2	482	Paraguay	1,979	10.2	413
Czech Republic	10,694	13.1	1,035	Peru	22,623	17.7	974
Dominican Republic	4,293	17.0	563	Philippines	39,302	21.9	586
Ecuador	14,955	22.1	1,332	Poland	42,160	14.3	1,094
Egypt	33,358	15.8	588	Romania	5,492	8.4	242
El Salvador	2,188	13.1	388	Russia	94,232	6.3	635
Estonia	186	..	124	Rwanda	954	14.7	123
Ethiopia	5,058	11.5	92	Senegal	3,678	14.9	445
Gabon	3,967	10.5	3,049	Sierra Leone	1,392	..	316
Gambia	419	14.4	388	Slovakia	4,067	9.3	761
Georgia	1,227	1.2	226	Slovenia	2,290	5.4	1,151
Ghana	5,389	24.8	324	Sri Lanka	7,811	8.7	437
Guatemala	3,017	11.4	292	Tajikistan	594	..	103
Guinea	3,104	14.2	483	Tanzania	7,441	20.5	258
Guinea-Bissau	816	15.2	782	Thailand	60,991	16.3	1,051
Haiti	712	1.2	102	Togo	1,455	7.8	363
Honduras	4,418	33.9	768	Trinidad and Tobago	2,218	31.6	1,713
Hungary	28,016	53.0	2,730	Tunisia	9,254	18.8	1,050
India	98,990	26.9	108	Turkey	66,332	33.4	1,090
Indonesia	96,500	32.4	507	Turkmenistan	418	4.2	95
Iran	22,712	22.5	363	Uganda	3,473	45.6	187
Jamaica	4,318	20.6	1,729	Ukraine	5,430	2.0	105
Jordan	7,051	12.4	1,747	Uruguay	5,099	16.1	1,612
Kazakhstan	2,704	1.9	161	Uzbekistan	1,156	3.2	52
Kenya	7,273	33.6	280	Venezuela	36,850	21.0	1,740
Korea, Rep.	54,542	7.0	1,227	Vietnam	25,115	6.1	349
Kyrgyzstan	441	4.8	99	Yemen	5,959	4.8	403
Laos	2,080	7.7	438	Zambia	6,573	31.5	714
Latvia	364	2.1	143	Zimbabwe	4,368	..	405

Source: World Development Report 1996, World Bank, New York, 1996.

Underestimating the debt problem

The joint analysis of the World Bank and the IMF is based on too narrow debt sustainability criteria, and overly optimistic macro-economic assumptions.

1.2 Eight countries have an unsustainable debt burden, and twelve more are 'possibly stressed', according to the joint IMF and World Bank analysis of the debt problems of Heavily Indebted Poor Countries (HIPCs). This estimate is based on 23 country studies and preliminary assessments of 16 more HIPCs, but many more countries are facing a debt problem.

The definition of debt sustainability

1.3 According to the Bank and the Fund, a debt burden is sustainable when "a country is able to meet its current and future external obligations in full, without recourse to relief or rescheduling of debts or the accumulation of arrears, and without unduly compromising economic growth".

1.4 To calculate debt sustainability, a certain time horizon is picked - mostly 10 to 20 years - and the growth rates of the main economic variables (such as GDP, exports, imports) are estimated, as well as the future inflows of foreign capital. At the end of the period, the country should have a sustainable debt level.

1.5 Based on empirical evidence, the "rule of thumb" ranges for a sustainable debt level are (a) a net present value (NPV) of debt to exports ratio between 200-250 per cent and (b) a debt service to export ratio between 20-25 per cent. It should be noted that the condition that a country should not need exceptional financing is abandoned in the joint analysis. If this condition were applied, 39 of the 41 HIPCs would be defined as 'unsustainable'.

The threshold ranges

1.6 The threshold ranges of 20-25 per cent and 200-250 per cent are arbitrary. In previous publications the Bank used ratio's of 20 per cent and 200 per cent. Moreover, the use of NPV is questionable. The discount rate used to determine this value is probably too high. Compared to a discount rate of 8 per cent, a rate of 6.5 per cent would increase a given stream of debt service payments by 10 per cent. Also, to measure the debt over-

hang, the nominal, and not the present value is relevant.

1.7 In addition, the NPV is calculated for a certain point in the future - 10 years for most countries in question, and five years for some. So it is not current but future sustainability which is calculated. For those countries which are given a clean bill of health, the projections thus do no more than indicate that sustainability can be achieved in ten years time, provided all the assumptions on which the projections are based are fulfilled.

Going beyond the present criteria

1.8 Eligibility criteria should go beyond the criteria presently used by the Bank and the Fund. Debt sustainability should

> ... debt indicators should be linked to indicators of social development, as the debt burden is an important cause of lack of sustainable social and human development in poor countries.

not only be based on indicators linked to exports and GNP:

• Debt indicators should be linked to budget ratios or payment capacity indicated as a percentage of the national budget, as such and in comparison to other expenditures, such as for health, education and defence.

• In particular, debt indicators should be linked to indicators of social development, as the debt burden is an important cause of lack of sustainable social and human development in poor countries. UNICEF has calculated that investments of around $9 billion in additional resources would be needed for sub-Saharan Africa to make investments which would save the lives of 21 million children and provide over 90 million girls with access to basic education. This is far less than the $13 billion currently spent on debt in the region.

• Because many HIPCs have large trade deficits, it would make more sense to link debt sustainability to 'net exports' (exports minus survival import needs) instead of gross exports.

• The amount of arrears should be taken into account.

• Fiscal indicators should be measured adequately.

1.9 If a wider range of more realistic debt indicators is assumed, around 32 countries could be classified as unsustainable. But even this amount may be too optimistic, as the joint analysis is also based on optimistic assumptions regarding export growth, new aid inflows, and private investments.

How many problem countries?

1.10 According to the Bank and the Fund, eight countries have an unsustainable debt burden: Mozambique, Sudan, Zaire, Zambia, Burundi, Guinea-Bissau, Nicaragua, and São Tomé and Principe. Twelve more countries are possibly stressed: Bolivia, Cameroon, Côte d'Ivoire, Tanzania, Uganda, Congo, Ethiopia, Guyana, Madagascar, Myanmar/Burma, Niger, and Rwanda. Total: 8+12.

1.11 The three countries that are not included in the analysis (Liberia, Somalia, and Nigeria) all have an unsustainable debt burden. Uganda, with an unsustainable debt-to-export ratio of 488 per cent in 2005 is mistakenly classified as possibly stressed. Total: 12+11.

1.12 If fiscal indicators are taken into account, Cameroon, Côte d'Ivoire, and Tanzania would move from possibly stressed to unsustainable; Benin, Kenya, and Togo from sustainable to possibly stressed; and Honduras and Senegal from sustainable to unsustainable. Total: 17+11.

1.13 Madagascar, Guyana, and Ethiopia would be classified as unsustainable instead of possibly stressed and Angola as possibly stressed if debt relief comparable with Naples Terms (or Trinidad) from Russia is not assumed. Total: 20+9.

1.14 Finally, if a shorter time frame (3-5 years) is taken, and using debt ratios of 200 per cent or 20 per cent, Burkina Faso, Kenya, Sierra Leone, Guinea (and Honduras) would move to the unsustainable group. In addition, the eleven original possibly stressed countries would all move to the unsustainable category. Total: 29+3.

1.15 In short, debt sustainability analyses should be based on more realistic assumptions and they should include a wider range of realistic criteria.

Source: Eurodad, 1996.

STRUCTURAL ADJUSTMENT

The World Bank and the IMF assert that the more a country integrates into the world economy - their formula for this process is "structural adjustment" - the better the standard of living will become. However evidence shows that these policies and the current form of globalization tend to favor speculative deals which exacerbate social inequalities worldwide.

Structural adjustment straitjacket

The economic "structural adjustment" policies guide the action of an increasing number of governments, including those of the so-called First World. Their main aims are to open up national markets to international influences and foreign capital; to suppress subsidies and price controls; to reduce budget commitments - particularly in education and social coverage - to reduce taxes on companies, capital and large incomes; to privatize public companies and deregulate salaries and working conditions.

2   According to the promoters of these policies, in particular the international organisations like the International Monetary Fund (IMF) or forums like the group of the seven richest countries in the world (G-7), these measures are absolutely indispensable to avoid phenomena like the "globalization" of the economies - particularly the increase in world trade - from contributing to the social inequalities in the poor countries or between the richer and poorer nations. In a communiqué released in the Lyons summit, in France, in June 1996, the G 7 wrote: "economic growth and progress are closely linked to the globalization process," which offers "an ever greater number of developing countries the possibility of improving their standards of living."

3   In its report "Global Economic Prospects and the Developing Countries," also from 1996, the World Bank wrote of accentuated imbalances between countries in relation to integration to the world market. According to this international organisation, the more foreign trade and direct foreign investment present in relation to the GDP of a nation, the better will be the percentage of manufactured

What are we talking about? Structural adjustment aims to open national markets, suppress subsidies and price controls, reduce budgets commitments, privatize public companies and deregulate salaries and working conditions.

goods in their exports and the more "secure" they will be considered for receiving credit ratings, and the faster will be their integration to the world market.

4   In general terms, the poorest states "integrate" at a slower rate, which, according to the World Bank, is partly due to the fact that they tend to resist the "structural adjustment" of their economies. This creates imbalances in sectors which are in full expansion, like foreign investment, as even though 38 per cent of this goes to the developing countries,

Argentina, Chile, Mexico, Ghana, the Czech Republic, Hungary, Poland and Turkey are the best pupils of the structural adjustment and receive more than one third of the foreign investment going to developing countries.

two-thirds of these investments are received by only eight of the 93 countries. The developing nations with faster "integration" include those which have applied the adjustment policies most rigorously: Argentina, Chile and Mexico in Latin America, Ghana in Africa, the Czech Republic, Hungary, Poland and Turkey in Europe.

5   In other words, the argument is that the more "adjustment" is applied, the better a nation will become integrated into the world economy, and the more it integrates, the better the standard of living will become. However, many experts and observers believe the adjustment policies and the current form of "globalization" tend to favor purely speculative deals accentuating the worldwide social inequalities. Also, in some cases, the adjustment policies themselves create such economic and social imbalances, that the achievements of years of "rigor" can be destroyed in a few minutes. The most obvious case of this is that of Mexico, whose "orthodoxy" allowed it to enter the Organization for Economic Cooperation and Development (OECD) - made up of the richest 25 nations in the world - only seven months before sudden, extensive capital flight left the nation bankrupt.

6   On the more global front speculative operations are on the up and up, encouraged by the liberalization of the currency markets amongst other elements. Daily currency deals can be estimated at around $1,500 billion and around 95 per cent of these transactions are purely speculative: there is nothing either produced, bought or sold. This has led specialists, like Nobel Prize winning scientist James Tobin, to call for a tax on currency deals. However, a measure of this kind would

have to be applied on all the currency markets of the world at the same time, as one of the characteristics of speculative capital is its great mobility, and were the taxes only applied on some markets, the others would undoubtedly see more action.

[7] According to experts opposed to adjustment, like US sociologist James Petras, the very volatility of this capital produces the opposite effects to those they are supposed to have on employment and social inequalities. This is related to the increasing application of policies similar to the structural adjustment plans in the developed countries, which tend to "Third Worldize" the nations, that is, they increase the number of poor people and concentrate wealth in the hands of an increasingly smaller social group.

[8] "Two basic tendencies appear to characterize the mutations currently suffered by the world economy: the explosion of speculative capital and the correlated increase of ever more precarious jobs," wrote Petras and Todd Cavaluzzi in "Le Monde Diplomatique," adding that, from 1969 to 1994 workers forced into part time employment (as they were unable to find full time jobs) in the United States rose from six percent to nearly 13 per cent of the working population. The number of "badly paid" workers, that is, those earning less than $15,000 per year, trebled, going from 8.4 per cent to 23.2 per cent, and the working poor, added to the unemployed, went from nearly 23 per cent to 38.5 per cent.

[9] For Petras and Cavaluzzi, "what capital gains in mobility, the working world loses in security. The increase in direct investments in nations with low salaries has two consequences. On one level, it creates a competent and competitive labor force, which pushes down salaries in the United States, and therefore in the other industrialized nations as well. On another, these investments cause an increase in exports towards the United States and the other "rich" countries. In the most extreme cases, well-paid indus-

trial jobs are replaced by poorly paid service occupations in supermarkets stocked with imported produce. At the same time, the adjustment policies have led to worsening working conditions in the Third World.

[10] In other words, phenomena like this lead to the most unfortunate sectors of the "poor" countries competing with the least favored of the "rich" nations for jobs which are ever more precarious and poorly paid. Similarly, the wealth distribution tends to be less equitable, even in the developed nations, which, on a social level, are becoming more like the so-called developing nations every day. Edward N. Luttwak, former adviser to the US State Department, wrote an essay entitled; "The Endangered American Dream: How to Stop the United States from Becoming a Third World Country," claiming that the United States was gradually doing so. The concentration of wealth is such that the richest Americans are appearing increasingly similar to the rich of the Third World. Luttwak argued that this type of wealth, characteristic of the poorest countries and the United States, is not only a result of underdevelopment, but also one of its causes.

[11] These phenomena also occur in the European nations, even though, up until now, the income concentration is less severe in Western Europe than in the United States. Thus, while 10 per cent of the richest people hold 70 per cent of the

wealth in the United States, this percentage is a more extensive 28 per cent in France. However, the concentration of wealth is also high in France, above all in land ownership, as 20 per cent of the population own nearly 70 per cent of the country.

[12] Former chief of "Le Monde Diplomatique" and author of "L'Empire Américain," Claude Julien, said that structural adjustment has led to "a true ideology of free trade, transmitted through the IMF and the World Bank, taking control of our souls." For Julien, no nation can modify this situation alone and he claimed that strikes and demonstrations, like those seen in France in December 1995, should be provided with global mechanisms, like, for example, union rights all over the world, in order to prevent wages from a continuous fall and to stop the world from becoming more like one gigantic underdeveloped nation every day.

Alternative policies

Once strictly the province of the International Monetary Fund, the set of policies known as Structural Adjustment Programs (SAPs for short) has, since the early 1980s, been used by the World Bank to place conditions on loans. By the end of 1992, 75 countries in Africa, Asia, Latin America, the Caribbean, and Eastern Europe had received SAPs from the IMF and World Bank, totaling more than $150 billion.

[1.2] According to Lisa McGowan of the Washington DC-based Development Group for Alternative Policies, SAPs are designed to boost a country's foreign-exchange earnings by promoting exports, reduce government deficits through cuts in spending, and improve the foreign-investment climate eliminating trade and investment regulations. Though SAPs are tailored slightly different-

ly from country to country, they usually include shifting domestic food production to production of food for export; devaluing the currency (to further encourage exports); cutting social spending (including health and education); restricting credit and suppressing wages; privatizing national industries; and, finally, liberalizing trade.

[1.3] The impact has been sharp and bitter. While small-scale producers suffer from cutbacks in credit and subsidies, large-scale agribusiness export for production has increased. Food production diminishes, and export crops take more and more scarce land and capital. The devaluation of currencies leads to dramatic price increases in food and other basic provisions, such as fuel.

Source: Brave New World Bank: 50 Years Is Enough! by Juliette Majot

AID

IN THE LAST FIVE YEARS the more industrialized countries (except the US) reaffirmed the commitment to bring development aid levels to 0.7 per cent of their GNP. However, in 1994, eight of the 21 donors cut their aid and only four maintained the contribution which they had pledged. The aid was an average 0.3 per cent of GNP, the lowest level in the last 20 years.

Preventive aid

The so-called aid for development consists of loans given to poor countries by the World Bank, the IMF and other international and government agencies from the OECD and OPEC countries, with the intended objective of eradicating poverty.

[2] This strategy started with the cold war, with the objective of "preventing the advance of Communism in the world", and contributed to the hastening of the process which hands over the control of the economy to transnational capital and leaves governments with the function to control their own people.

[3] The granting of credits, which gave political and commercial advantages to the elites, led to an increase of social contradictions instead of eliminating them. Market economies did not solve demand for basic food, education and the provision of health. Development, human rights and protection of the environment were no more than mere rhetoric. Social differences as well as technological and financial dependence among the "beneficiaries", widened.

[4] On the other hand, the creditors were benefited from a financial point of view in two ways: by having a bigger access to the flow of information from Third World countries and by purchases from these which were a by-product of the original loans.

[5] With the end of the cold war, the end of globalization and the internal decay of countries which contributed to the aid, the amount of loans were reduced and their destinations changed.

[6] In the last five years, each one of the donor OECD countries, except the US, reaffirmed their intention to contribute with 0.7 per cent of their GNP. However, in 1994, eight of the 21 donors cut their aid and only four maintained the contribution which they had pledged. The aid was an average 0.3 per cent of GDP, the lowest level in the last 20 years.

> In 1994, the less developed countries owed $1.9 billion. They paid $169 billion annually to service the debt, while receiving $56.7 billion in aid.

[7] Meanwhile, emergency situations gained priority over development. Expenditures in this field grew from $4.5 billion to $6.0 billion from 1993 to 1994. With respect to bilateral aid, the increase in emergencies was from 1.5 per cent to 8.5 per cent.

[8] In 1994, the less developed countries owed $1,921 billion. They paid $169 billion annually to service the debt, while receiving $56.7 billion in aid.

[9] Aid is increasingly being diverted to service debts. The major share of such aid is being spent on supporting debt service payments. About one-quarter of bilateral aid is being spent on supporting debt service payments to the multilater-

> A 1995 report estimated that the supply of extra spare parts and other equipment and services directly related to Danish development projects was worth about 25 per cent of the projects' value.

al institutions. Also, in 1993/94, for every three dollars of loans, two went straight back to the World Bank to pay debts. Part of the remaining dollar was used to repay credits to the IMF. In fact, many countries can only service their external debt when the donor community provides the resources to do so.

[10] Commercial pressures on aid are as strong as ever. During 1995 the Belgian press highlighted cases in Cape Verde, Benin, Ethiopia, Kenya and Indonesia of aid closely linked to the needs of commercial interests in Belgium; aid which had been inappropriate or had never worked. At the heart of these aid abuses was a lack of clarity in departmental responsibilities, strong commercial and political interests, and the absence of clear priorities.

[11] Tied aid locks developing countries into purchasing goods and services from a donor, often from a particular producer. By reducing competition, the Development Assistance Committee (DAC) estimates that tying increases costs by 15 per cent. With fully tied aid amounting to nearly $15 billion in 1994, overpricing is in the region of $2.0 billion, larger than the Swiss and Australian aid programmes combined.

[12] A 1995 report estimated that the supply of extra spare parts and other equipment and services directly related to Danish development projects was worth about 25 per cent of the projects' value. In addition, Danish companies expected to generate 75 per cent of project value in export contracts not directly related to the project.

[13] In Spain, tied aid credits, absorbing 70 per cent of bilateral Official Development Assistance (ODA), are still characterized by an exclusively commercial focus, a lack of transparency, and a total absence of final evaluations of the development benefits of the projects they finance.

Throwing down the gauntlet

At the 1995 Summit for Social Development in Copenhagen, Non–Governmental Organizations (NGOs) from around the world pressed the view that the aid regime promoted by the international financial institutions - the World Bank, the IMF and the regional development banks - had failed to create employment, had deepened social inequality and poverty, and thereby fed social disintegration.

1.2 For more than a decade, NGOs have challenged the harsh economic measures of Structural Adjustment Programmes (SAPs), designed by these institutions and accepted by all the major donors, and have been public voices for alternative measures. NGOs and affected populations point to unsustainable resource exploitation, massive environmental destruction, population displacement, the undermining of food security and human centred development strategies, as among the impacts of SAPs and many of the projects undertaken by the IFIs (International Financial Institutions).

1.3 The Banks have recognized the importance of poverty reduction, environmental protection, citizens' participation in the development process, as well as institutional transparency and accountability, as core values and policies that are intended to shape the impact of their programmes. The World Bank has opened access to information and set out an appeal process for those affected by Bank projects. NGOs in recent years have continued to press for improvements in Bank accountability and to extend these policies to other arms of the Bank (for example, the International Financial Corporation) and to other institutions.

1.4 But policies alone, in the absence of fundamental institutional reform and linkage to the UN Charter and Treaties, will not be sufficient to ensure a systematic integration of the principles of sustainable and human development into economic and lending decisions. Populations affected by World Bank-supported infrastructural projects continue to press their case, with support from NGOs arround the world. NGOs which actively lobby the IFIs are conscious of the need for more effective monitoring networks which will reveal the true impacts of IDA (International Development Aid) programs, recognizing positive language in IDA 10 and IDA 11 replenishment agreements which aim to include poverty alleviation and reduction goals in all lending programs.

1.5 Donors are a long way from adopting, let alone implementing, NGOs' recommendations that 50 per cent of aid should go to social sectors. In eight out of 18 cases for which figures were available, developing countries' spending on total health and education exceeds 20 per cent. In not one case does the proportion of aid allocated to these sectors exceed 20 per cent.

1.6 Actually, the NGOs (and not the banks) were the institutions which showed more creativity and flexibility when fighting against poverty.

Aid in figures

Official Development Assistance from OECD and OPEC members

	Millions $	As a percentage of donors' GNP
	1993	1993
OECD		
Aotearoa/NZ	98	0.25
Australia	953	0.35
Austria	544	0.30
Belgium	808	0.39
Canada	2,373	0.45
Denmark	1,340	1.03
Finland	355	0.46
France	7,915	0.63
Germany	6,954	0.37
Ireland	81	0.20
Italy	3,043	0.31
Japan	11,259	0.26
Netherlands	2,525	0.82
Norway	1,014	1.01
Sweden	1,769	0.98
Switzerland	793	0.33
United Kingdom	2,908	0.31
United States	9,721	0.15
Total OECD	**54,453**	..
OPEC		
Algeria	7	0.01
Kuwait	381	1.30
Libya	27	0.12
Qatar	1	0.02
Saudi Arabia	811	0.70
United Arab Emirates	236	0.66
Total OPEC	**1,463**	..

Source: World Development Report, World Bank, 1995.

Official aid: receipts

Net disbursement of ODA from all sources

	Millions of dollars 1993	Per capita dollars 1993	As % of GNP 1993		Millions of dollars 1993	Per capita dollars 1993	As % of GNP 1993		Millions of dollars 1993	Per capita dollars 1993	As % of GNP 1993
Albania	194	57.3	..	Guatemala	212	21.1	1.9	Niger	347	40.5	15.6
Algeria	359	13.4	0.7	Guinea	414	65.6	13.0	Nigeria	284	2.7	0.9
Argentina	283	8.4	0.1	Guinea-Bissau	97	94.6	40.3	Oman	1,071	538.8	9.2
Bangladesh	1,386	12.0	5.8	Honduras	324	60.7	9.7	Pakistan	1,065	8.7	2.1
Benin	267	52.4	12.5	India	1,503	1.7	0.6	Panama	79	31.3	1.2
Bolivia	570	80.6	10.6	Indonesia	2,026	10.8	1.4	Papua New Guinea	303	73.7	6.0
Botswana	127	90.4	3.3	Iran	141	2.2	..	Paraguay	137	29.1	2.0
Brazil	238	1.5	0.0	Israel	1,266	242.5	1.8	Peru	560	24.5	1.4
Burkina Faso	457	46.8	16.2	Jamaica	109	45.0	2.8	Philippines	1,490	23.0	2.8
Burundi	244	40.6	25.8	Jordan	245	59.7	4.4	Rwanda	361	47.7	24.1
Cameroon	547	43.7	4.9	Kenya	894	35.3	16.1	Saudi Arabia	35	2.0	..
Central African Rep.	174	55.0	14.1	Korea Rep.	965	21.9	0.3	Senegal	508	64.3	8.8
Chad	229	38.1	19.1	Kuwait	3	1.5	0.0	Sierra Leone	1,204	269.4	164.4
Chile	184	13.3	0.4	Laos	199	43.2	14.9	Singapore	24	8.5	0.0
China	3,273	2.8	0.8	Lesotho	128	65.7	16.8	Sri Lanka	551	30.8	5.3
China-Hong Kong	30	5.2	0.0	Madagascar	370	26.7	11.0	Tanzania	949	33.9	40.0
Colombia	109	3.0	0.2	Malawi	503	47.8	25.5	Thailand	614	10.6	0.5
Congo	129	52.9	5.2	Malaysia	100	5.2	0.2	Togo	101	25.9	8.1
Costa Rica	99	30.1	1.3	Mali	360	35.5	13.5	Trinidad and Tobago	3	2.0	0.1
Côte d'Ivoire	766	57.5	8.2	Mauritania	331	153.2	34.9	Tunisia	250	28.9	1.7
Dominican Republic	2	0.2	0.0	Mauritius	27	24.3	0.8	Turkey	461	7.7	0.3
Ecuador	240	21.9	1.7	Mexico	402	4.5	0.1	Uganda	616	34.2	19.0
Egypt	2,304	40.8	5.9	Mongolia	113	48.6	10.3	Uruguay	121	38.5	0.9
El Salvador	405	73.4	5.3	Morocco	751	29.0	2.8	Venezuela	50	2.4	0.1
Ethiopia	1,087	21.0	..	Mozambique	1,162	77.0	79.2	Vietnam	319	4.5	2.5
Gabon	102	100.9	1.9	Myanmar/Burma	102	2.3	..	Yemen, Rep.	309	23.4	..
Gambia	92	88.0	25.5	Namibia	154	105.6	6.2	Zambia	870	97.3	23.6
Ghana	633	38.5	10.4	Nepal	364	17.5	9.7	Zimbabwe	460	42.8	8.1
Greece	44	4.2	0.1	Nicaragua	323	78.5	17.9				

Source: World Development Report, World Bank, 1995.

DRUGS IS ONE OF THE MOST debated subjects in today's economical, political, judicial and medical spheres. Some plants like coca, important to some traditional cultures of the South, have been the target of the ultimate Western crusade.

The drugs crusade

The coca plant embodies a clear cut example of how truth, even scientific truth, responds to an interest or a point of view. A detailed description of its leaf concludes that, due to its richness in aminoacids, acids, and vitamins, the coca plant is earth's most complete plant in non proteic nitrogen. This kind of nitrogen eliminates toxins and pathogens from the human body, also hydrating and regulating the nervous system.

2 The plant is cultivated in warm and humid valleys, known in the Aymara language as *yungas*. Andean peasants chew it while working, resting, and even treat their guests with it. The habit of chewing - not only accepted but widely spread among millions of inhabitants in countries such as Colombia, Ecuador, Peru, Bolivia, Argentina and Chile - has an economic basis. For peasants, coca is a most beneficial crop because of its ability to yield three to four harvests a year, in non-arable soils. Peasants also praise coca for its better profitability, in comparison to other crops. Its very specific farming technique is very well adapted to the valleys through the construction of stone or walled ground platforms.

3 Raising coca in the Andean valley is not only a practical solution but also an ancestral custom. Since about 2000 BC, the leaf was intertwined with the Andean life. Andeans not only utilized it for conveying friendship, repaying services or simply as a coin, but also considered it sacred. Besides discovering its medicinal powers, they employed the leaf, mixed with certain oils, to soften rocks. When the Incas politically centralized the area, plantations were located all along the empire in order to maintain a stable production, the Inca being the sole proprietor of the sacred harvest. Later on, once the Spaniards imposed themselves in the area, the Spanish Crown distributed these plantations among some *colonos* under the *encomiendas* regime, and the payment with coca leaves was authorized.

4 Once the Spaniards discovered its energising properties, they encouraged the leaf consumption in order to increase the productivity of the natives forced to work in the Potosi mines. This inclined many Spaniards to the coca trade, which became a very important revenue source for the Crown, second only to mines exploitation. Tithes on coca contributed almost entirely to cover the Andean Catholic Church revenues.

5 Coca entered the market economy and the colonial society adopted the plant, fully incorporating it to its habits and

> Cocaine's desirability has launched a fabulous business - more lucrative than oil commerce and only second to the warfare business.

manners to the extent that physicians employed it as a medicine for asthma, hemorrhages, toothache, vomiting and diarrhoea.

The satanic bush

6 Nevertheless, despite its early assimilation by colonial society, Spaniards were not reluctant to blame the natives' ritual use of the leaf for delaying their conversion to Christianity, thus beginning to fight its consumption. When decolonization brought independent states in the region, the plant was once again accused, this time of blocking the natives' assimilation into "White" society.

7 However, it was the emergence of cocaine - one of the 14 alkaloids of the plant - that ignited the black history of this bush. Soon after being isolated in 1884, cocaine was applied as anesthesic in surgery, while Sigmund Freud recommended cocaine as a relief for nervous stress and fatigue. Towards the end of the 19th century, cocaine consumption extended through the upper classes and the artistic circles both of Europe and the US. By then Vin Mariani, a tonic based on the coca extract, was prescribed by every physician as a cure for several diseases, similar to the origins of Coca-Cola - patented in 1895 as a stimulant and headache reliever.

8 It was in 1906 that the US authorities declared cocaine illegal. In 1922, the US Congress expanded the trial by officially determining that cocaine was a narcotic and then prohibiting its import, together with the coca leaves. In spite of the prohibition - or eventually because of it - all through the century cocaine has become highly appreciated and consumed. UN Convention for Narcotics placed cocaine on its toxic drugs first page, listing it as "psychotropic" in 1961. But the truth is that its rocketing price makes cocaine one of the most profitable businesses on Earth. In financial, artistic and political milieus from Western Europe and the US, cocaine is regarded as synonous with opulence and distinction, also being consumed in Japan, Eastern Europe and Latin America, though to a lesser degree.

9 Cocaine's desirability has launched a fabulous business - more lucrative than oil and only second only to the warfare business - known as narcotraffic. The word defines the entire process of illegal production, transportation and selling of illegal or controlled drugs. In this transnational game, each one plays its role. The USA, Europe or France sustain a strong demand, while Andean countries like Peru, Bolivia or Colombia

supply the product. In these countries, coca consumption still differs from the one developed in the North. While cocaine paste expands among the young floating population, the natives and peasants - while disliking the paste - still preserve the habit of daily chewing. The coca producing regions have apparently transformed into developing zones, due to the fact that the drug cartels extend credit and insurance to the groups that produce cocaine. Coca planting peasants have increased their incomes: raising the leaf means much more profit than raising any other crop.

[10] In Bolivia, coca and its by-products generate a revenue of $600 million a year, giving occupation to 20 per cent of the adult labour force. In Peru, this sector occupies 15 per cent of the active labour force and reports a year income of $1 billion. In Colombia the drugs trade gives a revenue of $1 billion, a sum higher than coffee exports. The main gain, however, belongs in the consumer countries, where the money laundering is undertaken, chemicals for cocaine production are supplied and weapons to sustain drug dealers are sold.

Traffic and repressive policies

[11] The basic point about this amazing business seems to be hypocrisy. In the US more than $100 billion has been spent on arrests, imprisonments, education and other action since President Ronald Reagan initiated the war against drugs in 1983. However, in the period from 1983 to 1993 the death by drug abuse rate doubled, while assasinations linked to drug trafficking trebled. Statistics reveal that in 1992, in the US, 12,000 people died from drug abuse and 2,000 more perished in drug related murders. The worst statistics for drugs casualties are for adults between 35 and 50 years old, who in 1983 accounted for 80 per cent of the total drug casualties. Ten years later, the risk of dying by drug abuse was 15 times greater for a person in their forties than for a university student.

[12] The US authorities in charge of the fight against drugs give no explanation for these figures. They just present statistics showing an increase in the relatively low rate of teenagers who smoke marijuana. In this respect, Lee Brown, the White House official spokesman for drug affairs, in 1996 classified marijuana as an "extremely dangerous drug" that leads youngsters to "fight for their life in hospital beds". The true fact is that only 26 - out of the 20,375 drug-abuse deaths occurring in 1993 - related to teenagers with marijuana. Furthermore, in these cases, marijuana was combined with alcohol or hard drugs like heroine and co-

Fall-out from GATT

The creation of the World Trade Organization (WTO) as a guardian and guarantor organization of a multilateral trade system, was one of the foundations of the General Agreement on Tariffs and Trade's Uruguay Round.

[1.2] Since the Uruguay Round was concluded in Marrakech in 1994, developing countries have faced several problems to be able to implement the decisions taken on that occasion.

[1.3] According to Magda Shahin, Egypt's delegate at the WTO, "the most significant reason for developing countries to sign the agreements was the fear of being left out, not the conviction that the agreements would somehow benefit them".

[1.4] "The WTO must shape a dynamic ambience to ensure commercial regulations are in accordance with the evolution of world economy and its multilateral system of trade", said Shahin.

[1.5] The issue of whether developing countries are currently more integrated to world economy will be pending until the WTO agreements are effectively applied. Most of Third World countries have adopted substantial obligations regarding access to markets, consolidating the results of their liberalization programmes, unilaterally imposed.

[1.6] The WTO's first Ministers' Meeting, anticipated for December 1996 in Singapore, meant the southern countries would "face a series of decisive stages", according to Bhagirath Lal Das, who was India's delegate at the GATT.

[1.7] Lal Das believes that "during the talks (of the Uruguay Round), major developing countries were constantly subject to overwhelming pressure by the large developed countries and were dealt with on a single basis. Southern countries could not act collectively on any issue, while developed countries joined in order to exert their pressure on them".

[1.8] For the meeting in Singapore, developed countries have proposed several issues: environment, social clauses, com-

petition, corruption and investment policies. Lal Das states that these proposals are "not only of dubious utility but may be potentially harmful for developing countries."

[1.9] India's former delegate recommends the southern countries' strategy should be to act together against the inclusion of new issues in the Organization, since "at the WTO, just one of the 120 signing countries is enough to formally prevent an agreement from being reached". He warns about the dangers entailed by a lack of information and considers that developing countries should "recognize their own potential as decisive actors in the international economic scenario".

[1.10] According to Chakravarthi Raghavan, "from the perspective of developing countries generally, and more so of their poor and disadvantaged sectors, the new trade order under the World Trade Organization has more negative than positive features; and while it could be beneficial as a rule-based system (depending on how the major industrialized countries implement it in letter and spirit), the rules in some areas of obligations for the majors are ambiguous and vague, while those relating to developing countries are quite onerous...".

[1.11] "The WTO order will foreclose many development options for the countries of the South, and could lock them into a new colonial-type economic relationship with the U.S., Europe and Japan".

[1.12] "But joining the WTO need not mean surrender either. The developing countries and the smaller trading nations of the North, seeking to preserve some autonomy in national policy and decision-making, and the academics and activists, as well as the development community and NGO's, must actively continue their fights, identify the changes needed, and work together to bring about these changes in the WTO agreements and its system" says Raghavan.

caine. Only in one case was the death directly attibuted to a marijuana overdose.

[13] Although the authorities refuse to mention it, DAWN reports warn that aspirin - and its substitutes, like Tylenol - is the most dangerous drug among adolescents. During 1993, 25,000 teenagers were hospitalized because of an aspirin or substitute overdose, four times the total for heroin, cocaine and marijuana consumption. This kind of manipulation hides other aspects, like the fact that an Afro-American teenager has only one fifth of the risk of dying from drugs, but

10 times more chance of being arrested for drugs than a middle-aged white adult. However, in 1993 more than 6,000 death-by-drugs cases were white people.

Ethnic catastrophe and growing problems

[14] At present, the US market almost entirely absorbs the Latin American drug production (and a third of the world's heroin plus 80 per cent of the marijuana). Drug consumers in that country amount to 20 million, but in order to solve this domestic problem, the US decided to

fight it abroad. This exclusively domestic issue of drugs consumption has turned into one of the favourite excuses for US intervention abroad, the creation of the Drug Enforcement Agency (DEA) in July 1973 being one of the fundamental steps to institutionalize it.

[15] The intervention in Bolivia in 1986, or the extradition treaties (like the one signed with Colombia in 1982) should be counted among US' clearest acts of interference. This line of interference has expanded into a total oblivion of other countries' sovereignty: the US Supreme Court resolution, legalizing in 1992 the kidnapping of suspects in other countries, carries within a very serious threat to Human Rights, and mocks International Law.

[16] This is another way of mis-representing the real problems. Drug consumption, an unsolved domestic issue, becomes the object of a crusade, projecting the evil onto the producer and not on the consumer - on to the 'other' and not on to oneself. US policy centered on the total eradication of coca will surely not solve the problem for US citizens. But it cerainly constitutes a catastrophe for the Andean populations where coca occupies a space that identifies participants with their families and culture. For the Andean natives, coca has the power to protect, to change luck or foresee the future; for them, the chewing of coca still constitutes today a deep, mythical and collective ritual, through which they connect the past into the present.

[17] During the Ronald Reagan and George Bush administrations, the US carried out direct military interventions, fumigating entire coca plantations with herbicides that destroy organisms and nutrients, but the only tangible outcome these "prophylactic" operations have brought is entire coca plantations growing nowadays in North America's southern estates - while Andean natives testify how their ancestral plants were destroyed and burnt.

[18] Today, many respectable voices can be heard proposing that drugs become legalized, as a first step to solving some of the problems created by the prohibitions - such as the high prices, which often lead to corruption and violence, or the bad quality of the final product, that endangers health. After all, was not this the only way to finish the socially and politically disturbing years of the (alcohol) Prohibition in the United States in the 1920s?

International Trade

	EXPORTS 1994 million $	IMPORTS 1994 millions $		EXPORTS 1994 million $	IMPORTS 1994 millions $
Albania	116	596	Laos	300	564
Algeria	8,594	8,000	Latvia	967	1,367
Aotearoa/NZ	12,200	11,900	Lithuania	1,892	2,210
Argentina	15,839	21,527	Macedonia, FYR	1,120	1,260
Armenia	209	401	Madagascar	277	434
Australia	47,538	53,400	Malawi	325	491
Austria	45,200	55,300	Malaysia	58,756	59,581
Azerbaijan	682	791	Mauritius	1,347	1,926
Bangladesh	2,661	4,701	Mexico	61,964	80,100
Belarus	3,134	3,857	Moldova	618	672
Belgium	137,394	125,762	Mongolia	324	223
Bolivia	1,032	1,209	Morocco	4,013	7,188
Botswana	1,845	1,638	Mozambique	..	1,000
Brazil	43,600	36,000	Myanmar/Burma	771	886
Bulgaria	4,165	4,160	Namibia	1,321	1,196
Burundi	106	224	Nepal	363	1,176
Cameroon	..	1,100	Netherlands	155,554	139,795
Canada	166,000	155,072	Nicaragua	352	824
Chile	11,539	11,800	Nigeria	9,378	6,511
China	121,047	115,681	Norway	34,700	27,300
China-Hong Kong	151,395	162,000	Oman	5,418	3,915
Colombia	8,399	11,883	Pakistan	7,370	8,890
Costa Rica	2,215	3,025	Panama	584	2,404
Côte d'Ivoire	..	2,000	Papua New Guinea	2,640	1,521
Croatia	4,259	5,231	Paraguay	817	2,370
Czech Republic	14,252	15,636	Peru	4,555	6,794
Denmark	41,417	34,800	Philippines	13,304	22,546
Dominican Republic	633	2,630	Poland	17,000	21,400
Ecuador	3,820	3,690	Portugal	17,540	26,680
Egypt	3,463	10,185	Romania	6,151	7,109
El Salvador	844	2,250	Russian Federation	53,000	41,000
Estonia	1,329	1,690	Saudi Arabia	38,600	22,796
Ethiopia	372	1,033	Sierra Leone	115	150
Finland	29,700	23,200	Singapore	96,800	103,000
France	235,905	230,203	Slovakia	6,587	6,823
Gambia	35	209	Slovenia	6,828	7,304
Georgia	381	744	South Africa	25,000	23,400
Germany	427,219	381,890	Spain	73,300	92,500
Greece	9,384	21,466	Sri Lanka	3,210	4,780
Guatemala	1,522	2,604	Sweden	61,292	51,800
Guinea-Bissau	32	63	Switzerland	66,200	64,100
Haiti	73	292	Tajikistan	531	619
Honduras	843	1,056	Tanzania	519	1,505
Hungary	10,733	14,438	Thailand	45,262	54,459
India	25,000	26,846	Trinidad and Tobago	1,867	1,131
Indonesia	40,054	31,985	Tunisia	4,660	6,580
Iran	13,900	20,000	Turkey	18,106	23,270
Ireland	34,370	25,508	Turkmenistan	2,176	1,690
Israel	16,881	25,237	Uganda	421	870
Italy	189,805	167,685	Ukraine	11,818	14,177
Jamaica	1,192	2,164	United Arab Emirates	19,700	21,100
Japan	397,000	275,000	United Kingdom	205,000	227,000
Jordan	1,424	3,382	United States	513,000	690,000
Kazakhstan	3,285	4,205	Uruguay	1,913	2,770
Kenya	1,609	2,156	Uzbekistan	3,543	3,243
Korea, Rep.	96,000	102,348	Venezuela	15,480	7,710
Kuwait	11,614	21,716	Vietnam	3,770	4,440
Kyrgyzstan	340	459			

Source: World Development Indicators 1996, World Bank, 1996.

ARMS

IN 1990 THIRD WORLD MILITARY expenditure accounted for 15 per cent of the world total. It was the main market for arms sales, with 55 per cent of the total purchases. Between 1965 and the mid-1980s this expenditure increased faster than that in Western countries and accounted for a greater proportion of gross domestic product, an average over four per cent.

Arms race and globalization

The growth in military expenditure in the South began in the 1950s and 1960s when the birth of new Third World states coincided with an unfolding Cold War between the superpowers. Throughout this period the US and the USSR transferred vast quantities of defence equipment to Third World regions.

2   Conflicts fought within and between developing countries became proxy battles between the superpowers. Politically, institutionally and technically, the armed forces of the recipient countries became locked into a dependent relationship with the supplier countries. Domestically, the military established an influential political base in many developing countries.

3   In the 1970s, after the first oil-price rise, an increase in credit to developing countries facilitated a leap in Third World arms imports. For OECD economies, arms export was a means of recycling petrodollars and maintaining industrial output despite the threat of recession. Arms sales to the Third World remained buoyant until the recession. At the end of the 1980s despite the occurrence or continuation of several major conflicts - southern Africa, Central America - arms sales to the Third World began to decline.

4   After the end of Cold War hostilities, industrialized countries reduced their total military spending from a peak of $838 billion in 1987 to $762 billion in 1990. The potential for resource transfers is high. A one per cent reduction in US military expenditure could have increased American overseas development aid by 40 per cent.

5   Although developing countries also reduced military expenditure from a peak of $155 billion in 1984 to $123 billion in 1990, they did so for different reasons. Faced with chronic recession, indebtedness and foreign exchange shortages they could no longer afford the rising cost of defence.

6   At any rate, official figures are always lower than real figures, if one takes into account the value of black market arms deals, and the fact that all countries attempt to conceal the exact amount spent on arms and arms research.

7   The Middle East became the target of the international arms trade - both overt and covert - a trade in which the United States is the undisputed leader. US policy greatly contributed to this situation. In February 1991, the Pentagon informed the Senate that the sales of American arms would reach a historic record of $33 million, half destined for the Middle East.

8   Since military expenditure, unlike that on education, health and infrastructure, is not 'wealth' producing, the economic burden is cumulative. Even a small but steady use of productive assets for 'unproductive' military hardware becomes a major loss over time. This aspect becomes more significant when taking into account that military requirements for land have risen steadily, owing to the increase in the size of standing armed forces and the rapid pace of technological ad-

vances in weaponry. During the World War II, an armored infantry batallion in the United States - with some 600 soldiers - required less than 16 sq km. A similiar unit today requires twenty times as much space.

Post-Cold War nuclear strategies

9   Over 50 years ago the nuclear bombs dropped on Hiroshima and Nagasaki started the nuclear arms race. Today we are witnessing a nuclear arms race in reverse and nuclear weapons become increasingly irrelevant. About 50 years ago the United Nations was established to be handicapped most of this time by the deadlock of the Cold War. Today this crippling effect is over and more problems are placed on the tables of the UN family of organizations than they have the capacity to handle. The nuclear arsenals of the US and Russia will be cut from some 65,000 warheads to a few thousand.

10   Hans Blix, Director General of the International Atomic Energy Agency (IAEA) stated in 1994 that "there are some very specific questions to focus on

Guinea pigs for the chemical industry

After two years of study, Pentagon scientists have concluded that the 80,000 men and women who came home from the Gulf War with respiratory problems, memory loss, debilitating fatigue, reproductive defects and other serious problems don't really have anything wrong with them, except maybe a bad case of "stress".

12   A much more likely cause is highly-toxic chemicals that Gulf War soldiers were given by the military - insecticide-sprayed uniforms, repellents rubbed on their skin and chemical detergents fed to them orally. All three compounds are in the deadly organophosphate family of chemicals, which affect the brain, nervous system and reproductive organs.

13   Three weeks after the Pentagon's whitewash came out, researchers from Duke University and the University of Texas reported that they had tested these same organophosphates on chickens. When the chickens were dosed with any of the three chemicals separately, no problem. But when they were given two or three of the chemicals - even in very low doses - the combination caused the same symptoms as Gulf War Syndrome, seriously sickening and even killing the chickens.

A = Military spending (% of GDP) 1994
B = Imports of conventional weapons ($ millions/prices 1990) 1994

	A	B		A	B		A	B		A	B
Algeria	2.7	20	Cyprus	5.4		Laos	7.9		Philippines	1.4	
Angola	8.7		Djibouti	6.2		Lebanon	4.4		Qatar	3.8	
Antigua	0.8		Dominican Rep.	1.1		Lesotho	3.2		Rwanda	7.7	
Argentina	1.7		Ecuador	3.2		Liberia	2.5		Saudi Arabia	11.2	1,602
Bahrain	5.5	8	Egypt	5.9	1,370	Libya	3.7		Senegal	2.2	
Bangladesh	1.8	75	El Salvador	1.9		Madagascar	0.8		Sierra Leone	4.4	
Barbados	0.5		Eq. Guinea	1.4		Malawi	1.1		Singapore	4.8	70
Belize	1.9		Ethiopia	2.6		Malaysia	3.9		South Africa	3.3	
Benin	1.5		Fiji	1.5		Mali	3.0		Sri Lanka	4.7	
Bolivia	1.4		Gabon	2.3		Mauritania	2.7		Sudan	35.0	
Botswana	4.6		Gambia	3.7		Mauritius	0.4		Suriname	2.8	
Brazil	1.6	217	Guatemala	1.1		Mexico	0.7		Syria	8.6	194
Brunei	4.5		Guinea	1.2		Mongolia	2.8		Tanzania	3.5	
Burkina Faso	1.6		Guinea-Bissau	3.3		Morocco	4.3	181	Thailand	2.6	
Burundi	3.0		Guyana	1.4		Mozambique	7.1		Togo	2.7	
Cambodia	2.3		Haiti	2.2		Myanmar/Burma	3.1	248	Trinidad and Tobago	1.4	
Cameroon	1.4		Honduras	1.3		Namibia	2.2		Tunisia	1.4	
Cape Verde	0.9		India	2.8	773	Nepal	1.1		Turkey	3.2	2,135
Cen. Afr. Rep.	2.0		Indonesia	1.4		Nicaragua	2.0		Uganda	2.4	
Chad	2.6		Iran	3.8	780	Niger	0.9		United Arab Emirates	5.7	389
Chile	3.5	263	Iraq	14.6		Nigeria	3.1		Uruguay	2.5	
China	5.6	2	Jamaica	0.9		Oman	15.2		Venezuela	1.6	147
Colombia	2.3		Jordan	7.1		Pakistan	6.9	819	Vietnam	5.7	
Congo	1.7		Kenya	2.2		Panama	1.2		Yemen, Rep.	5.2	
Costa Rica	0.5		Korea Dem. Rep.	26.6	13	Papua New Guinea	1.1		Zaire	1.9	
Côte d'Ivoire	0.8		Korea Rep.	3.6	613	Paraguay	1.4		Zambia	1.0	
Cuba	2.7		Kuwait	12.2	80	Peru	1.8		Zimbabwe	3.5	

Source: Human Development Report 1996; UNDP, 1996.

in the process of nuclear disarmament and arms control:

1. What will happen to the quantities of enriched uranium and plutonium that will be recovered as the warheads are dismantled?

2. What is the existing capacity in some states, including the five declared nuclear-weapons states, to produce nuclear material for weapons use?

3. Whether a complete ban on the testing of nuclear weapons might be within reach, as the US, UK, France and Russia are presently observing declared moratoria;

4. Whether the present international situation provides a proper climate to prevent the spreading of nuclear weapons to further countries - what is generally termed the non-proliferation regime, formalized above all in the Non-Proliferation Treaty (NPT) and the Treaty of Tlatelolco.

[11] The Security Council shows an increasing determination to prevent a spread of nuclear weapons to further countries. In the statement made at the summit-level meeting in 1992 the members declared that "the proliferation of all weapons of mass destruction constitutes a threat to international peace and security".

[12] Given that Article 39 of the UN Charter enables the Council to decide on enforcement action when it identifies a threat to peace, the statement quoted seems to be a stern warning that the Council will feel free to take severe action to prevent proliferation.

The new risks of spreading

[13] Some new risks of a further spread of nuclear weapons have arisen. The disintegration of the Soviet Union gave rise to three new independent States with nuclear weapons on their territory: Ukraine, Belarus and Kazakhstan. However, today all the three are non-nuclear-weapon States parties to the Non-Proliferation Treaty and all the nuclear weapons on their territory have either been sent to Russia or are to be sent there.

[14] Evidence of another risk arising in the former Soviet Union and Eastern Europe has been seen in the increased incidence of trafficking in nuclear materials. Steps have now been taken to strengthen the co-operation between European customs and police authorities to stop this trade which is linked to widespread criminality.

[15] The IAEA is actively "helping" the countries in question to strengthen their control over facilities which are authorized to have nuclear material, and thereby to prevent any such material being taken illegally.

[16] A more serious risk of nuclear proliferation, however, was identified with the discovery after the Gulf War that Iraq, a party to the NPT and a State accepting IAEA safeguards, had a secret programme aimed at the production of highly enriched uranium and nuclear weapons. Despite a billion-dollar effort, some years of work would have remained before Iraq could have made a bomb.

[17] However, for the IAEA director the case raised several disturbing questions. When the NPT was concluded, the proliferation concern was chiefly about advanced industrialised countries - like Sweden, Germany, Switzerland, Italy, Japan - going for nuclear weapons. These countries seem later to have securely settled for a non-nuclear-weapon status. However, the case of Iraq showed that a growing number of developing countries might be attaining a technological level at which nuclear weapons might - with money and time - be within reach.

[18] The other disturbing matter for IAEA was that Iraq had been able to hide its large programme from the IAEA, from satellite observation and from intelligence. The case of Iraq seems a minor threat when considering that countries such as Pakistan or Israel are known to have developed full atomic bombs. Or, even more significantly, when remembering that De Klerk's South Africa officially announced that it had secretly developed atomic devices - though the country denied the military use of atomic energy.

[19] After the discoveries in Iraq a number of measures have been taken to strengthen the safeguards system and additional measures are being developed.

CARGILL - probably the world's largest trader - exemplifies the successful transnational corporation, with an immense network of companies influencing agriculture and agribusiness worldwide. The corporation's success has relied as much on its ability to shape public policy, capture government subsidies and key political actors as on its financial and business acumen.

An invisible giant: Cargill

The transnational corporations are those associations which possess and control means of production or services outside the country in which they were established. This was the UN Center for Transnational Corporations (CTC) definition. The office, charged with studying the environmental and social impact of transnationals, was closed in 1991.

[2] In one of its latest reports, UNCTC disclosed that half of the gases responsible for the greenhouse effect were created by transnationals linked with production of energy, oil, mining, agribusiness, construction of highways, and chlorofluorocarbons - compounds used mainly in aerosol sprays and refrigerators.

[3] Transnationals are also responsible for most of the world's ecological disasters, including over 2,000 deaths at Bhopal; the oil spill caused by the Exxon Valdez Tanker in the Prince William Canal (Alaska) and the pollution of the Rhine.

[4] To hold on and increase their power, transnationals have shown great ability in adapting to new situations. The increase in the number and range of their activities is ever greater.

[5] There are many invisible giants. Cargill - probably the world's largest trader - exemplifies the successful transnational corporation, its worldwide network of companies exerting an immense influence over global agriculture and agribusiness to the detriment of small producers and the environment. The corporation's success has relied as much on its ability to shape public policy, capture government subsidies and key political actors as on its financial and business acumen.

[6] It employs some 72,700 people worldwide in 800 locations in 60 countries, in more than 50 leading lines of business including corn, salt, peanuts, cotton, coffee, road transport, river-canal shipping, molasses, livestock feed, steel, hybrid seeds, rice milling, rubber, citrus, chicken, fresh fruits and vegetables, beef, pork, turkey and flour milling. Cargill is the world's largest producer of malting barley; the largest oilseed processor; and the second largest producer of phosphate fertilizer.

[7] Such wide-ranging, vertically-integrated operations have brought Cargill huge benefits. Acting as input supplier, banker, buyer of finished products and wholesaler allows the company to make profits at every stage in the production, distribution and consumption of the commodities in which it trades.

[8] The case of frozen concentrated orange juice provides a good illustration of the company's ability to use both the processing and trading of an actual commodity and the trading of imaginary commodities, such as futures contracts and derivatives, to influence, if not control, global markets. Thus, every five days, Cargill's custom-designed and built bulk tankers leave the Brazilian port of Santos, loaded with frozen concentrated orange juice from the company's orange processing plants inland, and head for New Jersey or Amsterdam where the juice is transferred to Cargill rail cars or trucks for delivery to processors and retail distributors.

[9] Cargill may well have supplied the orange trees to the farmers, told them how to grow the fruit, supplied the requisite chemical fertilizers and pesticides, hired the labour to pick the oranges, and even provided credit for the farmers to buy Cargill inputs. Meanwhile, for the five days that the frozen concentrate is in transit, Cargill, through its Financial Markets Division, has the opportunity to trade futures and derivatives based on the commodity up to 19 times.

[10] Indeed, Cargill's "invisibility" lies not only in the fact that its name seldom, if ever, appears on the retail product, but also in that most of its actual trading activities are in the non-existent, invisible commodities of futures contracts and derivatives. With its private financial system and resources to call upon, Cargill is in a position to exert immense leverage over the production and prices of food commodities, including where and by whom they will be produced.

[11] But what has proved beneficial and profitable for Cargill has proved the opposite for thousands of small farmers, livestock ranchers and meat processors in the countries in which the company operates. In 1989, for example, Cargill opened a $55 million beef packing plant in Alberta, Canada, with the help of $4 million from the Alberta government. It drove competitors out of business by simply paying more than others for cattle and selling them for less until it had gained the predominant market share. The company also drove down wage rates in packing houses across the country by $3 an hour simply by announcing, before its plant had even opened, the maximum rate it was prepared to pay.

[12] Cargill has a full array of highly sophisticated lobbying styles to manipulate government policy and programmes to its advantage. Its reputation in the grain trade for doing so is extensive.

[13] A prime mechanism is the revolving door of public service: usually senior Cargill executives take leave of Cargill for a stint in government advisory and policy positions, returning to the company when their mission is accomplished. The career of William Pearce, who retired as Cargill's vice-chair in 1993, is illustrative. In 1973, Pearce left Cargill to join the Nixon administration as deputy special representative for trade negotiations, steering a trade bill through

Congress that, in Cargill's own words, shaped international trade policy. Pearce rejoined Cargill a year later.

[14] Cargill employees or ex-employees have taken up key posts in the US Department of Agriculture (USDA) and in the US negotiating team for the recent GATT Uruguay round. Such is the extent to which Cargill employees have rotated through positions at the USDA that one government investigator has called the practice "structural corruption".

[15] The next level of lobbying activity takes place through the myriad trade associations that represent a commodity or processsing interest, such as turkey growers, flour millers, soybean processors, peanut growers or the feed industry.

[16] Cargill's approach to starting a business in a new country has its origins in military strategy. Historic product-line beachheads for the company have been hybrid seeds, commodity export market-

ing and animal feed milling. The strategy has been: create the beachhead with inputs of capital, technology and management nucleus; get the cash flow positive; re-invest the cash flow and expand the beachhead. The company generally insists on majority ownership in beachhead companies because it needs to be clear who is responsible for the management of an individual company.

The biggest 100 (1994)

Industrial corporation name	Sales $ millions	Employees	Country	Industrial corporation name	Sales $ millions	Employees	Country
Mitsubishi	175,835.6	36,000	Japan	Meiji Mutual Life Insurance	36,343.7	49,050	Japan
Mitsui	171,490.5	80,000	Japan	Daewoo	35,706.6	80,600	South Korea
Itochu	167,824.7	7,345	Japan	E.I. Du Pont De Nemours	34,968.0	107,000	US
Sumitomo	162,475.9	22,000	Japan	Union Des Assur. De Paris	34,597.0	50,448	France
General Motors	154,951.2	692,800	US	Mitsubishi Motors	34,369.9	28,742	Japan
Marubeni	150,187.4	9,911	Japan	Kmart	34,313.0	335,000	US
Ford Motor	128,439.0	337,778	US	Texaco	33,768.0	30,042	US
Exxon	101,459.0	86,000	US	Philips Electronics	33,516.7	253,032	Netherlands
Nissho Iwai	100,875.5	17,008	Japan	Électricité de France	33,466.6	117,575	France
Royal Dutch/Shell Group	94,881.3	106,000	UK/Neth.	Deutsche Bank	33,069.2	73,450	Germany
Toyota Motor	88,158.6	110,534	Japan	Fujitsu	32,795.1	164,364	Japan
Wal-Mart Stores	83,412.4	600,000	US	Mitsubishi Electric	32,726.4	110,573	Japan
Hitachi	76,430.9	331,673	Japan	Eni	32,565.9	91,544	Italy
Nippon Life Insurance	75,350.4	90,132	Japan	Renault	32,188.0	138,279	France
AT&T	75,094.0	304,500	US	Daiei	32,062.3	49,682	Japan
Nippon Telegraph & Telephone	70,843.6	194,700	Japan	Citicorp	31,650.0	82,600	US
Matsushita Electric Industrial	69,946.7	265,397	Japan	Industrial Bank Of Japan	31,072.3	5,433	Japan
Tomen	69,901.5	3,192	Japan	Chevron	31,064.0	45,758	US
General Electric	64,687.0	221,000	US	Hoechst	30,604.2	165,671	Germany
Daimler-Benz	64,168.6	330,551	Germany	Procter & Gamble	30,296.0	96,500	US
Intl. Business Machines (IBM)	64,052.0	243,039	US	Alcatel Alsthom	30,223.9	196,900	France
Mobil	59,621.0	58,500	US	Peugeot	30,112.3	139,800	France
Nissan Motor	58,731.8	145,582	Japan	Fuji Bank	30,103.3	16,252	Japan
Nichimen	56,202.6	2,591	Japan	Mitsubishi Bank	29,990.9	15,701	Japan
Kanematsu	55,856.1	8,431	Japan	Abb Asea Brown Boveri	29,718.0	207,557	Switzerland
Dai-Ichi Mutual Life Insurance	54,900.4	71,797	Japan	Sumitomo Bank	29,620.6	17,247	Japan
Sears Roebuck	54,825.0	360,000	US	Nippon Steel	29,003.8	96,800	Japan
Philip Morris	53,776.0	165,000	US	Sanwa Bank	28,799.0	14,909	Japan
Chrysler	52,224.0	121,000	US	Mitsubishi Heavy Industries	28,676.0	53,048	Japan
Siemens	51,054.9	382,000	Germany	Ito-Yokado	28,631.5	66,710	Japan
British Petroleum (BP)	50,736.9	60,000	Britain	Rwe Group	28,628.3	117,958	Germany
Tokyo Electric Power	50,359.4	43,115	Japan	Pepsico	28,472.4	471,000	US
U.S. Postal Service	49,383.4	728,944	US	PEMEX (Petróleos Mexicanos)	28,194.7	119,928	Mexico
Volkswagen	49,350.1	242,318	Germany	CIE Générale des Eaux	28,153.1	215,281	France
Sumitomo Life Insurance	49,063.1	70,911	Japan	Crédit Agricole	27,753.1	72,500	France
Toshiba	48,228.4	190,000	Japan	Amoco	26,953.0	43,205	US
Unilever	45,451.2	304,000	UK/Neth.	Basf	26,927.7	106,266	Germany
Iri	45,388.5	292,695	Italy	Ing Group	26,926.3	46,975	Netherlands
Nestlé	41,625.7	212,687	Switz.	Bayer	26,771.1	146,700	Germany
Deutsche Telekom	41,071.2	223,000	Germany	Asahi Mutual Life Insurance	26,505.5	39,499	Japan
Fiat	40,851.4	248,810	Italy	Dai-Ichi Kangyo Bank	26,500.2	19,061	Japan
Allianz Holding	40,415.2	69,859	Germany	Crédit Lyonnais	26,388.1	68,291	France
Sony	40,101.1	138,000	Japan	Sakura Bank	26,069.0	21,600	Japan
Veba Group	40,071.9	126,875	Germany	BMW (Bayerische Motoren Werke)	25,972.6	109,362	Germany
Honda Motor	39,927.2	92,800	Japan	France Télécom	25,706.3	152,886	France
Elf Aquitaine	39,459.1	89,500	France	Kansai Electric Power	25,585.3	26,707	Japan
State Farm Group	38,850.1	68,353	US	Hewlett-Packard	24,991.0	98,400	US
NEC	37,945.9	151,069	Japan	Total	24,653.0	51,803	France
Prudential Ins. Co. of America	36,945.7	99,000	US	East Japan Railway	24,643.4	79,709	Japan
Öesterreichische Post	36,766.0	56,983	Austria	Long-Term Credit Bank of Japan	24,605.1	3.877	Japan

Source: Fortune, 1995.

DEFORESTATION

MORE ANIMAL AND PLANT SPECIES are becoming extinct at an unprecedented rate as a result of human activity, according to a September 1996 report by the United Nations Environment Programme. Over 1,500 scientific experts from all over the world contributed to the preparation of the report, called the Global Biodiversity Assessment.

Losing diversity

The Biodiversity Convention is concerned with the conservation of the Earth's biodiversity, the sustainable use of its components, and the fair and equitable sharing of the benefits arising out of the use of genetic resources.

[2] According to the Global Biodiversity Assessment, which was released at the Biodiversity Convention talks held in Jakarta, almost three times as many species (112) became extinct between 1810 to the present as between 1600 and 1810, when 38 species became extinct. Between five and 20 per cent of some groups of animal and plant species are threatened.

[3] The assessment estimates that there are about 13 to 14 million different species on Earth, and only 1.75 million, or 13 per cent, have been scientifically described.

[4] The loss of biodiversity can have major repercussions on humans. Reuben Olembo, Deputy Executive of UNEP said, "The adverse effects on the Earth's biodiversity are increasing dramatically and we are threatening the very foundation of sustainable development."

[5] The report finds that biodiversity loss threatens food supplies, sources of wood, medicines and energy, and interferes with essential ecological functions such as regulation of water runoff, control of soil erosion, assimilation of wastes, purification of water and the cycling of carbon and nutrients.

[6] Human activity is the major cause of biodiversity loss worldwide, caused in large part by increasing demand of resources due to population growth, economic development, and over-consumption.

[7] Other reasons cited are the failure of people to consider the long-term consequences of their actions, the use of inappropriate technology, the failure of markets to recognize the value of biodiversity, and the failure of government policies to address the overuse of biological resources.

Source: From "Species becoming extinct at a record rate" by Daniel J. Shepard, Earth Times News Service.

What's at stake in deforestation?

Biological diversity: It is estimated that over half of all the plant and animal species on earth are found in tropical forests, which cover less than seven per cent of the world's land surface. Deforestation is driving many of these species to extinction. The unique ecosystems and species of temperate and boreal (coniferous) forest are also threatened. For example, over one thousand species in Scandinavia are endangered by industrial forestry. At a global level, this devastation is propelling the planet towards an ecological crisis. Species are being lost which could provide new pest-resistant crops or cures for diseases such as cancer and AIDS.

[1.2] **Environmental balance:** The large amount of rainfall characteristic of tropical forests is filtered through thick forest cover, regulating its impact on the local environment. Forest destruction leads to flooding and the loss of thousands of tonnes of top soil, often resulting in landslides. The siltation of rivers and streams endangers fish populations, threatening both the food of local people, as well as commercial fishing operations. Clearing forest for cattle-ranching, agriculture and logging also leads to the release of millions of tonnes of carbon dioxide, causing climatic instability and contributing to global warming.

[1.3] **Indigenous peoples:** Worldwide, more than 100 million indigenous people live in and depend on the forests for their livelihood, food and medicine. In the Amazon, for example, Indian lands are being invaded by logging companies to extract mahogany for foreign markets.

[1.4] **Economic options:** Over a billion people who do not actually live in the forest rely on its valuable resources. Harvests of fruit, nuts, medicine and rubber from intact forests offer viable economic options for local and indigenous communities. Foods we now take for granted such as cocoa, palm oil, coffee and corn originate from tropical forests. The potential for finding new foods is enormous; a wild species of caffeine-free coffee has only recently been discovered in the Comoros Island forests off eastern Africa.

Source: Greenpeace

African rainforest and the World Bank

Documents in possession of the British NGO Friends of the Earth showed, in September 1996, how World Bank staff misled their own board of directors in order to force through a project that could lead to new logging in some of Africa's most spectacular tropical rainforests.[1]

2.2 Internal memoranda relating to the Cameroon Transport Sector Project noted that roads paid for by the World Bank will "give access to protected forest areas where logging is likely to occur" and said that "most of what the Bank is supporting (or wants to support)... will increase incentive for logging".

2.3 World Bank staff noted that "Some serious flak is coming up around the Cameroon transport sector project" but instead of recommending the scheme be shelved, the documents showed that they misled the board of directors to get the project through.[2]

2.4 Tony Juniper, deputy campaigns director at Friends of the Earth, said: "The World Bank is a smoke screen behind which environmental and social destruction is carried forward in the name of 'development'. World Bank policies to protect forests and the poor are not worth the paper they're written on."

2.5 Aspects of this project were considered for funding by the much less stringent African Development Bank (AfDB) in 1991 but were turned down after protests from campaigners and an official environmental assessment (EA). These aspects appeared again in the World Bank project. Indeed, the World Bank documents said that "One of the roads on the map is a pure and simple logging road that AfDB rejected on environmental grounds, following an EA." The World Bank went on to say that "it appears that the World Bank may be funding road maintenance that AfDB rejected and without our own EA".

2.6 The project was approved by the Bank's board of executive directors on May 30 1996. The UK's board member was initially against the project but ultimately voted in favour. The United States' board member was the only one to vote against: he is also the only one required by national law to make his voting position public.

NOTES:

[1] The ecologically sensitive areas include the Dja wildlife reserve. According to the International Union for the Conservation of Nature (IUCN), "The Dja reserve is one of the largest, biologically richest and best preserved moist forest areas in Africa". It will be made further accessible to logging companies by a new road.

[2] The largest component of the AfDB project was the construction of a 131km engineered road between Abong-Mbang and Lomie in southeastern Cameroon to replace the existing narrow dirt track. This road, now part of the World Bank project, will pass in close proximity to the Dja wildlife reserve, one of Africa's key wildlife protection areas. The road will increase access for the poorly regulated logging industry which is widely accepted to operate to low and environmentally damaging standards.

Source: Friends of the Earth

Human impact on desertification

Desertification was the subject of a conference in June 1994. The document proposed ways to combat this phenomenon, often called "the advance of the desert," working alongside the local population and encouraging the decentralization of decision-making.

3.2 The developed nations, for their part, committed themselves to financing some action programs and to encouraging technology transfer. These programs were implemented both on a national and regional level within four priority zones: Africa, Asia, Latin America and the Caribbean and the northern Mediterranean.

3.3 However, the objectives of the document seem to be too general, at least partly due to the extreme diversity of the causes and mechanisms of desertification. Besides, the convention did not develop any independent funding mechanism, which, added to its complexity, makes it difficult to implement.

3.4 Desertification is not always related to drought, as appeared to be shown by the advance of the desert in the Sahel between 1968 and 1986, coinciding with a persistent drought which affected a zone which stretched for some 5,000 km from east to west, and some hundreds of kilometers from north to south. In some cases, desertification - which takes many very different forms - is not the result of any climatic change, and this is shown in the case of the Maghreb. In this North African region, the torrential rains on strongly inclined ground cause far more serious damage than in the Sahel, which has flatter surfaces.

Furthermore, if the soil is degraded by human activity, the damage tends to be irreversible: in the Maghreb the soil is replaced very slowly, and the continuous presence of human populations does not give them a chance to recover.

3.5 We are dealing here with an example of the incidence of human action and the complexity of the issue: the livestock raisers intensified grazing on the less fertile lands - as they tended to grow cereals on the better areas. Nowadays, when they found a stretch of pasture, the animals are taken there in lorries at top speed. The regeneration of the grazing circuits became very difficult or impossible .

3.6 In reality, in the Maghreb, the climatic risks matter less than planning errors in land use. Desertification does not progress on fronts, as in the Sahel - the much quoted "advancing desert", but by aureoles or patches spreading out from overpopulated rural areas, urban centers or other ecologically fragile surroundings. As a result, its increase is not as spectacular as in the Sahel, where it is far easier to chart or to show on the television news.

3.7 Having essentially socio-economic causes, the phenomenon should be easier to combat but, under the current conditions, nothing seems to be in the track of changes. The small importance the convention gave to the demographic problems is surprising, as the Maghreb - unlike the north of the Mediterranean - is a region where large demographic growth is an essential cause of desertification.

Source: Pierre Rognon in Le Monde Diplomatique.

GLOBAL WARMING

GLOBAL WARMING might be a sign of profound changes experienced by the earth's climate as a result of human activity. Rising sea levels - which in the medium term are likely to result in flooding some populated areas - should be counted among global warming's many possible consequences.

Climate change / global warming: not quite the same thing

Recent accounts of the scientific debate on climate frequently misrepresent what is being argued about. They suggest that scientists are still discussing whether or not the climate is changing in response to greenhouse gas (GHG) emissions, as if there were a simple yes/no answer. So if a scientist questions the adequacy of present climate models, or fails to find conclusive evidence for global warming in a particular data-set, he or she is often reported as claiming that "there isn't really a problem". However, in most scientific circles the issue is no longer whether or not GHG-induced climate change is a potentially serious problem. Rather, it is how the problem will develop, what its effects will be, and how these can best be detected.

[2] The confusion arises from the popular impression that "the enhanced greenhouse effect", "climate change", and "global warming" are simply three ways of saying the same thing. They are not. No one disputes the basic physics of the "greenhouse effect". However, some of the consequences of the basic physics, including higher average temperatures due to "global warming", are less certain (although highly probable). This is because the fundamental problem concerns the way GHG emissions affect the flow of energy through the climate system, and temperature is just one of many forms energy takes.

[3] In the long term, the earth must shed energy into space at the same rate at which it absorbs energy from the sun. Solar energy arrives in the form of short-wavelength radiation; some of this radiation is reflected away but, on a clear day, most of it passes straight through the atmosphere to warm the earth's surface. The earth gets rid of energy in the form of long-wavelength, infra-red radiation. But most of the

> By determining how the air absorbs and emits radiation, greenhouse gases play a vital role in preserving the balance between incoming and outgoing energy.

infra-red radiation emitted by the earth's surface is absorbed in the atmosphere by water vapour, carbon dioxide, and other naturally occurring "greenhouse gases", making it difficult for the surface to radiate energy directly to space. Instead, many interacting processes (including radiation, air currents, evaporation, cloud-formation, and rainfall) transport energy high into the atmosphere to levels where it radiates away into space. This is fortunate for us, because if the surface could radiate energy into space unhindered, the earth would be more than 30°C colder than it is today: a bleak and barren planet, rather like Mars.

[4] By determining how the air absorbs and emits radiation, greenhouse gases play a vital role in preserving the balance between incoming and outgoing energy. Man-made emissions disturb this equilibrium. A doubling of the concentration of long-lived greenhouse gases (which is projected to occur early in the next century) would, if nothing else changed, reduce the rate at which the planet can shed energy to space by about 2 per cent. Because it would not

Antarctic ozone depletion

Global warming and ozone depletion are two different problems, even if sometimes they are presented as similar ones. However, they could be related, or rather accentuated one by the other.

[1.2] In his article "Antarctica and Global Change", Australian Antarctic researcher Bill Budd suggested a feedback link between climate change and Antarctic ozone depletion. As stratospheric ozone has been lost, upper atmospheric temperatures have fallen. The decrease in stratospheric temperatures over the Antarctic has been coincident with a general increase in lower atmospheric temperatures.

[1.3] Budd stated that "This type of reaction would be expected to result from reduced stratospheric ozone but it is also a feature which can result from the CO_2 warming in the troposphere... these changes were strongest in the Antarctic, pointing to the dominance of the ozone effect, although no doubt both processes may be acting together and reinforcing each other". He went on to refer to "... the strength of the Antarctic stratospheric vortex which may be expected to increase and last longer with the reduction in Antarctic stratospheric temperatures. This could isolate the region of ozone reduction even more, and possibly lead to further reduction extending longer into summer... ".

[1.4] Later, J. Austin, N. Butchart and K. Shine, in their article "Possibility of an Arctic ozone hole in a doubled-CO_2 climate" demonstrated the link between global warming and ozone depletion in a modelling study of the Arctic, in which cooling of the stratosphere enhanced ozone loss.

Source: Greenpeace

affect the rate at which energy from the sun is absorbed, an imbalance would be created between incoming and outgoing energy. Two per cent may not sound like much, but over the entire earth it would amount to trapping the energy content of some three million tons of oil every minute.

5 The climate will somehow have to adjust to get rid of the extra energy trapped by human-made greenhouse gases. Because there is a strong link between infra-red radiation and temperature, one probable adjustment would be a warming of the surface and the lower atmosphere. But it is important to realize that a warmer climate is not the only possible change, nor even necessarily the most important one. The reason for this is that radiation is not the only energy transport mechanism within the lower atmosphere (although it does play a vital, controlling role). Instead, the surface energy balance is maintained, and surface temperatures are controlled, by that complex web of interacting processes which transport energy up through the atmosphere. In contrast to the case of radiation, it is very difficult to predict how processes like cloud-formation will respond to greenhouse gas emissions.

6 Global warming is a symptom of climate change, but it is not the problem itself. It may be the clearest symptom we have to look for, but it is important not to confuse the symptom with the disease. The fundamental problem is that human activity is changing the way the atmosphere absorbs and emits energy. Some of the potential consequences of this change, such as sea-level rise, will depend directly on how the surface temperature responds. But many of the most important effects, such as changes in rainfall and soil moisture, may take place well before there is any detectable warming.

7 If a scientist argues that the warming may not be as large or as fast as models predict, he or she is not suggesting that the problem of climate change should be ignored. The point is only that this particular symptom - global average temperature - may be unreliable. We know that the air's radiative properties are changing, and we know that the climatic effects of this change will be profound. All climate models indicate that the most significant change will be a global warming. But even if it isn't, other effects, equally profound, are inevitable. We are altering the energy source of the climate system. Something has to change to absorb the shock.

Source: Information Unit
on Climate Change (IUCC)

Nuclear energy in Europe

Unlike the Western European states, which were "obliged" to adopt policies of (relative) energy saving, following the so-called oil crisis of 1973, the Eastern European nations did not reduce their energy consumption, and this was provided at a low cost by the State. The Chernobyl accident, in 1986, did not make them question the "principle" of the production and consumption of large amounts of energy, and this is still the pattern even today - seven years after the fall of the Berlin Wall.

2.2 Up until the disappearance of the USSR, energy sales represented half the exports of the European nations of the Soviet bloc and, according to Perline and Mycle Schneider in "Le Monde Diplomatique," the money made here covered 60 per cent of their imports. This encouraged the maintenance of sectors with high levels of energy consumption, like the steel industry. Even today this sector uses more than half the total energy consumed in Eastern Europe, a figure which is running at around a third in Western Europe.

2.3 On another front, according to the Schneiders, there is a correlation between the energy consumption in the former socialist bloc and the exports of the old communist States to the 25 "rich" OECD countries, which include many of the Western European nations. The more energy the Eastern European countries consume, the more they export to the developed world. "In other words, the OECD tends to transfer the cost of the exhaustion of non-renewable resources and environmental contamination (to the former socialist bloc), which amounts to passing on the risks. In this way, in the Eastern nations, the energetic intensity - that is, the relation between the consumption of primary energy (coal, oil) or its secondary equivalent (electricity) and the GDP (the relation between the consumption of energy and the wealth produced) - is between two and three times higher than in the European Union (EU)," they explain.

2.4 Also, concomitantly with the strong fall in production of recent years in Eastern Europe, the intense energy use has continued to increase, amongst other reasons, because the factories are running at only partial capacity - hence with a low production to energy consumption ratio - and because the number of private cars has increased dramatically.

2.5 Seven of the 15 EU nations have "peaceful" nuclear programs, which generate from 2 per cent of the energy in the Low Countries and 75 per cent in France. Six of the 15 states in Central Europe have nuclear reactors, which produce from 20 per cent of the electricity in the Czech Republic, to 87 per cent in Lithuania. In the Commonwealth of Independent States (CIS), only two of the 12 countries use nuclear installations: Ukraine and Russia. These installations generate between 13 per cent and 33 per cent of their electricity in this way, according to data from 1993. Half of these reactors are of the RBMK model used in Chernobyl, two of which are still in action and produce nearly 5 per cent of the electricity in Ukraine.

Small island states

Virtually every nation in the world will ultimately have to face the implications of climatic change related to global warming. However, none so soon, and with as a great a level of foreboding as those nations classified by the United Nations as "small island states."

3.2 Global warming trends projected over the next century may result in sea level rises of one meter or more. This could result in several small island states, many of which are only 1.5-2 meters above sea level, being wholly submerged within the next century. These nations include the Marshall Islands, Tuvalu, Tokelau and the Maldives.

3.3 Additionally, rising ocean levels could severely denigrate fresh water supplies in such nations, devastate agriculture, and threaten coral reef habitats which provide a home for many species of fish critical to the subsistence of inhabitants in many small island states.

3.4 Finally, climatic change may substantially increase the incidence and severity of hurricanes and tropical storms, wreaking havoc on the economic infrastructure of many of these nations.

Source: William C. Burns, Pacific Center For International Studies, USA

WATER

OVER THE NEXT YEARS water scarcity may become a source of international conflicts and major shifts in national societies, affecting prospects for peace - in unstable regions like the Middle East - as well as food security, people's ways of life, the present growth of cities and the location of industries.

Use more, save more

Any discussion on the emerging global freshwater crisis can take as a starting point two undeniable statistics: planet earth has little freshwater, and it is decreasing on a per capita basis as human populations continue to grow. Freshwater is a major, everyday problem for more than one billion people.

[2] More than 50 countries are unable to provide safe and adequate freshwater for domestic use. Yet most of the freshwater discussion tends to be based on the assumption that water is being overused and misused on a per capita basis. The crisis, now becoming apparent in many parts of the world, is not due to a decline in the resource as such, in absolute or per capita terms.

[3] More than 50 developing countries, particularly in Africa, use less than one per cent of their annual renewable freshwater resources. In some of these countries, per capita freshwater consumption is as low as 50 cubic metres, against a world average of 650 m³, 1,200 m³ in industrialized countries, and 2,100 m³ in the US. At least nine countries in Africa, including Burkina Faso, Congo, Burundi, Guinea Bissau and Zaire, use less than 25 m³ per capita (domestic animals included). National average domestic water consumption represents less than 15 litres per person a day. These countries will have to increase their freshwater consumption in the future.

[4] Similarly, many developing countries, again particularly in Africa, need to supply electricity to their populations. Many households still lack electric light. At the same time most of these countries use but a fraction of their potential hydroelectricity.

[5] Africa is estimated to have 1,621,800 megawatts of known exploitable hydroelectric potential. At present the total installed capacity reaches only 1.2 per cent. In absolute terms the 54 African countries produce 53,000 GWh (million kilowatt hours) of hydroelectricity.

[6] By comparison, a small country like Switzerland produces 37,000 GWh (almost 95 per cent of potential and 60 per cent of total electricity) and the US 310,000 GWh of hydroelectricity. By 1975, all Western European countries had already constructed dams to produce more than 75 per cent of their potential hydroelectricity.

[7] Between 1992 and 1993 the OECD countries increased their hydroelectricity production by 70,000 GWh, more than the total produced by the entire African continent. Moreover, Africa's total electricity production (329,000 GWh of which South Africa alone produces 169,000 GWh) is less than 10 per cent of that generated by the United States.

[8] Any attempt to increase hydroelectricity production in Africa is bound to raise the issue of wildlife and habitat conservation. Recent proposals by Namibia to build a dam on the Cunene river at Epupa falls, and by Nigeria to build Kafin Zaki dam have raised valid social and environmental arguments.

[9] But what is the acceptable level of use of the hydroelectric potential of a country? And how can it be done without damaging ecosystems? Poorer countries, using less of their water resources, are home for much of the planet's remaining tropical forest and wildlife.

[10] The future of biodiversity in Africa clearly depends on how the continent uses its freshwater resources in the incoming decades. Freshwater use is related to food security, social order, and political stability. The challenge facing the world is to derive ways and means to increase freshwater use and hydroelectric production while conserving biodiversity. Any further delay in this direction will cause serious and irreparable damage to natural, biodiverse systems.

[11] This is not a new issue. It has been said many times in the past. But the conservation agenda today is dominated by the North and elites in the South who hold different views, especially towards nature and wildlife.

[12] The world as a whole may have to save water on a per capita basis and reduce the number of dams. While some countries will have to increase their use of freshwater and produce more hydroelectricity, others will seriously have to consider reducing water use and remove some dams on rivers to restore valuable wetlands and re-establish free-flowing rivers.

Source: Bishkam Gujja, from People & the Planet 1996

Current cost and projected future cost of supplying water to urban areas

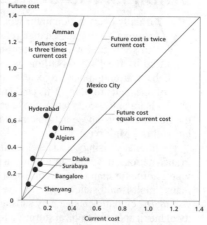

NOTE: Cost excludes treatment and distribution. Current cost refers to cost at the time data were gathered. Future cost is a projection of cost under a new water development project. Source: World Bank 1992.

Conflicts

Globally, water use has more than tripled since 1950, and the answer to this rising demand generally has been building more and bigger water supply projects - especially dams and river diversions. Around the world, the number of large dams (those more than 15 metres high) has climbed from just over 5,000 in 1950 to about 38,000 today. More than 85 per cent of the large dams now standing have been built during the last 35 years. This is a massive change in the global aquatic environment in a very short period of time.

1.2 Just south of the US-Mexico border, the Cucap, or "people of the river," have fished and farmed in the delta of the Colorado river for some 2,000 years. Today, they are a culture at risk of extinction. Just 40 to 50 families remain in the delta region. There is little work for the younger tribal members, and many have migrated to the cities. Traditionally, they ate fish three times a day, but now they are fortunate to have it once a week. In addition, their water is too salty to grow melons, squash, and other traditional crops.

1.3 The reason for the Cucaps' precarious state lies in the neon lights of Las Vegas, the cotton fields of Arizona, and the swimming pools of Los Angeles. The Colorado river, the Cucaps' lifeblood, has been so heavily dammed and diverted in the western United States that it literally disappears into the desert before it reaches the sea. Seventy-five years ago, the American naturalist Aldo Leopold described the Colorado delta as a "milk-and-honey wilderness" teeming with wildlife.

1.4 Once the planet's fourth largest lake, the Aral has lost half its area and three-fourths of its volume because of excessive diversions of its two major sources of inflow - the Amu Dar'ya and Syr Dar'ya - in order to grow cotton in the desert. Prior to 1960, the two rivers poured 55 billion cubic metres of water a year into the Aral. Between 1981 and 1990, their combined flow into the sea dropped to an average of 7 billion cubic metres, just 6 per cent of their total annual flow. Wetlands in the Aral basin river deltas have shrunk by 85 per cent. Twenty of the 24 native fish species have disappeared, and the fish catch, which totalled 44,000 tons a year in the 1950s and supported some 60,000 jobs, has dropped to zero.

1.5 Three principal forces conspire to make water scarcity a potential source of conflict: the depletion or degradation of the resource, population growth and unequal distribution or access. River basins most likely to be hot spots for hostility are those in which the river is shared by at least two countries. In some cases, water is insufficient to meet all projected demands, and there is no recognized treaty governing the allocation of water among all basin countries. Examples of such potential hot spots include the Ganges, the Nile, the Jordan, the Tigris-Euphrates, and the Amu Dar'ya and Syr Dar'ya.

1.6 At the moment, international law offers little concrete help in resolving water disputes between nations, and says virtually nothing about protection of water ecosystems.

Source: Sandra Postel, from People & the Planet 1996.

Growing demand for water in selected countries, 1955-2025

Annual renewable fresh water per person (cubic meters)

Source: Engelman and LeRoy 1993.

The international policy framework

The Copenhagen (1991) and Dublin (1992) principles

2.1 The basis for the present policy framework for integrated water resources management (IWRM) was laid at an international consultation held in Copenhagen in 1991. Two important principles were generally approved: (1) Water and land resources should be managed at the lowest appropriate level; that is, as close as possible to the water and land users, and; (2) Water should be considered as an economic good, with a value reflecting its most valuable potential use.

2.2 In 1992, these principles were increased to four at a pre-UNCED conference held in Dublin: (1) Water is a limited and vulnerable resource, necessary to maintain life, development and environment. Effective management should link all land and water use across the whole of a catchment area or groundwater aquifer; (2) Water development and management should be based on a participatory approach involving planners and users at all levels, and decisions should be taken at the lowest appropriate level; (3) There is a need to address the often forgotten fact that women play a crucial role in the provision, management and safeguarding of water, and; (4) Water has an economic value in all its competing uses and should be recognized as an economic good.

UNCED (1992) and after

2.3 Water issues were treated in a special chapter of Agenda 21, adopted at the UNCED conference in Río de Janeiro in 1992. The four principles mentioned above became the guidelines for this chapter, which also included recommendations concerning further action. Since 1992, several international high-level meetings have been held to further develop policies and strengthen joint actions within the field of integrated water resources management. During two conferences summoned by OECD/DAC in 1994 and 1995, IWRM was confirmed as the operative concept and donors were asked to prepare themselves for long-term commitments regarding development cooperation within this area. In 1997, the results will be evaluated at an international follow-up conference in New York, organized by the Commission for Sustainable Development.

United Nations

New York, USA
Geneva, Switzerland.
The United Nations is the highest level intergovernmental organization in the world. Founded in 1945 by 51 representatives of the Allied Nations engaged in World War II, the UN is the legal successor of the League of Nations which had emerged after World War I. The UN is an association of nations which signed the United Nations Charter, pledging to maintain international peace and security and to cooperate internationally to create the political, economic and social conditions to achieve peace and security. The charter does not authorize the organization to intervene in matters which are strictly the national concern of any state.

 ## SECURITY COUNCIL

The Security Council is the UN's principal organ for the enforcement of the organization's aims and principles. It is made up of five permanent members (China, France, Russia, United Kingdom and the United States) and ten members elected by the General Assembly for two-year terms, who may not be re-elected for consecutive periods. The Security Council's major function is to maintain world peace and security. It may urge UN members to apply economic and other sanctions and may take military action against an aggressor state. Decisions on procedural matters are made by the affirmative vote of any nine members. For decisions on other matters the required nine affirmative votes of all five permanent members, or at least their abstention.

 ## THE SECRETARIAT

The Secretariat comprises the Secretary-General, who is the UN chief administrative officer (appointed for five years by the General Assembly on the recommendation of the Security Council) and by a staff of over 4,000 specialists recruited from all member nations, in proportion to each nation's financial contribution to the UN.

 ## INTERNATIONAL COURT OF JUSTICE

The Hague, Netherlands.
Also known as the Hague Court, the International Court of Justice is the principal judicial body of the UN. All UN members have signed the court's statute which is an integral part of the UN Charter. The court is composed of 15 judges elected by the General Assembly and the Security Council.

 ## TRUSTEESHIP COUNCIL

The Trusteeship Council was created to protect the interests of those peoples who live in non-autonomous territories, that is, colonies placed under the trusteeship of an administering state. These territories are generally former colonies of countries that were defeated in either World War. Of the 12 territories originally under this system, only one remains today: the Pacific Islands, administered by the United States and split up in four associated states, a system similar to that applied in Puerto Rico. The remaining 11 either became independent or joined other independent countries. They are: British Togo (joined Gold Coast to form Ghana in 1967); French Cameroon (became Cameroon in 1960); Italian Somalia (joined British Somalia to form Somalia in 1960) French Togo (became Togo in 1960); Northern British Cameroon in 1961); Tanganyika (gained independence in 1961 an later joined Zanzibar to form Tanzania); West Samoa (gained independence in 1968); New Guinea (became independent Papua New Guinea in 1975); the Pacific Trust Territories (gave rise to Republic of Palau; North Marianas Community; Federated States of Micronesia and the Marshall Islands).

GENERAL ASSEMBLY

UN Plaza, New York, NY, 10017, USA
The General Assembly consists of all members of the UN and has a large Third World majority. It is the major world forum for the discussion and adoption of recommendations on issues affecting international peace and security. It does not, however, have the power to enforce its decisions.

ECONOMIC AND SOCIAL COUNCIL

The ECOSOC is made up of 54 members elected by the General Assembly for three-year terms. It is the major policy-making and co-ordinating organ for UN efforts on economic, social, cultural, educational and health questions and for the enforcement of human rights. Each ECOSOC member has one vote and decisions are made by any majority of votes. ECOSOC performs its functions with the help of several commissions. The are five regional economic commissions: the Economic Commissions for Latin America and the Caribbean (ECLAC) in Santiago, Chile; the Economic and Social Commission for Asia and the Pacific (ESCAP) in Bangkok, Thailand; the Economic Commission for Africa (ECA) in Addis Ababa, Ethiopia; the Economic Commission for Western Asia (ECWA) in Beirut, Lebanon; and the Economic Commission for Europe (ECE) in Geneva, Switzerland.

UNEF - United Nations Emergency Force.
UNMOGIP - United Nations Military Observation Group in India and Pakistan.
UNPROFOR - United Nations Protection Force.
Other UN peace-keeping and observation bodies.

Committee of Major States.

Disarmament Commission.

Main Commissions
CHR - Commission on Human Rights.
CND - Commission on Narcotic Drugs.
CSD - Commission for Social Development.
CSWP - Commission on the Status of Women and Population.
CSC - Statistical Commission.
CSTD - Commission on Science and Techonology for Development.
CSD - Commission on Sustainable Development.
CCPCJ - Commission on Crime Prevention and Criminal Justice.
CTC - Commission on Trasnational Corporations.

Permanent and procedural commissions and committees.
Other subsidiary organs of the General Assembly.

IAEA - International Atomic Energy Agency.

Habitat Commission on Human Settlements.
UNRWA - United Nations Relief and Works Agency for Palestine refugees in the near-east.
UNCTAD - United Nations Conference on Trade and Development.
UNICEF - United Nations Children's Fund.
UNHCR - United Nations High Commissioner for Refugees.
WFP - World Food Program.
UNITAR - United Nations Institute for Training And Research.
UNDP - United Nations Development Program.
UNEP - United Nations Environment Program.
UNFPA - United Nations Population Fund.
UNU - United Nations University.
UNSF - United Nations Special Fund.

Regional Commissions .
Organic Commissions.
Permanent, Special and Temporary committees.

GATT - General Agreement of Tariffs and Trade (now WTO).
ILO - International Labour Organization.
FAO - Food and Agriculture Organization.
UNESCO - United Nations Education Science and Culture Organization.
UNIDO - United Nations Industrial Development Organization.
WHO - World Health Organization.
IMF - International Monetary Fund.
IDA - International Development Association.
IMO - International Maritime Organization.
IBRD - International Bank for Reconstruction and Development.
IFC - International Finance Corporation.
ICAO - International Civil Aviation Organization.
UPU - Universal Postal Union.
ITU - International Telecommunications Union.
WMO - World Meteorological Organization.
WIPO - World Intellectual Property Organization.
WTO - World Trade Organization.

INTERNATIONAL ORGANIZATIONS

Directory

State and/or governmental organizations

Amazon Pact

[1]   Founded July 3, 1978. Member countries: Bolivia, Brazil, Colombia, Ecuador, Guyana, Peru, Surinam and Venezuela. Designed to halt the "internationalization" of the Amazon Basin and to restrict its exploration and economic exploitation to those countries sharing its territory.

Andean Pact

Casilla Postal 3237, Avda. Paseo de la República 3895, San Isidro, Perú.

[2]   Founded May 25, 1969. Member countries: Bolivia, Ecuador, Peru, Venezuela (founding members), and Colombia (1973). Associate member: Panama. (Chile was a charter member but dropped out in 1976). A sub-regional economic integration plan, the agreement establishes uniform customs tariffs for all the countries in the area. However, due to the state of crisis which besieged the region during the 1980s, the timetable for trade deregulation and introduction of a common external tariff has not been complied with.

Arab League

[3]   Founded in March 22, 1945. Member countries: Saudi Arabia, Algeria, Bahrain, United Arab Emirates, Egypt, Yemen, Iraq, Jordan, Kuwait, Lebanon, Libya, Morocco, Mauritania, Oman, Palestine, Qatar, Syria, Somalia, Sudan, Tunisia, and Djibouti. The league's objectives include closer cooperation between member States and the coordination of their political action. It promotes close mutual understanding in economic, financial, trade, aviation, postal, telecommunication, sanitation, cultural and legal matters.

Association of Southeast Asian Nations (ASEAN)

Box 2072, Jakarta, Indonesia.

[4]   Founded in 1967. Member countries: Brunei, Indonesia, Malaysia, Philippines, Singapore and Thailand. Its aim is "to accelerate economic progress and social and cultural development in the region."

Asia Pacific Economic Cooperation (APEC)

[5]   Founded in November 7, 1989, to promote trade and investment in the Pacific basin. Its members are all the ASEAN members (Brunei, Indonesia, Malaysia, Philippines, Singapore, Thailand), plus Aotearoa, Australia, Canada, China, Hong Kong, Japan, South, Korea, Papua New Guinea, Taiwan, USA.

British Commonwealth

Marlborough House, Pall Mall, London, United Kingdom.

[6]   Founded on December 31, 1931. The British Commonwealth originated with the breakup of the British Empire. Member countries: the United Kingdom and virtually all of its former colonies. The Commonwealth is not a federation of states, nor does it have any rigid contractual obligations. It operates through informal meetings of heads of state and government, at which resolutions are reached by consensus. Queen Elizabeth II is seen as a symbol of association by all member countries and as the Commonwealth's highest authority, though she wields no legal power among them. She is also acknowledged as the formal head of state in Aotearoa (New Zealand), Australia, the Bahamas, Barbados, Canada, Fiji, Grenada, Jamaica, Kiribati, Mauritius, Papua New Guinea, the Solomon Islands and Tuvalu.

Caribbean Community and Common Market (CARICOM)

Bank of Guyana Building, P.O. Box 607, Georgetown, Guyana.

[7]   Founded in July 4 1973, to replace the Caribbean Free Trade Association (CARIFTA), founded in September 1966. Member countries: Antigua, Barbados and Guyana (charter members), Jamaica, Trinidad and Tobago, St Kitts-Nevis-Anguilla, Dominica, Grenada, Montserrat, St Lucia, St Vincent and the Bahamas. The free-trade zone was converted into a regional common market, on August 1, 1973. Although CARICOM managed to increase intraregional trade, the insular nature of the member States' economies, as well as their bilateral relations with countries outside the region, have prevented further progress.

Central American Common Market (CACM)

4a. Avenida 10-25, Zona 14, Apto. Postal 1237, Guatemala City, Guatemala.

[8]   Introduced in 1958 under the Multilateral Treaty of Free Trade and Economic Integration, CACM was institutionalized in December 13, 1960, under the General Treaty of Central American Economic Integration. It provides for the elimination of tariffs between member countries and the establishment of common tariffs for trade between this region and the rest of the world.

Colombo Plan

Melbourne Avenue, P.O. Box 596, Colombo 4, Sri Lanka.

[9]   Founded in 1950 by the foreign ministers of the Commonwealth countries meeting in Colombo, the plan envisages cooperative economic and social development in Asia and the Pacific. At present it has 26 member countries, including Canada, United States and United Kingdom.

Common Afro-Mauritian Organization (OCAM)

BP 965, Bangui, Central African Republic.

[10]   Founded April 28, 1966, replacing the Afro-Malagasy Union for Economic Cooperation (UAMCE). Member countries: Burkina Faso, Benin, Côte d'Ivoire, Mauritius, Niger, Central African Republic Rwanda, Senegal and Togo.

Commonwealth of Independent States (CIS)

Minsk, Belarus

[11]   Founded in December 1991. Member states: Armenia, Azerbaijan, Belarus, Kazakhstan, Kyrgyzstan, Moldova, Russia, Tajikistan, Turkmenistan, Ukraine and Uzbekistan. Its aims are to maintain on a basis the economic, political and military relations of the republics of the former Soviet Union. In March 1992, the presidents of the parliaments of Armenia, Belarus, Kazakhstan, Kyrgyzstan, Russia, Tajikistan and Uzbekistan decided to create an inter-parliamentary assembly.

Customs and Economic Union of Central Africa (Union douanière et économique de L'Afrique Central-UDEAC).

96, Bangui, Central African Republic.

[12]   Founded in December 1964 to replace the union douanière équatoriale (founded in 1959); it began operating in January 1966. Member countries: Cameroon, Central African Republic, Zaire and Gabon.

Economic Community of West African States (ECOWAS) (Communauté Economique de L'Afrique de l' Ouest - CEAO)

Ouagadougou, Burkina Faso.

[13] Founded May 30, 1975. Member countries: Benin, Burkina Faso, Cape Verde, Côte d'Ivoire, Gambia, Ghana, Guinea-Bissau, Liberia, Mali, Mauritania, Niger, Nigeria, Senegal, Sierra Leone, Togo. Its aim is to lay the foundations for the economic integration of its member States. Entente Council (Conseil de l'entente) BP 20824 Abidjan, Côte d'Ivoire.

[14] Founded in May 1959. Member countries: Burkina Faso, Benin, Côte d'Ivoire, Niger and Togo. The Entente's main concerns are the integration and economic development of its member countries.

European Free Trade Association (EFTA)

Geneva, Switzerland

[15] Founded in 1959. In 1961 its member countries were: Austria, Denmark, Finland, United Kingdom, Norway, Portugal, Sweden and Switzerland. In 1973, United Kingdom and Denmark joined the EC and Portugal in 1986. The seven remaining members of EFTA negotiated an agreement with the EC for a reciprocal elimination of tariffs on industrial goods by January 1984. With the admission of Austria and Finland to the European Union, EFTA was due to merge with the European Union.

European Union

Headquarters: Rue de la Loi 200, B-1049, Brussels, Belgium.

[16] The idea of a united Europe first emerged in 1950. The European Union went into effect on January 1, 1995, replacing the European Community which had been set up in 1967. Its key institutions continue to be the Council of Ministers, the European Commission, the European Parliament, the Court of Justice and the European Council. It is made up of the Coal and Steel Community, founded in 1951; the Economic Community (EC, the "Common Market"); and the Atomic Community (EURATOM), both dating from the year 1957. Member countries: Belgium, France, Germany, Holland, Italy and Luxembourg (charter members); Denmark, Ireland and the United Kingdom (since 1973); Greece (since 1981); Spain and Portugal (since 1986). In February 1993,

in Maastricht, the Netherlands, the heads of state and of government of what was then known as the European Community ("The Twelve") signed treaties for political and monetary union (European Union), which were ratified by plebiscite or by the parliaments of all the members of the EC, a process which was completed in mid-1994. That same year, treaties were signed admitting four countries: Austria, Finland, Norway and Sweden, but Norwegians rejected membership in a referendum. New enlargement negotiations began in 1996 with some other countries. Turkey, Malta, Cyprus, Hungary and Poland have all applied for membership officially. Bulgaria, the Czech Republic, Estonia, Hungary, Latvia, Lithuania, Poland, Romania and Slovakia are invited as new candidate members to one meeting of the European Council every year.

Group of Eight

[17] Member countries: Uruguay, Colombia, Venezuela, Peru, Argentina, Mexico, and Brazil. (Panama was a founding member but its membership was suspended in 1989.) An organization set up to improve regional cooperation.

Group of 15 (G15)

[18] Founded in 1990 in Kuala Lumpur. Member countries: Algeria, Argentina, Brazil, Egypt, India, Indonesia, Jamaica, Malaysia, Mexico, Nigeria, Peru, Senegal, Venezuela, Yugoslavia, and Zimbabwe. Its aim is to promote cooperation and consultation among all the countries of the Third World, in order to confront common problems and devise a common strategy in dealing with international organizations and development agencies.

Group of Seven (G-7)

[19] Founded in 1975, it brings together the world's seven most industrialized countries: Germany, Canada, United States, France, Italy, Japan and the United Kingdom. Its initial goal was to coordinate economic, financial and trade policies, but in actual fact, G7 summit conferences dealt with all major international issues (like arms, energy, foreign debt, technology). In 1994, Russia was invited to participate in the Group of 7's meetings, thus broadening the original scope of the organization.

Group of 77 (G-77)

[20] Established in October 1967. The name of this organization goes back to the first UNCTAD

UN
Reforms and disagreements

The approval of the UN's reform project anticipated along with the organization's 50th anniversary in 1995, remains to be concluded. The international organization's material situation further complicates matters since the debt owed by important contributors such as the United States have left it on the brink of bankruptcy. Several countries refuse to pay and their debts amount to approximately $4 billion.

[2] In fact there are several reform projects. There is a general agreement on the modification of the Security Council's integration - increasing the number of members - and its transformation into an organisation where the countries of the world are more "equally" represented. Supporters of this radical reform explain that, unlike institutions such as the International Monetary Fund - also created at the end of World War II and whose organization was based on a number of votes equal to each country's financial contribution - the UN was designed according to the "principle of equal sovereignty among all of its members". However, the work group in charge of the reform project has only been able to find "significant differences" among the various proposals. Furthermore, no reform of the UN's Charter can be approved without the agreement of two thirds of the member countries, including all of the Security Council's permanent members.

[3] In other words, the reform's approval still depends on the agreement of the four countries considered to have won World War II - United States, Russia, United Kingdom, France - and China. These five "powers" seem willing to accept the incorporation of two other economically powerful countries as permanent members of the Council, Germany and Japan - considering them to be "eager to prevent conflicts and adequately equipped for it" - and to increase the number of non-permanent members from 10 to 13.

[4] The "powers" proposal has divided the other countries into a minority - formed by Colombia and Malaysia, among others - which criticizes the "confirmation of inequality" and requests the suppression of the permanent member category, and a majority which seeks less radical changes. Among these last the Canadian proposal to create a semi–permanent member status stands out; the elimination of the right to veto, sought by Iran, Mexico, Honduras and Cuba; or its restriction to certain decisions, as proposed by Australia, Spain and Zimbabwe.

[5] Disagreements also took place regarding the succession of UN Secretary-General since 1991, Boutros Boutros-Ghali. The former Egyptian Minister of Foreign Affairs proposed his own re-election in mid-1996, provoking hostility from the United States which threatened to veto his nomination. This led to speculation on his possible successors - among others, Ghana's Kofi Annan, head of the UN's peace maintenance operations division and Norway's former prime minister, Gro Harlem Brundland.

Conference, held in Geneva in 1964, when the 77 attending Third World countries approved a resolution to challenge the industrialized nations to be able to participate more equitably in world trade, dominated by the wealthy countries. In 1991, this association had 129 members.

Gulf Cooperation Council (GCC)

Riyadh, Saudi Arabia

[21] Founded in 1981. Member countries: Saudi Arabia, Bahrain, the United Arab Emirates, Kuwait, Qatar and Oman. The group was designed to neutralize the effects of Iran's Islamic revolution, although it is not the explicit objective of the group. The council coordinates the member countries' economic, social, educational and defence policies.

Inter-Parliamentary Union

Place du Petit-Saconnex, 1209 Geneva, Switzerland.

[22] Founded in 1889 to promote personal contact between members of the world's parliaments. It holds a yearly Inter-Parliamentary Conference with representatives from the 77 national parliamentary groups affiliated with the union.

Latin American Economic System (SELA)

Caracas, Venezuela.

[23] Founded in 1975. Member countries: 26 Latin American states. Its aims are to coordinate economic action to improve integration between member countries and address common problems in the financial, commercial and customs spheres. SELA is the only regional integration organization which includes Cuba.

Latin American Energy Organization (OLADE)

Quito, Ecuador.

[24] Founded November 2, 1973. Member countries: 24 Latin American countries. The organization is designed to promote Latin American integration in the area of energy. In 1981 the Latin American Program for Energy Cooperation (PLACE) was established as an instrument for the execution of projects oriented towards the study, planning and development of the region's energy resources.

Latin American Integration Association (ALADI)

Cebollatí 1461, Casilla de Correo 577, Montevideo, Uruguay.

[25] Founded in 1980 to replace the Latin American Free Trade Association (ALALC), created in 1962. Member countries: Argentina, Bolivia, Brazil, Colombia, Chile, Ecuador, Mexico, Paraguay, Peru, Uruguay and Venezuela. ALADI establishes a regional, preferential tariff system and provides mechanisms for agreements to be made on a regional basis. It also establishes differential treatment on the basis of three categories (according to level of development). The first group includes Bolivia, Paraguay and Ecuador; the second, Colombia, Peru, Chile, Uruguay and Venezuela; and the third, Brazil, Mexico and Argentina.

Latin American Parliament (Parlamento Latinoamericano)

Casilla 6401, Lima, Peru.

[26] Founded in 1965. Its purpose is to "promote, harmonize and channel the movement towards the economic, political and cultural integration of Latin America". It has an organizational structure similar to that of the Inter-Parliamentary Union. In the period of the region's military dictatorships, the Latin American Parliament became a forum for reports of violations, and granted membership to representatives of parliaments dissolved by these dictatorships. Member parliaments: Argentina, Aruba, Bolivia, Brazil, Chile, Colombia, Costa Rica, Cuba, Dominican Republic, Ecuador, El Salvador, Guatemala, Honduras, Mexico, Netherlands Antilles, Nicaragua, Panama, Paraguay, Suriname, Uruguay and Venezuela.

Southern African Development Community (SADC)

[27] Established 17 August 1992. Evolved from the Southern African Development Coordination Conference (SADCC), its aim is to promote regional economic development and integration. Members: Angola, Botswana, Lesotho, Malawi, Mozambique, Namibia, Swaziland, Tanzania, Zambia, and Zimbabwe.

South Centre

Chemin. du Champ d'Anier 17, 1211 Geneva 19, Switzerland.

[28] Established July 30, 1995. Evolved from the South Commission, its aim is to give support to Southern countries' cooperation and to strenghten their capacity in North-South negotiations. Members: Algeria, Angola, Benin, Bolivia, Brazil, Burundi, Cambodia, Cape Verde, China, Colombia, Côte d'Ivoire, Cuba, North Korea, Egypt, Ghana, Guyana, Honduras, India, Indo-nesia, Iran, Jamaica, Jordan, Libya, Malawi, Malaysia, Mali, Micronesia, Morocco, Mozambique, Namibia, Nigeria, Pakistan, Panama, Philippines, Seychelles, Sierra Leone, South Africa, Sri Lanka, Sudan, Surinam, Uganda, Tanzania, Vietnam. Yugoslavia and Zimbabwe.

The Lomé Conventions

Maison des ACP, Av. Georges Henri 451, 1040 Brussels, Belgium

[29] The Lomé Conventions are agreements between the European Economic Community (EEC) and 46 African, Caribbean and Pacific (ACP) states on trade and financial and economic cooperation, signed on February 28, 1975 and July 31, 1979 in Lomé, Togo. For the 22 participating African countries, the Lomé Conventions replace the Arusha Agreement of 1968 and the Yaoundé Conventions of 1969. Unlike other similar agreements, the Lomé Conventions do not compel ACP countries to adhere to mutual preferences in their trade relations, requiring only that they apply the "most favored nation" clause. Under the Lomé Conventions, a free trade zone was established for all industrial products and for 96 per cent of all agricultural products. In addition, the agreements established a system for stabilizing the exchange of exports, financial aid and technical cooperation, between members. The 1990 Lomé Accord IV was signed by 69 ACP countries.

MERCOSUR (Common Market of the South)

[30] Founded March 26, 1991 by Argentina, Brazil, Paraguay and Uruguay, with the aim of establishing an economically integrated zone in which goods and services will circulate freely. In mid-1994 the four governments readjusted the timetable announced in 1991 in order for the common external tariff to go into effect in 2001. Uruguay and Paraguay will have an additional 5 years to adapt their tariffs to the 14 per cent which had been agreed upon. On October 1st 1996, Chile made effective an agreement with the Mercosur, though not joining it as a full member.

Non-Aligned Movement (NAM)

[31] Founded in Belgrade, Yugoslavia in 1961, by countries not identified with the blocs involved in the East-West conflict, to promote decolonization, disarmament and self-determination, and to deter the use of force to resolve international controversies. By the time the 9th Summit Conference was held in 1989 in Belgrade, the Movement included 102 nations, mostly from Asia, Africa and Latin America. At the 10th Summit, held in 1992 in Jakarta, Indonesia, the Non-Aligned Movement initiated a reformulation of its objectives, as a result of the new international situation. New concerns included the defense of the environment, the promotion of more cooperative dialogue between the North and the South, and the democratization of the dealings of the United Nations. A series of conflicts between some member States, as well as the withdrawal of other countries, such as Argentina, have debilitated the Movement.

North American Free Trade Association (NAFTA)

[32] In February 1991, the governments of Canada, Mexico and the United States signed an agreement known as the North American Free Trade Association, in order to create a free trade and economic integration zone in North America. There was opposition to the agreement in both labor union and ecological quarters, especially in the United States, which faced the threat of the loss of millions of jobs, due to the wage differential between that country and Mexico. The three countries created the North American Development Bank with funds totalling more than $3 billion. This bank joins other existing international credit agencies.

Organization for Economic Cooperation and Development (OECD)

Chateau de la Muette, 2 rue André Pascal, 75775 Paris Cedex 16, France.

[33] Founded December 14, 1960, to replace the Organization for European Economic Cooperation (OEEC), established in 1948. Member countries: Aotearoa, Australia, Austria, Belgium, Canada, Denmark, Finland, France, Germany, Greece, Iceland, Ireland, Italy, Japan, Luxembourg, Mexico, Netherlands, Norway, Portugal, Spain, Sweden, Switzerland, Turkey, United Kingdom and United States. Its aim is to assist member governments in the formulation of their economic and social policies and coordinate the joint positions of Northern capitalist nations.

Organization of African Unity (OAU)

Box 3243, Addis Ababa, Ethiopia.

[34] Founded in May, 1963. Mem-

bers: 51 African countries, including the Arab Democratic Republic of Sahara, represented by the Polisario movement. Its aim is to address regional issues and encourage pacific solution to conflicts.

Organization of American States (OAS)

17th Street and Constitution Avenue, NW, Washington DC 20006, USA.

[35] Founded in May 1948, its charter went into effect in December 1951. The OAS replaced the Pan-American Union created in the 19th century. The charter was signed by 21 countries; Cuba was expelled in 1962. In the 1970s, six additional Caribbean and Central American countries joined. The OAS was created under the aegis of the US, which managed to mold the organization to suit Washington's policies in Latin American.

Rio de la Plata Basin Treaty

[36] Founded February 27, 1967. Member countries: Argentina, Bolivia, Brazil, Paraguay and Uruguay. "It is the decision of these Governments to carry out a complete study of the Rio de la Plata basin, in order for multinational, bilateral and national construction projects to be set up, which will contribute to the progress of this region". The body meets regularly to deal with problems concerning navigation and transportation links.

Financial Organizations

African Development Bank (AfDB)

1387, Abidjan, Côte d'Ivoire.

[37] Founded in November 1964. Member countries: all independent countries in the African continent. The bank was created on the initiative of the UN Economic Commission for Africa with the support of the Conference of African Heads of State and Government held in Addis Ababa in 1963. Since 1979, States outside the African continent may also become members of the bank. Capital base: $19.8 billion.

Arab Bank for Economic Development in Africa (ABEDA)

Sharaa el Baladia, P.O. Box 2640, Khartoum, Sudan.

[38] Founded in 1973 (became op-

erational in 1975). Created by the Arab League, its purpose is to finance development projects in sub-Saharan Africa.

Arab Monetary Fund (AMF)

Box 2818, Abu Dhabi, United Arab Emirates.

[39] Founded April 27, 1976. Member countries: 21 members. Created by the Arab League, the AMF is a regional financial institution with independent legal status. Its aim is to assist member countries in the elimination of trade restrictions and in developing Arab capital markets.

Asian Development Bank (AsDB)

Headquarters: Manila, Philippines.

[40] Founded in 1965. It is currently made up of 32 countries of the region, as well as 15 non-Asian states. Japan exerts a dominant influence within the Bank. Caribbean Development Bank (CDB) P.O. Box 408, Wildey, St. Michael, Barbados Founded in January 1970. Member countries: CARICOM countries and the Bahamas, Belize, Virgin Islands, Cayman, Turks and Caicos Islands, Colombia, Mexico, Venezuela and France.

Franc Zone (FZ)

[41] Member countries: France and its present or former colonies whose currencies are linked to the French franc and who agree to hold their reserves mainly in the form of French francs. This monetary union is based on bilateral agreements between France and each member country. Member countries: Benin, Burkina Faso, Cameroon, Central African Republican, Chad, Congo, Côte d'Ivoire, Equatorial Guinea, Gabon, Mali, Niger and Senegal.

Inter-American Development Bank (IDB)

17th Street, N.W., Washington, DC 20577, USA.

[42] Founded in December 1959. Member countries: all member states of the Organization of American States (OAS). The IDB's primary aim is to provide credits for infrastructure development projects and technical assistance in member countries. The US government is the main contributor to the bank's funds. Since 1976, non-American States may also be members of the IDB.

International Monetary Fund (IMF)

Washington DC, USA

[43] A specialized agency of the UN, the IMF was created in 1944 by the Bretton Woods Accords. Its aim is to improve the international system of payments. It offers lines of credit to governments, subject to certain commitments in their economic and fiscal policies. Voting power in the IMF's decision-making process is determined by the member country's contribution of capital, which means that its policies are controlled by the world's richest nations. The IMF currently has 165 member States, among them Russia and 11 other countries that made up the former Soviet Union (members since June 1, 1992).

Islamic Development Bank (ISDB)

Box 5925, Jeddah, S. Arabia.

[44] Founded in 1974. Set up by the Islamic Conference, the bank aims at encouraging economic development and social advancement in member countries and Muslim communities, in accordance with the principles of the Koran.

World Bank (WB-IBRD)

Washington DC, USA

[45] A specialized agency of the UN, the International Bank for Reconstruction and Development was created in 1944 by the Bretton Woods Accords. Its goals are to promote international trade and economic development by means of loans and technical assistance. Voting power (in the Bank's decisions) is determined by each country's contribution of capital to the Bank's funds, which in essence means that the Bank is controlled by the world's richest countries. In 1971 the WB had 171 members, including Russia and 12 other republics belonging to the former Soviet Union (which joined on June 1st, 1992). The WB Group also includes three complementary organizations: the International Development Association (IDA), founded in 1960; the International Financial Corporation (IFC), created in 1956; and the Multilateral Investment Guarantee Agency (MIGA), which was established in 1985 and is autonomous with relation to the IBRD.

World Trade Organization (WTO)

Headquarters: Centre William Rappard, 154 rue de Lausanne, CH 1211, Geneva, Switzerland.

[46] Created in 1994 to replace the General Agreement on Tariffs and

Trade (GATT). Established on 1 January 1995, is the legal and institutional foundation of the multilateral trading system. WTO members have agreed to enter into "reciprocal and mutually advantageous arrangements directed to the substantial reduction of tariffs and other barriers to trade and to the elimination of discriminatory treatment in international trade relations."

Military Organizations

Australia-New Zealand (Aotearoa)-United States Security Treaty (ANZUS)

[47] Signed September 1, 1951, the treaty went into effect on April 29, 1952. It serves as an "umbrella" group for the chain of military organizations which has been under United States leadership since the end of WWII: NATO, CENTO and SEATO. Since 1954, ANZUS has formed part of the Southeast Asian Treaty. Australia has decided to reduce its commitments with ANZUS and Aotearoa suspended its security obligations on August 11, 1986.

Central American Defence Council (CONDECA)

[48] Founded in 1964. Member countries: Guatemala, Honduras and El Salvador; Nicaragua is a former member. When the signatories to the treaty met in August 1979 in Guatemala, only two member countries were present: Guatemala and El Salvador. The treaty was not enforced again until October 1983, when the United States reactivated CONDECA as an instrument against Sandinistas in Nicaragua.

Inter-American Treaty of Reciprocal Assistance Tratado Interamericano de Asistencia Recíproca (TIAR)

[49] Founded in 1947 by the governments of all American countries, it is also known as the Treaty of Rio de Janeiro. Its original aim was to combat "external threats", although it became increasingly concerned with halting "subversion." During the Malvinas war, Argentina attempted to invoke the treaty against the United Kingdom, but the initiative was vetoed by the US.

North Atlantic Treaty Organization (NATO)

Brussels, Belgium

[50] Created April 4, 1949. Its current members are: Belgium, Canada, Denmark, France, Germany, Greece, Iceland, Italy, Luxembourg, the Netherlands, Norway, Portugal, Spain, Turkey, UK and the United States. The armed forces of France and Spain are not integrated into the NATO command structure; likewise, Iceland does not have armed forces of its own. NATO is currently promoting the association of the countries of eastern and southeastern Europe - members of the defunct Warsaw Pact - although not as full members. With the withdrawal of the major part of US troops from Germany, Franco-German military integration has taken on a greater importance. NATO, and especially the armed forces of the United States, France and UK, has intervened in increasingly open fashion, in conflicts beyond the North Atlantic region which NATO was originally created to defend (eg, Bosnia and Herzegovina).

Western European Union (WEU)

London, United Kingdom.

[51] Created in 1948 in Brussels, Belgium. Charter members: Belgium, France, Holland, Luxembourg and the United Kingdom. Germany and Italy joined in 1954, Spain and Portugal in 1988. Its original aim was to promote economic, political, cultural and military cooperation. In 1950, the WEU transferred some of its defence-related functions to NATO, and in 1954, the EC assumed those related to economic, political and cultural cooperation. In 1984, the WEU decided to expand its role, which at the time was restricted to solely military functions. In 1991, Greece demanded that it be admitted to the WEU.

Non Governmental Organizations

Amnesty International (AI)

1 Easton Street, London, WC1X 8DJ, United Kingdom

[52] Founded in 1961 to monitor the Universal Declaration of Human Rights and other internationally recognized human rights agreements. In 1996 AI had an active worldwide membership with more than 1,100,000 individual members, subscribers and supporters in over 160 countries and territories.

Alliance of the Indigenous-Tribal Peoples of the Tropical Forests

[53] Founded in February 1992, in Penang, Malaysia, by 29 representatives of indigenous peoples of Asia, Africa and Latin America, with the aim of combating the destruction of their ancestral lands, defending their rights and presenting their case before international agencies involved in drawing up policies related to the tropical forests.

Amazon Basin Indigenous Organizations Coordinating Committee (COICA)

[54] Founded in Lima Peru in March 1984 by national indigenous organizations in Peru, Bolivia, Ecuador, Brazil and Colombia. Its aim is to strengthen local organizations and ties among indigenous peoples, promote the exchange of experiences and the union of indigenous peoples in their struggle to defend their rights and their native lands. COICA is authorized to represent its members in international forums, before governments and intergovernmental organizations.

Womens' World Banking

New York, United States

[55] Established in 1980, it coordinates the work of over 45 financial institutions managed by women. Its aim is to increase the participation of women in the corporate world and in economic development.

Ecological Organizations

American Ecological Action Pact

[56] A plan of action approved in 1989, in Las Vertientes, Chile by representatives of different regional ecological organizations and open to the national ecological networks and other community-based organizations in Latin America. Its aims are to achieve sustained development by changing current development models, and the existing international economic order, to put an end to their destructive effects upon nature and society.

Asia-Pacific Peoples' Environment Network (APPEN)

43, Salween Road, 10050 Penang, Malaysia

[57] See also Third World Network below. Enda Tiers-Monde Dakar, Senegal.

[58] ENDA, or Environment and Development in the Third World, was established in 1972 with the support of the UN Environment Program. The group does practical field work as well as research, training, advising, publication and communication. The underlying focus of its work is the relationship between environment, poverty and democracy. As ENDA's main office is in Senegal, more than half of its programs are carried out in West Africa. However, it also has branches elsewhere in Africa and in Latin America, the Caribbean and India.

Environment Liaison Centre (ELC)

[59] Founded in 1975, the ELC seeks to support non-governmental organizations (NGOs), especially those in the Third World, which deal with ecology and sustainable development. It provides information, develops ties and coordinates activities with agencies and NGOs, and offers consultation for development projects. Some 250 organizations in 70 countries belong to the ELC, which also maintains links with many other groups.

Friends of the Earth (FoE)

London, United Kingdom

[60] Originally a Dutch organization, FoE coordinates 33 autonomous national branch organizations. Its aim is to support policies and develop courses of action in defense of the environment, persuading - whenever necessary - governments, corporations or international agencies to change their programs and their plans or operations.

Greenpeace International

Keizersgracht 176, 1016 DW Amsterdam, Netherlands.

[61] Founded in 1971. Its aims are to identify human activities which can affect the ecological balance, and to carry out campaigns in defence of the environment. It is a staunch defender of Antarctica, and has proposed that an "international park" be created on this continent. It operates on an international level, and has 43 offices in 30 countries, complemented by autonomous local organizations.

International Federation of Environmental Journalists (IFEJ)

14, rue de la Pièrre Levé F - 75011 Paris, France

[62] Founded in November 1994. Bringing together journalists, authors, and information professionals of all countries and modes of expression, the federation's goal is the diffusion, through all forms, of exact information free from any pressures, in the fields of ecology, environmental management, nature conservation and sustainable development.

Third World Network

[63] In addition to its activities which are specifically linked to ecology (see "Other Organizations"), the Third World Network and the Asia-Pacific Peoples' Environment Network (APPEN) jointly make up the International Secretariat of the World Rainforest Movement.

World Conservation Union (IUCN)

Rue Mauverney 28, CH-1196 Gland, Switzerland

[64] Formerly known as the International Union for the Protection of Nature and Natural Resources (IUPN), which was founded in 1948, the IUCN promotes the protection and rational use of natural resources. It is made up of nearly 500 national governments, state authorities, private entities and ecological groups, in 114 countries.

World Rainforest Movement (RAN)

Headquarters: 450 Sansome Street, Suite 700, San Francisco, CA 94111, USA.

[65] Established in 1986, it is an association of groups and individuals committed to actions aimed at protecting the planet's tropical rainforests. Likewise, it supports the indigenous peoples of these regions, who are fighting against the destruction of the jungle by commercial and mining interests (among others). Its International Secretariat is located in Penang, Malaysia, with branch offices in Brazil, India, southeast Asia, Europe, North America, Japan and Australia.

Worldwide Fund for Nature (WWF)

[66] The World Wildlife Fund changed its name to the Worldwide Fund for Nature in 1989.

Originally concerned with the preservation of wildlife and natural habitats, the WWF is currently involved in promoting development projects, controlling pollution and curbing wasteful consumption. From its headquarters in Switzerland, it coordinates 23 national branch offices on five continents.

Labor Organizations

Commonwealth Trade Union Council (CTUC)

Congress House, Great Russell Street, London WC1B 3LS, England

[67] Created March 1, 1980 to "promote the interests of the workers in Commonwealth countries through cooperation between national trade union federations", it groups together 25 million workers, affiliated with labour unions and federations in forty countries.

Latin American Federation of Journalists (FELAP)

Nuevo León 144, ap. 101, Mexico City, DF, Mexico.

[68] Founded in June 1976, the Latin American Federation of Journalists serves as a forum for the continent's journalists and media workers' organizations, identifying with national liberation struggles and the opposing transnational corporations and US imperialism.

International Confederation of Free Trade Unions (ICFTU)

Rue Montagne aux Herbes Potageres, 1000 Brussels, Belgium.

[69] Founded in 1949. Set up by trade federations which had withdrawn from the World Federation of Trade Unions (WFTU), ICFTU has national affiliates in 88 countries and regional offices in Asia, Africa and Latin America: the Asian Regional Organization (ARO), based in India; the African Regional Organization (AFRO), based in Liberia; and the Inter-American Regional Organization of Workers (ORIT), with headquarters in Mexico.

International Organization of Journalists

Parizska 9, 11001, Prague 1, Czech Republic.

[70] Founded in 1946, its purpose is to "defend the freedom of the press and of journalists and to promote their material welfare". The IOJ's activities include the publication of several magazines; it also maintains training and recreation centers for journalists.

World Confederation of Labor (WCL)

Rue Joseph 11, Brussels 1040, Belgium.

[71] Founded in 1920 as the International Federation of Christian Trade Unions; reorganized under its present name in 1968. The confederation brings together international trade union federations which are sympathetic to the Christian Democratic political philosophy. The WCL has regional offices in Latin American (Latin American Confederation of Trade Unions, with headquarters in Caracas, Venezuela), Asia (BATU, with headquarters in Manila, Philippines), and Canada.

Organizations of Political Parties and Other Organizations

Afro-Asian People's Solidarity Organization (AAPSO)

Abdel Aziz Al Saoud St., Marial, Cairo, Egypt.

[72] Founded in May 1954. Set up at the Heads of Government Conference held in Colombo, Ceylon (now Sri Lanka) with the participation of Burma/Myanmar, Sri Lanka, India, Indonesia and Pakistan, the AAPSO was confirmed at the Bandung Conference in April 1955. The organization's goals are to unite and coordinate the struggle of Asian and African peoples against imperialism and colonialism, assist in the liberation struggle of these peoples and ensure their economic, social and cultural development.

Christian Democratic World Union (UMDC)

Via del Plebiscito 107, 00186 Rome, Italy.

[73] Founded in 1956. Members: some 50 parties on four continents. It coordinates the activities of political parties of Christian Democratic inspiration throughout the world. Sponsored by the Konrad Adenauer Foundation, an organization linked to the German Christian Democrats.

Independent Committee on Development Problems (The Brandt Commission)

[74] Created in 1977 in an attempt to revive the unsuccessful North-South dialogue, the committee was made up of prominent individuals under the chairmanship of West Germany's former Prime Minister Willy Brandt.

International Association for Peace Research

PO Box 70, Tampere, Finland.

[75] Founded in 1965 to "carry out interdisciplinary research into the conditions of peace and the causes of war".

International Foundation for Development Alternatives (IFDA)

Place du Marché, 1260 Nyon, Switzerland

[76] Founded in 1976, its purpose is to promote innovative research and action in an effort to promote forms of development which imply the reorganization of international relations in the direction suggested by the New Economic Order. The foundation includes individuals from the Third World and certain industrial nations in the spirit of what the United Nations call the Third System nongovernmental organizations working side-by-side with governments and the UN international system.

International Organization of Consumer Unions (IOCU)

Emmastraat 9, 2595 EG The Hague, Netherlands

[77] Founded in 1960, dedicated to the promotion and defense of consumer interests, it involves over 170 organizations in 62 different countries.

Permanent Conference of Latin American Political Parties (COPPPAL)

Félix Parra 170, San José Insurgentes, Mexico 19 DF, Mexico.

[78] Founded in October 1979, COPPPAL brings many of the region's nationalist and center-leftist parties into contact with each other.

Socialist International

88A St. John's Wood High Street, London, NWS, United Kingdom.

[79] Reorganized in 1951, the Socialist International coordinates the activities of Socialist and Social Democratic parties throughout the world. In recent years, it has tried to increase its presence in Africa and Latin America.

Stockholm Institute for Peace Research and Investigation (SIPRII)

[80] Founded in 1966, the SIPRI specializes in research on disarmament, the arms race, and arms control. Its scientific activity has become the basis for other projects such as those implemented by the independent Commission for Disarmament, commonly called the Palme Commission.

Third World Network

International Secretariat:
228 Macalister Road, Penang, 10400, Malaysia.
African Secretariat:
PO Box 8604, Accra-North, Ghana.
Latin American Secretariat:
Juan D. Jackson 1136, 11200 Montevideo, Uruguay.

[81] Founded in 1984. The Third World Network consists of numerous Third World non-governmental organizations (NGOs) - ranging from research institutes to rural or urban activist groups - and includes leading figures from both the North and South. Its aims are to defend the rights and satisfy the needs of the peoples of the Third World, through a more equitable distribution of the planet's resources and forms of development which are ecologically sound. The Third World Network also runs a news service (in English and Spanish) on a wide variety of subjects, including health, politics, economics, culture, ecology, the environment, women, science and technology.

THE UNITED NATIONS' INDIGENOUS RIGHTS document has remained in draft form since 1982. Many states questioned the draft's recognition of self-determination for indigenous peoples, as well as the concept of group or community rights.

Pending indigenous rights

The word "Peoples" in the human rights context, especially with regard to Indigenous Peoples, has been criticized since the beginning: hence the name of the "Working Group On Indigenous Populations".The states resisted the word "peoples," in 1982, calling it the Working Group on Indigenous "Populations." Brazil, for example, from the beginning raised the issue of the word "peoples" and wanted to drop the "s" on Peoples.
[2] Some states are afraid that adopting the Declaration will lead to their break-up. They fear that indigenous peoples living within their borders will use the declaration to justify their independence. Another fear is that if indigenous peoples had self-determination, they could use this right to hamper the states' wish to determine development projects on Indigenous lands. The states also fear that they would be obliged to share with Indigenous Peoples the benefits of development. Furthermore, there are states afraid of losing control over indigenous lands. Other states voiced the fear of having the obligation to give back the land.
[3] During the discussion of the United Nations' document, the United States and others presented the position that all human rights are individual rights and that there are no human rights for groups or communities. The US representatives explained that, although the US recognizes group or community rights in its domestic law - as it applies to Native Americans - the US does not recognize these rights internationally.

Neo-liberalism and the land

One of Neo-liberalism's impacts on indigenous communities has been its hampering of their access to land. In North America, for instance, NAFTA has provided a rationale for new enclosures of indigenous lands for purposes of commercial exploitation. In Canada, indigenous land claims fall on deaf ears while in the United States privatization of public lands and the commercialization of indigenous lands furthers this problem. In Mexico the desire for oil, wood and grazing land results in the privatization of *ejidos*' lands. On a world scale, the GATT has provided a rationale for the enclosure of all public and indigenous lands.
[2.2] The material foundations of indigenous community survival and cultural development have eroded through these and many other neo-liberal policies. Indigenous peoples from all over the world have been forced off their lands and from their communities' cultural foundations. As a result, they and often have to work in the cities or in large farms or ranches.

A complicated trial
[2.3] Since the autumn of 1990 a trial in the district court in Sveg Sweden against four local Sami communities has been ongoing. The trial has questioned the Sami's rights to use certain parts of privately-owned forest area as grazing land for reindeer during the wintertime.
[2.4] Even though the highlands are sufficient for the needs during the summer they are not sufficient to provide food for the reindeer year round. Because of this the trial has been of vital importance for the survival of these communities.
[2.5] This trial was unique since it not only questioned the indigenous rights for a few individuals but the landrights for four communities affecting the Samis in a whole province. By February 21 1996 the court reached a decision, which is that the Samis have no right whatsoever to use privately owned areas in the forest land for winter grazing.

Inuits and whales

Subsistence hunting of the bowhead whale has been a vital element of a millennia-old Inuit culture. Such hunting posed no danger to the bowhead when its population was at a high level, but the current number of bowhead whales is very scarce. If its levels remain low, the species will be vulnerable to extinction by over-hunting, oil pollution, or the spontaneous "crashes" which occur in small populations.
[1.2] The Inupiat - a distinct ethnic grouping of the Inuit peoples who inhabit the Arctic regions of Greenland, Canada, Alaska, and Siberia - call themselves the "People of the Whales", since they have hunted the arctic stocks for at least 2,500 years.

INSTITUTO DEL TERCER MUNDO

Countries
of the World

Afghanistan

Afghanestan

Population: 22,789,000 (1994)
Area: 652,090 SQ KM
Capital: Kabul

Currency: Afghanis
Language: Pushtu, Dari

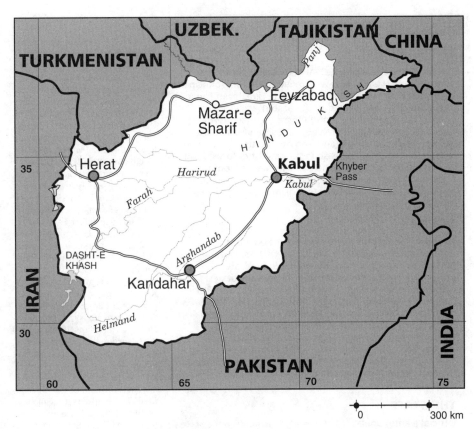

The group of mountains currently known as Hindu Kush (Caucasus by the Greeks and Paropamisos by the Persians) were sparsely populated until the agricultural revolution. During this period the region was a passageway for frequent migrations of displaced peoples, and the Khyber Pass became the gateway to northern India.

[2] The Hindu Kush was incorporated into the Persian Empire of Cyrus the Great in the sixth century BC, and 300 years later became part of the Hellenic world, as a result of the military campaigns of Alexander of Macedonia who founded Alexandropolis (present-day Kandahar). An armistice between the Greeks and Indians made it a province of the Mauryan Empire, and unified all of northern India.

[3] Between the first century BC and the third century AD, an invading nation of Scythian origin (speaking an Indo-European language) founded the state of Kusana, on the trade line between Rome, India and China, known as the "silk route". Following this trail, along the Tarim River basin, Buddhism found its way into China. In 240, Kusana was annexed to the new Sassanian Persian Empire until the beginning of the 8th century and the succession of Caliph Walid, who extended his rule, and the Islamic faith, as far as the Indus.

[4] Towards the 13th century, the Mongol invasion caused great commotion in the Old World. Afghanistan became part of the empire ruled by Genghis Khan from Karakoram. In 1360, the empire disintegrated from constant dynastic strife and the Afghan region was ruled by Tamerlane, whose descendants governed until the beginning of the 16th century.

[5] With the rise of the third Persian Shiite Empire (1502) and the Empire of the Great Mogul in India (1526), the region became the scene of constant battles between the Mongols of India who dominated Kabul, the Saffavid Persians who controlled the southern region, and the Uzbek descendants of Tamerlane who ruled the northwest.

[6] These battles and political upheavals led to the unification of the country in 1747, when an assembly of local chieftains elected Shah Ahmad Durrani, a military commander who previously served Persian sovereigns. Ruling by military might, the new shah consolidated the national borders. Nevertheless, due to its geopolitical relevance, the region would be prone to attacks and the frontiers would suffer severe modifications.

[7] Since the beginning of the 18th century the Russians had arrived in the area, eager to gain access to the sea ports of the Persian Gulf and the Arabian sea to maintain a closer watch on their main enemy there, the Ottoman Empire. On the other hand, the British were intent on controlling the Indus Valley area, where nomadic migrations and sanctuaries sheltering anti-British Indian rebels threatened their incipient colonial domain.

[8] Consequently, geopolitical conditions in the area were similar to those that now exist in Afghanistan, Iran and Pakistan. These regions held the geographic key to the colonial designs of both Russia and Britain. However, the two countries used divergent tactics: in Afghanistan, the Russians relied on diplomacy and bribery and the British used force.

[9] The British defeat in the first British-Afghan War (1839-1842) reinforced Dost Muhammad Shah's slightly pro-Russian sympathies. He increased his influence in northern India by encouraging anti-British movements. His son Sher Ali Shah continued this policy, to which the British responded by invading the country once again.

[10] As a result of the second British-Afghan War (1878-1880), the Durrani dynasty was overthrown, Afghanistan lost its territories south of the Khyber (including the Pass itself) and under the rule of the British-imposed Emir, lost control of foreign relations. Eventually, the British granted Afghanistan a narrow strip of land extending Afghan territory as far as the Chinese border. This closed off a common frontier between Russian and British territories.

[11] Afghanistan's borders, established artificially to create a buffer state between two empires, were never recognized by the peoples they were supposed to divide, who routinely crossed with their herds of cattle in search of greener pastures. In 1919, after a third British-Afghan war, lasting four months, the country was free from British "protection". Independence leader Emir Amanullah Khan (the heir and grandson of the British-imposed ruler) came to power, modernized the country and became the first head of state to recognize the revolutionary government of the Soviet Union. He was overthrown in 1929 by the Mohammedzai clan (descendants of the dynasty ousted in 1879) which crowned Mohamed Nadir Shah. Constitutional guarantees in 1931 recognized the autonomy of local leaders and created a system which remained unchallenged until 1953.

[12] In that year, Muhammad Daud Khan, the shah's cousin and brother-in-law, became prime minister

WORKERS

1994

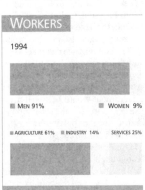

■ MEN 91%   ■ WOMEN 9%

■ AGRICULTURE 61% ■ INDUSTRY 14%   SERVICES 25%

PROFILE

ENVIRONMENT

The country consists of a system of highland plains and plateaus, separated by east-west mountain ranges (principally the Hindu Kush) which converge on the Himalayan Pamir. The main cities are located in the eastern valleys. The country is dry and rocky though there are many fertile lowlands and valleys where cotton, fruit and grain are grown. Coal, natural gas and iron ore are the main mineral resources. The rapid increase in the rate of deforestation (which relates to desertification and soil degradation) constitutes the main environmental problem. The shortage of drinking water has contributed to the increase in infectious diseases.

SOCIETY

Peoples: The Pushtus (Pathans) are a majority of the population, the Tajiks one third; also Uzbeks, Hazaras and nomads of Mongol origin. **Religions:** 99% of the population are Muslim (Sunni and Shiite). **Languages:** Dari is the national language. Pushtu is widely used; also Turkic and dozens of other languages. **Political Parties:** The three main Islamic parties are: the Jamiat-i-Islami, moderate Tajik, led by Ahmed Shah Massud; the Hezb-i-Islami, led by Gulbuddin Hekhmatyar, Integralist Pushtu, which supports the creation of an Islamic state; and the Harakat Islami, of Asef Mohseini. The People's Democratic Party of Afghanistan (Marxist-Leninist) changed its name to the Watan (Fatherland) Party.

THE STATE

Official Name: Doulat i Islami-ye Afghánistan (Islamic State of Afghanistan). **Administrative divisions:** 31 provinces. **Capital:** Kabul, 700,000 inhab. (1988). **Other cities:** Kandahar, 225,000 inhab.; Herat, 177,000 inhab.; Mazar-e-Sharif, 130,000 inhab. (1988). **Government:** Mohammad Rabbani head of the Supreme Council. **National Holiday:** May 27, Independence Day (1919).

---

and launched a new modernization process: he nationalized utilities, built roads, irrigation systems, schools and hydroelectric facilities (with US funding); he abolished the obligatory use of the *chador* (the veil) by women; reorganized the armed forces (with Soviet assistance); and maintained neutrality throughout the cold war. Unfortunately, land reform was never attempted and he was forced to abdicate in 1963.

[13] The following period was dominated by traditional forces. The People's Democratic Party of Afghanistan (PDPA), founded as an underground organization, staged its first anti-government demonstrations in 1965, as Zahir Shah and his council were drawing up a new constitution. The PDPA split into two factions: the Khalk advocated

revolution through a worker-peasant alliance, and the Parcham, who sought to find a broad-based front involving the intellectuals, national bourgeoisie, urban middle class and military. The military finally overthrew the monarchy and Zahir Shah in 1973, installing Muhammad Daud as president with PDPA support.

[14] In April 1977, the murder of PDPA leader Mir Akhbar Khyber triggered a popular uprising. The Parcham military reacted by deposing President Daud, replacing him with Nur Muhammad Taraki. Hafizulah Amin and Babrak Karmal were appointed vice premiers but conflicts arose and in April 1979, Amin was appointed prime minister, a position previously vacant. In September, Amin succeeded in a conspiracy to overthrow the presi-

dent, who was subsequently murdered.

[15] As the new leader, Amin introduced reforms: a compulsory literacy campaign; abolition of the dowry system and other traditional customs; and radical land reforms. All of which brought predictable opposition from traditional local leaders and religious authorities. In February 1979, the US ambassador to Kabul was kidnapped and murdered; consequently the US withdrew economic assistance and increased hostilities towards what they considered a pro-Soviet government. The period between April 1978 and September 1979 saw 25 different cabinets. Amin survived several assasination attempts until in late 1979 he was finally murdered.

[16] He was replaced by Babrak Karmal, who had been placed in power by the Soviet troops that had entered the country in December 1979. Karmal himself had legitimized their presence, citing a treaty of friendship and cooperation between the two states. Soviet intervention served as a pretext for the intensification of the "Second Cold War", initiated several months earlier with the US decision to freeze SALT, the Strategic Arms Limitation Treaty. The presence of the Soviets also triggered a sense of solidarity among Islamic fundamentalists, who travelled into Afghan territory to fight "Satan", in volunteer expeditions financed by Saudi Arabia. Afghan peasants who supported the mujahedin, or Islamic guerrillas, migrated to the cities or to neighboring Pakistan or Iran.

[17] In April 1986, heavy fighting near the border with Pakistan left 2,000 insurgents and 200 Pakistani soldiers dead (although no official figures were available), and a combined Afghan-Soviet force captured the main opposition base. On May 4, President Babrak Karmal returned from several weeks of medical treatment in Moscow and requested to be relieved of his party post.

[18] New PDPA Secretary-General Najibullah, a young doctor (born in 1947) of Pushtu origin was made president. He announced a unilateral ceasefire in January 1987, with guarantees for guerrilla leaders willing to negotiate with the government, an amnesty for rebel prisoners, and the promise of a prompt withdrawal of Soviet troops.

[19] In 1988, after six years of negotiations, an Afghan-Pakistani accord was signed in Geneva, guaranteed by both the US and USSR. The agreement set the terms of mutual relations specifying the principles of non-interference and non-intervention, and guaranteeing the voluntary return of refugees. Another document, signed by Afghanistan and the USSR, provided for the withdrawal of Soviet troops a month later.

[20] Three thousand Jamiat-i-Islami fighters announced they would take advantage of the amnesty law, but turned the disarmament ceremony into an ambush, killing several members of the Kabul military high command. The PDPA was renamed the Watan Party, the Party of the Homeland.

[21] In September 1991, the US and USSR stopped sending arms to the Afghan guerrillas. The US-USSR pact liberated the confrontation between Saudi Arabia and Iran, countries that financed the Afghan "mujahedin" groups. The Kabul regime had no foreign support since the USSR had disappeared.

[22] President Najibullah was replaced in mid-April, after he took refuge in the UN headquarters in Kabul. The government remained in the hands of the four vice-presidents. The authorities immediately announced the government's willingness to negotiate with the rebel groups and met commander Ahmed Shah Massud, of the Jamiat-i-Islami, at the city gates. Massud was a Tadjik leader, known as the "Lion of Panjshir", due to the fierce battles that he had waged in the northern region against the army and the Soviet occupation troops. Massud's presence in Kabul triggered protest demonstrations among the guerrilla groups who belong to the Pushtu majority in the south and east of the country. From Pakistan, Gulbuddin Hekhmatyar, the head of fundamenstalist group Hezb-i-Islami, threatened to start bombing the capital if Massud did not surrender to the government. Provisional president Abdul Rahim Hatif stated that the government would be transferred to a coalition of all the rebel groups and to none in particular. In the days that followed, the forces of Massud and Hekhmatyar fought in Kabul itself.

[23] In March, the president of the Central Bank fled the country, tak-

ing with him all the country's foreign reserves.

24 An interim government under the leadership of Sibgatullah Mojadidi took power, toward the end of April. The alliance of moderate Muslim groups, under the leadership of Ahmed Shah Massud - the new minister of defence - gained control of the capital, expelling the Islamic fundamentalists led by Gulbuddin Hekhmatyar. Pakistan, Iran, Turkey and Russia were the first countries to recognize the new Afghan government.

25 On May 6 1992, the Interim Council formally dissolved the communist Watan Party, whose various factions had governed the country since 1978. The Council set up a special court to try the former communist officials who had violated either Islamic or national laws. The KHAD, the country's secret police, and the National Assembly were also dissolved.

26 Some of the changes which were carried out showed the government's intention of imposing Islamic law: the sale of alcohol was outlawed, and an attempt was made to enforce new rules requiring women to cover their heads and use the traditional Islamic dress.

27 Toward the end of May, most of the rebel Afghan groups, including the Hezb-i-Islami and the Jamiat-i-Islami, announced a peace agreement. The main points were the decision to hold elections within a year, and the withdrawal of both the militia loyal to Minister of Defence Ahmed Shah Massud and the Uzbek militia loyal to Abdul Rashid Dostam, from Kabul.

28 A few days after the agreement, interim president Modjadidi miraculously escaped an attempt on his life. On May 31, the truce between the two main factions of guerrillas was broken. During the first few days of June, the Afghan capital once again became a battleground for Hezb-i-Islami and Jamiat-i-Islami troops. In a week of fierce fighting, the death toll reached 5,000, and Kabul took on the look of a city devastated by war.

29 On June 28, Mojadidi turned over the presidency to the leader of Jamiat-i-Islami, Burhanuddin Rabbani. The latter declared, upon assuming the presidency, "We have only one condition in our program: unity. We will not take a single step without (having) consensus".

30 Hekhmatyar continued fighting against Kabul, demanding the with-

drawal of Massud, as well as the militia loyal to Abdel Rashid Dostam. The latter had been a member of the communist government, but had defected in order to join the Muslim guerrillas that took power.

31 The UN announced a $10 million aid program, to provide food and medication to the civilian population that had been forced to leave Kabul because of the violence. As a result of the war, the country's economy had come to a standstill and 60 percent of its productive structure had been destroyed. Afghanistan had also become the world's largest producer of opium.

32 The Pakistani government, long term supplier of the mujahedin, decided to put a stop to the arms and food contraband across its border with Afghanistan, in order to weaken Hekhmatyar, whom it accused of being responsible for a deterioration in relations between the two countries. The United Nations High Commissioner for Refugees declared the existence of 4.5 million Afghan war refugees. Almost 3 million of these were in Iran, and the Iranian government announced its desire to expel the refugees. Most of these were "shiites", an Afghan minority.

33 In March 1993, the leaders of eight rival factions announced the

signing of a peace agreement in Islamabad, Pakistan. The accord, sponsored by Pakistan's prime minister, resulted in Nawaz Sharif-Rabbani and Hekhmatyar agreeing to share power for an 18-month period, until elections could be held. Rabbani was to continue as president, while Hekhmatyar was to be prime minister. Abdul Rashid Dostam, the powerful general whose militia controlled most of the northern part of the country, did not take part in the peace conference.

34 In June, Hekhmatyar became prime minister and Massud resigned to the Ministry of Defence. In September, Russian and Tadjik government forces confronted Tadjik rebels allied to Afghan fighters along the border with Tadjikistan. In spite of Moscow's and Dushanbe's accusations, Afghan authorities denied any participation in the conflict and demanded the withdrawal of Russian forces from their territory.

35 In January 1994, Dostam's militias, allied to Prime Minister Hekhmatyar, launched an offensive against the capital. The fight between both groups added to the central state's disintegration. Kabul remained divided in zones controlled by rival groups, while 75 per cent of the capital's 2 million population

took refuge in other regions. In June, Rabbani refused to hand over the government at the end of his term, which was finally extended by the supreme court.

36 Afghanistan's division is in part an outcome of rivalries between several countries in the region, including Iran, Saudi Arabia, Uzbekistan, Pakistan or Russia, which directly or indirectly intervened in the civil war. Other countries, less involved, are concerned with the growth of Islamic fundamentalism in the country due to the influence it could have in Muslim regions under their control. Thus, China fears the spread of fundamentalism in Sinkiang and India in Kashmir.

37 In 1995, the appearance of an armed group Taliban ("students" in Persian) in southern Afghanistan, modified the war's progress. These guerrillas, trained in Pakistan, sought to establish a united Islamic government in Afghanistan. They have extended support among wide sectors of society, especially in regions inhabited by Pushtus. This support and foreign aid - which might come from Pakistani secret services - enabled them to conquer Kandahar and several neighbouring provinces.

38 In February, Taliban took over Hekhmatyar's general headquarters in the center of the country. Meanwhile, Dostam kept strengthening his positions in the northwest. After Kabul's almost complete destruction, approximately two thirds of the population moved to areas controlled by the Uzbeki military chief.

39 In mid 1996, when no solution was foreseen to the civil war, 8,000 Taliban guerrillas began to shell Kabul's center from the suburbs, reinforcing the capital's siege and hindering the search for an agreement between the factions. In late September, Kabul fell and the leader of the Taliban movement, Mohammad Rabbani (no relation to ex-president Burhanuddin), took office as head of a six-man Supreme Council. Russian Foreign Minister Evgeni Primakov and other regional diplomats expressed concern about the situation, while Pakistan's government began to deliver aid to the new regime.

Albania

Shqipërí

Population: 3,202,000 (1994)
Area: 28,750 SQ KM
Capital: Tirana

Currency: New Lek
Language: Albanian

The Albanians are descendants of the ancient Illirians, an Indo-European people who migrated southward from Central Europe to the north of Greece by the beginning of the Iron Age. The southern Illirians were much in contact with Greek colonies, meanwhile the northern tribes of Albanians were united at various times under local kings. The most important of them was Argon, whose kingdom (second half of the 3rd century BC) expanded from Dalmatia in the north to the Vijose river in the south. By the year 168 BC the Romans conquered all of Illiria and then the Albanians became part of the prosperous Roman province of Illyricum. With the decline of the Roman empire, after the year 395 AD, the area was connected administratively to Constantinople. Despite the Hun incursions during the 3rd to 5th centuries, and the Slavic invasions during the 6th and 7th centuries, the Albanians were one of the few peoples in the Balkans who kept their own language and customs. When the Turks invaded in 1431, the Albanians put up stiff resistance, finally being occupied 47 years later. In the years which followed, the Ottomans imposed Islam upon the country. A popular uprising ended foreign domination in 1912, and independence was officially proclaimed by the agreement that put an end to World War I. In 1927, Ahmed Zogu, who had become president in 1925, signed a treaty with Mussolini, turning the country into a virtual Italian protectorate. A year later, Zogu proclaimed the country a monarchy.

[2] In April 1939, Italy occupied and formally annexed Albania to the kingdom of Victor Emmanuel III. The communists, under the leadership of Enver Hoxha, organized guerrilla resistance and a wide anti-Fascist front, receiving support from the allies during World War II. The occupation forces withdrew on November 29, 1944, and the People's Republic was declared on January 11, 1945.

[3] When the neighbouring Yugoslavian leader Tito and the Soviet Union's Josef Stalin split in 1948 (see Yugoslavia), the Albanian Workers' Party sided with the KOMINFORM - communist parties allied with the Soviet Union. When the Soviet Union started "destalinization" in the 1960s, Albania broke with Moscow and established close ties with the People's Republic of China, with which it eventually broke ties in 1981, when the Cultural Revolution came to an end and the Maoists fell from power. This break with China became official at the Eighth Workers' Party Congress, where a party line was put forth "against US imperialism, Soviet socialist-imperialism, Chinese and Yugoslavian revisionism, Eurocommunism and social democracy". At the same time condemning the policies of non-alignment and European detente, as set forth in the Helsinki accords, this party line virtually isolated the country from any international alliance.

[4] Until Albania's liberation from Italian Fascist occupation in 1944, 85 per cent of the population lived in the country, and 53 per cent lacked even a place to grow their own vegetables. In 1967 collective farming was established. In1977, Albania proclaimed self-sufficiency in wheat. According to official figures, between 1939 and 1992 industrial output has grown by 12,500 per cent, building materials by 26,200 per cent, and electricity by 32,200.

[5] In 1989, Ramiz Alia, the Albanian head of state since Hoxha's death in April 1985, initiated a reorganization process aimed at improving the economy, and breaking the country's international isolation.

[6] The pace of the reforms picked up during 1989 and 1990. Border immigration procedures were simplified to encourage tourism; talks were initiated with an aim toward resuming relations with both the US and the USSR; freedom of religion was declared; capital punishment was abolished for women; the number of crimes punishable by death was reduced from 34 to 11 and guidelines were established in the area of civil rights.

[7] Other innovations included the right to own a home, the opening of the economy to foreign investment and the election of workplace managers by employees through secret ballot. In February 1990, amnesty was granted to political and common prisoners. In November of the same year, Alia also announced that the constitution, which had been in effect since 1976, would be revised. In December, independent political parties were authorized.

[8] That same month free, direct elections were announced, the first to be held in Albania in 46 years. Although originally planned for February 1991, the election had to be postponed until March 31 due to the instability produced by a mass exodus of thousands of Albanians to Italy. Almost 2 million voters were faced with a choice of more than a 1,000 candidates from 11 political parties, under the scrutiny of about a 100 international observers, and more than 250 foreign journalists.

[9] Despite the overwhelming defeat of Ramiz Alia in Tirana - 18 of the 19 posts to be filled went to the hitherto unknown engineer, Franko Karogi, of the Democratic Party - the Communists managed to obtain 156 of the 250 parliamentary seats. The election, in which 95 percent of all registered voters participated, gave the Democratic Party (of social democratic orientation) 67 seats. The rest of the offices were won by small political groups like the Association for the Defence of the Greek Minority (OMONIA). Former King Leka I issued a statement from Paris, denouncing electoral fraud and opposition leader Gramoz Pashko stated that evidence was being collected which would prove the illegitimacy of the electoral proceedings.

[10] In May 1991, more than 300,000 workers went on strike, demanding the resignation of the communist government, as well as a 50 per cent salary increase. Prime Minister Fatos Nano dissolved his cabinet to form an alliance with the opposition. This, in turn, provoked angry protests from orthodox communists; at the party conference held in June, President Alia was accused of being a traitor.

[11] After deliberating the issue, the Albanian Workers' Party decided to change its name to the Socialist Party, naming Nano as president and Spiro Dede - a liberal - vice-president. A new government was formed with a communist majority, led by Ylli Bufi, with the opposition holding key positions, such as the defence ministry.

[12] On June 19, 1991, the Conference on Security and Cooperation in Europe announced the admission of Albania, the last European country remaining outside this group. In 1991, Albania received $77 million in aid from the European Community. Albania pays for its food imports by exporting chrome, agricultural goods, oil and copper. It also provides Greece and Yugoslavia with electricity.

[13] At the end of 1991 the Democratic Party accused the government of paralyzing the reforms and withdrew its ministers from the cabinet. The Socialist Party could not handle the crisis. In the aftermath of a collapse of the economy, the parliamentary elections of March 1992 gave the Democratic Party a spectacular success

PROFILE

ENVIRONMENT

A Balkan state on the Adriatic Sea, Albania's seacoast is comprised of distinct regions. Alluvial plains, which become partially swampy in the winter, extend from the Yugoslavian border to the Bay of Vlorë. Further to the south, the coast is surrounded by mountains and has a Mediterranean climate. The soil of the mountainous inner region is very poor. Cattle-raising predominates there, while cotton, tobacco and corn are grown on the plains. Irrigated valleys produce rice, olives, grapes and wheat. The country has large areas of forest and is rich in mineral resources, including oil deposits.

SOCIETY

Peoples: Albanians are a homogeneous ethnic group; there is a Greek minority. **Languages:** Albanian (official) and Albanian dialects. **Religions:** Freedom of worship was authorized in 1989, having been banned since 1967. Pre-1967 estimates placed Islam as the largest religion (70%), with the Albanian Orthodox Church in second place (20%) and the Catholic Church (10%) in third. **Political Parties:** A multi-party system was established by the constitution which has been in effect since April 30, 1991. The Democratic Party (PDA), a liberal democratic party, in favor of a free market economy; currently in power. Socialist Party (PS), opposition, formerly communist; today, it advocates a democratic socialism within a free market economy. **Social Organizations:** The Central Council of Albanian Trade Unions has 610,000 members.

THE STATE

Official Name: Repúblika e Shqip üsü. **Administrative divisions:** 26 districts. **Capital:** Tirana, 244,200 inhab. (1988). **Other cities:** Durrës, 85,400 inhab.; Elbasan, 83,300 inhab.; Vlöre, 73,800 inhab. (1988). **Government:** Parliamentary republic. Sali Berisha, President and Head of State (elected April 1992). Alexander Meksi, Prime Minister and Head of Government (since April 1992). Single-chamber legislature: People's Assembly, made up of 140 deputies elected by universal suffrage every four years. **National Holiday:** November 29, Liberation Day (1944). **Armed Forces:** 73,000 (22,400 conscripts). **Paramilitaries:** 16,000; International Security Force: 5,000; People's Militia: 3,500.

791

STATISTICS

DEMOGRAPHY

Urban: 37% (1995). **Annual growth:** 1.8% (1991-99). **Estimate for year 2000:** 4,000,000. **Children per woman:** 2.9 (1992).

HEALTH

Under-five mortality: 41 per 1,000 (1995). **Calorie consumption:** 102% of required intake (1995).

EDUCATION

School enrolment: Primary (1993): 97% fem., 97% male. University: 10% (1993). **Primary school teachers:** one for every 19 students (1992).

COMMUNICATIONS

97 **newspapers** (1995), 99 **TV sets** (1995) and 92 **radio sets** per 1,000 households (1995). 1.4 **telephones** per 100 inhabitants (1993). **Books:** 102 new titles per 1,000,000 inhabitants in 1995.

ECONOMY

Per capita GNP: $380 (1994). **Annual inflation:** 32.70% (1984-94). **Consumer price index:** 100 in 1990; 226.7 in 1994. **Currency:** 95 new leks = 1$ (1993). **Cereal imports:** 647,000 metric tons (1993). **Fertilizer use:** 338 kgs per ha. (1992-93). **Imports:** $596 million (1994). **Exports:** $116 million (1994). **External debt:** $925 million (1994), $289 per capita (1994). **Debt service:** 2.5% of exports (1994).

ENERGY

Consumption: 422 kgs of Oil Equivalent per capita yearly (1994), 28% imported (1994).

(65.6 per cent) over the Socialist Party (22.6 per cent). 80 per cent of the voters abstained.

[14] On April 4 1992 Sali Berisha, leader of the Democratic party, replaced Ramiz Alia as President of Albania. He won the parliamentary vote 96 to 35, becoming the first non-Marxist president since the end of World War II.

[15] By the middle of 1993 the trials of the main political figures of the previous regime began. Naxhmija Hoxha, widow of the historic communist leader Enver Hoxha, the ex-President Ramiz Alia, and the ex-Premier Fatos Nano among others, were sent to prison convicted of misuse of public funds.

[16] During 1994 and 1995 laws about private property were sanctioned, and this stimulated national and foreign investors. The gross internal income grew by an average of 7 per cent due to the development of the building industry, services and agriculture. Inflation and public deficit were reduced. Despite the fact that half a million Albanians live and work outside the country, unemployment remained as high as 18 per cent. Foreign debt was refinanced by the Western banks and it was reduced from $700 to $300 million.

[17] Concerning foreign politics, Albania was accepted as the 36th member of the European Council, and relations with neighboring countries improved. An agreement of military cooperation with the United States was signed, which included the establishment of air bases on Albanian territory for intelligence flights over Bosnia-Herzegovina.

[18] A state of conflict prevailed through 1995 and 1996. The press denounced the censorship and practices of intimidation, and the relationships among the political parties remained difficult. The legislation approved during this period reflects those tensions. The so-called "Verification Bill" made possible for the government to decide on the accessibility to public office for journalists and other persons according to the reports of Sigurimi, ex-secret police of the communist regime. The bill about "Genocide and Communist Crimes" forbade notorious ex-political

leaders access to public office until the year 2002, and the new electoral law forbade coalitions among the smaller parties.

[19] On May 26, 1996 parliamentary elections took place, in which the main leaders of the opposition were not allowed to participate. The Democratic Party won 122 out of 140 places. On the day of the elections the Socialist Party, the Democratic Alliance (center right), the Social Democratic Party, and the Human Rights Party withdrew their candidacies. The Socialist Party accused the government of controlling the voting through the police and party groups, and demanded the annulment of the elections. President Berisha, on the other hand, accused the socialists of organizing terrorist gangs. Observers sent by Washington verified the existence of pressure from above.

[20] During the following months, the questionable legitimacy of the elections was made the target of continuous criticism on the part of the US and some European countries, who are interested in investing in Albania. As an answer to a recommendation of the Eu-

ropean Council, the parliament created an ad hoc committee in order to investigate possible irregularities and violence against members of the opposition. In this committee some members of the opposition participated as a minority.

Algeria

Algerie

Population: 27,422,000 (1994)
Area: 2,381,740 SQ KM
Capital: Algiers (Alger)
Currency: Dinar
Language: Arabic

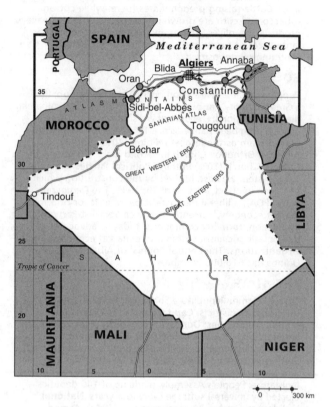

From ancient times, the states that arose in what is now Algeria developed links with the two power nuclei in that part of the world: Tunisia, beginning in the Carthaginian period (10th century BC), and Morocco, after the conquest of the Iberian Peninsula by the Arabs (711 AD). Those who resented the fact that long-time Muslims and recent converts were taxed differently, settled in this area. These "dissidents" would later embrace the Jariyite sect, a religious group which was characterized by its egalitarian principles. It maintained, for example, that caliphs did not have to be descended from Mohammed or his relatives; thus, any Moslem could become caliph regardless of race, color or social position. This tenet fitted in well with the Berbers' social reality, given their subordinate position in relation to the Arabs (although the latter were a minority, they nevertheless exercised greater political power). Variations of this doctrine spread widely among the Berbers, contributing to the formation of many North African empires (see Morocco: Almoravids and Almohads).

2   With the downfall of the Almohad Empire, Yaglimorossen ibn Ziane founded a new state on the Algerian coast. Its borders were consolidated as economic prosperity and cultural development led nomadic peoples to settle down. Ziane and his successors governed the country between 1235 and 1518. After the Christians had put an end to seven centuries of Moslem domination, in 1492, the Zianids were confronted with a series of Spanish military incursions in which various strategic sites, like Oran, were taken.

3   Algeria and Tunisia both became part of the Ottoman Empire in the 16th century. The Arroudj and Kheireddine brothers drove the Spanish from the Algerian coast and expanded the state's authority over a sizeable territory. Its mighty fleet won respect for the nation, and its sovereignty was acknowledged in a series of treaties (with the Low Countries in 1663, France in 1670, Britain in 1681, and the US in 1815).

4   Wheat production gradually increased until it became an export crop once again, for the first time since the Hilalian invasion (see Mauritania). Wheat exports would eventually be the indirect cause of European intervention as at the end of the 18th century, the French revolutionary government bought large amounts of wheat from Algeria but failed to pay. Napoleon, and later the Restoration monarchy, delayed payment until the Dey of Algiers demanded that the debt be paid. Reacting to further excuses and delays, he slapped a perplexed French official in the face - a show of temper that would cost the Turkish Pasha dearly. 36,000 French soldiers disembarked to "avenge the offence", a pretext used by the French to carry out a long-standing project: to reestablish a colony on the African coast opposite their own shores. However, the French encountered heavy resistance, and were defeated.

5   In 1840, disembarking with 115,000 troops, the French set out once again to conquer Algeria. Successive rebellions were launched against the French. In the South, however, the nomadic groups remained virtually independent and fought the French until well into the 20th century (see Sahara). In 1873, France decided to expropriate land for French settlers who wished to live in the colonies (500,000 by 1900, and over a million after WW II). As a result, the French pieds-noirs came to monopolize the fertile land, and the country's economy was restructured to meet French interests.

6   Nationalist resistance grew stronger from the 1920s onwards, until in 1945 it exploded when the celebration of the victory over Nazi-Fascism turned into a popular rebellion. The French forces tried to put down the rebellion and according to official French reports, 45,000 Algerians and 108 Europeans were killed in the ensuing massacre.

7   Shortly thereafter, the Algerian People's Party, founded in 1937, was restructured as the Movement for the Triumph of Democratic Liberties (MTLD), which was to participate in the 1948 and 1951 elections called by the colonialists.

8   Convinced of the futility of elections under colonial control, nine leaders of the OS (Special Organization, the military branch of the MTLD) founded the Revolutionary Committee for Unity and Action (CRUA). In November 1954, this committee became the National Liberation Front (FLN) that led the armed rebellion. Franz Fanon, a doctor from Martinique, who had fought in the liberation of France during World War II, joined the FLN and came to exert great theoretical influence not only in Algeria, but also throughout the Sub-Saharan region.

9   In order to maintain "French Algeria" and the pieds-noirs, the French colonial system destroyed 8,000 villages, eliminated more than a million civilians, made systematic use of torture and deployed more than 500,000 troops. Right-wing French residents of Algeria formed the feared Secret Army Organization (OAS), a terrorist group which blended Neo-Fascism with the demands of the French colonials, who resented the growing power of the

PUBLIC EXPENDITURE

% of GNP — 1992

MILITARY 2.7
EDUCATION 9.1
HEALTH 5.4

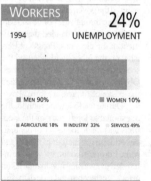

WORKERS

1994 — **24%** UNEMPLOYMENT

MEN 90%   WOMEN 10%

AGRICULTURE 18%   INDUSTRY 33%   SERVICES 49%

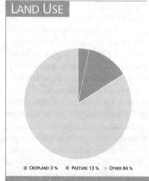

LAND USE

CROPLAND 3 %   PASTURE 13 %   OTHER 84 %

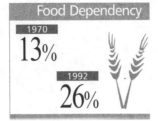

Food Dependency
1970
13%
1992
26%

External Debt
1994
PER CAPITA
$1,090

Illiteracy
1995
38%

Algerians. Finally, on March 18, 1962, De Gaulle signed the Evian Agreement, agreeing to a cease-fire and a plebiscite on the proposal for self-determination.

[10] Independence was declared on July 5, 1962, and a Constituent Assembly was elected later that year. Ahmed Ben Bella was named prime minister. Almost all foreign companies were nationalized, 600,000 French nationals abandoned the country taking everything they could with them, and 500,000 Algerians returned, to share the lot of 150,000 landless and hungry peasants. A system of locally undertaking agricultural and industrial operations was introduced by the new government.

[11] However, Ben Bella's self-management program came up against the reality of a government with little administrative expertise. In June 1965 a revolutionary council headed by Houari Boumedienne took power and jailed Ben Bella. A new emphasis on organization, centralization, and state power began to prevail over Ben Bella's self-management notions. Under Boumedienne, there were further nationalizations and a program of rapid industrialization based on revenues from oil and liquid natural gas. The country entered a period of economic expansion, which was not reflected in the countryside. Population grew more rapidly than agricultural production, and Algeria went from exporting to importing food.

[12] Houari Boumedienne died in December 1978, after a long illness, just as the country's political institutions were beginning to consolidate. In 1976, a new National Charter was approved and, in 1977, the new members of the National People's Assembly were elected, with Colonel Chadli Ben Jedid being designated president.

[13] Ben Jedid received almost unanimous support at the fourth FLN congress in January 1979, and was sworn in with full presidential powers.

[14] The new president initiated a policy of reconciliation by releasing Ahmed Ben Bella, who had been imprisoned for 14 years. Restrictions on travel abroad were lifted, taxes reduced and prohibitions lifted on private housing. Overlarge public enterprises gradually shrank and the restructuring of inefficient public enterprises gave an impulse to private companies.

ENVIRONMENT

South of the fertile lands on the Mediterranean coast lie the Tellian and Saharan Atlas mountain ranges, with a plateau extending between them. Further south is the Sahara desert, rich in oil, natural gas and iron deposits. Different altitudes and climates in the north make for agricultural diversity, with Mediterranean-type crops (vines, citrus fruits, olives, etc.) predominating. The country's flora and fauna are seriously threatened. More than 30 mammal, 8 reptile and 70 bird species are in danger of extinction. Desertification affects primarily the pre-Saharan regions. However, erosion also poses a serious threat to 45% of all agricultural land (12 million hectares).

SOCIETY

Peoples: Algerians are mostly Arab (80%) and Berber (17%). Nomadic groups linked to the Tuareg of Nigeria and Mali live in the south. Nearly a million Algerians live in France. **Religion:** Islam

Languages: Arabic (official), and Berber in some areas. Many people speak French, but Arabic has gradually been replacing it in education and public administration. **Political Parties:** National Liberation Front (FLN), Vanguard Socialist Party (PAGS), Front for Socialist Forces (FFS), Movement for Democracy in Algeria (MDA), Pro-Democracy and Culture Group (RDC), Islamic Salvation Front (FIS), Hamas. **Social Organizations:** General Union of Algerian Workers (UGTA), National Union of Algerian Peasants, National Union of Algerian Women, National Youth Union.

THE STATE

Official Name: Al-Jumjuriya al-Jazairia ash-Shaabiya **Capital:** Algiers (Alger), 1,721,607 inhab. **Other cities:** Oran, 663,504 inhab.; Constantine, 448,578 inhab.; Annaba, 348,322 inhab. **Government:** Lamine Zeroual, President. **National Holiday:** November 1st, Anniversary of the Revolution (1954). **Armed Forces:** 121,700 (65,000 conscripts). **Paramilitaries:** 180,000 (Gendarmerie, National Security Forces, Republican Guard)

[15] Ben Jedid was reelected in January 1984. In October 1988, a wave of protests broke out in several cities due to the lack of water and basic consumer goods; the legitimacy of the FLN and the military was called into question. Among the main groups participating in these mass protests were militant Muslim fundamentalists. Some mosques - above all, those in poorer neighborhoods - became the site of political demonstrations, particularly on Friday afternoons, when mosque prayers ended in political declarations voicing economic and social demands.

[16] Some sectors of the most radical forms of Islam, influenced by Iran, began sending volunteers to fight in Afghanistan, to carry out the jihad or "holy war" against the Kabul regime, supported by the USSR. In mid-1989, against a background of protest and upheaval, Bendjedid presented a new constitution which introduced a modified multiparty system, breaking the monopoly which the FLN had held.

[17] More than 20 opposition groups - including Muslims - openly expressed their views. The most significant were the Islamic Salvation Front (FIS), the Da'wa Islamic League, the communist Socialist Avant Garde Party (PAGS) and the strongly Kabbyle (ethnic minority of Berber origin) Pro-Democracy and Culture Group (RDC). Mouloud Hamrouche, a leading reformer, was appointed prime minister. In the first multiparty elections since Algeria's independence from France in 1962, the FIS defeated the FLN in the June elections.

[18] Hamrouche and his cabinet resigned in June 1991, against a background of social agitation promoted by the mosques. In early June, a state of siege was declared throughout the country, in the face of massive protests pitting the country's military and police forces against FIS agitators who demanded that presidential elections be held ahead of schedule and that an Islamic state be proclaimed. Sid Ahmed Ghozali, an oil technician who had been prime minister during the previous government, was designated as the new prime minister. Legislative and presidential elections were scheduled for later in the year, and the FIS suspended its campaign of social and political agitation.

[19] The country turned to the IMF for loans, to offset the effects of the fluctuation in the price of oil. Ghozali submitted a series of reforms to the National Assembly, to ensure the fairness of the election. However, these reforms - which included doing away with a man's right to vote in his wife's name - were rejected by the FLN, which had a parliamentary majority.

[20] In the December elections, 40 per cent of the country's 13 million voters abstained. The FIS won the first round of voting, winning 188 of the 430 seats contested in Parliament, with the support of 3.2 million voters. The Socialist Forces Front (FFS), a secular party dominated by Berbers, obtained 25 seats, while the FLN obtained 15.

[21] Anti-fundamentalists - led by the General Union of Algerian Workers (UGTA) and the FFS - were alarmed by the FIS victory in the first round of voting. They called a massive demonstration in the center of Algiers, which was attended by 100,000 people. Women's groups, professionals and intellectuals also participated in the demonstration.

[22] President Chadli Ben Jedid resigned, under strong pressure from the military and from political figures fearing a FIS victory. A State

Maternal Mortality
1989-95

PER 100.000
LIVE BIRTH

140

Foreign Trade
MILLIONS $ 1994

IMPORTS
8,000
EXPORTS
8,594

Security Panel, made up of 3 military officers and the prime minister, took over. Shortly afterwards, a 5-member Council of State was named, presided over by Mohamed Budiaf, who had been in exile since 1964. All FIS leaders who had not gone into hiding were subsequently arrested.

[23] On February 9, the High Council of State declared a state of emergency throughout the country, to remain in effect for a year. The army was opposed to any form of power-sharing with the FIS. Intellectuals, professionals, women and labor unions sided with the military.

[24] In March 1992, the FIS was dissolved, with the declaration of a state of emergency. In early April, Ghozali's government dissolved nearly 400 city and town councils which had been controlled by FIS supporters since the June 1990 municipal elections.

[25] In late April 1992, the Supreme Court ratified the declaration which banned the Islamic Salvation Front (FIS). Given the fact that the Court is the final resort in the Algerian legal system, its decision could not be appealed.

[26] The violence of both factions - government and Muslim fundamentalists - kept rising. In June, at a public rally, Budiaf was assassinated by one of his bodyguards while delivering a speech. The death of Budiaf - replaced by Ali Kafi - and Ghozali's resignation indicated growing government inflexibility towards Islamic opposition.

[27] In September, the government of Prime Minister Belaid Abdelsalam decreed a series of "anti-terrorist" measures, including the extension of the death penalty to several crimes. Amnesty International estimates that over 2,000 people died in the civil war's first year. In February 1993, the High State Council extended the state of emergency indefinitely, established a curfew in Algiers and five provinces, and dissolved all FIS linked organizations.

[28] From this moment on, the military hesitated whether to find a negotiated solution or to seek a military defeat of Muslim armed groups. In late 1993, the FIS claimed it was willing to initiate talks with the government, which responded by freeing 60 Muslim fundamentalist prisoners and called a conference to begin negotiations.

[29] However, the attempt was to be the first of a long series of failures

STATISTICS

DEMOGRAPHY

Urban: 54% (1995). **Annual growth:** 1.4% (1992-2000). **Estimate for year 2000:** 31,000,000. **Children per woman:** 4.3 (1992).

HEALTH

One **physician** for every 1,064 inhab. (1988-91). **Under-five mortality:** 65 per 1,000 (1995). **Calorie consumption:** 104% of required intake (1995). **Safe water:** 79% of the population has access (1990-95).

EDUCATION

Illiteracy: 38% (1995). **School enrolment:** Primary (1993): 96% fem., 96% male. Secondary (1993): 55% fem., 66% male. University: 11% (1993). **Primary school teachers:** one for every 27 students (1992).

COMMUNICATIONS

101 **newspapers** (1995), 100 **TV sets** (1995) and 102 **radio sets** per 1,000 households (1995). 4.0 **telephones** per 100 inhabitants (1993). **Books:** 87 new titles per 1,000,000 inhabitants in 1995.

ECONOMY

Per capita GNP: $1,650 (1994). **Annual growth:** -2.50% (1985-94). **Annual inflation:** 22.00% (1984-94). **Consumer price index:** 100 in 1990; 316.3 in 1994. **Currency:** 43 dinars = 1$ (1994). **Cereal imports:** 5,821,000 metric tons (1993). **Fertilizer use:** 123 kgs per ha. (1992-93). **Imports:** $8,000 million (1994). **Exports:** $8,594 million (1994). **External debt:** $29,898 million (1994), $1,090 per capita (1994). **Debt service:** 56% of exports (1994). **Development aid received:** $359 million (1993); $13 per capita; 0.7% of GNP.

ENERGY

Consumption: 1,030 kgs of Oil Equivalent per capita yearly (1994), -273% imported (1994).

caused by disagreements between opposing political sectors. The government refused to invite certain Muslim sectors, and the FLN and Hocine Ait Ahmed's Socialist Forces Front (FFS) stated that any president appointed by the conference would be illegitimate. After this failed attempt, the government appointed defence minister Lamine Zeroual as president of the country for three years.

[30] The fall of Abdelsalam, who was opposed to massive privatizations, enabled the signature of an agreement between the IMF and Redha Malek's new government in early 1994. Foreign debt growth coincided with a rise of unemployment, which affected 22 per cent of the economically active population. That year, Malek was replaced by another supporter of economic lib-

eralization, Mokdad Sifi, regarded as more lenient toward a dialogue with Muslim fundamentalists.

[31] In what became the beginning of a growing division among all political sectors, Islamic guerrillas broke off into the Armed Islamic Group (GIA) and Armed Islamic Movement. The violence of both factions - government and Muslim fundamentalists - kept growing throughout the year, without any of them in condition to obtain a military victory. In one of their most spectacular measures, fundamentalists enabled the escape of 1,000 prisoners from Tazoult high-security prison.

[32] The civil war continued throughout 1995 and, in spite of a certain military advantage obtained by the government, the defeat of Islamic opposition did not seem

possible. Political attempts to solve the conflict were also fruitless. In early 1995, after having met in Rome, the FIS, FLN, FFS and Hamas' moderate fundamentalists proposed to the government an end of violence, the liberation of political prisoners and the formation of a national unity government which would hold elections. In spite of support from Spain, the United States, France and Italy, Zeroual did not accept the proposal.

[33] Algeria's president continued waging the war against fundamentalists and held presidential elections, which led to his re-election in November 1995. The FIS, FLN and FFS boycotted the elections, won by Zeroul with 61 per cent of the vote against moderate fundamentalist Mahfoud Nahnah's 25 per cent. However, in spite of the presence of international observers, the election's credibility is strongly doubted.

[34] In the first months of 1996, Zeroual's government, which seemed to have the support of the FLN's new leading group, achieved significant military victories and continued its plan of structural adjustment recommended by the IMF, which increased growing impoverishment of a large part of the middle class and the underprivileged sectors.

Angola

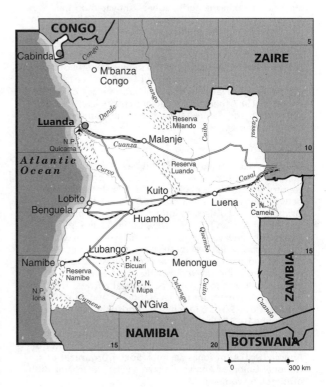

Population: 10,442,000 (1994)
Area: 1,246,700 SQ KM
Capital: Luanda

Currency: New Kwanza
Language: Portuguese

Angola

The original inhabitants of what is now Angola were probably Khoisan-speaking hunters and gatherers. During the 1st millennium AD, large-scale Bantu-speaking migrations (possibly originating in central Africa) into southern Africa absorbed the older population and expanded iron technology and cereal cultivation in the area. Eventually, they came to constitute the dominant ethnolinguistic group of southern Africa. Their occupation of the area that is now Angola was complete by 1600. The most significant Bantu kingdom in Angola was the Kongo, with its capital at Mbanza Kongo (called by the Portuguese São Salvador do Congo). This kingdom was a highly centralized and complex state that was divided into six provinces. South of the Kongo, the Mbundu people emplaced the Ndogno, an equally centralized kingdom that gave Angola its name from the title of its king, the ngola.

[2] In 1483 Portuguese explorers reached Angola, Christianized the ruling family of the Kongo, and engaged in trade and missionary work. However, increasing Portuguese involvement in the slave trade eventually turned the Kongo against Portugal, which in turn led to Portugal's destruction of the Ndongo kingdom to the south. Portugal's colonization of the area was slow, but its slave trade flourished, so much so that by the early 17th century some 5,000 to 10,000 slaves were being exported annually from Luanda. All the kingdoms in the area tenaciously resisted foreign occupation until the mid-18th century. War and slavery decimated the population: it dropped from 18 million in 1450, to barely 8 million in 1850. Even so, Angolan resistance to Portuguese colonization continued, led by figures like Ngo-

la Kiluange, Nzinga Mbandi, Ngola Kanini, Mandume and others.

[3] Portugal intensified its military incursions following the 1884 Berlin Conference which divided Africa among the European colonial powers. Nonetheless, it took 30 years of military campaigns (1890-1921) to "pacify" the colony. From then on, Portuguese settlers arrived in ever-increasing numbers. In 1900, there were an estimated 10,000; in 1950, 50,000; and in 1974, less than a year before independence, 350,000, only one percent living on farms inland. The colonial economy was parasitic, built upon the exploitation of mineral and agricultural wealth (diamonds and coffee), with the bulk of profits going to Portuguese middlemen.

[4] On December 10, 1956, several small nationalist groups (PLUA, MINA and MIA) fused to form the Popular Movement for the Liberation of Angola (MPLA). Their aim was to pressure the Portuguese government into recognizing the Angolan people's right to self-determination and independence. When Britain and France began to withdraw from their overseas colonies in the 1960s, Portugal did not follow the example and frustrated all Angolan attempts to win independence by peaceful means.

[5] On February 4, 1961, a group of MPLA militants from the most underprivileged classes stormed Luanda's prisons and other strategic points in the capital. This acted as a spur to resistance in other Portuguese colonies. Their manifesto was clear: they were fighting not just colonialism but also the international power system which sustained it. In addition, they were fighting racism and tribal chauvinism.

[6] In the years that followed, other independence movements with different regional origins sprang up: the National Front for the Liberation of Angola (FNLA) led by Holden Roberto, the Cabinda Liberation Front (FLEC), and the Union for the Total Independence of Angola (UNITA) led by Jonas Savimbi.

[7] Under the direction of Agostinho Neto, the MPLA militants held a conference in January 1964 to discuss and define their strategy of "prolonged peoples' war."

[8] Portugal's domestic problems, coupled with military setbacks in Angola, Mozambique and Guinea-Bissau and repeated shows of inter-

national solidarity with the independence fighters, dashed Portuguese army hopes of a military solution. An uprising led by the Armed Forces Movement (MFA) overthrew the Portuguese regime of Oliveira Salazar and Marcelo Caetano on April 25, 1974. The MFA expressly recognized the African colonies' right to self determination and independence.

[9] The MFA immediately invited the MPLA, FNLA and UNITA to participate with Portugal in a transitional government for Angola in the interim period, the mechanisms of which were established in the Alvor Accords, signed in January 1975. By this time, political and ideological divergences among the three groups had become irreconcilable; the FNLA was directly assisted by US intelligence services and received military aid from Zaire; UNITA received overt backing from South Africa and Portuguese settlers; while the MPLA was aligned ideologically with the socialist countries, and the accords were never implemented.

[10] The FNLA and UNITA unleashed a series of attacks on MPLA strongholds in Luanda, and a bloody battle for control of the capital ensued. Between September and October 1975, Angola was attacked on all sides: Zaire invaded

from the north while South Africa, with the complicity of UNITA, attacked from the south.

[11] On November 11, the date agreed upon to end colonial rule, the MPLA unilaterally declared independence in Luanda, preempting the formal transfer of sovereignty. Some 15,000 Cuban troops aided the new government in fighting off the South African invasion. In 1976, the United Nations recognized the MPLA government as the legitimate representative of Angola.

[12] The Angolan economy was severely debilitated. The war had paralyzed production in the extreme north and south of the country. The Europeans had emigrated en masse, taking all that they could with them and leaving production facilities unserviceable.

[13] Under these circumstances, the Angolan government began to restore the chief production centers, and to train the largely unskilled and illiterate workforce. Thus, a large public sector emerged which was to become the economy's driving force, and banking and strategic activities were nationalized.

[14] In May 1977, Nito Alves led a faction of the MPLA committed to "active revolt", in a coup attempt. Six leading MPLA members were killed but the conspiracy was suc-

WORKERS

1994

- MEN 62%
- WOMEN 38%

- AGRICULTURE 73%
- INDUSTRY 10%
- SERVICES 17%

ENVIRONMENT

The 150 kilometer-wide strip of coastal plains is fertile and dry. The extensive inland plateaus, higher to the west, are covered by tropical rainforests in the north, grasslands at the center and dry plains in the south. In the more densely populated areas (the north and central west) diversified subsistence farming is practiced. Coffee, the main export crop, is grown in the north; sisal is cultivated on the Benguela and Huambo plateaus; sugar-cane and oil palm along the coast. The country has abundant mineral reserves: diamonds in Lunda, petroleum in Cabinda and Luanda, iron ore in Cassinga and Cassala. The port of Lobito is linked by railway to the mining centers of Zaire and Zambia. There are a number of environmental problems aggravated by the civil war, the lack of drinking water (in 1987 there was a cholera epidemic in Luanda), soil erosion and deforestation as a result of the export of valuable timber.

SOCIETY

Peoples: As a consequence of centuries of slave trade the population density is very low. To maintain control over the country, the Portuguese colonizers fostered local divisions between the various ethnic groups; Bakondo, Kimbundu, Ovimbundu and others. **Religions:** The majority profess traditional African religions. 38% Catholic, and 15% Protestant. However, there are forms of syncretism which make it impossible to establish strict boundaries between one religion and another. **Languages:** Portuguese (official) and African languages derived from Bantu; Ovidumbo, Kimbundu, Kikongo and others. **Political Parties:** The People's Movement for the Liberation of Angola (MPLA), of President dos Santos, founded by Agostinho Neto on December 10, 1956; the National Front for the Liberation of Angola (FNLA) of Holden Roberto; the main opposition force is the National Union for the Total Independence of Angola (UNITA) of Jonas Savimbi. **Social Organizations:** National Union of Angolan Workers (UNTA); Organization of Angolan Women (OMA).

THE STATE

Official Name: República Popular de Angola. **Administrative Divisions:** 18 Districts. **Capital:** Luanda, 1,134,000 inhab. **Other cities:** Huambo (Nova Lisboa), 203,000 inhab., Lobito, 150,000 inhab., Benguela, 155,000 inhab. **Government:** José Eduardo dos Santos, President; Fernando Franca Van Dunem, Prime Minister. Unicameral Legislature. 223-member National Assembly, elected by direct popular vote. **National Holiday:** November 11, Independence Day (1975). **Armed Forces:** 82,000. **Paramilitaries:** 20,000 Internal Security Police.

---

cessfully put down within hours. Seven months later, at its first congress, the MPLA declared itself Marxist-Leninist and adopted the name of MPLA-Labor Party. In 1978, closer political and economic links were established with the countries of the socialist Council for Mutual Economic Assistance.

[15] The country's first president, Agostinho Neto, died of cancer in Moscow on September 10, 1979 and was succeeded by Planning Minister José Eduardo dos Santos.

[16] In August 1981, South Africa launched "Operation Smokeshell", in which 15,000 soldiers, with tanks and air support, advanced 200 kilometers into Cunene province. Pretoria justified this aggression as an operation against guerrilla bases of the Namibian liberation movement, the South West African Peoples Organization (SWAPO). The undeclared aim seems to have been to establish a "liberated zone" where UNITA could install a parallel government inside Angolan territory capable of obtaining some degree of international recognition.

[17] This incursion and successive attacks in the years that followed were contained by effective Angolan and Cuban military resistance. The cost of the war, plus international pressure and the mounting anti-apartheid campaign at home, obliged South Africans to resume diplomatic discussions with the MPLA government. In December 1988, Angola, South Africa and Cuba signed a Tripartite Accord in

---

African cultures before

U ntil a comparatively short time ago, Europeans knew little about Africa. Scholars of the 15th century summed up their knowledge of that continent writing on the maps "Ibi sunt leones" (Here there are lions). The most encyclopedic intellectual of the 19th century, the philosopher Hegel, described blacks as a childish race, and Africa as "the unhistorical spirit, the non-developed spirit, still submerged in the conditions of the natural..., situated on the threshold of the history of the world".

[2] Such notions corresponded with a lack of knowledge about the continent, and the subordination of its peoples to subjugation and slavery. But decolonization in the second half of the 20th century opened a period in which African peoples were confronted with the complex tasks of social and cultural change and development. Part of it consists in a reassessment and a revaluing of the African past. The history of Africa - Ki Zerbo has said - is the history of the awakening of its consciousness.

[3] Nevertheless, the knowledge that the West has accumulated about Africa is full of myths. Among them, the myth of the impossibility of a scientific history, or the myth of the inaccessibility of its past, or the myth of the absence of written testimonies and writing itself, or the myth of the necessary stagnation of the black peoples - which reduced historical interest about Africa to its connections with the Mediterranean world and to isolated subjects, such as Egypt, the Magreb, or Christian Ethiopia.

[4] If these myths are not overcome, the scientific history of Africa cannot be born. The task involves the double operation of deconstructing traditional African history - written under the light of the prejudices of a domineering Eurocentric perspective - and of critically investigating an ample and ignored past.

[5] In any case, the question is not to wonder about the fate of Africa if colonialism had not existed, but of considering the autonomous developmental factors of that continent, that is, its contribution to the history of human culture and its promise for future creations.

[6] Charles Darwin wrote: "It is likely that our first fathers had lived in Africa more than anywhere else." This intuition is apparently being confirmed by contemporary science. Present research supports the view that there were favorable conditions there for the development of the essential phases of the process of evolution to human form. Africa supposedly holds the most complete spectrum of prehistorical human remains.

[7] Archeological discoveries indicate the pre-eminence of African prehistory over the prehistory of other civilizations. Advanced techniques of elaboration of tools - quarries and workshops, manufacture of double-edged axes, usage of stones as heat accumulators, and pottery making - became differentiated as they extended over wider ecological areas, but "the initiative, the great tradition, and the 'fashion' came from Africa".

[8] The Neolithic era began in Africa three thousand years earlier than in Europe. Not in Egypt, but in the Sahara, which was then an attractive area, of wide rivers and abundant vegetation. There, the exchange of techniques between different communities made possible diversified agricultural activities. Corn, barley, sorghum, millet, palms, textile plants and so on, were cultivated. Cattle-breeding saw a modest development. These practices evolved in an autonomous and parallel way to those of other Asian and Indo-American peoples.

[9] African peoples of Neolithic Sahara created one of the first technical revolutions through their agricultural practices. This allowed them to lead stable lives and to carry on technical and cultural exchanges with peoples from other areas. The development came from the South upwards through the communities living in the Sahara and then reached the North of Africa. The civilization which flourished later in the Nile valley cannot be accounted for exclusively by reason of the changes that came about there as a result of the extraordinary fertility of the land, and the demographic concentration - which finally led to the desertification

---

heir entry into the Third World

of the Sahara. The cultural riches created and transmitted by the black peoples of the South was also a factor in this process. The Egyptians recognized that those peoples shared their ancestry.

[10] The Egyptians organized an agricultural civilization of strong craftspeople and a consistent governing and military structure which triumphed over the neighboring populations, among them the kingdoms of Nubia. They developed commercial relations along the river, and exported towards the South bronze manufactures and other products. They also invented their own writing. On their papyrus they drew hieroglyphs, which evolved and became partly alphabetical.

[11] Women had a relevant role in Egyptian civilization. The mother figure is foremost in their culture. There was also a female clergy. Landed property was inherited by women, and a dowry was given to the parents of the wife. Their imposing works of art had a profound religious inspiration. Their worldview did not underline the value of progress but rather of equilibrium and peace in opposition to the forces of decay and chaos.

[12] In the centuries immediately before and after the first century AD, migrations and fusions took place among the peoples established south of the Sahara. Several languages were born then, and the cultural achievements include an autochtonous iron culture which was as im-

> **Their worldview did not underline the value of progress but rather of equilibrium and peace in opposition to the forces of decay and chaos.**

portant a revolution as the neolithic one for the African continent. Iron was abundant in the region. The mastering of the techniques of production of iron tools, as well as the handling of these, augmented the productive abilities and the military capabilities of those peoples, and allowed their expansion and subjugation of less advanced groups. Ironsmiths and craftspeople gained the uppermost positions in these societies. This is the origin of a tradition which recognizes the king-smiths as the African ancestors. From the point of view of social organization, those centuries represent the passage from clan organization to the formation of kingdoms.

[13] Between the 7th and the 12th centuries the majority of the bigger African kingdoms were founded. The Arab-Muslim conquest of North Africa began in the 7th century. From then on inter-continental commerce through the Sahara became easier, and also from the coasts towards the kingdoms of sub-Saharan Africa. The Arabs were the intermediaries.

[14] The empires of Ghana and Awdaghost to the West, and those of Nubia and Aksum to the North-East dominated several other kingdoms and won territories rich in gold mines. Thanks to the commerce in gold and slaves they acquired considerable economic power. Those kingdoms created cities as market centers of agricultural products and cattle. Courts headed by black kings admitted civilian councils which, more frequently than not, included Muslim ministers, because of their technical competence. Powerful armies were organized, which supported campaigns of expansion which were either successful, or else heralded the decline of those kingdoms.

[15] It is today recognized that between the 12th and the 16th centuries were the "great centuries" of Black Africa. Its countries developed strongly then. They reached well-balanced positions, were socio-po-

litically well integrated and developed strong economies. The strongest were the empires of Mali, which at the time of Mahmud Ali had 400 cities in what is now west Sudan; the states of Hausa, Yoruba and Benin, respectively owards the south-east and the south-west of Nigeria; the Bantu Kingdom and the Kongo in Central Africa; and the Zimbabwe and Monomotapa in the South.

[16] The revenues of those empires derived from taxes on the harvests and the cattle, from tributes, from customs houses, from gold nuggets (which were by law the property of the government), and from war booty. They integrated politically different peoples through a government compounded of high officers whose competence was either functional - such as the ministers - or territorial - the chiefs who controlled the provinces - and whose mandates were revocable, not hereditary. Almost all of them had professional armies. Their lands were worked by peasants, compelled to pay a tribute according to the number of workers, families and hamlets, and by slaves. There was a distinction between war slaves and house slaves, who served in the court and in the households, and who had certain civil rights, and were eligible for eventual emancipation. But in some communities and ethnic groups, such as those of Equatorial Africa, there were no slaves.

[17] The cities included within a certain empire had centers of religious studies which evolved into universities of sorts, where religious and other studies developed, as in Timbuktu. The empires fostered artistic manifestations, as well as architecture, such as the prodigious buildings of Great Zimbabwe, which were heard about early in Europe through reports from the Portuguese colonists.

[18] From the 16th century onwards, exogenous events eroded the independent life of the African empires. The Muslim states were not happy with being mere intermediaries anymore, and they launched expansionist policies which demolished the inner empires. In this they were gradually replaced by the Europeans who, after exploring the continent, organized slave traffic in order to satisfy the needs of the New World with cheap manual labor. The commerce of slaves was organized from coastal stations where the captured slaves were concentrated. The operation involved some 10-15 million slaves. Some African kings and leaders collaborated.

[19] The economic involvement of Europe was not limited to the slave trade. During those centuries, an economy based on the export of grain was also developed, but it did not include, at least at the beginning, a direct political control, or the loss of sovereignty on the part of the African states.

[20] These decisive historical changes entrenched a situation of general dependency and contributed to the later under-development of Africa. In the 20 years between 1890 and 1910, the European powers conquered, occupied and finally subjugated a continent whose territory, or rather 80 per cent of it, was previously governed by autonomous leaders. Most of them resisted this imposition, and manifested their determination to defend their sovereignty and independence, their religions, and their traditional lifestyles.

[21] Despite this resistance, colonization destroyed the authentic life forms of those countries, it fractured their social and cultural balance, and established relationships of dependence. European capital and a world economy and commerce robbed African peoples of their resources through mining firms, commercial and financial institutions, compelling them to work not for themselves any longer, but for European development.

[22] However, these cultures did not disappear. Furthermore, African traditions and habits have not only helped to shape the cultural development of the European countries that once subjugated them but also have been revitalized in their own native soil over the last several years.

New York which put an end to the war between Luanda and Pretoria, and provided for the independence of Namibia and withdrawal of South African and Cuban troops from Angola.

[18] In June 1989, UNITA signed a truce in the presence of 20 African heads of state at Gbadolite, Zaire. However, the ceasefire was broken barely two months later. At the end of April 1990, Angolan authorities announced in Lisbon, Portugal, that direct negotiations with UNITA would be resumed to achieve a lasting ceasefire. A month later, UNITA leader Jonas Savimbi officially recognized José Eduardo dos Santos as head of state.

[19] Toward the end of 1990, the MPLA announced the introduction of reforms geared toward democratic socialism. On May 11, 1991, a law on political parties was published, which brought one-party rule (under the MPLA) to an end. In addition, the law banned political participation by active members of the armed forces, the police or the judicial branch. On May 23, the last Cubans left Angola. On the 31st of the same month, after 16 years of civil war, a peace settlement was signed by the Angolan government and UNITA, in Estoril, Portugal. This agreement included an immediate cease-fire, as well as a promise to hold democratic elections in 1992 and the creation of a Joint Politico-Military Commission (CCPM), charged with establishing a national army made up of soldiers from both opposition groups. The governments of Portugal, the United States and the Soviet Union were involved in the discussion and drawing up of the agreement, as was the United Nations, which was put in charge of supervising compliance with the terms of the peace agreement. From November 14 Supreme Command of the Armed Forces was shared between Joao de Matos (MPLA) and Ahilo Camulata Numa (UNITA).

[20] Holden Roberto, leader of the National Front for the Liberation of Angola (FNLA) and Jonas Savimbi, president of UNITA, returned to Luanda in August and September of 1991, respectively, after 15 years of exile, in order to launch their election campaigns. The United States continued to support UNITA, and as a result there was increasing polarization as they moved towards the 1992 elections.

[21] The Organization of Angolan

STATISTICS

DEMOGRAPHY

Annual growth: 2.3% (1992-2000). **Children per woman:** 6.6 (1992).

HEALTH

One **physician** for every 25,000 inhab. (1988-91). **Under-five mortality:** 292 per 1,000 (1995). **Calorie consumption:** 85% of required intake (1995). **Safe water:** 32% of the population has access (1990-95).

EDUCATION

Primary school teachers: one for every 32 students (1990).

COMMUNICATIONS

85 **newspapers** (1995), 83 **TV sets** (1995) and 81 **radio sets** per 1,000 households (1995). 0.5 **telephones** per 100 inhabitants (1993). **Books:** 81 new titles per 1,000,000 inhabitants in 1995.

ECONOMY

Annual growth: -6.80% (1985-94).
Consumer price index: 100 in 1990; 61,982 in 1994.
Currency: 30 new kwanzas = 1$ (1990).
Cereal imports: 272,000 metric tons (1990).

Women (OMA), was founded in 1961, as a branch of the MPLA. In response to the new political and social situation, in which there is an attempt to create a more open and more democratic society, the delegates of OMA separated from the MPLA, becoming a non governmental organization (NGO).

[22] In the first OMA Extraordinary Congress, held in August 1991, a platform was drawn up, seeking common ground between church groups, intellectuals and professional organizations. The government plans to carry out budget cuts were cause for concern among the women, as these measures included the elimination of subsidies and the devaluation of the currency, in order to promote free trade and comply with International Monetary Fund conditions. Angola is a country where women die from malnutrition and illegal abortions, and where a great many are subject to physical and psychological abuse. OMA's struggle to improve the situation of women is a difficult and uphill struggle.

[23] Beleaguered by a foreign debt of over $6 billion, the government's intensive pacification efforts have coincided with an economic policy which seeks to meet the basic needs of the population.

[24] Despite President José Eduardo dos Santos' appeals to the international community, the United States refused to call off its economic and diplomatic embargo, stating that Angola was a Marxist nation and announcing that it would not extend diplomatic recognition until after the 1992 election. As a result of this classification, US companies doing business in Angola were unable to receive new loans from US banks.

[25] The attempted reunification process broke down soon after it was initiated as a result of organization and infrastructure problems. Food and medicine could not be delivered to the more isolated villages because of the lack of an adequate road network.

[26] In the meantime, although the State maintains a monopoly on communication, health and educational services, the industrial sector has been opened up to capital investment.

[27] After intense negotiations between the government and UNITA, it was agreed elections would be held in September 1992. The MPLA, the party in power, obtained almost 50 per cent of the vote followed by UNITA's 40 per cent. Savimbi did not recognize his defeat and ordered hostilities

to be resumed. As UNITA troops advanced, they occupied the countryside's diamond mines, leaving oil as the only source of stable income for the government (netting between $1.6 and 1.7 billion per year).

[28] In November 1993, peace talks were continued in Lusaka, capital of Zambia, where a year later, in November 1994, a peace agreement was signed between both factions. In practice, the agreement - which was to establish a truce and constitutional amendments to enable Savimbi's appointment as vice-president - was not enforced until late 1995 and the war went on.

[29] During 1996, progress was made in the application of the Lusaka agreement. An amnesty law was passed in May and UNITA military members began to join Angola's Armed Forces. Furthermore, Savimbi's organization withdrew a large number of troops (as was stated in the agreements) and began to hand over some of its weapons.

[30] The social and economic crisis which affects the country - after years of civil war - is considered the most serious in its history. Social conditions led dos Santos to replace Prime Minister Marcolino Moco with Fernando Franca Van Dunem. The adoption of liberalization measures in the economy recommended by international organisms like the IMF and the World Bank have not changed the situation and acute problems, like Luanda's lack of food, are continuously experienced.

Anguilla

Population: 8,000 (1994)
Area: 96 SQ KM
Capital: The Valley

Currency: E.C. Dollar
Language: English

Anguilla

Anguilla is the most northerly of the Leeward Islands. Its small size, barely 96 sqkm including the neighboring Sombrero Island, and the lack of fresh water for agriculture, made the island unattractive to the British Empire. From 1816 to 1871 Anguilla, St Kitts-Nevis and the Virgin Islands were administered together as one colony. The Virgin Islands were split off in 1871, leaving the others as a colonial unit ruled from St. Kitts.

[2] Colonialism lasted until the group became one of the five Caribbean "States in Association with the United Kingdom". Anguillans opposed the agreement and rebelled against the St. Kitts government. Under the leadership of local businessman Ronald Webster, Anguilla demanded a separate constitution. In March 1969, British troops disembarked on the island to supervise the installation of its colonial Commissioner, but the separatist movement continued its struggle.

[3] In 1976 a new constitution was approved by Britain, establishing a parliamentary system of government under the patronage of the British Commissioner. But it was not until 1980 that Anguilla was able to formally withdraw from the Associated State arrangement with St Kitts-Nevis, gaining the status of "British Dependent Territory".

[4] The 1976 constitution provided for a Governor appointed by the British Crown, responsible for defence, foreign relations, internal security (including the police), utilities, justice and the public audit.

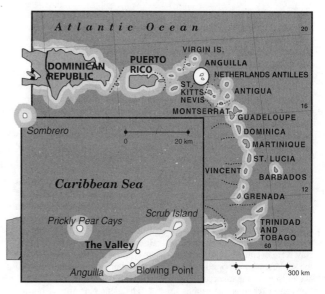

Atlantic Ocean

VIRGIN IS.
PUERTO RICO
DOMINICAN REPUBLIC
ANGUILLA
NETHERLANDS ANTILLES
ST. KITTS-NEVIS
ANTIGUA
MONTSERRAT
GUADELOUPE
DOMINICA
MARTINIQUE
ST. LUCIA
VINCENT
BARBADOS
GRENADA
TRINIDAD AND TOBAGO

Sombrero

Caribbean Sea

0   20 km

Scrub Island
Prickly Pear Cays
The Valley
Anguilla   Blowing Point

0   300 km

He presides over the Executive Council.

[5] The first legislative elections held in March 1976 voted in the Anguilla United Party leader, Ronald Webster, as Chief Minister. A year later he failed to win a vote of confidence and was replaced by opposition leader Emile Gumbs. In the 1980 general elections, the United Party of Anguilla, led by Webster, won by a landslide majority, obtaining six out of the seven seats. A year later, the government suffered internal divisions and Webster created the Popular Party of Anguilla (PPA). In the elections of June 1981, he won 5 seats, while the remaining 2 went to the Anguilla National Alliance, led by Emile Gumbs.

[6] In September, the constitution was changed leading to home rule on internal matters.

[7] On March 9 1984 in accordance with the new constitution of 1982 which requires a parliamentary government, elections were held and Emile Gumbs became Chief Minister once more. The ANA recieved 53.8 percent of the vote.

[8] The new ANA government requested greater power for the Executive Council and an increase in British investment in the island's economic infrastructure. After the Council called for constitutional changes, particularly with reference to the situation of women and people born outside the island who had relatives there, the Governor appointed a Committee to review the Constitution.

[9] Through the 1980s the building industry for tourism reduced unemployment from 26 per cent to 1 per cent, but at the beginning of the 1990s this trend stopped and the USA recession brought a reduction of tourists.

[10] In 1991 the Prime Minister Gumbs intended a closer cooperation between the British possessions and the Organization of East Caribbean States.

[11] The income generated by cattle-breeding, salt production, lobster fishing, boat building and the money sent by nationals working abroad (located mainly in the US), lost its former predominant role to the profit made by the building industry, tourism, and financial offshore services. In 1992 legislation gave Governor Alan W. Shave the task of granting permits to foreign firms, in an effort to control them more tightly.

[12] In March 1994 a victory was won in the parliamentary elections by Hubert Hughes, of the United Party of Anguilla. In November, the new legislature sanctioned a crucial package which regulates the activity of international firms and trusts, and includes legislation about fraud. In 1995 the registration of firms was computerized with the purpose of facilitating the activities of agents from all over the world 24 hours a day all year round.

Antigua

Antigua & Barbuda

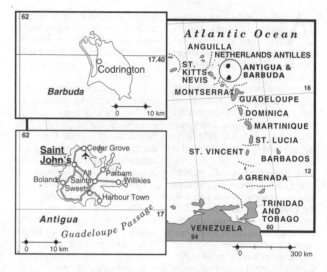

Population: 67,000 (1994)
Area: 422 SQ KM
Capital: St. Johns

Currency: E.C. Dollar
Language: English

The Caribs inhabited most of the islands in the ocean which took their name, but abandoned many of them, including Antigua, in the 16th century due to the lack of fresh water.

[2] In 1493 the name Antigua was given to one of the Antilles by Christopher Columbus in honor of Santa María la Antigua, a church in Seville. Other Europeans settled later (the Spanish in 1520, the French in 1629), but again left because of the scarcity of water. However, a few English were able to settle by using appropriate techniques to store rainwater.

[3] By 1640, the number of English families on the island had increased to 30. The few indians who dared to stay were eventually murdered by the settlers, who imported African slaves to work the tobacco plantations and later sugar plantations.

[4] Antigua felt the repercussions of the frequent armed conflicts between the great powers: in 1666, when war broke out between France and England, the governor of Martinique invaded the island, kidnapping all the African slaves. When England regained control in 1676, a rich colonist from Barbados, Colonel Cedrington, acquired large quantities of land and brought new African slaves. Thus sugar production - of excellent quality - was reinstated on the island.

[5] In the mid-8th century, European planters rebelled against the tyranny of governor Daniel Park, who in spite of representing the Crown in Antigua, was turned over to the slaves who immediately lynched him. In 1779, drought forced the settlers to import water from neighboring islands at the price of gold.

[6] Slavery was abolished in the British colonies in 1838. Antigua was administrated *in absentia* by the governor of the Leeward Islands, a group of islands which also included St Kitts, Nevis, Anguilla, Montserrat and the British Virgin Islands. The workers' situation did not noticeably change and quasi-slavery continued for several decades, until the early 20th century when the first trade unions appeared.

[7] The first trade union, led by Vere Bird, was formed on January 16 1939. Trade unions still play an important role in Antigua. The Labor Party of Antigua, the first political party, originated within a trade union. It was also led by Vere Bird, who had a life-long commitment to trade unions and politics.

[8] In the elections of April 1960, Bird's party won and he became Prime Minister. At the same time discontentment grew regarding the colonial statute. In 1966, a new constitution introduced self-government with a Parliament elected by Antiguans and Barbudans; Britain remained responsible for defence and foreign relations. Bird was again successful in the 1967 elections.

[9] The opposition Progressive Labour Movement (PLM) gained office for the first time in 1971, with 13 of 17 Parliament seats. George Walter, leader of the PLM, replaced Bird; in 1976, Bird returned to the government and two years later announced plans to negotiate for full independence of the archipelago.

[10] That same year Antigua ratified the transformation of the Caribbean Free Trade Association (CARIFTA) into the Caribbean Economic Community (CARICOM) and joined the new organization with 11 other islands.

[11] Bird's initiatives faced opposition disapproval who feared he would imitate Grenada's Eric Gairy and use power for his own self-interest. In 1979, opposition leader George Walter charged the government with human rights abuses, citing brutal repression of a teachers' strike as an example, and claimed repression could only become more so under independence. Bird was accused of being authoritarian by the progressive forces, and of using the island as a springboard for arms sales to Rhodesia (now Zimbabwe) and South Africa. The prime minister reacted by charging leading figures of the Labor-Progressive cabinet with corruption.

[12] Major opposition now came from the Barbudans demanding greater financial and administrative autonomy; creation of a separate police force for the island; and respect for the local system of land-ownership, conditions which the Antiguans judged inadmissible. As alternatives, Barbuda suggested either the continuation of colonial rule or the island's own independence. Britain rejected the proposal, fearing creation of a mini-state would involve too great a burden in financial assistance to achieve viability.

[13] Finally, on November 1 1981, Antigua and Barbuda became independent as a "sovereign, democratic and united state" and were admitted into the United Nations and the Caribbean Community (Caricom). Independence also gave the islands the right to become indebted to the IMF and the World Bank, a "privi-

> **Barbuda suggested either the continuation of colonial rule or the island's own independence. Britain rejected the proposal, fearing creation of a mini-state would involve too great a burden in financial assistance to achieve viability.**

lege" which they had not enjoyed as an associate state of Britain.

[14] The foreign debt grew dramatically and reached almost half of the GDP by the end of 1981. Tax exemptions granted to transnational companies led to rapid growth in the manufacturing sector, especially in the field of electronics. Neverthelessthe foreign trade defieit continued to grow.

[15] The IMF proposed its well-known prescription: 40 per cent cuts in the civil service, a wage-freeze and the elimination of a state program aimed at creating jobs for young people. The social impact of these measures would have been politically disastrous for the governing party. Yielding to public opinion, Bird refused to heed the Fund's recipes.

[16] In foreign affairs the prime minister still maintains close relations with the US, who use part of the island's territory for military purposes and pay an annual rent. Also, regional antennas of the "Voice of America" and the BBC are installed on the island.

[17] During 1982-83, the government strengthened ties with the US through incentives for the banking industry. Re-opening of the West Indies refinery encouraged promotion of other US companies, in particular for food and clothing manufacture. Antigua imports oil from Mexico and Venezuela, and exports oil by-products to other islands.

[18] The alliance with the US was further consolidated in 1983, when Antigua participated in the American invasion of Grenada. Parliamentary elections were held in April 1984, a year ahead of schedule in an effort to convince foreign investors of political stability according to Bird. He was in fact re-elected and despite accusations during the electoral campaign, his victory was attributed to approval of his stance on the Grenada issue.

[19] Three years later after a landslide victory, the ALP led by Vere Bird encountered serious internal problems. Adolphus Freeland, a minister transferred during the

ENVIRONMENT

The islands - Antigua with 280 sqkm and its dependencies Barbuda with 160 sq km and Redonda with 2 sq km - belong to the Leeward group of the Lesser Antilles. Antigua is endowed with beautiful coral reefs and large dunes. Its wide bays distinguish it from the rest of the Caribbean because they provide safe havens. Barbuda is a coral island with a large lagoon on the west side. It consists of a small volcano joined to a calcareous plain. Redonda is a small uninhabited rocky island, and is now a flora and fauna reserve. Sugar cane and cotton are grown along with tropical fruits; sea foods are exported. The reduction of habitats due to the reforestation of native forests with imported species is the main environmental problem of most of the Caribbean islands.

SOCIETY

Peoples: The majority of Antiguans and Barbudans are of African origin; they are descendants of slaves or immigrants from other Antilles islands. **Religions:** 80% Anglican and the remainder Catholic, Adventist and Methodist. **Languages:** English is the official language, but in daily life a local dialect is spoken. **Political Parties:** The Antigua Labour Party (ALP), led by Premier Lester Bird. The Antigua Caribbean Liberation Movement (ACLM). The United National Democratic Party (UNDP). The United Progressive Party (UPP). **Social Organizations:** The labor movement is divided into two groups: the Antigua Workers' Union, linked to the UNDP, and the Antigua Trades and Labour Union, with ALP leadership.

THE STATE

Official Name: Associated State of Antigua and Barbuda. **Capital:** St. Johns, 21,500 inhab. (1987). **Other cities:** Parham, Liberta. **Government:** James Carlisle, Governor-General, representative of Queen Elizabeth II, who is formally the Head of State; Prime Minister, Lester Bird. The British-style bicameral legislature is composed of a 17-member Senate appointed by the Governor-General and a 17-member House of Representatives elected by universal suffrage for a 5-year term. **National Holiday:** November 1, Independence Day (1981). **Land use:** Forested 11%; meadows and pastures 9%; agricultural and under permanent cultivation 18%; other 62%.

DEMOGRAPHY

Annual growth: 0.3% (1992-2000). **Children per woman:** 1.7 (1992).

COMMUNICATIONS

28.9 **telephones** per 100 inhabitants (1993).

ECONOMY

Consumer price index: 100 in 1990; 105.7 in 1991. **Currency:** 3 E.C. dollars = 1$ (1994).

---

gave rise to this debt were used to expand hotel capacity by 20 percent. The FAO, an agency of the UN, granted loans for installation of an irrigation system for local farming.

[23] In March 1991, three ministers including vice-premier Lester Bird demanded the resignation of Vere Bird, who responded by reshuffling the cabinet and removing his son,

> The IMF proposed its well-known prescription: 40 per cent cuts in the civil service, a wage-freeze and the elimination of a state program aimed at creating jobs for young people.

Lester. In August several financial measures were announced including the suspension of investments, the suspension of public hirings and the reorganization of the fiscal system.

[24] The 1992 budget did not create new taxes. Tourism from Europe grew 44 per cent while tourism from the US dropped 11 per cent. Aged 83, Bird confirmed he would complete his term in office. In late 1992, the Organization of Western Caribbean States, formed by St Kitts-Nevis, Antigua, Dominica, St Lucia, St Vincent, Grenada and Montserrat, announced it would open an embassy in Brussels, Belgium, in order to intensify trade relations with the European Community.

[25] In 1993, the government announced it would strengthen the suppression of drug trafficking in order to prevent Antigua from becoming a "transfer" point. A debate took place in August about corruption in the state due to accusations contained in a book by US writer Robert Coram.

[26] In 1994, Vere Bird retired from public life and his son Lester replaced him at the head of ALP. In the March elections, Labour lost 4 seats but retained an absolute majority at the Chamber of Representatives, with 11 seats out of 17. The United Progressive Party, a coalition of 3 opposition parties, received 5 seats and the People's Movement of Barbuda obtained one.

[27] In January 1995 a series of demonstrations began against new taxes and the rise of public utility rates. In September, hurricane Louis caused losses which amounted to $300 million, damaging 60 per cent of the country's buildings, whether partially or completely, including several main hotels, seriously affecting the country's tourism infrastructure.

---

cabinet shuffle of 1987 stated that the ALP needed new leadership. Bird, active for 50 years in politics, claimed that selfish and irresponsible attitudes of some ministers were responsible for the party disputes.

[20] Opposition leaders demanded the prime minister's resignation, citing as one of the reasons the misappropriation of funds in the St Johns Airport modernization project. At the same time an opposition alliance, led by Tim Hector and Ivor Heath, was formed. The "Herald" - a pro-government newspaper - accused the US of trying to "subvert order on the island" in its attempt to use the "Information Exchange Treaty" as a means of gaining access to bank accounts, in its search for "drug dollars".

[21] In spite of problems, the ALP won 15 of the 17 parliamentary seats during the March, 1989 elections, with one seat to an independent candidate, and one to the opposition, the United Democratic National Party.

[22] The foreign debt, in the meantime, went from 50 to 90 million Caribbean dollars. Loans which

Aotearoa

(New Zealand)

Population: 3,631,200 (1996)
Area: 270,028 SQ KM
Capital: Wellington

Currency: NZ Dollar
Language: English and Maori

Aotearoa, "the land of the long white cloud", was settled around the 9th century by the Maori, who arrived there from Polynesia. Over the years a distinct culture developed, based on tribal organisation and a strong affinity with the land. The Maori appointed themselves as guardians of the land for future generations.

[2] In 1642 Abel Tasman, from Holland, reached the South Island, the larger of Aotearoa's two main islands. However, the resistance of the indigenous population prevented him from going ashore. It was not until 1769 that James Cook, a Briton, surveyed the shores of the two most important islands, thus opening the door for a growing colonization of the country which Abel Tasman had named "New Zealand". Whalers, sealers and traders, along with a few deserters from the navy and fugitives from Australian jails, established themselves on the islands.

[3] In the early 19th century, colonization increased with the arrival of British immigrants and missionaries. They brought with them new diseases, values, and beliefs, which affected the traditional Maori way of life. Christianity weakened the traditions which gave substance and cohesion to the tribal society, and the introduction of commercial practices by the Europeans seriously damaged the material basis of the Maori lifestyle.

[4] In 1840 the territory of New Zealand was formally annexed by the British crown as a colony. The two larger islands were occupied under different legal schemes: the South Island was incorporated by virtue of the right of "discovery", and the North Island through the Treaty of Waitangi, signed in 1840 by the Maori chiefs and representatives of the British government. According to the text of the treaty - which is somewhat confusing and different in its English and Maori versions - Maori chiefs accepted the presence of British settlers and the establishment of a government by the Crown to rule the settlers. In exchange, the Maori were assured absolute respect of their national sovereignty.

[5] However, soon after the signing of the treaty an extremely violent process of expropriation of Maori lands began. The so-called "land wars" between the Maori and the

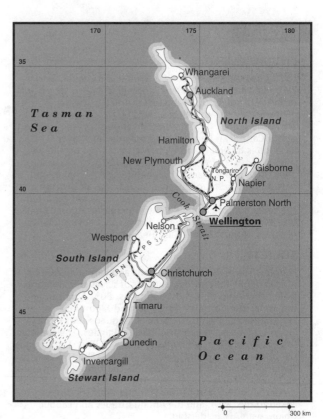

PROFILE

ENVIRONMENT

Aotearoa is situated in Oceania. Its two principal islands are relatively mountainous. The North Island is volcanic and has plateaus and geysers. The South Island is crossed by the Southern Alps, a mountain chain with peaks of over 3,000 metres. The climate is moderately rainy, with temperatures cooler in the south. The native rainforests have been devastated by early colonisation, replaced by agricultural land and pastures. The economic base of the country is agricultural, with advanced techniques being used in pastoral and agricultural by-products. The main exports are meat, dairy products, timber, seafood and wool.

SOCIETY

Peoples: Much of the population is descended from European settlers. 12.9% are Maori (a figure rapidly increasing), and 3.5% are from Pacific Islands descent. **Religions:** Anglican and other Protestant denominations dominate. There are also Catholic and various Maori church minorities. **Languages:** English and Maori are the official languages, with English dominant. **Political Parties:** A multi-party

system in change due to the adoption of a proportional representation voting system. The major parties are National, Labour, New Zealand First and the Alliance. **Social Organizations:** The Trade Union Federation and the Council of Trade Unions.

THE STATE

Official Name: New Zealand. **Administrative divisions:** Divided into 15 town and 58 district authorities. **Capital:** Wellington, 325,682 (1991). **Other cities:** Auckland, 885,571 inhab.; Christchurch, 307,179 inhab.; Hamilton, 148,625 inhab. (1991). **Government:** A parliamentary monarchy and member of the British Commonwealth. Governor-General: Michael Hardie Boys. Prime Minister: National Party Leader Jim Bolger (since September 1996). Aotearoa has a parliamentary system, with a unicameral legislative body of 120 members. The elected members of the government form a cabinet chaired by the Prime Minister. On occasions this cabinet acts as the Executive Council presided over by the Governor-General. The Governor-General appoints cabinet on the recommendations of the Prime Minister, and has the power to dissolve Parliament. **National holiday:** February 6, Waitangi Day. **Armed Forces:** 9,958 (1995).

Europeans were essentially for sovereignty and guaranteed rights to the lands, forests, fisheries and other *taonga* (treasures).

[6] Massive immigration led to the gradual annexation of Maori land. The magnitude of their loss becomes clearer if one considers that, out of the 27 million hectares they owned in 1840, they now have a little over a million left.

[7] While the north of the country was involved in a series of wars, the South Island settlers went through a period of prosperity because of the discovery of gold. This discovery brought a massive flow of British, Chinese and Australian immigrants, which energized the region's economy.

[9] Maori opposition to Pakeha colonization found new expression by the end of the 19th century. They organized petitions, delegations and submitted their claims before local courts and even before the British Crown itself, demanding compliance with the Treaty of Waitangi. These efforts were fruitless. The lands that had once belonged to the Maori were now used for farming, which had started to play a central role not only in the life of the settlers but also in the

DEMOGRAPHY

Urban: 86% (1995). **Annual growth:** 0.8% (1991-99).
Estimate for year 2000: 4,000,000. **Children per woman:** 1.95 (1995).

HEALTH

Under-five mortality: 7.3 per 1,000 (1995). **Calorie consumption:** 113% of required intake (1995). **Safe water:** 97% of the population has access (1990-95).

EDUCATION

School enrolment: Primary (1993): 101% fem., 101% male. Secondary (1993): 104% fem., 103% male. **University:** 2986 students per 100,000 inhabitants (1995). **Primary school teachers:** one for every 22 students (1995).

COMMUNICATIONS

28 daily **newspapers** (1995), 443 **TV sets** (1995) and 108 **radio sets** per 1,000 households (1995). 465 **telephones** per 1000 inhabitants (1993). **Books:** 107 new titles per 1,000,000 inhabitants in 1995.

ECONOMY

Per capita GNP: $15,646 (1995). **Annual growth:** 0.7% (1985-94). **Annual inflation:** 2.2% (1995). **Consumer price index:** 1000 in 1993; 1058 in 1995. **Currency:** $NZ1 = $US0.71 (November 1996). **Cereal imports:** 282,000 metric tons (1993). **Fertilizer use:** 2 million tonnes (1995). **Imports:** US$ 11,900 million (1995). **Exports:** US$ 12,200 million (1995).

ENERGY

Consumption: 4,352 kgs of Oil Equivalent per capita yearly (1994), 5% imported (1994).

Cook Islands

Cook Islands

Population: 18,500 (1993)
Area: 237 SQKM
Capital: Avarua
Currency: NZ Dollar
Language: English

The islands, which had already been explored and settled by Polynesians and Spaniards, received their name from the British navigator Captain James Cook, who drew up the first map of the archipelago in 1770.

[2] In 1821, Tahitian missionaries were sent to the islands by the London Missionary Society; a Protestant theocracy was established and all "pagan" structures were destroyed, as were many of the traditional forms of social organization. The islands were declared a British Protectorate in 1888 and became part of Aotearoa/New Zealand in 1901. The land rights of the native Maori were recognized, and the sale of real estate to foreigners was prohibited. In 1965, the United Nations promoted and supervised a plebiscite, and the population voted against independence, and in favour of maintaining its ties to Aotearoa.

[3] Prime Minister Geoffrey Henry governed the country with an iron hand for 15 years before being replaced in 1978 by Thomas Davis, of the Democratic Party, who gave incentives to private fruit planters. However he was dismissed from office by Parliament in 1987, being replaced by Pupuke Robati, also of the DP. In the 1990 elections, Geoffrey Henry's Cook Islands Party was returned to office, and re-elected in 1994.

[4] From 1991 on, when Aotearoa's economic injections to the Cook Islands fell to 17 per cent of the local budget, both governments decided the auditors in the Cook Islands would take charge of the State finances in place of the Aotearoan office.

[5] The Constitution allows for the Cook Islands to declare independence unilaterally; but in the meantime it will maintain its status as "free associate state."

ENVIRONMENT

Archipelago located in the South Pacific 2,700 km northeast of Aotearoa/New Zealand, made up of 15 islands which extend over an ocean area of 2 million sq km. They are divided into 2 groups. The northern group is made up of 6 small coral atolls, which are low and arid, with a total area of 25.5 sq km. The southern group comprises 8 larger and more fertile volcanic islands (211 sqkm). The capital, Avarua, is located on Rarotonga, the largest of the islands. Every 5 years, for the past 2 decades, the islands have suffered terrible droughts.

PEOPLES

Most inhabitants are Maori. **Religions:** Christian. **Languages:** English (official). Language and traditions similar to the Maori in Aotearoa/New Zealand. **Political Parties:** The Cook Islands Party (CIP); the Democratic Party (DP), the main opposition force; the Tumu Democratic Party; the Alliance Party.

THE STATE

Official Name: The Cook Islands. **Capital:** Avarua. **Government:** Autonomous associated state. Tim Caughley, Representative of Aotearoa. Geoffrey Henry, Prime Minister since January 1990. Single-chamber legislature; there is a Legislative Assembly, with 25 members elected by direct vote every five years.

DEMOGRAPHY

Population: 18,500 (1993).

EDUCATION

Primary school teachers: one for every 17 students (1988).

COMMUNICATIONS

25.1 **telephones** per 100 inhabitants (1993).

ECONOMY

Consumer price index: 100 in 1990; 120.6 in 1994. **Currency:** NZ dollars.

whole economy. New markets for dairy and meat products opened up in the 1880s, with the appearance of cold-storage systems which made long-distance shipping possible. These products constituted the basis of the country's economic development.

[10] For the Maori, British rule also meant the beginning of a process of cultural extermination, through the arbitrary imposition of European language, religion, and customs. Thus, the Maori language which was in theory protected by the Treaty of Waitangi, was given less and less importance.

[11] At the end of the century, the country's political scene was dominated by the Liberal government. They were the first in the world to grant the vote to women (1893) and to establish measures to protect the rights of industrial workers, a group which was growing parallel to the development of cities and manufacturing industries.

[12] In the 20th century new political movements appeared to oppose the power of the Liberals who had become a coherent, organized political party after 20 years in power. The Labour Party, with ample working-class and urban middle-class support, became one of the country's main political forces.

[13] During the 1930s, Maori demands acquired new strength thanks to an alliance between the groups representing Maori interests and the Labour Party, which came to power for the first time in 1935 with Maori support. But there was still no legislative recognition of the Treaty of Waitangi.

[14] World War II marked the beginning of a new era for the country. During the war Britain's inability to guarantee the security of its former colony led to closer ties with the US. Through a series of political and military alliances, US presence was consolidated in the region. In the 1950s and 1960s Aotearoa had to pay the price for this relationship, particularly when it found itself involved in the Vietnam war, a conflict which touched the political life of the country very deeply.

[15] Legislation in the 1950s forced many Maori from their land and the 1960s saw their increasing urbanisation. This brought them into greater contact with the Pakeha lifestyle and institutions, and the Maori gained better access to the Pakeha education system.

[16] In the 1970s, Aotearoa attempted, with limited success, to diversify its production and gain access to markets other than Britain and the US. Unemployment rose and inflation reached unprecedented levels due to the failure of the diversifica-

Foreign Trade

MILLIONS $ 1994

IMPORTS
11,900

EXPORTS
12,200

tion scheme, the rise in the price of oil and massive external loans to finance "think big" projects. 1975 saw a drastic fall in the purchasing power earned from primary exports. This fall, combined with the foreign loans, provoked a period of rising foreign debt. After 12 years in opposition, the Labour Party took power from 1972 until it was defeated by the conservative National Party in the 1975 elections. The new government closed the country's doors to immigration, which it used as a scapegoat for unemployment

[17] In 1975, growing Maori activism led to the formation of the Waitangi Tribunal to investigate Treaty claims. In 1986 the Labour government gave it the power to hear claims dating back to the 1840 signing. The Tribunal has no binding powers over the Crown, so that settlement of claims for compensation has been slow.

[18] In the mid-1980s, the Labour Party regained office, and introduced a monetarist economic policy which included privatisation of a number of public enterprises. These policies alienated many traditional Labour supporters.

[19] A law passed in 1987 banned the entry of nuclear weapons or nuclear powered ships into the country's ports, and questioned the military presence of France and the US which had used the South Pacific as a testing ground for such weapons. This decision led to the cessation of defence agreements between the US and Aotearoa. In 1996, prime minister Jim Bolger tried to secure the weakened links with the US.

[20] The National Party won the October 1990 elections but the shift of the ruling power did not alter the country's economic model. Privatizations increased, protectionism was further dismantled and there were significant cutbacks on health, education and social security benefits. The outcome was a fall of inflation along with a rise in unemployment.

[21] The National government's decline in popularity was not reflected in the 1993 electoral results, when the National Party (NP) obtained 50 of Parliament's 99 seats and Jim Bolger was re-elected prime minister. The Labour Party (LP), main opposition force, won 45 seats and the recently created Alliance (left-of-center coalition which joins the Democratic Party of New Zealand, the NewLabour Party, the Green Party of Aotearoa, Mana Motuhake o Aotearoa (Maori) and the Liberal Party) obtained 2 seats. The newly formed New Zealand First Party also won 2 seats. Deregulation policies continued and brought about macroeconomic improvements but at a high social cost: continuing unemployment, crises in the health sector, worsening conditions for beneficiaries, students, and low-paid workers, and the weakening of the trade unions.

[22] In October, Aotearoa achieved its first budget surplus in years; the currency increased in value, unemployment was reduced and inflation settled at 2 per cent.

[23] Between 1994 and 1995 the Bolger administration signed agreements with the Tainui from the North Island who claimed their rights to lands colonized during the last century, compensating them economically and with some 15,400 hectares of land. These agreements were ratified by Queen Elizabeth II of England who even apologized for lives lost during the islands' colonization.

[24] Although general unemployment rates continued to fall towards 6 per cent during 1995, according to government statistics, the rates for the young, Maori and immigrants from other South Pacific islands are still high. Almost 10 per cent of the population receive government subsidies and prior to the October 1996 elections, when the first parliament with proportional representation was elected, the campaign was focused on the loss of funds for health and education sectors during the last years and not on economic growth.

[25] The government's firm opposition to French nuclear tests on the Mururoa Atoll has helped to strengthen Prime Minister Bolger's situation. The new system of government forced the NP to seek electoral alliances, faced with the growth of the opposition parties.

Niue

Niue

Population: 2.321 (1994)
Area: 259 SQ KM
Capital: Alofi
Currency: NZ Dollar
Language: English

Niue was settled by Samoans and Tongans. Captain James Cook named it "Savage Island" when he visited it in 1774. The indigenous peoples' reputation for fierceness kept missionaries away (the island's first permanent mission dates from 1861), as well as the slave traders who caused much suffering in other areas of the Pacific. Emigration to the phosphate mines on other islands of the area initiated an outgoing stream which has since continued to the present. In 1900 the island was declared a British protectorate and was annexed by Aotearoa /New Zealand in 1901. It was administered along with the Cook Islands until 1904, when it separated to form a separate possession. In 1974 it became an Associated Territory of Aotearoa. The United Nations recognized this as a legitimate decision and eliminated the "Niue case" from the Decolonization Committee agenda. As well as economic help from Aotearoa, Niue also receives help with defence and international affairs.

[2] In April 1989, questions arose over the management of economic aid received from Aotearoa by Prime Minister Robert Rex's government. Despite this, the opposition lacked the votes necessary to approve a parliamentary motion to censure Rex's administration.

[3] In 1991, Aotearoa announced a reduction in financial aid. Diplomatic contacts were established with Australia in 1992. The prime minister, Robert Rex, died on December 12, and was replaced by Young Vivian, who was subsequently defeated in the elections and succeeded by Frank Lui, in March 1993.

ENVIRONMENT

Located in the South Pacific in southern Polynesia, 2,300 km southeast of Aotearoa, west of the Cook Islands and east of the Tonga Islands. Of coral origin, the island is flat and the soil relatively fertile. Its rainy, tropical climate is tempered by sea winds.

SOCIETY

Peoples: The people of Niue are of Polynesian origin.
Religions: Protestant.
Languages: English (official), local Niue language (national).
Political Parties: Niue People's Action Party (NPAP).

THE STATE

Official name: Niue.
Capital: Alofi.
Government: Autonomous associated state. Kurt Meyer, Representative of Aotearoa. Frank F. Lui, Prime Minister (since March 9, 1993). Single-chamber legislature - National Assembly, made up of 20 members (14 village representatives). Since October 1974, the island has had an autonomous local government. There is a 20-member Legislative Assembly which is headed by the Prime Minister. Aotearoa/New Zealand controls defence and foreign affairs.

DEMOGRAPHY

Population: 2,321 (1994). 14,500 Nieuans resident in Aotearoa/New Zealand.

EDUCATION

Primary school teachers: one for every 14 students (1988).

ECONOMY

Consumer price index: 100 in 1990; 111.6 in 1993.
Currency: NZ dollars.

Tokelau

Population:	2,000 (1994)
Area:	10 SQ KM
Capital:	Fakaofo
Currency:	NZ Dollar
Language:	Samoan

Tokelau Islands

From ancient times, the islands were inhabited by Polynesian peoples who lived by subsistence fishing and farming. The first European to set foot on the islands was English explorer John Byron, who arrived in 1765. However, the absence of great riches meant that the islands did not arouse the colonial interests of the British crown.

[2] In 1877, these islands became a British protectorate. Britain annexed them to the Crown in 1916, and included them as part of the colonial territory of the Gilbert and Ellice Islands (now Kiribati and Tuvalu). In 1925, Britain transferred administrative control of the islands to Aotearoa (New Zealand). In 1946, the group was officially designated the Tokelau Islands, and in 1958 full sovereignty was passed to Aotearoa.

[3] Like the Cook Islands, the Tokelau Islands were claimed by the US until 1980, when the United States signed a Friendship Treaty with Aotearoa and dropped its claims.

[4] The government of Aotearoa adopted policies intended to maintain traditional customs, institutions and communal relations in Tokelau. The population lives in relative isolation; with a single ship reaching the islands from Apia (Samoa) every two or three months. The ocean and lagoons provide fish and seafood which are the islanders' staple subsistence foods.

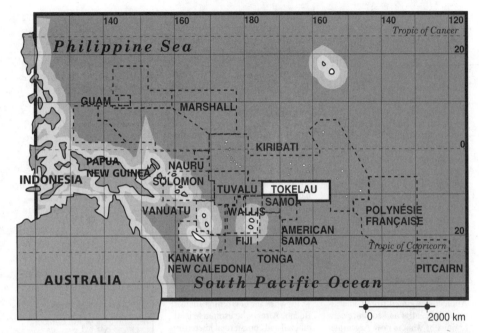

[5] In the 1980s, farming underwent a crisis due to a number of adverse climatic conditions. As a consequence, emigration to Aotearoa rose considerably. The government of Aotearoa tried to encourage Tokelau immigrants to return to their homeland, but its policies have not been successful.

[6] In 1976 and 1981 the UN sent delegations to Tokelau. On both occasions the envoys reported that the inhabitants of the islands did not wish to change their relations with Aotearoa.

[7] In December 1984, the UN Assembly discussed the situation of Tokelau and decided that Aotearoa should continue to report on the state of the islands. However, in the report submitted before the UN Special Committee in June 1987, Tokelau expressed a wish to achieve greater political autonomy, while maintaining unchanged relations with Aotearoa.

[8] In 1989, Tokelau's motion to apply sanctions on those States that fish in its territorial waters was supported by the South Pacific Forum. In November of that year, Aotearoa banned fishing in a 200-nautical-mile exclusion zone, in order to protect the fishing reserves which constitute the islands' main wealth.

[9] In 1989, a UN report on the consequences of the greenhouse effect included Tokelau among the islands which could disappear under the sea in the 21st century, unless drastic measures are taken to halt pollution.

[10] In February 1990, the three groups of islands which make up the country's territory were devastated by hurricane Ofa, which destroyed all the banana trees and 80 per cent of the coconut plantations, as well as hospitals, schools, houses, and bridges. As a result, emigration to Aotearoa increased once again.

[11] Aotearoa established the first regular maritime transport service between the three atolls during 1991.

[12] In 1993, the Samoan government announced that it had agreed to the transfer of the Office of Tokelau Affairs, from Apia, Western Samoa, to the islands themselves. Tokelau's administrator in 1996 was Brian Absolum, who took office in 1994.

Argentina

Population: 34,194,000 (1994)
Area: 2,791,810 SQ KM
Capital: Buenos Aires

Currency: Peso
Language: Spanish

Argentina

Two large groups populated what is now Argentina at the beginning of the 16th century: the Patagonians and the Andeans. Patagonians included Tehuelch, Rehuelche, Rampa, Mataco and Guaycure peoples. The latter two constituted stable, agrarian civilizations, while the others were nomadic hunters and gatherers. They settled in the south, center and north of the country. The principal Andean groups were the Ancient Rehuenche, Algarrobero Rehuelche, Huerpe, Diaguita, Capayane, Omahuaca and Patama. Through contact with the Incas, they perfected their agricultural system introducing terracing and artificial irrigation. They raised llamas, and traded in the northwest and west of the country.

[2] In the 17th and 18th centuries during the Spanish conquest, the Araucans migrated from Chile toward the central and southwestern regions of what is now Argentina, with the resulting "Araucanization" of the region's inhabitants.

[3] Explorations organized by Spain in the 16th century led Amerigo Vespucci, in 1502, and Juan Díaz de Solís, in 1516, to enter the estuary which they named Río de la Plata (River Plate, meaning silver river) in honor of the much sought after precious metal. In 1526, Sebastián Cabot built a fort on the banks of the Carcarañá River, the first settlement of present-day Argentina.

[4] To check Portugal's advance, Spain sent Pedro de Mendoza to the region with stipulated terms of political and economic advantages for the conqueror. In 1536, Mendoza founded Santa María del Buen Aire (Buenos Aires), but this small town was abandoned in 1541, after attacks by the native population became too numerous.

[5] Using Asunción (in today's Paraguay) as the center for colonization, the Spaniards founded several cities in Argentina (Santiago del Estero, Córdoba, Santa Fé), refounding Buenos Aires in 1580. This port soon became Spain's strategic, political and commercial base in the region.

[6] In 1776 the Viceroyalty of Río de la Plata was created - including today's Chile, Paraguay, Argentina, Uruguay and part of Bolivia - with Buenos Aires as its capital. A strong commercial bourgeosie, based in the port area, became the initiators of the 1810 revolutionary movement. They created the United Provinces of Río de la Plata and ousted the Viceroy, accusing him of disloyalty to Spain, then occupied by Napoleonic troops. Independence of the United Provinces of Río de la Plata was first initiated in 1816, when the Governing Junta of Buenos Aires, who aspired to institutional and commercial liberalism, found the ideas of the restored Spanish monarch Fernando VII and his servile party incompatible with their aims.

[7] General José de San Martín organized the armies that defeated the royalists and contributed decisively to the independence of Chile and Peru. During the same period, the United Provinces fell under the influence of Britain, as an exporter of hides, and import market of manufactured goods. In two decades this ruined the craftspeople of the upland provinces and increased benefits to the Buenos Aires port oligarchy who controlled import trade. In 1829, when Juan Manuel de Rosas came to power, he reconciled the interests of both parties by creating customs laws and other restrictions primarily to resist French and British penetration. Both powers resorted to military

aggression when they blockaded the port of Buenos Aires in an attempt to overcome the government; their efforts were, however, defeated.These first attempts to bring Argentina into the international market were successful after the "Great War", (waged in Argentina and Uruguay 1839-52), involving Argentina, Uruguay and Brazil, with direct intervention from Britain and France.

[8] In 1833 UK occupied the Malvinas Islands, with the blessing of the United States. After Rosas' defeat, Argentina became an importer of British manufactured goods and capital, and exporter of beef and cereals. In the name of economic and political liberalism, president Bartolomé Mitre - together with the Brazilian Empire of

Pedro II and Venancio Flores' Uruguay - waged the War of the Triple Alliance (1865-1870) against Paraguay. This war of extermination against their neighbor's prosperous economic experiment began in 1865 and ended with the death of Paraguayan president Francisco Solano López in 1870. (See box "The Triple Alliance"). At the time of "victory", Faustino Domingo Sarmiento had already replaced Mitre.

[9] After the war, technological upgrading was necessary to meet new world market requirements. Railroads and ports were built to transfer the country's wealth to London, while the indigenous peoples of the Patagonian and Chaco regions were exterminated to make way for new settlers. There was massive immi-

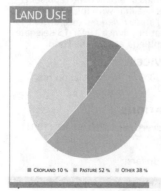

WORKERS

10%

1994   UNEMPLOYMENT

- MEN 71%
- WOMEN 29%
- AGRICULTURE 13%   INDUSTRY 34%   SERVICES 53%

LAND USE

- CROPLAND 10 %
- PASTURE 52 %
- OTHER 38 %

gration of European labor, mainly Italian and Spanish.

[10] In 1916, Radical Civic Union leader Hipólito Irigoyen was elected president with the vote of the middle class and the emerging proletariat. This election ended a succession of governments dominated by landed interests. Favorable conditions during WWI - in which Irigoyen maintained a position of active neutrality - generated some industrial development and brought new social groups and conflicts of interest to the fore. These built up to the "Tragic Week" of 1919, when troops machine-gunned striking workers in the streets, and extreme rightist opposition vigilantes ransacked the Jewish quarter in search of "Bolsheviks".

[11] The inadequacy of the agro-export model was revealed during the great depression of 1929. Irigoyen was overthrown by the ruling classes, who drew closer to Britain and consolidated their own economic privileges.

[12] World War II stimulated rapid industrial development based on import substitution. An alliance between the bourgeoisie, the working class and the Armed Forces dominated the political scene. Mass uprisings on October 17 1945 sparked events that carried the then Colonel Juan Domingo Perón to the presidency in 1946. Perón nationalized foreign trade, banks, railroads, oil and telephones, expanded the shipping fleet and set up an airline, increased the workers' share of national income by 50 per cent, and passed progressive social legislation. He also organized workers' and employers' associations into national confederations where economic and social policies were negotiated. In foreign relations, he argued for a "third position", independent of the contending blocs in the Cold War. Perón's wife, Evita, cultivated a charismatic relationship with the workers, and used her influence to secure female suffrage.

[13] US interests triggered a coup d'etat in 1955. Military dictator Pedro Aramburu repudiated Perón's "third position" and embraced the National Security doctrine. Regional defence against "the enemies of democracy" was delegated entirely to the United States. Aramburu's grossly misnamed "Liberating Revolution" was an attempted reversion to the 1940s, with the United States playing Britain's former role. The dictatorial re-

PROFILE

ENVIRONMENT

Argentina claims sovereignty over the Malvinas Islands and a 1,250,000 sq km portion of Antarctica. There are four major geographical regions. The Andes mountain range marks the country's western limits. The sub-Andean region consists of a series of irrigated enclaves where sugarcane, citrus fruits (in the north) and grapes (central) are grown. A system of plains extends east of the Andes: in the north, the Chaco plain with sub-tropical vegetation and cotton farms; in the center, the Pampa with deep, fertile soil and a mild climate where cattle and sheep are raised, and wheat, corn, forage and soybeans are grown, and to the south stretches Patagonia, a low, arid, cold plateau with steppe vegetation where sheep are extensively raised and oil is extracted. Untreated domestic waste has raised the levels of contamination of many rivers, especially the Matanza-Riachuelo in Buenos Aires. Another problem is increasing soil erosion, especially in the damp northern Pampas.

SOCIETY

Peoples: Most Argentinians are descendants of European immigrants (mostly Spaniards and Italians) who arrived in large migrations between 1870 and 1950. Among them is found the largest Jewish community in Latin America. According to unofficial figures, the indigenous population of 447,300, is made up of 15 indigenous and 3 mestizo peoples mainly in the north and southeast of the country, and in the marginal settlements around the major cities. Mapuche, Kolla, and Toba constitute the largest ethnic groups. The indigenous peoples in the east, center and southernmost tip are in decline. Religions: Catholic (92%, official); Protestant,

Evangelical, Jewish and Islamic minorities. Languages: Spanish. Minor groups maintain their indigenous languages: Quechua, Guarani and others. Political Parties: "Justicialista" Party (PJ, "Peronist" party), in power; Radical Civic Union (UCR), led by former president Raúl Alfonsín; the Frepaso is a coalition made up of former communists, socialists, independents, members of the Intransigent Party and former Peronists, including the head of the alliance, José Octavio Bordón; Socialist Unity; People's Socialist Party; Modin (nationalist); the Center Alliance, liberal, made up of the Progressive Democratic, Autonomist, Federal, Democratic and Union of the Democratic Center (UCeDé) parties; Republican Force, linked to the last military dictatorship. Social Organizations: The General Labor Confederation (CGT), Peronist in orientation, founded in 1930. In reaction to the current government's economic and labor policies, the confederation has split into three factions.

THE STATE

Official Name: República Argentina. Administrative divisions: 5 Regions with 22 Provinces, Federal District of Buenos Aires, National Territory of Tierra del Fuego. Capital: Greater Buenos Aires, 12,582,300 inhab. (1991). Other cities: Cordoba, 1,179,100 inhab.; Rosario, 1,078.400 inhab.; Mendoza, 801,900; La Plata, 542,600 inhab. (1991) Government: Presidential system; Carlos Saúl Menem, President since July 8 1989. Legislature: made up of the Chamber of Deputies and the Senate. Each province has 2 seats in the 48-member Senate. National Holidays: May 25, Revolution (1810); July 9, Independence Day (1816). Armed Forces: 67,300: army 60%, navy 26.8%, air force 13.2% (18,100 conscripts). Paramilitaries: (Gendarmes) 18,000.

gime imposed a reactionary income redistribution. Military and civilian opposition members faced firing squads in the 1956 Operation Massacre, a precursor of decades of political violence.

[14] Though outlawed, Peronism helped Arturo Frondizi win the 1958 presidential election. An advocate of development theories, Frondizi brought multinational oil and automobile concerns into the country and followed a model of economic growth and income concentration causing subsequent social confrontations. In 1962, Fron-

dizi was forced out of office. Two periods of violent struggle between the "blue" and "red" factions of the Armed Forces, saw General Juan Carlos Onganía emerge as the new leading force.

[15] With Peronism once again banned, the Radical Party's Arturo Illía was elected. For the first time in four decades Argentina was not subject to the "state of siege" or other extraordinary measures of repression or cultural censorship.

[16] Illía revoked oil contracts signed by Frondizi and paid indemnity to the affected foreign compa-

nies. Several good harvests and trade with socialist markets helped to overcome the economic crisis bequeathed by the military interregnum. The Illía administration refused to send Argentinian troops to assist the US invasion of the Dominican Republic. Confrontations with Peronist-led labor unions, which staged strikes, factory takeovers, and anti-government campaigns, continued to plague Illía's administration.

[17] In June 1966, Illía was overthrown by the "Argentinian Revolution" of Onganía. This new au-

thoritarian plan, clerical and corporatist in its politics, advocated outright economic liberalism and pursued a foreign policy based on the defence of "ideological borders". Onganía consecrated the country to the holy spirit, banned political parties, and intervened in universities and the Trade Union Confederation (CGT). Control of the economy passed from national companies to US, British and German consortiums with the ability to buy bankrupt Argentinian firms, very cheaply.

[18] The ensuing economic stagnation led to serious social unrest, particularly in 1969, when violent demonstrations and rioting paralyzed Córdoba, the nation's second largest city. The uprising, nicknamed "El Cordobazo", was organized by the Argentinian CGT, led by Raimundo Ongaro (Peronist), Antonio Scipione (Radical) and Agustín Tosco (Marxist). The Cordobazo and other outbreaks of guerrilla warfare, ended Onganía's plans for an "Argentinian Revolution". General Agustín Lanusse came to power in 1971. In order to pacify the country, Lanusse called elections but banned the candidacy of Perón, then in exile in Madrid.

[19] Héctor Cámpora, the "Justicialista" Liberation Front candidate - a Peronist electoral coalition - received 49 per cent of the vote in the March 1973 elections, taking office in May 1973, and resigning two months later to allow new elections after Perón's return to the country. On September 23, Perón was reelected in a new contest with 62 per cent of the vote, and the vice-presidency went to María Estela ("Isabelita") Martínez de Perón, his wife. He resumed diplomatic relations with Cuba, proposed a reorganization of the Organization of American States (OAS) to serve Latin America's interests, promoted Argentina's participation in the Non-Aligned Movement and increased trade with socialist countries. After the Ezeiza massacre on June 23 1973, the day Perón returned to Argentina, friction grew among various Peronist factions and a situation of open warfare broke out between the old-time labor leaders and the "special squads", as Perón used to call the guerrillas. When Perón died, his widow took office. Her Social Welfare Minister (Perón's former secretary), José López Rega, set up the "Triple A" (Argentine Anticommunist Alliance), a paramilitary hit squad that murdered Marxist opponents and left-wing Peronists.

[20] On March 25 1976, a military coup put an end to "Isabelita's" inefficient and corrupt administration. A military junta led by General Jorge Videla suspended all civil liberties and set in motion a cycle of kidnapping, torture, betrayal and murder. The term "missing person" became ominously commonplace, and government priorities were dictated by the newly adopted "National Security" doctrine. Human rights organizations drew up a list of over 7,000 missing persons, 80 per cent of whom had been arrested by the police in front of witnesses. Their fate has never been determined with certainty, but a 1983 Armed Forces communiqué stated that the *desaparecidos* (missing persons) should be considered "killed in action" with counter-insurgency troops.

[21] The Junta encouraged imports to the point of liquidating a third of the country's productive capacity. Fifty years of labor gains were wiped out, real wages lost half of their purchasing power, and regional economies were choked by high interest rates. The country's cattle herds decreased by 10 million head and the foreign debt climbed to $60 billion, a quarter of which had been spent on arms. It was the era of what Argentines called the "patria financiera" (financial fatherland), when governmental economic policies encouraged most of the country's productive sector to turn to speculation. In 1980, a wave of bankruptcies hit banks and financial institutions.

[22] In 1981, General Leopoldo Galtieri became president and Army Commander-in-Chief. He agreed on Argentinian military participation in the US intervention in Central America. Thinking that this was enough to secure US president Reagan's unconditional support,

STATISTICS

DEMOGRAPHY

Urban: 87% (1995). **Annual growth:** 0.3% (1992-2000). **Estimate for year 2000:** 36,000,000. **Children per woman:** 2.8 (1992).

HEALTH

Under-five mortality: 27 per 1,000 (1994). **Calorie consumption:** 109% of required intake (1995).

EDUCATION

School enrolment: Primary (1993): 107% fem., 107% male. University: 41% (1993). **Primary school teachers:** one for every 18 students (1992).

COMMUNICATIONS

101 **newspapers** (1995), 103 **TV sets** (1995) and 108 **radio sets** per 1,000 households (1995). 12.3 **telephones** per 100 inhabitants (1993). **Books:** 101 new titles per 1,000,000 inhabitants in 1995.

ECONOMY

Per capita GNP: $1,8110 (1994). **Annual inflation:** 317.70% (1984-94). **Consumer price index:** 100 in 1990; 310,967 in 1994. **Currency:** 1 peso = 1$ (1994). **Cereal imports:** 8,000 metric tons (1993). **Fertilizer use:** 78 kgs per ha. (1992-93). **Exports:** $15,839 million (1994). **Imports:** $21,527 million (1994). **External debt:** $77,388 million (1994), $2,263 per capita (1994). **Debt service:** 35.1% of exports (1994).

ENERGY

Consumption: 1,399 kgs of Oil Equivalent per capita yearly (1994).

Galtieri decided to ward off the domestic crisis by recovering the Malvinas Islands, and troops landed there on April 2 1982. Galtieri's error soon became evident. The UN Security Council opposed the invasion, Britain deployed a powerful fleet that included nuclear submarines, and the United States firmly supported its North Atlantic ally. After 45 days of fighting Argentina surrendered on June 15; 750 Argentinians and 250 British soldiers had been killed during the conflict. Two days later Galtieri was forced to resign from both his military and presidential posts.

[23] Seventeen member countries in the OAS acknowledged Argentina's right to the Malvinas. They also voted against Washington for having violated the Interamerican Reciprocal Assistance Treaty by supporting Britain, a non-American nation, in acts of aggression against American territory.

[24] Soon after Galtieri resigned, the upper echelons of the Armed Forces underwent a series of purges. The junta set elections for October 30, 1983 and the Army, without consulting the Navy and the Air Force, appointed retired General Reynaldo Bignone acting president until January 30, 1984.

[25] The Radical Civic Union's new leader, Raul Alfonsín, won the election with 52 per cent of the vote, well ahead of the Peronists' 40 per cent. During his campaign, Alfonsín denounced the existence of a pact between the military and trade unions, and he offered himself as the candidate of law and life, standing against tyranny and death. This won him the election, even in the Greater Buenos Aires industrial belt, a traditional Peronist stronghold.

[26] Economic efforts concentrated on inflation, where forecasts were gloomy, 688 per cent by the end of 1984. For 1985, it seemed reasonable to expect a monthly rate of 25 per cent. Government expenditure was cut drastically and the military budget shrank to 18 per cent of the total as compared to a previous 30 per cent. The budget deficit spiralled to an annual $70 million. Meanwhile, speculative investments yielded the highest interests paid anywhere in the world.

[27] Workers discovered that their share of national income had decreased from 48 to 35 per cent, and the Peronist CGT wavered towards a variety of mixed political alliances. It reached the point of endorsing

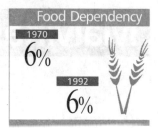

Food Dependency

1970
6%

1992
6%

External Debt

1994

PER CAPITA
$2,263

a stockbreeders' strike in exchange for an increase in rural wages and support for union control over social spending, especially on public health and domestic tourism.

[28] In mid-June 1985, the government introduced the Austral Plan which froze prices, service tariffs and a new currency, the Austral, was put into circulation, linked to the US dollar.

[29] Relations with Chile improved, after having been extremely tense due to a border dispute in the South (over the Beagle Canal). A treaty was signed, following a plebiscite in which a Vatican peace proposal, backed by the Argentine government, was supported by 80 per cent of the voters.

[30] CONADEP, the National Commission on Missing Persons, continued to reveal horrendous findings at the trials of the nine Commanders-in-Chief of the dictatorship, accused of crimes during that period. The indictment of several high-ranking military officers (including former president Videla) and the subsequent extension of the trials to lower-ranking officers created tension within the military. During Easter 1987, some units occupied their barracks to demand an end to the trials, and amnesty for those who had been indicted. Although the rebels represented a minority, many people doubted the willingness of the "loyal" troops to contain the rebellion.

[31] In response to President Alfonsín's appeal, more than a million people flocked to the Plaza de Mayo in defence of democracy. The crisis was dispelled on Easter Sunday when Alfonsín personally went to the Campo de Mayo barracks to meet with the rebels, who agreed to submit to the constitution. Consequently, several military officers were relieved of their commands, and the president sent Congress a bill on "due obedience", exempting the majority of the military personnel from prosecution, on the grounds that they had been following orders.

[32] In the years that followed, the majority of the population saw their real earnings steadily eroded by inflation. Resistance to the government's economic strategy was led by trade unions of the General Labor Confederation (CGT). Between December 1983 and April 1989, thousands of jobs were lost, wages came to be just one more "adjustable variable" and recession aggravated the already difficult situation of small and medium sized businesses. In this period there were 14 general strikes.

[33] The social cost of this process spiralled to the point where it is estimated that some 10 million people, nearly 10 per cent of the total population, lived outside the consumer market.

[34] In the summer of 1988 there was a second, but hardly significant, military uprising at Monte Caseros in Corrientes province. Its leader, Lieutenant Colonel Aldo Rico, was tried and subsequently expelled from the army. In the winter of that same year, Colonel Mohamed Ali Seineldin led a third insurrection with the declared aim of settling an internal army matter. The forces were split into "nationalists" loyal to Seineldin, and "liberals" who followed army orders. Even though the country survived both episodes, the internal divisions continued to exist.

[35] In February 1989, an armed group from the All for Our Country "movement", which had acted legally until then, attacked an Army barracks, allegedly in order to prevent a military coup. The response was disproportionately violent and those fighters who survived received heavy sentences. It is claimed that some prisoners were summarily executed after surrendering, and others "disappeared".

[36] The final 3 months of the Alfonsín government were marked by hyperinflation of between 100 and 200 per cent a month. The economy was wracked by recession and the president's prestige demolished. Stores in suburbs of the capital and in the interior were sacked once again, with looters stealing food and basic consumer goods.

[37] In the 1989 presidential elections the Peronist governor of La Rioja, Carlos Saúl Menem, received 47.6 per cent of the vote, and the radical candidate, Eduardo César Angeloz, 38.4 per cent. The economic collapse was so drastic that the new president was invited to take office a couple of months earlier than scheduled. Menem passed a State Reform Law in August 1989 starting the privatization of state enterprises, the deregulation of markets, and the decentralization of public administration. Within the year the main oilfields, communications media, the national telecommunications company and the state airline were all privatized.

[38] Menem re-established diplomatic relations with Britain, leaving the key issue of the Malvinas under a dubious "protective um-

brella", and in two stages, pardoned all the military responsible for human rights violations during the "dirty war".

[39] The persistent economic and social crisis caused Menem to change the traditional Peronist doctrine, wreaking havoc on the country's main institutions. The CGT split into two sectors: one pro-government and the other, opposition. The Army and political parties were also divided, the Church criticized the political leadership, whose public prestige waned.

[40] In December 1990, the warrant officers rose in arms in Buenos Aires. The uprising lasted for 24 hours and claimed the lives of several civilians and military. Colonel Seineldin, in custody at the time, accepted full responsibility for these events.

[41] During the Gulf War, the Argentinian government sent troops to join the US-commanded anti-Iraq coalition, without waiting for parliamentary approval. In 1991, Argentina announced its withdrawal from the Non-Aligned Movement, and in early 1992, it joined the United States in denouncing the human rights situation in Cuba, thereby breaking with its traditional non-interference policy.

[42] From the start, Menem's government was beset by scandal. Several of his top advisors had to resign, accused of drug money laundering or accepting bribes to favor companies in bids or privatization take-overs.

[43] In spite of the corruption, Menem succeeded in keeping his image intact and in the 1991 elections the Justicialist Party retained 13 of the 23 provinces, while 4 were won by the Radicals and 6 by local parties. Voters throughout the country gave 40 per cent of the vote to the ruling party, only 7 per cent less than in 1989.

[44] This political victory has been attributed to economic stability attained by the conversion plan which drastically reduced public spending and set a new currency, the peso, equivalent to one US dollar. Inflation fell, reaching record lows, standing at 7.4 per cent in 1993.

[45] In December 1992, Argentina, Brazil, Paraguay and Uruguay signed the agreement leading to Mercosur, a regional common market scheduled to go into effect in January 1995.

[46] In the 1993 parliamentary elections, the Peronists obtained their

fourth consecutive electoral victory with 43.2 per cent of the vote, while the Radicals received 30 per cent. Afterwards, the Radicals agreed to support a constitutional reform to enable presidential re-election - which would directly benefit Menem - in exchange for the reduction of the government term from six to four years.

[47] In the economic field, the Gross Domestic Product increased significantly. However, industrial activity remained below 1987 levels. Inequality of wealth distribution between the country's different regions widened. By the end of the year, civil servants in several provinces staged protests over unpaid back wages and in some cases, they set fire to public buildings and looted the homes of politicians and government officials.

[48] In the 1994 constitutional elections, the great surprise was the outcome of the left-wing "Frente Grande" (Large Front) coalition, which became the country's third most important political force, winning the election in the Federal Capital with 37.6 per cent of the vote. It also won in the southern province of Neuquen. In Buenos Aires, it became the second most important political force with 16.4 per cent of the vote. However, Peronists and Radicales obtained a sufficient majority to ensure the reform.

[49] The Mexican crisis of the beginning of 1995 spread shadows upon Argentinian economic stability. Economy minister Domingo Cavallo hastened to carry out budget cutbacks and apply austere measures. In his May 1 speech President Menem promised to reduce unemployment by half through a 5-year public investment plan which would create 300,000 jobs.

[50] In the May 14 1995 elections, Menem was re-elected with 50 per cent of the vote, while Frepaso candidate - and former Peronist - José Bordón received 29 per cent and Radical Horacio Massaccesi 17 per cent.

[51] In mid-1996, over 2 million had become unemployed and the number of underemployed rose to 1.5 million, according to official statistics. Unemployment reached a national average of 17.1 per cent and in some provinces exceeded 20 per cent. Economists from the government party admitted unemployment had settled itself in the country.

[52] By this time, foreign debt had grown 57 per cent since 1991, when the radical privatizations plan and the successful "convertibility" anti-inflation plan were initiated by Economy minister Domingo Cavallo. A man trusted by international financial organisation, Cavallo became a reference point for regional liberalization processes of the economy until in the first quarter of 1996, public debt reached the maximum expected for the whole year and the model began to show signs of exhaustion. After a long series of political disagreements, Menem removed minister Cavallo and replaced him with Roque Fernández, with a Doctorate from the University of Chicago, who set out to maintain his predecessor's policy, overcome the fiscal deficit, unemployment, recession, and renegotiate goals with the IMF.

Malvinas

(Falkland Is.)

Population:	2,000
	(1994)
Area:	12,170 SQ KM
Capital:	Port Stanley
	(Puerto Argentino)
Currency:	Pound Sterling
Language:	English

When the islands were sighted for the first time in 1520 by a Spanish ship, they were uninhabited. In the 18th century, they were baptized the "Malouines" in honor of Saint Malo, port of origin of the French fisherfolk and seal hunters who settled there. In 1764, Louis Antoine Bougainville founded Port Louis on Soledad Island. This move brought Spanish protests, and France recognized Spain's prior claim. That same year the British, who since 1690 had called the islands "Falkland" (after the treasurer of the British Navy), founded Port Egmont. This was later returned to Spain in exchange for £24,000, and renamed Puerto Soledad.

[2] In 1820, shortly after independence, Argentina appointed Daniel Jewit as first governor of the Malvinas. In 1831, governor Vernet impounded two North American ships on charges of illegal fishing. A North American fleet that was visiting South America avenged this act of "piracy" by destroying houses and military facilities at Port Soledad. On January 3, 1833, the English corvette Clio landed a contingent of settlers which the small local force was unable to repel.

[3] After WWII, the United Nations Decolonization Committee included the Malvinas and their dependencies on the list of "non-autonomous" territories and established that, as the inhabitants were British, the principle of self-determination was not applicable. The only juridically valid solution to the problem was to recognize Argentinian sovereignty.

[4] On April 2, 1982, Argentinian forces occupied the Malvinas. Two months later, at the cost of more than 1,000 lives, the Union Jack was once again hoisted over the islands.

[5] Diplomatic relations between Britain and Argentina were placed on hold. At the beginning of his administration, Menem renewed relations with London, but the intractable issue of the Malvinas' sovereignty was left pending. During his first official visit to Malvinas, British Foreign Minister Douglas Hurd emphasized London's determination to maintain the islands under British sovereignty. In November 1991 Britain authorized

After World War II, the United Nations Decolonization Committee included the Malvinas and their dependencies on the list of "non-autonomous" territories.

the governor of the Malvinas to award contracts, for the exploration for and exploitation of possible underwater oil deposits, around the islands.

[6] In March 1994, the Argentinian Minister of Defence reported that in 1982, in the Malvinas, at least 9 Argentinian soldiers had died at the hands of British forces, in circumstances that violated the Geneva Convention on the treatment of prisoners of war. There are still some 15,000 live mines scattered across the islands.

PROFILE

ENVIRONMENT

An archipelago with nearly 100 islands located in the South Atlantic. It includes Georgia, Sandwich and Shetland Islands. There are two main islands - Soledad to the east and Gran Malvina to the west - separated by the San Carlos Channel. The coast is rough and mountainous. More than half of the population lives in the capital on Soledad. The main economic activity is sheep raising. The islands' territorial waters are believed to contain oil reserves. There is also hope that krill (a microscopic crustacean rich in protein) can be marketed. Its proximity to Antarctica also gives the archipelago strategic importance.

SOCIETY

Peoples: Since June 1982, the British have maintained a 4,000-strong garrison on the islands. **Language:** English.

THE STATE

Capital: Port Stanley (Puerto Argentino), 1,329 inhab. (1989). **Government:** William H. Fullerton, Governor, appointed by UK.

DEMOGRAPHY

Population: 1,620. **Density:** 0.157 inhab./sq km. **Annual growth:** -0.7%

EDUCATION

Primary school teachers: one for every 15 students (1980).

COMMUNICATIONS

505 **radio receivers** per 1,000 inhab. (1991).

ECONOMY

Currency: pounds sterling. **Imports:** $8 million (1991). **Exports:** $10 million (1991).

ENERGY

Consumption: 9,106 kgs of Coal Equivalent per capita yearly, 71% imported. **Major source:** coal (1990).

Armenia

Population: 3,748,000 (1994)
Area: 29,800 SQ KM
Capital: Yerevan (Jerevan)

Currency: Dram
Language: Armenian

Hayastán

The first historical reference to the country "Armina" (Armenia) was made in the cuneiform writings from the era of King Darius I of Persia (6th-5th centuries BC). But the name Hayk, as the Armenians are called, comes from the name of the country, Hayasa, mentioned in the Hittite ceramic writings from the 12th century BC. The Urartians, direct ancestors of the Armenians, founded a powerful state in the 9th to 6th centuries BC; its capital was the city of Tushpa (today Van, Turkey). In the year 782 BC, they founded the fortress of Erebuni, in the north of the country (today Yerevan, capital of Armenia).

[2] With the collapse of Ur, the ancient Kingdom of Armenia emerged in its territory. The first rulers of Armenia were the *satraps* (viceroys) of the shahs of Persia. This period was recorded in the works of Xenophon and Herodotus. In "Anabasis", Xenophon described how the Armenians turned back ten thousand Greek mercenaries in 400 BC. His writings also describe Armenia's prosperous production, its wheat, fruit and delicious wines.

[3] After the expeditions of Alexander the Great and the rise of the Seleucid Empire, Armenia came under extensive Greek influence which had an important effect upon the cultural life of the country. The Seleucid State fell into the hands of the Romans in 190 BC, and Armenia became independent. The local government named Artashes (Artaxias) King of Greater Armenia.

[4] Armenia reached the pinnacle of its prosperity during the reign of Tigranes the Great (95-55 BC). King Tigranes united all Armenian-speaking regions and annexed several neighboring regions. Armenia's borders extended as far as the Mediterranean to the south, the Black Sea to the north and the Caspian Sea to the east. Tigranes' empire soon fell, and Armenia was proclaimed "friend and ally of the

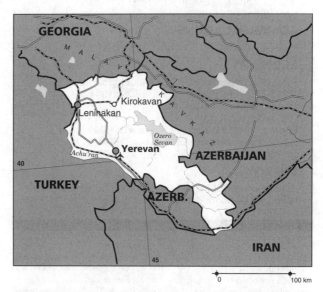

Roman people", a euphemism for the vassals of Rome.

[5] Armenia began to disintegrate over a 400-year period until it finally disappeared as a state in the year 428, when the Roman Empire and the new kingdom of Persia divided it up between them.

[6] In 301, Armenia became the first country in the world officially to adopt Christianity as a state religion At that time, St Gregory the Illuminator, the first Armenian patriarch, founded the monastery at Echmiadzin, still extant as the headquarters for the patriarchs of the Armenian Church. Ever since the 4th century, the Church has been identified with Armenian national feeling as religion made it possible to maintain the unity of the people over the long period during which the country lacked any actual state organization.

[7] In 405, a monk, Mesrop Mashtots, devised an alphabet which formed the basis of the Armenian writing system. The characters of this alphabet have remained unchanged, achieving a continuity which spans the centuries and links ancient, medieval and modern cultures. The 5th century was the golden age of religious and secular literature and of Armenian historiography; the natural sciences also developed during a later period. In the 7th century, Ananias Shirakatsi wrote that the world was round and formulated the hypothesis that there were several worlds inhabited by beings endowed with some form of intelligence.

[8] In the 5th and 6th centuries, Armenia was divided between Byzantium and Persia. The Persians tried to stamp out all traces of Christianity in the eastern Armenian regions triggering a massive rebellion. Prince Vartan Mamikonian, commander of the Armenian army, assumed the leadership of the rebellion. In the year 452, he led an army of 60,000 troops into battle against a vastly superior Persian force, in the Avaraev valley. The Armenians were defeated and Prince Vartan was killed, but the Persians also suffered heavy losses, and subsequently gave up all attempts to convert the Armenians to Islam. All those who lost their lives in this battle were later canonized by the Armenian Church.

[9] In the 7th century, Arab forces invaded Persia, bringing about the collapse of the Persian Empire. The new Muslim leaders also established control over the Armenian regions. The people resisted, fighting for their independence until the late 9th century, when Prince Ashot Bagratuni was named king of Armenia, and established an independent government. The epic novel "David of Sasun" describes this long struggle for independence.

[10] The prosperity of the Bagratids' reign was short-lived, for in the 11th century, the Byzantines and Seleucids began bearing down on the Transcaucasian region, from Central Asia. Many Armenian princes ceded their lands to the Byzantine emperor, in exchange for lands in Cilicia. The inhabitants of other Armenian regions began

DEMOGRAPHY

Urban: 68% (1995). **Annual growth:** 1.4% (1991-99). **Estimate for year 2000:** 4,000,000. **Children per woman:** 2.8 (1992).

EDUCATION

School enrolment: Primary (1993): 93% fem., 93% male. Secondary (1993): 90% fem., 80% male. University: 49% (1993).

COMMUNICATIONS

15.6 **telephones** per 100 inhabitants (1993).

ECONOMY

Per capita GNP: $680 (1994). **Annual growth:** -13.00% (1985-94). **Annual inflation:** 138.60% (1984-94). **Currency:** dram. **Cereal imports:** 350,000 metric tons (1993). **Fertilizer use:** 436 kgs per ha. (1992-93). **Imports:** $401 million (1994). **Exports:** $209 million (1994). **External debt:** $214 million (1994), $57 per capita (1994). **Debt service:** 1.7% of exports (1994).

ENERGY

Consumption: 667 kgs of Oil Equivalent per capita yearly (1994), 87% imported (1994).

PER 100,000 LIVE BIRTHS

35

flocking to Cilicia, fleeing the Turkish raids.

[11] At the end of the 11th century, the Rubenid dynasty founded a new Armenian State in Cilicia, which lasted 300 years. Cilicia had close ties to the western European states; Armenian troops took part in the Crusades, and intermarriage with other ruling dynasties introduced the Rubenids to the circle of European rulers. In 1375, Armenian Cilicia fell to the Mamelukes of Egypt, who retained Cilician science, culture and literature. In the meantime, the region that had originally been Armenia was devastated by invasions and wars.

[12] In the 13th century the Ottoman Turks replaced the Seleucids and began their conquest of Asia Minor. In 1453, they took Constantinople and marched eastwards, invading Persia. Armenia was the scene of numerous wars between Turkey and Persia, until the 17th century, when the country was divided between the two Islamic empires. During this period, the Church carried out secular functions and also called the attention of fellow-Christian European countries to the plight of the Armenians, who were forced to emigrate and settle far from their home country. Many of these expatriate Armenian "colonies" exist to this day.

[13] In 1722, Russian troops carried out an expedition to Transcaucasia, occupying the city of Baku and other territories belonging to Persia. Armenian princes in Nagorno-Karabakh and other neighboring areas seized the opportunity to join forces with the Russians, and organized a revolution against the Persians. The uprising was led by the Armenian national hero David-bek. However, Armenian hopes were dashed when the Russian czar,

Peter the Great, died; he had promised to support the Armenians, but on his death, Russia signed a peace treaty with Persia. Another war between Russia and Persia, a hundred years later, ended in 1813, with the Treaty of Gulistan. According to the terms of this peace treaty, Karabakh and other territories which had historically belonged to Armenia became part of the Russian Empire.

[14] Russia was at war with either Turkey or Persia during the major part of the 19th century; with each war, Russia annexed more and more Armenian territory. Finally, - home to more than 2 million Armenians - was swallowed up. However, the major part of Armenia's historic lands - with a population of more than 4 million - belonged to Turkey.

[15] Protected by Russia against wars and invasions, Eastern Armenia prospered, while within the Ottoman Empire the Armenians were the object of abuse and persecution. There were frequent disturbances and riots, which were cruelly suppressed by the Turks. During World War I, citing the Armenians' pro-Russian sympathies, the Young Turk government perpetrated the massacre of more than a million Armenians. While the men were executed in the villages, the women and children were sent to the Syrian deserts, where they starved to death. Survivors of these atrocities sought refuge in Armenian expatriate communities.

[16] Upon the fall of the Russian Empire, Armenia's independence was proclaimed in Yerevan. Turkey attacked Armenia in 1918 and again in 1920. In spite of some resounding victories on the part of the Armenian troops, the young republic's economy suffered and it also lost a significant part of its

territory. In late 1920, a coalition of communists and nationalists proclaimed the Soviet Republic of Armenia. The nationalists were eased out of power and, in February 1921, the communist government was brought down. However, with the help of the Red Army - which came into Armenia from Azerbaijan - the communists were back in power after 3 months of fighting.

[17] In 1922, Armenia, Georgia and Azerbaijan formed the Transcaucasian Soviet Federated Socialist Republic, which became a part of the USSR at the end of the year. In order to avoid ethnic friction between Christian Armenians and Muslim Azeris, the Soviet regime adopted the policy of the separation of nationalities into different political/administrative entities, which implied the relocation of large segments of the population. In 1923, the Nakhichevan (Nachicevan) Autonomous Soviet Socialist Republic was created as a dependency of Azerbaijan, from which the entire Armenian population had been removed. Azerbaijan was also given Upper Nagorno-Karabakh, a region which had historically been Armenian and which Azerbaijan had previously relinquished in 1920. In 1936, the Transcaucasian Federation was dissolved, and the republics joined the Soviet Union as separate constituent republics.

[18] In 1965, Armenians around the world commemorated the 1915 genocide for the first time. In the Armenian capital, demonstrators clamored for the return of their lands, referring to the region of Upper Karabakh. The first petition for the reunification of Nagorno-Karabakh and Armenia - signed by 2,500 inhabitants of the former - was submitted to the president of the USSR, Nikita Khrushchev, in May 1963. Ever since that time, there have been two diametrically opposed positions: Armenia, in favor of reunification, and Azerbaijan, against. In 1968, fighting broke out between Armenians and Azeris in Stepanakert, the capital of Nagorno-Karabakh.

[19] In February 1988, within the framework of *perestroika* initiated in the USSR, Nagorno-Karabakh Armenians (80 per cent of the local population) decided to join Armenia. Karabakh's Regional *Soviet* (Parliament) approved the resolution and in Armenia, the Karabakh petition for reunification was received enthusiastically. Moscow reacted violently and sent

in Russian troops to crush the demonstrations in Yerevan and Stepanakert.

[20] In the Armenian elections of August 1990, the National Pan-Armenian Movement won, determined to achieve independence by legal means. In the September 1991 referendum, 99.3 per cent of the electorate voted for separation from the USSR. The Armenian *Soviet* proclaimed independence and in October Levon Ter-Petrosian was elected president with 83 per cent of the vote.

[21] Nagorno-Karabakh also declared independence after 90 per cent of the voters approved the separation. Azerbaijan responded with an economic and military blockade around Nagorno-Karabakh, causing war between these republics. In December 1991 Armenia joined the Commonwealth of Independent States (CIS) and in February 1992 was admitted to the UN.

[22] From the beginning of 1993, while the pro-Armenian armed forces won important victories on the Nagorno Karabakh front, Yerevan began, at least officially, to withdraw unconditional support of it. In September, Azerbaijan imposed an economic blocade of Armenia, effecting a sharp decline in the standard of living and a severe energy crisis at the beginning of 1994.

[23] The Armenian economy grew again through 1994 and 1995, due in part to good relations with Iran, which allowed it to alleviate the consequences of the embargo imposed by Azerbaijan. The transit of merchandise through Iranian territory determined in 1995 a gigantic increase of Armenian exports to Iran, 40 times greater to those of 1992.

[24] In May 1994 - when, according to Azerbaijan, Armenian forces had conquered 12,000 sq km of disputed territory - Russian pressure made possible a cease-fire in High Karabakh. Through 1995 and the first months of 1996 negotiations went on among the contending parties. The US and Russia made efforts to maintain Nagorno-Karabakh under Azerbaijani sovereignty, but they could not compel the Armenians to an immediate agreement, and negotiations ended in an impasse.

Aruba

Population: 77,000 (1994)
Area: 193 SQ KM
Capital: Oranjestad

Currency: Florin
Language: Dutch

Aruba

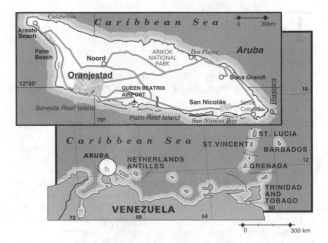

The Caiquetios were the first inhabitants of Aruba, an island of the Lesser Antilles. In 1634, some European colonists settled on the island to raise horses, the only economically feasible activity in an archipelago that had been declared worthless by Spain. The Caiquetios were conquered by the Spaniards and sold to Hispaniola as slaves.

[2] According to the Treaty of Westfalia (1648), the Netherlands ruled over Aruba, Curaçao and Bonaire. Aruba required little labor to support its horse and cattle-raising industry, which explains why only 12 per cent of the population was of African origin by the time slavery was abolished.

[3] The installation of large oil refineries on Aruba at the end of the

PROFILE

ENVIRONMENT

Located off the coast of Venezuela, the island of Aruba was until January 1986 one of the "ABC Islands," otherwise known as the Netherlands Antilles, together with Curaçao and Bonaire. The climate is tropical, moderated by ocean currents. Oil refineries, where Venezuelan petroleum is processed, and tourism are the main economic activities.

SOCIETY

Peoples: The island's ethnic composition differs from that of the rest of the Netherlands Antilles, given the fact that its African population is very small. Predominantly of European origin, its inhabitants intermingled with Latin American and North American immigrants. **Religions:** Mainly Catholic (82%). There is also a Protestant (8%) minority and small Jewish, Muslim and Hindu communities. **Languages:** Dutch (official). The most widely spoken language, as on Curaçao and Bonaire, is *Papiamento*, a local dialect based on Spanish with elements of Dutch, Portuguese (spoken by the Jewish community), English and some African languages. **Political Parties:** The People's Electoral Movement (MEP); the Aruba People's Party (AVP) formed with the Democratic Party of Curaçao, which dominated politics in the Antilles in the 1950s and 1960s; National Democratic Action (ADN); Aruban Patriotic Party (PPA); Aruban Liberal Organization (OLA). **Social Organizations:** The Aruba Workers Federation, linked to the MEP, is the largest labor organization with 4,800 members.

THE STATE

Official Name: Aruba. **Capital:** Oranjestad, 17,000 inhab. (1981). **Other city:** St. Nicolaas, 17,000 inhab. (1981). **Government:** Olindo Koolman, Governor designated by Holland since 1992. The Prime Minister and Minister of General Affairs is Nelson Odaber since 1989. Holland continues to be in charge of defence and foreign relations. Single-chamber legislature; Parliament made up of 21 members elected for a 4-year term.

DEMOGRAPHY

Density: 379 inhab./sq.km.

COMMUNICATIONS

30.4 **telephones** per 100 inhabitants (1993).

ECONOMY

Imports: $1,300 millions (1993). **Exports:** $1,600 millions (1993). **External debt (1987):** $81 millions ($1,350 per capita). **Consumer price index:** 100 (1990); 122.7 (1994). **Currency:** 2 florin = 1$ (1994).

1920s generated a wave of immigration of specialized, high income workers, particularly from the US. Many Arubans began to resent the administrative dominance of Curaçao.

[4] In 1971, the Aruba People's Party (AVP), led by Prime Minister Henny Eman, an anti-Dutch, anti-Curaçao nationalist, split into two separate parties. One was the Aruban Patriotic Party (PPA), which opposed the separation of Aruba. The other was the People's Electoral Movement (MEP), radical wing, led by Gilberto "Betico" Croes, which supported each island's freedom to choose its own constitution and even become an autonomous republic. After the Netherlands' refusal to accept this proposal, Croes threatened a unilateral declaration of independence.

[5] In 1979 in Curaçao, the Antiyas Nobo Movement (MAN), active on 16 islands, won a significant electoral victory. With the MEP and the Bonaire Patriotic Union (UPB), it established a left of center coalition government. But while MAN favored a federation with broad autonomy for each island, the MEP insisted on the secession of Aruba. These differences led to the disintegration of the governmental alliance in 1981.

[6] In 1983, the Dutch government granted Aruba separate status, effective from January 1986, establishing a governor appointed by the Netherlands' queen, a single-chamber parliament with 21 members and its own national symbols. In the 1985 election, 8 seats were won by the MEP, 7 by the AVP and the remaining 6 by three minor parties who entered a coalition with the AVP. Henny Eman was prime minister. Croes died in 1986 in suspicious circumstances.

[7] In the 1989 elections the MEP won 10 seats, the AVP won 8, and each of the 3 remaining parties, National Democratic Action (ADN), New Patriotic Party (PPN), and PPA, gained one seat. The new Prime Minister Nelson Odaber, defended continued association with the Netherlands.

[8] The economy of Aruba has the support of Dutch credit. It exports rum and tobacco, but 98.9 per cent of its exports come from refining Venezuelan oil. In 1985 Exxon, the owners of the refinery, withdrew from the country, leaving the island's economy in serious difficulties. With unemployment at 20 per cent, the government turned to the tourist industry, which has provided 35 per cent of GDP. In recent years the authorities have tried to reduce the dependence on tourism by promoting industrial activities.

[9] In 1986 Aruba split off from the Netherland Antilles as a first step towards what was to be total independence in 1996. The Netherlands revoked this resolution in 1990.

[10] During 1990 the hotel capacity of Aruba doubled and the refinery was reopened in 1990-91. Unemployment fell to 0.6 per cent in 1992. However, the financial crisis led to the bankruptcy of many tourism related foreign companies. The MEP won the elections again in 1993 and started to implement an economic diversification policy, but the coalition faltered and was replaced in 1994 by the AVP.

Australia

Population: 17,843,000 (1994)
Area: 7,686,850 SQ KM
Capital: Canberra

Currency: Australian Dollar
Language: English

Australia

The first inhabitants of Australia came from Southeast Asia some 50,000 years ago, but it is not possible to determine precisely whether at the time of their arrival they constituted one homogeneous ethnic group.

[2] These peoples, named "aborigines" by the Europeans, spoke over 260 different languages and had distinctive cultures, though they were all semi-nomadic hunters and gatherers. Their daily life was linked to a universal scheme through rituals which guaranteed the ordered changing of the seasons and the community's food supply.

[3] Land was held in common, and there was no social stratification, other than the prestige gained by practising religious rites.

[4] In 1606, the Spaniard Vaez de Torres explored the strait which today bears his name, located between Australia and Papua-New Guinea. Three years later, Fernández de Quirós, also Spanish, became the first European to set eyes on Australian territory. However, colonization did not begin until after the arrival of Captain James Cook, in the late 18th century.

[5] In 1788, the first British settlers arrived, and established a penal colony. This was the beginning of an assault on the way of life and customs of the Aborigines. The two fundamental objectives were: to take over the best lands and to obtain a supply of labor.

[6] At first, the British settled in coastal areas, but the process of inland expansion soon began, with the settlers acting as though the land were uninhabited. While British and international law recognized the prior rights of indigenous peoples, Australia was simply declared *terra nullius* (uninhabited).

[7] The wars to gain control of the land claimed the lives of 80 per cent of the Aborigines. The "Aborigine problem" was dealt with by genocide. Entire families were poisoned and others were forcibly removed from their lands onto reservations administered by the British.

[8] In 1830 more than 58,000 convicts were "transported" to Australia, most of them petty thieves, deserters from the Royal Navy and members of Irish opposition groups. Known by the British as "the human refuse of the Thames", most of them served out their sentences by providing inexpensive labor for European landowners.

[9] Australia served as a safety valve for social tensions generated by Britain's rapid industrialization. Most of the wealth accumulated in the colony returned to England, representing an important influx of capital. Finally, Australia's strategic location was important for British trade, in order to control its world-wide maritime network.

[10] Australian society was marked by a strong authoritarian streak, a legacy of its original function as a penal colony. Local government was normally limited to a high-ranking military officer, appointed by the Crown to oversee the prison population and take charge of the defence against possible attack by other European powers.

[11] The Aboriginal religion, based on a strong spiritual link between each human being and the earth, was given no credence by the settlers. As a result, the religion and language were debased and the people persecuted. The Aborigines lost lands rich in natural resources, fishing areas and any land that the settlers considered suitable for cultivation or grazing. Once the land was in the hands of settlers, it was stocked with sheep to provide the growing British textile industry with both an abundant supply of raw materials, and a cheap source of food for the industry's workforce.

[12] The development of the cattle industry and the subsequent discovery of gold and other precious metals, boosted economic development between 1830 and 1860. The possibility of getting rich quickly led to increased European settlement. As resistance was broken the Aborigines

LAND USE

■ CROPLAND 6 % ■ PASTURE 54 % ■ OTHER 39 %

Christmas Island

Population: 2,278 (1994)
Area: 130 SQ.KM
Capital: Flying Fish Cove
Currency: Australian Dollars
Language: English

The island, formerly a dependency of the British colony of Singapore, was transferred to Australia on October 1, 1958. Nearly all the island's population work in the Phosphate Mining Commission.

[2] As of 1981, the residents of the island were granted the right to become Australian citizens. In 1984, the Australian government extended social security, health and education benefits to the island, also granting the citizens political rights. There has been an income tax since 1985.

[3] In 1987, the authorities closed the mine, which was reopened in 1990 by private investors but under strict guidelines for the preservation of the environment. In 1991 investments were made to develop the island's tourist potential.

ENVIRONMENT

An island in the Indian Ocean, 2,500 km northeast of Perth, Australia, and 380 km south of Java, Indonesia. Mountainous and arid, it has a dry climate.

SOCIETY

Peoples: Nearly two-thirds of the population are of Chinese origin. There are some Malays and a minority of Australians. There are no native inhabitants. **Religions:** Christianity (Protestants), Confucianism and Taoism. **Languages:** English, Malay, Mandarin and Cantonese. **Capital:** Flying Fish Cove. **Government:** A D Taylor, Administrator appointed by the Australian government.

DEMOGRAPHY

Density: 24 inhab./sq.km.
Economy
Currency: Australian dollars.
Land use rate: Almost totally dedicated to mining.

Foreign Trade

MILLIONS $ 1994

IMPORTS
53,400
EXPORTS
47,538

PROFILE

ENVIRONMENT

Australia occupies the continental part of Oceania and the island of Tasmania, and has a predominantly flat terrain. The Great Dividing Range runs along the eastern coast. Inland lies the Central Basin, a desert plateau surrounded by plains and savannahs. The desert region runs west to the huge Western Plateau. Rainfall is greatest in the north where the climate is tropical, with dense rainforests. 75% of the population is concentrated in the southeast, which has a subtropical climate and year-round rainfall. Oats, rye, sugar cane and wheat are grown there; Australia is one of the world's largest producers of the latter. Australia has the largest number of sheep of any country in the world, located in the inland steppes and savannahs, and is also the world's largest exporter of wool. It also exports meat and dairy products and is one of the world's largest producers of minerals: iron ore, bauxite, coal, lead, zinc, copper, nickel and uranium. It is self-sufficient in oil and has a large industrial zone concentrated in the southeast. The soil has undergone increasing salinization. Many unique species, both plant and animal, are in danger of extinction due to the destruction of their habitats.

SOCIETY

Peoples: When the British "discovered" Australia in 1788, there were 250,000 Aborigines in approximately 500 different tribes. In 1901, only 66,000 of their descendents were still alive. Today, there are 200,000, representing just 1% of the national population. Descendents of British immigrants make up two-thirds of the population. The rest are immigrants from Asia, Europe and Latin America, a process which continues today. **Religions:** Christians 74% (Catholic 27%, Anglican 24%, Uniting Church 8% and others). Buddhist, Muslim, Confucian and other groups, 13%. **Languages:** English. **Political Parties:** Liberal (conservative) Party, a right-of-centre power (aligned with the National Party, which represents the interests of pastoralists), in power since 1996; Labor Party, a centre-left group; Australian Democrats (moderate); Party of the New Left, founded in 1989 after the dissolution of the Communist Party; Socialist Party; Marxist-Leninist Communist Party and Green Party. **Social Organizations:** The Australian Council of Trade Unions (ACTU) is the largest labour confederation, with 133 union affiliates.

THE STATE

Official Name: Commonwealth of Australia. **Administrative divisions:** 6 states and 2 territories. **Capital:** Canberra, 328,000 inhab. (1994). **Other cities:** Sydney 3,738,500 inhab.; Melbourne 3,198,200 inhab.; Brisbane 1,454,800 inhab.; Perth 1,239,400 inhab.; Adelaide 1,076,400 inhab.(1994). **Government:** Parliamentary monarchy. Sir William Deane, Governor General, appointed by the Queen of England. John Howard, Prime Minister since March 1996. **National Holiday:** January 26, Australia Day. **Armed Forces:** 56,100 (1995).

countries. Australia was the first country to grant women the right to vote in 1902.

[21] World War II loosened the ties between Britain and Australia, as the former colonial capital was unable to guarantee the security of its former colonies, in the face of the threat of Japanese attack. The US assumed the role of guaranteeing the region's security, as it has done ever since.

[22] The Korean War triggered a sharp increase in the price of wool on the world market, reinforcing Australia's economic growth. It also helped to avoid creating a gap between urban prosperity and the relative stagnation of the rural areas.

[23] With the onset of the Cold War, the ANZUS military assistance treaty was signed in 1951 by Australia, Aotearoa/New Zealand and the US. The aim of the treaty was to guarantee the security of US and allied interests in the region. This military alliance also committed the Australians to participate in the Vietnam War, which damaged the treaty's image internationally, and triggered an important anti-war movement.

[24] Japan became the main market for Australian minerals, financing coal exploration in Australia in the 60s. The large coal deposits discovered have supplied local industry ever since, and have also met Japan's huge industrial needs.

[25] In the 1960s, the Labour Party became the platform for Aborigine groups, helping to publicize and sensitize public opinion to their grievances and demands.

[26] A plebiscite held in 1967 - supported by the Labour Party - granted Australian Aborigines full citizenship rights. The Aboriginal issue was also placed under the jurisdiction of the federal government. However, the Aborigines still did not gain recognition of their land rights.

[27] There are currently 200,000 Aborigines in Australia, many retaining their original languages. Two-thirds of them no longer live in tribes, living instead in areas surrounding the major cities. A minority continues to live in areas which Europeans consider to be inhospitable, like the central desert or the wetlands which lie in the northern part of the country, where they have managed to keep

were progressively relegated to Australia's desert areas.

[13] For many years, rural settlers occasionally had to deal with armed Aboriginal resistance. Driven by their desire to free land from the Europeans, they fought the settlers and tried to disrupt their farming.

[14] Once suppressed, Aborigines were forced to sign "work contracts" written in English. These contracts committed them to the status of unpaid workers, household slaves or concubines, often subject to extremely severe disciplinary measures.

[15] In the meantime, a dynamic labor movement was beginning to appear in the cities, and it soon had a significant following. From the mid-19th century onwards, the unions obtained important victories

and concessions which Europe's working classes were still a long way from obtaining.

[16] In the latter part of the 19th century, Australia's growing urbanization process went hand-in-hand with an accelerated rate of industrial development, especially in Sydney and Melbourne, which took on the characteristics of any large urban centre.

[17] Successive waves of immigration also helped change the face of Australian society. The low cost of land and a demand for Australia's products on the world market, provided large numbers of mostly British people with the means for achieving rapid social mobility.

[18] A vast middle class developed alongside a wealthy urban industrial bourgeoisie, utterly transforming Australian life. The liberal

governments which dominated the country's political scene between 1860 and 1900 accelerated this political, social and economic transformation.

[19] After an uneasy period of consolidation, in 1901 the six British colonies (New South Wales, Victoria, South Australia, Western Australia, Queensland and Tasmania) became independent states united in the "Commonwealth of Australia". The Northern Territory and the capital did not join until 1911.

[20] A prolonged period of economic prosperity made it possible for the country to finance a series of social reforms. Australia was a relatively open society, in which social mobility was possible for certain sectors of the population, and there was protective legislation for workers before most European

Norfolk Island

Population: 3,000 (1994)
Area: 40 SQKM
Capital: Kingston
Currency: Australian Dollars
Language: English

There are no records on the existence of a native population before the arrival of the Europeans. Discovered by Cook in 1774, it was used as a prison site between 1825 and 1855. In 1913, the island was transferred to Australia as an overseas territory. In November 1976, two thirds of the Norfolk electorate opposed annexation to Australia. Since 1979, the island has had internal autonomy.

[2] In December 1991, the local population rejected a new proposal to become part of Australia's federal electorate.

ENVIRONMENT

The island is located in Southern Melanesia, northwest of the Great Island of Aotearoa The subtropical climate is tempered by sea winds.

SOCIETY

Peoples: 3,000 inhabitants in 1994. A large part of the population descend from the mutineers of the British vessel HMS. Bounty, who came from Pitcairn Island in 1856. **Religions:** Protestant. **Languages:** English (official).

THE STATE

Official name: Norfolk Island. **Capital:** Kingston. **Government:** Cdr. John A. Matthew, Administrator appointed by the Governor-General of Australia. The office of prime minister was replaced by that of President of the Legislative Assembly. In the May 1992 elections, David Ernest Buffett was elected to this new office.

DEMOGRAPHY

Density: 59 inhab./sq.km.

COMMUNICATIONS

1,000 **radio receivers** per 1,000 inhab. (1991).

ECONOMY

Currency: Australian dollars. **Land use rate:** arable, 0%; pasture, 25%; forest, 0% and other, 75%.

---

their own religious and social traditions alive.

[28] In 1983, powerful transnational corporations interested in developing mineral resources launched a campaign aimed at convincing the population that the Aborigines' defence of their territorial claims actually compromised the country's economic growth. As a result of this campaign, the government's promise of passing federal legislation on territorial rights of the Aborigines was shelved.

[29] In many cases the Aborigines found themselves forced to give in to pressure from mining companies, and sign land use agreements, even though they know that it may well adversely affect their sacred sites, the environment, and their traditional life style.

[30] Although some aspects of the Aborigines' situation are uniform throughout the country, their legal status nevertheless varies from one state to another. The state of Northern Territories, for example, has relatively advanced legislation, with regard to the recognition of the rights of Aborigines. On the other hand, other states, like Tasmania, have only recently begun to study proposals aimed at safeguarding some of the Aborigines' rights, as well as ways of combating the discrimination which they are subjected to.

[31] According to local and international human rights organizations, in 1987, the death rate of Aboriginal prisoners was two per month; putting it in terms of the total prison population, this would be the equivalent of 100 deaths of prisoners of European descent, in the same amount of time.

[32] Australian Aborigines are probably the people with the largest prison population in the world. In 1981, the rate was 775 per 100,000 inhabitants. Among the population of European descent, the proportion is 67 per 100,000. In Queensland, in the northeast of the country, Aborigines make up only 2 percent of the population, yet account for 35 percent of the prison population. In Western Australia, where they also make up 2 percent of the population, 44 percent of the prison population are Aborigines.

[33] Given this situation, the federal government created a Royal Commission charged with investigating the deaths of Aborigines in prison. After several years of work and research, the commission presented a provisional report which did not clearly establish what the actual causes of these events were, nor who was responsible for them. This report was harshly criticized by spokespersons of different groups, who called it a "sham" for all the country's citizens.

[34] In the area of health, the Ab-

origines are also at a clear disadvantage with relation to the rest of the population. Thus, they continue to come down with diseases which have almost disappeared among the European population, and for which total immunization is available.

[35] Although Australia agreed with Aotearoa in opposing British and French nuclear tests in the region, the virtual withdrawal of Aotearoa from ANZUS has made it necessary for the Australian government to redefine its role within this alliance, and within the region. Thus, the United States is in the process of negotiating new agreements allowing the maintenance of troops and a telecommunications centre in Australian territory.

[36] In 1989, in order to meet this new international reality head-on, Australia proposed the creation of the Asian Pacific Economic Cooperation (APEC). The project foresees the formation of a common market among the countries of the

region, with Australia taking a leading role. At the same time, Australia hopes to become the representative of food-exporting countries. To achieve this goal, it proposed the creation of the so-called "Cairns Group", which includes Argentina, Colombia, Thailand, Uruguay and other countries within the Uruguay Round of the GATT.

[37] In 1989, data supplied by Australian National University research damaged the Labour government's image, but above all it was a letdown to public opinion, which wanted to see itself as a model egalitarian society. The figures pointed to the fact that 1 percent of all Australians owns more than 20 percent of the total national wealth, and 13 percent of all Australian homes live below the poverty line.

[38] The social and economic crisis worsened, with unemployment figures totalling more than a million in 1991. This put pressure on Bob Hawke's Labour government, which found itself under growing

Coral Sea

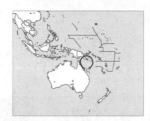

Created in 1969 as a separate administrative entity, the territory consists of several islets located east of Queensland (Eastern Australia). The major ones are Cato and Chilcott in the Coringa group, and the Willis archipelago. With the exception of a weather station, on one of the Willis islands, the rest of the islands are uninhabited.

[2] The Constitutional Act by which the territory was created did not provide for Australian administration of the islands, but only for control over foreign visitors by the Canberra government. However, this situation may change in view of the possible exploitation of hydrocarbons and the good prospects for fisheries development.

---

opposition from other political parties, and from within the party as well.

[39] Hawke had been elected in 1983, after eleven years as president of ACTU (the country's largest Labor union), a fact which had given him a huge amount of popular support at the beginning of his term.

[40] Toward the end of the 1980s, with the support of the more conservative elements within his party, Hawke embarked upon a series of economic reforms aimed at liberalizing the economy, including an accelerated program for privatizing public enterprises.

[41] There was growing discontent among the population at large, and opposition to the government began to emerge within the Labour Party itself. In addition, a power struggle developed between Hawke and his former Treasurer (Economics Minister), Paul Keating. In December 1991, the Labor government held internal elections in order to put these conflicts to rest, once and for all, and also, to open up the possibility of improving the government's image. Keating, representing the Labor Party, obtained a slim victory which made him Prime Minister.

[42] Shortly after assuming his post as head of the government, Keating announced that further measures would go into effect to liberalize the economy. Keating himself had initiated these measures between 1983 and June 1991, when he had been in charge of economic policy for Hawke's government. These measures were firmly rejected by the unions and the left wing of the Labor Party.

[43] In 1992, unemployment reached 11.1 per cent. The government approved legislation aimed at increasing employment, by reducing immigration by 27 per cent.

[44] In the 1993 general elections, Labor won by a narrow margin, with 50.5 per cent of the vote. One of Keating's campaign promises was to give the rights of indigenous Australians a high priority.

[45] An important decision on native land rights was made by the Australian High Court in June 1993. Known as the Mabo decision, it effectively overturned the notion of *terra nullius*, which had assumed that when the British colonists arrived in 1788, the Aborigines had neither right of law nor ownership of the land. Mabo was one of Australia's most significant legal decisions since federation in 1901. The decision drew on the history of the Murray Islands north of Queensland and to their Meriam people, whose rules of inheritance were crucial in the final judgment. Following this Mabo decision, the federal Labor government passed revised Native Title legislation. So far only one claim by indigenous Australians has been successful - that of the Dunghutti people in northern NSW in October 1996. There are also concerns over other matters affecting Aborigines and Torres Strait Islanders, including infant mortality, health, education and the country's terrible record of black deaths in custody. Progress Australia had been making in multiculturalism has also slowed, despite it being one of the most ethnically diverse countries in the world.

[46] With the defeat of the Labor government in early 1996, the debate as to whether or not Australia would become a republic lost considerable momentum. The previous prime minister, Paul Keating, had hoped the year 2001 (the centenary of Australian federation) might be the focus for the inauguration of the republic. However, with the significant electoral victory of the conservative Liberal government which is likely to hold power until the eve

Cocos

Population:	1,000 (1994)
Area:	10 SQKM
Capital:	West Island
Currency:	Australian Dollars
Language:	English

The Clunies Ross Company, founded by John Clunies Ross in the early 1800s, was the real owner of the islands, despite their formal status as a British (1857) and later Australian (after 1955) colony. Ross brought in Malayan labourers to work the coconut groves.

[2] In 1978, after many years of negotiations, Australia bought the islands from the company. However, the company kept a monopoly on the production and commercialization of copra. The transfer was designed to give Home Island residents formal ownership of their plots of land, in order to alleviate social tensions. On West Island, Australia also has a military base, purchased in 1951.

[3] In a 1984 referendum, the population voted in favour of Australian nationality for the islanders and full annexation of the territory to Australia. In December of that year the UN General Assembly validated the results of the plebiscite and Australia was freed from the obligation of reporting to the Decolonisation Committee.

ENVIRONMENT

A group of coral atolls in the Indian Ocean, southwest of Java, Indonesia. Of the 27 islets, only two are inhabited, West and Home. The climate is tropical and rainy, and the land is flat and covered with the coconut groves that give the islands their name.

SOCIETY

Peoples: The inhabitants of Home Island descend from Malayan workers. Australians predominate on West Island.
Religions: Mainly Protestant and some Sunni Muslims.
Languages: English.

THE STATE

Capital: West Island, 208 inhab. (1984). **Government:** B. Cunningham, administrator appointed by the Australian government.

DEMOGRAPHY

Density: 45 inhab./sq km.

ECONOMY

Currency: Australian dollars.
Land use rate: Arable, 0%; pasture, 0%; forest, 0% and other, 100%.

---

of that date, such a transition seems unlikely.

[47] Privatisation has continued, as initiatives both of the previous Labor and current Liberal governments. Significant sales include Qantas airlines and the Commonwealth Bank. The controversial proposed sale of Telstra, the national telecommunications agency, is still being pursued. The government continues to consolidate its trade and economic interests in the East- and South-Asian region.

[48] The government started to develop links with the Asian Pacific nations, especially the "Tahi-Laos Friendship Bridge," built and financed by Australia on the Mekong river. Their approaches to Cambodia, North Korea, Vietnam, Indonesia, Malaysia and China advanced with greater difficulty.

[49] The French nuclear tests on Mururoa atoll caused public protests throughout Australia. The Keating and Chirac governments were at loggerheads and broke off diplomatic relations. The Australian delegation which went to Europe with the Pacific Islands' minister of affairs did not gain the support of the British government, which refused to confront France. The British decision strengthened Australia's pro-independence stance. With the French nuclear testing firmly in mind, Australia successfully renegotiated the nuclear test-ban treaty at the United Nations in September 1996.

[50] Australian prime minister, John Howard, came to power in March 1996, undertaking to introduce a series of major industrial relations reforms, strongly opposed by the unions.

Austria

Österreich

Population:	8,028,000 (1994)
Area:	83,850 SQ KM
Capital:	Vienna (Wien)
Currency:	Schilling
Language:	German

1 The first traces of human settlement in the Austrian Republic date back to the Early Paleolithic age. From then on, the territory was occupied by various ethnic groups. The Austrian region of Hallstat gave its name to the main culture in the Iron Age, from 800-450 BC.

2 Celtic tribes moved into the Eastern Alps in the 5th century BC, and it is thought that they founded the kingdom of Noricum. Later, the Romans also settled here, attracted by the iron in the area and by its strategic military importance.

3 After peaceful penetration, the Roman troops conquered the whole of the country around the year 15 BC. Raetia, Noricum and Pannonia became Roman provinces, extending the Empire's dominions as far as the Danube. The provinces were subdivided into municipalities and a vast network of roads was built.

4 The Pax Romana ended with the arrival of the Germanic tribes between 166 and 180 AD. This invasion was repulsed, but the region did not recover its prosperity. Between the 4th and 6th centuries, the Huns and the Germans raided the area pushing back the Empire's borders on the Danube.

5 According to the only written records of the time, the Germanic tribes of the Rugii, the Goths, the Heruli and the Langobardi settled on Austrian territory. In the year 488 part of the population of the devastated province of Noricum was forced to emigrate to Italy.

6 After a succession of battles between the Germans and the Slavs, the Bavarian lands reached the border with the Avars in the 7th century. Upon the death of King Dagobert I, the Bavarian dukes became virtually independent. Christianity spread in the region, and it was practised by the Romans who had remained in the territory.

7 Bavarian military expansion and wealth were central to the advance of Christianity. Under their protection, the churches of Salzburg and Passau opposed the missions of the East, led by the Slav apostles, Cyril and Methodius.

8 During the 9th century the Magyars raided the area and ruled over Austrian territory until they were expelled in 960. The earliest evidence of a developed legal system was found in the region of Burchard, east of the river Enns, in this period. Later on, this legislation was extended as far as the forests of Vienna.

9 On several occasions Austria was the object of investiture disputes between the Pope and the Holy Roman Emperor, who were struggling over control of the German Church. Meanwhile, the reformists gained ground, with the foundation of the monasteries of Gottmelg, Lambach, and Admont in Styria.

10 Between the 10th and 13th centuries the dispute over the territories continued and new peoples settled in Germanic and non-Germanic areas. Apart from short interruptions, the Babenbergs kept their sovereignty over the duchies of Austria and Styria and expanded them both north and south.

11 Upon the death of Frederick II, the Babenberg dominions were coveted by their neighbors. The main beneficiary was Premysl Otakar II, from Bohemia, until Rudolf IV of Hapsburg occupied the German throne in 1273, using Hungarian help to displace him.

12 At first the Hapsburgs were rejected by the local nobility and their neighbors, but apart from a few minor defeats they succeeded in keeping control over their dominions. During the 14th and 15th centuries the Austrian territories were ruled by the Hapsburgs.

13 In 1490, after several crises, the Hapsburgs lost everything except Lower Austria. Upon the death of Frederick II, Maximilian I inherited the House of Austria and the German Empire; his son Philip I married Infanta Juana and became king of Spain.

14 By the end of the Middle Ages, the Hapsburg monarchy held an extent of Alpine land comparable in size to modern Austria. The House of Hapsburg had always wanted to rule over Bohemia and Hungary, an aspiration that Maximilian rekindled.

15 After unification, territories maintained their individuality and their own legal codes. Cities prospered, while rural settlement regressed, particularly in Lower Austria, because of the growing interest in mining.

16 Luther's ideas penetrated Austria, supported by the noble families, especially in the south and centre of the country. In 1521, Protestant pamphlets were printed in Vienna, and bans on their dissemination had no practical effect.

17 There were peasant revolts in Tyrol, Salzburg and Innerösterreich. Many peasants were joining the Anabaptists rather than the Lutherans. Their name came from the Greek meaning "to baptize again", as they considered the christening of children ineffectual, subjecting followers to a second christening. As they were more radical, and had no support from the powerful, they suffered greater persecution from the outset.

18 In 1528, in Vienna, Balthasar Hubmaier, leader of the Anabaptists in the Danube valley and south of Moravia, was burned at the stake. In 1536, in Innsbruck, Jakob Hutter, a Tyrolian, was sentenced to the same fate after he had led his followers into Moravia.

19 When King Jagiellon of Bohemia and Hungary died, Vienna saw an opportunity to broaden the power of the Hapsburgs. Ferdinand I was proclaimed king of Bohemia in 1526, but his troops were repulsed with the aid of the Turks, when he tried to impose himself over the Hungarians.

20 At the end of the dispute in 1562, the treaty of Constantinople divided Hungary into three regions: the north and the west went to the Hapsburgs; the centre, to the Turks; and Transylvania, with its surrounding territories, to Hungarian Hanos Zapolya and his successors.

21 Maximilian II succeeded Ferdinand I as emperor of Bohemia, which was a part of Hungary and the Austrian Danube. The heir had Protestant leanings, although he promised his father that he would uphold the Catholic faith. The Counter-reformation started with the Jesuits, who were strong in Vienna, Graz and Innsbruck.

22 Maximilian's successor, Rudolf II, had been strictly educated in Spain, in the Catholic faith. He removed all the Protestants who served at court and entrusted the conversion of cities and markets to Melchior Klesl, the Apostolic Administrator of Vienna, later to be bishop and cardinal.

23 The Catholic-Protestant controversy had ups and downs in the House of Hapsburg until the death of Emperor Matthias, in 1619, and the accession of Ferdinand II, who was the least involved with the Counter-reformation. Bohemia and Moravia rebelled, and war became inevitable.

24 The conflict went beyond the empire's borders: Spain, Bavaria and Saxony joined Ferdinand II; Bohemia was subjected, and Protestants emigrated to Germany in massive numbers. Germany was invaded, and the Protestant backlash came from Sweden in 1630, causing the Thirty Years War that followed.

25 Between the 17th and 19th centuries, the Hapsburgs were involved

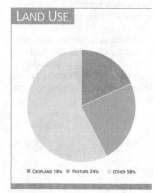

in all the European conflicts. The Napoleonic Wars virtually dismantled the Austrian Empire, and it was only after Napoleon's abdication in 1814 that the House of Austria recovered most of its territory.

[26] The Austrian chancellor Clemens Metternich was the creator of the Holy Alliance of the European powers in 1815. It upheld the principles of Christian authoritarianism and foreign intervention against the liberal and revolutionary movements of the time.

[27] In 1848, the repercussions of the Paris Commune reached Austria. A revolt broke out in Vienna, led by crowds demanding the liberalization of the regime. Metternich's resignation, rather than bringing peace, unleashed a revolution throughout the empire.

[28] In spite of their active presence in the street fights of Vienna, the workers had little significance in the society of the time, due to the country's lack of industrial development. The most important social consequence of the revolution was the liberation of peasants, sanctioned by the Emperor in 1849.

[29] At the same time the liberal government of Hungary demanded independence. In Germany the revolution installed a National Assembly in Frankfurt, which incorporated Austrian-German liberals and conservatives interested in separation from the Hapsburg Empire.

[30] The Emperor accepted the Budapest petitions, except on two key points: budget and military autonomy. The Hungarian Parliament declared the power of the Hapsburgs null and void, and proclaimed a republic in April 1849. The revolution was crushed four months later.

[31] Counter-revolution also annulled the Frankfurt Assembly, but the Austrian-Prussian dispute persisted. From then on, the Hapsburg Empire weakened inexorably. It lost, ceded and decentralized its dominions until its final disintegration in 1918, at the end of World War I.

[32] In October of that year, a National Assembly declared German Austria an independent state. In November, after the abdication of the Emperor, it proclaimed a republic. Socialist Karl Renner led the first coalition government of the main parties.

[33] Amidst the economic chaos and the hunger resulting from the war, the Austrian Social Democrats decided to use their strength to con-

front social unrest and growing Communist activism, inspired by the Soviet revolution in Russia in 1917 and in Hungary in 1919.

[34] Unlike the German Socialists, who made alliances with the old order, Austrian Socialists organised a People's Guard, which thwarted two coup attempts. Otto Bauer and Frederick Adler consolidated their strength when they defeated the communists in the workers' and soldiers' councils.

[35] Once political and social order were re-established in Vienna, the problem moved to the interior, where some states were demanding secession. The 1920 Constitution guaranteed a loose federal system, whereby the capital was governed by the Socialists and the rest of the country by the Conservatives.

[36] The new Constitution granted women the vote, something which had been guaranteed from the moment the republic was proclaimed in 1918.

[37] The Social Democrats had an unquestionable majority in Vienna, where one-third of the population lived. Social Christians had guaranteed support among the peasants and the Conservatives, while German nationalism fed on popular discontent, with the support of the urban middle classes.

[38] The League of Nations supported Austrian economic recovery during the postwar period, on the condition that the country would remain independent and would not join Germany. In 1922, the government was granted a loan which enabled complete stabilization of the country's finances.

[39] The great depression of 1929 brought the Austrian economy to the verge of collapse. The government tried to arrange a customs agreement with Germany which was harshly opposed by the rest of Europe. Together with the rise of Nazism, German nationalism in

Austria was showing signs of strengthening.

[40] In 1932, the Social Christian government of Engelbert Dollfuss attempted to take an authoritarian stance against the Social Democrats and the Nazis simultaneously. The Social Democrats rebelled and were declared illegal, and in 1934, the Nazis murdered Dollfuss in a failed coup d'etat.

[41] Taking advantage of the internal crisis and the government's weakness, German troops invaded Austria in 1938, unresisted by the European powers. A plebiscite carried out that same year in greater Germany recorded a vote of more than 99 per cent in favour of Hitler.

[42] In 1945, after Hitler's defeat, Austria was divided into 4 zones, occupied by US, French, British and Soviet troops. In the November elections, out of the 165 seats in the National Council, the conservatives obtained 85 and the Social Democrats 76 seats.

[43] The Austrian economy recovered with difficulty after the war, with the aid of the United Nations and the United States, through the implementation of the Marshall Plan. Heavy industry and banking were nationalized in 1946, and inflation was controlled by price and salary agreements.

[44] Conservatives and Socialists shared the government of the Second Austrian Republic, which only recovered full independence in 1955, with the Treaty of State and the withdrawal of the allied troops. The coalition held until 1966, when the People's Party were elected to govern alone.

[45] During the postwar period, Austria did not become a member of international organizations. During the Cold War, the country was liberal in the acceptance of political refugees from Poland, and it was a transit station for Soviet Jewish émigrés.

[46] Austria became a member of the United Nations in 1955 and of the Council of Europe in 1956. Since then, Austrian foreign policy has centered on the dispute with Italy over Sudtirol (Bolzano), resolved in 1969, and its association with the European Economic Community (EEC).

[47] In 1958, Vienna joined the European Free Trade Association, establishing a special agreement with the EEC. It established negotiations with the neighboring countries belonging to the Common Economic

IMPORTS
55,300
EXPORTS
45,200

STATISTICS

DEMOGRAPHY

Urban: 55% (1995). **Annual growth:** 0.7% (1991-99). **Estimate for year 2000:** 8,000,000. **Children per woman:** 1.6 (1992).

HEALTH

Under-five mortality: 7 per 1,000 (1995). **Calorie consumption:** 114% of required intake (1995).

EDUCATION

School enrolment: Primary (1993): 103% fem., 103% male. Secondary (1993): 104% fem., 109% male. University: 43% (1993). **Primary school teachers:** one for every 11 students (1992).

COMMUNICATIONS

106 **newspapers** (1995), 105 **TV sets** (1995) and 102 **radio sets** per 1,000 households (1995). 45.1 **telephones** per 100 inhabitants (1993). **Books:** 108 new titles per 1,000,000 inhabitants in 1995.

ECONOMY

Per capita GNP: $24,630 (1994). **Annual growth:** 2.00% (1985-94). **Annual inflation:** 3.20% (1984-94). **Consumer price index:** 100 in 1990; 114.7 in 1994. **Currency:** 11 schillings = 1$ (1994). **Cereal imports:** 184,000 metric tons (1993). **Fertilizer use:** 1,773 kgs per ha. (1992-93). **Imports:** $55,300 million (1994). **Exports:** $45,200 million (1994).

Assistance Council. In this way it remainied a neutral nation.

[48] The Socialist Party (SPÖ) obtained a close victory in 1970 forming a minority government led by Bruno Kreisky. In 1971 and 1975, the SPÖ obtained an absolute majority and monopolized government, supported by great economic stability and a policy of moderate social reforms.

[49] The Austrian government was a guest at the meetings of the Movement of Non-Aligned Countries and acted as a bridge between the Palestine Liberation Organization (PLO) and Western Europe. Vienna was energetic in its condemnation of the Israeli invasion of Lebanon in 1982.

[50] In 1978, the government lost a plebiscite over the installation of a nuclear power plant, but the SPÖ confirmed its support of the prime minister and held power until the 1979 elections. Kreisky resigned after 1983, when the SPÖ lost its majority and had to form a coalition government.

[51] The fall of the SPÖ was attributed to the rise of two green parties,

which gathered more than 3 per cent of the vote but did not obtain any parliamentary seats. In a coalition with the Freedom Party (FPÖ), the SPÖ maintained its policies of social welfare and active neutrality.

[52] In the presidential election of 1986, the SPÖ was defeated by the former Secretary General of the United Nations, Kurt Waldheim, who was an independent candidate with the support of the Austrian People's Party (ÖVP, Social Christian). However, the socialists maintained their majority in parliament until 1990.

[53] During the electoral campaign Waldheim was accused of having been an officer in the Nazi army. After taking office the questioning continued, when evidence of his involvement in the German campaign against the Yugoslav guerrillas was disclosed. This fact had negative repercussions on the country's relations with both the United States and Israel.

[54] The governor of the southern province of Carincia, Joerg Haider, leader of the ultranationalist Liberal Party (FPÖ), was deposed in

June 1991 after he praised the full employment policy of the Third Reich. In the November 1991 municipal elections, after a campaign in which he accused foreigners living in Austria of "stealing" Austrians' jobs, the FPÖ obtained 22.6 per cent of the vote becoming the second largest political power in Vienna.

[55] In 1992, amidst growing attacks against foreign residents, the government passed a law aimed at punishing neo-Nazi activities. In May, Thomas Klestil, from the ÖVP, was elected president with almost 57 per cent of the vote, after Waldheim decided not to stand for re-election and thus put an end to six years of Austrian international isolation.

[56] The changes which took place in Europe distorted Austrian economy. For example, in its first two years, German reunification caused a deficit in Austria's trade balance.

[57] In 1993, Haider intensified his xenophobic rhetoric blaming the rise in crime and unemployment on "uncontrolled" immigration. In February, 200,000 people marched through the streets of Vienna in opposition to racism. A motion by Haider, aimed at reducing the number of foreigners and limiting their rights, did not receive sufficient signatures for its approval.

[58] In a 1994 referendum, Austrians accepted to join the European Union which, in theory, does not affect the country's constitutional neutrality. In the regional elections in March, once again the FPÖ improved its results. In October's parliamentary elections, Haider's party received 23 per cent of the vote, while the SPÖ lost 7 points, with 35 per cent and the ÖVP lost 4 points, with almost 28 per cent.

[59] The growing number of foreigners living in Austria - approximately 300,000 in mid 1994 - coincided with new terrorist attacks against immigrants or their supposed Austrian "accomplices", like Maria Loley. On October 16 1995 this 71 year-old Austrian woman received a letter which exploded when she opened it, severing her left hand. Previously, in February, 4 gypsies had died in an explosion.

[60] Officially, Haider condemned these attacks and ridiculed the "teutonic mania" of those who carried them out. However, in a widely read book, Hans Henning Scharsach showed the similarities between Haider and Hitler, who "was also a populist at the beginning". The rise

of the FPÖ's leader coincides with a debate on Austrians' responsibilities during World War II.

[61] On October 12 1995 the coalition between the SPÖ and the ÖVP was broken and Vranitzky called for new elections. Two months later the SPÖ obtained 38 per cent of the vote, the ÖVP 28 per cent and Haider had his first electoral setback since 1986, with 22 per cent. After the elections, Vranitzky formed another coalition government with the ÖVP. In the country's first European elections in October 1996, the SPÖ obtained 29.1 per cent of the vote, the ÖVP 29.6 per cent and Haider came a close third place with 27.6, with 22 per cent.

[62] In spite of economic growth, higher employment and a rate of inflation lowered to 2-3 per cent, the budget deficit became even greater in 1995. The growth of the fiscal deficit and state indebtedness might jeopardize Austrian participation in the monetary unification scheduled by the European Union.

Azerbaijan

Azerbaidzhan

Population: 7,459,000
(1994)
Area: 86,000 SQ KM
Capital: Baku

Currency: Manat
Language: Azeri

The Azeris came from the mix of ancient peoples of eastern Caucasus. In the 9th century BC, the States of Mana, Media, Caucasian Albania and Atropatene (the name "Azerbaijan" being derived from the latter) emerged in the area corresponding to modern Azerbaijan. General Atropates proclaimed the independence of this province in the year 328 BC, when Persia was conquered by Alexander the Great.

2   Later, these states were incorporated to the Persian Arsacid and Sassanid kingdoms. There were a number of anti-Sassanid revolts, such as the Mazdokite rebellion. In the year 642, the Arab caliphate conquered Azerbaijan, which was still inhabited by tribes of different ethnic groups. The Arabs unified the country under Shiite Islam, despite a certain degree of resistance. Between 816 and 837, an anti-Arab revolt was led by Babek.

3   Between the 7th and 10th centuries, an important trade route passed through Azerbaijan, uniting the Near East with Eastern Europe. From the 11th to the 14th centuries the Seleucid Turks occupied Transcaucasia and the north of Persia. All the peoples of the region adopted the Turkish language, and the Azeris' ethnic identity was forged during this period. In the 15th and 16th centuries, the region of Sirvan, in the north of Azerbaijan, became an independent State.

4   Between the 15th and 16th centuries, the Setevid State emerged. Shah Ishmael I, founder of the dynasty, was supported by the nomadic Azeri tribes, who became the main power behind the state. In the late 16th century, Azeri nobility transferred its support to the Iranians.

5   In the 16th to 18th centuries, East Transcaucasia was the scene of Iranian-Turkish rivalry. In the 18th century, the Russian Empire began

its expansion toward Azerbaijan and by the middle of this century more than 15 Azerbaijani Khanates were dependent upon Iran. After several Russian wars against Turkey and Persia, the peace treaties of Gulistan (1813) and Turkmenchai (1828) were signed, granting Russia Northern Azerbaijan (the provinces of Baku and Yelisavetpol, corresponding to modern Gyanja).

6   The peasant reform carried out in Russia in 1879 accelerated the development of Azerbaijan, which was attractive because of the abundance of oil in the region. After the Russian Revolution of 1905, the nationalist *Musavat* (Equality) Party was founded in Baku in 1911. Its base was the country's bourgeoisie, and had a pan-Turkish, and pan-Islamic platform. After the triumph of the Bolsheviks in October 1917, Soviet power was established in Azerbaijan by the Commune of Baku.

7   In the summer of 1918, joint Turkish-British intervention ousted the Commune of Baku and brought the Musavatists to power. However, two years later the Red Army reconquered Baku and re-established Soviet power throughout Azerbaijan. It was proclaimed the Soviet Socialist Republic (SSR) of Azerbaijan and in March 1922, became part of the Transcaucasian Federation of Soviet Socialist Republics, along with Armenia and Georgia.

8   In the early 1920s, in an attempt to ease inter-ethnic tensions, Moscow decided to incorporate the regions of Upper Nagorno-Karabakh and Nakhicevan to Azerbaijan. These had previously belonged to Armenia as the ancient khanates of Karabakh and Nakhichevan. In July 1923, the Autonomous Region of Nagorno-Karabakh was founded, and in February of the following year, the Autonomous Region of Nakhichevan; both had formerly belonged to the SSR of Azerbaijan. In December 1936, the Transcaucasian Federation was dissolved, and the SSR of Azerbaijan joined the USSR on its own. In 1929 an attempt was made to substitute the Latin alphabet for Azeri script, which used Arabic characters. In January 1940, the Cyrillic alphabet was introduced.

9   Between 1969 and 1982, the leadership of the Communist Party of Azerbaijan was in the hands of Gueidar Aliev, a former KGB (Soviet secret police) agent who was

known and trusted by the Secretary of the Soviet Communist Party, Leonid Brezhnev.

10   By the end of the 1980s, socioeconomic, political and ethnic problems exacerbated the feeling of discontent among the Azeris. In 1986, the new Soviet leader Mikhail Gorbachev initiated a period of economic reforms *(perestroika)* and openness in the administration of the country *(glasnost)* which channeled popular discontent throughout the Soviet Union.

11   In the republic of Azerbaijan, there was a wave of strikes, political rallies and demonstrations. New political movements came into being, like the leading People's Front of Azerbaijan (PFA) with a platform stressing civil rights, free elections and political and economic independence for the country. The PFA opposed the long-standing aspiration of the Armenian population of Nagorno-Karabakh (90 per cent of that region's inhabitants) to rejoin the Armenian republic.

12   On September 25, 1989, Azerbaijan was proclaimed a sovereign state within the USSR. The worsening of ethnic conflicts between Azeris and Armenians led to the formation of extremist groups. In 1989, Armenians were massacred in Sumgait and in 1990, in Baku. After the events in Baku, the communist government decreed a state of emergency and called in troops from the USSR to re-establish order; a total of 100 people were killed in the ensuing violence.

13   The Nagorno-Karabakh issue increased friction with Armenia,

triggering an escalation of armed conflicts between Armenian and Azerbaijani guerrilla groups. The Autonomous Region of Nagorno-Karabakh proclaimed its independence from Azerbaijan. In December 1989, Armenia's Soviet approved reunification with Nagorno-Karabakh and Azerbaijan denounced this decision as interference in its internal affairs. In September 1991, the Republic of Nagorno-Karabakh declared independence from both Azerbaijan and Armenia. In November 1991, the Azerbaijani Soviet annulled the status of Nagorno-Karabakh as an "autonomous region".

14   On August 30, 1991, the USSR's Soviet Presidium approved Azerbaijan's declaration of independence. The state of emergency was lifted in Baku and on September 8, presidential elections were held. The Communist Party of Azerbaijan was dissolved. Former Azeri communist leader, Ayaz Mutalibov, who had supported the aborted coup against Gorbachev, was the only candidate, and was elected president. The PFA called the elections "undemocratic" because they were held under a state of emergency, and withdrew their candidate. The Autonomous Republic of Nakhichevan, with Gueidar Aliev, abstained from the voting.

15   On December 22, 1991, Azerbaijan joined the new Commonwealth of Independent States (CIS) which replaced the Soviet Union, and on February 2, 1992, it was admitted to the UN as a new member.

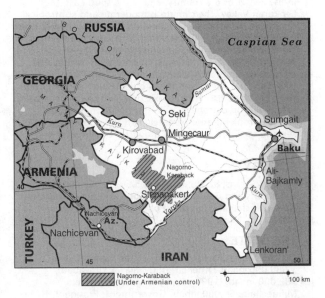

LAND USE

CROPLAND 23 %   PASTURE 26 %   OTHER 51 %

ENVIRONMENT

Located in the eastern part of Transcaucasia, Azerbaijan is bordered by Iran to the south, Armenia to the west, Georgia to the northwest, Dagestan (an autonomous region of the Russian Federation) to the north, and the Caspian Sea to the east. The mountains of the Caucasus Range occupy half of its territory; the Kura-Araks valley lies in the center of the country; and in the southeast, the Lenkoran Valley. The climate is moderate and subtropical, dry in the mountains and humid on the plains. Its principal rivers are the Kura and the Araks. The vegetation ranges from arid steppes and semi-deserts to Alpine-like meadows. The mountains are covered with forests. The country has important deposits of oil, copper and iron. Like Armenia, Azerbaijan has problems with soil pollution from pesticide use. Highly toxic defoliants have been used extensively on cotton crops. Water pollution is another serious problem: approximately half the population lacks sewage facilities, while only a quarter of all water is treated.

SOCIETY

Peoples: Azeris, 82.7%; Russians, 5.6%; Armenians, 5.6%; Ukrainians and Georgians. **Religions:** Muslim. **Languages:** Azeri (official) and Russian. **Political Parties:** New Azerbaijan, Musavat.

THE STATE

Official Name: Azerbacan Respublikasi. **Capital:** Baku, . 1,080,000 inhab. **Other cities:** Gyanja (ex Kirovabad), 282,000 inhab. (1990); Sumgait, 236,000 inhab.; Mingechaur, 90,000 inhab. **Government:** President, Heydar Aliev; Fuad Guliyev, Prime Minister. **National Holiday:** May 28, Independence (1918). **Armed Forces:** 86,700. **Paramilitaries:** Militia (Ministry of Internal Affairs): 20,000; Popular Front: Karabakh People's Defence: up to 12,000 claimed.

DEMOGRAPHY

Urban: 55% (1995). **Annual growth:** 1.4% (1991-99). **Estimate for year 2000:** 8,000,000. **Children per woman:** 2.7 (1992).

EDUCATION

School enrolment: Primary (1993): 87% fem., 87% male. Secondary (1993): 88% fem., 89% male. University: 26% (1993).

COMMUNICATIONS

8.7 **telephones** per 100 inhabitants (1993).

ECONOMY

Per capita GNP: $500 (1994). **Annual growth:** -12.20% (1985-94). **Annual inflation:** 122.80% (1984-94). **Consumer price index:** 100 in 1990; 453,423 in 1994. **Currency:** manat. **Cereal imports:** 480,000 metric tons (1993). **Fertilizer use:** 395 kgs per ha. (1992-93). **Imports:** $791 million (1994). **Exports:** $682 million (1994). **External debt:** $113 million (1994), $15 per capita (1994). **Debt service:** 0% of exports (1994).

ENERGY

Consumption: 1,414 kgs of Oil Equivalent per capita yearly (1994), -41% imported (1994).

[16] On December 10, 1991, besieged by Azeri troops and with bombs going off intermittently, the people of Nagorno-Karabakh participated in a plebiscite in which 99 per cent of the electorate voted pro-independence. The Azeri minority abstained from voting. CIS troops began withdrawing from Nagorno-Karabakh, but fighting between Azeri and Armenian guerrillas intensified.

[17] The People's Front of Azerbaijan accused the government of ineptitude in handling the Karabakh issue, and demanded that the government speed up the creation of an Azeri national army. Baku, in turn, accused CIS troops of facilitating the union of Armenia and Nagorno-Karabakh. On March 6, 1992, Ayaz Mutalibov resigned and Yuri Mamedov, the acting president of the Soviet, assumed presidential powers. At the same time, Azeri prime minister, Gusain Aga Sadikov, announced in New York that he would ask the UN Security Council to send a peace-keeping force in order to find a solution to the armed conflict in Armenia.

[18] In mid-March 1992, Turkish prime minister, Suleyman Demirel, pledged that his army would not intervene in the conflict, issuing a call to avoid seeing the conflict as a religious war between Muslims (Azeris) and Christians (Armenians).

[19] In April, fighting between Armenians and Azeris extended to Nakhichevan. Turkey warned that it would not accept any changes in its borders, which have been in effect since 1921.

[20] On May 14, 1992, Azerbaijan's Parliament, controlled by former communists, reinstated deposed President Ayaz Mutalibov. He subsequently suspended the June 7 presidential elections, banned all political activity, and imposed censorship of the press and a curfew. The leaders of the Popular Front, with the support of most of the leaders of the national militia, seized the Parliament building in Baku, declaring the reinstatement of Mutalibov illegal. After two days of street demonstrations and clashes in the capital between supporters of both factions, the opposition consolidated its position and ratified Mamedov as president. The Popular Front assumed key posts in the new government, such as the directorship of the security services and of the official media. The Popular Front's main leader and presidential candidate, Abulfaz Elchibei, confirmed that elections would be held as scheduled.

[21] On December 21, 1992, combat operations of unprecedented proportions were launched in Nagorno-Karabakh. Azeri troops suffered defeat and Parliament was besieged by the Popular Front; President Mutalibov, who was still ensconced within, was forced to resign. In the weeks which followed, the Armenian offensive intensified, and on May 7, 1993, an attack was launched upon Stepanakert, the administrative center of Nagorno-Karabakh.

[22] In June Albufaz Elchibei, leader of the Popular Front, won presidential elections, making possible the return of Aliev to public office. He was made Prime Minister. Nevertheless, the new and successful Armenian offensive in Nagorno-Karabakh was followed by a military coup. Headed by Colonel Guseinov and starting from Giandzha, the second city of Azerbaijan, the rebellious military took five regions of the country. When an offensive against Baku was started, Elchibei ran away, leaving Aliev as provisional president.

[23] In September 1993 the Soviet accepted the incorporation of Azerbaijan to the Community of Independent States and in October Aliev, ex-leader of the Communist Party of the Soviet Union, won in the new presidential elections and launched a successfull offensive in Karabakh. In May 1994, under the pressure of Russia, a cease-fire was agreed upon, and negotiations reissued among the conflicting parties. Contacts were pursued through 1995 and the first months of 1996, but no substantial progress was made in order to reach a final peace agreement.

[24] In October 1994 supporters of the Prime Minister Surat Husseynov occupied several cities of the country, but were quickly controlled by forces loyal to Aliev. In Baku, strong demonstrations in support of Aliev took place. He accused Husseynov of treason and dismissed him as Prime Minister. He decreed two months of martial law, and several members of the government, supposedly involved in the coup attempt, were arrested.

[25] In November 1995 Aliev's Party - New Azerbaijan - triumphed in parliamentary elections. These elections took place without the participation of the important Musavat Party, communist formations or pro-Islam groups, whose participation was forbidden by the government.

[26] Concerning the economy, the fall of the national internal income in 1995 was 20 per cent, equivalent to its fall the previous year, while inflation diminished from 880 per cent in 1994 to 790 per cent in 1995.

Bahamas

Population:	272,000 (1994)
Area:	13,880 SQ KM
Capital:	Nassau
Currency:	Bahaman Dollar
Language:	English

Bahamas

The Bahamian archipelago was one of the few areas of the Caribbean from which the Arawak Indians were not displaced by the Caribs. These Americans were probably the first to "discover" the lost European navigators, on October 12 1492.

[2] Indeed Columbus probably first trod American soil on the Bahamian island of Guanahani, or San Salvador, although he thought he was in Asia. "The Arawaks opened their hearts to us...", wrote the navigator to his Catholic sovereigns...we have become great friends...". Soon enough, the initial enthusiasm cooled to give way to less idealistic concerns: "from here, in the name of the Holy Trinity we can send all the slaves that can be sold... If Your Majesties so commanded, the entire population could be shipped to Castile or be enslaved on the island... as these people are totally ignorant of warfare...".

[3] Only 300 out of 500 natives lived through the first voyage to Spain. Moreover, most of the survivors died within a few years for they lacked immunity to European diseases. But Columbus had also set out to find gold for himself and the royal family. As Spanish historian Francisco de Gomara has pointed out, "over a period of 20 years, the Spaniards enslaved 40,000 natives who were sent to work in mines on the other islands, like, Santo Domingo".

[4] The Spaniards did not colonize the islands that lacked mineral resources. Instead, British privateers and pirates sought refuge in these islands after seizing "Spanish" gold (or rather, gold appropriated by the Spanish with the heavy cost of native lives in different American territories). Both the Bahamas and the Bermudas were located in a dangerous Caribbean area frequently hit by tropical storms, a fact that made

them particularly safe for buccaneers. However from 1640 the British began to settle the Bahamas. Sugar cane and other tropical crops were grown in plantations worked by African enslaved laborers whose descendants currently make up most of the local population. In 1873, the Treaty of Madrid settled the dispute over control of the Bahamas in favor of the British.

[5] The British refused to accept the independence of this strategic archipelago, and it was not until 1973 that the Bahamas proclaimed their independence within the British Commonwealth. This change actually meant little to the islanders because, in the meantime, the country had become increasingly dependent on the United States.

[6] Most of the tourists who now visit the Bahamas every year, drawn by its beaches and casinos, come from the US. The transnationals that use the Bahamas as their formal headquarters are also North American, taking advantage of the exemptions that make the country a "tax-haven". Also, US citizens are the chief buyers of lottery tickets, a source of fiscal revenues that contributes heavily to the state budget. In 1942, the US installed a naval base at Freeport, which helps control traffic from the Gulf of Mexico to the Atlantic, via the Florida Strait.

[7] The country's current situation closely reflects the prevailing conditions of the area. But while other Caribbean nations seek ways to bring about regional integration, the Bahamas prefers to take advantage of its proximity to the US, turning its back on regional cooperation. The islands have not joined any regional organization.

[8] The first political party - the Progressive Liberal Party (PLP) - was formed in 1953, and in 1958 the United Party of the Bahamas was founded. In the 1954 elections the PLP obtained six seats out of a total 29. In 1956 the Assembly passed an anti-discriminatory resolution, which was aimed at promoting ethnic equality. Thus, the Afro-Caribbean population was given access to places where they had never before been admitted.

[9] In the 1962 general elections the PLP won 8 of the 29 seats. Two years later the new Constitution was passed and a ministerial form of government was established. The number of representatives to the Assembly was increased to 38.

Sir Roland Symonette was elected prime minister and Lynden Pindling became the leader of the opposition. In the January 10, 1967 election, each party obtained 18 seats, and the representatives who won the 2 remaining seats joined the PLP, thus enabling this party to form a government. Lynden Pindling became Prime Minister.

[10] In 1977, when the economic and social crisis started to be felt, the government decided to give even greater incentives to foreign capital. It promoted the use the Bahamian flag by foreign ships as a "flag of convenience", to compete with Panama and Liberia. In an attempt to diminish the massive unemployment which threatened to create social tensions and changes in the archipelago, the government opened an industrial estate of 1,200 hectares near a deep water port in Grand Bahama. It was meant to be used as a storage point for merchandise which was later to be re-exported after minimal local processing.

[11] During the 1977 electoral campaign the opposition parties - the Free National Movement, the Democratic Party of Bahamas, and the Vanguard Party - accused the government of corruption and squandering public funds. Pindling prom-

ised to "Bahamize" the economy, that is, give more participation to domestic capital, emphasizing that he would not damage the country's image as a tax haven. Both left and right criticized the government's policy towards transnationals, the former because they considered it complacent, the latter because they thought that excessive taxation was driving foreign investment away. However, Pindling again won by a landslide and promised to lower unemployment. He opened the country's coasts to the "seven sisters" (the seven companies that control the world oil industry) which started oil prospecting in 1979.

[12] In 1984, the political scene was further upset when the US NBC Network News directly charged Prime Minister Pindling with receiving large sums of money for authorizing drug traffic through Bahamian territory. An investigation immediately confirmed that government officials were involved in smuggling but cleared Pindling of any responsibility in the affair.

[13] In 1987, the unemployment rate was estimated to be above 18 per cent, with the under-25s and 35 per cent of women affected. With these figures in mind, public protests against the recruiting of Haitian refugees to the labor force

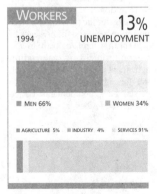

WORKERS	13%
1994	UNEMPLOYMENT

■ MEN 66% ■ WOMEN 34%

■ AGRICULTURE 5% ■ INDUSTRY 4% ■ SERVICES 91%

ENVIRONMENT

The territory comprises over 750 islands, only 30 of them inhabited. The most important are: New Providence (where the capital is located), Grand Bahama, and Andros. These islands, composed of limestone and coral reefs, built up over a long period of time from the ocean floor. Despite the subtropical climate, the lack of rivers has prevented the favorable climatic conditions being fully exploited for agriculture. Farming is limited to small crops of cotton and sisal. The main economic activity is tourism, centered on New Providence. The effects of industrial development beyond their borders, the warming of the seas, the increasing frequency and intensity of tropical hurricanes and coastal erosion all pose a growing threat to the islands' environment.

SOCIETY

Peoples: Descendants of enslaved African workers, 85 per cent, plus North Americans, Canadians and British. **Religions:** Non-Anglican Protestants 55%, of which 32% are Baptists 6%, Methodist and 6% of the Church of God; Anglicans 20%; Catholics 19%. **Political Parties:** National Free Movement, party in power. Its leader is Hubert Ingraham. Progressive Liberal Party (PLP) founded in 1953; bringing together the population of African origin, under a pro-independence banner. The Vanguard Party, socialist. The People's Democratic Force. **Social Movements:** There is a Trade Union Congress, and other trade unions such as the Bahamas Hotel Catering and Allied Workers Union, the Bahamas Union of Teachers, the Bahamas Public Service Union, the Airport Allied Workers Union, the Musicians amp; Entertainers Union, the Taxi Cabs Union. **Languages:** English (official) and Creole.

THE STATE

Official Name: The Commonwealth of the Bahamas **Capital:** Nassau, 172,200 inhab. (1990). **Other cities:** Adelaide, Freeport/Lucaya, Marsh Harbour. **Government:** Queen Elizabeth II is the head of State, and has been represented by Clifford Darling since 1995. Prime Minister Hubert Ingraham. There is a bicameral Legislative Power, with a 16-member Senate and a 49-member Assembly. **National Holiday:** July 10, Independence (1973). **Paramilitaries:** 2,550: Police (1,700); Defence Force (850).

DEMOGRAPHY

Annual growth: 0.6% (1992-2000). **Children per woman:** 2.1 (1992).

HEALTH

Under-five mortality: 10 per 1,000 (1995). **Calorie consumption:** 103% of required intake (1995).

EDUCATION

Illiteracy: 2% (1995).

COMMUNICATIONS

102 **newspapers** (1995), 103 **TV sets** (1995) and 104 **radio sets** per 1,000 households (1995). 30.3 **telephones** per 100 inhabitants (1993).

ECONOMY

Per capita GNP: $11,800 (1994). **Annual growth:** -0.80% (1985-94). **Consumer price index:** 100 in 1990; 118.0 in 1994. **Currency:** 1 Bahaman dollar = 1$ (1994).

---

became more understandable. Some local people felt that the Haitians were constantly landing on the coast and obtaining work permits from the government. Although they strongly supported the government in the past, labor unions now criticized the authorities for the lack of constructive, long-term programs aimed at solving the grave unemployment problem.

[14] After 20 years in office, Lynden Pindling won his sixth consecutive election on June 19 1987. His Progressive Liberal Party obtained 31 out of 49 seats for a 5-year term. The opposition Free National Movement (FNM), led by Cecil Wallace Whitfield (after lawyer Kendal Isaac's resignation following the electoral defeat), has repeatedly asked Pindling to step down, charging him with fraud, corruption, and being "soft" towards drug dealers.

[15] In 1991, there was a sharp decrease in the number of tourists, estimated at 3 million per annum. This was attributed to the fact that there had been a rise in the crime rate, compounded by the country's extremely high cocaine consumption rate. Although there had been a slow-down in inflation and a slight fiscal surplus, this did not prevent the people from taking to the streets, to voice discontent over economic conditions.

[16] US officials accused Pindling of allowing the Bahamas to be used as a transfer point for drugs. Testimony from the trial of the former Panamanian General, Manuel Antonio Noriega, alleged that Pindling had received at least $5 million in payments for this service.

[17] Tourism had generated over 65 per cent of the country's GDP, but

In 1991, there was a sharp decrease in the number of tourists, estimated at 3 million per annum.

in 1992 this fell by 10 per cent in relation to 1990. Meanwhile the banking system lost customers to its competitors in the Cayman Islands. In addition, the government was unable to increase agricultural production, making it necessary for the country to import 80 per cent of its food.

[18] Pindling's 25 years in office ended in 1992, when Hubert Ingraham, a former protege of Pindling's and leader of the National Free Movement, won the parliamentary elections with 55 per cent of the vote. The new government aimed to reduce unemployment through the liberalization of foreign investment laws and the re-establishment of the Bahamas as a major tourist destination.

[19] In February 1994, one of the prominent figures in Pindling's time, lawyer Nigel Bowe, was incarcerated in Miami for drug trafficking. That year, the government appointed a commission to investigate Pindling, accused of having used the Hotel Corporation's funds to increase his wealth. Pindling denied the charges and sought protection in bank secrecy, which limited the Commission's progress.

[20] In 1995, foreign investors carried out the privatization of several hotels belonging to the Hotel Corporation. The most significant problems for the government were unemployment and the presence of several thousands of Haitian and Cuban refugees. A repatriation agreement with Haiti was achieved, but the Cuban refugees refused to be sent back, and demanded to be transferred to the United States.

Bahrain

Bahrayn

Population: 557,000 (1994)
Area: 680 SQ KM
Capital: Manama (Al-Manamah)

Currency: Dinar
Language: Arabic

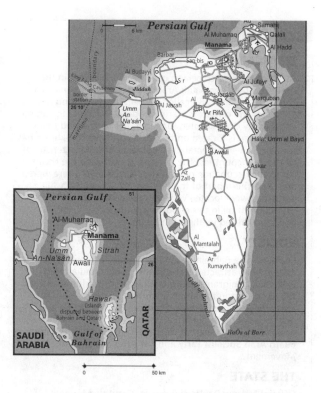

It is believed that the first Sumerians left Bahrain for Mesopotamia. From that time on, the country was central to the intense maritime trade between Mesopotamia and India. This trade became particularly prosperous between the 11th and 15th centuries, as Islamic civilization expanded over all the territory from the Atlantic Ocean to the South Pacific.

[2] Portuguese sailors occupied the island in 1507 and stayed there for a century until the Persians expelled them. Iran's claim to sovereignty over this part of the Persian or Arab Gulf dates from this period. Sheikh al-Khalifah took power in 1782, displacing the Persians the following year; his descendants are still in power. Independence lasted until 1861, when another Khalifah, afraid of Persian annexation, agreed to declare a "protectorate" under the British.

[3] During the two World Wars Bahrain was an important British military base. In 1932 the first oil wells were opened. Nationalist movements demanding labor rights, democracy and independence grew during the 1950s, as they did in other parts of the Arab world.

[4] In 1954, a strike broke out in the oil fields, and, in 1956, colonial administrative offices were attacked. British troops were sent to quash the rebellion, and opposition leaders were arrested and exiled. Slowly, some reforms were carried out and local participation in public administration increased.

[5] Finally, beginning in the early 1970s, the British decided to withdraw from their last colonies "east of Suez", though they maintained their economic and strategic interests in Bahrain. Bahrain and Qatar refused to join the United Arab Emirates, so in 1971, the country became independent under Sheikh Isa ibn-Sulman al-Khalifah.

[6] The new nation authorized Washington to set up naval bases in its ports. These were dismantled in 1973 following the Arab-Israeli conflict. Local elections were held the same year and the National Assembly came under the control of progressive candidates calling for freedom to organize political parties and greater electoral representation. The British felt their interests were being threatened, and, in August 1975, they backed al-Khalifah's decision to dissolve Parliament.

[7] In the 1970s, Iran managed to eclipse Saudi control of the Emirates, forcing a virtual protectorate over them. Meanwhile, heavy migration towards Bahrain threatened to create an Iranian majority, or at least a large Iranian minority on the islands.

[8] Shortly before his assassination, King Faisal reacted against growing Iranian influence by embarking on a diplomatic campaign that his successors continued to implement. Riyadh, the Saudi capital, put pressure on Qatar, the only Emirate that has traditionally been faithful to the Saudis, to solve their dispute with Bahrain over territorial waters. He appealed to the "Arab sentiment" that supposedly united them (Iran is Moslem, but not Arab). This attempt at solidarity lost ground in 1976 when Saudi Arabia increased its oil prices less than the other OPEC countries, including Iran.

[9] The fall of the Shah of Iran in 1979 made matters worse, as the unofficial representatives of the new Islamic Republic declared publicly that Iran maintained its claim to the Gulf islands. The Emir retaliated by cracking down even harder on both Iranian (or Shiite) immigrants and on all progressive movements in general. At the same time he drew closer to the other Arab governments, rejecting the Camp David agreement and signing mutual defence treaties with Kuwait and Saudi Arabia.

[10] Bahrain is the "least rich" oil producer in the Gulf. Its known total reserves (300 million barrels) will soon be exhausted if it continues the present rate of production. Aware that dry oil wells, and competition to Bahraini pearl-diving from Japanese oyster farms, may leave this desert country with fishing as its only resource, the government decided to slow down oil production in 1980 taking advantage of the islands' strategic location to turn them into a trade and financial center.

[11] In May 1975, Bahrain obtained a controlling interest in its oil industry, and in 1978 announced that all concessions would be cancelled.

[12] The state also promoted the establishment of industries, especially for copper and aluminium production. Generous tax exemptions and unrestricted facilities for profit remittance encouraged trans-national corporations to set up subsidiaries. The country is becoming a base for the re-export of all kinds of goods to other ports in the area. Bahrain also houses the second largest refinery in the Middle East, BAPCO, which processes local as well as a large part of Saudi crude oil.

[13] In 1981, the country joined the Gulf Cooperation Council. This organization was set up, with US help, to guarantee military and political control over the area, to counteract spreading Iranian influence, and to keep an eye on opposition groups in the member states.

[14] One of the most closely watched movements was the Bahrain National Liberation Front, supported mainly by oil workers, students and professionals. This group was a member of the Gulf Liberation Front until 1981.

[15] Toward the end of 1986, Bahrain bought ground-to-air missiles and warplanes from the United States to defend itself in the eventuality of Iranian aggression. In exchange, the US won authorization for the construction of a military airport and the right to use Bahrainian naval bases for its Gulf fleet, which "patrolled" the Gulf during the last few months of the Iran-Iraq war.

[16] In November 1986, a superhighway was opened between Saudi Arabia and Bahrain. As a result, Bahrain is, in practice, no longer an island. During the first year alone, the highway was used by more than a million vehicles. Such grandiose building projects may be a thing of the past, as the fall in world oil prices has plunged the country into a serious crisis.

[17] In 1989, Bahrain was forced to seek credit in order to balance its budget. Kuwait and Saudi Arabia contribute $100 million per year to ensure the stability of the Manama government.

[18] In March 1991, after the Iraqi

WORKERS

1994

- MEN 88%
- WOMEN 12%

- AGRICULTURE 3%
- INDUSTRY 14%
- SERVICES 83%

ENVIRONMENT

This flat, sandy archipelago consists of 33 islands in the Arab Gulf between Saudi Arabia and the Qatar Peninsula. The largest island, also called Bahrain, is 48 km long and 15 km wide. The climate is warm, moderately humid in summer and slightly dry in winter. Manama, the capital and main trade center, is located on the island of Bahrain. Bahrain has the same environmental problems that are characteristic of all the Gulf countries. The need to industrialize the country led to the occupation of what few fertile lands there were, in the northern part of the main island. In addition, industries were located very near residential areas. The extraction of oil in this region accounts for 4.7% of all pollution caused by the oil industry.

SOCIETY

Peoples: The Bahrainis are Arab people. The petroleum industry has attracted a number of Iranian, Indian and Pakistani immigrants. **Religions:** Islamic, predominantly of the Sunni sect in the urban areas and Shiite in the rural areas. **Languages:** Arabic. **Political Parties:** There are neither legal political parties nor trade unions. Some opposition forces are the Liberation Front of Bahrain, the Baath Arab Socialist Party and the Arab Nationalist Movement.

THE STATE

Official Name: Daulat al-Bahrayn. **Capital:** Manama (Al-Manamah) 140,000 inhab. (1992). **Other cities:** Al-Muharraq, 45,000 inhab. (1991). **Government:** Isa ibn-Sulman al-Khalifah, Emir since 1961; Khalifah ibn-Sulman al-Khalifah, Prime Minister since1970, assisted by an 11-member cabinet. The National Assembly, partially elected by popular vote, was dissolved in August 1975. **National Holidays:** August 15, Independence (1971); December 16, National Day. **Armed Forces:** 10,700 **Paramilitaries:** (Ministry of Interior): Coast Guard: 400; Police: 9,000.

DEMOGRAPHY

Annual growth: 0.5% (1992-2000). **Children per woman:** 3.7 (1992).

HEALTH

One **physician** for every 775 inhab. (1988-91). **Under-five mortality:** 20 per 1,000 (1995).

EDUCATION

Illiteracy: 15% (1995).

COMMUNICATIONS

103 **newspapers** (1995), 114 **TV sets** (1995) and 112 **radio sets** per 1,000 households (1995). 22.9 **telephones** per 100 inhabitants (1993). **Books:** 100 new titles per 1,000,000 inhabitants in 1995.

ECONOMY

Per capita GNP: $7,460 (1994). **Annual growth:** -0.70% (1985-94). **Consumer price index:** 100 in 1990; 103.6 in 1994. **Currency:** 0,4 dinars = 1$ (1994).

---

defeat in the Gulf War, the foreign ministers of Egypt, Syria and the six Arab member States of the Council for Cooperation signed an agreement with the United States in Riyadh in order to "preserve the regional security".

[19] In July 1991, the government announced that it would allow foreign companies to have full control of the local businesses they bought. Previously, all business interests on the island had to have a minimum of 51 percent of national capital. The change in policy was aimed at attracting foreign investment to compensate for economic losses caused by the Gulf War.

[20] After Kuwait, Bahrain was the emirate most affected by the conflict. The country was threatened by the oil slick, and Bahrain's skies were blackened for several months by Kuwait's burning oil wells. In the economic field, the effects of Kuwait's invasion and the subsequent war against Iraq were also felt. For instance, private bank deposits were reduced by 30 to 40 percent, and foreign banks, which had been 75 in 1981, were reduced to 51 by the end of 1991.

[21] As from 1992, foreign capital slowly began to return to Bahrain. Also that year, the country played an important role in the Middle East Peace Conference held in Moscow, which was attended by ten Arab nations, as well as Israel.

[22] In October 1994, Israeli Minister Yoosi Sarid's visit to Manama - the first high level public contact between Israel and a Gulf State - was a clear indication of the eco-

nomic liberalization policy of the regime.

[23] In December 1994, Shiite leader Sheik Al-Jamri was arrested, after signing a claim for the restauration of the Constitution and the Parliament, dissolved in 1975. His arrest provoked anti-governmental demonstrations, in which two students and a police officer died. In April 1995, Emir Isa ibn-Sulman al-Khalifa met with 20 opposition leaders in an attempt to put an end to growing violence. In August, both parties reached an agreement which concluded with the liberation of 1,000 political prisoners.

[24] Two months after the political agreement, Al-Jamri and six other opposition leaders declared that the government "had not honoured" its word and staged a ten day hunger strike ending on November 1. Some 80,000 people gathered in front of Sheik Al-Jamri to listen to a statement declaring the continuation of peaceful opposition for restoring the rule of constitutional law to the country. The Government reacted by closing down mosques, which led to further clashes and the re-arrest of Al-Jamri and his colleagues.

[25] In the first anniversary of the hunger strike new demonstrations shook Manama, a situation repeated in 1996. From October 23 to

November 1, 1996, popular demonstrations, strikes and switching-off were staged by the opposition, reaching various cities and the island of Nabih Saleh.

[26] The government turned to the death sentence to punish the identified "responsibles" for the uprisings and the Cassation Court validated the decision, arguing that under the provisions of the State Security Law, those sentenced by security courts have no right of appeal.

[27] International organizations expressed their concern about the human rights record of the country during recent years, when torture, deportation, death sentences and street repression became every day's news. The United Nations Working Group on Arbitrary Detention adopted three decisions on 20 September, 1996 "concerning the numerous persons, including children and minors" detained in Bahrain prisons. International agencies have been prevented from visiting the prisoners of conscience who are suffering from inhumane treatment and severe prison conditions.

[28] Due to the social unrest, some finance companies, like the US CoreStates Bank N.A., abandoned the country to settle themselves in Dubai, seeking political stability.

Bangladesh

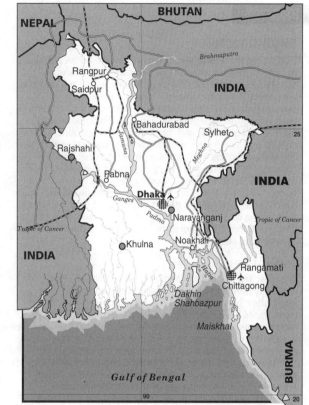

Population: 117,941,000 (1994)
Area: 144,000 SQ KM
Capital: Dhaka (Dacca)

Currency: Taka
Language: Bengali

Bangladesh

Though Bangladesh is a newly formed state, it is an old nation, with roots in the ancient state of Banga. The country has a written history of several thousand years, dating back to the ancient epic, the Mahabharat.

2   The British East India Company called it "Bengal" in the 17th century. It was renamed East Bengal after the partition of Bengal and India in 1947, and then East Pakistan when it joined the newly created state of Pakistan in 1956. It emerged as an independent state in 1971, earning its current name (meaning "the land of the Bengali speaking people") from the national language.

3   Hindu beliefs flourished in the region with the arrival of the Indo-Aryan peoples between 3,000 and 4,000 years ago. The country was subsequently occupied by Muslims during the 13th century, paving the way for a series of Afghan dynasties, which cut the ties with the ruling Muslim dynasties in Delhi. The country was periodically annexed by the Delhi empire, but these occupations were relatively brief. The Mughals conquered the territory in the 16th century and declared it an autonomous region. Later under Mughal rule, the country became virtually independent.

4   The Portuguese, Armenians, French and British came to Bangladesh in succession during the 17th and 18th centuries, establishing military and trade outposts. The British purchased 3 villages - Colikata, Sutanoti and Govindapur - from the landowners, and soon challenged the local authorities on trade and military issues. In June 1757, in the Battle of Plassy, 500 British soldiers led by Robert Clive defeated a much larger enemy force under the command of local ruler Nawab Siraj-ud-Dwola. The defeat of the local army amounted to the British conquering the country. This was the beginning of 190 years of British rule.

5   When the British partitioned India, transforming part of its territory into Pakistan in August 1947, Bengal was also partitioned. The Muslim majority areas, known as East Bengal, became part of Pakistan, and the Hindu majority areas became part of India. The people of East Pakistan became impatient when they became aware that massive amounts of resources were being transfered from their part of the country to West Pakistan. They also lost patience with the bureaucratic military oligarchy in West Pakistan, and began to demand regional autonomy.

6   As a result, East Pakistan elected its own political leadership in 1970. The Pakistani government refused to recognize this secession; the military blocked all attempts to form a government and began campaigns of aggression against East Pakistani civilians on March 25, 1971.

7   The people of East Pakistan declared independence and launched an armed resistance movement. They formed a government-in-exile in India with Sheikh Mujibur Rahman as president, and finally suceeded in ousting the occupation forces. About 3 million lives were lost during the war and 10 million refugees were reported to have crossed the border into India.

8   On December 16, 1972, Bangladesh adopted a constitution providing for parliamentary democracy. Large industries, banks and insurance companies were nationalized. Democracy, secularism, socialism and nationalism were declared as the state's basic principles.

9   But the task of recuperating the problematic war-torn economy proved too great a challenge for the inexperienced leadership of the ruling party, the Awami League. Nationalist fervor soon died down, giving way to widespread feelings of disappointment. Bangladeshi politics degenerated into confusion as insurgency and armed political movements erupted throughout the country.

10   In December 1974, the government declared a state of National Emergency, and in January 1975 fundamental civil rights were suspended. All opposition political parties and trade unions were banned leaving a single party, Bakshal, mainly made up of Awami League members and pro-Moscow communists. Newspapers were banned - except for four national dailies controlled by the government - and a new press and publication law suppressed all opposition views. On August 15, 1975, against this backdrop, a group of active and retired army officers assassinated President Sheikh Mujibur Rahman and his family, and declared martial law.

11   On November 7, 1975, General Zia ur Rahman emerged as the leading force in the opposition, following a series of coups and counter-coups. He founded the Bangladesh Nationalist Party (BNP), which included many members who had opposed Bangladesh's initial independence in 1971. The BNP won the 1978 and 1979 elections, and the Awami League went into opposition.

12   The country went through a brief period of relative stability. But in May 1981, Zia ur Rahman was killed in an abortive coup attempt.

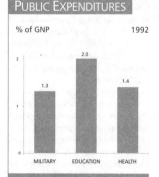

PUBLIC EXPENDITURES

% of GNP — 1992

MILITARY 1.3
EDUCATION 2.0
HEALTH 1.4

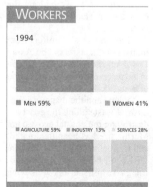

WORKERS

1994

■ MEN 59%   ■ WOMEN 41%

■ AGRICULTURE 59%   ■ INDUSTRY 13%   ■ SERVICES 28%

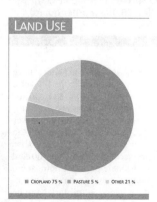

LAND USE

■ CROPLAND 75 %   ■ PASTURE 5 %   ■ OTHER 21 %

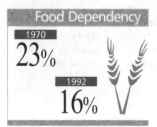

Food Dependency

1970
23%

1992
16%

External Debt

1994

PER CAPITA
$140

Illiteracy

1995

62%

PROFILE

ENVIRONMENT

Located on the Padma River Delta, formed by the confluence of the Meghna with the Ganges and the Brahmaputra, Bangladesh is a fertile, alluvial plain where rice, tea and jute are grown. There are vast rain forests and swamps. A tropical monsoon climate predominates, with heavy summer rains from June to September generally accompanied by hurricanes and floods with catastrophic consequences. Low-quality coal and natural gas are the only mineral resources. The increase in sea level poses a growing threat to the country. The coastal area along the Bay of Bengal has been severely affected by the discharge of sewage and industrial waste. This pollution, together with indiscriminate fishing to supply both internal and export markets, is leading to the destruction of one of the country's main resources with sea-borne pollution threatening to irreversibly damage the coastal ecosystem.

SOCIETY

Peoples: The people of Bangladesh are ethnically and culturally homogeneous, as a result of 25 centuries of integration between the local Bengali population and immigrants from Central Asia. There are small Urdu and Indian minorities. Of the present ethnic groups, the Chakma are known to have migrated from the hills of Chittagong to border states of India in 1981. **Religions:** Mostly Islamic (83%) and Hindu (16%), with Buddhist and Christian minorities. **Languages:** Bengali. **Political Parties:** Awami League (AL), in favor of a socialist economy with the intervention of the private sector; Bangladesh Nationalist Party, right of center (Bangladesh Jatiyatabadi Dal, BNP); National Party (coalition), an Islamic-inspired alliance of five nationalist parties; Islamic Assembly; Jamaat-I-Islami, Islamic fundamentalists; Communist Party and other minor parties.

THE STATE

Official Name: Gana Prajatantri Bangladesh. **Capital:** Dhaka (Dacca), 6,105,160 inhab. (1991). **Other cities:** Chittagong 2,040,663 inhab.; Khulna 877,388 inhab.; Rajshahi 517,136 inhab. (1991). **Government:** Parliamentary republic. Shahabuddin Ahmed was sworn in as president in Octuber, 1996. Sheikh Hasina Wajed, Prime Minister since June and head of government since June 23, 1996. Single-chamber legislature: Parliament made up of 330 members (300 elected by direct vote and 30 reserved for women, nominated by Parliament, for 5-year terms). **National Holidays:** March 26, Independence Day (1971); December 16, Victory Day (1971). **Armed Forces:** 115,500 (1995) **Paramilitaries:** Bangladesh Rifles: 30,000 (border guard); Ansars (Security Guards): 20,000.

Almost a year later, on March 24 1982, the army engineered a successful coup, placing General Ershad in the presidency.

[13] In 1985, General Ershad formed the Jatiya Party (JP), going on to win the 1986 general election with what the opposition saw as widespread fraud. As a result of this, massive popular uprisings forced Ershad to dissolve the parlia-

ment. Confidence in political integrity was so low that the 1988 elections, were boycotted by all the major political parties and a large percentage of the electorate as well.

[14] Since Sheikh Mujibur Rahman's death in 1975, socialism and secularism have been discarded as state policy. Public sector budgets have gradually been reduced and the military have be-

come a strong contender for political power.

[15] Bangladesh is one of the poorest and most densely populated countries in the world and has suffered many famines, an occurance that seems hardly explicable in a country with extremely rich soil and plentiful water and sun, all the elements necessary for agricultural production sufficient to feed the population. However, as the majority of the quality land is concentrated in the hands of a few large landowners, the low productivity and enormous privations faced by the rural population are not altogether surprising.

[16] Prior to the "green revolution" of the 1970s and 1980s, over 7,000 varieties of rice were commonly cultivated in Bangladesh. Now only one is under cultivation. This new high-yield variety promised to meet all the country's rice needs, but the only real achievement was to open a new market for the manufacturers of chemical fertilizers, upon which the production of this new variety of rice depends.

[17] In 1982, "Administrative decentralization" was initiated, leading to huge unproductive spending in rural areas and a growing alliance between bureaucrats and the rural elite to control public resources.

[18] Economic performance has been very poor, complicated by indiscriminate privatization, increasing military expenditure, low growth rates in manufacturing, falling agricultural production and the material damage caused by recurrent floods.

[19] On April 30, 1991, the worst storm to hit Bangladesh since 1970 killed nearly 100,000 people. Prime Minister Khaleda Zia reported that it had caused several million dollars' worth of damage and left millions of people homeless. The most devastated areas were along the coast and on islands of the Ganges delta.

[20] Bangladesh also suffers from man-made disasters. The Asian Development Bank has $5 million in aid reserved to end the pollution of Bangladesh's waters. Leaks from oil tankers and the disposal of industrial waste along the coast are destroying the coastal ecosystem. Marine life has been decimated, and it is believed that the sea here will soon be completely dead. Bangladeshi fisherfolk are having to go increasingly far out to sea in

order to obtain a satisfactory catch. Fishing accounts for 6 per cent of the GNP, and the export of fish products earns over $150 million per year.

[21] The district of Khulna is among the world's most polluted areas, with the destruction in recent years of forest worth over $110 million. There are more than a thousand offending industries, known by the government, which continue to operate in the knowledge that they cannot be closed without further aggravating an already serious unemployment situation.

[22] Bangladesh has always been a predominantly rural society. Agriculture accounts for about half of the gross domestic product, compared to just 10 per cent from manufacturing. About four-fifths of the labor force is employed in agriculture or related activities. Participation by women in the formal labor market is small: only seven per cent of the work force is female.

[23] Women's social status is very low, and under the fundamentalist inspired constitution, a woman may inherit only half as much as her brother, and in practice, this fraction goes to her husband or is saved as a dowry. Bengali feminists complain that women are treated as goods rather than individuals as they belong to their fathers in childhood, to their husbands in marriage (most marry at 13), and to their sons in old age. Their work in the home and the fields is not included in official production statistics, and divorce, a male prerogative under Islamic law, can be easily obtained if a woman's productivity drops. Unfortunately this frequently occurs as many women are seriously undernourished.

[24] Bangladesh joined the UN in 1974 after two earlier attempts were blocked by Chinese vetoes. It is a founding member of the Non-Aligned Movement and also of the Organization of the Islamic Conference. A gradual Islamization of politics has been evident in recent years. A 1989 constitutional amendment declared Islam the official religion, a move opposed by secular forces.

[25] In 1991, Ershad was deposed, and elections were called, in which the Awami League (AL) opposed the Bangladesh Nationalist Party (BNP). Both political alliances launched women as their major candidates; in both cases the widow of the leader of the alliance.

[26] On March 2 1991, Begum Khaleda Zia of the BNP was elected prime minister. Khaleda Zia declared that she supported the establishment of a parliamentary regime, thereby stripping the AL of one of its traditional platforms. Five months after her triumph in the general election, and with the unanimous approval of the legislative representatives of the 2 major parties, the Congress of Bangladesh replaced the country's presidential system with a parliamentary system.

[27] The previous president, Hussain Mohammad Ershad, had been sentenced in 1991 to 10 years in prison for abuse of power, unlawful wealth and possession of illegal weapons. However, he was able to run for the 1996 elections after appealing against the sentence.

[28] In mid-August 1991, the political reform was approved when, in a climate of great tension, President Shahabuddin Ahmed threatened to resign unless the system was modified.

[29] Bangladesh continues to rely heavily upon international aid; 95 per cent of its development programs are financed from abroad. The United States, Japan, the Asian Development Bank (ADB) and the World Bank contributed $2.3 billion in 1991, 10 million more than the previous year. 95 per cent of the country's budget is taken up in interest payments on its foreign debt.

[30] Bangladeshis have an average per capita annual income of $170. Fifty per cent of housing is of mud brick construction and has no plumbing: one toilet may be shared by as many as 50 families. More than 13 million people live in the shanty towns surrounding the cities, which continue to grow at a rapid rate.

[31] Among its measures to shore up the fragile economy, the government declared all-out war against contraband, in December 1991. Contraband is a big business in Bangladesh, and involves important financial institutions, the nouveau riche, and corrupt political and government officials.

[32] The IMF- and World Bank-launched privatization scheme, which included the sale of 42 public enterprises, came to a stand-still in 1991. Since then, Minister of Finance Saifur Rahman has faced the opposition of organized labor, especially in the jute industry (one of the main sources of export earnings), as well as textile and railway unions that fear the loss of jobs. General strikes - demanding a national minimum wage - have become commonplace.

[33] The Gulf War (1990-1991) provoked the need for Bangladesh to reaffirm its identity as a Muslim community. This feeling was intensified in 1992, when the repatriation of Muslim refugees began. Early in that year, Bangladesh received several groups of some 250,000 Bihari Muslims, who had supported Pakistan in 1971. In June, another 270,000 Rohingya Muslims arrived, fleeing persecution in Burma, a predominantly Buddhist country. The repatriation agreements signed between the two countries in 1992 failed to stem the flow of refugees.

[34] In 1994, writer Taslima Nasrin became renowned for her book

In 1982, "Administrative decentralization" was initiated, leading to huge unproductive spending in rural areas and a growing alliance between bureaucrats and the rural elite to control public resources.

"Shame" which denounced Muslim oppression over Hindus and other minorities. Nasrin was arrested for demanding that Islamic law grant more rights to women, which implied criticism of the Koran.

[35] Violence erupted in the Chittagong district when the Shanti Bahini guerrilla movement decided that the limited autonomy offered them in 1989 was not sufficient. This group comes from a Buddhist minority on the border between India and Burma, who are waging war on the authorities in Dhaka. As a result of this, and friction over the construction of the Farrakka dam on the Indian Ganges, relations with New Delhi have grown increasingly tense since 1993. The dam deprives Bangladesh of water for irrigation and river transportation during the dry season, and causes uncontrollable floods during the rainy season.

STATISTICS

DEMOGRAPHY

Urban: 17% (1995). **Annual growth:** 3% (1992-2000). **Estimate for year 2000:** 132,000,000. **Children per woman:** 4 (1992).

HEALTH

One **physician** for every 12,500 inhab. (1988-91). **Under-five mortality:** 117 per 1,000 (1995). **Calorie consumption:** 87% of required intake (1995). **Safe water:** 97% of the population has access (1990-95).

EDUCATION

Illiteracy: 62% (1995). **School enrolment:** Primary (1993): 105% fem., 105% male. Secondary (1993): 12% fem., 26% male. **Primary school teachers:** one for every 63 students (1992).

COMMUNICATIONS

84 **newspapers** (1995), 83 **TV sets** (1995) and 80 **radio sets** per 1,000 households (1995). 0.2 **telephones** per 100 inhabitants (1993). **Books:** 83 new titles per 1,000,000 inhabitants in 1995.

ECONOMY

Per capita GNP: $220 (1994). **Annual growth:** 2.00% (1985-94). **Annual inflation:** 6.60% (1984-94). **Consumer price index:** 100 in 1990; 115.8 in 1994. **Currency:** 40 taka = 1$ (1994). **Cereal imports:** 1,175,000 metric tons (1993). **Fertilizer use:** 1,032 kgs per ha. (1992-93). **Imports:** $4,701 million (1994). **Exports:** $2,661 million (1994). **External debt:** $16,569 million (1994), $140 per capita (1994). **Debt service:** 15.8% of exports (1994). **Development aid received:** $1,386 million (1993); $12 per capita; 6% of GNP.

ENERGY

Consumption: 65 kgs of Oil Equivalent per capita yearly (1994), 31% imported (1994).

[36] Throughout 1994, 154 MPs opposed to Khaleda Zia's government boycotted Parliament's activities until they jointly resigned in December. Sheikh Hasina, leader of the Awami League (AL), led peasant demonstrations in 1994 and 1995 demanding elections under a neutral administration.

[37] On February 15, 1996 - with the prospect of women voting for the first time - elections were held. Prime Minister Khaleda Zia from the BNP retained her office but elections - held under army surveillance - were considered a fraud.

[38] Confrontations between police and opposition activists added to the country's standstill encouraged by the AL. Violence did not stop with Zia's fall on March 30. On June 23, former Supreme Court of Justice president Mohammad Habibur Rahman - who had headed the neutral government prior to the elections held on June 12 to 19 - handed over the government to prime minister Sheikh Hasina, chosen in the second elections in four months.

Barbados

Barbados

Population: 260,000 (1994)
Area: 430 SQ KM
Capital: Bridgetown

Currency: Barbadian Dollar
Language: English

The peaceful and nomadic Arawak people expanded throughout the Caribbean region, and although they were dislodged from many islands by the Caribs, they remained on others, as in the case of Barbados.

[2] The Spanish landed on the island early in the 16th century and christened it the "island of the bearded fig tree". Satisfied that the island lacked natural wealth, they withdrew, but not before massacring the native population, taking a few survivors with them to amuse the Spanish court. In 1625 the British arrived and found a fertile, uninhabited territory.

[3] Around 1640 the island had close to 30,000 inhabitants. The majority were farmers and their families, and some were political and religious dissidents from Britain. The settlers grew tobacco, cotton, pepper and fruit on small plots of land, raising cattle, pigs and poultry.

[4] Sugar cane was introduced, with the support of the British, and caused extensive social change: plantation owners purchased large plots of land which the new crop required in order to be profitable, and small landowners, most of them in debt, sold off their plots to the plantation owners. The importation of slaves from Africa to work the sugar plantations began during this period.

[5] In 1667, 12,000 farmers emigrated to the 13 colonies of North America. Nonetheless, the island had a commercial fleet of 600 vessels and, as recorded by a French traveller in 1696, was "the most powerful island colony in America".

[6] Towards the end of the 18th century, the island was one huge sugar-producing complex which included 745 plantations and over 80,000 African slaves. By this time there was no woodland left on an island described in a 16th century account as "entirely covered with trees". Its ecological balance seriously impaired, the island fell victim to drought in the early 19th century and parts of it suffered soil exhaustion.

[7] The pursuit of increased profitability and an economy oriented toward foreign trade resulted in under-development in Barbados. A different approach might have led to development along the lines of the other North American colonies.

[8] Slavery was abolished in 1834 but the plantation economy still dominates the island's fiscal system. European landowners dominated local politics until well into the 20th century.

[9] In 1938, following the gradual extension of political rights, the Barbados Labor Party (BLP), led by Grantley Adams, developed from within the existing labor unions. Universal suffrage was declared in 1951, and Adams became leader of the local government.

In 1667, 12,000 farmers emigrated to other Caribbean islands or to the 13 colonies of North America. Nonetheless, the island had a commercial fleet of 600 vessels and, as recorded by a French traveller in 1696, was "the most powerful island colony in America".

[10] In 1961, internal autonomy was granted, and in 1966 independence was proclaimed within the British Commonwealth. Errol Barrow was elected prime minister. Unlike the rest of the West Indies, Barbados never severed its links with the colonial capital in spite of its political independence.

[11] After 1966, Errol Barrow's Democratic Labor Party (DLP) contributed to the creation of the Caribbean Free Trade Association which, in 1973, became CARICOM, involving 12 islands of the region. Barrow showed great interest in the Non-Aligned Countries movement.

[12] Free education and new electoral laws were not followed by any significant change in DLP policies towards the owners of sugar refining plants. The rise in unemployment diminished DLP support and resulted in the party's electoral defeat. In 1970, the country became a member of the International Monetary Fund (IMF).

[13] In 1976, the BLP won 17 of the 24 available seats. Tom Adams, Grantley Adams' son, was elected prime minister. He promised to fight corruption and described himself as a social democrat (the BLP became a member of the Socialist International in 1978). However, the government protected the interest of investors in sugar and tourism transnationals, while encouraging foreign investment.

[14] Adams pressured Washington to withdraw from its naval base in St Lucia, which they did in 1979. In 1981, Adams was re-elected and consolidated relations with Washington. Barbados supported the US invasion of Grenada.

[15] In order to attract foreign capital, the government passed new tax exemption laws and liberalized ship registration. In 1986, an agreement between the United States and Barbados led to 650 new companies registering in the off-shore sector. Unemployment and inflation within Barbados continued to grow.

[16] The 1986 election was won by the DLP and Prime Minister Errol Barrow committed the government to changing the foreign policy toward the US which had been followed by the BLP government in the preceeding 10 years. In 1987, Barrow died of a sudden heart attack and was succeeded as prime minister by Erskine Sandiford. Two years later, as a result of a split within the DLP, a new opposition group was founded, the National Democratic Party, under the leadership of Richard Heynes.

[17] In 1989, steadily falling sugar prices and rising interest rates fuelled inflation. The government's resolve to cut its budget deficit aroused fears of further unemployment and there

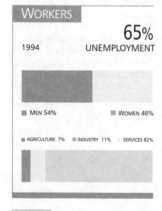

WORKERS

65%
1994 — UNEMPLOYMENT

- MEN 54%
- WOMEN 46%

- AGRICULTURE 7%
- INDUSTRY 11%
- SERVICES 82%

PROFILE

ENVIRONMENT

Of volcanic origin, it is the eastern-most island of the Lesser Antilles. The fertile soil and rainy tropical climate favor intensive farming of sugar cane, rotated with cotton and corn. The most serious environmental problems are the disposal of sewage and refuse, water consumption, and soil and coastal erosion. From a regional perspective the Caribbean Sea is increasingly contaminated and its resources are over-exploited.

SOCIETY

Peoples: Most are of African origin, with a minority of Europeans and mestizos. It is one of the most densely populated countries in the world, with an average of 1,700 inhabs. per sq. km. **Religions:** 70% Anglican. There are also Catholics, Methodists and Moravians. **Languages:** English (official). Creole is also spoken. **Political Parties:** The Barbados Labor Party (BLP), in power since 1994. The Democratic Labor Party (DLP); the right-wing National Democratic Party, founded in 1989 after a split with the DLP; the left-wing Popular Political Alliance (PPA); the Marxist Workers' Party of Barbados (WPB), founded in 1985. **Social Organizations:** Barbados Workers' Trade Union with 30,000 members, is the largest labor organization. Also strong are the unions of school teachers and civil servants.

THE STATE

Official Name: Barbados. **Capital:** Bridgetown, 85,000 inhab. (1980). **Other cities:** Speightstown; Holetown; Bathsheba. **Government:** Owen Arthur, Prime Minister since September 7 1994; Dame Nita Barrow, Governor-General, appointed in1990, died on December 19,1995. The bicameral Parliament has been shaped on the British model. The Legislative power is constituted by a Parliament, with a Low Chamber, the Senate and the post of Governor General. The Executive power is made up of a cabinet, with a Prime Minister and other government ministers. The Judiciary is independent from the other powers. Barbados has had one of the most stable governments in the Caribbean. **National Holiday:** November 30, Independence Day (1966).

STATISTICS

DEMOGRAPHY

Annual growth: 1.2% (1992-2000). **Children per woman:** 1.8 (1992).

HEALTH

Under-five mortality: 114 per 1,000 (1995). **Calorie consumption:** 110% of required intake (1995).

EDUCATION

Illiteracy: 3% (1995). **Primary school teachers:** one for every 17 students (1991).

COMMUNICATIONS

104 **newspapers** (1995), 103 **TV sets** (1995) and 111 **radio sets** per 1,000 households (1995). 31.8 **telephones** per 100 inhabitants (1993). **Books:** 102 new titles per 1,000,000 inhabitants in 1995.

ECONOMY

Per capita GNP: $6,560 (1994). **Annual growth:** 0.00% (1985-94). **Consumer price index:** 100 in 1990; 114.1 in 1994. **Currency:** 2 Barbadian dollars = 1$ (1994).

---

were several strikes. Nonetheless, industry recovered, exchange reserves increased, and about a hundred Panamanian offshore concerns expressed interest in Barbados. Japan signed agreements to invest in tourism on the island.

[18] In 1989, Barbados requested a review of the CARICOM charter as it was felt that the constraints were too limiting. In order to favor local agriculture, government import restrictions were introduced, particularly on vegetables from Guyana.

[19] In 1990, serious controversies arose over sugar cane prices between the sugar mills and the government; demands from workers threatened to end in strike action. The tourist industry employs 15 per cent of the workforce of the island and 450,000 tourists visited Barbados in 1988.

[20] In 1990 the government endorsed the use of solar power as electricity is very expensive on the islands.

[21] According to the Inter-American Development Bank, Barbados' economy has registered a negative growth rate since 1990, although inflation has been reduced. While overall unemployment remained constant, there was an increase in unemployment among female workers in the garment, electronics and tourism industries, the country's main source of foreign exchange. In the tourist sector, there was a 6.3 per cent decline in the number of tourists, as well as a reduction in the length of their stay on the island.

[22] In 1991, the DLP was re-elected with 49 per cent of the vote although it lost two of its twenty seats. In November, riots and protests due to an 8 per cent cutback on civil servants' wages, led to a general strike. In spite of having been questioned, Sandiford managed to remain in office.

[23] In 1992 the government obtained an IMF loan amounting to $64.9 million and the prime minister announced further wage reductions in the public sector, a rise of interest rates and cutbacks in the social system, as well as the privatization of oil and cement production and the tourism industry.

[24] After two successive finance ministers resigned, Sandiford himself assumed the post in late 1992, attempting to transform the country into a financial center. Tourist activity has flourished since 1993. That year, tourism revenues reached $500 million, equivalent to one third of the country's GDP.

[25] In June 1994, the Barbados Labor Party (BLP) withdrew their support for the prime minister and won the elections held on September 7. Economist Owen Arthur, the new prime minister, cancelled planned wage reductions, which was passed as law in February 1995.

[26] Arthur introduced a deficit budget in April, arguing expenditure was needed to tackle unemployment, which had reached 21.2 per cent. The IMF had advised caution in the management of public funds, but Arthur chose to overlook these prescriptions.

[27] The island's Governor General, Dame Nita Barrow, sister of former prime minister Errol Barrow, died on December 19, 1995. She had held the post since June 1990, when she had been nominated by the Queen of England.

Belarus

Belarus

Population: 10,356,000 (1994)
Area: 207,600 SQ KM
Capital: Minsk (Mensk)

Currency: Ruble
Language: Belarusian

The territory known today as Belarus (formerly Belorussia) was inhabited as of the 1st century AD by the Krivich, Radimich and Dregovich Slavic peoples. The first principalities were those of Polotsk and Turov-Pinsk. In the 9th century, Eastern Slav tribes belonging to the Western Rus joined other Eastern Slav peoples to form the Kiev Rus, the ancient Russian State (see history of Russia) which gave rise to modern Russia, Ukraine and Belarus. The major cities of the area, Polotsk, Turov, Brest, Vitebsk, Orsha, Pinsk and Minsk date from the 9th and 10th centuries.

2   In the 14th century, Lithuanian principalities annexed themselves to the Western Rus. As the Lithuanians lacked their own writing system, the language of the Belarusians (which was the only Russian language, at the time) became the official language of the State.

3   Between the 14th and 16th centuries, the Belarusian culture began to differentiate itself from that of the Russians and the Ukrainians.

4   In 1569, according to the terms of the Union Treaty between Poland and Lithuania, Belarus became a part of Poland. The Belarusians resisted domination by the Catholic Poles, and maintained their linguistic and cultural identity.

5   With the first partition of Poland in 1772, Russia kept the eastern part of Belarus. Between 1793-95, the rest of Belarus became part of the Russian Empire. The abolition of serfdom in Russia in 1861 accelerated the development of Belarus. Important railways developed, including the Moscow-Brest and the Libava-Romni lines. By the end of the 19th century, Belarus had Russia's second most important railway system.

6   The Russian czar's abolition of the serf system also put an end to the peasants' feudal bondage to the landed nobility. However, this failed to solve the land tenure problem, as in Belarus only 35 per cent of the land was handed over to the peasants. This triggered a number of uprisings, the most important of which took place in 1863 under the leadership of Kalinovski.

7   In March 1899, the First Congress of the Social Democratic Workers' Party of Russia (SDWPR) was held secretly in Minsk. This group was inspired by Marxist socialism and was determined to bring down the czar. Other sectors of the intelligentsia took their cue from the peasants' discontent, and founded an autonomous Belarusian movement in 1902, which sought to revive the nation's culture. During this period, the czarist regime deported Russian Jews to Belarus, where they came to account for a fifth of the local population.

8   During World War I (1914-1918), part of Belarus was occupied by German troops. After the Russian Revolution of February 1917, councils (soviets) made up of workers' representatives were formed in Minsk, Gomel, Vitebsk, Bobruisk and Orsha.

9   Soviet power began to be established at the end of 1917, and in February 1918 the large land holdings were nationalized. A few months later, land began to be distributed to the peasants. At the insistence of the Bolsheviks (socialist revolutionaries of the SDWPR), the first collective farms (koljoses) were set up. In reality, however, the land gradually went into State control.

10   The 6th Party Conference of the Russian Communists (Bolsheviks), the former SDWPR, held in Smolensk, approved the decision to found the Soviet Socialist Republic of Belarus (SSRB). In January 1918, the decision to join Lithuania was made at the 1st Congress of the SSRB Soviets. The Congress of the Soviets of Lithuania endorsed the same motion.

11   On February 28 1919 in the city of Vilno (today Vilnius, the capital of Lithuania) the government of the Soviet Socialist Republic of Lithuania and Belarus was elected, with Mickevicius-Kapsukas as head of state.

12   In February 1919, Poland, which had become independent from Russia after the Bolsheviks had seized power in St Petersburg, occupied a large part of Belarus. According to the Treaty of Riga, signed between Soviet Russia and Poland (1921), Western Belarus became part of Poland.

13   On August 1 1920 the Assembly of representatives of the Lithuanian and Belarusian Communist Parties and of labor organizations in Minsk and its surrounding areas, approved the declaration of the foundation of the independent Belarusian republic.

14   On December 30, 1922, Belarus joined the Union of Soviet Socialist Republics (USSR), as one of its official founders, along with the Russian Federation, Ukraine and the Transcaucasian Federation (Armenia, Georgia and Azerbaijan).

15   The industrialization and collectivization of agriculture began in the second half of the 1920s. On February 19 1937 the 12th Congress of the Soviets of Belarus approved the new constitution. In November 1939, as a result of the Molotov-Ribbentrop Pact - signed between the USSR and Germany - Western Belarus was reincorporated into the Soviet Union.

16   In June 1941 Belarus became the first of the Soviet republics to suffer Hitler's aggression against the Soviet Union. The fortress at Brest offered fierce resistance, becoming a symbol of Belarus's heroic defence. Guerrilla warfare extended throughout the country for the duration of the war, a conflict in which more than 2 million Belarusians lost their lives.

17   After Germany's defeat in 1945, Belarus' current borders were established. It became a charter member of the United Nations that same year; like Ukraine, it had its own delegation, independent of the USSR.

18   It took Belarus almost two decades to restore its economy, becoming in the process an important producer of heavy trucks, electrical appliances, radios and television sets. Between the 1920s and the 1980s, Belarus ceased to be a rural country - in which 90 per cent of the population lived from traditional farming and livestock raising - becoming an urban, industrialized country with nearly 70 per cent of its population living in cities.

19   Belarus is one of the few republics that has managed to maintain an ample majority of its local population - more than 80 per cent - with slight influence from Russian or other immigration.

20   Until 1985, the Communist Party and the Belarusian government followed the course established by the Communist Party of the Soviet Union. Among the outstanding communist leaders of this period was Piotr Masherov.

21   Because of the policies of *glasnost* (openness) and *perestroika* (restructuring) initiated by President

LAND USE

CROPLAND 30 %   PASTURE 15 %   OTHER 55 %

ENVIRONMENT

Belarus is located between the Dnepr (Dnieper), Western Dvina, Niemen and Western Bug rivers. It is bounded in the west by Poland, in the northwest by Latvia and Lithuania, in the northeast by Russia and in the south by Ukraine. It is a flat country, with many swamps and lakes, and with forests covering a third of its territory. Its climate is continental and cool, with an average summer temperature of 17-19 degrees, and 4-7 degrees below zero in winter. Due to its geographical proximity to Ukraine, Belarus received intense radioactive fall-out following the disaster at the Chernobyl nuclear plant in 1986. A quarter of the country's arable land shows signs of chemical pollution from the overuse of pesticides.

SOCIETY

Peoples: Belarusians, 77.9%; Russians, 13.2%; Poles, 4.1%; Ukrainians, 4.1%; Jews, 1.1% (1989). **Religions:** Christian Orthodox; in the west, Catholic. **Languages:** Belarusian (official); Russian (spoken as a second language by most of the population) and Polish. **Political Parties:** Communist Party, Belarusian Popular Front, Agrarian Party. **Social Organizations:** Labor Union Federation of Belarus; Free Union.

THE STATE

Official name: Respublika Belarus. **Administrative divisions:** Six regions (Brest, Gomel, Grodno, Minsk, Moguiliov and Vitebsk). **Capital:** Minsk (Mensk), 1,671,000 inhab. **Other cities:** Gomel, 506,000 inhab.; Mogilov, 363,000 inhab.; Vitebsk, 373,000 inhab.; Grodno, 291,000 inhab.; Brest, 269,000 inhab. **Government:** Executive power: President, Aleksander Lukashenka; Prime Minister, Mikhail Chigir. The Supreme Soviet is the maximum legislative body; Myachaslau Hryb, President. **National Holiday:** June 27, Independence (1991). **Armed Forces:** 102,600. **Paramilitaries:** Border Guards (Ministry of Interior): 8,000. **Land use:** Forested 33.7%; meadows and pastures 15.1%; agricultural and under permanent cultivation 30.1%; other 21.1%.

Foreign Trade

MILLIONS $ 1994

IMPORTS
3,857

EXPORTS
3,134

Maternal Mortality

1989-95

PER 100,000 LIVE BIRTHS
25

DEMOGRAPHY

Urban: 69% (1995). **Annual growth:** 0.4% (1991-99). **Estimate for year 2000:** 10,000,000. **Children per woman:** 1.9 (1992).

EDUCATION

School enrolment: Primary (1993): 95% fem., 95% male. Secondary (1993): 96% fem., 89% male. University: 44% (1993).

COMMUNICATIONS

17.6 **telephones** per 100 inhabitants (1993).

ECONOMY

Per capita GNP: $2,160 (1994). **Annual growth:** -1.90% (1985-94). **Annual inflation:** 136.70% (1984-94). **Consumer price index:** 100 in 1990; 622,203 in 1994. **Currency:** ruble. **Cereal imports:** 1,250,000 metric tons (1993). **Fertilizer use:** 2,228 kgs per ha. (1992-93). **Imports:** $3,857 million (1994). **Exports:** $3,134 million (1994). **External debt:** $1,272 million (1994), $123 per capita (1994). **Debt service:** 4.3% of exports (1994).

ENERGY

Consumption: 2,692 kgs of Oil Equivalent per capita yearly (1994), 89% imported (1994).

Mikhail Gorbachev, there was no strong pressure within Belarus to secede immediately from the USSR, although there were movements favoring a multi-party political system, as well as protest demonstrations against high food prices.

[22] Belarus was the region most affected by the 1986 catastrophe at the Chernobyl, Ukraine nuclear plant. According to foreign researchers, in the years following the accident there was a rapid increase in cases of cancer, leukemia, and birth defects in the areas affected by the disaster: Ukraine, Belarus and western Russia.

[23] A report of the UN International Atomic Energy Agency (IAEA) released in May 1991 stated that some health problems attributed to the accident were nervous disorders and not the result of exposure to atomic radiation. The IAEA study added that safety measures adopted by Soviet authorities went beyond international guidelines.

[24] In June 1991 Belarus declared independence and in October signed an economic integration agreement with Kazakhstan and Uzbekistan. On December 8 1991 the presidents of the Russian Federation and Ukraine, together with the president of the Belarusian Parliament, signed a historic agreement ending the USSR.

[25] Opposing President Gorbachev's idea of signing a new Union Treaty, the presidents decided to found an association of sovereign states. On the 21st of that same month, in Alma-Ata (Kazakhstan), 11 republics signed an accord creating the Commonwealth of Independent States (CIS) whose members requested admission to the UN as separate countries.

[26] At the end of 1991, Belarus was among the eight CIS republics that assumed their share of responsibility for the former USSR's foreign debt, and agreed to submit to the demands of their creditors and the International Monetary Fund, in order to continue being eligible for loans from western banks. The CIS countries agreed to reduce their budget deficits, lift price controls, liberalize currency exchange, and carry out the structural reforms necessary to achieving a market economy.

[27] In February 1992, Russia began to withdraw its nuclear weapons. A situation developed between Sushkevich, who defended the need to maintain economic links exclusively with Russia, and Prime Minister Kebich, who was opposed to this position.

[28] In September 1993 the opposition between the president of the Parliament and the prime minister reached such an intensity that there were rumors of a coup. In January 1994, Sushkevich was replaced by Mecheslav Grib. In March, the Parliament approved the new constitution, which replaced the 1977 one, formulated when Belarus was still part of the Soviet Union.

[29] The country became a presidential Republic with a Parliament of 260 members and Alexander Lukashenka became prime minister. Lukashenka won only 45 per cent of the votes in the first round on June 23 1994, followed by Kebich, with 17 per cent. On July 10 Lukashenka, after a campaign in which he attacked and denounced the "corruption" of his predecessors, obtained 80 per cent of the votes of the second round.

[30] Despite having criticized it before the elections, the leader followed a policy of rapprochement towards Moscow, and agreed to the currency unification with Russia, which took place in April 1994. In August he met the Russian president, Boris Yeltsin, and in November decreed a state of emergency with the avowed purpose of stopping inflation.

[31] In February 1995 Minsk and Moscow signed a treaty of friendship and cooperation, which allows Russia to continue its deployment of military forces in Belarusian territory. In the parliamentary elections in May, whose date Lukashenka arranged in order to coincide with the referendum concerning his policy of rapprochement with Moscow, the low participation did not allow all the members of parliament to be chosen, determining a new partial election to take place in November 1995.

[32] During the first months of 1996 the opposition to Lukashenka and his "pro-Russian" policy, as well as the pressure he exerted against opposing political groups, such as the Popular Front, incited strong demonstrations. In March, at the commemoration of the anniversary of the independent Republic created in Belarus in 1917, tens of thousands demonstrated against the union with Russia.

Belgium

België/Belgique

Population: 10,131,000 (1995)
Area: 30,528 SQ KM
Capital: Brussels (Bruxelles/Brussel)
Currency: Belgian Franc
Language: French, Flemish and German

B elgium, Holland, Luxembourg and a part of northern France make up the Low Countries, which had a common history until 1579 (see the Netherlands). The linguistic separation which took place between the Roman and Germanic languages coincided with the borders of the Holy Roman Empire, which divided the Low Countries in two.

[2] In 1579, the creation of the Union of Arras, achieved by joining together the Catholic provinces of Artois and Hainaut, led Spain to resume its war against the Dutch Protestants. According to the terms of the Treaty of Utrecht, signed in 1713, the southern provinces and Luxembourg were left in the hands of Charles VI, head of the Holy Roman Empire and of the Austrian branch of the Hapsburgs.

[3] The region's economy was based on the production of linen and textiles. Industrialization was facilitated by the fact that manufacturers and landowners were often one and the same, and that textile mills were concentrated in the hands of a few owners. Factories sprang up in Ghent, Antwerp, and Tournai, with over 100 low-paid workers.

[4] Fearing the spread of the French Revolution, the House of Austria attempted some progressive reforms, but the conservative revolt of 1789 put an end to these. After French Republican troops invaded in 1794, France did away with the region's autonomy and the privileges of the local aristocracy, while encouraging the industrial revolution. After the fall of Napoleon, in 1814, the European powers enforced reunification with the north. The southern provinces had already forged an identity of their own, and were unwilling to accept Dutch authority.

[5] In 1830, the Belgian bourgeoisie took up arms against the Dutch authorities. When the conflict spread and proved substantial, European powers recognized the independence of the southern provinces, which were henceforth known as Belgium. Congress adopted a parliamentary monarchy, with an electing body made up of property owners. The duchies of Luxembourg and Limburg were divided between Belgium and Holland.

[6] By the end of the 19th century, workers forced the government to pass laws aimed at providing housing for working class people, and improving working conditions, especially for women and children. At the same time, Parliament changed the Constitution and in 1893, established limited male suffrage.

[7] Between 1880 and 1885, Leopold II financed international expeditions to the Congo, making it effectively his 'private' colony. Economic mismanagement and protests from several European nations against the severe repression and exploitation occurring in the Congo, forced the Belgian government to take it as a colony in 1908. With this, the worst excesses ended, but the Belgian government, businesses and church increased their influence over the following years.

[8] During World War I, Belgium was invaded by Germany. The Treaty of Versailles returned the territories of Eupen and Malmedy to Belgium. In 1920 Belgium signed a military assistance treaty with France, and the following year, it formed an economic alliance with Luxembourg. In Africa, Belgium colonized the former German colonies of Rwanda and Burundi after the League of Nations granted them a mandate.

[9] After the war, the government established universal male suffrage. Parliament approved an eight-hour workday, a graduated income tax, and a retirement pension plan. From 1930, increased pressure was put on the government to tackle discrimination against the Flemish people and the Dutch language. This resulted in several "linguistic laws" and two regions: Flanders and Wallonia. In the former, Dutch is the official language, while in the latter it is French.

[10] During World War II Belgium remained neutral, but it was occupied by Germany in 1940, until liberation in 1944. The return of the king, held prisoner by the Germans, unleashed a major controversy. In a plebiscite, 57 per cent of the population voted in favor of his return to the throne, but tension in Wallonia forced Leopold III to abdicate in favor of his son Baudouin (1950).

[11] In 1947, Belgium, the Netherlands, and Luxembourg formed an economic association known as the Benelux, still extant within the European Union. Belgium also became a member of NATO in 1949. In 1960 the Belgian Congo became independent, but Belgium and the western powers continued to intervene in the former colony (now Zaire). In 1962 Rwanda and Burundi became independent.

[12] Female suffrage was granted in 1949. In 1975, Belgian women gained the right to equal pay. During the 1960s and 1970s the questions of language and the autonomy of the regions became increasingly important. In 1980, a new federal structure was approved by Parliament, made up of Flanders, Wallonia, and Brussels as the capital district. Wallonia and Flanders have been autonomous economic and cultural entities since 1970.

[13] In 1983, the installation of NATO atomic missiles on Belgian soil unleashed a national controversy. The missiles were withdrawn in 1988, after an arms reduction agreement was signed between the United States and the USSR.

[14] In April 1990, despite the opposition of Christian Democrats within the coalition government, Parliament approved an abortion law. In order to avoid having to sign the proposed legislation, King Baudouin found it necessary to step down temporarily from the throne.

[15] The largest Belgian transnational corporation is Petrofina, with 23,800 workers, annual sales of $17.3 billion, and branches in Europe and Africa. Solvay follows in second place, with 45,600 workers, annual sales of $7.6 billion and subsidiaries in countries like Brazil and Mexico.

[16] Between 1983 and 1988, Belgium had an average unemployment rate of 11.3 per cent, the third highest in Western Europe, after Spain and Ireland. The economic situation improved in 1989, but after falling significantly in 1990, unemployment began to rise again in 1991. In October 1996 it stood at 13.8 per cent.

[17] In the November 1991 elections, the "Vlaams Blok" ("Flemish Bloc"), a xenophobic group which had geared its campaign around the expulsion of foreign residents, obtained 6.6 per cent of the vote and increased its seats in the house of commons from 2 to 12. The Green parties went from 9 to 17 seats (9.9 per cent of the vote). The gains made by the far right and the Greens marked an important change in Belgian politics.

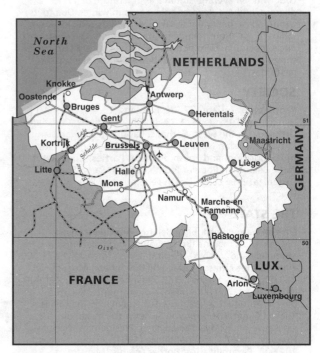

LAND USE

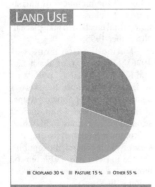

CROPLAND 30 % PASTURE 15 % OTHER 55 %

Foreign Trade

MILLIONS $ 1994

IMPORTS
125,762

EXPORTS
137,394

ENVIRONMENT

Northwestern Belgium is a lowland, the Plains of Flanders, composed of sand and clay deposited by its rivers. In southern Belgium the southern highlands rise to 700 m on the Ardennes Plateau. One of the most densely populated European countries, its prosperity rests on trade, helped by its geography and by the transport network covering the northern plains, converging at the port of Antwerp. It has highly intensive agriculture, and fosters a major industrial center. Heavy industry was located near the coal fields of the Sambre-Meuse valley, and textiles were traditionally concentrated in Flanders, where new industries have developed since the late 1960s. Water protection laws were approved in 1971. However, the Meuse River, supplying 5 million people with drinking water, is contaminated by industrial waste. Pollution, by the excessive use of both animal wastes and chemical fertilizers, has increased the nitrate concentration in many rivers and accounts for the presence of algae blooms. Belgian smokestack industries contribute to air pollution in Europe, and are responsible for acid rain in neighboring countries.

SOCIETY

Peoples: The country's two major language-based groups are the Flemish (55%) and the Walloons (44%). There is also a German minority (0.69%). Many industrial workers are immigrants (Italian, Moroccan and, in lesser numbers, Turkish and African). Foreigners represent about 9% of the whole population. **Religions:** Mainly Catholic. **Languages:** French, Dutch and German are the official languages. French is the main language spoken in the south and east, and Dutch in the north and west, while German is spoken by 0.69% of the population. Brussels is bilingual. **Political Parties:** Flemish Christian People's Party (CVP); French-speaking Social Christians (PSC); French-speaking Socialists (PS); Flemish Socialists (SP) Flemish Liberal Party (VLD); French-speaking Liberal Party (PRL); Green parties (AGALEV/ECOLO); Flemish Nationalists (VU). **Social Organizations:** The General Labor Federation of Belgium (ABVV/FGTB), Confederation of Christian Labor Unions of Belgium (ACV/CSC).

THE STATE

Official Name: Royaume de Belgique/ Koninkrijk België. **Capital:** Brussels (Bruxelles/Brussel) 949,000 inhab. (1994). Brussels is also the capital of the European Union. **Other cities:** Antwerp 463,000 inhab.; Liège, 196,000 inhab.; Ghent 228,000 inhab. (1994). **Government:** Parliamentary monarchy. The Head of State is King Albert II, since August 1, 1993. Prime Minister, Jean-Luc Dehaene, Christian Democrat (CVP), since March 1992. **National Holiday:** July 21, National Day (ascension of King Leopold to the throne); (1831). **Armed Forces:** 53,000. Conscription was abolished in 1994 and there is now only the professional army.

DEMOGRAPHY

Urban: 97% (1995). **Annual growth:** 0.36% (1991-99). **Estimate for year 2000:** 10,280,000. **Children per woman:** 1.6 (1992).

HEALTH

Under-five mortality: 11.5 per 1,000 (1995). **Calorie consumption:** 120% of required intake (1995).

EDUCATION

School enrolment: Primary (1993): 100% fem., 100% male. Secondary (1993): 104% fem., 103% male. **Primary school teachers:** one for every 13 students (1992).

COMMUNICATIONS

103 **newspapers** (1995), 105 **TV sets** (1995) and 104 **radio sets** per 1,000 households (1995). 43.7 **telephones** per 100 inhabitants (1993). **Books:** 13,913 new titles in 1991.

ECONOMY

Per capita GNP: $22,870 (1994). **Annual growth:** 2.30% (1985-94). **Annual inflation:** 3.20% (1984-94). **Consumer price index:** 100 in 1988; 120.3 in 1995. **Currency:** 32 Belgian francs = 1$ (1994). **Cereal imports:** 5,308,000 metric tons (1993). **Fertilizer use:** 4,246 kgs per ha. (1992-93). **Imports:** $125,762 million (1994). **Exports:** $137,394 million (1994).

ENERGY

Consumption: 5,091 kgs of Oil Equivalent per capita yearly (1994), 77% imported (1994).

He was succeeded by his brother Albert, who was crowned as Albert II. At the same time, Dehaene began negotiations with unions to produce a "comprehensive plan" against unemployment, which reached 14.1 per cent rate in September 1994. Charges of corruption led three leading socialist politicians to resign from their positions as ministers.

[20] The 1994 European elections reflected the population's lack of confidence towards the governing coalition. French-speaking socialists went from 38.5 per cent of the vote in 1989 to 30.4 per cent in 1994. In the municipal elections of October, the socialists lost votes once again, while the Vlaams Blok became the second party in Antwerp, by winning 18 out of the 55 seats at stake on the city council.

[21] On May 21 1995, Belgium organized its first legislative elections after the approval of the Constitution that turned the country into a federal state. Together with the members of parliament, voters chose for the first time 75 members for the three new regional assemblies of Brussels, Flanders and Wallonia. The government alliance led by Premier Dehaene attained a clear victory in the national elections, allowing him to go on as head of government.

[22] The 1996 budget, which attempted to reduce the fiscal deficit to 3 per cent of GDP in line with the conditions of the Maastricht Treaty for the political and financial unification of the EU, was the target of heated discussions between the government and the unions, which considered the curbs on public spending excessive.

[23] In autumn 1996, after the discovery of the bodies of 4 missing children, a wave of protest arose against the ineffective administration of justice. Names of ministers, judges and top industrialists were leaked to the press. In October one of the biggest-ever demonstrations took place in Brussels. Belgium went into the deepest constitutional crisis since the postwar controversy over the return of King Leopold III.

[18] Wilfried Martens, who had been premier for 10 years, was succeeded by his party colleague Jean-Luc Dehaene, who formed a new government with Christian Democratrs and socialists in March 1992. Dehaene immediately announced a plan to reduce the fiscal deficit, including drastic curbs on public spending.

[19] On July 31 1993, King Baudouin died after a reign of 42 years.

Belize

Population: 211,000 (1994)
Area: 22,960 SQ KM
Capital: Belmopan

Currency: Belize Dollar
Language: English

Belize

A native American people known as the Itzae were the original occupants of what is now Belize (formerly British Honduras). Belize, together with Guatemala and southern Mexico, formed part of the Mayan empire. In Belize, the Mayas built the cities of Lubaatún, Pusilhá, and a third which archeologists call San José (its original name is unknown).

2 In 1504 Columbus sailed into the bay naming it the Bay of Honduras. Spain was nominally the colonial power in the region, but it never pushed further into Belize, because it encountered tough resistance from the natives. According to the terms of the Treaty of Paris (1763) Spain allowed the British to start exploiting timber in the area. This authorization was later confirmed in the Treaty of Versailles (1783). From the Captaincy General of Yucatan (now Mexico) the Spaniards tried on several occasions to drive out the British, many of whom were involved in piracy. In 1798 the British gained control of the colony, although Spain retained sovereignty over it until it became a British colony in 1862. As of 1871 the Crown took charge of local government with the territory being administered by the governor of Jamaica until 1884.

3 By the second half of the 17th century, some British entrepreneurs attracted in particular by cedar, campeche wood, and logwood, started establishing themselves in the depopulated coastal areas, importing African slaves to work their estates. Shortly afterwards, slaves outnumbered Europeans, and in 1784 only 10 per cent of the population was of European extraction, a ratio which has been gradually decreasing until the present.

4 In 1950 the People's United Party (PUP) was founded, led by George Price. First organized as a "people's committee" to fight against arbitrary treatment by the colonial administration, the PUP won its first elections by a landslide majority. In 1954 universal suffrage was granted, and the majority of the legislative representatives were elected directly. In 1961 a ministerial system of government was established, and in 1964 the country was granted internal autonomy, with George Price becoming Prime Minister. On June 1, 1973 the country changed its name from British Honduras to Belize.

5 Guatemala claimed to have inherited sovereignty over Belize from Spain, and did not recognize the Guatemala-Belize border. In March 1981, Guatemala and Britain signed a 16-point agreement. Britain assured the future independence of Belize in exchange for some concessions to the Guatemalan regime, such as free and permanent access to the Atlantic, joint exploitation of the marine resources, the building of a pipeline, and an "anti-terrorist" agreement.

6 Tourism brought in $4 million in 1983, but even if this source of income were to expand rapidly, it is unlikely that it would reach the $36 million earned by sugar.

7 Price and the PUP were accused of partiality towards Cuba and Nicaragua, an issue exploited by the right-wing opposition, the United Democratic Party (UDP), which won 21 out of 28 seats in the December 1984 elections.

8 The new Prime Minister, Manuel Esquivel, a US-educated physics professor, adopted a liberal economic policy and supported the private import-export sector, which was in the hands of inexperienced family business ventures. He also encouraged foreign investment and sought to attract US, Jamaican, and Mexican investments in tourism, energy, and agriculture. The previous government had bequeathed Belize one of the most liberal Third World legislations concerning foreign investment.

9 Sugar-cane production generates 50 per cent of the country's revenue, but a fall in international prices badly affected the economy. While the market value stood at 4 cents per pound, a local subsidiary of the British company, Tate and Lyle, claimed its production costs reached 11 cents. The industry only survives because of import quotas guaranteed by the US and EEC markets, which take 60 per cent of the sugar output, with the remainder being sold at a loss.

10 In March 1986, Prime Minister Esquivel proposed a plan for the sale of Belizean citizenship, aimed primarily at Hong Kong businessmen (as Hong Kong will revert to China in 1997). Anyone investing $25,000 in government bonds - really worth only $12,500 - will immediately be granted citizenship in Belize.

11 Meanwhile the government tried to resettle the thousands of illegal refugees living on the Guatemalan border, in order to legalize their status. During the 1980s, the country received approximately 40,000 Salvadoran, Guatemalan, Honduran and Nicaraguan refugees.

12 In spite of official tolerance some officials have started to blame immigrants for the dramatic rise in marijuana trafficking and crime.

13 When Vinicio Cerezo became president of Guatemala, relations between the countries changed substantially. In December 1986, Cerezo's government re-established diplomatic relations with Britain, broken over two decades before, due to Guatemalan claims to Belizean territory. A Permanent Joint Commission was formed, with Belizean, Guatemalan and British representatives, to find a peaceful solution to the issue.

14 In spite of the existence of the commission, Belize fears a Guatemalan invasion and keeps a standing army of 1,800 British troops. However Esquivel declared in Mexico that he "would not allow the installation of US military bases", stressing that he did not wish to become involved in the Central American crisis.

15 However, the number of US Embassy personnel has grown six-fold since independence, and the number of Peace Corps volunteers is ten times higher. Dean Barrow, Minister of Foreign Affairs and Economic Development, admitted that he was aware of the country's dependence upon the US. But he also stated that Belize can defend its territorial integrity through the Non-Aligned Movement.

16 Drug-trafficking, or the sale of "Belize's breeze", as marijuana is known, has shown spectacular growth. According to some foreign economists, it has become the country's main export, and it is estimated that some 700 tons, worth at least $100 million have been brought into the US, where the resale price is at least ten times this sum. The area under cultivation has increased by at least 20 per cent, in spite of a US-sponsored herbicide spreading campaign carried out with Mexican helicopters.

17 In the September 6 1989 general elections George Price was re-

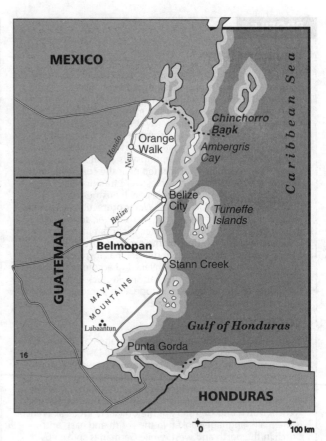

ENVIRONMENT

Belize covers the southeastern tip of the Yucatan Peninsula. The land is low, and the climate warm and rainy in the north. In southern Belize, the hillsides sustain a variety of crops. The northern coastline is marshy and flanked by low islands. In the south there are excellent natural harbors between reefs. Significant oil and gas deposits are believed to exist off the coast. The country shares many of the problems of the Caribbean region, including the deterioration of its water-quality and soil erosion. Close to the cities, the surface of the water is covered with residual wastes and by-products of sugar cane production. Industrial waste, dumped in the water, has generated public health problems and killed fish stocks.

SOCIETY

Peoples: The historical majority, the English-speaking population of African origin (37%), have been outnumbered by Spanish-speaking mestizos (44%). 11% of the population are of Mayan descent. There are also European, Indian, Chinese, and Arab minority groups (1991). **Religions:** 58% are Catholic, most of the rest are Protestant (34%). **Languages:** English (official); Spanish; Mayan dialects are also spoken. **Political Parties:** The United Democratic Party (UDP), led by Manuel Esquivel. The People's United Party (PUP), led by Prime Minister George Price, founded in 1950. The National Alliance for the Rights of Belize, founded in 1992 by members of the UDP opposed to compromise with Guatemala. **Social Organizations:** The General Workers' Union, the Christian Workers' Union (CWU) and the Public Service Union of Belize.

THE STATE

Official Name: Belize. **Administrative divisions:** 6 districts. **Capital:** Belmopan 3,800 inhab. (est. 1993). **Other cities:** Belize City 47,700; Orange Walk 11,900 inhab.; San Ignacio/ Santa Elena 9,700 inhab. (1993). **Government:** Parliamentary Monarchy, Queen Elizabeth II of England is Head of State. Colville Young, Governor-General since 1995. Manuel Esquivel, Prime Minister since June 1993. Legislative power lies with the House of Representatives, which has 28 members and is elected by universal suffrage, and an 8-member Senate appointed by the Governor General. **National Holidays:** September 10, National Day; September 21, Independence Day (1981).

DEMOGRAPHY

Annual growth: -0.1% (1992-2000). **Children per woman:** 4.5 (1992).

HEALTH

Calorie consumption: 77.8% of required intake (1988-90).

EDUCATION

Primary school teachers: one for every 25 students (1989).

COMMUNICATIONS

14.0 **telephones** per 100 inhabitants (1993).

ECONOMY

Per capita GNP: $2,530 (1994).
Annual growth: 5.00% (1985-94).
Consumer price index: 100 in 1990; 106.3 in 1994.
Currency: 2 Belize dollars = 1$ (1994).

[20] A 1992 census revealed that the Spanish-speaking mestizo population had outgrown the English-speaking population of African origin who, together with the European descendants, had historically been the dominant group.

[21] Banana, sugar cane and citrus production has increased, representing 40 per cent of the country's

> In spite of official tolerance some officials have started to blame immigrants for the dramatic rise in marijuana trafficking and crime.

total production and four-fifths of its exports. Tourism has become the sector with the greatest potential. Several environmental organizations have cooperated in protecting nature reserves and attempts to improve the quality of the Caribbean waters. Recent Mayan archeological discoveries have attracted many visitors.

[22] Hours after President Serrano of Guatemala was ousted, on June 1 1993, Price brought the election date forward 15 months, counting on a new PUP victory to ratify agreements made with this neighbor. His opponent Manuel Esquivel questioned the validity of Price's concessions to Guatemala, and proposed that these be legitimized by a referendum.

[23] In the June 1993 elections the PUP was pushed out by the UDP, led by Esquivel and dominated by the mestizos. In March 1994, the UDP also won the elections for the nation's 7 local councils. The government formed a group of economic advisors to look into establishing facilities for the capital and tourist markets, and the development of one or more free zones. The fear of new territorial demands from Guatemala and the withdrawal of British troops led to an increase in defence spending and new recruitment for the armed forces.

[24] In January 1995 Prime Minister Esquivel reshuffled his cabinet in an attempt to control the recession and the crime wave. The government founded a committee to renew the economic citizenship programme closed in June 1994, in order to attract investors and reduce the foreign debt. On another front it froze public sector salaries for the 95/96 period to reduce the fiscal deficit.

turned to power when the PUP won 15 of the 28 seats. The 13 remaining seats went to Esquivel's UDP.

[18] In September 1991 Guatemalan president Jorge Serrano Elias officially recognized Belize's sovereignty and right to self-determination, 10 years after the country had unilaterally declared its independence. The Government of Belize, in turn, gave Guatemala free access to the Gulf of Honduras, thereby reducing its own territorial waters.

[19] Within the past few years, a large number of Haitians have come to Belize as agricultural laborers, attracted by higher wages than at home. Lower inflation rates and more equal income distribution were incentives for immigration; immigrants also had access to health and education services. The social services consequently became overloaded, triggering a backlash from the Belizeans, who demanded repatriation measures.

Benin

Population: 5,325,000 (1994)
Area: 112,620 SQ KM
Capital: Porto Novo

Currency: CFA Franc
Language: French

Benin

Benin (known as Dahomey until 1975) is considered one of the poorest countries in the world. It lies in the region of the Yoruban culture, which developed at the ancient city of Ife. It was here that the Ewe peoples, who stemmed from the same linguistic family, developed into two distinct kingdoms during the 17th century: the Hogbonu (today known as Porto Novo) and the better-known Abomey, further inland. These states developed around the booming slave trade, serving as intermediaries.

[2] The traditional rulers of Abomey, the Fon, built a centralized state that extended east and west beyond Benin's present-day frontiers. A well-disciplined army, with European rifles and a large contingent of female soldiers, enabled them to end the tutelage of the Alafin of Oyo (Nigeria) and capture various Yoruban cities. After the 17th century, Ouidah became the main port for British, French and Portuguese slave traders receiving their human cargo.

[3] The ruling group of Abomey suffered a setback in 1818 when Britain banned the slave trade, although Ghezo, who ruled between 1818 and 1856, maintained a thriving clandestine traffic with Brazil and Cuba. He also promoted the development of agriculture and established a strict state monopoly on foreign trade.

[4] In 1889, Ghezo's grandson Benhanzin inherited a prosperous state, but one already threatened by colonialism. In 1891, Fon troops resisted the French invasion only to be defeated a year later. The king and his army retreated to the forests, where they held out until 1894. Benhanzin, who became a symbol of anti-colonial resistance, died in exile in Martinique in 1906.

[5] The colonists destroyed the centralized political structure of the ancient Fon state. Traditional Fon society was dismantled and replaced with a system based on the exploitation of farm labor. The French also declared a monopoly on the palm-oil trade and ruined families that had resisted foreign penetration for nearly a century.

[6] By the beginning of the 20th century, the colony of Dahomey (as the French called it), was no longer self-sufficient. When it gained independence in August 1960, oilseed exports stood at 1850 levels, while the population had tripled.

> In 1980, a new Revolutionary Assembly was elected through direct vote. The government switched to a more pragmatic foreign policy and diplomatic relations with France were resumed.

[7] Independence came as a direct consequence both of France's weakness by the end of World War II and activities of European-educated nationalists, led by Louis Hunkanrin, who waged a stubborn 20-year struggle against compulsory labor imposed by the French. All forms of political organization were banned and in retaliation, Hunkanrin created the Human Rights League. A period of ruthless repression followed: hundreds of villages were burned down, nearly 5,000 people were killed, and Hunkanrin took refuge in Mauritania.

[8] By 1960, Dahomey had become an unbearable economic burden, and France agreed to independence. The new government inherited a bankrupt economy and a corrupt infrastructure. A series of 12 military and civilian governments marked a 16-year period of instability.

[9] The neocolonial elite collapsed when then-Major Mathieu Kerekou headed a coup by a group of young officers opposed to political corruption and official despotism. Two years later, a Marxist-Leninist state was proclaimed and its name changed to Benin, with a communitarian political and economic system: all foreign property was nationalized, and a single-party system was introduced with the creation of the People's Revolutionary Party.

[10] The revolutionary government became the target of several conspiracies plotted abroad. There was an unsuccessful invasion in January 1977 with the participation of French mercenaries and the support of Gabon and Morocco.

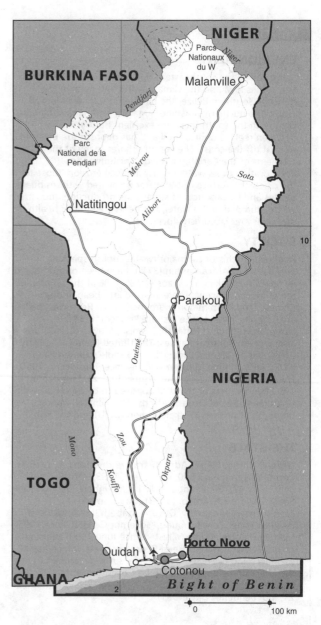

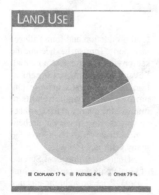

WORKERS
1994
65% UNEMPLOYMENT

MEN 53%   WOMEN 47%

AGRICULTURE 70%   INDUSTRY 7%   SERVICES 23%

LAND USE

CROPLAND 17 %   PASTURE 4 %   OTHER 79 %

Maternal Mortality
1989-95

PER 100,000 LIVE BIRTHS
2,500

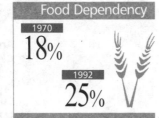

Food Dependency
1970
18%
1992
25%

External Debt
1994

PER CAPITA
$304

Illiteracy
1995

63%

PROFILE

ENVIRONMENT

Benin is a narrow strip of land which extends north from the Gulf of Guinea. Its 120km sandy coast lacks natural ports. Several natural regions cut across the country from south to north: the coastal belt where oil palms are cultivated; the tropical wooded lowlands; and the plateau which rises gradually towards the headwaters of the Queme, Mekrou, Alibori and Pendjari rivers, in a region of tropical hills. One of the major environmental problems is that of desertification, which has been aggravated in recent years by a significant decrease in rainfall.

SOCIETY

Peoples: The Beninese stem from 60 ethnic groups. The Fon (47%), Adja, Yoruba and Bariba groups are the most numerous and before French colonization they had already developed very stable political institutions. **Religions:** 70% of the Beninese practice traditional African religions, 15% are Muslim and 15% Catholic. **Languages:** French (official). Other widely spoken languages are Fon, Fulani, Mine, Yoruba and Massi. **Political parties:** Democratic Renewal Party (PRD); Benin Renaissance Party (RPB). **Social organizations:** The Benin Workers' National Trade Union (UNSTB).

THE STATE

Official name: République Populaire du Bénin. **Capital:** Porto Novo, 177,000 inhab. (1992). **Other cities:** Cotonou 533,000 inhab.; Djougou 132,000 inhab.; Abomey-Calavi 125,000 inhab.; Parakou 106,000 inhab. (1992). **Government:** Presidential republic. Nicéphore Soglo, President. Single chamber parliament. **National Holidays:** July 1, Independence (1960); November 30, Revolution Day (1974). **Armed Forces:** 4,800. **Paramilitaries:** Gendarmerie: 2,000; People's Militia: 1,500-2,000.

STATISTICS

DEMOGRAPHY

Urban: 30% (1995). **Annual growth:** 1.5% (1992-2000). **Estimate for year 2000:** 6,000,000. **Children per woman:** 6.2 (1992).

HEALTH

One **physician** for every 14,286 inhab. (1988-91). **Calorie consumption:** 93% of required intake (1995). **Safe water:** 50% of the population has access (1990-95).

EDUCATION

Illiteracy: 63% (1995). **School enrolment:** Primary (1993): 44% fem., 44% male. Secondary (1993): 7% fem., 17% male. **Primary school teachers:** one for every 35 students (1990).

COMMUNICATIONS

82 **newspapers** (1995), 83 **TV sets** (1995) and 85 **radio sets** per 1,000 households (1995). 0.4 **telephones** per 100 inhabitants (1993). **Books:** 81 new titles per 1,000,000 inhabitants in 1995.

ECONOMY

Per capita GNP: $370 (1994). **Annual growth:** -0.80% (1985-94). **Annual inflation:** 2.90% (1984-94). **Currency:** 535 CFA francs = 1$ (1994). **Cereal imports:** 134,000 metric tons (1993). **Fertilizer use:** 82 kgs per ha. (1992-93). **External debt:** $1,619 million (1994), $304 per capita (1994). **Debt service:** 10.1% of exports (1994). **Development aid received:** $267 million (1993); $52 per capita; 13% of GNP.

ENERGY

Consumption: 18 kgs of Oil Equivalent per capita yearly (1994), -239% imported (1994).

[11] In 1980, a new Revolutionary Assembly was elected through direct vote. The government switched to a more pragmatic foreign policy and diplomatic relations with France were resumed. Although palm-oil production continued to drop as the trees grew older, cotton and sugar sales rose. High unemployment continued, but in 1982 offshore oil was discovered thus guaranteeing energy self-sufficiency. In addition large phosphate deposits were discovered in the Mekrou region.

[12] Hopes of recovery dimmed as drought reached the northern provinces. Desertification was exacerbated, and the region was unable to supply its own food even in traditional subsistence crops such as manioc, yams, corn and sorghum.

[13] To make matters worse, Nigeria cracked down on immigrants, expelling thousands back to Benin and closing the border.

[14] The economic crisis forced the government to accept the terms of the International Monetary Fund, including a 10 per cent income tax and a 50 per cent reduction in non-wage social benefits. In January 1989 students, teachers and other educational bodies staged a strike demanding back pay and scholarship payments which had not been honored.

[15] On December 8 1989, disappointed with results and besieged by street demonstrations, President Kerekou announced that he was abandoning his Marxist-Leninism. A new constitution was drawn up, providing for a series of political and economic reforms, especially the promotion of free enterprise.

[16] On March 24 1991 Prime Minister Nicéphore Soglo defeated President Kerekou with 68 per cent of the votes in the country's first presidential election in 30 years. In 1992, former president Kerekou, prosecuted for his activities following the 1972 coup d'etat, was granted an amnesty, and political prisoners were released.

[17] Soglo continued the economic liberalization and privatization policy initiated by Kerekou in 1986. Debt servicing still represented a high percentage of the resources annually obtained (in 1992 debt service was equivalent to 27 per cent of the country's income).

[18] The 100 per cent devaluation of the CFA franc decreed by France in January 1994 provoked contradictory effects in Benin's economy. GDP continued to grow at a 4 per cent annual rate and cotton exports increased. However, public expenditure retrenchment - aimed among other things at curbing inflation after the devaluation - brought about drastic cuts in social spending schemes.

[19] In July 1994, Soglo assumed the leadership of the Renaissance Party of Benin (RPB), founded by his wife Rosine in 1992. The RPB was defeated in the legislative and municipal elections of March 1995 by the opposition Democratic Renewal Party (DRP). In May 1995 the government banned food exports to reduce local sale prices and curb inflation, at the same time that it refused to increase salaries.

[20] In December 1995 the Summit of French-speaking countries, held in Cotonou, once again noted that Benin is "an example of democratic openness" in the region.

Bermuda

Population:	63,000 (1994)
Area:	50 SQ KM
Capital:	Hamilton
Currency:	Bermuda Dollar
Language:	English

Bermuda

The Bermuda archipelago was the first colony of the British Empire. Sighted by the Spanish navigator Juan Bermúdez in 1503, it was settled in 1609 when British emigrants on their way to America were shipwrecked nearby.

[2] From 1612, the colony welcomed religious and political dissidents. In 1684, it began to be administrated by the British Crown, and the first parliament was installed. Even though African slaves formed the majority, only plantation owners could elect representatives.

[3] Agriculture almost disappeared in the 20th century, being replaced by tourism, gambling and transnational corporations lured by numerous tax exemptions. During the Prohibition Laws (1919-1933) traffickers living in Bermuda smuggled rum into the United States. Since 1941, Washington set up air and naval bases on the islands and completely replaced British military presence since 1957.

[4] After the creation of the Bermuda Industrial Union (BIU), in 1963 workers founded the Progessive Labor Party (PLP), which favors the country's total independence and the introduction of an income tax. The following year, the Right organized the United Bermuda Party (UBP).

[5] In 1968, Britain granted the islands greater administrative autonomy, and the majority party was given the right to name the prime minister. The period leading up to the elections was marked by racial and political violence. The assassi-

nation of the governor led to intervention by British troops. When the election was held, the UBP won by a large margin.

[6] In the 1976 election, the PLP increased its number of seats. From its position in opposition, the PLP continued to demand greater autonomy for the island.

[7] In 1977, two members of the "Black Cadre" were sentenced to death, for participating in armed anticolonial activities. The execution unleashed a wave of protests, with British troops intervening once again. The Minister of Communal (Race) Relations was withdrawn from his post, and his duties were delegated to the Bermuda Regiment, responsible for putting down protest and discontent.

[8] In spite of charges of racism, the European population supported the government, while the black population opted for the PLP or more radical organizations.

[9] The 1979 Bermuda Constitutional Convention, failed to reach a consensus on representation and the minimum voting age. However, a reduction in the number of non-Bermudan voters went into effect in that December.

[10] The UBP won both the 1980 and 1983 elections, although the opposition increased its representation. With the UBP in power, the country continues without self-determination as the government claims that the majority is against this.

[11] The islands' strategic location along North Atlantic routes explains the current presence of British, Canadian and US troops.

[12] In 1989 Prime Minister John Swan of the UBP was re-elected and his party gained 23 seats . The PLP obtained 15 seats; one seat went to the National Liberal Party (NLP), a center party, and the Environmental Party won one seat.

[13] In 1989 and 1990, unemployment rose by 0.5 per cent and 2 per cent, respectively. Employment within the industrial sector was 3.1 per cent. The US recession continued to affect tourism in the first few months of 1992. London rejected plans for independence.

[14] In 1992, Sir John Waddington was nominated Governor-General.

[15] Intense drug trafficking produced a massive diversion of investment, in 1993. That year, measures also went into effect to control offshore banking operations.

[16] In the October 5, 1993 elections, J. Irving Pearman was elected vice-premier. The UBP retained its majority at the Assembly with 22 seats while the PLP, led by Frederick Wade, confirmed its growth with 18 seats.

[17] In August 1995, a plebiscite was held regarding the prime minister's proposal on independence. The result was negative and so Bermuda will continue being a British colony.

Bhutan

Population: 675,000 (1994)
Area: 47,000 SQKM
Capital: Thimbu

Currency: Ngultrum
Language: Dzongkha

Druk Yul

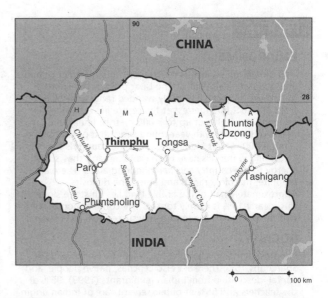

Bhutan lies in the heart of the vast Himalayan mountains. Early explorers and envoys of the British colonial government called it Bootan, land of the Booteas, or sometimes Bhotan. Wedged between giant neighbors, China and India, and secluded by some of the world's highest peaks, it was little known to the rest of the world. Even today, the origin of its name remains unknown. Perhaps it came from the Sanskrit Bhot ante ("the end of Tibet") or from another Sanskrit word Bhu'nthan ("high land"). To the people of Bhutan it is Druk or Druk Yul, land of the Thunder Dragon. It was here that the Drukpa Kargyud sect of Buddhism flourished in the 7th century.

[2] From the 12th century, the Drukpa Kargyud tradition became dominant. After a long period of rivalry among various groups, the country was united by a Drukpa Kargyud Lama named Ngawang Namgyal in the 17th century. Druk, the country's original name, is derived from the Kargyud Sect of Mahayana Buddhism (Drukpa), now the state religion. Namgyal, popularly known as Shabdrung ("at whose feet one submits"), was both the country's spiritual and secular ruler. Factionalism gradually eroded the power of the subsequent Shabdrungs. On December 17 1907 Ugen Wangchuck united the country and established Bhutan's first hereditary monarchy.

[3] The British colonial administration in India signed important treaties with Bhutan in 1774 and 1865. The 1910 Treaty of Punakha stipulated that the British would not interfere in Bhutan's internal affairs, but made the country a British protectorate in terms of external relations. Similar provisions were included in the 1949 treaty signed between Bhutan and independent India.

[4] Bhutan emerged from its self-imposed isolation in the 1960s, by joining the Colombo Plan in 1962 and the Universal Postal Union in 1969. A Department of Foreign Affairs established in 1970 was upgraded to ministerial rank in 1971. That same year, Bhutan became a member of the United Nations, opened a diplomatic mission in New Delhi and a permanent mission to the UN in New York; it also became a member of the Non-Aligned Movement.

[5] Present King Jigme Singye Wangchuk, fourth in line to the throne, was crowned in 1972. His father, the late King Jigme Dorji Wangchuk, was considered the architect of modern Bhutan. Under his rule, the political and administrative machinery were restructured and modernized, and successive 5-year development plans were initiated.

[6] Agriculture and animal husbandry are the predominant means of livelihood in the country. As Bhutan opened up, the development of an economic infrastructure became a priority. A road network, particularly imperative with the closing of the Tibetan border in 1960, is now in place across the country, with important arterial routes heading south toward India.

[7] Many rivers and water falls have been exploited as sources of hydroelectric energy. The massive Chukha project lies on the main route between Phuntsholing and Thimphu. With a potential of 8,000 megawatts, it supplies electricity not only within Bhutan, but also to adjacent parts of India; parts of southern Bhutan have been linked with the Indian grid system. 80 per cent of Bhutan's population lack electricity.

[8] The first chemical industry, Bhutan Carbide and Chemicals Ltd, began production in June 1988. The Bhutan Development Finance Corporation (BDFC) was also established in 1988 to encourage rapid expansion of the private sector. A rural credit scheme introduced in 1982 under the Royal Monetary Authority has been brought under the auspices of the BDFC.

[9] The semblance of political stability, which the government attempts to project abroad, has been seriously affected by the campaigns of the various opposition groups, who clamor for the democratization of the country. The main agitator among these movements is a group representing the Nepalese minority which has been extremely active in recent years.

[10] Ethnic friction has been on the increase since the 1988 census, when the king stepped up his "Bhutanization" campaign throughout the kingdom, decreeing the obligatory public use of the national dress - the *ko* for men and the *kira* for women - as well as the mandatory use of the Dzongkha language. Teaching in Nepalese was prohibited, tourists were barred from sacred sites, and television signals from India were blocked. Work visas for foreigners were also outlawed.

[11] These measures infuriated the Nepalese minority (25 per cent of the population), who immediately rejected them. However, their demands were met with absolute intransigence on the part of the monarchy, on various occasions.

[12] In September 1990, demonstrations and street protests by the banned Bhutan People's Party were repressed by government forces in the city of Thimbu. Opposition groups claimed that 300 people were killed in the incidents.

[13] The BPP was founded by a group of students of Nepalese origin, to represent the Nepalese minority and fight for an end to Drukpa domination, to achieve a more democratic constitution, and legalization of the political parties.

[14] Since September 1990, social agitation has increased: Nepalese commandos have set fire to schools and destroyed bridges, they have also carried out kidnappings to raise money, and the government has blamed them for no fewer than thirty violent deaths.

[15] Bhutan depends upon India for its imports of consumer goods, fuel, grain, machinery, replacement parts, vehicles and exports; with 93 per cent of its produce sold to India. The main Bhutanese exports are hydroelectric energy, cement, handcrafts, fruits and spices such as cardamoms.

[16] In the past few years, Bhutan has tried to diversify foreign trade. At the same time, a number of development programs are in effect with financing from India, Norway, Kuwait, Japan and Switzerland amongst other countries.

[17] Attempts to introduce a market economy system constantly come up against the fact that Bhutan is among the world's 42 least developed countries. In addition, the mountains make it difficult to build highways and providing an adequate infrastructure is both costly and difficult. Peasants are forced to work in infrastructure projects without remuneration.

[18] Despite these problems, the country's 5-year plans have been followed and their timetables respected. As a result, the country's GNP grew in real terms by an annual average of 7.5 per cent between 1980 and 1990. Since 1992, the 7th plan has been implemented aiming to increase exports, conserve the environment, decentralize and promote women's rights. To achieve

WORKERS

1994

- MEN 68%
- WOMEN 32%

- AGRICULTURE 92% INDUSTRY 3% SERVICES 5%

PROFILE

ENVIRONMENT

This Himalayan country is made up of three distinct climatic and geographical regions: the Duar plain in the south, humid and tropical, is densely wooded ranging in height from 300 to 2,000 metres; at the centre lies a temperate region with heights of up to 3,000 metres; finally, the great northern heights, with year-round snow and peaks that climb 8,000 metres high. Forests are the country's mainstay. Rivers and waterfalls have been exploited and they supply energy to the country and neighbouring areas. There are graphite, marble, granite and limestone deposits. Approximately half of the arable land lies on steep slopes; of this, nearly 15% is merely topsoil. The combination of these two factors makes the terrain very susceptible to erosion.

SOCIETY

Peoples: Bhutia (Ngalops) 50%; Nepalese (Gurung) 35%; Sharchops 15%. There are also Lepchas, native people, and Santal, descended from Indian immigrants (1993). 95% of the teachers and 55% of public servants are of Indian origin. **Religions:** Buddhist 69.6%; Hindu 24.6%; Muslim 5%; other 0.8% (1980). **Languages:** Dzongkha (official). Nepalese and other dialects are also spoken. **Political Parties:** The Bhutan People's Party (BPP), founded in Nepal in 1990; the National Bhutanese Congress founded in 1992, in Nepal and India; the People's Forum of Human Rights (PFHR), arose in 1990, supervising five refugee camps in Nepal; the United People's Liberation Front (UPLF), also founded in 1990.

THE STATE

Official Name: Druk Yul. **Administrative Divisions:** 18 Districts. **Capital:** Thimbu, 30.400 inhab. (1993) **Government:** Jigme Singye Wangchuk, king since July 21 1972. The hereditary monarch is assisted by a council of 9 members, of whom 5 are elected by the people, 2 are appointed by the autocratic ruler and 2 by the Buddhist religious dignitaries whose 6,000 lamas (monks) are headed by the Je Khempo. There is also a consultative assembly (Tsogdu) of 150 members 101 of whom are elected and 49 appointed, most being Buddhist monks. The ruler also dispenses justice. **National Holidays:** August 2, Buddhist Lent; October 30, end of Buddhist Lent.

STATISTICS

DEMOGRAPHY

Annual growth: 3.8% (1992-2000). **Estimate for year 2000:** 2,000,000. **Children per woman:** 5.9 (1992).

HEALTH

One **physician** for every 11,111 inhab. (1988-91). **Under-five mortality:** 193 per 1,000 (1995). **Calorie consumption:** 98% of required intake (1995).

EDUCATION

Illiteracy: 58% (1995). **Primary school teachers:** one for every 37 students (1988).

COMMUNICATIONS

83 **newspapers** (1995), 82 **TV sets** (1995) and 77 **radio sets** per 1,000 households (1995). 0.2 **telephones** per 100 inhabitants (1993).

ECONOMY

Per capita GNP: $400 (1994). **Annual growth:** 4.40% (1985-94). **Currency:** 31 Ngultrum = 1$ (1994). **Cereal imports:** 37,000 metric tons (1992). **Fertilizer use:** 8 kgs per ha. (1992). **Development aid received:** $64 million (1991); $44 per capita; 25% of GNP.

---

these aims the government has encouraged foreign investment, receiving technical assistance from India. Likewise a cautious privatization plan was introduced, with shares in state enterprises offered to the general public.

[19] Drastic cuts in public spending, the drive against nepotism and the voluntary reduction of the king's personal wealth - including moving from the royal palace into a modest home - encouraged international agencies to provide financing for development projects.

Dependence upon foreign aid was consequently reduced from 50 per cent in 1982, to 20 per cent in 1992. The budget for 1994-95 reveals a new growth of international financial support: 41.6 per cent were contributions from the United Nations and other agencies and 21.2 per cent came from the government of India.

[20] Bhutan has long maintained border disputes with China and India. In 1992, talks continued with China over the region of Arunachal Pradesh. As for India, the 2 countries were attempting to establish conclusive borders in a small area between Sarbhang and Gueyilegfug.

[21] In a country where neither political parties nor unions are permitted, there is nevertheless a National Association of Women dedicated to improving conditions for Bhutanese women. They have encouraged women to leave their traditional agricultural activity, offering them training in the production of woven handcrafts.

[22] In 1992 and 1993, conflict in the south between the population of Nepalese origin and the authorities, all members of the northern Drukpa majority, continued. In May 1993 the Nepalese government condemned the death of some 500 Nepalese immigrants as a result of hazardous conditions in the temporary camps where they were being housed by the Bhutanese authorities.

[23] Until 1995 the Nepalese government had held 6 fruitless meetings with the authorities to solve the problem of over 100,000 immigrants - Nepalese and Bhutanese of Nepalese background - persecuted or expelled from Bhutan, living in refugee camps. Since 1990 Bhutan has taken the view that the refugees, called *hotsampas*, do not have a nationality.

[24] Nepalese activists point out that in 1995, over half the population of Bhutan was Nepalese but the government estimates they make up only one third of the population. Bhutanese authorities try to control Nepalese minorities so they do not organize and take part in political activity. Most civil servants are Hindu. Using scholarships granted by other countries, the government selects youngsters - 30,000 in 1994 - who are trained abroad.

[25] In 1996, the European Parliament condemned the human rights situation of Thimbu's government. Up to that time, Nepal and Bhutan had not resolved how to face the refugee problem. Nepal proposed a survey to reveal the *hotsampas'* nationality but Bhutan did not accept.

Bolivia

Population: 7,237,000 (1994)
Area: 1,098,580 SQ KM
Capital: La Paz

Currency: Boliviano
Language: Spanish

Bolivia

[1] In 2000 BC the region of modern-day Bolivia was inhabited by farmers in the Andes, and forest hunter-gatherers in the east. Their terrain ranged from high mountain areas, *puna*, to hot valleys and forests. They produced livestock, potatoes, cotton, maize, and coca; they also fished and mined. This region, rich in natural resources, sustained several kingdoms and fiefs around Lake Titicaca, with the Tiahuanaco civilization at their center.

[2] The basic social unit was the *ayllu* kinship group, in which there was no private land ownership. The society was stratified into farmers, artisans and the ruling *ayllu*, of priests and warriors, who appointed the *malku* (chief).

[3] By 800 AD, Tiahuanaco formed the first Pan-Andean empire. By 1100, the *ayllu* of the Incas, from the Cuzco Valley in Peru, had colonized the other Andean peoples and formed a confederation of states called the Tahuantinsuyu. Also known as the Inca empire, it adopted elements of Tiahuanaco culture, technology, religion and economics, particularly the *ayllu* social unit.

[4] Through the *mita*, a mutual cooperation commitment between the different *ayllus*, each worker would render service to the centralized state; this system was later cruelly exploited by the Spaniards. The social structure was rigid; at the top was the Inca (son of the Sun), followed by the nobility and the priests. Then came the *capac*, governors of the regions into which the empire was divided, and lastly came the *curacas* (leaders of the *ayllus*) and the farmers. Social organization was based on self-sufficient, communal production.

[5] When the Spaniards arrived at the beginning of the 16th century, the Tahuantinsuyu extended from southern Ecuador, through Peru, to northern Chile, and from Lake Titicaca and the *altiplano* highlands to northern Argentina, embracing the mountain valleys and the eastern plains. One million people are estimated to have been living within the area of the present Bolivia, and from two to three million in the Tahuantinsuyu as a whole, making it the most densely populated area of South America. This society included a number of ethnic groups, predominantly Aymara (around Lake Titicaca) and Quechua. The eastern plains were inhabited by dispersed groups of Tupí and Guaraní, with no central nucleus. To this day, Aymara and Quechua are the most widespread languages in Bolivia.

[6] In 1545 the Spanish discovered the Inca mines at Potosí. They extracted immense quantities of silver which contributed to the capital accumulation of several European countries. Hundreds of thousands of native Americans died there, barbarically exploited to the point of exhaustion. Potosí was one of the three largest cities of the 17th century, growing up at the foot of the hill. It became the economic nerve center for vast regions of Chile and Argentina, and nurtured a rich mining bourgeoisie, guilty of corruption, ostentation and squandering.

[7] Decades of popular struggle against the Spaniards reached their peak in the successive failed rebellions of Tupac Katari (1780-82), the Protective Board of La Paz (1809), headed by *mestizo* Pedro Domingo Murillo, and the "independence guerrillas". The pro-independence movement was subsequently taken up by the creoles (Spaniards born in America), who distorted it by advocating social, economic and political systems based, not on native models, but on those of emerging European capitalist powers. A British blockade interrupted the supply of mercury, essential for treating the silver, and the Bolivian mining industry went into decline. The Buenos Aires based trading bourgeoisie soon lost interest in "Upper Peru" (Bolivia) and offered little resistance when it fell under the influence of the independence leader Simon Bolívar. The country was renamed after him when the Assembly of Representatives proclaimed independence in Chuquisaca, in 1825.

[8] Peru exerted great influence over the independent Bolivia until 1841. Bolivian president Marshall Andrés de Santa Cruz, tried to modernize Bolivia, founding universities and the Supreme Court of Justice, and compiling law codes.

[9] A mine-owning oligarchy, including Patiño, Aramayo, and Hochschild, worked with the politicians and generals, who were their associates, handling the Bolivian republic as if it were part of the tin business. British imperialist interest, initially in the saltpeter at Antofagasta and later in Bolivia's southern oil reserves, triggered two wars in South America: the Pacific War (Chile against Bolivia and Peru from 1879-83), and the Chaco War (Paraguay and Bolivia, from 1932-35). As a result of these conflicts, Bolivia lost its sea-coast and three quarters of its territory in the Chaco region. The ceding of Amazonian Acre to Brazil, in 1904, completed the country's dismemberment.

[10] On July 21 1946 Bolivian president Gualberto Villaroel was overthrown, assassinated, and his body hung from a lamp-post in downtown La Paz. His predecessor, German Busch, from the military generation of the Chaco War, had nationalized Bolivia's oil reserves, and he was accused of being a Nazi by the State Department propaganda.

[11] Nationwide frustration at these humiliations gave way to a powerful current of reformism and anti-imperialism. The Nationalist Revolutionary Movement (MNR) grew up alongside progressive labor and peasant movements. After several

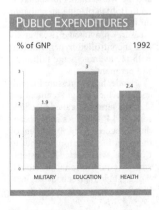

PUBLIC EXPENDITURES

% of GNP — 1992

- MILITARY: 1.9
- EDUCATION: 3
- HEALTH: 2.4

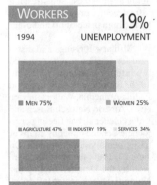

WORKERS

19% UNEMPLOYMENT

1994

- MEN 75%
- WOMEN 25%

- AGRICULTURE 47%
- INDUSTRY 19%
- SERVICES 34%

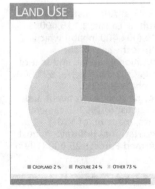

LAND USE

- CROPLAND 2 %
- PASTURE 24 %
- OTHER 73 %

Foreign Trade

MILLIONS $ 1994

IMPORTS
1,209

EXPORTS
1,032

Maternal Mortality

1989-95

PER 100,000
LIVE BIRTHS
373

PROFILE

ENVIRONMENT

A landlocked country with three geographic regions. 70% of the population live in the cold, dry climate of the *altiplano*, an Andean highland plateau with an average altitude of 4,000 meters. This region holds the country's mineral resources: tin (second largest producer in the world), silver, zinc, lead and copper. The subtropical valleys (*yungas*) of the eastern slopes of the Andes form the country's main farming area, where coffee, cocoa, sugar cane, coca and bananas are grown. The tropical plains of the East and North, a region of jungles and grasslands, produce cattle, rice, corn, and sugar cane. The area is also rich in oil. Bolivia is made up of three drainage basins which empty into Lake Titicaca, the Amazon and the Rio de la Plata. Unconstrained exploitation of lumber threatens the country's forest resources, as well as water supplies and fauna. In "El Alto" (a high district of La Paz), air pollution has risen in direct proportion to the number of motor vehicles.

SOCIETY

Peoples: 48% of Bolivians belong to the Quechua and Aymara nations. *Mestizos* and *Cholos* account for 31% of the population. A minority of European descent has ruled the country since the Spanish conquest. The Tupí and Guaraní peoples live in the eastern forests.
Religions: Mainly Catholic. Freedom of religion.
Languages: Spanish (official), spoken by only 40% of the population; Quechua and Aymara; there are 33 ethnic-linguistic groups. **Political Parties:** The parties with parliamentary representation are the Nationalist Revolutionary Movement, of President Gonzalo Sánchez de Losada; the new Patriotic Conscience party (CONDEPA), directed by media businessman Carlos Palenque; the Solidary Civic Union (UCS); the Nationalist Democratic Action (ADN), headed by General Hugo Bánzer Suárez; the Free Bolivia Movement; the Socialist Democratic Group (ASD); the Revolutionary Movement of the Left (MIR); the Pachacuti-Axis, supported by *campesinos* engaged in coca cultivation; and the MRTK-L (the political wing - albeit small in number - of the Tupac Katari Movement). **Social Organizations:** The Bolivian Workers' Confederation (COB); the Sole Labor Union Confederation of Farm Workers of Bolivia (CSUTCB). There are also important ethnic and environmental organizations like the Indigenous Confederation of the Eastern region, Chaco and Bolivian Amazon Area (CIDOB); the Guaraní People's Assembly (APG); the Federation of Campesino Women; the Federation of Neighborhood Commissions; and the Bolivian Forum on Environment and Development (FOBOMADE).

THE STATE

Official Name: República de Bolivia. **Administrative Divisions:** 9 Departments. **Capital:** La Paz, 1,115,400 inhab. (1992) - including El Alto (404,400 inhab.) which became a separate city in 1988 - is the seat of the government. Sucre is the constitutional capital, and seat of the judiciary. **Other cities:** Santa Cruz de la Sierra, 694,600 inhab.; Cochabamba, 404,100 inhab. (1992)
Government: Gonzalo Sánchez de Losada, President and Head of the government since August 6, 1993. Bicameral legislature: Chamber of Deputies, made up of 130 members; Senate, 27 members. **National Holiday:** August 6, Independence Day (1825). **Armed Forces:** 25,000 (1993). **Paramilitaries:** 23,000 Police.

uprisings, and a 1951 electoral victory which was not honored, the MNR led a popular insurrection in 1952. Civilians defeated seven regular army regiments in the streets, and they carried first Victor Paz Estenssoro, and then Hernán Siles Zuazo, to the presidency. The Bolivian revolution nationalized the tin mines, carried out agrarian reform and proclaimed universal suffrage. Workers' and peasant militias were organized and, together with the Bolivian Workers' Confederation (COB), formed a "co-government" with the MNR. The army was eliminated after the insurrection, but was later reorganized under US pressure.

[12] Troubled by internal divisions, the MNR gradually lost its drive and was defeated in November 1964 by a military junta led by René Barrientos. Ernesto "Che" Guevara tried to establish a guerrilla nucleus in the Andes to spread revolutionary war throughout South America, but he was caught by US-trained counter-insurgency troops, and assassinated on October 8, 1967.

[13] Division within the army, coupled with pressure from the grassroots level, led to an anti-imperialistic faction taking over government in 1969, with General Juan José Torres as the leader. During his short time in office, there was a significant increase in the number of grassroots organizations. The People's Assembly was formed, with links to the COB and the parties of the left. In August 1971 he was ousted by Colonel Hugo Bánzer Suárez, who formed a government with MNR support. The civilian-military coalition government remained in power until July 1978, with an authoritarian but development-oriented administration encouraging agribusiness and stressing infrastructure projects, bolstered by the high price of oil and other minerals.

[14] Military uprisings and disregard for election results occurred repeatedly between 1978 and 1980. Juan Pereda Asbún, David Padilla Arancibia, Walter Guevara Arce, Alberto Natusch Busch and Lidia Gueiler all had fleeting presidencies, ended either by coups or parliamentary appointments. On June 29, 1980, the elections were won by the Democratic Popular Union (UDP), a center-left coalition whose candidate, Hernán Siles Zuazo, was prevented from taking office by another bloody coup engineered by General Luis García Meza.

[15] According to reports received by Amnesty International, thousands of people were killed or tortured, including Socialist leader Marcelo Quiroga Santa Cruz, Trotskyist parliamentary deputy Carlos Flores Bedregal, and miners' leader Gualberto Vega.

[16] By 1982, the army was seething with internal dissension, government links with paramilitary groups and drug trafficking were eroding its international prestige, and it was facing dogged popular resistance led by the Bolivian Workers' Confederation (COB), leading to its destruction. In September, the military command decided to convene the government which had been democratically elected in 1980. On October 10, after 18 years of military rule, a civilian, Hernán Siles Zuazo took office, a legal constitutional stage that has continued ever since. Siles Zuazo, with a populist/nationalistic outlook, gave administration of the state-owned mines to the labor unions. He also announced the "non-payment" of Bolivia's foreign debt. The labor movement and the peasants exerted pressure on the government through demonstrations, and several laws passed by his administration allowed these groups to participate in the economic policy decisions of large businesses, and in local committees dealing with food, health and education issues. The Peasant Agricultural Corporation took partial control of the markets and set up collective machinery pools and work teams. In response to these measures, creditor banks, the IMF and the World Bank blocked credits to Bolivia placing an embargo on its international trade. These measures provoked a fiscal crisis and uncontrolled hyperinflation, with the average wage falling to $13 per month.

[17] Under heavy pressure from all social sectors, the government cut short its term and called elections for July 14, 1985. As neither candidate secured over 50 per cent of the vote, the decision was put to Congress, who elected Paz Estenssoro as constitutional president, though Hugo Bánzer had marginally more votes.

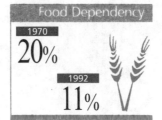

Food Dependency

1970
20%

1992
11%

External Debt
1994

PER CAPITA
$656

Illiteracy
1995

17%

[18] The Estenssoro government decreed a program of neo-liberal measures to end subsidies and close down state enterprises. It eliminated price controls and the official listing of the dollar against the local currency. Mines were closed down, and others rented out, leaving thousands of miners jobless.

[19] There was a drastic cut in real wages, with massive redundancies, while investment ground to a standstill. Inflation increased to four figures.

[20] In the 1989 national elections, Jaime Paz Zamora's Revolutionary Movement of the Left proved itself a potent new political force, coming in third with 19 per cent of the votes (double that of previous elections). The MNR candidate, Gonzalo Sánchez de Lozada, obtained 23 per cent and Bánzer's Nationalist Democratic Action, 22.6 per cent. As no party had achieved a majority, an agreement between the MIR and the ADN, known as the "Patriotic Accord" made it possible for Paz Zamora to be nominated president by the National Congress.

[21] The so-called "Patriotic Accord" (AP) continued Paz Estenssoro's neo-liberal economic policy. In 1991, the government embarked upon a transformation of the state, which entailed initiating privatization. Congress passed a law permitting the state to sell off 22 of the existing 64 public enterprises, though the Supreme Court ruled it unconstitutional. The government promoted joint ventures between the state-owned Mining Corporation (COMIBOL) and private companies. The Federation of Mining Workers (FSTMB) launched a series of hunger strikes and threatened to occupy the mines, defending the principle of continued state ownership.

[22] In April 1991, Bolivia's Congress authorized US military officers to come to Bolivia to train local personnel in the war against drugs. Despite military action and the policy of crop substitution - a program called "development for coca" - the area under coca cultivation has increased. It has been estimated that in 1992 there were some 463,000 people involved in coca-cocaine production, and that the national net earnings from this crop reached $950 million per year.

[23] The loss of power in the workers' movement was compensated for by new organizations of indigenous peoples and communities. In October 1991, the 9th Congress of the Confederation of Indigenous Peoples of the Eastern Region, the Chaco and the Bolivian Amazon Region (CIDOB) met in Santa Cruz. 340 representatives approved a series of demands, including claims to lands lost during the Spanish invasion, and the use of their native languages in the educational system.

[24] The population of Bolivia's eastern region includes 250,000 people of 10 linguistic groups and 35 ethnic groups. In September 1990, a group of peoples from this region carried out a 750 km march from the east to the capital with the slogan "Land and Dignity". Paz Zamora's government approved a "National Plan for the Defence and Development of the Indigenous Peoples", and in August 1991, recognized the Santa Ana de Horachi Mosetana Community's right to 8,000 hectares of land, which they consider collective property. These resolutions have been opposed by companies which have been exploiting the region's vast forestry resources.

[25] In January 1992, Presidents Paz Zamora and Alberto Fujimori (of Peru) signed an agreement where Peru ceded an area of 327 hectares to Bolivia, for it to develop a free zone at the port of Ilo, in southern Peru, with which Bolivia attains a "way out to the sea", that is, a free port for its international trade.

[26] The MNR won the national elections of June 1993, with 36 per cent of the vote. Its presidential candidate was Gonzalo Sánchez Losada, and the vice-presidential candidate was Victor Hugo Cárdenas, an Aymara sociologist and leader of the Tupac Katari Movement. While the ADN and MIR suffered significant losses, the new populist and nationalist movements, such as CONDEPA and UCS, led by Carlos Palenque and Max Fernández retained their support among mestizos and the poorer neighborhoods.

[27] In its first year the government established the right to education in the native languages (Aymara, Quecha, Guarani). The Capitalization Bill aimed to privatize 50 per cent of the main public industries (telecommunications, electricity, oil, gas, railways, airlines) on the basis of transferring half the shares to the Bolivian

citizens as pension funds. The aim was to attract foreign investment, reduce unemployment and increase the GDP.

[28] Many local movements reacted against the importing of toxic waste from Europe and the environmental deterioration caused by mining, while others protested against the closing of these same mines and the consequent job losses. The burning of coca plantations by the United States led to continuous confrontations between the rural workers and the military. Meanwhile, the World Bank reported that 97 per cent of the rural population live in poverty.

[29] In 1994, the YPBF oil corporation, on the brink of privatization, was made even more attractive by an agreement with Brazil for the construction of a $2 billion gas pipeline to join Santa Cruz de la Sierra with São Paulo. The Capitalization bill, unpopular amongst workers fearful of losing their jobs, led to a series of strikes in 1995. The government declared a state of emergency on two occasions, granting the police special powers and imposing a curfew. Public meetings were banned and more than 100 leaders were imprisoned.

Bosnia and Herzegovina

Population: 4,383,000
(1994)
Area: 51,129 SQ KM
Capital: Sarajevo

Currency: Dinar
Language: Serbian and Croatian

Bosnia I Hercegovina

The earliest inhabitants of what is now Bosnia-Herzegovina were Illyrian tribes. After the Roman conquest in the mid-2nd century BC, this region became a part of the province of Illyria, although the Sava river basin remained within the province of Pannonia. Serbs settled in the area in the 7th century AD. Except for a brief interregnum (1081-1101), Bosnia remained politically separate from Serbia.

[2] Hungarian control over Bosnia began in the mid-12th century. The combined efforts of the papacy and of the Hungarians to impose their religious authority over Bosnia gave rise to strong national resistance. Although war was avoided, the region came under the jurisdiction of the Hungarian archbishop of Kalocsa.

[3] During this period, Bosnia was looked upon as a bastion of the Bogomils (or Cathari), one of southern Europe's main heretical movements, who found support among the local population, and from the State, as well. Neighboring Christians - both Orthodox Serbs and Catholic Croats - organized several crusades against this heresy. The Bogomils defeated the crusaders, but some parts of Bosnia were Christianized.

[4] Ban Prijezda founded the Kotromanic dynasty (1254-1395), under which Bosnia conquered the province of Hum (Herzegovina). In 1377, Tvrtko crowned himself king of Serbia, Bosnia and the coastlands. With the Turkish invasion, in 1386, the Serbs were defeated at Kosovo (1389), but Tvrtko carried out further conquests in the west, and in 1390 was crowned king of Rashka, Bosnia, Dalmatia, Croatia and the coastlands.

[5] The Ottoman Empire occupied Serbia in 1459 and Bosnia became a province of the Ottoman Empire in 1463. Hum resisted longer, but in 1482, the port of Novi fell (now called Hercegovini), and Herzegovina also became a province of the Ottoman Empire.

[6] After the Turkish conquest, a massive campaign began to convert Bosnian Bogomils to Islam. Thus, in addition to Catholic Slavs and Christian Orthodox Slavs, there were now Muslim Slavs. Ever since, relations between the three communities have been strained, and religion became the decisive social factor in the region.

[7] At first, the Turkish governor (*Pasha*) had his headquarters in Banja Luka, but these were later transferred to Sarajevo. In 1580, Bosnia was divided into 8 *sanjaks* (sub-regions), under the jurisdiction of 48 hereditary Kapetans, who exercised a feudal power over their territories. Mining and trade began to decline; only the manufacture of weapons and wrought metals survived.

[8] In the 16th and 17th centuries, Bosnia played an important role in the Turkish wars against Austria and Venice. In 1697, Prince Eugene of Savoy captured Sarajevo. By the Treaty of Karlowitz (1699) the Sava River, which formed Bosnia's northern border, also became the northern boundary of the Ottoman Empire. Herzegovina and the part of Bosnia east of the Una River were ceded to Austria in 1718, and returned to Turkey in 1739.

[9] In the 19th century, Bosnia's nobility resisted Turkish interference. In 1837, Herzegovina's regent declared independence. Uprisings became chronic, bringing Christians and Muslims together, despite their differences, against the bureaucracy and corruption of the empire.

[10] In 1875, a local Herzegovinian conflict unleashed a rebellion, which spilled over into Bosnia. Austria, Russia and Germany tried unsuccessfully to mediate between Turkey and the rebels. The Sultan's promise to reduce taxes, grant religious freedom and install a provincial assembly was also rejected.

[11] By a secret agreement, in 1877, Russia authorized Austria-Hungary to occupy Bosnia-Herzegovina, in exchange for its neutrality in Russia's upcoming war against Turkey. After the Russo-Turkish War of 1877-78, the Congress of Berlin disregarded Serbian wishes, and assigned Bosnia-Herzegovina to Austria-Hungary (although nominally they continued to be under Turkish control).

[12] In 1878, Vienna put down armed resistance by Bosnia-Herzegovina. The revolution launched by the Young Turks, in 1908, brought on a crisis within the Ottoman Empire.

[13] The Turkish government asked Bosnia-Herzegovina to participate in the new parliament at Istanbul but Austria-Hungary annexed the two provinces in 1908, with Russian consent. Vienna established a provincial assembly (*sabor*) and special laws of association, without representation in Vienna or Budapest.

[14] The 1910 constitution consolidated social and religious differences by establishing three electoral colleges - Orthodox, Catholic and Muslim - each with a fixed number of seats in the *sabor*.

[15] The influence of the Young Bosnia (Mlada Bosna) movement and other revolutionary groups led the empire's authorities to close Bosnia's *sabor*, and dissolve several Serbian political groups. The assassination of Archduke Francis Ferdinand and his wife, the Duchess of Hohenberg, in Sarajevo, in 1914, by a Bosnian Serbian student, triggered World War I.

[16] After the collapse of the Austro-Hungarian empire in the Balkans, in 1918, Bosnia-Herzegovina became a part of Serbia.

[17] During the Nazi occupation, Bosnia-Herzegovina was subjected to the puppet administration established in Croatia. In the two provinces, Croat Ustashes (Fascists) massacred the Serbs. The rivalry which had always existed between Muslim, Serb and Croat degenerated into deep hostility.

[18] After the war, Bosnia-Herzegovina became a Yugoslav republic. The slogan of the Yugoslav federated socialists was "Brotherhood and Unity", but ethnic confrontation was visible in the arts, the humanities and literature.

[19] In April 1990 the Yugoslav Communist League rescinded their total control, and in the first multi-party legislative elections to be held since the war, the nationalist parties elected 73 Serbs and 44 Croats, a fact which resulted in heavy losses for the Democratic Reform Party (ex-Communist) candidates and the liberal technocrats.

[20] The Muslims were represented by the Democratic Action Party (DAP) and its leader, Alija Izetbegovic, was elected president of the republic. The DAP held an absolute majority in 37 communities (administrative units) and a relative majority in 52 others. The Croatian Democratic Community, based in Zagreb, controlled 13 local communities and held a relative majority in 17 others.

[21] In October 1991, Bosnia-Herzegovina's *sabor* approved a declaration of independence from the Yugoslav federation. In January 1992, the parliament agreed to hold a plebiscite on the issue of separation. The Serbian communities of Bosnia wanted to remain within a Yugoslav federation, while the Croat and Muslim communities clamored for independence.

[22] Bosnian president Izetbe-

govic wanted to maintain the unity and integrity of the republic, promising that Bosnia-Herzegovina would not become a Muslim State, and guaranteed that the rights of all nationalities would be respected. In early March, an all-out conflict broke out, when the referendum was supported by 99.4 per cent of the Muslims and Croats, who ratified the republic's independence.

[23] On April 7, the European Community and the US recognized the independence of Bosnia-Herzegovina. The Bosnian republic was accepted as a member state in the Conference of Security and Cooperation in Europe; in May it became a member of the UN.

[24] At the same time, the Serbian community proclaimed the independence of the "Serbian Republic of Bosnia-Herzegovina" in the areas under Serbian control (Bosnian Krajina, with its center at Banja Luka). The conflict quickly extended throughout the entire region.

[25] Local Croat forces also controlled certain areas of the republic; there were sporadic confrontations with Bosnian government troops. In late July, Croatia and Bosnia signed a mutual recognition pact.

[26] In January 1993, Serbian troops killed Bosnian deputy prime minister, Hakija Turajlic, in Sarajevo. In March the UN decreed a cease-fire in that city, which was under siege. At this point in the war, there were numerous reports on the existence of Serbian concentration camps, as well as an "ethnic cleansing" campaign. The latter involved the expulsion of all members of rival ethnic groups, especially from the smaller villages.

[27] According to Amnesty International, thousands of civilians, as well as soldiers that had been captured or wounded, were executed. In addition, prisoners were systematically submitted to torture and harsh treatment. According to UN figures, a total of 40,000 women had been raped. Although excesses were committed by all sides, it was the Serbians who bore the major responsibility, while the Muslims were the main victims.

[28] The UN Protection Forces (UNPROFOR) sent in some 20,000 peacekeeping troops. The

PROFILE

ENVIRONMENT

Bosnia and Herzegovina has a 20km coastline on the Adriatic Sea. In the west it is bounded by Croatia, Montenegro and Serbia. The major part of the country lies in the Dinaric Alps, with elevations of around 4,265 meters, making overland communication difficult. The country is drained by the Sava and Neretva rivers and their tributaries. The territory takes its name from the Bosna River, a tributary of the Sava. The main crops are grains, vegetables and grapes; there is also livestock rearing. There is a wealth of mineral resources, including coal, iron, copper and manganese. Because of air pollution, respiratory ailments are very common in urban areas. Barely half of the region's water supplies are considered to be safe, the Sava River being the most polluted of all.

THE SOCIETY

Peoples: Muslim-Slavs 49.2%, Christian-Orthodox Serbs 31.3%, Catholic-Croats 17.3%. Ethnic differences stem from historical and religious factors. Serbs make up the majority in northeastern Bosnia, living in or around Banja Luka; Croats form the majority in western Herzegovina, Mostar being their main urban center. It is impossible to draw an "ethnic dividing line" in other regions. In the capital, Sarajevo, there is a Muslim majority, and Serbs and Croats. There were 1,200 Jews until 1992. **Religions:** The majority is Muslim. Other religions: Christian Orthodox and Roman Catholic. **Languages:** Serbian and Croatian. **Political Parties:** Party of Democratic Action (Muslim), currently in power; the Serbian Democratic Party; the Serbian Renaissance Movement; the Croatian Democratic Community; the Democratic Reform Party (formerly the Communist Party). **Social Organizations:** Currently in the process of being reorganized.

THE STATE

Official Name: Republika Bosna I Hercegovina. **Administrative Division:** 50 Districts. **Capital:** Sarajevo, 415,600 inhab. (1991), reduced to less than 50,000 in September 1995. **Other cities:** Banja Luka, 142,600 inhab.; Tuzla 142,644 inhab.; Mostar 110,377 inhab. (1991) **Government:** As of the September 1996 elections, Bosnia has a three-member Presidency. Bosnia's Muslim President, Alija Izetbegovic, is first chairman; the Serb nationalist Momcilo Krajisnik came second and the Croat leader, Kresimir Zubak was third. **National Holiday:** March 1st, Independence (1992). **Armed Forces:** approximately 60,000 troops. Reserves: 120,000. (1993). After the conflict, the United Nations sent in several thousand peace-implementation troops, and this is soon to be replaced by a new multinational force. There are more than 150,000 troops belonging to the Federal Yugoslavian Army, the Bosno-Serbian Army and the Serbian Nationalist Militia, among others.

COMMUNICATIONS

13.7 **telephones** per 100 inhabitants (1993).

US refused to send troops to Bosnia, despite pressure from the UN and European countries. Several security zones were decreed in such cities as Tuzla, Zepa, Gorazde, Bihac and Sarajevo. However, these were not always observed, just as successive cease-fires were consistently violated.

[29] In October, Serbian occupation of Bosanki Brod opened up a passageway between Serbia and Bosnian Krajina. Serbians controlled 70 per cent of the territory, as a result of their superior artillery and armored vehicles, as well as their control of the bridges over the Drina River - on the border between Serbia and Bosnia - which allowed them to receive arms and other supplies from the Yugoslav Federation. As a result of their support of the Serbians, the UN called for an economic blockade of the Federation, and an arms embargo aimed at Bosnians and Croats.

[30] The Muslims found themselves cornered in Sarajevo and a few other minor sites, receiving what little financial and moral support they could from a few Islamic countries.

[31] In June 1993, Serbian president Slobodan Milosevic and his Croatian counterpart, Franjo Tudjman, announced the partition of Bosnia into three ethnic entities (Serbian, Croatian and Muslim), within the framework of a federal state. Their scheme coincided with a UN-EU peace proposal which provided for a division into semi-autonomous provinces, controlled by each of the ethnic groups.

[32] The Croats, faced with the partition of Bosnia, sought to gain an upper hand, so as to negotiate from a position of strength. In July they launched an offensive against Mostar, capital of Herzegovina. The UN Human Rights Commission reported that 10,000 Muslims had been held in Croatian concentration camps, where they were submitted to torture and in some cases, to summary executions.

[33] In the meantime, the situation in Sarajevo grew progressively worse: there was a lack of electricity, water and food. Besieged by a series of epidemics, the city's 300,000 inhabitants managed to survive on minimal rations, while international aid

agencies tried to reach the city under mounting difficulties. In early 1994, the UN appealed to the Serbians to stop their attacks upon Sarajevo and withdraw the heavy artillery surrounding the city. Despite resistance from the Russians, NATO threatened to bomb Serbian positions. When the formal deadline expired, the Serbians withdrew.

[34] The peace conference sponsored by the UN and the EU proposed the territory be divided into Bosnian, Muslim, and Croatian areas as ethnically homogeneous republics. This option called for the transfer of population from one sector to another. The UN's human rights commission for the former Yugoslavia criticized the creation of etchnic boundaries and defended a reform of the democratic system.

[35] In 1994, the US and Russia exerted growing pressure upon the Serbs to accept the proposal. Croatians and Muslims approved a federal agreement between the two communities: 51 per cent of the territory would remain among Bosnians and Croatians while Serbs would be in control of 49 per cent, without the need of separating Bosnia into three ethnically distinct states.

[36] With support from the EU, Washington and Moscow, the federal agreement was signed by presidents Franjo Tudjman of Croatia and Alija Izetbegovic of Bosnia, but the Serbs rejected it. Negotiations were hampered because Yugoslav president Slobodan Milosevic - who was the Serbs' diplomatic representative - stated he had no authority over the self-proclaimed Republic of Sprska.

[37] In 1995, the Serbian-Bosnians held several UN troops hostage and took Bihac. In August, the situation was radically changed by NATO's bombing of Serb-Bosnian positions in the siege on Sarajevo. Almost at the same time, Croatia expelled Serb-Croat forces from the Eastern side of the country forcing their delegates to negotiate.

[38] According to the Dayton (Ohio, US) agreements, elections would be held in September 1996, hoping to promote more tolerant leaders among each of the nationalities in conflict. The agreement was carried out under US military pressure, with the underlying objective of influencing President Clinton's campaign for re-election in November. The presence of American troops forced a peaceful settlement and, at the same time, froze the political situation.

[39] The Dayton agreements acknowledge the existence of two ethnically pure mini-states due to the physical elimination or expulsion of ethnic minorities: the Serb-Bosnian Republic (Sprska) and the Croatian-Muslim Federation.

[40] The International Criminal Tribunal for the Former Yugoslavia at the Hague convicted Radovan Karadzic, leader of the *Sprska Republic* and his military commander Ratko Mladic, for genocide. In spite of being convicted, neither was incarcerated, having great influence in the republic's political life even though the Dayton agreements had banned electoral participation of individuals accused of committing war crimes.

[41] In June 1996, public power, water and transport services began to be reinstated and Sarajevo regained part of its former vitality. The environment of political tranquility considered necessary for the run-up to the September elections, was hindered by the refugees' inability to return to their homes unless their fellow countrymen ruled over the area, the limitations of journalists' movements and the difficulties imposed upon the activities of several NGO's.

[42] Since the September 1996 elections, Bosnia has a three-member Presidency. Bosnia's Muslim President, Alija Izetbegovic, is first chairman; the Serb nationalist Momcilo Krajisnik came second and the Croat leader, Kresimir Zubak was third. These three were the most nationalist among the candidates in each federated republic.

THE DAYTON PEACE PROCESS

Steps toward unity?

After four years of conflagration, the Bosnia and Herzegovina war ended in 1995. Witnessed by the European Union Special Negotiator and representatives of the five Contact Group nations (United States, Russia, United Kingdom, Germany and France), the presidents Alija Izetbegovic of the Republic of Bosnia and Herzegovina, Franjo Tudjman, Republic of Croatia, Slobodan Milosevic representing both the Federal Republic of Yugoslavia and the *Republika Srpska*, and President Zubak, for the Federation of Bosnia and Herzegovina, signed in Paris on December 14th, 1995, the final agreement for peace in Bosnia and Herzegovina.

[2] The peace talks had begun earlier at Wright-Patterson Air Force Base in Dayton, Ohio, on November 21st, under US Assistant Secretary of State Richard Holbrooke's initiative. The general framework agreement, signed in the military base, established that Bosnia and Herzegovina, the Federal Republic of Yugoslavia and Croatia would respect one another's sovereignty.

[3] The agreement established a federated country maintaining the ancient boundaries of the former Yugoslavian republic, but acknowledged the existence of three communities (Muslims, Croats and Serbs) and two political entities (the Muslim-Croat Federation of Bosnia and Herzegovina and the Serb *Republika Srpska*). Each community should elect a president to join the collective executive branch of the federal goverment, which would be headed by the candidate who obtained the relative majority of the total ballot of the three communities.

[4] A multinational military Implementation Force, the IFOR, was invited to monitor and guarantee compliance with the military aspects, also fulfilling supporting tasks that included the right to the use of force. Airspace control and protection of the different populations' migrations, following the two administrative divisions, were also responsibilities of the IFOR.

[5] Some problems arose during 1996, for example when the Tribunal on Human Rights banned the *Republik Srpska*'s president, Radovan Karadzic, from acting in politics. Finally Karadzic resigned and the elections were held as scheduled. Each of the three different communities elected their respective authorities and named a representative to enter the collective presidency of the federative republic.

[6] The partition of the country, though ignored in the papers, continued to be the main concern of the American and European negotiators. The election's outcome, favouring the most nationalist candidates in each community, spread shadows on the possibilities of Bosnia and Herzegovina remaining a united country.

Botswana

Botswana

Population: 1,443,000 (1994)
Area: 581,730 SQ KM
Capital: Gaborone

Currency: Pula
Language: Setswana

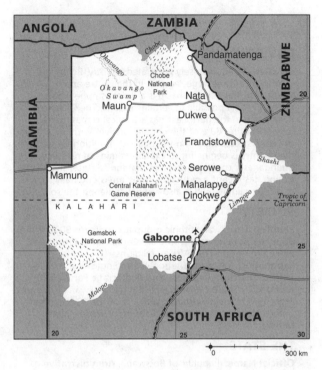

In the course of its history the "Bechuanaland", (English corruption of *ba'tswana*, "Tswana people") now Botswana, was also known as the "fatal crossroads" and the "eye of the storm", on account of its position in the heart of southern Africa, providing a passageway for British, Dutch and Portuguese colonizers. The British used the "missionary route" from the south of Africa to Sudan and Egypt, whereas the Portuguese sought to unite their colonies of Angola and Mozambique by controlling "Bechuanaland".

[2] Thus, since the 18th century this land has been a crossroads for various strategic colonial interests, as well as for the Tswana who had lived in the area since the 17th century. Around 1830, Botswana was penetrated by Boer colonists of Dutch origin, fleeing northward from Cape Town in South Africa to escape the British. These farmers fought with the natives for possession of the scant fertile lands, while the Tswana also clashed with Zulus who had been driven from their lands in South Africa by the Boer settlers (see South Africa: "Zulu expansion"). In 1894, leaders of Botswana's three major ethnic groups travelled to London to seek support in their fight against Boer settlers. The British promptly complied, and Botswana soon became a "protectorate".

[3] British trusteeship prevented political absorption by South Africa but paved the way for Afrikaner economic supremacy. Consequently, when the first nationalist movements arose they aimed at putting an end to both situations.

[4] Although Botswana has a large semi-arid area, it came to be one of southern Africa's major cattle and meat exporters. At the beginning of this century, 97 per cent of the pop-

ulation lived in rural areas, every family owned at least 2 cows, and the richest had oxen to plow their fields. By the 1960s, 15 per cent of the population had migrated to the cities and at least 40 per cent of the rural population had lost their cattle. Concentration of land ownership enabled Afrikaners to gain control of agricultural production and to supply 60 per cent of all meat exports.

[5] The struggle for independence became entangled with the wedding of Seretse Khama, a leader of the Bamangwalo ethnic group, who went to England to study law and married Ruth Williams, a European white-collar worker. This upset both the British and the Afrikaners, who prevented Seretse from returning home. He withstood every pressure including offers of money from the British and, firmly supported by his people, he retained leadership of the country's main ethnic group. Seretse finally returned home in 1956, 9 years before the general elections in which his Botswana Democratic Party (BDP) polled 80 per cent of the vote.

[6] In September 1966, Botswana gained independence and Seretse was elected first president. A year later he was knighted by the British. The BDP pursued a conciliatory policy towards Boer descendants, who owned 80 per cent of the country's economic resources, and Botswana continued to be highly dependent upon South Africa. Almost all its imports came through Cape Town, and 60 per cent of its exports were purchased by South Africa. Nevertheless, in the political arena, Seretse maintained his distance from South Africa, and supported the region's anti-racist movements. Botswana was one of the "frontline nations" fighting apartheid (see South Africa), and a

member of the SADCC (see Organizations) seeking to end the economic dependence on South Africa of the nine southernmost African countries.

[7] Seretse Khama died of cancer in July 1980 and was succeeded by the vice president, Quett Masire. Strong pressures were exerted upon Masire by revolutionary socialist groups to limit the concentration of arable land in the hands of citizens of European origin and to increase the area allotted to cooperatives. The rural poor accused large landowners of overgrazing, causing a deterioration in the quality of the land, and threatening their future. A movement also arose, demanding the nationalization of rich

diamond, iron, copper and nickel deposits exploited by South African companies.

[8] In 1982, the country faced balance of payments problems resulting from a drought that affected cattle farming and exports. By early 1983, and the end of the drought, the difficulties had subsided. Government austerity measures also contributed to increasing farm production and mineral exports grew. The deficit became a surplus in the first semester of 1983, enabling the government to slacken the austerity program and grant wage increases and tax cuts.

[9] However, 60 per cent of the population still depend on subsistence crops or engage in informal economic activities (occupations which escape official records, taxation, and the trade circuit). The informal labor market includes approximately 15,000 nomads.

[10] Botswana is the world's third largest producer of diamonds, after Australia and Zaire. In recent years, the government has associated itself with South African mining companies in order to prospect for gold, platinum, and other non-precious metals such as nickel. This effort produced record annual economic growth of 12 per cent between 1978 and 1988.

[11] In 1985, South Africa with-

PUBLIC EXPENDITURES

% of GNP   1992

MILITARY 3.1   EDUCATION 8.4   HEALTH 1.5

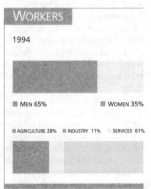

WORKERS

1994

MEN 65%   WOMEN 35%

AGRICULTURE 28%   INDUSTRY 11%   SERVICES 61%

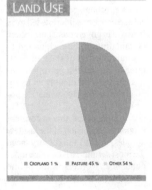

LAND USE

CROPLAND 1 %   PASTURE 45 %   OTHER 54 %

PROFILE

ENVIRONMENT

An extensive and sparsely populated country, Botswana is divided into three large regions. In the centre and southwest, the Kalahari basin is a desert steppe where grazing is only possible in certain seasons. The Okavango River basin in the northeast has a tropical climate suitable for agriculture. 80% of the population lives within a strip in the east, stretching along the railroad. Traditionally pastoral, the country is beginning to exploit its mineral resources (manganese, copper, nickel and diamonds). Intensive cattle-raising has meant a reduction in the areas originally set aside for wildlife, and is also rapidly depleting the soil.

SOCIETY

Peoples: The Tswana ethnic group accounts for 90% of the population. **Religions:** Some are Catholic and Protestant, while the rest practise African religions.
Languages: Setswana (national) and English (official).
Political Parties: The Botswana Democratic Party (BDP), founded by Seretse Khama in 1961; the Botswana National Front (BNF). **Social Organizations:** 5 of the 13 local trade unions form the Botswana Federation of Trade Unions, founded in 1976.

THE STATE

Official Name: Republic of Botswana. **Administrative Divisions:** 4 Districts. **Capital:** Gaborone, 133,000 inhab. (1991). **Other cities:** Francistown, 65,000 inhab.; Selebi-Phikwe, 40,000 inhab. (1991). **Government:** Sir Quett Masire, President since July 18, 1980. The single chamber National Assembly has 46 members. **National Holiday:** September 30, Independence Day (1966). **Armed Forces:** 7,500 (1994). **Paramilitaries:** 1,000 Transport Police.

STATISTICS

DEMOGRAPHY

Urban: 26% (1995). **Annual growth:** 3.5% (1992-2000). **Estimate for year 2000:** 2,000,000. **Children per woman:** 4.7 (1992).

HEALTH

One **physician** for every 4,762 inhab. (1988-91). **Under-five mortality:** 101 per 1,000 (1995). **Calorie consumption:** 93% of required intake (1995). **Safe water:** 93% of the population has access (1990-95).

EDUCATION

Illiteracy: 30% (1995).
School enrolment: Primary (1993): 120% fem., 120% male. Secondary (1993): 55% fem., 49% male. **University:** 3% (1993). **Primary school teachers:** one for every 29 students (1992).

COMMUNICATIONS

92 **newspapers** (1995), 85 **TV sets** (1995) and 90 **radio sets** per 1,000 households (1995). 3.1 **telephones** per 100 inhabitants (1993). **Books:** 101 new titles in 1995.

ECONOMY

Per capita GNP: $2,800 (1994). **Annual growth:** 6.6% (1985-94). **Annual inflation:** 11.7% (1984-94). **Consumer price index:** 100 in 1990; 164.2 in 1994. **Currency:** 3 pula = 1$ (1994). **Cereal imports:** 133,000 metric tons (1993). **Fertilizer use:** 9 kgs per ha. (1992-93). **Imports:** $1,638 million (1994). **Exports:** $1,845 million (1994). **External debt:** $691 million (1994), $479 per capita (1994). **Debt service:** 4.3% of exports (1994). **Development aid received:** $127 million (1993); $90 per capita; 3% of GNP.

ENERGY

Consumption: 380 kgs of Oil Equivalent per capita yearly (1994), 55% imported (1994).

---

drew claims for a non-aggression treaty with Botswana. There had been repeated skirmishes in the border areas due to the Botswana government's support of the African National Congress in its struggle against apartheid. In 1987 South Africa put pressure on the country by blocking roads leading to Gaborone, the capital of Botswana. In March, 1989 nine South Africans were expelled from the country for "security reasons", and in May, five ANC members were arrested by Botswana troops, accused of illegal possession of arms.

[12] In 1989 elections were held for the National Assembly. They were won by the Botswana Democratic Party, which obtained 31 of the 34 seats in dispute, with the Botswana National Front obtaining 3 seats. Quett Masire was consequently re-elected president.

[13] In the 1988-89 period growth reached 8.9 per cent, but the following year fell to 4.8 per cent. Mean-

while, unemployment hit 35 per cent, while in the private sector a large number of jobs were held by foreigners, in particular at the managerial and technical levels.

[14] In spite of the economic crisis, the country was not eligible for loans from the African Development Bank, due to the fact that its national currency, the pula, was considered to be one of the strongest in the region.

[15] A profound conflict of political and economic interests led to an increase in government corruption, culminating in the resignation of several ministers of state. In 1991, three of the seven opposition parties created a political front to oppose the ruling Botswana Democratic Party.

[16] In spite of the almost constant economic expansion of the last few decades, in 1991 the country faced the most serious strikes since independence. Workers in the public sector demanded a salary rise of

154 per cent, and the government responded by dismissing 18,000 public employees.

[17] In 1992, unemployment came close to 25 per cent. In an attempt to increase employment and raise the GDP, the government launched a series of incentives for opening industries not linked to the mining sector. As a result of the drought, a state of emergency was declared by government authorities. Public spending was drastically reduced, which meant that a third of the manpower employed directly or indirectly by the state was laid off.

[18] Meanwhile, Botswana started a policy of rapprochement with the United States and intensified negotiations with the South African De Beers group to try and reduce this company's control over 80 per cent of diamond exploitation. In Octo-

ber 1994, the BDP won the legislative elections - despite losing 9 seats - winning 26 of the 40 contested positions.

[19] In February 1995, the pardon granted the three perpetrators of the "ritual sacrifice" of a girl in Mochudi provoked violent demonstrations. For the BNF, the social inequalities - the greatest in the world according to some studies - and unemployment were the real causes of the outburst. Meanwhile, in 1996 the nation continued to depend on the market price of export minerals to a large extent, as these represent 47 per cent of its income.

Brazil

Brasil

Population: 159,128,000 (1994)
Area: 8,511,970 SQ KM
Capital: Brasilia

Currency: Real
Language: Portuguese

[1] It is estimated that in the early 16th century Brazil had between 2 and 5 million inhabitants, belonging mainly to the Tupi, Guarani, Carib and Arawak linguistic groups. Some were hunter-gatherers, others farmed, and they were often skilled weavers and sensitive artists. When the Europeans arrived, many of these "Indian" cultures had already reached their apogee and were gradually declining.

[2] It is clear that the Portuguese intended to expand their domain well before Pedro Alvarez Cabral landed at All Saints' Bay in 1500. Under the 1494 Tordesillas Treaty between Spain and Portugal, the line dividing the future colonial empires was shifted further west, assuring Portugal a greater share of the land in the New World.

[3] The colonists dominated the indigenous population and established a succession of trading posts along the coast where they traded for brazilwood, which was used as a dye by the European textile industry. Portuguese settlement was spurred on by the fear that the French might seize Brazilian territory. Extraction of brazilwood was succeeded by sugar cane farming, initially employing native labor.

[4] The indigenous societies were non-hierarchical, the appropriation of each other's labor was unknown and they systematically resisted slavery. They were also extremely vulnerable to European diseases. This led to the Portuguese decision to import African slave laborers to the colony.

[5] It is calculated that more than 3.5 million African slaves were shipped to Brazil between 1532 and 1855, with most of the profit from this horrific trade going to Britain.

[6] Thousands of Africans escaped slavery, fleeing from the coastal plantations to seek refuge in the dense forests. There Africans, indigenous Americans and their mixed descendants joined forces in constant war against the colonial military expeditions. They also founded stable settlements called *quilombos* or *mokambos*, the most famous of which was at Palmares in northern Brazil, from 1630-1695, where the legendary Zumbi led the struggle to preserve an African identity. The Brazilian anti-racist movement still commemorates the date of Zumbi's death, November 20, as Negro Consciousness Day.

[7] Apart from determining Brazil's ethnic make-up, Africans also played an important role in building the country. Their presence and availability did not always alleviate the burden of the indigenous peoples. Some landowners who were unable to purchase slaves organized groups of *bandeirantes*. They attacked Spanish enclaves, particularly the Guaira Jesuit missions, where they could capture natives already resistant to common European diseases and trained in plantation agriculture. *Bandeirante* manhunts were so predatory that the missions were gradually pushed further and further South until they reached the "Seven Towns", currently in the state of Rio Grande do Sul.

[8] When Portugal was annexed by Spain in 1580, there were major consequences for Brazil. All of South America became Spanish territory and the Tordesillas border lost effect, encouraging *bandeirante* penetration to the interior. Brazil also found itself involved in a war of independence between the Netherlands and Spain; Flanders and the Netherlands were inherited territo-

PUBLIC EXPENDITURE

% of GNP — 1992

MILITARY 0.7 · EDUCATION 4.6 · HEALTH 2.8

WORKERS

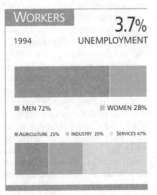

1994 — **3.7%** UNEMPLOYMENT

MEN 72% · WOMEN 28%

AGRICULTURE 25% · INDUSTRY 25% · SERVICES 47%

LAND USE

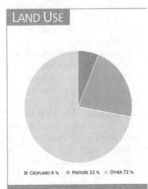

CROPLAND 6% · PASTURE 22% · OTHER 72%

ries of the Spanish crown. Between 1630 and 1654 the Dutch took Pernambuco after a failed attempt to conquer Bahia, which was repulsed by the joint efforts of native Americans, Africans and Portuguese.

[9] When Spain and Portugal subsequently separated, their colonial borders had shifted and the Tordesillas line could not be reinstated. A crisis in the sugar trade triggered the search for other sources of wealth and, in 1696, a *bandeirante* expedition struck the first gold in today's state of Minas Gerais. Gold mining peaked in the 18th century and the "sugar cycle" was superseded by the "gold cycle". The local ruling class began

> From 1939 to 1954, the economic policy of Getulio Vargas virtually dominated Brazilian life, producing the transition from agricultural to industrial economy.

to benefit from an expanding export economy, and soon expressed a growing desire (and later a political platform) to dispense with the Portuguese role as brokers in their trade with Europe. The first moves towards independence came in the late 18th century and were rapidly crushed by the colonial power. The most well remembered figure of the Brazilian liberation movement was Ensign Tiradentes, Joaquim José da Silva Xavier, who was executed in 1792 as a result of his leading role in the Minas Conspiracy of 1789. This Conspiracy became the symbolic beginning of the Brazilian independence movement.

[10] When Napoleon invaded the Spanish peninsula in 1808 the Portuguese King Dom João VI had to transfer his court to Brazil, making the country semi-independent. Portugal ceased to be an intermediary, and Brazil dealt directly with its main customer, UK. British prime minister William Pitt said "the South American empire and Great Britain will be tied forever by bonds of exclusive trade." So, the Brazilian trade bourgeoisie

prospered at the expense of other sectors linked to the Portuguese monopoly; the 1821 Oporto revolution in Portugal was an attempt to reinstate their trade monopoly. When the king returned to Lisbon, the local trade bourgeoisie, determined to secure their gains, declared independence with British blessing. Brazil became an empire and Pedro I, formerly prince-regent, became its emperor.

[11] The Empire lasted from 1822-1889. During this time Brazil consolidated its national unity and extended its borders over the areas settled by *bandeirantes* in the 17th and 18th centuries. This period of territorial expansion included a failed attempt to annex the Cisplatine Province (now Uruguay) that ended in the creation of an independent counntry between Brazil and Argentina in 1828; the war of the Triple Alliance, in which Brazil annexed 90,000 square kilometers of Paraguayan territory (see The Triple Alliance); and, towards the end of the century, the annexation of Bolivia's Acre to Amazonia.

[12] Brazil's economy continued to depend on large estates, exports of tropical fruit and vegetables, and slave labour. The abolition of slavery in 1888 accelerated the fall of the monarchy, but brought little change in the social conditions of former enslaved workers. As illiterates, they were denied the vote and so excluded from political representation.

[13] For many years the main economic activity was coffee production, which was the power base of the ruling oligarchy. The advent of the Republic did little to modify this situation even though it was faced with continuous uprisings and armed revolutions. The rebellions in the South, fought by the Rio Grande *gaúchos*, were helped by fighters from the River Plate countries. This assistance occurred in the Farrapos War (1835-1845), the 1893 federalist revolt and the Santo Angelo uprising.

[14] But it was only with the Revolution of 1930 - a new stage in the bourgeois revolution in Brazil which was unleashed towards the end of the 19th century - that the oligarchy was to lose its predominance, already corroded by the world crisis which had wiped out the coffee economy. Within this context there was an acceleration of the country's industrialization.

[15] From 1939 to 1954, the eco-

nomic policy of Getulio Vargas virtually dominated Brazilian life, producing the transition from agricultural to industrial economy. His "import substitution" model gave priority to national industrial development, especially in iron and steel during World War II. Vargas held power from 1937 to 1945 as dictator of the New State and, in 1950, was elected constitutional president. Nationalism and reformist defence of workers' rights were the two outstanding features of his government.

[16] In 1953, during Vargas's last term, the Petrobras state oil monopoly was created, and social security laws were passed. In August 1954, he committed suicide, leaving a letter accusing imperialism and its internal accomplices of blocking his efforts to govern according to popular and national aspirations.

[17] With Vargas' death, the economic policies of preceding decades were called into question. As a result, progress in nationalist proposals and plans for independent industrialization slackened. Juscelino Kubitschek's administration (1956-1961) promoted development and admitted transnationals to the Brazilian market, granting them exceptional privileges. Brasilia was built during his term in office to mark a new era in the country's economic development. In 1960, the federal capital, previously Rio de Janeiro, was transferred to the new city. Kubitschek's successor, Janio Quadros, initiated some changes in foreign policy, but resigned in peculiar circumstances barely seven months after taking office.

[18] In September 1961, Vice-President João Goulart, the Labor Party leader and Getulio Vargas' political heir, occupied the presidency. High-ranking military officers opposed his appointment, but the new president was backed by a civilian/military movement which wanted legal government, led by Leonel Brizola, governor of Rio Grande do Sul. A compromise parliamentary solution was adopted, with Tancredo Neves becoming Prime Minister. The presidential system was instituted in January 1963 after a national plebiscite. Goulart then tried to introduce a series of "basic reforms", including agrarian reform and legislation to regulate profit transfers abroad by foreign companies. A US-backed military coup deposed him on April 1, 1964.

The military coup followed the doctrine of national security, a widespread political philosophy of the time which was adopted by the Southern Cone countries, underpinning a succession of military dictatorships in Uruguay, Chile and Argentina.

[19] The new government passed Institutional Act No. 1, which repealed the 1946 liberal Constitution, allowing the revocation of parliamentary mandates and the suspension of political rights. A string of arrests all over the country forced major political leaders into exile: Joao Goulart, Leonel Brizola, Miguel Arraes and, later even Juscelino Kubitschek, left the coun-

> Goulart tried to introduce a series of "basic reforms", including agrarian reform and legislation to regulate profit transfers abroad by foreign companies.

try. Others, like Luis Carlos Prestes and Carlos Marighela kept up their fight underground.

[20] The Supreme "Revolutionary" Command appointed army general Humberto de Alencar Castello Branco head of government for the remainder of the constitutional period, but his mandate was later conveniently extended until 1967. In the October 1965 state elections, opposition candidates won in Rio de Janeiro and Minas Gerais. The military retaliated with Institutional Act No. 2, stating that the president would henceforth be appointed by an electoral college, under military control, and banning existing political parties. A two-party system was created guaranteeing a majority to the pro-government National Renewal Alliance (ARENA), excluding the opposition Brazilian Democratic Movement (MDB) from power.

[21] In January 1967, new authoritarian Constitution came into effect, and two months later army general Arthur Da Costa e Silva was appointed president. In December of the following year, to confront mounting popular opposition, Insti-

ENVIRONMENT

There are five major regions in Brazil. The Amazon Basin, in the North, is the largest tropical rainforest in the world. It consists of lowlands covered with rain forest, criss-crossed by rivers. The Carajás mountain range contains one of the world's largest mineral reserves, rich in iron, manganese, copper, nickel and bauxite. The economy is mainly extractive. The Northeastern "sertao" consists of rocky plateaus with a semiarid climate and scrub vegetation. Cattle raising is the main economic activity. The more humid coastal strip, situated on the "Serra do mar"(Coastal Sierra), has numerous sugar cane and cacao plantations. In the Southeast, the terrain consists of huge plateaus bordered in the East by the Serra do Mar mountain range. The main crops are coffee, cotton, corn and sugar cane. The southern plateau, with sub-tropical climate, is the country's main agricultural region, where coffee, soybeans, corn and wheat are grown. In the far South, on the Rio Grande do Sul plains, cattle raising is the main economic activity. Finally, the mid-West region is made up of vast plains where cattle raising predominates. With the devastation of the Amazon region, caused by unrestrained felling of trees, the habitats of many species have been destroyed.

THE SOCIETY

Peoples: Brazilians come from the ethnic and cultural integration of native inhabitants (mainly Guaraní), African slaves and European (mostly Portuguese) immigrants. Arab and Japanese minorities have also settled in the Rio-Sao Paulo area. Contrary to what is commonly admitted, racial discrimination does exist and the groups fighting it are rapidly gaining strength. **Religion:** Most are baptized Catholic; but there is considerable merging into syncretic Afro-Brazilian cults (macumba and umbanda).
Language: Portuguese is the official and predominant language.

THE STATE

Official Name: República Federativa do Brasil. **Administrative Divisions:** 26 States, 1 Federal District. **Capital:** Brasilia, 1,492,500 inhab. (1991). **Other cities:** São Paulo 9,393,700 inhab.; Rio de Janeiro 5,473,900 inhab.; Salvador 2,070,300; Belo Horizonte 1,529,600 inhab.; Recife 1,297,000 inhab.; Porto Alegre 1,237,200 inhab.; Manaus 1,005,600 inhab. (1991). **Government:** Fernando Henrique Cardoso, President since 1995. Bicameral legislature. **National Holiday:** Sept. 7, Independence Day (1822). **Political Parties:** Brazilian Social Democratic Party (PSDB), led by Senator Artur da Távora and President Fernando Henrique Cardoso; Party of the Liberal Alliance (PFL), led by Jorge Bornhausen; Party of the Brazilian Democratic Movement (PMDB), led by Luis Henrique da Silveira; Worker's Party (PT), led by Rui Falcao; People's Reform Party (PPR), led by senator Espiridião Amin; Democratic Labor Party (PDT), led by Leonel Brizola; Brazilian Labor Party (PTB), led by Jose E. de Andrade Vieira; the Communist Party split into two groups: orthodox communists (PC), led by Oscar Niemeyer, and the People's Socialist Party, led by senator Roberto Freire. In addition, there are countless other minor national or provincial organizations. **Social Organizations:** Workers are grouped together primarily in the Sole Group of Labor Unions (CUT), the General Confederation of Workers (CGT) and the Labor Union Forum. Many labor unions do not belong to any of these, preferring to remain independent. Movement of the Landless (MST), an association of workers without land whose agenda is agrarian reform in rural areas, and land for the construction of housing, in urban areas. National Union of Indigenous Peoples (UNI), an association of Brazil's different indigenous groups. Pastoral Commission of the Earth (CPT) and Indigenous Missionary Council (CIMI), pastoral groups of the Catholic Church involved in social action in these areas. Defence Network of the Human Race (REDEH), an eco-feminist organization. "Torture Never More", state groups committed to the defence of human rights. **Armed Forces:** 295,000 troops (1995). **Paramilitaries:** 243,000 Public Security Forces.

---

tutional Act No. 5 was passed, granting full autocratic powers to the military regime. In August 1969, Costa e Silva was replaced by a Military Junta that remained in power until the following October, when another army general, Emilio Garrastazú Médici, former head of the National Information Services (SNI), was brought into presidential office. The Médici administration was marked by extremely violent repression of both the legal and illegal opposition, while his economy ministers fed middle class consumerism. In 1974, army general Ernesto Geisel, was appointed president by the Junta. He put an end to the state oil monopoly, signed a controversial nuclear agreement with West Germany and granted further prerogatives to foreign investors. Arms are an important source of earnings, and Brazil is the world's fifth-largest producer.

[22] Under the Geisel administration, there had been a "slow and gradual relaxation" of political controls, which the ruling armed forces expected would institutionalize the regime while preserving its basically authoritarian nature. In 1974 and 1978, despite media censorship, the MDB achieved significant victories at the polls. At the end of his mandate, Geisel delivered the reins of government to army general and SNI member, João Baptista Figueiredo.

[23] Figueiredo came to power in March 1979, announcing that he intended to complete the process of political relaxation. A month later, a strike by 180,000 metalworkers in Sao Paulo, led by Luis Inácio "Lula" da Silva, was settled without violence, as a result of negotiations between the Labor Minister and the union. In November, the Brazilian Congress passed a bill granting a far wider amnesty to political opponents than the Executive had originally intended. As a result, political prisoners were released and exiles began to return.

[24] The after-effects of the monetary policies applied by successive military administrations were felt during the Figueiredo administration. The foreign debt spiralled, and in the early 1980s, Brazil had to change from being an importer to an exporter in its search for funds to pay off the interest on its $100 billion debt.

[25] Social inequalities became more acute. In 1985, official statistics registered six million unemployed and thirteen million under-employed, in cities alone. In July 1985 Ministry of Labor technicians declared that "not even 7 per cent growth per year for two consecutive decades would be enough to improve these peoples' living conditions".

[26] This explosive situation spelled the military regime's downfall. The opposition's electoral victory in November 1983 reflected enormous popular discontent. The government held only 12 states; the opposition won 10, including the economically decisive states of São Paulo, Rio de Janeiro and Minas Gerais which, between them, account for 59 per cent of the population and 75 per cent of the country's GDP. Of these ten, the PMBD (Brazilian Democratic Movement Party, formed in 1980 from moderate MDB members and the Popular

Party) won nine state governments and the PDT (Democratic Workers Party), one; Rio de Janeiro.

[27] The military's fate was sealed. Tancredo de Almeida Neves, governor of Minas Gerais, was the chief coordinator of the opposition front. A popular campaign for direct elections in 1984 failed, but the opposition took the Electoral College. Tancredo Neves was elected president, and José Sarney, who two weeks earlier had been president of the government party, became vice-president. Tancredo announced plans for a new social order: the New Republic.

[28] Neves' intentions were to lay the foundations of a solid political democracy and to correct distortions in Brazil's recent development. Effective modernization and democratization depended on agrarian reform, re-negotiation of the foreign debt, and economic growth. On March 14, 1985, the day before his inauguration, Tancredo Neves was hospitalized and rushed into surgery. José Sarney was sworn in as interim president. On April 21,1985, Tancredo died. Sarney legalized the communist party and other left-wing organizations, some after 20 to 40 years of illegal existence. "Redemocratization" began to take place with the approval of direct elections for presidents and mayors of state capitals. A national assembly was convened for January 1987 to draft a new constitution, and illiterates, over 21 million people, were granted the vote.

[29] In 1986 Sarney declared a moratorium on the foreign debt and launched the "Cruzado Plan" to fight inflation. The plan produced impressive short-term results: a boom in consumption and economic growth. This startling prosperity coincided with the November 1987 parliamentary and state governmental elections. Not surprisingly, the PMDB won a landslide victory. The new congress faced the task of drafting a new constitution which would formally inaugurate the return to democratic government. It set Sarney's term at five years, resisting considerable pressure to ignore the groups who wanted Tancredo's promise of reducing the term to four years

[30] The Cruzado Plan could not be maintained without fighting financial speculation and halting pressure from the financial sector. Two days after the election, the price

freeze came to an end and inflation leapt to a monthly rate in double figures. The planned agrarian reform was gradually reduced.

[31] In the 1988 municipal elections, the left-wing and center PDT, PT (Workers Party) and PSDB (Brazilian Democratic Socialist Party) parties began to make inroads on the PMDB's flagging popularity.

[32] Violence against grassroots social movements went on once democracy was reinstated. In December 1988, the murder of Chico Mendes made this situation internationally known. Mendes had led the "seringueiro" (rubber tappers) movement and the Amazonian indigenous peoples, in their struggle to secure the right to live and work in the forest, and protect its future.

[33] Direct presidential elections, the first in 29 years, were held in November and December 1989. Nearly 80 million electors voted, and the final round was fought between conservative candidate, Fernando Collor de Mello, and Workers' Party leader, Luiz Inacio "Lula" da Silva. Collor de Mello, a young politician who had founded his career under the military regime, won the second round with 42.75 per cent to his opponent's 37.86 per cent of the vote.

[34] On taking office, on March 16, 1990, Collor de Mello announced the New Brazil Plan, an attempt to stem inflation by confiscating 80 per cent of the country's available financial assets.

[35] Collor adopted the neoliberal model of opening up the economy, with the privatization of state enterprises and the reduction of tariff barriers controlling the importation of foreign products. However, he failed to control inflation, to hold off a recession and to reduce unemployment.

[36] In the complex economic scenario, Collor's government was criticized for the social situation and an escalation of violence. In 1991, in Rio de Janeiro, more than 350 street children were murdered. The parliamentary commission created to investigate this atrocity estimated that over 5,000 children have been killed in this way in the past three years. The same commission announced that the persecution of homeless children; seven million - according to estimates of the Brazilian Center for Childhood and Adolescence - is carried out by paramilitary groups financed by shop owners.

[37] The indigenous peoples, meanwhile, are under the constant threat of "progress". They have suffered epidemics, the degradation or loss of natural resources, pollution, and a systematic fall in their standard of living. The Brazilian indigenous peoples are being decimated by diseases unknown to them until now, by people looking for metals and by the police who attack and murder them. Many are refraining from having children as they feel too threatened.

[38] The disappearance of indigenous cultures is linked to the accelerated destruction of the tropical forest, by large landowners to exploit its mining and logging wealth or to turn it into grassland. This predatory policy prevents the regeneration of the forest, which in turn causes soil exhaustion. Several local and international bodies such as the UN Economic and Social Council have called the Brazilian government's attention to the need to adopt measures to guarantee the preservation of the tropical forest and the rights of its traditional inhabitants. Although progress has been made in this direction, the desired results have yet to be achieved.

[39] In September 1991, thousands of people belonging to the Landless Movement of Brazil (MST) staged a march in the State of Rio Grande do Sul, where there are 150,000 families of landless farmers alongside 9 million hectares of land not in production. The protest demanded that the 4,700 million cruzeiros earmarked for agrarian reform should be spent. Up till then, only 800 million had actually been used.

MERCOSUR

On October 29, 1991, the Treaty of Asunción was ratified by Argentina, Brazil, Paraguay and Uruguay. It had previously been signed in the Paraguayan capital on March 26 of the same year, by presidents Carlos Menem, Fernando Collor, Andrés Rodríguez and Luis A. Lacalle. This was the first step in the creation of the Southern Common Market (Mercosur), an economic integration zone effective since January 1, 1995, when goods and services began to circulate almost freely.

[2] The four countries, with a total population of almost 200 million inhabitants and a combined GDP of almost US$550 billion, committed themselves to a gradual elimination of tariffs (taxes on imported goods) for merchandise from Mercosur countries. In principle, internal barriers were to be totally eliminated; free circulation of goods and services was scheduled within four years. Exceptions were made for the less powerful members (Uruguay and Paraguay).

[3] The Mercosur countries have also agreed to eliminate non-tariff barriers, such as laws protecting national products or protectionist tax breaks to certain industries, which create unfair competition for imported goods. The aim of this measure - according to Mercosur's four member governments - is to give Mercosur greater power in negotiating with other countries or economic blocs, as well ensuring more favorable prices for national consumers.

[4] The treaty also establishes a common external tariff for the importation of products from non-member countries, as well as the possibility of dealing as a united economic bloc with other countries, economic communities and even world negotiation entities such as the General Agreement on Trade and Tariffs (GATT).

[5] Two administrative levels operate within Mercosur: the Common Market Council and the Common Market Group. The former is the more important of the two, and is made up of the ministers of Foreign Affairs and of the Economy, from each of the four countries. The latter is the executive branch and is made up of representatives of the aforementioned ministries and of the Central Banks of each of the four countries.

[6] For Mercosur's smaller members it is difficult to carry out the reconversion of industry which is essential for the scheme to work, given the fact that the two larger nations are in a position to place more competitive products throughout the Mercosur region. Uruguay's labor union group (PIT-CNT) called for meetings to be held with labor union organizations from all Mercosur, with the aim of seeking solutions to the real possibility of Argentine and Brazilian products "flooding" Uruguayan and Paraguayan markets, and thereby affecting their economies.

[7] Given the different stages of relative development, in mid-1994 the four governments decided to readjust the timetable which had been announced in 1991, in order for the common external tariff to go into effect in 2001. Uruguay and Paraguay will have an additional five years to adapt their tariffs to the 14 per cent which had been agreed upon. At the time of this renegotiation, the latter two countries maintained certain advantages with regard to their free zones and trade agreements with their major trading partners, which in some cases will remain in effect until the year 2001, and in others, until 2006.

[8] Mercosur, though, is far from being a single market. Some experts are concerned that disputes may increase now that most tariffs are down to zero. Vulnerable areas include Argentina's sugar industry, which would have a hard time competing against Brazil's low-wage sugar producers, and Brazil's wine and fruit industries, which lag behind Argentina's in quality and productivity. Brazil's reduction of car supplies imports from Argentina in 1995 brought some clashes between the two biggest partners of the customs union.

[9] But despite these differences, the four countries began to put in practice their plan to enlarge the Mercosur. Chile, Bolivia and Venezuela were the first targets for this plan. As for the rest of the world, the Mercosur countries have set preferences: to establish agreements with the United States and the European Union are the main objectives.

[10] In June Chile signed the first document stating its decision to join the group. Two months later, in the 10th presidential meeting of the alliance held in San Luis, Argentina, the four presidents asked their collegues of Chile and Bolivia, Eduardo Frei Ruiz and Gonzalo Sánchez de Lozada to participate in the debates. They also signed a statement of "democratic guarantee", committing the Mercosur to the democratic system of government, and the respect of human and civil rights.

[11] On October 1 1996, Chile made effective an agreement with the Mercosur, though not joining it as a full member. Custom fares were reduced 40 per cent and both Mercosur and Chile increased their trade possibilities, thus paving the way to reach Pacific and Atlantic markets from within the five countries' boundaries. Chilean exports to Mercosur in 1995 totalled US$1.77 billion, a 31 per cent jump over the previous year and representing 11 percent of total exports. Seventy per cent of Chilean foreign investment takes place in the Mercosur countries. This reached US$7.53 billion as of June 30, 1996.

40 Late in September 1991, the government devalued the Brazilian currency by 20 per cent over 2 days, accelerating the spiralling inflation. That year prices rose by 400 per cent. The increase in bank interest rates, near to a thousand per cent a year, triggered massive redundancies in the industrial sector. Over a million people lost their jobs in the industrial city of Sao Paulo alone.

41 In May 1992, a Parliamentary Investigating Commission was formed to study claims made by President Collor's brother. According to his statements, the former treasurer of Collor's electoral campaign, Paulo César Farías, was in charge of official corruption within the government. This took the form of influence-peddling, in exchange for deposits made to the president's personal account.

42 Protest demonstrations against corruption and the discovery of evidence implicating other government figures in these schemes led all the country's political parties to vote for the president's impeachment.

43 In the month of September, Congress voted to relieve the president of his duties, so that he could be tried. Vice-president Itamar Franco became interim president. In December 1992, Collor was found guilty by the Senate of "criminal responsibility"; his sentence included the loss of his presidential mandate and the suspension of his political rights until the year 2000. Franco officially assumed the presidency until the end of the presidential term in progress.

44 Franco was supported by a wide range of parties, cutting across partisan lines. The president tried to project an image of austerity and ethics upon the management of public affairs. After several changes in the ministry of economics, Fernando Henrique Cardoso was appointed to this cabinet position.

45 Nevertheless, the changes launched by the government failed to alter Brazil's social situation, marked by misery and growing violence. According to figures provided by the Pastoral Commission of the Earth, in 1992 there were 15,042 rural slaves, triple the number recorded the previous year. According to data gathered by the Federal Bureau of Statistics, nearly 4 million people living in rural areas work under conditions of virtual slavery. Collor's administration had designated 20 million hectares as

new indigenous zones, a notion which alleviated but did not fully solve the dire situation of the indigenous peoples.

46 In the cities, violence reached unprecedented levels. The police have been discredited, as an institution, as many policemen are accomplices or perpetrators of organized crime.

47 Three incidents, of the many that have taken place in recent years, serve to illustrate the proportions which this violence has taken. Toward the end of 1992, 111 inmates of the prison at Carandiru, São Paulo, were killed by members of the Military Police, who had entered the premises to deal with a dispute between rival groups of prisoners. Postmortem examinations found that 85 of the prisoners had been executed after having surrendered. In July 1993, members of the Military Police murdered 8 street children while they were asleep on the doorsteps of the Church of the Candelaria, in Rio de Janeiro. The following month, 50 men entered the *favela* (shantytown) of Vigario Geral and killed 21 innocent people, to avenge the death of four military police who had been killed in a confrontation with drug dealers.

48 The general outcry over these incidents did lead to some official measures. In Rio de Janeiro, for the first time ever, a judge handed down prison sentences to the leaders of "jogo do bicho", an illegal lottery which has been in existence for decades and is intrinsically linked to organized crime. In São

Paulo, a police investigator received the maximum prison sentence (516 years) for causing the death of 18 prisoners, who died of asphyxiation in a police station cell in 1992.

49 However, the best examples of projects confronting the prevalent poverty come from the non-gov-

> The Brazilian indigenous peoples are being decimated by diseases unknown to them until now, by people looking for metals and by the police who attack and murder them.

ernmental sector. The "Citizens' Action against Hunger and for Life", founded in 1993 by Herbert de Souza, organized tens of thousands of committees throughout the country to collect and distribute food, and obtain more jobs by putting pressure on both businessmen and the government to cooperate. Two million people - mostly tenants, housewives, religious and labor union groups - belong to this grassroots movement. By August 1994, over 4 million families had received food aid.

50 Towards the end of 1993, the finance minister Cardoso presented

the Plan Real, an economic stabilization project, which ended index-linking and created a new currency, the real, in July 1994. The success of this anti-inflationary policy rapidly made Cardoso the most popular candidate for the October elections. In the first round he beat the previuos favourite Luis Ignacio "Lula" da Silva, of the WP.

51 The Cardoso administration began with the privatization of large State-owned companies like Petrobras and the telecommunications network, but the economic recession caused increasing unemployment, union conflicts and crime. Land occupations were carried out by the Landless Movement. On many occasions these ended with many people killed by the paramilitaries and the impunity of the guilty parties. In April 1996, police officers rounded up, tortured and killed 19 rural workers on the outskirts of Eldorado de Carajas, causing so much public outcry that President Cardoso created a Ministry of Agrarian Reform and announced some plans for land redistribution. Half way through the year, the government announced that around 40,000 families had already benefited from the official land distribution plans.

In the 1996 municipal elections in Pilar, a village in the northeastern state of Alagoas, Frederico the goat - better known for having been the local mascot which represented the Brazilian football team in the Atlanta '96 Olympic Games - was nominated for Mayor. The alternative political formula was completed with Juliete Maria, a fake transvestite played by a radio show host. On September 5, when the "candidates" were being led through the streets of Pilar in a demonstration, the carriage of Frederico and Juliete Maria was hit by a burst of machine-gun fire. There were no injuries. The person responsible for the attack was not identified.

Brunei

Population: 280,000 (1994)
Area: 5,770 SQ KM
Capital: Bandar Seri Begawan

Currency: Brunei Dollar
Language: Bahasa Malaysia

Brunei

Today's Sultanate of Brunei is what remains of a 13th century Islamic empire that once covered most of the island of Borneo, from which it derives its name. Portuguese, Spanish and Dutch presence in the area occurred virtually uninterrupted from the 16th century. The colonial powers only attempted full occupation of this huge island at the beginning of the 19th century. While the Dutch were making inroads from the south, the Sultan of Brunei sought British support, and succeeded in remaining autonomous.

[2] In 1841, as payment for help in quelling a two year long rebellion, the sultan had to turn the province of Sarawak over to James Brooke, a strange figure who became a European "Rajah" over a Malayan state. In 1846 the British annexed the strategic island of Labuan, and in following years paved the way for the secession of the province of Sabah. In 1888, the British consolidated their position and established separate protectorates over Brunei, Sarawak and Sabah.

[3] Following World War II, despite Britain's efforts, the island began the decolonization process. An agreement signed with Rajah Brooke in 1946 made Sarawak and Sabah into British colonies while Kalimantan (former Dutch Borneo) gained independence in 1954 as part of Indonesia.

[4] All that was left of the British protectorate was the Sultanate of Brunei, reduced to a tiny territorial enclave between two Malaysian provinces, scarcely 40 km from the border with Indonesia. In 1929, the transnational company Shell discovered oil deposits in the area. In the following decades, drilling for oil and natural gas began, reaching current production rates of 175,000 barrels a day.

[5] In 1962, Sultan Omar Ali Saiffudin accepted a proposal from Malaysian Premier Abdul Rahman to join the Federation of Malaysia, which at the time included Sabah, Sarawak, Singapore, and the provinces of the Malayan peninsula. The Brunei People's Party, (Rakyat) which held 16 seats in the 33-member Legislative Council, was not keen on the idea and opposed the move, proposing instead the creation of a unified state comprising Northern Borneo, Sarawak and Sabah, but excluding peninsular Malaya.

[6] A mass uprising broke out in December 1962, staged by the *Rakyat*, backed by the *Barisan Sosialis* of Singapore, with support from the anticolonialist Sukarno regime in Indonesia. The rebels opposed integration to the Federation of Malaysia, demanding participation in administration and the end of the autocratic regime. The rebellion was rapidly stifled, the party outlawed and the leaders arrested or forced into exile.

[7] Finally, in spite of ethnic, historical, and cultural ties with Malaya, Sultan Omar decided to keep his sultanate out of the Federation.

> **Although popular discontent has not broken out again since the 1962 rebellion, observers are not particularly convinced of the sultanate's stability.**

He was not satisfied with arrangements for power-sharing with the other Malayan rulers and least of all with Federation hopes of a share in his territory's oil resources.

[8] In 1976, with Malaysian prompting and UN support, an opportunity arose to renegotiate the anachronistic colonial statute, when the newly-elected Malaysian prime minister Datuk Hussain Onn promised to respect Brunei's independence. In 1977, Brunei finally accepted independence, but postponed implementation until January 1, 1984.

[9] Power was formally transferred on schedule but celebrations were postponed until February 23, 1985, so that foreign guests could attend. One month later independence was formally proclaimed, and Hassanal Bokiah, son of Sultan Omar, who had abdicated in his favor in 1967, dissolved the Legislative Council and went on to govern by decree.

[10] Brunei embarked on independent existence in particularly favorable conditions for a Third World country. It has a relatively small population, a high per capita income, low unemployment, a generous social security system and considerable foreign exchange reserves. These reserves stood at $14 billion in 1984.

[11] Although popular discontent has not broken out again since the 1962 rebellion, observers are not particularly convinced of the sultanate's stability. The main sources of tension now are the power struggle within the ruling family and the presence of foreigners in all the key positions of public office, the economy and the armed forces.

[12] Other potential problems are perhaps even more worrying. Problems on the domestic front include poor basic education. In spite of advanced legislation for education, including a student transport allowance and free accommodation, the illiteracy rate keeps climbing; it was estimated at 45 per cent in 1982, posing a serious obstacle to filling civil service posts with Brunei nationals. There are also large ethnic minorities within the state;

Chinese influence is especially significant, as they constitute 25 per cent of the population and because Brunei has strong ties with neighboring Chinese communities which control some key areas of regional trade and industry.

[13] The Sultan is aware that the country depends on a non-renewable resource, and Brunei currently imports nearly 80 per cent of the foodstuffs it consumes. Consequently, with a view to achieving self-sufficiency in food production, he has attempted to diversify the economy and promote a new landowning class. Currently, only 10 per cent of the arable land is cultivated, and small farmers, especially rubber-tree growers, tend to emigrate to the city where they believe a higher standard of living awaits them.

[14] Economically, Brunei depends upon the complex interplay of transnational interests. The government's partnership in exploiting natural gas reserves with Brunei Shell Petroleum and Mitsubishi, a shipping contract with Royal Dutch

PROFILE

ENVIRONMENT

Brunei comprises two tracts of land located on the northwestern coast of Borneo, in the Indonesian Archipelago. It has a tropical, rainy climate slightly tempered by the sea. Rubber is harvested in the dense forests. There are major petroleum deposits along the coast. The country is one of the world's main exporters of liquid gas.

SOCIETY

Peoples: Malay 67%; Chinese 15%; Indian and other 18%.
Religions: Islam (official), Buddhism and Christianity.
Languages: Bahasa Malaysia (official), Chinese and English.
Political Parties: Brunei Solidarity National Party

THE STATE

Official Name: Islamic Sultanate of Brunei. **Administrative divisions:** 4 districts. **Capital:** Bandar Seri Begawan, 46,000 inhab. (1991). **Other Cities:** Kuala Belait 21,200 inhab.; Seria 21,000 inhab.; Tutong 13,000 inhab. **Government:** Hassanal Bolkiah Muizzaddin Waddaulah, Sultan and Prime Minister. **National Holiday:** January 1, Independence Day; July 16, the Sultan's birthday. **Armed Forces:** 4,900.

STATISTICS

DEMOGRAPHY

Annual growth: 0.3% (1992-2000). **Children per woman:** 3.1 (1992).

HEALTH

Under-five mortality: 10 per 1,000 (1995). **Calorie consumption:** 104% of required intake (1995).

EDUCATION

Illiteracy: 12% (1995). **Primary school teachers:** one for every 15 students (1991).

COMMUNICATIONS

104 **newspapers** (1995), 106 **TV sets** (1995) and 101 **radio sets** per 1,000 households (1995). 19.7 **telephones** per 100 inhabitants (1993). **Books:** 98 new titles per 1,000,000 inhabitants in 1995.

ECONOMY

Per capita GNP: $14,240 (1994). **Consumer price index:** 100 in 1990; 109.9 in 1994. **Currency:** Brunei dollars.

Shell and recent oil field concessions to Woods Petroleum and Sunray Borneo, introduce powerful new parties into the process of national decision-making.

[15] In March 1985, the government announced the creation of an Energy Control Board to supervise the activities of the Brunei Shell Petroleum Company, a company funded equally by the government and Shell.

[16] Membership of ASEAN and the UN, obtained in 1984, meant greater international support for the new nation.

[17] In Brunei, 20 per cent of the population lives below the poverty line of $500 a month. Also, 90 per cent of all consumer goods, including basic foodstuffs, are imported, as the country produces little other than oil. The vast amount of imports explains the extremely high cost of living.

[18] In early 1987, it was reported that a request from US colonel Oliver North for "non-lethal" aid to the Nicaraguan contras led the Sultan of Brunei to deposit a $10 million donation in a Swiss bank account.

[19] The state of emergency which was declared during the 1962 rebellion is still in force, and the country's military dependence on Britain and the US has increased. Under economic restructuring policies applied in recent years, only the military budget has grown.

[20] While foreign investments in other areas have remained unchanged or fallen, interests in the military sector have increased. In August 1988, 3,000 British soldiers and Hong Kong Gurkhas carried out joint maneuvers in the sultanate's tropical forests, and the US has begun to show an interest in installing military bases in Brunei, should its forces have to leave the Philippines.

[21] In early 1991, Sultan Hassanal Bolkiah freed 6 political prisoners who had been detained after the failed 1962 revolt. The liberation was ascribed to political pressure by the British government. This same year Brunei signed a contract for almost $150 million with the United Kingdom in order to modernize its army.

[22] In mid-1992, Brunei joined the Non-Aligned Movement, together with Vietnam and India. The commemoration of the 25th anniversary of Hassanal Bolkiah's reign stressed the role of the monarchy in affirming Brunei's national identity. Brunei's sultan, the only absolute monarch in the Far East, is also the richest person in the world.

[23] In September 1992, Brunei signed with other members of the Association of South East Asian

> The state of emergency which was declared during the 1962 rebellion is still in force, and the country's military dependence on Britain and the US has increased.

Nations (ASEAN) - Indonesia, Singapore, Thailand, Malaysia and Philippines - an agreement to create the first integrated market in Asia in 2007. This project stipulates the creation of "growth triangles" - association between some ASEAN members to deregulate trade in certain economic sectors, as a preparation for the overall liberalization planned for the year 2007. Thus, in March 1994, Brunei created - together with Philippines, Malaysia and Indonesia - a sub-regional market to intensify trade in toursim, fishing, and transport by sea and air.

[24] In February 1995, the Solidarity National Party of Brunei - the sole political organization in the country - organized its opening congress and declared its total support to the Sultan. Its leader, Abdul Latif Chuchu also spoke about a royal decree of 1984 which asserted that Brunei's national building is based on the principle of a democratic Malayan-Muslim monarchy. The Sultan's announcement that the state of emergency would be extended was interpreted as a sign of "openness" by the monarch. In October 1995, Brunei joined the World Bank and the IMF.

Bulgaria

Bulgaria

Population:	8,435,000 (1994)
Area:	110,910 SQ KM
Capital:	Sofia (Sofija)
Currency:	Leva
Language:	Bulgarian

B ulgaria was occupied by the Slav people who spread across the area between the Danube and the Aegean Sea in the 7th and 6th centuries BC. Some of the old Thraco-Illyrian population were expelled, and the rest were assimilated by the invaders. The Slavs worked the land in small organized communities.

2   The Bulgarians were among the non-European tribes who followed Attila. They were fierce warriors living from warfare and plunder. They first appeared in the region towards the end of the 5th century AD when they settled temporarily on the steppes north of the Black Sea and northeast of the Danube.

3   Some tribes disappeared and others were enslaved by the Turks. Those tribes led by Kubrat remained in this region until the mid-7th century when the Kazars bore down on them, forcing them into crossing the Danube. They reached as far as Moesia, at that time a province of the Byzantine Empire. Emperor Constantine IV formally recognized the State of Bulgaria in the year 681.

4   Constant invasions by foreign tribes, led to the new State consolidating its power in the center and southwest. Bulgarian leaders adopted the Slav language and culture. Prince Boris was initially baptized Roman Catholic, but in 870 he turned to the Orthodox Church.

5   While the Patriarch of Byzantium recognized the independence of the Bulgarian Church Rome refused to appoint a national patriarch. This support helped to consolidate the kingdom's power. Under Simeon (893-927), the Bulgarian State extended its domain as far as the Adriatic, subduing the Serbs, and becoming the most powerful kingdom of Eastern Europe.

6   Bulgaria's power declined after Simeon's death. Internal disputes among the nobility, the opposition of the peasants and renewed attacks from abroad led to its downfall. In 1014, Bulgaria lost all of its territory to the Byzantine Empire, which kept control for more than 150 years. After a massive uprising in 1185, the northern part of Bulgaria recovered its independence.

7   Under the reign of Ivan Asen II (1218-41) Bulgaria regained power and territory, including Albania, Epirus, Macedonia and Thracia. However, none of his successors ever managed to impose a central authority over these diverse areas where feudalism was the norm. By 1393 the whole of Bulgaria had fallen under Ottoman rule.

8   In the 17th and 18th centuries, after the wars with Austria and the unsuccessful siege of Vienna, the Ottoman Empire began to decline, though it still retained much of its territory. In the early 19th century, Bulgaria was still unheard of in Europe, despite the participation of Bulgarian volunteers in the Serbian and Greek uprisings. The former Bulgarian State was twice invaded by Russia in 1810 and 1828.

9   Throughout the invasion period the Bulgarians maintained their cultural identity; keeping their language, music and folklore alive. Under Turkish domination, the Greek Orthodox Church assumed religious leadership, suppressing the independent patriarch. Hence, Bulgarian monks were among the precursors of the national liberation movement.

10   The Bulgarian church fought for 40 years to recover its independence. In 1870, the Sultan gave permission for the church to create an exarchate with 15 dioceses under its jurisdiction. The first exarch (deputy patriarch) and his successors were declared schismatic and were excommunicated by the Greek partriarch; this further strengthened Bulgarian nationalism.

11   From 1876 onwards, a series of Bulgarian revolts was cruelly put down. Some Bulgarian volunteers joined the ranks of the Serbian and Russian armies, who went to war against the empire. One of Moscow's conditions for the Treaty of Saint Stefano, was the creation of a Bulgarian State. The European powers feared the creation of a Russian satellite in the Balkans so they blocked the motion.

12   In the 1878 Berlin Congress, the "autonomous province" of Rumelia was created in the south, and the State of Bulgaria in the north. Rumelia was nominally under the sultan's control, and Macedonia was to continue as part of the Ottoman Empire. The legal framework and the election of a governor for the new state remained in the hands of an assembly of selected people.

13   The assembly approved a liberal Constitution, establishing a constitutional monarchy. Prince Alexander of Battenberg, grandson of Alexander II of Russia, was subsequently elected. He assumed the Bulgarian throne in July 1878, swearing to uphold the Constitution. He suspended it two years later.

14   The prince set up a dictatorship, headed by the Russian general, LN Sobolev and other conservatives. The Russian emperor's death modified Alexander's behavior, making him more attuned to Bulgarian issues. In 1885, he supported the liberal rebellion in Rumelia, the governor there was replaced and union with Bulgaria was proclaimed.

15   The treaties of Bucharest and Top-Khane, signed in 1886, recognized Prince Alexander as the ruler of Rumelia and Bulgaria. However, he was subsequently taken to Russia against his will and forced to abdicate. Looking for someone who would be acceptable to Russia, as well as the rest of Europe, the Bulgarians finally appointed Ferdinand of Saxe-Coburg-Gotha as his replacement.

16   Although initially distrusted, Ferdinand was able to gain the support of Vienna, London, Rome and Russia. He then concentrated on the re-unification of the Bulgarians whose territory had been divided under the terms of the Treaty of Berlin. The prince proclaimed Bulgaria's independence in 1908, thereby breaking its nominal dependence upon Turkey.

17   In 1912, Bulgaria signed secret military agreements with Greece and Serbia. In October of that year, Montenegro, accompanied by its Balkan allies, declared war on Istanbul. In May, Turkey ceded its European dominions, north of a line between Enos, in the Aegean Sea, and Midia, on the Black Sea.

18   The allies did not agree with the distribution; Bulgaria confronted Greece and Serbia over the issue and Romania joined Serbia as its ally. The Second Balkan War quickly ended in Bulgarian defeat. In Bucharest, in August 1913, Macedonia was divided between Greece and Serbia, and Romania gained an area of northern Bulgaria, rich in natural resources.

19   In 1913, the Bulgarian government abandoned its traditionally pro-Russian stance, seeking closer ties with Germany. When World War I broke out the Bulgarian people and the army disapproved of the official policy, even though Serbia was beaten. Ferdinand surrendered to the Allies in 1918 and abdicated in favor of his son Boris.

20   Having lost much territory, Bulgaria was disarmed and forced to pay extensive war damages. With the restoration of the 1878 Constitution, elections were held in 1920. The anti-war reaction gave the Agrarian Party a wide margin. Working on a Soviet model, the government started radical agrarian reforms. The government was not pro-soviet, however, and local communists were persecuted.

21   Bulgaria joined the League of

LAND USE

- Cropland 39 %
- Pasture 17 %
- Other 44 %

PROFILE

ENVIRONMENT

Located on the Balkan Peninsula, Bulgaria comprises four different natural regions. The fertile Danubian plains in the north are wheat and corn-producing areas. South of these plains are the wooded Balkan Mountains, where cereals and potatoes are cultivated, cattle and sheep are raised, and the country's major mineral resources - iron ore, zinc and copper, are found. Cattle are also raised on the Rhodope Mountains in southern Bulgaria. South of the Balkan mountains there is a region of grassland, crossed by the Maritza River, where tobacco, cotton, rice, flowers and grapes are cultivated. Heavy metals, nitrates, petroleum derivatives and detergents have damaged the lower and mid-tributaries of the most important rivers. These flow into the Black Sea, which is highly polluted. A fourth of the country's woodland is currently suffering the effects of air pollution.

SOCIETY

Peoples: Most of the Bulgarian population is of Slav origin, but there are also people of Turkish, Armenian and Greek origins. **Religions:** Orthodox Christian and Muslim. **Languages:** Bulgarian (official); Turkish, Greek, and Armenian are also spoken. **Political Parties:** Union of Democratic Forces (a coalition made up of 16 parties); Bulgarian Socialist Party (formerly the Communist Party); Movement for Freedom and Human Rights (of the Turkish minority); Bulgarian Agrarian Party.

THE STATE

Official Name: Narodna Republika Balgaria. **Capital:** Sofia (Sofija), 1,114,000 inhab. (1993). **Other cities:** Plovdiv, 345,000 inhab.; Varna, 307,000 inhab. (1993). **Government:** Yelio Yelev, President; Zhan Videnov, Prime Minister **National Holiday:** March 3, National Liberation Day. **Armed Forces:** 102,000 **Paramilitaries:** Border Guards (Ministry of Interior): 12,000; Security Police: 4,000; Railway and Construction Troops: 18,000.

Nations and followed a conciliatory line of diplomacy for some time. However, its territorial losses and the pressure exerted by Bulgarians living abroad soon led to new tensions with its neighbors. Aleksandur Stamboliyski, the leader of the Agrarian Party and head of the government, was ousted from power and assassinated by a conspiracy of Macedonians and opposition figures in 1923.

[22] Aleksandur Tsankov assumed control of the government, heading a multi-party alliance which excluded the Liberal, Communist and Agrarian parties. Uprisings and armed activity by the opposition led to hundreds of executions and assassinatons. The government declared martial law and reinforced the army in order to avoid outright rebellion.

[23] In 1926, Tsankov resigned in favor of Andrei Liapchev, leader of the Democratic Party, to make way for more liberal policies. The government approved a partial amnesty, and permitted a reorganization of the Agrarian Party. In 1932, Liapchev's cabinet was made up of members of the Democratic, Liberal and Agrarian parties, known as the National Bloc.

[24] In 1934, fearing the effects of the worldwide economic depression and taking a cue from his neighbors, King Boris III supported the action of the conservative group, Zveno, who deposed Liapchev and set up a dictatorship. All political parties were proscribed, there was censorship of the press, the universities were closed and an ultra-right youth movement was established.

[25] Tension with Turkey eased, and in 1937 a peace and friendship treaty was signed with Yugoslavia. The following year, Bulgaria signed a non-aggression pact with the Balkan alliance, in exchange for the rearmament of the Bulgarian army. While the king was once again seeking rapprochement with Germany, the Bulgarian dream of re-establishing its former borders was gathering strength.

[26] In 1940, Germany made Romania return the parts of Bulgaria which it had won in the second Balkan War. Bulgaria signed the Antikomintern pact and German troops set up bases aimed at Greece and Yugoslavia on Bulgarian soil. In exchange, Bulgarian troops were allowed to occupy the part of Thracia belonging to Greece, and the part of Macedonia belonging to Yugoslavia, and part of Serbia.

[27] When Bulgaria refused to declare war on the Soviet Union, King Boris was assassinated, and a new pro-German government was formed. Growing anti-Nazi resistance, led by the communists, contributed to the formation of the Patriotic Front in 1942. The Republicans, left-wing Agrarians, Democrats and independents all subsequently joined.

[28] In May 1944, paralyzed by the civil war, the pro-German Boshilov resigned, and was replaced by Bagrianov. While Soviet troops advanced toward the Danube, Bagrianov sought an agreement with the Allies. In August, Bulgaria proclaimed its neutrality, and ordered the disarmament of the German troops on its soil.

[29] Ignoring the Sofia Declaration, the Red Army entered Bulgarian territory. The resistance fanned local insurrection. On September 8, General Kyril Stanchev's troops took the capital, and the Patriotic Front formed a government headed by the Republican Kimon Georgiev.

[30] Sofia signed a treaty with the allies in October 1944. Bulgarian troops, under Soviet command, collaborated in the defeat of the German forces in Hungary, Yugoslavia and Austria. During 1945, war trials led to the imprisonment of 6,870 people, and the execution of 2,680 others.

[31] In March 1945, communist leader Giorgi Dimitrov returned to Bulgaria after several years at the Komintern headquarters, in Moscow. A few months later a crisis arose when Georgiev failed to have the electoral lists drawn up in time, postponing the approaching election.

[32] In a referendum held in September 1946, 92 per cent of the electorate approved of the creation of the Republic of Bulgaria. In the October election, the Patriotic Front obtained 364 seats with 70.8 per cent of the vote, 277 seats went to the Communist Party (BCP). In November, Dimitrov became prime minister of the new government.

[33] In 1947, Britain and the US recognized the government, the National Assembly ratified the peace treaty with the Allies, the new Constitution went into effect and, at the end of the year, the Soviet troops withdrew from the country. After joining the opposition, some of the Patriotic Front's former leaders were arrested and sentenced to death for conspiracy.

[34] Under the leadership of the BCP, the Bulgarian State adopted the Soviet socio-economic model. A process of accelerated economic growth was set into motion, without taking into account the lack of raw materials or the technical preparation of the labor force. In the countryside, collective farming was enforced.

[35] Dimitrov resigned from the government in March 1949, and died a few months later. Vulko Chervenkov succeeded him, first as leader of the government, and then as leader of the party. In March 1954, Todor Zhivkov was named First Secretary of the BCP, becoming prime minister in 1962. Bulgaria was the USSR's closest ally among the Warsaw Pact countries, and in 1968 Bulgarian troops accompanied Soviet troops in the invasion of Czechoslovakia.

[36] In the 1980s Bulgaria was accused of imposing a policy of forced assimilation, upon the country's 10 per cent Turkish minority. In 1986, Sofia refuted Amnesty International's accusation that more than 250 Turks had been detained or imprisoned for refusing to accept new identity cards.

[37] In 1988, Bulgaria and Turkey signed a protocol governing bilateral economic relations. The dialogue was interrupted in the following year, when it was revealed that the Bulgarian militia had used violence to put down a protest by 30,000 Turks, who had been demonstrating against the government's policy of assimilation.

[38] In June 1989, more than 80,000 Turks were expelled from Bulgaria.

Foreign Trade

MILLIONS $ 1994

IMPORTS
4,160

EXPORTS
4,165

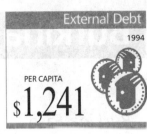

External Debt

1994

PER CAPITA
$1,241

Turkey promised to accommodate them all, but towards the end of August, after receiving 310,000 Bulgarian Turks, it closed its borders. 30,000 of these refugees subsequently returned to Bulgaria.

[39] In November, the largest protest meeting since the war was held in front of the National Assembly. A group known as "Eco-glasnost" demonstrated against a proposed nuclear power plant on an island in the Danube, citing the fact that it lay in an earthquake-zone. They also protested against the construction of a reservoir in one of the country's largest nature reserves.

[40] In December, some 6,000 members of the Pomak community - a Muslim minority group with an estimated 300,000 members - demanded religious and cultural freedom.

[41] The Central Committee of the BCP replaced Zhivkov as Secretary General, a post he had held for 35 years. He was succeeded by Petur Mladenov, who had been prime minister since 1971 and was considered to be a proponent of liberalization of the regime. In December, Mladenov also replaced Zhivkov as president of the State Council.

[42] The National Assembly lifted the ban against anti-government demonstrations, and granted amnesty to political prisoners. There was an immediate increase in the number of political demonstrations demanding reforms and elections.

[43] In March 1990, the Union of Democratic Forces (UDF), made up of 16 opposition parties, and the official party agreed to the election of a Constitutional Assembly. In the July election, the Bulgarian Socialist Party, BSP (formerly BCP) won 206 seats, and the UDF 136, out of a total of 400. In October, a new government was formed by the BSP, led by Yelio Yelev, a dissident during the 1970s and leader of the Social Democratic wing of the UDF.

[44] The new coalition government adopted a program of economic reforms in consultation with the IMF and the World Bank. They reached an agreement with the labor unions for a 200-day "social peace" until the reforms could go into effect.

[45] In July 1991, the new Constitution was approved, establishing a parliamentary system of government, allowing personal property, and freedom of opinion. After the October election, Parliament named Filip Dimitrov as prime minister. Virtually alienated from

the Social Democrats and the "Greens", co-founders of the UDF, Dimitrov was chosen because of the support he had from the right-wing of the opposition coalition and the Movement for Freedom and Human Rights (of the Turkish minority).

[46] In the January 1992 election, in which 75 per cent of registered voters participated, Yelev was elected president of the republic. As leader of the political transition period and author of the rapprochement toward the West, he received 54.4 per cent of the vote in the second round of voting, defeating the BSP candidate, Velko Valkanov.

[47] With a general strike protesting low wages and unemployment, Podkrepa - the national labor union - and the government parted company. The unemployment rate reached 17 per cent, and retirement pensions came to an average of $25 per month.

[48] In June 1992, Bulgaria joined the Council of Europe. During this year, former Communist leader Todor Zhivkov, three former prime ministers and another former member of governments prior to 1991, were arrested charged with corruption in the exercise of their duties. The economic situation led the MDL to withdraw its support of Dimitrov, which caused the downfall of his cabinet.

[49] Dimitrov was substituted by Liuben Berov from the MDL. The new prime minister stated he was willing to restore the lands confiscated by the communists to the Turkish minority. Transition from a centralized planned economy to a free market economy continued to be difficult and caused paradoxical situations. Thus, when agricultural cooperatives were dismantled, two million heads of cattle were slaughtered as they could not be managed by the new private farms.

[50] When former USSR stopped buying two thirds of Bulgarian exports, foreign trade was significantly reduced. Furthermore, UN sanctions imposed on neighbouring ex-Yugoslavia caused losses of 1.5 million to Bulgaria. In 1993, Berov went on with the reforms to install a market economy at a rate considered excessively slow by the IMF, which caused a certain amount of tension between Sofia and the international organisation.

[51] However, Berov held on to power until mid 1994, thanks to the support of the MDL, the BSP and

STATISTICS

DEMOGRAPHY
Urban: 70% (1995). **Annual growth:** -0.8% (1991-99). **Estimate for year 2000:** 8,000,000. **Children per woman:** 1.5 (1992).

HEALTH
Under-five mortality: 19 per 1,000 (1995). **Calorie consumption:** 117% of required intake (1995).

EDUCATION
School enrolment: Primary (1993): 84% fem., 84% male. Secondary (1993): 70% fem., 66% male. University: 32% (1993). **Primary school teachers:** one for every 14 students (1992).

COMMUNICATIONS
111 **newspapers** (1995), 102 **TV sets** (1995) and 101 **radio sets** per 1,000 households (1995). 26.3 **telephones** per 100 inhabitants (1993). **Books:** 103 new titles per 1,000,000 inhabitants in 1995.

ECONOMY
Per capita GNP: $1,250 (1994). **Annual growth:** -2.70% (1985-94). **Annual inflation:** 42.20% (1984-94). **Consumer price index:** 100 in 1990; 2,841.7 in 1994. **Currency:** 65 leva = 1$ (1994). **Cereal imports:** 241,000 metric tons (1993). **Fertilizer use:** 663 kgs per ha. (1992-93). **Imports:** $4,160 million (1994). **Exports:** $4,165 million (1994). **External debt:** $10,468 million (1994), $1,241 per capita (1994). **Debt service:** 14% of exports (1994).

ENERGY
Consumption: 2,786 kgs of Oil Equivalent per capita yearly (1994), 63% imported (1994).

the New Union for Democracy, a breakaway group from the UDF. Members of this opposing coalition frequently accused the prime minister of wanting to "restore socialism". Disagreements between Berov and the MDL in matters related to human rights and the social condition of Bulgarians of Turkish background, further weakened the government.

[52] In June 1994, Berov was able to pass the privatization law and in September he presented his resignation. After three failed attempts to form a new government, President Yelev dissolved Parliament and called new elections. The BSP won an absolute majority in the National Assembly - 125 seats over a total of 240 - in the December elections, while the anti-communist UDF gained 69 seats.

[53] In January 1995, socialist leader Zhan Videnov formed a new government which included members from the BSP, the Bulgarian

Agrarian National Union and the Eco-glasnost Political Club. His cabinet was the first in the history of post-communist Bulgaria with an absolute majority at the National Assembly. Differences between the new government and Yelev were frequent. In July, the president criticized the government because he believed reforms leading toward a market economy were not being implemented swiftly, suggesting the BSP was "genetically connected" to organized crime and that the Videnov administration was incapable of eliminating it.

Burkina Faso

Population: 10,118,000 (1994)
Area: 274,200 SQ KM
Capital: Ouagadougou

Currency: CFA Franc
Language: French

Burkina Faso

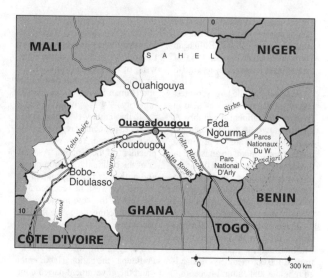

During the 11th century, the Mossi martial caste subdued the neighboring peoples in the region where the Volta rivers (Volta Noire, Volta Blanche and Volta Rouge) rise. In the following two centuries, they established a series of highly organized kingdoms, the most notable being Yatenga and Ougadougou. In the latter group, rulers were chosen from the royal family by four officials with ministerial rank, whose role was to seek a balance between the Mossi aristocracy and the Mande base. This electoral system persisted into the 20th century.

In the first year of Sankara's government, Upper Volta became Burkina Faso "land of the incorruptible"; the national anthem was sung in African languages; land reform was carried out and popular courts were set up to dispense justice.

[2] The Mossi and Mande tenaciously resisted attempts at annexation by the Mali and Songhai empires (see Mali and Guinea) and remained independent throughout the Fulah invasions of the 18th and 19th centuries. In a series of military incursions between 1896 and 1904, the French laid waste to the central plains, burning houses, and slaughtering humans and animals. The ensuing reign of terror finally sparked off an insurrection in 1916, which met with such violent repression that millions of Burkinabes were forced to emigrate, mostly to Ghana.

[3] By then the country was incorporated into the colony of Haute Sénégal-Niger. It was reconstituted as the separate colony of Upper Volta in 1947.

[4] Upper Volta had no directly productive role in the international division of labor. Its task was to produce people: forced labor on plantations in the Côte d'Ivoire, or cannon fodder for French armies at war in Europe or the colonies.

[5] Finally, in line with French neocolonial strategy, the country was declared independent in 1960. The 1960 elections were won by the Voltese Democratic Union (UDV), a party backed by landowners and private enterprise. Maurice Yameogo was elected president and was re-elected in 1965 amidst intense trade union agitation, which occurred in response to economic crisis, administrative chaos and austerity measures. Yameogo was overthrown a year later in a military coup led by Army Chief-of-Staff, Colonel Sangoule Lamizana.

[6] The 1970s witnessed a succession of elections, military coups, and more or less fraudulent re-elections orchestrated by Colonel Lamizana. Starvation was widespread, herds were dwindling and an estimated quarter of the population had emigrated to neighboring states. Major Jean-Baptiste Ouedraogo, a doctor, succeeded Zerbo in 1982 by another coup and was in turn ousted by Thomas Sankara. This young army captain was popular among soldiers and the rural poor as he brought in an anti-corruption campaign and organized brigades to assist victims of the prolonged drought and deforestation.

[7] In the first year of Sankara's government, Upper Volta became Burkina Faso - "land of the incorruptible"; the national anthem was sung in African languages; land reform was carried out and popular courts were set up to dispense justice.

[8] At the head of the National Revolutionary Council, Sankara set a target of two meals and 10 liters of water per inhabitant per day.

[9] The implementation of such measures presented a considerable challenge in a country where 82 per cent of adults are illiterate and absolute poverty is the norm. There are practically no bank accounts held within the country and kerosene is the energy source for lighting. Wood is the main fuel and cutting for personal use contributes some 50,000 to 100,000 hectares to the total annual deforestation. Encroachment by the Sahel desert, due partly to neocolonial policies favoring export crops over subsistence farming, further aggravates the lack of firewood.

[10] On December 15, 1987, Sankara was overthrown, tried and executed along with 12 of his supporters in a coup led by his second-in-command, Blaise Compaoré, who pledged to continue the Popular Front and "rectify" the regime internally.

[11] Contrasting with the image of Sankara, a man of austere habits who shunned personal property, Compaoré built a government palace and purchased a presidential plane. His economic strategy included encouragement of private enterprise and foreign capital, and consideration of denationalization, deregulation and agreements with international lending institutions.

[12] In September 1989, four high-ranking officers were accused of plotting to kill the head of state, and were shot.

[13] The 4th Republic adopted a new constitution, which was approved by referendum on June 2, 1991 (90 per cent voting in favor, with 50 per cent of registered voters participating). The new constitution, which was put into effect by de facto President Compaoré, re-established a multi-party system and the division of powers, while reducing presidential powers. The office of prime minister was created, and general elections were to be held every seven years.

[14] In the May 1992 elections, the official party won 78 of the 107 National Assembly seats. The main opposition force, con-

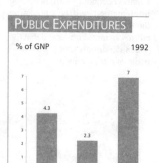

PUBLIC EXPENDITURES

% of GNP — 1992

MILITARY 4.3
EDUCATION 2.3
HEALTH 7

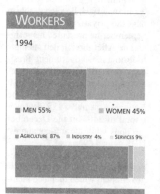

WORKERS

1994

MEN 55% · WOMEN 45%

AGRICULTURE 87% · INDUSTRY 4% · SERVICES 9%

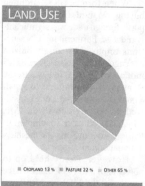

LAND USE

CROPLAND 13% · PASTURE 22% · OTHER 65%

Maternal Mortality 1989-95

PER 100,000 LIVE BIRTHS

939

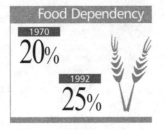

Food Dependency

1970 **20%**

1992 **25%**

External Debt 1994

PER CAPITA

$111

Illiteracy 1995

81%

PROFILE

ENVIRONMENT

A landlocked country, Burkina Faso is one of the most densely populated areas of the African Sahel (the semi-arid southern rim of the Sahara). The Mossi plateau slopes gently southwards and is traversed by the valleys of the three Volta Rivers (Black Volta, White Volta and Red Volta). Though more fertile, these lands are plagued with tse tse flies. As in the whole Sahel region, this nation suffers from mounting aridity caused by inappropriate farming techniques introduced to produce peanuts and cotton for export. Wood, a popular energy source, is becoming scarce as a result of deforestation.

SOCIETY

Peoples: Over half of the population are Mossi. Pouhi herders and the Tamajek clans with their vassals, the Bellah, amount to around 20 per cent. Djula peasants and traders are indigenous minorities. The language of these three population groups is the linguistic bridge among the various regions. Senufo and Bobo-Fing cultures inhabit the western plains where the savannah merges into the forest. In the south, the predominant Lebi, Bobo-Ule, Gurunsi and Bisa cultures extend into several states. The savannah to the east holds the great Gurmantehé civilization, while the Sampo, Rurumba and Marko cultures live in the desert regions to the north and northeast. **Religions:** The majority of the population profess traditional African religions; 30% are Muslim and 10% are Christian. **Languages:** French (official); Mossi, Bobo and Gurma are the most widely spoken languages. **Political Parties:** Organization for the People's Democracy-Workers' Movement (ODP/MT); National Convention of Progressive Patriots-Social Democrats (CNPP-PSD); Burkinabe Socialist Party (PSB). **Social organizations:** Voltanese Free Unions Organization, Confederation of Voltanese Unions.

THE STATE

Official Name: République de Burkina Faso. **Administrative division:** 30 provinces, 300 departments and 7200 villages. **Capital:** Ouagadougou, 500,000 inhab. **Other cities:** Bobo-Dioulasso, 228,668 inhab.; Koudougou, 51,925 inhab.; Ouahigouya, 38,902 inhab. **Government:** Republican parliamentary system, with a strong head of state, Blaise Compaoré, President. Single-chamber legislature: National Assembly (107 members elected by direct vote for a 7-year term). Judiciary: judges responsible to a Superior Council, under the authority of the president of the republic. **National Holiday:** August 5 (independence from France) (1960). **Armed Forces:** 5,800 (1995). **Paramilitaries:** Gendarmerie: 1,500; Security Company (CRG): 250; People's Militia (R): 45,000 trained.

STATISTICS

DEMOGRAPHY

Urban: 23% (1995). **Annual growth:** 6.9% (1992-2000). **Estimate for year 2000:** 12,000,000. **Children per woman:** 6.9 (1992).

HEALTH

One **physician** for every 33,333 inhab. (1988-91). **Under-five mortality:** 169 per 1,000 (1995). **Calorie consumption:** 91% of required intake (1995). **Safe water:** 78% of the population has access (1990-95).

EDUCATION

Illiteracy: 81% (1995). **School enrolment:** Primary (1993): 30% fem., 30% male. Secondary (1993): 6% fem., 11% male. **Primary school teachers:** one for every 60 students (1992).

COMMUNICATIONS

81 **newspapers** (1995), 83 **TV sets** (1995) and 78 **radio sets** per 1,000 households (1995). 0.2 **telephones** per 100 inhabitants (1993). **Books:** 78 new titles per 1,000,000 inhabitants in 1995.

ECONOMY

Per capita GNP: $300 (1994). **Annual growth:** -0.10% (1985-94). **Annual inflation:** 1.60% (1984-94). **Consumer price index:** 100 in 1990; 126.2 in 1994. **Currency:** 535 CFA francs = 1$ (1994). **Cereal imports:** 121,000 metric tons (1993). **Fertilizer use:** 60 kgs per ha. (1992-93). **Development aid received:** $457 million (1993); $47 per capita; 16% of GNP.

ENERGY

Consumption: 16 kgs of Oil Equivalent per capita yearly (1994), 100% imported (1994).

centrated in the National Convention of Progressive Patriots-Social Democratic Party (CNPP-PSD) won 13 seats. Blaise Compaoré, who ran unopposed, became the first president of the 4th Republic, with 21.6 per cent of the vote.

[15] Prime Minister Yussuf Ouedraogo promised to prioritize agriculture and provide incentives for the creation of small and medium sized enterprises in order to create new jobs. International pressure made Ouedraogo pledge to reduce the country's bureaucracy; Burkina Faso had 36,000 public employees, most of them on pensions.

[16] In December 1992 and March 1993 there were two general strikes against the implementation of the structural adjustment plan recommended by the IMF. Devaluation of the CFA franc in January 1994 led to Ouedraogo's resignation and his replacement by Marc-Christian Kabore in March that year. In July, in spite of protests from the opposition, Parliament passed a law which allowed the government to privatize 19 national companies.

[17] In February 1995, the opposition boycotted municipal elections since it considered they were being organized by a non-independent electoral commission. The ruling Organization for the People's Democracy-Worker's Movement won in 26 of the country's largest 33 cities. At a social level, conditions remained tense since indirect consequences from the 1994 devaluation led the cost of living to rise 30 per cent in the first semester of 1995.

[18] The government continued its policy of economic liberalization and officially entered Burkina Faso into the World Trade Organization in June 1995. A project to restructure the recently privatized railways, including 500 lay-offs, caused a new strike which led the country once again toward a tense situation, typical of the early 1990s.

Burundi

Population:	6,183,000 (1994)
Area:	27,830 SQ KM
Capital:	Bujumbura
Currency:	Burundi Franc
Languages:	Rundi Kirundi and French

Burundi

Burundi is one of the poorest countries in the world and one of the most densely populated in Africa. For five centuries it has been torn by ethnic and economic struggles involving the Hutu majority and the Tutsi (or Watusi, as the Germans dubbed them). At least five major massacres affecting one or the other community have occurred since the 15th century.

[2] The Hutu, descendants of the Bantu, were communal farmers. Five hundred years ago they were overrun by the Tutsi, who migrated south from Uganda and Ethiopia in search of fertile grazing land for their cattle. The Tutsi had more sophisticated weapons and were thus able to subdue the Hutu and enslave their people. Tutsi rulers held considerable power during the 17th and 18th centuries, but rivalry between the various local groups weakened the central power structure in the 19th century. This paved the way for German colonization around 1890. The Europeans upheld Tutsi domination and in turn forced the traditional ruler (Mwami) to accept German tutelage. In 1899, Burundi and Rwanda (see Rwanda) were merged into a single colony, Rwanda-Urundi, which became famous for its ivory exports controlled by a German trade monopoly.

[3] Following the defeat of Germany in World War I, Belgium took over the colony and separated Burundi from Rwanda again only to merge the former with Zaire. The Belgian system of indirect rule assigned Tutsi elites a privileged role and gave rise to nationalist movements in the 1950s. This process bred the Union for National Progress (UNAPRO) led by Louis Rwagasore, elected prime minister

in 1960. The Belgians feared that Rwagasore might become a Burundian Lumumba (see Zaire) and he was therefore assassinated before independence was granted. Burundi gained independence on July 1 1962, governed by a puppet Tutsi ruler under Belgian control.

[4] Violence marked the first four years of autonomous rule under five different prime ministers. Instability predominated until November 1966, when the prime minister, Captain Michael Micombero, staged a coup and proclaimed the Republic of Burundi. The new president carried out a massive purge of all Hutu government officials. In 1971, 350,000 Hutus were killed through government repression and an additional 70,000 went into exile.

The high rate of deforestation is in part due to the transformation of forests into farming lands for landless peasants.

[5] In 1976, Lieutenant-Colonel Jean Baptiste Bagaza took power promising to end ethnic persecution and set up a reformist government. Bagaza broadened the UNAPRO, put a land reform program into practice, and rehabilitated labor unions, all in blatant defiance of the Tutsi elite and their foreign capitalist allies. In foreign policy, the new government drew closer to Tanzania, and China sent aid to develop Burundi's mineral resources.

[6] In 1979, the first post-independence congress of UNAPRO was held and a new constitution drafted, effective in 1981. The new constitution prevented the exploitation of the Hutu majority by the Tutsi minority, prompted the modernization of the political structure, and gave women equal rights. Reforms stirred up strong feelings between the government and the conservative Catholic hierarchy, leading to confiscation of properties and deportation of 63 missionaries.

[7] Elections held in 1982, 20 years after independence, upheld President Bagaza's policies and brought about a swift stabilization of the young nation's political situation.

[8] Despite Burundi's established political independence, the country faced serious problems due mostly to economic dependence and its geographical location. Being a landlocked country raised the prices of both imports and exports. Population density is high and unevenly distributed geographically; 70 per cent live in the north where overuse of the land has caused massive soil erosion. Wood is the main source of domestic fuel and unfortunately the high demand outweighs the forest production of the country. 90 per cent of agricultural products are consumed by the domestic market. Burundi's economic hopes lie in the rich nickel deposits and in its as yet unexploited hydroelectric potential. Belgian and US companies are currently developing the Musongati mines

which also have large cobalt and uranium deposits.

[9] In September 1987, while attending a summit meeting of French-speaking countries in Quebec, Bagaza was overthrown by army major Pierre Buyoya in a bloodless coup.

[10] In August 1988, strife between Hutus and Tutsis broke out in Dtega and Marangana, in the north of the country, leaving several thousand dead, most of them Hutus. The rebellion against Tutsi landowners was cruelly dealt with by an army made up mostly of Tutsis. Over 60,000 Hutus sought refuge in Rwanda. The government responded by designating a Hutu prime minister, Adrien Sibomana, with a new cabinet in which half the ministers were Hutus.

[11] In 1989, most refugees returned and there was reconciliation between the military government and the Catholic Church, who had regained their property. A structural adjustment program was initiated which included the privatization of public enterprises and the creation of a "watchdog" tribunal to combat corruption.

[12] Burundi depended largely on the export of its main crop, coffee, and on its fluctuating prices. Poverty and high population density added to the environment's degradation. The high rate of deforestation is in part due to the transformation of forests into farming lands for landless peasants.

[13] In 1992, Buyoya enacted a

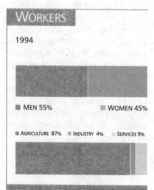

WORKERS

1994

- MEN 55%
- WOMEN 45%
- AGRICULTURE 87%
- INDUSTRY 4%
- SERVICES 9%

LAND USE

- CROPLAND 13 %
- PASTURE 22 %
- OTHER 65 %

Maternal Mortality
1989-95

PER 100,000
LIVE BIRTHS

1,327

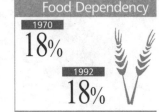

Food Dependency

1970
18%

1992
18%

External Debt
1994

PER CAPITA
$182

Illiteracy
1995

65%

PROFILE

ENVIRONMENT

Most of the land is made up of flat plateaus and relatively low hills covered with natural pastures. Located in the African zone of the Great Lakes (Tanganyika, Victoria), the Ruvubu River valley stretches through the country from north to south. Tropical forests are found in the low, western regions. Most inhabitants engage in subsistence agriculture (corn, cassava, sorghum, beans). Coffee is the main export product. Internal communications are hampered by natural barriers and the nearest sea outlet is 1,400 km beyond the border, a fact which makes foreign trade even more difficult. The forests are fundamental in maintaining climatic balance, and deforestation is one of the country's worst problems, aggravated by the growing incidence of farming on unsuitable lands.

SOCIETY

Peoples: Most Burundians (86%) belong to the Hutu ethnic group, an agricultural people of Bantu origin. They were traditionally dominated by the Tutsi or Watusi (13%), shepherds of Hamitic descent. There is a small minority (1%) of Twa pygmies. **Religions:** 67% Christian; 32% follow traditional African religions and 1% are Muslim. **Languages:** Rundi, Kirundi and French (official), with Swahili the business language. **Political Parties:** Burundi Democratic Front (DEFROBU); Unity for National Progress (UNAPRO), led by President Pierre Buyoya. **Social Organizations:** Workers' Union of Burundi (UTB)

THE STATE

Official Name: Républika y'Uburundi. **Administrative divisions:** 15 Provinces. **Capital:** Bujumbura, 272,000 inhab. (1992). **Other cities:** Muhinga; Gitega, 21,000 inhab.(1990). **Government:** Pierre Buyoya, President. **National Holiday:** July 1, Independence (1962). **Armed Forces:** 12,600.

STATISTICS

DEMOGRAPHY
Urban: 7% (1995). **Annual growth:** 3.5% (1992-2000). **Estimate for year 2000:** 7,000,000. **Children per woman:** 6.8 (1992).

HEALTH
One **physician** for every 16,667 inhab. (1988-91). **Under-five mortality:** 176 per 1,000 (1995). **Calorie consumption:** 88% of required intake (1995). **Safe water:** 70% of the population has access (1990-95).

EDUCATION
Illiteracy: 65% (1995). **School enrolment:** Primary (1993): 63% fem., 63% male. Secondary (1993): 5% fem., 9% male. **University:** 1% (1993). **Primary school teachers:** one for every 63 students (1992).

COMMUNICATIONS
82 **newspapers** (1995), 82 **TV sets** (1995) and 81 **radio sets** per 1,000 households (1995). 0.3 **telephones** per 100 inhabitants (1993).

ECONOMY
Per capita GNP: $160 (1994). **Annual growth:** -0.70% (1985-94). **Annual inflation:** 5.40% (1984-94). **Consumer price index:** 100 in 1990; 128.3 in 1994. **Currency:** 247 Burundi francs = 1$ (1994). **Cereal imports:** 22,000 metric tons (1993). **Fertilizer use:** 34 kgs per ha. (1992-93). **Imports:** $224 million (1994). **Exports:** $106 million (1994). **External debt:** $1,125 million (1994), $182 per capita (1994). **Debt service:** 41.7% of exports (1994). **Development aid received:** $244 million (1993); $41 per capita; 26% of GNP.

ENERGY
Consumption: 23 kgs of Oil Equivalent per capita yearly (1994), 97% imported (1994).

---

multi-party Constitution and called elections to be held in 1993. Buyoya, leading the Union for National Progress (UNAPRO) - with a majority of Tutsi leaders - was defeated by Melchior Ndadaye, from the opposing Democratic Front of Burundi (DEFROBU), composed of a Hutu majority.

[14] On October 24 1993, three months after he was elected, Ndadaye was assassinated during an attempted coup d'etat. Prime Minister Sylvie Kinigi, who sought asylum in the French embassy, managed to keep the situation under control. The leaders of the rebellion were either arrested or fled to Zaire. Cyprien Ntaryamira - a Hutu like Ndadaye - was appointed president by the Parliament.

[15] In spite of the failed coup d'etat, the assassination of Ndadaye led to one of the worst massacres in Burundi's history. Supporters of the ex-president attacked UNAPRO members - Tutsis or Hutus - causing the death of tens of thousands of people and the flight of some 700,000. The so-called "extremist armed militias" - hostile to cohabiting with the other ethnic group - were consolidated by this time, as were the "Undefeated" Tutsis or the "Intagohekas" ("those who never sleep") Hutus. Violence began to spread.

[16] On April 6 1994, Ntaryamira died along with Rwandan president Juvenal Habyarimana when their airplane was attacked in Kigali, capital of Rwanda. Another Hutu, Sylvestre Ntibantunganya, replaced the assassinated president. Violence was intensified, especially between militias in favour of "Hutu power" and the army, controlled by Tutsi officers.

[17] In February 1995, the UNAPRO left the government to force prime minister Anatole Kanyenki-ko to resign. His resignation enabled the nomination of Tutsi Antoine Nduwayo and the return of UNAPRO to the coalition government which it had been part of along with DEFROBU.

[18] The number of "domestic refugees" continued on the rise in 1996. Many Tutsis fled to the cities, seeking protection from the army, while thousands of Hutu peasants - willingly or not - left with the guerrillas toward mountain or jungle regions. At 47, Pierre Buyoya led a new and successful coup d'etat in July 1996 to become Burundi's new president.

[19] Buyoya, a Tutsi military officer, dissolved the National Assembly and banned strikes and concentrations. Afterwards, he formed a government with Hutu and Tutsi ministers and promised a new "democratic beginning".

Foreign Trade

MILLIONS $ 1994

IMPORTS
224

EXPORTS
106

Cambodia

Cambodia

Population: 9,951,000 (1994)
Area: 181,040 SQ KM
Capital: Phnom Penh (Phnum Pénh)

Currency: Riels
Language: Khmer

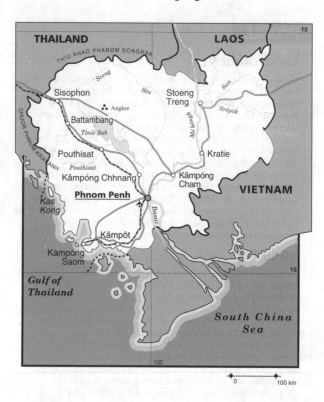

The Khmer civilization once occupied virtually all of Indochina, reaching its height, between the 9th and 13th centuries AD. It was gradually confined to the limits of today's Cambodia as a result of several territorial partitions, especially in Vietnam and Thailand. The country withstood colonization attempts from the Portuguese, Dutch and British before becoming a French "protectorate" in 1863. In 1930 the Indo-Chinese Communist Party was founded, which soon succeeded in uniting the anti-imperialist forces of Vietnam, Laos and Cambodia.

[2] During World War II, the Japanese occupied Cambodia. On March 12 1945, when the invaders withdrew, King Norodom Sihanouk proclaimed independence but the French invaded again. Instead of waging a new liberation war, as Vietnam did, the king negotiated autonomy.

[3] In 1947, a new constitution kept King Sihanouk on the throne. However his power was limited by parliament. In 1949, negotiations with France resulted in Cambodian independence "within the French union", with Paris retaining control over military and foreign affairs. In 1953, the Vietnamese offensive forced French troops to withdraw from Cambodia, leaving the country fully independent. In 1955, Sihanouk abdicated in favor of his father, changing his name to Prince Sihanouk so that he could participate in political life. From 1960 onwards, Sihanouk became head of state.

[4] During the first years of US aggression against Vietnam, the Cambodian government sought to maintain the country's political and territorial neutrality. In 1965, however, following air raids by US-controlled South Vietnamese forces, Cambodia broke relations with the US. In 1968 the US began launching regular air raids on the so-called "Ho Chi Minh trail" through Cambodian territory, allegedly supplying the Viet Cong from North Vietnam.

[5] While taking diplomatic action abroad to protect his country's sovereignty, Sihanouk was ousted in a coup staged by the US Central Intelligence Agency and replaced by General Lon Nol. During his five-year rule, Lon Nol received $1.6 billion in aid from Washington. During that period, US forces intervened in Cambodia to fight the Communist Khmer Rouge guerrillas. Armed clashes left nearly 100,000 Cambodians dead.

[6] From his exile in Beijing, supported by the Khmer Rouge, Sihanouk organized the National United Front of Kampuchea (NUFK) using the name of the country in Khmer language. By 1972, the NUFK was in control of 85 per cent of Cambodian territory, and its forces closed in on the capital, Phnom Penh (Phnum Pénh).

[7] In 1975, the Khmer Rouge achieved a resounding victory. Lon Nol was replaced by the Democratic Republic of Kampuchea, and in 1976, while a new Constitution was being approved, the People's Congress confirmed Sihanouk and Khieu Sampham as head of state and head of government, respectively. However, on his return from exile, Sihanouk was forced to resign and kept under house arrest, while Pol Pot rose to the fore as the regime's new leading force.

[8] Under Pol Pot, the country was completely sealed off from contact with the outside world. Borders were closed and diplomats were not allowed to leave their embassies. Determined to build an unprecedented social system, Pol Pot eliminated money and transferred large numbers of the urban population to the countryside, enforcing the return to an agricultural lifestyle. Mass purges and executions, hunger and illness, left at least a million people dead.

[9] The Pol Pot regime strengthened ties with China, broke off diplomatic relations with Hanoi, refused to recognize borders "established by colonialism" and sent Khmer Rouge troops to invade Vietnamese territory in late 1977. The disputed borders had been established in 1973 by agreement between Vietnam and Sihanouk, as

historical accounts could provide no exact basis for establishing boundaries.

[10] In December 1978, when the border war was on the verge of being won by Vietnam, the United Front for the Salvation of Kampuchea was created under the presidency of General Heng Samrin. On January 11 1979, the Vietnamese forces and those of the United Front entered Phnom Penh and proclaimed the People's Republic of Kampuchea. A People's Revolutionary Council was established, and the government began the enormous task of rebuilding a hungry and devastated nation.

[11] Heng Samrin's government defined its policy as being one of "independence, non-alignment and democracy directed toward socialism." Pol Pot was tried in absentia for war crimes, and his forces turned to guerrilla warfare. With the support of China and the United States, the Khmer Rouge managed to maintain UN recognition as the country's legitimate government. In 1982, Sihanouk and the Khmer Serei, led by Son Sann, joined Khieu Samphan's opposition government-in-exile.

[12] One of the main goals of Heng Samrim's government was to reactivate the industrial, agricultural and transportation sectors. Soviet, and other socialist countries gave aid which was focused on the reconstruction of power sources and the formation of a very basic industrial infrastructure.

[13] In March and April 1981 municipal and legislative elections were held, with many candidates for each post. The Non-Aligned Movement summit meeting at the time in New Delhi did not support any of the contending groups. Faced with a set of eclectic electoral options, Cambodia's seat in the movement was left unfilled.

[14] In July 1985, the Vietnamese announced that their 150,000 troops would gradually be withdrawn from Cambodia over the following five years.

[15] Difficulties stemming from the war affected the Phnom Penh government, and it was only officially recognized by some 30 countries. Still, it managed to bring about a steady increase in the production of rice, cattle, pigs and poultry.

[16] In 1986, the government began overtures toward Sihanouk within the context of a general agreement, offering him the position of head of state in a government from which the Khmer Rouge would

WORKERS

1994

MEN 59%	WOMEN 41%

| AGRICULTURE 74% | INDUSTRY 7% | SERVICES 19% |

ENVIRONMENT

The territory is mainly a plain surrounded by mountains, like the Cardamour Mountains in the southwest. In the north, the Dangrek range rises abruptly from the plain. The central basin of the country, occupied by the Tonle Sap (Great Lake) depression, is the point of confluence into the Mekong, one of the largest rivers in Asia. The country has a sub-tropical climate with monsoon rains. The population is concentrated in the central basin where rice, the mainstay of the diet and principal export crop, is grown. The country also exports rubber. The mineral reserves: phosphate, iron ore and limestone, are as yet unexploited. The main environmental problem is deforestation, caused by defoliants and the explosion of bombs during the Vietnam War.

SOCIETY

Peoples: Mostly Khmers. The Khmers inhabit an area that extends beyond the present boundaries of Cambodia. There are Vietnamese and Chinese minorities. **Religions:** In 1986 Buddhism - the religion of the majority - became the country's official religion. There is an Islamic minority (Cham). **Languages:** Khmer, official and predominant. **Political Parties:** United National Front for an Independent, Neutral, Peaceful and Cooperative Cambodia (Funcinpec); Cambodian People's Party; Liberal Democratic Buddhist Party; National Liberation Movement of Cambodia.

THE STATE

Official Name: Preach Reach Ana Pak Kampuchea. **Administrative divisions:** 22 provinces. **Capital:** Phnom Penh (Phnum Pénh), 900,000 inhab. (1987). **Other cities:** Battambang 45,000 inhab., Kompong Cham 33,000 inhab. (1987). **Government:** Parliamentary monarchy. Norodom Sihanouk, King. Norodom Ranariddh, First Prime Minister; Hun Sen, Second Prime Minister. The legislature is made up of a 120-member National Assembly, in which representatives hold 5-year terms. **National Holiday:** November 9, Independence Day (1953). **Armed Forces:** 88,500.

DEMOGRAPHY

Annual growth: 3.1% (1992-2000). **Children per woman:** 4.5 (1992).

HEALTH

Under-five mortality: 177 per 1,000 (1995). **Calorie consumption:** 90% of required intake (1995). **Safe water:** 36% of the population has access (1990-95).

COMMUNICATIONS

82 **newspapers** (1995), 84 **TV sets** (1995) and 88 **radio sets** per 1,000 households (1995). 0.1 **telephones** per 100 inhabitants (1993).

ECONOMY

Currency: riels. **Cereal imports:** 20,000 metric tons (1990).

---

be excluded. From the end of 1987 to mid-1989, Hun Sen, who had become prime minister, met with Sihanouk on six different occasions. The main disagreement each time was whether the Khmer Rouge would be admitted to a new provisional government or not.

[17] During the course of 1989, a curfew, which had been in effect since 1979, was lifted and private ownership of some "non-strategic" enterprises was permitted. In addition, collective farms passed into the hands of peasant families, with the assurance that the land could henceforth be passed on from generation to generation, and transportation services were privatized. On June 1 1989, to complete the diplomatic offensive and improve its image, the country changed its name to the State of Cambodia. After a meeting between Hun Sen and the new Thai prime minister, Chatichai Choonhavan, the possibilities of regional agreement and economic relations between the countries improved.

[18] The International Conference for Peace in Cambodia, held in Paris in July 1989, ended in failure as the three factions of the armed opposition could not come to any agreement. The two main points of disagreement were UN monitoring of the Vietnamese withdrawal and Khmer Rouge participation in government. Even without this agreement, Vietnamese troops withdrew completely in September, as planned, followed by an advance in the position of rebel forces.

[19] In April 1990, after a military offensive by Heng Samrin and Hun Sen's government, opposition forces retreated to the Thai border. With this development, the Thai Premier, Chatichai Choonhavan, sponsored a new meeting of Sihanouk and Hun Sen. In Bangkok, both leaders agreed to the installation of a supernational agency which would symbolize the sovereignty and national unity of Cambodia. An "adequate" UN presence was also agreed to by both parties.

[20] In July 1990, the US announced they would recognize the State of Cambodia and would initiate negotiations with Vietnam over a peace settlement.

[21] In October 1991, a peace treaty was signed. The Supreme National Council was created, with representatives of the Phnom Penh government and part of the opposition, it was chaired by Sihanouk and was to govern the country until elections in 1993. In 1992, the UN sent a peace-keeping force consisting of 20,000 troops to enforce the ceasefire and organize the elections.

[22] In the constituent elections of May 1993, boycotted by the Khmer Rouge, Funcinpec supporters, led by Sihanouk's son Norodom Ranariddh, won 58 seats, the former Cambodian People's Communist Party 51, the Liberal Democratic Party 10, and the National Liberation Movement of Cambodia, one. In the new government, Ranariddh and Hun Sen share the post of prime minister.

[23] In September 1993, a new Constitution was enacted. It turned the National Assembly into Parliament and established a parliamentary monarchy. Sihanouk was appointed king, "independent" of all political parties. But, despite he spent part of 1994 in China receiving cancer treatment, he still advocates a "national reconciliation" government, with the inclusion of the Khmer Rouge.

[24] In 1995, fighting intensified between the official army and the Khmer Rouge, which controlled about 15 per cent of the territory. The International Monetary Fund expressed its faith in the country's economy, which grew 7 per cent in 1995 according to the government.

[25] In 1996, the Khmer Rouge began to show signs of weakness. The diverse versions about Pol Pot's death in June were construed as an indication of the existence of significant internal divisions in the guerrilla movement. Many defections - like Ieng Sary's, another well known leader of the armed group, in August - suggested that the government policy aimed to split the Khmer Rouge was proving successful.

Cameroon

Cameroun

Population: 12,986,000 (1994)
Area: 475,440 SQ KM

Capital: Yaoundé

Currency: CFA franc
Language: French and English

Cameroon is the birthplace and original homeland of the Bantu ethnic group, which after 200 BC migrated east and south, spreading new crop varieties and methods for working iron. In 1472 Fernando Po named a river Camarones (Spanish for shrimp) because of the crustaceans swarming at its mouth, a name that eventually became Cameroon.

[2] The Fulah migration (see Niger: Routes of the Sahara; also, Senegal: The Fulah States) helped transform the local economies into part of a regional circuit which later gave rise to the Emirate of Adamaua in the north central region.

[3] German penetration began in June 1884, when German envoy Gustav Nachtigal signed an agreement with the ruler of the Doualas, a coastal group, making it a protectorate. A year later, the Berlin Conference awarded Cameroon to Germany. But the Emirate of Adamaua, which the British wanted, was not given to the Germans until 1894.

[4] This protectorate did not survive very long; the Doualas dominated trade that the Germans wanted to control between the coast and Yaoundé, the trade center between Adamaua and the south. In 1897, the Doualas resisted German interference, and a bloody four-year war ensued.

[5] The Germans appropriated the most fertile lands, and the Africans began to die of hunger by the thousands, in an area where stable agriculture had existed for centuries. In 1918, France and Britain invaded Cameroon: the French taking three-fourths, and the English taking the rest. The pro-independence organizations found their cause was facilitated by the tensions among the colonial powers. In 1945, the People's Union of Cameroon (UPC) was founded, headed by Rubem Um Nyobé. The UPC earned popular support and launched a series of legal campaigns between 1948 and 1956, when it was outlawed.

[6] Nationalist leaders fled to the British held western sector and organized a guerrilla movement. The UPC established liberated zones in the southern forests, where they set up their own administration, a first for sub-Saharan Africa. The efficiency of the guerrillas made it possible for them to resist constant French attacks, until 1960. Two years later, Rubem Um Nyobé died, but the revolution continued.

[7] UPC resistance forced the French to adopt a new strategy, adding political maneuvers to their repressive tactics. Paris created the National Union of Cameroon (UNC), merging two conservative, predominantly Islamic, northern-based parties. UNC leader Alhaji Ahmadou Ahidjo went into power after the French granted the country independence, remaining so after a British plebiscite approved reunification. UPC nationalists were unable to help shape their long-held dream of an independent Cameroon as they were mostly underground or in exile.

[8] Ahidjo stepped up persecution of the opposition in the 1960s, developing one of sub-Saharan Africa's most efficient repressive systems. European human rights groups revealed the existence of thousands of political prisoners in the country. In 1982, Ahidjo suddenly resigned, and was succeeded by his former prime minister, Paul Biya.

[9] Biya maintained his predecessor's political and economic policies, though many of Adhidjo's followers supported a coup attempt by a group of military officers. Young people, especially students, resisted the coup attempt by taking to the streets.

[10] Growing unemployment and food shortages undermined Cameroon's traditionally prosperous image. Biya attempted to reinforce his control by calling early elections for April 1984. With democratic organizations prohibited, he was re-elected. Nevertheless, the overall instability of the country led to a new coup attempt, followed by a series of bloody incidents in which 200 people were killed.

[11] In the 1980s, the UNC changed its name to "Democratic Group of the People of Cameroon" (RDPC), but its political line remained the same.

[12] The UPC later adopted a more flexible stance in order to broaden its social scope, and Biya created new northern provinces to give the Muslims greater economic and political power. Disagreements over oil revenues aggravated inter-ethnic and inter-regional friction. Economic problems stemming from a drop in world commodity prices for coffee, rubber and cotton were aggravated by Cameroon's dependence on French companies which control almost 44 per cent of the export market.

[13] In 1988, Biya was re-elected. Censorship increased, journalists were arrested and the post of prime minister was eliminated.

[14] Faced with the fall in oil prices on the world market and foreign debt payments, the government sought World Bank and IMF support for a structural adjustment program to stabilize its finances. By reducing imports and state expenditure, privatizing public enterprises and reorganizing the banking system, the country's debts were renegotiated with the Paris Club.

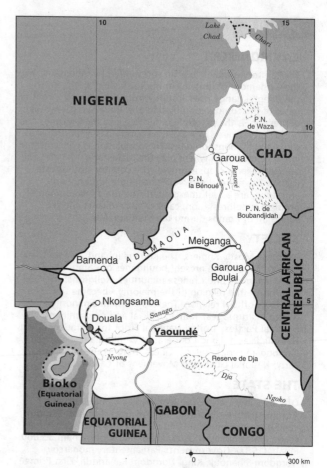

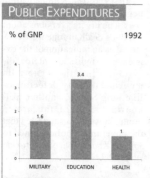

PUBLIC EXPENDITURES

% of GNP — 1992

- MILITARY: 1.6
- EDUCATION: 3.4
- HEALTH: 1

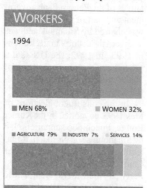

WORKERS

1994

- MEN 68%
- WOMEN 32%
- AGRICULTURE 79%
- INDUSTRY 7%
- SERVICES 14%

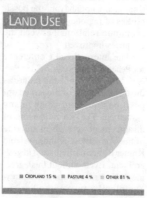

LAND USE

- CROPLAND 15%
- PASTURE 4%
- OTHER 81%

ENVIRONMENT

The country is divided into three regions: the plains of the Lake Chad basin in the northern region (savannas where cattle are raised and corn and cotton are grown); the central part, made up of humid grasslands; and the southern part, an area with rich volcanic soil where the main cash crops (coffee, bananas, cocoa and palm oil) are grown. It is in the latter that most of the population lives. Drought and desertification are the main concerns in the southern region, which covers 25% of the nation's land area, and houses over a quarter of the population.

SOCIETY

Peoples: There are some 200 ethnic groups, the main ones being those of Bantu origin: the Dualas, Bamilekes, Tikars and Bamauns in the south; the Euondos and Fulbes in the west, and the Fulanis in the north. In the southeast live the Baka Pygmies, who live by hunting and fishing. Neither their rites, their language nor other cultural features show Bantu influence. **Religions:** Half the population practise traditional African cults. Christians are a majority in the south while Muslims predominate in the north. **Languages:** French and English are the official languages. There are nearly one hundred African languages and dialects. **Political Parties:** Democratic Alliance of the People of Cameroon (RDPC); Movement for the Defence of the Republic; Social Democratic Front (FSD); National Union for Democracy and Progress (UNDP); Front of Allies for Change (FAC). **Social Organizations:** National Workers' Union of Cameroon.

THE STATE

Official Name: République du Cameroun.
Capital: Yaoundé, 650,000 inhab. (1987). **Other cities:** Douala, 810,000 inhab.; Garoua 142,000 inhab. (1987). **Government:** Parliamentary republic. Paul Biya, President since 1982. **National Holidays:** January 1, Independence Day (1960); October 1, Reunification (1961), May 20, proclamation of the Republic (1972). **Armed Forces:** 14,600.

Maternal Mortality
1989-95

PER 100,000 LIVE BIRTHS
511

Illiteracy
1995

37%

DEMOGRAPHY

Urban: 43% (1995). **Annual growth:** 1.9% (1992-2000). **Estimate for year 2000:** 16,000,000. **Children per woman:** 5.8 (1992).

HEALTH

One **physician** for every 12,500 inhab. (1988-91). **Under-five mortality:** 95 per 1,000 (1995). **Calorie consumption:** 91% of required intake (1995). **Safe water:** 50% of the population has access (1990-95).

EDUCATION

Illiteracy: 37% (1995).
School enrolment: University: 2% (1993). **Primary school teachers:** one for every 51 students (1992).

COMMUNICATIONS

86 **newspapers** (1995), 87 **TV sets** (1995) and 90 **radio sets** per 1,000 households (1995). 0.5 **telephones** per 100 inhabitants (1993). **Books:** 81 new titles per 1,000,000 inhabitants in 1995.

ECONOMY

Per capita GNP: $680 (1994). **Annual growth:** -6.90% (1985-94). **Annual inflation:** 1.30% (1984-94). **Consumer price index:** 100 in 1990; 103.0 in 1994. **Currency:** 535 CFA francs = 1$ (1994). **Cereal imports:** 281,000 metric tons (1993). **Fertilizer use:** 30 kgs per ha. (1992-93). **External debt:** $7,275 million (1994), $560 per capita (1994). **Debt service:** 16.7% of exports (1994). **Development aid received:** $547 million (1993); $44 per capita; 5% of GNP.

ENERGY

Consumption: 83 kgs of Oil Equivalent per capita yearly (1994), -525% imported (1994).

[15] In 1990, social and political organizations reported an increase in government repression of hundreds of citizens.

[16] A few months later, the government authorized the creation of political parties: some 70 political organizations were granted legal recognition.

[17] In November 1991, opposition leaders demanded constitutional reform and a return to the federal system abolished in 1972. The political crisis deepened, strikes and pro-democracy demonstrations proliferated, 40 demonstrators were killed by the police.

[18] In December, the government set the legislative elections, for February 16, 1992. The opposition wanted the constitution and electoral law to be modified prior to the elections, nonetheless, several parties participated in the elections on March 1, 1992.

[19] The RDPC won 88 of the 180 parliamentary seats. The National Union for Democracy and Progress (UNDP) came in second, with 68 representatives. According to the official report, 61 per cent of the four million registered electorate voted.

[20] Biya advanced the presidential election, to October 11, 1992. Seven parties fielded candidates: among them, the Social Democratic Front (FSD), led by John Fru Ndi, the main opposition party representing the English-speaking community. Both the FSD and the Democratic Union of Cameroon, led by Adamu Ndam Njoya, had boycotted the March elections.

[21] Amid widespread accusations of fraud, the government's victory, with 39.98 per cent over the FSD's 35.97, provoked incidents in the English-speaking Northwestern Province, Fru Ndi's territory.

[22] International observers confirmed the fraud, but the Supreme Court refused to annull the election. Fru Ndi proclaimed himself president, and the government decreed a state of emergency in the Northwest. Fru Ndi and his supporters were immediately placed under house arrest.

[23] In November a "coalition government" of the RDPC and minor groups was formed. Government repression intensified, bringing international condemnation. In December, four members of the Bar Association were arrested for leading a protest, one dying after

being tortured. When the US suspended foreign aid, Biya cancelled the state of emergency and freed Fru Ndi.

[24] In January 1993, Benjamin Menga, the UDC vice president was assassinated. Several political leaders were imprisoned and press censorship was stepped up. In 1994, the split in the opposition facilitated Biya's "hard handed" policy. In April, Fru Ndi's supporters could not reach agreement during a national conference which called together various groups representative of the English speaking people of Cameroon. The disagreements over the future federal system and the delegation which would represent the English speakers in any constitutional conference weakened the position of the opposition leader.

[25] On the economic front, Biya ignored the opposition requests and kept the economic free zone open, despite the 100 per cent devaluation

of the CFA decided in Paris. After promising to make 20,000 civil servants redundant and to go ahead with the privatizations, Cameroon's government received funds from the IMF and the Paris Club.

[26] In 1995, the opposition continued to split. In May, Sigo Assanga, former secretary-general of the Fru Ndi Social Democratic Front, abandoned this political group and founded the Social Democratic Movement. Meanwhile, the UNDP also broke into factions after the expulsion of two members who had participated in Biya's governments.

[27] In spite of the opposition movements, the government imposed its economic policy and achieved 5 per cent economic growth during 1995.

Canada

Population: 29,248,000
(1994)
Area: 9,976,140 SQ KM
Capital: Ottawa

Currency: Canadian Dollar
Language: English and French

Canada

The first inhabitants of present-day Canada were Inuit (wrongly known as Eskimos) and other people that had come from the Asian continent across the Bering Straits. The European colonists who came to North America in the 16th century estimated the indigenous population of the entire continent to be between ten and twelve million.

[2] Between the 17th and 19th centuries, these territories were colonized by Britain and France. Some areas changed hands several times until 1763, when the Peace of Paris, which ended the Seven Years' War, granted Canada to Britain. Colonization increased, thanks to the profitable fur trade, with the local population growing to almost half a million by the end of the 19th century.

[3] The British North America Act of 1867 determined that the Canadian constitution would be similar to Britain's, with executive power vested in the King and delegated to a Governor General and Council. The legislative function would, meanwhile, be carried out by a Parliament composed of a Senate and a House of Commons.

[4] In 1931, the Statute of Westminster released Britain's dominions from the colonial laws under which they had been governed, giving Canada legislative autonomy. That same year, Norway recognized Canadian sovereignty over the Arctic regions to the north of the main part of its territory.

[5] In 1981, the Canadian government reached an agreement with the British Parliament over the initiation of a constitutional transition. The following year, the 1867 Constitution was replaced by the Act of Canada, which granted Canada the autonomy to reform the constitution.

[6] The Constitution Act of 1982 included a Charter of Rights and Freedoms, which recognized the country's pluralistic heritage and the rights of its indigenous peoples. It set forth the principle of equal benefits among the country's ten provinces, and the sovereignty of each province over its own natural resources, though Quebec did not sign the agreement.

[7] When Quebec was occupied by English Protestants in 1760, its predominantly French population sought to maintain a separate Quebecois identity, backing the cause with both religious and politically nationalistic forces. Thus, Quebec assumed a special role as guardian of the Catholic faith, the French language and the French heritage in North America. It was a role which it performed well, given the fact that from a French population of 6,000 in 1769, the number of "Quebecois" had increased to 6 million by 1960.

[8] In Quebec, four-fifths of the population speak French as a first language and are fiercely proud of their cultural identity, and provincial autonomy has always been a delicate issue. In 1977, the government of the separatist Parti Quebecois (PQ), led by René Lévèsque, adopted French as the official language of education, business and local public administration.

[9] Lévèsque discarded the possibility of a unilateral separation, proposing instead a "sovereign association", with a monetary and customs union. However, voters rejected this proposal by 59.5 to 40.5 per cent, in a plebiscite held in 1980. Fifteen years later, in a new plebiscite favoured by the Parti Quebecois, the margin was narrowed down: 50.6 per cent of voters opposed separation and 49.6 per cent approved it.

[10] Antiquated British law governed relations between the sexes in North America for a long time, and it was not until October 1929 that Canadian women obtained full legal rights, and it was only under the 1982 Constitution that true legal sexual equality was established.

[11] In addition to the federal law, each province has an equal rights law, guaranteeing access to housing, jobs, services and other facilities, without discrimination on the basis of race, religion, age, nationality or sex, though Quebec is the only province which prohibits discrimination on the basis of "sexual inversion", their terminology for homosexuality.

[12] Nevertheless, women are still the victims of much discrimination. Not all professions are open to them, and in a great many professional fields they must accept lower salaries. Between 1969 and 1979, there was a 62 per cent increase in the number of women working outside the home. By 1981, this figure represented 49 per cent of all women, and 39 per cent of the economically active population.

[13] Liberal governments, led by Pierre Trudeau, were elected in 1968, 1972, 1974 and again in 1980, after a brief Conservative interlude. Trudeau loosened Canada's traditional ties with Western Europe and the United States, and strengthened those with the Far East, Africa and Latin America. In addition, he refused to participate in the economic blockade against Cuba.

[14] Economic difficulties stemming from the worldwide recession triggered a sharp drop in the Liberal Party's popularity, in favour of the Conservative Party. In 1983, Conservative leader Brian Mulroney, a labour lawyer and businessman from Quebec became prime minister, replacing John Turner, who had succeeded Trudeau as head of the Liberal Party.

[15] Mulroney re-established a "special relationship" between Canada and the United States, with the initiation of negotiations for a free trade agreement in 1985. This agreement, which went into effect in January 1989, provoked criticism from the part of the Liberals and other members of the opposition, who claimed that the terms of the agreement were overly favourable to the US. Nevertheless, it received majority support from the voters in the 1988 election.

[16] The Conservative victory was made possible by the Quebecois votes, reflecting the growing economic importance of French-speaking voters. This victory also reflected the impact of the Meech Lake Accord, initiated by Mulroney and signed in 1987, in which the federal government ceded important powers to the provinces, granting Quebec recognition of its unique cultural status, for the first time in order for it to accept the constitution and Charter of Rights and Freedoms.

[17] To go into effect, the Meech Lake Accord had to be ratified by the unanimous consent of each of the provinces, and it was blocked by Newfoundland and Manitoba. In Manitoba the one vote against came from a Native Indian who would not accept the "distinct society" clause for Quebec.

[18] The Northwest Territories (NWT) - which make up one-third of the country's land area with only 52,000 inhabitants, half of whom are Inuit or members of other indigenous groups - were considering the possibility of splitting into two

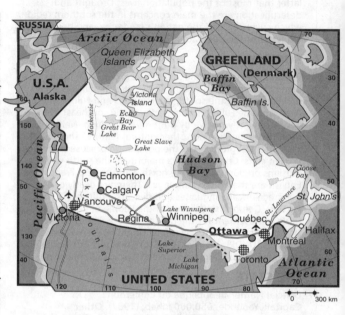

LAND USE

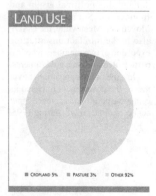

■ CROPLAND 5%   ■ PASTURE 3%   ■ OTHER 92%

regions, Nunavut and Denendeh, each with an autonomous government. While this plan was approved by the provincial legislature in 1987, it is subject to ratification by a plebiscite of its residents, and the Federal Government.

[19] The Federal Government declared its willingness to accept the right of the Inuit and other native peoples of the NWT to self-government, incorporating this right into the Constitution if approved by the provincial governments and by organizations representing the interests of these peoples. In 1988, it transferred 673,000 sq km to the native peoples and the Inuit. In 1992, representatives from the Federal and provincial governments signed an agreement on the land claims of 17,500 native peoples from the Northwest Territories. The way was paved for the creation of "Nunavut" ("Our Land" in Inuit), a project approved in a plebiscite held on April 4. Nunavut, to the north of Hudson Bay and the Arctic archipelago, covers an area of 2.2 million sq km, equal to one fifth of the country. The agreement also included the transference of 350,000 sq km to the Inuits and the payment of $580 million as a counterpart for the waiver of future land claims. "Our Land" did not achieve autonomy, but it provided the population with the right to take part in the exploitation of their lands' natural resources.

[20] The Inuit from northwestern Canada have been working with other indigenous groups of Greenland and Alaska to deal with ecological, cultural, social and political problems which affect their nation. The Fifth Circumpolar Conference of the Inuit (CCI) in 1989 reaffirmed Inuit rights to lands which Canada considers government property, and leases out to large mining interests.

[21] Canada's first trading partner is the United States, and vice versa, but while the volume of Canada's sales to the US accounts for one-fifth of its total production, American exports to Canada represent less than 3 per cent. No other major Western economy maintains a trade imbalance of such magnitude; only comparable to the dependence of Southern countries with relation to the industrialized countries.

[22] With the free trade agreement which went into effect in 1989, Canada's integration with the American economy has been ac-

PROFILE

ENVIRONMENT

Canada is the second largest country in the world in land area, divided into five natural regions. The Maritime Provinces along the Atlantic coast are a mixture of rich agricultural land and forests. The Canadian Shield is a rocky region which is covered with woods and is rich in minerals. To the south, along the shores of the Great Lakes and the St. Lawrence River, there is a large plain with fertile farmlands, where over 60% of the population is concentrated, and the major urban centres are located. Farming (wheat, oats and rye), is the mainstay of the Central Prairie Provinces. The Pacific Coast is a mountainous region with vast forests. The "Great North" is almost uninhabited, with very cold climate and tundra. There are ten provinces, and two territories; the Yukon and Northwest. These are working towards provincial status, but this is complicated by Native Land claims. Canada has immense mineral resources; it is the world's largest producer of asbestos, nickel, zinc and silver and the second largest of uranium. There are also major lead, copper, gold, iron ore, gas and oil deposits. Its industry is highly developed. Industrial emissions from Canadian and US plants contribute to the "acid rain" which has affected thousands of lakes. The construction of a number of hydroelectric plants in Quebec is threatening to destroy the lands and livelihood of the indigenous population.

SOCIETY

Peoples: There are about 800,000 Native Americans, "Metis" (mixed race) and Inuit ("Eskimos") ranging from highly acculturated city-dwellers to traditional hunters and trappers living in isolated northern communities. There are six distinct culture areas and ten language families; many native languages such as Cree and Ojibwa are still widely spoken. About 350,000 native people are classified as such: that is, they belong to one of 573 registered groups and can live on a federally protected reserve (though only about 70% actually do so). "Metis" and those who do not have official status as "native people" have historically enjoyed no separate legal recognition, but attempts are now being made to secure them special rights under the law. **Languages:** English and French, official. 13% of the population is bilingual, 67% speak only English, 18% only French, and 2% speak other languages (Italian, German and Ukranian). **Religions:** Roman Catholic 45.7%; Protestant 36.3%; Eastern Orthodox 1.5%; Jewish 1.2%; Muslim 1.0%; Buddhist 0.7%; Hindu 0.6%; nonreligious 12.4%; other 0.6%. **Political Parties:** The Liberal Party; the Conservative Party; Bloque Quebecois; the New Democratic Party. **Social Organizations:** The Canadian Labour Congress, with over two million members, is the largest trade union; The Assembly of First Nations.

THE STATE

Official Name: Canada. **Administrative Divisions:** 10 Provinces and 2 Territories. **Capital:** Ottawa, 920,857 inhab. (1991). **Other cities:** Toronto, 3,893,046 inhab., Montreal, 3,127,242 inhab., Vancouver, 1,602,502 inhab. (1991). **Government:** Ramon J. Hnatyshyn, Governor-General. Jean Chrétien, Prime Minister and head of government. Canada is a federation of ten provinces and a member of the British Commonwealth. The system of government is parliamentary. **National Holiday:** July 1, Canada Day (1867). **Paramilitaries:** 6,400 Coast Guard.

centuated. Through trade, the extension of credit, and investments in Canada, the US has secured ever greater control over Canadian natural resources, and majority control over shares in Canadian industry.

[23] A recession in the US means a recession in Canada; Ottawa's monetary policy is essentially defined by its neighbour's. This degree of dependence has been labelled "colonial", despite the fact that Canada is the world's eighth largest industrial power, and that its standard of living is the tenth highest in the world, according to OECD figures.

[24] The Canadian Armed Forces (CF) are charged with protecting national interests both inside and outside the country, with defending North America in cooperation with the United States, and complying with its NATO commitments; in addition, it has participated in the UN Peacekeeping Forces.

[25] In addition to these functions, the Canadian Forces Europe (CFE) supplied the NATO Supreme Command with troops which were permanently stationed in Germany, including 4 Mobile Command Brigades and an air division, all of which maintained maximum operational readiness. They also participated in the United States intervention in Haiti in 1994.

[26] Relations between Canada and the United States became tense in 1985, when a US Coast Guard vessel passed through the Northwest Passage without Canadian authorization. The US recognizes Canadian sovereignty over the Arctic islands but not over its waters; a similar dispute exists with relation to the waters surrounding the French islands of St Pierre and Miquelon.

[27] Canada has made little headway in getting the US to control industrial gas emissions, which drift over Canadian territory, bringing with them acid rain.

[28] The Canadian government, has however, made a commitment to reduce its own industrial emissions by half. This program, which has already gone into effect, foresees a reduction of industrial pollution of 20 per cent by the year 2005.

[29] In 1991, Quebec's Premier Robert Bourassa initiated the second phase of the huge James Bay Power Project, which involves damming or diverting nine rivers which flow along 350,000 km of the northwestern part of the

MILLIONS $ 1994

IMPORTS
155,072
EXPORTS
166,000

province, emptying into the James and Hudson Bays.

[30] The first phase, called "La Grande", is capable of generating 10,282 megawatts of hydro-electric power, and once the project is completed, it will reach a capacity of 27,000 megawatts, far greater than the output of Itaipu Dam, in Brazil, or that of Three Gorges Dam in China.

[31] The region affected by the projects is located in the hunting and fishing grounds of approximately 11,000 Cree and 7,000 Inuit, peoples who have lived in the area for over 5,000 years. This region contains the largest beluga whale population in the world, as well as the largest caribou herds and most of the world's fresh-water seals. Vast numbers of migratory birds make the extensive wetlands their home for several months each year.

[32] In order to divert four of the rivers and build the dams and reservoirs, 13,000 sq km of land has been submerged. In addition, in 1984, 10,000 caribou died when water was released from one of La Grande's reservoirs.

[33] The completion of the second phase of the project will dam the rivers which empty into the aforementioned bays and will flood an additional 10,000 sq km of land. The project will allow Canada to sell electricity to the United States, set up plants for aluminium production in Quebec and guarantee an energy supply for Canadian consumption, which is among the highest in the world. The ecological consequences of this project on the regional ecosystem are unforeseeable.

[34] In 1973, hunting and fishing provided two-thirds of the local population's food, with this percentage dropping to one-fourth, twenty years later. A 1975 treaty allowed local people to retain a small part of their lands, granting some economic compensation for the loss of the rest of their territory. Even so, the chances of the local indigenous culture surviving the nationalistic ecocide of Quebec are severely limited.

[35] The Inuit, who inhabit the Belcher Islands in the Hudson Bay and navigate the frozen waters in search of fish and game, were not taken into account in these agreements, and they are very concerned about what effect these major changes will have upon the ecosystem, and their way of life.

[36] Canada currently has 326,000 indigenous people belonging to 577 different groups, in addition to 25,000 Inuit. There are also at least 100,000 Métis and indigenous peoples who have been assimilated into the population of European descent (4 per cent of the total population, according to official figures, with actual numbers possibly reaching 850,000 people).

[37] Canada's original inhabitants have been gradually becoming more organized and, between the 1960s and 1970s, these efforts culminated in the creation of the National Indian Brotherhood (NIB). The aim of this organization is to represent the indigenous peoples point of view to the general public, and also in making their grievances known to the federal government. Recently, the NIB was replaced by the Assembly of First Nations.

[38] The indigenous peoples are fighting for government respect of the treaties which affirm their rights to their land and resources. Those who live from hunting and fishing, like the Cree, the Dene, the Innu, the Haida and the Iroquois, are especially vulnerable, facing multiple threats to their lands and traditional way of life.

[39] After World War II, the government tried to restrict the Innu to small reservations, and by the 1970s, most of them were living in areas designated for this purpose. Although Canada officially recognizes the claims of indigenous peoples over their lands, it has sought to reach agreements with individual groups, to set them up in smaller areas in exchange for economic compensation.

[40] The experience that has occurred in one generation of the Innu people has demonstrated that it is a mistake to accept these government propositions. A 1984 report revealed that the suicide rate among the communities in the North of Labrador is 5 times the national average. In 1988, in retribution, the Innu invaded the base at Goose Bay, setting up a "peace camp". In spite of the fact that many were arrested, they managed to disrupt that year's test flight program.

[41] Four Innu charged with invading the base were unexpectedly released under the argument that, since land ownership is a concept which is alien to Canada's original inhabitants, it was not reasonable for the court to consider evidence

North American Free Trade Agreement

On January 1, 1994, the governments of Canada Mexico and the United States signed the document creating NAFTA, the North American Free Trade Agreement. According to the terms of the agreement, the three countries will eliminate all customs and tariff restrictions between them, forming a common economic zone.

[2] The three partners, with a joint population of 370 million and a combined GNP of more than $6 trillion per year, pledged to eliminate some trade tariffs immediately, gradually doing away with the rest over a 15-year period. The three countries also decided that the United States' and Mexico's most vulnerable industries would be eligible for protection for a longer period of time. According to the treaty, any of the members may withdraw from NAFTA, after giving notice six months in advance.

[3] The process leading up to the signing of the treaty was marked by deep resistance on the part of the US labor unions, fearful of losing jobs. There was a split vote in Congress by President Bill Clinton's Democratic Party, with unions belonging to the AFL-CIO accusing the president of allowing the causes of unemployment to continue to grow. The majority of the Republican congressmen supported the agreement, which passed by a narrow margin in Congress.

[4] The unions' fears are based on the fact that mean wages are 15 per cent lower in Mexico than in the United States. The large transnational companies will receive enormous benefits by taking advantage of a labor force capable of operating First World technology at Third World wages.

[5] Since the treaty went into effect, several Latin American countries have asked to join NAFTA: Chile, Colombia and Argentina have repeatedly offered to sign bilateral agreements with the new common market.

[6] Within NAFTA's projected framework, the three countries agreed to create the North American Development Bank. With funds totalling more than $3 billion, it will join other existing international credit agencies.

[7] The unification of an economic area like NAFTA signals a reordering of priorities in US relations with Latin America. Just as the European Community and Japan maintain clear, almost exclusive areas of commercial interest, the treaty will give the United States privileged access to Mexico's mineral and oil resources.

[8] A large number of environmental organizations have called attention to the lack of controls on industrial production in Mexico. A massive transfer of industries to that country could cause considerable damage to the environment in the United.States. According to official sources, the US government has already earmarked $225 million, to be released over a four-year period following the signing of the treaty, to finance an environmental clean-up along the US-Mexican border.

based on British or Canadian legal norms. The Canadian government filed an immediate injunction, calling into question the validity of the judge's ruling, when it became obvious that this decision could set a precedent for land claims of the region's indigenous groups.

[42] In late 1990, Parliament approved restrictive legislation with regard to the granting of refugee status, effective as of January 1991. The new legislation establishes that people fleeing persecution can now be returned to their country of origin if that country is considered to be "safe". Both the opposition and humanitarian organizations have accused the government of going "from one extreme to the other". The government claimed that it had not closed its borders completely, as it authorized the selection of 13,000 refugees by its embassies abroad and will also allow 10,000 family members of foreign residents to come to Canada.

[43] In mid-1991, the International Task Force for Indian Affairs declared that the Canadian government was not respecting the rights of the Mohawk people to practise their religion, and denounced specific acts of cultural aggression such as the construction of a golf course on their communal lands.

[44] Toward the end of 1991, in a conference organized by the "Indigenous Woman 500" Committee of Canada, delegates from 22 American countries rejected the domination and discrimination to which they are subjected. The participants in the meeting decided to take steps to recover the leadership role which women held in Indian society before the arrival of the European colonists.

[45] In October 1991, Ottawa announced its decision to earmark some $700 million for the agricultural sector; however, farmers were not satisfied with this amount, considering it insufficient to compensate for the losses of the previous year. Farm subsidies by the governments of the United States and European countries have left Canadian farmers at a distinct disadvantage, as present policy is to pursue liberalized trade under the GATT agreement.

[46] In December, the Canadian Minister of Finance stated that the country had registered its first trade

DEMOGRAPHY

Urban: 77% (1995). **Annual growth:** 1.3% (1991-99). **Estimate for year 2000:** 30,000,000. **Children per woman:** 1.9 (1992).

HEALTH

Under-five mortality: 8 per 1,000 (1995). **Calorie consumption:** 113% of required intake (1995).

EDUCATION

School enrolment: Primary (1993): 104% fem., 104% male. Secondary (1993): 103% fem., 104% male. University: 103% (1993). **Primary school teachers:** one for every 17 students (1992).

COMMUNICATIONS

108 **newspapers** (1995), 109 **TV sets** (1995) and 110 **radio sets** per 1,000 households (1995). 59.2 **telephones** per 100 inhabitants (1993). **Books:** 103 new titles per 1,000,000 inhabitants in 1995.

ECONOMY

Per capita GNP: $19,510 (1994). **Annual growth:** 0.30% (1985-94). **Annual inflation:** 3.10% (1984-94). **Consumer price index:** 100 in 1990; 109.4 in 1994. **Currency:** 1,4 Canadian dollars = 1$ (1994). **Cereal imports:** 1,095,000 metric tons (1993). **Fertilizer use:** 479 kgs per ha. (1992-93). **Imports:** $155,072 million (1994). **Exports:** $166,000 million (1994).

ENERGY

Consumption: 7,795 kgs of Oil Equivalent per capita yearly (1994), -46% imported (1994).

deficit in 15 years (in September, the trade deficit was $275 million). 1991 was characterized by economic recession, high taxes, corruption scandals, increasing separatism and a decline in Mulroney's popularity.

[47] In the first few months of 1992, there was an intense debate over Quebec's growing demands for autonomy. In August, this province rejected the government's scheme for granting it "special" status, as the terms were considered "insufficient".

[48] On August 12, 1992, after 18 months of negotiations, Canada, the United States and Mexico signed an agreement creating a free trade zone (North American Free Trade Agreement, NAFTA), (see Box).

[49] In February 1993, Prime Minister Brian Mulroney, decided to resign from the Conservative Party (and therefore from the government) with the lowest popularity ratings ever. Earlier, in June, he had tried to take advantage of his 10-seat parliamentary majority to obtain ratification of NAFTA. But his resignation gave credit to the opposition's argument that elections should be anticipated, and that no key issues should be decided on before new elections.

[50] Mulroney's resignation led to the Minister of Defence, Kim Campbell, becoming the 19th prime minister, and the first woman to occupy this post. Her first decision, after assuming office, was to reduce the cabinet to 25 members, 10 fewer than before.

[51] In June of the same year, Parliament ratified the NAFTA treaty. Two months later, the new prime minister travelled all over the country promoting the adoption of austerity measures to combat the budget deficit. In the meantime, her rival, Liberal Jean Chrétien, promised to give high priority to the creation of new jobs.

[52] In the October 1993 general elections, the Liberal Party (opposition) regained power after nine years, with a landslide victory over the Conservative Party, whose representation in the House of Commons dropped from 155 seats to 2. The Liberals won 178 seats, against 79 in the previous legislature. The separatist Bloque Quebecois (54 seats) and the right-of-center Reform Party (52 seats) both made significant gains. This was the worst defeat for a governing party in Canada's 126-year history.

[53] The new prime minister, Jean Chrétien, took office on November 4, 1993. The following month, former prime minister Kim Campbell resigned as head of the Conservative Party.

[54] The reduction of public debt and state expenditures, the rise of unemployment to over 10 per cent and Quebec's separatist efforts were the government's main concerns during 1994 and 1995. In May 1994, inflation fell for the first time in 40 years. With a family income at over $45,000 per year, Canada continues to be one of the wealthiest countries in the world.

[55] The possibility of Quebec's separation from Canada was raised once again in 1994 by the separatist Parti Quebecois (PQ). The popularity of Prime Minister Chrétien had grown 13 per cent since the 1993 elections but the PQ won in the province of Quebec -which Chretien had recognized as "distinct". The new Quebec premier, Jacques Parizeau, promised to do everything possible to turn Quebec into a sovereign state.

[56] Due to the geopolitical context imposed by the NAFTA, Quebec's case was closely followed by the agreement's member countries since it was concluded on January 1, 1994, even though the PQ's project stated the province would abide by the obligations assumed by Canada. Separation was rejected by 50.6 per cent of voters on October 30, 1995. The PQ claims it will hold a new plebiscite in case of winning the 1988 legislative elections.

Cape Verde

Cabo Verde

Population: 372,000
(1994)
Area: 4,030 SQ KM
Capital: Praia

Currency: Escudo
Language: Portuguese

When the Portuguese settled on the Cape Verde archipelago in the 15th century, the islands were deserving of their descriptive name. They were covered by lush tropical vegetation that stood out against the black volcanic rock and the blue sea. Some 400 years later, Portuguese colonization had wrought devastation and transformed the islands into a "floating desert". Most of the population emigrated to escape starvation, while those who stayed depended upon foreign aid.

[2] Cape Verde was a very important 16th century port of call for ships carrying slaves to America. Incursions by French, British and Dutch pirates were so frequent that Portugal brought farmers to the islands from the Alentejo region to ensure a more permanent presence. The agriculture practised by the new settlers quickly eroded the thin layer of fertile soil and periodic droughts have struck the country since the 18th century.

[3] Falling farm production caused massive emigration of Cape Verdeans; many went to Guinea-Bissau, another former Portuguese colony with close ties to the archipelago. In later years, there was further emigration to Angola, Mozambique, Senegal, Brazil, and the US.

[4] During the independence struggle, Cape Verde developed closer ties to Guinea-Bissau (see Guinea-Bissau). One major reason was the formation of the African Party for the Independence of Guinea and Cape Verde (PAIGC) in 1956, supported by both colonies. Amilcar Cabral, the PAIGC's founder and ideologist, planned that the two economically similar countries, could fight together for freedom and development, once independence was achieved.

[5] In 1961, guerrilla warfare broke out on the continent, and hundreds of Cape Verdean patriots joined the fight. Portugal's colonial government was overthrown in 1974, a transition government was installed, and in 1975, the islands were proclaimed independent. The PAIGC set an unusual precedent, as for the first time ever one political party took office in two different countries at the same time. Aristides Pereira was elected president and commander Pedro Pires became prime minister. The PAIGC leadership took the first steps toward establishing a federation between Cape Verde and Guinea-Bissau, and the national assemblies of the two countries sat together as the Council of the Union.

[6] In 1968 the Cape Verde government faced the devastating effects of a drought that left 80 per cent of the population of the archipelago without food. However, the prudent action of civilian organizations supplemented by foreign assistance averted a major catastrophe.

[7] From 1975 onwards, the forested areas of Cape Verde increased from 3,000 to 45,000 hectares. The government predicted that in the next 10 years a further 75,000 hectares would be planted making the islands self-sufficient in firewood. Early in the rainy season, the population of Cape Verde voluntarily spend a week planting trees.

[8] The government implemented land reform, giving priority to subsistence farming to replace the export crops of the colonial period, when barely 5 per cent of food needs was met by local production. The drastic decline in agricultural production led the government to invest in fisheries.

[9] Cape Verde supported Angola during its "second liberation war" (see Angola) by allowing Cuban planes to land on the archipelago during the airlift that helped defeat the invasion of Angola by Zaire and South Africa. Cape Verde adopted a policy of non-alignment, declaring that no foreign military bases would be allowed in their territory.

[10] In 1981, while the PAIGC was negotiating a new constitution for Guinea-Bissau and Cape Verde, Guinea-Bissau's president Luis Cabral was overthrown. Joao Bernardo Vieira took office, and was initially hostile to integration with Cape Verde. In January 1981, an emergency meeting of PAIGC members in Cape Verde discussed the political developments in Guinea-Bissau. This reaffirmation of support for the principles of Amílcar Cabral led the party congress to adopt a new name "the African Party for the Independence of Cape Verde" (PAICV) stressing their independence from Guinea.

[11] Relations between the two governments became tense, but

Cape Verde supported Angola during its "second liberation war" by allowing Cuban planes to land on the archipelago during the airlift that helped defeat the invasion of Angola by Zaire and South Africa.

mediation efforts, particularly by Angola and Mozambique paid off. Reconciliation came in August 1982, in Maputo, when President Machel brought Aristides Pereira and Joao Bernardo Vieira together. Further progress was made at the Conference of Former Portuguese Colonies in Africa, held in Cape Verde in November 1982. In Praia, President Joao Bernardo Vieira met with colleagues from Angola, Mozambique, Cape Verde, and Sao Tomé. Diplomatic relations returned to normal, but plans for reunification were abandoned.

[12] Political stability, defiance and perseverance helped Cape Verde avoid famine and the other drastic consequences of the drought which afflicted Africa. In 1984, crop yields fell by 25 per cent, the trade deficit stood at $70 million and the foreign debt reached $98 million. However, the food distribution system, and efficient state and resource management have prevented famine, although the country has been ravaged by droughts for 19 years. Despite these positive factors, some sectors of the population are still severely undernourished.

[13] The paucity of resources made Cape Verde dependent upon foreign assistance, submitting to a number of projects which were doomed to failure, including the nation's First Development Plan.

[14] The Second Development

WORKERS

1994

■ MEN 68%   ■ WOMEN 32%

■ AGRICULTURE 31%   ■ INDUSTRY 6%   ■ SERVICES 63%

ENVIRONMENT

An archipelago of volcanic origin, composed of Windward Islands Santo Antão, São Vicente, São Nicolau, Santa Luzia, Sal, Boa Vista, Branco and Raso, and Leeward Islands Fogo, Santiago, Maio, Rombo and Brava. The islands are mountainous (heights of up to 2,800 meters) without permanent rivers. The climate is arid, influenced by the cold Canary Islands current. Agriculture is poor; nonetheless it employs most of the population. Cape Verde lies within the Sahel region, which is undergoing increased desertification, with periodic droughts. This phenomenon is aggravated by the islands' small size, topography of rolling hills and the prevalence of high winds.

SOCIETY

Peoples: Cape Verdeans are Africans of Bantu origin with strong European influence. **Religions:** Roman Catholic 93.2%; Protestant and other 6.8%. **Languages:** Portuguese is the official language, but the national language is Creole, based on old Portuguese with African vocabulary and structures. **Political Parties:** African Party for the Independence of Cape Verde (PAICV); Movement for Democracy (MPD). **Social Organizations:** Organized workers belong to the National Cape Verde Workers' Union-Central Trade Union Committee (UNTC-CS).

THE STATE

Official Name: República do Cabo Verde. **Administrative divisions:** 9 islands and 14 counties. **Capital:** Praia, 61,000 inhab. (1990). **Other cities:** Mindelo, 47,000 (1990). **Government:** Parliamentary republic. Antonio Mascarenhas Monteiro, President of the republic. Carlos Alberto de Carvalho Veiga, Prime Minister. **National Holiday:** July 5, Independence Day (1975). **Armed Forces:** 1,100.

DEMOGRAPHY

Annual growth: 3.1% (1992-2000). **Children per woman:** 4.3 (1992).

HEALTH

Under-five mortality: 73 per 1,000 (1995). **Calorie consumption:** 102% of required intake (1995). **Safe water:** 97% of the population has access (1990-95).

EDUCATION

Illiteracy: 28% (1995). **Primary school teachers:** one for every 33 students (1989).

COMMUNICATIONS

84 **newspapers** (1995), 86 **TV sets** (1995) and 92 **radio sets** per 1,000 households (1995). 3.8 **telephones** per 100 inhabitants (1993). **Books:** 87 new titles per 1,000,000 inhabitants in 1995.

ECONOMY

Per capita GNP: $930 (1994). **Annual growth:** 2.00% (1985-94). **Consumer price index:** 100 in 1990; 109.8 in 1994. **Currency:** 81 escudos = 1$ (1994).

Plan, launched in 1986, prioritized private enterprise, especially in the informal sector. In agriculture, concentrated efforts were made in the fight against desertification. In 1990, the plan was to recover more than 500 square kilometers, and establish a centralized system of administering water reserves for the whole country. During the first phase of the plan, more than 15,000 dams were built to store rainwater, and 231 square kilometers were forested.

[15] Despite adverse climatic conditions, farm production has gradually increased to the point where the country is practically self-sufficient in meat and vegetable requirements.

[16] Amidst the political decline linked to the crisis of socialism worldwide, the PAICV lost the first pluralist parliamentary elections, held in February 1991. The Movement for Democracy (MPD), led by Carlos Veiga, won the elections with over 65 per cent of the vote, though 38 per cent of the electorate abstained from voting. Antonio Mascarenhas Monteiro, a legal expert educated at the University of Louvain, Belgium, and President of the Supreme Court of Cape Verde during the previous decade, was elected president.

[17] In September 1992, the new constitution went into effect. It ratified a multi-party system and banned the formation of parties on the basis of religious doctrines, tribal or regional divisions.

[18] The new government began the transition to a free-market economy with the privatization of insurance companies, fishing and banking, in line with the requirements of international agencies on which the country greatly depended on. Foreign aid accounted for 46 per cent of the GDP, while money remittances from the 700,000 Cape Verdians residing abroad, constituted another 15 per cent of it.

[19] The liberal government, faced with a 25 per cent unemployment rate, declared its goal to restructure the State. In the first quarter

> **Despite adverse climatic conditions, farm production has gradually increased to the point where the country is practically self-sufficient in meat and vegetable requirements.**

of 1993, government authorities announced a 50 per cent reduction of the country's ual deregulation of prices was initiated.

[20] The 1994 budget, despite establishing a retrenchment of the public expenditure, included an increase of public investment, from $80 million in 1993 to $138 million in 1994. The priority sectors for state investment were transport, telecommunications and rural development.

[21] In January 1995, Prime Minister Carlos Veiga introduced significant changes in government in order to "facilitate the country's transition to a free-market economy". One of the most important changes was the merger of the Ministries of Finance, Economic Coordination and Tourism, and Industry and Commerce into a sole Ministry of Economic Coordination. Inflation in 1995 was 6 per cent and Cape Verdian economy is still greatly dependent upon foreign aid, particularly from the European Union.

Cayman Islands

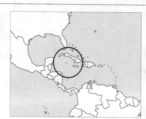

Population: 33,000 (1994)
Area: 260 SQ KM
Capital: George Town

Currency: Cayman Dollar
Language: English

Cayman Islands

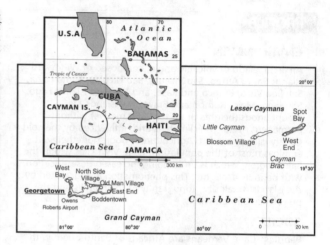

In 1503, Christopher Columbus sighted the Cayman Islands and named them the "Turtles" because of the vast numbers of turtles, iguanas and alligators, which were their sole inhabitants. The Cayman Islands include Grand Cayman, which is the largest, Cayman Brac and Little Cayman, and all three are surrounded by coral reefs.

2 Shortly afterwards, the islands came under French control, or more precisely, buccaneers and privateers "the fathers of the French colonies in the West Indies". For a good part of the 17th century, the Caymans were the pirate headquarters of the Antilles.

3 The first permanent colonial settlement was established in Grand Cayman, and most of the settlers came from Jamaica; it was not until around the year 1833 that they started to live on the other two islands. The Antilles were all affected by power struggles between the European colonialists. When Jamaica was ceded to the British Crown in 1670, under the Treaty of Madrid, the Cayman Islands became a Jamaican dependency. In 1959 they were given the status of British dependent territory.

4 In 1972, a new Constitution was passed granting some local autonomy on domestic matters. The governor is appointed by the Crown and is responsible for defence, foreign affairs, internal security, and some social services. There is also an Executive Council and Legislative Assembly. The new Executive Council, created after the 1988 general elections, has two new members and two who were already appointed.

5 The Cayman Islands have few natural resources other than the sea and sand, which make them popular tourist resorts. With the exception of turtle farming, local industry and agriculture only meet domestic needs. The Caymanese are renowned schooner builders and sailors. One of the country's main sources of revenue is the money sent home by sailors.

6 Tax exemption has attracted a growing number of offshore banking and trust companies. In 1987, there were 515 banks in the Cayman Islands.

7 A quarter of the economically active population works in tourism, there are hotels throughout the islands, with a total of 5,200 beds. The influx of visitors has produced considerable immigration, which currently accounts for 35 per cent of the population. In 1989, the Cayman Islands were visited by 618,000 tourists. Fishing declined seriously in the 1970s, and has not recovered. Turtle farming is being developed to replace traditional turtle fishing.

8 The Cayman issue is considered regularly by the UN Decolonization Committee, which asserts the legitimate right of the inhabitants to self-determination and independence. On the islands, however, the people do not seem anxious to sever links with Britain. In 1982, when the Malvinas War broke out, the islands made a contribution of $1 million to the British forces.

9 In August 1991, McKeeva Bush, a deputy member of the Assembly, founded the first political party in the Caymans. The DPP (Democratic Progressive Party) aimed to change the legal status of the islands, which are presently considered British dependencies. The DPP promotes constitutional reform, which, would bring in the creation of a party system, a rise in the number of members of the Executive Council, the creation of the post of prime minister, and an increase in the number of seats in the Legislative Assembly. The governor would become the president of the extended Executive Council.

10 The Bank of Credit and Commerce International (BCCI) scandal disclosed the peculiar banking and speculation conditions of the Cayman Islands. In 1991 five countries took legal steps to close the BCCI, which they accused of systematic fraud. When the fraud was revealed, the Bank of England explained that the irregularities included the concealment of losses, the drawing up of false balance sheets and other serious alterations to documents. The two most important offices of the BCCI are in Luxembourg and on the Cayman Islands; both of which provide for tax exemptions and bank secrecy.

ENVIRONMENT

Located west of Jamaica and south of Cuba, this small archipelago is part of the Greater Antilles. It comprises the Grand Cayman islands, where most of the population live, Little Cayman, and Cayman Brac. The archipelago is of volcanic origin, and has rocky hills and considerable coral formations. The climate is tropical and rainy, tempered by oceanic influences.

SOCIETY

Peoples: Over half the population are *mestizos*; one third are of European descent, and the rest are of African origin. **Religions:** Protestant. **Languages:** English. **Political parties:** Until August 1991 there were no formally constituted political parties. McKeeva Bush, a deputy member of the Legislative Assembly, founded the Democratic Progressive Party (DPP), which aims to change the relationship with Britain. Groups of independent citizens are formed to elect the local government.

THE STATE

Official Name: Cayman Islands. **Administrative divisions:** There are eight districts (Creek, Eastern, Midland, South Rown, Spot Bay, Stake Bay, West End, Western). **Capital:** George Town, 12,921 inhab. (1991). **Other cities:** West Bay, 5,362 inhab., Bodden Town 3,407 inhab. **Government:** Michael E.J. Gore, Governor since September 4, 1992. There is an 18 member Legislative Assembly (15 elected by direct, popular vote and 3 named by the governor).

COMMUNICATIONS

44.9 **telephones** per 100 inhabitants (1993).

ECONOMY

Consumer price index: 100 in 1990; 111.2 in 1992.
Currency: Cayman dollars.

Central African Republic

Population: 3,234,000 (1994)
Area: 622,980 SQ KM
Capital: Bangui

Currency: CFA Dollar
Language: French

République Centrafricaine

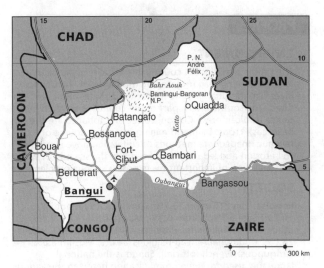

1 The old Ubangui-Chari empire was associated with other neighboring countries, such as Chad, Gabon and Congo during the colonial period. It has been ruled by tribal chiefs, a sultan, colonial traders, French civil servants, a president, a military dictator, an emperor, and a corrupt politician. This process of unstable leadership started late in the last century and has continued since the Central African Republic proclaimed independence, in August 1960.

2 The region suffered one of the most savage and devastating colonial regimes ever imposed by European powers in Africa. France took an interest in the Central African highlands in the late 19th century, when the coast had already been colonized. Occupation of the interior was begun by three military expeditions directed against Sultan Rabah, the governor of the territory presently known as Sudan. The occupation provided an opening for about 40 trading companies which took over rubber and ivory operations in the region, later adding coffee, cotton and diamond mining to their production. This inaugurated a period of nearly 30 years in which the enslavement of native peoples, killings and persecution were widely used by the companies which controlled 70 per cent of the colonized area, renamed Ubangui-Chari in 1908.

3 After World War II, France was short of the financial resources necessary for maintaining a complete colonial administration, so it granted partial autonomy to the Bangui government. The Europeanized local elite became an instrument of neocolonialism, by which the French retained control of foreign trade, defence and tax collection.

4 Violence and oppression were

on the increase in 1949 when Barthelemy Boganda, a former minister, founded the Movement for the Social Evolution of Black Africa (MESAN) to fight for the independence of the Central African Republic. Boganda's prestige grew enormously despite the French attempts to discredit him. In 1956, the French secret service infiltrated MESAN in an attempt to bribe two of Boganda's main advisers, David Dacko and Abel Goumba. Boganda died a year before independence in a suspicious airplane crash, in March 1959.

5 David Dacko took over the post vacated by MESAN's founder, using the position to unleash a purge of the party's more progressive elements. He drew increasingly closer to the French as the US began to show an interest in exploiting the country's uranium and cobalt reserves. Dacko lost the support he had inherited from Boganda, and his government became increasingly corrupt, while the country plunged into deep economic crisis.

6 In December 1965, Dacko was ousted in a coup led by his cousin, Colonel Jean Bedel Bokassa who was pro-France, having served in the French Army for 22 years. Bokassa belonged to a small bourgeois group of M'Baka landowners from Lobaye, where France had recruited agents for its colonial administration. In 1972, he proclaimed himself president-for-life, later taking the rank of field marshall, and finally renaming the country the Central African Empire, crowning himself emperor. The coronation ceremony in December 1977 cost $28 million and was financed by France, Israel and South Africa.

7 In 1978, Bokassa gave 30,000 sq km rich in diamonds to the Israe-

li army, hiring a notorious international arms dealer as his military adviser. He sent troops to Zaire to help Mobutu put down a rebellion in the province of Shaba (see Zaire).

8 Popular discontent finally erupted in a series of workers' and student rebellions, stifled with the support of Zairean troops. In April 1979, the students refused to comply with regulations specifying that uniforms had to be bought in stores owned by the emperor, and in the ensuing street demonstrations, 100 youths were arrested and taken to the infamous Ngaragba prison, where they were tortured and most of them killed under Bokassa's personal supervision.

9 Resorting to neocolonial manipulations, France decided to depose the emperor, attempting to erase the negative image it had created in supporting his crowning. On September 20 1979, while Jean Bedel Bokassa was travelling in Libya, France sent a military plane

to bring former president David Dacko back to Bangui. He took power, making no effort to hide his status as a colonial envoy, protected by a thousand French troops. Dacko dissolved the empire and reinstated a republic, granting France a ten-year lease on the enormous Bouar air base.

10 Dacko's return amounted to no more than a change of name as corruption and repression continued. In 1980, the Central African government broke relations with the USSR and Libya, expelling all their technical advisers and diplomats. Political opposition was ruthlessly repressed, and almost all opposition leaders were sent to prison or exiled. Amidst countless conspiracies, Dacko was overthrown by another military coup in September 1981, bringing General Kolingba to power.

11 The new government's first measures were to ask the French to pay the salaries of 24,000 civil servants, and to grant new economic concessions to the US, allowing them to exploit uranium resources. Meanwhile, the return to normality, initially set for 1982, was postponed until 1986. On November 21 of that year, in an election barred to the opposition, Kolingba was elected president with the support of his Central African Party of Democratic Recovery (PCRD). A constitution was passed establishing a one-party system, and on July 21, 1987, the members of the General Assembly were chosen from PCRD members.

12 The government launched a structural adjustment plan, follow-

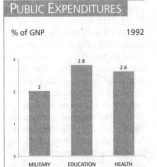

PUBLIC EXPENDITURES

% of GNP   1992

MILITARY 2   EDUCATION 2.8   HEALTH 2.6

WORKERS

1994

■ MEN 55%   ■ WOMEN 45%

■ AGRICULTURE 81%   ■ INDUSTRY 3%   SERVICES 16%

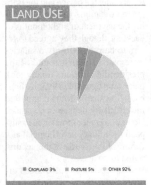

LAND USE

■ CROPLAND 3%   ■ PASTURE 5%   ■ OTHER 92%

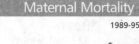

Maternal Mortality
1989-95

PER 100,000
LIVE BIRTHS
649

Food Dependency
1970
17%
1992
19%
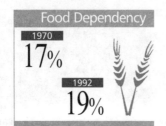

External Debt
1994

PER CAPITA
$276
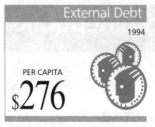

Illiteracy
1995

40%

PROFILE

ENVIRONMENT

The CAR is a landlocked country in the heart of Africa; it is located on a plateau irrigated by tributaries of the Congo River, like the Ubangui, the main export route, and Lake Chad. The southwestern part of the country is covered by a dense tropical forest. Cotton, coffee and tobacco are the basic cash crops. Diamond mining is a major source of revenue. Inadequate methods of cultivation have increased soil erosion and led to a loss of fertility. The scarcity of water and pollution of important rivers pose a serious problem.

SOCIETY

Peoples: Baya (Gbaya) 23.7%; Banda 23.4%; Mandjia 14.7%; Sara 6.5%; Mbum 6.3%; Mbaka 4.3%; Kare 2.4%; other 18.6%. **Religions:** Protestant 40.0%; Roman Catholic 28.0%; traditional African religions 24.0%; Muslim 8.0%. **Languages:** French (official); Sango is the national language, used for intercommunication between the various ethnic groups. **Political Parties:** Central African People's Liberation Party; Democratic Movement for the Rebirth and Evolution of the Central African Republic (MDRERC).

THE STATE

Official Name: République Centrafricaine. **Administrative Divisions:** 16 Prefectures, 52 Sub-prefectures.
Capital: Bangui, 452,000 inhab. (1988) **Other cities:** Bouar 40,000, Berberati 42,000 inhab. (1988).
Government: Ange-Félix Patassé, President and Head of State, elected in September 1993. Bicameral legislature.
National Holiday: December 1, Independence Day (1960).
Armed Forces: 2,650 (1995). **Paramilitaries:** 2,700 Gendarmes.

STATISTICS

DEMOGRAPHY

Urban: 39% (1995). **Annual growth:** 1% (1992-2000). **Estimate for year 2000:** 4,000,000. **Children per woman:** 5.8 (1992).

HEALTH

One **physician** for every 25,000 inhab. (1988-91). **Under-five mortality:** 175 per 1,000 (1995). **Calorie consumption:** 86% of required intake (1995). **Safe water:** 18% of the population has access (1990-95).

EDUCATION

Illiteracy: 40% (1995). **School enrolment:** Primary (1993): 51% fem., 51% male. University: 2% (1993). **Primary school teachers:** one for every 90 students (1989).

COMMUNICATIONS

81 **newspapers** (1995), 82 **TV sets** (1995) and 81 **radio sets** per 1,000 households (1995). 0.2 **telephones** per 100 inhabitants (1993). **Books:** new titles per 1,000,000 inhabitants in 1995.

ECONOMY

Per capita GNP: $370 (1994). **Annual growth:** -2.70% (1985-94). **Annual inflation:** 2.60% (1984-94). **Consumer price index:** 100 in 1990; 116.6 in 1994. **Currency:** 535 CFA dollars = 1$ (1994). **Cereal imports:** 32,000 metric tons (1993). **Fertilizer use:** 5 kgs per ha. (1992-93). **External debt:** $891 million (1994), $276 per capita (1994). **Debt service:** 12.9% of exports (1994). **Development aid received:** $174 million (1993); $55 per capita; 14% of GNP.

ENERGY

Consumption: 29 kgs of Oil Equivalent per capita yearly (1994), 76% imported (1994).

---

ing IMF guidelines. Nevertheless, the economy as a whole showed no signs of recovery.

[13] In 1986 Bokassa returned from exile in France. Sentenced to death, during his absence, the former "emperor" was imprisoned.

[14] Within the framework of the democratization process, direct municipal elections were held in May 1988, with universal suffrage. It was hoped that this would help to improve trade relations with the developed countries. Seeking popular support for a return to democratic institutions, the government tried to address the question of self-sufficiency in food production, although the country still figures among the world's poorest.

[15] Senegal, the Côte d'Ivoire, Gabon, Djibouti and the Central African Republic are the key areas of French foreign policy in Africa.

[16] In July 1991 a plan was approved for Constitutional reform and for the adoption of a multiparty system. Elections were sched-

uled for October 1992, but shortly after the voting started, Kolingba annulled the process claiming there had been irregularities. After the annulment a Provisional National Council of the Republic was formed, made up of the five presidential candidates. This council announced new elections for April 1993, but these were also postponed on several occasions.

[17] In May, a group of soldiers took the presidential palace and the Bangui radio station by force, demanding the payment of eight months of backpay, though they agreed to return to barracks after two salaries were paid. The first round of the presidential elections finally took place on August 11, 1993. Kolingba, who came in fourth, annulled the result by decree once again. However, France threatened to suspend its military and financial aid, forcing him to allow the second round to go ahead.

[18] On September 1, Kolingba ordered the political prisoners to be

freed, including former emperor Bokassa, who had been accused of cannibalism, murder and the misappropriation of public funds. On September 19, Ange-Felix Patassé, Bokassa's former prime minister, was returned as president with 52.47 per cent of the vote.

[19] In 1994, Patassé continued the rapprochement with Paris, which had supported the candidacy of former president Dacko in 1993. The economy continued to give signs of weakness. The payment of civil servants returned to normal, but the government continued to owe large sums of backpay. In August 1994, a parliamentary delegation from Kuwait visited Bangui to thank the Central African Republic for its support during the Gulf War.

[20] On December 28, 1994, a new Consitition was approved by referendum. However, the opposition

considered this a defeat for Patassé, as only 46 per cent of the electorate turned out to vote. The increase in international prices for cotton and diamonds produced economic growth, estimated at 7 per cent for 1995. Nonetheless, inflation caused by the French imposed 100 per cent devaluation of the CFA continued to affect a large part of the population.

[21] In May 1996, Patassé asked France for military intervention for the second time in two months in order to put down a new group of mutinous troops. The direct participation of French soldiers in street fighting provoked a series of demonstrations against the intervention. Following the mutinies and the French military action, Bangui was left partially destroyed by the fires and looting, and the Central African economy became even more dependent on France.

Chad

Tchad

Population: 6,288,000 (1994)
Area: 1,284,000 SQ KM
Capital: N'Djamena

Currency: CFA Dollar
Language: Arabic and French

The Sahel region, of which the Republic of Chad forms part, has been inhabited from time immemorial. Artifacts dating from 4,900 BC have been found in several tombs in the Tibesti mountain region, in the northern part of the country. The southern region around Chad has been under settlement since about 500 BC. In the 8th century AD, increased desiccation of Saharan borterlands stimulated Berber migrations into the area. At that time, the kingdom of Kanem was founded; it converted to Islam in 1085. By the early 1200s the borders of Kanem had been expanded to Fezzan in the north, Quaddaï in the west, and north of Lake Chad into the Bornu kingdom. The new kingdom of Kanem-Bornu was established through merger, its power and prosperity peaking in the sixteenth century because of its command of the southern terminus of the trans-Sahara route to Tripoli.

[2] At this point the rival kingdoms of Baguirmi and Ouaddaï evolved in the south. At the height of their power in the 17th century, The Baguirmi made an unsuccesful attempt to expand their territory into Kanem-Bornu. Eventually, in the years 1883-93, all three kingdoms (Ouaddaï, Kanem-Bornu and Baguirmi) were to fall to the Sudanese adventurer Rabih az-Zubayr during 1883-1893.

[3] Towards the end of the 18th century, European missionaries converted some of the southern Sara to Christianity and gave them a European-style education. These converts subsequently sided with the Europeans against the native peoples of the north and by 1900 Rabih az-Zubayr was overthrown. The French "acquired" Chad in 1885 during the Berlin Conference, but did not establish themselves in the territory until a 1920 invasion by the notorious French Foreign Legion, which defeated the northern Muslim groups.

[4] The colonizers introduced cotton farming in 1930, and small parcels of land were distributed to peasants to grow this crop, while the French monopolized its commercialization. In the south, forced changes in the land tenure system replaced a communal practice of self-sufficient agriculture with large-scale farming. The result was a cotton surplus and a food shortage, accompanied by famine.

[5] In August 1960, when France granted Chad its independence, southern leaders, who had been negotiating with the colonizers since 1956, assumed power. However, Chad's first president, François Tombalbaye, leader of the Chad Progressive Party, was unable to unite the country whose frontiers reflected the arbitrary colonial divisions.

[6] The Chad Liberation Front (FROLINAT), was founded in 1966 and heavily put down by the French troops. The front became fully active about the time that southern peasants rebelled against a cotton marketing system introduced by the French company Cotonfran, which controlled the cotton industry. Northerners provided FROLINAT's main political support.

[7] In 1970, the front controlled two-thirds of the national territory, and by 1972 FROLINAT guerrillas were within range of N'Djamena, the capital. In 1975 Tombalbaye was ousted and killed in a French-staged coup which put General Felix Malloum in power. The guerrilla offensive continued, and France began to take advantage of the split within the Front, which had, at this time, divided into more than 10 groups.

[8] Paris supported Hissene Habré's faction because he opposed Goukouni Oueddei, FROLINAT's president, who was receiving support from Muammar Khaddafi's government in Libya. At the time, Habré was at the helm of the Armed Forces of the North (AFN), and Oueddei was the leader of the People's Armed Forces (PAF). Two years later, France dispatched an additional force of 3,000 soldiers, along with fighter planes.

[9] The French announced that their troops would maintain neutrality in the struggle between FROLINAT factions, while Hissene Habré accepted the post of prime minister in Malloum's government.

[10] Towards the end of 1979, Chad's 11 major political groups formed a Provisional Government of National Unity (PGNU). Habré was named Minister of Defence. However, the French were displeased with the make-up of the Cabinet because of their strategic interest in Chad - linked to the Maghreb and to the uranium and oil discoveries in the 1960's. In March 1980 Habré resigned, broke the alliance, and unleashed civil war that enabled him to consolidate his power in the south.

[11] The ruling coalition split into three factions. In May 1980, Oueddei requested military aid from Libya, and Khaddafi sent 2,000 troops.

[12] That same year, over 100,000 refugees fled the country, after the

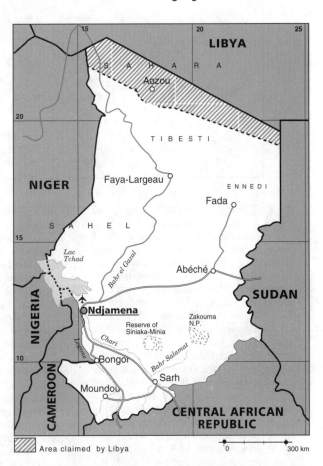

Area claimed by Libya
0   300 km

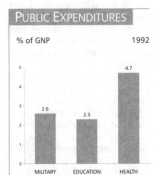

PUBLIC EXPENDITURES

% of GNP   1992

MILITARY 2.6 EDUCATION 2.3 HEALTH 4.7

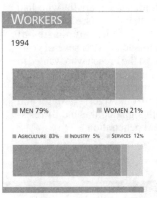

WORKERS

1994

MEN 79%   WOMEN 21%

AGRICULTURE 83%   INDUSTRY 5%   SERVICES 12%

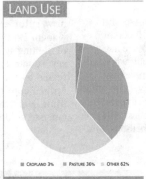

LAND USE

CROPLAND 3%   PASTURE 36%   OTHER 62%

Kanem-Bornu

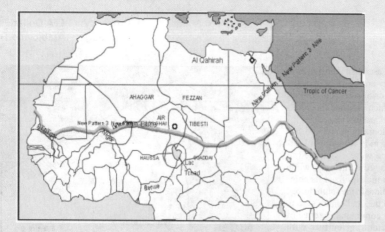

Where the Bilma trade route neared Lake Chad, there lived according to oral tradition communities of "small red men" who must have given rise, between the first and fifth centuries of our era, to the so-called Civilization of Chad, known for its high-quality artistic and metallurgical production. The Sao probably arrived towards the 10th century. They were a Nilotic nation, noted for their great size. They exterminated the "little red men" and established a confederation of groups to the southeast of the lake. On the other side of the lake, communities of shepherds related to the Tibu (autochthonous nomads from the Tibesti mountains) had settled around the eighth century. These shepherds were the Kanuri who, after subduing the local population, set up the state of Kanema, ruled by a military caste. Converted to Islam in the 11th century, they cultivated close ties with the Arab world, based on the slave trade.

2 The capture of slaves frequently took the Kanuri to Sao territory which they eventually conquered. Thereafter, the Saif (Kanuri ruling family) came to be called "the rulers of Kanem, the masters of Bornu" and their nation was known by this compound name.

3 The most outstanding period was during Dunama Dibalim's.rule. Between 1220 and 1260 Dibalim ruled over the territory that extended from the banks of Lake Chad to Fezzan (Libya) in the north, and from the Haussa states in the west to Ouaddai (border of present-day Sudan) in the east.

4 The flourishing traffic allowed him to maintain a standing army of 30,000 horsemen. He also built a large residence in Cairo (Al Qahirah) for the young Kanuri who studied at the famous Al Azhar University.

5 However total dominion over the other nations did not occur: the Sao, Tibu and Bulala (to the southeast) shook the 14th and 15th centuries with frequent rebellions, forcing the Saifs to leave Kanem and to move their capital.

6 Idris Alaoma (1571-1603) decided to solve these problems drastically; he bought muskets, hired instructors in Tunis and modernized the army to confront his enemies one by one.

7 The Sao, long considered rebellious, were completely exterminated in a campaign that razed all their towns. The Tibu were forced to abandon their mountains and settle in Bornu where they were easier to control. To subdue the Bulala he had to conquer the north of what is now Cameroon.

8 Paradoxically, this "putting things in order" did not benefit the Kanuri economically. Although the Haussa merchants yielded to Idris' military drive, and eventually paid tribute to the Kanem-Bornu rulers, they succeeded in gaining control of the slave trade, thus benefitting from the "peace" obtained with so much blood.

9 This situation remained more or less stable until the 19th century, when the Fulani invasion forced the king of Bornu to turn for help to Mahamad Al-Kanemi, a military chieftain. Al-Kanemi repulsed the invaders, but stayed and became the real power behing the throne, even though the crisis was over. This dual situation lasted until 1846, when the last of the Saifs, Ibrahim, was executed by Omar, al-Kanemi's son.

10 In 1893, Omar's successor was deposed by Rabah, a Sudanese guerrilla who had fought alongside the Mahdi (see Sudan). The new ruler extended the boundaries of Kanem-Bornu to Sudan. After several years of fighting against the French who wanted to seize control of the territory, Rabah was defeated by the joint action of three colonial armies coming from Algeria, Congo and Mali. His death signaled the end of resistance to colonial penetration.

bombing of N'Djamena by Habré's forces. In October, Libyan troops reached the capital, mediation efforts by the Organization for African Unity (OAU) failed, and the defeated Habré fled to Cameroon in December 1980.

13 France unleashed an international campaign against Libyan expansionism in Africa, with support from the United States, Egypt, Sudan, and other African countries, fearful that Khaddafi's revolutionary thrust would eventually "infect" poor Islamic populations of the Sahel region, south of the Sahara.

14 Habré was accused of being opportunistic and corrupt; however, Libyan support made it possible for France to divide Oueddei's allies. In April 1981, backed by France, Habré reorganized his followers in Sudan. In July, in Nairobi, the OAU decided to send a peace-keeping force to Chad, with the help of the French and troops from six African countries. Oueddei, yielding to foreign pressure and tensions within the PGNU, requested the withdrawal of Khaddafi's troops in November.

15 After his defeat by Habré's forces in June 1982, in October the exiled Oueddei set up a Provisional National Salvation Government, with eight of the 11 groups that had opposed Habré in the civil war. The new civil war practically split the country in two: northern Chad, under the control of a recently formed National Liberation Council with Libyan logistical support; and southern Chad, with the Habré government dependent on French military support.

16 When Habré seized power, bringing this phase of the civil war to an end, the country was in ruins. The population of the capital, N'Djamena, had dropped to 40,000, and half of its businesses and small enterprises had closed. Outside the cities, 2,000 wells and all the country's water towers had been destroyed, and the health and educational infrastructures were practically nonexistent.

17 An intense drought triggered widespread famine, despite French and American food aid. As hunger spread, massive protests and peasant uprisings occurred in the south. Repression by the Habré government forced 25,000 Chadians to flee across the border into the Central African Republic.

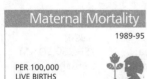
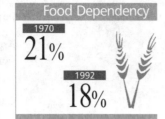

PROFILE

ENVIRONMENT

The northern part of Chad is desert, and 40% of its territory is part of the Sahara Desert, with the great Tibesti volcanic highlands. The central region, the Sahel, which stretches to the banks of Lake Chad, is a transition plain where nomadic pastoralism is common. The lake, only half of which lies within Chad's borders, is shallow and mostly covered with swamps. Thought to be the remnant of an ancient inland sea, its waters are fed by two rivers, the Logane and the Chari, which descend from the plateau that separates this basin from that of the Congo River. The banks of these rivers in southern Chad, fertilized by flooding, contain the country's richest agricultural lands and are the most densely populated areas. In colonial times, economic activity was concentrated there. Cotton is the main export product but subsistence agriculture, though hampered by droughts, still predominates. In recent years, mineral reserves of uranium, tungsten and oil have attracted the attention of the transnationals. Desertification and drought are endemic to the region and affect all aspects of daily life.

SOCIETY

Peoples: Northern Chadians are basically nomadic shepherds of Berber and Tuareg origin; Tubu, Quadainee, while southerners; Sara, Massa, Mundani, Hakka, are mainly traditional farmers. Famines have periodically driven people from the Sahel, in the north to the fertile southern regions. **Religions:** More than 44% of all Chadians are Muslim, 23% practice traditional African religions and 33% are Christian. **Languages:** Arabic and French are the official languages. There are many local languages, the most widely spoken being Sara. **Political Parties:** Patriotic Salvation Movement; National Union for Democracy and Socialism; Chad National Liberation Front (FROLINAT); National Front of Chad (FNT); Movement for Democracy and Development (MDD). **Social Organizations:** Chad Federation of Labor Unions.

THE STATE

Official Name: République du Tchad. **Administrative Divisions:** 14 Prefectures. **Capital:** N'Djamena, 530,000 inhab. (1993). **Other cities:** Sarh 198,000 inhab., Moundou 281,000 inhab., Abéché 187,000 inhab .(1993). **Government:** President Idriss Deby. **Armed Forces:** 25,200 **Paramilitaries:** 4,500 Gendarmes.

STATISTICS

DEMOGRAPHY

Urban: 21% (1995). **Annual growth:** 1.1% (1992-2000). **Estimate for year 2000:** 7,000,000. **Children per woman:** 5.9 (1992).

HEALTH

One **physician** for every 33,333 inhab. (1988-91). **Under-five mortality:** 202 per 1,000 (1995). **Calorie consumption:** 81% of required intake (1995). **Safe water:** 24% of the population has access (1990-95).

EDUCATION

Illiteracy: 52% (1995). **School enrolment:** Primary (1993): 38% fem., 38% male. University: 1% (1993). **Primary school teachers:** one for every 64 students (1992).

COMMUNICATIONS

81 **newspapers** (1995), 82 **TV sets** (1995) and 94 **radio sets** per 1,000 households (1995). 0.1 **telephones** per 100 inhabitants (1993).

ECONOMY

Per capita GNP: $180 (1994). **Annual growth:** 0.70% (1985-94). **Annual inflation:** 1.70% (1984-94). **Currency:** 535 CFA dollars = 1$ (1994). **Cereal imports:** 59,000 metric tons (1993). **Fertilizer use:** 26 kgs per ha. (1992-93). **External debt:** $816 million (1994), $130 per capita (1994). **Debt service:** 8.1% of exports (1994). **Development aid received:** $229 million (1993); $38 per capita; 19% of GNP.

ENERGY

Consumption: 16 kgs of Oil Equivalent per capita yearly (1994), 100% imported (1994).

[18] In 1987, the southern forces, supported by France, announced they had taken Fada, Faya Largeau and the frontier strip of Aouzou, claimed by Libya. In 1989, Chad and Libya signed an agreement on this 114,000 sq km territory, which included the return of prisoners and the presentation of a territorial lawsuit before the Interational Court in The Hague.

[19] In 1990, Idriss Deby, leader of the Patriotic Salvation Movement, supported by France, defeated Habré after a three month military campaign. During the deposed president's term in office in the eighties, some 40,000 people were executed or "disappeared."

[20] Habré fled to Senegal, where he began to plan a new insurrection. Some 5,000 of his rebel supporters assembled in the N'Guigmi region, near Lake Chad, from where they launched several offensives. In 1992, 400 rebels died and 100 were captured when government forces, with French support, put down an uprising by troops loyal to Habré.

[21] In 1993, Deby officially inaugurated a national conference to "democratize" Chad, with the participation of some 40 opposition parties, another 20 organizations and six armed rebel groups. Adoum Maurice El-Bongo was designated president of the conference, and this meeting called a special court to judge Habré and chose Fidele Moungar to take the role of prime minister during the transition period.

[22] In February 1994, the International Court in The Hague ruled the Aouzou strip belonged to Chad. In April, after Chad and the IMF signed an agreement to implement an annual growth programme, the Transition High Council postponed the elections for one more year. In May, Libya officially returned the Aouzou strip to N'Djamena.

[23] In March 1995, The Transition High Council delayed the elections yet again for another 12 months. Also, the measures taken by the government to limit the impact of the devaluation of the CFA, decreed by France in 1994, were considered inadequate. In the presidential elections, finally carried out in June and July 1996, Deby was elected constitutional president.

Chile

Population: 13,994,000 (1994)
Area: 756,950 SQ KM
Capital: Santiago

Currency: Peso
Language: Spanish

Chile

At the beginning of the 16th century, the north of Chile formed the southernmost part of the Inca empire (see: Peru, Bolivia and Ecuador). The area between Copiapo to the north and Puerto Montt to the south, was populated by the Mapuche, later called Araucanians by the Europeans. Further south lived the fishing peoples: the Yamana and the Alacalufe.

[2] Diego de Almagro set off from Peru in 1536, to begin the conquest of Chile. With an expeditionary force of Spanish soldiers and enslaved native Americans, he covered nearly 2,500 km, reaching central Chile when a mutiny in Lima forced him to turn back.

[3] Between 1540 and 1558, Pedro de Valdivia settled in what is now the port of Valparaiso, and founded several cities, including Santiago. The Mapuche, headed by their traditional leader Lautaro, put up fierce resistance. They defeated the invaders on several occasions, using war strategies learned from the European enemy. Valdivia died in one of these battles, and his successor, Francisco de Villagran, caught and killed Lautaro in 1557. Spanish governor Garcia de Hurtado sought out the Mapuche, who were now led by an old warrior, Caupolican, and defeated them. The rebellions continued off and on, and Spanish-Creole domination did not extend over all of the territory until the middle of the 19th century.

[4] In 1811, the Governing Junta headed by Jose Miguel Carrera instigated the independence process. General Bernardo O'Higgins, the son of a former viceroy in Peru, joined this movement. War broke out between the independence army and the royalist forces, with their strongholds in Valdivia and Concepción. Helped by the army of José de San Martín, which crossed the Andes to fight the royalists, the independence army finally defeated the Europeans on April 5, 1818 in the Battle of Maipú.

[5] In 1817, O'Higgins was designated Supreme Head of State, while the royalist troops still maintained pockets of resistance. He lay the political foundations of the country, which were re-inforced in the 1833 Constitution, during the presidential term of Diego Portales. This "aristocratic republic", denied all forms of political expression to the new urban sectors, the middle class and the rising proletariat. English companies, in alliance with the creole oligarchy, organized an export economy based on the rich saltpeter deposits of the north. These deposits extended along the maritime coast of Bolivia and reached as far as the southernmost regions of Peru. Soon the English controlled 49 per cent of Chile's foreign trade. Chilean and British capital owned 33 per cent of Peru's saltpeter, but they wanted total control.

[6] The "Nitrate War", or the Pacific War,1879-1884, was caused by this Chilean-British alliance. Chilean territory increased by a third and left Bolivia in its present landlocked state. The victory brought about the rapid growth of the saltpeter industry and its labor force.

[7] Jose Manuel Balmaceda, was elected president in 1886, and he tried to break the oligarchic order. The nationalism fostered by war, economic growth, social diversification and the encyclopedic education of the wealthy helped to gain him support. He encouraged protectionism to develop national industry. The oligarchy reacted violently, supported by the English. The army defeated the President's partisans, and Balmaceda committed suicide in the Argentine embassy in 1891.

[8] In 1900, the first union was founded in Iquique; in 1904, 15 unions with 20,000 members, joined to form a federation; the National Convention. That year the unions came into conflict with military in Valparaíso, and three years later, automatic weapons supplied by the US, were used to massacre 2,500 workers and their families in a school in Iquique

[9] Housing, railroads and new mines continued to expand, stimulating trade, new services, and public administration. In 1920, populist politician Arturo Alessandri became the leader of the new social factions that sought to subvert the oligarchic order and achieve representation in politics. Alessandri's government promoted constitutional reform and welfare legislation with electoral rights for literate men over 21, direct presidential election, an eight hour working day, social security, and labor regulations.

[10] Chile's economy, based on farm and mineral exports, was severely affected by the 1929-30 depression. Recovery did not come until the Chilean bourgeoisie was able to impose an industrialization pro-

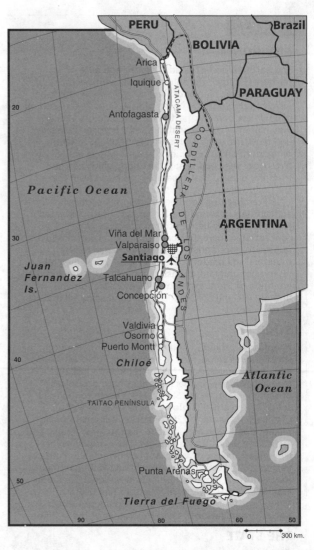

PUBLIC EXPENDITURES

% of GNP — 1992

- MILITARY: 2.7
- EDUCATION: 3.7
- HEALTH: 3.4

WORKERS

1994 — **5.9%** UNEMPLOYMENT

- MEN 71%
- WOMEN 29%

- AGRICULTURE 19%
- INDUSTRY 26%
- SERVICES 55%

LAND USE

- CROPLAND 6%
- PASTURE 18%
- OTHER 76%

gram to produce previously imported goods. Their proposal served as a base for the 1936 Popular Front, which marshalled support from Communists and Socialists. The front was led by Pedro Aguirre Cerda, who later became president. The armed forces were purged and withdrew from the political scene for almost forty years. Although the oligarchy was weakened, it made a pact with the government on agrarian policies, and Aguirre Cerda never allowed land reform or the formation of rural workers' unions.

[11] The alliance worked out in the thirties broke down during González Videla's term of office, 1946-1952. The atmosphere created by the Cold War legitimized the

> English companies, in alliance with the creole oligarchy, organized an export economy based on the rich saltpeter deposits of the north.

Law for the Permanent Defence of Democracy which banned the Communist Party and deprived its members of the vote. On January 9, 1949, to compensate for these measures, the government brought in female suffrage. The Radical Party's repressive and de-nationalizing policies, and the disabling of the left enabled the Populist Carlos Ibáñez to win the 1952 election.

[12] Economic deterioration rapidly eroded the strength of the populist government. In 1957, offshoots from the National Falange (populist) and from the old Conservative Party (oligarchic), founded the Christian Democratic Party. The Communist Party was legalized and the left rebuilt its alliances, forming the Popular Action Front. The people wanted change, but, sensitive to an aggressive anti-communist campaign, voted for Eduardo Frei's "revolution in freedom", which initiated agrarian reform in 1964.

[13] The UP (Popular Unity), a coalition formed by the Socialist Party, the Communist Party, the United Popular Action Movement (MAPU), and the Christian Left, led by Salvador Allende, won the

1970 election, obtaining 35 per cent of the vote, while the rest of the electorate was split between the Christian Democrats and the conservative parties. The following year the UP won almost 50 per cent of the vote in the legislative

elections, leading the right to fear a definitive loss of its majority.

[14] Allende nationalized copper and other strategic sectors, together with private banks and foreign trade. He increased land reform, promoted collective production and

PROFILE

ENVIRONMENT

Flanked by the Andes mountain range in the east and the Pacific in the west, the country is a thin strip of land 3,500 km long and never wider than 402 km. Its length explains its variety of climates and regions. Due to the cold ocean currents, the northern territory is a desert. The central region has a mild climate which makes it good for agriculture. The southern part of the country is colder and heavily wooded. The major salt and copper mines are located in northern Chile. 65% of the population live in the central valleys. During the military regime, some 40 thousand hectares of native forests were cut down each year, and replaced by other tree species, causing the displacement and death of wildlife.

SOCIETY

Peoples: Chileans result from ethnic and cultural integration between the native American population and European immigrants. 300,000 Mapuche live mainly in southern Chile. **Religions:** Mainly Catholic (77%), Protestant 13%. **Languages:** Spanish. **Political Parties:** The Christian Democratic Party (PDC); the Democratic Party (PPD);the Socialist Party (PS) and the Radical Party (RP). The National Renovation Party (RN) and the Independent Democratic Union (UDI). Democratic Alliance; Union of Center Center (UCC); National Party (PN); Communist Party of Chile (PCCh); Revolutionary Movement of the Left (MIR); the Manuel Rodríguez Patriotic Movement (MPMR); Humanistic Party (PH); the Green Party. **Social Organizations:** The Central Workers' Union (CUT), the main labor organization until 1973, had been banned during the dictatorship, and became legal again in 1990. Other trade unions are: the Copper Workers' Confederation (CTC) led by Rodolfo Seguel; the Confederation of Civil Servants of Chile (CEPCH), led by Federico Mujica; the National Association of Civil Servants; the United Workers' Front (FUT); the National Labor Co-ordination Board (CNS); the National Federation of Taxi Drivers, and unions representing artists, petroleum workers professionals and truck drivers; the Democratic Workers' Union (UDT), and the National Union of Civil Servants.

THE STATE

Official Name: República de Chile. **Administrative divisions:** 13 Regions and the Metropolitan Area of Santiago. **Capital:** Santiago, 4,628,300 inhab. (1993). **Other cities:** Viña del Mar 319,400 inhab.; Concepción 318,100 inhab.; Valparaiso 301,700 inhab.; Temuco 262,600 inhab. (1992). **Government:** Eduardo Frei, President since March 11, 1994. **National Holiday:** September 18, Independence Day (1810). **Armed Forces:** 93,000 troops (1994). **Paramilitaries:** Carabineros, 31,000.

created a "social sector" in the economy, managed by workers.

[15] The traditional elite, now displaced from government, conspired with the Pentagon, the CIA and transnational corporations, particularly ITT, to topple the government. The Christian Democrats were indecisive, and finally supported the coup. Inflation, a shortage of goods and internal differences within Popular Unity contributed to the climate of instability.

[16] On September 11, 1973 General Augusto Pinochet led a coup d'etat. The Presidential Palace at La Moneda was bombed by the air force, and President Allende died during the fighting. Violent repression ensued: people were shot without trial, sent to concentration camps, tortured, or simply "disappeared".

[17] The Chilean military dictatorship was one of many that ravaged the Southern Cone of South America during the 1970s, inspired by the National Security Doctrine. This was supported by the Chilean oligarchy and the middle classes, as well as by transnational corporations who recovered the companies which had escaped their control.

[18] After the coup, Chile's economic policy started to be based on neoliberal doctrines. Inflation dropped below 10 per cent per year, unemployment practically disappeared and imported manufactured goods flooded the market. Alongside this ran a loss of earning power in workers' salaries and an overall impoverishment of the poorer classes.

[19] The deteriorating influence of these neo-liberal economic policies became evident in 1983, two years after a constitution containing a mandate for "continuity" was approved in a plebiscite by 60 per cent of the voters. Unemployment reached 30 per cent, 60 per cent of the population were malnourished, protein consumption had dropped by 30 per cent, real salaries had been reduced by 22 per cent in just two years and 55 per cent of all families were living below the poverty line. This situation provided the background for the violent popular uprising which took place in November 1983, led by the National Labor Co-ordination Board and the National Workers' Command.

[20] In 1984 the Church started talks. To participate in these talks, the opposition formed the Democratic Alliance, led by the PDC. The talks with Home Minister Ser-

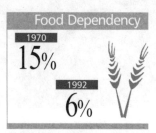

Food Dependency

1970
15%

1992
6%

External Debt

1994

PER CAPITA

$1,639

STATISTICS

DEMOGRAPHY

Urban: 84% (1995). **Annual growth:** 0.2% (1992-2000). **Estimate for year 2000:** 15,000,000. **Children per woman:** 2.7 (1992).

HEALTH

One **physician** for every 943 inhab. (1988-91). **Under-five mortality:** 15 per 1,000 (1995). **Calorie consumption:** 99% of required intake (1995). **Safe water:** 85% of the population has access (1990-95).

EDUCATION

Illiteracy: 5% (1995). **School enrolment:** Primary (1993): 98% fem., 98% male. Secondary (1993): 70% fem., 65% male. University: 27% (1993). **Primary school teachers:** one for every 25 students (1992).

COMMUNICATIONS

103 **newspapers** (1995), 103 **TV sets** (1995) and 102 **radio sets** per 1,000 households (1995). 11.0 **telephones** per 100 inhabitants (1993). **Books:** 101 new titles per 1,000,000 inhabitants in 1995.

ECONOMY

Per capita GNP: $3,520 (1994). **Annual growth:** 6.50% (1985-94). **Annual inflation:** 18.50% (1984-94). **Consumer price index:** 100 in 1990; 176.6 in 1994. **Currency:** 403 pesos = 1$ (1994). **Cereal imports:** 983,000 metric tons (1993). **Fertilizer use:** 849 kgs per ha. (1992-93). **Imports:** $11,539 million (1994). **Exports:** $11,800 million (1994). **External debt:** $22,939 million (1994), $1,639 per capita (1994). **Debt service:** 20.3% of exports (1994). **Development aid received:** $184 million (1993); $13 per capita; 0.4% of GNP.

ENERGY

Consumption: 943 kgs of Oil Equivalent per capita yearly (1994), 66% imported (1994).

---

gio Onofre Jarpa failed, and from Seprmber 1984, the break with the Church became evident. After this date, the Vicarate for Solidarity of the Archbishopric of Santiago played a key role in defending human rights.

[21] The political left joined together under the Popular Democratic Movement (MDP), vindicating all forms of struggle against the dictatorship. One faction of the left formed the "Manuel Rodriguez Patriotic Movement", an armed group which carried out numerous attacks, the most important being an attempted assassination of Pinochet on September 7, 1986.

[22] The international isolation of Chilean government during the Carter administration eased a little with the elections of Ronald Re-

agan, in the US, and Margaret Thatcher, in the UK. In August 1985 Chile authorized American space shuttle landings on Easter Island. At the beginning of 1986 an American delegation proposed that Chile be condemned before the UN Human Rights Commission, thereby avoiding having to take more drastic action against Chile.

[23] On October 5, 1988, an eight-year extension of Pinochet's rule was submitted to a plebiscite, and a wide opposition alliance ended his government, bringing elections the following year.

[24] Facing up to the fact that change was inevitable, Pinochet negotiated Constitutional reform. Proposed changes included further restrictions of the power of future governments, an increase in the number of

senators, shortening the presidential term from eight years to four and a liberalization of the proscription of left-wing parties. The reform was approved by referendum on July 30, 1989.

[25] Elections were held on December 14, 1989, and Patricio Aylwin, the leader of the Christian Democrats, obtained 55.2 per cent of the vote, taking office on March 11, 1990.

[26] In April, Aylwin appointed a

> The Chilean military dictatorship was one of many that ravaged the Southern Cone of South America during the 1970s, inspired by the National Security Doctrine.

"Truth and Reconciliation" commission to investigate the issue of missing people. The commission confirmed that there were at least 2,229 missing people, who were assumed dead. It also made a detailed study of repression during the dictatorship. When the facts were made public in March 1991 the president asked the Nation for forgiveness in the name of the State, he announced that judicial procedures would follow, and he requested the co-operation of the armed forces in these proceedings

[27] Official recognition virtually brought about an institutional crisis. The armed forces and the Supreme Court reaffirmed their conduct during the Pinochet dictatorship, and denied the validity of the government report, thus discrediting the president. The crisis was defused when the government accepted the pre-requisites of "Chilean transition", preserving the judicial system of the previous regime and guaranteeing impunity from human rights violations.

[28] On April 2, senator Jaime Guzman, former advisor and ideologue of the military regime, was murdered and the assassination was attributed to radical left wing groups, enabling the right-wing to raise the

issue of terrorism once more. Political life was slowly brought back to normal. On April 23, the MRPM announced its decision to abandon armed struggle.

[29] The Chilean economy maintained a ten year expansion, with annual growth rates of over 6 per cent, mainly due to high levels of investment (especially fixed capital), and the expansion of the external sector. Direct foreign investment remained at significant levels: $3.5 and $2.9 billion in 1992 and 1993, respectively.

[30] During Aylwin's presidency, social indicators improved. In 1993, real salaries increased 5 five per cent, the unemployment rate fell to 4.5 per cent, and social spending increased 14 per cent.

[31] In August 1993, the Special Commission on Indigenous Peoples, a government agency, proposed introducing indigenous language instruction in Mapuche, Aymara and Rapa Nui at primary school level. This measure was considered to be of utmost importance by Mapuche educators, to reduce the loss of cultural identity indigenous children experience when thrown into a school system where Spanish is the sole medium of instruction.

[32] In the 1993 elections, Eduardo Frei, candidate for the the Christian Democratic Party and the "Concertacion" coalition, won the presidency with 58 per cent of the vote. However, he did not achieve the Parliamentary majority needed to do away with authoritarian habits. Eight "designated" seats remained in the Senate which made opposition to the necessary reforms a majority.

[33] Shortly before handing over the presidency, president Aylwin pardoned four MRPM activists, sentenced to death for the assassination attempt against Pinochet in 1986.

[34] The Frei government considered social programs a top priority and announced a plan to tackle poverty affecting nearly one fourth of Chile's population. In May 1995, the minimum wage was raised by 13 per cent. The sale of cigarettes and automobiles was taxed to finance a 10 per cent increase in the lowest pensions and a 5 per cent increase in education spending. In June, the government requested an association with the Mercosur in order to conclude bilateral agreements with its members. It also negotiated its incorporation to the NAFTA.

[35] In May 1995, Brigadier General Pedro Espinoza and retired general Manuel Contreras were sentenced to prison for their participation in the assassination of former foreign minister Orlando Letelier, committed in Washington DC in 1976. Contreras rebelled and the institutional crisis was disguised by his confinement under custody in a military hospital. Pinochet supported military solidarity with the accused, but then requested respect for civilian authorities. At the same time, the government suspended investigations on the former dictator's son for corruption. Some observers suspect there was an agreement between both sides for this reason.

[36] The stand taken by young military officers who were trained and active during Pinochet's regime, openly disrespectful of democracy and civilian power, has caused concern in relation to the lack of power which will be left in the military leadership when the former dictator retires in 1998.

[37] The Chamber of Deputies approved by a large majority of 76 to 26 in August the trade agreement with the countries of the Southern Common Market (Mercosur). Two right wing parties (UDI and RN) voted against the treaty. The government signed the free trade agreement June 25, forging a so-called "four-

> The government accepted the prerequisites of "Chilean transition", preserving the judicial system of the previous regime and guaranteeing impunity from human rights violations.

plus-one" partnership with Mercosur countries Argentina, Brazil, Uruguay and Paraguay. Chile's trade arrangement with Mercosur came into effect in October 1, 1996.

[38] A report by the National Agricultural Society (SNA) concluded that the trade agreement will result in annual losses of $460 million for the Chilean agricultural sector, but

City lights over northern observatories

Chile's pristine northern skies in Region IV, so clear and starry that they have attracted multi-million dollar investments in world-class observatories, are now being threatened by light coming from the urban sprawl of cities like La Serena, Coquimbo and Vicuña.

[2] Marc Phillips, observatory director at Cerro Tololo, says, "We have been aware of this growing problem with city light for the past 25 years. We need to assure that these areas in the north remain dark for at least the next fifty years. It is already possible to appreciate the growing light contamination." While the problem is not yet critical, Phillips says action needs to be taken in view of the many new observatories that are due to be constructed soon at an estimated cost of $500 million. These include the Gemini project on Cerro Pachon (Region IV) with two eight meter telescopes; the ESO project with four eight meter telescopes in Paranal (Region II) and the LCO project with two 6.5 meter telescopes in Las Campanas (Region III).

[3] The existing observatory at Tololo appears to be the most affected thus far. Officials there estimate that night-time luminosity is about 10 per cent greater than "normal," with the problem expected to double by the year 2020, and grow to 37 per cent by the year 2045. Government officials have taken notice, and legislation to assure continued darkness in the northern skies is currently under consideration.

[4] In the meantime, some measures have already been adopted. City authorities are already working to assure that all light bulbs are directed towards the ground, and not the sky, and that existing incandescent mercury lighting is replaced with sodium lighting. In the city of La Serena some 5,442 lights have been so changed, while another 1,534 have been changed in Vicuna.

[5] Jorge May, of the University of Chile's astronomy department, says the future of Chile's observatories is at stake. "If Chile is able to enact legislation to assure that the northern skies will remain dark, the country will continue being a very important center for astronomy activity. If it does not take measures, we may have a catastrophe on our hands," says May.

Source: El Mercurio

the country's economy would be integrated to a market of more than 200 million people at lower tariffs, while prices will drop for goods imported from Mercosur countries.

[39] The October municipal elections gave a majority of the vote to the governing Concertación Nacional but the opposition on the right also claimed victory. The Christian Democrats (PDC), Socialist Party (PS), Party for Democracy (PPD), Radical Social Democrats (PR) increased their representation on municipal councils. However, the number of Concertacion party mayors declined

as compared to the previous elections of 1992. Within the Concertacion the popularity of the dominant Christian Democratic party diminished, while the PPD and PS gained council seats and mayorships, bolstering their strength within the alliance and thus breeding the Socialist's aspirations to head the presidential ticket in 1997.

China

Zhonghua

Population: 1,190,918,000 (1994)
Area: 9,596,960 SQ KM
Capital: Beijing

Currency: Yuan renminbi
Language: Chinese

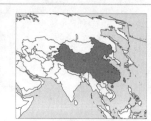

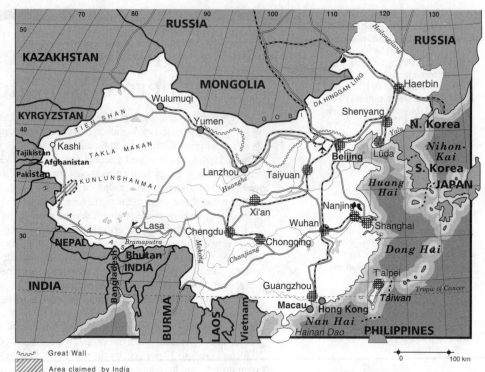

ㅅㅅㅅㅅ Great Wall

▨▨▨ Area claimed by India

0 —— 100 km

China has one of the world's oldest civilizations, which achieved an unparalleled degree of cultural and social homogeneity several thousand years before most modern nations were formed.

2 Although the nation disintegrated into warring kingdoms at various points in its history, the development of Chinese civilization can be traced back five thousand years to its emergence in the Yellow River (Huanghe) basin.

3 Territory was consolidated early on, and a single writing system was established by emperor Qin Shihuang some 2,200 years ago, in 221-206 BC; the same system is still in use today. In his determination to consolidate Chinese thinking he also ordered the burning of all previous books, and the execution of many scholars.

4 By the time of the Tang Dynasty, 618-907AD, China had developed a cultured civilization, and during the Song Dynasty, 960-1279 AD, the nation was more developed than Europe in every field.

5 Construction of the 5,000-kilometer Great Wall of China was started in the third century BC, in an attempt to repel invaders from the north. The wall ultimately failed, and China fell to Mongolian invaders; Genghis Khan and his grandson Kublai Khan ruled China from 1276 to 1368.

6 One of the most influential figures in Chinese culture was the scholar Confucius who lived about 2,500 years ago. As a philosopher, he emphasized respect for one's elders, the importance of unquestioning loyalty and responsibility to superiors in the social hierarchy and the central role of the family. Other philosophers such as Lao Zi emphasized a more imaginative role for individuals in society, but Confucian thinking is most accurately reflected in the partially surviving social structures of China, Korea, Japan and Vietnam.

7 With their highly developed culture, the Chinese soon developed a disdain for the "barbarians" beyond their borders. This attitude was often combined with the corruption of imperial dynasties, leading to eras of isolationism and social stagnation.

8 Regular contact with Europeans began when the Portuguese opened up maritime routes in the 15th century. The Chinese allowed Europeans to use a limited number of trading ports under strict conditions and granted the tiny enclave of Macau to the Portuguese in 1557. But the Chinese still refused to take foreign powers seriously.

9 The British eventually found a commodity that could be traded with China to pay for the Chinese silk, tea and porcelain products so fashionable in Europe. They began to import opium from India. When China tried to outlaw the drug trade, and later, to stop smuggling, Britain declared war in the name of free trade.

10 Modern European weapons easily defeated the Chinese imperial armies in the First Opium War, 1839-42. The victorious British demanded that five ports be opened to their trade, that low customs duties be imposed, and that the territory of Hong Kong be given to them.

11 The tottering Chinese Empire suffered another type of defeat with the Taiping rebellion in 1853. The rebel empire controlled much of southern China for 11 years before being crushed with the help of Western troops.

12 Massive imports of opium were now being paid for in Chinese silver, impoverishing the nation as it was weakened by widespread drug addiction. Another war from 1856-1860, this time against Anglo-French forces, ended with the capture of Beijing and a new round of concessions, including the admission of missionaries.

13 In 1895, China suffered an even more humiliating defeat at the hands of its former tributary state Japan, which overran part of the Korean peninsula and the island of Taiwan.

14 An anti-western rebellion broke out in 1898, led by the Boxers, and a joint British, Russian, German, French, Japanese and US expedition intervened to put down the uprising. The victorious armies divided the country into "zones of

PUBLIC EXPENDITURES

% of GNP · 1992

MILITARY 2.7 · EDUCATION 3.7 · HEALTH 3.4

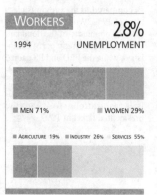

WORKERS

2.8% UNEMPLOYMENT

1994

MEN 71% · WOMEN 29%

AGRICULTURE 19% · INDUSTRY 26% · SERVICES 55%

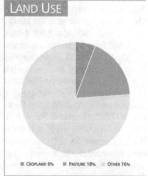

LAND USE

CROPLAND 6% · PASTURE 18% · OTHER 76%

ENVIRONMENT

The terrain of the country is divided into three large areas. Central Asian China, comprising Lower Mongolia, Sin Kiang and Tibet, is made up of high plateaus, snow-covered in winter but sporting steppe and prairie vegetation in summer. North China has vast plains, such as those of Manchuria and Hoang-Ho, which boast large wheat, barley, sorghum, soybean and cotton plantations, and coal and iron ore deposits. Manchuria is the country's main metallurgical center. South China is a hilly area crossed by the Yangtse Kiang and Si-kiang rivers, this area has a hot, humid monsoon climate, and most of the nation's rice plantations. The country possesses great mineral wealth: coal, oil, iron ore and non-ferrous metals. The use of coal as the main source of energy produces acid rain. Less than 13% of the land is covered by forest. The construction of the Three Gorges dam is considered a potential environmental disaster.

SOCIETY

Peoples: There are a total of 56 different nationalities, from the Luoba, with 2,300 people, to the Han, with 1.04 billion. (see box) **Religions:** nonreligious 59.2%; Chinese folk-religionist 20.1%; atheist 12.0%; Buddhist 6.0%; Muslim 2.4%; Christian 0.2%; other 0.1%. **Languages:** Chinese (official), is a modernized version of northern Mandarin. Variant dialects can be found in the rest of the country, the most widespread being Cantonese, in the south. Ethnic minorities speak their own languages. **Political Parties:** The Chinese constitution states that the Communist Party is "the leading nucleus of all the Chinese people".

THE STATE

Official name: Zhonghua Renmin Gongheguo. **Administrative divisions:** 23 provinces, 5 autonomous regions and 3 municipalities. **Capital:** Beijing, 5,700,000 inhab. (1990). **Other cities:** Shanghai, 7,496,000 inhab.; Shenyang, 3,600,000 inhab. (1990). **Government:** Jiang Semin, president and general secretary of the Communist Party. Li Peng, Prime Minister. The 3,000-member National People's Congress (NPC) sits for about two weeks a year to ratify laws. NPC delegates are drawn from various geographical areas and social sectors (army, minorities, women, religious). **National Holiday:** October 1 and 2, proclamation of the People's Republic. **Armed Forces:** 2,930,000 (1994) **Paramilitaries:** 1,200,000 Armed Peoples' Police, Defence Department.

influence" and demanded that the Chinese pay huge war reparations.

[15] The most spectacular example of foreign involvement in China was in Shanghai, where the port was developed for trading, and foreign companies invested in hundreds of factories, located within foreign "concessions", to take advantage of cheap Chinese labor.

[16] Foreign exploitation also brought new ideas and concepts to China. The government belatedly allowed small groups of students to study overseas. Nationalist groups emerged to focus and articulate the massive anti-foreign sentiment. In 1911, nationalists led by Sun Yat-Sen, won the support of several imperial generals. A number of garrisons mutinied, and a Republic was proclaimed.

[17] Most of the generals who had joined the uprising wanted power and had no intention of instituting democratic reforms. Simmering rivalries soon burst into the open and China was thrust into an era of civil war as local "warlords," backed by various foreign powers, fought over the fragmented country.

[18] In 1921, 13 people from all over China gathered in a small house in the French concession of Shanghai to hold the first national congress of the Chinese Communist Party (CCP).

[19] The new party prioritized the organization of workers and it enjoyed considerable success. By 1926 it was involved with 700 trade unions representing 1.24 million members. They formed an alliance with Chiang Kai-Shek's Kuomintang (KMT) nationalist party to oppose the warlords. In 1927, the KMT turned on their allies and massacred some 40,000 Communist labor leaders.

[20] In the face of such a defeat, a Communist leader, Mao Zedong (Mao Tse Tung), argued that the Chinese peasantry should be mobilized as a revolutionary force. He led most of the Communist Party leadership into rural China, where they carried out political organizing and raised a peasant-based army.

[21] When the KMT armies tried to surround the CCP's remote bases in 1934, the communist army, their leaders and supporters began a march in search of an alternative site. They travelled for over a year, their numbers dwindling under frequent attacks and severe conditions, but they were eventually able to regroup. This legendary journey became known as the "long march."

[22] In 1937 a full-scale Japanese invasion of China forced the CCP and the KMT to set aside their differences and form a second United Front against the common enemy. While fighting against the Japanese, the Communists were also able to extend their influence among the workers in KMT-controlled areas. Chiang Kai-shek showed more interest in fighting the Communists than the Japanese and broke the pact on several occasions. On one occasion, his own generals arrested him and forced him to negotiate with the CCP to continue joint resistance of the Japanese.

[23] When the allies defeated the Japanese in 1945, the CCP-KMT front collapsed into civil war. The corrupt and autocratic KMT proved no match for the highly-motivated CCP armies and their supporters. On October 1,1949, the communist leadership declared the foundation of the People's Republic of China. An era of foreign control and humiliation went out and an era of social and economic reform came in.

[24] The remnants of the KMT government and army fled to the island of Taiwan. There, under US protection, they claimed to represent all of China and laid plans to eventually "reconquer" the mainland.

[25] The Chinese Communists instituted far-reaching land reforms, adding to those already carried out in their own liberated zones before the final victory. They also nationalized all foreign-controlled property and initiated widespread education and health programs.

[26] The first five-year development plan incorporated many aspects of the Soviet model; the emphasis was on industrial investment rather than consumption, and heavy industry rather than light, all under a centralized economic plan. The creation of a strong industrial base could only be achieved by extracting a large surplus from rural areas, but by 1957 agricultural production was badly stagnating.

[27] In the name of open debate, and in search of solutions to the nation's problems, the party launched a campaign to "let a hundred flowers bloom, let a hundred schools of thought contend," inviting people to criticize the system and suggest alternatives. When citizens began to complain about the lack of democracy and question party rule, the leadership turned about-face and, cracked down on those who had spoken out.

[28] In 1958, Mao launched the Great Leap Forward, aimed at accelerating rural collectivization and urban industrialization. This heavy-handed, dogmatic and inflexible plan led the country down a disastrous path, sparking widespread famine in

Foreign Trade

MILLIONS $ — 1994

IMPORTS
115,681

EXPORTS
121,047

Maternal Mortality

1989-95

PER 100,000
LIVE BIRTHS
158

STATISTICS

DEMOGRAPHY

Urban: 29% (1995). **Annual growth:** 2.7% (1992-2000). **Estimate for year 2000:** 1,255,000,000. **Children per woman:** 2 (1992).

HEALTH

Under-five mortality: 43 per 1,000 (1995). **Calorie consumption:** 102% of required intake (1995). **Safe water:** 67% of the population has access (1990-95).

EDUCATION

Illiteracy: 19% (1995). **School enrolment:** Primary (1993): 116% fem., 116% male. Secondary (1993): 51% fem., 60% male. University: 4% (1993). **Primary school teachers:** one for every 22 students (1992).

COMMUNICATIONS

96 **newspapers** (1995), 89 **TV sets** (1995) and 92 **radio sets** per 1,000 households (1995). 1.5 **telephones** per 100 inhabitants (1993). **Books:** 96 new titles per 1,000,000 inhabitants in 1995.

ECONOMY

Per capita GNP: $530 (1994). **Annual growth:** 7.80% (1985-94). **Annual inflation:** 8.40% (1984-94). **Consumer price index:** 100 in 1990; 165.6 in 1994. **Currency:** 8 yuan renminbi = 1$ (1994). **Cereal imports:** 7,332,000 metric tons (1993). **Fertilizer use:** 3,005 kgs per ha. (1992-93). **Imports:** $115,681 million (1994). **Exports:** $121,047 million (1994). **External debt:** $100,536 million (1994), $84 per capita (1994). **Debt service:** 9.3% of exports (1994). **Development aid received:** $3,273 million (1993); $3 per capita; 0.8% of GNP.

ENERGY

Consumption: 647 kgs of Oil Equivalent per capita yearly (1994), -1% imported (1994).

National minorities

Mongol, 4.8 million. Their main livelihood is cattle-raising and agriculture. Their language is Mongol, and Lamaism, their religion.

Hui, 8.6 million. Most are engaged in agriculture, some in cattle-raising, handcrafts and commercial activity. They are famous for their restaurants. Their religion, Islam, is a very important part of their daily life.

Tibetan, 4.59 million. Their main livelihood continues to be animal husbandry. Their religion is Lamaism, and their language, **Tibetan**.

Uighur, 7.21 million. Involved in agricultural activities, and have raised cotton for many generations.

Miao, 7.39 million. Live primarily from agriculture; their main crops are rice, corn, potatoes and legumes. In 1956, a unified Miao language was created.

Yi, 6.57 million. Mostly involved in agriculture and cattle-raising. Apart from advances in these areas, they have excelled in astronomy, meteorology and medicine, and developed their own calendar.

Zhuang, 15.48 million, mostly agricultural. They speak the Zhuang language, for which an alphabet was created in 1956.

Buoyci, 2.54 million. Blessed with fertile soil and a temperate climate, they have their own language. As for religion, they are polytheistic and worship their ancestors, although some have converted to Catholicism.

Korean, 1.92 million. They are farmers, known for their rice crops despite the fact that they live in the cold northern region. They speak their own language.

Manchu, 9.82 million. Engaged in agriculture, although those who live in cities work mostly in the industrial sector. They have their own language, although nowadays they also speak Mandarin Chinese. A minority practice shamanism.

Dong, 2.51, mostly involved in agriculture.

Yao, 2.13 million. Although the majority live from agriculture or forestry, some are also hunters.

Bai, 1.59 million. They are good farmers, known for high productivity.

Tujia, 5.7 million. One of the ancient nationalities of south central China, they are mostly involved in agriculture. They have their own spoken language, but no writing system. Many speak and use the Han language.

Hani, 1.25 million. Engaged in agriculture, many practice terrace farming. They have their own calendar.

Kazakh, 1.11 million. Except for a few non-nomadic farmers, most are involved in cattle-raising and wander over extensive areas in search of pasture with the changing seasons. They are Muslim and have their own dialect.

Dai, 1.02 million. They are farmers, and their religion is Hinayana, a form of Buddhism. They have their own spoken and written language.

Li, 1.1 million. Mostly farmers, they have contributed a great deal to the development of Hainan Island. Their language is Li.

Lisu, 574,856, most are farmers. They have their own spoken and written language.

Sho, 630,378. They are farmers; their principal crops are rice, tea, sugar cane and ramie. They have their own spoken language.

Va, 354,974. They are mostly farmers. Some are Hinayana Buddhist, others are Christian.

Gaoshan , 400,000, mostly farmers and hunters. They have their own spoken language.

Lebu, 411,476. They raise rice,

---

the countryside. The official Chinese figure for deaths between 1959 and 1961 is 20 million; one of the greatest human tragedies of the century, and one of the least widely reported.

[29] In 1962, Mao was forced to deliver a public self-criticism of his economic policies and was replaced by Liu Shaoqi as state chairman. Mao still enjoyed support among many radicals and, most importantly, from the People's Liberation Army.

[30] During this period, relations with the Soviet Union were souring due to ideological differences. This was highlighted in a 1956 speech by Soviet leader Nikita Khruschev. The split came in 1963, and USSR advisors left China taking plans and blueprints with them.

[31] In 1966, the army and young people, the "Red Guards", armed with compilations of Mao's thoughts, initiated campaigns throughout China attacking bureaucrats and university professors as reactionaries and "capitalist followers." The Great Proletarian Cultural Revolution had begun.

[32] There are many interpretations of the Cultural Revolution, and the upheaval occurred on many levels.

[33] At some stages during the Cultural Revolution, with the internal "struggle between two lines" attaining civil war proportions, production suffered severely with many factories and universities closing altogether. An estimated 10 million people died during the years of turmoil, including Liu Shaoqi, Mao's chief opponent.

[34] At the same time, China took a leading role in founding the Non-Aligned Movement, and sent large numbers of workers abroad on high-profile development projects such as the Tanzam railway, linking Zambia to the coast via Tanzania in Africa. Liberation movements in many countries were inspired by China, adopting the peasant war strategies outlined in the works of Mao Zedong and Lin Piao.

[35] In the early 1970s, Prime Minister Zhou Enlai started talks with the US, and Liu Shaoqi's protégé, Deng Xiaoping, was returned to power.

[36] In 1971 the mainland Communist government won sufficient support to replace Taiwan as the representative of China in the United Nations. Unlike earlier occasions, this time the US refrained from exercising its veto because it saw distinct advantages in improving relations with China. In 1976, the US and China re-established diplomatic relations.

[37] Internationally, China seemed to oppose all Soviet allies. This led to Chinese support for movements like UNITA in Angola and the PAC in South Africa. It also caused a growing rift with Vietnam after it defeated the US in 1975. In 1979, China invaded Vietnam to "teach it a

Han
Mongolian
Manchu
Korean
Hui
Uygur
Yi
Buly
Miao
Tibetan
Kazak
Other National Minorities
Zhuang
Tajik
Oroquen

tea, tobacco, hemp, sisal and other cash crops. They have their own spoken language and recently created a written system.

Shui, 345,993. They live in small agricultural communities. They speak the Shui dialect, although they also speak Han.

Dongxiang, 373,872. They are an agricultural people who raise potatoes and corn, as well as a few cash crops. They are Muslim. They have their own spoken language, but no writing system.

Naxi, 278,000. Farming is their main occupation. They have their own oral and written language. Their religion is called «Dongha», although some are Buddhist, Taoist or Christian.

Jingpo, 119,200. Their main crops are rice and millet. They are Buddhist.

Tu, 191,624. Originally shepherds and goatherds, after the Ming dynasty they began to turn to farming. They are Lamaist.

Kirghize, 141,549. In ancient times, they were nomadic shepherds; today, they still raise animals, although they also do some farming. They have their own oral and written language. Most are Muslim, although some are Lamaist.

Daur, 121,357. Their main livelihood is farming, which they supplement through hunting and cattle-raising. They are shamanist.

Mulam, 159,328. They are farmers, and have their own oral and written language.

Qiang, 198,252. They are farmers. The Qiang dialect has no writing system.

Blang, 82,280. Their main crop is rice, although they also raise such cash crops as cotten and tea (one specific variety).

Salar, 87,697. They are Muslim. Mostly farmers, some raise cattle or weave textiles.

Maonan, 71,968. They are farmers and are Taoist or Christian.

Gelao, 473,997. They live in scattered communities and are mostly farmers. Their language has no writing system.

Xibe, 172,847, half of which live in the northeast, and the rest in Xinjiang. Originally they were a nomadic people, although later they were displaced from their region and settled down to farming. They have their own oral and written language.

Achang, 27,708. They are famous for their rice crops, as well as their metallurgical products.

Tadzhik, 33,538. Basically cattle-farmers, they supplement this activity with farming. They are Muslims. Although they have their own language, they generally speak Uighur.

Nu, 27,123. Although they are farmers, hunting and gathering constitute an important part of their lives. They speak a dialect which has no writing system. They worship nature, although some members of this people are Christian or Lamaist.

Pumi, 29,657. They are farmers or raise cattle. Although they have their own language, the majority use Mandarin. Most are polytheistic and worship their ancestors, although some are Lamaist.

Uzbek, 14,502. Most live in urban areas, scattered throughout different cities and towns of Xinjiang province. They earn their living as tradesmen-especially silk and handcrafts-although some are still farmers. They practice Islam.

Russian, 13,504. Most live in cities and work in the service or transport sectors, or with handicrafts. They speak Russian and belong to the Eastern Orthodox Church.

Ewenki, 26,315. This nationality is unique in China as they are the only ones who herd reindeer. Half raise animals and the rest are engaged in agriculture or farming. They have no written language. Most are shamanist.

De'ang, 15,462. They are considered to be the most ancient tea planters. They have their own spoken language, but no written language, and are Hinayana Buddhist.

Bonan, 12,212. Most are farmers, although some raise cattle or make handcrafts. They have their own spoken language, although they use Mandarin to express themselves in written form. They are Muslim.

Yugur, 12.297. They have their own language, and use Mandarin as a written language. They are shamanist or Lamaist.

Jing, 18,915. This people, formerly known as the "Yue" nationality, changed their name in 1958. It is the only ethnic group entirely devoted to fishing along the Chinese coastline. They have accumulated a vast knowledge of the sea and fishing. They are polytheistic, worshiping not only their ancestors

but also gods connected to the sea. Their have their own language.

Tatar, 4,873. A high percentage of this group are intellectuals. They have their own language, although they use Uighur as a lingua franca. Most are Muslim.

Drung, 5,816. Their language has no written system. They are Christian.

Oroqen, 6,965. For generations, they lived in forests; they are hunters, well-versed in the habits and characteristics of native animals. They now lead more settled lives and have developed a diversified economy. They are shamanist.

Mazher, 4,245. Native to northern China, they are the only nationality whose livelihood is exclusively fishing, and the only one that uses dogsleds. Poverty and oppression reduced their numbers to 300 during the 1940s, but they have since grown significantly. Their language has no written system. They are shamanist.

Momba, 7,475. This nationality has close cultural ties to the Tibetans, and like the latter, its members are Lamaists. Many speak and write Tibetan, although they also have their own dialect. Today their main economic activity is farming, especially rice cultivation.

Lhoba, 2,312. Most are farmers, although they also engage in hunting, gathering and fishing. They have their own dialect, and are Lamaist.

Jino, 18,021. They raise different crops, especially tea. Their language has no writing system.

(Based on official data supplied by the government of China at the beginning of the 90's).

---

lesson" over its military contribution to the Khmer Rouge defeat in Cambodia.

38 In 1975, Zhou, revolutionary veteran Zhu De, and finally Mao died in succession. The pragmatists and reformists prevailed, taking advantage of support among party bureaucrats.

39 The leaders of the Maoist faction; his widow Jiang Qing, Zhang Chunqiao, Yao Wenyuan and Wang Hongwen, dubbed the "Gang of Four", and were arrested and charged with plotting to usurp state and party power. They became scapegoats for the failures and excesses of the Cultural Revolution. In

1981 they were put on public trial as a final symbolic gesture that the era of "revolutionary struggle" was over.

40 Deng Xiaoping was rehabilitated by the new leadership, and he was universally acknowledged as its leading authority.

41 The new leadership announced an ambitious economic development program. It envisioned significant advances in agriculture, industry, defense and science and technology.

42 1978 and 1979, were marked by the government's acceptance of greater popular expression and criticism. They knew that this would be directed against the tur-

moil of the Cultural Revolution and the radicals responsible for it. The "Beijing Spring" was centered on the "Democracy Wall", where common citizens could display posters expressing their opinions. The wall also served as a spot where dissidents distributed unofficial magazines. When criticism began to focus on the country's core political structure, the leadership stopped the movement. The Democracy Wall, a continuing institution throughout the Cultural Revolution, was discontinued. One dissident, Wei Jingshen, who published a magazine calling for a "Fifth Modernization" - democracy - was

tried and sentenced to a 15-year prison sentence, which he is still serving.

43 In December 1978, at the 3rd plenary of the 11th Party Central Committee, Deng was fully rehabilitated and sweeping economic reforms were announced.

44 In the countryside, the People's Communes were disbanded and land was redistributed into family-size units leased from the state. Production quotas were replaced by taxes, and peasants were permitted to sell their surpluses for cash in the towns and cities.

45 China also announced a dramatic new openness to foreign

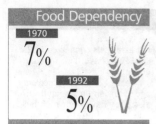

Food Dependency

1970
7%

1992
5%

External Debt

1994

PER CAPITA
$84

Illiteracy

1995

19%

trade, investment and borrowing. To attract foreign factories, Special Economic Zones were established near Hong Kong and Macau, offering a range of incentives; such as tax breaks, and cheap land and labor, similar to free trade zones elsewhere.

[46] In industry, a good deal of decision-making power was passed from the ministries to individual plant managers, who were now free to plan production and distribution, and choose the sources of their raw materials. Young workers were offered a contract system of employment instead of lifetime assignment to a production unit. Individuals and families were given permission to start small businesses such as restaurants and shops.

[47] The government began a phased program of removing price subsidies on consumer goods, allowing the market to determine the price of such basics as food and clothing, to spur economic growth and encourage consumption.

[48] As the changes took hold, more goods and foodstuffs appeared in shops, and wages increased for several years. The long-standing problem of unemployment and underemployment diminished as more people became self-employed. Restrictions on the freedom of travel inside and outside China were loosened, greater artistic diversity and expression were tolerated and a greater diversity of information was made available.

[49] These economic reforms began to accelerate inflation, especially on such basic items as food and clothing, and in the late 1980s, workers found that their buying power was falling.

[50] In the countryside, peasants with easy access to urban centers benefited as they were able to supply large markets. Those in more remote areas fell behind. The use of chemical pesticides and fertilizers soared, and production initially rose as well. But production levels began to fall, prompting farmers to apply greater quantities of chemicals, approaching danger levels. Pesticide and fertilizer prices rose sharply.

[51] Workers' welfare deteriorated, and the new employment systems generated insecurity. While managers were given more pow-

er, even to hire and fire workers and set production targets, trade unions were not granted any corresponding freedom to act.

[52] High-level debate over the scale and pace of economic reforms included discussion of a clearer division between party and state.

[53] In 1986 students demonstrated in Shanghai, calling for freedom of the press and political reform. As a result, Hu Yaobang, then a relatively young party general secretary, was forced to step down, and a campaign against "bourgeois liberalism" was launched by hard-liners. This aimed to root out what they called "western" ideas; most importantly political pluralism, though consumerism and corruption were also mentioned. Many viewed this campaign as a veiled attack on Deng's economic reform policies, given that Hu Yaobang was one of Deng's protégés.

[54] In March 1989, police in what is known as Central Asian China fired on Tibetan demonstrators who were protesting against cultural and religious persecution and demanding greater political rights. The shooting triggered three days of widespread rioting which resulted in the imposition of Martial Law, lifted only in April 1990. Tibetan exiles reported the detention of many dissidents and several executions. Tibet was officially annexed to China in 1950 and converted into an autonomous region in 1965.

[55] In April 1989, the death of Hu Yaobang served as a pretext for thousands of students to assemble in Tiananmen Square in downtown Beijing. In addition to mourning the death of the man who to them symbolized liberal reform, the students called for an end to corruption and changes in both society and the party.

[56] The students camped in the square and their numbers grew, buoyed by the anniversary of the student demonstrations of May 4, 1919. Hundreds of thousands of citizens and workers in Beijing began to demonstrate support for the reform movement, bringing food and other supplies to the students, marching, and blockading intersections when it seemed likely that troops would move against the students.

[57] Students, citizens and workers began similar demonstrations in other cities, and the banners of groups of workers from the official media, trade unions, and various party sections began to appear in the daily demonstrations. Various autonomous bodies were set up by demonstrators, challenging four decades of complete party control over all social organizations.

[58] Finally, on June 4, 1989, the government moved decisively and violently, sending troops into downtown Beijing to clear out the citizens and students. Many hundreds were killed and thousands wounded. Many Beijing residents and students were killed as the troops forced their way past barricades in armored vehicles and opened fire on the crowds. The pro-democracy demonstrations were smashed. The assault on Tiananmen was followed by a nationwide hunt for leaders of the autonomous student and workers movements, thousands were arrested and badly beaten; many others were executed. There were important political changes, and Li Peng became prime minister.

[59] The first Western government to send representatives to the capital after the Tiananmen massacre was Britain, in September 1991. They signed an agreement with Beijing on the construction of a new airport in Hong Kong, within the constraints of the negotiations to return the British colony to China, in 1997.

[60] With one-third of all public enterprises operating at that time at a deficit, and less and less possibility of balancing the State budget, structural reforms seemed all but inevitable in state enterprises. The Chinese Communist Party was paving the way for a return to the economic liberalization program initiated by Deng Xiaoping, who had reappeared on the political scene at the age of 87. At the same time, there were other indicators of changing attitudes: Jiang Zemin, Deng's successor as Communist Party Secretary, called for "free thinking", and charges against former Secretary Zhao Zhiyang were reduced. Shortly afterwards, however, he was blamed for events leading to the Tiananmen massacre, rendering impossible his chances of rehabilitation.

[61] The Paris accords on Cambodia (see Cambodia), signed in October 1991, and the end of the Soviet Union accelerated a rapprochement between China and Vietnam. In November 1991, diplomatic relations between the two countries were re-established.

[62] In late November 1991, the government freed the student leaders from the Tiananmen demonstrations. However, there were an estimated 70,000 political prisoners in various provinces of the the Xinjiang and Qinghai region.

[63] In 1992, China signed the Nuclear Non-Proliferation Treaty. The CCP secretary general, Jiang Semin, was designated president of the republic, becoming the first person since Mao to combine the functions of head of state, the party and commander of the armed forces. Prime Minister Li Peng was confirmed in his post.

[64] In April 1992, the decision was made to build the giant Three Gorges dam on the Yangtse river. This controversial project would be completed in 2009 and would flood lands inhabited by 1.3 million people, drowning ten cities and more than 800 towns. Ecologists immediately opposed the project, saying it would destroy the habitats of endangered species and would leave millions of people exposed to the danger of earthquakes, avalanches and floods. They also considered the dam would not be profitable.

[65] In September, the government said the pro-independence action in Tibet would be "implacably repressed". On the economic front, an austerity plan was launched in the State apparatus and taxes on the rural population were increased. However, a series of protests and demonstrations forced the government to remove the new charge from a sector of the population which totalled around 800 million people.

[66] The GDP increased by 12.8 per cent in 1992, a figure unprecedented in the previous decade. However, this growth (the economy nearly doubled since 1990) had its first undesirable effects in 1993, when inflation hit 20 per cent in the first half year. In March 1994, Li Peng proposed limiting the economic expansion to 9 per cent in order to lim-

Tibet: high pendulum

Tibet was invaded by China in 1950 and today its population is roughly half Chinese and Tibetan with some Hui (Chinese Muslims), Hu, Monba, and other minority nationalities. Most of the inhabitants of this high plateau at the east of the Himalayas bear the same ethnic origin and have traditionally practised the same religion, also sharing the same language.

2   The first known religion in the region was Bon, which combined a belief in gods, demons, and ancestral spirits who were responsive to priests, or shamans. Chinese Buddhism was introduced in ancient times - the first scriptures belonging to the 3rd century - but the mainstream of Buddhist teachings came to Tibet from India in the 7th century. The syncretic blend of both produced the particular Lamaist Buddhism of the region and its many different sects. The principal abbots of the many monasteries spread over the country became also temporal rulers and began to compete against each other. The Mongols and the Chinese in different times took advantage of these conflicts, favouring one or another according to their own convenience, setting a pendular policy that lasted centuries.

3   The Tibetans developed a phonetic alphabet about AD 600 and, after centuries of rivalries, a theocratic kind of feudal state was established in the early 10th century. Both political and religious power was conferred on the lamas, and monasteries who, along with the ruling class, controlled the servants and the produce of the land.

4   In 1240, the Mongols marched on central Tibet and attacked many important monasteries. In 1247, Köden, the younger brother of Güyük Khan, symbolically invested the Sa-skya lama with temporal authority over Tibet. Kublai Khan, grandson of Genghis Khan, appointed the lama Phags-pa as his "Imperial preceptor". The political-religious relationship between Tibet and the Mongol Empire was stated as a personal bond between the emperor as patron and the lama as priest. During a century, many Sa-skya lamas, living at the Mongol court, became viceroys of Tibet on behalf of the Mongol emperors.

5   The Dre-lugs-pa hierarchies set a new alliance with the Mongols in 1578, re-establishing the relationship that shared religious authority - assigned to Tibetan lamas - and political dominance - lying on Mongolian leaders. Atlan conferred the Bsod-nams-rgya-mtsho - third in the lamas hierarchy - the title of Dalai ("oceanwide") Lama. In 1640, the Mongols invaded Tibet and two years later the Güüshi Khan enthroned the fourth Dalai Lama as ruler of Tibet, keeping for himself the military protection of the region.

6   The assasination of two high Chinese commissioners in 1751 - instigated by the Dalai Lama - brought the immediate reaction of the Manchu dynasty in power in Beijing: in retaliaton the emperor sent a bloody military expedition to Lhasa. Since then, the Dalai Lama's relationship with China became more difficult than the Panchens', the other leaders of the religious hierarchy. Worldly competition between the two heads of the Lamaist Buddhism often fostered discrepancies and sectarianism.

7   In the 17th century, the fifth Dalai Lama declared that his tutor, Blobzang chos-kyi-rgyal-mtshan (1570-1662), who was the current Panchen Lama, would be reincarnated in a child.

8   In 1904 a British military expedition headed by Colonel Younghusband invaded Tibet and forced the Tibetans into making a treaty.

9   In 1910 Chinese troops entered Lhasa. The Dalai Lama appealed to the British to help expel them - and was refused. But the Chinese empire was in its death throes and fell to the Nationalists in 1911. Tibetans seized the chance to expel the invaders and in June 1912 the Dalai Lama proclaimed Tibetan independance. About ten years later, disagreements between the Dalai Lama and the Panchen Lama ended in the flight of the latter to Beijing. A boy born of Tibetan parents about 1938 in Tsinghai province, China, Bskal-bzang Tshe-brtan, was recognized as his successor by the Chinese government and brought to Tibet in 1952. Eventually, he entered Lhasa under communist military escort and was enthroned as head abbot of the Tashilhunpo Monastery.

10   In 1949 the communists came to power in China and the following year the government declared its intention to 'liberate' Tibet from feudalism. Some 84,000 troops penetrated Tibet's eastern province of Kham and met resistance from Tibetan Khampa horseriders but this resistance was crushed and in 1951 the present Dalai Lama, Tenzin Gyatso, then aged 15, signed an agreement with China.

11   In 1959 Tibetan resistance exploded into an uprising which was crushed by the Chinese. Chinese figures record 87,000 deaths; Tibetan sources suggest as many as 430,000 were killed. When the 1959 revolt collapsed, the Panchen Lama remained in Tibet, while the Dalai Lama fled into exile. The Panchen Lama's refusal to denounce the Dalai Lama as a traitor brought him into disfavor with the Chinese government, which imprisoned him in Peking in 1964. He was released in the late 1970s and died in 1989.

12   The Dalai Lama and the Tibetan government-in-exile, established in India in 1959, continue to fight for the independence of Tibet and the return to traditional society. In 1994 the Chinese declared a five-year-old boy to be the next Panchen Lama instead of the boy found by Tibetan religious authorities. The Dalai Lama and Tibetan government-in-exile has rejected the legitimacy of the nomination.

it inflation despite protests from the coastal provinces like Guandong, the main beneficiaries of the Chinese "boom."

67   The social inequalities between the new rich in the cities and the vast majority of the workers and rural population continued to increase along with the migration of millions of people from the countryside to the cities. This led the government to be prudent in closing and privatizing unprofitable State enterprises, as a sharp rise in unemployment would only worsen an already tense and precarious social situation.

68   However, another project which planned to limit the social effects of the economic reforms was postponed. The plan was to establish a redundancy payment for workers laid off from closed state companies, but the total lack of a state-run social security system in the nation - without so much as unemployment benefit - made the government back off to prevent them from launching excessively radical reforms.

69   In 1995, Jiang Semin consolidated his power even further, putting himself in the best possible position to continue ruling the nation in the "post-Deng" era. The authorities continued to be concerned about the social effects of the reforms, hence they maintained the large subsidies for State enterprises.

70   In that year, inflation reached only 13 per cent, and the 1996-2000 five year plan forecast annual growth of "only" 8-9 per cent. The amount of corruption scandals increased dramatically. The CCP first secretary in Beijing, Chen Xitong, was forced to leave his post when it became clear he had been misappropriating funds along with important communal leaders and a large municipal metalurgical company. In April, the vice-mayor of Beijing, Wang Daosen, accused of having embezzled $37 million of government funds, committed suicide.

71   As in previous years, trade relations with the United States strengthened, despite the constant public disagreements between the nations, like the situation provoked by military manoeuvres in the waters around Taiwan (see Taiwan) or the nuclear tests carried out by Beijing. In November, Jiang made the first ever visit of a Chinese president to South Korea, in another demonstration of the commercial and political rapprochement of China with countries it had classed as anticommunist during the Cold War.

72   In 1996, the Chinese and Western authorities continued to be concerned about a possible "explosion" of the country due to the separatist tendencies, even though they did not consider this probable. In May, Amnesty International (AI) condemned the repression of Buddhist monks in Tibet by the Chinese authorities. According to AI, 80 monks were injured for refusing to respect a ban on the public exhibition of pictures of the Dalai Lama.

Hong Kong: the return

Hong Kong on the south coast of China is the third largest financial center in the world, after New York and London.

2 The territory was an integral part of the ancient and well-organized Chinese administrative system. Britain took control of Hong Kong in three stages, each the result of military threats or aggression towards the ailing imperial Chinese regime. The island of Hong Kong was ceded to Britain in perpetuity in 1842 when the British defeated China in the First Opium War. Eighteen years later, the British gained the rights to Kowloon, the mainland peninsula opposite Hong Kong Island. In 1898, the British completed their colonization by forcing the Chinese to grant them a 99-year lease on a rural area north of Kowloon known as the New Territories.

3 Initially Hong Kong developed as a trading center, providing an entry-point to China. But when the US and Britain imposed a boycott on trade with China in the 1950s, after the communist victory, Hong Kong rapidly developed a light manufacturing base, and became a major exporter of textiles and garments, plastics and electronic products.

4 The growth in trading and manufacturing for export provided the opportunity for Hong Kong also to develop as a financial, communications and transport center. This was encouraged by the government's laissez faire approach which allowed low tax rates, minimal customs duties, confidentiality and ease of capital movement.

5 When China announced a new era of openness to foreign trade and investment in the late 1970s, Hong Kong, with its sophisticated international trade and financial systems, large, modern container terminals and great natural harbour, was in a position to take full advantage. It is estimated that between 30 and 50 per cent of China's foreign exchange is generated through Hong Kong, and 90 per cent of foreign investment in Kwangtung province comes from there.

6 With its unique combination of historical and geographical circumstances, Hong Kong has enjoyed consistently high growth-rates since 1975, raising the per-capita GNP to the level of Aotearoa (New Zealand) and Spain by 1987.

7 Hong Kong is one of the Four Tigers (newly-industrialized East Asian states), along with Singapore, Taiwan and South Korea. Yet the Hong Kong "government" has never intervened actively in the economy to promote specific industrial sectors.

8 Until the 1980s, most political activity took the form of pressure groups and lobby groups campaigning on specific issues.

9 Nearly all of the inhabitants of Hong Kong are Chinese, migrants from China during this century, fleeing famines and political turmoil, mostly after the communists took power in 1949. First or second generation refugees from China perceived themselves as "guests" arriving in a relatively prosperous enclave from a huge poor neighboring socialist country.

10 While Hong Kong has seen the rise of a large, visible middle class, and periods of near full employment, many workers face difficulties with such basic needs as housing, unemployment benefits, pensions and welfare provision for old age.

11 Hong Kong is the most densely-populated city in the world. Massive high-rise public housing projects now hold sixty per cent of the population. But demand exceeds supply, and the waiting list for such accommodation can run to five or ten years.

12 1980s, further pressure on housing and other services was created by a continual influx of refugees from China, providing a constant supply of cheap labor for Hong Kong employers.

> In the early 1980s, the British and Chinese governments began formal talks over the future of Hong Kong, as the 99-year lease on most of the land area expires in 1997.

13 Vietnamese refugees have been arriving since 1975 creating some social pressure. The Hong Kong government has maintained a policy of granting these arrivals "first asylum" refuge on humanitarian grounds. Hong Kong has increasingly placed these arrivals in closed detention centers in the hope that this would discourage further groups from leaving Vietnam bound for Hong Kong. It has also started screening arrivals to categorize them as political refugees or economic migrants. Resentment over the situation has grown among the Hong Kong population as they witness illegal Chinese forcibly returned across the border each day, while the Vietnamese are taken in. In late 1989, in a move which drew international criticism, most notably from the United States, the Hong Kong government forcibly returned the first group of Vietnamese arrivals to be declared economic migrants.

14 The human rights group Asia Watch has called for a stop to forced repatriation. In October, 1991, Britain and Vietnam signed an agreement stating that Hong Kong can send 222 refugees back to Vietnam.

15 The rapid, unregulated industrialization of Hong Kong has resulted in severe water, land and air pollution throughout the territory. Many of the sea areas around the colony contain human bacteria and toxic waste levels many times higher than WHO standards. Government action on these problems has been slow and ineffective.

16 Hong Kong's Governor, appointed by the British government, used to wield absolute power over the daily running of the territory. He was assisted by executive and legislative councils of members largely representing business, finance and professional groups.

17 In the early 1980s, the British and Chinese governments began formal talks over the future of Hong Kong, as the 99-year lease on most of the land area expires in 1997. Hong Kong people were not represented at these talks.

18 Before the talks with China began, special immigration laws were passed in Britain which downgraded the status of 3.25 million Hong Kong-born residents who had received British passports and citizenship. The changes removed any right of abode in the UK and the right to pass nationality on to descendants, resulting in a newly-created British Dependent Territories Citizen passport which does not confer effective citizenship of any nation.

19 In 1984, the Sino-British talks produced an agreement on the future of Hong Kong. Under the agreement, the entire territory will revert to Chinese sovereignty in 1997, but will enjoy a "high degree of autonomy" as a Special Administrative Region of China. The agreement states that Hong Kong will retain its current "social and economic systems" for at least fifty years after 1997; it will have a separate executive, legislature and judiciary, issue its own currency, and remain a separate immigration and customs area. The agreement hands responsibility for Hong Kong's foreign affairs and security to China.

20 China announced that the agreement contained an important new concept of "one country, two systems" which they also applied to the Portuguese colony of Macau and have frequently hinted could be a basis for the reunification of China and Taiwan.

21 A new "Basic Law" for Hong Kong containing working details for the operation of Hong Kong after 1997 in line with the agreement, will serve as a constitution.

22 When Britain and China began talks on the future of the colony the Hong Kong population made their voices heard. Many in Hong Kong worry that China will impose its communist style of rule on Hong Kong, or will try to regain Hong Kong as "a colony in working order", and rule by decree just as the British have.

23 Over one fifth of Hong Kong's population, some one and a quarter million people, took to the streets to demonstrate support of the Chinese democracy movement in May and June 1989. The events in China produced a wave of cultural, political and emotional identification with China among Hong Kong people, and a renewed consciousness that they shared a common destiny.

24 In September 1991, for the first time in 150 years, the inhabitants of Hong Kong elected the Legislative Council. Most of the seats were won by the candidates of the United Democrats coalition (UDHK), who favor democratization and have been critical of the Hong Kong colonial government and of China.

25 In 1992, repatriation agreements with the Vietnamese government eliminated immigration from Vietnam. The transfer of industries to southeast China caused increased unemployment and a drop in industrial wages.

26 British governor Chris Patten reformed the electoral system, establishing a total separation of the Legislative and Executive branches. This was criticized by Beijing, claiming the reform contradicted the principles set out in the Basic Statutes. In 1993, China appointed a Preparatory Committee for the Administrative Region of Hong Kong, chaired by Foreign Minister Qian Qichen and made up of prominent citizens of Hong Kong - mostly businesspeople - and consultants representing the colonial government.

27 In 1994 Patten launched a plan to increase the number of voters in the 1995 elections which led to a new conflict with Beijing. In September 1995, the Democratic Party, opposed to Chinese official interpretation of the Basic Statutes, won the elections to choose a

Legislative Council. Qian informed of Beijing's intention to dissolve the Council in 1997 since "it does not take into account the interests of all social sectors" in Hong Kong.

28 Patten publicly reproached Chinese authorities of being ill-informed of what takes place in Hong Kong since, according to this member of the British Conservative Party, they only talk "with multi-millionaires whose only concern is to continue being multi-millionaires". The "dictator sent by the Queen", as Chinese leaders once alluded to Patten, was referring to an alliance established in 1996 between Hong Kong and Beijing's main business community.

29 This alliance has a solid economic and commercial base: Hong Kong investments in the People's Republic of China are estimated in $80 billion, which will help to employ four million people and form 20 per cent of China's GDP.

Profile

ENVIRONMENT

Comprises an area of the southern Chinese mainland, plus several nearby islands to the east of the Pearl River estuary on the South China Sea. The city centers on the harbor of the Hong Kong Island and Kowloon peninsula, but several new towns have been developed in the traditionally rural New Territories to the north. Although almost 70% of the land is mountainous, Hong Kong has an agricultural sector providing 50% of its fresh vegetables and poultry. Rice production ceased in the territory in the 1970s. Hong Kong is not self-sufficient in fresh water, and has to buy half of its supplies from China.There are high levels of air and water pollution, as well as of the soil. Noise pollution is also a serious problem in the city, due to non-stop construction and high volumes of traffic.

SOCIETY

Peoples: 98% are Chinese, the great majority of these being Cantonese from southern China. Europeans, North Americans and other Asians make up small minorities like the Vietnamese. **Religions:** Religious practices are not widespread, but folk religious rituals and offerings predominate at festivals, dedications and funerals. There are also Muslim, Buddhist and Christian minorities. **Languages:** Chinese; English. **Political Parties:** United Democrats of Hong Kong; Association for Democracy and Support of the People; Democratic Alliance for the Betterment of Hong Kong; Liberal Party; the New Society of Hong Kong; Communist Party of China; Kuomintang.

THE STATE

Government: Christopher Patten, Governor until July 1997. Legislative Council elected in 1995. The territory will revert to Chinese sovereignty in july 1997, but will enjoy a "high degree of autonomy" as a Special Administrative Region of China.

DEMOGRAPHY

Urban: 95% (1995). **Annual growth:** 0.2% (1992-2000). **Estimate for year 2000:** 6,000,000. **Children per woman:** 1.4 (1992).

HEALTH

Under-five mortality: 116 per 1,000 (1995). **Calorie consumption:** 105% of required intake (1995). **Safe water:** 100% of the population has access (1990-95).

EDUCATION

Illiteracy: 8% (1995). **Primary school teachers:** one for every 27 students (1987).

COMMUNICATIONS

117 **newspapers** (1995), 103 **TV sets** (1995) and 105 **radio sets** per 1,000 households (1995). 51.0 **telephones** per 100 inhabitants (1993). **Books:** 108 new titles per 1,000,000 inhabitants in 1995.

ECONOMY

Per capita GNP: $21,650 (1994). **Annual growth:** 5.30% (1985-94). **Annual inflation:** 9% (1984-94). **Consumer price index:** 100 in 1990; 143.2 in 1994. **Currency:** 8 Hong Kong dollars = 1$ (1994). **Cereal imports:** 640,000 metric tons (1993). **Imports:** $162,000 million (1994). **Exports:** $151,395 million (1994). **Development aid received:** $30 million (1993); $5 per capita; 0% of GNP.

ENERGY

Consumption: 2,280 kgs of Oil Equivalent per capita yearly (1994), 100% imported (1994).

Macau

Population: 444,000 (1994)
Area: 16 SQ KM
Capital: Macau

Currency: Pataca
Language: Portuguese and Cantonese

Macau

The tiny Portuguese enclave of Macau is located across the Pearl River estuary from Hong Kong. The Portuguese set up Macau in 1557 as an important link in their chain of trading ports which stretched from Europe, along the coasts of Africa and India, past Melaka, and on to Nagasaki in Japan. Rent was paid for the territory until 1849 when Portugal declared it independent. China accepted this in 1887, after Portugal pledged never to alienate Macau and its dependencies without agreement with China. It was declared an Overseas Province of Portugal in 1951.

[2] For several hundred years, Macau represented the main point of contact for economic and cultural links between Europeans and the vast Chinese empire. With the development of Hong Kong next-door, and the decline of Portugal as an International colonial power, Macau lost its importance.

[3] During the height of the Cultural Revolution in China, Macau experienced serious anti-Portuguese rioting, and the government reportedly offered to abandon the territory within a month, but China refused the offer.

[4] In 1974, after the fall of the regime of Antonio de Oliveira Salazar, the Portuguese government again offered to return the colony to China. Anxious not to alarm Hong Kong or Taiwan, China refused. The Portuguese government unilaterally declared Macau to be Chinese territory under Portuguese administration.

[5] In 1985, once negotiations between China and Britain on the future of Hong Kong had been successfully completed, Beijing turned its attention to Macau, and negotiated a settlement with the Portuguese under which Macau will be returned to China in 1999, with arrangements for retaining some local autonomy like in Hong Kong.

[6] Macau is an impractically small enclave, isolated from a central government thousands of miles away and with an administration reluctantly serving out its time. Not surprisingly, corruption is widespread. In 1988 senior officials were implicated in a scandal over privatization of the colony's television station.

[7] Much of Macau is separated from China by only a narrow river, preventing the authorities from stopping the large flow of illegal immigrants into the city. These peo-ple arrive with no legal means of support and often turn to crime.

[8] In 1988 the government began to allow workers from China into Macau. The move was criticized by labor groups as a method of keeping wages down.

[9] Macau has few natural resources: for some time, China has provided part of its water, and since 1984, part of its electrical energy. However, as of 1989, more than 90 per cent of the electricity used has been produced within the territory. In the past, the colony depended upon tourism, a certain amount of trade with China and light industry (especially toys and textiles). Recently, however, since China began its economic liberalization program and created a Special Economic Zone on the other side of the border, construction was begun on a hotel and a new airport.

[10] Between 1985 and 1989, the GNP grew at an annual rate of 8.1 per cent. Most of Macau's tourists come from Hong Kong. The casinos are the main tourist attraction, and account for 62 per cent of the government's income. The construction of the airport has attracted investment from Taiwan and other countries, who have benefited from the low cost of labor.

[11] In 1989, Portuguese passports were issued to around a hundred thousand Macau residents of Chinese origin. Macau's "old-time" residents were always considered to be Portuguese citizens, enjoying all the rights of citizens, including automatic entry to Portugal and as of 1992, the rest of the European Community.

[12] In 1991, governor Vasco Rocha Vieira allowed Chinese citizens to work as civil servants and declared Cantonese an official language along with Portuguese. In the September 1992 legislative elections, pro-Chinese organizations won a majority of seats.

[13] In 1993, the final draft of the Basic Statutes was passed which will become the territory's Constitution as of 1999, when it will become a Special Administrative Region of the People's Republic of China. The draft was immediately ratified by Beijing. The document assures a continuation of the capitalist system, a ban on Chinese taxes, the continued operation of the casinos, the election of a Head of State by a local electoral college and the formation of a Legislative Council with a mandate until the year 2001.

[14] In 1994, the apparently illegal handing over - by Beijing authorities- of an Australian citizen with a Chinese background to Macau police provoked strong international criticism. In 1995, the Supreme Court of Macau refused to extradite three individuals to Beijing, charged with committing crimes in China. A report from the US State Department in March 1996 stated that the self-censorship of Macau's press will continue or increase toward 1999, especially on matters deemed sensitive by the People's Republic of China.

> **For several hundred years, Macau represented the main point of contact for economic and cultural links between Europeans and the vast Chinese empire.**

PROFILE

ENVIRONMENT

Macau is an enclave located on the southern coast of China on the South China Sea, lying to the west of the Pearl River estuary at the mouth of the Si-kiang river. The territory consists of a peninsula and two small islands, Taipa and Coloane, attached to the mainland by a bridge and a causeway.

SOCIETY

Peoples: Most of the population are Cantonese from southern China, although a sizeable proportion of Mecanese of mixed Chinese and European ancestry have developed during the city's long history. **Religions:** Folk religious rituals predominate at festivals, dedications and funerals. There is also a high proportion of Christians, mostly Catholic. **Languages:** Portuguese and Cantonese. **Political Parties:** Though officially there are no political parties, their functions are performed by civic associations such as: Association for the Defence of Macau's Interests (ADIM), a conservative group; the reformist Democratic Center of Macau (CDM); and the Macau Community Development Studies Group (GEMC), regarded as a moderate, center organization; Union to Promote Progress, pro-China.

THE STATE

Capital: Macau, 356,000 inhab. (1991). **Government:** The current governor is Vasco Rocha Vieira, named by Portugal in April 1991. The Legislative Assembly is made up of 23 members: seven named by the governor, eight elected by direct vote and eight by indirect vote.

COMMUNICATIONS

34.4 telephones per 100 inhabitants (1993).

ECONOMY

Consumer price index: 100 in 1990; 133.8 in 1994. **Currency:** patacas.

Taiwan

Taiwan

Population: 21,268,000 (1994)
Area: 35,981 SQ KM
Capital: Taipei

Currency: New Dollar
Language: Chinese (Mandarin)

The island of Taiwan has a government which insists that it represents all of China, which calls itself the Republic of China, and which for four decades has pledged itself to the cause of recovering the mainland, from which it is separated by the 200 kilometer-wide Taiwan Strait.

[2] China's Qing (Manchu) emperors incorporated Taiwan into the empire in 1683, and the island was proclaimed a separate province of China in 1887. During this time, the political and administrative systems of mainland China were extended to Taiwan, and many people migrated there from the mainland.

[3] After China's defeat in the 1895 Sino-Japanese war, Taiwan became a Japanese colony, but was returned to China after the defeat of Japan at the end of World War II. At first, the people of Taiwan rejoiced at the end of Japanese colonialism, but they soon discovered that life under the ruthlessly authoritarian, corrupt and vehemently anti-communist Kuomintang (KMT) party led by Chiang Kai-shek resembled nothing less than different kind of colonialism.

[4] On February 28, 1947, there was a major demonstration against the KMT authorities. The KMT reacted at first by lifting martial law, and inviting the opposition to form a Settlement Committee of politicians, trade unionists and student groups to discuss possible political reforms. Meanwhile, they drafted in 13,000 additional troops, and when the opposition came forward, the KMT massacred large numbers of them, imprisoning others.

[5] In 1949, the entire KMT government, the remnants of its armies, and their relatives and supporters, fled to Taiwan after losing the mainland civil war to the Communist armies. From its refuge, backed by the United States, the KMT declared Taiwan to be the temporary base of the Republic of China pending recovery of the mainland. Most Western nations continued to recognize the KMT as the representatives of all China.

[6] When the Korean War erupted, with China supporting the North Koreans, the United States redoubled its military and economic commitment to Taiwan, protecting it as a front-line state in the battle to defend the "free world".

[7] Democracy vanished. Human rights were violated, demonstrations, strikes and political par-

ties banned, and martial law imposed, all in the name of the battle to reconquer the mainland. The KMT set up a governmental system which claimed to represent the whole of China, with legislators representing each mainland province.

[8] Taiwan's remarkable industrialization began in the 1960s, when World Bank and US technocrats helped the government apply an export-oriented development strategy. The era of Japanese colonialism had left Taiwan with only partially-developed transport and an education system. The US granted ideologically-motivated financial, trade and aid advantages to bolster an authoritarian political regime which, in turn, hectored Taiwan's disenfranchised, and politically disorganized workforce.

[9] Taiwanese output grew at an average of 8.6 per cent each year between 1953 and 1985, and became one of the four newly-industrialized dragons of East Asia. Growth was entirely export-oriented, and the island developed the world's second-largest trade surplus with the US, following Japan. Today, Taiwan has highly developed plastics, chemical, shipbuilding, clothing and electronics industries.

[10] Sweatshop conditions were common in the 1960s and 70s, and many workers put in long shifts, often exposed to toxic substances.

[11] In 1971, the United States decided to seek closer ties with China, no longer vetoing the latter's admission to the UN. Taiwan therefore lost its representation in that world body. A few years later, in 1979, the United States officially broke diplomatic relations with Taiwan.

[12] The country found itself at an economic crossroads: the island was over-dependent on a few export markets and labor-intensive industries, importing nations were pressuring for more balanced trade and Taiwan's labor was no longer as cheap as that of many of its Asian neighbors. Technocrats argued that, in order to compete, Taiwan should develop a more open political system. The KMT also faced an internal succession crisis as aging politicians from the civil war era clung precariously to power.

[13] In this climate, social movements were proliferating, with ecological groups protesting against pollution and nuclear power, farmers' groups demanding higher pric-

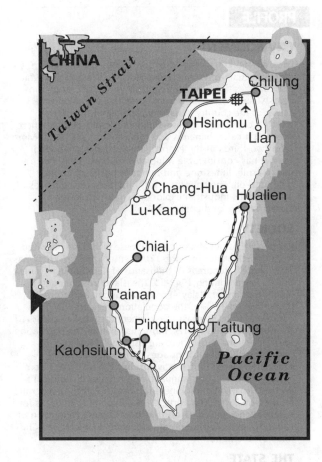

es, and students calling for more academic freedom and an end to human rights abuses.

[14] In addition, a significant portion of the population does not consider their country to be a part of

China, and rejects both the authoritarian policies of the KMT and Deng Xiaoping's unification proposal of "one nation, two systems"

with Taiwan becoming a Chinese dependency.

[15] In September 1986 the Democratic Progress Party (DPP) was formed as the first opposition party to challenge the KMT's political stranglehold. Although technically illegal, the party was allowed to survive, and Martial Law was formally lifted on July 15, 1987.

[16] 1987 witnessed a huge upsurge in union activity, strong independent sections were formed in the trade union system and supporters founded parallel political structures in the Labor Party and Workers' Party.

[17] In December 1989, elections were held. The KMT won 53 per cent of the vote, against the DPP's 38 per cent. The DPP also fared well in municipal and council elections, winning the mayorship of Taipei.

[18] While the KMT is resolutely opposed to independence, some DPP members formed the New Wave group proposing that Taiwan declare self-rule, and other groups advocate a referendum for self-determination.

> After China's defeat in the 1895 Sino-Japanese war, Taiwan became a Japanese colony, but was returned to China after the defeat of Japan at the end of World War II.

PROFILE

ENVIRONMENT

Located 160 km southeast of continental China, Taiwan is part of a chain of volcanic islands in the West Pacific which also includes the Japanese islands. A mountain range runs across Taiwan stretching from north to south along the center of the country. A narrow plain along the island's western coast constitutes its main agricultural area where rice, sugar cane, bananas and tobacco are cultivated. More than two-thirds of the island's area is densely wooded. Taiwan has considerable mineral resources: coal, natural gas, marble, limestone and minor deposits of copper, gold, and oil. The country is suffering the consequences of its enormous industrial explosion, with high levels of water, air and land pollution.

SOCIETY

Peoples: Most inhabitants are Chinese who have migrated from the mainland since the 17th. century and are known as "Taiwanese". Hundreds of thousands of Kuomintang Chinese fled to Taiwan during 1949-50. The island's indigenous inhabitants are of Malayo-Polynesian origin, and currently do not exceed 1,7% of the population. They are concentrated on the east coast, where they make up 25% of the population. **Religions:** More than half of the population are Chinese Buddhist. There are also small Muslim and Christian minorities. **Languages:** Chinese (Mandarin), official. Taiwanese, a derivative of the Chinese dialect from Fujian province, is the language of the majority. **Political Parties:** Kuomintang (Chinese Nationalist Party); Democratic Progressive Party; New Party; World Union of Formosans for Independence. **Social Organizations:** Chinese Federation of Labor; National Federation of Independent Unions; Tao-Chu-Miao Brotherhood.

THE STATE

Official name: Republic of China. **Administrative divisions:** 7 municipalities and 16 counties. **Capital:** Taipei, 2,650,000 inhab. (1994). **Other cities:** Kaohsiung, 1,416,000 inhab., T'aitung, 836,000 inhab., T'ainan 702,000 inhab. (1994). **Government:** President, Lee Teng-hui. **National holiday:** January 1st, Day of the Republic. **Armed Forces:** 425,000. **Paramilitaries:** Militarized police, 25,000.

STATISTICS

HEALTH

Calorie consumption: 105% of required intake (1995).

COMMUNICATIONS

105 **newspapers** (1995), 106 **TV sets** (1995) and 109 **radio sets** per 1,000 households (1995). **Books:** 104 new titles per 1,000,000 inhabitants in 1995.

ECONOMY

Currency: new dollars.

---

[19] The KMT achieved yet another victory over the opposition DPP in the National Assembly elections in December 1991. The results were regarded as a plebiscite on the independence issue. The DPP's electoral platform, favoring a definitive separation from China, was supported by a mere 21 per cent of the electorate, against the KMT's 71 per cent.

[20] In 1992, fourteen members of the World Union of Formosans for Independence - a party which had been banned by the government- were arrested. In addition, diplomatic ties were broken with South Korea, which established relations with China. In December, in the first open elections since 1949, the PDP obtained 31 per cent of the vote, and 50 of the 161 legislative seats.

[21] In early 1993, and as a direct consequence of the electoral results, there were two significant resignations by members of the KMT leadership: that of Prime Minister Hau Pei-tsu, and that of Secretary General James Sung, both of the conservative wing of the party. Lien Chan, the first government leader to have been born in Taiwan, was named prime minister. Shortly afterwards, the governing party experienced its first major division. Thirty deputies broke with the KMT to found the Party of the New Nationalist Alliance.

[22] Starting from 1994, distinctive voices are making themselves heard demanding an independent way for Taiwan, leaving behind the assumption of being a government representative of all Chinese. However, due to Beijing's opposition to any measure which could further lead the island toward independence, Taipei's efforts to be accepted in the UNO were vain.

[23] The first multi-party municipal elections were held in December. Most of the vote went to the ruling Kuomintang and the recently formed Democratic Progress Party (DPP). Meanwhile, in spite of opposition from environmentalists and anti-nuclear activists and DPP objections, the Kuomintang sponsored the approval for the construction of a fourth nuclear plant on the island.

[24] In 1995, economic relations with Beijing were intensified. Taiwan became the second "foreign" investor in the People's Republic after Hong Kong, placing approximately $22 billion in China. In spite of this exchange, political relations between the two countries were deteriorated after a private visit from Taiwanese president Lee Teng to the US in June. Heedless of US warnings, Beijing carried out a series of missile launchings in July and August on waters located 140 kilometers from Taiwan.

[25] In spite of losing ground from previous elections, the Kuomintang won the December legislative elections with 46 per cent of the vote and 33 per cent by the DPP. In the campaign prior to the March 1996 presidential elections, new Chinese military manoeuvres near the coasts of Taiwan led Washington to send warships in defence of Taiwan's alleged threatened territorial integrity.

[26] On March 20, Lee Teng won the first presidential elections with direct suffrage in the island's history.

> Starting from 1994, distinctive voices are making themselves heard demanding an independent way for Taiwan, leaving behind the assumption of being a government representative of all Chinese.

Colombia

Population: 36,330,000 (1994)
Area: 1,138,910 SQ KM
Capital: Santa Fé de Bogotá

Currency: Peso
Language: Spanish

Colombia

The best known of Colombia's indigenous cultures is that of the Chibchas or "Muiscas", as they called themselves. They lived in the northern Colombia and Panama, farming and mining.

[2] In America, colonization resulted in the plundering of native wealth and the population was subjected to thinly disguised forms of slavery. After 300 years of colonialism, a large part of the indigenous population had disappeared.

[3] Spain conquered Colombia between 1536 and 1539. Gonzalo Giménez de Quesada decimated the Chibchas and founded the city of Santa Fe de Bogotá which became the center of the Viceroyalty of New Granada in 1718.

[4] Extensive, export-oriented agriculture (coffee, bananas, cotton and tobacco) replaced traditional crops (potatoes, cassava, corn, wood and medicinal plants), with African slaves replacing the more rebellious indigenous population.

[5] The Revolt of the *Comuneros* started the process leading to the declaration of independence in Cundinamarca, in 1813. The road to independence was marked by constant struggles between the advocates of centralized government and the federalists, headed by Camilo Torres. Antonio Nariño (who had drafted the declaration of independence) represented the urban bourgeoisie, linked to European interests, while Torres presided over the Congress of the United Provinces, representing the less privileged.

[6] In 1816, Pablo Morillo reconquered this territory, executing Torres. Three years later, Simon Bolívar counter-attacked from Venezuela, liberated Colombia and founded the Republic of Greater Colombia, including Venezuela, Ecuador and the province of Panama. Regionalism and strong British pressure brought about the secession of Venezuela and Ecuador in 1829-30. The Republic of New Granada was then proclaimed and in 1886 Colombia adopted its current name.

[7] From 1830 to the early 20th century, the country went through civil wars, local wars (including two with Ecuador), three military uprisings and 11 constitutions. The Liberals and Conservatives have kept a permanent hold on the Colombian political situation, separated by a mutual hatred passed down from generation to generation, despite having similar platforms for governing the country.

[8] Between 1921 and 1957, overproduction of the country's oil reserves led to a depletion of this resource, leaving American oil companies with a profit of $1.137 billion, and Colombia with no oil. American companies control 80 to 90 per cent of banana production and mining, and 98 per cent of energy production.

[9] In 1948, in Bogotá, Mayor Jorge Eliecer Gaitan of the Liberal Party was assassinated, and public indignation over his death triggered widespread riots, known as El Bogotazo. That same year, a Liberal mayor organized a guerrilla group, the first of 36 which were active during the presidencies of Ospina Pérez, Laureano Gómez and Rojas Pinilla. In 1957, a constitutional reform ensured the alternation of Liberals and Conservatives in the government, every 12 years.

[10] The Revolutionary Armed Forces of Colombia (FARC), led by Manuel *"Tiro Fijo"* (Sure Shot) Marulanda and Jacobo Arenas, appeared on the scene in 1964. Guerrilla strategists included Camilo Torres Restrepo, a priest and co-founder of the National Liberation Army (ELN), who was killed in combat in 1965.

[11] Large landowners organized, armed and paid "self-defence" groups to fight these rural-based guerrilla movements. These were supported by members of the army and, in some cases, by foreign mercenaries. Closed out of official circles, the army also created paramilitary groups, later condemned by Amnesty International.

[12] From 1974, President Alfonso López Michelsen, a Liberal, tried to give greater attention to popular demands, but vested economic interests led to the failure of this policy. Figures for 1978 reveal that only 30 per cent of industrial workers and 11 per cent of the rural workforce had social security benefits. Colombia is dependent on international coffee prices on the American and German markets, for its foreign exchange, as these countries consume 56 per cent of the Colombian product.

[13] Guerrilla movements, particularly the FARC and the April 19 Revolutionary Movement (M-19), continued their activities into the late 1970s. Military repression grew during the government of president Julio C Turbay Ayala (1978-82).

[14] In 1982, a divided Liberal Party nominated two candidates, thus handing over the victory to the Conservative Party's Belisario Betancur, a journalist, poet and humanist, who had actively participated in the peaceseeking process in Central America. Betancur proposed that Colombia join the Non-Aligned Movement and reaffirmed the right of debtor nations to negotiate collectively with creditor banks. Also, in 1983 he entered into peace talks with leaders of the M-19.

[15] The M-19 had initiated the peace process in 1980, when guerrilla leader Jaime Bateman proposed a high-level meeting in Panama. Bateman subsequently died in a suspicious airline accident and talks were suspended. In the meantime, the FARC and the government reached an agreement which led to a ceasefire between them and to the adoption of political, social and economic reforms.

[16] Large landowners fiercely opposed talks between the government and the guerrillas. The rural oligarchy, holding 67 per cent of the country's productive land, de-

PUBLIC EXPENDITURES

% of GNP — 1992

- MILITARY: 2.4
- EDUCATION: 2.9
- HEALTH: 1.8

WORKERS

1994

- MEN 77%
- WOMEN 23%
- AGRICULTURE 10%
- INDUSTRY 24%
- SERVICES 66%

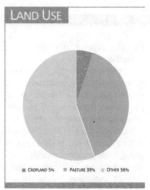

LAND USE

- CROPLAND 5%
- PASTURE 39%
- OTHER 56%

Foreign Trade

MILLIONS $ 1994

IMPORTS
11,883
EXPORTS
8,399

Maternal Mortality

1989-95

PER 100,000
LIVE BIRTHS
107

PROFILE

ENVIRONMENT

The Andes cross the country from north to south, in three ranges: the western range on the Pacific coast, and further inland, the central and eastern ranges, separated by the large valleys of the Cauca and Magdalena Rivers. North of the Andes, the swampy delta of the Magdalena River opens up, leaving flat coastal lowlands to the west - along the Pacific coast - and to the east, plains covered by jungle and savannas extend downward to the Orinoco and Amazon Rivers. This diversity results in great climatic variety, from perpetual snows on Andean peaks to tropical Amazon rain forests. The country's Andean region houses most of the population. Coffee is the main export item, followed by bananas. Abundant mineral resources include petroleum, coal, gold, platinum, silver and emeralds. Intensive agriculture and mining have contributed to soil depletion. There is significant deforestation, and two-thirds of the country's bird species are in danger of extinction.

SOCIETY

Peoples: Colombians are the result of cultural and ethnic integration among three groups: native Americans, Africans and Europeans. **Religions:** 93% of Colombians are Catholic. Although this is the country's official religion, there is religious freedom. **Language:** Spanish (official). **Political Parties:** The Liberal Party (LP); the Social Conservative Party (SCP); the New Democratic Force (NFD), led by Andrés Pastrana, conservative; M-19 Democratic Alliance (ADM-19); National Salvation Movement (MSN), which broke away from the PSC; Patriotic Union, formed by the military. **Social Organizations:** There are four major labor organizations: the Confederation of Colombian Workers; the Confederation of Workers' Unions of Colombia; and the General Labor Confederation. The Colombian United Labor Federation (CUT), founded on 26 September 1986, and 80% of all salaried workers affiliated to trade unions are members.

THE STATE

Official Name: República de Colombia. **Administrative divisions:** 32 departments and the capital district. **Capital:** Santa Fé de Bogotá, 5,237,600 inhab. (1995). **Other cities:** Cali 1,718,900 inhab.; Medellín 1,621,400 inhab.; Barranquilla1,064,300 inhab; Cartagena745,700 inhab. (1995). **Government:** Ernesto Samper Pizano, president since August 7, 1994. **National Holiday:** July 20, Independence Day (1810). **Armed Forces:** 146,400 troops (1994). **Paramilitaries:** National Police Force, with 85,000 members. Coast Guard: 1,500 (1993).

increase exports and reduce the $2 billion fiscal deficit by 30 per cent.

[18] Rejected by the labor unions and the parties of the left, the plan also failed to satisfy the country's creditors. A commission of 14 banks, presided over by the Chemical Bank, stated that the government would have to sign a letter of intent and reach a formal agreement with the IMF, two steps which Betancur had wanted to avoid.

[19] According to the Human Rights Commission, 80 prisoners had disappeared in one year, political prisoners had been tortured and 300 clandestine executions were confirmed. The number of political activists who had disappeared rose to 325.

[20] On November 6, 1985, 35 guerrillas from M-19 took over the Palace of Justice in Bogotá. The army attacked, causing a massacre. All the guerrillas were killed, along with 53 civilian victims, including magistrates and employees. A guerrilla commander known as Alonso insisted that those who had been killed "were deliberately sacrificed by the army".

[21] Taking advantage of this panorama of violence, the drug lords, and their traffickers, have created a secure power base for themselves.

[22] Over 2,000 left wing activists were killed by terrorists, and in 1987, Jaime Pardo Leal, a member of the Patriotic Union, was assassinated. Liberal senator and presidential candidate in the 1990 election, Luis Carlos Galán - who had promised to dismantle the paramilitary groups and fight drugs - was assassinated in August 1989. War broke out, between the government and the drug mafia. In March 1990, Bernardo Jaramillo, the Patriotic unions' presidential candidate, was assassinated, with Carlos Pizarro (replacing Jaramillo) being killed 20 days later.

[23] Official figures indicate that there are more than 140 paramilitary groups in Colombia, most financed by the drug mafia. Meanwhile, as part of its "War on Drugs", the United States Drug Enforcement Administration is allegedly bombarding coca plantations with chemical herbicides.

[24] Despite the huge profits generated by the planting, processing and export of drug products (which have given rise to an underground economy), agriculture continues to be the backbone of Colombia's legal economy. Despite using only 5 per cent

of the land it represents 23 per cent of the GNP, and in 1981, it employed 21 per cent of the workforce.

[25] The presidential elections of May 27, 1990 were won by Liberal Party candidate Cesar Gaviria, who received 48 per cent of the vote in an election where the abstention rate was 58 per cent. The Movement for National Salvation obtained 23.7 per cent of the vote, the Democratic Alliance of the M-19 (ADM-19) 12.56 per cent, and the Conservative Party, 11.90 per cent.

[26] In December 1990, elections were held to form a constituent assembly with a 65 per cent abstention rate. The number of votes received by ADM-19 was significant, as they won 19 seats, 4 fewer than the Liberal Party (the party in power).

[27] In June 1991, members of Gaviria's government met in Caracas with representatives of the Revolutionary Armed Forces of Colombia (FARC), the National Liberation Army (ELN), the People's Liberation Army (EPL) and members of the Simon Bolívar Guerrilla Coordinating Committee, who controlled 35 per cent of the country. The "long and difficult" negotiations dealt with demobilization of the guerrillas, constitutional guarantees, the subordination of the Armed Forces to civilian authority, the dismantling of paramilitary groups and the reinsertion of guerrilla fighters into areas where they can exert political influence.

[28] On July 5 1991, the new Colombian constitution went into effect. In addition to creating the office of the vice-presidency, it eliminated the possibility of presidential reelection. It also included the following gains: civil divorce for Catholic marriages, direct election of local authorities, autonomy for indigenous peoples, and the democratic tools of referendum and grassroots legislative initiatives.

[29] Despite only a token presence of women in the constituent assembly, the constitution created by this assembly guarantees equal opportunities for women. Its provisions guarantee special support from the State where women are heads of families, as well as subsidies for unemployed pregnant women or those who have recently given birth.

[30] Once approved, the constitution met with criticism from the left for not introducing measures to control the Armed Forces, for failing to give civilian courts jurisdic-

nounced the pacification process as "a concession to subversion" and proposed the creation of private armies. Paramilitary action started up again; subsequent investigations revealed the hand of the MAS, "Death to the Kidnappers", which had opposed the withdrawal of the army from guerrilla-controlled areas. A one-year truce went into effect; but, M-19 withdrew five months later, claiming

that the army had violated the cease-fire.

[17] In January 1985, the government passed a series of unpopular economic measures: drastic cuts in public spending, a salary freeze for civil servants, substantial fuel and transport price rises, increased taxes on over 200 products including some basic consumer goods, limited salary increases which lowered real purchasing power, and a currency devaluation. The object was to

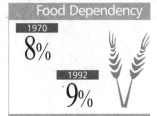

tion over members of the military who have committed crimes against civilians and for granting judicial powers to State security agencies.

[31] On October 27, a pact between the president and the three major political forces participating in the constituent assembly reduced the number of representatives; after the dissolution of Congress, parliamentary elections were held. Abstention remained unmodified, and 60 per cent of the vote went to the Liberal Party, which had now split into several groups. The percentage of votes going to the ADM-19 dropped to 10 per cent.

[32] On October 30, in Caracas, the new government met with the guerrillas. Two months later, a radical branch of the EPL began fighting once again.

[33] The peace process reached a low point in 1992. After talks had been discontinued, the government promoted its so-called "Integral War", which in addition to military action advocated intervention in civilian organizations with suspected links to rebel groups. Among those targeted for "intervention" were human rights groups, development projects, and certain media and social research institutes.

[34] The Simón Bolívar Coordinating Committee resisted the army offensive continuing its campaign. In the meantime, paramilitary groups resumed their activities, primarily in the Mid-Magdalena (River Valley) region, Boyacá and the Medellín. The violence spread from the areas where the conflict had been concentrated, to other regions in the country's interior.

[35] In the March 1992 municipal elections - marked by a 70 percent abstention rate among voters - the Liberal and Conservative parties maintained their majorities (including candidates considered vital by President Gaviria's government), while the ADM-19 continued to weaken as a political force.

[36] In November the government decreed a state of emergency when Pablo Escobar Gaviria, head of the Medellín Cartel, a powerful drug-trafficking ring, escaped from prison in mid-1992 stepping up the cartel's violent action. In January 1993, a group known as "PEPES" or People Persecuted by Pablo Escobar, appeared; within a two-month period, they killed 30 cartel members, destroyed several of Escobar's properties and harassed

members of his family. The confrontation reached serious proportions, with the detonation of several car-bombs causing dozens of deaths.

[37] Finally, on December 2, Escobar was killed by the police forces in a shoot-out in central Medellín. Although his death was a heavy blow to the political and social power of the Medellín Cartel, the drug-trafficking mafia still had many other holds, including the more discreet Cali Cartel, which emerged stronger.

[38] The Supreme Court of Justice decriminalized the use of marihuana, cocaine and other drugs which brought about the radical opposition of vast political and religious sectors headed by president Gaviria.

[39] The coffee market crisis, the 1993 drought and the reduction of its banana quota by the European Community affected exports. However, with an annual 2 billion dollars in revenues from drug traffic and the finding of oil in the province of Casanare, the country achieved a sustained rate of growth of 2.8 per cent per capita. Construction increased by 8 per cent in 1993, trade and transport grew 5 per cent each. Unemployment fell to less than 9 per cent in the country's seven main cities and wages continued on a positive trend. Despite this, 45 per cent of the population continue to live in extreme poverty.

[40] President Gaviria held on to his popularity especially at an international level. He was appointed Secretary General of the OAS with the decisive support of the USA and added to the victory of his party's candidate, Ernesto Samper, in the 1994 elections.

[41] In the second electoral round and with almost 50 per cent of the vote, Samper defeated conservative Andres Pastrana who obtained 48.6 per cent. In the first round the ADM-19 declined to 4 per cent of the vote while abstentionism was slightly reduced to 65 per cent.

[42] Samper began his government with a succession of victories against drug traffic but a political scandal broke out in September 1995 when a spokesman for the Cali cartel revealed details of that organization's contributions to the Samper and Pastrana electoral campaigns, which was not confirmed. Defence minister Fernando Botero, former director of Samper's campaign, was sent to prison on charges of unlawful wealth.

[43] In August, Samper declared a state of emergency to control the wave of violence and kidnappings. This was seen as an attempt to protect himself from drug-related scandals. Violence went on with the murder of opposition leaders and the continued activity of FARC and ELN guerrillas, which attacked high power lines, oil pipelines and police and military facilities. With almost 100 active war fronts, these two groups control growingly extended and economically powerful regions in the coffee-growing areas, the Caribbean and even Bogota or Medellin vicinities, where they are in charge of government duties.

[44] Efforts to eradicate coca and poppy crops as well as armed operations against the cartels' operative bases continued. Some of the main leaders of the Cali cartel - which holds 70 per cent of the world traffic of cocaine - turned themselves

in voluntarily. Although the cartel was weakened, other groups which aimed their activities at neighbouring countries came out strengthened. In March 1996, the US took Colombia off its list of countries which cooperate in the fight against drugs, thus depriving it from bilateral aid and denying it access to foreign financial sources. In July, Washington announced it would refuse to issue Samper's visa, in a diplomatic attempt to corner the Colombian president.

Comoros

Population: 485,000 (1994)
Area: 2,238 SQ KM
Capital: Moroni

Currency: CFA Franc
Language: Arabic and French

Comoros

C omoros was populated around the 5th century by one of the last Indonesian migrations (see Madagascar). The Comoros Islands remained isolated from the continent until the 12th century when Moslem traders from Kilua settled on the islands, founded ports and reproduced the civilization of the eastern African coast (see Tanzania: The Zandj Culture in East Africa).

[2] In the 16th century the Comoros had a prosperous economy but the Portuguese seized the islands and destroyed their active trade. When the Sultan of Oman finally drove the Portuguese from the region, the Comoros came under the influence of Zanzibar. The Bantu population, originally brought from the continent to work on the islands, increased considerably as a result of the slave trade.

[3] In the 19th century, Zanzibar split from the sultanate of Oman and the French occupied Mayotte in 1843. Colonial domination eventually spread to the entire archipelago as, the islands' position on the Cape route gave them strategic value.

[4] The Comoros National Liberation Movement (MOLINACO) was created in the context of successful anti-colonial struggles throughout Africa. It joined forces with the local Socialist Party (PASOCO) to form the United National Front (FNU), which pressured the French government into holding a plebiscite in 1974. 154,182 people voted in favor of the islands' independence, and only 8,854 voted against.

[5] Most of those who wished to remain under French rule were Mayotte residents (63 per cent of its voters). France has air and naval bases on Mayotte and the economy is controlled by a few dozen Catholic families, sympathetic to France and politically represented by the "Mahorés People's Movement", led by Marcel Henry.

[6] Ahmed Abdallah, the archipelago's leading rice exporter and prime minister of the semi-autonomous local government, proclaimed the independence of Comoros in July 1975, before the French announced the result of the referendum. Abdallah was afraid that his Udzima (Unity) Party would lose out to the FNU in a future assembly to draft a new constitution. The Mahorés People's Movement took advantage of the situation to declare that Mayotte would continue under French rule. Paris supported the secession in order to maintain its military presence in the Indian Ocean, violating its previous commitment to respect the territorial integrity of Comoros and the result of the referendum. France did not oppose the islands' membership in the UN but vetoed specific Security Council resolutions to reincorporate Mayotte into the archipelago.

[7] Less than a month before the declaration of independence, a small group of FNU youths seized the national palace in Moroni and appointed their leader, the socialist Ali Soilih, as president, in place of Ahmed Abdallah. France immediately reacted by sending a task force of three warships and 10,000 soldiers to Mayotte; one soldier for every three inhabitants.

[8] In May 1978, a mercenary force, under the command of Ahmed Abdallah, in exile in Paris, landed on Grand Comoro overthrowing Ali Soilih, who was assassinated three days later. At the head of the operation was the notorious French mercenary Bob Denard, who had been tried in 1977 for mercenary acts of war against the government of Benin. His presence in Comoros triggered international protests, to the point of the Comoros Islands' delegation being expelled from a ministerial meeting of the Organization for African Unity in Khartoum.

[9] From then on, Denard became a key figure in the archipelago's politics, and Abdallah's control of the government came to depend on support from Denard and his 650 troops.

[10] On November 26, 1989, a coup d'état, led by Bob Denard succeeded in ousting President Ahmed Abdallah, who was killed in the fighting.

[11] The European mercenaries were financed by South Africa, and the Comoros were also said to have

WORKERS

1994

MEN 62% WOMEN 38%

AGRICULTURE 83% INDUSTRY 6% SERVICES 11%

PROFILE

ENVIRONMENT

The Comoros Islands are located at the entrance of the strategic Mozambique Channel, on the oil tanker route between the Arab Gulf and western consumer nations. The four major islands of this volcanic archipelago are: Njazidja, formerly Grand Comoro; Nzwani, formerly Anjouan; Mwali, formerly Moheli; and Mahore, also known by its former name of Mayotte. Njazidja has an active volcano, Karthala, 2,500 meters in altitude. The mountainous island is covered by tropical forests. Only 37% of the cultivated land is used to grow cash crops; vanilla and other spices, the rest is devoted to subsistence farming which is carried out without permanent rivers. The rainwater, which is stored in reservoirs, is easily polluted.

SOCIETY

Peoples: The original Malay-Polynesian inhabitants were absorbed by waves of Bantu and Arab migrations. Today, the latter groups predominate, co-existing with minor Indian and Malagasay communities. **Religions:** Islam (official). There is one mosque for every 500 inhabitants. **Languages:** Arabic and French are official. Most people speak Comoran, a Swahili dialect, and some groups speak Malagasay. **Political Parties:** Comoran Union for Progress, formerly "Udzima"; National Union for Democracy in Comoros; Democratic Alliance; People's Democratic Movement; Comoros Party for Democracy and Progress; Socialist Party. **Social Organizations:** The main organization is the Comoran Workers' Union.

THE STATE

Official Name: République Fédéral et Islamique des Comores. **Government:** Parliamentary republic. Mohamed Taki Abdoulkarim, President. Tajidine Ben Said Massomde, Prime Minister since 1996. **National Holiday:** July 6, Independence Day (1975).

served as a supply base for RENA-MO, the Mozambican SA-backed rebel group.

[12] In the days following this coup, the French government suspended all economic aid to the islands and initiated negotiations designed to oust Denard and his mercenaries from the island. The mercenaries left the islands in mid-December, heading for South Africa, which has reported political and economic links to the former Comoran presidential guard. Prior to his departure, Denard is said to have turned the responsibility for the island's armed forces over to a French army contingent. The interim president of the Comoros permitted the French troops to stay on the islands for a further year or two.

[13] Legal restrictions on the formation of political parties ceased after Abdallah's death, and a number of opposition groups returned from exile.

[14] The IMF demanded a reduction of the number of civil servants. This instruction was carried out by the government as of 1989 and has been extended in the last years. France continues to be the main source of international credit, with Japan close behind.

[15] In August 1991, the Supreme Court of the Islamic Federal Republic of the Comoros found President Said Mohamed Yohar unfit to govern and guilty of serious negligence.

[16] An alliance of all the political parties of the Island of Mwali (25,000 inhabitants) demanded that central government divide civil service jobs and economic benefits, more fairly among the islands.

Mayotte

The Eighth Conference of Non Aligned Countries, held in Harare (Zimbabwe), issued the following statement on France's failure to honor its commitment to respect Comoran territorial integrity: "The Heads of State or Government of the Non Aligned Countries reaffirmed the fact that the Comoran island of Mayotte, still under French occupation, is an integral part of the sovereign territory of the Federal Islamic Republic of the Comoros. They regret that the Government of France, despite repeated promises, has not, to date, adopted any measure or initiative that could lead to an acceptable solution of the problem of the Comoran Island of Mayotte. In October 1991, the UN General Assembly reaffirmed the sovereignty of the Comoros over Mayotte by an overwhelming majority.

ENVIRONMENT

A mountainous island with tropical climate and heavy rainfall throughout the year. The island is of volcanic origin, has dense vegetation and is located at the entrance to the Mozambique Channel.

SOCIETY

Peoples: Native Mayotte inhabitants are of the same origin as Comorans. There is an influential community of French origin that controls the island's commerce. **Religions:** 98% Moslem. **Languages:** Arabic, Swahili, Comoran, and French (official). **Political Parties:** the Mayotte Federation for the Organization of the Republic (RPR); the Mahorais People's Movement (MPM); the Party for the Democratic Organization of the Mahorais (PRDM), which seeks unification with the Comoros; the Union for French Democracy (UDF); the Social Democratic Center (CDS).

THE STATE

Capital: Dzaoudzi, 5,865 inhab. (1985).
Other cities: Mamoutzou, 12,026 inhab (1985).
Government: Jean-Jaques Debacq, prefect, appointed by the French government.

ECONOMY

Currency: CFA francs.

[17] After a failed attempt to oust the president, in late 1991, the political parties and Yohar signed a national reconciliation pact. In January a new government was formed, headed by Mohammed Taki, leader of the National Union for Democracy in Comoros. At the same time, a National Conference drafted a new constitution, which was approved by referendum, in June 1992.

[18] In July, President Yohar dismissed Taki, after accusing him of appointing a former French mercenary to his cabinet.

[19] In September, there was a failed coup d'état against Yohar, led by two of former president Ahmed Abdallah's sons. Government forces put down the coup, and arrested Abdallah's sons. A series of arrests of opposition leaders accused of conspiring against the government began, and at least six civilians were killed.

[20] In November, the country's first legislative elections were held. A number of irregularities meant that voting was not completed until December.

[21] The opposition obtained 25 seats in the Assembly, while the official party won 17. In June 1993, Ahmed Ben Cheikh Attoumane became prime minister.

[22] In January 1994, the main opposition parties requested Yohar's resignation and denounced the "brutality" of democratic transition. In early 1995, the prime minister was forced to leave office due to mounting tensions which took place during the "Activities for Democracy and Renewal". President Yohar asked Finance minister Mohamed Caabi el-Yachroutu to reshuffle the government. This was to be the 14th. administration in five years.

[23] Denard attempted a new coup d'etat in September but was arrested along with his nearly 1,000 followers by French forces who arrived from neighbouring Mayotte. Since October, when 80 year old Yohar travelled to Reunion to receive medical treatment, Yachroutu declared himself "interim" president, but refused to return the post once Yohar announced his comeback.

[24] Yohar recovered the presidency in January 1996 and between March 6 and 17, elections were held to be won by Mohamed Taki Abdoulkarim, from the National Union for Democracy in Comoros.

Congo

Population: 2,577,000
(1994)
Area: 342,000 SQ.KM
Capital: Brazzaville

Currency: CFA Franc
Language: French

Congo

Today's Congo was originally populated by Pygmies and Bushmen. By the 16th century it comprised the Bantu states of Luango and Kacongo, for many years ruled by the Manicongo (see p.200: The Bantu States of the Congo). These nations managed to stave off early Portuguese attempts at colonization. Instead, for three centuries, under the Batekes of Anzico, they acted as suppliers and intermediaries of British and French slave dealers. Towards the end of the 19th century, trade in rubber and palm oil replaced slave traffic. The new trade brought with it French colonization.

[2] In 1880 French troops led by Savorgnan de Brazza began to colonize the Congo by force, and by the 1920s two-thirds of the native population had been killed. This genocide, along with the use of forced labor to build the Brazzaville-Pointe Noire railroad, resulted in the early emergence of semi-religious independence movements, under the leadership of Matswa. After World War II, labor and student movements inspired by socialist ideas formed the backbone of resistance.

[3] Sponsored by the French, Friar Fulbert Youlou led the Democratic Union for the Defence of African Interests and was the first president of independent Congo in 1960. Mass participation grew, however, and Youlu's neo-colonialist policies were rejected. A wave of demonstrations against corruption and the banning of trade unions ended in a popular uprising during the "three glorious days" (August 13-15) of 1963.

[4] Youlou resigned and the president of the National Assembly, socialist Alphonse Massemba-Debat, took office and forced the French troops stationed in the country to withdraw. He then founded the National Movement of the Revolution (MNR) as the country's sole party.

[5] This force could not coexist with the neo-colonial army, equipped and trained by the French, and the consequent crisis led Massemba-Debat to resign on January 1, 1969. He was replaced by a young major, Marien N'Gouabi, backed by leftwing army officers. N'Gouabi founded a Marxist-Leninist party called the Congolese Worker's Party (PCT). In 1973, a new constitution was approved, proclaiming the Congo a People's Republic, and the state took control of energy, water, major industrial firms, and petroleum.

[6] In December 1975, N'Gouabi made a public self-appraisal, issued a call "to extend the revolution," and launched a general review of party structures, the state apparatus and mass organizations.

[7] To afford greater public participation in the running of the state, an ambitious educational reform was begun and the structure of the government was reformed. Until then all the school text-books had come from France and were therefore Euro-centric.

[8] When Angola gained independence in 1975, Congo lost no time recognizing Agostinho Neto's government. His clear position on the problem of Cabinda was decisive in frustrating attempts to have this oil-rich province break away from Angola, a scheme which was being promoted by transnational companies with interests in the Congo.

[9] On March 18, 1977, N'Gouabi was assassinated by followers of ex-President Massemba-Debat. The conspirators failed in their efforts to seize power, however, and Massemba-Debat was executed.

[10] The new president, Colonel Joachim Yombi Opango, did not maintain N'Gouabi's austere style and was forced to resign on February 6, 1979, charged with corruption and abuse of power. Denis Sassou N'Guesso replaced him.

[11] N'Guesso launched a campaign against corruption in the public sector, as well as a broad administrative and ministerial reform.

[12] Toward the end of 1981, the Congo faced foreign trade difficulties directly linked to inefficiency in state enterprises which the government refused to close. In 1982, the Congo's finances improved. Part of the recovery can be attributed to oil exports, which financed 49.44 per cent of the budget.

[13] Congolese oil is exploited in partnership with French, US and Italian firms. Annual exports approached eight million tons of crude in 1988.

[14] President N'Guesso's foreign policy has been pragmatic; he sought closer relations with Eastern Europe, while maintaining commercial ties with the United States and France. He played a leading role in negotiations between Angola, South Africa and Cuba, which culminated in the signing of a peace settlement for the region in December 1988, in Brazzaville, paving the way for Namibia's independence.

[15] The fall of the Berlin Wall and the demise of Soviet perestroika precipitated political and economic changes, and in December 1990, the country adopted a multi-party system.

[16] In July 1991, within the framework of these innovations, Andre Milongo became prime minister for a transitional period, until presidential elections could be held.

[17] That same year, the World Bank decided not to grant any new credits to the country, until it met interest payments on loans already received.

[18] In January 1992, an attempt by Prime Minister Milongo to break the control held by a small circle of military officers close to President N'Guesso, triggered a military rebellion. After weeks of uncertainty, N'Guesso took over as command-

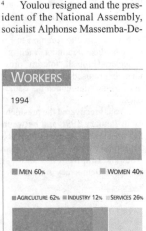

WORKERS

1994

■ MEN 60%   ■ WOMEN 40%

■ AGRICULTURE 62% ■ INDUSTRY 12% ■ SERVICES 26%

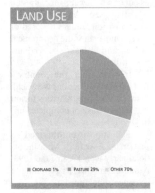

LAND USE

■ CROPLAND 1%   ■ PASTURE 29%   ■ OTHER 70%

PROFILE

ENVIRONMENT

The country comprises four distinct regions: coastal plains; a central plateau (separated from the coast by a range of mountains rising to 800 meters); the Congo River basin in the northeast; and a large area of marshland. The central region is covered with dense rain forests and is sparsely populated. Two-thirds of the population live in the south, along the Brazzaville-Pointe Noire railroad. Lumber and agriculture employ more than one third of the population. The country has considerable mineral resources including oil, lead, gold, zinc, copper and diamonds. The environmental situation is characterized by haphazard urban development, the accumulation of refuse, a lack of sewage facilities, the proliferation of contagious diseases, pollution, deforestation and the disappearance of native fauna.

SOCIETY

Peoples: The Congolese are of Bantu origin; the Bakongo prevail in the South, the Teke (or Bateke) at the center; the Sanga and Vilil in the North. **Religions:** Catholic 54%; Protestant 25%; African Christian 14%; Traditional beliefs 5%. **Languages:** French (official), Kongo and various local dialects. **Political Parties:** Pan-African Union for Social Democracy; Congolese Workers' Party (PCT); Congolese Movement for Democracy and Integral Development. **Social Organizations:** The four existing labor organizations merged in 1964 to form the Congolese Labor Confederation (CSC).

THE STATE

Official Name: République du Congo. **Administrative divisions:** 8 regions and 6 communes.
Capital: Brazzaville, 937,600 inhab. (1992). **Other cities:** Pointe Noire, 576,200 inhab.; Loubomo, 83,600 inhab. (1992). **Government:** Parliamentary republic. Pascal Lissouba, president since August 1992. Joachim Yombi Opango, prime minister since June 1993. **National Holiday:** August 15, Independence Day (1960). **Armed Forces:** 10,000.
Paramilitaries: 6,100: Gendarmerie: 1,400; People's Militia: 4,700.

STATISTICS

DEMOGRAPHY

Urban: 57% (1995). **Annual growth:** 1.6% (1992-2000). **Estimate for year 2000:** 3,000,000. **Children per woman:** 6.6 (1992).

HEALTH

One **physician** for every 3,571 inhab. (1988-91). **Under-five mortality:** 109 per 1,000 (1995). **Calorie consumption:** 98% of required intake (1995). **Safe water:** 38% of the population has access (1990-95).

EDUCATION

Illiteracy: 25% (1995). **Primary school teachers:** one for every 66 students (1992).

COMMUNICATIONS

87 **newspapers** (1995), 83 **TV sets** (1995) and 86 **radio sets** per 1,000 households (1995). 0.8 **telephones** per 100 inhabitants (1993). **Books:** 82 new titles per 1,000,000 inhabitants in 1995.

ECONOMY

Per capita GNP: $620 (1994). **Annual growth:** -2.90% (1985-94). **Annual inflation:** -0.30% (1984-94). **Consumer price index:** 100 in 1990; 141.1 in 1994. **Currency:** 535 CFA francs = 1$ (1994). **Cereal imports:** 148,000 metric tons (1993). **Fertilizer use:** 118 kgs per ha. (1992-93). **External debt:** $5,275 million (1994), $2,047 per capita (1994). **Debt service:** 51.5% of exports (1994). **Development aid received:** $129 million (1993); $53 per capita; 5% of GNP.

ENERGY

Consumption: 147 kgs of Oil Equivalent per capita yearly (1994), -2,492% imported (1994).

---

er-in-chief of the armed forces, effectively crushing the coup attempt.

[19] In March, a plebiscite was held on the text of a proposed constitution. In the August presidential elections, Pascal Lissouba - representing the Pan-African Union for Social Democracy (UPADS) - was elected president in the second ballot, with 61 per cent of the vote. Bernard Kolelas, of the Congolese Movement for Democracy and Integral Development (MCDDI) followed, with 37 per cent of the vote. Both Sassou N'Guesso, of the PCT, and Milongo, leader of the Forces for Change, were eliminated in the first ballot.

[20] In the National Assembly, however, the absolute majority went to a coalition of the PCT, the MCDDI, and another five parties. The Assembly passed a vote of no confidence in Prime Minister Stephane Maurice Bongo - Nuarra, who had been nominated by Lissouba. The prime minister was forced to resign and in mid-November,

Lissouba dissolved the Assembly.

[21] This decision caused fighting in the streets and the formation of a transition government with the participation of the armed forces. In the legislative elections, brought forward to May 1993, the ruling party won 62 seats to the opposition coalition's 49. The opposition accused the government of fraud and further clashes between demonstrators and troops left six people dead.

[22] Lissouba appointed Jacques Yhombi-Opango, a retired army officer, as prime minister, which caused the opposition to form a shadow government led by Bernard Kolelas. In July and December, new disturbances caused the death of some 80 people. Meanwhile, the government continued to be bankrupt and only paid civil servants seven of their 12 salary cheques for 1993.

[23] In January 1994, the army called on the artillery to counteract attacks from armed opposition groups. The disturbances caused the death of more than 100 people.

An agreement between the opposition and the government in mid-March marked the beginning of a cease-fire. Later, in July, the election of Kolelas as mayor of Brazzaville calmed the situation even further, allowing for a public national reconciliation ceremony the following month.

[24] Meanwhile, Lissouba accepted the IMF conditions for reestablishing aid for the economic structural adjustment, which included reducing the number of civil servants. On another front, the competition between several oil multinationals forced the French company Elf to increase the amount of profits reinvested in the nation from 17 per cent to 31 per cent in order to maintain its dominant position.

[25] 1995 was dominated by the quest for a definitive agreement to disarm the anti-government urban militias, and their eventual integra-

tion into the official army. On February 19, a general strike was started calling for the payment of outstanding salaries. On March 1, an agreement was made, but the civil servants rejected it, as it proposed a reduction in salaries in return for a shorter working week. The tension was largely due to the government desire to apply the IMF imposed structural adjustment programme, which had already reduced the number of State employees from 80,000 to 50,000.

[26] Despite the reductions in civil service salaries and the intensive exploitation of petrol reserves, the government continued in bankruptcy and the country was still one of the poorest and most heavily indebted in the world.

The Bantu States of the Congo

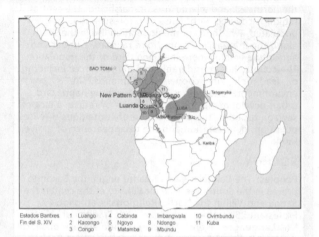

Estados Bantxes
Fin del S. XIV

1 Luango   4 Cabinda   7 Imbangwala   10 Ovimbundu
2 Kacongo   5 Ngoyo   8 Ndongo   11 Kuba
3 Congo   6 Matamba   9 Mbundu

The Congo River valley (which includes present-day Zaire, Angola and Congo as well as parts of Gabon and Zambia) was populated by pygmies who lived in the heart of the jungle region and by San people who inhabited the savannas, where their characteristic rock paintings have been found. At the beginning of the Christian era both groups were displaced by Bantu migrants. Some were driven further and further into the jungle, while others were pushed southward.

[2] When the Bantu reached the shores of Lake Kariba, which now marks the border between Zambia and Zimbabwe, they rapidly developed a trade economy - probably using their links with the trade centers on the east coast and with Zimbabwe. Hindu and Chinese objects from the 8th and 10th centuries that have been found in the area are an indication of the extent of such trade.

[3] Further northwest, in Katanga, the region around the current borders of Zaire, Angola and Zambia, the Luba (or Baluba) ethnic group engaged in trading, and also developed copper mining. Studies show that between the 10th and 15th centuries, at least 100,000 tons of copper were extracted. This Luba proto-state extended its influence over neighboring Bantu groups - the Lunda of Kasai and the Kimbundo of Cuanza - and penetrated westward and northward as far as the area around the mouth of the Congo River.

[4] By the end of the 14th century the region was home to a series of states with different degrees of integration. Luango (along the coast of present-day Gabon and Congo), Kacongo, Cabinda, Ngoyo (further inland), Matamba and Imbangwala in the Kuango River valley (Angola, Zaire), Ndongo and Mbundu in the Cuanza Valley (Angola), Ovimbundu (between the Cuanza River and the coast) and Kuba, on the Kasai, formed a constellation of satellites around the major state of the region, Congo.

[5] The Manicongo - master of the Congo - ruled this territory. The neighboring nations paid him tribute, and he exercised a kind of protectorate over them, arbitrating their disputes, regulating trade and eventually demanding homage and, if necessary, military assistance. Apparently, his power was not based on force but rather was gradually recognized by formerly autonomous units which needed someone to settle their disputes. Thus, the influence of the Manicongo reached beyond his "protectorate" as other nations began to respect his decisions.

[6] In 1482, when Portuguese explorers from Diogo Cao reached the region, they visited Manicongo Nzanga Nkuwu in his capital, Mbanza (present-day M'Banza Congo in Angola) and established friendly relations. Trade and support in the form of Portuguese weapons consolidated the already existing power structure and contributed to a substantial intensification of slavery-the practice was common locally but slaves had not previously been sold internationally. Moreover, cultural and religious conflict emerged when one of the successive leaders converted to Christianity and changed his name to John I. His successors adopted European customs and organized their "court" along the Lisbon model; the son of one was named a bishop and was consecrated in Rome by the Pope.

[7] In 1512, Portugal determined that the Portuguese king and the "king" of the Congo would have the monopoly of the slave trade, which was exporting nearly 5,000 people a year. Traders in São Tomé, who had been acting as middlemen, resented the move and began to encourage anti-European sentiment. They hoped to destabilize the Manicongo's rule and take advantage of the resulting crisis. Another source of conflict was the meddling of Jesuit missionaries in the affairs of native groups. This reinforced anti-Christian sentiments. Moreover, Congolese "King" Alfonse I defied the virtual religious monopoly of the Portuguese when he demanded direct access to the Pope.

[8] In 1544 his successor, Diego I, openly broke with the Portuguese, reaffirmed his native name, Nzanga Mbemba, and launched a "re-Africanization" campaign with broad popular support. During this period, Portuguese business shrank to semi-clandestine transactions between São Tomé and coastal cities not under the Manicongo's control. In 1561, Portugal's attempts to impose Nzanga Mbemba's successor led to a massacre of Portuguese residents. It was only in 1568, with Alvaro I, that the European power managed to recover its influence.

[9] By then, northern nations were trying to push southward. While the Congolese state successfully repelled the invaders, it was left weakened. Political and economic power swiftly passed to the neighboring Ndongo nation (the name Angola, would later be derived from the Ndongo ruler who was addressed as "Ngola").

[10] As a result, Portuguese attention shifted increasingly to Ndongo and Luanda, the major slave shipping port. In 1575, the Crown proposed to create a colony with settlers in Luanda, as it had in Brazil. However, there was fierce resistance, and the Portuguese suffered severe defeats, one of the best known occurring in Ngoleme in 1590. European maneuvers finally managed to oust Nzinga, the woman who ruled Ndongo and the leading organizer of the resistance, replacing her with a figurehead. Nzinga failed to relent, however; she seized power in a neighboring nation and used it as a base for her battle against the invaders. A century of almost constant warfare, with battles ranging from Matamba to the Congo and back again to Ndongo, frustrated Portuguese plans to set up a colony and resume coastal trade.

[11] One of the consequences of this turbulent period along the coast was a shift in regional power. Luango assumed the leading role, in the Gabon-Congo region, now linked to the British and French slave trade. Further south, the center of power moved inland. The Balunda, already possessing firearms by the 18th century, extended their rule over the Katanga plateau and the surrounding areas (southern Zaire, eastern Angola, northern Zambia). The Mwata Yambo (traditional leaders) of theLunda gained influence over such a vast territory (from Bié in central Angola to Lake Tanganyika) that they had to rule through delegate local leaders (Mwata Kazembes). The decline of the slave trade in the 19th century undermined the power of the Mwata Yambo and the Kazembe became increasingly independent. Toward the end of the century, encroaching Porutuguese colonialism irrevocably changed this state of affairs.

Costa Rica

Population: 3,304,000 (1994)
Area: 51,100 SQ KM
Capital: San José

Currency: Colón
Language: Spanish

Costa Rica

U nlike the rest of Central America, where the Spanish conquistadores found advanced cultures upon their arrival, the territory corresponding to what is now Costa Rica was less densely populated with natives belonging to the Chibcha family (Muisca: see Colombia). Columbus landed on the shores of Costa Rica on his last voyage in 1502.

[2] Gaspar de Espinosa, Hernán Ponce de León and Juan de Castañeda explored its coast between 1514 and 1516. Between 1560 and 1564, the territory continued to be explored by Juan de Cavallón, Father Juan de Estrada Rabago and Juan Vázquez de Coronado.

[3] The conquest of the territory was concluded in the second half of the 16th century. The colonists remained isolated from Spain for a long period of time, and were unable to establish the system of *encomiendas* through which native labor was virtually enslaved in other Spanish colonies.

[4] The territory was annexed to the Captaincy-General of Guatemala. A patriarchal society of small landowners was formed, without the powerful landowning oligarchies which rose in the neighboring countries. Instead of becoming a country eternally dominated by military dictatorships, modern Costa Rica became the "Switzerland of America", a country with no army and with more teachers than policemen.

[5] In 1821, Costa Rica joined the newly-formed independent United Provinces of Central America and remained a part of this federation until its dissolution in 1840. A persistent opponent to the Balkanization induced by British imperialism, its territory was used as a base of operations by Francisco Morazán - an advocate of Central American unity - until 1848, when Costa Rica became an independent state.

[6] Only two occasions of brief unrest occurred to disturb the peaceful Costa Rican tradition: in the mid-19th century American adventurer William Walker, who had previously taken possession of Nicaragua, tried to extend his domination southwards and was defeated during the administration of president Juan Rafael Mora. One hundred years later, in 1948, an electoral fraud led to civil war finally won by the opposition forces led by José Figueres.

[7] Between 1940 and 1948 the government was backed by coffee plantation owners and bankers. However, in 1948, opposition leader Otilio Ulate, nominated by the National Unity Party, won a presidential election which was annulled by Congress. This unleashed civil war which ended with the appointment of a junta, presided over by José Figueres. The government thus formed issued a call for the election of new representatives, who in turn confirmed Ulate's victory. A year later, a new constitution was ratified, establishing a presidential system of government and prohibiting the formation of armies.

[8] The populist revolution led by Figueres spread anti-dictatorial ideas throughout Central America.

[9] José Figueres was elected president in 1954 and during his administration Costa Rica became a strongly anti-communist welfare state. In 1958, the conservatives defeated Figueres and imposed an import-substitution development model.

[10] Henceforth, the traditional antagonism between liberals and conservatives gave way to new tensions between the National Liberation Party (LN), led by Figueres, and a heterogeneous group consisting of various small parties. In 1966, the opposition managed to form an electoral coalition, the United National Opposition that elected José Joaquín Trejos to the presidency.

[11] After the 1970 election, the LN returned to power with Figueres, who remained in office until 1974, when Daniel Odúber Quirós, co-founder of the party in 1950, was elected president.

[12] Odúber tried to restore unity to the Central American Common Market, left in a critical situation after the 1969 war between El Salvador and Honduras. However, his obvious stance in favor of democracy did not meet with the approval of Somoza's regime in Nicaragua. Costa Rica was constantly harassed by its neighbor and became the nearest safe place for thousands of political refugees.

[13] On the domestic scene, favorable conditions in 1975 increased wages, when transnational oil companies were nationalized and coffee prices rose on the world market

[14] In 1978, contrary to all expectations, presidential elections gave the victory to a conservative coalition, who had been critical of the previous government's administration. The leftist influence, grouped in the United People's Coalition, increased considerably and became the third most important political force; however, they remain isolated and unable to propose social changes within the democratic and pluralist system favored by the majority.

[15] The new president, Rodrigo Carazo Odio, imposed an unpopular economic policy prescribed by the IMF, which resulted in growing confrontation with labor and leftist forces. However, in 1979, encouraged by popular sympathy towards the Sandinista rebels, and under threats of invasion by neighboring dictator Anastasio Somoza, the Costa Rican government decided to actively support the Nicaraguan Sandinistas.

[16] A radically different attitude was taken in 1980 with regard to El Salvador's insurgency. In spite of grave and continuous human rights violations in that country, the San José government supported the Salvadoran military junta. In 1981,

PUBLIC EXPENDITURES

% of GNP — 1992

MILITARY 0.9 — EDUCATION 4.6 — HEALTH 3

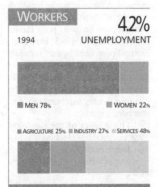

WORKERS — 1994

4.2% UNEMPLOYMENT

MEN 78% — WOMEN 22%

AGRICULTURE 25% — INDUSTRY 27% — SERVICES 48%

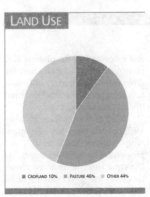

LAND USE

CROPLAND 10% — PASTURE 46% — OTHER 44%

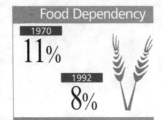

IMPORTS
3,025
EXPORTS
2,215

11%
1992
8%

PER CAPITA
$1,163

5%

PROFILE

ENVIRONMENT

A mountain range with major volcanic peaks stretches across the country from northwest to southeast. Costa Rica has the highest rural population density in Latin America, with small and medium sized farmers who use modern agricultural techniques. Coffee is the main export crop. The lowlands along the Pacific and the Caribbean have different climate and vegetation. Along the Caribbean coast there is dense, tropical rainforest vegetation. Cocoa is grown in that region. The Pacific side is drier; extensive cattle raising is practised along with artificially irrigated sugar cane and rice plantations. Deforestation has been partially responsible for soil erosion and reduced fertility.

SOCIETY

Peoples: Costa Ricans, usually called 'ticos' in Central America, are the result of ethnic integration between native American inhabitants and sizeable European migration, mainly Spanish. African descendants, brought in from Jamaica, make up 3% of the population and are concentrated along the eastern coast. Indigenous peoples, 1%. **Religions:** 80 % of the population are Catholic; Evangelical Protestant 15% (1992). **Languages:** Spanish is the official language, spoken by the majority. Mekaiteliu, a language derived from English, is spoken in the province of Limón (east coast). Several indigenous languages.
Political Parties: the National Liberation Party, social democrat; the Democratic Force, centre left coalition; the Social Christian Unity Party (PUSC); the "Cargagines" Agricultural Union, regional: the Agrarian Party. **Social Organizations:** The Unitary Workers' Union (CUT); the 50,000-member CUT groups both workers and peasants from the National Federation of Civil Servants, the National Peasant Federation; the Federation of Industrial Workers, regional federations and smaller unions. 375 cooperatives.

THE STATE

Official Name: República de Costa Rica. **Administrative Divisions:** 7 Provinces. **Capital:** San José, 280,600 inhab. (1992). **Other cities:** Limón, 51,000 inhab.; Alajuela, 45,400 inhab. and Puntarenas, 38,300 inhab. (1992). **Government:** José María Figueres, President since May 8, 1994; member of the National Liberation Party (PLN), of social democratic orientation. Legislature, single-chamber Assembly, made up of 57 members. **National Holiday:** September 15, Independence Day (1821). **Armed Forces:** Abolished in 1949. **Paramilitaries:** 7,500 Civil Guard, Frontier Guard and Rural Guard (1994).

STATISTICS

DEMOGRAPHY

Urban: 49% (1995). **Annual growth:** 1.1% (1992-2000). **Estimate for year 2000:** 4,000,000. **Children per woman:** 3.1 (1992).

HEALTH

One **physician** for every 1,136 inhab. (1988-91). **Under-five mortality:** 16 per 1,000 (1995). **Calorie consumption:** 103% of required intake (1995).

EDUCATION

Illiteracy: 5% (1995). **School enrolment:** Primary (1993): 105% fem., 105% male. Secondary (1993): 49% fem., 45% male. University: 30% (1993). **Primary school teachers:** one for every 32 students (1992).

COMMUNICATIONS

102 **newspapers** (1995), 102 **TV sets** (1995) and 101 **radio sets** per 1,000 households (1995). 11.1 **telephones** per 100 inhabitants (1993). **Books:** 98 new titles per 1,000,000 inhabitants in 1995.

ECONOMY

Per capita GNP: $2,400 (1994). **Annual growth:** 2.80% (1985-94). **Annual inflation:** 18.20% (1984-94). **Consumer price index:** 100 in 1990; 195.4 in 1994. **Currency:** 165 colones = 1$ (1994). **Cereal imports:** 535,000 metric tons (1993). **Fertilizer use:** 2,354 kgs per ha. (1992-93). **Imports:** $3,025 million (1994). **Exports:** $2,215 million (1994). **External debt:** $3,843 million (1994), $1,163 per capita (1994). **Debt service:** 15% of exports (1994). **Development aid received:** $99 million (1993); $30 per capita; 1% of GNP.

ENERGY

Consumption: 558 kgs of Oil Equivalent per capita yearly (1994), 41% imported (1994).

Rodrigo Carazo broke diplomatic relations with Cuba.
[17] Furthermore, in January 1982, the US-backed Central American Democratic Community was set up in San José with the main objective of isolating revolutionary Nicaragua.
[18] In February 1982, Luis Alberto Monge, a right-wing LN candidate, became president with 57.8 per cent of the votes. The following month, Monge visited Israel, and ignoring UN Resolution 478, raised the Costa Rican flag over the diplomatic headquarters in Jerusalem. The president-elect declared this an "act of sovereignty".
[19] Monge took office in May 1982, proclaimed his alignment with "western democracies", and announced austerity measures. At the same time, he fostered closer ties with the governments of El Salvador, Guatemala and Honduras, thus aggravating relations with Nicaragua.
[20] Costa Rica's hostile attitude towards its neighbor showed clearly when the US declared a commercial embargo on the revolutionary government. A series of border incidents brought relations between the two countries to a breaking point during July and August 1985. However, prompt action taken by the Contadora Group checked mounting tension in the area; both governments agreed to place neutral observers along the common frontier to arbitrate any further border clashes.
[21] The winner of the February 1986 presidential election was Social Democrat Oscar Arias, who won a tight victory with 52 per cent of the vote.
[22] Arias devoted himself to the task of designing a policy which would break both the logic of war and the escalating tension within the region. In August 1987 he presented a peace plan at a summit meeting held in Esquipulas, Guatemala, accepted and signed by the presidents of El Salvador, Nicaragua, Guatemala, and Honduras. The focal points of the plan were: a simultaneous cease-fire in Nicaragua and El Salvador, an immediate end to American aid to the Nicaraguan "contras", a democratization timetable for Nicaragua which included holding free elections and putting an end to the use of foreign territory as supply/attack bases.
[23] The signing of this peace plan, known as "Esquipulas II", earned Costa Rica a special place in international relations, and constituted a

CONTROL AND VIOLENCE:
The situation of women in Latin America

In the early 1990s, the situation of women and women's rights have become the subject of public discussion and concern. Sexual violence, sexual abuse and reproductive rights have become important issues.

2 In Costa Rica, domestic violence reached the headlines when two young women were killed by their partners in early 1994. In 1993, an office within the Ministry of Justice registered 5,500 appeals for help from women, while the Office for the Defence of Women dealt with 2,000 such cases.

3 In Brazil, according to information provided by the 119 Offices of Women's Defence in the State of São Paulo, the number of cases of sexual abuse against women in the first semester of 1994 was 30 per cent higher than the corresponding figure for 1993. Around 35 per cent of all cases of sexual abuse were perpetrated by the victim's father or another male relative.

4 Despite the fact that preventive and/or punitive measures are beginning to be taken, the socio-economic situation characterized by high unemployment, underemployment and job instability throughout the continent, does not make it easy to address this situation.

5 Violence against women is predominantly an expression of male dominance depriving the woman of the most basic rights, robbing her of her self-esteem and leaving her feeling shamed and guilty. In the majority of cases women fail to report the crime.

6 Women's rights are also an issue in the public domain. Women continue to hold only a small minority of political positions. Therefore, in the political arena, where decisions are made on issues that are intimately linked with women's prospects for a better life, such as the right to prevent or interrupt an unwanted pregnancy, are decided by men.

7 In April, during the the Third Preparatory Commission for the UN International Conference on Population and Development, to be held in Cairo in September 1994, abortion was one of the most controversial issues. The governments of the United States, India, China and most of Western Europe defended women's access to institutions for safe and sanitary abortion. However, the vast majority of Latin American governments were opposed to any interruption of pregnancy.

8 An estimated 500,000 illegal abortions are performed in Argentina each year, and the figure for Uruguay is estimated at 250,000.

9 The issue of sterilization without knowledge or consent is another hotly disputed topic in Latin America. In 1994, NGOs in Mexico denounced the government promotion of fallopian tube-tying through a mass disinformation campaign. Aimed at reducing the country's population growth from the current 1.8 per cent, to 1 per cent per year, the measure was primarily aimed at indigenous women.

10 In early August, the US Agency for International Development (AID) denied involvement in a massive sterilization campaign being carried out by the Honduran government, through the use of a "mini-pill" allegedly supplied by AID. In addition to having an abortive effect, this pill has been singled out as being responsible for sterilization in nursing mothers, as well as problems in sexual hormone development in children.

personal triumph for President Arias, who received the Nobel Peace Prize in recognition of his efforts, in October 1987.

24 During his term, Arias instituted the first phase, and later the second, of a structural adjustment program, with World Bank support. The objective of this program was the transformation of the productive sector, through technological modernization, increased efficiency and greater productivity. Neoliberal formulas were applied, following the counsel of international financial organizations. However, according to the labor unions, no planning for these "prescriptions" took the social effects and their repercussions into account.

25 In July 1989, an investigation into drug trafficking was carried out by a parliamentary commission. The report produced by the commission found both of the main political parties, the PNL and the PUSC, guilty of receiving drug money during the 1986 electoral campaign. At the same time, another scandal broke out over the financing of electoral campaigns, with accusations that both parties - and Oscar Arias was singled out, by name - had received money from Panamanian General Noriega in 1986.

26 Between March and July, 1988, the structural adjustment policy of the Arias government led to public discontent and demonstrations by many sectors of the population: public employees were against the concessions made to the IMF and the World Bank; the peasant unions, gathered in the UNSA, staged a one-week strike to protest against the agrarian policy, and farmers went on strike several times.

27 In 1989 this situation worsened, and in August a coalition of regional trade union federations, professional guilds, and citizens' groups called for a strike in the province of Limón. The strike paralyzed sea traffic on the Caribbean coast for four days. Some sectors on the Atlantic coast also joined the protests. In September, teachers staged a nationwide strike.

28 In 1990, 29.9 per cent of the economically active population were women. However, the percentage on the informal job market was 41 per cent. In the political arena - both the Executive and Legislative branches - women hold only 15 per cent of posts. Women's organizations have observed an increase in teenage prostitution through prostitution rings, which have become multi-million-dollar enterprises for their owners.

29 Under the slogan of "change", and focusing specifically on low-income groups with lower educational levels, the social Christian candidate Rafael Angel Calderón won the election held in February 1990, by a wide margin; in addition, he obtained an absolute majority in the legislature, with one seat to spare.

30 The application of a severe economic adjustment program led to a reduction in the state apparatus as well as in the fiscal deficit, which had reached 3.3 per cent of the GDP. As a result of these cuts, there was a rise in unemployment and an increase in popular discontent.

31 In the 1994 elections, social-democrat candidate Jose Maria Figueres defeated government candidate Miguel Rodriguez by a slight margin, after a campaign with few differences in the political platforms but with notably harsh speeches from the rival parties.

32 In January 1995, a free trade agreement was signed with Mexico. However, the deterioration of economy, inflation and a fiscal deficit led the government to increase taxes in order to balance the budget. The World Bank rejected the economic plan and refused to finance the structural adjustment schedule.

33 In April, the PLN was forced to accept the banking system's liberalization and the privatization of insurance, hydrocarbon and telecommunications sectors proposed by the Christian opposition in exchange for the approval of its fiscal adjustment. Labour unions set up strikes against these measures, especially against the discharge of thousands of civil servants, but privatizations were carried out.

Côte d´Ivoire

Population: 13,841,000 (1994)
Area: 322,460 SQ KM
Capital: Abidjan

Currency: CFA Franc
Language: French

Côte d´Ivoire

According to Baulé tradition, in 1730 Queen Aura Poka (sister of a defeated pretender to the Ashanti throne) emigrated westwards with her people. They founded a new state in the center of a territory known to Europeans since the 15th century as the Ivory Coast (or Côte d'Ivoire) because of the active trade in elephant tusks. This Ashanti state soon grew and became a threat to the small states of Aigini, on the coast, and Atokpora, inland. In 1843, these states requested French protection, thus enabling France to obtain exclusive rights over the coastal trade centers.

[2] 50 years later, French troops moved up the coast aiming to join Côte d'Ivoire to Guinea, Mali and Senegal. However, they met with stiff resistance from Samori Touré, a Fulani leader who had risen through the social ranks because of outstanding leadership ability. He had set up a state in the heart of the region which the Europeans were planning to unify under a central colonial administration (See Guinea).

[3] After three decades of bloody fighting (1870-1898), Touré was defeated and the leaders of the dominant Fulani groups were forced to sign colonial agreements with the French.

[4] In 1895 the area became known as French West Africa, comprising Senegal, French Sudan (now Mali), Guinea and Côte d'Ivoire. Later, what are now Chad, Burkina Faso and Mauritania were also annexed. The French hoped to balance the poorer regions of Chad and Burkina Faso with the better off Senegal and Côte d'Ivoire. This centralized administration failed after independence.

[5] Several attempts at economic integration were made, with different degrees of success, but real progress was made only recently with the creation of the Economic Community of West Africa, which includes 15 former colonies of France, Britain and Portugal.

[6] In French West Africa, modern political activities began in 1946 with the creation of the African Democratic Union (ADU), a political party with branches in Senegal, Mali and Guinea. The ADU promoted independence and unity for French colonies in the area. Felix Houphouet-Boigny, hereditary leader of the Baulé ethnic group, and also a physician and well-to-do farmer, was appointed party president, based on his experience leading a farmers' association which had fought against colonial policies.

[7] Tactically allied to the French Communist Party, the only anti-colonial French party at the time, the ADU staged strikes, demonstrations and boycotts of European businesses. The nationalist cause was savagely suppressed, leaving dozens of activists dead and thousands imprisoned, giving Boigny grounds for ending his alliance with the French communists in 1950. He soon reached a new agreement with François Mitterrand, then Minister for Overseas Territories. This move undermined the ADU's standing, and Boigny was only just able to maintain his prestige in his native land.

[8] Between 1958 and 1960 all of French West Africa became independent and the newborn states joined the UN. Aware of their meager economic prospects, the political leaders proposed to form a federation. Boigny, however, confident in the relative prosperity of his country and his privileged neo-colonial relations with France, opposed the idea. "We are not saying 'goodbye' to France, but 'see you later'", he had declared when announcing independence.

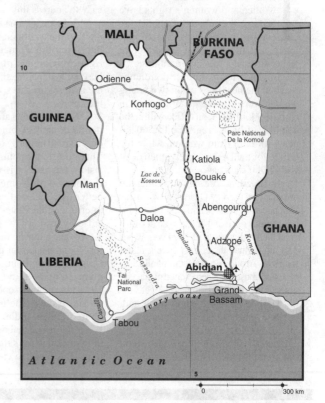

[9] As a major producer of cocoa, coffee, rubber and diamonds, Côte d'Ivoire was able to attract transnational investors. In addition to the political stability produced by Boigny's paternalistic, authoritarian rule, Côte d'Ivoire offered extremely cheap labor, mostly supplied by neighboring countries. In the ten-year period (1966-76) the economic growth rate remained between 8 and 10 per cent a year. But prosperity withered as the West went into a recession after 1979. Agricultural exports fell from $4 billion to barely $1 billion between 1980 and 1983. Half the industries set up between 1966 and 1976 closed down, pushing unemployment figures up to 45 per cent of the active population. By 1985, the foreign debt was five times that of 1981.

[10] Following International Monetary Fund recommendations, President Boigny made severe budget cuts, paralyzed nearly all public projects and slashed food subsidies. These decisions did not spark social unrest immediately, but only because of the weakness of the labor unions and the lack of fresh political leadership. The only groups which retained potential for opposition were students and liberal professionals, and they organized mass demonstrations and strikes.

[11] In May 1984, following negotiations with the Paris Club and other sources of financing, Côte d'Ivoire was allowed to restructure its foreign debt. The economy has continued to deteriorate, however, particularly as coffee and cocoa export prices have fallen steadily.

[12] In October 1985, the 8th PDCI party congress issued a resolution nominating octogenarian Houphouet-Boigny for a sixth presidential term. The appointment was confirmed that same month by 99 per cent of the electorate. In November, however, elections for the General Assembly were marked by a very low turnout of 25 per cent.

[13] The Boigny regime's grandiose building schemes stand in stark contrast to African realities. In the face of severe economic hardship, he is currently building Africa's largest cathedral, a project offensive to the 23 per cent of the population which is Moslem.

[14] Côte d'Ivoire is the world's largest producer of cocoa and, like all Third World producers of raw materials, it has suffered from falling commodity prices. From July to October 1987, the price of cocoa dropped 50 per cent on the world market.

[15] In 1990, a government austerity plan went into effect with IMF and World Bank support. These

WORKERS

1994

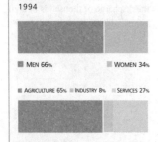

- ■ MEN 66%
- ■ WOMEN 34%
- ■ AGRICULTURE 65% INDUSTRY 8% ■ SERVICES 27%

LAND USE

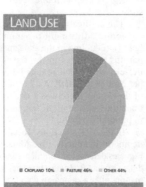

- ■ CROPLAND 10% ■ PASTURE 46% ■ OTHER 44%

Maternal Mortality 1989-95

PER 100.000
LIVE BIRTH

822

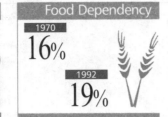

Food Dependency

1970 **16%**

1992 **19%**

External Debt 1994

PER CAPITA

$1,333

Illiteracy 1995

60%

PROFILE

ENVIRONMENT

Located on the Gulf of Guinea, the country is divided into two major natural regions: the South, with heavy rainfall and lush rain forests, where foreign investors have large plantations of cash crops like coffee, cocoa and bananas; and the North, a granite plain characterized by its savannas, where small landowners raise sorghum, corn and peanuts. Côte d'Ivoire has one of the fastest rates of deforestation in the world.

SOCIETY

Peoples: The population includes five major ethnic groups, the Kru, Akan, Volta, Mande and Malinke, some from the savannas and others from the rain forests, sub-divided into approximately 80 smaller groups. **Religions:** Nearly two thirds of the population practice traditional African religions, 23% are Moslem and 12% Christian. **Languages:** French (official). There are as many languages as ethnic groups; the most widely spoken are Diula, in the North, Baule, center and West and Bete, Southeast. **Political parties:** Côte d'Ivoire Democratic Party (PDCI), in power since independence in 1960; D'Ivoire People's Alliance (FPI), the major opposition force. **Social Organizations:** General Workers' Union.

THE STATE

Official Name: République de Côte d'Ivoire. **Administrative divisions:** 49 departments. **Capital:** Abidjan, 2,170,000 inhab. (1990). **Other cities:** Bouaké, 330,000 inhab.; Daloa 121,842 inhab.; Korhogo 109,445 inhab.; Yamoussoukro 106,786 inhab. (1988) **Government:** Henry Konan Bédié, President since December 8, 1993. Daniel Kablan Duncan, Prime Minister since December 11, 1993. Single-chamber legislature: National Assembly with 175 members, elected by direct popular vote every 5 years. **National Holiday:** August 7, Independence (1960). **Armed Forces:** 8,400. **Paramilitaries:** 7,800: Presidential Guard: 1,100; Gendarmerie: 4,400; Militia: 1,500; Military Fire Service: 800.

STATISTICS

DEMOGRAPHY

Urban: 42% (1995). **Annual growth:** 1.4% (1992-2000). **Estimate for year 2000:** 17,000,000. **Children per woman:** 6.6 (1992).

HEALTH

One **physician** for every 11,111 inhab. (1988-91). **Under-five mortality:** 150 per 1,000 (1995). **Calorie consumption:** 99% of required intake (1995). **Safe water:** 72% of the population has access (1990-95).

EDUCATION

Illiteracy: 60% (1995). **School enrolment:** Primary (1993): 58% fem., 58% male. Secondary (1993): 17% fem., 33% male. **Primary school teachers:** one for every 37 students (1992).

COMMUNICATIONS

85 newspapers (1995), 93 **TV sets** (1995) and 88 **radio sets** per 1,000 households (1995). 0.7 **telephones** per 100 inhabitants (1993). **Books:** 82 new titles per 1,000,000 inhabitants in 1995.

ECONOMY

Per capita GNP: $610 (1994). **Annual growth:** -4.60% (1985-94). **Annual inflation:** 0.20% (1984-94). **Consumer price index:** 100 in 1990; 101.6 in 1993. **Currency:** 535 CFA francs = 1$ (1994). **Cereal imports:** 590,000 metric tons (1993). **Fertilizer use:** 132 kgs per ha. (1992-93). **External debt:** $18,452 million (1994), $1,333 per capita (1994). **Debt service:** 40.1% of exports (1994). **Development aid received:** $766 million (1993); $58 per capita; 8% of GNP.

ENERGY

Consumption: 170 kgs of Oil Equivalent per capita yearly (1994), 82% imported (1994).

---

economic measures brought about decreases in worker salaries in both public and private sectors.

[16] Public reaction, particularly by students, brought about a reduction in the price of staple goods. In the meantime, the government decided not to carry out a planned 11 per cent wage cut, calling the measure a "contribution to national solidarity".

[17] In 1990, a multiparty system was established and presidential elections were held. Houphouet-Boigny was re-elected with 81 per cent of the vote. The post of prime minister was created, and Alassan Oattara was nominated. The opposition won 11 of the 175 seats in the National Assembly.

[18] In 1991, the opposition requested a national conference to begin a transition toward a multi-party system. The government's decision to reduce the number of civil servants by 25 per cent led to a series of strikes and demonstrations of dissent. In spite of these austerity measures, the government could not fully apply its three-year program to put its finances in order, partially privatize state companies and encourage the "competitiveness" of private corporations.

[19] In mid 1993, a group of military officers led several revolts, demanding higher wages and social benefits, ending in talks between Houphouet-Boigny and delegates from the Republican Guard. The president's death in December 1993 triggered a succession quarrel. He was replaced by Henri Konan-Bédié, National Assembly chairman, who managed to consolidate his power within the ruling Democratic Party, in spite of former Prime Minister Ouattara's opposition.

[20] In 1994, the government faced a strong labour union mobilization demanding compensation measures after the CFA franc's 100 per cent devaluation in January. In what was seen as a reward for having accepted the devaluation - decided by France and the IMF - half of Côte d'Ivoire's debt with the Paris Club was cancelled. However, the country continued having the world's largest per capita foreign debt.

[21] In October 1995, Bédié won a controversial presidential election, boycotted by the opposition. For the first time since the country's independence, residing foreigners were not allowed to vote, neither were those whose parents were not Ivorian. This allowed Bédié to dispose of Ouattara, whose father was from Burkina Faso. In spite of protests from the opposition, Bédié refused to modify these regulations and elections were held in a violent environment.

[22] The rise of coffee and cocoa prices in the international market significantly improved Ivory Coast's trade balance. Thus, in 1995, Abidjan's bilateral trade with France had a surplus for the first time in ten years.

Croatia

Hrvatska

Population: 4,778,000 (1994)
Area: 56,538 SQ KM
Capital: Zagreb

Currency: Dinar
Language: Serbo-Croatian

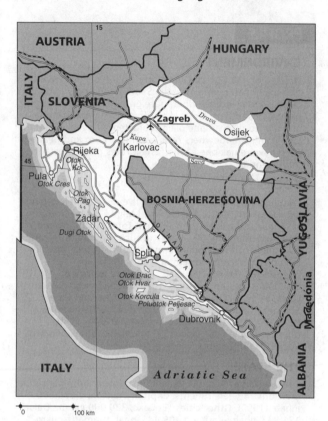

The Croats migrated in the 6th century AD, from White Croatia, a region that now belongs to Ukraine, to the lower Danube Valley. From there they continued moving toward the Adriatic, where they conquered the Roman stronghold Salona in the year 614. After settling in the former Roman provinces of Pannonia and Dalmatia, the Croats freed themselves from the Avars and began developing independently.

[2] The Croatians were a farming people, and maintained their old way of life, uniting under their tribal chieftains. In the 7th century, the Croats were converted to Roman Catholicism; a bishopric was established at Nin for this region, and the Croats obtained the right to use their own language in their religious services.

[3] In the 8th century, Croatian tribes organized themselves into larger units, made up of two duchies, Pannonia and Dalmatia. By the Peace of 812 between the Franks and the Byzantines, the former became a part of the Frankish Empire, and the latter, of Dalmatia. Both duchies freed themselves in the mid-9th century, uniting to form the first independent Croatian kingdom toward the end of the century.

[4] Tomislav and his heirs had to defend themselves against both the Bulgarian Empire in Pannonia, and Venetian expansion along the Dalmatian coast. The Byzantine Empire helped Stjepan Drzislav (969-97) defend himself against Venice but in exchange, the Byzantines were allowed to reestablish its influence in the Adriatic. Peter Kresimir (1058-74) broke off relations with Byzantium, and strengthened his links with the papacy. During this period, Croatia reached the peak of its power, as well as expanding its territory.

[5] However, during Kresimir's government, the country split in two, with a Latin group favoring the king, and an opposition national group, backed up by broad popular support. When the Pope invited Dimitrije Zvonimir to involve the kingdom in a war against the Seljuq Turks, the opposition accused him of being the pope's vassal and assassinated him in 1089. The civil war which was then unleashed marked the beginning of the decline of the Croatian kingdom.

[6] The Byzantines recovered Dalmatia, and in the meantime, Lazlo I of Hungary conquered Pannonia in 1091, laying claim to the Croatian crown. Lazlo founded a bishopric at Zagreb in 1094, which became the center of the Church's power in the region. Petar Svacic was crowned king by the Dalmatians, but the Pope considered him a rebel, and turned to King Kalman of Hungary to unseat him. Kalman invaded the country; Svacic - the last king of Croat blood - fell in 1097.

[7] After an extended war Kalman signed a treaty, called the Pacta Conventa, with the Croat representatives. Only Bosnia, then a part of the Croatian kingdom, refused to submit to a foreign monarch. For the next eight centuries, Croatia was linked to Hungary. In the 14th century, Dalmatia became a part of Venice, which ruled over it for 400 years.

[8] After the defeat of the Croatian and Hungarian forces in the battles of Krbavsko Polje (1493) and Mohacs (1526), most of Pannonia and Hungary fell into Turkish hands.

[9] Turkish domination altered the ethnic composition of Pannonia, as many Croats migrated northwards, some even going into Austria. In the meantime, the Turks brought in German and Hungarian settlers, and gave incentives for Serbians fleeing the Balkans to settle in the Vojna Krajina.

[10] When the Turks were driven back, in the 17th century, Austria tried to limit Croatia and Hungary's state rights, to make them mere provinces of the empire. The Croatian and Hungarian nobility conspired together to organize an independence movement, which failed. The Croatian leaders were executed and their lands were distributed among foreign nobles.

[11] After the annexation of Rijeka (Fiume) in the 1770s, Hungary tried to impose its language, but this triggered a nationalistic reaction among the Croatians. In the meantime, the French Revolution and the Napoleonic Wars which had incorporated Dalmatia, Pannonia and the area south of the Sava river to the French Empire, further stimulated Croatian nationalism. Upon the fall of Napoleon, relations between Hungary and Croatia rapidly deteriorated.

[12] In April 1848, the Hungarian Parliament adopted a series of measures severely limiting Croatian autonomy. The Croatian Diet declared its separation from Hungary, abolishing serfdom and approving equal rights for all its citizens. Hungary's troops were weakened by this conflict, making it easier for the Hapsburgs to put down the Hungarian Revolt and regain power later that year.

[13] The Croatian Diet was dissolved in 1865, and two years later, with the division of the crown, Germany and Hungary became the major nations of the Austro-Hungarian Empire. In 1868, Hungary accepted the union of Croatia, Slavonia and Dalmatia as a separate political entity, though Austria refused to relinquish its claim to Dalmatia.

[14] In the early 20th century, Croatian nationalist groups intensified their activity. An alliance of Croatian and Serbian leaders adopted the "Rijeka resolution", a plan of action which enabled them to win the 1906 elections. In the meantime, the Croatian Peasant Party began political activity among the peasants. The Crown responded by increasing repression.

[15] In 1915, Croatian, Serbian and Slovenian leaders organized the Yugoslav Committee in Paris, in favor of separation from the empire and union with an independent Serbia. Austria-Hungary's defeat accelerated the creation of the Yugoslav kingdom, in 1918.

[16] The Serbian dynasty's policy of amalgamating the regions immediately came into conflict with Croatian desires for independence. Croatia demanded the creation of a Yugoslav federation. As of 1920, the Peasant Party, led by Stjepan Radic, headed the Croatian opposition. The assassination of Radic and other members of the opposition, in 1928, provoked a serious crisis.

[17] When World War II broke out, Yugoslavia was divided internally, so it was easily occupied by Hitler in 1941. The German army set up a puppet regime made up of Croatia, Slavonia, parts of Dalmatia and Bosnia-Herzegovina. A racist campaign was launched in which Serbians, Jews, Gypsies and Croats opposed to Fascism were massacred

LAND USE

■ CROPLAND 25% ■ PASTURE 22% ◻ OTHER 52%

ENVIRONMENT

Croatia is bounded to the north by Slovenia and Hungary and to the east by Serbia. The Dalmatian coast - a 1,778-kilometer coastline on the Adriatic, with numerous ports and seaside resorts - lies in southern and western Croatia. The territory is made up of three regions with different landscapes: rolling hills in the north, around Zagreb; rocky mountains along the Adriatic coast and the inland valleys of the Pannonian Basin. The Pannonian Basin, under the influence of masses of continental air, is colder and not as rainy as the coast. The climate along the Dalmatian coast is Mediterranean, with an average temperature of 4°C in winter, and 24°C in summer. The country's traditional economic activities have been agriculture and cattle raising. After World War II, light industry was developed, and the discovery of rich deposits of oil changed the country's economy. The government is in the process of reconstructing the country's productive apparatus, as well as vast areas of the country which were devastated by the conflict. Likewise, it has begun to assess the impact of the civil war upon the environment.

SOCIETY

Languages: Serbo-Croatian (official). **Religions:** Mainly Roman Catholic **Political Parties:** The Croatian Democratic Union, right-of-center, currently in power; the Croat Socio-Liberal Party (CSLP); the Croat Peasant Party. **Social Organizations:** These are in the process of being reorganized, after recent political and institutional changes.

THE STATE

Official Name: Republika Hrvatska. **Administrative Divisions:** 102 Districts. **Capital:** Zagreb, 706,770 inhab. (1991). **Government:** Franjo Tudjman, President since April,1990. Zlatko Matesa, Prime Minister since November 4, 1995. The bicameral parliament (called Sabor) is one of Europe's oldest: the Chamber of Deputies has 127 members and the Chamber of Districts has 68 members, elected for a 4 years term. **National Holiday:** June 25, Independence (1991). **Armed Forces:** 103,300 (1993). **Paramilitaries:** 40,000 Police (1,000 deployed in Bosnia). **Serbian** Krajina **Forces,** 40 - 50,000. Paramilitaries: 24,000. UN **Forces:** 15,000.

External Debt
1994

PER CAPITA

$482

Foreign Trade
MILLIONS $ 1994

IMPORTS
5,231

EXPORTS
4,259

in concentration camps or forced into exile.

[18] After the anti-Nazi guerrillas had occupied Zagreb, in May 1945, the anti-Fascist Council of National Liberation of Croatia assumed control of the government. By the end of the year, Croatia was one of the constituent republics of the new People's Federated Republic of Yugoslavia.

[19] Within the Yugoslav socialist system (see Yugoslavia), Croatia maintained and strengthened its national independence, although this worked against its ethnic minorities. In 1972, a purge within the Croatian League of Communists led to the suspension of the Matica Hrvatska, the Croatian cultural organization.

[20] At the end of the 1980s, Yugoslavia modified its political system; the Yugoslav League of Communists (YLC) renounced the monopoly and leading role assigned to it by the constitution, and in April 1990, the first multi-party elections since World War II were held in the different regions.

[21] In Croatia, parliamentary elections gave a majority to the Croatian Democratic Union (CDU), a center-right party which favored making Yugoslavia a confederation of sovereign states. The leader of the CDU, retired general Franjo Tudjman, was elected president of the republic. In December, Parliament ratified a new constitution which endorsed the right to separate from the federation.

[22] In June 1991, Croatia proclaimed its independence from the federation, while the Serbs in Krajina declared their intention of separating from Croatia. The United States and the European Community held back from recognizing Croatia and Slovenia, which had also declared their independence.

[23] The federal army, whose high command answered primarily to Serbia, intervened in Croatia and Slovenia, claiming that separation constituted a threat to Yugoslavia's integrity. War broke out, with many victims on both sides.

[24] In November 1991, Tudjman's government arrested the leaders of the ultra-right-wing Croatian Ustasha (Fascist) Party (CUP), and dissolved its militia, under the accusation of conspiring against the country's legal authorities. Founded in February 1990, the CUP adopted the name of a nationalist party of the past, whose leader had been the leader of the Croatian puppet government during Nazi occupation in World War II.

[25] In December 1991, Germany recognized Croatia's independence, accelerating the process whereby the EC would do the same. A few days earlier, the Serbs of Krajina (in southern Croatia), Slavonia, Baranja and western Serm (to the east) proclaimed the republic of Krajina, as a new federated Yugoslav entity, an alliance of Serbian States.

[26] In January 1992, through EC mediation, Serbia and Croatia accepted a peace plan, and agreed to allow the UN to station 15,000 troops in the war zone; however, Krajina's representatives objected to the agreement. Almost a third of Croatia's territory was controlled by federal troops and Serbian irregulars. Both forces accused each other of violating the cease-fire, and of committing war crimes.

[27] In February 1992, after overcoming Krajina's opposition, the UN Security Council unanimously ratified sending a multinational force to Croatia. UN troops are charged with keeping the peace and protecting Croatia's Serbian minority, which maintains civilian guerrilla groups. All parties agreed to the UN peacekeeping force remaining in the region for a one-year period.

[28] In May, the General Assembly of the UN, meeting in New York, formally admitted the former Yugoslav republics of Croatia, Slovenia and Bosnia-Herzegovina, increasing the number of its members to 178.

[29] After his re-election on August 2, 1992, President Franjo Tudjman managed to reopen the main highway link between Zagreb and Belgrade, through an agreement with the New Yugoslav Federation (Serbia and Montenegro). This agreement was engineered through UN mediation, with former US Secretary of State Cyrus Vance acting as chief negotiator.

[30] In Geneva, in October 1992, presidents Tudjman and Cosic, of the Yugoslav Federation, announced their opposition to "ethnic cleansing". They agreed to provide more humane treatment of refugees.

[31] In November 1992, Radovan Karadzic, leader of the Bosnian Serbs, announced the formation of a confederation between the "Serbian Republic of Bosnia and Herzegovina" and the "Serbian Republic of Krajina", in Croatia.

[32] According to official sources, 99 percent of the Krajina electorate favored unification with the Bos-nian Serbs at the time; however, a year went by following Karadzic's announcement and, at the end of 1993, nothing had yet been done to bring about such a union.

[33] This situation intensified tensions in Croatia. Tudjman, concerned about avoiding the proliferation of separatist feelings, offered an amnesty and relative autonomy to the Serbs of Croatian Krajina. Likewise, the latter allowed the bridge at Maslenica - which links the northern and southern parts of the country - to be rebuilt.

[34] However, subsequent discrepancies led to a resumption of hostilities by the Serbs; not even mediation by the UN was able to ensure security of the bridge.

[35] As for the conflict in Bosnia, Tudjman supported Bosnian-Croat claims in regions where they constitute a majority. However, he stated off the record that an understanding has been reached with President Slobodan Milosevic (Yugoslav Federation) whereby, in case of an eventual partition of Bosnia, a mere 15 per cent of the territory would be earmarked for Muslim Bosnians (who make up 44 per cent of the population).

[36] Tudjman has refused to comment on the massacre of Bosnian Muslims at the hands of Bosnian-Croats in the city of Mostar, during the first half of 1993. During the events in question, Bosnian-Croat forces adopted the emblem used by the Ustashes, Croats who sided with the Nazis during World War II.

[37] As a direct result of this, at the UN in February 1993, Moscow defended the extension of international sanctions to Zagreb.

[38] That same month, the UN created an international tribunal to judge war crimes committed in the former Yugoslavia.

[39] Amnesty International (AI) has reported that thousands of people, most of them Serbs, were accused of armed rebellion or of constituting a threat to the territorial unity of Croatia. Among them were those whom AI considered to be prisoners of conscience, given the fact that they were arrested because of their ethnic origin. AI has also received reports of summary executions perpetrated by members of the Croatian army.

[40] The conflict in Croatia, although less bloody than that of Bosnia-Herzegovina, has caused thousands of deaths and injuries among the civilian population.

[41] Throughout the Balkan region

EDUCATION

School enrolment: Primary (1993): 87% fem., 87% male. Secondary (1993): 86% fem., 80% male. University: 27% (1993).

COMMUNICATIONS

21.5 **telephones** per 100 inhabitants (1993).

ECONOMY

Per capita GNP: $2,560 (1994). **Consumer price index:** 100 in 1990; 53,818 in 1994. **Currency:** 6 dinars = 1$ (1994). **Imports:** $5,231 million (1994). **Exports:** $4,259 million (1994). **External debt:** $2,304 million (1994), $482 per capita (1994). **Debt service:** 4.2% of exports (1994).

ENERGY

Consumption: 1,057 kgs of Oil Equivalent per capita yearly (1994), 28% imported (1994).

there are some 750,000 refugees, including 450,000 refugees from Bosnia and Herzegovina.

[42] The outbreak of war rocked the republic's economic structures, producing growing unemployment (20 per cent by the end of 1994) and inflation (2 per cent per day). In March a general strike brought about the fall of Prime Minister Hrvoje Sarnic, who was succeeded by Nikica Valentic. The new prime minister was an active member of the party in power, the Croatian Democratic Union, and president of INA, the state-owned petroleum company.

[43] Toward the end of 1993, Tudjman and Izetbegovic, the president of Bosnia-Herzegovina, signed a cease-fire agreement which also provided for the dismantling of the prisoner of war camps.

[44] In January 1994, Croatia and Serbia agreed in Geneva to a full restoration of transportation and communication links between the two republics. Likewise, offices were opened in Zagreb and Belgrade which are to function as diplomatic representations, although with lesser rank than embassies.

[45] However, there has been no solution to the situation of the enclave of Krajina, still under Serbian control. Croatia attempted to win Serbian recognition of its territorial demands, but Slobodan Milosevic opposed it.

[46] In any case, UN and US mediators Thorvald Stoltenberg and Lord Owen welcomed the understanding, which diffused the tension between the two former Yugoslav republics.

[47] Towards the end of January, Croatia made an unsuccessful bid to recover the Krajina enclave by force, thereby violating the ceasefire with Serbia. In February, the UN Security Council unanimously approved an extension of the UN peacekeeping force mandate in Croatia.

[48] The need for economic resources led Croatia to turn to the European Reconstruction and Development Bank, to obtain investment funds, principally in the areas of tourism, agriculture and energy projects. However, that agency declined, stating that "it cannot finance projects in Croatia as long as Croatia remains at war".

[49] The hyper-inflationary spiral forced the government to adopt austerity and stabilization measures in October 1993. The adoption of these measures also seems to have been a signal to the international community.

[50] Russia tried mediation to get Serbians to respect peace agreements in Serbia, Bosnia and Croatia. Vitali Churkin, Russian vice-minister of Foreign Affairs, has proposed that the rebel Serbs of Krajina put their "republic" under the sovereignty of Zagreb, in exchange for a high degree of autonomy. Likewise, he recommended that Croatians form a federal State in which Serbian rebels would be willing to live, "because Europe has adopted the path toward integration".

[51] Both Russia and the United States agreed that the problem of the violent disintegration of Yugoslavia demands a global solution acceptable to the three major groups: Serbians, Croats and Muslim.

[52] The course of action suggested by the US government implies a union of Bosnian-Croats and Muslims, to form a confederation between the Bosnia which would result from that union, and Croatia. This would, in turn, reduce the risk of Islamic fundamentalism given that Croatian Catholics would constitute an ample majority within the new State.

[53] In March 1994, Bosnian president Alija Izetbegovic and his Croatian counterpart, Franjo Tudjman, signed an agreement in Washington on the principles of a Confederation between the two

republics which emerged from the former Yugoslavia (see Bosnia and Herzegovina).

[54] This projected confederation will elect a president and vice-president in democratic elections. Both will have a 4-year term, and the presidency will alternate each year between a Muslim and a Croat; the same will happen with the the head of Government, whose "second in command" will also be a member of a different ethnic group.

[55] In June 1994, Croatia launched the "kuna", the currency used in that country during World War II by the government which had been installed by Hitler's troops. The kuna replaced the dinar, which had been adopted after independence.

[56] The new currency was rejected by the opposition. However, Tudjman pointed out that the "kuna", which in Croatian means "marten" (the animal), dates back to the 11th century, when the fur of the marten was used as currency.

[57] The opposition also questioned the government's decision to change the names of streets, schools and plaques dedicated to the memory of notorious anti-Fascist war heroes.

[58] Pope John Paul II's visit to Croatia on September 10 and 11, helped to raise the international profile of the nation. The Serb rebels and the Croat government reached an agreement on opening the Zagreb-Belgrade road, the Adria oil pipeline and for the supply of water and electricity to Serbs in the occupied territories.

[59] Croatia dominated inflation in 1994, though industrial production fell and unemployment remained at around 20 per cent. The emigration of university professionals and qualified workers increased significantly during this year.

[60] In January 1995, Croatia informed the UN that authoritization for the peace maintaining troops would not be extended beyond March 31, the date the agreement ended. According to Tudjman, the 12,000 UN blue helmets legitimated the Serb presence in Krajina, occupying 27 per cent of Croat territory. The Serb and Croat forces took offensive action. The UN said the withdrawal of its troops could cause the most dangerous situation in Europe since 1945.

[61] On April 13, Serb forces bombarded Dubrovnik airport and ten days later blocked the road between Zagreb and Belgrade, in eastern Slavonia, again. The Croat army

regained control in this area in a swift operation in May. Zagreb was immediately shelled by Serb troops.

[62] President of the Yugoslav Federation, Slobodan Milosevic said that, if the UN would lift its sanctions on Yugoslavia, peace would return to the Balkans in a matter of months. An agreement between Franjo Tudjman and the United States allowed the international force to remain but to withdraw gradually, with help from NATO troops.

[63] In early August, the Serbs suffered their worst defeat since the break-up of the former Yugoslavia. The Krajina region, in central Croatia, was taken by the Croats, who forced Serb troops and civilians to leave. According to Richard Holbrooke, the US mediator, "Tudjman drew new lines on the map, which they had not been able to achieve via diplomatic means".

[64] Allegations of torture and pillage were lodged against the Croats who occupied Krajina. Some 250,000 Serbs joined the mass of more than 700,000 refugees already uprooted in the Balkans. The Croat occupation of Krajina was considered the biggest ethnic cleansing operation of the Balkan war. By May 1996, only eight Croats had been tried in their absence by the the International Criminal Tribunal at the Hague

[65] In October, representatives of the Croat government and the Serb leaders of eastern Slavonia agreed the basis for the peaceful restructuring of the region. On October 29, Tudjman won the elections with 45.23 per cent of the vote. He did not receive the two thirds majority necessary for reforming the Constitution and increasing presidential powers.

[66] The signing of an agreement between the presidents of Croatia, Serbia and Bosnia and Herzegovina at an airbase near Daytona, in the United States, marked the end of hostilities, though civilians maintained the aggression towards UN and NATO troops during 1996.

[67] In August 1996, a schedule of meetings between Serb and Croat authorities was established. The return of the refugees and preparations for elections were the main issues on the agenda. Tudjman, accused of hindering the return of the refugees, allowed a limited number of Serbs to settle in the Croat territory.

Cuba

Population:	10,978,000 (1994)
Area:	114,524 SQ KM
Capital:	Havana (La Habana)
Currency:	Peso
Language:	Spanish

Cuba

Until the 16th century the island of Cuba was inhabited by the Taino and Ciboney. In 1492, the Spanish emissary to the Indies arrived, but it was not until 1509 that a voyage was made around the coast proving that Cuba was an island. In 1514, the conquest begun in 1510 was completed. This was a violent period in the island's history, with a number of uprisings, like those led by native chieftains Hatuey and Guama.

[2] In fact, the Spanish expeditions which would subsequently conquer a large part of the Caribbean, Mexico and Central America, departed from Cuba, which a mid-16th century document described as "high and mountainous" with small rivers "rich in gold and fish". In 1511, colonists from Santo Domingo started mining Cuban gold. It was a short-lived economic cycle, probably because the native population was rapidly exterminated. As the number of African enslaved workers on the island was insufficient, economic life soon declined and did not recover until the end of the 16th century, with the advent of sugar production.

[3] As early as the 17th century, economic diversification was achieved through shipbuilding and the developing leather and copper industries. The economic center of the island gradually moved from Santiago, on the southern coast, to Havana, in the north, a port of great importance in the mid-17th century.

[4] Sugar cane plantations in colonial America used slave labor and caused economic dependence. There were periods, during 1840 for example, when slave labor represented 77 per cent of the total Cuban workforce, and there is also evidence that there were *palenques*, settlements of escaped slaves (see Brazil, paragraph 6), on the island,

and there was an abortive slave revolt in 1843. It took a long, complex process to change this situation, comparable only with that of Brazil, the last American country to abolish slavery. British pressure against the slave trade, and the contribution of the enslaved workers during the Ten Year War, contributed decisively to the abolition of slavery in 1886.

[5] The struggle for independence began in the first few decades of the 19th century, but the decisive battle, with José Martí, Antonio Maceo and Máximo Gómez was not fought until 1895. In 1898, aware that the victory of the Cuban patriots was inevitable, the US declared war on Spain and invaded Cuba to guarantee their continued influence in the Caribbean.

[6] US occupation forces ruled the country from 1899 to 1902, imposing a constitution including the so-called "Platt Amendment". This secured the US rights to intervene in Cuba and to retain a portion of its territory. Thus, Guantánamo became a powerful air and naval base that is still maintained. The "right" to intervene has been exercised on various occasions, with US Marines remaining on Cuban soil for extended periods of time.

[7] With the fall of the dictator, Machado, in 1933 Carlos Manuel de Céspedes came to power. Several months later there was a military coup and the rebelling sergeants established a short-lived "Pentarchy". A few days later, Grau formed the "Government of a Hundred Days" with Batista and Guiteras, and several anti-imperialistic measures were implemented. In January 1934, Batista led a reactionary counter-coup, and Guiteras was assassinated. Elections were held, ushering in a series of governments presided over by Batista's Authentic Party, dominating a peri-

od characterized by corruption and gangsterism under the auspices of the United States. On March 10, 1952, Batista engineered yet another coup, establishing a dictatorial regime which was responsible for the death of 20,000 Cubans.

> ## The "right" to intervene was exercised on various occasions, with US Marines remaining on Cuban soil for extended periods of time.

[8] On July 26, 1953, Fidel Castro and a group of revolutionaries attacked the Moncada Army Base in Santiago de Cuba. Although the attack itself failed, it marked the beginning of the revolution. After a short period of imprisonment, Castro went into exile in Mexico only to return in December 1956. Castro's revolutionary program had been defined during his trial after the failed initial uprising, ending with his well-known words: "History will absolve me".

[9] On December 31, 1958, Batista fled from Cuba as guerrilla columns led by Ernesto "Che" Guevara and Camilo Cienfuegos, closed on Havana. They were the vanguard of the Rebel Army. In just over two years, the guerrillas of the "26th of July" Movement, organized by Fidel Castro, broke the morale of Batista's corrupt army.

[10] In 1961, counter-revolutionaries disembarked at Playa Giron (Giron Beach), on the Bay of Pigs,

in an attempt to bring down the regime which had carried out Agrarian Reforms and expropriated various American enterprises. They had counted on a popular uprising against the revolutionary government, but this did not materialize. After 72 hours of fierce fighting, the Bay of Pigs invasion ended in the defeat of the invading forces. Two days before the invasion, on April 15, while the victims of the Havana Airport bombing were being buried, Castro proclaimed the socialist nature of the revolution.

[11] Also in that year, all the pro-government organizations joined together in a common structure. This was initially known as the Integrated Revolutionary Organizations, and later became the United Party of the Socialist Cuban Revolution (PURSC). The United States had Cuba excluded from the Organization of American States (OAS). They also put pressure on other countries to break diplomatic relations with Cuba, engineering an economic blockade of the island. That year the October Crisis made the possibility of the next World War, into a near-probability. The United States photographed Soviet nuclear missile launching sites on the island, and began preparations for another invasion. However, a peaceful solution was reached as the Soviet Union and the United States came to an agreement. The US pledged not to invade Cuba, but disregarded the Cuban government's demands for an end to the blockade, the withdrawal of US troops from Guantanamo and an end to US-managed terrorist activities.

[12] The literacy campaign during these years soon bore fruit, and by 1964, Cuba was free of illiteracy. Improvements in health were also one of the government's priorities. In October 1965, the People's

PUBLIC EXPENDITURES

% of GNP   1992

MILITARY 5   EDUCATION 6.6   HEALTH 3

WORKERS

1994

MEN 67%   WOMEN 33%

AGRICULTURE 24%   INDUSTRY 29%   SERVICES 47%

ENVIRONMENT

The Cuban archipelago includes the island of Cuba, the Isle of Youth (formerly the Isle of Pines) and about 1,600 nearby keys and islets. Cuba, the largest island of the Antilles, has rainy, tropical climate. With the exception of the southeastern Sierra Maestra highlands, wide and fertile plains predominate in the country. Sugar cane farming takes up over 60% of the cultivated land, particularly in the northern plains. Nickel is the main mineral resource.

SOCIETY

Peoples: Cubans call themselves "Afro-Latin Americans" on account of their mixed Afro-European-American ethnic background. There are also approximately 30,000 Asian natives, mainly Chinese. **Religions:** Catholic, Protestants, Afrocuban syncretists. **Languages:** Spanish. **Political Parties:** The Cuban Communist Party is defined by the constitution as the "supreme leading force of society and the state". Founded in October 1965, it replaced the United Party of the Socialist Cuban Revolution (PURSC). This group had been in existence since 1962, when the Integrated Revolutionary Organizations (ORI) were reorganized and renamed. The ORI, a political front formed in 1961, resulted from a merger of the Revolutionary Movement July 26, the Revolutionary Leadership March 13, and the People's Socialist Party. **Social Organizations:** Cuban Workers' Union (CTC). The CTC had 2,984,393 members in 1988, and represented more than 80% of the Cuban labor force; the National Association of Small Farmers, 167.461 members (1988) with 3,500 grass roots organizations; the Federation of Cuban Women (FMC), 2,420,000 members; the University Student Federation (FEU) and the Federation of Secondary School Students (FEEM), 450,000 student members; the José Martí Pioneers' Union, 2,200,000 children and young people; the Revolution Defense Commitees.

THE STATE

Official Name: República de Cuba. **Administrative Divisions:** 14 Provinces, 169 Municipalities including the special municipality of Isla de Pinos. **Capital:** Havana (la Habana), 2,176,000 inhab. (1993). **Other cities:** Santiago de Cuba 4,440,100 inhab.; Camaguey 294,000 inhab.; Holguín 242,100 inhab.; Guantánamo 207,800 inhab. (1993). **Government:** Fidel Castro Ruz, president of the Council of State and the Council of Ministers, Cuban Communist Party secretary-general and Commander-in-Chief of the Revolutionary Armed Forces. The constitution, approved in a plebiscite on November 15, 1976, states that "all the power belongs to the working people and is exercised through the Assemblies of People's Power". The local Assemblies delegate power to successively more encompassing representative bodies until a pyramid is formed which peaks in the National Assembly. Representatives are subject to recall by the voters. **National Holidays:** January 1st, Liberation Day (1959); July 26, Assault on the Moncada military barracks (1953). **Armed Forces:** 105,000 (1995) **Paramilitaries:** 1,369,000 Civil Defense Force, Territorial Militia, State Security, Border Guard.

Socialist Party, the Revolutionary Directory and the 26th of July Movement (led by Castro himself) dropped the name PURSC and created the Cuban Communist Party (PCC). This was just one more step closer to the Soviet Union, a relationship which had been strengthened by the economic blockade.

[13] With the consolidation of the socialist regime, Cuba began lending technical assistance to like-minded peoples and governments of the Third World. It sent troops to countries like Ethiopia and Angola, who requested help to resist invading forces.

[14] The revolution began to be institutionalized after the first PCC congress, in 1975. A new constitution was approved in 1976 and there were subsequent elections of representatives for the governing bodies at municipal, provincial and national levels. In 1979, Castro and the leaders of the PCC launched a campaign of revolutionary requirements to correct weaknesses in the administrative and political management areas of the revolutionary process.

[15] Tension between Cuba and the US had increased when Ronald Reagan came to the White House in 1979.

[16] In the mid-1980s, some 120,000 people left Mariel bound for Florida taking advantage of a clause offering automatic U.S. residency to any Cuban arriving in the United States - brought in when emigration from Cuba was practically impossible.

[17] Cuba's relations with many Latin American countries improved as a consequence of the Cuban attitude during the Malvinas war. Anxious to put a halt to Cuban-Latin American rapprochement, the United States took a series of measures against Havana. In August 1982 the US Senate passed a resolution allowing the government to use the armed forces to halt "Marxist-Leninist subversion" in Central America.

[18] In the midst of mounting tension in Central America, Fidel Castro proposed to remove all foreign troops and advisors from the region. Meanwhile on the domestic scene, he outlined a new development strategy aimed at achieving greater economic autonomy. He aimed to make the island less dependent upon external factors and less vulnerable to manipulation by its creditors in the socialist bloc.

[19] These proposals were largely due to open threats of military aggression from the US during Reagan's second term. Consequently, defence had top priority, and efforts to strengthen the economy were made in order to fund the defense budget. Unavoidable imports were to be financed through exports.

[20] Cuba depends on capitalist countries for 15 percent of its imports, and has foreign debts of some $3 billion, to non-socialist countries and private banks. It must therefore be able to guarantee the availability of hard currency. To meet the debt payments there are two choices open. One is, to save on oil imports, and the other is, to produce more and better goods, both for export and to replace imported goods in the domestic market.

[21] As economic difficulties increased, the upper echelons of the PCC and the government proposed a series of new measures. Several party leaders were removed from their posts for failing to adapt to the new situation; a new Office for Religious Affairs was charged with strengthening links between the churches and the State, and state control over the media was relaxed.

[22] The results of the Third Congress were formalized in April 1986 and the so-called "process of the rectification of errors and negative tendencies" was initiated. This coincided with the changes which were taking place in the Soviet Union, but the Cubans avoided adopting the eastern European model.

[23] In June 1989, a high-ranking group of Army officers and officials of the Ministry of the Interior were brought to trial and executed for being involved in drug trafficking.

[24] From the time George Bush took office there was an increase in American pressure on Cuba. In the first few months of 1990 there were important military maneuvers at Guantanamo Base and in the Caribbean. The US also violated Cuban television airspace with transmissions by "Televisión Martí", which was supported by "The voice of America". However, this broadcast was only picked up on the island for one day, after which it was jammed.

[25] At this time, the last of Cuba's troops returned home from the Congo and Angola. More than 300,000 Cubans had served in Angola, with the loss of 2,016 lives.

[26] After a seven-month delay, the 4th PCC congress took place in October 1991. It was decided to

reform the constitution so that members of the National Assembly could be elected directly, a major ratification of the single-party system. People with religious beliefs were also admitted, and the government recognized the need for joint ventures, especially with Latin American investors.

27 With the changes in Eastern Europe, and the disappearance of its former CMEA allies, some of Cuba's basic supplies dropped to critical levels. To ease this crisis, the government has strengthened its ties with China, Vietnam and North Korea, and is looking for ways to capitalize on its recent technological advances.

28 Biotechnological research has made Cuba a Third World leader in this area. (They recently developed an anti-meningitis vaccine).

29 The government decided to ration fuel when the former USSR cut oil shipments by 25 percent between 1989 and 1991. They also decided to strengthen the tourist industry. The Cuban government have initiated joint ventures with Spanish companies. The investors maintain 50 percent ownership of the enterprise once it enters operation.

30 The loss of its main trading partners, and the re-enforcement of the US blockade led to the creation of a Special Plan, aimed at distributing scarce resources equitably. In 1990, bread was rationed at 100 grams per person per day, three newspapers ceased to be published, and the official newspaper, "Granma", was forced to cut its circulation by more than two thirds. Alternative proposals include the possibility of offering more sugar, tobacco and coffee on the international market, an overall strengthening of the agricultural sector, and the possibility of oil purchase agreements with Mexico and Venezuela.

31 The severe shortage of food forced the authorities to carry out an emergency food program. Over the last two years, this scheme has sought to increase agricultural production rapidly, maximizing sugar production so that sugar cane derivatives can be used as animal feed. The emergency program also aims to create new sources of employment.

32 With the break-up of the Soviet Union, the United States urged Russia to take a tougher stand in its relations with the island, and this trend became evident during Mikhail Gorbachev's last months in power. Gorbachev's decision to withdraw three thousand Soviet soldiers, stationed in Cienfuegos and Havana, took the Cubans completely by surprise as the issue had never even been discussed.

33 In August 1991, through its representative at the United Nations, Cuba presented a motion that the UN should debate the necessity of ending the American economic blockade. The Cubans quickly withdrew the motion, when they realized what they were up against; the US had unleashed pressures and threats against all the other delegations.

34 In early 1993, National Assembly (parliament) elections were held, and deputies were elected directly for the first time. Ricardo Alarcón, foreign minister, was designated president of the National Assembly, and Roberto Robaina, secretary general of the Communist Youth organization, became foreign minister.

35 On July 26, 1993, in his speech for the commemoration of the 40th anniversary of the attack on Moncada Army Post, Castro explained the difficult situation which the country was facing, and the need to adopt special measures. He announced that it would henceforth be legal for Cubans to possess and use foreign currency, and to be self-employed.

36 Toward the end of the same year, the National Assembly met to discuss the country's critical situation, especially its internal finances. The Assembly decided to take no measures for the time being, but to discuss the issue and collect opinions and ideas in workplaces throughout the country. These "Workers' Parliaments" were held during the first quarter of 1994.

37 Towards the end of thiat year, the National Assembly agreed to organize a debate on the internal financial crisis in each work centre. These Workers Parliaments, which met in 1994, produced information on the state of the work centres and their economic management.

38 In April 1994, a meeting entitled "The Nation and Emigration" was held in Havana on the initiative of Foreign Minister Roberto Robaina, and it was attended by some groups of Cuban emigrés living abroad. Shortly afterwards, there were a series of incidents in Havana involving people who wanted to leave the country illegally in rickety boats. When Cuba announced it would not stop the "raft people" from leaving, the United States started official negotiations to regulate the illegal departure of the immigrants.

39 In July, Cuba entered the Association of Caribbean States as a full member. Cuban participation in the ACS, a group which is emerging as a new economic bloc, will encourage greater integration of the Cuban economy in the region, offering tariff benefits and trade facilities.

40 In 1995, the fiscal deficit fell for the third year running, through the reduction of the public payroll and the withdrawal of subsidies. The peso was put on a par with the dollar and the possession of US currency was legalized. The Cuban Parliament approved a new investment law which allowed for the establishment of 100% foreign funded companies, including businesses run by Cuban expatriates. The United States Congress responded with the Helms-Burton bill, which aimed to penalise companies dealing with Cuba via third nations. The international community, especially the European Union, severely criticized this measure as a breach of the World Trade Organization agreements and GATT.

41 In February 1996, the Cuban air force shot down two small planes belonging to a group of exiles - called "Brothers to the rescue"- resident in Miami, who habitually violated Cuban airspace to distribute anti-government pamphlets. The incident caused further confrontations with the United States. Amnesty International denounced the imprisonment of several people linked to Cuban Conciliation, an organization which according to AI includes 140 groups of opposition journalists, professionals and union leaders, whom the government links with US interference.

Cyprus

Kipros Kibris

Population: 726,000 (1994)
Area: 9,250 SQ KM
Capital: Nicosia (Levkosía)
Currency: Pound
Language: Greek and Turkish

The first civilization on Cyprus may be 3,000 years old. Hittites, Phoenicians, Greeks, Assyrians, Persians, Egyptians, Romans, Arabs and Turks trooped through its valleys and over its hills until 1878, when the British negotiated the island with Turkey in exchange for protection against Czarist Russia. Cyprus became a bridgehead of the British expansion eastward.

[2] In 1930, movements favoring enosis (annexation) by Greece, gained ground stimulated by the incorporation of Crete to Greece in 1913. Enosis was promoted by the Greek Orthodox Church, the spiritual guide of Greek-Cypriots, who make up the majority of the island; population. But Turkey, Greece's perennial rival, was averse to being completely surrounded by a hostile neighbour on its Mediterranean coast. The British sent many Greek-Cypriot priests into exile. Nevertheless, after World War II the country; most important political figure was Archbishop Vaneziz Makarios, who led the Cypriot anticolonial movement from exile.

[3] In 1959, representatives of the Greek and Turkish communities, of Makarios; Democratic Party and of British colonial interests, approved a plan whereby Cyprus was to become an independent republic with constitutional guarantees for the Turkish minority, and British sovereignty over the island; military bases. Independence was proclaimed on August 16, 1960, and Makarios took office as president. An active supporter of anti-colonialism, the president played an important role in the Movement of Non-Aligned Countries. Makarios was re-elected in 1968 and 1973. Tensions between Greece and Turkey persisted and had frequent repercussions in Cyprus where conflicts erupted between the Greek and Turkish communities.

[4] The US government was not friendly toward Makarios, either, considering him the "Fidel Castro of the Mediterranean". In 1963 there was an unsuccessful coup attempt by members of the radical right, who subscribed to enosis. On July 15, 1974, the crisis came to a head when the Cypriot National Guard, under the command of Greek army officers, ousted the president-Archbishop, who fled to Britain, and appointed Nikos Sampson, who favored annexation by Greece. Five days later, Turkey invaded northern Cyprus, bombed Nicosia, and drove 200,000 Greek Cypriots southward, under the pretext of protecting the Turkish minority.

[5] Sampson turned the presidency over to the president of the Lower House, Glafkos Clerides, on July 23. That same day, the dictatorial junta of colonels that governed Greece also stepped down, unable to face the prospect of a war with Turkey, domestic opposition, and world-wide repudiation.

[6] The Turkish forces, who were occupying 40 per cent of the island, refused to return to the situation which had prevailed prior to the coup, continuing to occupy the north of the country. On August 16, 1974 a Turkish -Cypriot Federal state was proclaimed in the northern part of the island - the part under Turkish control- under the presidency of Rauf Denktash.

[7] Makarios returned to Cyprus in December 1974 and held the presidential office until his death in 1977. Spyros Kyprianou succeeded Makarios one month later and followed the same policy, refusing to recognize the division of Cyprus. He also remained a member of the Movement of Non Aligned Countries.

[8] A pact between Makarios and Denktash had set four basic conditions for a negotiated settlement: the establishment of communal, non-aligned, independent, and federal republic; an exact delimitation of the territories that each community would administrate; the discussion of internal restrictions on travelling, ownership rights and other important issues within the frame of a federal system with equal rights for both communities; and sufficient federal power to ensure unity.

[9] The Turkish refusal to withdraw the troops, meant that no substantial progress was made. On November 15, 1983, the Turkish Republic of North Cyprus was proclaimed, but only Turkey recognized the new state.

[10] In the 1980s, Cyprus enjoyed a period of economic prosperity brought by tourism, foreign aid, and international businesses that made the island a financial centre, replacing the city of Beirut, whose money markets had been paralyzed by the Lebanese civil war. The main beneficiary of this was the Greek-Cypriot bourgeoisie. In the northern part of the island there had been a large influx of 40,000 new Turkish immigrants. Added to the 35,000 Turkish soldiers stationed in the area, and the migration of 20,000 Turkish-Cypriots from the south, the area population profile was substantially altered, as the Turkish population in the area tripled.

[11] On May 5, 1985, in northern Cyprus, a constitution for the Turkish-Cypriot Republic was submitted to a referendum. The turnout was 70 per cent, of which 65 per cent voted affirmatively. The low turnout was a political setback for the legitimacy of the government.

[12] In February 1987, the Greek premier cancelled a visit to the US, because the US had favorable relations with Turkey and installing US arms in the Turkish part of Cyprus.

[13] In 1988, Giorgis Vassiliu was elected president of Cyprus. He reestablished negotiations with Denktash, which had been suspended in 1985. Political leaders on both sides and representatives of the 350,000 Cypriots in exile have called for constructive flexibility on the part of both leaders. The Greek community -backed by their economic prosperity and the fact that they are numerically in the majority - want independence to be guaranteed by the United Nations; they also seek freedom of movement and property owner-ship through-out the island. The Turkish-Cypriots, on the other hand - basing their position on the status quo and the superiority of the Turkish army - demand a binational federation, under Ankara protection.

TURKEY

Mediterranean Sea

Morfou Bay · Kyrenia · Trikomo · Famagusta Bay · Ammókhostos (Famagusta) · Levkosia (Nicosia) · Lárnax (Larnaca) · U.K. Sovereign Base Area · TRODOS · Limassol · Paphos · U.K. Sovereign Base Area

///// TERRITORY UNDER TURKISH CONTROL   0   50 km

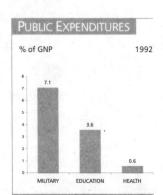

PUBLIC EXPENDITURES

% of GNP   1992

MILITARY	EDUCATION	HEALTH
7.1	3.6	0.6

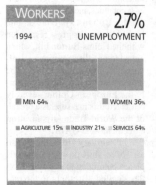

WORKERS   **2.7%**

1994   UNEMPLOYMENT

MEN 64%   WOMEN 36%

AGRICULTURE 15% INDUSTRY 21% SERVICES 64%

ENVIRONMENT

Once a part of continental Europe, the island of Cyprus is located in the eastern Mediterranean, close to Turkey. Two mountain ranges - the Troodos in the southwestern region, and the Kyrenia in northern Cyprus - enclose a fertile central plain. The temperate Mediterranean climate, with hot, dry summers and mild, rainy winters, is good for agriculture. The growing levels of atmospheric pollution rise even higher during certain "peak" hours of urban activity, with the combined effect of industrial pollution and traffic. However, though the quality of the air has been affected by pollution, it still lies within acceptable limits. Soil degradation is the result of natural causes and human abuse, particularly where there has been an excessive use of agrochemicals. The coastal ecosystem has felt the effects of tourism, particularly the construction of large hotels and high-rise apartment blocks.

SOCIETY

Peoples: Cypriots are divided into Greek (80%) and Turkish (18%) communities, within which they have separate political, cultural and religious organizations. **Religions:** Greek Orthodox and Muslim. **Languages:** Greek and Turkish (official). **Political Parties:** Democratic Party (Greek sector) and National Unity (Turkish sector), are the main parties. **Social Organizations:** There are two major labor organizations, the Pancyprian Federation of Labour, and the Cyprus Worker's Confederation. They group approximately 30 unions.

THE STATE

Official Name: Kypriaki Dimokratia-Kibris Cæmhuriyeti. **Capital:** Nicosia (Levkosia), 177,000 inhab. (1992). **Other cities:** Limassol (Lemesú), 137,000; Larnaca (Lçrnax), 61,000 inhab. (1992) **Government:** Greek sector: presidential republic, constitution in effect since August 16, 1960. Glafkos Klerides, Head of State and of the government since January 28, 1993. Single-chamber legislature: House of Representatives with 71 members (56 from the Greek community and 15 from the Turkish community). Turkish sector: parliamentary republic, constitution in effect since May 5, 1985. Rauf Denktash Head of State since February 13, 1975. Single-chamber legislature: 50-member Assembly, elected every five years. Rauf Denktash has occupied the presidency of the "Turkish Federated State of Cyprus" in the north of the island, since 1975. This government, based upon Turkish occupation and recognized only by Turkey, exercises an independent administration, including a judicial system and a Legislative Assembly. **National Holiday:** October 1, Independence Day (1960). **Armed Forces:** 10,000 **Paramilitaries:** Armed Police: 3,700.

DEMOGRAPHY

Annual growth: 1% (1992-2000). **Children per woman:** 2.4 (1992).

HEALTH

One **physician** for every 585 inhab. (1988-91). **Under-five mortality:** 10 per 1,000 (1995). **Calorie consumption:** 107% of required intake (1995).

EDUCATION

Primary school teachers: one for every 18 students (1991).

COMMUNICATIONS

102 **newspapers** (1995), 101 **TV sets** (1995) and 97 **radio sets** per 1,000 households (1995). 49.4 **telephones** per 100 inhabitants (1993). **Books:** 107 new titles per 1,000,000 inhabitants in 1995.

ECONOMY

Per capita GNP: $10,260 (1994).
Annual growth: 4.60% (1985-94).
Consumer price index: 100 in 1990; 122.8 in 1994.
Currency: 0,5 pounds = 1$ (1994).
Cereal imports: 498,000 metric tons (1990).

[14] The economy of the island grew by 6.9 per cent in 1989. In April 1990, Denktash was re-elected as president of the "Turkish Republic", in the north of the island. In 1991, with the support of US President George Bush, Turkey proposed a summit with representatives from Ankara, Athens and both Cypriot communities. Washington's support for the initiative was considered compensation for the help given to Turkey during the Gulf War by Greece and the Greek Cypriots.

[15] Nicosia demanded participation in the UN and the European Community, while rejecting the presence of Turkish Cypriot representatives. In April 1992, the UN declared Cyprus a bi-communitarian and bi-regional country, with equal political rights for both communities. In early 1993, Glafkos Klerides defeated Giorgis Vassiliu in the presidential elections.

[16] In the north of the island, Denktash formed a new coalition government in December 1993. In 1994, the Cypriot courts ratified the "legitimacy" of the British military bases on the island, while the European Court ordered an embargo on Turkish Cypriot exports. Despite help from Ankara, this embargo seriously damaged the economy in the north of the island.

[17] In 1995, the idea of Cyprus joining the European Union began to be considered as a way of overcoming the division on the island. Meanwhile, the island received more than two million tourists and its economy continued to grow. Massive investments from Eastern Europe led to accusations of money laundering.

[18] In August 1996, a series of demonstrations on the island's internal frontier left several people dead. Amnesty International accused the Turkish Cypriot police of being directly or indirectly responsible for at least two of the deaths.

Czech Republic

Population: 10,333,000 (1994)
Area: 78,864 SQ KM
Capital: Prague

Currency: Koruny
Language: Czech

Ceska Republika

B ohemia and Moravia, the present Czech Republic, and Slovakia share a common history until 1993. No one knows the exact origin of the first inhabitants of the Mid-Danube region, apart from the signs left by the Boii, a Celtic people whose name gave rise to the Latin name "Bohemia". The Celts were displaced, without major conflict, by Germanic tribes and later, in the 6th century, by the Slavs, while the former continued their migration southward.

[2] The inhabitants of mountain and forest areas had the natural protection of these areas. However, the lowlands, which extended north of the Danube, were repeatedly invaded by the Avars. The Slavs were able to repel these invasions when they had leaders strong enough to unite the tribes. In the 8th century, calm was restored to Bohemia with the defeat of the Avars at the hands of Charlemagne.

[3] In the early 9th century, three potential political centers emerged: the plains of Nitra, the Lower Morava Basin and Central Bohemia. The Slavs of Bohemia, known as Czechs, gained the upper hand in most of the region, which remained peaceful (and beyond the missionaries' sphere of influence), except for Charlemagne's invasion in the year 805.

[4] The Morava River Basin, an important trade route between the Baltic and Adriatic Seas, was controlled by Czech princes. The first of these was Mojmir I, who extended his realm as far as Nitra. His successor, Rostislav I, institutionalized the State and consolidated political relations with the Eastern Frankish empire, as a way of maintaining his own sovereignty.

[5] The Christian Franks organized the first missions at Nitra and in Bohemia, but Rostislav would not allow Latin to be taught, and asked the Byzantine emperor to send preachers who spoke the Slav language. Religious texts were also translated to Slav. Constantine and Methodius arrived in 863, leading a group of Greek missionaries.

[6] Methodius won recognition from Rome for his work in Moravia and in Panonia, which became an ecclesiastical province linked to the Archbishop of Sirmium. The Franks came to consider Methodius an enemy; he was captured and kept prisoner until 873, when he returned to Moravia.

[7] Rostislav was the founder of Great Moravia, uniting the territories inhabited by the Slavs of the region for the first time. Slovakia, bordered by the northern ring of the Carpathian Mountains and by the Morava River, was a natural member of Great Moravia. In addition to having common Slav roots, the Czech and Slav peoples were also linked by strong family ties.

[8] Religious rivalry between the Latin and Slav languages continued throughout the period of expansion of the Bohemian principality. The death of Methodius, in 885, strengthened the position of the Frankish Bishop Wiching, his lifelong enemy. Wiching displaced Methodius's disciples, and convinced the new Pope to outlaw the Slavic liturgy.

[9] Some years later, Czech expansion collided with the kingdom of Germania. King Arnulf sent a military expedition to Moravia in 892, allying himself later with the Magyars to defeat the principality. Between 905 and 908, Great Moravia went through several foreign occupations, until an agreement was reached between Mojmir II and Arnulf.

[10] In the 10th century, between the strengthening of Germania and the restoration of the Holy Roman Empire, Bohemia lost the major part of its possessions. When Bretislav I ascended to the throne in 1034, the principality recovered part of Moravia and invaded Poland in 1039. However, King Henry III of Germania forced a retreat, and the Hungarian Crown kept Slovakia.

[11] In order to maintain its independence, Bohemia found it necessary to be actively involved in the campaigns of the Holy Roman Empire. Thus, the situation arose that the Bohemian princes were being crowned king by both of the rival powers. The ups and downs in Bohemia's relationship with the empire finally came to an end in 1212 with the signing of the Golden Bull of Sicily.

[12] In the early 13th century the Church separated from the State, and the feudal lords began demanding greater political participation. At the same time, Germanic immigration was responsible for an increase in population. A series of incentives to build new urban centers and develop mineral resources gave rise to a new class of tradesmen and entrepreneurs.

[13] Under the Przemysl dynasty, which lasted until 1306, Bohemia controlled part of Austria and the Alps; at one point there was even a single king ruling over Bohemia and Poland. This dynasty was succeeded by the Luxemburgs in 1310. With the coronation of Emperor Charles I in 1455, Bohemia and the Holy Roman Empire were joined together. As the capital of the kingdom and the empire, this was Prague's greatest moment.

[14] The high level of corruption among the clergy triggered a religious reform movement in the 14th century, which became even more radical under the influence of Father John Huss, of Prague's Bethlehem Chapel. Excommunicated by the Pope, Huss was later tried for heresy and sedition by the Council of Constance, and was burned alive in 1415 after refusing to recant.

[15] The anger which followed Huss's execution marked the birth of the Hussite movement in Bohemia and Moravia. The Germanic peoples remained faithful to Rome, however, and in addition to these religious differences, the ethnic issue remained, triggering political conflict between them. The Holy Roman Empire, allied with the German princes, launched several military campaigns in Bohemia, but they were repulsed by the Hussites.

[16] Religious differences prevented political union between Bohemia and its former possessions for many years. Vladislav II reigned over Bohemia from 1471 on, but Moravia, Silesia and Lusacia were ruled by Mathias, of Hungary. Only when Vladislav II was elected king of Hungary, upon Mathias's death in 1490, were the territories reunited.

[17] When the kings' residence was in Budapest, the Diet - made up of the nobility and the wealthy urban tradesmen - acquired an importance comparable to that of a royal court. The noblemen, supported by the lesser nobility, held a majority on this council, and could therefore block attempts to limit the privileges of the burghs or royal districts, or measures taken against the lower classes.

[18] The death of Vladislav II's son, Louis II, in 1526, paved the way for the rise of the Hapsburgs. Ferdinand I, Louis' brother-in-law, became king by currying the favor of the nobility. Austria's victory over the Protestant Society of Schmalkaldica in 1547 permitted Ferdinand to impose the right of hereditary succession to the throne upon Bohemia and its states.

[19] The Hapsburgs strengthened the Counter-reformation throughout the region. Slovakia remained

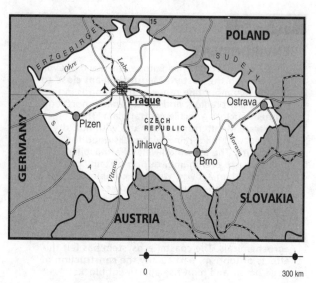

LAND USE

CROPLAND 43%   PASTURE 11%   OTHER 46%

within its realm because the Hapsburgs had retained it when Hungary was invaded by the Ottoman Empire, in 1526. Most Slovaks had turned to Calvinism and, like the Czech Protestants, used Slavic in their liturgy.

[20] Rudolf II (1576-1612) transferred the seat of the empire to Prague, making it once again one of the continent's most important political and cultural centers. Many key positions of the kingdom were filled by Catholics during Rudolf's reign, as he himself was Catholic. However, this triggered a rebellion by the non-Catholic (Reformed church) majority, and the king was deposed in 1611.

[21] After a stormy succession, Ferdinand II of Styria, with the support of Maximilian I of Bavaria, defeated the Protestants and ruled with a strong hand. The Diet lost its power to initiate legislation, being reduced to merely approving the king's petitions. The Germanic language was added to the traditional use of Czech, and only the Catholic religion was authorized.

[22] Unlike Bohemia, Moravia did not become involved in the fight against the Hapsburgs, and therefore, did not suffer the effects of civil and religious strife as severely. In Moravia, despite minor friction between the Germanic people and the Slavs, there was religious tolerance, which permitted the growth of Protestantism in the state, which remained separate from the Austrian Crown until 1848.·

[23] Despite the hegemony of the Germans and the prohibition of political activity, the Czechs conserved their ethnic identity, their language and their culture. Something similar had also occurred in the Hungarian counties inhabited by Slovaks. This set the scene for a resurgence of nationalism in the early 19th century, which strengthened the traditional ties between these two peoples.

[24] Czechs and Slavs, together with the inhabitants of the German republics, helped put a stop to the absolutist doctrine, amidst a revolutionary wave which swept Europe in 1848. In 1867, the empire split in two: Austria, where ethnic Germans out-numbered the Czechs, Poles and other nationalities; and Hungary, where the Magyars subdued the Slavs.

[25] In Bohemia, toward the end of the century the "Young Czechs" - a radical dissident nationalist group -

ENVIRONMENT

The Bohemian massif occupies the western region, bounded to the southeast by the Moravian plains. Cereals and sugar beets are cultivated in the lowlands of Bohemia and Moravia, where cattle and pigs are also raised. Rye and potatoes are grown in the Bohemian valleys. The country has rich mineral deposits: coal, lignite, graphite and uranium in the Bohemian massif; coal in Moravia. Sulphur dioxide emissions-produced in the generation of electrical energy are very high, causing acid rain. Air pollution has destroyed or damaged large areas of forest. Approximately three-fourths of all the country's trees show a high degree of defoliation. Water pollution levels are also very high, especially in rural areas. Waste from industry, mining and intensive farms threaten the purity of the water, both above and below the ground.

SOCIETY

Peoples: Czechs, 81.2%, Moravians, 13.2%, Slovaks, 3% **Religions:** Mainly Catholic. **Languages:** Czech (official). **Political parties:** Civic Democratic Party; Communist Party of Bohemia and Moravia; Social Democratic Party, Socialist Party; Agrarian Party, Green Party, Christian Democratic Union-Czechoslovakian People's Party, Association for the Republic-Republican Party, Democratic Civic Alliance; Movement for Autonomous Democracy of Moravia and Silesia. **Social Organizations:** Central Trade Union Council.

THE STATE

Official Name: Ceska Republika. **Administrative divisions:** 8 regions subdivided into municipalities. **Capital:** Prague, 1,217,000 inhab. (1994). **Other cities:** Ostrava, 326,000 inhab.; Brno 390,000 inhab.; Olomouc 106,000 inhab. **Government:** Parliamentary republic. Vaclav Havel, President since January 1993. Vaclav Klaus, Prime Minister since July 1992. **National holiday:** October 28, Independence Day. **Armed Forces:** 92,900 in 1994. **Paramilitaries:** Border Guards: 7,000; Internal Security Forces: 2,000; Civil Defence Troops: 2,000.

created hope for change, which was frustrated by German resistance. From then onward, the electorate began to make ideological choices: the workers and urban population voted for the Social Democratic Party, while the rural areas supported the Agrarian or Republican Party.

[26] With World War I the fall of the Austro-Hungarian Empire finally brought about the recognition of the Republic of Czechoslovakia. The new state's borders established in 1919 by the victorious powers included parts of Poland, Hungary and the Sudeten land, all home to ethnic minorities, and all sources of potential conflict.

[27] Czech and Slovak leaders charged the National Assembly

with drawing up a Constitution. The assembly opted for a strict parliamentary system, in which the president and his cabinet would be responsible to two legislative chambers. Women's right to vote or to be elected was granted for the first time.

[28] In 1920, the Social Democratic Party lost the majority which it had held in previous elections. In 1922, the left wing of the party split off, creating the Communist Party, which was affiliated with Komintern. The Republican Party won a majority in subsequent governments, while the German minority in the Sudeten land organized itself to defend its own rights.

[29] The 1930s world-wide depres-

sion affected the Sudeten land intensely, as it was a highly industrialized region. It also accentuated nationalistic feeling among the German people there, who developed a separatist movement alongside Hitler's rise to power in 1933. Britain, France and Italy negotiated the cession of the Sudeten land to Germany in 1938, thus paving the way for the German occupation of Czechoslovakia in 1939.

[30] During World War II, Nazi SS troops imposed a harsh regime upon the Czechs, though toward the end they encountered strong national resistance from communist led groups. After Hitler's defeat, Czechoslovakia recovered its 1919 borders, while the German population was almost entirely expelled from the country.

[31] The Communist Party (CKC) obtained 38 per cent of the vote in the 1946 election, increasing to 51 per cent in 1948. In June, a People's Republic was proclaimed, and the CKC applied the economic model in effect at the time in the USSR. Czechoslovakia joined the Council for Mutual Economic Assistance (CMEA) and the Warsaw Pact.

[32] In 1960, the People's Republic of Czechoslovakia added "socialist" to its name. During the 1960s, intellectuals and artists started demonstrating increasingly openly against the regime.

[33] In early 1968, the election of Alexander Dubcek as Secretary of the CKC, and of Ludwik Svoboda as the country's president led to the implementation of a program to decentralize the economy, and affirm national sovereignty, against a background of broad popular support.

[34] The decision to change political course was the culmination of a process which had begun early in the decade, when Slovak leaders, expelled from the party in the 1950s, were rehabilitated. The Slovak struggle for autonomy (which had been even further restricted by the new socialist constitution) together with the 1967 student strikes brought an end to Antony Novotny's leadership of the CKC.

[35] Moscow viewed the possibility of Czechoslovakian reforms as a threat to the integrity of the socialist camp; in August 1968, it used Warsaw Pact Forces to intervene in the country and crush the movement. The leaders of the "Prague Spring" were expelled from the CKC and political alignment with the USSR was reestablished.

DEMOGRAPHY

Urban: 65% (1995). **Estimate for year 2000:** 11,000,000. **Children per woman:** 1.9 (1992).

HEALTH

Under-five mortality: 10 per 1,000 (1995). **Calorie consumption:** 115% of required intake (1995).

EDUCATION

School enrolment: Primary (1993): 100% fem., 100% male. Secondary (1993): 88% fem., 85% male. University: 16% (1993). **Primary school teachers:** one for every 18 students (1992).

COMMUNICATIONS

106 **newspapers** (1995), 105 **TV sets** (1995) and 103 **radio sets** per 1,000 households (1995). 19.0 **telephones** per 100 inhabitants (1993). **Books:** 104 new titles per 1,000,000 inhabitants in 1995.

ECONOMY

Per capita GNP: $3,200 (1994). **Annual growth:** -2.10% (1985-94). **Annual inflation:** 11.80% (1984-94). **Consumer price index:** 100 in 1990; 147.7 in 1994. **Currency:** 29 koruny = 1$ (1994). **Cereal imports:** 519,000 metric tons (1993). **Imports:** $15,636 million (1994). **Exports:** $14,252 million (1994). **External debt:** $10,694 million (1994), $1,035 per capita (1994). **Debt service:** 13.1% of exports (1994).

ENERGY

Consumption: 3,902 kgs of Oil Equivalent per capita yearly (1994), 13% imported (1994).

Foreign Trade

MILLIONS $ 1994

IMPORTS
15,636

EXPORTS
14,252

External Debt

1994

PER CAPITA
$1,035

---

the Civic Movement. In Slovakia, the Public against Violence Party divided into two opposing groups, the larger of which (PVC-for a Democratic Slovakia) urged the electorate to "fight against the swing to the right" that the country was taking.

[43] In the June legislative elections, the Czech Civic Democratic Party and the Slovak group, Democratic Slovakia, won in their respective republics.

[44] When negotiations over the statutes of the new federation came to an impasse, Czech and Slovak leaders meeting in the Moravian city of Brno admitted that separation was inevitable.

[45] On July 17, 1992, Czechoslovakian president Vaclav Havel announced his resignation after the declaration of sovereignty in the Slovakian National Assembly.

[46] Ex-president Havel spoke of the separation as "a peaceful and democratic decision" of the Czech and Slovak peoples, in his farewell speech. Czechoslovakia disappeared from the world map, replaced by the Czech Republic, with its capital Prague (Praha), and the Republic of Slovakia, with its capital Bratislava.

[47] As soon as Slovakia became independent, in January 1993, the country began a rapid process of integration with Western Europe. On January 26, Vaclav Havel, the former president of Czechoslovakia, was elected president of the Czech Republic, with the support of a parliamentary majority led by Prime Minister Vaclav Klaus.

[48] The country adopted the market economy with less negative results than the other former socialist republics. As for the rest of the region, the freeing of prices and the introduction of consumer taxes had limited inflationary effects: the price increase was 17 per cent in 1993. Unemployment remained at 2.5 per cent and per capita income continued to be one of the hightest in Eastern Europe.

[49] The Czech Republic became the eleventh signatory of the Peace Association in March 1994, an organism which developed out of NATO to progressively include the nations of the former socialist bloc in the alliance.

[50] On an internal front, Klaus went ahead with his policy of privatizing vast sectors of the Czech economy. The nation's debt continued lower than that of the other re-

gional former communist states, while foreign investment was one of the highest. Meanwhile, some specialists claimed the unemployment rate - 3.5 per cent of the active population and exceptionally low for the region - could increase markedly with Klaus's projects to restructure heavy industry.

[51] In 1995, the Liberal prime minister introduced paid registration in the universities. The banking system was affected by an excess of deposits, many from illegal economic activities. Also, the former director of the privatization agency, Jaroslav Lizner, who had been arrested in 1994, accused of receiving "hidden commissions," was sentenced to seven years imprisonment. In the same year, the Czech Republic became the first former Communist OECD member. In June 1996, the large growth in electoral support for the Social Democrats left Klaus without a majority in Parliament, though he managed to remain in the post and form a new government.

[52] Meanwhile, several sources indicated an increase in racial violence against the gypsies. According to the Ministry of the Interior, the attacks on the gypsies were not more frequent than before, but they were far more violent and were openly supported by a high percentage of the population.

---

[36] In 1977, a manifesto known as Charter 77, denouncing the lack of civil rights in Czechoslovakia, was divulged outside the country. Despite repressive measures, the reform movement grew, and in 1985 it called for the dissolution of NATO and the Warsaw Pact, along with the withdrawal of Soviet missiles and troops from Czech territory.

[37] From the time Mikhail Gorbachev was named Communist Party Secretary in the USSR, the reform process there brought about changes in Czechoslovakia, with the CKC trying to keep its hold on power. In 1989, despite violent repression, anti-government protests continued, precipitating a crisis within the regime.

[38] The government was forced to negotiate with the Civic Forum, an alliance of several opposition groups. Among other reforms, parliament approved the elimination of the CKC's leadership role, issuing a condemnation of the 1968 Soviet intervention. In late 1989, a provisional government was formed, with a non-communist majority.

[39] In December 1989, the Civic Forum declared that the CKC had redistributed cabinet positions in such a way as to keep its own people in the key positions. 200,000 people gathered in Prague to de-

mand a greater opposition representation in the cabinet. Gustav Husak resigned the presidency of the federation and was replaced by Vaclav Havel, who immediately granted amnesty to all political prisoners and called elections for June 1990.

[40] In late 1989, the Soviet and Czech governments agreed on a gradual withdrawal of the 70,000 Red Army troops stationed within the country. The previous year the USSR had started removing its nuclear warheads from Czechoslovakia. In October 1991, after the dissolution of the Warsaw Pact, President Havel and his Hungarian and Polish counterparts requested some kind of association with the Atlantic alliance.

[41] In the June 1990 elections 46.6 per cent of the votes went to the Civic Forum, against 13.6 per cent for the Communist Party. Havel was confirmed in the presidency, and the Czech and Slovak Federal Republic was proclaimed. In September, the republic joined the World Bank and the IMF.

[42] After the elections in the republics, the Forum divided into the Civic Democratic Party, self-proclaimed as "a party of the right, with a conservative program", and

Denmark

Danmark

Population: 5,205,000 (1994)
Area: 43,093 SQ KM
Capital: Copenhagen

Currency: Kroner
Language: Danish

The first hunters established themselves in Denmark in about 10,000 BC as the neolithic period drew to a close. This was followed by a flourishing Bronze Age civilization, about 1000 BC Around the year 500 AD, northern Germanic peoples began settling on the islands becoming fishermen and navigators. Certain place names bear witness to the worship of Scandinavian gods, such as Odin, Thor and Frey.

[2] The first evidence of hierarchical society in Denmark comes from the Viking age, mostly from cemeteries and settlement sites. The Vikings were Scandinavian farmers, navigators, merchants and, above all, raiders, who ruled the northern seas between the 8th and 10th centuries AD.

[3] Archeological remains indicate that Roskilde on the island of Sjaelland, Hedeby, to the north of Jutland, and Jelling, to the south, were the most intensely populated areas. After Danish victory over the Germans the Eider River became the final southern border. A huge wall was built to the south and west of Hedeby.

[4] In the 10th century, after continuous conflict with rival kingdoms, the center of the kingdom's power was transferred to Jelling, where Gorm became king of Jutland. His son, Harald Bluetooth (Blatand), is credited with uniting Denmark and conquering parts of Norway.

[5] Subsequent Viking reigns extended Danish possessions as far as modern-day England and Sweden. In 1397 King Margrethe managed to unite Denmark, Norway, Iceland, Greenland, Sweden and Finland in the "Union of Kalmar".

[6] The introduction and spread of Christianity and strengthening of the Hanseatic League went hand-in-hand with the weakening of Denmark's military power.

[7] The Danish kings were involved in successive wars between themselves, campaigns which were

interspersed with peasant rebellions and bourgeois revolts; a large and powerful middle class had developed as a result of growing mercantile activity. These conflicts ceased in the 17th century, when a weakened nobility gave the king power as absolute sovereign. He was then able to create laws to be imposed throughout the land.

[8] During the 18th century, Denmark colonized the Virgin Islands. The Danish colonists organized local production using African slave labor and in 1917 the islands and their population were sold to the United States. (See "American" Virgin Islands)

[9] Peace existed in Denmark and Norway from 1720 until the Napoleonic Wars. After Napoleon's defeat, Sweden attacked Denmark and, under the Kiel Peace Treaty, annexed Norway in 1814.

[10] The loss of Norway, combined with British trading impositions, brought on an economic collapse which worsened as a result of low wheat prices. The ensuing agricultural crisis forced land reform to a standstill. The situation later improved when agricultural prices stabilized, trade increased and industrialization began. In 1814 an educational reform made schooling obligatory.

[11] After the European revolutions of 1848, King Frederick VII called an assembly. A parliamentary monarchy was established and absolutism was abolished. The 1849 Constitution guaranteed freedom of the press, of religion and of association,

as well as the right to hold public meetings. The main trends of the period were nationalism and liberalism.

[12] A territorial dispute with Germany over the Duchies of Schleswig and Holstein reinforced nationalistic sentiment. In 1864, when Denmark was defeated by Prussia and Austria, it lost its claim to these lands and the national-liberal government was brought down.

[13] The 1866 Constitution maintained the monarchy. In 1871, Louis Pio, a former military officer, attempted to form a socialist party. A series of strikes and demonstrations organized by the socialists was put down by the army, and Pio was deported to the United States. The Social Democratic Party, mainly supported by intellectuals and workers, dates back to 1876.

[14] The peasants and emerging middle class weakened the monar-

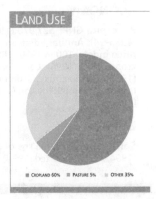

LAND USE

■ CROPLAND 60%   ■ PASTURE 5%   ■ OTHER 35%

IMPORTS
34,800
EXPORTS
41,417

chy on three fronts: the farm cooperative movement, a liberal bourgeois party (popularly referred to as "leftist") and the social democrat party.

[15] In 1901 the United Left (Liberal) came to power establishing a new government. The emergence of the UL and the Social Democrats as a leading force at the turn of the century was the result of agrarian reform, industrialization and the development of railroads. The growth of urbanization and overseas trade (accelerating the formation of labor unions throughout the country and the rise of co-operatives in the countryside) was the main reason for these changes.

[16] In 1915, the Constitution was revised. The voting age of 35 was maintained, but the right to vote was extended to women, servants and farm hands. There were judicial reforms, bringing in trial by jury, and land distributions in the largest states.

[17] Also in 1915 women were granted the right to stand for election, a right they had not enjoyed since viking times. The first woman was appointed to the cabinet in 1924.

[18] After the 1870-71 Franco-German War, Denmark adopted a neutral international stance. World War I offered Copenhagen favorable trading opportunities with the warring nations, but at the same time affected its supplies. During World War II Denmark was invaded by Germany, al-

STATISTICS

DEMOGRAPHY

Urban: 85% (1995). **Annual growth:** 0.2% (1991-99). **Estimate for year 2000:** 5,000,000. **Children per woman:** 1.8 (1992).

HEALTH

Under-five mortality: 7 per 1,000 (1995). **Calorie consumption:** 116% of required intake (1995).

EDUCATION

School enrolment: Primary (1993): 98% fem., 98% male. Secondary (1993): 115% fem., 112% male. University: 41% (1993). **Primary school teachers:** one for every 11 students (1992).

COMMUNICATIONS

104 **newspapers** (1995), 103 **TV sets** (1995) and 103 **radio sets** per 1,000 households (1995). 58.9 **telephones** per 100 inhabitants (1993). **Books:** 118 new titles per 1,000,000 inhabitants in 1995.

ECONOMY

Per capita GNP: $27,970 (1994). **Annual growth:** 1.30% (1985-94). **Annual inflation:** 2.90% (1984-94). **Consumer price index:** 100 in 1990; 108.0 in 1994. **Currency:** 6 kroner = 1$ (1994). **Cereal imports:** 579,000 metric tons (1993). **Fertilizer use:** 2,088 kgs per ha. (1992-93). **Imports:** $34,800 million (1994). **Exports:** $41,417 million (1994).

ENERGY

Consumption: 3,996 kgs of Oil Equivalent per capita yearly (1994), 27% imported (1994).

Faeroe

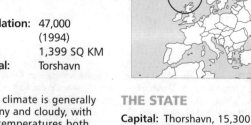

Population: 47,000 (1994)
Area: 1,399 SQ KM
Capital: Torshavn

The climate is generally rainy and cloudy, with mild temperatures both in summer and winter. Only 6% of the land is cultivated. Most of the production consists of vegetables as the land is not suitable for grain. Sheep are raised throughout the islands and there is mining in Suderoy. 20% of the population are employed in handicraft production and 21% in fisheries, which provide 90% of the islands' exports.

SOCIETY

Peoples: The population is of Scandinavian origin. **Religions:** Lutheran. There are a large number of Baptists and a small Catholic community. **Languages:** Danish (official) and the Faeroese dialect. **Political Parties:** Social Democratic and Union (liberal) parties, which favor ties with Denmark; the Republican "Left", Popular and Progressive parties, favor independence.

THE STATE

Capital: Thorshavn, 15,300 inhab. (1988).
Government: The parliament (Løgting) has up to 30 members, elected on a basis of proportional representation. Parliament names a national cabinet (Landstyret) of 6 ministers. After the 1990 elections, the latter was a coalition between the Social Democratic and Popular parties. A high commissioner (Rigsombudsman) represents the Crown. The islands elect 2 representatives to the Danish Parliament.
Diplomacy: In January 1974, the Logtinget decided not to join the EEC. To protect the fishing industry, a 200- mile fishing zone was established in 1977. Modernization of the islands' fishing fleet and methods was financed through a series of foreign loans, which totaled $839 million in 1990, that is, $18,000 per capita.

COMMUNICATIONS

49.5 **telephones** per 100 inhabitants (1993).

ECONOMY

Consumer price index: 100 in 1990; 110.6 in 1993.

though it officially maintained its independence until 1943.

[19] When Hitler attacked the USSR, Denmark created a volunteer army and outlawed communist activity. The 1943 election was an anti-Nazi plebiscite, with the electorate throwing its support behind the democratic parties. Resistance to the Nazi regime, strikes, and the government's refusal to enforce Nazi rule, led the German occupation forces to declare a state of emergency dissolving Denmark's police and armed forces.

[20] In September 1943, the Danish Freedom Council was created to co-ordinate the anti-Nazi opposition. When Germany finally surrendered, a transition government was formed, with representatives from the Council and the traditional parties. The 1945 election was won by the liberals.

[21] The Faeroe Islands, under Danish control since 1380, were occupied by Britain during the war and subsequently returned to Denmark. The 1948 Constitution gave the islands greater autonomy, though the Danish Parliament retained control of defence and foreign affairs. Denmark is currently disputing the control of resources under the sea bed in island waters.

[22] In Greenland, also under Danish control since 1380, Denmark was responsible for foreign policy and justice. In 1979, the island obtained the right to have its own legislative assembly (Landsting), which has power over internal affairs. Other areas are dealt with by

the Danish Parliament which has two representatives from the island.
[23] Since 1947, there has always been at least one woman in the cabinet. That same year, Protestant women achieved the right to be pastors.
[24] In the post-war period, a proposed Nordic Defence Alliance did not come about, as Norway did not agree with Denmark and Sweden. Denmark joined NATO in 1949, increasing its military power with the help of the United States. A US proposal for setting up air bases on Danish soil was, however, turned down.
[25] In 1953, the constitution was amended reducing the Legislature

> **Both divorce and the option to live together - rather than marry - are on the increase. Marriage between people of the same sex has been authorized by law since 1989.**

to a single chamber (*Folketing*). In 1954 the post of 'ombudsman' was created to ensure that municipal and national government complied with the law, and to protect "ordinary" citizens from the misuse of power on the part of government officials or agencies.
[26] In 1959, Denmark became the founding member of the European Free Trade Association (EFTA). Denmark joined the European Economic Community (EEC) in 1972 when 63.7 per cent of the electorate approved the move. In 1973, the oil crisis seriously affected the Danish economy and unemployment rose to over 14 per cent.
[27] In the 1973 election, the traditional parties were ousted in favour of two new parties; the extreme right Progressive Party and the Democratic Center, the right-wing of the Social Democratic Party. In 1975, the Social Democrats were voted back into power. They lost a significant number of votes in subsequent elections because of high unemployment and unpopular economic measures. The Socialists

were voted out in 1982, and the Conservatives won office for the first time since 1894. Their victory was due to their alliance with the Liberals, the Democratic Center and the Christian People's Party.
[28] In the 80s, 23 per cent of the seats in Parliament were held by women. In 1982, the government appointed two female ministers responsible for Labor and Religious affairs. In that year 43.9 per cent of the economically active population were women, and they were protected by specific legislation. A women can interrupt her career for up to two years to raise her children. The average Danish family has four members.
[29] With regard to ecological issues, in 1986 Parliament approved a strict environmental protection law which has proved costly for both the industrial and agricultural sectors.
[30] Despite its progressive tradition, and unlike other European nations, Parliament did not ban trade with South Africa - in protest against apartheid - until 1988.
[31] Denmark is one of the few countries that has complied with United Nations recommendations to contribute at least 1 per cent of its GDP to development. It's ecological record has been good: Denmark recycles 72 per cent of all the paper it uses, and recovers more than 100,000 tons of industrial waste per year.
[32] Both divorce and the option to live together - rather than marry - are on the increase. Marriage between people of the same sex has been authorized by law since 1989.
[33] Within the last few years, there has been an increase in poverty levels in the cities. In 1993, the unemployment rate reached 12.5 per cent. The right-wing Progress Party reacted by calling for an end to immigration. Immigrants currently make up 3.6 per cent of the population, including 30,000 Turks, 30,000 Europeans from Community nations, 22,000 Scandinavians and other Europeans, Pakistanis and Sri Lankans.
[34] There has traditionally been a colonial attitude toward Greenland, where Danes and other Europeans changed the environment known by the Inuit, also introducing alcohol, previously unknown human diseases and canine rabies.
[35] In 1993, the people of Denmark

Population: 55,000 (1994)
Area: 2,175,600 SQ.KM
Capital: Nuuk (ex-Godhaab)
Currency: Kroner
Language: Greenlandic Danish and Inuit

Located in the Arctic Ocean, the island is the second largest tract of frozen land on the planet. Nearly four fifths of its surface is covered by an ice cap. In the month of June, soon after the rapid thaw, moss and lichen vegetation appears on certain parts of the coast. Most of the population is concentrated in the western region, where the climate is less severe. Fishing forms the basis of the economy; whale oil and salted or frozen fish are exported. The country has lead, zinc, and tungsten deposits. Cryolite from the large reserves in Ivigtut is also exported.

SOCIETY

Peoples: 80% are Inuit. The remaining 20% are Danish or other short-term European residents.
Religions: Lutheran. The Greenlandic Church comes under the jurisdiction of the Bishop of Copenhagen and the minister of ecclesiastical affairs.
Languages: Greenlandic, Danish and Inuit **Political Parties:** Siumut, Social

Democratic; Atassut, liberal-conservative; Inuit Atgatigiit, nationalist federation.

THE STATE

Capital: Nuuk (ex-Godhaab), 12,426 inhab. (1989).
Government: Greenland elects two representatives to the Danish *Folketing*, and has one representative on the Nordic Council.
Diplomacy: On January 1, 1985 the island withdrew from the European Economic Community. However, EEC countries can still fish in Greenland's territorial waters upon payment of US$ 20 million.

DEMOGRAPHY

Density: 0 inhab./sq km
Annual growth: 1.1% (1970-86).

COMMUNICATIONS

198 **TV sets** and 411 **radio receivers** per 1,000 inhab. (1991).
Consumption: 4,630 kgs of Coal Equivalent per capita yearly, 100% imported. (1990).

finally approved the Maastricht Treaty and the European Union, after rejecting it the previous year.
[36] The Social Democrat prime minister, Poul Nyrup Rasmussen was re-elected in 1994, after being appointed to the post the year before, following the scandalous fall of the Liberal-Conservative government for the violation of refugee laws. The former justice minister Erik Ninn Hansen was found guilty of having prevented Sri Lankan Tamil refugee families from being reunited.
[37] Rasmussen achieved a parliamentary majority through alliances with centre and left-wing parties. The nation's economic growth hit a high in 1994 with a 4.4 per cent increase in GDP and 2 per cent inflation; growth slowed in 1995 and 1996. The reduction of the exten-

sive social welfare benefits allowed for unemployment to be reduced from 12.5 per cent to 10 per cent in two years.
[38] In March 1995, Copenhagen hosted the United Nations Social Development Summit, which discussed aims for the erradication of poverty, job creation and personal security. Some 20,000 people from 180 countries attended the summit.

Djibouti

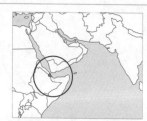

Population: 603,000 (1994)
Area: 23,200 SQ KM
Capital: Djibouti

Currency: Franc
Language: French

Djibouti

In about the 3rd century BC Ablé immigrants came from Arabia and settled generally in the north and partially in the south. The Afars, or Danakil, are descendants of these peoples. Later the Somali Issas pushed the Afars out of the south and settled in the coastal regions. In AD 825 Islam was brought to the area by missionaries. Arabs controlled the trade in this region until the 16th century, when the Portuguese competed for it. On 1862, Tadjoura, one of the Sultanates on the Somalian coast (see Somalia) sold the port of Obock and adjoining lands to the French for 52,000 francs and in1888 French Somaliland (Côte Française des Somalis) was established.

[2] Djibouti became the official capital of this French territory in 1892. A treaty with Ethiopia in 1897 reduced the territory somewhat in size. A railway was built to connect Djibouti with the Ethiopian hinterland, reaching Dire Dawa in 1903 and Addis Ababa in 1917. The interior of the area was effectively opened between 1924 and 1934 by the construction of roads and administrative posts. After World War II Djibouti port lost trade to the Ethiopian port of Aseb (now in Eritrea). In 1946 French Somaliland acquired the status of an overseas territory (from 1967 called the French Territory of the Afars and Issas), and in 1958 it voted to become an overseas territorial member of the French Community under the Fifth Republic.

[3] Independence and the reunification of neighboring Somalia stimulated the emergence of anticolonialist movements such as the Somaliland Liberation Front and the African League for Independence, both of which used the legal and armed branches.

[4] During the 1970s, renewed resistance forced acting governor Ali Aref to resign. France called a plebiscite on May 8 1977, and 85 per cent of the population voted for independence. Hassan Gouled Aptidon, main leader of the African League for Independence, became president of the newborn Republic.

[5] The government of the new state had to deal with tension between the Afar and Issa peoples and with refugees from the war zones in Ethiopia and Somalia. In an attempt to overcome old ethnic divisions, Gouled granted governmental participation to various groups. He even appointed several Afar ministers. Though French remained the official language, Djibouti was admitted into the Arab League.

[6] Ethiopia and Somalia, its two neighbors, both had territorial designs on Djibouti. Ethiopia's interest in Djibouti was basically geopolitical. In the event of Eritrean nationalists achieving independence, Ethiopia would become a landlocked territory - if an agreement can not be reached for the use of Eritrean ports, Djibouti would be their only available harbor. For Somalia it was mainly a historical aim to unify the Somali nation.

[7] In mid-1979, president Hassan Gouled resumed relations with Ethiopia and Somalia, signing trade and transportation agreements with them. The participation of Afars in the government and in the newly-formed army was encouraged as a way of securing some semblance of national unity.

[8] A ruling party was organized and municipal administrations were created to encourage political participation. Foreign aid was basically used for irrigation works and to improve the situation of refugees from the Ogaden war.

[9] The Gouled administration

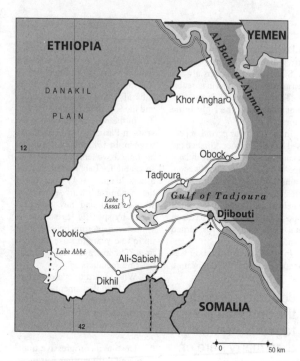

> ## In October 1981, President Gouled amended the constitution to introduce a single-party system.

received aid from Saudi Arabia, Kuwait, Iraq and Libya. So far, Gouled has skillfully administrated his country, playing upon its only valuable resource: its strategic position at the mouth of the Red Sea. Contrary to expectations which foretold a rapid annexation by Ethiopia or Somalia, Djibouti has stressed its determination to remain independent.

[10] In spite of a successful performance in diplomatic relations, Gouled faced serious domestic problems, the most pressing of which result from ethnic rivalries between Afars and Issas. Afars, who account for 35 per cent of the population, complain of political and economic discrimination. The Issas, who constitute 60 per cent of the population and hold key positions in the government, refute all accusations of group favoritism, and support Gouled's radical policies to neutralize the opposition. After the prohibition of the Popular Liberation Movement in 1979, the Afars tried to reorganize as the Djibouti Popular Party in 1981, but this was also banned.

[11] In October 1981, President Gouled amended the constitution to introduce a single-party system. The official Popular Association for Progress (RPP) became the sole legal party and the other groups were banned. It was argued that they had racial or religious aims and in consequence were potentially harmful to national unity.

[12] During 1983, the first steps of a project planned to radically change Djibouti's economy, were taken. Incentives were to be given to transform the country into a financial center and free trading port, a sort of Middle Eastern Hong Kong. Six foreign banks opened offices in Djibouti, mainly attracted by a solid national currency backed by dollar deposits in the US.

[13] By the end of 1984, the first results were not encouraging: the number of passengers and goods in transit to Ethiopia and Somalia had dropped considerably and, therefore, customs revenues and bank activities had decreased. The continuing conflict in the area has been seen as the major cause for the withdrawal of European capital.

[14] With the support of the UN High Commissioner, the government of Hassam Gouled resumed the voluntary repatriation of more than a hundred thousand refugees, a process which had been interrupted in 1983.

[15] At the same time, bilateral agreements were signed with Ethiopia to combat contraband and promote peace in the border areas. These had been closed in 1977, at the outset of the conflict with Somalia over the Ogaden region.

[16] In August 1987, foreign military presence in Djibouti grew as French bases started to be used by US and British forces participating in maneuvers in the Persian Gulf. Relations with the former colonial power were reinforced by President François Mitterand's visit, in December.

[17] Djibouti depends on French economic aid to a large extent, estimated at $200 million per year, a figure which is several times the total of its exports. This dependence on foreign aid worried local authorities, who were making every effort to diversify the country's commercial relations. With this in mind, Prime Minister Barkad Gourad Hamadou signed an economic and

ENVIRONMENT

Located in the Afar triangle, facing Yemen, it is one of the hottest countries in the world (average annual temperature: 30°C). The land is mostly desert, its only green area is found in the basalt ranges of the northern region. Agriculture, confined to oases and to a few spots along the coast, satisfies only 25% of the domestic food demand. Extensive cattle raising is practised by nomads. Economic activity is concentrated around the port.

SOCIETY

Peoples: Djiboutians are divided into two major ethnic groups of equal size; the Afar, scattered throughout the country, and the Issa, of Somalian origin, who populate most of the southern territory and predominate in the capital. **Religions:** Muslims. There is a small Christian minority (5%). **Languages:** Afar and Issa (Somali), French (official) and Arabic (religious). **Political Parties:** People's Assembly for Progress (RPP), which emerged in 1979 from the African People's League for Independence, founded in 1975 by Hassan Gouled. Democratic Renovation Party (PRD); Democratic Front for the Liberation of Djibouti; Movement for the Liberation of Djibouti and the Front for the Restoration of Unity and Democracy (FRUD).

THE STATE

Official Name: République de Djibouti. **Administrative divisions:** 5 districts. **Capital:** Djibouti, 450,000 inhab. (1989). **Other cities:** Ali-Sabieh 4,000 inhab.; Tadjoura 3,500 inhab.; Dikhil 3,000 inhab. (1989) **Government:** Hassan Gouled Aptidon, President since June 1977, reelected in the May 1993 elections. Barkad Gourad Hamadou, Prime Minister. The prime minister is traditionally Afar to counterbalance the power of Gouled, who is Issa. **National Holiday:** June 27, Independence Day (1977). **Armed Forces:** some 3,900 **Paramilitaries:** Gendarmerie (Ministry of Defence): 600; National Security Force (Ministry of Interior): 3,000.

DEMOGRAPHY

Annual growth: 0.5% (1992-2000). **Children per woman:** 5.8 (1992).

HEALTH

Under-five mortality: 158 per 1,000 (1995). **Safe water:** 95% of the population has access (1990-95).

EDUCATION

Illiteracy: 54% (1995). **Primary school teachers:** one for every 43 students (1990).

COMMUNICATIONS

87 **newspapers** (1995), 95 **TV sets** (1995) and 85 **radio sets** per 1,000 households (1995). 1.3 **telephones** per 100 inhabitants (1993).

ECONOMY

Currency: 178 francs = 1$ (1994). **Cereal imports:** 44,000 metric tons (1990).

Ougoureh Kible. In June, Gouled and the FRUD jointly decided to end the two and a half year war.
[22] In June, a demonstration of mostly Afar residents from the Arhiba district, opposed to the demolition of their homes for "security" reasons, was quelled by government forces. Police intervention left an outcome of four dead, 20 injured and 300 arrests, including that of Muhammad Ahmad Issa, United Opposition Front president.
[23] In October, dissent within the FRUD had grown and the move-

after the Constitution was revised, a section of the FRUD formed an alliance with the ruling party. Ahmad Dini Ahmad said this alliance amounted to "treason".
[24] In mid 1995 and under pressure from the IMF, the government considerably reduced public spending and took a series of measures to increase fiscal revenues.

technical aid agreement with Turkey in June 1989.
[18] In 1990 France wrote off Djibouti's debt by granting it 40 million dollars as a way of public development aid. In 1991, confrontations between the government and the Front for the Restoration of Unity and Democracy (FRUD) guerrillas were renewed. In November, Amnesty International accused the government of Djibouti with the torture of 300 prisoners. Due to mounting strains, France pressed Gouled to begin talks with the opposition.
[19] In September 1992, amid harsh combat against the FRUD, the government introduced a constitutional reform which established a multi-party system. In the May 1993 presidential elections, Gouled was once again re-elected with over

60.8 per cent of the vote issued. However, encouraged by the FRUD, half of the electorate abstained from voting and the opposition considered the elections a "fraud".
[20] Armed confrontations between government troops and guerrillas escalated in the weeks following the elections, which led thousands to seek refuge in Ethiopia. Acting as mediator, the French government sought a cease-fire and negotiations began. Meanwhile, Gouled believed the rebellion was part of a plan orchestrated by Ethiopia.
[21] In March 1994, the FRUD was divided due to a possible agreement with the government. The movement's Politburo, led by Ahmad Dini Ahmad, who opposed the agreement, was replaced by an executive council headed by Ahmad

Armed confrontations between government troops and guerrillas escalated in the weeks following the elections, which led thousands to seek refuge in Ethiopia

ment's leaders banned Ahmad Dini Ahmad and Muhammad Adoyta Yussuf, another breakaway leader, from holding any "activity or responsibility" in the FRUD. In 1995,

Dominica

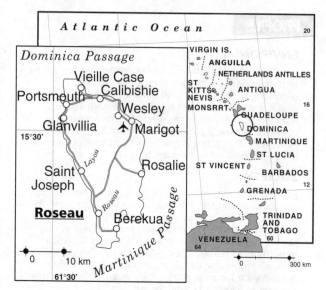

Population:	72,000 (1994)
Area:	750 SQ KM
Capital:	Roseau
Currency:	E.C. Dollar
Language:	English

Dominica

On Sunday, November 3, 1493, Spanish-Italian navigator Christopher Columbus reached an island which he named Dominica. He planted a cross to claim Spanish sovereignty over the island, claiming the newly discovered territory for Queen Isabella and King Ferdinand of Spain.

[2] This previously unclaimed territory was actually already fully populated by Carib peoples, peaceful and friendly indigenous groups. They watched the ceremony from a distance, but could not have understood its significance. Soon afterwards, the Spanish returned in a more aggressive frame of mind, searching for the vast stores of gold that they believed the Caribs were hiding from them. They rapidly exterminated all but a few of the legitimate owners of the island; a pattern of destruction and despoilment that would be repeated throughout the colonial possessions in the new continent.

[3] According to reliable historical sources, by 1632 there were only approximately one thousand surviving Caribs on the island. These few people had only managed to continue their existence through perpetual brave resistance. Today, there are only five hundred of their descendents left living in the reservations.

[4] The landscape of present day Dominica has changed greatly since their ancestor's days. In colonial times, the forests were entirely felled to clear the land for extensive sugar cane plantations. The Carib peoples could not provide a sufficiently sturdy and reliable workforce, so the plantations were worked by thousands of slave laborers transported from Africa.

[5] In the 17th century, the Spanish withdrew and the French took their place. They introduced cotton and coffee production to complement the sugar trade. In the following two centuries, frequent British attempts to seize the island from French hands culminated in Dominica becoming a British colony in 1805.

[6] French influence, however, still persists. The Catholic religion predominates despite strong competition from well-financed Protestant sects, and "Creole", the language spoken by the people, is based on French mixed with African languages.

[7] After five centuries of colonization, the Dominicans inherited a barely developed agrarian economy based on the monocultivation of bananas for export. They are the only significant export item, because most of the land went over to banana production when sugar cane became unprofitable at the end of the 19th century.

[8] The political system adopted on achieving internal autonomy in 1967, was a copy of the British model. The constitution established the "free association" of the Associated States of the West Indies with Great Britain. Britain retained responsibility for defence and foreign relations and each member island elected its own individual state government. The seat of the federal government was in Barbados.

[9] From 1975, independence was negotiated separately by each state. In the case of Dominica the negotiations were conducted by Labor Party premier, Patrick John. In 1978, the British Parliament approved a new statute for the island and on November 3, 1978, exactly 485 years after the arrival of Columbus, Dominica became an independent country once more.

[10] The protest rallies that took place during the celebrations were an indication of the difficulties that lay ahead. Political division immediately became evident: a conservative group of veteran politicians opposed independence; and a progressive sector supported by young people disagreed with the Labor government which had done the negotiating.

[11] Since the island lacks the natural resources to attract transnationals and tourism, Dominica cannot compete with its neighbors. Vast numbers of young people emigrate to neighboring Guadeloupe, while others, who are more radical, find political inspiration in the "black power" ideology of US black activists.

[12] In May 1979, the police fired at demonstrators protesting against two decrees limiting union activities and press freedom. At the same time, secret agreements between John and the government of South Africa were revealed; the Dominican premier had helped to plan a mercenary attack on Barbados and intended to supply Pretoria with oil refined in Dominica. Patrick John was forced to resign. His successor, Oliver Seraphine, from the progressive wing of the labor party, called elections and strengthened Dominica's ties with progressive neighboring governments. In September 1979, Hurricane David devastated the island. The conservative opposition skillfully associated the catastrophe with the Labor administration and gained electoral support.

[13] In the July 1980 election, Mary Eugenia Charles won a landslide victory and became the first woman to head a Caribbean government. She was the daughter of a wealthy and influential landowner and she describes herself as "liberal, democratic and anti-communist". Since her election, her government has survived two hurricanes and an attempted *coup d'état*.

[14] In the 1985 election, Charles' party (DFP) obtained 15 of the 21 Parliamentry seats; the Labour Party, led by Michael Douglas, won only 5 seats and the Dominican Unified Labour Party won the remaining seat for its leader. During the electoral campaign, the opposition called for an investigation into a $100,000 donation made to the government. It was believed that the donation had been made by the CIA to confirm Dominican collaboration in the US invasion of Grenada.

[15] In August 1986, Air Force commander Frederick Newton was sentenced to death by hanging for his participation in a plot, to topple the conservative government in 1981. Former prime minister Patrick John and Michael Reid, captain of the Defense Forces, were found guilty of similar charges, and of involvement in a minor Ku Kux Klan conspiracy. They were sentenced to 12 years in prison.

[16] After the May 28, 1990 parliamentary election, the third government since independence took office. Of the 21 seats up for election, Charles's DFP (Dominica Freedom Party) won by a slim majority of 11 seats, while the Labor Party (led by Douglas) won 4 and the United Labor Party, a newcomer in the electoral process, led by Edison James, won 6 seats. Leonard Baptist's Progressive Force did not win any seats.

[17] The day after the election, which gave the 71-year-old lawyer Charles her third consecutive term,

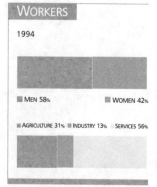

WORKERS

1994

MEN 58%   WOMEN 42%

AGRICULTURE 31%   INDUSTRY 13%   SERVICES 56%

ENVIRONMENT

The largest Windward Island of the Lesser Antilles, located between Guadeloupe to the north and Martinique to the south, Dominica is a volcanic island. Its highest peak is Morne Diablotin, at 2,000 meters. The climate is tropical with heavy summer rains. The volcanic soil allows for some agricultural activity, especially banana and cocoa plantations.

SOCIETY

Languages: English (official). A local French patois, with African elements, is widely spoken. **Religions:** Roman Catholic 70.1%; six largest Protestant groups 17.2%, of which Seventh-day Adventist 4.6%, Pentecostal 4.3%, Methodist 4.2%; other 8.9%; nonreligious 2.9%. **Political parties:** Dominica Revolutionary Party (PRD), led by José Francisco Peña Gómez; Social Christian Reformist Party (PRSC), led by Jacinto Peynado; Dominican Liberation Party (PLD), led by Leonel Hernández; Democratic Unity (UD).

THE STATE

Official Name: Commonwealth of Dominica. **Administrative divisions:** 10 parishes. **Capital:** Rosean **Other cities:** Portsmouth, 3,621 inhab., Marigot, 2,372 inhab. (1991). **National Holidays:** November 3, Independence Day (1978) and Discovery by Christopher Columbus (1493).

DEMOGRAPHY

Children per woman: 2.5 (1992).

HEALTH

Calorie consumption: 105% of required intake (1988-90).

EDUCATION

Primary school teachers: one for every 29 students (1991).

COMMUNICATIONS

19.1 **telephones** per 100 inhabitants (1993).

ECONOMY

Per capita GNP: $2,800 (1994). **Annual growth:** 4.30% (1985-94). **Consumer price index:** 100 in 1990; 114.9 in 1994. **Currency:** 3 E.C. dollars = 1$ (1994).

Patrick John and Michael Reid were freed.

[18] In 1990, prime minister Eugenia Charles signed an agreement with her counterparts James Mitchell, of Saint Vincent, John Compton, of Saint Lucia, and Nicholas Braithwaite, of Grenada, for the four islands to form a new state.

[19] In the same year the Regional Constitutional Assembly of the Eastern Caribbean, of government officials, religious authorities, and representatives from social organizations, started to operate. In a second meeting, in Saint Lucia in April 1991, the Assembly studied various proposals for the establishment of the future State.

[20] This integration project is in agreement with the Chaguaramas Treaty, which aims at the unification of the entire English-speaking Caribbean community, although the three islands of Kitts, Montserrat, and Antigua have refused to commit themselves to the process. The opposition parties of Grenada and Lucia have also stated that they were against it. The seven mini-States of the Eastern Caribbean have a common Central Bank, which coins their shared currency.

[21] In 1991, the government decided on a series of measures to stimulate the national economy, centering its efforts on the development of agriculture and communications.

[22] In April, the prime minister narrowly escaped censure, through a motion presented by the United Workers' Party. At the end of 1991, Charles attempted to enact legislation making it illegal for civil servants to protest against the government.

[23] In 1993, Prime Minister Charles raised the rates on several services, suspended government investment and proposed a wage freeze, in an attempt to deal with the budget deficit. Thanks to her total support of US policies, the island obtained American foreign aid, earmarked for the revitalization of its agriculture.

[24] Charles proposed offering Dominican passports to 1,200 Asians, on the condition that each one invest at least $35,000 in the country, but resistance to this idea increased during 1992-3. In mid-1993, the prime minister modified the project, raising the amount of the investment to $60,000 and imposing restrictions on the voting rights of the new citizens.

[25] In April 1994, the government decision to increase the number of transport vehicle licenses caused protests and public disorder in Roseau. The prime minister accused the opposition of trying to bring forth early elections, and the crisis

> In 1990, prime minister Eugenia Charles signed an agreement with her counterparts James Mitchell, of Vincent, John Compton, of Lucia, and Nicholas Braithwaite, of Grenada, for the four islands to form a new state.

continued until an agreement was signed on May 6.

[26] In the same year, Dominica decided to vote against the establishment of a whale sanctuary in the South Atlantic. This attitude earnt them a threatened tourism ban from the International Wildlife Coalition. This measure, which would have seriously affected a national economy heavily dependent on tourism, was not put into practice.

[27] In June 1995, the United Dominican Labour Party (UDLP) won the election, taking 11 of the 21 seats in Parliament. The Freedom Party and Liberal Party took five seats each. The new prime minister, Edison James, decided to promote the banana industry and privatize State enterprises in order to invest in the social infrastructure.

[28] In August and September 1995, the government plans were severely shaken. A succession of hurricanes and tropical storms destroyed the plantations and export projects. Houses, bridges, roads, hotels and public facilities were also destroyed and had to be rebuilt, absorbing vast amounts of resources.

Dominican Republic

República Dominicana

Population: 7,622,000 (1994)
Area: 48,730 SQ KM
Capital: Santo Domingo

Currency: Peso
Language: Spanish

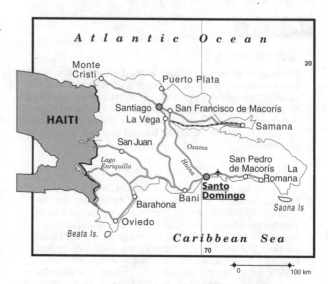

The island of Quisqueya was made up of two present-day countries: Haiti and the Dominican Republic. The first inhabitants belonged to several ethnic groups; the Lucayo, the Ciguayo, the Taino and the Carib. They were communities of fishers and gatherers who practiced rudimentary agriculture. There was always a great deal of contact between the Caribbean islands and trade between the tribes.

[2] In December 1492, Christopher Columbus reached the island of Quisqueya, which he renamed Hispaniola. With the wood from one of his vessels he built a fort; initiating the European colonization of America. Within a few years, the Europeans had appropriated the whole island becoming owners of the land and the native Carib population. The terrible living and working conditions imposed by the Spaniards nearly exterminated the natives. Faced with this shameful situation, Bishop Bartolomé de Las Casas proposed that they replace native slave labor with Africans, millions of whom were distributed throughout the American continent.

[3] Dominican historical records show that in 1523 a group of rebel African slave workers, founded the first *quilombo* (former slave settlement) on the island. Subsequent rebel groups, in 1537 and 1548, set up their own quilombos. The replacement of native American labor with African labor, accompanied a change in economic focus, from panning for gold to plantations of sugar cane and extensive cattle raising. As historian Pierre Vilar points out, the gold cycle in Hispaniola was "destructive, not of raw materials, but rather, of the labor force". During the colonial period the extraordinary economic potential of the Dominican Republic was comparable only to that of Brazil. The island was successively the greatest gold producer in the Antilles, one of the largest producers of sugar in the New World, between 1570 and 1630, and finally, such an important cattle producer that, there were 40 cattle per person on the island.

[4] As major sugar producer with a key position on the trade route from Mexico and Peru to Spain, Hispaniola was coveted by the other colonial powers. In 1586, the English buccaneer Francis Drake raided the capital and in 1697 the French occupied the island's western half. When they were given official ownership under the Treaty of Ryswick, they renamed it Haiti. Later, the whole island fell under French rule but was partially recovered by Spain in 1809, after the first Afro-American republic had been established.

[5] Haiti's native government regained control over the whole island in 1822. The Spanish descendants', or *criollo*, resistance came to a head after an uprising in Santo Domingo. The independence of the Dominican Republic was proclaimed, but in 1861 the government asked Spain to renew colonial status, in an attempt to gain support for the *criollos*, whose dominance was threatened by the negro and mulatto majorities.

[6] However, Spain did not defend its colony effectively and the Dominican Republic became independent again in 1865 after a mulatto uprising. The economic system remained unchanged.

[7] By that time the US, fully recovered from the civil war, began to gain influence in the West Indies. In 1907 the US imposed an economic and political treaty on the Dominican Republic prefiguring "dollar diplomacy". This treaty helped the US to invade the Dominican Republic and impose a protectorate that lasted until 1924.

[8] In 1930, when the country was autonomous again, Rafael Leónidas Trujillo seized power. He was Head of Staff of the National Guard, elected and trained by the American occupation forces. He set up a dictatorial regime, with US backing, without nominally occupying the presidency. His crimes were so numerous and so monstrous that he finally became too embarrassing, even for the US, and the CIA planned his assassination which was carried out in May 1961.

[9] One extreme example of his crimes was the case of Jesús de Galíndez, a Spanish Republican professor, labor leader, and member of the Basque Nationalist Party. He was kidnapped by assassins, on Trujillo's payroll, and was subsequently killed, although his body was never found. Trujillo had the gunmen killed some time later which proved to be a fatal mistake; among the assassins was an American with relatives in the CIA, and they settled the score by killing Trujillo. On his death, Trujillo owned 71 per cent of the country's arable land and 90 per cent of its industry.

[10] In 1963, following a popular rebellion, the first democratic elections were held and writer Juan Bosch was elected president. Seven months later, he was overthrown by military officers from the Trujillo regime. In April 1965, Colonel Francisco Caamaño Deñó led a constitutionalist armed uprising. Accusing the nationalists of having "Pro-Castro/communist" sympathies, the US intervened once again, sending in 35,000 Marines who supressed the insurgency.

[11] Before leaving the country, the US occupation force paved the way for an unconditional Trujillo supporter, Joaquín Balaguer, to rise to power. In return, he opened the country to transnationals; especially Gulf and Western. The sugar industry fell under Gulf control, and the corporation also bought shares in local banking, agro-industry, hotels and cattle raising, consequently, becoming very influential in Dominican Republic.

[12] The nationalist opposition kept up its resistance and in 1973 Francisco Caamaño was killed while leading a guerrilla group. The Dominican Revolutionary Party, originally led by Juan Bosch, split, with the right wing, of landowner Antonio Guzmán, eliminating the main reformist measures from its program. This action made it acceptable to the State Department, and in 1978, when the PRD won the elections, the US used its influence, in the name of human rights, to make sure that the results would be respected.

[13] The PRD program promised to re-establish democratic freedoms and to follow an economic policy of income redistribution favoring the majority. The first promise was fulfilled and popular organizations took advantage of the new situation to reorganize their weakened structures after decades of harsh repression.

[14] New presidential elections were held on May 16, 1981, and Salvador Jorge Blanco became president in PRD's second successive victory. José Francisco Peña Gómez, one of the Latin American leaders in the Socialist International, was elected mayor of Santo Domingo. On July 4, the departing president, Antonio Guzmán, killed himself, generating political tension, which ended with the announcement of the electoral result.

[15] When president Blanco took office the trade deficit amounted to $562 million. In 1982, sugar prices fell and oil prices rose causing for-

WORKERS

1994

- MEN 85%
- WOMEN 15%

- AGRICULTURE 46%
- INDUSTRY 15%
- SERVICES 39%

MILLIONS $ 1994

IMPORTS
2,630

EXPORTS
633

1970
18%

1992
6%

ENVIRONMENT

The Dominican Republic comprises the eastern part of the island of Hispaniola, the second largest of the Antilles. The Cordillera Central, the central mountain range, crosses the territory from northwest to south east. Between the central and northern ranges, lies the fertile region of the Cibao. Sea winds and ocean currents contribute to the tropical, rainy climate. Between 1962 and 1990, the country lost a significant portion of its woodlands. Coral reefs are suffering the effects of pollution, which has harmed marine habitats and reduced fish populations.

SOCIETY

Peoples: Most Dominicans are of Spanish and African descent, with a small native American component. About one-fifth of the population is considered genuinely Afro-American but at least 75% have some African blood. There are also European, North American and Asian minority groups. **Languages:** Spanish. **Religions:** Roman Catholic 91.2%; other 8.8%. **Political Parties:** Social Christian Reformist Party; Dominican Revolutionary Party; Dominican Liberation Party. **Social Organizations:** Most workers are represented by the General Workers' Union (CGT) and the Unity Workers' Union (CUT). In March 1991, 4 major labor groups, 57 federations and 366 labor unions merged within the CUT.

THE STATE

Official Name: República Dominicana. **Administrative Divisions:** 29 Provinces, 1 National District. **National Holiday:** November 27, Independence (1844). **Paramilitaries:** National Police, 15,000.

eign debt to climb to $2 billion. Unemployment affected 25 per cent of the active population.

[16] Blanco tried to tackle the situation by applying IMF-tailored austerity measures. But during 1983, the international price of sugar fell 50 per cent below the cost of production, and sugar accounted for 44 per cent of Dominican exports. In 1984, the government withdrew the subsidies on several products, and imposed a 200 per cent price increase on staple and medical goods. These measures brought about protest rallies led by leftist organizations and labor unions. In return, the Union headquarters were occupied by soldiers leaving 100 dead, 400 injured, and over 5,000 imprisoned.

[17] In 1985, the US reduced their Dominican sugar quota again, causing another decrease in exports. Unemployment rose abruptly, provoking yet another wave of social unrest. The government continued

to toe the IMF line, harshly repressing all the strikes and protests against its policies.

[18] Internal struggles at the PRD convention led to armed confrontations in which two people were killed; Jacobo Majluta was finally designated party candidate for the 1986 election.

[19] In this year, the seriousness of the economic crisis became evident. The judges went on strike and paralyzed the courts for three months. In May, the powerful Dominican Medical Association began a new round of conflicts, accompanied by the nurses of the Social Security system, the Association of Agriculture Professionals and the Veterinary Association. In August, the Association of Engineers followed suit, supported by the architects and surveyors, who demanded a minimum wage of $260. The Dominican Association of Economists wrote that impoverishment, since 1980, had been "initially limited to the low-income

groups, (but) had spread to the point of affecting the middle classes".

[20] In the lower income brackets, the situation was untenable. Committees for Popular Struggle began to appear and grass roots movements organized to resist price rises on basic goods and services. In April, dockworkers found 28 young Dominican girls asphyxiated in a ship's container. They could not find work and had expected to find a way of making a living on another island. It was disclosed that every two weeks a "cargo" of Dominican girls left for the Franco-Dutch island of Saint Martin, where a brothel manager sold them to other Caribbean islands for between 800 and 1,000 German marks.

[21] On May 16, the national election was held with three candidates: Jacobo Majluta for the PRD, Joaquín Balaguer for the Social Christian Reform Party (PRSC) and Juan Bosch for the Dominican Liberation Party (PLD). The post-election climate was marked by confusion and the vote count was interrupted. Eventually Joaquín Balaguer's PRSC won with 41 per cent of the votes. The PRD had 40 per cent and the PLD, 19.

[22] The conditions surrounding Balaguer's inauguration were very difficult. Clearly a conservative, without a parliamentary majority and faced with highly organized social opposition, his hands were virtually tied. He did not even have the necessary power to impose a more restrictive economic program to get new loans from the IMF.

[23] The situation worsened when the US further restricted the Dominican sugar quota. This led to 17,000 redundancies, in an administration that was already suffering 27 percent unemployment.

[24] In July 1988, former president Jorge Blanco was found guilty of illegal dealings. Despite judicial restrictions prohibiting such activities, he had sold weapons to the armed forces at inflated prices. This incident is just one of many, which have made people lose faith in political organizations, over the past two decades.

[25] The May 16, 1990, general election went to President Joaquín Balaguer. His rival, Juan Bosch, leader of the Marxist PLD, accused him of an electoral fraud of monumental proportions. In the last opinion poll carried out before the election, Bosch was leading, with 36 per cent of surveyed voters saying

that they intended to vote for him, as opposed to the 26 per cent committed to Balaguer. The third candidate, PRD's Peña Gómez, suggested a recount which was held at the polling stations. This recount confirmed Balaguer's victory, with 35 per cent, to Bosch's 34 and Peña Gómez's 23.

[26] Although constituting almost the same proportion of the electorate as men, there were only 88 women running for Congress, out of a total of 6,240 candidates. This deficiency of female candidates was not reflected in the tone of campaign programs and speeches, however, which gave major attention to women's issues for the first time.

[27] In 1991 two of the country's main political leaders, Joaquín Balaguer, aged 85, and Juan Bosch, aged 82, announced their resignations from politics.

[28] In the meantime, the social and economic situation continued to deteriorate. The recession which had begun in 1990 continued throughout the following year: unemployment rose and inflation soared, while agricultural production dropped (6 per cent below 1990 levels) as did mineral production (11 per cent down). The combined effect of a fall in the price of Dominican goods on the world market, and an increase in the price of imported oil (caused by the Gulf War), led to a worsening of the country's foreign trade situation.

[29] From 1990 to 1991 there were hundreds of thousands of Haitian immigrants working as sugar cane cutters in the Dominican sugar industry. In June 1991, Balaguer expelled illegal immigrants.

[30] According to several national and international studies, about half of the country's population, some 3,300,000 people, live in poverty. Of these, 1,105,000 suffer serious deprivation. In 1984, those classified as "poor" represented 31 per cent of the country's population, and the "seriously deprived" made up 16 per cent. In late 1991, the Economic and Social Council of the United Nations reprimanded the Dominican government, as it had done on a previous occasion, for allowing the authorities to evict hundreds of poor families. Between 1986 and 1991, more than 100,000 people lost their homes.

[31] Between 1990 and 1991, labor conflict reached one of the highest levels registered in recent years.

PER CAPITA
$563

18%

STATISTICS

DEMOGRAPHY

Urban: 63% (1995). **Annual growth:** 1.2% (1992-2000). **Estimate for year 2000:** 8,000,000. **Children per woman:** 3 (1992).

HEALTH

One **physician** for every 935 inhab. (1988-91). **Under-five mortality:** 21 per 1,000 (1995). **Calorie consumption:** 95% of required intake (1995). **Safe water:** 76% of the population has access (1990-95).

EDUCATION

Illiteracy: 18% (1995). **School enrolment:** Primary (1993): 99% fem., 99% male. Secondary (1993): 43% fem., 30% male. **Primary school teachers:** one for every 47 students (1989).

COMMUNICATIONS

97 **newspapers** (1995), 100 **TV sets** (1995) and 93 **radio sets** per 1,000 households (1995). 7.4 **telephones** per 100 inhabitants (1993). **Books:** 101 new titles per 1,000,000 inhabitants in 1995.

ECONOMY

Per capita GNP: $1,330 (1994). **Annual growth:** 2.20% (1985-94). **Annual inflation:** 28.90% (1984-94). **Consumer price index:** 100 (1990); 100 (1989). **Currency:** 13 pesos = 1$ (1994). **Cereal imports:** 961,000 metric tons (1993). **Fertilizer use:** 694 kgs per ha. (1992-93). **Imports:** $2,630 million (1994). **Exports:** $633 million (1994). **External debt:** $4,293 million (1994), $563 per capita (1994). **Debt service:** 17% of exports (1994). **Development aid received:** $2 million (1993); $0 per capita; 0% of GNP.

ENERGY

Consumption: 340 kgs of Oil Equivalent per capita yearly (1994), 89% imported (1994).

The government was faced with a series of general strikes, which were brutally repressed; in 1990, 15 people died and thousands were detained.

[32] Negotiations carried out by Balaguer made it possible to refinance the foreign debt. In the meantime, tourism has grown, making the country the fourth most important tourist destination in the Caribbean, in 1993.

[33] In the past few years, as a result of the political crisis in neighboring Haiti, the Dominican Republic has supplied contraband across the Haitian border. This practice has foiled the international embargo against the Haitian regime, and oil is the main product smuggled into the country.

[34] Economic problems cause hundreds of Dominicans to leave the country each year with forged visas and documents. Many sail in unseaworthy vessels, bound for Puerto Rico, normally as a "stopover" on their way to New York.

[35] More than a million Dominicans live in the United States, and half are illegal. There are also an estimated 20,000 in Spain, half of them illegal. The majority are women who work as domestic helpers, while another 25,000 women work as prostitutes throughout Europe, most of them lured by false promises of employment.

[36] Although he had announced his retirement, Balaguer sought re-election. His perennial opponent

Juan Bosch also ran in the May 16, 1994 election. According to opinion polls, José Francisco Peña Gómez, of the Dominican Revolutionary Party (PDR) held a slight lead over Balaguer. Intent upon avoiding another case of electoral fraud, four of the five participating parties signed a "civility pact", with the Catholic Church acting as warrantor. In spite of this, the electoral campaign turned violent, with hundreds of people injured and 12 people killed.

[37] International observers were

> According to several national and international studies, about half of the country's population, some 3,300,000 people, live in poverty. Of these, 1,105,000 suffer serious deprivation.

called in to supervise the elections, denouncing irregularities, especially in the interior. The PRD impugned the election, stating that some 200,000 voters were remained unable to vote as a result of official party manipulation. According to official figures, Balaguer obtained 43 per cent of the vote, leading Peña Gómez by 1.5 per cent, while Juan Bosch obtained 31.2 per cent.

[38] The US government, anxious to win Balaguer's support in enforcing the embargo on Haiti, gave assurances that it would approve the final decision of the special commission set up to clarify the electoral results. Possible solutions being considered by the commission are: repeating the election in certain districts, or even repeating the entire electoral process. However, 3 months after the August 1994 election, the special commission had still not reached a verdict. Finally, the Central Electoral Committee declared Balaguer the winner.

[39] But in mid-August, Peña Gomez and Balaguer agreed to hold general elections on November 16, 1995. Both leaders decided to reform the constitution in order to bar

presidential re-election. In the meantime, Balaguer was proclaimed president. Official returns gave him a 22,281 vote lead.

[40] During 1995, the drought affected the country's electricity supply. The power utility reported that 40 per cent of electricity consumed was not paid for; moreover, a fourth of power production was lost due to technical faults.

[41] Growth and inflation control goals set by the government were endangered by the power crisis. In March, public transport companies increased their fares by 50 per cent. This led to violent protests and the government decreed the raise was illegal. However, a new raise was authorized in June. Several individuals died in further confrontations between demonstrators and police.

[42] In 1995, opposing political parties prepared for the first elections in several decades with a definite chance of replacing Balaguer in power, who had reached 88 by then. An agreement between the PRSC and the Dominican Liberation Party (PLD) enabled elections to be postponed until May 16, 1996.

[43] Jose Francisco Peña, candidate for the Dominican Revolutionary Party (PRD), was the most voted but did not obtain an absolute majority. The government's candidate, Jacinto Peinado, did not reach the second electoral round. On June 30, 1996, Leonel Fernandez Reyna from the PLD won the second round of elections. On August 18 he took office, succeeding Joaquin Balaguer, who had held the country's presidency for 7 terms.

East Timor

Timor Leste

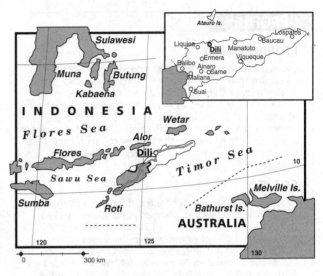

Population: 821,000 (1994)
Area: 14,870 SQ KM
Capital: Dili
Language: Tetum

L ong before the arrival of Vasco da Gama, the Chinese and Arabs knew Timor as an "inexhaustible" source of precious woods which were exchanged for axes, pottery, lead and other goods of use to the local inhabitants.

[2] Timor's traditional Maubere society consisted of five major categories: the *Liurari* (kings and chiefs), the *Dato* (lesser nobles and warriors), the *Ema-reino* (freemen) the *Ata* (slaves) and the *Lutum* (nomadic shepherds).

[3] In 1859, Portugal and the Netherlands agreed to divide the territory between them. The Portuguese kept the eastern part, under an accord ratified in 1904. Resistance to colonialism included armed insurrections in 1719, 1895 and 1959, all of which were put down. Passive resistance by the Maubere enabled their culture to survive five centuries of colonialism. It fared better than the forests of precious woods: species like sandalwood were exhausted very early, or replaced with coffee plantations, which are still Timor's economic mainstay.

[4] The independence movement began later than in other Portuguese colonies in Africa, but in the mid-1970s a national liberation front was formed, bringing together nationalist forces and all sectors of society.

[5] In April, 1974, when the clandestine struggle against colonial rule had already grown and gained broad support, the "Carnation Revolution" took place in Lisbon. With the fall of the fascist colonial regime in the metropolis the political scene in Timor changed and the patriotic movement was legalized. In September, the Revolutionary Front for the Independence of East Timor (FRETILIN) was created.

[6] The new Portuguese government promised independence but the colonial administration favored the creation of the Timor Democratic Union (UDT) which supported the colonial status quo and "federation" with Portugal. At the same time, the Indonesian consulate in Dili, Timor's capital, encouraged a group of Timorese to organize the Timor Popular Democratic Association (APODETI) which wanted fulll independence from Portugal, and supported integration with Indonesia.

[7] A period of conflict ensued between Portuguese neo-colonialist interests, Indonesian annexationists and the independence movement.

In August, the UDT attempted a coup d'état, causing FRETILIN to issue a call for general armed insurrection, and the Portuguese administration withdrew from the country. FRETILIN attained territorial control and declared independence on November 28, 1975, proclaiming the Democratic Republic of East Timor. To this day, however, Portugal's withholding of official recognition continues to have important diplomatic and political implications.

[8] On December 7, 1975, Indonesia invaded the new republic. A few hours earlier, US president Gerald Ford had visited Jakarta where he had probably learned of, and endorsed, Suharto's expansionist plans. FRETILIN was forced to withdraw from the capital, Dili, and from the major ports. On June 2, 1976, a so-called "People's Assembly", made up of UDT and APODETI members, approved Timor's annexation as a province of Indonesia. However, this illegal resolution was not recognized by the United Nations Decolonization Committee, which still regards Portugal as the colonial power responsible for Timor.

[9] Meanwhile, the Democratic Republic of East Timor established diplomatic relations with numerous former Portuguese colonies and socialist states.

[10] In December 1978, FRETILIN president Nicolau dos Reis Lobato died in combat. Despite this great loss, the liberation movement reorganized and continued resistance. According to reliable sources, Indonesia has adopted a policy of extermination on the island, to date killing off nearly 20 per cent of the population.

[11] Tactics used in the war have varied. In 1978, in a peculiar decision, the Front organized the massive surrender of civilians, who then moved into the Indonesian-controlled cities. Young men were armed and trained by the Indonesian army in an attempt to "Timorize" the war and set the Maubere against each other. But acting on FRETILIN instructions, now well-armed and equipped recruits, rebelled and re-joined the revolutionary forces. The Front became active in both the countryside and the cities.

[12] Indonesia's responses to the the war are erratic. In 1983, FRETILIN commander-in-chief, Xanana Gusmao, signed a cease-fire with the chief of the expeditionary force, Colonel Purwanto. However, President Suharto objected to this agreement, and the guerrilla war continued.

[13] The region is well reknowned for its mineral reserves, and in April, 1985, an international consortium was formed to explore oil and natural gas reserves in Timor territorial waters off the coast of Australia.

[14] FRETILIN sought closer ties with the Timor Democratic Union (UDT), and in 1978 established a

> ### The independence movement began later than in other Portuguese colonies in Africa.

coordinating body, the National Convergence. This union helped Portugal to actively resume its role. The internal structure of the nationalist movement was reorganized, to become politically independent and more locally based. Gusmao was confirmed as commander-in-chief of Timor's liberation army.

[15] Portugal won an important diplomatic victory when the European Parliament and the European Commission (the executive organ of the European Economic Community) adopted a position on the East Timor issue. They defended the Maubere people's right to self-determination, recognized the need for a negotiated settlement, and condemned the Indonesian occupation.

[16] In October 1989, the United Nations Human Rights Sub-commission passed a motion condemning Indonesian occupation and repression in East Timor. The disturbances had spread throughout the island, above all in Dili where students took to the streets, burned cars and destroyed the houses of several Indonesian officers.

[17] Repression increased after this. The island has long forbidden entrance to foreign correspondents, isolating Dili from the rest of the world. There are no telephone lines out of the country; nor are there any diplomatic representatives in the capital.

[18] When Pope John Paul II visited Dili in October 1989, a group of young people unfurled a FRETILIN banner twenty meters from the altar where mass was celebrated. Of the 80,000 people who attended the ceremony, 13,000 are calculated to have been members of the Indonesian security forces. The young protesters shouted independence slogans while the army waded in to prevent them. Reporters accompanying the Pope, some of whom had their cameras confiscated, witnessed at first hand the political oppresion operating in East Timor, and were able to inform the rest of the world.

[19] Repression has however, continued: every family is required to hang a list on their door naming the family members living in the house, and this list can be

ENVIRONMENT

Located between Australia and Indonesia, East Timor comprises the eastern portion of Timor Island, the dependency of Oecusse, located on the northwestern part of the island, the island of Atauro to the North, and the islet of Yaco to the East. Of volcanic origin, the island is mountainous and covered with dense rainforest. The climate is tropical with heavy rainfall, which accounts for the extensive river system. The southern region is flat and suitable for farming.

SOCIETY

Peoples: The Maubere people result from the integration between Melanesian and Malayan populations. In 1975 there was a Chinese minority of 20,000 as well as 4,000 Portuguese, 3,000 of whom were in the armed forces. 200,000 people are estimated to have died as a result of the Indonesian occupation. There are 6,000 Maubere refugees in Australia and 1,500 in Portugal. **Religions:** Most of the population profess traditional religions. 30% are Catholic. **Languages:** Tetum is the national language. There are several dialects. Indonesian occupation has banned the use of these languages in education, and virtually all the teaching is done in Bahasa, the main Indonesian language. A minority also speaks Portuguese. **Political Parties:** On 31 December 1988, FRETILIN and the UDT formed the National Convergence.

THE STATE

Official Name: República de Timor Leste. **Capital:** Dili, 67,000 inhab. (1980). **Government:** The country has been occupied by Indonesia since its independence. Xanana Gusmao, head of the National Liberation Armed Forces for Convergence, currently in prison in Djakarta, is responsible for the reorganization of the structures of the revolutionary movement. Konis Santana is taking his place. **National Holiday:** November 28, Independence Day (1975).

[22] Opposition leaders accused the US, Australia, the Netherlands, Japan and other countries with important economic interests in Indonesia, of cooperating with Jakarta in its attempt to play down the genocide and silence the international press.

[23] Late in 1991, Portugal reported that Jakarta and Canberra had signed a contract with twelve companies to extract around a billion barrels of oil from the sea around Timor. The list of companies is led by Royal Dutch Shell (British and Dutch capital), and Chevron (US). They are followed by six Australian companies, Nippon Oil (Japan), and transnational corporations; Phillips Petroleum, Marathon, and the Enterprise Oil Company. Meanwhile, representatives of the Australian government have announced that they will not support the sanctions against Jakarta.

[24] In Timor, the leaders of Nationalist Convergence urged Portugal to break negotiations with Indonesia and to take the harsher measures needed for a diplo-

> In October 1989, the United Nations Human Rights Subcommission passed a motion condemning Indonesian occupation and repression in East Timor.

matic solution to the conflict, through UN intervention. Nationalist Convergence expects that conditions to put pressure on the Indonesian government will be more favorable in 1992, when Portugal takes up the EEC presidency, both at EEC, and UN, level. The Timorese demand compliance with the UN resolutions: the withdrawal of occupation troops, and a referendum to decide on the country's political future.

[25] In March 1992, the ship "Lusitania Expresso" left Port Darwin, in Australia. On board

were human rights activists from over 23 countries, and well-known Portuguese politicians, including president Ramalho Eanes. "The Peace Boat", left with the aim of commemorating the Dili massacre of November 1991. The Indonesian authorities immediately announced that the ship would be diverted from its course to a nearby island; where the government would decide which members of the committee would be authorized to enter Timor.

[26] According to reports published in early 1994, every Saturday the mothers of the dead, imprisoned without trial or "disappeared" gather in Dili's main square to pray and voice their protest.

[27] Reports from the island in early 1994 indicated that the families of the dead, "disappeared" and prisoners met in the main square of Dili every Saturday to pray and protest. In November, some 100 people were arrested after a series of public demonstrations. The tension continued, and a year later, in October 1995, between 50 and 100 people were arrested following three days of disturbances in Dili.

[28] In early 1996, shortly before the seventh meeting between the Indonesian and Portuguese foreign ministers to seek a solution to the situation in East Timor, Amnesty International asked for free access for human rights observers to the occupied nation. Other specialist reports insist there are social inequalities between the occupied and occupiers, despite the well-publicized investments made by the Indonesian government in sectors like education. In October, Roman Catholic Bishop Carlos Filipe Ximenes Belo and activist José Ramos-Horta recieved the Nobel Peace Prize for their work towards reaching a peaceful resolution of East Timor's conflict, bringing the country back to the international spotlight.

checked at any time by occupation forces. Thousands of Maubere women have been compulsorily sterilized, and tetum, the national language, is banned from schools. The transmigration policy enforced by the Indonesian authorities is reducing the Maubere people to the status of a minority within their own country. Also mass graves have been discovered in different parts of the country, containing corpses which showed signs of having suffered mass execution at the hands of the occupation forces.

[20] In early November, during a massive peaceful funeral procession accompanying the body of a young man who had been murdered, the army opened fire on the crowd, 200 people were killed, and countless more were injured. Media coverage of this event, caused the Portuguese government to appeal to the EEC countries to break trading relations with Indonesia. Indonesia has a preferential trading agreement with the EEC, being a member of the six countries of the Association of South East Asian Nations (ASEAN). Portugal also requested a meeting of the UN Security Council, criticizing them harshly because they did not react to Indonesia as they had to Iran on the invasion of Kuwait in August 1990.

[21] A visit to the island by Portuguese members of parliament, scheduled for the first days of November 1991, was cancelled after the Indonesians refused to allow entry to an Australian journalist accompanying the delegation.

Ecuador

Population:	11,227,000 (1994)
Area:	283,560 SQ KM
Capital:	Quito
Currency:	Sucre
Language:	Spanish

Ecuador

In 1478, Inca Tupac Yupanqui united the agricultural peoples who had inhabited the territory that is now Ecuador since 2,000 BC, within the Tahuantinsuyu (Inca empire). Within a few years the northern region of the Quechua territory, of which Quito was the center, acquired great economic importance and became the commercial and cultural center of a great civilization. But the rivalry for succession between Atahualpa (from Quito) and Huascar (from Cuzco) weakened the power of the Empire.

[2] The Spanish conquistadors commanded by Sebastián de Benalcázar took advantage of the situation to capture Quito in 1534. Indigenous South Americans, who had been free people working the land and living in communities characterized by a high level of social organization and mutual help (*ayllus*) were enslaved and cruelly exploited.

[3] At first, the country formed part of the Viceroyalty of Peru as an administrative dependency known as the *Real Audiencia de* Quito (Royal District of Quito). In 1717, Quito was transferred by the Bourbons to the Viceroyalty of Nueva Granada, which also comprised present-day Colombia, Venezuela and Panama.

[4] In 1822, General Antonio José de Sucre, Bolívar's deputy, defeated the royalist forces in the Battle of Pichincha, thus ending Spanish domination and incorporating Ecuador into Bolívar's Greater Colombia scheme.

[5] The Royal District of Quito seceded from Greater Colombia in 1830 and adopted the name Republic of Ecuador. The change in name probably meant little to the peasant majority, as nothing changed with regard to the land tenure system.

[6] In 1895, the Liberal Revolution led by Eloy Alfaro rekindled the people's hopes for real land reform. Church property was nationalized although large landowners were not affected. Alfaro was killed and the country, like other parts of the continent, came under the influence of British imperialism.

[7] Ecuador successively lost one part of its territory after another, due to the ruling oligarchy's lack of political capacity. Under Spanish domination, Ecuador had covered 1,038,000 sq km, shrinking to 283,560 sq km by 1942 through successive agreements with external powers.

[8] In 1944 a popular revolt ousted President Carlos Alberto Arroyo, ushering in a populist government headed by José María Velazco Ibarra and comprising conservatives, communists and socialists, under the common banner of the Democratic Alliance. The Cold War made it impossible for this alliance to continue, and the left began to be the object of repression. In 1962, under pressure from the United States, President Carlos Arosemena broke off diplomatic relations with Cuba.

[9] In the early 1970s, bananas, coffee and cocoa exports were replaced by crude oil. That year the political situation also changed. Veteran populist leader, José María Velasco Ibarra was ousted for the fourth time by the armed forces. Under the government of General Guillermo Rodríguez Lara, the country joined the OPEC. The government purchased 25 per cent of the shares of Texaco-Gulf and affirmed its rights over 200 miles of territorial waters, and the "tuna war" ensued when this claim was disputed by US fishing interests.

[10] Jaime Roldós, nominated by the Concentration of People's Forces (CPF) and the People's Democratic Party, became president in August 1979. Ecuador renewed diplomatic relations with Cuba, China and Albania. The government also initiated a program aimed at integrating marginalized rural and urban populations into the economy. However, it encountered a hostile Congress, as well as opposition from the United States, which did not take kindly to its human rights policies and its antagonism to the dictatorships in the southern cone.

[11] Toward the end of January 1981, the Five Day War broke out between Ecuador and Peru, with skirmishes along borders which had not been clearly delineated by the 1942 Protocol.

[12] In May of the same year, Roldós died in a suspicious airplane accident, whereupon the vice president, Osvaldo Hurtado, took office. The following year was marked by the most severe social crisis since the military had left power. It was triggered, on the one hand, by the application of IMF formulas, and on the other, by an overt drive to build up military power to achieve parity with the Peruvian armed forces.

[13] The 1984 election was won by conservative León Febres Cordero, of the Social Christian Party. For the most part, Febres Cordero was able to fulfill his campaign promises: to encourage free enterprise, develop agriculture and mining, attract foreign investment and establish relations with the IMF. The agreement with 400 creditor banks entailed assigning 34 per cent of estimated export revenues to meet new commitments.

[14] Febres was an ardent defender of US president Reagan's Central American policy. In October 1985 he broke off diplomatic relations with Nicaragua, and on several occasions helped finance Nicaraguan "contra" leaders' travel and expenses.

[15] The May 1988 election was won by Social Democrat Rodrigo Borja, inaugurated in August of that year with the support of a coalition made up of the Democratic

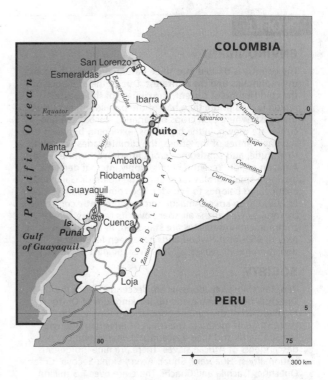

PUBLIC EXPENDITURES

% of GNP — 1992

- MILITARY 2.2
- EDUCATION 2.8
- HEALTH 0.4

WORKERS

1994 — **7.1%** UNEMPLOYMENT

- MEN 81%
- WOMEN 19%
- AGRICULTURE 33%
- INDUSTRY 19%
- SERVICES 48%

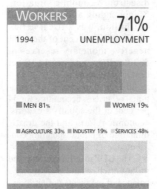

LAND USE

- CROPLAND 11%
- PASTURE 8%
- OTHER 82%

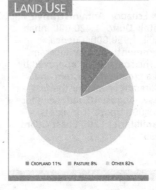

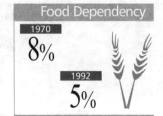

PROFILE

ENVIRONMENT

The country is divided into three natural regions: the coast, the mountains and the rainforest. Due to the influence of the cold Humboldt current, the climate of the coastal region is mild. More than half of the population lives along the coast, where cash crops of bananas, cocoa, rice and coffee are grown. In the highlands, extending between two separate ranges of the Andes, the climate varies according to altitude, and subsistence crops are grown. In the eastern Amazon region, recent oil finds supply internal demand, leaving a small surplus for export. The Colón or Galapagos archipelago belongs to Ecuador. The country also lays claim to 200,000 sq km of Amazon territory presently controlled by Peru, as well as the air space over its territory, where communication satellites are stationed. In the coastal region, 95% of the woodlands have been felled. Soil depletion has increased by 30% over the past 25 years.

SOCIETY

Peoples: Most Ecuadoreans are descended from the Quechua people who made up the kingdom of Quito. A break-down of the current population shows a high percentage of mestizos, the result of intermarriage with the Spaniards and their descendents, in addition to descendents of African slaves. There are nine indigenous nationalities: Huaorani, Shuar, Achar, Siona-Secoya, Cofan, Quechua, Tsachila and Chachi. There are over 1.5 million Quechua living in the inter-Andean valley.
Religions: Mainly Catholic. **Languages:** Spanish (official), although 40% of the population speaks Quechua. **Political Parties:** The Social Christian Party of Jaime Nebot; the United Republican Party, conservative; the Ecuadorean Roldosista Party; the Democratic Left, social democratic, affiliated to the Socialist International; the Conservative Party; People's Democracy; the Christian Democratic Union; the Democratic People's Movement; the Ecuadorean Socialist Party; the Concentration of Popular Forces; the Radical Liberal Party; the Broad Front of the Left; the Radical Alfarista Front; the National Patriotic Front, of Frank Vargas; the Ecuadorean People's Revolutionary Action; the National Liberation Party. **Social Organizations:** The major union federations are the Ecuadorean Central Organization of Class Unions (CEDOC), and the Central Organization of Ecuadorean Workers (CTE) coordinated with the United Worker's Front (FUT). In the last few years the Confederation of Indigenous Nationalities of Ecuador (CONAIE) has gained importance; the National Federation of Small Producers (FNPA).

THE STATE

Official Name: República del Ecuador. **Administrative divisions:** 21 Provinces. **Capital:** Quito, 1,420,000 inhab. (1992). **Other cities:** Guayaquil, 1,950,000; Cuenca, 325,000 inhab. (1992). **Government:** Abdala Bucaram, President since August 1996. The constitution approved by plebiscite in 1978 established a presidential system and granted the right to vote to illiterate people for the first time in 1984. **National Holiday:** August 10, Independence Day (1809). **Armed Forces:** 58,000 troops (conscripts). 100,000 reserves (1993). **Paramilitaries:** 200 Coast Guard and 6 coastal patrol units (1993).

STATISTICS

DEMOGRAPHY

Urban: 57% (1995). **Annual growth:** 1.2% (1992-2000). **Estimate for year 2000:** 13,000,000. **Children per woman:** 3.5 (1992).

HEALTH

One **physician** for every 671 inhab. (1988-91). **Under-five mortality:** 57 per 1,000 (1995). **Calorie consumption:** 94% of required intake (1995). **Safe water:** 71% of the population has access (1990-95).

EDUCATION

Illiteracy: 10% (1995). **School enrolment:** Primary (1993): 122% fem., 122% male. Secondary (1993): 56% fem., 54% male. **Primary school teachers:** one for every 29 students (1988).

COMMUNICATIONS

102 **newspapers** (1995), 100 **TV sets** (1995) and 102 **radio sets** per 1,000 households (1995). 5.3 **telephones** per 100 inhabitants (1993). **Books:** 96 new titles per 1,000,000 inhabitants in 1995.

ECONOMY

Per capita GNP: $1,280 (1994). **Annual growth:** 0.90% (1985-94). **Annual inflation:** 47.50% (1984-94). **Consumer price index:** 100 in 1990; 424.4 in 1994. **Currency:** 2.269 sucres = 1$ (1994). **Cereal imports:** 428,000 metric tons (1993). **Fertilizer use:** 380 kgs per ha. (1992-93). **Imports:** $3,690 million (1994). **Exports:** $3,820 million (1994). **External debt:** $14,955 million (1994), $1,332 per capita (1994). **Debt service:** 22.1% of exports (1994). **Development aid received:** $240 million (1993); $22 per capita; 2% of GNP.

ENERGY

Consumption: 517 kgs of Oil Equivalent per capita yearly (1994), -223% imported (1994).

Left, Osvaldo Hurtado's Christian Democrats and a dozen parties of the left.
[16] From the time Borja took office, his government was confronted by high rates of inflation and a disastrous economic situation. Compounding a foreign debt of $11 billion, was a fiscal deficit equalling 17 per cent of the GDP, negative monetary reserves of $330 million, and nearly 15 per cent unemployment.
[17] In 1990, the increase in international oil prices, accounting for 54 per cent of the country's exports, economic reform and the drastic limitation of public spending contributed to a mild economic recovery. The GDP rose 1.5 per cent, and inflation fell, as did the country's balance of payments deficit; however, real wages declined. The foreign debt burden clearly had a negative effect upon Ecuador's economic recovery.
[18] Under Borja, important gains were made on the domestic front: the Taura Commandos, who had kidnapped Febres Cordero in January 1987, were deactivated and the guerrilla movement Alfaro Vive ("Alfaro Lives") was integrated into the political mainstream.
[19] On the international scene, under Borja, Ecuador participated actively in the different groups involved in subregional integration. In addition, it supported the Group of Eight, which replaced the Contadora Group (charged with mediating the Central American political crisis). In addition, it was a member of the Rio Group, it rejoined the

The cultural rebirth of the aboriginal peoples

The indigenous uprising which rocked the country in 1990 marked a decisive moment for the consolidation of two processes: the cultural rebirth of indigenous peoples, and the revitalization of their land rights struggle.

[2] When the Spanish arrived in this part of the continent, America's indigenous peoples were living in advanced unified nations, based on communities, ethnic castes, confederations and the embryonic Inca nation.

[3] Spanish colonization broke the foundations of this process, but was unable to do away with the community as an organizational unit. The indigenous people constantly fell back on the community, working from this towards the different phases of national unification which they were trying to achieve. The end of the colonial period was marked by rebellions sparked by the confiscation of community lands in order to form haciendas. These rebellions re-ignited the national unification process, becoming national liberation movements.

[4] In 1860, the uprising led by Fernando Daquilema against the government of García Moreno was the last of this period of indigenous movements. Subsequently the hacienda system was consolidated, the communities lost their autonomy, becoming appendages to the hacienda and the national economy. After becoming serfs, many Indians lost their community ties. The idea of "national liberation" bowed to the land issue, and those taking up the struggle were no longer community members, but the serfs - the silent, dispossessed partners - from within the confines of the haciendas.

[5] The US-promoted transformation of the land tenure system in the 1960s and 1970s was an expression of the need to remove the obstacles to economic development, and to hold back the specter of armed insurrection by campesinos, prompted by the example of Cuba's successful revolution.

[6] Agrarian reform was carried out with impeccable logic. Its aim was to defuse the time bomb of the hacienda system by launching the modernization of the old patriarchal system of haciendas located on the best lands and close to markets. At the same time, it channelled the campesino struggle onto the least economically productive haciendas belonging to a lower echelon of landowners, supporting these moves with a legal and institutional State framework. Thus, an agrarian bourgeoisie was consolidated, and the rest of the countryside was fragmented into thousands of tiny properties, condemned to increasing differentiation and competition among themselves. The poorest of these would provide cheap labor for industry, particularly the construction sector.

[7] The plan was carried out exactly according to the blueprint. It was a perfect plan, apart from one major flaw: it did not take the indigenous peoples themselves into account.

[8] In fact, when the agrarian reform freed the Indians from servitude, it led to the recomposition of the indigenous community and a resurgence of a national movement, further stimulated by the presence of ethnic groups from the Amazon region. The community came back into being, even though the extremely fragmented property units had subdued it for generations.

[9] What is being seen now is a rebirth of indigenous culture: music, dance, poetry, medical knowledge, and the unfolding of conceptions of nature, being, time and death, a cosmic vision which has been very influential on the development of contemporary ecological thought. The inter-marriage between indigenous and Spanish peoples has weakened considerably, and with it, the pressure on indigenous people to adopt Western culture. Thus the indigenous peoples preserve their identity in the cities including Guayaquil, and the coastal region which is not a native region. The indigenous population has currently begun to increase for the first time since the Conquest, and semi-urban areas in the highlands have been "seized".

[10] Many activists have been working for the consolidation of the indigenous peoples. There are nine nationalities: in the east, the Huaoranik, Shuar, Achuiar, Siona, Secoya, Cofán, Quechua, and in the highland and coastal areas, the Tsachila and Chachi. The Quechua alone, inhabiting the inter-Andean valley, currently total more than a million and a half people.

[11] Ethnic cohesion created favorable conditions for a new phase in the land struggle. A new generation descried the scarcity of land, and the crisis aggravated pre-existing poverty and misery, with the 1990 uprising proving the culmination of this process. The land struggle is also the struggle for territory; not "territory" as geographical space, but rather as a historical entity and a natural-cultural reality - the campesino struggle for land therefore has parity with the national struggle of the indigenous peoples.

[12] As yet there is no economic or territorial space designated for the indigenous people to practise their alternative forms of economic and social relations, to make a foundation for territorial sovereignty. Quite the contrary, their agrarian and craft organisations are more than ever linked to the national and even the international market. To make matters worse, this external control makes them even more vulnerable to the current economic crisis.

[13] One of the most evident causes of the 1990 uprising was the pauperization of the Indians, through the economic crisis and structural adjustment policies. The issue of the indigenous peoples is further complicated by a deeply rooted paradox, for while they comprise a people who despite the affirmation of their culture tend to be autonomous; on the other, they make up a set of classes and social strata already integrated into the country's dominant society.

[14] The most recent Indian movement has led to the creation of a national organization, the Confederation of Indigenous Nationalities (CONAIE), and has defined a general political and economic program which is characterized by the issues of autonomy and integration. The central thesis behind all their activity is, fundamentally, the creation of a multinational state which recognizes the autonomy and political rights of all the nine nationalities in the country. They want an understanding of "territoriality" as cultural space and not political sovereignty. The organization's second demand is for land and agrarian reform. The national indigenous movement and CONAIE are becoming an influential force in contemporary Ecuador.

External Debt
1994

PER CAPITA
$1,332

Illiteracy
1995

10%

Movement of Non-Aligned Countries, and became the site of numerous high-level international meetings and forums.

[20] In May 1990, the presidents of Bolivia, Colombia, Peru and Venezuela agreed to begin eliminating tariff barriers between their countries as of January 1, 1992, as a first step toward the creation of an Andean Common Market by 1995. In addition, the Andean leaders underscored the importance of collective cooperation with the United States in the war against drugs.

[21] On May 28 1990, a group of Indians from the coast took over Santo Domingo Church in Quito, demanding that they be allowed to own land and that their human rights be respected. Shortly thereafter, several highways were closed by other indigenous groups from the highlands and the eastern part of the country.

[22] Dozens of middle-sized cities in the Andean region were symbolically taken over by tens of thousands of indigenous peoples from neighboring villages. Subsequently, native peoples from the Amazon region marched on Quito. In the ensuing police action, one person was killed and several injured, although the large-scale violence which had marked previous incidents was avoided.

[23] In the June 1990 legislative election, President Borja's party was badly defeated, with the winning share going to the conservatives, represented by the Social Christian Party, and the left, by the Socialist Party.

[24] On May 28 1991, more than a thousand Indians peacefully occupied the assembly room of the National Congress, demanding amnesty for some thousand Indians who had been tried in connection with the 1990 uprising.

[25] In the April 1992 elections, the Social Christian Party and the United Republican Party were the most successful participants. They obtained 19 and 13 of the 77 seats in Congress, respectively. The ex-president Febres Cordero was elected mayor of Guayaquil with 70 per cent of the returned votes. The turnout for the election was 73 per cent, six points higher than traditional level. None of the presidential candidates managed to achieve 50 per cent of the vote, so the election went to a second round between the two most popular nominees. In the second ballot on July 5,

Sixto Durán Ballén of the PUR gained 56 per cent of the vote over Jaime Nebot of the PSC who achieved 43 per cent.

[26] The left-of-center and the left contributed to the electoral victory of Sixto Durán, considered the "lesser of two evils", when compared with Nebot's authoritarianism.

[27] The new government proposed a program based on the "modernization of the State", including a plan for privatizing state enterprises and a rigorous structural adjustment plan. This meant eliminating subsidies, increasing or floating the prices of basic goods (including gasoline and other hydrocarbon derivatives) and keeping wage adjustments below the accumulated inflation rate.

[28] After a year in office, Vice-President Dahik, considered the real power behind the throne, proclaimed the government's adjustment policies to be a success. They had achieved a 60 per cent reduction of inflation in 1992 and 32 per cent in 1993, an increase in monetary reserves and a reduction of the fiscal deficit, especially in the public sector, and reduction of public spending to a mere 26 per cent of the GDP in 1993.

[29] The opposition, the Confederation of Indigenous Nationalities of Ecuador (CONAIE) and the United Workers' Front (FUT) questioned these policies, citing increased poverty and unemployment, and the total absence of social policies. In actual fact, the mean urban wage had continued to fall being worth only a fifth of the average 1980 value in 1993.

[30] The main conflict stemmed from privatization of the social security system, electricity, telecommunications and oil. The opposition was strengthened by the creation of a National Sovereignty Defense alliance, bringing together strategic labor unions, CONAIE and other social organizations.

[31] The legislative package authorizing the privatizations underwent several modifications. The attitude of the Social Christian Party, the main parliamentary force, proved decisive. Although the party was ideologically in favor of privatization policies, it had to operate with an eye on its electoral prospects for 1996, so the government proposals were eventually passed in much reduced versions to avoid having to pay a high political cost.

[32] Throughout this period the gov-

ernment committed a number of faux pas, including the Ingenio Azucarero Aztra (a sugar mill) and Ecuatoriana de Aviación (the national airline) scandals. Both of these companies were allowed to go bankrupt before the privatization process began.

[33] The damage to the government was extensive. In the May 1994 elections, the PUR did not obtain a single parliamentary seat, and the Conservative Party only won 6 of a total 77. Together, they obtained less than 10 per cent of the vote. The main beneficiary was the Social Christian Party, with 22 deputies, 25 per cent of the vote, and a candidate, Jaime Nebot, as the main contender for presidential succession.

[34] The two left and left-of-center parties, the Democratic Left and Christian Democrats suffered serious setbacks, winning fewer than 15 legislative seats and 15 per cent of the vote. The surprise element was the increase in strength of the MPD - a party with a strong base in the educational sector - and the triumph in Quito of General Frank Vargas Pazzos, whose platform was marked by strong nationalism.

[35] Meanwhile, the economy went through a period of deep recession. The PBI and exports decreased, specially banana exports, due to the restrictions of the European Community. The building industry and oil exports were the exceptions. The government gave new licences for oil prospecting and planned the building of an oil pipeline. The measure was heavily criticised by the opposition, worried, so they said, about the possible exhaustion of the deposits.

[36] In April 1994 a fire in Isabela, on the Galapagos Islands, destroyed 6,000 hectares of woods and vegetation and endangered the life and habitat of the giant turtles, a fact which prompted a decree about control of the tourist flux, immigration, and illegal fishing.

[37] By mid-year the government closed an involved process of renegotiation of the foreign debt, in which the actual reduction did not go over 45 per cent of it. The interest rate payments were not considered in this calculation.

[38] In June the President Durán Ballén eliminated the agrarian reforms of the years 1964 and 1973. These reforms had finished with the big haciendas or land properties, and had given land, for the first time, to the Indians and the small

peasants. As an answer the CONAIE blocked access to several towns and cities. Following which the government decreed a state of emergency and sent the army to control the situation. In Ecuador 48 per cent of rural land is in the hands of peasant communities, mostly Indian ones, and 41 per cent belongs to private citizens.

[39] At the beginning of 1995 new armed conflicts took place against Peru at Cordillera del Condor, where the frontier had never been determined with precision, and where there are prospects of finding gold, uranium, and oil deposits. Durán Ballén decreed a state of emergency once again and called the conscripts. Despite international mediation, there were a few dozen casualties, mostly on the Peruvian side.

[40] By the year's end several ministers were questioned by the Parliament about misuse of public funds. Some of them were subsequently arrested, and the Vice-President Alberto Dahik fled the country.

[41] After much debating, the Congress approved the sale of 35 per cent of the state telephone company. The government had envisaged achieving this first important privatization before the end of its mandate in August 1996.

[42] The right wing populist leader Abala Bucaram won the elections with 54 per cent of the votes over the Social Christian Party Jaime Nebot, who got 46 per cent of the votes. The first task of the government was to assuage the fears of the business and financial groups about the possible fulfilling of Mr Bucaram's campaign promises to the poor. Journalistic versions forecasting a military coup in case of a Bucaram victory proved false.

[43] During a bilateral meeting with the Chilean president Eduardo Frei in September 1996, Bucaram expressed his admiration for Chilean policies concerning social security and lodging. He also invited Domingo Cavallo, ex-Minister of the Economy of Argentina, to enter his governing team.

Egypt

Misr

Population: 56,767,000 (1994)
Area: 1,001,450 SQ KM
Capital: Cairo (Al-Qahirah)

Currency: Pound
Language: Arabic

Six thousand years ago, the inhabitants of the Nile valley; Nahr-an Nil formed a civilization that eventually developed into a centralized state. While struggling to control the periodic flooding of the Nile, the Egyptians built the pyramids and created a culture which was used as an inspiration for the later "western civilization". They succeeded in feeding a large population for such a small area of land, and became a busy center of economic, diplomatic and cultural relations. In the last millennium BC, during the decline of this remarkable civilization, the country was ruled by Libyan and Sudanese Pharaohs and then directly by the Assyrian, Persian, Greek and Roman empires.

[2] During the period of Greco-Roman domination, Alexandria (Al-Iskandariyah) was one of the most influential cultural centers in the classical world, and its famous library was the largest in existence until it was burned down under the roman emperor Aurelian in the late 3rd century AD. It brought together the most outstanding philosophers, scientists and scholars of the era. In 642, when the Arabs conquered the country, little remained of its highly developed past and the Egyptians adopted Islam and the Arabian language.

[3] Three centuries later, under the government of the Fatimid Caliphs, the new capital, Cairo (Al-Qahirah), became one of the major intellectual centers in the Islamic world and scholars, particularly African Muslims, were attracted by its university.

[4] Between the 10th and 15th centuries, Egypt benefited from its geographic location, becoming the trade center between Asia and the Mediterranean. The Venetians and

Genoese came to trade here, and even the constant warfare provoked by the European Crusades in Palestine, in the 11th to 13th centuries, did not stop active trade.

[5] Once the Crusaders were driven out, it seemed that Egypt would naturally become the center of the ancient Arab empire, but the sultanate of the Ottoman Turks was the rising power in the Islamic world at the beginning of the 16th century, and they soon conquered Egypt. The opening of the sea route between Europe and the Far East had already put an end to Egypt's previous trade monopoly, reliant on its dominion over the Red Sea, and Egypt had begun its economic decline.

[6] Until the 19th century, Turkish domination was little more than nominal and the real power lay in the hands of Mameluke leaders. In 1805, Muhammad Ali, an Albanian military leader, took power. He forcibly eliminated local Mameluke leaders, established a centralized regime, reorganized the army, declared a state monopoly on foreign trade of sugar cane and cotton and achieved increasing autonomy from the Sultan of Istanbul laying the foundations of a modern economy.

[7] The economy was poorly managed by Muhammad Ali's successors, the crisis deepened and dependency on Europe increased. In 1874, Egypt was forced to sell all its shares in the Suez Canal, built as a joint Egyptian-French project between 1860 and 1870, to pay its debts to the British.

[8] The situation continued to deteriorate, loans piled up and, in 1879, the creditors imposed a Bureau of Public Debt formed by three ministers, one English, one French and one Egyptian. This bureau assumed the management of the country's finances.

[9] This degree of interference awakened an intense nationalistic reaction, supported by the army. That year, the military overthrew Muhammad Ali's successor Khedive Ismail, forcing his son, Tawfiq, to expel the foreign ministers and appoint a nationalist cabinet. The colonial power reacted promptly: in 1882 an Anglo-French fleet landed English troops in Alexandria, seizing military control of the country.

[10] The occupation was "legalized" in 1914, when Egypt was formally declared a protectorate. This situation continued until 1922, when an Egyptian committee negotiated independence in London. This independence was negotiated in such a way that the resulting conditions

really meant the continuation of the protectorate.

[11] During World War II, Egypt was used as a British military base again. Anticolonial feeling reached its height in 1948, when the state of Israel was created in Palestine.

[12] Egypt and other Arab nations launched an unsuccessful war against the new state, and the frustration of defeat brought about massive demonstrations against the royal government.

[13] Against a background of widespread government corruption, a nationalist group known as the "Free Officials" was formed within the Egyptian Army, led by General Mohamed Naguib and Colonel Gamal Abdel Nasser.

[14] On July 23, 1952 this group ousted King Faruk and, in June 1953, proclaimed a Republic. Three years later, Nasser became president.

[15] The new regime declared itself nationalist and socialist, deciding to improve the living conditions of the *fellahin*, the country's most impoverished peasants. Land reform was started, limiting the landowners' monopoly of the majority of the land.

[16] Nasser gave priority to the construction of the Aswan dam, one of the world's largest dam projects. The construction was carried out by the Soviet Union, after the western

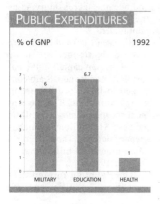

PUBLIC EXPENDITURES

% of GNP — 1992

MILITARY 6 · EDUCATION 6.7 · HEALTH 1

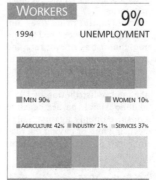

WORKERS

1994 — **9%** UNEMPLOYMENT

MEN 90% · WOMEN 10%

AGRICULTURE 42% · INDUSTRY 21% · SERVICES 37%

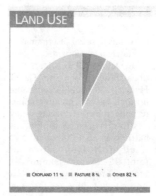

LAND USE

CROPLAND 11% · PASTURE 8% · OTHER 82%

Foreign Trade

MILLIONS $ 1994

IMPORTS
10,185
EXPORTS
3,463

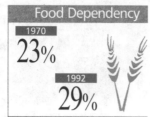
Food Dependency

1970
23%

1992
29%

PROFILE

ENVIRONMENT

99% of the population live in the Nile valley and the delta although this constitutes only 30% of the land. The remaining land is covered by deserts except for a few isolated oases. The floods of the Nile set a pattern for the country's economic life thousands of years ago. There are several dams, the major one being the Aswan dam in the south. This has enabled farmers to add new crops such as cotton and sugar cane to the traditional crops; wheat, rice and corn. The hydroelectric power supply, together with the northeastern oil wells, in the Sinai Peninsula, favored industrial development and the disproportionate growth of the cities. Problems which remain unsolved include the unchecked growth of the cities, which has swallowed up fertile lands; erosion of the soil and the use of fertilizers; and water pollution.

SOCIETY

Peoples: Egyptians are Arabs of Hamite origin with East Asian communities in the north and central regions and African groups in the Upper Nile area. **Religions:** Muslim, majority Sunni. There is a 7% Coptic minority and other smaller Christian groups. **Languages:** Arabic (official), English and French in business, Nubian and Oromo in daily use. **Political Parties:** The National Democratic Party (NDP), founded in 1978 by Anwar Sadat, had its origins in the Arab Socialist Union, the single party formed by Nasser in 1954; the New Wafd Party (NWP), heir of the Wafd Movement of 1919; the Muslim Brotherhood; the Liberal Socialist Party; the Socialist Labour Party; the National Union Progressive Party (NPUP). **Social Organizations:** The Egyptian Labor Federation is the only central labor organization, and the Union of Egyptian Students represents the university students.

THE STATE

Official Name: Al-jumhuriat Misr al-Arabia. **Administrative divisions:** 26 provinces. **Capital:** Cairo (Al-Qahirah), 6,850,000 inhab. (1994). **Other cities:** Alexandria (Al-Iskandariyah) 3,382,000 inhab; El Giza (Al-Jizah), 2,144,000 inhab. (1994) **Government:** Hosni Mubarak, President. The Majlis (single-chamber legislature) has 454 members, of which 10 are named by the president. **National Holiday:** July 23, Revolution Day (1952). **Armed Forces:** 430,000 (incl. 270,000 conscripts). **Paramilitaries:** Coast Guard, National Guard and Border Guards, 74,000.

---

powers had refused to take it on. The dam, which had been hailed as the key to the country's industrialization and "development", in fact caused serious environmental disruption.

[17] In 1955, Nasser was one of the leading organizers of the Bandung Conference, a forerunner neutralist Afro-Asian movement which preceded the Movement of Non-Aligned Countries. Twenty-nine Afro-Asian countries condemned colonialism, racial discrimination and nuclear armament.

[18] In October 1956, after the nationalization of the Suez Canal, French, British and Israeli troops invaded Egypt. The government responded by distributing weapons to civilians. A diplomatic battle was also launched; as a result of UN intervention and joint US-Soviet disapproval, France, Britain and Israel were forced to withdraw and the canal finally came under Egyptian control.

[19] On February 1, 1958, the union of Egypt and Syria was officially announced, under the name of the United Arab Republic (UAR). This lasted until September 1961, when Syria decided to separate from Egypt, though Egypt continued to call itself the United Arab Republic.

[20] After Nasser's re-election in 1965, Egypt gave high priority to the conflict with Israel. However, its attempt to economically paralyse Israel by blockading the Gulf of Aqaba failed during the Arab-Israeli conflict, the "Six-Day War", of June 1967. This ended in another defeat of the Arab countries - Egypt, Syria and Jordan - when Israeli forces occupied the Sinai Peninsula, the Gaza strip, the West Bank, and the Syrian Golan Heights. The cost of the war aggravated Egypt's financial problems and only Soviet aid prevented its total collapse.

[21] Nasser died in 1970, and Vice-President Anwar Sadat, a member of the right wing of Nasser's Arab Socialist Party, took his place. Sadat put the *infitah* into practice; a government plan which meant an opening to western influence, the de-nationalization of the Egyptian economy and the end of the single-party system. Furthermore, the new government broke relations with the Soviet Union and US economic and military aid flowed into Egypt.

[22] In 1973, Egyptian troops crossed the Suez Canal, beginning the fourth Arab-Israeli war. The Egyptians were not defeated, and their success resulted in OPEC substantially increasing oil prices. This move did not produce the desired effect, and Israel retained the rest of the occupied territories.

[23] Substantial price rises and unemployment worsened the living conditions of workers and resulted in massive anti-government demonstrations in 1976 and 1977. Peasants rebelled against the land redistribution of 1952, and the Islamic parties began to conspire openly against Sadat, accusing him of paving the road for a new period of foreign domination.

[24] Sadat's visit to Jerusalem in November 1977 raised a wave of protest in the Arab world. The process of rapprochement with Israel reached its apex in March 1979, with the signing of the Camp David Agreement, wherein the US negotiated the return of Sinai to Egypt. From then on, Egypt became the main beneficiary of US military aid, aimed at turning the country into the new US watchdog in the Arab World, as Shah Pahlevi of Iran had recently been deposed.

[25] In October 1981, Sadat was killed in a conspiracy organized by certain sectors of the military opposed to *infitah* and the repression of fundamentalist Islamic movements. Vice-president Hosni Mubarak became president on October 14.

[26] Repression, corruption and increasing poverty led to widespread popular discontent. To distract public attention Mubarak ordered an inquiry into the wealth accumulated by the Sadat family. He also extended further concessions to foreign companies.

[27] There were some improvements in Egyptian foreign affairs during 1984. Egyptian diplomacy managed to overcome the most adverse reactions to the Camp David agreements. Their new position on the Palestinian question argued that any fair settlement of the Middle East crisis had to contemplate the rights of the Palestinian people and that Arab solidarity was "the only way to recover the usurped rights".

[28] From early 1985 the economic situation became more difficult as revenues from the four economic pillars: oil, emigrants' repatriated pay, canal fees and tourism shrank considerably. Islamic Fundamentalism was very powerful in opposition and became increasingly so as the government's popularity plummeted.

[29] Between 1980 and 1986 there was a dramatic increase in the role of foreign capital in the national economy and US aid continued to be a very important source of income. The government receives almost $3 billion per year, $1.3 billion of which is spent on defence. The IMF granted a further $1.5 billion loan in October 1986.

[30] The Egyptian foreign debt went from $2.4 billion in 1970,

External Debt
1994

Illiteracy
1995

PER CAPITA
$588

49%

to $35 billion in 1986; the figure for military aid alone multiplying seven times. Military expenditure and losses caused by war had understandably negative effects upon the economy.

[31] Parliamentary elections, originally scheduled for 1989, were called two years early, and in April 1987 the National Democratic Party (NDP) was elected with 75 per cent of the vote.

[32] In September 1989, in the United Nations General Assembly, Mubarak, representing contact between the Arab countries and the United States, proposed arranging an Israeli-Palestinian dialog, with no prior conditions. In October of that year relations with Libya were renewed.

[33] In the 1980s, Egyptian emigration to the Gulf countries reached very high levels. In some villages in the province of Sohag, in Upper Egypt, up to 60 per cent of the men left their homes, mainly to seek work in Kuwait and Lybia. This situation, which continues today, has forced Egytian women to take sole responsibility for their families, generating conflict and tension between the mothers and their children. Emigrants manage to earn large amounts, but at the expense of weakening the links with and unity of their families.

[34] In August 1990 Iraqi troops invaded Kuwait; Egypt was among the first Arab countries to condemn the action, sending troops to the Gulf immediately.

[35] When the land offensive started in January 1991, the US announced the cancellation of the Egyptian military debt, which amounted to $7 billion.

[36] Egypt's alignment with the West in the war against Iraq was not supported by the majority of the country's population. In February 1991 a demonstration to end the hostilities and seek a peaceful solution to the war in the Persian Gulf was staged in Cairo.

[37] There was great opposition to war against Iraq, despite the fact that Iraqi president Saddam Hussein had expelled nearly two million migrant workers from Iraq and Kuwait during the two previous years, and had attempted to forcibly recruit Egyptians to the army.

[38] In 1990, the foreign debt reached a record $40 billion. Per capita income averaged some $600 per year, with over a third of the population living below the poverty line.

[39] In May 1991, the IMF approved a stand-by loan of $372 million to Egypt, conditional on an economic "structural adjustment plan". Cairo committed itself to privatizing State-run companies, to eliminating controls on production and investment and to reducing the current fiscal deficit from 21 per cent to 6.5 per cent of the GNP. In order to achieve these goals the government decided to cut back subsidies on food and other staples and to reduce the program of aid for the needy.

[40] On May 15, 1991, foreign minister and deputy minister Esmat Abdel Meguid was named the new secretary general of the Arab League. Coming upon the heels of the return of Arab League headquarters to Cairo (from Tunis) this appointment signified Egypt's recovery of its leadership role within the Arab world.

[41] The escalation of violence from Islamic fundamentalists led the government to enact an anti-terrorist law and extend for another three years the state of emergency which had been in effect for the past ten years. According to official statistics, this violence caused 175 deaths between February 1992 and August 1993. Thousands of Islamic followers were arrested and in June and July 1993, 15 people were executed.

[42] The government continued its economic liberalization policies by facilitating foreign banks' operations. In March 1993, the IMF supported a privatization plan with the cancellation of $3 billion in foreign debt. In October, after being re-elected in a plebiscite, Mubarak continued his iron hand policy with Islamic fundamentalists. However, attempts against the lives of foreign tourists multiplied in 1994.

[43] In April, lawyers' associations denounced the suspicious death in a police station of an Islamic activists' defender. After a week of demonstrations the movement concluded with a general strike which revealed the influence had by Islamic fundamentalists among lawyers' associations. In October, the number of dead since March 1992 when the rebellion of Islamic fundamentalists against Mubarak began, rose to 460.

[44] In May, the president formed a committee to organize political talks between the government and the opposition with the exclusion of communists, the Muslim Brotherhood and groups representing

the Coptic minority. Simultaneously, relations with the IMF were hampered due to an alleged government slowness in implementing the planned economic liberalization.

[45] In the international sphere, Egypt recovered its main role in the Middle East peace talks and in political exchanges between Arab countries. This was shown by a meeting held in Alexandria in December with the participation of leaders from Egypt, Saudi Arabia and Syria. In February 1995, a summit joined leaders in El Cairo from Egypt, Jordan, Israel and Palestine.

[46] In 1995, Mubarak was unable to find a solution to the confrontation with Islamic fundamentalists. In January, secretary al-Alfi attended a meeting of Arab coun-

tries' interior ministers to attempt to coordinate the fight against violent Islamic movements.

[47] In November, the ruling National Democratic Party won the parliamentary elections with the participation of all the parties acknowledged by the government. Elections, held amid a violent atmosphere, granted 416 of the 444 seats at stake to the ruling party which provoked several accusations of fraud. In January 1996, Mubarak appointed Kamal al-Ganzouri as prime minister, replacing Atef Sedki.

STATISTICS

DEMOGRAPHY

Urban: 44% (1995). **Annual growth:** 0.6% (1992-2000). **Estimate for year 2000:** 63,000,000. **Children per woman:** 3.8 (1992).

HEALTH

Under-five mortality: 52 per 1,000 (1995). **Calorie consumption:** 112% of required intake (1995). **Safe water:** 80% of the population has access (1990-95).

EDUCATION

Illiteracy: 49% (1995). **School enrolment:** Primary (1993): 89% fem., 89% male. Secondary (1993): 69% fem., 81% male. University: 17% (1993). **Primary school teachers:** one for every 26 students (1992).

COMMUNICATIONS

100 **newspapers** (1995), 100 **TV sets** (1995) and 102 **radio sets** per 1,000 households (1995). 4.3 **telephones** per 100 inhabitants (1993). **Books:** 88 new titles per 1,000,000 inhabitants in 1995.

ECONOMY

Per capita GNP: $720 (1994). **Annual growth:** 1.30% (1985-94). **Annual inflation:** 16.40% (1984-94). **Consumer price index:** 100 in 1990; 164.9 in 1994. **Currency:** 3 pounds = 1$ (1994). **Cereal imports:** 7,206,000 metric tons (1993). **Fertilizer use:** 3,392 kgs per ha. (1992-93). **Imports:** $10,185 million (1994). **Exports:** $3,463 million (1994). **External debt:** $33,358 million (1994), $588 per capita (1994). **Debt service:** 15.8% of exports (1994). **Development aid received:** $2,304 million (1993); $41 per capita; 6% of GNP.

ENERGY

Consumption: 608 kgs of Oil Equivalent per capita yearly (1994), -67% imported (1994).

El Salvador

Population: 5,635,000 (1994)
Area: 21,040 SQ KM
Capital: San Salvador

Currency: Colón
Language: Spanish

El Salvador

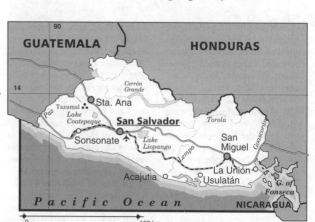

The region of El Salvador was inhabited from early times by Chibcha people (*muisca*; see Colombia), most of whom were Pipile and Lenca. The Maya also lived within the region, but they were not very influential here.

[2] The Spaniards subdued the Aztecs in Mexico, and subsequently began the conquest of Central America under the leadership of Pedro de Alvarado. In 1525, Alvarado founded the city of El Salvador de Cuscatlan. The territory formed part of the Captaincy General of Guatemala, a dependency of the Viceroyalty of Mexico. Central America became independent from Spain in 1821 and organized itself into a federation.

[3] In 1827, internal rivalries between "imperialists" and "republicans" led to civil war. In 1839, General Francisco Morazan, president of the Central American Republic, tried to prevent the break-up of the federation. From El Salvador, Morazan struggled to preserve the union with the support of some Nonulco natives, headed by Anastasio Aquino. Morazan was defeated in 1840 and sent into exile, and Anastasio Aquino was arrested and executed.

[4] When the federation was dissolved, Britain took advantage of the situation dominating the isthmus. In 1848, President Doroteo Vasconcelos refused to bow under British pressure, and the British blockaded Salvadoran ports. Over the last three decades of the 19th century, the US began to push the British out.

[5] At the end of the century, the invention of artificial coloring destroyed the demand for indigo, El Salvador's principal export product, and its price fell to rock bottom. Indigo was replaced with coffee. Coffee required larger and more extensive farming areas, and the Liberal Revolution of 1880 drove thousands of peasants from their communal lands, forming a rural working class and a countryside full of anger and conflict. The coffee plantation owners became the dominant oligarchy and the ruling class of El Salvador.

[6] The 1929 financial crash caused the coffee market to collapse, crops were left unharvested, and thousands of sharecroppers and poor peasants starved. This led to a mass uprising headed by the Communist Party of El Salvador and Farabundo Martí, a former secretary of Augusto C. Sandino in his campaign against the US invasion of Nicaragua.

[7] The January 1932 rebellion was ruthlessly crushed, around 30,000 Salvadorans were massacred by the troops of General Maximiliano Hernández Martínez, who had taken power in 1931. This started a series of military regimes which lasted half a century. Twelve thousand people died as a consequence of the repression.

[8] In 1960, the Alliance for Progress sponsored an industrialization program, within the Central American Common Market. High economic growth rates were attained without reducing the rampant unemployment which had caused 300,000 landless peasants to emigrate to neighboring Honduras. Population growth and competition between the local industrial bourgeoisies, led to war between El Salvador and Honduras in June 1969. The regional common market collapsed after the 100-hour conflict, severely damaging Salvadoran industry.

[9] In the early 1970s unions and other civilian movements took on new life. Guerrilla fighters appeared in El Salvador and the legal opposition parties joined into a national front; the UNO, formed by the Christian Democrats (PDC), the Communists (UDN) and the Social Democrats (MNR). Colonel Arturo Molina, presidential candidate for the official National Conciliation Party, defeated the UNO's candidate Napoleón Duarte in a fixed election in 1972.

[10] In 1977, another fraudulent election made General Carlos Humberto Romero president. Mass riots broke out in protest but social unrest was suppressed leaving 7,000 dead.

[11] The elimination of the legal political opposition encouraged the growth of guerrilla organizations who began to co-ordinate their armed action with the democratic opposition. In response, the US State Department, afraid of a repeat of the Nicaraguan situation, supported a coup by reformist military officers.

[12] On October 15 1979, a civilian/military junta seized power. It included representatives of the Social Democracy and the Christian Democracy. The Junta lacked real power and had no control over the ruthless repression campaigns carried out by police and military forces. Civilian members resigned and were replaced by right-wing Christian Democrats from Duarte's party.

[13] On March 24 1980, the Archbishop of San Salvador, Monsignor Oscar A. Romero, was assassinated while performing mass in a clear reprisal for his constant defence of human rights. The leading guerrilla organizations, democratic parties and mass organizations united under the policy of a common program of "popular revolutionary democracy".

[14] In October 1980, the five anti-regime political-military organizations agreed to form the Farabundo Martí Front for National Liberation (FMLN). On January 10 1981 the FMLN launched a "general offensive" throughout most of the country. However, the Front was unable to seize power.

[15] In August 1981, the Mexican and French governments signed a joint agreement recognizing the FMLN and the Democratic Revolutionary Front (FDR) as "a representative political force".

[16] The US administration, led by President Ronald Reagan, saw the situation in El Salvador as a national security issue. The US became directly involved in the political and social conflict, and was the military and economic mainstay of the "counter-insurgency" war which the Salvadoran Army was unsuccessfully waging.

[17] On March 28, 1982, as instructed by Washington, the regime held an election for a Constituent Assembly. In response, the rebels launched an offensive ending in a

PUBLIC EXPENDITURES

% of GNP — 1992

- MILITARY: 1.7
- EDUCATION: 1.8
- HEALTH: 2.6

WORKERS

7.9% UNEMPLOYMENT

1994

- MEN 72%
- WOMEN 28%
- AGRICULTURE 11%
- INDUSTRY 23%
- SERVICES 66%

LAND USE

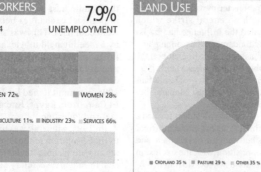

- CROPLAND 35%
- PASTURE 29%
- OTHER 35%

ENVIRONMENT

It is the smallest and most densely populated country in Central America. A chain of volcanos runs across the country from east to west and the altitude makes the climate mild. Coffee is the main cash crop in the highlands. Subsistence crops such as corn, beans and rice are also grown. Along the Pacific Coast, where the weather is warmer, there are sugar cane plantations. It is the country with the greatest problems of deforestation in Latin America.

SOCIETY

Peoples: 89% of the Salvadoran population are mixed descendants of American natives and Spanish colonizers, 10% are indigenous peoples, and 1% are European. **Religions:** Mainly Catholic (75%), Protestant, Mormon, Jehovah's Witness. **Languages:** Spanish is the official and predominant language. Indigenous minority groups speak Nahuatl. **Political Parties:** The Nationalist Republican Alliance (ARENA); the Christian Democratic Party; the left-wing Democratic Convergence, the Party of National Reconciliation (PCN). The Farabundo Martí National Liberation Front (FMLN), founded in October 1980, is made up of five political-military organizations; the Farabundo Martí People's Liberation Forces (FPL), the El Salvador Communist Party (PCS), the National Resistance Armed Forces (FARN), the People's Revolutionary Army (ERP) and the Central American Workers' Revolutionary Party (PRTC). **Social Organizations:** the National Union of Salvadoran Workers (UNTS), the United Union and Guild Movement (USIGES) and the National Confederation against Hunger and Repression (CNHR) created in 1988.

THE STATE

Official Name: República de El Salvador. **Administrative division:** 14 Departments. **Capital:** San Salvador, 422,600 inhab.; 1,522,100 metropolitan area (1992). **Other cities:** Soyapango 251,800 inhab; Santa Ana 202,300 inhab.; San Miguel, 182,800 inhab.; Mejicanos 145,000 (1992). **Government:** Armando Calderón Sol, President since May 1994. **National Holiday:** September 15, Independence Day (1821). **Armed Forces:** 30,700 troops (1994). **Paramilitaries:** National Civilian Police, made up of former guerrillas, soldiers and police.

External Debt
1994
PER CAPITA
$388

Illiteracy
1995
29%

one-week siege of Usulatán, a provincial capital.

[18] After continuous internal tussling for power, the presidency of the constitutional convention went to Roberto D'Aubuisson, the main leader of the ultra-right-wing Nationalist Republican Alliance (ARENA) and the power behind the assassination of Monsignor Romero.

[19] Against a background of an upsurge in fighting, general elections were held on March 25, 1984. These were boycotted by the FDR-FMLN; the abstention rate by voters was 51 per cent. Ostensibly supported by the US, the PDC - led by Napoleon Duarte - obtained 43 per cent of the vote, against the 30 per cent won by ultra-right-wing candidate Major Roberto D'Aubuisson.

[20] The extreme right parties disputed the election, but quick responses from the Minister of Defence and the High Command, in support of Duarte, quashed any further reaction. It was the first time that the Armed Forces had publicly supported the reformists. The only contacts between the two parties took place in La Palma and Ayagualo, in 1984.

[21] In October 1986, a relatively strong earthquake brought about a virtual cease-fire, which eventually led to the renewal of negotiations in October 1987. These talks took place within the new framework of regional peace making; the Central American governments had signed the Esquipulas agreements in August 1987, agreeing to strive for peace.

[22] Between 1987 and 1989, the US supported the Duarte administration's search for a political solution to the conflict with the FDR-FMLN. Internal disputes, and pressure from the ultra-right-wing sectors and the Armed Forces, blocked such a solution.

[23] Elections were held in October 1989; these were boycotted by some of the guerrillas, but civilian sectors of the FDR (members of the social democratic and Social Chris-

tian parties) participated, with Guillermo Ungo as their presidential candidate. Alfredo Cristiani, the ARENA (right-wing) party's candidate, won the election.

[24] In November 1989, the FMLN launched an offensive occupying several areas of the capital and surrounding regions. The government responded by bombing several densely populated areas of the capital. Six Jesuits, including the rector of the University of Central America, Ignacio Ellacuria, were tortured and killed by heavily armed soldiers. This provoked a world-wide outcry, especially from the Catholic Church, and American economic aid was threatened.

[25] According to the El Salvador Human Rights Commission, a non-governmental organization, women, especially students and members of labor unions, are the people who have suffered most from repression. During the last 12 years, the human rights movement has been led by mothers, wives, daughters and relatives of the thousands of victims of repression, and by the National Union of Salvadoran Workers (UNTS). It has challenged the harsh military regime and has denounced the continuous human rights violations.

[26] On March 10 1991, the legislative and local elections reflected a new spirit of negotiation. For the first time in 10 years the FMLN did not call for the boycott of the elections, instead they decreed a 3-day unilateral truce. Abstention was still above 50 per cent, and there were acts of paramilitary violence immediately prior to the polls. The voters narrowly elected the ruling party with 43 out of 84 seats. On March 12 the fighting resumed.

[27] In Mexico on April 4 1991 delegates of the Cristiani government and the FMLN started negotiations for a cease fire agreement. On April 19, 10,000 demonstrators, from 70 social organizations, gathered in the Permanent Committee for National Debate (CPDN), demanding that the Constitution be reformed. This was two weeks before both the end of the parliamentary term and the deadline given by the FMLN for the signing of the peace agreement.

[28] On April 27, after several attempts, representatives of the government and the Farabundo Marti Front signed the "Mexico Agreements" restricting the function of the Armed Forces to the defence of national sovereignty and territorial

integrity. The formation of paramilitary groups was banned, and it was agreed to reform article 83 of the constitution to say that sovereignty "resides in the people, and that it is from the people that public power emerges". In New York , in June, another agreement was reached; the Salvadoran government committed itself to dismantling the National Guard and the Rural Police (Policia de Hacienda), replacing it with Civilian Police including FMLN-members.

[29] On November 16, new talks began in the UN headquarters. This time, the FMLN declared an indefinite unilateral truce until a new, definite, cease-fire was signed. Then, a Spanish parliamentary delegation wrote a report on the murders of the six Spanish Jesuits from the Central American University (UCA). The report, submitted to the Spanish, European, Salvadoran, and American parliaments, accused the Salvadoran government and the army of concealing evidence which could help to clarify the facts.

[30] On January 1 1992, after 21 weeks of negotiation and 12 years of civil war, both parties met in New York to sign agreements and covenants establishing peace in El Salvador. The war left 75,000 people dead, 8,000 missing, and nearly 1 million in exile. The two parties fixed a cessation of hostilities period from February 1 until October 3 1992, for all armed conflict to cease, and for an atmosphere to develop that would be propitious for the agreements to go into effect, and for UN and OAS-sponsored negotiations to continue.

[31] The final agreements were signed in the Mexican city of Chapultepec on January 16 1992. They include substantial modifications to the Constitution and to the structure, organization, regulation, and form of the Armed Forces. They guarantee to change rural land tenure and to alter the terms of employee participation in the privatization of State companies; they establish the creation of bodies for the protection of human rights, and guarantee the legal status of the FMLN.

[32] According to the peace accords the government would have to reduce its troops by half by 1994, bringing the number down to 30,000; in addition, it would have to disband its intelligence service. As of March 3, a new civilian police would be created, made up in

Foreign Trade

MILLIONS $ 1994

IMPORTS
2,250

EXPORTS
844

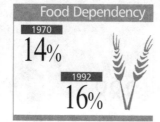

Food Dependency

1970
14%

1992
16%

STATISTICS

DEMOGRAPHY

Urban: 45% (1995). **Annual growth:** 0.6% (1992-2000). **Estimate for year 2000:** 6,000,000. **Children per woman:** 3.8 (1992).

HEALTH

One **physician** for every 1,563 inhab. (1988-91). **Under-five mortality:** 56 per 1,000 (1994). **Calorie consumption:** 93% of required intake (1995). **Safe water:** 55% of the population has access (1990-95).

EDUCATION

Illiteracy: 29% (1995). **School enrolment:** Primary (1993): 80% fem., 80% male. Secondary (1993): 30% fem., 27% male. University: 15% (1993). **Primary school teachers:** one for every 44 students (1992).

COMMUNICATIONS

101 **newspapers** (1995), 100 **TV sets** (1995) and 105 **radio sets** per 1,000 households (1995). 3.2 **telephones** per 100 inhabitants (1993). **Books:** 81 new titles per 1,000,000 inhabitants in 1995.

ECONOMY

Per capita GNP: $1,360 (1994). **Annual growth:** 2.20% (1985-94). **Annual inflation:** 15.50% (1984-94). **Consumer price index:** 100 in 1990; 166.8 in 1994. **Currency:** 9 colones = 1$ (1994). **Cereal imports:** 286,000 metric tons (1993). **Fertilizer use:** 1,073 kgs per ha. (1992-93). **Imports:** $2,250 million (1994). **Exports:** $844 million (1994). **External debt:** $2,188 million (1994), $388 per capita (1994). **Debt service:** 13.1% of exports (1994). **Development aid received:** $405 million (1993); $73 per capita; 5% of GNP.

ENERGY

Consumption: 219 kgs of Oil Equivalent per capita yearly (1994), 58% imported (1994).

---

part by former members of the FMLN. In January 1992, according to the terms of the Law of National Reconciliation, amnesty was granted to all political prisoners.

[33] In addition, the government pledged to turn over lands to the combatants and provide assistance to campesinos belonging to both bands. The FMLN became a political party as of April 30 1991, and held its first public meeting on February 1 1992. After years of being underground, it was presided over by guerrilla commanders Shafick Handal, Joaquin Villalobos, Fernan Cienfuegos, Francisco Jovel and Leonel Gonzalez; they called for the unification of all opposition forces for the 1994 elections.

[34] In early March 1992, the first implementation difficulties began

to be seen. Several leaders of the National Union of Salvadoran Workers accused the government of violating the accords, and launching a propaganda campaign against grassroots organizations.

[35] On February 15 1993, the last 1,700 armed rebels turned over their weapons in a ceremony which was attended by several Central American heads of State and by UN Secretary-General Boutros Boutros-Ghali. The National Civil Police was created, as well as a Human Rights Defence Commission and a Supreme Electoral Court.

[36] The result of the investigation of human rights violations, carried out by the Truth Commission created by the UN, led to the resignation of Defence Minister General Rene

Emilio Ponce, singled out in that investigation as being the one who ordered the assassination of six Jesuits at the University of San Salvador in 1989. According to the Commission's final document, the military, the death squads linked to these and the State were responsible for 85 percent of the civil rights violations committed during the war.

[37] The Truth Commission recommended the dismissal of 102 military leaders and that some former guerrilla leaders be deprived of their political rights. President Cristiani proposed a general amnesty for cases where excesses had been committed; this proposal was approved on March 20, 1993, only 5 days after the document drawn up by the Truth Commission had been made public. With this measure, the most serious crimes committed during the war met with total impunity.

[38] A year later on March 20 1994, the first elections since the civil war were held. The candidate of the left coalition, Democratic Convergence - made up of the FMLN and other groups - won 25.5 per cent in the first round of voting, against 49.2 per cent for the right candidate, Armando Calderón Sol, from the ARENA party. Although the left declared the existence of fraud, UN observers in the country (ONUSAL) insisted that the voting had been fair. After the elections, the FMLN faced an internal crisis triggered by discrepancies between the groups that make up the alliance.

[39] According to ONUSAL, the peace accords did not bring an end to the violence. In addition to the existence of intelligence activities within the Armed Forces, members of the military were linked to organized crime. Likewise, the fact that nothing was done to create viable employment opportunities for discharged troops (from both bands) led to an increase in petty crime.

[40] The prisons were overflowing. Built for a total capacity of 3,000 prisoners, they were holding twice that number. In addition, the abominable prison conditions and the delays in bringing cases to trial generated discontent which vented itself in multiple prison riots, in which 70 people were killed and more than a hundred were injured.

[41] The long-promised turn-over of lands to demobilized fighters was slow and inefficient. By mid-1994, only one-third of the potential ben-

eficiaries -12,000 of a total 37,000 former members of the army or the guerrilla - had obtained their plots. The rest remained inactive, living in sub-standard temporary housing and often drifting into organized crime.

[42] On the regional scene, Calderón Sol's administration intensified talks with neighboring Honduras, to resolve longstanding border disputes.

[43] The government managed to raise the country's annual economic growth to 5 per cent, although 29 per cent of the population live in poverty and 15 per cent of children under 5 years old showed signs of malnutrition, according to official figures.

[44] Official studies revealed that 90 per cent of indigenous vegetation has disappeared and that two thirds of farmlands suffer from erosion. 90 per cent of the rivers are polluted with chemical waste and almost half the population does not have access to drinking water.

[45] In May 1995, an agreement between ARENA and the Democratic Party - a breakaway from the former FMLN - allowed for the raise of the valued added tax from 10 to 13 per cent. This raise was based on the need to obtain funds for land, electoral and judicial reforms as well as infrastructure repairs, harmed in the war.

[46] National Civilian Police was unable to control criminals, who according to the local press committed an offence each hour. The government sent 5,000 soldiers to reinforce police patrolling of highways and rural areas. Death squads were accused of dozens of crimes. Three policemen were found among the members of the so-called "Black Shadow". Meanwhile, a raid revealed a modern and sophisticated arsenal belonging to "The Benedicts", whose leader was linked to a network of Central American criminal organizations.

[47] Having started in April, the UNO's supervision of the observance of the 1992 agreements between the guerrillas and the government concluded in October that year. The slowness of the process, which included the transfer of lands and compensation for war veterans, led in August to the occupation of streets and public buildings by those involved.

Equatorial Guinea

Guinea Equatorial

Population: 386,000 (1994)
Area: 28,050 SQ KM
Capital: Malabo

Currency: CFA Franc
Language: English

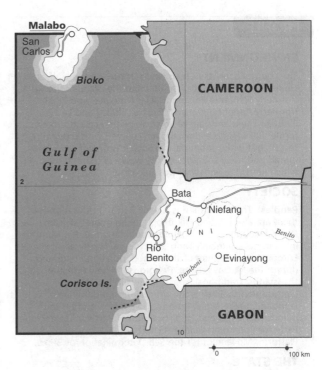

Around the 13th century, Fang and Ndowe people settled in the Rio Muni area, south of Gabon, subduing the Bayele pygmy population, who now only exist in a few isolated groups. From the continental coast, these nations expanded onto the nearby islands which were described as "densely populated" in the 15th century. In the colonial division of Africa, the Rio Muni area and the islands received the name of Equatorial Guinea.

[2] The Ndowe became allies and intermediaries of Portuguese, Spanish, Dutch and English slave traders. The Fang, whose social organization did not include slavery, withdrew into the forests, convinced that the white men were cannibals.

[3] The kings of Portugal, proclaiming themselves the lords of Guinea, ceded the entire "District of Biafra" to Spain under the treaties of San Ildefonso and Pardo in 1777 and 1778, in exchange for Spanish territory in southern Brazil. In 1778, an expedition sailed out from Montevideo to occupy the islands. They lost their commanding officer, Argelejos, in a battle against the Annoboneuses, and the survivors, under their new leader, Lieutenant Primo de Rivera, turned back. The French and British gradually took over sections of the territory, with the British finally occupying it, founding the first settlements. They used it as a base for their conquest of Nigeria and turned the freed slaves, or "Fernandinos", into their agents, creating a ruling group that, in many aspects, still exists today.

[4] Between 1843 and 1858, the District was militarily reconquered by Spain, re-establishing their "rights" over the area. At that time, economic activity centered on cacao, coffee and timber, but the territory was ineffectively controlled from a distance. The Spanish colonists, led by Fernando Po, supported Franco in the 1936 civil war and were later granted almost full powers over the archipelago. Admiral Carrero Blanco, Franco's Prime Minister, owned the colony's largest cacao plantations.

[5] In 1963, the colony obtained a degree of internal autonomy, and three legal political organizations were formed: MONALIGE (the Equatorial Guinean National Liberation Movement), MUNGE (the Equatorial Guinean National Unity Movement), and IPGE (the Equatorial Guinean Popular Ideal). Mounting international pressure forced Spain to grant the colony independence and this was officially proclaimed on October 12, 1968.

[6] Francisco Macías Nguema assumed the presidency with the support of Atanasio Ndongo, of MONALIGE and Edmundo Bosio, leader of the Bubi Union of Fernando Po. He headed a coalition government formed by IPGE and sizeable dissident groups from MUNGE and MONALIGE; movements that had emerged during the struggle for independence.

[7] Within a year, on the pretext of foiling an alleged coup attempt, Macías began a violent campaign to eliminate the opposition killing thousands of political prisoners, murdering opponents, causing disappearances and exiling 160,000 people. Amnesty International reported that two thirds of the National Assembly had mysteriously vanished.

[8] Proceeding with the repression, Macías Nguema dissolved all political parties, and created the Sole Traditional Workers' Party (PUNT), proclaiming himself president for life and "grand master of popular education, science and culture". His reign of terror was kept going by the "Youth on the March with Macías" organization, who extended persecution to Catholic priests and Protestant missionaries.

[9] In August 1979, the rule of Macías Nguema ended in a coup led by his nephew, Lieutenant-Colonel Teodoro Obiang Mba Nzago. Macías was arrested, tried and executed for crimes against humanity.

[10] Persecution, indiscriminate arrests and corruption were soon rampant, reminiscent of the previous regime. In August 1982, legislative elections were held but the list of candidates was drawn up by the official party. The National Alliance for the Recovery of Democracy formed in exile, sought to negotiate with Obiang but accomplished little.

[11] A coup attempt in June 1986 sparked off another wave of arrests. Among those detained was vice-prime minister, Fructuoso Mba Onana.

[12] In September 1984, Obiang joined the Customs and Economic Union of Central Africa (UDEAC), linking the Ekwele (the national currency) to the central banks of the region. Stronger ties with the West resulted in a drastic reduction of facilities offered by the Soviet Union and East Germany. The US, France and Spain dominated the economy, controlling the extraction of oil, iron ore and lumber. Meanwhile, Morocco sent troops to guarantee the stability of the new government.

[13] Teodoro Obiang was re-elected in June 1988 and he continued to apply IMF prescriptions, thus obtaining a $16 million loan to improve public investments, restructure the banking system, and accelerate growth. At a meeting of international loan agencies in November 1988, the regime obtained additional loans of $58 million to be spread over three years.

[14] Teodoro Obiang's visit to France in September 1988 produced closer ties between the two countries, and Equatorial Guinea later called for the integration of all French-speaking countries. At the same time, he obtained an amnesty for two-thirds of his country's foreign debt to Spain.

[15] There was no easing of the repression. Two members of the armed forces were executed in September 1988 on the charge of

> The Ndowe became allies and intermediaries of Portuguese, Spanish, Dutch and English slave traders. The Fang, whose social organization did not include slavery, withdrew into the forests, convinced that the white men were cannibals.

WORKERS

1994

MEN 60% WOMEN 40%

AGRICULTURE 77% INDUSTRY 2% SERVICES 21%

PROFILE

ENVIRONMENT

The country consists of mainland territory on the Gulf of Guinea (Rio Muni, 26,017 sq km) and the islands of Bioko (formerly Fernando Po, and Macías Nguema) and Pigalu (formerly Annobon, Corisco, Greater Elobey and Lesser Elobey). The islands are of volcanic origin and extremely fertile; Rio Muni is a coastal plain covered with tropical rain-forests but without natural harbors. It is one of the most humid and rainy countries of the world, a characteristic which limits the variety of possible crops.

SOCIETY

Peoples: The population is mostly Bantu. In the islands there are also Ibo and Efik peoples who migrated from Nigeria, subduing the local Bubi population. In Rio Muni the inhabitants are mainly Fang and Ndowe. Nearly all of the Europeans and a third of the local population emigrated during the Macías regime. **Religions:** Mainly Christian on the islands; traditional African beliefs in Rio Muni. **Languages:** Spanish is the official and predominant language. In Rio Muni, Fang is also spoken, and on the islands, Bubi, Ibo and English. **Political Parties:** Democratic Party for Equatorial Guinea (PDGE), Joint Opposition Platform, Movement for the Self-Determination of Bioko.

THE STATE

Official Name: República de Guinea Ecuatorial. **Administrative divisions:** 4 continental and 3 island regions. **Capital:** Malabo, 30,000 inhab. (1983). **Other city:** Bata, 24,300 inhab. (1983). **Government:** Obiang Nguema Mba Nzago, President since August 25, 1979. **National Holiday:** October 12, Independence Day (1968). **Armed Forces:** 1,300.

STATISTICS

DEMOGRAPHY

Annual growth: 2.8% (1992-2000). **Children per woman:** 5.5 (1992).

HEALTH

Under-five mortality: 177 per 1,000 (1994). **Safe water:** 91% of the population has access (1990-95).

COMMUNICATIONS

82 **newspapers** (1995), 83 **TV sets** (1995) and 102 **radio sets** per 1,000 households (1995). 0.3 **telephones** per 100 inhabitants (1993). **Books:** 92 new titles per 1,000,000 inhabitants in 1995.

ECONOMY

Per capita GNP: $430 (1994). **Annual growth:** 2.20% (1985-94). **Currency:** 535 CFA francs = 1$ (1994). **Cereal imports:** 11,000 metric tons (1990).

---

conspiring against the government, and around the same time, an opposition leader, José Luis Jones, was arrested on his return to the country, and condemned to a 17-year prison term. He was eventually pardoned.

[16] In June 1989, Obiang was re-elected in elections where the Equatorial Guinea Democratic Party, was, once again, the only legal party. Governmental policy continued unchanged. A new agreement was signed with the EEC countries granting them fishing rights from 1989-1992 for ECU6 million (European currency units).

[17] In early 1991, the Democratic Coordinating Committee of Opposition Parties was founded in Gabon, and in May, Feliciano Moto, one of the main opposition leaders, was assassinated. In mid- 1991, the ban on all political parties and censorship of the press were lifted. Toward the end of the year, a hu-

man rights commission was formed to investigate violations that had occurred. The new constitution, approved in a November referendum, was rejected by the opposition because it set no limit on presidential re-election.

[18] At the annual meeting of the UN Human Rights Commission, held in Geneva in March 1992, the Malabo government was censured for human rights violations.

[19] In October 1991, the inhabitants of the Pigalu island, which belongs to Equatorial Guinea, made an appeal to the international community to prevent the dumping of European industrial and toxic waste on the island. The island, which had hitherto been unspoilt, is, with Obiang's consent, rapidly becoming a waste dump. The first symptoms of pollution-related illnesses are already appearing.

[20] In early 1992, the government announced a general amnesty. The

law regulating political parties banned those based on tribal, regional or provincial loyalties, and required political groups to pay nearly $160,000 in order to be legally recognized as parties. In addition, it established a ten-year residence requirement for

> In October 1991, the inhabitants of the Pigalu island, which belongs to Equatorial Guinea, made an appeal to the international community to prevent the dumping of European industrial and toxic waste on the island.

presidential candidates, thereby disqualifying all those in exile.

[21] Despite these obstacles, in early 1993, ten political parties were legalized. At the same time, the UN released a report denouncing the systematic viola-

tion of the civil liberties of political opponents.

[22] The credibility of legislative elections held in late 1993, was questioned by international observers. According to official returns, the ruling Democratic Party won 68 in 80 seats. After Spain's general consul in Bata was expelled in January 1994, that country reduced its aid to Equatorial Guinea by half.

[23] In March 1995, Severo Moto, leader of the opposing Progressive Party, was sentenced to two and a half years in prison for a supposed bribe offered to a police officer and for having "harmed the reputation" of Obiang. In April, Moto was sentenced once again to 28 years in prison charged with treason and conspiracy, which caused Western countries' reproval, especially Spain's.

[24] In July, several members from the Movement for Bioko's Self-Determination (Bioko is one of Equatorial Guinea's islands) were put under arrest. In midst of accusations of arbitrary arrests and tortures, the government decided to hold presidential elections in early 1996. In February that year, shortly before the elections, the government banned the Joint Platform Opposition and arrested several of its members. On February 25, Obiang won the controversial vote.

Eritrea

Eritrea

Population: 3,482,000 (1994)
Area: 124,320 SQ KM
Capital: Asmara

Currency: Ethiopian Birr
Language: English

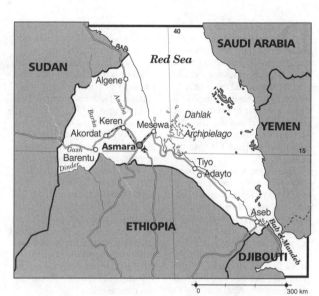

As the site of the most important ports of the Aksumite empire (flourished 4th-6th century AD), Eritrea was linked to the beginnings of the Ethiopian kindom, but it retained much of its independence until it fell under Ottoman rule in the 16th century. From the 17th to the 19th century control over the territory was disputed among Ethiopia, the Ottomans, the kingdom of Tigray, Egypt, and Italy. In 1890, the treaty of Wichale between Italy and Menilek II of Ethiopia recognized Italian possessions on the Red Sea, and the colony, created on January I, 1890, was named by the Italians for the Mare Erythraeum ("Red Sea" in Latin) of the Romans.

[2] Eritrea was used as the main base for the Italian invasions of Ethiopia (1896 and 1935-36) and it became one of the six provinces of Italian East Africa. In 1941 the area came under British administration and remained so until Eritrea was federated as an autonomous unit to Ethiopia in 1952.

[3] The common struggle against the Italians had brought a reasonable degree of unity to almost one million Eritreans. On December 2, 1950, United Nations Resolution 390A declared that Eritrea should become a federated state within the Ethiopian Empire. The resolution rejected Ethiopian demands for outright incorporation, but also left the process of Eritrean self-determination undefined.

[4] In Eritrea, a national assembly was elected which enjoyed some autonomy until 1962, when Haile Selassie forced a group of Eritrean congressmen to vote for its complete incorporation into Ethiopia. The decision was contested by nationalist groups and immediately sparked a rebellion.

[5] The oldest anti-Ethiopian resistance movement is the Eritrean Liberation Front (ELF) founded in 1958, in Cairo, by journalist and union leader Idris Mohamed Adem. It began guerrilla activities in September 1961. In 1966, a split produced the Eritrean Popular Liberation Front (FPLE). In 1974, with Sudanese mediation, the two groups agreed to coordinate their actions and in the next few years, the FPLE imposed its leadership upon the rebel movement.

[6] After the pro-Soviet Mengistu government was installed in Ethiopia, relations between Addis Ababa and the Eritrean rebels did not improve. The Eritreans felt that the changes in Addis Ababa did not bring their self-determination closer, so they had no reason to stop fighting. The war against Ethiopia caused thousands of victims on both sides.

[7] In 1989, US government attempts at mediating the conflict were unsuccessful. Negotiations were held in Alabama between delegations representing the Ethiopian government and the FPLE, aiming to put an end to the conflict.

[8] In February 1990, the FPLE captured the port of Massewa, and nearly all of the Eritrean territory. The former Ethiopian president suddenly fled to Zimbabwe allowing the FPLE rebels to enter Asmara on May 25, and the port of Aseb on the following day. The highway from Aseb to Addis Ababa is the only land supply route to the Ethiopian capital.

[9] At the end of May, the FPLE announced the formation of a Provisional Government, and in July, in the Ethiopian capital, an agreement was reached to hold a UN supervised referendum within a two-year period.

[10] Since 1991, Asmara and Addis-Ababa have begun to deal with each other on a diplomatic level. After an agreement between the interim Ethiopian president and the FPLE, the ports on the Red Sea were reopened to let international aid into the country. At about the same time, it rained in Eritrea, ending the two year drought.

[11] 99.8 per cent of voters favoured independence in a plebiscite held in April 1993. The de facto ruling EPLF, formed a provisional government - that would draft a new constitution within four years and would call multi-party election - led by Isaias Afwerki.

[12] In 1993, Eritrea joined the UN and the provisional National Assembly announced multi-party elections would be held in 1997. In February 1994, the EPLF held its third congress and became a political party: the People's Front for Democracy and Justice (PFDJ). Eritrea joined the IMF and the World Bank, which praised its "realistic" development policy.

[13] In December 1994, relations with Sudan deteriorated, hampering the repatriation of 500,000 Eritrean refugees from that neighbouring country. In June 1995, Eritrea hosted a meeting of Sudanese opposition movements and in October, Afwerki stated his government would supply weapons to any group willing to overthrow Khartoum's regime.

[14] The country's reorganization after 30 years of war and its economic plans are problems difficult to solve for the provisional government. In 1996, the planned presentation of lands to foreign investors was hindered by several disputes regarding the property of those lands. According to official sources, these disputes "may be solved in 1997".

PROFILE

ENVIRONMENT

Eritrea is in the horn of Africa. The northeast coast, all 1,000 km of it, is bordered by the Red Sea; to the northeast lies Sudan; to the south, Ethiopia and to the southeast, Djibouti. The dry plains and extremely hot desert steppes are inhabited by pastoralists. Deforestation and the consequent erosion are partly responsible for the frequent droughts.

SOCIETY

Peoples: The nine ethnic groups are the Tigrinya, Tigre, Bilen, Afar, Saho, Kunama, Nara, Hidareb and Rashaida. The majority are pastoralists or farmers; 20% are urban workers. Half a million Eritrean refugees live in Sudan, 40,000 in Europe and 14,000 in the US. **Religions:** Almost half of all Eritreans are Coptic Christians; most of the rest are Muslim, although there are Catholic and Protestant minorities. **Languages:** Tigrinya and at least nine local languages. **Political Parties:** People's Front for Democracy and Justice (former Eritrean People's Liberation Front).

THE STATE

Capital: Asmara, 400,000 inhab. (1992). **Other cities:** Asseb 50,000 inhab.; Keren 40,000 inhab.; Massaua (Mesewa) 40,000 inhab. in 1992. **Government:** Isaias Afwerki, president. **National holiday:** May 24, Independence (1993).

COMMUNICATIONS

0.6 **telephones** per 100 inhabitants (1993).

Estonia

Eesti

Population: 1,499,000 (1994)
Area: 45,274 SQ KM
Capital: Tallinn

Currency: Krona
Language: Estonian

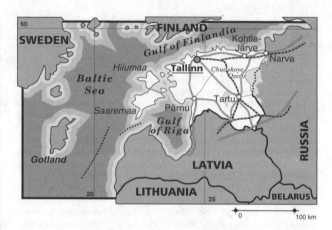

Despite historical and political links with their southern neighbors, Lithuania and Latvia, the Estonian people have always been known for their spiritual and cultural independence. Belonging to a branch of the Finno-Ugric nations, Estonians have greater cultural and linguistic ties with the Finns to the north than with the Indo-European Balts to the south.

2   The region was settled some 6,000 years ago. Around the year 400 AD, the semi-nomadic peoples' hunting and fishing activity began to be replaced by agriculture and cattle raising. At the same time, navigation and trade with neighboring countries along the Baltic Sea intensified. In the 11th and 12th centuries, combined Estonian forces successfully repelled Russia's attempts to invade the territory.

3   The Germans, Russians and Danes, invading Estonia in the 13th century, found a federation of states with a high level of social development and a strong sense of independence, keeping them united against foreign conquerors.

4   In the 13th century, the Knights of the Sword, a Germanic order which was created during the Crusades, conquered the southern part of Estonia and the north of Latvia, creating the kingdom of Livonia and Christianizing the inhabitants of this area. German traders and landowners brought the Protestant Reformation to Estonia in the first half of the 16th century.

5   The northern part of Estonia remained under Danish control. Livonia was much disputed between 1558 and 1583; it was repeatedly attacked by Russia before being dismembered in 1561. Poland conquered Livonia in 1569; a hundred years later, it ceded the major part of the kingdom to Sweden. In the Nordic Wars (1700-

1721), Russia took Livonia away from Sweden, and kept these lands under the Treaty of Nystad.

6   Russia received the Polish part of Livonia in 1772, with the first partition of Poland. The former kingdom of Livonia became a Russian province in 1783. Power was shared between the czar of Russia and local German nobles, who owned most of the lands and the peasants lived in serfdom.

7   The abolition of serfdom in Russia and peasant landownership rights (1804), strengthened the Estonian national consciousness. During the second half of the 19th century, the Society of Estonian Literati developed the written language, and the preservation of national folklore, enabling the people to withstand the czar's Russification campaigns.

8   In 1904, Estonian nationalists seized control of Tallinn, ousting the German-Baltic rulers. After the fall of the czar in February 1917, a demonstration by 40,000 Estonians in Petrograd forced the Provisional Government to grant them autonomy, maintained even after the Bolshevik Revolution.

9   In November 1917, with the election of a constituent assembly, the Estonian Bolsheviks only obtained 35.5 per cent of the votes. On February 24 1918, Estonia declared its independence from the Soviet Union and set up a provisional government. The following day, German troops occupied Tallinn and the Estonian government was forced to go into exile.

10   After World War I, the Estonians sucessfully fought both the Red Army and the Germans. On February 2 1920, the Soviet Union recognized Estonia's independence.

11   Estonia, Latvia and Lithuania, joined the League of Nations in 1921. Following the Swiss model, the Estonian constitution established a parliamentary democracy. The government began the reconstruction of the country and initiated agrarian reform. In the 1920s, Estonia established the world's first shale-oil distillery.

12   Estonia passed legislation guaranteeing the rights of minorities, and ensuring that all ethnic groups had access to schools in their own languages. The economic depression of the 1930s led to a modification of the Estonian constitution; in 1933, it became a virtual dictator-

ship before adopting a presidential-parliamentary system in 1937.

13   The secret protocols of the Molotov-Ribbentrop pact, signed in 1939 determined that Estonia - like its two Baltic neighbors - would remain within the Soviet sphere of influence. At the same time, Tallinn signed a mutual assistance treaty with Moscow giving the USSR the right to install naval bases on Estonian soil.

14   In June 1940, after demanding the right for his troops to enter Estonian territory under the pretext of searching for missing soldiers, Stalin deposed the Tallinn government and replaced it with members of the local Communist Party. Elections were held during the Soviet occupation, after which the Communist Party seized power.

15   Following the examples of Latvia and Lithuania, the new government adopted the name "Soviet Socialist Republic of Estonia", joining the USSR. According to the official record, all three Baltic States voluntarily became a part of the USSR. In 1941, around 60,000 Estonians were deported to Siberia.

16   When the German offensive against the USSR began in 1941, Nazi troops invaded Estonia, establishing a reign of terror which was especially directed against Jews, gypsies and Estonian nationalists. The USSR recovered the Baltic States in 1944; Estonian groups began pro-independence guerrilla activity, but had little impact against Soviet military might.

17   Industrialization was established by the Soviet regime, and collective farming was forced upon the national population. Some 80,000 Estonians emigrated to the West, while Russian colonization gradually altered the traditional eth-

nic composition of the population. The history of Estonia was rewritten, and its national symbols replaced, with Russian language and culture replacing those of Estonia.

18   Around 20,000 Estonians were deported between 1945 and 1946. The third wave of mass deportations took place in 1949, when another 40,000 Estonians were sent to the farthest regions of the USSR, most of them farmers who refused to accept collectivization of the land, imposed by Soviet authorities.

19   In 1982-83, toward the end of the Brezhnev era, a number of underground publications appeared in Estonia, including The Democrat. In addition, several new movements emerged, including the Democratic Movement, the Estonian National Front and others; members of these movements were arrested and sent to forced labor camps.

20   The reforms set in motion in 1985 by Soviet president, Mikhail Gorbachev stimulated social and political activity within Estonia. In August 1987, a demonstration in Tallinn demanded the publication of the Molotov-Ribbentrop pact. Latvians and Lithuanians also asked that the contents of the protocols be revealed.

21   In January 1988, former Estonian political prisoners founded the Estonian Independence Party, defending the country's right to self-determination. In addition, this group called for the re-establishment of multi-party democracy, and the restoration of Estonian as the country's official language. Another group, the Estonian Heritage Society, began trying to locate and restore the country's historical monuments.

22   The Popular Front of Estonia

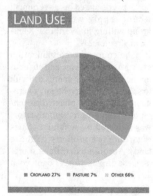

ENVIRONMENT

Located on the northeastern coast of the Baltic Sea, Estonia is bounded by the Gulf of Finland in the north, the Russian Federation in the east and Latvia in the south. The Estonian landscape - with its numerous lakes and river - bears witness to its glacial origin. There are more than 1,500 lakes, the largest of which are Lake Peipsi (Europe's fourth largest) and Lake Vorts. Forests make up 38% of Estonia's territory; the highest elevation is Mount Suur Muna Magi (317 m). The climate is temperate, with average temperatures of 28°C in summer and around 0°C in winter. The Baltic coast is 1,240 km long and includes a number of fjords. Some 1,500 islands close to the Baltic coast make up around one-tenth of Estonia's territory. The Gulf of Finland has numerous ice-free bays, of which Tallinn is the largest. Estonia's most important mineral resources are bituminous shale (which meets most of Estonia's energy needs), and phosphates. Estonia also has an important fishing industry, although its fleet has had to begin fishing far from its coastal waters due to the polluted state of the Baltic Sea, which contains toxic waste dumped by industries in several countries with Baltic coastlines. Neither Tallinn nor other minor cities have satisfactory sewerage systems.

SOCIETY

Peoples: Estonians, 61.5%; Russians, 30.3%; Ukranians, 3.1%; Belarusians, 1.8% Finnish, and others 3.3%. **Religions:** Lutheran (majority), Orthodox, Baptist. **Languages:** Estonian (official); Russian and others. **Political Parties:** Country Coalition Party; Royalist Party of Estonia; Rural Union of Estonia; Greens of Estonia; Citizen's Union of Estonia; Reform Party; Pro Patria Party.

THE STATE

Official Name: Eesti Vabariik. **Administrative divisions:** 15 districts and 6 cities. **Capital:** Tallinn, 443,000 inhab. (1994). **Other cities:** Tartu, 106,000 inhab.; Narva, 79,000 inhab.; Kohtla-Jarve, 73,000 inhab.; Pärnu, 52,000 inhab. (1994). **Government:** Parliamentary republic. Lennart Meri, President and Head of State since October 1992. **Legislature:** single-chamber assembly (*Riigikogu*), with 101 members, elected every 4 years. **National Holiday:** February 24, Independence (1918). **Armed Forces:** 2,500 (1993). **Paramilitaries:** 2,000 (Coast Guard).

DEMOGRAPHY

Urban: 73% (1995). **Annual growth:** 0.1% (1991-99). **Estimate for year 2000:** 2,000,000. **Children per woman:** 1.8 (1992).

EDUCATION

School enrolment: Primary (1993): 83% fem., 83% male. Secondary (1993): 96% fem., 87% male. University: 38% (1993). **Primary school teachers:** one for every 25 students (1992).

COMMUNICATIONS

23.2 **telephones** per 100 inhabitants (1993).

ECONOMY

Per capita GNP: $2,820 (1994). **Annual growth:** -6.10% (1985-94). **Annual inflation:** 77.30% (1984-94). **Consumer price index:** 100 in 1990; 9,960 in 1994. **Currency:** 14 krona = 1$ (1993). **Cereal imports:** 46,000 metric tons (1993). **Fertilizer use:** 1,512 kgs per ha. (1992-93). **Imports:** $1,690 million (1994). **Exports:** $1,329 million (1994).

ENERGY

Consumption: 3,552 kgs of Oil Equivalent per capita yearly (1994), 42% imported (1994).

(FPE), founded in April by nationalists and communists, organized a rally in June 1988 which was attended by 150,000 people. The outlawed Estonian flag was displayed on this occasion. In September, some 300,000 Estonians held another rally; a few days later, the ban on the flag was lifted.

[23] In October, the first FPE congress reaffirmed Estonia's demand for autonomy, and asked Moscow for an admission of the fact that Estonia had been occupied against its will in 1940. The following month, Estonia's soviet (parliament) declared the country's sovereignty, and affirmed its right to veto laws imposed by Moscow without consent.

[24] In August 1989, some 2 million Estonians, Latvians and Lithuanians formed a 560-km human chain from Tallinn to Vilnius to demand the independence of the Baltic States. In February 1990, a convention of Estonian representatives approved the Declaration of Independence, based on the 1920 Peace Treaty of Tartu between the Soviet Union and the newly independent Estonian republic.

[25] In the May 1990 elections, the FPE and other nationalist groups won an ample majority within parliament. Moderate nationalist leader Edgar Savisaar was named as the leader of the first elected government since 1940. In August, parliament proclaimed the independence of Estonia, but Moscow did not consider it to be valid.

[26] In late 1990-early 1991, Moscow threatened to impede Estonia's separation from the USSR, by force, and skirmishes took place between Soviet troops and nationalist groups. In September, the USSR recognized the independence of the three Baltic States, and that same month, they were admitted to the United Nations.

[27] In January, Savisaar and his government resigned in the face of growing criticism over his economic policy. Parliament named former transportation minister Tiit Vahi to head the new government. Estonia has had to ration food and fuel ever since the Russian Federation began restricting and raising the price of its products.

[28] The Russian Federation, which is Estonia's main trading partner and its chief supplier of fuel, began charging international prices for its exports of crude oil. At the same time, given the limitations on its own consumption, Russia reduced its wheat exports and its importation of clothing and electrical appliances from Estonia.

[29] On June 20 1992, the new constitution (based on the 1938 Constitution) was ratified by referendum.

[30] In September, the Rligikogu (parliament) of the 7th Legislature was elected. On October 7, it issued a declaration announcing the end of the transitional government, and reinvoking the constitution.

[31] The following month, Estonia initiated negotiations with the Russian government for the withdrawal of former Red Army troops. In addition, the government has territorial claims over part of the Russian region of Pskov, with a land area of 2,000 sq km.

[32] Lennart Meri, of the National Country Coalition Party, was elected president of Estonia by parliament on October 5 by 59 votes to 31.

[33] In 1992, Estonia became the first Eastern European country to abandon the rouble, and create its own currency, the krona.

[34] The privatization process came to a halt in December 1992 as a result of the resignation of the official in charge, who had been accused of negligence and fraud. In early 1993, privatizations resumed, although these basically entailed returning properties and goods confiscated during the Communist regime to their former owners.

[35] In June 1993, a blatantly nationalistic law was approved, aimed at foreigners, especially those of Russian origin who make up 39 per cent of the total population. The law obliges foreigners to apply for a discretionary residence permit.

[36] Russian troops concluded their withdrawal from Estonia in August 1994 while Tallinn had an active participation in NATO's Partnership for Peace. On the domestic side, a former Communist Party leader, Indrek Toome, was arrested two months later charged with alleged corruption.

[37] The March 1995 elections led to the defeat of the coalition which had ruled Estonia since the former Soviet republic broke away from the USSR. The new prime minister, Tiit Vahi, caused a controversy when he named a supposed "disproportionate" number of former communist ministers in his government. In October his cabinet was forced to resign due to corruption charges against the minister of the Interior. The new government was formed with the inclusion of Reform Party members.

[38] On September 20 1996, and after a tight vote, Lennart Meri was re-elected as president of Estonia by an electoral college.

Ethiopia

Yaitopya

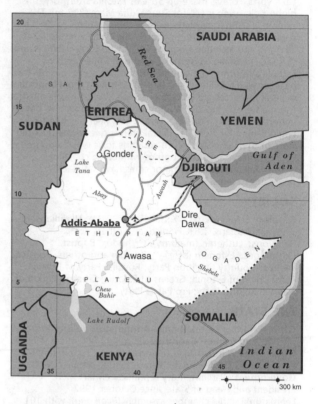

Population: 54,890.000 (1994)
Area: 1,221,900 SQ KM
Capital: Addis-Ababa (Adis-Abeba)

Currency: Birr
Language: Amharic

In ancient times, the Greek term "Ethiopian", meaning "burnt face" in Greek, was applied indiscriminately to all Africans. Ethiopia's other name, Abyssinia, came from the Arabic "Habbashat", which was the name of one of the groups that emigrated from Yemen to Africa around 2000 BC.

2 Axum, in the north of present-day Ethiopia, was the center of trade between the Upper Nile valley and the Red Sea ports which traded with Arabia and India, it reached its height in the first centuries AD. Ethiopia was a rich and prosperous state, which was able to subdue present-day Yemen, but which went through a crisis in the 7th century. Trade routes moved as Arab unification and expansion dominated the area, conquering Egypt. The Ethiopian ruling elite had converted to Christianity in the 4th century, further contributing to their isolation. Expansion towards the south, excessive growth of the clergy, and declining trade led to a process of social and economic stratification similar to that in feudal western Europe. By the 16th century, one third of the land belonged to the "king of kings"; another third belonged to the monasteries and the rest was divided among the nobility and the rest of the population.

3 The Muslim population that had developed a powerful trade economy on the coast of the Red Sea (see Tanzania: "The Zandj Culture of East Africa"), instigated an insurrection, leading Ethiopia to resume its relations with Europe to request assistance. The aid took almost a century to arrive, but the Portuguese fleet, when it finally arrived in 1541, was decisive in destroying the Sultanate of Adal (See Somalia).

4 For 150 years, Ethiopian emperors focused their efforts on the coast, giving the Galla (a nation akin to the Haussa) a chance to gradually penetrate from the west until they outnumbered the local population. Their influence grew so great that a Galla became emperor, between 1755 and 1769; though the Ahmara ruling elite took great pains to oust him.

5 This state of affairs continued until 1889 when Menelik II came to power. Designated heir to the throne in 1869, he spent the next 20 years training an army (with British and Italian assistance) and organizing the administration of his own territory, the state of Choa. His efficiency was fortunate; in 1895 his former allies, the Italians, invaded the country claiming that previous commitments had not been honored. The final battle was fought in Adua in 1896 where, 4,000 of the 10,000 Italian soldiers were killed. It was the most devastating defeat suffered by European troops on African soil until the Algerian War. In the diplomatic negotiations that followed their defeat, the Italians succeeded in obtaining two territories that Ethiopia did not really control: Eritrea and the southern Somalian coast. In 1906, the world powers recognized the independence and territorial integrity of what was then known as Abyssinia, in exchange for certain economic privileges.

6 This arrangement saved Ethiopia from direct colonization until 1936, when Italian Fascist dictator Benito Mussolini invaded the country, taking advantage of internal strife among Menelik's would-be successors. Despite his pleas to the League of Nations, Haile Selassie, heir to the throne, got no concrete help. During the five-year occupation, several basic industries and coffee plantations were started, and a system of racial discrimination was installed, similar to that of apartheid in South Africa.

7 In 1948, Ethiopians won their autonomy back from Britain, which had taken over the country after Mussolini's defeat in 1941. When Selassie returned to the throne, his country was floundering in unprecedented crisis: foreign occupation had disrupted production; nationalist political movements had strengthened in the struggle for autonomy and rejected a return to feudalism; and poverty in the interior had grown considerably.

8 Selassie denounced colonialism, favored non-alignment and supported the creation of the Organization for African Unity, which finally set up headquarters in Addis-Ababa. He also maintained close links with Israel. On the domestic front, the crisis deepened as his government was dominated by a corrupt oligarchy and the Orthodox Church, which between them held 80 percent of the country's fertile lands. The domestic crisis came to a head in 1956 when Eritrean separatist rebels intensified their attacks. In 1974, after a series of strikes, student rallies and widespread protests against absolutism and food shortages, Haile Selassie was overthrown.

9 An Armed Forces Coordination Committee, the Dergue ("committee" in Amharic), headed by General Aman Andom abolished the monarchy and proclaimed a republic, suspending the Constitution and dissolving Parliament

10 Popular movements became more radical, leading to a series of disputes among the military, and a subsequent crisis. Colonel Mengistu Haile Mariam rose to power in December 1977. He managed to

PUBLIC EXPENDITURES

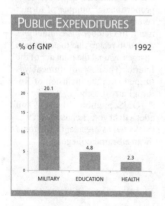

% of GNP   1992

- MILITARY 20.1
- EDUCATION 4.8
- HEALTH 2.3

WORKERS

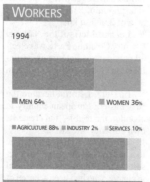

1994

- MEN 64%
- WOMEN 36%
- AGRICULTURE 88%
- INDUSTRY 2%
- SERVICES 10%

LAND USE

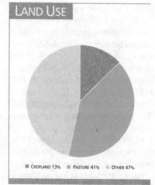

- CROPLAND 13%
- PASTURE 41%
- OTHER 47%

hold the Dergue together and put an end to the military's internal struggles.

[11] The military government managed to carry out agrarian reform and to nationalize foreign banks and heavy industry. US military bases were closed down. Right-wing groups carried out assassinations and bombings. Left-wing organizations retaliated by killing conspirators and monarchists on sight. Between December 1977 and April 1978, the violence left over 5,000 people dead.

[12] Once the military crisis was resolved, a struggle broke out for the control of grassroots organizations. At first the Dergue supported the Pan-Ethiopian Socialist Movement (Meison), which set up neighborhood committees (kebeles) as a basis for mass organization. The Meison grew stronger and proposed a system of self-controlled civilian units. This set them on a collision course with the Dergue, which rejected the creation of autonomous military groups.

[13] Mengistu declared the Meison illegal and backed the Revolutionary Flame, a group of officers and civilians who had studied in socialist Europe and were loyal to the president of the Dergue. After settling this latest crisis, the government was able to confront the two separatist movements which had been gaining strength since 1977, in Eritrea and the Ogaden desert. The Eritrean rebels considered that their independence struggle was valid, regardless of the existence of a progressive government in Addis-Ababa, and the Somali population of the Ogaden desert (claimed by Somalia) took advantage of the domestic crisis in Addis-Ababa to further their own separatist cause.

[14] Mengistu rejected Eritrean separatist demands. He alleged that their struggle had only been reasonable when they had confronted the feudal monarchy, and not when they had opposed the pro-soviet regime.

[15] Faced with the Somalian intention of annexing the Ogaden, the Soviet Union broke its military treaties with Siad Barre. In this modern armed war, Soviet and Cuban support was decisive in his defeat. With the situation resolved on both battle fronts, Mengistu turned his attention back to domestic policy.

[16] In 1979 the government set up the Ethiopian Workers' Party Organization Committee (COPWE), and their first general congress was held in 1980. In that year, there was a 15 per cent increase in cultivated land, which raised the GNP by 6 per cent.

[17] In 1984, the country was struggling under the effects of a prolonged drought which had begun in 1982, causing thousands of deaths from starvation. The drought affected twelve provinces, threatened five million lives and killed over half a million people.

[18] The Ethiopian Workers' Party (PWE) held their founding congress in 1984, approving a program to transform the country into a socialist state.

[19] The newly-elected Assembly (the *Shengo*, or parliament) proclaimed the People's Democratic Republic of Ethiopia on September 12, ratifying Mengistu Haile Mariam as head of state. Separatists extended operations in Eritrea and Tigre, as well as in Wollo, Gondar and Oromo in the south.

[20] The new constitution provided for the creation of five autonomous regions and 25 administrative regions. Eritrea was able to legislate on all matters except defence, national security, foreign relations and its legal status in relation to central government. The separatists rejected the proposal, calling it "colonial".

[21] In December 1987, after a bloody battle, the Eritreans took Af Abed. 18,000 Ethiopian troops, three Soviet advisors and the army commander of the northern region were taken prisoner.

[22] The Tigre People's Liberation Front (Tigray) captured important cities like Wukro, a distribution center for international aid to drought victims. Building on these victories, the rebels from Eritrea and Tigre signed cooperation agreements and began to plan joint military strategies.

[23] Increased rebel activity took a heavy toll on the Ethiopian army and, in 1989, 12 divisions (150,000 troops) stationed in the front line attempted a coup. Mengistu returned hastily from West Germany and managed to abort the coup. He executed dozens of officers, among them Generals Amha Desta, airforce commander-in-chief, and Merrid Negusie, chief-of-staff of the armed forces.

[24] By the end of the 1980s, the war was consuming 60 per cent of Ethiopia's national budget, and agricultural production was slumping. In 1989, former US president Jimmy Carter's mediation brought together delegations from the government and FPLE in Atlanta, in the US. The FPLE condensed its demands into one: a plebiscite for Eritrea's future. Mengistu refused.

[25] In September 1989, the last Cuban soldiers withdrew from Ethiopia. The government had signed a peace agreement with Somalia in April 1988, and no longer needed their services.

[26] Ethiopia re-established contact with Israel, as relations had been interrupted after the 1973 war. Mengistu looked to Tel Aviv for substantial military aid and facilities for 17,000 Ethiopian Jews to emigrate to Israel. Meanwhile, in Eritrea and Tigre, where the drought had reduced the grain harvest by nearly 80 per cent, the situation of millions steadily worsened. International agencies warned of the danger of yet another famine.

[27] In 1990, the Central Committee of the Ethiopian Workers' Party decided to restructure the party and change its name to the Democratic Ethiopian Unity Party (PDUE). While excluding the possibility of a multi-party system, the changes sought to lay the basis of "a party of all Ethiopians", open to "opposition groups". The Marxist-Leninist tag was dropped. The government established a mixed economy, including state enterprises, cooperatives and private businesses.

[28] Within one month, the former USSR withdrew its military advisors from Eritrea, and the FPLE took the port of Massawa, thus controlling all of Eritrea, except Asma-

PROFILE

ENVIRONMENT

A mountainous country with altitudes of over 4,000 meters. Ethiopia is isolated from neighboring regions by its geography. In the North, the Eritrean plain, a desert steppe, extends alongside the Red Sea. In the mountains and plateaus, vegetation and climate vary with altitude. The Dega are cool, rainy highlands, above 2,500 meters, where grain is grown and cattle raised. The deep valleys which traverse the highlands are warm and rainy with tropical vegetation, known as the Kolla (up to 1,500 m). The drier, cooler, medium-range plateaus where coffee and cotton are grown,(1,500 to 2,500 m) are the most densely populated parts of the country. To the East lies the Ogaden, a semi-desert plateau inhabited by nomadic shepherds of Somali origin. Many regions which were once rich in vegetation are now rocky, desert areas. Desertification and erosion have increased within the past decade.

SOCIETY

Peoples: Two thirds are of Amhara or Oromo descent. There are many other ethnic groups including the Tigre, Gurage, Niloti, Somali and Danakil. **Religions:** Most Ethiopians are Coptic Christians. Traditional African religions are also practiced. **Languages:** Amharic; Arabic; and at least 100 local languages. **Political Parties:** Ethiopian People's Revolutionary Democratic Front; All Amhara People's Organization; Oromo Liberation Front; Ogaden National Liberation Front.

THE STATE

Official Name: Itiopia. **Administrative Divisions:** 25 Administrative Regions, 5 Regions. **Capital:** Addis-Ababa (Adis-Abeba), 1,673,100 inhab. (1988). **Other city:** Dire Dawa 117,700 inhab. (1988). **Government:** Negasso Gidada, President; Meles Zenawi, Prime Minister. **National Holidays:** March 21, Proclamation of the Republic (1975). **Armed Forces:** 120,000 (1995).

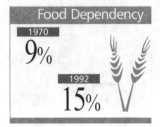

Food Dependency
1970
9%
1992
15%

External Debt
1994
PER CAPITA
$92

Illiteracy
1995
65%

STATISTICS

DEMOGRAPHY

Urban: 13% (1995). **Annual growth:** 1.9% (1992-2000). **Estimate for year 2000:** 67,000,000. **Children per woman:** 7.5 (1992).

HEALTH

One **physician** for every 33,333 inhab. (1988-91). **Under-five mortality:** 200 per 1,000 (1994). **Calorie consumption:** 81% of required intake (1995). **Safe water:** 25% of the population has access (1990-95).

EDUCATION

Illiteracy: 65% (1995). **School enrolment:** Primary (1993): 19% fem., 19% male. Secondary (1993): 11% fem., 12% male. University: 1% (1993). **Primary school teachers:** one for every 27 students (1992).

COMMUNICATIONS

81 **newspapers** (1995), 82 **TV sets** (1995) and 93 **radio sets** per 1,000 households (1995). 0.2 **telephones** per 100 inhabitants (1993). **Books:** 82 new titles per 1,000,000 inhabitants in 1995.

ECONOMY

Per capita GNP: $100 (1994). **Consumer price index:** 100 in 1990; 167.1 (1994). **Currency:** 6 birr = 1$ (1994). **Cereal imports:** 1,045,000 metric tons (1992). **Fertilizer use:** 95 kgs per ha. (1992-93). **Imports:** $1,033 million (1994). **Exports:** $372 million (1994). **External debt:** $5,058 million (1994), $92 per capita (1994). **Debt service:** 11.5% of exports (1994).

ENERGY

Consumption: 21 kgs of Oil Equivalent per capita yearly (1994), 86% imported (1994).

---

ra, which was totally cut off from the rest of Ethiopia by occupied land.

[29] In May 1991, Mengistu Mariam unexpectedly fled the country, overwhelmed by the repeated guerrilla victories in the north. The government was left in the hands of Vice-president Tesfaye Gabre Kidane, considered a moderate, who initiated his transitional government by negotiating a cease-fire with the Eritrean rebels.

[30] Kidane's government took part in peace talks, in London, presided over by the US, with the participation of the most important rebel groups. They aimed at reaching an agreement which would stave off civil war. Kidane resigned in late May, when the US advised the forces of the Revolutionary Democratic Front of the People of Ethiopia (FRDPE) to take control of Addis-Ababa.

[31] Ata Meles Zenawi, the 36-year-old leader of the FRDPE became interim president, until a multi-party conference could be held. He promised to bring the civil war to an end, re-establish democracy and conquer the problem of hunger. Three months later, upon reopening Parliament and passing a new Constitution, Zenawi pledged to hold elections within a year.

[32] The interim president also promised to honor economic commitments assumed by the deposed government, and to respect the will of the Eritrean people. The Eritreans had announced the formation of an autonomous government, and scheduled a referendum to decide the issue of independence (see Eritrea).

[33] The new government confronted two practically insoluble problems: reconciling the various guerrilla groups, and hunger. Ethiopia has become one of the poorest countries in the world. In the past decade, it has been simultaneously confronted with severe drought and civil war on at least four fronts.

[34] It is estimated that within the past 20 years, more than a million Ethiopians have died of starvation, and another million have become refugees in neighboring countries. After an agreement was signed with the Eritreans, international aid was once more transported through the ports of Aseb and Massawa on the Red Sea.

[35] On July 9, for the first time in 17 years, residents of Addis-Ababa elected their representatives to the kebeles (neighborhood committees) which are beyond the control of any political organization. The voters' main concerns were to achieve peace and security within the capital, and to elect a local administration.

[36] In March 1992, the minister of Insurance and Rehabilitation declared that if international aid did not arrive quickly, the transition to democracy would be jeopardized by the instability caused by the extreme poverty of a large number of Ethiopians. Population displacements and ethnic conflicts worsen the situation, events which are reminiscent of the 1984-85 famine.

[37] The transition government promised to promote the market economy, stimulate agricultural production and reduce poverty, within a five year programme coordinated by United Nations organisms and the World Bank. In January 1993, large-scale student demonstrations during a visit from the UN Secretary-General Boutros Boutros-Ghali, once again reflected the tense social situation in the nation.

[38] In 1994, the delivery of the $1.2 billion in five years planned in the economic programme, slowed down considerably as the international organisms decided the Ethiopian government was privatizing too slowly. In a criticism of the structural adjustment plan proposed by the IMF and the World Bank, humanitarian organizations supporting Ethiopia said there should be greater investment in seeds, tools and livestock in order to fight famine.

[39] Ten years after the 1984 famine, the situation became critical again in the first six months of 1994, with 5,000 deaths in the Wolayata district in the south of the country. In May, the Council of Representatives, a temporary 87-member body, approved a draft Constitution which created the Federal Democratic Republic of Ethiopia. This draft was based on the "ethnic federalism" doctrine, which put an end to the previous official unitarian vision of the nation. According to the approved document, the "sovereignty resides in the nations, nationalities and peoples of Ethiopia" and not the people as a whole.

[40] In July, elections were held to elect a Constituent Assembly, though these were boycotted by the majority of opposition parties, like the Oromo Liberation Front and the National Ogaden Liberation Front. In September, the police carried out mass arrests in the west of the country, an area mainly settled by Oromos. Several human rights organizations, like Amnesty International, expressed their concern at the situation.

[41] In May and June 1995, there were parliamentary elections, which were also boycotted by the majority of opposition parties. The new federal republic was officially established in August, when Negasso Gidada, a christian Oromo from the Welega region in west Ethiopia, became president. The outgoing president, Meles Zenawi, took the post of prime minister and the 17 members of government were carefully selected to reflect "the ethnic balance" of the nation.

[42] The trials of officials from the old Mengistu regime were postponed for the second time and attempts to extradite the former president from asylum in Zimbabwe, came to nothing. Amnesty International condemned the Ethiopian government once again, following the arrest of five political opponents.

[43] The government went ahead with the privatization of state companies - 144 were sold off in 1995 - and the annual grain deficit reached around a million tons.

Andorra

Population: 65,000
(1994)
Area: 453 SQ KM
Capital: Andorra la Vella
Language: Catalan

Located in the eastern Pyrenees, the co-principality of Andorra is made up of deep ravines and narrow valleys surrounded by mountains with altitudes varying between 1,000 to 3,000 meters. The Valira de Ordino and Valira de Carrillo rivers join in Andorran territory under the name of the Valira river. Wheat is grown in the valleys, but livestock (especially sheep-raising) has given way to tourism as the primary economic activity. More than 12 million people visited the country in 1995.

SOCIETY

Peoples: Spanish 46.4%; Andorran 28.3%; Portuguese 11.1%; French 7.6%; British 1.8%; German 0.5%; other 4.3% (1993).
Religions: Catholic. **Languages:** Catalan (official), French and Spanish. **Political Parties:** Liberal Union Party (Unió Liberal, UL), actually in offfice. It formed a coalition with the Liberal Group (Grup Liberal), the National Andorran Coalition (Coalició Nacional Andorrana), and the Canillo-La Massana Grouping (Agrupació Canillo-La Massana). National Democratic Grouping of former chief of State, Marc Forné. **Social Organizations:** There are no organized trade unions. Many workers have joined French trade unions.

THE STATE

Official Name: Principat d'Andorra.
Administrative divisions: 18 provinces.
Capital: Andorra la Vella, 22,387 inhab. (1993).
Other cities: Les Escaldes, 13,177 inhab.; Encamp, 9,654 inhab. (1993).
Government: Parliamentary republic. Constitution in effect since March 14, 1993. The Bishop of Urgel (Spanish jurisdiction) and the President of France, represented by Franceso Badia-Batalla and Louis Deblé, respectively, are "co-princes" of the territory. Oscar Ribas is president of the Executive Council and head of Government. Single-chamber legislature (General Council), with 28 members elected by direct popular vote, every four years. At some point in the future, total independence from France and the Spanish bishopric is foreseen.

COMMUNICATIONS

42.2 **telephones** per 100 inhabitants (1993). 154 **TV sets** and 222 **radio receivers** per 1,000 inhab. (1991). **Books:** 49 new titles in 1991.

Liechtenstein

Population: 31,000
(1994)
Area: 160 SQ KM
Capital: Vaduz
Language: German

This small principality lies between Switzerland and Austria, in the Rhine valley. Wheat, oats, rye, corn, grapes and fruit are produced. 38% of the land is pasture. In recent years, the principality has transformed into a highly industrialized country, producing textiles, pharmaceutical products, precision instruments, refrigerators and false teeth.

SOCIETY

Peoples: Liechtensteiner 61.6%; Swiss 15.6%; Austrian 7.2%; German 3.6%; other 12.0%.
Religions: Catholic and Protestant.
Languages: German (official). **Political Parties:** Multiparty system. The Fatherland Union (VU), 13 seats. The Progressive Citizens' Party (FBP), 11 seats; Free List, green,1 seat. **Social Organizations:** The Trades Union Association, and the Agricultural Union.

THE STATE

Official Name: Fürstentum Liechtenstein.
Administrative divisions: 11 communes.
Capital: 5,067 inhab. (1995). **Other cities:** Schaan, 5,143 inhab.; Balzers, 3,752 inhab.; Triesen, 3,586 inhab. (1995).
Government: Liechtenstein is a constitutional monarchy. Prince Hans Adam II, head of State since November 13, 1989. Mario Frick, Prime Minister, head of government. Government functions are carried out by a FBP and VU coalition. Cornelia Gassner (FBP), Secretary of Construction, the first woman to be named to a cabinet post. Single-chamber legislature: Parliament with 25 members elected every four years. **Diplomacy:** A member of the European Council, Liechtenstein has a customs and monetary alliance with Switzerland, which is its representative abroad. When Liechtenstein decided to join the European Economic Area (EEA) - despite Switzerland not being a member of this organization - both governments maintained their alliance, keeping their economic links untouched. **National holiday:** February 14.

COMMUNICATIONS

62.4 **telephones** per 100 inhabitants (1993). 307 **newspapers**, 345 **TV sets** and 719 **radio receivers** per 1,000 inhab. (1991).

Monaco

Population: 33,000
(1994)
Area: 1 SQ KM
Currency: French Francs
Language: French

A small principality on the Mediterranean coast, with a 3.5 km coastline, surrounded by the French department of the Maritime Alps. The territory is made up of three urban centers: Monaco-Ville, the capital, built upon an isolated cliff; Condamine, a residential center; and Montecarlo, to the northeast of the port of Monaco, home of the famous casino and consequently the most often visited location. Olives and citrus fruits are cultivated in the narrow inland strip. Tourism is the country's basic source of revenue. In recent years support has been given to infrastructure projects and the modernization of tourist facilitites.

SOCIETY

Peoples: 47% French, 17% Monacan, 16% Italian, 4% English, 2% Belgian and 1% Swiss.
Religions: Catholic. **Languages:** French (official) and Monegasque. **Political Parties:** There are no political parties. Candidates form lists to run for election in the National Council (CN). Until 1992, the CN was dominated by the National Democratic Union. In 1993, the main groups were the Cámpora List (15 seats) and the Médicin List (2 seats). Last CN election: January 1993. Paul Dijoud is the Minister of State since 1994.

THE STATE

Official Name: Principauté de Monaco.
Administrative divisions: 1 commune.
Government: Parliamentary monarchy. Constitution in effect since December 17, 1962. Prince Rainier III, sovereign since May 9 1949, head of State and of the government. Single-chamber legislature: National Council, with 18 members elected every 5 years.

EDUCATION

Primary school teachers: one for every 20 students (1991).

COMMUNICATIONS

10.6 **telephones** per 100 inhabitants (1993). 296 **newspapers**, 819 **TV sets** and 1,126 **radio receivers** per 1,000 inhab. (1991). **Books:** 41 new titles in 1990.

ECONOMY

Currency: French francs.

San Marino

The Republic of San Marino, founded in 1866 and located between the Italian provinces of Romana and Marca, is the world's smallest republican state. Its mountainous terrain is part of Mount Titano, an eastern branch of the Apennines; the La Rocca peak, 749 m, is the highest point in the city-state. Grapes and grain are grown in the farmlands, and sheep are also raised. Tourism is the country's major source of income. In 1994, more than 3 million people visited the country. The small republic is linked to the port of Rimini, on the Adriatic.

SOCIETY

Peoples: Sammarinesi 76.8%; Italian 22.%; other 1.2%. **Religions:** Mainly Catholic.
Languages: Italian (official) and a local dialect.
Political Parties: Multiparty system. The Christian Democrat Party (PDC), 26 seats; Socialist Party (PS), 14 seats; Progressive Democratic Party (PDP) formerly communist, 11 seats; Popular Alliance (AP), 4 seats; Democratic Movement, (MD), 3 seats; Communist Refoundation (RC), 2 seats. **Social Organizations:** The Unitarian Trade Union Central, the General Democratic Confederation of Workers and the General Confederation of Labor.

THE STATE

Official Name: Serenissima Repubblica di San Marino. **Capital:** San Marino, 2,315 inhab.

Area: 60 SQ KM
Capital: San Marino
Currency: Lire
Language: Italian

(1990). **Other cities:** Serravalle/Dogano 4,709 inhab.; Borgo Maggiore 2,367 inhab.
Government: Presidential republic. The Executive of the Council of State, with ten members elected every six months (dominated by the PS and the PDC) is presided over by two regent Captains: Patricia Busignani and Salvatore Tonelli, head of State and head of Government, respectively, elected in April and September, 1993. Single-chamber legislature of the General Grand Council, with 60 members elected every five years by direct popular vote. **Diplomacy:** At the beginning of 1992, San Marino joined the United Nations and the IMF. It has had a Friendship Treaty with Italy since 1862. As a neutral European country, it is a Non-Aligned Movement observer.

EDUCATION

Primary school teachers: one for every 6 students (1991).

COMMUNICATIONS

61.3 **telephones** per 100 inhabitants (1993). 87 **newspapers**, 351 **TV sets** and 589 **radio receivers** per 1,000 inhab. (1991).

ECONOMY

Consumer price index: 100 in 1990; 109.5 in 1994. **Currency:** lire.

Vatican City

Located within the city of Rome, not far from the banks of the Tiber river, the Vatican comprises St Peter's square and basilica, the palaces and gardens of the Vatican itself, the palace and basilica of St John Lateran, the papal villa at Castelgandolfo and another 13 off-limit buildings which enjoy diplomatic privileges because they house congregations of the Catholic Church. The Vatican State dates back to the 9th century when Charlemagne, emperor of the Franks, legalized its existence. The Vatican ceased to exist in 1870 when the first king of Italy, Victor Emmanuel of Savoy, seeking national unification, formally occupied the territory and proclaimed Rome its capital. The 1871 Guarantees Law established the inviolability of the Pope and recognized his ownership of the Vatican, but the Roman pontiffs did not accept the situation until February 11 1929, when the Holy See signed the Treaty of Letran with Benito Mussolini, establishing the current State's borders and privileges.

SOCIETY

Peoples: Church members whose functions require residence, become Vatican citizens. Most permanent members are Italian, but there is also a large number of Swiss residents, and some other nationalities. **Religions:** Catholic.
Languages: Italian (official language of the State) and Latin (official language of the Church).

THE STATE

Official Name: Stato della Cittá del Vaticano.
Administrative divisions: Two parallel administrations: Holy See (supreme body of the Catholic Church) and Vatican City (site of the Church). **Capital:** Vatican City, 1,000 (1995).
Government: Elected lifelong monarchy.
Monarch: John Paul II (Karol Wojtyla, of Polish origin), elected on October 16 1978 by the cardinals' conclave (secret meeting), the first non-

Area: 1 SQ KM
Language: Italian and Latin

Italian Pope in 456 years. Secretary of State (a position equivalent to prime minister): Angelo Cardinal Sodano. The state government is exercised by a 5-member pontifical commission appointed by the Pope and headed by the Secretary of State. The Pope is also the Bishop of Rome and the supreme ruler of the Catholic Church. In Church government, he is assisted by the College of Cardinals and the Synods of Bishops, which meet when so instructed by the Pope. The Church's administrative departments comprise 9 holy congregations, 3 Secretariats and several commissions, mayoralties and tribunals, known collectively as the Roman Curia. **Economy** Originally, the state's revenues derived from the financial investment of 1.75 billion lire indemnity established in the Treaty of Lateran as payment for the territories lost in 1870. Later they were increased by contributions and donations from all over the world, especially from the United States and West Germany. At present, these resources are administered by the Institute for Religious Works - known popularly as Bank of Vatican - re-organized following banking scandals originated in the bankruptcy of the Ambrosiano Bank. Allegedly the Institute has reserves exceeding $11 billion, participates in numerous other banks and enterprises, and own countless real-estate properties throughout the world. In 1993, the estimated budget deficit ($91.7 million) surpassed the projected deficit ($86.1 million), and there were two unprecedented events. Firstly, Vatican workers organized the first ever protest, demanding wage increases and pensions, and secondly the commercialization of the Pope's image was authorized, during his visit to the United States in June, allowing the sale of posters and T-shirts. Two years later, the Vatican produced a compact disc of prayers recited by the Pope which sold more than 20 million copies. In 1995, the Vatican economy went back into the black, leaving the financial troubles behind.

Fiji

Population: 767,000
(1994)
Area: 18,270 SQ KM
Capital: Suva City

Currency: Fiji Dollar
Language: English

Fiji

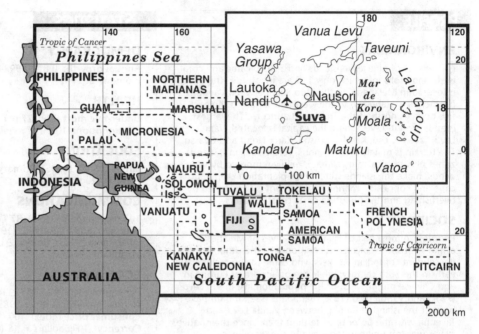

Four thousand years ago the Fiji archipelago was already populated. Melanesian migrations first reached the islands in the 6th century BC, and the Fijians had one of the leading Pacific cultures. In 1789, British Captain William Bligh visited the islands, writing the first detailed account of Fijian life.

2   Social life on the islands was organized in families and clans which gradually formed larger communities. One of these, ruled by traditional leader Na Ulivau, extended its influence from Ngau over the rest of the islands, achieving unification.

3   Early in the 19th century, British adventurers and merchants exploited the forests of Fiji, rich in sandalwood and other aromatic species highly prized in China. To settle their rivalries, Europeans sought support from local leaders. In 1830, the first Christian missionaries arrived from Tonga and 24 years later achieved a major victory on christening ruler Thakombau, son of Na Ulivau. Contemporary accounts report a "refined, courteous and honorable gentleman", and this "king", formerly reported to be a "cannibal" became an enthusiastic admirer of the western world, offering to annex Fiji to the US, though this may have been more due to the Fijian ruler's heavy debts after prolonged wars with rival leaders. The White House, caught up in the turmoil of the Secession War, missed the opportunity to tack another star on the flag. The British, faced with a shortage of sugar, "discovered" the archipelago's potential for growing cane and, with Thakombau's consent, officially annexed the islands on October 10, 1874.

4   Fijian officials raised no objections to the purchase or expropria-tion of large tracts of land for the new crops, though the peasants were unwilling to leave their communal estates to work on the plantations. Consequently, a massive influx of bond workers were brought in, first from the Solomon islands, and then from India. Following the abolition of slavery, the British used this same system of transferring Indian workers to Africa and the Caribbean to curb the labor shortage and ease the population squeeze in India.

5   At the end of their contracts, the workers brought their families to Fiji and became small shopowners, craftsmen or bureaucrats in the colonial administration, and though far from India, they retained their language, religion and caste system. Colonial regulations meant that they were restricted from buying land from native Fijians; a law ostensibly designed to prevent interethnic conflict.

6   Initial moves toward local autonomy resulted in a complicated electoral system securing political control by the "natives" who were in the minority. Prime Minister Sir Kamisese Mara initiated an elaborate scheme of representation, where each electoral category was allotted a quota for governmental representation: Fijians, 22, Indians, 22 and General, 8. A peculiar system, that nonetheless proved useful in preventing major disruptions during the transition to indepen-dence in 1970. His Alliance Party claimed to be multiracial, and marshalled support from some Indians and other minor groups, in addition to the Fijian electorate.

7   In 1976, the government turned down a constitutional reform project which would have required multi-group cooperation in the country's administration.

8   Interethnic co-existence has been successful except for two fierce clashes in 1959 and 1968; however, potential conflict situations could still erupt. The recent "back-to-the-roots" trend, centered mainly around the University of the Pacific in Suva, has encouraged young people to honor their Melanesian, Polynesian, Hindu, Islamic or Chinese cultural heritage, thus widening the existing gap between co-existent but and miscible cultures. Indians predominate in trade and the liberal professions. Fijians, in turn, own the arable land and lease it on very advantageous terms to poor Indians.

9   In the 1970s economic difficulties forced thousands of Fijians to emigrate to Aotearoa in search of job opportunities. Tourism, possibly the single industry which could improve the deficit balance of payments, requires Melanesians to remain "genuine" to satisfy Australian and Aotearoa's visitors' thirst for exoticism. Individualism and private land ownership were encouraged, seeking to bolster the economic situation of "natives" over that of the Indians. However, under these conditions, the rural workers' gained few improvements; urban income increased by 3.5 per cent in 1978, but rural income increased only 0.3 per cent.

10   The difficult economic situation swelled the number of "racist" groups, such as the Nationalist Party, who militate against Indian majority, though Indians have been on the islands for four or five generations.

11   Former prime minister Kamisese Mara's conservative administration was deeply marked by racial and ideological trends. His government was the only one in the Commonwealth to maintain relations with the racist Rhodesian and South African regimes and to welcome Chilean dictator Augusto Pinochet on an official visit in early 1980. Mass demonstrations during Pinochet's visit demonstrated the growing progressive movement in Fiji.

12   In the 1982 election, the Alliance Party, in power, gained victory by a small margin over the opposition NFP, who had unified its two rival factions despite growing conflict within the party.

13   In the April 11, election, 1987, the Indian majority won and Timoci Bavadra became premier, ending 16 years of Polynesian government. In May, a few short weeks after the election, Bavadra's labor

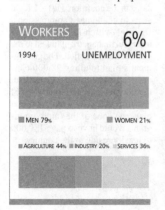

WORKERS

6%

1994   UNEMPLOYMENT

■ MEN 79%   ■ WOMEN 21%

■ AGRICULTURE 44% ■ INDUSTRY 20% ■ SERVICES 36%

ENVIRONMENT

Fiji consists of nine large islands and 300 volcanic and coral islets and atolls, of which only 100 are inhabited. The group is located in Melanesia, in the Koro Sea, between Vanuatu (former New Hebrides) to the west and Tonga to the east, five degrees north of the Tropic of Capricorn. The largest islands are Viti Levu, where the capital is located, Vanua Levu, Taveuni, Lau, Kandavu, Asua, Karo, Ngau and Ovalau. The terrain is mainly mountainous. Fertile soils in flatland zones plus tropical rainy climate, mildly tempered by sea winds, make the islands suitable for plantation crops, sugar cane and copra. The marine environment is threatened by overfishing and the pollution of its coastal waters.

SOCIETY

Peoples: Almost half of Fijians are natives of Melanesian origin with some Polynesian components, the other half are descendants of Indian workers who came to the archipelago early in the 20th century; while the others are of European and Chinese origin. Most of the population live in Viti Levu (73%) and Vanua Levu (18%). Banabans (see Kiribati) have bought the island of Rambi, between Vanua Levu and Taveuni, with the purpose of settling there since their native island was left uninhabitable by phosphate mining.
Religions: 53% of Fijians are Christian (mainly methodist and other protestant sects 38% are Hindu and 8% Muslim.
Languages: English (official), Urdu, Hindi, Fijian, Chinese.
Political Parties: Fijian Political Party and United National Front, representing the Fijians; National Federation Party and Fiji Labour Party, depending on the Indo-Fijian support; the General Vote Party representing the remaining minorities. **Social Organizations:** The Association of Fijian Young People and Students.

THE STATE

Official Name: Sovereign Democratic Republic of Fiji.
Administrative Divisions: 5 Regions divided into 15 Provinces. **Capital:** Suva City, 69.700 inhab.(1986). **Other cities:** Lautoka 28,700 inhab.; Lami 8,600; Nadi 6,900 inhab.(1986). **Government:** Kamisese Mara, Head of State since December 16, 1993; Sitiveni Rabuka, Head of Government since June 1992. Legislature: bicameral Parliament: the Chamber of Representatives has 70 members; 37 Fijian, 27 Hindi, 1 from Rotuma Island and 5 others. The Senate is made up of 34 members; 24 named by the Grand Council of the Chieftains, 9 by the president and 1 by the council of Rotuma. **National Holiday:** October 10, Independence (1970).

DEMOGRAPHY

Annual growth: 0.8% (1992-2000). **Children per woman:** 3 (1992).

HEALTH

Under-five mortality: 27 per 1,000 (1994). **Calorie consumption:** 103% of required intake (1995).

EDUCATION

Illiteracy: 8% (1995). **Primary school teachers:** one for every 31 students (1991).

COMMUNICATIONS

103 **newspapers** (1995), 87 **TV sets** (1995) and 112 **radio sets** per 1,000 households (1995). 7.1 **telephones** per 100 inhabitants (1993). **Books:** 91 new titles per 1,000,000 inhabitants in 1995.

ECONOMY

Per capita GNP: $2,250 (1994).
Annual growth: 2.40% (1985-94).
Consumer price index: 100 (1990); 119.0 (1994).
Currency: 1 Fiji dollars = 1$ (1994).

October 6, 1987, Rabuka retaliated proclaiming a Republic, a move intended to disavow the authority of the head of state, the British appointed governor.
[15] In December 1987, Rabuka resigned as Head of State, in an attempt to create an image of a joint civilian-military government, aimed at improving its foreign image. Penaia Ganilau was named president and Camisese Mara, prime minister, a regime never submitted to the approval of the electorate.
[16] In July 1990 a new constitutional decree based on apartheid went into effect, assuring the native 37 of the Chamber of Representatives' 70 seats and 24 of the Senate's 34 seats. A constitutional referendum announced for 1992 was finally cancelled. The following year, the apartheid was denounced by the UNO's General Assembly, Mauritius and India. In November, the Association of Fijian Young People and Students organized a public burning of the new constitution's text. Demonstrators were harshly subdued.
[17] Rabuka founded the Fijian Political Party and in 1992, amid growing political and social strains, a military officer was named Prime Minister.
[18] Fiji has not been able to overcome its chronic balance of pay-

ments deficit: almost all of its fuel and manufactured products are imported and its main sources of income - sugar exports and tourism - are not enough to balance the budget.
[19] In November 1993 six FPP members voted together with the opposition against the budget, forcing general elections. Rabuka retained power with 31 of the 37 Fijian seats and support from independent and General Vote Party members. Dissidents formed a Fijian Association obtaining only 5 seats.
[20] In November 1994, the government began a timid revision of the racist constitution. In 1995, Rabuka had to reorganize his Cabinet several times due to internal divisions in the coalition. He took legal steps to challenge a commission's findings which involved him in unlawful management linked to the Central Bank's deficit. The government's plan to allow 28,000 Chinese from Hong Kong to settle in the country also led to a wave of criticism. According to the official plan, each one of the future immigrants would have to pay $130,000 to legally reside in the country.

government was overthrown by a military coup led by Colonel Sitiveni Rabuka. The colonel justified himself as "attempting to solve the ethnic problem", though the real objective seemed to be the removal of the Indian government, who believed in an independent foreign policy, and planned to join the treaty of Rarotonga. The treaty promoted regional denuclearization; endorsed by Australia and Aotearoa, but criticized by Britain and the US.
[14] In 1987, Bavadra, imprisoned by the military during the coup, was freed. He rejected the idea of new elections proposed by the military for the end of 1988, ostensibly to draw up a new constitution. The main Indian and Melanesian political parties reached an agreement, with added pressure from the Commonwealth, which appeared to appease the military. However, on

Finland

Population: 5,089,000 (1994)
Area: 338,130 SQ KM
Capital: Helsinki

Currency: Markkaa
Language: Finnish

Süomi

Finland was inhabited from 7500 BC onwards, and the first settlers were the Sami (later called "Lapps" by other peoples). The Sami were essentially hunters, fishermen and gatherers. Over the centuries they began domesticating animals and practicing subsistence agriculture, remaining isolated from the rest of Europe. Much later - in the 1st century AD - descendants of the Finno-Ugrics entered from the south, and they pushed the Sami to the north, taking control of the major part of the territory. By the year 1000 AD, the Finns had established many settlements in the south.

[2] The ancestors of the *tavastlanders* came from the southwest across the Gulf of Finland, and the Carelians arrived from the southeast. Scandinavian peoples occupied the west coast, the archipelagoes and also the Aland Islands (Ahvenanmaa).

[3] From the 12th century onwards, Finland was claimed by both the Russian and Swedish empires. In 1172, the Pope advised the Swedes to control the Finns, to avoid their being proselytized by the Russian Orthodox Church. With the Protestant Reformation, Lutheranism became Sweden's official religion.

[4] In 1809 Finland became a grand duchy of Imperial Russia, although it was allowed to maintain its own parliament, army and judicial system.

[5] In 1889 the Social Democratic Party was founded, a party which was to play a leading role in Finnish political life from that moment onward. The presence of Lenin and other Bolshevik exiles also helped to strengthen the country's socialist leanings. In 1906 a single-chamber parliament was created, and universal suffrage was established. Finland became the third country in the world to recognize female suffrage.

[6] Finland declared independence in December 1917. Spurred on by the example of the Russian Revolution, Finnish Socialists tried to seize power by means of a revolution. In January 1918, the Social Democratic Party (SDP) took Helsinki and the major industrial centers. The government counterattacked, backed by the White Army and German troops under the command of General Mannerheim. In May, the civil war came to an end and the Socialist leaders were tried and given harsh sentences.

[7] Finland became a monarchy, and German Prince Frederick Charles of Hessen was elected king. After Germany's defeat in World War I, the monarchy collapsed and General Mannerheim was appointed regent, while new republican institutions were established.

[8] The 1919 Constitution established a parliamentary system with a strong presidential figure; in addition, it made the prime minister the head of the government, and the president, the head of state. Although the constitution recognized both Finnish and Swedish as official languages, nationalist youth groups later demanded that Finnish be given priority. This led to the creation of the Swedish People's Party. Subsequent constitutional reforms maintained the right of the Swedish minority to its own language.

[9] In 1918, 70 per cent of all Finns depended upon agriculture and forestry for a living. The importance of this segment of the population led to the formation of the Rural Party, which became the Center Party.

[10] The country has been governed almost uninterruptedly since its independence by alliances between the Social Democrats and the Centrists. From 1920 to 1930, these two political forces backed an ambitious program of social and economic reforms. Agricultural production was modernized, and the exploitation of the country's lumber resources was intensified. Likewise, the government fostered industrialization in the cities, and sponsored progressive legislation, aimed at protecting the emerging working class and agricultural workers.

[11] The growth of the Communist Party (FCP) gave rise to the ultra-right-wing Lapua (*Lappo*) Movement, whose terrorist acts and mass protests were supported by conservative groups and by some Rural Party members. In 1930, the FCP was banned by law, with the support of these parties. In 1931 the Lappo movement began to carry out attacks against the Social Democrats. After an attempted coup d'etat the following year, they were finally arrested by the government.

[12] After the German invasion of Poland, the USSR demanded part of the Karelian Islands, a naval base on the Hanko peninsula and other islands of the Gulf of Finland, as a means of defending themselves against imminent German attack. Given the refusal of the Finns to cede the lands, the USSR seized these areas by force, giving rise to a brief conflict and the loss of 80,000 Finnish lives. The 1940 Treaty of Moscow forced Finland to cede the aforementioned territories.

[13] Finland had declared its neutrality at the beginning of World War II, but since it had become involved in a conflict with the USSR, it permitted German troops to use its territory to launch an attack again their mutual enemy. Thus, Finland was briefly able to recover the territories lost in 1940.

[14] The 1944 Soviet counteroffensive and subsequent German retreat strengthened the Finnish peace movement. The Karelian Islands once again came under the control of the Red Army, causing the resignation of President Ryti, the architect of the alliance with Germany. He was succeeded by General Mannerheim, who negotiated an armistice with Moscow, recognized the 1940 Treaty of Moscow and organized the fight to expel German troops from the country.

[15] When the war ended, the country's political life was once again controlled by the Social Democratic-Finnish Rural Party alliance and the leader of the Finnish Rural Party, Urho Kekkonen, occupied the presidency from 1956-1982. The office of prime minister was also held by SDP figures for most of that same period.

[16] During the post war period, Finland was faced with a critical social and economic situation. Production had come to a standstill, there were thousands of unemployed and more than 300,000 refugees from the Karelian Islands.

[17] Neutrality became the cornerstone of Finnish foreign policy. In 1948, Helsinki and Moscow signed

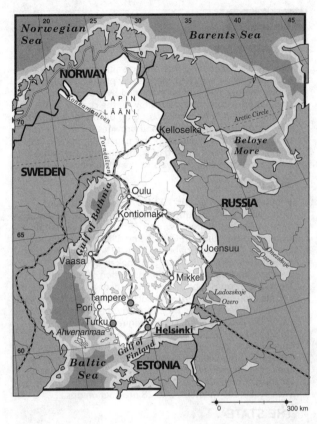

LAND USE

▪ CROPLAND 8%   ▪ PASTURE 1%   ▪ OTHER 91%

IMPORTS
23,200
EXPORTS
29,700

PROFILE

ENVIRONMENT

Finland is a flat country (the average altitude is 150 meters above sea level) with vast marine clay plains, low plateaus and numerous hills and lakes formed by glaciers. The population is concentrated mainly on the coastal plains, the country's main farming area. Logging of coniferous forests is the central economic activity and provides the main export products: lumber, paper pulp and paper. Sulphur dioxide emissions and the dumping of contaminated water into the Baltic Sea pose serious threats to the environment.

SOCIETY

Peoples: Ethnic make-up: 92.1% of the population is Finnish and 7.5% Swedish. There are Romany (gypsy) and Sami minorities. **Religions:** over 86% of the population belongs to the Lutheran Church and 1% to the Finnish Orthodox Church (official churches). 12% are non religious. **Languages:** Finnish (official and spoken by 93.2 %). Swedish, spoken by 6% of the population, is also an official language. Lapp is spoken by a minority of around 1,700 people. **Political Parties:** The Social Democratic Party (SDP), center-left; Center Party (KP), formerly the Finnish Rural Party; the People's Democratic League; the National Coalition Party (KOK), moderate conservative; the People's Liberal Party; the People's Swedish Party, liberal, represents interests of the Swedish-speaking minority; the Green Party; the Christian League of Finland; the Rural Party. **Social Organizations:** Central Organization of Finnish Unions, with 1,086,000 members and 28 member unions.

THE STATE

Official Name: Suomen Tasavalta. **Administrative divisions:** 12 provinces. **Capital:** Helsinki, 515,800 inhab.; metropolitan area 874.900, (1994). **Other cities:** Espoo 186,500 inhab., Tampere 179,200 inhab.; Vantaa 164,400 inhab.; Turku (Abo) 162,400 inhab. (1994). **Government:** Unicameral parliamentary democracy. Martti Ahtisaari, President since March 1, 1994; Paavo Lipponen, Prime Minister since April 1993. **National holiday:** December 6 (Proclamation of independence). **Armed Forces:** Total: 31,200 (1994). **Paramilitaries:** Border guards: 4,400.

STATISTICS

DEMOGRAPHY

Urban: 62% (1995). **Annual growth:** 0.4% (1991-99). **Estimate for year 2000:** 5,000,000. **Children per woman:** 1.9 (1992).

HEALTH

Under-five mortality: 5 per 1,000 (1994). **Calorie consumption:** 108% of required intake (1995).

EDUCATION

School enrolment: Primary (1993): 100% fem., 100% male. Secondary (1993): 130% fem., 110% male. University: 63% (1993). **Primary school teachers:** one for every 14 students (1980).

COMMUNICATIONS

110 **newspapers** (1995), 105 **TV sets** (1995) and 107 **radio sets** per 1,000 households (1995). 54.4 **telephones** per 100 inhabitants (1993). **Books:** 117 new titles per 1,000,000 inhabitants in 1995.

ECONOMY

Per capita GNP: $18,850 (1994). **Annual growth:** -0.30% (1985-94). **Annual inflation:** 4.20% (1984-94). **Consumer price index:** 100 in 1990; 211 in 1994. **Currency:** 5 markkaa = 1$ (1994). **Cereal imports:** 108,000 metric tons (1993). **Fertilizer use:** 1,363 kgs per ha. (1992-93). **Imports:** $23,200 million (1994). **Exports:** $29,700 million (1994).

ENERGY

Consumption: 5,954 kgs of Oil Equivalent per capita yearly (1994), 62% imported (1994).

---

the Finno-Soviet Pact of Friendship, Cooperation and Mutual Assistance. Seven years later, the USSR returned the naval base at Porkkala and the treaty was again ratified in 1970 and 1983. In 1955, Finland joined the European Free Trade Association (EFTA), the United Nations and the Nordic Council, but blocked the creation of a customs union within the framework of the Council. Finland was the only member of the OECD which simultaneously maintained relations with the EC and the socialist bloc CMEA. This situation explains the country's relatively rapid economic recovery.

[18] The Finnish government offered incentives to develop the lumber industry, axis of the economy. It promoted industrialization and took advantage of the profits generated by its lucrative foreign trade.

[19] However, with the disintegration of the USSR, which had provided the market for over 25 per cent of its exports, economic growth came to a standstill.

[20] In 1991, Prime Minister Esko Aho, politically moderate, launched a drastic structural adjustment program which was resisted by workers. Economic stagnation caused a decline in Finnish standards of living, and unemployment went from 3.5 to over 20 per cent in December 1993.

[21] Finland requested EU membership in 1992. A plebiscite in 1994 supported the initiative, which was confirmed in early 1995. The internal impact of this measure was softened by special subsidies for the farming sector, disadvantaged compared with its community partners, because of the cold climate. Part of the State monopoly on public health and the sale of alcoholic beverages was also maintained.

[22] President Ahtisaari said that despite the resolution, Finland would not join the Western European Organization, which could be a first step towards NATO membership. The Finnish government considered a common defence policy should be based on consensus, and that their traditional policy of neutrality had become obsolete since the fall of the iron curtain. He also warned the West about the risk of isolating Russia and the environmental danger represented by the nuclear plants and industries of this nation.

[23] In the March elections, the Social Democrat Party, led by Paavo Lipponen, pushed out the rural based Center Party, which had been the most important group in Parliament. In October, the government proposed a plan to reduce unemployment, which was running at around 17 per cent.

[24] On an environmental front, the nation managed to cut sulphur dioxide emissions from the paper industry, but more than 50 per cent of the air pollution comes from the neighboring nations.

France

Population: 57,928,000 (1994)
Area: 551,500 SQ KM
Capital: Paris

Currency: French Franc
Language: French

France

The two regions occupied by the Celts were known to the Romans as Gaul. The region between the Alps and Rome was Cisalpine Gaul, and beyond the Alps was Transalpine Gaul. With natural borders on all sides; the Alps, the Pyrénées, the Atlantic Ocean and the Rhine, Gaul occupied not only what is now France, but also Belgium, Switzerland and the western banks of the Rhine.

[2] Gallic society was essentially agricultural, with almost no urban life. The few cities, were used as fortresses, where the peasants sought refuge when under attack. Society was divided into the nobles (who were also warriors), the people and the Druids, keepers of Celtic wisdom and religious traditions.

[3] The Romans came to Gaul in 125 BC. They conquered the area along the Mediterranean, the Rhone valley and Languedoc, calling the combined area "Provincia". Caesar divided Gaul into two regions; Provincia and Free Gaul. Free Gaul was subdivided into ; Belgian Gaul in the north, between the Rhine and the Seine; Celtic Gaul in the center, between the Seine, the Garonne and the lower Rhine; and Aquitaine, in the southwest.

[4] In 27 BC, Augustus Caesar set up administrative centers in Gaul, to manage Rome's affairs encouraging urbanization. Bridges and an extensive road network were built throughout the region, facilitating an increase in trade. Wheat production was increased and vineyards were planted, with wine replacing beer as the traditional beverage. After a series of invasions by the Visigoths in the south and the Burgundians along the Saone and the Rhone, the northern Gauls conquered the rest of Gaul under the leadership of Clovis, adopting the name, "Franks".

[5] Between the 5th and the 9th centuries France appeared, as the Merovingian and Carolingian dynasties brought the entire region under the influence of Christianity. With the Islamic expansion and the fall of the Roman Empire, trade ceased, urban civilization was almost completely wiped out, population decreased and the culture fell into decadence.

[6] By the 9th century, feudalism had become firmly established. Centralized authority practically disappeared, as local people were unable to repel the Scandinavians, Hungarians, Saracens, and other invaders of this era. By the end of the century, the previously united land was a conglomerate of more than 300 independent counties.

[7] From the 10th century onward, the royal dynasties slowly recovered their power. They established hereditary succession to the throne, they shared power with the Church and became the main feudal landowners.

[8] In the 13th century, an increase in commercial activity led to a remarkable rebirth of the cities, and agricultural techniques were improved, as the population increased. The Crusades led to a greater circulation of people and goods, and the gradual disappearance of serfdom gave rise to greater social mobility. This was the "Golden Age" of the French Middle Ages, when France had great power over, and influence upon Western civilization.

[9] Paris was one of Europe's most important cities, and the prestige of its University was linked to its cultural pre-eminence. The University trained lawyers in Roman law, and their influence helped form a new concept of the State where the king was no longer a feudal lord, but rather the embodiment of the law. Over a period of time, nationalistic feelings began to develop.

[10] Before the 18th century, France suffered from the Hundred Years' War with England, the Thirty Years' War with Spain several wars with Italy, over a hundred revolts, wars between Catholics and Protestants, and the Black Death which scoured the country in the 14th and 15th centuries. Over the following century, the social and economic structures of the country changed, making conditions ripe for revolution.

[11] All forms of servitude disappeared and many Feudal lords had to sell their property. The country's mercantile structures stabilized with the rise of manufacturing and trade, triggering population growth and urbanization.

[12] Louis XIV, the "Sun King", personified the concept of absolute monarchy. He came to the throne in 1661, and established the "Divine Right of Kings". He consolidated the unity of France, giving rise to the concept of the modern State. During his reign, French cultural influence reached its apogee.

[14] The 1789 Revolution opened up a new era in the history of France. The National Assembly, convened in July of that year, replaced the absolute monarch with a constitutional monarchy. The fall of the Bastille, on July 14, and the Declaration of the Rights of Man, on August 27, brought the ancien regime to an end, opening the way for the bourgeoisie - the prevailing class of the burghs - whose reforms came into direct conflict with the Church and the King. Finally, the Assembly overthrew the monarchy and proclaimed the First French Republic.

[15] The rest of Europe joined forces against revolutionary France. Danton and Robespierre declared

France - dependencies

OVERSEAS DEPARTMENTS:

Guadeloupe; Martinique; French Guyana; Reunion (see each section).

OVERSEAS TERRITORIES:

Mayotte (see Comoros), Saint Pierre and Miquelon (text follows); New Caledonia (see Kanaky); Wallis and Futuna (text follows); French Polynesia (see corresponding section).

SOUTHERN AND ANTARCTIC TERRITORIES:

Comprising two archipelagos: Kerguelen (7,000 sq km, with 80 inhab. in Port-Aux-Français) and Crozet (500 sq km, with 20 inhab.); two islands: New Amsterdam (60 sq km, with 35 inhab.) and Saint Paul (7 sq km, uninhabited), located in the southern Indian Ocean; the Land of Adélie (500,000 sq km with 27 inhab. at the Dumont Durville Base), in Antarctica. The territories are governed by an Administrator-General advised by a seven-member Consultative Council appointed by the French government. Technical personnel at weather stations are the only inhabitants.

ENVIRONMENT

In the north is the Paris Basin, which spreads out into fields and plains. The Massif Central, in the center of France, is made up of vast plateaus. In the southeast, the Alps rise up. The southern region includes the Mediterranean coast, with mountain ranges, like the Pyrénées, and plains. Grain farming is the main agricultural activity; wheat is grown all over the country, especially in the north. Grapes are grown in the Mediterranean region for wine exports. The main mineral resources are coal, iron ore and bauxite. Dependence on nuclear energy poses a serious problem. Atomic reactors operating in the country generate three quarters of the national consumption of electricity, making France the second largest producer of nuclear energy, after the United States. There is also a nuclear reprocessing plant, as well as one which generates plutonium.

SOCIETY

Peoples: Most of the population stem from the integration of three basic European groups: Nordic, Alpine and Mediterranean. There are 3.6 million immigrants from northern Africa - Algerians, Tunisians and Moroccans - other former African colonies, Portugal, Spain, Italy.
Religions: Mainly Catholic; There are over 2.5 million Muslims (French Muslims as well as immigrants), making Islam the country's second largest religion; close to 2% are Protestant, and 1%, Jewish. **Languages:** French is the official and predominant language. There are also regional languages: Breton in Brittany, a German dialect in Alsace and Lorraine, Flemish in the northeast, Catalan and Basque in the Southwest, Provençal in the South-East, Corsican on the island of Corsica. Immigrants speak their own languages, particularly Portuguese, Arab, Berber, Spanish, Italian and diverse African languages. **Political Parties:** Socialist Party; Rally for the Republic (RPR); Union for French Democracy (UDF); National Front; Communist Party; the Greens; Worker's Struggle (LO). **Social Organizations:** General Labor Confederation (CGT); French Democratic Labor Confederation (CFDT); Workers' Force (FO); French Confederation of Christian Workers (CFTC). France has the lowest level of unionization in the European Community (10%).

THE STATE

Official Name: République Française. **Administrative divisions:** 22 Regions with 96 Departments in France and 5 possessions, called "Overseas Departments".
Capital: Paris, 2,150,000 inhab.; Greater Paris, 9,000,000 (1990). **Other cities:** Marseilles, 1,230,000 inhab.; Lyon, 1,260,000 inhab.; Toulouse, 608,000 inhab.; Nice 475,000 inhab. **Government:** President: Jacques Chirac, since May 1995. Alain Juppé, prime minister since May 1995. Bicameral legislature: Chamber of Deputies, made up of 577 members, including 22 from "overseas departments"; Senate, made up of 317 members, 13 from overseas and 12 representing French nationals who reside abroad. **Armed Forces:** 409,000 (1995) **Paramilitaries:** 96,300 Gendarmes.

DEMOGRAPHY

Urban: 73% (1995). **Annual growth:** 0.6% (1991-99). **Estimate for year 2000:** 59,000,000. **Children per woman:** 1.8 (1992).

HEALTH

Under-five mortality: 9 per 1,000 (1994). **Calorie consumption:** 114% of required intake (1995).

EDUCATION

School enrolment: Primary (1993): 105% fem., 105% male. Secondary (1993): 107% fem., 104% male. University: 50% (1993). **Primary school teachers:** one for every 12 students (1992).

COMMUNICATIONS

102 **newspapers** (1995), 104 **TV sets** (1995) and 106 **radio sets** per 1,000 households (1995). 53.6 **telephones** per 100 inhabitants (1993). **Books:** 105 new titles per 1,000,000 inhabitants in 1995.

ECONOMY

Per capita GNP: $23,420 (1994). **Annual growth:** 1.60% (1985-94). **Annual inflation:** 2.90% (1984-94). **Consumer price index:** 100 in 1990; 109.7 in 1994. **Currency:** 5 french francs = 1$ (1994). **Cereal imports:** 1,188,000 metric tons (1993). **Fertilizer use:** 2,354 kgs per ha. (1992-93). **Imports:** $230,203 million (1994). **Exports:** $235,905 million (1994).

ENERGY

Consumption: 3,839 kgs of Oil Equivalent per capita yearly (1994), 47% imported (1994).

the nation "to be in peril" and formed a citizen army. This Committee of Public Salvation was able to forestall foreign invasion but internal confrontations resulted in the "Reign of Terror". Robespierre and his companions were overthrown and executed by the liberal, more moderate bourgeoisie, in July 1794.
[16] For the next five years, the revolutionaries tried to regain control of the country, which had fallen victim to corruption, internal strife and instability. Napoleon Bonaparte's coup d'etat (1789) brought an end to the dying regime. Seizing power, he had himself named Consul for Life, in 1802, and then Emperor, in 1804.
[17] Although Napoleon's reign represented a return to absolutism, it preserved the main achievements of the Revolution. Legal, administrative, religious, financial and educational reorganization changed the country irrevocably. Napoleon strove hard to bring the rest of Europe under his control and his armies occupied the whole of the continent from Madrid to the outskirts of Moscow. Finally, exhausted by war, France was defeated at Waterloo, in 1815.
[18] 19th century France was rocked by rebellions in 1830, 1848 and 1871, and constant disturbance. In spite of this, the Industrial Revolution brought factories, railroads, large companies and credit institutions to France. The Third Republic, beginning in 1870, was to be France's longest-lasting regime in almost a century and a half.
[19] With the establishment of universal male suffrage in 1848, the peasants and the urban middle class had the greatest electoral power. The government managed to win their support by protectionism and the establishment of free, secular and mandatory primary education, which raised aspirations of greater social mobility.
[20] Under the Republic, France had

a period of colonial expansion, beginning with the conquest of Algeria in 1830, and continuing with other territories in Africa and the Far East. A large empire was built, with colonies in the Caribbean, Africa, the Middle East, the Indochinese peninsula and the Pacific.

[21] World War I enabled France to recover Alsace and Lorraine, which had been annexed by Germany in 1870. The war left France devastated. More than a million and a half young people had been killed, property damage coupled with the internal and foreign debt added up to more than 150 billion gold francs, and the nation's currency lost its traditional stability. Neither the right-wing bloc of parties, nor the radical party was able to bring the economic and political situation under control.

[22] The world-wide recession reached France in 1931. In 1936, the parties of the left, who had joined together to form the Popular Front, had won the legislative elections, and they had carried out important social reforms, such as paid vacations and the 40-hour working week, but they were unable to hold back unemployment or the looming economic crisis. Government became polarized between adherents of Italian Fascism and those of Communism. This split led to an impasse until 1939, when Germany invaded Poland.

[23] Germany went on to occupy almost one-third of France. In 1940, Marshall Petain signed an armistice proclaiming the "national revolution" transforming non-occupied France into a satellite of Berlin. From London, General de Gaulle urged the French people to continue the fight.

[24] France was liberated in 1944. The French participated in the liberation and the invasion of Germany, with the "maquis" (especially the communists and socialists), the Free French Forces (liberal nationalists) and French troops fighting alongside the Allies. In October 1946, the Fourth Republic was proclaimed. This period was characterized by political instability and bitter opposition between the French Communist Party (PCF) and General de Gaulle's party.

[25] Funds from the US Marshall plan, led to the economic and social reconstruction of the country. Production reached a 6 percent annual growth rate; per capita income increased 47 per cent between 1949 and 1959; women were granted the right to vote; the banks were nationalized and a social security program went into effect. In 1957 France was one of the six signatories of the Treaty of Rome, which created the European Economic Community (EEC).

[26] After 1945, France was unable to reestablish its pre-war control over its colonies. The spirit of democracy and increasing awareness of universal human rights, was consolidated by the anti-Fascist alliance and the creation of the United Nations. Colonization had not allowed local identity to influence policy, so decolonization took place through fierce pro-independence movements, with little room for negotiation.

[27] In 1945, Syria and Libya were the first countries to become independent, followed by Morocco, Tunisia and Madagascar. Vietnam, Laos and Cambodia became independent only in 1954, after a long and bloody war. In May 1958, four years after the Algerian revolution began, the pieds-noirs - Frenchmen residing in the colony - revolted. They made the army promise not to leave the country. This dealt a mortal blow to the Fourth Republic, and the government called in General de Gaulle, to deal with the crisis.

[28] The establishment of the Fifth Republic, in 1958, and the decision to elect the president by direct universal suffrage, in 1962, laid the foundation for a regime with strong presidential powers. After the independence of Algeria and the last remaining African colonies, the country sought to achieve greater stability. With national independence as the primary aim of their foreign policy, France made use of "disuasive" atomic power and withdrew from the military structure of the North Atlantic Treaty Organization (NATO), in 1966.

[29] France continued to be a member of the Atlantic alliance, and maintained enormous economic, cultural and political influence over its former African colonies, south of the Sahara. Diplomatic relations with Algeria, Vietnam, and other countries that had fought bloody wars for independence were not restored until 1982.

[30] In May 1968, the greatest social and political crisis of the Fifth Republic took place. Huge student protests and labor strikes throughout the country were brought on by the

Wallis and Futuna

Population: 14,000 (1994)
Area: 265 SQ KM
Capital: Mata-Utu
Currency: CFP Franc
Language: French

Wallis was named after Samuel Wallis, a navigator who "discovered" it in 1767. Marist missionaries arrived in the archipelago in 1837 and converted the inhabitants to Catholicism. It became a French protectorate in 1888, and in December 1959, after a referendum, the country adopted the status of French Overseas Territory.

[2] Unlike other French dependencies in the Pacific, there are no pro-independence movements on the islands. In 1983, the two kingdoms of Futuna achieved separation from Wallis, but maintaining their relationship with France. The islands' economic prospects are poor: in addition to the devastating effect of cyclones which periodically hit the islands, the only bank on the islands was closed. Wallis & Futuna received FF55 million in aid from France in 1987. Approximately 50 per cent of the economically active population has had to emigrate to other parts of Polynesia in search of work; their remittances, together with public works projects, constitute the main source of income for the islands.

[3] In the 1992 elections for the Territorial Assembly, the left managed to defeat the neo-Gaullist RPR, a party of the right, for the first time in twenty years.

ENVIRONMENT

The territory consists of the Wallis archipelago (159 sq.km), formed by Uvea Island - where the capital is located - and 22 islets, plus the Futuna (64 sq km) and Alofi Islands (51 sq km). This group is located in western Polynesia, surrounded by Tuvalu to the north, Fiji to the south and the Samoa archipelago to the east. With a rainy and tropical climate, the major commercial activities are copra and fishing.

SOCIETY

Peoples: Of Polynesian origin. Approximately two thirds of the population live on Wallis and the rest on Futuna. Nearly 12,000 inhabitants live in Kanaky/New Caledonia and Vanuatu. **Religions:** Catholic. **Languages:** French (official) and Polynesian languages.

THE STATE

Capital: Mata-Utu, (located on Uvea) 815 inhab. (1983). **Government:** Overseas territory administered by a French-appointed Chief Administrator, Robert Pommies since 1990, assisted by a 20-member Assembly elected for a 5-year term. The kingdoms of Wallis and Futuna (in Sigave and Alo) from which the country was formed, have very limited powers. They send one deputy to the French National Assembly, and another to the Senate.

COMMUNICATIONS

6.3 **telephones** per 100 inhabitants (1993).

ECONOMY

Currency: CFP francs.

regime's growing authoritarianism in the educational and social sectors. For a whole month, the government seemed to be seriously threatened. Apart from the student leaders, however, there were no other political forces capable of toppling the government, and the general strike was called off when a salary increase was promised.

[31] The years which followed saw the birth of other groups based around social issues, such as the feminist, ecological and antinuclear movements. In 1972 the Socialist Party and the Communist Party created the Union of the Left, and François Mitterand, the Socialist candidate, was elected president in 1981 because of the Communist votes; his was the first left-wing cabinet since 1958.

[32] The new government carried out reforms: the nationalization of important industrial and banking groups, new labor rights for workers, the 39-hour working week, an increase in social benefits, retirement at the age of 60 and a decentralization of power. However, unemployment, the economic crisis and an increase in imports led the government to enforce a harsh economic policy and to carry out restructuring of the industrial sector. This led communist ministers to resign from the government, in 1984.

[33] That year, Prime Minister Pierre Mauroy, a symbol of the left's union, was replaced by Laurent Fabius, a young technocrat, considered a loyal friend of Mitterrand's. Fabius formed a more "centrist" government than Mauroy's, in which Communists refused to take part.

[34] In 1985, relations between France and several South Pacific countries like Australia and New Zealand were deteriorated when it was discovered the French were responsible of sinking a Greenpeace ship which caused the death of a militant ecologist when it was headed for the Mururoa atoll to demonstrate against French nuclear tests.

[35] In 1986, expecting a defeat of the left, Mitterrand substituted the majority electoral system by a proportional system which gave more seats to the losers than the previous one. In March, a right-wing coalition made up mainly of the Group for the Republic (RPR), a party led by neo-Gaullist Jacques Chirac and by former president (1974-1981) Valery Giscard D'Estaing's Union for the French Democracy (UDF), defeated the left in the legislative elections. The elections confirmed a clear progress of Jean Marie Le Pen's ultra-right National Front. Chirac had to form a new government and for two years the country lived its first cohabitation experience between a left-wing president and a conservative council of ministers.

[36] Chirac's government wiped out some of the 1981 and 1982 reforms with the privatization of several companies nationalized by the left but kept most of the social gains. In the meantime, it continued a liberalization policy of the economy and finances begun in the socialist government's last years. In the field of individual liberties, the strong hand of Interior minister Charles Pasqua, especially regarding legislation concerning foreigners living in France, was criticized by several humanitarian organizations.

[37] In the 1988 presidential elections, Mitterrand defeated Chirac among other things, thanks to the vote of many Jean Marie Le Pen followers who had almost 15 per cent in the first round. Mitterrand's re-election allowed for the socialists' return to power as well as the incorporation to the government of several turncoats from the right. New Prime Minister Michel Rocard's Cabinet had the participation of well-known conservative figures like former Giscardians Jean Pierre Soisson and Lionel Stoleru.

[38] Socialist economic policy did not differ substantially from the right's. Unemployment kept growing, reaching 9.5 per cent in 1991. That year Rocard, who was opposed to Mitterrand within the Socialist Party, was replaced by the president's loyal friend Edith Cresson who became the first female head of government in contemporary France. However, only nine months after Cresson came to office, Mitterrand replaced her with former Economy minister Pierre Beregovoy who was widely supported among the industry and finance spheres.

[39] In 1993, the left lost legislative elections once again and Mitterrand nominated conservative Edouard Balladur as Prime Minister. The majority electoral system, reinstated by the right in 1986, prevented the National Front and the Greens from obtaining seats at the National Assembly. Scandals due to accusations of corruption, whose main targets had been Socialist leaders, shifted their aim and affected well-known right-wing figures, leading to the resignation of three of Balladur's ministers in 1994.

[40] In 1995, Mitterrand ended his term amid harsh criticism for what was considered by many to be the left's betrayal, when it supposedly applied conservative policies in its ten years in power, and for his alleged right-wing past and support of marshal Petain during the German occupation (1940-1944).

[41] In the first round of elections in April 1995, Socialist Lionel Jospin and Chirac (with 23 per cent and 20.5 per cent of the vote) defeated their rivals. Ultra right-wing and xenophobe Le Pen obtained over 15 per cent of the vote and extreme left-wing Arlette Laguiller had over 5 per cent, which was considered a sign of strong discontent. In the second round Chirac was elected president of France with almost 53 per cent of the vote.

[42] In December, the largest civil servant strike to erupt since 1968, paralyzed the country for over three weeks. The social situation continued to be strained in 1996, which some observers linked not only to growing unemployment but also with the differentiated income distribution - 20 per cent of the population receive 44 per cent of total personal income- and property distribution since 20 per cent own 69 per cent of the national estate.

Saint Pierre and Miquelon

Population: 6,000 (1994)
Area: 240 SQ.KM
Capital: St. Pierre
Currency: French Franc
Language: French

An archipelago of 8 small islands, near the Canadian coast, in the north Atlantic, economically dependent on fishing.
Capital: St Pierre, 5,663 inhab.(1990) Other towns: Miquelon, 709 inhab.(1990)

SOCIETY

The People: The majority are descendants of French settlers. Religions: Catholic. Language: French.

THE STATE

Government: A French overseas department administered by a 19-member Conseil Général elected a 6-year term, and a Commissioner appointed by the French government. In the French National Assembly, the area is represented by one deputy and has one senator and one representative in the European Parliament.

DEMOGRAPHY

Density: 26 inhab./sq.km. Annual growth: 1.5% (1970-86). Education Primary school teachers: one for every 15 students (1986).

COMMUNICATIONS:

652 **TV sets** and 698 radio receivers per 1,000 inhab. (1991). 57.0 **telephones** per 100 inhabitants (1993).

ECONOMY

Currency: French francs. Energy Consumption: 7,000 kgs of Coal Equivalent per capita yearly, 98% imported. (1990).

French Guyana

Guyane Français

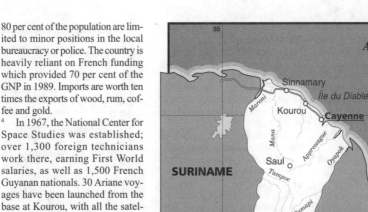

Population: 141,000 (1994)
Area: 91,000 SQ KM
Capital: Cayenne

Currency: French Franc
Language: French

The Arawaks were short, copper-skinned people with straight, black hair. They grew corn, cotton, yams and sweet potatoes. They built round huts with thatched, cone-shaped roofs, and slept in "hammocks", an Arawak word which survived the culture which gave rise to it. The Caribs displaced them from the area and later resisted the Spaniards who began to arrive toward the beginning of the 16th century.

[2] In 1604 the French occupied Guyana, despite Carib resistance. The colony passed successively into Dutch, English and Portuguese hands, until the beginning of French domination in 1676. Toward the end of the 18th century, France sent more than 3,000 colonists to settle the interior. Few survived the tropical diseases, but those who did sought refuge in a group of islands off the coast which they named "Health Islands". After a brief period of prosperity brought about by the discovery of gold in the Appranage River basin, the colony gradually fell into decline.

[3] In 1946, Guyana became a French "Overseas Department". Nine-tenths of the country is covered by forests, and although the country is fertile, most of the food is imported. The creole (African and mestizo) 80 per cent of the population are limited to minor positions in the local bureaucracy or police. The country is heavily reliant on French funding which provided 70 per cent of the GNP in 1989. Imports are worth ten times the exports of wood, rum, coffee and gold.

[4] In 1967, the National Center for Space Studies was established; over 1,300 foreign technicians work there, earning First World salaries, as well as 1,500 French Guyanan nationals. 30 Ariane voyages have been launched from the base at Kourou, with all the satellites sent into space financed by European consortia.

[5] During the 1970s, the autonomist Socialist Party of Guyana (PSG) became the majority party at the local level. Early in the 1980s, armed groups attacked "colonialist" targets, but tensions were defused with the victory in France of the French Socialist Party in 1981.

[6] In 1986, Guyana's representation in the French National Assembly increased to two members. In the 1989 municipal elections, Cayenne and 12 other districts, out of the 19 at stake, were won by the left. Georges Othily, a PSG dissident was elected for the French Senate.

[7] In 1992, the country's persistent economic crisis triggered a week-long general strike, called by trade unions and two entrepreneurial federations. Paris agreed to finance an infrastructure and education improvement program. Also in that year almost half of the 6,000 Surinamese refugees were repatriated. In 1994, Cayenne joined the Association of Caribbean States in the capacity of associated member.

[8] The present social and economic situation of Guyanans is less encouraging than that suggested by figures. The per capita income is the highest in South America, but the Guyanan economy is highly dependent on its base at Kourou, French subsidies and food and energy imports. Also the high unemployment rate, particularly among young people, reflects a difficult social situation which some observers consider explosive.

PROFILE

ENVIRONMENT

Guyana is located slightly north of the Equator. Due to its hot and rainy climate, only the coastal alluvial lowlands are suitable for agriculture (cocoa, bananas, sugar cane, rice and corn). The hinterland mountains are covered with rainforests. The country has large reserves of bauxite and gold.

SOCIETY

Peoples: Mostly from native and European roots. There are Carib and Tupi Guarani ethnic groups, and Cimarrones, descendants of African slaves. In the cities there are Chinese, Indian and French minorities. **Religions:** Mainly Catholic, also Hindu. **Languages:** French (official), Creole. **Political Parties:** The Union for the Republic (RPR); the Socialist Party of Guyana (PSG); the De-Colonization Movement of Guyana; the Union for French Democracy (UDF).

THE STATE

Official Name: Département d'Outre-Mer de la Guyane française. **Administrative Divisions:** 2 Districts. **Capital:** Cayenne, 41,667 inhab. (1990). **Other Cities:** Kourou, 13,873 inhab., St Laurent-du-Maroni, 13,606 inhab.(1990). **Government:** Jean-Franáois Cordet, prefect designated by France. Regional parliament (consultative body): General Council, made up of 19 members, and Regional Council, with 31 members. The department has 2 representatives in the National Assembly and 2 in the French Senate. **National Holiday:** All French holidays. **Armed Forces:** 8,400 French troops.

DEMOGRAPHY

Annual growth: 2.7% (1970-86).

EDUCATION

Literacy: 84% male, 82% female (1982) **University:** 374 students per 100,000 inhab. **Primary school teachers:** one for every 23 students (1983).

HEALTH SERVICES

Calorie consumption: 96% of required intake (1988-90).

COMMUNICATIONS

10 **newspapers**, 213 **TV sets** and 759 **radio receivers** per 1,000 inhab. (1991).

ECONOMY

Consumer price index: 100 in 1980, 216 in Jul.93. **Currency:** French francs. **Imports:** $769 million (1991). **Exports:** $70 million (1991). **Major export products:** shellfish, 46%; rice, 13%; gold, 5%; wood, 4%; fresh fish, 4%. **Major markets:** EEC, 78%; Latin America, 20%; EFTA, 1%.

ENERGY

Consumption: 2,590 kgs of Coal Equivalent per capita yearly, 100% imported. (1990).

French Polynesia

Polynésie Français

Population: 219,000 (1994)
Area: 4,000 SQ KM
Capital: Papeete

Currency: CFP Franc
Language: French

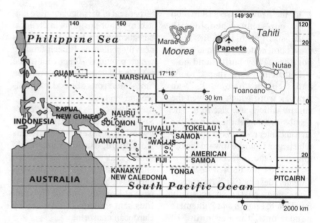

Polynesia's first inhabitants either came from Latin America or Indonesia, but the evidence is inconclusive.

[2] In 1840 the islands were occupied by France and in 1880, despite native resistance, the archipelago was officially made a colony under the name of "the French Establishments of Oceania". In 1958, they became an Overseas Territory.

[3] Except for some concessions on the domestic front, France continues to maintain its control over the islands, with a hard-line policy because of the strategic location of the islands and the atomic tests which have taken place on the Mururoa atoll since 1966. In 1975 France also carried out tests on the Fangataufa atoll for tests, despite strong opposition to the growing militarization of the islands from the population of the territory and other countries in the area, especially New Zealand,.

[4] In the face of French intransigence, there has been increasing resistance from separatist groups, who have taken their case to the UN Decolonization Committee, though few concrete resolutions have been made.

[5] The "nuclearization" of the area has rapidly resulted in the destruction of Tahiti's traditional economic base, as the island is currently dependent upon the French military budget. In less than one generation, the economy has gone from self-sufficiency to dependence upon imported goods. By the end of 1980, 80 per cent of basic foodstuffs were imported.

[6] Both the environment and the health of the islands' inhabitants have deteriorated over the last two decades, and there have been dramatic increases in the incidence of brain tumors, leukemia and thyroid cancer. The scale of these illnesses is hard to evaluate, as the French government has refused to divulge medical statistics in recent years.

[7] In 1985, the explosion of a bomb planted by two French secret agents killed an ecology activist and destroyed a Greenpeace ship that was heading for Mururoa in order to protest against nuclear testing. In 1992, French president François Mitterrand decided a temporary suspension of nuclear tests and Paris began negotiations with Papeete for an economic plan to be carried out after the permanent closure of the experiment centers in Mururoa and Fangataufa.

[8] The "Progress Pact" between Paris and "French" Polynesia is based in assumptions similar to those in structural adjustment programs advocated by the IMF, ie economic liberalization, privatizations and "major balancing" in public finance. For the 1994-1998 period, Paris has planned a slight increase in financial contributions for the pact to attain its goals. However, the main pro-autonomy political party, Tavini Uiraatira (that means "serving the people" in the Polynesian language) led by Oscar Temaru and representing 15 per cent of the voters, proposes nationalizations and the extension of free services.

[9] French President Chirac, after another controversial set of six tests, announced the permanent suspension of nuclear tests in January, 1996. After the first one of these explosions, there were disturbances and fighting in Tahiti.

PROFILE

ENVIRONMENT

The territory is located in the southeast portion of Polynesia. Most of the French Polynesian islands are of volcanic origin but they also have major coral formations. The islands are largely mountainous, with tropical climate and heavy rainfall. Relatively fertile soils favor agricultural development. The Windward and Leeward Islands form the Society Archipelago. Nuclear tests carried out by France over a 26-year period have caused damage to the environment and human beings, that is difficult to evaluate.

SOCIETY

Peoples: Polynesian 78%, Chinese 12%, French descendants 6%. **Religions:** Mostly Christian, 55% Protestant, 32% Catholic. **Languages:** French (official), Tahitian (national). **Political Parties:** Tahoeraa Huiraatira (RPR), conservative; Pupu Here Ai'a, pro autonomy, led by Jean Juventin; Ai'a Api, a coalition led by Emile Vernaudon; Tavini Huiraatira, an antinuclear and independence movement led by Oscar Temaru. **Social Organizations:** The Pacific Christian Workers Central (CTCP), the Federation of French Polynesian Unions (FSPF) and the Territorial Union of the General Labor Confederation "Workers' Force" (UTSCGT).

THE STATE

Official Name: Territoire d'outre-mer de la Polynésie française. **Administrative Divisions:** Windward Islands, which include Tahiti, Murea, Maio; Papeete constitutes the center of the district. Leeward Islands, with the capital in Utoroa on Ralatea island, also includes the islands of Huahine, Tahaa, Bora–Bora and Maupiti. Tuamotu and Gambier Archipelagoes, the Austral Islands and the Marquesas Islands. Capital: Papeete, 170,000 inhab. (1988). **Government:** The French government is represented by a high commissioner, who controls the areas of defense, foreign relations and justice. A local 41-member Territorial Assembly is elected by universal suffrage for a 5-year term. This assembly, in turn, elects the president of the Council of Ministers, an executive body which is made up of 5 to 10 members of the Territorial Assembly, selected by the president. President of the Council of Ministers: Alexandre Leontieff. President of the Territorial Assembly and Mayor of Papeete: Jean Juventin.

EDUCATION

Primary school teachers: one for every 14 students (1990).

ECONOMY

Consumer price index: 100 (1990); 105.6 (1994). **Currency:** CFP francs.

Gabon

Population: 1,301,000 (1994)
Area: 267,670 SQ KM
Capital: Libreville
Currency: CFA Franc
Language: French

Gabon

Tools found in the Gabon forests indicate that this region has been inhabited since the paleolithic age. Nothing else is known about life in the area until the 16th century, when the migrations that triggered the crisis in the ancient state of Congo (see "Congo: The Bantu states of Congo") brought the Myene and, in the 17th century, the Fang to Gabon. Thereafter, they monopolized the slave and ivory trade together with the Europeans.

[2] The Portuguese arrived in 1472. Around the middle of the 19th century, the French, Dutch and British established a permanent trade of ivory, precious woods and slaves. In 1849, Libreville was founded and established as a settlement for freed slaves from other French colonies. The territory was of little economic interest to the French, who used it as a base for expeditions into the heart of the continent.

[3] The quest for independence was relatively uneventful because the two local parties (the Joint Mixed Gabonese Movement of Leon M'Ba and the Democratic and Social Union of Jean-Hillaire Aubame) were willing to accept neo-colonial tutelage.

[4] Gabon has abundant resources: iron ore, uranium, manganese, timber and oil. The transnational oil company Shell recently discovered oil in Rabikuna, near Port Gentil. These oil reserves contain enough crude to last for 50 years and the oil wells are already in operation. Shell holds an 80 per cent interest in the wells already in operation and ELF the remaining 20 per cent; Amoco (an American company) and Braspetro (Brazilian) have been authorized to continue oil exploration in the Port Gentil area.

[5] Until recently, timber mills were the only local industry of any size, but a few years ago, French and US transnationals discovered that Gabon could serve as a spearhead to penetrate the markets of central Africa, and they initiated its industrialization. This "development", relying on foreign capital has only exacerbated social conflicts and the promise of jobs in the cities has encouraged urban migration. Gabon's social structure is changing as independent producers become suppliers of cheap labor for transnational industries. Economically, the breakdown of the rural economy has completely blocked the way to food self-sufficiency, destroying the only social sector which remained independent of foreign capital.

[6] The neo-colonial system has been guaranteed by a military treaty signed between Libreville and Paris in 1960. Its effectiveness was proved in 1964 when M'Ba was deposed by a group of progressive officers.

[7] When M'Ba died in 1967, he was succeeded by Omar Bongo, head of cabinet, who faithfully followed, and even refined his predecessor's style, becoming Giscard D'Estaing's "privileged spokesman" in Africa. Applying the US thesis of "sub-imperialisms" to French interests, Bongo became its "Watchdog" in central Africa, a base for aggression against neighboring progressive regimes. Thus, in January 1977, Gabon provided the planes and arms used by a mercenary group in an unsuccessful attack on the People's Republic of Benin.

[8] Bongo's foreign policy continued to evolve and he maintains good relations with several states in the region, without altering the country's privileged relationship with France. Like Senegal, Ivory Coast, Chad, and the Central African Republic, Gabon also has French troops on its soil.

[9] In 1979 and again in 1986, Bongó was reelected with 99 per cent of the vote, in presidential elections in which he was the only candidate.

[10] His outrageous squandering led to violent protests from the Gabonese in the early 1980s. The revolt spread to the police who, in 1982, organized an unheard-of demonstration, demanding an increase in wages and the withdrawal of French advisors. The protests were brutally repressed by Gabon's secret police, officially known as the "Documentation Center".

[11] The government repressed the National Reorientation Movement (MORENA), formed by intellectuals, workers, students and nationalist politicians. The movement was accused of having expropriated 30 tons of weapons in October 1982. At that time, the Bongó family and French military installations were the target of armed attacks. At least 28 top MORENA leaders were sentenced to 15 years' imprisonment. The French Socialist Party criticized the rulings, thus seriously upsetting Bongo's relations with president François Mitterrand.

[12] This episode did not stop Bongo from visiting France in March 1984. The friendly greeting at the presidential palace received as much criticism as the decision to allow the French government to build a nuclear plant in Gabon.

[13] In late 1989 there were signs of democratic participation for Gabon's opposition forces. After violent confrontations in the streets the president had the constitution amended, introducing a multiparty system and lifting censorship of the press. In the meantime, he invited six opposition leaders to join the cabinet.

[14] There was a period of calm after all this social upheaval. However, in May, Joseph Redjambe, president of the Gabonese Progressive Party was murdered in a Libreville hotel. His death provoked strong reactions against the government, thought to be associated with the murder. A state of rebellion through the entire Port Gentil region lasted ten days and the French government evacuated 5,000 French residents, charging the presidential guard with re-establishing order. According to information confirmed by Amnesty International, the incidents left a toll of six dead and a hundred wounded.

[15] Once calm had been restored, the political and social groups met on June 6, in a National Conference and an agreement was reached whereby presidential elections would be held. This was undoubtedly a major victory for the op-

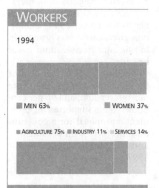

WORKERS

1994

- MEN 63%
- WOMEN 37%
- AGRICULTURE 75% INDUSTRY 11% SERVICES 14%

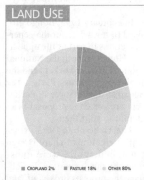

LAND USE

- CROPLAND 2% PASTURE 18% OTHER 80%

ENVIRONMENT

Irrigated by the Ogooue River basin, the country has an equatorial climate with year-round rainfall. The land is covered by dense rainforest and a large part of the population work in the local timber industry. The country has manganese, uranium, iron and petroleum reserves. Deforestation is one of the most serious environmental problems, together with the depredation of the country's wildlife.

SOCIETY

Peoples: Gabon was populated by pygmy hunters. In the 16th century it was invaded by Myene and other ethnic groups. Today over half of the population are Bantu, divided into more than 40 different groups including Galoa, Nkomi, and Irungu. One third of the population are Fang and Kwele, while in the northern and southern parts of Gabon there are Punu and Nzabi minorities. **Religions:** Mainly Christian. More than one third practice traditional African religions and there is a small Muslim minority.
Languages: French (official). Bantu languages are spoken along the coast; Fang is used in northern Gabon; local languages and dialects are used in the rest of the country.
Political Parties: Gabonese Democratic Party (PDG), pro-government. Gabonese Progress Party and National Union of Timberworkers, in the opposition.

THE STATE

Official Name: République Gabonaise. **Administrative divisions:** 9 provinces and 37 prefectures.
Capital: Libreville, 419,600 inhab. (1993). **Other cities:** Port Gentil 178,200 inhab. (1993); Franceville 75,000 inhab. in1988. **Government:** Parliamentary republic with strong head of State. Member of the African Financial Community (CFA). Omar (Albert-Bernard) Bongo, President and head of State since December 1967. Paulin Obame-Nguema, Head of Government and Prime Minister since November 1994. Single-chamber legislature: National Assembly, with 120 members elected every five years. **National Holiday:** July 17, Independence Day (1960). **Armed Forces:** 4,700 (1995). **Paramilitaries:** Coast Guard: 2,800; Gendarmerie: 2,000.

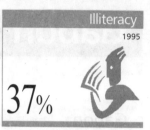

Maternal Mortality 1989-95
PER 100.000 LIVE BIRTHS
438

Illiteracy 1995
37%

DEMOGRAPHY

Urban: 48% (1995). **Annual growth:** 1.5% (1992-2000). **Estimate for year 2000:** 2,000,000. **Children per woman:** 5.9 (1992).

HEALTH

One **physician** for every 2,500 inhab. (1988-91). **Under-five mortality:** 151 per 1,000 (1995). **Calorie consumption:** 98% of required intake (1995). **Safe water:** 68% of the population has access (1990-95).

EDUCATION

Illiteracy: 37% (1995). **School enrolment:** Primary (1993): 136% fem., 136% male. University: 3% (1993). **Primary school teachers:** one for every 44 students (1992).

COMMUNICATIONS

86 **newspapers** (1995), 88 **TV sets** (1995) and 87 **radio sets** per 1,000 households (1995). 2.4 **telephones** per 100 inhabitants (1993). **Books:** new titles per 1,000,000 inhabitants in 1995.

ECONOMY

Per capita GNP: $3,880 (1994). **Annual growth:** -3.70% (1985-94). **Annual inflation:** 3.30% (1984-94). **Consumer price index:** 100 in 1990; 123.8 in 1994. **Currency:** 535 CFA francs = 1$ (1994). **Cereal imports:** 77,000 metric tons (1993). **Fertilizer use:** 11 kgs per ha. (1992-93). **External debt:** $3,967 million (1994), $3,049 per capita (1994). **Debt service:** 10.5% of exports (1994). **Development aid received:** $102 million (1993); $101 per capita; 2% of GNP.

ENERGY

Consumption: 520 kgs of Oil Equivalent per capita yearly (1994), -2,268% imported (1994).

---

position, who extracted promises of multiparty elections from the president. But Bongo advanced the election date from 1992 to late 1990, a tactic meaning that he could use the power base he had built up over the past 20 years in office to carry out his campaign, while the opposition had no time to organize itself adequately or overcome its internal divisions.

[16] In September 1990, President Bongo's Gabonese Democratic Party (PDG), obtained a majority in the National Assembly. The new constitution, approved in March 1991, formally established a multiparty system.

[17] In 1991, the country was rocked by a new outbreak of political and social violence. The economic crisis continued to worsen and the economic stabilization program launched in September under IMF guidelines, and the structural adjustment plan promoted by the World Bank failed to generate much hope.

[18] The absence of an electoral timetable and student agitation demanding greater resources for the university increased tension even further. The president closed the university, called off the elections and banned all political meetings. The opposition called a general strike, which paralyzed Port Gentil, the center of the oil industry. Concerned by the extent of the protests, Bongo reopened the university and lifted the ban on demonstrations.

[19] Gabon has played an important role in the mediation of several regional disputes, like the border dispute between Libya and Chad and the resuming of talks between the US and Angola. However, some of the government's international policies have generated friction. In October 1992, Gabon announced the expulsion of 10,000 Nigerians, saying that their status in the country was "irregular", angering authorities in Abuja.

[20] In February 1993, President Bongó proposed that a date be set for general elections, restricting access to only part of the opposi-

tion. Abessole, leader of the National Union of Timberworkers, and Pierre Louis Agondjo, leader of the Gabonese Progress Party did not respond to the proposal, seeing it as a mere ploy to divide the opposition. These parties formed the second most influential bloc in parliament, and had maintained contact with Bongo.

[21] In February 1993, Bongo proposed fixing a date for the general elections to part of the opposition. However, both the National Union of Timberworkers and the Gabonese Progress Party - the second strongest force in Parliament - rejected the proposal as an attempt to split them.

[22] The December 1993 presidential elections, where Bongo triumphed again, were questioned by the opposition. In February the protests broke out, and

were harshly put down, leaving 30 people dead. In May, a strike was called at the Omar Bongo University, which was closed by the president for three months in June.

[23] Rumours of rapprochement between military sectors and the opposition made Bongo more prepared to negotiate. In meetings held in Paris from September to November 1994, a coalition government was brought in to rule until free elections could be called.

[24] In July 1995, 63 per cent of the electorate turned out to vote in a referendum, and 96 per cent of the voters supported the president's proposal for a constitutional reform with presidential and legislative elections for 1997.

Gambia

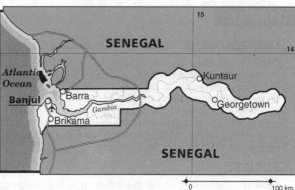

Population: 1,079,000 (1994)
Area: 11,300 SQ KM
Capital: Banjul

Currency: Dalasi
Language: English

Gambia

The earliest settlers of the Gambia river valley came from what is now Senegal. Attracted by Gambia's coast, which lent itself to trade and navigation, they settled along the river, carrying out subsistence farming.

2 In the 15th century, the region was colonized by the Mandingo who, together with the Mali Empire (see Mali: "The Mali Empire" and Senegal: "The Fulah States"), established their authority in the Gambia Valley. They also founded several kingdoms in the area, which controlled coastal trade and enabled them to develop economically and culturally.

3 With the arrival of the Portuguese in 1455, most of the region's domestic trade was displaced toward the Atlantic coast, bringing about the decline of the local kingdoms which had prospered under this trade. For the Portuguese, Gambia became the point of departure for a large quantity of precious metals.

4 Gambia was also important to them as a prosperous port of call on their route to the Orient. However, in 1618 the Portuguese Crown sold its commercial and territorial rights to the British Empire, which at the height of its naval prowess, was trying to assert its dominance as a colonial power by acquiring a foothold in Africa.

5 At that time, a war began between Britain and France (controlling all of what is now Senegal), which was to last for more than two hundred years. From 1644, the British used this coastal area as a source of slaves, setting up alliances with inland tribal princes to provide them with human merchandise.

6 Throughout the 17th century, Gambia was nothing more than a source of slave labor for Britain's colonies and for its slave trade with other colonial powers. The British therefore limited themselves to establishing a rudimentary trading post in the area, founded in 1660. Border disputes between the British and French increased during the 18th century.

7 Throughout the 19th century a series of religious wars in the interior resulted in the complete Islamization of the country, with an increase in Muslim immigration from other parts of Africa. In the meantime, the region lost its international economic significance when the slave trade was abolished in Britain.

8 It did however gain strategic importance by being a British enclave inserted into the heart of Senegal, a region instrumental in France's designs on sub-Saharan Africa. Slavery actually continued to exist within the colony until the 20th century, not being outlawed until 1906.

9 In 1889, France and Britain reached an agreement as to the boundaries of their respective colonies, ensuring peace in the region and the formal recognition of British sovereignty over Gambia by other European powers.

10 Gambia's status as a British colony remained unchanged throughout the first half of the 20th century. Although the decolonization process in Africa began after World War II - resulting in the creation of a number of independent states in what had formerly been European colonies - Gambia did not receive administrative autonomy from Britain until 1963. Two years later, Gambia obtained full independence and joined the British Commonwealth. At the time of its independence, Gambia did not seem to constitute a nation as such because of its ethnic, cultural and economic complexity.

11 After independence, the territory's social and economic structures remained unchanged. Exports continued to be based on a single crop, peanuts, while traditional social structures remained so unassailable that they were finally legitimized by the 1970 Constitution. Thus, although some legislative seats were determined by the election of deputies, others were assigned to the 5 regional leaders.

12 Dawda Jawara, founder of the People's Progressive Party (PPP), has dominated Gambian politics since the 1960s. He won the 1962 elections, but failed to assume office because of a vote of "no-confidence" from the opposition. However, he yet again won the election in 1970 when the country was proclaimed a republic, with a presidential system of government.

13 Around 1975, the success of Alex Haley's book "Roots" turned Gambia into an important tourism center; but alongside this, prostitution and drug trafficking also increased while organized Islamic opposition emerged.

14 The lack of border controls in Gambia led to it becoming a sort of paradise for west African smuggling and a large part of Senegal's agricultural produce is illegally shipped through the port of Banjul.

15 The close economic relationship between the countries led the government of Dawda Jawara to accept a project for union in 1973.

16 In July 1981, Muslim dissidents attempted to overthrow Jawara, aiming to end official corruption through the establishment of a revolutionary Islamic regime. The revolt was crushed by Senegalese troops who entered Gambia at the request of President Dawda, who was in London at the time.

17 The proposed union with Senegal had been planned for 1982, but the coup attempt against Gambia's government accelerated plans for creating the Senegambian federation.

18 The 1980s were marked by a worsening of the country's economic situation, and a severe drought led to a sudden drop in the production of agricultural exports. The most obvious consequences were an increase in unemployment, migration from rural areas to the capital and increasing foreign debt to finance food imports.

19 Senegambia was officially established in February 1982 with Abdou Diouf of Senegal as its first president, assisted by a council of ministers and a bi-national parliament. The federation did not totally unite Senegal and Gambia, as both states retained their autonomous character and internal organization. The treaty ensured that Dawda Jawara gained protection against internal rebellions and Senegal gained greater control over the leak of export tax revenues through smuggling.

20 Gambia became unhappy with this alliance and its dissatisfaction became apparent in mid-1985, with their reluctance to sign the treaties aimed at strengthening ties with Senegal. More recently Jawara failed to

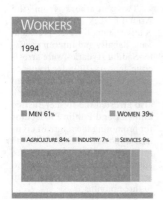

WORKERS

1994

MEN 61%   WOMEN 39%

AGRICULTURE 84%   INDUSTRY 7%   SERVICES 9%

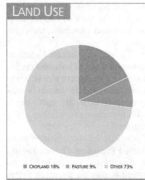

LAND USE

CROPLAND 18%   PASTURE 9%   OTHER 73%

MILLIONS $ 1994

IMPORTS
209

EXPORTS
35

PER 100,000
LIVE BIRTHS
1,050

External Debt 1994

PER CAPITA
$388

Illiteracy 1995

61%

PROFILE

ENVIRONMENT

One of the smallest African countries, Gambia stretches 320 km along the Gambia River, which is fully navigable and one of the main waterways in the area. The climate is tropical; there are rainforests along the river-banks and wooded savannah further inland. The economy is based on peanut exports and tourism. Because of the use of firewood as a main energy source and the production of export crops large areas of forest have been felled.

SOCIETY

Peoples: The majority of Gambians belong to the Mandingo ethnic group, 40%; 14% are Fulah, 13% are Wolof who belong to the same root and 7% are Diula. There are minor ethnic groups inland (Serahuilis, Akus). Five to ten thousand laborers migrate from Mali, Senegal and Guinea-Bissau every year only to return home after the harvest. **Religions:** The majority of Gambians practise the Islamic faith; a minority practise traditional religions and Christianity, mostly Protestant. **Languages:** English (official); the most widespread local languages are Mandingo, Fulani and Wolof (the Wolof form a majority in the main trade center). **Political Parties:** The People's Progressive Party (PPP), founded by Dawda Jawara in the 1960s; the National Convention Party (NCP) from a PPP split in the 1970s; the People's Party of Gambia (PPG). **Social Organizations:** There are three trade union federations: the Gambia Workers' Union (linked to the PPP); the Gambia Labor Union; and the Gambia Salesmen's and Merchants' Union.

THE STATE

Official Name: Republic of The Gambia. **Administrative Divisions:** 5 Provinces and the Capital. **Capital:** Banjul, 42,300 inhab. (1986). **Other cities:** Brikama 24,300 inhab.; Bakau 23,600 inhab. (1986). **Government:** Presidential republic. Lt Yaya Jammeh, President of the Provisional Military Council. Legislature. Single-chamber Parliament: Chamber of Deputies, with 50 members, 36 of whom are elected by universal suffrage, 9 special members and 5 tribal chieftains. **National Holiday:** February 18, Independence (1965). **Armed Forces:** 800 troops (1994).

STATISTICS

DEMOGRAPHY

Urban: 24% (1995). **Annual growth:** 2.4% (1992-2000). **Children per woman:** 6.5 (1992).

HEALTH

Under-five mortality: 213 per 1,000 (1995). **Calorie consumption:** 96% of required intake (1995). **Safe water:** 95% of the population has access (1990-95).

EDUCATION

Illiteracy: 61% (1995). **School enrolment:** Primary (1993): 61% fem., 61% male. Secondary (1993): 13% fem., 25% male. **Primary school teachers:** one for every 30 students (1992).

COMMUNICATIONS

83 **newspapers** (1995), 81 **TV sets** (1995) and 92 **radio sets** per 1,000 households (1995). 1.6 **telephones** per 100 inhabitants (1993). **Books:** 93 new titles per 1,000,000 inhabitants in 1995.

ECONOMY

Per capita GNP: $330 (1994). **Annual growth:** 0.50% (1985-94). **Annual inflation:** 10.10% (1984-94). **Consumer price index:** 100 in 1990; 128.8 in 1994. **Currency:** 10 dalasis = 1$ (1994). **Cereal imports:** 87,000 metric tons (1993). **Fertilizer use:** 44 kgs per ha. (1992-93). **Imports:** $209 million (1994). **Exports:** $35 million (1994). **External debt:** $419 million (1994), $388 per capita (1994). **Debt service:** 14.4% of exports (1994). **Development aid received:** $92 million (1993); $88 per capita; 26% of GNP.

ENERGY

Consumption: 56 kgs of Oil Equivalent per capita yearly (1994), 100% imported (1994).

comply with the terms of the agreements promising Senegal military and diplomatic support in case of external or internal conflicts.
[21] Within the framework of the 1991-92 budget, the Minister of Finance announced a 6 per cent increase in public employees' salaries. He also reinstated a series of benefits eliminated when the structural adjustment program went into effect. Gambia has one of the world's highest infant mortality rates at 143 per 1,000 live births, and the effects of the adjustment program, particu-

larly within the health sector, have been devastating.
[22] In May 1991, Gambia and Senegal took the first steps toward reconciliation, by signing a good-will and cooperation treaty. The agreement foresees an annual meeting between the two heads of State, and the creation of a joint commission presided over by the foreign ministers of both countries.
[23] In January 1992, Jawara announced his decision not to seek re-election in the April 7 elections. However he did in fact stand, obtaining 58.4 per cent of the vote, against the 22

per cent obtained by Sherif Mustafa Dibba of the NCP.
[24] In 1993 both agriculture and tourism were hit by ripples from the European financial crisis. Trade with Senegal was harmed by a decision from the Central Bank of the Western African States to discontinue financing of business deals based upon the African franc (CFA) concluded outside the area of the member countries of this monetary system.
[25] The government took steps to initiate a stage of national reconciliation. The death penalty was abolished and an amnesty was granted to guerrilla movements which sought to overthrow the regime.
[26] In July 1994, a military coup ousted President Dawda Jawara who requested asylum in Senegal after his stay on a United States

warship which was visiting the country. The ship's presence in Banjul suggested US complicity with the military. The coup, led by Lieutenant Yahya Jammeh, was triggered by protests from soldiers demanding payment for their peace services carried out in Liberia. There were no deaths.
[27] Two members from the Armed Forces' Provisional Executive Council - Vice-president Sana Sabally and Interior minister Sadibu Hydara - were arrested in January 1995 for trying to return the government to civilian leadership. In March, Jammeh also arrested the former Justice minister and Public Prosecutor for promoting the return of civilians to power. In November the military Council extended the authority of the security forces to arrest opposition suspects for up to three months.

Georgia

Sakartvelo

Population: 5,418,000 (1994)
Area: 69,700 SQ KM
Capital: Tbilisi

Currency: Kupon
Language: Georgian

The peoples of the Caucasian Isthmus, considered to be the creators of metallurgy, entered the Bronze Age around 2000 BC. The ancestors of modern Georgians are thought to have formed the first tribes in that region about this time. The kingdom of Kolkhida - which dates back to the 6th century BC - is mentioned in Homer's poems and in traditional Greek mythology.

[2] Two centuries later, a legendary chieftain, Farnavaz, created the kingdom of Iberia, in what is now eastern Georgia. These two kingdoms were the first Georgian States, the result of a fusion of ancient agricultural and iron-smelting tribes of the region.

[3] Kolkhida and Iberia were subdued by Greece and then by Rome, as the result of Pompei's campaigns, in the 1st century BC. In the year 337 AD, King Mirian of Iberia adopted Christianity, making it the country's official religion. Like the Armenian church, the Georgian church separated from Rome in 506, to form the national Church of St. George, with its headquarters at Tiflis (today Tbilisi).

[4] From a minor city, Tiflis became first the capital of Iberia, and later of all Georgia, after the unification of the 8th and 9th centuries. Both kingdoms fell under Iranian (Sasanid), then Byzantine and finally Arab control, between the 6th and 10th centuries. Feudal Georgia prospered under King David "the Builder", 1089-1125 and Queen Tamara, 1184-1213.

[5] In the 13th and 14th centuries Georgia was invaded by the Tatars and Tamerlane. In the 15th century it dissolved into small fiefdoms, becoming the object of territorial disputes between Turkey and Iran from the 16th to the 18th centuries and there were a number of anti-Turkish and anti-Iranian revolts.

LAND USE

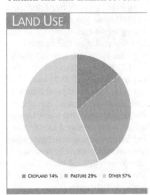

CROPLAND 14%   PASTURE 29%   OTHER 57%

[6] Through the Treaty of Georgievsk (1783) Russia established its protection over Eastern Georgia, annexing it in 1801. In the second half of the 19th century, Western Georgia (Tiblisi and Kutaisi) met the same fate. The Georgian provinces were incorporated into Transcaucasia, where the government was in the hands of a viceroy designated by the czar.

[7] Annexation to Russia had a number of negative effects on Georgian life. The native language was eliminated from administrative documents, and it was replaced by Russian in literature and in schools. The Catholic Church of St George was outlawed, its patriarchs deported, being replaced by Russian Orthodox bishops and Russians were encouraged to come and settle in all of Georgia's cities, a fact which explains why the revolutionary movement was so strong in Georgia, in all of its Marxist, nationalist, populist and social democratic forms. It was here that Josif Dzhugashvili, henceforth "Stalin", began his political career. After frequent rebellions during this period, Georgians played an important part in the revolution which shook the empire in 1905.

[8] When World War I broke out, the Caucasus saw fighting between Russia and Turkey. The Georgians formed a legion to fight alongside the Turks. In February 1917, a new wave of uprisings was unleashed by war policies and by hunger, and the peoples of the Caucasus brought down the czarist regime.

[9] In November, after the Bolshevik triumph in Petrograd, power in Transcaucasia fell into the hands of the Mensheviks (who wanted gradual change rather than revolution). Confronted with this situation, Stalin's group allowed the Caucasus to join the Bolsheviks led by Lenin. In April 1918, in Tbilisi, the United Government of Transcaucasia announced its separation from Soviet Russia. On May 26, 1918, Georgia proclaimed independence, which was recognized by Moscow two years later.

[10] Between 1918 and 1920, German, Turkish and British troops entered Georgia, all intent upon bringing down the socialist regime. In February 1921, the Red Army occupied Georgia and established Soviet power. On February 25, 1921, the Soviet Socialist Republic of Georgia was proclaimed, with the Abkhaz Autonomous Region as an administrative division of this republic. In March 1921, a peace treaty between the Russian Federation and Turkey was finalized, with the latter ceding Batumi and the northern part of Adzharistan, which became an autonomous republic within the Georgian SSR.

[11] In March 1922, Georgia, Azerbaijan and Armenia were reorganized as the Transcaucasian Federation. The following month, the South Ossetian Autonomous Region was formed, as a part of Georgia. On December 5, 1936, the federation was dissolved and Georgia became one of the 15 republics of the Soviet Union (USSR).

[12] Stalin became secretary general of the Communist Party of the Soviet Union in 1922, and after Lenin's death he became party leader. From 1928, he was leader of the USSR. Georgia witnessed many purges and suffered the effects of centralization and the repressiveness of Stalinism, although to a lesser degree than other republics. In 1924, the Red Army intervened to put down a peasant uprising against forced land collectivization.

[13] In the following years the country benefitted from the cultural and economic development promoted by the Soviet Union. Many Georgians - apart from Stalin - were in the leadership circle of the Soviet Union's CP, including Sergo Ordzhonikidzegv, the USSR's Minister of Heavy Industry, and Lavrenti Beria, head of the secret police (NKVD) who was executed after Stalin's death, in 1953. When Krushchev denounced his predecessor's crimes at the 20th Party Congress in 1956, it hurt Georgian national pride.

[14] In 1953, Moscow made Eduard Shevardnadze chief of police in Tbilisi; a Georgian, he had been head of the Communist Youth (Konsomol) during Vasili Mzhavanadze's period as first secretary of the Georgian Communist Party. Implicated in cases of corruption and fraud, he was replaced by Shevardnadze in 1972. A new nationalist feeling began to emerge, declarations in defense of the Georgian language and violent acts of sabotage were made.

[15] In 1978, the new Constitution of the USSR triggered a wave of protest, as it made Russian the official language of the Soviet Union. However, Shevardnadze managed to have the measure abolished and allowed an anti-Stalin movie, "Repent", to be screened. This marked the development of glasnost and perestroika process in Georgia.

[16] Shevardnadze remained at the head of Georgia's Communist Party and government until 1985, when he was made Foreign Minister of the Soviet Union. After the political changes made by Mikhail Gorbachev, several movements and parties emerged in Georgia, clamoring for the independence which Georgia claimed to have lost in 1921, with the establishment of Soviet power. On April 9, 1989, in Tbilisi, Red Army units dispersed a crowd meeting in the main square to demand the secession of Georgia, citing Article 72 of the Constitution of the USSR. Nineteen people were killed, most of them women or adolescents.

[17] In the mid-80s, a Green Party emerged in Georgia, amid the first ever open debates on the country's ecological problems. Despite the fact that it is a traditional summer

ENVIRONMENT:

Located in the central western part of Transcaucasia, Georgia is bounded in the north by Russia, in the east by Azerbaijan and in the south by Armenia and Turkey, with the Black Sea to the west. Most of its territory is occupied by mountains. Between the Little and the Great Caucasus lies the Kolkhida lowland and the Kartalinian Plain; the Alazan Valley lies to the east. Subtropical climate in the west, moderate in the east. Principal rivers: Kura, Rioni. Heavy precipitation in the western area, along the shores of the Black Sea. 40% of the republic is covered by forests. Georgia is well-known for its wine. Bacterial pollution of 70% of the Black Sea constitutes a serious problem. Only 18% of the residual waters in the main port of Batumi undergo adequate treatment. The abuse of pesticides has resulted a high level of chemical toxicity in the soil.

SOCIETY

Peoples: Georgians, 70%; Armenians, 8.1%; Russians, 6.3%; Azeris, 5.7%; Ossetes, Abkhazians and Adzharians. **Religions:** Georgian Orthodox (65%), Muslims (11%), Russian Orthodox (10%), and Armenian Orthodox (8%). **Languages:** Georgian (official), Russian and Abkhazian. **Political Parties:** Union of Citizens of Georgia, Georgian Communist Party, National Democratic Party, Aidguilara (national movement of the Abkhazians), Admon Nyjas (national movement of the South Ossetians).

THE STATE

Capital: Tbilisi, 1,270,000 inhab. (1993). **Other cities:** Batumi, 137,000 inhab.; Sukhumi, 112,000 inhab., Kutaisi, 240,000 inhab.; Rustavi, 158,000 inhab. (1993) **Official Name:** Sakartvelos Respublika. **Government:** President Eduard Shevardnadze; prime minister Niko Lekishvili. **National Holiday:** May 26, Independence (1918). **Armed Forces:** 4,500 **Foreign Forces:** Russia, 22,000.

Foreign Trade

MILLIONS $ 1994

IMPORTS
744

EXPORTS
381

Maternal Mortality

1989-95

PER 100,000
LIVE BIRTHS
55

DEMOGRAPHY

Urban: 58% (1995). **Annual growth:** 0.4% (1991-99). **Estimate for year 2000:** 5,000,000. **Children per woman:** 2.2 (1992).

COMMUNICATIONS

10.5 **telephones** per 100 inhabitants (1993).

ECONOMY

Annual inflation: 228.30% (1984-94). **Currency:** kupon. **Cereal imports:** 500,000 metric tons (1993). **Fertilizer use:** 680 kgs per ha. (1992-93). **Imports:** $744 million (1994). **Exports:** $381 million (1994). **External debt:** $1,227 million (1994), $226 per capita (1994). **Debt service:** 1.2% of exports (1994).

ENERGY

Consumption: 572 kgs of Oil Equivalent per capita yearly (1994), 81% imported (1994).

---

resort, swimming in the Black Sea has been banned due to the high level of chemical and biological pollution. The country also suffers serious erosion and deforestation. The "Greens" have combined their concern for the environment with a defense of non-violence, democracy and human rights. The Georgian Green Party has tried to coordinate action with similar organizations in other countries on the Black Sea. Georgia is the second former Soviet republic - after Estonia - to join the European "Green" Parliament.
[18] On October 28, 1990, elections were held for the Soviet (Parliament) of the Georgian SSR, with the "Free Georgia Round Table" coalition victorious. Zviad Gamsakhurdia, a well-known political dissident who had been proscribed during the Soviet regime, became the leader of the coalition. The Georgian Soviet decided to change the country's name to "the Republic of Georgia".
[19] Ethnic tensions increased, and the South Ossetian Autonomous Region claimed republican status. Georgia's Soviet annulled this decision, declared a state of emergency in South Ossetia and organized a blockade of this region. In April 1991, after fighting broke out be-

tween Ossetian guerrillas and Georgian troops, the Soviet approved the region's independence.
[20] On May 27, Zviad Gamsakhurdia was elected president. In September 1991, a struggle began between President Gamsakhurdia's supporters and the opposition, led by Dzhaba Ioseliani and Tenguiz Kitovani, commander of the National Guard. Repeated protest meetings and demonstrations ended in violence on January 6, 1992, when government forces (following Kitovani's orders) stormed government headquarters. Gamsakhurdia fled first to Armenia, and then to the Chechen Autonomous Republic, whose president, General Dzhokhar Dudaev, was seeking secession from the Russian Federation.
[21] A Military Council led by Kitovani seized power in Tbilisi and suspended the constitution. Gamsakhurdia loyalists continued to fight against the new regime's forces in western Georgia. In late January 1992, attempts to put an end to the civil war failed. In March, former Foreign Minister, Eduard Shevardnadze, returned to Georgia to take charge of the presidency of the Council of State, the most powerful governing body. Chosen by the

Military Council, he will hold this post until democratic elections are held. Shevardnadze declared that his country will normalize relations with the international community, giving preference to relations with the former USSR republics, though for the moment not joining the Commonwealth of Independent States (CIS).
[22] On June 28 of the same year, a cease-fire was called in South Ossetia, supervised by a peace-keeping force of Russians, Georgians and Ossetians. From their capital in Sukhumi, the Abkhazian authorities resolved to limit the jurisdiction of Georgia's central government within the autonomous region. On August 14, government troops entered Abkhazia and occupied Sukhumi. Local authorities fled to the city of Gudauta, which became a center of resistance.
[23] In early October, with the support of Russian volunteers and Army regulars, they managed to extend their control over a large part of Abkhazia, coming close to Sukhumi.
[24] In early 1993, the Russian Air Force began regular bombing of Sukhumi, and by mid-March fighting had reached the outskirts of the city. On June 28, a Russian mediated armistice went into effect, with both parties agreeing to demobilize their troops.
[25] In mid-September 1993, a strong Abkhazian offensive began, taking the Georgians by surprise and culminating in the capture of

Sukhumi by the Abkhazian army. Eduard Shevardnadze remained in Sukhumi until the last moment, leaving on the last flight out of the city.
[26] In November, supporters of former President Gamsajurdia launched a broad offensive, but were defeated with the help of Russian troops. In the same month, Georgia entered the CIS. In early 1994, Gamsajurdia died in a suicide attempt, according to official reports. In February, Georgia signed a friendship treaty with Russia, and in April a peace treaty was signed with the Abkhazian rebels in Moscow.
[27] Despite the rapprochement with Russia, Tbilisi entered the Partnership for Peace program promoted by NATO. The economic situation continued to worsen: in the first 10 months of the year, industrial production fell by 42 per cent. The inability to repay a debt to the Turkmenistan government led to a reduction in gas sales -the main Georgian source of energy. In late November, Tbilisi had neither heating, electricity nor running water.
[28] In 1995, the economic situation improved. In the first 11 months of the year, industrial production increased 20 per cent. In August, the Parliament approved a new Constitution, which defined Georgia as a presidential republic. In November, Shevardnadze won the presidential elections, while his party, the Citizens' Alliance, won 124 of the 235 seats in Parliament.

Germany

Population: 81,538,604
(1994)
Area: 356,978 SQ KM
Capital: Berlin

Currency: Mark
Language: German

Deutschland

The first reference to the Germanic tribes comes from between 58 and 51 BC when the Romans, under Julius Caesar, invaded a part of Gaul north of the Alps and west of the Rhine. The Germans descended from the Teutoni and the Cimbri, who had in turn descended from the Danes. Between 113 and 101 BC. Teutoni and Cimbri tribes invaded the Mediterranean regions including the Italian peninsula

[2] In 9 AD Arminius, a Germanic chief and the earliest recognised national hero, led a revolt which defeated three Roman legions in the Teutoburger Wald. By the second half of the 3rd century, the Germanic tribes had fused into larger political units; the Saxons, Franks and Germans.

[3] The Romans found it increasingly difficult to keep their dominion over the region and withdrew gradually. In the 4th century Attila's Huns invaded, and a long period of Migrations followed throughout Europe. In 455, the Germans defeated the Mongols at the battle of the river Nedao and the Mongol Empire subsequently collapsed. Clovis and Charlemagne later subjected the Germanic and Saxon tribes to the Frankish Empire.

[4] After the Treaty of Verdun and the division of the Carolingian empire in 843, the first entirely Germanic kingdom was established under Louis the German. Under Otto I, crowned in Rome in 936, Germania became Europe's most powerful kingdom.

[5] During the 12th and 13th centuries Germany experienced a rapid growth in population, and continuous expansion.

[6] The instability of royal dynasties encouraged the strengthening of secular and religious principalities. Princes were free to build fortresses, exploit natural resources and administer justice in their dominions.

[7] In 1356, the authority of the king in relation to the papacy was consolidated in the Golden Bull of Charles IV (1346-1378) establishing the right to designate the king without the Pope's approval, and strengthened the hand of the principalities, on whose support the king depended.

[8] During the 15th and 16th centuries internal instability persisted. In 1517 Martin Luther led the Protestant Reformation, which coupled with political rivalries frustrated reunification efforts.

[9] The Reformation provided a forum for criticisms of secularisation and corruption in the German Church, which had become an increasingly prosperous economic and financial institution; it owned a third of the land in some districts and profited from selling indulgences. It was the scandal over indulgences that finally led to its downfall. After a succession of internal wars, including peasant uprisings, the Peace of Augsburg was reached in 1555, giving equal rights to both Catholics and Lutherans.

[10] Political and religious factions combined their strengths; in 1608 the Protestant Union was created, and the Catholic League a year later. The Bohemian rebellion triggered the Thirty Year War (1618-1648), which spread to the entire continent, and ultimately reduced the population of central Europe by around 30 per cent, ending with the Peace of Westfalia (1648).

[11] In the 18th century, the Kingdom of Prussia emerged as a dynamic economic and political unit, responsible for creating tensions among the German states. Napoleon's victories over Prussia in 1806 and the formation of the Confederation of the Rhine put an end to the Holy Roman Empire.

[12] In Central Europe during the 18th century, culture formed an outlet for intellectual energies which could not be expressed through politics, as political life itself was dominated by the autocratic rulers. Philosophers like Kant and Herder, and writers like Goethe and Schiller, expressed the idealism and spiritualism which characterized German art and literature of this time.

[13] When Napoleon fell, the German princes created a confederation of 39 states, which were independent except for external policy. Austrian and Prussian opposition to broader forms of representation increased popular unrest, leading to the 1830 revolts and increased repression.

[14] In 1834, Prussia threw its growing economic weight into the political realm, with the establishment of the German Customs Union, from which Austria was excluded. The main consequences of the Union were the duplication of trade among its members over the next ten years, as well as the formation of industrial centres and the emergence of a working class. Due to the rapid growth of the urban population, the supply of manpower greatly exceeded the demand. The resulting impoverishment of industrial workers and artisans served as a breeding ground for the rebellions of subsequent years, culminating in the revolutionary wave of 1848-49.

[15] A National Assembly first met in Frankfurt on May 18, 1848; its representatives belonged mostly to liberal democratic sectors. They campaigned for German unity, and guarantees of political freedom. However, internal divisions facilitated the recomposition of the forces of the former regime, leading finally to the dissolution of Parliament in June 1849 and the repression of opposition organizations.

[16] With the revolutionary tendencies crushed, Austria and Prussia were free to dispute their respective roles in German unification. The issue was settled in 1866 with Prussia's victory in the Seven Weeks' War. The union was forged around the North German Confederation, a creation of the Prussian chancellor, Otto von Bismarck, that aimed to halt liberalism. The Parliament (Reichstag) was inaugurated in February 1867.

[17] Three years later, war broke out with France. Prussia's victory in 1871 was the last step in Bismarck's scheme to unify Germany under a single monarch, and Prussian domination.

[18] The empire had to deal with opposing internal forces, namely, the Church and Social Democracy. Bismarck passed the May Laws, secularising education and

LAND USE

CROPLAND 34%   PASTURE 15%   OTHER 51%

PROFILE

ENVIRONMENT

The northern part of the country is a vast plain. The Baltic coast is jagged, with deep, narrow gulfs. The center of the country is made up of very old mountain ranges, plateaus and sedimentary river basins. Of the ancient massifs, the most important are the Black Forest region and the Rhineland. The southern region begins in the Danube Valley, and is made up of plateaus (the Bavarian Plateau), bordered to the south by the Bavarian Alps. There are large deposits of coal and lignite along the banks of the Ruhr and Ens rivers, which provided the backbone of Germany's industrial development. Heavy industry is concentrated in the Ruhr Valley, mid-Rhineland and Lower Saxony. The south of the former German Democratic Republic is rich in coal, lignite, lead, tin, silver and uranium deposits. The chemical, electrochemical, metallurgical and steel industries are concentrated there. This region suffered severe air pollution as a result of the carbon output of the industries. The emission of sulphur dioxide in eastern Germany was 15 times that in the west, compounding the problem of acid rain. Untreated industrial effluents carrying heavy metals and toxic chemicals have contaminated many east German rivers, ending up in the highly polluted Baltic Sea.

SOCIETY

Peoples: German 91.2%; Turkish 2.5%; former Yugoslav 1%; Italian 0.7%; Greek 0.4%; Bosnian 0.4%; Polish 0.3%; Croation 0.2%; others 3.3% (1995).
Religions: Christian; a Catholic majority in the South and Protestant in the North. There are approximately 28 million Protestants and the same number of Catholics. Before re-unification, Catholics constituted a majority in West Germany. **Languages:** German (official) and local dialects which, in spite of restrictions, are regaining popularity. Turkish, Kurdish. **Political Parties:** Christian Democratic Union (CDU); Christian Social Union (CSU), Liberal Party (FDP), Social Democratic Party (SPD); the Greens/Alliance-90; Democratic Socialist Party (PDS) , the Republicans (REP). **Social Organizations:** Workers' Federation (DGB).

THE STATE

Official Name: Bundesrepublik Deutschland.
Administrative divisions: Federal parliamentary State made up of 16 *Länder* (federated states), as of October 3, 1990. Eleven *Länder* made up what was formerly West Germany (Schleswig-Holstein, Hamburg, Bremen, Niedersachsen, Nordrhein-Westfalen, Hessen, Rheinland-Pfalz, Saarland, Baden-Wättemberg, Bayern and Berlin) while the former German Democratic Republic was divided into five *Länder* (Mecklemburg, Brandemburg, Sachsen-Anhalt, Sachsen and Thuringen). **Capital:** Berlin, 3,471,418 inhab. (1995). **Other cities:** Hamburg, 1,707,901 inhab.; München (Munich), 1,244,676 inhab.; Köln (Cologne), 963,817 inhab.; Leipzig, 481,112 inhab.; Dresden, 500,000 inhab.; Frankfurt/Main, 652,412 inhab.; (1995).
Government: Roman Herzog is president; Helmut Kohl is chancellor, appointed by the federal parliament.
Armed Forces: 338,000 **Paramilitaries:** Federal Border Guard (Ministry of Interior): 28,000.

some other activities. Bismarck was later to reverse himself on this, securing the Church as an ally against socialism. Alarmed at the growth of Social Democracy, the regime used repression and social reform to neutralize the latter's potential.

[19] Bismarck's government introduced commercial protectionism to increase domestic income and foster national industry and the German economy grew substantially, especially in heavy industry and production. The creation of the Triple Alliance with Austria and Italy, and the acquisition of colonies in Africa and Asia after 1884, placed the German Empire as a leading world power.

[20] However, Germany's imperial venture was not highly successful, either in its African possessions (Tanganyika, Cameroun, Togo and Southwestern Africa) or those in Asia (New Guinea, the Carolina, Mariana, Marshall and Samoan Islands). At the end of World War I, these were all transferred to the victorious powers.

[21] German rivalry with France and Britain in the west, and Russia and Serbia in the east, triggered the First World War. The capitulation of the Austro-Hungarian Empire and Turkey in November 1918, led to Germany's final defeat. The crisis was aggravated by an internal revolution which led to the abdication of the Emperor. Government was handed over to the socialist Friedrich Ebert, who was to call a National Constituent Assembly.

[22] German Social Democracy split into a moderate tendency favouring a gradual evolution to socialism, and a radical tendency favouring revolutionary change. The radical group, "the Spartacists" , headed by Karl Liebknecht and Rosa Luxemburg, identified with the Russian revolution of October 1917 and wanted to set up a system similar to that of the Soviets.

[23] In December 1918, the government called constituent elections for a month later. The army was called in to put down a radical socialist coup, and in January 1919 the Spartacist leaders were arrested and executed. A few days later, the electorate returned a moderate socialist majority to power.

[24] In February 1919, in Weimar, the Assembly appointed Ebert as the first president of the Republic. In June of that year, after some hesitation, the government of socialist

Gustav Bauer signed the Treaty of Versailles.

[25] In 1921, the victorious powers fixed German reparations at 132,000 million marks, an amount that the country was in no position to pay.

[26] Germany was declared bankrupt by the Commission of Reparations of the victorious powers, and in January 1923, Belgian and French troops occupied the industrial Ruhr region. Economic and financial instability caused hyperinflation, affecting most of the *petite bourgeoisie*. A fiscal reform in late 1924, managed to stabilize the country's finances.

[27] When Ebert died in 1925, Marshall Paul von Hindenburg was elected President. His main concern was to re-establish Germany's position among the great powers. He supported the ancient landed oligarchy's desire to return to power.

[28] The German economy expanded until the Wall Street crash of 1929, when a lack of foreign credit threw it into recession. In 1930 Hindenburg dissolved Parliament, and that year the vote for both Communists and National Socialists (Nazis) rose sharply. The Nazi Party, led by Adolf Hitler, became the second largest political force behind the Social Democrats.

[29] The middle classes and the impoverished masses became increasingly attracted to the Nazi promise of rebuilding the Great Germany. Humiliated by the postwar treaties they seized on the campaign that blamed Jews and Communists for the economic crisis. The Communists and Socialists held a majority in Parliament, but due to differences they did not join forces against Hitler. By the time they realized the danger he embodied it was too late to stop him.

[30] In the 1932 elections, amidst violent clashes with the Communists, the Nazi Party doubled its support, obtaining 37 per cent of the total vote. The president, who had banned the Nazis from making public demonstrations, offered Hitler participation in a coalition government led by the right-wing Franz von Papen. He did not yet accept Hitler's demand to be head of government.

[31] Between 1932 and 1933, in international negotiations, Germany attempted to make up for war losses and recover its right to rearm. On the domestic scene,

one crisis followed another and the number of unemployed exceeded 6 million.

[32] In January 1933, at the urging of representatives of the upper bourgeoisie, the president turned over the government to Hitler, who dissolved Parliament in February. He also called elections, in which the NSDAP obtained an absolute majority.

[33] In Potsdam in March 1933, the new Parliament granted Hitler the power to issue decrees outside the Constitution without the approval of the Legislative body or the President. For four years Hitler had the power to fix the annual budget, request loans and sign agreements with other countries, re-organize both his cabinet and the supreme ranks of the armed forces, and proclaim martial law.

[34] In July 1933 Hitler abolished the German Federation and set up absolute central power. He outlawed all parties except his own, as well as trade unions and strikes. Germany also withdrew from the Disarmament Conference and the League of Nations.

[35] The Hitler regime installed its first concentration camps to confine thousands of political opponents, gypsies (Sinti and Roma ethnic groups) and homosexuals. In addition, it launched a scheme to eliminate the handicapped. Protests from the churches, whose support was still important to the regime, put a stop to these policies, at least for the time being.

[36] After the death of President Hindenburg in August 1934, the cabinet was forced to swear personal allegiance to the chancellor, Adolf Hitler. In 1935, in violation of the Treaty of Versailles, he began to re-arm Germany. The European powers protested, but were unable to stop him.

[37] With the 1935 "Nuremberg Laws", the regime provided a legal framework for its racist ideology, laying the foundation for its subsequent policies of ethnic and religious extermination.

[38] In October 1936 Germany and Italy signed a co-operation agreement, including support of General Franco in the Spanish Civil War. The following month Germany and Japan (the Axis powers) agreed to set up a military exchange, and in November 1937 Germany, Italy and Japan signed the Anti-communist Pact in Rome.

[39] In March 1938, German troops invaded Austria, and Hitler proclaimed its annexation. That same year, the pressure of Hitler and German nationalism on the Sudetenland forced the European powers to cede this Czechoslovakian region to Germany. On "Kristallnacht", November 9-10, the government carried out a systematic destruction of Jewish commercial property and religious and cultural institutions.

[40] In 1939, taking advantage of the disagreements between Czechs and Slovaks, German troops advanced into Prague; Bohemia, Moravia, and Slovakia became protectorates.

[41] Britain assured Poland, Rumania, Greece and Turkey that it would protect their independence, and Britain and France attempted to establish an alliance with the Soviet Union. In August 1939, however, the USSR signed a nonaggression treaty with Germany. On September 1, German troops invaded Poland. Britain and France issued Germany an ultimatum and World War II broke out.

[42] By 1940, Germany had invaded Norway, Denmark, Belgium, the Netherlands, Luxembourg, and France. In 1941 Hitler started his offensive against the USSR, but German troops were halted a few miles outside Moscow. They were finally defeated after the siege of Stalingrad (Volgograd) in 1943

[43] From the outset, German aggression against its neighbours was accompanied by the systematic extermination of the Jewish population in concentration camps, primarily in Poland. Over six million Jews, and a million other people were killed.

[44] The advance of the Red Army on the eastern front-culminating in the capture of Berlin-and the 1944 Allied landing in Normandy on the western front forced Germany to surrender in May 1945.

[45] Four million Germans from neighbouring countries and from the territories annexed by Poland and the USSR, were forced to move to the four zones Germany was divided into, while it re-

mained occupied by the United States, France, Britain and the USSR. In 1949, discord between the ex-allies over the future governing of Germany led to the creation of the Federal German Republic, in the West, and the German Democratic Republic, in the East. The issue of the two Germanies became a bone of contention in postwar relations between the USSR and the United States.

[46] In 1955, the sovereignty of both Germanies was recognized by their occupying forces. During the Cold War, West Germany became a member of NATO (the North Atlantic Treaty Organization), while East Germany joined the Warsaw Pact. Foreign forces continued to be based in their territories and the two republics were still subject to limitations on their armed forces and a ban on nuclear weapons.

[47] The Unified Socialist Party (SED), formed from the union of Communists and Social Democrats in 1946, took over the government of East Germany. The USSR partly compensated them for war losses with money, equipment and cattle, and a social system similar to that of USSR was set up. In 1953, the political and economic situation of the German Democratic Republic (GDR) led to a series of protests, which were put down by Soviet troops. In the meantime, emigration to the Federal Republic of Germany (FRG) increased.

[48] Between the state's creation and 1961, when the East German government forbade all emigrations to the west, some three million East Germans emigrated to West Germany. To enforce their resolution, the East closed its borders and built the Berlin Wall between the eastern and the western sections of the city. In 1971, Erich Honecker took over the leadership of the SED party, and later the GDR government.

[49] Between 1949 and 1963, Chancellor Konrad Adenauer, a conservative Christian Democrat, oversaw the reconstruction of the FRG, under the slogan "[establishing] a social market economy". With US support (the Marshall Plan) and huge amounts of foreign capital, the FRG became one of the most developed capitalist economies, playing a key role in the founding of the European Community (EC).

50 With the victory of the Social Democratic Party (SPD) in the 1969 elections, the government of Chancellor Willy Brandt launched a policy of rapprochement toward Eastern Europe and the German Democratic Republic. In 1970 the first formal talks between the FRG and the GDR began, and in 1971, the occupying powers agreed to free access of FRG citizens, to the GDR. A Basic Treaty of bilateral relations was signed by both Germanies in 1973; in September they were admitted to the United Nations.

51 In the 1970s, the number of power stations producing nuclear energy increased. In response to this, a strong environmental movement was formed, with a network of hundreds of grassroots groups throughout the country.

52 In 1974, after the discovery that his private secretary was an East German spy, Brandt resigned as head of the Government, being succeeded by Helmut Schmidt. The modernization of Soviet medium-range missiles in the GDR, and the December 1979 NATO decision to do the same with its arsenal in the FRG, gave a strong impetus to the anti-nuclear movement in both States, with massive protests in the FRG.

53 In 1982, the SPD had to leave power when the Liberal Party withdrew from the government alliance after 13 years. It was succeeded by the liberal-conservative alliance (CDU/CSU, FDP) currently in power, led by Chancellor Helmut Kohl (CDU). In the 1983 elections, the environmental party (the "Greens") won parliamentary representation on a national level for the first time. Approximately half of its members of Parliament were women.

54 In mid-1989, Hungary liberalized regulations regarding transit across the border with Austria; within a few weeks, some 350,000 Germans from the GDR had emigrated to the FRG. Street demonstrations demanding change brought on a crisis in the GDR. In August, Honecker resigned and was replaced by Egon Krenz. On November 9, the GDR opened up its borders and the Berlin Wall fell.

55 Kohl immediately proposed the creation of a confederation of East and West Germany. In December 1989, during his first visit to East Germany, Kohl insisted on this proposal and committees were set up to develop bilateral relations. In the meantime, demonstrators began clamouring for reunification of the GDR and the FRG.

56 In February 1990, the East German government agreed to German unification and the withdrawal of all foreign troops from its territory. Kohl supported the plan and proposed starting negotiations after the March elections in the East. After the Christian Democrat victory, promising rapid re-unification, monetary union was achieved in July. The exchange rate was very favourable to East German investors.

57 The fusion of the two states was officially recognised in August 1990, as the Federal Republic of Germany. Political union was made possible when the USSR accepted the East German entry into NATO. An important difference between the two Germanies is the Eastern preservation of a much more liberal abortion law.

58 Industrial plants in the East, working with obsolete technology and generating serious pollution, were closed down by the end of 1990, triggering several strikes. At the time of re-unification, 50 per cent of women in the West worked outside the home, as compared to 83 per cent in the East.

59 In the first parliamentary elections of the new Federal Republic, in December 1990, the governing coalition of Social Christians and Liberals obtained 54 per cent of the vote, Social Democrats had 35 per cent, the Greens 5 per cent, while the Democratic Socialists, descendants of the old Democratic Republic Socialist Party, also obtained 5 per cent. Women make up 20.54 per cent of members of parliament, 4.5 per cent more than in the former FRG and 12.5 per cent less than in the former GDR.

60 In 1991, the extreme right made important gains at a local level. In Bremen, they increased their share of the vote by 7 per cent. The elections took place against a background of racist demonstrations and increasing xenophobia. The victims of this violence were mainly citizens of Third World countries, but some were immigrants from the old European socialist bloc. These eastern European people had been coming to Germany in large numbers since 1989, with over 100,000 arriving in 1991 alone.

61 During 1992, there were 2,280 attacks on foreigners and Jewish monuments, which left 17 people dead. After one incident which caused the death of a Turkish woman and two children, the government outlawed three neonazi organisations.

62 Massive demonstrations against racist violence and anti-semitism were held all over the country. The deliberate burning of a house in the western city of Solingen, which killed two Turkish women and three small girls once again demonstrated the strength of the ultra-right in Germany.

63 In May 1993, the Supreme Court annulled the law to liberalize abortion, one of the few dispositions inherited from the GDR. The closing of most of the industry in this nation, the economic recession - the biggest since 1945 - and the increased productivity caused a constant increase in unemployment. In March, 1994, more than four million people were out of work.

64 This same month, an attack on the synagogue in the city of Lübeck showed that antisemitic violence was far from extinct. On May 23, the Conservative Roman Herzog, with the support of Kohl, was designated president of Germany by a special electoral assembly, after defeating the Social Democrat Johannes Rau.

65 In the second part of the year a cycle of economic expansion began, allowing for growth of 2.8 per cent in 1994. By the end of the year, unemployment had already fallen to 3.5 million .

66 Kohl triumphed again in the October general elections, though his parliamentary majority was reduced to 10 seats out of the 672 in play (under the previous legislature, the difference between the government and opposition was 134 seats). The conservative alliance which supported him, the CDU-CSU, won 41.5 per cent of the vote, with the additional 6.9 per cent of their liberal FDP allies. Meanwhile, the social democratic SPD won 36.4 per cent, the ecologists 7.3 and the East German former Communists PDS 4.4 per cent, due to good support in several regions of the former German Democratic Republic.

67 In 1995, the constant weakening of the Liberal Party in various local elections prompted the resignation of foreign minister Klaus Kinkel as chair of the Liberal Party. On the social front, parliament adopted a new law which, despite re-authorizing abortion during the first 12 weeks of pregnancy, was, in fact, more restrictive than that annulled in 1993, as it obliged women to be "assessed" before intervention occurs.

68 The economic expansion continued, with growth evaluated at 1.9 per cent in 1995, which led to the reappearance of pay disputes from the unions. On June 16, 1996, a gigantic demonstration was held in Bonn, the former capital of East Germany, to oppose Kohl's economic "rationalization" plans, calling for a larger share of the "fruits of growth" to reach the salaried workers.

70 On the international front, the rapprochement between Berlin and Moscow was disturbed by the war in Chechenia, even though it appeared to be one of the most stable points of German foreign policy. Meanwhile, the union movement questioned the de facto imposition of the German monetary policy -increasingly criticized within its own frontiers on the European Union allies.

Ghana

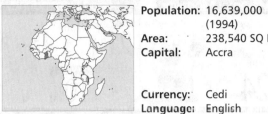

Population: 16,639,000 (1994)
Area: 238,540 SQ KM
Capital: Accra

Currency: Cedi
Language: English

Ghana

n about 1300, the Akans or Ashantis moved into Ghana from the north. The powerful and organized Fanti State of Denkyira was already established on the coast so the Akans settled in the inland jungles. Here, they founded a series of small cities that paid tribute to the coastal nation.

2   In the 15th century the Ashantis began to trade in the markets of Begho, on the border of present-day Côte d'Ivoire. They traded slaves and gold for fabrics and other goods from North Africa and further afield. An ornate pitcher that had once belonged to Richard II, king of England between 1367 and 1400, was discovered in the treasure of a Kumasi ruler.

3   In the 17th century, new waves of migration threatened the existence of the small jungle states. This crisis forced the Akans to unite in confronting the Doma invaders. The Doma were defeated.

4   With the decline of the Songhai, the Moroccan incursions and the beginning of the Fulah expansion, the North African trade network collapsed with serious repercussions for the inland economy. The Ashanti were also deprived of access to the shoreline trading points as these were in Denkyira Territory. Instead of paying taxes, they declared war on the Denkyira and defeated them. They then organized a centralized state, ruled by the "Ashantihene" (leader of the Ashanti nation) endowed with a powerful army. By 1700, the Ashanti had control of the slave trade to the coast and the flow of European goods to the interior.

5   When the British stopped trading in slaves, the Ashanti reacted by attempting to seize the shoreline from the Fanti, who still retained a significant share of the coastal trade. English backing of the Fanti led to the Anglo-Ashanti wars, 1806-1816, 1825-28 and 1874. After the last war, the British turned the Fanti territory into a colony. In 1895 under the pretext of defending the region from Samori Turé (see Guinea) they also proclaimed a protectorate over the northern territories.

6   While the coastal and northern regions were under British rule, the central region belonged to the autonomous Ashanti nation. It did not take long for friction to build up and another Anglo-Ashanti war broke out in 1896. The capital, Kumasi, was razed by cannon fire, and the ruler was deposed and exiled. The Ashanti were then told that they owed 50,000 ounces of gold in compensation for "war damages". An attempt to collect the debt coupled with the British governor's culturally insensitive desire to sit on the gold throne, led to a general rebellion four years later. This was immediately put down with a great loss of lives. In 1902 the Ashanti state was formally annexed to the Gold Coast Colony.

> In the 17th century, new waves of migration threatened the existence of the small jungle states.

7   During the first half of the 20th century, a strong nationalist current developed in spite of the ethnic and religious differences within Ghana. Economic opposition to the British also grew at this time both in the north, where the traditional structures had remained intact, and in the south, where a westernized middle class and a relatively significant working class had developed.

8   Popular pressure on the colonial administration led to political concessions. In 1946, London admitted a few Africans into the colonial administration and in 1949 Kwame Nkrumah formed the Convention People's Party (CCP) to campain for greater reform.

9   Nkrumah is known as one of the founding fathers of Pan-Africanism and African nationalism. He established a solid rural and urban party structure and, in 1952, he became prime minister of the colony. In his inaugural speech, he proclaimed himself a "socialist, Marxist and Christian", promising to fight against imperialism.

10   Nkrumah represented Ghana at the Bandung Conference in 1955. This conference marked the birth of the Movement of Non-aligned Parties, and together with Tito, Nasser, Nehru and Sukarno, Nkrumah played an important part (See Non-Aligned Countries). In 1957 he achieved a major victory and Ghana became the first country in West Africa to gain independence. Known as Osagyefo (redeemer), Nkrumah enthusiastically embraced the Pan-African anti-colonialism. He initiated a series of internal changes based on industrialization, agrarian reform and socialist education.

11   These new measures, however, did not suit everybody. Traditional and neo-colonial interests con-

---

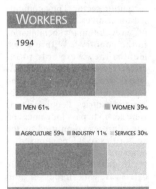

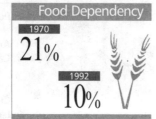
PROFILE

ENVIRONMENT

The southern region is covered with dense rainforest, partially cleared to plant cocoa, coffee, banana and oil palm trees. Wide savannahs extend to the north. The rest of the territory is low-lying, with a few high points near the border with Togo. The Volta, Ghana's main river, has an artificial lake formed by the Akossombo dam. The climate is tropical, with summer rains. The subsoil is rich in gold, diamonds, manganese and bauxite. Desertification in the northwest and deforestation are the main environmental problems.

SOCIETY

Peoples: Ghanaians come from six main ethnic groups: the Akan (Ashanti and Fanti), 44%, located in the mid southern part of the country; the Ewe, 13%, and Ga-Adangbe, 8%, on both sides of the Volta in southern and southeastern Ghana; the Mole-Dagbane,16%, in the northern savannahs; the Guan, 4%, and the Gurma, 3%, in the valleys and plateaus of the northeastern territory. **Religions:** 50% Christian, 32% traditional religions, 13% Muslim. **Languages:** English (official); Ga is the main native language, Fanti, Haussa, Fantéewe, Gaadanhe, Akan, Dagbandim and Mampusi are also spoken. **Political Parties:** Progressive Alliance, coalition made up of the National Democratic Conference (NDC); the National Convention Party (NCP); and the Every Ghanaian Living Everywhere (EGLE); People's National Convention; National Independence Party (NIP); New Patriotic Party (NPP); People's Heritage Party (PHP). **Social Organizations:** Ghana Trade Union Congress.

THE STATE

Official Name: Republic of Ghana. **Administrative divisions:** 10 regions, subdivided into 110 districts. **Capital:** Accra, 950,000 inhab. (1988). **Other cities:** Kumasi, 385,000 inhab.; Tamale, 151,000 inhab. (1988). **Government:** Jerry Rawlings, President. Legislature: Parliament (single-chamber), with 200 members elected by direct popular vote for 4-year terms. **National Holiday:** March 6, independence (1957). **Armed Forces:** 7,000 **Paramilitaries:** People's Militia: 5,000.

spired against the government and Nkrumah was overthrown. The coup leaders drew up a parliamentary constitution and in 1969 held elections for a civilian government. The CPP was not allowed to participate. Nkrumah died in exile in Bucharest in 1972.

[12] Also in 1969, Colonel Ignatius Acheampong led a new coup replacing the government of Dr Kofi Busia. Acheampong dropped the ambitious industrialization and development plans, substituting an essentially agrarian policy which favored the owners of large cocoa plantations.

[13] Acheampong survived eight coup attempts in five years, but his economic policy was not as successful. In 1977, Ghana had an inflation rate of 36 per cent, an oppressive foreign debt, a devalued currency and hundreds of imprisoned intellectuals and students, condemned for questioning government policy.

[14] In July 1977, the so-called "revolt of the middle class" occurred. The period of social unrest came to a head in July 1978 and Acheampong resigned. A new military regime was established, lead by General William Frederick Akuffo. The opposition claimed this was simply a continuation of the previous government under a different figurehead. On June 4, 1979, a coup led by lieutenant Jerry Rawlings overthrew Akuffo and called fresh elections. The People's National Party, which included Nkrumah's follow-ers, won a large majority. A transitional government went into power promising an eventual return to a constitutional system.

[15] On October 1, 1979, with the acquiescence of the Revolutionary Council of the Armed Forces, Hilla Limann, a PNP leader, took over the presidency. Limann abandonned many of Nkrumah's nationalistic economic policies, replacing them with International Monetary Fund policies, in an attempt to reduce the fiscal deficit. To attract foreign investors and overcome the sharp fall in cocoa export revenues, the Ghanaian government drastically reduced all imports, including food. As a result, food prices escalated wildly while the purchasing power of workers fell drastically. This inevitably led to a series of strikes in 1980 and 1981.

[16] Support for Rawlings remained high among the poor. He accused the government of being indecisive and preoccupied with foreign capital. Inflation exceeded 140 per cent and the unemployment rate was more than 25 per cent. This created an unstable situation which culminated in a further coup, led by Rawlings, on January 1, 1982.

[17] The first concern of the new group of officers was to launch a campaign against corruption in the public sector. They committed themselves to nothing short of "a revolution for social justice in the country". Within a few months they were able to increase the amount of taxes collected, and greatly reduce cacao smuggling (*kalabule*) to neighboring countries. People's courts were set up to pass judgement on irregularities committed by the authorities of the previous government.

[18] To deal with the economic crisis it was necessary to straighten out economic policy,

> In 1900 there were only 8 urban centers in the whole of the country, but by 1984 the number had risen to 180.

and look for sources of foreign capital. Rawlings had already expressed his willingness to carry out an austerity program dictated by the IMF. He intensified relations with them in 1983, and generous loans were granted to Ghana, on the basis that its prescriptions would be followed religiously and that Ghana would serve as a model for other countries in the region.

[19] To increase national income and bring contraband under control Ghana devalued its currency (the cedi). This went from 2.74 per $1 in 1982, to 183 per $1 in 1988. Taxes were increased, subsidies were removed, the government salaries budget was reduced, financing for inefficient private enterprises was eliminated and the printing of currency was virtually suspended.

[20] However,there were some positive results. Inflation fell from 200 to 25 per cent, the banking system improved and better prices were obtained for cocoa producers.

[21] By applying this economic policy Ghana obtained loans on extremely generous terms; $500 million to be repaid in two instalments, the first in 1994, at 5 per cent annual interest, and the second in 1999, at 0.5 per cent. The major drawback of this scheme is that Ghana's finances are under rigid IMF control. Ghana's foreign debt is approaching $4 billion and the servicing of the debt uses up two-thirds of the country's export earnings.

[22] The social cost of this readjustment program has been high. Consumer prices rose by some 30 per cent between 1983 and 1987, 45,000 public employees have lived under the constant threat of losing their jobs and there has been a general loss of earning power leading to an increase in hunger, infant mortality and illiteracy.

[23] In 1900 there were only 8 urban centers in the whole of the country, but by 1984 the number had risen to 180. In the past few years there has been a migration towards the major cities, resulting in the creation of shantytowns with no drinking water or sanitation. In the poorest quarters of Takoradi there are only 16 public toilets per 3,250 people. If current rates of rural exodus continue, by the year 2020 over half

External Debt
1994

PER CAPITA

$324

Illiteracy
1995

36%

the population of Ghana will be living in urban centers. In order to palliate this situation a Program of Actions to Mitigate the Social Costs of Adjustment (PAMSCAD) was established, and also a program to transfer 12,000 people per year to rural areas was put into practice.

24 The economic upheaval has had environmental costs as well. Tropical forest used to cover 34 per cent of the country's surface. This forest is now only a quarter of its original size; and 42 per cent of the area which is officially considered " forest ", is in fact covered with timber plantations, secondary vegetation, or immature trees. For the people who live in the country, and for most of those who live in the cities, forest plants constitute the basis of their traditional medicine. 75 per cent of the population rely on bush meat for their basic source of protein and forests also provide firewood, a basic household necessity.

25 All of these activities are ignored by official plans, which see the forest merely as a source of timber for export. In Ghana at present some 70,000 people are employed in the timber industry.

26 A National Forestry Administration Program (NFAP) has been established, supported by international bodies like AID and countries such as Canada and the UK. The FAO also supports the plans. These programs blame deforestation, desertification and soil degradation on the poor population and their search for new areas of farmland. Thus, the poor are blamed for environmental degradation, when in fact they are the victims of an economic model which encourages exports at any price.

27 Ghana has considerably increased its production of cocoa, gold, wood and bauxite, but the fall in the world market price of cocoa generated a loss of $200 million, and many industries were forced to close down. In meetings at the Paris Club in January and February of 1989, further loans of $900 million were granted to Ghana.

28 In 1990 in, spite of monetary infusions, the hopes of growth collapsed. The GDP growth fell from 6.1 to 2.7 per cent. Inflation at 25 per cent in 1989, went up to 37 per cent. The current account

deficit almost doubled and earnings for exports were lower than in 1988. Between 1988 and 1989, around 120 industries closed down because their products could not compete with cheaper and higher-quality goods from China, South Korea, and Taiwan.

29 In December 1991, the World Bank announced that Ghana would be granted new loans. The country would receive two more loans, totalling $155 million, for the government to reform the fi-

> The economic upheaval has had environmental costs as well. Tropical forest used to cover 34 per cent of the country's surface. This forest is now only a quarter of its original size.

nancial system and improve the rural road network.

30 This latest crisis has forced the Rawlings government to initiate a democratization program and a devolution plan, favoring local administrations. In June 1991, an Advisory Assembly of 260 members was elected to draft a new Constitution. The National Council for Women and Development, a body with ministerial status, won 10 seats and the Organization of Graduate Nurses and Midwives is also represented.

31 In December, 1991, Amnesty International denounced the Ghanaian policy of silencing or intimidating its opponents. In view of such serious and constant human rights violations, the Ghana Committee for Human and Popular Rights was created in January 1992, the first such organization in the country.

32 In this year, a new Constitution was approved and elections were held. Rawlings took 58.3 per cent of the vote in the November presidential elections and the three parties who supported him took 197 of the 200

seats in the December legislative elections.

33 In spite of supposed irregularities, 80 countries and international organizations attended Rawlings' investiture in January 1993. This foreign support was attributed to the rigorous application of the IMF imposed structural adjustment plans in the previous decade and the prompt payment of servicing on the foreign debt, which represented some 35 per cent of export earnings.

34 In February 1994, land ownership confrontations between members of different ethnic groups killed more than 1,000 people and the migration of 150,000 more. The government declared a state of emergency, and in June, representatives of both sides signed an agreement. In May, Rawlings reshuffled his cabi-

net and formed a committee to fix the salaries of high-ranking executives in the State enterprises, to avoid them fixing their own levels.

35 The 1994 budget had a surplus of $80 million, following two years of deficit, and predictions for 1995 forecast 5 per cent growth. In May, demonstrations against the new value added tax left five people dead. The minister of economy, Kwesi Botchwey, withdrew the new tax from the budget and, after 13 years in the post - coinciding with the implementation of the IMF plans - resigned his office.

STATISTICS

DEMOGRAPHY

Urban: 35% (1995). **Annual growth:** 1.4% (1992-2000). **Estimate for year 2000:** 20,000,000. **Children per woman:** 6.1 (1992).

HEALTH

One **physician** for every 25,000 inhab. (1988-91). **Under-five mortality:** 131 per 1,000 (1995). **Calorie consumption:** 88% of required intake (1995). **Safe water:** 56% of the population has access (1990-95).

EDUCATION

Illiteracy: 36% (1995). **School enrolment:** Primary (1993): 70% fem., 70% male. Secondary (1993): 28% fem., 44% male. **Primary school teachers:** one for every 29 students (1992).

COMMUNICATIONS

96 **newspapers** (1995), 85 **TV sets** (1995) and 101 **radio sets** per 1,000 households (1995). 0.3 **telephones** per 100 inhabitants (1993). **Books:** 88 new titles per 1,000,000 inhabitants in 1995.

ECONOMY

Per capita GNP: $410 (1994). **Annual growth:** 1.40% (1985-94). **Annual inflation:** 28.60% (1984-94). **Consumer price index:** 100 in 1990; 202.7 in 1994. **Currency:** 1,053 cedis = 1$ (1994). **Cereal imports:** 396,000 metric tons (1993). **Fertilizer use:** 38 kgs per ha. (1992-93). **External debt:** $5,389 million (1994), $324 per capita (1994). **Debt service:** 24.8% of exports (1994). **Development aid received:** $633 million (1993); $39 per capita; 10% of GNP.

ENERGY

Consumption: 91 kgs of Oil Equivalent per capita yearly (1994), 64% imported (1994).

Greece

Ellás

Population: 10,426,000 (1994)
Area: 131,990 SQ KM
Capital: Athens (Athínai)

Currency: Drachma
Language: Greek

B etween 5,000 and 3,000 BC, the Greek peninsula was inhabited by maritime peoples from Asia, and in Thessaly, central Greece and Crete traces have been found of previous inhabitants, nomadic and agricultural peoples. From the second milennium BC, the Achaeans, an Indo-European people, started to extend throughout the peninsula.

2 The Achaeans founded Mycenae, Tyrins and Argos, later conquering Athens and the eastern Peloponnese, invading Crete and raiding Troy. They were warriors, with an economy based on agriculture and cattle-breeding. The hierarchical social organization had kings, noblemen, and warriors dominating farmers, artisans, and peasants.

3 By the first milennium BC, Mycenian civilization had succumbed to Dorian invasions. The Dorians used iron weapons, unknown to the Achaeans, and they soon mixed with the subject population, bringing linguistic unity to the region.

4 The peninsula's topography favored the appearance of city-states (*polis*), ruled by a king or tyrant, supported by the military aristocracy. The peasants were forced to pay tributes in kind, and if they could not produce sufficient crops they became serfs or their whole family was sold into slavery.

5 Despite the social differentiation, the Greeks had an original concept of the human being. All previous civilizations had seen human beings as a mere instruments of the will of the gods or the monarchs, whereas in Greek philosophy they acquired the value of individuals. The concept of a male citizen, as an individual who belongs to the *polis*, regardless of his nobility, was one Greek culture's key contributions to Western civilization.

6 The cities alternately established alliances and fought each other. However, Hellenic groups started to feel they belonged to a common nationality bound by elements like the Olympic games, and a common religion and language.

7 In the 8th century BC most city-states went through crises, due to the decline in power of the monarchs (who were progressively replaced by magistrates appointed by the nobility), to the lack of fertile soil, and to demographic growth. This produced great social tension spurring the Greeks to colonize the Mediterranean, creating active trade routes and expanding the use of Greek as the language of commerce.

8 Around 760 BC the Greeks established colonies in southern Italy, in the bay of Naples and in Sicily. The Phoenicians and the Etruscans prevented them from ruling the whole of Sicily and the south of Italy, but their culture was deeply influential in later developments on the Italian peninsula.

9 When colonization started, the social and political structure of the polis underwent transformations. Merchants, wealthy as a result of maritime expansion, were not ready to leave government to the nobility. They joined the peasants in pressing for political participation. Between the 7th and 6th centuries BC, Athens, one of the most prosperous cities in the peninsula, started a process of political transformation which led to the gradual democratization of its governmental structures. In 594 BC, a reformer called Solon took the first steps in this direction with the establishment of written law, a court of justice and an assembly of 400 members, elected from among the wealthy, which was in charge of legislating on city matters.

10 Sparta, the other great city-state in the region, evolved in a completely different way. It consolidated an oligarchic state, with a strict social and political structure. Spartan society was completely militarized because of the central importance of the army, which had been a determining factor in expansion and annexation of neighboring territories.

11 In 540 BC the Persians started advancing on Asia Minor, conquering some Greek cities. These cities revolted, backed first by Athens and then by Sparta, resulting in several wars until the Persians were

defeated around 449 BC. These wars served to consolidate Athenian power in the region, exerting its political and economic influence over the other city-states through the League of Delos.

> Around 760 BC the Greeks established colonies in southern Italy, in the bay of Naples and in Sicily. The Phoenicians and the Etruscans prevented them from ruling the whole of Sicily and the south of Italy, but their culture was deeply influential in later developments on the Italian peninsula.

12 The Athenian triremes played a key role in the wars against the Persians. The oarsmen, who belonged to the lowest strata of Athenian society, became an indispensable tool in the defence of Athens,

and were thus in a position to demand improvements in their living conditions and increased political rights. In 508 BC, after a period when the Athenian oligarchy had succeeded in recovering its political power, a reformer called Cleisthenes raised the number of assembly members to 500, turning this body into the main governmental structure. Participation in the Assembly was open to all free citizens in the polis. However, these constituted a minority of the population; Athenian prosperity was actually based on the exploitation of huge numbers of slaves. Athenian society of this period is therefore referred to as a "slave holding democracy" by historians.

13 In 446 BC the *archon* or governor of Athens, Pericles, agreed the Thirty Years' Peace with Sparta, recognizing each city's areas of influence in the Athenian League and the League of the Peloponnese.

14 Under Pericles' rule, Athens became the trading, political, and cultural centre of the region. The domination of maritime trade and the prosperity this brought about enabled Pericles to start new democratic reforms. This was the period of scholars such as Anaxagoras, of dramatists such as Sophocles, Aeschylus, Euripides, Aristophanes, and of sculptors like Phidias. The Greeks achieved great scientific knowledge, much of which, particularly in the fields of medicine or astronomy, has now

LAND USE

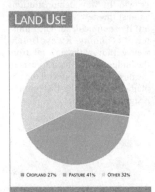

■ CROPLAND 27% ■ PASTURE 41% ■ OTHER 32%

been altered, but their contributions to geometry and mathematics are still indispensable for most of present-day science.

[15] A long period of continuous struggles over who should control the region ensued between the Spartans and Athenians. This mutual attrition enabled the Macedonians to conquer this dominion under the rule of Philip II, 359-336 BC. Philip's son, Alexander the Great, 336-323 BC, continued to spread Hellenic influence over the north of Africa and the Arabian Peninsula, through Mesopotamia, and as far as India. This empire, built over eleven years, contributed to the dissemination of Greek culture in the East. During the years of the conquest, many trading cities were founded. Alexander promoted the fusion of Greek culture with that of the conquered peoples, giving rise to what is known as Hellenism. Upon his death, the Macedonian Empire collapsed and wars and rebellions continued to shake the peninsula.

[16] These internal disputes resulted in a decline in the Greek civilization, leading to subsequent devastation and impoverishment enabling the Romans to move in. The Macedonian wars lasted from 215 to 168 BC, ending with the Romans establishing their rule over Greece by 146 BC.

[17] Under the Romans, Greece was acquainted with Christianity in the 3rd century, and it became a more rural society which was subjected to several invasions. From 395 to 1204, it belonged to the Eastern Empire, whose domination ended with the formation of the Eastern Roman Empire and the region's division into feudal possessions. In 1504, as a consequence of the Schism of the Roman Church, Greek Christians pledged obedience to the Orthodox Church in Constantinople.

[18] The Turks invaded and conquered in 1460, dividing Greece into six provinces which were forced to pay taxes. The Turks kept an occupation army in the country for 400 years. External attempts to expel the Turks, in particular incursions led by Venice - eager to acquire this strategic territory for trade with the East - and internal revolts all failed. The Peace of Passarowitz incorporated Greece into the Ottoman Empire in 1718.

[19] In 1821 a Greek uprising succeeded in freeing Tripolitza, and a

national assembly drafted a constitution, declaring independence. In 1825 the attempt ended in a bloodbath when the Turks, aided by Egypt, regained control of the city.

[20] In 1827 Russia, France and Britain, eager to keep the Turks away from their borders, signed the Treaty of London, which demanded Greek autonomy. The Turks refused it, and that year the allied fleet defeated the Turkish-Egyptian navy. In 1830, the London Convention declared the total independence of Greece, although the country had to sacrifice the region of Thessaly.

[21] In later years the European powers fought to control the peninsula, intervening in its domestic affairs and supporting kings who complied with their interests. Otto of Bavaria, 1831-1862, who favored the Russians, was followed by George I, 1864-1913, who was backed by the British.

[22] In 1910, a coup led by Eleuthe-

rios Vinizelos resulted in the enactment of a Constitution (1911) establishing a parliamentary monarchy. Throughout the two World Wars and the interwar years, subsequent military coups brought sympathizers of each side alternately to power.

[23] Once the German occupation was ended in 1944, an important part of the country remained under the control of communist guerrillas. They were led by Markos Vafiades, and had been crucial in Nazi resistance. The British and Americans supported the government in repressing the guerrillas until they were totally destroyed in 1949.

[24] Greece remained under American influence, becoming a member of the Council of Europe in 1949 and of NATO in 1951. In the 1956 elections women voted for the first time. After the war, Greece existed in a state of almost perpetual political instability.

[25] In April 1967, a group of colo-

nels staged a coup. Martial law was enforced, the Constitution was suspended, democratic movements were harshly repressed, and socialist leader Andreas Papandreou was sentenced to nine years' imprisonment. In December the king attempted to oust the military junta, but he failed and fled to Rome. The military appointed General Zoitakis president, and Papadopoulos prime minister.

[26] The "colonels'" regime, as it was known, was supported by the United States and by great tycoons such as Onassis and Niarchos. Attempts were made to mask the dictatorship behind a unicameral Parliament in 1968, but in reality, the military junta ruled by decree.

[27] Between 1973 and 1974 the military government grew weaker. In November, student demonstrations in the Polytechnic University of Athens were harshly repressed, leaving hundreds of casualties, and attracting international condemnation.

In the 1981 parliamentary elections, the Pasok (Pan-Hellenic Socialist Movement), led by Papandreou, gained an absolute majority, and the first socialist government in the history of Greece took office. The country also became the 10th member of the European Economic Community (EEC).

[28] In July 1974, the Greek military junta promoted a coup in the state of Cyprus, by collaborating with the Cypriot National Guard. The coup succeeded in deposing President Archbishop Vaneziz Makarios, who was forced into exile in London. A minister in favor of Greek annexation was appointed. The Turkish army immediately invaded Cyprus, allegedly defending the Turkish minority in the country.

MILLIONS $ 1994

IMPORTS
21,466
EXPORTS
9,384

STATISTICS

DEMOGRAPHY

Urban: 64% (1995). **Annual growth:** 0.5% (1991-99). **Estimate for year 2000:** 11,000,000. **Children per woman:** 1.4 (1992).

HEALTH

Under-five mortality: 10 per 1,000 (1994). **Calorie consumption:** 118% of required intake (1995).

EDUCATION

Primary school teachers: one for every 19 students (1992).

COMMUNICATIONS

101 **newspapers** (1995), 101 **TV sets** (1995) and 101 **radio sets** per 1,000 households (1995). 45.7 **telephones** per 100 inhabitants (1993). **Books:** 103 new titles per 1,000,000 inhabitants in 1995.

ECONOMY

Per capita GNP: $7,700 (1994). **Annual growth:** 1.30% (1985-94). **Annual inflation:** 15.50% (1984-94). **Consumer price index:** 100 in 1990; 175.7 in 1994. **Currency:** 240 drachmas = 1$ (1994). **Cereal imports:** 708,000 metric tons (1993). **Fertilizer use:** 1,309 kgs per ha. (1992-93). **Imports:** $21,466 million (1994). **Exports:** $9,384 million (1994). **Development aid received:** $44 million (1993); $4 per capita; 0.1% of GNP.

ENERGY

Consumption: 2,235 kgs of Oil Equivalent per capita yearly (1994), 63% imported (1994).

The Greek military government became even more discredited and internationally condemned, and they relinquished power immediately at the prospect of war with Turkey.

[29] In that month, former conservative prime minister Konstantin Karamanlis returned from exile and took over government. In the 1974 elections his party won a parliamentary majority and a later referendum abolished the monarchy. In June 1975, Parliament adopted a new Constitution and Konstantinos Tsatsos, a Karamanlis partisan, was elected first president of the republic.

[30] From 1974 onwards Greece decided not to take part in NATO military exercises, after conflict with Turkey, another member of the organization.

[31] In the 1981 parliamentary elections, the Pasok (Pan-Hellenic Socialist Movement), led by Papandreou, gained an absolute majority, and the first socialist government in

the history of Greece took office. The country also became the 10th member of the European Economic Community (EEC).

[32] The socialist government approached Third World countries, in particular the Arab nations, recognized the PLO and led a worldwide campaign in favor of handing back works of art that had been stolen during colonial domination.

[33] In 1983, salaries were frozen, provoking a wave of protests and strikes. Trade unions were granted greater participation in the public sector, but their right to strike was curtailed.

[34] The 1983 census revealed that women constituted one-third of the active population. Most of them occupied the service sector and their pay was lower than that of their male counterparts. Peasant women accounted for 40 per cent of the active female population, not counting the 400,000 women who worked unpaid on family plots.

[35] In the 1984 elections the Pasok

won again, by an even greater margin than it had held in 1981. The 1986 constitutional amendment gave more power to the parliament at the expense of the president. Successive austerity plans and salary freezes fuelled new protests and strikes.

[36] In November 1988, details of embezzlement in the Bank of Crete, involving several government members were disclosed. This scandal caused several ministerial crises. In June of the following year, the Greek Left and the Communist Party formed the Left Alliance.

[37] In the 1989 elections the Pasok lost its majority, and the conservative New Democracy party received a large part of the vote. As there was no adequate parliamentary majority and no agreement to form a government, the presidency went to the leader of the Left-wing coalition, the communist Charilaos Florakis. He formed a temporary government with the New Democracy, aiming to investigate the financial scandals.

[38] In September of that year Parliament announced that Papandreou and several of his ministers would be tried by a special court.

[39] The November 1989 election results, and those of other later elections were inconclusive, making it necessary to form a coalition government.

[40] Between 1983 and 1989, Greece and the United States signed several cooperation agreements. The terms included maintaining the four US military bases on Greek territory, in exchange for economic and military assistance, and US support in Greece's disputes with Turkey, in particular over Cyprus.

[41] In January 1990 Washington and Athens publicized a new agreement, establishing the closure of two military bases as part of an American plan to reduce its military presence in the area.

[42] On March 7 1990, legislation was passed on collective work agreements for the private sector, and for public companies and services. The new law establishes the freedom of negotiation between workers and employers, ending 50 years of State intervention. It includes regulations on the organization of committees in companies and trade unions, and on the participation of workers in company decision-making. An office for medi-

ation and arbitration was also created; it will appoint mediators and arbitrators for three-year periods.

[43] Following Karamanlis' triumph in the April 1990 presidential elections, a new government was formed headed by the Conservative Constantinos Mitsotakis. In 1991, Mitsotakis brought in a policy to reduce public spending, free prices and privatize.

[44] The social cost of these measures was reflected in the defeat of the Conservative government in the 1993 legislative elections. On October 12 of that year, Papandreou's PASOK gained the support of nearly 47 per cent of the electorate - compared with the 40 per cent of New Democracy - and the absolute majority in Parliament.

[45] The public debt and pressure from the European Union for a "rigorous" economic policy almost immediately complicated Papandreou's leadership. In the European elections of 1994, both PASOK and, to a greater extent, New Democracy, lost votes to the smaller parties like the Political Spring, the Communist Party and the Progressive Left Coalition.

[46] In 1995, amidst permanent rumours of Papandreou's retirement on health grounds and new accusations of the misappropriation of public funds against him, the finance minister, Alexandros Papadopoulos, carried out unpopular fiscal reforms and launched a policy of "strict (budgetary) rigor."

[47] Ill and increasingly criticized, even within his own party, Papandreou stood down in January 1996. The historic leader of Greek Socialism was replaced by his former minister of industry, Kostas Simitis. Papandreou died in June.

[48] In the September elections, the Socialist PASOK Party got over 45.5 per cent of the vote and 162 seats in Parliament. Conservative New Democracy got 38.15 per cent and 108 seats, the Communist Party came third with 5.60 per cent and 11 seats, Coalition of the Left (Sinaspismos) won 5.10 per cent of the vote and 10 seats and Democratic Social Movement received 4.43 per cent and 9 seats.

Grenada

Population: 92,000 (1994)
Area: 340 SQ KM
Capital: St. George's

Currency: E.C. Dollar
Language: English

Grenada

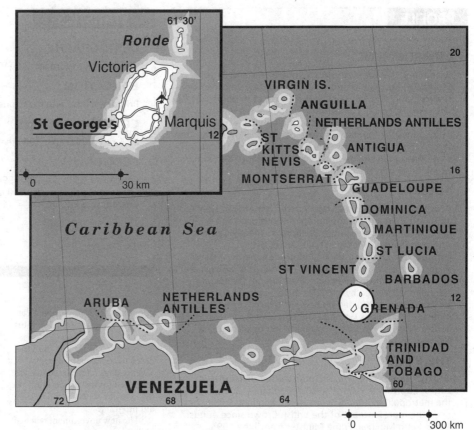

The Carib Indians inhabited Grenada when Christopher Columbus arrived, around 1498, and named the island "Concepción". This European visit did not disrupt the island's peace, but two centuries later, in 1650, the governor of the French possession of Martinique, Du Parquet, decided to occupy the island. By 1674, France had already established control over Grenada, despite fierce resistance from the Caribs.

2   In 1753, French settlers from Martinique had around a 100 sugar mills and 12,000 slaves on Grenada. The native population had been exterminated.

3   The British took control of the island towards the end of the 18th century and introduced the cultivation of cacao, cotton and nutmeg, using slave labor. In 1788 there were 24,000 slaves, a number which remained stable until the abolition of slavery in the following century.

4   The severe living conditions of workers resulted in the creation of the first union in the mid-20th century, the Grenada Manual and Metal Workers Union. In 1951, a strike broke out and the labor struggle won considerable wage increases. Eric Matthew Gairy, a young adventurer who had lived away from the island most of his life, formed the first local political party, the Grenada United Labor Party (GULP) which favored independence. In 1951, GULP won a legislative election and Gairy became leader of the assembly.

5   In 1958, Grenada joined the Federation of the British West Indies, which was dissolved in 1962. The country became part of the Associated State of the British Antilles in 1967. That year, Gairy was appointed prime minister when his party won the August elections. Their main objective was total independence from Britain.

6   GULP soon obtained semi-independence which gradually led to full independence. By that time, left-wing groups such as the New Jewel Movement (NJM) had appeared on the island, opposing separation from Britain. Though apparently paradoxical, many Grenadians considered that Gairy was seeking independence for his personal benefit, manipulating a politically unprepared population.

7   In January 1974, an "anti-independence" strike broke out to prevent Gairy from seizing power.

After some weeks of total paralysis of the country "Mongoose squads", similar to the Haitian Tonton-Macoutes appeared. They were on the prime minister's payroll and brutal repression was used to end the strike. Independence was proclaimed the following week.

8   As feared, Gairy exploited power for his own personal benefit. He distributed government jobs among the members of his party, while promoting the paramilitary squad to the level of a "Defence Force", making it the only military body on the island. The "Mangoose Squad" received military training from Chilean advisers and increased its numbers considerably by incorporating ex-inmates of St George's Prison.

9   In the December 1976 election, the People's Alliance, made up of the NJM, the National Party of Grenada and the United Popular Party, gained new ground in Parliament, moving from one to six representatives in a total of 15 parliamentary seats. On March 13 1979, while Gairy was out of the country, the opposition carried out a bloodless coup and seized power. Thanks

to widespread popular support, they were able to establish a provisional revolutionary government, under the leadership of Maurice Bishop, a lawyer.

10   In four years, the People's Revolutionary Government stimulated the formation of grass roots organizations, and they created a mixed economy, expanding the public sector through agro-industries and state farms. Private enterprise was encouraged making Grenada more compatible with global economic policy.

11   However, because of the government's socialist ideology, the NJM was under constant harassment from the United States and Grenada's conservative neighbors, who did everything possible to destabilize the Grenadian government. Claiming that Grenada's modern airport at Point Saline might be used as a stop-over for Cuban troops en route to Africa, the United States began a coordinated campaign to crush Grenada's economy.

12   The government based its foreign policy on the principles of anti-imperialism and non-align-

ment. Special attention was given to the development of ties with the socialist world. Cuba agreed to collaborate in the construction of an international airport. The project was conceived as a way to stimulate tourism, which employed 25 per cent of the country's work force.

13   Prime Minister Bishop faced constant pressure from the NJM's extreme left wing. This faction, led by the assistant prime minister, was intent upon radicalizing the political process. On October 10, 1983, upon his return from a short State visit to Hungary, Czechoslovakia and Cuba, Bishop was placed under house arrest, while General Hudson Austin, head of the army, seized power. Bishop was freed by a crowd of sympathizers, only to be shot dead by troops. A number of others were killed along with him, including his wife Jacqueline Creft - Minister of Education, the Foreign and Housing Ministers, two union leaders and 13 members of the crowd.

14   The US then began to take definite steps towards military intervention, a move which had been decided upon and planned for over

ENVIRONMENT

Grenada is the southernmost Windward island of the Lesser Antilles. The island is almost entirely volcanic. Lake Grand Etang and Lake Antoine are extinct volcanic craters. The rainy, tropical, climate, tempered by sea winds, is fit for agriculture which constitutes the country's major source of income. It is famous for its spices, and is known as "the Spice Island of the West". The territory includes the islands of Carriacou (34 sq km) and Petite Martinique (2 sq km), which belong to the Grenadines.

SOCIETY

Peoples: Grenadians are descended from African slaves, although there are a large number of mestizos and a minority of European origin. **Religions:** Mainly Catholic. Anglican, Baptist, Methodist and Adventist.
Languages: English (official and predominant). A patois dialect derived from French, and another from English are also spoken. **Political Parties:** The National Democratic Congress (NDC), led by Nicholas Braithwaite. The New National Party (NNP), led by Keith Mitchell. Former Prime Minister Eric Gairy's Grenada United Labor Party (GULP). The Maurice Bishop Patriotic Movement, led by Terrence Marryshow.

THE STATE

Official Name: Grenada. **Administrative divisions:** 7 zones. **Capital:** St George's, 30,000 inhab. (1990), 29,500 in the metropolitan area. **Government:** Reginald Palmer has been the representative of the British Crown since August 1992. Keith Mitchell, Prime Minister from June 1995. **Legislature:** the bicameral Parliament is made up of the 15-member Chamber of Deputies and the 13-member Senate. Deputies are elected by direct popular vote, while 7 of the senators are appointed by the governor, 3 by the prime minister and 3 by the leader of the opposition. **National holiday:** February 7, Independence (1974).

DEMOGRAPHY
Children per woman: 2.9 (1992).

EDUCATION
Primary school teachers: one for every 27 students (1989).

COMMUNICATIONS
22.1 **telephones** per 100 inhabitants (1993).

ECONOMY
Per capita GNP: $2,630 (1994).
Consumer price index: 100 in 1990; 112.3 in 1994.
Currency: 3 E.C. dollars = 1$ (1994).

a year; troop-landing had been rehearsed in maneuvers. Early in the morning of October 25 1983, 5,000 marines and Green Berets landed on the island. They were followed several hours later by a symbolic contingent of 300 policemen from six Caribbean countries; Antigua, Barbados, Dominica, Jamaica, St Lucia and St Vincent, who joined the farce of a "multi-national intervention for humanitarian reasons".
[15] Resistance from the Grenadian militia and some Cuban technicians and workers, meant that the operation lasted much longer than expected. The US suffered combat casualties, and the press was barred from entering Grenada until all resistance had been eliminated. This made it impossible to verify how many civilians had been killed in attacks on a psychiatric hospital and other non-military targets.
[16] While strict US military control continued, Sir Paul Scoon, official British crown representative in Grenada, assumed the leadership of an interim government with the task of organizing an election within a period of 6 to 11 months. Under the discreet surveillance of the invaders, voting was held on December 3, 1984 to elect the members of a unicameral parliament. In turn, the representatives appointed Herbert Blaize prime minister. Blaize led a coalition of parties that was presented to public opinion as the New National Party (NNP) and received support from the US.
[17] Neither NATO nor the OAS dared to condone the aggression. Twenty days later, Barbados was rewarded for its "cooperation" in the invasion with a $18.5 million aid program.
[18] The new government reached a classic agreement with the International Monetary Fund including a reduction of the civil service, a wage-freeze and incentives to private enterprise.
[19] The System of Regional Security (SSR), permitting the prime minister to invoke the presence of troops from neighboring Caribbean islands if he feels threatened, was established by Blaize in December 1986, under the pretext that the trial of those involved in the 1983 coup was coming to an end. Bernard Coard, his wife Phyllis, and former army commander Hudson Austin were sentenced to death, along with 11 soldiers. Three others were tried, receiving prison sentences of 30 to 45 years.
[20] During his first years in government, Blaize achieved considerable economic growth, at between 5 and 6 per cent per year, basically from tourism. However, youth unemployment continued to increase, along with crime and drug addiction.
[21] Blaize died in 1989. He was replaced by Ben Jones of the National Party, who was backed by big business and the landowners.

On March 13 1990 - the anniversary of the coup which ousted Eric Gairy in 1979 - the general elections were won by Nicholas Braithwaite, former interim head of government after the invasion, and warmly looked upon by the United States.
[22] Studies carried out in 1994 revealed unemployment of 30 per cent and a marked pattern of emigration. The fall in international prices of bananas, coconuts, wood and nutmeg influenced this trend. Between March and April 1995, there were strikes in key sectors of the economy, like the hotels, and sugar and cocoa production. Although tourism is the second most important source of income for the nation, Grenada cannot compete with the other Caribbean islands on this front.
[23] In the June 1995 general elections, the governing NDC was pushed out by Keith Mitchell, former mathematics professor at the Howard University in Washington and leader of the NNP. His victory was attributed to his promise to remove income tax, which had been revoked in 1986 and reimposed by the NDC in 1994.
[24] Despite the agreement made between Indonesia and Grenada - the leading world nutmeg producer - in 1994, the international prices continued to tumble in 1995.

Guadeloupe

Population:	421,000 (1994)
Area:	1,710 SQ KM
Capital:	Basse Terre
Currency:	French Franc
Language:	French

Guadeloupe

The entire archipelago of present-day Guadeloupe was inhabited by the Caribs. They originally came from South America, and they dispersed throughout the islands after overpowering the Arawak people, resisting the Spanish invasion in 1493, but were defeated by the French two centuries later.

[2] The French colonizers built the first sugar mill on the island in 1633 and began importing African slave laborers. By the end of the 17th century, Guadeloupe had become one of the world's main sugar producers. The colonists killed the last of the archipelago's surviving Caribs in the early 18th century.

[3] With the abolition of the slave trade in 1815, France restructured its formal links with the Caribbean islands, giving them the status of colonies. In 1946, with the initia-

PROFILE

ENVIRONMENT

Includes the dependencies of Marie Galante, La Désirade, Les Saintes, Petite-Terre, Saint Berthélemy and the French section of Saint Martin forming part of the Windward Islands of the Lesser Antilles. The tropical, rainy climate is tempered by sea winds, and sugar cane is grown.

SOCIETY

Peoples: A majority of descendants of native Africans with a small European minority. **Religions:** Mainly Catholic. **Languages:** French (official). **Political Parties:** Guadeloupe Federation of the Socialist Party (PS), Guadeloupe Federation of the Union for the Republic (RPR), and Guadeloupe Federation of the Union in French Democracy (UDF), are the local branches of the colonial parties; the Communist Party of Guadeloupe (PCG); the Guadeloupe Democrats; the Popular Union for the Liberation of Guadeloupe (UPLG), the largest of the radical pro-independence groups; the Revolutionary Caribbean Alliance (ARC), founded in Point-a-Pitre in 1983, has carried out several armed attacks. **Social Organizations:** The Guadeloupe General Labor Confederation, with 15,000 members; the Department Union of Christian Workers' French Confederation, with 3.500 members; the Department Organization of Trade Unions(CGT-FO), with 1,500 members.

THE STATE

Official Name: Département d'Outre-Mer de la Guadeloupe. **Administrative Divisions:** 3 Arondissements, 36 Cantons **Capital:** Basse-Terre, 14,000 Inhab. (1991). **Government:** Franck Perriez, commissioner appointed by the French government. Legislative power: A 42-member General Council, and 41-member regional Council. Guadeloupe has 4 deputies and 2 senators in the French parliament.

EDUCATION

Primary school teachers: one for every 19 students (1991).

COMMUNICATIONS

36.7 **telephones** per 100 inhabitants (1993).

ECONOMY

Consumer price index: 100 in 1990; 110.1 in 1994. **Currency:** french francs.

tion of the new French Constitution, Guadeloupe achieved greater political autonomy, as an Overseas Department.

[4] French government subsidies increased per capita income and the consumption of imported goods, but ruined the local economy.

[5] In 1981, in a country where 90 per cent of the population is of African or racially mixed origin, racism has made the social climate more tense.

[6] In 1985 there were violent clashes between pro-independence demonstrations and the police. Several disruptive actions led to the imprisonment of members of the Revolutionary Caribbean Alliance (ARC).

[7] In the 1988 election, François Mitterrand received 70 per cent of the vote on Guadeloupe, although two-thirds of the voters abstained. In the regional elections, of October 1988, the left, already in the majority within the General Council, was further strengthened, gaining another seat.

[8] In 1989, ARC members sentenced in 1985 were granted amnesty. The ARC took part in the elections for the first time, and their failure was taken as a show of support for the present situation.

[9] In 1989 hurricane Hugo left 12,000 people homeless. The French government announced that it would freeze the interest on the foreign debt, and would send $5.4 million in aid to the island.

[10] In 1992 the European Community decision to reduce banana import quotas from the "French" Antilles to benefit imports from Africa and Latin America triggered strong protests in Guadeloupe. In the elections for the Regional Council - one of the two main assemblies in the island - the Guadeloupe Federation of the Union for the Republic (RPR) won. The RPR is a conservative, neo-Gaullist party, led by current French president Jacques Chirac. The RPR obtained 15 of the 41 seats, followed by the Socialist Party with 9.

[11] In 1993, the reduction in tourism aggravated the economic situation. The unemployment rate reached 24 per cent of the active population, while the informal sector of the economy increased.

[12] Despite that fact that banana production has replaced sugar cane to a great extent, many islanders still consider sugar cane essential for Guadeloupe. Indeed, it not only employs 1,000 workers and accounts for 20 per cent of the country's GDP but is also considered vital to preserve the local culture and identity. Pointe-a-Pitre government authorities planned a production increase to 800,000 tons by the year 2000. It had gone down to 250,000 tons in 1994.

[13] In 1995 and 1996, as further integration with the European Union drew closer, several observers highlighted the growing "social unrest". "Unemployment increase, the spread of social conflicts, the marginalization of youngsters who no longer find reasons to live, are the symptoms of a deep crisis", was the opinion published by the French monthly journal "Le Monde Diplomatique".

Guam

Population: 146,000 (1994)
Area: 549 SQ KM
Capital: Agaña

Currency: US Dollar
Language: English

Guam

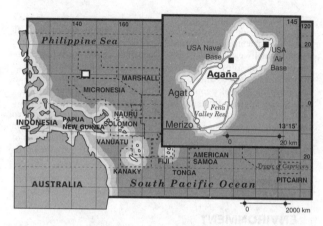

Guam shares a common history with the rest of the Micronesian archipelago (see Micronesia). The native population, who settled on the island thousands of years ago, became the victim of extermination campaigns at the hands of Spanish colonizers, between 1668 and 1695.

[2] As a result of armed aggression and epidemics (the natives lacked immunity against European illnesses), the population declined from 100,000 at the beginning of the 17th century to fewer than 5,000 in 1741.

[3] The few survivors inter-married with Spanish and Filipino immigrants, producing the Chamorro people who presently populate the island.

[4] For three centuries, Guam was a port of call on the Spanish galleon route between the Philippines and Acapulco (Mexico), a major depot on the trade route to Spain.

[5] Under the terms of the Treaty of Paris of 1898, Guam changed hands from Spain to the United States, together with the Philippines. The island continued to serve as a stop-over place, until it was invaded by the Japanese in 1941. Recovered in 1944, it became a US military base.

[6] Since 1973, the United Nations have unsuccessfully urged Washington to permit the islanders to exercise their right to self-determination. In January 1982, a plebiscite on a new statute did not achieve the majority required.

[7] In a self-determination plebiscite, 75 per cent of the electorate favored a system of association with the United States. As a result of this the UN General Assembly recommended the US to implement Guam's decolonization in December 1984. The UN also reiterated its conviction that military bases are a major obstacle to self-determination and independence, two principles established in the UN Charter.

[8] In February 1987, former governor Ricardo Borballo, who had been elected in 1984, was found guilty of bribery, extortion and conspiracy.

[9] In November 1987, the High Commission initiated a self-determination referendum for Guam, and voters supported negotiation of a new relationship with the US. In February 1988 Guam received US economic aid, or more precisely, disaster relief, to repair the damage done by the typhoon in January.

[10] The economy of Guam is based on tourism, fish and craft exports, and fundamentally on the presence of thousands of US armed forces personnel billeted in air and naval bases on an island the US considers a leading geo-strategical enclave. In 1995, Carl TC Gutiérrez was elected governor.

[11] In early September 1996, the island was used as the base for the US bombers which carried out a "limited attack" against Iraq. Three hundred Iraqi Kurds, employed by the US government for aid and intelligence work in northern Iraq traveled to Guam with their families after the Kurdistan Democratic Party - allied with Saddam Hussein - conquered the north of the country. The Kurds, who worked under the protection of the pro-Iranian Patriotic Union of Kurdistan were evacuated to Guam on their way to asylum in the United States.

PROFILE

ENVIRONMENT

Guam is the southernmost island of the Marianas archipelago, located east of the Philippines and south of Japan. Of volcanic origin, its contours are mountainous except for the coastal plain in the northern region. The climate is tropical, rainy from June to November (over 300 mm a month) and drier and colder from December to May. It has rainforest vegetation. One-third of the island is occupied by military installations.

SOCIETY

Peoples: Chamorro natives account for approximately 60% of the population. There are 20,000 US troops and dependents. **Religions:** 95% of all Guamanians are Catholic. **Languages:** English (official), Chamorro (a dialect derived from Indonesian), and Japanese. **Political Parties:** Republican and Democrat, as in the US.

THE STATE

Official Name: Territory of Guam. **Capital:** Agaña, 50,000 inhabitants. **Government:** Joseph Ada, Governor, re-elected on November 6 1990. His office responds to the US Interior Department and actually has less autonomy than the local military commander, as one-third of the island is under control of the US Navy and Air force. Although Guamanians formally possess US citizenship they have no representation in the US Congress nor do they participate in the US presidential elections.

DEMOGRAPHY

Density: 219 inhab./sq km. **Annual growth:** 1.8% (1970-86). **Life expectancy:** 73 years.

EDUCATION

Literacy: 99% male, 99% female (1990). **Primary school teachers:** One for every 20 students (1988).

COMMUNICATIONS

161 **newspapers**, 658 **TV sets** and 1,397 **radio receivers** per 1,000 inhab. (1991).

ECONOMY

Consumer price index: 100 in 1980, 277 in June 93. **Currency:** US dollars. **Imports:** $420 million (1991). **Exports:** $62 million (1991).

ENERGY CONSUMPTION:

5,261 kgs of Coal Equivalent per capita yearly, 100% imported. (1990).

Guatemala

Guatemala

Population: 10,322,000 (1994)
Area: 108,890 SQ KM
Capital: Guatemala

Currency: Quetzal
Language: Spanish

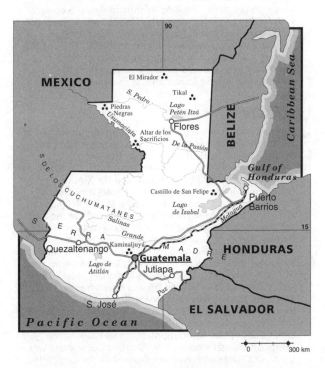

The Maya civilization flourished during the first ten centuries AD in what is now Guatemala and parts of Mexico, Honduras and El Salvador.

[2] Spanish troops, under the command of Pedro de Alvarado, entered the country in 1524, founding the city of Guatemala and gaining total control over the country two years later. This process was facilitated by the fact that the country was undergoing a gradual transition and readjustment among its various ethnic groups - the K'iche', Kaqchi', Mam, Q'eqchi', Poqomchi', Q'anjob'al, Tz'utujiil and others - all of which stemmed from a common Mayan ancestry. Still, though the situation favored the invaders, there were nevertheless frequent incidents in which they faced stiff resistance.

[3] On September 15, 1821, large landowners and local businessmen joined forces with colonial officials, to peacefully proclaim the independence of the Viceroyalty of New Spain, including the five countries of Central America.

[4] The new political and administrative union, called the Federation of Central American States, was dissolved in 1839 as a result of internal struggles prompted by the British policy of "divide and conquer".

[5] In 1831, under tremendous debt pressure, the government yielded large portions of territory to Britain for lumbering. This later became British Honduras, now the independent nation of Belize.

[6] The discovery of synthetic dyes in Europe in the mid-18th century caused a serious economic crisis in Guatemala, where the main export products were vegetable dyes. Coffee replaced dyes as the country's main cash crop; while plantation owners deprived the natives of the better part of their communal lands through the Liberal Reform of 1871. During the late 19th century, Guatemalan politics were dominated by the antagonism between liberals and conservatives. German settlers in Guatemala developed ties with German companies and initiated an import-export business which damaged the interest of the incipient national bourgeoisie.

[7] Toward the end of the 19th century, Manuel Estrada Cabrera rose to power and governed Guatemala until 1920. He began an "open door policy" for US transnationals, who eventually owned the railroads, ports, hydroelectric plants, shipping, international mailing services and the enormous banana plantations of the United Fruit Company.

[8] General Jorge Ubico Castañeda, the last of a generation of military leaders, was elected president in 1931. However, a popular uprising known as the "October Revolution" toppled him in 1944 calling fresh elections. The winner was reformist Juan José Arévalo, who launched a process of economic and social reform.

[9] Arévalo's government heralded a climate of political and economic liberalization. In 1945, literate women were granted the right to vote. That same year the first campesino labor union was formed. The land reform program, under which extensive tracts of unused United Fruit Company land were expropriated, were considered "a threat to US interests" by Washington. An aggressive anti-communist campaign was launched, with the sole aim of harassing Arévalo and his successor, president Jacobo Arbenz.

[10] John Foster Dulles, secretary of state, but also a United Fruit Company (UFCO) share-holder and company lawyer pressured the OAS to condemn the reforms being carried out by Jacobo Arbenz's government. Allen Dulles, director of the CIA and former president of the company was the one who organized the invasion carried out by Castillo Armas in June 1954. With the fall of Arbenz, UFCO managed to get back its lands, and subsequently changed its name to United Brands.

[11] Two decades of military regimes followed. Guatemalans are unanimously of the opinion that the four elections which followed, in 1970, 1974, 1978 and 1982, were rigged with the top military candidates invariably elected.

[12] This kind of political atmosphere bred armed insurgency. The Rebel Armed Forces (FAR) entered the scene in 1962, followed by the Poor People's Guerrilla Army (EGP) and the Organization of the People in Arms (ORPA) in 1975 and 1979, respectively.

[13] According to estimates from several humanitarian organizations, official repression took some 80,000 lives between 1954 and 1982. The Guatemalan National Revolutionary Unity (URNG) was founded in February 1982, uniting the Poor People's Guerrilla Army, the Rebel Armed Forces, the Organization of the People in Arms, and the Guatemalan Labor Party (PGT, National Leadership Committee).

[14] On March 23, 1982, only a few days after a rigged election, a military coup ousted Lucas García from office and replaced him with General Efraín Ríos Montt.

[15] Ríos Montt's counter-insurgency campaign surpassed that of his predecessor's in terms of aggressiveness. Over 15,000 Guatemalans were killed in the first year of Ríos Montt's administration. 70,000 took refuge in neighboring countries, especially Mexico, and 500,000 fled to the mountains to escape the army. Hundreds of villages were razed, while the number

PUBLIC EXPENDITURES

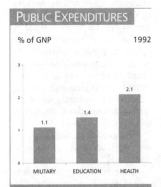

% of GNP — 1992

- MILITARY 1.1
- EDUCATION 1.4
- HEALTH 2.1

WORKERS

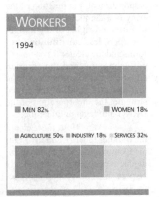

1994

- MEN 82% — WOMEN 18%
- AGRICULTURE 50% — INDUSTRY 18% — SERVICES 32%

LAND USE

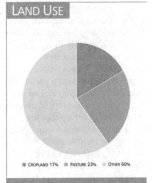

- CROPLAND 17% — PASTURE 23% — OTHER 60%

ENVIRONMENT

The Sierra Madre and the Cuchumatanes Mountains cross the country from east to west, and these are the areas of volcanic activity and earthquakes. Between the mountain ranges there is a high plateau with sandy soil and easily eroded slopes. Although the plateau occupies only 26% of the country's territory, 53% of the population is concentrated there. The long Atlantic coastline is covered with forests and is less populated. In the valleys along the Caribbean coast and in the Pacific lowlands there are banana and sugar cane plantations. In 1980, 41.9% of the country's land area was covered by forests; in 1990, that figure had been reduced to 33.8%, threatening the exceptionally diverse ecosystems, with a great variety of species.

SOCIETY

Peoples: Approximately 90% are of Mayan descent. Amid the country's great cultural and linguistic diversity, four major peoples can be distinguished: the Ladino (descendants of Amerindians and Spaniards), the Maya, the Garifuna (of the Caribbean region) and the Xinca. **Religions:** Mainly Catholic. In recent years a number of Protestant groups have appeared. The Mayan Religion has also survived. **Languages:** Spanish is official but most of the population speak one of the 22 Maya dialects. **Political Parties:** National Advancement Party, conservative; Guatemalan Republican Alliance (FRG), ultra-right-wing led by former dictator Efraín Ríos Montt; Christian Democracy of Guatemala (DCG); Union of the National Center (UCN); New Guatemala Democratic Front, a new left wing coalition; Solidarity Action Movement (MAS). The National Revolutionary Unity of Guatemala (URNG) is a coalition of different armed groups. **Social Organizations:** Union of Labor and Popular Associations; National Labor Union Alliance; National Workers' Coordinating Committee; Labor Union of Guatemalan Workers; Committee for Campesino Unity; National Coordinating Committee of Indigenous Campesinos; Communities of Peoples Resistance; Permanent Commissions of Refugees in Mexico; National Commission of Widows of Guatemala; Mutual Support Group; "Runujel Junam" Council of Ethnic Communities; Office of Human Rights of the Archbishopric; Coordinating Committee of Mayan Women; Academy of Mayan Languages of Guatemala; "Majawil Q'uij" Coordinating Committee and Council of Mayan Organizations of Guatemala.

THE STATE

Official Name: República de Guatemala. **Administrative Divisions:** 22 Departments. **Capital:** Guatemala, 1,942,953 inhab. (1993). **Other cities:** Quezaltenango 542,556 inhab; Escuintla 526,249 inhab.; Puerto Barrios 116,217 inhab. (1993) **Government:** Alvaro Arzu, President since January 1996. **National Holiday:** September 15, Independence Day (1821). **Armed Forces:** 44,200 troops (1994). **Paramilitarles:** 10,000 National Police, 2,500 Hacienda Guard; 500,000 Territorial Militia.

ist Democratic Party and called for elections in November 1985.

[19] The election, boycotted by the URNG, gave a clear victory to the Christian Democratic candidate, Vinicio Cerezo. One of the first measures was the "total and definitive" suspension of secret police activities.

[20] In October 1987, representatives of the URGN and Vinicio Cerezo's government met in Madrid, the first direct negotiations between the government and guerrilla forces in 27 years of conflict. That year, the National Reconciliation Commission (CNR) played a decisive role in the rapprochement process. The commission was created as a result of the Esquipulas II peace plan for Central America, signed by 6 countries in the region.

[21] On March 30, in Oslo, guerrillas and government agreed on an operational pattern for the meetings and the role of CNR and UN mediators. Despite the persistence of political persecution and assassination, on June 1, 1990 a basic agreement was signed in Madrid by the National Commission for Reconciliation, the political parties and the URGN. The overall aim of the accord was to continue the search for peace in Guatemala.

[22] During the last few months of 1990, negotiations came to a standstill and a high degree of skepticism developed among voters, which led to a 70 per cent abstention rate in the November 11, 1990 presidential elections. During the second round of the elections, held on January 6, 1991, Jorge Serrano Elías, of the Solidary Action Movement (MAS), was elected president.

[23] The Serrano government and the URNG decided to take up peace negotiations. A fortnight later, in Cuernavaca, Mexico, a three-day meeting was held. After three decades of violence, during which over 100,000 people were murdered and 50,000 went missing, the government and the guerrillas committed themselves to the negotiation process, attended by top level delegates and with the aim of achieving a firm and lasting peace agreement in the shortest possible time. The agenda included topics such as: democratization, human rights, the strengthening of civil empowerment, the function of the army in a democratic society, the identity and rights of the indigenous peoples, constitutional reforms and the electoral system, socio-economics and the agrarian situation, the resettlement of the landless as a result of the armed struggle, the incorporation of the URNG into legal political life.

[24] Late in May, 1991, the leader of the Runujel Junam Council of Ethnic Communities disappeared. Indigenous peoples constitute a mere 16 per cent of the economically ac-

tive population and they are press-ganged into joining the army.

[25] In July, the US Senate suspended military aid to Guatemala. The URNG demanded that human rights violations cease immediately. Human rights organizations found that in the first 9 months of Serrano's rule there had been 1,760 human rights violations, 650 of them executions without trial. Several deaths of street children below the age of six, were also reported.

[26] Also in September 1991, the Guatemalan president recognized the sovereignty and self-determination of Belize, the ex-British colony which proclaimed its independence in 1981. The announcement caused the resignation of chancellor Alvaro Arzú, the leader of the National Advancement Party (PAN), and one of the ruling party's main allies.

[27] In 1992, a national debate began on the existence of government armed civilian groups, such as the Civilian Self-Defence Patrols (PAC). The Catholic Church criticized the government's economic policy and spoke out in favor of agrarian reform. In the meantime, organizations representing the indigenous peoples demanded the ratification of ILO Agreement 169, dealing with indigenous and tribal peoples.

[28] The government created the "Hunapú" force, made up of the Army, the National Police and the "Hacienda Guard", replacing the former PACs. In April, members of the "Hunapú" provoked an incident during a student demonstration demanding improvements in the education policy. One student was killed and 7 injured. The World Bank, the US government and the European Parliament urged the Guatemalan government to end political violence.

[29] In October, while the quincentenary of Columbus's arrival on the continent was celebrated, Rigoberta Menchú Tum, an indigenous leader from the Quiché ethnic group, won the Nobel Peace Prize. Menchú has traveled world-wide, denouncing the situation of her country's indigenous peoples.

[30] On May 25, 1993, President Serrano, backed by a group of military officers, carried out an "auto-coup", revoking several articles of the constitution and dissolving Congress and the Supreme Court. On June 1, national and international rejection of this measure - including pressure from the United States - made Serrano abandon office. On June 6, after several days of uncertainty, former Human Rights attorney Ramiro De León Carpio was elected head of the Executive, to finish Serrano's term.

[31] De León Carpio began by purging the officers that had supported Serrano, changing five military commands. Shortly afterwards, Jorge Carpio Nicolle, leader of the

of "model hamlets" increased systematically. Peasants were taken by force to these hamlets, where they were required to produce cash crops for export, rather than growing subsistence crops.

[16] In August 1983, another coup staged by the CIA deposed Ríos Montt and General Oscar Mejía Víctores came into power, promising a quick return to a democratic system.

[17] An 88-seat Constituent Assembly was elected on July 18, 1984, to replace the legal framework in effect since 1965, and annulled after the 1982 coup. The Assembly offered new constitutional guarantees, habeas corpus and electoral regulations. Seventeen parties ran candidates; however, lack of guarantees forced the left to abstain from running yet another time.

[18] The Constituent Assembly approved the right to strike for civil servants, authorized the return from exile of leaders of the Social-

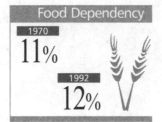

Food Dependency

1970
11%

1992
12%

Maternal Mortality
1989-95

PER 100,000
LIVE BIRTHS
464

Illiteracy
1995

44%

Union of the National Center and the president's cousin, was assassinated.

[32] Despite the intense campaign against the PACs and compulsory military service, President De León Carpio stated that he would maintain both institutions as long as the situation of armed conflict persisted, a U-turn on his previously stated opinion. On August 5, the government announced that the "Files" - records kept on citizens considered a "danger" to State security - had disappeared. However, the loss of these papers also implied the elimination of evidence against those responsible for human rights violations. In November, excavations in several clandestine cemeteries uncovered the remains of 177 women and children assassinated by the military in the 1982 "Rio Negro Massacre".

[33] The 1994-95 Government Plan, presented in August, reaffirmed the structural adjustment program already in effect, prioritizing the end of state intervention in the economy, along with fiscal reform and the privatization of enterprises.

[34] De León Carpio's stated goal was to fight corruption in the public sector. On August 26, the president called for the resignation of legislative deputies and members of the Supreme Court, causing a confrontation between the president and Congress. This led to a clash of economic and political interests, culminating in the Executive and Congress agreeing on constitutional reform.

[35] A plebiscite was scheduled for January 30, 1994, for nation's voters to decide on reforms to public administration and in the constitution.

[36] Following a 22-day occupation of the local OAS office by members of the Committee for Campesino Unity and the National Commission of Widows of Guatemala, some 5,000 members of indigenous groups carried out a march demanding the dissolution of the PACs.

[37] In January 1994, the government and guerrilla signed agreements for the resettlement of the population displaced by the armed conflict, without a mediated cease fire. As a result of the accords 870 people were able to settle in the zones of Chacula, Nenton and Huehuetenango, but the majority of the resettlement areas were still under army control.

[38] Around the same date, a referendum for constitutional reform registered an abstention rate of 85 per cent. Of those who did vote, 69 per cent supported the moves for a new Congress and a new Supreme Court. In general elections with a similar abstention rate, the Guatemalan Republican Alliance led by former dictator Efrain Rios Montt took the majority. The National Vanguard Party came second.

[39] Monica Pinto, the UN Human Rights observer, recommended the demilitarization of society through the gradual reduction in size of the army, the disbanding of the PACs and the Presidential Staff, along with the creation of a Truth Commission.

[40] Following a prolonged silence in early 1994, the foreign minister recognized Belize as an independent State, but maintained its territorial claim, meaning no frontier could be established.

[41] In March, the government and the URNG ratified an agreement related to the disbanding of the PACs and the international verification of human rights by the UN. Three days later the president of the Court of Constitutionality, Epaminondas González, was murdered.

[42] In April, the police evicted 300 rural workers' families from an estate they had occupied in Escuintla. In May a new contingent of nearly 2,000 refugees returned, heading for the Quiche area, which was still occupied by the military.

[43] The URNG and the government signed a draft agreement for the "Resettlement of the People Uprooted by the Armed Confrontation," in Oslo, Norway, in June. The Communities of Peoples were recognized as non fighting civilians and the vital importance of land for these uprooted populations was explicitly stated. The second agreement in Oslo enshrined the principle of not individualizing responsibility for human rights violations as a way of neutralizing the action of sectors opposed to a negotiated outcome.

[44] Income from tourism fell by $100 million due to the hunger strike by a US citizen protesting for the liberty of her husband, who had been detained in 1992 for alleged links with left-wing organizations, and

because another US citizen, suspected of involvement in a baby trafficking network, was brutally flogged before a crowd.

[45] The peace negotiations between the guerrillas and the government reached stalemate during 1995, due to the elections and the disinterest of the army and landowners, and the government weakness in face of them. The UN mission reported that impunity continued to be the main obstacle to the respect for human rights, and described hundreds of cases of torture, illegal detentions and extrajudicial executions.

[46] In August 1995, Congress leader Efrain Rios Montt and another two members of the FRG

lost their legal immunity and were tried by the Supreme Court on charges of bugging telephones, falsifying documents and abuse of power.

[47] In the November general elections, Alvaro Arzu, the National Vanguard Party candidate, took 42 per cent of the vote, while his closest rival, Alfonso Portillo Cabrera, of the FRG, achieved only half the amount. In a second round on January 7 1996, Arzu was elected president. Abstention reached the record level of 63 per cent.

STATISTICS

DEMOGRAPHY

Urban: 41% (1995). **Annual growth:** 1.1% (1992-2000). **Estimate for year 2000:** 12,000,000. **Children per woman:** 5.1 (1992).

HEALTH

One **physician** for every 4,000 inhab. (1988-91). **Under-five mortality:** 70 per 1,000 (1994). **Calorie consumption:** 93% of required intake (1995). **Safe water:** 62% of the population has access (1990-95).

EDUCATION

Illiteracy: 44% (1995). **School enrolment:** Primary (1993): 78% fem., 78% male. Secondary (1993): 23% fem., 25% male. **Primary school teachers:** one for every 34 students (1992).

COMMUNICATIONS

89 **newspapers** (1995), 93 **TV sets** (1995) and 83 **radio sets** per 1,000 households (1995). 2.3 **telephones** per 100 inhabitants (1993). **Books:** 92 new titles per 1,000,000 inhabitants in 1995.

ECONOMY

Per capita GNP: $1,200 (1994). **Annual growth:** 0.90% (1985-94). **Annual inflation:** 19.50% (1984-94). **Consumer price index:** 100 in 1990; 181.7 in 1994. **Currency:** 6 quetzales = 1$ (1994). **Cereal imports:** 486,000 metric tons (1993). **Fertilizer use:** 833 kgs per ha. (1992-93). **Imports:** $2,604 million (1994). **Exports:** $1,522 million (1994). **External debt:** $3,017 million (1994), $292 per capita (1994). **Debt service:** 11.4% of exports (1994). **Development aid received:** $212 million (1993); $21 per capita; 2% of GNP.

ENERGY

Consumption: 186 kgs of Oil Equivalent per capita yearly (1994), 70% imported (1994).

Guinea

Guinée

Population: 6,425,000 (1994)
Area: 245,860 SQ KM
Capital: Conakry

Currency: Franc
Language: French

Guinea was initially inhabited by pygmies who were driven towards the more inhospitable regions by the Mande peoples. The country was originally connected to the great states of Sudan and their trade system. The Bambuk gold mines that fed the Mediterranean economy for centuries were located on the ridges of the Futa Dyalon massif (see Mali and Morocco). The same region witnessed the rise and fall of the Fulah states between the 16th and 19th centuries (see Senegal: "The Fulah States").

[2] In 1870, against this background, Samor became the local political and spiritual leader (Almamy) of a state that included the greater part of Guinea and parts of present-day Mali and Ivory Coast. He had first confronted the French troops in 1886, as they were advancing from Senegal, and he had continued to fight constantly until 1898 when he was taken prisoner and exiled to Gabon. He died there two years later.

[3] This episode had been kept alive in the people's memory until 1947 when a small group of activists led by Ahmed Sekou Touré founded the Guinea Democratic Party (GDP). Supported by labor unions, the GDP soon became a very influential political organization.

[4] Overwhelmed by the loss of Indochina in 1954, the independence of Tunisia and Morocco in 1956, and the uprising of Algeria which began in 1954, French President de Gaulle attempted to salvage his interests in West Africa by creating the French Community in 1958. Through agreements and covenants he attempted a neo-colonialist project, to guarantee continued French hegemony in the area; in particular through its large monopolies. When the population of Guinea was consulted, 1,200,000 opposed de Gaulle's proposal for a French Community and only 57,000 supported it.

[5] Four days later, the country proclaimed independence. Sekou Touré stated: "We'd rather be poor and free than rich and enslaved". It was the first time that such a thing had happened in "French" west Africa, and in retaliation, Paris withdrew its technical advisers, paralyzed the few industries that there were, and blocked Guinean trade.

[6] In 1959, the state took over the main economic activities and Guinea immediately created its own currency, becoming independent from the French franc. Industry and farming were diversified in an attempt to attain self-sufficiency, and aluminium production surpassed one million tons per year.

[7] French aggression mounted. In 1965, Guinea's bank accounts in Paris were blocked, large-scale smuggling was fostered and counterfeit currency introduced. In 1970, Portuguese mercenaries invaded the country in an attempt to overthrow the government and destroy the PAIGC, which was fighting for the independence of neighboring "Portuguese" Guinea. In response to the ever more frequent destabilization campaigns, Local Revolutionary Powers (LRPs) were created in every neighborhood or village with the aim of fighting against corruption, mismanagement and contraband.

[8] In November 1978, the Congress decided to change the country's name to the People's and Revolutionary Republic of Guinea, and to re-open relations with France. After a long period of isolation,

> In November 1978, the Congress decided to change the country's name to the People's and Revolutionary Republic of Guinea, and to re-open relations with France.

Sekou Touré visited African and Arab countries in search of new sources of foreign capital to exploit the country's mineral wealth and to attract investments which might help to pay the foreign debt.

[9] The rapprochement between France and Guinea brought new agreements. The exploitation of the Mount Nimba iron ore reserves was granted to a French firm while the French Oil Company started offshore oil prospecting and the exploitation of rich bauxite deposits discovered in the 1970s. Guinea became the world's second largest bauxite producer. At that time Guinea and Mali signed an agreement envisaging the merger of the two neighbors into an economic and political federation.

[10] At the end of March 1984, Sekou Touré died in a US hospital while undergoing treatment for an old ailment. A week later, Colonel Lansana Conté headed a military coup which overthrew interim president Louis Beavogui.

[11] The new regime dismantled the political structure, the grassroots organizations and the Assembly; the Constitution was abolished too. The Democratic Party of Guinea was banned and the name of the country was changed, eliminating the words "People's" and "Revolutionary". The military government prompted private enterprise, eliminated mixed companies and applied for economic aid to France, the US and the African States.

[12] Under the burden of an $800 million foreign debt, the government decreed a 100 per cent devaluation of the *syli* (the national currency), and reduced public expenditure, a pre-condition of Guinea's entry into the French franc monetary scheme.

[13] In December 1984, the number of cabinet members was reduced, as was the number serving on the Reconstruction Comittee. In addition, Colonel Conté consolidated power by simultaneously taking on the positions of head of state, prime minister and minister of defence.

[14] In December 1985, an Economic Recovery program was introduced, where agriculture, especially the cultivation of rice, was given top priority. Industrial and agricultural enterprises were also to be privatized. Self-

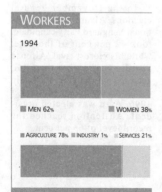

WORKERS

1994

MEN 62%   WOMEN 38%

AGRICULTURE 78%   INDUSTRY 1%   SERVICES 21%

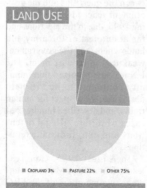

LAND USE

CROPLAND 3%   PASTURE 22%   OTHER 75%

The Songhai empire

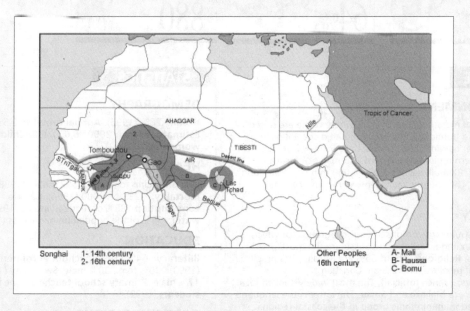

Songhai 1- 14th century
        2- 16th century

Other Peoples   A- Mali
16th century   B- Haussa
                C- Bornu

[1] The Songhai, who were a nation of Nilotic origin coming from the east, settled on the banks of the Niger river at the beginning of the eighth century.
[2] According to tradition, they rapidly achieved political unity under two leaders of Berber origin.
[3] As early as the 10th century, Arab historians wrote about the nation of Kugha and its traditional leaders, the Dia family, reporting a state nearly as remarkable as that of Ghana.
[4] In the subsequent period they founded various city-states along the Niger. These states were autonomous, but they were related through kinship, as all the traditional leaders came from the Dia family.
[5] The most outstanding of these city-states was Gao, located where the Niger joins the desert trade route (see Niger - box). As a result of its geographical position it rapidly grew into an important trade centre.
[6] Although the leaders of this ever more prosperous trade city were at one time subject to the Sultans of Mali, by the 15th century they had regained their autonomy and extended their power over a far wider area.
[7] In 1468 "Sonni" (traditional leader) Ali conquered Timbuktu (Tombouctou). This city had formerly been ruled by the Mali sovereign, but had fallen to the Tuareg attackers. From this date until his death, "Sonni" Ali Ber (the Great) concentrated his energies in expanding his domain until it comprised almost all of Mali, the

north of Burkina Faso and Benin, and southwest Niger.
[8] In spite of his Muslim origin, Ali Ber was a stern advocate of the local African native traditions in which he educated his sons. He confronted the Muslim clergy and fiercely persecuted the Fulani nations because he considered them lay propagators of Islam.
[9] His death in 1492 raised the question of Islamic domination once again, and this soon led to a confrontation. When the new Sonni refused to be converted, the head of the army, Muhammad Turé of Senegal, deposed him and took his place under the name Askia Muhammad, thereby imposing Islamic leadership.
[10] The Songhai empire reached its greatest extension and its best political organization under the Askias. Divided into four administrative units, each under a Sonni delegate, the state branched into numerous provinces with local governors, allowing for efficient administration in which the Fulani again acted as intermediaries, while a professional army and a fleet on the Niger guaranteed peace and trade throughout the vast empire.
[11] The Tuareg resumed paying taxes, cultural and economic development again centered in Timbuktu, and the city flourished on the profits of trans-Saharan trade.
[12] Leon the African, author of historical records from this period, reported that the most profitable items on the Songhai market were books, because they were in such high demand, as the wealth of the traders of Timbuktu was measured by the number of manuscripts in their libraries.
[13] However, the splendor of the Songhai

empire did not even last a century. Gold production in Bambuk and Bure slowly declined, removing a major source of income, and an ever-growing proportion of the gold that was produced was channelled towards the Atlantic, bound for Europe.
[14] The shortage of gold in northern Africa led the Sultan of Morocco, Mulay Ahmad the Victorious, to send an expedition armed with artillery to conquer the Songhai nation and its mines.
[15] In 1591, the Moroccans (mainly Spanish renegades in the Sultan's service) conquered Timbuktu, Gao and the major cities of the mid-Niger. They were looking for the rumored large gold reserves, which the Sultan expected them to deliver to him. They could not find them and they were unable to return empty-handed, so they were left to fend for themselves.
[16] Formally, there continued to be "Moroccan" pashas in Timbuktu until 1770, but in this period they were really only tax-payers contributing to the Tuareg treasury; the Spanish renegades had soon mixed in with the local population.
[17] The collapse of the empire made way for other nations, particularly the Bambar of Mali. In the 17th century, they founded the states of Ségou and Kaarta along the upper Niger, which held until the Fulani expansion in the 19th century.

PER CAPITA
$483

64%

PER 100,000
LIVE BIRTHS
880

PROFILE

ENVIRONMENT

The central massif of Futa-Dyalon, where cattle are raised, separates a humid and densely populated coastal plain, where rice, bananas and coconuts are grown, from a dryer northeastern region, where corn and manioc/cassava are cultivated. Rainfall reaches 3,000-4,000 mm per year along the coast. The country has extensive iron and bauxite deposits.

SOCIETY

Peoples: Guineans comprise 16 ethnic groups, of which Fulah, Mandingo, Malinke and Sussu are the most numerous. **Religions:** About 85% are Muslim, 5% practise traditional religions, and 2% are Christian.
Languages: French (official). The most widely-spoken local languages are Malinke and Sussu. The latter is the language of the predominant ethnic group in the coastal region.
Political Parties: Party for Unity and Progress; Guinean People's Union; Union for National Prosperity; Democratic Party of Guinea/African Democratic Union. **Social Organizations:** National Confederation of Guinean Workers.

THE STATE

Official Name: République de Guinée. **Administrative divisions:** 33 regions. **Capital:** Conakry, 700,000 inhab. (1989). **Other Cities:** Kankan, 55,000 inhab.; N'Zerekore, 45,000 inhab. (1983). **Government:** General Lansana Conté, President. **Armed Forces:** 9,700.
Paramilitaries: People's Militia: 7,000; Gendarmerie: 1,000; Republican Guard: 1,600.

STATISTICS

DEMOGRAPHY

Urban: 28% (1995). **Annual growth:** 2.5% (1992-2000). **Estimate for year 2000:** 8,000,000. **Children per woman:** 6.5 (1992).

HEALTH

One **physician** for every 7,692 inhab. (1988-91). **Under-five mortality:** 223 per 1,000 (1994). **Calorie consumption:** 88% of required intake (1995). **Safe water:** 55% of the population has access (1990-95).

EDUCATION

Illiteracy: 64% (1995). **School enrolment:** Primary (1993): 30% fem., 30% male. Secondary (1993): 6% fem., 17% male. **Primary school teachers:** one for every 49 students (1992).

COMMUNICATIONS

81 **newspapers** (1995), 83 **TV sets** (1995) and 79 **radio sets** per 1,000 households (1995). 0.2 **telephones** per 100 inhabitants (1993). **Books:** new titles per 1,000,000 inhabitants in 1995.

ECONOMY

Per capita GNP: $520 (1994). **Annual growth:** 1.30% (1985-94). **Annual inflation:** 18.60% (1984-94). **Consumer price index:** 100 in 1990; 155.6 in 1994. **Currency:** 981 francs = 1$ (1994). **Cereal imports:** 335,000 metric tons (1993). **Fertilizer use:** 47 kgs per ha. (1992-93). **External debt:** $3,104 million (1994), $483 per capita (1994). **Debt service:** 14.2% of exports (1994). **Development aid received:** $414 million (1993); $66 per capita; 13% of GNP.

ENERGY

Consumption: 65 kgs of Oil Equivalent per capita yearly (1994), 87% imported (1994).

---

sufficiency in food is still an unattained goal.

[15] To meet the country's serious economic problems head-on, in 1987 there were mass dismissals of public officials. In 1988, Conté moved a group of military officers away from the capital, as their discontent over low salaries constituted a threat to his government. In January of the same year, the 80 per cent increase in all salaries caused a "price explosion" with prices tripling over a few days. The demonstrations which took place at the time forced the government to freeze the prices of consumer good and rents.

[16] In February 1990 the government announced an amnesty for political prisoners and exiles. In December, a referendum approved by a vast majority modified the Constitution and created a Temporary Council for National Development (CTDN) in charge of leading the transition to democracy. President Conté announced that legislative elections would be held in 1992, with the participation of several parties.

[17] The drop in the international price of bauxite, which accounts for 60 per cent of the country's earnings and 96 per cent of its exports, aggravated the state of the economy. According to international credit agencies, official corruption was so widespread that a structural-adjustment program would be impossible to implement without first eradicating these practices. In 1992, the World Bank demanded that the government fire 40,000 civil servants.

[18] The country's economic problems are reflected in the quality of life of its inhabitants. Nutritional deficiencies and diseases result in a short life expectancy of 42 years, as well as a high infant mortality rate. The most serious environmental problems are water pollution and the lack of sanitation, which result in the spreading of parasitic diseases like ancylotomiasis and amebiasis.

[19] In 1992, 650,000 refugees crossed the border from Sierra Leone and Liberia, setting up makeshift camps. The government increased the number of peacekeeping troops sent to both countries by the Southwest African Economic Community, to patrol the borders and prevent more refugees from entering the country.

[20] When the multi-party system was approved, opposition leader Alpha Conde returned from exile and sponsored the creation of the National Democratic Forum (FND) which enclosed 30 opposition groups. However, tension and political persecution continued.

[21] Lansana Conté was re-elected in the December 1993 presidential elections with almost 51 per cent of the vote. Alpha Conde accused him of having staged a coup d'etat which led to confrontations between the police and opposition followers. In January 1994, dozens of people died in riots in Macenta near the border with Liberia.

[22] In June 1995, the ruling parties won 76 seats out of 114 in parliamentary elections which the opposition deemed fraudulent. Estimated annual GDP growth reached almost 5 per cent and, in recognition of what it considered to be good economic results, the Paris Club cancelled $85 million of Guinea's foreign debt, refinancing another $85 million still owed by Conakry.

Guinea-Bissau

Population: 1,044,000 (1994)
Area: 36,120 SQ KM
Capital: Bissau

Currency: Peso
Language: Portuguese

Guinea-Bissau

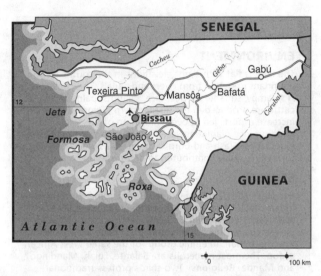

Guinea-Bissau was the first Portuguese colony to gain independence in Africa. This was achieved even before the fall of the Portuguese dictatorship, in a successful political and military struggle led by Amilcar Cabral's, African Party for the Independence of Guinea and Cape Verde (PAIGC).

[2] After belonging to the Mali and Songhai empires (see box in Guinea), the peoples in the Geba river valley became independent. This independence was soon threatened by the Portuguese, who had been settled on the coast from the end of the 15th century, and by the Fulah, coming from the interior in the 16th century. Inland, the state of Gabu remained autonomous until the 19th century (see Senegal: "The Fulah States") while the coastal population suffered the consequences of the slave trade and forced displacement to the Cape Verde Islands.

[3] Resistance against the European colonists began in 1500 when the Portuguese arrived in Guinea. At that time, the country was inhabited by several different groups, immigrants from the state of Mali along with the Fulah and Mandingo groups, who lived in organized autocratic societies in the savannahs. During the 17th century Guineans made their first contact with the inhabitants of the Cape Verde islands, a mandatory stopover for the ships carrying slaves to Brazil.

[4] As the country was small and poor the monopoly on trade and agriculture was dealt with by a private company, the Uniao Fabril. Guineans were forced to cultivate export crops while massively reducing the acreage available for subsistence farming. In the 1950s, infant mortality reached the remarkable rate of 600 deaths per 1,000 births. There were only 11 doctors in the country and only one percent of the rural population was literate. In the early 1960s, only 11 Guineans had completed secondary education

The PAIGC government diversified agriculture giving priority to feeding the population and reducing the emphasis on export crops. Foreign trade was nationalized, agrarian reform was implemented together with a mass literacy campaign.

[5] It was against this setting of misery and exploitation that Amilcar Cabral founded the Athletics and Recreational Association, in 1954. This organization developed into the African Party for the Independence of Guinea and Cape Verde, two years later. The party called on all Guineans and inhabitants of Cape Verde to unite in anti-colonial resistance, regardless of color, race or religion. In September 1959, after trying fruitlessly to engage the Portuguese in negotiations for three years, the PAIGC embarked upon guerrilla warfare. The fighting spread quickly and by 1968 the Portuguese were confined to the capital, Bissau, and a few coastal strongholds. Four years later, the PAIGC had taken two-thirds of the nation's territory under their control. In September and October 1972, the first free elections were held in the liberated areas. A Popular National Assembly was elected and a year later, on September 24 1973, the "democratic, anti-imperialist and anti-colonialist republic of Guinea" was proclaimed. Two months later the UN General Assembly recognized the independent state.

[6] Amilcar Cabral was assassinated in Conakry in February 1973, by Portuguese agents. He left many books and studies on the struggles for freedom in the African colonies. His successor, Luis Cabral, set up the Government Council of Guinea-Bissau in a small village called Madina do Boé, in the heart of the liberated area.

[7] The impact of the unilateral independence of Guinea-Bissau and its immediate recognition by the UN shook the infrastructure of Portuguese colonialism. General Spinola, commander of the 55,000 colonial soldiers,came to the realization that his army could not be successful in the war against the PAIGC and argued for political changes in Portugal. It was as a result of these conditions in Bissau that the Captain's Movement was born. This movement later became the Armed Forces' Movement, the group which was responsible for the coup that overthrew the dictatorial regime in Portugal on April 25, 1974. Four months after the coup, Portugal recognized the independence of Guinea-Bissau.

[8] The PAIGC government diversified agriculture giving priority to feeding the population and reducing the emphasis on export crops. Foreign trade was nationalized, agrarian reform was implemented together with a mass literacy campaign. In foreign relations, the new government opted for non-alignment and unconditional support for the struggle against apartheid and colonialism in Africa. The PAIGC congress also gave top priority to economic integration with the archipelago of Cape Verde, with a view towards uniting the two countries.

[9] In 1980, a military conspiracy led by Joao Bernardino (Niño) Vieira, a former guerrilla commander, overthrew Luis Cabral replacing all the government bodies with a centralized Revolutionary Council, headed by Bernardino Vieira.

[10] Talks with Cape Verde were abruptly cut off while the two countries were discussing a unified constitution (See Cape Verde). The new government in Bissau was immediately recognized by the neighboring Republic of Guinea, which had been at loggerheads with the former president Cabral in a dispute over offshore oil rights in an area presumed to be rich in petroleum deposits.

[11] Contact between the two Guineas was intensified in September 1982, and in February 1983 diplomatic missions were exchanged.

[12] The first development plan of 1983-86 proposed an initial investment of $118.6 million, of which 75 per cent would be financed by international funds. In 1984, the construction of five ports was started at an estimated cost of $40 million, and the construction of the Bisalan-

Workers

1994

■ MEN 61%   ■ WOMEN 39%

■ AGRICULTURE 82% ■ INDUSTRY 4% ■ SERVICES 14%

Land Use

■ CROPLAND 12%   ■ PASTURE 38%   ■ OTHER 50%

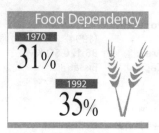

Food Dependency

1970
31%

1992
35%

External Debt

1994

PER CAPITA
$782

Illiteracy

1995

45%

Foreign Trade

MILLIONS $ 1994

IMPORTS
63

EXPORTS
32

PROFILE

ENVIRONMENT

The land is flat with slight elevations in the southeast and abundant irrigation from rivers and canals. The coastal area is swampy, suitable for rice. Rice, peanuts, palm oil and cattle are produced in the drier eastern region. The need to increase exports led to over-cultivation of the soil; in addition, rice plantations are replacing part of the coastal woodlands. Slash-and-burn techniques, as well as numerous forest fires, have contributed to deforestation.

SOCIETY

Peoples: The people of Guinea-Bissau come from three major roots: *mestiços* are descendants of Cape Verdeans, *assimilados* come from GuineaConakry immigrants, and *indigenas* are the native inhabitants. *Indigenas* are divided into more than 40 ethnic groups of the same West African group. The most numerous are Balante, Fulah, Mandingo and Mande. **Religions:** Two-thirds profess traditional African religions; nearly one-third are Muslim and there is a small Catholic minority. **Languages:** Portuguese (official). The *crioulo* dialect, a mixture of Portuguese and African languages, is used as the lingua franca. The most widely spoken native languages are Mande and Fulah. **Political Parties:** The African Party for the Independence of Guinea and Cape Verde (PAIGC) party in power and the country's sole political party until the end of 1991; Party for Social Renovation.

THE STATE

Official Name: República da Guiné-Bissau. **Administrative Divisions:** 8 Regions and 1 Autonomous Sector. **Capital:** Bissau, 125,000 inhab. (1988) **Other cities:** Bafatá, 13,429 inhab.; Gabú, 7,803 inhab.; Mansôa, 5,390 inhab. (1979) **Government:** Gen. João Bernardino (Niño) Vieira, president. National People's Assembly, parliament of 100 seats. **National Holiday:** September 24, Independence Day (1973). **Armed Forces:** 7,200 (1995). **Paramilitaries:** 2,000 Gendarmes.

STATISTICS

DEMOGRAPHY

Urban: 21% (1995). **Annual growth:** 2.4% (1992-2000). **Estimate for year 2000:** 1,000,000. **Children per woman:** 6 (1992).

HEALTH

Under-five mortality: 231 per 1,000 (1994). **Calorie consumption:** 95% of required intake (1995). **Safe water:** 53% of the population has access (1990-95).

EDUCATION

Illiteracy: 45% (1995). **Primary school teachers:** one for every 25 students (1987).

COMMUNICATIONS

83 **newspapers** (1995), 81 **TV sets** (1995) and 79 **radio sets** per 1,000 households (1995). 0.8 **telephones** per 100 inhabitants (1993). **Books:** new titles per 1,000,000 inhabitants in 1995.

ECONOMY

Per capita GNP: $240 (1994). **Annual growth:** 2.20% (1985-94). **Annual inflation:** 65.70% (1984-94). **Currency:** 15.369 pesos = 1$ (1994). **Cereal imports:** 70,000 metric tons (1993). **Fertilizer use:** 10 kgs per ha. (1992-93). **Imports:** $63 million (1994). **Exports:** $32 million (1994). **External debt:** $816 million (1994), $782 per capita (1994). **Debt service:** 15.2% of exports (1994). **Development aid received:** $97 million (1993); $95 per capita; 40% of GNP.

ENERGY

Consumption: 37 kgs of Oil Equivalent per capita yearly (1994), 100% imported (1994).

ca Airport was completed. The government started a campaign against corruption and inefficiency in public administration, and as a result, in 1984, Vice-president Victor Saude Maria was asked to resign. Shortly afterwards, the Popular Assembly eliminated the position of prime minister, and the Revolutionary Council became the Council of State.

[13] In November 1984, the vice-president of the Council of State, Colonel Paulo Correia, led an unsuccessful coup against President Vieira. Paulo Correia was executed on July 21, 1986.

[14] The 1984 stabilization plan failed, causing further deterioration of the economic and financial situation. The international economic crisis increased, the price of Guinea-Bissau's oil exports lost market value. The government adopted a "corrective" policy, including freezing salaries and reducing public investment. This was an attempt to bring it into line with IMF conditions for refinancing the servicing of the foreign debt. The economy was subsequently opened up to foreign capital in the hope of attracting resources from Portugal and France, particularly in the high-priority areas of telecommunications. In 1989, Portugal participated in the creation of a commercial bank.

[15] In 1986, the government granted an amnesty to political prisoners. Two years later, in June 1988, the Central Committee of the PAIGC began discussing a proposal for separating the executive and legislative powers. At that time Guinea-Bissau spent more on its military budget than on education and health combined.

[16] In February 1991, the PAIGC approved a political reform which anticipated elections for 1992. The IMF cancelled several loans granted to Guinea-Bissau after the resignation of Economy minister Manuel dos Santos. Meanwhile, the World Bank deferred the payment of $6.5 million belonging to the structural adjustment program begun in 1987, to force Bissau to pay its foreign debt on time.

[17] Dependence on agricultural exports - especially on peanuts, whose price had fallen abruptly - caused significant social and economic imbalances along with political tensions in 1992 and 1993. The 1992 elections were not held due to a government decision and after the murder in March 1993 of a high military commander they were delayed once more.

[18] Finally, in 1994 Joao Bernardo Vieira defeated Kumba Iala of the Party for Social Renovation. After he obtained 46 per cent of the vote in the first round in July - Iala had less than 22 per cent - Vieira obtained 52 per cent in the second round and was elected president. During the campaign, Iala accused him of supporting tribalism and racism. In the parliamentary elections, Vieira's PAIGC conquered 64 of the 100 seats at stake. Believing that the ruling party had "bought" votes, Iala refused to take part in a national unity government.

[19] In January 1995, the IMF granted a new $14 million loan in support of economic reforms. In June, the visit of Senegalese president Abdou Diouf led to a rapprochement with Dakar after a period of relative hostility. Both countries agreed to exploit joint energy and mineral resources. In August, Iala denounced Vieira's coming to terms with France, the rise in the price of basic items like rice and human rights violations by the government.

Guyana

Guyana

Population: 826,000
(1994)
Area: 214,970 SQ KM
Capital: Georgetown

Currency: Guyana Dollar
Language: English

The original inhabitants of what is now Guyana were the Arawaks, but they did not increase substantially in number due to the humid lands and the mangrove-covered coasts, which were mostly swamps. The Arawaks were displaced from the area by the Caribs, warriors who dominated the region before moving on to the islands in the sea, which was later called after them.

2   Both the Arawaks and the Caribs were nomads. Organized into families of 15 to 20 people, they lived by fishing and hunting. There are thought to have been half a million inhabitants at the time of the arrival of Europeans in Guyana; there are presently 45,000 Indians, divided into nine ethnic groups, of which seven maintain their cultural identity and traditions.

3   Led on by the legend of El Dorado, in 1616, the Dutch built the first fort there. Guyana was made up of three colonies: Demerara, Berbice and Essequibo. But in 1796, the Dutch colony was taken over by the Britlish, who had already begun a massive introduction of slaves. A slave, Cuffy, led a rebellion in 1763, which was brutally put down. To this day, Cuffy is considered a national hero.

4   Those slaves that escaped from the plantations went into the forests to live with the indigenous peoples, giving rise to the "bush blacks". The English brought in Chinese, Javanese and Indian workers as cheap labor. In the second half of the 20th century, Guyana's population managed to channel independence feelings into a single movement, the People's Progressive Party (PPP), with policies of national independence and social improvements, with long term aims for a socialist country. Cheddi Jagan stayed prime minister of the colony for three successive terms.

5   After years of struggle and periods of great violence, Britain rcognized Guyana's independence within the Commonwealth on May 26, 1966. By that time, the PPP had split; the Afro-Guyanese population joined the People's National Congress (CNP), while indigenous remained loyal to Jagan. Forbes Burnham, leader of the CNP took office, supported by other ethnic minorities.

6   This process was influenced by ethnic conflict and by foreign interests, particularly from the US, which felt its hegemony in the Caribbean threatened by Jagan's socialism.

7   Even though Burnham came to power with Washington's blessing, he kept his distance. He declared himself in favor of non-alignment and proclaimed a Cooperative Republic in 1970. The bauxite, lumber, and sugar industries were nationalized in the first half of the 1970s, and by 1976, the state controlled 75 per cent of the country's economy. At the same time, regional integration was implemented through CARICOM, the Latin American Economic System (SELA), and the Caribbean Merchant Fleet. During the first decade after independence, Burnham and Jagan defended the same political platform.

8   In May 1976, Cheddi Jagan stated the need to "achieve national anti-imperialist unity", when disputes broke out with Brazil over the border. The PPP representatives returned to parliament, from which they had withdrawn three years earlier to protest over electoral corruption. Shortly afterwards, Burnham announced the creation of a Popular Militia.

9   Elections were postponed in order to hold a constitutional referendum, with Parliament drawing up a new constitution. This gave way to harsh criticism by the PPP, which withdrew from legislative activity for the second time. In 1980 Burnham was elected president. However, according to international observers, the election had been plagued by fraud. In 1980 Burnham granted authorization for transnational corporations to carry out oil and uranium operations. In addition, he turned to the IMF to obtain credit.

10   In June of the same year, Walter Rodney, a world renowned intellectual and founder of the opposition "Working People's Alliance" (WPA) was killed by a car bomb. The culprits were never found.

11   In the post-election period, border disputes escalated. Venezuela claimed the Essequibo region, approximately 159,000 sq km (three fourths) of Guyanese territory, arguing that British imperialism illegally deprived Venezuela of that area in the 19th century.

12   In 1983, both countries turned to the UN. In 1985 direct negotiations started again to settle the dispute "within a framework of cordiality and goodwill". Negotiations were focused on an outlet to the Atlantic Ocean for Venezuela.

13   While financial difficulties increased during 1984 and the government faced a new crisis in its relations with labor unions, Burnham resumed contacts with the IMF to obtain a $150 million loan. Burnham considered the conditions on the loan "unacceptable". The American invasion of Grenada - and Guyana's criticism of this action - led to a deterioration of the relations between the two countries. Guyana made overtures to the socialist countries.

14   Burnham died in August 1985 and was replaced by Desmond Hoyte. In general elections held in 1985, the PNC won, with 78 per cent of the votes, and once again the opposition levelled accusations of fraud. In 1986, five of the six opposition parties joined together to form the PCD (Patriotic Coalition for Democracy) but decided not to

> In four years, the People's Revolutionary Government stimulated the formation of grass roots organizations, and they created a mixed economy, expanding the public sector through agro-industries and state farms.

WORKERS

1994

MEN 74%   WOMEN 26%

AGRICULTURE 27%   INDUSTRY 26%   SERVICES 47%

ENVIRONMENT

90% of the population and most of the country's agriculture are concentrated on the coastal plane which ranges between 15 and 90 km in width. Rice and sugar cane are the main crops. As most of the shore is below sea level, dams and canals have been built to prevent flooding. The inner land consists of a 150 km-wide rainforest where the country's mineral resources are concentrated (bauxite, gold, and diamonds). To the west and south, the rest of the country is occupied by an ancient geological formation, the Guyana mountain range. Guyana is a native word meaning "land of waters". There are many rivers, as a result of the tropical climate and year-round rains. In comparison with world-wide deforestation, the Guyanas have suffered little, and until 1990, only a small fraction of the extensive forests had been felled. However, foreign agencies are pressing for the intensification of lumber exploitation. In some areas, reforestation after logging has not been carrried out, leading to soil erosion.

SOCIETY

Peoples: Half the population descend from Indian workers, one-third from African natives and the rest are native Americans, mestizos, Chinese and Europeans.
Religions: Half of the population are Christian, one third are Hindu and there is a Muslim minority.
Languages: English is official, Hindi and Urdu are used in religious ceremonies. **Political Parties:** People's Progressive Party; People's National Congress (PNC). **Social Organizations:** Trade Union Congress (TUC).

THE STATE

Official Name: Cooperative Republic of Guyana.
Administrative Divisions: 10 Regions.
Capital: Georgetown, 248,000 inhab. (1992). Other cities: Linden 27,200 inhab.; New Amsterdam 17,700 inhab.
Government: Cheddi Jagan, President, Sam Hinds, Prime Minister. Parliament is composed of a unicameral assembly with 65 members, of which 12 are regional representatives and 53 are elected through direct vote and proportional representation. **National Holiday:** February 23, Proclamation of the Republic (1970). **Armed Forces:** 1,600 (1995). **Paramilitaries:** 4,500 People's Militia, national service.

DEMOGRAPHY

Annual growth: 1.6% (1992-2000). **Children per woman:** 2.6 (1992).

HEALTH

Under-five mortality: 61 per 1,000 (1994). **Calorie consumption:** 99% of required intake (1995).

EDUCATION

Illiteracy: 2% (1995). **Primary school teachers:** one for every 36 students (1988).

COMMUNICATIONS

102 **newspapers** (1995), 89 **TV sets** (1995) and 105 **radio sets** per 1,000 households (1995). 5.1 **telephones** per 100 inhabitants (1993). **Books:** 91 new titles per 1,000,000 inhabitants in 1995.

ECONOMY

Per capita GNP: $530 (1994).
Annual growth: 0.40% (1985-94).
Consumer price index: 100 in 1990; 167.2 in 1989.
Currency: 143 Guyana dollars = 1$ (1994).
Cereal imports: 53,000 metric tons (1990).

---

participate in the municipal elections the same year; therefore, all offices up for election remained in PNC hands. In January 1987, President-elect Hoyte announced that his government intended to return to Cooperative Socialism.

[15] Guyanan women have formed movements, in conjunction with the Department of Social Sciences of the university, to denounce sexual discrimination, and cases of abuse or battering of women.

[16] Parliament gathered on December 3, 1991, five days after the government declared a state of emergency, in order to postpone the elections scheduled for the 16th of the month. The government disregarded the objections of the opposition and extended the state of emergency until June 1992.

[17] On October 5, 1992, Cheddi Jagan defeated President Desmond Hoyte by 54 to 41 per cent, in the general elections. The Progressive People's Party (PPP) won 32 seats in the National Assembly, while the National Congress of the People (NCP) won 31.

[18] One of the first Latin American leaders to carry the Marxist banner in the 1950s, Jagan's return to power put an end to 28 years of NCP government.

[19] Toward the beginning of 1993, President Cheddi Jagan surprised observers by going against the policy of neighboring countries, including Brazil, permitting US troops to carry out military training in Guyana.

[20] Brazil expressed concern that US military bases in Panama might eventually be transferred to Guyana. Jagan denied this possibility, but admitted there would be collaboration with US military forces in combatting drug trafficking and in developing sanitation facilities in the country's interior.

[21] Jagan intended to modify the structural adjustment program launched by his predecessor, Desmond Hoyte, in line with the IMF. Jagan favored the use of "non-conventional" methods to solve land distribution, transportation, health, housing and education problems.

[22] He proposed a free market strategy for solving the problem of poverty, which affected 80 per cent of the population. The size of this problem was reflected in the fact that migration outpaced demographic growth, with the population falling from 1,020,000 in 1989 to 808,000 in 1992.

[23] In March 1994, the government and the country's workers jointly resisted IMF pressure to bring wage increases for the public sector in line with inflation.

[24] In June 1994, Jagan rejected the US candidate for ambassador, accusing him of "subversive activities" during the last few years of British colonial administration. In 1995, four million cubic meters of cyanide-contaminated waste fell into the Omai river, which flows into the Essequibo, the main Guyanese river, creating the worst environmental accident in the history of the nation.

[25] In February 1996, the human rights organization Amnesty International denounced the use, for the first time since 1990, of the death sentence by hanging in Guyana.

Haïti

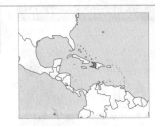

Population: 7,008,000 (1994)
Area: 27,750 SQ KM
Capital: Port-au-Prince

Currency: Gourd
Language: French

Haïti

When Quizqueya, as the local population named it, was "discovered" by Christopher Columbus in 1492, the island was inhabited by numerous Arawak peoples. The contact with Europeans was devastating for the Arawaks, who almost entirely disapeared form the island over the next few decades. Spanish colonists, supported by the Dominican Christianization project, called the islands after their patron saint, St Dominic.

2   The island was later colonized by the French and other European settlers who were attracted by the sugar plantations. Disputes arose among the Europeans and in 1697 Spain ceded the west of the island to France, under the Treaty of Ryswick.

3   After gaining control of the island, France began to exploit it, introducing about 20,000 African slaves per year, leading to rapid racial mixing. Sugar soon became the principal export product of the region, and during the 18th century Haiti became the most important French possession in the Americas.

4   This prosperity was based on African slave labor. By 1789, the number of African slaves in the colony had reached 480,000, there were 60,000 mulattos and free "colored" people, while the rich landowning Europeans constituted a minority of no more than 20,000. The Haitians were influenced by the revolutionary movement that had started in the colonial capitals and they waged a revolutionary war, led by former slave Toussaint L'Ouverture, lasting for 12 years, from 1791 to 1803, ending in the proclamation of the first black republic in the world.

5   Haiti's struggle for independence went through several stages. Initially, the large landowners joined the revolutionary movement, along with the slaves, merchants and poor Europeans (called *petit blancs*), forming a local Assembly to demand an end to colonial rule. In the second stage, the free mulattos supported the French revolution, under the assumption that it would give equal rights to all people, regardless of the color of their skin. However, in 1790 the European planters fiercely rejected the demands of the freed slaves, leaving them no alternative but to ally themselves with the *marrons,* two groups of rebel slaves, a year later.

6   It was L'Ouverture who gave the *marrons* their direction, rallying them to the call of "general freedom for all", transforming the different groups into a disciplined army. On February 4 1794, taking advantage of the fissures in the French colonial system, he succeeded in getting the French National Convention to ratify a decree abolishing slavery in Santo Domingo and appointing L'Ouverture as a General.

7   After the coup of the Brumaire 18, Napoleon Bonaparte sent a colossal military expedition to reconquer the colony and re-establish slavery. L'Ouverture responded with a general uprising, but he was imprisoned and died in exile in France in 1803.

8   Jean-Jacques Dessalines, took over leadership of the war of independence, aided by Henri Christophe and mulatto Alexandre Pétion, who together radicalized L'Ouvertures's legacy. They succeeded in unifying the Africans and mulattos and after a series of heroic campaigns, they forced the French troops to capitulate. Independence was proclaimed on November 28 1803, and Haiti became the first independent state in Latin America.

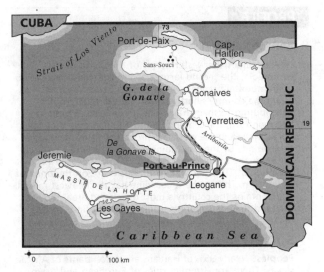

9   Dessalines gave his government a strong nationalist direction, trying to consolidate his personal power by creating an autocratic State, similar to the one emerging in France. Like Napoleon, Dessalines proclaimed himself emperor, calling himself Jacques I. Pétion and Christophe immediately started to plot against him, and he was murdered during a revolt in 1806.

10   The east of the island was recovered by the Spaniards under the Treaty of Paris in 1814, while in the west Christophe and Pétion fought for the leadership, dividing the territory. Henri Christophe established a republic in the north, later making it a kingdom with himself as King Henri I from 1811 to 1820. In the south, Alexandre Pétion ruled over a separate republic, from 1808 to 1818, supporting Simon Bolívar with weapons and funding. Pétion was convinced that only total American independence could guarantee that of Haiti, as the country was being harassed by European powers and the US.

11   In 1818, JP Boyer was elected president instead of Pétion. Boyer recovered the north of the country in 1820, putting an end to Christophe's monarchic experiment. Two years later he conquered Santo Domingo in the east of the island, thus achieving a fragile reunification, which lasted for a quarter of a century. In 1843, a revolution led by the Santo Domingan Creoles divided the island into two definite, independent States: the Dominican Republic in the east, and the Republic of Haiti in the west.

12   In spite of the permanent political violence throughout the 19th century, foreign investors gained access to the Haitian market by building ports and railways and by buying profitable plantations. The new transport and communications led to a faster and more effective extraction of the island's riches, with the profits going to the industrialized countries. The unfavorable trading exchange led the country into debt, until it became wholly dependent on its creditors, who were mostly American.

13   From 1867, a bloody civil war launched a period of political instability and economic crisis which lasted until 1915, when the country was occupied by US marines for non-compliance with "its commitments". A year later the US invaded the Dominican Republic, gaining control of the whole island.

14   The invasion of Haiti was heroically resisted by Charlemagne Péralte's "Revolutionary Army". Péralte was treacherously murdered

PUBLIC EXPENDITURE

% of GNP   1992

MILITARY	EDUCATION	HEALTH
2.1	1.8	3.2

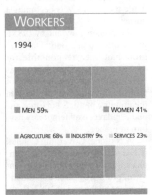

WORKERS

1994

MEN 59%   WOMEN 41%

AGRICULTURE 68%   INDUSTRY 9%   SERVICES 23%

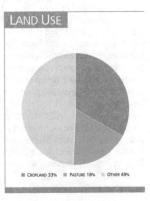

LAND USE

CROPLAND 33%   PASTURE 18%   OTHER 49%

MILLIONS $ 1994

IMPORTS
292

EXPORTS
73

1989-95

PER 100,000
LIVE BIRTHS
600

PROFILE

ENVIRONMENT

Haiti occupies the western third of the island of Hispaniola, the second largest of the Greater Antilles. Two main mountain ranges run from east to west, extending along the country's northern and southern peninsulas, contributing to their shape. The hills and river basins in between form the center of Haiti. The flatlands that open up to the sea in the west, are protected from the humid trade winds by the mountains to the north and east. Coffee is the main export product. Copper ceased to be exploited in 1976 and bauxite deposits are on the brink of exhaustion. The northern coastline has the heaviest rainfall and contains the country's most developed area, though the land there is suffering from serious erosion. Forests make up less than 2% of the land area.

SOCIETY

Peoples: Nearly 95% of Haitians are descendants of African slaves. There are also minorities of European and Asian origin, and integration has produced a small mestizo group. Thousands of Haitians have emigrated in recent years, especially to Colombia, Venezuela and the United States. **Religions:** Voodoo, a mixture of Christianity and various African beliefs. Catholicism. **Languages:** French (official), spoken by less than 20% of the population. Most people speak creole, a local dialect of combined Spanish, English, French and African languages. **Political Parties:** National Front for Change and Democracy (FNCD); National Alliance for Democracy and Progress (ANDP); National Committee of the Democratic Movements (Konakom); National People's Assembly (APN); Agricultural Industrial Party (PAIN); Unified Party of Haitian Communists (PUCH) ; Union for National Reconciliation (UNR); Lavalas Political Organization. **Social Organizations:** The "Grassroots" Church groups, of Catholic origin; Movement in Favor of the Creole Language; Solidarity of the Woman of Haiti (SOFA).

THE STATE

Official name: Repiblik Dayti. **Administrative divisions:** 5 departments. **Capital:** Port-au-Prince, 1,255,000 inhab. (1992). **Other Cities:** Carrefour, 241,000 inhab.; Delmas, 200,000 inhab. (1992). **Government:** President René Préval. **National Holiday:** January 1, Independence Day (1804). **Armed Forces:** 1,500.

geoisie, the ecclesiastic authorities, the state bureaucracy, and the US State Department used Duvalier to control the country for over 30 years. With Washington's backing, Duvalier or "Papa Doc" proclaimed himself "President-for-life" in 1964, passing on the title to his son Jean-Claude or "Baby Doc" on his death in 1971.

[18] In the meantime Haiti fell from being a rich sugar and coffee producer to being the only Latin American country to rank among the 25 poorest nations in the world.

[19] Assailed on the international level by continual condemnations of human rights violations, and on the domestic level by active opposition, Duvalier's government

> After gaining control of the island, France introduced about 20,000 African slaves per year. Sugar soon became the principal export product of the region, and during the 18th century Haiti became the most important French possession in the Americas.

called elections in 1984. 61 per cent of the population abstained. Organizations opposed to Duvalier renewed their underground activities and the opposition grew, organizing itself into parties and trade unions, while the regime was becoming a burden to the US.

[20] Repression grew, and by 1985 it was estimated that Baby Doc's regime had been responsible for 40,000 murders. The country was enveloped by a growing wave of protests and strikes. Duvalier fled the country in a US air force airplane and received temporary asylum in France.

[21] A National Governing Council (CNG) led by General Henri Namphu assumed control of the government, promising "free and direct" elections by the end of 1987. How-

ever, the opposition leaders denounced the continuation of the regime, which was confirmed by Duvalier upon his arrival in France.
[22] The dictator's flight did not end the mobilization of the people. Mass lynching of Tonton-Macoutes forced the National Governing Council to dissolve the repressive force and to free all the prisoners.
[23] In October 1986, the CNG called elections to elect a Constituent Assembly to draw up a new constitution. Less than 10 per cent of the 3 million Haitians participated in the election, and, on March 29 1987 a referendum approved the new constitution with 99.8 per cent of the vote. The new constitution limits the presidential term to 5 years, prohibits re-election and divides the power with a prime minister chosen by Parliament.
[24] The elections were to be held in November 1987, but a few hours after the polling stations were opened, they were sabotaged by factions of the armed forces and by ex-Tonton Macoutes, and the elections were suspended, finally taking place in January 1988. In a very volatile atmosphere, Leslie Manigat, the "official" candidate, was elected, only to be deposed in June, in a coup led by General Namphu.
[25] In September 1988, a movement of sergeants and soldiers deposed General Namphu, putting Prosper Avril, the *eminence grise* of the Duvalier period, into power.
[26] In March 1990, during a period of intense popular protest, General Avril was ousted by General Abraham, who relinquished control to a provisional civilian government headed by Judge Ertha Pascal-Trouillot, the first woman to occupy the presidency in Haiti. The provisional government created suitable conditions to put the Constitution into practice, and called elections for December 1990.
[27] These elections were won by a priest, Jean-Bertrand Aristide, who obtained 67 per cent of the vote as the leader of the National Front for Change and Democracy. He was voted in mostly by the poor urban sectors, and he took office on February 7 1991.
[28] Aristide, a Liberation Theology activist, was censured in 1988 by the church authorities and expelled from the Salesian order. His governmental program was based on a war against corruption and drug trafficking, including a thorough literacy campaign, and a project to

in 1919. The US troops finally defeated the resistance and controlled the country until 1934, turning it into a virtual colony. That year, President Vincent succeeded in getting the US troops to withdraw from the island, but he could not eliminate US influence from the country's domestic affairs.
[15] The national army, or Garde d'Haiti, took a central role in national politics, staging coups against presidents Lescot, 1941 to 1946, Estime, 1946 to 1950, and Magloire, 1950 to 1957. In 1957, François Duvalier, a middle-class

doctor, seized power supported by the army and the US.
[16] The elite mestizo group which was in power repressed the African-based culture until François Duvalier recognized the strategic power he could achieve by manipulating it. Over time he succeeded in building two great supports for his domination. The first was Voodoo, the syncretic and distinctive religion of Haiti that combines African and Catholic rites. The second was the Tonton-Macoutes, a group of some 300,000 "national security volunteers".
[17] The army, the commercial bour-

move from "extreme poverty to poverty with dignity".

29 On September 30, General Raoul Cédras staged a bloody coup. In protest the Organization of American States (OAS), declared a trade embargo, starting diplomatic negotiations in the region and in the UN. Meanwhile, the rebels tried to avoid international isolation by officially recognizing the sovereignty and operation of parliament.

30 In February 1992, OAS representatives, Haitian members of parliament, and the deposed Aristide signed an agreement in Washington to re-establish democracy and reinstate the former president. By the end of April 1992 there had been over 4,000 reports of political assassinations, and thousands were leaving the country for fear of further repression.

31 Since Aristide's fall, more than 40,000 Haitians were picked up at sea by the US Navy. In May, the US government ordered that the Coast Guard stop all boats coming from Haiti, and return them to their country of origin. Two months later, a US Court of Appeal upheld the refugees' claims, declaring that this policy of interdiction violated the 1980 Refugees Act.

32 In June 1992, the military government named Marc Bazin as prime minister. He was the founder of the liberal Movement for the Restoration of Democracy.

33 President Clinton announced in January 1993 that his government would continue the interdiction policy. A US federal court ordered that 158 Haitian refugees be released from Guantanamo Base, in order to receive treatment for AIDS. The court called the refugee camp "an HIV prison", with conditions "similar to those reserved for spies and assassins".

34 UN and OAS Special Envoy Dante Caputo initiated a diplomatic offensive to bring about Aristide's return to office. The points for negotiation included amnesty for military officers implicated in the September 1991 coup, and on the economic front, the implementation of a "development plan" drawn up by the World Bank.

35 Two human rights organizations, Americas Watch and National Coalition for Haitian Refugees, opposed the proposed amnesty and indicated, in a joint letter to Dante Caputo, that it violated international law.

36 In January, the de facto government carried out parliamentary elections which were considered illegitimate, designed as they were to replace only a part of Parliament. Less than 3 per cent of those registered to vote took part in the elections. A few months later, Marc Bazin resigned as prime minister.

37 On June 27 1993, talks began on Governor's Island, New York, between General Raoul Cédras and deposed president Aristide. In the meantime, the UN Security Council imposed a financial, oil and arms embargo upon Haiti. In July, Aristide and Cedras signed an agreement which confirmed the president's return and guaranteed an amnesty for all military leaders involved in the coup. In accordance with the terms of the agreement, Aristide named Robert Malval prime minister, in August. The Security Council announced a conditional suspension of the embargo.

38 However, a new wave of violence soon broke out in Haiti, to prevent the agreement from going into effect. In October, a US warship patrolled the coast near Haiti's capital. An armed mob threatened the ship's troops, and port facilities were closed. President Clinton ordered the ship back to Guantanamo Base, while the Security Council reinstated the naval embargo.

39 Malval's Minister of Justice, Guy Malary, was assassinated. The attack was carried out in almost the same place where Antoine Izmery, a pro-Aristide businessman, had been killed a month earlier. Those responsible for the killings were members of the pro-military Front for the Advancement and Progress of Haiti, who use the same tactics as the Tonton-Macoutes.

40 In the meantime, the embargo against Haiti began to have the opposite effect of what had been intended, destroying the formal economy and facilitating blackmarket dealings of the military in power. International sanctions increased poverty although they did not substantially affect the lifestyle of the country's wealthiest families.

41 In early 1994, Aristide called Clinton's policy of interdiction "a floating Berlin Wall". According to UN reports, between May 1993 and February 1994, 426 supporters of deposed president Aristide were assassinated.

42 The military junta surrendered on September 19, in the face of an imminent invasion by a multinational force led by the United States

STATISTICS

DEMOGRAPHY

Annual growth: 1.9% (1992-2000). Children per woman: 4.7 (1992).

HEALTH

One physician for every 7,143 inhab. (1988-91). Under-five mortality: 127 per 1,000 (1994). Calorie consumption: 86% of required intake (1995). Safe water: 28% of the population has access (1990-95).

EDUCATION

Illiteracy: 55% (1995). Primary school teachers: one for every 29 students (1990).

COMMUNICATIONS

83 newspapers (1995), 82 TV sets (1995) and 80 radio sets per 1,000 households (1995). 0.7 telephones per 100 inhabitants (1993). Books: 91 new titles per 1,000,000 inhabitants in 1995.

ECONOMY

Per capita GNP: $230 (1994). Annual growth: -5.00% (1985-94). Annual inflation: 13.20% (1984-94). Consumer price index: 100 in 1990; 248.7 in 1994. Currency: 13 gourdes = 1$ (1994). Cereal imports: 236,000 metric tons (1990). Imports: $292 million (1994). Exports: $73 million (1994). External debt: $712 million (1994), $102 per capita (1994). Debt service: 1.2% of exports (1994).

ENERGY

Consumption: 47 kgs of Oil Equivalent per capita yearly (1994), 70% imported (1994).

and supported by the UN Security Council. Aristide returned to the country and resumed the presidency on October 15, after the coup leaders had gone into exile. In December, troops were demobilized, with a view to creating a new national police force.

43 In January 1995, the Security Council approved a motion to send a contingent of UN troops to Haiti, to replace the multinational force before March 31. The legislative and municipal elections of April 28 were considered significant, as a barometer of the country's new political situation. On February 7 1996, Jean-Bertrand Aristide finally took his post as president in a country where 80 per cent of the population were living below the poverty line.

Honduras

Population:	5,750 (1994)
Area:	112,090 SQ.KM
Capital:	Tegucigalpa
Currency:	Lempira
Language:	Spanish

Honduras

Before the arrival of the Spaniards, the region of present-day Honduras was inhabited by the Chibcha (see Colombia), the Lenca and the Maya. In the north of Honduras lies the city of Copan, which belonged to the ancient Mayan empire. Copan's splendor lasted until the 9th century AD, ending with the downfall of the Mayan Empire.

[2] The first European to reach Honduras appears to have been Amerigo Vespuccio in 1498, but Pedro de Alvarado was in charge of the final Spanish conquest of Honduran territory, joining it to the Captaincy General of Guatemala, despite strong resistance from the Indians under Lempira's command.

[3] In 1821, Honduras gained independence from Spain. Together with the other Central American provinces it joined the short-lived Mexican Empire of Iturbide, which collapsed two years later. Francisco Morazán, and other Honduran leaders of the last century, sought in vain to set up an independent Central American federation. Their efforts were no match for Britain's "Balkanization" tactics.

[4] With the liberal reform of 1880, mining became the backbone of the economy, and to encourage the development of this sector, the country was opened to foreign investment and technology. In 1898, another "empire" managed to penetrate Honduras: the notorious US United Fruit Company (Unifruco). The Company took over vast tracts of land, it produced almost the entire fruit output of the country, it ruled railroads, ships and ports, and dictated many key political decisions.

[5] US Marines invaded Honduras in 1924, imposing a formal democracy, allowing Unifruco to establish a monopoly in banana production by buying out its main competitor, the Cuyamel Fruit Company. Washington eventually handed power over to Tiburcio Carias Andino, and he governed Honduras from 1933 to 1949.

[6] Border disputes with Guatemala led to US arbitration in 1930. In 1969, friction developed with El Salvador when the number of Salvadoran peasants emigrating to Honduras reached a critical level. This led to a further war, which was triggered by a soccer game, and was finally ended with mediation by the Organization of American States (see El Salvador).

[7] In 1971, nationalists and liberals signed the Unity Pact. General Osvaldo López Arellano, in power since 1963, permitted elections and Ramón Ernesto Cruz, of the National Party, was elected president.

[8] However, in 1972 López Arellano overthrew the Cruz administration. He demonstrated his sensitivity to peasant demands for land reform and began to impose controls upon United Brands as Unifruco was now called. This resulted in López Arellano's resignation and his replacement by Colonel Juan A. Melgar Castro.

[9] The Army Commander-in-Chief, General Policarpo Paz García took power in August 1978. This regime became closely allied to that of the dictator Anastasio Somoza, in neighboring Nicaragua. The Sandinista Revolution hastened the election of a constituent assembly which promptly ratified Paz García as president. General elections were held in 1981, but leftist parties were banned. Liberal Party candidate, Roberto Suazo Córdova, won the presidency and was inaugurated in January 1982.

[10] The price of consumer goods immediately rose and an anti-terrorist law was passed forbidding strikes as intrinsically subversive. Death squads acted with impunity and opposition political figures disappeared daily.

[11] Honduras tolerated the presence of US troops and the installation of Nicaraguan "contras" in its territory. It was estimated that the Pentagon had 1,200 soldiers there in 1983. They participated in some military operations, gave military instruction and logistical support, and established a military infrastructure. The "contras", had some 15,000 soldiers within Honduran territory, alongside "Nica" camps, holding around 30,000 refugees.

[12] In April 1985, 7,000 US soldiers were on maneuvers near the Nicaraguan border. Washington offered the Honduran air force the renewal of all its combat planes. With almost $300 million in military aid, the Honduran army doubled its number of troops and renewed the Air Force combat fleet.

[13] Jose Azcona Hoyo, of the Liberal Party, won the 1985 elections.

The new president requested help from Washington so that the "contras" could leave the country; he also tried to stimulate foreign investment.

[14] Poor management of US aid unleashed a wave of corruption, especially within the armed forces. The government's privatization plan failed, as did its scheme for reducing public spending; in the agricultural sector, seasonal unemployment reached 90 per cent.

[15] In 1989, Rafael Callejas, the National Party candidate, won by a wide margin, although the elections were marked by fraud. Supported by the US and by the business community, Callejas launched an overall liberalization of the economy.

[16] In 1990, after the Sandinista defeat in neighboring Nicaragua's elections, US President Bush's administration made significant cuts in its economic aid to Honduras. Callejas sought closer ties with the armed forces, in the hopes of keeping growing social discontent under control.

[17] In early 1990, the government raised taxes, increased fuel prices by 50 per cent and devalued the national currency.

[18] In December, the government extended an amnesty to all political prisoners or victims of political persecution. The anti-terrorism law was abolished and a forum for political discussion was set up, with no group excluded.

[19] On January 12 1991, after eight years in exile, four leftist political leaders returned to Honduras, announcing the end of their armed struggle. In October 1991, the

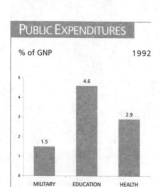

PUBLIC EXPENDITURES

% of GNP — 1992

- MILITARY: 1.5
- EDUCATION: 4.6
- HEALTH: 2.9

WORKERS

1994

- MEN 79%
- WOMEN 21%
- AGRICULTURE 38%
- INDUSTRY 15%
- SERVICES 47%

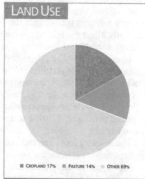

LAND USE

- CROPLAND 17%
- PASTURE 14%
- OTHER 69%

ENVIRONMENT

80% of the land is covered by mountains and rainforests. Both population and economic acvtivity are concentrated along the Caribbean coast and in the southern highlands, close to the border with El Salvador. The coastal plains have the largest banana plantations in Central America. Coffee, tobacco and corn are grown in the southern part of the country. Deforestation, misuse of the soil and uncontrolled development have led to soil depletion.

THE SOCIETY

Peoples: Honduran people are of mixed Mayan and Spanish descent. There is a 10% minority of native Americans; 2% are Afro-American. Garifunas, descendants of fugitive slaves and native Americans, live along the Caribbean coast and on the nearby islands, maintaining their traditional lifestyles. **Languages:** Spanish is the official and most widely spoken language. Some native groups maintain their own languages. **Religions:** Roman Catholic 85%; Protestant (mostly fundamentalist, Moravian, and Methodist) 10%; other 5%. **Political Parties:** People's Liberal Alliance (ALIPO), National Party, conservative; Liberal Party; "Rodista" National Movement (MNR), Revolutionary Democratic Movement; Innovation and Unity Party (PINU); Christian Democratic Party; Socialist Action Party of Honduras (PASOH); Marxist-Leninist Communist Party (PCML); Unified National Direction (DNU) Alliance for the Liberation of Honduras.

THE STATE

Official Name: República de Honduras. **Administrative divisions:** 18 Departments. **Capital:** Tegucigalpa, 738,500 inhab. (1993). **Other cities:** San Pedro Sula, 353.800 inhab.; La Ceiba, 82.900 inhab.; El Progreso, 77.300 inhab.; Choluteca, 69.400 inhab. (1993). **Government:** Carlos Roberto Reina, President since January 27, 1994. **National Holiday:** September 15, Independence Day (1821). **Armed Forces:** 16,000 troops, 13,200 conscripts (1994). **Paramilitaries:** 5,500 members of the Public Security Force.

DEMOGRAPHY

Urban: 43% (1995). **Annual growth:** 1.5% (1992-2000). **Estimate for year 2000:** 7,000,000. **Children per woman:** 4.9 (1992).

HEALTH

One **physician** for every 1266 inhab. (1988-91). **Under-five mortality:** 54 per 1,000 (1995). **Calorie consumption:** 91% of required intake (1995). **Safe water:** 65% of the population has access (1990-95).

EDUCATION

Illiteracy: 27% (1995). **School enrolment:** Primary (1993): 112% fem., 112% male. Secondary (1993): 37% fem., 29% male. University: 9% (1993). **Primary school teachers:** one for every 38 students (1992).

COMMUNICATIONS

99 **newspapers** (1995), 99 **TV sets** (1995) and 106 **radio sets** per 1,000 households (1995). 2.1 **telephones** per 100 inhabitants (1993). **Books:** 106 new titles per 1,000,000 inhabitants in 1995.

ECONOMY

Per capita GNP: $600 (1994). **Annual growth:** 0.50% (1985-94). **Annual inflation:** 13.00% (1984-94). **Consumer price index:** 100 in 1990; 196.4 in 1994. **Currency:** 9 lempiras = 1$ (1994). **Cereal imports:** 197,000 metric tons (1993). **Fertilizer use:** 210 kgs per ha. (1992-93). **Imports:** $1,056 million (1994). **Exports:** $843 million (1994). **External debt:** $4,418 million (1994), $768 per capita (1994). **Debt service:** 33.9% of exports (1994). **Development aid received:** $324 million (1993); $61 per capita; 10% of GNP.

ENERGY

Consumption: 169 kgs of Oil Equivalent per capita yearly (1994), 71% imported (1994).

"Lorenzo Zelaya" People's Revolutionary Forces accepted the government's peace declaration and renounced armed conflict.

[20] Armed forces commander, General Arnulfo Cantarera, accused of human rights violations, was relieved of his command and replaced by General Luis Discua, who favored increased intervention by the military in the country's political life. Political assassination and other abuses committed by the military were denounced by the Honduran Committee for the Defense of Human Rights.

[21] The Callejas administration's structural adjustment program permitted the renegotiation of the country's $3.6 billion foreign debt. New agrarian laws authorized the sale of expropriated lands, which led to speculation by transnational agribusiness.

[22] The increase in military power and the country's overall political instability compounded the weakness of the economy, which was suffering the effects of losing dollars previously provided by US military aid.

[23] Voters registered their discontent in the ballots, bringing about the triumph of opposition candidate Carlos Roberto Reina, a social democrat, in the November, 1993 elections.

[24] One of the new government's first decisions was the abolition of the mandatory military service. This measure was approved by Parliament in May 1994 and ratified by the legislature in April 1995, with 125 votes for, and 3 against it. However, in August 1994 giving in to pressure from the military, the government agreed to a temporary draft call to fill 7,000 openings in the armed forces. In mid 1994, the government dissolved the infamous National Bureau of Investigations, which had once been the torture section of the armed forces.

[25] As a result of the fall in banana exports to the EC, the Tela Railroad Company (formerly part of the United Fruit Company) closed four plantations, claiming that they were no longer profitable. They also laid off 3,000 employees for three months. Union workers went on strike and after several days of tension, the government called off the strike and forced the company to reopen the banana plantations and rehire the workers who had been laid off. After negotiating directly with the union - SITRATERCO - the company reinstated 1,200 workers, though nearly 1,000 female employees were not re-employed.

[26] As a result of the drought which hit the country during the first half of 1994, 90 villages throughout the country lost over 60 per cent of their subsistence crops. Given the danger of famine, which threatened more than 1.5 million people, the government requested aid from the UN's food agency, FAO. The latter provided $900 million for agricultural development of the declared emergency areas. However, 73 per cent of the population of Honduras still lives in conditions of poverty or extreme poverty.

[27] The drought also affected the forests: the green cover area went down from 36 per cent in 1980 to less than 27 per cent of the territory in 1995. The rapid deforestation rate has worsened soil erosion.

[28] While the armed forces continued with police assignments in the cities, the Legislative Assembly began to work on amendments to the constitution that handed control of public security forces over to civilians. In January 1995 the new Crime Investigations Unit began to act, led by civilians, replacing the secret police that had been dismantled the previous year. The new body, initially of 1,500 agents, was trained by the Israeli police and the FBI from the US. At that time, more than 50 people were murdered each day in the country.

[29] Senior government officials were arrested in 1995 because of their links with the trafficking of official passports. The Supreme Court of Justice revoked former President Callejas' immunity to allow him to testify in court on false documents and misappropriation of public funds. President Reina himself was found out by his own offensive against corruption. The president was investigated for using state funds in the solution of a personal labor dispute.

Hungary

Population: 10,261,000 (1994)
Area: 93,030 SQ KM
Capital: Budapest

Currency: Forint
Language: Hungarian

Magyarország

The territory of Hungary belonged to the Roman provinces of Pannonia and Dacia. By the end of the 4th century Rome had lost Pannonia, which had been occupied by tribes of Germans, Slavs, Huns and Avars. The latter ruled over the Danube basin during the 7th and 8th centuries, until they were conquered by Charlemagne.

[2] Charlemagne's successors set up a series of duchies in the western and northern parts of the basin, while the southern and eastern parts were under the sphere of influence of the Byzantine Empire and Bulgaria. The duchy of Croatia became independent in 869 and Moravia put up stiff resistance to the Carolingians until the appearance of the Magyars.

[3] The Magyars had organized a federation of tribes west of the lower Don. These were made up by several clans led by a hereditary chieftain. The federation was called On-Ogur (Ten Arrows); "Hungarian" is a derivation of this term in the Slav language. In 982 Emperor Rudolf turned to the Magyars to break Moravian resistance.

[4] Led by Arpad, the Magyars crossed the Carpathians and conquered the inhabitants of the central plateau. Moravia was defeated in 906 and Pannonia a year later. The Hungarians then expanded northwards and repeatedly raided the rest of Europe.

[5] The German emperor Otto I defeated Arpad in 955 and halted Magyar expansion. Arpad's heirs reunified the tribes and adopted Western Christianity. Stephen I was crowned by Rome, subsequently laying the foundations of the Hungarian State.

[6] The question of succession upon Stephen's death triggered two centuries of instability, but Hungary consolidated its dominions as far as the Carpathian mountains and Transylvania in the north, and the region between the Sava and Drava rivers in the south. In addition, it ruled over Croatia, Bosnia, and Northern Dalmatia (although the latter remained a separate state).

[7] With the Mongol invasion of the 13th century, Hungary lost half of its population. This prompted the kingdom to reorganize and open its doors to new settlers; however, it had to make important concessions to the Cuman overlords and immigrants, who further weakened the

kingdom. Finally, the succesion of Charles Robert of Anjou - a foreigner - to the throne, stabilized the country.

[8] As the struggles between the Holy Roman Empire and the Papacy did not involve Hungary, the 14th century was the country's golden age. The kingdom established friendly relations with Austria, Bohemia and Poland, and strong links with Bosnia. However, during this period the country seized Dalmatia from Venice, and other territories from Serbia.

[9] Sigismund of Luxembourg (1387-1437) was both a German and Czech king. His long absences and arbitrary rule enabled the Hungarian Diet (parliament), which was made up of nobles, to expand its power. The Diet's consent was required in order for laws to be passed. Taxes were continually being exacted from the peasants, who started revolts in the north and in Transylvania.

[10] After another controversial succession, Matthias Corvinus of Prague was elected king of Hungary in 1458. He ruled over his country with an iron fist. With the help of the Black Army, an army of mercenaries, Matthias subdued his enemies inside the country and expanded his dominions over Bosnia, Serbia, Walachia, and Moldova. He also carried out a series of campaigns against Bohemia and Austria.

[11] Upon Matthias' death in 1490 the magnates appointed Vladislav II, who was king of Bohemia and whose weak personality was well known. The Black Army was disbanded but the oppressed peasants rebelled again in 1514; the rebellion was ruthlessly put down. Meanwhile, Austria recovered the southern provinces and established its authority over Hungary.

[12] Hungary was conquered by the Ottoman empire in 1526. At first the sultan supported Zapolya of Hapsburg to succeed the king who had been killed in battle. However, upon Zapoyla's death he occupied Budapest himself, and annexed a large portion to the south and centre of the country. Croatia and the western and northern strip of the country remained under the rule of Ferdinand of Hapsburg, who had to pay tribute to the Turkish empire.

[13] During the 17th and 18th centuries Hungary was under two empires whose only interest in the country lay in the tributes it paid.

The situation became more volatile when the majority of the population embraced the Reformation, while Vienna attempted to re-establish Catholicism. Even the nobles began to react against absolutism, peasant oppression and stagnation.

[14] During the 1848 revolution, the Hungarian Diet passed the April Laws, which brought changes in agricultural and fiscal matters. They reorganized on a more representative basis and proposed the reunification of the country and the creation of a separate administration in Budapest. The reform was met with distrust on the part of large landowners and Serbian, Romanian, and Croatian minorities.

[15] When the revolution was defeated, Austria, aided by Russia, annulled all reforms and regained control over Hungary. When Austria was defeated by Prussia in 1866, Vienna decided to subdivide its empire and accepted Hungary's April Laws. A "Nationalities Law" guaranteed respect for the rights of minorities, giving way to the establishment of the Austro-Hungarian Empire in 1867.

[16] In the early 20th century, Hungarian politics was still dominated by conservative landowners, as most businesspeople were foreigners, Jewish or German, and the majority of the population remained excluded from political activity. Minorities continued to be oppressed and the country lacked a strong sense of autonomy.

[17] With the collapse of the Hapsburg empire during World War I, a provisional government took power and proclaimed the Republic of

Hungary. But Serbians, Czechs and Romanians seized two-thirds of the country and the central government was paralyzed. In 1919 a communist rebellion was followed by the formation of a "Soviet" republic.

[18] Bela Kun's Bolsheviks expected Moscow to support them, but they were forced to flee when the Romanian troops took over the capital city. The European powers pushed Romania into withdrawing and installed a provisional government. The 1920 Parliament restored monarchy and appointed Admiral Milkos Horthy as provisional ruler.

[19] In the Trianon Treaty, the victorious allies recognized Hungarian independence, but Yugoslavia, Romania, and Czechoslovakia remained in possession of most of the country's territory and 60 per cent of its population. Austria, Poland, and Italy also benefitted from the partition of Hungary.

[20] Hungary was forced to pay heavy reparations; its industry and production were left in a shambles. Unemployment rose to unprecedented figures and nearly 400,000 refugees arrived from the territories it had lost. The middle classes and the refugees set up right-wing armed groups, blaming the left for their ruin.

[21] Funds granted by the League of Nations, followed by private investments and credits, alleviated domestic tension, but the depression of the 1930s had serious effects on Hungary. Horthy formed an extreme-right government which sided with Germany, while anti-Semitism grew within the country.

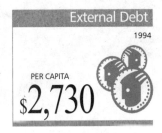

[22] The alliance with Berlin enabled Budapest to recover part of Slovakia, Ruthenia, and the north of Transylvania. Hungary cooperated in the German attacks on Romania, Yugoslavia and the USSR, but nothing could stop the Red Army counter-offensive. In the Treaty of Paris, Hungary was forced to retreat to the borders fixed in the Treaty of Trianon.

[23] In 1944 a Provisional Assembly formed a coalition government and took up a program proposed by communist leaders and backed by the USSR. This program included the expropriation of large estates, the nationalization of banking and heavy industry, as well as guarantees for small landowners and private initiative, democratic rights and liberties.

[24] The Communists, at that time represented by the Workers' Party, assumed control of the government. In 1946 the Constitution of the People´s Republic of Hungary was established.

[25] In 1948, the communist government forcibly collectivized agriculture and started a series of development plans which stressed the development of heavy industry. Upon Stalin's death in 1953, Matyas Rakosi, head of government, was replaced by Imre Nagy, who promised political changes.

[26] In 1955, Nagy was deposed and expelled from the party. Rakosi returned for a short period, until 1956, when he handed power over to Erno Gero, who had similar political views to his own.

[27] Students staged a demonstration in Budapest, and were joined by the rest of the population. Gero reacted harshly: the police were instructed to shoot into the crowd, and the demonstration turned into a popular revolution, backed by the Army. The government was handed back to Nagy, heading a broad national coalition.

[28] In November 1956, Nagy announced the withdrawal of Hungary from the Warsaw Pact, and requested that the United Nations recognize its neutrality. The Red Army, which had withdrawn during the revolution, occupied the country again and reinstated the communist government, led by Janos Kadar.

[29] Kadar followed Soviet guidelines, where foreign policy was concerned. As for domestic policy, central planning was partly liberalized from 1968 onwards, the standard of living rose, but bureaucracy and corruption increased.

[30] Sexual discrimination continued on all fronts. In 1981, while women made up 45 per cent of all workers, they were systematically paid less than men and their job opportunities were restricted to specific fields of traditionally female labor.

[31] Kadar was elected First Secretary of the Hungarian Workers' Socialist Party and head of government, and he ruled without serious opposition until March 1986, when a youth protest march was heavily crushed. Two years later some 10,000 people demanded freedom and reforms in a demonstration in Budapest.

[32] The government relaxed press censorship and permitted the formation of trade unions and independent political groups - such as the newly constituted Hungarian Democratic Forum - to be formed.

[33] Between 1986 and 1988 Hungarian and Austrian environmentalists protested against the construction of a dam on the Danube; part of a Hungarian-Czechoslovakian project supported by Austria. The Budapest government resisted the pressure but in the end the project was shelved.

[34] In January 1989 Parliament passed a law legalizing strikes, public demonstrations and political associations. Meanwhile, it adopted an austerity plan which reduced subsidies and devalued the currency. Unemployment was expected to rise to 100,000 and annual inflation to 30 per cent.

[35] The Workers' Party also approved the elimination of the single party system and agreed to celebrate Independence Day on March 15, the date of the 1848 revolt against Austria. In March 1989, 100,000 people demonstrated in Budapest, demanding elections and the withdrawal of Soviet troops.

[36] The opposition candidates won the provincial elections held in 1989. In August, two million workers went on strike, protesting against price increases. In October, after an agreement between the Workers' Party and the opposition, Parliament proclaimed the Republic of Hungary and eliminated the single-party system.

[37] Hungary was the first country in the Warsaw Pact to break its Cold War alliances. Within two years, Budapest established relations with Israel, South Korea, and South Africa.

[38] By the end of the year, the Workers' Party had become the Hungarian Socialist Party, with one faction deciding to keep its old name. In the 1990 elections, the Hungarian Democratic Front obtained 43 per cent of the vote, forming a coalition government with two smaller parties: the Socialist Party and the Workers' party. These received 10.3 and 3.5 per cent of the vote, respectively.

[39] In late 1991, the Hungarian prime minister met with the Czechoslovakian and Polish presidents in the city of Krakow, Poland. The three leaders expressed their wish to establish formal relations with NATO and the West European Union (WEU; see Greece). The Krakow declaration also stressed the importance of associating with the European Community.

PROFILE

ENVIRONMENT

The country is a vast plain with a maximum altitude of 1,000 m partially ringed by the Carpathian Mountains. The mountainous region has abundant mineral resources (manganese, bauxite, coal). Between the Danube and its tributary, the Tisza, lies a very fertile plain, site of most of the country's farming activity. Cattle are raised on the grasslands east of the Tisza. In the past 30 years there has been rapid industrial expansion, particularly in the production of steel, non-ferrous metals, chemicals and railway equipment. Oil and natural gas deposits have been found in the Szegia and Zala river basins. 41% of the population is exposed to sulphur dioxide and nitrogen dioxide in the air. The sulphurous emissions are greater than in most eastern European countries.

SOCIETY

Peoples: Magyar 92%; Gypsy 3%; German 1%; Slovak 1%; Jewish 1%; Southern Slav 1%; other 1%.
Religions: Christian 92.9%, of which Roman Catholic 67.8%, Protestant 25.1%; atheist and nonreligious 4.8%; other 2.3%. **Languages:** Hungarian (official) **Political Parties:** Hungarian Socialist Party (HSP); Hungarian Democratic Forum (MDF); Alliance of Free Democrats (SZDSZ); Smallholders' Party; Federation of Young Democrats (FIDESZ); Popular Christian Democratic Party; Social Democratic Party of Hungary; Agrarian Party; Hungarian Socialist Workers' Party (communist). **Social Organizations:** Central Council of Hungarian Trade Unions (SZOT); Hungarian Women's Network.

THE STATE

Official Name: Magyar Köztársaság. **Administrative Divisions:** 19 Counties and the Capital.
Capital: Budapest,1,995,000 inhab. (1994). **Other cities:** Miskolc, 189,000; Debrecen, 217,000; Szeged, 179,000. (1994). **Government:** Arpad Göncz, President; Gyula Horn, Prime Minister since June 1994. The single-chamber National Assembly (352 representatives elected for five-year terms) is the supreme authority in the Republic. Eight seats in the Assembly are reserved for each of the country's minorities. **National Holiday:** March 15, Independence (1848), April 4, Anniversary of the Liberation (1945). **Armed Forces:** 70,500. **Paramilitaries:** Border Guard, 15,900. Civil Defence Troops, 2,000. Internal Security Troops, 1,500.

[40] Hungary had a foreign debt of $21 billion, the largest in Europe. Its agricultural and industrial production fell by 10 per cent - a situation aggravated by IMF policies-preventing domestic capital accumulation and favoring foreign investors. In 1990, inflation reached 30 per cent, with the average family spending 75 per cent of its income on basic consumer goods. The following year, the figure was 90 per cent, for the same goods. Of its 10 million inhabitants, 2 million were living below the poverty line.

[41] In this new era in Hungary, women's representation in politics has declined. In the 1990 elections, women won 7.5 per cent of the seats, while in 1985 they had held 21per cent. Women constitute 46 per cent of the active workforce, but this is also beginning to change with the increasingly popular ideological position of restoring the "natural order". The idea of women staying at home clashes with the new economic reality of the country, where two salaries are needed to cover even the most basic needs of a nuclear family.

[42] The womens' struggle is also heavily involved in the abortion issue. Over 4 million legal abortions have been carried out in Hungary over the past 25 years. However, nearly all the current political parties, except for the Federation of Young Democrats and the Free Democrats, have come out against the practice. Meanwhile, the Hungarian Women´s Network is working on campaigns against the illegalization of abortion and in defence of the living standards, working rights and health of women.

[43] In January 1992, the government announced its decision to reduce the public deficit, which stood at a level in excess of $900 million in 1991. They hoped to achieve this by imposing a 4 per cent reduction in public spending. Growing popular discontent, which was reflected in a reduced turn-out for elections and in repeated union strikes, have forced Joszef Lantall's right-of-center government to take a more cautious position.

[44] The economic crisis encouraged nationalistic and xenophobic demonstrations, including the notorious publication of a manifesto signed by the writer and vice-president of the Democratic Forum, Istvçn Csurka, blaming the ex-communists, the "westernized" liberals, the gypsies and the Jews for "the deterioration of the social climate".

[45] The liberal wing of the MDF immediately repudiated the declaration, and a few days later 70,000 people from various political groups took part in a demonstration, more precisely, a "march for democracy" through the city, to "chase the ghost of radicalism out of the country". After this event 48 skinheads were brought to face charges in court.

[46] Shortly afterwards, Csurka renounced his vice-presidency of the MDF. The internal difficulties of the Forum, based on differences between the radical nationalist and more liberal tendencies, led to various changes of ministers in 1993. In July, the National Assembly approved a law guaranteeing full rights to the ethnic minorities living in Hungary. The nation has made great efforts to cultivate an image of tolerance with its neighboring countries, where 3 million Hungarians live.

[47] In October, the prime minister, Jozef Antall, was hospitalized when he became seriously ill. Minister of the Interior Peter Boross replaced him, later being confirmed in the post upon Antall's death the following December.

[48] In the winter of 1993, a television report declared high levels of air pollution in Budapest over a ten day period, recommending that children should not be allowed to play outside. In the last ten years the number of vehicles in Budapest has doubled and few people use public transport. Traffic moves through the city center at under 10 km per hour.

[49] The government and the opposition were unable to agree on a law defining the situation of the state radio and television stations, which virtually monopolize the airwaves. Their budgets were directly administrated from the prime minister's office.

[50] In March 1994, in the middle of the electoral campaign, the government sacked 129 radio reporters. This action was condemned by international press and human rights organizations. This measure meant the end of 30 cultural and political programs in the three state radio stations.

[51] In the general elections of May 1994, the Hungarian Socialist Party, led by Gyula Horn, won in the second round. The ex-communists won 54 per cent of the vote and 208 of the 386 seats. In second place came the Free Democrats, with 70 seats compared with 90 in the 1990 election. The Democratic Forum was the largest loser, only retaining 37 of its previous 165 seats.

[52] Horn assumed the post of prime minister and, by the end of the year, local elections confirmed the electoral advantage of the former communists. In March 1995, Horn's "romance" with his voters came to an end upon the adoption of an unpopular package of economic measures. In order to reduce the fiscal deficit, the government drastically reduced the budget for education, unemployment insurance funds and maternity allowances, among other measures. A survey released in 1996 showed that half of Hungarian youngsters considered their life conditions had worsened since 1989.

[53] In the international field, Horn maintained his policy of drawing closer to Western powers, and negotiation with neighboring countries, in an attempt to help Hungarian minorities of these states in preserving their "cultural identity".

Iceland

Population:	266.000 (1994)
Area:	103,000 SQ.KM
Capital:	Reykjavik
Currency:	Kronur
Language:	Icelandic

Island

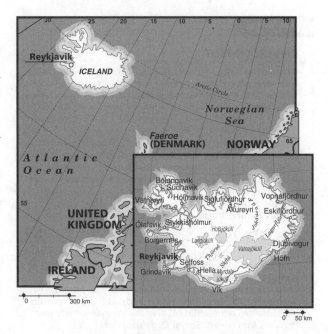

In 874, immigrants of European origin began settling in Iceland. The population is made up of descendants of Norwegians, Scots and Irish. Iceland was an independent state between 930 and 1264, but under the terms of the "Old Treaty" of 1263, the country became part of the kingdom of Norway.

[2] In 1381, Iceland and Norway were conquered by Denmark, but when Norway separated from the Danish Crown in 1814, Iceland remained under its protection. In 1918, Iceland became an associated state of Denmark, until it recovered its independence and a republic was proclaimed in June 1944.

[3] Until the end of the 19th century, Iceland had no roads or bridges in the interior, and trade was entirely in the hands of Danish companies.

[4] In 1915, Icelandic women acquired equal rights with men, being granted the right to vote and to hold elected office. This was attained through the same legislation which granted these rights to Danish women, as at the time, the country was a Danish colony.

[5] After World War I, local agriculture was revitalized through a series of laws aimed at the agrarian sector, and the introduction of modern equipment. From 1920 onwards, the fishing industry grew steadily. In addition, communication throughout the island improved with the building of a new network of inland roads and bridges, and the creation of a coast guard.

[6] Iceland joined the Council of Europe and NATO in 1949; it has no army or navy, and the United States provides the country with defence forces, within NATO's strategic framework. Iceland only has three coast guard ships, one airplane and two helicopters to protect itself against illegal fishing in its territorial waters.

[7] In 1952, Iceland founded the Nordic Council, including Sweden, Norway, Finland and Denmark.

[8] Due to the importance of fishing for Iceland, and the fear of overfishing near the island by foreign fishing fleets, Reykjavik extended its territorial waters to 12 nautical miles in 1964, and again in 1972, to 50 miles. This triggered two serious conflicts known as the "cod wars" with the United Kingdom. They were resolved by a treaty signed in 1973.

[9] In October 1975, Iceland announced the extension of its territorial waters to 200 nautical miles, citing the protection of the environment and of its economic interests. The failure of the 1973 treaty with the UK, and the impossibility of reaching a new agreement led to the third and most serious "cod war".

[10] In February 1976, Iceland briefly broke off relations with the United Kingdom, the first breaking of diplomatic ties to occur between NATO members. In June, the two countries reached an agreement and in December, English trawlers withdrew from Icelandic waters. In July 1979, Iceland re-affirmed its 200 mile claim.

[11] NATO membership became a controversial issue in Iceland in the 1970s. In the 1978 election, the joint victory of the Popular Alliance, who favored renouncing the treaty, and the Social Democratic Party (SDP), who were more moderate, caused a two-month delay in the formation of a new government, which depended upon the alliance of the two political groups.

[12] According to 1980 figures, women made up 43.4 per cent of the country's economically active population.

[13] In 1980, Vigdis Finnbogadottir, an independent candidate campaigning against the retention of American military bases on the island, won the presidential election with the support of the left, making her country's first female head of state. This post did not confer the power for her significantly to alter government policy, so she was unable to carry out her campaign commitment. In 1984 she was re-elected with no opposition candidates.

[14] In 1983, a strict austerity plan was implemented, aimed at reducing inflation. Trade unions accepted the conditions, and the inflation rate consequently fell from 130 per cent at the beginning of the year, to 27 per cent by December.

[15] In May 1985, Parliament approved a resolution declaring Iceland a "nuclear-free zone", thereby prohibiting nuclear weapons from being brought into the country. Iceland was chosen as the site for the important summit meeting between United States president Ronald Reagan and his Soviet counterpart Mikhail Gorbachev in October 1986.

[16] Although Iceland's continuing NATO membership has not really been threatened, the air base at Keflavik continued to cause controversy. With 2,200 troops, the base is a part of the US "advance warning" system. In September 1988, Iceland's executive power formally declared that there would be no new military projects initiated in the country.

[17] That same year, relations between Reykjavik and Washington became tense. The US claimed that Iceland's decision to catch 80 fin whales and 40 sei whales violated the moratorium imposed by the International Whaling Commission. Iceland replied that it was a scientific program and did not modify its decision.

[18] The United States threatened to boycott Icelandic fish products and two Icelandic whaleboats were sunk by ecological activists in Reykjavik Bay. In 1987, Iceland announced the reduction of its catch to 20 whales. The threats of sanctions continued, and the following year it reduced this quota even further.

[19] In June 1988, President Finnbogadottir was re-elected for her third four-year term, with the backing of the main parties and a margin of over 90 per cent of the electorate. This victory was obtained in spite of the existance of an electoral rival, who had proposed increasing presidential power during the campaign.

[20] "President Vigdis", as she is popularly known in her country, led the reforestation process on the island, through the active encouragement of tree planting. She was a founding member and sponsor of the "Save the Children" (Barnaheill) foundation, which aims to develop education. Finnbogadottir has also encouraged the development of national drama.

[21] In September 1988, Steingrimur Hermannsson became prime minister, as head of a coalition including center and left wing groups, made up of the SDP and the Popular Alliance. The new government promised to initiate an austerity program for Iceland's economy, which was suffering from high inflation and recession, accompanied by currency devaluations.

[22] In August 1989, as a result of growing international pressure, including an international boycott of Icelandic products organized by Greenpeace, Reykjavik announced a two-year suspension of its whale hunting.

[23] In 1989 Iceland faced economic difficulties, like its Arctic neighbors, the Faeroe Islands, Greenland and northern Norway, affected by a sudden reduction in the international price of fish, due to over-production.

[24] As a result of these economic difficulties, the country's labor unions signed a wage agreement for 1990, accepting an 8 per cent drop in purchasing power. One indication of the reduction in consumer demand was the fact that the total number of registered motor vehicles - the world's largest in propor-

Iceland

PROFILE

ENVIRONMENT

Located between the North Atlantic and the Arctic Ocean, Iceland is an enormous plateau with an average altitude of 500 meters. A mountain range crosses the country from east to west passing through an extensive ice-covered region; source of the main rivers. The coastline falls sharply from the plateau forming fjords. Most of the population live in the coastal area in the south and west, where ocean currents temper the climate. Reykjavík, the capital and economic center, is located in a fertile plain where the largest cities are found. The northern coast is much colder as a result of Arctic Ocean currents. Geysers and active volcanoes are used as a source of energy. Agricultural output is poor; fishing accounts for 80% of all exports.

SOCIETY

Peoples: Most Icelanders are descendents of Norwegian, Scottish and Irish immigrants. **Religions:** Mainly Lutheran. **Languages:** Icelandic. **Political Parties:** Independence Party, conservative; Popular Alliance, socialist; Progress Party, left-of-center); Social Democratic Party. Other parties with parliamentry representation are: the Women's Party and the People's Movement. **Social Organizations** : Icelandic Federation of Workers, with 47,000 members.

THE STATE

Official Name: Lydhveldidh Island. **Administrative divisions:** 7 districts. **Capital:** Reykjavík, 103,036 inhab. (1994). **Other cities:** Kópavogur, 17,431 inhab.; Hafnarfjördhur, 17,238 inhab. (1994). **Government:** Olafur Ragnar Grimsson, from the People's Alliance party, President and Head of State since August 1st,1996. David Oddsson, Prime Minister and Head of Government. Unicameral parliament composed of 63 members, elected by popular vote for a 4 years term. **National Holiday:** June, 17. **Paramilitares:** 130 coast guard (1994); 2,200 NATO troops.

STATISTICS

DEMOGRAPHY

Annual growth: 1.2% (1991-99). **Children per woman:** 2.2 (1992).

HEALTH

Under-five mortality: 5 per 1,000 (1994). **Calorie consumption:** 115% of required intake (1995).

COMMUNICATIONS

111 **newspapers** (1995), 103 **TV sets** (1995) and 105 **radio sets** per 1,000 households (1995). 54.4 **telephones** per 100 inhabitants (1993). **Books:** 119 new titles per 1,000,000 inhabitants in 1995.

ECONOMY

Per capita GNP: $24,630 (1994). **Annual growth:** 0.30% (1985-94). **Consumer price index:** 100 in 1990; 117.0 in 1994. **Currency:** 68 kronur = 1$ (1994).

---

tion to its population - decreased that year for the first time ever.

[25] In 1990 the government created the new post of environment minister, and Julius Solnes was appointed.

[26] In the parliamentary elections of April 1991, the majority again went to the center-left coalition of Prime Minister Hermannsson. His party held 32 of the 63 seats in the *Althing*, the Icelandic parliament. The conservative Independence Party, led by David Oddsson, the mayor of Reykjavik, came second, with 38 per cent of the vote and 26 seats, eight more than in the previous election. The coalition groups came in with the following results: the Progress Party, 13 seats (18.9 per cent); the SDP, 10 seats (15.5 per cent); the Popular Alliance, 9 seats (14.4 per cent), and the feminist Women's Alliance, 5 seats,

though its electorate dropped by 1.83 per cent.

[27] From November 1992, discussion of the potential benefits of joining the European market, with its 380 million consumers, divided the political scene. The opposition Popular Alliance, including socialists and dissident Social Democrats, was against joining. Jon Baldvin Hannibalsson, foreign minister and minister of foreign trade, on the other hand, claimed that joining the EC would bring a flow of capital and new foreign investment to the island.

[28] Fishing, the most important source of employment, went into crisis as reserves of certain species, especially cod, suffered from overfishing. This situation led to new laws restricting the foreign access to territorial waters, causing friction

in commercial relations with other countries.

[29] Iceland produces abundant electric power and in the early 1990s it had hoped to develop its non-ferrous metal industry. This aim has had to face stiff competition from Russia, which has depressed prices on the international market.

[30] The tourist industry is expanding and there is hope that income from this will entail economic recovery for the island. A series of laws was passed in 1993 to protect polar bears which were being systematically exterminated.

[31] Iceland's entry into the European Union was approved by the *Althing* on January 1 1993. The preparations for unification caused excitement during that year and 1994. There was a feeling that unification would not be attained and a debate was triggered on whether joining the European Union was convenient or not.

[32] In 1994 a number of Iceland's fishing boats entered an area of the Barents sea near Russian and Norwegian territorial waters, causing friction in relations with these two countries. Both nations had undertaken a process to recover fish reserves in the area by reducing capture. By the end of 1995, diplomatic negotiations were considered the only solution for the difference.

[33] In the parliamentary elections of April 8 1995, the Independence Party won 25 of the 63 seats and

was forced to create an alliance with the Progress Party - with 15 seats - in order to reach a majority in the *Althing*. The representation of the Popular Alliance and the Women's Alliance dropped, while the new Popular Movement obtained 4 seats.

[34] Iceland's economy managed a slight recovery in 1995 after a slow growth in the previous years. The gross domestic product increased 3 per cent and inflation did not exceed 2 per cent. Unemployment - aggravated by redundancies at the NATO base - reached 5 per cent.

[35] After governing the country for 16 years, "President Vigdis" did not run for the elections held on June 29 1996. Olafur Ragnar Grimsson, former minister of finance and a member of the Popular Alliance, won the election with 41.4 per cent of the vote and assumed office on August 1 of the same year.

India

Bharat

Population: 913,600,000
(1994)
Area: 3,287,590 SQ.KM
Capital: New Delhi

Currency: Rupee
Language: Hindi

Around 3000 BC, the Dravidian inhabitants of the Indus Valley, in present-day Pakistan, built about a hundred cities, erected huge temples in larger urban centers, like Harappa and Mohenjo Daro, created a written language, that has yet to be deciphered, and carved cylindrical seals of rare perfection. With their irrigated agriculture they developed a prosperous economy and maintained active trade from the Indian Ocean to the spurs of the Himalayas, using the Indus River as their main means of communication. Little is known about their culture, their political organization or development except that, after five centuries of existence, invaders devastated the whole region, exterminating the population and destroying their civilization.

[2] Towards 1600 BC, waves of Indo-Europeans arrived from Afghanistan, gradually conquering the sub-continent. Armed with iron weapons, protected by armor and using war chariots, they subdued the local population and established numerous states. The civilization they created, later called Vedic, was based on a rigid caste system in which the conquerors constituted the dominant nobility. They were called the *ariana* or *ayriana*, the nobility, originating the term Aryan which was later generically used to designate all Indo-Europeans.

[3] The Iranian-Greek invasions of the 6th-4th centuries BC (see Iran) did not reach Magadha in the Ganges River valley, the most powerful state in India. Under the rule of Ashoka, 274-232 BC, Magadha occupied the entire sub-continent, except for the extreme south. Indian civilization proper dates from this period. Ashoka and his descendants were the driving force behind

a cultural unification that included the organization and diffusion of Buddhism, based on the preaching of Gautama Siddhartha, who lived from 563 to 483 BC, and was later known as Buddha. Between the 1st and 3rd centuries AD, this civilization began to break up, fragmented by the development of the Seythian Kusana (see Afghanistan) and Ksatrapa states in the northeast.

[4] When the Guptas of Magadha seized power in the 3rd-6th centuries AD, a new process of unification ushered in one of the most brilliant periods of Indian culture. The 8th century spread of Islam failed to take hold in India, but it was more successful four centuries later, when the Turks of Mahmud of Ghazni re-introduced the faith. Successive groups of Islamic migrants from central Asia invaded the sub-continent, ending with the Tartars of Timur Lenk (Tamburlaine). Between 1505 and 1525, one of their descendants, Babur, founded what later came to be known as the Empire of the Grand Mogul, with its capital in Delhi.

[5] Babur's descendants consolidated Islam, particularly in the northwest and northeast (see Pakistan and Bangladesh). Culture and the arts developed remarkably, the Taj Mahal was built around 1650, but the European presence, that had been limited to coastal trading posts, began to be felt. In 1687, the British East India Company settled in Bombay. In 1696, it built Fort William in Calcutta, and throughout the 18th century, the Company's private army waged war against the French competition, emerging victorious in 1784. From 1798, Company troops led by a brother of the Duke of Wellington methodically conquered Indian territory in various campaigns. By

1820, the English were in control of almost all of India, except for the Punjab, Kashmir and Peshawar, which were governed by their Sikh ally, Ranjit Singh. After his death in 1849, the British annexed these territories. The "loyal allies" retained nominal autonomy and were allowed to keep their courts, great palaces and immoderate luxury, much to the satisfaction of European visitors.

[6] The Indian economy was completely dismantled. Its textile industry, whose exports of high-quality cloth had reached half the globe,

was an obstacle to the growth of British weaving.

[7] The ruin of this industry, based on individual weavers, brought widespread impoverishment to the countryside. Peasants were also hard hit by the reorganization of agriculture for export crops. The early results of British domination were lower incomes and greater unemployment. Public accounts were conveniently arranged. All military spending, including the campaigns in Afghanistan, Burma and Malaya, was covered by the Indian treasury, 70 per cent of whose budget was earmarked for these "defence expenses". All British spending, however remotely connected with India, was entered as expenditures of the "Indian Empire".

[8] "Divide and rule" was a motto of British domination. Mercenaries recruited in one region were used to subdue others. Such was the case with Nepalese Gurkhas and Punjabi Sikhs. Religious strife was also fomented; an electoral reform at the beginning of the 20th century stated that Muslims, Hindus and Buddhists could each vote only for candidates of their own faiths. Throughout the colonial period all this manipulation generated innu-

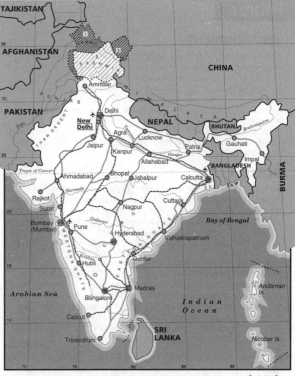

1 Territory under Pakistan control 2 Territory under China's control 3 Territory claimed by Pakistan

0   300 km

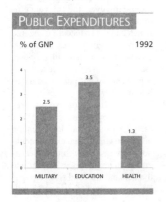

PUBLIC EXPENDITURES

% of GNP   1992

- MILITARY: 2.5
- EDUCATION: 3.5
- HEALTH: 1.3

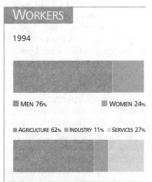

WORKERS

1994

- MEN 76%   WOMEN 24%
- AGRICULTURE 62%   INDUSTRY 11%   SERVICES 27%

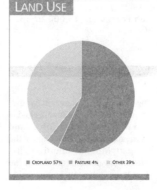

LAND USE

CROPLAND 57%   PASTURE 4%   OTHER 39%

Foreign Trade

MILLIONS $ 1994

IMPORTS
26,846
EXPORTS
25,000

Maternal Mortality

1989-95

PER 100,000
LIVE BIRTHS
437

Food Dependency

1970
21%
1992
5%

PROFILE

ENVIRONMENT

The nation is divided into three large natural regions: the Himalayas, along the northern border; the fertile, densely populated Ganges plain immediatly to the south, and the Deccan plateau in the center and south. The Himalayas shelter the country from cold north winds. The climate is subject to monsoon influence; hot and dry for eight months of the year, and raining heavily from June to September. Rice farming is widespread. Coal and iron ore are the main mineral resources. There is a long-standing territorial dispute with Pakistan over Kashmir, in the northwest.

SOCIETY

Peoples: The population of India contains a multitude of racial, cultural and ethnic groups. Most are descendants of the Aryan peoples who developed the Vedic civilization and created a caste system so robust that it has survived until the present. The influence of Arab invasions, in the 7th and 12th centuries, and Mongolian incursions, in the 13th century, is still felt in the north. Peoples of Dravidian origin still predominate on the Deccan plateau in the centre and south of India.
Religions: 80.3% Hindu, 11% Muslim, 2.5% Sikh. There are Christian and Buddhist minorities. **Languages:** There are 18 officially recognized languages of which Hindi is the most widely spoken. English is an associated language, widely spoken for official purposes. **Political Parties:** The United Front, former National Front-Left Front, currently in office. The Congress Party, which fought for independence under Gandhi and Nehru became the party of government for much of the post-war period. In the wake of Indira Gandhi the party split. The Bharatiya Janata Party (BJP) is a rightwing nationalist party which supposedly has connections with the Rashtriya Swayan Sevak Sangh (RSSS), a paramilitary communal organization. The Communist Party of India-Marxist (CPI-M). The Lok Dal claims to be the party of the poor peasants. Its main support comes from the middle castes. Amongst the most important regional parties are the all-India Anna Dravida Munetva Kazhagam (AIA-DMK), the National Conference of Jammu and Kashmir, the Asom Ghana Parishad, ruling in Assam, the Telegu Desam, and the Akali Dal which leads a struggle for greater Punjabi autonomy.
Social Organizations: The most important are the Indian National Trade Union Congress (INTUC), with over 4 million members; the Bharatiya Mazdoor Sangh, with approximately 2 million members; the All-India Trade Union Congress (AITUC), with 1.5 million members; and the Center of Indian Trade Unions (CITU), with 1 million members.

THE STATE

Capital: Delhi and New Delhi, 8,720,381 inhab. (1995). **Other cities:** Greater Bombay, 12,596,243 inhab.; Calcutta, 11,021,915 inhab; Madras, 3,841,396 inhab; Bangalore, 3,302,296 inhab; Hyderabad, 3,145,939 inhab.
Government: Shankar Dayal Sharma, President since July 1992. H. D. Deve Gowda, Prime Minister since June 1, 1996, after resignation of Atal Bihari Vajpayee. **National Holidays:** August 15, Independence Day (1947); January 26, Day of the Republic (1950). **Armed Forces:** 1,145,000 troops (1995). **Paramilitaries:** 1,421,800.

of administering such a large domain and led the British crown, after violent repression, to assume direct government of India.

[10] The educational system was based on the classic British model and was conceived to train "natives" for colonial administration in the civil service. However, it did not exactly fulfill this purpose. What it did do was create an intellectual elite fully conversant with European culture and thinking. The British had certainly never planned that the first association of civil servants in India, created in 1876 by Surendranath Banerjee, would take the Italian revolutionary Giuseppe Mazzini as its patron and inspiration and not just quietly follow the government line. Years later in 1885, it was this intelligentsia that formed the Indian National Congress which included British liberals and, for a long time, limited itself to proposing superficial reforms to improve British administration.

[11] When Mohandas K. Gandhi, a lawyer educated in England with a good knowledge of colonial methods in South Africa, returned to India in 1915, he became aware of the need to break out of the straitjacket of Anglo-Indian "cooperation". Gandhi tried to win Muslims over to the autonomist cause. He reintroduced Hindu teachings that Ram Mohan Roy had reinterpreted in the 19th century, giving particular importance to mass mobilization. His ties with the Indian National Congress strengthened the movement's most radical wing where young Jawaharlal Nehru was an activist. In 1919, the Amritsar massacre occurred; a demonstration was savagely repressed leaving, according to British sources, 380 dead and 1200 wounded. In 1920, in response to the Amritsar massacre, at Gandhi's urging, the Indian Congress launched a campaign which showed the effectiveness of unarmed civilian opposition. The campaign tactics included the boycotting of colonial institutions; non-participation in elections or administrative bodies, non-attendance at British schools, refusal to consume British products, and passive acceptance of the ensuing legal consequences. The movement spread nationwide at all levels. Gandhi came to be called *Mahatma* (Great Soul) in recognition of his leadership.

[12] A new campaign between 1930 and 1934 aimed to attain full inde-

pendence and denounce the state salt monopoly. This demonstrated Gandhi's ability to combine a key political goal with a specific demand affecting all the poor; one they would understand and support. For the first time the British saw women flocking to demonstrations. Jails overflowed with prisoners who did not resist arrest posing an immense problem to the colonial authorities. It was impossible not to negotiate with Gandhi, and after World War II the British were left with no option but to rapidly grant independence.

[13] With the British withdrawal in 1947, the sub-continent was divided into two states; the Indian Union, and Pakistan, which was created to concentrate the Muslim population into one area (see Pakistan and Bangladesh). This "Partition" as it was called was a painful and often violent separation. The Indian Union brought together an enormous diversity of ethnic, linguistic, and cultural groups in a single federated state, consolidating the sentiment of national unity forged in the independence struggle that the British had never managed to stifle. The excitement of independence was clouded by the assassination of Gandhi less than a year later, on 30 January 1948.

[14] After independence, Prime Minister Jawaharlal Nehru, along with Sukarno of Indonesia, Gamal Abdel Nasser of Egypt and Tito of Yugoslavia, advanced the concept of political non-alignment for newly de-colonized countries. In India, he applied development policies based on the notion that the industrialization of the society would bring prosperity.

[15] In a few decades India made rapid technological progress, which enabled it to place satellites in orbit and, in 1974, to detonate an atom bomb, making India the first nuclear power in the non-aligned movement. However, the relevance of this kind of project to a country which had yet to feed all its people was widely questioned. Pakistan's civil war and conflict with India over East Pakistan eventually led to the independence of this portion as Bangladesh, in 1971.

[16] The Indian economy was severely hit by the oil crisis of the early 1970s as it was dependent on oil imports. Industrial exports could not grow fast enough to offset increases in import prices, nor to meet the demand for food from a population growing at a rate of 15 mil-

merable uprisings, both large and small, on local and national levels.

[9] The most serious of these were the 1857-1858 rebellions by sepoys, Indian soldiers in the British army. These began as a barracks movement, eventually incorporat-

ing a range of grievances and growing into a nationwide revolt. Hindus and Muslims joined forces and even proposed the restoration of the ancient Moghul empire. This movement demonstrated that the East India Company was incapable

lion per year. In 1975, the economic crisis, and popular resistance to the government's mass sterilization campaigns, led Indira Gandhi (Nehru's daughter, who had taken over as prime minister when her father died in 1966) to declare a state of emergency and impose press censorship.

[17] Abandoning the Congress Party's traditional populist policies, Mrs Gandhi followed World Bank economic guidelines, losing mass support for the government, without winning wholehearted backing from business sectors (particularly those linked to foreign capital), which demanded even greater concessions. The British tradition of respect for democratic freedoms instilled in mass organizations, big business and the middle class, led them (for different reasons) to close ranks in opposition and force the government to call parliamentary elections for March 1977. The Congress Party was roundly defeated by the Janata Party, a heterogeneous coalition formed by a splinter group of rightwing Congress Party members, the Socialist Party headed by trade union leader George Fernandes, and the Congress for Democracy, led by Jagjivan Ram, leader of the "untouchables" and former minister in Indira Gandhi's cabinet.

[18] India's foreign policy of non-alignment remained basically unchanged under aging prime minister Morarji Desai. In his time in power he was also unable to fulfill his promises of full employment and economic improvement. By mid-1979, disagreements within the party caused groups led by Charan Singh and Jagjivan Ram to split from the Janata Party. Desai had to resign and, unable to form a cabinet with a stable parliamentary majority, the prime minister called early elections, which returned Indira Gandhi to power in January 1980.

[19] Indira's second term was marked by a growing concentration of power, and accusations of unnecessary bureaucracy and corruption in the government, which gradually blemished her image. The problems in the Punjab, where the government faced increasingly strident demands from Sikh separatists, exemplified regionalism at its worst. Small groups of militant Sikhs harassed Hindus, to drive them out of the Punjab and create an absolute Sikh majority in the province. After that, the next step would be se-

cession and the formation of independent "Khalistan". Indira accused "forces from abroad" (namely Pakistan and the US) of destabilizing the country.

[20] On June 6 1984, the prime minister called in the army to evict hundreds of militant Sikhs from their most sacred shrine, the Golden Temple at Amritsar, which they had transformed into a command post for their separatist war. Hundreds of Sikh extremists were killed within the temple, including prominent Sikh leader, Jarnail Singh Bhindranwale. This bloody confrontation left moderate Sikh leaders with no room for negotiation. The atmosphere which prevailed throughout the region after the incident led to Indira Gandhi's assassination by two of her Sikh bodyguards, on October 31 1984.

[21] After the assassination, thousands of Sikhs fell victim to indiscriminate retaliation by Hindu paramilitary groups. Bypassing party and institutional formalities, Indira's son Rajiv was rapidly promoted to the office of prime minister and leader of the Congress Party.

[22] Elections in January 1985 gave him overwhelming support: he obtained 401 of Parliament's 508 seats. None of the other parties reached the 50 seats needed to be officially recognized as the opposition. Despite the landslide victory, some strongly regionalist areas such as the states of Karnataka and Andhra Pradesh turned their backs on Gandhi, influenced by Rama Rao, a charismatic former actor. Gandhi also lost in Sikkim, a Himalayan kingdom annexed to India in the 1970s, where the separatist Sikkim Sangram Parishad party won.

[23] The new premier took several steps to deal with Punjab's problems. He appointed a conciliatory figure as governor of the area, released political prisoners including opposition leaders, and directed that militants within his own party who had participated in the anti-Sikh violence be tried and punished. These measures paved the way for dialogue with Akali Dal, the regional majority party of Sikhs and other dissident groups. In spite of radically different stands, progress was made. The Punjab autonomists proposed the Indian central government should maintain its responsibility for defence and foreign affairs, the issuing of currency, mail, highways and telecommunications. Meanwhile, local

STATISTICS

DEMOGRAPHY

Urban: 26% (1995). **Annual growth:** 1.1% (1992-2000). **Estimate for year 2000:** 1,016,000,000. **Children per woman:** 3.7 (1992).

HEALTH

One **physician** for every 2,439 inhab. (1988-91). **Under-five mortality:** 119 per 1,000 (1995). **Calorie consumption:** 92% of required intake (1995). **Safe water:** 81% of the population has access (1990-95).

EDUCATION

Illiteracy: 48% (1995). **School enrolment:** Primary (1993): 91% fem., 91% male. Secondary (1993): 38% fem., 59% male. **Primary school teachers:** one for every 63 students (1992).

COMMUNICATIONS

94 **newspapers** (1995), 89 **TV sets** (1995) and 85 **radio sets** per 1,000 households (1995). 0.9 **telephones** per 100 inhabitants (1993). **Books:** 84 new titles per 1,000,000 inhabitants in 1995.

ECONOMY

Per capita GNP: $320 (1994). **Annual growth:** 2.90% (1985-94). **Annual inflation:** 9.70% (1984-94). **Consumer price index:** 100 in 1990; 157.0 in 1994. **Currency:** 31 rupees = 1$ (1994). **Cereal imports:** 694,000 metric tons (1993). **Fertilizer use:** 720 kgs per ha. (1992-93). **Imports:** $26,846 million (1994). **Exports:** $25,000 million (1994). **External debt:** $98,990 million (1994), $108 per capita (1994). **Debt service:** 26.9% of exports (1994). **Development aid received:** $1,503 million (1993); $2 per capita; 1% of GNP.

ENERGY

Consumption: 243 kgs of Oil Equivalent per capita yearly (1994), 20% imported (1994).

government would have greater autonomy than in India's other states.

[24] In 1987, India intervened in the conflict in Sri Lanka, pressing for a cease-fire and an agreement between Sinhalese and Tamils by sending in troops. Three years later, the Indian Peace Forces had to discreetly withdraw, having suffered many casualties.

[25] Rajiv Gandhi was an active participant in the group of six neutral countries (Argentina, India, Sweden, Tanzania, Mexico and Greece) which appealed to the superpowers to curtail the arms race. The prime minister announced India would not relinquish nuclear weapons in the future if Pakistan persisted in making the atomic bomb.

[26] India's foreign policy remained loyal to non-alignment but some domestic changes were announced. Rajiv made the personal computer

the symbol of his swift "modernization" policy. He promised the private sector he would lift restrictions on imports and on the purchase of foreign technology, while relaxing fiscal controls. Labor unions, however, feared that sophisticated technology would lead to unemployment and that Indian industry would not survive the competition from foreign products.

[30] Elections in November 1989 were held amidst a violent atmosphere which left more than 100 dead and revealed the opposition's progress. The Congress Party's representation went down to 192 seats while Janata Dal's went up to 141. Although a minority, this party succeeded in forming a government coalition with the National Front which appointed Vishwanath Pratap Singh as prime minister. This was possible because of the unusual simultaneous support from the right-wing Bharatiya Janata Party (BJP) and several left-wing groups.

[31] In March 1990, tensions between India and Pakistan mounted as Pakistan increased its support for the Kashmir independence movements. There were fears that, if both countries went to war, one of them could resort to nuclear weapons. In November, confrontations between Hindus and Muslims escalated amidst a general worsening of the economic crisis. Prime Minister Singh was replaced by Chandra Shekhar, also from the Janata Dal.

[32] An election campaign which claimed over 280 lives gave way to parliamentary elections which started on May 20 1991. The following day, they were adjourned following the assassination of Rajiv Gandhi, attacked by the Tamil liberation movement. A week later, Narasimha Rao was appointed Gandhi's successor as leader of the Congress Party. The elections, in which only 53 per cent of those registered voted and which were to become the most violent in India's independent history, were resumed between June 12 and 16. The Congress Party obtained 225 seats followed by the Bharatiya Janata Party with 119.

[33] In August 1991, the new government announced a drastic shift toward liberalism that would change the economic policy in force since independence. This change gave rise to criticism and protests. Already in September of the previous year, 70,000 representatives of tribal groups had met to halt dam constructions on the holy Narmada river, which threatened to flood ancient temples and ancestral lands belonging to the region's peasants. Over 200,000 people driven from their lands by the dams demonstrated their disagreement in 1991.

[34] This and other infrastructure projects formed part of a tradition of industrial modernization without weighing possible ecological consequences. Safety problems of the country's seven nuclear reactors remain unsolved and controls over companies which use or emit toxic substances are insufficient.

[35] Numerous acts of violence took place in 1992 by Hindu fundamentalists against the Islamic population in the northern cities of Bombay and Ayodhya. According to Hindu tradition, in the latter - which had a mosque that was built in 1528 - the god Rama had been born. In December, Lal Krishnan Advani, leader of the BJP, ordered his followers to destroy the temple. The following clashes between the two communities left approximately 1,300 dead in several Indian cities reaching neighboring countries such as Pakistan and Bangladesh. In 1995, the Supreme Court was consulted about the possible existence of a Hindu temple in Ayodhya, where the Babri mosque was to be built afterwards and then destroyed in December 1992 by Hindu fundamentalists. The members of the Court unanimously declined to issue an opinion on the matter.

[36] The reforms planned by Prime Minister Rao, implemented by finance minister Manmohan Singh, included opening up India's market to foreign investment. The state gradually reduced its intervention in the economy, leaving the rupee to fluctuate freely against the dollar, and removing import controls.

[37] The outcome was reflected in the reduction of inflation (below 10 per cent in 1995) and the fiscal deficit while exports have grown steadily since 1992. The country continues to bear the burden of a foreign debt equal to 38 per cent of the GDP while the per capita income of its over 935 million inhabitants is one of Asia's lowest.

[38] Economic reforms have triggered protests from several sectors, especially from agriculture. Resistance to multinationals involved in fertilizers and seeds was very strong. As part of the "green revolution" and a capital-intensive agriculture, the World Bank had granted significant loans during the 1960s and 1970s for the purchase of hybrid seeds while the government subsidized farmers.

[39] The government proceeded to eliminate those subsidies following instructions from the World Bank's structural adjustment plan. The Farmers' Association of the State of Karnataka (KKRS) headed rural protests which were responsible for a number of direct attacks against representatives of multinational corporations since 1991.

[40] At the International Conference on Rights of Third World Farmers held in Bangalore on October 3 and 4, 1993, farmers declared that the "seeds, plants, biological material and wealth of the Third World form part of the Collective Intellectual Property of the peoples of the Third World". They pledged to develop these rights in the face of the private patenting system which encourages the spread of monoculture and threatens biodiversity.

[41] Despite the violence among religious communities, protests against the economic reforms and natural disasters - an earthquake killed over 10,000 people in September 1993 - the government was optimistic in early 1994. However, differences between the states were growing. In the north, Utta Pradesh, home to over 140 million people, showed figures of social development far below the national average. Furthermore, the male population outnumbered that of females (882 women for every 1,000 men in 1995). In the southern state of Kerala, beggars are seldom seen, 90 per cent of the population can read and write and infant mortality (17 per 1,000) resembles that of Washington DC instead of the rest of India (79 per 1,000).

[42] In 1994, the Supreme Court of India declared that state governments could lose some powers if they did not enforce religious freedom, which was stipulated in the Constitution. Conflicts between Hindus and Muslims continued in Karnataka. In December, three Cabinet members resigned after being involved in corruption cases related to public security.

[43] That year India signed two agreements with China: the first one stated a reduction of military troops stationed along the 4,000 km of common borders between the two countries while the other encouraged trade relations. When former Pakistani prime minister Nawaz Sharif claimed his country had nuclear weapons, relations between New Delhi and Islamabad were strained. Pakistan closed its consulate in Bombay.

[44] In 1995, Prime Minister Narasimha Rao changed his cabinet three times. Elections in several states revealed a weakened Congress Party, with several breakaway groups, and a growing BJP. The new BJP government in the state of Maharashtra decided to change the name of the city of Bombay to Mumbai.

[45] In Bombay, New Delhi and Baroda, over 200 people were prosecuted for the killings of Sikhs after Indira Gandhi's death and Hindu-Muslim confrontations in 1993. The widow of Rajiv Gandhi complained about the slow investigation of her husband's assassination.

[46] Economic stability and assistance schemes announced by Rao, which included a plan of school meals for 110 million children and the construction of 10 million rural homes, were not enough to keep him in power. On May 10 1996, after his party's decisive defeat at the general elections held between April 27 and May 7, Rao resigned. The BJP, in alliance with other minority parties, obtained 187 seats in the Chamber of Representatives, followed by the Congress Party with 138 and the National Front-Left Front - which was to become the United Front - with 117.

[47] The BJP did not achieve a parliamentary majority which would grant it the necessary support. The new prime minister Atal Bihari Vajpayee resigned on May 28, twelve days after he had taken office. President Shankar Dayal Sharma appointed H D Deve Gowda, leader of the United Front, as head of the new government.

Indesnia

Indonesia

Population: 190,388,992 (1994)
Area: 1,904,570 SQ.KM
Capital: Jakarta

Currency: Rupiah
Language: Bahasa Indonesia

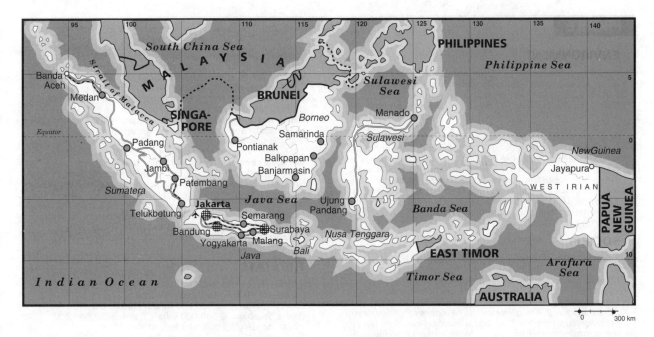

Indonesia has some of the world's earliest homo sapiens remains. The ancestors of the present-day population were Malay immigrants who arrived on the islands of Java and Sumatra around 400 BC. They brought with them the cultural and religious influences of India. Indonesian civilization reached its height in the 15th century when the state of Mojopahit extended far east, beyond Java, Bali, Sumatra, and Borneo, becoming commercially and culturally linked with China.

[2] At the end of the 13th century, through contact with Arab traders, Islamic beliefs were introduced, rapidly flourishing in the archipelago. These beliefs were not imposed through conversion campaigns, but were freely adopted because they represented a simple, egalitarian faith suited to local conditions.

[3] When the Muslim traders sent Indonesian spices to Europe, the Europeans' interest was aroused, and they set their colonial sights on the region. In 1511, the Portuguese arrived in Melaka; in 1521, the Spanish reached the Moluccas and in 1595, private Dutch businessmen organized their first expedition. The Netherlands had just won its independence from Spain and wanted to secure its own supply of spices. In 1602, Dutch merchants founded the Dutch East India Company, obtaining a trade monopoly, with a colonial mandate from the governor of the region.

[4] During the 17th and 18th centuries, Indonesia was fought over by Spain, Portugal, the Netherlands and Britain, the latter creating another private company. The Netherlands gained the upper hand and became the colonial power. Cash crops, coffee and sugar, were introduced, yielding excellent profits for the European investors but seriously upsetting the local socio-economic organization, which had remained intact until then but was virtually destroyed by the export economy. Anti-colonial revolts broke out. Towards the end of the 19th century, rubber, palm oil and tin became the main export products, though industry only began to develop during World War II, when the Netherlands was unable to meet its production needs at home.

[5] In December 1916, nationalist pressure caused the formation of the *Volksraad* (People's Council), a body designed to defend the rights of the local population. Although its proposals were largely ignored, the council encouraged political participation among the local population.

[6] In 1939, eight nationalist organizations formed a coalition called the Gabusan Politick Indonesia (GAPI), demanding democracy, autonomy and national unity within the framework of the antifascist struggle. GAPI adopted the red and white flag and Bahasa Indonesia as the national language.

[7] After the outbreak of World War II, the Netherlands was invaded by Germany, and Indonesia by Japan in 1942. The Japanese, who claimed to be the "Asiatic brothers" of the Indonesians, freed nationalist leaders like Ahmed Sukarno and Muhammad Hatta who had been imprisoned under Dutch rule. On August 11 1945, just four days prior to the Japanese surrender, they invested Sukarno and Hatta with full powers to establish an autonomous Indonesian government.

[8] Indonesia proclaimed its independence on August 17 1945. The Dutch tried to recover the archipelago, forcing the Indonesians to fight back. The Arab and Indian communities actively supported the pro-independence guerrillas, while Britain backed the Netherlands. The US pressed for a negotiated settlement.

[9] Unable to recover military control, the Netherlands were forced to surrender in 1949, although they retained partial control through a Dutch-Indonesian confederation.

[10] In 1954, the Dutch-Indonesian Union, which was never fully implemented, was renounced by the Sukarno government, and the archipelago became fully independent. Sukarno's government soon began its own colonial policies in the region. In 1963, reacting to the Hague's refusal to withdraw from the island of New Guinea, Indonesia occupied West Irian (Irian Jaya), the former Dutch colony whose territory took up half of the island. The Indonesian independence process offered an important example for the other Third World countries. Together with other major events of

WORKERS

1994

- MEN 69%
- WOMEN 31%

- AGRICULTURE 56% — INDUSTRY 14% — SERVICES 30%

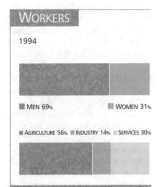

LAND USE

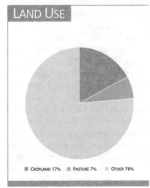

CROPLAND 17%   PASTURE 7%   OTHER 76%

Foreign Trade

MILLIONS $ 1994

IMPORTS
31,985
EXPORTS
40,054

Food Dependency

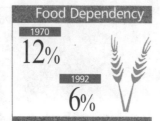

1970
12%

1992
6%

PROFILE

ENVIRONMENT

Indonesia is the largest archipelago-state in the world, made up of approximately 3,000 islands. The most important are Borneo (Kalimantan), Sumatra, Java, Celebes, Bali, the Moluccas, Western New Guinea and Timor. Lying on both sides of the equator, the island group has a tropical, rainy climate and dense rainforest vegetation. The population, the fourth largest in the world, is unevenly distributed: Java has one of the highest population densities in the world, 640 inhab. per sq km, while Borneo has fewer than 10 inhab. per sq km. Cash crops, mainly coffee, tea, rubber and palm oil are cultivated, along with subsistence crops, especially rice. Indonesia is the tenth largest oil producer and the third largest tin producer in the world. Like neighboring Malaysia, Indonesia has suffered deforestation due to the expansion of the paper and lumber-exporting industries.

SOCIETY

Peoples: 90% of Indonesians are of Malay origin, half of whom belong to the Javanese ethnic group. There are also Chinese and Indian minorities. **Religions:** Mainly Muslim; about 10% are Christian, around 3% are Buddhist and Hindu. **Languages:** Bahasa Indonesia (official), similar to Malay, is the official language. The governments of Indonesia and Malaysia have agreed to gradual language unification based on Melayu, the common mother tongue. Javanese, native language to 60 million of the country's inhabitants. English, the language of business and commerce. 250 regional languages. **Political Parties:** A presidential decree of January 1960 allows the president to dissolve any party whose ideology runs counter to official policies. The governing party, the Sekretariat Bersama Golongan Karya (Sekber Golkar Party), was created in 1971. Supported by the armed forces, the Sekber Golkar is a coalition of various professional and interest groups. President Suharto is chairman and president of the party's Advisory Board. Civil servants are mostly affiliated to the National Association of Civil Servants (KDRPR), part of Sekber Golkar. Opposition parties organized by the government in 1973 include the United Development Party (PPP), conservative and Muslim-based, and the Indonesian Democratic Party (PDI), which caters to Christians and old nationalistic groups from the Sukarno era. The Communist Party of Indonesia (PKI) which, until the 1965 coup, was the third largest Communist party in the world, has been outlawed and since then has gone underground. In 1990, the Free Aceh Movement and the Aceh National Liberation Front, made up of insurgents from the northern region of the island of Sumatra, started attacks on the government, but they were virtually crushed by the Army. Repression has also characterized the response to liberation movements in East Timor and West Irian. **Social Organizations:** All-Indonesian Union of Workers (SPSI), founded in 1973 and renamed in 1985.

THE STATE

Official Name: Republik Indonesia. **Administrative Divisions:** 26 Provinces (excluding East Timor). **Capital:** Jakarta, 10,000,000 inhab. (1992). **Other cities:** Surabaya, 2,500,000 inhab; Bandung, 2,000,000 inhab.; Medan, 1,700,000 inhab. (1990). **Government:** General Suharto, President since March 1968, re-elected for the sixth time in March 1993. Legislature, single-chamber, made up of 500 members, 400 elected by direct popular vote, and 100 designated by the president. **National Holiday:** August 17, Independence Day (1945). **Armed Forces:** 276,000 (1994). **Paramilitaries:** Police, 215,000, Auxiliary Police, (*Karma*), 1.5 million, Auxiliary Armed Force, in regional commands (*Wanra*).

---

the period - Indian and Pakistani independence, the Cuban Revolution, the nationalization of the Suez Canal in Egypt, and the French defeats in Vietnam and Algeria - it heralded the arrival of the South onto the international political scene. Sukarno was one of the main leaders of the pro-Third World movement, and in 1955, the Indonesian city of Bandung played host to a meeting of heads of government, giving rise to the Non-Aligned Movement.

[11] The three-million strong Communist Party, the second most powerful in Asia after the Chinese, supported Sukarno. He launched nationalist development programs, aimed at raising the living standards of a population with one of the world's lowest per capita incomes. The country had sizeable petroleum reserves, which were banked on to spur economic development. Sukarno created a state petroleum company, PERTAMINA to break the domination of the Anglo-Dutch transnational, Royal Dutch Shell.

[12] In 1965, these Indonesian oil deposits were nationalized. In October of that year, a small force of soldiers led by General Suharto seized power under the pretext of stemming the "communist penetration."

[13] This bloody coup left nearly 700,000 dead, and some 200,000 political activists were imprisoned. Though deprived of any real power, Sukarno remained the nominal president until 1967, when Suharto was officially named head of state. Suharto opened the doors to foreign oil exploiting companies that wanted drilling rights. But with the rise in oil prices, the influx of capital, and a liberal economic policy, the government also widened the gap between those with the highest and lowest incomes. Living conditions did not improve for millions of rural dwellers, many of whom had to leave their lands, joining the shanty-towns in the country's large cities.

[14] In 1971, defying repression, students took to the streets in opposition to the "corrupt generals, Chinese merchants, and Japanese investors" controlling the nation. In an attempt to divert the attention of young officers and neutralize local dissent, Suharto invaded East Timor, on that country's independence from Portugal in 1975. US President Gerald Ford visited Jakarta a few hours before the invasion and apparently gave it his blessing. However, from the point of view of the Maubere people of Timor the Indonesians were not seen as liberators but as new colonists. The unyielding resistance on the island only deepened Indonesia's internal problems.

[15] In May 1977, when elections were held for a portion of the seats in the House of Representatives, discontent surfaced once again. Despite repression, the banning of leftist parties and press censorship, the official Sekber Golkar Party lost in Jakarta to a Muslim coalition which had campaigned against the rampant corruption. The governing party also lost ground in rural areas where it had always previously maintained effective political control.

[16] In the light of this setback, and to ensure victory in the election five years later, the regime clamped down on political activity and re-organized the electoral system, making it dependent on the Ministry of Home Affairs.

[17] Sekber Golkar was the predictable winner in May 1982, obtaining 244 of 460 legislative seats. The United Development Party (PPP) gained 96 representatives, while those of the Democratic Indonesian Party (PDI) went down from 29 to 25. The 96 remaining seats, appointed directly by Suharto, were mostly occupied by officers of the armed forces.

[18] On March 10 1983, in spite of growing opposition, the People's Consultative Assembly unanimously re-elected Suharto for a fourth 5-year presidential term.

[19] Indonesia has adopted a birth control policy which has led to a reduction in the growth rate of its population. While figures for the 1984 census revealed an annual increase of 2.34 per cent, the average rate of growth between 1980

and 1990 was only 1.8 per cent. Even so, demographic pressure, particularly on the island of Java, together with the radical reorientation of the country's economy and commerce toward a world market, coupled with rapid industrialization, have all led to a deterioration in the quality of the environment and the depletion of agricultural lands. From 1979, the government reacted to this situation with a population transfer project known as "Transmigrasi", which involved transferring 2.5 million Javanese to other less populated islands, causing great social tension, and destroying the lifestyles of the peoples already resident on these islands.

[20] In the past few years over 300 ethnic groups in the area have seen their standards of living drop sharply. The most energetic protests have been those of the inhabitants of Irian Jaya (West Irian) who demand self-determination and freedom of movement to and from the neighboring territory of Papua New Guinea, with which they have a high degree of cultural and historical affinity.

[21] The continued Indonesian occupation of East Timor (see East Timor) has seriously damaged the Indonesian international image. In 1988 it caused the rejection of Indonesia's candidacy for the chair of the Non-Aligned Movement. Meanwhile, the European Parliament recognized East Timor's right to self-determination.

[22] Suharto was re-elected in 1988, in an election marked by unprecedented discussion of issues in the press.

[23] In 1991 the fighting between the army and the Adeh liberation movements in Sumatra became more acute when the commander of the army called for the annihilation of the insurgents. In November in Dili, capital city of East Timor, Indonesian troops fired against 1,000 protesters, 185 of whom died. At the beginning of March, 1992, an armed offensive of several separatist groups was started in the Irian Jaya province. At the same time the US Government asked the Congress to approve $2,300,000 for the training of the official security forces of Indonesia.

[24] Regarding the parliamentary elections of 1992 PPP, the main opposition party reached an agreement with Muslim leaders of great rural influence. Meanwhile, the PDI had Sukarno's children appear on the political scene, among them Guruh, a well-known author of popular songs. But Golkar once again won a victory, with 68 per cent of the votes. In March 1993 the Popular Consultive Assembly chose Suharto for the sixth time as president.

[25] In the 1990s the exports of gas, oil, wood and the new industrial goods, as well as tourism, contributed to an annual rate of growth of 6 per cent to 8 per cent. Nevertheless the foreign debt put 27 per cent of Indonesians under conditions of extreme poverty. The compulsive cutting of timber and the large infrastructural works resulted in the loss of fertile land and displacement of the peasant population.

[26] The Muslim 90 per cent of the population increasingly protested against the privileges of Chinese and Christian minorities alike. In March 1994 a scandal exploded which implicated Chinese business people and official politicians. In June, the government closed an opposition paper and two weekly magazines. The protests and demonstrations by artists and intellectuals did not receive media coverage.

[27] During 1995 the technocrat middle class manifested a growing interest in political openess. But it is thought that Suharto will win the presidential elections of 1998. The controversial minister of research and technology, JB Habibie, appears as a possible successor to Suharto. He promoted the young among the presidential allies, thus displacing the old generals.

[28] In August, Suharto liberated three political prisoners held in prison for 30 years for participating in the rebellion that eliminated Sukarno, and promised to remove the stigma of the identity cards of more than 1.3 million ex-political detainees. In May, he announced a plan for 1997 to reduce the seats assigned to military men in Parliament.

[29] Two years before the campaign for the parliamentary elections of mid-1997, Suharto's minister of information, Harmoko, toured Indonesia in support of Golkar, the most important political organization of the country. The opposition accused the government of interfering in its activities. Particularly vocal was the PDI, headed by Megawati Sukarnoputri, daughter of Sukarno, the still-respected independence leader.

[30] In March 1995 Harmoko arrested several members of the Alliance of Independent Journalists, a group formed after the closure of three publications. In September, the Government censored the publication of the memoirs of an ex-Sukarno-aide, as well as public appearances of several opposition leaders. In the ex-colony of East Timor the separatist movement was repressed by military forces. The armed forces investigated the death of civilians at the hands of its troops.

[31] In 1996 the pre-election debates reached a new high when complaints about the illegal enrichment of the Suharto family and its circle grew. The military saw Islamic groups as well as Sukarnoputri and the PDI as the main threats to Suharto's power. The change of political climate is due partly to a wider circulation of information via the Internet as well as campaigning journalism, and vigorous protest about arms sales to Indonesia.

STATISTICS

DEMOGRAPHY

Urban: 33% (1995). **Annual growth:** 2.7% (1992-2000). **Estimate for year 2000:** 206,000,000. **Children per woman:** 2.9 (1992).

HEALTH

One **physician** for every 7,143 inhab. (1988-91). **Under-five mortality:** 111 per 1,000 (1995). **Calorie consumption:** 101% of required intake (1995). **Safe water:** 62% of the population has access (1990-95).

EDUCATION

Illiteracy: 16% (1995). **School enrolment:** Primary (1993): 112% fem., 112% male. Secondary (1993): 39% fem., 48% male. University: 10% (1993). **Primary school teachers:** one for every 23 students (1992).

COMMUNICATIONS

91 **newspapers** (1995), 93 **TV sets** (1995) and 90 **radio sets** per 1,000 households (1995). 0.9 **telephones** per 100 inhabitants (1993). **Books:** 83 new titles per 1,000,000 inhabitants in 1995.

ECONOMY

Per capita GNP: $880 (1994). **Annual growth:** 6.00% (1985-94). **Annual inflation:** 8.90% (1984-94). **Consumer price index:** 100 in 1990; 139.9 in 1994. **Currency:** 2.200 rupiahs = 1$ (1994). **Cereal imports:** 3,105,000 metric tons (1993). **Fertilizer use:** 1,147 kgs per ha. (1992-93). **Imports:** $31,985 million (1994). **Exports:** $40,054 million (1994). **External debt:** $96,500 million (1994), $507 per capita (1994). **Debt service:** 32.4% of exports (1994). **Development aid received:** $2,026 million (1993); $11 per capita; 1% of GNP.

ENERGY

Consumption: 393 kgs of Oil Equivalent per capita yearly (1994), -101% imported (1994).

Iran

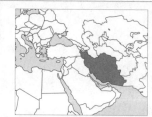

Population: 62,550,000 (1994)
Area: 1,648,000 SQ.KM
Capital: Teheran (Tehran)

Currency: Rial
Language: Persian

Iran

Shortly before the 18th century BC, nations of the Indo-European area reached the plains of Iran, subduing the shepherds who inhabited the region. More of these people arrived up until the 10th century, contributing to the Mesopotamian cultural mix. They later became known by different names: Medes, from the name of the ruling group; Iranians, the name they adopted in Persia and India (from Sanskrit *"ayriana"* meaning "nobles"); Persians (a Greco-Latin term alluding to Perseus, the mythological ancestor that the Greeks foisted on Iranians), with its corrupted forms: Parsis, Farsis, Fars, or Parthians (according to the time and source). Whatever their names, they first won control of the mountainous region, then conquered the Mesopotamian plains under the reign of Ciaxares. During the rule of Ciro the Great (559-530 BC), this wave of expansion reached as far west as Asia Minor, and as far east as present-day Afghanistan. These borders were later extended as far as Greece, Egypt, Turkestan and part of India.

2 Towards the end of the 4th century BC, this vast empire fell into the hands of Alexander of Macedonia. Alexander's successors, the Seleucides and Romans (see Syria) lost their hold on the eastern part of the empire to the Persians, who recovered their independence with the Arsacid dynasty (2nd century BC to 3rd AD). They remained independent under the Sassanids until the 7th century, though constantly at war with the Romans and Byzantines.

3 After the Arab conquest reached the region in 641 (see Saudi Arabia), Islamic thought and practices became dominant. Unlike the people in most other provinces of the Arab Empire, they retained their own language and distinctive styles in arts and literature. With the fall of the Caliphate of Baghdad, Persia attained virtual independence, first under the descendants of Tahir, the last Arab viceroy, and later under the Seleucidian Turks and the Persian dynasties. Despite political restlessness, the period was remarkably rich in cultural and scientific progress, personified by the poet, mathematician, philosopher and astronomer, Ummar al-Khayyam.

4 The Mongol invasion led by Hulagu Khan, that began in 1258, was an altogether different matter. Three centuries of Mongol domination brought dynastic strife between the descendants of Timur Lenk (Tamburlaine) and the Ottomans. The dispute paved the way for Persian Ismail Shah whose grandson Abbas I (1587-1629) succeeded in unifying the country. He expelled the Turks from the west, the Portuguese from the Ormuz region, and also conquered part of Afghanistan. For a short time Iran ruled a region extending from India to Syria.

5 The weakness of the Persian ruler during frequent Anglo-Russian interventions in Iran and Afghanistan encouraged a strong nationalist movement, ideologically influenced by Syrian Pan-Islamic intellectuals.

6 A 1909 treaty divided the country into two areas - one of Russian and the other of British economic influence. A British firm was given the opportunity to exploit Iranian petroleum fields. Military occupation by the two powers during World War I, in addition to government corruption and inefficiency, led to the 1921 revolution headed by journalist Sayyid Tabatabai and Reza Khan, commander of the national guard.

1 Neutral Zone   Kurdistan

0   300 km

7 Reza, the revolution's war minister, became prime minister in 1923. Two years later, the National Assembly dismissed Tabatabai, and Reza ousted the shah, occupying the throne himself. Reza repealed all treaties granting extraterritorial rights to foreign powers, abolished the obligatory use of the veil by women, reformed the education and health systems, and cancelled oil concessions that favored the British.

8 His attempt to establish a militarily strong, internationally neutral modern state, that became known as Iran in 1935, met with strenuous resistance from European powers. Reza Shah insisted in maintaining strict neutrality, rejecting a July 1941 ultimatum demanding passage for allied arms to the Soviet Union through Iranian territory. In August, British and Soviet forces invaded, forcing the Iranian army to surrender in September. The shah

was overthrown and sent into exile after abdicating in favor of his son, Muhammad Reza Pahlevi. Educated in Europe and more amenable to European interests, the new ruler governed under Anglo-Soviet tutelage until the end of the war.

9 A 1949 constitution curtailed the shah's authority, and progressive, nationalist forces won seats in parliament. With their support, Prime Minister Muhammad Mossadegh attempted to nationalize oil reserves and expropriate the Anglo-Iranian Oil Company.

10 Mossadegh maintained that it was "better to be independent and produce only one ton of oil a year than produce 32 million tons and continue as slaves of England", despite the fact that Britain had other sources of supply in the Arab states, Venezuela and the US, while Iran had no alternative market. Mossadegh's audacity was met in 1953 by an economic blockade and a coup backed by the US Central Intelligence Agency (CIA) which returned almost absolute power to the shah. This chain of events was accompanied by the killing of nationalists and left-wing activists. Thousands were imprisoned. Mossadegh remained in prison until his death in 1967.

11 The shah's power depended principally on the oil economy. He encouraged multinational penetration of Iran, citing "modernization" which effectively consisted of promoting Western consumer habits.

12 Westernization was resisted by Islamic clergy who feared encroaching

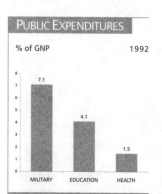
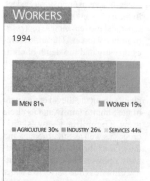
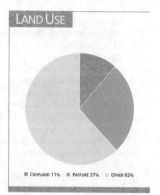

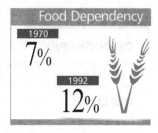

Foreign Trade

MILLIONS $ 1994

IMPORTS
20,000

EXPORTS
13,900

Food Dependency

1970
7%

1992
12%

secularism, and by those social sectors most affected, notably small farmers and the urban poor. By the end of the 1970s, the growing power of foreign enterprises in the domestic market, plus rapidly changing consumption patterns, had lost the shah the sympathies of the powerful commercial and industrial elites known locally as the "bazaar".

[13] Opposition groups included the National Front founded by Mossadegh, the Tudeh Communist Party, and Fedayin (Marxist) and Mujahedin (Islamic) guerrillas - each using a different strategy. The most effective opposition was centered around the Muslim leader Ayatollah Ruhollah Khomeini, exiled in 1964. Khomeini was resident in Iraq for many years until the shah, irritated by political ferment in Iranian mosques, pressured the Iraqi government into deporting him to France.

[14] The recordings of Khomeini's preaching in Paris became well-known and encouraged the masses to organize. During 1977, demonstrations began in secondary schools and by 1978 became generalized. The shah was forced to flee in January 1979 and Khomeini made a triumphant return to Iran. On February 11 1979 crowds invaded the imperial palace; the shah's prime minister resigned, and a divided army accepted the new state of affairs. The Islamic Revolution, seen as a successful alternative to both socialist and capitalist Western models, was welcomed enthusiastically, not only in Iran but throughout the Muslim world.

[15] Prime Minister Mehdi Bazargan of the National Front sought to reconcile Muslim traditions with a model of progressive development permitting broad participation. However he was supported neither by the revolutionary left nor by the Muslim fundamentalists. The latter, bolstered by the "revolutionary guards" and Khomeini's popularity, first excluded their former left-wing and right-wing allies and then proceeded to suppress them harshly.

[16] In early November 1979, a student group stormed the US embassy in Tehran, taking the diplomatic personnel hostage. Documents found inside provided evidence of CIA involvement in Iranian politics. In April 1980, a US attempt to rescue the hostages was unsuccessful.

[17] On July 17 1980, the shah died in Egypt and lengthy negotiations began for release of the hostages.

This was finally accomplished on January 20 1981.

[18] In September 1980, Iran and Iraq went to war which lasted for eight years (see box in Iraq).

[19] The Islamic Revolutionary Party (IRP) won the presidential elections with more than 90 per cent of the vote in 1981. On August 30, violent incidents and bombings caused the death of 72 political leaders, among them President Muhammad Ali Rajai and Prime Minister Muhammad Javad Bahonar. A new president, Ali Khamenei, former secretary-general of the IRP, was elected.

[20] The clergy held a favored position, and the government made widespread use of imprisonment, exile and execution for both political opponents and common criminals. Government sources registered 500 executions in 1989 - mostly for drug trafficking.

[21] In 1985, the country made great commercial profits, despite the possibility of war with its neighbour. Their main clients were West Germany, Japan, Switzerland, Sweden, Italy and the Arab Emirates, with petroleum being the main export product, constituting 98 per cent of exports.

[22] Contrary to Western expectations, the death of Ayatollah Khomeini on June 3 1989 did not lead to widespread chaos and instability. As more than eight million, mostly poor mourners, gathered to bury Khomeini, President Ali Sayed Khameini was appointed Iran's spiritual leader by a vote of the Assembly of Experts.

[23] In August, new elections gave a landslide victory to Hojatoleslam Ali Akbar Hashemi Rafsanjani, ex-speaker of the parliament, and now president. The new constitution enhanced the president's role and Rafsanjani took office as a fully empowered president rather than a ceremonial one. Iran's international image deteriorated when Khameini sentenced Indo-British writer, Salman Rushdie, to death, stating that the writer's book "The Satanic Verses" was blasphemous. Rushdie went into hiding, but in late 1990, he made a few public appearances, re-confirmed his Islamic faith and became reconciled with the main Islamic authorities. However, the Iranian clergy have not lifted the death sentence (*fatwa*), and Rushdie remains in hiding.

[24] In 1990, Iran condemned the Iraqi invasion of Kuwait, taking advantage of the situation to negotiate previous border disputes. Iraq with-

ENVIRONMENT

Central Iran is a steppe-like plateau with a hostile climate, surrounded by deserts and mountains (the Zagros on the western border and the Elburz to the north). Underground water irrigates the oases where several varieties of grain and fruit trees are cultivated. The shores of the Caspian Sea are more humid and suited to tropical and subtropical crops (cotton, sugar cane and rice). Modern industries (petrochemicals, textiles, building) expanded during the time of Shah Pahlevi, but the industry of hand-made rugs and other materials remain an important activity. Deforestation is occuring in the last remaining forested areas along the Caspian Sea and in the western mountain region, the sea was heavily polluted during the Gulf War, threatening marine life, and there is air pollution from automobile exhaust emissions, industry and oil refineries, especially in urban areas.

SOCIETY

Peoples: The Persians, or Farsis, are a Indo-European people (unlike the Arabs, who are Semitic), whose culture was integrated like the rest of Central Asia by Islam by the 7th century. Over one third of the population consists of different minorities (Kurds, Arabs, Turks, Armenians and some nomadic groups of Afghani and Pakistani ancestry).
Religions: Mainly Islamic, with the Shi'ites clearly predominant. Iran's Islamic constitution permits other Moslem groups, as well as Catholics, Jews and Zoroastrians to practise their religions. Members of the Bah'ai sect suffered persecution in the early years of the Islamic revolution. **Languages:** Persian (official), and languages of ethnic minorities.
Political Parties: Movement for the Liberation of Iran; Communist Party (Toudeh); National Front. The Revolutionary Council of the Resistance, founded in Paris in 1981, is a coalition made up of supporters of former president Abolhassan Bani Sadr, the People's Mujahedin Movement, the Movement for the Independence of Kurdistan, and others.

THE STATE

Official Name: Dshumhurije Islámlje Irân. **Administrative divisions:** 23 provinces, 472 towns and 499 municipalities. **Capital:** Tehran, 6,475,000 inhab. (1991). **Other cities:** Mashhad, 1, 760,000 inhab.; Isfahan, 1,128,000 inhab.; Tabriz, 1,090,000 inhab. (1991). **Government:** Presidential republic. Ali Akbar Hashemi Rafsanjani, President since 1989. Religious power is exercised by Ali Sayed Khamenei. Legislature: Islamic Consultative Assembly (single-chamber), with 270 members elected by direct popular vote every four years. The members of parliament are elected by secret, universal ballot. **National Holiday:** February 11, Revolution Day (1979). **Armed Forces:** 513,000 (1994). **Paramilitaries:** BASIJ 'Popular Mobilization Army' volunteers, mostly youths; strength has been as high as 1 million during periods of offensive operations; Gendarmerie: (45,000 incl border guards).

DEMOGRAPHY

Urban: 58% (1995). **Annual growth:** 1% (1992-2000). **Estimate for year 2000:** 75,000,000. **Children per woman:** 5.5 (1992).

HEALTH

Under-five mortality: 51 per 1,000 (1995). **Calorie consumption:** 110% of required intake (1995). **Safe water:** 84% of the population has access (1990-95).

EDUCATION

Illiteracy: 28% (1995). **School enrolment:** Primary (1993): 101% fem., 101% male. Secondary (1993): 58% fem., 74% male. University: 15% (1993). **Primary school teachers:** one for every 32 students (1992).

COMMUNICATIONS

90 **newspapers** (1995), 96 **TV sets** (1995) and 100 **radio sets** per 1,000 households (1995). 5.9 **telephones** per 100 inhabitants (1993). **Books:** 100 new titles per 1,000,000 inhabitants in 1995.

ECONOMY

Annual inflation: 23.40% (1984-94). **Consumer price index:** 100 in 1990; 236.0 in 1994. **Currency:** 1.736 rials = 1$ (1994). **Cereal imports:** 4,840,000 metric tons (1993). **Fertilizer use:** 755 kgs per ha. (1992-93). **Imports:** $20,000 million (1994). **Exports:** $13,900 million (1994). **External debt:** $22,712 million (1994), $363 per capita (1994). **Debt service:** 22.5% of exports (1994). **Development aid received:** $141 million (1993); $2 per capita; 0.2% of GNP.

ENERGY

Consumption: 1,565 kgs of Oil Equivalent per capita yearly (1994), -127% imported (1994).

External Debt 1994
PER CAPITA $363

Illiteracy 1995
28%

drew from 2,600 sq km of occupied Iranian territory, war prisoners were exchanged, and the countries' sovereignty over the Shatt-al Arab canal was divided. Iran remained neutral when hostilities broke out in 1991.

[25] The election of Rafsanjani to the country's presidency meant a strengthening of the "liberal" wing in government. According to the Iranian constitution, religious and lay authorities share power. But, whereas his predecessor Ruhola Khomeini had held the title of Great Ayatollah, the present supreme leader, Ali Khameini, does not. It is the old ayatollah, Araki, who bears that responsibility.

[26] The radical sectors were pre-eminent, but did not have a two-thirds majority in the Majlis (Parliament), therefore having no power to change presidential resolutions. The Revolutionary Guard meanwhile were merged into the army and acquired military ranks, depriving radical fundamentalists of an important pressure group.

[27] Iranian neutrality during the Gulf War not only brought advantages from Iraq, but represented an effort to gain acceptance in regional and international diplomacy. When the hostilities began, Iraq sent its air force for safekeeping in Iran, and the Iranian government subsequently decided to keep the planes as compensation for the Iran-Iraq War in the 1980s. Diplomatic relations with Britain were re-established in 1990, and with

Saudi Arabia in March 1991 during the Gulf War.

[28] This rapprochement with the West and the refusal openly to support the other Islamic states during the US offensive affected relations with radical groups in the area. Iran's lack of support to the Shiite rebellion in Iraq, at the end of the war, disillusioned those sectors which saw Tehran as the capital of Shiite expansion.

[29] In 1990, a five-year plan was approved to organize the country's economy and diversify its sources of foreign currency. Foreign investors were, however, wary of the revolutionary leadership and no foreign capital could be attracted to Iran. The government also announced that foreign debt accounted for 12 per cent of the GNP in 1990.

[30] The much desired re-establishment of normal relations between Iran and the West has been hindered by the terrorist groups linked to Tehran. In April 1990, the brother of the Mujahedin guerrilla movement leader was murdered in Switzerland; in April 1991, the secretary of Chaput Bakhtiar, Shah Reza Pahlevi's former prime minister, was assassinated in Paris; and in May Bakhtiar himself was killed. Relations between Switzerland and Iran were on the brink of collapse in December 1991, as a result of using diplomatic immunity for espionage and terrorist agents.

[31] With the disintegration of the former Soviet Union, a new area of

influence lay open to Iran. In 1991, Tehran opened embassies in the Islamic republics of the Caucasus and the Middle East, for trading and cultural agreements, opening new roads of communication with its neighbors.

[32] Azerbaijan seems to have been the first target for Iranian diplomacy, partly because the Azeri population lives on both sides of the border between the two countries. It is also clear that US support of Armenia in 1991, during the conflict over Nagorno Karabakh, caused other Muslims and nationalists to favor Iran.

[33] Two main problems with the expansion of Iranian influence in the two republics of the Commonwealth of Independent States are that Islam has little significance in everyday life in these two countries, and Sunni, not Shiite Islam is the predominant religion.

[34] In 1991, Iran's most successful move in its rapprochement towards the West was the release of ten hostages held by pro-Iranian Lebanese forces. However, early in 1992 the upper echelons of the Revolutionary Guard, sent to Lebanon to organize the Hezbollah (Party of God), stated that Iran would not cease to fight against their great common enemy, the US, confirming existing conflicts within the regime.

[35] Also in 1992, Rafsanjani's government announced plans to privatize some large companies, previously nationalized during the 1979 Islamic Revolution. This was seen as an obvious attempt to attract foreign capital or Iranian investors abroad.

[36] On April 10 1992, new legislative elections were held, with the moderates who supported president Rafsanjani winning a clear victory over the radical candidates.

[37] In July, Khameini, Iran's spiritual leader, launched a campaign to "eradicate Western influence", clashing with Rafsanjani and his more moderate vision of Islam to such an extent that the latter threatened to resign.

[38] However in June 1993 Rafsan-

jani was confirmed in his post by 63 per cent of voters. The high abstention rate - 41 per cent of the 29 million voters - was interpreted as a display of discontent with the corruption and nepotism within the leadership.

[39] In February 1994 - during a time of great tension between the various tendencies of this hierarchy - Rafsanjani escaped unharmed from an attack in Tehran during the celebration of the 15th anniversary of the Islamic Revolution. Disagreements with Saudi Arabia related to the annual Muslim pilgrimage to Mecca - the main sacred site of Islam, which lies within Saudi territory - damaged relations with the government of King Fahd and once again highlighted the fight between the two countries for the world leadership of Islamic nations.

[40] On the economic front, the planned results were not forthcoming. Oil sales from March 1994 to March 1995 brought in $10.5 billion, compared with the $17.7 billion of the year before. At the same time, the government withdrew subsidies on 23 imported products - mainly food and medicine - causing subsequent price increases within Iran.

[41] Despite improving trade relations with the United States and consolidating his political power, Rafsanjani continued to make less economic progress than planned during 1995. Crude oil sales from March 1995 to March 1996 raised $15 billion - an improvement on the previous year, but still less than planned and less than two years previously. Inflation and the loss of buying power created increasing discontent amongst much of the population.

[42] On the international front, Iran continued its fruitless attempts at rapprochement with the Western nations. On a regional scale, Tehran kept up its opposition to the Palestine-Israeli peace agreement and reiterated the importance of not endangering the territorial integrity of Iraq.

Iraq

Population:	20,356,000 (1994)
Area:	438,320 SQ.KM
Capital:	Baghdad
Currency:	Dinars
Language:	Arabic

Iraq

The territory of present-day Iraq was the cradle of the Sumerian civilization around 4000 BC. Over the centuries it was the home of a series of prestigious civilizations; the Akkad, Babylonian, and Assyrian or Chaldean. Known as Mesopotamia, from the Greek, "between rivers", it was a crossroads for innumerable migrations and conquest expeditions; Hittites, Mitannians, Persians, Greeks, Romans and Byzantines all passed through the territory, leaving behind more destruction than cultural heritage.

[2] In the 7th century the Arab conquest transformed Mesopotamia into the center of an enormous empire (see Saudi Arabia). A century later, the new Abbas dynasty decided to move the capital east from Damascus. Caliph al-Mansur built the new capital, Baghdad, on the banks of the Tigris and for three centuries the city of the "thousand and one nights" was the center of a new culture.

[3] This culture led to the greatest flourishing of the arts and sciences in the Mediterranean region since the days of the Greeks. The empire began to fall apart after the death of Harun al-Raschid. The African provinces were lost (see Tunisia: "Islam in North Africa"), and the region north and east of Persia won independence under the Tahiris (the Kingdom of Khorasan). The caliphs depended increasingly on armies of slaves or mercenaries (Sudanese or Turks) to retain their grip on an ever-shrinking empire. When the Mongols assassinated the last caliph in Baghdad in 1258, the title had already lost its political meaning. The conquests of Genghis Khan devastated the region's agricultural economy, and the region was subsequently ruled in whole or part by Seleucids, Otto-

mans, Turks, Mongols, Turkomans, Tartars, and Kurds. The movement of steppe peoples (see Afghanistan), brought great instability to the fertile crescent, which finally achieved unification under the Ottoman Turks in the 16th century, having repelled an attack by Timur Lenk (Tamburlaine) in the 14th century.

[4] A subsequent period of relative calm gave the people a chance to reorganize the economy; irrigation systems were rebuilt and arable areas were expanded.

[5] In the early 20th century, Arab Renaissance movements were active in Iraq, paving the way for the rebellion that rocked the Turkish realm during World War I (see Saudi Arabia, Jordan and Syria). The British already had a foothold in the region and were keen to expand their influence. With the defeat of the Turks, Iraq entertained hopes of independence. These however were dashed when the revolutionary Soviet government revealed the existence of the secret Sykes-Picot treaty (signed in 1916), whereby France and Britain divided the Arab territories between themselves. Faisal, son of the shereef Hussein, had proclaimed himself king of Syria and occupied Damascus. However, since this territory "belonged" to the French, who had promised nothing to the Arabs, (see Saudi Arabia, Jordan and Syria), he was forcibly evicted. In 1920, Britain was awarded a mandate over Mesopotamia by the League of Nations, triggering a pro-independence rebellion.

[6] In 1921, Emir Faisal ibn Hussain was appointed king of Iraq, in compensation for his previous bad treatment. In 1930, general Nuri as-Said was appointed prime minister, signing a treaty of alliance with the British, under which the country would become nominally independent on October 3 1932.

[7] That year, the Baghdad Pact was founded, making Iraq part of a military alliance with Turkey, Pakistan, Iran, Britain and the US. This pact was resisted by all Iraqi nationalists. In July 1958, anti-imperialist agitation resulted in a military coup led by Abdul Karim Kassim, bringing about the execution of the royal family.

[8] In 1959 the regime tried to forge a union with Syria, but the Communist Party - one of the most important in the East - and the democrats, whose model was the European

1 Neutral Zone ///// Kurdistan
0 — 300 km

parliamentary system, opposed the move. In July, in an attempt to consolidate the regime, Kassim banned all political parties and proclaimed that the emirate of Kuwait belonged to Iraq. The Arab League, dominated by Egypt, authorized the deployment of British troops to protect the oil rich enclave.

[9] Kassim's overstated ties with the Soviet Union and China fomented predictions that Iraq could become "a new Cuba". In the summer of 1960 the country suddenly moved towards the West. Steps toward economic change were taken, a land reform program was implemented, and the profits of the Iraq Petroleum Company were severely restricted. In 1963, Kassim was deposed by pan-Arabian sectors within the army. Several unstable governments ensued until July 17 1968, when a military coup placed the Baath party in power.

[10] Founded in 1947, the Arab Baath Socialist Party (baath meaning "renaissance") was inspired by the ideal of Pan-Arabianism, regarding the Arab World as an "indivisible political and economic unit" where no country "can be self-sufficient". The Baathists proclaim that "socialism is a need which emerges from the very core of Arab nationalism". It is organized on a "national" (Arab) level, having several "regional" leaders in each country.

[11] Iraq nationalized foreign com-

panies, and Baghdad defended the use of oil as a "political weapon in the struggle against imperialism and Zionism". It insisted on protected prices and the consolidation of OPEC as an organization which would support the struggle of the Third World for the recovery and enhancement of its natural resources. A land reform program was decreed, and ambitious development plans encouraged the reinvestment of oil money into national industrialization.

[12] In 1970, the Baghdad government gave the Kurdish language official status, and granted Kurdistan domestic autonomy. However, abetted by the Shah of Iran, and fearful of land reform, the traditional leaders rose in armed confrontation. In March 1975, the Iran-Iraq border agreement deprived the Kurds of their main foreign support and the rebels were defeated. The Baghdad government decreed the teaching of Kurdish in local schools, greater state investment in the region, and the appointment of Kurds to key administrative positions.

[13] On July 16 1979, President Ahmed Hassan al-Bakr resigned because of ill-health. He was replaced by Vice-President Saddam Hussein.

[14] Saddam Hussein has since tried to establish himself as leader of the Arab world. He was one of the most outspoken critics of the 1979 Camp David agreements between Egypt,

WORKERS

1994

MEN 78%   WOMEN 22%

AGRICULTURE 14%   INDUSTRY 19%   SERVICES 67%

ENVIRONMENT

The Mesopotamian region, between the rivers Tigris and Euphrates (Al-Furat) in the center of the country, is suitable for agriculture, and contains most of the population. In the mountainous areas in the north, in Kurdistan, there are important oil deposits. In Lower Mesopotamia, on the Shatt-al-Arab channel, where the Tigris and the Euphrates merge, 15 million palm trees produce 80% of the world's dates which were sold worldwide, before the blockade. The war's devastation included the destruction of the major part of the country's infrastructure. Tank and troop movements caused profound damage to road surfaces and soil, especially in the environmentally sensitive area along the Saudi Arabian border.

SOCIETY

Peoples: Iraqis are mostly Arab; 20% belong to a Kurdish minority living in the north. **Religions:** Mainly Islamic. Most of the Shiite Muslims live in the south. In the center, Sunni Arabs predominate, and they share a common religion with the Kurds in the north. **Languages:** Arabic (official and predominant); in Kurdistan it is regarded as a second language, after Kurdish. **Political Parties:** The Arab Baath Socialist Party has been in power since 1968; the Kurdistan Democratic Party; the Kurdistan Revolutionary Party; the Communist Party; Ad-Da'wa al-Islamaya (the Voice of Islam); Umma Party. **Social Organizations:** The General Federation of Trade Unions of Iraq.

THE STATE

Official Name: Al-Jumhuriyah al-'Iraqiyah. **Administrative divisions:** 15 provinces and 3 autonomous regions. **Capital:** Baghdad, 4,044,000 inhab. (1990). **Other cities:** Diyala, 961,073 inhab.; as-Sulaymaniyah, 951,723 inhab.; Irbil, 770,439 inhab.; Mosul, 664,221 inhab. **Government:** Saddam Hussein, President of the Revolutionary Command Council since 1979. **National Holidays:** July 14, Proclamation of the Republic (1958); July 17, Revolution Day (1968). **Armed Forces:** 382,000 (1985).

STATISTICS

DEMOGRAPHY

Annual growth: 0.7% (1992-2000). **Children per woman:** 5.7 (1992).

HEALTH

One **physician** for every 1,667 inhab. (1988-91). **Under-five mortality:** 71 per 1,000 (1995). **Calorie consumption:** 106% of required intake (1995). **Safe water:** 44% of the population has access (1990-95).

EDUCATION

Illiteracy: 42% (1995). **Primary school teachers:** one for every 25 students (1990).

COMMUNICATIONS

99 **newspapers** (1995), 100 **TV sets** (1995) and 101 **radio sets** per 1,000 households (1995). 3.4 **telephones** per 100 inhabitants (1993). **Books:** 82 new titles per 1,000,000 inhabitants in 1995.

ECONOMY

Consumer price index: 100 in 1990; 286.5 in 1991. **Currency:** 0,3 dinars = 1$ (1994). **Cereal imports:** 2,834,000 metric tons (1990).

Israel and the US, but Iraqi relations with other Arab countries still worsened. A branch of the Baath party took over power in Syria in 1970, but its discrepancies with Baghdad led to rivalry and some border disputes.
[15] These issues encouraged the Iraqi forces to wage attacks on Iranian positions. They were confident of a quick victory but in the event the war lasted for eight years (see box).
[16] In 1981, claiming that Iraq intended to produce atomic weapons, Israeli planes bombed the nuclear power station at Tamuz, which had been built with French assistance.
[17] During the war, Saudis and Kuwaitis, who had benefited from Iraq acting as a bulwark against Iranian fundamentalism, granted Baghdad many loans, which were used both in the conflict and for strengthening the country's infrastructure. An oil pipeline was built through Turkey as an alternative to the one which crossed Syria to the Mediterranean; Syria had closed this in sympathy with Iran. The roads to Jordan were also improved.

[18] In November 1984, 17 years after they had broken diplomatic relations, official links with the US were re-established. Despite American declarations of neutrality in the Iran-Iraq conflict, events disclosed the superpower's doublespeak. The "Iran-contra" scandal made this clear (see box).
[19] The 1988 armistice meant that Iraq retained 2,600 sq kms of Iranian territory with its powerful and skillful army. Apart from refusing to establish export quotas, Kuwait also extracted more oil than it was allowed to from deposits under the border. As the US declarations seemed to state that they would remain neutral in the event of conflict, Baghdad thought that it would be able to retaliate against the West by occupying the neighboring territory and exploiting its wealth. On August 2, Iraq invaded Kuwait and took thousands of foreign hostages.
[20] Four days later the UN decided on a total economic and military blockade until Iraq unconditionally retreated from the occupied territory. Withdrawal was rejected, but a proposal for an international conference to discuss the Middle East

issue was submitted. When Iraq started to release the hostages and to make new attempts at negotiating, the US refused to talk and demanded an unconditional surrender.
[21] On January 17 1991, an alliance of 32 countries led by the US started attacks on Iraq. When the land offensive began in March, Saddam Hussein had already announced his unconditional withdrawal. The Iraqi army did not resist the offensive and hardly attempted to stage an organized withdrawal, yet it suffered great losses. The war ended early in March, with the total defeat of the Iraqis (see box "The Gulf War", in Kuwait).
[22] At the end of the offensive, the US encouraged an internal revolt of the southern Shiites and of the northern Kurds so that Saddam Hussein would be deposed. However, the political differences between these factions made an alliance impossible, and Washington abandoned the rebels to their own devices, whereupon they were crushed by the still powerful Iraqi army. Over one million Kurds sought refuge in Iran and Turkey to escape the Baghdad forces, and

thousands starved or froze to death when winter came.
[23] Between 150,000 and 200,000 people, mostly civilians, died in the war. As a result of the blockade - which is still in effect - some 70,000 more may have died in the first year, among them 20,000 children. At the end of 1991, both the Turkish and Iraqi armies were continuing to harass the Kurds in the border area.
[24] According to documents captured after the war and publicized by human rights organization Middle East Watch, the Iraqi government used chemical warfare against the Kurds in an attempt to exterminate them toward the end of the 1980s. Specific reference was made to the attack on Halabja in March 1988, in which 5,000 Kurdish civilians were killed.
[25] The conditions stipulated for lifting the blockade became even more demanding with the increased determination on the part of the US government to bring down Hussein. In addition, The New York Times and the Sunday Telegraph (London) reported that the United States had introduced huge amounts of counterfeit dinars (Iraq's currency), smuggled across the Jordanian, Saudi Arabian, Turk-

ish and Iranian borders. Baghdad established the death penalty for anyone participating in these operations.

[26] Toward the end of 1991, the Iraqi government authorized UN inspections of military centers. In 1992, Iraq was found to have a uranium enrichment project, which had been developed using German technology. UN inspection teams destroyed 460 x 122mm warheads armed with sarin, a poisonous gas. They also dismantled the nuclear complex at al-Athir, the uranium enrichment installations at Ash-Sharqat and Tarmiuah, and the chemical weapons plant at Muthana.

[27] In 1992 and 1993, the United States carried out several missile attacks on military targets and factories near Baghdad. They also bombed Iraqi troops along the border with Kuwait, recovering weapons abandoned by the retreating Iraqis. When Bill Clinton assumed the presidency of the United States, the Iraqi government asked for a cease-fire. They put no conditions on further UN inspections.

[28] During 1994, the pressure on Iraq continued in an attempt to force Baghdad to recognize the new frontier with Kuwait, amongst other things, but according to President Saddam Hussein this awarded a small part of Iraqi territory to the neighboring country.

[29] This same year a frontier crossing was opened with Turkey to allow certain UN authorized foodstuffs and medicines to enter the country - the only exceptions to the trade embargo. However, a few months later, in March 1995, Turkish troops invaded Iraqi Kurdistan - under the military protection of "allied," basically US, troops - to repress members of the Kurdish Workers' Party (PKK) which was launching attacks from there on Turkish troops stationed in Turkish Kurdistan.

[30] On another front, the international isolation of Baghdad became even more serious as Jordan's improving relations with Kuwait and Saudi Arabia distanced it from Saddam's government.

[31] Iraq once again became the target of military action in 1996. In September, following fighting between members of the Democratic Party of Kurdistan, supported by Iraq, and the Patriotic Union of Kurdistan, supported by Iran, the United States launched new missile attacks on Iraqi positions. As on previous occasions - and even in the Gulf War itself - it was hard to tell how many civilians were killed.

[32] The continued sanctions began to have serious consequences for a large part of the Iraqi population. According to estimates, 500,000 children under five years old died between 1990 and 1996 as a direct or indirect consequence of the embargo.

[33] On another front, some observers noted strong pressure on the United States from countries like Saudi Arabia not to lift the embargo. In fact, Saudi oil replaced the Iraqi product in many markets and an increase in the volume of sales of crude oil caused a new fall in prices, which was opposed by the governments of the other exporting nations.

The Iran - Iraq war

The 1980-1988 war between Iran and Iraq was rooted in long-standing rivalries, a legacy of the disputes between the Persian and Ottoman Empires. Upon achieving independence, Iraq retained full sovereignty over the Shatt-al-Arab Canal where the Tigris and Euphrates Rivers meet - its only access to the sea.

[2] Iranian Shah Reza Pahlevi denounced this sovereignty in 1969 and, seeking to destabilize the Iraqi government he actively supported the Kurdish separatist movement. Pahlevi and Saddam Hussein (then Iraq's vice-president) signed the Algiers Agreement on March 6 1975. It contained three main points: the recognition of the land borders established in 1914; the division of the Shatt-al-Arab Canal along the deepest course of the river bed, granting Iran independent access to the Persian Gulf for its Abadan refineries; and border security to prevent armed incursions.

[3] In 1978, the Iranian government demanded that Ayatollah Ruhollah Khomeini, exiled in Baghdad, cease political activities directed against the shah. The religious leader was obliged to seek refuge in France.

[4] When the shah was deposed in 1979, Iran no longer felt bound by the previous agreement. Border incidents multiplied. The Baghdad government claimed that as the Algiers Agreement was no longer in force, they had exclusive sovereignty over the Shatt-al-Arab Canal, leaving Iran landlocked.

[5] In September 1980, Iraqi troops crossed the canal, precipitating the war. Iraq gained the support of Saudi Arabia and Jordan, creating a united front which hoped to prevent the Iranian revolution spreading throughout the Gulf region. Syria, an old opponent of Iraq, and Libya supported Iran.

[6] Over the years, the military situation stagnated. Mediation attempts by the United Nations and multiple calls for a ceasefire got nowhere. The war caused more than a million casualties, two-thirds of whom were Iranian.

[7] In early 1986, US warships were sent to "protect" shipping and ensure petroleum supplies for its Western allies. Iran mined the Strait of Hormuz and used speedboats to carry out sporadic attacks, but they could not prevent the US from gaining control of navigation routes in the region.

[8] The US, which sold arms to Saudi Arabia, Jordan and Kuwait, supplied intelligence information to the Iraqi government and secretly sold arms to Iran to obtain the release of hostages. These moves were aimed either at strengthening moderate factions within the Iranian government, or simply at prolonging the conflict. Israel, an intermediary in the deals, had a particular interest in prolonging the conflict, as this war kept the two largest Islamic armies, after that of Egypt, fully occupied.

[9] On August 20 1988, an armistice was announced to end the war. Negotiations mediated by the United Nations led to a ceasefire - but a definite peace accord could not be signed because of disagreements over the border, the key issue in the conflict.

[10] A final peace agreement was signed at the end of 1990, after the Iraqi invasion of Kuwait. The Baghdad government gave in to all the Iranian claims thereby securing Iranian neutrality during the Iraqi conflict with the US.

Ireland

Population: 3,571,000 (1994)
Area: 70,280 SQ.KM
Capital: Dublin

Currency: Pound
Language: Irish and English

Eire

The religious tensions in Irish history date back to the 17th century when a large number of British Protestants (in particular, Scottish Presbyterians) were encouraged to settle in the province of Ulster in the northern part of the island by the English colonial rulers. They subjugated the native Catholic population, Celtic peoples who had been converted to Christianity in the early 5th century.

[2] By virtue of the 1800 Act of Union, Ireland was incorporated into the United Kingdom. During the 19th century most of the population outside the Protestant-dominated North-east supported independence, which led to the formation of a strong nationalist movement.

[3] In 1916, the republican Easter Rising in Dublin was crushed by the occupying forces but it marked the foundation of the Irish Republican Army (IRA) and the final stage of the long struggle for freedom.

[4] The IRA's campaign of guerrilla warfare forced the British in 1920 to grant independence to the 26 counties with Catholic majorities. Southern Ireland became a self-governing region within the UK. The remaining six north-eastern counties became Northern Ireland, with a devolved government in Belfast and representation in the British parliament in Westminster. In 1922 Southern Ireland had declared independence and become the Irish Free State, a dominion under the British crown. In 1949 the Free State became the Irish Republic, formally breaking its final links with the British commonwealth.

[5] The 1937 Irish Constitution considers Ireland to be a single country, where all the inhabitants - North and South - have citizenship rights.

[6] Supported by the Protestant majority, the Ulster Unionist Party always managed to retain control of the Parliament in Belfast. For 50 years, Northern Ireland was ruled exclusively by Unionists, led by the Provincial Prime Minister and a Governor acting as a representative of the British Crown.

[7] The Catholic minority in Northern Ireland was thus excluded from domestic political affairs, and many other fields. This led to the creation of an active civil rights movement in the 1960s.

[8] Although it was non-violent, the extremist Protestant groups considered the civil rights movement a threat to the region's status and their dominant position and they reacted to it with violence.

[9] In April 1969, amidst growing disturbances, the Northern Ireland government requested British troops to protect the region's strategic installations. In August, Belfast and London agreed that all the

> In April 1969, amidst growing disturbances, the Northern Ireland government requested British troops to protect the region's strategic installations. In August, Belfast and London agreed that all the province's security forces should come under British command.

Province's security forces should come under British command.

[10] As a counterpart to the Provisional IRA, the "loyalists" (loyal to the British crown) formed a number of paramilitary organisations, including the Ulster Volunteer Force and the Ulster Defence Association. Between 1969 and the middle of 1994, more than 3,100 people died at the hands of the Protestant and Catholic paramilitaries, the British army and the Ulster police force, the Royal Ulster Constabulary (RUC). In 1972 political status was given to paramilitary prisoners, but this amendment was abolished in 1976.

[11] The growing violence provoked London to take over full responsibility for the maintenance of law and order in Northern Ireland. The government in Belfast was abolished and a system of "direct rule" from Westminster installed. In a plebiscite held in March 1973, 60 per cent of the population of Northern Ireland voted in favour of union with Britain.

[12] At the end of 1973, a Northern Irish Assembly and Executive was created in Belfast in which Protestant and Catholic representatives were supposed to share power.

[13] In December, the London and Dublin governments agreed on the establishment of an Irish Council. Both this agreement and the new power-sharing Assembly were bitterly opposed by most Protestants in Ulster. In 1974, the Protestant Ulster Workers' Council declared a general strike in Northern Ireland. This resulted in the resignation of the Executive and London took over direct government once more.

[14] In the 1973 election in the Irish Republic, Fianna Fáil - a conservative nationalist party which had been in power for 44 years - was defeated. A coalition between the conservative Fine Gael and the Labour Party took office. This new government was committed to power-sharing between the two communities in Ulster, but it rejected the immediate withdrawal of British troops from the region.

[15] In 1976, after the IRA murdered the British ambassador in Dublin, the Irish government took stricter anti-terrorist measures. Fianna Fáil was elected to government in 1977 and it maintained the friendly relations with London established by the previous administration. Prime Minister Jack Lynch supported the creation of a delegated government in the North, instead of total unification.

[16] In August 1979, Dublin agreed to increase border security after the murders, on the same day, of Lord Mountbatten (a prominent British public figure linked to the Royal Family) in the Irish Republic, and 18 British soldiers in Warrenpoint, Northern Ireland. In December,

LAND USE

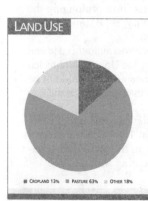

CROPLAND 13% PASTURE 63% OTHER 18%

Lynch resigned and was replaced by former prime minister Charles Haughey, who went back to the old idea of reunification with some form of autonomy in Ulster.

[17] The Irish Council of Womens' Affairs was created in 1972. The weight of Catholic tradition has been a hindrance to women's liberation. Women in both parts of Ireland had to travel to England to have an abortion, and in the Irish Republic the sale of contraceptives was limited to licensed outlets.

[18] In 1983, a controversial referendum in the Republic approved the outlawing of abortion under the Irish constitution. The law against divorce, in force for 60 years, was maintained in spite of the government's proposal to revoke it.

[19] In 1980 there were talks between the heads of the Irish and British governments. This led to the formation of the Anglo-Irish Intergovernmental Council a year later, with the aim of holding ministerial level meetings.

[20] Discussions between the British and Irish governments finally led to the signing of the Anglo-Irish Agreement in 1985, under which the Dublin government has some influence on the political, legal, security and border affairs of Northern Ireland. A majority of Protestants remain strongly opposed to the Agreement, although it guarantees that the constitutional position of Northern Ireland is subject to the will of the majority of its population.

[21] In 1988, Dublin and London quarrelled over the British government's refusal to investigate members of the RUC accused of shooting at terrorist suspects before attempting to arrest them.

[22] At the same time, an inquest was started in the Ulster Defence Regiment (UDR), a locally-recruited, part-time British Army regiment, which is dominated by Protestants, over the death of six IRA members in 1982.

[23] During the first months of 1989 a new wave of violence broke out. In April, three members of the paramilitary Ulster Defence Association were arrested in Paris together with a South African diplomat. Pretoria was supposedly supplying them with arms in exchange for secret British missile technology stolen from a Belfast arms production plant.

[24] In September 1989, the Irish government demanded a complete

review of the UDR, in particular of their links with the Protestant paramilitaries and the leaking of intelligence documents on Republican suspects. The United Kingdom accepted the Irish demands and as a result 94 UDR soldiers were arrested.

[25] In the Republic during the 1980s, unemployment ran high - an average of 16.4 per cent between 1983 and 1988 - and so did emigration. The country also faced spiralling inflation and industrial recession. A strict austerity program, implemented since 1987, has brought about economic growth, and a significant decrease in inflation.

[26] In May 1987, Ireland reaffirmed its EEC commitments through a national referendum. The Single Europe Act was approved by 69.9 per cent of the electorate.

27 Since joining the EEC in 1973, Community aid has played an important role in the economy of the Irish Republic, one of the poorest EEC members.

[28] Irish women accounted for 38.5 per cent of the labour force in 1987; only one country in Western Europe had a smaller proportion of women in paid employment.

[29] In 1990, the General Synod of the Church of Ireland approved the ordination of female priests in the country. This decision placed the Church of Ireland ahead of most other Anglican churches.

[30] In November 1990, academic lawyer Mary Robinson was elected President of the Republic, the first-ever woman to be elected to this post. An independent candidate, supported by the Labour Party, the Political Association of Women and the trade unions, Robinson gained 52 per cent of the vote.

[31] In a country that is predominantly Catholic and overwhelmingly conservative, Mary Robinson has spoken out in defence of gay and womens' rights and the legal recognition of illegitimate children.

[32] In April 1991, multi-party negotiations began in Belfast, in another attempt to define Northern Ireland's political future. These were the first important talks to be held since 1974. Representatives of Ulster's main "constitutional" parties participated (Protestant unionists, Catholic nationalists and the small non-sectarian Alliance Party), as did representatives of the government in London. Sinn Féin, the political wing of the Provisional IRA, which supports the immediate withdrawal of British troops, the disarming of the RUC and the reunification of Ireland, were excluded from the talks due to their refusal to condemn the IRA's campaign of violence.

[33] Opinion polls released in London at the time of the peace talks showed that most Britons favor the withdrawal of British troops from Ulster.

[34] In January 1992, the IRA launched one of its major offensives, in Ulster and in England. In County Tyrone, Northern Ireland, the IRA killed 7 Protestant construction workers, stating that, as they were working on British Army barracks, they were considered to be "collaborating with the occupiers". Republicans also carried out a series of arson attacks in Ulster on commercial establishments.

[35] From late 1991, there was a sharp increase in attacks on Catholics by Protestant paramilitary groups. These groups vowed to retaliate for each IRA attack with "an eye for an eye".

[36] In response to the demands of Protestant politicians and in order to confront the escalating violence, London sent several hundred military reinforcements to Ulster. They joined the 11,000 British troops already stationed there, as well as the 6,000 volunteers of the Ulster Defence Regiment, and the 12,000-strong police force.

[37] In March 1992, the issue of

STATISTICS

DEMOGRAPHY

Urban: 57% (1995). **Estimate for year 2000:** 4,000,000.
Children per woman: 2 (1992).

HEALTH

Under-five mortality: 7 per 1,000 (1995). **Calorie consumption:** 119% of required intake (1995).

EDUCATION

School enrolment: Primary (1993): 103% fem., 103% male. Secondary (1993): 110% fem., 101% male. University: 34% (1993). **Primary school teachers:** one for every 25 students (1992).

COMMUNICATIONS

108 **newspapers** (1995), 104 **TV sets** (1995) and 106 **radio sets** per 1,000 households (1995). 32.8 **telephones** per 100 inhabitants (1993). **Books:** 105 new titles per 1,000,000 inhabitants in 1995.

ECONOMY

Per capita GNP: $13,530 (1994). **Annual growth:** 5.00% (1985-94). **Annual inflation:** 2.00% (1984-94). **Consumer price index:** 100 in 1990; 110.5 in 1994. **Currency:** 0,6 pounds = 1$ (1994). **Cereal imports:** 409,000 metric tons (1993). **Fertilizer use:** 6,391 kgs per ha. (1992-93). **Imports:** $25,508 million (1994). **Exports:** $34,370 million (1994).

ENERGY

Consumption: 3,136 kgs of Oil Equivalent per capita yearly (1994), 70% imported (1994).

---

right to have access to information on methods for interrupting pregnancy, as well as the right to travel abroad for an abortion. However, the electorate voted against legalizing abortion in cases where the mother's life is in danger.

[41] In 1993, with the reduction in workers in the rural and traditional industry sectors, unemployment reached 20 per cent, while the GDP grew due to an increase in exports, especially high-tech goods.

[42] The government lifted broadcasting restrictions on the Sinn Fein in early 1994. In August, the IRA declared a complete cessation of military operations, a gesture which enabled political talks to take place on the island.

[43] In November, the Reynolds administration collapsed after the Labour Party withdrew its support following a controversy over long delays in the extradition - to Northern Ireland - of a paedophile priest. Reynolds was succeeded as prime minister by John Bruton from Fine

Irish women accounted for 38.5 per cent of the labour force in 1987; only one country in Western Europe has a lesser proportion of women in paid employment.

rized doctors to inform women about abortion clinics abroad. In a close referendum, the Irish approved (with 50 per cent to 49 per cent of the vote) a constitutional reform which allows the divorce of couples who have been separated for a period of 4 years.

[47] Bruton established an excellent relationship with British prime minister John Major, while also having close ties with Ulster's John Hume, SDLP leader, and Gerry Adams, president of Sinn Fein. Prince Charles' mid-year visit enhanced the need for links between England and Ireland. The opposition limited itself to staging peaceful demonstrations. Late in the year, the visit of United States president Bill Clinton had an encouraging effect on the parties to find a formula which would solve the armed conflict.

[48] The prime minister took part in a round of talks in Belfast in June 1996. Bruton supported Britain's initiative to exclude Sinn Fein representatives from the negotiations since both governments believed IRA behaviour - which in February had ended the unilateral truce observed since August 1994 with bombings in mainland UK - had made no contributions to peace.

---

abortion hit the headlines. A 14 year-old girl who became pregnant as the result of a rape was forbidden by the High Court from travelling to England for an abortion. It is estimated that up to 4,000 Irish women per year have abortions in Britain as the operation is forbidden under the Irish constitution. The case of the 14 year-old girl went before the Supreme Court, provoking demonstrations from both pro-choice and anti-abortion groups, and calls for another referendum on abortion. The Supreme Court eventually overturned the High Court decision, and the girl was allowed out of the country. The case raised important issues with regards to the relationship between the Irish constitution and European Community law, which guarantees the freedom of movement of EC citizens within member nations.

[38] In mid -July 1992, 69 per cent of the Irish population voted in favour of ratifying the Treaty of Maastricht and the further integration of the European Community. This figure exceeded all expectations, with 57 per cent of the electorate participating in the vote.

[39] In January 1993, Albert Reynolds, of Fianna Fail, was confirmed as head of government, despite the fact that his party and the Fine Gael had lost seats in the last election. In 1993, Fianna Fail had 77 representatives in Parliament, while Fine Gael had 45 and Labour 33. This led to the formation of a coalition government of Fianna Fail and the Labour Party.

[40] In the meantime, a plebiscite was held on the abortion issue. Two-thirds of the voters pronounced themselves in favor of the

Gael, with support from the Labour Party and the democratic left.

[45] The government coalition achieved political and social stability in 1995. The February budget was advantageous for civil servants, small business and low-waged workers while economic growth amounted to 5.25 per cent.

[46] The Catholic Church continued to be the target of accusations regarding priests who molested children. In Dublin, the archbishop publicly confessed he had used part of the diocese's funds to pay-off one of the clergy's victims. Meanwhile, the Supreme Court autho-

Israel

Yisra´el

Population: 5,383,000 (1994)
Area: 20,770 SQ.KM
Capital: Jerusalem/Tel Aviv

Currency: New Shekel
Language: Hebrew and Arabic

In 1896 Viennese journalist Theodor Herzl published a book entitled The Jewish State. Influenced by the European nationalism which had engendered the unification of Germany and the Italian Risorgimento, Herzl envisaged a Jewish nation-state which would put an end to anti-Semitic acts, like the Russian pogroms and the Dreyfus Affair - a spy case in France.

2   This Jewish state was to be established in Palestine, then a Turkish colony. The name "Palestine" inspired the suffering Jews of Eastern Europe, who dreamed of the return to Zion, the land of the ancient kingdoms of Israel. Zion is the name of a hill in Jerusalem; by extension it became a synonym for Jerusalem itself and then for the whole of Palestine. But the Zionists, as Herzl's followers began to be called, preferred to ignore the fact that half a million Arabs had been living there for over 1,000 years, naturally becoming attached to their land and traditions.

3   Zionism firmly pursued a policy of forming alliances with the great capitalist powers, denying the peoples of Palestine their national identity - the precise right the Jews claimed for themselves. This ideology was born out of the same Eurocentric womb that produced the English colonialist, explorer, and magnate Cecil Rhodes; French statesman and colonialist Jules Ferry; and Chancellor Otto von Bismarck, the godfather of German imperialism. Zionism has never been able to rid itself of this heritage, although it has always contained left wing and socialist sectors.

4   During World War I, Britain and France agreed to divide up the remains of the Ottoman Empire in the Middle East. In 1917, British foreign secretary Arthur Balfour declared his support for the establishment of a national homeland for the Jewish people in Palestine, although as prime minister in 1905 he had opposed Jewish immigration into Britain. The Balfour Declaration stated "that nothing shall be done which may prejudice the civil and religious rights of existing non-Jewish communities in Palestine". These groups made up 90 per cent of the population.

5   At the end of World War II, the French colonies of Syria and Lebanon and the British colonies of Iraq and Transjordan obtained their independence. But Britain retained control of Palestine, based on its commitment under the Balfour Declaration.

6   Zionists regarded Jews the world over as exiles, and they organized migration to Israel from all corners of the globe. At the beginning of this century there were 500,000 Arabs and 50,000 Jews living in Palestine. By the 1930s, the number of Jews had risen to 300,000. The anti-Semitic persecutions in Nazi Germany raised immigration above the legally permitted quotas, alarming the British, who saw their power in Palestine threatened.

7   In 1939 London declared that its aim was not to set up a Jewish state but an independent, binational Palestinian state with both peoples sharing government. Ships bringing refugees from Hitler's Europe were turned back from Palestinian ports. The Zionists organized acts of sabotage and terrorism, using force to hold Britain to its promise.

8   Using donations from Jews all over the world, the Zionists purchased Arab lands from wealthy absentee owners living in Beirut or Paris, who cared little about the fate of their tenants, the Palestinian *fellahin*. The Jews then arrived, deeds in hand, to expel peasant families whose ancestors had lived there for untold generations. They set up agricultural colonies, kibbutzim, defended by armed Zionist militia.

9   In February 1947, in view of intensified anti-British attacks, London submitted the Palestinian problem to the United Nations. A special committee recommended partition of the territory into two independent states; one Arab, the other Jewish. Jerusalem would remain under international administration.

10   The Soviet Union preferred a Jewish state to continued British military presence, and its support was decisive in the creation of Israel. At the same time, London and Washington considered the partition unfeasible. US Secretary of Defense James Forrestal proposed that President Harry Truman send troops to impose a UN trusteeship over the territory of Palestine.

11   The UN General Assembly finally approved the partition plan in a 33 to 13 vote, there were 10 abstentions and the Arab countries and India voted against partition. Armed Zionist organizations seized control of major cities and towns and began to expel Palestinians en masse, alleging an imminent attack by Arab armies. This policy culminated in a massacre at the village of Deir Yasin in April 1948, when its entire population was murdered by the Irgun, an extremist group led by Menachem Begin.

12   On May 14 1948, the British high commissioner withdrew from Palestine, and David Ben Gurion proclaimed the state of Israel. The Jordanian, Egyptian, Syrian, Iraqi and Lebanese armies attacked immediately, and the governments of those countries urged the Palestinians to abandon their possessions and take refuge.

13   The war ended with an armistice in January 1949, granting Israel 40 per cent more territory than

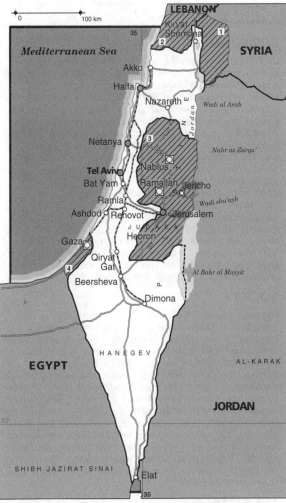

OCCUPIED BY ISRAEL
1   Syria - Golan Heights - occupied in 1967
2   Southern Lebanon - occupied in 1983
3   West Bank - occupied in 1967
4   Gaza Strip - occupied in 1967
◎   Recovered by Palestine from 1994: Gaza Strip, Jericho, Nablus, Ramallah

LAND USE

■ CROPLAND 21%   ■ PASTURE 7%   ■ OTHER 72%

PROFILE

ENVIRONMENT

20,770 sq km, within the pre-1967 borders (see Palestine). These territories are currently occupied, both militarily and with settlements of Jewish colonists: Golan Heights (Syria) 1,150 sq km, West Bank 5,879 sq km, Gaza Strip 378 sq.km, Greater Jerusalem 70 sq km The land comprises four natural regions: the coastal plains, with a Mediterranean climate, the country's agricultural center; a central hilly and mountainous region, stretching from Galilee to Judea; the western lowlands, bound on the north by the Jordan River, which flows into the Dead Sea; and the Negev Desert, to the south, which covers half of the total territory. The main agricultural products are citrus fruits for export, grapes, vegetables, cotton, beet, potatoes and wheat. There is considerable livestock production. Industrial production is growing rapidly and there are serious difficulties with water; the country has over 2,000 sq km of irrigated land -part of the Negev has been reclaimed through irrigation. Pollution, caused by both industrial and domestic waste, as well as pesticides, is also an important problem.

SOCIETY

Peoples: Jewish, 81.4%; Arab and other, 18.6%.
Religions: Judaism. Arabs are Muslim and Christian.
Languages: Hebrew and Arabic are official. English is used for business. Native languages of immigrants (notably Yiddish and Russian) are also spoken. **Political Parties:** The Zionist left is made up of the Labor Party, and the Meretz (a front which includes three parties). The non-Zionist left is made up of Hadash (formerly Communist) and the Democratic Arab Party. The Zionist right is formed by Likud; the Tehiya (Renaissance); the Tsomet and the Moledet. The religious parties are: the Shas; the Agudat-Israel (the League of Israel); the National Religious Party and the Degel Hatora. **Social Organizations:** The Histadrut Haoudim Haleumit (National Labor Federation) is the main trade union.

THE STATE

Official Name: Medinat Yisra'el (Hebrew); Isra'il (Arabic).
Administrative divisions: 6 districts, 31 municipalities, 115 local councils and 49 regional councils. **Capital:** In 1980, Jerusalem, 567,100 inhab. (1991) was proclaimed the "sole and indivisible" capital of Israel. The UN condemned this decision and most diplomatic missions are in Tel Aviv. **Other cities:** Haifa, 245,900 inhab.; Tel Aviv-Jaffa, 357,400 inhab.; Holon, 162,700 inhab. (1991). **Government:** Esser Weizman, President. Binyamin Netanyahu, Prime Minister. Parliamentary system of government. **National holiday:** May 7, Independence (1948). **Armed Forces:** 172,000. **Paramilitaries:** Border Police: 6,000; Coast Guard: 50.

it was due under the partition plan. Soviet weapons and aircraft, purchased in Czechoslovakia, proved decisive in the Israeli victory.

[14] Israel's government changed sides during the Cold War and formed a permanent strategic alliance with the United States. This al-

liance was not forged easily nor immediately.

[15] In 1956, when Israeli troops, backed by France and Britain, invaded Egypt in response to the nationalization of the Suez Canal, the United States and the Soviet Union both firmly opposed the move, and

the expeditionary force had to be withdrawn.

[16] The diaries of former Prime Minister Moshe Sharett, published posthumously, revealed that Israel used Zionist agents to sabotage Western targets in Cairo, forcing US President Dwight D. Eisenhower to opt for Tel Aviv and not Cairo as its principal Middle East ally. The episode is known historically as the Lavon affair.

[17] In the 1960s, Egyptian president Gamal Abdel Nasser was deprived of US credit and arms, and he turned to the Soviet Union. This was enough to produce closer relations between Israel and Washington.

[18] In the 1967 Six Day War, Israel carried out a lightning strike, using North American weaponry, and seized the whole of Palestine, the

> Using donations from Jews all over the world, the Zionists purchased Arab lands from wealthy absentee owners living in Beirut or Paris, who cared little about the fate of their tenants, the Palestinian *fellahin*. The Jews then arrived, deeds in hand, to expel peasant families whose ancestors had lived there for untold generations.

Syrian Golan Heights and Egyptian Sinai .

[19] Paradoxically, this military victory undermined Israel's moral strength; the newborn state was transformed into an expansionist power in control of areas housing thousands of Palestinians.

[20] Resolution 242 of the UN Security Council, passed on November 22 1967, called on Israel to withdraw from the occupied territories. Israel refused, claiming that it need-

ed "secure borders" in order to resist Arab threats to its existence.

[21] Armed conflict broke out again in 1973, when Egyptian troops crossed the Suez Canal and ended the myth of Israeli military invincibility. In truth, by the time armistice was decreed, battlefield positions had hardly changed.

[22] In 1977, Menachem Begin was elected prime minister, breaking with the political and ideological continuity of the previous governments. For the first time in the history of the Israeli state, the Labor Party was not in power.

[23] Begin's victory resulted partly from growing social tension between Ashkenazim (economically and culturally privileged Jews of European origin who are traditionally Labor supporters) and Sephardim (Eastern Jews from Arab countries).

[24] The Labor establishment was recruited from and supported by the kibbutzim. Although members of kibbutzim account for only about three per cent of the population, they have enormous political and economic influence. Originally intended as egalitarian agricultural co-operatives, they have branched out into industrial production, employing mostly Arab and oriental Jewish laborers who do not share membership privileges. The oriental Jews, who culturally have more in common with the Arabs than with European Jews, make up 55 per cent of the population, but their average income is just two-thirds that of the Ashkenazis. Their resentment of the Ashkenazim, aggravated to the point of fanaticism, provided the electoral basis for the political right.

[25] Begin completely rejected the idea of negotiating with the Palestine Liberation Organization (PLO) and expressed his intention to annex the West Bank. Meanwhile in 1977, the US persuaded Egyptian President Anwar Sadat to sign the Camp David Agreement, leading to peace between Cairo and Tel Aviv and the return of Sinai to Egypt.

[26] From then on, international attention gradually shifted from the issue of secure borders for Israel to the plight of the Palestinians. Changing tactics, the PLO gained sympathy and allies, so that when Jerusalem was declared the capital of Israel in 1980, the move was condemned as an arrogant gesture by many of Israel's supporters.

[27] In June 1982, the Israelis launched "Operation Peace for Galilee," invading Lebanon and

devastating Beirut, under the pretext that these were maneuvers to stop infiltration by Palestinian guerrilla groups. PLO combatants agreed to evacuate the city in exchange for the deployment of a joint force of Italians, French, and North Americans to guarantee the security of Palestinian civilians.

[28] In spite of the agreement, hundreds of Palestinians in the Sabra and Shatila refugee camps, in Israeli-controlled areas, were murdered by right-wing militia with Israeli complicity. The invasion of Lebanon fomented dissent in Israel, and 400,000 people attended a demonstration sponsored by the group Peace Now. Begin was forced to appoint a commission of inquiry which found Defense Minister Ariel Sharon and other military leaders "indirectly responsible" for the massacres.

[29] After the September 1984 elections, the Labor Party, headed by Shimon Peres, was able to form a government, though lacking an absolute majority. Its coalition included Likud, and the agreement stipulated that the leader of that party, Yitzhak Shamir, would assume the prime minister's post in 1986.

[30] At the end of 1987, several young Palestinians were killed in a clash with Israeli military patrols in the occupied West Bank. The funerals turned into protest demonstrations and led to further confrontations, deaths, and general strikes - civil protests where people refuse to leave their homes. This marked the beginning of the *intifada* or insurrection, and Middle Eastern politics were deeply disturbed from the least expected quarter; from the unarmed grassroots. In 1989, during the last National Palestine Congress, the PLO recognized the State of Israel, accepting UN resolutions 242 and 338.

[31] Images of young Palestinians, armed with slingshots, confronting the mighty Israeli army were televised worldwide, ruining Israel's international standing. This development came as quite a shock for a state founded on the basis of international solidarity.

[32] The most apparent outcome of the questioning of Israeli credibility is that 250,000 young Israelis leave the country each year. Negative net migration throughout the 1980s was balanced only in 1989 with the massive immigration of Jews from the Soviet Union. Since its creation as a solution to the Jew-

STATISTICS

DEMOGRAPHY

Urban: 90% (1995). **Annual growth:** 2.9% (1991-99). **Estimate for year 2000:** 6,000,000. **Children per woman:** 2.7 (1992).

HEALTH

Under-five mortality: 9 per 1,000 (1995). **Calorie consumption:** 109% of required intake (1995).

EDUCATION

School enrolment: Primary (1993): 96% fem., 96% male. Secondary (1993): 91% fem., 84% male. University: 35% (1993). **Primary school teachers:** one for every 16 students (1992).

COMMUNICATIONS

106 **newspapers** (1995), 103 **TV sets** (1995) and 103 **radio sets** per 1,000 households (1995). 37.1 **telephones** per 100 inhabitants (1993). **Books:** 104 new titles per 1,000,000 inhabitants in 1995.

ECONOMY

Per capita GNP: $14,530 (1994). **Annual growth:** 2.30% (1985-94). **Annual inflation:** 18.00% (1984-94). **Consumer price index:** 100 in 1990; 166.1 in 1994. **Currency:** 3 new shekels = 1$ (1994). **Cereal imports:** 2,293,000 metric tons (1993). **Fertilizer use:** 2,253 kgs per ha. (1992-93). **Imports:** $25,237 million (1994). **Exports:** $16,881 million (1994). **Development aid received:** $1,266 million (1993); $243 per capita; 2% of GNP.

ENERGY

Consumption: 2,815 kgs of Oil Equivalent per capita yearly (1994), 96% imported (1994).

ish issue, the state of Israel has managed to attract only 20 per cent of the world's Jews.

[33] Life in the kibbutzim also changed. The generations raised there tended to be more individualistic, and more consumer oriented. They rejected the cooperatives' rigid laws, which forced them to select their careers according to the needs of the kibbutz. These phenomena explain the decline of the kibbutzims' influence upon the country's economy.

[34] Economically, Israel is dependent on US support to balance its trade deficit. Egypt is a growing, regional market for Israeli goods though it is unable to absorb all Israeli production.

[35] According to US sources, their aid to Israel between 1949 and 1991 totalled $53 billion. In the 1979 Camp David Agreements, the figure reached $40 billion, 21.5 per cent of all US foreign aid.

[36] In Israel military service is mandatory for both women (two years) and men (three years). All men up to the age of 51, and unmarried women up to age 24, belong to the reserve. The militarization of society also results in a high level of state interference in the economy and in an enormous weight of war industry on the country's production figures.

[37] In its search for markets, Israel has become an arms exporter involved in scandalous deals like the "Iran-Contras" scheme, supplying arms to Iran. It has also provided military training for Colombian drug militias.

[38] Land belongs mostly to the state and its sale or privatization is restricted, especially in the case of Arab citizens or Arab capital, as land ownership and control are considered to be matters of strategic importance.

[39] Demographically, the Israeli idea of a stable Jewish nation-state is contradicted by the government's

expansionist practices. Successive territorial annexations have incorporated some two million Arabs into the present territory of Israel. By simple demographic progression, without including the Palestinians in exile, Arabs will outnumber the Jewish population by the end of the century.

[40] In March 1990, these fundamental problems combined with other differences to produce a split in the Labor-Likud alliance.

[41] Everything changed when Iraq invaded Kuwait, on August 2 1990. Israel was excluded from the anti-Iraq coalition led by the United States, because if it had participated, the Arab countries would not have joined. Israel made preparations for war regardless, and a state of siege was declared in the occupied territories.

[42] On January 17, in answer to the initiation of Allied bombing, Iraq launched several missile attacks against Israel. Their aim was to try and force Israel to enter the war. However, Israel did not join the war, leaving defence to Patriot anti-missile units manned by American troops. Israel's defence was out of its own hands for the first time in its history.

[43] The Iraqis could not establish a direct link between a solution to their conflict and the larger Arab-Israeli question, but in March a joint Soviet-US statement expressed hope of finding a negotiated solution to the Arab-Israeli conflict. The terms of the statement were vague, however, and Israel harshly rejected it.

[44] When the war ended in March 1991, the US presented diplomatic circles with a "land for peace" proposal. Two months later in Damascus, Syria and Lebanon signed a "brotherhood, co-operation and coordination" treaty. To Israel, this treaty constituted a military threat to the "water tank" of Israel; a region in the north containing many important water sources.

[45] Israel asked the United States to approve loans of $10 billion to ease its economic difficulties, caused by the resettling of between 250,000 and 400,000 Soviet Jews from 1989 to 1991.

[46] The government began transferring immigrants to new settlements on the West Bank where, in 1991, unemployment reached 11 per cent, despite efforts to create new jobs for immigrants.

[47] Wishing to launch peace nego-

tiations in the region, the US government imposed the condition that loans granted would not be invested in settlements in occupied territories.

[48] The creation of Jewish settlements and the construction of new housing for Soviet Jews on the West Bank, became a double-edged sword for Shamir's government. The US loans and new settlements were vitally important, but Palestinians and other Arabs demanded that resettlement cease, in order for the peace talks to continue.

[49] On October 30 1991, a Middle-eastern Peace Conference was held in Madrid, sponsored by the US and the former USSR. Shortly afterwards, hundreds of thousands of Israelis demonstrated, calling upon their government to maintain a dialogue with the Palestinians and Israel's Arab neighbors. Delegations from Jordan, Lebanon, Syria and Israel attended the conference. The Palestinians formed part of the Jordanian delegation, as Israel refused to negotiate directly with Palestinian representatives.

[50] The Arab countries supported the "land for peace" scheme, though it was rejected by Israel. Israel, in turn, showed itself willing to discuss limited self government by the Palestinian population in the occupied areas, as a provisional solution for a 5-year period. This offer triggered a crisis in the Likud's coalition government, when the Tehiya and Moladet (ultra-right-wing parties) walked out of the coalition, leaving Shamir without a parliamentary majority.

[51] At the joint request of Israel and the US, the UN General Assembly did away with a 1975 resolution defining Zionism as "a form of racism and racial discrimination". In February 1992, Israel bombed Palestinian refugee camps in the south of Lebanon, killing Abas Musawi, the leader of Hezbollah, a pro-Iranian group.

[52] This unleashed a new wave of violence along the northern border. Israeli soliders received orders to shoot without warning at any Palestinian bearing weapons. The appearance of "death squads" in the occupied territories was attributed to the modification of the Penal Code, which granted immunity to members of these groups.

[53] On June 23 1992, Labor won a decisive victory in the general elections and Yitzhak Rabin was nominated prime minister. Construction of housing in the occupied territo-

ries came to a sudden standstill, and as a result, the United States removed the embargo on loan guarantees for Israel.

[54] According to United Nations sources, since the beginning of the *intifada*, 62 children under the age of 13, and 170 between the ages of 13 and 17, have been killed. Amnesty International reported that some 25,000 Palestinians were arrested in 1992 for security reasons. That same year, 120 died at the hands of Israeli forces. The intifada began using fire arms for its attacks, killing 20 Israeli civilians, 18 soldiers and police, and more than 200 Palestinians suspected of "collaborating" with Israeli authorities.

[55] In December 1992, the Israeli government charged 415 Palestinians with belonging to "Hamas", a

Images of young Palestinians, armed with slingshots, confronting the mighty Israeli army were televised worldwide, ruining Israel's international standing.

guerrilla organization, deporteding them to southern Lebanon. When Lebanon failed to admit them, they were left in no man's land. International pressure eventually led to their repatriation.

[56] The Israeli government proceeded to "settle accounts", attacking Hezbollah and PLO bases in southern Lebanon by land, sea and air. In six days, 130 people were killed, 520 wounded, 15 towns destroyed and hundreds of thousands were left homeless.

[57] On September 13, 1993, after months of secret negotiations in Oslo, Norway, Israeli authorities and PLO leaders signed the Declaration of the Principles of the Interim Self-government Agreement, in Washington. This agreement foresaw the installation of a limited autonomy system for Palestinians in the Gaza Strip and the city of Jericho, for a five-year period. This agreement would subsequently be extended to include all of the West Bank.

[58] Difficulties arose in the negotiations which followed in Cairo, due to differences of opinion over border station control and water access

rights. In addition, the PLO demanded the release of 1,000 political prisoners.

[59] Hamas and the pro-Iranian Hezbollah questioned the agreement. Meanwhile, Jewish settlers in the occupied territories - who had been supplied with arms by the government - rejected the agreement out of hand, as it provided for the withdrawal of settlers and the Israeli army.

[60] It was not long before this situation erupted in violence. On February 25 1994, Baruch Goldstein, a member of Kahane's Kach, an ultra-right-wing Jewish movement, entered the Tomb of the Patriarchs in Hebron, armed with an M-16 rifle and massacred several dozen Palestinians who were at prayer. He was subsequently beaten to death by Palestinians.

[61] The official investigation of the Hebron massacre revealed that soldiers in the occupied territories had been under orders not to shoot against armed settlers under any circumstances. According to the investigation, several soldiers fired into the crowd, and doors were closed blocking the victims' escape. The PLO withdrew from the negotiations demanding that the settlers turn over their arms immediately, and that the United Nations issue an official condemnation. The latter approved a statement which categorically condemned the massacre and voted for a peacekeeping force to be sent to the occupied territories.

[62] The Israeli government ordered several dozen militant settlers turn in their rifles. They also cracked down on Kach, and began the release of Palestinian prisoners. There was an outbreak of violence between Palestinians and Israeli troops.

[63] In early May 1994, Israeli prime minister Yitzhak Rabin and PLO chairman Yasser Arafat signed an agreement in El Cairo granting autonomy to Gaza and Jericho. Late that month, the Israeli army withdrew from Gaza, ending 27 years of occupation (see Palestine).

[64] Negotiations with several Arab states progressed, especially with Jordan, which signed an agreement with Tel Aviv in July ending with the "state of aggression" between the two countries. However, disagreements with Syria carried on, among other things because Israel refused to accept a full

evacuation of the Golan Heights, occupied since 1967.

[65] The new situation favoured economic expansion: in 1994, the GDP grew 7 per cent and investments 20 per cent while unemployment went from 11 per cent to 7.6 per cent. The following year, economic growth continued at 7 per cent and inflation fell from 14.5 per cent to 8.5 per cent.

[66] 1995 was marked by a growing division of Israeli society regarding the peace process with the Palestinians. Anti-Rabin demonstrations were frequent and led to the prime minister's assassination by a young Israeli member of the far right. Shimon Peres, also from the Labour Party, took Rabin's place but was defeated in the general elections of May 1996 by Likud leader Binyamin Netanyahu.

[67] The Conservatives' return to power hindered negotiations with Palestine and put the country on the verge of a new war in September. The government allowed the opening of a tunnel below the Al-Aqsa mosque, Islam's third holy site, breaking the status quo held since the signing of the peace agreements and provoking a Palestinian reaction. Several Israeli soldiers and dozens of Palestinians died during the riots.

Italy

Italia

Population: 57,120,000 (1994)
Area: 301,270 SQ.KM
Capital: Rome (Roma)
Currency: Lire
Language: Itallan

I n the 13th century BC, ancient central European peoples occupied the northern part of what is now Italy.

2 Upon the fall of the Hittite empire, around the year 900 BC, the Etruscans established themselves to the north of the Tiber River. Their influence extended throughout the Po valley until the end of the 6th century, when the Celts bore down on them, destroying their territorial unity.

3 According to legend, Romulus founded the city of Rome upon the Palatine hill in the year 753 BC; during the following century, this settlement was united with those on the Quirinal, Capitoline and Esquiline hills. The first form of government was an elective monarchy. Its powers were limited by a Senate and a people's assembly of clans which held the power of *imperium*, or mandate to govern.

4 There were two social classes: the patricians, who could belong to the Senate, and the plebeians, who had to band together to protect themselves from the abuses of the large landowners.

5 Under King Tarquinius Priscus (616-578), Rome entered the Latin League. The poverty of the plebeians and the system of debt-induced slavery led to the expulsion of the kings in 509. In the 5th century, the traditional laws were written down. This "Law of the Twelve Tables" extended to the plebeians, who after a lengthy struggle had managed to win some rights.

6 The Punic Wars against Carthage in the 3rd century allowed Rome to increase its possessions once again; in the early 2nd century, after displacing the Macedonians, Greece became a "protectorate". Within a few years, Asia Minor, the northeast of Gaul, Spain, Macedonia and Carthage (including the northern part of Africa) had fallen into Roman hands.

7 Toward the end of the 2nd century, the Gracchi brothers, Tiberius and Gaius - both Roman representatives - were assassinated by the nobles, along with 3,000 followers, for supporting the plebeians.

8 The Roman Empire controlled the land from the Rhine in Germany to the north of Africa, and also included the entire Iberian Peninsula, France, Great Britain, Central Europe and the Middle East as far as Armenia. The 2nd century brought internal disputes which plunged Rome into chaos.

9 In 330, the Emperor Constantine transferred the capital of the empire to Byzantium - called New Rome - and converted to Christianity. In 364, the empire split into two parts: the Western and Eastern Roman Empires.

10 The end of the 5th century was marked by the invasions of the Mongols and other northern tribes, and by the attempts of the Byzantine Empire to recover its lost territories. In the mid-6th century, Italy became a province once again, but the Lombards conquered the northern part of the peninsula.

11 When the capital of the empire had been transferred to Byzantium, the bishops of Rome had presented themselves as an alternative to Byzantine power as a separate power base in Rome. When the Lombard kings began taking up arms in defence of Christianity against Rome's enemies, the bishops broke the alliance, in order to maintain their temporal power.

12 In 754, Pope Steven II asked for help from Pepin the Short, and in exchange, crowned him king of the Franks. After the defeat of the Lombards, Pepin turned over the center of the peninsula to the Pope. Charlemagne, Pepin's son, was crowned king and emperor of Rome in 800, but the Muslim invasions which took place mid-century once again left the region without government.

13 Between the 9th and 10th centuries, the Church formed Pontifical States in the central region, including Rome itself. In the 12th century, self-government arose in some cities because of the lack of a centralized power.

14 In the 14th century, when the struggle intensified between the Guelfs (those who favored the Pope) and the Ghibellines (the defenders of the German empire), the Holy See was transferred to Avignon, where it remained for the next 7 papacies. Two centuries later, the prosperity and stability of cities like Venice, Genoa, Florence and Milan produced an intellectual and artistic movement known as the Renaissance.

15 In the early 16th century, the peninsula was attacked by the French, the Spanish and the Austrians, who all craved control of Italy. In 1794, Napoleon Bonaparte entered the country expelling the Austrians. Four years later, he occupied Rome and created the Roman Republic and the Parthenopean Republic, in Naples. Only the two Italian states of Sicily and Sardinia were not under Napoleon's control, as they were governed by Victor Emmanuel I. The French emperor rescinded the temporal power of the popes and deported Pius VII to Savona.

16 Before the fall of Napoleon in 1815, Victor Emmanuel II named Camillo Benso di Cavour president of the council of ministers. Cavour was to be the architect of Italian unification, forging a single kingdom of Italy from those of Sardinia and Piedmont, with only Rome and Venice remaining outside the realm. In 1870, the Italians invaded Rome and, given Pope Pius IX's refusal to renounce his temporal power, they confined him to the Vatican, where his successors would remain until 1929. In 1878, the king, Humberto I brought Italy into the Triple Alliance with Austria-Hungary and Germany. Italy's colonial conquest of Eritrea and Somalia, in eastern Africa, also began.

17 In 1872, influenced by the events of the Paris Commune, Italy's first socialist organization was formed, giving rise in 1892 to the Socialist Party (PSI). The encyclical Rerum Novarum (1891) oriented Catholics toward militant unionization and the union movement grew tremendously.

18 When World War I broke out, Italy proclaimed its neutrality; however, in the face of growing pressure from nationalist groups on the left, it ended up declaring war against its former allies of the Triple Alliance.

19 Benito Mussolini, who had been expelled from the PSI for supporting Italy's entry into the war, was able to manipulate resentment over the poor outcome. Through a blend of nationalism and pragmatism, he called for the unions to work toward collaboration between capital and labor, in the name of "the interests of the nation". In 1921, a group headed by Amadeo Bordiga and Antonio Gramsci split

LAND USE

■ CROPLAND 41% ■ PASTURE 15% ■ OTHER 45%

ENVIRONMENT

The northern region of the country consists of the Po River plains, which extend as far as the Alps. It is the center of the country's economic activity, having the main concentration of industry and farming. Cattle are raised throughout the peninsula; important crops include olives and grapes, with vineyards extending along the southern coastal strip. The country includes not only the peninsula - which is divided by the Apennines - but also the islands of Sicily and Sardinia.

SOCIETY

Peoples: Italians 96%. Others, particularly German on Alto Adige and immigrants from Africa, 4%.
Religions: Predominantly Catholic. Muslims are about 700,000. **Languages:** Italian (official). Several regional languages, like Neapolitan and Sicilian, are widely spoken. French is spoken in Val d'Aosta and German in Alto Adige. Immigrants speak their own languages, particularly African languages. **Political Parties:** Forza Italia; Democratic Party of the Left (PDS); National Alliance; League of the North; Reformed Communist Party; Popular Italian Party (PPI); Liberal Party (PLI); Republican Party (PRI); Southern Tyrolean Popular Party; National Federation for the Green List, ecologists. **Social Organizations:** Three central unions: the CGIL; the CISL and the UIL.

THE STATE

Official Name: Repubblica Italiana. **Administrative Divisions:** 20 Regions divided into 95 Provinces. **Capital:** Rome (Roma), 2,723,000 inhab. (1993). **Other cities:** Milan (Milano), 1,358,000 inhab; Naples (Napoli), 1,070,000 inhab.; Palermo 697,000 inhab.; Turin (Torino), 953,000 inhab. (1993). **Government:** Oscar Luigi Scalfaro, President. Romano Prodi, Prime Minister. Bicameral parliamentary system. **National Holiday:** June 2, Anniversary of the Republic (1946). **Armed Forces:** 322,000 (1994). **Paramilitaries:** 111,800 Carabinieri.

Foreign Trade

MILLIONS $ 1994

IMPORTS
167,685
EXPORTS
189,805

off from the PSI to form the Communist Party (PCI), leaving the PSI without its radical wing.

[20] Having confronted one government crisis after another, and following an impressive march on Rome, Victor Emmanuel III turned over the government to Mussolini. An electoral reform, giving Mussolini's Fascist Party a majority, was denounced by socialist leader Giaccomo Matteotti, who was subsequently assassinated by followers of "Il Duce" Mussolini in 1924. A new constitution established censorship of the press; in 1929 the Pact of Letran was signed with the Vatican, re-establishing the temporal power of the popes, and thereby gaining Catholic support for the government.

[21] Mussolini's foreign policy was directed almost exclusively toward the acquisition of colonies. In 1936, Italy invaded Ethiopia, and a year later the Italian East African Empire was formed. During the Spanish Civil War, closer ties developed with Hitler's Germany, forming the basis for what was to become the Rome-Berlin axis. In April 1939, Italian troops took Albania.

[22] During World War II, Italy declared war on France and Britain. Not only did it lose its colonies at the end of the war, but the entire peninsula became a combat zone for Germans and Allies alike. In 1943, the latter defeated Hitler's troops, and Humberto II became king.

[23] Following a referendum at the end of the war (June 1946), the monarchy was abolished and the republic of Italy was formed. Under the leadership of Alcide de Gasperi, the Christian Democracy (DC) won a relative majority and formed a government. These first elections marked the beginning of the Christian Democrats' hold on power. In May 1948, Luigi Einaudi, also of the DC, was elected Italy's first president.

[24] The 1946 International Conference authorized Italy to continue administrating Somalia (known as "Italian Somaliland"), a situation which continued until 1960. During the 1950s, Italy participated in the reconstruction of Europe. In 1957, it became one of the charter members of the Common Market.

[25] At the same time that the socialists were seeking closer links with the Christian Democrats, the Communist Party (PCI) embarked upon a revision of its political base under the leadership of Palmiro Togliatti. The PCI reaffirmed its belief in a democratic means toward a socialist aim. During this period, the party's influence extended into the labor unions, shaping the power structure of the General Union of Italian Workers (CGIL). The PCI, the country's second strongest electoral force, as well as the second largest Communist Party in the Western world, defined itself in terms of "Eurocommunism", that is, communism with a European face.

[26] After successive victories at the polls, in 1961 the DC began "opening up toward the left", seeking alliances with the Socialists and Social Democrats. The powerful Communist Party, in spite of its strong electoral presence, was permanently excluded from the cabinet. The economic and institutional crises which took place during that decade led radical groups from the right and left alike to turn to violence as a means of bringing about change. The far right organized bombings to draw attention to its demands, while the Red Brigades of the left used political kidnapping as their main tool. In 1978, the kidnapping and assassination of former prime minister Aldo Moro sealed their isolation from mainstream politics.

[27] Between 1952 and 1962, the average income of Italians doubled, as a result of the development of industry, which had come to employ 38 percent of the national workforce. At the same time, agricultural employment dropped by 11 per cent, triggering migrations from the countryside to the cities, and from the south to the north. The industrial triangle of Milan, Turin and Genoa attracted a concentration of millions of people, living in overcrowded conditions inferior to those of the rest of Europe.

[28] According to the 1948 constitution the president of the republic, who is elected every seven years, is the head of state. One of the duties is to select a head of the government, the prime minister, who will have the support of the bicameral parliament. Until 1978, when socialist Sandro Pertini was elected, all presidents had belonged to the DC Francesco Cossiga, elected in 1985, returned Italy to the tradition of Christian Democrat presidents.

[29] Cabinets which take lengthy negotiations to form generally only manage to last a few months. Charges against Christian Democrat Arnaldo Forlani's government - linking his supporters and allies to an organization called "Propaganda Due"- brought the government down in May 1981, giving way to just over a year of government led by Republican Giovanni Spadolini.

[30] The instability of the government, coupled with astronomical fiscal deficits and the political influence of the Mafia and the "Camorra" have led many to consider the need for constitutional reform.

[31] In February 1991, the last secretary-general of the PCI proposed that the old name be changed to the Democratic Party of the Left (PDS); immediately afterwards, the PDS sought admission into the Socialist International. The PDS also called for reunification with the PSI, hoping to heal the split which had taken place 70 years earlier. In December 1991, dissenters from the official party line decided to create the "Reformed" Communist Party (PRC), using the red flag, hammer and sickle as its symbols.

[32] By the end of 1991, confrontations between DC's Cossiga and the PDS leader, Achille Occhetto wors-

ened. In February 1992, the president and the prime minister dissolved parliament, bringing the legislative elections forward.

[33] In the April 1992 elections, the DC did not achieve a parliamentary majority, the first defeat of a Christian-Democratic government since 1946. Days afterwards, Prime Minister Giulio Andreotti announced the dissolution of his government and President Cossiga resigned early. At the end of May, Judge Giovanni Falcone, enemy number one of the Mafia, was killed in Sicily. This allowed the election of former president of the chamber of deputies, Oscar Scalfaro.

[34] Two months after the assassination of Judge Falcone, the Mafia killed Paolo Borsellino, who had inherited the investigation against organized crime upon Falcone's death.

[35] In October 1992, the government presented a series of economic adjustment measures. These gave rise to a general strike in protest against the measures, with 10 million workers participating.

[36] Public prosecutor Antonio di Pietro launched an investigation which revealed a complex system of illegal operations involving politicians of all political persuasions and business people. As a result, more than a thousand political and business leaders were sent to prison in 1993. Well known public figures, such as former prime ministers Bettino Craxi and Giulio Andreotti, were brought to trial in Operation "Clean Hands".

[37] Between 1980 and 1992, corruption deprived state coffers of some $20 billion. Because of bribes to government officials and politicians, Italian public investments were 25 per cent higher than in the rest of the European Community.

[38] In April 1993, former president of the Central Bank, Carlo Azeglio Ciampi, was named prime minister. A few days later, a plebiscite was held on an electoral reform, to elect three-quarters of the seats of the two chambers - the Senate and Chamber of Deputies - by simple majority in the electoral districts. The remaining 25 per cent would be allotted by proportional representation.

[39] The increasingly tarnished image of the traditional parties led to the defeat of the Socialists and Christian Democrats in the municipal elections, with gains registered by the left-wing PDS, the neo-Fascist extreme-right-wing of the Italian Social Movement, and the federalist Lombard League. In total, the left won 72 municipal elections, and the right 16.

[40] In 1993, the budget deficit reached 10 per cent of the GDP. Ciampi's government reformed public administration, privatized companies and reduced public spending. The unemployment rate rose to 11 per cent, and large corporations began drastic reductions of their personnel. In November, the Fiat Automotive Division announced the elimination of 5,000 jobs.

[41] For the March 1994 legislative elections, the PDS, which had been less affected by corruption scandals, was the favorite. However, within a few months, media magnate Silvio Berlusconi set up a new political party, the Forza Italia, which allied with the Northern League of Umberto Bossi and the neo-Fascist National Alliance, of Gianfranco Fini and attained an unexpected victory.

[42] The new alliance led by Berlusconi, the "Freedom Poll", with 34.4 percent, obtained an absolute majority in Parliament. The left coalition led by the PDS obtained only 26 per cent and the Popular Party, heir to the Christian Democrats, 10.7 per cent. Berlusconi, who had promised a new "economic miracle" for Italy by means of liberalization, transparency and fiscal balance without tax increase, was appointed prime minister.

[43] In June Berlusconi consolidated his popularity, when Forza Italia took 30 per cent of the vote in the European election, while Bossi obtained 6.6 per cent and Fini 12.5 per cent. The results meant 15 points for the "Freedom Poll" as a whole. However, the relations with the Northern League, which continued criticizing Berlusconi and the National Alliance "Fascists", began to complicate the governmental action.

[44] In October, the trade unions opposed the retirement reform proposed by the prime minister, because according to them, it constrained benefits, increased contributions and "privatized" part of the system. After a massive demonstration in Rome, which gathered more than one million people, Berlusconi withdrew his reform project. In December, Bossi - who, despite his scarce electoral weight accounted for almost a fifth of the Chamber of Deputies - withdrew his support for the government and Berlusconi resigned.

[45] Scalfaro refused to hold a new election and appointed Berlusconi's former minister of the economy Lamberto Dini as prime minister. With more support among international financial agencies than his predecessor - he had been an IMF official some years before - Dini formed a supposedly "apolitical" government of technocrats.

[46] The new prime minister managed to amend the retirement system and curb the fiscal deficit, thanks to the support of the Left. As it had happened with Ciampi - another "technocratic" prime minister who had governed the country in 1993 and 1994 - Dini forged an alliance with the PDS in order to impose economic austerity measures. Berlusconi's weakening, under charges of corruption and increasing criticism from the right, prepared the field for the electoral victory of the left.

[47] On April 21 1996, the "El Olivo" coalition, led by ex-Christian Democrat Romano Prodi and supported by the PDS, obtained the victory in the legislative election. Prodi was appointed prime minister and formed a government with the participation of prominent PDS leaders, as well as well-known conservative independent figures like Ciampi and Dini himself. The distinctive feature of the new governmental period was the entry of former communists into the cabinet and the support provided by the Reformed Communist Party.

Jamaica

Population: 2,497,000 (1994)
Area: 10,990 SQ.KM
Capital: Kingston

Currency: Jamaica Dollar
Language: English

Jamaica

Ethnological studies report that the name Xamayca was given to this Caribbean island by the Arawaks. The Arawaks had pushed out the Guanahatabey the original inhabitants, who had come from North America. The name Xamayca means "land of springs", derived from the abundant natural irrigation of its luxuriant forests. The Arawaks lived in villages of houses made from palm branches and carried out farming and fishing activities.

2 Columbus reached Jamaica on his second voyage to the New World in May 1494, but it was his son, Diego Colon, who conquered the island in 1509. From then on, the number of Arawaks decreased dramatically. In the first half of the 16th century, around 1545, Spanish historian Francisco López de Gomara wrote that "Jamaica resembles Haiti in all respects - here the Indians have also been wiped out." Some sources believe that prior to the Spanish Conquest, there may have been as many as 60,000 Arawaks.

3 The Spanish, now absolute rulers of the island, began to plant sugar cane and cotton, and to raise cattle. There were incursions by the British in 1596 and 1636, and in 1655, 6,500 British soldiers under the command of William Penn dislodged the 1,500 Spaniards and Portuguese. Jamaica rapidly became a haven for pirates who ravaged Spanish trade in the Caribbean. The last important enemies that the English had to face on the island were the enclaves of rebel slaves or "quilombos", hidden in remote areas like the Blue Mountains. In 1760 a general rebellion in the colony was put down, and in 1795 a further revolution shook the island.

4 By the late the 19th century there were approximately 800 sugar mills and more than 1,000 cattle ranches in Jamaica. The economy was built on the labor of 200,000 slaves brought from Africa. The anti-slavery and anti-colonial rebellions of the 18th and 19th centuries were followed by labor union struggles in the first few decades of the 20th century. The two large contemporary political parties, the Labor Party and the People's National Party both grew out of workers' organizations. Independence was proclaimed in 1962, but successive Labor governments failed to rescue the economy from foreign hands.

5 In 1942 rich deposits of bauxite were discovered, and the aluminum transnationals; ALCOA, ALCAN, Reynolds and Kaiser, quickly established themselves on the island. The sugar industry was replaced by bauxite exploitation.

6 The transnationals exploited Jamaica's bauxite by shipping the raw metal out of the country, making all the decisions on production, and paying minimal customs duties.

7 After independence, a few plants were built to transform bauxite into aluminum, but the bulk of the mineral extracted continued to be shipped unprocessed to the US. In 1973, Jamaica was the second largest producer of bauxite in the world. Bauxite accounted for half of the country's exports but employed only one per cent of the labor force. Transnationals' earnings were not reinvested in the country. Export and import duties were reduced as the transportation of the mineral was classified as an internal transfer, with preferential taxation rates.

8 The People's National Party (PNP) won the 1972 elections and in 1974 the new Prime Minister, Michael Manley, raised the bauxite export tax. In addition, he began negotiations with the foreign companies to recover the third of the nation owned by them, and to assert greater control over their activities.

9 At the same time, the PNP government strongly supported Caribbean integration. A bi-national bauxite marketing company was created with Venezuela. Jamaica became a member of the Caribbean Multinational Merchant Fleet with Cuba and Costa Rica. Together, the associates devised ways to prevent penetration by US capital.

10 These progressive measures were resisted by the mining companies and by local conservative forces who sought to upset the December 1976 elections. The PNP won anyway, with an overwhelming victory at the polls and an increased majority in parliament. Manley advocated socialism within the existing constitutional structure.

11 Jamaica took active part in the Movement of Non-Aligned Countries. It also supported anticolonial positions at all world forums and proclaimed its unconditional solidarity with all liberation movements, particularly those of southern Africa.

12 This stand strained relations with the US. As a result, the transnational mining companies reduced their production, transferring operations to Guinea. Export revenues fell, and with them, funding for government social programs.

13 In 1979, the government was forced to seek loans from the IMF, which imposed harsh conditions. In February 1980, Manley announced that he was suspending negotiations with the Fund since meeting its requirements would mean a drastic reduction in living standards for the population. That year elections were held ahead of schedule, in a climate of destabilization generated by the right-wing opposition, which finally won a landslide victory. The new Labor government, headed by Edward Seaga, expelled the Cuban ambassador and imposed neoliberal economic policies, opening the country to unconditional foreign investment and even suggesting that it might request Jamaica's admittance to the United States. The results were counterproductive: between 1981 and 1983, unemployment increased, bauxite production continued to decline, inflation rose and the foreign debt doubled.

14 In November 1983, Jamaica was one of the small group of Caribbean countries that gave diplomatic support and symbolic military assistance to the invasion of Grenada. A month later, taking advantage of the favorable political atmosphere, Seaga decided to call early parliamentary elections. The PNP boycotted the elections, accusing the government of not having fulfilled a previous commitment to update the electoral register and reorganize the system of voter identification, so as to avoid rigging. Only the governing party nominated candidates, thereby winning all 60 seats.

15 In 1984 the need to negotiate new loans with the IMF and the consequent cuts in public spending aggravated an already critical situation. Inflation soared from 4.7 per cent in 1981 to 32 per cent in 1984. In addition, in 1985 revenues from bauxite production fell and there was a 50 per cent decrease in tourism. Union federations announced that workers' living conditions were being allowed to deteriorate deliberately in order to attract transnationals to the Kingston "free

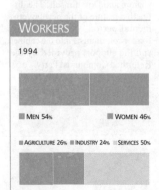

WORKERS

1994

MEN 54%   WOMEN 46%

AGRICULTURE 26%   INDUSTRY 24%   SERVICES 50%

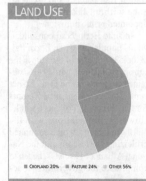

LAND USE

CROPLAND 20%   PASTURE 24%   OTHER 56%

Foreign Trade

MILLIONS $ 1994

IMPORTS
2,164
EXPORTS
1,192

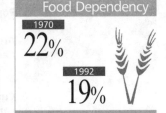

Food Dependency

1970
22%
1992
19%

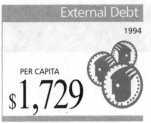

External Debt

1994

PER CAPITA
$1,729

Illiteracy

1995

15%

PROFILE

ENVIRONMENT

Jamaica is the third largest of the Greater Antilles. A mountain range, occupying one third of the land area, runs across the island from East to West. A limestone plateau covered with tropical vegetation extends to the West. The plains are good for farming and the subsoil is rich in bauxite. The climate is rainy, tropical at sea level and temperate in the eastern highlands. Soil loss from deforestation and erosion reaches 80 million tons per year. In some metropolitan areas, the lack of sewerage and the dumping of industrial wastes have polluted drinking water supplies, threatening the urban population.

SOCIETY

Peoples: Most Jamaicans are of African descent. There are small Chinese, Indian, Arab and European minorities. **Religions:** Mainly Christian (Anglicans and Catholics). **Languages:** English (official). A dialect based on English is also spoken. **Political Parties:** People's National Party (PNP); Jamaican Labor Party (JLP); National Democratic Movement; Workers' Party of Jamaica (JWP) . **Social Organizations:** National Workers Union of Jamaica (NWUJ); Bustamante Industrial Trade Union (BITU).

THE STATE

Official Name: Jamaica. **Administrative Divisions:** 3 parishes. **Capital:** Kingston, 104,000 inhab. (1991). **Other cities:** Montego Bay, 83,500 inhab.; Spanish Town, 92,000 inhab. (1991). **Government:** Parliamentary monarchy. Head of State: Queen Elizabeth II of Britain. Prime Minister since March 1992: Percival J. Patterson. Bicameral Legislature: 21 member Senate, designated by the Governor General; Chamber of Deputies, 60 elected by direct popular vote every 5 years. **National Holiday:** First Monday in August, Independence Day (1962). **Armed Forces:** 3,350 troops.

STATISTICS

DEMOGRAPHY

Urban: 53% (1995). **Annual growth:** 0.9% (1992-2000). **Estimate for year 2000:** 3,000,000. **Children per woman:** 2.7 (1992).

HEALTH

One **physician** for every 7,143 inhab. (1988-91). **Under-five mortality:** 13 per 1,000 (1995). **Calorie consumption:** 100% of required intake (1995). **Safe water:** 86% of the population has access (1990-95).

EDUCATION

Illiteracy: 15% (1995). **School enrolment:** Primary (1993): 108% fem., 108% male. Secondary (1993): 70% fem., 62% male. University: 6% (1993). **Primary school teachers:** one for every 38 students (1992).

COMMUNICATIONS

100 **newspapers** (1995), 101 **TV sets** (1995) and 103 **radio sets** per 1,000 households (1995). 10.6 **telephones** per 100 inhabitants (1993). **Books:** 88 new titles per 1,000,000 inhabitants in 1995.

ECONOMY

Per capita GNP: $1,540 (1994). **Annual growth:** 3.90% (1985-94). **Annual inflation:** 27.60% (1984-94). **Consumer price index:** 100 in 1990; 267.9 in 1992. **Currency:** 33 Jamaica dollars = 1$ (1994). **Cereal imports:** 429,000 metric tons (1993). **Fertilizer use:** 973 kgs per ha. (1992-93). **Imports:** $2,164 million (1994). **Exports:** $1,192 million (1994). **External debt:** $4,318 million (1994), $1,729 per capita (1994). **Debt service:** 20.6% of exports (1994). **Development aid received:** $109 million (1993); $45 per capita; 3% of GNP.

ENERGY

Consumption: 1,112 kgs of Oil Equivalent per capita yearly (1994), 100% imported (1994).

---

zone" where workers' rights are not respected.

[16] Ninety-six per cent of the work force in the "free zone" are women, putting in long working hours in unhealthy conditions, without social security and with low pay. This indirectly affects the children of Kingston, as 52 per cent of all heads of households are women.

[17] In June 1986, popular discontent over rates increases and the removal of price controls was rife, the cost of living had doubled, and Seaga called municipal elections. Michael Manley's Popular National Party defeated the Labor Party with 57 per cent of the vote, winning 140 of the 187 seats.

[18] Between 1985 and 1987, in agreement with the United States, the Labor government adopted an anti-drug program. Marijuana trafficking from Jamaica to the United States was estimated to earn $750 million per year.

[19] In 1989 Michael Manley's PNP returned to power with a program substantially different from that of 1976: it was based on promoting free enterprise and fostering good relations with the US. The new government announced the re-establishment of diplomatic relations with Cuba. It also stated that the agreements with the IMF would be respected.

[20] In April, 1992, Percival Patterson replaced Manley as prime minister. Manley's long illness also led to his resignation as chair of the PNP. Patterson was a well-known politician, a former senator and minister of several PNP govern-

ments, and he was elected president of the party.

[21] In 1992 a privatization process was started in some 300 state companies and public services deemed to be "unproductive", including the sugar industry.

[22] In March 1993, the PNP took more than 60 per cent of the vote and 52 of the 60 seats in parliament, which led to the confirmation of Patterson as prime minister. The Labor Party - led by the increasingly criticized Edward Seaga - refused to participate in the partial elections held in 1994 in order to show their disagreement with the electoral system and to ask for it to be changed before the 1998 general elections.

[23] In October 1995, a group of Seaga's opponents within the Labor Party abandoned the group to found the National Democratic Movement. Meanwhile, Prime Minister Patterson continued with the economic and financial liberalization which has, amongst other things, made the financial establishment the most prosperous sector in Jamaica.

Japan

Nihon

Population: 124,961,000 (1994)
Area: 377,800 SQ.KM
Capital: Tokyo

Currency: Yen
Language: Japanese

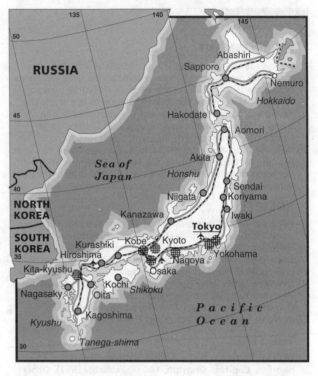

E arliest culture in Japan dates back 10,000 to 30,000 years ago to the Paleolithic Age, known as the Pre-Cambrian era. This was followed by the Jomon Neolithic culture, which lasted until 200-300 BC and extended throughout the Japanese archipelago.

[2] The Jomon culture was altered by the arrival of the Yayoi, who probably arrived from the continent at a time when Japan was still linked by land bridges across the straits of Korea, Tsushima, Soya and Tsugaru. The Yayoi introduced rice cultivation, horses and cows, the potter's wheel, weaving, and iron tools.

[3] According to Chinese chronicles, at the beginning of the Christian era the Wo region (Wa in Japanese) was divided into more than 100 states. During Himito's reign, some 30 of them were grouped together. The Wa people were divided into social classes and paid taxes, advanced building techniques were known and used, there were large markets and correspondence was exchanged with the continent.

[4] From the end of the civil war until Yamato's consolidation as emperor (266-413), the Wa territory was isolated. This led to the unification of the nation by the middle of the 4th century, a prerequisite for further expansion. In the year 369, Yamato subdued the Korean kingdoms of Paekche, Kaya and Sila and established military headquarters from which he could control the region.

[5] The Yamato empire suffered a rapid decline, due to both the resistance of its Korean subjects and to internal fighting within the court. During that period, between the years 538 and 552, Buddhism was introduced into the country, though initially it was merely an object of

curiosity and admiration, due to its majestic temples and the magical powers that were attributed to it.

[6] The most important traces of Chinese presence in Japanese culture are the grid system for dividing the land, which dates back 1,500 years and is still visible today, the Chinese characters used in the writing system and the Buddhist religion.

[7] However, this cultural heritage underwent successive adaptations to the local weather, language, and habits, especially during the 17th and 18th centuries. This was particularly apparent in the architecture and the language, each of which shows evidence of very different local influences.

[8] Japan's first permanent capital was Nara, established in the year 710. In the 9th century, tribal chieftains were replaced by a permanent court with a hereditary right to title. The aristocracy made Buddhism a controlling force, which served to reinforce the power of the state. After several conflicts, Kammu (781-806) re-established the empire's independence and transferred the capital to Heian (Kyoto).

[9] In the new capital, the power of the Fujiwara family was consolidated. The Fujiwaras established the regent as the ruling figure, above and beyond the power of the emperor. With imperial approval, two new Buddhist sects - Tendai and Shingon - developed in Heian. They were both seen to be more closely identified with Japanese culture than the previous Buddhist sects, a fact which brought an end to Nara's religious hegemony.

[10] The imperial land tenure system also fell increasingly into private hands. The members of the aristocracy and the religious institutions began taking over large extensions of tax-free land (shoen). The nobility organized private armies for themselves, and a rural warrior class - the samurai - emerged.

[11] The Taira and Minamoto clans, who were prominent families and local leaders, became involved in a power struggle which led to a number of military confrontations between these two warlords. The Taira were in power from 1156 until their defeat in the Gempei War (1180-85). The shogun (general), Minamoto Yoritomo, founded the Kamakura shogunate, the first of a series of military regimes which ruled Japan until 1868.

[12] The Kamakura were put to the

test during the Mongol invasions of 1274 and 1281. Aided by providential storms, which were called kamikaze (divine winds), the Japanese defeated the invaders. During this period, several new Buddhist sects emerged, such as Pure Land Buddhism, True Pure Land and Lotus.

[13] In the early 14th century, the Kamakura shogunate was destroyed, and Emperor Go-Daigo re-established his authority over the warlords in the Kemmu Restoration. However, a short while later he was expelled from Kyoto and replaced by a puppet emperor, who was controlled by the military clans. Go-Diago established his court in Yoshino, and for 56 years there were two imperial courts operating alongside each other.

[14] The Onin War (1467-77), over succession within the Ashikaga shogunate, became a civil war which lasted a hundred years. New military chiefs emerged, independent of imperial or shogun authority. They established themselves and their vassals within fortified cities, leaving the surrounding villages to run themselves, and to pay tribute.

[15] In the cities, trade and manufacturing ushered in a new way of life. Portugal began trading with Japan in 1545, and the missionary Francis

Xavier introduced Catholicism to the Japanese in 1549. However, Christianity caused conflicts with feudal loyalties, and so it was proscribed in 1639. All of the Europeans in Japan, except for the Dutch, were banished from the country.

[16] Toward the end of the 16th century, the Japanese warlords isolated themselves from the rest of society, pacifying and uniting the country around a single national authority. To achieve these objectives, the warlords made use of firearms and of military fortresses; in addition, they disarmed the peasants and achieved greater control over the land.

[17] During the 17th century, the Tokugawa clan gained supremacy over the entire country. From the city-fortress of Edo (Tokyo), the Tokugawa shogunate governed Japan until 1867. A careful distribution of the land among their relatives and the local chieftains guaranteed them the control of the largest cities, Kyoto, Osaka and Nagasaki, as well as of the most important mines.

[18] Local chieftains were compelled to spend half their time on the shogun's affairs while their families remained behind, as hostages. Transformed into military bureaucrats, the samurai were the highest level of a four-class system,

LAND USE

CROPLAND 12%   PASTURE 2%   OTHER 86%

ENVIRONMENT

The country is an archipelago made up of 3,400 islands, the most important being Hokkaido, Honshu and Kyushu. The terrain is mountainous, dominated by the so-called Japanese Alps, which are of volcanic origin. Since 85% of the land is taken up by high, uninhabitable mountains, 40% of the population lives on only 1% of the land area, in the narrow Pacific coastal plains, where demographic density exceeds 1,000 inhabitants per sq km. The climate is subtropical in the south, temperate in the center and cold in the north. Located where cold and warm ocean currents converge, Japanese waters have excellent fishing, and this activity is important to the country's economy. Japan's intensive and highly mechanized farming is concentrated along the coastal plains (rice, soybeans and vegetables). There are few mineral resources. Highly industrialized, the country's economy revolves around foreign trade, exporting manufactured products and importing raw materials. The major environmental problems are air pollution, especially in the major urban areas of Tokyo, Osaka and Yokohama, and acid rain in many parts of the country. One of the world's largest heavy industries has polluted many coastal areas.

SOCIETY

Peoples: The Japanese are culturally and ethnically homogeneous, having their origin in the migration of peoples from the Asian continent. There are Korean, Chinese, Ainu and Brazilian minorities. **Religions:** Buddhism and Shintoism. **Languages:** Japanese. **Political Parties:** The government coalition is made up of the Social Democratic Party of Japan (SPDJ), the Liberal Democratic Party (LDP) and the New Party Sakigake. The Shinshinto (Japan Renewal Party) has been in the opposition since 1994. **Social Organizations:** The General Council of Japanese Trade Unions has 4,500,000 members.

THE STATE

Official Name: Nihon. **Capital:** Tokyo, 8,021,943 inhab. (1994). **Other cities:** Yokohama, 3,300,513 inhab., Osaka, 2,575,042 inhab.; Nagoya, 2,153,293 inhab.; Sapporo, 1,744,806 inhab. (1994). **Government:** Parliamentary constitutional monarchy. The Diet (Legislature) is bicameral: House of Representatives, made up of 512 members; House of Counsellors, with 252 members, elected by direct popular vote every 4 and 6 years, respectively. Emperor Akihito has been Head of State since 1989, although his official coronation did not take place until November 12, 1990. Ryutaro Hashimoto (Liberal Democratic Party) Prime Minister and head of the government since January 11, 1996. **National Holiday:** January 1, New Year's Day. February 11, Founding of the Country (1889). **Armed Forces:** 239,500 (including 8,000 women). **Paramilitaries:** 12,000 (non-combat Coast Guard, under the jurisdiction of the Ministry of Transport).

---

followed by the peasants, artisans and traders.

[19] A national market arose for textiles, food, handcrafts, books and other products. As of 1639, the Tokugawas implemented a policy of almost total isolation from the outside world. Nagasaki was the only exception; here, the Chinese and the Dutch were allowed to open trading posts, although the latter were restricted to a nearby island.

[20] In the 19th century, the old economic and social order went into a state of collapse. Peasant revolts became more and more frequent, and the samurai and local chieftains found themselves heavily in debt with the traders. In 1840, the government tried to carry out a series of reforms, but these failed and weakness allowed the United States to prize open its ports.

[21] Japan was forced out of its isolation by Commodore Matthew Perry, using cannon to persuade the Japanese to cooperate with him. The signing of unfavorable trade agreements with the United States and several European countries simply deepened the crisis within the *shogunate*. The samurai carried out several attacks against the foreigners and then turned against the *shogun*, forcing him to resign in 1867.

[22] Imperial authority was restored with the young Meiji emperor, in 1868. During the Meiji Restoration, Japan's modernization process began, following the Western model. The United States exerted its influence, as did England, France and Germany, in the fields of education,

the sciences, communication and Japanese cultural expression.

[23] Within less than 50 years, Japan was transformed from a closed, feudal society into an industrialized world power. Western advisers and technology were brought in, for education, trade and industry. An army based on the draft replaced the military authority of the samurai, who were defeated when they tried to rebel, in 1877.

[24] In 1889, succumbing to internal political pressure, the emperor approved a constitution which established a constitutional monarchy, with a bicameral legislature (Diet). However, only one per cent of the population was eligible for office, and the prime minister and his cabinet were responsible to the emperor who continued to be seen as a divine figure.

[25] Japan defeated China in the war of 1894-95, and maintained control over Korea. Japan's victory in the Russo-Japanese War (1904-05) enabled it to annex the Sajalin Peninsula, with Korea being annexed a few years later (1910). Japan entered World War I as a British ally, as a treaty had been signed to that effect in 1902.

[26] The war allowed Japan to gain control of several German possessions in East Asia, including the Chinese territory of Kiaochow. In 1915, Japan forced China to accept the extension of its influence over Manchuria and Inner Mongolia. In 1918, Hara Takashi became the head of the first government to have a parliamentary majority.

[27] In 1921-22 Japan signed a naval arms limitation treaty with the US in Washington, replacing an agreement with Britain, and establishing a new balance of power in the Pacific.

[28] The economic difficulties caused by the international depression of the 1930s gave the militarists the excuse they were seeking to attack the government. They proposed that the country's problems could only be solved by expanding its military power, and through the conquest of new markets for their products and as sources of raw materials.

[29] Within this context, Japanese officers occupied Manchuria in 1931, without government authorization. Unable to deter the military, the government accepted the creation of the puppet state of Manchukuo, in February 1932. Three months later, the country's political leaders were forced to turn the government over to the militarists, who retained power until 1945.

[30] In 1940, Japan invaded Indochina hoping to open up a passage through to Southeast Asia. The United States and Britain reacted by imposing a total embargo upon Japanese merchandise. The Japanese attack on Pearl Harbor, in Hawaii, and of the Philippines, Hong Kong and Malaysia, unleashed the war with the United States and opened up a new phase of World War II.

[31] Japan surrendered on August 14 1945, after the US had dropped two atomic bombs on Hiroshima and Nagasaki on August 6 and 9. Japan was subsequently occupied by US troops who remained in the country between 1945 and 1952, and was governed by the Supreme Command of the Allied Powers (SCAP) under the leadership of General Douglas MacArthur. SCAP forced Japan to abandon the Meiji institutions, to renounce the emperor's claim to divinity, and transfer the government to a parliament, which was charged with electing the prime minister and to establishing an independent judiciary.

[32] Although imposed upon the Japanese from the outside, the principles laid down in the 1947 constitution were accepted by all sectors of society and in 1952 the country recovered its independence. Japanese sovereignty was restored over the Tokara archipelago in 1951, over the Amami islands in 1953, over the Bonin islands in 1968, and over the rest of the Ryukyu, including Okinawa, in 1972.

[33] SCAP also took other measures to weaken the hierarchical model of the Meiji family-state. These ranged from giving tenants the right to purchase the land they lived on, and laws aimed at strengthening free trade and preventing the return of monopolies. However, the Japanese financial system remained intact and provided the basis for economic recovery at the end of the occupation.

[34] In 1955, opposing the country's conservative and nationalist sectors, which had supported the war policy, the Liberal Democratic Party (LDP) was formed. It was a center-right party, holding a majority in Parliament and has governed the country from its foundation to the present.

[35] The 1947 Constitution established restrictions upon the devel-

opment of Japanese military power. During the postwar period, Japan bowed to US strategy for the region forming alliances with Taiwan and South Korea. In 1956 it joined the UN and re-established relations with the USSR.

[36] The return to independence found the Japanese economy in a state of growth and change. Farmers became unable to survive with the traditional methods of small-scale production and left the land in droves, leading to great urban migration. Industrialization and full employment triggered the need for technological innovation in the countryside to increase the food supply.

[37] During the 1960s, Japan specialized in the production of high technology products, which made it necessary to establish stable trade relations with more industrialized countries instead of its previous Asian partners. The oil crisis of 1973 did not halt the growth of the Japanese industry, which led the world in steel, ship building, electronics, and automobile manufacturing.

[38] Although Prime Minister Kakuei Tanaka's visit to Beijing in 1972 signalled Japanese recognition of the People's Republic of China, it damaged the country's relations with Taiwan. The scandal following the disclosure of the fact that Tanaka had been bribed by the Marubeni Corporation (a representative of the US Lockheed Aircraft Corporation), adversely affected the LDP's popularity and in 1976, for the first time in its history, it lost its absolute majority in parliament.

[39] During the 1960s and 1970s, there was a large trade surplus in Japan's favor in its trade with the US. During this period, Japan began ranking first or second with all its trading partners. With direct investment and the establishment of subsidiaries of Japanese companies, Japan expanded worldwide.

[40] The Japanese corporate world is dominated by the Sogo-Shosha system, huge conglomerates which commercialize virtually all kinds of raw material in almost every country in the world, by means of state-of-the-art information systems capable of supplying data for decision making instantaneously.

[41] The organization of Japanese corporations still maintains some principles which are the legacy of Japan's medieval tradition. The worker is bound to the corporation by an allegiance similar to that which bound the medieval peasant to his land and the local warlord. This often results in great company discipline and production efficiency.

[42] The impressive development of the Japanese economy is due not only to this efficiency but also to a policy of foreign investment in projects which quickly deplete non-renewable natural resources. This policy has caused irreversible damage to rainforests and serious alterations to the Third World ecosystem.

[43] Japan imports over 16 million cubic metres of tropical timber every year. This has caused a massive deforestation in Malaysia, Thailand, Indonesia, the Philippines, and Papua New Guinea. Japan has also become a consumer of endangered species and products derived from them, such as elephant tusks and tortoise shells.

[44] Japan makes direct investments in Third World countries and gives credits through its agencies to aid in building roads and in scientific research projects. As with all international aid this assistance is geared toward its own interest in gaining cheap, easy access to the raw materials of those countries.

[45] Throughout modern Japan, the pursuit of economic success has become people's main objective. Family, leisure, and individual ideals are sacrificed to the factory or the company. Within this system, women play a very subservient role as the pillar of the home and of the children's education.

[46] In Japan there is great social pressure for women to get married: 80 per cent of women are married by the age of 30 and 98 per cent are married, widowed, or divorced by the age of 50. Women may work outside the home, generally in second-rate, badly paid jobs.

[47] In spite of this age-old discrimination - a legacy of the country's traditional cultures - Japanese women are active in local movements against the pollution caused by industry and nuclear power plants. They also actively defend the quality of life and of consumer goods.

[48] The traditional full-employment situation in Japan was threatened by two factors: demographic growth and technological modernization which created a labor shortage. Japanese resistance to foreign immigrants and the progressively ageing profile of the population only intensified the problem. In June 1990, an immigration law went into effect opening up the Japanese labor market to foreign workers, for the first time ever.

[49] Although the United States continues to give military support to Japan, Washington started exerting pressure in 1982 so that the country would increase its military expenditure (around 0.9 per cent of its GDP) and assume greater responsibility in regional security in the Western Pacific.

[50] Since the early 1950s, Japan began demanding that the USSR return four small islands which Japan claimed belonged to the Kurile archipelago. Relations between the two countries entered a new era in 1986 with Minister of Foreign Affairs Eduard Shevardnadze's visit to Tokyo. This visit resulted in the decision to hold regular ministerial consultations, and to increase trade.

[51] Prime Minister Yasuhiro Nakasone travelled to Eastern Europe in 1987, in the first such visit by a Japanese head of government. Three years later, Prime Minister Toshiki Kaifu travelled to several European countries and declared his support of the liberalization of Eastern Europe.

[52] The economic and social stability of southeast Asia is of utmost importance for Japan because a growing part of its investments and a third of its foreign trade depends on this region, which provides the country with raw materials vital to its industry.

[53] The death of Emperor Hirohito in January 1989 brought to an end the Showa era, which had begun in 1926. The coronation of his successor, Akihito, in the traditional Japanese style, marked the beginning of the Heisei era (achievement of universal peace). The coronation ceremony was attended by more heads of state than had ever gathered together for any such event. The expense incurred generated internal protests.

[54] Japanese relations with the EC and the US have been strained at times, due to problems of protectionism and a trade imbalance (in Japan's favor). In 1987, Washington protested at Japan's sale of sophisticated submarine technology to the USSR, between 1982 and 1984. Nevertheless in 1989 Naboru Takeshita was the first head of government to be received by newly-elected US president, George Bush. Takeshita resigned in April, as a result of a real estate scandal. He was succeeded by Sosuke Uno, who was forced to resign after less than three months in office after admitting having sexual relations with a geisha woman.

[55] At the end of the Cold War, Japan emerged as one of the three main world economic powers, together with the US and the EC. At present, it is the country with the largest overseas investment. It is a key participant in the world financial system, and is influential in the exchange of Third World debt funds.

[56] On November 5 1991, Prime Minister Toshiki Kaifu presented his resignation, and was replaced by 72-year-old Kiichi Miyazawa who had been elected president of the Liberal Democratic Party nine days before. Kaifu had been elected in August 1989 because he was one of very few party members who had never been involved in a corruption scandal. His political career rapidly came to an end when he lost the support of the Takeshita clan, the most influential of the official party's five factions, and the one which had backed his election.

[57] The Takeshita clan also backed the next prime minister, Miyazawa, even though he had been forced to resign as minister of finance in December 1988, because of his involvement in the Recruit scandal. In his inaugural speech, Miyazawa outlined the objectives he had set for his administration: to expand relations with, and aid to China; to negotiate with the United States and to normalize relations with the USSR. In addition, he announced his willingness to liberalize the rice market, making concessions similar to those already made by the EC and the United States, to ward off a failure of the Uruguay Round of the GATT.

[58] The traditional leader of the PLD, 77-year-old Shin Kanemaru stated that despite his personal aversion to him, Miyazawa ap-

Foreign Trade

MILLIONS $ 1994

IMPORTS
275,000

EXPORTS
397,000

DEMOGRAPHY

Urban: 77% (1995). **Annual growth:** 0.4% (1991-99). **Estimate for year 2000:** 127,000,000. **Children per woman:** 1.5 (1992).

HEALTH

Under-five mortality: 6 per 1,000 (1995). **Calorie consumption:** 105% of required intake (1995). **Safe water:** 97% of the population has access (1990-95).

EDUCATION

School enrolment: Primary (1993): 102% fem., 102% male. Secondary (1993): 97% fem., 95% male. University: 30% (1993). **Primary school teachers:** one for every 20 students (1992).

COMMUNICATIONS

119 **newspapers** (1995), 119 **TV sets** (1995) and 119 **radio sets** per 1,000 households (1995). 46.8 **telephones** per 100 inhabitants (1993). **Books:** 102 new titles per 1,000,000 inhabitants in 1995.

ECONOMY

Per capita GNP: $34,630 (1994). **Annual growth:** 3.20% (1985-94). **Annual inflation:** 1.30% (1984-94). **Consumer price index:** 100 in 1990; 107.1 in 1994. **Currency:** 100 yen = 1$ (1994). **Cereal imports:** 28,035,000 metric tons (1993). **Fertilizer use:** 3,951 kgs per ha. (1992-93). **Imports:** $275,000 million (1994). **Exports:** $397,000 million (1994).

ENERGY

Consumption: 3,825 kgs of Oil Equivalent per capita yearly (1994), 82% imported (1994).

peared to be the only politician which the party could back at the moment.

[59] On November 23, US secretary of defense Richard Cheney called on Japan to play a more active role in the military and political aspects of world affairs. US military forces currently in Japan consist of three air bases, an aircraft carrier and some 56,000 troops, including a division of Marines.

[60] On February 9 1992, Yukihise Yoshida, an opposition candidate supported by all four opposition parties, defeated the official candidate by 51 per cent to 37 per cent, in the western district of Nara. This outcome was attributed to popular disenchantment with the party in power, because endemic corruption was not only extant, but was very obviously on the increase. Prime Minister Kiichi Miyazawa himself appeared to be entirely surrounded by people with a record of corruption.

[61] On February 10, 1992, it was revealed that the minister of postal services, Hideo Watanabe, had admitted to receiving a bribe greater than the $40,000 he had previously admitted to during the Recruit scandal of 1988. On February 14, four people were arrested with relation to another financial scandal, which involved more than 100 members of Miyazawa's government, as well as the mafia group known as Inagawagumi. It was revealed that billions of yen had been siphoned off from party donations making the new case, known as Kyubin, an even more serious breach of conduct than the Recruit scandal.

[62] Toward the middle of the year, after heated debate, a law was passed authorizing troops to be sent abroad for the first time since World War II.

[63] The LDP won the July elections for the Upper House, although it failed to win a majority in the Diet. Keizo Obuchi assumed the leadership of the Takeshita faction of the LDP, after its leader resigned on corruption charges.

[64] Japanese foreign relations continued to improve. In September, Japanese troops were sent to Cambodia as members of the UN peacekeeping force, and Emperor Akihito signed bilateral trade agreements with China during his end-of-year visit.

[65] In early 1993, Miyazawa promised to broaden the scope of Japa-

nese forces, both in terms of funding and personnel, within the framework of the UN peacekeeping missions.

[66] Also in 1993 the prime minister was censured by the Diet for failing to carry out electoral reforms needed to end the endemic corruption in Japanese political life. The official party split, losing 54 seats, and elections were moved up to July.

[67] The two dissident groups, led by Tsutomu Hata and Masayoshi Takemura, formed new parties; the Reformation Party and the Pioneer Party.

[68] Debate intensified in April over whether Japanese troops should be sent abroad, after a civilian and a member of the Japanese police force were killed in Cambodia as members of the UN forces. In May, another contingent of 75 police and 600 soldiers left for Cambodia, and that same month, more soldiers were sent to Mozambique.

[69] Diplomatic relations with the Russian government improved slightly when Japan announced that, while not reliquishing its claim to the Kuril Islands, it would not allow its claim to the islands to prevent economic aid to Russia.

[70] On July 18, general elections drew the lowest rate of voter participation since the war (67.3 per cent), a clear sign of the electorate's dissatisfaction with the corrupt state of Japan's political scene.

[71] The LDP, in power since 1955, lost its majority in the Kokkai (Diet), obtaining only 228 of the 512 disputed seats in the House of Representatives, compared with 275 in the 1990 election. This result was significant enough to alter the balance of power which had been in effect since World War II.

[72] Miyazawa, who had announced his decision to remain in office regardless of the election results, changed his mind. A few days later, he resigned the party presidency, and assumed responsibility for the LDP's defeat. Yohei Kono was named party president.

[73] In July, the "new majority" of the Socialist, Reformation, Komeito (Buddhist), Democratic Socialist, Unified Democratic Socialist and Pioneer parties, agreed to form a new government based upon a limited platform which would not attempt any profound changes.

[74] Hosokawa, the former governer of the province of Kunamoto, was elected prime minister in August. Upon assuming office, he an-

nounced a far-reaching political-reform bill, aimed at fighting corruption, putting an end to the recession and modernizing the pension and health systems.

[75] In his first speech before the Diet, Hosokawa referred to the aggression with which Japan had treated its Asian neighbors, from the 1930s until the end of World War II. On August 15, the 48th anniversary of Japan's surrender, Hosokawa offered condolences and apologies to the victims of Japanese colonialism.

[76] In December 1993, the Japanese government opened up the rice market, allowing up to 4 per cent of internal consumption, amounting to 10 million tons, to be imported. Many families tend to hoard domestic rice, and shopkeepers speculate with prices. The "Rice Lord", or *kome*, is a sacred part in Japanese tradition, and is linked to the earth and the ancestors.

[77] Japan's enormous trade surplus has been a source of constant tension on the international market, particularly with the United States. This surplus reached $107 billion in 1992, and $150 billion in 1993.

[78] Beyond the problems caused by the surplus, the Japanese economy has also suffered the effects of world recession. The most apparent consequence being that of unemployment, which has been on the

increase since mid-1992 and reached its peak in 1996 with 3.2 per cent.

[79] A corruption scandal among top government circles had been revealed in 1992 when Shin Kanemaru, historic LDP leader, admitted he had received bribes. Kanemaru was arrested in March 1993 and was granted an amnesty shortly afterward. When Hosokawa succeeded Miyazawa, the Confederation of Industries announced it would discontinue its "contributions" to the LDP. According to local press versions, these had amounted to $1 billion per year.

[80] In August, 165 political leaders and 445 business people were arrested for irregularities committed during the July 1993 election campaign. 5,500 cases were brought to court to deal with electoral fraud. In October, Shinji Kiyoyama - president of Kajima, the country's second largest construction firm - was arrested for paying a bribe in exchange for a building permit.

[81] Hosokawa's "romance" with public opinion, after reaching record levels of popularity, started to cool off in early 1994 as press reports became critical, accusing him of having received huge amounts of money in exchange for favors.

[82] Already in February 1994, the prime minister had been forced to aban-

American troops out of Okinawa!

After a year of strong protests, the inhabitants of Okinawa, 2,000 km south of Tokyo, voted for the dismantling of US bases on the island. The proportion of votes against the military presence of the US was 10 to 1. The results of the September 1996 referendum coincided with the repeated refusals of the governor of Okinawa, Masahide Ota, to cooperate with the government in the negotiations to maintain the bases.

[2] Criticism against the US military grew on September 29, 1995, when three soldiers raped a 12 year-old Japanese girl. The US Secretary of Defense, William Perry, formally apologized while calling the troops to a "day of reflection". Ota asked the Government to revise the agreement with the US, but the then prime minister, Tomiichi Murayama refused to consider the protests.

[3] After the results of the referendum were known, Ryutaro Hashimoto - successor of Murayama - recognized that Tokyo had ignored the complaints of the population of Okinawa. Hashimoto announced that in April 1997, he would initiate talks with Washington in order to close, or to move, 9 of the 40 military sites.

[4] The island of Okinawa was colonized by Japan in 1879, and in 1945 half a million allied soldiers landed on its coasts. The invasion left a toll of 200,000 casualties, half of them civilians. After the Japanese surrender, the US kept the island, settling there the biggest and most complex US air base in the world.

[5] 75 per cent of US troops in Japan are based in Okinawa, which represents only 1 per cent of the country's territory. 10.8 per cent of the island's population - 1,230,000 people - are US military personnel and their families. A fifth of the territory of Okinawa - a subtropical island which is a tourist center for the Japanese population - is surrounded by barbed wire. Besides the military installations, the bases have fast food restaurants, fun resorts, schools, and shops selling US imported goods.

[6] In addition to the continuous roar of war planes and shooting practice, the inhabitants of Okinawa have to share space with 28,000 soldiers who sometimes cause trouble. A week before the referendum, two marines assaulted and robbed a 56-year-old woman. Soldiers are also accused of introducing forbidden drugs, and of using the island for the traffic of turtle shells.

[7] The US president Bill Clinton justified the presence of his troops in Japan. Washington's main regional concern is focused on North Korea and China. The former has a million soldiers and also intends to develop nuclear technology, while the latter has raised its military budget and undertook intimidating navy maneuvers near Taiwan at the beginning of 1996.

[8] A study prepared by US analysts in 1996 advised the withdrawal of 18,000 army personnel of the Third Navy Division, arguing that the Taiwan Strait is sufficiently defended by sea and air, and that this body would not be able to act with due celerity in the case of South Korea being invaded. Shunji Taoka, analyst on defence issues for the influential magazine Asahi Shimbun, stressed that the sole purpose for the Third Division to remain in Okinawa is "to ensure its officers' jobs".

don a scheme which he himself had sponsored, aimed at keeping the government coalition intact. Worn out by the pressure, he had to issue a self-critique - unpleasant for a head of government - explaining the reckless manner in which he had handled the situation.

[83] A new $140 billion plan to reactivate the economy also drew criticism. Its feasibility was questioned and the press stressed the plan did not address the problems posed by a rapidly ageing population.

[84] In March 1994, a summit meeting between Japan and the United States ended in failure. Although Hosokawa showed a willingness to accept opening up the Japanese automobile, telecommunications, medicine and insurance markets, he rejected the US scheme of mandatory quotas. Japanese businessmen disqualified Hosokawa and felt inclined to wage a commercial war with the US before yielding to its pressures. The automobile industry employs 11 per cent of Japan's labor and equals 30 per cent of the GDP.

[85] A new corruption case involved LDP member of parliament Naka-mura, who was arrested for accepting bribes from Kajima and other construction firms. Hosokawa was unable to shake off accusations from the opposition about his own involvement in illicit business deals. In April he resigned and asked "sincerely for forgiveness from the people of Japan".

[86] Tsutomu Hata was nominated prime minister and on April 28, Japan's first minority government in four decades was formed. The socialists had walked out of the government coalition leaving the government with only 182 of the 512 seats in the lower house.

[87] Hata made official visits to Europe aiming to establish closer trade links with the European Community. He also admitted that Japan's trade surplus had caused the trade crisis with the United States. After an initial agreement was reached between the two countries, Hata launched a scheme to promote economic deregulation.

[88] Socialist Tomiichi Murayama was elected prime minister on June 29, 1994 and took office on July 18. His party, the Social Democratic Party of Japan (SDPJ) did not obtain a majority in parliament but formed an alliance with its traditional rival, the LDP, and with a new party, the Sakigake.

[89] In a visit to Seoul in July, Murayama apologized for Japan's actions carried out against Korea in the past. The opening of Kansai airport, located on an artificial island, led to a blossoming of island-city projects. The overpopulation of Japan's large cities has led to a permanent real estate development, taking up what little space is available for construction.

[90] On January 17 1995, an earthquake hit the area of Hanshin. Over 6,000 died, 100,000 buildings were destroyed in the city of Kobe and over 300,000 were left homeless. The government's delayed response to the disaster was harshly criticized. The lower house added $ 10 billion to the budget for the area's recovery.

[91] In March, a series of attacks with poisonous Sarin gas killed 12 people and affected 12,000 others in Tokyo's subways. A similar attack had taken seven lives in Matsumoto in June 1994. Shoko Asa-hara, the leader of a religious sect called Aum Shinriyko (Supreme Truth), was charged with the attack and arrested along with 16 other leaders from the movement.

[92] Murayama was defeated when Independent Yukio Aoshima was elected governor of Tokyo in the April 9 1995 elections. The Socialists did not obtain favorable returns at the July elections either, which renewed half of the upper house's seats and in which opposition party Sakigake had a significant growth.

[93] Protests from Okinawa residents for the decision to deploy 29,000 US troops on the island were added to the trade frictions between Japan and the United States. When three US soldiers were found guilty of having raped a Japanese girl, pressures on the two governments were intensified, but Murayama favoured an extension of the authorization for the troops to remain on the island.

Jordan

Urdunn

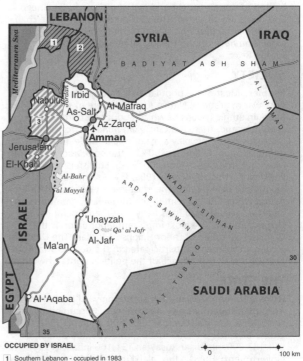

Population: 4,035,000 (1994)
Area: 89,210 SQ.KM
Capital: Amman
Currency: Dinar
Language: Arabic

OCCUPIED BY ISRAEL

[1] Southern Lebanon - occupied in 1983
[2] Syria - Golan Heights - occupied in 1967
[3] West Bank - occupied in 1967 (some individual towns now have autonomy)

0   100 km

In biblical times, when it was divided among the Semitic nations of Gilead, Moab and Edom, the territory of present day Jordan was central to the development of the region's history. In classical times, the state of Petra was one of the "desert sentries" allies of the Roman Empire. In the 7th century it was the site of the battle of Yarmuk, in which the Arabs fought Byzantine Emperor Heraclius, winning access to the fertile crescent (see Saudi Arabia), and during the Crusades the western part of the territory served as the operational base to initiate warfare against the European strongholds.

[2] In the 16th century, Turkish domination made the territory part of the district of Damascus, and it remained so until the beginning of World War I in 1914.

[3] The Jordanians participated actively in the Arab rebellion against the Turkish Ottoman Empire. The secret 1916 Sykes-Picot Treaty between France and Britain, later formalized by the League of Nations, gave the French control of Lebanon and Syria and Britain a mandate over Iraq and Palestine (which included present-day Jordan). The British had promised Shereef Hussein of Mecca a unified Arab nation, including these territories, and the Arabian peninsula.

[4] The conflict with Faisal, the shereef's son, in Syria (see Syria) and his expulsion by the French in 1920, led his brother Prince Abdullah to organize a support force of Jordanian Bedouin. The British persuaded Abdullah that he would be better off accepting the government of Trans-Jordan.

[5] With this, and the later creation of the state of Iraq, which was granted to Faisal as compensation, Britain put Shereef Hussein in an awkward position; if he insisted on his Greater Arabia idea, his sons would lose their recently acquired positions as heads of state. Conversely, as Hussein controlled Hidjaz, three parts of the proposed nation were already in the hands of the family, making subsequent unification a possibility. This was not to happen, however. In 1924 the shereef proclaimed himself Caliph. Emir Ibn Saud, king of Nejd, viewed this as a threat, so he invaded Hidjaz and drove Hussein out of Mecca (see Saudi Arabia).

[6] The emirate of Trans-Jordan remained under British mandate until 1928, when the borders with Palestine were established. Abdullah and his heirs were given legislative and administrative powers. Foreign and military affairs remained in British hands until May 1946, when the emirate became the "Hashemite Kingdom of Trans-Jordan". After the Arab-Israeli War of 1948, King Abdullah annexed Palestinian territories on the West Bank of the Jordan River, calling the new nation Jordan. Palestinian refugees, the legal status of Jerusalem, and the doubling in length of the border with Israel were problems he had to face.

[7] In 1951, Abdullah was assassinated and his son Talal succeeded him. Talal appeared to harbor anti-British sentiments and promised a progressive government, but a year later he was deposed. In 1953, his 17-year-old son Hussein occupied the throne.

[8] King Hussein has managed to remain in power because of firm US support and the personal devotion of Bedouins, both essential to keep the army under control.

[9] For over 40 years, and particularly after Israel occupied the West Bank, Palestinian immigration has been large. First came those driven from their lands by Israeli occupation troops in 1948, then those expelled from refugee camps in 1967. Palestinians currently make up two-thirds of Jordan's population. Demographic pressure and Hussein's ambition to replace the PLO as spokesperson for the Palestinians, were detonating factors in "Black September", the 1970 Jordanian massacre of Palestinians.

[10] After the 1973 Arab-Israeli war, there was a startling change in Jordanian policy and the king re-established relations with the PLO in 1979.

[11] In 1984, Jordan reasserted its position against unilateral negotiations with Israel though closer links with Egypt began. In 1985, King Hussein and PLO leader Yasser Arafat announced a common diplomatic initiative for peace in the Middle East; this failed as Israel refused to negotiate.

[12] In July 1988, Hussein relinquished his claim on the West Bank, and made the PLO legally responsible for the territories under Israeli occupation.

[13] Most international currency came in remittances from migrant workers and the national budget depended on financial aid from Arab countries, estimated at $1 billion in 1989.

[14] A backlog of repayments, and a foreign debt that reached $6 billion in 1989, led the government to turn to the IMF. Price increases in consumer goods caused a popular revolt in April 1989.

[15] King Hussein gave parliament power over the monarchy, he freed political prisoners and called for democratic elections.

[16] The outbreak of the Gulf crisis, when Iraq invaded Kuwait in August 1990, found the king at the height of popularity, but his position was very difficult. He was surrounded by Israel and Iraq, most of his Palestinian subjects supported the Iraqi leader, Saddam Hussein.

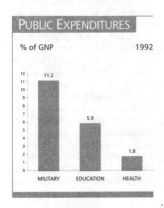

PUBLIC EXPENDITURES

% of GNP   1992

MILITARY 11.2
EDUCATION 5.9
HEALTH 1.8

WORKERS

1994

■ MEN 89%   ■ WOMEN 11%

■ AGRICULTURE 10% ■ INDUSTRY 26% ■ SERVICES 64%

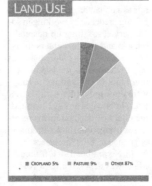

LAND USE

■ CROPLAND 5%   ■ PASTURE 9%   ■ OTHER 87%

ENVIRONMENT

75% of the country is a desert plateau, 600 to 900 m. in altitude. The western part of this plateau has a series of cleavages at the beginning of the great Rift Fault, which crosses the Red Sea and stretches into east Africa. In the past, these fissures widened the Jordan River valley and formed the steep depression which is now the Dead Sea. As most of the country is made up of dry steppes, farming is limited to cereals (wheat and rye) and citrus fruits. Sheep and goats are also bred. The shortage of water is the chief environmental problem. Desertification and urban expansion have caused the loss of arable land near the Jordan River.

SOCIETY

Peoples: Most of the population is Palestinian, from Israeli post-war migrations. Native Jordanians are of the 20 large Bedouin ethnic groups of which about one third are still semi-nomadic. There is a Circassian minority from the Caucasus, who now play a major role in trade and administration. **Religions:** 90% of the population are Sunni Muslim. **Languages:** Arabic (official). English is often used. **Political Parties:** By March 1993, the government had recognized nine parties. The most important are: the Islamic Action Front, Shiite oriented; the Jordanian Communist Party and the Socialist Arab Baas Party of Jordan. **Social Organizations:** The most important labor union is the General Federation of Labor Unions of Jordan. The Union of Jordanian Women has participated in the democratization process and in the defence of the political rights of women.

THE STATE

Official Name: al-Mamlakah al-Urdunniya al-Hashimiyah. **Administrative divisions:** 8 provinces, 3 occupied by Israel since 1967. **Capital:** Amman, 963,500 inhab. (1994). **Other cities:** az-Zarqa 344,500 inhab.; Irbid 208,200 inhab.; as-Salt 187,000 inhab.; ar-Rusayfah 131,100 inhab. (1994). **Government:** Hussein Ibn Talal, King since 1952. Abdul Karim Al-Kabariti, Prime Minister since February 1996. **Legislative branch:** National Assembly (bicameral), with a 40 member Senate appointed by the king, and an 80 member Chamber of Deputies elected by direct popular vote; the latter can be dissolved by the king. **National Holiday:** May 25, Independence Day (1946). **Armed Forces:** 100,600 troops (1993). **Paramilitaries:** 6,000 soldiers under the authority of the Department of Public Security; 200,000 militia in the "People's Army"; 3,000 Palestinians in the Palestinian Liberation Army, under the supervision of the Jordanian Army.

DEMOGRAPHY

Urban: 70% (1995). **Annual growth:** 0.7% (1992-2000). **Estimate for year 2000:** 5,000,000. **Children per woman:** 5.2 (1992).

HEALTH

One **physician** for every 649 inhab. (1988-91). **Under-five mortality:** 25 per 1,000 (1995). **Calorie consumption:** 104% of required intake (1995). **Safe water:** 89% of the population has access (1990-95).

EDUCATION

Illiteracy: 13% (1995). **School enrolment:** Primary (1993): 95% fem., 95% male. Secondary (1993): 54% fem., 52% male. University: 19% (1993). **Primary school teachers:** one for every 22 students (1992).

COMMUNICATIONS

101 **newspapers** (1995), 101 **TV sets** (1995) and 102 **radio sets** per 1,000 households (1995). 7.0 **telephones** per 100 inhabitants (1993).

ECONOMY

Per capita GNP: $1,440 (1994). **Annual growth:** -5.60% (1985-94). **Annual inflation:** 9.20% (1984-94). **Consumer price index:** 100 in 1990; 120.4 in 1994. **Currency:** 0,7 dinars = 1$ (1994). **Cereal imports:** 1,596,000 metric tons (1993). **Fertilizer use:** 398 kgs per ha. (1992-93). **Imports:** $3,382 million (1994). **Exports:** $1,424 million (1994). **External debt:** $7,051 million (1994), $1,747 per capita (1994). **Debt service:** 12.4% of exports (1994). **Development aid received:** $245 million (1993); $60 per capita; 4% of GNP.

ENERGY

Consumption: 997 kgs of Oil Equivalent per capita yearly (1994), 97% imported (1994).

Jordan depended on Saudi Arabia financially, and on Baghdad for oil.

[17] Jordan joined in the commercial sanctions against Iraq, but opposed the use of military force to enforce the resolutions of the UN Security Council.

[18] The country lost $570 million as a result of the blockade. It also received 40,000 Kurdish refugees, and some 300,000 Jordanians of Palestinian origin, repatriated from Kuwait, in retaliation for Jordan's support of Iraq.

[19] On June 9 1991, a new Constitution was signed, legalizing political parties, extending political rights to women, and ending press censorship.

[20] That same month, Taher Al Masri replaced Mudar Badram as prime minister. During his short term, Al Marsi favored Jordan's participation in the Middle East Peace Conference and he carried out a policy of rapprochement with the Bush administration.

[21] Fundamentalism met with less repression here than in other countries of the region, and the Muslim Brotherhood was authorized to function as a philanthropic organization, gaining credence by working in social services; setting up hospitals, schools and several centers for Islamic studies. The political arm of the Brotherhood is the Islamic Action Front (IAF), and both groups were opposed to the Arab-Israeli talks. In November 1992, two IAF members of parliament were released from prison, under an amnesty granted to 1,480 prisoners.

[22] In November 1991, Taher Al Masri was deposed after a vote of no confidence from the fundamentalist bloc in Jordan's parliament. He was replaced by Sharif Zeid Bin Shaker.

[23] The king's frail health and his choice of his brother, Hassan Ibm Tall as heir, helped to add momentum to the fundamentalist campaign for a theocratic state.

[24] In May 1993, Abdul Salam Madjali, leader of the Jordanian delegation in the peace talks with Israel, was nominated prime minister. In November, the first parliamentary elections were held. The IAF was the leading minority and Tuyan Faisal, the television announcer, became the first woman elected to Parliament in the country.

[25] The difficult relations between the Islamic groups and the throne blocked the negotiations with Israel until the Israel-PLO agreement was signed in September 1993. It took until July 1994 for Jordan and Israel to establish a schedule, in Washington, to end the 46 year war, including arrangements to repatriate nearly 60,000 Palestinian refugees from Jordan.

[26] King Hussein and the Israeli prime minister, Yitzhak Rabin, signed a bilateral peace agreement on October 26 1994, in the presence of US President Bill Clinton and 5,000 guests in an open air ceremony on the border. Israel handed over 300 sq km of desert to Jordan and the frontier was established, but far more significance was attributed to King Hussein being given custody of the Muslim sacred sites in Jerusalem.

[27] This concession was questioned by the Palestinians, as Israel and the PLO had left the issue open just the year before. However, the Jordanian government and the Palestine National Authority reached a cooperation agreement in January. In November, King Hussein visited Jerusalem for the first time since Israel occupied the city in 1967 to attend the funeral of Yitzhak Rabin.

[28] The Jordanian public was surprised by the speed with which the peace agreement was made and by the meagre economic benefits for their country. In January, King Hussein designated Sharif Zaid ibn Shaker his prime minister, and in May, the new premier banned an opposition conference planned by the Islamic Action Front. In December, the Islamic dissident Leith Shubailat, critic of the peace agreements, was arrested. In February 1996 the prime minister was replaced by Abdul Karim Al-Kabariti, who also headed the foreign and defence ministries.

Kanaky / New Caledonia

Kanaky

Population: 178,000 (1994)
Area: 19,080 SQ.KM
Capital: Noumea

Currency: CFP Franc
Language: French

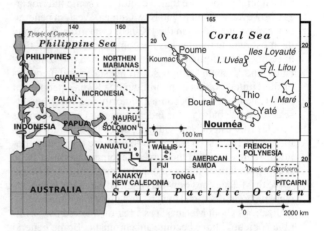

New Caledonia (Kanaky) was populated by Melanesians (Kanaks) three thousand years ago. The islands were named by Captain Cook in 1774, as the tree-covered hills reminded him of the Scottish - Caledonian landscape.

[2] In 1853, the main island was occupied by the French Navy which organized a local guard to suppress frequent indigenous uprisings. Nickel and chrome mining attracted thousands of French settlers. The colonizers pushed out the original inhabitants, and traditional religions, crafts and social organizations were obliterated, and many landless natives were confined to "reservations", and the system of terraced fields were trodden over by cattle. The last armed rebellion, stifled in 1917, only accelerated European land appropriation.

[3] After Algerian independence, in July 1962, colonization increased with the arrival of pieds-noirs, the former French colonists in Algeria. By 1946, New Caledonia had become a French Overseas Territory, but the resulting political autonomy did not favor the Kanaks, now reduced to a minority group, in relation to the *caldoches* (descendants of Europeans who settled a century ago).

[4] The election of President Mitterrand in 1981 rekindled the hopes of the pro-independence parties. The French socialist leader was supported by most Kanaks, who saw independence as a way to end the unfair income distribution on the island. This stood at $7,000 per capita (the highest in the Pacific except for Nauru) but the vast majority of the money was concentrated in the hands of European - mostly French - business people, the *métros*, who enjoyed incredible fiscal benefits, and the *caldoches* who monopolized the most important official positions.

[5] Most of the Kanaks actively supported independence, the bulk of them lived in poverty, with high unemployment rates, and suffered educational discrimination.

[6] In the 1970s, discontent with the economic situation, produced by colonial domination, caused strikes, land invasions, experiments in cooperative work, and a powerful campaign to restore traditional lands to the local groups. These had been totally occupied by settlers and used mostly as cattle pastures. The rescue of *coutume* (cultural traditions) and the Kanak identity became a priority, and the proposed luxury tourist camps organized by the "Club Méditerranée" were firmly rejected.

[7] Kanak claims were supported by other independent Melanesian countries (Fiji, Solomon, Papua New Guinea and above all Vanuatu), and were put forward at the South Pacific Forum in August 1981. A month later, pro-independence leader Pierre Declercq, a Catholic of European origin, was murdered at his home by right-wing extremists, changing the malaise to a fully-blown political crisis.

[8] Another strong reason why France is hesitant to grant Kanaky independence is that it has the world's second largest nickel deposits, and extensive reserves of other minerals including chrome, iron, cobalt, manganese, and polymetallic nodules, discovered recently on the ocean floor within territorial waters.

[9] Furthermore, the islands' strategic position is of great military value. Its ports, facilities and bases house 6,000 troops and a small war fleet (including a nuclear submarine), considered by the military command as a "vital point of support" for the French nuclear-testing site on Mururoa atoll.

[10] In July 1984, the French National Assembly passed special bills concerning the colony's autonomy, though it rejected amendments submitted by pro-independence parties, confirming Kanak fears that the socialist government of France had no intention of granting independence. In November, the main opposition force, the Socialist Kanak National Liberation Front (FLNKS) called for a boycott of local Territorial Assembly elections, which were sure to endorse the French government plan of postponing Kanak independence indefinitely.

[11] In December 1984, local government became fully controlled by the *caldoches* with no indigenous Kanak representation, therefore, the FLNKS unilaterally declared

PROFILE

ENVIRONMENT

The territory consists of the island of New Caledonia (16,700 sq.km), the Loyalty Islands (Ouvea, Lifou, Maré and Walpole), the archipelagos of Chesterfield, Avon, Huon, Belep, and the island of Noumea. The whole group is located in southern Melanesia, between the New Hebrides (Vanuatu) to the East and Australia to the West. Of volcanic origin, the islands are mountainous with coastal reefs. The climate is rainy, tropical, and suitable for agriculture. The vegetation is dense and the subsoil is rich in nickel deposits.

SOCIETY

Peoples: Indigenous New Caledonians are of Melanesian origin (the Kanaka group), 44.8%; there are French and descendants of French (known as *caldoches*), 33.6%; as well as Wallisian, 8.6%, Vietnamese, Indonesian, Chinese and Polynesian minorities. **Religions:** Roughly 60% Catholic, 16% Protestant and around 5% Muslim. **Languages:** French (official), Melanesian and Polynesian languages. **Political Parties:** Rally for Caledonia within the Republic (RPCR); Kanak Liberation Party (PALIKA); Kanak Socialist Liberation Front (FLNKS). **Social Organizations:** The Caledonian Workers Confederation (CTC); the Federation of New Caledonian Miners' Unions (FSMNC); the New Caledonian Federation of Laborers' and Employees' Unions (USOENC); and the Union of Exploited Kanak Workers (USTKE).

THE STATE

Official Name: Territoire d'outre-mer de la nouvelle-calédonie. **Administrative divisions:** Three provinces: Southern, Northern and Islands. **Capital:** Noumea, 74,000 inhab. **Other Cities:** Mont-Doré, 16,370 inhab.; Dumbéa, 10,052 inhab.; Poindimié; Koné. **Government:** High Commissioner named by Paris. 54-member Territorial Assembly. **Armed Forces:** French troops; 3,700 (1993).

HEALTH

Calorie consumption: 570% of required intake (1988-90).

EDUCATION

Primary school teachers: one for every 20 students (1991).

ECONOMY

Currency: CFP francs.

Melanesian Revivalism

The indigenous peoples of much of the Melanesian region are now part of the world economic system and are subject to pressures of Christianization and Westernization. In some areas such forces have operated for more than a century. In some interior areas, however, particularly in New Guinea, Western penetration came much later - in the 1930s or even after. But today the most remote regions have become accessible, and they have been transformed. Papua New Guinea, Solomon Islands, and Vanuatu are now sovereign states and members of the United Nations; the indigenous Kanak peoples of New Caledonia have become a minority in a French overseas territory, battling for independence. The indigenous peoples of western New Guinea , since incorporation into Indonesia, have been subjected to massive disruption, political repression, and the forced accommodation of large settler populations under national transmigration policies.

[2] With Melanesians now serving as diplomats, businessmen, bishops, doctors, lawyers, and professors, many generalizations about Melanesia as a region marked by "primitive societies" have become anachronistic. Some general observations still seem possible, however. One is that classless societies have become class-stratified societies, with politicians, public servants, and entrepreneurs constituting an emerging elite. Moreover, at least in the English-speaking areas, the elites increasingly share a common (Westernized and consumerist) culture and common political and economic interests and ideologies that cut across boundaries not only between cultures and language groups but also between nations.

[3] The countries of modern Melanesia show increasing polarization between metropolitan centres and village hinterlands. Squatter settlements on urban peripheries and movement into towns are increasingly found, and both serve as links between the villages and urban life. The more remote villages have little access to the educational, medical, and economic services of the state. It is in the marginal areas that the traditional culture tends to be the most resilient.

[4] Various modes of capitalist enterprise, as well as dependence on imported goods, have penetrated ever farther into the Melanesian village hinterlands, with some areas attaining a measure of prosperity by Western standards through the production of high-value crops. Roads and airfields now connect once-isolated hinterlands to regional networks. Among the new elite, cultural nationalist ideologies have tended to focus on "custom" and "the Melanesian way"; cultural revivalism has become a prominent theme. Art festivals, cultural centres, and ideologies of kastom have cast in a more positive light the traditional cultural elements, such as ceremonial exchange, dance and music, and oral traditions, that had long been suppressed by the more conservative and evangelistic forms of Christianity. The emphasis on traditional cultures as sources of identity has been expressed in the perpetuation or revival of old genres of exchange. In Papua New Guinea , for example, the kula exchange system of arm shells and necklaces continues in the Massim, carried on through the medium of air travel and among politicians, professionals, and public servants as well as by villagers in canoes. Members of the new elite conspicuously pay bridewealth in shell valuables.

New Caledonia independent, proclaiming a Kanak state. The resulting election boycott involved 80 per cent of the Kanak population, forcing the government to call off the election, and prepare for negotiations.

[12] On December 5 1984 - immediately after the French government announced its willingness to talk with the FLNKS - ten Kanak political activists were brutally murdered by right-wing caldoches. This incident led to widespread violence which continued throughout the following year, leaving a toll, among Kanak activists, of almost 40 dead and thousands wounded. Security forces sent in by the French government, far from controlling the situation, were in fact linked to several of the crimes and violent incidents carried out against the independence movement.

[13] In December 1986, the United Nations General Assembly proclaimed the right of the Kanak people to self-determination and independence, proposing that the FLNKS be recognized as their legitimate representative.

[14] One year later a referendum was held to determine whether or not ties with France should be maintained. Voting was open to all

residents of the island, even Europeans and immigrants who arrived as recently as three years and, for this reason, the FLNKS boycotted the referendum. According to the opposition and the Australian and New Zealand/Aotearoan governments, the high abstention rate of around 41.5 per cent invalidated any claim to legitimacy for continued colonial domination.

[15] In May 1988, the FLNKS captured 22 French gendarmes and held them hostage on the island of Ouvea. Their objective was to negotiate their freedom in exchange for a post-electoral agreement with the French government.

[16] When all attempts at negotiation failed for the Kanaks, the French attacked the island of Ouvéa, killing 19 people, most of whom were apparently executed rather than killed in combat.

[17] In June 1988, the FLNKS leader Jean-Marie Tjibaou and the main caldoche political leader, Jaques Lafleur, signed the Matignon accords in Paris, supported by the French prime minister, Michel Rocard. The territory was divided into three regions, two of them with a majority of Kanak voters. One of the objectives of this administrative division was to create a Melanesian

political and financial "elite", separating the army from power on behalf of the pro-independence groups in the majority of the territory.

[18] Other dispensations of the agreements provided for greater financial support from Paris over the next ten years, a first referendum in 1988 to ratify the accords and another in 1998 to decide on independence. In May 1989, Tjibaou and another leading independence leader who supported the Matignon accords were assassinated in Ouvea.

[19] In 1991, the trade balance was affected by the falling international nickel and fish prices. The effects of the Matignon accords began to be felt. In the two provinces controlled by the pro-indpendence groups a generation of new leaders appeared, but the situation for most of the Melanesian population worsened. The imbalance of income between Kanaks became more pronounced and, moreover, a greater access to consumption distanced many Melanesians from their community structures and traditions.

[20] In the caldoche sector, mainly covering the capital Nouméa, the social inequalities also became more serious. Not only because of the arrival of Melanesian farmers

who built shanty towns on the outskirts of the city, but also because of the impoverishment of some caldoches. Against a background of increasing social tension, street protests like those of March 1992 multiplied.

[21] The political repercussions of these new social contradictions were seen in the 1995 provincial elections. The Balika, one of the FLNKS members, registered separately, criticizing the leadership of the Front representatives in the two provinces controlled by the pro-independence groups. Both political sectors achieved similar results.

[22] On the union front, the increasing inequalities led many Caldoches to join the Union Exploited Kanak Workers (USTKE). Furthermore, this union, the main opponent of the business owners groups, voiced its criticisms of the "technocrats" leading the pro-independence provinces.

[23] The political future continued to look uncertain, partly because an increasing number of pro-independence leaders were ever more keen on immediate independence in 1998, which could mean a fall in living standards.

Kazakhstan

Kazajstan

Population: 16,811,000
(1994)
Area: 2,717,300 SQ.KM
Capital: Alma Ata

Currency: Tengue
Language: Kazakh

In the Bronze Age (about 2000 BC), the territory of Kazakhstan was inhabited by tribes who lived by farming and raising livestock. Around 500 BC, an alliance was formed among the Saka peoples and in the 3rd century BC, the Usune and Kangli tribes - who lived near the Uighur, Chechen and Alan - subdued the other tribes in the area. The region was subsequently occupied by Attila's Huns, until they were expelled by the Turks.

2   In the mid-4th century, the Turkish Kaganate (*khanate* or kingdom) was formed and later divided into Eastern and Western parts. In the 8th century two states emerged: a Turkish and a Karluk *kaganate*. The Turkish conquerors built mosques and tried to impose Islam upon the local population. Over the following 300 years scholarship flourished in the area.

3   Between the 9th and 12th centuries, the region was occupied by the Oghuz, Kimak, Kipchak and Karajanid tribes. The Kipchaks never achieved political unity and remained outside the realm of Islamic influence, which was concentrated in the cities along the Caspian Sea. Until the 13th century, successive waves of Seleucid, Kidan and Tatar invasions swept across the great steppes. Kipchak chiefs and Muscovite princes joined together to resist foreign domination, but did not achieve independence until the fall of the Mongols.

4   Most of these peoples were nomads but gradually, settled groups of farmers and artisans were organized, and cities like Otrar, Suyab, Balasagun, Yanguikent, Sauran and Kulan arose. The Silk Road, uniting Byzantium, Iran and China, passed through Kazakhstan. Trade relations developed between the no-

mads of the steppes and the inhabitants of the oases, which extended as far as Western Europe, Asia Minor and the Far East.

5   By the late 15th century, the khanate of Kazakh had been formed, disintegrating into three loosely allied but like-minded *yuzos* (hordes). By the 16th century, an ethnic identity had been forged among the Kazakhs. The khans of the Kazakh *yuzos* passed on their power to their heirs, who thought of themselves as descendants of Juchi, the eldest son of Genghis Khan. Below them were the sultans, with administrative and judicial power, who governed through the *biy* and the local chieftains.

6   In the 17th century, the *khanate* of Dzhungar carried out successive raids in the Kazakh region, sometimes looting and other times remaining and occupying the area. Russian colonial expansion from the north began in the 18th century. The Russians built a line of forts and then began working their way southward, creating a line of defence against the Dzhungars. The two smaller hordes or *yuzos* fell under Russian protection eliminating their autonomy in the 1820s because of the frequent rebellions. With the defeat of the Great Horde, annexation of Kazakhstan to the Empire was completed.

7   Russia installed its government institutions, collected taxes, established areas closed to the Kazakhs and built new cities, declaring the entire territory property of the State. As of 1868, there were six provinces under the control of governors general; the sultans and the *biy* became mere state officials. The conquest of Kazakhstan was a long process of wars against local tribes. The Cossack regiments were the vanguard of the Russian army and they overcame the khans of Khiva, Boukhara and Kokand, one by one, defeating the last of these in 1880.

8   The Kazakhs were registered in the censuses as citizens of the State of Russia and were incorporated into the Russian army's foreign expeditions. Kazakhstan became a place to which Russian deportees were sent, including the Decembrists (aristocrats who had conspired against Czar Nikolai I in 1825), Polish and Ukrainian revolutionaries (including the Ukrainian hero, Taras Shevchenko) and the members of the St Petersburg "Pe-

trashevski circle", which included the writer Feodor Dostoyevsky.

9   In the late 19th century and early 20th century, Russia built huge railroads (the Trans-Caspian, the Trans-Aralian and the Trans-Siberian), which crossed the region, uniting it with distant urban centers and facilitating the exploitation of Kazakhstan's fabulous mineral wealth. A third of Russia's coal reserves; half its copper, lead and zinc reserves; strategic metals like tungsten and molybdenum; iron, and oil were found in Kazakhstan. Agriculture remained stagnant and attempts to resettle farmers, after the abolition of feudal serfdom in Russia, met with limited success, although there was some development of cotton, wool and traditional Kazakh leather production.

10   At the beginning of this century, a small nationalist movement emerged in Kazakhstan, and after the Russian Revolution of 1905 the Kazakhs had their own representatives to the first and second Duma (parliament) convened by the czar. In 1916, when the Czarist regime ordered the mobilization of all men between the age of 19 and 43 for auxiliary military service, the Kazakhs rebelled, led by Abdulghaffar and Amangeldy Imanov. The revolt was brutally crushed, but in November 1917, after the triumph of the Soviet revolution in Petrograd, the Kazakh nationalists demanded total autonomy for their country. In the early decades of this century, Kazakhstan received massive waves of Ukrainian, Belorusian, German, Bulgarian, Polish, Jewish and Tatar immigrants. A nationalist

government was installed in Alma Ata in 1918, but the country soon became a battleground.

11   Fighting between the Red Army and the White Russians - the latter defending the government that had been overthrown - lasted until 1920, when the counter-revolution was defeated. The Autonomous Soviet Socialist Republic (ASSR) of Kirghiz was formed, as a part of the Russian Federation; it later became known as the ASSR of Kazakhstan. In 1925, the revision of the borders in Soviet central Asia was completed, and all Kazakh lands were unified. In 1936, Kazakhstan became one of the 15 republics of the USSR and the following year, the local Communist Party was founded.

12   In addition to developing its industrial potential, the Soviet regime increased the amount of land under cultivation. Previously considered not very fertile (in 1913, only 4.2 million hectares were under cultivation), Kazakhstan increased the number of hectares of tilled land to 35.3 million, 15 per cent of all agricultural land in the USSR. Production included wheat, tobacco, mustard, fruit and cattle. Bringing virgin territory under cultivation was an achievement associated with Leonid Brezhnev, during his period as head of the Communist Party of Kazakhstan. Brezhnev replaced Nikita Khrushchev in 1964 as head of the Soviet Party until his death in 1982. Although he proclaimed an era of "developed socialism", the country's economic and political problems

LAND USE

- CROPLAND 13%
- PASTURE 70%
- OTHER 17%

Foreign Trade

MILLIONS $ 1994

IMPORTS
4,205

EXPORTS
3,285

Maternal Mortality

1989-95

PER 100.000
LIVE BIRTHS
53

DEMOGRAPHY

Urban: 59% (1995). **Annual growth:** 1% (1991-99). **Estimate for year 2000:** 18,000,000. **Children per woman:** 2.7 (1992).

EDUCATION

School enrolment: Primary (1993): 86% fem., 86% male. Secondary (1993): 91% fem., 89% male. University: 42% (1993).

COMMUNICATIONS

9.1 **telephones** per 100 inhabitants (1993).

ECONOMY

Per capita GNP: $1,160 (1994). **Annual growth:** -6.50% (1985-94). **Annual inflation:** 150.20% (1984-94). **Consumer price index:** 100 in 1990; 2,265.0 in 1993. **Currency:** tengue. **Cereal imports:** 100,000 metric tons (1993). **Fertilizer use:** 134 kgs per ha. (1992-93). **Imports:** $4,205 million (1994). **Exports:** $3,285 million (1994). **External debt:** $2,704 million (1994), $161 per capita (1994). **Debt service:** 1.9% of exports (1994).

ENERGY

Consumption: 3,710 kgs of Oil Equivalent per capita yearly (1994), -16% imported (1994).

PROFILE

ENVIRONMENT

Kazakhstan is bordered to the southeast by China; to the south by Kyrgyzstan; and to the north by the Russian Federation. In the western part of the country lie the Caspian and Turan plains; in the center, the Kazakh plateau; and in the eastern and southeastern regions, the Altai, Tarbagatay, Dzhungarian Alatau and Tien Shan mountains. It has a continental climate, with average January temperatures of -18° in the north, and -3° in the south. In July, the temperature varies from 19° in the north, to 28° in the south. Important rivers include the Ural, Irtysh, Syr Dar'ya, Chu and Ili. There is also Lake Balkhash and the Caspian and Aral Seas. The vegetation is characteristic of the steppes, but vast areas have come under cultivation (wheat, tobacco, etc) or are used for cattle-raising. The region's abundant mineral wealth includes coal, copper, semi-precious stones and gold.

SOCIETY

Peoples: Kazakhs, 39.7%, Russians, 37.8%, Ukrainians, 5.4%, various minorities descended from early 20th century immigrants. **Religions:** Muslim and Christian Orthodox. **Languages:** Kazakh (official), Russian, German, Ugric, Korean, Tatar. **Political Parties:** Union of National Unity, President Nazarbayev's party, moderate; Socialist Party (which replaced the Communist Party); People's Congress; Republican Party, a nationalist party. **Social Organizations:** Independent labor unions are in the process of being formed. Birlik Movement, Zheltokso and Semipalatinsk-Nevada, an anti-nuclear movement.

THE STATE

Official Name: Republika Kazajstán. **Administrative divisions:** 19 regions and 2 cities. **Capital:** Alma Ata, 1,147,100 inhab. (1990). **Other cities:** Karaganda, 633,000 inhab., Semipalatinsk, 320,000 inhab., Pavlodar, 330,000 inhab., Ust'-Kamenogorsk, 321,000 inhab., Tselinograd, 248,000 inhab., Kokchetav, 127,000 inhab. **Government:** Nursultan Nazarbayev, President since December 1, 1991. Unicameral legislative power, with 360 members. **National Holiday:** December 16, Independence (1991). **Armed Forces:** 2 regiments of Russian Air Defence.

actually worsened during his administration.

[13] Until 1985, the person who wielded the power in Kazakhstan was Dinmujamed Kunaev, a member of the Politburo of the Soviet Communist Party Central Committee. In 1989, the forced resignation of Kunaev triggered student disturbances; the army resorted to violence in dealing with the demonstrators. That year, in mid-summer, a water shortage in the oil-producing city of Novy-Ouzen, on the Caspian Sea, led groups of city residents to launch an attack on the water distribution company. The official version attributed the incident to a group of trouble-makers but other sources claimed that the

disturbances were more serious, and that they had religious and nationalistic connotations.

[14] After the transformations set in motion in the USSR by Mikhail Gorbachev, the republic of Kazakhstan declared independence. During this period two social movements arose - Birlik and Zheltoksan - as well as the anti-nuclear movement Semipalatinsk-Nevada. The main test-sites for Soviet nuclear weapons were situated in Kazakhstan, as is the Baykonur cosmodrome for launching Soviet space vehicles. After the failed coup d'etat against the Soviet president in August 1991, Nursultan Nazarbayev resigned as head of the Soviet Communist Party, of which he

had been a member, in his capacity as president of a Soviet republic.

[15] In September, Kazakhstan presented a seven-point plan for the creation of a new union treaty, which was approved by Gorbachev and ten republics. On December 1 1991, Nazarbayev was elected the first president of an independent Kazakhstan. The Communist Party became the Socialist Party of Kazakhstan. On December 21, in Alma Ata, 11 republics signed an agreement which formally dissolved the USSR and created the new Commonwealth of Independent States, whose members will apply separately for admission to the UN. President Nazarbayev's foreign policy envisions a privileged alliance with Russia and the Islamic republics of the region, as well as closer links with the West.

[16] In 1992, Nazarbayev let the Russian president Boris Yeltsin know that he would not allow the nuclear missiles - installed when Kazakhstan was still part of the USSR - to remain under the exclusive control of Moscow. In 1993, Alma Ata promised to dismantle these missiles in return for financial aid from the United States.

[17] This same year several political entities were created, like the Socialist Party, the Peoples Congress Party and the ruling Union of National Unity. Nazarbayev also undertook to bring in swiftly a series of privatizations and to stimulate foreign investment.

[18] In March 1994 the first multiparty legislative elections were held, and they were won by Nazarbayev's party. Shortly after the victory, a series of scandals over alleged corruption forced the prime

minister Sergey Tereschenko to resign. He was replaced by Akezhan Kazhegeldin. The new head of government immediately announced he would accelerate the economic liberalization process.

[19] In April, Alma Ata launched a vast privatization plan, including 3,500 state enterprises, that is, 70 per cent of the public companies. The rapid introduction of the market economy and the natural riches of the country attracted a large number of foreign investors in 1995, but the economic liberalization also led to a fall in the standard of living for many Kazakhs.

[20] The Constitutional Court annulled the previous year's elections and the president said he would rule by decree until new elections could be held. As a consequence of popular discontent, and faced with the risk of not being re-elected in a contest between several candidates, Nazarbayev made the most of the political crisis to propose his term in office be extended to the year 2000 by referendum. The official results of this poll, which took place in April, gave the President almost unanimous support.

[21] In April 1996, amidst increasing accusations of authoritarianism and human rights violations against Nazarbayev and his government, the interior minister accepted that the Kazakh prisons were overcrowded, that the medical care was deficient and that the funds received for the prison budget in 1995 were less than half of what was needed. In June, the State announced that 20,000 prisoners accused of non-violent crimes had been released.

Kenya

Population: 26,017,000 (1994)
Area: 580,370 SQ.KM
Capital: Nairobi
Currency: Shilling
Language: English and Swahili

Kenya

The prosperous city of Malindi was founded in the 10th century on the coast of what is now Kenya. It was the center of a rich African Arab culture (see Tanzania: The Zandj Culture), which was destroyed by a Portuguese armed occupation in the 16th century. When they were forced to withdraw in 1698, the Portuguese left behind only a few abandoned forts and economic ruin, ideal conditions for the slave and ivory trades managed by Shirazi merchants from Zanzibar.

[2] The inland peoples were mainly Bantu with Nilotic and Somalian groups also present. They did not develop material cultures comparable to those of the coast, nor did they evolve into organized states. It was only in the 19th century that the Masai, Nilotic shepherds, succeeded in establishing a certain degree of authority over other groups in the region. A few decades later when a bovine plague annihilated almost all of their herds and deprived them of their economic mainstay, their power base collapsed.

[3] At the end of the 19th century, the Berlin Conference and subsequent German-British agreements delimited spheres of European influence in East Africa. Zanzibar, Kenya and Uganda were assigned to the English who had already settled in Uganda, and who decided to build a railway line from this colony to the coast.

[4] White settlers occupied the lands which "became available" along the railway line after half the local population died in an epidemic of smallpox brought in with Indian laborers. As late as 1948, 4,200 sq km of the approximately 5,000 sq km of fertile land were held between 5,000 European planters, while one million Kikuyu occupied fewer than 1,000 sq km.

There was no indemnity or compensation for this theft.

[5] In 1944, the Kenya African Union (KAU) was created to defend Kikuyu interests. Under the leadership of Jomo Kenyatta, KAU organized strikes, farmers' rallies and mass demonstrations.

[6] At almost the same time the Mau-Mau, a political and religious group organized as a secret society, launched its first offensive. The Mau-Mau program was both political and cultural. They demanded self-government, restitution of lands, and wage parity, while rejecting Christianity and other European influences in favor of traditional customs and beliefs.

[7] In 1952, in response to increased Mau-Mau attacks on settlers' lives and property, the colonial administration declared a state of emergency, arrested nationalist leaders (Kenyatta among them), dissolved political parties and imprisoned thousands of Kikuyu in concentration camps.

[8] After years of bitter repression and indiscriminate killing by colonial governments, KAU was legalized in 1960 as the Kenya African National Union (KANU).

[9] KANU found electoral support among leading urban and ethnic communities, and, with British encouragement, was able to defeat tribal-based political movements. In 1961, Kenya was freed and in 1962 the Legislative Council was elected. After the May 1963 elections Kenya was granted autonomy. It achieved independence within the Commonwealth on December 12 1963. Kenyatta became Prime Minister, and when his country achieved full independence on December 12 1964, he was elected first president.

[10] However, Jomo Kenyatta, a *nom de guerre* meaning "Kenya's

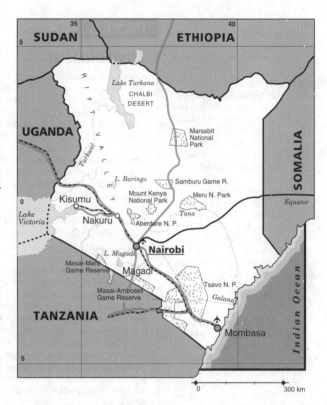

flaming spear", soon forgot his commitment to the nation as a whole, and began to favor his own ethnic group, the Kikuyu. His government encouraged private enterprise and transnational concerns. Farmers who had won back their lands lost them again under the burden of debt.

[11] A black bourgeoisie, largely Kenyatta's family and friends, took over where the former colonists had left off. KANU, once a model for other African parties in the struggle against foreign domination, succumbed to neocolonialism. It even went so far as to admit British military forces to the port of Mombasa

and allowed its own military installations to be used for the notorious Israeli raid on Entebbe, Uganda.

[12] This led to a breach of relations between Kenya and Uganda, and tension with Tanzania grew over differences in economic policies. The three countries were unable to consolidate 1967 plans for an East African Economic Community, and this ambitious integration project was finally dropped in 1977.

[13] Kenyatta died at the age of 85 in September 1978. He was succeeded by his vice president, Daniel Arap Moi, a member of the smaller Kalenjin ethnic group.

[14] This might have soothed ethnic conflict in Kenya, but penetration by transnationals produced structural imbalances, worsening an already difficult economic situation and aggravating social tensions. Even before Kenyatta's death there was growing distrust between the urban consumer bourgeoisie and the larger rural population. The country people's whole way of life was being steadily degraded, as the market for cash crops undermined traditional communal self-sufficiency.

[15] Arap Moi was interim president from September 1978 until the general election of November. KANU

PUBLIC EXPENDITURES

% of GNP — 1992

MILITARY 2.8
EDUCATION 6.8
HEALTH 2.7

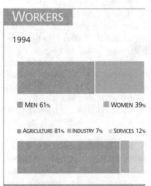

WORKERS

1994

MEN 61%   WOMEN 39%

AGRICULTURE 81%   INDUSTRY 7%   SERVICES 12%

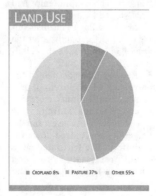

LAND USE

CROPLAND 8%   PASTURE 37%   OTHER 55%

was the only party authorized to run candidates, and Moi was confirmed as president.

[16] In early 1979, President Moi declared an amnesty for political prisoners and launched a campaign against corruption. The first measures of the new administration revealed a technocratic approach which deflated hopes of radical change. One of the outstanding technocrats was Charles Njonjo, appointed minister of home affairs in June 1980.

[17] Njonjo played a key role in dissolving the ethnic-based organizations that arose in the early 1970s. The largest of these, GEMA (the Gikuyo, Embu and Meru Association), was led by a millionaire businessman, becoming a powerful pressure group for tribal leaders who were growing rich from business with British and US companies.

[18] GEMA opposed Moi's designation, and after the death of Jomo Kenyatta had been indirectly involved in a conspiracy to assassinate all top government officials.

[19] Droughts and crop changes led to a drop in the production of corn and other basic consumer goods. Credits supplied by multinational companies encouraged Kenyan farmers to grow flowers and to plant sugarcane, coffee and tea for export under the control of large North American and British firms. As a result, the government had to import huge quantities of corn and wheat from the US and South Africa.

[20] These difficulties led President Moi to seek reconciliation with old political rivals who had been excluded from public life. It was a maneuver devised to neutralize potential opposition at a moment of great instability. Former vice-president Oginga Odinga, a veteran nationalist politician who had broken with Kenyatta in 1971 in disagreement with KANU's conciliatory policy, was one of the major beneficiaries of this relaxation of the regime. Odinga had lost his political rights after founding a dissident party, the Kenya Popular Union (KAPU). After eleven years of ostracism, he returned to parliament at the end of 1981.

[21] Despite all this reconciliation, the crisis erupted into violence in August 1982, in a military conspiracy that unleashed widespread rioting and looting of

PROFILE

ENVIRONMENT

Kenya is located on the east coast of central Africa. There are four main regions, from east to west: the coastal plains with regular rainfall and tropical vegetation; a sparsely populated inland strip with little rainfall which extends towards the north and northwest; a mountainous zone linked to the eastern end of the Rift valley, with a climate tempered by altitude, and volcanic soil fit for agriculture (most of the population and the main economic activities are concentrated here); and the west which is covered by an arid plateau, part of which benefits from the moderating influence of Lake Victoria. The principal environmental problems are soil exhaustion, erosion and desertification; deforestation; pollution of drinking water supplies, especially near such large cities as Nairobi and Mombasa.

SOCIETY

Peoples: Kenyans are descended from the main African ethnic groups: Bantu, Nilo-Hamitic, Sudanese and Cushitic. Numerically and culturally, the most significant groups are the Kikuyu, the Luhya and the Luo. Others include the Baluya, Kamba, Meru, Kissi and Embu. There are Indian and Arab minorities. **Religions:** 73% of the population are Christian. 6% are Muslim and 20% practise traditional religions. **Languages:** English and Swahili are the official ones. The latter is the national language. Kikuyu and other local languages are also spoken. **Political Parties:** Kenya African National Union (KANU); Forum for the Restoration of Democracy (FORD), split into two branches in 1992 (FORD-Asili and FORD-Kenya); Democratic Party (DP), Kenya Social Congress (KSC); Kenya National Congress (KNC). **Social Organizations:** Central Organization of Trade Unions (COTU).

THE STATE

Official Name: Jamhuri ya Kenya.
Capital: Nairobi, 1,500,000 inhab. (1989). **Other cities:** Mombasa, 465,000 inhab. (1989).
Government: Daniel Arap Moi, President; succeeded Jomo Kenyatta upon his death in September 1978.
Legislature: National Assembly (single-chamber), with 202 members, elected every five years. **National Holiday:** December 12, Independence Day (1963). **Armed Forces:** 24,200 troops. **Paramilitaries:** 5,000.

shops and public buildings in Nairobi. The coup attempt had been staged by members of the air force, leading to the disbanding of all airforce units after the army had suppressed the rebellion.

[22] Repression also affected the university, where dozens of professors and students were detained. Oginga Odinga was placed under house arrest and the university was closed down indefinitely. The frustrated coup d'état precipitated profound changes in Kenyan politics and

sowed distrust among the different political elements in KANU.

[23] In May 1983, President Moi denounced a conspiracy involving minister Charles Njonjo, supported by Israel and South Africa. In the midst of the confusion, President Moi decided to call general elections, in which his followers won a landslide victory, while Njonjo and his supporters were overwhelmingly defeated.

[24] Moi's administration reopened the border with Tanzania in November 1983, after a summit meeting in Arusha with Tan-

zania's Julius Nyerere and Uganda's Milton Obote. The meeting was a point of departure for gradually renewing economic co-operation between Kenya, Tanzania and Uganda after the failure of the the East African Economic Community in 1977.

[25] President Arap Moi had extended the executive's jurisdiction to the detriment of the parliament. He made it compulsory for civil servants to join KANU, and replaced secret voting in internal party elections with a public vote making intimidation easier. The moderate women's opposition group, Mandeleo Ya Wanawake, was taken over by the government in 1986.

[26] Reports backed by Amnesty International and other groups implicated the government in torture and the murder of opposition members, especially the "subversive" Mwakenya group.

[27] During October and November 1987, Muslim demonstrations in Mombasa served to justify a further wave of repression in which Nairobi University was once again closed. In December, a series of frontier skirmishes with Uganda, plus the expulsion of Libyan diplomats accused of encouraging civil disorder, seemed to herald new disturbances.

[28] Despite harsh "restructuring" policies imposed by the IMF and World Bank, elections in March 1988 consolidated the position of Moi's followers within both KANU and the government. By August, Moi had completed his authoritarian reorganization by placing the judiciary directly under his command. He prolonged the period for which detainees may be held without notifying a judge from 24 hours to 14 days.

[29] In the years that followed, cases of corruption and human rights violations became more widespread. In April 1989 vice-president Josephat Karanja lost the confidence of parliament, accused of furthering his interests and those of his own tribe. After he resigned, Karanja was expelled from KANU.

[30] Robert Ouko, the foreign affairs minister, and a harsh critic of corruption in the cabinet, was murdered in February, 1990. An inquest carried out by Scotland Yard disclosed that the culprits

were close advisors of the president. This triggered a new wave of popular anti-government protests.

[31] Through the active preaching of the Catholic and Protestant churches, democratization demands spread throughout the country. The suspicious circumstances surrounding the death of a Protestant minister involved in this democratization movement renewed street protests. The registration of new political parties was also prohibited.

[32] Kenya's diplomatic isolation became more marked during the 1980s, but it decreased after the Persian Gulf war. Kenya's strict alignment with the interests of the US-led coalition enabled the country to receive economic aid from Britain, and military support from Washington.

[33] The permanent deterioration in the human rights situation, however, led to the severing of diplomatic relations with Norway in 1991, and relations with Sudan, Ethiopia, and Uganda also became tense. The governments of Kenya and Sudan accused each other of protecting rebel groups, hostile to Nairobi and Khartoum, operating in the neighboring country. Political differences between Uganda and Kenya have caused permanent conflict since 1986.

[34] The economy has been incapable of achieving growth rates like those of the past decade. According to official estimates, inflation in 1990 amounted to 15 per cent, but all independent sources indicate it was very near to 30 per cent. As a result of the Gulf war oil import expenditure was a great deal higher than expected, and this had a negative effect on the trade balance.

[35] Tourism has dropped sharply in the last few years, and the servicing of the foreign debt accounted for 30 per cent of 1990 exports. Like other countries in the region, Kenya has implemented cuts in public expenditure, which include privatizing state-run companies and not filling vacancies in the public sector. Kenya's economy is heavily dependent on foreign aid.

[36] The KANU called the party Council to discuss the introduction of democratic reforms, including authorization for several national opposition parties. Lobbying groups such as the Forum for the Restoration of Democracy (FORD), led by Oginga Odin-

STATISTICS

DEMOGRAPHY

Urban: 26% (1995). **Annual growth:** 2.8% (1992-2000). **Estimate for year 2000:** 31,000,000. **Children per woman:** 5.1 (1992).

HEALTH

One **physician** for every 20,000 inhab. (1988-91). **Under-five mortality:** 90 per 1,000 (1995). **Calorie consumption:** 89% of required intake (1995). **Safe water:** 53% of the population has access (1990-95).

EDUCATION

Illiteracy: 22% (1995). **School enrolment:** Primary (1993): 91% fem., 91% male. Secondary (1993): 23% fem., 28% male. **Primary school teachers:** one for every 31 students (1992).

COMMUNICATIONS

88 **newspapers** (1995), 84 **TV sets** (1995) and 90 **radio sets** per 1,000 households (1995). 0.8 **telephones** per 100 inhabitants (1993). **Books:** 87 new titles per 1,000,000 inhabitants in 1995.

ECONOMY

Per capita GNP: $250 (1994). **Annual growth:** 0.00% (1985-94). **Annual inflation:** 11.70% (1984-94). **Consumer price index:** 100 in 1990; 290.6 in 1994. **Currency:** 45 shillings = 1$ (1994). **Cereal imports:** 569,000 metric tons (1993). **Fertilizer use:** 410 kgs per ha. (1992-93). **Imports:** $2,156 million (1994). **Exports:** $1,609 million (1994). **External debt:** $7,273 million (1994), $280 per capita (1994). **Debt service:** 33.6% of exports (1994). **Development aid received:** $894 million (1993); $35 per capita; 16% of GNP.

ENERGY

Consumption: 107 kgs of Oil Equivalent per capita yearly (1994), 82% imported (1994).

ga, and the Moral Alliance for Peace (MAP), became fully-fledged parties.

[37] In order to keep the situation under control, the government continued to imprison opposition leaders. Early in 1992, lawyer James Orengo and environmentalist Wangari Maathai were arrested and accused of "spreading malicious rumours" that President Moi planned to interrupt the democratization process started in 1991.

[38] In February, 1992, the Democratic Party (PD) was created. It was a new opposition group which proposed the creation of a multiparty democratic system. Meanwhile, women's groups have demanded greater

participation in politics; they constitute 53 per cent of the electoral roll and 80 per cent of the work force in agriculture, the country's main production sector. That same month, in Nairobi, a march organized by FORD rallied over 100,000 people demanding the end of repression and press censorship, and a definite date for elections. This was the first authorized anti-government march in the country's 22 years of independent life.

[39] Several government ministers had resigned in January to form new political parties, and in March, a general strike called by wives of political prisoners was crushed by the government. Further violence occurred in April, when clashes

broke out between various tribes in the western part of the country. The government banned all political meetings and began censoring the press. Nevertheless, the general strike called by the opposition constituted a setback for President Arap Moi.

[40] General elections were held in December. President Arap Moi was re-elected amidst accusations of fraud. His party obtained 95 seats with 36 per cent of the vote, while the opposition had only 88 seats with 60 per cent of the returns.

[41] In February 1993, the IMF considered Kenya's plan to privatize and liberalize foreign trade was insufficient. In April, Sheik Khaled Balala from the opposition's Islamic Party was arrested on charges for the murder of three policemen. In November, international finance organizations lifted the boycott imposed in February, since the government promised to fight against corruption and remove several restrictions on free trade.

[42] Liberalization continued in 1994: Nairobi eliminated exchange controls to attract private investors from Kenya and abroad. An extended drought hit several provinces especially in the eastern side of the country and in the Rift Valley which led the government to grant urgent assistance to the affected regions. The forced relocation of some 2,000 Kikuyu farmers from the Rift Valley caused serious riots and confrontation with police.

[43] In December, financial organisations and Kenya's creditor countries expressed their satisfaction with the country's economic policy and the introduction of a multi-party system by Nairobi. In 1995, the government announced the partial privatization of the national airline and other major state companies. However, despite some isolated progress in the field, organizations such as Amnesty International continued to accuse Arap Moi of human rights violations.

Kiribati

Population: 78,000 (1994)
Area: 728 SQ.KM
Capital: Bairiki on the island of Tarawa

Currency: Australian Dollar
Language: English and Gilbertan

Kiribati

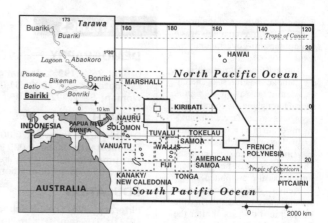

The islands that make up the Republic of Kiribati (formerly known as Gilbert Islands) are inhabited by Micronesian people. In 1764 these islands were visited by and named after British explorer Gilbert.

[2] Missionaries arrived in 1857, and three years later, trade in palm oil and copra began. In 1892 the island became a British protectorate. In 1915 the islands were annexed to the neighboring Ellice archipelago (now Tuvalu), to form the colony of the Gilbert and Ellice Islands.

[3] In 1916 Banaba Island became part of the colony. At that time, there were large deposits of guano on the island which were exploited by the British Phosphate Commission from 1920, and exported to Australia and New Zealand. The Banabans were evacuated during World War II and resettled on the island of Rabi, 2,600 km from Fiji. They were unable to return to their island because the open-cast mining of guano had made the island uninhabitable.

[4] After discussing different alternatives in 1981 the Banabans obtained an indemnity of £19 million from the British government. In 1957 the British government, as part of its nuclear armament program, detonated three hydrogen bombs near Christmas Island.

[5] The Polynesian population of the Ellice Islands obtained administrative separation in 1975, arguing that the ethnic, historical and cultural differences between them and the Melanesian majority of the Gilbert Islands made secession necessary. Under the name of the Territory of Tuvalu, these islands gained independence, in 1978.

[6] The inhabitants of the Gilbert Islands proclaimed independence on July 12 1979 adopting the name: Republic of Kiribati (the equivalent of "Gilbert" in Gilbertan).

[7] As the soil is not suitable for large-scale cultivation, copra and fish are the main exports. Copra production is in the hands of small landowners, while the exportation of the product is handled exclusively by the national trading company. Problems facing copra production include abrupt variations in price. Fishing is carried out primarily under agreements with Japanese, US, Korean and Taiwanese fishing fleets.

[8] Kiribati also hopes to exploit deep-sea mineral deposits. The manganese discovered is considered the highest-grade deposit of its kind in the world.

[9] In 1986 Kiribati began negotiations with the IMF and was recognized by the UN as one of the world's poorest countries, a fact which gives it access to certain credit and trade advantages.

[10] A 1989 UN report on global warming and the possible rise in sea level said that Kiribati could disappear under the rising sea.

[11] In the 1991 elections, Teatao Teannaki won with 46 per cent of the vote, producing the first change in president since independence.

[12] In a regional meeting attended by representatives of the Asian-Pacific nations in 1993, the Bairiki delegate asked for detailed studies of global warming. Australia promised aid to create a natural disasters prevention centre in Kiribati.

[13] In May 1994, the government of President Teannaki - accused of poor administration of public funds - lost a vote of confidence in Parliament and was forced to stand down. In July, the opposition coalition Maneaba Te Mauri won a parliamentary majority and in the September presidential elections, Teburoro Tito took over as national leader.

[14] The new president abandoned the privatization of State enterprises started by his predecessors, and in 1995 he announced his intention to increase the prices paid to the copra producers and the salaries of civil servants. The Tito government also expressed concern at the passage of Japanese plutonium-loaded vessels through the region and criticized French nuclear tests in the Pacific.

PROFILE

ENVIRONMENT

33 islands and coral atolls scattered over 5 thousand million sq km in Micronesia. The climate is tropical and rainy, tempered by the effect of sea winds. The country's large phosphate deposits are now virtually exhausted. Fishing and underwater mineral deposits make up Kiribati's major economic potential.

SOCIETY

Peoples: The population is Micronesian, with a small British minority. **Religions:** Christianity. **Languages:** English and Gilbertan. **Political Parties:** National Progressive Party and the Maneaba Party. **Social Organizations:** Kiribati General Labor Confederation.

THE STATE

Official Name: Republic of Kiribati. **Capital:** Bairiki on the island of Tarawa, 25,154 inhab. (1990). **Other cities:** Abaiang, 5,314 inhab.; Tarawa, 28,802 inhab.; Tabiteuea, 4,600 inhab. **Government:** Teburoro Tito, President. Parliament is made up of 39 members, elected by direct popular vote, plus one representative of Banaba Island. **National Holiday:** July 12, Independence Day (1979).

DEMOGRAPHY

Annual growth: 2% (1991-99).

HEALTH

Calorie consumption: 276% of required intake (1988-90).

EDUCATION

Primary school teachers: one for every 29 students (1992).

COMMUNICATIONS

2.3 **telephones** per 100 inhabitants (1993).

ECONOMY

Per capita GNP: $740 (1994).
Consumer price index: 100 in 1990; 227 in Oct.93.
Currency: 809 won = 1$ (Oct.93).

> As the soil is not suitable for large-scale cultivation, copra and fish are the main exports.

Korea

The Korean peninsula lies between China and Japan, and this position has shaped the nation's history, and the character of its people. The territory has frequently been the arena of struggles between armies from China, Mongolia and Japan.

2   Korea was first unified by the Silla tribe in 667, with its capital at Kyonju. This period saw the flowering of the Korean culture. The Koryo family, from which the nation took its name, came to power in the 10th century.

3   The Yi dynasty ruled Korea for six centuries between 1392 and 1920, with invasions by the Mongols, Manchus and Japanese, plus internal power struggles. The Manchurians dominated Korea for more than two centuries, a period when the country was isolated from the rest of the world and became known as "the hermit kingdom".

4   Emboldened by a victory over China in 1895, Japan occupied Korea after defeating the Russians in 1905, and in 1910 the country was formally annexed. Under the Japanese, Korea was exploited as a supplier of foodstuffs and as a source of cheap labor. Japanese landlords and factory owners settled in Korea, with a developed infrastructure merely to extract the wealth of the country. In the 1930s, the northern part of Korea saw industrial development of war materials, to supply Japan's goal of expansion into China.

5   With the defeat of Japan in World War II, Korean hopes for a united, independent country seemed to be on the verge of being realized. However, the country was once again thrown into a complex struggle in which powerful foreign interests were involved: the Korean peninsula was divided into two zones on each side of the 38th parallel. The northern part was to be occupied by Soviet troops, and the southern, under US control.

6   Negotiations were initiated to reunite the country, but these failed. In 1948 two separate States were created: in the north the People's Democratic Republic of Korea was established, and in the south, the Republic of Korea. Thus, what was occurring at the time with Germany in Europe was duplicated in Asia. Most of the foreign troops stationed in Korea left the following year. The US remained in the south, sanctioned by the United Nations,

which was US-dominated at that time.

7   In June 1950, North Korea launched a carefully planned offensive against South Korea. The United Nations convened all of its members to put a stop to the invasion. In the meantime, US President Truman called out his army to assist South Korea, without asking Congress to declare war.

8   Likewise, Truman failed to seek UN permission before sending the US Fleet to the Strait of Formosa to protect one of the US Army's flanks, and also to come to the assistance of Chiang Kai Shek's anti-communist Chinese regime, which had established itself on the island of Formosa (Taiwan). This, in turn, gave rise to fears in Peking of a nationalist invasion of "continental" China.

9   In the meantime, the military campaign on the Korean peninsula was turning into a disaster for the

> **With the defeat of Japan in World War II, Korean hopes for a united, independent country seemed to be on the verge of being realized. However, the country was once again thrown into a complex struggle in which powerful foreign interests were involved.**

South Koreans and US troops, who had arrived hurriedly and poorly equipped, and were forced to retreat nearly as far as Pusan, in the south.

10   Defeat was narrowly avoided by the action of General Douglas MacArthur, who landed some 160 km south of the 38th parallel and managed to divide and defeat North Korean troops.

11   China, concerned over the advancing allied troops, warned that the presence of the US in North Korea would force it to enter the war. MacArthur ignored the warning and in November launched his "Home by Christmas" offensive.

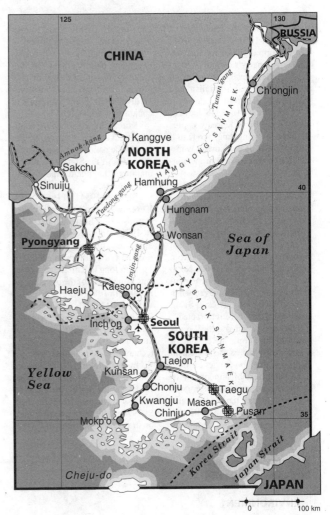

12   However, China sent 180,000 troops to Korea and by mid-December it had driven US troops back, south of the 38th parallel. On December 31 1951, China launched a second offensive against South Korea, subsequently taking up positions along the former border.

13   After differences of opinion on military strategy, MacArthur was relieved of his command by Truman. It was later revealed that MacArthur had outlined plans to use nuclear weapons against Chinese cities, advocating full-scale war with China.

14   The battle continued up and down the peninsula and the city of Seoul changed hands several times. The conflict lasted for 17 months and left four million dead.

15   After renewed fighting in June 1953, an armistice went into effect

the following month. From that time on, Korea was officially separated by the 38th parallel.

16   On August 8 1990, the UN Security Council unanimously approved the admission of both Koreas. On December 13 1991, Prime Ministers Yon Kyong Muk (North Korea) and Chong Won Shik (South Korea) signed a "Reconciliation, Non-Aggression, Exchange, and Cooperation Accord". This has been the most important step towards reunification since 1972. The Korean peninsula is in a strategic region located between the three large powers of Russia, China and Japan. There are currently intense negotiations underway to solve the nuclear question.

North Korea

Population: 23,448,000 (1994)
Area: 120,540 SQ.KM
Capital: Pyongyang
Currency: Won
Language: Korean

Choson Minjujuui In´min

The Korean Workers' Party, under the leadership of Kim Il Sung, has enjoyed uninterrupted power for four decades. The party's central philosophy is called *juche*, a strong blend of self-reliance, nationalism and centralized control.

[2] On February 16, 1948, the People's Republic of Korea was proclaimed in Pyongyang, and Kim Il Sung, leader of the Korean Workers' Party, was elected prime minister.

[3] With the war behind it (1950-53, see Korea), and armistice signed in Panmunjon, North Korea devoted all its efforts to the reconstruction of the country, which had been devastated by the war. When Syngman Rhee was ousted from office in South Korea (1960), North Korea attempted rapprochement with its neighbor. However, when the military took power in Seoul these overtures were interrupted.

[4] Despite its geographical proximity to the People's Republic of China, North Korea tried to remain neutral in the Sino-Soviet conflict.

[5] With the establishment of a socialist regime, agrarian reform was carried out, with part of the country's agriculture being collectivized. Industrialization had begun during Japanese occupation, and large textile, chemical and hydroelectric plants were established.

[6] The development of industry, expanding rapidly after 1958, was aided by the availability of rich mineral deposits. The Yalu River Hydroelectric Plant was built in conjunction with China.

[7] Korea adopted a socialist economy, with central planning; 90 per cent of the national industry was in the hands of the State with the remaining 10 per cent organized in cooperatives.

[8] Between 1954 and 1961, North Korea signed military assistance treaties with China and the USSR. In 1972, a new constitution was approved, making Kim Il Sung president as well as prime minister.

[9] Kim Jong Il, Kim Il Sung's son, was named head of the government in 1980. In 1984, North Korea provided relief to flood victims in South Korea. That same year, the formation of mixed enterprises was authorized in the construction, technology and tourist sectors.

[10] In 1988, North Korea began "ideological rectification" in direct contrast to the Soviet liberalization process introduced by Gorbachev. In 1990, there were massive demonstrations in support of the regime, on Kim Il Sung's 78th birthday; he was re-elected in May 1991.

[11] According to North Korea, any possibility of reunification must be preceded by a formal peace treaty with the US and removal of their troops from Korean soil. In 1990, North Korea submitted a proposal for denuclearization to the International Agency for Atomic Energy, subject to guarantees from the US that it would not use the 1,000 nuclear bombs it has in South Korea against them.

[12] In 1991, the USSR cut Korea's oil quota by half. Signs of the regime's growing economic difficulties began to show. As of 1992, Kim Jong Il assumed responsibility for the formulation of the country's foreign policy.

[13] Kim Il Sung's death at 82 - officially announced on July 7 1994 - once again complicated talks with the United States and deferred the summit planned between the two Koreas. Kim Jong Il succeeded his father without having the same real power as the historical leader of North Korean communism, which gave way to a fight for power among leading cadres.

[14] In 1995, extensive floods affected some five million people and worsened the food shortage. It is estimated the losses amounted to 1.9 million tons of crops which led the North Korean government to make an uncommon call for foreign aid. Japan, one of its main capitalist trade partners, contributed with 300,000 tons of rice and South Korea with 150,000 tons.

[15] The United States decided to lift its trade embargo and a delegation of American investors visited the country while Pyongang declared it no longer opposed the presence of US troops in South Korea.

[16] In May 1996, foreign contributions received after the floods amounted to $1,790,000, approximately one-fifth of the funds requested. UNICEF claimed in July that the ration of rice for North Korean children was limited to a daily 100 grams, adding that food shortage continued and that malnutrition among children would be extended for several months.

PROFILE

ENVIRONMENT

North Korea comprises the northern portion of the peninsula of Korea, east of China, between the Sea of Japan and the Yellow Sea. The territory is mountainous with wooded ranges to the East, along the coast of the Sea of Japan. Rice, the country's main agricultural product, is cultivated on the plains, 90% of the land being worked under a cooperative system. Abundant mineral resources (coal, iron, zinc, copper, lead and manganese).

SOCIETY

Peoples: Both north and south Koreans, probable descendants of the Tungu people, were influenced by their Chinese and Mongolian conquerors over many centuries. Homogeneous ethnic and cultural features in Korea stand out in sharp contrast with those of most other Asian countries. There are no distinct minorities. **Religions:** Religious practises are frowned upon. Buddhism, Confucianism, Chondokio (which combines Buddhist and Christian elements) are practised throughout the country, while traditional Shamanistic cults prevail in the interior. **Languages:** Korean (official). **Political Parties:** The Korean Workers' Party (KWP), founded in 1945 with Marxist-Leninist orientation, is the dominant political organization. The Korean Social Democratic Party and the Chondokio Chong-u party, the first being affiliated to the KWP in a National Front for the Reunification of Korea, created in June 1945. **Social Organizations:** The General Federation of Trade Unions is the only workers' organization, while the Union of Agricultural Workers is a peasants' association.

THE STATE

Official Name: The People's Democratic Republic of Korea. **Administrative Divisions:** 9 Provinces, 1 District. **Capital:** Pyongyang 2,000,000 inhab. (1986). **Other cities:** Hamhung, 670,000 inhab., Chongjin, 530,000 inhab., Sinudji, 330,000 inhab., Kaesong, 310,000 inhab. (1986). **Government:** Kim Jong Il, President of the Republic since July 1994, of the People's Assembly, and secretary general of the Korean Workers' Party. The supreme body of the state is the People's Assembly, with 541 members. **National Holiday:** September 9, Republic Day (1948). **Armed Forces:** 1,127,000. **Paramilitaries:** 3,800,000 Peasant Red Guard, 115,000 Security Troops, of the Ministry of Public Security.

COMMUNICATIONS

106 **newspapers** (1995), 85 **TV sets** (1995) and 88 **radio sets** per 1,000 households (1995). 4.8 **telephones** per 100 inhabitants (1993).

ECONOMY

Consumer price index: 100 in 1990; 129.3 in 1994. **Currency:** 789 won = 1$ (1994). **Cereal imports:** 11,271,000 metric tons (1993). **Fertilizer use:** 4,656 kgs per ha. (1992-93). **Development aid received:** $965 million (1993); $22 per capita; 0.3% of GNP.

South Korea

Population:	44,453,000 (1994)
Area:	99,020 SQ.KM
Capital:	Seoul
Currency:	Won
Language:	Korean

Taehaen – Min `Guk

On August 15, 1946, a Republic was established in the southern part of the peninsula, with its capital in Seoul. Syngman Rhee was named its first president and supported by US military, economic and political advisors, he ruled the southern portion of Korea for 14 years. Through constitutional laws giving him indefinite power, he ruled as a constitutional dictator. All opposition forces were controlled by charging them with cooperating with the North at the first signs of dissent.

[2] The US advisors convinced the Rhee regime of the need for a thorough land reform program. They were very much aware of events in China, where the nationalist Kuomintang had been defeated by the communists who promised rural peoples "land to the tiller". The US scheme in South Korea allowed for compensation, and land redistribution limited to three hectares per person; former Japanese-owned land was also redistributed. Many of the previous land-owners moved to the cities.

[3] Despite the introduction of a draconian National Security Law in 1958, when many political dissidents were jailed, the regime was unable to suppress opposition completely. Rhee was re-elected in 1952, 1956 and in 1960. The results of the last election were not accepted by the opposition, who claimed there had been fraud. There were protests in Seoul, which were repressed harshly by government forces. Growing opposition and the danger of a revolution extending to the rest of the country, led Syngman Rhee to resign on April 27 1960. He was succeeded by Huh Chung Tok as interim president.

[4] New elections were held, and Po Sun Yun, a member of the Democratic Party, was elected president. John Chang, who had been named head of the government, attempted to lead the country toward effective economic development and put an end to the corruption and waste which had characterized the previous regime. Some members of his cabinet were in favor of warmer relations with North Korea.

[5] In May 1961 a military coup ousted Chang; a military junta was installed, presided over by General Chan Yung. In July, General Chung Hee Park took command of the junta and proceeded to suspend all democratic freedoms and imprison all members of the previous regime.

[6] The new regime initiated a National Reconciliation policy, including a planned strategy against communism and corruption, and promised a return to civilian government after completion of these "revolutionary tasks".

[7] In March 1962, Park took the presidency from Po Sun Yun. In late 1963, Park was confident of power and held elections. He won by a mere 1.4 per cent despite suppression of the opposition. The protests that followed led him to declare martial law, the military retained their political domination and a long, harsh dictatorship ensued.

[8] The military regime established strong centralized economic planning and, with the help of western technocrats, Korea became a full example of export-oriented development. Imports and exports, domestic prices and access to credit were used as levers to control and guide the economy. Restrictive laws were established governing the right to strike, organize or allow collective bargaining.

[9] From a war-torn, mainly rural, indebted economy, South Korea developed into an industrial economy, dominated by large, Korean-owned transnational corporations producing steel, ships, cars and electronics goods.

[10] Villagers, affected by low grain prices, were forced into cities and the urban labor pool, contributing to the "Korean miracle". Some of the world's lowest wages, longest hours and most unsafe working conditions are the norm for Korean workers.

[11] Eighteen years after taking power, Park was still president, having won four fraudulent elections.

[12] In October 1979, Park was killed; shot by the director of the hated Korean Central Intelligence Agency, Kim Jae Kiu, in unclarified circumstances.

[13] Before the end of 1979, General Chun Doo Hwan, head of military intelligence, decided to secure his position and arrested rival generals; on May 17 1980, martial law was once again established and civilian politicians were arrested.

[14] The following day, citizens in the southern city of Kwangju took control in protest against the military regime, and particularly the arrest of Kim Daw Jung, a leading opposition figure and a native of Kwangju. The Korean army suppressed the citizens with extreme brutality and "hundreds lost their lives" with the knowledge and agreement of US military commanders stationed in the country. Although the protest had been carried out as a public reaction to the arrest of Kim Daw Jung, this fact was ignored and Daw Jung was sentenced to life imprisonment, charged with "instigating" the uprising. The Kwangju incident was used by the military as justification for seizing power and questions were raised as to whether the whole event had been deliberately provoked by the military.

[15] Chun launched a "purification campaign" of public and private sectors, even setting quotas for singling out the corrupt and "subversives". In the same manner as his predecessor, Chun Doo Hwan held elections in an attempt to legitimize and civilianize his rule; he won the elections in 1981.

[16] In October 1983, several members of the South Korean cabinet were killed by a bomb at the Martyrs' Mausoleum in Burma during a state visit; arriving late, President Chun escaped the bomb blast. The Burmese, claiming proof of North Korean involvement, broke off diplomatic relations and rescinded

PROFILE

ENVIRONMENT

The country is located in the southern part of the Korean Peninsula, east of China, between the sea of Japan and the Yellow Sea. The terrain is more level than North Korea's and the arable land area, mainly used for rice farming, is larger. Atmospheric pollution first appeared in the 1970s, as did pollution of water supplies, rivers and the sea . The western coast, on the Yellow Sea, is seriously contaminated.

SOCIETY

Peoples: North and South Koreans belong to the same ethnic group and have a common origin. There are no distinct minorities. **Religions:** Buddhism, Confucianism (a moral code rather than a religion), Christian groups and Chondokio. **Languages:** Korean (official). **Political Parties:** Democratic Liberal Party (DLP), in power; the Democratic Party (DP), fomerly Party for Peace and Democracy, in opposition; the National Unification Party, opposition; the United People's Party (PP), or Minjung Party constitutes the only left-wing party. **Social Organizations:** Legally, all unions must belong to the government-controlled Federation of Korean Trade Unions (FKTU) which has a persistent reputation for intervening against workers to end industrial action, sometimes violently. In January 1990, democratic trade unions formed by workers since 1987 announced the formation of the Korean Alliance of Genuine Trade Unions (Chonnohyop) claiming affiliation of 600 unions with 190,000 members, according to organization figures.

THE STATE

Official Name: Republic of Korea. **Administrative Divisions:** 9 Provinces. **Capital:** Seoul 10,628,000 inhab. (1990). **Other cities:** Pusan, 3,798,000 inhab., Taegu, 2,229,000 inhab. (1990). **Government:** Kim Young-Sam, President and Head of State, elected in February 1993. Li Hoi-Chang, Prime Minister and head of government, named in December 1993. Legislature, single-chamber: National Assembly, with 209 members, elected every 4 years. **National Holiday:** August 15, Liberation Day (1945). **Armed Forces:** 633,000 (1993). **Paramilitaries:** 3,500, 000 Civil Defence Corps, 4,500 Coast Guard.

their recognition of the North Korean government.

[17] Opposition to Chun's regime continued to grow as repression, reminiscent of Park's worst excesses, escalated. The US withdrew their support of the Marcos regime in the Philippines, after claims of election fraud and human rights abuses, and a worried Korean regime instituted some reforms. Censorship was lifted somewhat and Kim Daw Jung released from house arrest.

[18] During 1987, hundreds of thousands of Korean workers joined in strikes and factory occupations in an unprecedented wave of protests. Korean workers demanded the right to form democratic unions independent of the government-run Federation of Korean Trade Unions, higher wages, an end to forced overtime and, in general, a larger share of the benefits of the nation's spectacular growth.

[19] In July 1987, Chun stepped down and designated Roh Tae Woo both as his successor and as president of the official Democratic Justice Party. Demonstrations followed amid protests that Roh would continue the dictatorship as a close military colleague of Chun. Demands that Chun face trial for his part in the Kwangju massacre were also voiced.

[20] Faced with the possibility of larger street demonstrations and concern over its international image (South Korea was host of the Olympic Games in 1988) political restrictions were eased during the December, 1987 election campaign.

[21] Polls showed government candidates trailing badly in late November when a Korean Airways passenger jet disappeared near Thailand with 118 passengers aboard. North Korea was again suspected, as in the Burma bombing, but Pyongyang denied any involvement.

[22] In the elections, the newly-freed opposition together polled a majority, but failed to unite the two factions, led by Kim Yong Sam of the Reunification Democratic Party and Kim Dae Jung of the Party for Peace and Democracy. The split enabled the incumbent government, under Roh Tae Woo, to secure power with a majority of 42 per cent.

[23] In January 1990, the most conservative opposition groupings formed a merger with the official Democratic Justice Party, and became the Democratic Liberal Party, controlling 220 seats in the 298-member National Assembly. The Party for Peace and Democracy remained the only real parliamentary opposition.

[24] In the spring of 1990, a new offensive against independent trade unions and labor rights resulted in the arrest of the movement's leaders. In April, police stormed the Hyundai shipyards and arrested over 600 union activists, ending a 72-hour worker occupation protesting the arrest of union leaders. A few days later, 400 striking workers occupying the Korean Broadcasting System's headquarters were also arrested. The resulting nationwide protests precipitated the biggest drop in the history of the Korean stock market.

[25] In September 1991, US President George Bush made the decision to withdraw tactical nuclear weapons from South Korea, and in November this was accomplished. This significant step met one of the requirements of North Korea and was necessary for inspections of their territory to be made.

[26] In December of that year, Seoul and Pyongyang signed a "Reconciliation, Non-aggression, Exchange and Cooperation Accord", affecting bilateral realations (see Korea).

[27] In the legislative elections of March 1992, both the pro-government groups and the left suffered a serious setback. The Democratic Liberal Party won 149 of the 299 seats, one short of the majority. The Democratic Party (DP) came in second with 30 per cent, followed by the Party of National Unification with 17.4 per cent and 31 seats, a new grouping led by a former corporate leader. The left-wing People's Party (PP) won 1.4 per cent of the vote and no seats.

[28] In May, President Roh Tae Woo named Kim Young Sam, who had obtained 41.4 per cent of the returns in the December presidential election, as his successor. Kim's election coincided with a weakening opposition, worsened by the resignation in February 1993 of Chung Ju-Yung, leader of the United People's Party, charged with having accepted illegal contributions from a major corporation during the electoral campaign.

[29] Corruption scandals also reached the government. Choi Ki Son, a close aide of president Kim, admitted he had embezzled $1 mil-

lion in public funds. Furthermore, Suh Eui Hyun, leader of the country's largest Buddhist order, was accused of accepting $10 million from a businessman to hand them over to Kim Young Sam. This accusation caused confrontations among Buddhist monks which led to 134 arrests.

[30] In 1995, former presidents Chun Doo Hwan (1980-1988) and Roh Tae Woo (1988-1993) were arrested for their participation in the coup d'etat which took Chun to power in December 1979. In June, the government was defeated in the first local and provincial elections to be held without government control since 1961. The collapse of Sampoong shopping mall in Seoul which caused the death of over 500 people, brought up once more the

question of corruption, when it became known that local authorities had approved the construction of the fifth story after receiving concealed commissions.

[31] In April 1996, the ruling New Korea Party obtained 139 of the 299 seats at stake. The National Congress for a New Policy obtained 79 seats and the United Liberal Democrats 50.

Kuwait

Kuwayt

Population: 1,620,000 (1994)
Area: 17,820 SQ.KM
Capital: Kuwait (Al-Kuwayt)

Currency: Dinar
Language: Arabic

1 Neutral Zone

0   100 km

In the history of navigation, the voyage of Nearco, one of Alexander of Macedonia's admirals, was a milestone; he sailed from the Indo River to the farthest end of the Gulf of Arabia, ending his voyage in the port of Diridotis, in present-day Kuwait. This was not, however, the first contact between India and Kuwait, which had already been an active trade center for almost 2,000 years. The fate of the country was always intimately linked to that of the Mesopotamian civilizations (see Iraq) but, beginning in the 13th century, after the Mongol invasion caused collapse of the caliphate, the region entered into a long period of isolation.

[2] After this, new Arab groups began to settle there. In the 18th century, although nominally subject to the Ottoman Empire, the settlements were virtually independent; they decided to elect a *shaij* (sheikh) to conduct sporadic negotiations with the Turks. In 1756, the leader of the Anaiza tribe, Abdul Rahim al-Sabah, founder of the reigning dynasty, was designated for this task.

[3] At that time, the place, formerly known as the Qurain (horn) started to be called Kuwait, a diminutive for al-Kout, which is a local Arabic term to describe the fortified houses on the coast.

[4] The weakness of Turkish sultans and the growing British influence in the region were two fundamental factors that encouraged the formation of proto-states, of which the emirate of Najd (see Saudi Arabia) was the most dynamic.

[5] In order to avoid absorption by the Wahabites, the emirs of Kuwait requested support from the British who had moved the East India Company's overland mail terminal from Basra to Kuwait in 1779. The British sent Indian troops to guar-antee the autonomy of the emirate, an initiative which the Turks were helpless to stop. For decades Kuwait was the most prosperous and peaceful part of a region peppered by constant disputes.

[6] In 1892, with Turkish support, the emir of Najd was deposed and was granted asylum in Kuwait. Sheikh Mubarak al Sabah allowed the territory to be used as a base for their raids against the pro-Turkish Rashidis. Since this implied great risks, a treaty was signed with Britain in 1899, whereby the sheikh undertook not to cede any part of the territory to another country without British consent; in exchange, Britain guaranteed Kuwait's territorial integrity. This treaty, and the subsequent British military presence in the region, frustrated Turkish attempts to expand the Berlin-Baghdad railway to reach the Gulf through Kuwaiti territory.

[7] At the end of World War I, France and Great Britain split the remains of the Ottoman Empire. Kuwait was now considered a British protectorate, separate from the newly-created kingdom of Iraq, which claimed it as a province, alleging the historical subjection of the region to the government of Baghdad.

[8] In 1938, oil began to flow from the Burgan wells. Ahmad Jabir al-Sabah, the crown prince, opposed the exploitation of the extensive deposits, arguing that a high income and high salaries resulting from the new activity would ruin pearl fishing, until then the country's main economic activity, employing 10,000 sailors as deep-sea divers. After the impasse created by World War II, the emir granted the concession to the Kuwait Oil Company, owned by British BP and US Gulf, with oil first exported in 1946. The large-scale exploitation of these reserves soon turned the small port into a large trading center.

[9] In 1961, independence was negotiated, within the British policy of gradual decolonization. Sheikh Sabah proclaimed himself emir and took up full powers. Iraq refused to recognize the new state, claiming that it was an artificial creation of the British to maintain access to oil. Consequently, the British troops remained to defend the emirate until they were replaced by the troops of the Arab League.

[10] In 1962 a Constitution was enforced creating a National Assem-bly of 50 members, elected individually by male citizens aged over 21, whose fathers or grandfathers had resided in Kuwait before 1920. None of the candidates belonged to a political party, as these had been made illegal. Thus, out of the 826,500 Kuwaitis who lived in the country in 1990, a minority of the population; only 85,000, would have been entitled to vote if elections had been called.

[11] In 1966, Kuwait and Saudi Arabia solved their ancient border disputes, and the "neutral zone" which existed between both countries was divided equally. In 1969 the Central Bank of Kuwait was created, and in 1976 a Social Security Law and a Law of Reserves for Future Generations were passed. The latter established that 10 per cent of the State revenue was to be earmarked for an investment fund.

[12] In a few years, oil entirely changed the country. The Bedouins replaced their camels with luxurious air-conditioned cars. Pearl fishing disappeared. The entire population settled in brand new cities, where the stylized mosque towers stood side-by-side with shopping centers which replaced the old souks (markets). The population's educational standards and life expectancy rose. All manual labor and work in the oil industry was done by immigrant workers. In 1970 the number of migrants equalled the local population, and by 1985 this figure had almost doubled.

[13] Kuwaiti rulers became con-cerned that despite so much prosperity in such a poor area its legitimacy could be questioned. In 1961 the Arab Fund for Economic Development was created, in order to channel "soft" loans and donations to Third World countries. When the Organization of Petroleum Exporting Countries (OPEC) succeeded in raising prices in 1973, Kuwait increased its revenue immensely.

[14] Most of the Third World countries supported OPEC, hoping that they would receive help to establish a "New International Economic Order", by asking for better prices for the other raw materials that they supply to the Northern countries. However, instead of investing their oil revenues in their own countries or in other Third World nations, the Gulf monarchs placed their fortunes in transnational banks. This further accentuated the excess of liquidity in transnational private banks, which started to grant loans to the Third World quite indiscriminately. This situation was one of the main factors that provoked the "debt crisis" in 1982.

[15] Within the Gulf area, however, Kuwait was generous with its wealth. By the end of the 1980s, Kuwait had the highest rate of official development assistance in the world, proportional to its gross national product.

[16] Unfortunately, prosperity did not prevent political conflicts. In August 1976, the National Assembly was dissolved by Sheikh Jabir al-Sabah, arguing that it had acted

LAND USE

■ CROPLAND 0%   ■ PASTURE 8%   ■ OTHER 92%

ENVIRONMENT

Nearly all of the land is flat, except for a few ranges of dunes. The inland is desert with only one oasis, the al-Jahrah. The coast is low and uniform. The city and port of Kuwait is located in the only deep-water harbor. The climate is tempered by ocean currents but the temperature is high in summer. Winters are warm with frequent dust and sand storms. Petroleum is the main economic resource, with three refineries in Shuaiba, Mina al-Ahmadi and Mina Abdulla. Tremendous environmental damage was caused by burning oil wells, as a result of the Gulf war in 1991. The country also suffers from a lack of water.

SOCIETY

Peoples: Kuwaitis, of Arab descent, account for less than half of the population, 60% of which is made up of immigrant workers from Palestine, Egypt, Iran, Pakistan, India, Bangladesh, the Philippines and other countries. **Religions:** Muslim 85%, of which Sunni 45%, Shi'ah 30%; other Muslim 10%; other (mostly Christian and Hindu) 15%. **Languages:** Arabic (official). **Political Parties:** Islamic Constitutional Movement, a moderate Sunni group; Kuwaiti Democratic Forum, liberal; Salafeen, a fundamentalist Sunni group. **Social Organizations:** Federation of Unions.

THE STATE

Official Name: Dawlat al-Kuwayt. **Administrative divisions:** 5 governances. **Capital:** Kuwait (Al-Kuwayt), 31,200 inhab. (1993). **Other cities:** al-Jahra 139,476 inhab.; as-Salimiyah 116,104 inhab.; Hawalli 84,478 inhab.; al-Farwaniyah 47,106 inhab.(1993) **Government:** Jabir al-Ahmad al Sabah, emir. The National Assembly has 50 members. **National Holiday:** February 25, Independence Day (1961). **Armed Forces:** 16,600. **Paramilitary:** The National Guard has 5,000 members.

---

against national interests. In December 1977, the emir died and the Crown of Kuwait passed to Jabir Al-Sabah, who called national elections in February 1981. Over 500 candidates ran for the 50 seats in the National Assembly and 40 were won by Sunni candidates, loyal to the ruling family. Shiite candidates won four seats, and Islamic fundamentalists won six. Only 6.4 per cent of the population was permitted to vote.

[17] When the Iran-Iraq war broke out in 1979, Kuwait officially remained neutral, but in fact supported Iraq with large donations and loans. Kuwait considered Iraq to be a "first line of defence" against the Iranian Islamic revolution.

[18] In 1985, the National Assembly began to disapprove of governmental measures, such as press control, increases in the prices of public services and educational reforms, echoing other accusations that corruption existed within the ruling family. In August 1986, the emir dissolved the Assembly and began to rule by decree.

[19] In the late 1980s, the Kuwait Investments Office (KIO) had capital assets outside the country estimated at $1 billion, which included hotels, art galleries, European and US real estate, and major shares in large transnational corporations: 10 per cent of British Petroleum, 23 per cent of Hoechst, 14 per cent of Daimler-Benz, and 11 per cent of the Midland Bank.

[20] In 1987, alleging that Iraq was using the port of Kuwait to export oil and import weapons, the Iranian navy attacked Kuwaiti merchant ships. In response, Kuwait requested and obtained permission from the major powers - US, France, Britain, and the USSR - to use their flags for the Kuwait merchant navy. The US and Britain sent their navies to protect Kuwaiti ships in the Gulf.

[21] Once the war between Iran and Iraq ended in 1988, tensions between Kuwait and Iraq started to mount. Kuwait demanded the payment of $15 billion on account of war loans, which Iraq refused to return, alleging that those sums had been used to protect Kuwait. Iraq accused Kuwait of "stealing" the country's oil, by extracting large amounts of oil from the common deposits which stretch along the border. Iraq claimed that the emirate was pumping out much more oil than its entitlement, and demanded $2.4 billion in compensation.

[22] In spite of repeated attempts at mediation by Palestinian leader Yasser Arafat, tension mounted. On August 2 1990, Iraq invaded Kuwait quickly and decisively. Emir al-Sabah and his family took refuge in Saudi Arabia. Nearly 300,000 Kuwaitis fled the country, to join the other 100,000 who were abroad on their summer vacation. The occupation forces encouraged the exodus, perhaps to "deKuwaitize" the country. A pro-Iraqi provisional government, led by Al Hussein Ali, requested the total fusion of Iraq

---

Islamic civilization: origins and foundations

The Arab peninsula, inhabited during the first centuries of the Christian Era by nomad or semi-sedentary Bedouins, was the geographical and human context where Islamic culture and civilization surged. It is said that it was in Mecca, a center of pilgrimage, a city of caravans, and a trade centre of the Medieval world, that Muhammad was born. There truth was revealed to him and there he began his preaching of Islam until in the year 622 - the origin of the Hejira - he sought refuge under menace of death in the city of Medina, and the 30 years that followed Muhammad's death (632-661, during which period the four orthodox caliphs, who accompanied him in his preaching, governed) are considered by the Muslims as the "golden age" of Islam. Sustained by the inner belief in its message and by the overwhelming strength of the Arabic armies, Islamic expansion defeated the Sasanid and Byzantine empires, as well as the ancient Roman West and built a Muslim Empire which headed world trade and structured a web of major cities.

Islamic city and religious institutions

The Islamic city is a community of people who practise the Islamic religion. It constitutes the "umma", or nation, in which each Muslim person is included, whether he lives alone or as part of a group, whether peasant or citizen, nomad or settled. A much quoted expression defines it as "Dar al-Islam", 'the home of Islam', and it limits it to the countries or urban groups which obey the Islamic canonical law, and where its traditional lifestyles are practised. Islam, which signifies 'submission to God', comprises three basic religious institutions: the Koran, the Tradition of the Prophet (sunna), and the written and oral teachings of the jurists. Through the double testimony of the faith - "There is no more God than the One and Only" (Allah); "Muhammad is the messenger of God" - which declaration confers the condition of Muslim to every person of good will, the Koran proclaims its essential message, "al-tawhid" or 'Divine Unity', which establishes the rights of the Creator above all contingencies of our earthly existence and it achieves itself in the individual life of everyone who places his/her thoughts and actions nearest to God. With that purpose people are urged to read the Koran, to invoke the names of God, and to fulfill the obligatory practices of prayer, fasting, alms giving, and pilgrimage to Mecca, at least once in a lifetime. The Prophet, providentially 'chosen' in order to communicate Muslim law ("sharia"), became the most exalted person in the Islamic world. The collection of his sayings and advice, his acts and gestures, were gathered during the third century of Hejira, in the "hadits" or traditions, with the purpose of making the transmission and knowledge of them easier for the community of followers. Neither the Koran nor the Sunna, nevertheless, were elaborated as bodies of law. It was a later task of the learned men of Islam to formulate the juridical system that rules and divides the actions of the believers and makes them either compelling, recommended, permissible, damnable or prohibited.

Society, community, and individual

The essential thing about the Islamic city is the "durable combination of the effort by every man in order to submit to the will of the divine legislator within the communal framework which is the help and support of his effort" (JL Michon, 1976). In Islam, the link between the individual and the social whole is so strong that the task of individual salvation "includes ipso facto the sacralization of the social". The salvation of a single person depends on the people who surround him or her, as well as on favorable or unfavorable circum-

---

stances. Tradition assumes that it was Muhammad himself who formulated the principle of "iyma" or consensus of believers, which becomes concrete in Muslim law through the form of a collective statute called the 'duty of sufficiency'. It would exonerate a Muslim person of any compelling legal duty provided a sufficient number of followers would exonerate him of it. The individual, nevertheless, is not lost in the communal. Islamic law supposes that a person compromises only him/herself through his/her behavior and that, on its own day, s/he alone will appear before the Supreme Judge to respond for her/his behavior. The fact of all people being equal before God, and being also equally dependent on Him/Her, and compelled to obey the law, has given place to the definition of the Muslim community as 'equable theocracy' (L Gardet, 1961). The strong sense of social cohesion which accompanied the high degree of integration of traditional Muslim societies is due in a great degree to the socio-religious values which guided the life of its individuals and communities.

Government and politics: the islamic community

The community established in Medina in the first century of Hejira (the 7th century AD), was the prime model of institutional religious organization which ruled in all traditional Muslim societies. Its name being Yatrib before that date, its new name, al-Madina ("the city by excellence"), designated its condition as a center of "umma" and a siege of authority and justice. The caliph or iman, a successor of the Prophet, united in his person the spiritual and the secular authority, and was the supreme chief of the city. He was in charge of creating the conditions for the application of Koranic law, of directing the Holy War (jihad), of organizing the army, and of assuring an orderly administration and the security of the countries under its dominion. The caliph designated also, in each city, the ministers or vizirs, the governors, the army chiefs, the tax collectors, and even the police corps ("surta") that maintained order and protected the city from enemies. Justice in the traditional Islamic society was derived from divine mandate. It refers to an original pact through which God designated the people who exercise authority. These must protect the followers, and the followers must obey the authorities. The Platonic ideal of justice and Islam are synonomous: the order decreed by God will only prevail where virtuous people rule, who are capable of uniting their deep knowledge of the divinity to a high moral quality. They have in their hands "the task of making people, in this life and in this milieu enjoy a maximum of happiness, and also of making them reach the joys of a future life through communal institutions based on justice and brotherhood" (al-Farabi, 4th century of Hejira). Despite the fact that the juridical order of traditional Islamic cities lacked the local and municipal autonomy enjoyed by medieval European cities, its institutions, guided by values that negated racial, religious, or social discrimination, nevertheless stimulated the high degree of integration shared by all cities of the Muslim world from al-Andalus to India. Muslim jurists accepted through the centuries, as source of legislation, the local customs and habits of the different cities. This shows their flexibility and democratic inclinations.

The economy in medieval society

The economy of traditional Muslim cities was ruled by a corporate system which integrated the people dedicated to production, distribution, and services, whether they were proprietors, employees, home servants, independent workers or government employees, whether they were "people of high or low condition, Muslim, Christian, or Jew, native or foreign residents, they all belonged to the corporate system" (Yusuf Ibish, 1976). In the corporations the urban population was gathered according to the profession of each individual; there were artisan, merchant, auctioneer, lender, musician, singer, storyteller, transport, and sailor groups. The members of each corporation were considered also members of the

community of believers to which service was specially credited the contribution of each profession and trade. These were learnt through hard work supervised by a master ("sayj") connected to a chain of corporation masters, linked as well to other corporations, to the Holy Patrons, and even to the Prophet. The corporations were structured according to a conceptual and ritual system transmitted orally from generation to generation and linked to the sufi orders (Islamic loggias) To the acceptance of a young person as an apprentice in a workshop there followed the recitation of the first "azora" (chapter) of the Koran before the masters of the corporation and a period of years of work with low remuneration or lack thereof, compensated by the notion that this was the way of learning and of becoming gradually integrated to the rest of the community. A branch of sweet basil was given on behalf of the master to the young apprentice, signalling the beginning of his initiation. The initiate entered his own corporation, and through it, the umma. After some years, the execution of a masterpiece as a proof of his refined abilities could elevate the artisan to the position of a master.

Education and religious instruction

Muslim education, initiated at the time of the Prophet in Mecca, was fundamentally given through the mosque, and its content was the "sharia" or Islamic law, whose learning was a 'duty of sufficiency' for the Islamic community. The highest distinction in Islam was to acquire the 'knowledge' - al-'ilm - or the knowledge of revealed law. Memory was a praised gift, and the title of 'hafiz' was granted to the person who knew the Koran by heart. Religious instruction was one of the key elements in order to assure the survival of Islamic civilization. A citizen of average education could be a consultant inside the community, as well as conduct the prayers, and practise Koranic law. At the beginning, religious instruction and overall education were the same, but later there were differences between the two. The first century of Hejira, dedicated to military conquest and to the establishment of political authority in Islam, did not produce a noticeable development of Islamic education. But from the second century onwards - during which the institution of the mosque developed among the conquered territories - and above all from the third century onwards - in which a generation of jurists, theologians, and linguists sought to preserve the language and the traditions of a civilization which had grown far through very diverse cultural areas - education became a foremost preoccupation. During the third and the fourth centuries the mosque was a virtual public university, a center of worship and also a center for social gatherings. But it was then also that the elementary school ("kultab") appeared, as well as the 'houses of wisdom' or of 'science', exclusively for academic activities. In the fifth century appeared the high school or "madrasa", fostered by the Government, which was from then on the highest centre of learning in the Muslim world. Towards the ninth century, it was necessary to have finished studies in a madrasa in order to occupy a position in government. It was not only the acquisition of knowledge - which is the means of distinguishing between the prohibited and the praiseworthy - but its transmission also, which became a religious obligation in Islam, anticipating thus the historical effort in order to democratize teaching. "Islamic society repudiates the 'alim' (learned person) who does not transmit his learning to others." Islam has defended the freedom of thought, and has recognized the limits of reason. It cannot doubt neither divine unity nor the truth of Mohammed's message. From the point of view of Islam, reason can be either inborn - when it is a divine gift , or acquired - when it is the result of individual effort and experience. A valuable contribution of Islam is its recognition of the practical nature of thought and education, evident through a tradition attributed to the Prophet: "You should acquire all the wisdom of which you are capable! But God will not reward you for all that you have learned until you translate it into actions!"

DEMOGRAPHY

Urban: 97% (1995). **Annual growth:** 0.4% (1992-2000). **Children per woman:** 3.7 (1992).

HEALTH

Under-five mortality: 71 per 1,000 (1995). **Calorie consumption:** 108% of required intake (1995).

EDUCATION

Illiteracy: 21% (1995). **School enrolment:** Primary (1993): 65% fem., 65% male. Secondary (1993): 60% fem., 60% male. University: 16% (1993). **Primary school teachers:** one for every 16 students (1992).

COMMUNICATIONS

112 **newspapers** (1995), 110 **TV sets** (1995) and 107 **radio sets** per 1,000 households (1995). 24.5 **telephones** per 100 inhabitants (1993). **Books:** 103 new titles per 1,000,000 inhabitants in 1995.

ECONOMY

Per capita GNP: $19,420 (1994). **Annual growth:** 1.10% (1985-94). **Consumer price index:** 100 in 1990; 4,947.9 in 1994. **Currency:** som. **Cereal imports:** 120,000 metric tons (1993). **Fertilizer use:** 242 kgs per ha. (1992-93). **Imports:** $21,716 million (1994). **Exports:** $11,614 million (1994).

ENERGY

Consumption: 7,615 kgs of Oil Equivalent per capita yearly (1994), -711% imported (1994).

Maternal Mortality

1989-95

PER 100,000 LIVE BIRTHS

18

and Kuwait. A few days later the emirate was declared an Iraqi province.

²³ Iraq expected understanding from the US, in exchange for assured oil supplies, but the US reacted very harshly to the invasion and promoted a series of extremely tough measures from the UN Security Council. On August 6, the trading, financial, and military boycott of Iraq was voted through, and on November 29, the use of force against Iraq was approved if the country refused to withdraw from Kuwait before January 15, 1991 (see box). Only Cuba and Yemen abstained from voting.

²⁴ When the emir went into exile in Saudi Arabia he was pressured by the US to promise democratic elections once the country was liberated.

²⁵ The war devastated the country, not so much in terms of lives lost, for most of the fighting was done on Iraqi territory, but the bombings and forced withdrawal of the occupation troops left most of the country's oil wells burning and the country was unable to produce any oil until 1992.

²⁶ The war also turned Kuwait into an environmental disaster area. The black cloud formed by some fifty burning oil wells, and gigantic oil spills along the coast produced profound degradation of the air, marine resources and soil. An enormous mass of oil lined the coasts, posing a threat to birds and other animals. The Gulf's ecosystem was seriously affected, especially fish species

which form a staple of the local diet.

²⁷ It has been estimated that the clean up and reconstruction will cost between $150 and 200 billion. On January 9 1991, Britain committed itself to grant a loan of $950 million earmarked for reconstruction. Kuwait's debts to allied countries, arising from war costs, rose to more than $22 billion.

²⁸ After the war was over more than 1,300 people were killed by mines which were laid during the conflict.

²⁹ On March 18, 1991, Amnesty International in London denounced the "arbitrary detention and torture" of Palestinians residing in Kuwait, carried out by Kuwaiti civilians and military.

³⁰ Also in March 1991, while US Secretary of State James Baker was visiting Kuwait, the prince and premier Sheikh Saad al-Abdallah promised to lead the emirate towards democracy and to reinstate parliament.

³¹ On April 1 1991 a group of Kuwaiti citizens demanded freedom of the press, independence of the judiciary, the legalization of political parties, and the implementation of measures to fight corruption.

³² On April 8, the emir announced that a new National Assembly would be elected "when circumstances permitted", but he failed to state a precise date. According to several reports, popular discontent was running high as a consequence of the monarch's delay in returning

to the country and the vagueness of his political plans. .

³³ In June 1991, Sheikh Jabir al-Sabah called a National Council to discuss the elections, and female and foreign suffrage. The opposition were still demanding the re-establishment of the 1962 Constitution and the formation of a democratically elected parliament, though no opposition leaders would be allowed in the Council. Over a thousand people demonstrated in the streets after the emir made his announcement. Although street demonstrations are illegal in Kuwait, the government made no attempt to stop the march.

³⁴ When the war ended, the emir launched the slogan of "reKuwaitizing" Kuwait. The plan included a drastic reduction in the number of foreigners in the country. Over 800,000 people will have to leave the country, to comply with the sheikh's measure. A small number of foreigners have been deported, but most are forced to leave when they fail to find jobs and schools for their children.

³⁵ In mid-December 1991, Saudi Arabia, the United Arab Emirates, Oman, Qatar, Bahrain, and Kuwait held a summit meeting aimed at creating a collective security mechanism, and establishing a new defencive framework for the region which holds 40 per cent of the world's oil reserves.

³⁶ After renewing oil production, Kuwait initially accepted the OPEC quota of 2 million barrels per day. However, in February, it confronted OPEC's other members and unilaterally decided to raise daily production to 2.16 million barrels. The Minister of Petroleum explained that the country had confirmed the existence of new deposits of crude oil in what was formally a "neutral zone" along the border with Saudi Arabia.

³⁷ In January, press censorship was lifted. Demonstrations were organized by women's groups demanding the right to vote. Elections for the National Assembly were finally held on October 5, with independent opposition groups winning 31 seats, the fundamentalists 19 and the liberals 12. For the first time ever, the cabinet contained several members of parliament elected by direct popular vote.

³⁸ In early 1993, a financial scandal broke involving the Kuwait Investment Office (KIO), and two members of the royal family were found to have been involved. The personal holdings of high-ranking KIO officials were frozen when $5 billion were discovered to be missing. $1 billion had been diverted to Torras, a Spanish business consortium.

³⁹ As a result, the government prohibited the publication of any information on government corruption, without prior official consent.

Months later, the British newspaper, "The Financial Times", accused the KIO of having used $300 million to win the support of several UN diplomats in the military operation against Iraq.

⁴⁰ In 1993 the UN finally determined the frontier with Iraq, despite protests by Baghdad. As a response to supposed incursions by Iraqi troops, the US bombed Iraq repeatedly (see Iraq). Washington decided to install missiles in Kuwait and it began building a wall 130 kms long equipped with 1.3 million mines along the new frontier.

⁴¹ In August the ex-chief of the Temporary Free Government imposed by Baghdad, Al-Hussein Ali, was condemned to death, as well as five Kuwaitis and 10 Jordanians accused of collaborating with the Iraqi occupation.

⁴² In 1994 several members of the Kuwaiti élite were accused of corruption. In April the Minister of Oil, Ali Ahmad al-Baghli, was expelled from the government, after which he denounced the corruption within the Kuwait Petroleum Corporation. In June, the Constitutional Court declared itself incompetent to consider the denunciation against the ex-Minister of Finance, Sheikh Ali Khalifah as-Sabah, formulated by the National Assembly.

⁴³ In the same month, 13 people were condemned - six of them to death - after having been accused of attempting the assassination of the ex-US President George Bush, in April 1993. In 1995 the Government, supported by many Kuwaiti investors, continued privatize the state enterprises, while Kuwait paid a considerable amount buying arms to defend itself against a possible Iraqi attack.

⁴⁴ In February 1996 Amnesty International strongly criticized the lack of impartiality shown by Kuwaiti authorities concerning the judgement of the "supposed collaborators" of the Iraqi occupation in 1990 and 1991. The organization for the defence of human rights reminded world opinion that, under the martial law adopted after the Gulf War, there were summary executions, and 70 people "disappeared". The victims of these actions were, in many cases, foreign residents. Amnesty denounced also the practice of torture and expulsions without trial.

Kyrgyzstan

Population: 4,473,000 (1994)
Area: 198,500 SQ.KM
Capital: Pishpek (Bishkek)

Currency: Som
Language: Kyrgyz

Kyrguyzstan

In the Bronze Age (2000 years BC), the present Kyrgyzstan was inhabited by tribes who farmed and raised livestock. Between the 7th and the 3rd centuries BC the Saks settled in the area, and between the 3rd and 2nd centuries BC, a group of Usun tribes.

[2] From the 1st century BC to the 4th century AD, Kyrgyzstan was a part of the state of Kushan, which disappeared after the nomadic Eufalites invaded. Toward the end of the 4th century, Kyrgyzstan became part of the western region of the Turkish Kaganate (khanate), governed by the brothers Tumin and Istemi.

[3] In the 7th century, Kyrgyz territory was occupied by the Turguesh (under the chieftain Moje-Dajan); between the 8th and 10th centuries it came under the Karluks who were replaced by the Karahanyds between the 10th and 12th centuries. From the 12th to the 14th century, the Turkish peoples of Kyrgyzstan were absorbed as part of the Mongol Empire, in the ulus (province) of the Chagatai khan.

[4] During this period, Kipchak-Kyrguish tribes in the area between the Irtysh and Yenisey rivers migrated to Tian Shan, where they mixed with the Mongols and local Turkish tribes. In the 15th and 16th centuries, the Kyrgyz people were formed. The Kyrgyz Khanate was established in Tian Shan by Khan Ajmet.

[5] The distinguishing feature of Kyrgyz social organization was the lack of a ruling class of princes or nobles, with authority exercised instead by the *manaps* (elders), whose leadership depended on their personal prestige. The tribes maintained a high degree of cohesion because they were constantly at war against neighbouring peoples.

[6] Between the 16th and 18th centuries, the Kyrgyz were subdued by the Oirat-Yungars. In the mid-18th century the Oirat-Yungar State was conquered by Chinese troops. Kyrgyzstan became a protectorate of the Tsin (Chin) Empire. In the early 19th century, Russia began to show interest in the region, which had been conquered by the khan Madali, of Kokand.

[7] In 1862, Russian troops supported the Kyrgyz rebellion against the Khanate of Kokand, and took Pishpek. The following year, Kyrgyz representatives from central Tian Shan surrendered to Russia, in the city of Verni (today Alma-Ata). In 1864, Northern Kyrgyzstan joined the Russian Empire, as part of the region of Semirechensk.

[8] In 1867, Czar Alexander II created the Regional Government of Turkistan. Russian colonists began arriving by the thousands, and the Kyrgyz saw their best lands confiscated, and turned over to immigrants for agricultural use.

[9] In 1875 there was an uprising against Khan Jodoyar of Kokand. The Russian Empire leapt in to annex the Khanate of Kokand and Southern Kyrgyzstan. Despite the fact that the major part of its population was Uzbek, the region of Fergana was formed as part of the Province of Turkistan. Kyrgyzstan was divided into 3 regions -Semirechensk, Syr Darya and Fergana- within the province of Turkistan, with Tashkent as its center.

[10] Friction between the various ethnic groups periodically flared up over the issue of land ownership and mandatory military service, finally triggering a major revolt by the Kyrgyz, which was brutally put down by the czarist regime.

[11] Soviet power was established in the district of Pishpek in December 1917, though there were strong opposition from guerrilla groups in the rest of the country. In 1921 and 1922, agrarian and water reforms were carried out.

[12] In 1924 the Autonomous Region of Kara-Kirghiz was formed, as a part of the Russian Federation. In 1925, its name was changed to the Autonomous Region of Kirghizia, becoming, in 1926, an "Autonomous Soviet Socialist Republic". In December 1936, Kyrgyzstan became a federated republic of the Soviet Union (USSR).

[13] Under socialism the Kyrgyz had to change their nomadic way of life radically, joining the new agricultural cooperatives or working in the industries created by the State (textile, leather, tobacco, lumber, metal and hydroelectric power). They also had to change "Pishpek" to "Frunze", the name of a famous Red Army general, and the USSR continued to recruit a major part of its soldiers from among the Kyrgyz.

[14] As a part of the pre-World War II five-year development plans, the USSR built large plants in Kyrgyzstan for the extraction of antimony, as well as the processing of its agricultural products. The first metal-processing plants and modern blast furnaces were set up during this period.

[15] During the war and in the decades which followed, the region's industry underwent further expansion, producing machinery, building materials and electric power.

[16] From 1961 until the end of the Brezhnev era, Turdakun Usubaliev was first secretary of the Kyrgyzstan Communist Party, which functioned as a branch of the CP of the Soviet Union. Kyrgyz writer Chingiz Aytmatov achieved international fame during the 1970s, as the author of a play dealing with the moral compromises made under Stalinism.

[17] From 1986, with the changes put in motion in the USSR by Mikhail Gorbachev, an independence movement developed in Kyrgyzstan. After the failed coup against Gorbachev in August 1991, the republic of Kyrgyzstan decided to separate from the USSR.

[18] On August 31, 1991, Kyrgyzstan's Soviet declared the new republic a democratic and independent state. On that occasion, a statement was made in recognition of the three new Baltic states of Estonia, Latvia and Lithuania.

[19] In Kyrgyzstan's first presidential elections, held in October of that year, Askar Akaev, former president of the Academy of Sciences and a physicist of international renown, ran unopposed and was elected. Akaev was among the few leaders - along with Russian president, Boris Yeltsin - who personally resisted the failed coup in August, at the Parliament in Moscow.

[20] Akaev was the first head of state in the former Soviet Union to ban the Communist Party and to proclaim the independence of his republic. In December, Kyrgyzstan signed the founding charter of the Commonwealth of Independent States (CIS) with 10 other ex-soviet republics

[21] In December 1991, President Akaev had to exercise his veto power over a law approved by the Parliament, establishing the exclusive right of the Kyrgyz to the land. He thus avoided inflaming aggression between the native population and the Uzbeks of Fergana, which could have turned into an even greater conflict with the neighboring republic of Uzbekistan. The previous year, in the region of Os, in the western part of the country, more than 200 Uzbeks were assassinated by bands of young Kyrgyz, manipulated, it was claimed at the time, by agents of the former Soviet secret police (KGB).

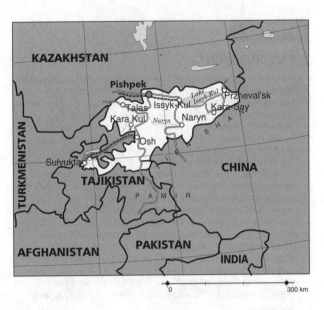

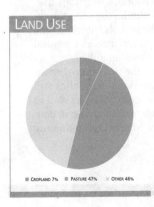

LAND USE

CROPLAND 7%   PASTURE 47%   OTHER 46%

PER CAPITA
$99

IMPORTS
459
EXPORTS
340

PER 100.000
LIVE BIRTH
43

PROFILE

ENVIRONMENT

Located in the northeastern part of Central Asia, Kyrgyzstan lies in the heart of the Tian Shan mountain range. It is bounded by China and Tajikistan in the south, Kazakhstan in the north, Uzbekistan in the west, the Pamir and Altai mountain ranges in the southwest and the Tian Shan range in the northeast. It has plateaus and valleys: in the north, the Chu and Talas valley; in the south, the Alai Valley; and in the southwest, the Fergana Valley. The climate is continental, with sharp contrasts between day and night temperatures. The eastern part of Tian Shan is dry, while the southwestern slopes of the Fergana range are rainy. The main rivers are the Narym and the Kara-Suu. Lake Issyk-Kul is the most important of the country's lakes. In the mountains there are forests and meadows, while desert and semi-desert vegetation abounds at lower altitudes. Metal deposits include lead and zinc; in addition, there are large coal reserves and some oil and natural gas deposits. Water pollution is a serious problem, since one third of the population obtains its water from rivers, streams, or wells.

SOCIETY

Peoples: Kyrgyz, 52.4%; Russians, 21.5%; Uzbeks, 12.9%; Ukrainians, 2.5%. **Religions:** Islam (Sunni) and Christian (Russian Orthodox Church). **Languages:** Kyrgyz (official), Russian, Uzbek. **Political Parties:** Democratic Movement of Kyrgyzstan, Erkin Kyrgyzstan Democratic Party, Communist Party.

THE STATE

Official Name: Respublika Kyrguyzstan. **Capital:** Pishpek (Bishkek), 631,000 inhab. (1990). **Other cities:** Os, 218,000 inhab.; Tokmok, 71,000 inhab.; Kara-Köl, 64.000 inhab. **Government:** President: Askar Akaev. **National Holiday:** December 15, Independence (1991). **Armed Forces:** 7,000.

DEMOGRAPHY

Urban: 39% (1995). **Estimate for year 2000:** 5,000,000.

COMMUNICATIONS

7.5 **telephones** per 100 inhabitants (1993).

ECONOMY

Consumer price index: 100 in 1990; 115.6 in 1993. **Currency:** 1 Australian dollars = 1$ (1994).

[22] Kyrgyzstan started to create a National Guard of a thousand troops, and asked CIS forces to remain in its territory. This military presence was justified, for "security reasons", until the republic was ready for full membership of the international community.

[23] In March 1992, Akaev reaffirmed his intention to maintain a strict neutrality, in spite of the attempts of Muslim states like Pakistan, Saudi Arabia or Iran to have closer ties. However, the president of Kyrgyzstan stated he was willing to cultivate closer relations with Turkey.

[24] In May 1993, a new constitution was approved while controversies and disagreements between Akaev and parliament continued, as parliament was opposed to his economic reform policies. A national currency, the som, was introduced that month. In the midst of growing tensions, many Kyrgyz of Russian origin, mainly intellectuals and engineers, began to flee the country.

[25] In 1994, Akaev continued his reforms policy recommended by the IMF. After the som's introduction, inflation fell from 40 and 50 per cent per month to a monthly 4 per cent, among the lowest in the former Soviet Union. At the same time, national production fell 19 per cent in 1992 and 16 per cent in 1993, imports -specially from Turkey and China- grew at an astounding pace and social inequality was increased.

[26] In February 1995, legislative elections did not significantly modify parliament's opposition to Akaev's economic liberalization policy. The president tried to extend his term through a referendum, but parliament decided to hold presidential elections in December, which were won by Akaev with 60 per cent of the vote.

Breastmilk, infant foods and pollution

Environmental degradation and the high rates of child mortality - which vary between 40 and 60 of every 1,000 live births in the Central Asian Republics - have led health professionals to investigate the link between environmental pollution and breastmilk.

[2] A 1993 report, based on samples taken in 1988 and 1989, identified three possible contaminants in human milk: agricultural pesticides, industrial chemicals and radioactive materials. However, the surveys carried out have not proven a link between infant mortality and breastfeeding.

[3] Working from available data, the report indicates that the most probable risk factors are: direct contact with pollutants in the air causing respiratory disorders, the lack of clean drinking water and adequate domestic and industrial waste treatment, causing high rates of infant diarrhea.

[4] Also, despite the fact that pollution might adversely affect the quality of breastmilk, the report highlights the benefits of breastfeeding, its nutritional and immunological value, its importance as a natural fertility control and its ecological qualities.

[5] The report concludes that the use of any breastmilk substitute in the Central Asian Republics will probably increase infant mortality rates, as well as levels of contamination (due to an increase in consumption of products from highly contaminating industries).

Laos

Lao

Population: 4,748,000 (1994)
Area: 236,800 SQ.KM
Capital: Vientiane

Currency: New Kip
Language: Lao

In the 14th century AD, Sam Sen Tal, a Burmese ruler unified the present northeast Thailand with most of the modern day territory of Laos, founding the flourishing state of Lang Xang, the "Land of a Million Elephants". In the late 18th century, the country split into three parts: Champassac, Vientiane and Luang Prabang.

2 By the 19th century, the Thais had established dominion over the land and Tiao Anuvong, prince of Vientiane, led an ill-fated nationalist rebellion in 1827. In 1892, the French invaded and by 1893, had established a protectorate over Luang Praband, with the rest of the country becoming part of French Indochina.

3 During World War II, Japan occupied Laos, and a pro-independence movement arose in Vientiane, led by three native princes, Petsarath, Suvana Fuma, leader of the National Progressive Party, and his half-brother, Tiao Sufanuvong, head of the Neo Lao Issara (Laos National United Front). In September 1945, Sufanuvong set up a provisional government, declaring independence and promulgating the constitution of "independent Laos", in spite of protests from King Sisavang Vong, reigning since 1904.

4 In early 1946, the country was again under French occupation. The Pathet Lao provisional government sought refuge in Bangkok, where the Pathet leaders organized the anti-colonialist struggle through the Lao Issara or "free Laos" movement.

5 On July 19, 1949, the Franco-Laotian Convention recognized Laotian independence "as part of the French Union". The Pathet Lao leaders saw this as mere formal independence and refused to recognize it. Suvana Fuma opted for negotiation, but the opposition coalition decided on active resistance; the military victories led to to a new treaty in 1953.

6 Differences between the Katay Sasorith, Suvana Fuma government, and the Pathet Lao were reconciled in November 1957. An agreement was reached, whereby the latter would participate in the political life of the country under the name "Neo Lao Haksat" (Laotian Patriotic Front), led by Tiao Sufanuvong.

7 The May 1958 elections were won by the left, worrying the US. Suvana Fuma's ruling party joined forces with the Independent Party to form the so-called Laotian People's Demonstration, together achieving a small majority. They formed a center-left coalition government, led by "neutralist" Suvana Fuma, with Sufanuvong as planning minister.

8 The US was strongly opposed and threatened to cut economic aid. This destabilized the government, and in August the leaders of the Committee for the Defence of National Interests took over. A military offensive against the Pathet Lao drove Sufanuvong from the capital, forcing him into guerrilla action in the forests.

9 In 1959, the army gained control, while the Pathet Lao controlled the strategic northern provinces and the central plain of Jars.

10 General Fumi Nosavang, the former defence minister, and his troops seized Vientiane on December 13, 1960, driving out Pathet Lao troops. A massive air raid left 1,500 dead and forced Vientiane to surrender.

11 The Thai and US supported Revolutionary Anti-Communist Committee, headed by Fumi Nosavang and Prince Bun Um claimed legitimacy to govern. On December 20 1960, however, Princes Suvana Fuma and Sufanuvong declared their intention of forming a national unity government.

12 The "neutralists" joined forces with the Pathet Lao and launched successful joint military campaigns.

13 In 1961 in Geneva, Britain and the Soviet Union initiated negotiations for a peaceful solution, and on January 19 1962, a final agreement was signed for a national unity government.

14 Growing US intervention resulted in the internationalization of the Vietnam war. Air raids on Laotian territory increased to 500 missions per day by 1970 and in nine years, Laos was subjected to a greater number of bombings than the whole of Europe during World War II.

15 The Pathet Lao declared an armistice in 1973, and the Vientiane government formed a new cabinet, including Pathet Lao members, with a Council of Ministers headed by Suvana Fuma. After the US defeat in Vietnam, right-wing groups lost the support they had been getting from the US. In 1975, the national unity cabinet gave way to a majority of Neo Lao Haksat ministers, and on December 1, a peaceful movement put an end to the monarchy. A People's Democratic Republic was proclaimed with Sufanuvong as president, led by the Lao People's Revolutionary Party (PPRL). Entrepreneurs and state bureaucrats left the country en masse, ruining the economy and crippling public administration.

16 The government nationalized banks and reorganized the public sector. Rice production rose from 700,000 tons in 1976 to 1,200,000 in 1981, when grain self-sufficiency was achieved. For access to the sea and to reduce dependence on Thailand, a road was constructed to the Vietnamese port of Danang, and an oil pipeline to Vietnam's refineries.

18 In 1986, Phumi Vengvichit became president. In November, Kaysone Phomvihane was re-elected secretary general of the PPRL, subsequently becoming prime minister.

19 A bloody border war with Thailand, at the end of 1987 resulted in a great loss of life, but ended rapidly in a cease-fire.

20 In 1988 diplomatic relations were renewed with China, and in early 1989 cooperation agreements were signed with the US, which granted a symbolic $10 million to combat the cultivation and trafficking of opium.

21 The changes in Eastern Europe resulted in the suspension of all Soviet economic aid to Laos, and trade fell by 50 per cent.

22 In 1991, the poor economic situation was exacerbated by floods

PUBLIC EXPENDITURES

% of GNP — 1992

MILITARY 6.1 · EDUCATION 1.1 · HEALTH 1

WORKERS

1994

MEN 55% · WOMEN 45%

AGRICULTURE 76% · INDUSTRY 7% · SERVICES 17%

ENVIRONMENT

Laos is the only landlocked country in Indochina. The territory is mountainous and covered with rainforests. The Mekong river valley, stretching down the country from north to south, is suited to agriculture, basically rice. It is estimated that 40% of the arable land was left barren as a result of the 25-year war. The climate is tropical and the lowlands are prone to disasters such as the 1978 drought and the 1988 flood. The main environmental problems are intense deforestation and as a result of the latter, dwindling water supplies and a loss of 70% of natural habitats.

SOCIETY

Peoples: 60% of Laotians are descendants of the Lao ethnic groups who inhabit the western valleys. The inhabitants of the mountains account for more than one-third of the total population, while 5% are of Chinese and Vietnamese origin.. **Religions:** Buddhist 57.8%; traditional religions 33.6%; Christian 1.8%, of which Roman Catholic 0.8%, Protestant 0.2%; Muslim 1.0%; atheist/no religion 4.8%; Chinese folk-religions 0.9%. **Languages:** Lao (official); minor ethnic group languages, and French. **Political Parties:** The Lao People's Revolutionary Party (PPRL), Marxist-Leninist, is in power. It originated in the old Neo Laotian Haksat. **Social Organizations:** The Patriotic Youth and the Association of Patriotic Women. The National Liberation Front, based in the Meo ethnic minority, wages guerrilla warfare and is accused of receiving Chinese support.

THE STATE

Official Name: Sathalanalat Paxathipatai Paxaxon Lao. **Administrative divisions:** 16 provinces, sub-divided into municipalities. **Capital:** Vientiane, 178,203 inhab. (1985). **Other cities:** Savannakhét 96,652 inhab.; Lovangphrabang 68,399 inhab. **Government:** Republic following the socialist model. Nouhak Phoumsavan, President since November 1992. Khamtai Siphandon, Prime Minister since August 1991. Legislature: Supreme Assembly of the People. **National Holiday:** December 2, Proclamation of the Republic 1975. **Armed Forces:** 37,000 troops (conscripts) (1994). **Paramilitaries:** 100,000 members of the Self-defence Militia Forces.

DEMOGRAPHY

Urban: 20% (1995). **Annual growth:** 2.9% (1992-2000). **Estimate for year 2000:** 6,000,000. **Children per woman:** 6.7 (1992).

HEALTH

One **physician** for every 4,545 inhab. (1988-91). **Under-five mortality:** 138 per 1,000 (1995). **Calorie consumption:** 97% of required intake (1995). **Safe water:** 45% of the population has access (1990-95).

EDUCATION

Illiteracy: 43% (1995). **School enrolment:** Primary (1993): 92% fem., 92% male. Secondary (1993): 19% fem., 31% male. University: 2% (1993). **Primary school teachers:** one for every 29 students (1992).

COMMUNICATIONS

83 **newspapers** (1995), 83 **TV sets** (1995) and 89 **radio sets** per 1,000 households (1995). 0.2 **telephones** per 100 inhabitants (1993). **Books:** 88 new titles per 1,000,000 inhabitants in 1995.

ECONOMY

Per capita GNP: $320 (1994). **Annual inflation:** 24.20% (1984-94). **Currency:** 716 new kips = 1$ (1993). **Cereal imports:** 8,000 metric tons (1993). **Fertilizer use:** 42 kgs per ha. (1992-93). **Imports:** $564 million (1994). **Exports:** $300 million (1994). **External debt:** $2,080 million (1994); $438 per capita (1994). **Debt service:** 7.7% of exports (1994). **Development aid received:** $199 million (1993); $43 per capita; 15% of GNP.

ENERGY

Consumption: 38 kgs of Oil Equivalent per capita yearly (1994), -19% imported (1994).

---

and pest infestations in a quarter of the country's farmlands. Consequently, over 200,000 tons of rice were imported to feed the population.

[23] Also in 1991, Thailand and Laos agreed to a repatriation plan for 60,000 Laotian refugees by 1994, also signing a Cooperation and Security Treaty. Thai investments were meanwhile concentrated in banking and trade. Laos has also put a lot of effort into establishing closer ties with China.

[24] In September 1991, 70-year-old Kaysone Phomvihane was named president. Upon his death a year later, Prime Minister Khamtay Sifandon assumed interim presidency of the country and the party (PPRL).

[25] In December 1992, legislative elections were called; only the PPRL and a few independent government-authorized candidates took part. This essentially represented a continuation of the one-party system. Anti-government demonstrations were banned, and scores of government opponents were imprisoned.

[26] The Supreme Assembly of the People met in February 1993, and named Nouhak Phumsavanh president of the republic, and Khamtai Siphandon leader of the PPRL.

[27] Major free-market oriented reforms were introduced, but without drastically changing the political system. In 1993, Prince Suvana Fuma was authorized to return to the country in an unofficial capacity to represent foreign companies.

[28] Following IMF-suggested strategies, the country managed to bring down the inflation rate - averaging 46 per cent in the 1980s - to an annual rate of 10 per cent, meanwhile the GDP grew by 7 per cent. It is estimated that part of that economic "growth" resulted from the introduction into the market of products and activities previously not accounted for, and not from an actual increase in production.

[29] The USAID (United States Agency for International Development) approved a $9.7 million loan to build roads in several regions of the country, arguing that access to markets would improve living conditions.

[30] Deforestation has become a serious environmental problem. The timber felled increased from 6,000 cubic meters in 1964 to over 600,000 in 1993. That year, the government began to restrict timber exports.

[31] As a result of this, there was an increase in the illegal felling of trees, and in 1993, half of all timber was logged illegally. In March 1994, the World Bank extended a loan given to Laos for reforestation, at an initial value of $8.7 million. A number of international environmental organizations have criticized the project, as it gives the funds directly to the government, with little or no participation from the local communities and for placing emphasis on commercial tree plantations, an activity that could further endanger the forests and the livelihood of its inhabitants.

[32] On April 8 1994, the "Friendship Bridge" over the Mekong was inaugurated, uniting Laos and Thailand. For Laos this reconciliation represented a distancing from Vietnam and marked the way for economic and cultural integration with Thailand - a more prosperous nation. The Thai banks, communication networks, transport companies and factories soon dominated investment in the Laos economy, which had now lost its socialist principles.

[33] A law approved in March of that year had established the rules for foreign investment, eliminating the remnants of the planned economic system. Another law guaranteed union rights and updated the labor legislation. Political liberalization, however, was not included in this reform: the government was determined to maintain its Communist identity.

[34] Laos surprised the Association of South East Asian Nations (ASEAN) in mid-1995, when it expressed its desire of becoming a full member within two years. The will to liberalize was also reflected in the ratification of a frontier treaty with Myanmar/Burma. The United States lifted a veto it had imposed on the nation since the Vietnam War limiting US economic aid.

[35] In recent years, tourism has increased considerably. The royal city of Lovangphrabang was declared national heritage and two-and-a-half million hectares of forests were converted into protected areas.

[36] From the macroeconomic point of view, the progressive weakening of the *kip* - the local currency - during 1995 and 1996 accentuated the monetary deficit of the nation. The price control imposed by the government produced inflation and discouraged investment. A small indication of recovery arrived in August when the Asian Development Bank granted a $20 million loan for urban infrastructure.

Latvia

Latvija

Population: 2,547,000 (1994)
Area: 66,547 SQ.KM
Capital: Riga

Currency: Lat
Language: Latvian

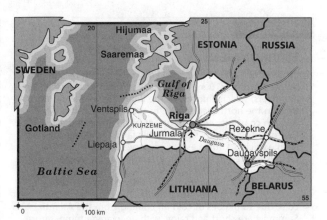

The first inhabitants of present-day Latvia were nomadic tribes of hunters, fishers and gatherers who migrated to the forests along the Baltic coast, after the last glaciers had retreated. Around 2,000 BC, these groups were replaced by the Baltic peoples, Indo-European tribes who began farming and established permanent settlements in Latvia, Lithuania and eastern Prussia.

[2] Latvians come from the main branches of the ancient Baltic peoples, who came into contact with the Roman Empire through the amber trade. This activity, which reached its peak during the first two centuries of the Christian era, was brought to a halt by Slav expansion toward central and eastern Europe. At this time, the Balts' trade and cultural relations turned northward, to their Scandinavian neighbors.

[3] The Danes used the Dvina and Dnepr rivers in their expansion toward the steppes north of the Black Sea, and therefore crossed Latvian territory. The Swedes and the Russians both claimed these lands during the 10th and 11th centuries and in the 12th century, German warriors and missionaries came to the Latvian coast. As it was inhabited at the time by the Livs, the Germans called it Livonia. In 1202, the bishop of the region, under authorization from Rome, established the Order of the Knights of the Sword.

[4] Before becoming the Knights of the Teutonic Order, in 1237, the Germans had subdued and Christianized the tribes of Latvia and Estonia. The Teutonic Knights created the so-called Livonian Confederation, consisting of areas controlled by the Church, free cities and regions governed by knights. In the mid-16th century, rivalries within Livonia became more pronounced with the expansion of Protestantism and discontent among the peasants.

[5] During this period, Latvians benefited from Riga's participation in the Hanseatic League, a German mercantile society which attained a high level of prosperity. Nevertheless, the Latvians were treated by the Germans like any vanquished people: the local nobility was done away with, and the peasants were forced to pay tithes and taxes, paying with labor when unable to pay otherwise. After being defeated by Lithuania and Poland, the power of the Teutonic Order declined. However, as the knights' power diminished, their exploitation of the Latvians actually increased.

[6] When Russia invaded the region in 1558, to halt Polish-Lithuanian expansion, the Order fell apart and Livonia was partitioned. At the end of the Livonian War, in 1583, Lithuania annexed the area north of the Dvina river, the south remained in Polish hands and Sweden kept the north of Estonia. In 1621, Sweden occupied Riga and Jelgava; Estonia and the northern part of Latvia were subsequently ceded to Sweden by the Truce of Altmark (1629).

[7] The region west of Riga, on the Baltic Sea, was organized into the Duchy of Courland, becoming a semi-independent vassal of Poland. In the mid-17th century, Courland became known as a major naval and trade center for northern Europe, and even had colonial aspirations. Duke Jacob led a brief Latvian occupation of Tobago, in the Caribbean, and of another island in the Gambia River delta in Africa. The name Great Courland Bay, in Tobago, dates back to this time.

[8] Sweden kept these territories until the Great Northern War, when it was forced to cede them to Russia, under the Peace of Nystad. In 1795, after the three partitions of Poland, Livonia was finally subdivided into three regions within Russia: Estonia (the northern part of Estonia); Livonia (the south of Estonia and north of Latvia) and Courland. The Russian Revolution of 1905 gave rise to the first instances of Latvian national reaffirmation.

[9] The peasants revolted against their German feudal lords, and the Russian rulers. Although the rebellion was put down by czarist troops, this uprising set the stage for the war of independence thirteen years later. After the Russian Revolution of 1917, the Latvian People's Council proclaimed the country's independence on November 18, 1918. A government led by the leader of the Farmers' Union, Karlis Ulmanis, was formed.

[10] Far from having its desire for independence and sovereignty respected, Latvia was attacked by German troops and by the Red Army. Only in 1920 was Latvia able to sign a peace treaty with the USSR, by means of which the latter renounced its territorial ambitions. In 1922, a constituent assembly established a parliamentary republic. The international economic crisis of the 1930s, and the polarization of socialists and Nazi sympathizers led to the collapse of the Latvian government. In 1934, Prime Minister Ulmanis suspended parliament and governed under a state of emergency until 1938.

[11] With the outbreak of World War II, according to the secret Russo-German pact, Latvia remained within the USSR's sphere of influence. In 1939, Latvia was forced to sign a treaty permitting the Soviets to install troops and bases on its soil. In 1940, it was invaded by the Red Army, and a new government was formed, which subsequently requested that the republic be admitted to the USSR.

[12] During the German offensive against the USSR, between 1940 and 1944, Latvia was annexed to the German province of Ostland, and its Jewish population was practically exterminated. The liberation of Latvia by the Red Army meant the re-establishment of Soviet government; before the Soviet forces arrived, 65,000 Latvians fled to Western Europe.

[13] In 1945 and 1946, about 105,000 Latvians were deported to Russia, and the far northeastern corner of Latvia - with its predominantly Russian population - was taken away from Latvia to form part of the USSR. In 1949, forced collectivization of agriculture triggered a mass deportation of Latvians, with about 70,000 being sent to Russia and Siberia. In 1959 the president of Latvia's Supreme Soviet, Karlis Ozolins, was dismissed because of his nationalist tendencies.

[14] Armed Latvian resistance to the Soviet regime was finally put down, in 1952. All symbols of Latvian independence - the national anthem, the flag, and national monuments and history - were banned or adapted to the new regime. Russian became the official language, and massive immigration of Russians and other nationalities began. To the nationalists, this was taken to be a deliberate colonization policy, aimed at diminishing the influence of the indigenous population.

[15] Until the 1980s, Latvian resistance was expressed in isolated actions by political and religious dissidents, which were systematically repressed by the regime; in addition, some nationalist campaigns were carried out by exiles. From 1987, the policy of glasnost (openess) initiated by Mikhail Gorbachev in the USSR, gave hope to Latvian aspirations, permitting public political demonstrations, and the reinstatement of the national symbols.

[16] In June 1987, some 5,000 people gathered in front of the Monument to Liberty, in Riga, to honor the victims of the 1941 Soviet deportations. This marked the beginning of increasingly important political agitation, in which independence began to be discussed openly once again. In 1988,

LAND USE

CROPLAND 28%   PASTURE 13%   OTHER 59%

IMPORTS
1,367
EXPORTS
967

PROFILE

ENVIRONMENT

Latvia's terrain is characterized by softly rolling hills (highest point: Gaizins, 310 m), and by the number of forests, lakes and rivers, which empty into the Baltic Sea and the Gulf of Riga, in the northeastern part of the country. Latvia has a 494 km coastline, with important ports and attractive beaches. The country's most fertile lands lie in the Zemgale Plain, which is known as the country's breadbasket. The plain is located in the south, extending as far as the Lithuanian border. The Highlands, which make up 40% of the land, lie in the western and northern parts of the country, crossing over into Estonia. The climate is humid and cold, due to masses of cold air coming from the Atlantic. Summers are short and rainy, with an average temperature of 17 degrees C; winters last from December to March, with temperatures below zero, sometimes as low as -40 degrees C. Two-thirds of all arable land is used for grain production, and the rest for pasture. The main industries are metal engineering (ships, automobiles, railway passenger cars and agricultural machinery), followed by the production of motorcycles, electrical appliances and scientific instruments. Industrial wastes and the pollution of the country's rivers and lakes are problems which remain to be solved.

SOCIETY

Peoples: Latvians and Lithuanians constitute the two main branches of the Baltic Indo-European peoples, with a distinct language and culture that sets them apart from the Germans and Slavs. Ethnic Latvians make up 52% of the country's population, followed by Russians, 34%; Poles, Belarusians, Ukrainians, Lithuanians and Estonians account for the remaining 14%. **Religions:** The majority is Protestant (Lutheran), followed by Catholics.
Languages: Latvian (official); Russian and Polish. **Political Parties:** National Bloc; National Conciliation Bloc.

THE STATE

Official Name: Latvijas Republika. **Capital:** Riga, 804,000 inhab. (1993). **Other cities:** Daugavpils 120,917 inhab.; Liepaja 95,046 inhab.; Jelgava 69,411 inhab.; Jurmala 55,256 inhab. **Government:** Parliamentary republic. Guntis Ulmanis, President. Single-chamber parliament (Saeima), made up of 201 members, elected by direct vote. **National holiday:** November 18, Independence (1918).

STATISTICS

DEMOGRAPHY

Urban: 72% (1995). **Annual growth:** -0.1% (1991-99). **Estimate for year 2000:** 3,000,000. **Children per woman:** 1.8 (1992).

EDUCATION

School enrolment: Primary (1993): 82% fem., 82% male. Secondary (1993): 90% fem., 84% male. University: 39% (1993).

COMMUNICATIONS

26.8 **telephones** per 100 inhabitants (1993).

ECONOMY

Per capita GNP: $2,320 (1994). **Annual growth:** -6.00% (1985-94). **Annual inflation:** 69.80% (1984-94). **Consumer price index:** 100 in 1990; 2,989.0 in 1994. **Currency:** lat. **Cereal imports:** 11,000 metric tons (1993). **Fertilizer use:** 982 kgs per ha. (1992-93). **Imports:** $1,367 million (1994). **Exports:** $967 million (1994). **External debt:** $364 million (1994), $143 per capita (1994). **Debt service:** 2.1% of exports (1994).

ENERGY

Consumption: 1,755 kgs of Oil Equivalent per capita yearly (1994), 88% imported (1994).

---

the National Independence Movement of Latvia (LNNK) demanded an end to the Russification of the country, and called for freedom of the press and for the formation of independent political parties.

[17] In September 1988, the Environmental Protection Club of Latvia (VAK) organized an ecology-awareness rally, in which 30,000 people joined hands all along a stretch of the Baltic Sea coast. The following month, the Latvian Communist Party renewed its leadership, incorporating politicians who were identified with the reforms and who had broad popu-

lar support. The national flag was legalized, and Latvian was adopted as the country's official language.

[18] In October 1988, close to 150,000 people gathered to celebrate the founding of the Popular Front of Latvia (LTF), which brought together all of Latvia's recently formed social and political groups, as well as militant communists. A month later, for the first time since Soviet occupation, hundreds of thousands of Latvians commemorated the anniversary of the 1918 declaration of independence. The LTF began to have influence with the local govern-

ment, and with the Moscow authorities.

[19] A year later, the LTF Congress endorsed the country's political and economic independence from the Soviet Union. Despite Moscow's resistance to Latvia's secession, the LTF's policy of carrying out changes peacefully met with widespread popular support, from citizens of Russian or other origins. The LTF program for non-violent change included public demonstrations, free elections and change through parliamentary procedure. Latvia's 1938 Constitution went into effect once again, for the first time since the 1940 Soviet occupation.

[20] May 4, 1990 marked the Declaration of the Re-establishment of Independence, as well as the reinstatement of the 1922 Constitution. In September 1991, the new Council of State of the USSR, in its inaugural session, formally recognized the independence of the Baltic republics. They were also immediately recognized by a number of countries, and admitted as new members to the UN in the General Assembly session held that same month. The three Baltic States initiated negotiations with the European Community (EC), in the areas of trade and financial assistance. In February

1992, Latvia and Russia signed an agreement on the withdrawal of the former USSR's troops, stationed within Latvian territory.

[21] In June 1993, a new parliament was elected which appointed Guntis Ulmanis as president of the country. The beginning of the economy's liberalization process caused a strong growth of unemployment. Foreign investments increased in 1994 but Latvian economy still depended on Russia, its main supplier of fuel and first exporting market. Furthermore, despite a massive privatization of state companies, the budget and the trade balance kept showing a deficit.

[22] In 1995, lack of controls over commercial banks enabled the bankruptcy of several financial agencies, which further enlarged the public deficit. The September legislative elections did not reveal a definite winner since nine parties obtained between 5 per cent and 16 per cent of the vote. A deal between the conservative National Bloc and two left-wing parties enabled the nomination of Prime Minister Andris Skele in December.

[23] In June 1996, Parliament re-elected Guntis Ulmanis as president with 53 votes out of 100.

Lebanon

Lubnan

Population: 3,930,000 (1994)
Area: 10,400 SQ.KM
Capital: Beirut (Bayrut)

Currency: Pound
Language: Arabic

Much of present-day Lebanon corresponds to the ancient land of the Phoenicians, who probably arrived in the region in about 3000 BC. Commercial and religious connections were established with Egypt after about 2613 BC and continued until the end of the Egyptian Old Kingdom and the invasion of Phoenicia by the Amorites (c. 2200 BC). Other groups invading and periodically controlling Phoenicia included the Hyksos (18th century BC), the Egyptians of the New Kingdom (16th century BC), and the Hittites (14th century BC). Seti I (1290-79 BC) of the New Kingdom reconquered most of Phoenicia, but Ramses III (1187-56 BC) lost it to invaders from Asia Minor and Europe. Between the withdrawal of Egyptian rule and the western advance of Assyria (10th century BC), the history of Phoenicia is primarily the history of Tyre. This city-state rose to hegemony among Phoenician states and founded colonies throughout the Mediterranean region. The Achaemenians, an Iranian dynasty under the leadership of Cyrus II, conquered the area in 538 BC. Sidon, 20 miles (32 km) north of Tyre, became a principal coastal city of this empire. In 332 BC Tyre capitulated to the army of Alexander the Great after resisting for eight months. This event marked the demise of Tyre as a great commercial city, as its inhabitants were sold into slavery. In 64 BC Phoenicia was incorporated into the Roman province of Syria.

[2] Emperors embracing Christianity protected the area during the later Roman and Byzantine periods (c. AD 300-634). A 6th-century Christian group fleeing persecution in Syria settled in what is now northern Lebanon, absorbed the native population, and founded the Maronite Church. In the following century, Arab tribes settled in southern Lebanon after the Muslim conquest of Syria. Four hundred years later, many of this Arab group coalesced their beliefs into the Druze faith. In the coastal towns the population became mainly Sunni Muslim. By the end of the 11th century Lebanon had become part of the crusader states, and it later became part of the Mamluk state of Syria and Egypt. Lebanon was able to evolve a social and political system of its own between the 15th and the 18th century. Throughout this period European, particularly French, influence was growing. In 1516 the Ottoman Turks replaced the Mamluks. The social system came under severe strain as the Christian population grew.

[3] Around 1831 the vigorous Egypt of Muhammed Ali extended its influence northwards undermining the decadent Turkish empire. The European powers did not want the Turkish empire to crumble until Europe was in a position to carry off the spoils. They conveniently decided that Christians anywhere in the world could be equated with Europeans, meriting their protection, and they began to support the Maronite Arab Christians against Egypt. Between 1831 and 1834, five powers intervened in the "Syrian question". Russia and Austria devoted their efforts exclusively to the Balkans, and France and Britain were left to dispute domination of the Arab countries, laboriously shunting Prussia aside. The Ottoman Turks ended the local rule of the Druze Shihab princes in 1842, exacerbating already poor relations between the Maronites and the Druze. These relations reached a low ebb with the massacre of Maronites by Druze in 1860.

[4] The French intervened on behalf of the Christians, forcing the Ottoman sultan to form an autonomous province within the Ottoman Empire for the mountainous Christian area, known as Mount Lebanon. In 1919, Syria and Lebanon became French protectorates, while Britain took Egypt, Jordan and Iraq. For administrative purposes, France separated Lebanon from Syria and when the army pulled out in 1947, they left behind two separate states.

[5] Camille Chamoun was elected president of Lebanon in 1952, following a clearly pro-Western foreign policy. The 1957 parliamentary elections were marked by Muslim demonstrations and riots, demanding closer alignment with Egypt and Syria, and rejecting reelection for the Maronite president.

[6] By the following year the rioting had grown into full scale insurrection, pitting Muslims against Christians in a bloody civil war. In July 1958, President Chamoun allowed 10,000 US Marines to disembark in order to "pacify" the country. The foreign troops remained in Beirut (Bayrut) and in other strategically important Lebanese cities until October of that year.

[7] By the early 1970s, the population had become mostly Muslim, and they began to question the traditional political system which required a Christian president and a Sunni Muslim prime minister in order to maintain a balance between the two communities. The Shiites had no specified role.

[8] Differences between the various religious and ethnic communities were combined with an eminently unjust social and economic system. Predatory colonial exploitation had exhausted the traditional cedars of Lebanon, leaving the land barren and farmers impoverished. The new distribution of labor and wealth kept trade control for the Maronites, while the Muslims worked as artisans, laborers and farm workers.

[9] These tensions erupted in civil war again, when the Christian right wing used the "Palestinian problem" as a pretext for conflict in 1975. Lebanon, which had remained neutral in the 1973 Arab-Israeli War, granted refuge to 300,000 Palestinians in the southern part of its territory. Israel used this fact to justify frequent incursions and bombings against civilians. Meanwhile, the Lebanese Christian Phalange militia unleashed its power on the Palestinian refugee camps.

[10] In 1976, Syrian troops, part of an Arab-League peace force, put a stop to the fighting and guaranteed national unity. The underlying causes of the civil war persisted, and Israeli attacks continued. In 1981, Israeli artillery, together with a unit of the former Lebanese commander Saad Haddad, bombarded

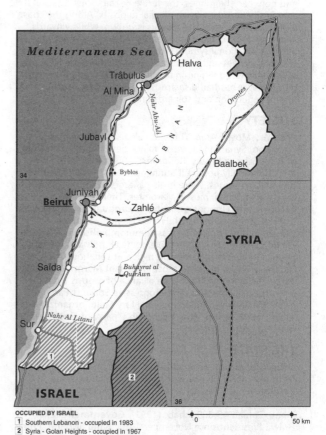

MEDITERRANEAN SEA · Trâbulus · Halva · Al Mina · Nahr Abu-Ali · Orontes · Jubayl · Baalbek · Byblos · Juniyah · **Beirut** · Zahlé · SYRIA · Saïda · Buhayrat al Qir'Awn · Nahr Al Litani · Sur · [1] · [2] · ISRAEL

OCCUPIED BY ISRAEL
[1] Southern Lebanon - occupied in 1983
[2] Syria - Golan Heights - occupied in 1967

0   50 km

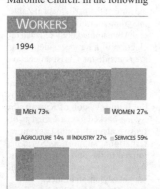

WORKERS

1994

MEN 73%	WOMEN 27%

AGRICULTURE 14%	INDUSTRY 27%	SERVICES 59%

ENVIRONMENT

One of the smallest Middle East countries, Lebanon has a fertile coastal plain where most of the population lives, with a mild Mediterranean climate making it rainy in the winter; two parallel mountain ranges, the Lebanon Mountains (whose highest peak is Sauda, at 3,083 m) and the Anti-Lebanon Mountains, with temperate forests on their slopes, and between the ranges, the fertile Bekaa Valley. The soil is very fertile because of rich alluvial deposits. Along the coast, a dry bush known as the *maquis* is grown, while vinyards, and wheat, olives, and oranges are cultivated. Cedar has become a national symbol. This wood was used to build the Phoenician fleet and temples. Nowadays there are little over 400 cedars left, ranging in age between 200 and 800 years old. The war has had disastrous consequences, causing a loss of vegetation and soil erosion.

SOCIETY

Peoples: Mostly Arab. There are also small Armenian, European, Syrian and Kurdish minorities, and a large group of Palestinian refugees. **Religions:** 75% are Muslim, with the traditionally powerful Sunni now being challenged by the militant Shiites. There is also a large Druze population. 25% are Christian, mostly Maronites. There are also Catholics and Orthodox Christians. **Languages:** Arabic (official and predominant). French is widely spoken, Armenian and English are less common. **Political Parties:** There are innumerable political, political-military, political-religious factions and militia in Lebanon. The Lebanese Phalange is the most important Maronite Christian party. The National Front is an alliance of groups from the left, especially Muslims. The Progressive Socialist Party is primarily Druzian. The political-military Shiite organization Amal ("Hope"), pro-Syria. Hezbollah ("party of God"), Shiite guerrilla group, pro-Iran.

THE STATE

Official Name: al-Jumhouriyah al-Lubnaniyah. **Administrative divisions:** 6 governmental divisions. **Capital:** Beirut (Bayrut) 1,100,000 inhab. (1991). **Other cities:** Tripoli (Tarabulus) 240,000 inhab.; Juniyah 100,000 inhab; Zahlah 45,000 inhab. (1991). **Government:** Elias Hrawi, President since November 1989. Rafiq al-Hariri, Prime Minister since May 16, 1992. **National Holiday:** November 22, Independence Day (1943). **Armed Forces:** 44,300 troops (1995). **Paramilitaries:** External Security Forces: UN peacekeeping forces 5,000; Syrian army 35,000. Civilian militias: Party of God 3,000; South Lebanese Army 2,500.

the cities of Tyre and Sidon (Saida), while Syrian troops installed anti-aircraft missiles in the Bekaa Valley. That July, Israeli jets launched a number of attacks against Palestinian positions, including a wave of air raids on western Beirut; they left 166 dead and 600 wounded.

[11] In June 1982, Israeli troops invaded Lebanon by land, sea and air. Tyre and Sidon were quickly overrun, and Nabataea and Tripoli (Tarabulus) were devastated by bombings. This marked the beginning of Israel's "Peace for Galilee" operation.

[12] Beirut, the most important political and cultural capital in the Arab world, was virtually destroyed. Thousands of civilians were killed. Bridges, oil pipelines, airports, hospitals, schools, major buildings and modest homes, factories and museums were razed.

Eight thousand Palestinians and Lebanese were captured and held in deplorable prison camps under suspicion of having belonged to the resistance. Israeli army officers and Mossad secret service agents ransacked the Palestine Liberation Organization's (PLO) Center for Palestinian Studies, where 80 scholars had spent 17 years studying Palestinian history and culture, and transferred the research material to an unknown destination.

[13] The PLO finally agreed to pull out of Beirut, providing that the evacuation was carried out under international supervision. US, French and Italian soldiers provided the necessary protection, and the people of western Beirut gave a heroes' farewell to the freedom fighters, who went to seven other Arab countries.

[14] On August 23 1982, what was left of the Lebanese Congress named Israel's nominee, Bashir Gemayel, the only candidate, to succeed President Elias Sarkis. The Maronite leader did not live to take office, however. He died on September 14 in a dynamite attack on the Phalange's eastern Beirut command headquarters. No one accepted responsibility for the act.

[15] The following day, Beirut was occupied by Israeli troops. On September 16 1982, a militia group led by former commander Haddad burst into the Palestinian refugee camps of Sabra and Chatila and assassinated hundreds of unarmed civilians, including old people, women and children.

[16] Later investigations left no doubt as to Israel's responsibility for the killings, as Israeli officers had encouraged the right-wing militia to act.

[17] While Israeli troops still held half of Lebanon, Amin Gemayel was elected president under the same conditions as his younger brother Bashir.

[18] The new president was unable to eliminate distrust among the various communities of Lebanon. A majority of the governing Falange party sought to replace the 1943 National Alliance with a new constitutional dictate to divide the state into districts under a central federal government. This raised the prospect of Lebanon's political division into several religious mini-states. Sunni and Shiite leaders favored administrative (but not political) decentralization, and the Druze community wanted greater autonomy.

[19] In May 1983 Israel and Lebanon signed an agreement to cease hostilities. Israel promised to withdraw from Lebanese territory, along with the other foreign troops. Lebanon in turn promised not to harbor any armed groups, organizations, bases, offices or structures with the aim of "carrying out raids on the other party's territory."

[20] In July 1984 the Lebanese currency, which had remained relatively stable since the beginning of the war in 1975, slumped abruptly, producing unprecedented inflation.

[21] Recession in the Gulf constituted the death blow to Beirut's economy which was already severely weakened by the withdrawal of foreign currency. This caused a balance of payments deficit of over $1.5 billion in 1984.

[22] Before the Israeli army official-ly withdrew from Lebanon in 1985, it ensured that the Christian militia had displaced the Muslims from southern Lebanon guaranteeing a "friendly" civilian population in the 10-km security zone they imposed.

[23] In September 1988, pro-Israeli Maronite General Michel Aoun took over the presidency, which had been vacant since Amin Gemayel's constitutional mandate was truncated in a palace coup. The country was now governed by two rival administrations - Aoun's, and that of Muslim Prime Minister Selim al-Hoss.

[24] In October 1989, Lebanon's single chamber parliament met, for the first time since independence, outside the country and under the auspices of the Arab League. The league includes Saudi Arabia, Algeria and Morocco, who acted as intermediaries in negotiations between the Lebanese factions. The meeting in Taif sealed the peace and laid down an alternative to the political system that had been in place since 1943. Under a new constitution, the next president would be elected indirectly by the parliament.

[25] On October 12, Christian and Muslim Lebanese legislators announced a national reconciliation agreement, granting greater power to Muslims and providing for the withdrawal of Syrian troops from Lebanon. General Aoun rejected the agreement because it was instituted by Saudi Arabia, and he considered it a "Syrian trap".

[26] On November 5 1989, René Moawad, a Maronite Christian sympathetic to Muslim causes, was unanimously elected president. Moawad was killed by a car-bomb scarcely 17 days before taking office.

[27] On November 24, 1989, Elias Hraoui, another Maronite Christian, was elected president by a meeting of the parliament at Zahle, in Syrian-controlled territory. General Aoun immediately rejected the election of a new president in an area outside the Christian enclave under his control.

[28] In October, 1990, taking advantage of the new situation created by the Iraqi invasion of Kuwait, the Syrian-backed forces started an offensive against Aoun. Lacking international support, as Syria belonged to the anti-Iraq coalition, Aoun was soon defeated and he sought asylum in France.

[29] In December 1990, a national

unity government was formed, for the first time since the beginning of the war. It incorporated the Lebanese Forces (Christian), the Amal (Shiite), the PSP (Druze), and the pro-Syrian parties.

[30] This government requested the UN to extend their mandate to the peace corps in the South (UNIFIL) in February 1990, and they demanded the withdrawal of Israeli troops from the borderline territories they occupied.

[31] On May 22 1991, the presidents of Lebanon and Syria signed a Brotherhood, Co-operation and Co-ordination Agreement in Damascus, the Syrian capital. Syria recognized Lebanon as a separate and independent state. The agreement established Syrian-Palestinian co-operation in military, security, cultural and economic matters, and it was ratified by the majority of the Lebanese parliament. The Israeli government, the Phalangist party and the Lebanese Forces militia (both Christian and traditionally anti-Syrian), opposed the treaty as they believed that it gave Syria control over Lebanon's domestic matters.

[32] Early in July, six thousand soldiers of the Lebanese army took over the territories occupied by the PLO around the port of Sidon, in the south of the country. This offensive forced the PLO out of its main operations base for campaigns against Israel. On July 7, in spite of the apparent defeat of the PLO, Israel stated that it would not withdraw its forces from the security zone.

[33] In November, Israel increased its offensive on the security zone, demanding that the Lebanese army should leave the territories within twelve hours. These actions caused the massive exodus of nearly 100,000 Shiites.

[34] In December, 1991, Lebanon was granted loans amounting to $700 million, to be used over the next three years for the rehabilitation and reconstruction of the country.

[35] The external and domestic war between 1975 and 1990, resulted in 94,000 civilian dead, 115,000 wounded, 10,000 handicapped and 20,000 missing. 800,000 people were displaced to other areas. The minimum wage dropped from $187 to $27 while the unemployment rate rose from 5.4 to 22 per cent. Deforestation accelerated soil erosion and there was significant damage to the country's archaeological and cultural heritage.

[36] On February 16 1992 - eight days before peace talks between Arabs and Israelis were due to resume - an Israeli air raid killed Sheikh Abbas Mussawi, head of the Hezbollah (Iranian-backed Islamic group). Six days later, the Shiite guerrillas suspended attacks against Israel as the result of an agreement between the Hezbollah, Amal (another Islamic grouping) and the representatives of Syrian and Lebanese forces. The Israeli defence minister said his country would retaliate for any attacks from that organization. The United Nations Security Council condemned the use of violence in the region, at the request of the Lebanese ambassador.

[37] In March, an intense Israeli mobilization at the border caused combats with the pro-Iranian Muslim guerrillas. In April, Israel shelled several Shiite Muslim villages in the Bekaa Valley.

[38] A general strike against the government's economic policy, accompanied by violent public demonstrations brought about the fall of the pro-Syrian Omar Karame government. President Elias Hraoui designated Rashid Al Sohl, a moderate Sunni, as the new head of government. He formed his cabinet with an equal number of Christians and Muslims. The August parliamentary elections were boycotted by the Christians. The new parliament included Hezbollah and Amal representatives. In October, Rafiq al-Hariri, a nationalized Saudi Arabian millionaire, was named prime minister.

[39] In 1993, the World Bank granted a loan for reconstruction and education projects. The currency regained 10 per cent of its value and the country began receiving more foreign investment. The gap between the rich and the poor grew.

[40] Israel continued to bomb Palestinian refugee camps and the bases of the Popular Front for the Liberation of Palestine (PLO). In August, Lebanese authorities rejected an Israeli proposal for total withdrawal since Israel demanded Hezbollah's total disarmament.

[41] Although political violence continued to destabilize domestic political consensus, progress toward a definite peace was made. The economy grew by 6 per cent and banks from the Netherlands, United Kingdom and France returned to Beirut after 20 years of absence.

[42] Among other terrorist incidents, a bomb killed a top Hezbollah leader, 21 Israeli soldiers died and - on the domestic side - Christian churches were shelled causing dozens of deaths. In March, the govern-

ment suspended private radio and television broadcasting until a new press law had been enacted while it reinstated the death penalty for political murders or crimes.

[43] During 1995, government attention was focused on finding a solution to the armed conflict and the reconstruction of Beirut. In January, the UN Security Council decided to extend its intervention in the country. At the beginning of talks between Israel and Palestine in Oslo, Norway, the Hezbollah and Southern Lebanon Army resumed their attacks in order to displace Israeli troops and defer negotiations with Syria.

[44] The Palestinians' fate in Lebanon was uncertain. The UN estimated the government had refused to acknowledge civilian rights to 338,000 Palestinians.

[45] On June 24, former Maronite leader Samir Geagea was sentenced to life imprisonment for the murders of his rival Dany Chamoun and his family. Political balance shifted toward the Muslims.

[46] Prime Minister al-Hariri promoted the "Horizon 2000" scheme to deal with Beirut's urban renovation and a constitutional reform which would extend the presidential term from 6 to 9 years to achieve the necessary stability to carry out his project. His goal of reinstating Lebanon as a central financial center in the Middle East, made progress. A plan to reconstruct Beirut's business and residential area was delayed by the discovery of archaeological remains, which led to a debate about their rescue and effect on the costs of urban renovation.

[47] Many people abstained from voting in the parliamentary elections held in five rounds between June and September 1996 and thousands of administrative irregularities were reported. The pro-government candidate list headed by al-Hariri - whose inclusion was influenced by the presence of Syrian troops - won the majority of the vote. In spite of everything, Hezbollah lost only one seat.

[48] Although peace had been settled by that time, there was no more talk of a "good border" but a "safe zone", and an implicit occupation had been accepted as such in the south.

STATISTICS

DEMOGRAPHY

Annual growth: 0.4% (1992-2000). **Children per woman:** 3.1 (1992).

HEALTH

One **physician** for every 413 inhab. (1988-91). **Under-five mortality:** 40 per 1,000 (1995). **Calorie consumption:** 109% of required intake (1995). **Safe water:** 94% of the population has access (1990-95).

EDUCATION

Illiteracy: 8% (1995).

COMMUNICATIONS

103 **newspapers** (1995), 108 **TV sets** (1995) and 116 **radio sets** per 1,000 households (1995). 9.3 **telephones** per 100 inhabitants (1993).

ECONOMY

Currency: 1.647 pounds = 1$ (1994). **Cereal imports:** 356,000 metric tons (1990).

Lesotho

Population: 1,942,000 (1994)
Area: 30,350 SQ.KM
Capital: Maseru

Currency: Maloti
Language: Sotho and English

Lesotho

For most African states, national unity is a task still to be accomplished. In some cases, however, the nation existed before the political state. Lesotho and Swaziland are both examples of this.

[2] The Zulu conquests, begun in 1818 by Shaka (see South Africa: "Zulu Expansion") affected a large number of Bantu nations, among them the North Sotho or Pedi who occupied a vast area in present-day Transvaal. While some withdrew northwards, the head of the Bakwena tribe, Moshoeshoe, brought other Sotho tribes and groups of dissident Zulus together under his command, retreating with them towards the Drakensberg mountains. The lengthy war of resistance fought first against the Zulus, and then against the expansionist Boers, consolidated the bonds between these groups of diverse origin who gave Moshoeshoe the title of "Great Leader of the Mountain", and called themselves Basothos.

[3] The Boers, the Dutch colonizers, tried to use the Basothos as labour. But they soon admitted defeat declaring that "these savages prefer liberty to slavery". Some Basothos refused to work on the Europeans' farms, saying "God created animals to feed men and not men to feed animals".

[4] Until 1867, Dutch colonization in South Africa was tenuous. Then diamonds were discovered. Soon afterwards gold was found, and along with its discovery the British came up from the Cape. In 1868, British missionaries persuaded the Basotho traditional leader Moshoeshoe, that only "protection" of the British Crown could save the people from subjection by the Boers. The territory became a protectorate, administered separately from South Africa even after the Boer War of

1899-1902, when the British took control of the whole country.

[5] Britain had promised the South African government that Basotholand (Lesotho), Bechuanaland (Botswana) and Swaziland, which were all in similar situations, would eventually be integrated into South Africa. However, when the South African Union broke all ties with London in 1961, consolidating its racist policy of apartheid, the British left those countries independent. In 1965, a constitution was promulgated in Basotholand and in 1966, the country proclaimed independence under the name of Lesotho.

[6] However, as an enclave within South Africa. Lesotho depended on the surrounding country as an outlet for its products of wheat, asbestos, cattle, and diamonds. The currency was the South African rand and South African companies controlled its economy and communications.

[7] Foreign trade was extremely unbalanced, with imports ten times higher than exports. The difference is offset by money that the migrant workers send home. Some 45 percent of the labor force works in the South African gold mines.

[8] The difficult economic situation enabled the opposition Congress Party to win a victory in the legislative elections of 1970. Prime Minister Leabua Jonathan engineered a coup, dissolved parliament, and sent King Moshoeshoe into exile. He was allowed to return on the promise that he would refrain from any political activity.

[9] After the student uprising which took place in Soweto, South Africa, in 1976, Lesotho opened its doors to thousands of young South Africans refugees despite the tremendous economic sacrifice this implied for such a poor country.

When South Africa began its "Bantustan" policy, Lesotho, in compliance with the UN resolution condemning this new form of apartheid, refused to recognize the puppet state in Transkei. In early 1977, South Africa closed the Lesotho border in retaliation. This economic aggression endangered the country's survival. Lesotho made dramatic appeals for international solidarity, requesting aid to resist the blockade.

[10] After Zimbabwean independence, Lesotho joined the economic integration project promoted by the Front Line states, tightening relations with Mozambique.

[11] South Africa retaliated by supporting of groups opposed to the Jonathan government. This led the leader to seek help from the United Nations and the European Economic Community. The official Basotho National Party (BNP) also had to face the opposition of groups linked to the founder of the Basotho Congress Party (BCP), led by Ntsu Mokhele.

[12] Most of the incidents were really caused by South African military groups interested in preventing the anti-apartheid refugees of the African National Congress (ANC) from organizing in Maseru (see South Africa). The most important of these attacks took place in Decem-

ber 1982, when a South African military air raid killed 45 people, 12 of them children, in the outskirts of Maseru. During the attack, three ANC leaders were killed, but the other victims were all bystanders.

[13] Since 1982, the government has also enforced an emergency law allowing the arrest of anyone suspected of illegal activities, without any previous judicial process. The army and the police were reinforced and a paramilitary group known as Koeko violently suppressed BCP supporters in the Drakensberg mountains.

[14] In March 1983, there was a border incident between Lesothan and South African troops, when saboteurs tried to enter the country to destroy the principal electric power plant. There was growing South African pressure to sign a non-aggression treaty with Pretoria, similar to the agreements that the apartheid regime had signed with Swaziland and Mozambique. The prime minister, Leabua Jonathan was against the treaty, but due to Lesotho's economic dependence on South Africa he was forced to be more flexible.

[15] Towards the end of 1984, the South African government began to prevent products from getting to Lesotho, particularly weapons purchased in Europe, and to delay the

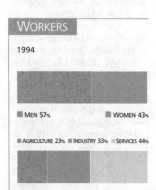

WORKERS

1994

- MEN 57%
- WOMEN 43%

- AGRICULTURE 23%
- INDUSTRY 33%
- SERVICES 44%

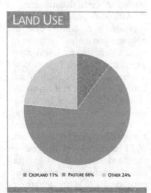

LAND USE

- CROPLAND 11%
- PASTURE 66%
- OTHER 24%

ENVIRONMENT

This small country is an enclave in the foothills of the Drakensberg mountains in southern Africa. Landlocked and mountainous, its only fertile land is located in the west where corn, sorghum and wheat are grown. In the rest of the country cattle are raised. Except for small diamond deposits, there are no mineral resources. Soil erosion is a serious problem.

SOCIETY

Peoples: Ethnically homogeneous, the country is inhabited by the Basotho (or Sotho) people, of Bantu origin. There are small communities of Asian and European origin.
Religions: Mainly Christian. Also traditional African beliefs are held. **Languages:** Sotho and English (official). **Political Parties:** The Basotho Congress Party (BCP), founded in 1952; the Basotho National Party (BNP), founded in 1958; the Lesotho Communist Party (LCP), founded in 1962; the Marematlou Liberation Party, founded in 1962; the National Independence Party (NIP), Christian; and the People's Front for Democracy (PFD), which emerged from a split in the LCP; the Kopanang of Basotho Party, founded in 1992, leading campaigns for women's rights. **Social Organizations:** The Lesotho General Workers' Union (LGWU), founded in 1954 and led by L. Hamatsoso, is the only central labor organization.

THE STATE

Official Name: Kingdom of Lesotho. **Administrative Divisions:** 10 Districts. **Capital:** Maseru, 109,000 inhab. (1990). **Other cities:** Leribe, Mafeteng. **Government:** Prime Minister Ntsu Mokhehle. King Letsic III. **National Holiday:** October 4, Independence Day (1966). **Armed Forces:** 2,000 troops (1995).

Illiteracy
1995

29%

Maternal Mortality
1989-95

PER 100.000
LIVE BIRTHS

598

STATISTICS

DEMOGRAPHY

Urban: 22% (1995). **Annual growth:** 3.2% (1992-2000). **Estimate for year 2000:** 2,000,000. **Children per woman:** 4.8 (1992).

HEALTH

One **physician** for every 25,000 inhab. (1988-91). **Under-five mortality:** 156 per 1,000 (1995). **Calorie consumption:** 92% of required intake (1995). **Safe water:** 52% of the population has access (1990-95).

EDUCATION

Illiteracy: 29% (1995). **School enrolment:** Primary (1993): 105% fem., 105% male. Secondary (1993): 31% fem., 21% male. University: 2% (1993). **Primary school teachers:** one for every 51 students (1992).

COMMUNICATIONS

89 **newspapers** (1995), 83 **TV sets** (1995) and 83 **radio sets** per 1,000 households (1995). 0.6 **telephones** per 100 inhabitants (1993).

ECONOMY

Per capita GNP: $720 (1994).
Annual growth: 0.60% (1985-94).
Annual inflation: 14.00% (1984-94).
Consumer price index: 100 in 1990; 168.9 in 1994.
Currency: 4 malotis = 1$ (1994).
Cereal imports: 131,000 metric tons (1993).
Fertilizer use: 178 kgs per ha. (1992-93).
External debt: $600 million (1994), $309 per capita (1994).
Debt service: 16.9% of exports (1994).
Development aid received: $128 million (1993); $66 per capita; 17% of GNP.

remittances sent by almost 400,000 migrant workers. They also delayed plans to build a dam on the Sengu River, on the border between the two countries. South African pressures were directed at intimidating the Lesotho electorate and strengthening opposition to the BNP, the conservative Basotho Democratic Party (BDA) and the (ANC-aligned) BCP.

[16] But even these tough measures did not satisfy Pretoria and on January 20, 1986, a military coup toppled Leabua Jonathan, and General Justin Lekhanya, head of Lesotho's paramilitary forces, led the Military Committee that replaced him.

[17] In 1988, workers residing in South Africa sent remittances totalling more than $350 million to Lesotho, that is 500 per cent of the total value of the country's exports.

[18] In March 1990, the military regime exiled King Moshoeshoe, accusing him of being a hindrance to the country's democratization program. His son Bereng Mohato Siisa replaced him as Letsie III. On April 30 1991, the armed forces went on strike for higher salaries, and a bloodless military coup deposed Lekhanya's government, setting up a Council chaired by colonel Elias P. Ramaema.

[19] The South African government blocked remittances from emigrant workers in 1991. In May, a demonstration against foreign interference in the economy ended with 34 people dead and 425 arrests.

[20] A new constitution designated the king head of state in 1993, without giving him either legal or executive powers. The July legislative elections gave all the seats to the BCP. In August the privatization of six state businesses was started with a loan from the IMF.

[21] Erosion affected 58 per cent of the low lying soils. Two thirds of all farming land belonged to emigrant workers, but was worked by their wives and families. Plans were developed to make use of the water from the highlands - the main natural resource of the country - through a hydroelectric project which will change the course of several rivers to feed water to South Africa in return for electricity.

[22] During 1994, the low salaries led to armed confrontations between rival military factions. The government plan to integrate the armed wing of the BCP into the army led to the kidnapping and murder of the finance minister by disgruntled soldiers.

[23] The king dissolved the government and Parliament. Protests outside the palace were put down by soldiers and police, with at least four people killed. The internal republican opposition and international pressure forced Letsie III to abdicate in favor of his father, King Moshoeshoe II, who was formally restored to the throne in January 1995.

[24] During 1995, the tunnel under the mountains, which will supply water to the Vaal river valley in South Africa, was completed. It will go into operation in 1997.

[25] In March 1995, members of the National Security Forces kidnapped several high-ranking officers, demanding their immediate withdrawal from the army for their part in assassinations and various sorts of corruption.

[26] In January 1996, the king died in a road accident. The Assembly designated his son Letsie III as his replacement.

[27] In a report published in June, Amnesty International addressed the civilian government and security forces of Lesotho calling for human rights guarantees, and for the clarification of numerous arbitrary arrests, torture cases and murders of prisoners. The victims included influential members of the government and police, members of parliament, and union and human rights campaigners.

The origins of the Yoruba culture

In the 5th century, iron working rural communities settled in the forest region of the coastal strip which stretches between Volta and Cameroon, establishing an organized agricultural economy with advanced and stable ways of life. Amongst the largest, and seemingly oldest of these were the Yoruba communities, found in the Ife, Ilesha and Ekiti regions.

In around the 13th century, a movement outwards by the groups which imposed economic and political supremacy on the weaker communities led to the creation of the Yoruba kingdoms. But the idea of a 'kingdom' associated with the traditional African societies is far from the Western view of, for example, the "kingdom of Louis XVI."

The Yoruba kingdoms or states

The Yoruba kingdoms - south of modern-day Nigeria - owe their linguistic and cultural homogeneity to this ethnic group, recognizing their roots in these ancestors. The splendor accomplished by the Ife and Oyo kingdoms spread to the traditions of the other areas and overshadowed their own origins. There is one belief structure which is based on the myth of Ife who created the world in the city of Ife-Ife, and another which attributes the creation to Oyo and says the Yoruba have their origins in a migration movement which came from the east. What is certain, however, is that the life of these kingdoms led to the spread of their institutions and practices amongst the local populations, and that the execution of complex functions - extensive agriculture, long-distance trade, tax systems, military expansion, citizen policies - were possible because of the well-planned and organized states.

Although each king aimed to leave his successor an even bigger kingdom, a policy of tolerance encouraged a certain level of interchange which led to the cultural enrichment of both the stronger communities and those they absorbed. It was this exchange which ultimately explained the heterogeneity of Yoruba civilization: a single culture expressed through differences.

The economic life of the Yoruba kingdoms

The Yoruba states were generally of a modest size, sometimes covering only a single city and the surrounding villages. The kingdom of Oyo was one important exception, spreading over a vast area and acquiring imperial status in the 17th century. The more common kingdoms consisted of a compact town surrounding the compound of the kings and elders in an area contained by a wall marking the edge of the kingdom. Resources produced from agriculture, a certain amount of mining and craftwork were taken to local markets organized on alternate days in order to prevent competition between markets. However, luxury articles - like the gold paid in tax in the courts, marble, art objects, nuts and others - were the main object of long distance trade, like that with the Hausa states of the eastern zones, to the benefit of the richest strata of society - the kings and their courts, officials, merchants and professionals. There was no slave trade amongst the early traditional Yoruba societies.

The Yoruba cosmovision

An ancient myth, revealing how mythical time is transposed over historical time, explains how the grandchildren of the mythical founder of Ife - the sacred city - spread out into the surrounding area establishing and naming the first generation of Yoruba states: Owu, Ketu, Benin, Illa, Dave, Popo and Oyo. The Yoruba belief system is based on the idea of a superior entity made up of three divinities, Olofi, Oloddumare and Olorun. The first of these created the world, which was initially only populated by orixas, or saints. The power, or *ache*, was later divided between the orixas, who from then on were empowered to intervene in human affairs and to represent people to Olofi through the mediation of the supreme judge or main messenger, Obbatala. As in most of the languages of Black Africa, "the power" is expressed amongst the Yoruba through the word "ache" which means "the force," not in the sense of violence, but as a vital energy which creates a multiplicity of process and determines everything from physical and moral integrity to luck.

The Yoruba cosmovision is prevalent in all the cultural creations of this group of people. As is generally the case with peoples where every action is carried out, interpreted and lived as part of an organic belief system which is not precisely religious, this cosmology includes the idea that the order of the cosmic forces can be upset by immoral actions which have the effect of unbalancing and damaging humanity, nature and the perpetrators themselves.

Community life

The community was of great value in the traditional Yoruba societies. It defined the conception of history - identified with the life of the group in continuous change - and time - conceived of as the social time, lived by the group, which transcends the time of the individual. It is, at the same time, the dimension where people can, and must, incessantly play out their fight against decadence and for the enrichment of their vital energy. The Yoruba believed that throughout the history lived by the group a certain *ache* was accumulated incarnated in objects. These objects were sent from the ancestors down through the successive generations via their patriarchs or kings, who were intermediaries between the transcendental and visible worlds, as gifts from the orixas.

Although they were headed by kings, the communities were led by governing councils made up of men of varied standing, where the elders enjoyed their deserved dignity. The Yoruba, as in the majority of traditional African societies, were societies of public opinion, where the conduct of the authorities was monitored, spied on, and the violations of the principles which ruled the community life were always denounced through persistent verbal criticism and rumors which were so wearing that, in time, the subject of these was forced either to explain their actions or stand down.

The Yoruba religion

The religion of the traditional Yoruba societies is characterized by the cult of God and a group of intermediary divinities, whose intervention and wills rule human life. The orixas were ancestors who accumulated power and knowledge over the forces of nature and humanity during their lifetimes, by virtue of which they one day changed from people into gods. Each one personifies certain forces of nature and is associated with a cult which obliges believers to offer food, sacrifices and prayers in order to escape their wrath and attract their favor.

The Yoruba religion is linked to the notion of family in the sense that each cult creates a religious community which comes directly from the orixa or a common ancestor, a group which includes both the living and the dead and goes beyond blood links.

The Yoruba gods and goddesses occasionally take possession of the faithful, and when this happens the god dances with his devotees in a friendly manner and sometimes speaks, offers advice or gives prophecies. The most well known orixas include Eleggua, the god who opens paths and is found behind the doors of Yoruba homes; Oggun, the inventor of the forge, god of the minerals and the mountains; Oxosi, the god of hunting; Xango, the god of fire and war; Oxun, the goddess of fresh water, love and all tenderness and Iemanya, the queen of the sea.

Yoruba art

The art in the most ancient Yoruba communities was distinguished by its sculptures, metalwork and ceramics. Bas-reliefs, woodworking, masks and human heads created by the lost-wax method were jealousy guarded as divine heirlooms.

However, Yoruba art is definitely dominated by music. Though this has an autonomous and profane meaning as an art form, music is inseparably united to the Yoruba religious cults and liturgy.

The most characteristic musical tradition is the predominance of the drums and especially the presence of the "bata" (family) drums, an exclusive creation of the Yoruba people. The sound and symphonic integrity of the bata (the vegetable sound of the wood of the drums, the animal sound of the skins, and the mineral for the accompanying bells and rattles), along with human voices, obey a magical criteria through which the Yoruba evoke the integrity of the cosmic powers.

Liberia

Liberia

Population: 2,719,000 (1994)
Area: 111,370 SQ.KM
Capital: Monrovia

Currency: Liberian Dollar
Language: English

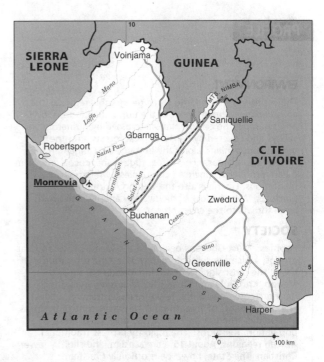

The present Liberia was formerly known as the Grain Coast, and was inhabited by 16 different ethnic groups. The Kru speakers lived in the southwest, and the Mande speaking peoples, including the Mandingo, lived in the east and northeast. After the arrival of the Portuguese, Mandingo traders and artisans played an important role as they spread throughout the territory, becoming the principal propagators of Islam.

2 Some people are unaware that the US also shared in the spoils of African colonization. Emancipated blacks posed a social problem to US southern slaveholders, long before US President Abraham Lincoln freed the slaves in 1865, during the US civil war. As a solution to the "problem", some were "repatriated." On the assumption that blacks would be at home in any part of Africa, it was planned to ship them to the British colony of Sierra Leone.

3 In 1821, the American Colonization Society purchased a portion of Sierra Leone and founded a city which was named Monrovia after James Monroe, president of the United States.

4 Only 20,000 US blacks returned to Africa. The native population distrusted these settlers whose language and religion were those of the colonizers. Supported by US Navy firepower, the newcomers settled on the coast and occupied the best lands. For a long time, they refused to mix with the "junglemals," whom they considered "savages." Even today only 15 per cent of the population speak English and practise Christianity.

5 In 1841, the US government approved a constitution for the African territory. It was written by Harvard academics, who called the country Liberia. Washington also appointed Liberia's first African governor: Joseph J. Roberts. In July 1847, a Liberian Congress, representing only the repatriates from the US, proclaimed independence. Roberts was appointed president and the Harvard-made constitution was kept, along with a flag which resembled that of the United States.

6 The emblem on the Liberian coat of arms reads: "Love of liberty brought us here." However, independence brought little freedom for the original population. For a long time, only landowners were able to vote. Today, the 45,000 descendants of the former US slaves form the core of the local ruling class and are closely linked with transnational capital. One of the principal exports, rubber, is controlled by Firestone and Goodrich under 99-year concessions granted in 1926. The same is true of oil, iron ore and diamonds. Resistance to this situation has been suppressed on several occasions by US Marine interventions to "defend democracy."

7 The discovery of extensive mineral deposits, and the use of the Liberian flag by US ships, fanned a period of economic growth beginning in 1960. This was instantly dubbed an "economic miracle", but this so-called miracle only reached the American-Liberian sector of the population, who secured significant increases in income during this period.

8 The political establishment was shaken in 1979, when the increase in the price of rice triggered demonstrations and unrest. A year later, Sergeant Samuel Doe overthrew the regime of William Tolbert, who was executed, along with other members of his government. These disturbances led to the suspension of the constitution and the banning of all political parties. In 1980, the beginning of a democratization process was announced, followed by the signing of the first agreement with the International Monetary Fund.

9 Falling exports, increasing unemployment, the reduction of salaries in both public and private sectors, and spiralling foreign debt, threw the country into a crisis of substantial proportions.

10 Popular discontent increased. Between 1980 and 1989 Doe's administration uncovered many new anti-government conspiracies.

11 Elections were held in 1985. With any viable political opposition banned, and accusations of fraud and imprisonment of opposition leaders abounding, Doe obtained 50.9 per cent of the vote. The Liberian People's Party (LPP) and United People's Party (UPP), which represented the major opposition forces, were not authorized to participate.

12 In 1987, most government financing came directly from the US, a fact related to the vast North American business interests in Liberia. These included $450 million of investments, military bases, a regional Voice of America station, and a communications center for all US diplomatic missions in Africa.

13 In May 1990, the National Patriotic Front of Liberia (NPFL) launched an attack against the city of Gbarnga, 120 km from the capital. The guerrilla movement rapidly took over several parts of the country.

14 In June, an NPFL victory, under the leadership of former civil servant Charles Taylor, seemed imminent. However, in July, when the battle to take Monrovia was just underway, the rebel front split, with one faction forming the Independent Patriotic Front (INPFL), led by Prince Johnson.

15 On July 31, 200 civilians, who had sought refuge in a Lutheran mission, were massacred by government soldiers.

16 In September 1990, President Samuel Doe was assassinated by Johnson's troops. In the mêlée which followed, Johnson, Taylor, Amos Sawyer (a civil servant in Doe's government) and Raleigh Seekie (head of the Presidential Guard) all proclaimed themselves "Interim President".

17 In November 1990, Sawyer formed a provisional government, with the support of Cote d'Ivoire, Gambia, Nigeria, Burkina Faso and Togo, the five west African coun-

> **The political establishment was shaken in 1979, when the increase in the price of rice triggered demonstrations and unrest. A year later, Sergeant Samuel Doe overthrew the regime of William Tolbert, who was executed, along with other members of his government.**

WORKERS

1994

■ MEN 71%　　■ WOMEN 29%

■ AGRICULTURE 75%　■ INDUSTRY 9%　■ SERVICES 16%

ENVIRONMENT

The country is divided into three geographic regions: the coastal plain contains most of the population, and is low and swampy; the central plateau, crossed by numerous valleys and covered by dense tropical forests; and the mountainous inland along the border with Guinea. In the fertile coastal areas, rice, coffee, sugar-cane, cocoa and palm oil are produced. American companies own large rubber plantations. Liberia is also the leading African iron ore producer. War, the loss of biodiversity and erosion are the main threats to the environment.

SOCIETY

Peoples: Most Liberians belong to the Mende, Kwa and Vai groups, which are split into nearly 30 ethnic sub groups. Of these, the most significant are the Mandingo, Kpelle, Mendo, Kru, Gola and Bassa (the Vai are renowned for having created one of the few African written languages). The descendants of "repatriated" US slaves control business and politics, though they constitute only 5% of the population. **Religions:** The majority profess traditional African religions. About 15% are Muslim, and slightly fewer Christian. The State, however, is officially Christian. **Languages:** English is the official language, though it is spoken by only 15% of the population. The rest speak local languages. **Political Parties:** National Patriotic Front of Liberia; United Liberation Movement of Liberia for Democracy; Liberia Peace Council; Ulimo-J and Ulimo-K (both born after the division of the United Liberian Independence Movement).

THE STATE

Official Name: Republic of Liberia. **Administrative divisions:** 11 Counties and 2 Territories. **Capital:** Monrovia, 400,000 inhab. in 1985. **Other cities:** Harbel 60,000; Gbarnga 30,000; Buchanan 25,000; Yekepa 16,000 (1985) **Government:** Interim Council of State headed by Ruth Perry **National Holiday:** August 26, Independence Day (1846). **Armed Forces:** the Armed Forces of Liberia (AFL), with a force of about 3,000, is confined to the capital city of Monrovia.

DEMOGRAPHY
Annual growth: 1.3% (1992-2000). **Children per woman:** 6.2 (1992).

HEALTH
Under-five mortality: 217 per 1,000 (1995). **Calorie consumption:** 94% of required intake (1995). **Safe water:** 46% of the population has access (1990-95).

COMMUNICATIONS
86 **newspapers** (1995), 85 **TV sets** (1995) and 98 **radio sets** per 1,000 households (1995). 0.2 **telephones** per 100 inhabitants (1993).

ECONOMY
Consumer price index: 100 in 1990; 108.1 in 1990. **Currency:** 1 Liberian dollars = 1$ (1994). **Cereal imports:** 70,000 metric tons (1990).

ment being signed in Geneva, on July 17, 1993. The two main armed forces and the provisional Sawyer government agreed a cease-fire in seven months and the calling of general elections.

> In 1821, the American Colonization Society purchased a portion of Sierra Leone and founded a city which was named Monrovia after James Monroe, president of the United States.

[23] The NPFL made several protests about ULIMO attacks. The civil war worsened. The UN embargo and the consequent food shortages caused a wave of guerrilla attacks against civilian targets.
[24] In August, following the timetable established in Geneva, a transitional Council of State was established with representatives of the NPFL, ULIMO and Sawyer's government.
[25] By 1993, the total of civil war victims had risen to 150,000. Nearly a million Liberians, out of a total population of 2.4 million, had been displaced to another part of the country, or were living as refugees in neighboring countries.
[26] The Council of State took power in March, 1994, but the setting up of a new government took until May due to disagreements between the NPFL, ULIMO and Sawyer's representatives. Meanwhile, fighting continued between rival armed groups and skirmishes of some of them with the recently deployed peace forces of the Economic Community of West African States (ECOWAS).
[27] In December the seven armed groups at war agreed a cease-fire. In 1995, negotiations continued, the Council of State increased its membership, Charles Taylor joined it, and a new government was formed.
[28] The civil war was triggered again in 1996, with violent clashes, particularly in the capital city, Monrovia, severely affecting the civil population. In September, Ruth Perry, the first woman head of state in Africa, took power as head of the Council of State, supported by ECOWAS. The war seriously affected the essential services, such as drinking water or food supplies.
[29] Some observers asserted that there were Nigerian interests acting in the Liberian conflict through ECOWAS. Others thought there were major deliveries of weapons to the different groups engaged in the conflict - in exchange for diamonds and wood - by foreign companies, particularly Belgian and French ones.

tries that had made up a peace-keeping force.
[18] In 1992, Taylor turned down the vice-presidency which Sawyer had offered him. The NPFL controlled most of the country, through the Government of the Patriotic National Assembly of Reconstruction. In August, 2,000 NPFL troops were killed in an attack upon Tubmanburg, north of Monrovi by United Liberian forces (ULIMO).
[19] ULIMO split into two factions in November. Alhaji Kromah, accused of being overly-friendly toward Muslim groups and the Libyan government, became the leader of the faction that established its

headquarters in Tubmanburg. Raleigh Seekie became the leader of the Sierra Leone faction.
[20] In June 1993, ULIMO attacked a refugee camp in Kata. According to the United Nations High Commissioner for Refugees, during the attack 450 people were killed and many of the victims' bodies were mutilated.
[21] US support for an increased UN role in the conflict, gave the Security Council the go-ahead to demand a cease-fire and decree an arms embargo, amongst other measures.
[22] The military stalemate and UN participation led to a peace agree-

Libya

Libiya

Population: 5,218,000 (1994)
Area: 1,759,540 SQ.KM
Capital: Tripoli (Tarabulus)

Currency: Dinar
Language: Arabic

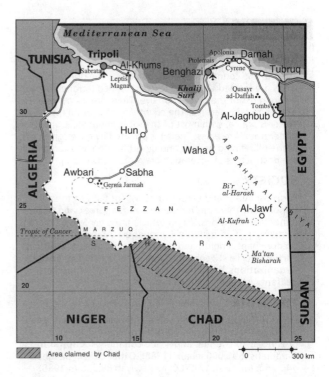

The Socialist People's Libyan Arab Jamahiriya (known as Libya) has always been torn between the different political and economic centers of North Africa. Bordering pharaonic Egypt, Libya shared its culture and two Libyan dynasties ruled Egypt between the 10th and 8th centuries BC. However such influences did not lead to a unified state. The Carthaginian and Roman empires on the western border further stressed this division. After the Arab conquest in the 7th century, Tunisia, Morocco and Egypt became the new power centers.

[2] The development of maritime trade and the ensuing piracy turned Tripoli (Tarabulus) into one of the major Mediterranean ports, bringing European intervention which caused further intervention by the Turkish Sultan. In 1551, Suleiman the Magnificent annexed the region to the Ottoman Empire. A weakened central authority gave increasing autonomy to the governors precipitating independence movements. The beginning of the 19th century saw piracy again on Libya's shores; this was used as a pretext for US military intervention and in 1804 US forces attacked Tripoli.

[3] In 1837, Muhammad al-Sanussi founded a clandestine Muslim brotherhood (the Sanussi religious sect) which promoted resistance to Turkish domination, though the Italians really posed a greater threat. With the decline of the Ottoman Empire, Italy declared war on Turkey in 1911 and seized the Libyan coast, the northernmost Turkish possession in Africa. With the outbreak of World War I, the Italian presence was confined to Tripoli and Homs (Al-Khums) while the rest of the territory remained autonomous. At the end of World War I

Italy attempted to recover control of the territory but faced resistance for twenty years by Sidi Omar al-Mukhtar's forces. In 1931, al-Mukhtar was captured and executed and the Italians formally annexed the territory.

[4] From Egypt and Tunisia, the Sanussi brotherhood remained active and cooperated with the Allies in World War II. Muhammad Idris al-Sanussi, leader of the brotherhood, was recognized as Emir of Cyrenaica by the British. At the end of the conflict, the country was divided into an British zone (Tripolitania and Cyrenaica) and a French (Fezzan) governed from Chad. In 1949, a UN resolution restored legitimate union to the region and established the independent nation of Libya, with Idris al-Sanussi as leader for his religious authority.

[5] Idris consolidated his position with support from the powerful Turkish-Libyan families, military advisers from the US and Britain and transnational oil companies. In 1960, foreigners settled in the country, as the oil began to flow in great quantities.

[6] In 1966, Muammar al-Khaddafi, the son of Bedouin nomads, founded the Union of Free Officials in London, where he was studying. He returned to Libya and, on September 1, 1969, he led an insurrection in Sebha overthrowing the king.

[7] Khaddafi's Revolutionary Council, proclaimed itself Muslim, Nasserist, and Socialist, beginning to eliminate all US and British military bases in Libya, and imposing severe limitations on transnationals operating in the country. The production of petroleum and its derivatives was placed under state control but the government kept some ties with the foreign companies.

[8] Khaddafi began an ambitious modernization program, with special emphasis on agricultural development. Each rural family was alloted 10 hectares of land, a tractor, a house, tools, and irrigation facilities. Over 1,500 artesian wells were drilled and two million hectares of desert were irrigated and turned into fertile farmland.

[9] Rapid growth meant that immigrant workers and experienced technicians from other Arab countries were needed. In 1973 following publication of his Green Book a complex structure of popular participation was created through people's committees and a People's General Congress.

[10] In the cities, a social security system was created, with free medical assistance and family allowances to encourage large families. Industrial workers were granted 25 per cent participation in the profits of the companies. Industrial investment was eleven times greater than during the monarchy and agricultural investment was 30 times greater. This massive oil-financed reform transformed Libya into the North African Nation with the highest per-capita income on the continent, at $4,000 a year.

[11] In 1977, the country changed its name to the Socialist People's Libyan Arab Jamahiriya (meaning mass state in Arabic). But while Khaddafi achieved ample positive results internally, external relations were dismal. Attempts to unite with Syria and Egypt met with failure, and overtures towards Tunisia came to nothing. Khaddafi became the main critic of the diplomatic rapprochement between Egypt and Israel; he clashed with the Saudi monarchy and the Emirates, and maintained his long-standing antagonism with King Hassan of Morocco. In OPEC, Libya opposed the moderate stand of Saudi Arabia and the Emirates on oil prices, and

firmly resisted the pressures and maneuvers of transnational corporations.

[12] From 1980, Libya became diplomatically active in Sub-Saharan Africa and Latin America. The government supported the Polisario Front and participated directly in the civil war in Chad, defending the Transitional Government of National Union, led by Goukouni Oueddei.

[13] US President Reagan undertook a huge international campaign to link the Libyan leader with world terrorism. In August 1981 in the Gulf of Sidra two Libyan planes were shot down by the US Sixth Fleet. Khaddafi skillfully avoided any violent response to the provocation, winning the sympathy of the conservative Arab regimes which had, until then, been hostile to his government.

[14] In 1983 Libya attempted a rapprochement with Morocco, a move which met with success in August 1984 when an agreement was signed. Most North African nations were surprised at the pact, for Moroccans and Libyans held opposite views on practically all political issues. However, the rapprochement could be regarded as a consequence of increasing cooperation between the Algerian, Tunisian and

WORKERS

1994

- MEN 90%
- WOMEN 10%

- AGRICULTURE 20% INDUSTRY 30% SERVICES 50%

ENVIRONMENT

Most of the country is covered by desert. The only fertile lands are located along the temperate Mediterranean coast, where most of the population live. There are no perennial rivers and rain is scarce. The country has major oil reserves. Water is scarce, with most of the supply pumped from underground reserves. The air is polluted by gases given off in the oil refining process and desertification, erosion and the destruction of vegetation are rapidly advancing.

SOCIETY

Peoples: Arabs and Berbers account for 89% of the population. There are Tunisian, Egyptian, Greek and Italian minorities. **Religions:** Islam (official), mainly Sunni. There is a small Christian minority. **Languages:** Arabic (official) is predominant. English and Italian are also spoken. **Political Parties:** Socialist People's Libyan Arab Jamahiriya. **Social Organizations:**There are mass organizations of workers, peasants, students and women.

THE STATE

Official Name: al-Jamahiriyah al-'Arabiyah al-Libiyah ash-Sha'biyah al-Ishtirakiyah **Administrative divisions:** 3 provinces, 10 counties and 1,500 communes. **Capital:**Tripoli (Tarabulus) 591,000 inhab. (1988). **Other cities:** Benghazi 446,250; Misratah 121,700; az-Zawiyah 89,338 in 1988. **Government:** The People's General Congress is the highest government body. Colonel Muammar Khaddafi, leader of the Revolution and commander-in-chief of the People's Armed Forces, is the Head of State. **National Holiday:** September 1, Revolution Day (1969). **Armed Forces:** 80,000 **Paramilitaries:** Revolutionary Guards, 3000.

DEMOGRAPHY

Annual growth: 0.6% (1992-2000). **Children per woman:** 6.4 (1992).

HEALTH

One **physician** for every 962 inhab. (1988-91). **Under-five mortality:** 95 per 1,000 (1995). **Calorie consumption:** 114% of required intake (1995). **Safe water:** 97% of the population has access (1990-95).

EDUCATION

Primary school teachers: one for every 12 students (1991).

COMMUNICATIONS

86 **newspapers** (1995), 101 **TV sets** (1995) and 100 **radio sets** per 1,000 households (1995). 4.8 **telephones** per 100 inhabitants (1993). **Books:** 90 new titles per 1,000,000 inhabitants in 1995.

ECONOMY

Currency: 0.4 dinars = 1$ (1994). **Cereal imports:** 2,290,000 metric tons (1990).

Mauritanian governments. Moreover, Morocco aimed at neutralizing Libyan support of the Polisario Front, while Libya sought to cut off Moroccan aid to Habre's regime in Chad.

[15] In January 1986, the US imposed an economic embargo on Libya, and on April 14 American war planes bombed Tripoli and Benghazi, leaving dozens of civilians dead. Subsequent information revealed the main objective had been to eliminate Colonel Khaddafi.

[16] In November, 1991, US and British courts blamed the Khaddafi government for the bombing of two airplanes: a Pan Am flight in Britain, with 270 deaths, 17 of whom were US citizens and a UTA flight in Nigeria, with 170 casualties. Interpol issued an international arrest warrant on two people accused of the bombings. In January, 1992, Libya expressed willingness to co-operate with the UN in the clarification of both attacks.

[17] Nevertheless Khaddafi rejected a UN extradition order, unsuccessfully proposing that the trial be held in Tripoli.

[18] Libyan inflexibility simply strengthened UN determination; in February and March the extradition of the accused Libyan agents was again demanded. The UN demanded that Libya explicitly renounce terrorism, setting April 15 as the deadline and threatening sanctions, a blockade and even military measures should the ultimatum not be met.

[19] Libya passed the deadline, invoking economic sanctions from the EC and the seven most industrialized countries. Khaddafi took the decision to the International Court of Justice.

[20] In August, when the embargo was renewed, Khaddafi decided on a change of foreign policy, and des-ignated as Chancellor a "moderate", able to negotiate Libyan and US positions. In 1993 Tripoli continued a policy of economic liberalization, initiated in 1989, when it broke relations with Iran.

[21] The isolation of Libya increased in 1994. The UN intensified the embargo despite Khadd-afi's concessions. In the end he accepted that the two accused men be judged in Scotland. Inside the country, the difficult situation made him more popular, since at least a part of public opinion made the US responsible for the scarcity of goods and related problems.

[22] Nevertheless, some concessions awoke resistance: in the southern region of Fezzan the pop-ulation protested when the area of Aouzou was given to Chad following a decision of the International Court of The Hague. On the other hand Tripoli made progress towards achieving an old project by signing an agreement for the building of an aqueduct that will allow water to come from other countries.

[23] In 1995 the country was still isolated despite the ongoing Libyan proposals for a dialogue with the West. This fact did not counter either the growth of the private sector or foreign investment. Several international firms were willing to participate in new projects to exploit oil ressources.

[24] On the other hand Tripoli decided on the expulsion of 10,000 Palestinians who were Libyan residents, and announced that the immigrants from neighbouring Arab countries and their families should also leave. The social situation went on deteriorating, to the apparent advantage of fundamentalist Islamic groups who have become more powerful.

> The production of petroleum and its derivatives was placed under state control but the government kept some ties with the foreign companies.

Lithuania

Lietuva

Population: 3,721,000 (1994)
Area: 65,200 SQ.KM
Capital: Vilnius

Currency: Lit
Language: Lithuanian

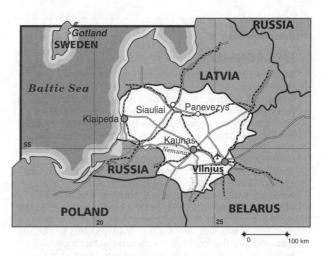

Lithuanians have lived along the shores of the Baltic Sea since long before the Christian era. Protected by the virgin forests, Lithuanian tribes fiercely resisted German efforts to subdue them in the 13th century, and united under the leadership of Mindaugas, who was crowned king by Pope Innocent IV in 1253.

2 In the 14th century, Lithuania began its eastward and southern expansion, going into Belarusian lands. Gediminas built the Grand Duchy of Lithuania, which extended from the Baltic Sea to the Black Sea, with its capital at Vilnius. In 1386, Jagiello, Gediminas's grandson, married the queen of Poland, thus uniting the two kingdoms.

3 Lithuania withstood a series of attacks from the Teutonic Order, who continued to combat the Lithuanian-Polish union - despite the fact that the latter were Christian - until the Battle of Tannenberg, in 1410, in which they suffered a crushing defeat. This defeat was a harsh blow to German supremacy in the Baltic region. A new pact between Lithuania and Poland, signed in 1413, reaffirmed the principle of union, while respecting the autonomy of both States.

4 With the coronation of Ivan III of Muscovy, as the sovereign of all Russia, a new and greater threat emerged for historic Lithuania. Nevertheless, the Lithuanian-Polish union reached its peak in the 16th century, when it was unrivalled in Europe as a political system (see Poland), only to fall in the 17th century, in the course of a series of devastating wars with Sweden, Russia and Turkey and the peasant rebellions within.

5 In the 1772 and 1793 partitions of Poland among Russia, Prussia and Austria, Russia kept only Belarus. But the Polish state disappeared in 1795, and all of Lithuania was in Russian hands by 1815. The Congress of Vienna granted the Russian emperor the additional titles of king of Poland and grand prince of Lithuania.

6 Lithuanians were harshly suppressed by the Russians. The czarist regime treated Lithuania as though it were a part of Russia, calling it the Northwest Territory after 1832. Between 1864 and 1905, Russification extended to all aspects of life: books which were printed in Lithuanian had to use the Cyrillic alphabet, and Catholics were persecuted.

7 With the Revolution of 1905, the peoples of the Russian Empire were granted freedom of speech. A congress with some 2,000 delegates called for the demarcation of Lithuania's borders, territorial autonomy and the election of a parliament by democratic means.

8 During World War I, Germany occupied a major part of Lithuania. In 1915, a congress - authorized by the occupying Germans - elected the 20-member Council of Lithuania. The 214 delegates to the congress called for the creation of an independent Lithuanian state within its "ethnic borders" and with Vilnius as its capital. On February 16, 1918, the Council declared Lithuania's independence and terminated all political ties with other nations.

9 In 1919, the Red Army entered Vilnius and formed a communist government, but the Allied powers forced it to withdraw. The new head of the Polish State, Josef Pilsudski, tried to re-establish the former union, but failed due to the resistance of the Lithuanians, Ukrainians and Belarusians. In the end, the League of Nations and the European powers agreed to the separation of Poland and Lithuania in 1923, but Lithuania refused to recognize the line of demarcation (with Poland) which had been established by these powers.

10 In 1926, Lithuania and the Soviet Union signed a non-aggression treaty. A treaty of good will and cooperation was signed in Geneva in 1934 by Lithuania, Latvia and Estonia. As of 1938, relations with Warsaw became tense, due to Poland's claim of sovereignty over Vilnius. Tension increased when a group of Nazis came to power in Klaipeda; they demanded the ceding of that city, which meant the loss of Lithuania's only port on the Baltic.

11 In September 1939, a secret German-Soviet non-aggression treaty brought Lithuania within the USSR's sphere of influence. In October, a mutual assistance treaty was signed in Moscow; according to its terms, Lithuania was forced to accept the installation of Soviet garrisons and air bases on its soil. In 1940, the Soviet Army occupied Lithuania and a number of local political leaders were arrested and deported, while others fled toward Western Europe.

12 The new prime minister, Justas Paleckis, and parliament asked to be admitted to the USSR, a request which was immediately granted by the Supreme Soviet. Lithuania thereby became a constituent republic of the USSR in August 1940. After German occupation in 1941, the Baltic Sates and Belarus became the German province of Ostland.

13 During German occupation, 190,000 Jews were sent to concentration camps. A hundred thousand residents of Vilnius - a third of that city's population, most of them Jews - were killed. Vilnius was known as the "Jerusalem of Lithuania" and had been considered to be one of the most important centers of Jewish culture in the world. Many non-Jewish Lithuanians throughout the rest of the country were killed, and tens of thousands of young people were sent to Germany to work.

14 Vilnius was reconquered by the Red Army in 1944. Lithuania was once again occupied by the Soviets and a new period of Sovietization began. This included religious persecution and massive deportations to northern Russia and Siberia, amid the forced collectivization of agriculture.

15 Religious persecution continued even after Stalin's death in 1953; thus, the resistance of the Catholic Church became identified with the nationalist movement. In 1972, the "Lithuanian Catholic Church Chronicle" (banned by the government) was published by the Lithuanian Movement for Human Rights. That year, a young man set fire to himself; his funeral triggered violent clashes in which 15 people were killed and 3,000 were arrested.

16 With the democratization process initiated by Mikhail Gorbachev in the USSR, Lithuania began a period of intense political agitation. In June, the Lithuanian Movement to Support Perestroika (restructuring) was founded; its Executive Committee adopted the name *Sejm* (the name of the Lithuanian parliament at the time of independence) and is known by the name "Sajudis". The Sajudis installed a kind of "shadow" government, and demanded a return to the peace treaties which recognized the country's independence, stating that Lithuania's admission to the USSR was the result of a secret agreement and, therefore, had no legal value.

17 In July, the Lithuanian Freedom League (LFL) emerged from underground activity. The LFL, which dates back to 1978, called for immediate withdrawal of Soviet troops from the country, and the independence of Lithuania. Police brutality toward participants of a demonstration sponsored by the LFL triggered joint protest actions by members of the League and the Saju-

LAND USE

CROPLAND 46%   PASTURE 7%   OTHER 47%

Foreign Trade

MILLIONS $ 1994

IMPORTS
2,210

EXPORTS
1,892

Maternal Mortality

1989-95

PER 100.000
LIVE BIRTHS
29

External Debt

1994

PER CAPITA
$118

PROFILE

ENVIRONMENT

On the eastern coast of the Baltic Sea, Lithuania is characterized by gently rolling hills and flat plains. The largest of the Baltic "mini-States", it has more than 700 rivers and streams, many forests and around 3,000 lakes. The Nemunas river, which crosses the country from east to west, is the country's largest river. It is lined with castles built centuries ago and it is still an important shipping route. Lithuania's climate is moderate because of its proximity to the sea: in summer, the average temperature is around 18 degrees C and there is heavy rainfall; in the winter, it is often cold and foggy, with the temperature often falling below zero. 49% of the land is arable; the chief crops are grain, potatoes and vegetables. There are large numbers of livestock. Among the country's most important industries are the food and machine manufacturing industries, as well as the export of energy. Most of the country's energy comes from Ignalina, a nuclear plant of the same type as that of Chernobyl, Ukraine, site of the nuclear accident in 1986. Since the 1980s, there has been an increase in pollution, especially bacterial pollution of rivers and lakes; the latter has been linked to the increase in infectious childhood diseases, particularly in the first few years of life

SOCIETY

Peoples: Close to 80% of the population is Lithuanian; the rest of the population includes Russians, 9%, Poles, 8%, Belorussians, 2% Germans, Jews, Latvians and Tatars. **Religion:** Catholic majority; there are Protestant minorities. **Languages:** Lithuanian (official); and Russian. **Political Parties:** Lithuanian Democratic Labor Party; Homeland Union.

THE STATE

Official Name: Lietuvos Respublika. **Administrative Division:** 44 Districts. **Capital:** Vilnius, 584,000 inhab. (1994). **Other cities:** Kaunas, 423,000 inhab., Klaipeda , 205,000 inhab., Siauliai, 149,000 inhab., Panevezys, 131,000 inhab. **Government:** Parliamentary republic, Algirdas Brazauskas, President, elected February 14, 1993 by direct vote. Single-chamber parliament (Seimas) made up of 141 members. **National Holiday:** February 16, Independence (1918). **Armed Forces:** 8,900. **Paramilitaries:** Coast Guard, 5,000.

STATISTICS

DEMOGRAPHY

Urban: 71% (1995). **Annual growth:** 0.7% (1991-99). **Estimate for year 2000:** 4,000,000. **Children per woman:** 2 (1992).

EDUCATION

School enrolment: Primary (1993): 90% fem., 90% male. Secondary (1993): 79% fem., 76% male. University: 39% (1993).

COMMUNICATIONS

22.9 **telephones** per 100 inhabitants (1993).

ECONOMY

Per capita GNP: $1,350 (1994). **Annual growth:** -8.00% (1985-94). **Annual inflation:** 102.30% (1984-94). **Consumer price index:** 100 in 1990; 31,156 in 1994. **Currency:** lit. **Cereal imports:** 415,000 metric tons (1992). **Fertilizer use:** 545 kgs per ha. (1992-93). **Imports:** $2,210 million (1994). **Exports:** $1,892 million (1994). **External debt:** $438 million (1994), $118 per capita (1994). **Debt service:** 2.8% of exports (1994).

ENERGY

Consumption: 2,194 kgs of Oil Equivalent per capita yearly (1994), 80% imported (1994).

dis, and led to a crisis within the Lithuanian Communist Party leadership.

[18] One of the main worries of Lithuanians, Latvians and Swedes was the nuclear plant at Ignalina with four reactors similar to those at Chernobyl. In 1988, after massive public protest, authorities closed down the second reactor, and suspended construction on the third and fourth reactors.

[19] The Lithuanian government opted not to follow the example set by Estonia's Supreme Soviet, which had made a unilateral decision on the question of sovereignty. Instead, Lithuanian authorities began making concessions to local movements, legalizing the use of the flag and the national anthem, designating Independence Day as a holiday and authorizing public commemoration of that day. In addition, Lithuanian was adopted as the country's official language, and the Vilnius Cathedral and other churches were reopened.

[20] In February 1989, the first secretary of the CP and the Sajudis attended the official commemoration of the country's 1918 independence, side by side. In December, Lithuania's Supreme Soviet did away with the article of the constitution which assigned the CP a leading role; the first decision of its kind within the USSR.

[21] In January 1990, Gorbachev announced in Vilnius that the details of a future relation with the Union would be established through legislation. In March, the Lithuanian Parliament proclaimed the nation's independence, to be effected immediately. In September 1991, the new Council of State of the USSR accepted the independence of the three Baltic States, which were immediately recognized by several countries and by the UN.

[22] In August 1991, after the failed coup d'etat against Gorbachev in the USSR, the Lithuanian Parliament banned the Communist Party, the Democratic Workers' Party and the Lithuanian Communist Youth organization. The following month, President Vytautas Landsbergis issued a call, before the United Nations, for the withdrawal of 50,000 Soviet troops from Lithuania; they had been stationed in Vilnius since January of that year.

[23] The new Constitution was approved by a referendum on October 25, 1992. That year, the GNP diminished by 35 per cent and inflation soared to 1,150 per cent.

[24] On February 14, 1993 the leader of the former Communist Party, Algirdas Brazauskas, was elected president with 60 per cent of the vote. The defeat of Stasys Lozoraitis, from the Sajudis, was caused largely by the growing difficult economic and social situation.

[25] The new president continued a transition policy towards a market economy, although at a somewhat slower pace. The Nationalist opposition collected 560.000 signatures to organize a referendum against the government's economic policy. However, on August 27, 1994, only 36.8 per cent of the electorate showed up to vote, which invalidated the event.

[26] In March 1995, government candidates obtained only 20 per cent of the vote in local elections as opposed to over 50 per cent for the Conservative opposition representatives. Once again, social conditions were among the reasons for the returns: it was estimated 80 per cent of Lithuanians were poor, 15 per cent middle class and 5 per cent wealthy.

[27] On February 8, 1996, Prime Minister Adolfas Slezevicius lost a vote of confidence in Parliament, after which he was replaced by Mindaugas Laurinas Stankevicius.

Luxembourg

Luxembourg

Population: 404,000 (1994)
Area: 2,586 SQ.KM
Capital: Luxembourg

Currency: Luxembourg Franc
Language: French and German

[1] Luxembourg, Belgium, the Netherlands, and part of northern France constitute the Low Countries, and until 1579 they shared a common history (see the Netherlands).

[2] In the war of the Low Countries against Spain, Luxembourg sided with the southern provinces, acknowledging the authority of Philip II. Luxembourg was conquered by France in 1684, but returned to Spain thirteen years later, under the Treaty of Rijswijk. In 1713, the Austrian Hapsburgs took control of the country until the Napoleonic invasion in 1795, when the country was annexed to the French empire.

[3] In 1815, after the defeat of Napoleon, the Congress of Vienna handed over the duchy of Luxembourg to William of Orange. William incorporated the duchy as his kingdom's eighteenth province. After the Belgian revolt in 1831, Luxembourg was divided again: the largest section given to Belgium, and the smallest was handed over to William as the Grand Duchy of Luxembourg, which he

finally accepted in 1839. Thereafter the duchy was administered independently until 1867. In 1866, the German Confederation was dissolved, the Treaty of London guaranteed the neutrality of the grand duchy giving control to the House of Nassau.

[4] Germany invaded the country twice, in 1914 and again during World War II. After the war productivity increased when the country formed an economic alliance with Belgium and the Netherlands called the Benelux, later becoming a member of the EEC.

[5] The Christian Social Party, a centre-right party, maintained a majority in Parliament from 1919 until 1974. Afterwards a centre-left coalition of the Socialist Workers' Party and the Democratic Party took office. The Christian Social Party recovered its majority in 1979, forming alliances with the Parti Socialist Workers' Party and the Democratic Party.

[6] In 1949 Luxembourg became a founder member of NATO. In 1986, women were admitted to the armed forces.

[7] Women were granted the right to vote in 1919. The female presence in the registered workforce was 25.2 per cent in 1979, increasing to 34 per cent by 1988. Luxembourgian women occupy 74.8 per cent of posts in education, the highest percentage in the European Community.

[8] In 1988, the metal, iron and steel industries accounted for 31 percent of the workforce and 32 percent of the country's GNP. The whole sector is dominated by the transnational corporation ARBED in Germany and Belgium.

[9] The June 1989 legislative elections reaffirmed the dominance of the Social Christians and Socialists. The PCS took 22 seats, the POS 18 and the Democratic Party 11. The Action Committee, a group created to defend the pension rights of private sector workers, won four seats.

[10] In 1990, border controls were abolished with Belgium, France, Germany and the Netherlands. In 1991, a new "financial scandal" hit the headlines when the International Bank of Credit and Commerce (BCCI) went bankrupt -an institu-

tion which was originally from Luxembourg with its headquarters in the Arab Emirates.

[11] In 1993, Josée Jacobs became the first woman to hold a cabinet position in the history of the nation, becoming minister of agriculture and viniculture. In 1994, the PCS and the POS, who had governed the country together since 1984, were returned to power in the general elections, allowing Jacques Santer to continue as prime minister.

[12] On an internal front, the hostility of many of the local people to foreign workers, who made up 50 per cent of the working population of the country in 1994, began to be expressed openly.

[13] In 1995, Santer became president of the European Union Commission and was replaced as prime minister by Jean-Claude Juncker. Luxembourg established itself as one of the main financial markets of the world, especially in the administration of social funds - like pension funds - with an estimated worth of $356 billion.

PROFILE

ENVIRONMENT

Located on the southeastern side of the Ardennes, Luxembourg has two natural regions: the north with valleys and woods, and a maximum altitude of 500 meters, is sparsely populated with potato and grain farming; the south (Gutland) is a low plain and the country's main demographic corridor, with most of the population and cities and major industries (iron, steel and mining), including the capital.

SOCIETY

Peoples: Luxemburger 69.7%; Portuguese 10.8%; Italian 5.0%; French 3.4%; Belgian 2.5%; German 2.2%; other 6.4%. **Religions:** Majority Catholic. (97%)
Languages: Luxemburgian, French, German, Portuguese. **Political Parties:** Christian Social Party (PCS); Socialist Workers' Party (POS); Liberal Party; Democratic Party; Communist Party; Green Alternative Party and Ecological Initiative. **Social Organizations:** Luxembourg National Workers' Confederation; National Council of Unions.

THE STATE

Official Name: Groussherzogtum Lëtzebuerg
Administrative divisions: 12 cantons.
Capital: Luxembourg, 76,000 inhab. (1991) **Other cities:** Esch-sur-Alzette 24,018; Dudelange 14,674; Differdange 8,520; Schifflange 6,870 (1991)
Government: Constitutional monarchy. Multi-party parliamentary system. Grand Duke Jean, head of State; Jean-Claude Juncker, Prime Minister. Single-chamber legislature: Chamber of Deputies, with 60 members elected by direct popular vote, every 5 years. **National holiday:** June 23 (National day) **Armed Forces:** 800. **Paramilitaries:** 560 (Gendarmes).

DEMOGRAPHY

Annual growth: 1% (1991-99). **Children per woman:** 1.7 (1992).

HEALTH

Under-five mortality: 9 per 1,000 (1995). **Calorie consumption:** 120% of required intake (1995).

EDUCATION

Primary school teachers: one for every 13 students (1990).

COMMUNICATIONS

105 **newspapers** (1995), 102 **TV sets** (1995) and 103 **radio sets** per 1,000 households (1995). 54.1 **telephones** per 100 inhabitants (1993). **Books:** 111 new titles per 1,000,000 inhabitants in 1995.

ECONOMY

Per capita GNP: $39,600 (1994). **Annual growth:** 1.20% (1985-94). **Consumer price index:** 100 in 1990; 112.6 in 1994. **Currency:** 32 Luxembourg francs = 1$ (1994).

Macedonia

Makedonija

Population: 2,100,000 (1994)
Area: 25,713 SQ.KM
Capital: Skopje

Currency: Denar
Language: Macedonian

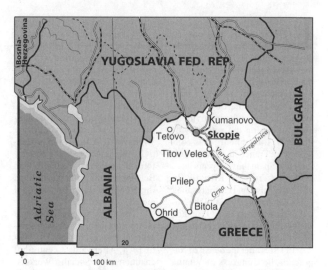

Macedonia's ancient cultural history is linked to that of Greece and Anatolia. According to archeological studies, the ancestors of the Macedonians can be identified in the early Bronze Age. In the year 700 BC, a people calling themselves "Macedonian" migrated toward the east from their native lands on the banks of the Aliakmon River. Aegae was the capital of the kingdom which, during the reign of Amyntas I, extended to the Axios River and beyond, as far as the Chalcidice Peninsula.

[2] Macedonia reached a position of power and influence within Greece during the reign of Philip II (359-336 BC). Warriors had the power to choose a new king, and also to try cases of high treason. Alexander "the Great", Philip's son and a student of Aristotle, defeated the Persian Empire and led the Macedonian armies to northern Africa and the Arabic peninsula, crossing Mesopotamia and reaching as far east as India.

[3] The Macedonian Empire, built up over eleven years, contributed to the propagation of Greek culture in the Orient. Alexander the Great founded a large number of cities and was responsible for the fusion of Greek culture with the cultures of the peoples he conquered, giving rise to what is known as Hellenism. His death in 323 BC was followed by a period of internal struggles, though Macedonia maintained the unity of the empire.

[4] Around the year 280 BC, groups of Galatian marauders invaded Macedonia, killing the king. Three years later, Antigonus II defeated the Galatians and was crowned king by an army from Macedonia. Most of the population were farmers, except for urban areas like Beroea and Pella, and the Greek settlements along the coast.

The king held exclusive rights to the mines and forests.

[5] During the reign of Philip V (221-179), Macedonia conquered Rome's client-states in Illyria and subsequently turned eastward and to the northeast, subduing the cities of Rhodes and Pergamum. Rome responded by going to war and defeated Philip in 197 adding Macedonia to its kingdom and taking Thessaly away from Macedonia. Philip collaborated with the Romans and consolidated his power. This was a prosperous period for Macedonia, which also managed to recover Thessaly.

[6] From 168 to 146 BC, Macedonia was a Roman province, with four independent administrative sections. Macedonia supervised the Greeks for Rome, keeping watch for rebellions and invasion attempts across the northern border. In the year 27, Macedonia became a senatorial province separate from Greece. By the 4th century AD, most Macedonians had adopted Christianity.

[7] The ethnic composition of the Macedonians was not significantly affected by the Goth, Hun and Avar invasions. However, when Slavs arrived in the Balkans, they established permanent settlements throughout Macedonia. Between the 7th and the 14th centuries, Macedonia was successively subdued by the Bulgarian, Byzantine and Latin empires until they were almost completely dominated by the Serbs, except in the Greek region of Salonika.

[8] Toward the end of the 14th century, Turkey began invading the Balkans. By 1371, it had conquered most of Macedonia and in 1389, inflicted a decisive defeat upon the Serbian Empire at Kosovo. The Ottomans seized the best lands for themselves and established a feudal system. Christian peasants either became vassals of Muslim lords, to whom they paid a tithe, or were driven onto the less fertile lands.

[9] In 1864, the Ottoman Empire divided Macedonia into three provinces: Salonika; Monastir, including parts of Albania; and Kosovo, which extended into "Old Serbia". In 1878, Russia forced Turkey into accepting the creation of Bulgaria, which included most of Macedonia, but the other European powers returned this territory to the Ottomans. During the ensuing years, Bulgaria, Serbia and Greece all continued to lay claim to Macedonia.

[10] Toward the end of the 19th century, a strong nationalist movement emerged in Macedonia. Macedonian Slavs created the VMRO (Vatreshna Makedonska Revolutsionna Organizatsia) in 1893, with the slogan "Macedonia for the Macedonians". At the same time, Bulgaria and Greece began sending guerrillas into Macedonia, provoking the Greco-Turkish War of 1897. Turkey supported the Serbs to offset the influence of the VMRO and Bulgarians.

[11] By 1903, the increase in opposition activities by Bulgarians, Greeks, Serbs and Macedonians led Russia and Austria-Hungary to demand that an inspector general be appointed, and that the police force be reorganized. Turkey agreed to meet the demands, but in August there was a general uprising, apparently initiated by Bulgaria. It was ruthlessly put down by the Turks, who destroyed 105 Macedonian-Slavonic villages.

[12] In 1908, after the fall of the Ottoman Empire, with the "Young Turks" rebellion, the clamour to divide up Ottoman-Turkish territories in the region culminated in the two Balkan Wars of 1912 and 1913. Bulgaria and Serbia signed a Mutual Assistance Treaty, with Greece and Montenegro subsequently joining the alliance. Russia supported the Balkan League, because of its interest in halting the southeasterly advance of the Austro-Hungarian Empire.

[13] After defeating Turkey in the first war, the allies turned against each other. Bulgaria provoked the second Balkan War, confronting both Greece and Serbia. Romania and Turkey then allied themselves with Greece and Serbia to defeat Bulgaria. The Treaty of Bucharest gave Greece Salonika and most of the Macedonian coastal area, while Serbia was given the center and the northern parts of the territory.

[14] When World War I broke out, Bulgaria saw a chance to reassert its territorial claims over Macedonia. Sofia joined the Central Powers (Austria-Hungary, Germany and Turkey), and occupied all of Serbian Macedonia and also a part of Serbia. The victorious allied powers left the Greco-Macedonian borders as they were, and Yugoslav Macedonia was incorporated into the new Serbian, Croat and Slovenian kingdom.

[15] In the inter-war period, Serbian domination deepened Yugoslav inter-ethnic conflicts. King Alexander, who assumed dictatorial powers in 1929, was assassinated in Marseilles in 1934 by Croatian nationalists. At the beginning of World War II, when Germany invaded Yugoslavia, these internal divisions meant that the invaders met little resistance.

[16] The Yugoslav nationalist struggle intensified during the following years. Guerrillas led by the Yugoslav Communist League (YCL) seized power in May 1945, later proclaiming the Socialist Federal Republic of Macedonia. Yugoslav Macedonia joined the new state as one of its six constituent republics.

[17] In 1947, Yugoslavia declared that Macedonia was the country's least developed region. The federal government subsequently began earmarking funds for industrializa-

LAND USE

■ CROPLAND 26% ■ PASTURE 25% ■ OTHER 49%

ENVIRONMENT

In the south-central part of the Balkan Peninsula, Macedonia, which has no maritime coast, is bounded in the north by Serbia and Kosovo, in the east by Bulgaria, in the south by Greece and in the west by Albania. Two mountain ranges cross the region, the Pindo (a continuation of the Alps) and the Rodope, in the center and the east. With a continental climate, the average temperature in the capital is 1° C in winter, and 24° C in summer. The country's main agricultural activity is centered in the Vardar River basin. In the mountain region, sheep and goats are raised. There are some copper, iron and lead deposits.

SOCIETY

Peoples: Macedonians 67%; Albanians 22,9%, Turcs 4,0%, Gypsies 2,3%, Serbs 2%, other 2,3% (1994) **Religions:** Christian Orthodox (majority); Muslim. **Languages:** Macedonian (official), Albanian. **Political Parties:** Social Democratic Alliance of Macedonia (former Communist League of Macedonia), currently in power, founded 1990. **Social Organizations:** In the process of being reorganized.

THE STATE

Official Name: Republika Makedonija. **Administrative divisions:** 30 districts. **Capital:** Skopje, 440.557 inhab. (1994) **Other cities:** Bitolj (Bitola) 75.386 inhab.; Prilep 67.371 inhab.; Kumanovo 66.237 inhab.; Tetovo 50.376 inhab. **Government:** President: Kiro Gligorov, re-elected on October, 1995. Prime Minister: Branko Crvenkovski. Unicameral parliament: an Assembly of 120 members, called Sobranje. **Political Parties:** Macedonias Alliance, a coalition of the Democrat Alliance of Macedonia, liberal, the Socialist Party and the Albanian Party for Democratic Prosperity. The Alliance was founded in 1994 and is in power since that date. **National Holiday:** September 8, Independence (1991). **Armed Forces:** 10.400 (8.000 conscripts) **Paramilitaries:** 7,500 (police).

DEMOGRAPHY

Urban: 59% (1995).

EDUCATION

School enrolment: Primary (1993): 87% fem., 87% male. Secondary (1993): 55% fem., 53% male. University: 16% (1993). **Primary school teachers:** one for every 20 students (1992).

COMMUNICATIONS

14.8 **telephones** per 100 inhabitants (1993).

ECONOMY

Per capita GNP: $820 (1994). **Consumer price index:** 100 in 1990; 35,826 in 1994. **Cereal imports:** 117,000 metric tons (1993). **Fertilizer use:** 248 kgs per ha. (1992-93). **Imports:** $1,260 million (1994). **Exports:** $1,120 million (1994). **External debt:** $924 million (1994), $440 per capita (1994). **Debt service:** 12.7% of exports (1994).

---

tion projects, especially in steel, chemical and textile production.

[18] In January, 1990, a special YCL Congress decided to adopt a multi-party system, and eliminate the leadership role which the constitution assigned to the party. However, the motion to grant greater autonomy to YCL branches in the republics was not accepted. After the Congress, the Communist Leagues of Slovenia, Croatia and Macedonia decided to separate from the YCL, and call themselves by a new name, the Democratic Renewal Party.

[19] On September 8, 1991, in the midst of fighting between the new separate republics of Croatia and Slovenia, and the federal army, a plebiscite was held in which Macedonians pronounced themselves in favor of separation from the former Yugoslav Federation. All Macedonian political parties, except for the Albanian ethnic minority, favored independence.

[20] In early 1992, when several countries recognized Croatia and Slovenia, Greece blocked recognition of Macedonia, considering the use of that name to be a "usurpation" of Greek history and culture.

[21] On January 20 1992, Greek and Bulgarian representatives held an informal meeting. Sofia had recognized Macedonia five days previously. From Skopje, the newspaper "Nova Makedonika" accused the Greek and Bulgarian governments of entertaining territorial designs upon their country.

[22] In a plebiscite held on January 12, Macedonia's Albanian minority voted in favor of having an independent State, but on April 3, 1992, the Independent Republic of Ilirida (the republic of the Albanian residents of Yugoslavia) was proclaimed, within Macedonian territory.

[23] The new Yugoslav Federation withdrew its troops from the country. In July, the cabinet resigned en masse and international recognition was not received. In August, Parliament rejected the EEC proposal to change the country's name.

[24] Social Democrat Branko Crvenkovski became prime minister that month, gaining official recognition of the new Macedonian state from Russia, Albania, Bugaria and Turkey.

[25] In April 1993, the UN Security Council recommended that the country be admitted to the General Assembly, with the provisional name of "The Former Yugoslavian Republic (TFYR) of Macedonia".

[26] Greece has withheld official recognition, fearing Macedonia's growing territorial ambitions on the Greek province of Macedonia.

[27] Athens did everything within its power to complicate Skopje's foreign relations, campaigning to change the country's official name, and to alter Macedonia's "expansionist" constitution. Athens also pressured Macedonia to adopt a new flag, as the current emblem is the Vergina star, associated with Alexander the Great, revered by the Greeks as a symbol of their own Hellenic culture.

[28] After Macedonia's admittance to the UN, negotiations with Greece began, mediated by Lord Owen (British) and Thornvald Stoltenberg (Swedish), both of whom were already conversant with Balkan politics. Andreas Papandreou's victory in the October 1993 Greek elections calmed the debate somewhat.

[29] The Balkan War seriously affected the Macedonian economy. Although it maintained international sanctions on the former Yugoslavia, Kiro Gligorov's government was forced to allow trucks to cross its territory, transporting goods to Belgrade. Likewise, the oil-transporting Athens-Belgrade railway continued to operate.

[30] In early 1994, the IMF and the World Bank approved a plan for economic reforms, backed by US and EU financial aid.

[31] On February 16 1994, two years and a half after the independence of Macedonia, Greece decided on a total economic blockade of the country. Greece did not recognize the adoption of the Greek name "Macedonia". The North frontier and the port of Salonica, through which 80 per cent of Macedonian exports, plus all of its oil, went abroad, were closed. The Greek Government was denounced before the European Court of Justice in April, but the blockade continued.

[32] On October 16 of that year Gligorov was re-elected, receiving 52,4 per cent of the votes. The Social Democratic Alliance of Macedonia (ASDM), headed by prime minister Branko Crvenkovski decided on a coalition government with the liberals, the Socialist Party, and the Party of Democratic Prosperity, the main Albanian political group. This coalition was called Alliance of Macedonia (AM).

[33] A year later, a bomb exploded in Gligorov's car. The president survived and during his recuperation the government was in charge of Stojan Andov, a representative of the Parliament. Simultaneously the negotiations between Macedonia and Greece for the lifting of the blockade, which were taking place in New York, made some progress.

[34] From the 15 of October on, Greece stopped the commercial embargo. Macedonia on its part agreed to withdraw the Vergina star from its flag, but the change of name was postponed. In that month, Macedonia was admitted as a member of the Organization for European Security and Cooperation (OSCE) under the name of the former Yugoslavian Republic of Macedonia. With the same name - approved by Greece - it joined the European Council.

Madagascar

Population: 13,100,000 (1994)
Area: 587,040 SQ.KM
Capital: Antananarivo

Currency: Franc
Language: Malagasy and French

Madagasikara

About 2,000 years ago, Malay-Polynesian navigators reached the African coast in canoes, voyage which was frequently repeated between the 1st and 5th centuries.

[2] Towards the 14th century, groups of Comoran traders established a series of ports in the northern region. These ports were destroyed by the Portuguese in 1506-1507.

[3] When the Portuguese found no gold, ivory, or spices, they lost interest in the territory. By this time the Europeans had introduced firearms to the island in exchange for slaves.

[4] In the 16th century the Sakalawas on the west coast and the Betsilios on the east coast established the first monarchies. In the 17th century the state of Merina or Imerina came into being on the eastern edge of the central plateau. A century later it was the Merinas, under their leader Nampoina, who initiated the process of unification which was completed later by Nampoina's son Radama I (1810-1828).

[5] Contact with the Arabs and Europeans became more frequent. Radama adopted the latin alphabet for the Malagasy language. He also used their help to create a modern army. However, the untimely death of the ruler and the ensuing conflicts over succession paved the way for European occupation of the island by the end of the century.

[6] The colonials cleared the virgin forests to make way for sugarcane, cotton, and coffee plantations. They seized the best lands and the peasants were forced to work in conditions of semi-slavery. The struggle for political rights and economic improvement led to a great uprising from 1947 to 1948, and this was ruthlessly put down by the French army with the loss of thousands of lives.

[7] The failure of the insurrection enabled the colonial administration to control the transition to autonomy. Independence was finally proclaimed in 1960. In the following September the country held it's first elections and the Social Democratic Party (PSD) won by a large margin. Its leader, Philibert Tsiranana, became the republic's first president, an office to which he was re-elected in 1965 and 1972.

[8] In May 1972, after a series of serious disturbances, Tsiranana was forced to resign; he turned over full presidential powers to General Ramanantsoa, who suspended the National Assembly and the Senate. He also eliminated the presidency and abolished the 1959 Constitution, giving a military government both executive and legislative power, and creating Institutional and National Councils. In October, he had these measures approved by referendum.

[9] A year later, France decided to withdraw its troops, and the succeeding three-years of instability ended in June 1975, when Commander D. Ratsiraka became prime minister. He adopted socialist policies, and called a referendum on December 21, 1975. His nomination as head of State was overwhelmingly approved, and the Charter of the Madagascan Socialist Revolution was adopted as the basis for a new constitution. On December 30, 1975 the State changed the country's name to the "Democratic Republic of Madagascar".

[10] In June 1976, the new power structure was put into action, with a 12 member Supreme Council of the Revolution and the Government, which was presided over by Colonel J. Rakotomalala. When he died in an airplane accident in July 1976, he was replaced by Justin Rakotoniaina. The legislative function was placed in the hands of a 144-member National Council.

[11] The progressive Malgache parties joined together to form the National Revolutionary Front. AREMA (the Malagasy Revolutionary Vanguard) was the leading party of the Front, and was created in 1975 in support of Ratsiraka's renewal program. The Supreme Council of the Revolution was made up primarily of AREMA members, with members of five other parties ranging in their views from Marxist/Leninism to Christian Democratism. In 1977, Désiré Rakotoarijaona became prime minister.

[12] In 1980, the fall in price of the main export commodities caused a crisis which forced Ratsiraka to impose austerity measures. The government acted repressively displeasing some social sectors and imprisoning opponents. At the same time, it attempted to diversify foreign trade, to reach reconciliation with France and to renew negotiations with the IMF. That year, the foreign debt reached $700 million.

[13] In 1982, the president was re-elected by 80 percent of the vote while the radical sector —represented by MONIMA— obtained the remaining 20 percent.

[14] Madagascar has unique ecological characteristics. Plant and animal species extinct in other parts of the world are found here. The island possesses 3 percent of the world's flora varieties, 53 percent of the bird species, and 80 percent of the reptile and amphibian families present on Earth. But this enormous biological reserve is in danger of extinction. Small-scale farmers, pressurized by the lack of land, add to the deforestation process by burning pieces of land for cultivation at a greater rate than ever before. Although this form of land-clearing has been practiced for centuries, the forest is now being destroyed faster than it can regenerate.

[15] In 1988, Lt. Colonel Victor Ramahatra succeeded Raotoarijaona as prime minister. AREMA, the party in power, won the 1989 elections; Didier Ratsiraka was reelected president, with 67.2% of the vote. His presidency was characterized by reform, with the restoration of a multi-party system and the inclusion in his cabinet of several members of the opposition by March 1990.

[16] In 1991, the opposition united around the Committee of Living Forces, made up of 16 organizations. A series of street demonstrations and the taking of the National Radio, led the government to declare a state of emergency. The Committee called for Ratsiraka's resignation and, in July, nominated a transition government.

[17] In August, following the detention of two ministers from the transition cabinet, 400,000 people took to the streets and repeated the demand for Ratsiraka's resignation.

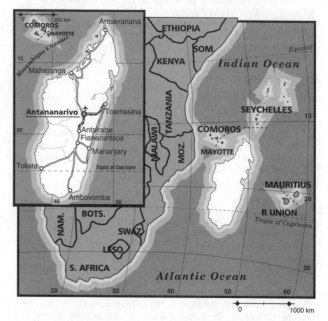

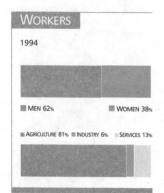

WORKERS

1994

MEN 62% | WOMEN 38%

AGRICULTURE 81% | INDUSTRY 6% | SERVICES 13%

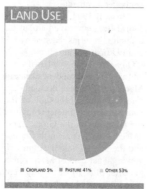

LAND USE

CROPLAND 5% | PASTURE 41% | OTHER 53%

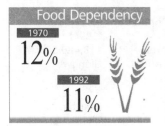

Food Dependency

1970
12%

1992
11%

Foreign Trade

MILLIONS $ 1994

IMPORTS
434

EXPORTS
277

Maternal Mortality

1989-95

PER 100.000
LIVE BIRTHS
660

External Debt

1994

PER CAPITA
$316

PROFILE

ENVIRONMENT

Madagascar is one of the world's largest islands, separated from the African continent by the Mozambique Channel. The island has an extensive central plateau of volcanic origin which overhangs the hot and humid coastal plains. These are covered with dense rainforest to the east and grasslands to the west. The eastern side of the island is very rainy, but the rest has a dry, tropical climate. The population is concentrated on the high central plateau. Rice and export products (sugar, coffee, bananas, and vanilla) are cultivated on the coast. Cattle-raising is also important throughout the island. The major mineral resources are graphite, chrome and phosphate. Deforestation is one of the major environmental problems, it is estimated that the destruction of forests affects up to 75% of the island. Only 10% of the rural population has assured access to drinking water. The lack of sewage facilities and concentration of organic wastes have led to the pollution of many watercourses.

SOCIETY

Peoples: The Malagasy people are believed to be descendants of Malay-Polynesian navigators who populated the island 2,000 years ago. The major groups are; the Merina who make up a quarter of the population, the Betsileo, the Sakalawa, the Antankarana, the Betsimisaraka and the Antasaka. From the 15th century, Arabs and Europeans introduced African slaves, mostly of Bantu origin, who mixed with the native population. There are also French, Indian, Chinese, and Comoran communities. **Religions:** Christian 51%, of which Roman Catholic 26%, Protestant 22.8%; traditional beliefs 47%; Muslim 1.7%; other 0.3%. **Languages:** Malagasy and French. Hovba and other local dialects are also spoken. **Political Parties:** Committee of Living Forces, an alliance of 16 groups; Militant pro-Socialist Movement. **Social Organizations:** Confederation of Malagasy Workers (FMM); Christian Confederation of Trade Unions (SEKRIMA); Federation of Autonomous Trade Unions of Madagascar (USAM); Federation of Workers' Unions of Madagascar (FISEMA).

THE STATE

Official Name: Repoblikan'i Madagasikara **Administrative Divisions:** 6 provinces, 10 districts, 1,252 sub-districts, and 11,333 towns. **Capital:** Antananarivo, 1,052,000 inhab. in 1993. **Other cities:** Toamasina, 127,000 inhab.; Mahajanga (Majunga), 100,000 inhab.; Fianarantsoa, 99,000 inhab.(1993) **Government:** Parliamentary republic. Albert Zafy, President ; Emmanuel Rakotovahiny, Prime Minister. Bicameral Legislature-Senate (2/3 selected by an Electoral College and 1/3 selected by the president) and a National Assembly, whose members are elected directly. **National Holiday:** July 26, Independence Day (1960). **Armed Forces:** 21,000. **Paramilitaries:** Gendarmerie: 7,500.

STATISTICS

DEMOGRAPHY

Urban: 26% (1995). **Annual growth:** 2.5% (1992-2000). **Estimate for year 2000:** 16,000,000. **Children per woman:** 6.1 (1992).

HEALTH

One **physician** for every 8,333 inhab. (1988-91). **Under-five mortality:** 164 per 1,000 (1995). **Calorie consumption:** 93% of required intake (1995). **Safe water:** 29% of the population has access (1990-95).

EDUCATION

School enrolment: Primary (1993): 72% fem., 72% male. Secondary (1993): 14% fem., 14% male. **University:** 4% (1993). **Primary school teachers:** one for every 38 students (1992).

COMMUNICATIONS

84 **newspapers** (1995), 85 **TV sets** (1995) and 94 **radio sets** per 1,000 households (1995). 0.3 **telephones** per 100 inhabitants (1993). **Books:** 84 new titles per 1,000,000 inhabitants in 1995.

ECONOMY

Per capita GNP: $200 (1994). **Annual growth:** -1.70% (1985-94). **Annual inflation:** 15.80% (1984-94). **Consumer price index:** 100 in 1990; 190.1 in 1994. **Currency:** 1.963 francs = 1$ (1993). **Cereal imports:** 111,000 metric tons (1993). **Fertilizer use:** 25 kgs per ha. (1992-93). **Imports:** $434 million (1994). **Exports:** $277 million (1994). **External debt:** $4,134 million (1994), $316 per capita (1994). **Debt service:** 9.5% of exports (1994). **Development aid received:** $370 million (1993); $27 per capita; 11% of GNP.

ENERGY

Consumption: 37 kgs of Oil Equivalent per capita yearly (1994), 83% imported (1994).

---

ation of the country, one of the poorest in the world, was considered disastrous. Per capita income barely increased between 1976 and 1992, going from $200 to $230 per year, while calorie consumption went from 108 per cent of those required in 1964-66, to 95 per cent in 1988-90.

[20] In March 1994, the government introduced a series of austerity measures, recommended by the International Monetary Fund, which increased the social tensions even further. At the end of this year, massive demonstrations were held in opposition to these policies. In January 1995, the governor of the Central Bank abandoned his post on request from the IMF and the World Bank.

[21] In September, the Malagasy

people approved increased powers for Zafy in a referendum. The debate over structural adjustment coincided with disagreements over the use of the nation's natural resources, when the mining transnational RTZ proposed opening a mine on the southern coast of the island to extract titanium dioxide. The project caused harsh protests by militant ecologists, convinced the project would destroy unique species of the Madagascar flora and fauna.

The demonstration was brutally put down, with 31 people killed and more than 200 injured.

[18] In March 1992, a multi-party forum was created to draw up a new constitution and presidential elec-

tions were set for August. In this month, the new document was approved and Albert Zafy was elected president with 66.2 per cent of the vote to Ratsiraka's 33.8.

[19] The economic and social situ-

Malawi

Malawi

Population: 9,532,000
(1994)
Area: 118,480 SQ.KM
Capital: Lilongwe

Currency: Kwacha
Language: Chewa

The state of Kitwara was part of a small country on the coast of Lake Malawi (previously Nyasa), and had been part of a series of nations related to gold production and ruled by the Monomotapa of Zimbabwe (see Zimbabwe). The decline of this power center allowed the Chewa to enlarge their territory, only to see it reduced again when the Changamira Rotsi restored the predominance of Zimbabwe. Around 1835, Zulu expansion (see South Africa) pushed the Ngoni-Ndwande to the shores of the lake, leading to sixty years of war between the Ngoni and the Chewa and Yao allies.

[2] The country was explored by Livingston in 1859 and it suffered a Portuguese attempt at colonization in 1890 which was ended by an ultimatum from the British government. Britain wanted to keep the territory which would eventually serve as a link in a continuous chain of colonies joining South Africa to Egypt. In 1891, Cecil Rhodes' British South African Company negotiated the Protectorate of what became Nyasaland.

[3] The British idea was to create a Central African federation embracing present-day Zimbabwe, Malawi and Zambia, regions linked by climatic similarities, plains, plateaus, dry forests and by common Bantu tribal ties. Politically this would have meant the extension of a Rhodesian-style white-racist domination to the entire federation.

[4] The Malawi Congress Party (MCP), and the UNIP of Zambia, favored a pro-independence stance. They were strongly influenced by Dr Hastings Kamuzu Banda, a doctor who had studied in the United States and who was presented as the "nation's saviour".

[5] Banda's demands for greater power within the party were granted to prevent internal division.

When the colony became independent in 1964, Banda seized control of the MCP and the country. The president established close economic and diplomatic relations with the racist governments of South Africa and Rhodesia, and with the colonial administration in Mozambique.

[6] South Africa became Malawi's main tea and tobacco purchaser while South African investors built roads, railways and a new capital city. South African managers took charge of the airline, information and development agencies and a large part of the state administration.

[7] In 1975, Mozambican independence radically changed Banda's situation. He had actively cooperated with the Portuguese in their struggle against FRELIMO. The closing of Mozambique's border with Rhodesia led to a drastic reduction in Malawi's trade with Rhodesia, depriving Ian Smith's racist government of an escape route from the international blockade.

[8] In June 1978, Banda held the first election in 17 years. All the candidates had to be MCP members and pass an English test, which immediately excluded the 90 percent of the population who do not speak English.

[9] After 1980, Zimbabwe's independence changed the economic and political situation of Malawi. Banda lost direct communication with South Africa, and was therefore deprived of major support. Consequently the government drew closer to the Front Line states, joining the SADCC association because of Malawi's dependence on the railway lines that run through Mozambique and Zimbabwe.

[10] The success of neighboring socialist governments strengthened

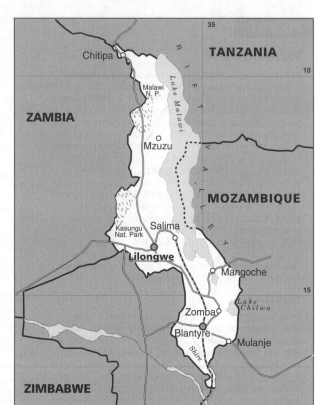

the Socialist League of Malawi (LESOMA) which favored breaking economic and political ties with South Africa, putting an end to Banda's dictatorship and fully redemocratizing the country. In 1980, this party created a guerrilla force. The Malawi Freedom Movement (MAFREMO), led by Orton Chirwa, also gained strength.

[11] In 1983, Chirwa and Attati Mpakati, a LESOMA leader, were accused of conspiracy and sentenced to death. Shortly thereafter,

Mpakati was assassinated by South African agents while visiting Harare. Chirwa and his wife were kidnapped in Zambia, where they lived as exiles, and imprisoned in Blantyre.

[12] Banda created a secret police force, called the Special Branch, with South African and Israeli advisers. The president-for-life also personally controlled the economy, owning 33 per cent of all businesses.

[13] Between 1987 and 1988, the country received 600,000 refugees from Mozambique, in whose civil war Malawi had supported the counter-revolutionaries of the National Resistance Movement. This support was discontinued in 1988, after the visit which Banda received from President Chissano of Mozambique.

[14] Also in 1988, Amnesty International denounced the imprisonment of well-known scholars and writers, among them Jack Mapanje, the country's foremost poet. The United States announced the cancellation of $40 million of foreign debt in November 1989.

PUBLIC EXPENDITURES

% of GNP   1992

3.4 EDUCATION
2.9 HEALTH
1.4 MILITARY

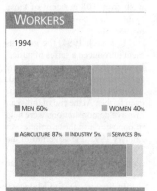

WORKERS

1994

MEN 60%   WOMEN 40%

AGRICULTURE 87%   INDUSTRY 5%   SERVICES 8%

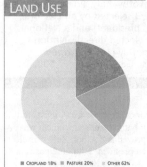

LAND USE

CROPLAND 18%   PASTURE 20%   OTHER 62%

PROFILE

ENVIRONMENT

The terrain and the climate are extremely varied. The major geological feature is the great Rift fault that runs through the country from north to south. Part of this large depression is filled by Lake Malawi, which takes up one fifth of the land area. The rest is made up of plateaus of varying altitudes. The most temperate region is the southern part, which is also the highest, containing most of the population and economic activities (basically farming). The lowlands receive heavy rainfall and are covered by grasslands, forests or rainforests, depending on the amount of rainfall they receive. The degradation of the soil and deforestation are the main environmental problems.

SOCIETY

Peoples: Maravi (including Nyanja, Chewa, Tonga, and Tumbuka) 58.3%; Lomwe 18.4%; Yao 13.2%; Ngoni 6.7%; other 3.4%. **Religions:** Many people follow traditional religions. There are significant Christian and Muslim communities. **Languages:** Chewa (official) and English; several Bantu languages - other than Chewa - are spoken by their respective ethnic groups. **Political Parties:** United Democratic Front (UDF); Alliance for Democracy (Aford); Malawi Congress Party (MCP). **Social Organizations:** Trade Union Congress of Malawi.

THE STATE

Official Name: Republic of Malawi. **Administrative divisions:** 24 districts. **Capital:** Lilongwe, 395,000 inhab. (1994). **Other cities:** Blantyre 446,800 inhab.; Mzuzu 62,700 inhab. (1994) **Government:** Presidential republic. Bakili Muluzi, President since May 1994. Legislature: single-chamber: National Assembly, made up of 177 members. **National Holiday:** July 6, Independence Day (1964). **Armed Forces:** 8,000 **Paramilitaries:** 1,500

Illiteracy
1995

44%

Maternal Mortality
1989-95

PER 100.000 LIVE BIRTHS

620

STATISTICS

DEMOGRAPHY

Urban: 13% (1995). **Annual growth:** 2.9% (1992-2000). **Estimate for year 2000:** 11,000,000. **Children per woman:** 6.7 (1992).

HEALTH

One **physician** for every 50,000 inhab. (1988-91). **Under-five mortality:** 221 per 1,000 (1995). **Calorie consumption:** 90% of required intake (1995). **Safe water:** 47% of the population has access (1990-95).

EDUCATION

Illiteracy: 44% (1995). **School enrolment:** Primary (1993): 77% fem., 77% male. Secondary (1993): 3% fem., 6% male. University: 1% (1993). **Primary school teachers:** one for every 68 students (1992).

COMMUNICATIONS

83 **newspapers** (1995), 82 **TV sets** (1995) and 95 **radio sets** per 1,000 households (1995). 0.4 **telephones** per 100 inhabitants (1993). **Books:** 84 new titles per 1,000,000 inhabitants in 1995.

ECONOMY

Per capita GNP: $170 (1994). **Annual growth:** -0.70% (1985-94). **Annual inflation:** 18.80% (1984-94). **Consumer price index:** 100 in 1990; 220.5 in 1994. **Currency:** 4 kwacha = 1$ (1993). **Cereal imports:** 514,000 metric tons (1993). **Fertilizer use:** 434 kgs per ha. (1992-93). **Imports:** $491 million (1994). **Exports:** $325 million (1994). **External debt:** $2,015 million (1994), $211 per capita (1994). **Debt service:** 17.4% of exports (1994). **Development aid received:** $503 million (1993); $48 per capita; 26% of GNP.

ENERGY

Consumption: 39 kgs of Oil Equivalent per capita yearly (1994), 59% imported (1994).

---

[15] The implementation of an IMF adjustment program led to a reduction in inflation and in the balance of payments deficit, as well as an increase in investments. However, it also aggravated the situation faced by the poorest sectors of the population.

[16] In 1990 and 1991, earthquakes and floods exacerbated food shortages among the rural population, 90 per cent of the country's total population. The privatization of the maize/corn market benefitted a few producers at the expense of the poorest rural sectors.

[17] Despite the constant denunciations of human rights violations, US vice-president Dan Quayle reaffirmed his government's support for the Banda regime.

[18] For the first time since independence, the Catholic Church wrote a pastoral letter criticizing the human rights situation, and calling for greater political freedom. This letter was followed by a popular uprising in Blantyre, which was harshly repressed by government forces.

[19] In April 1992, opposition leader Chafuka Chihana, of the Alliance for Democracy, was caught trying to return to the country, and imprisoned. An international campaign prevented him from being executed.

[20] During the Cold War, Banda was a staunch ally of the West, but the constant human rights violations of his government led to several countries cutting aid to Malawi.

[21] In May, a general strike called by textile workers was brutally put down, with 38 deaths and hundreds of injuries. In reprisal, the World Bank discontinued part of its financial aid.

[22] The ruling Congress Party of Malawi, the only party to participate in the June general elections, obtained 114 seats in the National Assembly.

[23] In late 1992, news was recieved of the death of Orton Chirwa of the Malawi Freedom Movement (MAFREMO). He had been in prison since 1983, and died under torture. To keep popular indignation from turning violent, Banda announced that a referendum would be held on opening up the political system.

[24] The referendum took place in June 1993. Two thirds of voters chose a multiparty system. That month, Banda promised presidential elections would be held in 1994 and released Vera Chirwa, widow of the assassinated dissident and who by that time had served the

longest prison term for political reasons in Africa.

[25] Opposition member Bakili Mukizi was elected president in May 1994. His party, the United Democratic Front (UDF), won 84 of the 177 seats at stake in the legislative elections. In September, having won only 55 seats, Banda decided to retire from political activity.

[26] Malawi suffered the consequences of an intensive drought, in 1994, which led to a food shortage. In the midst of a growing difficult social situation, the government went ahead with its IMF-sponsored policy to cut back public spending. In January 1995, ex-president Banda was arrested, charged with the murder of three former ministers. The alleged vi-

olation of human rights by Muluzi's government was frequently denounced by opposition members throughout the year.

[27] In May 1996, the president significantly modified his government's composition. Three key ministers and the country's second vice president were replaced. Among other changes, Foreign Minister Edward Bwanali - appointed Minister of Irrigation - was replaced by then transport minister George Ntafu.

Malaysia

Malaysia

Population: 19,669,000 (1994)
Area: 329,750 SQ.KM
Capital: Kuala Lumpur

Currency: Ringgit
Language: Malay

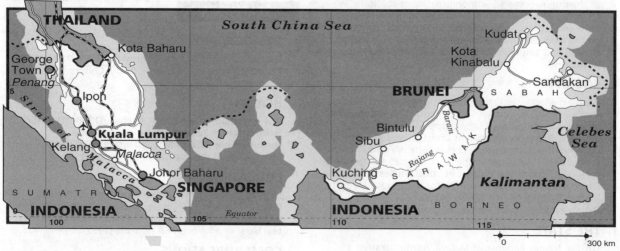

T he Malay Peninsula and the Borneo states of Sarawak and Sabah were originally inhabited by native aboriginal peoples, living in the forests.

[2] In the second millennium BC, there was migration from the south of present-day China to present-day Malaysia, Indonesia and the Philippines. Over the millenia, metal-working techniques and agriculture were introduced. Rice farming was not developed until the first millennium AD. Indian influence was all pervasive, bringing religion, political systems and the Sanskrit language.

[3] The "Indianized" kingdom of Funan was founded on the Mekong in the 1st century AD and Buddhist states eventually developed in the east, trading with China. In the 15th century, the port of Melaka (Malacca) was founded; its rulers were the first in the region to convert to Islam. Trade with Islamic merchants brought prosperity to Melaka. The new faith spread across the rest of present-day Malaysia and Indonesia, replacing Buddhism. At the be-

ginning of the 16th century, Melaka, attracted the Portuguese, who were competing with Arab merchants for the Indian Ocean trade routes.

[4] In 1511 the Portuguese viceroy of India, Alfonso de Albuquerque, seized the port by force. It was of vital strategic importance in the Portuguese struggle to maintain their monopoly on the spice-trade from the Moluccas Islands. These spices were exchanged for Indian textiles and Chinese silk and porcelain. In the 17th century, the Dutch formed an alliance with the Sultan of Johor to drive the Portuguese out of Melaka. This alliance between Johor and the Dutch was established in Batavia (present-day Jakarta) and it succeeded in eliminating European and Asian competition for a hundred years.

[5] Meanwhile, the British began to set up back-up points for their trade with China in northern Borneo (Kalimantan), and in 1786 founded the port of Georgetown, on the island of Penang (Pulau/Pinang), opposite the western coast of

the Malay peninsula. The British model of free trade proved more successful than the Dutch trade monopoly and Penang attracted a cosmopolitan population of Malays, Sumatrans, Indians and Chinese. In 1819, the British founded Singapore, but at the time they were more interested in safeguarding navigation than in the local spice trade, as their imports from China were being paid for with opium from India. Nevertheless, the Dutch and the British found it difficult to co-exist in the region; a treaty drawn up in 1824, granted control of Indonesia to the Dutch, while Malaya was left in British hands.

[6] The colonies of Penang, Melaka and Singapore became the key points of the British colony. The British encouraged Chinese (and to a lesser degree, Indian) immigration to the ports along the straits, and large numbers of Chinese arrived to work the tin mines and service the urban ports. The Malay peasants and fishing people continued their traditional activities. From 1870, the British began to sign

"protectorate" agreements with the sultans and in 1895 they encouraged them to form a federation, with Kuala Lumpur as its capital. The sultanates of northern Borneo (Brunei, Sabah and Sarawak, the latter ruled by James Brooke, and his heirs) also became British protectorates, administered from Singapore, but without any formal ties with the peninsula.

[7] Towards the end of the 19th century, the British introduced rubber farming with heveas seeds smuggled from Brazil, putting an end to the "rubber boom" in the South American Amazon. They promoted Tamil immigration, from southern India, to get workers for the plantations which faced growing demands from the incipient automobile industries.

[8] The colonial administration set up a three band education system that differed for Malays, Indians and Chinese. In the economy, the Malays stayed mainly in rural agriculture, the Chinese worked in tin mines and the urban service sectors, and the Indians worked on the rubber estates. In the first decades of the 20th century, the Malays joined the Islamic reform movements of the Middle East, the Indians supported the struggles of Mahatma Gandhi, and the Chinese were ideologically influenced first by the nationalism of Sun Yat-Sen and then by the Communist Party.

[9] In 1942, during World War II, the country was occupied by the Japanese. The Japanese tried to form alliances with local Southeast Asian nationalist movements to gain support against the European powers. The greatest resistance came from the Chinese, especially

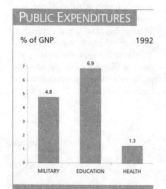

PUBLIC EXPENDITURES

% of GNP — 1992

MILITARY 4.8 — EDUCATION 6.9 — HEALTH 1.3

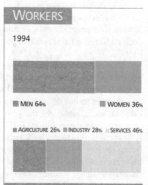

WORKERS

1994

MEN 64% — WOMEN 36%

AGRICULTURE 26% — INDUSTRY 28% — SERVICES 46%

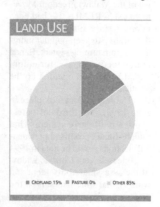

LAND USE

CROPLAND 15% — PASTURE 0% — OTHER 85%

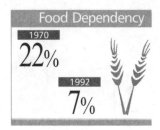

Food Dependency

1970
22%

1992
7%

Foreign Trade

MILLIONS $ 1994

IMPORTS
59,581

EXPORTS
58,756

PROFILE

ENVIRONMENT

The Federation of Malaysia is made up of Peninsular Malaysia (131,588 sq km), and the states of Sarawak (124,450 sq km) and Sabah (73,711 sq km) in northern Borneo (Kalimantan), 640 km from the peninsula in the Indonesian archipelago. Thick tropical forests cover more than 70% of the mainland area, and a mountain range stretches from north to south across the peninsula. Coastal plains border the hills on both sides. In Sabah and Sarawak, coastal plains ascend to the mountainous interior. There is heavy annual rainfall. Malaysia's economy is export orientated. Tin and rubber, the traditional export products have recently been replaced by petroleum and manufactured goods. Indiscriminate logging and the use of highly toxic herbicides worries ecologists who fear irreparable damage to native species, and the destruction of the few surviving native cultures.

SOCIETY

Peoples: Three groups of people have co-existed in the Malaysian peninsula for centuries, each maintaining their cultural identity. The Malays, or Bumiputeras, are estimated to make up over 50% of the population; the Chinese account for a little over 33%, and Indians make up the remaining percentage. There are minor ethnic groups in the northern part of the peninsula and in the insular provinces. Traditionally, Malays are farmers, civil servants and army recruits. The Chinese are mostly employed in trade, industry and mining, and the Indians tend to work in rubber farming or the liberal professions. **Religions:** Islam, the official religion, is practiced by the Malays, 54%. The Chinese are mainly Buddhist, 18%, and Taoist, 22%, and the Indians Hindu, 7%. There is a Christian minority, 7%.
Languages: Malay is the official language, the common language, and the main language of education. Chinese dialects, Tamil and English are also used. **Political Parties:** For the past two decades the government has been controlled by the Barisan Nasional (National Front), a coalition of 10 parties, most of which were formed to represent ethnic interests. The

main components are; the United Malays National Organization (UMNO); the Malaysian Chinese Association (MCA); the Malaysian Indian Congress (MIC); and the United Peoples' Party (PBB). The Front controls over two-thirds the parliamentary seats. The opposition coalition is fragmenting and weak. They include the Chinese-based Democratic Action Party (DAP); the PAS (the Islamic party); and Semangat 46 which broke off from UMNO. The People's Party (Parti Rakyat) attracts a small amount of support from low-income groups. Politics in Malaysia revolve mainly around ethnic issues, with ideological issues on the margins. Political rallies are forbidden, even during elections, and public meetings and street demonstrations require a police permit. **Social Organizations:** The leading labor organization is the Congress of Malaysia's Unions. Despite restrictions, grassroots organizations are very active in human rights, defense of the environment, and consumer rights. The Consumers' Association of Penang and Sahabat Alam (Friends of the Earth) Malaysia, are well-known for their influence on public opinion.

THE STATE

Official Name: Malaysia. **Capital:** Kuala Lumpur, 1,145,075 inhab. (1991): **Other cities:** Ipoh, 382,633 inhab., Johor Baharu, 328.646 inhab., Melaka, 295,999 inhab., Petaling Jaya, 254,849 inhab., Georgetown, 250,000 inhab. (1991) **Government:** Constitutional, parliamentary and federal monarchy. Tuanku Ja'afar ibni al-Marhum Tuanku Abdul Rahman is the current king or *yang di-pertuan agong* (sovereign). The sovereign is elected every five years from among the nine regents (sultans), and only the sultans can vote. Datuk Seri Mahathir bin Mohamad, Prime Minister Central bicameral parliament, 69-member senate, 192-member chamber of deputies, with a constitution and a legislative assembly for every state. **National Holiday:** August 31, Independence Day (1957). **Armed Forces:** 114,500 (1995) **Paramilitaries:** 21,500 Police, Regular, Sea and Air, Auxiliaries.

---

the Malayan Communist Party, which organized guerrillas to oppose the invading forces.
[10] At the end of the war it was clear that British domination could not persist without changes, but the difficulties of diverse ethnic interests, "protected" sultanates and ports under direct colonial administration, made it difficult to find a suitable political system. The British proposed a Malayan Union with equal citizenship for all. This threatened the position of the Malays, and Malay nationalists gathered around the symbolic figure of the sultans founding the United Malays National Organization (UMNO), controlled by the dominant class but with popular support.
[11] In 1948, there was a communist led insurrection that was sup-

pressed by British forces. Failure was partly due to the view of many impoverished Malays and Indians, that the revolt was a Chinese effort and not the action of a unifying, anti-colonial progressive movement. The Marxist parties were outlawed and their leaders jailed. From 1948 until 1960, the Communist Party waged guerrilla warfare in the northern Malayan Peninsula and in Borneo.
[12] Dato Onn made a new attempt at creating a pan-ethnic party, he left UMNO to found the Malayan Independence Party. He was defeated in the 1952 municipal elections by an alliance of UMNO and the Malayan Chinese Association. The alliance was later broadened to include the Indian Malayan Congress, winning nationwide elec-

tions, except for Singapore, where Lee Kuan Yew's socialists were victorious.
[13] Faced with the threat of an armed communist insurrection, the British decided to negotiate. This resulted in Malayan independence in 1957. Tunku Abdul Rahman, a prince who led the independence movement, became the first prime minister. A federation of 11 states was established with a parliamentary system and a monarch chosen every five years from among nine state sultans. A constitutional "bargain" was struck by the three communities; citizenship was granted to the non-Malays, but the Malays were recognized as indigenous people; they were accorded special privileges in education and public-sector employment and Malay

would be the official language. The country adopted the free market system and a reliance on foreign capital, which had been established under colonialism, remained dominant. Malaya was politically pro-Western.
[14] In 1963, the British colonial states of Singapore (south of Malaya), Sabah and Sarawak obtained independence and joined Malaya to form the Federation of Malaysia. There were major disagreements over ethnic policy and Singapore was expelled from the federation in 1965 becoming an independent republic.
[15] The conservative Alliance Party ruled with a large majority from 1957, but in the 1969 elections it lost many seats to the Islamic PAS Party, Gerakan and, mostly, to the

Illiteracy
1995
17%

Maternal Mortality
1989-95
PER 100.000 LIVE BIRTHS
34

External Debt
1994
PER CAPITA
$1,259

STATISTICS

DEMOGRAPHY

Urban: 52% (1995). **Annual growth:** 1.4% (1992-2000). **Estimate for year 2000:** 22,000,000. **Children per woman:** 3.5 (1992).

HEALTH

One **physician** for every 2,562 inhab. (1988-91). **Under-five mortality:** 15 per 1,000 (1995). **Calorie consumption:** 120% of required intake (1988-90). **Safe water:** 78% of the population has access (1990-95).

EDUCATION

Illiteracy: 17% (1995). **School enrolment:** Primary (1993): 93% fem., 93% male. Secondary (1993): 61% fem., 56% male. **Primary school teachers:** one for every 20 students (1992).

COMMUNICATIONS

106 **newspapers** (1995), 102 **TV sets** (1995) and 105 **radio sets** per 1,000 households (1995). 12.6 **telephones** per 100 inhabitants (1993). **Books:** 101 new titles per 1,000,000 inhabitants in 1995.

ECONOMY

Per capita GNP: $3,480 (1994). **Annual growth:** 5.60% (1985-94). **Annual inflation:** 3.10% (1984-94). **Consumer price index:** 100 in 1990; 117.4 in 1994. **Currency:** 3 ringgits = 1$ (1994). **Cereal imports:** 3,288,000 metric tons (1993). **Fertilizer use:** 1,977 kgs per ha. (1992-93). **Imports:** $59,581million (1994). **Exports:** $58,756 million (1994). **External debt:** $24,767 million (1994), $1,259 per capita (1994). **Debt service:** 7.9% of exports (1994). **Development aid received:** $100 million (1993); $5 per capita; 0% of GNP.

ENERGY

Consumption: 1,711 kgs of Oil Equivalent per capita yearly (1994), -66% imported (1994).

Chinese-based Democratic Action Party. Tensions escalated into widespread riots and unrest, largely blamed by the authorities on 'Communist Terrorists'. The parliamentary system was suspended and the country was ruled for two years by a National Operations Council. The Malay population were dissatisfied with their share of corporate equity, because in 1970 they made up over 50 per cent of the population, but had only one percent of the income. A new economic policy was formulated to deal with this. It set targets to increase the Malay and other "Bumiputera" (sons-of-the-soil or indigenous groups) share to 30 per cent by 1990, while the foreign share would drop from 70 to 30 per cent and the non-Bumiputera (mainly Chinese and Indian) share would be 40 per cent.

[16] This policy dominated internal economic policy for the next two decades. The Bumiputera equity share rose to almost 20 percent by 1989 whilst foreign ownership fell below 40 per cent. Because of these policies, Malaysians of Chinese descent claimed that they faced discrimination at work, in education and wherever economic opportunities are concerned.

[17] Communist guerrilla warfare gradually died out, and the National Front won over two-thirds of the parliamentary seats in elections during the 1970s and 1980s. The present prime minister, Mahathir Mohamad, took office in 1981, developing industry. In the late 1980s he faced increasing challenges to his leadership from UMNO (the Front's main party). Some of his opponents left to form a new opposition party, Semangat 46. In 1990 they formed a loose opposition coalition with the Islamic PAS, the Chinese-based Democratic Action Party (DAP) and the small left-wing People's Party.

[18] An Internal Security Act allows detention without trial for two years. Under Mahathir, the number of political prisoners declined from over 1,000 to a few hundred. In 1987 a major crackdown led to 150 more detentions, including opposition politicians and leaders of social groups, all of whom were released by 1990.

[19] In 1988, the chief judge and three other Supreme Court judges were sacked following a dispute between the judiciary and the executive. Lawyers and critics accused the government of threatening the judiciary's independence. There are currently serious limits imposed on the freedom of press and of assembly; permits are required for both newspapers and public meetings.

[20] Malaysian foreign policy has moved from its pro-western position in the 1960s through non-alignment in the 1970s to a pro-Third World stance in the 1980s. In 1990, Kuala Lumpur hosted the inaugural summit meeting of 15 Third World countries (including India, Brazil, Indonesia, Mexico, Venezuela, Tanzania and Senegal) aimed at fostering concrete South-South co-operation projects and aiming to reduce Northern dominance of Southern economies. It played a major role in the South Commission and actively supports the Palestine Liberation Organization and South Africa's African National Congress. Mahathir was internationally recognized as a pro-South leader.

[21] There has been a growth of activity and influence of civic organizations, including consumer, environmental and human rights groups. The Consumers Association of Penang has spearheaded actions on food safety, tenancy rights and business malpractices.

[22] The *Sahabat Alam* (Friends of the Earth) Malaysia assisted indigenous groups in Sarawak to defend their forest from loggers in a widely-publicized campaign. The natives (especially nomadic Penan) staged blockades on timber roads to prevent logs leaving the forest. In August 1991 a judge in Sarawak condemned eight protest leaders to nine months in prison.

[23] Malaysia is the major world exporter of tropical wood with a growing demand in the industrialized countries, especially Japan. In 1989 its timber exports earned $ 2.6 billion. More than 80 per cent of this timber comes from Sarawak and Sabah, in Borneo. It is estimated that out of the previous 305,000 sq km of tropical forests, only 157,000 sq km are left, and the country loses 5,000 sq km of tropical forest each year. Environmental organizations accuse several companies of destroying vast areas of forest, and of causing huge fires to create new areas for cultivation. Alternative development models based on satisfying basic needs and living in harmony with nature are increasingly advocated by social and ecological groups.

[24] In 1991, construction plans were approved for a hydroelectric dam on the Pergau River, north of Kelantan, the country's poorest state. The project was financed by Britain at a cost of $350 million and was tied to British arms purchases amounting to $1.5 billion. The dam would cause a damage to the environment which has met with opposition. A similar project in 1995 was also criticized for its anticipated social and ecological impact. Thousands of peasants were forced to leave forested areas which would later be flooded.

[25] The rapid growth of high-tech industries has led to a shortage of skilled and semi-skilled labour. Wages in the industrial sector have had a significant raise which led to increase social differences.

[26] According to Amnesty International, several prisoners of conscience have been jailed under the Internal Security Law. In Malaysia, prisoners are physically punished and the death penalty is instituted.

[27] In April 1994, Tuanku Ja'afar ibni al-Marhum Tuanku Abdul Rahman was elected king of Malaysia.

[28] The support had by prime minister Datuk was confirmed in 1995 when his coalition, the National Front, won 162 of the 192 seats in the Chamber of Representatives with 84 per cent of the vote.

[29] In the last years, economic expansion has carried on. The GDP has grown between 8 per cent and 9 per cent, based on the manufacturing industry and cheap labour. The Proton automobile has turned into a symbol of Malaysian economy. Inflation in 1995 reached 3.5 per cent which worried the government and led the prime minister to announce there would be no further price increases.

Maldives

Population: 246,000
(1994)
Area: 300 SQ.KM
Capital: Male

Currency: Rufiyaa
Language: Dhivehi

Maldives

The Republic of the Maldives consists of an archipelago of more than 1,000 coral islands in the Indian Ocean, southwest of India and Sri Lanka, of which only 192 are permanently inhabited. Maldivans are excellent sailors and fisherfolk and have always had close contact with the mainland of Asia. It was through this connection that Arab and Muslim influence reached them in the 12th century; the islanders converted to Islam and their ruler took the title of Sultan.

[2] European colonizers came to the Maldives quite early on, as it was a compulsory stop-over on the way to the Far East. The natives put up tenacious resistance to foreign domination, forcing the Portuguese to seek alternative harbors; hence the foundation of Goa on the west coast of India.

[3] Eventually, the sultan of the Maldives gave in to the enticing offers of the agents of British imperialism and, in 1887, he agreed to place his islands under British "protection". The economy was makeshift based on the production of coconut oil, fishing and the cultivation of tropical fruits. On the other hand, the Maldives possessed great strategic value, which was enhanced on the opening of the Suez Canal.

[4] A naval base was set up on Gan Island, on the equator, which became a link in the safety chain that protected navigation from Gibraltar to Hong Kong, via Aden and Singapore.

[5] The local population was of no interest to the authorities even as a source of labor, so very little was done for their education, health or welfare. Even today the country has only one teacher for every 2,000 inhabitants and one doctor for every 15,000, it also has one of the world's lowest per capita incomes.

[6] This neglect stimulated rebellion against the sultan, whose position as an intermediary between his people and the colonial capital made him the only person to benefit from this contact. In 1952, a popular revolt overthrew the ruler and a republic was proclaimed. British troops intervened to "restore order", putting the sultan back on the throne two years later. In 1957, Britain requested permission to enlarge the Gan naval base and install facilities for fighter planes to land there. The proposal aroused fierce opposition and pro-British prime minister Ibrahim Ali Didi was forced to resign. His successor, Amir Ibrahim Nasir, refused permission to build the new facilities claiming that such a project would violate the country's neutrality.

[7] In 1959, rebellion broke out in the southern Maldives, which decided to break away under the name of the Republic of Suvadiva. This autonomy was short-lived, and in 1960, the 20,000 Suvadivan republicans were restored to the sultanate with British help. The colonialists took the opportunity of signing a new agreement with the sultan extending the protectorate, and maintaining and enlarging the bases.

[8] This time the British paid nothing towards them, only the pay due to the British soldiers and the cost of the ammunition used to stifle the rebellion in the south.

[9] But the British Empire was losing power. During the 1960s, the British finally decided to withdraw from their strategic positions "east of Suez", making sure that their interests would continue to be defended by the United States. In 1965, the Maldives chose to become independent, receiving immediate recognition from the United Nations.

[10] The sultan was unable to survive the withdrawal of foreign support and, in 1968, a plebiscite was held which favored the establishment of a republic. Amir Ibrahim Nasir, the prime minister, now became president.

[11] The Gan naval base remained in British hands until 1975, when the building of modern US military installations on the neighboring island of Diego Garcia rendered it obsolete.

[12] In March 1975, President Nasir announced that he had uncovered a conspiracy led by Prime Minister Ahmed Zaki; Zaki was exiled to a desert island along with some of his supporters. Nasir's proposal to rent the unused installations on Gan Island to transnational enterprises was rejected by the Majilis (legislative council). Realizing that he had lost support, Nasir ended his second presidential term in 1978, without standing for re-election. The Majilis appointed Maumoon Gayoom, an Islamic intellectual of international prestige, to replace him.

[13] The new president concerned himself with the serious difficulties of the fishermen, and set up a state fisheries corporation to control the country's main economic resource from catch to marketing. Gayoom founded schools in 19 of the major atolls and opened up the archipelago to the rest of the world, personally travelling to Europe, the Middle East and the Sixth Summit Conference of the Non-Aligned Countries.

[14] In May 1980, he officially announced that his government had crushed an attempted mercenary invasion of the islands, organized from abroad by Ibrahim Nasir. Nasir's extradition from Singapore was immediately requested.

[15] The Republic of Maldives supports Islamic solidarity, non-alignment, Palestine, the New International Economic Order, disarmament, African solidarity, and demilitarization of the Indian Ocean. In spite of its size, population and resources, its strategic location secures considerable bargaining power for the country and an independent position in foreign policy. In addition, non-aligned foreign policy has provided the basis for assistance and support from all, without sacrificing too much freedom of action.

[16] In April 1980, the Maldives signed a scientific and technological cooperation agreement with the Soviet Union, and in April 1981, a trade agreement with India, which is the country's closest co-operator in the region. In July, 1981, Chinese foreign minister Huang Hua visited Male and signed a technical cooperation agreement.

[17] In July 1982, the government achieved a major diplomatic victory when the Maldives became the 47th member of the British Com-

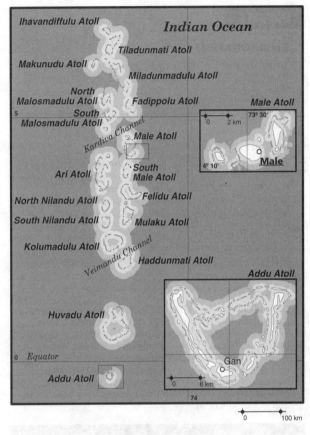

Ihavandiffulu Atoll / Indian Ocean / Tiladunmati Atoll / Makunudu Atoll / Miladunmadulu Atoll / North Malosmadulu Atoll / Fadippolu Atoll / Male Atoll / South Malosmadulu Atoll / Kardiva Channel / Male Atoll / 73° 30' / Ari Atoll / South Male Atoll / Male / 4° 10' / North Nilandu Atoll / Felidu Atoll / South Nilandu Atoll / Mulaku Atoll / Kolumadulu Atoll / Veimandu Channel / Haddunmati Atoll / Addu Atoll / Huvadu Atoll / Equator / Gan / Addu Atoll / 100 km

WORKERS

1994

MEN 78% WOMEN 22%

AGRICULTURE 25% INDUSTRY 32% SERVICES 43%

ENVIRONMENT

There are nearly 1,200 coral islets in this archipelago, where the land is never more than 3.5 meters above sea level. Vegetation is sparse except for coconut palms which are plentiful. A tropical monsoon climate prevails. There are no mineral or energy resources. Fishing is the economic mainstay.

SOCIETY

Peoples: The population of the archipelago came from migrations of Dravidian, Indo-Arian and Sinhalese peoples from India, later followed by Arab peoples.
Religions: Islam (official and proclaimed universal).
Languages: Dhivehi (official), an Indo-Arian language related to Sinhala, English and Arabic are also spoken.
Political Parties: There are no political parties. The Majilis, or parliament, is elected by direct vote, and in turn proposes a president whose candidacy is submitted to a plebiscite.

THE STATE

Official Name: Divehi Jumhuriyya (Republic of Maldives).
Administrative divisions: 20 districts. **Capital:** Male, 55,130 inhab. (1990). **Government:** Presidential republic. Maumoon Abdul Gayoom, President and Prime Minister re-elected in 1993. Legislative Power; the Citizens' Council, with 48 members, 8 of whom are elected by the President.
National Holiday: July 26, Independence Day (1965).
Armed forces: About 700-1,000. It performs both army and police functions.

DEMOGRAPHY

Annual growth: 0.9% (1992-2000). **Children per woman:** 6 (1992).

HEALTH

Under-five mortality: 78 per 1,000 (1995). **Calorie consumption:** 93% of required intake (1995). **Safe water:** 95% of the population has access (1990-95).

EDUCATION

Illiteracy: 7% (1995).

COMMUNICATIONS

88 **newspapers** (1995), 88 **TV sets** (1995) and 90 **radio sets** per 1,000 households (1995). 4.2 **telephones** per 100 inhabitants (1993). **Books:** 85 new titles per 1,000,000 inhabitants in 1995.

ECONOMY

Per capita GNP: $950 (1994).
Annual growth: 7.70% (1985-94).
Currency: 12 rufiyaa = 1$ (1994).

---

monwealth. This was the second major achievement of Gayoom's government since 1979, when the Maldives had joined the Non-aligned Movement. The Maldives have, however, refused to join ASEAN as this organization tends to act on bloc decisions agreed by a majority.

[17] Being the country closest to the US naval base at Diego Garcia in the Indian Ocean, Maldivan developments have attracted the interest of other countries. In a referendum held in 1983, Gayoom was elected for another five-year period, with 95.5 per cent of the vote.

[18] In August 1988, three months before being elected president for the third time, Maumoon Gayoom frustrated another coup attempt, allegedly promoted by Amir Nasir. Seventy-five people were arrested, most of them Sri Lankans, of whom 16 were sentenced to death, and 59 were given prison sentences.

[19] In 1985, there was a 37 per cent increase in tourism over the 1984 figure, coinciding with the virtual closing down of the Maldi-

van docks, as a result of the drastic reduction of its imports and the loss of important international markets. In 1986, tourism maintained its economic importance, generating $42 million per year, close to 17 per cent of the GDP, and in 1991, the Maldives had 200,000 tourists, who brought $80 million into the country. Fishing, however, continued to be the main activity, employing 45 per cent of the country's manpower and generating 24 per cent of the GDP.

[20] The local environmental effects of industrial activity in developed countries, is a major issue to Maldivans. In March, 1990, thousands of school children demonstrated in the streets of the capital, voicing their concern over the rising sea levels around their tiny archipelago. This phenomenon is attributed to the "greenhouse effect" which is producing global warming, that will lead to the melting of the polar ice caps, causing the sea level to rise. According to scientific predictions, if present trends continue most of the islands will

disappear over the next hundred years.

[21] To counteract the sea rise, large retaining walls were built around the islands. In 1987, gigantic waves flooded two-thirds of the capital, Male, causing damage estimated at $40 million. In June 1991, heavy rainfalls destroyed the homes of over 10 per cent of the population, causing damage for $30 million worth of damage.

[22] Twenty-eight tons of garbage are collected daily in Male, generating serious environmental problems, given the island's small surface area. A UN body determined that the garbage should not be recycled, but should be taken instead to a neighboring island, to be used as filling material to gain land from the sea.

[23] The government of Maldives denied Amnesty International access to prisoners held in the Dhoonidhoo detention center and the Gamadhoo prison, where there had been reports of inhumane treatment. Most political prisoners were sent there under the Prevention of Terrorism Act which, among other measures, authorizes detention without trial for up to 45 days.

[24] In 1993 and 1994, the Maldives went through an economic crisis caused by the 1991 monsoons and a chronic trade deficit

dating from 1992. The government sold 25 per cent of the Bank of Maldives' shares and applied a free market economy to all sectors except frozen fish exports.

[25] President Maumoon Abdul Gayoom began his fourth term in government in 1993 and remained as head of the ministries of defense, national security and finance. 27 per cent of the budget was turned over to education, health and social security. The fishing sector received 30 per cent since it employs one fourth of the working population.

[26] $113 million in tourism revenues from 1994 helped in the recovery of the archipelago's economy in 1995.

[27] Although there are no political parties or an organized opposition, young Maldivans educated abroad have shown signs of rejecting the system imposed by the president. The spread of religious radicalism throughout the islands is another reason for concern of Maumoon Abdul Gayoom's government.

Mali

Mali

Population: 9,524,000 (1994)
Area: 1,240,190 SQ.KM
Capital: Bamako

Currency: CFA Franc
Language: French

The empire of Mali, one of the great cultural and commercial centers of Africa (see box) was occupied by the French in 1850. Together with what are now Burkina Faso (Upper Volta), Benin and Senegal it was renamed the French Sudan, and later, French West Africa.

[2] Domination brought about considerable changes. Trade, which traditionally flowed towards the Mediterranean, was turned back towards the Atlantic where Dakar had become the official trade center of the colony. This in turn caused the decline of the trans-Saharan trade routes. After 1945 the anti-colonial feelings of French Sudan were expressed in the formation of the African Democratic Assembly (RDA), at a conference in Bamako encouraged by the democratic atmosphere that prevailed after World War II. Under the impact of the Dien Bien Phu defeat of French colonialism in Vietnam and the Algerian revolution, Paris embarked on a policy of gradual concessions that would lead to Mali's independence in 1960.

[3] Aware of their limitations the newly formed states formed the Federation of Mali, but their differences of interest soon caused the federation to collapse. French Sudan then severed its last remaining ties with France, becoming the Republic of Mali, with Modibo Keita as its first president.

[4] Together with Senghor from Senegal and Houphouet-Boigny from Cote d'Ivoire, Keita belonged to a generation of African leaders educated in France and inspired by social democracy. Unlike his neighbors, the president of Mali did not accept neocolonialism. Instead he pushed for reforms and economic development under the banner of African socialism; moving closer to the views held by Guinea's Sekou Touré and Nkrumah of Ghana.

[5] The economy was nationalized, industrialization stimulated and the proportion of children in school rose from 4 to 20 per cent. His struggle for Pan-Africanism, non-alignment and independent foreign policy earned Keita a good reputation among progressive forces throughout the continent. He was, however, unable to build a strong political structure and there was a successful coup against him in November 1968.

[6] The Military Committee for National Liberation (CMLN) took over, headed by Moussa Traoré. The committee promised to straighten out the economy and fight corruption, but the results were exactly the opposite. Grain, which had been produced in sufficient quantities to provide a surplus for export, began to be rationed in the 1970s.

[7] Economic dependence and agricultural intensification geared towards a world market earned Mali a large foreign debt during this decade.

[8] In 1974 seeking to gain political support, the CMLN held a referendum on a new constitution. With the opposition banned and the followers of Keita imprisoned, the government won a 99.8 per cent majority, too large to be believed. On May 16 1977 Modibo Keita died somewhere in the desert where he had been held prisoner since 1968. The official cause of death was "food poisoning", but this was commonly re-interpreted as simply "poisoning". In the largest mass demonstration ever seen in Bamoko, the people followed Keita's body to the cemetery in open defiance of the military regime.

[9] In 1979, President Moussa Traoré embarked upon an austerity program drawn up by the IMF and the international creditor banks. This naturally stirred up social unrest. In November 1979 students and teachers took to the streets staging rallies and demonstrations which lasted for over a month. Teachers also went on strike. The government reacted violently. Thirteen students were tortured to death and around 100 were arrested.

[10] After 1983 Mali also drew closer to France. At the same time Traoré distanced himself from the Soviet Union after having recieved economic and educational aid during the 1970s.

[11] Key sectors, such as agriculture, continued to decline. A prolonged drought drastically reduced Mali's livestock.

[12] In June 1985, President Traoré was re-elected for a further six years. As the official party's only candidate he recieved 99.94 per cent of the vote.

[13] In 1988 students, teachers and public employees took to the streets once again. The government was re-organised and a Social and Economic Council was created, headed by General Amadou Baba Diarra.

[14] The government faced a foreign debt which amounted to 125 per cent of its GDP, with the servicing of the debt exceeding a quarter of the income earned from exports. Negotiations for an economic adjustment program began with the IMF in 1988. The privatization of Mali's banking system was initiated, financed by an 8,000,000 CFA franc loan from France. At the same time the government announced a reduction in their number of employees, and the decision to sell off state enterprises.

[15] On April 10, 1991, a popular and military revolt against the Traoré regime carried Lt. Colonel Amadou Toumani Touré to power. He was the leader of the Council of Transition for the Salvation of the People (CTSP), which promised to transfer government to civilians early in 1992.

[16] The CTSP contained many cadres from the previous military dictatorship. Popular distrust of the military led to violent action by student organizations, as buildings associated with the old dictatorial regime were raided.

[17] Old political parties re-organized and new ones appeared. So-

PUBLIC EXPENDITURES

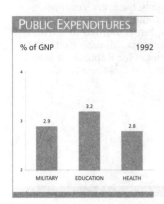

% of GNP — 1992

MILITARY 2.9 · EDUCATION 3.2 · HEALTH 2.8

WORKERS

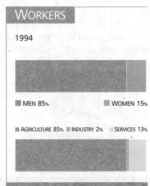

1994

MEN 85% — WOMEN 15%

AGRICULTURE 85% · INDUSTRY 2% · SERVICES 13%

LAND USE

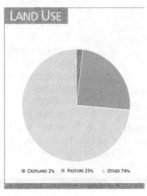

CROPLAND 2% · PASTURE 25% · OTHER 74%

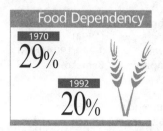

Food Dependency
1970 **29%**
1992 **20%**

Illiteracy
1995 **69%**

Maternal Mortality
1989-95
PER 100.000 LIVE BIRTHS **1,249**

External Debt
1994
PER CAPITA **$292**

The Malian Empire

According to tradition, during a drought in Niani the traditional ruler, Allakoi Keita, converted to Islam. Rain began to fall immediately and Keita made a pilgrimmage to Mecca to express his gratitude. By the time he returned to Niani in 1050 he had become a Sultan.

[2] This small state aroused the ambition of its neighbors, who hoped to control Niani's goldmines. As a result it was invaded in 1230 by the troops of Samanguru Konte, a Sosso ruler. In an effort to consolidate his power, Konte had the entire ruling family of Niani beheaded. All, that is, except the young invalid prince, Sundiata. Consolidation was not achieved as other nations realized that they might be subject to similar treatment. The only surviving Keita devoted the next few years to building a confederation of Malinke local groups. In 1235 his followers confronted the Sosso in an epic battle still recalled by Malian folk singers. The victory not only confirmed Sundiata's leadership, but also left him in a strong position to expand his territory. He was nicknamed "Mali Dajata", "the Lion of Mali". In 1240 he annexed the ancient state of Ghana. Upon his death in 1255 his dominions extended from the Atlantic to the curve of the Niger, and from the equatorial forests to the Sahara desert (see map). This area included present-day Senegal, Gambia, Guinea-Bissau, northern Guinea (Conakry), half of Mauritania, southern Algeria and most of Mali except for the eastern tip which was added in 1325.

[3] Niani in the south, Timbuktu in the north and Djene in between, became important trade centers. Timbuktu was especially prominant, dominating the southern end of the trans-Saharan trade route (see map in Mauritania). Among Sundiata's successors, Sultan Kankan Musa (1312-1337), is often cited by Arab and Venetian historians for his legendary wealth and extravagance. On one pilgrimmage to Mecca he distributed so much gold that the price of ingots fell in Cairo, and it took twelve years for

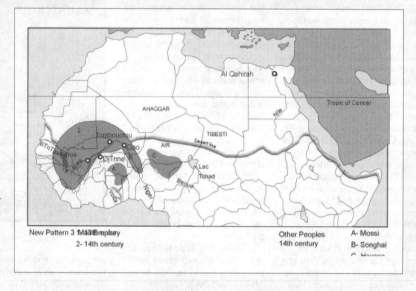

New Pattern 3 Mali Empire
1- 13th century
2- 14th century
Other Peoples
14th century
A- Mossi
B- Songhai
C- Haussa

the market price to recover its former level. The intensity of commercial and cultural interchange with the Arab world, and in particular with Egypt and Arabia, led to the development of an Islamic culture rich in local elements, and gave international prestige to the recently created University of Timbuktu.

[4] Thomas Hodgkin, the English historian, wrote that a 14th century student at Timbuktu University would have been just as at home in 14th century Oxford.

[5] Trade with the Arab world was intense. Each year a total of 12,000 camels crossed the desert between Mali and Cairo. The goldmines of Bambuk and Boure were the main suppliers of the Arab and European markets. The extent of trade was such that even sculpted gravestones were imported through Morocco to Arab dominated Spain. The period was well documented by Arab historians like Ibn Jaldun and Ibn Battuta. They wrote with admiration of the West-African empire of this period, and of the political skills of the Sultans who reconciled their muslim orthodoxy with the traditional beliefs of the majority of their subjects.

[6] Towards the end of the 15th century, however, this flourishing nation began to decline. External aggression - Fulah from the west, Tuareg from the

north, Mossi from the south and particularly the Songhai of Gao from the east-contributed to the downfall of the Malian empire. By far the most important reason for Mali's decline, however, was the arrival of the Portuguese on the Atlantic coast. A large proportion of the gold produced in Bambuk and Boure was diverted towards the coast where it was exchanged for European goods. According to Portuguese records, the West-African gold gained amounted to 720 kilos each year. This new market broke the trade monopoly of the Malian Sultans and greatly debilitated trans-Saharan traffic.

[7] The development of rival neighboring states contributed to the decline of the Malian empire. In much reduced circumstances Mali formally survived as a nation until the French occupation in the 19th century. The Keita dynasty still exists, but is limited to a tribal group in Kangara. Modibo Keita, the first president of the Republic of Mali, was a member of this same family.

ENVIRONMENT

There are three distinct regions the northern region which is part of the Sahara desert; the central region which consists of the Sahel grasslands, subject to desertification, and finally, the southern region, with humid savannah vegetation, most of the population, and the two major rivers: the Senegal and the Niger. The shortage of water and the over use of forest resources have given rise to further aridity.

SOCIETY

Peoples: There are many ethnic groups, the largest being the Bambara. Other important groups are the Malinke, Songhai, Peul-Fulani, Dogon, Tuareg and Moor. 10% of the population are nomads; they are the most affected by drought.
Religions: The majority, 70%, is Muslim, Nearly a third, 27% practise traditional African religions and there is a small Christian minority. **Languages:** French (official). Of the African languages, Bambara is the most widely spoken. Arabic and Tuareg are also spoken in the north. **Political Parties:** Alliance for Democracy in Mali (ADEMA); National Committee for Democratic Initiative (CNID); Sudanese Union-African Democratic Group (US-RDA); National Renaissance Party (PRN); Unified Movements and Fronts of Azawad (MFUA). **Social Organizations:** Workers' Union of Mali; Union of Malian Women (UNFM); National Union of Malian Youth (UNJM); Association of Pupils and Students of Mali (AEEM).

THE STATE

Official Name: République du Mali **Administrative divisions:** 8 regions and the district of Bamako.
Capital: Bamako, 650,000 inhab. (1987). **Other cities:** Ségou 88,877 inhab.; Mopti 73,979 inhab.; Sikasso 73,050 inhab.; Gao 54,874 inhab. (1987) **Government:** Parliamentary republic, with strong head of state. Alpha Oumar Konaré, President. Legislature: single-chamber: National Assembly, with 129 members, elected every 5 years. **National Holiday:** September 22, Independence Day (1960). **Armed Forces:** 7,350. **Paramilitaries:** 7,800 (Gendarmes, Republican Guard, Militia and National Police).

DEMOGRAPHY

Urban: 26% (1995). **Annual growth:** 2.4% (1992-2000). **Estimate for year 2000:** 12,000,000. **Children per woman:** 7.1 (1992).

HEALTH

One **physician** for every 20,000 inhab. (1988-91). **Under-five mortality:** 214 per 1,000 (1995). **Calorie consumption:** 91% of required intake (1995). **Safe water:** 37% of the population has access (1990-95).

EDUCATION

Illiteracy: 69% (1995). **School enrolment:** Primary (1993): 24% fem., 24% male. Secondary (1993): 6% fem., 12% male. **Primary school teachers:** one for every 47 students (1992).

COMMUNICATIONS

83 **newspapers** (1995), 82 **TV sets** (1995) and 80 **radio sets** per 1,000 households (1995). 0.1 **telephones** per 100 inhabitants (1993). **Books:** 87 new titles per 1,000,000 inhabitants in 1995.

ECONOMY

Per capita GNP: $250 (1994). **Annual growth:** 1.00% (1985-94). **Annual inflation:** 3.40% (1984-94). **Consumer price index:** 100 in 1990; 94.9 in 1993. **Currency:** 535 CFA francs = 1$ (1994). **Cereal imports:** 83,000 metric tons (1993). **Fertilizer use:** 103 kgs per ha. (1992-93). **External debt:** $2,781 million (1994), $292 per capita (1994). **Debt service:** 27.5% of exports (1994). **Development aid received:** $360 million (1993); $36 per capita; 14% of GNP.

ENERGY

Consumption: 22 kgs of Oil Equivalent per capita yearly (1994), 80% imported (1994).

---

cial organizations demonstrated in favour of several issues, including pay increases, and against the privatization of public companies in the telecommunication, railroad, textile, pharmaceutical and cement sectors.

[18] In June, the rising of the Tuaregs in the north and the Moors in the east increased the state of social unrest. The new president met with his colleagues in Algeria and Mauritania to deal with the Tuareg question.

[19] On July 14, an attempted coup by a sector of the armed forces led Toumani Touré to grant a salary increase of 70 per cent to the armed forces and civil servants in an effort to avert unrest.

[20] In December 1991, talks were renewed between Mali's authorities and the Tuareg rebels, who accepted a cease-fire in order for prisoners to be exchanged.

[21] In March 1992, with Algerian mediation, the government of Mali and the Tuaregs reached a new peace agreement. In April a final peace settlement was signed with the Azawad Unified Front and Movements, an organization of four Tuareg opposition groups.

[22] In the March 1992 municipal and legislative elections, the Alliance for Democracy in Mali (ADEMA) won 76 of the 116 seats. The rest were divided up among other political alliances and representatives of the expatriate Mali community.

[23] In April 1992, the Malian government and the Tuareg rebels in the north ratified a "national peace pact", putting an end to almost two years of armed conflict.

[24] On April 26, Alpha Oumar Konaré, leader of ADEMA, was elected president in the first multi-party elections since the country's independence in 1960. His main rival was Tieoule Mamadou Konaté, head of the Sudanese Union-African Democratic Group (US-RDA).

[25] Konaré, a teacher by profession, had played a key role in ousting Moussa Traoré's regime, in 1991.

[26] Konaré and Konaté called for a second ballot, as neither had won an absolute majority in the initial round of voting on April 12, although seven candidates were eliminated in this first round. Both rounds of voting were characterized by a very low voter turnout (only 21 per cent of the voters participated in the second ballot).

[27] Most political parties boycotted the parliamentary election and challenged the electoral procedures, this perhaps accounted for the high rate of absenteeism.

[28] In its annual report for 1992, Amnesty International condemned the imprisonment of several opposition figures without charges or trial. Fourteen death sentences had also been passed, although these were later commuted to life sentences. Finally, according to the report, several dozen members of the Tuareg community had been executed without trial by the army.

[29] A group of students occupied the state radio station in March 1993, and massive demonstrations were held against the government's economic policy in April. However, with support from international financial organizations, Konare consolidated the economic liberalization begun by Traore. The president reformed the tax system, reduced public spending, privatized state companies and eliminated price controls.

[30] In February 1994, new demonstrations were staged against the government. On the 15th, all the education centers except primary schools were closed. Some observers indicated the appearance of underground armed groups determined to attack properties belonging to Mali's major creditor countries.

[31] The government signed an agreement with the MFUA, one of the main groups representing Tuareg rebels. But violent incidents continued to take place and on one occasion, some 200 people, including many women and children, died due to police repression in the cities of Gao and Beher.

[32] In 1995, the government pursued negotiations with other Tuareg rebel groups - likewise with neighbouring countries - to organize the relocation of 120,000 Tuareg refugees from Algeria, Burkina Faso, Mauritania and Niger. A three year repatriation scheme was launched in October.

[33] The IMF approved the third annual structural adjustment plan after it considered Mali's economic results were good. At the same time, many foreign investors showed a renewed interest in Mali after new gold fields were found in the south.

Malta

Population: 368,000 (1994)
Area: 320 SQ.KM
Capital: Valleta

Currency: Liri
Language: Maltese and English

Malta

Malta's most valuable asset is probably its geographic location, hence becoming the historical focus of every conflict for domination of the Mediterranean. In ancient times, Phoenicians, Greeks, Carthaginians, Romans and Saracens successively occupied the island.

[2] In 1090, the Normans conquered the island and 300 years later it fell to the Spanish Kingdom of Aragon. In the 16th century, the defence of the island was entrusted to the ancient order of the Knights of St John of the Hospital. Renamed the Knights of Malta, they remained there for another three centuries, until they were driven out by the French in 1798. The Congress of Vienna granted Britain sovereignty over the island, in 1815.

[3] In 1921, after a popular rebellion, London agreed to a certain degree of internal autonomy for the islands, revoked at the beginning of World War II.

[4] During World War II, Malta suffered greatly; it was used as a base for the allied counter-offensive against Italy. The heroic struggle of the Maltese raised their national consciousness and in 1947, London restored autonomy to the island.

[5] Independence was formally declared on September 8 1964, but Britain assumed responsibility for defence and financial support. In 1971, Dominic Mintoff's govern-

ment established relations with Italy, Tunisia, the USSR and Libya, from whom it received substantial financial assistance. NATO forces left in 1971, and Malta joined the Movement of Non-Aligned Countries two years later. In December 1974, the Republic of Malta was proclaimed.

[6] In 1980, tensions developed in Malta's relations with Libya, primarily over oil explorations. The conflict was eventually turned over to the International Court of Justice in the Hague, and relations returned to normal.

[7] From the beginning of the Labor administration, measures to reduce the Church's power had been introduced. The bishops owned 80 per cent of all real estate on the island and virtually controlled education. In July 1983, the government expropriated all Church properties and made secular education obligatory in primary schools. In 1985, the government and the Church signed an agreement providing for a gradual transition to secular education at the secondary level.

[8] In May 1987, the Liberal prime minister, Dr Fenech Adami, initiated a policy of rapprochement towards US policies, favoring EEC membership.

[9] Forecasts of a falling birthrate and the increase in people aged over 60 have forced the government to draw up plans to deal with

the problems of the elderly. At present 7 per cent of those over 65 live in hospitals.

[10] There was a certain amount of tension with the United States in 1993, when President Bill Clinton accused Malta of violating the UN trade embargo on the Yugoslav Federation.

[11] Tourism is the country's chief economic activity, with one-third of the economically active population in this sector and more than one million visitors per year..

[12] In 1993, the EU declared that Malta's democratic stability and its human rights policy were acceptable for its incorporation into the Union. However, its economic

structures were considered "archaic" and the implementation of "fundamental economic reforms" was advised. On January 1 1995 a Value Added Tax came into effect, under criticism from labour unions and shopkeepers.

[13] Six months later, the EC granted economic aid for Malta's reforms and it was announced that in late 1996 negotiations would begin for the country's formal incorporation into the bloc.

[14] Libya has financed the construction of gigantic port installations designed as world trade facilities.

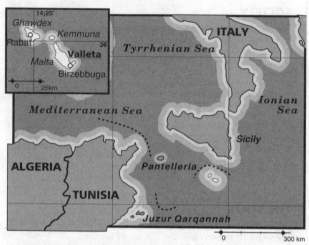

PROFILE

ENVIRONMENT

Includes five islands, two of which are uninhabited. The inhabited islands are the three largest ones: Malta, where the capital is located, 246 sq km; Gozo, 67 sq km; and Comino, 3 sq.km. The archipelago is located in the Central Mediterranean Sea, south of Sicily, east of Tunis and north of Libya. The coast is high and rocky, with excellent natural harbors. The islands' environmental problems are concentrated precisely in these areas. The problems are caused by the encroachment of civilization, the development of tourism, the gradual abandonment of lands devoted to agriculture and the increase in waste. The coastal areas are also polluted by industrial activity.

SOCIETY

Peoples: The Maltese come from numerous ethnic combinations, with strong Phoenician, Arab, Italian and British roots. **Religions:** mainly Catholic. **Languages:** Maltese and English, both official. **Political Parties:** The Nationalist Party (NP), with 34 seats; the Labor Party (LP), opposition party with 31 seats, Democratic Alternative, founded by Labor Party dissidents; its platform is the defense of the environment and of human rights. **Social Organizations:** The Confederation of Trade Unions: 14 unions with more than 50,000 members.

THE STATE

Official Name: Repubblika ta'Malta. **Capital:** Valleta, 9,144

inhab. (1994) **Other cities:** Birkirkara, 21,770 inhab.; Qormi 19,904 inhab.; Hamrun 13,654 inhab.; Sliema inhab.13,514 (1994)
Government: Ugo Mifsud Bonnici, president. Edward Fenech Adami, prime minister, re-elected in February 1992. **National Holiday:** Sept. 8, Independence day (1964). **Armed Forces:** 1,850 troops (1995)

DEMOGRAPHY

Annual growth: 0.7% (1991-99). **Children per woman:** 2.1 (1992).

HEALTH

Under-five mortality: 114 per 1,000 (1995). **Calorie consumption:** 109% of required intake (1995).

EDUCATION

Primary school teachers: one for every 21 students (1990).

COMMUNICATIONS

103 **newspapers** (1995), 113 **TV sets** (1995) and 104 **radio sets** per 1,000 households (1995). 43.0 **telephones** per 100 inhabitants (1993). **Books:** 110 new titles per 1,000,000 inhabitants in 1995.

ECONOMY

Annual growth: 5.10% (1985-94). **Consumer price index:** 100 in 1990; 110.2 in 1994. **Currency:** 0,4 liri = 1$ (1994).

Marshall Islands

Marshall

Population: 54,000 (1994)
Area: 180 SQ.KM
Capital: Majuro

Currency: US Dollars
Language: English

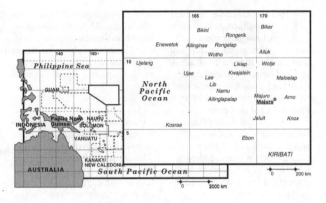

The Kwajalein and Bikini atolls of the Marshall Islands were introduced into modern history in February 1944 when heavy bombing by combined US naval and air force units hit the islands. This was followed by a prolonged and bloody battle which ended with the defeat of the Japanese and the occupation of the islands. Countless lives were lost in the attempt to control the small coral formations which, surround large lagoons. However, the military command considered it was a fair price to pay for the islands given their strategic position.

2   In 1979, the US proposed to make an Associated Free State with the four administrative units of the region; the Kwajalein and Bikini atolls, plus the Mariana and Caroline islands which were also trust territories. The Marshall Islands were granted jurisdiction over local and foreign affairs but the US declared the islands military territory for specific use. These terms formalized the situation of the Bikini and Kwajalein atolls; intense nuclear testing from 1946 to 1958 had turned the islands into the most radioactively contaminated area in the world.

3   In 1961, Kwajalein became the Pacific experimental missile target area, especially for intercontinental ballistic missiles launched from California, and early in the 1980s the US chose the atoll as a testing site for its new MX missiles. The local population was totally evicted and entry was forbidden to civilians.

4   The case of the Bikini atoll is well known. Twenty-three nuclear tests were performed there between 1946 and 1958, including the detonation of the first H-bomb. Despite the legacy of cancer, thyroidism and leukemia left behind by the bomb, the inhabitants of the atoll insisted on returning to their homeland after having been transferred to the Rogenrik atoll. In 1979, testing revealed that 139 of the total 600 inhabitants living on the Bikini atoll had extremely dangerous levels of plutonium in their bodies.

5   The inhabitants of Bikini, together with those living on Rongelap (also exposed to the radioactivity of the H-bomb dropped on Bikini) sued the US government for $450 million. Abnormalities already existed in children under 10 years of age and could not be eliminated. The charges were filed together with reports compiled by US government agencies proving that natives had been intentionally exposed to radioactivity in 1954 in order to study the effects of the bomb on human beings. Reports revealed by the US government in 1995 proved the dangers of exposure were known but this had never been passed on to the Marshall population.

6   In April 1990, the US announced that it would use the area to destroy all the nerve gas supplies which had been installed in Europe to date. Environmentalists denounced plans to dispose of 25 million tons of toxic waste on the atoll, between 1989 and 1994.

7   In October 1986, the US and the Marshall Islands signed a Free Association Pact, whereby the latter ceased to be a US trusteeship and became a Free Associated State, responsible for its own internal politics. According to the pact, the US will be responsible for the defence of the new state for a period of 15 years, which enables the US to have an air base on the island, and the island is also entitled to US financial aid.

8   In the first elections of the new state, held in 1986, Amata Kabua was voted president.

9   In the years following independence, the Marshall Islands attempted to consolidate diplomatic and commercial relations with most of the neighboring states. In 1988 they were admitted in the South Pacific Agreement on Economic Cooperation and Regional Trade, and in 1989 set up diplomatic relations with Japan and Taiwan. During the first half of 1993, embassies were established in the Philippines, Israel, Barbados, the United States and Britain.

10   On September 17 1990, during the 46th UN General Assembly, held in New York, the Marshall Islands were accepted as a member state of the organization.

11   Foreign affairs minister Tony de Brun founded the Ralik Ratak Democratic Party in June 1991, after becoming estranged from president Kabua. In the November elections, Kabua was re-elected for his fourth consecutive term.

12   That year, a Hawaiian court ordered the suspension of missile testing until the operations' effects on the environment had been assessed. Two years later, local environmentalists protested against an "alternative energy" project based on the burning of discarded tires, considering it a source of atmospheric contamination.

13   The controversy over the use of the Islands resurfaced in 1995 when the government announced plans to install a nuclear waste dump on Bikini, even though atomic explosions between 1946 and 1958 had made it uninhabitable for a period of 10,000 years. The government saw this as a chance for the country's economic salvation.

PROFILE

ENVIRONMENT

A total of 1,152 islands grouped in 34 atolls and 870 reefs. The land area covers 180 sq km, but the islands are scattered over a million sq km in the Pacific. The atolls of Mili, Majuro, Maloelap, Wotje and Likiep lie to the north-east. The south-west atolls are: Jaluit, Kwajalein, Rongelap, Bikini and Enewetak among others. The northern islands receive less rainfall than the southern atolls.

SOCIETY

Peoples: 60% of the population lives on Majuro and Kwajalein. A large part of Kwajalein's residents consists of US military personnel. **Religions:** Protestant 90.1%; Roman Catholic 8.5%; other 1.4%. **Languages:** English. **Political Parties:** Ralik Ratak Democratic Party.

THE STATE

Official Name: Major (Republic of the Marshall Islands) **Administrative divisions:** 25 districts. **Capital:** Majuro, 14,649 inhab. (1988). **Other cities:** Ebeye, 8,324 inhab. (1988). **Government:** Parliamentary republic. Amata Kabua, President of the local government. A High Commissioner (Janet McCoy) is the US government delegate and handles security, defence and foreign affairs. The constitution provides for a 33-member parliament which appoints the local president. **National holiday:** December 22.

DEMOGRAPHY

Annual growth: 3.9% (1991-99).

COMMUNICATIONS

4.4 **telephones** per 100 inhabitants (1993).

ECONOMY

Consumer price index: 100 in 1990; 104.8 in 1993. **Currency:** US dollars.

Martinique

Population: 383,000 (1994)
Area: 1,100 SQ.KM
Capital: Fort-de-France

Currency: French Francs
Language: French

Martinique

The Carib indians migrated throughout the central and northern parts of the South American continent, and from there to the nearby islands in the Caribbean. They called the largest of the Lesser Antilles "Madinina", becoming known as Martinique in the colonial period.

[2] The Caribs were primarily hunter-gatherers and banded together to form a small group on the island. The French occupied the island in 1635, but economic development did not occur until a century and a half later.

[3] Late in the 17th century, sugar cane cultivation transformed the natural and economic landscape of the island, ending the indians' lifestyle. The system of production changed on the sugar cane planta-

tions. The indigenous population was replaced by African slaves, brought to the island by slave-traders. This monoculture has since dominated Martinique's economy, reinforcing the colonial ties with France.

[4] A minority of 12,000 European land, sugar mill and business owners dominated 93,000 slave laborers, an imbalance that gave rise to countless uprisings, known as *Marronuage* - meaning collective rebellions of slaves. Accounts written in those days recall the existence of *quilombos* (see "Quilombos" in Brazil) on the island in 1811, 1822 and 1833; years when rebellions shook Martinique

[5] In the early 19th century, unprecedented social disturbances ocurred as the traditional plantation

PROFILE

ENVIRONMENT

Martinique is a volcanic island, one of the Windward Islands of the Lesser Antilles. Mount Pelée, whose eruption in 1902 destroyed the city of Saint Pierre is a dominant feature of the mountainous terrain. The fertile land is suitable for agriculture, especially sugar cane. The tropical, humid climate is tempered by sea winds.

SOCIETY

Peoples: Descendants of former enslaved Africans (mostly mestizos). There is a minority made up of people of European origin. **Religions:** Catholic (predominant). **Languages:** French (official), and Creole. **Political Parties:** Communist Party of Martinique, (PCM) founded in 1957 by Leopold Bissol; Progressive Party of Martinique, (PPM), a splinter group of the 1957 Martinican communist movement is now an independent left-wing party; Union of Martinican Democrats, (UDM); Socialist Revolution Group; Martinican Independentist Movement (MIM).

THE STATE

Official Name: Département d'outre-mer de la Martinique. **Capital:** Fort-de-France, 101,540 inhab. (1990). **Other cities:** Le Marin; La Trinité. **Government:** Jean François Cordet French appointed commissioner. General Council president. The Council's 36 members are elected through universal suffrage for 6-year terms. President of the 41-member Regional Council.

COMMUNICATIONS

40.5 **telephones** per 100 inhabitants (1993).

ECONOMY

Consumer price index: 100 in 1990; 113.3 in 1994.
Currency: French francs.

system was abandoned without investment in industrialization.

[6] In 1937 the formation of Martinique's Central Labor Union, prompted social mobilization.

[7] In 1946, the French government revised its approach to colonial relations, creating the Overseas Departments in 1948. At first, the Martinican middle class had hoped to have the same rights as French citizens, but lack of consensus within the anti-colonial movement hampered the acievement of this goal.

[8] In the March 1986 elections, the left-wing parties obtained 21 of the 41 seats on the Regional Council. Aimé Césaire, representing the leftist coalition, was re-elected president of the Council. Rodolphe Désiré became Martinique's first left-wing party (PPM) representative in the French Senate.

[9] In 1992, the establishment of a single European market worsened the island's economic situation. In the elections that year there was a tied vote between the Right on the one side and the pro-independence and Left parties on the other. In the plebiscite to ratify the Maastricht Treaty, the abstention rate was 75 per cent, showing hostility or at least a passive approach toward a deeper integration of the island into the European Union (EU).

[10] Other signs of hostility toward the EU were the demonstrations of banana producers, who protested against the opening of the European market to imports from Africa and Latin America, where fruits are

cheaper than in the "French" Antilles. The crisis worsened in 1994 with a 20 per cent increase in unemployment and a noticeable drop in corporate investments.

[11] One of the main issues in France's 1995 presidential campaign in Martinique was "social equality" with France, ie an increase in social benefits, minimum wage - currently lower than "metropolitan" ones. "Social equality" supporters also highlighted the fact that prices in Martinique were higher than in France.

[12] The island's future is still giving rise to disagreements between Martinican political parties. In the economic field, the PPM proposed to levy a tax for development, and will probably face growing pressure from the EU for more privatization and liberalization of the economy.

[13] In the political field, while the Communist Party of Martinique (PCM) speaks about "attaining responsibility", which suggests the idea of greater autonomy without reaching independence, the Martinican Independence Movement (MIM) still advocates "a negotiated process of access to independence".

Mauritania

Mauritanie

Population: 2,215,000 (1994)
Area: 1,025,520 SQ.KM
Capital: Nouakchott

Currency: Ouguiya
Language: Arabic and French

The process of desertification that turned fertile plains into the Sahara desert, would have separated southern Berbers from the Mediterranean coast if camels had not been introduced into the area. The groups that migrated southward in search of pastures kept in touch with their native culture, and later shared the benefits of the Islamic civilization, which flourished along the Mediterranean coast.

2   The southern part of present-day Mauritania was the setting for one of the most peculiar African civilizations (see "Ghana-Uagadu"). But the Almoravid conquest and later the Fulah migrations (see Senegal: "The Fulah States") stimulated the integration and unification of the population.

3   In the 14th century, the Beni Hilal, who had invaded North Africa three centuries before, reached Mauritania. For over 200 years, they plundered the region and warred with the Berbers throughout an area including present-day southern Algeria and Sahara. In 1644, all the Berber groups in the region joined to fight the Arabs, but the resulting conflict, the Cherr Baba War, ended 30 years later with the defeat of the Berbers. The Arabs became a warrior caste, known as the *Hassani*, monopolizing the use of weapons, while trade, education and other civilian activities were left in the hands of the local population. Beneath these two groups came the *haratan*, African shepherds from the south, kept as semi-serfs. Though this rigid social stratification weakened with time, it is still intact among the Arab-Berbers, the Fulah and Soninke.

4   Towards the end of the 17th century various emirates arose. These were unable to achieve the political organization of the country, due to their internal rivalries and dynastic quarrels. Nevertheless, they provided a minimal degree of order within the region which led, in turn, to a small revival of trade caravans. This efforts were supported by the cultural unification being carried out by the *zuaias*. These Berber-Marabouts devised a simple system of Arabic writing, propagating it along with their religious teachings.

5   In the 19th century, growing trade coincided with a French project to transfer Sudanese commercial activities to Senegal (see Mali), requiring the elimination of trans-Saharan trade and the frequent robberies in Senegal. The French therefore invaded Mauritania in 1858, under General Faidherbe, and the fighting continued until the 20th century. The resistance initially met in the emirates of Trarza and Brakna continued with the Sheikh Ma al-Aini (see Sahara), with his sons and later his cousin, Muhammad al-Mamun, the emir of Adrar. Pursued by the French almost a 1,000 km into the Sahara, Muhammad al-Mamun died in combat in 1934.

6   After World War II, Mauritania became a French Overseas Province, sending deputies to the French parliament. Ten years later internal autonomy was granted and in 1960 independence was gained. Lacking infrastructure and administrative capacities, the potential government officials were five university graduates and fifteen students.

7   A French transnational company, MIFERMA, was more powerful than the whole government. MIFERMA's iron ore mines supplied 80 per cent of the country's exports and employed one in four salaried workers.

8   The progressive PPM wing gradually gained ground and created a desire for true independence. In 1965 Mauritania withdrew from OCAM (the Common Afro-Mauritanian Organization), created to maintain French control over their former colonies.

9   In 1966, the government set up a state company, SOMITEX, with a monopoly on sugar, rice and tea imports, and a campaign was launched to rekindle the Arab culture. Mauritania set up its own customs system independent from Senegal; the Arab-Mauritanian Bank was granted the monopoly on foreign trade and the country began to issue its own currency.

10   In 1974 the iron mines were nationalized. The country started to replace French influence by a greater rapprochement with Muslim countries, and finally became a member of the Arab League. Saudi Arabia, Kuwait, and Morocco supplied economic assistance.

11   In 1975 Mauritania, fulfilling an old ambition to annex part of Sahara, joined Morocco in an attempt to divide the Spanish possession. With logistic and military support from France, 3,000 Mauritanian soldiers and 10,000 Moroccan troops occupied Sahara.

12   Ould Haidalla's government paid an extremely high price for this aggression. Mauritania soon became the target of violent reprisals by the Polisario Front (see Sahara) and was virtually occupied by Moroccan troops.

13   Economic hardships aroused widespread popular unrest leading to protest rallies and street fighting. Opposition groups condemned Mauritanian interference in the Saharan liberation war.

14   The crisis exploded in 1978 and over the next six years there were five coups. An attempt was made to Arabize the whole population, ignoring the African population (20 per cent), living in the southern part of the country.

15   Ould Haidalla's government ended the Arabization process of total and also relinquished Mauritania's claim to the Sahara, making peace with the Polisario Front in August 1979. With the support of young officers and intellectuals from the left, a decree was signed abolishing slavery in Mauritania.

16   A coup occurred on December 11, 1984. Maawiya Ould Sid'Ahmed Taya, an army colonel and chief of staff, ousted Haidallah. The first measures of the new administration were the recognition of the Democratic Saharaui Arab Republic and steps to dismantle a growing "clandestine economy" (only 50 per cent of the firms operating in the country kept legal accounting records).

17   In April 1985 the IMF announced that a $12 million stand-by credit had been granted to Mauritania under an agreement to restructure foreing debt. Conditions were

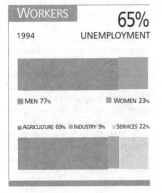

WORKERS

65% UNEMPLOYMENT

1994

MEN 77%   WOMEN 23%

AGRICULTURE 69%   INDUSTRY 9%   SERVICES 22%

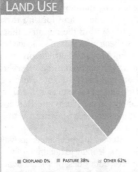

LAND USE

CROPLAND 0%   PASTURE 38%   OTHER 62%

Ghana-Uagadu

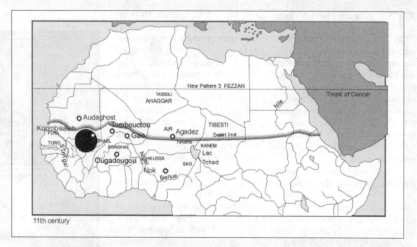

11th century

There are two sources of confusion regarding the name of this nation. First Arab historians used the word Ghan-a, a title used by traditional rulers (meaning war-chief), to address the nation and its capital, and handed it down to posterity. Actually, the country was called Uagadu in the Soninke language, and the capital was called Koumbisaleh by its people. The other source of confusion is contemporary. The modern nation of Ghana, adopted this name because of its historical prestige, but has no relation - either geographic cultural or historical - with the ancient Uagadu.

[2] This nation was founded in the 5th century AD by a group of Berbers who exercised a certain degree of hegemony over the Soinke nation, of Mande origin. This Berber group lived in the southern part of present-day Mauritania and Mali Tradition says that the first 44 rulers were Berber; that integration between the two nations produced the ethnic group called Sarakole today; and that in the eighth century, a family belonging to this group, the Cisse, ousted the Berbers and took over power until the state was dismembered.

[3] The power of rulers entitled them to a monopoly on gold production. Mining was carried out by local leading groups who kept the gold-dust and handed the nuggets over to the public treasury. Gold sales were reserved for the ruler too (who also held the title of "Kaya Ma Ghan", meaning chief of gold). In addition, he collected taxes on salt trading and on the commercialization of other products from the north. These goods were exchanged for slaves and local products such as the resin used in the making of Gum Arabic. Some records give an idea of the importance this trade achieved: the Arab historian bin Haukal saw, in a 977 letter of credit signed by a Moroccan merchant for 42,000 dinars of gold, the equivalent of almost $3 million today.

[4] Historians describe Koumbisaleh at the height of its splendor (950-1075) as a beautiful city of stone houses, with plenty of gardens and ample market places where Egyptian wheat, cloth and weapons from Damascus, jewels from Spain, fruits and enslaved women from the Mediterranean, could be purchased with gold dust. The city's trade quarter housed about 20,000 inhabitants and had 12 mosques. The city housing the ruling family and retinue as well as the religious dignitaries was overlooked by a huge castle decorated with stained-glass windows and sculptures. This flourishing culture was overrun in 1076 by the Almoravid invasion. The Almoravid ruled only for 10 years but destroyed the foundations of the nation. The state that survived barely controlled the area surrounding the capital.

particularly harsh for a country with a depleted agrarian economy affected by a persistent process of desertification and an annual grain deficit of 12,000 tons.

[18] Mauritania's geography is dominated by extensive desert areas, with a hot, dry tropical climate and limited precipitation along the Atlantic coast. The economic and social outlook is deteriorating; the southern farming and grazing lands are being reduced as the desert expands, and the impoverished nomadic people have been driven to the cities.

[19] In the fishing sector the government has adopted a long-term strategy, with two objectives: on the one hand, to save resources, and on the other, to integrate the fishing industry into the rest of the economy. No new licences are being given to foreign fishing fleets.

[20] Since 1987, a number of incidents have occurred between farmers and cattleherders in the border area along the Senegal River. In Nouakchott, many angry Mauritanians attacked hundreds of unarmed Senegalese with sticks and stones in 1989. In Dakar, Senegalese returning from Mauritania reported assassinations and mutilations of their compatriots in Mauritania. Scores of furious Senegalese reacted by murdering Mauritanian shop-keepers, and looting their shops. Within the country, there have been instances of violence between African Mauritanians from the south and Arabs and Berbers from the rest of the country. Throughout these events, the army supported the Arabs and Berbers, killing hundreds of African Mauritanians.

[21] Repeated border incidents with Senegal, leading to hundreds of victims on both sides, prompted Senegal to break diplomatic relations in August 1989. Thousands of refugees returned to their country of origin: 170,000 returned to Mauritania.

[22] In July 1991, the Mauritanians approved a referendum that recognized a multi-party system and established a democratic government. The opposition, concentrated in the United Democratic Front (FDU) exerted pressure on the government to achieve this objective, but by early 1991 the only response from the government had been increased repression. In spite of the constitutional reform, social tensions caused by the crisis did not de-

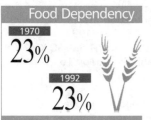

Food Dependency

1970
23%

1992
23%

Illiteracy
1995

62%

Maternal Mortality
1989-95

PER 100.000
LIVE BIRTHS

800

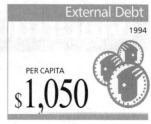

External Debt
1994

PER CAPITA

$1,050

PROFILE

ENVIRONMENT

Two thirds of the country are occupied by the Sahara Desert. The terrain consists of rocky, dry plateaus and vast expanses of dunes where extremely dry climate prevails. The desert region reaches into the southern part of the country, the Sahel grasslands, with some rainfall and sparse vegetation. The southwestern region receives slightly more rain and is irrigated by a tributary of the Senegal river. This region holds most of the population and the main economic activities. Nomadic shepherds are scattered throughout the country. Desertification is a serious problem for Mauritania.

SOCIETY

Peoples: Three-fourths of all Mauritanians are descendants of the Moors; a mixture of Arab, Berber and African peoples, nomadic shepherds of north-west Africa. The other 25% are from African groups in the south, the most important being the Peul and the Soninke. There are small groups of Wolof and Bambara. **Religions:** Muslim (official and predominant). **Languages:** Arabic and French (official). The Moors speak Bassanya, an Arab dialect. In southern Mauritania, Peul-Fulani and Sarakole (of the Soninke) are also spoken. **Political Parties:** Democratic and Social Republican Party; Union of Democratic Forces-New Era (UDF-EN); Union for Democracy and Progress (UDP); Movement of Independent Democrats. **Social Organizations:** Workers' Union of Mauritania

THE STATE

Official Name: al-Jumhuriyah al-Islamiyah al-Muritaniyah **Administrative divisions:** 12 regions and the district of Nouakchott. **Capital:** Nouakchott, 480,000 inhab. in 1992 **Other cities:** Nouadhibou 72,305; Kaédi 35,241; Kiffa 29,300; Rosso 27,783 (1992) **Government:**Maawiya Ould Sid'Ahmed Taya, president. **National Holiday:** November 28, Independence Day (1960). **Armed Forces:** 15,600. **Paramilitaries:** 6,000 (National Guard and Gendarmes).

STATISTICS

DEMOGRAPHY

Urban: 51% (1995). **Annual growth:** 2.1% (1992-2000). **Estimate for year 2000:** 3,000,000. **Children per woman:** 6.8 (1992).

HEALTH

One **physician** for every 16,667 inhab. (1988-91). **Under-five mortality:** 199 per 1,000 (1995). **Calorie consumption:** 96% of required intake (1995). **Safe water:** 66% of the population has access (1990-95).

EDUCATION

Illiteracy: 62% (1995). **School enrolment:** Primary (1993): 62% fem., 62% male. Secondary (1993): 11% fem., 19% male. University: 4% (1993). **Primary school teachers:** one for every 51 students (1992).

COMMUNICATIONS

81 **newspapers** (1995), 89 **TV sets** (1995) and 90 **radio sets** per 1,000 households (1995). 0.4 **telephones** per 100 inhabitants (1993). **Books:** 84 new titles per 1,000,000 inhabitants in 1995.

ECONOMY

Per capita GNP: $480 (1994). **Annual growth:** 0.20% (1985-94). **Annual inflation:** 7.20% (1984-94). **Currency:** 128 ouguiyas = 1$ (1994). **Cereal imports:** 286,000 metric tons (1993). **Fertilizer use:** 82 kgs per ha. (1992-93). **External debt:** $2,326 million (1994), $1,050 per capita (1994). **Debt service:** 23.3% of exports (1994). **Development aid received:** $331 million (1993); $153 per capita; 35% of GNP.

ENERGY

Consumption: 103 kgs of Oil Equivalent per capita yearly (1994), 100% imported (1994).

---

crease. A period of great social conflict ensued, and the majority of the population went on strike.

[23] Six months later, in January 1992, President Maawiya Ould Sid Ahmed Taya was reelected, in the first multiparty elections to be held. The opposition and numerous international observers suspected a fraud. Taya defeated his main rival, Ahmed Ould Daddah, winning 63 percent of the vote to Dadda's 33. In March, after two rounds of voting in the legislative election, the DSRP (the official party) achieved an absolute majority in the National Assembly, winning 67 of the 79 disputed seats.

[24] Six opposition groups boycotted the voting and denounced there had been a fraud in the elections. In April, Mauritania reestablished diplomatic relations with Senegal and Mali. This allowed to resume negotiations on pending border disputes and to find a possible solution for the situation of Mauritanian refugees in both neighbouring countries.

[25] Foreign aid granted by China and France gave new impetus to Taya's discredited government, accused of electoral fraud and of being responsible for the country's difficult social conditions. The application of an IMF-sponsored adjustment plan further aggravated mounting tensions, peaking in October when the prices of basic consumer goods were increased.

[26] Protest demonstrations were widespread and harshly quelled by the police. In early 1993, Interior minister Ba Aliu Ibra Hasni Uld Dudi was forced to resign and was replaced by a "moderate" official.

In January 1994, the ruling party won the local elections, also considered a fraud by opposition members. In May, the government tried to prevent Le Calame newspaper from being issued since it included a report on Mauritania by the Human Rights International Association.

[27] Sixty Islamic leaders were arrested in October, on charges of "creating an atmosphere of fear". Afterward, the government ordered all "fundamentalist" militants to cease political activity.

[28] In January 1995, thousands of demonstrators rocked Nouakchott, setting cars on fire and looting shops after a new valued added tax caused a rise in prices. In June, Mauritania's creditors accepted a renegotiation of the public debt which was then partially cancelled.

[29] Exactly one year later, president Taya appointed Fishing Minister El Afia Ould Mohamed Khouna as Prime Minister.

Mauritius

Population: 1,115,000 (1994)
Area: 1,860 SQ.KM
Capital: Port Louis

Currency: Rupee
Language: English

Mauritius

According to Portuguese accounts, the island of Mauritius was deserted when they first explored it in the 16th century. The island was immediately coveted by the imperial powers, who fought over it ceaselessly by both diplomatic and military means. Between 1598 and 1710, the Dutch, attracted by the island's ebony, took up occupation and named it Mauritius. In 1721, it was recolonized by the French Bourbons, who renamed it Ile de France. When rivalry between France and Britain recurred over the control of India, the island of Mauritius constituted an important strategic base. The island was used in trade with India, and treaties were signed with Madagascar and Mozambique to use their coasts and expand sugar cane plantations.

[2] During the French Revolution, Mauritius acquired a certain degree of autonomy, but in 1810, it fell into British hands. In 1815, after Napoleon's defeat, the Treaty of Paris recognized the status of the island as a British colony. It was the British themselves who introduced sugar cane, which became the island's main economic resource right up until the present.

[3] In the 19th century, sugar cane plantations expanded considerably, but in 1835 the emancipation of slaves, who constituted 70 per cent of the population, caused a serious labor shortage. This emancipation was opposed by the European landowners, who tried to alleviate the situation by importing over 450,000 Indian indentured "hired" workers over the next 100 years. With the passage of time, the Indians became an increasingly important sector of the Mauritian social and economic structure, and at present, they constitute the majori-

ty of the Mauritian population. The local culture however still bears the mark of French influence.

[4] In 1936, the Labor Party was organized but a series of strikes were brutally repressed, causing the death, imprisonment or exile of most party leaders. Seewooagur Ramgoolam, a doctor, started his political career as the leader of the Advance group. In the late 1940s, encouraged by the colonial administration, he succeeded in defeating the leadership of the Labor Party of Mauritius (PLM).

[5] During World War II, the British were unable to guarantee the security of their colonial dominions, which encouraged the emergence of claims for independence. At the same time, US political and military influence increased. The inhabitants of Mauritius fought for and achieved representation within British colonial government, and in 1957, a new government structure was created, giving Mauritius its own prime minister. In 1959, the first elections with universal suffrage brought the PLM to power, making Seewooagur Ramgoolam prime minister in 1961.

[6] From Mauritius, the British administration ran the islands of Rodrigues, Cargados-Carajos and the Chagos archipelago. In 1965, with the approval of Ramgoolam, Chagos and other islands became the British Indian Ocean Territories, and the United States established a major naval base on one of its islands, Diego García.

[7] The local population was secretly transferred to Mauritius, a move that caused a scandal in Congress when it was later made public. However, the island was not returned to Mauritius and its inhabitants were not granted permission to return home.

[8] Mauritius became independent in 1968, after a long decolonization process. Britain expected the process to end with the granting of a very limited autonomy for the island, allowing it merely to solve its domestic affairs. The 1958 Constitution, which is still in force, was created by the Colonial Office following the usual model used in the Commonwealth. Mauritius is a monarchy, ruled by Elizabeth II, and the sovereign is represented by a Governor-General and a High Commissioner. In 1964, the Council of Ministers and the Legislative Assembly were created. The latter appoints the prime minister (head of government) and the rest of the ministerial cabinet.

[9] On independence there was an established two-party system in the country. The Labor Party, supported by the Indian population, was the majority party until 1958. That year, the French-Mauritians and Creoles gathered in the Social Democratic Party of Mauritius (PSDM), and Muslim groups founded the Muslim Action Committee. During the 1960s, the common struggle for independence drew the Indian Labor groups and Muslims together, and in 1967 they formed an alliance which took them to power. Sir Seegwooagur Ramgoolam was re-elected prime minister, and Gaetan Duval, the leader of the PSDM, led the group which opposed independence. In 1969, a

coalition government was formed with the Labor Party and the PSDM. Duval, the representative of the French elite and the main opponent to independence, became foreign minister, showing that the British-granted independence was a mere formality.

[10] Internally, the government was tainted by electoral fraud and trade union repression, and in foreign affairs, it established close links with Israel and South Africa. Pretoria had a free zone in Port Louis which enabled it to trade with the EEC, thus evading international sanctions.

[11] In the 1970s a new opposition group emerged, called the Militant Mauritian Movement (MMM). It denounced the alliance between the Labor Party and the former French settlers. The growth of the MMM and the general deterioration of the political environment led to the legislative elections scheduled for 1972, being postponed until 1976. The MMM was the best supported party in the 1976 elections, but the Labor party succeeded in forming weak alliances with certain minority groups, managing to stay in power. The workers' demonstrations and the social unrest caused by unemployment increased, reaching their peak in 1979.

[12] In the 1982 elections, the MMM, allied with the Socialist Party of Mauritius (PSM) and won with a landslide majority. This gave

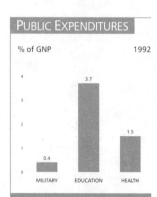

PUBLIC EXPENDITURES

% of GNP   1992

- MILITARY: 0.4
- EDUCATION: 3.7
- HEALTH: 1.5

LAND USE

■ CROPLAND 52% ■ PASTURE 3% ■ OTHER 44%

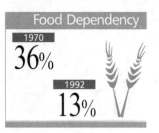

Food Dependency

1970
36%

1992
13%

Foreign Trade

MILLIONS $ 1994

IMPORTS
1,926

EXPORTS
1,347

PROFILE

ENVIRONMENT

This island, located in the Indian Ocean, 800 sq km east of Madagascar, is 53 km wide and 72 km long. Mauritius also controls several dependencies: Rodrigues Island, 104 sq km with 23,000 inhabitants who farm and fish; Agalega, 69 sq km with 400 inhabitants, producing copra; and St Brandon, 22 small islands inhabited by people who fish and collect guano. The islands are of volcanic origin, surrounded by coral reefs. The terrain climbs from a coastal lowland to a central plain enclosed by mountains. Heavy rainfall contributes to the fertility of the red tropical soils. Sugar cane is the main crop. The degradation of the soil, due to monoculture and use of pesticides, and water pollution are the most serious environmental problems.

SOCIETY

Peoples: Indo-Pakistani 68%; Creole (mixed Caucasian, Indo-Pakistani, and African) 27%; Chinese 3%. **Religions:** 51% Hindu; 30% Christian (mostly Catholic, with an Anglican minority); 17% Muslim; 2% other. **Languages:** English (official) and Creole. Hindi, Bhojpuri (a Hindi dialect), Urdu, Hakka, French, Chinese, Tamil, Arabic, Marati, Telegu, and other languages are also spoken. **Political Parties:** Mauritian Socialist Movement (MSM); Militant Mauritian Movement (MMM); Mauritian Militant Renaissance (RMM); Mauritian Social Democratic Party (PMSD). **Social Organizations:** General Workers' Federation (GWF); Mauritian Women's Committee.

THE STATE

Official Name: Republic of Mauritius. **Administrative divisions:** 4 islands and 9 districts. **Capital:** Port Louis, 142,000 inhab. in 1992. **Other cities:** Beau-Bassin-Rose Hill, 94,000 inhab.; Curepipe, 75,000 inhab.; Quatre Bornes, 72,000 inhab. (1992). **Government:** Parliamentary republic. Cassam Uteem, president. Legislature: single-chamber: National Assembly, with 72 members, elected every 5 years. **National Holiday:** March 12, Independence Day (1968). **Armed Forces:** non-existent. **Paramilitaries:** 1,800 (includes 500 Coast Guards).

STATISTICS

DEMOGRAPHY

Urban: 41% (1995). **Annual growth:** 0.3% (1992-2000). **Estimate for year 2000:** 1,000,000. **Children per woman:** 2 (1992).

HEALTH

One **physician** for every 1,176 inhab. (1988-91). **Under-five mortality:** 23 per 1,000 (1995). **Calorie consumption:** 104% of required intake (1995). **Safe water:** 99% of the population has access (1990-95).

EDUCATION

Illiteracy: 17% (1995). **School enrolment:** Primary (1993): 106% fem., 106% male. Secondary (1993): 60% fem., 58% male. University: 4% (1993). **Primary school teachers:** one for every 21 students (1992).

COMMUNICATIONS

101 **newspapers** (1995), 104 **TV sets** (1995) and 103 **radio sets** per 1,000 households (1995). 9.6 **telephones** per 100 inhabitants (1993). **Books:** 98 new titles per 1,000,000 inhabitants in 1995.

ECONOMY

Per capita GNP: $3,150 (1994). **Annual growth:** 5.80% (1985-94). **Annual inflation:** 8.80% (1984-94). **Consumer price index:** 100 in 1990; 132.8 in 1994. **Currency:** 18 rupees = 1$ (1994). **Cereal imports:** 240,000 metric tons (1993). **Fertilizer use:** 2,512 kgs per ha. (1992-93). **Imports:** $1,926 million (1994). **Exports:** $1,347 million (1994). **External debt:** $1,355 million (1994), $1,215 per capita (1994). **Debt service:** 7.3% of exports (1994). **Development aid received:** $27 million (1993); $24 per capita; 1% of GNP.

ENERGY

Consumption: 387 kgs of Oil Equivalent per capita yearly (1994), 92% imported (1994).

---

them control of 62 of the 66 seats and full control of the government. Aneerood Jugnauth became prime minister. The government program of the MMM-PSM alliance promised to increase job opportunities and salaries, nationalize key sectors in the economy, reduce economic links with South Africa, and demand the return of Diego Garcia from the US.

[13] The increase in the price of oil, coupled with the drop in the world price of sugar, caused a deficit in the balance of payments equal to 12 percent of the gross domestic product. The government was forced to turn to the IMF which approved five stand-by loans between 1979 and 1985.

[14] To receive these loans, the government had to agree to a series of conditions imposed by the IMF; to adopt austerity measures, to postpone part of the planned salary increase and job creation programs, to relinquish part of its control over public expenditure, to cut subsidies on basic foods and to devalue the Mauritian rupee.

[15] All these measures led to clashes between the MMM and the Socialists, a situation which eventually led to a call for early elections, in August of that year. The MMM was defeated and a new coalition was formed including labor, socialists and Duval's social-democrats. However, this apparent political stability was very weak and was followed by a series of governmental alliances which did not consolidate its position. Majority parties underwent serious divisions, giving rise to new political groups such as the Mauritius Socialist Movement (MSM), a splinter of the PSM, led by the current president Aneerood Jugnauth.

[16] In the past 12 years, the country's average annual economic growth has been 7 per cent. Traditionally, sugar was the main export, but now it only accounts for 40 per cent because of the diversification of sources of foreign currency. The main economic potential lies in manufacture, in particular in textiles. The industrial free zones employ 90,000 people, around 10 per cent of the population, and the per capita income of the population in Mauritius is three or four times higher than the average for other African countries, though inflation is increasing at an annual rate of 16 per cent. The lack of skilled labor on Mauritius could bring a new migratory wave from Madagascar, India, and Kenya.

[17] In 1988, in agreement with the Organization of African Unity (OAU), Mauritius again demanded the devolution of the island of Tromelin, administered by France, and of the Chagos archipelago, along with the demilitarization of the Indian Ocean, which is used for maneuvers by the superpowers. Their claims are supported by environmentalist groups, because of the proliferation of nuclear weapons in the claimed territories.

[18] In the legislative elections of

17%

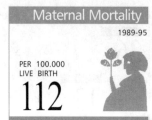

PER 100.000 LIVE BIRTH

112

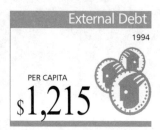

PER CAPITA

$1,215

December, 1987, the coalition of MSM, PSDM and Labor won a very narrow victory, giving it a majority in parliament. Aneerood Jugnauth was re-elected prime minister, and the MMM became the main opposition force.

[19] The headlines have announced scandals linking political leaders with drug deals and money laundering. There have also been several attempts to murder the prime minister, at least one of which has been attributed to drug dealers.

[20] In the legislative elections held in September 1991 the governing MSM succeeded in maintaining Jugnauth in the post of prime min-ister by reinforcing the traditional alliance with the MMM. In March 1992, Mauritius ceased to be a constitutional monarchy to become a republic, and, in June, Cassam Uteem became the first president of the country.

[21] In August 1993, the foreign minister Paul Bérenger, of the MMM, withdrew from the cabinet, leaving Jugnauth without an absolute majority. In 1994, the island's economic results were still regarded as satisfactory by international agencies such as the World Bank. The foreign debt accounted for 25 per cent of the GDP and the per capita income was $2,740.

[22] In January 1995 Jugnauth included in his government representatives of the right-wing Mauritian Social Democratic Party. In the December elections, a coalition of the opposition led by Bérenger and Nuvin Ramgoolam obtained two thirds of the seats.

[23] The economic shifts in the island were reflected in exports, since sales of textiles abroad accounted for 52 per cent of total sales, as compared with 28 per cent in sales of sugar, the traditional crop in Mauritius.

Diego García

Population: 1,000 (1994)
Area: 52 SQ.KM

The settlement of Diego García began in 1776, when the French Viscount de Souillac sent a ship there from Mauritius island, trying to establish French presence before the British could get there. French entrepreneurs obtained permission to exploit all of the island's riches; coconuts (for coconut oil), giant tortoises, fish and birds, and they established a leper colony on the island. With the defeat of Napoleon in 1815, the island passed into the hands of the British Crown, together with Mauritius' other dependencies. During the 19th century, many workers came from India and Africa. The risk of catching leprosy, especially among men, and the demanding nature of their work, gave rise to a matriarchal family organization.

[2] Around 1900, there were approximately 500 inhabitants, but the population increased radically over the next decades, with the arrival of Africans, Madagascans and Indians. Once they had settled, they developed a culture of their own. They spoke creole, a mixture of their native languages and they took part in the Tamul rituals (of Madagascan origin) even though they were mostly Roman Catholic. The indigenous Chagos community, the Ilois, had lived in their traditional manner, and this remained practically unchanged until the 1960s.

[3] In 1965, Britain decided to remove Diego García from Mauritian jurisdiction, annexing it to the BIOT (British Indian Ocean Territory). Although this change was condemned by the UN on December 14, 1960, the British and Mauritian governments made a deal where a large sum changed hands. Two years later, Britain ceded the island to the United States for 50 years, in exchange for a discount on its purchase of nuclear arms. To make way for the construction of important air and naval bases, the Ilois were deported to Mauritius in the early 1970s.

[4] The 2,000 or so Ilois were abandoned as soon as they arrived in the port of Mauritius, finding themselves in a situation of total indigence; they had not been allowed to keep their belongings and they suffered the effects of being uprooted. Many died of despair or hunger, and to this day they lack the basic human right of having a nationality. The British government does not recognize them as subjects, and the Mauritian government does not consider them to be citizens. Ilois attempts to return to their land became increasingly numerous, as did their demands for a solution to their problems, and they began to receive the support of international opinion. The details of this situation reached the United States Senate in 1975, but to this day, both the US and British governments have continued to ignore the real problem, each blaming the other for the situation and both spending intermittent sums of money which do nothing more than temporarily alleviate the severity of the problem. A report issued by the Mauritian government in 1981 revealed that 77 per cent of the Ilois wished to return to their homeland.

[5] In the meantime, the govern-ment of Mauritius has decided to stop shipping foodstuffs and supplies to Diego García, by way of reiterating its desire to recover the islands.

[6] Between 1992 and 1993, the government issued repeated claims to the Chagos archipelago, including Diego García, to the UN and the International Court of Justice. As a reprisal, Britain cut economic aid to the country.

PROFILE

ENVIRONMENT

The Chagos Archipelago is located some 1,600 km southwest of India, in the middle of the Indian Ocean. The main islands are Diego García (8 km long by 6 km wide), Chagos, Peros, Banhaus and Solomon. The islands are made up of coral formations and are flat; they have a large number of coconut trees, which grow well here because of the tropical climate and year-round rain.

SOCIETY

Peoples: All of the inhabitants are American or British military personnel.

THE STATE

Government: It is nominally exercised by an English commissioner from the Foreign Office in London. Britain is represented on the islands by a 25-man Royal Navy detachment.

DEMOGRAPHY

Density: 20 inhab./sq.km.

Mexico

México

Population: 88,543,000 (1994)
Area: 1,958,200 SQ.KM
Capital: Mexico City

Currency: New Peso
Language: Spanish

The 20,000-year history of the present Mexican territory includes only 2,000 years of urban life. Over this period, native meso-American peoples developed advanced civilizations such as the Olmeca, Teotihuacan, Maya and Mexica. These cultures had complex political and social organizations, and they had advanced artistic, scientific, and technological skills.

[2] When the Spaniards arrived, Moctezuma II, Emperor of the Aztecs, ruled over an empire the size of modern Italy, with the capital, Tenochtitlan, under the present Mexico City. The conquest was completed in 1521, when Spanish explorer Hernán Cortés took advantage of internal strife between the ruling Aztecs and other native peoples who paid them tribute. Cortés succeeded in imposing Spanish domination over the local armies and, from then on, the colonials initiated Christianization and Hispanization of the indigenous inhabitants.

[3] In the 17th century, the major economic structures of the so-called "New Spain" were laid down. The *hacienda*, a landed estate, emerged as the basic production unit, and mining became the basis of a colonial economy conceived to meet the gold and silver needs of the Spanish motherland. The native American population was over-exploited and decimated by hard labor and disease. By1800, Mexico had become one of the world's richest countries, a nation of "great wealth and great poverty".

[4] After almost three centuries of colonial domination, in 1810 the struggle for independence began, led by *criollos* (Mexicans of Spanish descent) such as Miguel Hidalgo and José María Morelos, two priests. The struggle became a broad-based national movement as natives and mestizos joined its ranks, but the rebels were soon crushed by the royal army. The liberal revolution in Spain radically changed the situation. Afraid to lose their privileges, the Spanish residents and the conservative clergy came to an agreement with the surviving revolutionaries. This pact became known as the Iguala Plan: trading independence for a guaranteed continuation of Spanish dominance. In 1821, General Iturbide proclaimed himself emperor, but was soon replaced by an unstable republican government.

[5] At this time, Mexico was Latin America's largest country; including the Central American provinces it covered 4.6 million sq km and was also beset by enormous economic, political, and social problems.

[6] Mexico sustained attacks from Spain, France and the US in its early years of independence. (see box.)

[7] Between 1821 and 1850, the country had 50 different governments. The first election did not produce stability. As with many other Latin American states, the Mexican bourgeoisie adhered to two political parties: the liberals and the conservatives. At the end, a brief and minor conflict between Mexico and France (known as the Pastry War) arose in 1838-39 from the claim of a French pastry cook living in Tacubaya, near Mexico City, that some Mexican army officers had damaged his restaurant. A number of foreign powers had pressed the Mexican government without success to pay for losses that some of their nationals claimed they had suffered during several years of civil disturbances. France decided to back up its demand for $600,000 by sending a fleet to Veracruz and after occuppying the city, the French won a guarantee of payment through the good offices of Britain and withdrew their fleet. The most important result of this conflict was the further enhancement of the prestige and influence of political dictator Antonio López de Santa Anna, who lost a leg in the fighting.

[8] The Liberal victory of 1857 led by Benito Juarez, tried to strengthen the republic, instituting a free market economy, prescribing individual rights and guarantees, and expropriating the wealth of the clergy. Juárez, a Zapotec indian educated as a middle-class liberal, and his party, sought to give a juridical base for their reforms by drafting a new constitution. The Constitution of 1857 prohibited slavery and restrictions on freedom of speech or the press. It abolished special courts and prohibited civil and ecclesiastical corporations from owning property, except buildings in use; it

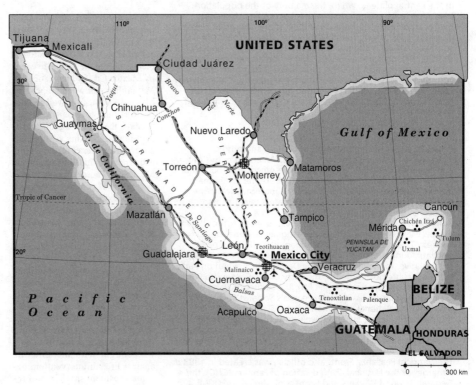

PUBLIC EXPENDITURES

% of GNP — 1992

MILITARY 0.5 — EDUCATION 4.1 — HEALTH 1.6

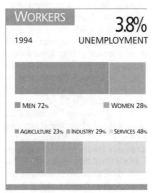

WORKERS

3.8% UNEMPLOYMENT

1994

MEN 72% — WOMEN 28%

AGRICULTURE 23% — INDUSTRY 29% — SERVICES 48%

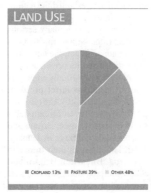

LAND USE

CROPLAND 13% — PASTURE 39% — OTHER 48%

ENVIRONMENT

The country occupies the southern portion of North America. It is mostly mountainous, with the Western Sierra Madre range on the Pacific side, the Eastern Sierra Madre on the Gulf of Mexico, the Southern Sierra Madre and the Sierra Neovolcánica Transversal along the central part of the country. The climate varies from dry desert wasteland conditions in the north to rainy tropical conditions in southeastern Mexico, with a mild climate in the central plateau, where the majority of the population lives. Due to its geological structure, Mexico has abundant hydrocarbon reserves, both on and off shore. The great climatic diversity leads to very varied vegetation. The rainforests in the southeastern zone and the temperate forests on the slopes of the Sierra Neovolcánica are strategic economic centers. The river system is scanty and unevenly distributed throughout the country. Air, water and soil pollution levels are high in industrial areas. Deforestation affects 6,000 sq km per year. Mexico City and its surrounding area suffers from high levels of smog and atmospheric pollution.

SOCIETY

Peoples: Mexicans are the result of cultural and ethnic integration between predominant Meso-American native groups and Spanish colonizers. Of the 56 native American groups that live in Mexico today, the most important are the Tarahumara, Nahua, Huichole, Purepecha, Mixteco, Zapoteca, Lacandone, Otomie, Totonaca and Maya. Approximately seven million Mexicans live in the south of the United States.
Religions: Mainly Catholic. **Languages:** Spanish (official). One million Mexicans speak indigenous languages. **Political Parties:** The Institutional Revolutionary Party (PRI), in power since its foundation in 1929; the National Action Party (PAN), conservative, founded in 1939; the Democratic Revolution Party, led by Cuauhtémoc Cárdenas; the Party of the Cardenista Front for National Reconstruction (PFCRN); the Authentic Party of the Mexican Revolution (PARM); the People's Socialist Party (PPS); the Zapatista National Liberation Front (FZLN) arose in 1994, in the state of Chiapas; Ejército Revolucionario Popular. **Social Organizations:** The Labor Congress is a grouping of 34 large union organizations serving as a basic labor support for the PRI government. Altogether they represent nearly 8 million workers. The National Peasants' Front, also official, was created in 1983 in a merger of the National Confederation of Farmers (CNC), the General Union of Workers and Farmers of Mexico (UGOCM), the National Confederation of Small Landowners (CNPP) and the Independent Farmers' Central Union (CCI).

THE STATE

Official Name: Estados Unidos Mexicanos. **Administrative Divisions:** 31 states and a federal district. **Capital:** Mexico City, 15,300,000 inhab. (1992). **Other cities:** Guadalajara, 1,650,000 inhab.; Ciudad de Netzahualcóyotl (neighbouring the Federal District) 1,255,500 inhab.; Monterrey, 1,069,000 inhab.; Puebla, 1,007,200 inhab.; Ciudad Juárez, 789,500 (1990). **Government:** Ernesto Zedillo, President since August 1994. The system is both federal and presidential. Re-election is not permitted. **National Holiday:** September 16, Independence Day (1810). **Armed Forces:** 175,000 troops, 60,000 conscripts, 300,000 reserves. **Paramilitaries:** 14,000 Rural Defense Militia.

Food Dependency

1970
7%

1992
11%

Foreign Trade

MILLIONS $ 1994

IMPORTS
80,000
EXPORTS
61,964

control it thorugh a puppet, Maximilian of Austria, they withdrew their troops. Despite a major defeat at Puebla on May 5 1862, reinforcements enabled the French to occupy Mexico City in June 1863, and Maximilian - crowned Emperor of Mexico - soon followed to take control of the government. Forced to leave the capital again, Juárez kept himself and his government alive by a long series of retreats that ended only at Ciudad Juárez, on the Mexican-US border. Early in 1867, as a result of continued Mexican resistance, increased US pressure and criticism at home, Napoleon decided to withdraw his troops. Soon afterward, Mexican forces captured Maximilian and executed him.

10 In 1871, Juan de Wata Rivera published the newspaper "The Socialist". On September 10, this newspaper published the general statutes of the 1st Marxist International for the first time in Latin America. General Porfirio Díaz, who fought with the Liberals against French intervention, became president in 1876 and remained so until 1911. During the 35 years of his dictatorship, the country opened its doors to foreign investment, the economy was modernized and social inequalities increased.

11 In 1910, Francisco Madero led the Mexican Revolution under the slogan "effective suffrage, no re-election". This was the first popular Latin American revolution of the century. In 1913, US ambassador Henry Lane Wilson participated in a conspiracy, resulting in the assassination of Madero. The people responded by taking up arms more vigorously than before. Peasants joined the revolt, led by Emiliano Zapata and Francisco (Pancho) Villa. The principles of this Revolution were set out in the Constitution of 1917, promulgated by Venustiano Carranza. It was the most socially-advanced constitution of its time and many of the principles are still in force today. However, conflict between the various revolutionary factions continued, resulting in the deaths of the major leaders.

12 In 1929, under President Plutarco Elías Calles, these factions joined to form the National Revolutionary Party. In 1934, General Lázaro Cardenas took office. He embodied the continuation of the revolutionary process, and was one of the main driving forces behind its accomplishments. The major reforms included land reform, the nationalization of oil (the founding PEMEX), the expropriation of oil refineries, incentives for new industries, and a national education system. The "No re-election" principle became inscribed forever in the Mexican Constitution. When Cardenas was asked to amend the constitution for his own re-election, he refused. The National Revolutionary Party became the Institutional Revolutionary Party (PRI) and the deeply rooted revolutionary socialist principles were gradually abandoned.

13 Conditions created by World War II accelerated the first phase of Mexico's industrialization which reached its peak during Miguel Aleman's term of office (1946-1952). The changes that took place during this period altered the former social balance. Mexico's population still remained predominantly rural, with only 40 per cent in the cities, but this rapid development was not able to absorb the quickly growing population. Communal land ownership, which had stimulated solidarity and revolutionary feeling among peasants in the 19th century, was gradually replaced by a new type of individual land tenure causing the formation of large estates or *latifundios*. The socialist Mexico planned by the revolutionary prophets of 1910,had thus developed into Latin America's most successful post-war capitalist nation.

14 During the following decade, Cardenas' successors, though not always loyal to his principles, stabilized the system by reinforcing those factors responsible for its success in a continent generally afflicted by underdevelopment and stagnation. Strong governmental influence, involving substantial public investment, kept the economy strong. A reasonable balance was maintained between heavy and light industry, tourism was encouraged and a stable government, seeming to satisfy the people's demands, was in power. The party was made up of peasants, workers, merchants, businessmen and technocrats; only the Church and traditionally conservative landowners were excluded.

15 Against the backdrop of the 1968 Mexico Olympics, the students' movement organized protests against the true social situation

eliminated monopolies and prescribed that Mexico was to be a representative, democratic, republican nation. Neither the religious community nor the military accepted this constitution, and both inveighed against the reform, calling for retention of "religión and fueros." The church excommunicated all civil officials who swore to support the constitutioin; civil war erupted and foreign powers became involved in the Mexican struggle. On April 6 1859, the United States recognized the Juárez government, permitted war supplies to be shipped to the liberal forces and encouraged its citizens to serve the liberal cause as volunteers; while

Spain, Britain and France favoured the conservatives.

9 President Benito Juárez was succesfull in re-establishing national unity in January, 1861. He was however faced with many serious problems: the opposition's forces still remained intact, the new Congress distrusted its president, and the treasury was virtually empty. As a solution to this latter problem Juárez decided in July 1861, to suspend payment on all foreign debts for two years. France, Britain and Spain invaded Mexico in January, 1862, landing with their troops in Veracruz. But when Britain and Spain realized that Napoleon III intended to conquer Mexico and

STATISTICS

DEMOGRAPHY

Urban: 74% (1995). **Annual growth:** 0.7% (1992-2000). **Estimate for year 2000:** 99,000,000. **Children per woman:** 3.2 (1992).

HEALTH

One **physician** for every 621 inhab. (1988-91). **Under-five mortality:** 32 per 1,000 (1995). **Calorie consumption:** 131% of required intake (1988-90). **Safe water:** 83% of the population has access (1990-95).

EDUCATION

Illiteracy: 10% (1995). **School enrolment:** Primary (1993): 110% fem., 110% male. Secondary (1993): 58% fem., 57% male. University: 14% (1993). **Primary school teachers:** one for every 30 students (1992).

COMMUNICATIONS

103 **newspapers** (1995), 102 **TV sets** (1995) and 100 **radio sets** per 1,000 households (1995). 8.8 **telephones** per 100 inhabitants (1993). **Books:** 91 new titles per 1,000,000 inhabitants in 1995.

ECONOMY

Per capita GNP: $4,180 (1994). **Annual growth:** 0.90% (1985-94). **Annual inflation:** 40.00% (1984-94). **Consumer price index:** 100 in 1990; 166.3 in 1994. **Currency:** 5 new pesos = 1$ (1994). **Cereal imports:** 6,223,000 metric tons (1993). **Fertilizer use:** 653 kgs per ha. (1992-93). **Imports:** $80,100 million (1994). **Exports:** $61,964 million (1994). **External debt:** $128,302 million (1994), $1,449 per capita (1994). **Debt service:** 35.4% of exports (1994). **Development aid received:** $402 million (1993); $5 per capita; 0.1% of GNP.

ENERGY

Consumption: 1,577 kgs of Oil Equivalent per capita yearly (1994), -55% imported (1994).

in the country. A student rally held on the *Plaza de las tres culturas* was dispersed by the army. The order to shoot to kill was given without warning, and hundreds of people died or were injured in the "Tlatelolco Massacre".

[16] During the presidency of José López Portillo (1976-1982), important oil deposits were discovered. This bound Mexico more closely to the US, as its prime oil supplier.

[17] In 1982, Miguel de la Madrid assumed the presidency implementing an IMF economic adjustment plan. Subsidy and public spending cuts, changes in the pattern of public investment, and a dual currency exchange rate, caused public discontent and the PRI's first electoral defeat since its foundation. In the 1983 municipal elections, the government lost two major cities and their seat in the capital.

[18] Under the pressure of foreign debt the trends of 1983 continued: growing inflation; losses in real wages; reductions of public spending; falls in production, and rising unemployment.

[19] In 1985, direct foreign investment increased by $1.5 billion, 66 per cent coming from American investors. However, export profits also increased, totaling $344 million in the first nine months. The $96 billion foreign debt continued to be the main problem, as it demanded an average annual servicing of $12 billion.

[20] The devastating earthquake of September 1986, which buried more than 20,000 people alive, further aggravated an already critical situation. De la Madrid continued a policy of economic growth through exports, to service the debt, but in spite of the fact that the exports increased by 3 per cent, the government was only able to meet payments by taking out new loans. The reduction in the oil quota by the US made it necessary to generate alternative sources of income. One of these was tourism, and another was the "maquiladoras"; foreign companies on the border with the US which are exempt from taxes and from paying their workers social insurance, disposing of the goods they produce on the vast domestic market.

[21] In 1986, debt servicing increased, while social discontent, inflation and unemployment continued to rise, and border tensions with the United States mounted. US Senator Helms led a defamation campaign against the Mexican government, accusing it of sponsoring illegal immigration into the United States, and of having links with the drug mafia.

[22] Although the trade surplus and the accumulation of reserves grew in 1987, industrial employment fell by 7 per cent during the first quarter, and inflation - which stood at 106 percent in 1986 - was 134 per cent higher by August 1987.

[23] The internal situation became more difficult for the PRI and it was accused of fraud in the 1986 municipal elections. The formation of an independent Labor Union Negotiation Board posed a threat to the labor wing of the PRI, which had traditionally formed a bloc within the Labor Congress, the largest co-ordinating board of labor federations.

[24] Political leaders Cuauhtémoc Cárdenas and Porfirio Muñoz Ledo challenged the mechanism for designating the PRI's presidential candidate. Both named their own alternative candidates. In response, they were barred from using party sites or the PRI emblem, and were also asked to leave the organization. The nomination ultimately went to Carlos Salinas de Gortari, minister of planning and budgets. However, the Left presented at least two candidates in the 1988 presidential elections; five organizations joined together to form a coalition Socialist Mexican Party, while the Workers' Revolutionary Party and six other minor organizations backed another candidate.

[25] On July 6 1988, in the national elections several parties gained a significant vote, something which had not happened since 1910. Carlos Salinas de Gortari, the PRI candidate, won the election with - according to official figures - 50 per cent of the vote (the lowest percentage in party history). The left, for the first time, emerged as a real alternative to the PRI. Cuauhtemoc Cardenas, the son of President Cardenas, led a coalition of groups operating as a single party, the FDN, representing the masses and opposing the Institutional Revolutionary Party. Cardenas took second place in an election marked by accusations of fraud and irregularities, obtaining more votes than the PAN (National Action Party), a conservative party. The abstention rate was 49.72 per cent.

[26] In 1989, the FDN split, and Cardenas founded the Democratic Revolution Party, made up of former members of the PRI, communists and members of smaller organizations. In the meantime, the People's Socialist Party (PPS), the Authentic Party of the Mexican Revolution (PARM) and the Party of the Cardenista Front for National Reconstruction (PFCRN) continued as autonomous organizations which voted with the PRI in Congress.

[27] The PRI administration resolved to open up the country to foreign investment, and also announced a series of measures aimed at controlling inflation. Both of these decisions were welcomed by the US government. Mexico made overtures to the US to sign a free trade agreement, coinciding with Mexico's entry into GATT and the legal authorization for foreign investment in Mexican enterprises to go above the previously stipulated 49 per cent. In May 1990, President Salinas de Gortari privatized the banking system, which had been nationalized eight years previously.

[28] On August 18 1991, elections were held for the following posts: 500 Congressional representatives, 32 senators, 6 state governors and 66 city council members (for Mexico City). Amid accusations, on the part of the opposition, of "widespread fraud", the PRI proclaimed itself the winner, with 61.4 per cent of the vote. This gave the PRI control of the Chamber of Deputies, and power to carry out constitutional reforms.

[29] On November 8 1991 an Agrarian Reform was presented by the president. Approved in December, it granted property rights to *campesinos* (peasants) who farm state lands known as *ejidos* (co-operative farms ceded by the Zapata Revolution in 1917). According to the PRI, the new system is designed to reduce the current dependence on an annual 10 million tons of imported food. According to the opposition, the reform - which allows the *campesinos* to sell their lands - will bring about a transfer of small landholdings to larger investors.

[30] In August 1992, PEMEX, the national oil company, intensified oil extraction operations in the state of Tabasco, in southeastern Mexico, affecting a 4,000 sq km strip of wetlands, the largest in Central America. Despite protests and alternative proposals from scien-

The war "against" the United States

Between 1836 and 1848, the United States stole more than half of Mexico. In 1836, North Americans occupied the Mexican territory of Texas, later requesting the recognition and protection of the United States. In 1846 the US army invaded this part of Mexico with the aim of defending the North American occupiers.

After the war, which lasted two years, some members of the United States government wanted to annex all of the area north of the Rio Bravo, while others wanted to seize all of Mexico. Finally, New Mexico, Alta California (what is now the North American state of California), and parts of the Mexican states of Tamaulipas, Coahuila, Chihuahua and Sonora were annexed. The United States paid Mexico $15,000,000 and assumed $3,250,000 in claims held by US citizens against Mexico.

tists and farming and fishing organizations, production increased to half a million barrels per day by August 1992.
[31] During that year, the government's liberal economic policies continued. In January, the constitutional guarantees that had long protected the Mexican oil industry were discontinued.
[32] On December 17, 1992, the Mexican, U S and Canadian governments signed the North American Free Trade Association (NAFTA) Treaty. (See box in Canada).
[33] During the Salinas administration, inflation was reduced from three-digit figures to a rate of 9 per cent, in 1993. Between the end of 1988 and mid-1993, the state received some $21 billion from the privatization of state interests. The private foreign debt increased by $11 billion in 1993, to an accumulated total of $34.265 billion.
[34] In the first quarter of 1994 - as the NAFTA treaty went into effect - trade with the US increased at an unprecedented rate. Mexican sales increased by 22.5 per cent, while US exports to Mexico were up 15.7 per cent on the preceding quarter.
[35] The high interest rates suffocated the small and middle-sized businesses. Unemployment and underemployment affected 5 and 12 million people respectively, mostly indigenous people working as street vendors and domestic servants.
[36] On January 1 1994, the previously unknown Zapatista National Liberation Army (EZLN) took three towns in the southern state of Chiapas, declaring them a "freed zone." Chiapas was the region where the paramilitary guards contracted by the landowners to expel the indigenous people from their land acted with most impunity. It is one of the states with the largest Maya population, mostly from the Tzotzil group. It also has the highest level of Spanish language illit-

eracy and the lowest incomes. However, it also has large oil and gas reserves - 21% of the nation's crude oil is extracted in Chiapas.
[37] The government initially refused to accept the importance of the uprising, which spread rapidly, increasing the size of the area under their control, with the rural public and workers of other areas taking up the cause. As a result, a fifth of the army was deployed to the region. When the number of dead rose above a thousand and the complaints of summary executions continued, the government, on the insistence of Catholic Bishop Samuel Ruiz, agreed to negotiate.
[38] In February, a cease-fire was agreed and negotiations began in San Cristobal de las Casas - the city in the center of the Zapatista area - with the rebels demanding electoral reforms, the creation of new municipal areas, ethnic representation in Congress and the government of Chiapas, schooling in the native languages, health and education infrastructure, along with modifications to the Penal Code and land ownership. The government accepted half the Zapatista claims, but after consulting with the indigenous communities, the rebels rejected the agreement.
[39] The PRI presidential candidate Luis Donaldo Colosio was murdered on March 23 in Tijuana. Three members of his bodyguard were involved. The PRI nominated Ernesto Zedillo as his replacement for the August 21 elections, where he was voted in by 49 per cent of the electorate. Both the PRD and the EZLN accused the government of electoral fraud.
[40] In September, Jose Ruiz Massieu, PRI secretary-general, was assassinated, and in November, his brother Mario resigned from his post as attorney general as he believed Party officials were blocking the criminal investigation. His brother's death increased suspi-

cions that leading PRI figures, and perhaps the drug mafia, were also involved in the murders.
[41] The economy continued to grow until on December 20, due to an acceleration of capital flight, the new government abandoned the policy of gradually depreciating the currency. By the end of the year, the peso had lost 42 per cent of its value, with its plunge causing the stock market to collapse in an event which came to be known as "the Tequila effect."
[42] The Zedillo government took until March 1995 to put an austerity plan into action, gaining the support of the United States and the IMF, which provided loans helping the nation to confront the financial crisis. The conditions attached to the international aid provoked a harsh recession.
[43] In order to alleviate the effects of the austerity plan on the poorer sectors, the government increased the minimum wage by 10 per cent and extended free health care for the unemployed from two to six months.
[44] During the course of the year, millions of Mexicans joined spontaneous protest movements, like "El Barzon". Many of these protests, by the middle classes and the small business sector, were prompted by the increase in debts caused by the devaluation. In August, President Zedillo approved a relief plan which benefited 7.5 million debtors.
[45] In January, Zedillo attempted to achieve a quick, peaceful solution to the situation in Chiapas. Following the failure of his initiative, in February, he launched a military offensive. The investigations of the Colosio and Ruiz Massieu murders had still got nowhere.
[46] Raul Salinas de Gortari, brother of the former president, was arrested in February accused of mas-

terminding the Ruiz Massieu killing. At the same time, the government called for the extradition of Mario Ruiz Massieu from the United States, on charges of hampering the investigation into his brother's death. Raul Salinas's wife was arrested in Switzerland in November when using false documents to carry out a banking transaction. Mrs Salinas was attempting to transfer funds - presumably from the laundering of drug money - from her husband's account. The former president Carlos Salinas said he was astonished by his brother's illegal wealth.
[47] The PRI and the two leading opposition parties agreed an electoral reform in December, which included establishing an independent control commission and limiting campaign expenditure.
[48] In September 1996, the Zapatistas accused the government of being insensitive to their claims, arrogant and racist, bringing 16 months of peace negotiations to a halt. Meanwhile, the interior minister declared there was no reason for suspending the talks.

Micronesia

Population: 104,000 (1994)
Area: 700 SQ.KM
Capital: Colonia (Kolonia)
Currency: US Dollar
Language: Kosrean, Yapese, Pohnpeian and Turkese

Micronesia

The name Micronesia is derived from the Greek, meaning small islands. It refers to the Marshall Islands, the Marianas (including Guam) and the Caroline Islands.

[2] In 1531, Ferdinand Magellan named them the Islands of Thieves, but they were later renamed after Queen Mariana of Austria, Spanish regent.

[3] In 1885, the Germans tried to impose a protectorate on the islands, but the Spaniards appealed to the Vatican and managed to keep the islands. Guam was annexed to the United States by the Treaty of Paris of December 12 1898, and the rest of the Marianas were sold to Germany.

[4] In 1914 Micronesia was occupied by the Japanese who were negotiating the demilitarization of the region with the US. The accord was broken in 1935, and the Japanese Empire launched the attack on Pearl Harbor from Micronesia, on December 7 1941, bringing World War II to the Pacific.

[5] In 1947, when the fate of the Japanese and German possessions was decided, an agreement between the United Nations Security Council and the US government assigned the islands to Washington as a trust territory. Though located 13,000 km from the coast of the United States, they were already under Washington's control.

[6] A UN mandate obliged the US to develop national consciousness and promote an economy enabling the native population to exercise their right to self determination.

[7] This 'Free Associated State' system allows Washington to keep military bases in such countries, while handling their defence and foreign affairs. In 1975, the Northern Marianas voted for Free Associated State status, and in February 1978, the islands became part of United States under a trusteeship agreement.

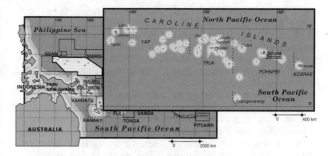

[8] In 1978, another plebiscite led to the creation of the Federated States of Micronesia. Four districts supported the motion, but Palau and the Marshall Islands decided to continue as autonomous states.

[9] Before World War II the major Micronesian economic activities were fishing and coconut plantations. After the war, the population increase made this economic base insufficient.

[10] Bikini and Eniwetok atolls; where the first hydrogen bomb was tested in 1954, were used as sites for nuclear experiments and the people displaced from these islands have not been able to return. In 1968 the Bikini atoll and 34 neighbouring islands were mistakenly declared suitable for human inhabitation, but tests carried out in 1977 revealed that water, fruit, and vegetables were still too radioactive for consumption.

[11] Several countries dump radioactive waste in the Pacific, turning the ocean into a lethal sewer for industries and nuclear power, stations. This situation has drawn strong protests from the local population, who have already had to suffer the consequences of nuclear waste, and displacement from atolls used for US navy and air force target practice.

[12] In October 1982, the US signed a Free Association agreement with the Marshall Islands and the Federated States of Micronesia. This means that all the islands, including the Northern Marianas, would manage their own internal affairs, while the US would be responsible for defence and security. In 1983 the agreement was approved in plebiscites, and in October 1986 President Reagan issued a proclamation, formally putting an end to the US administration of Micronesia. However, the US still control Micronesian foreign affairs.

[13] On September 17 1990, the 46th UN General Assembly approved the admission of Micronesia and other six countries, increasing the number of member states from 159 to 166.

[14] Toward the end of the year, Hurricane Owen hit the islands, leaving 4,500 people homeless and destroying 90 per cent of subsistence crops.

[15] In the March 1991 legislative elections, President John Haglelgam was defeated, and in May, Bailey Olter was elected president.

[16] The government decreed a state of emergency in April, 1992 to try to counteract the consequences of a severe drought. In June 1993 Micronesia joined the IMF which according to Olter was part of a scheme to solve the country's problems caused by the drought.

[17] The president called a national conference in 1994 to analyze the long-term possibilities of the Micronesian economy. One of the matters touched on was the country's dependence on the United States, whose aid equals about two thirds of the GDP. The participants also referred to problems dealing with fishing, tourism and the country's small industrial sector.

[18] In March 1995 Olter won the presidential elections. The president began his second term taking no heed of the pressures in favour of establishing a rotating presidency which would allow representatives from different regions of the country to become president.

[19] One month after taking office, Olter officially denounced Japan's plans to continue carrying plutonium on ships crossing the Pacific ocean.

Moldova

Moldova

Population: 4,350,000 (1994)
Area: 33,700 SQ.KM
Capital: Kishinev

Currency: Lei
Language: Moldovan

Moldovans are descended from the peoples of the southern part of Eastern Europe who had been subdued, and culturally influenced, by the Roman Empire. The Vlachs were mentioned in Byzantine chronicler, John Skilitsa's, writings (976 AD) as the forebears of the Moldovan people. In the mid-14th century, Vlach tribes from the northeast formed their own state, independent of the Hungarian Kingdom, in the territory of South Bukovina. The first *gospodar* (governor) of the Moldovan Principality was Bogdan (1356-1374), although according to legend, it was Dragos who founded the principality.

2 During the second half of the 14th century, the Moldovans freed themselves from Hungarian domination, and of the Tatar khans. By the early 15th century, Moldova's borders were the Dnestr River (to the the west), the Black Sea and the Danube (to the south) and the Carpathian mountains (to the west).

3 This small principality found itself subject to the influence and interest of the larger states: Hungary, Poland, the Grand Principality of Lithuania, Turkey and the Crimean Khanate. The principality was alternately a vassal of Hungary, Poland and the Ottoman Empire. The Christian Orthodox Church was the official church, and the country's language - known as Ecclesiastical Slav - was used for church liturgy, for official documents and education. The principality's first capitals were Baya, Stret and Suchava. The main economic activities were livestock herding and agriculture, especially wheat and vinyards.

4 The Moldovan Principality achieved its greatest political and economic success under the *gospodars* Alexander the Good (1400-

1432) and Stephen the Great (1457-1504). During this period, Moldova warred against Hungary, Poland and the Crimean Khanate, but its main threat came from the Turks. Turkish expansionism posed a constant danger, Moldova also had to pay tribute to the Porte (the Ottoman government in Turkey). In 1475, the Turkish army invaded Moldova, suffering a resounding defeat at the Battle of Vaslui. Nevertheless, the Moldovans were no match for the Turks numerically, and in 1484 the Turks stripped Moldova of its territories surrounding the fortresses at Kilia and Belgorod giving them the Turkish name Akkerman, and the *raya* - enclaves ruled over by the Turks - were created.

5 In the early 16th century, Moldova lost its independence as a state, being forced to recognize the power of the Turkish sultan (although still maintaining considerable autonomy within the Ottoman Empire). Turkish domination over Bukovina lasted until 1775, when the latter was incorporated into the Austrian Empire. Bessarabia was under Turkish control until 1812, and the rest of the Moldovan Principality until 1878. Turkey seized one Moldovan territory after another; by the mid-18th century, Moldova had lost half of its lands between the Prut and the Dnestr rivers.

6 Anti-Turkish sentiment increased because of Moldova's territorial losses, by the increase in the tributes paid to the sultans, and invasions by Turkish and Tatar troops, who devastated Moldovan cities and towns. Gospodars Petra Rares (1527-38, 1541-46), Ioann Voda Liuti (1572-74) and Dmitri Kantemir (1710-11) all turned against Turkey. In 1711, Dmitri Liuti and his Moldovan army joined forces with Russian czar, Peter the Great. In its struggle against the Ottomans, Moldova was forced to ally itself with the large powers which had opposed Turkey, signing pacts with Hungary, Austria and Poland. From the end of the 18th century, Moldova began developing closer relations with Russia.

7 All the wars between Russia and Turkey in the 18th and 19th centuries included Moldova in some way. The Prut expedition undertaken by Peter the Great in 1711 was a failure for Russia and its ally, Moldova, but it demonstrated

Moldova's intention of achieving independence by relying upon Russian aid. Distrusting the Moldovans, the Ottomans began putting Greek Phanariotes (from Phanar, a suburb of Istanbul), on the Moldovan throne. They continued to rule over Bessarabia until the beginning of the 19th century, and over the rest of the Moldovan Principality until 1821. There were bloody Russo-Turkish wars on Moldovan soil (1735-39; 1768-74; 1787-91), with a considerable number of Moldovan volunteers fighting against the Turks in Russian ranks.

8 Under the Peace of Jassy (1791), the Russians obtained the left bank of the Dnestr, south of the Yagolik River. During the second partition of Poland between Russia, Prussia and Austria in 1793, Russia obtained the other part of the left bank of the Dnestr. After the Russo-Turkish War of 1806-1812 and the Peace of Bucharest, Russia seized the territory between the Prut and the Dnestr rivers calling it all Bessarabia. The Muslim population was deported, thus putting an end to the Turkish invasions of Bessarabia. At first, this territory was an autonomous region within Russia, with Kishinev (Kisinov) as its capital. However, in 1873 it became a Russian province, subject to all the laws of the Russian Empire.

9 During the 19th century, the population of Bessarabia grew

from 250,000 to 2,500,000, and by the end of the century, Moldovans made up half of the province's population. There were also a significant number of Ukrainians and Russians, as well as Bulgarians, Germans, Jews and Gagauz (Muslims). In 1812, Kishinev had 7,000 inhabitants; by 1897, it had 109,000. During the Russian-Turkish wars of 1828-29, 1877-78 and the Crimean War (1853-56), Bessarabia acted a rearguard for the Russian army. Under the Treaty of Paris (1856), the part of Southern Bessarabia next to the Danube and the Black Sea was incorporated into the Moldovan Principality, which joined Walachia in 1859 to form the State of Romania. In 1878, the Treaty of Berlin returned this territory to Russia.

10 Moldovan schools began to be closed in the 1840s, and as of 1866 the Moldovan language was no longer taught. After the Russian Revolution of 1905, the teaching of Moldovan was once again authorized. On December 2, 1917, the People's Republic of Moldova was proclaimed. Romanian troops entered Bessarabia and ousted local Soviet authorities. Between December 1917 and January 1918, first the Soviets and then the Romanians gained control over Moldova. Toward the end of January, the independent Moldovan Republic was proclaimed.

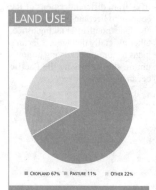

LAND USE

■ CROPLAND 67% ■ PASTURE 11% ■ OTHER 22%

Foreign Trade

MILLIONS $ 1994

IMPORTS
672
EXPORTS
618

Maternal Mortality

1989-95

PER 100.000
LIVE BIRTHS
34

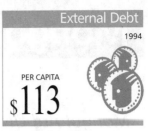

External Debt

1994

PER CAPITA
$113

PROFILE

ENVIRONMENT

Located south of Russia, Moldova is bounded in the west by Romania and the east by Ukraine. Moldova lies at the foot of the Carpathian Mountains, and is made up of plateaus of relatively low altitude. The region is drained by the Dnestr and the Prut rivers. The soil is black and very fertile. Average temperature in summer is 19-22°C, and in winter, -3 to -5°C. The country's main economic activities are livestock raising and agriculture -especially vinyards, sugar beets, fruits and vegetables. 40% of the underground waters are contaminated with bacteria and 45% of watercourses and lakes are contaminated with chemicals.

SOCIETY

Peoples: Moldovans, 63.9%; Russians, 14.2%; Ukrainians, 12.8%; Gagauz, 3.5%; Bulgarians, 2%. **Religions:** Christian Orthodox. **Languages:** Moldovan (official), Russian, Ukrainian, Gagauz. **Political Parties:** Agrarian Democratic Party; Popular Front; Congress for Intellectuals.

THE STATE

Official Name: Republica Moldova. **Capital:** Kishinev 753,000 inhab. (1991). **Other cities:** Tiraspol 186,000 inhab.; Balti 165,000 inhab.; Tighina 141,000 inhab. **Government:** President, Mircea Snegur, since December 1991. Prime Minister, Andrei Sangheli. **National Holiday:** August 31 ,"Mother Tongue" Day. **Armed Forces:** 11,800 **Paramilitaries:** Internal Troops (Ministry of Interior): 2,500; OPON (riot police) ((Ministry of Interior): 900.

STATISTICS

DEMOGRAPHY

Urban: 50% (1995). **Annual growth:** 0.5% (1991-99). **Estimate for year 2000:** 4,000,000. **Children per woman:** 2.3 (1992).

COMMUNICATIONS

12.0 **telephones** per 100 inhabitants (1993).

ECONOMY

Consumer price index: 100 in 1990; 11,176 in 1994. **Currency:** 4 lei = 1$ (1994). **Cereal imports:** 200,000 metric tons (1993). **Fertilizer use:** 612 kgs per ha. (1992-93).

[11] In the 1920s and 1930s, the territory of modern Moldova was divided into two unequal parts. Bessarabia was part of the Romanian Kingdom, while the left bank of the Dnestr belonged to the USSR. On October 12 1924, the Autonomous Republic of Moldova was formed, in Ukraine. Its first capital was Balta, and after 1929, Tiraspol, Moldova made certain gains in industrial and cultural development, while Bessarabia, as part of Romania, remained at a standstill.

[12] On June 28 1940 - with World War II already underway - the Soviet government issued an ultimatum whereby Romania was forced to accept Soviet annexation of Bessarabia. On August 2, the Federated Republic of Moldova was founded as a part of the USSR, uniting the central part of Bessarabia and the Autonomous Republic of Moldova. The northern and southern parts of Bessarabia, and the eastern region of the Autonomous Republic of Moldova, remained within Ukraine. In June 1941, Nazi troops invaded the USSR. Romania made an alliance with Hitler and re- covered all of Bessarabia, as far as the Dnestr and Odessa. Three years later, the Red Army took on a weakened Germany, recovering Bessarabia and northern Bucovina.

[13] Leonid Brezhnev's political career began in Moldova, where he was leader of the local Communist Party. He was later to become Secretary General of the Communist Party of the Soviet Union (CPSU) and President of the USSR, positions which he held simultaneously until 1983.

[14] After the liberalization process initiated by Soviet president, Mikhail Gorbachev, in 1985, political and ethnic problems began to emerge in Moldova. In 1988, the Democratic Movement in Support of Perestroika (restructuring) began demanding the return to the use of the Latin alphabet instead of the Cyrillic alphabet for writing the Moldovan language. Moldovan nationalists called for an end to the political and economic privileges which Russian residents enjoyed, stating that if they could not do without them, they should return to their native land. In July 1989, a violent confrontation was barely avoided between Moldovan nationalists and Russians, who make up 14.2 per cent of the population. In August, some 300,000 Moldovans carrying Romanian flags, demonstrated in Kishinev in favor of Moldovan independence.

[15] On November 10 1989, Parliament approved the Official Language Law, which established Moldovan as the country's official language for political, economic, social and cultural affairs, with Russian to be used only in the press and other means of communication. There was an outbreak of nationalist protests, and a strike by some 80,000 Russian workers. In the meantime, separatist activity increased in Dnestr, where a large proportion of the population is Russian, and the Gagauz region.

[16] On August 27 1991, Moldova declared its independence from the USSR. A month later, Dnepr and Gagauz - opposing both Moldova's independence and the possibility of union with Romania declared themselves independent republics.

[17] After the failed coup against Gorbachev in Moscow in August 1991, the Moldovan government arrested the leaders of the local separatist movements. In December, the country's first presidential elections were held, with Mircea Snegur winning the election. A few days later, 13 people were killed in violent clashes between Moldovan soldiers and members of the Russian minority, in Dnestr. On March 2 1992, Moldova was admitted to the UN as a new member. Given the continued fighting in Dnestr, President Snegur declared a state of emergency on March 16, and ordered the elimination of the opposition forces fighting in the region.

[18] In June 1992, fighting broke out between the pro-Russian Pridnestrovie government in the Dniester River Basin and the Moldovan army. Pridnestrovie managed to maintain its independence.

[19] In Moldova, the government's refusal to use force to resolve the conflict marked an important change in policy. The political scene was divided in two camps, one pro-unification with Romania, and the other pro-sovereignty and independence.

[20] The pro-independence parties - which were opposed to unification with Romania - won by a large majority in the parliamentary elections, confirming Prime Minister Andrei Sangheli in his post. In August, the new constitution came into operation, defining Moldova as an "independent" and "democratic" state, and in October, an agreement was signed with Moscow to begin the withdrawal of Russia's troops from this former soviet republic.

[21] In 1995, Kishinev accelerated its privatization plans and facilitated the participation of foreign capital in the process by authorizing investors to acquire up to 60 per cent of the shares in any company. In early 1996, the situation in the Dnestr valley remained confused, after a referendum in late 1995 showed a wide majority were in favor of independence for this Russian-speaking region.

Mongolia

Population: 2,363,000 (1994)
Area: 1,565,000 SQ.KM
Capital: Ulaanbaatar (Ulan Bator)

Currency: Tughrik
Language: Khalkha Mongolian

Mongol Ard Uls

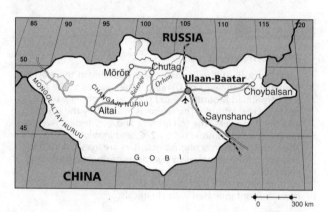

1 The Mongols constitute one of the principal ethnic groups of northern and eastern Asia, linked by cultural ties and a common language. Dialects vary from one part of the region to another, but few cannot be understood by a Mongolian.

2 Direct lineage from a male ancestor gives the family or clan its name, though there was an earlier tradition of female lineages. Intermarriage between members of the same clan was forbidden; so there was great need for establishing alliances between clans, who formed tribes.

3 The Mongols were mostly nomadic, with the movement of livestock and campsites determined by pasturage needs throughout the year. Animals were individual property, while the grazing lands were collective property.

4 The most powerful clans tended to control the tribe's activities. The weakest families maintained their own authority and their ownership of the animals, but they were obliged to pay tribute to the dominant clan. They moved, camped, grazed their livestock and went to war under that clan's orders.

5 Political and military organization was adapted to the needs of each clan or tribe. A man capable of handling a weapon could be a chief or a soldier, according to the needs of the moment. Capturing livestock, women or prisoners from other tribes was a common means of acquiring wealth.

6 When a tribe became very powerful, as with that of Genghis Khan in the 13th century, they organized themselves in groups of 10, 100, 1,000 or 10,000 soldiers. The leaders of large units were assigned a territory where they could collect tribute and recruit warriors for the supreme leader.

7 The Siung-nu, or Huns, were the earliest inhabitants of the Selenga valleys, joining Siberia to the heart of Asia, and they are thought to have been settled in this region by 400 BC.

8 The Huns created a great tribal empire in Mongolia when China was undergoing unification as an imperial State under the Ch'in and Han dynasties (221 BC-220 AD). The Hunnish empire warred against China for centuries, until it disintegrated - perhaps due to internal conflicts - around the 4th century.

9 Some of the southern tribes surrendered to China and settled in Chinese territory, where they were eventually absorbed by the Chinese, while others migrated westward. In the 5th century, Attila's Huns conquered almost all of Europe, reaching Gaul and the Italian peninsula.

10 The Huns were subsequently displaced by the Turks, who established themselves throughout the region. The major Hun chieftains set up general headquarters, surrounded by cultivated lands; where they bred larger, stronger horses, capable of carrying a warrior and his armor.

11 This led to a differentiation between aristocrats and traditional tribal archers, who rode smaller horses. Agriculture also became more important to the economy. The Uigurs, who came to occupy the Orhon valley after the Turks, developed a settlement around an oasis, where agriculture was possible.

12 The term "Mongol" first appeared in records of different tribes which were written during the T'ang Chinese dynasty. It then disappeared until the 11th century, when the Kidan became the rulers of Manchuria and northern China,

controlling almost all of present day Mongolia.

13 The Kidan established the Liao dynasty in China (907-1125) and ruled Mongolia, fostering division between the different tribes. Historical records mention the existence of a single Mongol nation, although it did not include all Mongolian-speaking peoples.

14 The Kidan were succeeded by the Juchen, who were in turn succeeded by the Tatars, before the era of Genghis Khan. Born in 1162 Temujin was the grandson of Qabul (Kublai Khan), who had been the Mongols' greatest leader to date. Temujin inherited several fiefdoms that had been seized from his family.

15 In 1206, because of his political and military prowess, Temujin was recognized as leader of all the Mongols, and given the title of Genghis Khan. His armies invaded northern China, reaching Beijing. By 1215, the Mongolian Empire extended as far as Tibet and Turkistan.

16 Upon Genghis Khan's death in 1227 disputes among his successors, caused the Mongolian Empire to disintegrate, until the Chinese throne was left in the hands of the Ming dynasty in 1368. China invaded Mongolia and destroyed Karakorum, the former imperial capital by fire, though it was unable to bring the territory under control.

17 In the 15th and 16th centuries, controlling the areas beyond the Great Wall of China demanded military mobilization. In addition, cities were needed, to act as centers of trade and serve as a market place for food produced by local peasants.

18 From far-away western Mongolia, the Oyrat began gaining control of the territory. They added their own mercantile and adminis-

trative expertise to the Mongols' tribal organization.

19 The separation of the Oyrat from the Jaljas began during this period, with the latter forming the core of what was later to become Outer Mongolia. A tribal league was formed between the Khalkhas in the north and the Chahars in the south, while the leadership passed over to the Ordos, during the reign of Altan Khan (1543-83).

20 To keep their hold on power, the Mongolian princes thought it useful to be backed up by a religious ideology. They adopted the Tibetan Buddhist religion as Tibet posed no cultural threat, and the Tibetan script was easy to use.

21 Altan Khan proceeded to invite a Tibetan prelate, whom the Mongols called "Dalai Lama" to lead the state religion. The merging of religious interests with those of the State was accomplished by claiming that an heir to the Khalkhas clan was the first "reincarnation" of the Living Buddha of Urga.

22 In 1644, after consolidating their power in Manchuria, the Manchus seized the Chinese throne, with the help of Mongolian tribes from the far east. Before occupying Beijing, the Manchus took control of southern Mongolia, which was henceforth known as Inner Mongolia.

23 It took China almost a century to conquer Outer Mongolia. Meanwhile, Inner Mongolia became a part of China, and the Khalkhas' desire to retain power in the south prevented the Oyrats from attaining reunification.

24 This was the final stage of the great wars among the Mongols; ending in their overall dispersal. Several groups of Khalkhas remained in the south; some Chahars settled in Sinkiang and the Oyrat

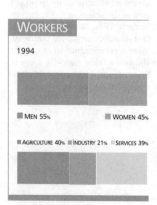

WORKERS

1994

MEN 55% WOMEN 45%

AGRICULTURE 40% INDUSTRY 21% SERVICES 39%

LAND USE

CROPLAND 1% PASTURE 80% OTHER 19%

Foreign Trade

MILLIONS $ 1994

IMPORTS
223
EXPORTS
324

Maternal Mortality

1989-95

PER 100.000
LIVE BIRTHS
240

dispersed in different directions, including czarist Russia.

[25] In the Russo-Japanese War of 1904-05, both armies used Mongolian troops and staff. This served Japanese interests well as a resurgence of Mongolian nationalism could weaken both Russia and China. At the end of the war, Russia secretly recognized Inner Mongolia as belonging to Japan's sphere of influence.

[26] With the outbreak of the Chinese Revolution in 1911, there was a pervading malaise in Mongolia. Until then, the region had been nothing more than an object of disputes between Russia and Japan. However, the Mongolians' social and political discontent was directed against the Manchus and the local government.

[27] Led by their Buddhist leader, Mongolia proclaimed independence from China and sought Russian support. However, because of its secret treaties with Japan and England, it could offer nothing more than mere "autonomy". After lengthy negotiations, this status was granted to Outer Mongolia.

[28] This situation continued until the Russian Revolution in 1917. China sent in troops and made the Mongolians sign a request for aid from Beijing. But the region was invaded by retreating czarist troops, who expelled the Chinese and mistreated the Mongolians.

[29] With the traditional leaders discredited because of their poor handling of the Chinese and White Russian interventions, some groups of Mongolian revolutionaries sought help from the Bolsheviks. Russian and Mongolian troops took the capital, Urga, in July 1921.

[30] This was the beginning of the republic, although the first monarchy had the Living Buddha as puppet king, only authorized to endorse the new regime's proposals. Upon his death in 1924, the People's Republic of Mongolia was proclaimed.

[31] The People's Revolutionary Party (PRP), made up of conservatives and revolutionary nationalists, wavered between Beijing and Moscow until the defeat of the Chinese Revolution, at the hands of Chiang Kai-shek. At this time, Mongolia began to fall increasingly under the influence of the USSR, and Joseph Stalin.

[32] The new republic proclaimed the right for women to vote. The efforts to impose socialism upon Mongolia, which was still feudal in many ways, were marked by even greater excesses than were committed in other republics.

[33] Following the Soviet model, the PRP government tried to collectivize the economy in order to break the power of the feudal lords and the Buddhist priests. Between 1936 and 1938, the Mongolian regime purged the party and the army, executing many leaders.

[34] In the 1930s, the PRP destroyed 750 monasteries and killed over a thousand monks. Mongolia's demographic level had been unchanged for hundreds of years, as a large part of the male population became Bhuddist monks.

[35] In 1939, Japan invaded northeastern Mongolia, along the Siberian border. With Soviet help Japan was defeated; a blow to the Axis powers in Berlin and Tokyo. Mongolia and the USSR fought together in the Inner Mongolian and Manchurian campaign, two weeks before the end of World War II.

[36] As part of the Yalta accords, Chiang Kai-shek agreed to hold a plebiscite in Mongolia. Although the result overwhelmingly favored independence, Mongolia failed to receive diplomatic recognition because of an unresolved territorial dispute. In 1961, Mongolia was admitted to the UN.

[37] In 1960, government officials in Ulan Bator accused the Chinese government of mistreating Mongolian citizens and of seeking territorial expansion, at Mongolia's expense. In the early 1970s, there were a number of incidents along the border between the two countries, and 2,000 Chinese immigrants were expelled from Mongolia

[38] Friction continued until 1986, when the Chinese Deputy Minister of the Council of Ministers visited Mongolia and re-established consular and commercial relations. In 1987, the Soviet government announced the withdrawal of part of its military forces, as a good-will gesture aimed at stabilizing the region.

[39] In March 1988, China and Mongolia signed the first treaty ever, aimed at defining the 4,655 km border between the two countries. A year later, during Mongolian premier Tserenpylium Gombasuren's visit, the first in 40 years, relations between the two countries were returned to normal.

[40] In 1989, within the framework

PROFILE

ENVIRONMENT

Comprises the northern section of Mongolia, also known as "Outer Mongolia" (the southern part, "Inner Mongolia", is part of the Chinese territory under the name of Inner Mongolian Autonomous Region). At the center of the country, lies the wide Gobi desert which is bordered north and south with steppes where there is extensive nomadic sheep, horses and camel raising. The Altai mountain region, in west Mongolia, is rich in mineral resources: copper, tin, phosphates, coal and oil. Desertification is aggravated by an arid climate and fragile soils. Water is a limited resource, especially in areas bordering the Gobi Desert.

SOCIETY

Peoples: Khalkha Mongol 78.8%; Kazakh 5.9%; Dörbed Mongol 2.7%; Bayad 1.9%; Buryat Mongol 1.7%; Dariganga Mongol 1.4%; other 7.6%.
Religions: Buddhism. **Languages:** Khalkha Mongolian
Political Parties: Mongolian People's Revolutionary Party (MPRP); Mongolian National Democratic Party (MNDP); Mongolian Social Democratic Party (SDP); Mongolian Democratic Renewal Party. **Social Organizations:** Central Council of Mongolian Unions; "Blue Mongolia"; Mongolian Confederation of Free Unions.

THE STATE

Official Name: Mongol Uls. **Administrative divisions:** 18 provinces and 1 municipality (Ulan Bator). **Capital:** Ulaanbaatar (Ulan Bator), 680,000 inhab. (1994). **Other cities:** Darhan 85,800 inhab.; Erdenet 63,000 inhab.; Choybalsan 46,000 inhab.; Ölgiy 29,400 inhab. (1994) **Government:** Parliamentary republic. Punsalmaagiyn Otchirbat, President ; Mendsayhany Enkhsaikhan, Prime Minister. Legislature: single-chamber Assembly with 76 members elected every 4 years. **National holiday:** July 11. **Armed Forces:** 21,100 (1995) **Paramilitaries:** 10,000 (internal security and border guards).

of Soviet perestroika (restructuring), Moscow announced that three-quarters of its troops would be withdrawn in 1990. Shortly afterwards, both governments agreed to the complete withdrawal of all Soviet military personnel and equipment from Mongolian territory, by the end of 1992.

[41] At the same time, the PRP leadership admitted that social and economic reforms were not having satisfactory results. The official party adopted changes in its internal elections, making them more democratic, and rehabilitated some figures who had been purged during the 1930s.

[42] In 1989 and 1990, several opposition groups emerged. One of the most active, the Democratic

Union of Mongolia, was officially recognized in January 1990. In March of that year, increasingly frequent public demonstrations against the government triggered a new crisis within the PRP.

[43] The National Assembly approved a constitutional amendment withdrawing the reference to the PRP as society's "prime moving force" and approving new electoral legislation; however, no changes were made with relation to political party activity.

[44] Toward the end of March, demonstrators in the streets of Ulan Bator clamored for the dissolution of the National Assembly. Leaders of the opposition stated that the changes which had been proposed were insufficient, and called for

External Debt
1994

PER CAPITA
$187

legislation allowing all parties to present candidates.

[45] Rich in oil, minerals, livestock, lumber and wool, what Mongolia currently lacks is the money, machinery and specialized manpower to exploit these resources, particularly after the loss of Soviet subsidies and withdrawal of 50,000 Soviet technicians and government advisers.

[46] The legendary figure of Genghis Khan, whose name was forbidden for many years, has begun to be rehabilitated as an authentic expression of Mongolian pride and tradition, sentiments which until recently were condemned as being an expression of a narrow-minded "nationalism".

[47] In spite of 65 years of Soviet aid, Mongolia has an economy that maintains vestiges of nomadism. Urbanization is just beginning, and half of Ulan Bator's population lives in tents, with rudimentary electric and water supplies.

[48] In the first trimester of 1991, Mongolia registered a substantial reduction in its foreign trade. Its balance of payments deficit reached $250 million. Shortages of food, medicine and fuel became more acute. The currency suffered devaluation, from 7.1 to 40 tughrik per US dollar, and government income declined sharply, while expenditures steadily increased.

[49] Prime Minister Dashiun Byambasuren announced a new economic policy which included incentives to attract foreign investment, the establishment of a national stock exchange, the sale of two-thirds of the state's capital goods, deregulation of prices and changes in the banking system. The government warned that the reform is very complex and will take time to implement.

[50] The life of the Mongolians, whose median income is $100 per year, has become even harder than before. The population is not only anxious to see the results of the changes announced by the government, but is also worried about the scandals which have affected this administration.

[51] After an official investigation, the president of the Central Bank of Mongolia, Zhargalsaikhan, together with a group of new investors, were arrested in December 1991, for a $82 million fraud, money which they had invested in business deals that went wrong. As a result, the country lost the major part of its reserves. At the same time, the deputy prime minister, Cabaadorjiyn Ganbold - the highest ranking reform figure in a government dominated by former communists - was accused of secretly authorizing the transfer of 4,400 kg of gold to a branch of Goldman Sachs - a British merchant bank. The gold was collateral for a $46 million loan, apparently earmarked for covering losses. The reform sector said that the minister of finance would try to cover the losses, and accused the former communists of using such denunciations to try to discredit the transition toward a market economy.

[52] In January 1992, the National Assembly approved a proposed constitutional reform presented to the government, adopting the official name "Republic of Mongolia", and dropping the term "People's". The reform also established a pluralistic democratic system, replacing the socialist system in effect until the present.

[53] The governing People's Revolutionary Party gained more than 70 of the 76 Mongolian parliamentary seats in the June 1992 letgislative elections. The opposition democratic coalition only obtained three or four seats according to the Electoral Commission which supervised the count. Over 90 per cent of the electorate voted.

[54] In October 1992, after being defeated in the June elections, the Mongolian National Democratic Party (MNDP) became an opposition party. The Social Democrats preferred to remain independent.

[55] In November 1992, the scheme implemented for privatizing 80 percent of State enterprises involved distributing bonds for the purchase of shares to all of the country's inhabitants. However, it seems that the scheme was not explained fully enough to the largely nomadic population. They ap-peared not to understand it, and most of the bonds were sold on the black market.

[56] In 1992, the withdrawal of Russian troops - begun in 1987 - was completed. Meanwhile, Otchirbat had a rapprochement with the NMDP and the SDP to prepare the presidential elections in June 1993. Thanks to the two groups, which had been previous members of the opposition, the president was re-elected with almost 58 per cent of the vote and announced the "westernization" of economy.

[57] However, disagreements between Otchirbat and the former communist majority in parliament were frequent. Poverty and unemployment kept growing and, according to official estimations, 26.5 per cent of the population did not earn the minimum income to subsist.

[58] Meanwhile, for the first time since the collapse of the Communist regime, Mongolia yielded a 2.5 per cent economic growth in 1994. Differences between the parliament majority and the opposition NMDP and SDP groups diminished, which enabled a deal in 1995 to modify the election system. Thus, it was decided that 24 of the 76 members of parliament would be elected by proportional vote and the majority system would continue for the remaining seats.

[59] In the economic field, several international organizations criticized Mongolia's supposed sluggishness in liberalizing its economy and promote the private sector's development.

[60] Parliamentary elections in June 1996 resulted in an opposition victory with 50 of the 76 seats, while the former communists went down from 70 to 25 representatives. After the new MPs assumed office, parliament appointed Mendsayhany Enkhsaikhan as prime minister.

Montserrat

Population: 11,000 (1994)
Area: 100 SQ.KM
Capital: Plymouth
Currency: E.C. Dollar
Language: English

Montserrat

Montserrat was first inhabited by the Caribs and after being sighted for the first time by Columbus in 1493, it shared the history of the other Lesser Antilles.

[2] The island was colonized by Irish people, driven out from neighboring St Kitts. Owing to its barren nature and unsuitable coastline, it was never really coveted by Europeans, and apart from sugar cane and cotton, no crops were grown in the Island. In the mid-19th century it had a population of about 10,000 people, including 9,000 slaves.

[3] Montserrat was a member of the Federation of the Leeward Islands and, later, of the West Indies Federation, until the latter was dissolved in 1962. Since 1960 Montserrat's colonial statute has ensured a rather autonomous political organization. The Progressive Democratic Party (PDP) founded by William Bramble and later headed by his son, Austin Bramble, refused to change this situation when the group of Associated States of the Antilles was formed in 1967 as a preliminary step towards total independence.

[4] Despite its isolation the island has felt the influence of the decolonization process. Bramble's party lost all its parliamentary seats in November 1973, driven out by the new People's Liberation Movement (PLM). In 1979, Montserrat's prime minister John Osborne announced that the island would attain full independence in 1982, a promise that failed to materialize.

[5] Barely a quarter of the country's surface area is agricultural land, and the population density on those lands is very high. Half of the island is suitable for cattle raising, which has been a major source of export for some time. Montserrat has one of the highest swimming-pool counts per capita in the world: 2 per inhabitant. Tourist activity has developed since 1979, largely because of US immigration. Another source of income is international radio stations; Radio Antilles, and the Voice of Germany; both broadcast from the island.

[6] In August, 1987, 11 months late, Osborne called elections. The PLM won four of the seven elective seats, the National Development Party (NDP) two and the PDP one.

[7] The influx of tourists and pay sent home from Montserrat workers abroad are insufficient to cover the enormous balance of payments deficit. At the moment, Britain makes up the balance, and this is the main sustaining argument of those who stand against independence.

[8] Relations between the local government and the United Kingdom are not good. By early 1989, Montserrat was the accommodation address for 347 offshore banks,

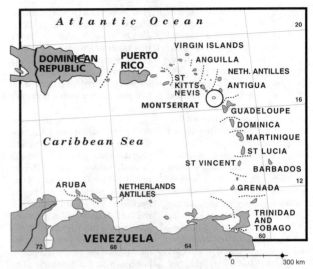

though the government only recieves about 5 per cent of the sale of licences. The lack of control over banking has given rise to accusations that the island has been used in fraud and in laundering drug money. Since 1990, the island's new governor, David Pendleton, has encouraged stiffer regulations for offshore companies.

[9] In 1989, hurricane Hugo ravaged Montserrat. The British granted considerable assistance; £17.5 million to rebuild the island. This fresh money refuelled the national economy: three new factories were opened in 1991, and tourism is on the increase.

[10] In 1991, new legislation made it possible to re-establish offshore banking. That same year, the government abolished the death penalty.

[11] Systematic investigations into banking operations carried out on the island led Britain to close 90 per cent of the offshore banks in 1992. These were found to be involved in illegal transactions, and the laundering of drug money.

[12] In February 1993, Frank Savage was appointed governor.

ENVIRONMENT

Montserrat is one of the Leeward islands, in the Lesser Antilles, located 400 km east of Puerto Rico, northwest of Guadeloupe. The terrain is volcanic in origin and quite mountainous, with altitudes of over 1,000 meters. The tropical, rainy climate is tempered by sea winds. The soil of the plains is relatively fertile and suitable for agriculture.

SOCIETY

Peoples: The population is predominantly the result of integration of African and European immigrants.
Religions: Mostly Christian; Anglicans, Catholics and Methodists predominate. **Languages:** English (official). Most people speak a local dialect. **Political Parties:** The People's Liberation Movement (PLM), of John Osborne, founded in 1975, is the majority party in parliament; the Progressive Democratic Party (PDP), led by the Bramble family, is conservative; the National Development Party (NDP) is the second largest force in the legislative.

THE STATE

Official Name: Colony of Montserrat. **Capital:** Plymouth, 1,478 inhab. (1980). **Government:** Frank Savage, Governor appointed by the British Queen, on February 1993. Reuben Meade, Chief Minister. The government has limited autonomy and consists of an Executive Council and a Legislative Council with seven elected members.

DEMOGRAPHY

Density: 119 inhab./sq.km. **Annual growth:** 1.8% (1970-86).

HEALTH SERVICES

Under-five mortality: 14 per 1,000 (1991). Communications 148 **TV sets** and 577 **radio receivers** per 1,000 inhab. (1991).

ECONOMY

Per capita GNP: $3,330 (1990). **Currency:** E.C. dollars. **Imports:** $10 million (1991). **Exports:** $1 million (1991).

ENERGY CONSUMPTION

1,450 kgs of Coal Equivalent per capita yearly, 106% imported. (1990).

COMMUNICATIONS

39.4 **telephones** per 100 inhabitants (1993).

Morocco

Magreb

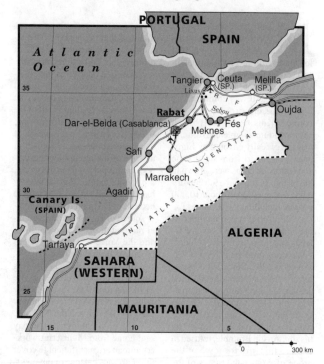

Population: 26,367,000 (1994)
Area: 446,550 SQ.KM
Capital: Rabat

Currency: Dirham
Language: Arabic

Morocco was the cradle of the two North African empires that dominated the Iberian peninsula (see "Almoravids and Almohads"). It became one of the power centers of the region because of its geographic location; it was close to Spain, and at the northern end of the trans-Saharan trade routes. Although neither Fez nor Marrakesh achieved the academic prestige of Cairo, their political influence was felt as far as Timbuktu and Valencia. Their close ties with Spain were culturally enriching during the Cordovan caliphate, but they brought negative consequences to bear on Morocco in the final stages of the "Reconquest". The war moved into Africa and the Spanish seized strongholds on the coast (Ceuta in 1415, and Tangiers in 1471). European naval dominance blocked Mediterranean and Atlantic routes to Morocco causing a decline in trade.

[2] Unlike Algeria and Tunisia, Morocco was not formally annexed to the Ottoman empire, but it did benefit from the presence of Turkish corsairs who hampered Luso-Spanish expansion. This precarious balance allowed the sultans to remain autonomous until the 20th century. France's policy of economic penetration meant that France was supervising Moroccan finances, aiming to guarantee repayment of the Morroccan foreign debt, while France argued with Germany over who should have political sway over the area. The French finally won, securing agreements with Spain over the borders of the Spanish Sahara, and sultan Muley Hafid ended his support of the Saharan rebels (see Sahara). In 1912, an agreement between France, Spain and Britain transformed Morocco into a French protectorate, giving Spain the Rif region, to the north (where Ceuta and Melilla are located), and the Ifni region to the south, near the Sahara. In exchange, Britain obtained French consent for its policies in Egypt and Sudan. The city of Tangier was declared an international free port and the sultan became a figurehead.

[3] The areas under Spanish control became sanctuaries for the nationalists who disagreed with European domination. In 1921, it was on Spanish territory that the Berber revolt led by Emir Abdel Krim (Abd al-Karim al-Khattab) began. Backed by the Third International and the Pan-Islamic Movement, he proclaimed the Republic of the Confederated Tribes of the Rif, induced the inland tribes to rebel, and put Spain on the defensive. The French intervened causing the rebellion to spread throughout the entire country. It took them until 1926 to force the Emir to surrender.

[4] In the south, Spanish rule was really only nominal, and French pressures to close down this "sanctuary" for Algerian, Moroccan, Saharan and Mauritanian rebels failed (see Sahara).

[5] During World War II, there was sustained nationalist agitation. Demands for liberation were so pressing that Sultan Muhammad V himself became the spokesperson for the cause. To calm the people, the French thought they could rely on the prestige of the elderly emir Abdel Krim who had been deported to Reunion. Unfortunately for them, the veteran fighter took advantage of a stopover in Egypt to escape from the ship seeking refuge in Cairo, where he died in 1963. Growing tension led the French to depose Muhammad V in 1953, only succeeding in making the nationalist movement more radical. They raised an army and fought until they achieved his return to power. In 1956, the French were forced to acknowledge the total independence of Morocco.

[6] On April 7 1956, Morocco recovered Tangier, as well as the "special zones" of Ceuta and Melilla, although the ports of these two cities still remain under Spanish control. Ifni was not returned to Morocco until 1969.

[7] The goal set by Muhammad V was "to move forward slowly" gradually modernizing the country's economic and political structures. But his son, Hassan II, who succeeded him in 1961, had more conservative ideas. His family came from the lineage of Muhammad, the prophet, and his theocratic regime, based on a paternalistic system of favors and duties prevented the development of authentic national businessmen. The king also encouraged foreign investment to exploit the nation's natural resources, especially from France.

[8] In 1965, Ben Barka, leader of the powerful National Union of Popular Forces (NUPF) was assassinated on the orders of Hassan II. The NUPF worked for the social and economic welfare of workers and peasants.

[9] The death of Ben Barka in Paris was followed by a crackdown on popular organizations. The NUPF split and Ben Barka's followers were banned. Meanwhile, another NUPF faction, headed by Abderrahim Bouabid, changed its name to the Socialist Union and traded Ben Barka's principles for a minority place in parliament. The Istiqlal, who were originally anti-colonialist, became supporters of right-wing nationalist expansionism, fully backing Hassan's dream of restoring "the Great Morocco" through the annexation of the western Sahara and, if possible, Mauritania. (See Sahara).

[10] In 1975, the conflicts underlying Moroccan society surfaced when King Hassan ordered the occupation of Sahara, unleashing a war that has brought about important political changes in north Africa.

[11] Funds for the military campaign, the fall in the price of phos-

PUBLIC EXPENDITURES

% of GNP — 1992

- MILITARY: 4
- EDUCATION: 5.5
- HEALTH: 0.9

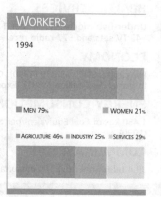

WORKERS

1994

- MEN 79% — WOMEN 21%
- AGRICULTURE 46% — INDUSTRY 25% — SERVICES 29%

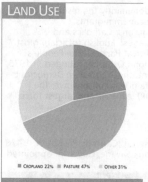

LAND USE

- CROPLAND 22% — PASTURE 47% — OTHER 31%

ENVIRONMENT

The country has a long, 800-km coastline influenced by the ocean. In the eastern part there are two mountain ranges (Atlas and Rif), covered with a barren steppe and inhabited by nomadic Berbers. In the foothills lie irrigated plains where citrus fruit, vegetables and grain are cultivated. On the western slopes of the Atlas Mountains (Grand Atlas and Anti-Atlas), cattle are raised and there are phosphate, zinc and lead mines. Along the coastal plains, grapes and citrus fruit are grown. Fishing is important, though mainly exploited by foreign fleets. In the cities, many run-down neighborhoods have been occupied by the homeless, turning them into shanty towns. Some 700,000 people live as squatters in 12 such shanty towns in historic zones. This type of housing has increased by 10 percent per annum between 1971 and 1982. The lack of any kind of planned development in the countryside-which has a high agricultural potential-has led to depletion of the soil.

SOCIETY

Peoples: Arab 70%; Berber 30%; other, less than 1%. **Religions:** Mainly Muslim. There are Christian and Jewish minorities. (Many members of the Jewish community have emigrated since 1975). **Languages:** Arabic (official), is spoken by most of the population. French, Spanish and various Berber dialects are also spoken. **Political Parties:** The Independence Party (Istiqlal) founded in 1943; the Constitutional Union (UC) founded in 1983 by former prime minister Maait Buabid. The Socialist Union of Popular Forces (USFP), progressive. The Independence National Union (RNI), moderate. The Popular Movement, Berberist. The National Union of Popular Forces (UNFP), founded by Mehdi Ben Barka. The Party of Progress and Socialism, communist. **Social Organizations:** Moroccan Workers' General Union (UGTM), founded in 1960; Moroccan Labor Union, founded in 1955; Democratic Labor Confederation, founded in 1979.

THE STATE

Official Name: al-Mamlakah al-Maghribiyah **Administrative divisions:** 30 provinces and two municipalities, Casablanca and Rabat. **Capital:** Rabat, 1,220,000 inhab. (1993). **Other cities:** Casablanca (Dar-el-Beida), 2,940,000 inhab.; Marrakesh, 1,517,000 inhab.; Fez, 565,000 inhab. **Government:** Hassan II, king since 1961. The September 1992 electoral reform granted greater power to Parliament. **National Holiday:** March 3, Independence Day (1956). **Armed Forces:** 195,500 (1995) **Paramilitaries:** 40,000: Gendarmerie Royale: 10,000; Force Auxiliaire: 30,000.

DEMOGRAPHY

Urban: 47% (1995). **Annual growth:** 1% (1992-2000). **Estimate for year 2000:** 30,000,000. **Children per woman:** 3.8 (1992).

HEALTH

Under-five mortality: 56 per 1,000 (1995). **Calorie consumption:** 105% of required intake (1995). **Safe water:** 55% of the population has access (1990-95).

EDUCATION

Illiteracy: 56% (1995). **School enrolment:** Primary (1993): 60% fem., 60% male. Secondary (1993): 29% fem., 40% male. University: 10% (1993). **Primary school teachers:** one for every 28 students (1992).

COMMUNICATIONS

86 **newspapers** (1995), 99 **TV sets** (1995) and 100 **radio sets** per 1,000 households (1995). 3.1 **telephones** per 100 inhabitants (1993).

ECONOMY

Per capita GNP: $1,140 (1994). **Annual growth:** 1.20% (1985-94). **Annual inflation:** 5.00% (1984-94). **Consumer price index:** 100 in 1990; 126.3 in 1994. **Currency:** 9 dirhams = 1$ (1994). **Cereal imports:** 3,653,000 metric tons (1993). **Fertilizer use:** 326 kgs per ha. (1992-93). **Imports:** $7,188 million (1994). **Exports:** $4,013 million (1994). **External debt:** $22,512 million (1994), $854 per capita (1994). **Debt service:** 33.3% of exports (1994). **Development aid received:** $751 million (1993); $29 per capita; 3% of GNP.

ENERGY

Consumption: 307 kgs of Oil Equivalent per capita yearly (1994), 95% imported (1994).

phates on the international market, and the loss of financial aid from Saudi Arabia - in retaliation for Hassan's support of the Camp David agreements between Israel and Egypt - deepened the economic crisis. The political consequences of the crisis soon appeared; in 1979, students and workers held massive street demonstrations. The with-

drawal of Mauritania from the Sahara war, in July 1980, constituted a serious blow to the Moroccan government, which now bore sole responsibility for carrying on the fight.
[12] Severe drought in 1980 and 1981 drastically reduced food supplies, forcing the government to increase food imports. This sent the

country's foreign debt soaring to intolerable levels. The IMF assisted the monarchy with emergency loans which carried a condition eliminating subsidies on food and housing, a measure which increased the hardships faced by the working classes. The government failed to achieve their ambitious development aims, and the export of unemployed workers was limited by french immigration restrictions.
[13] The situation worsened when several moderate opposition parties decided to break the tacit political truce. The Socialist Union of Popular Forces (USFP) staged anti-government street demonstrations. In June 1981, Casablanca was the scene of bloody repression which left 60 dead, according to official reports, or 637, according to the opposition. Two thousand people were arrested. This"Casablanca

massacre", led to an open controversy between the king and the leftist parties over the high cost of the Sahara war; it was swallowing more than $1 million a day, and the opposition refused to keep on paying.
[14] The stalemate on the battlefield caused conflict within the Moroccan armed forces, and in early 1983, they began to show signs of internal dissent. This tension became evident with the assassination of General Ahmed Dlimi, the supreme commander of the Royal Armed Forces, who was killed in mysterious circumstances after having secret liasons with Europe over ending the Sahara war.
[15] On June 10 1983, the Moroccans went to the polls to elect their 15,492 municipal councillors. The results were very controversial as international observers believed that figures had been tampered with

Almoravids and Almohads

Around 1030 AD, as frequently occurred in Islamic history, a religious campaign with hidden political aims was launched. The campaign, led by Abdallah Ibn Yassin and some Moroccan Berbers was really intended to recover the influence the Berbers had lost as Arab-influenced Moroccans had gained ground. Finding little support in their own country, they emigrated to what they called "Bilad as-Sudan" (the land of the Negro) setting up a religious center on an island of the Senegal river. By 1042 they had recruited thousands of followers, Berber from the south and other Islamicized Africans. With this following they launched a holy war or "Jihad". 20 years later they had won control of all the territory between Senegal and the Mediterranean. This empire, with its capital in Marrakesh, lasted for a hundred years and was known as Almoravid - derived from al-Murabitum, meaning the predestined or devout.

[2] In 1076 the Berbers, headed by Ysuf ibn Tashfin, conquered the ancient Sudanese state of Ghana-Uagadu (see box in Mauritania), introducing Islam to central Africa. They did not succeed in dominating the gold producing region and only made small profits from the gold trade. In 1086 the Almoravids left Ghana, concentrating their forces on the invasion of Spain. They succeeded at this, claiming to be allies of "the Taifa kingdoms", the states which had emerged after the fragmentation of the Cordovan Caliphate. Within a few years, the Almoravids had managed to dominate all the Muslim states except Valencia, which was conquered by "El Cid", Lord Rodrigo. They added the European possessions to their African empire.

[3] This created a joint Spanish-Arab-African culture, which proved to be extremely rich and creative, in spite of intense political activity. In 1125 the Almohad, the "unitarian" Berber of the central Atlas region, raised an army under the leadership of Muhammad Ibn Tumart, who proclaimed himself "Mahdi". He accused the Almoravid of drifting towards polytheism and neglecting the Koran. The conflict lasted for 20 years, until the last Almoravid sultan was defeated and killed near Oran.

[4] Abd al-Munin, the first Almohad sultan, consolidated his power over the African part of the empire, including Algeria, Tunisia, an eastern section of Libya and part of southern Mauritania. In 1165, his successor crossed the Mediterranean easily dominating the Almoravid emirs who failed to present a united defence. At its height, the empire extended from Senegal to the Ebro river, and from the Atlantic Ocean to Libya. Its "achilles heel" was the Spanish section, where Christian troops were constantly pressing during the "reconquest" campaign. In

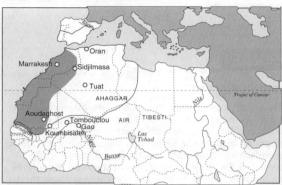

Almoravid Empire ■ 11th century
Almohad Empire □ 12th century

1212, in Navas de Tolosa, the Almohad suffered their first important defeat. The need to concentrate troops on the European front weakened their African rearguard, enabling other local groups to grow powerful and dispute their supremacy. As a result they gradually lost ground on both continents. In 1269 the Marini - eastern Berbers - occupied Marrakesh, delivering the coup de grace to the empire and ending the unification of the Magreb.

[5] Ten years later, the death of the Berber leader, and the need to strengthen the invasion force in Spain forced the Berbers to retreat. They left Islam in their wake, far into Central Africa, giving the region distinct cultural characteristics.

[6] The Almoravids had succeeded in controlling the gold bearing region of Bambuk, at the foot of the Futa Djalon massif, north of present-day Guinea (see Mali-map), separating it from its trade center. They had also siezed control of the western desert route (Sidjilmasa-Audaghost-Koumbisaleh) interrupting the gold and salt trade, causing it to be re-organized along a more central route running from Tuat (southern Algeria) to Gao or Timbuktu, on the Niger river, passing through the Tilemsi valley (see Niger box/map).

[7] The small states that arose, or rather re-emerged, in the region: Diara, in southwestern Mali; Tekrour, valley of Senegal; Ghana, southern Mauritania ; Djénne, on the Bani, a branch of the Niger River; Sosso, southeastern Mali; and Niani, northern Guinea, including the Bambuk, were too small to keep the trading territory together, so trade was suspended for the following 150 years.

---

by the government. The opposition denounced the electoral fraud, and accused King Hassan II of ignoring the will of the people.

[16] In 1984, the Saharan Arab Democratic Republic (RASD), proclaimed by the Polisario Front's fighters in the territory of the former Spanish Saharan colony, was recognized as a full member of the Organization of African Unity (OAU). Morocco, reacted by withdrawing from the pan-African organization.

[17] In his capacity as religious leader, the Moroccan king became worried about the current of Islamic fundamentalism on the increase

throughout the Arab world. For this reason, Hassan II improved administrative measures for reinforcing the power of the "ulemas" and other representatives of the religious sector.

[18] In 1987, the Moroccan monarch suggested to King Juan Carlos of Spain that both governments form a "study group" designed to consider the future of Ceuta and Melilla (see sections at the end of this chapter). The proposal was not well received in Spain, as there is an insistence on the "historical nature" of Spain's presence in Ceuta and Melilla.

[19] In May 1988, after 12 years of

tension, Morocco and Algeria re-established diplomatic relations, through the mediation of Saudi Arabia and Tunisia. The cause of the disruption in their relations had been the war in the Sahara, as Algeria had openly supported the Saharan nationalists from the very beginning.

[20] Better relations between Algeria and Morocco meant that a gas pipeline was built across the Strait of Gibraltar joining the two countries to Europe. From 1995, an Algerian-Moroccan firm based in Rabat will transport between 10 and 15 billion cubic meters of Algerian gas via the line.

[21] Unemployment in Morocco is on the increase. Urban migration of a million people per year has exacerbated the urban housing situation, putting a strain on sanitation, water and other services. In October 1992, 800 Moroccans were arrested in Tarifa, in the south of Spain, trying to enter the country illegally, while another 35 failed to make it to Spain, drowning in the Mediterranean.

[22] In March 1992, the Paris Club of creditor nations signed an agreement with Morocco, renegotiating its foreign debt of $21.3 billion. Development credits were renegotiated for 20 years, while other

types of credit were extended to 15 years. The country's economy has been undergoing adjustment. The state deficit was reduced from 10 per cent in the early 1980s to 3.2 per cent in 1992. The balance of payments has improved and the country's net reserves have increased.

[23] The United Nations International Council for the Control of Narcotics, condemned the fact that many farmers in countries like Morocco have been pressurized into cultivating opium and coca, raw materials for the production of heroin and cocaine, respectively.

[24] In Sahara, cooperation between Algeria and Morocco allowed a negotiated settlement to the conflict. A UN supervised referendum offered the inhabitants of the RASD the choice of being either an independent country, or a part of Morocco. In the meantime, the Moroccan government has given full support to the plebiscite policy, confident of a definitive defeat of the Polisario Front (see Sahara).

[25] Torture and disappearances are common both in Sahara and in Morocco itself. In February 1993, the Moroccan Human Rights Association announced the existence of 750 political prisoners - Nubier Amauí, secretary general of the Democratic Confederation of Labor, was sentenced to two years imprisonment, for 'defaming' the regime. He was released a few months later, after the opposition triumph in Parliament.

[26] In August 1992, King Hassan II dismissed Prime Minister Azedine Laraki and his fellow ministers, replacing the entire executive branch. Mohamed Karim Lamrani became prime minister; having already held that office from 1971 to 1983. A government referendum approved a new constitution designed to extend parliament's powers, though the king retained the power to choose the prime minister.

[27] The opposition triumphed in the first parliamentary elections following the reform, in June 1993, winning 99 of the 222 seats, while the ruling party won only 74. Two months later, Hassan held a spectacular inauguration ceremony in the biggest mosque in the world, built in Casablanca at a cost of $536 million.

[28] Despite the constitutional reform, the king continued to dominate national politics, and in May 1994, he designated one of his relations by marriage, Abd al-Latif Filali as prime minister. In August, the king issued a surprise call for the "integration of Berber culture an language into national life."

[29] The economic situation in the country became abruptly worse in 1995, mainly due to the lack of rains, which meant the harvests were only a sixth as big as the previous year. The GDP, which had increased 12 per cent in 1994, fell to 4 per cent in 1995.

[30] After tough negotiations, Morocco and the European Union signed a new association agreement in November. However, several observers stressed the need to respect the transition clauses of this accord, as an overly hasty liberalization of the Moroccan market could destroy 60 per cent of the nation's industrial sector.

Ceuta

This is an enclave on the Mediterranean coast of Morocco facing Gibraltar and Spain. The climate is Mediterranean with hot summers and moderate winters. There is sparse rainfall in winter.

[2] Occupied by troops of Portuguese king John in 1415, Ceuta was transferred to Spain in 1688 and retained by the Spanish after Moroccan independence in 1956. Despite many UN-backed claims and negotiations, before the United Nations Decolonization Committee, this port is still in Spanish hands. It is used as a harbor, but two thirds of the territory is reserved exclusively for military purposes.

SOCIETY

Peoples: 80% of the population is Spanish . Minorities include Arabs, 10% and others (according to November 1991 census taken by the Ceuta municipal goverment).
Religions: Catholic (majority), Muslim and others.
Languages: Arabic and Spanish.

THE STATE

Government: Civil authority is exercised by a representative of the Spanish Home Affairs Ministry, who administers it as part of Cádiz Province. Military authority lies in the hands of a General Commander. The territory has one representative in the Spanish parliament.

DEMOGRAPHY

Population: 71,190 (1986).
Density: 3,747 inhab./sq.km.

ECONOMY

Currency: pesetas.

Melilla

Melilla is a small peninsula on the Mediterranean coast of Morocco with two adjacent island groups. The climate is similar to that of Ceuta. Melilla is an ancient walled town built upon a hill with a modern European-style city on the plain. It is an important port which, like Ceuta, hopes to be a tourist attraction.

[2] Founded by Phoenicians and successively held by Romans, Goths and Arabs, Melilla was occupied by Spain in 1495. It was repeatedly besieged by the Rifs, a Berber group that opposed French and Spanish domination, most recently in 1921. Like Ceuta, it has been claimed by Morocco and it also is a port and a military base; more than half of the territory is used for military purposes.

[3] On the northern Moroccan coast the Spanish also hold Peñón (rock) de Vélez de la Gomera, which had 60 inhabitants in 1982; Peñón de Alhucemas with 61 inhabitants in 1982; (with the islets of Mar and Tierra) and the Chafarinas Archipelago (Islands of Congreso, Isabel II and Rey), 1 sq.km. in area and with a population of 191 inhabitants in 1982.

SOCIETY

Peoples, languages, religions and other features of the population are similar to those of Ceuta.

THE STATE

Government: A representative of the Spanish government is responsible for administrating the territory's civilian affairs. There is a military command in charge of military affairs. Like Ceuta, Melilla keeps a representative in Cortes, the Spanish parliament.

DEMOGRAPHY

Population: 58,460 (1986). Density: 4,872 inhab./sq km.

ECONOMY

Currency: pesetas.

Mozambique

Moçambique

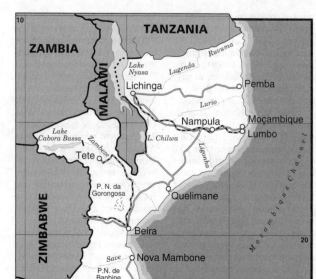

Population: 15,463,000
(1994)
Area: 801,590 SQ.KM
Capital: Maputo

Currency: Meticais
Language: Portuguese

The city of Sofala was founded by Shirazis towards the end of the 10th century. It became a point of contact between two of the most flourishing developed cultures in Africa: the commercial, Muslim cultures of the east coast (see box in Tanzania: "The Zandj Culture") and the metallurgical, animist culture of Zimbabwe (see box in Zimbabwe).

[2] As with the rest of the civilizations of the continent, the Portuguese presence on the coast of present-day Mozambique was fatal; they planned to seize control of the Eastern trade which had nourished the two civilizations for centuries. This led to the destruction of the ports and the stifling of Zimbabwean gold exports.

[3] The Portuguese were never able to re-establish this trade to their own benefit, nor were they able to achieve their other goals. The Monomotapa (leaders of the Karanga) became Portuguese subjects in 1629, but these authorities were insignificant figureheads on the coast, and the way to the gold mines was still closed by the Changamiras of Zimbabwe.

[4] When Zanzibar expelled the Portuguese from their area of influence, the colonists turned to the slave trade as a profitable business. Attempts at connecting Mozambique and Angola by land failed repeatedly and European control was confined to a coastal strip where their entire "administration" was limited to granting *prazos*, concessions of huge areas of land, to Portuguese and Indian adventurers who either plundered the land or searched for natives to enslave. These *prazeiros* became virtually independent of the Portuguese authorities.

[5] In 1890, the English questioned Portuguese control over

these lands and threatened to occupy them. The Portuguese government underwent a long, hard struggle to forcibly subdue the *prazeiros*, and prove their authority over the region. The conquest of the interior, however, was only completed around 1920 when they finally defeated the ruler, Mokombe, in the Tete region.

[6] Mozambique started to supply South African gold mines with migrant workers (up to one million every year) and its ports were open to South African and Rhodesian foreign trade.

> **When Zanzibar expelled the Portuguese from their area of influence, the colonists turned to the slave trade as a profitable business.**

[7] Portuguese colonialism controlled the country as an "Overseas Province", encouraging local group rivalries to prevent nationalist feelings from developing. Split into several movements, the nationalists staged strikes and demonstrations in their struggle for independence.

[8] In 1960, a spontaneous and peaceful demonstration in Mueda was fiercely repressed, leaving over 500 people dead. This convinced many Mozambicans that any peaceful negotiations with colonialism were doomed to failure.

[9] The following year, Eduardo Mondlane, then a United Nations official, visited his home country and persuaded the struggling pro-

independence groups that their unity was essential. This was finally achieved on July 25 1962, in Tanzania, with the creation of FRELIMO (the Front for the Liberation of Mozambique). FRELIMO was made up of individuals and organizations from all regions and ethnic groups of Mozambique.

[10] On December 25 1964, after two years of underground activity, FRELIMO started guerrilla warfare to win "total and complete independence". By the end of 1965, FRELIMO controlled some areas in Mozambique and by 1969, one-

fifth of the country's territory was under their authority.

[11] In February of that year, Mondlane was assassinated by colonialist agents. Differences of opinion developed within FRELIMO about the desired form of independence; some wanting a mere "Africanization" of the established system and others seeking to create a new popular democratic society.

[12] The second faction dominated FRELIMO's second congress, held in the liberated areas, and Samora Machel was elected president of the organization. From then on, the fighting intensified and spread to other areas. The impossibility of winning the colonial wars in Africa led to a military uprising in Lisbon on April 25 1974, ending Salazar's and Caetano's regime.

[13] A transitional government was established in Mozambique and on June 25, 1975, the Popular Republic of Mozambique was formed, with Samora Machel announcing that "the struggle will continue" in solidarity with the freedom fighters in Zimbabwe and South Africa.

[14] On the domestic front, the FRELIMO government nationalized education, health care, foreign

PUBLIC EXPENDITURES

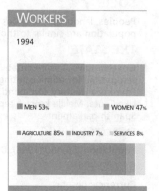

% of GNP — 1992

MILITARY 10.2 · EDUCATION 6.3 · HEALTH 4.4

WORKERS

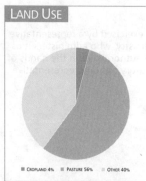

1994

MEN 53% — WOMEN 47%

AGRICULTURE 85% · INDUSTRY 7% · SERVICES 8%

LAND USE

CROPLAND 4% · PASTURE 56% · OTHER 40%

Illiteracy
1995

60%

Maternal Mortality
1989-95

PER 100.000
LIVE BIRTH

1,512

PROFILE

ENVIRONMENT

The wide coastal plain, wider in the south, gradually rises to relatively low inland plateaus. The Tropic of Capricorn runs across the country and the climate is hot and dry. Two major rivers cross the country: the Zambezi in the center and the Limpopo in the south. Due to its geographic location, the country's ports are the natural ocean outlets for Malawi, Zimbabwe and part of South Africa. However, trade has been hampered by wars during the past two decades. Mineral resources are important though scarcely exploited. The war devastated the country's entire productive structure, especially in the agricultural sector. The use of its mangrove forests for firewood has caused deforestation.

SOCIETY

Peoples: The Mozambican population is made up of a variety of ethnic groups, mainly of Bantu origin. The main groups are: Tsonga and Changone in the south; and Sera and Macondo in the northeast. There are also minorities of European and Asian origin. **Religions:** The rural population practise traditional religions. Most of the urban population is Christian or Muslim. Islam prevails in the north.
Languages: Portuguese (official). Most of the population speak Bantu languages, the main ones being Swahili and Macoa-Lomne. **Political Parties:** Mozambique Liberation Front (FRELIMO); National Resistance of Mozambique (RENAMO); Liberal and Democratic Party of Mozambique (PALMO); National Union of Mozambique (UNAMO). **Social Organizations:** Organization of Mozambican Women; Mozambican Youth Organization was formed.

THE STATE

Official Name: República de Moçambique. **Administrative divisions:** 10 provinces and 1 city (Maputo).
Capital: Maputo 931,000 inhab. (1991).
Government: Joaquim Chissano, President. **National Holiday:** June 25, Independence Day (1975). **Armed Forces:** 50,000 (government); 20,000 (RENAMO).

plement a more rational distribution of political leaders between the government and the Front.

[19] All of these projects were affected by the increasing deterioration of relations with South Africa. In 1981 they invaded Mozambican territory, attacking the Maputo suburb of Matola. They also backed the anti-government Movement of National Resistance (RENAMO), made up of former Salazar followers and mercenaries. South Africa's terrorism was aimed at their anti-racist refugees living in Mozambique, while RENAMO aimed to sabotage economic objectives and intimidate the rural population.

[20] At the end of 1982, the government cracked down on the black market and launched a major offensive against RENAMO in the Gorongoza region, where RENAMO's armed bands were attacking the communal villages of the interior.

[21] In the fourth congress of FRELIMO, in April 1983, deep chang-

> **1985 marked the beginning of a critical period for Mozambique, with RENAMO terrorist attacks on the one hand and a severe drought, on the other.**

es in the government's economic program were discussed. One of the main subjects under debate was a proposed reduction in the emphasis given to large agricultural projects to provide greater incentives for smaller projects.

[22] Debate of the "eight points" of the fourth congress resulted in the wider participation of party members who were called in to supervise the behaviour of government officials more directly. This initiative to re-furbish the links between the party leadership and its rank and file, brought about a complete change in the constitution of the delegation attending the congress. The overwhelming majority became rural workers and the number of women delegates had more than doubled since the third FRELIMO congress in 1977.

[23] The emphasis given to small-

scale agriculture and industry was the result of a major reappraisal of all the large state-run farms. These were charged with excessive centralization, bureaucracy and economic inefficiency.

[24] 1985 marked the beginning of a critical period for Mozambique, with RENAMO terrorist attacks on the one hand and a severe drought, on the other. The drought decimated cattle, causing a 70 per cent drop in production, and was also responsible for a 25 per cent reduction in grain production.

[25] President Samora Machel denounced South Africa's covert support of RENAMO, saying that it was a violation of the Nkomati agreements of March 1984, when the two countries had signed a treaty of non-aggression.

[26] Already tense economic and defence situations were compounded by the death of President Samora Machel on October 19, 1986; his plane crashed while returning from a high-level conference in Zambia. At that conference, presidents Kenneth Kaunda of Zambia, Mobutu Sese Seko of Zaire, José Eduardo Dos Santos of Angola and Machel himself, had debated joining forces to confront South Africa's growing aggression toward the independent countries of southern Africa. They wanted to end South African support of UNITA and RENAMO in Angola and Mozambique. It has never yet been established whether the plane crash in which Machel died was an accident or an act of sabotage.

[27] On November 3, the Central Committee of the FRELIMO Party met in a special session where they elected Joaquim Chissano (minister of foreign relations) as the new president and commander-in-chief of the armed forces.

[28] In the following year, the government began to reconsider the economic strategy it had adopted on independence in 1975. A more flexible foreign investment policy was adopted and local producers were encouraged to invest more. This was the first step toward the establishment of a mixed economy, a concept adopted by the FRELIMO Congress in July 1989. The party dropped all references to its Marxist-Leninist orientation.

[29] Peace negotiations between RENAMO and the government in Maputo began in 1990. These negotiations were made easier as the new constitution had recently

banks and several transnational corporations. The government promoted communal villages, bringing together the scattered rural population and organizing collective production methods.

[15] In 1977, FRELIMO held its third congress in Maputo. Marxist-Leninism was announced as the Front's ideology.

[16] Mozambique supported Zimbabwe's struggle, blockading the imports and exports of Ian Smith's racist regime despite severe repercussions on the Mozambican economy. Zimbabwean freedom fighters were given permission to set up bases within Mozambican territory. The white minority regimes retaliated with air raids and invasions.

[17] Zimbabwe finally gained independence in 1980, altering the po-

litical outlook of the region, tightening the circle around apartheid and allowing Mozambique to revitalize its economy through the economic integration of the Front-Line nations (Tanzania, Zambia, Angola and Botswana) and with Zimbabwe, Malawi, Lesotho and Swaziland.

[18] In March 1980 President Samora Machel launched a political campaign aimed at eliminating corruption, inefficiency and the increasing bureaucracy of state agencies and companies. A ten-year program for economic development was implemented, calling for investments of $1 billion in agriculture, transport and industry, and the political organization of the country was improved by consolidating FRELIMO, which decided to im-

PER CAPITA
$355

STATISTICS

DEMOGRAPHY

Urban: 31% (1995). **Annual growth:** 3.9% (1992-2000). **Estimate for year 2000:** 20,000,000. **Children per woman:** 6.5 (1992).

HEALTH

One **physician** for every 33,333 inhab. (1988-91). **Under-five mortality:** 277 per 1,000 (1995). **Calorie consumption:** 80% of required intake (1995). **Safe water:** 33% of the population has access (1990-95).

EDUCATION

Illiteracy: 60% (1995). **School enrolment:** Primary (1993): 51% fem., 51% male. Secondary (1993): 6% fem., 9% male. **Primary school teachers:** one for every 53 students (1992).

COMMUNICATIONS

83 **newspapers** (1995), 82 **TV sets** (1995) and 79 **radio sets** per 1,000 households (1995). 0.4 **telephones** per 100 inhabitants (1993). **Books:** 82 new titles per 1,000,000 inhabitants in 1995.

ECONOMY

Per capita GNP: $90 (1994). **Annual growth:** 3.80% (1985-94). **Annual inflation:** 53.20% (1984-94). **Currency:** 6.651 meticais = 1$ (1994). **Cereal imports:** 507,000 metric tons (1993). **Fertilizer use:** 15 kgs per ha. (1992-93). **External debt:** $5,491 million (1994), $355 per capita (1994). **Debt service:** 23% of exports (1994). **Development aid received:** $1,162 million (1993); $77 per capita; 79% of GNP.

ENERGY

Consumption: 40 kgs of Oil Equivalent per capita yearly (1994), 74% imported (1994).

---

brought in a multi-party system. The continuation of the single-party system had been one of the arguments used by the rebels to justify their terrorist activity.

[30] In October 1991 the authorities of the Manica province, one of the most fertile regions of the country, declared a state of emergency because of the drought which hit this part of the country destroying most of the crops. It was considered the worst drought to have affected the area in the last 40 years causing enormous shortages for the 300,000 local inhabitants. Faced with this situation, Joaquím Chissano's government requested international aid of more than a million tons of food to prevent starvation of a large part of the population.

[31] In Rome, in November 1991, a peace protocol was signed by the government of Mozambique and RENAMO. It foresaw the recognition of the rebel movement as a legal political party. This protocol was considered the forerunner to a peace agreement.

[32] In addition to the new laws regulating the political parties, this agreement guaranteed freedom of information, expression and association, and pledged that elections would be held.

[33] The refinancing of the $1.6 billion foreign debt was contingent upon the success of the peace accord. Prime Minister Machungo explained that his country was suffering badly from the effects of the cessation of aid from the former USSR and the Eastern European countries.

[34] In early 1992, the new political party regulations became one of the major hindrances to peace negotiations in Mozambique. The Chissano government initially offered RENAMO a special status, guaranteeing their members political rights but the rebels turned down the offer. The armed opposition also refused to accept the terms which stated that they would have to have a minimum of a hundred registered members in each province, as well as in the capital, to qualify as a *bona fide* political party.

[35] According to World Bank estimates, the cost of the civil war in Mozambique could reach $20 billion. The government requested $1 billion from the Paris Club to carry out a post-war reconstruction plan. However, everything depended upon the outcome of the peace talks. When RENAMO activities continued, the elections scheduled for 1991 were postponed.

[36] In May 1991, a new opposition party, the Liberal and Democratic Party of Mozambique, was created.

> According to World Bank estimates, the cost of the civil war in Mozambique could reach $20 billion.

The following month, a coup attempt by the sector opposed to peace negotiations ended in failure.

[37] In August, Chissano was re-elected during FRELIMO's 6th Congress, and Feliciano Salamao was named secretary general.

[38] On October 4 1992, with Italy as mediator, Chissano and Alfonso Dhlakama (of the RENAMO) signed a peace agreement in Rome, putting an end to 16 years of conflict which had caused over a million deaths, and five million refugees.

[39] In October 1992, the government accused RENAMO of violating the agreement by occupying four cities. However, despite this situation, the UN Commission of Supervision and Control arrived in Maputo and set up working commissions to solve sectorial problems.

[40] According to the terms of the agreement, RENAMO and government troops were confined to pre-established areas, and weapons were to be turned over to UN soldiers charged with disarming both sides within a six-month period.

[41] Zimbabwean troops, who controlled the corridors linking that country with Mozambique's ports, were to be withdrawn. The agreement also provided for the creation of an army of both government and guerilla forces.

[42] Discrepancies between the two parties led to direct UN Security Council intervention, in December 1992. The Peace Plan approved civilian observers and 7,500 troops being sent into Mozambique. The first contingent, under the command of a Brazilian general, reached Maputo in February 1993, and the remaining 4,700 troops arrived in May, including troops from Bangladesh, Italy, Uruguay, Argentina, India, Japan and Portugal.

[43] In February and March 1993, FRELIMO participated in joint military manoeuvres with the United States. This change of attitude was interpreted as an attempt at rapprochement with the West, a source from which Mozambique hopes to obtain foreign aid.

[44] The peace process nearly collapsed toward the end of 1993, when Dhlakama refused to lay down arms and demanded that the government cede five of the country's ten provinces to him.

[45] The UN decided to postpone elections until October 1994, hoping to overcome the stalemate in which the peace process had fallen.

[46] After several comings and goings, the RENAMO agreed to take part in the elections. Chissano was re-elected with over 53 per cent of the vote and in parliamentary elections, FRELIMO won with 44.3 per cent followed by the RENAMO with 37.7 per cent. In spite of his defeat, Dhlakama announced he was willing to cooperate with the government.

[47] In March 1995, the Paris Club promised to give Maputo $780 billion for the country's reconstruction. Social conditions after the civil war were disastrous, among other things because Mozambique, one of the poorest countries in the world, came out of the strife with a dismantled agriculture sector and a large part of its fields ridden with mines.

Myanmar/Burma

Population: 45,581,000 (1994)
Area: 676,550 SQ.KM
Capital: Yangon (Rangoon)
Currency: Kyat
Language: Burmese

Myanmar/Burma

Inscriptions dating from the 6th century BC testify to the very early establishment of advanced civilizations in Burma. Migrations occurred frequently from north to south and from the mountains to the coast. The people intermingled and occasionally engaged in battles until the 11th century when Burmese conquered the southern Hmong and the northern Kadu, establishing the state of Pagan.

[2] The following two centuries were a veritable "golden age" in Burmese thought and architecture. The Mongols attacked from the North, with aid from the Great Khan in Beijing. In 1283, the Mongol invasion ended the Pagan state and the Mongols remained in power until 1301. Marco Polo, in the service of Kublai Khan, participated in the invasion and is thought to be the first European to visit the country.

[3] Burma remained divided into small ethnic states until the 16th century when Toungoo local leaders reunified the territory. The second of these rulers, Bayinnaung, extended his domain to parts of present-day Laos and Thailand.

[4] Extravagance undermined the agricultural foundations of the economy, resulting in an exodus of peasants to neighboring states. The process of fragmentation was hastened by the presence of early European traders and their consequent rivalries.

[5] In 1740, a Toungoo ruler again achieved unification, with the help of the British. But when his successors continued the project of national reconstruction, they clashed with British interests in Assam, India, and a confrontation resulted with their former European allies. The Burmese fought three wars against the British throughout the 19th century, in 1820-26, 1852-53 and 1885-86. During the last war, King Thibaw was taken prisoner and Burma was annexed to the British viceroyalty of India.

[6] The 1930s began with a rising tide of nationalist movements; that of the Buddhist monk, U Ottama, inspired by Gandhi; Saya San's attempt to restore the monarchy; and uprisings organized from the University of Rangoon, bringing together Buddhists and Marxists. In 1936, a student demonstration became an anti-British national protest, led by Aung San.

[7] The anti-colonial movement was not restricted to the urban elites. Heavy taxes and the collapse of the world rice market in 1930 led thousands of small farmers into debt and ruin at the hands of British banks and Indian moneylenders. Discontent expressed itself as a generalized xenophobia, leading to popular rebellions in 1938 and 1939.

[8] When WWII broke out, a group of militant anti-colonialists in Bangkok, known as "the 30 comrades", including members from the newly-created Communist Party, formed the Burma Independence Army (BIA). They joined the Japanese against the British and invaded the capital on 7 March 1942. Minority groups of Karen, Kachin and Chin organized guerilla groups to combat both the BIA and the Japanese.

[9] The Japanese granted Burma independence on August 1 1943, appointing Ba Maw head of state. The "national" army was placed under the command of Ne Win. However, friction soon developed between the Japanese and the socialist wing of the "30 comrades". On March 27 1945, the BIA declared war on Japan and was recognized by the British as the Patriotic Burmese Forces. On May 30, they captured Rangoon, this time with the help of the British. Aung San organized a transition government, and in 1947, a constitution was drafted. On July 19, a military commando assassinated Aung San and several aides in the palace, and U Nu stepped in as premier. On January 4 1948, independence was proclaimed.

[10] Several challenges faced the new government: ethnic minorities rebelled; recently-defeated Chinese Kuomintang forces moved into Shan state, where they were also involved in drug trafficking, and Aung San's army, renamed the People's Volunteer Organization and linked to the Communist Party, mounted a further armed insurrection.

[11] On May 2 1962, General Ne Win overthrew U Nu, who had been very successful in the 1960 elections. Burma had peacefully settled its border conflicts with India and China, and throughout the war in southeast Asia Rangoon maintained a policy of non-alignment. Ne Win nationalized the banks, the rice industry (which accounted for 70 per cent of foreign earnings), and the largely Indian controlled trade.

[12] In 1972, a new constitution confirmed the governing Burma Socialist Program Party (BSSP) as the only legal political organization.

[13] Ne Win's regime declined after the 1973 economic crisis and opponents emphasized the ambiguity of his "Burmese Socialism".

[14] In 1979, Burma withdrew from the Non-Aligned Movement. In 1981, the National Congress named San Yu as successor to Ne Win, who resigned the presidency but continued as party chairperson and thus maintained control of the country.

[15] At the end of 1987, the social and economic situation worsened. In August, Ne Win admitted making mistakes in the economic policy of the previous 25 years. A BSPP ruling congress appointed Sein Lwin as head of state, triggering a new wave of protests. Hundreds of students and Buddhist monks died in the streets, and Lwin

WORKERS

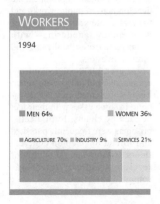

1994

- MEN 64%
- WOMEN 36%
- AGRICULTURE 70% INDUSTRY 9% SERVICES 21%

LAND USE

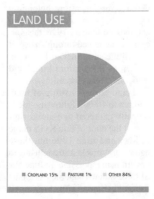

- CROPLAND 15% PASTURE 1% OTHER 84%

ENVIRONMENT

The country lies between the Tibetan plateau and the Malayan peninsula. Mountain ranges to the east, north and west surround a central valley where the Irrawaddy, Sittang and Salween rivers flow. Most of the population is concentrated in this area where rice is grown. The climate is tropical with monsoon rains between May and October. Rainforests extend over most of the country. Deforestation has been responsible for the destruction of two-thirds of the country's tropical forest.

SOCIETY

Peoples: Three-fourths of the population are Burmese. The Shan, Karen, Rakhine, Mon, Chin, Kachin and other ethnic groups have traditionally resisted political domination by the Burmese majority. Other minority groups are Chinese, Tamil and Indian. **Religions:** Bhuddist, 89%, Christian, Muslim and Hindu. **Languages:** Burmese (official) and the languages of the ethnic groups. **Political Parties:** The single party system was abolished by Saw Maung's military government in 1988. The opposition National League for Democracy (NLD), formed in 1988 and led by Aung San Suu Kyi. Shan Nationalities League for Democracy; Arakan League for Democracy; the National Unity Party (NUP), the new name of the former Burma Socialist Program Party (BSPP). There are small armed groups linked to the different ethnic minorities, the Mong Thai being the sole relevant among them. **Social Organizations:** The Workers' Organization of Burma and the Peasants' Organization, both founded in 1977, are BSPP mass organizations.

THE STATE

Official Name: Pyidaungzu Myanma Naingngnandaw. **Capital:** Rangoon 2,513,000 inhab. (1983). **Other cities:** Mandalay, 533,000 inhab.; Moulmein, 220,000 inhab; Pegu, 150,000; Bassein, 144,100 inhab. (1983). **Government:** Military regime. General Than Shwe, President, Chairman of the State Law and Order Restoration Council (SLORC) since April 1992. Legislature: suspended since 1988. **National Holiday:** January 4, Independence Day (1948). **Armed Forces:** 286,000 (1995). **Paramilitaries:** 85,000 (People's Police, People's Militia).

Illiteracy
1995

17%

Maternal Mortality
1989-95

PER 100.000 LIVE BIRTHS

517

DEMOGRAPHY

Urban: 26% (1995). **Annual growth:** 1.4% (1992-2000). **Estimate for year 2000:** 52,000,000. **Children per woman:** 4.2 (1992).

HEALTH

One **physician** for every 12,500 inhab. (1988-91). **Under-five mortality:** 109 per 1,000 (1995). **Calorie consumption:** 114% of required intake (1988-90). **Safe water:** 38% of the population has access (1990-95).

EDUCATION

Illiteracy: 17% (1995). **Primary school teachers:** one for every 35 students (1989).

COMMUNICATIONS

85 **newspapers** (1995), 82 **TV sets** (1995) and 85 **radio sets** per 1,000 households (1995). **Books:** 85 new titles per 1,000,000 inhabitants in 1995.

ECONOMY

Annual inflation: 26.50% (1984-94). **Consumer price index:** 100 in 1990; 263.9 in 1994. **Currency:** 6 kyats = 1$ (1994). **Cereal imports:** 21,000 metric tons (1992). **Fertilizer use:** 69 kgs per ha. (1992-93). **Imports:** $886 million (1994). **Exports:** $771 million (1994). **External debt:** $6,502 million (1994), $143 per capita (1994). **Debt service:** 15.4% of exports (1994).

---

was forced to resign after 17 days, replaced by Maung Maung.

[16] The opposition organized for the multi-party elections scheduled for May 1990. The government changed the country's name to Union of Myanmar and dropped the term "Socialist".

[17] The National League for Democracy (NLD) won 80 percent of the vote, while the ruling National Unity Party (ex-BSPP) retained only 10 of the 485 seats. The election results were ignored by the government who banned opposition activities, imprisoned or banished its leaders, and harshly repressed street demonstrations.

[18] In July 1989, the leader of the NLD, Aung San Suu Kyi, the daughter of anti-colonial hero Aung San, was sentenced to house arrest and held incommunicado. She received the Nobel Peace Prize in 1991.

[19] The opposition was reinforced by agreements between students, Buddhist monks and some minorities. In March 1992, the UN High Commission for Refugees denounced the massacres carried out against ethnic minorities. All political parties were dissolved or banned.

[20] In April, General Than Swe took power releasing 200 dissidents and permitting 31 universities and schools to reopen. Myanmar returned to the Movement of Non-Aligned Countries. In September, martial law was suspended, but Amnesty International reported the continued use of torture.

[21] In January 1993, the military invoked a National Convention to draw up a new constitution. At the end of that year, Amnesty International denounced the imprisonment of over 1,550 opposition figures.

[22] An article of the new 1994 constitution stipulated presidential candidates could neither be married to foreigners nor bear children under foreign citizenship and they should have been residing in Myanmar for the last 20 consecutive years. The regulation was custom-made for Suu Kyi who is married to a British subject and lived several years abroad. The military junta met with her in September 1994 for the first time since she was arrested. No agreement was reached regarding the new constitution.

[23] In July 1995, Suu Kyi was released from house arrest and called on the State Law and Order Restoration Council (SLORC) to hold a dialogue. The SLORC refused, jailed dozens of dissidents and maintained the ban on political debates.

[24] The headquarters of the rebel minority in Manerplaw was taken by the SLORC in January. The fall of Manerplaw was an important defeat for the opposition since it was also an important base for undercover organizations belonging to the rebel armies and democrat activists. In January 1996, the government achieved through a secret accord the surrender of Khun Sa, known as the "opium king" and leader of a Shan force.

[25] Huge crowds - some of them with up to 10,000 followers - periodically gathered to express support at the door of Suu Kyi's home in May and June 1996 but these demonstrations lost strength due to government repression. In June, the SLORC banned statements against the government and threatened with a complete prohibition of the NLD's activities and the arrest of its members for unlawful association. In July, Amnesty International denounced the government of Myanmar for arresting hundreds of NLD members, in some cases for watching videos of Suu Kyi's speeches in their own homes.

[26] Taking advantage of the NLD's weakness, the military held monthly press conferences both for the local and foreign press, which shortly before they had considered biased. The SLORC declared 1996 as year of tourism. It promoted the development of public works and the real estate sector - a base for drug money laundering - with strong participation by the state and foreign investors.

[27] At least 1,000 political prisoners remained in jail in September. Many were sentenced to forced labour, in some cases with their feet bonded by chains. In tiny cells, prisoners were frequently punished and forced to sleep on concrete floors, without any possibility of receiving visitors. Torture was commonly practised.

Namibia

Nammibia

Population: 1,508,000 (1994)
Area: 824,290 SQ.KM
Capital: Windhoek

Currency: Rand
Language: English

Southwestern Africa was occupied by Khoikhoi and San peoples and the Bantu-speaking Herero. In the late 1480s Portuguese navigators explored the coastal regions and during the 17th, 18th and early 19th centuries the Dutch and English explored the coast and limited inland areas. In the 1840s the German connection with the territory began with the arrival of the Rhenish Missionary Society. In the 1860s tribal wars prompted missionaries, settlers, and the German government to offer the British territorial sovereignty in exchange for protection. Each request was denied by the British, until, in 1876, the Cape Colony secured treaties with the tribal chiefs that brought the territory under British colonial rule. The British government agreed only to annex Walvis Bay and some adjacent territory.

[2] Finally, in 1884, Germany´s chancellor, Otto von Bismarck took the initiative of offering complete German protection which developed into full-fledged annexation to Germany. In their determination to take full possesion of the territory the Germans massacred the Hereros and Nama.

[3] The country acquired greater strategic value when rich deposits of iron ore, lead, zinc, copper and diamonds were discovered. More recently metals with military applications including manganese, tungsten, vanadium, cadmium and large quantities of uranium were discovered.

[4] When conflicts between imperialist powers led to the outbreak of World War I, the British invaded Namibia from South Africa.

[5] At the end of the war, Namibia became a League of Nations trust territory, assigned to the Union of South African as "a sacred trust in the name of civilization" to "promote to the utmost the material and moral well-being of its inhabitants."

[6] After overcoming their initial differences, the South African *Boers* and the German colonists of Namibia soon discovered a common interest, namely the brazen exploitation of the native population whose well-being they were supposed to be promoting.

[7] In 1947, after World War II, South Africa formally announced its intention to annex the territory to the United Nations. The UN, which had inherited responsibility for League of Nations trust territories, opposed the plan, arguing that "the African inhabitants of South-West Africa have not yet achieved political autonomy". Until 1961, the UN insisted on this point, year after year, and was systematically ignored by South Africa's apartheid regime.

[8] Between 1961 and 1968, the UN tried to annul its trusteeship and establish Namibian independence. Legal pressure was ineffective and the Namibian people, led by the South West African People's Organization (SWAPO) chose to fight for their freedom. The first clashes occurred on August 26 1966.

[9] In 1968, the UN finally declared the South African occupation of the country, now internationally known as Namibia, to be illegal. A UN Council for Namibia was established to give the territory legal representation until its people achieved sovereignty. However, efforts by the majority of the UN General Assembly to reinforce this condemnation with economic sanctions were routinely vetoed by the Western powers.

[10] Angolan independence, in 1975, radically changed the Namibian struggle for freedom, as it provided SWAPO with a friendly rearguard. The guerrilla war was stepped up in spite of South African pressure on Angola, and the Western powers began to urge Pretoria to seek a "moderate" settlement to prevent a revolutionary regime from coming to power.

[11] In December 1978, South Africa held elections in Namibia to establish an autonomous government the first step towards independence. But these elections lacked credibility as United Nations observers were denied access, apartheid troops were occupying the territory and SWAPO was excluded.

[12] South African economic concerns wield great influence in Namibia. Foreign interests are concentrated in the three key sectors of the Namibian economy; mining, agriculture and fishing.

[13] Transnationals increased their presence in the economy through investments in mining, which is controlled by British, North American, German and South African companies. The Consolidated Diamond Mines company, a subsidiary of the transnational AngloAmerican, contributed 40 per cent of South Africa's administrative budget for Namibia in mining taxes.

[14] In Swakopmund, the Rossing uranium mine, the largest open-pit mine in the world, produces over 5,000 tons of uranium dioxide a year. This mine is controlled by a British transnational, Rio Tinto Zinc corporation. Two other transnationals, the French Elf-Aquitaine and Anglo-American (South Africa), also hold uranium concessions in Namibia.

[15] US firms AMEX and Newmont are associated with the Tsumeb Mining Company, which operates copper and pyrites mines in Otjihase. All of these companies benefitted from the extremely generous facilities granted to foreign firms by the South African administration in Namibia.

[16] The South African-controlled administration in Windhoek operated along traditional colonial lines. The country produced what it did not consume, and imported everything it needed, especially food. Namibia still exports maize, meat and fish, especially to South Africa, and imports rice and wheat. It exports minerals and raw materials, and, having no industries, imports manufactured goods. Ninety per cent of consumer goods in Namibia come from South Africa.

[17] Until 1990, the privilege system continued. Indigenous Africans, 90 per cent of the population, consumed 12.8 per cent of the GDP while the immigrants, 10 per cent of the population, consumed 81.5 per cent of the GDP. When independence was declared, three-quar-

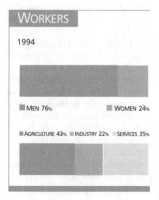

WORKERS

1994

- MEN 76%
- WOMEN 24%

- AGRICULTURE 43%
- INDUSTRY 22%
- SERVICES 35%

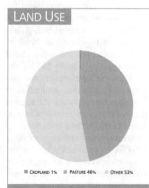

LAND USE

- CROPLAND 1%
- PASTURE 46%
- OTHER 53%

Foreign Trade

MILLIONS $ 1994

IMPORTS
1,196
EXPORTS
1,321

PROFILE

ENVIRONMENT

Mainly made up of plateaus in the desert region along the Tropic of Capricorn. The Namib desert, along the coast, contains rich diamond deposits and is only populated because of mining activities. To the east, the country shares the Kalahari desert with Botswana, an area populated by shepherds and hunters. The population is densest in the north, and in the central plateau, where rainfall is heaviest. There is a coastal fishing industry which, together with cattle raising, was the mainstay of the economy before mining began in the 1960s. The country has important reserves of copper, lead, zinc, cadmium and uranium.

SOCIETY

Peoples: Ovambo 47.4%; Kavango 8.8%; Herero 7.1%; Damara 7.1%; Nama 4.6%; other 25%. **Religions:** Many people practise traditional African religions although there are a large number of Christians. **Languages:** English (official), Khoisan, Bantu, German and Afrikaans. **Political Parties:** Southwest African People's Organization (SWAPO); Southwest African People's Organization for Justice (SWAPO for Justice); Democratic Turnhalle Alliance (DTA); United Democratic Front; National Christian Action. **Social Organizations:** National Union of Workers of Namibia.

THE STATE

Official Name: Republic of Namibia. **Administrative divisions:** 13 districts. **Capital:** Windhoek, 125,000 inhab. (1990). **Other cities:** Swakopmund, 15,500 inhab.; Rundu 15,000 inhab.; Rehoboth 15,000 inhab.; Keetmanshoop 14,000 inhab. **Government:** Sam Nujoma, President. **Armed Forces:** 8,100.

ters of agrarian and livestock production belonged to farmers of European descent.

[18] Average per capita income in Namibia is one of the highest in Africa: $1,410, but this figure conceals enormous inequalities. While the average salary of a person of European origin is $1,880, an African will not earn more than $108.

[19] The independence of Namibia was estimated to cost South Africa $240 million in lost exports and an additional $144 million in importing foreign products.

[20] As a result, South Africa not only refused to seek a negotiated settlement for the independence of Namibia, but also increased its military budget by 30 per cent to pay for a campaign against the Popular Liberation Army of Namibia, an affiliate of SWAPO.

[21] During late 1982 and early 1983, fighting intensified along the border between Namibia and An-

gola. It also spread to the south, with SWAPO forces staging daring attacks within the "iron triangle", close to the city of Grootfontein, where the main South African military forces were concentrated.

[22] On the diplomatic front, the failure of the five-point plan proposed by the Contact Group (France, United States, Britain, West Germany and Canada) brought the issue of Namibian independence to a deadlock. South Africa bound any accord to the withdrawal of Cuban troops from Angola and demanded guarantees that its investments in Namibia would not be affected.

[23] SWAPO refused to agree to special benefits for the European population, nor would it accept limitations, suggested by the Contact Group, for constitutional change after independence. Pretoria also balked at SWAPO demands for elections with UN supervision.

[24] In 1983, the interim Democratic Turnhalle Alliance (DTA) government in Namibia was ousted by the South African government when its leaders began to take critical stands on apartheid issues despite the party's strong links with Pretoria.

[25] This attempt to gain some popular support failed, and the government of Namibia was handed over to a colonial administrator appointed directly by Prime Minister Botha. France immediately withdrew from the Contact Group arguing that its efforts had been frustrated by Pretoria's delaying tactics.

[26] In February, 1984, Angolan and South African government representatives met in the Zambian capital, Lusaka, for peace negotiations. They agreed on a South African troop withdrawal from southern Angola in exchange for a ceasefire.

[27] In May of the same year, SWAPO delegates and other Namibian political party leaders met with a South African representative, but Pretoria's uncompromising attitude dashed the hopes of any agreement. South African troops failed to withdraw from Namibia on schedule and the deadlock continued.

[28] In June, Botha appointed an "interim government" based on a Multiparty Conference (MPC) of six multi-racial and white groups, including a section of the ruling National Party. The new regime was to have limited authority in domestic matters while South Africa would control foreign affairs and military matters. SWAPO was excluded.

[29] From 1987, all the states directly or indirectly involved in the conflict began to recognize the need to end this long drawn out confrontation, though negotiations were laborious due to the complexity of the economic and political interests at stake.

[30] There were many reasons for this sudden turnaround; after 14 years of uninterrupted war, Angola's economy was on the brink of collapse, and the war is calculated to have cost the country $13 billion.

[31] Pretoria's permanent harassment of Angola and occupation of Namibia were costing the regime dearly both economically and diplomatically. Its final drive to set up its counter-revolutionary protégé UNITA (National Union for the Total Independence of Angola) in government over the "liberated territory" was defeated by joint Ango-

lan and Cuban forces in the battle of Cuito-Cananale (April 1988).

[32] In December, 1988, after prolonged US-mediated negotiations, South Africa, Angola and Cuba reached an agreement whereby South African troops would leave Namibia, and Cuba would gradually withdraw its 50,000 soldiers.

[33] Throughout 1989, thousands of exiles began to flood back into the country and many political prisoners were released. In September, 60-year-old SWAPO leader Sam Nujoma returned from 30 years in exile to lead his organization into the elections two months later. SWAPO, which had not been party to the peace accord and did not feel bound by it, attempted to infiltrate from Angolan territory with large numbers of troops. These were detected and neutralized by South African troops, causing heavy losses.

[34] The liberation movement in exile was shaken by two scandals. It was discovered that political commissars at refugee camps had tortured alleged dissidents, and that the number of exiles had been wildly exaggerated in order to obtain greater international food assistance.

[35] In November 1989 more than 710,000 Namibians voted for the members of the National Assembly who would draft the country's first constitution. Ten parties contested the elections, supervised throughout by the UN. SWAPO won a resounding victory with over 60 per cent of the vote. This ensured control of the constituent assembly and the appointment of Sam Nujoma as Namibia's first president.

[36] Independence was proclaimed on March 21 1990. The presidential guard of honor included SWAPO and the South West Africa Territorial Force fighters. The latter had protected the South African-backed regime and was to be integrated into the new National Army of Namibia.

[37] The government of independent Namibia has had to deal with many social injustices and inequalities inherent in the apartheid system imposed by South African domination, especially in the health and education spheres. In the territories which used to be reserved for whites there is an acceptable infrastructure for education and good sanitation, but in the areas where the blacks were forced to live, in particular in the north of the coun-

try, there are inadequate schooling and health facilities and few well-qualified staff.

38 On independence, Namibia adopted English as its official language, replacing Afrikaans. The government has set up an informal department of education to teach women how to read. The body's initial work, which had UNICEF support, showed that Namibian women would have to be taught how to read in their own languages, before they would be able to learn English. As for health, the main concern of the government is the fate of some 40,000 disabled people, mostly victims of the 23-year-long guerrilla warfare.

39 The UN's Development Program considered land reform to be Namibia's most pressing issue. This belief was based on the fact that 65 per cent of the land is owned by people of European origin, many of whom were absentee landlords.

40 In mid-February 1991, the civil servants, the town hall workers, and the Namibian Parastatal Corporation went on strike demanding higher salaries, labor legislation, and protesting against alleged government racism. The government responded by accusing the workers of attempted economic sabotage of the administration.

41 The inequality in Namibian incomes has led economic analysts from various international organizations to call the country a "dual economy nation". Seventy per cent of the inhabitants live in Third World conditions, 25 per cent live in a transition economy, and only 5 percent benefit from the economic conditions of development.

42 Its economy is heavily dependent on mining and imports of manufactured consumer goods. Over 70 per cent of the population lives directly or indirectly from agriculture, but this sector represents only 10 per cent of Namibia's GNP.

43 Namibia has been expanding its fishing industry, but one of the main problems with this is that its territorial waters are frequently invaded by fishing boats from other countries. This friction over territorial waters has aggravated a dispute with South Africa over the sovereignty of territorial waters, a dispute that is difficult to solve, because Namibia lacks military power.

44 In 1991, the Nujoma government reached an agreement with

the US to start a military education and training program, under which Namibian Defence Force officers will receive free professional military training. In December of that

Throughout 1989, thousands of exiles began to flood back into the country and many political prisoners were released.

same year, the defence minister, Peter Muesiniange, travelled to Brazil to be advised on the establishment of a modern air force.

45 Early in 1991, the Namibian president, ready to play a leading role in southern Africa, donated a million rand to the African National Congress (ANC), then the main African opposition to the minority government in South Africa. The money was given to ANC leader Nelson Mandela, when he visited Namibia on January 31. Democracy in South Africa and lasting peace in Angola, which shares an economic identity with Namibia along their common border, would enable Namibia to achieve significant economic development.

46 In September, an agreement was reached with South Africa over joint administration of the port at Walvis Bay. The agreement also established the middle of the Orange River as the country's southern border, which should reduce armed confrontations between the two countries in that region.

47 The following month, Namibia announced the opening of a commercial office in Pretoria. In November the government set up control posts along the border with Zambia to fight smuggling and drug trafficking.

48 That same month, the government set up a commission to investigate charges that anti-independence Namibian parties were being financed by Pretoria. The ensuing scandal made it impossible for the opposition Turnhalle Democratic Alliance, which was under suspicion in the affair, to become a unified party under the leadership of Mishae Muyongo.

49 On December 11 1991, the first legal SWAPO congress established the need to promote political and economic relations with the world community, particularly with Third World countries. During the congress, Nujoma was reelected president of SWAPO.

50 In 1992, Namibia and South Africa agreed to return the port of Walvis Bay to Windhoek in 1994. A year later, the National Council, the upper chamber of parliament, went into operation. In 1994, the government approved a land law which aimed, amongst other things, to limit the concentration of property, as it was already estimated that 1 per cent of the population owned 75 per cent of the national land area.

51 In December of that year, Sam Nujoma was reelected president with 70 per cent of the vote. However, in May 1995, SWAPO - the president's party - split, leading to the creation of SWAPO for Justice.

52 After achieving 5 per cent growth in 1994, the government predicted economic growth of 3 per cent for 1995, while unemployment affected some 38 per cent of the active population.

Nauru

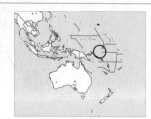

Population: 10,000
(1994)
Area: 21 SQ.KM

Currency: Australian Dollar
Language: English

Nauru

The island, which lies in the central Pacific, was populated by migrating Polynesian, Micronesian and Melanesian people. The British sailor John Fearn was the first European to visit the island, which he named Pleasant Island. Whaling ships often arrived at Nauru during the 19th century, until it was annexed by the German Second Reich in 1888.

[2] In 1899, Nauru was found to have phosphate-rich rock which made it one of the wealthiest countries in the world. In 1905, an Anglo-German company began mining on the island.

[3] At the beginning of World War I, Nauru came under Australian trusteeship and during World War II, under Japanese control with

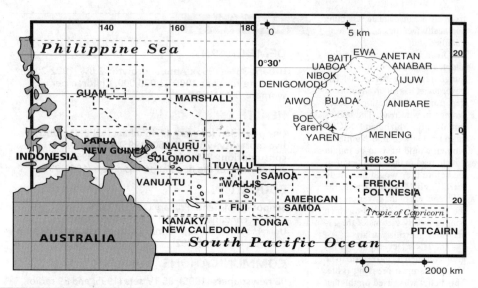

thousands of Nauruans sent to forced labor camps on the island of Truk (in present-day Micronesia). At the end of the war, only 700 came back alive. Australia recovered the island, and mining was continued by British, Australian and Aotearoan transnational consortium.

[4] The Nauruans did not adjust well to mining and were soon replaced by immigrants, mostly Chinese. Immigrant workers became so numerous that in 1964, the Australian government officially suggested that they should accept resettlement on a different island or elsewhere in Australia. Nauruans rejected this, deciding to stay on their island and seek autonomy.

[5] The mining sector was nationalized in 1967, which meant a large increase in per capita income. In 1992, the minimum annual salary was $6,500. Nauru declared independence on January 31 1968, rejecting Australian attempts at domination. It is still the smallest republic in the world. A year later it became a member of the Commonwealth.

[6] Government expenses and public investment are paid out of a huge legal fund fed by phosphate revenues. Education, medical care and housing are free.

[7] Hammer de Roburt dominated the political scene, and as the first president, he governed continuously until 1976 when he lost the majority to Bernard Dowiyogo, of the Nauru Party. Two years later, De Roburt once again obtained the majority, holding office from 1978 to 1986. In the general election that

year a woman was elected to parliament for the first time since independence.

[8] Bernard Dowiyogo returned to the presidency in 1989. His government started a court case against the Australian state for the indiscriminate exploitation of the phosphate mines during the last 50 years. In 1991, Australia recognized Nauru's right to claim compensation. Britain and Aotearoa promised to collaborate with Australia in the payment of $100 million to Nauru.

[9] This year the economy of Nauru registered a budget deficit for the first time in its history. It is estimated that if mining continues at the same rate until the year 2000, the phosphate resources of the island could be exhausted. The current dependence on phosphate exports has forced islanders to think of conversion. The only apparent viable alternative is fishing, as the land is not fit for agriculture. The production of vegetables and tropical fruits is enough to meet domestic needs.

[10] The government, headed by Lagumot Harris since 1995, has started to invest resources in developing the nation's air and fishing fleets.

Nepal

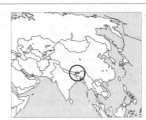

Population: 20,885,000 (1994)
Area: 140,800 SQ.KM
Capital: Kathmandu

Currency: Rupee
Language: Nepali

Nepal

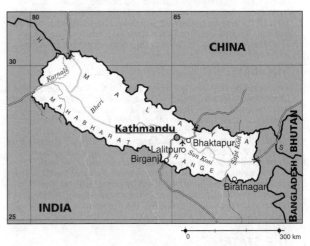

Nepal is a landlocked country in the central Himalayas between two of the world's most densely populated countries; India and China. References to Nepal Valley and Nepal's lower hill areas are found in the ancient Indian classics, suggesting that the Central Himalayan hills were closely related culturally and politically to the Gangetic Plain at least 2500 hundred years ago. Lumbini, Gautama Buddha's birthplace in southern Nepal, and Nepal Valley also figure prominently in Buddhist accounts. There is substantial archaeological evidence of an early Buddhist influence in Nepal, including a famous column inscribed by Ashoka (emperor of India, 3rd century BC) at Lumbini and several shrines in the valley.

[2] A coherent dynastic history for Nepal Valley becomes possible, though with large gaps, with the rise of the Licchavi dynasty in the 4th or 5th century AD. Although the earlier Kirati dynasty had claimed the status of the Kshatriya caste of rulers and warriors, the Licchavis were probably the first ruling family in that area of plain Indian origin. This set a precedent for what became the normal pattern thereafter -Hindu kings claiming high-caste Indian origin ruling over a population much of which was neither Indo-Aryan nor Hindu.

[3] The Licchavi dynastic chronicles, supplemented by numerous stone inscriptions, are particullarly full from AD 500 to 700; a powerful, unified kingdom also emerged in Tibet during this period, and the Himalayan passes to the north of the valley were opened. Extensive cultural, trade and political relations developed across the Himalayas, transforming the valley from a relatively remote backwater into the major intellectual and commercial centre between South and Central Asia. Nepal's contacts with China began in the mid-7th century with the exchange of several missions. But intermittent warfare between Tibet and China terminated this relationship; and while there were briefly renewed contacts in subsequent centuries, these were reestablished on a continuing basis only in the late 18th century.

[4] The middle period in Nepalese history is usually considered synonymous with the rule of the Malla dynasty (10th-18th century) in Nepal Valley and surrounding areas. Although most of the Licchavi kings were devout Hindus, they did not impose Brahmanic social codes or values on their non-Hindu subjects; the Mallas perceived their responsibilities differently, however, and the great Malla ruler Jaya Sthiti (reigned c. 1832-95) introduced the first legal and social code strongly influenced by contemporary Hindu principles. His successor Yaksa Malla (reigned 1429-1482) divided his kingdom among his three sons, thus creating the independent principalities of Kathmandu, Patan, and Bhaktpur (Bhagdaon) in the valley. Each of these states controlled territory in the surrounding hill areas, with particular importance attached to the trade routes northward to Tibet and southward to India that were vital to the valley's economy. There were also numerous small principalities in the western and eastern hill areas, whose independence was sustained through a delicate balance of power based upon traditional inter-relationships and, in some cases, common ancestral origins among the ruling families.

[5] In the 16th century virtually all these principalities were ruled by dynasties claiming high-caste Indian origin whose members had left the hills in the wake of Muslim invasions of Northern India. In the early 18th century the principality of Gorkha (also spelled Gurkha) began to assert a predominant role in the hills and even to pose a challenge to Nepal Valley. The Mallas, weakened by familial dissent and widespread social and economic discontent, were no match for the great Gorkha ruler Prithvi Narayan Shah, who conquered the valley in 1769 and moved his capital to Kathmandu shortly thereafter, providing the foundation of the modern state of Nepal.

[6] The Shah (or Sah) rulers faced tremendous and persistent problems in trying to centralize an area long characterized by extreme diversity and ethnic and regional parochialism. They absorbed dominant regional and local elites into the central administration of Kathmandu, thus neutralizing potentially disintegrative political forces and involving them in national politics. However, this also severely limited the centre's authority in outlying areas because local administration was based upon a compromise division of responsabilities between the local elites and the central administration.

[7] The British conquest of India in the 19th century posed a serious threat to Nepal and left the country with no real alternative but to seek an accommodation with the British to preserve its independence. This was accomplished by the Rana family regime after 1860. Under this de facto alliance, Kathmandu permitted the recruitment of highly valued Gurkha units for the British Indian Army and also accepted British "guidance" on foreign policiy. In exchange, the British guaranteed the Rana regime protection against both foreign and domestic enemies and allowed it virtual autonomy in domestic affairs. Nepal, was also careful to maintain a friendly relationship with China and Tibet, for economic reasons.

[8] When the British withdrew from India in 1947, the Ranas were deprived of a vital external source of support and the regime was exposed to new dangers. Anti-Rana forces, composed mainly of Nepalese residents in India who had served their political apprenticeship in the Indian nationalist movement, formed an alliance with the Nepalese royal family, led by the king. A tripartite agreement between Nepal, India and the UK was signed in Kathmandu in November 1947, and Gurkha troops were used by India in the war against China, in 1961 to 62; Pakistan in 1965 and 1971, and by the UK against Argentina in 1982.

[9] In February 1951, the Nepalese Congress succeeded in overthrowing the Rana regime with support from King Tribhuvan Bir Bikram Shah Deva. From this time to December 1960 a series of democratic experiments took place in Nepal; political parties were allowed to form governments and a general election was held in 1959, based on the Constitution approved by King Mahendra Bir Bikran Shah Deva. On December 15 1960, the first elected prime minister, B P Koirala, was arrested, the parliament was dissolved and most of the provisions of the constitution were suspended. Political parties were outlawed and a tutelary democracy under a non-party system or Panchayat was introduced in December 1962.

[10] Nepal became a member of the United Nations in 1955, and it has been an active member of the Non-Aligned Movement since the days of the Bandung Conference. By July 1986, some 75 countries, including the major powers, endorsed a proposal by King Birendra Bir Birkram Shah Dev that Nepal should be declared a peace zone. All of its neighbors except India and Bhutan endorsed the proposal.

[11] India is Nepal's major trading partner, and they have signed several treaties on trade and transit since 1950. The last treaty expired in 1989, and the two countries became locked in an undeclared trade war. Nepal had earlier antagonized

---

WORKERS

1994

MEN 68% | WOMEN 32%

AGRICULTURE 93% | INDUSTRY 1% | SERVICES 6%

ENVIRONMENT

Nepal is a landlocked country in the Himalayas with three distinct geographical regions: the fertile, tropical plains of Terai, the central plateaus, covered with rainforest, and the Himalayan mountains, where the world's highest peaks are located. The climate varies according to altitude, from rainy and tropical, to cold in the high mountains. This diversity allows for the cultivation of rice, sugarcane, tobacco, jute and cereals. Livestock is also important; sheep and buffalo. Mineral and hydroelectric resources are as yet unexploited. Wood provides 90% of the energy consumed, resulting in deforestation and soil erosion. Air and water pollution especially affect urban areas. The lack of sewerage systems in the most important cities has also contributed to the deterioration of the environment.

SOCIETY

Peoples: The Nepalese are descended from Indian, Tibetan and Mongolian migrants. Nepalese 53.2%; Bihari (including Maithili and Bhojpuri) 18.4%; Tharu 4.8%; Tamang 4.7%; Newar 3.4%; Magar 2.2%; Abadhi 1.7%; other 11.6% **Religions:** Hinduism 86,2% (official), Buddhism 7,8%, Muslim 3,8%, Christian 0.2%; Jain 0.1%; other 1.9%. **Languages:** Nepali (official) is spoken by only half the population; many other languages are spoken, corresponding to the different cultural communities, with Tibetan being the second most common. **Political Parties:** The Nepalese Congress Party (NCP), led by prime minister Sher Bahadur Deuba; United Communist Party of Nepal (UCPN), National Democratic Party (NDM), the Communist Party of Nepal-Maoist (CPNM) and a people's front, the Sumukta Jan Morcha (SJM)

THE STATE

Official Name: Sri Nepála Sarkár. **Administrative divisions:** 75 districts. **Capital:** Kathmandu, 419,073 inhab. (1991). **Other cities:** Biratnagar, 130,129 inhab.; Lalitpur, 117,203 inhab.; Pokhara, 95,311 inhab.; Birganj, 68,764 inhab. (1991) **Government:** Birendra Bir Bikram Shah Deva, King since 1972. Sher Bahadur Deuba, leader of the Nepali Congress Party (NCP), Prime Minister since September 12, 1995. The House of Representatives has 205 members. **National Holidays:** February 15 (Constitution) and February 18 (Homeland) **Armed Forces:** 35,000 (1994) **Paramilitaries:** 28.000 (Police Forces).

DEMOGRAPHY

Urban: 13% (1995). **Annual growth:** 4.1% (1992-2000). **Estimate for year 2000:** 24,000,000. **Children per woman:** 5.5 (1992).

HEALTH

One **physician** for every 16,667 inhab. (1988-91). **Under-five mortality:** 118 per 1,000 (1995). **Calorie consumption:** 89% of required intake (1995). **Safe water:** 46% of the population has access (1990-95).

EDUCATION

Illiteracy: 73% (1995). **School enrolment:** Primary (1993): 85% fem., 85% male. Secondary (1993): 23% fem., 46% male. University: 3% (1993). **Primary school teachers:** one for every 39 students (1992).

COMMUNICATIONS

84 **newspapers** (1995), 82 **TV sets** (1995) and 80 **radio sets** per 1,000 households (1995). 0.4 **telephones** per 100 inhabitants (1993). **Books:** 82 new titles per 1,000,000 inhabitants in 1995.

ECONOMY

Per capita GNP: $200 (1994). **Annual growth:** 2.30% (1985-94). **Annual inflation:** 12.10% (1984-94). **Consumer price index:** 100 in 1990; 157.6 in 1994. **Currency:** 50 rupees = 1$ (1994). **Cereal imports:** 27,000 metric tons (1993). **Fertilizer use:** 391 kgs per ha. (1992-93). **Imports:** $1,176 million (1994). **Exports:** $363 million (1994). **External debt:** $2,320 million (1994), $111 per capita (1994). **Debt service:** 7.9% of exports (1994). **Development aid received:** $364 million (1993); $18 per capita; 10% of GNP.

ENERGY

Consumption: 23 kgs of Oil Equivalent per capita yearly (1994), 84% imported (1994).

India by importing arms from China, and India had suspended trade with Nepal, closing 19 of the 21 transit routes from March 1989, seriously affecting the Nepalese economy. Nepal claimed that transit was a basic right for a landlocked country and that India had accepted trade as a separate issue in previous treaties they had agreed on. India threatened to withdraw all special trade facilities and suspended the supply of petroleum products and other essentials.

[12] In 1979, student protest movements emerged in Kathmandu and other cities, challenging the system. King Birendra responded by holding a plebiscite to choose between a multiparty system or a reformed Panchayat. The 1980 referendum frustrated the hopes of a return to a multiparty system as 55 per cent of the electorate voted for continued, reformed, Panchayat rule. The transition to parliamentary democracy with a constitutional monarchy was proclaimed by royal decree in April 1990.

[13] The Birendra monarchy was accused of having one of Asia's worst human rights records. In mid-1989 more than 300 political prisoners were held in Nepalese jails. The king has even passed laws that contradict the human rights provisions of the country's 1962 constitution.

[14] On April 12 1991, the first free elections were held in Nepal after 32 years of semi-monarchic rule. The Communist Party of Nepal and the Nepalese Congress allied for the elections, and other groups joined them. The communists obtained 4 of the 5 seats in the capital, but the national majority went to the Nepalese Congress Party.

[15] In 1991, the Nepali and Indian governments signed two treaties on trade and transit, both subject to parliamentary approval. The major criticism, from the left and the right alike, was the fact that the 1950 Friendship Treaty was not revised or abrogated. Critics considered it to be a threat to the country's sovereignty, in that it expressly forbid the acquisition of military equipment from countries other than India.

[16] Escaping from the ethnic conflict in Bhutan, 22,000 Bhutanese people of Nepalese origin entered the country in 1992. One month later, a general strike against corruption and price increases paralyzed the nation. The repression of the demonstrations left many people dead.

[17] Girija Prasad Koirala, prime minister and secretary general of the Congress Party, announced the convertibility of the rupee to stimulate foreign investment, and new trade agreements were signed with India. Two thirds of the Nepalese budget was funded by foreign aid.

[18] From 1994 onwards, the government decided to charge $50,000 to any expedition planning to climb Mount Everest. The rivers flowing from the Himalayas potentially make Nepal one of the world's richest countries; however, the full exploitation of this resource for energy production, flood control and irrigation has been deterred by the lack of agreement between Nepal and India on sharing dam construction costs.

[19] Koirala was unsuccessful in his fight against poverty and illiteracy. The loss of parliamentary support and infighting in the monarchist Nepalese Congress Party (NCP) forced him to resign on July 10 1994. In the November elections, the United Communist Party took 88 seats in the Chamber of Representatives, surpassing the NCP who had 83. Man Mohan Adhikari was nominated prime minister. When unable to gain majority political support for his leadership in September 1995, the Communist prime minister handed over control to NCP leader, Sher Bahadur Deuba.

[20] According to the World Bank, Nepal is the eighth poorest country in the world. The annual per capita income stood at $180 in 1995, infant mortality was running at 10 per cent, and 71 per cent of the population lived under the poverty line. Living conditions in Kathmandu and the provinces are very different, for while life expectancy is 71 years in the capital, in rural areas the average drops to 34.

[21] In early 1996, masked guerrilla groups identified with the two Maoist parties went into action proclaiming they were eradicating feudalism. Some confrontations occurred, where police officers, guerrillas were killed.

Netherlands

Nederland

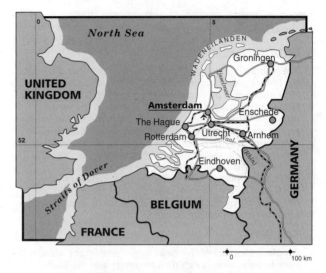

Population: 15,381,000 (1994)
Area: 37,330 SQ.KM
Capital: Amsterdam

Currency: Guilder
Language: Dutch

The Netherlands, Belgium, Luxembourg and the northern part of France constitute the region known as the Low Countries. The first inhabitants of these lands arrived at the end of the last Ice Age. They changed, over thousands of years, from hunting and gathering groups to the more elaborate and hierarchical cultures which the Romans encountered.

[2] When the Romans reached the area, in the 1st century BC, it was inhabited by Celts and Germanic people. The Empire never managed to occupy the land of the Frisians, in the north above the Rhine, so the Romans settled in the southern delta, where they created the provinces of Belgica and lesser Germanica.

[3] The Frisians lived by fishing and raising cattle, while in the south, agriculture was practiced around the villages and towns. By the second half of the 3rd century, the encroaching sea had drastically altered the economic basis of the region.

[4] The strengthening of the Germanic tribes forced Rome to grant them custody of the Empire's borders, as it did with the Franks in Toxandria and Brabant. The line between the Romance and Germanic languages ran across the middle of the Low Countries, coinciding with the borders of the Roman Empire.

[5] The Frisians remained independent until the 7th century, when the Franks and the Catholic Church started a strong offensive. By the end of the century, the region had been subdued by the Franks, under the Pepin and Carolingian dynasties.

[6] The decline of the Carolingian Empire led to a period of instability during the 10th century. Several principalities were formed, with feudal ties to the kingdoms of Germany and France. The Frisians still remained free of sovereign authorities.

[7] The secular principalities of Flanders (Vlaanderen), Hainaut, Namur, Loon, Holland, Zeeland, Guelders, and the duchies of Brabant and Limburg were formed. In the principalities of Utrecht and Liege, the secular and ecclesiastic authorities shared power.

[8] The principalities worked towards greater freedom from royal authority. Flanders was the first to establish an efficient administration, followed by Brabant, Hainaut, and Namur. The designation of their own bishops marked the end of German influence, and the establishment of closer links between the principalities.

[9] France tried to subdue Flanders but was defeated in the Battle of the Spurs in 1302. In general, France and Britain kept a balance of power, a situation that contributed to preserving the autonomy of the region.

[10] Population pressure led to the creation of new farmlands. On the coast, the Cistercian and Premonstratensian monks constructed many dikes. These were initially built to defend people from the high tides, but they later served to reclaim land.

[11] From the 11th century onwards, the Frisians developed a drainage system which removed the sea-water, providing land for pasturage and arable fields. During the 12th and 13th centuries, a vast peat bog in the Netherlands and Utrecht was reclaimed for agriculture.

[12] This area of polders on the coast of Flanders and Friesland, became economically very significant. Between the 12th and 14th centuries, the struggle against the sea and inland water became so important that water authorities were created to organize the construction of dikes and the use of water.

[13] This increase in arable land and population brought about growth, not only in agriculture, but also in industry and commerce. The new towns gave birth to new social classes who sought autonomy. Merchants in the towns had to swear an oath of co-operation; promising that they would keep law and order.

[14] The towns gradually became independent centers, having the power to sign commercial, political, and military agreements, with other towns or with the prince. The town owned the land within its boundaries and the inhabitants did not rely on any external authorities.

[15] During the second half of the 14th century, the dukes of Burgundy, from the French royal house of Valois, ruled over most of the Low Countries where they tried to create a centralized state. There was a movement to prevent this centralization, in 1477, but the accession of the Hapsburgs interrupted it.

[16] The fate of the Low Countries depended on the outcome of the House of Austria's struggle for European domination. Centralization was on the increase and the Church was no exception. A Papal Bull was issued, creating a Rome-based administration, with three archbishops and 15 bishops. This was very much resented by the local nobility.

[17] The Lutheran and Anabaptist faiths had difficulty gaining converts in the Low Countries, while Calvinism quickly gained acceptance among the lower classes and the intellectuals. Repression drove many Calvinists into exile, but they still had influence in the 1567 anti-absolutism rebellion.

[18] Popular unrest and the nobility and urban gentry's desire for autonomy led to a successful rebellion in Holland. After the defeat of the Spanish troops, the rebellion extended to all the provinces, ending with the Pacification of Ghent in 1576.

[19] Three years later, for geographic, economic, political, and religious reasons, this union dissolved. The union of Arras was founded in the south, and the union of Utrecht in the north, both of these being within the boundaries of a larger unit led by the States-General.

[20] In 1581, the States-General passed an Act of Abjuration making the Union of Utrecht into a State, seizing sovereignty from Phillip II. The Netherlands was the largest economic and political power within the union, so all of the union's inhabitants became known as Netherlanders.

[21] During the Twelve Years' Truce with Spain, 1609-21, controversies within the Union grew. The collaboration between the province of Holland and the House of Orange during the war gave way to a growing rivalry. This situation was antagonized by the dispute over Church-State relations.

[22] In 1618, Maurice of Orange, supported by the States-General, executed the leader of the main party in the Netherlands. When war against Spain resumed in 1621, the French and Dutch rivals were forced to re-unite until the signing of the Peace of Utrecht in 1713.

[23] The 17th century is called the Golden Age in Dutch history, because the country was the center of attention, and it established relations with the great powers of the time. This was viewed as the birth of a great nation, a fact that was only called into question at the end of the 17th century.

[24] Dutch prosperity was not only the result of continental trade, but also of its colonial power. In 1602, the East India Company was created, with bases in Ceylon, India and Indonesia. The Netherlands had sovereign powers in these colonies,

LAND USE

- CROPLAND 27%
- PASTURE 31%
- OTHER 41%

Foreign Trade

MILLIONS $ 1994

IMPORTS
139,795
EXPORTS
155,554

PROFILE

ENVIRONMENT

The country is a vast plain and 38% of its territory is below sea level. Intensive agriculture and cattle raising produce high priced milk products and crops (particularly flowers). Population density is amongst the highest in the world. Highly industrialized, the country is the world's third largest producer of natural gas, and is a dominant influence in petroleum activity. It has large refineries in the Antilles and Rotterdam, the world center of the free, or "spot", crude oil market. As a result of the massive use of pesticides, underground water supplies contain high levels of nitrates. Similarly, the country's main rivers are filled will all kinds of organic and industrial wastes, which often originate in other European countries.

SOCIETY

Peoples: Most are of Germanic origin. Immigrant minorities from former colonies include Surinamese and Antilleans. Immigrant workers; mainly Turks and Moroccans, making up about 5% of the economically active population. The port of Rotterdam has one of the largest overseas communities of Cape Verde nationals. **Religions:** 32% Catholic, Dutch Reformed Church 15%; Calvinist 7%; Muslim 3.7%; other 2.3%; no religion 40%. **Languages:** Dutch. **Political Parties:** Christian Democratic Appeal, center rightist; People's Party for Freedom and Democracy, liberal (right wing); Democrats 66, liberal (left wing); Labour Party, affiliated to the Socialist International; among the minor parties are three religious (protestant) parties; Green Left, the Centrum Democrats (ultra right) and the General Union of the Elderly . **Social Organizations:** There are two large central unions: the Federation of Dutch Trade Unions and the Christian National Trade Union.

THE STATE

Official Name: Koninkrijk der Nederlanden. **Administrative Divisions:** 12 Provinces. **Capital:** Amsterdam, 724,096 inhab. (1994). Although the government has its seat in The Hague, Amsterdam is still considered the capital. **Other cities:** Rotterdam, 589,521 inhab.; The Hague (s'-Gravenhage), 445,279 inhab.; Utrecht, 234,106 inhab., Eindhoven, 196,130 (1994). **Government:** Pluralist, constitutional and parliamentary monarchy. Queen Beatrix is Head of State; Wim Kok has been Prime Minister since August 22, 1994. Bicameral parliament, with 75 members elected by regional parliaments, 15 member second chamber. **National Holiday:** April 30. **Armed Forces:** 70,900 (1994). **Paramilitaries:** 3,600 Royal Military Corps. **Dependencies:** See "Netherlands" Antilles and Aruba.

as did the British and French in theirs.

[25] At first the company only held coastal trading bases, but later on, it started to occupy the hinterland controlling the region and certain commodities. The colonial administration was autonomous, as the Dutch preferred governing through agreements with local leaders.

[26] In 1621, the West Indies Company was founded. It made the bulk of its profits from the slave trade and privateering, it also operated outside Zeeland, especially against Spanish ships. The Netherlands dominated the slave trade during the 17th century.

[27] In 1648, the Dutch had three large settlements in the Americas: one in the north, for the fur trade; another on the Atlantic coast, with outposts for the slave trade and smuggling with Spanish colonies; and another in what is now Brazil and Suriname, formerly Dutch Guyana. By 1700, they were left with only the trading outposts in Curaçao, St Eustacius and St. Martin, the plantations in Guyana, and the slave port of Elmina.

[28] Dutch maritime power weakened in the 18th century, especially after the war against Britain (1708-84). The country used up a portion of its capital to buy bonds of foreign governments. The bankers in Amsterdam were among the most powerful in Europe.

[29] The 1750s saw the emergence of the Patriot movement, a group of factions who had been overlooked in governmental policies: they ranged from bankers and simple artisans, to dissident Protestants and Catholics, unhappy with monarchic abuse.

[30] The Patriots had to go into exile during the Prussian invasion of 1786. Their hopes were rekindled with the French Revolution, but it

> **The colonial administration was autonomous, as the Dutch preferred governing through agreements with local leaders.**

was not until 1794 that France managed effective opposition to Britain and Prussia, who supported William V. They proclaimed the Batavian Republic and started the process of political modernization.

[31] The new republic declared equality among its citizens and changed the institutional framework. It replaced the assembly of States-General for a National Assembly with direct electoral representation, and executive power was separated from the legislative and the judicial.

[32] In 1806, France annexed the state to the Empire, under the name of the Napoleonic Kingdom of the Netherlands. Five years later, Bonaparte incorporated the Netherlands into France, until the fall of the empire. In 1814, King William I of Orange was chosen by the Dutch leaders, and he restored the monarchy.

[33] In the Congress of Vienna, the victorious powers gave William I sovereignty over the whole of the Low Countries. The Belgian revolution of 1830 broke out during his reign. His successors, William II and William III made the final move towards parliamentary monarchy.

[34] Universal male suffrage was approved in 1917; women were granted the vote in 1922. After decades of debate over the school system, Protestants and Catholics allied themselves against the liberals and, in 1888, the first private denominational schools were opened.

[35] New political parties were founded, based on the religious ideas and ideologies of the time. To the Liberal, Protestant and Catholic parties were added the Protestant Conservative, the Socialist and the Communist parties. As none could obtain a majority, coalitions became commonplace.

[36] During World War I, the Netherlands declared its neutrality, and political parties agreed on a truce in order to devote themselves to the domestic economy and trade. The merchant navy had recovered and industry grew, in particular textiles, electronics, and chemicals.

[37] During the postwar period, the Netherlands was a member of the League of Nations, but it re-affirmed its neutrality. Symbolically, The Hague became the seat of the International Court of Justice. During the Versailles negotiations, Belgium tried unsuccessfully to revive an old territorial claim against the Netherlands.

[38] During World War II, Hitler attacked France through the Netherlands, and Queen Wilhelmina formed a government in exile in London. All political factions took part in anti-Nazi resistance. German repression was harsh, and by the end of the war the country was on the verge of famine.

[39] In 1945, an agreement was signed by the government, companies and trade unions. It lasted 20 years and was aimed at controlling prices and salaries. Indonesia became independent soon afterwards, while Suriname had to wait until 1975. The Netherlands underwent a rapid industrialization process, in particular in the fields of steel production, electronics, and petrochemicals.

[40] Dutch capital controls some

large transnational corporations, such as Royal Dutch Shell, the oil company (jointly a British company). Unilever, the world's largest food and soap manufacturer, has branches all over the world and its products are used by companies everywhere. Other Dutch transnationals include Philips electronics, AKZO, a chemical manufacturer, DSM, Hoogoven Groep and Heineken.

[41] In the postwar period, the government was composed of Labor party (the former socialists) and Catholic coalitions. The Netherlands gave up neutrality, becoming a member of NATO and of the European Economic Community. Also, together with Belgium and Luxembourg it formed an economic alliance known as the Benelux.

[42] During the 1960s, youth demonstrations developed into violent riots, the royal marriages also became the subject of public controversy and ideological and religious questioning occurred in all Dutch institutions.

[43] During the 1970s, voters favored the center and the left, while the government reformed the tax system and redistributed income. The most controversial subjects were the defence budget and the installation of NATO atomic missiles within the country.

[44] The Netherlands is one of the OECD countries giving over 0.7 per cent of its GNP for aid to the Third World. It has also maintained a coherent policy on the defence of human rights and against apartheid in South Africa, though links with Israel have tended to estrange it from some Arab countries.

[45] In 1989 the government approved a 0.6 per cent increase in the defence budget for 1990 and 1991, to be followed by a budget suspension until 1995 and the withdrawal of 750 Dutch soldiers posted in West Germany. This Dutch resolution caused much malaise within NATO.

[46] In the Netherlands there are some 100,000 farms, with around 4,500,000 head of cattle. The cost to the country of raising and feeding each cow is $2,000 per year, a sum which is higher than the per capita income of many Third World countries. The Dutch export dairy products and flowers, but also import a great amount of raw materials for their agro-industry. Fifteen million hectares of land in other countries (including 5 million in the

Third World) is under production for Dutch needs. This is mainly because much of the soya and tapioca that their farm animals eat comes from other countries, mainly Brazil and Thailand. Only 20 per cent of the vegetable protein fed to animals is converted to animal protein. The rest produces 110 million tons of manure a year, of which only half is used in agriculture. The other half ends up polluting the soil, drinking water and the air, where it adds to industrial emissions to cause acid rain. Over 40 per cent of Holland's trees have been irreversibly damaged by acid rain and the country has exceptionally high levels of contamination.

[47] Dutch agriculture is also characterized by a high energy consumption (in heated greenhouses to produce summer crops in winter) and the intensive use of pesticides, at a rate of about 20 kg per hectare per year. These seep into underground water reserves, contaminating drinking water supplies. Likewise, the high level of industrialization and population density have given rise to the presence of heavy metals, nitrates and other organic

> Over 40 per cent of Holland's trees have been irreversibly damaged by acid rain and the country has exceptionally high levels of contamination.

wastes in the Maas, Rhine and Waal rivers.

[48] Due to the high level of consumption which characterizes Dutch society, each person produces an estimated 3,000 kg of refuse per year. In Amsterdam and other cities many apartment buildings are inhabited by people who live alone, a fact which increases the demand for housing.

[49] Taking advantage of the fact that the Netherlands was occupying the EC presidency at the time, the Dutch Parliament exhorted EC heads of state and governments, to condemn all forms of racism in Maastricht in December 1991. In addition, they urged the adoption of legislation prohibiting xenophobic acts throughout Europe.

[50] Despite this official position, the country was beset by growing racism, although to a lesser degree than other European countries. In September 1993 the Netherlands adopted legislation restricting the admission of immigrants from outside the European Union.

[51] Governmental agencies devised economic indicators capable of assessing environmental damage, and methods for calculating national income. In 1992, the Central Office of Statistics announced the "green gross national product" (GGNP). This indicator evaluates losses in natural resources, according to the capacity for regeneration, and the effect on local communities.

[52] The parties in the coalition government (the Christian Democratic Appeal and the Labor Party) lost their majority in the March 1994 local elections and the growth of right-wing parties - some of which are xenophobic - became apparent. On August 22, Wim Kok took office as Prime Minister. Unemployment increases and the cost of social benefits were some of the problems that the new government had to deal with. Kok went on with the reduction and restructuring of the armed forces and the policy to strengthen educational quality.

[53] For the second consecutive year, in 1995, the complex hydraulic Dutch system showed flaws. The repairs of dikes, channels and damages cost $1 billion.

Netherlands Antilles

Population: 198,000 (1994)
Area: 960 SQ.KM
Capital: Willemstad

Currency: NA Dollar
Language: Dutch

The Caiqueti were the original inhabitants of what today are the islands of Aruba, Bonaire and Curaçao. Soon after the arrival of the Spaniard Alonso de Ojeda, in 1499, these local people were enslaved and taken to Hispaniola (present-day Haiti and Dominican Republic). The natives that Columbus found on the "S Islands" (the Windward Islands) in 1493 met a similar fate.

[2] The conquistadors attached little importance to these islands because they lacked readily exploitable natural resources. However, the Windward Islands (particularly St Martin Island) acquired strategic importance as a port of entry into the Caribbean and also became important for their salt deposits. Ports on the Iberian Peninsula were closed to Holland after the war with Spain and Portugal, so the Dutch were forced to seek alternative sources of salt in the Antilles.

[3] In 1606, the Spanish Crown decided to put a halt to this intense traffic, and prohibited Dutch shipping in the Antilles. Holland retaliated by creating the West Indies Company, which was charged with establishing, managing and defending the colonies. Thus the theft of Spanish ships became one of the main sources of income.

[4] Although salt and Brazilwood were important products, the slave trade became the principal economic activity. In 1634, the stockholders of the Company in Amsterdam decided to invade Curaçao. The Spaniards offered no resistance and Curaçao became a major international slave-trading center.

[5] In 1648, after three centuries of conflict, the Treaty of Westfalia was signed, granting Holland control over these small islands in the Antilles. The native populations - particularly those of Curaçao and Bonaire - were replaced by African slaves to meet the needs of new agricultural operations, and thus became a minority on these islands. As in the rest of the Antilles, slave rebellions were frequent, culminating in the 18th century with a massacre at the hands of colonial troops.

[6] During the Napoleonic wars, at the beginning of the 19th century, the three islands passed into British hands on two different occasions; however, the change failed to produce any improvement in the lives of the local inhabitants. Although the slave trade had been prohibited

in 1814, it was not abolished in the Dutch colonies until 1863. Even then, although the slaves were "free", their lives did not change substantially. Many migrated to the Dominican Republic, Panama, Venezuela and Cuba.

[7] Once slavery was abolished, interest in the islands lagged. In

> The issue of independence has long been at the center of local political life.

1876, the Dutch parliament proposed to sell them to Venezuela, but negotiations fell through. With the rise of the oil industry at the beginning of the 20th century, refineries were installed here because of the proximity of the islands to Lake Maracaibo. From the 1920s onward, this new activity radically changed the colony. The plantations were abandoned and the demand for labour attracted thousands of immigrants from

Venezuela, Suriname and the British West Indies.

[8] However, this boom dwindled over time, as automation significantly reduced the number of jobs. On May 30, 1969, when unemployment stood at 20 percent, police broke up a labor demonstration. This sparked large-scale riots which were quelled only by the joint intervention of 300 Dutch marines and some US marines from the American fleet, which happened to be anchored in the archipelago. As a result of these disturbances, parliament was dissolved.

[9] The islands have been politically active since 1937 when the first local parties were founded. However, they have only existed as states since 1948, when the new post-war Dutch constitution renamed what had been known as "Curaçao and dependencies" as the "Netherlands Antilles"

[10] The issue of independence has long been at the center of local political life. In 1954, a new law established the islands' autonomy over their internal affairs. Nevertheless, the People's Electoral Movement (PEM), founded in 1971 in Aruba, maintains that, as the islands are no longer "Dutch", they have nothing in common (not

even a name, because "Antilles" is the generic designation of the whole region) and each island should be free to choose its own constitution and set itself up as an autonomous republic. Secessionist feeling was strong in Aruba, where a referendum carried out in 1977 for consultation purposes only (i.e., not politically binding), showed that the majority wished to separate from the other islands.

[11] The Netherlands however preferred to maintain the political unity of the islands, arguing that this would ensure better economic prospects for the whole group and, above all, more political stability in the turbulent Caribbean region. This difference of opinion, together with the diversity of views held by the local parties on the matter, delayed negotiations for independence.

[12] The delay radicalized the electorate and in 1979 the New Antilles Movement (MAN) achieved a parliamentary majority in Curaçao (7 out of 12 seats). In coalition with the PEM and the UPB (Bonaire Patriotic Union) it formed the first left-of-center government in the history of the islands. The MAN favoured a federal formula with considerable autonomy for each of

Illiteracy
1995

34%

External Debt
1994

PER CAPITA
$2,651

PROFILE

ENVIRONMENT

The Netherlands Antilles are made up of two Caribbean island groups. The main group is composed of Bonaire (288 sq km) and Curaçao (444 sq km). Located near the coast of Venezuela, they are known (together with Aruba) as the "Dutch Leeward Islands" or the "ABC Islands". The smaller group is made up of three small islands of volcanic origin: St Eustatius (21 sq km), Saba (13 sq km) and Saint Martin (34 sq km - the southern part of which is a dependency of the French island of Guadeloupe). These are known as the "S Islands" or the "Dutch Windward Islands", although in fact they are part of the Leeward group of the Lesser Antilles. In general the climate is tropical, moderated by ocean currents. There is little farming. One of the largest oil refineries in the world is located on Curaçao; it is leased to a Venezuelan oil company. The coastal areas of the islands have suffered the consequences of economic development. The soil has been polluted, and waste disposal poses a serious problem. Curaçao is the most polluted of the islands.

SOCIETY

Peoples: Most of the population are descended from African slaves. **Religions:** Mainly Catholic.
Languages: Dutch (official). The most widely spoken language in Curaçao and Bonaire is *papiamento*, a local dialect based on Spanish with elements of Dutch, Portuguese, English and some African languages. In St Eustatius, Saba and St Martin, English is the main language. Spanish is also spoken and many claim that it should be taught as a second language (instead of English) to facilitate integration with Latin America. **Political Parties:** National People's Party (NVP), a group which broke from the Catholic Party in 1948, is linked to the World Christian Democratic Union. New Antilles Movement (MAN), founded in 1971 by left-wing intellectuals who split from the Workers' Front, is led by former premier Domenico "Don" Martina, of Curaçao.

In 1981 the party joined the Socialist International. Workers' Liberation Front (FOL). Patriotic Union of Bonaire (UPB), the island's major party. Democratic Party (DP), the only party with supporters on all six islands. **Social Organizations:** The largest labor confederation in Curaçao is the National Confederation of Curaçao Trade Unions (AVVC), with 13,000 members. Next in importance are the General Workers' Union of Curaçao (CGTC) with 5,000 members, linked to the social-Christian Latin American Workers' Confederation (CLAT); the Workers' Federation of Aruba (FTA) with 4,800 members, linked politically to the MEP; the General Federation of Bonaire Workers (AFBW) and the Bonaire Labor Federation (FEDEBON), each with 500 members.

THE STATE

Capital: Willemstad, 50,000 inhab. (1981).
Government: Jaime M. Saleh, governor appointed by the Dutch government. Miguel Pourier, prime minister since 1994. There is an Advisory Council and a Council of Ministers, responsible to the Staten (Legislature) composed of 22 elected members. Defense and foreign relations are handled by the Dutch crown. **National Holiday:** May 30. Anti-colonial Movement (1969).

COMMUNICATIONS

25.5 **telephones** per 100 inhabitants (1993).

ECONOMY

Consumer price index: 100 in 1990; 109.5 in 1994.
Currency: 2 NA dollars = 1$ (1994).

---

the islands but the PEM insisted on the total separation of Aruba.

[13] In 1980 it was agreed to set independence for 1990, on the condition that each of the six islands submit the issue to a plebiscite as soon as possible. Aruba chose to become an individual associated state, breaking away from the federation in January 1986 (see Aruba).

[14] The Netherlands Antilles have become modern trading posts, totally dependent on trans-national oil companies. Exxon and Royal Dutch Shell (an Anglo-Dutch consortium), maintain a monopoly on all oil refining and processing. They also control the petrochemical fertilizer industry and handle all transportation and sales. These powerful companies, linked financially and commercially to branches of approximately 2,500 foreign firms registered on the islands, have

an impact on 85 per cent of the total imports, 99 per cent of the exports and 50 per cent of the islands' net income.

[15] Despite modest industrial growth and a booming tourist industry, the economy is fragile. This was clearly demonstrated toward the end of 1984, when the transnational oil companies announced their withdrawal from the islands and spread panic among the people and local political leaders. The conflict was partially resolved in October 1985, with the purchase of the Curaçao refinery by the Antilles government, which in turn rented the facililty to Petroleos de Venezuela S.A., an oil company owned by the Venezuelan government. The deal did not include the Exxon plant on Aruba, which closed.

[16] The separation of Aruba favored the 1984 victory of a right-of-center coalition headed by Maria

Liberia Peters of the National People's Party (PNP). This coalition was unable to sustain the minimum consensus necessary to stay in power. Therefore in January 1986, MAN leader, Domenico Martina, became prime minister once again. Two years later, Maria Liberia Peters returned to power.

[17] In 1990 the government renewed its contracts with the Venezuelan oil company and introduced a series of austerity measures designed to cover the deficit generated by the Aruban withdrawal from the federation.

[18] The election of the Island Councils in 1991 was marked by the defeat of the Democratic Party as a result of internal divisions and financial irregularities committed by the administration.

[19] The Netherlands requested that each island make a separate proposal for constitutional reform in 1993. Accordingly, Curaçao received special status, St Martin was made independent, while Bonaire, Saba and St Eustatius continued under Dutch control. The constitution also dictated that the Treaty of Strasburg, which predicted complete independence for 1999, would not be applied in the islands.

[20] In a plebiscite held in 1994, the constituent members of the Netherlands Antilles voted to preserve their federation.

Nicaragua

Population: 4,156,000 (1994)
Area: 130,000 SQ.KM
Capital: Managua

Currency: Córdoba
Language: Spanish

Nicaragua

What today is Nicaragua was in pre-Colombian times an area of influence for the Chibcha (see Colombia) and the Maya. The Caribbean coast was inhabited by the Miskito and was visited by Christopher Columbus in 1502. By converting local leaders Nicoya and Nicarao to Christianity and crushing the resistance of Diriangen's armies, the conquistadors Gil Gonzales Dávila and Andrés Nino consolidated Spain's hold over the territory. In 1544 it was incorporated into the Captaincy-General of Guatemala.

[2] In 1821 Nicaragua became independent, together with the rest of Central America, joining the Mexican Empire but subsequently withdrawing in 1824 to form the Federation of the United Provinces of the Center of America.

[3] Nicaragua left the Federation in 1839, declaring itself an independent state. The country was divided between two groups: the coffee and sugar oligarchies, and the artisans and small landowners. The former would become the conservatives, and the latter, the liberals who favored free trade.

[4] In 1856, 120 men landed in Nicaragua under the command of William Walker, an American mercenary. With Washington's unspoken but tacit support, he proclaimed himself president of Nicaragua. His purpose was to find new territories for slavery which was on the point of being abolished in the Union. Walker was defeated by the allied armies of Central America in 1857, and later executed. Nicaragua's ports were occupied by Germany in 1875 and Britain in 1895. Britain commandeered Nicaragua's Customs in order to collect on unpaid debts.

[5] After 30 years of conservative rule the Liberal Party came to power in 1893, and José Santos Zelaya became president. The liberals refused to comply with demands made by the United States so in 1912 the US invaded. After killing the Liberal Party leader, Benjamin Zeledón, the US Marine force remained in Nicaragua until 1925. In 1926 they returned to protect president Adolfo Díaz, who was about to be overthrown.

[6] This second US occupation was resisted by Augusto C. Sandino, who raised and commanded a popular army of about 3,000 troops. For more than six years they held out against 12,000 US Marines who were supported by the airforce and ground troops of the local oligarchy. Sandino promised that he would lay down arms when the last marine left Nicaragua. He carried out his promise in 1933, but was assassinated by US-backed Anastasio Somoza García who, after seizing power, ruled despotically until he was killed by the patriot Rigoberto López Pérez in 1956.

[7] During two decades of dictatorship, Somoza had achieved almost absolute control of the nation's economy. He was succeeded by his son, engineer Luis Somoza Debayle, who in turn handed the government on to his son, Anastasio, a West Point graduate.

[8] Anastasio outlawed all trade unions, massacred members of peasant movements and banned all opposition parties. In the 1960s the Sandinista National Liberation Front was founded. This organized and developed guerrilla warfare which lasted for 17 years. When opposition leader Pedro Joaquín Chamorro, editor of the daily La Prensa, was assassinated on Somoza's orders in January 1978, a nationwide strike was called and there were massive protest demonstrations.

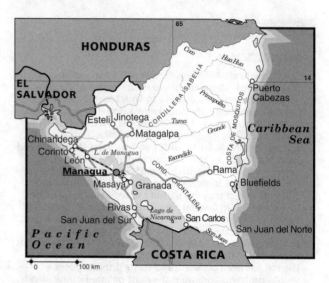

[9] By March 1979, the Sandinista Front had unified its three factions and began to lead the political opposition, the 'Patriotic Front'. In May 1979 the Front launched the "final offensive", combining a general strike, a popular uprising, armed combat, and intense diplomatic activity abroad. On July 17, Somoza fled the country, bringing an end to a dynasty that had killed 50,000 people. The Junta for National Reconstruction, created a few weeks before in Costa Rica, was installed in Managua two days later.

[10] The victorious revolutionaries nationalized the Somozas' lands and industrial properties, which constituted 40 per cent of the economic resources of Nicaragua. They also replaced the defeated National Guard with the Sandinista Popular Army. The revolutionary government implemented a literacy campaign and began the reconstruction of the devastated economy.

[11] In May 1980 two non-Sandinista members of the Junta, Violeta Barrios de Chamorro and Alfonso Robelo, resigned. The government avoided a crisis by replacing them with Rafael Córdoba and Arturo Cruz, two "moderate" anti-Somoza activists. Other important policies were; non-alignment, a mixed economy, political pluralism, and respect for individual rights and liberties.

[12] In 1981, US President Ronald Reagan announced his aim of destroying the Nicaraguan Sandinistas. Between April and July 1982 deputy interior minister, Edén Pas-tora ("Commander Zero"), deserted and announced from Costa Rica that he would drive the Sandinista Front's National Directorate out "at gunpoint". 2,500 former National Guardsmen, supported by the US, invaded Nicaragua from Honduras. From then on Nicaragua was harassed without respite, forcing the authorities to extend the state of emergency, to institute compulsory military service and to ban pro-American political declarations.

[13] In 1983, President Reagan admitted the existence of secret funds destined for covert CIA operations against Nicaragua. The funds were also used to aid counter-revolutionaries or *contras* operating from Honduran territory. Reagan referred to the *contras* as "freedom fighters".

[14] Concerned at the serious threat of a war that might escalate throughout Central America, the governments of Colombia, Mexico, Panama and Venezuela began to seek a negotiated settlement to the conflict. As the "Contadora Group", these countries' foreign ministers advanced peace plans which won great diplomatic support and prevented an imminent invasion by US forces.

[15] *Contra* attacks intensifed steadily with open US backing. Elections were held in November 1984. Candidates were drawn from the Sandinista National Liberation Front (FSLN), the Democratic Conservative Party, the Independent Liberal Party, the Popular Social Christian Party, the Communist Party, the Socialist Party and the Marxist-Leninist People's Action Movement. 82 per

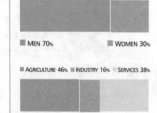

WORKERS

14% UNEMPLOYMENT

1994

MEN 70%   WOMEN 30%

AGRICULTURE 46%   INDUSTRY 16%   SERVICES 38%

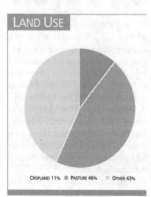

LAND USE

CROPLAND 11%   PASTURE 46%   OTHER 43%

ENVIRONMENT

Nicaragua has both Pacific and Caribbean coastlines. It is crossed by two important mountain ranges: the Central American Andes, running from northwest to southeast, and a volcanic chain with several active volcanoes along the western coast. The Managua and Nicaragua lakes lie between the two ranges. On the eastern slopes, the climate is tropical with abundant rainfall, while it is drier on the western side where the population is concentrated. Cotton is the main cash crop in the mountain area, while bananas are grown along the Atlantic coast. Changes in water and soil have affected approximately 40% of the country's territory.

SOCIETY

Peoples: Over 70% of Nicaraguans are mixed descendents of American natives and Spanish colonizers; there are minorities of Europeans, Miskitos, Sunos and Ramas, and African descendants. **Religion:** Mainly Catholic. **Languages:** Spanish (official and predominant). Miskito, Suno and English are spoken on the Atlantic coast. **Political Parties:** Sandinista Front for National Liberation (FSLN); Political Opposition Alliance, formerly the National Union of the Opposition (UNO), made up of: Democratic Party of National Confidence (PDCN), National Conservative Party (PNC), Conservative Popular Alliance Party (PAPC), Independent Liberal Party (PLI), Social Democratic Party (PSD), Liberal Constitutionalist Party (PLC). Other parties include: Nicaraguan Socialist Party (PSN); Communist Party of Nicaragua (PCdeN); Neoliberal Party (PAL); Nicaraguan Democratic Movement (MDN); National Action Party (PAN); Central American Integrationist Party (PIAC); National Conservative Alliance Party (PANC); People's Social Christian Party (PPSC); Yatama Indigenous Movement; Yatama Social Christian Party (PSC); United Revolutionary Movement (MRU); Party of the Nicaraguan Resistance (PRN). **Social Organizations:** The Sandinista Labor Confederation (CST); the General Labor Confederation (Independent) (CGT-I); the Nicaraguan Labor Confederation (CTN); the Labor Unity and Action Confederation (CAUS); the Rural Workers' Association (ATC); the Workers' Front (FO); the Unified Labor Confederation (CUS); the National Employees Union (UNE); the National Confederation of Professionals (CONAPRO).

THE STATE

Official Name: República de Nicaragua. **Administrative Division:** 16 Departments. **Capital:** Managua, 973,800 inhab. (1992). **Other cities:** León, 172,000 inhab.; Masaya, 101,900; Chinandega ,101,600; Matagalpa, 95,300; Granada, 91,900 (1992). **Government:** President, Arnoldo Aleman, elected in October 1996. **National Holidays:** September 15, Independence Day (1821); July 19, Sandinista Revolution Day (1979). **Armed Forces:** 15,200 troops (1994).

---

cent of Nicaragua's 1.5 million registered voters went to the polls and the Sandinista Front government led by Daniel Ortega obtained 67 per cent of the vote. In November Reagan was re-elected and in April 1985 he declared a trade embargo against Nicaragua and seized its assets.

[16] In 1986, the Assembly discussed the details of a new Constitution, which went into effect in January 1987. It provided for a presidential system, with a president elected by direct vote for a six-year term. Legislators would be elected on the basis of proportional representation.

[17] In 1987, with UN and OAS participation, the Central American presidents met for negotiations in Esquipulas, Guatemala. The Esquipulas II accords stipulated an end to external support for armed opposition groups; the opening of internal dialogue in each of the countries, mediated by the Catholic Church; and an amnesty for those who lay down their arms, with guarantees of political representation.

[18] In Nicaragua a National Reconciliation Commission was formed. Among its most spectacular achievements was the return of *contra* leader Fernando Chamorro from exile. He was granted an amnesty after he renounced violence. Press censorship was lifted and Violeta Chamorro's opposition daily *La Prensa* re-appeared. On October 7, a unilateral cease-fire went into effect in several parts of the country, although *contra* leaders announced that they would continue hostilities.

[19] Throughout 1988, as US pressure increased , the economic situation worsened. A monetary reform and a 10 per cent reduction in the government budget in February were insufficient to halt spiralling inflation.

[20] In July 1988, the US ambassador to Managua was expelled on the accusation of encouraging anti-Sandinista activities. The US government responded by expelling Nicaragua's representative in Washington.

[21] The Esquipulas II accords seemed to be doomed to oblivion, but when the five Central American presidents met at Costa del Sol, El Salvador, in February 1989, President Daniel Ortega embarked on fresh negotiations. The Sandinista proposal was to bring the elections forward to February 1990 and to accept proposed modifications to the 1988 electoral law. The condition was that the *contras* dismantle their bases in Honduras within three months of an agreement. The US however insisted that the *contras* continue in Honduras, and George Bush persuaded Congress to award them $40 million in "humanitarian aid".

[22] Daniel Ortega was the FSLN presidential candidate. The National Opposition Union (UNO), a 14-party coalition, presented Pedro Joaquin Chamorro's widow, Violeta Barrios de Chamorro. Unexpectedly, the UNO won the elections with 55 per cent of the vote against the FSLN's 41 per cent. The Sandinistas accepted defeat and ascribed it to the Nicaraguan people's desperate desire for peace, the dire state of the economy and FSLN overconfidence that victory was assured.

[23] On April 25, before she took office, the President and the FSLN signed a "Transition Protocol". This included respecting the standing Constitution and the social achievements of the revolution, and supporting disarmament of the *contras*. The new president announced that she would personally assume the Defense portfolio and maintain the Sandinista General Humberto Ortega as commander of the armed forces. She also indefinitely suspended compulsory military service.

[24] The UNO's vice-president, Virgilio Godoy, and other members of the coalition withdrew from the government accusing Chamorro of betraying pre-election agreements by keeping Humberto Ortega in office.

[25] In May 1990, public employees went on strike for wage increases of up to 200 per cent. The government declared the strike illegal, and revoked the civil service law (under which civil servants cannot be fired without just cause) as well as the agrarian reform law passed by the Sandinista government. Workers responded by extending the strike over the whole country. After a week, the government partially gave in to the workers' demands, and the strike ended.

[26] In mid-1990, the government received several offers from international consortia interested in carrying out projects in northern Nicaragua. This area comprises some 270,000 hectares of tropical rain forests, occupying more than half of the country's total land area. The

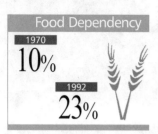

proposals ranged from the creation of landfills for toxic waste, to the exploitation of the region's vast fishing, mineral and forestry resources.

[27] When the government was accused of carrying out secret negotiations with a Taiwanese enterprise, the existence of large mineral deposits was inadvertantly revealed. These include gold, silver, copper, tungsten and Central America's largest deposits of calcium carbonate, a raw material used in cement production.

[28] A partial disarming of the *contras* was carried out, at the same time that significant reductions in

Inflation fell from 7,000 per cent in 1990 to 3.8 per cent in 1992 due to an IMF and World Bank-sponsored adjustment program. Productive investments and spending in education and health were reduced. Unemployment rose to 60 per cent.

army personnel were announced. In October 1991, a number of former *contra* commanders founded the Nicaraguan Resistance Party. This new political group rejected the return to violence by other *contras* - some 600, in all - in the northern part of the country, where civilians were being killed, and farms as well as farm cooperatives were being burned.

[29] In 1991, President Chamorro agreed with the FSLN to recognize agrarian reform and to set aside for the workers at least 25 per cent of shares in state enterprises slated for privatization.

[30] Inflation fell from 7,000 per cent in 1990 to 3.8 per cent in 1992 due to an IMF and World Bank-sponsored adjustment program. Productive investments and spending in education and health were re-

STATISTICS

DEMOGRAPHY

Urban: 62% (1995). **Annual growth:** 0.9% (1992-2000). **Estimate for year 2000:** 5,000,000. **Children per woman:** 4.4 (1992).

HEALTH

One **physician** for every 2,000 inhab. (1988-91). **Under-five mortality:** 68 per 1,000 (1995). **Calorie consumption:** 95% of required intake (1995). **Safe water:** 58% of the population has access (1990-95).

EDUCATION

Illiteracy: 34% (1995). **School enrolment:** Primary (1993): 105% fem., 105% male. Secondary (1993): 44% fem., 39% male. University: 9% (1993). **Primary school teachers:** one for every 37 students (1992).

COMMUNICATIONS

101 **newspapers** (1995), 100 **TV sets** (1995) and 103 **radio sets** per 1,000 households (1995). 1.7 **telephones** per 100 inhabitants (1993). **Books:** 83 new titles per 1,000,000 inhabitants in 1995.

ECONOMY

Per capita GNP: $340 (1994). **Annual growth:** -6.10% (1985-94). **Annual inflation:** 1.311.20% (1984-94). **Consumer price index:** 100 in 1990; 3,767.4 in 1992. **Currency:** 7 córdobas = 1$ (1994). **Cereal imports:** 125,000 metric tons (1993). **Fertilizer use:** 246 kgs per ha. (1992-93). **Imports:** $824 million (1994). **Exports:** $352 million (1994). **External debt:** $11,019 million (1994), $2,651 per capita (1994). **Debt service:** 38% of exports (1994). **Development aid received:** $323 million (1993); $79 per capita; 18% of GNP.

ENERGY

Consumption: 241 kgs of Oil Equivalent per capita yearly (1994), 84% imported (1994).

duced. Unemployment rose to 60 per cent.

[31] Differences between the president and the UNO led them to break in 1993 after which Chamorro received support from the Sandinistas and the UNO's Center Group. The following month, the UNO expelled that group and changed its name to Political Opposition Alliance (APO).

[32] Parliamentary debate on constitutional reform caused the FSLN's orthodox sector, led by former president Daniel Ortega, to exclude former vice-president Sergio Ramirez, head of the Sandinista parliamentary bloc, from the Front. By-passing the party's leadership, the parliamentary bloc presented its own bill against nepotism which banned presidential re-election and prohibited relatives of the current president from running for presi-

dent. This clause put an end to the political aspirations of Chamorro's son-in-law, minister Antonio Lacayo.

[33] The economic crisis was intensified by a drought which led to the loss of 80,000 hectares of crops and left 200,000 farmers without food. Child malnutrition has affected 300,000 children and some have been blinded by a lack of vitamin A.

[34] In January 1994, the UNO, with less than half its founders and unable to obtain the support to set up a constituent assembly, put an end to a year of boycotting the National Assembly. Violence continued between the army, gangs of criminals and small guerrilla groups.

[35] General Humberto Ortega confirmed he would resign after a new military law had been passed. In

August the Assembly passed the law which aimed at eliminating political involvement by the Sandinista Popular Army and increasing its dependence on civilian authority, although the power was actually left in the hands of a military council.

[36] The government signed a 3-year agreement with the IMF which opened the possibility of renegotiating public debt. Unemployment was estimated between 43 per cent and 60 per cent. The per capita GDP dropped for the eleventh year in a row.

[37] Debate on the constitutional reform prevailed in the political scene in 1995. In February, the Assembly proposed to change the army's name, ban compulsory military service and grant guarantees to private property. These measures were supported by President Chamorro but she did not agree with the shift of power from the executive to the legislative branch, eg regarding the right to raise taxes. The Assembly published the reforms unilaterally in February and began to implement them.

[38] In June, an agreement was reached on a general law for constitutional reforms which stated they had to be supported by a majority of 60 per cent in the Assembly before being signed by the president, who concluded the agreement in July. The National Assembly ratified its choice of judges for the Supreme Court of Justice and a new Supreme Electoral Tribunal was appointed.

[39] The approval of the nepotism law was deferred. President Chamorro's son-in-law remained in the government, although he announced his resignation to launch his presidential campaign.

[40] Conservative Arnoldo Aleman, ex-mayor of Managua, obtained 49 per cent of the vote in the October 1996 elections. The electoral law establishes that if any of the candidates obtains more than 45 per cent of the vote, there is no need of a second vote, thus leaving Sandinista ex president Daniel Ortega (with 39 per cent) out of the race. The FSLN denounced a fraud but the Supreme Electoral Court named Aleman president.

Niger

Population: 8,730,000 (1994)
Area: 1,267,000 SQ.KM
Capital: Niamey

Currency: CFA Franc
Language: French

Niger

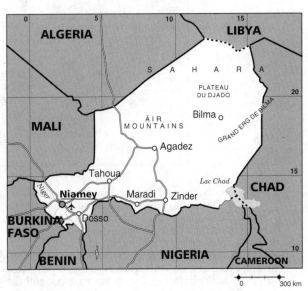

Fossil remains found in the Niger region indicate that it was inhabited in prehistoric times and later populated by many different nomadic and trading peoples. The Nok Empire, reaching its peak in present day Nigeria between the 15th century BC and the 5th century AD (see Nigeria) left its mark on this neighboring region.

During the 7th century AD, the western part of the country became a part of the Songhai Empire created by the Berber (see box, Guinea), who were important propagators of Islam from the 11th century onwards.

Between the 14th and 19th centuries the eastern part of the territory belonged to the Kanem-Bornu State, which had been founded by the Kurani in the 8th century (see box, Chad). Meanwhile, during the 19th century, the Haussa states flourished in the south (see box, Nigeria); until conquered by the Fulani.

2 Throughout the 19th century, the French colonized the territory, making Niger an official French colony in 1922. Traditional subsistence crops were replaced by cash crops, like peanuts and cotton, which were grown for export, causing endemic food shortages in Niger.

3 In the 1950s, Niger launched its own independence movement, led by Hamani Diori. In 1960, the country's first constitution was approved and Niger became an independent state.

4 When it broke its colonial ties, Niger was the poorest country in French West Africa, with 80 per cent of the population living in rural areas with persistent drought, soil erosion and population pressure that threaten the country's agriculture and ecology.

5 In the new republic's first election, Hamani Diori, the Progressive Party's candidate, was elected president. The new government maintained deep economic and political ties with France, to the point of allowing French troops to remain within its territory. In the first years of his presidency, Diori banned the Sawaba party, forcing its leader, Djibo Bakari into exile. The government was accused of corruption and of the harsh repression of the growing political opposition.

6 In the early 1970s, the drought which hit the Sahel region meant that the army was charged with distributing food among the peasants. This brought them into contact with the needs of the rural people for the first time. On April 13 1974, a Supreme Military Committee took power, naming Seyni Kuntch'e president. Price controls were set on agricultural products, salaries were raised, nepotism was eliminated, investments were reoriented and education and health services were planned. An attempt was made to access underground water sources and incentives were given to set up farming cooperatives.

7 The new government tried out new ways of organizing the youth to establish the political base which the country lacked. It also expelled Djibo Bakari, who had returned from exile to lend his support to the regime, and signed bilateral agreements with France. These agreements allowed neocolonialism to focus greater attention on the exploitation of Niger's mineral wealth than on traditional colonial products.

8 During the 1970s the country experienced an economic boom, based on an increase in the international price of uranium. This mineral accounted for 90 per cent of the country's exports in 1980, when the so-called "miracle" came to an abrupt end.

9 The foreign debt was $207 million in 1977, and by 1983 it had increased to $1 billion, forcing Kuntch'e to introduce an IMF supervised structural adjustment program, in an attempt to turn the economy around. The hoped-for return of favorable uranium prices did not materialize and, between 1984 and 1985, the perennial drought in the Sahel region worsened. The government also confronted political challenges on different fronts: in 1983, it put down an attempted coup by former members of the secret police and it also engaged in combat with the Tuareg, an ethnic group which had resorted to guerrilla warfare.

10 The economic crisis worsened when Nigeria closed its borders between April 1984 and March 1986 halting traffic between the two countries, including the transportation of cattle and basic foodstuffs. The decision to close the borders was made by authorities in Lagos to carry out a readjustment of their monetary system. An exception was made in 1985, allowing fuel to be transported across the border.

11 In 1985 there was no change in the situation: the price of uranium fell an additional four points, there was a deficit of 400,000 tons of grain, a growing foreign debt and the increasing cost of servicing this debt

12 In 1986 the government tried changing its policies, but General Kuntch'e had a cerebral hemorrhage, dying in a Paris military hospital eleven months later. The Supreme Military Committee designated Ali Seibou as his successor. Among his first measures as president was the appointment of 10 new ministers, and the declaration of an amnesty which provided for the return of political exiles, including Djibo Bakari and Hamani Diori, who were personally welcomed by the president and were allowed to return to political activity. In April 1989, Diori died in Rabat, Morocco.

13 In 1988, Seibou was faced with 3,000 protesting students, who boycotted classes for 22 days, until their demands were finally met.

14 On August 2, the National Movement for a Developing Society (MNSD) was formed as the only government authorized party, and the National Development Council drew up a new constitution, put to the vote and approved by plebiscite in 1989. In December, Seibou was elected president of the Republic by universal suffrage, in the first elections to be held since independence in 1960.

15 The drought in the Sahel forced the government to give special attention to agricultural production and to the inhabitants of the coun-

WORKERS

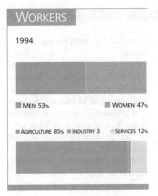

1994

MEN 53%   WOMEN 47%

AGRICULTURE 85%   INDUSTRY 3   SERVICES 12%

LAND USE

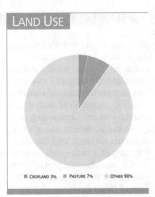

CROPLAND 3%   PASTURE 7%   OTHER 90%

Routes of the Sahara

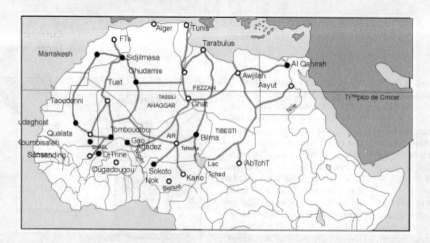

It is generally believed that the Sahara has always existed or that it was always a natural boundary between the nations of the Mediterranean coast and the peoples inhabiting the savannahs and tropical rainforests. These suppositions are incorrect.

2   Seven thousand years ago, the desert of Tassili was covered with cypresses; the arid region of Ahaggar was a vast savannah where giraffes and ostriches roamed, and the desert of Ténéré (one of the most desolate regions of the Sahara) was an enormous lake that fed many fishing villages. The last vestige of this vast water reservoir is today's Lake Chad.

3   The progressive desiccation of the area throughout the centuries forced innumerable human migrations towards the coast or towards the green savannahs of the south. In the Sahel area, a transitional barren plain between the desert and the savannahs, there was also human transition; groups of shepherds, farmers and fisherfolk; populations coming from the Nile or the Atlantic coast, from the tropical rainforests or from the Libyan beaches, all intermingled there.

4   The great desert was never an insurmountable barrier between this region and the Mediterranean shores. Even in the 5th century BC, Greek historian Herodotus wrote about the routes that led into the Sahara and were used by "Garamants" (Berber) in their trade with "Ethiopians". These routes are marked on rupestrian paintings and decorations showing carts with four wheels being used to carry goods. From Greco-Roman times, and even before, traders travelled to the south in search of gold and slaves that were exchanged for salt and manufactured goods. For centuries these routes were used continuously in both directions.

5   The western route, beginning in Sidjilmasa, a salt production center in Morocco, allowed traders to reach the capital of the Ghanaian empire in two and a half months. In 1913,

the salt caravan that covered the Bilma-Chad route twice a year was made up of 25,000 loaded camels.

6   As groups of often aggressive nomadic shepherds (Berber, Tuareg, Fulani, Tibu and Nilotic), settled in the region the existence of organized states which provided a minimum of security in the trade zone, became a key element in enabling trade to develop. A series of states and empires succeeded one another throughout the Sahel, guaranteeing the continuance of local commerce with their protection. Outstanding examples were the state of Ghana-Uagadu (see Mauritania); the Almoravid Empire (see Morocco); the Empire of Mali (see Mali); the Songhai Empire (see Guinea); the Haussa states (see Nigeria); Kanem-Bornu (see Chad) and the Fulani states (see Senegal). In spite of their strength and influence, it is impossible to establish a direct relationship between these states and the first known sub-Saharan civilizations. Nok, a nation that flourished in present-day Nigeria between the 10th century BC and the fifth century AD, was one of these independent civilizations, which developed metal forging techniques influencial over the rest of Africa.

7   Wars, conflicts and climatic changes had an influence on trade routes: they appeared, and fell into disuse, causing trade centers to rise and decline, bringing about the downfall of some of the political centers that had made exchange possible. Mediterranean civilizations would not have flourished as they did without the Sudanese gold which they depended upon, and the populations of the savannahs and rainforests would have developed differently without the salt produced north of the Sahara. Thus, the development of a vast region was historically dependent on the successive city-states of this region.

---

try's rural areas. During the 1980-90 period they invested 32 per cent of the national budget in agriculture. A fundamental part of the environmental problem is that 98 per cent of people use firewood for cooking. In a country facing deforestation and almost permanent drought, this almost exclusive use of wood for household energy is dangerously destructive. Facing the problem head-on, the government had solar and wind-powered gener-

ators installed, developed electric energy and sponsored the manufacture of low-cost energy-saving stoves. With wood consumption at close to 2 million tons per year, the government are proposing six-fold increases in the cost of wood-cutting permits, in the hope that this will put an end to further deforestation.

16   The new president has a favorable economic outlook on his side, as there was a grain surplus of

200,000 tons in 1989. He promised that he would initiate a true democratization process, but this did not occur and throughout 1990 there was intense opposition from the political sector, the labor unions and the students, demanding salary increases, the enactment of educational reforms, and the establishment of a multi-party system. Strikes and massive protest demonstrations were organized in support of these demands and these

were harshly suppressed by the police.

17   In 1990, the country's agricultural production dropped by more than 70 per cent, while the population continued to increase at 3 per cent per year. Despite doubling the amount of arable land, food production has risen by only one per cent, and the land has lost its fertility through phosphate and nitrogen depletion.

18   To confront the crisis the gov-

Food Dependency	Illiteracy	Maternal Mortality	External Debt

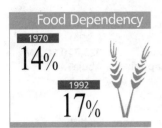

Food Dependency

1970
14%

1992
17%

Illiteracy

1995

86%

Maternal Mortality

1989-95

PER 100.000
LIVE BIRTHS

593

External Debt

1994

PER CAPITA

$180

PROFILE

ENVIRONMENT

Most of this land-locked country is made up of a plateau at an average altitude of 350 m. The north is covered by the Sahara desert and the south by savannas. There are uranium, iron, coal and tin deposits, and possibly oil. 80% of the population lives in rural areas. There are nomadic shepherds in the center of the country, and peanut, rice and cotton farming in the south. 85% of all energy is provided by firewood. This region is affected by marked desertification. There is also considerable erosion, produced by strong winds. There is air and water pollution primarily in densely populated urban areas.

SOCIETY

Peoples: Among the shepherds of the central steppe, ethnic origins vary from the Berber Tuareg, to the Fulah, including the Tibu (Tubu). In the south there are western African ethnic groups; the Haussa, Djerma (Zarma), Songhai and Kamuri. **Religions:** mostly Muslim; in the south there are traditional African religions and a Christian minority. **Languages:** French (official) and several local languages. **Political Parties:** National Movement for a Developing Society (MNSD); Alliance for Democracy and Progress (AFC). **Social Organizations:** Nigerien National Workers Union (UNTN)

THE STATE

Official Name: República du Niger. **Administrative Divisions:** 7 Departments. **Capital:** Niamey, 392,000 inhab. (1988). **Other cities:** Zinder, 120,000 inhab. (1981); Maradi, 104,000 inhab.; Tahoua, 50,000 inhab. (1988). **Government:** Ibrahim Bare Mainassara, President. **National Holiday:** August 3, Independence (1960). **Armed Forces:** 5,300. **Paramilitaries:** 5,400 Gendarmes, Republican Guards and National Police.

STATISTICS

DEMOGRAPHY

Urban: 16% (1995). **Annual growth:** 2.3% (1992-2000). **Estimate for year 2000:** 11,000,000. **Children per woman:** 7.4 (1992).

HEALTH

One **physician** for every 50,000 inhab. (1988-91). **Under-five mortality:** 320 per 1,000 (1995). **Calorie consumption:** 94% of required intake (1995). **Safe water:** 54% of the population has access (1990-95).

EDUCATION

Illiteracy: 86% (1995). **School enrolment:** Primary (1993): 21% fem., 21% male. Secondary (1993): 4% fem., 9% male. University: 1% (1993). **Primary school teachers:** one for every 38 students (1992).

COMMUNICATIONS

82 **newspapers** (1995), 83 **TV sets** (1995) and 83 **radio sets** per 1,000 households (1995). 0.1 **telephones** per 100 inhabitants (1993). **Books:** 101 new titles per 1,000,000 inhabitants in 1995.

ECONOMY

Per capita GNP: $230 (1994). **Annual growth:** -2.10% (1985-94). **Annual inflation:** 0.20% (1984-94). **Consumer price index:** 100 in 1990; 98.3 in 1992. **Currency:** 535 CFA francs = 1$ (1994). **Cereal imports:** 136,000 metric tons (1993). **Fertilizer use:** 4 kgs per ha. (1992-93). **External debt:** $1,569 million (1994), $180 per capita (1994). **Debt service:** 26.1% of exports (1994). **Development aid received:** $347 million (1993); $41 per capita; 16% of GNP.

ENERGY

Consumption: 37 kgs of Oil Equivalent per capita yearly (1994), 83% imported (1994).

---

ernment launched another structural adjustment plan, imposed by the World Bank and the IMF. A two-year freeze on the salaries of public employees was announced as part of the plan. Workers and students reacted by calling a new series of strikes and holding more demonstrations. In late 1990, Seibou publicly announced his commitment to leading the country toward a multi-party democratic system and he created the National Conference, which was charged with organizing the political transformation.

[19] After four months, the National Conference decided to form a transitional government, headed by a new prime minister, Amadou Cheffou. André Salifou was named president of the High Council of the Republic, the body holding legislative power during the transition period. However, the country's situation had never been more critical.

The state was bankrupt; there was no money to pay the salaries of the public employees, and students did not receive money for their grants.
[20] The economic disaster and its accompanying social crisis can be traced back to the end of the uranium boom; in 1989, this mineral cost Fr30,000,000 per kilo, dropping to Fr19,000,000 per kilo in 1991.
[21] In February 1992, Tuareg guerrillas rose up in arms against the government once more. In December, the new Constitution was approved with nearly 90 per cent of the votes in a referendum, but two months later, in February 1993, the ruling party was defeated in the legislative elections.
[22] In April, Mahamane Ousmane became the first president of Niger, with 55.4 per cent of the vote in the second round. The efforts to reach an agreement to end the insurrection of the Tuareg guerrillas in the north of the country continued

throughout the year and lasted for most of 1994.
[23] The fighting continued - in May 1994, 40 people were killed in confrontations between the rebel and government forces - but the main guerrilla group, the Coordination of Armed Resistance and the government managed to seal an agreement. Thus, in June, Niamey granted independence to part of the country inhabited by some 750,000 Tuaregs.
[24] Amidst student protests calling for, amongst other things, the payment of money owed in grants, the government arrested 91 members of the opposition. In September, Prime Minister Mahamadou Issoufou resigned after his party, the Nigerien Party for Democracy and Socialism, withdrew from the government coalition, leaving it without an absolute majority in Parliament.
[25] In January 1995, an opposition

coalition triumphed in the legislative elections and immediately replaced the prime minister, Boubacar Cissé Amadou with Hama Amadou. The latter announced that his first move would be to bring in an economic austerity plan, reaching an agreement for settling the payment of overdue civil service salaries.
[26] The situation between the new government and the president became increasingly tense. In January 1996, a military coup ousted Ousmane, who was replaced by a National Salvation Council under the control of Ibrahim Bare Mainassara. In July, Mainassara was elected president with 52 per cent of the vote.

Nigeria

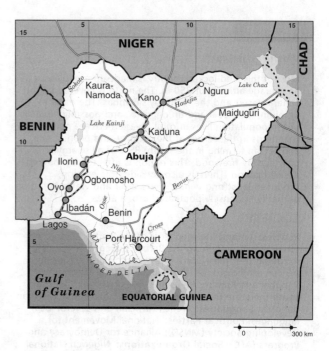

Population: 108,014,000 (1994)
Area: 923,770 SQ.KM
Capital: Abuja

Currency: Naira
Language: English

Nigeria

As heirs to the ancient Nok civilization (see Niger: "Civilizations of Sudan"), the Yoruba lived in walled cities with broad avenues. As early as the 11th century they developed a democratic system of urban administration, also producing beautiful ceramics and bronze sculptures.

[2] Between the 10th and 11th centuries, Ife, Oyo, Ilorin and Benin (not the present-day nation) were loosely confederated city-states extending their influence from the Niger River in the east to present-day Togo.

[3] Ife always enjoyed a reputation as the main religious center of the nation, and the Oni of Ife is still the High Priest of all Yorubans, whether Nigerian or not.

[4] The city of Oyo, strengthened since the 16th century by the slave trade, had a higher status as political and economic center, through the Alafin (ruler). The dependence on slavery caused its downfall when that institution was abolished.

[5] The northern part of the country held the Haussa states and was the center of a different culture (see "Haussa States"). A similar group of active Ibo traders emerged in the southeast, but they did not develop urban civilizations like the Yoruba.

[6] The British colonial system disregarded these differences, forcing all the territories under a single administration in 1914, creating an artificial state that has not attained national unity. British interests focused on the exploitation of tin and agricultural and timber resources.

[7] British administration in northern Nigeria was indirect, based on traditional Muslim emirs who acted as middlemen. Consequently this area, populated by Haussa and Peul, enjoyed greater political autonomy than the other regions.

[8] Independence in 1960 brought the Northern People's Congress to power, in an alliance with the National Council of Nigerian Citizens, an Ibo organization. The country had been divided into a federal structure of four states, and a two-chamber parliament was established, based on the British model.

[9] These political imports proved to be poorly suited to local conditions. Regional governors acquired more power than the President Nnambi Azikiwe. Progressive parties were pushed aside in a succession of electoral frauds, while political leaders lost their national outlook, encouraging ethnic rivalries.

[10] After months of infighting, the military chose General Yacuba Gowon as president. About this time the oil industry began to develop, just as France was inciting the separatist movement among the Ibo, provoking a three-year secessionist civil war in Biafra.

[11] Nigeria became the world's 8th largest oil producer, and Gowon expropriated 55 per cent of the transnational petroleum operations, creating financing for local entrepreneurs.

[12] Real power was vested in the nationalist Supreme Military Council, with different leaders acting as president. The council closed US military and espionage installations. During Olusegun Obasanjo's presidency, Barclays Bank and British Petroleum assets were nationalized as these companies were violating economic sanctions against South Africa.

[13] In 1978, constitutional reform was proposed, calling for elections and a return to civilian government. The Federal Election Commission authorized only five parties, all representing the traditional financial and political elite. Parties with socialist or revolutionary perspectives were barred from the electoral process, under the pretext of avoiding political fragmentation.

[14] The National Party of Nigeria (NPN) won the election, with 25 per cent of the vote, and the Unity Party of Nigeria (UPN) came second, with 20 per cent.

[15] The inauguration of Shagari, who lacked majorities in both legislative chambers, ended 13 years of military rule. He launched a capitalist development plan based exclusively on petrodollars, to transform Nigeria into the development hub of Subsaharian Africa. His promises included constructing a new capital, doubling primary school attendance and achieving self-sufficiency in food production via controversial "green revolution" methods.

[16] None of these proposals came to anything. Contraband, large urban concentrations of immigrants and poor peasants, unemployment and poverty all increased.

[17] The International Monetary Fund demanded the refinancing of the foreign debt putting the country under further pressure. Shagari announced new elections, and took part as the NPN candidate. He was re-elected, amidst accusations of electoral fraud and military conspiracies.

[18] On January 1 1984, Muhamad Buhari staged the fourth coup in the country's republican history. The new leaders accused their predecessors of corruption in the petroleum sector, accounting for 95 per cent of export earnings. Hundreds of people were arrested, and all civilian government officials were replaced by military personnel.

[19] However, the crisis was not checked, and the price of rice quadrupled in a single year. Repression extended to foreigners and 600,000 were expelled, classed as illegal immigrants. These events set the stage for another coup, the country's sixth, on August 26 1985, with General Ibrahim Babangida becoming the new president. That year, the foreign debt reached $15 billion.

[20] In December 1987, local elections were held with 15,000 unaffiliated candidates taking part. A National Election Commission had been appointed to oversee the voting and ensure that the election was clean and free of coercion. The lack of proper preparation for the voting itself, however, led to acts of violence, confusion and subsequent accusations of fraud. The election was finally annulled.

[21] On December 7 1989, the military government announced that the elections, originally scheduled for the end of the month, would be postponed until December 1990. Six months later, President Babangida announced that the ban on political activism had been lifted, in an attempt to monitor the transition from military to civilian government in 1992.

[22] In May 1990, Babangida visited Britain; one outcome of the visit was a treaty between the two countries granting Nigeria $100 million in aid. Most of its trade is with the United States, the United Kingdom and France, though it has sought more balanced international relations by maintaining relations with all the countries in the world.

[23] Nigeria has seen its fortunes rise and fall with the fluctuations in oil prices. In 1990 the foreign debt reached $30 billion, and to service its debt, the government has had to

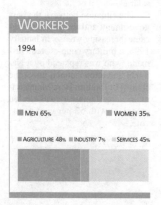

WORKERS

1994

■ MEN 65%   ■ WOMEN 35%

■ AGRICULTURE 48%   ■ INDUSTRY 7%   ■ SERVICES 45%

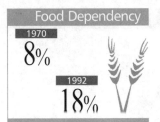

Food Dependency

1970
8%

1992
18%

Foreign Trade

MILLIONS $ 1994

IMPORTS
6,511
EXPORTS
9,378

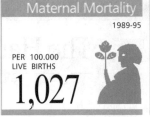

Maternal Mortality

1989-95

PER 100.000
LIVE BIRTHS
1,027

External Debt

1994

PER CAPITA
$310

PROFILE

ENVIRONMENT

The country's extensive river system includes the Niger and its main tributary, the Benue. In the north, the "harmattan", a dry wind from the Sahara, creates a drier region made up of plateaus and grasslands where cotton and peanuts are grown for export. The central plains are also covered by grasslands, and are sparsely populated. The southern lowlands receive more rainfall, they have dense tropical forests and most of the country's population lives there. Cocoa and oil-palms are grown in this area. The massive delta of the Niger River divides the coast into two separate regions. In the east, oil production is concentrated around Port Harcourt, the homeland of the Ibo, who converted to Christianity and fought to establish an independent Biafra. To the west, the industrial area is concentrated around Lagos and Ibadan. Yoruba are the predominant western ethnic group, and some of them have converted to Islam. Nigeria has lost between 70 and 80% of its original forests.

SOCIETY

Peoples: Nigeria has the largest population in Africa. The 250 or so ethnic groups can be divided into four main groups: the Haussa and Fulani in the north; the Yoruba in the southwest; and the Ibo in the southeast.
Religions: The north is predominantly Muslim, while Christians form the majority in the southeast; Muslims, Christians and followers of traditional African religions can be found in the southwest. **Languages:** English (official). Each region has a main language depending on the predominant ethnic group, Hausa, Ibo or Yoruba. **Political Parties:** National Democratic Coalition; Movement for the Survival of the Ogoni People (MOSOP). **Social Organizations:** Unified Federation of Nigerian Unions

THE STATE

Official Name: Federal Republic of Nigeria. **Administrative Divisions:** 30 States. **Capital:** Abuja, 100,000 inhabitants (1991) **Other cities:** Lagos, 1,347,000 inhab.; Ibadan, 1,295,000 inhab.; Ogbomosho, 660,000 inhab.; Kano, 700,000 inhab. (1992). **Government:** General Sani Abacha declared himself absolute ruler in November 1993, after a coup against Ernest Shonekan. **National Holiday:** October 1, Independence Day (1960). **Armed Forces:** 77,100 troops (1995) **Paramilitaries:** 7,000 National Guard, 2,000 Port Security Police.

STATISTICS

DEMOGRAPHY

Urban: 38% (1995). **Annual growth:** 2% (1992-2000). **Estimate for year 2000:** 128,000,000. **Children per woman:** 5.9 (1992).

HEALTH

One **physician** for every 5,882 inhab. (1988-91). **Under-five mortality:** 191 per 1,000 (1995). **Calorie consumption:** 91% of required intake (1995). **Safe water:** 40% of the population has access (1990-95).

EDUCATION

Illiteracy: 43% (1995). **School enrolment:** Primary (1993): 82% fem., 82% male. Secondary (1993): 27% fem., 32% male. **Primary school teachers:** one for every 39 students (1992).

COMMUNICATIONS

85 **newspapers** (1995), 88 **TV sets** (1995) and 92 **radio sets** per 1,000 households (1995). 0.3 **telephones** per 100 inhabitants (1993). **Books:** 84 new titles per 1,000,000 inhabitants in 1995.

ECONOMY

Per capita GNP: $280 (1994). **Annual growth:** 1.20% (1985-94). **Annual inflation:** 29.60% (1984-94). **Consumer price index:** 100 in 1990; 403.2 in 1994. **Currency:** 22 nairas = 1$ (1994). **Cereal imports:** 1,584,000 metric tons (1993). **Fertilizer use:** 175 kgs per ha. (1992-93). **Imports:** $6,511 million (1994). **Exports:** $9,378 million (1994). **External debt:** $33,485 million (1994), $310 per capita (1994). **Debt service:** 18.5% of exports (1994). **Development aid received:** $284 million (1993); $3 per capita; 1% of GNP.

ENERGY

Consumption: 162 kgs of Oil Equivalent per capita yearly (1994), -484% imported (1994).

carry out constant negotiations with international agencies.
[24] In April and June 1990, news services announced a failed coup attempt, put down by forces loyal to the government.
[25] Between August and October, the creation of nine new States to separate hostile ethnic groups generated protests which were repressed by the army, causing some 300 deaths according to unofficial figures. The government imposed a curfew.
[26] Toward the end of 1991, the government invalidated the allegedly fraudulent internal party elections, held to select governmental candidates. In November, a new census eliminated 20 million non-existent voters from the electoral register. On December 14, governmental elections were held; the Social Democratic Party (SDP) won in 16 states, and the National Republican Convention (NRC) in 14.
[27] The opposition were granted a general amnesty, and 11 dissidents were freed. A law prohibiting former government officials from running for office was revoked.
[28] In early 1992, the imprisonment of 263 Muslim militants, caused protests in the State of Katsina. During this period, there was also an escalation of inter-ethnic conflict between Haussa and Kataj in the State of Kaduna, and territorial conflict between Tiv and Jukin in the Taraba.
[29] On December 5, Nigeria recovered its position in the International Court of Justice, after the UN Security Council and the General Assembly voted to fill the vacancy left by the death of Nigerian Judge Taslim Elias. The Nigerian general prosecutor and minister of justice, Ola Ajibola, won over the Ghanaian, Kenyan, and Ugandan candidates.
[30] Nigeria received $240 million in aid from the European Community to be used in agricultural, health, telecommunication, road building, and conservation projects.
[31] In July 1992, legislative elections were held, but the National Assembly did not convene until December. The SDP obtained 52 seats in the Senate, and 314 in the Chamber of Representatives. The National Republican Convention obtained 37 seats in the Senate, and 275 in the Chamber of Representatives. Nigeria subsequently began the transition to civilian government, after 23 years of military regimes.
[32] In October, presidential primary elections were held, with candidates from the two main parties, the SDP (left-of-center) and the RNC (right-of-center). Both parties were created by the regime, with nearly indistinguishable manifestos. An-

The Haussa states

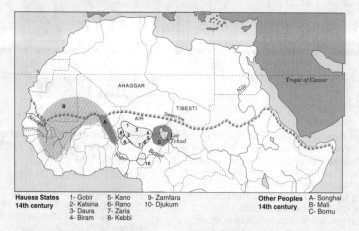

Haussa States 14th century
1- Gobir
2- Katsina
3- Daura
4- Biram
5- Kano
6- Rano
7- Zaria
8- Kebbi
9- Zamfara
10- Djukum

Other Peoples 14th century
A- Songhai
B- Mali
C- Bornu

The Haussa nations established between the Niger river and Lake Chad - today's Nigeria - followed the example set by their western neighbors, the Songhai in the fifth century, beginning to form city-states with kinship ties between the rulers. The term Haussa does not really identify a nation, but rather, groups of different origins and constitutions who achieved cultural unity through the common Haussa language, which developed from a Semitic root.

2   The first group of cities, in present-day Niger and northern Nigeria, were Kano, Daura, Gobir, Katsina, Biram, Zaria and Rano; all located in the valleys of Sokoto, a tributary of the Niger, and Hadejia, a tributary of Lake Chad. The cities of Djukum, on the Benue, Kebbi and Zamfara developed further south later on.

3   Through contact with Mali between the 13th and 14th centuries, the Haussa gradually converted to Islam and integrated into the trade circuits, specializing in the slave business. They soon became the main suppliers of eunuchs - castrated men - heavily in demand as harem guards in the Islamic world. Although the harem abduct people from the forest villages further south, they more often sold members of their own Haussa groups as slaves. Their victims were highly valued on the Arab market because they had cultural and linguistic similarities to the Arabs.

4   There was never a serious attempt to unify the city-states. They constantly fought amongst themselves, but the wars were not very bloody. In fact these wars were more commercial forays aimed at obtaining the largest possible number of healthy prisoners, and not at annihilating the enemy or destroying its political and economic structures.

5   The principal Haussa sovereigns had the chronicles of their states recorded, and these would have been invaluable sources of information on Central African history, as the Haussa had links with all the major states of the area; Mali, Songhai and Kanem-Bornu all of which eventually ruled different Haussa cities. Unfortunately, these chronicles were destroyed after the Fulani conquest in the 19th century.

---

other 23 candidates also took part. President Babangida invalidated the election, claiming fraud, stripped the candidates of their authority and ousted the leadership of the SDP and RNC. Later that month, the military government suspended all political activity. These developments halted the democratic transition process, jeopardizing the possibility of a transfer to civilian government in early 1993.

[33] In November Babangida announced that elections slated for January 1993 had been postponed until June. He also ratified the proscription of the 23 1992 presidential candidates. Finally, he postponed the transfer to a democratic regime until August.

[34] On June 12, the first presidential elections since 1983 were held. The military government did not divulge election results until it had concluded an investigation of alleged fraud. The main contest was between the NCR and SDP candidates, who had been authorized to take part in the election, although the original candidates continued in proscription.

[35] Babangida invalidated the election results on June 23, accusing SDP and NCR candidates of "buying votes". Abiola, a wealthy Muslim who was the SDP candidate and who had apparently won the June election, left for London to campaign for international condemnation of the Babangida regime.

[36] The United States and the United Kingdom suspended their economic aid and training of Nigerian military personnel. Abiola launched a civil disobedience campaign and massive protests broke out in the streets of Lagos, where at least 25 people were killed by federal troops.

[37] Clashes continued, and on August 26, Babangida resigned, leaving the provisional command of the country in the hands of Ernest Shonekan, who promised to hold new elections.

[38] The following month, Abiola returned to London and labor unions called a general strike, demanding he be recognized as Nigerian president. Toward the end of 1993 the minister of Defence, General Sani Abacha, overthrew Shonekan, dissolved Parliament and banned all political activity.

[39] This new "strongman" had been a very influential member of the previous military regime and a key figure in the military coup that had ousted the government in 1983. In one of his first statements, Abacha announced he would abandon some liberal economic reforms adopted in the 1980s.

[40] The interest rates went down and a new foreign exchange control was established, at a time in which any possibility of reaching an agreement with the IMF was increasingly remote. Popular support for Abiola mounted; in June 1994, the political leader was arrested, which triggered a 10-day-strike by oil workers, the most important sector in the country.

[41] Abacha did not give in despite all the pressures for Abiola's freedom. In April 1995, Archbishop Desmond Tutu - on behalf of South African President Nelson Mandela - travelled to Nigeria in support of Abiola's liberation. Despite Abiola's agreement to accept the annulment of the 1993 election, the Nigerian president refused to free him.

[42] In November, the execution of nine members of the Movement for the Survival of the Ogoni People, resulted in the isolation of the military regime. Several countries, including the United States, withdrew their ambassadors from Nigeria.

[43] In February 1996, the organization for the defence of human rights, Amnesty International, stated its concern for the situation of other Ogoni activists arrested.

Northern Marianas

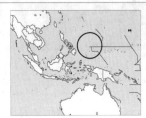

Population: 47,000 (1994)
Area: 404 SQ.KM
Capital: Saipan

Currency: US Dollar
Language: English

Northern Mariana Islands

In Saipan, the largest island of the Mariana archipelago, evidence has been found of human habitation from 1500 BC.

[2] During his first expedition around the world, the Portuguese navigator Ferdinand Magellan first sighted the islands in 1521 and claimed them for the Spanish Crown. They were held by the Spanish until being ceded to Germany in 1899 as the Spanish empire declined.

[3] During World War I the islands came under the control of Japan, which had an alliance with the United Kingdom since 1902. The Japanese occupied the islands until World War II.

[4] In June 1944, after fierce fighting, the US took Saipan and Tinian from the Japanese due to their strategic location in the North Pacific, on the route between Hawaii and the Philippines. They finally came to form part of the Trust Territory of the Pacific Islands in 1947.

[5] The islands continued in this situation until they became a Free Associated State under a referendum held in 1975. In 1977, US President Jimmy Carter approved the constitution of the Northern Mariana islands. At the end of the year this new constitution brought about elections for posts in the bicameral legislative.

[6] In 1978, the islands began to be administered autonomously. Washington developed a project to turn two-thirds of Tinian into a military air base and an alternative center for the storage of nuclear weapons. Negotiations with Tinian landowners have been under way since 1984.

[7] When news leaked out that cement deposits containing radioactive waste from Japanese nuclear plants - cobalt 60, strontium 90 and cesium 137 - had been dumped in this part of the Pacific, the alarm was sounded on similar US projects that would directly affect the Mariana Islands.

[8] In 1984 US President Ronald Reagan granted some civil and political rights to the islands' residents, such as equal employment opportunities in the federal government, the civil service and the US armed forces.

[9] The Northern Marianas were formally admitted to the Commonwealth of the United States in 1986. Mariana inhabitants were granted US citizenship but not the right to vote in presidential elections. They have a representative in the US Congress with no voting rights.

[10] The main economic activities of the country are fishing, agriculture - concentrated in smallholdings - and tourism which employs about 10 per cent of the workforce. Some of these activities have been affected often by the typhoons which the islands suffer in the rainy season: in January 1988 Rota Island was devastated by a typhoon which forced the US to declare a state of emergency there. Two years later, the Koryn typhoon struck the archipelago in January 1990.

[11] In the local elections held in 1989, Republicans retained the governorship, ousting the Democrat representative from Washington. Larry Guerrero was elected governor after Pedro Tenorio had decided to stand down.

[12] The termination of UN Security Council trusteeship of the islands was approved in 1990. Thus, the Northern Marianas became an independent state, associated to the United States.

[13] The Republican Party won the November 1991 legislative elections by a wide margin. Given this majority, it hoped to modify the islands' status with relation to the United States, in order to guarantee control over a 200-mile exclusive economic zone.

[14] The United States Supreme Court ratified the land tenure system, restricting land ownership to local people. Froilan C Tenorio was elected governor in 1994.

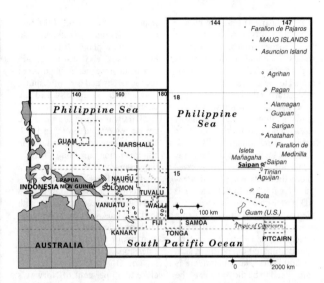

> In 1978, Washington developed a project to turn two-thirds of Tinian into a military air base and an alternative center for the storage of nuclear weapons.

Norway

Norge

Population: 4,337,000 (1994)
Area: 386,958 SQ.KM
Capital: Oslo

Currency: Kroner
Language: Norwegian

The earliest traces of human life in Norway correspond to the period between the 9th and the 7th millennia BC. Germanic tribes are thought to have emigrated to these areas when glaciers receded from the northern European coasts and mountains. Cave drawings show that navigation was already known to the peoples of the time, as were something like skis, designed for gliding over the snow.

2   Historians believe that Norwegian nationality and Christianization began between 800 and 1030 AD. These dates also correspond to the rise of the Vikings, the name given to Scandinavian seafarers, merchants, and above all, raiders, who dominated the northern seas for 200 years. Harald Harfagre is considered to be the founder of the nation, a feat which he accomplished after defeating his rivals in a naval battle at Hafrsfjord, near the city of Stavanger. After this victory, a large part of the country came under his control.

3   Viking expeditions expanded the Norwegian empire, reaching Greenland to the west, and Ireland to the south. According to official history, in 1002, the Norwegian Leif Erikson and his men were the first Europeans to cross the Atlantic and reach North America, which they named Vinland.

4   At the end of the Viking era, Norway was an independent kingdom in which four regional peasant assemblies (*lagting*) elected the monarch. Legitimate and illegitimate children of the king had equal rights to succession before the *lagtings*. In the 10th and 12th centuries it was common for two kings to govern simultaneously without any conflict arising between them.

5   King Magnus III Barfot (1093-1103) conquered the Scottish Orkney and Hebrides Islands. His three sons governed together: they

imposed a tithe, founded the first monasteries and built cathedrals. At the beginning of the 12th century, a hundred-year civil war broke out, as a result of the increased power of the monarchy and also of disputes between the monarchy and the church.

6   This war continued until the coronation of Haakon IV in 1217. The new king reorganized public administration, imposing a hereditary monarchy. Haakon signed a treaty with Russia over the country's northern border. Greenland and Iceland agreed to a union with the king; thus, with the Scottish islands and the Faeroes, the Norwegian empire attained its maximum extension.

7   The Black Death killed off close to 50 per cent of Norway's population between 1349 and 1350. The upper classes were decimated; Danes and Swedes were hired to fill the positions left vacant in the higher levels of the government and the church. However, the king lost control over his dominions and isolated regions organized autonomous administrations.

8   The ascent of Queen Margaret of Denmark to the throne, in 1387, paved the way for the union of the Scandinavian countries. In 1389, she was crowned Queen of Sweden and in 1397, her adopted nephew, Erik, was elected king of all Scandinavia in Kalmar, Sweden. With the Kalmar union, Norway was gradually subordinated, ultimately becoming a province of Denmark, a situation which lasted for more than 400 years.

9   After 1523, Norway's administrative council tried to obtain greater independence from Denmark. However, the fact that power lay in the hands of the Catholic bishops made it difficult to obtain Swedish support. At the end of the civil war, between 1533 and 1536, the council was abolished. In 1537, the Danish king made the Lutheran religion the country's official religion; the Norwegian Church has been a State church ever since.

10   During this period, social conditions in Norway were better than in Denmark. There was a class of rich landlords in the countryside which exploited the regional timber resources, as well as a large group of rural wage-earners. Most of Norway's population were peasants and fishermen and cities were limited to fewer than 15,000 inhabitants.

11   At the end of the Napoleonic wars, Denmark unilaterally surrendered control of Norway to Sweden. In 1814, Norway's constituent assembly proclaimed national independence. Sweden re-established its dominance over Norway by force, but this union was dissolved once again without bloodshed in 1905, with Norway regaining its sovereignty and with its full previous territories intact.

12   In spite of its subordination to Sweden, the majority of the laws enacted in 1814 remained in force during this period. The Norwegian constitution is one of the oldest in the world, second only to that of the United States. It is based on the principles of national sovereignty, the separation of powers and the inviolability of human rights.

13   Under a constitutional amendment in 1884, Norway adopted a parliamentary monarchy as its system of government. The Danish prince Carl was elected king of Norway, under the name Haakon VII, in 1905. Up until 1914, the country experienced rapid economic expansion, with the hydroelectric

wealth of the region allowing large-scale industrial development.

14   The sale of a large number of Norwegian water courses with hydroelectric power-generating potential, to foreigners, caused great concern among the population. In 1906, 75 per cent of Norway's hydroelectric dams belonged to foreign investors. In 1909, Parliament passed laws for the protection of the country's natural resources.

15   Universal suffrage, a term which applied to men only when it was passed in 1898, was extended to women by reforms approved in 1907 and 1913. One consequence of industrialization and universal suffrage was the growth of the Labor Party (LP).

16   During World War I, Norway tried to remain neutral, but was obliged by the other powers to cut trade with Germany. Anti-German feeling was strong, particularly because of the various accidents caused by German war submarines. Price increases provoked by the conflict hit the workers hardest.

17   Unlike other western European social democracies, Norway's LP

LAND USE

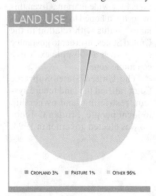

CROPLAND 3%   PASTURE 1%   OTHER 96%

ENVIRONMENT

The Scandinavian mountain range runs north-south along the coast of the country. On the western side, glacier erosion has gouged out deep valleys that are way below actual sea level, resulting in the famous "fjords", narrow, deep inlets walled in by steep cliffs. Maritime currents produce humid, mild winters and cool summers. The population is concentratred in the south, especially round Oslo. Nine tenths of the territory is uninhabited.

SOCIETY

Peoples: Mainly Nordic, Alpine and Baltic Germanic peoples and 40,000 Sami, an indigenous people who live chiefly in the northern province of Finnmark. There are also 12,000 Finns and 195,000 other immigrants. **Religions:** 88% of the population belongs to the Church of Norway (Lutheran), there are Evangelical and Catholic minorities.
Languages: Two forms of Norwegian are officially recognized; 80% of school children learn the old form "Bokmal", and 20% learn the neo-Norwegian "Landsmal". In the north, the Sami speak their own language. **Political Parties:** Labor Party; Conservative Party; Progress Party; Party of the Socialist Left; Christian Democratic Party, Center Party, Liberal Party. **Social Organizations:** Norwegian Federation of Unions (LO); Organization of Academics (AF); Organization of Trades (YS).

THE STATE

Official Name: Kongeriket Norge **Administrative divisions:** 19 provinces (*Fylker*). **Capital:** Oslo, 482,000 inhab. (1995). **Other cities:** Bergen, 221,000 inhab.; Trondheim, 143,000 inhab.; Stavanger 103,000 inhab. (1995). **Government:** Parliamentary constitutional monarchy since 1884; Harald V, king since January 17, 1991. Prime Minister: Thorbjørn Jagland, of the Labor Party, since October 25, 1996. Legislative power resides in the *Storting*, the 165-member unicameral parliament. **National Holiday:** May 17, Constitution Day (1814). **Armed Forces:** 32,500 **Paramilitaries:** Coast Guard: 680.

DEMOGRAPHY

Urban: 73% (1995). **Annual growth:** 0.5% (1991-99). **Estimate for year 2000:** 4,000,000. **Children per woman:** 1.9 (1992).

HEALTH

Under-five mortality: 8 per 1,000 (1995). **Calorie consumption:** 111% of required intake (1995).

EDUCATION

School enrolment: Primary (1993): 99% fem., 99% male. Secondary (1993): 114% fem., 118% male. **University:** 54% (1993). **Primary school teachers:** one for every 6 students (1992).

COMMUNICATIONS

106 **newspapers** (1995), 103 **TV sets** (1995) and 102 **radio sets** per 1,000 households (1995). 54.2 **telephones** per 100 inhabitants (1993). **Books:** 109 new titles per 1,000,000 inhabitants in 1995.

ECONOMY

Per capita GNP: $26,390 (1994). **Annual growth:** 1.40% (1985-94). **Annual inflation:** 3.00% (1984-94). **Consumer price index:** 100 in 1990; 109.8 in 1994. **Currency:** 7 kroner = 1$ (1994). **Cereal imports:** 302,000 metric tons (1993). **Fertilizer use:** 2,276 kgs per ha. (1992-93). **Imports:** $27,300 million (1994). **Exports:** $34,700 million (1994).

ENERGY

Consumption: 5,326 kgs of Oil Equivalent per capita yearly (1994), -636% imported (1994).

(in which the left-wing formed the majority) decided to join the Third Communist International in 1918. However, the Norwegian LP could not agree with the centralization applied by the Soviet Communist Party, and cut its ties with the Comintern in 1923.

[18] Despite economic difficulties and serious labor conflicts (with unemployment reaching 20 per cent in 1938), Norway underwent vigorous industrial expansion in the inter-war years. Moreover, the government extended social legislation to include old-age pensions, mandatory leave for workers and unemployment benefits.

[19] In 1940, at the beginning of World War II, Norway was invaded by Germany, which seized control of the country after two months of fighting. King Haakon and the government went into exile in London, coordinating the resistance from there. Norway was liberated in 1945. Haakon died in 1957 and was succeeded by his son Olaf V.

[20] The LP governed continuously between 1935 and 1965, except for a brief one-month period in 1963. In 1965, the LP lost its parliamentary majority and Per Borten, the leader of the Center Party, was named Prime Minister. However, he resigned in 1971 when it was revealed that he had leaked confidential information during EEC negotiations.

[21] Since World War II, the Norwegian National Health Service has provided free medical care, including hospitalization and most medical services.

[22] After the war, Norway abandoned its neutrality policy, joining NATO in 1949. This membership was encouraged by the fear of Soviet expansionism and the unsuccessful attempt to establish a Scandinavian military alliance between Norway, Denmark and Sweden. However, in 1952, the Scandinavian countries established the Nordic Council to deal with their common interests. Norway has been a member of the Nordic Council since 1952, and of EFTA since 1960. By 1972, the idea of joining the EEC had been proposed. The issue divided the population and a plebiscite was held to decide the question; 53 per cent voted against, while 47 per cent were in favor.

[23] Trygve Bratteli, the Labor prime minister resigned and a new government was formed by Lars Korvald of the Christian People's Party, to negotiate a trade agreement with the EEC. After this the Labor Party formed a minority cabinet under Oddvar Nordli, who served until 1981, when the LP appointed Gro Harlem Brundland, the first female prime minister.

[24] In 1977, Norway extended its territorial waters to 200 miles and designated a protected fishing zone in its territory of Svalbard.

[25] In 1986, the Labor Party formed a minority government, once more under Gro Harlem Brundland, who appointed 8 women to her 18-member cabinet.

[26] Trade conflicts between Norway and the US have frequently occurred; in 1987, the state-owned Kongsberg Vapenfabrik exported advanced automated machinery to

Foreign Trade

MILLIONS $ 1994

IMPORTS
27,300
EXPORTS
34,700

Whale-hunting and politics

Norway continues whaling and resumed, in 1995, the killing of baby seals for "scientific research", despite international bans on these activities. Moreover, errors discovered in the software Oslo used to provide estimates of the North Atlantic minke whale population weakened its arguments justifying whaling.

[2] The whaling moratorium went into effect in 1985 and was especially questioned by the Japanese and Norwegian governments, who argued that the minke whale population was large enough for a total of 50 examples to be hunted each year. According to the terms of the ban, 400 whales may be hunted per year, but only for "scientific purposes".

[3] The Norwegian government's disagreement with the IWC's anti-hunting policies has a series of antecedents. In June 1992, Norway and Iceland announced that they would not observe the moratorium, and Iceland left the IWC. On September 15 1992, Norway, Iceland, the Faeroe Islands and Greenland founded the North Atlantic Maritime Mammals Commission (Nammco) with the aim of promoting a sustainable policy for the commercial hunting of whales. A Japanese delegate attended as an observer.

the Soviet Union, violating NATO sales restrictions to Warsaw Pact and Third World countries.

[27] There were also problems with 'heavy water' exports. Romania and West Germany, had re-sold some to India, but later pledged not to re-sell the material for weapons production without Norwegian authorization. The dispute was not properly resolved until Norway prohibited all 'heavy water' exports in 1988.

[28] In 1988, US threats to impose sanctions were dropped after Norway agreed to limit its whaling quota to scientific studies. In 1990, however, the government announced its intention to return to the whaling tradition.

[29] In 1986, the pollution of rivers and lakes in southern Norway was attributed to 'acid-rain' coming from the United Kingdom. The British government announced its intention to reduce emissions by 14 per cent by 1997, but the Norwegians considered this measure to be insufficient. A 60 per cent reduction of emissions within the country since the early 1980s has been insufficient to stop the environmental damage.

[30] The vulnerability of the Norwegian environment to events beyond national borders was underscored in April 1986 when a fire broke out in the Chernobyl nuclear power plant, in the Ukraine. The high levels of radioactivity released into the atmosphere most seriously affected the Sami population in the northern part of the country, and over 70 per cent of all reindeer meat had to be destroyed.

[31] Since the early 1970s, Norway and the former USSR have been at odds over rights to large parts of the Barents Sea. A temporary agreement signed in 1978 defined a 'grey area' for the joint management of fishing. The unresolved questions continued to create tension until early 1989, when the USSR proposed that they begin negotiations.

[32] Protests have been triggered by acid rain emanating from industries on the Kola Peninsula to the east of Norway. Since the fall of the communist regime this situation has changed, and there is growing trade across the border. Norway, Sweden and Finland have joined forces to help Russia solve the immense environmental problems in the area.

[33] Toward the end of 1986, Norway suffered a recession, immediately following a veritable consumer boom which had transformed Norway into the Scandinavian country with the highest standard of living. There was a large devaluation that year, tax reforms were carried out, and unemployment increased. The government responded by implementing a harsh austerity plan.

[34] Norway is the world's sixth largest producer of natural gas and crude oil, and the third largest oil exporter. Operating outside the OPEC system, oil concessions have always been handled as an economic and political weapon. Although drilling rights are granted to foreign companies, the state looks after its own interests by requiring 50 per

> A 60 per cent reduction of industrial emissions within the country since the early 1980s has been insufficient to stop the environmental damage.

cent participation by Statoil, the government oil company, in all operations.

[35] In January 1992, Norway's mission in South Africa was upgraded from a consulate to an embassy, and in the course of 1993, Norwegian diplomacy gained a great deal of prestige when the minister of foreign affairs, Johan Juergen Holst served as an intermediary in the Israeli-PLO negotiations.

[36] In 1993 the economy picked up, although unemployment remained at 6 per cent. Oil, fishing (with 25,000 employed), the merchant marine (with 1,059 ships and a capacity of 36 million tons) and agriculture play important roles in the country's economy. The Norwegian agricultural system is one of the most heavily subsidized in Europe, but while the country must import fruit, vegetables and cereals, it is self-sufficient in animal products.

[37] Despite serious disagreements between the various sectors of Norwegian society, Oslo sought entry to the current European Union (EU) in late 1992. The unions considered joining the former European Community would threaten national sovereignty, while many entrepreneurs - particularly in the export sector - wanted full access to the EU market.

[38] However, for the entry to be effective it had to be approved by a referendum, and a date was set for November 1994. The campaign for this vote largely dominated political life for two years and revolved around three main issues: oil exploitation, the regional policy and the fishing policy.

[39] The social democrats and conservatives supported joining the EU, while all the other opposed the move, leading to an extremely close outcome: 52.4 per cent of the electorate voted against integration, blocking Norway's membership of the bloc.

[40] At the same time, the Norwegian economy showed high levels of growth compared with the EU nations. Economic expansion was measured at 5 per cent in 1994 and was estimated at around 4 per cent for 1995 and 1996. Even excluding oil, the main sector of the economy, there was 3 per cent growth in 1994, while it was estimated that unemployment, which stood at 4.8 per cent in 1995, will fall to 3.5 per cent by the year 2000.

Oman

´Uman

Population: 2,098,000 (1994)
Area: 212,460 SQ.KM
Capital: Muscat (Masqat)

Currency: Rials Omani
Language: Arabic

Sumerian clay tablets from the third century BC mention Oman as one of the outstanding markets in the economy of the Mesopotamian cities. Omani navigators became the lords of the Indian Ocean, connecting the Gulf to India, Indonesia and Indochina. In the 7th century they also played a major role in the peaceful propagation of Islam (see Saudi Arabia). Around 690 AD Abd Al-Malik decided to control the expansion of dissident sects. Consequently, some defeated leaders were forced to abandon the country. One of them, Prince Hamza, emigrated to Africa where he founded Zanzibar (see Tanzania: "The Zandj Culture") beginning a relationship between Oman and the African coast which would last until the 19th century.

[2] Persecution only served to further reinforce Sharite "heresy". Towards 751, Oman took advantage of the dynastic strife in Damascus to elect an imam who gradually evolved from a spiritual leader to a temporal sovereign. Omani wealth gave the imam considerable power in the entire Gulf region, but also made Oman the object of successive invasions by the caliphs of Baghdad, the Persians, the Mongols and the local groups of central Arabia. The Portuguese arrived from 1507, destroying the fleet and coastal fortifications, opening the way for the occupation of the principal cities and the control of the Straits of Hormuz. The Portuguese held control of the region for almost 150 years, suffocating the Gulf's trade.

[3] In 1630, Imam Nasir bin Murshid launched an inland struggle against the invaders. His son, Said, concluded this endeavour in 1650 with the expulsion of the Portuguese from Muscat (Musqat) and the recovery of Zanzibar and the

African coast of Mombasa in 1698. Thus a powerful state was created which obtained the political unification of the African and Asian territories where a common culture and economy had developed.

[4] Sultan Said expanded the African territories and moved the capital to Zanzibar in 1832. At the time of his death in 1856, the British presence was already being strongly felt on both continents. Said's sons argued and as a result, the African and Asian parts of the state were separated: the elder son, Thuwaini, kept the sultanate of Oman while his brother Majid took control of Zanzibar. The 1891 Canning Agreement virtually made Oman a British protectorate.

[5] In 1913, inland peoples elected their own imam in opposition to the sultan's hereditary rule. The struggle came to an end only in 1920, when a treaty was signed acknowl-

In 1913, inland peoples elected their own imam in opposition to the Sultan's hereditary rule.

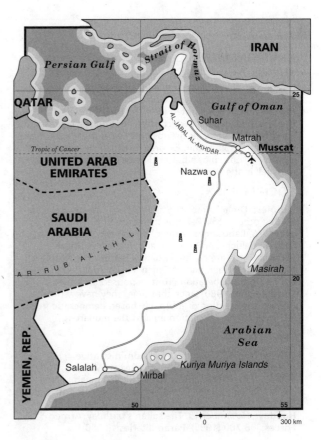

edging the country's division in two: the Sultanate of Muscat and the Imamate of Oman.

[6] Muscat is an extremely poor country, where arable lands account for less than one per cent of the total territory. Between 1932 and 1970, it suffered the despotic rule of Sultan Said bin Taimur, who fanatically opposed any foreign influence in the country, even in education and health. This did not prevent him from granting to Royal Dutch

Shell control over the country's oil deposits.

[7] Imam Ghaleb bin Alim, elected in 1954, proclaimed independence and announced his intention to join the Arab League. In 1955 the British reunified Oman. Since then, a liberation movement has stubbornly put up resistance, especially in the southern province of Dhofar.

[8] Taimur was overthrown by his own son Qaboos on July 23 1970. Those who expected the young, Oxford-educated monarch to introduce modernizing changes soon realized that British domination was only being replaced by US domination. The US became a net importer of oil and began to develop an active interest in the area.

[9] Oil has been produced commercially since 1967. It provides more than half of the Gross Domestic Product. However more than half of Oman's labour force remain involved in agriculture, in the thin coastal strip that contains the country's only arable land.

[10] With US assistance, Qaboos organized a mercenary army but when this force proved incapable of

smashing the Popular Front for the Liberation of the Gulf, he signed an agreement with Shah Reza Pahlevi to secure Iranian intervention in the conflict.

[11] The guerrilla fighters were forced to retreat under the superior firepower of Iranian troops. Iranians also placed the Strait of Hormuz under their jurisdiction.

[12] Upon the Shah's downfall, Iranian soldiers were quickly replaced by Egyptian commandos and troops. The sultan decided to give the US the Masira island air base, and later the air bases at Ihamrit and Sib and the naval bases at Matrah and Salalah. Two-thirds of the national budget are committed to defence, while the people continue to suffer acute poverty and widespread illiteracy.

[13] Oman's strategic importance, with a geographic position giving the country full control over oil routes, led the US to concentrate on the sultanate in efforts to settle in the area.

[14] The country's isolation gradually came to an end during the first part of Qaboos' rule, but a massive

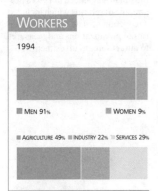

WORKERS

1994

MEN 91% WOMEN 9%

AGRICULTURE 49% INDUSTRY 22% SERVICES 29%

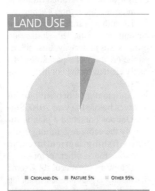

LAND USE

CROPLAND 0% PASTURE 5% OTHER 95%

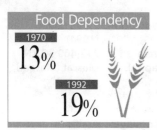

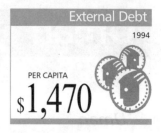

Food Dependency	Foreign Trade	Maternal Mortality	External Debt
1970 13% **1992** 19%	MILLIONS $ 1994 **IMPORTS** 3,915 **EXPORTS** 5,418	1989-95 PER 100.000 LIVE BIRTHS 184	1994 PER CAPITA $1,470

PROFILE

ENVIRONMENT

With its 2,600 km coastline, Oman occupies a strategic position on the southeastern edge of the Arabian Peninsula, flanking the Gulf of Oman, where oil tankers leave the Persian Gulf. It is separated from the rest of the peninsula by the Rub al Khali desert which stretches into the center of the country. Local nomadic groups now live with petroleum and natural gas exploitation. Favored by ocean currents, the coastal regions enjoy a better climate. Monsoon summer rains fall in the north.

SOCIETY

Peoples: Omani Arab 73.5%; Pakistani (mostly Baluchi) 21.0%; other 5.5%. **Religions:** Muslim 86%; Hindu 13%; other 1%. **Languages:** Arabic, official and predominant, English, Baluchi and Urdu are also spoken. **Political Parties:** There are no legal political parties. The People's Front for the Liberation of Oman (FPLO), founded in 1965 as the Dhofar Liberation Front, espoused an armed struggle until 1980. Since that time, they have worked toward the creation of a broadly based democratic front, to mobilize the population against the monarchy.

THE STATE

Official Name: Saltanat 'Uman **Administrative divisions:** 59 Districts. **Capital:** Muscat (Masqat), 100,000 inhab.(1990). **Other cities:** Nizwa 62,880 inhab.; Sama`il 44,721 inhab.; Salalah 10,000 inhab.. **Government:** Qaboos ibn Said, Sultan, in power since July, 1970. **National Holiday:** November 19. The sultan's birthday. **Armed Forces:** 36,700 (1995). **Paramilitaries:** 3,900.

STATISTICS

DEMOGRAPHY

Urban: 12% (1995). **Annual growth:** 3.4% (1992-2000). **Estimate for year 2000:** 2,000,000. **Children per woman:** 7.2 (1992).

HEALTH

Under-five mortality: 27 per 1,000 (1995). **Safe water:** 63% of the population has access (1990-95).

EDUCATION

School enrolment: Primary (1993): 82% fem., 82% male. Secondary (1993): 57% fem., 64% male. University: 5% (1993). **Primary school teachers:** one for every 27 students (1992).

COMMUNICATIONS

94 **newspapers** (1995), 116 **TV sets** (1995) and 107 **radio sets** per 1,000 households (1995). 8.6 **telephones** per 100 inhabitants (1993). **Books:** new titles per 1,000,000 inhabitants in 1995.

ECONOMY

Per capita GNP: $5,140 (1994). **Annual growth:** 0.50% (1985-94). **Annual inflation:** 0.10% (1984-94). **Consumer price index:** 100 in 1990; 106.6 in 1993. **Currency:** 0,4 rials Omani = 1$ (1994). **Cereal imports:** 369,000 metric tons (1993). **Fertilizer use:** 1,270 kgs per ha. (1992-93). **Imports:** $3,915 million (1994). **Exports:** $5,418 million (1994). **Development aid received:** $1,071 million (1993); $539 per capita; 9.2% of GNP.

ENERGY

Consumption: 2,347 kgs of Oil Equivalent per capita yearly (1994), -801% imported (1994).

"opening" towards foreign capital, followed by insignificant action in the education and cultural fields, did not end the extremely authoritarian features of the traditional social system. The only real opposition is the People's Front for the Liberation of Oman (PFLO), which dropped the armed struggle, to create a broad based democratic front. The organization's objectives were to form an alliance with the Arab nationalists, especially the Palestinians, and to expel the foreign troops.

[15] During the Iran-Iraq war, US presence grew in Oman; 10,000 soldiers were stationed in different bases, mainly on the Masira Island base, equipped with atomic weapons and where the Rapid Intervention Force has operated since 1984 despite violent protests. Subsequently Oman purchased two F-16 combat aircraft squads with sophisticated equipment.

[16] In June 1989, the Petroleum Development of Oman announced the discovery of the most important natural gas deposits found within the last 20 years.

[17] In March, Oman adopted a con-

> During the summer of 1994, some 500 dissidents were arrested charged with trying to topple the goverment "using Islam as a cover".

ciliation policy toward Iran, which included an economic cooperation agreement, though this aid was contingent upon efforts to achieve political stability in the Gulf region.

[18] In March 1991, after the Iraqi invasion of Kuwait, a member of the Gulf Cooperation Council, Oman suspended all aid to Jordan and the PLO.

[19] In 1991, the Foreign Ministers of Egypt, Syria and the six Arab member States of the Cooperation Council signed an agreement with the United States in Riyadh, the capital of Saudi Arabia, aimed at maintaining the region's security.

[20] Later that year, the government announced that the democratization process was underway; this included the creation of a parliament directly elected by the country's citizens.

[21] Anticipating a depletion of oil reserves before the year 2010, Sultan Qaboos launched a plan to diversify the economy, aiming to develop fishing, agriculture and tourism, among other sectors.

[22] During the summer of 1994, some 500 dissidents were arrested, including several senior officials and well-known business people, charged with trying to topple the government "using Islam as a cover". The fiscal deficits accumulated by the government since 1981, regarded as excessive, led the World Bank to warn that the level of State expenditure was "unsustainable".

[23] Taking heed of the international financial organization's stand, the sultan announced in 1995 a programme of reforms which included a reduction of state spending, a series of privatizations and measures to attract foreign investment.

Pakistan

Pakistan

Population: 126,284,000 (1994)
Area: 796,100 SQ.KM
Capital: Islamabad

Currency: Rupee
Language: Urdu

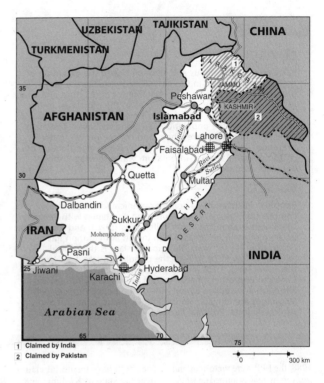

1 Claimed by India
2 Claimed by Pakistan

0   300 km

P akistan means the land of the pure, as it was religion (Islam) that bound together the people of different ethnic communities and languages. Poet-philosopher Mohammad Iqbal articulated the concept of Pakistan in its basic form in 1931 when he proposed a separate state for the Muslims in India.

2   The first Muslims to arrive in the Indian sub-continent were the traders from Arabia and Persia. A permanent Muslim foothold was achieved with Mohammad bin Qasim's conquest of Sind in 711 AD. It was in the early 13th century that the foundations of Muslim rule in India were laid, establishing borders and a capital in Delhi. The region including the present territory of Pakistan was subsequently ruled by several Muslim dynasties, finishing with the Mughals.

3   The question of Muslim identity emerged alongside the decline of Muslim power and the rise of the Hindu middle class during British colonialism. In the early years of the 20th century, Muslim leaders became convinced of the need for effective political organization. A delegation of Muslim leaders met the viceroy (the chief representative of the British imperial government in India) in October 1906, demanding the status of a separate electorate for the Muslims. The All India Muslim League (ML) was founded in Dhaka that year with the objective of defending the political rights and interests of Indian Muslims. The British conceded the status of a separate electorate in the Government of India Act of 1909, confirming ML's status as the representative organization of Indian Muslims.

4   In the 1930s there was a growing awareness of a separate Muslim identity and a greater desire to pre-serve it within separate territorial boundaries. Under the leadership of Mohammad Ali Jinnah, ML continued its campaign for Pakistan; a separate homeland in British India. After the general election of April 1946, ML called a convention of the newly-elected ML parliamentarians in Delhi. A motion by Hussain Shaheed Suhrawardi, then chief minister of Bengal, reiterated the demand for Pakistan in no uncertain terms.

5   The Hindu-Muslim relationship had been seriously affected by communal tensions and riots in different parts of India. This convinced the leadership of the Indian National Congress (representing mainly the nationalists) to accept Pakistan as a solution for the communal problem. A Partition Plan for transfer of power was announced on June 3 1947. Both the ML and the Congress accepted the Plan and on August 14, 1947, the new state of Pakistan was born comprising West Punjab, Sind, Baluchistan, North-West Frontier Province and East Bengal.

6   Pakistan has suffered numerous political crises. The first constitution was adopted on March 23 1956, but the civilian government was deposed in a coup d'etat on October 7 1958. Martial law was proclaimed and the constitution abrogated. On October 27 1958, General Ayub Khan emerged as the new leading force. He introduced "basic democracy," a system of local self-government and indirect presidential elections. Martial law was withdrawn in 1962 and a new constitution, granting absolute power to the president, was instituted, with Pakistan being declared an Islamic Republic. Ayub Khan was forced to resign on March 25, 1969, following a popular uprising. Martial law was promulgated again and General Yahya Khan became president.

7   A general election was held at the end of 1970, the first ever in Pakistan under neutral administration. Two political parties, the Awami League (AL) and the Pakistan People's Party (PPP) emerged victorious in East and West Pakistan, respectively. However, the AL won an absolute majority in the parliamentary election on an all-Pakistan basis, forming the federal government. However, the parliamentary session was postponed, and the people of East Pakistan began the movement for an independent Bangladesh under the leadership of the AL in March 1971. The AL was banned and its leader Sheikh Mujibur Rahman was arrested. A civil war broke out, leading the AL to form a government in exile in India. The Indian army intervened and on December 16 1971, Bangladesh became independent.

8   Zulfiqar Ali Bhutto, the leader of the PPP, formed a civilian government in 1972 following the resignation of General Yahya Khan. He encouraged strong public sector participation in the economy, followed a non-aligned foreign policy and approved radical land reforms. The PPP was victorious again in the general election of 1977, but the results were rejected by opposition political parties who accused the PPP of vote-rigging. Against this backdrop of political unrest, General Zia-ul Haq ousted the Bhutto government and proclaimed martial law. Bhutto was arrested and later sentenced to death on charge of conspiring to murder an opposition political leader.

9   General Zia accelerated the process of "Islamicization" in all spheres of political and social life. Many political opponents were harassed and detained. A general election on a non-party basis was held in February 1985 under martial law, and a pro-Zia government was formed. Zia was killed in a mysterious air crash in August 1988. In the general election of November

PUBLIC EXPENDITURES

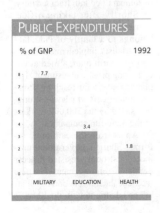

% of GNP   1992

MILITARY 7.7
EDUCATION 3.4
HEALTH 1.8

WORKERS

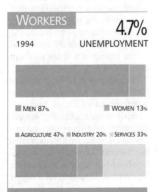

4.7%
1994   UNEMPLOYMENT

MEN 87%   WOMEN 13%

AGRICULTURE 47% INDUSTRY 20% SERVICES 33%

LAND USE

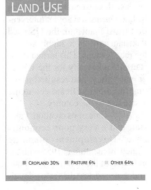

CROPLAND 30% PASTURE 6% OTHER 64%

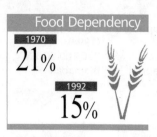

PROFILE

ENVIRONMENT

Pakistan is mountainous and semi-arid, with the exception of the Indus River basin in the east. This is virtually the only irrigated zone in the country, suitable for agriculture and vital to the local economy. The Indus rises in the Himalayas in the disputed province of Kashmir and flows into the Arabian Sea. The majority of the population live along its banks. The main agricultural products are wheat and cotton, grown under irrigation. The country's main industry is the manufacture of cotton goods. Pakistan suffers from water shortage, soil depletion and deforestation.

SOCIETY

Peoples: Pakistan has a complex ethnic and cultural composition, basically Indo-European combined with Persian, Greek and Arab in the Indus Valley, plus Turkish and Mongolian in the mountainous areas. Much Indian immigration has occurred in recent years. **Religions:** Islam is the official religion followed by 97% of the population (most belong to the Sunni sect); 1.6% are Hindu; the remainder belong to Christian and other smaller sects. **Languages:** Urdu (official, although it is spoken by only 9% of the population). Other languages are Punjabi, Sindhi, Pashto, Baluchi, English and several dialects. **Political Parties:** The Pakistan People's Party

(PPP) of former president Zulfiqar Ali Bhutto, currently led by his daughter, Prime Minister Benazir Bhutto; the Pakistan Muslim League (IJI or IDA) led by Nawaz Sharif, formed a coalition including minor parties and independents. In the elections of October 1993, the PPP won 86 out of the total 217 seats in the National Assembly, against 72 obtained by Sharif. The Mohajir Qaumi Movement, boycotted the 1993 elections. There are pro-independence parties in Azad Jammu and Kashmir. **Social Organizations:** The National Pakistan Federation of Unions.

THE STATE

Official Name: Islam-i Jamhuriya-e Pakistan. **Administrative divisions:** 4 Provinces. **Capital:** Islamabad, 204,364 inhab. (1981). **Other cities:** Karachi, 5,208,132 inhab.; Lahore, 2,952,689 inhab.; Faisalabad, 1,104,209 inhab. (1981). **Government:** Faruk Ahmed Leghari, President and Head of State since November 1993. Legislature: bicameral. 87-member Senate; 217-member National Assembly, elected every 6 and 5 years, respectively. **National Holidays:** August 14, Independence Day (1947); March 23, Proclamation of the Republic (1956). **Armed Forces:** 587,000 (1995). **Paramilitaries:** 275,000 (National Guard, Border Corps, Maritime Security, Mounted Police).

1988, the PPP were victorious and a government was formed under the leadership of Bhutto's daughter Benazir Bhutto. The democratic system was brought back after an 11 year absence.

[10] Bhutto became the first woman to serve as head of state of a predominantly Islamic country. One of her first moves was to release all female prisoners charged with crimes other than murder. Many of these had been imprisoned under discriminatory "Black Laws" passed during Zia's term in office. They include the law of evidence, under which the declaration of one man is given more weight than that of two women in legal proceedings. Zia left a legacy of such laws in the form of constitutional amendments, which required a virtually politically-impossible two-thirds majority to overturn.

[11] The army has always been a strong institution in Pakistan. The country became a member of the Southeast Asian Treaty Organization (SEATO) in 1954 and the Central Treaty Organization (CENTO) in 1955, two strong military alliances led by the United States. Although Pakistan later withdrew from these alliances, bilateral relations with the US have been fluent. Pakistan has received military and economic aid from the US.

[12] Relations with India have always been strained. The two coun-

tries both have territorial claims over the state of Kashmir. India treats it as an integral part of the country, while Pakistan has been demanding a plebiscite there to allow the people of Kashmir to decide their own fate. After wars in 1948, 1965 and 1971, the two states agreed on a cease-fire zone on both sides of the Kashmir border. This meant not only a division of land, but also a separation of the local population. Since then, Kashmiri nationalist groups have demanded the creation of an independent state in the region.

[13] Pakistan was also very vocal in its opposition to the Soviet intervention in Afghanistan. About three million Afghan refugees arrived in Pakistan in mid-1990. Pakistan also supported the Mujahedin, Afghan resistance groups based in Pakistan, in their fight against the pro-Soviet regime in Kabul. During the Soviet intervention in Afghanistan, the US used Pakistani territory to supply arms to the rebel groups. This turned Pakistan into a key ally for the US regional policy, and resulted in the granting of significant economic assistance to the country.

[14] At the time of its creation, Pakistan had a very poor economic base. Development planning gained momentum in 1955 with the initiation of the first five-year plan, and a strong textile sector devel-

oped in the 1960s. Several large-scale hydroelectric projects have been commissioned since the 1960s. Pakistan has registered consistent growth in national income over the years. A country that once suffered from huge food deficits has been converted into one with a modest surplus.

[15] Pakistan's nuclear program has created suspicion among many countries including India, which accused Pakistan of developing nuclear weapons. Pakistan has denied the allegations stating their nuclear program is for peaceful purposes.

[16] Pakistan's political stability and the prospects for constitutional rule depend on the relationship between the political regime and the army. The declared policy of the army leadership is that it will remain loyal to the political regime. However, in recent years, political stability has been threatened by increasing ethnic riots between the locals and Indian immigrants in the southern province of Sind.

[17] On August 6 1990, President Ghulam Ishaq Khan dissolved the government of Prime Minister Benazir Bhutto, charging her administration with nepotism and corruption. The president suspended the National Assembly and named Ghulam Mustafa Jatoi, leader of the Combined Opposition Parties

(COP) coalition, head of the interim government.

[18] On October 24, elections were held. Nawaz Sharif was elected prime minister, with the support of the Muslim League, the main party in the coalition opposing Bhutto. Benazir Bhutto also submitted her candidacy, but her People's Party of Pakistan claimed the elections had been a fraud and started an intense opposition campaign.

[19] After the Iraqi invasion of Kuwait when the war broke out in the Persian Gulf, Pakistan quickly aligned with the US and sent troops to Saudi Arabia. Surveys showed that the population had strong pro-Iraqi tendencies, but the government announced that the country's forces would only defend Islamic holy places, and would not take part in combat or go into Iraqi territory.

[20] Shortly after taking office, Prime Minister Nawaz Sharif, who belongs to a family of Pakistani industrialists, implemented an economic reform plan, aimed at encouraging private investment. The plan includes a far-reaching privatization process which was strongly resisted by the 300,000 workers in the State-run companies.

[21] Apart from the economic reforms, Sharif's government promoted a strongly-resisted process of Islamic reinforcement. The plan included the introduction of "Sharia" or Islamic law. In the Sharif's

DEMOGRAPHY

Urban: 34% (1995). **Annual growth:** 1.7% (1992-2000). **Estimate for year 2000:** 148,000,000. **Children per woman:** 5.6 (1992).

HEALTH

One **physician** for every 2,000 inhab. (1988-91). **Under-five mortality:** 137 per 1,000 (1995). **Calorie consumption:** 92% of required intake (1995). **Safe water:** 79% of the population has access (1990-95).

EDUCATION

Illiteracy: 62% (1995). **Primary school teachers:** one for every 41 students (1992).

COMMUNICATIONS

96 **newspapers** (1995), 86 **TV sets** (1995) and 88 **radio sets** per 1,000 households (1995). 1.3 **telephones** per 100 inhabitants (1993). **Books:** 85 new titles per 1,000,000 inhabitants in 1995.

ECONOMY

Per capita GNP: $430 (1994). **Annual growth:** 1.30% (1985-94). **Annual inflation:** 8.80% (1984-94). **Consumer price index:** 100 in 1990; 150.6 in 1994. **Currency:** 31 rupees = 1$ (1994). **Cereal imports:** 2,893,000 metric tons (1993). **Fertilizer use:** 1,015 kgs per ha. (1992-93). **Imports:** $8,890 million (1994). **Exports:** $7,370 million (1994). **External debt:** $29,579 million (1994), $234 per capita (1994). **Debt service:** 35.1% of exports (1994). **Development aid received:** $1,065 million (1993); $9 per capita; 2% of GNP.

ENERGY

Consumption: 255 kgs of Oil Equivalent per capita yearly (1994), 38% imported (1994).

Illiteracy 1995

62%

External Debt 1994

PER CAPITA $234

opinion, Benazir Bhutto's term was a period of religious regression. The enforcement of Sharia immediately resulted in obvious setbacks of women's social and legal status. The government has banned the media from making any reference to a woman's right to divorce. The main leaders of the women's movements, who demand the end of discrimination against women, have been harassed by security forces, and some have been kidnapped and raped.

[22] On November 22 1990, a serious political and administrative scandal broke out. The opposition accused Nawaz Sharif of embezzling public funds. Sharif was responsible for the bankruptcy of several cooperative credit institutions. Only the unconditional support of President Ishaq Khan prevented the fall of the prime minister and the matter was referred to the judiciary. The scandal triggered a wave of demonstrations led by the PPP, and the government responded with increased repression.

[23] In 1991, there was an increase in the accusations of human rights violations levelled against the Pakistani government. In November, Amnesty International demanded the suspension of the sentences on twenty people on death row. According to Amnesty International, the people sentenced to capital punishment were denied the most basic legal rights.

[24] Apart from the disputes between the government and the opposition, Pakistan has been periodically shaken by inter-ethnic violence, in particular between the Sindhis and the Muhajirs (former Indian refugees). Violence has particularly affected the Sind region and the city of Karachi, which is the country's main industrial center.

[25] The end of Soviet intervention in Afghanistan placed Pakistan in a difficult position. From the US viewpoint, Pakistan had lost the key position that it had held during the years of Soviet military intervention. The Islamicization process started by Sharif was also unacceptable to the US, because of the implication of greater cooperation with other Islamic regimes in the area.

[26] In February 1992, the ancient dispute over the border territory of Kashmir brought Pakistan and India to the brink of a new armed conflict. The Jammu and Kashmir Liberation Front, a Muslim group which demands the creation of an independent state in this borderline region, staged a protest march along the line that divides the country between Pakistan and India. The organizers encouraged thousands of demonstrators on both sides to cross the border. The Pakistani government ordered the army to shoot at the demonstrators in order to stop the separatist march. Five people died and fifty were injured.

[27] Amidst the tension created by the Kashmir border crisis, the differences between Washington and Islamabad became more obvious. When Pakistan announced their nuclear weapon construction project was well under way, the US began to exert pressure on the country. This resulted in the virtual suspension of US economic assistance and arms sales to Pakistan. As a response to the US measures, Pakistan announced that it had guaranteed Chinese economic and technological support to continue its nuclear research program.

[28] In November 1992, Benazir Bhutto led an opposition march from Rawalpindi to Islamabad, demanding Sharif's resignation. The government arrested 1,600 people, including Bhutto, who was held for 30 days in a Karachi prison.

[29] President Ishaq Khan accused Prime Minister Mohammed Nawaz Sharif of poor administration, corruption and nepotism, forcing him to resign in April 1993. The National Assembly was dissolved and elections were announced for July. Sharif appealed to the Supreme Court and was reinstated in May. The dissolution of the Assembly was revoked and the call for elections cancelled.

[30] Sharif's return aggravated the conflict with President Ishaq Khan. Both leaders resigned in July. The interim president, Moin Kureishi, a former World Bank and IMF official, took unprecedented measures: he abolished the old system whereby members of parliament had received funds for investment in their own districts, a source of endless corruption, and also taxed large property units.

[31] Traditionally, powerful landowners had dominated Pakistan's economy and political system and had exempted themselves from paying taxes. Only one per cent of the country's 130 million inhabitants paid their taxes, but charges of tax evasion had never been filed.

[32] Benazir Bhutto returned to power with the October 6 elections. The PPP won 86 of the Assembly's 217 seats, against 72 obtained by Sharif. Bhutto received the support of minor parties and was confirmed with 121 votes. Her position was strengthened with the election of Faruk Ahmed Leghari (also from the PPP) as president in November.

[33] The prime minister tried to activate the economy and avoided confrontation with the conservative clergy. However, Islamic fundamentalists put a price on the head of the justice minister, who supported a proposal to modify Pakistan's laws on blasphemy. The modification included making false charges of blasphemy as an offence.

[34] Bhutto's efforts to bring democracy and equal rights to her country were tainted by political and ethnic violence in 1994 and 1995, the bloodiest years since 1971 when Bangladesh separated. Karachi and the northern separatist areas were the center of disputes. Over 3,500 people died in confrontations between August 1994 and October 1995.

[35] An attempted coup d'etat led by fundamentalist military officials was aborted by the government in late 1995.

[36] The Supreme Court of Justice introduced several changes to its office during 1996. The most significant was the abolition of designations according to political background. In June, hundreds of people protesting against the proscription of pro-independence candidates in local elections in Azad Jammu and Kashmir were arrested.

[37] One of Benazir Bhutto's main political rivals, her brother Murtaza, died in a confrontation with police on September 20 1996. Murtaza commanded a guerrilla group demanding the prime minister's resignation. The Bhuttos had been rivals since their father was overthrown in 1977.

[38] In November 1996 Benazir was fired by President Laghari, and was put under house arrest. She faced charges of corruption and incompetence. New elections were held in February 1997.

Palau

Palau

Population: 16,000
(1994)
Area: 494 SQ.KM
Capital: Koror
Language: English

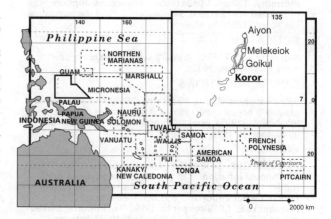

Five thousand years ago, sailors from Formosa and China peopled the islands of Micronesia, forming highly stratified societies where age, sex and military prowess defined rank, and wealth.

[2] European colonization in the 19th century, did not totally eliminate the indigenous peoples of Palau as it had on other islands.

[3] In 1914, the Japanese took the islands from Germany, who had bought them from Spain in 1899. During World War II, Japan installed its main naval base in Palau and it soon became the scene of fierce combat when the US recognized its strategic position in relation to the Philippines.

[4] By the end of World War II, the indigenous population of Palau had been reduced from 45,000 to 6,000. Micronesia became a US trust territory. The US used the islands as nuclear testing grounds and reneged on promises of self-government, instead fostering economic dependence.

[5] In 1978, at the beginning of the transition to self-government, the archipelago opted for separation from Micronesia. In January 1979, the new Constitution banned all nuclear weapons installations, nuclear waste storage, and foreign land ownership. A further amendment established a 200-mile area of territorial waters under the UN-approved Law of the Sea, which the US tried to veto.

[6] The new constitution received majority approval of assembly members, but the US put pressure on the local parliament to change provisions which would frustrate their attempts to use Palau as a military base. In July 1979, a referendum showed 92 per cent support for an unamended constitution. Bending to US pressure, the local Supreme Court annulled the referendum and entrusted Palau's parliament to draft a "revised" constitution. In October 1979, this document was rejected by 70 per cent of the voters, and In July 1980, a third referendum ratified the original document by a 78 per cent majority.

[7] US Ambassador Rosemblat declared the Constitution as "incompatible with the system of Free Association" in effect since November 1980 "because it impedes (the US) from exercising its responsibility of defending the territory.

[8] In 1981, International Power Systems (IPESCO) scheduled the installation of a 16-megawatt power plant in Palau; the scheme was to be financed by the US, subject to amendments to the anti-nuclear constitution.

[9] President Haruo Remliik was persuaded to hold another plebiscite to decide on the IPESCO plan and a "Compact of Free Association" with the US. The Compact received majority support, but failed to achieve the 75 per cent needed to override the constitution. In 1983, despite this ambiguous ballot, the president signed agreements for a loan totalling $37.5 million. UN intervention was immediately called for by the oppositon.

[10] In 1984, the power plant was built and the following year, Remliik refused to call a third plebiscite when the US proposed to revoke the historic anti-nuclear constitution. A few days later he was assassinated.

[11] The US insisted on Palau adopting the status of "Free Associated State", whereby Palau would have limited sovereignty and financial assistance from the US. In turn, the US required guaranteed access to ports and military bases for 50 years. Clauses in the new statute gave permission for nuclear vessels to enter Palau's territorial waters. In 1986, this proposal, which carried an attractive offer of $421 million in aid, was approved by 72 per cent of the electorate, still not enough to override the constitution.

[12] The Palau Congress amended the constitution by a referendum requiring a simple majority, and on August 21 1987, a further plebiscite approved the Compact. However, in April 1988, the constitutional amendment was invalidated by the Supreme Court, stating the 75 per cent majority requirement.

[13] In August 1988, then President Salii was found dead with a bullet in his head and a pistol at his side. His death was officially described as suicide.

[14] A new referendum held in February 1989 again did not gain the necessary majority for approval. Moreover, to restrain constant violations of its territorial waters by Indonesian vessels, in early 1992 Palau requested Japanese assistance in monitoring the area, in exchange for fishing rights.

[15] Towards the end of that year, Kuniwo Nakamura was elected president by a narrow margin. At the same time, a plebiscite was supported by 62 per cent of the electorate to reduce the percentage needed to approve the status of Free Associated State of the US to a simple majority in a new referendum.

[16] This eighth referendum on the same issue was held in July 1993. No special majority was required, and therefore the Free Associated State system was approved by the favorable vote of 68 per cent of the electorate.

[17] Palau became a Free Associated State as of October 1994, when the new status - so many times postponed due to the controversial storage of nuclear materials in the island - came into effect.

[18] Nakamura succeeded in becoming closer to Japan and increasing commercial relations and tourism between both countries, for which he visited Tokyo in April 1995. Palau's president also visited Taiwan, the occasion on which he established diplomatic relations with Taipei.

PROFILE

ENVIRONMENT

A barrier reef to the west of Palau forms a large lagoon dotted with small islands. Coral formations and marine life in this lagoon are among the richest in the world with around 1,500 species of tropical fish and 700 types of coral and anemones.

SOCIETY

Peoples: Most are of Polynesian origin.
Religions: Catholic. **Languages:** English. **Political Parties:** The Coalition for Open, Honest and Fair Government is opposed to the proliferation of nuclear arms and toxic wastes in the region and rejects the status of "Free Associated State"; the Palau Party favors "Free Association".

THE STATE

Official Name: Republic of Palau. **Administrative Divisions:** 16 States. **Capital:** Koror, 10,486 inhab. (1990). **Government:** Stella Guerra is the High Commissioner appointed by the President of the United States, working under US jurisdiction. Kumino Nakamura, President since January 1 1993.

Palestine

Capital: Jerusalem
(Al-Quds
Ash Sharif)

Language: Arabic

Palestine

A round 4000 BC, the Canaanites, a Semitic people from the inner Arabian peninsula, settled in the land which became known as Canaan and later, Palestine. The Jebusites, one of the Canaanite tribes, built a settlement which they called Urusalim (Jerusalem), meaning "the city of peace".

[2] The Egyptian Pharaohs occupied part of Canaan in 3,200 BC, building fortresses to protect their trade routes, but the country kept its independence. Around 2000 BC, another Semitic people, Abraham's Hebrews, passed through Palestine on their way south. Seven centuries later, twelve Hebrew tribes returned from Egypt, following Moses. There was fierce fighting over possession of the land. The Bible records that "The sons of Judah were unable to exterminate the Jebusites that dwell in Jerusalem" (Joshua 15, 63).

[3] Four centuries later, Isaac's son David managed to defeat the Jebusites and unite the Jewish nation. After the death of his son Solomon the Hebrews split into two states, Israel and Judah. These later fell into the hands of the Assyrians, in 721 BC, and Chaldeans, in 587 BC. It was in 587 that Nebuchadnezzar destroyed Jerusalem and took the Jews into captivity in Babylon.

[4] In 332 BC, Alexander the Great conquered Palestine, but the territory returned to the Egyptian Empire of the Ptolemies soon after his death. The country was subdued by the Seleucids from Syria before a rebellion, headed by Judas Maccabeus, restored the Jewish state in 67 BC.

[5] In 63 BC, the Roman Empire seized Jerusalem, placing the city under its domination. Maccabeans, Zealots and other Jewish tribes resisted the invaders but were fiercely subdued. Solomon's temple was demolished around 70 AD, and the Jews were expelled from Jerusalem around 135 AD.

[6] The Romans gave Palestine its present name, and Roman domination was followed by that of the Byzantine Empire (the Roman Empire in the East), which lasted until 611 when the province was invaded by the Persians. The Arabs, a Semitic people from the inner peninsula, conquered Palestine in 634, and according to legend, it was in Jerusalem that the prophet Muhammad rose to the heavens. As a result, the city became a holy

place for all three monotheistic religions.

[7] The Islamic faith and the Arabic language united all the Semitic peoples except for the Jews. With short intervals of partial domination by the Christian Crusaders and the Mongols in the 11th, 12th and 13th centuries, Palestine had Arab rulers for almost 1,000 years and Islamic governments for 15 centuries.

[8] In 1516, Jerusalem was conquered by the Ottoman Empire which maintained power until the end of World War I. During this conflict, the British promised Shereef Hussein the independence of the Arab lands in exchange for his cooperation in the struggle against the Turks. At the same time, in 1917, British Foreign Secretary Lord Balfour promised the Zionist Movement the establishment of a "Jewish National Homeland" in Palestine.

[9] Britain had no power at all over the area, either de facto or de jure, but it soon obtained this right by defeating the Turks, with the help of Arab allies, with a League of Nations mandate in 1922. Massive immigration raised the Jewish population of Palestine from 50,000 at

the beginning of the century to 300,000 prior to World War II (see Israel).

[10] The Palestinians staged a general strike in April 1936, in protest against this immigration, which they saw as a threat to their rights. The British put forward a plan for the partition of Palestine into three states: Jewish in the north, Arab in the south, and a third section under British administration in the Jerusalem-Jaffa (Tel Aviv) corridor. The Arabs rejected the plan and rebellion broke out, lasting until 1939, when London gave up the idea and set limits to immigration.

[11] Once World War II was over, Britain handed the problem over to the newly-established United Nations.

[12] When the UN General Assembly approved a new partition plan (1947) 749,000 Arabs and 9,250 Jews lived in the territory where the Arab State would be set up, while 497,000 Arabs and 498,000 Jews lived in the part which was to become the Jewish state.

[13] To drive the Palestinians from their land, a detachment of the Irgun organization commanded by Menahem Begin raided the village

of Deir Yasin on April 9 1948, killing 254 civilians. 10,000 terrified Palestinians left the country.

[14] On May 14 1948, Israel unilaterally proclaimed itself an independent country. Neighboring Arab armies immediately attacked, but were unable to prevent the consolidation of the Jewish State. On the contrary, the latter emerged from the 1949 war with a land area larger than that proposed by the United Nations.

[15]

[16] In the eyes of the United Nations, and therefore, of international law, the Palestinians were not a people, but simply refugees, a "problem" to be solved.

[17] Political decisions about the Palestinian cause were left entirely in the hands of the Arab governments, who even had the right to appoint the Palestinian representative to the Arab League. At the 1964 Arab summit, Egyptian leader, Gamal Abdel Nasser, asked the League to take on the task of forging a unified Palestinian organization.

[18] In Jerusalem, on May 27, the Palestine National Council met for the first time. There were 422 partic-

Map legends

Jewish State, UN Plan, 1947

Arab State, UN Plan, 1947

[1] West Bank

Jewish State, UN Plan, 1947

Annexed to Israel in the 1948 war

[2] Gaza Strip

Israel's pre-1967 borders

Occupied by Israel in 1967

Recovered by Palestine from 1994:
Gaza Strip, Jericho, Nablus, Ramallah

ipants, including personalities, business leaders, representatives of the refugee camps, the trade-union organizations, and the young people's and women's groups; they founded the Palestine Liberation Organization (PLO).

[19] Palestinian groups already operating secretly, such as Al Fatah, were wary of this Arab promoted organization as they distrusted its emphasis on using diplomatic channels for its struggle. They were convinced that their land could only be recovered by military force.

[20] On January 1 1965, the first armed operation took place in Israel. The attacks intensified during the following months, until the outbreak of the Six Day War in 1967, when Israel occupied all of Jerusalem, Syria's Golan Heights, Egypt's Sinai Peninsula, the Palestinian territories of the West Bank and the Gaza strip. The defeat of the regular Arab armies strengthened the conviction that guerrilla warfare was the only path.

[21] In March 1968, during a battle in the village of Al-Karameh, Palestinians forced the Israelis to withdraw. The event passed into folk history as the first victory of the Palestinian force.

[22] With their prestige thus restored, the various armed groups joined the PLO and obtained the support of the Arab governments. In February 1969, Yasser Arafat was elected chairman of the organization.

[23] The growing political and military strength of the Palestinians was seen as a threat by King Hussein of Jordan, who had acted as their representative and spokesman. Tension mounted between the king and the Palestinians eventually reaching explosive proportions. In September 1970, after much bloody fighting, the PLO was expelled from Jordan to set up its headquarters in Beirut.

[24] This new exile reduced the possibility of armed attacks on targets inside Israel, and new radical groups such as "Black September" directed their efforts towards Israeli institutions and businesses in Europe and other parts of the world. Palestinians, until then regarded by world opinion purely as refugees, quickly came to be identified by some as terrorists.

[25] PLO leaders promptly realized the need to change their tactics and, without abandoning armed struggle, launched a large-scale diplomatic

ENVIRONMENT

In terms of international law "Palestine" is the 27,000 sq km. territory west of the Jordan River which the League of Nations handed over to Britain's "mandatory" power in 1918. This territory comprises: the area occupied by Israel before 1967, 20,073 sq km; Jerusalem and its surroundings, 70 sq km; the West Bank area, 5,879 sq km; and the Gaza Strip, 378 sq km. It is a land of temperate Mediterranean climate, fertile on the coast and in the Jordan Valley. It is surrounded in the south and the northeast by the Sinai and Syrian deserts respectively. The Gaza region suffers from a severe scarcity of water. The accumulation of waste waters and refuse convert the refugee camps into highly contaminated areas. Soil erosion and deforestation are also serious problems.

SOCIETY

Peoples: Palestinians are an Arab people. There are 700,000 in Israel (1967 frontiers); 1,500,000 on the West Bank; 800,000 in the Gaza Strip, and 33% of the inhabitants of the occupied territories live in refugee camps. There are large Palestinian populations in other Arab countries, the US, Chile and Brazil. **Religions:** Most are Muslim, but there are a large number of Christians, mostly Eastern-rite Catholic. **Languages:** Palestinians speak Arabic and often also use Hebrew in the occupied territories. **Political Parties:** The main political organization represented in the PLO is the Al-Fatah National Liberation Movement, founded in 1965 by Yasser Arafat. The Popular Front for the Liberation of Palestine (PFLP); the Popular Front for the Liberation of Palestine-General Command (PFLP-General Command); the Democratic Front for the Liberation of Palestine (DFLP); Al-Saiqa; the Arab Liberation Front; Hamas (Islamic Resistance Movement); Islamic Jihad; Hezbollah.

THE STATE

Official name: State of Palestine. **Capital:** Jerusalem has traditionally been the capital of Palestine; the Palestine National Authority (PNA) is in Jericho, 16,000 inhab. (1993). **Government:** Yasser Arafat, President. The Autonomous Council acts as a Parliament.

PLO leaders promptly realized the need to change their tactics and, without abandoning armed struggle, launched a large-scale diplomatic offensive.

offensive, starting to devote much of their energy to consolidating Palestinian unity and identity. The Algiers Conference of Non-Aligned Countries (1973) identified the Palestine problem, and not Arab-Israeli rivalry, as the key to the conflict in the Middle East for the first time.

[26] In 1974, an Arab League summit conference recognized the PLO as "the only legitimate representative of the Palestinian people". In October of the same year the PLO was granted observer status in the UN General Assembly, which recognized the right of the Palestinian people to self-determination and independence, and condemned Zionism as "a form of racism".

[27] The PLO program proposes to set up "a secular and independent state in the whole of the Palestinian territory, where Muslims, Christians and Jews can live in peace, enjoying the same rights and duties". This necessarily implies the end of the present state of Israel. Without giving up this ultimate goal, the PLO has gradually come to accept the "temporary solution" of setting up an independent Palestinian state "in any part of the territory that might be liberated by force of arms, or from which Israel may withdraw".

[28] In 1980, the Likud prime minister, Menahem Begin and Egyptian president Anwar Sadat signed a peace accord at Camp David, with US mediation. Shortly afterwards, Begin began officially to annex the Arab part of Jerusalem, proclaiming it the "sole and indivisible capital" of Israel. Jewish settlements on the West Bank multiplied, using Palestinian lands and increasing tension in the occupied territories. Successive United Nations votes against these measures, or for any action against Israel, were stripped of any practical value by the US using its veto in the Security Council.

[29] In July 1982, in an attempted "final settlement" of the Palestine issue, Israeli forces invaded Lebanon. The intention, as it later became clear, was to destroy the PLO's military structure, capture the greatest possible number of its leaders and combatants, annex the southern part of Lebanon and set up a puppet government in Beirut. Surrounded in Beirut, the Palestinian forces only agreed to withdraw after receiving guarantees of protection for civilians under a French-Italian-North American international peace-keeping force.

[30] The massacres that took place at the refugee camps of Sabra and Chatila showed the ineffectiveness of international protection, but the PLO managed to transform what seemed a final defeat into a political and diplomatic victory. The headquarters of the organization were moved to Tunis and Yasser Arafat toured Europe receiving the honors due a head of state in various countries, most notably in the Vatican.

[31] The PLO quietly initiated talks with Israeli leaders receptive to a negotiated settlement with the Palestinians. With the invasion of Lebanon, small but active peace groups emerged in Israel, demanding the initiation of a dialogue with the PLO. Palestinian radicals called these overtures into question, breaking with Yasser Arafat's pol-

Steps to statehood

In its first stage, the new Palestinian territory includes the Gaza Strip and the region of Jericho, even though these two zones are separated by Israeli territory. Jericho is the seat of the new Palestine National Authority (PNA).

2   Eight hundred thousand Palestinians live cramped in refugee camps in the Gaza Strip, a 378 sq km band of territory on the Mediterranean. There are also approximately 5,000 Jewish colonists living in 17 different settlements covering 20 per cent of the Gaza area.

3   Between April and July 1994, many thousands of Palestinian police and members of the Palestine Liberation Army moved into the region. Nevertheless, the Israeli army continued to patrol the main roads of the Strip, guarding the Jewish settlements.

4   Until self-government goes into effect, the Palestinians in Gaza have restricted access to the scarce water resources in the area. The Israeli state established a water distribution system which is preferential to the Jewish colonists. It is precisely this issue of the control and use of water resources that is the main bone of contention in the Palestinian-Israeli negotiations.

5   Infrastructure in Gaza is non-existent. There are no drains or sewerage systems, nor is there any refuse collection. Several miles of the desert zone around the camps is covered by waste. The zone has high levels of infant malnutrition and maternal anaemia, and the one hospital has only 900 beds.

6   The ancient oasis city of Jericho is close to the border with Jordan, on the West Bank of the Jordan river.

7   In 1948, the population of Jericho numbered approximately 2,000 inhabitants. More recently, within a short period of time some 70,000 Palestinian refugees settled in temporary camps surrounding the city. As a result, the city underwent a rapid economic expansion, sustained by the large source of cheap labor in the temporary camps.

8   With the Six Day War, the bulk of the refugees were expelled to Jordan, reducing the present population of Jericho to 16,000, including several thousand people still living in the camps. The regional economy is largely agricultural though the input from tourism is currently increasing.

9   In contrast to the mainstays of the local economy, approximately 62 per cent of the Palestinian population in the Occupied Territories are salaried workers, with large numbers of them working in Israel. The Palestinians fill the posts the Israelis will not take, working for half the salary of an Israeli worker. The non-laboring classes of Palestinian society are mostly found outside the area, in Lebanon and other Gulf countries. These are the professional classes; directors of oil companies and banks, university professors and government functionaries in the region.

10   Some 800,000 ex-patriate Palestinian students are to be found studying in universities throughout the Arab world and beyond, while few Palestinian residents of the Occupied Territories manage to reach university level education.

11   The geographical separation of the Palestinian social classes is reflected in Palestinian politics, where the different social sectors express their different interests, interests that do not always coincide.For example, in Gaza groups like Hamas have great influence on the population, whereas in Jericho where the economic situation is better, Fatah holds the majority of popular support, as Abdul Kareem, leader of the local PLO office, affirmed "we are a moderate city in all senses".

12   The economy of Gaza, meanwhile, is at the level of its infrastructure. Between 1986 and 1990 the GNP fell continuously, agriculture is modest and industry nearly non-existent. The biggest business concern in the region is a bottling company, employing no more than 100 workers. The underground economy, in contrast, is on the increase. The majority of small factories employ undeclared workers on low pay and without social security, invoices and accounts are rare. Remittances sent by workers in the other Gulf countries, international aid and the PLO themselves contribute to making daily life a little less difficult for the Palestinian settlers.

13   The new Palestinian territory of Gaza and the region of Jericho, now has to implement a basic infrastructure plan for water and energy distribution, roads, port and airports. There is a desperate need for new housing, approximately 125,000 homes need to be constructed over the next few years, the refugees and returned exiles need to be housed, and thousands of jobs need to be created.

14   Japan, the European countries and the US have all announced their intentions to give aid and offer assistance, but the promised funds are rarely forthcoming. The participation of expatriate Palestinians in the construction of the new economy is a fundamental necessity, and their investments are desperately needed. At the beginning of the Gulf War, Palestinian deposits in the Kuwaiti banks exceeded $9 billion.

15   The PLO leaders are aware that total economic separation from Israel is neither possible nor desirable. However, the terms of the mutual economic relationship remain undefined especially as far as tariffs and financial matters are concerned, key elements in the Palestine National Authority search for investments. One possible option would be Palestine's integration into a regional market, a process that might also strengthen the peace process throughout the region.

---

icies. This division of the PLO put its factions at odds with each other, sometimes causing violent confrontations.

32   In 1987, after several years of internal difficulties, the Palestine National Council met in Algiers, with representatives from all Palestinian organizations, except those groups that favored direct action, and the internal structure of the PLO was rebuilt.

33   In November 1987, several Palestinian laborers were run over by a military truck in the Gaza Strip. In protest, Palestinian businesses closed down and the people took to the streets.

34   The official answer to the Arab protests was to increase the repression. But unlike what had happened on other occasions, this time the military intervention only managed to increase the number of women, elderly people and children taking part in the demonstrations. The more civilian casualties there were, the greater the hatred grew and the more demonstrations, strikes, and closures occurred. Funerals transformed into acts of open political defiance. This marked the beginning of the *intifada* or rebellion.

35   During the first few months of 1988 a great many Palestinians with Israeli citizenship participated in the strikes called by the so-called "Unified Direction of the Popular Uprising in the Occupied Territories". This was the first instance of their joint political expression with the Palestinians of the occupied territories.

36   In July 1988, King Hussein of Jordan announced that all economic and political links were being broken with the inhabitants of the West Bank. From that moment on,

the PLO assumed sole responsibility for the territory's inhabitants.

[37] At a meeting in Algeria on November 15 1988, the Palestine National Council proclaimed an independent Palestinian state in the occupied territories, citing Jerusalem as its capital. It also approved UN resolutions 181 and 242, which in effect meant accepting Israel's right to exist. Within the next 10 days, 54 countries around the world recognized the new state.

[38] Arafat, elected president, was received in Geneva by the UN General Assembly, which had called a special session in order to hear him. The Palestinian leader repudiated terrorism, accepted the existence of Israel and asked that international forces be sent to the occupied territories. As a result of his speech, US president Reagan decided to initiate talks with the PLO.

[39] The Security Council met in Geneva, at the bidding of the Arab countries, to listen to Arafat discuss violence in the occupied territories. The Palestinian leader exhorted the UN to call an urgent international conference on peace in the Middle East.

[40] When tensions began between Iraq and Kuwait, in the second half of 1990, Arafat tried unsuccessfully to start negotiations between the countries. After the invasion, the Palestinian position seemed to strengthen when a parrallel could be drawn between Kuwait and Palestine; if Iraq could be forced to submit to UN resolutions, then so could Israel.

[41] When the war broke out, it was clear that the Palestine people were pro-Iraqi. This support deprived the PLO of the financial support of the rich Gulf monarchies, who were opposed to the Iraqi regime. When Iraq was defeated in March 1991, the situation in the occupied territories was very volatile.

[42] On a diplomatic level, the US-Soviet declaration expressed the hope that a peace agreement would be achieved. This showed that a wedge had been driven between US and Israel.

[43] In September 1991, in the closing session of the Palestine National Council, Yasser Arafat was confirmed as President of Palestine and of the PLO. The body accepted the resignation of Abu Abbas, the leader of the Palestine Liberation Front. Abbas had been given a life sentence in absentia by an Italian tribu-

nal, for the hijacking of the "Achille Lauro" liner in 1985.

[44] Between October 30 and November 4 1991, the first Peace Conference for the Middle East was held in Madrid, with support from the US and the former USSR.

[45] The Arab delegations unanimously demanded that the negotiations should be based on resolutions 242 and 338 of the UN Security Council. These resolutions forbid the acquisition of territories by force and recommend the granting of territories in exchange for peace agreements.

[46] Three issues were put forward, concerning the possible autonomy of the Palestinian territories: the end of Jewish settlements in the occupied territories, the future legal status of Jerusalem and the procedures for Palestinian self-determination.

[47] The Conference for the Middle East continued in Washington in December.

[48] No progress was made as far as the Palestinian issue was concerned, as Israel reaffirmed the validity of its own interpretations of the UN resolutions. At the end of the Conference, the Israeli delegation left satisfied because UN resolution 3379, defining Zionism as a form of racism, had been eliminated.

[49] Following the Israeli elections of June 1992, the Labor leader and new prime minister Yitzak Rabin froze the settlement of new colonies in the Gaza strip and on the West Bank. However, it was difficult to restart negotiations which had been interrupted by the expulsion of 415 Palestinians from the Hamas group to Lebanon.

[50] Secret negotiations between the PLO and Israel, with the active participation of Norwegian diplomats, resulted in mutual recognition in September 1993. The Declaration of Principles on the autonomy of the occupied territories granted limited autonomy to Palestinians in the Gaza strip and the city of Jericho in the West Bank. This autonomy will be extended to the rest of the West Bank and, five years later, a definite status will be negotiated for the occupied territories and the part of Jerusalem occupied by Israel since 1967.

[51] Hamas and Hezbollah on the Palestinian side, as well as settlers in the occupied territories and far-right parties on the Israeli side, opposed the agreement. In a climate of hostility, Israeli military withdrawal from Gaza and Jericho an-

ticipated for December 13 was postponed.

[52] After the killing by Israeli troops of Salim Muafi, commander of "Al Fatah Hawks" armed group, Al Fatah proclaimed a general strike that paralyzed the Gaza Strip for three days. In May, Rabin and Arafat signed the "Gaza and Jericho first" autonomy agreement, while Israeli withdrawal continued, enabling the return of several contingents of the Palestinian Liberation Army exiled in Egypt, Yemen, Libya, Jordan or Algeria.

[53] Arafat arrived in Gaza in July and took office as head of the Palestine National Authority's Executive Council. The struggle between the PLO leader and his fundamentalist rivals became increasingly

> The "Gaza and Jericho first" autonomy agreement enabled the return of several contingents of the Palestinian Liberation Army exiled in arab countries.

violent. After the death of Islamic Jihad leader Hani Abed, allegedly by Israeli secret services, three Israeli soldiers were murdered in November 1994. A week later, the recently created Palestinian police shot a group of people outside a mosque regarded as a sacred place for fundamentalist activists. 13 people were killed.

[54] Once again, Gaza was on the brink of civil war in April 1995 when an explosion destroyed a building, killing 7 people, including Kamal Kaheil, one of the leaders of the Ezzedin-El-Kassam brigades. In retaliation, suicide attacks by Hamas and Islamic Jihad caused the death of 7 Israeli soldiers and a US tourist, leaving 40 wounded. The Jihad's military wing said the attempt was a "heroic suicide operation" and a "gift for the soul of the criminal massacre's martyrs" - referring to the building's explosion.

[55] Tension continued, as well as

negotiations between Islamic fundamentalists and PLO leaders. Among other things, Arafat wanted Hamas to participate in the Palestinian general elections of January 1996, which would have legitimated his leadership. After comings and goings, the fundamentalists decided to boycott the elections. Arafat was elected president with 87 per cent of the vote and government candidates won 66 out of the 88 seats.

[56] The confrontation between Arafat and the fundamentalist opposition did not stop. Approximately 1,000 people were arrested in March - due to new suicide attempts, mostly in Jerusalem - and the Palestinian president placed all of Palestine's mosques under his direct authority.

[57] Benyamin Netanyahu's election as Israeli prime minister (see Israel) in May 1996 aggravated tension between both countries. This led to new confrontations such as the controversial opening of a tunnel below El-Aqsa mosque in Jerusalem by Israeli authorities in September. Dozens of Palestinians and Israelis died in the riots and the critical situation had to be discussed in a summit meeting between Arafat and Netanyahu with the participation of United States president Bill Clinton.

Panama

Panama

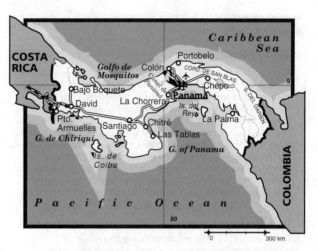

Population: 2,612,000 (1994)
Area: 77,080 SQ.KM
Capital: Panama City

Currency: Balboa
Language: Spanish

The Chibcha civilization (see Colombia: muisca) was one of America's great cultures, which developed in the Isthmus of Panama. They had a highly stratified society, developed elaborate architecture, crafted gold and had a wide scientific knowledge.

[2] In 1508, Diego de Nicuesa was given the task of colonizing what was known as the Gold Coast; present day Panama and Costa Rica. The enterprise ended in complete failure. In 1513, Vasco Nuñez de Balboa, was sent to look for what was assumed to be a "South Sea", and on the 25th of September he found the Pacific.

[3] The isthmus soon acquired great geopolitical importance. Panama became an important commercial center with merchandise distributed throughout America from San Francisco to Santiago. The Peruvian gold followed the opposite route, and the Ecuadorian straw hats are still called Panama hats because they were transported from there. The concentration of riches attracted English pirates and buccaneers; Francis Drake razed Portobelo in 1596 and Henry Morgan set fire to Panama in 1671.

[4] Panama was dependent upon the viceroyalty of Peru until 1717, when the Bourbons transferred it to the new viceroyalty of Granada, this was later to be a part of Greater Colombia when the country became independent from Spain in 1821.

[5] In 1826, Panama was selected by Simón Bolívar as the site of the Congressional Assembly which was to seal the continent's unity. But the economic decadence of the end of the 18th century, coupled with the change in commercial routes meant that Panama did not maintain its strategic importance after breaking with Spain, and did not become an independent nation with the disintegration of Greater Colombia in 1830.

[6] In 1831 Panama seceded from New Granada for a year, with the intention of forming a Colombian Confederation while maintaining autonomy. The state of Panama was created in 1855 within the Federation of New Granada (present-day Colombia).

[7] The first direct reference to the United States' "right" of military intervention in Panama, is the Mallarino-Bidlak treaty of 1864 signed by the governments of Washington and Bogota. The document authorized the United States to obtain a faster means of uniting the east coast with the west and to build a railroad across the isthmus; the Atlantic terminal was the Island of Manzanillo in the Bay of Limón. With the railway the US tried to offset British presence in the area, especially in Nicaragua.

[8] On January 1 1880, a French company started to build the Panama Canal. In 1891, the company was accused of fraud in its dealings, causing its eventual bankruptcy, though 33 km of the project had already been completed.

[9] In 1844 the New Panama Canal Company was formed to complete the canal project.

[10] In 1902, the United States bought out the French company, and in January 1903, the Hay-Herran Treaty was signed with a representative of the Colombian government. The treaty spelled out the terms of the construction and administration of the canal, granting the United States the right to rent a 9.5 km-wide strip across the isthmus in perpetuity. The Colombian Senate rejected the treaty unanimously, considering it improper and an affront to Colombia's sovereignty, and only a revolution allowed the United States to remain. The "revolutionaries", supported by US marines, declared Panama's independence in November 1903, and the US recognized the new state within three days. While Theodore Roosevelt was president, the "Big Stick" policy of sending troops into Central American states was common practice.

[11] A new treaty, the Hay-Buneau Varilla treaty, granted the United States full authority over a 16 km-wide strip and the waters at either end of the canal, in perpetuity. Buneau Varilla, a former shareholder of the canal company and a French citizen, signed as the official representative of Panama. He received payment for his services in Washington, and did not return to Panama. The canal, covering a distance of 82 km, was officially dedicated on August 15 1914, and from then on was administered and governed by the United States.

[12] The Canal Zone brought incalculable wealth to the United States, not so much in toll fees but in time and distance saved by vessels travelling between California and the East coast. US military bases in Panama functioned as an effective means of control over Latin America. Against the backdrop of the Cold War, American military instructors lectured Latin American military officers on the National Security Doctrine, a politico-military system which guaranteed loyalty to the US, including when necessary ousting legally constituted governments and replacing them with military dictators. Also, the financial center created in the isthmus became an initial foothold for the expansion of US transnational corporations and money laundering.

[13] In January 1964, 21 students died in an attempt to raise Panama's flag in the Canal Zone, under US jurisdiction The sacrifice of these young Panamanian lives transformed them into national martyrs.

[14] The demand for full sovereignty over the Zone was taken up by the government of General Torrijos. He rose to power in 1969, upon the dissolution of a 3 member Military Junta which had overthrown President Arnulfo Arias in 1968. The diplomatic battle against the colonial enclave was waged in all the international forums and gained the support of the Latin American countries, the Movement of Non-Aligned Countries and the United Nations.

[15] The struggle for sovereignty drew Panamanians closer together stimulating nationalistic feelings submerged by decades of foreign cultural penetration. At the same time, the Torrijos government initiated a process of transformation aimed at establishing a more equi-

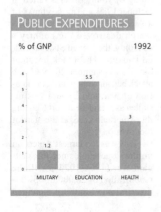

PUBLIC EXPENDITURES

% of GNP — 1992

MILITARY 1.2 / EDUCATION 5.5 / HEALTH 3

WORKERS — 1994

14% UNEMPLOYMENT

MEN 72% — WOMEN 28%

AGRICULTURE 27% — INDUSTRY 14% — SERVICES 59%

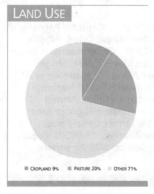

LAND USE

CROPLAND 9% — PASTURE 20% — OTHER 71%

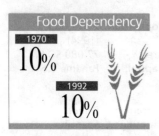

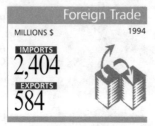

ENVIRONMENT

The country is bordered by the Caribbean in the north and the Pacific in the south. A high mountain range splits the country into two plains, a narrow one covered by rain-forests along the Atlantic slopes and a wider one with forests on the Pacific slopes. Panamanians generally say that their main resource is their geographic location, since the canal and the trade activities connected to it constitute the main economic resource of the country. Tropical products are cultivated and copper is mined at the large Cerro Colorado mines. Air and water pollution in both urban and rural areas is significant. 40 million tons of raw sewage are pumped into Panama Bay each year.

SOCIETY

Peoples: 64% of the inhabitants are the result of integration between native Americans and European colonist immigrants. 14% are African descendants and mulattos. The three main native groups are the Cunas on the island of San Blas in the Caribbean, the Chocoles in the province of Darien and the Guaymies, in the provinces of Chiriqui, Veraguas and Bocas del Toro. There are also minorities of Jamaican, Chinese and East Indian origin, whose ancestors worked on the construction of the Canal. **Religions:** 80% Catholic, 10% Protestant, 5% Muslim. **Languages:** Spanish, official and spoken by the majority. English is also spoken. **Political Parties:** The most important, in the last general elections held in May 1994, were: the United People's Alliance, a social democratic coalition made up of the Democratic Revolutionary Party (PRD), founded by Omar Torrijos, the liberal Republican Party and the Labor Party; the Democratic Alliance of Civilian Opposition (ADOC), a three-party coalition (Christian Democrats, Authentic Liberals and the Nationalist Republican Liberal Movement) with Mireya Moscoso - widow of former President Arnulfo Arias - as its candidate; the "Papa Egoro" Movement, led by Rubén Blades, founded in 1991; the "Change '94" Alliance, neo-liberal, led by Rubén Darío Carlés; the Christian Democratic Party, led by Eduardo Vallarino. **Social Organizations:** The National Council of Workers' Organizations (CONATO) combines the three largest labor unions: the leftist National Workers' Central Union of Panama (CNTP), the social-democratic Workers' Confederation of the Republic of Panama (CTRP), and the Authentic Central Union of Independent Workers (CAT), of Christian left tendency. The Christian-Democratic Workers' Central Union (CIT) does not belong to CONATO, and is the smallest of the four trade union organizations.

THE STATE

Official Name: República de Panamá. **Capital:** Panama City, 450,705 inhab. (1990). **Other cities:** Colón, 54,654 inhab.; David, 50,000 inhab. (1990). **Government:** Ernesto Pérez Balladares, President since May 8, 1994. Single-chamber parliament: Legislative Assembly, made up of 67 members, elected every 5 years by direct vote. **National Holiday:** November 3, Independence (1903). **Armed Forces:** The National Guard was declared illegal in June 1991. US forces stationed in the former Canal Zone amount to 7,100 (1994). **Paramilitaries:** 11,000 National Police, 400 National Maritime Service, 400 Air service (1993).

table social order. The outstanding reforms included agriculture, education, and the exploitation of copper on a nationalized basis. The "banana war" was also waged against transnational fruit companies like the United Fruit Company to obtain fairer prices.

[16] The US finally agreed to open negotiations in favour of a new canal treaty, as the Panama issue was damaging its image in Latin America. The 1977 Torrijos-Carter Treaty abrogated the previous one and provided for a totally Panamanian canal from the year 2,000. Amendments introduced by the US Senate, however, added provisions to the treaty that were contrary to Panamanian sovereignty. The United States retained the right to intervene "in defence of the Canal" even after expiration of the treaty, scheduled for December 31 1999.

[17] On July 31 1981 General Omar Torrijos died in a suspicious airplane accident. Unconfirmed reports suggested that the plane's instruments were interfered with from the ground. President Aristides Royo, who succeeded Torrijos in 1978, lost the support of the National Guard and was forced to resign by his new commander-in-chief, Ruben Paredes. He started realigning the country's policies, adopting a pro-US stance.

[18] The role of the United States in the Malvinas (Falklands) War and the Contadora Group, (a meeting of South American heads of states seeking peace in Central America) of which Panama was the first host, led to new friction in the relations between the nations. The island of Contadora had already gained a certain notoriety as the hiding place chosen by the US for the exiled shah of Iran, Reza Pahlevi, after he was overthrown.

[19] In 1983, Paredes was replaced as commander-in-chief of the National Guard by General Manuel Noriega. Presidential and legislative elections were narrowly won by Nicolas Barletta, the candidate of the Democratic Revolutionary Party, which was founded by Torrijos and supported by the armed forces. The opposition, led by veteran politician Arnulfo Arias brought accusations of fraud. Barletta encountered growing opposition to his economic policies, and resigned toward the end of 1985.

[20] He was succeeded by Eric del Valle, but the leading force continued to be General Noriega, who

was singled out as a target for the United States to oust. As an old protegé of the American government, they could not forgive his lack of collaboration with its plans to invade Nicaragua. A "settling of accounts" began, in which Noriega was accused of links with drug traffickers and other crimes. The opposition united behind the National Civil Crusade, made up of parties of the right and center with broad support of the business community.

[21] In 1987, the United States withdrew its economic and military aid. In 1988, it froze Panama's assets in the United States and imposed economic sanctions, including the cessation of payments for Canal operations. In March all the Panamanian banks closed for several weeks, provoking a financial crisis. American military presence increased. Del Valle ousted Noriega, but the National Assembly backed the commander-in-chief and removed the president, replacing him with the Minister of Education, Manuel Solis Palma.

[22] Elections were called for May 5 1989. The official candidate was Carlos Duque and the opposition, the so-called Democratic Alliance of Civilian Opposition, put forward Guillermo Endara. Amid interference from the White House, which discredited the electoral process and its results even before it had taken place, the results of the ballot were kept secret for several days and, because they favored the opposition, the election was declared null and void.

[23] Solis declared that the United States's objective was to set up a puppet government regardless of the election results and retain control of the Canal Zone, going back on the commitments it had assumed through the Torrijos-Carter treaty.

[24] An anti-Noriega uprising by a group of young officers failed in October 1989, and as the climax of a series of economic sanctions which destroyed the country's economy, the United States invaded Panama. The attack began at dawn on December 20 1989, the attack began, without prior declaration of war, and Endara was installed as president at Fort Clayton, the American base, at the start of the invasion.

[25] This was the largest American military operation since the Vietnam War (1964-1973), with the mobilization of 26,000 troops. Indiscriminate bombing damaged

·Illiteracy
1995

9%

External Debt
1994

PER CAPITA

$2,721

STATISTICS

DEMOGRAPHY

Urban: 53% (1995). **Annual growth:** 0.7% (1992-2000). **Estimate for year 2000:** 3,000,000. **Children per woman:** 2.9 (1992).

HEALTH

One **physician** for every 562 inhab. (1988-91). **Under-five mortality:** 20 per 1,000 (1995). **Calorie consumption:** 97% of required intake (1995). **Safe water:** 83% of the population has access (1990-95).

EDUCATION

School enrolment: University: 23% (1993). **Primary school teachers:** one for every 23 students (1992).

COMMUNICATIONS

100 **newspapers** (1995), 102 **TV sets** (1995) and 96 **radio sets** per 1,000 households (1995). 10.2 **telephones** per 100 inhabitants (1993). **Books:** 95 new titles per 1,000,000 inhabitants in 1995.

ECONOMY

Consumer price index: 100 in 1990; 104.9 in 1994. **Currency:** 1 balboas = 1$ (1994). **Cereal imports:** 159,000 metric tons (1993). **Fertilizer use:** 476 kgs per ha. (1992-93). **Development aid received:** $79 million (1993); $31 per capita; 1% of GNP.

heavily-populated neighborhoods. The invaders withheld all information concerning the number of people killed or injured, as well as figures for property damage. The number of dead is estimated at between 4,000 and 10,000, and the Chamber of Commerce estimated losses at more than $2 billion.

[26] Panamanian resistance proved to be stronger than the invaders had anticipated, and prolonged the military undertaking. Noriega, who took refuge within the Vatican Embassy, was later arrested and transfered to the United States. About 5,000 Panamanians were arrested and temporarily imprisoned.

[27] The new government of Guillermo Endara eliminated the National Defence Force and replaced it with a minor police agency, called the Public Force. In order to disarm the population, $150 was paid for each weapon that was turned in. American economic aid, which the new government had counted on, did not materialize and Endara himself began a hunger strike in order to get it. The government had to accept the presence of US "supervisors" in the ministries and the actions of the South Commando troops outside the canal zone in order to fight drug trafficking and Colombian guerrilla warfare.

[28] Washington's interest in the region waned after the defeat of the Sandinistas in Nicaragua, so the economic crisis in Panama at the time of the invasion was never overcome. Independent sources put unemployment at 20 per cent in 1991.

[29] The Organization of American States called the invasion "deplorable" and called for a vote on the withdrawal of troops, there were 20 votes in favor, one against (the United States) and six abstentions. Britain supported the invasion and France vetoed the UN Security Council denunciation. In Latin America, only El Salvador supported the US invasion.

[30] In spite of this, control of the canal was given to Panama. In March 1991, a Panamanian took over the administration of the canal for the first time.

[31] In April 1991, Endara announced the end of his alliance with the Christian Democrat Party (PDC), dismissing five of their ministers. The government's precarious stability was shaken by five putschist attempts during its first two years. The Democratic Revolutionary Party (PRD), which won a significant victory in the by-elections of January 1991, called for general elections.

[32] In the course of General Noriega's trial, which started late in 1991, in Miami, it was disclosed that the former leader had close connections with the Drug Enforcement Agency (DEA) and the CIA. That same year it was discovered that President Endara's legal office had connections with 14 companies which laundered drug money. The DEA also disclosed that drug dealings had increased since the American invasion. In June 1992, Noriega was sentenced to 40 years in prison.

[33] The government suffered a serious setback when, in a plebiscite on constitutional reform on November 15 1992, the people rejected the formal abolition of the Defence Force by 63.5 to 31.5 per cent. The government was further weakened when the deputy vice-president, Christian Democrat Ricardo Arias Calderón, resigned on December 17 after accusing Endara of failing to deal with the social crisis.

[34] Foreign Minister Julio Linares was forced to resign in August 1993 as he had been involved in the sale of weapons to Serbian forces in Bosnia through the consulate in Barcelona.

[35] Economist Ernesto Perez Balladares, fomer minister and devoted admirer of Omar Torrijos, was elected president in 1994 with 34 per cent of the vote. Balladares ran as candidate for the social democratic alliance Pueblo Unido (United People), made up of the PRD, the Liberal Republican Party and the Labour Party. Mireya Moscoso, from the right-wing Democratic Alliance, had 29 per cent and salsa singer Ruben Blades, from the Papa Egoro Movement, had 17 per cent. These were the first general elections held after the US invasion.

[36] In June 1994, the Kuna and Embera indigenous groups rejected the construction of a road through their autonomous territory. They were supported by environmentalist groups and the Catholic Church. The highway was to have crossed the 550,000 hectare rainforest called "Darien's Stopper", declared a universal heritage by the UNESCO. The 108-km highway would link Panama with Colombia. For years, this forest has endured illegal logging of oak, cedar and mahogany. Likewise, the area has been devoted to coca cultivation and arms smuggling.

[37] A plan to assassinate Pérez Balladares and several cabinet members was uncovered in January 1995. Ten National Police members were arrested on charges of conspiracy but the investigation was shelved for lack of evidence.

[38] The country continued to have an active role in arms and drug traffic as well as money laundering. The explosion of a package during a routine drug inspection killed three officials and injured 25. The explosives, grenades and ammunitions, were being sent to Ecuador, supposedly to guerrilla groups. Two arms deposits were found in the capital, belonging to a Colombian citizen.

[39] The reform of the labor code, aimed at attracting foreign investment, led to an atmosphere of social violence and strikes, since it would reduce labor security and the freedom to unionize and negotiate collectively. Confrontations between workers and students with the police left 4 dead and 86 injured in August. However, the law was passed.

[40] Talks with the United States over the sovereignty of the Canal as of 1999 seem to point to the possibility that Howard navy and air bases might remain under US control.

Papua New Guinea

Population: 4,197,000 (1994)
Area: 462,840 SQ.KM
Capital: Port Moresby

Currency: Kina
Language: English

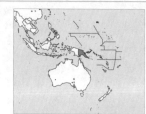

Papua New Guinea

Papua New Guinea (PNG) has been inhabited by Melanesian peoples since 3,000-2,000 BC. Traditionally, its population lived in groups scattered throughout the dense tropical jungle, cut off from the outside world. As a result, over 700 dialects are spoken on the New Guinean island. One third of the population lives in the highlands, and many communities there knew neither the wheel, iron implements nor Europeans until the 1930s.

[2] In a short time, PNG has passed from the stone age into the plastic age, and it currently faces an economic boom based on the exploitation of non-renewable natural resources (copper, gold, oil and hardwoods).

[3] The present state of Papua New Guinea was formed when the Territory of Papua, a British protectorate, under Australian administration since 1906, was joined with New Guinea, a German colony until World War I, later administered by Australia under mandate, first from the League of Nations and then the United Nations.

[4] The process leading to self-government really began in 1964, and culminated in the proclamation of independence in 1975. That same year, there was an unsuccessful attempt at secession by the island of Bougainville. This island has rich deposits of copper and gold and the population is ethnically closer to the Solomon Islanders than the New-Guineans. Australia maintains a strong presence in the young republic, providing considerable investment in business, and military and financial assistance.

[5] Papua New Guinea's system of government is currently parliamentary. There have been four prime ministers since independence, backed by shifting party alliances.

Each government enjoys a six-month truce period after which it has to submit its actions to parliamentary approval. Political parties form around personalities and regional groups rather than ideological differences. Although this might suggest a tendency to instability, the fact that no single group is dominant has produced a government of political consensus, which in itself has a stabilizing influence.

[6] Papua New Guinea has strong links with Australia and, in turn, exercises a degree of leadership among South Pacific states. Its military contributed to supressing a secession attempt by the island of Espiritu Santo (Vanuatu) in 1979, led by Jemmy Stevens, and backed by French authorities and US economic interests.

[7] In May 1988, Papua New Guinea, together with Vanuatu and the Solomon Islands, signed an agreement to defend and preserve traditional Melanesian cultures. It also expressed support for Kanaky's (New Caledonia) independence. Relations with the government of Indonesia, which occupies the western portion of the island, Irian Jaya, are troubled. The Free Papua independence movement operates in this province, and Indonesian military operations here in 1984 led 12,000 inhabitants to seek refuge in Papua-New Guinea.

[8] The country's peculiar personality has been caricatured by a North American economist who said "Papua New Guinea is a mixture between a political system *a la italiana*, an economic policy worthy of a West German bundesbank (for its fiscal and monetary discipline), and African infrastructure". The labor market is much like Australia's, with minimum wages renegotiated every three years, and indexed to inflation. Indigenous

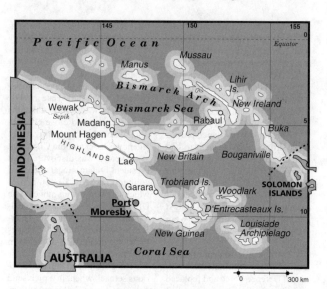

The Bougainville rebelion

Bougainville is an island located 800 km from Port Moresby, the capital of Papua New Guinea. One of the world's largest deposits of copper is found on the island. Since it began, Australian-financed exploitation of the copper mine has been plagued by conflict. In 1969, local people organized a movement called Napidokae Navitu to protest against the poisoning and flooding of their rivers caused by effluence from the mine workings.

[2] Violent confrontations between the Masioi, the Australian enterprises and the Papuan government have occurred constantly since the beginning of mining activity. Agreements based on financial compensation for the islanders have only achieved brief periods of peace.

[3] In May 1990, the island of Bougainville declared its independence from Papua New Guinea and sought recognition from the rest of the world without much success. Armed conflict between the rebels and the Papuan government caused more than 1,500 deaths one year.

[4] In September 1994 the Papua New Guinea government recaptured the Bougainville copper mine. The Bougainville Revolutionary Army continued fighting, in spite of talks on a cease-fire at a meeting in Cairns, Australia. In November 1995, Papua New Guinea's prime minister, Julius Chan, announced that the talks were being abandoned.

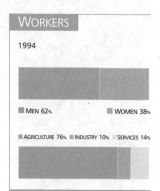

WORKERS

1994

■ MEN 62%   ■ WOMEN 38%

■ AGRICULTURE 76% ■ INDUSTRY 10% ■ SERVICES 14%

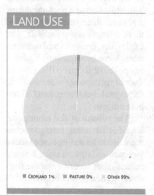

LAND USE

■ CROPLAND 1%   ■ PASTURE 0%   ■ OTHER 99%

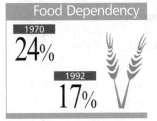

Food Dependency	Foreign Trade	Illiteracy	Maternal Mortality
1970 **24%** / 1992 **17%**	MILLIONS $ 1994 / IMPORTS **1,521** / EXPORTS **2,640**	1995 **28%**	1989-95 / PER 100.000 LIVE BIRTHS **700**

PROFILE

ENVIRONMENT

Located east of Indonesia, just south of the Equator, the country is made up of the eastern portion of the island of New Guinea (the western part is the Indonesian territory of Irian Jaya) plus a series of smaller islands; New Britain, New Ireland and Manus, in the Bismarck archipelago; Bougainville, Buka and Nissau, which form the northern part of the Solomon Islands; the Louisiade and Entrecasteaux archipelagos; and the islands of Trobriand and Woodlark, southeast of New Guinea. The terrain is volcanic and mountainous, except for the narrow coastal plains. The climate is tropical and the vegetation is equatorial rain forest. The country suffers from deforestation, due to large-scale indiscriminate felling.

SOCIETY

Peoples: The inhabitants are mostly Melanesian. There are Micronesian, Polynesian and "Negritoid" groups (the latter being an ethnic group from the Malaysian archipelago). There are 40,000 inhabitants of European descent, mostly Australians. **Religions:** Many people practise local traditional religions but they also belong to Catholic and Protestant communities. **Languages:** English (official). A local Pidgin, with many English words and Melanesian grammar is widely spoken, as well as 700 other local languages. **Political Parties:** Popular Democratic Movement (PDM); Pangu Pati (PP); Popular Action Party; Popular Progressive Party (PPP).

THE STATE

Official Name: Independent State of Papua New Guinea. **Administrative Divisions:** 20 Provinces. **Capital:** Port Moresby, 200,000 inhab. (1990). **Other cities:** Lae, 80,655 inhab.; Madang, 27,057 inhab.; Wewak, 23,224 inhab.; Goroka, 17,855 inhab. (1990). **Government:** Queen Elizabeth II of the United Kingdom, Head of State, represented by a Governor General. Julius Chan, Prime Minister and Head of the Government. Single-chamber legislature: Parliament, with 109 members. **National Holiday:** September 16, Independence Day (1975). **Armed Forces:** 3,800 troops (1995).

STATISTICS

DEMOGRAPHY

Urban: 16% (1995). **Annual growth:** 1.6% (1992-2000). **Estimate for year 2000:** 5,000,000. **Children per woman:** 4.9 (1992).

HEALTH

Under-five mortality: 95 per 1,000 (1995). **Calorie consumption:** 94% of required intake (1995). **Safe water:** 28% of the population has access (1990-95).

EDUCATION

School enrolment: Primary (1993): 67% fem., 67% male. Secondary (1993): 10% fem., 15% male. **Primary school teachers:** one for every 31 students (1992).

COMMUNICATIONS

87 **newspapers** (1995), 82 **TV sets** (1995) and 82 **radio sets** per 1,000 households (1995). 1.0 **telephones** per 100 inhabitants (1993).

ECONOMY

Consumer price index: 100 in 1990; 120.4 in 1994. **Currency:** 1 kina = 1$ (1994). **Cereal imports:** 227,000 metric tons (1993). **Fertilizer use:** 308 kgs per ha. (1992-93). **External debt:** $2,878 million (1994), $686 per capita (1994). **Debt service:** 30% of exports (1994). **Development aid received:** $303 million (1993); $74 per capita; 6% of GNP.

---

communities own 98 per cent of the land, but as the economy is increasingly open to foreign penetration, the exploitation of natural resources has reached proportions that are seriously damaging the environment.

[9] Two thirds of the population are subsistence agriculturalists, and the 0.3 per cent that work with mining concerns generate 66 per cent of export earnings. Mines at Bougainville and Ok Tedi boast the largest copper deposits in the world and also produce gold. The Porgera gold mine went into production in 1990, and will soon be the largest in the world outside South Africa. Another on the island of Lihir may

be even larger. The mines are exploited by transnationals, some of them Australian. Sizeable oil deposits have also been found and will be exploited only after a 175-mile oil pipeline is built, at a cost of $1 billion.

[10] In 1989, Papua New Guinea strengthened its ties with Southeast Asia, signing a friendship and co-operation treaty with ASEAN and opening negotiations with Malaysia. At the same time, it granted the Soviets permission to open an embassy in Port Moresby. In December, a maritime agreement was signed with the Federated States of Micronesia which allowed PNG to take advantage of an existing mul-

tilateral fishing agreement between the South Pacific and the US.

[11] Parliament reinstated the death penalty in August 1991 34 years after it had been abolished. This move was denounced by Christian movements, feminists and human rights organizations. The National Council of Women feared that the reinstatement of the death penalty might lead to a return of 'bounty' payments in assassination cases, as was traditional in Papua New Guinean culture.

[12] Prime Minister Palas Wingti launched a fight against corruption in July 1992, through a public accountability campaign for those holding public office. Several governors were forced to resign.

[13] In June, the Justice Department brought corruption charges against former prime minister Namaliu and his minister of Finance, Paul Pora.

[14] Wingti has pledged to increase the country's participation in joint ventures with foreign capital. This triggered an angry reaction from PJV, an Anglo-Austro-Canadian consortium which controls 90 per

cent of Porgera, the country's largest gold mine.

[15] Wingti was replaced in August 1994 by Julius Chan. That year, the government recovered the copper mine on Bougainville island (see box), but an explosion in the Porgera gold mine interrupted extraction works.

[16] Chan harshly denounced the new series of French nuclear tests in the Pacific. After the second atomic test was carried out in October 1995, the prime minister of Papua New Guinea suspended talks between France and the South Pacific Forum, acting as president of the regional organization.

Paraguay

Paraguay

Population: 4,788.000 (1994)
Area: 406,750 SQ KM
Capital: Asunción

Currency: Guaraní
Language: Spanish

There were three ethnic groups in the Paraguay River basin in the 16th century: the Guaranís, in the center, and the Guaycurús and Payaguás, in the Chaco region to the south. The Guaranís cultivated cassava, squash, sweet potatoes and corn. They were settled agriculturalists, organized in villages headed by a *tubichá* (chief).

2 The Payaguás and Guaycurús were nomadic hunters and fishers. They periodically attacked the Guaraní plantations, leading the Guaranís to aid the Spaniards in the conquest of the Chaco area.

3 Spanish conquistadors sailed up the Paraná River in the mid-16th century, searching for the mythical silver mountains that gave the Rio de la Plata (silver river) region its name. In 1537, Juan de Salazar founded the fort of Nuestra Señora de Asunción. This tiny settlement grew and became capital of the Rio de la Plata Province for over a century.

4 Having failed to find precious metals in these "worthless lands", it was only much later that Spanish colonizers were attracted by the wild cattle herds that prospered on the fertile plains of the Banda Oriental (today's Uruguay) and the humid Argentinian Pampas. A powerful oligarchy of middlemen developed in the ports of Buenos Aires and Montevideo, and the former provincial capital, Asunción declined.

5 A pre-capitalist system of large landed estates grew up around Buenos Aires and Montevideo. Meanwhile, in what is now Paraguay, the Jesuits organized a system of agricultural colonies in which the indigenous population worked the land and produced handicrafts on a communal basis. Private interests and the government both resented this system, and the Society of Jesus was finally expelled in 1767. The natives either became slaves on Brazilian sugar cane plantations or peasant laborers on large cattle ranches.

6 Although the Paraguayan province did not initially join the 1810 Liberal Revolution, the Asunción oligarchy ousted the Spanish governor Velazco on May 14 1811, demanding free trade for *yerba mate* (a local tea) and tobacco.

7 This rebellion began a process which concentrated all the power in the hands of Gaspar Rodríguez de Francia. He was supported by the peasants, owners of small and medium-sized plots who wanted order and feared the "anarchy" that was affecting the other provinces of the old viceroyalty. The oligarchy in Asunción, the new system's only enemy, was eliminated, and Paraguay withdrew into itself. Stimulated by public investment, industry began to develop, laying the foundations for Paraguay's future economic strength.

8 Paraguay continued to develop under the patriarchal governments of Gaspar Rodríguez de Francia (*El Supremo*), Carlos Antonio López and his son, Francisco Solano López. It remained cut off from the rest of the world for decades and was untouched by the British influence that was so greatly felt in the other newly independent Rio de la Plata provinces. Left to develop alone, the state assumed an important role in the country's economy, controlling agricultural production (*yerba mate* and high quality timber), and the first railroads, telegraph system and steel furnaces in South America.

9 The leaders of Brazil, Argentina, and Uruguay - Pedro II, Mitre and Venancio Flores respectively - signed the Triple Alliance. In 1865, with the support of the British Empire and the Baring Brothers' Bank, they began a war to eliminate Paraguay on the pretext of frontier disputes (see "Triple Alliance").

10 Sixty-two years later, Paraguay fought a second fratricidal war, this time against Bolivia, orchestrated by rival oil multinationals of the "Seven Sisters" group. The Chaco War (1932-1935) was won, but it cost Paraguay 50,000 lives.

11 In 31 years, there were 22 presidents, until the 1954 coup d'état which brought General Alfredo Stroessner to power. Descended from Germans, and an admirer of Nazism, he gave refuge to war criminals fleeing Europe. Stroessner had himself re-elected seven times, most recently in February 1988. He allowed only a legal "opposition", whose leaders he often appointed himself.

12 By the mid-1980s, the Stroessner regime was showing clear signs of weakening. Democratization processes elsewhere in Latin America doubtless contributed to this. Internal opposition was growing in a variety of forms: an Inter-Union Labor Movement; a Permanent Assembly of Landless Rural Workers (APCT); indigenous peoples' organizations; and the Rural Womens' Co-ordination Group. The Catholic church, through the Paraguayan clergy, began to voice the need for change.

13 On February 3 1989, General Stroessner was overthrown by a coup headed by his son's father-in-law, army commander General Andrés Rodríguez, who immediately called free elections for the following May, to elect a government and legislators to see out Stroessner's term.

14 The elections, held on May 1 1989, were open to all political parties except the still-banned Communist Party. General Rodríguez was elected president with 68 per cent of the vote. The principal opposition group, the Authentic Radical Liberal Party, led by Domingo Laíno, took 21 per cent. Despite the presence of foreign observers, voting was plagued by countless irregularities, attributed to the Colorado Party which was leaving power.

15 Nevertheless, opposition groups still considered the election to be the start of a democratization process, which was to be completed by 1993, with a Constituent Assembly convened, and a new constitution in place.

16 Transnationals have become involved in the production of soybeans and cotton. At present, 2 per cent of all landholders have 85 per cent of the land. Most of the companies and foreign settlers congregate along the border with Brazil on

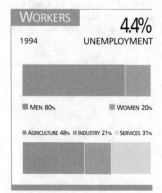

WORKERS

4.4%
1994 **UNEMPLOYMENT**

MEN 80%   WOMEN 20%

AGRICULTURE 48% INDUSTRY 21% SERVICES 31%

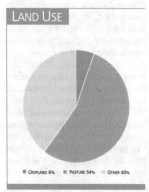

LAND USE

CROPLAND 6%   PASTURE 54%   OTHER 40%

ENVIRONMENT

A landlocked country in the heart of the Río de la Plata basin, Paraguay is divided into two distinct regions by the Paraguay River. To the east lie fertile plains irrigated by the tributaries of the Paraguay and Paraná rivers, and covered with rainforest; this is the main farming area, producing soybeans (the major export crop), wheat, corn and tobacco. The western region or Northern Chaco is dry savannahs, producing cotton and cattle. Farmers, especially those linked to large agricultural operations, have felled large areas of forest, destroying habitats in the process. Indiscriminate hunting, combined with deforestation, have resulted in 14 species of mammals, 11 birds and 2 reptiles now being in danger of extinction. Large amounts of industrial and domestic waste are dumped into watercourses, and untreated sewage from the capital city flows directly into the Bay of Asunción.

SOCIETY

Peoples: 90% of Paraguayans are of mixed descent from indigenous peoples and Spanish colonizers. The native Guaraní people currently account for only 5% of the population and are threatened with the loss of their cultural identity. There are German, Italian, Argentinian and Brazilian minorities. This last group is expanding along the border between the two countries. One million Paraguayans live abroad, some 200,000 having migrated for political reasons. **Religions:** Mainly Catholic, official. **Languages:** Spanish (official) and Guaraní (national); most Paraguayans are bi-lingual. **Political Parties:** Colorado Party; Authentic Radical Liberal Party, linked to the "International Liberal Network"; National Encounter, a new electoral coalition made up of independent sectors, dissidents from the traditional parties, the "Febrerista" Revolutionary Party, and the "Constitution for All" Movement; Colorado Popular Movement (MOPOCO) - which grew from a split within the Colorado Party, in 1959; Christian Democratic Party; Communist Party. **Social Organizations:** The Paraguayan Labor Confederation; the United Workers' Center (CUT) founded in 1990, uniting labor unions and the powerful rural workers' movement; the Paraguayan Women's Union.

THE STATE

Official Name: República del Paraguay. **Administrative Divisions:** 19 departments. **Capital:** Asunción, 502,400 inhab. (1992). **Other cities:** Ciudad del Este, 133,900 inhab. (1995); Pedro Juan Caballero, 80,000 inhab.; Encarnación, 31,445 inhab.; Pilar, 26,352 inhab.; Concepción, 25,607 inhab. (1984).
Government: Presidential republic. President Juan Carlos Wasmosy, since August 15, 1993. **National Holiday:** May 14, Independence Day (1811). **Armed Forces:** 20,300 (1995).

Food Dependency

1970
19%

1992
13%

Foreign Trade

MILLIONS $ 1994

IMPORTS
2,370
EXPORTS
813

DEMOGRAPHY

Urban: 51% (1995). **Annual growth:** 1.4% (1992-2000). **Estimate for year 2000:** 6,000,000. **Children per woman:** 4.6 (1992).

HEALTH

One **physician** for every 1,587 inhab. (1988-91). **Under-five mortality:** 34 per 1,000 (1995). **Calorie consumption:** 103% of required intake (1995). **Safe water:** 35% of the population has access (1990-95).

EDUCATION

Illiteracy: 8% (1995). **School enrolment:** Primary (1993): 110% fem., 110% male. Secondary (1993): 38% fem., 36% male. University: 10% (1993). **Primary school teachers:** one for every 23 students (1992).

COMMUNICATIONS

93 **newspapers** (1995), 94 **TV sets** (1995) and 83 **radio sets** per 1,000 households (1995). 3.1 **telephones** per 100 inhabitants (1993).

ECONOMY

Per capita GNP: $1,580 (1994). **Annual growth:** 1.00% (1985-94). **Annual inflation:** 26.20% (1984-94). **Consumer price index:** 100 in 1990; 204.0 in 1994. **Currency:** 1.940 guaraní = 1$ (1994). **Cereal imports:** 82,000 metric tons (1993). **Fertilizer use:** 96 kgs per ha. (1992-93). **Imports:** $2,370 million (1994). **Exports:** $817 million (1994). **External debt:** $1,979 million (1994), $413 per capita (1994). **Debt service:** 10.2% of exports (1994). **Development aid received:** $137 million (1993); $29 per capita; 2% of GNP.

ENERGY

Consumption: 261 kgs of Oil Equivalent per capita yearly (1994), -141% imported (1994).

---

a strip of land 2000 km long and 65 km wide. The main language is Portuguese, and the Brazilian cruzeiro is common currency. Between 300,000 and 400,000 Brazilians live in the border districts.
[17] The World Bank-supported Caazapa and Caaguazu rural development projects involve the settlement and ranching of indigenous lands in eastern Paraguay. The Indians find themselves restricted to tiny pockets of their former territories. Twelve indigenous communities asked the Unitary Workers' Center (CUT) to represent them in the National Constitutional Assembly. In December 1991, in the first

round of elections to designate representatives to this Assembly, the CUT gave its support to the "Constitution for All" Movement.
[18] When the huge Itaipú hydroelectric plant, on the border with Brazil, went into operation, economic ties between the two countries increased.
[19] The first free municipal elections, held in June 1991, were marked by the emergence of the independent Asuncion for All movement in the capital. Led by Carlos Filissola, a 31 year-old doctor, the movement counted on the support of the United Workers Center and various social groups, and

was elected with more than 35 per cent of the vote.
[70] In the elections for the National Constituent Assembly in December, the Colorado Party took 60 per cent of the vote. The Authentic Liberal Radical Party (PLRA) of Domingo Laíno came second with 29 per cent and the Constitution for All Movement, led by Filizzola, came third.
[21] The Constitution of June 1992 replaced the one brought in by Stroessner in 1967. The presidents of the Republic, the Supreme Court and Congress did not take part in the public act to announce the new Constitution, which included broad dispensations for the protection of human rights and banned the death penalty for common crimes.
[22] The repeated accusations of widespread corruption within the

armed forces, including drug trafficking, led to investigations being undertaken. Warrants were issued for the arrest of leading generals, including the commander in chief of the army.
[23] In December 1992, Paraguay signed an agreement for the constitution of a common market, the Mercosur, which would come into operation in 1995, with Argentina, Brazil and Uruguay (see box in Brazil).
[24] In an atmosphere tempered by the danger of fraud and declarations from the army that it would keep the Colorado Party in power, the ruling party candidate, Juan Carlos Wasmosy took 40 per cent of the vote in the May 1993 elections. Some 75 per cent of the 1.7 million eligible adults voted. However, troops positioned on the borders

prevented the entry of citizens resident in Argentina and Brazil. The PLRA took 33 per cent of the vote, and the National Encounter, a pro-business coalition, 25 per cent.

[25] When Wasmosy took over the presidency in August, he became the first elected civilian president in the 182 years since national independence. Congress granted civilian power the control over the military.

[26] In September, the World Jewish Congress asked President Wasmosy for access to the government archives in order to locate Nazis who took refuge in Paraguay following the World War II.

[27] Parliament voted in a law banning members of the military from party politics in May. The government and the high command immediately started action to declare this law unconstitutional. This attitude led the opposition to abandon the cooperation pact with the Colorados.

[28] Under pressure from the United States, Wasmosy nominated General Ramón Rozas Rodríguez leader of the war on drugs. Rozas was assassinated in 1994, when he should have presented a report on illicit behavior involving the military hierarchy, including General Lino Oviedo.

[29] Violent confrontations between rural people and the police affected several regions of the nation. A hundred demonstrators blocking the access to Asunción were injured by rubber bullets. The farmers were supported by church organizations, unions and opposition parties. A general strike was declared in May to press for pay rises and protest against the privatization of State companies. The government admitted that purchasing power had fallen by 42 per cent in the last five years and upped salaries by 35 per cent. In December, all timber exports were suspended with the aim of halting deforestation.

[30] In January 1995, the National Human Rights Commission said many of the crimes from the time of Stroessner had gone unpunished and that rural people continued to be murdered. Wasmosy forced eight high-ranking officers to retire. The Colorado Party and the opposition finally agreed to ban the military from politics in May.

[31] When presented with the order for his retirement by President Wasmosy, General Oviedo dug himself in with many of the new officers promoted in April 1996. Despite the support of thousands of demonstrators, Wasmosy took refuge in the US embassy. In order to overcome the impasse, he nominated Oviedo Minister of Defence, a measure he revoked the following day with the support of the international community. The crisis became associated with the imminent selection of the Colorado Party leader, which would determine the identity of the 1998 presidential candidate. Luis María Argaña, one of Stroessner's closest allies, was chosen.

Illiteracy 1995

8%

Maternal Mortality 1989-95

PER 100.000 LIVE BIRTHS

180

The triple alliance war

I n 1865, Brazil, Argentina and Uruguay (represented by Pedro II, Mitre and Venancio Flores, respectively) signed a treaty known as the Triple Alliance with British consent. The war broke out in 1865 and continued until 1870 and only ended with the death of Paraguay's president, Francisco Solano López, destroying Paraguay and wiping out between 60 and 80 per cent of the population The only males left alive were babies - which the victors sought out eagerly to kidnap and sell as slaves in Brazil - and very old men.

[2] After the fight for independence from Spanish domination, the power struggle in the ex-Río de la Plata Viceroyalty ended with Buenos Aires becoming the new center. Provincial efforts to resist Buenos Aires's centralizing and unifying efforts, including the *montonera* guerrilla movements of federalist *caudillos* (leaders) like Felipe Varela, Peñaloza, Quiroga, Urquiza and Artigas, all ended in defeat. As a result, the role of "economic exploiter" simply moved, from the colonial power to the port city of Buenos Aires. During the war, the *porteña* oligarchy was headed by the "Unitarian" Bartolomé Mitre from 1862-1868, and by Faustino Sarmiento between 1868 and 1874.

[3] In the early 1860s, the Balkanization used by British diplomacy between 1820 and 1830, had not yet been imposed. It was therefore possible for Mitre to implement a double strategy ridding himself of two undesirable elements at once: Uruguay - a hot bed of followers of Artigas' ideas on integration, and Paraguay - an example of successful and independent development.

[4] A similar situation of revolt against central power had developed in Brazil. Pedro II's empire faced rebellion by the *farrapo* movement in the South which wavered between aggressive autonomy and almost total de facto independence. In short, both Argentina and Brazil suffered from acute internal turmoil.

[5] Paraguay resisted the excessive port taxes levied by Buenos Aires while suffering constant pressure from Brazil, which demanded the right to navigate freely on Paraguayan rivers.

[6] Uruguay, governed by Bernardo Prudencio Berro of the Blanco Party, was on friendly terms with Paraguay and had historical links with the Argentinian federalist cause. Mitre and Pedro II agreed to finance and support a coup which, carried Venancio Flores to power in 1865. With the fall of Berro, Paraguay's isolation was complete.

[7] The Paraguayan political model was offensive to the Argentinian and Brazilian governments. Also, more importantly, it was a thorn in the side of the British empire. It was a totally independent system in which 90 per cent of land was state-owned. Practically all foreign trade was nationalized. The state monopolized hardwood and *maté* (tea) exports, and held a large share in tobacco exports. It had blast furnaces (the basis for a steel industry), its own shipyards, national railways, and a telegraph system. Paraguay coined its own currency in a national mint built with foreign technical assistance. What was more, it had no foreign debt.

[8] As president of Paraguay, Field Marshal Francisco Solano López led his nation on the battlefield for five years. He died defending his country, on May 1 1870, at Cerro Cora. Besides the enormous number of Paraguayan dead, defeat was crippling in other ways. Paraguay lost 160,000 sq km of territory. Free navigation on its rivers was exacted by force. It paid 900 million pesos in war reparations to Brazil, 400 million to Argentina and 90 million to Uruguay. Its lands, industrial facilities and services were denationalized. Its blast furnaces were dynamited, its weapons confiscated and its fortifications demolished. The British banking firm, Baring Brothers, lent Paraguay £300,000, against a collateral of 300,000 hectares of Paraguayan soil, and demanded repayment of £1.5 million. When Paraguay could not pay the debt, the bank finally refinanced it for £ 3.2 million. By 1907, the debt had reached £7.5 million.

Peru

Perú

Population: 23,238,000 (1994)
Area: 1,285,220 SQ.KM
Capital: Lima

Currency: New Sol
Language: Spanish and Quechua

The *Tahyantinsuyu* (Inca empire) with its center in Cuzco developed in the Peruvian highlands during the 12th century. This socially developed empire united various advanced urban cultures, adeptly resolving ecological, communication, administration and distribution problems. Irrigated terraces were built to grow crops on the mountain slopes, roads joined the main urban centers, and Inca officials coordinated government activities from Argentina to Quito.

[2] A rivalry over succession coincided with the arrival of the Spanish in Perú in 1524. Francisco Pizarro manipulated the situation to achieve a quick military victory. Spain colonized the country, decimating the native population through hard labor, wars and epidemics of diseases brought from Europe. It took until 1975 for the Peruvian population to regain the level of 15 million inhabitants, that it had had in the 16th century.

[3] The viceroyalty of Peru dates from 1542, when the region became important for the mining of precious metals. Resistance to colonial rule never ceased over the next two centuries, but the opposition ended in 1780 with the defeat of Tupac Amaru's rebellion.

[4] The powerful mine and landowning oligarchy, which had developed in Peru during colonial times, firmly resisted change. The country was the last on the continent to become independent when the combined armies of San Martín and Bolívar defeated the Spanish forces at Ayacucho (1824). The first years of Peru's independence witnessed continuous fighting between the liberals and the conservative oligarchies who yearned for the good times they had had under Spanish rule. The 1827 war with

Colombia took place against this background, and unification with Bolivia, attempted in 1835 by Bolivian president Andrés Santa Cruz, proved to be a failure, both economically and socially.

[5] Marshal Ramón Castilla, who ruled from 1845 to 1862, shaped the modern Peruvian state, after abolishing slavery and proclaiming a constitution.

[6] In 1864 Spain tried to re-establish a presence in the Pacific coast. Peru, Chile, Bolivia and Ecuador declared war on Spain, which bombed Valparaíso, Chile and Callao, Peru, before its fleet was defeated in 1866.

[7] Once Peru's silver mines had been depleted, guano - fertilizer from bird droppings - became Peru's main export commodity. When the guano "boom" was over, it was replaced by saltpeter, which was found in the southern desert. The exploitation of this natural resource brought about the War of the Pacific (1879-1883) between Peru, Bolivia and Chile (the latter backed by British companies). Peru and Bolivia lost the war, and with it, their provinces of Arica, Tarapacá and Antofagasta.

[8] The first few decades of the 20th century marked the beginning of large-scale copper mining, especially by the Cerro de Pasco Copper Corporation, an American mining company. Foreign capital was also involved in oil, sugar and cotton production.

[9] Within this context, the American Popular Revolutionary Alliance (APRA), Marxist-inspired and commited to Latin Americanism, gained wide popular support. Successive military coups prevented APRA from coming to power.

[10] APRA dissidents and members of revolutionary leftist groups

turned to guerrilla warfare in the 1960s, without success.

[11] In 1968, a military faction headed by General Juan Velasco Alvarado ousted President Fernando Belaúnde Terry. The oil and fishing industries were nationalized, a cooperative-based land reform program was launched,and other important social reforms were initiated. The military government followed an independent non-aligned foreign policy.

[12] Velasco fell ill and gradually lost control of the situation. In August 1975, he was replaced by his prime minister, General Francisco Morales Bermúdez. Beset by financial pressure from the IMF and political harassment from an oligarchy anxious to regain power, Morales began to prepare a return to civilian rule. In the 1978 elections ,the populist APRA and the conservative Christian People's Party, each gained an almost equal number of seats in a Constituent Assembly.

[13] Belaúnde Terry's Acción Popular, which had boycotted the Constituent Assembly elections, won the presidency in 1980. Belaúnde began his term of office by implementing an economic policy in line

with IMF prescriptions. The economic results were disastrous: real unemployment rose and runaway inflation led to a devaluation of the currency. There was dizzying growth of the "underground" sector of the Peruvian economy, work carried out beyond the reaches of all legal regulation or social benefits. In 1980 there was a new outburst of armed violence, led by the Peruvian Communist Party for the "Shining Path" or *Sendero Luminoso*; and in 1984 the Tupac Amaru Revolutionary Movement (MRTA) also rose up in arms.

[14] The United Left, a coalition of left-wing parties, won the mayoralty of Lima in the 1982 municipal elections. The government's answer to the increase in guerrilla warfare was to put those provinces affected by the violence under military control. In 1983, given the growing social discontent, a state of emergency was decreed over the entire country.

[15] In the 1985 election the APRA candidate Alan García Pérez won, with 46 per cent of the vote. The United Left (IU), with 22 per cent of the vote, was consolidated as the country's second political force.

WORKERS

8.9%
UNEMPLOYMENT

1994

MEN 76% — WOMEN 24%

AGRICULTURE 35% — INDUSTRY 12% — SERVICES 53%

LAND USE

CROPLAND 3% — PASTURE 21% — OTHER 76%

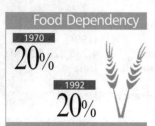

PROFILE

ENVIRONMENT

The Andes divide the country into three regions. The desert coastal area, with large artificially irrigated plantations and some natural valleys has historically been the most modern and westernized. Half of the populaton live in the *Sierra* (highlands), between two ranges of the Andes. Numerous peasants here are still organized into *ayllus* (communities) with Incan roots. Subsistence farming of corn and potatoes is practised, the traditional raising of llamas and alpacas having been forced onto the higher slopes due to the incursions of mining and sheep rearing. The eastern region, comprising the Amazon lowlands, with tropical climate and rainforests, is sparsely populated. The country suffers from soil depletion, on soils which are poor to begin with. Indiscriminate fishing has endangered some species. The coastal area has been polluted by both industrial and household waste.

SOCIETY

Peoples: Nearly half of all Peruvians are of indigenous American origin, mostly from the Quechua and Aymara ethnic groups living on the *Sierra*. Along the coast, most of the population are mixed descendents of native Indians and Spaniards. There are several indigenous groups in the East Amazon jungle. **Religions:** Catholic (official). **Languages:** Spanish and Quechua (official). Aymara is also spoken. **Political Parties:** The Change '90-New Majority, founded in 1989 by Alberto Fujimori; Popular Action (AP), led by Fernando Belaúnde Terry; the Peruvian "Aprista" Party, founded in 1930; the Christian Popular Party; the United Left (IU) is an alliance of the National Workers' and Campesinos' (Farmers') Front, the Peruvian Communist Party and the Union of the Revolutionary Left; the Socialist Left, of Alfonso Barrantes, withdrew from the IU in the 1990 elections. **Social Organizations:** The Peruvian Workers', General Central Union (CGTP), founded in 1928 and predominantly Communist. The Peruvian Workers' Central Union (CTP), founded in 1944, linked to the APRA. The independent National Worker's Confederation (CNT), founded in 1971. The Workers' Central of the Peruvian Revolution, founded in 1972 by Velasco backers. In March, 1991, the four trade union federations (CGTP, CTP, CTRP, and CNT) created the national Coordinating Body of Trade Union Federations, as a step towards the creation of a single federation. There are two rural organizations: the National Agrarian Confederation (CNA), founded in 1972, and the Peruvian *Campesino* Confederation (CCP), founded in 1974.

THE STATE

Official Name: República del Perú. **Administrative divisions:** 25 Departments, 155 Provinces and 1,586 Districts. **Capital:** Lima, metropolitan 5,706,100 inhab. (1993). **Other cities:** Arequipa, 619,200 inhab.; Callao, 615,000 inhab.; Trujillo, 509,300 inhab.; Chiclayo, 411,500 inhab. (1993). **Government:** Alberto Fujimori, President since July 28 1990, dissolved Congress on April 5 1992, and was re-elected in 1995. 120-seat Congress. **National Holiday:** July 28, Independence (1821). **Armed Forces:** 115,000 troops (65,000 conscripts). 188,000 reserves. **Paramilitaries:** National Police: 60,000 members. Coast Guard: 600 members. Campesino Rounds: the campesino self-defense forces, made up of 2,000 groups mobilized within emergency zones.

Acción Popular (AP) received only 5 per cent. When García took office, in July 1985, Peru's foreign debt stood at \$14 billion, with annual servicing of the debt costing \$3.5 billion. The president announced that he would limit the payment of the foreign debt to 10 per cent of the country's annual export earnings, and would negotiate directly with Peru's creditors. Until 1987, the APRA's economic policy revived the economy: urban employment and salaries increased for two consecutive years. In October 1987 banks, financial institutions and private insurance companies were nationalized.

[16] From 1986 on, there were armed attacks against APRA local party offices, mayors and military personnel, in areas under the state of emergency. In June 1986, the suppression of mutinies by *Sendero* prisoners in El Frontón, Lurigancho and Santa Barbara prisons left around 300 prisoners dead or missing.

[17] Having used up the country's international reserves, García's government put a new economic policy into effect amid growing inflation and recession. Imports were drastically reduced, including both medicine and food. Exports, on the other hand, increased by more than \$1 billion between 1988 and 1989. In 1989, underemployment in Lima reached the highest level ever, at 73.5 per cent. Inflation reached 2,000 per cent in 1988 and continued to grow. The underground economy displaced legally constituted businesses as the prime moving force in the economy.

[18] Parties participating in the general election included APRA, two factions of the left, the Democratic Front (FREDEMO) - which fielded writer Mario Vargas Llosa as its candidate - and a new group, Change '90 (Cambio 90), who had the agricultural engineer Alberto Fujimori as its candidate. The latter, the son of Japanese immigrants and without previous political experi-

ence wâs elected president by a majority of 56.4 per cent.

[19] The election also implied a defeat for *Sendero Luminoso*, which had called for abstention from voting. On July 28 1990, Alberto Fujimori was inaugurated president and he immediately adopted a strict counter-inflationary policy.

[20] The last months of 1990 were marked by labor and social unrest, and in March 1991, the four trade union federations combined, creating the National Coordinating Body of Trade Union Federations. However, popular organizations and political parties have not succeeded in uniting against the government.

[21] Fujimori was forced to fight the opposition of the political parties and Congress, who tried to disempower him. The president began governing by decree, issuing 126 legislative decrees in 1991.

[22] Fujimori's economic policy accentuated the impoverishment of the Peruvian population. Out of a

total of 22 million inhabitants, 12 million live in extreme poverty.

[23] In 1991-2, a cholera epidemic ravaged the country, affecting some 270,000 people, and leaving 2,540 dead. Against this backdrop of crisis, coca production thrives. With 200,000 hectares under cultivation, the export of semi-processed cocaine brings in export-earnings of \$2 billion annually. In May 1991, the Peruvian government signed an agreement with the US where the US pledged to fund the replacement of coca with alternative crops, though this has not been honored.

[24] According to the reports of international agencies, the number of missing people rose from 200, before Fujimori took office, to a total of 3,000. The president has admitted the existence of paramilitary groups and the impunity of the armed forces in fighting the guerrillas.

[25] On April 5 1992, backed by the armed forces and police, Fujimori announced the temporary dissolu-

Illiteracy
1995

11%

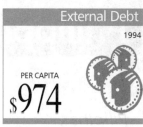

External Debt
1994

PER CAPITA
$974

tion of parliament and the reorganization of the judicial system, claiming that the reconstruction of the country was being hindered by the incompetence and corruption of both powers. A few days later, he passed a law granting himself full power to legislate and modify the constitution, and to bring back morality to public administration and justice, by fighting against drug trafficking, corruption, and subversion. A survey carried out by the polling agency Peruana de Opinion Publica, reported over 70 per cent support for Fujimori's measures, although all Peruvian social and political organizations condemned the coup d'etat.

[26] Reactions in Latin America ranged from the condemnation of the coup d'etat, to the attitude of the OAS, which deplored the situation without condemning it. Venezuela broke off relations, and Brazil and Chile suspended bilateral negotiations which were in progress at the time.

[27] The US announced the suspension of all economic and military aid to Peru, explaining that it would not recognize San Roman, Fujimori's appointed president but would continue to negotiate with the government. Late in April, the IMF announced in Washington that it approved of the program of economic and structural reforms being carried out by the Peruvian government.

[28] The government's greatest political triumph was in its fight against rebel movements. Prior to Fujimori's presidency, *Sendero Luminoso* ("the Shining Path") had waged a civil war with a toll of 25,000 deaths. This group had up to 5,000 members and had been fighting for 13 years.

[29] In September 1992, the government publicly displayed Abimael Guzmán, founder and leader of *Sendero*, in jail. The arrests of Guzman and other *Sendero* leaders pressed the organization into holding peace talks.

[30] The government's need to re-enter the international financial system and the president's confidence in his popularity led to elections in November. At the Democratic Constitutional Congress, Fujimori's Change 90-New Majority party won an absolute majority. The traditional parties did not stand for election.

[31] The opposition rejected the government's proposed constitu-

tional reform as it restricted political and individual liberties and established the death penalty for terrorists. In the October referendum, 53 per cent of voters supported the reform while 47 per cent opposed it.

[32] In mid-1993, a professor and nine students from La Cantuta university were murdered. A congressional investigation implicated the commander of the army and a military adviser to Fujimori, who was reticent to submit them to civilian trial.

[33] In 1993, the State sold two mining companies to a state enterprise in China, which became the country's fourth largest foreign investor. Aeroperu (the national airline) and Petroleos del Mar (which provided service for oil tankers) were also sold. Fujimori promoted the country's integration into the Pacific Basin. He travelled frequently to Japan, which contributed with loans to Peru where there is a large Japanese community. The economy had a 6.5 per cent growth rate due to the fishing, construction and manufacturing sectors and to the dynamic undercover economy. Peru's unwillingness to reduce tariff barriers obstructed the formation of an Andean common market with Colombia, Venezuela and Bolivia anticipated for 1995.

[34] Fujimori successfully carried out a military offensive against *Sendero Luminoso* and launched a propaganda campaign to convince its followers to surrender the cause. Human rights organizations accused the government of murdering, kidnapping and torturing innocent civilians.

[35] Security in the Andean region enabled mining activity to resume, with the participation of foreign prospectors and investors. The GDP had a 12 per cent growth rate during 1994.

[36] In early 1995, Peru and Ecuador held military confrontations at their border in the Condor mountain range. The hostilities caused 200 dead and millions of dollars in losses for both countries. Peace talks were held under the patronage of the Protocol of Rio with Argentina, Brazil, Chile and United States acting as guarantors.

[37] Fujimori had an overwhelming victory in the April presidential elections. His main rival, former secretary general of the UN Javier Pérez de Cuellar, did not obtain the votes needed for a second round.

Fujimori's wife, Susana Higuchi could not run for office and her party, 21st. Century Harmony, did not get any seats in the Congress due to irregularities in its lists. In July, the president and his wife were divorced.

[38] Boosted by his triumph in the elections, Fujimori withdrew the autonomy from San Marcos and La Cantuta universities and granted amnesty to military and police members prosecuted for human rights violations in their fight against the guerrillas since 1980. It is estimated that between 1992 and late 1995, the "faceless judges" had sentenced 2,000 people to death.

[39] The decrease of terrorist activity led tourism to return to pre-1991 levels.

STATISTICS

DEMOGRAPHY

Urban: 71% (1995). **Annual growth:** 0.6% (1992-2000). **Estimate for year 2000:** 26,000,000. **Children per woman:** 3.3 (1992).

HEALTH

One **physician** for every 1,031 inhab. (1988-91). **Under-five mortality:** 58 per 1,000 (1995). **Calorie consumption:** 90% of required intake (1995). **Safe water:** 71% of the population has access (1990-95).

EDUCATION

Illiteracy: 11% (1995). **School enrolment:** University: 40% (1993). **Primary school teachers:** one for every 28 students (1990).

COMMUNICATIONS

101 **newspapers** (1995), 100 **TV sets** (1995) and 100 **radio sets** per 1,000 households (1995). 2.9 **telephones** per 100 inhabitants (1993). **Books:** 86 new titles per 1,000,000 inhabitants in 1995.

ECONOMY

Per capita GNP: $2,110 (1994). **Annual growth:** -2.00% (1985-94). **Annual inflation:** 492.20% (1984-94). **Consumer price index:** 100 in 1990; 1,625.5 in 1994. **Currency:** 2.180 new soles = 1$ (1994). **Cereal imports:** 1,920,000 metric tons (1993). **Fertilizer use:** 216 kgs per ha. (1992-93). **Imports:** $6,794 million (1994). **Exports:** $4,555 million (1994). **External debt:** $22,623 million (1994), $974 per capita (1994). **Debt service:** 17.7% of exports (1994). **Development aid received:** $560 million (1993); $25 per capita; 1% of GNP.

ENERGY

Consumption: 351 kgs of Oil Equivalent per capita yearly (1994), 1% imported (1994).

Philippines

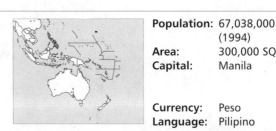

Population: 67,038,000 (1994)
Area: 300,000 SQ.KM
Capital: Manila

Currency: Peso
Language: Pilipino

Pilipinas

The Philippine archipelago was first inhabited in paleolithic times, and the Neolithic culture on the islands began around 900 BC with metal working beginning in about the 15th century. Native peoples, such as the Aeta and the Igorot, probably subsisted without being assimilated by the later groups of migrants.

[2] Between the 2nd and 15th centuries AD, migrants from Indonesia and Malaysia settled on the islands, gathering in clans. Unlike the rest of the Malays, they were virtually uninfluenced by the classical Indian culture. Between the 11th and 13th centuries the coastal areas were raided by Muslim, Japanese, and Chinese merchant ships, bringing traders and craftsmen to the islands. The southern islands adopted Islam and sultanates soon appeared.

[3] The archipelago was only "discovered" by Ferdinand Magellan in 1521 when it was named after the Spanish king Felipe. The explorer was killed on one of its beaches and Spanish possession of the islands, which were also coveted by the British and the Dutch, was not secured until 1564. The Igorot of the Cordillera region and the Islamic population of Mindanao were never fully incorporated by European colonization, and most of the rural population preserved their subsistence economy, never even paying tribute to the Europeans. Several uprisings by these communities and the Chinese were repressed by the Spaniards.

[4] Spanish colonization in the Philippines followed similar patterns to that in America. However, the Philippines had two distinguishing features; they were located on the oceanic trading routes, in a position that received merchandise from all over southeast Asia on

its way to Europe, and they were ruled by the Viceroyalty of Mexico.

[5] Late in the 19th century, a local independence movement developed, led by the native bourgeoisie, who wanted the political power which was barred to them. They were soon followed by the other oppressed sectors. Anti-colonial revolution erupted in 1896 and independence was proclaimed on July 12th. However, the US was attracted by the archipelago's strategic position and vulnerability, and it immediately stepped in to sieze control. This resulted in the signing of the Treaty of Paris, on December 10 1898, ending the Spanish-US war (see history of Cuba and Puerto Rico). In meetings, which barred Filipino delegates, Spain decided to cede the archipelago to the US in exchange for compensation.

[6] Between 1899 and 1911, more than 500,000 Filipinos died in the struggle against occupying troops commanded by US general Douglas MacArthur. During World War II, the archipelago was occupied by Japan and the Huk Movement, peasant-based and socialist inspired, emerged during the struggle against the invaders. In 1946, after the war had ended, occupation troops returned led by the son of the first Douglas MacArthur. The archipelago was eventually granted formal independence but the Philippines have remained under US economic domination ever since.

[7] Nor did independence bring about any social changes. The *hacienda* system still persists in the country, where large estates are farmed by sharecroppers. More than half the population are peasants, and 20 per cent of the population own 60 per cent of the land. Although the sharecropper is supposed to receive half of the harvest, most of the peasant's actual income

goes to paying off the debts incurred with the *cacique*, the landowner. There is almost 9 per cent unemployment and the country suffers from the consequences of a balance of trade deficit, typical of a producer of agricultural commodities.

[8] The Nationalist Party, a conservative party of landowners, stayed in power until 1972 when Ferdinand Marcos, president since 1965, declared Martial Law. In 1986 a coalition of opposition forces rebelled against the continuous abuses of Ferdinand Marcos. During his

presidency repression grew both against armed movements (the Muslim independent groups of Mindanao and the New People's Army led by the Maoist Communist Party), and against political and trade union opposition. This repression often had the military support of the US.

[9] Due to its long presence, the Catholic Church is deeply rooted in Filipino society here. This is reflected in the fact that 75 per cent of Filipinos over the age of 10 have learnt to read in institutions dependent on the Church. The Church played an active role in the denunciation of fraud when the 1976 referendum supported the invocation of martial law. Five years later, 45 political and trade union organizations united to boycott the fraudulent and unconstitutional elections that Marcos used to stay in power. In September 1981 thousands of people demonstrated in Manila, demanding the end of dictatorship and the withdrawal of the US military bases. Since the 19th century, the US army has had its two largest foreign bases in the Philippines.

[10] On August 21 1983, opposition leader Benigno Aquino was mur-

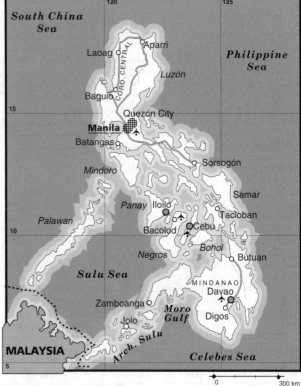

PUBLIC EXPENDITURES

% of GNP — 1992

MILITARY 2.2 · EDUCATION 2.9 · HEALTH 1

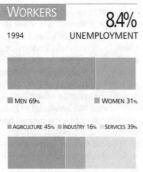

WORKERS — 1994

8.4% UNEMPLOYMENT

MEN 69% · WOMEN 31%

AGRICULTURE 45% · INDUSTRY 16% · SERVICES 39%

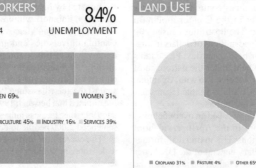

LAND USE

CROPLAND 31% · PASTURE 4% · OTHER 65%

Food Dependency

1970
11%

1992
218

Foreign Trade

MILLIONS $ 1994

IMPORTS
22,546

EXPORTS
13,304

PROFILE

ENVIRONMENT

Of the 7,000 islands that make up the archipelago, spread over 1,600 km. from north to south, eleven account for 94% of the total area and house most of the population. Luzon and Mindanao are the most important. The archipelago is of volcanic origin, forming part of the "Ring of Fire of the Pacific". The terrain is mountainous with large coastal plains where sugar cane, hemp, copra and tobacco are grown. The climate is tropical with heavy rainfall and dense rainforests. The country is the main producer of iron ore in Southeast Asia, there are chrome, copper, nickel, cobalt, silver and gold deposits. Like other countries in the region, it is suffering the effects of rapid deforestation.

SOCIETY

Peoples: Successive waves of immigration account for the country's ethnic heterogeneity. Most of the population migrated from Malaysia and Indonesia between the 24th century BC and the 11th. century AD. However, it is probable that native peoples such as the Aetas and the Igorots subsisted without being assimilated to the other ethnic groups. In the 15th century, when Islamic communities arrived from Borneo, some 200,000 Chinese traders were already established on the islands. The former resisted the evangelization carried out by the Spanish conquerors who arrived in 1521, whereas the rest of the population gave in to Spanish influence. Some peoples of Malay extraction, who were at a different stage of societal development, were not christianized either. Finally, from 1898 onwards, US colonization was extremely influential on Filipino culture and society. **Religions:** Catholics 83%; Protestants 8%, Muslims 5%, Anglipayans (Independent Filippino Church), 3%; Animists and Buddhists. **Languages:** The Philippines presents a complex linguistic map. 55% of the population speaks Pilipino (official), based on Tagalog, a language of Malaysian origin. English, spoken by 45%, is obligatory in schools. However, 90% of the population speaks one of the following languages on an everyday basis: Cebuano (6 million); Hiligayano (3 million); Bikolano (2 million); Waray-Waray (1 million). Spanish and Chinese are minority languages. **Political Parties:** Lakas ñg Bansa (The People's Struggle), of Fidel Ramos; the Laban Party (The People's Power Movement), founded in 1988 by Benigno Aquino, the Democratic Christian Party; the BISIG movement, socialist of the Tagalo speakers; the People's reform Party, of Miriam Defensor Santiago, founded in 1991; Eduardo Cojuangco, leader of a broad right-wing coalition; Imelda Marcos, widow of the ex-dictator Ferdinand Marcos; the Liberal Party, led by Jovito Salonga; the Nationalist Party, right-wing, of ex-vice-president Salvador Laurel. The main leftist opposition force is the National Democratic Front (NDF) which includes different mass organizations (workers, peasants, youth and women), religious (Christians for National Liberation) and cultural-based groups (artists, writers, teachers) led by the Philippine Communist Party (PKP) and its military wing, the New People's Army (NPA). In the Cordillera region, an NPA splinter group, the Cordillera People's Liberation Army (CPLA) organized the Cordillera Bodong Association (CBA) which demands full autonomy for the north. In the southern Muslim areas, the Moro National Liberation Front wages guerrilla warfare also demanding autonomy but, recently broke into three factions as a result of differences in opinion on how to deal with the new democratic government; Kababaihan Para Sa Inang Bayan (Women for the Motherland), created in 1986, the first party exclusively for women; the Mindanao Alliance, regional. **Social Organizations:** Labor is divided between the left-wing Kihusan Mayo Uno (May Day Confederation) and the Trade Union Congress of the Philippines (TUCP), affiliated with AFL-CIO. The Philippines have 720 voluntary non governmental organizations, popular organizations, and church groups, gathered in the "Green Forum".

THE STATE

Official Name: Republika ñg Pilipinas. **Administrative Divisions:** 3 regions, 73 provinces. **Capital:** Manila, 1,894,660 inhab. (1991). **Other cities:** Quezon City, 1,627,900 inhab.; Davao, 867,800 inhab.; Cebu 641,000 inhab.; Calaocan, 629,500 inhab.; Zamboanga, 453,200 (1991). **Government:** Presidential republic, Fidel Ramos, president since June 30, 1992. Bicameral Legislative: Chamber of Deputies (Congress) 250 members, 200 elected by general vote, 50 chosen by the president, and a 24 - member Senate. **National Holiday:** July 12th., Independence Day (1946). **Armed Forces:** 106,500 troops (1995) **Paramilitaries:** 40,500 National Police (Ministry of the Interior), 2,000 Coastguards.

---

dered at the Manila airport as he stepped down from the commercial flight that had brought him back to his country after prolonged exile in the US, a murder which was attributed to Marcos. More than 500,000 mourners followed his coffin to the cemetery, blaming Marcos for the crime. The crowds flooded the streets refusing to go home until the dictator was ousted.

[11] In the May 14 1984 by-elections, the opposition made important progress; for though it lost the elections, it won 78 of the 183 parliamentary seats. In April 1985, the "Reform of the Army Movement" (RAM) was formed by officers who were opposed to Marcos' interventions in the army.

[12] Amidst a scenario of increased violence and repression, compared by the Church with the dirty war of Argentina in the 1970s, a large sec- tion of the population put pressure on Marcos, demanding early elec- tions in 1986, and supporting the candidacy of Corazón Aquino, widow of the assassinated leader.

[13] Elections were held in February 1986, but widespread fraud pre- vented Cory Aquino from winning, and she subsequently called for civ- il disobedience. Marcos's minister of defence, Juan Ponce Enrile, at- tempted a coup against the dictator, but failed. A million supportive ci- vilians surrounded the rebels in the field where they had taken refuge. The air force refused to bombard their comrades-in-arms and the people who supported them, so Marcos opted for exile and Corazón Aquino assumed the presidency with Enrile as her minister of de- fence.

[14] Aquino faced various attempts at ousting her; the most threatening of which took place in November 1986 and September 1987, both ending in cabinet reshuffles. After the first attempt, Enrile resigned from his post. The breach with the more conservative figures had to be balanced by the removal of several first-line advisors who were direct- ly involved in grass-roots move- ments and the defense of human rights.

[15] The new constitution was ap- proved by a large majority in the Feb- ruary 1987 plebiscite. The charter granted autonomy to the Mindanao and Cordillera regions, thus paving the way for a truce with guerrilla groups operating in those areas. The New People's Army negotiators soon left the negotiating table, after sever- al acts of provocation against mass organizations and attempts on the lives of civilian leaders. Agrarian re- form, which should have been the cornerstone of the government's plan for social transformation, was dilut- ed after going through a legislature where many of the members are land- owners. The future of the Clark and Subic Bay American military bases began to be debated in April 1988, as the contracts were due to expire in 1991.

[16] In 1990, 39 per cent of the Filipi- no population was under the age of 14, and in spite of NGO efforts to al- leviate their suffering, only a third of their needs are met. A number of gov- ernment "internal refugee" camps were created to provide basic assis- tance to some 1,250,000 homeless people.

[17] The UN has recorded a total of 110 regional and ethnic groups in the Philippines. During 1991 in- creasing pressure from these groups, the urgent need for better land and wealth distribution and,

Illiteracy
1995

5%

Maternal Mortality
1989-95

PER 100.000
LIVE BIRTHS

208

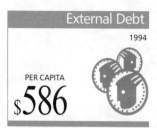

External Debt
1994

PER CAPITA

$586

STATISTICS

DEMOGRAPHY

Urban: 52% (1995). **Annual growth:** 1.8% (1992-2000).
Estimate for year 2000: 77,000,000. **Children per
woman:** 4.1 (1992).

HEALTH

One **physician** for every 8,333 inhab. (1988-91). **Under-five
mortality:** 57 per 1,000 (1995). **Calorie consumption:** 94%
of required intake (1995). **Safe water:** 85% of the
population has access (1990-95).

EDUCATION

Illiteracy: 5% (1995). **School enrolment:** University: 26%
(1993). **Primary school teachers:** one for every 36 students
(1992).

COMMUNICATIONS

100 **newspapers** (1995), 94 **TV sets** (1995) and 92 **radio
sets** per 1,000 households (1995). 1.3 **telephones** per 100
inhabitants (1993). **Books:** 85 new titles titles per 1,000,000
inhabitants in 1995.

ECONOMY

Per capita GNP: $950 (1994). **Annual growth:** 1.70%
(1985-94). **Annual inflation:** 10.00% (1984-94). **Consumer
price index:** 100 in 1990; 151.7 in 1994. **Currency:** 24
pesos = 1$ (1994). **Cereal imports:** 2,036,000 metric tons
(1993). **Fertilizer use:** 540 kgs per ha. (1992-93).
Imports: $22,546 million (1994). **Exports:** $13,304 million
(1994). **External debt:** $39,302 million (1994), $586 per
capita (1994). **Debt service:** 21.9% of exports (1994).
Development aid received: $1,490 million (1993); $23 per
capita; 3% of GNP.

ENERGY

Consumption: 364 kgs of Oil Equivalent per capita yearly
(1994), 70% imported (1994).

[21] The issue of closing the Subic
Bay naval base, with the relocation
or loss of around 12,000 personnel,
resulted in several vacillations in
the Filipino Senate due to the economic implications. In October, the
Senate voted not to renew the
bases' agreement.

[22] The evacuation of the Clark
base and its relocation on a nearby
island outside Filipino sovereignty
worried local and international environmental movements. They are
concerned that the constant circulation of nuclear weapons may pose
a threat to the area.

[23] The forest, which covered 75
per cent of the land in the 1950s,
was reduced to 42 per cent by 1990.
The effect of natural disasters like
typhoon Uring which killed 8,000
in 1991, was magnified by the lack
of trees.

[24] Twenty-five of the thirty-two
million Filipinos entitled to vote
took part in the May 1992 elections,
considered the calmest and cleanest
in the country's history. The winner
was Fidel Ramos, former minister
of defence of the Aquino government. In September, the United
States left its naval base at Subic
Bay with an infrastructure worth $8
billion and thousands of rural wom-
en who worked as prostitutes were
left unemployed.

[25] In 1993 the government and
Congress tightened their links with
Muslim secessionists and the guerrillas in northern Luzon. At the
same time, the authorities were
charged with torture and forced
"disappearances".

[26] In 1994, the Ramos government
had to turn to the opposition's support to control evasion of the 10 per
cent VAT tax. This measure won
him the IMF's support - including
a loan - and enabled a 5 per cent
growth of the GNP. The campaign
against crime, now headed by vice-
president Estrada, implicated 2 per
cent of police forces - who were
discharged - in criminal activies,
while another 5 per cent was kept
under investigation. The NPA communist guerrillas lost strength due
to an amnesty for its members and
to intestine conflicts regarding the
amnesty.

[27] In 1995 Imelda Marcos won the
elections in the Chamber of Deputies, in spite of being accused of
many corruption charges. Swiss
banks returned to the country $475
million deposited by her husband
during the dictatorship, but the gov-

ernment is convinced billions remain in other accounts.

[28] The elimination of restrictions
to investments, the reduction in
customs barriers and the presence
of educated and cheap labour, attracted investors which led to a 6
per cent growth of the GNP. $2 bil-
lion belonging to remittances from
4.2 million workers living abroad -
mainly housemaids - came into the
country in 1995.

[29] Due to negotiations between the
Moro National Liberation Front
(MNLF), the radical Islamic Moro
Liberation Front armed 3,000
troops in the jungle and fundamen-
talist group Abu Sayyaf attacked a
Christian village - killing 47 - loot-
ed banks and set buildings on fire.

[30] In late 1995, an unprecedented
food crisis took place with a 70 per
cent increase of the price of rice at
a time when 70 per cent of the population lived below the poverty level. Farmers' organizations blamed
the government for an incoherent
and corrupt agrarian policy and demanded an effective land reform,
with rural industrialization, food
self-sufficiency and protection of
the environment.

[31] In spite of Christians' heated
protests, the government continued
negotiating with the Muslims. They
were offered membership of the
Congress, the Supreme Court of
Justice and the Cabinet while 5,000
guerrillas would be absorbed by the
Armed Forces. The MLNF set up a
peace and development council in
the southern region with a 2 million
population, seeking to hold a plebiscite within three years regarding
the administration's limits. The
government's main reason to negotiate was the need for peace in order
to establish economic reforms and
put the country into competition
with its Association of South East
Asian Nations (ASEAN) partners,
which are mostly Muslim.

possibly, the approaching presidential elections of May 1992, led
Corazón Aquino to create a Bureau
of Northern Communities. The bureau was concerned with the mountain tribes and ethnic groups, particularly in Luzon. There is also a
Bureau of Southern Cultural Communities, excluding the Muslims.
The staff of those bureaus are recruited from among the communities concerned.

[18] In the Mindanao region there
has been a Bureau of Muslim Affairs since 1987. Potential conflict
is brewing there, as the 20,000-
strong Moro Liberation Front considers that the government has not
respected conditions set out in the
1987 Constitution.

[19] In June 1991, the eruption of
Mount Pinatubo shook the country,

claiming the lives of over 700 Filipinos, flattening entire villages,
forcing the evacuation of over
300,000 people, and completely
burying the evacuated Clark US
airforce base under the ashes.

[20] With its airbase completely un-
usable, and swamped with difficult
negotiations as the contract approached their expiry dates, the US
opted to abandon the base of its
own accord. On November 26,
1991, the Clark airbase was formally and definitely abandoned. It had
employed over 6,000 US military,
and over 40,000 Filipinos, most of
whom had carried out the menial
work. The Aetan people claimed
the territory they had been deprived
of when the base was installed, and
the Filipino government made
plans to build an airport there.

Pitcairn

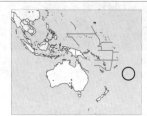

Area:	5 SQ.KM
Capital:	Adamstown
Currency:	NZ Dollar
Language:	English

Pitcairn

Like most of the islands of the region, Pitcairn's first settlers were Polynesians. The first European to visit the island was the English seaman Robert Pitcairn, who sailed along its coasts in 1767.

[2] In 1789, part of the crew of HMS Bounty mutinied, on their return to Britain from six months in Tahiti. The captain and the rest of the crew were given a small boat and the other men returned to Tahiti. They stayed there a short time and then transferred to Pitcairn Island.

[3] The group, led by Fletcher Christian, was made up of eight crew members, six Tahitian men and twelve women. Ten years later, only one of the mutineers, John Evans, was still alive, with eleven women and 23 children. Purportedly guided by "apparitions", Adams christened the children and peopled Pitcairn, which later became a British colonial dependency.

[4] The population reached 200 inhabitants in 1937, but decreased in recent times as young islanders emigrated to Aotearoa/New Zealand in search of work.

[5] Pitcairn was under the jurisdiction of the governor of Fiji between 1952 and 1970. At this time, it became a dependency of the British High Commissioner in Aotearoa, who assumed the functions of a governor, in consultation with the island's Administrative Council.

[6] The island's communications with the outside world are limited to a radio and boats that occasionally anchor off the coast.

[7] Despite the lack of communications, education in the islands is important and primary school is compulsory for all children between the ages of 5 and 15. A New Zealand teacher is appointed for a period of 2 years, and is also responsible for publishing the Pitcairn Miscellany, a four-page bulletin.

[8] The island's Administrative Council is made up of 11 representatives, only 5 of which are chosen from inhabitants over the age of 18 who have lived on the island at least three years.

[9] The only national celebration is the Queen's birthday, which is celebrated on the second Saturday in June. The Queen is the head of State.

[10] The inhabitants work almost exclusively at subsistence fishing and farming. The fertile valleys produce a wide variety of fruit and vegetables, including citrus, sugar cane, watermelons, bananas, potatoes, and beans. However, the island's main source of income is the export of postage stamps for sale to stamp collectors.

[11] The reforestation plan carried out in 1963 increased plantations of the miro tree, which provides a kind of wood that can be used for many kinds of handcrafts.

[12] Since 1987 the Japanese Tuna Fisheries Cooperative Association fleet has been allowed to operate within Pitcairn's 200-mile Exclusive Economic Zone, which ended in 1990.

[13] That same year, the British High Commissioner in Fiji - representing Pitcairn, the United Kingdom's last possession in the South Pacific - met with representatives of the United States, France, Aotearoa/New Zealand, and six other States of the region, to sign a Regional Convention of the South Pacific for the Protection of the Environment. The aim of this agreement was to put an end to the dumping of nuclear waste in the region.

[14] In 1989, Henderson island, located 68 kms northeast of Pitcairn, was listed among the UK property to be preserved as a natural bird reserve. It is the home of five unique species. In early 1992, deposits of manganese, iron, copper, zinc, silver and gold were identified in underwater volcanos within the island's territorial waters.

[15] In 1995, Robert John Alston became the new governor, replacing David Moss.

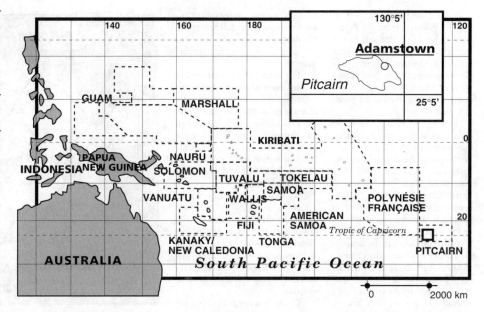

PROFILE

ENVIRONMENT

Pitcairn consists of four islands of volcanic origin, of which only Pitcairn is inhabited. The others are: Henderson, Ducie and Oeno. The group is located in the eastern extreme of Polynesia, slightly south of the Tropic of Capricorn, east of French Polynesia. The economy is based on subsistence agriculture, fishing, hand crafts, and there are very few export products. The rainy, tropical climate is tempered by sea winds. The islands are subject to typhoons between November and March.

SOCIETY

Peoples: The population consists of descendants of mutineers from HMS Bounty and Polynesian women from Tahiti. **Religions:** Seventh Day Adventist. **Languages:** English (official). **Political Parties:** None.

THE STATE

Official Name: Pitcairn, Henderson, Ducie and Oeno Islands. **Capital:** Adamstown. **Government:** Robert John Alston, British High Comissioner in Aotearoa, Governor of the islands since 1995. Jay Warren, magistrate and president of the Council since 1993. The Council is made up of ten members. In addition to the magistrate who presides over it, one member functions as the Council's secretary. Five members are elected every year and three are named by the government for a one-year period.

ECONOMY

Currency: NZ dollars.

Poland

Polska

Population: 38,544,000 (1994)
Area: 312,680 SQ.KM
Capital: Warsaw (Warzawa)

Currency: Zloty
Language: Polish

The name Poland comes from a Polanian tribe, the Poles, who tilled the soil in the Warta basin region. A region later known as Greater Poland. Their chiefs belonged to the Piast dynasty, descendants of a legendary ancestor, and they lived in the burgh of Gniezno.

2 During the 10th century, the Polanians subdued the Kujavians, the Mazovians, the Ledzians, the Pomeranians, the Vistulans, and the Silesians. Mieszko I (960-992) duke of the Piast, founded the first Polish State when he united neighboring tribes under a common state structure. Mieszko was christened in the year 966 establishing the Christian influence in the area.

3 Poland was a hereditary monarchy until the 12th century. The duke and a group of noblemen held power, supported by an army of elite warriors. Free peasants, who constituted the largest and poorest group, were mobilised as they were needed, and paid taxes to support the system.

4 The greatest dilemma for Poland was what stance to take towards the German Empire and the Pope. As the empire expanded, Mieszko accepted submission, in exchange for an acknowledgment of his sovereignty. As compensation, he appealed for Papal protection and in 1,000 AD he founded the first Polish ecclesiastical city state.

5 From then on, the Roman Catholic Church became a crucial element in the political structure of the Polish State. Poland alternated phases of independence and annexation until the 12th century, when the State started to split up.

6 During the feudal period, Poland was subdivided into several duchies, led by the Piasts and some 20 overlords. The nobility and the

Church became increasingly autonomous as the central power and the dukes lost strength. This was also a period of great demographic growth.

7 The arrival of German settlers created a new ethnic situation in the country, for up to then the population had been of Slavic stock. From the 13th century, the population of the towns became increasingly German and Jewish. Immigrants brought in their own legal systems, their capital, their crafts and their agricultural skills.

8 Under the reign of Casimir the Great (1333-70), Poland became a monarchy divided into estates, with the king acting as an arbiter between the nobility, the clergy, the bourgeoisie, and the peasants. The State ceased to be royal property and in 1399 the monarchy became elective.

9 Through a royal marriage in 1386, Poland joined with Lithuania, and they adopted an organization which respected the differences between the two countries. In 1409, the Teutonic Order attempted to stop Polish-Lithuanian expansion, but it was defeated at Grunwald. The Peace of Torun, in 1411, did not stop the conflict, and the power of the Order suffered a setback.

10 In 1466, after a new victory over the Teutons, Poland recovered Pomerania of Gdansk and Malbork, Elblag, and the Land of Chelm; it also gained the territory of Warmia. In recognition of their assistance during the war, Poland granted autonomy to Pomerania and some privileges to the towns. A period of economic prosperity and cultural renaissance began.

11 During the 15th century the General Diet (parliament) of Poland and Lithuania was created. It had two houses: a lower house, constituted by members of the nobility, and an upper house, or royal council, presided over by the king. The two states shared the king, the diet, and the management of foreign affairs, while administration, justice, finance and the army remained separate.

12 The 16th or "Golden" century is also known as the period of the Royal Republic, for the king had to consult the noblemen before fixing taxes or declaring war. The reduction of the rights of the bourgeoisie and the peasantry in favor of the nobility and the clergy modified the

original monarchy of divided estates.

13 In 1573, with the extinction of the Jagiellon dynasty, the Diet approved the free election of the king and guaranteed religious tolerance, which was exceptional in a time when Europe was being shaken by religious wars. King Stephen Bathory (1576-86) gave up arbitration, and the nobility started to elect their own courts.

14 During the 17th century, international developments were unfavorable for Poland and Lithuania. Sweden fought against Poland over the control of the Baltic, and Russia entered into conflict with Lithuania. Turkish and Austrian ambitions in central Europe also exerted pressure on Poland.

15 On the lower Dnepr, on the border with the Ukraine, free peasants and impoverished noblemen became the first cossacks, warriors who lived by pillaging. In 1648, the cossacks and the Ukranian peasants started a national revolt. The king made unsuccessful attempts to reach an agreement with the rebels, whose victories weakened the Polish republic.

16 The cossacks formed occasional alliances with the Turks and the Russians. In 1654, Russian troops entered Polish territory, and Sweden invaded the rest of the country a year later. King John Casimir fled to Silesia, and Poland was aided by Austria, while the peasants in the

provinces organised the first large-scale armed resistance.

17 The Swedes and the Turks were expelled from the country and the cossacks were defeated. Russia kept Smolensk and the Ukraine, the left bank of the Dnepr and the city of Kiev. The wars devastated the land, decimated the population and split the republic.

18 In 1772, Russia, Prussia, and Austria agreed on the territorial partition of Poland. A second partition followed in 1793, after a fresh Russian invasion annulled the 1791 Constitution and put an end to a new Polish attempt to reorganise the State.

19 A patriotic insurrection was crushed in 1794, and was followed a year later by the third partition of Poland. The Polish State disappeared from the map, but the Polish people retained a feeling of national identity.

20 During the 19th century, the Poles made several attempts to free their motherland. National conscience and Catholicism, both under persecution, became stronger. New political parties appeared; a peasants', a workers', and a national party, and resistance was expressed through culture.

21 The consolidation of the Russian revolution in 1917 brought Poland the support of western powers. In 1918 a provisional government was created, led by Jozef Pilsudski. It established an eight-hour

LAND USE

■ CROPLAND 48%   ■ PASTURE 13%   ■ OTHER 38%

working day and equal rights for men and women.

22   Boundaries could not be moved back, as a national sense of identity had emerged in the Ukraine, Lithuania and Belarus. The creation of a federation failed with the counteroffensive of the Soviet army. The Peace of Riga, signed in 1921, guaranteed the independence of the Baltic states and fixed the eastern border of Poland at Zbrucz.

23   The 1921 Constitution adopted a parliamentary system. During the years that followed, the government had to resort to coalitions, generally between the National and Social Christian parties, which were both conservative, or the Peasant s'party, which was moderate.

24   Political instability, a tariff war against Germany, unemployment and social unrest benefited the Communist Party, which had been banned in 1923. Their main military leader, Jozef Pilsudski, staged a coup d'etat in 1926, but the ensuing prosperity was ended by the effects of the 1929 Wall Street crash.

25   The growing power of Germany and the USSR became a threat to Poland, and Britain and France only gave the country formal support. In a secret agreement, the USSR and Germany divided the Polish territory once more. On September 1 1939, Germany invaded Poland.

26   Britain and France declared war on Germany. On September 17, the USSR also invaded Poland. In the occupied territories millions of Poles died, especially Jews, some of whom were taken to German concentration camps, many others starved or were executed.

27   The Polish government-in-exile led the resistance. A military contingent fought on the western front, while the Home Army staged subversive acts. After the German invasion, the USSR accepted the creation of a Polish army under its jurisdiction.

28   The Soviet counter-offensive modified relations with Poland. The government-in-exile demanded an inquiry into the murders of Polish officers, and the USSR broke diplomatic relations. The Red Army invaded Poland and re-established military occupation.

29   After the defeat of Germany, the allies gathered at Yalta and agreed on a Provisional Polish

PROFILE

ENVIRONMENT

On the extensive northern plains, crossed by the Vistula (Wisla), Warta and Oder (Odra) rivers, there are coniferous woodlands, rye, potato and linen plantations. The fertile soil of Central Poland's plains and highlands yield a considerable agricultural production of beet and cereals. The southern region, on the northern slopes of the Carpathian Mountains, is less fertile. Poland has large mineral resources: coal in Silesia (the world's fourth largest producer); sulphur in Tarnobrzeskie (the world's second largest producer); copper; zinc and lead. Major industries are steel, chemicals and shipbuilding. The country has high levels of air pollution; due to its geographical location in the center of Europe, it absorbs polluted water and air "in transit" from other countries. Soil depletion has caused extensive erosion, due to the excessive deforestation that has accompanied intensive agriculture.

SOCIETY

Peoples: Polish, 98.7%; Ukrainian, 0.6%; other (Belarusian, German), 0.7%. **Religions:** Catholic, 90%; Orthodox, 1.5%. **Languages:** Polish. **Political Parties:** Democratic Left Alliance (SLD); Polish Peasant Party (PPP); Democratic Union (UD); Solidarity; Work Union; Reform Support Block; Confederation for an Independent Poland. **Social Organizations:** Central Council of Trade Unions, Central Union of Agricultural Groups, Solidarity.

THE STATE

Official Name: Rzeczpospolita Polska. **Administrative Divisions:** 49 provinces. **Capital:** Warsaw (Warzawa), 1,643,000 inhab. (1994). **Other cities:** Lódz, 834,000 inhab.; Kraków, 745,000 inhab.; Wroclaw, 643,600 inhab.; Poznan, 589,700 inhab. **Government:** Parliamentary republic. Aleksander Kwasniewski, Head of State and President. Wlodzimierz Csimoszewicz, Prime Minister. Legislature, bicameral: Senate and *Sejm*, whose members are elected every four years. **National Holiday:** July 22. **Armed Forces:** 283,600 (1993). **Paramilitaries:** 23,400 Border Guard, Police, Coast Guard.

Government of National Unity (made up by representatives from pro-Soviet and exile groups) which was to call elections. The government was dominated by the Polish Workers' Party.

30   In 1945 the provisional government and the USSR signed an agreement establishing the Polish eastern border, along the Curzon line. The same year, the allies fixed the eastern border, along the Oder-Neisse line of Lusetia.

31   The Polish Workers' Party and the Socialist Party of Poland constituted the Polish United Workers' Party (PUWP). The Polish Peasants' Party disintegrated, and elections were postponed.

32   The PUWP governed the country, modelling themselves on the Soviet Communist Party (CPSU) in the USSR. Industry and commerce were nationalised, the State built great steel and metal works, and forcibly collectivized agriculture. Women were incorporated to the workforce.

33   The denunciation of Stalin's crimes during the 20th Congress of the CPSU, in 1956, had repercussions on the PUWP. In November of that year Wladyslaw Gomulka was elected party first secretary and promised to take a "Polish path towards socialism". Gomulka freed Cardinal Stefan Wyszynski, who was the head of the Catholic

Church, and stirred up popular expectations.

34   In 1968, action taken by an anti-Semitic group within the PUWP forced Jewish groups to leave the country. In a treaty signed in 1970, West Germany recognized the Polish borders established after the war. East Germany had done so in 1950.

35   In 1970, strikes broke out because of a staggering increase in prices. The government ordered the army to open fire on the workers and started another crisis within the PUWP. Gomulka was replaced by Edward Gierek, but the regime underwent renewed crises, over corruption charges and internal fights within the party.

36   In 1976, new strikes broke out, and this time they were repressed not by the use of firearms but by long prison sentences. In 1979, the Polish pope John Paul II visited his native land rekindling the hopes of the population, who welcomed him at massive gatherings.

37   The 1980 strike at Gdansk's Lenin Dockyard was led by Lech Walesa. It turned into a general strike and the government had to negotiate with the strikers. Two months later, the government was forced to recognise Solidarity, a workers' union with 10 million members. Rural Solidarity was created, to represent three million peasants.

38   The PUWP suffered other crises that led to the appointment of Wojciech Jaruzelski, the then prime minister, to the post of party first secretary. In December 1981, Jaruzelski declared martial law, Solidarity was banned and its leaders went underground.

39   Martial law was lifted in 1983, but the Constitution was modified to include a state of emergency. With the Catholic Church acting as a mediator, the government and Solidarity representatives went back to negotiations in 1989, while the USSR was starting its "perestroika" (restructuring).

40   In the elections in June of that year, the PUWP only obtained the fixed number of representatives that had been agreed on during the negotiations with the opposition. Solidarity rejected General Jaruzelski's proposal to share government with the PUWP. Tadensz Mazowiecki, a journalist and a moderate member of Solidarity, was appointed the first president of a non-com-

Foreign Trade

MILLIONS $ 1994

IMPORTS
21,400
EXPORTS
17,000

External Debt

1994

PER CAPITA
$1,094

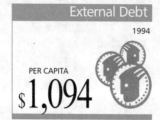

STATISTICS

DEMOGRAPHY

Urban: 64% (1995). **Annual growth:** 0.4% (1991-99). **Estimate for year 2000:** 39,000,000. **Children per woman:** 1.9 (1992).

HEALTH

Under-five mortality: 16 per 1,000 (1995). **Calorie consumption:** 112% of required intake (1995).

EDUCATION

School enrolment: Primary (1993): 97% fem., 97% male. Secondary (1993): 87% fem., 82% male. University: 26% (1993). **Primary school teachers:** one for every 17 students (1992).

COMMUNICATIONS

103 **newspapers** (1995), 104 **TV sets** (1995) and 102 **radio sets** per 1,000 households (1995). 11.5 **telephones** per 100 inhabitants (1993). **Books:** 102 new titles per 1,000,000 inhabitants in 1995.

ECONOMY

Per capita GNP: $2,410 (1994). **Annual growth:** 0.80% (1985-94). **Annual inflation:** 97.80% (1984-94). **Consumer price index:** 100 in 1990; 435.6 in 1994. **Currency:** 2 zlotys = 1$ (1994). **Cereal imports:** 3,142,000 metric tons (1993). **Fertilizer use:** 811 kgs per ha. (1992-93). **Imports:** $21,400 million (1994). **Exports:** $17,000 million (1994). **External debt:** $42,160 million (1994), $1,094 per capita (1994). **Debt service:** 14.3% of exports (1994).

ENERGY

Consumption: 2,563 kgs of Oil Equivalent per capita yearly (1994), 5% imported (1994).

munist government in the East European bloc.

[41] Poland re-established diplomatic relations with the Vatican and with Israel. The United States and Germany committed themselves to financial assistance. German reunification caused some alarm, but the Four-Plus-Two negotiations (see Germany) ratified the postwar Polish borders.

[42] In December 1989, the National Assembly approved the reinstating of the name, the Republic of Poland. In January 1990, the PUWP was dissolved to create another party, but it later split into the Social Democracy of the Republic of Poland and the Polish Social Democratic Union.

[43] In January 1990, the government started an economic adjustment programme agreed to with the International Monetary Fund. Poland requested its incorporation to the Council of Europe and established relations with the European Economic Community. The United States and the Council of Europe made entry to NATO dependent on the results of the economic reforms under way.

[44] On May 10 1990, the first strike was held against Mazowiecki's government. This exercise of power hastened the fragmentation of Solidarity. The union became a political party in July 1990, and later divided into several factions.

[45] In December 1990, the first direct presidential elections were held, and Lech Walesa was elected head of State with 75 per cent of the vote. In August 1991, new prime minister Jan Krysztof Bielecki resigned, upsetting the precarious balance of political transition. The former Communist party and a small peasant party were ready to accept his resignation, but Walesa backed the prime minister and insisted on giving him special powers, by threatening to dissolve the Diet.

[46] In the parliamentary elections of October 1991, 60 per cent of registered voters abstained, and the rest of the vote was dispersed. Former prime minister Tadeusz Mazowiecki obtained the largest number of votes, with 13.6 per cent, followed by the former communists with 12.9 per cent. Solidarity only secured 5.3 per cent, and prime minister Bielecki's Liberal Congress had 7.5 per cent.

[47] In December 1991, the new Diet appointed Jan Olszewski as prime minister. He was accepted by President Walesa three days later, but the cabinet was not ratified by the Diet until December 23, and then only by a narrow margin of 17 votes.

[48] As of November 1991, Poland has been the 26th member country of the European Council, a Western European organization which also includes Turkey, former Czechoslovakia, and Hungary.

[49] On January 13 1992, there was a general strike; it lasted for one hour and was the first since the non-Communist government took office. Solidarity called the strike to protest against the non-negotiated rises in the price of electricity, gas, and municipal heating.

[50] On February 17 1992, the finance minister Karol Lutkowski resigned as he disagreed with the policy of issuing banknotes to control recession, and he also disagreed with Olszewski's proposal to increase the budget deficit.

[51] In the first days of March, the majority of the *Sjem* rejected the government's economic plan. In the following months the crisis was aggravated by confrontations between Walesa and Olszewski over decisions concerning the army, provoking the resignation of the Ministry of Defence.

[52] At the end of May, President Walesa asked the *Sjem* to form a new government because of a lack of confidence in Olszewski. In mid-July parliament accepted Walesa's proposal and appointed Hanna Suchocka prime minister, supported by a 7-party coalition.

[53] Suchocka applied strict monetary controls and promoted the privatization of approximately 600 state enterprises with a privatizations act in August 1992. Meanwhile, in February 1993 Walesa made the provocative decision to revoke, under pressure from the Church, the law which instituted the right to abortion.

[54] Dissatisfied with Suchocka's social policies, Solidarity sponsored a motion censuring the government, finally passed by a one-vote difference, forcing the prime minister to call early legislative elections.

[55] Elections were held in September and led to the return to power of the political sectors which had supported the communist regime: the Democratic Left Alliance (SLD) and the Polish Peasants' Party (PSL), winning 73 of the 100 seats.

[56] Initially refusing to acknowledge the defeat, Walesa appointed PSL leader Waldemar Pawlack as prime minister. Although 1994 featured constant conflicts between the president and the government, the latter did not significantly modify the economic liberalization policy, despite its attempt to slow the pace of some reforms in order to reduce its social impact.

[57] The former communists' return to power was concluded in November 1995 when Aleksander Kwasniewski defeated Walesa in the second round of presidential elections, with 52 per cent of the vote. The outgoing president had based his election campaign on anti-communism and the support of "Christian values", insisting on the need to maintain the ban on abortion.

[58] At first, the government of prime minister Josef Olesky - Pawlack's successor - had no modifications. However, in January 1996 Olesky resigned and was replaced by Wlodzimierz Csismoszewicz the following month.

Portugal

Population: 9,902,000 (1994)
Area: 92,390 SQ.KM
Capital: Lisbon

Currency: Escudo
Language: Portuguese

Portugal

In ancient times Portugal was inhabited by Lusitanians, an Iberian tribe whose cultural influence extended over a vast area including the whole western shore of the Iberian Peninsula. This area was successively conquered by several Middle Eastern peoples, who only occupied the coastal areas.

2 During the 2nd century BC the Romans settled in the territory, ruling over it until the fall of the Empire around the 5th century AD. Like the rest of Europe, Portugal was invaded by Northern European peoples, generically called Barbarians, who raided the Roman dominions. Among these peoples were the Visigoths. They had a developed culture and settled on the Iberian Peninsula dividing the territory into various kingdoms, and spreading the Christian faith. Their domination over the whole region lasted for nearly six centuries.

3 In the 8th century AD Arab peoples invaded the region imposing their political and cultural domination in spite of resistance from the earlier inhabitants.

4 During the 11th century the reconquest of the Lusitanian territory started, ending with the expulsion of the Arabs a hundred years later. With Muslim domination over, the territory was politically unified and Portugal entered a period of great economic prosperity. This reached its height during the 15th and 16th centuries, with great maritime expeditions and conquests of vast territories in America, Africa, and the Far East.

5 Its maritime superiority enabled Portugal to develop active worldwide trade and achieve a privileged economic position within Europe. A long time elapsed before other nations like Britain and the Netherlands were in a position to threaten Portugal's naval supremacy.

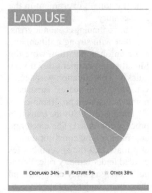

LAND USE

CROPLAND 34%   PASTURE 9%   OTHER 38%

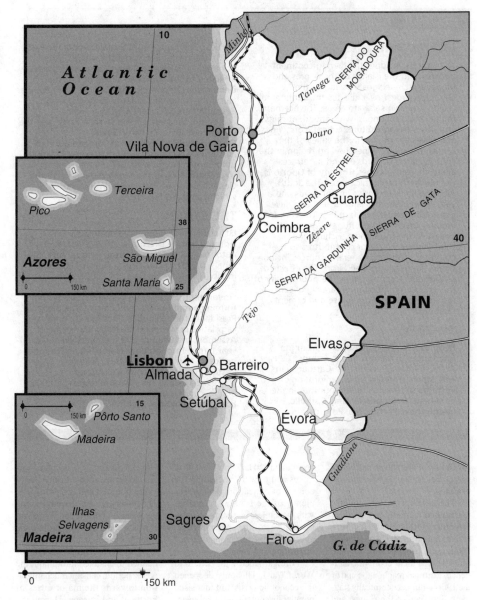

Atlantic Ocean

Porto
Vila Nova de Gaia

Terceira
Pico
38
Azores
São Miguel
Santa Maria 25
0   150 km

Minho
Tamega
SERRA DO MOGADOURA
Douro
SERRA DA ESTRELA
Guarda
Coimbra
Zêzere
SERRA DE GATA
40
SIERRA DE GATA
SERRA DA GARDUNHA

SPAIN

Tejo
Elvas

Lisbon
Almada
Barreiro
Setúbal
Évora

15
0   150 km   Pôrto Santo
Madeira

Guadiana

Ilhas Selvagens
Madeira 30

Sagres
Faro
G. de Cádiz

0   150 km

6 Following a series of dynastic stuggles, the country was made subject to Philip II, king of Spain. The two kingdoms remained united until 1688 when Portugal succeeded in having its independence recognized in the treaty of Lisbon. Unity with Spain brought the decline of Portugal's power. Most of the maritime empire collapsed, besieged by the British and the Dutch, who started to control most of the trading routes and outposts.

7 By the time Portugal recovered its independence, it had been devastated by three decades of war against Spain. The country was forced to look on while the new maritime powers siezed most of its colonies in Africa and Asia. Brazil, meanwhile, remained under Portuguese rule. The position of Britain as the leading maritime power became painfully obvious when Portugal was forced to sign the Treaty of Methuen, which established Portugal's political and economic dependence on the British. Pombal, an advisor of Jose I, carried out economic reforms. Like the Spanish Bourbons, Pombal had been influenced by the ideas of the French Enlightment and he changed colonial management. The discovery and exploitation of gold mines in Brazil enabled the country to enjoy a period of great economic prosperity. But in spite of Pombal, Portugal finally went into decline.

8 Dependence on Britain was further consolidated when Portugal was forced to seek support to end Napoleonic occupation, which lasted from 1807 to 1811, and French domination led to the independence of Brazil. The Portuguese court had fled there in exile, and Brazil had enjoyed a significant expansion in trade, in particular with Britain. At the end of the Napoleonic period in Europe, the rising Brazilian bourgeoisie was not ready to be displaced, so in 1821 Brazil declared independence. Meanwhile a civil war broke out in Portugal between those who wanted the

Foreign Trade

MILLIONS $ 1994

IMPORTS
26,680
EXPORTS
17,540

PROFILE

ENVIRONMENT

The country includes the Iberian continental territory and the islands of the Azores and Madeira archipelagos. The Tagus, the country's largest river, divides the continental region into two separate areas. The northern region is mountainous, with abundant rains and intensive agriculture: wheat, corn, vines and olives are grown. In the valley of the Douro, the major wine-growing region in the country, large vinyards extend in terraces along the valley slopes. The city of Oporto is the northern economic center. The South, Alentejo, with extensive low plateaus and a very dry climate, has large wheat and olive plantations and sheep farming. The cork tree woods, which made Portugal a great cork producer, are found here. Fishing and shipbuilding are major contributors to the country's economy. Mineral resources include pyrite, tungsten, coal and iron. The effects of erosion are accentuated by the poor quality of the soil. Air pollution levels are significant in urban areas, or near cellulose and cement factories.

SOCIETY

Peoples: The Portuguese came from the integration of various ethnic groups: Celts, Arabs, Berbers, Phoenicians, Carthaginians and others. There is a great migration of Portuguese towards richer countries in the continent. **Religions:** Mainly Catholic. **Languages:** Portuguese. **Political**

Parties: The center-right Social Democratic Party (PSD); the center-left Socialist Party (PS); the Communist Party; the Center Democratic Party, affiliated to the European Union of Christian Democrats. Other minor parties: the right Popular Party; the National Solidarity Party, the Revolutionary Socialist Party; the Greens, ecological; the People's Monarchist Party; and the Democratic Renewal Party. **Social Organizations:** The General Confederation of Portuguese Workers (CGTP), a nation-wide multi-union organization with 287 union members (represents 80% of the organized workers); the General Union of Portuguese Workers (UGTP), which combines 50 unions.

THE STATE

Official Name: República Portuguesa.
Administrative Divisions: 18 districts, 2 autonomous regions. **Capital:** Lisbon, 681,100 inhab. (1991). **Other cities:** Oporto, 309,500 inhab.; Vila Nova de Gaia, 247,500 inhab.; Amadora, 176,100 inhab. (1991).
Government: There is a parliamentary system forming a multi-party democracy. Jorge Sampaio, President of the Republic since March 1996. Legislative power is exercised by a unicameral Assembly, made up of 230 members elected for 4-year terms, through universal suffrage. **National holiday:** April 25, Liberty Day. December 1, Restoration of Independence. **Armed Forces:** 54,200 (1995). **Paramilitaries:** 20,900 Republican National Guard; 20,000 Public Security Police; 8,900 Border Security Guard.

restoration of absolutism and liberal groups preferring greater political participation.
[9] The economy of the country maintained its traditional agrarian structure. Other countries were embarking on an accelerated industrialization process which would quickly give them economic eminence. Portugal reached the end of the 19th century economically stagnant, deprived of the richest and largest part of its colonial empire, and suffering from acute internal political crisis.
[10] The monarchy was incapable of ensuring the stability needed to start economic recovery, and was definitively overthrown by liberal opposition forces in 1910. This started the Republican period. Once they had attained their objective, the alliance of Liberal and Republican groups started to fragment and internal differences prevented them from achieving a common governmental agenda. One of the few things they shared was active opposition to the Church, which had been a traditional ally of the *ancien*

regime and had had important privileges and powers, including the control of education. The inefficiency of the Liberals, together with the ruthless persecution of representatives of the ancien regime, encouraged the formation of a vast opposition movement.
[11] Portugal sided with Britain in World War I. This only deepened the economic crisis and increased popular discontent. Political instability and economic stagnation are the most salient features of the period. In 1926 this led to a coup bringing a right-wing military group to power. They set up an authoritarian corporatist regime which they called the "New State". With a few changes, it was to rule the country for over 40 years. Political opposition was proscribed, with major figures imprisoned or exiled. Trade unions were dissolved and replaced by corporatist organizations similar to those in fascist Italy.
[12] The most significant figure of this period and the true ruler behind the military was economist Antonio de

Oliveira Salazar, who occupied various positions and dominated Portuguese political and economic life. The country remained neutral during the Spanish Civil War and World War II which could have jeopardized the country's barely stable economy.
[13] The agricultural system remained unchanged throughout the whole period, causing much migration towards the major cities of Portugal and Europe. During the 1950s the decolonization movement started, and the country was faced with the possibility of losing its last dominions in Africa. From that moment on, Salazar's regime fought against the liberation movements emerging in the Portuguese colonies, becoming increasingly isolated from other countries and unpopular at home.
[14] The Portuguese government was adamant in its refusal to recognize the independence of Mozambique, Angola, and the other members of its former colonial empire. This prevented the country from being accepted as a member of the United Nations until 1955, earning

it the condemnation of the UN General Assembly on several occasions.
[15] The human and economic cost of these colonial wars accelerated the internal attrition of Salazar's government. Repressive measures had to be increased to halt the growing opposition. Salazar's death in 1970 and the deepening of the economic crisis showed that the end of the regime was close at hand.
[16] In 1974, amidst the opposition of many social groups and political parties, a significant number of dissatisfied army officers gathered under the Armed Forces Movement (Movimento das Forças Armadas–MFA). In April, they staged a coup intending to end the wars in Africa and start the democratization process.
[17] The military government which emerged from the "Revolution of the Carnations" or "Captains' Revolution" had vast popular support. This support came out as active encouragment of the progressive tendencies within the MFA. The new government quickly decolonized Angola, Mozambique, and Guinea-Bissau. Meanwhile it actively sought international recognition and attempted to improve the country's image abroad. It legalized left-wing political parties, decreed an amnesty for political prisoners, and passed a series of land laws aiming at breaking up large rural estates and modernizing agricultural production.
[18] The Socialist and Communist parties were the main supporters of the new regime, but after a year in office, their alliance began to break up. This slowed down the political democratization of the country which had been started in April 1974. In the 1976 general elections, the Socialist Party led by Mario Soares won the majority vote to become Portugal's first democratic constitutional government in the 20th century.
[19] The continuing economic crisis, together with strong political and trade union opposition, wore the Socialist government down very quickly. This situation was aggravated by Mario Soares' harsh economic adjustment programme.
[20] During the 1980s the transition process continued. The electorate approved a new constitution and eliminated all the special bodies created under military rule. Portugal's foreign policy changed, and the country underwent speedy eco-

nomic and political integration with Europe, including incorporation in NATO and the EEC, in 1986. That year, the Socialist Party lost power again, this time to its one-time ally the centre-left Social Democratic Party (PSD).

[21] By the end of the 1980s the country was experiencing significant economic growth, but still far below the average level for the rest of Europe. Changes accelerated after the electoral victory of the PSD, which used its ample parliamentary majority to liberalize the economy. The new economic policy received strong opposition, particularly from the workers in the public sector, who make up 5 per cent of the country's active labour force and who saw the PSD's reforms as a threat to their jobs.

[22] The trade union movement brought the country to a complete or partial standstill on several occasions. It opposed the privatization of public companies, the elimination of labor legislation enacted during the 1974 revolution, and the attempt to repeal the 1974 land reform legislation. In 1984 an extreme left-wing group, called the Popular Forces of April 25 (FP-25), also started to take action against these measures. The group, which had been very active in recent years, demanded that the parties in power respect the achievements of the "Revolution of the Carnations".

[23] In April 1987, the governments of Portugal and the People's Republic of China signed an agreement charging Portugal with the administration of Macau until 1999. Sovereignty will then be transferred to China, under the "one country, two systems" principle (see Macau).

[24] In 1988, after over a month of negotiations, the PSD and the Socialist Party agreed to modify the constitution to allow the re-privatization of various companies nationalized during the "Revolution of the Carnations" and to further reduce presidential powers. Mario Soares, former leader of the Socialist Party and current president of the Republic, opposed these reforms. This led to his split with the current party leadership and a permanent rivalry with Prime Minister Aníbal Cavaco Silva.

[25] Portuguese politics became polarized between the ruling PSD and the PS. The latter was a more viable left-wing alternative after the collapse of real socialism. However-

STATISTICS

DEMOGRAPHY

Urban: 35% (1995). **Annual growth:** -0.6% (1991-99). **Estimate for year 2000:** 10,000,000. **Children per woman:** 1.5 (1992).

HEALTH

Under-five mortality: 11 per 1,000 (1995). **Calorie consumption:** 111% of required intake (1995).

EDUCATION

School enrolment: Primary (1993): 118% fem., 118% male. University: 23% (1993). **Primary school teachers:** one for every 14 students (1992).

COMMUNICATIONS

105 **newspapers** (1995), 102 **TV sets** (1995) and 93 **radio sets** per 1,000 households (1995). 31.1 **telephones** per 100 inhabitants (1993). **Books:** 105 new titles per 1,000,000 inhabitants in 1995.

ECONOMY

Per capita GNP: $9,320 (1994). **Annual growth:** 4.00% (1985-94). **Annual inflation:** 12.00% (1984-94). **Consumer price index:** 100 in 1990; 122.1 in 1994. **Currency:** 159 escudos = 1$ (1994). **Cereal imports:** 2,147,000 metric tons (1993). **Fertilizer use:** 813 kgs per ha. (1992-93). **Imports:** $26,680 million (1994). **Exports:** $17,540 million (1994).

ENERGY

Consumption: 1,828 kgs of Oil Equivalent per capita yearly (1994), 90% imported (1994).

er, in the October 1991 parliamentary elections, the PSD won over 50 per cent of the vote while the PS did not reach 30 per cent. The political victory of Cavaco Silva was due to the social democratic message with

The country remained neutral during the Spanish Civil War and World War II which could have jeopardized the country's barely stable economy.

which he disguised his orthodox liberal economic orientation. The right-wing Social Democratic Center received a mere 4 per cent which led its leader Diego Freitas do Amaral to resign. Support for the

Communist Party dropped 3 per cent since 1987.

[26] In January 1992, Portugal took over the rotational presidency of the European Community. The new president, Luis Mira de Amaral, Portuguese minister of industry and energy, announced he would promote industrial cooperation with Latin America, Africa and central Europe. Another of his priorities was the signing of the Maastricht Treaty, which entailed political, economic and monetary union between the members of the Community.

[27] In August 1993, the Assembly restricted the right to seek asylum and enabled the expulsion of foreigners from the country. The legislation was based on the defence of the job market and was opposed by President Soares. Unemployment reached 8 per cent but it was set to increase with the privatization or closing of state-run airlines, shipyards and steel industries.

[28] Meanwhile a plan financed by the European Union was approved for the 1993-1997 period for the poorest members, including Portugal, providing investment in education, transport, industrial reconversion and job creation.

[29] Portugal's population numbered 10 million in 1993, with 4 million workers abroad.

[30] Portuguese politics were rocked by intense student protests against the cost and quality of education, and by strikes for higher wages in the public sector. The 50 per cent increase in the toll charged at Lisbon's access bridge cause several blockades by transport workers. The government justified the raise citing the need to finance a new bridge for the last world Expo of the century, to be held in 1998.

[31] The equilibrium between socialist president Mario Soares and centre-right prime minister Cavaco Silva was broken. Soares denounced the excessive power in the hands of the prime minister and warned about the dangers of a "dictatorship of the majority". Prior to the elections, the PSD privatized 28 per cent of Portugal Telecom and 40 per cent of Portucel Industrial.

[32] The October 1995 general elections were won by the Socialist Party which gained an absolute majority at the Assembly. Antonio Guerres was appointed prime minister, replacing Anibal Cavaco. After 10 years of PSD dominance, oriented toward European integration and economic liberalism, the PS capitalized on domestic discontent with education and health and assured the financial market it would not interfere with the goals regarding monetary union and privatization.

[33] Socialist candidate Jorge Sampaio won the 1996 presidential elections with 54 per cent of the vote followed by 46 per cent from former prime minister Cavaco Silva. He took office in March.

Puerto Rico

Population: 3,651,000 (1994)
Area: 8,900 SQ.KM
Capital: San Juan

Currency: US Dollar
Language: Spanish and English

Puerto Rico

In 1508, fifteen years after Christopher Columbus had arrived on the Caribbean island of Borinquen, that territory came under colonial domination, a status which remains unchanged to this day. Due to its strategic location at the entrance of the Caribbean - Puerto Rico is the easternmost of the Greater Antilles - the island suffered 400 years of Spanish rule, as well as repeated attacks of pirates and regular naval forces, whether British, Dutch or French, plus US administration after the SpanishAmerican War of 1898.

[2] In the first few decades of the 16th century, a sugar-based economy began to take shape. According to a document of the period, by 1560 there were already 15,000 slaves on the island. This was however a short-lived period in the island's history: in the latter part of the century leather became the principal product, a repetition of what had occurred some time before in Santo Domingo.

[3] As in neighboring islands, the native Taino people were exterminated by war, disease and overwork. African slaves were brought in to take their place in the fields where most food supplies for Spanish expeditions into the mainland were produced. Thus, Puerto Rican culture is a blend of its African and Spanish heritage.

[4] Spanish rule was continually challenged by external attacks and rebellions by both the Tainos and enslaved African workers. The latter rebelled successively in 1822, 1826, 1843, and 1848. The struggle for independence in the rest of Latin America had its counterpart in Puerto Rico's struggle for administrative reform (1812-1840), but Spanish troops ruthlessly stifled the uprising.

[5] In 1868, five years before slavery was finally abolished, a group of patriots led by Ramón Emeterio Betances moved things a step nearer to liberation. In the town of Lares they proclaimed Puerto Rico's independence and took up arms to free the island. Despite their defeat, the Lares revolt signalled the birth of the Puerto Rican nation.

[6] The independence movement continued to gain strength in the following years. In 1897, the Cubans were already up in arms led by José Martí in a movement that reached Puerto Rico. US military intervention in the war against Spain, in 1898, hastened European

defeat, but for Puerto Rico it only meant the imposition of a new ruler.

[7] US colonial administrations, first military and then civilian, imposed English as the official language and attempted to turn the island into a sugar plantation and a military base. Puerto Ricans were made US citizens in 1917, though they were given no participation in the island's government. As a result, resistance to colonial rule continued to increase in strength. In 1922 the pro-independence Nationalist Party (PN) was founded. PN-led uprisings in 1930 and 1950 were fiercely put down.

[8] The PN leader from 1930 until his death in 1965 was Pedro Albizu Campos, who suffered exile and imprisonment for his anticolonial activities.

[9] In 1947, intense internal and international pressure forced the US to allow Puerto Rico to elect its own governor. The 1948 elections gave the post to Luis Muñoz Marín, leader of the Popular Democratic Party (PPD), who favored turning the country into a self-governing, commonwealth, or free associated state. The US government authorized the drafting of a constitution

in 1959, which was approved by a plebiscite and later ratified by the US. Muñoz Marín's program was thus sanctioned.

[10] Commonwealth status, still in effect today, leaves defence, financial affairs and foreign relations to Washington, while maintaining common citizenship and currency, as well as free access to the US for Puerto Ricans and vice versa.

[11] With the institution of Commonwealth status, US administrations were freed from the obligation of reporting on Puerto Rico's status to the UN Decolonization Committee. Moreover, in this way the UN tacitly endorsed the arrangement declaring the "end" of colonial rule. Nevertheless, in September 1978, the Decolonization Commission reconsidered the situation, and in December of the same year a UN General Assembly resolution defined Puerto Rico as a colony once again and demanded self-determination for its people.

[12] During World War II, Puerto Rico was again turned into a military garrison for controlling the Caribbean. The US built seven bases on the island.

[13] Muñoz Marín promoted industrialization on the island, through

massive US private investment enticed by government tax incentives. During the 50s, Puerto Rican agriculture was destroyed and farmers uprooted by an influx of US products, resulting in over 50 per cent of the island's food consumption being imported. The newly-formed labor reserve was more than enough to supply cheap hands for the growing US corporate community, and Puerto Ricans soon began migrating en masse to the US, and especially to New York, in search of work. The 1980 US census registered over two million Puerto Ricans living in the United States.

[14] With the great social upheavals of the 1960s, the struggle for independence flowered anew on the island. Despite the revival of the independence movement, a 1967 plebiscite confirmed the Commonwealth. Moreover, elections the following year gave the governorship to the New Progressive Party (PNP) which favored making the island the 51st state of the US. Nevertheless after the 1972 elections the PPD returned to power led by Rafael Hernández Colón. In 1976, supporters of the statehood option returned to the governorship. Carlos Romero Barceló, announced that if

ENVIRONMENT

The smallest and easternmost island of the Greater Antilles. A central mountain range, covered with rainforests, runs across the island. In the highlands, subsistence crops are grown (corn, manioc); on the western slopes there are large coffee plantations and in the central region small tobacco farms. On the northern slopes citrus and pineapples are grown, exported to the United States. The main crop is sugar cane which uses the best farmlands along the coastline. The islands of Vieques (43 sq km), Mona (40 sq km) and Culebra also belong to Puerto Rico.

SOCIETY

Peoples: Most Puerto Ricans are descendants of African slave workers and Spanish colonizers. In addition, about 3 million Puerto Ricans have emigrated to the United States to escape unemployment and poverty. **Religions:** Mainly Catholic. **Languages:** Spanish and English; in all relations with the United States, English is used. Compulsory introduction of English in education, administration and communications was supported when in 1993 it was declared joint official language together with Spanish. **Political Parties:** the New Progressive Party (PNP) advocates total integration to the United States as the 51st state of the Union; the Democratic People's Party (PPD), founded in 1938 by Luis Muñoz Marín, supports the current "Commonwealth" status; the Puerto Rican Renewal Party (PRP), a split of the PNP; the Puerto Rican Independence Party (PI), founded in 1946 from a breakup of the PPD; the Communist Party (PC), founded in 1934.

THE STATE

Official Name: Estado Libre Asociado de Puerto Rico. **Capital:** San Juan, 1,085,000 inhab. (1980). **Other cities:** Ponce, 189,046 inhab.; Bayamon, 196,206 inhab.; Caguas, 117,959 inhab. (1980). **Government:** Pedro Rosselló, Governor, since January 2, 1993. **National Holiday:** September 23, the Battle of Grito de Lares (start of the anti-colonial armed revolt in 1868).

DEMOGRAPHY

Urban: 73% (1995). **Estimate for year 2000:** 4,000,000.

COMMUNICATIONS

103 **newspapers** (1995), 104 **TV sets** (1995) and 108 **radio sets** per 1,000 households (1995). 33.5 **telephones** per 100 inhabitants (1993).

ECONOMY

Annual growth: 1.60% (1985-94).
Consumer price index: 100 in 1990; 113.0 in 1994.
Currency: US dollars.

---

he was re-elected for a second term, he would call a pro-statehood referendum. Romero Barceló went on to win in 1980, but by such a slim margin that plans for a statehood plebiscite were abandoned, despite encouragement from President Reagan.

[15] Puerto Rico has one representative in the US Congress, but with no voting rights other than in committees. US citizenship only gave Puerto Ricans the right to participate in the 1980 presidential elections, although residents in the US are able to vote in all elections.

[16] Rafael Hernández Colón was elected president on November 6, 1984. He promised a "four-year term of struggle against corruption and unemployment". He renewed Puerto Rico's Commonwealth status thus rejecting his predecessor's intention to integrate into the Union.

[17] Hernández Colón was re-elected in November 1988, with 48.7 per cent of the vote, against 45.8 per cent for those in favor of annexation by the United States, and 5.3 per cent for those who favor independence.

[18] In 1989 the Special United Nations Decolonization Committee expressed its wish that the people of Puerto Rico exercise their right to self-determination and independence. The resolution stressed the "clearly Latin American character and identity of Puerto Rico's people and culture".

[19] In April 1991, governor Hernández Colón passed a law granting official status for the Spanish language. A few weeks later, the Puerto Rican people were granted the Prince of Asturias award by the Spanish crown, "in recognition of the country's efforts to defend the Spanish language".

[20] In the plebiscite carried out late in 1991, various strategies were proposed to promote development on the island. The governor succeeded in rallying moderate nation-

> Despite the revival of the independence movement, a 1967 plebiscite confirmed the Commonwealth.

alists and supporters of independence, who campaigned together. They were in favor of self-determination, the end of subjection to US jurisdiction, the affirmation of Puerto Rican identity, regardless of any future referendum decisions, and the maintenance of US citizenship. However, all this effort came to nothing in the polls, when 55 per cent of the voters supported the PNP's position that a break with Washington had to be avoided.

[21] Given this result, Hernández's position within his party was weak-

ened, strengthening the hand of more pro-independence sectors and Hernández finally resigned the Popular Democratic Party leadership.

[22] Pedro Rosselló, a supporter of Puerto Rico's annexation by the United States, was elected governor in 1992. His plan to transform English into the only official language on the island - replacing Spanish - caused massive protest demonstrations. Finally in 1993 English was made an official language alongside Spanish.

[23] In November, a referendum was held to decide on the political future of the island. Those wanting to maintain the "free associated state" status won a narrow victory, with 48.4 per cent of the vote, while the group supporting the transformation of Puerto Rico into the 51st US state took 46.2 per cent. Independence supporters had a mere 4 per cent.

[24] Some pro-independence supporters had agreed to collaborate - in sectors like culture - with the "pro-annexation" cabinet led by the present governor. However, when Rosselló sacked the director of the Puerto Rico Institute of Culture, Awilda Palau, in May 1995 the possibilities of collaboration were reduced.

Qatar

Population: 610,000 (1994)
Area: 11,000 SQ.KM
Capital: Doha (Ad-Dawhah)

Currency: Riyals
Language: Arabic

Qatar

In a small desert country without a single river, 400 farms are now producing almost all the foodstuffs required for domestic consumption. The miracle that changed the desert into a kitchen garden was the advent of oil revenues. Half a million barrels of oil per day, and the successful policies of Amir Khalifa bin Ath-Thani's government, have helped bring about these achievements.

[2] Like the neighboring island of Bahrain, the Qatar peninsula has participated from ancient times in the active Gulf trade between Mesopotamia and India. Islamized in the 7th century (see Saudi Arabia), at the time of the caliphate of Baghdad, Qatar had already obtained autonomy which was maintained until 1076 when it was conquered by the Emir of Bahrain.

[3] From the 16th century, after a brief period of Portuguese occupation, the country lived in great prosperity due to the development of pearl fishing attracting immigrants. Settled on the coasts, under the leadership of the Al-Thani family, these settlers succeeded in politically unifying the country in the 18th century, though it remained subject to Bahrain's sovereignty. The process of independence, begun in 1815 by Sheikh Muhammad and his son Jassim, culminated in 1868 with the mediation of the English; the Al-Thanis agreed to end the war in exchange for guaranteed territorial integrity.

[4] The Turkish sultans, nominal sovereigns of the entire Arabian peninsula since the 16th century, did not look favorably upon increasing British penetration in the Gulf. Consequently, they named the reigning sheikh (Jassim Al-Thani) governor of the "province" of Qatar as a pretext to establishing a small military garrison in Dawhah

(Doha). Neither the Qataris nor the British concerned themselves over this formal affirmation of sovereignty and the garrison remained until World War I without the slightest effect on British influence in the region.

[5] In 1930, the price of pearls dropped with the Japanese flooding the market with a cheaper version of cultivated pearls.Consequently Sheikh Abdullah sold all the country's oil prospecting and exploitation rights, and granted a 75-year lease on its territorial waters, for £ 400,000. The Anglo-Iranian Oil Co. discovered oil in 1939 but actual production only began after World War II, attracting other companies who purchased parts of the original concession. The immense wealth obtained by the Royal Dutch Shell did not seem to concern Sheikh Ahmad bin Ali Al-Thani as oil and tariff revenues increased his personal fortune by £ 15 million.

[6] Shortly thereafter, Ahmad was ousted by his own family, who replaced him with his cousin Khalifa and giving him the task of "removing any elements which are opposed to progress and modernization".

[7] Sheik Khalifa created a Council of Ministers and an Advisory Council to share the responsibilities of his absolute power and promised "a new era of enlightened government, social justice and stability". Redistributing the oil revenues, he exempted all inhabitants from taxation and provided free education and medical attention. His greatest achievement has perhaps been the subsidizing and promotion of productive activities not connected with oil prospecting, activities which were symbolically expressed by the first tomato exports at the beginning of his mandate.

[8] To reduce Qatar's dependency on a single product, the fishing industry was revitalized, industrialization was accelerated with new cement and fertilizer plants, plus iron and steel mills being built. The country took advantage of its strategic position to provide commercial and financial services to the economies of the entire region. In addition, a large part of the country's financial surplus was invested abroad (in Europe and the US). In 1980, estimated income from this "exportation of capital" will eventually equal all oil revenues. In this way Qatar seeks to ensure its future when the oil wells run dry.

[9] As an indispensable element of control, a state oil company, the Qatar Petroleum Producing Authority (QPPA), was set up in 1972, and by February 1977, all foreign oil installations had been expropriated.

[10] Qatar's economic expansion required the large-scale immigration of foreign technical experts and workers (the former being European and American; the latter Iranian, Pakistani, Indian and Palestinian). Estimates as to the actual number of immigrant workers vary widely, since surveys are not taken, but the figure is believed to be approximately 150,000 people, or about 60 percent of the total population.

[11] To avoid any profound transformation of the local culture, the government prefer and promote immigration from other Arab countries. Nevertheless, the advanced systems of social security also protect foreign workers. The local Iranian community lives in better conditions than in any other Gulf emirate and, because of this, Qatar has had fewer conflicts with the Khomeini

regime than any other country in the region.

[12] Adhering to OPEC policy, in 1982 the country decreased crude oil production by 25 percent. Consequently, exports decreased, reflected in a considerable reduction in volume of petrodollars invested in the West, in public spending and in industrial expansion. Nevertheless, Qatar has the infrastructure to face the change. In the industrial center of Umm Said the iron and steel plant was producing 450,000 tons a year at the beginning of the decade. A liquid gas plant is also in full operation and the government decided to go ahead with a six-billion-dollar natural gas project for use in its energy and desalinization programs.

[13] Qatar's geographical position has brought the country both economic benefits and geopolitical worries. Israeli aggression against Lebanon, Syria and Iraq, the lack of solutions to the Palestinian problem, the war between Iran and Iraq and the Reagan administration's frightening interest in the region's "military security" all

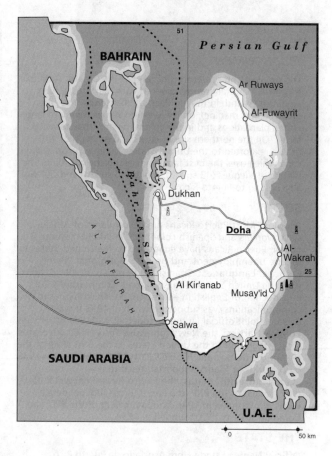

WORKERS

1994

- ■ MEN 93%   ■ WOMEN 7%

- ■ AGRICULTURE 3%   ■ INDUSTRY 28%   ■ SERVICES 69%

PROFILE

ENVIRONMENT

The country consists of the Qatar Peninsula, on the eastern coast of the Arabian Peninsula in the Persian Gulf. The land is flat and the climate is hot and dry. Farming is possible only along the coastal strip. The country's main resource is its fabulous oil wealth on the western coast. Qatar suffered the negative effects of the burning oil wells during the Gulf War.

SOCIETY

Peoples: The native Qatari Arab inhabitants account for 20% of the population. Arabs, though, are a majority thanks to Palestinian, Egyptian and Yemenite immigrants, who account for 25% of the population. The remaining 55% are immigrants, mostly from Pakistan, India and Iran.
Religions: Muslim (official and predominant). The majority are Sunni, with Iranian immigrants mainly Shiite. There are also Christian and Hindu minorities. **Languages:** Arabic (official and predominant). Urdu is spoken by Pakistani immigrants, and Farsi by Iranians. English is the business language. **Political Parties:** There are no organized political parties.

THE STATE

Official Name: Dawlat Qatar. **Capital:** Doha (Ad-Dawhah), 236,000 inhab. (1987). **Other cities:** Rayyan, 100,00 inhab., Wakrah, 26,000 inhab., Umm Salal, 12,100 inhab. (1987). **Government:** Hamad ibn Khalifa ath-Thani, Emir and Head of State since june 1995. Legislative Power: a Consultative Council with 30 members appointed by the Emir to 3 year-terms. **National Holiday:** September 3, Independence Day (1971). **Armed Forces:** 11,100 (1995)

STATISTICS

DEMOGRAPHY

Annual growth: 0.4% (1992-2000). **Children per woman:** 4 (1992).

HEALTH

Under-five mortality: 25 per 1,000 (1994).

EDUCATION

Illiteracy: 21% (1995). **Primary school teachers:** one for every 11 students (1991).

COMMUNICATIONS

104 **newspapers** (1995), 114 **TV sets** (1995) and 108 **radio sets** per 1,000 households (1995). 21.4 **telephones** per 100 inhabitants (1993). **Books:** 110 new titles per 1,000,000 inhabitants in 1995.

ECONOMY

Per capita GNP: $12,820 (1994).
Annual growth: -2.40% (1985-94).
Consumer price index: 100 in 1990; 104.4 in 1991.
Currency: 4 riyals = 1$ (1994).

helped to convince Qatar's rulers of the need to enter new alliances with their neighbors on the Arabian Gulf.
[14] Since 1981, together with Bahrain, Kuwait, Oman, the United Arab Emirates and Saudi Arabia, Qatar has participated in the Gulf Cooperation Council, an organization designed to coordinate the area's policies in political, economic, social, cultural and military issues. Qatar strongly supported Saudi Arabia's stand regarding the need to reduce and, if possible, eliminate US troops stationed in the region, mainly in Oman.
[15] In April 1986, tensions flared between Bahrain and Qatar, over the artificial island of Fasht ad-Dibal. This conflict was resolved through negotiations sponsored by members of the Gulf Cooperation Council. In November 1987, the government renewed diplomatic relations with Egypt.
[16] In March 1991, after the Iraqi invasion of Kuwait, the Gulf Cooperation Council, to which Kuwait and Qatar belong, suspended all economic aid to Jordan and the PLO.
[17] Also in March 1991, the Foreign Affairs ministers of Egypt, Syria, and the six Arab countries which belong of the Cooperation Council gathered in Riyadh, the capital of Saudi Arabia, to sign an agreement with the US to preserve security in the region. The plan included four points; a common military strategy between the US and the Arab countries in the anti-Iraqi coalition; mechanisms to avoid arms proliferation; acceptance of a peace treaty by Israel; and a new economic program for the development of the region.
[18] During the Gulf War, oil wells set ablaze in Kuwait and Saudi Arabia filled the air over Qatar with thick smoke. Estimated pollution problems include acid rain and climatic disturbances in the area. Health experts stated that burning oil wells cause a high concentration of sulphur dioxide and soot particles that cause asthma and bronchitis, and increase the risk of cancer.
[19] In September 1991, private European companies began exploiting huge gas deposits along the coasts of Qatar.
[20] In December, a group of 53 people, among them some government officials, drew up and signed a petition addressed to the Emir of Qatar, asking for free parliamentary elections, a written constitution and increased personal and political freedoms. As a result, several citizens were arrested and detained in the Doha Central Prison; they were eventually released several months later.
[21] Toward the end of 1992, Qatar and Bahrain became involved in another territorial dispute over several small islands off the coast of the two emirates: Howard Island, and more especially, underground rights to Dibval and Qitat, both potentially rich in oil. Growing tension between the two countries prompted Saudi Arabia to step in and mediate the conflict.
[22] In 1993, the fall in the price of oil in the international market triggered an almost 20 percent decrease in fiscal income. Furthering his program to develop alternative income sources, in 1994 Doha negotiated new agreements with several Asian companies for the exploitation of natural gas.
[23] In June 1995, the prince's heir Hamad ibn Khalifa ath-Thani overthrew his father in an unbloody coup and became the new Emir of Qatar. The new sovereign promised to intensify his efforts to solve the territorial disputes with Saudi Arabia and Bahrain, and -after deciding a sudden withdrawal of his country representation from the Gulf Cooperation Council in December- he cast some doubts on a possible withdrawal of Qatar from that organization.

Réunion

Population: 640,000
(1994)
Area: 2,510 SQ KM
Capital: Saint Denis

Currency: French Franc
Language: French

Réunion

The island of Réunion was uninhabited until the beginning of the 17th century when Arab explorers arrived and called it Diva Margabin. The Portuguese renamed it Ilha Santa Apolonia and the French settlers called it Bourbon, and after the French revolution it was given its current name: Réunion.

[2] The colonial regime replaced subsistence farming with the export commodities of sugar cane and coffee, and the island's fishing resources were depleted because of the over-exploitation of several species.

[3] The island's four renamings reflect the struggle between the colonial powers for possession of this strategically important point in the Indian Ocean. After decades of fighting, the French finally retained colonial rights over Réunion

[4] Since March 1946, Réunion has been considered an Overseas Department by the French government. The local middle-class and the French colonists supported the integration of the island into French territory. Later, conservative political changes in France brought realignment of political forces on the island.

[5] While sectors tied to colonial interests defended the Department status, the Réunion Communist Party (PCR) changed its stance, and in 1959 campaigned for partial autonomy. The PCR suffered the consequences of its about face, and for ten years was the sole supporter of gradual independence.

[6] In July 1978, the UN Decolonization Committee issued a pronouncement in favor of full independence for Réunion The question of independence was now transformed into an international issue.

[7] In 1978, the independence movement made important international contacts in a meeting with other anti imperialist and anti colonialist organizations in the Indian Ocean area. At this meeting, a Permanent Liaison Committee was formed in a joint effort against foreign domination. In June 1979, during elections for the European Parliament, Paul Vergés raised the issue of double colonialism, charging that Réunion Island was not only dominated by the French, but forced to serve the interests of the entire European Community.

[8] The economy deteriorated. There was a crisis in production, unemployment increased, and small and medium-scale industries were forced to cut back because of the general situation and a sluggish domestic market. The situation was aggravated by the island;183;s population explosion: in the past ten years the population increased by 20 per cent.

[9] In 1989 there was a record number of exports, but the balance of trade deficit increased, and unemployment hit 37 per cent of the active population. French subsidies, essential for economic survival, also weaken independence movements.

[10] Street riots became frequent as social conditions continued to deteriorate. 8 people died in 1991 when they were taking part of demonstrations against the banning of a pirate television channel, Tele Free-DOM, headed by popular Camille Sudre. To oppose the closing, Sudre - a Frenchman residing in Réunion - formed a political list in a record time and won the 1992 regional elections. The CPR's support enabled him to become the region's president, one of the island's two principal executive authorities.

[11] However, the French top administrative tribunal invalidated the elections due to alleged irregularities and banned Sudre's future nomination. The Free-DOM list - headed by Marguerite Sudre, Camille's wife - won in new regional elections in 1993. The new president had a rapprochement with the French right-wing and was appointed minister of Edouard Balladur's (1993-1995) conservative cabinet in Paris.

[12] Social tension kept mounting on the island, where it is estimated unemployment affects over 40 per cent of the active population. Amoung young people, the rate is even higher.

> The island's four renamings reflect the struggle between the colonial powers for possession of this strategically important point in the Indian Ocean.

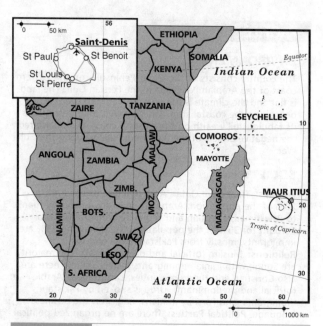

PROFILE

ENVIRONMENT

Located in the Indian Ocean, 700 km east of Madagascar, Réunion is a volcanic, mountainous island with a tropical climate, heavy rainfall and numerous rivers. These conditions favor the growth of sugar cane, the main economic activity.

SOCIETY

Peoples: Mostly of African descent: 20% are of French, Malayan, Indians and Chinese origin. **Religion:** Mainly Catholic. Muslim, Hindu and Buddhist minorities. **Languages:** French (official) and Creole. **Political Parties:** Free-DOM Movement; Communist Party of Reunion; Union for the Republic (RPR); Union for French Democracy (UDF); Socialist Party.

THE STATE

Official Name: Département d'outre-mer de la Réunion. **Administrative Divisions:** 5 Arrondisements. **Capital:** Saint Denis, 122,000 inhab. (1990). **Other cities:** St Paul, 72,000 inhab., St Pierre 59,000 inhab. (1990). **Government:** Prefect appointed by the French government. There are two local councils: the 47-member General Council, and the 45-member Regional Council. The island has 5 representatives and 3 senators in the French parliament. **National Holiday:** December 20, Abolition of slavery (1848). **Armed Forces:** 4,000 French Troops.

COMMUNICATIONS

32.0 **telephones** per 100 inhabitants (1993).

ECONOMY

Consumer price index: 100 in 1990; 113.5 in 1994. **Currency:** french francs.

Romania

România

Population: 22,731,000 (1994)
Area: 237,500 SQ KM
Capital: Bucharest
Currency: Lei
Language: Romanian

The earliest inhabitants of Romania included the Thracians, whose descendants, known as the Getae, established contact with Greek colonies that appeared on the shore of the Black Sea in the 7th century BC. Together with the Dacians, a related people living in the Carpathian Mountains and in Transylvania, the Getae established a distinct society by the 4th century BC.

[2] In the first century BC, the Geto-Dacians bitterly resisted conquest by the Romans, who were interested in the region's mineral wealth. Rome finally triumphed over the powerful Dacian kingdom in 106 AD, putting its inhabitants to death, or expelling them to the north. According to some Roman sources, most of the males of the conquered area who could not run were put to death or brought to Rome as slaves. However, many Dacians ran away from the center of the former kingdom, into vast areas that belonged to the so-called free Dacians, at north, east and north-east of the Roman province of Dacia.

[3] The province was then subdivided into Dacia Superior, Dacia Inferior and Dacia Porolissensis. Emperor Marcus Aurelius abandoned Dacia Superior and Inferior during the period between 271-275. Unwilling to recognize before the Senate that he withdrew from such important provinces, he reorganized the province of Moesia Superior at the south of Danube into Dacia Ripensis and Dacia Mediterranea (to keep the name Dacia). The center and east of present Walachia, southern Moldova and Dobrudja were at that time part of the Roman province of Moesia Inferior, which he still occupied. The Roman rule left an enduring legacy in the Romanian language, which is derived from Latin; this rule continued on the territory of present Romania until the 7th century, when the Byzantin Empire lost its last northern Danubian srongholds in favour of the migratory peoples.

[4] Between the 3rd to the 12th century, the region underwent successive invasions by Germanic peoples, Slavs, Avars, and others. Roman Christianity was brought by Romans during the 2nd and 3rd centuries and was generalized on the whole region by the 4th century. The Romanian-Bulgarian Empire, which lasted over two hundred years, introduced Greek Orthodox Christianity through the north of the Danube. The Romanian Church adopted the Slavonic language due to the Bulgarian influence. Towards the end of the 9th century, the Bulgarians were expelled by the Magyars from the Pannonic plains (nowadays Hungary) though they remained in what is present-day Romania.

[5] Transylvania, the cradle of the Romanian nation, was conquered by Hungary during the 10th to the 12th century. According to Hungarian sources (Gesta Hungarorum) the existing feudal states in Transylvania put up fierce resistance. The Vlachs, from Transylvania, reappeared in the 13th century to the south of the Carpathian Mountains, in two separate regions, Walachia and Moldova; inlanders inmediately accepted the newcomers, as people with the same origins. The Tatar-Mongol invasion of 1241 produced big losses among the Romanians.

[6] The first Romanian state, Walachia, was established south of the Carpathians during the early 14th century, and a second, Moldova, was founded in 1349 east of the Carpathians in the Spruth River valley.

[7] The principalities of Walachia and Moldova fought for their independence from Hungary, which after conquering Transylvania, tried unsuccesfully to conquer them also. This struggle ended in the first half of the 14th century, when Hungarian invaders were defeated both in Walachia and Moldova. By the 15th century the Ottoman Empire began to be a greater threat.

[8] After defeating the Serbs in Kosovo in 1389, the Ottoman Empire began closing in on Walachia, with pressure intensifying after the fall of Bulgaria in 1393. Walachia became a vassal state of Sultan Mehmed I, in 1417, though prince Mircea maintained the claim to the throne and the Christian religion remained intact. In 1455 Moldova also became a vassal state.

[9] King Mircea's death in 1418 was followed by a rapid succession of princes, until the Turks appointed a Romanian prince of their choice to the throne. Walachia put up resistance, but after the Hungarian defeat at the battle of Mohacs in 1526, Turkish domination became inevitable. The Hungarian kingdom disapeared and was transformed by the Ottomans into a Pashalic. The three Romanian principalities, Walachia, Moldova and Transylvania entered into the sphere of influence of the Ottomans, but kept some indepence by paying a tribute in money.

[10] In 1594, the Turkish inhabitants of Walachia were massacred by Prince Michael, in alliance with Moldova. He went on to invade Turkish territory, taking several key sites along the banks of the Danube. Faced with the collapse of his counter-offensive, the sultan had no choice but to recognize the sovereignty of Walachia, which subsequently became linked to Transylvania.

[11] Five years later, in 1599, Segismund Bathory of Transylvania abdicated and Michael dethroned Andreas, Segismund's successor. The Vlach peasants in Transylvania rose up against Hungary. Walachia, not wishing to break the alliance, helped put down the rebellion.

[12] In 1600, Michael conquered Moldova, proclaiming himself regent. The Austrian Emperor Rudolf II recognized the claim, although he later tried to take over Transylvania and Moldova. Michael the Brave is a national hero because he was the first and the last to unite the Romanians until the modern reunification in 1918.

[13] In the 17th and 18th centuries, Walachia and Moldova fell under Turkish rule again. The Sultans did not trust the native Romanian princes, so they named either Greek princes from Romania, or Romanian princes from the Greek quarter in Istanbul, Fanar, to rule Walachia and Moldova. Russia occupied the region in 1769, but Austria forced Russia to return the principalities to the sultan in 1774.

[14] Russia's power over the area gradually increased, until it included the right to designate the princes. The sultan ousted these princes, and in 1806, Russia retaliated by invading the region. Under the Peace of Bucharest in 1812, Russia retained the southeastern part of Moldova - Bessarabia.

[15] The Walachian prince, Ion Caragea, was linked to "Philiki Etaireia", a Greek revolutionary movement sponsored by Russia. Alexander Ypsilantis, Prince Konstantinos's son and the czar's aide entered Moldova in 1821, leading the Etaireians, and provoking the Turks.

[16] The sultan managed to divide the Romanians and the Greeks by allowing the principalities to pass laws in their own languages and elect native-born princes. Ion Sandu Sturza and Grigore IV Ghica assumed control of the government of Moldova and Walachia, respectively, and both moved towards Greece and Russia.

LAND USE

CROPLAND 43%   PASTURE 21%   OTHER 36%

ENVIRONMENT

The country is crossed from north to center by the Carpathian Mountains, the western part of which are known as the Transylvanian Alps. The Transylvanian plateau is contained within the arc formed by the Carpathian Mountains. The Moldavian plains extend to the east, while the Walachian plains stretch to the south, crossed by the Danube, which flows into a large delta on the Black Sea. The mountain forests supply raw material for a well developed timber industry. With abundant mineral resources (oil, natural gas, coal, iron ore and bauxite), Romania has begun extensive industrial development. Its economy, still depends to a great extent on the export of raw materials and agricultural products. Romania defines itself as a "developing country", and is attuned with Third World demands. It is also one of Europe's largest oil producers. Copsa Mica, in the center of Romania, is considered to be one of the areas with the highest levels of industrial pollution in Europe.

SOCIETY

Peoples: Mostly Romanian, the population also includes sizeable minorities of Hungarian, 9%, Romany, 7%, German, Ukrainian, Turkish, Greek and Croatian origin. **Religions:** Mainly Greek Orthodox. There are Catholic and Protestant minorities. **Languages:** Romanian (official language, spoken by the majority); ethnic minorities often speak their own languages, particularly Hungarian and Romany. **Political Parties:** Democratic Front for National Salvation; Hungarian Democratic Union of Romania (HDUR); National Liberal Party; National Unity Party (PUNR). **Social Organizations:** General Confederation of Trade Unions.

THE STATE

Official Name: Románia. **Capital:** Bucharest, 2,066,000 inhab. (1993). **Other cities:** Timisoara, 325,000 inahb.; Constanza, 350,000 inhab.; Iasi, 338,000 inhab. (1993). **Government:** Ion Iliescu, President and Head of State. Nicolae Vacaroiu, Prime Minister and Head of Government. Legislature, bicameral: Senate, 143 members; Deputies, 341 members, 13 representing ethnic minorities. **Administrative Divisions:** 40 Districts and the Municipality of Bucharest. **Armed Forces:** 217,400 (1995). **Paramilitaries:** 43,000 Border Guard, Gendarmes, Construction Troops.

[17] Another war broke out between Russia and Turkey in 1828. The following year, the Peace of Adrianopolis maintained the principalities as tributaries of the sultan, but under Russian occupation. Russian troops remained in the region, and the princes began to be named for life.

[18] The local nobility drew up a constitution known as the "Reglement organique", which was passed in Walachia in 1831 and in Moldova, in 1832. This established administrative and legislative bodies made up of natives of the principalities. In 1834, after the sultan's approval of the "Reglement", Russia withdrew.

[19] During the European revolutions of 1848, nationalist sentiment in Moldova and Walachia was stimulated by peasant rebellions. These reached a climax in May with the protests at Blaj, which were put down by Turkish and Russian troops, restoring the "Reglement organique". The revolutions were defeated with foreign aid, Ottoman in Walachia, Russian in Moldova, and Austrian and Russian in Transylvania.

[20] During the Crimean War, the three Romanian principalities were occupied alternately by Ottoman, Russian and Austrian troops. The Treaty of Paris, 1856, maintained the ancient statutes of the principalities until its revision by a European commission in Bucharest in 1857, with delegates representing the sultan and both principalities.

[21] The local delegates made the following proposals: that the provinces be autonomous, joining together under the name "Romania"; that a foreign king be elected, with the right to hereditary succession; and that the country be neutral. In August 1858, despite the sultan's opposition, the Treaty of Paris created a commission to carry out the unification process.

[22] In 1859, the principalities elected a single prince, Alexandru Ion Cuza, who was recognized by the major powers and by the sultan in 1861. The Constitution of 1864 established a bicameral legislative body, granting property holders greater electoral power.

[23] When the war between Russia and Turkey resumed in 1877, Romania authorized the transit of Russian troops through its territory in April. Moscow declared war on Turkey in May. After suffering heavy losses at Pevna, the Russians asked Prince Carol I of Romania for military support. Romanian troops contributed to the Russian victory against the Turks.

[24] The 1878 Treaty of Berlin respected Romania's independence, but failed to return Bessarabia to Romania, instead giving it Dobrudja - without its southern part - and the Danube Delta. In 1881 Romania became a kingdom.

[25] When the Balkan War broke out in 1912, tension from territorial disputes in past wars, persisted between Romania and its neighbors. After the first few battles, Bucharest demanded a ratification of its borders in Dobrudja. The St Petersburg Conference of 1913 gave Romania Silistra, much to Bulgaria's displeasure.

[26] Romania took advantage of the second Balkan War of 1913 to shore up its position. The Treaty of Bucharest gave Romania the southern part of Dobrudja, which was occupied by Romanian troops. At the beginning of World War I, Romania wavered between taking Bessarabia or Transylvania, finally opting for the latter.

[27] In 1916, Romania allied itself with Britain, France, Russia and Italy, declaring war on Austria and Hungary. After the occupation of Bucharest, King Ferdinand, and the Romanian government and army took refuge in Moldova. The defeat of the Central Powers in 1918 made it possible for Romania to double its size, since Transylvania, Bessarabia, Bucovina and Banat expressed their will to belong to Romania through referendums.

[28] In 1918 the king approved electoral reforms making voting obligatory for men over the age of 21, and introducing the secret ballot. At the time the peasants made up 80 per cent of the population, and they had their own political party. Conservatives found their position weakened because of having supported Germany, liberals included only people linked to professional and commercial activity, and the socialists carried little weight.

[29] Social upheaval, and the landowners' fear of having their lands expropriated, led General Averescu (the hero of two wars and now the head of the government) to take harsh measures. The general strike of 1920 was put down and the Communist Party was declared illegal in 1924.

[30] In the 1928 election, the National Peasant Party (NPP) obtained 349 of 387 seats. The government abolished martial law and press censorship, also decentralizing public administration, an issue supported by the ethnic minorities. It also authorized the sale of land and foreign investment in the country.

[31] The council of regency, which had been formed upon Ferdinand's death, was dissolved when King Carol assumed the throne in 1930, a succession agreed on by the major political parties. The economic crisis led to the displacement of the NPP, and the king took advantage of the emergence of the Iron Guard, a fascist Moldovan group similar to those that existed in Germany and Italy, to weaken the traditional parties.

[32] In 1938, after a fraudulent plebiscite, Carol passed a new corporative constitution. Seeking closer ties with Germany, he met Hitler in November 1938. Upon his return, he had 13 officials of the Iron Guard assassinated, along with its leader, and Carol founded his own party, the National Renaissance Front (NRF).

[33] Carol assured the other powers that he was acting under pressure from Hitler, obtaining French and German assurances of the country's territorial integrity. When Germany and the USSR invaded Poland in 1939, Romania not only renounced the mutual defence treaty which it had signed with Warsaw, but also detained Polish authorities as they fled across Romanian territory.

[34] Between June and September of 1940, Romania was forced to turn Bessarabia and Bucovina over to the USSR, north-east Transylvania over to Hungary, and southern Dobrudja to Bulgaria. Because of his disastrous foreign policy, Carol had no choice but to abdicate in September, leaving his son Michael on the throne and turning over the government of the country to General Antonescu. Romania was occupied by 500,000 German troops, and was proclaimed a "national legionary state".

[35] Romanian troops cooperated with the abortive German offensive against the USSR, but when the counter-offensive was mounted by the Red Army, Bucharest abruptly changed sides. In June 1944, the Peasant, Liberal, Social Democratic and Communist Parties created the National Bloc and in August, King Michael ousted Antonescu and declared war on Germany.

[36] In September, with the major part of its territory occupied since mid-August by Soviet troops, Romania signed an armistice with the Allies. After three short-lived military governments, the USSR intervened in the naming of a new prime minister, Petru Groza, leader of the Ploughers' Front, a leftist party which was not a part of the National Bloc.

[37] At the Potsdam Conference, the Allies decided to resume relations with Romania, provided its government was "recognized and democratic". The USSR granted it immediate recognition while the United States and Britain adopted a "wait-and-see" attitude. In the 1946 election, the party in power was re-elected, with 71 per cent of the vote.

[38] The government initiated a series of detentions, summary trials and life sentences against the leaders and members of the Social Democratic, the National Liberal and National Peasant parties. In

1947, King Michael was forced to abdicate.

[39] In 1948, the Communists and some Social Democrats formed the Romanian Workers' Party (RWP) which joined the Ploughers' Front and the Hungarian People's Union to form the People's Democratic Front (PDF). In the March election that year, the PDF won 405 of the 414 seats of the National Assembly.

[40] On December 30 1947, the People's Republic of Romania was proclaimed, and a soviet style socialist constitution was adopted. In April, the government adopted centralized economic planning.

[41] Between 1948 and 1949, Bucharest signed friendship and cooperation treaties with the European socialist bloc, and joined the Council for Mutual Economic Assistance (CMEA). In 1955, Romania joined the Warsaw Pact, but in 1963 it began to drift away from the Soviet fold.

[42] In 1951 the First Five Year Plan was started, aimed at socialist industrialization of steel, coal, and oil. In 1952, the new regime started to consolidate under President Groza and Prime Minister Gheorghiu-Dej, head of state from 1961.

[43] Gheorghe Maurer was appointed prime minister in 1961, while Gheorghiu-Dej was president of the republic. Maurer tried to achieve greater economic and political independence, so Romania did not take sides in the Sino-Soviet dispute. In 1962 the economy underwent a series of reforms. The land collectivization policy was finished, and trade with the US, France, and Germany began.

[44] In 1965, when Gheorghiu-Dej died, Nicolae Ceausescu was elected First Secretary of the RWP, which subsequently changed its name to the "Romanian Communist Party" (RCP). The constitution was reformed, and the name of the country was changed to "the Socialist Republic of Romania", by the National Assembly. In 1967, Ceausescu was elected president of the State Council.

[45] On the diplomatic front, Ceausescu placed some distance between his country and the USSR. In 1966, he affirmed that his country was continuing its struggle for independence. Romania established diplomatic relations with West Germany and, unlike other members of the Warsaw Pact, it did not break relations with Israel in 1967, nor intervene over the Soviet invasions of Czechoslovakia in 1968.

[46] Reaffirming Romania's independence from the USSR but not renouncing communism, Ceausescu modified the structure of the RCP and the State.

[47] In the 1970s and early 1980s, Ceausescu was re-elected several times, in apparently free elections, as secretary general of the RCP and as president of the country. Despite numerous government reorganizations,

economic hardships, accentuated by administrative corruption, led to growing discontent.

[48] From 1987, difficult living and labor conditions triggered marches and strikes. These were put down by the security forces. In 1988 and 1989, several government scandals broke out; various cabinet ministers and government authorities were subsequently tried and dismissed.

[49] Toward the end of 1989, confrontations between civilians and the army in Timisoara left many dead or injured, the international press spoke of hundreds of deaths, and the news had strong repercussions within Romania.

[50] The government declared a state of emergency, but a faction within the regime carried out a coup with massive popular support. Accused of "genocide, corruption and destruction of the economy", Ceausescu and his wife were secretly executed by army soldiers. The National Salvation Front (NSF) assumed control of the government.

[51] The leaders of the NSF came under suspicion. Resistance towards the new government increased, leading to violent confrontations in the streets. In the May 1990 election, the NSF claimed 85 per cent of the vote; but charges of election rigging were confirmed by international observers.

[52] In December 1991, 77 per cent of the electorate approved the new constitution which turned Romania into a multi-party presidentialist democracy. In Transylvania, however, the new constitution received scant support.

[53] On January 1 1992, the cooperative farming system created by the Ceausescu regime ceased to exist legally, but in fact it was still operating. In February 1992, in the municipal elections, incidents occurred in Transylvania when the Magyar Democratic Union candidate was barred from participation.

[54] One month before the elections, a new opposition party, the Democratic Convention, was created. It obtained 24.3 per cent of the vote, against the 33.6 per cent of the FSN. Opposed to the First Minister Petre Roman's reelection as party president, Ion Iliescu's faction withdrew from the FSN in March, founding the FSN/22nd of December Group, changing its name in April to the Democratic Front for National Salvation (FDSN).

[55] Also in January, Romania and Germany signed a Friendship and Cooperation Treaty. In the meantime, the United States included Romania on the list of "Most Favored Nations" (MFN), granting preferential treatment in trade. The state privatized 6,000 enterprises in June, selling 30 per cent of the shares to the general public.

[56] The World Bank and the IMF granted Romania loans of $400 million for economic reforms. Iliescu won the September and October

general elections, but in Parliament, the FDSN obtained a mere 28 per cent of the seats.

[57] Western countries and international financial organizations continued to voice their discontent to the government for its alleged sluggishness to implement economic reforms. One of the features criticized was land ownership, since 30 per cent still belonged to the state in 1994.

[58] 1995 was also dominated by disagreement between Bucharest and some of its trade partners, such as the European Union, on the pace of the reforms. Finally, after two years of deliberation, Bucharest passed in June the law to privatize state enterprises.

[59] Human rights organizations, such as Amnesty International, also criticized Romania especially for the violence and discrimination often practised against the approximately 500,000 to 2,000,000 Romanian gypsies.

[60] After years of negotiations, Bucharest and Budapest signed a treaty in September 1996 regarding the 1.6 million Hungarians living in Romania. Hungary was satisfied with Romania's pledge to "guarantee the minority's rights", shelving its request of "autonomy" for Transylvanian Hungarians.

Foreign Trade

MILLIONS $ 1994

IMPORTS
7,109

EXPORTS
6,151

External Debt

1994

PER CAPITA
$242

STATISTICS

DEMOGRAPHY

Urban: 55% (1995). **Estimate for year 2000:** 23,000,000. **Children per woman:** 1.5 (1992).

HEALTH

Under-five mortality: 29 per 1,000 (1994). **Calorie consumption:** 110% of required intake (1995).

EDUCATION

School enrolment: Primary (1993): 86% fem., 86% male. Secondary (1993): 82% fem., 83% male. University: 12% (1993). **Primary school teachers:** one for every 21 students (1992).

COMMUNICATIONS

102 **newspapers** (1995), 101 **TV sets** (1995) and 90 **radio sets** per 1,000 households (1995). 11.5 **telephones** per 100 inhabitants (1993). **Books:** 100 new titles per 1,000,000 inhabitants in 1995.

ECONOMY

Per capita GNP: $1,270 (1994). **Annual growth:** -4.50% (1985-94). **Annual inflation:** 62.00% (1984-94). **Consumer price index:** 100 in 1990; 2.617 in 1994. **Currency:** 1.767 lei = 1$ (1994). **Cereal imports:** 2,649,000 metric tons (1993). **Fertilizer use:** 423 kgs per ha. (1992-93). **Imports:** $7,109 million (1994). **Exports:** $6,151 million (1994). **External debt:** $5,492 million (1994), $242 per capita (1994). **Debt service:** 8.4% of exports (1994).

ENERGY

Consumption: 1,750 kgs of Oil Equivalent per capita yearly (1994), 27% imported (1994).

Russia

Rossiya

Population: 148,350.000
(1994)
Area: 17,075,400 SQ KM
Capital: Moscow (Moskva)

Currency: Ruble
Language: Russian

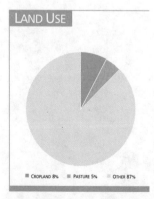

■ CROPLAND 8%   ■ PASTURE 5%   OTHER 87%

Before the Slavs appeared on the historical scene, the European territory of the present Russian Federation, Belarus and Ukraine was inhabited by different peoples, and underwent successive invasions by the Huns, Avars, Goths and Magyars. The first mention of the Slavs dates back to the 6th century. Byzantine writers wrote of several Slavic peoples: the Polians (based in Kiev), Drevlians, Dregoviches, Kriviches, Viatiches, Meria and others.

[2] In the 9th century, the first Russian State was formed, known as the "Ancient Rus" or "Kievan Rus". The latter emerged from the struggle against the Khazars of the south, and the Varangians (Scandinavians) of the north. In the 9th century, the route between the Baltic and Black seas, down the Dnepr River, "the route of the Varangians and the Greeks", became important for European trade.

[3] In 882, Prince Oleg of Novgorod conquered Kiev and transferred the center of the Russian State there. In the year 907, Oleg signed a treaty which was to be beneficial for the Rus. During the reign of Sviatoslav, Oleg's grandson, the struggle between Byzantium and Bulgaria intensified. Vladimir (980-1015), Sviatoslav's son and successor, consolidated the judicial, dynastic and territorial organization of the Russian State. To overcome the isolation of the "pagan" Rus vis-á-vis a monotheistic Europe, Vladimir made Christianity the state religion in 988, adopting the ornate Byzantine ritual.

[4] At the end of Vladimir's reign, strong separatist tendencies were evident within the Principality of Novgorod. Svyatopolk, Vladimir's successor, killed three of his brothers in order to consolidate his own power, but Yaroslav, the fourth brother and prince of Novgorod, ousted Svyatopolk and assumed power in Kiev, granting Novgorod several benefices. After his death, the feudal republic of Novgorod was formed, along with the principalites of Vladimir-Suzdal, Galich-Volin and others. Muscovy (the principality of Rostov-Suzdal) is first mentioned in historical chronicles in 1147.

[5] In 1237, the troops of the Tatar khan Batu, the grandson of Genghis Khan, invaded the principalities of Riazan and Vladimir, taking Moscow and other Russian cities. From 1239-40, the conquest of Russian principalities continued, followed by two and half centuries of Tatar rule. Apart from the Mongols in the East, the Teutons and the Swedes posed an additional danger in the West. In 1242, Prince Alexander of Novgorod defeated the Teutons in the famous "battle on ice" on Lake Chudskoye, near the Neva River, obtaining the title "Prince of Nevsky" as a result of this victory.

[6] The Mongols governed through local princes or Turkish chieftains and Muslim merchants, who were given the authority (*yarlik*) to govern. In the early 14th century, Tver, Moscow, Riazan and Novgorod were the main principalities. Dmitry, grand prince of Moscow began uniting forces in order to expel the Tatars, but ran up against the opposition of the princes of Tver, Nizhni-Novgorod and Riazan. In 1378, the Mamai Khan led an expedition aimed at conquering Russia, but was defeated.

[7] In 1380, Dmitry defeated Mamai in the battle of Kulikov, near the Don river, marking the beginning of Russian liberation from the Tatars. The struggle for liberation lasted a century, and ended in 1480, when Ajmat, the last of the Golden Horde's khans, retreated from a confrontation with the troops of Prince Ivan III, at the Ugra river.

[8] In 1547, Ivan "the Terrible" came to the throne. In 1552, Ivan IV took Kazan, and annexed the territory of the middle course of the Volga river, inhabited by Tatars, Chuvashes, Mari, Morduins and Udmurts. In 1556, he occupied Astrakhan, while the west continued to war against the Polish-Lithuanian State, trying to gain an outlet to the Baltic Sea. Serfdom was established, with peasants losing the right to leave without the feudal lord's permission. Ivan IV established absolute power by eliminating several upper nobility, *boyar* clans.

[9] Upon the death of Ivan IV, he was succeeded by his mentally ill

ENVIRONMENT

The largest country in the world, Russia is divided into five vast regions: the European region, the Ural area, Siberia, Caucasia and the Central Asian region. The European region is the richest of the Russian Federation, and lies between Russia's western border and the Ural Mountains (the conventional boundary between Europe and Asia); it is a vast plain crossed by the Volga, Don and Dnepr rivers. The Urals, which extend from north to south, have important mineral and oil deposits in their outlying areas. The third region, Siberia, lies between the Urals and the Pacific coast. It is rich in natural resources, but sparsely populated because of its rigorous climate. Caucasia is an enormous steppe which extends northward from the mountains of the same name, between the Black and Caspian Seas. Finally, the Central Asian region is an enormous depression of land made up of deserts, steppes and mountains. Grain, potatoes and sugar beet are grown on the plains; cotton and fruit in Central Asia; tea, grapes and citrus fruit in the subtropical Caucasian and Black Sea regions. The country's vast mineral resources include oil, coal, iron, copper, zinc, lead, bauxite, manganese and tin, found in the Urals, Caucasia and Central Siberia. Chelyabinsk, a city which lies south of the Ural Mountains, has high levels of radioactivity, due to leaks in its plutonium plant. Pollution, caused by the dumping of industrial wastes, threatens Lake Baikal. Heavy industry and mining have contributed to increasing contamination of the country's main rivers, air and soil. Other contributing factors include the dependence upon coal in electrical generating plants, defects in nuclear reactors and the abuse of agrochemical products. Deforestation and soil erosion also threaten large areas of the countryside.

SOCIETY

Peoples: Russians, 82.6%; Tatars, 3.5%; Ukrainians, 3%; over 100 other nationalities.
Religions: Christian Orthodoxy is the main one. There are also Muslim, Protestant and Jewish minorities. **Languages:** Russian (official). Minorities also speak their own languages. **Political Parties:** Our Home is Russia; Liberal Democratic Party; Communist Party; Agrarian Party; Yabloko; Congress of Russian Communities. **Social Organizations:** Federation of Independent Labor Unions of Russia; Sotsprof Labor Union Association

THE STATE

Official Name: Rossiyskaya Federatsiya.
Administrative divisions: The federation is made up of 26 autonomous republics: the republics of Bashkortostan (formerly Bashkir), Chechen-Ingush, Chuvash, Dagestan, Kabardino-Balkar, Kalmykia, Komi, Mari, North Ossetia, Tatar, Tuva and Saja (formerly Yakut); the regions of Adigueya, Gorno-Altai, Hebrea and Karachai-Cherkessk; and the territories of Buryat-Aguin, Buryat-Ust-Ordin, Chukchi, Dolgano-Neneos of Taimir, Evenkos, Janti and Mansi, Koriakos, Neneos and Yamalo-Neneos. **Capital:** Moscow (Moskva), 8,570,000 inhab. (1994). **Other cities:** St Petersburg, 4,320,900 inhab.; Nizhny-Novgorod, 1,424,600 inhab.; Novosibirsk, 1,418,200 inhab.; Yekaterinburg, 1,347,000 inhab.; Samara, 1,222,500 inhab.; Omsk, 1,161,200 inhab.; Chelyabinsk, 1,124,500 inhab.; Kazan, 1,092,300 inhab.; Ufa, 1,091,800 inhab.; Perm, 1,086,100 inhab.; Rostov-na-Donu, 1,023,200 inhab. (1994)
Government: Boris Yeltsin, President; Victor Chernomyrdin, Prime Minister. Parliamentary republic. Legislature, bicameral: Congress of People's Deputies, with 1,068 members elected by direct popular vote; Supreme Soviet, with 252 members elected by the members of Congress. **Armed Forces:** 1,520,000 (1995).

son Fyodor in 1584 and power passed to the *boyar* Boris Godunov. Godunov conducted a short war against Sweden, signed an alliance with Georgia - which became a Russian protectorate - and annexed the principality of Siberia. On Fyodor's death in 1598, the Council of Territories (*Zemski Sodor*) elected Boris Godunov as czar. However, the *boyar* clans of older lineage felt they had a greater claim to the throne, and a period of intrigue and conflict over succession began.

[10] In 1601-02, in the Ukrainian territory - which belonged to Poland - an impostor calling himself "Prince Dmitry" appeared, claiming to have escaped an attempt on his life by Boris Godunov. Amassing an army, he moved toward Moscow, managing to gain a following among the disaffected members of the population, and eventually seizing the throne. In 1606, the *boyars* killed the "False Dmitry" but in 1607 another impostor - a second False Dmitry - appeared, supported by Poles, Lithuanians and Swedes. Polish troops occupied Moscow with the help of turncoat *boyar*. The Poles were finally expelled from Moscow, and in 1613 the Zemsky Sodor elected Michael Romanov as the new czar. Between 1654 and 1667, Russia warred against the kingdoms of Sweden and Poland, managing in the process to annex eastern Ukraine.

[11] Under the Romanovs, the Russian State became an absolute monarchy, administrated by an efficient bureaucracy and an oligarchy (made up of noblemen, merchants and bishops) which was integrated into the government structure. The church was reformed and the bible translated into cyrillic script, causing a schism in the Russian church. During the 17th century the economy grew rapidly, as a result of territorial expansion, and the exploitation of Siberia's natural resources. A market also developed for Russia's forest products and semi-manufactured goods, primarily in Britain and Holland.

[12] In 1694, after the accession of Peter I, the Muscovite kingdom became known as the Russian Empire. Peter turned to the West to secure its scientific and technical advances, especially in order to develop the Russian bavy. Allied with Denmark and Poland, Russia successfully intervened in the Great Northern War (1700-21). In 1703,

he founded St Petersburg, transferring the capital of his empire there and organizing the government along a set of strict regulations.

[13] Peter established what amounted to a caste system, as well as an espionage network within his own administration, essential for maintaining his strict autocracy. He put down the *boyar* in Moscow, and had his own son, Alexei, tortured and executed for joining them.

[14] In 1721, by the Treaty of Nystad, Russia obtained control of the Gulf of Finland and the provinces along the east coast of the Baltic Sea. After winning the war against Persia, Peter extended Russia's southern borders as far as the Caspian Sea. The territorial, economic and commercial expansion which characterized this period made Russia one of the major European powers but also created a mosaic of ethnic and cultural groups which could not easily be assimilated into a single unit.

[15] Peter's sudden death in 1725 ushered in a period of instability, until the accession of Catherine II, in 1762. The empire's conquests continued, with the occupation of Belarus and the part of Ukraine east of the Dnepr; the partition of Poland between Russia and Prussia; and the annexation of Lithuania and Crimea. Russia also gained control of the northern coast of the Black Sea, penetrating the steppes beyond the Urals and all along the coast, they began exercising an ever-increasing influence over the Balkans.

[16] In the meantime, the impoverishment of the peasants increased. The military democracy of the Ukrainian Cossacks was abolished, and the nationalist sentiment of the subdued peoples began to cause friction.

[17] Toward the end of the 18th century the French Revolution and the fight against absolutism influenced the Russian intelligentsia, which

began rebelling against existing social conditions. The accession of Alexander I to the throne brought with it radical changes in the policies of the Russian empire.

[18] Although Alexander sought peace, Napoleon declared war in 1805, and defeated Russia at Austerlitz. In 1812, Napoleon's troops invaded Russia. The "War for the Motherland", in which peasant guerrillas were also involved, ended with the triumph of the Russian army commanded by Marshall Kutuzov. This victory transformed Russia into the continent's major power. In December 1825, after the death of Alexander I, a group of aristocrats later known as "the Decembrists" carried out a failed coup attempt in the Senate Square in St Petersburg. When the 1848 revolution shook Europe, Russia - governed by Nikolai I - remained untouched and used its army to subdue the Hungarians in Transylvania, though it was defeated in the

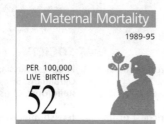

Crimean War against Britain and France (1853-56).

[19] In 1861, Czar Alexander II abolished serfdom. However, freedom was not "free", as the peasants had to pay their former landlords for the land which they farmed. A system of elected local assemblies was also established, with peasant representation, though the landowners maintained control.

[20] In the 1860s-70s, radical groups emerged whose aims ranged from demanding a constituent assembly to calling for insurrection. Socialist ideas influenced students and intellectuals, who saw in the peasants a revolutionary class. In 1861-62, several different revolutionary groups created an underground association in St. Petersburg, called "Land and Freedom", which existed until 1864. The Polish uprising of 1863 and a failed attempt on the czar's life in 1866 deepened government repression.

[21] The intellectuals reached the conclusion that they would have to "go to the people" in order to explain to them who their enemies were. In 1876, another "Peace and Freedom" underground movement was established. Three years later the group split into three, including the "People's Will", which assassinated Alexander II in 1881, and whose main leaders were hanged. Alexander III did away with his predecessor's reforms, and increased the power of the autocracy.

[22] In the early 20th century, Russian socialism had two main currents. The socialist revolutionaries promoted the socialization of the land in peasant communities, while the social democrats based socialism upon industrialization and the working class. Within the Social Democratic Worker's Party of Russia (SDWPR), the Bolsheviks, led by Vladimir Ilich Ulyanov (known as "Lenin") defended the idea of a workers' revolt, while the Mensheviks shared the idea of an evolutionary socialism, put forward by Europe's other social democratic parties.

[23] Russian expansionism in Eastern Asia led to a war with Japan in 1904. The violent repression of a demonstration in Moscow in 1905 unleashed a revolt in both capitals, which was only brought under control by the October 17 Manifesto: the czar's promise to convene a national parliament (Duma). The Bolsheviks boycotted the elections. The first Duma's majority was in the hands of constitutional democrats who were moderate liberals.

[24] The Duma's demands for agrarian reform, equal rights for members of all religions (Russian Orthodoxy was the official religion), an amnesty for political prisoners, and autonomy for Poland - were unacceptable to the czar. The Duma was dissolved and the regime put down the revolt after a long and bloody struggle. In 1907, the second Duma was elected, with the presence of the SDWPR. Both the left and the right wings were reinforced, and the main political issue continued to be land. Prime Minister Stolipin promoted an agrarian reform, in order to create a a class of landowning peasants and put an end to the traditional communal use of the land. However, he was unsuccessful and was assassinated. The second Duma was also dissolved. An electoral reform guaranteed that the Duma committees which followed also had a conservative majority.

[25] Russia's entry into World War I precipitated a crisis within the regime. The losses brought about by war, and the lack of food, simply deepened popular discontent. In January 1917, in Petrograd, a Council (Soviet) elected by workers, soldiers and members of the Duma formed a government. In February, Nicholas II abdicated and the Duma committee established a new Provisional Government, while local soviets multiplied. The German government authorized a group of Bolsheviks, led by Lenin, to return to Russia in a sealed railway car, to destabilize the enemy's internal situation. The influence of the Bolsheviks within the soviets began to grow, and a power split emerged between the councils and Alexander Kerensky's government.

[26] The Bolsheviks launched the slogans "Peace, land and bread" and "All power to the soviets", exhorting the people to "turn the world war into a civil war". On November 7, Lenin led the uprising which brought down the government, and the first socialist republic was established. In early 1918, the Bolsheviks dissolved the Constituent Assembly, in which the revolutionary socialists held a majority.

[27] The Soviet government (Council of People's Commissars) approved a peace "without annexation or indemnities", the abolition of private ownership of the land and its being turned over to the peasants and the nationalization of the banking system. Other measures were approved, including the control of factories by their workers, the creation of a militia and of revolutionary tribunals, the abolition of the privileges associated with class and of the right of inheritance, the separation of Church and State and equal rights for men and women.

[28] Following Russia's unilateral peace settlement in 1918, Germany, France and Britain sent expeditionary forces to Russia (their former ally), to support the "White Russians" - the former regime's army forces - and bring down the revolutionary government. This foreign intervention force was ousted in 1920, and two years later the civil war ceased, with the triumph of the Red Army. During this period, the Soviet government's policy of "war communism" resulted in a maximum centralization of power and a near-collapse of the monetary system.

[29] In December 1922, the Russian Federation, Ukraine, Belarus and the Transcaucasian Federation (Azerbaijan, Armenia and Georgia) established the Union of Soviet Socialist Republics (USSR). During the civil war, the regime had become the official preserve of the Russian (Bolshevik) Communist Party, or R(B)CP.

[30] In 1921, the R(B)CP - faced with the danger of an imminent economic collapse- was forced to do away with its "war communism" policy, and adopt a New Economic Policy (NEP). This consisted of a return to the laws of a market system and private ownership of small business, while the State took care of infrastructure, heavy industry and general planning. In 1924, the NEP was interrupted by Lenin's death; internal disputes within the R(B)CP came to an end with the triumph of Joseph Stalin, who had become secretary general of the Party in 1922. Trotsky (Lenin's close associate) went into exile and was assassinated under Stalins' orders, in Mexico in 1940.

[31] Stalin re-established the system of centralized planning of the economy, and forcibly imposed the collectivization of agriculture. Opposition figures and dissidents were eliminated through summary trials, like the Great Purge of 1935-38. Official plans cited huge figures for grain harvests as well as the production of electricity and steel, while the secret police held almost absolute power.

[32] At the beginning of World War II, by means of a secret agreement with Germany (the Molotov-Ribbentrop Pact), the USSR occupied part of Poland, as well as Romania, Estonia, Latvia and Lithuania. In 1941, Hitler launched a massive attack against Moscow, sending in thousands of troops, as well as German air power. At a cost of some 20 million lives, the Soviets were able to repel the German attack, and the Red Army liberated several countries, finally taking Berlin, in May 1945.

[33] In 1945, at the Yalta conference, the Western powers and the USSR "carved up" their respective areas of influence. In those countries occupied by the Red Army (Bulgaria, Hungary, Romania, Czechoslovakia, Poland and East Germany), the Communists took power and proclaimed first "people's republics", then socialist republics, following the model of the Communist Party of the Soviet Union (CPSU).

[34] In 1956, at the 20th Party Congress of the CPSU, Nikita Krushchev began a de-Stalinization process, which abruptly came to an end when Leonid Brezhnev ousted Krushchev in October 1964. Despite difficulties of implementation, "developed socialism" was proclaimed.

[35] The strategy of the cold war devised by the United States in the post-war period fuelled the arms race. A by-product of the cold war was the creation of the Warsaw Pact, between the USSR and its Eastern European allies in 1955. The East-West confrontation eventually included nuclear weapons and the control of outer space, where the United States and the USSR actively pursued their own space programs in the 1960s and 70s, with neither one actually taking a lead.

[36] In 1985, Mikhail Gorbachev became Secretary General of the CPSU and initiated drastic changes to avert a social and economic crisis. Glasnost (transparency) and perestroika (restructuring) -geared to carrying out a series of transformations within the country - unleashed forces which had long been repressed, and brought out into the open the issue of the autonomy of different ethnic groups and nationalities. Changes within the USSR set in motion similar processes in

Foreign Trade

MILLIONS $ 1994

IMPORTS
41,000
EXPORTS
53,000

STATISTICS

DEMOGRAPHY

Urban: 75% (1995). **Annual growth:** 0.4% (1991-99). **Estimate for year 2000:** 150,000,000. **Children per woman:** 1.7 (1992).

EDUCATION

School enrolment: Primary (1993): 107% fem., 107% male. Secondary (1993): 91% fem., 84% male. University: 45% (1993).

COMMUNICATIONS

15.8 **telephones** per 100 inhabitants (1993).

ECONOMY

Per capita GNP: $2,650 (1994). **Annual growth:** -4.10% (1985-94). **Annual inflation:** 124.30% (1984-94). **Consumer price index:** 100 in 1990; 64,688 in 1994. **Currency:** 1,200 rubles = 1$ (1993). **Cereal imports:** 11,238,000 metric tons (1993). **Fertilizer use:** 417 kgs per ha. (1992-93). **Imports:** $41,000 million (1994). **Exports:** $53,000 million (1994). **External debt:** $94,232 million (1994), $635 per capita (1994). **Debt service:** 6.3% of exports (1994).

ENERGY

Consumption: 4,038 kgs of Oil Equivalent per capita yearly (1994), -52% imported (1994).

other nations throughout Eastern Europe.

[37] In April 1986, there was a serious accident in one of the reactors of the nuclear power plant at Chernobyl in the Ukraine. 135,000 people were evacuated and by 1993 7,000 people had died (see Ukraine).

[38] Gorbachev initiated internal reforms and also made other initiatives in the area of foreign policy, such as the withdrawal of Soviet troops from Afghanistan (where the Soviet army had been involved in a war since 1979). In addition, he was responsible for accords leading to the reduction of nuclear weapons in Europe, and also accepted the reunification of Germany. In 1990, he proposed that the Warsaw Pact and NATO be phased out gradually.

[39] Economic reform, the opening of the country to foreign capital and a return to a free-market economy were all slow in being implemented due to the resistance of the CPSU leadership. In June 1991, Boris Yeltsin was elected president of Russia. After a failed coup, in August, the CPSU was dissolved after 70 years in power.

[40] Trouble broke out in the Chechen-Ingush region (Northern Caucasia). Toward the end of October, parliamentary and presidential elections were held there. General Dzhojar Dudaev, leader of the Chechen nationalist movement, seized power. In early November he proclaimed the independence of the Chechen Republic. An economic embargo was promptly announced by Moscow.

[41] On October 6, 1991, Yegor Gaidar was named deputy prime minister, and he launched liberal economic reforms, dubbed "shock therapy". On December 8, Boris Yeltsin (Russia), Stanislav Shushkevich (Belarus) and Leonid Kravchuk (Ukraine) revoked the 1922 treaty under which the USSR had been founded, proclaiming the Commonwealth of Independent States (CIS) to take its place (see box on CIS). Russia assumed the formal representation of the former USSR in foreign affairs. Latvia, Estonia and Lithuania withdrew and were recognized as separate countries by the UN.

[42] In early 1992, there was a growing rivalry between the two most important members of the CIS, Russia and Ukraine, revolving around the issue of who controlled the nuclear weapons and navy which had belonged to the former USSR. Relations between them were further complicated by the issue of Crimea's sovereignty. President Yeltsin declared that the United States was no longer a "strategic rival", and continued the reform of the economy that Gorbachev had begun, including the liberalization of prices, and the privatization of industry, agriculture and trade.

[43] On February 23, 1992, violent confrontations broke out between communist and nationalist protesters, and the militia, with several people being killed by the police.

[44] On March 13, the autonomous republics associated to Russia signed the Federation Treaty, except for the Tatar and Chechen republics. A week later, in a referendum in the Tatar republic, the population voted in favor of state sovereignty and the termination of the bilateral treaty with Moscow.

[45] In April 1992, Yeltsin managed to maintain control of his reformist government, despite attacks from the 6th Congress of People's Deputies. In June he threatened to dissolve parliament, which opposed his reforms. In the meantime, the Constitutional Court declared that Yeltsin's decree banning the Communist Party was illegal. On June 12 and 22 there were further clashes between opposition members and the police.

[46] On October 31 a territorial dispute broke out in Northern Caucasia between Ossetia and Ingushetia over part of North Ossetia which had previously been part of Ingushetia. Moscow declared a state of emergency in the region, and after several days, the rebellion was put down.

[47] In early December 1992, the 7th Congress of People's Deputies refused to ratify Boris Yeltsin's choice for prime minister, Yegor Gaidar. After exchanging recriminations, the president and Congress agreed to ratify Victor Chernomyrdin in the position.

[48] The "arm wrestling" between the president and parliament continued. In March 1993, Congress opposed a referendum proposed by Yeltsin and tried unsuccessfully to limit the powers of the leader. Following further disagreements, Yeltsin took all the power from parliament, which ousted the president and substituted him with Alexandr Rutskoi. On the same day, September 22, the police surrounded the seat of the legislative power.

[49] The tension continued to increase and on October 4 parliament was taken by force after having been attacked with tanks. Several opposition leaders, like Rutskoi, vice-president of Congress, and its leader, Ruslan Khasbulatov, were arrested. A few days later, Yeltsin called for new elections and organized a referendum to increase his own powers.

[50] The December elections marked the defeat of those sectors faithful to Yeltsin, but 60 per cent of the voters approved the constitutional reform which granted him greater powers. In February 1994, a bilateral agreement was signed with the Russian republic of Tartarstan and a similar document was expected to be signed with Chechnya. However, the tension between Moscow and the pro-independence groups of this mostly Muslim republic, which had declared independence in 1991, became more serious and in December 1994, Yeltsin ordered military intervention.

[51] Despite protests both within Russia and abroad, the president maintained the military attacks on Grozny, the Chechen capital, which was almost totally destroyed in 1995. In December of this year, the Communist Party of Zyuganov won the legislative elections with 22.3 per cent, followed by the ultra-right and xenophobic Liberal Democratic Party of Vladimir Zhirinovski, with 11.8 per cent and the Russia Our Home of Prime Minister Victor Chernomyrdin, with 10.1 per cent.

[52] The communist victory made Yeltsin fearful of defeat in the July 1996 presidential elections. The president tried to modify his policies, stopping the privatizations and nominating Evguenni Primakov - a diplomat from the Soviet era who was an ally of Gorbachev - as foreign minister, amongst other things. All the opposition candidates from Gorbachev to the Communists, criticized the unlimited financial speculation, corruption and "clanishness" of Yeltsin and his allies.

[53] However, in the second round of elections, in July, Yeltsin took 53.8 per cent of the votes and Siuganov only 40 per cent (4.8 per cent of voters voted against both candidates). The outgoing president managed to triumph thanks to an unexpected alliance with Alexandr Lebed, an opposition candidate who had taken 11 million votes, who hoped to replace Yeltsin if he became more ill.

[54] Lebed was nominated state security advisor and immediately began negotiations to try and end the war in Chechnya, which - he said in September - had caused the death of 80,000 people, mostly civilians. When it was announced that Yeltsin would undergo surgery, Lebed and Prime Minister Chernomyrdin became the two possible - and opposed - alternatives to fill any eventual power vacuum.

Rwanda

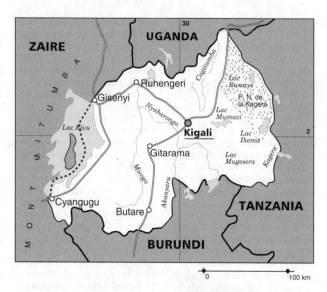

Population: 7,755,000 (1994)
Area: 26,340 SQ KM
Capital: Kigali

Currency: Franc
Language: Rwanda and French

Rwanda

Inhabited since ancient times by the Hutu (Bantus) and Twa (pygmy) ethnic groups, Rwanda's highlands were invaded in the 15th century by the Tutsis (or Watutsis) from Ethiopia. After subduing the local population, the Tutsis set up a stratified society. This social and political organization remained basically unchanged even after German colonization started in the region in 1897. The territory became part of German East Africa, which also included Burundi.

[2] After World War I, the territory named Rwanda-Urundi was placed under Belgian custody and administered by them from the Congo (today's Zaire). Of the three groups which inhabit the region, the minority - made up of the pastoralist and warrior Tutsis - consolidated its control over the agricultural Hutu and artisan Twa. In 1959, farm workers organized by the Parmehutu (Party of the Hutu Emancipation Movement) revolted against Tutsi rule. A bloody civil war ensued and the Belgian colonial government chose to abandon the territory. The Parmehutu won elections supervised by the United Nations in 1961 and proclaimed a Republic in 1962, autonomous from neighboring Burundi.

[3] The power structure favoring Tutsi rulers was abolished and land distributed on the capitalist basis of private ownership. However, this did not lead to true national unity, nor did it settle controversies between ethnic groups. The following year, civil war broke out again. Approximately 20,000 people died and 160,000 Tutsis were expelled from the country.

[4] Lacking other patterns, the Parmehutu re-organized society according to ethnic group interests: the Twa (40,000 in all) were assigned the crafts industry; the Tutsis, cattle-raising, and land ownership was reserved for the Hutus. The system virtually ignored the possibility of urban development, a fact evidenced by the tiny size of the capital city of Kigali. In order to set an example, President Gregory Kayibanda worked his own piece of land. As a result agriculture was almost entirely reduced to subsistence farming, with little surplus left over for the market.

[5] In the late 1960s, attempts to increase market production saw a revival of the coffee plantations, burnt down in 1959, in protest against colonial despotism. The new policy, however, did not solve Rwanda's economic problems but rather exacerbated social tensions as a new group of landowners arose, adding complexity to existing ethnic conflicts, and giving rise to new and violent outbreaks.

[6] Faced with the threat of a new civil war, Colonel Juvenal Habyarimana overthrew Kayibanda on July 5 1973. Habyarimana, previously defence minister, dissolved the Parmehutu, jailed Kayibanda (who died shortly afterwards) and launched a diplomatic offensive successfully dealing with conflicts which had arisen between Rwanda and neighboring countries during the previous regime.

[7] Since 1959 the country has been following a liberal economic model. The oligarchy controls farming, commerce and banking.

[8] The close relationship between Habyarimana and France, and with Zaire's President Mobutu caused disagreements within the governing party, the National Revolutionary Movement for Development (MRND). Progressive factions criticized the situation and Habyarimana's liberal policies. During the 1980 party congress they were expelled. Their leader, labor minister Alexis Kanyarengwe, went into exile in Tanzania to escape prison. Resolutions by the party congress endorsed "planned liberalism", which resulted in an open-door policy for foreign investments.

[9] In 1982, Uganda expelled large numbers of Rwandan exiles, who were not allowed back into their own country. Whole towns were burned down by Ugandan troops leaving at least 10,000 people homeless and without food.

[10] In 1986, the new president of Uganda, Museveni, announced that Rwandans who had lived in the country for more than ten years would automatically receive citizenship. Relations improved between the two countries; in 1988, both presidents signed a declaration confirming progress in their international policies.

[11] President Habyarimana changed his policies and took firm steps towards establishing a democracy. Initially he set up a National Development Council in an attempt to draw up national policies taking into consideration the country's different social and economic realities. Later, he executed an austerity program, which yielded some positive economic results, he released more than 1,000 political prisoners, and implemented measures to ensure respect for human rights in the prisons.

[12] Habyarimana was re-elected in 1988. On December 26 that year, elections were held to renew one-third of the 70 deputies. Of these, eleven women, but only two Tutsis, were elected. 60,000 Hutu refugees from Burundi, entered the country in the course of the year.

[13] On September 30 1990, Fred Rwigyema, an important government official belonging to the Tutsi ethnic group, led an uprising, entering Rwanda from Uganda. President Habyarimana requested help from Belgium, France and Zaire, whose troops played a decisive role in repelling the rebel offensive, although there was also fighting in the capital. In October, a cease-fire was reached between the two parties, through the intervention of Belgian prime minister, Wilfried Martens.

[14] At the end of January 1991, some 600 troops belonging to the Rwandan People's Front entered the country from Uguanda. In March, a cease-fire was signed. Among its provisions were the liberation of political prisoners and prisoners of war, and a commitment by President Habyarimana to initiate negotiations establishing a

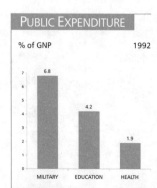

PUBLIC EXPENDITURE

% of GNP — 1992

- MILITARY: 6.8
- EDUCATION: 4.2
- HEALTH: 1.9

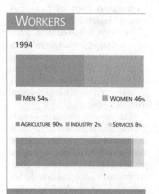

WORKERS

1994

- MEN 54%
- WOMEN 46%

- AGRICULTURE 90%
- INDUSTRY 2%
- SERVICES 8%

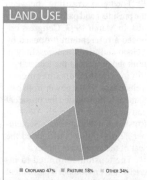

LAND USE

- CROPLAND 47%
- PASTURE 18%
- OTHER 34%

External Debt
1994

PER CAPITA
$123

Illiteracy
1995

40%

PROFILE

ENVIRONMENT

Known as the "country of a thousand hills" on account of its geographical location between two mountain ranges, Rwanda lies in the heart of the African continent. The terrain is mountainous and well irrigated by numerous rivers and lakes supporting varied wild life. The population is concentrated in the highlands where the economic mainstay is subsistence agriculture. The lowlands have been eroded and their natural vegetation is disappearing as a result of excessive grazing. 90% of the energy consumed by Rwandans is derived from natural resources, consequently leading to deforestation and erosion.

SOCIETY

Peoples: Rwanda's ethnic composition is the result of close integration between successive migrations and the original pygmy population. Today, 84% are Hutu (a branch of Bantu people), 15% Tutsi (Hamitic) and 1% Twa (descendants of the pygmies). There is a minority of European origin most of whom are Belgian. **Religions:** Most of the population (69%) practice traditional African religions, 20% are Catholic, 10% Protestant and 1% Muslim. **Languages:** Rwandan and French (both official). **Political Parties:** Rwandan Patriotic Front (FPR); National Revolutionary Movement for Development (MRND); Republican Democratic Movement (MDR); Liberal Party (PL).

THE STATE

Official Name: Repubulika y'u Rwanda. **Capital:** Kigali, 237,000 inhab. (1991). **Other cities:** Butare, 29,000 inhab.; Ruhengeri, 30,000 inhab.; Gisenyi, 22,000 inhab. (1991). **Government:** Pasteur Bizimungu, President. Pierre-Célestin Rwigema, Prime Minister. **National Holiday:** July 1, Independence Day (1962). **Armed Forces:** 5,200.

STATISTICS

DEMOGRAPHY

Urban: 6% (1995). **Annual growth:** 1.9% (1992-2000). **Estimate for year 2000:** 9,000,000. **Children per woman:** 6.2 (1992).

HEALTH

One **physician** for every 25,000 inhab. (1988-91). **Under-five mortality:** 139 per 1,000 (1994). **Calorie consumption:** 85% of required intake (1995). **Safe water:** 66% of the population has access (1990-95).

EDUCATION

Illiteracy: 40% (1995). **School enrolment:** Primary (1993): 50% fem., 50% male. Secondary (1993): 9% fem., 11% male. **Primary school teachers:** one for every 58 students (1992).

COMMUNICATIONS

84 **newspapers** (1995), 81 **TV sets** (1995) and 82 **radio sets** per 1,000 households (1995). 0.2 **telephones** per 100 inhabitants (1993). **Books:** 88 new titles per 1,000,000 inhabitants in 1995.

ECONOMY

Per capita GNP: $80 (1994). **Annual growth:** -6.60% (1985-94). **Annual inflation:** 4.50% (1984-94). **Consumer price index:** 100 in 1990; 147.2 in 1993. **Currency:** 146 francs = 1$ (1993). **Cereal imports:** 115,000 metric tons (1993). **Fertilizer use:** 6 kgs per ha. (1992-93). **External debt:** $954 million (1994), $123 per capita (1994). **Debt service:** 14.7% of exports (1994). **Development aid received:** $361 million (1993); $48 per capita; 24% of GNP.

ENERGY

Consumption: 27 kgs of Oil Equivalent per capita yearly (1994), 78% imported (1994).

---

more open political atmostphere. In June of this year, the president signed the new constitution, which provided for a multi-party system, creating the post of prime minister, guaranteeing freedom of the press, limiting the presidential term to two five-year periods with no further re-election, and establishing a separation of the powers of the state.

[15] In March 1992, at least 300 members of the Tutsi ethnic minority were assassinated, and another 15,000 were forcibly relocated to the region of Bugesera. The leaders of the two principal opposition parties, the Republican Democratic Movement (MDR) and the Liberal Party (PL), blamed the government for the violence, and especially a militia made up of young Hutus belonging to the MRND.

[16] In early 1993, President Habyarimana rejected the agreement signed in Arusha between the Rwandan delegation - led by Prime Minister Dismas Nsengiyaremye, of the Republican Democratic Movement - and the rebels of the Rwandan Patriotic Front (FPR), led by Alex Kanyarengue, made up of Tutsi exiles who had been questioning the Hutu domination of their community.

[17] The president refused any form of power-sharing with the FPR, which had demanded five cabinet positions, the incorporation of its soldiers into the regular army and the repatriation of Tutsi refugees from Uganda and Tanzania.

[18] Although both bands had respected the cease-fire which had been in effect since August 1992, the FPR broke off negotiations in February, launching a new offensive which enabled them to gain control over most of the country's territory, and prepare an advance upon the capital.

[19] Rwandan president, Juvenal Habyarimana, and Burundian president Cyprien Nytaryamira were killed in a mortar attack against the airplane in which they were travelling to Kigali, after attending a peace conference in Tanzania on April 6, 1994. The deaths of the two presidents - both Hutus - led to a new bloodbath.

[20] In the face of renewed civil war, France, the United States and Belgium decided to send in troops in April, in order to "guarantee the safety of foreigners, and evacuate them". French troops seized control of Kigali airport, in order to ensure the repatriation of the 600 French nationals in the capital.

[21] Four weeks after the death of the presidents of Rwanda and Burundi, almost half a million Rwandans, most of them members of the Tutsi minority, had been killed. During the final offensive upon the capital, thousands of civilians found themselves trapped between the lines of fire, while members of the government fled southwards. The WHO issued reports that on May 30, 50,000 Rwanda corpses had been seen floating in Lake Victoria. According to the government of Tanzania, there were some 3 million Rwandan refugees in that country.

[22] In mid-1994, reports issued by the International Federation of Human Rights (FIDH) on the activities of the "death squads" during Habyarimana's government, implicated France as one of the countries which had maintained the Rwandan regime in power, in the full knowledge that it systematically eliminated its opponents. Between 1990 and 1993, some 10,000-15,000 people had already been assassinated. Nevertheless, according to the report, President Mitterand re-

ceived Habyarimana in Paris with all the honors accorded to a head of State.

[23] With the endorsement of the UN Security Council, encouragement from Washington, and a "green light" from its closest European allies, in late June Paris initiated its participation in the armed conflict, with the stated aim of offering humanitarian aid.

[24] The death toll from the violence passed a million. In addition, there were approximately 2.4 million Rwandan refugees in neighboring African countries.

[25] On July 4, the FPR captured the city of Kigali, the government's last bastion in the southern part of the country, while French troops had received orders to stop the advancing rebels.

[26] On August 7, Faustin Twagiramungu, a Rwandan politician and leader of the Democratic Republic Movement (MDR), accepted the request by the guerrillas to head a government of national unity based on the Arusha peace accords. FPR vice-president Patrick Mazimpaka stated that the guerrillas were not an ethnic movement, and that more than 40 per cent of his combat troops belonged to the other two communities, Hutu and Twa.

[27] The new government tried to reactivate Rwanda's dismantled economy and set up trials against those who had committed what the UN described as a "genocide".

However, the government did not receive the expected assistance from Western countries and many militia members guilty of carrying out massacres kept operating from Zaire, where they had taken refuge.

[28] During 1995 and 1996, common graves continued to be found almost weekly, while violence amongst the new Rwandan army, made up of FPR fighters and the militias, led to new deaths. Many of the victims were simple witnesses of what had happened in 1994, as well as Hutu civil servants, accused of "collaborationism".

[29] The disappearance of several witnesses complicated even more the task of preparing the trials held against those guilty of genocide. A law discussed by the National Assembly in July 1996 proposed to separate those who had given orders to kill from those who had executed them. The former would be sentenced to death, while the latter would serve minor sentences. However, two years after the massacre, several Western countries - such as France, whose influence in the region was reduced when the government of Habyarimana fell - continued to refuse financial collaboration to facilitate fair trials.

[30] In November 1996 many Rwandan refugees returned home from camps in eastern Zaire.

Siblings, ethnic groups and classes

The peoples which traditionally inhabited the territory encompassing what is now Rwanda and its neighboring equatorial African countries have a common ancestry. The Banyarwanda included the Hutu, the Tutsi and the Twa. They speak Kinyarwanda, and are related to the Bayakole and the Bakiga of Uganda, and the Barundi of Burundi.

[2] In addition to a common language, these groups also share the same territory, the same traditional political institutions and physical characteristics.

[3] Up until the fall of the monarchy, the kingdom of Rwanda was a highly organized society, with a rigid system of social stratification. The posts of nobility and upper echelons of the army - and nearly all the country's livestock - were almost exclusively in hands of the Tutsis, while the Twa were traditionally hunters and potters, and the Hutus agriculturalists.

[4] By the mid-20th century, the population break-down was as follows: Hutus 84 per cent; Tutsis 14 per cent, and Twa 1 per cent. However, contrary to popular belief, not all Tutsis were members of the governing or privileged classes.

[5] At the time of independence, all 43 chieftains and 549 of the 559 sub-chieftains in Rwanda-Burundi were Tutsi, and around 80 per cent of positions in the judicial branch of the government were held by members of that ethnic group.

[6] The hold which the Tutsis were able to maintain over the Hutu majority until the dawn of independence was based upon their almost exclusive monopoly of livestock ownership. In traditional times, the distribution of cattle was carried out through a social mechanism known as "ubuhake". This system meant that a person of lower prestige and wealth - generally a Hutu - offered his services to another - generally a Tutsi - in exchange for access to the produce of one or more animals.

[7] The Tutsis also maintained their esprit de corps and their sense of superiority by dominating the military, membership of which was barred to members of other ethnic groups.

[8] Social cohesion was guaranteed by a series of hierarchical structures (territorial, military, mutual solidarity, etc.) to which every inhabitant belonged, with the king overseeing them all.

[9] European colonization resulted in massive changes in these ethnic relationships, and began undermining some aspects of the country's social structure. While the Rwanda-Burundi region was still under Belgian control, a movement giving voice to Hutu demands began to emerge in 1957. Two years later, the mysterious death of King Mutara, a reformist who lived in Bujumbura -now the capital of Burundi - marked the beginning of a bloody civil war. By 1962, 22,000 Tutsis had been displaced, within the country's borders.

[10] The Hutus, who came to power in Rwanda through the electoral process, never managed to establish ethnic harmony between the different groups which share the same geographical space. Violent rivalries erupted periodically, affecting both Burundi (which was governed until only a few years ago mostly by Tutsis) and its neighbors to the north.

[11] Fearing for their lives, close to 60 per cent of Rwanda's Tutsis left the country between 1959 and 1964, heading mostly for Burundi. In 1966, ethnic violence once again took a heavy toll. In Burundi, in turn, violence between the two groups erupted in 1965, 1966, 1969, 1972 and 1988.

[12] The last episodes in this chain of violence began in Rwanda in the early 1990s when the government headed a movement appealing to Hutu primacy. This campaign had its peak during 1994s genocide, which cost at least 500,000 lives and affected mostly but not only Tutsis. Confrontations between both groups continued in Burundi, and the 1996 coup made thousands of Burundans seek asylum in Zaire.

Western Sahara

Sahara Occidental

Population: 1,000.000
Area: 266,000 SQ KM (1994)
Capital: El Aaiú

Language: Arabic and Spanish

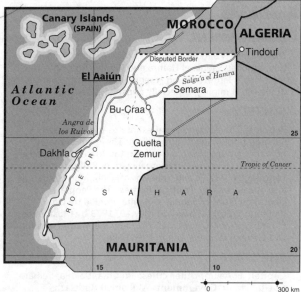

From the 5th century, the far west of the Sahara has been populated by Moors, Tuaregs and Tubus. These peoples evolved from migrations triggered by unbroken periods of drought which have affected the region since the Neolithic period.

[2] The presence of these peoples is documented by the Tassili stone carvings and other sources. In the 7th century, waves of migrants from Yemen integrated with people already in the area. Four hundred years later, the first Saharan confederation of peoples appeared.

[3] Spanish occupation of the Saharan coast was carried out mainly for strategic reasons: to cover the flank of the Canary Islands. The occupation was practically limited to Villa Cisneros (present-day Dakhla) until 1886, when, following the Berlin Conference (see appendix to Zaire), Madrid was determined that this "empty space" would not fall to another power.

[4] However, once the borders of the Spanish Sahara were established by agreement with France in 1904 the situation reverted to its former state. The colonial powers split the territory four ways, and the nomads continued to live in total independence, ignorant of the agreements reached in European offices.

[5] In 1895, Sheikh Ma al-Aini founded the Smara citadel and with the support of the sultan of Morocco, continued fighting the Franco-Spanish presence until 1910. Under French pressure, the Sultan of Morocco finally suspended assistance to the rebels, who enlarged their field of action to include Morocco and even threatened Marrakesh. A French counter-attack invaded "Spanish" territory and occupied Semara in 1913, though resistance continued until 1920.

[6] The French pressured Spain for increased control over the territory and in 1932 the city of Aaiún was founded. In 1933 Muhammad al Mamún, the Emir of Adrad and Ma al-Aini's cousin, defeated the French, forcing a change of tactics. France occupied the rebel base at the Tinduf oasis and advanced into Algeria, Mauritania and Morocco, while Spanish troops took Semara, overcoming the rebels in 1934.

[7] When the French deposed Sultan Muhammad V, in Morocco the National Liberation Army (ALN) was created and its Southern Division cooperated closely with Sahar-

an groups. However, Moroccan independence in 1956 and the dissolution of the ALN-South left Saharans alone to face the Spanish troops, and by 1958 the resistance was faltering.

[8] During this period the exploitation of phosphate deposits began at Bu-Craa. The Franco government, in close association with transnational companies, invested more than $160 million in Sahara, transforming the country, particularly the population distribution. In 1959, Al-Aiún had 6,000 inhabitants; in 1974 there were 28,000, while the per centage of nomads decreased from 90 to 16 per cent.

[9] With the progressive abandonment of the traditional nomadic way of life, tribal ties and relationships began to weaken in spite of the colonial administration maintaining tribal divisions through the political recognition of the *shiuj* (heads of clans) and nobles of the different tribal groups. The national identity card system also contributed to this, as these specified which tribe or faction the holder belonged to. Over the years, however, a new national identity has slowly been forged, transcending traditional divisions.

[10] In 1967, the Saharans founded the Al Muslim movement and, a year later, the Saharan Liberation Front. In 1973, the revolutionary

> In the 7th century, waves of migrants from Yemen integrated with people already in the area. Four hundred years later, the first Saharan confederation of peoples appeared.

leadership opted for armed struggle, and the Polisario Front was created. It was led by Uali Mustafa Seyid. The events of the war and UN resolutions in favor of independence for Sahara, led the Franco government to recognize the right of self-determination. Nevertheless, control was not relinquished, and instead the United National Saharan Party (PUNS) was formed.

[11] Spain withdrew from the area,

in exchange for the right to keep 33.7 per cent of the phosphate mines, and to continue fishing in Saharan waters for the next ten years. Spain also persuaded Morocco to relinquish its claim on Ceuta and Melilla.

[12] In 1974, the World Bank labelled the Western Sahara the richest territory in the Maghreb region because of its fishing resources, the richest in the world, and its phosphate deposits. These deposits are estimated at 1.7 billion tons in the Bu-Craa area, but there are thought to be another 10 billion tons in other parts of the region.

[13] In 1975 Morocco claimed sovereignty over the area; a claim rejected by the International Court of Justice in The Hague and replaced by requests for decolonization.

[14] Hassan II, the King of Morocco, organized the so-called "Green March" through the Sahara, a propaganda move which mobilized 350,000 Moroccans southward across the border to press for a solution and protest the Court's decision.

[15] The Spanish government ceded Sahara to Morocco and Mauritania in 1975, but the following year, on February 27, the Saharans proclaimed the Democratic Arab Republic of Sahara (RASD).

[16] The new African republic was created in Bir Lahlu, in the desert area of Saguia El Hamra, just a few kilometers from the Mauritanian border. A few hours earlier, in El Aaiún, the last representative of the

colonial administration had officially announced the end of Spain's presence.

[17] The new nation was recognized by several countries, but war still broke out with Morocco and Mauritania. In 1979, Mauritania, on the verge of bankruptcy, sought an end to the conflict and signed a peace treaty with the Polisario Front. Hassan's troops, however, intensified their attacks, with French and American support.

[18] The Polisario Front's military victories led to a diplomatic triumph at the June 1980 conference of the Organization for African Unity (OAU) in Freetown. Twenty-six African nations officially recognized the RASD as the legitimate title for the Sahara, and four months later the UN issued a resolution requesting Moroccan withdrawal.

[19] Between 1980 and 1981, 50 countries maintained diplomatic ties with the RASD. The military offensive continued and in March 1981 the city of Guelta Zemur was taken.

[20] The war continued, the Polisario Front controlled the coasts and levied fishing charges on the 1,200 boats operating in its territorial waters. Spain alone paid Morocco some $500,000 in fishing rights, per year. As part of its Development Aid Fund, Spain also granted Morocco 25 billion pesetas of aid, money which was in fact used for arms purchases.

[21] The RASD was accepted as a full member of the OAU in Novem-

ENVIRONMENT

The country is almost completely desert and is divided into two regions: Saguia el Hamra and Rio de Oro. It has one of the world's largest fishing reserves, but the principal source of wealth is mining, especially phosphate deposits.

SOCIETY

Peoples: The Polisario Front estimates the dispersed Saharan population at one million. These are primarily nomadic groups which differ from the Tuaregs and Berbers in their social and cultural organization. **Religions:** Muslim. **Languages:** Arabic and Spanish (official). Many Saharans also speak Hassani. **Political Parties:** The People's Liberation Front of Saguia al-Hamra and Rio de Oro (Polisario Front) founded on May 10, 1973 by Mustafa Seyid El-Uali. **Social Movements:** The Saguia el Hamra and Rio de Oro General Workers' Union (UGTSARIO).

THE STATE

Capital: El Aaiún. **Other cities:** Bu Craa, Dakhla, Semara, Guelta Zemur. **Government:** Mohamed Abdelaziz, President of the republic, secretary-general of the Polisario Front and president of the Superior Council of the Revolution. The Superior Council of the Revolution is the main governmental body of the Saharan Arab Democratic Republic (RASD). **National Holiday:** February 27, Proclamation of the Republic (1976).

ber 1984, and this was quickly followed by Morocco's withdrawal from the organization. On November 14 1985, the United Nations Decolonization Committee recognized the Saharan people's right to self-determination.

[22] Since the war began, the UN has acted as mediator between the Polisario Front and Hassan II. With Algeria's support, UN-sponsored talks were held and eventually a peace strategy was defined.

[23] On March 26 1988, the peace plan received a boost from the United States, when Under-Secretary of State Richard Murphy declared before the Senate that the US did not recognize Moroccan sovereignty over Sahara. He added that a negotiated settlement to the conflict should be sought. The Polisario leaders and Morocco agreed to hold a referendum on Sahara allowing the local population to choose between independence or annexation to Morocco.

[24] In July 1990, in Geneva, representatives of Morocco and the Polisario Front debated a code of procedure for the referendum. One of the major difficulties was deciding who had voter eligibility. The latest census of the Sahara dates from 1974 and, moreover, Morocco's administrative personnel in the area expect to have voting privileges. The dual agreement was that the

The Spanish government ceded Sahara to Morocco and Mauritania in 1975, but the following year, on February 27, the Saharans proclaimed the Democratic Arab Republic of Sahara (RASD).

UN should exercise "sole and exclusive" control over voting centers, counting of votes, and announcement of the result. The two parties also agreed to cease hostilities so that the referendum could be completed, and that the 160,000 Moroccan troops in the area would be gradually reduced to around 25,000. The troops would then withdraw in the 24 hour-period following announcements of the plebiscite's final results. Polisario Front forces, for their part, were to leave the area of Tindouf in Algeria.

[25] On April 29, 1991, the UN Security Council approved a peace plan establishing a cease-fire enforced on September 6, and the creation of a Mission for the Organiza-

The Moroccan government settled thousands of Moroccan citizens in the Saharan territory, to ensure their voter eligibility; furthermore, the cease-fire was ignored.

tion of a Referendum in the Western Sahara (MINURSO). The referendum was scheduled for January 1992.

[26] In the following months, the Moroccan government settled thousands of Moroccan citizens in the Saharan territory, to ensure their voter eligibility; furthermore, the cease-fire was ignored, and journalists were hindered. They also banned international observers as repression against the Saharans mounted. A propaganda campaign was carried out during this period where the referendum was presented as a confirmation of Morocco's ownership of the Sahara.

[27] In an unexpected change of foreign policy, King Hassan re-established diplomatic relations with Algeria, broken off in March 1976. This gave rise to speculation that Algeria might discontinue support of the Polisario Front. The renewal of relations with Algeria had a positive effect upon the implementation of the peace plan, as Algeria continued to promote direct dialogue between Morocco and the Polisario Front.

[28] The MINURSO was in charge of drawing up the electoral register, based on the 1974 census. This meant that a number of Saharans would not be able to vote, but, in exchange, all Moroccan immigrants who arrived after 1976 would also be excluded from the referendum. The Saharans of voting age living in the refugee camps in Algeria were to be transferred from the camps to their home towns.

[29] In December 1991, Johannes Manz, special representative of the UN secretary general in the Western Sahara, resigned, giving a clear indication of the difficulties experienced in the decolonization plan. In January 1992, the date scheduled for the referendum, the MINURSO were nowhere near completing the program of voter identification and the Saharan repatriation plan had not been completed. Meanwhile, 60,000 Moroccan soldiers were still stationed in the Sahara.

[30] In 1992, Brahim Hakim, a member of the Polisario Front leadership, returned to Morocco ending his exile in Algeria. Hakim announced that the armed struggle had become useless, and urged his followers to stop fighting. Within recent years, increasing numbers have been leaving the Polisario Front, because of numerous military setbacks.

[31] Algeria, which formerly supported Polisario's cause, also changed its policies, cutting off financial support and urging Polisario to seek a settlement to the conflict through UN negotiations.

[32] The Polisario Front accused Morocco of violating the UN mediated cease-fire. The capital, El Aaiún, and Smara had been virtually surrounded by the Moroccan army.

[33] Morocco continued trying to block a referendum, convinced - according to a number of observers - that stalling would favor Moroccan authority. After postponing the plebiscite several times during the following years, the Polisario Front threatened to resume war. At the present pace, the program for the identification of voters will be finished by 2005, according to the secretary-general of the Polisario Front, Muhammad Abdelaziz in May 1996.

[34] Meanwhile, 170,000 Saharans remain in refugee camps in the Tindouf region, a desert area of Algeria, near Western Sahara, where the Polisario Front has tried to organize schools and other services.

St Helena

Saint Helena

Population: 8,000 (1994)
Area: 122 SQ KM
Capital: Jamestown

Currency: Pound Sterling
Language: English

St Helena was uninhabited when Portuguese navigators arrived in 1502. In 1659 an outpost of the British East India Company was established on the islands, and since then St Helena has been an English colony. Of scant economic interest, the island acquired notoriety as the location of Napoleon's second exile, from 1815 until his death in 1821. The Malvinas War put it back on the map, a century and a half later.

[2] An English representative stated: "It was only with the help of Ascension Island and the labor force provided by St Helena, that we could recover the Falklands". This may justify the expensive maintenance of this British enclave through the Overseas Development Administration. The British could have a shared base like Ascension which is a US base and a bridgehead of the Royal Air Force, or they could simply transfer the territory to the US, as they did with Diego Garcia.

[3] The United Nations supports all nations' rights to self-determination and independence, and in December 1984, the UN General Assembly urged Britain to bolster the fishing industry, handcrafts and reforestation on the island and to foster awareness of the right to independence. Washington and London both voted against the resolution. The UN also questioned the existence of the military base in Ascension since there should not be any base in non-autonomous territories.

[4] On January 1 1989 a new constitution was instituted conferring greater powers on the members of the Legislative Assembly and enabling civil servants to become election candidates with the approval of the governor. The new constitution also reduced the voting age to 18.

[5] The island's only export is fish, but in the last few years, there has been a sudden decline in the total catch: 27.2 tons in 1985, and a mere 9.2 tons in 1990.

[6] St Helena is of scientific interest because of the its rare flora and fauna. The island has some 40 plant species unknown in the rest of the world.

PROFILE

ENVIRONMENT

Located in the South Atlantic some 2,000 km from the coast of Angola, slightly west of the Greenwich meridian. Of volcanic origin, the island is mountainous but has no mineral reserves.

SOCIETY

Peoples: The island's population is largely of mixed European (mostly British), Asian, and African descent. Some 6,000 natives of St. Helena have emigrated to South Africa and Britain. **Religions:** Protestant. **Languages:** English (official). **Political Parties:** St. Helena Progressive Party (SHPP), the St Helena Labor Party (SHLP), both inactive since 1976.

THE STATE

Official Name: St Helena Colony with Dependencies. **Capital:** Jamestown, 1,744 inhab. (1989). **Other cities:** Longwood (where Napoleon Bonaparte was exiled). **Government:** Alan Hoole, governor appointed by the British crown. There is an Executive Council and a Legislative Council of 12 members.

COMMUNICATIONS

21.9 **telephones** per 100 inhabitants (1993).

ECONOMY

Consumer price index: 100 in 1990; 122.8 in 1994. **Currency:** pounds sterling.

Tristan da Cunha

The most important of a group of South Atlantic islands 2,400 km west of Cape Town, South Africa, under the administration of St Helena. The islands total 201 sq km (Tristan da Cunha 98 sq km; Inaccessible island 10 sq km, 32 km west of Cunha; Nightingale Islands 25 sq km, 32 km south of Cunha, and Diego Alvarez or Gough Island 91 sq km, 350 km south of Cunha).

[2] The 306 inhabitants (1988) are concentrated in Tristan da Cunha, the majority employed by the government and in a lobster processing factory. There were volcanic eruptions in 1961 and the island was evacuated, though the population returned in 1963.

[3] On Diego Alvarez there is a small weather station run by the South African government. The main religion is Protestant and the official language is English. There is an administrator, R. Perrys, who represents the government of St Helena and an advisory council with executive and legislative duties, comprising 8 elected and 3 appointed members.

Ascension

Wideawake Island

Ascension Island is of volcanic origin, covering 88 sq km. Its importance derives from its strategic location in the South Atlantic, 1,200 km northwest of St Helena. It is a communication relay center between South Africa and Europe, and the United States maintains a missile tracking station there under an accord with Britain. The island's naval installation and airbase were vital to Britain during the Malvinas war (April–June 1982) and afterwards as a base for the ships and planes that continue to supply the British troops occupying the islands which are claimed by Argentina.

[2] There is no native population on Ascension, and the majority are employees of the St Helena government. In 1988, of 1,099 inhabitants, 765 were from St Helena, 222 were British, 102 were American, and 10 were of other nationalities. These figures do not include British military personnel. The main religion is Protestant and the official language is English. The island's administrator, Brian Connelly, represents the government of St Helena.

St Kitts-Nevis

Population: 41,000
(1994)
Area: 267 SQ KM
Capital: Basseterre

Currency: E.C. Dollar
Language: English

St Kitts-Nevis

The island of Liamuiga, or "fertile land" in the language of the Carib Indians who lived there, was renamed St Christopher by Columbus on his second voyage to America, in 1493. It was not colonized by Europeans until 1623, when adventurer Thomas Walker established the first English settlement in the Caribbean. The neighboring island of Nevis was colonized five years later. After the rapid extermination of the Caribs, the English started to grow plantation crops, especially sugar cane, for which they used slaves brought over from their African colonies.

[2] After many disputes between France and Britain, St Kitts and Nevis became a British colony under the treaty of Versailles in 1793, together with some other Caribbean islands.

[3] In the 20th century, the decolonization process following World War II, gave the islands total internal autonomy, while foreign relations and defence were left to the colonial capitals. These islands joined the Associate States of the West Indies. In 1980, Anguilla formally separated from St Kitts and Nevis (see Anguilla), and the islands became governed by a prime minister and a parliament, both elected by universal suffrage.

[4] The Labor Party had been in office since 1967, but suffered a major defeat in the 1980 election, at the hands of an opposition coalition of the People's Action Movement (PAM) and the Nevis Reformation Party (NRP), making Kennedy Alphonse Simmonds prime minister. The opposition victory meant that independence, planned for June 1980, had to be postponed, as the NRP was opposed to a post-independence federation with St Kitts. The 1976 plebiscite showed that 99.4 per cent of the population of Nevis favored separation, but the secessionist unrest on Nevis appeared to ease since the NRP became involved in government.

[5] In March 1983, the new constitution was approved by parliament despite labor opposition which objected to a clause granting Nevis the right to secede. On September 19, 1983, St Kitts became an independent federated state.

[6] In the 1984 elections, Kennedy Simmonds and his government increased their parliamentary representation. Simmonds was re-elected in March 1989, in line with US interests in the region.

[7] In 1990, strikes broke among agricultural workers who had been denied a 10 per cent wage increase. Sugar companies responded by hiring close to 1,000 workers from St Vincent and the Grenadines, to cut sugar cane. Nevis prime minister Daniel Simeon promised that Nevis would secede from St Kitts by the end of 1992.

[8] In June 1992, the Concerned Citizens' Movement, in opposition, won the election in Nevis, obtaining 3 seats in the Nevis Assembly, ousting the Nevis Reformation Party (NRP) led by Daniel Simeon; together with the People's Action Movement, the NRP made up the main coalition. Weston Paris, Governor General Sir Clement Athelston's representative on Nevis, died in unclarified circumstances.

[9] The election of November 1993 was not conclusive and new elections were held in July, 1995. The People's Action Movement (PAM) was defeated by the St Kitts-Nevis Labor Party (SKNLP), led by Denzil Douglas. The SKNLP won 7 out of 11 seats. The PAM only retained one seat and Prime Minister Simmonds lost his seat.

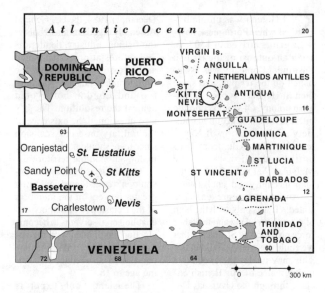

PROFILE

ENVIRONMENT

Divided between St Kitts, 168.4 sq km, and Nevis, 93.2 sq km. The two islands are in the Windward Islands of the Lesser Antilles. They are of volcanic origin, hilly, with a rainy tropical climate, tempered by sea winds which make the land fit for plantation crops, especially sugar cane.

SOCIETY

Peoples: Most of the population are mestizo, resulting from the integration of African enslaved laborers and European colonizers. There is a British minority. **Religions:** Mainly Protestant. **Languages:** English (official). **Political Parties:** the People's Action Movement (PAM) of Prime Minister Kennedy Simmonds; St Kitts-Nevis Labour Party led by Denzil Douglas; the Nevis Reformation Party (NRP) led by Simeon Daniel; the Labour Party (Workers' League); the United National Movement (UNM), headed by Eugene Walwyn; and the Concerned Citizens' Movement.

THE STATE

Official Name: Federation of St Christopher (St Kitts) and Nevis. **Capital:** Basseterre, 15,000 inhab. (1989). **Government:** Denzil Douglas, Prime Minister; Clement Athelston Arrindell, Governor-General appointed by Great Britain. There is a Parliament with 11 members chosen by universal suffrage (8 representatives of St Kitts and 3 of Nevis), and 3 appointed senators. **National Holiday:** September 19, Independence (1983).

DEMOGRAPHY

Annual growth: 1.4% (1992-2000). **Children per woman:** 2.6 (1992).

EDUCATION

Primary school teachers: one for every 21 students (1991).

COMMUNICATIONS

29.6 **telephones** per 100 inhabitants (1993).

ECONOMY

Consumer price index: 100 in 1990; 107.2 in 1992. **Currency:** 3 E.C. dollars = 1$ (1994).

St Lucia

Population: 160,000 (1994)
Area: 620 SQ KM
Capital: Castries

Currency: E.C. Dollar
Language: English

Saint Lucia

Before Christopher Columbus named it Santa Lucia in 1502, this island had already been conquered by the Caribs who had expelled the Arawaks.

[2] Neither the Spaniards nor the British succeeded in defeating local resistance, and in 1660 the French settled on the island, starting a dispute with Britain which was to last for 150 years. Over this period, the flag of St Lucia changed 14 times.

[3] In 1814, the Treaty of Paris transferred the island from France to Britain, which kept it until 1978. France left the legacy of *patois*, a pidgin language of mixed African and French.

[4] Under British rule, St Lucia became one big sugarcane plantation populated by African slave laborers. Agriculture is still the main economic resource but 20 years ago sugar gave way to banana cultivation.

[5] The island was part of the Colony of the Windward Islands, and between 1959 and 1962 St Lucia belonged to the West Indies Federation. In 1967 the island became more autonomous and was granted a new constitution as one of the Federate States of the Antilles.

[6] In the first elections held as an independent nation, in July 1979, Prime Minister John G. M. Compton and his United Workers Party (UWP), who had governed the is-

land since 1964, were beaten. The progressive St Lucia Labor Party (SLP) won.

[7] The new prime minister, Allen

> Neither the Spaniards nor the British succeeded in defeating local resistance, and in 1660 the French settled on the island, starting a dispute with Britain which was to last for 150 years.

Louisy, promised to help workers and peasants and to encourage small business as a means of curbing unemployment. George Odlum was deputy prime minister and minister of trade, industry, tourism and foreign affairs. He was also leader of the predominant SLP's "new left" wing, and he promoted the country's entry to the Non-Aligned Movement and established diplomatic relations with Cuba and North Korea. He also adopted a

policy of close collaboration with the neighboring island of Grenada, which had launched its own revolutionary process some months before.

[8] After repeated political crisis, the United Workers' Party won the 1982 and 1987 elections. Compton returned to power with a conservative platform, favouring a market economy and proposing adjustment measures recommended by the IMF. The increase in exports and

tourism revenues was not enough to leave economic crisis behind, which continued through the 1990s.

[9] The years 1994 and 1995 were marked by protests from banana plantation workers - the island's main export - and also by dock employees, demanding higher wages. In 1996, Vaughn Allen Lewis was elected president.

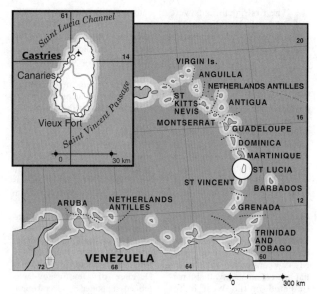

ENVIRONMENT

One of the volcanic Windward Islands of the Lesser Antilles, south of Martinique and north of St. Vincent. The climate is tropical with heavy rainfall, tempered by ocean currents. The soil is fertile, bananas, cocoa, sugar cane and coconuts are grown.

SOCIETY

Peoples: Most inhabitants are mestizos, from the integration of African slave laborers and European colonists. **Religions:** Roman Catholic 79%; Protestant 15.5%, of which Seventh-day Adventist 6.5%, Pentecostal 3 %; other 5.5%. **Languages:** English (official) and a local dialect (Patois) derived from French and African elements. **Political Parties:** The St Lucia Labor Party, currently in office. The United Workers' Party (UWP), conservative, led by former prime minister John Compton. George Odlum's Progressive Labor Party (PLP), which calls itself "the new left". The Citizens' Democratic Party, founded by a group of businessmen in 1995.

THE STATE

Capital: Castries, 52,862 inhab. (1987)
Government: Stanislaus James, Governor-General; Vaughn Allen Lewis, prime minister of the parliamentary British-style governmental system since April 2, 1996. **National Holiday:** 13 December, Independence Day (1978), and Discovery by Christopher Columbus.

DEMOGRAPHY

Annual growth: 1% (1992-2000). **Children per woman:** 3.2 (1992).

EDUCATION

Primary school teachers: one for every 29 students (1989).

COMMUNICATIONS

15.4 **telephones** per 100 inhabitants (1993).

ECONOMY

Consumer price index: 100 in 1990; 115.3 in 1994.
Currency: 3 E.C. dollars = 1$ (1994).

St Vincent

Saint Vincent

Population: 110,000 (1994)
Area: 388 SQ KM
Capital: Kingstown

Currency: E.C. Dollar
Language: English

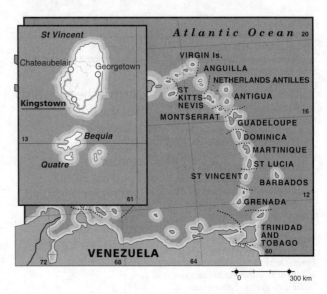

The first inhabitants were Arawaks later displaced by the Caribs, who were on the islands when Columbus arrived in 1498.

[2] In 1783, St Vincent became a British colony. However, the inhabitants resisted European conquest; the former slaves who had rebelled on the neighboring islands and those who took refuge in St Vincent joined the Caribs to oppose the invaders. In 1796, they were defeated and exterminated or deported.

[3] A plantation economy developed, using slave labor, the chief crops being sugar cane, cotton, coffee and cocoa. In 1833, St Vincent became part of the Windward Islands colony. In 1960, together with the Grenadine Islands, it was granted a new constitution with substantial internal autonomy. It also participated in the West Indies Federation until its dissolution in 1962.

[4] St Vincent became a self-governing state in association with the UK in 1969. The post of head minister - equivalent to that of prime minister, but with more limited powers - was held then by Milton Cato, together with the pro US St Vincent Labor Party (SVLP). Defence and foreign relations continued to be controlled by Britain, but independence was declared in October 1979.

[5] The elections in December 1979 reinforced the predominance of the SVLP, and the opposition PPP received only 2.4 percent of the vote.

[6] The new government faced an armed rebellion of Rastafarians led by Lennox "Bumba" Charles on Union Island. This rebellion was quickly put down by troops from Barbados.

[7] In the early 1980s, the government faced a serious socio-economic crisis, enabling the popular movements to gain ground. In May 1981, the national Committee in Defence of Democracy was formed, supported by several opposition parties, the labor unions and other organizations. Several days later, the government attempted to impose repressive legislation designed to maintain "public order", triggering mass protests.

[8] The economy of the archipelago is in the hands of transnationals that also control tourism, the banana plantations and the financial center. Another major source of income is remittances sent by migrant workers.

[9] Cato's government supported the US invasion in Grenada and sent a police detachment to join the occupation forces (see Grenada). Cato called early elections, but the economic crisis and the SVLP's tax policies meant the New Democratic Party (NDP) won the election, while the left found its vote reduced by half, to 7.2 per cent.

[10] In the May 1989 election, James Mitchell (NDP) was re-elected, going on to sign an agreement with the prime ministers of Dominica, St Lucia, and Grenada to create a new state of the four islands in 1990 (see Dominica).

[11] In February 1992, the government of St Vincent adhered to the Tlatelolco treaty to ban nuclear weapons in Latin America and the Caribbean.

[12] Given the Grenadines' secessionist feelings, which had already erupted in violence in 1980, Mitchell created a Ministry of Grenadine Affairs and named Herbert Young, a Grenadine, as Minister of Foreign Affairs.

[13] The SVLP took three of the fifteen seats in Parliament in the early elections held in February 1994, becoming the official opposition to the NDP government. In August, Mitchell survived a motion of no confidence. The opposition blamed him for the failure to fight the crisis and the fall in banana exports. This product represents 80 per cent of the income of St Vincent.

[14] Three men accused of murder were hanged in February 1995. The suspension of the moratorium on hanging, in place for many years, was condemned by Amnesty International.

PROFILE

ENVIRONMENT

Comprises the island of St Vincent (345 sq km.) and the northern part of the Grenadines (43 sq km) including Bequia, Canouan, Mustique, Matreau, Quatre, Savan and Union. They are part of the Windward islands of the Lesser Antilles. Of volcanic origin, the islands have fertile rolling hills. The climate, tropical with heavy rainfall and tempered by ocean currents, is fit for plantation crops. St Vincent is a leading arrowroot producer, a plant with starch-rich rhizomes, used in the manufacture of a type of paper employed in electronics. The population is concentrated on the island of St Vincent.

SOCIETY

Peoples: Descendants of African slaves 82%, mixed 14%; there are also European, Asian, and Indigenous minorities.
Religions: Anglican, Catholic and other Protestant **Languages:** English (offficial); also a local dialect. **Political Parties:** The New Democratic Party (NDP), holds 12 seats in Parliament. The St Vincent Labour Party (SVLP), 2. The National Unity Movement (MNV), which has one seat, and the United People's Movement (UPM), make up a Marxist coalition. The National Reformist Party (NRP), led by Joel Miguel.

THE STATE

Official name: Commonwealth of St Vincent and the Grenadines **Capital:** Kingstown, 15,824 inhab. in 1993. **Government:** David Jack, Governor-General appointed by Britain, since 1989. James Fitz-Allen Mitchell, Chief Minister and head of state since 1984. There are 13 elected representatives in Parliament and a Senate with 6 members. **National Holiday:** October 27, Independence Day (1979).

DEMOGRAPHY

Annual growth: 2.3% (1992-2000). **Children per woman:** 2.5 (1992).

EDUCATION

Primary school teachers: one for every 20 students (1990).

COMMUNICATIONS

14.8 **telephones** per 100 inhabitants (1993).

ECONOMY

Consumer price index: 100 in 1990; 133 in 1994. **Currency:** 3 E.C. dollars = 1$ (1994).

Samoa

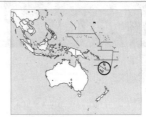

Population: 164,000 (1994)
Area: 2,840 SQ KM
Capital: Apia

Currency: Tala
Language: Samoan and English

Samoa

The archipelago of Samoa has been inhabited since at least 1,000 BC. The people developed a complex social structure. Four of these local groups still hold a privileged position: the Malietoa, the Tamasese, the Mataafa and the Tuimalealiifano.

[2] The Dutch were the first Europeans to visit the islands in 1722, but colonization did not begin until the end of the 19th century.

[3] In 1855, Germany finally occupied the islands. German merchants bought copra with Bolivian and Chilean currency valued at 10 times its real worth. In 1889 a new treaty recognized the US rights over the part of Samoa located east of meridian 171; "rights" which the US still retains. The western half remained under German rule.

[4] In 1914, upon the outbreak of World War I, New Zealand occupied the German part of the island, which was later granted to Aotearoa (New Zealand) by the League of Nations as a trust territory.

[5] In 1920, an influenza epidemic killed 25 per cent of the population. At about the same time, the 'Mau' movement began to spread throughout the archipelago, preaching resistance to the foreign governments. Samoans carried out a nine-year-long civil disobedience campaign, which eventually became a

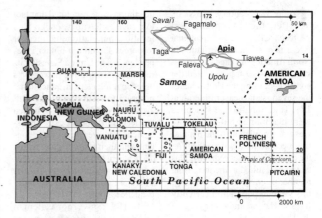

vigorous pro-independence movement.

[6] In 1961 after intense protests and pressure from the UN, a plebiscite was held for Samoans to vote on independence. This was achieved the following year, with a Constitution based largely on the traditional social structure and the executive power in the hands of two rulers, Tupua Tamasase Meoble and Malietoa Tanumafili. Only the leading group (*matai*), around 8,500 in all, are eligible to vote.

[7] After being elected prime minister in 1970, Tupua Tamasese Lealofi launched a battle against the *matai*. He also backed the establishment of foreign corporations on the archipelago, despite strong opposition.

[8] The 1976 elections were won by the opposition. Tupuola Tais became prime minister. In 1979 he retained office by only one vote in parliament. The increase in oil prices and a drop in exports led his government to place strict controls on certain basic consumer items such as rice, meat and poultry, which are largely imported. The price of cocoa and copra fell, further aggravating the economic and financial situation, making it necessary for 2,000 Samoans to emigrate each year.

[9] In February 1982, Va'al Kolone, leader of the Party for the Protection of Human Rights, came to power. In September he was removed from government, under accusations of corruption and abuse of power.

[10] In April 1988, Tofilau Eti Alesana came to power; his political party won an absolute majority in the legislative in the 1985 elections, when it obtained 31 of the 47 seats. In spite of the political changes, the country's economic situation continued to be critical.

[11] Malnutrition is on the increase; one in six preschool children and 11 per cent of all primary school children are under-nourished. The islands have 200,000 hectares of arable land but less than a third of this is cultivated.

[12] The 1991 constitutional reform extended the parliamentary term from three to five years, and increased the number of seats from 47 to 49. Fiame Naomi became minister of education that year, the first woman to be appointed to the cabinet.

[13] As well as the traditional crops of copra, cocoa and fruit, new export products have appeared in the last decade: quality woods, livestock and textiles. However, in 1993 and 1994, exports suffered a serious slump, provoking a trade imbalance. In 1994, the country exported barely 4 per cent of the value of its imports. The money sent home by Samoan workers abroad was an important source of income.

[14] Following advice from the United States, Samoa opened up to foreign investment. There are large tourist development projects underway, which imply the construction of hotels and airports.

[15] A petition of more than 80,000 signatures was presented to the government in 1995 asking for the tax on goods and services to be annulled. The chief auditor and state treasurer had to give up their posts when an investigation proved they were involved in corruption cases.

PROFILE

ENVIRONMENT

Includes the islands of Savai'i, 1,690 sq.km, 40,000 inhab.; Upolu, 1,100 sq.km, 110,000 inhab., Manono and Apolina, in Polynesia, northeast of the Fiji Island. The eastern portion of the Samoa archipelago is under US administration. The islands are of volcanic origin, mountainous, with fertile soil in the lowland areas. The climate is tropical, tempered by sea winds.

SOCIETY

Peoples: Samoans are mostly Polynesian. "Euronesians" (a result of European and Polynesian integration) make up 10% of the population. There are Europeans and Pacific islanders.
Religions: Congregational 47.2%; Roman Catholic, 22.3%; Methodist, 15.1%; Mormon, 8.6%; other, 6.8%.
Languages: Samoan and English are the official languages.
Political Parties: The Human Rights Protection Party (HRPP) holds 26 seats in Parliament since April 1996; the Samoa National Development Party has 13 and independent candidates won the other 10.

THE STATE

Official Name: Samoa i Sisifo. **Capital:** Apia, 34,126.inhab. (1991). **Government:** Malietoa Tanumafili II, Head of State for life. Tofilau Eti Alesana, Prime Minister. By the 1991 constitutional reform, 2 of the 49 representatives are elected by universal suffrage while the rest are chosen by the Heads of the Samoan clans and families. After the death of Tanumafili II, the president will be elected by direct vote. **National Holiday:** June 10, Independence Day (1962).

DEMOGRAPHY

Annual growth: 0.5% (1992-2000). **Children per woman:** 4.5 (1992).

EDUCATION

Primary school teachers: one for every 27 students (1986).

ECONOMY

Per capita GNP: $1,000 (1994).
Annual growth: -0.30% (1985-94).
Consumer price index: 100 in 1990; 128.9 in 1994.
Currency: 3 tala = 1$ (Oct.93).

American Samoa

Population: 55,000 (1994)
Area: 200 SQ KM
Capital: Pago Pago

Currency: US Dollar
Language: Samoan and English

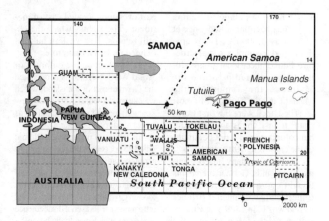

American Samoa

In the 18th century, 150 years after the Europeans had reached the islands, the colonial interests of Germany, Britain and North America disputed the possession. The 1899 treaty settled the conflict, granting the United States the seven islands east of Meridian 171. Traditional social structures were maintained but agriculture was not stimulated so the population became totally dependent on the external colonial economy. This situation resulted in the increasing number of emigrants; over half of the Samoan population currently lives in Hawaii and other parts of the United States.

[2] On December 5, 1984, the UN General Assembly considered Eastern Samoa's right to self-determination and independence. A unanimous vote reiterated that factors such as territory, geographic location, population and meager resources should not hinder independence. The United States, in its role as administrative power, was urged to implement an educational program to assure Samoans' full awareness of their rights, to hasten the decolonization process. The islanders, however, seem to be content with their present state, which allows them to emigrate to the United States without restrictions. There are no organized pro-independence groups.

[3] In 1984, governor Coleman (elected in the first elections for governor held in 1977) submitted proposals for a new Constitution in American Samoa for ratification by the US Congress. The proposals were withdrawn in May of the same year, because it was feared that they would be harmful to the interests of US citizens. In November, A.P. Lutali was elected governor and Faleomavaega Eni Hunkin became vice-governor.

[4] In July 1988, the delegate to the US house of representatives, Fofo Sunia, announced that he would not stand for re-election because he was going to be subjected to official investigations after accusations of financial mismanagement. In October, he was sentenced to a 5 to 15 week term in prison for fraud. Hence, Eni Hunkin replaced Sunia. In November, Coleman was re-elected for his third term and Galeani Poumele replaced Hunkin as vice-governor.

[5] In the second half of 1986, the governments of American Samoa and Samoa signed an agreement to create a permanent committee for the development of both countries in tourism, transport, and fishing.

[6] Economic development is dependent on the 90 per cent of Samoan foreign trade going to the US. Tuna fishing and tuna processing plants are the backbone of the private sector economy, with the canneries, as the second-largest employers, exceeded only by the government.

[7] Despite reforms to the 1967 constitution during the 1980s, proposed changes were not ratified by the United States Congress. According to the constitution currently in effect, in addition to a governor, who is elected for a four-year term, there is also a legislature, or "Fono" with an 18-member Senate elected every four years by the Matai, or clan heads. There is also a 20-member House of Representatives elected by direct popular vote for two-year terms. Women do not have the right to vote.

[8] After his re-election for governor in 1992, Lutali took measures to cut public spending, especially by reducing the number of government employees. The projected social security reform in the United States and its dependencies led to a debate in the second half of 1996 about the consequences for the inhabitants of American Samoa.

[9] Figures from the archipelago expressed the concern for the reform's effects on American Samoan residents - regarded as US "nationals" but with fewer rights than the "citizens" of that country - and on immigrants, specifically those from other Polynesian territories.

> Economic development is dependent on the 90 per cent of Samoan foreign trade going to the US.

PROFILE

ENVIRONMENT

The island occupies 197 sq.km of the eastern part of the Samoan archipelago, located in Polynesia, slightly to the east of the International Date Line, northwest of the Fiji islands. The most important islands are Tutuila (where the capital is located), Tau, Olosega, Ofu, Annuu, Rose and Swains. Of volcanic origin, the islands are mountainous with fertile soil on the plains. The climate is rainy and tropical, tempered by sea winds. There is dense, woody vegetation and major streams of shallow waters. The main export is fish, especially tuna.

SOCIETY

Peoples: 46,800, including the military personnel stationed in the area. Samoans are mostly Polynesians; there are Euronesians (a result of European and Polynesian integration) and a European minority.
Religions: Christian; Protestant and Catholic.
Languages: Samoan, predominant, and English are the official languages.

THE STATE

Official Name: Territory of American Samoa. **Capital:** Pago Pago, on Tutuila, 3,050 inhab. (1980). **Other city:** Fagatogo, 1,340 inhab. (1970). **Government:** A.P. Lutali, governor. There is a bicameral Legislative Power, chaired by Letuli Toloa. Samoans are considered US "nationals", but without the right to vote in presidential elections while living on the islands.

EDUCATION

Primary school teachers: one for every 15 students (1991).

COMMUNICATIONS

15.4 **telephones** per 100 inhabitants (1993).

ECONOMY

Consumer price index: 100 in 1990; 109.2 in 1993.
Currency: US dollars.

São Tome and Principe

Population: 125,000 (1994)
Area: 960 SQ KM
Capital: São Tomé

Currency: The dobra
Language: Portuguese

São Tomé e Príncipe

S ão Tomé island was probably uninhabited when it was first visited by European navigators in the 1470s. Thereafter, the Portuguese began to settle convicts and exiled Jews on the island and established sugar plantations there, using slave labor from the African mainland. Strategically located 300 kms off the African coast, the islands' natural ports were used by the Portuguese as "supply stops for ships" in the 15th century. Dutch, French, Spanish, British and Portuguese slave traders bought enslaved African laborers to be sold in the American colonies. Some of those slaves remained on the islands, which later became the leading African producers of sugarcane

[2] Rebellion soon broke out and after the failure of an insurrection headed by Yoan Gato, a slave named Amador led a revolt that succeeded in controlling two-thirds of the island of São Tomé, where he proclaimed himself ruler.

[3] Soon defeated, the rebels hid in *quilombos* (guerrilla shelters in the forest) after burning their crops. The landlords who moved to Brazil with their enslaved workers took with them the seed of insurrection which quickly produced *quilombos* in Brazil, some of which, such as the Palmares *quilombo*, became true republics and held out for nearly a century.

[4] In São Tomé and Príncipe, agriculture disappeared for three centuries. In the 17th century the island was held briefly by the Dutch. After a period of decline, the colony recovered its prosperity in the late 19th century with the cultivation of cocoa.

[5] Even after abolition was declared in 1869, slavery continued in a disguised manner ("free" workers signed contracts for nine years at fixed salaries), leading to revolts and an international boycott against the "cocoa slavery" of the Portuguese colony in the early 20th century.

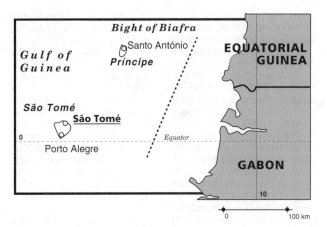

[6] This neo-slave system continued until the mid-1900s. A Society for Immigration of São Tomé organized the modern slave trade, "hiring" plantation workers in other Portuguese colonies: Angola, Cape Verde, Guinea and Mozambique. This flow "re-Africanized" the country as the "filhos da terra" (sons of the earth), the result of several centuries of integration between the natives and the Portuguese, mixed with the African immigrants. During the colonial regimes of Salazar and Caetano, repression was particularly harsh. In February 1953, over 1,000 people were killed in Batepá in less than a week.

[7] This massacre revealed the need for the rebels to join forces, and in 1969 the Movement for the Liberation of São Tomé and Príncipe (MLSTP) was founded, with two main objectives: independence and land reform.

> Foreign companies owned 90 per cent of the land of São Tomé and despite the fertile soil, most food was imported due to the island's monoculture policy.

[8] Foreign companies owned 90 per cent of the land of São Tomé and despite the fertile soil, most food was imported due to the island's monoculture policy. Rural workers were and still are one of the major pillars of the MLSTP, as shown by a 24-hour strike in August 1963, which paralyzed all the plantations.

PROFILE

ENVIRONMENT

The country comprises the islands of Sao Tomé (857 sq km) and Príncipe (114 sq km), and the smaller islands of Rólas, Cabras, Bombom and Bone de Joquei in the Bay of Biafra of the Gulf of Guinea, facing the coast of Gabon. The islands are mountainous, of volcanic origin, with dense rainforests, a tropical climate and heavy rainfall. Cocoa, copra and coffee are the main export crops.

SOCIETY

Peoples: Most are Africans of Bantu origin traditionally classified in five groups formed as a result of different migratory waves: the *Filhos da terra* (sons of the earth), descendants of the first enslaved workers brought to the islands and intermingled with the Portuguese; the *Angolares*, believed to descend from Angolans who came to the islands in the 16th Century; the *Forros,* descendants of freed slaves when slavery was abolished; the *Serviçais,* migrant workers from Mozambique, Angola and Cape Verde; and the native *Tongas*. Since independence, these categories have begun to disappear. In 1975, approximately 3,000 Portuguese colonists returned to Portugal. **Religions:** Roman Catholic, about 80.8%; remainder mostly Protestant, predominantly Seventh-day Adventist and an indigenous Evangelical Church. **Languages:** Portuguese (official). Creole, a dialect with Portuguese and African elements, is widely spoken. **Political Parties:** Movement for the Liberation of São Tomé and Príncipe (MLSTP), founded in 1972 from the Committee for the Liberation of São Tomé and Príncipe (CLSTP). Party of Democratic Convergence, in government; Democratic Coalition, and Christian Democratic Front, all emerged after the democratization process started in 1990. **Social Organizations:** Women's, Youth and Pioneers Organizations linked to the MLSTP.

THE STATE

Official Name: República Democrática de Sao Tomé e Príncipe. **Capital:** São Tomé, 43,420 inhab. (1991). **Other cities:** Trindade, 11,388 inhab.; Santana, 6,190 inhab.; Neves, 5,919 inhab.; Santo Amaro, 5,878 inhab. (1991) **Government:** Miguel Trovoada, President, re-elected on July 21, 1996; Armindo Vaz D'Almeida, Prime Minister, from December 31, 1995 to September 20, 1996, when he was censured by Parliament. Legislative: 55-member National People's Assembly. **National Holiday:** July 12, Independence Day (1975). **Administrative Divisions:** 7 Districts.

[9] As the island's terrain did not favor guerrilla warfare, the MLSTP launched an intensive underground political campaign that resulted in its recognition by the OAU and the Non-Aligned Nations. Together with the MPLA of Angola, the PAIGC of Guinea and Cape Verde, and FRELIMO of Mozambique, the MLSTP joined the Conference of National Organizations of the Portuguese Colonies. It was the only legitimate group in existence when, after the 1974 revolution, Portugal began to free its colonies.

[10] Rejecting reactionary claims that the islands were not colonies but overseas provinces, the MLSTP joined a transition government in 1974 and in the following year declared independence. Thereafter, accomplishments were impressive: banks and farms were nationalized, medicine was socialized, a national currency was created, a major administrative reform was launched to reorganize public administration, and numerous "centers of popular culture", based on the culture-building educational method of Brazilian Paulo Freire, were created as part of a literacy campaign.

[11] Opposing these reforms was a rightist faction led by Health Minister Carlos da Craça, who fled to Gabon to plot a mercenary invasion of the islands in early 1978. The thwarted conspiracy consolidated the unity of the MLSTP, which held its first congress in August 1978. Mass organizations to defend the revolution were fostered and a People's Militia was formed. Miguel Trovoada, prime minister since independence, was removed, thus strengthening the more progressive trends of the Party Coordinating Committee.

[12] On the economic front, the government turned its attention to cocoa, the islands' main export item. Cocoa production reached its zenith early in this century but fell steadily due to soil impoverishment, obsolete farming methods, pests and, after independence, emigration of almost all the Portuguese technicians.

[13] In March 1986, two opposition groups based outside the country - the São Tomé e Príncipe Independent Democratic Union (UDISTP) and the more radical São Tomé e Príncipe National Resistance Front (FRNSTP), founded by Carlos da Graça - announced the formation of an alliance called the Democratic Opposition Coalition. The aim of this coalition was to put pressure on the government to hold free elections. A month later, a fishing vessel with 76 members of the FRN-STP on board arrived in Walvis Bay, the South African enclave in Namibia. They asked the Pretoria government to supply military aid needed to destabilize the São Tomé government. These events led to Carlos de Graça's resignation as president of the FRNSTP. In May, he announced his willingness to cooperate with the government, on the condition that Cuban and Angolan troops stationed in the country be withdrawn.

[14] In 1985, in the midst of the worst drought in the country's history, the government sought to open up the economy: new legislation was designed to promote foreign investment and privatize the so-called "people's stores". Gradually, the state relinquished economic control, previously heavily dependent upon such key products as cocoa, coffee and bananas. Almost 70 per cent of these plantations

> Even as economic reforms were being implemented, the government took the first steps toward reforming the political structure created by the revolution.

were nationalized after independence. The government sought ways to attract foreign capital to the agricultural, fishing and tourist sectors. In 1989 there was a 20 per cent devaluation of the nation's currency (the dobra) which triggered a steep increase in basic commodity prices.

[15] Even as economic reforms were being implemented, the government took the first steps toward reforming the political structure created by the revolution. In October 1987, the Central Committee of the MLSTP announced a constitutional reform including the election by universal suffrage of the president of the republic and of the members of the National People's Assembly. This reform would enable independent citizens to be elected to the National People's Assembly; however, existing mechanisms whereby the MLSTP authorities designate presidential candidates, would remain in place. The MLSTP Central Committee also proposed that the post of prime minister be reestablished, and questioned the continuing existence of a single-party system. Concurrent with the open economic policy, there was a trend toward more liberal politics.

[16] Late in 1989, the MLSTP leadership discussed reform of the party statutes and of the national constitution. In March 1990, the People's National Assembly approved amendments to the Charter, later to be submitted to a referendum. These changes made possible a move to a multi-party system. Independent candidates were admitted in the legislative elections, and the tenure of the President was limited to no longer than two five-year terms.

[17] The first parliamentary elections after independence were held in January 1991. The opposition Democratic Convergence of Leonel d'Alva was voted into power. In March, the former prime minister Miguel Trovoada returned from exile and found no opposition in the presidential elections.

[18] The heads of state of São Tomé and Principe, Cape Verde, Guinea-Bissau, Mozambique and Angola - all former Portuguese colonies - met in February 1992. After being organized under single party systems, the five countries experienced rapid political change and economic liberalization processes.

[19] The social and economic situation of the country worsened in recent years as the result of an IMF and World Bank imposed austerity plan. Public sector salaries were frozen, a third of all civil servants were dismissed and the local currency was devalued by 80 per cent. While inflation fell, the price of basic foodstuffs quadrupled and unemployment reached 30 per cent.

[20] The island of Principe declared independence on April 29 1995 and established a five-member regional government.

[21] In August, a group of army officers took power in a bloodless coup. Negotiations led to the immediate re-establishment of the legal government.

[22] The General Assembly censured Prime Minister Armindo Vaz D'Almeida, in power since December 31 of the previous year, for "bad management, inefficiency, incompetence and corruption."

Saudi Arabia

´Arabiyah as-Sa´udiyah

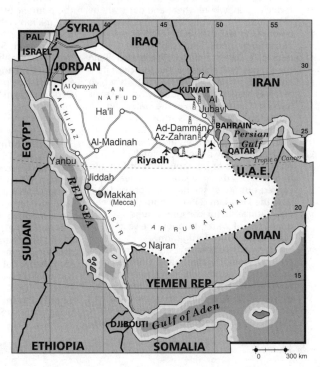

Population: 17,985,000 (1994)
Area: 2,149,690 SQ KM
Capital: Riyadh (Ar-Riyad) (royal capital)
Currency: Saudi Arabian Riyal
Language: Arabic

Arabia was drawn into the orbit of western Asiatic civilization toward the end of the 3rd millennium BC; caravan trade between south Arabia and the Fertile Crescent began about the middle of the 2nd millennium BC. The domestication of the camel around the 12th century BC made desert travel easier and gave rise to a flourishing society in South Arabia, centred around the state of Saba (Sheba). In eastern Arabia the island of Dilmun (Bahrain) had become a thriving entrepót between Mesopotamia, South Arabia, and India as early as the 24th century BC. The discovery by the Mediterranean peoples of the monsoon winds in the Indian Ocean made possible flourishing Roman and Byzantine seaborne trade between the northern Red Sea ports and South Arabia, extending to India and beyond. In the 5th and 6th centuries AD, successive invasions of the Christian Ethiopians and the counter-invasion of the Sasanian kings disrupted the states of South Arabia.

2 In the 6th century Quraysh - the noble and holy house of the confederation of the Hejaz controlling the sacred enclave of Mecca - contrived a series of agreements with the northern and southern tribes. Under this aegis, caravans moved freely from the southern Yemen coast to Mecca and thence northward to Byzantium or eastward to Iraq. Furthermore, members of the Quraysh house of 'Abd Manaf concluded pacts with Byzantium, Persia, and rulers of Yemen and Ethiopia, promoting commerce outside Arabia. The 'Abd Manaf house could effect such agreements because of Quraysh's superior position with the tribes. Quraysh had some sanctity as lords of the Meccan temple (the Ka'bah) and were

themselves known as the Protected Neighbors of Allah; the tribes on pilgrimage to Mecca were called the Guests of Allah.

3 In its enclave Quraysh was secure from attack; it arbitrated in tribal disputes, attaining thereby at least a local pre-eminence and seemingly a kind of loose hegemony over many Arabian tribes. The Ka'bah, through the additions of other cults, developed into a pantheon, the cult of other gods perhaps being linked with political agreements between Quraysh - worshippers of Allah - and the tribes.

4 Muhammad was born in 570 of the Hashimite branch of the noble house of 'Abd Manaf; though orphaned at an early age and in consequence with little influence, he never lacked protection by his clan. Marriage to a wealthy widow improved his position as a merchant, but he began to make his mark in Mecca by preaching the oneness of Allah. Rejected by the Quraysh lords, Muhammad sought affiliation with other tribes; he was unsuccessful until he managed to negotiate a pact with the tribal chiefs of Medina, whereby he obtained their protection and became theocratic head and arbiter of the Medinan tribal confederation (ummah). Those Quraysh who joined him there were known as *muhajirun* (refugees or emigrants), while his Medinan allies were called *ansar* (supporters). The Muslim era dates from the hijrah (hegira) - Muhammad's move to Medina in AD 622.

5 Muhammad's men attacked a Quraysh caravan in AD 624, thus breaking the vital security system established by the 'Abd Manaf house, and hostilities broke out against his Meccan kinsmen. In Medina two problems confronted him - the necessity to enforce his

role as arbiter and to raise supplies for his moves against Quraysh. He overcame internal opposition, removing in the process three Jewish tribes, whose properties he distributed among his followers. Externally, his rising power was demonstrated following Quraysh's failure to overrun Medina, when he declared it his own sacred enclave. Muhammad foiled Quraysh offensives and marched back to Mecca. After taking Mecca in AD 630 he became lord of the two sacred enclaves; however, even though he broke the power of some Quraysh lords, his policy thenceforth was to conciliate his Quraysh kinsmen.

6 After Muhammad's entry into Mecca the tribes linked with

Quraysh came to negotiate with him and to accept Islam; this meant little more than giving up their local deities and worshipping Allah alone. They had to pay the tax, but this was not novel because the tribal chiefs had already been taxed to protect the Meccan enclave. Many tribesmen probably waited to join the winner. From then on Islam was destined for a world role. Under Muhammad's successors the expansionist urge of the tribes, temporarily united around the nucleus of the two sacred enclaves, coincided with the weakness of Byzantium and Sasanian Persia. Tribes summoned to the banners of Islam launched a career of conquest that promised to satisfy the mandate of their new faith as well as the desire for booty and lands.

7 As the conquests far beyond Arabia poured loot into the Holy Cities (Mecca and Medina), they became wealthy centres of a sophisticated Arabian culture. Medina became a centre for Qur'anic (Koran) study, the evolution of Islamic law, and historical record. Under the caliphs - Muhammad's successor - Islam began to assume its characteristic shape. Paradoxically, outside the cities it made little difference to Arabian life for centuries. After the prophet's death, Omar, the second caliph, led the

PUBLIC EXPENDITURE

% of GNP — 1992

MILITARY 11.8, EDUCATION 6.2, HEALTH 3.1

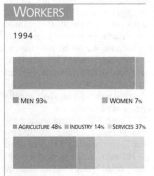

WORKERS

1994

MEN 93% WOMEN 7%

AGRICULTURE 48% INDUSTRY 14% SERVICES 37%

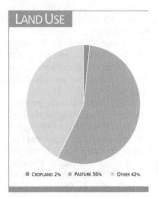

LAND USE

CROPLAND 2% PASTURE 56% OTHER 42%

ENVIRONMENT

The land occupies up to 80% of the Arabian peninsula. There are two major desert regions: the An Nefud in the north, and the Rub al-Khali in the south. Between the two lie the Nejd massif, of volcanic origin, and the plain of El Hasa, the country's only fertile region, where wheat and dates are cultivated. The country's cultivated land amounts to less than 0.3%; 90% of the agricultural products consumed, are imported. Oil extraction, concentrated along the Persian Gulf shores, is the source of its enormous wealth. The intense activity which revolves around the oil industry has raised levels of contamination in the water and along the coastline. (This situation was exacerbated during the Persian Gulf War: 640 km of beaches and swampland were affected by oil spillages totalling 4.5 million barrels of oil; thousands of fish and migratory birds were killed; the clean-up, carried out under the direction of the UN, cost the kingdom $450 million). The agricultural sector consumes huge quantities of water, through the vast system of artificial irrigation. Environmentalists have issued warnings on the shortage of water, made more acute by the slow rate at which water levels are replenished. Reports on this issue have estimated that, should the current rate of water use continue, this resource will be completely depleted in one or two decades.

SOCIETY

Peoples: Saudis are of Arab origin with a trace of African influence from the import of slave labor. In recent years there has been large scale immigration of Iranians, Pakistanis and Palestinians who have settled in the new eastern industrial areas, bringing the number of foreigners to an estimated five million (1992). However, it is believed that official demographic estimates are exaggerated. **Religions:** Islam, mostly Sunni orthodox Wahhabism (official); Christian minorities. **Languages:** Arabic. **Political Parties:** Not permitted. **Social Organizations:** Not permitted.

THE STATE

Official Name: al-Mamlakah al-'Arabiyah as-Sa'udiyah. **Administrative Divisions:** 5 regions and 14 districts. **Capital:** Riyadh (Ar-Riyad) (royal capital), 1,380,000 inhab. (1986 census). **Other cities:** Jeddah (administrative center),1,210,000 inhab.; Mecca (Makkah - religious center), 550.000 inhab. **Government:** Absolute monarchy. King Fahd ibn Abd al-Aziz as-Saud, Head of State and of the Government since June 13, 1982. Advisory Council, with advisory role to the king. **National Holiday:** September 23, National Unification (1932). **Armed Forces:** 105,000 (plus 57,000 active National Guard). **Paramilitaries:** 10,500: Coast Guard, 4,500; Special Security Force, 500.

Arab conquest. Within ten years the Arabs occupied Syria, Palestine, Egypt and Persia. With Muawiya, the caliphate became hereditary in the family of the Ummaias and the Arabs became a privileged caste which ruled over the conquered nations.

[8] In the 8th century, the borders of the Arabian Empire reached from North Africa and Spain to the west, to Pakistan and Afghanistan in the east. Upon moving the capital to Damascus, Syria became the cultural, political and economic center of the Empire, and it was in Damascus that the foundations of a new culture were laid. Greco-Roman, Persian and Indian components blended into an original combination in which science played an important role. Contrary to Muhammad's expectations, the Arabian peninsula was to remain on the sidelines within the enormous empire, except in religious matters. Mecca, although failing to match Baghdad or Damascus in socio-economic and cultural importance, continued to be the center of Islam and the destination of multitudinous pilgrimages that came from all over the world.

[9] This situation remained unchanged for centuries; the empire split up; the capital moved to Baghdad and the power of the caliphs passed over to the viziers, while culturally Arabic civilization at-

tained the highest standards in all fields of knowledge and artistic creation. Arabic became the language of scholars, from Portugal to India. But nothing changed in the land that gave birth to all this civilization; nomadic groups continued to shepherd their flocks; the settled population kept up their commerce, and rivalries between the two were frequently settled through war. As in Muhammad's time, demographic growth was channelled towards conquest, with the emigration of whole communities, such as the Beni Hilals in the 11th century. Trading caravans carrying supplies to Mecca and stopping points along the way, became much more frequent, while the ports became more active as a result of trade with Africa (see box in Tanzania: "The Zandj culture"). The peninsula was governed from Egypt, first by Saladin and later by the Mamelukes. The Turks ruled from the 16th century to the 20th century, without introducing any major changes in the socio-economic pattern of the Arab nation.

[10] Under Turkish rule the provinces of Hidjaz and Asir, on the Red Sea, had a reasonable degree of autonomy due to the religious prestige of the shereefs of Mecca, descendants of Muhammad. The interior, with Riyadh as the main urban center, became the Emirate of Najd

at the end of the 18th century, through the efforts of the Saud family supported by the Wahabite sect (known as the "Islamic Puritans"). In the 19th century, with Turkish assistance, the Rachidi clan forced Abd al-Rahman ibn Saud out of power; the ousted leader sought exile in Kuwait. In 1902, his son Abd al-Aziz, again with the support of the Wahabites, organized a religious-military sect, the Ikhwan, in which he enlisted nearly 50,000 Bedouins in order to reconquer Najd. Twelve years later the Saudis defeated the Rachidis and added the Al-Hasa region on the Gulf until then under direct control of the Turks, against whom the Saudi forces fought during World War I. At the end of the conflict, Britain, the predominant power in the area, faced a difficult situation. In exchange for Abd al-Aziz's continued anti-Turk campaign, Britain had promised to guarantee the integrity of his state. But for the same reason, it had also promised to make Hussein ibn Ali (the shereef of Mecca) king of a nation that would encompass Palestine, Jordan, Iraq and the Arabian peninsula.

[11] The Emir of Najd waited for some time. It seemed clear to him that Britain would not keep its word to Hussein: such a powerful kingdom, ruled by the Prophet's family with the capital in the holy city,

would alter the regional balance of power. But in 1924 Hussein proclaimed himself caliph (see Jordan). Abd al-Aziz invaded his territory immediately, despite English opposition and in January 1926 was declared King of Hidjaz and Sultan of Nadj in the great mosque of Mecca. Six years later the "Kingdom of Hidjaz, of Nedj and its dependencies" was formally unified under the name of Saudi Arabia.

[12] In 1930, the monarch gave US companies permission to drill for oil. When he died in 1953, his son Saud squandered all the revenues paid into the kingdom's treasury by the Arab-American Oil Company (Aramco), on palaces, harems, fancy cars and casinos on the French Riviera. In 1964 the country was on the verge of bankruptcy when Saud was ousted by his brother Faisal, an able diplomat who had also proved a valiant soldier in the wars. Monogamous, deeply religious and austere to the point of asceticism, Faisal gave new life to the country's economy and began to invest petrodollars in ambitious development programs, though maintaining the traditional feudal structure headed by the autocratic ruler.

[13] Faisal's religious views led him to bluntly reject the Soviet Union and any other system linked with atheism, including Nasser's nationalism in Egypt, as well as Iraqi or Syrian Ba'athism. His strategic alliance with the United States was seen as "natural", but was disturbed by US support of Israel after the war of 1967 (Faisal used to say that his only dream was to be able to pray one day in Omar's mosque, in a freed Jerusalem) and rivalries with neighboring Iran, which under the Shah Pahlevi also played watchdog for Washington's interests in the Gulf.

[14] During the 1973 Arab-Israeli war, Faisal supported an oil embargo on the countries backing Israel, including the US. The sudden oil shortage allowed OPEC countries to increase oil prices in a short space of time augering a new era in international relations. On March 25, 1975, Faisal was murdered by an apparently insane nephew. His brother Khaled, almost paralyzed by a rheumatic illness, was named as his successor. However, due to his paralysis, the governing duties in an era of superabundance fell to his brother, Crown Prince Fahd ibn Abdul Aziz.

[15] The oil revenue, which amounted to $500 million a year when Faisal was crowned in 1964, had gone up to almost $30 billion when he died. That same year the Bank of America closed its balance with assets of $57.5 billion, and West Germany's exports amounted to $90 billion. These figures reveal that, when compared with other major economic powers, Saudi economic power has been overrat-

ed. These are still staggering figures for the Third World: three times the income for Egypt, whose population is eight times bigger.

[16] New cities, universities, hospitals, freeways and mosques sprouted up everywhere. Yet there was surplus money available. Instead of planning oil production to meet the country's needs, which would have avoided the fall in prices and the weakening of OPEC during the 1980s, fortunes accumulated in western banks. Thus, Saudi Arabia tied its future more closely with the industrialized capitalist world. Besides, it created a surplus of money in circulation, which the banks lent to Third World countries in some irresponsible ventures. This gave rise to the 1984-85 foreign debt crisis, with a rise in interest rates.

[17] The Muslim fundamentalist groups denounced the Saudi dynasty for allegedly betraying Islam, leading to several violent confrontations with government forces. The theological basis which legitimized the ruling autocracy may be undermined by the progressive rise to power of new members of the ruling family, trained in European and US universities and military academies, rather than in the traditional desert-tent Koranic schools.

[18] Since the overthrow of the Shah of Iran, the Saudi government drew closer to the US. After King Khaled's death, on June 13 1982, his brother Fahd, the creator of Saudi Arabia's foreign policy, became king, without dispute.

[19] The year before, the new monarch had created a peace plan for the Middle East which had been approved by several Arab countries, the PLO and the United States, though the plan collapsed in August, 1981 due to Israeli opposition. This plan proposed the creation of a Palestine State, with Jerusalem as its capital, the withdrawal of Israel from the occupied Arab territories, and the dismantling of the Jewish colonies set up in 1967, as well as the recognition of the right of every State in the area to "live peacefully". On September 26, 1982, during the Haj or pilgrimage to Mecca, the king condemned Israeli intervention in Lebanon, which he called "criminal aggression". He also criticised Soviet intervention in Afghanistan, accused the Ayatollah Khomeini of attempting to destabilize the regime through sabotage, and generally condemned the "influence of great powers" in the Muslim world.

[20] Naval bases in Jubail and Jeddah were built under the supervision of the US Army Corps of Engineers. Also, investments and bank deposits by the Saudi State and nobility are closely linked with the performance of the US economy: two-thirds of the huge amount of Saudi petrodollars invested abroad have gone into corporation

stocks, treasury bonds and bank deposits in the US.

[21] The official development plans established an industrial policy to reduce dependency from oil resources. Priority was given to the petrochemical industry, and there was a proliferation of steel mills, mechanical manufacturing plants, heavy industry, and others.

[22] The industrial infrastructure is concentrated in two new cities: Jubail, on the shores of the Gulf, and Yanbu, on the Red Sea. They are connected with the eastern hydrocarbon deposits through gas and oil pipelines.

[23] The 1985-89 five-year plan promised a "wide income redistribution". Yet, that intention coincided with the first signs of economic trouble in the kingdom. In 1984, due to another drop in the price of crude, the official budget closed with a deficit for the first time. The Minister of Industry, Ghazi Al Gosaibi, was forced to resign after writing a poem which made reference to corruption.

[24] Million-dollar investments in all kinds of mainly US weapons continued, aiming to secure the Gulf coast against the potential Iranian aggression and satisfying the ambitions of an increasingly powerful and influential army.

[25] During the Iran-Iraq war, Saudi Arabia backed Iraq financially, afraid that the Iranian Islamic revolution might spread over the Gulf. Fahd changed his title of King for that of Guardian of the Holy Sites, but every year the pious pilgrims to Mecca protested more loudly, angry about the alliance between Riyadh and Washington and by what they consider a commercialization and westernization of the sacred places, surrounded today by shopping centers, highways and other symbols of transnational culture. On August 1, 1987, a march by Iranian women and maimed war veterans in Mecca was fired upon by Saudi police. Hundreds of pilgrims were killed. In retaliation, the embassies of Saudi Arabia and Kuwait in Tehran were attacked and burned, and the relations between both countries grew very tense.

[26] After the Iraqi invasion of Kuwait, in August 1990, Saudi Arabia was the scene of a huge military deployment by a multinational coalition led by the United States. Apart from the loss of human lives, the war caused a huge ecological disaster. Since the end of the war, Saudi Arabia has spent some $14 billion on US weapons.

[27] In March 1992, King Fahd issued several decrees aimed at decentralizing political power. This legislation, of 83 articles, is called "The Basic System of Government". Amongst other amendments, it established an Advisory Council, which has the right to review all matters of national policy.

DEMOGRAPHY

Urban: 79% (1995). **Annual growth:** 0.5% (1992-2000). **Estimate for year 2000:** 22,000,000. **Children per woman:** 6.4 (1992).

HEALTH

One **physician** for every 704 inhab. (1988-91). **Under-five mortality:** 36 per 1,000 (1994). **Calorie consumption:** 106% of required intake (1995). **Safe water:** 95% of the population has access (1990-95).

EDUCATION

Illiteracy: 37% (1995). **School enrolment:** Primary (1993): 73% fem., 73% male. Secondary (1993): 43% fem., 54% male. University: 14% (1993). **Primary school teachers:** one for every 14 students (1992).

COMMUNICATIONS

101 **newspapers** (1995), 108 **TV sets** (1995) and 104 **radio sets** per 1,000 households (1995). 9.1 **telephones** per 100 inhabitants (1993). **Books:** 85 new titles per 1,000,000 inhabitants in 1995.

ECONOMY

Per capita GNP: $7,050 (1994). **Annual growth:** -1.70% (1985-94). **Annual inflation:** 2.80% (1984-94). **Consumer price index:** 100 in 1990; 105.6 in 1994. **Currency:** 4 Saudi Arabian riyals = 1$ (1994). **Cereal imports:** 5,186,000 metric tons (1993). **Fertilizer use:** 1,438 kgs per ha. (1992-93). **Imports:** $22,796 million (1994). **Exports:** $38,600 million (1994).

ENERGY

Consumption: 4,744 kgs of Oil Equivalent per capita yearly (1994), -435% imported (1994).

However, the ultimate decision-making power always belongs to the king.

[28] The royal decrees also created the "mutawein", or religious police, whose mandate is to ensure the observance of Islamic customs. The home is defined as a sacred place, and therefore the state (the religious police) may enter without legal authorization. This situation has led to the detention of several alleged violators of the law.

[29] In the past few years, and especially since 1991, both the powerful business community and the Islamic fundamentalists have been agitating, with the aim of extending their political influence and forcing the removal of a number of government officials accused of corruption.

[30] The Iraqi invasion of Kuwait, and the subsequent crisis in the Persian Gulf, exposed the kingdom's political and military dependence upon the United States. It also served as a catalyst for political reform within Saudi Arabia itself.

[31] In 1992, Fahd confirmed his brother Abdullah Ibn Abdul Aziz al-Saud as heir to the throne. Human rights organizations, such as Amnesty International, expressed their concern throughout 1993 over the arbitrary arrest of mainly Christians and Shiite Muslims, but also Sunnis.

[32] In the economic field, Fahd announced in May 1994 his goal to implement a program to privatize large state enterprises, such as the national airline and telephone companies. Despite a small rise in the price of oil in 1995, the King decreed significant reductions of public spending, which amounted to 20 per cent in the education sector.

[33] According to several observers, growing social differentiation aggravated tensions among Saudis themselves, regarded as privileged compared to the millions of foreign workers living in the country. The attempt against the offices of US advisers in Riyad, in November, killing at least 6 people, was seen as proof of those tensions.

[34] King Fahd became increasingly ill late in the year and was temporarily replaced by Abdullah, amidst rumours of fights for power within the royal family and of the future king's hostility toward the United States and his sympathy for radical Islamic sectors. However, Fahd had fully recovered and was back in office by the second half of 1996.

Senegal

Sénégal

Population: 8,263,000 (1994)
Area: 196,720 SQ KM
Capital: Dakar

Currency: CFA Franc
Language: French

The banks of the Senegal River have been inhabited for many centuries by peoples who had been Islamicized through contact with neighboring Arab countries. This group of countries make up the region known as the Sahel. The Wolof (who constitute more than a third of the population), Fulani, Pulaar and other peoples, all lived within the area of modern Senegal.

[2] When the French occupied the area during the 17th century, they incorporated it into the triangular world trade pattern of the time: European manufactured goods were exchanged for slaves, sold in the Antilles, where colonial European powers produced rum and sugar, which were sold back in Europe.

[3] After slavery was abolished by the French revolution of 1848, the Senegalese became "second class citizens" of the French Empire, with one political representative in Paris. At this time Senegal exported thousands of tons of peanuts per year and supplied the French army with soldiers.

[4] During the second half of the 19th century, there were frequent rebellions among the Muslim leaders, and it was only in 1892 that the French managed to fully "pacify" the country. (see "The Fulah States").

[5] The pan-African movement inspired Senegalese Leopold Sédar Senghor and Martinique's Aimé Césaire, and together in Paris, they created the concept of négritude, an idea first presented to public opinion in 1933. Négritude stood against the imposition of French cultural patterns, which were enforced throughout the colonies. The concept was designed to serve as "an effective liberation tool". In Senghor's view, it had to escape "peculiarism" and find a place within contemporary movements

of solidarity, that is, to achieve a definite political profile.

[6] Senghor had been a member of the French maquis during World War II, he had fought against fascism and was elected representative to the French National Assembly in 1945. Three years later, he brought about the foundation of the Senegalese Progressive Union, which demanded greater autonomy for the colonies, though not independence.

[7] Finally, on April 4, 1960, Senegal declared independence, and on September 5, adopted a republican system. Senghor was elected president and through successive re-elections he stayed in power for two decades. He was finally replaced by Abdou Diouf on January 1 1981.

[8] The ideology of "African socialism" promoted by Senghor was based on the premise that traditional African agrarian society had always been essentially collectivist. Senghor believed that socialism already existed on the continent and therefore did not need to be imposed. This "collectivism" actually mainly served to provide cheap labor for the export-oriented production of peanuts and cotton. 82 per cent of the nation's industry was French-controlled.

[9] In an effort to keep up with political changes in France, Senghor requested membership of the Socialist International. To this end he pushed through a constitutional reform restricting political representation to three political parties: a "liberal democratic" party (the Senegalese Democratic Party), a Marxist-Leninist party (the African Party for Independence) and a "social democratic" party (his own, renamed the Senegalese Socialist Party).

[10] In the 1980s Diouf brought flexibility to the rigid political system he inherited from his predeces-

sor. This prevented the major political adversary Abdoulaye Wade's Democratic Party of Senegal (PDS), from uniting all the government opposition forces. In 1982, in the southern province of Casamance, on the border with Guinea-Bissau, there was a resurgence of separatist activity, leading to the creation of the Casamance Movement of Democratic Forces (MFDC) led by an abbot, Austin Diamacoune Senghor. The people of Casamance are primarly from the Diole ethnic group, animists who historically pride themselves on their independence and resistance to the Islamic hierarchical societies of the north. In December 1983, after successive confrontations with the Senegalese police forces, many leaders of the MFDC were detained, among them Abbot Diamacoune.

[11] The country has faced a series of serious droughts over the last 20 years. These are seen as part of the desertification process caused by climatic conditions and the French-imposed substitution of export crops for traditional food crops,. In the early 1980s the government gave in to pressure from the US, France, the World Bank, and the IMF, meaning the application of a structural adjustment plan. The elimination of agricultural subsidies triggered a rise in production costs and the prices of basic consumer goods.

[12] The drought afflicting the country since 1983 ruined peanut production (peanuts represent over one third of total exports), and the food

deficit required the import of 400,000 tons of rice.

[13] In 1988 the Socialist Party (PS) won another victory in the national elections, obtaining 72.3 per cent of the vote, while the Senegalese Democratic Party (PDS) won only 25.8 per cent. After the legitimacy of the election was called into question by various groups, government troops were mobilized to keep the peace. In the aftermath, hundreds of people were arrested, and the PDS candidate, Abdoulaye Wade, was forced to go into exile, only returning to Senegal in March 1989.

[14] A month later a border conflict broke out with Mauritania triggered by violent altercations between peasants and farmers, causing hundreds of casualties and forcing some 70,000 refugees to enter Senegal.

[16] In January 1991, Amnesty International accused the Senegalese government of human rights violations, on account of the tortures and murders committed in the Casamance region. Late in May, the government announced its intention to free political prisoners, including MFDC leader, Abbot Diamacoune Senghor as part of an amnesty settlement.

[16] This conflict affected the planned Senegambian integration, as Diouf criticized Gambian tolerance of the Casamance guerrillas and the signing of a mutual defence treaty with Nigeria (see box in Gambia).

[17] In July 1991, Diouf was elected president of the ECOWAS (Economic Community of the Western African

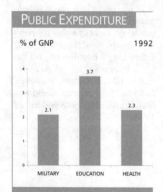

PUBLIC EXPENDITURE

% of GNP — 1992

	MILITARY	EDUCATION	HEALTH
	2.1	3.7	2.3

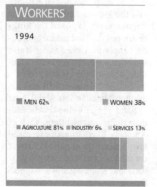

WORKERS

1994

MEN 62% · WOMEN 38%

AGRICULTURE 81% · INDUSTRY 6% · SERVICES 13%

The Fulah states

This African nation of shepherds (according to different sources, known as Fulah, Fulani, Peul, and also "nations of the western group"), were nomads in some cases and migrants in others. Throughout their history these peoples developed certain traits that explain the peculiar role they played in the area known as Sahel -south of the Sahara desert.

2   Their nomadic condition allowed them to interact with the sophisticated cultures of the north as well as the rudimentary local societies of the south, giving the Fulani a particular cultural flexibility. Pastoralists in hostile lands, they quickly became excellent warriors. Their traditional social system, hierarchically graded to an extent reminiscent of feudal echelons, favored a rapid emergence and concentration of centralized power among the Fulani.

3   In the ninth century, emigration led people out of present-day Mauritania, towards the south and the east, in search of better pastures. Most of their tribes were early converts to Islam; the Fulani became true travelling religious promoters making a significant contribution to the spread of Islam.

4   Futa Toro (present-day north Senegal) was the first state they established in the 10th century. Initially they paid tribute to Ghana but later became allies of the Almoravid and eventually extended their domination over Tekrour, central Senegal. The Fulani successively fell under Mali and Songhai rule but following the crisis of this empire, the Koli dynasty was extended to Futa Djalon (north Guinea), where it remained in control until the 18th century. In the 15th century, Fulah groups settled in Massina, in the Niger valley, north of Djénne, and their leader was appointed local ruler by the Sultan of Mali. Since the Fulani knew the area, and other nations had no connection with the sedentary population, Fulah leaders frequently served as local administrators and intermediaries of established states. Whether under Mali, Songhai or Bambar sovereignty, which they accepted equally, the Fulani of Massina maintained their local autonomy until the 19th century.

5   In the 16th century, other groups of the same nation settled in Liptako, in similar conditions. The need to defend the territory from the bold, neighboring Mossi (inhabitants of present-day Burkina Faso) allowed leader Ibrahima Faidu to claim greater autonomy. Though formally a subject of the Songhai empire, he set up a practically independent state that survived until 1810. Later, other Fulah groups settled among the Haussa in present-day Nigeria and in the north-central part of what is now Cameroon.

6   In the 18th century a revival of religious fervor among the Fulani, within the context of an Islamic rebirth, led to the rise of Mahdism (see Sudan) and had major political consequences throughout the area.

7   In Tekrour and Futa Toro, religious leader Abdel Kader Torodo succeeded in displacing the Koli dynasty, and in 1776 adopted the title of Almamy (variant of the Arabic al-Imam which means prayer leader), establishing a virtual theocracy.

8   A similar process occurred in Futa Djalon, led by two religious leaders Ibrahima Sory and Karamoko Alfa; however this situation was threatened by ensuing rivalries. In 1784, the two families came up with the Solomonic and original solution of "alternate theocracy". In effect, an Almamy from the Sory family ruled the country for two years and then "switched" power over to an Almamy from the Alfa family who undertook to return office to the Sory family two years later; a system that was in effect until 1848.

9   In the south of present day Niger, consolidation of the Fulah community was stimulated by Osman Dan Fodio's preachings. Born in 1754 and with a widespread reputation for holiness, he spoke in favor of the regeneration of Islam. The Haussa rulers (see Nigeria box) became aware of the danger and tried to control the movement. Osman called for a holy war, seized the state of Gobir in 1801 and proclaimed himself Sheik and then Commander of the Believers. Within a short time, he had captured all the Haussa states except Bornu (see Chad-box), where Ali Kaneime's troops fought off the Islamic crusaders. From the capital in Sokoto, Dan Fodio ruled over a large portion of today's Niger and Nigeria until his death in 1818. His successors held power as emirs of Sokoto until 1917.

10   One of Osman's disciples, Adama, assumed the leadership of the Fulani of the Benue Valley in north central Cameroon. In 1810, this relationship matured leading to the creation of the Emirate of Adamaua ruled by Adama's descendents until the German invasion of 1910.

11   In Massina, Hamadu Bari preached similarly and in 1816 his followers succeeded in overthrowing the Fulah local officials and the Bambar ruler of Ségou. Bari's Sheikdom gained recognition from Osman Dan Fodio, considered by the Fulani as their spiritual leader. His troops went on to seize Djene and Timbuktu (Tombouctou), and founded a state that survived until the 1860s.

12   Omar Saidu Tall (1797-1864), known as El Hadj Omar for his pilgrimage to Mecca, was Dan Fodio's last disciple. In 1848 Omar organized an army in Futa Djalon and set out to enlarge its boundaries. His troops overran the Bambar domain of Kaarta but failed in their attempt to recover Futa Toro, occupied by the French in 1860. Turning to the east, Omar's army seized all the area from Massina to Timbuktu and the emir (grandson of Hamadu Seku) was executed for his leniency with Europeans. Omar Saidu Tall died in combat in 1864. His son Ahmadu, who succeeded "El Hadj", was left simultaneously confronting both the heirs of the emir and the French who gradually seized control of the region between 1889 and 1893.

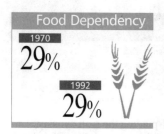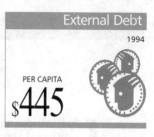
PROFILE

ENVIRONMENT

Located on the west coast of Africa, its northern border formed by the Senegal River, the country's population is concentrated in the less arid western part, close to Dakar. The Senegal Valley is still under-populated as a result of slave trade which was very intense in this region. Senegal is undergoing an acute deforestation and desertification process. A project which is currently underway for building a hydroelectric dam in the Senegal River valley in the north poses a threat to the local environment.

SOCIETY

Peoples: Wolof, 3.7%; Peul- (Fulani-) Tukulor, 23.2%; Serer, 14%; Diola, 5.5%; Malinke (Mandingo), 4.6%; other, 9%. **Religions:** 94% Muslim, 5% Christian, traditional African religions and others 1%. **Languages:** French (official). The most widely spoken indigenous languages are Wolof, Peul and Ful. **Political Parties:** Socialist Party; Senegalese Democratic Party. There are several other minor parties and a coalition of opposition groups, the Coordination of Democratic Forces. **Social Organizations:** National Federation of Senegalese Workers (CNTS); Union of Free Senegalese Workers (UTLS).

THE STATE

Official Name: République du Sénégal. **Administrative Divisions:** 10 Districts. **Capital:** Dakar,1,729,000 inhab. (1992). **Other cities:** Thies, 201,350 inhab.; Kaolack, 179,900 inhab.; Ziguinchor, 148,800 inhab.; Saint-Louis, 125,800 inhab. (1992). **Government:** Abdou Diouf, President since January 1, 1981. 120-member single-chamber Legislative Power. Prime Minister Habib Thiam, since April 1991. **National Holiday:** April 4, Independence Day (1960). **Armed Forces:** 13,350, and 1,200 French troops.

STATISTICS

DEMOGRAPHY

Urban: 41% (1995). **Annual growth:** 1.2% (1992-2000). **Estimate for year 2000:** 10,000,000. **Children per woman:** 5.9 (1992).

HEALTH

One **physician** for every 16,667 inhab. (1988-91). **Under-five mortality:** 115 per 1,000 (1994). **Calorie consumption:** 94% of required intake (1995). **Safe water:** 52% of the population has access (1990-95).

EDUCATION

Illiteracy: 67% (1995). **School enrolment:** Primary (1993): 50% fem., 50% male. Secondary (1993): 11% fem., 21% male. University: 3% (1993). **Primary school teachers:** one for every 59 students (1992).

COMMUNICATIONS

84 **newspapers** (1995), 89 **TV sets** (1995) and 87 **radio sets** per 1,000 households (1995). 0.8 **telephones** per 100 inhabitants (1993). **Books:** 82 new titles per 1,000,000 inhabitants in 1995.

ECONOMY

Per capita GNP: $600 (1994). **Annual growth:** -0.70% (1985-94). **Annual inflation:** 2.90% (1984-94). **Consumer price index:** 100 in 1990; 128.9 in 1994. **Currency:** 535 CFA francs = 1$ (1994). **Cereal imports:** 579,000 metric tons (1993). **Fertilizer use:** 72 kgs per ha. (1992-93). **External debt:** $3,678 million (1994), $445 per capita (1994). **Debt service:** 14.9% of exports (1994). **Development aid received:** $508 million (1993); $64 per capita; 9% of GNP.

ENERGY

Consumption: 102 kgs of Oil Equivalent per capita yearly (1994), 100% imported (1994).

States), which included 16 countries of the region.

[18] In September, the US government cancelled $42 million of the Senegalese debt as a recognition of the country's support to the allies during the Gulf War and to its contribution to the "peace troops" stationed in Liberia. The IMF approved a fresh $57.2 million loan.

[19] In 1991, a series of political and institutional changes was launched. The prime ministership was restored and members of opposition parties were named to two cabinet posts. In September 1991, after three months of negotiations with opposition parties, a consensus was reached for reforming the electoral laws. The conflict with the Casamance separatist movement (MFDC) resumed in 1992, after a truce which had been in effect since February 21.

[20] On February 21, 1993, amid denunciations of widespread fraud, President Diouf won the presidential elections on the first ballot, with 58.4 per cent of the vote. In the May legislative elections, his party, the PSS, maintained control of the National Assembly, with 84 of the 120 seats. The main opposition party, the PDS, won 27.

[21] The economic and financial situation of Senegal became more difficult during 1993, amongst other things because the international prices of the Senegalese export products fell considerably. This made the trade deficit worse, which was compensated by new loans which took the foreign debt to $3.5 billion, more than 60 per cent of the GDP.

[22] The 100 per cent devaluation of the CFA franc decided by France and the IMF accentuated the social tensions in early 1994. The opposi-tion Coordination of Democratic Forces organized an anti-government demonstration on February 16, which ended in confrontations with the police. Six police officers died and dozens of people were injured.

[23] Some 180 people were arrested with supposed responsibility for the disturbances, including the opposition leaders, Abdoulaye Wade and Landing Savané, who were declared innocent and freed in July. The repressive policy of the Senegalese government was criticized by the human rights organizations, the European Parliament and US Congress.

[24] Wade began negotiations with the government, which led to him entering Diouf's cabinet in March 1995. The president received the clear support of the international funding organizations, when he announced legislative modifi-cations to encourage foreign investment and accelerate privatization.

[25] Despite support from Paris and Washington, the government offensives did not achieve any decisive advantage over the Casamance guerrillas. According to some observers, the popularity of the MFDC amongst the Casamance youth, along with the geography of the region, ruled out any possibility of military victory for Dakar. The adoption of a "regionalization" project in the nation, in 1996, did not produce any political solution to the conflict either.

Seychelles

Population: 72,000 (1994)
Area: 280 SQ KM
Capital: Victoria

Currency: Rupee
Language: English and French

Seychelles

During the 18th century, the French and British colonists fought violently over this Indian Ocean colony. After expelling the French in 1794, the British paid little attention to it, and it was governed from the island of Mauritius until 1903. During the world wars, the Seychelles archipelago gained strategic significance.

[2] The Seychelles People's United Party (SPUP) founded in 1964 gave the local population - most of them descendants of enslaved Africans and Indian workers - a new sense of nationalism. The party proved its strength during the general strikes of 1965 and 1966, and in the mass demonstrations of 1972. The colonial interests were opposed to the SPUP and independence, and they organized themselves into the Seychelles' Taxpayers Association. It was later renamed the Seychelles Democratic Party, led by James Mancham.

[3] In the legislative elections of

the first president of the Republic of Seychelles. Shortly before independence, he had agreed to "return" the strategic British Indian Ocean Territory (BIOT) islands to Britain; they had been administered from Mahé since 1967. The British in turn, passed them on to the United States which built the important Diego Garcia naval base there.

[4] Aware that the people would not accept this deal, Mancham postponed the elections until 1979, arguing that they were not necessary as all of the parties were in favor of independence.

[5] Mancham's foreign policy was directed at cementing a strong alliance with South Africa, the source of most of the tourists, while domestic policy destroyed tea and coconut plantations to make room for new five-star hotels, owned by foreign companies. Entire islands were sold off to foreigners like Harry Oppenheimer, the South African gold magnate, and the actor Peter Sellers.

[6] In 1977, while Mancham was maneuvering to postpone the elections yet again, the SPUP, took over the country "with the collaboration of the local police force" while Mancham was out of the country. Accused of "leading a wasteful life while his people worked hard", Mancham was replaced by SPUP leader Albert René.

[7] René renewed his support for the Non-Aligned Movement, which had recognized the SPUP as the country's sole legitimate liberation movement prior to independence. He also strengthened the Seychelles' ties with left-wing countries and movements around the Indian Ocean. The new government turned to socialism, promising to reorganize tourism, give priority to self-sufficiency in agriculture and fishing, increase education, and cut the high level of unemployment; which was affecting nearly half of the working population.

[8] In mid-1978, in order to meet the new political situation, the SPUP became the Seychelles People's Progressive Front (SPPF). In June 1979, the SPPF won the national elections with 98 per cent of the vote, in elections where the abstention rate was a mere 5 per cent. After the victory, President René announced his decision to close down the American satellite tracking station on the archipelago and demanded that the US base on Di-

ego Garcia be closed, with the island returning to Mauritius.

[9] In August 1978, he launched a land reform program calling for the expropriation of all uncultivated land. He also nationalized the water and electricity services, the con-

struction industry and transportation. Controls were placed on food prices and an obligatory literacy program was introduced. These measures led to a rapid economic recovery in the Seychelles, and by 1979-80, on account of mineral and petroleum wealth, they had the largest per capita income of any of the islands in the region. The national currency was also revalued, something virtually unheard of in the Third World.

[10] This economic growth was encouraged and supported by tourism and effective administration. The Seychelles receive an average of 80,000 tourists a year. Under the Mancham government, most of them were South Africans, but after René took over more Europeans came, especially British. The Seychelles received funding for the expansion of several quays from the African Development Bank, aiming to increase the country's fishing activity.

[11] The opponents of Albert René, mainly the South African government, did not give up their efforts to overthrow the socialist government of the Seychelles. In November 1981, a group of 45 mercenaries led by former colonel Mike Hoare tried to invade the island and oust the government. The plot had been hatched in London by former president Mancham, with South African assistance. The attempted coup failed and the mercenaries had to hijack an Indian airliner to escape to South Africa. After the unsuccessful invasion, the Seychelles' government declared a state of emergency and imposed a curfew. The popular militia proved effective in neutralizing conspiracies and tightening controls on all foreigners.

[12] The thwarted invasion and the economic recession in Europe caused tourism to decline by approximately 10 per cent. In August 1982, the situation deteriorated fur-

> In the 1970s, This economic growth was encouraged and supported by tourism and effective administration. The Seychelles receive an average of 80,000 tourists a year. Under the Mancham government, most of them were South Africans, but after René took over more Europeans came, especially British.

April 1974, the SPUP won 47.6 per cent of the vote. However, the peculiar colonial "democratic" system awarded the party only 2 of the 15 seats and Mancham remained the prime minister. In spite of this, it was too late to stem the tide of nationalism and, in 1976, Mancham agreed to the British Foreign Office's suggestion that he become

> In 1986, the Seychelles and the United States renegotiated their treaty, originally signed in 1976 and renewed in 1981, allowing a US satellite tracking station to be stationed on Mahé.

21%

ENVIRONMENT

An archipelago composed of 92 islands. Mahé, Praslim, and La Digue are the largest islands, and are granite islands; the rest are coralloid. The climate is tropical with plentiful vegetation and heavy rainfall. Only the largest islands are inhabited, but some economic use is made of the other islands.

SOCIETY

Peoples: Most residents of the Seychelles are mestizos; descendants of mixed African and European origin and most of them live on the island of Mahé. There are minorities of European, Chinese and Indian origin. **Religions:** Roman Catholic, 88.6%; other Christian (mostly Anglican), 8.5%; Hindu, 0.4%; other, 2.5% (1987). **Languages:** English and French (official); most of the population speak Creole, a local dialect with European and African influences. **Political Parties:** The People's Progressive Front of Seychelles (FPPS), in power. Democratic Party (PD), opposition. Seychelle 3 for Democracy. Christian Democratic Party/Encounter of the People of Seychelles for Democracy. **Social Organizations:** The National Workers' Union, is the only labor union and is part of the SPPF.

THE STATE

Official Name: Repiblik Sesel. **Capital:** Victoria, 24,325 inhab. (1987). **Government:** Albert René, President since 1977. Legislature, single-chamber: National Assembly, with 33 members. **National Holiday:** June 28, Independence Day (1976). **Armed Forces:** 800 (1994). **Paramilitaries:** 1,300 National Guard, Coast Guard.

DEMOGRAPHY

Annual growth: 1.7% (1992-2000). **Children per woman:** 2.7 (1992).

EDUCATION

Illiteracy: 21% (1995). **Primary school teachers:** one for every 19 students (1989).

COMMUNICATIONS

16.2 **telephones** per 100 inhabitants (1993).

ECONOMY

Per capita GNP: $6,680 (1994).
Annual growth: 4.80% (1985-94).
Consumer price index: 100 in 1990; 108.7 in 1994.
Currency: 5 rupees = 1$ (1994).

---

ther when military sectors plotted an unsuccessful rebellion, strengthening rumors of another approaching mercenary conspiracy, and European conservative groups organized a smear campaign to support these rumours. In June 1984, Albert René was re-elected with 93 per cent of the vote, and the last political prisoner was released immediately after the elections. The country's greater political stability allowed the president to adopt a more pragmatic policy.

[13] In 1986, the Seychelles and the United States renegotiated their treaty, originally signed in 1976 and renewed in 1981, allowing a US satellite tracking station to be stationed on Mahé.

[14] In September 1986, there was another military coup attempt. Albert René, who was in Zimbabwe participating in a Non-Aligned Countries summit conference, returned immediately and put down the rebellion. Most of those responsible for the coup were arrested, among them the main leader of the

conspiracy, Minister of Defence Colonel Ogilvy Berlouis.

[15] The Seychelles' government has been promoting the creation of a peace zone in the Indian Ocean. It requires all warships wanting to call at its ports to be "nuclear-free", and for this reason, the United States and United Kingdom naval fleets have stopped calling at the archipelago. The Seychelles established diplomatic relations with Comoros and Mauritius in 1988, and with Morocco and Côte d'Ivoire in 1989.

[16] In the presidential elections of June 1989, Albert René was the only candidate and was re-elected for a third term, with 96 per cent of the vote, according to final election returns. After the election, the president announced that he was reorganizing his cabinet. He personally took charge of the ministry of industry and amalgamated the ministries of planning, foreign affairs, finance and tourism into a single cabinet post.

[17] In September 1989, Indian

president Ramaswamy Venkataraman carried out a three-day visit to Victoria, signing a cultural exchange agreement. In January 1990, the Seychelles established diplomatic relations with Kenya.

[21] The opposition forced President René to accept a multi-party system. The first experiment was car-

> The opposition forced President René to accept a multi-party system. The first experiment was carried out in July 1992, when a commission was elected to draw up the new Constitution.

ried out in July 1992, when a commission was elected to draw up the new constitution. The People's Progressive Front of Seychelles (FPPS), the ruling party, took 13 of the 23 seats. The Democratic Party (PD) of former president Mancham won eight seats.

[22] The government continued its authoritarian ways. The PD soon withdrew from the commission and the first draft constitution was re-

jected. The project was suspended until the PD returned to the commission in June 1993, and the text was written with support from both parties. 73.6 per cent of the voters approved the new constitution.

[23] This document made the multi-party system, the 33-member National Assembly and a five year presidential term official. The president can be re-elected twice. René won the July 1993 elections and his party took the absolute majority in the National Assembly.

[24] Tourism continued to be the main source of income for the nation. The annual number of visitors was higher than the national population in 1993 and 1994. Oil derivatives and tinned tuna made up more than 80 per cent of the islands' exports during this period. Despite unemployment running at 22 per cent, the gross national product of the Seychelles reached $5,480 per head in 1994. 99 per cent of the population have access to drinking water and 100 per cent can meet their basic nutritional needs.

[25] Seychelles participates in the Indian Ocean Commission which is responsible for the preservation of regional biodiversity, amongst other things.

Sierra Leone

Population: 4,399,000 (1994)
Area: 71,740 SQ KM
Capital: Freetown

Currency: León
Language: English

Sierra Leone

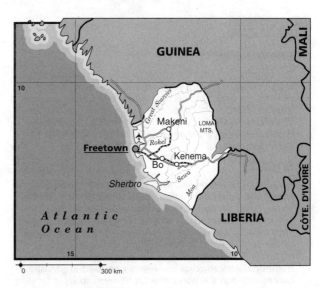

The Temne and Mende ethnic groups constitute the majority of the population of Sierra Leone. Minority groups of Lokko, Sherbo, Limba, Sussu Fulani, Kono, and Krio also exist.

2   When Britain faced a "demographic problem" with runaway slaves in the 1800s, the government decided to "return" them to Africa. They came to London hoping to benefit from a court ruling that had abolished slavery in the city of London. Abolitionist leader Granville Sharp purchased an area of 250 sq km for £60 from the local rulers, organizing an agricultural society based on democratic principles. However, this society was quickly transformed into a colonial company involved in the British conquest of the entire country.

3   Tossed into a land where they had no roots, the "creoles" sought to copy European culture, and considered themselves superior to the "savages" of the inland, acting as middlemen for British colonialism.

4   Yet, the "savages" resisted strongly and in 1898, Bai Buré, rallied most of the inland people, already angered by a British tax on their dwellings. However, after nearly a year of resistance, British military superiority overwhelmed and stifled the opposition.

5   In 1960, the British wanted to withdraw from Sierra Leone, and an agreement was negotiated with the traditional leaders to protect their interests. In 1961, Sir Milton Margai, secretary general of Sierra Leone People's Party (SLPP), became the first prime minister of an independent Sierra Leone.

6   The creoles, together with British and Syrian-Lebanese merchants, held the colony's economic power. Although removed from political power, they maintained significant influence over the Margai government.

7   When Margai died in 1964, he was succeeded by his brother Albert and the situation quickly deteriorated. Corruption and gambling became so widespread that it was compared to the Batista period in Cuba. The diamond trade gave birth to a chain of smuggling operations and crime became a main source of income.

8   In 1967 the All People's Congress (APC), party led by Syaka Stevens won the elections, signalling the possibility for change; however, the conservative creoles, the traditional leaders, and the neocolonialist British, united to prevent Stevens from taking control of the country as they considered him "dangerously progressive". As a result, Stevens was overthrown by a military coup and forced into exile in Conakry (Guinea).

9   In April 1968, a group of low-ranking officers took power through the so-called Sergeants' Revolt. They brought back Stevens who in 1971, broke all ties with Britain and declared Sierra Leone a republic, becoming its first president.

10   Important goals were achieved in the first years: the lumber industry was nationalized and the state seized majority control of the mining firm, which controlled the diamond trade. To protect the prices of iron ore and bauxite, Sierra Leone joined the associations of iron and bauxite producing countries.

11   In 1978, Stevens called for a plebiscite to establish a one-party system, in an attempt to appease the opposition. Stevens' proposal was approved and the APC incorporated prominent members of the SLPP within its ranks, granting them posts within the government.

12   In 1979, President Stevens' popularity began to ebb, as a result of worsening living conditions, growing inflation, authoritarian measures, government corruption, a fall in exports and the increasing foreign debt.

13   Popular unrest reached its climax in September 1981, when the Sierra Leone Labor Congress (SLLC) declared a general strike, demanding immediate changes in economic policy. The strike spread throughout the country and nearly toppled the government, which had to make several concessions.

14   In urban areas, the scarcity of staple foods, particularly rice, as well as periodic fuel shortages became chronic. Inflation and the high cost of living reduced the purchasing power of salaried workers by 60 per cent, and the late payment of salaries became normal. Members of parliament were paid with sacks of rice which were later sold on the black market at a large profit.

15   The powerful Lebanese trade clique controlled the unofficial market, and more than 70 per cent of the country's exports. Gold and diamond smuggling was estimated at nearly $150 million per year while official exports dropped from $80 million to barely $14 million between 1980 and 1984.

16   After several postponements and an alleged coup attempt, the 1982 elections took place amid serious outbreaks of violence. The Liberian government was forced to accept refugees, and the following year they announced the presence of 4,000 refugees from Sierra Leone, living in their country.

17   In November 1985, Syaka Stevens handed over the presidency to Joseph Momoh, a member of his cabinet, changing nothing in the political and social condition of the country.

18   The dimensions of the crisis culminated in the government declaration of a state of economic emergency in 1987. This measure included granting the state the sole right to commercialize gold and diamonds, a 15 per cent surcharge on imports, and the reduction of salaries of public employees.

19   In March 1991, rebel forces operating from Liberia occupied two border towns. Guerrilla groups from Burkina Faso, Liberia, and Sierra Leone joined them, occupying one third of the country.

20   On April 13 1991, President Joseph Momoh announced that Nigeria and Guinea had deployed troops in his country to help repel the rebel incursions from Liberia. Sierra Leone, Guinea, and Nigeria are members of the West African Economic Community, an organization which had signed a peace agreement with Liberia in August 1990.

21   In August 1991, a referendum approved a new Constitution establishing a multi-party system. However, the economic crisis continued alongside ever increasing corruption.

22   In 1992, the government launched a structural adjustment program which had been imposed by the International Monetary Fund. James Funa, a former World Bank executive, was named minister of finance. He implemented monetary control, incentives for natural resource exploration by foreign companies, widespread priva-

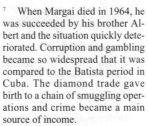

WORKERS

1994

MEN 68%   WOMEN 32%

AGRICULTURE 70%   INDUSTRY 14%   SERVICES 16%

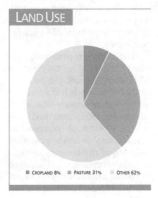

LAND USE

CROPLAND 8%   PASTURE 31%   OTHER 62%

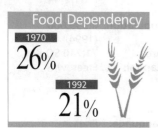

Food Dependency
1970 **26%**
1992 **21%**

Foreign Trade
MILLIONS $ 1994
IMPORTS **150**
EXPORTS **115**

Illiteracy
1995
69%

Maternal Mortality
1989-95
PER 100,000 LIVE BIRTHS
800

PROFILE

ENVIRONMENT

The country is divided into three regions. The coastal strip, with a width of nearly 100 km is a swampy plain that includes the island of Sherbro. Several rivers drain the inner tropical forest, mostly cleared for agricultural exploitation. The eastern plateau contains the country's diamond mines. Deforestation is very severe, 85% of all natural habitat has been destroyed.

SOCIETY

Peoples: The Temne and Mende ethnic groups account for nearly one-third of the population. Lokko, Sherbo, Limba, Sussu, Fulah, Kono and Krio are other important groups. The Krio are descendants of African slaves freed in the 19th century who settled in Freetown (Krio is a corruption of the English word "creole"). There is also an Arab minority. **Religions:** Most of the people profess traditional African religions; nearly one third are Muslims, concentrated in the north; the Christian minority is located in the capital. **Languages:** English (official). The most widely spoken native languages are Temne, Mende and Krio. The latter serves as the commercial language in the capital. **Political Parties:** Sierra Leone's People Party (SLPP); All Peoples' Congress (APC); United Political Movements Front; Revolutionary United Front (RUF). **Social Organizations:** Sierra Leone Labor Congress.

THE STATE

Official Name: Republic of Sierra Leone.
Capital: Freetown, 469,776 inhab. (1985). **Other cities:** Koidu, 80,000 inhab.; Bo, 26,000 inhab; Kenema, 13,000 inhab. (1985). **Government:** Ahmad Tejan Kabbah, President. **National Holiday:** April 19, Republic Day (1971). **Armed Forces:** 13,000 troops (1995).

STATISTICS

DEMOGRAPHY

Urban: 35% (1995). **Annual growth:** 2.1% (1992-2000). **Estimate for year 2000:** 5,000,000. **Children per woman:** 6.5 (1992).

HEALTH

Under-five mortality: 284 per 1,000 (1994). **Calorie consumption:** 84% of required intake (1995). **Safe water:** 34% of the population has access (1990-95).

EDUCATION

Illiteracy: 69% (1995). **Primary school teachers:** one for every 34 students (1990).

COMMUNICATIONS

83 **newspapers** (1995), 84 **TV sets** (1995) and 97 **radio sets** per 1,000 households (1995). 0.2 **telephones** per 100 inhabitants (1993). **Books:** 82 new titles per 1,000,000 inhabitants in 1995.

ECONOMY

Per capita GNP: $160 (1994). **Annual growth:** -0.40% (1985-94). **Annual inflation:** 67.30% (1984-94). **Consumer price index:** 100 in 1990; 258.5 in 1992. **Currency:** 613 leones = 1$ (1994). **Cereal imports:** 136,000 metric tons (1993). **Fertilizer use:** 26 kgs per ha. (1992-93). **Imports:** $150 million (1994). **Exports:** $115 million (1994). **Development aid received:** $1,204 million (1993); $269 per capita; 164% of GNP.

ENERGY

Consumption: 73 kgs of Oil Equivalent per capita yearly (1994), 100% imported (1994).

tization, an overhaul of the State apparatus, and a purge on the inherent corruption.

[23] On April 29, Captain Valentine Strasser seized power through a coup and, after suspending the constitution, created the National Provisional Governing Council, prohibiting all political activity and confirmed the Minister of Finance in his post. In June, all civilian members of the Governing Council were expelled, and its name was changed to "Supreme Council of State". Press censorship began.

[24] The eastern part of the country was occupied at the time by the United Movement for Liberian Liberation and Democracy, which used Sierra Leone as a base from which to carry out attacks against Charles Taylor's forces (see Liberia). In the meantime, the Revolutionary United Front of Sierra Leone was operating in the southeast.

[25] The regions affected by warfare also happen to be the country's richest, where gold and diamond deposits and the main agricultural production area are located. Guerrilla activity led to an abrupt decline in mining activity; the per centage represented by diamonds in the country's exports dropped from 54.7 per cent in 1987 to a mere 7 per cent in in 1990. The per capita GDP dropped from $320 in 1980 to $210 in 1991.

[26] More than a million people were displaced as a result of the war. Some fled to more secure areas within the country, while others sought refuge in neighboring Guinea and Liberia.

[27] On December 1, 1993, British pressure forced the government to decree a unilateral cease-fire, promising to grant an amnesty to all rebels who lay down their arms. The RUF refused to accept the proposal.

[28] Extreme poverty gave rise to the trafficking of children and young people. According to the terms of an unusual sort of contract, Lebanese traders in Sierra Leone took the daughters of poor families to Lebanon to work as household servants for five-year periods. However, the traders frequently failed to keep their part of the bargain, and the girls never returned.

[29] Those who managed to escape from the system have described the verbal and physical abuse to which they have been subjected. Although it is estimated that hundreds of girls and children have been involved, no government action was taken against the influential Lebanese community.

[30] The government promised to organize elections did not convince the RUF, which continued with the armed struggle. The rebels had a series of military victories in early 1994, which left a total of 150 civilians dead.

[31] The World Bank expressed its "satisfaction" with the economic policies of the country, which, amongst other things, had managed to reduce inflation from 120 per cent in 1991 to 15 per cent in 1994.

[33] Presidential elections were held in March 1996, and Ahmed Tejan Kabbah, of the Sierra Leone People's Party, was elected by nearly 60 per cent of the vote.

[34] According to various estimates, the war caused some 10,000 deaths between 1991 and 1996, forcing two million people to flee the conflict zones and take refuge in neighboring countries and other areas of Sierra Leone.

[32] In early 1995 the war extended over nearly all the country. The governmental forces recovered the Sierra Rutile titanium mine, which produced 50 per cent of Freetown's foreign trade. However, despite dedicating 75 per cent of the national budget to the war, and having increased the size of the army to 13,000 troops, the government did not appear in a position to defeat the guerrillas.

Singapore

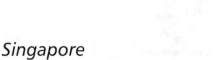

Population: 2,930,000
(1994)
Area: 620 SQ KM
Capital: Singapore

Currency: Singapur Dollar
Language: Malay, English,

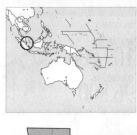

Singapore

Singapore's original name was Temasek, meaning "people of the sea" in Sanskrit. Later on, the city was named Singa-Pura, the "city of the lion", after a visiting prince experienced a vision.

[2] Singapore was founded in 1297 and destroyed a 100 years later. From 1819 onwards, it became an extremely important base for the British; Sir Thomas Stamford Raffles established the local headquarters of the British East India Company there.

[3] In 1824, the island of Singapore and the small adjacent islets were purchased as a single lot by Raffles from the Sultan of Johore (see Malaysia). The Company appointed Prince Hussein as the new ruler of Singapore. In gratitude, he granted the Company royal authorization to improve the port. Chinese immigrants soon constituted the majority of the local population.

[4] Singapore was part of the British colony called "the Straits Settlements", together with the ports of Penang and Melaka (Malacca). In 1946, Penang and Melaka joined the Malayan Union and Singapore became a crown colony.

[5] During World War II the Japanese at one point dominated the whole of southeast Asia. They conquered Singapore in 1942 but were later defeated. Their defeat was partly due to a strong internal resistance group, organized by a revolutionary movement and led by the Communist Party of Malaya (CPM). The name Malaya included both Singapore and the Malay peninsula; the separation of the latter was always questioned by the left.

[6] Once the war was over, the Malay sultanates and the former Straits Settlements attempted to form a union or federation, with a view to attaining independence for the territory. However, the conflicting interests of the Chinese and Malayan communities, and the conservative and progressive forces, made progress difficult. In January 1946, the Singapore Labor Union declared a general strike and in 1948 the Communist Party led an anti-colonial uprising which failed, as it failed to gain the support of the Malays and the poorer sectors of the Indian population. Marxist parties were outlawed and had to move into the forests, where they resorted to guerrilla warfare.

[7] As the first step towards the self-government of the city-state, municipal elections were held in 1949. Only English-speaking people were allowed to vote until 1954, when the People's Action Party (PAP) was founded. The anti-imperialism advocated by the PAP brought citizens of British and Chinese backgrounds together for the first time. In 1959, the Chinese were allowed to vote; full internal autonomy was granted, and the PAP obtained an overwhelming victory. Lee Kuan Yew, founder of the party, became prime minister, campaigning on a platform of social reforms and independence. He planned a federation with Malaya, which had been independent since 1957.

[8] The PAP split into a socialist faction, led by Lim Chin Siong, and the "moderates" of Lee Kuan Yew, who encouraged the promotion of private enterprise and foreign investment

[9] In 1961, the left wing of the PAP founded Barisan Sosialis (the Socialist Front), which opposed the prospect of uniting Singapore and Malaya under British control. In September 1963, the Federation of Malaysia, consisting of Singapore, the Malay peninsula, Sarawak and Sabah (both to the north and northeast of Borneo) was set up, after those opposed to the federation had been conveniently purged.

[10] As a Federation member, Singapore depended heavily on the peninsula; even its water supply came from there. Integration at that time was not feasible because of profound disagreements between Singapore's Chinese community and the Malay community of the rest of the Federation. On August 9, 1965, after several ethnic conflicts, Prime Minister Lee decided to withdraw from the Federation.

[11] In 1965, serious internal conflicts were caused by the mistreatment of the Malayan population and other ethnic minorities. There were also intense struggles with the left-wing opposition, which the government classified as "communist subversives". The island became an independent republic, and a Commonwealth member. A mutual assistance and defence treaty was signed with Malaysia, and on October 15 of the same year Singapore became a member of the United Nations.

[12] The first years of independence witnessed substantial economic growth. The island was turned into an "enclave" for the export of products manufactured by transnationals, and into an international financial center which controlled the regional economy.

[13] The situation turned gloomy in 1974, when the oil crisis upset Singapore's export scheme; Singapore is the fourth largest port in the world; with the second highest per capita income in Asia, after Japan. The ensuing economic deterioration brought about public demonstrations by students and workers. These protests were fiercely controlled, reaching such proportions that the Socialist International expelled the PAP from its ranks in 1975.

[14] In order to placate criticism from the opposition, the regime approved a reform which allowed the entering of two representatives of the Worker's Party and the Democratic Party of Singapore, respectively.

[15] Despite the sustained economic growth since 1987, the government expelled thousands of Thai and Filipino workers blaming them for taking over the jobs of the natural citizens.

[16] The decision to prohibit the circulation of foreign publications judged detrimental by the government, the imprisonment of opposition members and a Security Law that allows indefinite imprisonment without trial for two renewable years each time, raised countless denunciations of violations of human rights.

[17] In the 1988 elections the opposition vote grew, but due to the mechanisms of the electoral system its parliamentary representation

PUBLIC EXPENDITURES

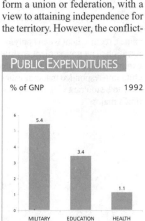

% of GNP   1992

MILITARY 5.4 · EDUCATION 3.4 · HEALTH 1.1

LAND USE

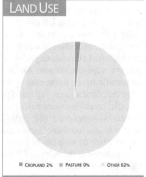

CROPLAND 2%   PASTURE 0%   OTHER 62%

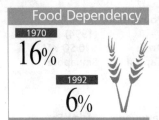

Food Dependency

1970
16%

1992
6%

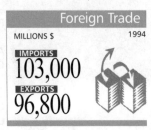

Foreign Trade

MILLIONS $ 1994

IMPORTS
103,000

EXPORTS
96,800

Illiteracy

1995

9%

PROFILE

ENVIRONMENT

Singapore consists of one large island and 54 smaller adjacent islets. The country is connected to Malaysia by a causeway across the Johore Strait. The terrain, covered by swampy lowlands, is not conducive to farming and the population traditionally works in business and trade. The climate is tropical with heavy rainfall. In its strategic geographical location, central to the trade routes between Africa, Asia and Europe, the island has become a flourishing commercial center. Economically, the main resources of the country have been its port, the British naval base and, more recently, industrial activity; textiles, electronic goods and oil refining. Industrialization has caused severe air and water contamination.

SOCIETY

Peoples: 76% of Singaporeans are of Chinese origin. Malaysians account for 15% and 6% are from India and Sri Lanka. **Religions:** Buddhism, Islam, Hinduism and Christianity. There are Sikh and Jewish minorities. **Languages:** Bahasa Malaysia, English, Chinese (Mandarin) and Tamil are the official languages. Bahasa Malaysia is viewed as the national language, but English is spoken in the public administration and serves as a unifying element for the communities. Various Chinese dialects are spoken, with Punjabi, Hindi, Bengali, Telegu and Malayalam in the Indian communities. **Political Parties:** The People's Action Party (PAP), founded in 1954 to fight for independence, held an anti-imperialist stance and was led by Prime Minister Lee Kuan Yew, who later welcomed international financial interests and businesses. Singapore Democratic Party; the Workers' Party (the only one which won parliamentary representation); Barisan Sosialis (the Socialist Front); the Singapore Justice Party; the United People's Front and the Singapore United Front. **Social Organizations:** The major national trade union is the Congress of Singapore Labor Unions.

THE STATE

Official Name: Republic of Singapore. **Capital:** Singapore, 2,703,703 inhab.(1990). **Government:** Parliamentary republic. Ong Tenc Cheong, President elected on August 29, 1993, by direct popular vote. Goh Chok Tong, Prime Minister since November 1991. Legislature, single-chamber, made up of 81 members, dominated since independence by the official Popular Action Party, which obtained 77 seats in 1991. **National Holiday:** August 9, Independence Day (1965). **Armed Forces:** 55,500, inc. 34,800 conscripts. **Paramilitaries:** 11,600 Police and Maritime Police (estimate). 100,000 Civil Defence Force.

STATISTICS

DEMOGRAPHY

Urban: 100% (1995). **Estimate for year 2000:** 3,000,000. **Children per woman:** 1.8 (1992).

HEALTH

One **physician** for every 725 inhab. (1988-91). **Under-five mortality:** 6 per 1,000 (1995). **Calorie consumption:** 107% of required intake (1995). **Safe water:** 100% of the population has access (1990-95).

EDUCATION

Illiteracy: 9% (1995). **Primary school teachers:** one for every 26 students (1989).

COMMUNICATIONS

107 **newspapers** (1995), 107 **TV sets** (1995) and 104 **radio sets** per 1,000 households (1995). 43.4 **telephones** per 100 inhabitants (1993). **Books:** 106 new titles per 1,000,000 inhabitants in 1995.

ECONOMY

Per capita GNP: $22,500 (1994). **Annual growth:** 6.10% (1985-94). **Annual inflation:** 3.90% (1984-94). **Consumer price index:** 100 in 1990; 111.5 in 1994. **Currency:** 1 Singapur dollars = 1$ (1994). **Cereal imports:** 798,000 metric tons (1993). **Fertilizer use:** 56,000 kgs per ha. (1992-93). **Imports:** $103,000 million (1994). **Exports:** $96,800 million (1994). **Development aid received:** $24 million (1993); $9 per capita.

ENERGY

Consumption: 6,556 kgs of Oil Equivalent per capita yearly (1994), 100% imported (1994).

---

diminished. In 1991, the PAP again won an overwhelming majority of seats. In November, Goh Chok Tong replaced Lee Kuan Yew as prime minister, but the latter nevertheless kept a considerable political weight. At his initiative, Singapore offered Washington the possibility of installing bases in the country when the Filipino Congress decided to close US military installations in the Philippines.

[18] As an exporter of high technology goods, industry was only marginally affected by the recession of its main partners, the US, Japan, and the European Union. In order to expand its activity, the government started a policy of "regionalization" increasing its investments in Indonesia and Malaysia. The investments in China surpassed $1 billion per year, and Singapore became the main commercial partner of Vietnam. From 1993 on, annual growth was over 8 per cent.

[19] Between 1994 and 1995 relations were strained between the government of Singapore and the US, Philippines, and Holland. The hanging of a Dutch engineer accused of heroin traffic, the condemning of a young US citizen to be hit with a rattan cane for vandalism, and the execution of a Filipino maid accused of murdering a fellow servant were the reasons for this diplomatic stress. In the case of the maid the crime was proved and diplomatic relations were re-established. Concerning the young American, the number of blows and the number of months in prison were reduced at the behest of President Bill Clinton.

[20] Concerning domestic affairs, an employee, two economists, and two journalists were sent to prison for having violated a law which protects state secrets. Their crime was to publish the economic growth forecast before its official publication.

[21] In June 1996, the "Speak Mandarin Campaign" was questioned by the ethnic minorities, concerned by the fact that this language became a requirement when applying for a job. The government - while exhorting the Chinese to have more children - responded that minorities have to be tolerant with the population's majority.

Slovakia

Slovenska

Population: 5,347,000 (1994)
Area: 43,035 SQ KM
Capital: Bratislava

Currency: Slovak Koruna
Language: Slovak

The region made up of Slovakia, Bohemia and Moravia had a common history until 1993 (see Czech Republic). In the Slovakian legislative elections of June 1992, the Movement for a Democratic Slovakia (MED) had returned 37.3 per cent of the vote: The Democratic Left Party, 14.7 per cent: the Christian Democrat Movement, 8.9 per cent and the Slovak National Party, 7.9 per cent. The Hungarian minority party Co-existence received 7.4 per cent, representing 600,000 people.

[2] On February 15 1993 Michal Kovac was elected president of the new Republic of Slovakia. Vladimir Meciar, leader of the MED and the architect of Czechoslovakian separation, was named prime minister.

[3] Meciar's administration was marked by controversy and accusations of authoritarianism, from the opposition. He re-nationalized the newspaper "Smena", and created a compulsory television slot for the broadcasting of government news and propaganda.

[4] Separation from the Czech Republic proved detrimental to the Slovakian economy, which lost nearly $1 billion in bilateral trade. The country suffered more severely than the Czech Republic, because of dependence upon the Socialist bloc.

[5] Economic reforms and privatizations came to an almost complete standstill. In order to reduce the budget deficit, which in the first eight months of 1993 came to some $280 million - 6 per cent of the GDP - the government cut subsidies and social programs.

[6] In April, the Confederation of Slovakian Labor Unions mobilized a demonstration of workers through the streets of the capital, to protest the lack of any concrete action on the part of government authorities.

[7] Pressure from environmental groups forced Hungarian authorities to abandon an important binational hydroelectric plant plan, a project which had been approved and signed in conjunction with the former Czechoslovakia. This project was to have constructed two large dams, diverting the Danube River. This decision led to a distinct cooling of relations between the two governments.

[8] Another cause of friction between these countries was the treatment of the Hungarian minority, which was demanding greater cultural autonomy. While in June the Slovakian Parliament approved a law allowing the Hungarian language to be taught in schools, Meciar also called for the enforcement of an old law, which had fallen into disuse, making it mandatory for Hungarian women to add the Slav feminine suffix "-ova" to their last names.

[9] By the end of 1993, more than half of all industries had recorded losses and the construction industry had suffered a 60 per cent reduction.

[10] Jozef Moravcik was named prime minister in March 1994, after long negotiations between the parties which had ousted "the dictatorial and corrupt regime of Vladimir Meciar". Moravcik announced the formation of a government of national salvation and urgent economic measures.

[11] The new government, with the participation of the Democratic Left Party, and Alternative for Real Democracy (both made up of former MED allies) and the right-wing Christian Democratic Movement was built upon two objectives: to keep Meciar out of power and to accept the general principles of European democracy. The internal differences in the coalition weakened the government and made it impracticable as a long-term alternative.

[12] Meciar won 35 per cent of the vote in the September 30 and October 1 elections of 1994. The coalition parties suffered a tough defeat and the parties of the Hungarian minorities formed a coalition which emerged as the third force in Parliament. Meciar became prime minister again and immediately put the privatization initiative, started by his predecessor, into reverse, stressing nationalism.

[13] President Kovac had played an active part in ousting Meciar, and as a result the prime minister was tireless in his efforts to topple him. The pressure from parliament was increased by the issue of a warrant for the arrest of Kovac's son by the German authorities in connection with a corruption scandal. In the autumn of 1995, the young Kovac was kidnapped by Slovak intelligence agents and abandoned in Austrian territory, where he was arrested and extradited to Germany.

[14] The Meciar government made many attempts to limit the rights of the Hungarians. A campaign to prevent Slovak being imposed as the only official language of the country did not receive wide enough support.

Slovenia

Slovenija

Population: 1,989,000 (1994)
Area: 20,251 SQ KM
Capital: Ljubljana

Currency: Tolar
Language: Slovene

One of the southern Slav groups, the Slovenes occupied what is now Slovenia, and the land to the north of this region, in the 6th century AD. Subdued by the Bavarians around the year 743, they were later incorporated into the Frankish Empire of the Carolingians.

[2] With the division of the empire in the 9th century, the region was given to the Germans. The Slovenes were reduced to serfdom and the region north of the Drava River was completely Germanized.

[3] The Slovene people preserved their cultural identity because of the educational efforts of their native intelligentsia who were mostly Catholic monks and priests. The House of Austria gradually established itself in the region, from the latter part of the 13th century onwards.

[4] Between the 15th and 16th centuries, the Slovenes participated in several peasant revolts - some, like the 1573 revolt, in conjunction with the Croats - leading the Hapsburgs to improve the system of land tenure.

[5] After 1809, a large part of Slovene territory fell within the Napoleonic empire's Illyrian provinces. After Napoleon's defeat in 1814, Hapsburg (House of Austria) rule was restored within the region. With the 1848 Revolution, the Slovenes called for the creation of a unified Slovene province within the Austrian Empire. The first glimmer of hope for a union of southern Slavs (Slavs, Serbs and Croats) emerged in the 1870s.

[6] In the 1890s, the Slovene People's Party (Catholic), and the Progressive (Liberal) and Socialist parties were formed. Members of the Catholic clergy also promoted a large-scale organization of peasants and artisans into cooperatives.

[7] In 1917 the Austrian Parliament, representing Slovenes and other southern Slav peoples, defended the unification of these territories into a single autonomous political entity, within the Hapsburg realm.

[8] At the end of World War I, amid widespread enthusiasm over the fall of the Austro-Hungarian Empire, Slovene leaders supported the creation of a kingdom of Serbs, Croats and Slovenes. In 1919, the new state adopted the name Yugoslavia (land of the southern Slavs). Nevertheless, at the Paris Peace Conference, the victorious powers handed Gorica to Italy despite the presence of a large Slovene population.

[9] The St-Germain Treaty, signed between the victorious powers and Austria, gave Yugoslavia only a small part of southern Carintia. Two plebiscites were announced, to define the future of the rest of Carintia. However, when the southern region opted to join Austria in 1920, the second plebiscite was not held, and both regions remained part of Austria.

[10] Serbian hegemony within the Yugoslav kingdom gave rise to a certain resentment among Slovenes. This feeling never reached the extremes that characterized Croat feelings and which led to a strong anti-Serbian movement there.

[11] In World War II, Slovenia was partitioned between Italy (the southwest), Germany (the northeast) and Hungary (a small area north of the Mura River). The most prominent group within the Slovene resistance movement was the Liberation Front, led by the communists.

[12] The communist guerrillas fought on two fronts at the same time: against the foreign invaders and against their internal enemies (especially groups belonging to the Slovene People's Party). The occupiers, in turn, organized anti-communist military units, with the participation of the local population. After the defeat of the Axis (Germany, Italy and Japan), the major part of old Slovenia was returned to Yugoslavia.

[13] Upon the foundation of the Federated People's Republic of Yugoslavia, in 1945, Slovenia became one of the federation's six republics, with its own governing and legislative bodies. Legislative power was made up of a republican council, elected by all citizens, and the council of producers, elected from among Slovenian industrial workers and officials.

[14] Although such entities did not add up to an autonomous government, Slovenia managed to maintain a high degree of cultural and economic independence through this self-management brand of socialism (led by the Yugoslav League of Communists). In 1974, changes in the Yugoslavian federal constitution made Slovenia a Socialist Republic.

[15] Slovenia became one of the most industrialized of the federation's republics, especially in the area of steel production and the production of heavy equipment. Yugoslavia's first nuclear power plant was completed in 1981 in Krsko, with the assistance of a private US firm.

[16] During the socialist period, Slovenia was first or second among the Yugoslav republics with regard to family income, leading the table for economically active population outside the rural sector. Slovenia produces both agricultural produce and cattle.

[17] In the late 1980s, influenced by the changes in Eastern Europe, Slovenia evolved toward a multi-party political system. In January 1989, the Slovene League of Social Democrats was founded, the country's first legal opposition party, and in October Slovenia's National Assembly approved a constitutional amendment permitting Slovenia to secede from Yugoslavia.

[18] The Slovene League of Communists left the Yugoslav League in January 1990, becoming the Democratic Renewal Party. In April, in the first multiparty elections to be held in Yugoslavia since World War II, the Demos coalition came out victorious. This coalition's members represent a broad political spectrum - including several communists - united in their aim of achieving Slovenian separation from the federation.

[19] Slovenia and Croatia declared their independence on June 25, 1991. In the hours which followed, central government tanks flocked to Slovenia's Austrian, Hungarian and Italian borders; 20,000 federal troops stationed within the republic were mobilized. After fierce fighting and the bombing of the Ljubljana airport, Belgrade announced that it controlled the federation's borders.

[20] A cease-fire went into effect on July 7 1991, following negotiations held on the Yugoslav island of Brioni between federal and Slovene authorities, with the mediation of the European Community. The agreement resulting from the negotiations reaffirmed the sovereignty of Yugoslav peoples and postponed Slovenian independence for a three-month period. At the same time, the federal president's authority over the army was recognized, and the Slovene police authority over Slovenia's borders.

[21] According to Brioni's compromise, customs tariffs along Slovenia's borders remained in the hands of Slovene police, although the income generated was to go into a joint account belonging to the various Yugoslav republics. The feder-

LAND USE

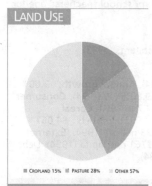

■ CROPLAND 15% ■ PASTURE 28% □ OTHER 57%

Foreign Trade		External Debt	
MILLIONS $	1994		1994
IMPORTS 7,304		PER CAPITA	
EXPORTS 6,828		$1,151	

PROFILE

ENVIRONMENT

Bounded in the west by Italy, in the north by Austria, in the northeast by Hungary and in the south and southeast by Croatia, Slovenia is characterized by mountains, forests and deep, fertile valleys. The Sava River flows from the Julian Alps (highest peak: Mt. Triglav, 2,864 meters), in the northwest of the country, to the southeast, crossing the coal-mining region. The Karavanke mountain range is located along the northern border. The region lying between the Mura, Drava, Savinja and Sava rivers is known for its vineyards and wine production. To the west and southwest of Ljubljana, all along the Soca river (known as "Isonzo" on the Italian side), the climate is less continental, and more Mediterranean. The capital has an average annual temperature of 9° C, with an average of 1° C in the winter and 19° C in summer. The country's main mineral resources are coal and mercury, which contribute to the country's high level of industrialization.

SOCIETY

Peoples: Slovene, 87.8%; Croat, 2.8%; Serb, 2.4%; Bosnian, 1.4%; Magyar, 0.4%; other, 5.2%. **Religions:** The vast majority is Catholic; there are Jewish and Muslim minorities. **Languages:** Slovene, official; Serbo-Croat. **Political Parties:** Demos, a right-of-center coalition, including ex-communists, currently in power. Slovene League of Social Democrats. Party of Democratic Renewal (formerly the Slovene League of Communists). **Social Organizations:** In the process of being reorganized after recent institutional and political changes.

THE STATE

Official Name: Republika Slovenija. **Administrative Divisions:** 62 Districts. **Capital:** Ljubljana, 276,133 inhab. (1991) **Other cities:** Maribor, 108,122 inhab.; Celje, 41,279 inhab.; Kranj, 37,318 inhab.; Velenje, 27,665 inhab. (1991). **Government:** President, Milan Kucan, since 8 April 1990. Janez Drnovsek, Prime Minister since April 22, 1992. Bicameral legislature: 90-member National Assembly, and 40-member National Council. **National Holiday:** June 25, Independence (1991). **Armed Forces:** 8,400 (1994). **Paramilitaries:** 4,500 Police (1994).

STATISTICS

DEMOGRAPHY

Urban: 62% (1995). **Estimate for year 2000:** 2,000,000.

EDUCATION

School enrolment: Primary (1993): 97% fem., 97% male. Secondary (1993): 90% fem., 88% male. University: 28% (1993). **Primary school teachers:** one for every 18 students (1992).

COMMUNICATIONS

25.9 **telephones** per 100 inhabitants (1993).

ECONOMY

Per capita GNP: $7,040 (1994). **Consumer price index:** 100 in 1990; 494.2 in 1994. **Currency:** 117 tolar = 1$ (1994). **Cereal imports:** 549,000 metric tons (1993). **Fertilizer use:** 2,306 kgs per ha. (1992-93). **Imports:** $7,304 million (1994). **Exports:** $6,828 million (1994). **External debt:** $2,290 million (1994), $1,151 per capita (1994). **Debt service:** 5.4% of exports (1994).

ENERGY

Consumption: 1,506 kgs of Oil Equivalent per capita yearly (1994), 19% imported (1994).

al army was to remain in the green area (free of border controls) during the three-month period established for jurisdiction to be turned over to Slovene authorities.

[22] In order for the cease-fire to be enforced, the agreement established an unconditional return of federal army units to army barracks, and the demobilization of the Slovene territorial defence forces. Likewise, directives were established for clearing the highways and freeing prisoners on both sides.

[23] Within the following three months, Slovenia strengthened its separatist resolve, while federal troops gradually withdrew from the territory. Only hours before the three-month period was to expire, Slovene authorities announced the creation of a new national currency, the *tolar*, which was to replace the Yugoslavian *dinar*. On October 8 1991, Slovenia's declaration of independence went into effect.

[24] When it was part of Yuglosavia, the population of Slovenia was only 8 per cent of the total (approx. 1.9 million inhabitants). Yet its industrial production accounted for 25 per cent.

[25] Between 1991 and January 1992, the EC countries recognized Slovenia and Croatia as independent states, although civil war continued in the latter. The EC threatened to apply economic sanctions to Belgrade unless fighting stopped. The homogeneity of the Slovenian population made its secession the least painful in the Yugoslavian dissolution process.

[26] Slovenia's recognition was also one of the most clear-cut cases for the international community, as it controlled its own borders, maintained its own armed forces, and had already issued its own currency. After the withdrawal of Yugoslavian troops, the government, headed by Milan Kucan, undertook the task of economic reconstruction, with practically no interference from the former Yugoslav government.

[27] Between 1991 and 1992, the independence process was marked by two years of steady decline in economic activity. During this period, inflation climbed to more than 200 per cent annually, and unemployment to over 13 per cent. By the beginning of 1993, the foreign debt totalled $1.764 billion. Even so, Slovenia maintained an annual per capita income of over $6,000 - higher than that of any other former Yugoslav republic.

[28] Slovenia began to define itself as a "European, not Balkan" State in 1993. Fluent trade relations held with the European Union and Slovenian authorities' willingness to open up its markets led the country to seek association into the bloc. The growth of tourism and the tolar's stability - convertible since September 1995 - enabled Slovenia's exports and gross domestic product to increase.

[29] Italy opposed Slovenia's association with the EU. Italian authorities demanded compensation for the nationalization of property belonging to 150,000 Italians between 1945 and 1972. The Catholic Church also demanded the return of property which had been nationalized by the communist regime.

[30] The European Commission approved Slovenia's request in May 1995 but Italian opposition continued. In late 1995, Croatia and Slovenia had not reached an agreement regarding the sovereignty of the waters of Piran bay. The destination of Croatian funds deposited in Slovenia's main bank, the Ljubljanska Banka, prior to the break-up of former Yugoslavia, had yet to be decided.

Solomon Islands

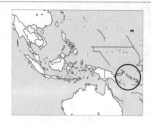

Population: 365,000 (1994)
Area: 28,900 SQ KM
Capital: Honiara

Currency: Solomon Is. Dollar
Language: English

Solomon Islands

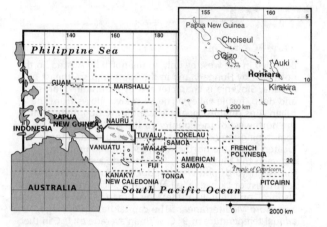

[1] The Solomon Islands have been inhabited by Melanesians since about 2000 BC. The social organization here was less complex than in other Melanesian archipelagos of the Pacific.

[2] The Spaniard Alvaro de Mendaña, arrived in 1567, searching for "El Dorado" - land of gold. In the 18th and 19th centuries, the islands were used as a source of slave labor for the sugar plantations of Fiji and Australia.

[3] After World War I, Britain colonized the archipelago, which was taken over by the Japanese in 1942. After World War II, the archipelago was recovered and divided in two. The eastern part, some 14,000 sq km, was placed under Australian administration, and later annexed to Papua New Guinea.

[4] On May 21 1975, Britain agreed to independence claims. The islands were granted internal autonomy in 1976, and independence was finally declared on July 7 1978.

[5] The financial system and the islands' land holdings, the main center of production, are under foreign control. Although efforts have been made at economic diversification, the Solomon Islands continue to depend almost entirely on external markets.

[6] Solomon Mamaloni was elected prime minister in August 1981, after the islands' parliament censured Kenilorea. He banned atomic-powered ships and planes, or those carrying nuclear arms, from the islands' territorial waters and air space.

[7] In the 1984 election, a coalition worked together to achieve the re-election of Kenilorea as prime minister. He patched up relations with the United States and in 1986, named Ezequiel Alebua as his successor. After taking office, Alebua followed Kenilorea's policies faithfully.

[8] In 1989, Mamaloni was re-elected prime minister. In 1991 the islands turned down a proposal by an American company to dispose of industrial waste on their territory. The environmental organization Greenpeace revealed that within the last 20 years, there had been close to 20 similar proposals.

[9] Mamaloni's government was subjected to parliamentary investigation in 1992 because of irregularities in the sale of the National Fishing Development Company, a Canadian enterprise, tarnishing the government image, though it was not found directly responsible. On another front, the government's development plan included the creation of 21 rural training centers to modernize agriculture, and incentives to increase tourism.

[10] The islands' relations with Papua New Guinea deteriorated as a result of the conflict in the secessionist island of Bougainville. Two Solomon Islanders had been killed in an armed confrontation there in September 1992. In April 1993, after attempts at defusing the situation, Papuan troops carried out several incursions into Solomon Island territory (see Papua New Guinea).

[11] Francis Billy Hilly was elected prime minister in June 1994 but was forced out of office on October 31 after being censored by the Parliament. A week later, Mamaloni was elected prime minister.

[12] Billy Hilly's government had announced that log exports would be banned in 1997. In spite of the wood industry's growth, the concern to preserve the levels of sustainability was seen as a top priority. With aid from Japan and searching for substitute activities, the government decided to sponsor fishing.

[13] Mamaloni did not continue his predecessor's preservationist policy and in April 1995 ordered the logging of all the trees on Pavuvu island. Despite the protests, Pavuvu's residents were relocated to other islands. (According to Billy Hilly, the timber industry was a source of government corruption.)

PROFILE

ENVIRONMENT

Comprises most of the island group of the same name, except for those in the northwest which belong to Papua New Guinea, the archipelago of Ontong Java (Lord Howe Atoll), the Rennell Islands and the Santa Cruz Islands. The Solomon Islands are part of Melanesia, east of New Guinea. The major islands, of volcanic origin, are: Guadalcanal (with the capital), Malaita, Florida, New Georgia, Choiseul, Santa Isabel and San Cristobal. The land is mountainous and there are several active volcanoes. Fishing and subsistence agriculture are the traditional economic activities. Deforestation is severe. Heavy rains cause soil erosion, particularly in exposed areas.

SOCIETY

Peoples: Most of the population are of Melanesian origin. There are also Polynesian, Micronesian, Chinese and European minorities. **Religions:** Christian, 96.7%, of which Protestant account for 77.5% and Roman Catholic for 19.2%; Baha'i, 0.4%; traditional beliefs, 0.2%; other and no religion, 2.7%. **Languages:** English (official), Pidgin (local dialect derived from English) and over 80 dialects. **Political Parties:** Group for National Unity and Reconciliation (formerly party for the Alliance of the People), of Solomon Mamaloni; People's Alliance Party; National Action Party of Solomon; Labor Party of Solomon; United Party. **Social Organizations:** The Solomon Islands Council of Trade Unions (SICTU), formed in 1986, made up of 6 trade unions.

THE STATE

Official Name: Solomon Islands.
Administrative Divisions: 8 Provinces and the Capital. **Capital:** Honiara, 35,288 inhab. (1990). **Other cities:** Gizo, 3,727 inhab.; Auki, 3,262 inhab.; Kira Kira, 2,585 inhab.; Buala, 1,913 inhab. **Government:** Moses Pitakaka, Governor-General since June 1994, appointed by the British government. Solomon Mamaloni, Prime Minister since November 7, 1994. Unicameral Parliament of 47 members. **National Holiday:** July 7, Independence Day (1978).

DEMOGRAPHY

Annual growth: 3% (1992-2000). **Children per woman:** 5.8 (1992).

HEALTH

Under-five mortality: 32 per 1,000 (1995). **Calorie consumption:** 92% of required intake (1995). **Safe water:** 96% of the population has access (1990-95).

EDUCATION

Primary school teachers: one for every 25 students (1991).

COMMUNICATIONS

88 **newspapers** (1995), 81 **TV sets** (1995) and 90 **radio sets** per 1,000 households (1995). 1.5 **telephones** per 100 inhabitants (1993).

ECONOMY

Per capita GNP: $810 (1994). **Annual growth:** 2.20% (1985-94). **Consumer price index:** 100 in 1990; 155.9 in 1994. **Currency:** 3 Solomon Is. dollars = 1$ (1993).

Somalia

Somaliya

Population: 8,775,000 (1994)
Area: 637,660 SQ KM
Capital: Mogadishu (Muqdisho)
Currency: Shilling
Language: Somali and Arabic

The region of Somalia, called "the land of Punt" by the Egyptians thousands of years ago, had active commercial relations with Egypt. Centuries later, the Romans called it "the land of aroma" because of the incense produced there.

[2] This ancient commercial tradition took on new dimensions from the eighth century when Arab refugees founded a series of commercial settlements on the coast.

[3] In the 13th century, Islamicized and led by Yemeni immigrants, Somalians founded a state which they called Ifat, with its principal center in Zeila. Tribute was initially paid to the Ethiopian empire but Ifat quickly entered into conflict with the Abyssinians and succeeded in consolidating its independence. New territories were annexed and a new name adopted: the Sultanate of Adal.

[4] Ties established with Arab markets and the southern part of the coast of Zandj contributed to develop intense commercial activity. At the same time, the sultans tried to enlarge their dominion, at the expense of the none too stable Ethiopian empire.

[5] However, it was not until 1541 that the Portuguese government, having acquired a clearer knowlege of Indian Ocean trade, sent its fleet. Backed up by the Ethiopian army, the latter razed the city of Zeila, going on to destroy Mogadishu (Muqdisho), Berbera and Brava.

[6] The Portuguese destroyed but did not occupy the area, though the presence of their armada hindered economic reconstruction. Adal declined, and was divided into a series of minor sultanates, the northern ones controlled by the Ottoman Empire, while the ones in the south accepted the sovereignty of the sul-

tan of Zanzibar after the expulsion of the Portuguese in 1698.

[7] The Suez Canal gave new strategic value to the region. In 1862, the French bought the port of Obock, leading to the creation of present-day Djibouti and in 1869 the Italians settled in Aseb and later extended their control over Eritrea.

[8] In 1885 the British, who had already occupied Aden on the Arabian peninsula, took the Egyptian settlements in Zeila and Berbera. In 1906, in compensation for their defeat in Ethiopia, the Italians obtained Somalia's southern coast.

[9] The British colony was the leading centre of resistance to foreign domination. Sheik Muhammad bin Abdullah Hassan organized an Islamic revolutionary movement, which defeated the British troops on four occasions between 1900 and 1904. The British finally gained control of the territory in 1920.

[10] On July 1 1960, with the start of decolonization in Africa, the British and Italian regions became independent and were merged as the Republic of Somalia. The new republic adopted a parliamentary system, until October 21 1969, when a group of officers led by General Said Barre seized power and proclaimed a socialist regime.

[11] In July 1976, the Front for the Liberation of Western Somalia initiated a military offensive in the Ogaden supported by the Somali government. The invasion was repelled by the Ethiopian army, supported by Cuban troops and backed by most African nations who opposed changes in colonial frontiers. The country broke off relations with Cuba and ended its military agreements with the Soviet Union.

[12] The war against Ethiopia caused serious difficulties to Somalia's fragile economy; prices of fuel and grain increased and a severe drought affected most of the country in 1978-79, bringing the Barre government to the brink of collapse. A group of army officers unsuccessfully attempted to overthrow the government, in April 1978. In October 1980 Barre declared a state of emergency and reinstated the Supreme Revolutionary Council which had ceased functioning in 1976.

[13] The northeastern province of Kenya has historically been claimed by Somalia, but relations with Kenya have improved since

1984, through commercial and technical cooperation agreements.

[14] Problems with Ethiopia recur sporadically because of the dispute over the Ogaden plains and the steady flow of Ethiopian refugees into Somalian camps. In 1988 alone the number of refugees was estimated at 840,000.

[15] Said Barre was re-elected in December 1986 by 99 per cent of the vote. In February 1989, he sent Prime Minister Ali Samater to London and Washington, to announce the granting of amnesty and commutation of sentences on 400 political prisoners.

[16] Western credits and investments were drastically reduced and exports to Saudi Arabia, the principal market for its cattle, stopped almost completely.

[17] In January 1991, the opposition formed the United Somali Congress (USC) and ousted the president, replacing him with Mohammed Ali Mahdi, leader of the USC and representing business interests.

[18] After bloody confrontations between the two factions of the USC, President Mahdi fled Mogadishu in November 1991. The capital remained in the hands of Gen-

eral Mohammed Farah Aidid, leader of the military wing of the USC.

[19] The two rival factions of the United Somali Congress were made up of members of different sub-groups of the Hawiye clan, led by Farah Aidid and Ali Mahdi, respectively.

[20] Since the beginning of the war between the rival clans, 300,000 Somali children have died, and 1.5 million inhabitants, a quarter of the total population, have left the country.

[21] On March 9, taking advantage of a cease-fire, 100,000 women and children held a peaceful march through Mogadishu. Nine tribal leaders of the Hiraan region met to discuss how to bring about peace in their country. It was the first meeting of "ugas" or kings in more than a hundred years.

[22] In 1993, after two years of civil war and total anarchy, Somalia continued to have no central governmental authorities. The traditional authorities, incapable of handling these unprecedented conditions, had abandoned their normal functions, leaving clans and sub-clans unrestricted. In some regions, though, the local population still re-

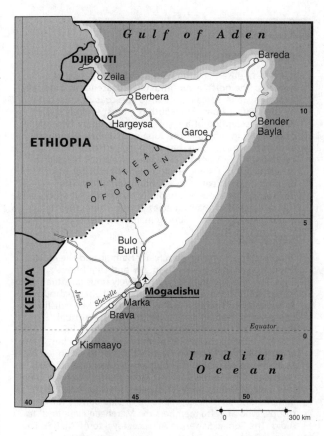

WORKERS

1994

MEN 62%   WOMEN 38%

AGRICULTURE 76%   INDUSTRY 8%   SERVICES 16%

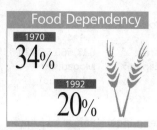

Food Dependency

1970
34%

1992
20%

ENVIRONMENT

It is a semi-desert country with a large population of nomads. The north is mountainous, descending gradually from the Galia-Somali plateau to the coastal strip bathed by the Gulf of Aden. The south is almost entirely desert with the exception of a fertile area crossed by two rivers originating in Ethiopia, the Juba and Shebeli. Drought, endemic in this region, has been exacerbated by overpasturing. A drastic increase in livestock has led to desertification. Fishing using explosive charges has damaged coral reefs and aquatic vegetation. The destruction of these fish habitats may have put stocks and future catches of many species at risk. There are 74 endangered species, including mammals, plants and birds.

SOCIETY

Peoples: The Somalian people are Hamitic and the most important ethnic groups (Isaq, Dir and Digil) have linguistic and cultural unity, a relatively unusual situation in Africa. **Religions:** Islamic (official), mostly its orthodox Sunni version. There are a few Catholics in Mogadishu. **Languages:** Somali and Arabic (official). **Political Parties:** the United Somali Congress, created in 1991: made up mainly of members of Hawiye clan. There is confrontation between rival factions of this clan led by Mohammed Ali Mahdi and Mohammed Farah Aidid (now deceased). The Somali Patriotic Movement, of General Omar Jess, part of United Somali Congress, loyal to Aidid. The Somali National Alliance, made up of Siad Barre's former troops, who are also members the Marehean clan, loyal to Aidid. The Somali Salvation Alliance, loyal to Ali Mahdi. The Somali Democratic Salvation Front, a party of the northeast, inhabited mainly by Darods (including Marehean and other sub-clans); this party engaged in a power struggle with the Islamic Union, a new political group. The National Somali Movement (founded in 1981), a party of the northwest, of the Issaq clan, whose sub-clans also have internal conflict. The Revolutionary Socialist Party of Somalia (SRSP), only legal party since 1991 coup d'etat. **Social Organizations:** The General Federation of Somali Trade Unions, created in 1977 and chaired by Mohamed Ali Almed.

THE STATE

Official Name: Jamhuriaydda Soomaliya. **Administrative divisions:** Somalia is divided into 18 regions or provinces. **Capital:** Mogadishu (Muqdisho), 570,000 inhab. (1984). **Government:** Neither the government nor the public administration has functioned since November 1991, when Mohammed Ali Mahdi fled the capital. **National Holiday:** July 1, Independence Day (1960). **Armed Forces:** Regular armed forces have not existed since rebel forces ousted the government in 1991.

DEMOGRAPHY

Annual growth: 1.4% (1992-2000). **Estimate for year 2000:** 10,000,000. **Children per woman:** 6.8 (1992).

HEALTH

Under-five mortality: 211 per 1,000 (1995). **Calorie consumption:** 86% of required intake (1995). **Safe water:** 37% of the population has access (1990-95).

COMMUNICATIONS

81 **newspapers** (1995), 84 **TV sets** (1995) and 80 **radio sets** per 1,000 households (1995). 0.2 **telephones** per 100 inhabitants (1993).

ECONOMY

Annual growth: -2.30% (1985-94). **Currency:** 930 shillings = 1$ (1989). **Cereal imports:** 296,000 metric tons (1992).

[24] Under pressure from the United States, the UN sent 28,000 troops to assist. This was the first time the world organization had intervened militarily in the internal affairs of a nation.

[25] On suggestions from the UN, military chiefs, councils of elders and prominent citizens agreed in March 1993, to create a provisional government and a National Transition Council. Aidid's forces confronted the UN "blue helmets" with losses on both sides. The elite US Rangers failed in their attempt to capture Aidid. In March 1994, the UN withdrew 10,000 European troops, leaving 19,000, mostly Africans and Indians. The peace efforts did not prosper, despite a meeting organized in Nairobi, Kenya, between the main opponents: Aidid and Ali Mahdi. The last contingent of UN troops left the country in March, protected by a force of 1,800 marines. And even though the regional conflicts continued, the civil war did not spread and economic life seemed to recover.

[26] The country remained divided during 1995. In the northeast, the Somaliland government maintained control over its territory. The Mogadishu region was split between the factions of Farah Aidid, the Somali National Alliance (SNA) and the Somali Salvation Front. The southern agricultural area around the port of Kismayu was in the hands of General Muhammad Said Hersi ("Morgan") - son-in-law of the former dictator Said Barre, who had died in Nigeria in January - and the interior was under the control of Colonel Ahmad Omar Jess. The plains between the Juba and Shebelle rivers had been settled by the group of Rahawayn clans.

[27] Aidid's position was weakened when his right-hand-man, the millionaire businessman Osman Hassan Ali ("Ato") left him to join Ali Mahdi. In June the United Somali Congress designated Aidid as president of the SNA, and his supporters elected him president of all Somalia. With the aim of consolidating his power, Aidid launched an offensive against the Rahawayn forces, who fought back. In Somaliland, President Muhammad Ibrahim Egal had to resist attacks from his predecessor Abd ar-Rahman Ahmad Ali ("Tur"), a supporter of Aidid.

[28] Representatives of all the parties met in Jeddah, Saudi Arabia, in September 1995. At the end of many peace conferences, which Aidid refused to attend, the decision was made to establish a national government. General Farah Aidid died on August 1 1996.

spected the traditional authority of the tribal elders. However, in the vacuum created by the absence of central administration, power remained in the hands of the warlords and of heavily armed bands of looters.

[23] The Republic of Somaliland, the former British colony which declared independence in 1992, holds 30 per cent of the national Somali territory but has not been internationally recognized. With the economy in ruins, it sheltered a million people displaced from the southern zone.

South Africa

Population: 40,539,000 (1994)
Area: 1,221,040 SQ KM
Capital: Pretoria
Currency: Rand
Language: Afrikaans, English, isi Ndebele, Sesotho sa Lebowa, etc.

South Africa

The first Dutch settlers arrived in Cape Town in 1652, more than 150 years after the voyage around the Cape of Good Hope by the Portuguese navigator Vasco da Gama. The Portuguese had focused their attention on India and showed little desire to challenge the Khoikhoi people, who had been established in the region for more than a thousand years, and were hostile to foreign navigators.

[2] Jan Van Riebeeck was the first Dutchman to challenge the Khoikhoi. In Cape Town he founded a colony where vegetables were cultivated and animals raised to supply the ships en route to Indonesia. By 1688, about 600 farmers had already settled in Cape Town, dividing their time between farming and fighting the war against the Khoikhoi. Being in such a tiny minority, the first Dutch settlers were fiercely united and aggressive, two characteristics which were to become striking features of the whole Boer ("farmer") society in South Africa.

[3] The Dutch increased the areas under their control to gain greater security, but they were not typical colonists. They were not allowed to trade with local people and they had to hand over everything they produced to the Company's ships. This arrangement finally led them into conflict with their overseas employers. In this dispute, the Boers slowly gained ground and by the end of the 17th century, the so-called free settlers, the burghers, were already in a majority. The population of European origin split into two groups; those linked to foreign trade and settled agriculture and those (pastoralists) who moved deeper inland in search of new territories.

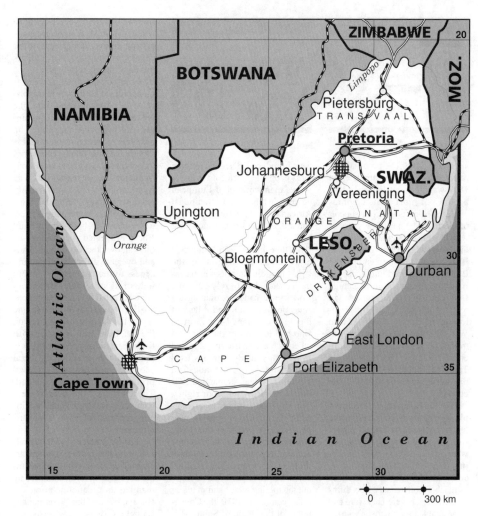

[4] Between 1770 and 1840, there were seven large wars against the Xhosa, or "kafirs" - a pejorative Boer term for the Africans. In 1806, with the weakening of the Dutch colonial empire, the British installed themselves in Cape Town. The British established a system which would enable them to exchange merchandise, incorporate African leaders as intermediaries and end slavery. This quickly led to a clash with the Boers, who began calling themselves Afrikaners to distinguish themselves from the more recently arrived colonists. In 1834 close to 14,000 Boers migrated inland, initiating the Great Trek which would take them to what is now Transvaal, the Free State and KwaZuluNatal. Objecting to outside interference, the Afrikaners established the states of Transvaal in 1852 and the Orange Free State in 1854. These regions were already occupied by African people who fiercely resisted the new settlers.

[5] The British recognized the independence of both regions, as European occupation of the new lands contributed to Cape Town's security. The Boers had no choice but to trade through the ports which were held by the British. Although official history minimizes the African resistance, it was extremely strong and organized and caused many casualties within the native population. In their expansion northward, the Afrikaners confronted the Xhosas and the Zulus. The latter had been led by Shaka, a military genius, who died in 1828. The resistance blocked the colonists' advance over a period of 50 years. Shaka had become the head of a huge empire which crumbled shortly before the Great Trek, not because of external pressure, but due to internal problems linked to the issue of royal succession.

[6] Conflict between the Boers and the British Crown erupted in the late 19th century when rich gold and diamond deposits were discovered in the interior. The realization that the area had great strategic importance led the British to propose a federation between the Cape Colony, the Transvaal (South African Republic), Natal and the Or-

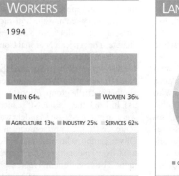

WORKERS

1994

■ MEN 64%   ■ WOMEN 36%

■ AGRICULTURE 13% ■ INDUSTRY 25% ■ SERVICES 62%

LAND USE

■ CROPLAND 11% ■ PASTURE 67% ■ OTHER 23%

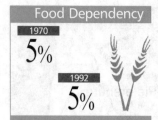

Food Dependency

1970	5%
1992	5%

Foreign Trade

MILLIONS $ 1994

IMPORTS
23,400

EXPORTS
25,000

ange Free State. The proposal was rejected by the Boers and later, in 1899, war broke out. Britain had the backing of most of its colonies while the Boers enjoyed the support of Germany. After three years of war, in which about 50,000 people died and 100,000 were confined to British concentration camps, the Boers surrendered and accepted British domination in exchange for certain concessions in regional autonomy. The British victory in the Boer or South African War signalled the end of the controlling power of landed Boer farmers in the Orange Free State and the Transvaal, and the beginning of the pre-eminence of mining in the economy.

[7] For the Boer settlers, local people were nothing but savages who had to be tamed by force. "White" supremacy and racial segregation were established to justify the subjugation of the black population, and to guarantee a supply of cheap farm labor from tenants. Boer farms in the interior were backward and unprofitable in comparison with the farming practised in the Cape and Natal.

[8] The British focus on trade and liberalism led them to consider slavery a restraint on the creation of consumer markets. This view did not prevent them from erecting rigid barriers to exclude South African black people from economic and social gains. Labor laws imposed severe controls on worker mobility. The 1843 Master and Servant Act, along with later decrees, made it a criminal act to break a work contract.

[9] In the 19th century the British also hired black workers in the territories now called Mozambique, Lesotho and Botswana, as well as Indians. No "imported" worker was allowed to bring any members of his or her family; they received only a minimum salary and were obliged to return to their own country if they lost their job. Both in the Cape and Natal hut and poll taxes

were introduced, designed to force local people to work for far lower wages than workers of European descent. In this way, the traditional African ways of life were destroyed and at the same time, wages were kept low.

[10] When the gold and diamond mines came into operation, the large European interests which controlled them had to rely on white workers who had some training and educational background. Many of them were former Boer farmers who had lost out in the war, and others came from Europe, attracted by the gold rush. Both groups, used to the workings of the industrial capitalist system, demanded economic rewards and labor rights. This situation was ably manipulated by the mining companies, who promised benefits to the white workers provided that they become accomplices in the exploitation of the black labor force. The so-called colour bar was established throughout the mining sector and in major urban centres. In 1910, the Constitution of the Union of South Africa (a federation of Cape Province, Natal, the Orange Free State and the Transvaal) deprived most black people of the right to vote or to own land.

[11] After 1910, segregationist legislation gradually increased. The Natives Land Act of 1913 set aside 7 per cent of the nation's land for blacks as reserves, which became home to 35-40 per cent of the country's population. Much of the remaining land was reserved for ownership by whites, who represented only 20 per cent of the population. On the overpopulated black reservations, subsistence agriculture predominated. The rest - belonging to whites - was intensively farmed with the reservations functioning as a permanent source of cheap labor. The 1923 Native Urban Act tightly regulated blacks' lives in cities which were considered to be white strongholds.

[12] Until 1984, political participa-

tion was limited very largely to whites. Most Black Africans, comprising almost 66 per cent of the country's population, were still deprived of the right to vote.

[13] From colonial times, power was monopolized by the white minority. At the outset of World War I, the white economy was based on mining and intensive farming. The post-war recession made it necessary for the large mining companies to hire blacks, leading to racial confrontations among the workers. The Rand strike in 1922, was harshly put down by the government. Most of the strikers were poor whites, descendants of both English and Boers. Frustrated over having lost the war, over the loss of their lands and over the difficulty of gaining access to the country's incipient industrial structure, Afrikaners were attracted by the ultra-nationalistic propaganda of the extreme right. The Nationalists, who won the 1924 election together with their English-speaking allies, broke with the tradition of liberal economic policies, and imposed protectionist measures instead. These affected the large mining companies and were aimed at initiating a process of internal industrialization.

[14] The emergence of state capitalism promoted by the Nationalists, enabled the country to achieve impressive growth rates, classified by many as an "economic miracle". But at the end of the 1920s, euphoria gave way to a new crisis: the Great Depression. The Nationalists were forced into another alliance in order to survive. This time they joined forces with traditionally despised foreign capital, keeping the system of racial segregation, which guaranteed a supply of cheap labor. The industrial development which followed brought with it an increase in the number of blacks employed in industry, and an increase in racial conflict. The Afrikaner BroederBond or Brotherhood emerged; it was a secret society, which became the bastion of white right-wing politics.

[15] The recession that followed World War II, again led to the "poor whites" rebelling when faced with the possibility of losing their jobs. Racism once again flourished, under the slogan *Gevaar* KKK (beware of blacks, Indians and communism - Kafir, Koelie, Komunism). Predictably enough, the Nationalists formed a govern-

ment by themselves in 1948 and imposed even harsher restrictions on black people.

[16] The first national political organization of South African blacks appeared in 1912. The African National Congress (ANC) was created by a group of former students from schools run by missionaries. Among its founders were several people who had studied or gained degrees in American or European universities. The first ANC leaders believed that the Afrikaners could be persuaded of the injustice of its segregationist laws. They also believed that Anglophile liberals would accept black co-participation in politics. These high hopes were dashed in 1920, when the black mine workers strike was crushed and segregation entrenched.

[17] In the 1940s the ANC adopted a strategy of non-violent resistance against segregationist laws. In 1955 the anti-racist front was broadened with the so-called Freedom Charter, proclaimed at a multi-racial gathering in Kliptown. The Charter contained a vehement denunciation of apartheid (separateness) and called for its abolition and a redistribution of wealth.

[18] Beginning in 1943, the Youth League branch of the ANC launched a more aggressive program. Its leaders, Nelson Mandela and Oliver Tambo, gradually attained positions of leadership within the ANC. In 1958, sectors of the ANC that did not agree with the multi-racial policy of the movement created the Pan-Africanist Congress (PAC). In 1960, the PAC organized a demonstration in the town of Sharpeville to protest against the pass law which restricted the movement of black workers in areas reserved for whites. The demonstration was brutally repressed by the police and 70 people were killed. Immediately after the "Sharpeville Massacre", the PAC, the ANC and the Communist Party were outlawed and the anti-racist struggle underwent a radical change. The African National Congress created an armed movement, *Umkhonto we Sizwe* (Spear of the Nation) while the PAC organized another guerrilla group, *Poqo* (Only us). In 1963, the main ANC leaders were arrested. Nelson Mandela was sentenced to life imprisonment and Oliver Tambo, who went into exile, took over leadership of the movement. The government's repressive violence, and the lack of supply

ENVIRONMENT

Located on the southern tip of the African continent, with a coastline on the Indian and Atlantic Oceans, South Africa has several geographic zones. A narrow strip of lowland lies along the east coast, with a hot and humid climate, large sugar cane plantations. In the Cape region are vineyards and *fynbos* vegetation. The vast semi-arid and arid Karoo, with cattle and sheep ranching, makes up over 40 per cent of the total territory, extending inland. The Highveld extends to the north and is the richest arable area. It surrounds the Witwatersrand, a mining area in the Transvaal where large cities and industries are found. The country's economic base lies in the exploitation of mineral resources: South Africa is the world's largest producer of gold and diamonds, the second largest producer of manganese, and the eighth largest producer of coal. The country's water resources are overtaxed, salinization is the main threat in dry areas. Soil erosion is a serious problem, particularly in the former Bantustans (homelands).

SOCIETY

Peoples: Over 76% of the population is of African origin. There are also Coloured (descendants of whites, slaves and Khoisan). European descendants account for 13% of the total. Asian groups, predominantly Hindu, make up less than 3% of the total population.
Religions: Christianity predominant (68%) including African independent churches, 22%; the Dutch Reformed Church, 12%; the Roman Catholic, 8%; Methodists, 6%. Traditional African religions are also followed. **Languages:** Afrikaans, English, isi Ndebele, Sepedi, Sesotho, siSwati, Xitsonga, Setswana, Tshivenda, isi Xhosa, isi Zulu. **Political Parties:** African National Congress (ANC); National Party (NP); Inkatha Freedom Party (IFP); Freedom Front (FF); Democratic Party (DP); Pan Africanist Congress (PAC); African Christian Democratic Party (ACDP); African Muslim Party (AMP); Dikwankwetla Party; Federal Party; African Democratic Movement; Ximoko Progressive Party (XPP). **Social Organizations:** South African Students' Congress (SASCO), the Congress of South African Trade Unions (COSATU), the National Congress of Trade Unions, the Federation of South African Labour Unions. South Africa has a large number of NGOs, estimates vary between 8,000-15,000.

THE STATE

Official Name: Republiek van Suid-Afrika, Republic of South Africa. **Administrative Divisions:** 9 Provinces. **Capital:** Pretoria (administrative capital), 1,080,200 inhab.; Cape Town (legislative capital), 2,350,200 inhab.; Bloemfontein (judicial capital) 300,150 (1991). **Other cities:** Johannesburg, 1,916,100 inhab.; Durban 1,137,400 inhab.; Port Elizabeth, 853,200 inhab. (1991). **Government:** Nelson Mandela, Pesident since May 10, 1994. According to the constitution which has been in effect since that date, the Legislature is made up of the Senate and the National Assembly. **National Holiday:** 27 April, Freedom Day (1994). **Armed Forces:** 136,900 (1995). **Paramilitaries:** 110,000 South African Police, 37,000 reserves.

---

bases in neighboring countries which were under regimes allied with the Afrikaners, kept the guerrillas from making the progress necessary to attract massive numbers of recruits.

[19] In 1964, any South African who was not white could not vote, had to use passes, could not leave certain areas, could be arrested arbitrarily, could not belong to a union nor support a strike, could not go to schools or universities with whites, could be transferred to other areas against their will, did not have access to public services, and could not participate in public demonstrations against segregation. The survival of the racist system depended, to a large extent, on international capitalist investment in the region, attracted by the availability of a large pool of cheap labor. Foreign investment, especially from America, quintupled in value between 1958 and 1967. The Afrikaners' protectionist policies created the infrastructure necessary for large industries to move there, with the long-term objective of creating an industrial development zone capable of supplying all southern Africa.

[20] The 1960s witnessed an increase in the migration of black farm workers to the cities. The massive exodus was fuelled by a number of factors: the misery of the "homelands", or Bantustans, with poor quality of the soil, which was incapable of producing very much, and the absence of social services. This migration affected the expectations of other urban sectors, like the coloured people, who saw their hopes of integration in the white economy dashed. In 1976 the black community in the suburbs of Johannesburg rebelled. The rebellion by the young people of SOWETO (South Western Township) made whites realize that the crisis had reached their own cities, where they had once felt safe. In 1970, 75 per cent of all those who worked in agriculture, mining or services were black, and the participation of non-whites in specialized jobs had tripled within the last 20 years, though blacks earned 5 to 10 times less than whites for equal work. The ruling white minority were dependent upon black labour, and they proposed a few reforms to apartheid, aimed at preventing new social crises among workers who had migrated to the cities. The Pretoria regime declared four Ban-

tustans, Transkei, Ciskei, Venda and Bophuthatswana, independent States, in an attempt to prevent further urban migration by the unemployed. Eight million people were thus deprived of their South African citizenship, though few countries in the world recognized these newly "independent" states.

[21] The situation in southern Africa was radically modified by the independence of Angola and Mozambique in 1975, and Zimbabwe in 1980. The African National Congress found the support bases it desperately needed, in these, and the other Front Line States of Botswana, Tanzania and Zambia. South Africa, with an economy three times greater than that of these independent countries combined, initiated a destabilisation campaign which included economic pressure, sabotage, support of rebel movements and invasion. All these measures were designed to force these governments to deny support to the anti-apartheid movement, and block attempts by newly independent countries to escape South African domination.

[22] One of the main scenes of conflict in southern Africa was Namibia, a former German colony which South Africa had occupied during World War I. In 1966 the UN determined that South Africa should grant Namibia its independence. This proposal was taken up by the OAU and the Front Line countries, despite the delaying tactics adopted by South Africa and the Western powers. The independence of Namibia did not actually occur until March 21 1990.

[23] For South Africa to be able to impose its power throughout the southern part of Africa, the support which it received from Europe and the United States was essential. Approximately 400 US companies had interests in South Africa; US capital and technology were vital for South African industrial development and for maintaining its military might. South Africa is also the world's major exporter of platinum, gold, manganese, chrome and vanadium; the second largest exporter of antimony, diamonds, fluoride and asbestos; and the third, of titanium, uranium and zirconium. Also, its geographical location is strategic: ships carrying 70 per cent of the oil destined for Europe, and 30 per cent of the oil destined for

the United States, navigate its waters.

[24] During the Reagan era, the United States used all its economic and military leverage to avoid revolutionary change in southern Africa. On the domestic front, South African Prime Minister PW Botha (in office 1978-1989) began reforming the segregationist system. Between 1982 and 1984, a constitutional reform went into effect whereby Indians and coloured people were given the right to vote through the creation of two more chambers in the legislature for these racial groups. Blacks continued to be excluded, limiting their participation to the local level. Many non-white citizens boycotted the reform, abstaining from voting. The South African economy went into recession in 1983, due to the fall in the international price of gold, its main export product. The economic difficulties affected the white middle class, which until then had been largely impervious to economic troubles. Given this situation, the racist parties of the extreme right attracted more votes.

[25] The gradual liberalization of apartheid promised by Botha found

---

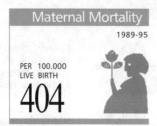
many opponents. Repression against blacks did not diminish; it was merely complicated by inter-tribal confrontations, and conflict between opposing parties. In July 1985 the government declared a state of emergency in 36 districts. By the end of 1986, more than 750 people had been killed and thousands of the government's opponents had been arrested.

[26] In the United States and western Europe the anti-apartheid campaign grew over the next few years. Public pressure forced Western governments and a growing number of companies and banks to limit their activities in South Africa. The US Congress lifted the veto imposed by President Reagan on economic sanctions, bringing about a change in the policy of constructive engagement in the area. Within South Africa, internal political opposition led to the creation of the United Democratic Front (UDF), which brought together more than 600 organizations working together within the law.

[27] From the beginning of 1988, the government showed greater intransigence toward the opposition, to the point of declaring all opposition illegal in February of that year. Religious leaders opposed to apartheid were briefly detained, among them black Archbishop Desmond Tutu, who had won the Nobel Peace Prize for his activity against racial segregation. Labour unions, led by the Congress of South African Trade Unions (COSATU), founded in 1985, were suppressed as subversive. In August 1989, Botha was faced with an internal crisis within his own party which had been in power for the past 41 years, and he resigned. He was succeeded by FW de Klerk, who was aware of the need to change the image of the South African government in the eyes of international public opinion, and he came out in favour of a change in the racist system. In 1990, South Africa had to refinance part of its $12 billion foreign debt, and changes in its system of segregation would help in its negotiations with international agencies.

[28] In September 1989, parliamentary elections were held within the state of emergency which had been in effect since 1986. The Mass Democratic Movement (MDM) an anti-apartheid coalition of legal organizations, called a general strike. Despite police round-ups and threats, three million South African blacks interrupted their work in the largest protest in the country's history. The growing momentum was accompanied by repression and killings. But by now an increasing number of the white minority was joining the protests, making the situation even more difficult to subdue.

[29] In February 1990, de Klerk legalized the African National Congress and other opposition groups. After several announcements, and an equal number of delays, Nelson Mandela, the country's most famous political prisoner was freed on the 11 of that month. He had served the longest prison sentence of all political prisoners; 27 years. A period of negotiation began. Upon being freed, Mandela, acclaimed by multitudes, took up his role as leader of the black majority.

Zulu expansion

The Zulus and Xhosas of southern Africa form the Nguni group of peoples. This group, along with the Shona of Zimbabwe, the Sotho of Lesotho, Ngoni in Malawi and Zambia, and the Tswana of Botswana, constitute the group of southeastern Bantu peoples. A great variety of southern African peoples were affected, directly or indirectly, by the unification process started in 1818, by a young Zulu leader called Shaka. He raised an army of all the able adult males and some women. He set up an efficient network of couriers, a system of espionage and counter-intelligence and a centralized supply network. With these resources he conquered an area of 500,000 sq km (in the present-day South African provinces of Gauteng and KwaZulu-Natal) which he endowed with a centralized organization, aiming to eliminate local power structures. All the conquered peoples were incorporated into the empire under the same conditions, regardless of their racial or geographical origins. The young men were recruited into the army, and when they married, after the age of thirty, a great many married outside their original ethnic groups.

[2] In 1828 Shaka was assassinated by Dingaan, his brother and successor, but his ten years of warfare had left important legacies across southern Africa. Dingaan went on to fight the Boers, beginning in 1835, at the start of the Boers "Great Trek" to the north, and continuing over the following four years. During this time, Dingaan inflicted many severe defeats upon the invaders. Zulu resistance was eventually put down in 1839 and Dingaan was executed by Andries Pretorious, leader of the trekkers.

[3] The Ndwande, defeated by Shaka in 1818, retreated northwards, reaching Malawi and Mozambique in 1835. They still form a minority there today. In 1823 some Sotho groups conquered by the Zulu settled in the southwest of present-day Zambia. They subjugated the Rotsi (1835) with whom they eventually divided the territory, known to the British as Barotseland.

[4] The rest of the Sotho unified under the leadership of Moshoeshoe, "the Great Chief of the Mountain". They resisted the Ndebele and the Zulus under Shaka and Dingaan maintaining their independence in what would become the nation of Lesotho.

[5] One of Shaka's lieutenants, Mzilikazi took a large contingent of warriors with him. He tried to transform them into a nation, giving them the name of Matabele. This group ended up occupying southern Zimbabwe, an area still known as Matabeleland.

[6] The Tswana, who inhabited the western region of Zulu territory, also felt compelled to unite. This union eventually led to the present-day republic of Botswana, although more people of Tswana descent live in South Africa than in Botswana.

[7] In northeastern Natal, other Bantu groups joined in resistance in the mountains. By the time he died in 1839, Sobhuza had unified these diverse peoples into one state, and one nation. They became known as the Swazi, after M'swazi, Sobhuza's son and successor.

[8] A substantial part of the political background of present-day South Africa was moulded by the Zulu in their reaction to European expansion. While Shaka had tried to eliminate ethnic differences and divisions within a large centralized empire, the rich ethnic mixture he left behind him has actually shaped itself into separate ethnic groups, like the Ndebele or nations like Lesotho and Botswana.

Being in a position of leadership is not altogether without its difficulties, however, one of these being the confrontation between the ANC and members of the Zulu nationalist organization "Inkatha", which had not been fighting apartheid.

[30] ANC leaders maintained that the conflict lay between those who want a united, democratic and nonracist South Africa, and those who would only accept changes based on the preservation of privileges. The ultra-right, playing on the fears of the poor whites, defended racial segregation. In May 1990, after three days of meetings between the ANC and the government, Mandela announced an agreement to put an end to violence and begin a normal political life. He called on the international community to maintain economic sanctions and other forms of pressure upon the South African government. Some opposition groups on the left were against talks with the government, claiming that these would only postpone the struggle and the radical change which they demanded. The police, whose terrible repressive habits had not been modified, were linked to the formation of death squads, adding to the uncertainty of a peaceful end to apartheid. In May 1990, the government renounced the policy of creating Bantustans (ten had been established), and abolished racial segregation in hospitals. In December, ANC president Oliver Tambo returned to the country, after more than 30 years in exile.

[31] In mid-April 1991, the European Community began studying the possibility of lifting the economic blockade, and set June 30 as a deadline for beginning the democratization of the country. At the same time, the government did away with the Population Registration Act and the Land Acts, whereby blacks were forbidden to own land. De Klerk promised to begin negotiations for a new constitution. The United States went ahead and lifted the blockade, while the EC, though intending to lift it, got caught up in internal debates. In the end, Denmark and Spain, which had just hosted a visit from Mandela, vetoed the move.

[32] While these developments were taking place, the "Inkathagate" scandal broke out. This was chiefly about state money being given to the Inkatha Freedom Party (IFP) for their activities, many of which were aimed at opposing the ANC. Given the fact that friction between the ANC and Inkatha had claimed 5,000 lives since 1986, these revelations cast doubt on the government's intentions. The ANC had suspended hostilities in August 1990, but the confrontations became more violent after June 1991. In late June, de Klerk replaced the ministers of Defence and of the Interior. In October, despite an agreement between the two groups and the government, signed the previous month, more than 50 deaths were reported.

[33] In addition to these confrontations, crime was on the rise, as a result of the economic recession. For the third consecutive year, the GDP dropped by 0.6 per cent in 1991, and unemployment reached almost 40 per cent of the economically active population.

[34] In August 1991, Mandela reiterated his demands: an interim government, the election of a Constituent Assembly, the release of all political prisoners and the principle of one person, one vote. In November, negotiations between the government and 20 political organizations led to the establishment of the Convention for Democracy in South Africa (CODESA), which would discuss the new constitution. Groups which excluded themselves by choice from these talks included the ultra-right white Conservative Party, the National Reform Party, the Afrikaner Resistance Movement (AWB) and black groups including: the Pan-Africanist Congress (PAC) and the extreme left Azanian People's Organization (Azapo). In December, CODESA met for the first time; but, Mandela insisted that the economic blockade be maintained.

[35] In February 1992, the National Party lost the by-election at Potchefstroom, a small university city located west of the capital. The Conservative Party took advantage of the election results to claim that de Klerk did not have a mandate to negotiate within CODESA. De Klerk immediately announced a referendum among South African whites, to decide whether the government should continue the negotiations. The ANC and the labor organization COSATU rejected the referendum, claiming that the country's destiny must not depend exclusively on the opinion of the whites. The conservatives, however, took up de Klerk's challenge. On March 17 almost 70 per cent of the white population voted "yes" to the transition envisioned by de Klerk.

[36] Death squad activity continued. In June, the massacre of 46 people in the township of Boipatong brought talks between the ANC and the government to a halt.

[37] In September 1992, with the resumption of the negotiations, a Protocol of Understanding was signed between President de Klerk and ANC leaders. This document established the guidelines for the transition, including the creation of a joint committee charged with preparing elections for a Constitutional Assembly, as well as a multiracial transitional government. In addition, it provided for the release of all political prisoners. The agreement was rejected by Inkatha and extreme right-wing groups. Armed confrontations broke out between Inkatha and the ANC, especially in Natal Province, where the government doubled the number of troops. The ANC's regional representative from Natal, Reggie Hadebe, was shot dead when he was returning from a meeting in which peace in the region was being discussed.

[38] Russia re-established diplomatic relations with Pretoria, followed shortly by Côte d'Ivoire. A few days later, de Klerk made an official visit to Nigeria, where he expressed South Africa's desire to join the Organization of African Unity (OAU).

[39] Difficulties continued to mount on the economic front. By the end of 1992, the nation's GNP had fallen by 5.7 per cent, the highest figure in recent years, due to a decrease in agricultural production, brought on in part by drought but also affected by the lack of investment and capital flight.

[40] The Executive Committee of the ANC, meeting on November 25, recognized the necessity of forming a "government of national unity". The government set the date for elections for April 1994. For

STATISTICS

DEMOGRAPHY

Urban: 50% (1995). **Annual growth:** 0.8% (1992-2000). **Estimate for year 2000:** 47,000,000. **Children per woman:** 4.1 (1992).

HEALTH

Under-five mortality: 68 per 1,000 (1995). **Calorie consumption:** 107% of required intake (1995). **Safe water:** 70% of the population has access (1990-95).

EDUCATION

Illiteracy: 18% (1995). **School enrolment:** Primary (1993): 110% fem., 110% male. Secondary (1993): 84% fem., 71% male. University: 13% (1993).

COMMUNICATIONS

97 **newspapers** (1995), 100 **TV sets** (1995) and 101 **radio sets** per 1,000 households (1995). 9.0 **telephones** per 100 inhabitants (1993). **Books:** 101 new titles per 1,000,000 inhabitants in 1995.

ECONOMY

Per capita GNP: $3,040 (1994). **Annual growth:** -1.30% (1985-94). **Annual inflation:** 14.30% (1984-94). **Consumer price index:** 100 in 1990; 157.0 in 1994. **Currency:** 4 rands = 1$ (1994). **Cereal imports:** 2,275,000 metric tons (1993). **Fertilizer use:** 596 kgs per ha. (1992-93). **Imports:** $23,400 million (1994). **Exports:** $25,000 million (1994).

ENERGY

Consumption: 2,253 kgs of Oil Equivalent per capita yearly (1994), -33% imported (1994).

their contributions to the dismantling of apartheid, the Swedish Academy gave the 1993 Nobel Peace Prize to Nelson Mandela and FW de Klerk.

[41] The progress made by the political leadership did not put an end to the violence. In January 1993, the Independent Human Rights Commission set the number of violent deaths of the previous year at 3,499, of which 299 were members of the security forces and 123 were civilians killed in police custody. The ANC denounced the formation of a new group of armed whites, made up of members and former members of the South African Defence Forces.

[42] In April 1993 talks resumed in a new forum - the Council of Multiparty Negotiation - which replaced the defunct CODESA, suspended in June 1992.

[43] However, yet another act of violence threatened to derail the talks definitively. On April 10, Chris Hani, Secretary General of the Communist Party, member of the Executive Committee of the ANC, was assassinated. He had a wide following and respect, especially among the most radical sections of the ANC, and was thought by many to be the likely successor to Mandela. A large number of ANC youth reacted to Hani's death by calling for a return to armed conflict. Several members of the Conservative Party were implicated in the conspiracy which had led to Hani's death. The weeks which followed were marked by numerous protest demonstrations.

[44] In May, another plot was uncovered - this time to assassinate Joe Slovo, former Secretary General of the Communist Party. In October, Janusz Walus, a South African citizen of Polish origin, and Clive Derby-Lewis, a former Conservative Party MP and former presidential adviser, were found guilty of assassinating Hani and sentenced to death.

[45] The mobilization of the ultra-right parties reached its peak during this period, when a conference of 21 racist groups was convened. These included the neo-Nazi Afrikaner Resistance Movement (AWB) led by Eugene Terreblanche, the Reformed National Party; the Afrikaner National Union, as well as white labor unions and organizations representing agribusiness. The main outcome of this meeting was the for-

mation of the Afrikaner National Front (NFA), led by General Viljoen.

[46] The NFA attempted to build an independent white state. In January, some 3,000 armed militants from the AWB deployed their forces at the World Trade Centre, where the negotiating talks were being held.

[47] A Human Rights Commission report registered 9,352 deaths in the period from 1990 to 1993 as a result of political violence.

[48] Maintaining a distance from the talks on the provisional constitution, the Inkatha Freedom Par ty, the NFA and the Conservative Party decided to boycott the electoral process. Bophuthatswana president Lucas Mangope declared he would join the boycott, in the midst of a civil servants' strike. Street protests immediately ensued. Simultaneously, neo-Nazi leader Terreblanche ordered his followers from the Afrikaner Resistance Movement to invade Bophuthatswana, and come to Mangope's aid. Resistance from black civilians and local forces made them withdraw. Mangope was deposed and the South African army seized control. 50 dead and 300 wounded was the outcome. A few days later, Terreblanche decided to participate in the electoral process through the recently created Freedom Front.

[49] In the meantime, Inkatha boycotted ANC activities and had severe confrontations with Mandela's followers, particularly in the Inkatha stronghold of KwaZulu-Natal. Buthelezi finally agreed to participate in the elections after the constitution finally recognized his nephew Goodwill Zwelethini as king of the Zulus.

[50] In October, the UN agreed to lift sanctions against the South African regime. The US immediately lifted its financial restrictions.

[51] The provisional Constitution formed a 400-member National Assembly and a 90-member Senate. The president, with less powers than the prime minister, would be elected by the Assembly for a 5-year term. The territory was divided differently, with nine provinces, each with its own governor and legislature; the ten bantustans were abolished as such and incorporated into the provinces.

[52] The South African National Defence Force (SANDF) was created which brought together the South Africa Defence Force and some ANC and PAC guerrillas.

[53] The first multi-ethnic elections in South Africa were held between April 26-29, 1994. 87 per cent of those registered to vote took part. The ANC obtained 63 per cent of the vote, de Klerk's NP 20 per cent and Inkatha 10 per cent. The far right, represented by the Freedom Front, won 2 per cent.

[54] The Government of National Unity included NP and IFP members as well as the ANC majority. The minister of finance and the governor of the South African Reserve Bank from the previous government remained in their posts.

[55] Although the barriers of apartheid have fallen, economic and cultural obstacles remain. Black workers earn far less than whites and unemployment is far higher respectively. The overall infant mortality rate for the country is 52 per 1,000 but this disguises the fact that the rate is much higher among the black population that the white.

[56] The May 1994 constitution was radical in tone. Among the measures to be implemented by his government, Mandela proposed free health care for children under 6 and pregnant women, the launching of a basic dietary plan for schoolchildren and provision of electricity for 350,000 homes. He pledged to create 2.5 million jobs and build a million homes for 1999.

[57] The implementation of the Reconstruction and Development Program progressed slowly in 1995 as a result of financial and bureaucratic limitations. A law set a new structure for education. In October it was announced 3.5 million would have access to water services in the following 18 months. The first budget completely elaborated by the Government of National Unity allotted 47 per cent to social services and 26 per cent to education, investment in housing was doubled and military expenditures were reduced. An ambitious agrarian reform program was put into effect by minister Derek Hanekom, a farmer. An act of labor relations was passed which guaranteed the right to strike and formed forums of discussion at workplaces. South Africa registered a lower number of strikes than in previous years.

[58] In January, the ANC withdrew the immunity it had guaranteed prior to the elections to two former Cabinet members and 3,500 police members, who were under investigation by the Truth and Reconciliation Commission. The trial of a former police colonel

charged with 121 murders, kidnappings and fraud brought about new evidence regarding police incitement of political violence during the previous regime. Prominent Inkatha leaders were implicated in payoffs to security police. A report by the Goldstone Commission sent to President de Klerk in 1994 was published, listing all these charges. In June an IFP under-secretary was arrested for murders committed in 1987.

[59] Strikes and accusations of racism among the police led to the resignation of the force's chief. His successor, George Fivaz, proposed their demilitarization, among other reforms. Although there was a growing concern over the increase of crime, the new Constitutional Court abolished the death penalty.

[60] The Government of National Unity had internal tensions, especially as the November local elections grew nearer. Inkatha withdrew from parliament and the constitutional assembly as a result of what it believed was the ANC's intention to establish complete dominance. The ANC accused the IFP of promoting KwaZulu-Natal secession, threatened to cut the supply of funds to the area and sent military and police forces to the province. All in all, political violence in Natal was reduced: in mid-1995 there were 70 deaths per month while prior to the elections the number amounted to 300.

[61] The local elections in November 1995 were a success for the ANC all over the country, except in the Western Cape and KwaZulu-Natal. In May 1996, the NP left the unity government to join the opposition.

Spain

España

Population: 39,143,000 (1994)
Area: 504,780 SQ.KM
Capital: Madrid

Currency: Peseta
Language: Spanish

Between the 9th and the 8th century BC Celtic peoples started to settle in the center and west of the Iberian peninsula. Later, during the 6th and 5th centuries BC, the Iberian culture developed in the south of the peninsula. The fusion of these two cultures gave rise to what is known as the Celtiberian peoples. The peninsula was colonized by the Phoenicians, the Greeks, and the Romans in succession.

[2] The fall of the Roman Empire coincided with the spread of Christianity and, more importantly, with the invasion of the northern peoples who raided Europe. The Iberian peninsula was occupied by the Visigoths who ruled over the area for 300 years.

[3] In 700 AD the peninsula was invaded by Arabs who defeated Rodrigo, the last Visigoth king, thus starting a period of Muslim domination. The descendants of the Visigoths lived in the north of the territory and set up kingdoms like Castille, Catalonia, Navarre, Aragon, Leon and Portugal. Over the centuries they underwent a slow unification process, which consolidated when they started fighting against the Arabs.

[4] The Arabs called the lands in the south of the Iberian peninsula Al-Andalus, a region that reached its peak during the 10th century. In contrast with the rest of impoverished rural Europe, its cities, and Cordoba in particular, prospered through active trade with the East. Religious tolerance enabled Muslims, Jews, and Christians to live side by side, and science, medicine, and philosophy developed. Copies and translations of the Greek classics were made, paving the way for the 15th century European Renaissance.

[5] In 1492, a triple process of national unification took place in Spain, through the marriage of Isabel of Castille and Fernando of Aragon, the expulsion of the Moors, and the conquest and subsequent colonization of the new American territories. The unification of the country's political power and the creation of the kingdom of Spain were carried out at the expense of the Jews (and members of other cultures), who were expelled from Spain after having lived there for many centuries. Both the Inquisition and centralized power were institutionalized under the new system, while the new American colonies supplied precious metals, sustaining three centuries of economic bonanza. The indigenous people in America had Christianity imposed upon them by the Crown, and the exploitation they were subject to through forced labor was responsible for decimating many of their populations.

[6] The economic prosperity provided by the colonies was reflected in a period of great cultural development in Spain. Literature in particular, developed extensively during the 16th and 17th centuries, which were dubbed the Spanish Golden Age.

[7] In the 18th century, the Bourbons came to the Spanish throne. They reorganized domestic and colonial administration, ruling in accordance with the principles of Enlightment, as the liberal ideas of the French Revolution spread throughout Europe and America. Combined with Napoleonic expansion, this view contributed to the disintegration of the Spanish colonies after the wars of independence.

[8] At the end of the Napoleonic era, there was great conflict between the liberal sectors seeking political and economic modernization, and the absolutists who wished to preserve the traditional order. The disputes between groups weakened the power of the empire, making way for revolutions in Spanish America.

[9] By the end of the 19th century Spain had renounced its last American territories and had come to terms with the loss of its privileges.

[10] At the beginning of the 20th century, Spain was plunged into a deep political, social, and economic crisis, exacerbated by World War I. Within an atmosphere of extreme polarization, Primo de Rivera's dictatorship, which resulted from the 1923 coup, attempted to halt any further increase in demands from workers or regional groups seeking autonomy. The government, closely resembling the Italian fascist model, retained power until 1931. Its eventual loss of power was a result of existing contradictions within the Church, the armed forces and industry, who were all fascist supporters, and not to the continuous opposition from political and labor organizations. The end of the dictatorship marked the end of the monarchy, and the dawn of a new republican era.

[11] The "Second Republic" was born in the midst of a series of extremely complex political and economic difficulties. In 1936, after two moderate governments, the People's Front, formed with socialists, republicans, communists and anarchists, gained a narrow electoral triumph, causing much friction with their political opponents.

[12] Immediately after the elections, the army, the church, and powerful sectors of the Spanish economy started to plot the overthrow of the government. The deep contradictions within the Front and the incessant opposition led to the rise of an important faction of the army, led by Francisco Franco, who provoked the Spanish Civil War lasting for three years. The republican government waited in vain for assistance from the European democracies, but they had decided not to intervene in the conflict. The Soviet Union was the only state which provided material support to the republican government, while a large number of volunteers from America and Europe also joined the republican army.

[13] Franco won in March 1939, assisted by internal divisions within the republican forces, the military superiority of the troops loyal to him, and German and Italian support.

[14] When the civil war ended, Franco became head of the new Spanish State. He set up an authoritarian regime along fascist lines, with a corporatist state, a personality cult, and extreme nationalism. Franco ruled over a deeply divided society and an economy that had been virtually devastated by the civil war.

[15] From the beginning of the Cold War, the US tried to secure Spanish support and Spain became a member of the United Nations in 1955, confirming a change in Franco's foreign policy, which was becoming preoccupied with improving the international image.

[16] In the 1960s, Franco opened parliament to other groups and movements. During those years key figures of the Opus Dei, an ultra-conservative Catholic movement, occupied key posts in government and changed the country's economic policies.

[17] Spain rid itself of economic isolation and liberalized its econo-

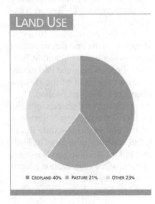

LAND USE

- ■ CROPLAND 40%
- ■ PASTURE 21%
- ■ OTHER 23%

PROFILE

ENVIRONMENT

Spain comprises 82% of the Iberian Peninsula and the Balearic and Canary Islands. The center of the country is a plateau which rises to the Pyrenees in the north, forming a natural border with France. The Betica mountain ranges extend to the South. In the inland the Central Sierras separate the plateaus of Nueva and Vieja Castilla. In the Ebro River basin, to the north, lie the prairies of Cataluña, Valencia, and Murcia, and the Guadalquivir River basin, to the south, forms the plains of Andalucía. The climate is moderate and humid in the north and northwest, where there are many woodlands. In the interior, the climate is dry in the south and east. 40% of the land is arable. Approximately 5% of the total land area is under environmental protection. Natural resources include coal, some oil and natural gas, uranium and mercury. Industry, which is undergoing a rapid rate of growth, is concentrated in Catalonia ("Cataluña") and the Basque Provinces; in the latter, there are blast furnaces and paper mills. Per capita emissions of air and water pollutants exceed Western European averages. Since 1970, the use of nitrogen fertilizers has doubled. Nitrate concentrations in the Guadalquivir River are 25% above 1975 levels. The percentage of the population serviced by sewage systems rose from 14% in 1975, to 48% in the late 1980's. This fact, together with industrial wastes from oil refining plants and natural gas production, has raised the level of contamination of the Mediterranean. The government has implemented a reforestation program to increase production of timber and prevent further erosion. Many seedlings have been lost in recent years to forest fires. An additional problem is the fact that the large-scale planting of single species, such as eucalyptus, does not take biodiversity into account. Approximately 22% of Spain's forests have suffered some degree of defoliation.
Capital: Madrid, 2,909,800 inhab. (1991).
Other cities: Barcelona, 1,623,500 inhab.; Valencia,752,500 inhab.; Sevilla, 659,100 inhab.; Zaragoza, 586,200 inhab. (1991).

SOCIETY

Peoples: The population is made up of Castillians, Asturians, Andalucians and Valencians, Catalans, Basques and Galicians, descended from a fusion of the Iberian people of the Mediterranean, the Celts of Central Europe and the Arabs of North Africa. Within the past 30 years, the urban population has grown from 56 to 78%. **Religion:** Catholicism is the religion of the vast majority of the Spanish people.
Languages: Spanish or Castillian (official national); there are also official regional languages, such as Basque, Catalan, Valencian and Galician. **Political Parties:** Spanish Socialist Workers' Party (PSOE); Peoples' Party, rightist; United Left; Convergence and Union; Basque Nationalist Party, as well as other regionalist parties. The ETA, an armed separatist Basque group. **Social Organizations:** There are several national workers' organizations: the Workers' Commissions (socialist tendency); the National Confederation of Labor (anarchist tendency); the Workers' Trade Union (of social democratic tendency).

THE STATE

Official Name: Estado Español. **Administrative Divisions:** Spain is divided into 17 autonomous regions: the Basque Country, Catalonia, Galicia, Andalucia, the Principality of Asturias, Cantabria, La Rioja, Murcia, Valencia, Aragon, Castilla, La Mancha, the Canaries, Navarre, Extremadura, the Balearic Islands, Madrid and Castilla León. Each one has its own institutional system, a local leader in executive power and a unicameral legislative. **Government:** Hereditary monarchy, with King Juan Carlos I de Borbón y Borbón as head of State. Prime Minister: José María Aznar since April 1996. The parliament, *las Cortes,* consists of two bodies: the Senate, with 255 members, and the Chamber of Deputies with 350 members elected by proportional representation.
Armed Forces: 206,000 (1995). **Paramilitaries:** Guardia Civil: 66,000 (3,000 conscripts); Guardia del Mar: 340.

my by eliminating many mechanisms of state control. The urban middle classes enjoyed a substantial improvement in their standard of living, which allowed a significant political relaxation. However, the peasants were still extremely poor and many emigrated to the major Spanish and European cities.
[18] Franco died in 1975, and power was handed over to his successor, the heir to the Spanish crown, Juan Carlos I of Bourbon. The new monarch immediately started negotiations with the political opposition to re-establish the democratic system overthrown in 1939.
[19] Between 1976 and 1981 Adolfo Suarez, the leader of the Center Democratic Union (UCD), was prime minister. In December

1978, while he was in office, a plebscite was held, turning Spain into a parliamentary monarchy, re-establishing political freedom, and guaranteeing the right of autonomy to some Spanish regions. Several politicians, intellectuals, and artists were able to return to the country, after up to forty years in exile.
[20] In February 1981, a group of Civil Guard officers took the *Cortes* (parliament) by force. The firm reaction of all the democratic political groups and in particular of King Juan Carlos, who had the support of the army, guaranteed the failure of the plan and the consolidation of the democratization process.
[21] The Spanish Socialist Workers' Party (PSOE) won the October 1982 elections, with a solid major-

ity in the *Cortes.* Felipe González became President of the Government, the equivalent of Prime Minister in other countries. He was subsequently re-elected in 1988 and 1993.
[22] Despite some setbacks, the parliamentary majority of the PSOE was able to enforce an ambitious adjustment and growth plan which has deeply transformed the Spanish economy and given vast social sectors access to unprecedented levels of consumption. However, this modernization resulted in high unemployment and tensions which led to a split between the government and the UGT, the trade union which had supported the PSOE.
[23] Spanish transition has also

been punctuated by radical nationalist activism, in particular that of the Basque separatist group ETA, which many times has used violent means to achieve its political goals. In recent years ETA has been extremely active, attacking government officials, military officers, shopping malls and entertainment arcades throughout Spain. In February 1990, the Basque Parliament declared the right of the Basque people to self-determination.
[24] ETA suffered serious setbacks in 1993. Cooperation between French and Spanish security forces led to the arrest of some of its leaders, and the discovery in Bayonne, France, of the organization's main weapons deposit.
[25] The Spanish government has

DEMOGRAPHY

Urban: 76% (1995). **Annual growth:** 0.2% (1991-99). **Estimate for year 2000:** 39,000,000. **Children per woman:** 1.2 (1992).

HEALTH

Under-five mortality: 9 per 1,000 (1994). **Calorie consumption:** 114% of required intake (1995).

EDUCATION

School enrolment: Primary (1993): 105% fem., 105% male. Secondary (1993): 120% fem., 107% male. University: 41% (1993). **Primary school teachers:** one for every 21 students (1992).

COMMUNICATIONS

101 **newspapers** (1995), 106 **TV sets** (1995) and 97 **radio sets** per 1,000 households (1995). 36.4 **telephones** per 100 inhabitants (1993). **Books:** 108 new titles per 1,000,000 inhabitants in 1995.

ECONOMY

Per capita GNP: $13,440 (1994). **Annual growth:** 2.80% (1985-94). **Annual inflation:** 6.50% (1984-94). **Consumer price index:** 100 in 1990; 122.9 in 1994. **Currency:** 132 pesetas = 1$ (1994). **Cereal imports:** 4,955,000 metric tons (1993). **Fertilizer use:** 769 kgs per ha. (1992-93). **Imports:** $92,500 million (1994). **Exports:** $73,300 million (1994).

ENERGY

Consumption: 2,414 kgs of Oil Equivalent per capita yearly (1994), 69% imported (1994).

Gibraltar

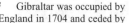
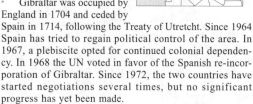

A peninsula on the southern coast of Spain, only 32 km from Morocco, Gibraltar's strategic position allows it to control maritime trade between the Mediterranean and the Atlantic ocean.
2   Gibraltar was occupied by England in 1704 and ceded by Spain in 1714, following the Treaty of Utretcht. Since 1964 Spain has tried to regain political control of the area. In 1967, a plebiscite opted for continued colonial dependency. In 1968 the UN voted in favor of the Spanish re-incorporation of Gibraltar. Since 1972, the two countries have started negotiations several times, but no significant progress has yet been made.

PROFILE

Peoples: Most of the permanent population is of British origin. The non-permanent population are mainly Spanish workers. **Religions:** Anglican and Catholic. **Languages:** English (official) and Spanish. **Official Name:** Gibraltar. **Government:** Non-autonomous territory, subject to UN control. Administrative power: United Kingdom. General Governor and commander-in-chief named by the British crown: Sir Derek Refell. The 15-member advisory council appoints the chief minister: Joe Bossano, leader of the Socialist Labor Party reelected in January 1992. Bossano obtained 73% of the vote, while the second political force, the Social Democrat Party, was supported by only 20% of the electorate. In February 1992, Bossano suggested transferring foreign relations to the EC. Britain would maintain nominal control of the colony, while it governed itself. Britain, however made Gibraltarian independence dependent upon an agreement with Spain. In September 1996, the NATO Headquarters aknowledged that in the future the military control over the peninsula would be exercised by Spain.

COMMUNICATIONS

129 **newspapers**, 316 **TV sets** and 1,171 **radio receivers** per 1,000 inhab. (1991).

ENERGY

Consumption: 894 kgs of Coal Equivalent per capita yearly, 275% imported. (1990).

---

been very active in international affairs, joining the European Economic Community and NATO in 1986. Membership of NATO had been opposed by the PSOE while it was in the opposition, but once in power the party defended the decision and confirmed it through consultation with the people. It has maintained a longstanding dispute with Britain over the possession of Gibraltar, which has been under British control since 1704; it maintains control of Ceuta and Melilla, which are claimed by Morocco.
26   In 1990, Felipe Gonzalez's government started negotiations with the US to reduce US military presence in Spain. Despite the agreements, some of the B-52 bombers that raided Iraqi territory during the Gulf War took off from the US airbases in Spain.
27   In 1992 Spain spent $10 billion on the celebration to mark the 500 years since the conquest of Latin America. At the same time the country made political moves towards Europe. The demands of the European policy resaulted in a weakening of Spains's traditional links with Latin America.
28   The will to align the country with European standards led the government to carry out cutbacks in defence, public spending and sub-

sidies for the industrial sector. In November, Parliament ratified the Maastricht treaty. The legal proposal to facilitate the dismissal and redeployment of workers caused a general strike in January 1994, but this was approved anyway with the argument that it was necessary to make the nation competitive. Unemployment reached 22 per cent, the highest level in the European Union.
29   Many of the Socialist Party leaders were implicated in fraud investigations. The former governor of the Bank of Spain was imprisoned along with the former director of the Madrid Exchange Reserve, and the financier Mario Conde, who had supervised the sale of the Spanish Credit Bank (Banesto).
30   Due to the numerous scandals, the socialist government lost a key part of its parliamentary support in 1995. The Catalan nationalist group Convergence and Union withdrew their support in July and voted in the budget with the right-wing People's Party (PP). Whilst defending his action and that of his government, Gonzalez moved the elections forward a year.
31   The imminent electoral success of the right led ETA to change their strategy: in January they assassinated the PP leader in the prov-

ince of Guipuzcoa, and in April tried to kill the PP candidate for prime minister, Jose Maria Aznar. In the May regional elections, the PP comfortably won 10 of the 13 areas at stake.
32   A former security chief and a socialist official were tried for their connections with the Anti-terrorist Liberation Groups (GALs), which legal investigations had linked to the security forces. One of them told the Court that Felipe Gonzalez had been responsible for the GALs, which had killed 27 people on the frontier with France in the so-called "dirty war" against ETA. A former interior minister was also investigated for paying French officials to collaborate in the persecution of the Basque separatists.
33   Fishing conflicts set Spain at odds with Canada and Morocco. Both ended with the European Union mediating in agreements

limiting their action in the territorial waters of the other nations. Spain took over the presidency of the European Union from July 1, 1995, and held several international summits. The opposition, however, questioned the government's ability to fulfill the necessary requirements for integration to European Monetary Union planned for 1997.
34   The People's Party won the March 1996 parliamentary elections with 38.9 per cent of the vote. The PSOE came second with 37.5 per cent. The IU (Izquierda Unida, United Left) took 10.6 per cent. On April 5, PP leader José Maria Aznar became prime minister.

Sri Lanka

Population: 17,865,000
(1994)
Area: 65,610 SQ.KM
Capital: Colombo

Currency: Rupees
Language: Sinhalese

Sri Lanka

The island of Ceylon was populated by the Vedda in ancient times, it was then successively invaded by the Sinhalese, Indo-Europeans and Tamils, who laid the foundations of an advanced civilization. When the Portuguese arrived in 1505, the island was divided into seven autonomous local societies.

[2] A century and a half later, the Dutch expelled the Portuguese from their coastal trading posts, but it was the British, who had already taken possession of neighboring India, who finally made the island a colony in 1796. Even then, it took them until 1815 to subdue all the local governments, as they fought hard to remain autonomous. The British then introduced new export crops such as coffee and tea, products which gave Ceylon a worldwide reputation, because of their excellent quality.

[3] In the 20th century, a strong nationalist movement developed in Ceylon. In 1948, independence was achieved and Ceylon became a member of the British Commonwealth. Under the leadership of Sir John Kotelawala and Prime Minister Bandaranaike, Ceylon pursued a vigorous anticolonial foreign policy. In August 1954, Bandaranaike met in Colombo with India's Nehru, Muhammad Ali of Pakistan, U Nu of Burma and Indonesia's Sastroamidjojo. The meeting was of great political importance as it led to the 1955 summit conference of Afro-Asian countries in Bandung, heralding the Movement of Non-Aligned Countries.

[4] In the late 1950s, the Tamil minority staged a series of secessionist uprisings and, in September 1959, the prime minister was assassinated. His widow, Sirimavo Bandaranaike, led the Sri Lanka Freedom Party to electoral victory at the beginning of 1960, although she had no previous political experience. Sirimavo Bandaranaike became the first woman in the world to head a government. Governing in coalition with the Communist and Trotskyist parties, in 1962 she nationalized various US oil and other companies. In 1965, she was defeated by a right-wing coalition but regained power in 1970, in a landslide election victory.

[5] She was faced with a "Guevarist" (after Che Guevara) guerrilla uprising which she had forcibly crushed. She never gave up her anti-imperialist stand and in 1972 declared Sri Lanka a republic, cutting all ties with the British empire.

[6] A land reform program nationalized British-owned plantations but had little effect on the standard of living of the bulk of the rural population.

[7] Conflict between the Sinhalese majority and the Tamil minority, who are descended from the Dravidians of South India, has persisted throughout most of the island's known history. The Sinhalese account for 74 per cent of the country's population, while the Tamils comprise 22 per cent, and are divided into two groups: the "Sri Lanka Tamils" and the "Indian Tamils". The "Sri Lanka Tamils" reached the island approximately 2,000 years ago and settled principally in the northern and eastern provinces, while the "Indian Tamils" are more recent immigrants. Both groups possess common ethnic characteristics and seek regional autonomy or even the formation of a separate Tamil nation. The Tamil United Liberation Front (TULF) founded on May 4 1972, resulted from the fusion of three Tamil parties: the Federal Party, the Tamil Congress and the pro-Indian Ceylon Workers Congress.

[8] Bandaranaike's main achievement was probably her appointment as president of the Non-Aligned Movement in 1976, at the Conference in Colombo. However, accusations of nepotism, and reaction from the country's intellectuals to the censorship of the press and to emergency security measures in force since 1971, caused Bandaranaike to lose the July 1977 election.

[9] The United National Party, led by Junius Jayewardene, won a comfortable parliamentary majority, with the Tamil Liberation Front coming in second. In spite of his "socialist" leanings, the new prime minister's economic policy exposed the country to transnational capital.

[10] A constitutional reform, which went into effect in 1978, made Jayewardene Sri Lanka's first president. In November 1980, a series of IMF approved economic measures started to be applied, with disastrous consequences for the country. A month earlier, Bandaranaike had been expelled from parliament and deprived of her political rights for seven years. A presidential commission had found her guilty of "abuse of authority" during her coalition government, between 1970 and July 1977.

[11] An International Monetary Fund "package" of measures went into effect. There was a general strike against the measures and the new economic policy, and 44,000 civil servants lost their jobs in the summer of 1980. In 1981 the government received substantial US aid, supplied with the alleged purpose of "stabilizing the region". The same argument was put forth to justify increasingly frequent visits by US naval ships to the port of Colombo.

[12] Sri Lanka's first presidential election, of October 1982, gave Jayewardene a clear victory, with 52.5 per cent of the vote. His campaign was backed by the state apparatus and benefitted from the division and internal strife of the Freedom Party. In some parts of the country, the political upheaval meant that elections had to be carried out under "state of emergency" conditions.

[13] The project to turn Sri Lanka into an "export center" like Hong Kong or Taiwan led to the creation of a free zone in Latunyabe where foreign investment has been steadily on the increase since 1978.

[14] In early 1982, in spite of earli-

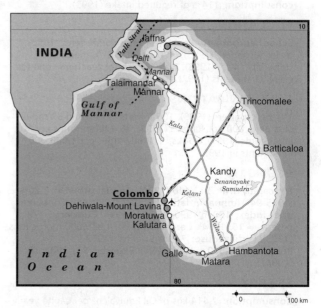

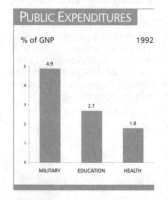

PUBLIC EXPENDITURES

% of GNP — 1992

- MILITARY 4.9
- EDUCATION 2.7
- HEALTH 1.8

WORKERS

13.6%
UNEMPLOYMENT

1994

MEN 73%   WOMEN 27%

AGRICULTURE 49%   INDUSTRY 21%   SERVICES 30%

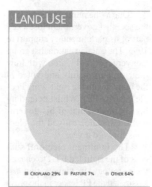

LAND USE

CROPLAND 29%   PASTURE 7%   OTHER 64%

ENVIRONMENT

An island in the Indian Ocean, southeast of India, separated from the continent by the Palk Channel. The land is flat, except for a central mountainous region. These mountains divide the island into two distinct regions and also block the monsoon winds which are responsible for the tropical climate. The southwest area receives abundant rainfall, and the remainder of the island is drier. Large tea plantations cover the southern mountain slopes, and other major crops are rice, for domestic consumption, and rubber, coconuts and cocoa, for export. Deforestation and soil erosion are important environmental problems. Likewise, air pollution is on the increase, due to industrial gas emissions.

SOCIETY

Peoples: 74% of Sri Lankans are Sinhalese; Tamils are the largest minority group, 22%, and there is a small Arab minority. **Religions:** 69% Buddhist; 15% Hindu; 8% Christian; 7% Muslim; and 1% other religions. **Languages:** Sinhalese (official). Also Tamil and English. **Political Parties:** The United National Party (Siri Kotha), founded in 1946, defined itself as "socialist" before taking power; the Sri Lanka Freedom Party, founded in 1951, led by Prime Minister Sirimavo Bandaranaike; the United Democratic National Front; the Tamil United Liberation Front (TULF): made up of the Tamil Congress - founded in 1944 - and the Federal Party - founded in 1949; the Communist Party, founded in 1943; the Equal Society Party (Lanka Sama Samaja), a Trotskyist group founded in 1935; the People's Democratic Party (Mahajana Prajathantra); the People's Liberation Front (Janata Vimuktui Peramuna), which backs the formation of a multi-ethnic workers' party and the Popular Alliance coalition. **Social Organizations:** The four trade unions are in the process of merging; the Workers' Congress, linked to the Social Democratic ICFTU; the Trade Union Federation, affiliated with the WFTU; the Trade Union Council; and the Labor Federation.

THE STATE

Official Name: Sri Lanka Prajathanthrika Samajavadi Janarajaya. **Administrative Divisions:** 9 Provinces, 24 Districts. **Capital:** Colombo, 615,000 inhab. (1990). **Other cities:** Dehiwala-Mount Lavina, 196,000 inhab.; Moratuwa, 170,000 inhab.; Jaffna, 129,000 inhab.; Kandy, 104,000 inhab.(1990). **Government:** Chandrika Kumaratunga, President and Head of State since November 1994. Sirimavo Bandaranaike, Prime Minister since 1994. Unicameral Parliament with 225 members. **National Holiday:** February 4, Independence Day (1948). **Armed Forces:** 126,000 (1994) **Paramilitaries:** 70,000 Police Force, National Guard, Private Guard.

---

er denials and the government's outspoken commitment to non-alignment, the US Navy was granted permission to use Sri Lanka's refuelling facilities in Trincomalee, a vital spot linking eastern and western sea routes through the Suez Canal.

[15] Early in 1983, the ethnic conflict worsened. It reached crisis point in July and August, leaving hundreds dead and wounded and thousands more homeless. More than 40,000 Sri Lankan Tamils fleeing the conflict sought refuge in Tamil Nadu. President Jayewardene's refusal to negotiate with the Tamil minority led the Indian government to withdraw its attempts at mediation.

[16] In 1984, Jayewardene made overtures toward the Israeli government, which had offered him support to bring the rebel minorities under control. A diplomatic mission was set up in Colombo with the support of the US embassy. However, Secretary of State Douglas Liyanage was forced to resign in September 1984, after violent protest demonstrations following a government visit to Tel Aviv. The protesters were mainly from the Muslim community.

[17] The "ethnic war", turned more bitter during 1985, discouraging foreign investors. Tourism is still one of the main sources of income of this "fiscal paradise".

[18] Military operations also seriously affected agriculture in the northern and eastern provinces, where rice farming covers more than 200,000 hectares. The government had to import 150,000 tons of Chinese rice to stabilize the prices.

[19] Peace negotiations between the Tamils and Sinhalese, took place throughout 1986. The efforts included a meeting of the leaders of seven southeast Asian countries in Bangalore, India, all of whom were concerned about the civil war that had shaken Sri Lanka. The Freedom Party, the main opposition group, demanded that elections be called in advance and that the government disclose how people felt about the ethnic conflict; the government chose to ignore them. In a gesture, aimed at diffusing internal opposition, Jayewardene promised to restore former prime minister Sirimavo Bandaranaike's political rights.

[20] In late July 1987, in Colombo, presidents Rajiv Gandhi (of India) and Junius Jayewardene finally signed an accord granting a certain autonomy to the Tamil minority of the northern and eastern provinces of Sri Lanka. The accord provided for the merger of the two provinces under a single government, and gave Tamil the status of a national language. India guaranteed the agreement and sent troops - the India Peace Keeping Force - to ensure that its terms were complied with. These terms were to be ratified by the parliaments of India and Sri Lanka, and approved by the main guerrilla group, the Liberation Tigers of Tamil Eelam (LTTE) known as the Tamil Tigers.

[21] These efforts did not manage to pacify the country, and the Indian military presence, far from guaranteeing peace, became a further cause for irritation and renewed confrontations. Inter-ethnic violence worsened as the Tamil Tigers' rejected the 1987 accords.

[22] In November 1988, the official party's hold on power was ratified by 50.4 per cent of the votes. Jayewardene, at the age of 82, ceded his post to Ranasinghe Premadasa, the former prime minister. Political violence was so pervasive that only 53 per cent of the electorate actually went to the polling stations to vote. Elections were boycotted, by both the Tamil guerrillas and the Popular Liberation Front (made up of Sinhalese who are violently opposed to any kind of concessions to ethnic minorities). Political opposition to the government increased, partly fueled by a strong student movement which the government eventually put down in early 1989, using extremely harsh repressive measures.

[23] In early 1990, the Indian government withdrew the last of its 60,000 peace keeping force, which had been stationed in Sri Lanka since 1987. More than 1,000 troops had died on the island. Amnesty International announced that, in 1990, the government had killed thousands of civilians in the region.

[24] In May 1991, the Tamil Tigers were accused of murdering Rajiv Gandhi, in a suicide mission. Rajiv had become an enemy of the Tigers after the Indian peace forces in Sri Lanka had attacked the rebels. The Tigers, however, denied any involvement in the attempt. Their guerrilla group is considered the most efficient and best armed in the world. It is financed by expatriate Tamils who make annual donations of millions of dollars. The guerrilla leader, Velupillai Prabhakaran, has his general headquarters in the north of the country, hidden in the jungle.

[25] The next stage of the war against the Tigers started in June 1991, when the Tigers broke off after 13 months of talks with the government. Since then, thousands more people have died in combat, and many civilians have taken refuge in the south or in India. In August, according to official estimates, over 2,000 guerrillas and 170 Sri Lankan soldiers died in an attack in the strategic "Pass of the Elephant", which was under government control. This was deemed to be the most important battle of the decade, and as yet, a peace agreement seems unlikely.

[26] In spite of the war, the country was visited by some 300,000 tourists in 1990. In some western countries the island is promoted for its countless brothels. The number of prostitutes in the capital is put at 50,000, with over 10,000 young boys and girls in the business. One reason for the extremely high number of prostitutes is that half of the population in the country live in extreme poverty.

[27] Nearly two decades after the large tea plantations were nationalized, the government decided to hand management over to private companies, while maintaining state ownership. It was estimated that the state companies which managed these plantations, employing over 400,000 people (mostly Tamil), had accumulated debts of $128.1 mil-

STATISTICS

DEMOGRAPHY

Urban: 22% (1995). **Annual growth:** 1.3% (1992-2000). **Estimate for year 2000:** 19,000,000. **Children per woman:** 2.5 (1992).

HEALTH

One **physician** for every 7,143 inhab. (1988-91). **Under-five mortality:** 19 per 1,000 (1994). **Calorie consumption:** 94% of required intake (1995). **Safe water:** 93% of the population has access (1990-95).

EDUCATION

Illiteracy: 10% (1995). **School enrolment:** Primary (1993): 105% fem., 105% male. Secondary (1993): 78% fem., 71% male. University: 6% (1993). **Primary school teachers:** one for every 29 students (1992).

COMMUNICATIONS

100 **newspapers** (1995), 90 **TV sets** (1995) and 96 **radio sets** per 1,000 households (1995). 0.0 **telephones** per 100 inhabitants (1993). **Books:** 100 new titles per 1,000,000 inhabitants in 1995.

ECONOMY

Per capita GNP: $640 (1994). **Annual growth:** 2.90% (1985-94). **Annual inflation:** 11.00% (1984-94). **Consumer price index:** 100 in 1990; 151.4 in 1994. **Currency:** 50 rupees = 1$ (1994). **Cereal imports:** 1,149,000 metric tons (1993). **Fertilizer use:** 964 kgs per ha. (1992-93). **Imports:** $4,780 million (1994). **Exports:** $3,210 million (1994). **External debt:** $7,811 million (1994), $437 per capita (1994). **Debt service:** 8.7% of exports (1994). **Development aid received:** $551 million (1993); $31 per capita; 5% of GNP.

ENERGY

Consumption: 111 kgs of Oil Equivalent per capita yearly (1994), 83% imported (1994).

Illiteracy 1995
10%

Maternal Mortality 1989-95
PER 100,000 LIVE BIRTHS
30

lion by the end of 1989. Agriculture continues to be the main source of revenue, with 36.3 per cent of the exports coming from tea, rubber, and coconuts.

[28] In August 1991, President Premadasa, who had been accused of violating the constitution and other crimes - such as treason, embezzlement, corruption, abuse of power, and moral depravity - suspended parliamentary sessions to avoid being censured. He also faced serious confrontations within his party and the hostility of the opposition, led by Sirimavo Bandaranaike, who demanded democratization and the end of the presidential system.

[29] In February 1992, in negotiations observed by the UN, the LTTE supremo, Velupillai Prabhakaran declared his desire for a peaceful end to the conflict. Prabhakaran stated that he wanted an agreement based on the recognition of a degree of autonomy for the northern and eastern provinces, where most of the Tamils live. President Premadasa expressed an interest in the proposed solution.

[30] In February 1993, the leader of the LTTE expressed his willingness to accept a negotiated settlement, in resolving the federalist issue. The government rejected Prabhakran's request for a lifting of the economic blockade of the Tamil region of Jaffna (decreed in February 1992 and subsequently extended in January 1993); nevertheless, parliament began discussing the terms of such an agreement.

[31] The issue of human rights violations by both sides complicated the negotiations. In October 1992, the Tamil Tigers had massacred 170 Muslims in the village of Polonnarua, in the northeastern part of the country. The following month, a car-bomb had killed a high-ranking naval officer in Colombo. In April 1993, the government required that all property owners in the capital provide a list of tenants, as a way of monitoring activist movement. That same month former cabinet minister Lalith Athulatmudali, president of the United National Democratic Front (split from the UNP in 1991) was assassinated.

[32] On May 1 1993, during the Labor Day parade in Colombo, President Ranasinghe Premadasa was assassinated by a suicide attacker, who threw himself upon the presidential entourage with a bomb tied to his body. The government accused the Tamil separatists of carrying out the assassination, but spokespersons for the rebels denied responsibility for the attack. Thousands of Tamils fled the capital, fearing government reprisals.

[33] On May 7, Prime Minister Dingiri Banda Wijetunge was named president. Wijetunge promised to continue the political agenda established by his predecessor. A few days later, police identified the presidential assassin as a Tamil, from Jaffna.

[34] The UNP won the May 17 elections, which took place in an atmosphere of relative calm, with a majority in six of the seven provinces. In August, Velupillai accused his vice-president of negotiating with the government and gave him the sack. The possibilities of a peace agreement appeared very remote.

[35] The confrontations left the economy of the north and eastern regions of the island devastated and thousands of people were forced to emigrate.

[36] In the November 1994 presidential elections, the Popular Alliance candidate, Chandrika Bandaranaike Kumaratunga, won with 63 per cent of the vote. Kumaratunga had been designated prime minister following the parliamentary elections in August and now became the first woman president of Sri Lanka. Her mother, Sirimavo Bandaranaike succeeded her as prime minister.

[37] The government and the Tigers agreed to start negotiations in January 1995, but the pact was broken by the Tamils. The guerrillas initiated a new series of attacks against the government forces in April.

[38] Kumaratunga presented a plan for state reform in August. The proposal, which was supported by the Tamil parliamentarians, included transforming Sri Lanka into a federation of eight regions. The central government was to keep control over defence, foreign relations and international economic relations.

[39] This project required the support of two-thirds of Parliament followed by approval by referendum before it could come into action. The negotiations between the various parties were slow and the increasing military action prevented it being completed.

[40] The city of Jaffna, on the peninsula of the same name, was at the center of the fighting from October 1995. The government forces occupied the city on December 5. Only 400 of the 140,000 people who lived there stayed throughout the fighting. By mid-1996, half of the Tamil population of the peninsula had returned to their homes.

[41] An attack on the Central Bank, in the center of Colombo, killed 84 people in February 1996. This led to fears of a more widespread inter-ethnic conflict. The Tigers' attacks continued in the streets and trains of Colombo. More than 50,000 people have died in 14 years of war on the island.

[42] A report by Amnesty International, released in August, said the government justified extrajudicial executions, disappearances, the killing of civilians and torture because of the war. Military spending ran at $2 million per day in 1995.

Sudan

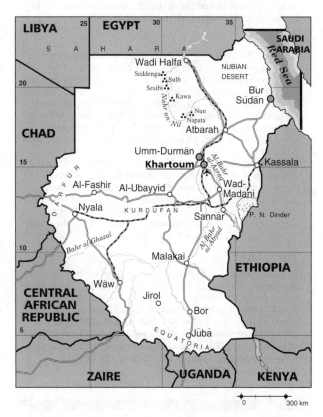

Population: 27,364,000 (1994)
Area: 2,505,813 SQ.KM
Capital: Khartoum (Al-Khartum)

Currency: Pound
Language: Arabic

Sudan

A round 3000 BC, the region called Kush by the Egyptians and Nubia by the Greeks, received its influence from Pharaonic Egypt, a fact which considerably delayed the formation of an organized state. Pharaohs favored disparate groups under their rule. Consequently, the state of Napata did not come into being until the eighth century when Egypt's decline gave foreign dynasties the opportunity to take over the country. The last of these was Sudanese. The rulers of Napata conquered Egypt in 730 BC and governed as pharaohs until the Assyrian conquest in 663 BC, when the ruling family was ousted. The country disintegrated, though it was not occupied. Shortly afterwards, the three "waterfall states", which would last over 20 centuries, rose in its place: Nobatia, Dongola and Alodia.

[2] While Egypt was ruled by Persians, Greeks, Romans and Arabs in succession, the waterfall states maintained their political and cultural autonomy. This was due to their position as trade intermediaries between the Mediterranean market and the sources of slaves, ivory, furs, and other goods in equatorial Africa. They were converted to Christianity in the 6th century under Ethiopian influence. A century later, the Arabs invaded forcing the ruler of Dongola to give financial aid to Arab merchants and to allow Muslim preaching in exchange for preservation of the territorial integrity of Dongola and Alodia. This treaty remained in effect for over 600 years.

[3] Egyptian Mamelukes destroyed Dongola and Alodia in the 14th century. Raids grew more frequent despite the emergence of new Islamic states: Sennar, on the Blue Nile; Kordofan to the west, and Darfur in the middle of the desert.

[4] Intent on exterminating Egyptian soldiers, Pasha Muhammad Ali (see: Egypt) entered Sudan in 1820. The Egyptian forces set up permanent quarters in Khartoum, holding control of the entire country by 1876. Foreign military occupation brought about radical changes: the country was unified, limiting the autonomy of various small local governments; the introduction of foreign religious rites (even within the predominant Orthodox Sunni religion) upset established religious communities. Under British pressure, slavery was abolished, undermining the influence of powerful slave dealers. Additionally, the burden of heavy taxes, particularly for farmers and cattle raisers, contributed to a general atmosphere of discontent.

[5] In 1881, Muhammad Ahmad proclaimed himself Mahdi, meaning saviour or redeemer, and launched a crusade to restore Islam. He rapidly gained support, especially among the Arabized population in the North. The British, having occupied Egypt in 1882, intervened militarily but were unable to change the course of the conflict. In 1885 Mahdi occupied Khartoum, defeated General Gordon's troops, and established the first national government, which was to last 13 years. The existence of the state threatened the British plan to unify all the territory from Cairo to Cape Town under their rule. In 1898, Egyptian troops moved in from Uganda and Kenya, attacking Mahdi from both fronts.

[6] The French, with their own east-west transcontinental designs, were also interested in Sudan, and sent troops into the area. This three-pronged attack defeated Mahdi in September 1898. The colonial armies met in Fachoda, with small skirmishes between the French and British, but France finally recognized British predominance over the Nile basin. In 1899, Sudan was jointly administered by Britain and Egypt.

[7] Egypt sought to achieve unity of the Nile region by joining Cairo and Khartoum politically, and Britain was determined to prevent this. The empire threatened to grant "federated independence" to the southern population, who were animists and Christians, but not to the Arabic and Muslim north. To make their threat effective, the British started a policy of "closed districts", which prevented any contact between north and south.

[8] A self-governing statute obtained by Sudan in 1953 was followed by the election of an all-Sudanese parliament in 1955 and the declaration of independence on January 1 1956. However, southerners charged that they had been politically marginalized, and five months before independence a civil war broke out, which continued for 16 years.

[9] In 1969, General Gaafar al-Nimeiry seized power through a coup d'etat. He dissolved parliament, proclaimed the creation of the Sudanese Democratic Republic, and set up a one-party system, the Sudanese Socialist Union. The new government promised to support reconstruction and the development of the southern territories, offering them a certain degree of administrative autonomy.

[10] After coming to power, Nimeiry changed course; he broke with the Sudanese Socialist Union and drew closer to conservative Arab regimes, never fulfilling his promise of greater autonomy for the South.

[11] Nimeiry was forced to sign an agreement with the guerrillas in 1972. This agreement granted autonomy to the southern provinces, while the guerrillas were incorporated into the regular army.

[12] In May 1977, Nimeiry was re-elected for a six-year term. Shortly afterwards the government announced a "national reconciliation" process which enabled exiled political leaders to return and paved the way for the rehabilitation of former opponents of the ruling Umma Party. However, the Sudanese Communist Party and the National Front, formed by former Finance Minister Shereef al-Hindi, were ignored.

[13] In 1976, Sudan signed a mutual defence pact with Egypt and initially the Nimeiry government backed the Camp David accords signed between Egypt, Israel and the US. However when it became clear that this position would isolate Sudan from the Arab world, Nimeiry distanced himself from Cairo and grew closer to Saudi Arabia. Nimeiry stressed the Islamic nature of his regime, arousing opposition in the non-Muslim south.

[14] In 1979, a wave of popular protests broke out against austerity measures dictated by the International Monetary Fund (IMF) and

---

WORKERS

1994

MEN 77% | WOMEN 23%

AGRICULTURE 72% | INDUSTRY 5% | SERVICES 23%

ENVIRONMENT

The largest African country, Sudan has three distinct geographic regions: the barren deserts in the north, the flatlands of the central region and the dense rainforests of the south. Most of the population lives along the Nile, where cotton is grown. Port Sudan handles all the country's foreign trade. Desertification has affected nearly 60 per cent of the territory. Industrial waste has contaminated coastal areas and some rivers.

SOCIETY

Peoples: The country has a wide ethnic spectrum with over 570 groups. Arabs, together with Nubians, account for nearly half of the population. In the South there are nearly 400,000 refugees from neighbouring nations (Chad, Uganda, Ethiopia and Eritrea). **Religions:** Islam is the predominant religion among Arabs and Nubians with a majority of Sunni Muslims. In the south, traditional African religions are practised and there are Christian communities in both north and south.
Languages: Arabic (official and spoken by most of the population); local languages. **Political parties:** Political parties were dissolved after the 1989 coup. The leading parties based on the 1986 electoral results were: the Umma Party; the Democratic Unionist Party (DUP); the National Islamic Front. Oppostion separatist movements in the south include the Sudan People's Liberation Movement (SPLM).

THE STATE

Official Name: Jumhuriyat as-Sudan. **Administrative Divisions:** 9 states, 66 Provinces and 281 Local Government Areas. **Capital:** Khartoum (Al-Khartum), 476,218 inhab. (1983). **Other cities:** Omdurman (Umm-Durman), 526,287 inhab. (1983); Port Sudan (Bur Sudan), 215,000 inhab. (1990). **Government:** General Omar al-Bashir, Head of State since June 30, 1989, when civilian government was deposed. **National Holiday:** January 1, Independence Day (1956). **Armed Forces:** 118,500 (1995). **Paramilitaries:** 30,000 - 50,000 People's Defence Force.

DEMOGRAPHY

Annual growth: 1.9% (1992-2000). **Estimate for year 2000:** 33,000,000. **Children per woman:** 6.1 (1992).

HEALTH

Under-five mortality: 122 per 1,000 (1994). **Calorie consumption:** 88% of required intake (1995). **Safe water:** 60% of the population has access (1990-95).

EDUCATION

Illiteracy: 54% (1995). **Primary school teachers:** one for every 34 students (1990).

COMMUNICATIONS

89 **newspapers** (1995), 95 **TV sets** (1995) and 100 **radio sets** per 1,000 households (1995). 0.2 **telephones** per 100 inhabitants (1993). **Books:** new titles per 1,000,000 inhabitants in 1995.

ECONOMY

Annual growth: -0.20% (1985-94). **Consumer price index:** 100 in 1990; 477.9 in 1992. **Currency:** 400 pounds = 1$ (1994). **Cereal imports:** 654,000 metric tons (1992). **Fertilizer use:** 72 kgs per ha. (1992).

against the consequences of the "green revolution" in agriculture.
[15] Nimeiry's weakness became apparent in March 1981 when he was barely able to stave off the 12th plot against his government. In October 1981, in an effort to neutralize his opponents and secure a new term in office, he dissolved the National Assembly and the Regional Assembly of the South, promising to hold new elections. This move coincided with growing unrest among southerners who staunchly opposed plans to locate refining operations exclusively in the north for oil extracted in the south. Southerners charged that Nimeiry made no effort to develop their region and, in fact, transferred the region's scant resources to the north.
[16] Nimeiry was re-elected to a third six-year term in 1983, amid widespread accusations of electoral fraud. In September, Nimeiry suddenly imposed Islamic law (the Sharia) on the entire national territory on the advice of Saudi Arabia, and in order to obtain economic assistance. This measure triggered a general protest among the animists and Christians in the south, and recharged the guerrilla movement.

[17] There was an instant increase in guerrilla activity. A revolt broke out in the city of Bor, where the Sudanese Popular Liberation Movement (SPLM) arose. This political and military organization gave the southern guerrilla movement a new ideological foundation, with the objectives of achieving national unity and establishing socialism, while respecting southern autonomy and religious freedom.
[18] In the north, Nimeiry was severely criticized by the Muslim Brotherhood and opposition parties for using Islamic law as a tool of repression against dissidents. The financial community also exerted pressure on the Sudanese president to limit the application of the Sharia in business as it stood in opposition to IMF policies.
[19] Sudan's external debt rapidly rose to $8 billion. Debt servicing was systematically delayed until early 1984, with the country considered bankrupt on at least two occasions.
[20] In January 1984, tensions became critical, extending into the north when the government sentenced the country's top Islamic republican leader, Mahomoud Taha, to death by hanging. The US

slowly withdrew government support, doubting the reliability of official control over nationalist military officers. In 1985 the US suspended all credit and the IMF forced the government to raise food prices. Rebellion broke out, quickly reaching to the capital. Nimeiry visited the US in search of backing, but was unable to return as Abdul Rahman Suwar al-Dahab, minister of defence and army chief of staff, had seized power in his absence.
[21] The coup did not have much of an effect on the political scenario. The Islamic bourgeoisie in the north made deals with the government, while the SPLM continued its activities in the south, as the political and economic discrimination against the region continued. Political parties were dissolved and the previous government's subdivision of the south was revoked. The new government promised to review the application of the Sharia, and Dahab promised elections in 1986.
[22] The elections were held in April 1986, and Sadiq al-Mahdi was elected prime minister. His Party of the People (UMMA), based on the Koran and on Islamic tradition, obtained 99 seats, while the Democratic Unionist Party (DUP) won 64. The UMMA had 8 ministers, the DUP, 6, and other minor parties, 4. A ruling coalition was thus established based on agreements which had led to the approval of a new Constitution in 1985.
[23] Thirty-seven representatives were not elected because of the war in the southern provinces. The SPLM's 12,000 guerrillas besieged government garrisons in the south-

ern provinces, practically splitting Sudan in two, and the south suffered from food shortages as a result of an insurgent blockade. However, the guerrillas agreed to let through airborne food and medical supplies sent by the UN to the besieged cities of Juba, Jirol and Waw. Meanwhile, unsettled economic, political and cultural disputes between the north and south prolonged the war.
[24] In June, 1989 the war between the SPLM and the army continued, the foreign debt reached $12 billion and the social tension increased, exacerbated by price increases. In February of that year, Major-General Omar al-Bashir ousted the president, blaming his government for the political and economic crisis. Bashir dissolved political parties and created a 15-member military junta, promising to bring the war to an end.
[25] Ten months after the military government took office, there was a coup attempt by a faction of army officials. Ethiopia, Kenya, Uganda, Zaire, and the US all attempted mediation and all failed. Peace negotiations failed, and southern people continued to suffer at the hands of government troops and Arab paramilitary groups financed by Bashir. These armed groups harassed the southern African peoples and forced them from their lands.
[26] According to UN figures, 7,100,000 people were threatened by famine in Sudan. In 1989 a food distribution program started in the devastated southern territories, but was suspended by the government in 1990. Meanwhile, the government continued its policy of selling agricultural reserves to finance the war.

Seven years of human rights violations

In mid-1996, as the Sudanese government was preparing to mark the seventh anniversary of the 30 June 1989 coup which brought it to power, Amnesty International (AI) condemned continuing human rights violations and called on United Nations member states to deploy human rights monitors.

2 One year after the human rights organization mounted an international campaign on human rights abuses in Sudan, it released a report detailing further atrocities committed by government forces, including the deliberate and arbitrary killings of villagers, the abduction of scores of children, torture and ill-treatment and incommunicado detention of suspected government opponents.

3 AI confirmed that during 1996 the government's security forces rounded up suspected opponents in the capital and attacked civilian targets in the war zones of the south - deliberately killing adults and forcibly abducting children as a tactic of war.

4 In March 1996 the government held elections - boycotted by the opposition - to become a civilian administration. But these elections have not made any difference to systematic repression as a method of securing control. In a typical incident in late April, government-controlled Popular Defence Force troops in the remote area of Udici abducted five children and shot two men dead. The children have not been seen since.

5 The abduction of children by government forces and allied militia is a growing feature of the civil war which began in 1983. Some of the children are held for ransom, some appear to be taken into domestic slavery, while others end up in government schools run like armed camps. Some children have reportedly been shot when trying to escape from these schools.

6 Suspected government opponents remain at risk of torture, Amnesty International said. Student demonstrators were severely beaten in September 1995 and in October 1995.

7 The report, "Sudan: Progress or public relations?", also documents the failure of the Sudan People's Liberation Army (SPLA) and rival armed opposition South Sudan Independence Army (SSIA), which in April concluded a peace deal with the government, to take practical steps to protect human rights.

8 "Sudan: Progress or public relations?" describes developments in the country since Amnesty International's 1995 campaign. Despite limited progress in some areas, such as the partial closure of a secret detention centre and the publicized release of 50 political prisoners in August 1995, the human rights situation remains grave.

9 The situation in the contested but largely SPLA-controlled region of Bahr al-Ghazal is especially grim. Cattle-owning civilians are the targets of several government-backed militia, including the Popular Defence Force (PDF), which is led by an army general. The government denies the PDF abducts children, describing attacks on villages as traditional' cattle raids. In reality the raids are part of a deliberate strategy to destabilize areas controlled by the SPLA.

10 Thousands of people have fled their lands with their cattle. They have then been hit by raids by troops from the SSIA. In early March 1996 SSIA troops were reported to be among a force which killed over 50 adults and children in attacks on cattle camps in the Makuac area. Boys were hacked to death with spears and large knives.

11 The SPLA has made important pledges to respect human rights, making a signed commitment to the Geneva Conventions and the UN Convention on the Rights of the Child in July 1995. In early May 1996 a conference of its political wing endorsed a Charter on human rights, saying that it is seeking to build a strong civil society in southern Sudan. However, the armed opposition group has yet to translate these pledges into action. For example, SPLA officials have denied a raid on Ganyliel in July 1995 in which over 200 people were killed was authorised, but have failed to provide any evidence of genuine investigations.

12 Despite the decision taken in 1995 by member states of the UN Commission on Human Rights to create a monitoring team to investigate human rights abuses in Sudan, the task was delayed. The Security Council voted in March 1996 for sanctions against Sudan for its alleged involvement in international terrorism but no references were made to the internal situation.

Source: Amnesty International

27 On February 4 1991, the government established a federal system in the country. According to the decree, Sudan would be divided into nine states, each administered by a governor and a ministerial cabinet. On January 31 of that year, the government of General Omar Bashir also signed a new criminal code based on the Sharia, only to be implemented in the north of the country, where Islam is predominant.

28 In March the army launched a military offensive against the People's Liberation Army (SPLA), supported by Ethiopia, Iran and Libya. Government forces recaptured the city of Bor, in the south, which had been a symbol to the rebel movement.

29 With the People's Liberation Army led by John Garang operating in the south, and the Nasir rebel group - a splinter group of the PLA led by Riek Masha - operating in the north, the government lost control. This led to violent government repression against its opponents in urban areas, and against rural communities which supported the guerrillas in the countryside. In March 1992, Amnesty International denounced the systematic extermination of Nubians, as well as the imprisonment of hundreds of alleged government opponents, among them dozens of people qualifying as prisoners of conscience. Amnesty International also disclosed that both factions were responsible for a number of homicides, both deliberate and accidental.

30 Negotiations in May 1992, sponsored by the president of Nigeria, in Abuja, Nigeria, ended in June with the issuing of an ambiguous communiqué. International pressure forced Omar Bashir to revoke the ban on airplane movements to the south of the country, which had kept food and medicine from being transported. It was estimated that some 6 million people were on the verge of starvation; this, in addition to the 60,000 people who had already died of hunger in Parayang, 800 km southeast of Khartoum.

31 In January 1993, disagreements among high-ranking members of the government led to a cabinet shake-up, aimed at bringing the government's policies into line with the IMF and World Bank objectives. However, both international agencies found the reforms insufficient; especially as the debt had not been serviced. The financing for infrastructure projects was consequently discontinued in April. In the meantime, $2 million a day were being spent on defense.

32 In February 1993, the PLA and the government resumed negotiations in Entebbe, Uganda. These were boycotted by Mashar, who was in Nairobi trying to speed up negotiations to merge with another PLA splinter group, led by William Nyuon. This alliance, signed in April, paved the way for a cease-fire agreement with the government, and the promise to carry on with the talks initiated in Nigeria. Disagreements between Garang and Bashir, in Abuja, over the devolution of power to the provinces, caused the talks to flounder in June 1993, and the impasse remained until the end of the year.

3 In May 1994, the government signed a protocol for humanitarian aid with both rebel factions in order to provide relief for villages cut off by the conflict. Their situation continued to worsen and humanitarian organizations voiced their denunciations (see box). In July 1995, African Rights charged Khartoum of being responsible for the "genocide" of the Nubian ethnic group.

34 The government held elections in March 1996. Bashir obtained almost 76 per cent of the vote, remaining as president. Some observers estimate that the possibilities of cohabitation between the "theocrats" from the country's north and the rebel leaders from the south are diminishing. The 12-year long war has caused the death of one million people, forcing 3 million more to become refugees.

Suriname

Population: 407,000 (1994)
Area: 163,270 SQ.KM
Capital: Paramaribo
Currency: Guilder
Language: Dutch and Spanish

Suriname

The Caribs, descendants of the Mongols who crossed the Bering Straits at the end of the last Ice Age, settled along the Atlantic coasts of Central and South America. The original groups gradually dispersed; some inland, and others to the islands that lie off the coast of the continent. They lived in small communities where they existed by hunting, fishing and small-scale farming. They were also able warriors.

[2] In the early 17th century Dutch colonists founded a settlement in Guyana, and began to trade in slaves.

[3] In 1863, when slavery was abolished in Dutch territories it was replaced by another source of cheap labor, Asian Indian and Javanese immigrants. This gave rise to a complex ethnic structure in the Guyana region. Asian Indians formed one of the largest groups, the most resistant to intermarriage and most strongly attached to their cultural heritage. Then came the creoles or Afro-Americans, the Javanese, the "bush people" whose ancestors had rebelled against slavery and fled the plantations, indigenous people and a small European minority.

[4] These ethnic, cultural and linguistic differences hindered the development of a national identity. Political movements tended to represent individual ethnic communities, especially as these broadly corresponded to social divisions. The creoles formed the NPK (National Party Combination), a coalition of four center-left parties, and they led the fight for independence after World War II. Jaggernauth Lachmon's Vatan Hitakarie represented the Indian population of shopowners and business people, and they sought to postpone independence.

[5] Since 1954, local government

had enjoyed a degree of autonomy and, in October 1973, the independence faction won the legislative elections. Liberal Hanck Arron, NPS (National Surinamese Party) leader was appointed prime minister.

[6] Arron and Lachmon reached an agreement, and independence was finally proclaimed in 1975. Many middle-income Surinamese took advantage of their status as Dutch citizens to emigrate to the ex-colonial power. Nearly a third of the population, mostly Asian Indian, left the country, causing a serious shortage of technical, professional and administrative personnel. The country lost much-needed qualified labor with only SURALCO and Billiton, the two transnationals which monopolize local bauxite mining, surviving the exodus. Economic activity decreased and agriculture declined to dangerous levels. In this difficult situation, Arron and his group of politicians were repeatedly accused of corruption and complicity with foreign companies.

[7] On February 25 1980, the premier was overthrown by a coup which had begun as a protest by the sergeants' union for better working conditions. The National Military Council (NMC) summoned opposition leaders to form a government and several leftwing politicians accepted cabinet posts.

[8] The Military Council, led by Lt-Colonel Desi Bouterse, took power from the politicians on February 4 1981, accusing the administration of corruption and underhand dealings with the Netherlands

WORKERS
1994   16.3% UNEMPLOYMENT
MEN 70%   WOMEN 30%
AGRICULTURE 20% INDUSTRY 20% SERVICES 60%

PROFILE

ENVIRONMENT

The coastal plain, low and subject to floods, is suitable for agriculture. Rice, sugar and other crops are grown. Land has been reclaimed from the sea through drainage and polders. Inland, the terrain is hilly with dense tropical vegetation, rich in bauxite deposits. Year-round heavy rainfall feeds an important system of rivers, some of which are used to generate hydro-electric power for the aluminum industry.

SOCIETY

Peoples: Suriname Creole 35%; Indo-Pakistani 33%; Javanese 16%; Bush Negro 10%; Amerindian 3%; other 3%.
Religions: Hindu 26.0%; Roman Catholic 21.6%; Muslim 18.6%; Protestant (mostly Moravian) 18.0%; other 15.8%.
Languages: Dutch and Spanish (official). English, Hindi, Javanese, and a form of Creole are spoken. The Creole, called either Taki-Taki or Senango-Tongo, is based on African languages mixed with

Dutch, Spanish and English. **Political Parties:** New Front for Democracy and Development (NF) comprised of four parties: Progressive Reform Party (VHP), Suriname National Party (NPS), Javanese Farmers' Party (KPTI) and the National Democratic Party NPD), Democratic Alternative, Jungle Commando, Angola, Mandela Liberation Movement, Tucaya. **Social Organizations:** Suriname Trade Union Federation; Central Organization of Civil Servants.

THE STATE

Official Name: Republiek Suriname.
Administrative Divisions: 9 Districts.
Capital: Paramaribo, 200,000 inhab. **Other cities:** Nieuw Nickerie 6,078 inhab.; Meerzorg 5,355 inhab.; Marienburg 3,633 inhab. **Government:** President, Jules Wijdenbosch since September 1996. 51-member unicameral parliament. **National Holiday:** November 25, Independence Day (1975)

and the United States. This new government established relations with Cuba, in the face of opposition from the major internal parties, from the United States and from the Netherlands. Brazil came to its aid on more than one occasion.

[9] In January 1983, Bouterse formed a new government, which appointed Errol Halibux, a nationalist and member of the Farmers' and Labor Union, as

> These ethnic, cultural and linguistic differences hindered the development of a national identity. Political movements tended to represent individual ethnic communities.

prime minister. After the American invasion of Grenada, the Suriname government did an about-face in its relations with Cuba, asking Havana to recall its ambassador and suspending all agreements for cooperation between the countries.

[10] In late 1983 and early 1984, there was a series of strikes by bauxite workers - a product which accounted for 80 per cent of the country's exports - who were joined by power, water and transport workers. Bouterse attributed it to errors by the prime minister and the entire cabinet automatically resigned. Bouterse proceeded to eliminate all fiscal restrictions by presidential decree, and appointed two union leaders to the new cabinet.

[11] An effort was made to reduce Suriname's isolation and diversify its foreign relations, which were excessively dependent on Holland. To accelerate this process, the government began to pursue a more aggressive foreign policy, joining CARICOM (Caribbean Community and Common Market) as an observer and establishing relations with Cuba, Grenada, Nicaragua, Brazil and Venezuela. It

also became a member of SELA (Latin American Economic System), the OAS (Organization of American States) and the Amazon Pact.

[12] In mid-1986, former soldier Ronnie Brunswijk, aided by foreign mercenaries and funded by the Netherlands, began a campaign of harassment from bases in French Guyana.

[13] In April 1987, the 33-member National Assembly (of whom 11 are labor union representatives) unanimously approved a draft Constitution providing for a return to institutional government. The draft was supported by the three main political parties.

[14] The coalition "United Front for Democracy and Development" won. On July 21, 1989, President Ramsewak Shankar granted an amnesty to the still-active guerrilla movement, allowing them to hold on to their weapons as long as they were in the forests. Bouterse and the NDP opposed this agreement, arguing that it legalized an autonomous military force and encouraged the division.

[15] In December 1990, another coup deposed Ramsewak Shankar, who had been the constitutional president since January 1988. Colonel Desi Bouterse, who had resigned a week prior to the coup, resumed his leadership of the army on December 30. The National Assembly, which had been set up in September 1987 for a period of five years appointed the NPS's Johan Draag as provisional president.

[16] During 1991, Dutch and US sources accused the top military echelons of Bouterse's regime of being involved in drug trafficking.

[17] The May elections were held to choose the National Assembly, and the New Front (NF) won; they are a broad coalition of civilian parties and ethnic groups who oppose the military government. The NF proposed the re-establishment of relations with the Dutch government and the tightening of economic and political links with the former colonizer.

[18] Fresh elections in May 1991 elected a new government to begin ruling in September. Spokespeople in The Hague and Washington independently announced their intentions of carrying out a military intervention in the country "should the new

STATISTICS

DEMOGRAPHY

Annual growth: 1.3% (1992-2000). **Children per woman:** 2.8 (1992).

HEALTH

Under-five mortality: 33 per 1,000 (1994). **Calorie consumption:** 101% of required intake (1995).

EDUCATION

Illiteracy: 7% (1995). **Primary school teachers:** one for every 23 students (1988).

COMMUNICATIONS

101 **newspapers** (1995), 101 **TV sets** (1995) and 107 **radio sets** per 1,000 households (1995). 11.6 **telephones** per 100 inhabitants (1993). **Books:** new titles per 1,000,000 inhabitants in 1995.

ECONOMY

Per capita GNP: $860 (1994).
Annual growth: 1.80% (1985-94).
Consumer price index: 100 in 1990; 741 in 1994.
Currency: 2 guilders = 1$ (1994).

government request it". International observers warned about the threat implicit in these announcements, in the light of the recent experiences in Grenada and Panama.

[19] In September, Ronald Venetiaan of the NF was elected presi-

> In the 1980s the government made efforts to reduce the country's isolation.

dent. The new head of state launched a 50 per cent cut in defence spending and a peace process, including a UN-sponsored guerrilla disarmament, under the supervision of Brazil and Guyana. In June 1992, a cooperation agreement was signed with the Netherlands, and a year later, a severe structural adjustment program was adopted, which gave rise to discontent among the population.

[20] Poverty and unemployment in the rural agricultural communities formed the background for the occupation of the

Afobakka dam, 100 kms south of Paramaribo, in March 1994. The rebels, who called for the resignation of the government, were expelled by government troops, after a four-day occupation. Another important movement took place in rural areas in 1995, when representatives of Indians and *cimarrones* (runaway slaves living in the forest in their own communities) gathered to protest for the environmental damage caused by a Canadian mining company and an Indonesian timber company.

[21] In September 1996, the Parliament elected Jules Wijdenbosch as president, thus blocking Venetiaan's re-election.

Swaziland

Population: 906,000 (1994)
Area: 17,360 SQ.KM
Capital: Mbabane

Currency: Emalangeni
Language: English

Swaziland

Like Lesotho, (see Lesotho) Swaziland achieved centralized administration when groups of very different origins joined to form a nation. The danger posed by Zulu expansion, (see South Africa) led Sobhuza, head of the Dlamini local groups, to bring together the remains of several groups split apart by Shaka, including Zulu deserters and Bushmen or San who remained in the region. They became a powerful force in the northeastern part of the present-day South African province of Natal. Sobhuza died shortly after the Zulu defeat by the Boers (1839) leaving his son, M'swazi, the task of keeping the nation together, in the face of constant threats from the Afrikaners. The nation was named after its king who led the nation in 30 years of resistance, allying with the British shortly before his death so as to avoid defeat by the Boers.

[2] In 1867, Swaziland formally became a British protectorate, like Basutoland, present-day Lesotho, and Bechuanaland, present-day Botswana.

[3] When Britain defeated the Boers and imposed its dominion over the whole of South Africa, these countries remained under separate colonial administrations, though South African settlers attempted to establish claims over the territories. In 1941, native authorities were formally recognized according to British criterion of making use of "local" intermediaries to facilitate colonial administration.

[4] In 1961, when the Union of South Africa broke relations with Britain and toughened racial segregation policies, London accelerated the decolonization process in the region. Swaziland was granted internal autonomy in 1967 and formal independence the following year.

Sobhuza II was recognized as head of state and governed with two legislative chambers. On April 12, 1973, he dissolved parliament, claiming that it contained "destructive elements" and proclaimed himself absolute monarch, banning all political parties and activities.

[5] In the event, Sobhuza II placed himself in the service of the South African colonizers against whom his grandfather had fought. Communications, the postal service, transport, currency and the Bank of Swaziland all became completely dependent on South Africa. South African companies exploited the country's asbestos and iron ore and South African experts ran the state's public administration - guaranteeing efficient control - and supervised agricultural production. Cotton, the main export product, was marketed by South African merchants and white South Africans opened brothels and cabarets which the strict "puritanical" official morality prohibited in their own country.

[6] Due to the lack of job opportu-

> **Sobhuza II was one of only three African rulers who never broke diplomatic relations with the government of Tel Aviv.**

nities, thousands of workers emigrate every year to work in the South African gold mines. The money they send home to their families accounts for 25 per cent of Swaziland's foreign currency earnings. In October 1977, teachers went on strike for better salaries and were supported by students. The following year, the Swaz Liberation Movement (Swalimo) was founded, led by Dr Swane, who managed to escape from prison in the capital, Mbabane. The strengthening of the opposition was accompanied by a rapid growth of the armed forces, from 1,000 in 1975 to more than 5,000 in 1979, and the police force was considerably reinforced.

[7] Mounting opposition inside the country was encouraged by the consolidation of a socialist regime in Mozambique, amongst other

things. Increasing tension in Swaziland also led the government to develop closer military relations with South Africa and Israel; Sobhuza II was one of only three African rulers who never broke diplomatic relations with the government of Tel Aviv.

[8] After the 1977 strikes, the government severely repressed all opposition, invoking a special law allowing detention without charges for up to 60 days.

[9] The Constitution was suspended in 1973 and a reformed version was introduced in 1978 without approval from or consultation with the electorate. Following tribal rules, the process of approval consisted of consultation with the heads of the 40 clans a fortnight before the enactment of the new constitutional draft. It banned the existence of opposition parties and established a parliament with little decision-making power.

[10] After 1980, the economic situation in Swaziland was affected by the world recession. The post-independence boom period came to an end with an increase in the prices of imported goods and a drop in prices for corn, sugar and wood exports. Inflation climbed, and a deficit appeared in the balance of payments, the rate of investment fell radically, and the GNP dropped by 4 per cent. Mineral exports fell from 40 per cent of total exports to

only 10 per cent due to the depletion of iron ore reserves. Various new coal deposits were discovered in 1980 but their exploitation has been slow due to legal difficulties and the lack of adequate transport facilities.

[11] In August 1982, King Sobhuza died at the age of 83 when his successor, Prince Makhosetive was only 15 years old. The lack of effective government sparked a power struggle between several princes and Prince Mabandla Dlamini was deposed from his position as prime minister. He was succeeded by Bhekimpi Alamini, a pro-South African conservative, who began his reign by persecuting South African refugees.

[12] In August 1983, Ntombi, one of Sobhuza's widows, overthrew reigning Queen Dzellue and took power strengthening the conservative faction. Two months later, 200,000 of the country's 760,000 inhabitants voted for a new parliament through a complicated indirect electoral system called *tinkhundla*. As expected, Prince Bhekimpi Dlamini was successful in an electoral contest where no political parties participated and the electorate did not know what they were voting for.

[13] The election resulted in closer ties with South Africa, giving it greater influence and causing rumors of the formation of a non-ag-

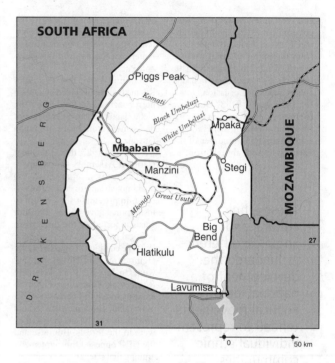

WORKERS

1994

MEN 59%   WOMEN 41%

AGRICULTURE 74%   INDUSTRY 9%   SERVICES 17%

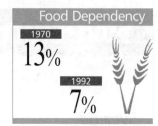
PROFILE

ENVIRONMENT

The country is divided into three distinct geographical regions known as the high, middle and low "veld" (fields) all approximately the same size. The western region is mountainous with a central plateau and flatlands to the east. The main crops are sugar cane, citrus fruits and rice (irrigated), cotton, corn (the basic foodstuff), sorghum and tobacco. In the lower veld, infections caused by polluted water are responsible for the high mortality rate. Wildlife was exterminated by European hunters in the first half of this century.

SOCIETY

Peoples: Close to 90% of the population belong to the Swazi ethnic group. Zulu account for 6%; Tonga and Shangaan another 3%; there is a European minority. **Religions:** 60% of the population are Christian, the rest follow traditional African religions. **Languages:** Swazi and English (official); ethnic minorities speak their own languages. **Political Parties:** Popular United Democratic Movement (PUDEMO); Swaziland Democratic Alliance; National Congress of Ngwane Liberation. **Social Organizations:** Swaziland Federation of Trade Unions (SFTU); Swaziland Youth Congress.

THE STATE

Official Name: Umbuso weSwatini **Administrative Divisions:** 210 Tribal Areas, including 40 traditional communities. **Capital:** Mbabane, 46,000 inhab. (1990). **Other cities:** Manzini, 53,000 inhab. (1990). **Government:** King Mswati III, crowned on April 25, 1986. Prime Minister: Barnabas Sibusiso Dlamini.

STATISTICS

DEMOGRAPHY

Annual growth: 3% (1992-2000). **Children per woman:** 6.6 (1992).

HEALTH

One **physician** for every 9,091 inhab. (1988-91). **Under-five mortality:** 107 per 1,000 (1994). **Calorie consumption:** 100% of required intake (1995).

EDUCATION

Illiteracy: 23% (1995). **Primary school teachers:** one for every 32 students (1991).

COMMUNICATIONS

90 **newspapers** (1995), 86 **TV sets** (1995) and 91 **radio sets** per 1,000 households (1995). 1.8 **telephones** per 100 inhabitants (1993).

ECONOMY

Per capita GNP: $1,100 (1994). **Annual growth:** -1.20% (1985-94). **Consumer price index:** 100 in 1990; 156.8 in 1994. **Currency:** 4 emalangeni = 1$ (1994).

---

gression pact. The repression of anti-apartheid militants increased in 1984, with detention and return to the Pretoria government becoming the norm.

[14] The power struggle within the royal family continued. The election had favored a financial elite, but by April 1984 this group was divided. During the second half of 1984, a student protest led the government to close down the university. Prime Minister Bhekimpi Dlamini's authoritarianism stimulated the revival of the Swazi Liberation Movement, which resumed its activities in January 1985. This movement was led by Prince Clement Dumisa Dlamini, who was the former secretary-general of the Progressive Party and a well-known nationalist leader. He was later exiled to England.

[15] On April 25 1986, Prince Makhosetive was crowned, taking the name King Mswati III. His inaugural address reflected his conservative stance. In May he dissolved the

Liqoqo (an assembly of local group leaders acting as Supreme State Council), consolidating his power and that of his ministers. Prime Minister Bhekimpi was relieved of his post and replaced by Sotsha Dlamini. In September 1987, King Mswati dissolved parliament, announcing elections for November, a year ahead of schedule. The 40 members of Parliament and 10 senators were elected by an electoral college. The king however disputed the election of the senators, demanding that the process be repeated in a more satisfactory manner.

[16] Relations with South Africa did not change with the change of king. The government condemned economic sanctions against the Pretoria regime and continued to harass anti-apartheid militants.

[17] Since the late 1980s the country's economic situation improved noticeably. The economy grew and foreign investment continued. A significant part of the food produced was sold to the European

Community. This improvement, a direct consequence of trade sanctions against South Africa, allowed the manufacturing sector to increase - contributting 20 per cent of the GDP by 1991 - and helped the country raise its economic growth rate to 3.5 per cent per year.

> Due to the lack of job opportunities, thousands of workers emigrate every year to work in the South African gold mines.

[18] In 1992, a significant part of the opposition joined the Popular United Democratic Movement (PUDEMO) and lobbied for the king to accept a multy-party system. Discontent mounted in 1993, as a prolonged drought destroyed the corn crop and generated further unemployment.

[19] According to some studies car-

ried out in 1994, even though calorie consumption is slightly higher than the necessary minimum, the country maintained a high child mortality rate - 107 per 1,000 - and only 30 per cent of the population had access to drinking water. Moreover, Swaziland was still dependent on South Africa, the source of 90 per cent of imports.

[20] Protests against Mswati III continued. After the deliberate fire of February 1995 in the seat of parliament, for which the Swaziland Youth Congress claimed responsibility, 40,000 people took part in a demonstration in support of a two-day general strike. South Africa pressed the king, and in February 1996 he expressed his willingness to authorize political parties.

Sweden

Sverige

Population: 8,781,000
(1994)
Area: 449,964 SQ.KM
Capital: Stockholm

Currency: Kronor
Language: Swedish

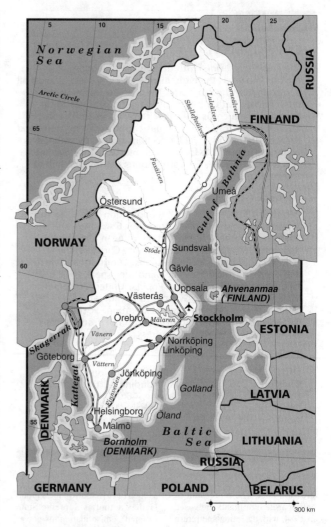

According to archeological research, the first inhabited area in Sweden is thought to have been the southern part of the country, with occupation dating back to 10,000 years BC. Between 8,000 and 6,000 BC, the region was inhabited by tribes who made a living by hunting and fishing, using simple stone tools. The Bronze Age (1,800- 500 BC) brought with it cultural development, reflected in particular in the richness of the tombs of that period.

[2] In 500 AD, in Lake Malaren valley, the Sveas created the first important center of political power. From the 6th century BC until the year 800 the population went through a migration period, later becoming settled, with agriculture becoming the basis of economic and social activities.

[3] Between the 9th and 11th centuries, the Swedish vikings, on trade expeditions, as well as pirate raids, reached the Baltic shores and also went as far as what is now Russia, reaching the Black and Caspian Seas. There they established relations with the Byzantine and Arab empires.

[4] Between the 9th and 11th centuries, Christian missions from the Carolingian empire (led by the missionary, Ansgar) converted most of Sweden. However, the gods of the ancient local mythologies survived into the 12th century. Sweden had its first archbishop in 1164.

[5] From the middle of the 12th century, Sverker and Erik's respective fiefdoms fought against each other to gain control of the Swedish kingdom, alternating in power between 1160 and 1250. The feudal chieftains remained relatively autonomous until the second half of the 13th century, when the king enforced nationwide laws and annexed Finland.

[6] The Black Death brought the country's growth to a standstill in 1350, and it remained so until the second half of the 15th century. In that period the foundries in the central region became important. During the 15th and 16th centuries the German Hanseatic League dominated Swedish commerce and encouraged the foundation of several cities.

[7] In 1397, the royal power of Norway, Sweden and Denmark was handed over to the Danish queen Margaret, who proclaimed the Union of Kalmar. The ensuing conflicts between the central Danish power and the rebellious Swedish nobility, townspeople and peasants ended in 1523, with the accession of Gustav Vasa to the throne of Sweden.

[8] Under the reign of Vasa, the monarchy ceased to be elected by the nobility and became hereditary. A German administrative model was adopted and the foundations were laid for a nation state. The possessions of the Church went to the state, with the Protestant Reformation following hot on its heels. From then on, Sweden aspired to becoming the main power in the Baltic region.

[9] In 1630, after intervening successfully in the Thirty Years' War, Sweden waged two more wars to conquer the Danish regions of Skane, Halland, Blekinge, and the Baltic island of Gotland, as well as the Norwegian islands of Bohuslan, Jamtland, and Harjedalen.

[10] Sweden thus became a great power in northern Europe, as it now ruled over Finland, several northern German provinces, and the Baltic provinces of Estonia, Latvia and Lithuania. However, the country was still basically rural and lacked resources to maintain that position indefinitely.

[11] After its defeat in the Great Northern War (1700-21), the Swedish empire lost most of the provinces to the south and east of the Gulf of Finland. It was reduced to the territories roughly corresponding to modern Sweden and Finland, with Finland being ceded to Russia during the Napoleonic Wars.

[12] In 1718, after the death of Charles XII, a Parliament (*Riksdag*) made up of nobles did away with the monarchy, assuming power itself. However, the new king Gustav II staged a coup in 1772, gradually claiming more and more power, and finally re-establishing full monarchic powers in 1789.

[13] In compensation for the losses incurred during the Napoleonic Wars, Norway was ceded to King Charles XIV John, who had become king in 1810. After a short war, it was forced to join Sweden in 1814. Nevertheless, after a series of conflicts, the union dissolved peacefully in 1905.

[14] In the second half of the 19th century, Sweden continued to be a poor country, with 90 per cent of the population engaged in agriculture. At this point, a great emigration movement began: one million out of a total of 5 million Swedes left, mainly for North America.

[15] During this period the liberal majority in parliament, supported by King Oscar I, established universal education (1842), the free enterprise system and the liberalization of foreign trade (1846). Legislation was also passed establishing sexual equality in inheritance law (1845), the rights of unmarried women (1858), and religious freedom (1860).

[16] Important social movements such as the temperance league, free churches, women's advancement, and especially the workers' movement emerged, growing with industrialization and influencing the government through social democracy.

[17] From 1890 onwards, with the support of foreign capital, the process of industrialization accelerated in Sweden. The country had one of the most thriving economies in that part of Europe. Finished products using Swedish technical innovations quickly became the country's main exports.

[18] Universal male suffrage has

LAND USE

- CROPLAND 7%
- PASTURE 1%
- OTHER 92%

been operating since 1908, a state old-age pension system was adopted in 1913, and in 1918 the 8-hour working day was established by law. That same year, the vote was extended to women.

[19] The parliamentary system has been in effect since 1918. The Social Democrats came to power in 1932; their policies fostered understanding between workers and industrialists, which was enshrined in the 1938 Pact of Salsboejaden.

[20] Sexual equality has been achieved to a greater extent in Sweden than in other industrialized countries, through the efforts of feminists and state campaigns against social and sexual prejudice.

[21] The country's educational system - from primary onwards - seeks to eliminate sexist conditioning by teaching both girls and boys how to keep house, take care of a newborn baby, do woodwork and metalwork.

[22] Abortion is available on demand to women up to the third month of pregnancy. The law concerning rape states that there are no extenuating circumstances to be considered in the man's case, and there are also stiff sentences for rape within marriage.

[23] Women can take on any profession, with the exception of the army. However, the female population is still concentrated in the lowest income groups. In general their earnings are 20 per cent lower than those of their male counterparts for similar jobs. In 1981, over 50 per cent of all married women had a professional activity outside the home.

[24] Within the civil service, there is an official position of Ombudsman, whose role is to investigate cases of abuse of authority, breach of ethics, and ethnic or sexual discrimination in state agencies and private companies.

[25] Since World War I, Sweden has refused to take part in peacetime alliances in order to stay neutral in times of war. This policy relies on Sweden's strong defence system and compulsory male national service.

[26] Trade routes were disrupted during the war, which caused serious food shortages in the country. For security reasons Sweden is almost self-sufficient, with up to 80 per cent of its agricultural produce produced internally. The government maintains a protectionist policy on agricultural imports, which

ENVIRONMENT

Sweden lies on the eastern part of the Scandinavian Peninsula. In the wooded northern area there are iron mines and paper manufacturing concerns. The central region, with fertile plateaus and plains, is the country's major industrial area, where the steel industry is located. The south is mainly agricultural: wheat, potatoes and sugar beet production, as well as cattle raising. This southern area is also the most densely populated in the country. Coastal areas along the North and Baltic Seas are highly polluted.

SOCIETY

Peoples: Swedes are of Germanic origin, with Finn and Lapp (Sami) minorities, and European and Latin American immigrants. **Religion:** Lutheran (official), Catholic and Jewish minorities. **Languages:** Swedish (official); there are Finnish and Lapp (Sami)-speaking minorities. **Political Parties:** The left-of-center Social Democratic Labor Party; the Center Party; the Moderate Party (conservative); the Green Party; the right-wing Liberal Party and the Communist Party. There are also the Christian Democratic Party (center-right) and New Democracy. **Social Organizations:** Confederation of Swedish Trade Unions, Confederation of Professional Associations, Central Wage Earners' Organization.

THE STATE

Official Name: Konungariket Sverige. **Administrative divisions:** 24 provinces. **Capital:** Stockholm, 704,000 inhab. in 1995. **Other cities:** Göteborg 444,553 inhab.; Malmö 242,706 inhab.; Uppsala 181,191 inhab.; Linköping 130,489 inhab. (1995). **Government:** Hereditary constitutional monarchy. Sovereign: Carl XVI Gustaf (since September 15, 1973). Prime Minister: Göran Persson, since March 1996. The single-chamber Parliament (*Riksdag*) has 349 members. **National Holiday:** June 6, Swedish Flag Day. **Armed Forces:** 64,000 (1994). **Paramilitaries:** Coast Guard: 600 (1993).

means that food prices are higher than those on the international market.

[27] Sweden favored a thaw in East-West relations during the Cold War, and worked actively for international disarmament. One of the cornerstones of its foreign policy is its support for the UN.

[28] During four decades in power, from 1932 to 1976, under the leadership of Tage Erlander, the Social Democrats established conditions which made it possible to combine rapid economic and industrial growth with a redistribution of wealth through direct taxation. Thus, a costly social welfare system was established, using the frame-

work of a capitalist economy with strong state intervention.

[29] There are a number of Swedish corporations among the 500 largest transnationals in the world, including Volvo, Electrolux, Stora Kopparbergs Bergslags, Svenska, SKF, Trelleborg, Cellulosa, Nobel Industries, KF Industry, Procordia, Mo Och Domsjo, Sandvik, Esselte, Saab, Svensk Stal.

[30] However, during the 1970s, there was a slow-down of economic growth, due partly to the increasing cost of oil imports to satisfy half the country's energy needs. Of the other 50 per cent 15 comes from nuclear reactors and 7 percent from coal and coke. After a plebiscite held in 1980, Parliament decided that until

the year 2010, the 12 nuclear reactors which were already in operation would remain closed.

[31] From 1981 onwards, relations between Sweden and the USSR were affected by the incursion of Soviet submarines into Swedish waters and the effects of the Chernobyl disaster.

[32] Atomic radiation from the Soviet Chernobyl nuclear plant affected the south of Lapland in 1986. The two main subsistence activities of the Sami people there, reindeer-herding and fishing, were badly affected by radiation, which contaminated all animal life in the area.

[33] Sweden was the first Western state to officially recognize North Vietnam (1969). The country's prime minister Olof Palme also acted as a mediator in the conflict between Iran and Iraq in 1982, and in 1988 Sweden sponsored a meeting between Palestinians and American Jews in 1988.

[34] After the 1976 elections, amid an incipient economic crisis, the moderate, liberal and conservative parties formed a coalition. Six years later, the Social Democrats returned to power. In 1986, charismatic leader Olof Palme was shot dead in a murder case that has still not been solved. Ingvar Carlsson succeeded him as prime minister.

[35] In the years that followed, parliament investigated an alleged case of bribery - involving the government and Bofors, an arms manufacturing company - with relation to the sale of weapons to the Middle East and India. The country's laws forbid such conduct, as well as trade with areas where there are military tensions or with countries that are at war.

[36] In spite of this scandal, the population was more concerned with the economy and the environment. Two accidents attributed to toxic substances destroyed the ecological balance along the west coast, and a virus killed off much of the fauna in the Baltic and North Seas.

[37] Gas emissions from Soviet industries increased the fall of acid rain and polluted snow in northern Scandinavia. In 1989, the USSR committed itself to a 40 per cent reduction in emissions before 1993.

[38] The social service and welfare system established in Sweden by the Social Democrats was aimed at ensuring a minimum level of assistance and a more egalitarian wealth redistribution, through graduated

Foreign Trade

MILLIONS $ 1994

IMPORTS
51,800

EXPORTS
61,292

STATISTICS

DEMOGRAPHY

Urban: 83% (1995). **Annual growth:** 0.6% (1991-99). **Estimate for year 2000:** 9,000,000. **Children per woman:** 2.1 (1992).

HEALTH

Under-five mortality: 5 per 1,000 (1994). **Calorie consumption:** 106% of required intake (1995).

EDUCATION

School enrolment: Primary (1993): 100% fem., 100% male. Secondary (1993): 100% fem., 99% male. University: 38% (1993). **Primary school teachers:** one for every 10 students (1992).

COMMUNICATIONS

106 **newspapers** (1995), 104 **TV sets** (1995) and 104 **radio sets** per 1,000 households (1995). 67.8 **telephones** per 100 inhabitants (1993). **Books:** 111 new titles per 1,000,000 inhabitants in 1995.

ECONOMY

Per capita GNP: $23,530 (1994). **Annual growth:** -0.10% (1985-94). **Annual inflation:** 5.80% (1984-94). **Consumer price index:** 100 in 1990; 119.7 in 1994. **Currency:** 7 kronor = 1$ (1994). **Cereal imports:** 202,000 metric tons (1993). **Fertilizer use:** 1,077 kgs per ha. (1992-93). **Imports:** $51,800 million (1994). **Exports:** $61,292 million (1994).

ENERGY

Consumption: 5,603 kgs of Oil Equivalent per capita yearly (1994), 36% imported (1994).

---

taxation of the highest income groups.

[39] In spite of the important benefits obtained by the Swedish population, the largest taxpayers managed to evade taxation. On Carlsson's initiative, parliament approved a fiscal reform in 1989, but it was considered too liberal and divided the Social Democrats.

[40] Under this new law, citizens with incomes below 160,000 krona a year (around 90 per cent of the population) would pay the same rate of tax, at 30 per cent. The rate for incomes above this limit went down from 72 to 50 per cent, while capital started to be taxed directly.

[41] Toward the end of the 1980s, the prospect of the European Union rekindled debate in Sweden over relations with its main trading partners. Until that time, Sweden had been a member of OECD, the Council of Europe and EFTA (European Free Trade Association). It was also on the Nordic Council, which had created a common labor

market and had achieved a high degree of uniformity in legislation among the five member countries.

[42] In 1990 negotiations were held in order to create a "European economic space" and in June 1991 Sweden - represented by the Carlsson government - applied for entry to the EC.

[43] In the September 1991 elections, the Social Democrats lost their parliamentary majority. A coalition of the Moderate, Liberal, Center, and Christian Democratic parties appointed Carl Bildt, the leader of the Moderates, as prime minister. The new government promised to speed up the union of Sweden with the EC, and the liberalization of the country's economy.

[44] In the last three elections, three new political parties have entered the Swedish Parliament: the Green Party, which had seats in 1988 but lost them in 1991; the Christian Democrats, in the new government coalition, and the New Democracy,

a right-wing party. Both of the latter gained their first seats in 1991.

[45] By the end of 1991, unemployment in Sweden hit a record 160,000 workers, doubling figures for the previous year. An additional 83,000 people were receiving vocational training, and 14,000 had been absorbed by government unemployment schemes.

[46] In 1991 and 92, the economic crisis served as the background for a wave of xenophobia and attacks perpetrated against foreigners, phenomena which had previously been non-existent in a country that had had liberal immigration and asylum policies throughout the 1960s and 1970s.

[47] The government budget for the 1992-93 fiscal year involved serious cutbacks in public spending. This refuelled the controversy over the Swedish model, whose main features are a strong public sector and heavy taxation. In late January, for the first time in 35 years, thousands of Swedish construction workers held demonstrations, demanding jobs.

[48] The president of the trade union confederation, Stig Malm, considered the government plan a declaration of war. However, because of the high unemployment rate, trade unions were in a weak position to use confrontational tactics. Bildt had the support of employers, but his parliamentary majority depended on the vote of the New Democracy, which was not its unconditional ally.

[49] On August 26 1993, the king opened a new parliament for the Sami population of Lapland. There are currently 17,000 Sami in Lapland, of a total population of 60,000 living in Norway (40,000), Finland, Russia and Sweden. Most of the people by herd reindeer, with 10 percent living by hunting deer. A major dispute with the government is in process over the abolition of the Sami's exclusive hunting rights over their lands.

[50] Sami children are given two weeks off in the autumn and two weeks in the spring, in order for them to learn about pasturing reindeer. In 1971, the law regulating reindeer-raising recognized special Sami rights over the lands and watercourses where this is practised.

[51] Unemployment rose significantly during 1993, and in 1994 it reached a record 13 per cent of the active population. Also in 1994, GDP increased for the first time

since 1990, while fiscal deficit remained higher than in many European countries: 13 per cent of the gross product.

[52] After a contentious electoral campaign, in a referendum held in November, 52 per cent of voters approved Sweden's entry into the European Union (EU). A significant part of the political establishment and a number of entrepreneurs supported entry, while left-wing groups and environmentalists were against it. There were also "geographical" disagreements, since the majority of southern urban population voted for it and rural and northern inhabitants voted against it.

[53] The referendum was an acid test for Social Democrats - who had returned to power that same year, after obtaining a small majority in parliament in the general elections - because their party did not succeed in reaching an agreement and had a divided opinion on the EU issue. Entry into the former European Community came into force on January 1 1995. In September, the disappointment of many Swedes became evident in the European elections. Groups against entry took 30 per cent of the vote, doubling the percentage of the general election of the previous year. The turnout for the elections was only 41 per cent.

[54] Relations with France, one of the 15 members of the EU, became complicated because Stockholm openly criticized Paris' decision to hold further nuclear tests in late 1995 and 1996. In the domestic field, Social Democrats continued their policy of reducing public deficit, by increasing taxes and reducing state expenditure, which brought about the confidence of the so-called "international financial markets".

[55] In May 1996, Minister of Finance Göran Persson replaced Carlsson and became head of a government that included 11 women and 10 men.

Switzerland

Schweiz / Confederation Helvetica

Population: 6,994,000 (1994)
Area: 41,290 SQ.KM
Capital: Bern
Currency: Swiss Franc
Language: German, French and Italian

Some Celtic tribes occupied the territory of Switzerland before Roman colonization. The most important of these tribes were the Helvetians, who settled in the Alps and the Jura mountains. The area was strategically important for Rome, with access to its dominions. Consequently, the Alpine valleys north of the Italian peninsula were conquered by Julius Caesar in 58 BC.

[2] The Germanic tribes north of the Rhine invaded from the year 260 onwards. Between the 5th and 6th centuries the Germans established permanent settlements in the region east of the Aar river, together with Burgundian and Frankish groups. By 639 they had founded the kingdoms that would later become France.

[3] The Christian survivors from Roman times had completely disappeared when St Columba and St Gall arrived in the 6th century. These missionaries created the dioceses of Chur, Sion, Basel, Constance and Lausanne. Monasteries were built, in Saint-Gall, Zurich, Disentis and Romainmotier.

[4] Until the partition of Verdun, in 843, these territories belonged to Charlemagne's empire. Thereafter, the region west of the Aar was allotted to Lothair, while the east remained in the hands of Louis the German. The French and German influence formed a peculiar blend with the Latin tradition of the Roman Catholic Church.

[5] Around 1033, for dynastic and political reasons, Helvetia became a part of the Holy Roman Empire, remaining so during the Middle Ages. In the 11th century the region was divided after the re-establishment of imperial authority and its disputes with the Papacy. Dukes, counts, and bishops exerted virtually autonomous local power.

[6] Walled cities served as administrative and commercial centres, and protected powerful families seeking to expand their possessions through wars against other lords and kingdoms. In the 13th century, Rudolf IV of Hapsburg conquered most of the territories of Kyburg and became the most powerful lord in the region.

[7] In the cities independence gradually developed in opposition to the nobility. Meanwhile the peasant communities, in the most inaccessible valleys, practised economic cooperation to survive the harsh conditions, rejecting forced labor and payment of tithes in cash or kind.

[8] In 1231 Uri fell under the authority of the Holy Roman Empire, and in 1240 Schwyz and Nidwald were subjected to Emperor Frederick II, although retaining the right to choose their own magistrates.

[9] The Hapsburg overlords questioned this freedom and uncertainty remained until Rudolf of Hapsburg was crowned king of Germany, in 1273. He exerted his imperial rights in Uri and inherited rights over Schwyz and Unterwald until his death in 1291. These regions, thereafter, constituted the Perpetual League.

[10] As with other circumstantial alliances among the regions, the Perpetual League constituted an agreement for dispute arbitration, putting law above armed strength. The honorary magistrates had to be residents of those cantons.

[11] The league of the Uri, Schwyz, and Unterwald cantons was joined by the city of Zurich, constituting the first historic antecedent of the Swiss Confederation. This confederation was consolidated with the victory of Margarten in 1315, defeating an army of knights sent to impose imperial law in the region by the Hapsburgs.

[12] The Confederation was supported by new alliances. In 1302 the League signed a pact with the city of Luzern, previously dependent on Vienna. In 1315 Zurich reaffirmed its union and in 1353 it was joined by Bern. The Glarus and Zug cantons joined later, forming the core of an independent state within the Germanic Empire.

[13] During the second half of the 14th century, the rural oligarchy was defeated, and their lands and laws given over to city councils. This democratic rural movement gave birth to the "Landesgemeinde", a sovereign assembly of canton inhabitants and a similar movement was led by the city guilds.

[14] The Confederation soon launched into territorial conquest. During the 15th century the union grew to thirteen cantons, it made alliances with other states, and a body known as the Diet was formed where each canton was represented by two seats and one vote.

[15] Disputes among the cantons verged on conflict but this was dispelled by the Stans agreement of 1481, where the confederates committed themselves to keeping the peace and repressing coups and domestic risings.

[16] In 1516, after the defeat of the Helvetians, the king of France forced a peace treaty with the cantons. In 1521 an alliance gave France the right to recruit Swiss soldiers. Only Zurich refused to sign this alliance, maintaining military and economic links with the Old Confederation until its end in 1798.

[17] The Reformation came to Switzerland with Huldrych Zwingli, a priest who preached against the mercenary service and the corruption and power of the clergy. Popular support for Zwingli strengthened the urban bourgeoisie. The Reformation became more radical in rural areas where harsh repression re-established the domination of cities over peasants.

[18] Zwingli's attempt to alter the federal alliance to benefit the reformed cities was frustrated by the military victory of the Catholic rural areas. The second national peace of Kappel, signed in 1531 gave the Catholic minority advantages over the Protestant majority.

[19] The areas where both religions co-existed were subject to constant tension, but cooperation was required to preserve the union of the federation. In Catholic regions agriculture prevailed, while in Protestant areas trade and industry flourished, aided by French, Italian and Dutch refugees.

[20] Ownership of real estate, trade, industry, and the recruitment of mercenary troops kept the wealth and power with a small group of families; especially in the cities. The majority of peasants were deprived of their rights and forced to work poor lands or sell their labor as rural workers.

[21] Popular consultation disappeared in the 17th century. City impositions caused uprisings, such as the great peasant revolt of 1653, which were harshly repressed. Three years later when a further war ensued the prerogatives of the Catholic cantons were re-established.

[22] During the European conflicts of the 17th and 18th centuries Swit-

LAND USE

CROPLAND 11%   PASTURE 32%   OTHER 56%

PROFILE

ENVIRONMENT

A small landlocked state in continental Europe, Switzerland is a mountainous country made up of three natural regions. To the northwest, on the French border, are the massive Jura mountains, an agricultural and industrial (watchmaking) area. Industry is concentrated in the Mitteland, a sub-Alpine depression between the Jura and the Alps, with numerous lakes of glacial origin. It is also an agricultural and cattle-raising region. The Alps cover more than half of the territory and extend in a west-east direction with peaks of over 4,000 meters. The main activities of this region are dairy farming and tourism. The country's ecosystem is currently threatened by acid rain.

SOCIETY

Peoples: Two-thirds of the population are of Germanic origin, while 18% and 13% are of French and Italian origin, respectively. 17.1% of the country's citizens or permanent residents are Italian, Yugoslav, Portuguese, German, Turkish or other nationalities.
Religions: 47.6% Catholic, 44.3% Protestant, 2.2% Muslim, 1% Orthodox Christian, 0.3% Jewish.
Languages: German, French and Italian. A very small minority in certain parts of the Grisons canton to the east speak Rhaeto-Romanic, of Latin origin. **Political Parties:** Major political parties are: the Social Democrats (54 seats), the center-

right Radical Party (45); the center-right Christian Democratic Party (34); the right-wing Swiss People´s Party (29), the center left Socialist Party; the Democratic Union of the Centre, the right-wing Liberal Party; the Independent Party, centre; and the leftist Labor Party. The ecologist Greens, formed recently (9). **Social Organizations:** The Swiss Federation of Trade Unions, with 480,000 members; the Confederation of Christian Trade Unions, with 110,000 members; the Federation of Swiss Employees' Societies with 160,000 members.

THE STATE

Official Name: Confederation Helvetica.
Administrative Divisions: 20 Cantons, 6 Sub-Cantons. **Capital:** Bern, 129,423 inhab. (1994). **Other cities:** Zürich, 343,045 inhab.; Basel, 176,220 inhab.; Geneva (Genève), 171,744 inhab.; Lausanne, 117,153 inhab. (1994) **Government:** Jean-Pascal Delamuraz, President for 1996 until 31 December. The president is elected annually from among the members of the Federal Council. Parliamentary republic with strong direct democracy. The legislative power is exercised by a federal assembly, with two bodies: the State Council, with 46 members, and the National Council, with 200 members. The executive power is in the hands of the Federal Council, with 7 members elected by parliament for 4-year terms. **National Holiday:** August 1. **Armed Forces:** 1,800 regular troops, 28,000 annual conscripts (15 week courses). **Paramilitaries:** 480,000 Armed Civil Defence.

Socialist Party was formed to give workers a political voice.
[30] In 1910, 15 per cent of the workers in Switzerland were foreign. Most of them were anarchists and socialists who had suffered persecution in their own countries and they consequently encouraged radical positions in the workers' movement. The Swiss Workers' Union took up a platform of "proletarian class struggle", and the Socialist Party was inspired by the Second International's Marxism.
[31] World War I brought great internal tensions to Switzerland, especially between the French- and German-speaking regions. Under the leadership of Ulrich Wile, the Swiss army cooperated with Germany. Tension only decreased after the French victory, when Switzerland formally approached the allies and became a member of the League of Nations.
[32] Growing clashes between trade unions and employers echoed the tensions between the different linguistic regions. The 1918 general strike, although lifted three days later under pressure from the armed forces, led the bourgeoisie to form an anti-Socialist bloc. That year proportional representation was introduced.
[33] The 1919 elections marked the end of the liberal hegemony, held since 1848. The Socialists obtained 20 per cent of the vote, leading liberals to ally with the peasants who had 14 per cent, while the conservatives became the second power in the Federal Council.
[34] These changes produced noticeable consequences during the following years. The 48-hour week was included in factory legislation, while in 1925 an article on old-age pensions was added to the constitution. Assistance to the unemployed improved and collective work contracts became more common.
[35] In the years before World War II, the Socialist Party was threatened by foreign and national fascism, and a sector split to form the Communist Party. The socialists were forced to include formal recognition of the state and national defence in their policies.
[36] During World War II, European powers recognized Swiss armed neutrality, but the country still suffered strong pressure from Nazi Germany. Throughout the war Switzerland maintained a delicate

zerland remained neutral due to its religious division and its mercenary armies. Neutrality became a condition for the Confederation's existence; with the policy of armed neutrality, which is still maintained, first formulated by the Diet in 1674.
[23] In 1712 the Protestant victory in the second battle of Villmergen ended religious struggles, ensuring the hegemony of cities which were undergoing industrial expansion. Switzerland became the most industrialized country in Europe. Industry was based on labor at home, completely transforming work in the countryside.
[24] Throughout the 18th century, a series of popular revolts against the urban oligarchy clamoured for the reform of the Swiss Constitution. In March 1798, the Old Confederation fell under pressure from Na-

poleon's army. The Helvetic Republic was proclaimed, "whole and indivisible", with sovereignty for the people.
[25] Between the unitary Republic and the 1848 Federal Constitution, Switzerland was shaken by coups, popular revolts, and civil wars. The new federal pact marked a final victory for liberalism in the country. Two legislative bodies were established guaranteeing the rights of the small Catholic cantons.
[26] A state monopoly was created for custom duties and money minting, while weights and measures were unified, thus satisfying the industrial and commercial bourgeoisie's economic requirements. The 1848 Constitution thus eliminated the hindrances to capitalist expansion.
[27] Nepotism and the concen-

tration of capital benefitted only the few and fuelled growing opposition to the institutional system. The 1874 Constitution partially addressed these issues, and introduced the mechanism of referendum as an element of direct democracy.
[28] Expansion of the home labor system delayed organization of the Swiss workers' movement in relation to the country's industrialization. The Swiss Workers' Federation, created in 1873 had only 3,000 members, and the Swiss Workers' Union, which replaced it in 1880, only exceeded this figure 10 years later.
[29] The first achievement of the workers' movement was factory legislation, passed by parliament in 1877. The working day was limited to 11 hours with improved working conditions. In 1888 the

balance between accepting Hitler's advances and defending its independence; a strategy that kept them out of the conflict.

[37] After the war the West was resentful of Switzerland's relations with Germany, and the USSR refused to re-establish diplomatic relations, broken in 1918. However, the country's financial power paved the way for return to the international community. During the cold war, Switzerland sided with the West, but did not join the UN, in order to preserve its neutrality.

[38] The Swiss economy expanded massively during the postwar period. The chemical, food, and machinery exporting industries became great transnational corporations. In 1973 Switzerland was placed fourth in direct foreign capital investments, after the US, France, and Britain.

[39] Switzerland's main transnational corporation, the food and babymilk manufacturer Nestlé AG, had 196,940 workers worldwide and sales worth $29.36 billion in 1989.

[40] The Swiss economic expansion attracted workers from Italy, Spain and other southern European countries. Between 1945 and 1974 the number of immigrants rose from 5 to 17 per cent. Several referenda called for an end to immigration and the 1974-76 crisis forced thousands of people back to their countries.

[41] Due to its political neutrality, Switzerland did not join the European Economic Community in 1957. However, it has been a member of EFTA (European Free Trade Association) since 1960. In 1983 Swiss development aid constituted only 5 per cent of the total capital invested in poor countries.

[42] In 1959, after 10 years of voluntary absence, the socialists joined the Federal Council with two representatives. The Executive consists of two Radicals, two Christian Democrats, two Socialists, and a peasant representative. Thus, 80 per cent of the electorate is represented in government.

[43] Women gained the right to vote in 1971, but some cantons retained male-only suffrage until 1985. In 1984, the first woman minister was elected - Elisabeth Kopp was made minister for justice and police.

[44] Switzerland is governed by consensus, and the population has increasingly abstained in referenda.

In the 1979 elections participation was lower than 50 per cent for the first time.

[45] New economic and social phenomena have recently occurred. The labor structure has become more differentiated and minorities (women, young and old people, immigrants) are still subject to discrimination, bringing the institutional system into question.

[46] During the 1980s new opposition groups appeared, some feminist, some opposed to nuclear plants, and some youth groups fighting against the consumer society. A referendum in 1981 added a clause on equal rights to the Constitution.

[47] According to 1981 statistics, male students outnumber females by three to one in Switzerland. Women accounted for no more than 32.5 per cent of the economically active population, while men made up 63.9 per cent.

[48] Since 1986 environmental problems have become more serious and the government has taken measures to curb pollution, especially "acid rain" and pollution of the Rhine. In 1987 France, Germany and the Netherlands received compensation for damage caused by an accident in the Swiss chemical industry.

[49] Another serious problem is the damage to the Alpine ecosystem. The rapid growth of cities in this area, and the increased transit of heavy trucks, have contributed to regional desertification. As a direct result, so-called "natural disasters", such as floods and avalanches of rock, mud and snow have increased; events that would be less likely to occur if the area still had the protection offered by its natural tree coverage.

[50] The Chernobyl disaster greatly concerned the Swiss population and in 1989, a series of demonstrations took place. A referendum proposed the gradual elimination of existing nuclear power plants, and the Federal Parliament cancelled the construction of a sixth nuclear station.

[51] Increasing social problems and the presence of immigrants have given encouragement to the Swiss extreme right. Although being small, the Swiss Democratic Party and the Party of Drivers, xenophobic and opposed to social security policies, have both gained support since the fall of the Berlin Wall.

[52] Switzerland's integration into the IMF was approved by a plebiscite in May 1992. In June 1993, parliament approved the incorporation of Swiss troops into the United Nations peace-keeping forces. This represented an important change in Switzerland's traditional policy of neutrality. However, most of the Swiss voted against this proposal in a referendum in 1994.

[53] One of the main obstacles to Swiss integration into the European Union was the objection to the free movement of workers between countries. A referendum in the same year approved an anti-racism law which punished discrimination while another granted the police powers to use greater "harshness" against illegal immigrants committing crimes within the country. This measure was widely criticized and considered a violation of the Swiss Constitution and the European Human Rights Convention.

[54] In the general elections of October 22 1995, there was a large increase in support for the Social Democrats, who took 54 seats, the Radicals had 45, the Christian Democrats 34, the Swiss People's Party took 29 and the Greens 9. Following the elections, a group of parliamentarians from various parties called for reform, the formulation of new policies and reconsideration of membership of the European Union.

[55] The government carried out a study of the future of unemployment, where it predicted a large reduction by the year 2000. According to their report, if the annual GDP growth rate does not fall from 2.25 per cent, by the year 2000 unemployment will affect 60,000 people, compared with 165,000 at present.

Syria

Population: 13,844,000 (1994)
Area: 185,180 SQ.KM
Capital: Damascus (Dimashq)

Currency: Pound
Language: Arabic

Suriyah

Syria was once the name for the entire region between the peninsulas of Anatolia (Turkey) and Sinai. Ancient civilizations coveted the territory: the Egyptians wanted it as a port, while the Persians considered the region a bridge to their plans of a universal empire.

[2] Between the 12th and 7th centuries BC, the Phoenician civilization developed on the central coastal stretch of the territory; a society of sailors and traders without expansionist aims. Phoenician cities were always independent, although some exercised temporary hegemony over others, and they developed the world's first commercial economy.

[3] The Phoenicians invented the alphabet, constructed ocean-going ships, practised large-scale ceramic and textile manufacturing, expanded and systematized geography, sailing around the coast of Africa. The propagation of these throughout the Mediterranean helped form what would be later called "Western civilization," of which the Greeks were the main exponents.

[4] The territory came under the control of Alexander of Macedonia (also known as Alexander the Great) during the 4th century BC. After Alexander's death, his vast empire was divided and Syria became the center of a Seleucidan state (named for Seleucus Nicator, one of Alexander's generals) that initially stretched as far as India. The eastern part was later lost to the Parthians. In the Roman era the province of Syria was a border zone constantly shaken by fierce local wars and resistance. One example of the latter was Queen Zenobia, whom the Romans finally managed to defeat in 272 AD.

[5] The Arabization of the territory was carried out by the Ummaia Caliphs, who, between the years 650 and 750, turned Damascus (Dimashq) into the capital of the empire (see Saudi Arabia), and fostered a strong national spirit. When the Ummaias were defeated by the Abbas, the capital was transferred to Baghdad, where the new caliphs enjoyed greater support. Although still economically and culturally important, the loss of political power proved significant: in the 11th century when Europeans invaded during the Crusades, the caliphs of Baghdad reacted with indifference. Local emirs were left to their own resources, and disagreements among them allowed a small Christian force to conquer the area, leading to 200-years of occupation.

[6] The Egyptians initiated the process to drive out the Europeans. One result was that Syria became a virtual Egyptian province and center stage for a confrontation with Mongol invaders. In the 16th century, the country became a part of the Ottoman Empire.

[7] The Crusaders left behind a significant Christian community, especially with the Maronites, serving as sufficient cause for European interference from the 17th century onwards. The Egyptian Khedive Muhammad (Mehemet) Ali conquered Syria in 1831, and heavy taxes and compulsory military service provoked revolt among both Christian and Muslim communities. The European powers, concerned with propping up the Turkish Empire to stave off the potential emergence of a more aggressive power like Egypt, used the repression of Christians as an excuse for intervention. Ali's offensive was suppressed and the "protection of Syrian Christians" was entrusted to the French. A withdrawal of Egyptian troops took place in 1840, along with the restoration of Ottoman domination and the establishment of Christian missions and schools subsidized by Europeans.

[8] However, the scene was set for confrontation and in 1858, Maronite Christians concentrated in the mountainous region between Damascus and Jerusalem rebelled against the ruling class and eliminated the traditional system of land ownership. Their Muslim neighbors, particularly the Druze, moved to repress the movement before it spread further, triggering a conflict that culminated in the deaths of a large number of Christians in June 1860.

[9] A month later, French troops disembarked in Beirut, ostensibly to protect the Christian community, and forcing the Turkish government to create a separate province called "Little Lebanon." This was to be governed by a Christian appointed by the Sultan but with the approval of the European powers, with its own police force, and traditional privileges were abolished in the territory. The social conflict thus became a confrontation between confessional groups with the Christians in "Little Lebanon" placed in a position of superiority over the local Muslim population, providing firm roots for the politico-religious conflict that has marked Lebanese history ever since.

[10] The proliferation of European educational institutions bred an intellectual elite, which readily absorbed the ideas of nationalist movements. Many of these intellectuals were at the forefront of Arab nationalism, centered in Syria.

[11] Consequently, when an Arab rebellion broke out during World War I (see Saudi Arabia, Jordan and Iraq), Emir Faisal (the son of the shereef of Mecca) hastened to lead his army of Bedouins to Damascus to take control of the main center of Arab political agitation. Faisal was proclaimed king of Syria. At the time French and British intentions were unknown, but the Sykes-Picot agreement divided the fertile crescent giving Syria (with Lebanon) to France, and Palestine (including Jordan) and Iraq to Britain.

[12] In 1920 France occupied Syria forcing Faisal to retreat. Two months later, Syria was divided into five states: Greater Lebanon (adding other regions to the province of "Little Lebanon"), Damascus, Aleppo, Djabal Druza and Alawis (Latakia). The latter four were reunified in 1924.

[13] Throughout 1932, Syria experienced a period of relative stability, and independence seemed imminent. A Syrian president and parliament were elected that year, but France made it clear that autonomy was unacceptable. This attitude engendered political agitation and confrontation, which only end-

> Phoenician cities were always independent, although some exercised temporary hegemony over others, and they developed the world's first commercial economy.

PUBLIC EXPENDITURES

% of GNP — 1992

	MILITARY	EDUCATION	HEALTH
	16.6	4.1	0.4

ed with a 1936 agreement with the French. France recognized certain Syrian demands, chiefly reunification with Lebanon. However, the French government never ratified the agreement, and this led to new waves of violence, which culminated in the 1939 resignation of the Syrian president and a French order to suspend the 1930 constitution that governed both Syria and Lebanon.

[14] In 1941, French and British troops occupied the region to flush out Nazi collaborators. In 1943 Chikri al-Quwatli was elected president of Syria, and Bechara al-Kuri, president of Lebanon. Bechara al-Kuri proposed elimination of the mandate provisions from the constitution; however, he and his cabinet members were imprisoned by the ever-present French troops. Violent demonstrations followed in both Lebanon and Syria, and the British pressed for withdrawal of the French. In March 1946, the United Nations finally ordered the European forces to withdraw. This was completed in 1947 and the end of the French mandate was officially declared.

[15] Independence brought social conflict; part of the legacy of the colonial period. In 1948 Syrian troops fought to prevent the partition of Palestine, and in 1956 joined Egypt in the battle against Israeli, French and British aggression. This aggression was the answer to Egyptian President Gamal Abdel Nasser's decision to nationalize the Suez Canal.

[16] In 1958, Syria joined Egypt in founding the United Arab Republic, but Nasser's ambitious integration project collapsed in 1961. Ten years later the scheme was reactivated, with greater flexibility and the Federation of Arab Republics was created including Libya.

[17] In 1963, after a revolution, the Ba'ath Arab Socialist Party came to power, with their main tenet holding the Arab countries to be merely "regions" of the larger Arab Nation. In November 1970, General Hafez al-Assad rose to the national presidency. He launched a modernization campaign, including a series of social and economic changes. The subsequent party congress named Assad party leader and proposed "accelerating the stages towards socialist transformation of different sectors." This guideline was adopted, and became part of

ENVIRONMENT

To the west, near the sea, lies the Lebanon mountain range. To the south there are semi-desert plateaus and to the north low plateaus along the basin of the Euphrates River. Farming -grains, grapes and fruit- is concentrated in the western lowlands that receive adequate rainfall. In the south, the volcanic plateaus of the Djebel Druze are extremely fertile farmlands, some as the oases surrounding the desert, the main one being that of Damascus. Cotton and wool are exported. Exploitation of oil fields is the country's chief industry. The dumping of toxic substances from the extraction of crude is responsible for the high levels of pollution of Syria's waters.

SOCIETY

Peoples: Syrians are mostly Arabs, with minority ethnic groups in the north: Kurds, Turks and Armenians. There are few statistics on the number of Palestinian refugees, and the Jewish population was authorized to emigrate in 1992. **Religions:** Mainly Muslims, mostly Sunni, followed by Alamites, Shiites and Ismailites. Also minor communities of eastern Christian religions.
Languages: Arabic (official). Minority groups speak their own languages. **Political Parties:** The Ba'ath Arab Socialist Party, founded in Damascus in 1947 by Michel Aflaq, including the National Progressive Front and other minor parties. The armed political-religious opposition is represented by the Muslim Brotherhood, a fundamentalist Sunni sect. **Social Organizations:** The General Federation of Labor Unions unites ten workers' federations representing different sectors of the economy.

THE STATE

Official Name: Al-Jumhouriya al Arabiya as-Suriya.
Administrative Divisions: 14 Districts.
Capital: Damascus (Dimashq), 1,549,930 inhab. (1994).
Other cities: Aleppo (Halab), 1,355,000 inhab., Homs (Hims), 481,000 inhab. (1990). **Government:** Hafez al-Assad, President since 1970, re-elected in 1971, 1978, 1985 and 1991; Mahmoud Az-Zoubi, Prime Minister. The People's Council (a unicameral legislature) has 195 members.
National Holidays: April 17, Independence Day (1946); November 16, Revolution Day (1978). **Armed Forces:** 423, 000 troops (1995). **Paramilitaries:** 8,000 Gendarmes, 4,500 Palestinian Liberation Army.

the new constitution which was approved in 1973.

[18] Syria took an active part in the Arab-Israeli wars of 1967 and 1973, during which Israeli troops occupied the Golan Heights. Syria also resisted US efforts to impose a "settlement" in the Middle East, together with Algeria, Iraq, Libya, Yemen and the Palestine Liberation Organization (PLO). They also opposed the Camp David agreement (see Egypt).

[19] Syrian troops formed a major part of the Arab Deterrent Force that intervened in Lebanon in 1976 to prevent partition of the country by right-wing Lebanese factions allied with Israel.

[20] In 1978, the Syrian and Iraqi branches of the Ba'ath Party drew closer, but negotiations for creation of a single state disintegrated. In late 1979, at the congress of the Syrian branch of the Ba'ath Party, the Muslim Brotherhood (a right-wing Islamic movement) was harshly censured and were labelled "zionist agents".

[21] The ensuing confrontations

between members of the fundamentalist Brotherhood and the Syrian government forces constituted one of the most serious internal problems for Assad's regime.

[22] In 1980, the government set a September deadline giving brotherhood members the opportunity to surrender before capital punishment was enforced for acts of sabotage. A reported 1,000 members of the movement surrendered delivering reports that the Iraqi and Jordanian governments had supplied them with weapons and training.

[23] Fearful that the Muslim Brotherhood could potentially establish a reactionary Islamic system more severe than the Assad regime, various Sunni Muslim factions joined together to form the United Islamic Front and declared support for democratic pluralism and respect for religious minorities.

[24] Brotherhood attacks continued. In 1982 the army launched a full offensive which resulted in thousands of casualties. The Syrian government blamed Iraq for arming the rebels, and closed the border between the two countries in April 1982. Iraq retaliated by closing the oil pipeline connecting Kirkuk, Iraq, to the Syrian port of Banias.

[25] The virtual alliance formed in 1980 between Saudi Arabia, Iraq and Jordan and tensions between those three countries and Syria were exacerbated by the outbreak of the Iran-Iraq War. Assad charged Iraq as the aggressor and diverted attention from what he called the major regional issue - the Palestinian question. Toward the end of the year, Syrian accusations of Jordanian support for the Brotherhood brought the two countries to the verge of war. The mediation of Saudi prince Abdalla Ibn Abdul-Aziz averted armed conflict.

[26] In 1981 the "missile crisis" broke out in Syria, when the Christian Phalangist Movement sought to extend their area of authority to include the area around the Lebanese city of Zahde. An Arab Deterrent Force, organized and commanded by Syria, attempted to prevent the advance. Syria installed Soviet surface-to-air SAM-9 missiles, triggering an Israeli reaction. The crisis was finally averted, but in 1982 Israel invaded Lebanon, and destroyed the bases of Syrian missiles.

[27] In mid-1983 there was a crisis between the Syrian authorities and the PLO leadership. This encour-

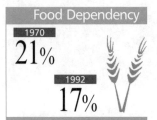

Food Dependency

1970 21%

1992 17%

STATISTICS

DEMOGRAPHY

Annual growth: 0.8% (1992-2000). **Estimate for year 2000:** 17,000,000. **Children per woman:** 6.1 (1992).

HEALTH

One **physician** for every 1,220 inhab. (1988-91). **Under-five mortality:** 38 per 1,000 (1994). **Calorie consumption:** 109% of required intake (1995). **Safe water:** 85% of the population has access (1990-95).

EDUCATION

Primary school teachers: one for every 25 students (1991).

COMMUNICATIONS

91 **newspapers** (1995), 97 **TV sets** (1995) and 102 **radio sets** per 1,000 households (1995). 4.1 **telephones** per 100 inhabitants (1993). **Books:** 83 new titles per 1,000,000 inhabitants in 1995.

ECONOMY

Annual growth: -2.10% (1985-94).
Consumer price index: 100 in 1990; 158 in 1994.
Currency: 11 pounds = 1$ (1994).
Cereal imports: 1,440,000 metric tons (1992).
Fertilizer use: 549 kgs per ha. (1992).

---

aged Syria to support the Palestinian groups opposed to Yasser Arafat's leadership, until the division ended in 1987. The fall in oil prices further aggravated the economic problems caused by the war and the government set up strict austerity measures in 1984: financial activities were tightly controlled, borders were closed to the illegal trafficking of merchandise, previously tolerated, and public expenditure was substantially reduced. Smuggling from Lebanon was reduced by 90 per cent, and the balance of payments deficit halted.
[28] In 1985, President Assad won a new seven-year term. In 1987, Prime Minister Abdul Rauf al-Kassem was forced to resign amid charges of corruption, following the resignation of four other ministers. In November 1987, Mahmoud Az-Zoubi, president of the People's Assembly, was chosen as prime minister.
[29] In April 1987, at a summit meeting of Arab countries, most nations sought alignment with Syria, in exchange for economic assistance, and condemned Iranian prolongation of the war. Syria maintained its position on Iran, and vetoed the motion for Egypt's re-entry into the Arab League.

[30] In October 1986, the British government accused Syria of supporting a terrorist attempt against an Israeli plane at London airport. In November most of the EEC countries broke diplomatic rela-

> The ensuing confrontations between members of the fundamentalist Brotherhood and the Syrian government forces constituted one of the most serious internal problems for Assad's regime.

tions with Syria, although there were disagreements from some countries. In July 1987 all the EEC countries except for Britain re-established relations with Syria, when the Syrian government withdrew support of the Al-Fatah Revolu-

tionary Council, a Palestinian organization which opposed Yasser Arafat. In May 1990, Syria finally re-established diplomatic relations with Egypt.
[31] When Iraq invaded Kuwait, Syria immediately sided with the anti-Iraqi alliance and sent troops to Saudi Arabia. Diplomatic relations with the US improved noticeably. During the crisis Syria increased its influence over Lebanon and strengthened the allied government in that country; they were also successful in disarming most of the autonomous militias.
[32] On April 29, 1991, Syrian president Hafez Assad met in Damascus with Iranian president Akbar Ashemi Rafsanjani, and announced his country's military presence would be maintained in Lebanon. He thereby gained the support of the Hezbollah (Party of God), and of the Revolutionary Guards (fundamentalist Iranians).
[33] In April 1991, Syria took part in the Middle East Peace Conference, held in Madrid with US and USSR support. Syria supported the proposal of negotiating with Israel on the basis of UN declarations 242 and 338, which condemned the occupation of territories by Israel. The Arab countries supported the motto "peace for territories", which was rejected by Israel.
[34] In May 1991 Syria and Lebanon signed a cooperation agreement whereby Syria recognized Lebanon as an independent and separate state, for the first time since both countries gained independence from France.
[35] In December 1991, al-Assad was re-elected for the fourth time, by 99.98 per cent of the vote, in elections which had him as sole candidate. That month, the government announced the pardon of 2,800 political prisoners, members of the Muslim Brotherhood.
[36] In 1992, the government abolished the death penalty and allowed the emigration of 4,000 Jews. In the economic level, the GNP grew 7 per cent and exports of oil and its by-products amounted to $2 billion. New legislation favored investments in the private sector, which had a significant growth between 1991 and 1993.
[37] Syria stayed away from the first stages of the regional peace process, which enabled the establishment of a limited autonomy for Palestine and the signing of agreements between Israel and Jordan in July 1994. In January, a "historic"

meeting took place between US president Bill Clinton and Assad in Geneva and in September, the Syrian minister of foreign affairs was interviewed for the first time on Israeli television. However, Damascus maintained its refusal to officially negotiate with Israel, due to the murder of 29 Palestinians in Hebron, committed by an Israeli settler in February.
[38] In the domestic field, the death of Basel al-Assad, eldest son and supposed successor of the Syrian president, increased uncertainty regarding the country's political future. In August, the ruling National Progressive Front won the general elections but only 49 per cent of registered voters took part.
[39] In June 1995, during official negotiations, Israel and Syria almost reached an agreement which would have returned the Golan Heights to Syria, occupied by Israeli forces since 1967. However, the agreement was not concluded since Israel wanted to keep a limited military presence indefinitely. In October, an ambush set by Hezbollah guerrillas on Israeli troops in southern Lebanon, complicated the talks between both countries. In mid 1996, President Assad took part with other regional heads of state in attempts to coordinate a common strategy for Arab countries in negotiations with Israel.
[40] Meanwhile, Assad tried to secure relations with Egypt and the Gulf countries, while he continued his policy of encouraging Syria's private sector of the economy, opening key sectors like power generation, production of cement or pharmaceuticals to private investment.

Tajikistan

Tadzhikistan

Population: 5,751,000 (1994)
Area: 143,100 SQ.KM
Capital: Dushanbe (Dusanbe)

Currency: Ruble
Language: Tajik

Around 500 BC, Central Asia and southern Siberia's first centers of sedentary civilization arose in what is now Tajikistan. The Bactrian State was located in the upper tributaries of the Amu Darya. Likewise, in the Zeravshan river basin and the Kashkadarya river valley lay the nucleus of another State, Sogdiana. The inhabitants of these early civilizations built villages, with adobe and stone houses, all along the rivers, which they used for irrigating their crops. These included wheat, barley and millet, and a variety of fruits. Navigation was well developed and the cities which lay along the route of the caravans uniting Persia, China and India became important trading centers.

2   In the 6th century BC these lands were annexed by the Persian Achaemenid empire. In the 4th century BC, Alexander the Great conquered Bactria and Sogdiana. With the fall of his empire in the 3rd century, the Greco-Bactrian State and the Kingdom of Kushan emerged, and subsequently fell to the onslaught of the advancing steppe Yuechzhi and Tojar tribes. In the 4th and 5th centuries AD, Sogdiana and its surrounding regions were invaded by the Eftalites, and in the 6th and 7th centuries, by Central Asian Turkish peoples. In the 7th century, Tajikistan became a part of the Arab Caliphate, along with other Central Asian regions. After its demise, the region was incorporated into the Tahirid and Samanid kingdoms. In the 9th and 10th centuries, the Tajik people emerged as an identifiable ethnic group.

3   From the 10th to the 13th centuries AD, Tajikistan formed part of the Gaznevid and Qarakhanid empires, as well as the realm of the Shah of Khwarezm. In the early 13th century, Tajikistan was conquered by Genghis Khan's Mongol Tatars. In 1238, Tarabi, a Tajik artisan, led a popular revolt. From the 14th to the 17th centuries, the Tajiks were under the control of the Timurids and the Uzbek Shaybanid dynasty. From the 17th to the 19th centuries, there were small fiefdoms throughout Tajikistan whose chieftains alternately submitted to or revolted against the khans of Bukhara.

4   In the 1860s and 1870s, the Russian Empire conquered Central Asia, and the northern part of Tajikistan was annexed by Russia. The khanate of Bukhara however maintained a relation of sovereignty rather than annexation, with Russia. The Tajik population of Kuliab, Guissar, Karateguin and Darvaz was incorporated into the khanate as the province of Eastern Bukhara. Oppression by the Russian bureaucracy and the local feudal lords triggered a wave of peasant revolts toward the end of the 19th century and beginning of the 20th century. The most important of these was the 1885 uprising led by Vose.

5   In 1916, during World War I, the native populations of Central Asia and Kazakhstan revolted over the mobilization of their people for rearguard duty with the retreating Russian army. After the triumph of the Bolshevik revolution in St Petersburg in October 1917, Soviet power was established in Northern Tajikistan. In April 1918, this territory became a part of the Autonomous Soviet Socialist Republic of Turkistan. Nevertheless, a massive number of Tajiks remained under the power of the Emirate of Bukhara, which existed until 1921. In early 1921, the Red Army took Dushanbe, but in February it was forced to withdraw from Eastern Bukhara.

6   Having broken Alim Khan's resistance in 1921-22, Soviet power was proclaimed throughout the entire territory of Tajikistan. In 1924, the Soviet Socialist Republic of Uzbekistan was formed, of which the Autonomous Soviet Socialist Republic of Tajikistan was a part. In January 1925, the Autonomous Region of Gorno-Badakhshan was established, high up in the Pamir range. On November 16 1929, Tajikistan became a federated republic of the Soviet Union. In the 1920s and 1930s, land and water reforms were carried out. The collectivization of agriculture was followed by industrialization, and the so-called "cultural revolution" campaign.

7   After World War II, the Soviet regime carried out a series of large construction projects, including a system of canals and reservoirs linking up with water projects in the neighboring republic of Uzbekistan. The purpose of this irrigation network was to develop the region's agricultural potential, especially for the cultivation of cotton. In the 1970s and 1980s, the effects of mismanagement and economic stagnation were felt in Tajikistan, one of the poorest regions in the USSR, with a high rate of unemployment, especially among the young, who make up most of the population.

8   From 1985 onwards, the changes promoted by President Mikhail Gorbachev made the democratization process possible , which in turn gave way to the expression of long-suppressed ethnic and religious friction in Tajikistan. In February 1990, there were violent incidents between Tajiks and Russians in the capital, with more than 30 people killed and scores injured. The government decreed a state of emergency, which remained in effect during that year's Supreme Soviet (Parliament) elections, in which the Communist Party won 90 per cent of the available seats. The opposition attributed the February events to provocation by the former Soviet secret police (KGB), aimed at discrediting Gorbachev's reforms and suppressing nationalist tendencies. Muslims and Democrats called for the disso-

lution of the Soviet, because the elections had been held while a state of emergency was in effect.

9   After the February 1990 events in Dushanbe, there was an exodus of Russian and Ukrainian technicians, which triggered the closure of half the country's hospitals, schools and factories, further aggravating its social and economic problems. In late August, Parliament passed a vote of no-confidence on President Majkamov, accusing him of supporting those responsible for the Moscow coup, and forcing him to resign. He was succeeded by Kadridin Aslonov as interim president of the republic.

10   Aslonov's decision to ban the Communist Party (CP) of Tajikistan, announced after the coup attempt against Gorbachev, was not supported by parliament. Dominated by hardline communists, they forced Aslonov to resign, naming Rajmon Nabiev to take his place as interim president. Nabiev had been First Secretary of the local CP until ousted by Gorbachev in 1985. In mid-September 1991, parliament approved the Declaration of Independence and the new constitution, decreeing a state of emergency and banning the Islamic Renaissance Party.

11   Tajikistan's religious revival has been stronger than that in the rest of the former USSR republics. In the early 1980s, there were 17 mosques; 10 years later, as a result of the liberalization promoted by Moscow, there were 128 large mosques, 2,800 places of prayer, an Islamic institute and five centers offering religious instruction. Ac-

Foreign Trade

MILLIONS $ 1994

IMPORTS
619

EXPORTS
531

Maternal Mortality

1989-95

PER 100.000
LIVE BIRTH
39

PROFILE

ENVIRONMENT

The Tian Shan, Guissaro-Alai and Pamir mountains occupy more than 90% of Tajhikistan's territory. This republic is located in the southeastern part of Central Asia between the Syr Darya River and the Fergana Valley in the north, the Pamir and Paropaniz mountain ranges in the south, the Karakul lake and the headwaters of the Murgab in the east, and the Guissar and Vakhsh valleys in the southwest. The Turkistan, Alai and Zeravshan mountains cross Tadzhikistan from north to south, joining the Pamir plateau. The highly cultivated valleys (lying at an altitude of .1000-2000 meters) have a warm, humid climate. Lower mountains and valleys, located in northern and southeastern Tadzhikistan, have an arid climate. It is bounded by Kyrgyzstan in the north, Uzbekistan in the west, China in the southeast and Afghanistan in the southwest. Abundant mineral resources include iron, lead, zinc, antimony and mercury, as well as important uranium deposits. Since 1960, land under irrigation has increased by 50%, but as in other countries where cotton is the sole crop, the salinity of the soil has increased. Due to the lack of adequate sewage systems, and insufficient supplies of drinking water, a high proportion of Tajik infants suffer from diarrhea, the main cause of the country's high infant mortality rate.

SOCIETY

Peoples: Tajik, 63.8%; Uzbek, 24.0%; Russian, 6.5%; Tatar, 1.4%; Kyrgyz, 1.3%; Ukrainian, 0.7%; German, 0.3%; other, 2.0%. **Religions:** Muslim (Sunni); in the Pamir region, the majority belongs to the Shiite (Ishmaelite) branch of Islam. **Languages:** Tajik (official), Uzbek, Russian. **Political Parties:** Communist Party; Socialist Party; Rastokhez People's Movement; Islamic Renaissance Party; Democratic Party. **Social Organizations:** Union Federation of Tajikistan.

THE STATE

Official Name: Jumhurii Tojikistan. **Administrative Divisions:** 3 Regions, 57 Districts. **Capital:** Dushanbe (Dusanbe), 602,000 inhab. (1990). **Other cities:** Khujand (formerly Leninabad) 164,500 inhab.; Kulob 79,300 inhab.; Qurghonteppa 58,400 inhab.; Urateppa 47,700 inhab. **Government:** Imamali Rajmonov, President. Yakhyo Azimov, Prime Minister. **National Holiday:** September 9, Independence (1991). **Armed Forces:** 6,000; 24,000 Russian troops (1994).

STATISTICS

DEMOGRAPHY

Urban: 32% (1995). **Annual growth:** 2.8% (1991-99). **Estimate for year 2000:** 7,000,000. **Children per woman:** 5.1 (1992).

EDUCATION

Primary school teachers: one for every 21 students (1992).

COMMUNICATIONS

4.6 **telephones** per 100 inhabitants (1993).

ECONOMY

Per capita GNP: $360 (1994). **Annual growth:** -11.40% (1985-94). **Annual inflation:** 104.30% (1984-94). **Currency:** ruble. **Cereal imports:** 450,000 metric tons (1993). **Fertilizer use:** 1,618 kgs per ha. (1992-93). **Imports:** $619 million (1994). **Exports:** $531 million (1994).

ENERGY

Consumption: 642 kgs of Oil Equivalent per capita yearly (1994), 55% imported (1994).

---

cording to unofficial estimates, between 60 and 80 per cent of the population are practising Muslims. The Islamic Renaissance Party, created within the past few years, has proposed that the state allow political and religious freedom, but that it be based on Islam, with *sharia* (Islamic law) as the law of the land. Since 1991, the Tajik government has officially instituted the celebration of several important Muslim holy days.

[12] The mobilization of pro-Islam-ic forces, on the one hand, and mediation by Moscow, on the other, led to the state of emergency and the Communist Party ban being lifted in early October. Two months later, in its first presidential elections in November 1991, Nabiev was confirmed in his post by 58 per cent of the voters. According to official tallies, Davlav Khudonazarov, a former member of the CP-Central Committee and president of the Soviet Film Actors Union, and a candidate supported by Muslims and democrats, came in second with 38 per cent of the vote.

[13] On December 21 Tajikistan entered the Commonwealth of Independent States (CIS). At the same time, the Autonomous Region of Gorno-Badakhshan, high in the Pamir mountains, asked to be granted the status of an autonomous republic. Like neighboring Afghanistan, the majority of this region's inhabitants are Shiite Muslims.

[14] During his visit to Dushanbe, in February 1992, US Secretary of State James Baker expressed the West's concern over the region's uranium deposits. Tajikistan, together with Turkmenistan and Uzbekistan, have important deposits of this mineral, which is crucial to the development of nuclear weapons.

[15] In April and May 1992, there were a number of confrontations, leading to the resignation of the president of parliament, Kendzhaev. On May 7, a coalition government was formed, with government and opposition representatives.

[16] Outside Dushanbe, however, there was violence from both pro-government and opposition forces, causing a conflict of civil-war proportions. Another protest demonstration against Nabiev led to his resignation on September 8. A private pro-communist army in the northern part of the country, far from accepting this state of affairs, launched a major offensive. The attack ended in December, with the fall of the capital and the formation of a new government controlled by the Popular Front.

[17] Some 300,000 people fled to other CIS republics and Afghanistan. As of February 1993, the persecution and massacre of members of the opposition began to diminish.

[18] The Russian government of President Boris Yeltsin was the first to recognize the new regime, led by Imamali Rajmonov, whom Yeltsin considered ideal for keeping the Islamic groups under control. In March, bombing started from Afghanistan, followed by incursions of opposition troops across the Russian-controlled frontier.

[19] Following nine months of confrontations, in April 1994, representatives of the government and the Islamic and liberal opposition met in Moscow, at a time when Tajikistan was more dependent than ever on Russian aid, both military and economic. Rajmonov proposed a new constitution, approved in November by referendum. The outgoing president was also re-elected in the presidential elections which accompanied the referendum, but the Islamic opposition accused the government of fraud.

[20] The exiled opposition representatives did not recognize the validity of the February 1995 parliamentary elections either, and sporadic armed confrontations continued to affect the country. On an economic front, the war-devastated Tajikistan could not convince Russia to form a monetary union between the two nations.

[21] In February 1996, Prime Minister Jamshed Karimov stood down and was succeeded by Yakhyo Azimov.

Tanzania

Population: 28,817,000 (1994)
Area: 945,090 SQ.KM
Capital: Dodoma

Currency: Shillings
Language: Swahili and English

Tanzania

The oldest fossils of the human race were found in the Olduvai gorge, in the north of Tanzania. These remains date from millions of years ago, yet little is known about life on most of modern Tanzania's mainland before the 7th century.

[2] A mercantile civilization, heavily influenced by Arab culture, flourished in this region between 695 and 1550 (See "The Zandj Culture in East Africa") but it was eventually destroyed by Portuguese invaders. One hundred and fifty years later, under the leadership of the sultan of Oman, the Arabs succeeded in driving out the remaining Lusitanians, but the rich cultural and commercial life of past times did not return. As the slave trade grew, Kilwa and Zanzibar became the leading trade centers.

[3] Between 1698 and 1830, Zanzibar and the coast were under the rule of the Oman Sultanate, with the sultan living in Zanzibar. His sons, under British pressure, later split the inherited domain, dividing the old Sultanate in two.

[4] At the end of the 19th century a German adventurer, claiming rights over Tanzania, based on "treaties" signed by local Tanganyika leaders set up a company which immediately received imperial endorsement, leasing out the mainland coastal strip to the sultan of Zanzibar. The British had also made a similar deal, so the Berlin Conference, where the European powers distributed Africa among themselves, was obliged to recognize the claims of both powers. The European powers eventually agreed upon the "cession of German rights" in favor of Britain.

[5] The areas of influence were delimited in 1886. Tanganyika, Rwanda and Burundi were recognized as German possessions while Zanzibar formally became a British

protectorate in 1890. German troops and British warships joined efforts to stifle a Muslim rebellion on the coast of Tanganyika in 1905.

[6] After the defeat of Germany in World War I, the League of Nations placed Tanganyika under British mandate, while Rwanda-Burundi was handed over to the Belgians.

[7] Nationalist feelings were channelled into TANU (Tanganyikan African National Union), a party founded in 1954 by Julius Nyerere, a primary school teacher known by the people as Mwalimu, the teacher.

[8] After seven years of organizing and fighting against racial discrimination and the appropriation of lands by European settlers, independence was achieved in 1961, and Nyerere became president, elected by an overwhelming majority.

[9] Meanwhile, in Zanzibar, two nationalist organizations, which had been active sporting and cultural institutions since the 30s, merged to form the Afro-Shirazi Party in February 1957. In December 1963, the British transferred power to the Arab minority, and a month later this government was overthrown by the Afro-Shirazi. In April, Tanganyika and Zanzibar formed the United Republic of Tanzania, though the island maintained some autonomy.

[10] Under Nyerere's leadership, Tanzania based its foreign policy on non-alignment, standing for African unity, providing unconditional support to liberation movements, particularly FRELIMO in neighboring Mozambique.

[11] In February 1967, TANU issued the Arusha Declaration proclaiming socialism as its objective. The declaration lays down the principle of self-sufficiency and gives top priority to the development of agriculture, on the basis of commu-

nal land ownership, a traditional system known in Swahili as *ujamaa*, which means family.

[12] Ten years later TANU and the Afro-Shirazi Party merged into the Chama Cha Mapinduzi (CCM), which officially incorporated the aim of "building socialism on the basis of self-sufficiency".

[13] In October 1978, Tanzania was invaded by Ugandan troops in an obvious attempt by dictator Idi Amin to distract attention from his internal problems. At the same time he sought to divert Tanzanian energies from supporting the liberation struggles in Southern Africa. The aggression was repelled in a few weeks, and Tanzanian troops cooperated closely with the Ugandan National Liberation Front to overthrow Idi Amin.

[14] The cost of mobilizing the army and maintaining troops to prevent chaos in Uganda weighed

heavily on the State budget, which was facing serious difficulties by the end of the 1970s. The falling price of Tanzania's main exports; coffee, spices, cotton, pyrethrum and cashew nuts, plus the growing cost of imported products, caused serious financial imbalances and forced cutbacks in the program to expand the *ujamaa* community villages.

[15] These villages were conceived as the nucleus of the Tanzanian economy and were basically designed to be self-sufficient. To achieve autonomy, they received sizeable initial investments from the government. Despite Nyerere's enormous efforts, various factors hindered the progress of *ujamaa* which grew slowly when compared with small and medium-sized private farms. They continued to depend upon imported foodstuffs and as the government aid was cut, many of them collapsed.

[16] At the end of 1982, the CCM held its second general congress. A new generation of young leaders, most of them strongly committed to the development of the socialist project were appointed to key posts within both the government and the party.

[17] Edward Sokoine was appointed prime minister and immediately launched a campaign against corruption while adopting a more flexible policy towards foreign investment. Sokoine also arranged for Tanzania and Kenya to resume re-

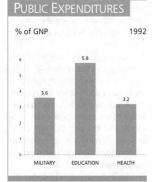

PUBLIC EXPENDITURES

% of GNP — 1992

MILITARY 3.6 — EDUCATION 5.8 — HEALTH 3.2

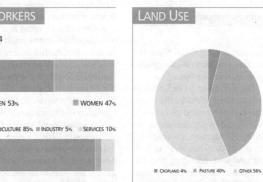

WORKERS

1994

MEN 53% — WOMEN 47%

AGRICULTURE 85% — INDUSTRY 5% — SERVICES 10%

LAND USE

CROPLAND 4% — PASTURE 40% — OTHER 56%

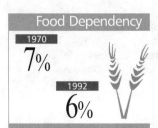

1970
7%

1992
6%

MILLIONS $ 1994

IMPORTS
1,505

EXPORTS
519

STATISTICS

DEMOGRAPHY

Urban: 23% (1995). **Annual growth:** 2.9% (1992-2000).
Estimate for year 2000: 33,000,000. **Children per
woman:** 6.3 (1992).

HEALTH

Under-five mortality: 159 per 1,000 (1994). **Calorie
consumption:** 92% of required intake (1995). **Safe
water:** 50% of the population has access (1990-95).

EDUCATION

Illiteracy: 32% (1995). **School enrolment:** Primary
(1993): 69% fem., 69% male. Secondary (1993): 5% fem.,
6% male. **Primary school teachers:** one for every 36
students (1992).

COMMUNICATIONS

85 **newspapers** (1995), 82 **TV sets** (1995) and 78 **radio sets**
per 1,000 households (1995). 0.3 **telephones** per 100
inhabitants (1993). **Books:** 83 new titles per 1,000,000
inhabitants in 1995.

ECONOMY

Per capita GNP: $140 (1994). **Annual growth:** 0.80%
(1985-94). **Annual inflation:** 33.30% (1984-94). **Consumer
price index:** 100 in 1990; 254.1 in 1992. **Currency:** 523
shillings = 1$ (1994). **Cereal imports:** 215,000 metric tons
(1993). **Fertilizer use:** 137 kgs per ha. (1992-93).
Imports: $1,505 million (1994). **Exports:** $519 million
(1994). **External debt:** $7,441 million (1994), $258 per
capita (1994). **Debt service:** 20.5% of exports (1994).
Development aid received: $949 million (1993); $34 per
capita; 40% of GNP.

ENERGY

Consumption: 34 kgs of Oil Equivalent per capita yearly
(1994), 83% imported (1994).

PROFILE

ENVIRONMENT

The country is made up of the former territory of
Tanganyika plus the islands of Zanzibar and Pemba. The
offshore islands are made of coral. The coastal belt, where
a large part of the population live, is a flat lowland along
the Indian Ocean with a tropical climate and heavy rainfall.
To the west lies the central plateau, dry and riddled with
tse-tse flies. The north is a mountainous region with slopes
suited for agriculture. Around Lake Victoria, a heavily
populated area, there are irrigated farmlands. Large
plantations of sisal and sugar cane stretch along the
coastal lowlands. Mount Kilimanjaro, the highest peak in
Africa at 6,000 m is located in the northern highlands. The
need to increase exports has led to intensification of
agricultural production in semi-arid areas, leading, in turn,
to further soil erosion. Indiscriminate cutting of forests has
continued.

SOCIETY

Peoples: Most Tanzanians are of Bantu origin, split into
around 120 ethnic groups. In the western region of the
mainland there are Nilo-Hamites; in Zanzibar there is a
Shirazi minority, of Persian origin. In both regions there are
small groups of Arabs, Asians and Europeans.
Religions: Muslim, 35%; animist, 35%; Christian, 30%.
Languages: Swahili and English are the official ones.
Several local languages are spoken. **Political
Parties:** Revolutionary Party of Tanzania (CCM-Chama Cha
Mapinduzi); Party for Democracy and Progress (CHADEMA);
Civic United Front (CUF); National Convention for
Construction and Reform-Mageuzi (NCCR-Mageuzi);
Movement for a Democratic Alternative (MDA), of
Zanzibar. **Social Organizations:** Union of Tanzanian
Workers (JUWATA), Youth League (VIJANZ), Union of
Women of Tanzania (UWT), Co-operative Union of Tanzania
(WASHIRICA).

THE STATE

Official Name: Jamhuri ya Muungano wa Tanzania.
Administrative Divisions: 25 Divisions. **Capital:** Dodoma,
203,000 inhab. (1988). **Other cities:** Dar es Salaam,
1,360,000 inhab.; Mwanza, 223,000 inhab.; Tanga, 187,000
inhab.; Zanzibar, 158,000 inhab. (1988).
Government: Benjamin Mkapa, President. Frederick
Sumaye, Prime Minister. Unicameral Legislature: 291-
member Assembly, 216 elected by direct popular vote.
National Holiday: April 26, Union Day (1964). **Armed
Forces:** 34,600 (1995). **Paramilitaries:** 1,400 Rural Police,
85,000 Militia.

lations which had broken down in
1977 when the East African Economic Community of Tanzania,
Kenya and Uganda was dissolved.
[18] After his death, his sucessor Salim
Ahmed Salim from Zanzibar continued
along the same lines.
[19] A national debate was organized to discuss constitutional reform to reorganize the executive
power, grant greater political participation to women, strengthen

democracy along CCM lines and
prohibit more than two successive
presidential re-elections.
[20] On November 5 1985, after 24
years as head of state, President
Julius Nyerere passed power on to
Ali Hassan Mwinyi, winner of the
October 27 election, gaining 92.2
per cent of the vote. Nyerere took
on a new post, presiding over the
South-South Commission, an organization based in Geneva, working

to strengthen Third World unity and
its negotiating capacity with the
North.
[21] In 1986, an economic recovery
plan went into effect. It was designed by the government of Ali
Hassan Mwinyi, following IMF
and World Bank guidelines. The
measures emphasized the reduction
of tariff barriers which stifle imports and included incentives for
private capital.
[22] Agricultural production improved and some industrial enterprises increased their profits.
[23] In the meantime, the crisis surrounding the ujamaa model intensified, partly due to its poor results
and partly to the growing resistance
of the population towards the resettlement of entire villages, especially when carried out by force. The
recovery of the economy as a whole
now depended on the credit promised by international agencies,
which would only be granted when
the required structural modifications have been introduced.
[24] The development of private
capital and incentives to create such
capital have caused new problems.
According to UNICEF studies, half
the children in the country are malnourished. Tanzania remained
among the 30 poorest countries in
the world, although it has managed

to escape the famine which has hit
other Central African nations.
[25] Agricultural tasks have traditionally been carried out chiefly by
women. While nearly half the work
force in the economy as a whole is
made up of women, 85 per cent of
all agricultural work is performed
by women. In the outlying areas
around the major cities, the female
population is increasingly opting
for work in the "underground"
economy.
[26] Economic data for 1989 revealed that the agricultural sector
accounts for 51 per cent of the GDP,
industry 10.2 per cent, trade 13.3
per cent and services 25 per cent.
With only 10 per cent of the employed work force, the latter is the
most competitive sector; agriculture, in turn, is the least capital-intensive but the most labor-intensive, with 85.6 per cent of the total
work force.
[27] At the beginning of 1990,
former president Julius Nyerere
said that he was not opposed to a
multi-party democracy in his country. According to Nyerere, the absence of an opposition party contributed to the fact that the CCM
had abandoned its program and its
commitments, straying from its
commitment to the masses.
[28] In February 1991, under the

External Debt	Illiteracy	Maternal Mortality
1994	1995	1989-95

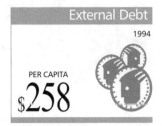

PER CAPITA
$258

32%

PER 100,000
LIVE BIRTHS
748

The Zandj culture

The eastern coast of Africa, south of the "horn" of Somalia, was known to Mediterranean navigators in ancient times but it was only between the 4th and 5th centuries AD that trade began to develop in the area. In the 7th century, this activity became extremely important when links were established with Arabia, and then with Persia.

[2] For political or religious reasons, small groups of Arabian dissidents subsequently left their homeland to settle in eastern Africa: in 695, Prince Hamza of Oman went to Zanzibar with a handful of followers. In 740, fugitives from Mecca founded Maqdishu (Mogadishu); in 834, the survivors of a frustrated rebellion in Basrah became prosperous pirates with headquarters in Socotra; and in 920, a group of Omanians conquered Mogadishu, driving its founders into the interior where they created trade caravans. Around 975, Ali bin Sultan al Hassan, prince of Shiraz was driven out of his country. When he reached Africa with his family he built the ports of Kilwa, Pemba, Manisa (later, Mombasa) and Sofala (near present-day Beira in Mozambique). His descendants, and by extension, the whole mestizo population of the coast, called themselves the "Shirazis", a generic term still in use today.

[3] The Arabs called the whole region "the country of the Zandj" or the country of the Blacks (Zandj- bar, meaning "coast of the Blacks") since the small white contingent soon merged totally with the Somalians and Bantus of the coast. Their contribution was formative, both economically and culturally. They brought a form of writing which supplemented the Swahili language (the most widespread of the Bantu languages), supplying an essential element for forming cultural unity from Mogadishu to Sofala. This gave the Zandj access to Arabian civilization and to the appropriate markets for their products.

[4] Direct trade was established, first with Arabia and Persia, and then with India, Siam (Thailand) and even China. In 1415, an official mission from China arrived in Zandj, escorted by a fleet headed by the first admiral of the Ming Empire. The main African export products included excellent steel from Malindi and Manisa later used in Syria (or India) to manufacture the famous swords of Damascus, and ivory of a quality that surpassed Indian ivory. They also exported furs from the savannas, gold from Zimbabwe (through Sofala) and, of course, slaves. All these were in high demand internationally.

[5] In exchange, the Zandj received cloth, books, jewels, pearls and porcelain; even today, archaeologists are amazed at the enormous amount of Chinese pottery found in the region.

[6] This chain of city-ports, though independent, maintained intense exchange and close cooperation between themselves. With houses built of stone or coral blocks, in the Arabian style, and large public spaces where poets and minstrels recited epics and poems in front of large audiences, these cities contributed to building a common culture. Though heavily influenced by Arabian culture, the newborn nation had genuine local roots which was perfectly evident at the most sophisticated levels of artistic and intellectual creation.

[7] In 1498, when the Portuguese reached Zandj on their way to India, they were impressed by the size and cleanliness of the cities, the quality of the houses, the luxurious good taste with which they were decorated and the beauty and elegance of the women who were active participants in society.

[8] However, since the main goal of the Portuguese was trading with India and later establishing a trade monopoly, they viewed Zandj cities as a frightening source of competition that had to be destroyed. In 1500 they attacked and ruined Mozambique, continuing their devastating work with such application that, within 50 years, they had destroyed all the cities on the eastern coast. Their target was to transfer all this thriving business to their own trading posts. Not only did they fail in their efforts, but their action caused long-lasting economic and cultural damage to all the peoples involved.

---

leadership of Abdullah Fundikira, a commission was formed to oversee the country's transition period.

[29] In its 1991 report, Amnesty International disclosed the existence of 120 prisoners on the island of Zanzibar, of which at least 40 are presumed to be political prisoners. Mwinyi's government denied that any political arrests had been made, and invited the human rights organization to prove its charges.

[30] In March 1991, the Tanzanian Workers' Organization (JUWATA) broke its ties with the CCM.

[31] In December, after a 23-year exile in England, opposition leader Oscar Kambona announced his plan to return to Tanzania and lead the fight for a multi-party system.

[32] The opposition leader announced the founding of the Democratic Alliance of Tanzania party. However, the national elections held in April 1993 confirmed yet again the predominance of the Chama cha Mapinduzi, the party in power, which obtained 89 per cent of the vote.

[33] The government pledged to the IMF it would implement a harsh program of economic adjustment which, among other measures, included the elimination of 20,000 public jobs and a reduction of the budget deficit. It thus reduced the education budget which went from 30 per cent in 1960 to 5 per cent of total public expenditure and in February 1994, authorized a 68 per cent raise in electricity bills and 233 per cent increase of several local taxes.

[34] In March, the World Bank praised Tanzania considering it the second-best African "student" after Ghana. The country's harsh social conditions were aggravated by a massive influx of Rwandan refugees, fleeing from massacres which killed over 500,000 in the neighboring state.

[35] 1995 was characterized by multi-party legislative and presidential elections held in October, won by the CCM once again due to Nyerere's support. Benjamin Mkapa became the new president in October and appointed Frederick Sumaye as prime minister.

Thailand

Prathet Thai

Population: 58,024,000
(1994)
Area: 513,120 SQ.KM
Capital: Bangkok
(Krung Thep)

Currency: Baht
Language: Thai or Siamese

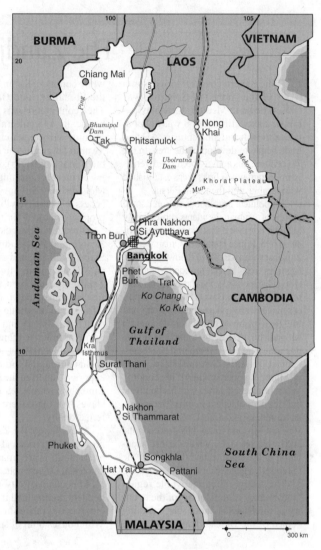

The Thais were originally from Central Asia, but the 6th century saw a migration south. In the 13th century, they founded the state of Siam, with its capital in Ayuthia. For 400 years they fought against the neighboring Khmers and Burmese before finally consolidating their present-day borders. In 1782, Rama I founded the dynasty that still rules the country, with its capital in Bangkok (Krung Thep).

[2] Throughout the 19th century, Siam was disputed by the French and the British. In 1896, the two powers agreed to leave the state formally independent, but continued to compete for control over flourishing agricultural resources. In World War I, Siam fought alongside the Allies and later joined the League of Nations.

[3] On June 24 1932, a coup imposed limits on the power of the monarchy, creating a parliament elected by universal suffrage. The democratic experience was short-lived and in 1941, during World War II, the Bangkok government allowed Japan to use its territory, becoming a virtual satellite of that country. During the conflict, Siamese troops occupied part of Malaysia, but was forced to abandon it in 1946, after the allied victory.

[4] In June of that year, King Ananda Mahidol was assassinated in unclarified circumstances. Maneuvers by the US were successful in putting his brother, Rama IX, into power. Born in the United States, the new king had never hidden his pro-US leanings. Since then, Thailand has remained under Washington's tutelage.

[5] US interests in Thailand are essentially strategic as its geographical location and plans for a projected channel through the Kra Isthmus made the country critical to US military policy in the area. In 1954, the Southeast Asian Treaty Organization (SEATO), a military pact designed to counterbalance the growing power of revolutionary forces in the region, established its headquarters in Bangkok.

[6] In 1961, large numbers of US troops entered the country in reaction to an insurrection in Laos. The US military maintained its presence in Thailand for 14 years, resulting in strong ties being forged between the Thai and US armed forces. In exchange for their participation in the anti-communist struggle, the Thai military enjoyed greater political influence and impunity from their corrupt activities, which included control of the drug trade from the famous "golden triangle" in the north.

[7] Between 1950 and 1975, the United States support of military regimes in Thailand cost them over $2 billion. On October 14 1973, a popular uprising by students brought down the government, and in 1975 as a consequence of this the first civilian government in 20 years was formed. Elections brought Prince Seni Pramoj to power, and he demanded the immediate withdrawal of US troops, dismantling of military bases and an improvement of relations with neighboring revolutionary governments.

[8] The domestic military disagreed with the new government, and in October 1976, Pramoj was overthrown in a bloody coup plotted by right-wing navy officers. Thousands of students and intellectuals joined a guerrilla struggle led by the Communist Party in the rural areas, and Thai relations with neighbors became extremely tense. Finally, in October 1977, a second coup brought the "civilized right wing" of the military to power, taking a liberal line in policy-making, as they were eager to attract new transnational investments. In 1979, Thailand granted asylum to Cambodian refugees. Refugee camps along the border became the rearguard of Cambodian strong man Pol Pot's Khmer Rouge guerrillas; a large part of international humanitarian aid was unwittingly channelled to them. Thailand thus served US interests in the region, as a base for new attacks against Vietnam.

[9] On April 1 1981, another military coup shook Bangkok, this time led by General Sant Chitpatima with the support of young middle-ranking officers demanding institutional democracy and social change. The king and prime minister General Prem had been asked to lead the coup and had apparently accepted. However, they finally opposed the uprising, and it was put down after three days of great tension. The young officers' revolt,

PUBLIC EXPENDITURES

% of GNP — 1992

- MILITARY 2.7
- EDUCATION 3.8
- HEALTH 1.1

WORKERS

2.7%

1994 — UNEMPLOYMENT

- MEN 56%
- WOMEN 44%
- AGRICULTURE 67%
- INDUSTRY 11%
- SERVICES 22%

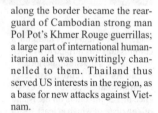

LAND USE

- CROPLAND 41%
- PASTURE 2%
- OTHER 58%

Food Dependency
1970
5%
1992
6%

Foreign Trade
MILLIONS $ 1994
IMPORTS
54,459
EXPORTS
45,262

rooted within the military establishment, was basically a reaction to government measures bringing in forced retirement for certain senior generals. Nevertheless, the short-lived coup gained unusually strong support from trade unions and student groups, an extraordinary state of affairs, coming as it did from the Thai military.

[10] At the same time, the left-wing opposition was severely weakened by an internal split in the Communist Party into the pro-Vietnamese and pro-Chinese factions.

[11] On April 20, 1983, elections were held and General Prem Tinsulanonda was appointed prime minister for a second successive term.

[12] Toward the end of 1984, the currency was sharply devalued, sparking discontent from General Arthit Kamlang-Ek and other hard-line generals, who threatened to withdraw military support for Prem. Some timely political maneuvering by the prime minister, offering incentives to the officers for their support, allowed Prem to isolate Arthit. After a further coup attempt in September 1985, Prem relieved Arthit of his command and designated General Chaovalit Yongchaiyut as the head of the armed forces.

[13] Political instability continued in Bangkok, with several cabinet changes and requests for early legislative elections on two separate occasions. In May 1986, the Democrat Party obtained enough of a majority to form a coalition, presided over once again by General Prem as prime minister. While General Chaovalit's stature continued to grow, encouraged by his harsh attacks on corruption, the government called for early elections in April 1988, to avoid the censure of Prem. The main criticisms of his administration were his questionable management of public funds and overall incompetence, particularly in the handling of the border war with Laos several months earlier, which had escalated from a dispute over the control of three villages.

[14] In the first general election since 1976, amid massive vote-buying campaigns (a practice considered normal by almost all candidates), the Thai Nation Party won the election. King Bhumibol Adulyahed asked General Chatichai Choonhavan to form a new government, which he did by arranging a six-party coalition.

[15] Thailand's new leading force and a successful businessman, Chatichai introduced important policy changes, breaking with the traditional focus on internal affairs and security. His main idea was to convert what had once been the Indochinese battlefield into a huge regional market. Tempted by the possibility of Cambodia and Laos opening up their markets, he invested considerably in those countries. This formula further fuelled the Thai economy, and in 1989 the growth rate exceeded 10 per cent. Thailand then challenged Western Europe and its subsidy policies for agricultural export products. There were also disagreements with the United States over Bangkok's refusal to accept North American trade criteria on intellectual property, especially with regard to computer programs.

[16] Chatichai's government made foreign policy one of its main priorities, while failing to pay sufficient attention to serious domestic problems such as the deterioration and pollution of the environment, and the extreme poverty of the majority of the population. In 1989, rapid deforestation contributed to serious flooding, as the regulatory effect of the forests on the water cycle was altered and trees had been removed from river banks, where they had formed natural dikes. To avoid more serious damage in the future, the government was forced to prohibit further logging. Peasants have also resisted the massive eucalyptus plantation projects promoted by the government and the World Bank. The natural forests have been transformed into private commercial plantations for urban industrial use without considering the local people and their traditional use of forest products.

[17] Charges of corruption and nepotism levelled against the Chatichai government grew as quickly as the economy. General Chaovalit, a political "rising star," was named minister of defence in 1990, in order to lend an image of honesty to the government.

[18] In March 1991, the military carried out another coup led by General Sunthorn Kongsompong, who presented King Bhumibol Adulyahed with a draft for a new constitution. The latter approved the draft, and justified the military coup on the grounds of "growing corruption" within the civilian government. The king also agreed with the

ENVIRONMENT

Thailand is located in central Indochina. From the mountain ranges in the northern and western zones, the Me Ping and Menan Rivers flow down to the central valley through extensive deltas, into the Gulf of Siam. The plains are fertile with large commercial rice plantations. The southern region occupies part of the Malay Peninsula. Severe deforestation of the area resulted in decreased production of rubber and timber, and has been responsible for migration of part of the native population. There has been an increase in air pollution in urban areas due to gas emissions, particularly from industry.

SOCIETY

Peoples: The Thai group constitutes the majority of the population. The most important minority groups are the Chinese, 12% and, in the south, the Malay, 13%. Other groups are Khmer, Karen, Indians and Vietnamese. **Religions:** Most people (94%) practise Buddhism. Muslims, concentrated in the south, make up about 4% of the population. There is a Christian minority. **Languages:** Thai or Siamese (official). Minority groups speak their own languages. **Political Parties:** Thai Nation Party, in power, formed a coalition with Chart Thai Party, New Aspiration Party, Righteous Force Party, and smaller parties after elections in July 1995. Democratic Party, liberal, opposition.

THE STATE

Official Name: Muang Thai, or Prathet Thai.
Capital: Bangkok (Krung Thep) 5,620,591 inhab. (1991).
Other cities: Nonthaburi 264,201 inhab.; Ratchasima 202,503 inhab.; Chiang Mai 161,541 inhab.; Khon Kaen 131,478 inhab.; Nakhon Pathom 37.200 (1991).
Government: Bhumibol Adulyahed, king since 1946. Banharn Silapa-archa, Prime Minister since July 13 1995.
National Holiday: December 5, the King's birthday (1927).
Administrative Divisions: 5 Regions and 73 Provinces.
Armed Forces: 256,000 (1994). **Paramilitaries:** 141,700 National Security Volunteer Force, Police and "Hunter Soldiers".

military on the need for calling new elections. As an indirect result of the military coup in Thailand, peace negotiations in neighboring Cambodia came to a standstill, with the Cambodian government denouncing Thailand for its renewed support of the Cambodian armed opposition.

[19] Throughout 1991, Thailand remained under the command of the Council for the Maintenance of National Peace (NPKC), a body of the military commanded by Sunthorn. In December, the king approved the new constitution, which stipulated that elections to replace the NPKC government would be held within 120 days. However, the military junta reserved the right to directly designate 270 senators, out of a total of 360, giving them total control over the new government.

[20] In elections held on March 22 1992, the majority of votes went to the opposition. There were 15 parties, with a total of 2,740 candidates, and 32 of 57 million Thais turned out to vote. In early April, General Suchinda Kraprayoon, commander in chief of the army at the time, became prime minister, backed by a small majority made up of five pro-military parties. On his 49-member cabinet, Suchinda included 11 former ministers who had been accused of embezzlement during Chatichai's government. A few weeks later, 50,000 people attended a demonstration held by four opposition parties, demanding that the head of the government

DEMOGRAPHY

Urban: 19% (1995). **Annual growth:** 1.6% (1992-2000). **Estimate for year 2000:** 65,000,000. **Children per woman:** 2.2 (1992).

HEALTH

One **physician** for every 4,762 inhab. (1988-91). **Under-five mortality:** 32 per 1,000 (1994). **Calorie consumption:** 94% of required intake (1995). **Safe water:** 86% of the population has access (1990-95).

EDUCATION

Illiteracy: 6% (1995). **School enrolment:** Primary (1993): 97% fem., 97% male. Secondary (1993): 37% fem., 38% male. University: 19% (1993). **Primary school teachers:** one for every 17 students (1992).

COMMUNICATIONS

102 **newspapers** (1995), 101 **TV sets** (1995) and 95 **radio sets** per 1,000 households (1995). 3.7 **telephones** per 100 inhabitants (1993). **Books:** 101 new titles per 1,000,000 inhabitants in 1995.

ECONOMY

Per capita GNP: $2,410 (1994). **Annual growth:** 8.60% (1985-94). **Annual inflation:** 5.00% (1984-94). **Consumer price index:** 100 in 1990; 119.5 in 1994. **Currency:** 25 baht = 1$ (1994). **Cereal imports:** 638,000 metric tons (1993). **Fertilizer use:** 544 kgs per ha. (1992-93). **Imports:** $54,459 million (1994). **Exports:** $45,262 million (1994). **External debt:** $60,991 million (1994), $1,051 per capita (1994). **Debt service:** 16.3% of exports (1994). **Development aid received:** $614 million (1993); $11 per capita; 0.5% of GNP.

ENERGY

Consumption: 770 kgs of Oil Equivalent per capita yearly (1994), 59% imported (1994).

Illiteracy 1995

6%

Maternal Mortality 1989-95

PER 100.000 LIVE BIRTH

155

---

resign. In the meantime, political leader Chamlong Srimuang and 42 militants from opposition parties began a hunger strike.

[21] At the end of May, what began as an anti-government demonstration ended in a massacre, with scores of people killed and hundreds injured. Army troops fired into a crowd gathered at the monument to democracy. The protests continued until an unexpected television appeal for national reconciliation from the Thai king. Suchinda, in the meantime, announced his support of a constitutional amendment whereby the prime minister would have to be an elected member of Parliament, a clause which would disqualify even Suchinda himself from holding the post. Srimuang was subsequently freed, and an amnesty announced for those who had been arrested. With a curfew still in effect in Bangkok, Parliament initiated discussion of the constitutional amendment and the king appointed General Pren Tinsulanonda to supervise the process.

[22] On May 24, Suchinda resigned and his vice president, Mitchai Ruchuphan, became interim president. Constitutional amendments approved in June reduced military participation in the government. In the first week of

June, after the amendment had been passed by Parliament, King Bhumibol named Anand Panyarachun prime minister. Panyarachun, who enjoyed a great deal of prestige in Thailand, had been prime minister after the 1991 military coup.

[23] Parliament accepted the king's nomination of Panyarachun, who named a number of technocrats to cabinet positions and requested the resignations of the twelve military officers responsible for the May massacre.

[24] On June 29, he dissolved parliament and announced elections for September. The Democratic Party (DP) won the election, obtaining 79 seats in the Chamber of Representatives. On September 23, a parliamentary majority made up of the DP, the Palang Dharma (PD) and the Party of New Aspirations emerged, holding 177 seats, and anti-military political leader Chuan Leekpai was elected prime minister. Shortly afterwards, the pro-military Social Action Party joined the coalition.

[25] Prime Minister Chuan Leekpai announced an economic program on January 19 1993, which aimed at winning back the foreign investment which had been scared off by political instability. Thailand's

main attraction is that it has the cheapest workforce in the region.

[26] Finance Minister Tarrin Nimmanahaeminda's plan included financial incentives to attract migration from other parts of the country to rural areas. In addition, it was designed to stimulate a return to the provinces by the urban unemployed, in an effort to check Bangkok's spiraling demographic growth. The 1993 budget deficit ($22 billion), equal to 0.9 per cent of the GDP, was justified by the government by the need to invest in infrastructure, given its goal of transforming Thailand into a "bridge" for companies investing in India or China.

[27] Nevertheless, many obstacles remain in Thailand, among the most important is that 15 per cent of the population lives in dire poverty. Also of major concern is the proliferation of AIDS in a country which has been a center for prostitution for many years. The opposition accused Prime Minister Chuan Leekpai of being slow and inefficient at a time when the country's problems demand urgent attention. Between June 10 and 12, 1993, a motion to censure Leekpai was put before parliament; his coalition managed to defeat the motion by 204 votes to 153.

[28] Although Leekpai's personal honesty was not called into question, some members of his cabinet were linked to illegal behaviour. The most upsetting element for the opposition was the interruption of the case against Montri Pongpanich. As a leader of the Social Action Party and a member of the government coalition, Pongpanich possessed legal immunity.

[29] Several violent incidents motivated by religious sensitivities took place in 1993. Concerned with his country's role in the region, Chuan reinforced ties with China and Indonesia and in October officially terminated the support which Thailand had provided to Cambodia's Khmer Rouge since 1979. Cambodian authorities accused Thailand of supporting the "Pol Potists" after 14 Thais were arrested during a thwarted coup attempt in Phnom

Penh. Toward the end of the year, Thailand and Brazil were involved in intense diplomatic negotiations to extradite Brazilian financier Paulo César Farias, arrested in Bangkok in November 1993. After the extradition, Farias was killed in mid-1996, in a Brazilian seaside resort.

[30] The government approved a reform to liberalize the banking system in January 1994. Among other things, the reform permitted the opening of branches of foreign banks in various Thai cities.

[31] A proposal for a constitutional reform generated a long debate in Parliament. The new project envisaged a reduction of the size and power of the Senate and lowered the voting age to 18 years. The subject remained unsettled due to mutual accusations by the different parties and the parliamentary changes which took place in 1995.

[32] Although in 1994 the economy of Thailand was among those that grew most, its economic stability did not reach the political sphere. The first parliamentary session of 1995 brought an exchange of accusations and the tense political atmosphere led to the dissolution of Parliament. On July 2 early elections took place. The Thai Nation Party obtained 25 per cent of the seats and the Democrats 22 per cent. Banham Silapa-archa, leader of the winning party, formed a government supported by several small parties.

[33] The cabinet designated by Silapa-archa was severely criticized by the press, since some of its members had been investigated for their links with the state coup of 1991. The prime minister's response to the military insinuations of a coup was made public through a radio program. He said that coups were "obsolete".

[34] In 1995, inflation reached 6.5 per cent. It was considered "manageable" by the Central Bank. With an annual population growth of 1.5 per cent and a growth in the economy of over 8 per cent, Thailand remained on the list of countries with sustained economic success.

Togo

Togo

Togo was peopled by the Ewe (of the same origin as the Ibo and Yoruba of Nigeria and the Ashanti of Ghana), who still make up a third of its population. The Ewe were a relatively poor and peaceful people with simple social structures like the Dagomba kingdoms in the north.

2 The Tvast slave trade gave the region the name of Slave Coast, and millions of human beings were shipped to slavery in the US between the 16th and 19th centuries.

3 Togo became a German colony in 1884 under "treaties" signed with local chieftains. Occupied by Anglo-French troops during World War I, the territory was divided between the two powers with the endorsement of the League of Nations. The western part was annexed to Ghana in 1956, after a British organized plebiscite, while the eastern part remained under French rule.

4 In 1958, Sylvanus Olympio, (a member of the Togolese Unity party) won the elections. His program was pro-independence, but moderate. The 1960 proclamation of independence seemed to overlook a contract signed in 1957 with the Benin Mines Company, in which the French consortium had seized control of Togo's phosphate reserves, its main natural resource, thus compromising independence. When Olympio confronted this situation in 1963, proposing a series of basic reforms - doubtlessly influenced by Nkrumah's radical pro-independence program in neighboring Ghana - he was assassinated in a military revolt, allegedly involving a young military officer, Eyadéma.

5 Olympio's successor, the neo-colonialist Nicholas Grunitzky, from the opposition Togolese Progress Party, was overthrown by a further coup, led by General Gnassingbé Eyadéma, who became the new head of state.

6 In 1969, Eyadéma's Togo Peoples' Group (RPT) brought in a one-party system. In 1972 a state law gave the state 35 per cent of shares of the mining company, which was raised to 51 per cent in 1975 and in 1976 the production and export of phosphate was nationalized.

7 On December 30 1979, a Constitution was approved and Eyadéma was re-elected president for a seven-year term. His government was confident of improving the economy, basing its optimism on tourism, oil and a rise in world phosphate prices. However in 1981 phosphate prices dropped by 50 per cent and a worldwide economic recession reduced the number of European tourists. A serious balance of payments deficit increased the foreign debt to $1 billion.

8 In June 1984 a refinancing agreement with stringent conditions, was reached with the Paris Club and the IMF. Consequently, there were salary freezes, reduction of government investments and high taxes levied to increase government funds, including a so-called "solidarity tax" which took 5 per cent of the population's incomes.

9 In January 1985 the Lomé III agreements were signed in Togo, regulating cooperation between ACP (Africa, Caribbean and Pacific) countries and the EEC.

10 In January 1986 Eyadéma was re-elected for 7 years, with 99.95 per cent of the vote. Consultation with the IMF resumed and new agreements were planned in 1988. In 1990, adjustment programs were initiated, including denationalization, on account of trade deficits and a public deficit of $1.27 billion. Meanwhile food production rose slightly.

11 In 1991, more than 10,000 peasants in the northern district of Keran Oti lost their lands to create an 80 sq km game reserve for hunting.

12 In early April 1991, there were protest demonstrations throughout the country, and on April 8 more than a thousand people took over one of Lomé's neighborhoods, setting up barricades and demanding Eyadéma's resignation. The police reacted violently, killing more than thirty people and injuring many more. However on April

> The creation of a 80 sq km game reserve for hunting in the northern district of Keran Oti in 1991, left more than 10,000 peasants without land.

Population: 4,007,000 (1994)
Area: 56,790 SQ. KM
Capital: Lomé
Currency: CFA Franc
Language: French

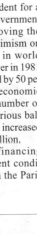

PUBLIC EXPENDITURES

% of GNP   1992

- MILITARY: 3.1
- EDUCATION: 5.7
- HEALTH: 2.5

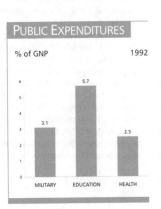

WORKERS

1994

- MEN 65%   WOMEN 35%
- AGRICULTURE 65%   INDUSTRY 6%   SERVICES 29%

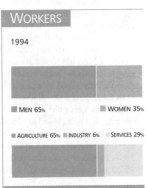

LAND USE

- CROPLAND 45%   PASTURE 4%   OTHER 52%

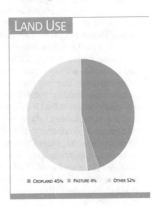

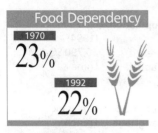

Food Dependency
1970
23%
1992
22%

Illiteracy
1995
48%

Maternal Mortality
1989-95
PER 100,000
LIVE BIRTHS
626

External Debt
1994
PER CAPITA
$363

PROFILE

ENVIRONMENT

The country is a long, narrow strip of land with distinct geographical regions. In the South, a low coastline with lakes, typical of the Gulf of Guinea; a highly populated plain where manioc, corn, banana and palm oil are grown. In the North, subsistence crops are gradually giving way to coffee and cocoa plantations. The Togo Mountains run through the country from Northeast to Southwest.

SOCIETY

Peoples: The prevailing ethnic groups are the Ewe, Kabye and Mina. The descendants of formerly enslaved Africans who returned to Togo from Brazil are called Brazilians. They form a caste with great economic and political influence. The small European minority is concentrated in the capital. **Religions:** The majority profess traditional African religions. There are Christian and Muslim minorities. **Languages:** French (official). The main local languages are Ewe, Kabye, Twi and Hausa. **Political Parties:** Rally of the Togolese People; Union for Justice and Democracy; Action Committee for Renewal. **Social Organizations:** National Confederation of Togolese Workers (CNT).

THE STATE

Official Name: République Togolaise. **Administrative Divisions:** 5 Regions and 21 Prefectures. **Capital:** Lomé, 600,000 inhab. (1989). **Other cities:** Sokodé, 48,098 inhab.; Kpalimé, 27,700 inhab. (1983). **Government:** Gnassingbé Eyadéma, President since January 1967; Kwassi Klutse, Prime Minister since 1996. Unicameral 79-member Legislative Assembly. **National Holiday:** April 27, Independence Day (1960). **Armed Forces:** 6,950 (1995). **Paramilitaries:** 750 Gendarmes.

STATISTICS

DEMOGRAPHY

Urban: 30% (1995). **Annual growth:** 1.6% (1992-2000). **Estimate for year 2000:** 5,000,000. **Children** per **woman:** 6.5 (1992).

HEALTH

One **physician** for every 11,000 inhab. (1988-91). **Under-five mortality:** 132 per 1,000 (1994). **Calorie consumption:** 91% of required intake (1995). **Safe water:** 63% of the population has access (1990-95).

EDUCATION

Illiteracy: 48% (1995). **School enrolment:** Primary (1993): 81% fem., 81% male. Secondary (1993): 12% fem., 34% male. University: 3% (1993). **Primary school teachers:** one for every 59 students (1992).

COMMUNICATIONS

82 **newspapers** (1995), 83 **TV sets** (1995) and 98 **radio sets** per 1,000 householdss (1995). 0.4 **telephones** per 100 inhabitants (1993).

ECONOMY

Per **capita GNP:** $320 (1994). **Annual growth:** -2.70% (1985-94). **Annual inflation:** 3.30% (1984-94). **Consumer price index:** 100 in 1990; 100.5 in 1991. **Currency:** 535 CFA francs = 1$ (1994). **Cereal imports:** 63,000 metric tons (1993). **Fertilizer use:** 183 kgs per ha. (1992-93). **External debt:** $1,455 million (1994), $363 per capita (1994). **Debt service:** 7.8% of exports (1994). **Development aid received:** $101 million (1993); $26 per capita; 8% of GNP.

ENERGY

Consumption: 46 kgs of Oil Equivalent per capita yearly (1994), 100% imported (1994).

12, opposition parties were legalized and Eyadéma announced an amnesty for political prisoners, and liberalization of the political system.

[13] On August 28 1991, a National Conference named Kokou Koffigoh - president of Togo's Bar Association and a recognized human rights leader - provisional prime minister. A legislative assembly was established which ousted Eyadéma as head of the armed forces and blocked his candidacy in the upcoming 1992 elections.

[14] The dissolution of the RPT on November 26 1991, caused a military coup. Two days later, the Armed Forces took over the government, dissolved the legislative assembly and kidnapped Koffigoh for several hours. The perpetrators of the coup claimed that the legislative assembly was made up of extremists, whose aim was to "disrupt the army and discredit the military regime and its achievements".

Political parties and other opposition organizations cooperated together within the framework of the National Resistance Committee, and encouraged people to resist a return to dictatorship.

[15] Togo's complicated political scenario is compounded by economic, social and tribal problems. Eyadéma belongs to the southern Kabye tribe, which provides the majority of the country's 12,000 troops. Meanwhile, Koffigoh is a member of the Ewe tribe. Negotiations have begun between those who favor the military regime and the opposition. The negotiations hoped to precipitate the transition to civilian government by 1992. That year, a timetable was established for a return to democracy, and all citizens were re-registered to vote.

[16] In June, opposition leader Tavio Komlavi was assassinated. After an attempt upon the life of Mining Minister Joseph Yao Amefia, Eyadéma took

advantage of the incident to have his powers restored by the High Council of the Republic (the interim legislature). Eyadéma proceeded to pack the cabinet with his supporters. In the face of widespread protests from the opposition, the government postponed elections indefinitely. The unions, in reprisal, called a long-term general strike.

[17] The opposition split from Koffigoh in January 1993, after which they met in Benin and nominated a government in exile. In this same month the presidential guard killed around a hundred demonstrators in Lomé, which led to thousands of people fleeing to Ghana and Benin.

[18] Eyadéma agreed to organize elections for August. In an atmosphere of civil war, he won with 96.5 per cent of the vote, in a poll classed as "fraudulent" by the opposition. The protests in the streets intensified, and in January 1994, Eyadéma was unharmed in a foiled assassination attempt - carried

out by around a hundred armed attackers - which left 67 people dead.

[19] The opposition won the February legislative elections, but Eyadéma did not allow one of its leaders, Edem Kodjo, to form a government without members of the ruling parties. The Action Committee for Renewal, which included the opposition groups, decided to boycott Parliament, paralyzing government for a large part of the year.

[20] On another front, widespread floods in Lomé left 150,000 people homeless in July. In September three weeks of heavy rains destroyed whole villages, roads and bridges, particularly in the northern and central regions of the nation, leaving 21,000 people destitute.

[21] In August 1996, Kodjo resigned from his post as prime minister and was replaced by Kwassi Klutse.

Tonga

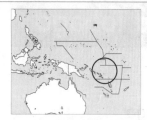

Population: 101,000 (1994)
Area: 699 SQ.KM
Capital: Nukualofa

Currency: Pa'anga
Language: Tongan and English

Tonga

The native inhabitants of Tonga, which means "south" in several Polynesian languages, immigrated from Fiji and Samoa over 1,000 years ago. Tongans developed a complex social organization, with a traditional leader in charge. According to oral tradition, the first Tui Tonga (ruler, Son of the Creator) of the islands was Aholitu in the second half of the 10th century. Towards the 15th century, religious and social roles were assigned to different leaders and this system of double leadership had continued until the Dutch came to the islands in 1616. English explorer Captain James Cook, visited them in 1775 and named them the "Friendly Islands".

[2] The archipelago only achieved political reunification in the mid-19th century, in a civil war, under the leadership of Taufa'ahau Tupou, ruler of the island of Haapai from 1820 onwards. Christianized and subsequently backed by European missionaries, Taufa'ahau Tupou seized Vavau and Tongatopu, securing control over the whole of the territory which he ruled under

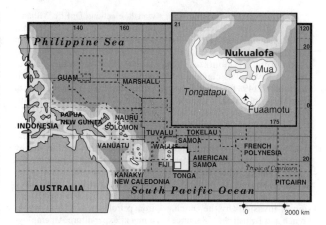

STATISTICS

ENVIRONMENT

This archipelago is also known as "Friendly Isles" and is located in western Polynesia, east of the Fiji Islands, slightly north of the Tropic of Capricorn. It comprises approximately 169 islands, only 36 of which are permanently inhabited. It includes three main groups of islands: Tongatapu, the southernmost group where more than half of the population live; Vavau, to the north, and Haapai in between. The volcanic islands are mountainous while the coral ones are flat. The climate is mild and rainy with very hot summers. The fertile soil is suitable for growing banana, copra and coconut trees.

SOCIETY

Peoples: Tongans are Polynesian people. It is estimated that 20% of them now live abroad. **Religions:** Free Wesleyan, 43%; Roman Catholic, 16%; Mormon, 12.1%; Free Church of Tonga, 11%; Church of Tonga, 7.3%; other, 10.6%. **Languages:** Tongan and English are official. **Political Parties:** Tonga Democratic Party.

THE STATE

Official Name: Pule'anga Fakatu'i 'o Tonga. **Administrative Divisions:** 23 Districts. **Capital:** Nukualofa, 21,400 inhab.(1986) **Other cities:** Neiafu 3,879 inhab.; Haveluloto 3,070 inhab.; Vaini 2,697 inhab.; Tofoa-Koloua 2,298 inhab. **Government:** Taufa'ahau Tupou IV, King; Baron Vaea, Prime Minister. **National Holiday:** June 4, Independence Day (1970).

DEMOGRAPHY

Annual growth: -0.3% (1991-99).

HEALTH

Calorie consumption: 118% of required intake (1988-90).

EDUCATION

Primary school teachers: one for every 24 students (1990).

COMMUNICATIONS

6.4 **telephones** per 100 inhabitants (1993).

ECONOMY

Per **capita GNP:** $1,590 (1994). **Annual growth:** 0.30% (1985-94). **Consumer price index:** 100 in 1990; 119.2 in 1993. **Currency:** 1 pa'anga = 1$ (1994).

the Christian name of George I until 1893. The king introduced a parliamentary system including the local leaders, and an agrarian reform which granted each adult male in the country 3.3 ha of arable land. This ensured social stability and agricultural self-sufficiency. This system is still used but since the population has grown very quickly, there are not so many fertile plots left and families have to share them.

[3] In 1889, Britain and Germany signed a treaty which gave Britain a free hand in its relations with Tonga. The following year the archipelago was made a "protectorate" of the British crown, though the monarchy was kept with limited powers.

[4] Queen Salote, great granddaughter of George I was crowned in 1918. In 1960, she gave women the right to vote in legislative elections.

[5] The British transformed the country's agriculture, which was re-oriented towards the export crops of copra and bananas.

[6] Present king Taufa'ahau Tupou IV was crowned in 1967, and three years later Tonga obtained independence. The high international market prices for copra and bananas helped Tonga to develop a wide-ranging social security system which included free education and medical services for all the inhabitants of the islands.

[7] In the mid-1970s welfare programs suffered because of the drop in fruit prices, and about 10,000 Tongans emigrated to Aotearoa/ New Zealand. Tonga's economy suffers from serious inflation, with a high level of unemployment, despite emigration, and a substantial trade deficit. It has also suffered from the whims of nature; in 1982 hurricane Isaac destroyed a number of plantations.

[8] In February 1990, the proponents of the reform were re-elected and they announced the creation of the first political party in the kingdom. The king accused the leader of the movement, Akalisi Pohiva, of Marxist tendencies.

[9] The UN questioned Tonga about the absence of trade unions in the country, in spite of there being over 3,000 civil servants. The government of Tonga explained that there were no trade unions because they were unnecessary. The UN dismissed the Tongan response.

[10] The crown prince and prime minister since 1965, Fatafehi Tu 'ipelehake, was replaced by Baron Vaea in 1991. In 1992, opposition groups founded the Pro-Democracy Movement which won 6 of the 9 seats in the National Assembly in the February 1993 elections.

[11] However, power continued to be controlled by the king, subjects named by him to high posts in the government and a small group belonging to the royalty. In September 1994, the Pro-Democracy Movement became the Tonga Democratic Party, led by 'Akilisi Pohiva.

[12] The king continued to pledge a multi-party system but did not modify his quasi-monopolistic grip on power. As tourism had a significant growth of 11.3 per cent in 1994, Tonga refused to authorize Tokyo's request to fish for whales in its territorial waters.

[13] Along with other countries from the South Pacific Forum, Nukualofa harshly denounced French nuclear testing in the Polynesian atolls of Mururoa and Fangataufa in 1995 and 1996.

Trinidad and Tobago

Population: 1,295,000 (1994)
Area: 5,130 SQ.KM
Capital: Port of Spain

Currency: Trinidad Dollar
Language: English

Trinidad and Tobago

Although Trinidad and Tobago currently form a single nation, the two islands have different histories. Trinidad, 12 km from the mouth of the Orinoco River, was claimed by Columbus in 1498 for Spain. Tobago had been inhabited by Carib Indians, but when the Dutch arrived in 1632, they found it uninhabited. Shortly thereafter, Spanish troops invaded to prevent the Dutch from using Tobago as a base for exploring the Orinoco, where there was thought to be gold.

[2] The islands suffered a succession of invasions by the Dutch, French and British. In 1783, three centuries after the arrival of the first Europeans, the local population comprised 126 people of European origin, 605 of African origin including 310 slaves, and 2,032 Amerindians. Trinidad finally became a British colony in 1802, and Tobago in 1814.

[3] The local economy was based on sugar. Once slavery was abolished in 1834, most Africans moved to the cities and were replaced on the plantations by Asian Indian workers and some Chinese. These changes defined the island's current ethnic and social composition. The Africans were mostly urban workers, while Asian Indians constituted the large rural working class. Nonetheless, some African workers remained in the countryside, where they survived by a system of mutual aid called *gayap*.

[4] In 1924, the first moves towards autonomy were made and the colonial administration was reformed, allowing elections for certain secondary positions although suffrage was limited. The newly-organized trade unions also raised the issue of independence. In 1950 internal autonomy was granted, and the People's National Movement (PNM)

won the elections. Dr Eric Williams became prime minister, a post which he held until his death in 1981. In 1962, after a brief period as part of the West Indies Federation (1958-62), Trinidad and Tobago became fully independent. On August 31, 1976, a new constitution proclaimed a republic.

[5] At the turn of the century sugar production began to decline and was gradually replaced by oil; by 1940, the latter had become the principal economic activity. Between 1972 and 1982, there was a five-fold increase in fiscal earnings from petroleum, as well as in government expenditure. Unemployment was reduced to less than 10 per cent and imports (primarily of luxury items) increased 11-fold. Trinidad had the highest per capita income in Central and South America, and began to take on the aspect of a "consumer society". In the meantime, President Eric Williams' Peoples' National Movement (PNM) government initiated nationalization of the oil industry and adapted price policies to those established by OPEC, while encouraging investment from transnational corporations.

[6] However, during the 1970s there was social unrest. In 1975, a number of strikes occurred and oil workers of African origin and Asian Indian plantation hands fought together for better conditions. The strike movement finally foundered when the PNM government called in the army and gave it the responsibility of gasoline distribution.

[7] The over-dependence on oil caused serious problems and instability from 1982. This was due to a number of factors: the international recession and the fall in oil prices, as well as factors linked specifically to the oil industry; the

decrease in the demand for heavy oil, competition from major refineries which had been installed along the southern and eastern coasts of the United States, and a downturn in production rates. At the current rate, oil reserves will be depleted by the end of the decade unless new deposits are discovered.

[8] In 1983, the economy took a downturn. A 26 per cent decrease in oil revenues was predicted, leading to an actual net loss of 5.2 per cent of the GNP. The government was able to control this situation by eliminating subsidies, a decreasing public investment and "containing" public salaries.

[9] In October 1983, the government led by George Chambers opposed the American invasion of Grenada and did not provide troops for the expeditionary force from six Caribbean countries. Chambers' government faced growing social

and political unrest. The opposition reformed as the National Alliance for Reconstruction (NAR) and won elections in December 1986, taking 33 of the 36 parliamentary seats to end 30 years of PNM government.

[10] Arthur Napoleon Robinson's new government proposed a 5-year "readjustment" plan, effective as of December 1990, designed to accelerate the process of integration with other CARICOM members (Caribbean Community and Common Market). This project entails eliminating protection for the textile industry, developing tourism and new industrial projects, such as methane and natural gas refineries.

[11] The government applied austerity measures agreed with the IMF, from whom it received $110 million in 1988, as well as a $128 million standby loan in 1989. There have been a number of strikes in the oil industry, including a general strike in March 1989.

[12] The Robinson administration committed itself to reducing inflation to 5 per cent a year, restructuring state-owned companies, limiting the number of public employees and further liberalizing the economy. To achieve these goals, the government abolished export licences, liberalized the national price system (while maintaining control over basic food products and pharmaceuticals), and decreed a 10 per cent reduction in the salaries in the public sector.

[13] On July 27 1990, Trinidad and Tobago suffered their first coup attempt since winning independence

PUBLIC EXPENDITURES

% of GNP — 1992

MILITARY	EDUCATION	HEALTH
1.3	4.1	1.7

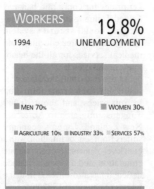

WORKERS

19.8% UNEMPLOYMENT — 1994

MEN 70% WOMEN 30%

AGRICULTURE 10% INDUSTRY 33% SERVICES 57%

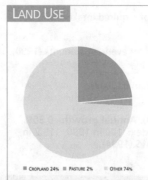

LAND USE

CROPLAND 24% PASTURE 2% OTHER 74%

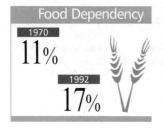

Food Dependency

1970 **11%**

1992 **17%**

Foreign Trade

MILLIONS $ 1994

IMPORTS **1,131**

EXPORTS **1,867**

Illiteracy

1995

2%

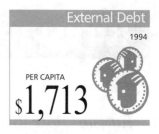

External Debt

1994

PER CAPITA

$1,713

PROFILE

ENVIRONMENT

The country is an archipelago located near the Orinoco River delta off the Venezuelan coast in the southern portion of the Lesser Antilles in the Caribbean. Trinidad, the largest island (4,828 sq km), is crossed from East to West by a mountain range which is an extension of the Andes. One-third of the island is covered with sugar and cocoa plantations. Petroleum and asphalt are also produced. Tobago, 300 sq km with a small, central, volcanic mountain range, is flanked by Little Tobago (1 sq km), the islet of Goat and the Bucco Reef. In the archipelago the prevailing climate is tropical with rains from June to December, but tempered by the sea and east trade winds. Rivers are scarce, but dense forest vegetation covers the mountains. Its geographical proximity to the main sea route linking the Caribbean to the Atlantic, has led to pollution of the islands' coasts.

SOCIETY

Peoples: There is a large minority of African origin (about 41%), and a slight majority descended from Asian Indians brought during the 19th century as contract workers (about 39%). European and Chinese groups make up a small minority. **Religions:** Mainly Christian, with a Catholic majority. About 25% are Hindu and 6% Muslim. **Languages:** English (official). Hindi, Urdu, French and Spanish. **Political Parties:** The Peoples' National Movement (PNM); National Alliance for Reconstruction (NAR); United National Congress (UNC); National Joint Action Committee (NJAC). **Social Organizations:** Trinidad and Tobago Labor Congress (TTLC); Jamaat-al-Muslimeen.

THE STATE

Official Name: Republic of Trinidad and Tobago. **Administrative Divisions:** 7 Counties, 4 Cities with own government, 1 semi-autonomous island, Tobago. **Capital:** Port of Spain, 50,900 inhab. (1990). **Other cities:** San Fernando, 30.000 inhab.; Arima, 29,700 inhab. (1990). **Government:** Noor Hassanali, President since March 1987. Basdeo Panday, Prime Minister since 1995. **National Holiday:** August 31, Independence Day (1962). **Armed Forces:** 2,600 (1994). **Paramilitaries:** 4, 800 Police.

STATISTICS

DEMOGRAPHY

Urban: 71% (1995). **Annual growth:** 0.7% (1992-2000). **Estimate for year 2000:** 1,000,000. **Children** per **woman:** 2.8 (1992).

HEALTH

One **physician** for every 1,370 inhab. (1988-91). **Under-five mortality:** 20 per 1,000 (1994). **Calorie consumption:** 106% of required intake (1995). **Safe water:** 97% of the population has access (1990-95).

EDUCATION

Illiteracy: 2% (1995). **School enrolment:** Primary (1993): 94% fem., 94% male. Secondary (1993): 78% fem., 74% male. University: 8% (1993). **Primary school teachers:** one for every 26 students (1992).

COMMUNICATIONS

103 **newspapers** (1995), 105 **TV sets** (1995) and 104 **radio sets** per 1,000 householdss (1995). 15.0 **telephones** per 100 inhabitants (1993). **Books:** 100 new titles per 1,000,000 inhabitants in 1995.

ECONOMY

Per **capita GNP:** $3,740 (1994). **Annual growth:** -2.30% (1985-94). **Annual inflation:** 6.50% (1984-94). **Consumer price index:** 100 in 1990; 122.5 in 1993. **Currency:** 6 Trinidad dollars = 1$ (1994). **Cereal imports:** 232,000 metric tons (1993). **Fertilizer use:** 801 kgs per ha. (1992-93). **Imports:** $1,131 million (1994). **Exports:** $1,867 million (1994). **External debt:** $2,218 million (1994), $1,713 per capita (1994). **Debt service:** 31.6% of exports (1994). **Development aid received:** $3 million (1993); $2 per capita; 0.1% of GNP.

ENERGY

Consumption: 4,549 kgs of Oil Equivalent per capita yearly (1994), -89% imported (1994).

from Britain in 1962. One hundred Muslims of African origin occupied parliament and took hostages. They demanded that Prime Minister Arthur Jay Robinson resign and call elections. On August 1, after freeing all the hostages, the group surrendered unconditionally.

[14] The 1990 balance showed a slight economic recovery, achieved through an increase in oil and petrochemical exports, which increased as a result of the Gulf crisis. This enabled the government to avoid enforcing the 10 per cent reduction in salaries.

[15] In December 1991, the PNM won the general elections with 46 per cent of the vote, and obtained 20 of the 36 parliamentary seats. The Unity National Congress (UNC) obtained 26 per cent and Robinson's NAR, 20 per cent. Abstentions were considerably higher than before.

[16] The main obstacles facing the new government, led by Patrick Manning, were a foreign debt amounting to $2.51 billion, and 24 per cent unemployment in a 460,000-strong workforce.

[17] The government's economic adjustment and privatization plans sparked protest demonstrations in January 1993. The civil servants also demanded the payment of overdue backpay. Manning called the army in to "control" the situation given the increasing agitation.

[18] Partial parliamentary elections in 1994 showed the increasing unpopularity of Manning, as the governing PNM lost two of the three debated seats.

[19] On another front, the amnesty decreed by then prime minister Robinson on the 114 members of the Jamaat-al-Muslimeen Islamic group responsible for the July 1990 coup attempt, was annulled. However, this was a symbolic measure, as the court decided the accused would be neither arrested nor taken to trial.

[20] Faced with what he considered a more favorable economic and political situation, Manning decided to call elections in November 1995, a year early. However, he had miscalculated, for the PNM took only 17 seats, as did the opposition UNC, led by Basdeo Panday.

[21] After making an alliance with the NAR, Panday became the first person of Asian Indian origin to be head of government in Trinidad and Tobago.

Tunisia

Tunisie

Population: 8,815,000 (1994)
Area: 163,610 SQ.KM
Capital: Tunis

Currency: Dinar
Language: Arabic

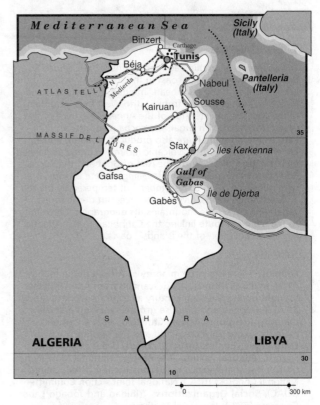

From the 12th century BC the Phoenicians had a series of trading posts and ports of call on the North African coast. Carthage was founded in the 8th century BC in the general vicinity of present-day Tunis, and by the 6th century the Carthaginian kingdom encompassed most of present-day Tunisia. Carthage became part of Rome's African province in 146 BC after the Punic Wars. Roman rule endured until the Muslim Arab invasions in the mid-7th century AD. In Tunisia, the Arabs met the strongest resistance to their westward advance, but this region eventually became one of the best cultivated and developed of their cultural centers; the city of Kairuan is associated with some of the most outstanding names in Islamic architecture, medicine and historiography. During the dissolution of the Almohad Empire (see appendix to Morocco: "Almoravids and Almohads") the region of Tunisia attained independence under the Berber dynasty of the Haffesides who, between the 13th century and the beginning of the 16th century, extended their power over the Algerian coast.

2 The development of European maritime trade attracted Turkish corsairs. The most famous, Khayr ad-Din (known as Red Beard), set up his headquarters in Tunisia, placing the Tunisian-Algerian coast under the domination of the Ottoman sultans. The inland regions, however, remained in the hands of the Berbers. The need to work with them, coupled with the distance from Istanbul, allowed the "bey" (designated governor) to attain a large degree of autonomy and in practice, become a hereditary ruler. The Murad family ruled between 1612 and 1702, and from 1705 until after independence in 1957, this role was filled by the Husseinite family.

3 After the French occupation of Algeria, European economic penetration became increasingly evident and with it came increasing debts. In 1869 the burden of the foreign debt forced the bey to allow an Anglo-Franco-Italian commission to supervise the country's finances. Foreign interference increased until 1882, when 30,000 French soldiers invaded the country, under an agreement whereby Britain, which had just occupied Egypt, "transferred its rights" to Tunisia, to compensate France for its loss of control over the Suez Canal.

4 In 1925, the Tunisians launched a campaign for a new Constitution (Destur 1), which would bring autonomy to the country. After the Second World War, the pro-independence Neo-Destur party grew considerably, and a series of demonstrations and anti-colonial uprisings led to armed struggle between 1952 and 1955. In March 1956, France recognized the sovereignty and independence of the country under the bey's regime. The new ruler was a direct heir of the sovereign under whom the country had become a French Protectorate in 1881. A year later, on July 25, 1957, the bey was deposed by a constituent assembly controlled by "Desturians". A republic was proclaimed and the leader of the party, Habib Bourguiba, was elected president. Bourguiba started an energetic campaign against the French presence at the Bizerta naval base, finally dislodging them in 1964. Also in 1964, the New Destur Party became the Destur Socialist Party (PSD) and, until 1981, was the only legal political organization.

5 Between 1963 and 1969, minister of finance Almed Ben Salah undertook a program of collectivization of small farms and trading companies, and nationalized foreign enterprises.

6 In 1969, following Ben Salah's arrest and subsequent deportation, government social and economic policy changed direction. The collectivization process was aborted and the economy thrown open to foreign investment. A 1972 law practically turned the whole country into a duty-free zone for export industries. Habib Bourguiba, the "Supreme Combatant", was appointed president for life.

7 Towards the end of the 1970s, the economy suffered the effects of declining phosphate exports, and protectionist measures by the EEC against Tunisia's substantial textile industry. In January 1978, the UGTT (General Union of Tunisian Workers) called a general strike against wage controls and the repression of unions. The government declared a state of emergency and street fighting left dozens dead. The union leaders, including the president Habib Achour, were arrested.

8 Appointed prime minister in 1980, Mohamed Mzali initiated a program of political liberalization. One year later, political parties were allowed to reorganize and trade union elections were held to renew the UGTT. General elections were held in November 1981 and, despite the vitality of the opposition parties, the government National Front won 94 per cent of the vote and all of the seats. Numerous irregularities were reported.

9 In January 1984, the government decided to end several food subsidies. The price of bread rose 115 per cent and violent demonstrations left more than 100 dead. President Bourguiba cancelled the price

PUBLIC EXPENDITURES

% of GNP — 1992

MILITARY 3.3 | EDUCATION 6.1 | HEALTH 3.3

WORKERS

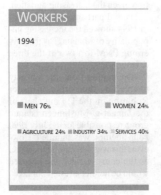

1994

MEN 76% | WOMEN 24%

AGRICULTURE 24% | INDUSTRY 34% | SERVICES 40%

LAND USE

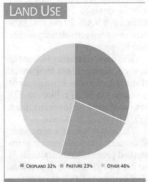

CROPLAND 32% | PASTURE 23% | OTHER 46%

33%

PER 100,000
LIVE BIRTHS
139

PROFILE

ENVIRONMENT

Tunisia is the northernmost African state. The eastern coastal plains are heavily populated and intensively cultivated with olives, citrus fruit and vinyards. The interior is dominated by the mountainous Tell and Aures regions populated by nomadic shepherds. The Sahara desert, in the south, has phosphate and iron deposits, while dates are cultivated in the oases.

SOCIETY

Peoples: 93% of Tunisians are Arab, 5% are Berber and 2% European. **Religions:** 99% Islam, Sunni. There are also Jewish and Catholic groups. **Languages:** Arabic (official). French, Berber. **Political Parties:** Constitutional Democratic Assembly; Communist Workers Party; Democratic Socialist Movement; Hezb Ennahda. **Social Organizations:** General Union of Tunisian Workers (UGTT).

THE STATE

Official Name: Al-Jumhuriyah at-Tunisiyah. **Administrative Divisions:** 25 Government Regions. **Capital:** Tunis, 675,000 inhab. (1994). **Other cities:** Safaqis (Sfax), 230,000 inhab.; Aryanah, 153,000 inhab.; Ettadhamen, 149,000 inhab.; Susah, 125,000 inhab. **Government:** Zine al-Abidine Ben Ali, President since 1987, re-elected in 1994. Hamed Karoui, Prime Minister since 1989. 163-member unicameral parliament. **National Holiday:** June 1, Independence (1959). **Armed Forces:** 35,500 (1993). **Paramilitaries:** 13,000, National Police; 10,000, National Guard.

STATISTICS

DEMOGRAPHY

Urban: 56% (1995). **Annual growth:** 0.9% (1992-2000). **Estimate for year 2000:** 10,000,000. **Children** per **woman:** 3.8 (1992).

HEALTH

One **physician** for every 1,852 inhab. (1988-91). **Under-five mortality:** 34 per 1,000 (1994). **Calorie consumption:** 107% of required intake (1995). **Safe water:** 99% of the population has access (1990-95).

EDUCATION

Illiteracy: 33% (1995). **School enrolment:** Primary (1993): 113% fem., 113% male. Secondary (1993): 49% fem., 55% male. University: 11% (1993). **Primary school teachers:** one for every 26 students (1992).

COMMUNICATIONS

97 **newspapers** (1995), 100 **TV sets** (1995) and 97 **radio sets** per 1,000 households (1995). 4.9 **telephones** per 100 inhabitants (1993). **Books:** 90 new titles per 1,000,000 inhabitants in 1995.

ECONOMY

Per **capita GNP:** $1,790 (1994). **Annual growth:** 2.10% (1985-94). **Annual inflation:** 6.30% (1984-94). **Consumer price index:** 100 in 1990; 124.6 in 1994. **Currency:** 1 dinars = 1$ (1994). **Cereal imports:** 1,044,000 metric tons (1993). **Fertilizer use:** 223 kgs per ha. (1992-93). **Imports:** $4,660 million (1994). **Exports:** $6,580 million (1994). **External debt:** $9,254 million (1994), $1,050 per capita (1994). **Debt service:** 18.8% of exports (1994). **Development aid received:** $250 million (1993); $29 per capita; 2% of GNP.

ENERGY

Consumption: 590 kgs of Oil Equivalent per capita yearly (1994), -7% imported (1994).

---

increases. In 1985, there was further trade unrest and the UGTT was placed under government control. There were also violent confrontations with the emerging, illegal, Islamic fundamentalist movement which ended with hundreds of arrests and some death sentences.

[10] During the period dominated by Bourguiba, Tunisia had developed as the most westernized of Arab societies. From 1986 onwards, the Fundamentalist reaction confronted the public with the issues of religion in the life of the individual and the community, and Tunisia's identity as an Arab-Muslim country. Female emancipation and European tourism were brought into question.

[11] In 1985, Colonel (later General) Zina El Abidine began his rise to power. In 1987, he was appointed prime minister and, in November, he dislodged president-for-life Bourguiba whom a medical board declared mentally and physically unfit to govern. This marked the start of a period of "national reconciliation", meaning greater press freedom and the liberation of hundreds of political prisoners and Islamic militants. The PSD was renamed the Constitutional Democratic Grouping (RSD) without losing political dominance.

[12] The presidential and legislative elections of April 2 1989, were considered by observers to be the "freest" since independence, even though 1,300,000 citizens were not registered to vote. The electorate split between the government RSD (with 80 per cent of the vote and all the seats) and the Hezb Ennahda Islamic movement, whose independent candidates won 15 per cent. Center and left wing opposition parties trailed as tiny minorities. President Ben Ali was re-elected by 99 per cent of voters.

[13] After distancing itself from the Arab League during the 1970s, Tunisia denounced the Camp David accords between Israel and Egypt, and became host to the PLO after its eviction from Beirut. Although President Ben Ali made overtures toward Islam during his first few years in office, he later unleashed a repressive campaign against the outlawed Hezb Ennahda movement and other opposition groups.

[14] In June 1990, Amnesty International published a report on Tunisia which included detailed denunciations of torture and mistreatment of prisoners, including solitary confinement. The organization called for the commutation of the sentences of two prisoners who had been sentenced to death during Ben Ali's presidency. The first of the two was executed in November of that year. In mid-1991, political figures urged their fellow-citizens to support the student movement, and asked the international community to offer the solidarity necessary to "sustain the Tunisian people in their struggle for liberation, democracy, progress and the respect for human rights".

[15] In reply to the mobilization, Ben Ali continued to reinforce repressive legislation. A restrictive Law of Association was passed in March 1992 and in July, members of the Islamic Hezb Ennahda were sentenced to life in prison. Unofficially, representatives from Western countries suggested the risk of expansion of Islamic fundamentalism in Tunis could justify Ben Ali's policy.

[16] In the meantime, human rights organizations continued to denounce Tunis for practising torture. In November 1993, Ben Ali further enacted a law restricting "fundamental liberties". In this context, the president was re-elected with 99 per cent of the vote in the March 1994 general elections, the ruling party having obtained the control of 88 per cent of parliamentary seats.

[17] The president pursued his policy of economic liberalization, wielding an "iron hand" in the political sphere: one of the main opposition leaders, Mohamed Moada, was sentenced to 11 years in prison in October 1995 for publishing a report on the curtailment of liberties in Tunis. Meanwhile, World Bank chairman James Wolfensohn visited Tunis in April 1996 and described this North African country as the "World Bank's best student in the region".

Turkey

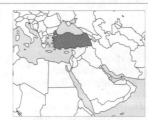

Population: 60,840,000 (1994)
Area: 779,450 SQ.KM
Capital: Ankara

Currency: Lire
Language: Turkish

Türkiye

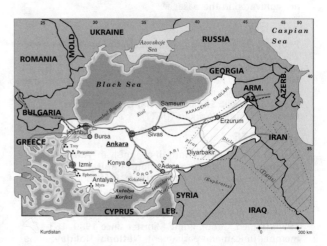

Kurdistan

The province of Anatolia has been inhabited for thousands of years. In the eighth century BC the Greeks founded the city of Byzantium on the strategic strait of Bosphorus, controlling traffic between the Black Sea and the Mediterranean. The city was conquered by the Romans in 96 AD and emperor Constantine had it rebuilt. Under the new name of Constantinople, it became the capital of the Eastern Roman Empire, which outlived the Barbarian destruction of Rome and the Western Roman Empire by a hundred years.

[2] Constantinople was seized by the Turks in 1453 and became the capital of the flourishing Ottoman Empire. In the 16th century, the domains of Suleyman the Magnificent reached from Algeria to the Caucasus and from Hungary to the southern end of the Arabian Peninsula, populated by 50 million people (ten times the population of contemporary England). Western visitors marvelled at the efficiency and prosperity of the government, and the respect shown for the rights of the peasants. This may explain why more than once, the Christian peasants of the Balkans fought alongside the "infidel" Muslim, against Christian noblemen and clergymen.

[3] From the 17th century, the empire fell behind the impressive technological advance of western Europe and its aggressive trade expansion. The route to the East round the Cape of Good Hope, opened by the Portuguese, slowly deprived the Ottoman Empire of its trade monopoly between Europe and eastern Asia. In fact, the very borders of the empire gave way to the thrust of European traders, whose companies introduced French, British and Dutch products

in the eastern Mediterranean region.

[4] The fate of Turkey foreshadowed the course of events that later affected the great civilizations of China and India. They all became peripheral markets where Europeans imposed the sale of their manufactured goods and collected the food and raw materials for their workforce and industries. Turkish artisans went bankrupt, the local textile industry failed to compete with British technology and the peasants grew increasingly poor,

> In World War I, Turkey sided with the second German empire and the Austro-Hungarian empire, and it was defeated along with them.

having to pay the price of imported goods.

[5] In the 19th century, a modernizing movement known as *Tanzimat* tried to impose European patterns and to build a centralized state, using modern technology, the telegraph and railways. New accords were signed with Britain and Germany but the influx of capital and goods further deepened Turkish dependency on Europe. One of the outstanding consequences was a large external debt which turned Turkey into "the sick man of Europe".

[6] At the beginning of the 20th century, a clandestine organization started to emerge from the discon-

tent with the autocratic government. The "Young Turks" formed within the universities and military academies. In July 1908, they led a rebellion in Macedonia, and within a few days, sultan Abdul Hamid yielded to their demand for a constitution limiting his power. The following year he was forced to abdicate in favor of Muhammad V, who bestowed real power on the Young Turks. They followed a nationalistic policy opposed to foreign suppliers, and various ethnic minorities, in particular the Greeks and Armenians.

[7] In World War I, Turkey sided with the second German empire and the Austro-Hungarian empire, and it was defeated along with them. The Ottoman Empire was torn apart and many small autocratic states arose on the Arabian Peninsula and in the Balkan region. The ethnic minority groups that remained within Turkish borders were ruthlessly repressed. In 1915 approximately 800,000 Armenians were killed in an episode that went down in history as "the first genocide of the 20th century".

[8] The conditions imposed on Turkey by the Treaty of Sevres (1920) were so humiliating that military leader Mustafa Kemal (later known as Kemal Ataturk) deposed sultan Muhammad VI, who had signed the accord, and launched a national liberation war. He finally succeeded in negotiating a new treaty, signed in Lausanne, in 1923. Turkey was relieved of paying war reparations, and the privileges enjoyed by foreign traders (capitulations) were annulled. In exchange, the straits were declared international waters, open to all ships regardless of their nationality, in times of peace.

[9] Ataturk proclaimed a republic and had a new constitution approved. The government initiated a process of rapid modernization; state and church separated, the Muslim Friday ceased to be the weekly holiday, replaced by Sunday, the Latin alphabet replaced Arabian notation, and women were urged to stop wearing veils.

[10] After the death of Ataturk, in 1938, the military remained very influential in Turkish politics. The government also crushed the leftist groups which had grown under the influence of the neighboring Soviet revolution and during the fight against neo-colonialism.

[11] After World War II, Turkey - now a US ally - became an anti-Soviet bastion. The US built major military bases in its territory and military doctrine was gradually replaced by the Pentagon's concept of "national security". With US influence Turkey adopted a multi-party system and gave incentives to foreign investors. Although the economy was still dependent upon agricultural exports, rural areas no longer provided enough jobs and young people emigrated to the cities or to western European countries, particularly West Germany.

[12] Turkish military intervention in Cyprus caused the island to split in 1974, provoking the resignation of prime minister Bulent Ecevit, a social democrat. Conservative Suleyman Demirel succeeded Ecevit but the rivalry between these two leaders prevented the formation of a stable government for the rest of the decade.

[13] Demirel was ousted by the military in 1980 and General Kenan

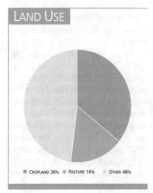

WORKERS
7.9%
1994 UNEMPLOYMENT

MEN 66%   WOMEN 34%

AGRICULTURE 47%   INDUSTRY 20%   SERVICES 33%

LAND USE

CROPLAND 36%   PASTURE 16%   OTHER 48%

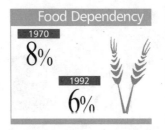

Evren became president. A new accord changed the conditions of the military bases, and Turkey started to receive $1 billion per year for leasing them. Labor unions and political parties were banned and the government was accused of systematic human rights violations.

[14] In 1983, a new constitution heralded a period of political liberalization designed to soothe the western European critics. The new government program aimed at gaining access to the European Economic Community, giving new impulse to modernization, moving away from the old nationalistic policies and embracing economic liberalism.

[15] In 1987, rivalries between the main opposition groups, formed by the followers of Demirel and Ecevit, enabled the Motherland Party to win the elections.

[16] Prime Minister Turgut Ozal embarked enthusiastically on a policy of privatization, economic liberalization and promotion of exports, following International Monetary Fund and World Bank guidelines. His willingness to adopt these policies was based on the assurance that they would lead to prosperity and to admittance into the European Community, an undeniably popular goal in a country where nearly all families have at least one relative working in Western Europe. However, political freedoms were not liberalized to the same extent and the Community postponed considering the Turkish bid until 1993, alleging that its high rates of inflation and unemployment, the lack of a concrete social policy and Turkey's conflict with Greece made its incorporation into the Community unviable.

[17] Accused of nepotism, corruption and a lack of sensitivity over the social impact of his economic policy, Ozal suffered a complete defeat in the March 1989 municipal elections. However, as a result of successful political maneuvering, he managed to persuade parliament to name him president in October of the same year, the first civilian in several decades to occupy the presidency. In 1990, his government was confronted by the growing activism of the Kurd separatists in the southeast of the country, where the (Kurdish Workers' Party) PKK was taking military action.

[18] In August 1990, when Iraq was blockaded after the invasion of

ENVIRONMENT

The country is made up of a European part, Eastern Thrace, and an Asiatic part, the peninsula of Anatolia and Turkish Armenia, separated by the Dardanelles, the Sea of Marmara and the Bosphorus. Eastern Thrace, located in the extreme southeast of the Balkan Peninsula, makes up less than one-thirtieth of the country's total land area including an arid steppe plateau, the Istranca mountains to the east, and a group of hills suitable for farming. Anatolia is a mountainous area with many lakes and wetlands. The Ponticas range in the north and the Taurus range in the south form the natural boundaries for the Anatolian plateau, which extends eastward to form the Armenian plateaus. The far east is filled by the Armenian Massif, centered around the lake region of Van, where there is a great deal of volcanic activity and occasional earthquakes. Parallel to the Taurus there are a number of ranges known as Antitaurus, which run along the border of the Georgia, together with the Armenian mountains. The country is mainly agricultural. The lack of natural resources, and absence of capital and appropriate infrastructure, have been major obstacles to industrialization. The air in the area surrounding Istanbul contains high levels of sulphur dioxide and the Marmara Sea is contaminated with mercury.

SOCIETY

Peoples: The inhabitants of present-day Turkey are descendants of ethnic groups from Central Asia that began to settle in Anatolia in the 11th century. There are Kurdish, 12 %, Arab, 1%, Jewish, Greek, Georgian and Armenian minorities, whose cultural autonomy, including teaching in their respective languages, is severely limited.
Religions: Mainly Islamic. Sunni Muslims represent about 80% of the population and Alevi (non-orthodox Shiites) almost 20%.
Languages: Turkish (official), Kurdish and other minority languages.
Political Parties: True Path Party (DYP); Motherland Party (ANAP); Social Democratic Populist Party (SHP); Party of the Democratic Left (DSP); Nationalist Action Party; Republican People's Party (CHP); Socialist Party; Welfare Party; Democracy Party; Kurdish Workers' Party (PKK). **Social Organizations:** Confederation of Turkish Trade Unions; Confederation of Progressive Trade Unions of Turkey.

THE STATE

Official Name: Türkiye Cumhuriyeti.
Administrative Divisions: 74 Provinces.
Capital: Ankara, 2,720,000 inhab. (1993)
Other cities: Istanbul, 7,330,000 inhab.; Izmir (Smyrna), 1,920,000 inhab. (1993).
Government: Suleyman Demirel, President since May 16, 1993. Necmettin Erbakan, Prime Minister since June 1996.
Armed Forces: 507,800 troops (1995).
Paramilitaries: 70,000 Gendarmes-National Guard, 50,000 Reserves.

Kuwait, Turkey interrupted the flow of Iraqi oil to the Mediterranean by blocking the pipeline through the country. Although it did not send troops, Turkey authorized the use of its military airports and US bases for the mass bombing of Iraq. The opposition questioned these measures, arguing that they would affect relations with a neighboring country in an area already made unstable by Kurdish independence movements.

[19] Ankara feared that the independence of Iraqi Kurdistan might affect the Turkish Kurds. The 19 million Kurds live in a territory divided between four countries; Turkey, Syria, Iran, and Iraq. They are the world's largest ethnic minority without a territory of their own, and in Turkey they have no right to their language and cultural identity.

[20] In October 1991, the Turkish

army entered northern Iraq in order to attack PKK bases, supported by aircraft and helicopters. Kurdish representatives accused the Turkish government of bombing the civilian population.

[21] The parliamentary elections of October 20, were won by Suleyman Demirel's True Path Party (DYP), which obtained 27 per cent of the vote, gaining 178 of the 450 parliamentary seats. This narrow majority forced Demirel to seek alliances with Erdal Inonu's Social Democratic Populist Party (SHP), which came in third with 21 per cent of the vote. The Motherland Party (ANAP), which obtained 24 per cent of the vote, formed the opposition.

[22] Demirel faced a budget deficit of $6 billion, a foreign debt of $44 billion, and an annual inflation rate of 70 per cent. Apart from the fun-

damentalist movements, all the other political parties agree that Turkey's main priority is to gain entrance to the European Economic Community.

[23] On November 14, a Filipino freighter, carrying thousands of sheep crashed into a Lebanese ship between the Black Sea and the Sea of Marmara, near Istanbul. The freighter "Madonna Lilli" sank, taking its cargo 29 meters under water. The decomposition of the animals on the sea-bed will generate methane gas and consume the water's oxygen. Experts have termed the process an "environmental time bomb". This episode brought attention to the pollution problems facing the most populated city in Turkey. It came as a reminder that the rivers that run through Istanbul are heavily pollut-

Illiteracy
1995

18%

Maternal Mortality
1989-95

PER 100,000
LIVE BIRTHS

186

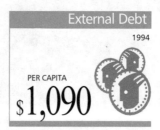

External Debt
1994

PER CAPITA

$1,090

STATISTICS

DEMOGRAPHY

Urban: 66% (1995). **Annual growth:** 1.9% (1992-2000). **Estimate for year 2000:** 68,000,000. **Children** per **woman:** 3.4 (1992).

HEALTH

Under-five mortality: 55 per 1,000 (1994). **Calorie consumption:** 110% of required intake (1995). **Safe water:** 80% of the population has access (1990-95).

EDUCATION

Illiteracy: 18% (1995). **School enrolment:** Primary (1993): 98% fem., 98% male. Secondary (1993): 48% fem., 74% male. University: 16% (1993). **Primary school teachers:** one for every 29 students (1992).

COMMUNICATIONS

101 **newspapers** (1995), 103 **TV sets** (1995) and 92 **radio sets** per 1,000 householdss (1995). 18.4 **telephones** per 100 inhabitants (1993). **Books:** 100 new titles per 1,000,000 inhabitants in 1995.

ECONOMY

Per **capita GNP:** $2,500 (1994). **Annual growth:** 1.40% (1985-94). **Annual inflation:** 65.80% (1984-94). **Consumer price index:** 100 in 1990; 967.0 in 1994. **Currency:** 38.726 lire = 1$ (1994). **Cereal imports:** 2,107,000 metric tons (1993). **Fertilizer use:** 702 kgs per ha. (1992-93). **Imports:** $23,270 million (1994). **Exports:** $18,106 million (1994). **External debt:** $66,332 million (1994), $1,090 per capita (1994). **Debt service:** 33.4% of exports (1994). **Development aid received:** $461 million (1993); $8 per capita; 0% of GNP.

ENERGY

Consumption: 955 kgs of Oil Equivalent per capita yearly (1994), 56% imported (1994).

Syrian government stated that it would close the PKK training camps and carry out stricter controls along its borders. [27] In 1992, the Council of Europe pressured the government to reduce repression against the Kurdish community. The Turkish authorities subsequently granted an amnesty to 5,000 political prisoners and authorized the circulation of two Kurdish-language newspapers. In November, the EC set 1996 as the date for Turkish admission to the European Customs Union, a first step toward eventual membership of the EU.

[28] Upon the death of President Turgut Ozal, in April 1993, Prime Minister Demirel was elected to be his successor. Tansu Ciller, minister of economic affairs, assumed the leadership of the DYP and was named prime minis-

> The signature of a military agreement with Israel in 1996 spoilt the Turkey's relationship with several arab countries. Shortly before the Muslim sacrifice festivities of Aid el Adha, Ankara closed the supply of water from its dams on the Euphrates, leaving Damascus under water shortage.

ter. Ciller, the first woman to become a government leader in Turkey, presented her program, which was approved by parliament in July. It included an accelerated privatization program - to be carried out by decree - as well as fiscal reform. In order to stem the growing budget deficit, which stood at $9.4 billion, there would be a moratorium on public investment. In July, public employees carried out strikes and demonstrations to protest the massive loss of jobs which would be

brought about by privatization. Some 700,000 people participated in the protests, which took place over a two-day period in Ankara, Istanbul and Izmir.

[29] The PKK ordered a cease-fire and offered to withdraw demands for the formation of an independent state in Kurdistan, in return for the initiation of formal negotiations with the government. At the end of May, government evasiveness led the guerrillas to declare an "all-out war" on Ankara and to take several actions in European cities, especially in Germany, accused of lending military support to Turkey.

[30] The army extended its offensive during 1994, forcing the inhabitants of hundreds of Kurdistan villages to leave their homes while it bombed Iraqi Kurdistan to destroy PKK bases.

[31] The growth of Islamic fundamentalism in Turkey was clearly visible in 1995, approaching the December elections. Its main representative, the Welfare Party, pledged during the election campaign to form Islamic organizations to compensate for the influence of NATO and the European Union. The party became Turkey's main political group winning 158 of the 550 seats.

[32] Prime Minister Ciller's DYP and the Motherland Party (ANAP), overcame their differences to prevent Islamic fundamentalists from coming to power and formed an unexpected government coalition, led by Mesul.Yilmaz from the ANAP, who took office in March 1996. However, the alliance was swiftly dissolved and the DYP chose to rule with the fundamentalists. Necmettin Erbakan became head of government in June 1996.

[33] At an international level, the signing of a military agreement with Israel in April further complicated Turkey's relations with several Arab countries, such as Syria. Tension mounted on the April 24, when Ankara decided to temporarily close for "technical reasons" the supply of water from its controversial dams on the Euphrates river, forcing Syrian authorities to ration water in Damascus, shortly before Muslim Aid el Adha (the sacrifice) festivities.

ed by sewage, and are lifeless 20 meters below sea level.

[24] The "Purple Roof Women's Shelter Foundation", a women's organization formed in 1980, conducted studies into violence against women. The results showed that 45 per cent of Turkish men agree that women should be punished when they disobey their husbands, and that one out of every four single women, and one out of every three married women are battered. The movement has grown in the past few years, and has helped women to speak out about their plight.

[25] In December 1991, prime minister Suleyman Demirel acknowledged the identity of the Kurdish people during a visit to the Kurdish provinces in the southeast of Turkey. In all the villages, the govern-

ment delegation was met by local residents demanding the respect of human rights, the end of aggression, and a halt to the torture of political prisoners and activists.

[26] In mid-March 1992, the banned Kurdish Workers' Party (PKK) announced the formation of a war government and a national assembly in the territory which they claim forms the core of Kurdistan, the internationally unrecognized Kurdish state. A few days later, coinciding with the Kurdish new year, there was an uprising in the southeastern provinces with violent action between the guerrillas and the Turkish security forces, especially in Cizre. In April, the Turkish interior minister, Ismet Sezgin, visited Damascus in Syria and Turkey announced an agreement to fight against the Kurdish "terrorist organizations". Under the agreement, the

Turkmenistan

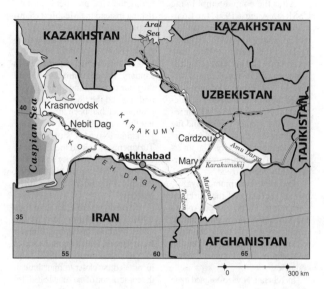

Population: 4,406,000 (1994)
Area: 448,100 SQ.KM
Capital: Ashkhabad (Aschabad)

Currency: Manat
Language: Turkmen

Turkmenistán

From 1000 BC onwards, the area now known as Turkmenistan was part of several large states: the Persian Empire (controlled by the Achaemenid dynasty) and later, the empire of Alexander the Great. In the third century AD, Turkmenistan was conquered by the Sassanids, an Iranian dynasty. Between the 5th and 8th centuries, there were successive invasions by the Eftalites, the Turks and the Arabs.

2   From the 6th to the 8th century, the entire area adjacent to the Caspian Sea was under the Arab Caliphate, and when the latter dissolved in the 9th and 10th centuries, the territory became a part of the Tahirid and Sassanid States. In the mid-11th century, the Seljuk Empire was formed in Turkmenistan, and towards the end of the 12th century it was conquered by the dynasty of the Shah of Khwarezm (Anushteguinidas).

3   During the Seljuk period the Turkmens - the most distinctive of the Turkish peoples of central Asia - formed as an ethnic group through the fusion of Oguz Turks and native tribes. This was a gradual process which was completed by the 15th century. In the early 13th century, Turkmenistan was invaded by Genghis Khan, whose heirs divided up the country. A large part of the territory was incorporated into the State of Hulagidas, and the northern regions were overrun by the Golden Horde of the Mongol Tatars. In the 14th and 15th centuries, the country fell under the Timurids, who were succeeded by the Uzbek khans of the Shaybani dynasty. From the 16th to the 18th century, Turkmenistan belonged to the khanates of Khiva and Bukhara, and the Iranian state of the Safavids successively.

4   In the 1880s, Turkmenia - as it was then known - was conquered by the Russian army. The major part of its territory lay within the Trans-Caspian region and the province of Turkistan, but the lands inhabited by the Turkmens passed into the hands of Khiva and Bukhara, which were Russian protectorates. There was strong resistance to Russian domination in Turkmenia, until the Battle of Geok-Tepe in 1881, in which the last of the rebels were defeated. The Turkmens were active participants in the 1916 uprising against the czar. The most violent revolt was in the city of Tedzhen, where several Russian residents and government officials were executed by the local population.

5   The fall of the czar in February 1917, led to the creation of *soviets* (councils of workers, peasants and soldiers) throughout the empire. The Trans-Caspian region fell under the control of the Russian Provisional Government. In December 1917, after the Bolshevik (socialists in favor of the revolution) take-over in St Petersburg, Soviet power was proclaimed. In July 1918, the English expeditionary forces re-established the Provisional Government in Trans-Caspia. After two years of civil war, Soviet power was rein-

stated in 1920 and on February 14 1924, the Soviet Socialist Republic of Turkmenistan was founded (joining the USSR, at the same time).

6   Up until that time, Turkmenistan had never known even nominal national political unity. Tribal organization was the only form of social organization, and most of the population was nomadic. For this reason, the Soviet regime resorted to repressive techniques in order to

impose industrialization and collective agriculture upon the local population.

7   After World War II, Turkmenistan entered a period of economic growth, accompanied by increases in oil and gas production. In addition, gains were made in the area of cotton farming. But during Leonid Brezhnev's administration, from 1963-83, political problems worsened and the economy entered a period of stagnation. In those republics where the Soviets had adopted a monocultural policy, as in the case of Turkmen-

LAND USE

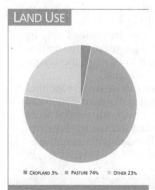

CROPLAND 3%   PASTURE 74%   OTHER 23%

PROFILE

ENVIRONMENT

Turkmenistan is in an arid zone, with a dry continental climate, located in the southeastern part of Central Asia between the Caspian Sea to the west, the Amu Darya River to the east, the Ustiurt mountains to the north and the Kopet-Dag and Paropamiz mountain ranges to the south. The land is flat and most of the territory (80%) lies within the Kara-Kum desert. Topographically, 90% of Turkmenistan is sandy plain. On the eastern shore of the Caspian Sea lie the Major and Minor Balkan ranges, of relatively low altitude. The Amu Darya river crosses Turkmenistan from east to west. The Kara-Kum Canal diverts the waters of the Amu Darya to the irrigation systems of the Murgab and Tedzhen oases, as well as those of the Mary and Ashkhabad areas. Turkmenistan is bounded by Kazakhstan to the northwest, Uzbekistan to the east; and Afghanistan and Iran, to the south. There are rich mineral deposits, including natural gas and oil.

SOCIETY

Peoples: Turkmenis, 73%; Russians, 10%; Uzbeks, 9%, Kazakhs, 2% (1992).
Religions: Mainly Sunni Muslim (Sufi).
Languages: Turkmen (official), Russian.
Political Parties: Socialist Party; Democratic Party; Agzibirlik (Unity).

THE STATE

Official Name: Türkmenistan Jumhuriyäti.
Administrative divisions: 3 provinces and 1 'Dependent Region' (Ashkhabad).
Capital: Ashkhabad (Aschabad), 416,000 inhab. (1991). **Other cities:** Chardzhou (Cardzou), 166,000 inhab.
Government: Presidential republic. Member of the (CIS) Commonwealth of Independent States. Saparmurad A. Niyazov, President, Head of State and of the government since October 27, 1990. **National Holiday:** March 21, Nauruz. **Armed Forces:** 25,000 Army (under joint control: Turkmenistan/Russia) (1995).

istan with the cultivation of cotton, agriculture was more vulnerable to the effects of the recession.

8 As of 1985, the changes promoted in the USSR by President Mikhail Gorbachev vitalized political life in Turkmenistan. There was also an Islamic religious rebirth, expressed through the construction of a number of mosques.

9 After the coup attempt in the USSR, in August 1991, the Communist Party of Turkmenistan lost its legitimacy to govern. Niyazov declared the need to restore state sovereignty and in October, a plebiscite brought in a presidential system. In November, Turkmenistan joined the Commonwealth of Independent States (CIS), which replaced the former USSR, and the Communist Party changed its name to the Democratic Party. The sovereignty of Turkmenistan was internationally recognized.

10 President Saparmurad Niyazov, former secretary of the Turkmen Communist Party, visited Ankara and Teheran. Niyazov stated that he would give priority to the country's relations with Turkey, given the cultural proximity between the two countries.

11 In Teheran, Niyazov signed trade agreements, giving joint top priority to exploration for natural gas and the construction of a transportation network. He also expressed the need to integrate the Turkish and Turkmen banking systems.

12 In early 1992, the new constitution was approved, granting Niyazov the dual powers of head of state and head of government, and abolishing the vice-presidency, establishing instead the Mechlis (Islamic parliament) and a People's Council. Niyazov rejected a free-market economy and the privatization of state-owned enterprises. In January, the government reduced the price of bread and other staples, and stopped the liberalization program which had been initiated during the Gorbachev era.

13 Under Niyazov's presidency, Turkmenistan joined the European Conference for Security and Cooperation (ECSC) and in March 1992, became a member of the United Nations.

14 Niyazov was re-elected in June 1992 with 99.5 per cent of the votes. Without distancing himself from Russia, which signed a series of trade agreements in 1993, Turkmenistan drew closer to Iran, through the construction of the Ashkhabad-Teheran railway, amongst other things. It also achieved the status of most favored nation with the United States.

15 Niyazov was reconfirmed in his post by 99.9 per cent of the vote in the January 1994 referendum. The President continued with his "hard handed" policy against the opposition, which led several opposition leaders to take refuge in Moscow.

16 In January 1995, Turkmenistan, Turkey, Iran, Kazakhstan and Russia signed an agreement to fund an oil pipeline which would allow Ashkhabad to export its natural gas to Eastern Europe through Iran and Turkey.

17 Despite attempts to reconcile its openness to Iran - and its reticence at becoming fully involved in the Community of Independent States - with a "balanced" policy towards Moscow, the Russian government was obviously uneasy about Niyazov's rapprochement with Teheran.

18 Russian observers claimed the July 1995 demonstrations against the Turkmeni president — where between 300 and 500 people protested against the "dictatorship" — were organized with Moscow's support.

The family legacy

After the disintegration of the USSR in December 1991, many problems arose among the ex-soviet republics. Although internal boundaries were declared untouched, the use of the rivers, internal seas, ports and highways of common use were matters of dispute. The control over the Black Sea Fleet was the most publicized of those problems, but the division and determination of the legal status of the Caspian Sea is still one of the unsolved legacies of the Soviet era.

2 Shared by five independent republics nowadays, the Caspian Sea (which can be considered a lake despite of its name, since it has no direct link to any ocean) used to be the common space of both Iranians and Russians. Soviet Russia made the first attempt to create an international legal regulation for the Sea by concluding a Treaty with Persia (now Iran) on 26 February, 1921.

3 The provisions of this treaty evolved further in the treaty between the USSR and Iran of 25 March 1940, according to which the line across the Sea from Astara to Gasankouli became the sea border between the USSR and Iran. Thus, the Caspian was a sea of two states.

4 After the collapse of the Soviet Union and the declarations of sovereignty by Azerbaijan, Kazakhstan, Russia and Turkmenistan a new geopolitical situation was created. Since Iran did not claim any modifications in the sea's status, the different ex-soviet states began discussions on the issue.

5 Over 42 billion barrels of oil reserves lie in the Caspian Sea region. Exploration is still in the early stages - some experts estimate as much as 200 billion barrels will ultimately be discovered. The Caspian Sea is likely to be the second largest depository of oil in the world after the Persian Gulf.

6 This wealth of oil makes Russia very cautious about the possibility of opening the sea as proposed by international regulation but its neighbors believe it is the only way to solve their isolation.

7 The three landlocked Caspian States (Azerbaijan, Kazakhstan, Turkmenistan) demand the signing of a treaty to guarantee their access to the high seas. As a result of this, agreement on the legal situation for the Volga delta, the river Volga, the Volga-Baltic and the Volga-Don canals should be included in the agenda.

8 The Russian Federation stated its general stance in an official document distributed at the UN in October, 1994, clearly asserting that it considered the Caspian Sea an internal lake. Claiming that the International Law of the Sea - which guarantees the coastal states free access to the high seas - does not apply, Russia tried to prevent the Caspian countries (including Iran) from having to the right to the use of the Volga system of waterways. The Kremlin's authorities also argued that the Caspian Sea is an indivisible reservoir which is a unified ecosystem, all the natural resources of which belong to all the Caspian states and may be utilized only by the mutual agreement of the parties, in order not to damage the Caspian's flora and fauna.

9 The four former soviet republics continued the talks during 1995 and 1996. However, no results were achieved and the sea (or lake) remained closed to other countries.

Turks and Caicos

Population: 14,000 (1994)
Area: 430 SQ.KM
Capital: Cockburntown

Currency: US Dollar
Language: English.

Turks and Caicos Islands

Turks and Caicos are two island chains separated by a deep water channel, lying approximately 150 km north of Haiti on the southernmost tip of the Bahamas chain. Like many of the smaller Caribbean islands, these were first inhabited by Arawak peoples. Some researchers claim it was on East Caicos or Grand Turk that Columbus first set foot on New World soil in 1492. The first Europeans to settle on the islands were saltrakers from Bermuda in 1678.

[2] During the following century, Turks and Caicos faced several invasions by both French and Spanish forces. The islands were a refuge during this period for both pirates and their merchant-vessel victims, Spanish galleons carrying American wealth to Europe. By 1787, colonial settlers had established cotton plantations and imported African slaves, and British domination was consolidated. Both archipelagos remain British colonies.

[3] Turks and Caicos were administered from the Bahamas until the Separation Act of 1848. After 1874, the islands were annexed to Jamaica, remaining a dependency until independence in 1962, when they again became a separate colony. During World War II, the United States built an airstrip on South Caicos and in 1951 the islands' authorities signed an agreement permitting the US to establish a missile base and a Navy base on Grand Turk island.

[4] After the Bahamas' independence in 1972, the islands received their own governor. Further autonomy achieved through the 1976 constitution, provided for a Governor, a Legislative Council, a Supreme Court and a Court of Appeals.

[5] In the 1976 elections, the pro-independence People's Democratic Movement (PDM), led by JAFS McCartney, won over the pro-US Progressive National Party (PNP). The PDM favored a new constitution which granted internal auton-

omy as a prior step towards eventual independence.

[6] In 1980, an election year, Britain showed an interest in granting independence to the islands, which receive $2 million a year in aid. The overwhelming electoral triumph on the part of the PNP, can be attributed to the failure of the PDM to solve the country's economic crisis, including a 30 per cent unemploy-

> During the 18th century the islands faced several invasions from both French and Spanish forces.

ment rate and the local population's fears that the economy could worsen with independence.

[7] The new head minister, businessman Norman Saunders, convinced Britain to shelve the idea of independence. He concentrated on bringing new business to the islands: tourism talks with the French Club Mediterranée; the development of light industry and off-shore banking; and finally an agreement with BCM Ltd, for construction of an oil refinery with a capacity of 125,000 barrels per day.

[8] Internal differences arose within the government with regard to economic policy. However, these differences were quickly forgotten when Saunders and Stafford Missik, his development minister and a key opposition figure within the government, were arrested in Miami. They were in the process of creating an international drug network, using the islands as a bridge between the United States and South America.

[9] Saunders was released after paying a supposedly exorbitant sum of bail money, and immediately resigned his post. His successor was Nathaniel Francis of the PDM, also forced to resign a year later, together with the entire cabinet, after accusations were made of "unconstitutional behavior and mismanagement". The British government suspended the ministerial system of government until 1988, when a general election was held. PDM's Oswald Skippings won the election, with 11 of the 13 parliamentary seats going to his party. The election marked the end of direct British government of the islands.

[10] The government declared its aim of establishing financial independence within 4 years, and political independence in no less than 10 years - by 1998 -. According to the government's announcement, this goal must be based upon a solid economy and upon the will of the people of Turks and Caicos Islands.

[11] Skippings left office in 1990. Washington Misick took his place in 1991 and was chief minister for four years. In 1995, Derek H. Taylor took office as head of government.

Tuvalu

Population:	13,000 (1994)
Area:	2 SQ.KM
Capital:	Fongafale
Currency:	Australian Dollar
Language:	Tuvaluan and English

Tuvalu

Around 30,000 BC, the peoples from Australia, ancestors of the aborigines, started their expansion toward the Pacific islands. By the 19th century AD, they had spread throughout practically all of Polynesia. The inhabitants of the archipelago then known as "Funafuti" came there from the islands of Samoa and Tonga.

[2] In the 16th century the Europeans reported seeing the islands but did not settle on them due to the lack of exploitable resources. Years later, a European named the atolls the Ellice Islands.

[3] Between 1850 and 1875, thousands of islanders were captured by slave traders and sent to the phosphate (guano) works in Peru and the saltpeter mines in Chile. Within few years, the population had shrunk from 20,000 to 3,000 inhabitants. As a source of slave labor, the Polynesian islands offered direct access to markets on the Pacific coast.

[4] A new wave of invasions began in 1865, with the arrival of British and North American missionaries. By 1892 the islands had become a British protectorate. In 1915, with their neighbors, the Gilbert Islands (see Kiribati), the British formed the Colony of the Gilbert and Ellice Islands.

[5] This arbitrary merger was decided on administrative grounds and had little to do with other factors. In a referendum of 1974, 90 per cent of the Ellice islanders voted in favor of separate administrations, after having attained relatively autonomous local government.

[6] The split became official in October 1975, and the first elections in independent Tuvalu were held in August 1977. Toaripi Lauti was named prime minister.

[7] The archipelago became independent on October 1 1978, adopting the name Tuvalu, which in the local language means "united eight", symbolizing the eight inhabited islands which make up the country. The islands became autonomous from London, but they fell under the economic influence of Australia, which had already started to hold considerable sway on the economy; the Australian dollar is used as the local currency.

[8] Under a friendship treaty of 1979, presently awaiting ratification by the US Senate, Washington relinquished its claim over the islands of Nurakita, Nukulaelae, Funafuti, and Nukufetau.

[9] On December 8 1981, Lauti was succeeded by Tomasi Puapua.

[10] In September 1985, general elections were held for the third time in the archipelago's history. Dr Tomasi Puapua was re-elected prime minister. In February 1986, the government showed its opposition to French nuclear weapon testing in Mururoa, Polynesia, by denying entry to a French warship which was on a "goodwill" mission.

[11] Tuvalu is the smallest of the Lesser Developed Countries, in both population and land area. One of its main sources of foreign exchange is the sale of stamps, which are valuable collectors' items, and the granting of fishing licences to foreign fishing fleets which earns some $100,000 per year.

[12] A United Nations report on pollution induced global warming and the subsequent rise in sea levels, stated that Tuvalu could be completely submerged, unless severe measures are taken.

[13] Faced with a lack of natural resources, a sparse population density, and an almost total lack of internal sources of savings or investment, the country depends heavily on foreign aid, of which Australia is the principal source, to finance its regular budget and its development budget. In 1989, the state was able to meet about 10 per cent of its expenses, most of the money sent home by the quarter of the islands' population, mostly young people, who reside in neighboring islands working in the phosphate mines.

[14] The situation of the archipelago has been worsening, as citizens returning from abroad look for jobs, competing with young people just finishing school. Only 30 per cent of school-leavers are successful in finding employment.

[15] In 1993, Kamuta Laatasi became prime minister. His government dismissed Governor-General Tomu Sione in 1994, arguing that he was a political appointee of the previous regime.

[16] At the annual Independence Day celebrations in October 1995, a new flag officially replaced the Union Jack. However, suggestions of a republican future for Tuvalu met with little enthusiasm.

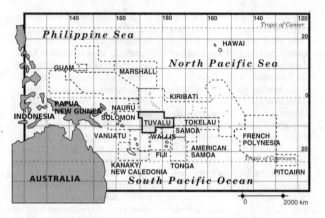

PROFILE

ENVIRONMENT

Tuvalu is distributed over the Funafuti, Nanumanga, Nanumea, Niutao, Nui, Nukufetau, Nukulaelae, Nurakita and Vaitupu coral atolls, known as the Ellice Islands, within an ocean area of 1,060,000 sq km. The archipelago is situated slightly south of the equator, 4,000 km northeast of Australia, south of the Gilbert Islands, between Micronesia and Melanesia. The climate is tropical with heavy rainfall. The islands are completely flat with a thin layer of topsoil suffering from severe erosion. Fishing and coconut farming are traditional activities.

SOCIETY

Peoples: Tuvaluans are mainly Polynesians; 2,250 of them work abroad. **Religion:** Church of Tuvalu (Congregational), 96.9%; Seventh-day Adventist, 1.4%; Baha'i, 1%; Roman Catholic, 0.2%; other, 0.5%. **Languages:** Tuvaluan and English. **Political Parties:** Prominent families control Tuvalu's politics. They are not formally organized in parties.

THE STATE

Official Name: Tuvalu. **Capital:** Fongafale, on Funafuti atoll, 2,810 inhab. (1985). **Other atolls:** Vaitupu, 1,231 inhab.; Niutao, 904 inhab.; Nanumea, 879 inhab.; Nukufetau, 694 inhab. (1985). **Government:** Governor-General, Tulaga Manuella; Prime Minister, Kamuta Laatasi. Tuvalu has a parliamentary government modelled on the British system. Parliament has 12 members. The Executive Branch is the governor general, who represents the British crown and is appointed on the recommendation of the prime minister. **National Holiday:** October 1, Independence Day (1978). Statistics Tonga

EDUCATION

Primary school teachers: one for every 21 students (1990).

COMMUNICATIONS

1.3 **telephones** per 100 inhabitants (1993).

ECONOMY

Consumer price index: 100 in 1990; 102.3 in 1993. **Currency:** Australian dollars.

Uganda

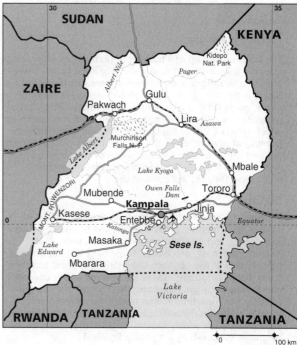

Population: 18,592,000 (1994)
Area: 235,880 SQ KM
Capital: Kampala

Currency: Shilling
Language: Swahili

Uganda

In present-day Uganda the large mud walls of Bigo state are evidence of urban civilizations dating from the 10th century.

[2] Various places in Uganda still bear the ruins of sizeable hilltop fortresses, overlooking the surrounding territory. Built with earthen walls, moats and trenches, these fortifications mark the lines of Bacwezi penetration. Around the 13th century, the Bacwezi, a nation of Nilotic herders, arrived from the north and subdued the Bantu inhabitants of the area. Their characteristic fortresses, in some cases up to 300 meters in diameter, were built to protect themselves and their cattle - their main source of wealth and status symbols. The fortresses were gradually abandoned, and the conquerors mixed with the conquered, adopted Bantu languages. The descendents of the Bacwezi, who preserved their nomadic herding lifestyle, intermingled less with the local population though they too adopted a Bantu language, and came to be called Bahima.

[3] The peoples of Bunyoro, Buganda, Busoga and Ankole were connected to the eastern coast and the Sudanese slave trade; they consolidated their national unity between the 17th and 18th centuries. A supremacy dispute arose between Bunyoro, supported by the Sudanese traders, and Buganda, linked to the "Shirazis" of Zanzibar. At the beginning of the 19th century, Bunyoro was already losing ground, and when some of their allies defected to form the independent state of Toro the predominance of Buganda became inevitable.

[4] Buganda was governed by Kabakas or traditional leaders, who were, in theory, absolute rulers but in practice were limited by the Lukiko, a council representing the higher castes. In the mid-19th century, Buganda maintained a standing army which was enough to guarantee their autonomy from the regional powers; Egypt and Zanzibar. It had a balanced society in which caste privileges were more honorary and political than economic and a solid agricultural economy which allowed Buganda to overcome the decline of the slave trade.

[5] The first contact with Europe occurred in 1862. The second contact had more far-reaching results. HM Stanley, adventurer and journalist, arrived in 1875. He vociferously denounced the spread of Islam in the region and announced an alleged "request" made by Kabaka Mutesa I asking Europe to send missionaries to check Egyptian-Sudanese religious infiltration.

[6] These missionaries soon arrived; English Protestants in 1877 and French Catholics in 1879. They quickly converted part of the Bugandese hierarchy, splitting the power elite into three parties. Two of these reflected the rivalry between missionaries, in local dialect the "Franza" and "Ingleza" parties, while the third took the role of defending national interests, as they were moderate and Islamic.

[7] The main consequence of this conflict was the consolidation of European presence. The missionaries succeeded in deposing the Moslem Kabaka, Mwanga in 1888, and swift on their heels came the Imperial British East Africa Company (IBEA), a typical colonial trading company, followed by the British government.

[8] The 1886 Anglo-German agreements had delimited areas of influence and the states of the lakes area went to the British who established a Protectorate over them in 1893.

[9] The other organized local groups did not have governmental institutions similar to those of Buganda's but these were imposed on them as the British thought that the Lukiko resembled their own parliamentary system.

[10] With the intention of developing a ruling elite to serve as intermediaries to the colonial power, the British undertook "land reform". This consisted of distributing communally-owned land to select individuals, depriving the rural population of their legitimate property to benefit the bureaucracy that sat in the Lukiko.

[11] The severe disruption of production caused by this confiscation was aggravated by the introduction of cash crops for export in the post World War II period. These crops were alien to the agricultural traditions of the region, and their production resulted in a deterioration in the living conditions of most of the inhabitants.

[12] There was no way of changing this state of affairs until the 1960s when the decolonization movement led to the independence of Uganda. Kabaka Mutesa II of Buganda was the first president of the new republic and the prime minister was Dr Milton Obote.

[13] In 1965, Obote succeeded in reforming the constitution and he assumed greater powers eliminating the federal system imposed by the British, which had created various relatively autonomous territorial sub-divisions. He also adopted a policy that favored the most impoverished sectors.

[14] This aroused fierce opposition among the Asian bourgeoisie; a minority of 40,000 people who controlled almost all the commercial activity in the country. Since the Asians held British passports, they had resisted total integration into the new nation.

[15] Obote decisively supported regional economic integration with Tanzania and Kenya to counterbal-

PUBLIC EXPENDITURES

% of GNP — 1992

- MILITARY 2.9
- EDUCATION 2.9
- HEALTH 1.6

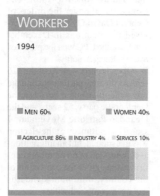

WORKERS

1994

MEN 60% — WOMEN 40%

AGRICULTURE 86% — INDUSTRY 4% — SERVICES 10%

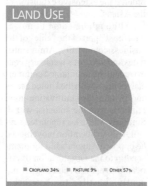

LAND USE

CROPLAND 34% — PASTURE 9% — OTHER 57%

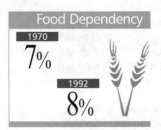

Food Dependency

1970
7%

1992
8%

Foreign Trade

MILLIONS $ 1994

IMPORTS
870

EXPORTS
421

Illiteracy

1995
38%

PROFILE

ENVIRONMENT

The land is made up of a number of plateaus, gently rolling towards the northwest where the Nile River flows. There are volcanic ranges and numerous rivers, the largest of which is the Nile. Nearly 18% of the territory is taken up by rivers, great lakes and swamps. The climate is tropical tempered by altitude. Lumber is taken from the rainforest which covers 6.2% of the land. In addition to subsistence farming of rice and corn, the cash crops are coffee, cotton, tea and tobacco. Lake Victoria is one of the largest fish reservoirs in the world. Swampland is being drained indiscriminately for agricultural use.

SOCIETY

Peoples: Most Ugandans come from the integration of various African ethnic groups, mainly the Baganda, Bunyoro and Batoro, and to a lesser extent, the Bushmen, Nile-Hamitic, Sudanese, and Bantu. Some have distinct physical traits typical of the northern Nile, Hamitic groups, but they speak Bantu languages. There are also minorities of Indian and European origin. **Religions:** Nearly half of the people are Christian, one-third follow traditional religions and there is a Muslim minority. **Languages:** Swahili (official) and Luganda are widely spoken. English, the other official language, is spoken by a minority. **Political Parties:** National Resistance Movement (NRM), led by President Museveni. All other parties have been disenfranchised by the government; the most important among them are the Ugandan People's Congress (UPC); the Democratic Party (DP), the Ugandan Freedom Movement (UFM); the Conservative Party (CP); the Nationalist Liberal Party; the Ugandan Democratic Alliance; the Ugandan Revolutionary Movement for Independence; the Ugandan United Nationalist Movement; and the Ugandan Patriotic Movement. **Social Organizations:** National Organization of Trade Unions (NOTU).

THE STATE

Official Name: Republic of Uganda. **Capital:** Kampala, 773,000 inhab.(1991). **Other cities:** Jinja, 61,000 inhab.; Mbale, 54,000 inhab. (1991). **Government:** Yoweri Museveni, President since January 30, 1986, re-elected May 1996. Kintu Musoke, Prime Minister since July 1996. **National Holiday:** October 9, Independence Day (1962). **Armed Forces:** 50,000 (1995).

ance the effects of Uganda's lack of access to the coast. This led to the formation of the East African Community.

[16] In January 1971, Obote was overthrown in a bloody coup led by former paratroop sergeant and boxing champion Idi Amin Dada. The economic base of the deposed government had been seriously shaken by a series of destabilization maneuvers by the Asian minority and interests connected to transnationals. Idi Amin quickly demonstrated his authoritarianism. In 1972 he ordered the extensive expulsion of the Asians.

[17] Although he came to power with the support of the "elite" of the business community, Amin's attitudes and measures were very controversial: he maintained commercial relations with the United States and Britain, also cultivating good relations with the socialist world. Similarly, while he supported several African liberation movements, he opposed Angola's bid for membership to the Organization of African Unity (OAU) and was permanently hostile to Nyerere's government in Tanzania. Finally, although he had been trained as a paratrooper in Israel, he expropriated the lands and other properties of members of the Jewish community and made overtures toward the Arab countries.

[18] Amin declared himself president-for-life. In 1978, he exacerbated tensions with Tanzania by annexing its northern territories. In April 1979, war between the two countries ended Amin's regime and he was forced to flee Kampala, after a joint offensive launched by Tanzanian troops and opposition activists unified in the Ugandan National Liberation Front (FNLU).

[19] The new government's main body was a National Advisory Council led by Yusuf Lule, a politically inexperienced university professor with conservative tendencies. 68 days later, Lule was replaced by the FNLU's powerful Godfrey Binaisa.

[20] The Front was a frail, eclectic movement, founded merely on joint efforts aimed at ending the terror imposed by Idi Amin. As expected, Binaisa was unable to conciliate the conflicting tendencies within his movement. He was even less capable of confronting the growing prestige of Milton Obote whose Uganda People's Congress party (UPC) continued to be very popular.

[21] The president held the elections scheduled for 1981 prematurely, and tried to ban Obote's candidacy. This fuelled a crisis that exploded in May 1980 when the army, on the pretext of preventing the government's policies to be continued, forced him to resign.

[22] The president was replaced by a Military Commission entrusted with maintaining the electoral schedule and enforcing the democratic principles of the movement that had overthrown Amin. The Commission, under the orders of General David Oyite Ojok, supervised the elections in December 1980, and the UPC won the predicted overwhelming majority.

[23] Obote's party won 73 of the 126 seats in the new parliament and the rest were distributed between the Democratic party, 52 seats, and the Ugandan Patriotic Movement, 1 seat. The legitimacy of the election was questioned, but an international commission of 60 members confirmed that, given the constraints of the country, minimum standards had been complied with.

[24] Obote inherited a bankrupt country. Copper mines had not been worked for several years and corruption and speculation were commonplace. Despite the reaffirmation of Obote's support in the elections, in 1981, the defeated conservative groups initiated an intense destabilization campaign that gradually became an openly anti-governmental guerrilla movement.

[25] In spite of the guerrilla presence, Obote authorized the return of Asian businesses, regulated the participation of foreign capital and undertook the reorganization of the economy, fighting corruption and speculation. Despite the intensification of political violence in 1985, Obote requested and obtained the withdrawal of Tanzanian troops who had been in Uganda since the fall of Amin.

[26] By early 1982, coffee exports had been resumed, international trade was returning to normal and negotiations to restructure the foreign debt; largely inherited from Amin's period, had been initiated with the IMF.

[27] In 1983, Obote and his party were re-elected, gaining 90 seats compared with the 35 of the Democratic party. Shortly before the elections, the UPC had achieved another major goal; the revival of the East African Community, dissolved in 1977. The Community slowly and steadily gained in strength.

[28] Between 1981 and July 1985, 16 major military offensives were launched against the main strongholds of opposition guerrillas. These groups included; the National Resistance Army (NRA), the military wing of the National Resistance Movement (NRM) founded by former president Yusuf Lule and led by Yoweri Museveni; the Uganda National Rescue Front, led by retired Brigadier Moses Ali, (one of Idi Amin's former associates); and Lt-Colonel George Nleawanga's Federal Democratic Movement (FEDEMU).

[29] Civil warfare displaced 100,000 people from their land. In 1981, the government took steps to prevent cattle smuggling by the nomadic peoples across the border, causing starvation among thousands in the province of Karamoja.

[30] In spite of the guerrilla activity, Uganda's economy had grown an average 5 per cent yearly since 1982, and, according to the World Bank, exports had increased 45 per cent since 1983. Food was not scarce and gasoline no longer rationed.

Maternal Mortality
1989-95

PER 100.000
LIVE BIRTH
550

External Debt
1994

PER CAPITA
$187

[31] Nevertheless, the floating exchange rate system introduced in 1981 accentuated inflation, the most serious problem, causing a 1,000 per cent devaluation of the Ugandan shilling, and the price of bread increased 5,000 per cent between 1979 and 1984.

[32] General elections were scheduled for December 1985, and a UPC victory was expected. However, in late July 1985, General Bazilio Olara Okello led a military coup putting an end to Obote's government. General Tito Okello, unrelated to the rebel leader, was appointed president, and within 12 months, he called elections to form a broad-based government. The Okellos belong to the northern Ajuli ethnic group, and Obote, member of the Langis, was accused of favoritism towards his own people.

[33] After the coup, the National Resistance Army intensified its action, occupying Kampala, the capital, in January 1986. After a bloody fight, Okello was ousted, and on January 30, NRA leader, Yoweri Museveni, assumed the presidency. The northern town of Golu, the last bastion of forces loyal to Okello, finally fell in March of that year.

[34] Museveni was faced with the reconstruction of a country virtually destroyed by a series of authoritarian regimes which had left almost a million dead, two million refugees, 600,000 injured and incalculable property damage.

[35] Resources are scarce in a country with a fertility rate of 7.1 children per woman, and a life expectancy of not more than 51 years. This difficult situation has been aggravated by an extremely high incidence of AIDS, which is said to have achieved epidemic proportions in some parts of the country.

[36] Uganda's foreign debt, rose to $1.2 billion in 1987. To solve this extremely serious problem, Museveni resorted to exchange arrangements with other African countries.

[37] The country has attempted to establish its economic independence, while evading the International Monetary Fund (IMF). This has led to some problems with Western countries, who do not approve of Uganda's relations with Cuba and Libya. The United States pressured Tanzania and Rwanda into ending the exchange operations which they had organized with Uganda.

[38] In 1991, the Ugandan government banned logging in the east of the country to try to prevent further environmental destruction. Although this measure has caused significant economic losses in the short term, the government announced that it would stop the timber exports.

[39] In various regions throughout the country, local elections were scheduled for March 1992. The partisans of former president Milton Obote announced their decision to boycott these.

[40] Meanwhile, President Museveni ordered the imprisonment of several journalists who reported human rights violations committed by the army in the north and northeast of the country. He also passed a law granting the government controlling power over the local press.

[41] In February 1992, local human rights organizations accused Museveni of harassing his political opponents and of not allowing the establishment of a multi-party democracy in the country. The government replied that it preferred to build a democracy based on traditional tribal structures, and that political parties were therefore unnecessary.

[42] But pressure from the opposition and some international agencies led President Museveni to authorize the election of a Constituent Assembly, which would be announced in February 1993, and established in 1995, charged with studying a draft of the new constitution. Nevertheless, the text drafted by the government was criticized by the opposition (Democratic Party and the UPC) for maintaining the partial ban on political parties for a seven-year period.

[43] In February, Pope John Paul II visited Uganda and urged the population to practice sexual abstinence, to prevent the spread of AIDS. This advice caused grave concern in a country where over 20 per cent of the population are HIV-carriers.

[44] As part of a policy aimed at winning the support of the Baganda ethnic group, Museveni authorized the restoration of the monarchy. During Prince Ronald Muenda Mutebi's coronation ceremony as kabaka, on July 31, authorities returned all royal property, which had been confiscated during former president Obote's administration.

[45] On another front, Museveni was accused by the opposition of having ordered the assassination of opposition leader Amon Bazira in Kenya in August. In the March 1994 elections the supporters of the Ugandan president took around half the seats, but the direct designation of some of the posts gave Museveni a broad majority in the new assembly.

[46] Continuing with his policy of restoring the local authorities, the leader once again authorized the Nioros, a people in the north of the country, to have their own kingdom, and this decision was consolidated in June.

[47] The discussion of a multi-party system continued throughout 1995. Museveni continued to claim the authorization of several parties would only make the "tribal divisions" more serious. On another front, the international funding organisations said they were satisfied with the economic performance of Uganda. Foreign investments grew, but budget cuts worsened the situation of most of the population as they already lived in poverty.

[48] On May 9 1996, Museveni was re-elected by more than 75 per cent of the electorate, with a 72.6 per cent turnout, defeating Paul Semogerere and Muhammad Mayanja. The president was victorious again in the May legislative elections, as his party members took 156 of the 196 seats at stake. The new government was appointed in July, with Kintu Musoke as prime minister.

Ukraine

Ukrayina

Population: 51,921,000
(1994)
Area: 603,700 SQ.KM
Capital: Kiev (Kijef)

Currency: Grivna
Language: Ukrainian

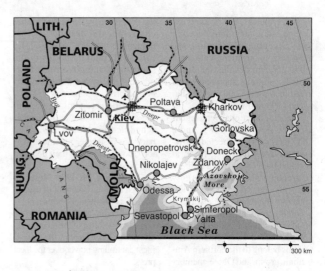

Between the 9th and 12th centuries AD, most of the present Ukraine belonged to the Kievan (Kijef) Rus, which grouped together several alliances of Eastern Slavic peoples. Its nucleus was the Russian alliance, with its capital at Kiev. The ancient Russian people gave rise to the three main eastern Slav nations: Russia, Ukraine and Belarus. In the 12th century, the Kievan Rus separated into the principalities of Kiev, Chernigov, Galich and Vladimir-Volynski, all in what is now Ukrainian territory. In the 14th century, the Grand Principality of Lithuania annexed the territories of Chernigov and Novgorod-Severski, Podolia, Kiev and a large part of Volin. The Khanate of Crimea emerged in the southern part of Ukraine and Crimea, and expanded into Galicia and Podolia. After the 11th century, Hungary began seizing the Transcarpathian territories.

2   The Ukrainians emerged, as an identifiable people, in the 15th century. Their name was derived from *krai*, a word meaning "border", which in 1213 was the name given to the territories along the Polish border. In the 16th century, the use of the name was extended to the entire Ukrainian region. Historically, there were close ties between the Ukrainians and the Russians, they had fought together against the Polish and Lithuanian feudal kingdoms and against the Tatars, in Crimea. The Ukrainian territories (Volin, eastern Podolia, Kiev and part of the left bank of the Dnepr) were incorporated into the *Rzecz Pospolita* (the union of Poland and Lithuania), which imposed Roman Catholicism.

3   During the first half of the 17th century, the struggle for independence from Poland and Lithuania intensified. The "Khmelnytsky insurrection" (1648-1654), under Bohdan Khmelnytsky, ended with the unification of Ukraine and Russia, approved by the Rada (council) of Pereyaslav. In March 1654, Ukrainian autonomy within the Russian Empire was ratified. The region along the right bank of the Dnepr and Galicia remained under Polish jurisdiction. Ukrainian colonization of what is now Kharkov (Char'cov) began in the 17th century.

4   In 1783, the khanate of Crimea - home of the Tatars - was annexed by Russia. After the partition of Poland among Russia, Prussia and Austria (1793-95), the right bank of the Dnepr became a part of Russia and Ukraine's autonomy was abolished at the end of the 18th century. In 1796, the "left-bank" of Ukraine became the Province of Malo-Rossiya (Little Russia).

5   After the fall of czarism, a dual system emerged in Ukraine, with power being divided between the Provisional Government of Saint Petersburg and the Ukrainian Central Rada (council), in Kiev. In December, after the Bolshevik Revolution, a Ukrainian Soviet government was formed in Kharkov. The Ukrainian Central Rada supported the Austro-German troops which invaded the country in the spring of 1918. In December, the Ukrainian Directorate, led by Symon Petlyura, seized power. Between 1918 and 1920, Ukraine was the scene of major fighting between the Soviets and their internal and external enemies. In December 1922, Ukraine attended the first All-Union Congress of the Soviets, held in Moscow, where the Treaty and Declaration of the Founding of the Union of Soviet Socialist Republics (USSR) was ratified.

6   In the period between the world wars, the Soviet government carried out rapid industrialization and collectivization of agriculture.

7   The secret clauses of the 1939 Soviet-German non-aggression pact incorporated western Ukraine into the USSR. In 1940, the Ukrainian Soviet Socialist Republic was enlarged through the addition of Bessarabia and Northern Bukovina. Germany attacked the Soviet Union in 1941, a strong guerrilla offensive began and by the end of World War II, all areas inhabited by ethnic Ukrainians became a part of the USSR. Ukraine participated in the founding of the UN as a charter member.

8   In 1954, Crimea - which had formerly belonged to the Russian Federation - was turned over to Ukraine, by the Soviet centralized authority. The leader of the Soviet Communist Party at the time was Nikita Krushchev, former first secretary of the Ukrainian CP.

9   By 1955 Ukraine's economy was second only to that of the Russian Federation.

10   On April 26 1986, the nuclear plant at Chernobyl - 130 km north of Kiev - was the scene of the worst nuclear accident in history, when one of its reactors exploded. The explosion affected an area inhabited by 600,000 people; 135,000 were evacuated and by 1993, 7,000 had died of radiation-related diseases. Six days after the explosion, military helicopters bombarded the reactor with 6,000 tons of sand, lead and boron, among other materials, in order to put out the fire. Later the site was covered with a thick layer of concrete forming a structure dubbed the "sarcophagus". The radioactive cloud emanating from Chernobyl affected Ukraine and neighboring Belarus most seriously, but its effects were felt as far away as Sweden. Within the next few years, foreign researchers found an increase in cases of cancer and other radioactivity related diseases.

11   In 1985 within the framework of the reforms in the USSR, Communist leaders and Ukrainian nationalists founded the Ukrainian People's Movement for Perestroika (restructuring) (RUKH), which demanded greater political and economic autonomy. In the March 1990 legislative elections, RUKH candidates received massive support from the population. On July 16 1990, the Ukrainian Supreme Soviet (Parliament) proclaimed the sovereignty of the republic. On August 24 1991, the Ukrainian Parliament approved the republic's independence, and convened a plebiscite to ratify or reject the decision.

12   On December 1 1991, 90 per cent of all Ukrainians voted in favor of ratifying the country's independence and Leonid Kravchuk, former first secretary of the Ukrainian Communist Party, was elected president, with 60 per cent of the vote. Russia, Canada, Poland and Hungary immediately recognized Ukrainian independence, underscoring the failure of Gorbachev's attempt to reach a new union treaty, within the USSR.

13   On December 8 1991, the presidents of Ukraine, the Russian Federation and Belarus pronounced the end of the USSR, founding the Commonwealth of Independent States (CIS). Ukraine declared itself a neutral and nuclear-free state, and stated its willingness to participate in the process of European integration. A few days later, seven former republics of the USSR joined the CIS, but differences among them kept them from defining the scope of the new alliance.

14   In late 1991, the Ukrainian government had claimed sovereign rights over the nuclear arms sited in their territory, affirming its commitment to destroy them within the seven years established in the Strategic Arms Reduction Treaty (START), signed between the US and the USSR in 1990. Kiev later agreed to turn these arms over to Russia, provided that they be destroyed.

15   In the first months of 1992, the Ukrainian government declared the liberation of prices; created a new currency, "karbovanets" and called for bids to build arms factories, announcing incentives for attracting foreign investment.

16   On May 5, Crimea declared independence, but it was vetoed by the Ukrainian Parliament. Crimea yielded, and withdrew the declaration. Russia reacted to the situation in June, annulling the 1954 decree

ENVIRONMENT

Ukraine is bounded by Poland, Czechoslovakia, Hungary, Romania and Moldavia in the west and southwest; by Belarus in the north and Russia in the east and northeast. The Black Sea and the Sea of Azov (Acovsko More) are located in the south. Ukraine is mostly made up of flat plains and plateaus, with the Carpathian Mountains (max. altitude, 2061 m) along the country's southwestern borders, and the Crimean Mountains (max. altitude, 1545m) in the south. The climate is moderate and mostly continental. There is black soil; both wooded and grassy steppes in the south. Much of the north is made up of mixed forest areas (such areas occupy 14% of the republic's total land surface). Pollution of the country's rivers and air has reached significant levels. Approximately 2,800,000 people currently live in areas contaminated by the Chernobyl disaster.

SOCIETY

Peoples: Ukrainians, 72.7%; Russians, 22.1%; Belarusians, 0.9%; Moldavians, 0.6%, Poles, 0.4%. **Religions:** Mainly Christian Orthodox. **Languages:** Ukrainian (official), Russian. **Political Parties:** The Christian Democratic Party; the Social Democratic Party; the National Party; the Republican Party; the People's Democratic Party; the Democratic Peasant Party; the Green Party; the Democratic Renaissance Party; the Democratic Party. **Social Organizations:** Unions are being reorganized after recent institutional changes.

THE STATE

Official Name: Ukrayina. **Administrative Divisions:** 25 Regions, the Republic of Crimea has special status as well as great internal autonomy. **Capital:** Kiev (Kijef), 2,645,000 inhab. (1994). **Other cities:** Kharkov (Char'cov), 1,622,800 inhab.; Dnipropetrovsk, 1,162,000 inhab.; Doneck, 1,115,000 inhab.; Odessa, 1,073,000 inhab. **Government:** Leonid Kuchma, President and Head of State since July 10, 1994. Legislature, single-chamber: Supreme Council, with 450 members. **National Holiday:** August 4, Independence (1991). **Armed Forces:** 517,000 (1994). **Paramilitaries:** 72,000 National Guard and Border Guard.

STATISTICS

DEMOGRAPHY

Urban: 69% (1995). **Annual growth:** 0.3% (1991-99). **Estimate for year 2000:** 52,000,000. **Children** per **woman:** 1.8 (1992).

EDUCATION

School enrolment: Primary (1993): 87% fem., 87% male. Secondary (1993): 95% fem., 65% male. University: 46% (1993). **Primary school teachers:** one for every 17 students (1991).

COMMUNICATIONS

15.0 **telephones** per 100 inhabitants (1993).

ECONOMY

Per **capita GNP:** $1,910 (1994). **Annual growth:** -8.00% (1985-94). **Annual inflation:** 297.00% (1984-94). **Currency:** 25.000 grivna = 1$ (1993). **Cereal imports:** 1,500,000 metric tons (1993). **Fertilizer use:** 841 kgs per ha. (1992-93). **Imports:** $14,177 million (1994). **Exports:** $11,818 million (1994). **External debt:** $5,430 million (1994), $105 per capita (1994). **Debt service:** 2% of exports (1994).

ENERGY

Consumption: 3,292 kgs of Oil Equivalent per capita yearly (1994), 43% imported (1994).

---

by which it had ceded Crimea to Ukraine, demanding that it be returned. Kiev refused, but granted Crimea economic autonomy. After the dissolution of the USSR, Crimea became one of the chief sources of conflict between the Kiev and Moscow governments.

[17] In March 1993, Ukraine suspended the transfer of tactical nuclear weapons to the Russian Federation, claiming there were no guarantees that these would actually be destroyed.

[18] The four CIS States with nuclear arms (Russia, Ukraine, Belarus and Kazakhstan) subsequently agreed to form an international commission to supervise the withdrawal and destruction of the nuclear weapons sited on Ukrainian soil.

[19] Tension surrounding the "ownership" of the Black Sea fleet came to a head in April, with both the Russian and Ukrainian governments claiming sovereignty over the powerful former-Soviet navy. Nevertheless, they subsequently agreed to initiate negotiations to settle the issue.

[20] Prime Minister Vitold Fokin resigned in September over the failure of his economic policies. He was replaced by Leonid Kuchma, former president of the Union of Industrialists and Entrepreneurs.

[21] The liberal policies of the new government and its privatization scheme soon came up against the dual obstacles of the Supreme Council - dominated by former communists- and worker resistance. Kuchma presented his resignation on May 21, but it was not accepted.

[22] In June, in a direct challenge to Kravchuk's moderate foreign policy, the Supreme Council announced the appropriation of the entire ex-USSR nuclear arsenal in Ukraine. With the disintegration of the Soviet Union, Ukraine became the world's third most important nuclear power.

[23] Finding himself politically vulnerable, in September 1993 Kravchuk ceded the part of the Black Sea fleet belonging to Ukraine to Russia, in compensation for debts incurred through oil and gas purchases from Moscow. In addition, he accepted help from Russia in dismantling the 46 powerful intercontinental SS24 missiles which Ukraine had wanted to keep, as a "last bastion" against any possible future expansionist schemes on the part of Russia. However, opposition in Kiev led to the invalidation of the agreement.

[24] In the meantime, the economy went out of control altogether - with inflation reaching 100 per cent per month and Kuchma resigned.

[25] In November 1993, the Council ratified the START-1 strategic weapons limitation treaty, and agreed to the gradual dismantling of 1656 nuclear warheads.

[26] The first presidential elections of the post-soviet era took place in June and July 1994. The former prime minister Leonid Kuchma defeated the outgoing president with 52 per cent of the vote, after which he declared his intention to tighten links with Russia and enter fully into the Commonwealth of Independent States (CIS).

[27] The economic and social situation continued to worsen in 1994 and 1995. The GDP fell by more than 20 per cent in 1994 and more than 12 per cent in 1995, while the constantly falling standard of living worsened poverty in the country.

[28] The tenth anniversary of the accident in the Chernobyl nuclear center in April 1996, led to an international campaign in favor of closing the many nuclear plants still functioning in the former socialist nations. According to recent studies, more than a million people were directly affected by the radiation. Ten years after the accident, 2.4 million people were living in regions which were affected by the radiation which had contaminated 12 per cent of the farming land of the nation.

United Arab Emirates

Ittihad al-Imarat al-´Arabiyah

Population: 1,877,000 (1994)
Area: 83,600 SQ KM
Capital: Abu Dhabi (Abu Zaby)
Currency: Dirham
Language: Arabic

In the southeastern corner of the Arab peninsula, between the Jabal Ajdar mountain range in Oman and the rocky steppe of Najd extending from the center of the peninsula to the Gulf shores (al-Hasa), is the Rub al-Khali desert taking up part of the territory of present-day Saudi Arabia and almost the entire United Arab Emirates.

[2] Toward the 6th century the oases which were spread over the land supplied enough water to the small stable population, enabling them to farm regular crops. The residents of the area spoke different Arabian dialects and shared diverse lifestyles. Some were grain growers, others were merchants or craftsmen from small villages and others still, usually known as "Bedouins", were nomads who raised camels, sheep and goats, making the most of the desert's scant water resources. Along with these activities, the coast alpeoples also fished in Gulf waters.

[3] Despite being a minority, the nomads - taking advantage of their condition as a mobile and armed group - and together with the merchants, dominated farmers and craftspeople. Gathered around relatively stable family chiefs, they had a typically tribal organization.

[4] Among shepherds and farmers, religion had become another type of social control but it appeared not to have a defined form. Local gods were identified with the heavenly bodies and could incarnate into rocks, trees or animals. Some families, claiming they had the power to understand the language of the gods, managed to control the others.

[5] Until the 7th century, Byzantine and Sassanid empires waged a long war with constant comings and goings which involved the peninsula, although not directly the land of the present-day Emirates. Likewise, such activity and opening of trade routes brought immigrants (mostly merchants, dealers and craftspeople) who shared their culture and knowledge of the foreign world.

[6] Islam was peacefully adopted in the area during Muhammad's lifetime. Tribal chiefs secured their power but the lifestyle of the few inhabitants was not significantly modified.

[7] Upon the prophet's death, various groups appeared which disputed his spiritual inheritance. One of them, the Ibadis (who claimed to descend straight from Muhammad), created the Uman (Oman) *imanate* in mid-8th century. It lasted up to the 9th century when it was suppressed by the Abassids, caliphs who claimed a universal authority and whose main capital was in Baghdad.

[8] After the 11th century, the Sunni form of Islam gradually went on from being the ruling groups' religion to reach the population at large. Likewise, the Ibadis communities continued to exist until the 15th century, exerting strong religious authority.

[9] The Gulf ports were already of significant importance for the traffic of commodities which came from China (textiles, glass, porcelain and spices), being transported to the Red Sea through an oasis chain.

[10] During the 17th and 18th centuries, the Ottoman Empire occupied a large area of the peninsula but had not reached the Gulf coast in the southeastern region. The Ibadis had re-instated their imanate under a Yaribi dinasty. Further to the north, Bahrain was under Iranian domain.

[11] Beyond the reach of the Ottomans (busy with constant fights in Europe, Africa and Asia), the southeastern region of the peninsula thrived on trade. Ruling families linked directly with merchants appeared and piracy developed, benefiting from the natural advantages of the indented coasts. The area came to be known as "Pirate coast".

[12] An important change was introduced when European fleets increasingly used the maritime route through the Cape. British influence grew gradually, since they used the Gulf ports as a stop on the way to India, and helped to combat piracy.

[13] At the beginning of the 19th century, Britain held complete control over the region, obtained through agreements reached with the local chiefs and small governors of the ports. Since then, the Pirate Coast was to be known as the "Trucial States", which included Abu Dhabi, Dubai and Sharja. Relations with Britain were conducted in the same way until the first decades of the 20th century.

[14] Around 1914, the Saudi state re-emerged as a great force in Central Arabia, becoming a threat even for the Ottoman power. Russia, France and Germany also sought to intensify their presence in the area. This led the British to formalize relations with the Trucial States of Bahrain, Oman and Kuwait, which let the government in London handle their affairs with the rest of the world.

[15] World War I, which led to the end of the Ottoman Empire and the independence of several Arab peninsula states, did not alter the relations between Britain and the Trucial States. London was the real power behind the puppet Abd al-Aziz in the new kingdom of Saudi Arabia, who controlled the southern and southeastern coasts of the peninsula. Furthermore, air routes were beginning to develop and the Gulf's airfields along with those in Egypt, Palestine and Iraq, had an important role in the Middle East.

[16] During the period between the world wars, England focused its attention onto Egypt and Iraq (where oil fields were already being exploited). After World War II, relations among Arab countries changed. The League of Arab States was formed in 1945 by those countries which had some form of independence.

[17] At the beginning of the 1960s, the Middle East's oil deposits were known to be among the largest in the world. The United States joined Britain in keeping their control over the Gulf States, which received almost 100 per cent of their revenues from oil.

[18] The growing influence of Gamal Abdel Nasser in the Arab world led Britain to allow greater local participation in the governments of several states of the protectorate. In 1968, it decided to withdraw all its military force from the region. That same year, the OPEAC was created - a branch of OPEC - formed exclusively by oil-exporting Arab states.

[19] In 1971, upon Britain's definite withdrawal, Abu Dhabi began a large-scale exploitation of its oil wells. Thus, the clear setting of borders within the territories became indispensable. Under British influence, the United Arab Emirates were created that year without participation of Qatar or Bahrain.

[20] Immediately after its formation,

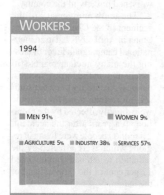

WORKERS

1994

MEN 91%   WOMEN 9%

AGRICULTURE 5%   INDUSTRY 38%   SERVICES 57%

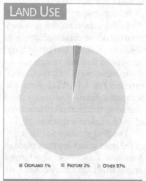

LAND USE

CROPLAND 1%   PASTURE 2%   OTHER 97%

ENVIRONMENT

Located in the southeastern part of the Arabic peninsula, stretching from the Qatar peninsula toward the strait of Hormuz, the land is mostly desert with few oases and wadis (dry and rocky river beds). The coastal areas are very hilly lowlands with coral islands offshore and sand dunes. Most of the oil fields are located in these areas.

SOCIETY:

Peoples: Population includes large immigrant component. From Bangladesh, India, Pakistan and Sri Lanka 45% (1993); Arabs 25%, of which 13% come from other Arab countries (mainly Egypt) and 12% from the Emirates themselves; Iranians 17%; other Asian and African countries 8%; Europeans and US 5%. 80% live in Dubai, attracted by the oil wealth. **Religions:** Muslim 94,9% (Sunni 80%, Shiite 20%); Christian 3,8%; others 1,3%. **Languages:** Arabic (official). English is used among immigrants and in business. **Political Parties:** There are neither political parties nor trade unions.

THE STATE

Official name: Daulat al-Imarat al-'Arabiyah al-Muttahidah. **Administrative divisions:** 7 emirates: Abu Dhabi, Dubai, Sharjah, Ras al-Khaimah, Ajman, Fujairah and Umm al-Qaiwain. **Capital:** Abu Dhabi (Abu Zaby), 363,432 inhab. **Other cities:** Dubai, 585,189 inhab.; Al-'Ayn 176,411 inhab.; Ash-Shariqah 125,000 inhab.; Ra's al-Khaymah 42,000 inhab.(1989). **Government:** Parliamentary Republic with a president of the federation as a head of state (Sheikh Zaid ibn Sultan al-Nuhayyan, Emir of Abu Dhabi, since 1971); a Vice President and Prime Minister (Sheikh Rashid ibn Said al-Maktum, Emir of Dubai, since 1990) and a National Federal Council with 40 members appointed by the emirs, serves as the legislative body. **National Holiday:** December 1. Proclamation of the Union (1971) **Armed Forces:** 70,000 (1995)

STATISTICS

DEMOGRAPHY

Urban: 83% (1995). **Annual growth:** 0.6% (1992-2000). **Estimate for year 2000:** 2,000,000. **Children per woman:** 4.5 (1992).

HEALTH

One **physician** for every 1,042 inhab. (1988-91). **Under-five mortality:** 20 per 1,000 (1994). **Calorie consumption:** 114% of required intake (1995). **Safe water:** 95% of the population has access (1990-95).

EDUCATION

Illiteracy: 21% (1995). **School enrolment:** Primary (1993): 108% fem., 108% male. Secondary (1993): 94% fem., 84% male. **University:** 11% (1993). **Primary school teachers:** one for every 17 students (1992).

COMMUNICATIONS

107 **newspapers** (1995), 102 **TV sets** (1995) and 105 **radio sets** per 1,000 householdss (1995). 32.1 **telephones** per 100 inhabitants (1993). **Books:** 100 new titles per 1,000,000 inhabitants in 1995.

ECONOMY

Annual growth: 0.40% (1985-94). **Currency:** 4 dirhams = 1$ (1994). **Cereal imports:** 483,000 metric tons (1993). **Fertilizer use:** 4,436 kgs per ha. (1992-93). **Imports:** $21,100 million (1994). **Exports:** $19,700 million (1994). **Development aid received:** $-9 million (1993); $-5 per capita; 0% of GNP.

ENERGY

Consumption: 12,795 kgs of Oil Equivalent per capita yearly (1994), -470% imported (1994).

---

the new state had to face a conflict with Iran which, claiming historical reasons, occupied the islands of Abu Mussa, Tunb al-Cubra and Tunb al-Sughra on the strait of Ormuz. During the first decade, oil production had constant growth, mainly in the three most important emirates: Abu Dhabi with 79 per cent of the total, Dubai and Sharjah. National participation in the control of oil exploitation also grew.

[21] When the oil-exporting countries, through OPEC, decided in 1973 to raise the price of the barrel by 70 per cent and reduce supply by 5 per cent, a new era in the relations with the industrialized world had begun. The results of this policy were explosive. From the rock-bottom price of $1.50 a barrel of crude oil in the early 1960s, it rose to $10 in 1973 and to $34 between 1979 and 1980. The Emirates' annual growth rate in the 1970s was over 10 per cent due to the revenues obtained.

[22] This led to two immediate effects: a rapid growth of cities with state-of-the-art highways, oil pipelines and banks, and a large influx of immigrants attracted by the region's economic possibilities. Little was left of the ancient peoples' pursuits of fishing or collecting of pearls in the coast.

[23] The 1980s began with the Iran-Iraq war. Although the Emirates maintained an apparently neutral stand, they gave economic support to Iraq to avoid a possible "Iranization" of the region. Once the conflict had ended, the Arab Emirates had become the Middle East's third biggest oil producer, after Saudi Arabia and Libya.

[24] Although since 1981 the government has tried to develop other industrial fields, technological differences with other countries and a limited domestic market make difficult a possible industrial diversification which would diminish dependence on oil production.

[25] The country has been part of the Non-Aligned Movement and has supported Palestinian claims. In late 1986, diplomatic relations were established with the Soviet Union and the People's Republic of Benin. In 1987, relations with Egypt were reinstated, broken after the Camp David agreements with Israel.

[26] During the Gulf War, the Emirates made financial contributions to the fight against Iraq. Once the conflict had ended, the policy to diversify the economy seemed to gather strength when the free port of Jabel Ali was opened to over 260 foreign companies.

[27] In March 1991, the Gulf Cooperation Council signed an agreement with the US to preserve security in the region, which includes a common military strategy and mechanisms to prevent arms proliferation, among other items.

[28] In 1992, with Syrian mediation, Iran restricted the claims it had over the islands of the Strait of Ormuz, occupied since the 1970s. Tehran claims sovereignty over Abu Mussa and has allowed expelled Arab residents to return. The conflict is, by pressure from the Emirates, under international arbitration.

[29] Successive immigration waves have formed a very heterogeneous population. According to 1993 figures, the Arab population in the country amounts to barely one quarter, half of it coming from other countries (mainly Egypt). The rest of the population is made up by immigrants from Bangladesh, Pakistan, Sri Lanka, Iran and other Asian or African states.

[30] Everything indicates that the influence of Islamic fundamentalism has grown in the Emirates between 1993 and 1996. Sheikh Zaid, president of the Union, has often made it understood - as in a speech at the end of Ramadan in 1993 - that official circles were concerned about the spread of Muslim "radicalism". In February 1994, Zaid decided to extend "Islamic law" to many criminal cases which had been dealt with by civil courts up to then.

[31] In 1995, new contacts were held with Iran to try to solve the dispute over the Strait of Ormuz islands but, as had happened in previous attempts, there was no positive outcome. Meanwhile, the US State Department reported in March 1996 that human rights continued to be curtailed in the Emirates.

United Kingdom

Population:	58,395,000 (1994)
Area:	244,880 SQ.KM
Capital:	London
Currency:	Pound Sterling
Language:	English

United Kingdom

The first known inhabitants of the island of Britain were paleolithic hunters, following herds of wild animals. After the final ice age, agriculturalists began to settle on the island. Over thousands of years, these people, and the many others who migrated from the continent, evolved increasingly complex social systems. By the final millennium BC, Britain was dominated by Celtic tribes who used iron tools and had extensive contact with the European mainland.

[2] In 44 AD the Romans invaded southern Britain. In 90 AD they created the province of Britannia, and between 70 and 100 AD, founded London. In the early 5th century, they abandoned the island, leaving it largely defenceless against the raids of Angles, Saxons and Jutes. These Germanic peoples pushed the Celts westwards, taking over the southern part of the island and establishing Anglo-Saxon kingdoms.

[3] During the 5th century, the inhabitants of Ireland and Wales adopted Christianity. In the 7th century, the British church came under the power of Rome.

[4] During the 7th and 9th centuries Danish invaders overran the eastern part of England. In the 11th century, the Normans, led by William the Conqueror, invaded England and secured the throne. Successive Anglo-Norman kings maintained their power by establishing various forms of vassalage over the feudal lords.

[5] The prestige which Richard the Lion-Heart (1189-99), one of the leaders of the Third Crusade, gained for the English Crown was lost under the reign of his successor, King John (1199-1216). Under John, England lost its French territories, and the barons, in alliance with the clergy, were able to restrict the power of the monarchy through the Magna Carta, signed in 1215.

[6] The Magna Carta laid the foundations for the British parliamentary system. It also marked the beginning of a continuous power struggle between the monarchy and the nobility. The growing power of the landowning class and later the bourgeoisie, eventually led to the consolidation of a parliamentary monarchy.

[7] Frequent dynastic conflicts, disputes over territories in France belonging to the English crown, commercial rivalry between England and France in Flanders, and French aid to Scotland in its wars with England paved the way for the Hundred Years' War (1337-1453), which culminated in the loss of the English possessions on the continent.

[8] The negative effects of the war increased the unpopularity of the monarchy, which was faced, at the same time, with an anti-Papal movement led by the followers of Wycliffe (a precursor of Luther) and a peasant rebellion. The peasants, led by Wat Tyler, rose up against the payment of tribute and the power of the feudal lords. In 1381, Tyler and his followers managed to enter London and negotiate directly with the king. The Peasants' Revolt was unsuccessful, however, and Tyler was later executed.

[9] The period following the Hundred Years' War was dominated by a long struggle for control of the throne between the Houses of Lancaster and of York. This finally led to the War of the Roses which ended with the coming to power of the Welsh House of Tudor in 1485. The Tudor period is considered the beginning of the modern British state. One of the Tudor kings, Henry VIII (1509-47) broke away from the Church of Rome, confiscating all its monasteries and founding the Anglican Church. The desire to extend English authority and the religious Reformation to Ireland led to the subjugation of Ulster by Henry's successor Elizabeth I (1558-1603). Tudor involvement in Ireland laid the foundations of centuries of religious and political conflict in the country.

[10] Under the reign of Elizabeth I (1558-1603), poetry and the theatre flourished with playwrights such as Ben Jonson, Marlowe, and Shakespeare. Industry and trade developed, and the country embarked upon its "colonial adventure", the beginning of its future empire. After defeating the Spanish fleet, the "Invincible Armada", in 1588, the British Navy "ruled the waves", with no other fleet capable of opposing it.

[11] British merchant ships involved in the slave trade, laden with colonists or belonging to pirates and privateers sailed the oceans freely. Markets multiplied, demand grew rapidly, and producers were forced to seek new techniques in order to accelerate production. It was a prelude to the industrial revolution which was to arise in the country at the beginning of the 18th century.

[12] In 1603, the crowning of James I (James VI of Scotland) put an end to the independent Scottish monarchy. The religious intolerance of James' son Charles I, led to a Scottish uprising and increasing discontent in England. The deteriorating political situation led to the Puritans forming an army supported by Parliament; led by Oliver Cromwell, they defeated the royal forces in 1642. In 1649, Parliament sentenced the king to death and proclaimed Cromwell "Lord Protector", establishing a republic known as the Commonwealth. After his death, in 1658, the monarchy was restored with Charles II.

[13] The priorities of the new regime were the colonization of North America and trade with America, the Far East and the Mediterranean. The slave trade, the kidnapping, trafficking and selling of slaves from Africa to buyers in America and other places, initiated in the 16th century, became one of the main sources of income for the empire.

[14] The absolutism of James II and his espousal of Catholicism were opposed by the Protestant Parliament which deposed James through the "Glorious Revolution". Parliament invited the Dutch prince William of Orange to assume the English throne. William was forced to sign the Declaration of Rights (1689), limiting royal powers and guaranteeing the supremacy of Parliament.

[15] In this period John Locke summarized revolutionary ideals, proposing that human beings have basic natural rights: to property, life,

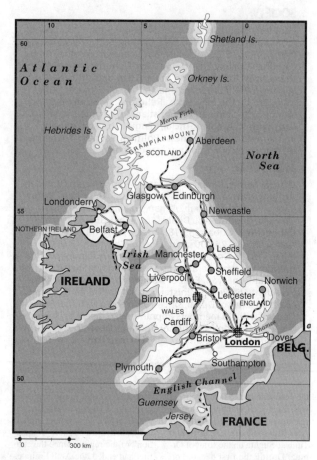

LAND USE

■ CROPLAND 27%   ■ PASTURE 46%   ■ OTHER 27%

liberty and personal security. Government, created by society to protect these rights, must fulfil its mission; if it fails to do so, the people have the right to resist its authority.

[16] In 1707, the parliaments of Scotland and England joined together, creating the United Kingdom of Great Britain. England intervened in the war of succession in Spain, obtaining Minorca, Gibraltar and Nova Scotia through the Treaty of Utrecht (1713). In 1765, increased taxes imposed by the Stamp Act triggered the rebellion and secession of the American colonies, who declared their independence in 1776.

[17] During this period, the two large traditional political parties were formed: the Conservatives (Tories), representing the interests of the large landowners, and the Liberals (Whigs), representing the merchant class. The ideas forming the basis of economic liberalism were developed at this time by Adam Smith. The liberal doctrine provided the political ideology for British imperialism, which used the concept of "free trade" as a justification for forcing open the ports and markets of the Third World, often with the use of naval force. Perhaps the most notorious examples of this were the Opium Wars fought against China in the mid-19th century.

[18] After the crushing of a nationalist rebellion in Ireland in 1798, the United Kingdom of Great Britain and Ireland was created in 1801 with the dissolution of the Irish Parliament.

[19] The 18th century gave rise to the agricultural "revolution", which introduced important innovations in farming techniques, as well as major changes in land tenure. The large landowners enclosed their properties, eliminating communal lands which had hitherto been used by small farmers, and introducing a more capitalist agricultural economy.

[20] At the same time, the industrial revolution began, with the textile manufacturers being the first to confront the problem of meeting a growing demand for fabric overseas. The introduction of machinery changed the way in which work was done, and the medieval shop was replaced by the factory. On the heels of the textile industry came mining and metallurgy. The mechanization process was consolidated with the invention of the steam engine, the use of coal as a fuel and

ENVIRONMENT

The country consists of the island of Great Britain (England, Scotland and Wales), Northern Ireland, and several smaller islands. The Grampian mountains are located in Scotland and the Cambrian mountains in Wales. The largest plains are in the southeast, around London. The climate is temperate. Farming is highly mechanized and is now a secondary activity. The huge coal and iron ore deposits, which made the Industrial Revolution possible, have nearly run out, but recent gas and oil finds in the North Sea have turned the United Kingdom into an exporter of these products. The country is highly industrialized and environmental pollution, especially air pollution, poses a serious problem. A nuclear reprocessing plant on the Irish Sea appears to be linked to the high incidence of leukemia in areas along both coastlines.

SOCIETY

Peoples: English, Scots, Welsh and Irish. **Minorities:** Pakistani, 0.9%, Indian, 1.5%, and Afro-Caribbean, 0.8%. **Religions:** Religious participation of about 8,400,000 active members in 1990 - Christian, 80% of which Roman Catholic, 21%; Anglican, 20%; Presbyterian, 14%; Methodist, 5%; Baptist, 3%; Muslim, 11%; Sikh, 4%; Hindu, 2%; Jewish, 1%; other, 2%. **Languages:** English (official), Welsh and Gaelic. **Political Parties:** Conservative Party, led by John Major, right-wing, in government; Labour Party, led by Tony Blair, affiliated to the Socialist International, in opposition; Liberal Democrat, "centre"; National Front, far-right; and Green Party. **Social Organizations:** Trade Union Congress (TUC).

THE STATE

Official Name: United Kingdom of Great Britain and Northern Ireland. **Administrative Divisions:** 39 Counties and 7 Metropolitan Districts. **Capital:** London, 6,933,000 inhab. (1993). **Other cities:** Birmingham, 1,012,400 inhab.; Leeds, 724,500 inhab.; Glasgow, 681,500 inhab.; Sheffield, 531,900 inhab.; Bradford, 488,000 inhab.; Liverpool, 477,000 inhab.; Edinburgh, 441,600 inhab.; Manchester, 432,000 inhab.; Bristol, 397,600 inhab. **Government:** Queen Elizabeth II, Head of State. John Major is Prime Minister. Parliamentary monarchy. Legislative Power: a House of Lords, with little effective power, and a House of Commons, with 651 members elected every five years. **Armed Forces:** 236,900 (1995). **Dependencies:** Anguilla, Cayman, Montserrat, Turks and Caicos and Virgin Islands (Caribbean), Malvinas-Falklands (Argentina), Gibraltar (Spain), Bermuda and St. Helena, Tristan da Cunha, Pitcairn (Pacific Ocean).

the substitution of first iron, then steel, for wood in construction.

[21] This period was characterized by population growth (going from 10,900,000 in 1801 to 21,000,000 in 1850), increasing demand and expanding trade, improvements in the transport system, capital accumulation, the creation of a vast colonial empire, scientific advances and the golden age of the bourgeoisie. England became the world's number one manufacturing nation. Its colonial policy helped to prevent competition with its factories; for example it established regulations which destroyed the Indian textile industry.

[22] The United Kingdom obtained new territories from its different wars with France, and its triumph over Napoleon at Waterloo (1815).

[23] One result of the industrial revolution was growing discontent among the rapidly increasing working class, due to low salaries, unhealthy working conditions, unsatisfactory housing, malnutrition, job insecurity and the long and tiring working days to which men, women and children were subjected. In many cases, popular uprisings were characterized by violence, and were met with equally violent repression.

[24] In the early stages of the industrial revolution, spontaneous movements arose, like the "Luddites", weavers who destroyed machinery to prevent it from destroying their trade. Trade unions began to appear later.

[25] In 1819 a demonstration in Manchester was violently put down, and repressive legislation followed, limiting the right of association and freedom of the press. Nevertheless, resistance movements continued their activity. One of the main movements of this period was the nationalist Irish Association led by Daniel O'Connell.

[26] The most important of the mass movements was the Charter Movement, made up primarily of workers. It took its name from the People's Charter, published in 1838 at a mass assembly in Glasgow, Scotland. This movement brought a number of issues to the forefront, both political; universal suffrage, use of the secret ballot, reform of voting registers, and social; better salaries and better working conditions. After its demonstrations and strikes, "Chartism" faded away; however, it had a far-reaching influence and its grievances were subsequently taken up by some members of parliament.

[27] Robert Owen (1771-1858), considered to be the founder of socialism and the English cooperative movement, argued that the predominance of individual interests led to the impoverishment of the masses. From 1830 on, he devoted himself to the establishment of co-operatives and the organization of labor into trade unions.

[28] During the long reign of Queen Victoria (1837-1901), the traditional nobility strengthened its alliance with the industrial and mercantile bourgeoisie. Trade unions were legalized in 1871, and shortly afterwards, some labor legislation was approved.

[29] Beginning in 1873, the rising population numbers led to a food shortage, making imports neces-

MILLIONS $ 1994

IMPORTS
227,000
EXPORTS
205,000

STATISTICS

DEMOGRAPHY

Urban: 89% (1995). **Annual growth:** 0.3% (1991-99). **Estimate for year 2000:** 59,000,000. **Children** per **woman:** 1.8 (1992).

HEALTH

Under-five mortality: 7 per 1,000 (1994). **Calorie consumption:** 110% of required intake (1995).

EDUCATION

School enrolment: Primary (1993): 113% fem., 113% male. Secondary (1993): 94% fem., 91% male. University: 37% (1993). **Primary school teachers:** one for every 20 students (1990).

COMMUNICATIONS

108 **newspapers** (1995), 105 **TV sets** (1995) and 110 **radio sets** per 1,000 households (1995). 49.4 **telephones** per 100 inhabitants (1993). **Books:** 109 new titles per 1,000,000 inhabitants in 1995.

ECONOMY

Per capita GNP: $18,340 (1994). **Annual growth:** 1.30% (1985-94). **Annual inflation:** 5.40% (1984-94). **Consumer price index:** 100 in 1990; 114.3 in 1994. **Currency:** 0,6 pounds sterling = 1$ (1994). **Cereal imports:** 3,534,000 metric tons (1993). **Fertilizer use:** 3,205 kgs per ha. (1992-93). **Imports:** $227,000 million (1994). **Exports:** $205,000 million (1994).

ENERGY

Consumption: 3,754 kgs of Oil Equivalent per capita yearly (1994), -9% imported (1994).

---

sary. At the same time industry began to feel the competition from the US and Germany. Britain increased its imperial activities in Africa, Asia and Oceania, not only for economic reasons, but also because of the political ambition to build a great empire. The Boer or South African War (1899-1902), fought to secure control over southern Africa, was the most expensive regional conflict of the 19th century.

[30] The first quarter of the 20th century saw the birth of the women's liberation movement. The militancy of the suffragettes led to some women obtaining the right to vote in 1917. The most famous example of their militancy was the suicide of Emily Davison, who threw herself in front of the king's horse during a race in 1913.

[31] In Ireland, the majority Catholic population were stripped of their lands, restricted in their civil rights because of their religion, and deprived of their political autonomy. Millions emigrated, and political unrest periodically resulted in violent uprisings. Not until 1867 were the privileges of the Anglican Church eliminated; at the same time, measures were taken to improve the situation of the peasants. The 1916 Easter Rising in Dublin was ruthlessly put down by the British, but the crown forces were unable to win the ensuing guerrilla war which began in 1918, and Britain finally granted Ireland independence in 1921. Six counties in the north-east, with Protestant majorities, remained under British control with a devolved administration in Belfast.

[32] Economic and political rivalry between the European powers led to the outbreak of World War I (1914-18). The Central Powers of Austro-Hungary and Germany, joined subsequently by Turkey and Bulgaria, fought against the Allied powers of France, Great Britain, Russia, Serbia and Belgium, with

Italy, Japan, Portugal, Romania, the United States and Greece joining during the course of the war.

[33] Despite its victory, Britain emerged from the war weakened. It had invested $40 billion in military expenditure, mobilized 7,500,000 troops, suffered a loss of 1,200,000 soldiers and acquired an enormous foreign debt. The deep economic depression in the post-War years led to renewed unrest among workers, which reached its height in the General Strike of 1926. The Conservative government declared the strike illegal, but did not take any measures to revive British industry. In the elections of 1929, the Labour Party came to power for the first time.

[34] The United Kingdom supported the US proposal to create the League of Nations. In 1931, the British Community of Nations (Commonwealth) was established under the Statute of Westminster. This formally recognized the independence of Canada, Australia, New Zealand and South Africa.

[35] On September 3 1939, Britain declared war on Germany, marking the beginning of its participation in World War II (1939-45). In May 1940 a coalition cabinet was formed, with Winston Churchill as prime minister. From 1939 to 1941, Britain and France were ranged against Germany, which was joined by Italy in 1940. Hungary, Romania, Bulgaria and Yugoslavia participated in the war as "lesser" allies of the Nazis.

[36] In 1941, the Soviet Union, Japan and the US entered the conflict. On May 8 1945, Germany surrendered. The UK, US and the USSR emerged as the major victors from the war. However, the British Empire was eclipsed by the rising power of the US, which became the undisputed economic, technological and military leader.

[37] In May 1945, the Labour government of Clement Attlee, who won the elections with the slogan "We won the war, now we will win the peace.", nationalized the coal mines, the Bank of England and the iron and steel industries.

[38] Pakistan was formed and India became independent in 1947, although both remained members of the British Commonwealth. During the following decade, most of Britain's overseas colonies obtained their independence. Britain was a founder member of NATO in 1949.

[39] The Franco-British military in-

tervention in the Suez Canal Zone in 1956, which failed due to a lack of US support, was met by strong criticism from both inside and outside Britain (see Egypt). The following year, the United Kingdom set off its first hydrogen bomb in the Pacific Ocean.

[40] The general election of 1964 was won by the Labour Party, under the leadership of Harold Wilson. His government faced serious problems, such as the declaration of independence by Southern Rhodesia (today Zimbabwe), and the breaking of diplomatic relations with nine other African countries.

[41] In 1967, having been denied entry to the Common Market, and faced with rapidly increasing unemployment, Wilson withdrew British troops from South Yemen, evacuated all bases located east of Suez except for Hong Kong, discontinued arms purchases from the US and implemented a savage austerity budget.

[42] In 1967 a law was passed legalizing the termination of pregnancy up until the third month. Three years later, sexual equality was guaranteed by law.

[43] In Northern Ireland in 1969, the latent conflict erupted into action, a number of people were killed and wounded in riots between Catholics and Protestants. The Catholics demanded equal political rights, and better access to housing, schools and social security. The Protestant-controlled Northern Irish government responded by sending in their armed police reserves against the Catholic demonstrators. The British government sent in their troops to separate the two sides and took control of police and reserve forces away from the Belfast government.

[44] In August 1971, the prime minister of Northern Ireland, Brian Faulkner, opened internment camps and authorized the detention of suspects without trial. Protests against these measures resulted in more than 25 deaths. On January 30 1972, "Bloody Sunday", British soldiers opened fire on a peaceful protest march in Derry (Londonderry), killing 13 Catholics and injuring hundreds more. The Irish Republican Army (IRA) responded with numerous assassinations.

[45] In March 1973, the people of Northern Ireland voted in a referendum to remain within the United Kingdom rather than join a united Ireland. Voting was characterized by a high rate of ab-

stentions of 41.4 per cent, and the protestant majority voted to stay in the UK.

[46] In the 1970s, social conflict in Britain intensified, and Edward Heath's Conservative government was faced with strikes in key public enterprises. The dockers, coal miners and railway workers all went on strike. Inability to deal with this labor unrest led to his resignation in 1974, and the Labour Party won the following elections. Amid a complex situation on the domestic front, in January 1973, a majority of the electorate voted in favor of entering the EEC. A policy of progressive integration with Europe began, as well as a search for new markets for ailing British industry.

[47] A new divorce law was passed in 1975. The same year, feminists campaigned successfully against restrictions being added to the 1967 abortion law.

[48] In 1979, voters in Scotland and Wales turned down autonomy for their regions in referenda organized by James Callaghan's Labour government.

[49] In May of that year, after the "Winter of discontent", characterized by strikes, the Conservative Party won the election, with Margaret Thatcher as its leader. The new prime minister brought in a severe monetarist policy to bring down inflation. She began to revert the nationalization process carried out under Labour, and returned to a free market policy.

[50] In 1981, a group of IRA prisoners began a hunger strike as part of their campaign to win recognition as political prisoners. The government refused to accept the prisoners demands. The strike resulted in 12 deaths.

[51] In April 1982, Thatcher sent a Royal Navy force, including aircraft carriers and nuclear submarines, to the Malvinas islands (Falklands) which had been occupied by troops from the military junta in Argentina. After 45 days of fighting the British recovered the islands for the Crown (see Argentina).

[52] In October 1983, the British government decided to withdraw its troops from Belize. The following year, in agreement with a treaty dating back to the First Opium War, Britain ceded sovereignty over Hong Kong to the People's Republic of China, to be effected in June 1997.

[53] During the Thatcher administration, the labour union movement suffered serious setbacks, with the loss of local affiliates in the industrial sector, itself in decline. The 1984/85 miners' strike culminated, after a year of violent internal strife and confrontation with the police, in defeat for the union.

[54] In 1987, Thatcher was elected to her third consecutive term in office. She continued her policies as before; radical economic liberalization, privatization of state corporations, and opposition to union demands. In foreign affairs, Britain opposed greater European Community integration and continued to align itself closely with the US.

[55] In mid-1989, after several years of growth, inflation reached high levels, unemployment continued to rise steeply, productive investments fell and the balance of payments deficit grew. At the same time, the introduction of a new Poll Tax and other projected reforms produced strong popular resistance.

[56] In February 1990, the United Kingdom and Argentina renewed diplomatic relations, and their representatives met in Madrid to negotiate the issue of the return of the Malvinas.

[57] In November 1990 Thatcher was replaced, as head of the government and Tory leader by her former minister, John Major. Upon taking office, Major declared himself to be in favor of capitalism with a human face, thus setting himself apart from the reality of the "Iron Lady's" harsh capitalism.

[58] In 1991, Major announced that the Poll Tax would be replaced, and promoted the adoption of a bill of rights for the sick, for working women, the consumer and the family, among other social categories. However, Major continued most of Thatcher's reforms, including that of the health system and the selling-off of nationalized industries such as the railways.

[59] In European affairs, the prime minister distanced himself from his predecessor. In 1991, London gave its backing to European agreements on monetary union. However, the fidelity which British diplomacy showed towards the United States, remained unaltered, as proved by British aid to the US in the Gulf War.

[60] The government managed to bring down inflation (which dropped from 10 per cent to 3.8 per cent between 1990 and 1991) and interest rates (which decreased from 15 per cent to 9.5 percent), but

economic activity remained stagnant. In 1991, industrial production declined and numerous small businesses failed. Unemployment, in turn, continued to rise (over 9 per cent by late 1991), and started affecting white-collar workers and professionals, groups which had supported Thatcher's neo-liberal policies.

[61] Within this context, Major's personal popularity (the highest in British history since Winston Churchill) has not translated itself into a similar level of support for his party. In the local elections held at the end of April 1991, the Tories lost 800 council seats and several of its traditional districts, while Labour gained more than 400 council seats. In these elections, the Liberal Democratic Party also made important gains.

[62] In September 1991, frustration with high unemployment and cuts in social spending erupted in a wave of urban violence not seen since 1976. Demonstrations - specially by young people - took place in cities as far apart as Cardiff, Newcastle, Birmingham and Oxford. At the same time, racist violence had significantly increased, according to Scotland Yard.

[63] In 1992, the Conservatives won parliamentary elections for the fourth consecutive time with 336 seats out of 651 giving them an overall majority. Shortly after the voting in which Gerry Adams, president of Sinn Fein - the political wing of the Irish Republican Army (IRA) - lost his seat, several high-powered bombs went off in London.

[64] As of 1993, the Conservatives suffered a series of setbacks at the polls in local elections, in the middle of an economic recession and high unemployment affecting some 3 million. On December 15, London signed a joint declaration with Dublin regarding the situation in Northern Ireland, which paved the way for peace talks (see Ireland).

[65] In spite of a more favorable economic context, the Conservative government was not able to recover its popularity during 1994. The GDP grew 3 per cent and unemployment went down to 2.5 million (9 per cent of the working population), but a series of scandals, such as the illegal financing of a dam in Malaysia, further tarnished the Tories' image. Meanwhile, parliament decided to lower the legal age of consent for homosexual intercourse from 21 to 18, refusing to put it on

the same level as heterosexual intercourse, where the age of consent is 16.

[66] The economic situation in 1995 - similar to the previous year although with a slightly lower growth and somewhat higher inflation, while unemployment affected 2.2 million - did not help to increase Major's popularity. Meanwhile the Labour Party, led by Tony Blair since July 1994, pursued its "modernization" program, eliminating from the party's constitution the decision to advance toward "the common ownership of the means of production, distribution and exchange".

[67] The Labour Party won new victories at the local election polls in 1996 and is hopeful of a return to power after the elections in 1997. According to Le Monde Diplomatique, the "US-style modernization" of Blair's party would coincide with its "ideological decay" and transform it into the representative for the British middle class rather than for the trade unions or for the "classes brutalized by 15 years of ultra-liberalism".

United States

United States

Population: 260,650 ,000 (1994)
Area: 9,372,610 SQ KM
Capital: Washington

Currency: Dollar
Language: English

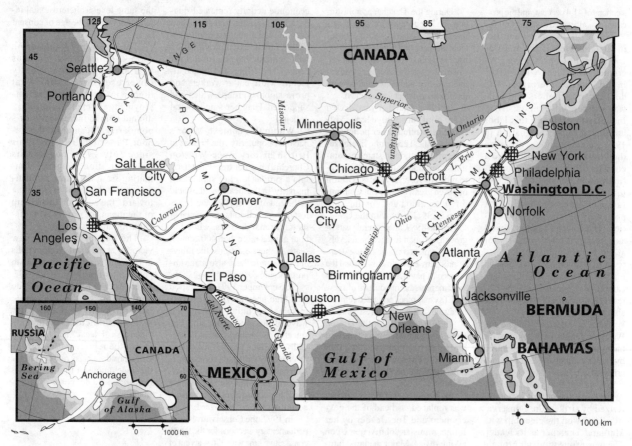

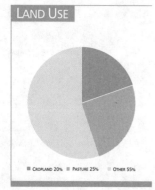

LAND USE

■ CROPLAND 20% ■ PASTURE 25% ■ OTHER 55%

The continental territory occupied by the United States was inhabited 30,000 years before the arrival of the Europeans, by peoples who came from the northwest, probably from Asia, across the Bering Strait. Among the most important peoples were the Apache, the Arapaho, the Cherokee, the Cheyenne, the Chippewa, the Crow, the Comanche, the Hopi, the Iroquois, the Lakota, the Navajo, the Nez Perce, the Oglala Sioux, the Pawnee, the Pueblo, the Seminole, the Shawnee, the Shoshone and the Ute.

2   In the deserts and on the plains, they were primarily hunters and gatherers, living in small tribes with a simple social structure. Where the lands were more fertile, agriculture developed and relatively large towns were established. The first and largest of these, Cahokia, close to what is now St Louis, is thought to have had a population of 40,000 in the year 1000 AD.

3   The religious beliefs of these peoples were rooted in a cosmic conception of the Earth, which is that as it belongs to the Universe, it belongs to no one. It is considered to be a living being, with both material and spiritual powers. Their religious leaders, or shamans, have the power to call upon the forces of this sacred Universe, either to foretell the future, to lead their people, or to heal the sick.

4   The first Europeans to come to America were Scandinavian navigators, but they did not settle permanently in the region. After the voyage of Christopher Columbus, in 1492, the Spaniards established the colonies of Saint Augustine in Florida, and Santa Fe in New Mexico; in addition, they explored both Texas and California. After the Spaniards came the British, French and Italian, all bent on territorial conquest.

5   In 1540, Hernando De Soto wrote in his journals of having found among the Cherokee an advanced agricultural society, linked to the peoples of the Ohio, the Mississippi and even the Aztecs. Europeans introduced not only firearms, but also the concept of killing (the enemy) as the objective of warfare: both significantly altered life among the American peoples, and ultimately established domination by the white man.

6   There were an estimated 1,500,000 Native Americans in the 15th century. Two centuries later, the large plantations of the South began buying slaves, and by 1760 there was a total of 90,000 Africans, twice the number of whites in that part of the country. The total number of British settlers on the Atlantic coast at the time was 300,000, greatly exceeding the French in the Mississippi Valley.

7   Most British immigrants left their country fleeing from poverty, religious persecution and political instability. Yet the birth of the colonies was marked by war against the native peoples, whose culture and way of life were systematically destroyed, and against other European colonists. By 1733, there were 13 English colonies, whose chief economic activities were agriculture, fishing and trade.

8   In 1763, with the 17th and 18th century European imperial wars behind them, France ceded its colonies east of the Mississippi to Britain, while its possessions west of that river went to Spain.

9   War broke out in 1775, and the Declaration of Independence, which marked the birth of the United States, was signed on July 4 1776. The war continued, but in the end the United States won (aided by its ally, France), with American sovereignty finally being recognized by England in 1783.

10   In 1787, the Philadelphia Constitutional Convention drew up the first federal constitution, which went into effect the following year.

ENVIRONMENT

There are four geo-economic regions. The East includes New England, the Appalachian Mountains and part of the Great Lakes and the Atlantic coast, a sedimentary plain which stretches from the mouth of the Hudson River to the peninsula of Florida. To the west are the Appalachian mountains, where mineral deposits (iron ore, and coal) abound. This is the most densely populated and industrialized area, where the country's largest steelmaking plants are located. A highly technological agriculture provides food for the large cities. The Midwest stretches from the western shores of Lake Erie to the Rocky Mountains, also including the middle Mississippi. Formed by the grasslands of the central plain, the Midwest is the country's largest agricultural area; horticulture and milk production predominate in the north, while wheat, corn and other cereals are cultivated in the south, side by side with cattle ranches where oxen and pigs are raised. Major industrial centers are located near the Great Lakes, near the area's agricultural production and large iron ore and coal deposits. The South is a subtropical flatland area, comprising the south of the Mississippi plain, the peninsula of Florida, Texas and Oklahoma. Large plantations, (cotton, sugarcane, rice) predominate here, while there is extensive cattle-raising in Texas. The region is also rich in mineral deposits (oil, coal, aluminum, etc.). The West is a mountainous, mineral-rich area (oil, copper, lead, zinc). A sizeable horticultural production exists along the fertile valleys of the Sacramento and San Joaquin rivers in California. Large industrial centers are located along the Pacific coast. In addition, the US has two states outside its original contiguous area: Alaska, on the continent's northwest where Mt. McKinley is located (Mt. "Denali", in the indigenous Atabasco language), the highest peak in North America, and Hawaii, an archipelago in the Pacific Ocean. The United States is the main producer of gases - principally carbon dioxide - which are responsible for the greenhouse effect.

SOCIETY

Peoples: The three largest ancestral groups are Germans, Irish and Italian, mixed with immigrants from all parts of the world. The largest ethnic minorities are of African origin, 11% of the total population; Latin American, 10%; and Asian, 8%. There are 1.9 million indigenous Americans, half of whom live on the 300 reservations. **Religions:** There is a Protestant majority, which encompasses several denominations; in addition, there are a considerable number of Catholics and the largest Jewish community in the world, approximately 5 million, and a Muslim minority. **Languages:** English (official); Spanish is spoken by a growing part of the population, and is a compulsory school subject in areas where Spanish-speaking people predominate. **Political Parties:** Though the formation of new parties is not banned, the US is effectively a two-party system. Republicans and Democrats have congressional representation. Independence Party, led by Texas billionaire Ross Perot. **Social Organizations:** The American Federation of Labor-Congress of Industrial Organizations (AFL-CIO) is the country's largest workers' organization, with 13,500,000 members. Many of the country's rural laborers, especially those of Mexican origin, are organized in the United Farm Workers (UFW) labor union, founded by César Chávez.

THE STATE

Official Name: United States of America. **Administrative Divisions:** Federal State, 50 States and 1 Federal District, Columbia. **Capital:** Washington, DC, 606,900 inhab. (1990). **Other cities:** New York, 7,333,300 inhab.; Los Angeles, 3,448,600 inhab.; Chicago, 2,731,700 inhab.; Houston, 1,702,100 inhab.; Philadelphia, 1,524,200 inhab. (1994). **Government:** Bill Clinton, President since 1992. Presidential government, federal system (50 states). There is a bicameral Congress where each state has two senators and a number of representatives proportional to its population. **National Holiday:** July 4, Independence (1776). **Armed Forces:** 1,547,300 (1995). · **Paramilitaries:** 68,000 Civil Air Control.

George Washington, commander of the Continental Army, was elected president in April 1789. In 1791, ten amendments dealing with individual freedoms and "states' rights" were added to the original constitution.

[11] The prime attraction of the West was the possibility of acquiring land and wealth without previous ownership, simply by staking a claim. The presidents in power at the time justified this "Empire of Freedom": it was the United States' "Manifest Destiny" to occupy the entire continent, in order to become a great nation.

[12] In 1803, the Louisiana purchase doubled the size of the Union. Between 1810 and 1819, the US went to war against Spain in order to annex Florida. In 1836, the Texans rebelled against Mexico and set up a republic, subsequently joining the Union in 1845. The United States declared war on Mexico, taking over vast areas of Mexican territory. California became a state in 1850, and Oregon in 1853.

[13] Westward expansion meant not only a change of "ownership", but also still another tragedy for the original inhabitants of the region, decimated by successive waves of land and gold fever. Occupation of their lands after the signing of treaties was systematically followed by an ignoring of the treaties, or the imposition of new treaties by force. In 1838, 14,000 Cherokee were forced off their lands by the Army, 4,000 perishing in the march to their new lands.

[14] The Civil War (or "War of Secession"), from 1861-65, revolved around the question of the preservation of slavery, but it was in fact a struggle between the two economic systems prevailing in the country. While the industrial North sought to free an important source of labor, the South's vital interest was to keep free access to its external markets.

[15] In 1850, of the 6 million inhabitants of European origin in the South, only 345,525 were slave owners; nevertheless, most whites were pro-slavery, remembering the slave rebellions which had taken place in South Carolina (1822) and Virginia (1800 and 1831).

[16] The election of Abraham Lincoln in 1860 precipitated the conflict: before he had even been inaugurated as president, the Southern states seceded. Committed to preserving the Union, and with a superior industrial base and superior weapons, the North finally triumphed over the South, although a million people were killed on both sides. Slavery was abolished, but ill-feeling between the 2 regions continued to exist.

[17] After the Civil War, the Native Americans of the Great Plains, especially the Sioux, launched repeated wars of defence. The treaties of 1851 and 1868, which had granted them sovereignty over their tribal lands, were ignored after the discovery of gold on these lands. The gradual occupation of their territory was completed by 1890, when the Sioux were defeated for the last time.

[18] In the 1880s, the remaining Native Americans were confined to reservations, most of which were on arid, barren lands. Years later, when uranium, coal, oil, natural gas and other minerals were discovered on some of the reservations, mining companies became interested in developing these mineral resources. This triggered the question of "rights" to the land on which the reservations were located.

[19] Between 1870 and 1920, the population of the United States grew from 38 to 106 million, and the number of states increased from 37 to 48. It was a period of rapid capitalist expansion, triggered by the growth of the railroads into huge companies. By the turn of the century, what had been essentially an agrarian country had been transformed into an industrial society.

[20] By the end of the 19th century, a two-party system of government had been established, with the Republican and Democratic parties alternately in power. Despite their different traditions, both parties have historically maintained a large degree of consensus on major national and international issues, a fact which led to a highly coherent foreign policy. The long struggle for women's suffrage began in 1889, being approved in 1920.

[21] Having dealt with its principal domestic problems, the United States ventured onto the international scene. The Spanish-American War in 1898, over Cuba and the Philippines, marked the beginning of an American imperial era. The occupation of Panama, and subsequent construction of the Panama Canal and a series of military bases, turned Central America - an area of

Foreign Trade

MILLIONS $ 1994

IMPORTS
690,000

EXPORTS
513,000

STATISTICS

DEMOGRAPHY

Urban: 76% (1995). **Annual growth:** 0.9% (1991-99).
Estimate for year 2000: 276,000,000. **Children** per
woman: 2.1 (1992).

HEALTH

Under-five mortality: 10 per 1,000 (1994). **Calorie
consumption:** 117% of required intake (1995).

EDUCATION

School enrolment: Primary (1993): 106% fem., 106%
male. Secondary (1993): 97% fem., 98% male.
University: 81% (1993).

COMMUNICATIONS

104 **newspapers** (1995), 111 **TV sets** (1995) and 123 **radio
sets** per 1,000 householdss (1995). 57.4 **telephones** per
100 inhabitants (1993). **Books:** 101 new titles per
1,000,000 inhabitants in 1995.

ECONOMY

Per **capita GNP:** $25,880 (1994). **Annual growth:** 1.30%
(1985-94). **Annual inflation:** 3.30% (1984-94). **Consumer
price index:** 100 in 1990; 113.4 in 1994. **Currency:** 1 dollar
= 1$ (1994). **Cereal imports:** 4,684,000 metric tons (1993).
Fertilizer use: 1,011 kgs per ha. (1992-93).
Imports: $690,000 million (1994). **Exports:** $513,000
million (1994).

ENERGY

Consumption: 7,905 kgs of Oil Equivalent per capita yearly
(1994), 19% imported (1994).

"vital security interest" to the US - into a kind of protectorate.

[22] The United States justified its interventions with the Monroe Doctrine, and former-President Monroe's slogan "America for the Americans". France was forced to withdraw its troops from Mexico (where they were protecting Emperor Maximilian), and Britain had to drop a territorial dispute with Venezuela. In 1890, the first Pan-American conference was held, paving the way for the inter-American system later set up.

[23] When World War I broke out, the United States declared its neutrality; though in 1917 it declared war on Germany, Austria and Turkey. In 1918, President Woodrow Wilson was one of the architects of the Treaty of Versailles, which established the framework for a new European peace. However, in 1920, US entry into the League of Nations was blocked by Congress.

[24] The United States emerged from the war more powerful than ever, but the Great Crash on Wall Street (the US Stock Exchange) shook the country in 1929. In its wake, many banks failed, industry and trade were severely affected and unemployment figures rose to 11 million. Under Franklin Roosevelt's presidency (1933-45), the government managed to bring the financial crisis under control.

[25] While Europe headed steadily toward another war, Congress passed a law in 1935 proclaiming US neutrality. However, when war broke out, Roosevelt introduced amendments to that law, in order to allow arms shipments to France and Britain. In 1941, the Japanese attack on the US military base at Pearl Harbor, in Hawaii, precipitated the entry of the US into World War II.

[26] The war acted as a dynamo for the entire American economy. With 15 million Americans at the front, employment grew from 45,500,000 to 53,000,000, because of the demands of the military industrial complex. To meet the demand for workers, 6 million people migrated from the countryside to the cities. On the labor front, in spite of a labor "truce" in support of the war effort, there were 15,000 strikes during the war, a fact which led Congress to pass a law limiting the right to strike.

[27] With Germany defeated on the European front, President Harry Truman (who had assumed the presidency upon Roosevelt's death) unleashed the United States' new military might upon Japan. On August 6 and 9, 1945, he gave the order for history's first atomic bombs to be dropped on the cities of Hiroshima and Nagasaki, wiping them out. That same year, at Yalta and Potsdam, Britain, the United States and the Soviet Union spelled out the terms for peace, and divided up the areas which would come under their respective spheres of influence.

[28] Truman presided over the opening of the United Nations in 1945, and was re-elected president in 1948. As the number one Western power, the United States devised a global strategy known as the "cold war", based on confrontation with the USSR and the socialist system. Within this framework, the Inter-American Treaty of Reciprocal Assistance was signed, and the North Atlantic Treaty Organization (NATO) was created.

[29] The United States took it upon itself to safeguard the global capitalist system, with the support of such international institutions as the World Bank and the IMF, as well as armed intervention by its troops all around the world.

[30] In 1952, General Dwight Eisenhower, commander-in-chief of American forces in Europe during the war and head of NATO, was elected president of the United States. The events of this period - the Korean War (1950-53), the partition of Germany, popular uprisings in Poland and Hungary, the delicate balance of nuclear weapons between the super-powers and the space race - all served to maintain the tension in US-Soviet relations. In 1960, a summit meeting between Eisenhower and Khrushchev had to be called off when an American U-2 espionage plane flying over Soviet territory was brought down.

[31] In the 1950s, the US economy grew steadily, as did its population, leading to a sense of over-all prosperity and contentment. The "American way of life" was exported to the rest of the world, hand-in-hand with American investments abroad. In 1956, the US offered South Vietnam's puppet government military support. Labor conflict increased; in 1959, Congress passed a law aimed at curbing corruption in the labor unions.

[32] The election of Democrat presidential candidate John F Kennedy in 1960 reflected the American people's desire for relief from domestic and foreign tensions. However, the president supported the failed Bay of Pigs invasion of Cuba in 1961, initiated the economic blockade of Cuba and presided over the Missile Crisis involving that Caribbean island. In 1963, the assassination of President Kennedy in Dallas, Texas, underscored the violence in American society.

[33] With regard to Latin America, President Kennedy launched the Alliance for Progress in Uruguay, in 1961, in an attempt to stem the influence of the Cuban Revolution throughout the hemisphere. However, the funds earmarked for this project were insufficient to effect real change, serving merely to expose the region's problems. Faced with growing guerrilla activity in the region, the United States decided to support official regional armies.

[34] Lyndon Johnson, elected in 1964, launched the "Great Society". However, the country was brought face to face with new frustrations. The war in Vietnam escalated, and came to a stalemate, unleashing a wave of protests throughout the US. This led the government to initiate a withdrawal from Vietnam, a blow to the American psyche. Racial segregation led to increasing racial confrontation in 1968, with violent protests in black neighborhoods in urban areas and the assassination of black Civil Rights leader, Revd Martin Luther King.

[35] That same year, the American Indian Movement (AIM) was founded by two Chippewa leaders. In 1969, AIM occupied the abandoned prison on Alcatraz Island in San Francisco to call attention to its demands and denounce the mistreatment of their people, which was followed by similar actions elsewhere in the country.

[36] Richard Nixon was re-elected in 1972 and visited Moscow and Beijing in 1973. In 1974, he signed the final withdrawal of US troops from Vietnam but was forced to resign that same year, after the discovery that Republicans had spied

U.S. Dependencies

Johnston:

Coral atoll of 2.6 sq km, made up of the Johnston island and the Sand, East and North islets, located approximately 1,150 km west-southwest of Honolulu (Hawaii). With an Air Force base the population (all military) is approximately 1,000. The atoll was uninhabited when it was discovered by British Captain Johnston in 1807. In 1858, Hawaii claimed sovereignty, but that same year US companies started exploitation of phosphate. Since 1934, the atoll has been administered by the US navy. Civilians are not admitted.

Midway:

A round atoll which comprises two islands: Sand and Eastern, with 5 sq km of total area and 2,300 inhabitants (1985) almost exclusively military. In 1867, they were annexed by the US and are currently administered by the Navy of that country. The islands are used for military purposes and operate as a refuelling base for trans-Pacific flights. In 1942, during World War II, an important sea battle took place here.

Wake:

Together with neighboring Wilkes and Peale, they make up an atoll with an area of 6.5 sq km and an estimated population of 1,600 inhabitants, mainly military personnel. Located between Midway and Guam, the island was seized by the United States during the 1898 war with Spain. It has a large airport which used to be a stopover for trans-Pacific flights, but nowadays it is no more used for business purposes. Since 1972 it has been administered by the Air Force which now uses it as a missile testing station.

Howland, Jarvis & Baker:

Located in central Polynesia, in the Equatorial Pacific they were occupied by the US in the middle of the 19th century. Uninhabited since the end of World War II as major phosphate deposits had been depleted. There is a lighthouse on Howland. The islands are administered by the US Fishing and Wild Life Service.

Palmira & Kingman:

The northernmost islands of the Line archipelago. Palmira is an atoll surrounded by more than 50 coral islets covered with exuberant tropical vegetation. It was annexed in 1898 during the war with Spain and is now a private property, dependent on the US Department of the Interior. Kingman is a reef, located north of Palmira, annexed by the US in 1922. The total area of both islands is 7 sq km. Although uninhabited, the US-Japanese project to turn them into deposits of radioactive waste raised protests throughout the Pacific region in the 1980s.

on Democrat headquarters located in the Watergate Hotel.

[37] In 1978, thousands of Native Americans demanding action on their grievances, carried out "The Longest March", descending on Washington DC from the official "Indian reservations" throughout the country. In 1982, AIM occupied the Black Hills in South Dakota, a site sacred to the Sioux, to protest against the mining operations of some 30 multinational companies, in that area.

[38] In the postwar period, the economy expanded through the transnational companies with affiliates all over the world: Ford and General Motors, with more than a million workers, the oil companies Exxon and Mobil Oil, International Business Machine (IBM), ITT, General Electric and Philip Morris, amongst others.

[39] The conservative government of Ronald Reagan (1980-88) reduced taxes on wealth and state social assistance, while increasing defence expenditure. The military-industrial complex dynamized the whole economy and partly made up for the lag behind Japan and Western Europe on other fronts.

[40] The population of Latin American origin increased in size by 53 per cent in the 1980s, reaching more than 22 million. However, the group which increased most per year was that of Asian origin.

[41] In the 1980s the United States emitted 25 per cent of the world total of carbon dioxide, (compared with the 13 per cent for the European Community which had a similar sized industrial sector) and refused to establish a deadline for reducing this. In 1991, the United States was the leading world producer of nuclear energy and liquid gas, coming second in coal, hydroelectric energy and natural gas, and third in oil. It was also the main consumer.

[42] In February 1991, the United States led the multinational force which confronted Iraq. The Gulf War demonstrated US military supremacy and allowed President George Bush to propose a new world order under the hegemony of his country.

[43] In 1991, 15 per cent of the population lived below the poverty line. Those most affected were the US citizens of African (33 per cent) and Latin American (29 per cent) origins.

[44] Bill Clinton, governor of Arkansas, was elected president in November 1992, and the Democrat Party also gained the majority in both houses of Congress. Clinton promised to concentrate on domestic problems, to move the greatest income tax burden on to the richer sectors, and to meet the health needs of all citizens. The health system reform, conducted by the nation's first lady Hillary Clinton in 1994, was rejected by Congress.

[45] The North American Free Trade Agreement with Mexico and Canada went into action in January 1994. The measures tending towards the globalization of the economy were criticized for giving other countries the jobs lost internally, both in the public and private sectors. The feeling of rejection grew amongst the immigrant workers. California state approved a law taking away the right to education from immigrants' children.

[46] Although the crime rate fell, the US citizens called for increasingly tougher punishment for offenders. The prisons overflowed as the number of inmates doubled and even tripled the figures of ten years earlier. In 1994, civilians were banned from using assault weapons, a measure which was questioned by arms consumers' associations.

[47] The summit between Clinton and Russian president, Boris Yeltsin in Moscow in January marked the end of the threat of nuclear war.

[48] The economy recovered and unemployment fell. However, workers were made to work an extra five hours per week. Consumption remained high despite increased credit card interest rates, only falling towards the end of the year. The social security reform included cuts in benefits and aggressive worker training programs.

[49] The Democrats lost the parliamentary elections of 1994, for the first time in 40 years. The discontent also showed in 1995 through acts of civil disobedience, like the right-wing paramilitary organizations who refused to recognize federal authority, and the extreme case of one Gulf War veteran who exploded a car bomb outside a federal office in Oklahoma.

[50] The 1996 political campaign saw President Clinton re-elected over Republican former senator, Bob Dole. In foreign policy, Clinton intervened militarily in Bosnia and Herzegovina and imposed his position on the Dayton accords of November 1995. In October 1996, he continued to promote Palestinian-Israeli peace talks in Washington in the midst of a serious crisis in Jerusalem and the West Bank, strengthening his image as a world leader.

Uruguay

Population:	3,163,000 (1994)
Area:	177,410 SQ.KM
Capital:	Montevideo
Currency:	Peso
Language:	Spanish

Uruguay

The range lands and forests of the eastern shore of the Uruguay River have been inhabited for at least ten thousand years. By the beginning of the 16th century there were three major cultural groups: the Charrúa, the Chaná and the Guaraní. The Charrúa were nomadic hunters, while the Chaná developed a rudimentary form of agriculture along the banks of the Uruguay River. Eventually, the Guaraní, developed some agricultural skills, mastered ceramics and navigation of the river in canoes.

[2] In 1527, Sebastián Cabot was the first European to sail up the rivers Paraná and Uruguay, where he established the first Spanish settlement in the territory. The settlers ignored the east bank of the Uruguay river for over a century, until cattle were brought in by the governor of Asunción, Hernando Arias de Saavedra (Hernandarias), in 1611.

[3] Good pasture and a moderate climate were conducive to large-scale reproduction of the cattle, which attracted the "faeneros", Río de la Plata equivalent of cowboys, from Brazil and Buenos Aires. Through intermarriage with the Indians (whose civilization was totally transformed by the introduction of beef in their diet, and the horse as a means of obtaining it) these "cowboys" (hunters of wild cattle, and not cattle tenders) gave rise to the "gaucho". In the 18th century, cattle-raising led to the extinction of some indigenous mammals, a reduction of plant diversity and an impoverishment of the soil. The Indians who could not adapt to this new life were displaced toward the Jesuit missions further north, or exterminated by the mid-19th century.

[4] In search of cattle and to gain control of the rivers which led to the interior regions of the Río de la Pla-

ta basin, the Portuguese advanced on the Banda Oriental (East Bank), an area approximately equivalent to that of present-day Uruguay, and founding "Colonia del Sacramento" on the shore opposite Buenos Aires, in 1680. This settlement caused constant friction between Portugal and Spain. In 1724, Spain ordered the Governor of Buenos Aires, Bruno Mauricio de Zabala, to cross the Río de la Plata and establish a fort in the Bay of Montevideo.

[5] Montevideo depended on the vice-royalty of Peru until 1776, as did Buenos Aires. That year, the Bourbonic reforms decreed the creation of the Viceroyalty of the Río de la Plata, with Buenos Aires as the capital, and Montevideo as a Naval Station.

[6] The "May Revolution" which broke out in Buenos Aires in 1810 was rejected by the Montevideans, who profited from the monopoly on trade with Spain, but found support in the countryside among small and medium-scale producers and the landless masses, led by the popular *caudillo* (leader) José Artigas. A "Federal League" was created with what is now Uruguay and the Argentine provinces of Cordoba, Corrientes, Entre Ríos, Misiones and Santa Fe.

[7] Uruguay was invaded by the Portuguese in 1816, with the tacit approval of Buenos Aires, alarmed by events termed as the "Artigan chaos". In 1820, Artigas fought his way to Paraguay, where he remained until his death in 1850. Portugal ruled over the Uruguayan territory until 1823, when Brazil became independent. Uruguay thus became the Brazilian Cisplatine Province. In 1825 the campaign against the Brazilian Empire was renewed, and one of its objectives was to return to the United Provinc-

es of the Río de la Plata. On August 25 of that year, independence was declared and the decision made to reunite with the territories of the former vice-royalty.

[8] The country was eventually granted its independence in 1828, through an agreement between Buenos Aires and Brazil, mediated by Britain. The 1830 Constitution created a republic for the elite: only independently wealthy males of European origin could vote or be eligible for office. The 19th century was shaken by civil wars and witnessed the formation of two political parties: the Blanco (White) or National Party, linked to Argentine leader Juan Manuel de Rosas, and the Colorado ("Red", but

not communist), linked to European capital and liberal ideas.

[9] In 1865, Colorado dictator Venancio Flores signed an agreement with Brazil and Argentina and created the Triple Alliance, breaking Paraguayan isolation, with European support, forcing the country to open its borders to foreign trade (see Paraguay).

[10] Between 1876 and 1879, under the dictatorship of Colonel Lorenzo Latorre, the countryside was enclosed by fences. The implications of this event were far-reaching: the land was completely appropriated by the private sector, production was geared to the capitalist market (over which Britain held undisputed control) and the countryside was "organized" leaving no space for the lifestyle of the "gaucho", who was transformed into a hired hand.

[11] In 1903 the Colorado José Batlle y Ordoñez assumed the presidency; the following year he fought and defeated Aparicio Saravia - the last rural *caudillo* of the Blancos. With the ideological influence of European social-democracy, and the support of immigrants, Batlle laid the foundations for the modern Uruguayan state. The high rate of productivity of extensive cattle-raising generated such a surplus that, without changing the land

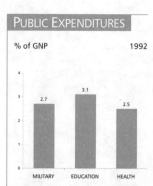

PUBLIC EXPENDITURES

% of GNP — 1992

MILITARY 2.7
EDUCATION 3.1
HEALTH 2.5

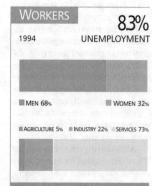

WORKERS

1994 — **8.3%** UNEMPLOYMENT

MEN 68%   WOMEN 32%

AGRICULTURE 5%   INDUSTRY 22%   SERVICES 73%

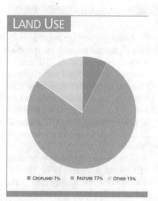

LAND USE

CROPLAND 7%   PASTURE 77%   OTHER 15%

ENVIRONMENT

Uruguay has a gently rolling terrain, crossed by characteristic low hills -an extension of Brazil's southern plateau- belonging to the old Guayanic-Brazilian massif. Its average altitude is 300 meters above sea level. This, together with its location and latitude, determines its temperate, subtropical, semi-humid weather, with rainfall throughout the year. The vegetation is made up almost entirely of natural grasslands, suitable for cattle and sheep raising. The territory is well irrigated by many rivers, and has over 1,100 km of navigable waterways, in particular on the rivers Negro and Uruguay, and on the Plata estuary. The coast is made up of many sandy beaches which attract tourists from the neighboring countries. There is increasing pollution in the northern departments as a result of emissions from a thermoelectric plant in Candiota (Brazil), while rivers and streams suffer from the use of poisonous agricultural chemicals. A growing loss of ecosystems on the plains and in the eastern wetlands has been observed, due to monoculture of wood and rice, respectively. Large infrastructure projects (e.g. the Buenos Aires-Colonia bridge, and connecting highway to Sao Paulo, Brazil) pose an additional threat to the environment.

SOCIETY

Peoples: Most Uruguayans are descendants of Spanish, Italian and other European immigrants. In spite of the genocide carried out against indigenous peoples in the mid-19th century, recent historical and genetic research has established part of the population is of Amerindian descent. The native population was annihilated in the mid-19th century. Afro-American descendants make up about 10% of today's population.
Religions: There is no predominant religion but a minority practice Catholicism. Afro-Brazilian cults are also practiced.
Languages: Spanish. **Political Parties:** An electoral system which allows for multiple presidential candidacies within each party made it possible for the two traditional

parties -the *Blancos* and the *Colorados*- to survive, in spite of their many internal differences. The two-party system has been shaken by the growth of previously marginal ideological parties. The Broad Front (*Frente Amplio*) is a coalition of the Communist and Socialist parties, the "Artiguista" Alliance, the Tupamaros-backed People's Participation Movement and other minor socialist organizations and individuals. The Broad Front (Alliance) joined together with important sectors which formerly belonged to the "traditional" parties and the Christian Democratic Party, to form the "Progressive Encounter", fielding a joint candidate for the November 1994 elections. The "New Space" (*Nuevo Espacio*) is the other party represented in Parliament. **Social Organizations:** PIT-CNT (Inter-Union Workers' Bureau - National Workers' Convention), is a confederation of all trade unions. FUCVAM (Uruguayan Federation of Housing Construction by Mutual Help). Students in the public sector are grouped in the Federation of University Students (FEUU), and the Federation of Secondary Education Students (FES). The National Association of Pensioners includes pensioners country-wide. There are several women's rights associations.

THE STATE

Official Name: República Oriental del Uruguay. **Administrative Divisions:** 19 Departments **Capital:** Montevideo, 1,378,707 inhab.(1996) **Other cities:** Salto, 81,600 inhab., Paysandú, 79,100 inhab., Las Piedras, 68,400 inhab., Rivera, 59,700 inhab., Maldonado, 40,700 inhab. **Government:** Julio María Sanguinetti, president-elect since November, 1994. Presidential system, with a bicameral legislative. A 99-member House of Representatives and a 30-member House of Senators, chosen by proportional representation. Vice-president Hugo Batalla chairs the Senate. **National Holidays:** August 25, Independence (1825); July 18, Constitution (1830). **Armed Forces:** 25,600 (1994). **Paramilitaries:** 700 Metropolitan Guard, 500 Republican Guard.

Areco's government, yielding to IMF prescriptions, curtailed the purchasing power of wage-earners and aimed to eliminate the bargaining power of trade unions. The National Workers' Convention and the student organizations stood in opposition to such policies, and parallel to this, the Tupamaro (MLN) guerrilla movement were active throughout the country. In 1971 the Broad Front (Frente Amplio), a left wing coalition, was founded. It promoted a progressive government program, and nominated a retired General, Liber Seregni, in the presidential elections of that year.
[18] Juan Maria Bordaberry, the Colorado and rural nominee, won the elections, which the National Party denounced as fraudulent. During that government, the military took up the fight against the guerrilla movement, with the backing of a parliamentary declaration of a state of civil war, and in 1972 the Tupamaros were defeated. A campaign including the systematic use of torture rapidly dismantled the clandestine organization. On June 27 1973, President Bordaberry and the armed forces staged a coup. Parliament was dissolved, and a civilian-military government was formed. Local trade unions resisted the coup with a 15-day strike.
[19] The Uruguayan military set up a regime based on the National Security Doctrine, a system learnt in US training academies. Economically, the neo-liberal theories of US professor Milton Friedman were applied, favoring the concentration of wealth in transnational corporations. Salaries lost 50 per cent of their purchasing power, while the foreign debt reached $5 billion. During this period, all public and private institutions, political parties and unions were banned, people disappeared and torture and arbitrary detentions were commonplace.
[20] The economic situation deteriorated and the civilian population systematically rebuffed efforts to legalize the police state, voting overwhelmingly against the draft of a new constitution that would support this. The Uruguayan generals had no alternative but to work out a timetable for reinstating an institutionally legitimate government.
[21] In April 1983, the Inter-union Plenary of Workers (PIT) was formed and May Day was celebrated for the first time since the coup. The three political parties recognized as legal by the military gov-

structure of large units in the hands of a few, the state was able to redistribute the wealth in the form of welfare, education, and protection of local industry oriented toward domestic consumption.
[12] The country underwent rapid urbanization. Commerce and services expanded quickly, with the state as the main employer. A large liberal middle-class developed, educated in state schools with Europeanized curricula.
[13] The Church and the state were separated and divorce was legalized. A collegiate system of government was introduced in 1917 to avoid power accumulation. Urban migration and European immigration enlarged the cities, but the birth-rate dropped. Abortion was

the major birth-control technique, becoming so common that it was legal between 1933 and 1935, although illegalized again after political negotiations with Catholic sectors. In 1934, women were granted the right to vote. This open legislation earned the country the title of "the Switzerland of South America".
[14] Uruguayan exports grew during both world war periods. Meat and its products were supplied first to the Allies, who were fighting against Nazism and Fascism in Europe, and later to US troops fighting in Korea.
[15] Trade balance surpluses secured the country large foreign currency reserves. The welfare policy of subsidies encouraged the emergence of relatively strong import-

substitution industries. Meanwhile, a prosperous building industry contributed to maintaining a high rate of employment.
[16] Instead of reinvesting their profits, land owners invested capital abroad, engaging in financial speculation and superfluous consumption. This inevitably led to inflation, corruption and the exacerbation of social conflict. In the mid-1950s, the industrial sector stagnated, a situation that proved impossible to reverse.
[17] The first Blanco government of the century, which came to power in 1959, accepted IMF economic guidelines which accelerated the recession causing a strong reaction among workers. The situation became critical in 1968, when Jorge Pacheco

DEMOGRAPHY

Urban: 90% (1995). **Annual growth:** 0.3% (1992-2000). **Estimate for year 2000:** 3,000,000. **Children** per woman: 2.3 (1992).

HEALTH

Under-five mortality: 21 per 1,000 (1994). **Calorie consumption:** 101% of required intake (1995). **Safe water:** 75% of the population has access (1990-95).

EDUCATION

Illiteracy: 3% (1995). **School enrolment:** Primary (1993): 108% fem., 108% male. University: 30% (1993). **Primary school teachers:** one for every 21 students (1992).

COMMUNICATIONS

104 **newspapers** (1995), 102 **TV sets** (1995) and 104 **radio sets** per 1,000 householdss (1995). 16.8 **telephones** per 100 inhabitants (1993). **Books:** 102 new titles per 1,000,000 inhabitants in 1995.

ECONOMY

Per **capita GNP:** $4,660 (1994). **Annual growth:** 2.90% (1985-94). **Annual inflation:** 73.80% (1984-94). **Consumer price index:** 100 in 1990; 758.9 in 1994. **Currency:** 6 pesos = 1$ (1994). **Cereal imports:** 110,000 metric tons (1993). **Fertilizer use:** 608 kgs per ha. (1992-93). **Imports:** $2,770 million (1994). **Exports:** $1,913 million (1994). **External debt:** $5,099 million (1994), $1,612 per capita (1994). **Debt service:** 16.1% of exports (1994). **Development aid received:** $121 million (1993); $39 per capita; 1% of GNP.

ENERGY

Consumption: 623 kgs of Oil Equivalent per capita yearly (1994), 68% imported (1994).

Foreign Trade

MILLIONS $ 1994

IMPORTS
2,770
EXPORTS
1,913

Maternal Mortality

1989-95

PER 100,000
LIVE BIRTHS
36

ernment took up the banner for the labor movement and the causes espoused prior to the dictatorship. They also demanded the release of all political prisoners. The Broad Front whose president had been in prison since 1973, was excluded from political activity.

[22] Protests against the military regime intensified as a result of the break-down in negotiations. Women's groups defied the government's prohibition of street demonstrations. After a successful 24-hour general strike called by the PIT, talks between the political parties and the military were re-initiated in January 1984, this time with the participation of representatives of the Broad Front.

[23] The National Party leader, Wilson Ferreira Aldunate, was arrested on his return to the country after an 11-year exile. He was not permitted to run as a candidate in the 1984 election. General Liber Seregni and hundreds of others were also proscribed as candidates.

[24] Under the slogan of "cambios en paz" (changes without conflict), the leader of the Colorado Party, Julio Maria Sanguinetti, won the national election, in 12 of the country's 19 departments, obtaining 40 per cent of the vote, against 34 per cent for the National Party or Blancos, 21 per cent for the Broad Front and 2 per cent for the Civic Union. During the new government's first

month in power, an amnesty law was approved whereby all political prisoners were freed.

[25] President Sanguinetti supported many international diplomatic moves in Latin America. The first two years of his term coincided with a favorable international situation, but he returned to the neo-liberal policies of the previous government, which were strongly resisted by the population.

[26] In 1986, under pressure from the president and the armed forces, Parliament approved an amnesty to military personnel accused of violations of human rights, with a ratio of 3 to 1. A referendum in April 1988 ratified the law, with 56 per cent in favor, with 42 per cent wanting to try the military for their crimes.

[27] In May, 1989, the government signed a secret structural adjustment agreement with the World Bank, in exchange for rescheduling Uruguay's international debt. The government committed itself to reducing expenditure on social security; to privatize bankrupt banks absorbed by the state; and to reform public companies, making them profitable and attractive for privatization. Six months before the election, the Party for Government by the People, which had supplied over 40 per cent of the Broad Front's votes, withdrew from the coalition.

[28] In the November 1989 election the

National Party, with Luis Alberto Lacalle, inflicted a severe electoral defeat upon the Colorado Party, taking 17 of the country's 19 departments. The Broad Front won in Montevideo with its candidate Tabaré Vázquez. The Left assumed responsibility for municipal administration, for the first time in the history of the country. Six women were elected to the Chamber of Deputies and another seven to the Municipal Council (legislative) of the Capital.

[29] The associations of pensioners succeeded in bringing about constitutional reform against the agreement with the World Bank, and pensions were readjusted, in line with the salaries of civil servants. This legislation was approved in a referendum.

[30] Lacalle carried out his neo-liberal policy: taxes were increased and the privatization of state-run companies was promoted. Salaries lost over 15 per cent of their purchasing power during the first year of his term, while inflation mounted from 90 to 130 per cent at the end of his term. In March 1991, Argentina, Brazil and Paraguay approved the Mercosur agreement (see box in Brazil).

[31] A commission convened by the labor union movement, constituted by members of several parties, presented a petition requesting a plebiscite on a privatization law. In December 1992, 72 per cent of the population voted to repeal the law. However, the government continued trying to privatize certain state enterprises, including the national airline, the gasworks and the sugar cane plantations in the north of the country. The latter caused several demonstrations, including a cane workers' march on Montevideo.

[32] The mayor of Montevideo had approval ratings higher than the electoral support which he had received during his first four years in office. He decentralized the departmental government into 18 zones, within which local voters could elect their own representatives.

[33] In the general and regional elections of November 27 1994, Colorado Party candidate Julio Maria Sanguinetti was elected once again by a small majority.

[34] The "Progressive Encounter" - made up of the Broad Front, the Christian Democrats and other sectors which had split from the traditional parties - considerably increased its electoral support, winning the mayoralty of the capital for the second consecutive time with over 43 per cent of the vote. Nationally, 31.2 per cent of the population voted for the Colorado Party, 30.03 for the Blanco Party and 29.82 per cent for the Progressive Encounter.

[35] Sanguinetti took office on March 1 1995, two months after the Mercosur came into force, and proposed political and economic reforms. The fall in inflation and the loss of jobs - mainly among the construction and industrial sectors - were the outcome of austerity measures implemented by Economy Minister Luis Mosca.

[36] In October, Uruguay became the first country to officially receive Cuban president Fidel Castro in two decades. Sanguinetti rejected Castro's invitation to visit Cuba in September 1996, considering the time was not right.

[37] The wife of former president Lacalle and several of his closest aides were involved in accusations of corruption which rocked the foundations of the political parties in mid-1996. A system of social security by savings accounts was implemented that year, in line with reforms previously carried out in Argentina and Chile.

[38] In late 1996, Parliament discussed a constitutional reform which entailed difficult political negotiations. The main goal of the reform was the electoral system. In his last trips abroad - Malaysia, Chile, Mexico - Sanguinetti travelled with businesspeople who promoted the country as a financial center fit for investments in the future capital of Mercosur.

Uzbekistan

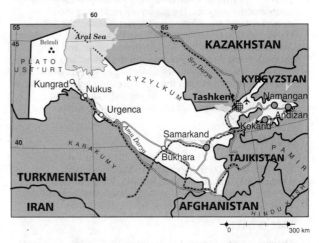

Population:	22,378,000 (1994)
Area:	497,400 SQ KM
Capital:	Tashkent
Currency:	Som-Kupon
Language:	Uzbek

Uzbekistan

Several ancient agricultural centers developed in what is now Uzbekistan, including Khwarezm, Ma Wara an-Nahr and the Fergana Valley. The population and the languages of the first states to emerge in the 10th century BC were Indo-European. Between the 6th and 7th centuries AD, these states formed part of the Persian Achaemenid empire, the empire of Alexander the Great, the Greco-Bactrian kingdoms, and that of Kusha, the white Hun or Eftalite state.

[2] Turkish nomads defeated the Eftalites and brought the major part of Central Asia into their empire, the Turkish *kaganate* (khanate), between the 6th and 8th centuries. During this period, Turkish-speaking peoples began coming to the region and inter-marrying with the native population. In the mid-8th century, the country was conquered by the Arabs. Islam spread quickly, particularly in the cities.

[3] When the power of the Arab caliphs declined, during the reign of the local Samanid and Karajanid dynasties, and under the shahs of Khwarezm (9th to early 13th centuries), an Islamic civilization developed. Agriculture was its main activity, made possible by a highly developed irrigation system; this culture was also noted for its artisans. The cities of Bukhara, Samarkand and Urgenca became prosperous trading centers for the caravans that crossed the Great Silk Route, from China to Byzantium.

[4] From 1219-1221, the state ruled over by the shahs of Khwarezm was overrun by the Tatars, and completely devastated. Khwarezm was turned over to the Golden Horde, under the leadership of Genghis Khan's oldest son. Ma Wara an-Nahr and Fergana went to the second son, Chagatai; the inhabitants of the area began calling themselves Chagatais. Turkish and Mongol tribes moved into the steppes, and in the second half of the 14th century, Timur, the head of one of these tribes, settled in Ma Wara an-Nahr, making Samarkand the capital of his empire. Later, Samarkand became the residence of Ulug Beg, the grandson of Timur Lenk (Tamburlaine), a khan and an astronomer.

[5] The union of nomad tribes who called themselves Uzbeks was achieved in the 15th century, in Central Kazakhstan. During the second half of the century, the great Uzbek poet and thinker Ali Shir Navai, who lived at the court of one of Timur's descendants, became a world-renowned figure. The leader of the Uzbeks, Muhammad Shaybani, conquered Ma Wara an-Nahr at the beginning of the 16th century, and the Uzbek immigrants gave their name to all the country's inhabitants. After dissolving the State of Shaybani, Uzbek khanates emerged in the region. In 1512, the Khanate of Khiva was formed; its military elite belonged to the Kungrats, an Uzbek people. In 1806 Muhammad Amin, head of the Kungrats, founded the dynasty which was to rule over Khiva until 1920.

[6] In the mid-16th century, the khanate of Bukhara was formed. Its military elite belonged to the Uzbek Manguite people. Muhammad Rajim, leader of the Manguites, founded his dynasty in 1753; it also lasted until 1920. The khanate of Bukhara reached its apogee at the time of Nasrula, a khan who ruled from 1826-1860. In the early 19th century, the emirs of Fergana (of the Ming dynasty) created the khanate of Kokand. These feudal theocracies were populated by Uzbeks, as well as Turkmens, Tajiks, Kyrgyz and Karakalpaks, who were in constant conflict.

[7] None of these States had definite boundaries, nor were they capable of exacting absolute fidelity from their chieftains. The emirs of Khiva and Bukhara exercised only nominal sovereignty over the Turkmen tribes of the Kara Kum Desert, who were slave traders in Iranian territory. Although the three Uzbek khanates reached their highest level of organization during this period, they were not prepared to face the imminent European expansion. In Central Asia, Russian and British colonial interests clashed over the cotton market.

[8] In 1860, the Russian offensive against the khanates began, though it was impeded by their geographical isolation, particularly in the case of Khiva, which was surrounded by desert. In 1867, the czar created the province of Turkistan, with its center in Tashkent. Although it belonged to Kokand, it was officially annexed in 1875. By the end of the 19th century, the province included the regions of Samarkand, Syr Darya and Fergana. On August 12 1873, the khan of Khiva signed a peace treaty with Russia, accepting protectorate status, as did the khan of Bukhara in September of the same year. The difficult living conditions imposed on the people by czarist Russia triggered several uprisings, like Andizhan in 1898, and Central Asia in 1916.

[9] With the fall of the czar in February 1917, power was shared by the committee of the Provisional Government, and the *Soviets* (councils) of workers and soldiers. After the triumph of the October Revolution in St. Petersburg power was concentrated in the hands of the Tashkent Soviet. In 1918, Red Army units crushed both the attempt to form an autonomous Muslim government in Kokand and an anti-Bolshevik uprising in Turkistan. The Red Army occupied Khiva in April 1920, and entered Bukhara in September. Land and water reform began in spring 1921, with military operations continuing until mid-1922, when the implementation of the reforms left the rebels without a cause.

[10] In 1924, the Soviet government reorganized Central Asian borders along ethnic lines, and proclaimed the Soviet Socialist Republic (SSR) of Uzbekistan. In May 1925, Uzbekistan became a federated republic of the Soviet Union. The Autonomous Soviet Socialist Republic of Tajikistan was a part of Uzbekistan until 1929, when it joined the USSR. In Moscow, during the Great Purge of 1937-38, a number of Uzbeks, including the prime minister, Fayzullah Khodzhayev and the first secretary of the Uzbek Communist Party, Akmal Ikramov, were sentenced to death. In 1953, both leaders were rehabilitated.

[11] The socialist reforms were aimed primarily at taking advantage of the region's agricultural potential. Huge mechanized irrigation systems, including the Great Fergana, North Fergana and Tashkent canals, were constructed by the State, as were reservoirs like the Kattakurgan on the Zeravshan River (the Uzbek Sea), the Kuyumazar and the Akhangaran (the Tashkent Sea). Between 1956 and 1983, when Sharaf Rashidov was first secretary of the Communist Party, the republic's economy revolved around cotton and the national industry was geared toward producing machines and heavy equipment for cotton cultivation, harvesting and processing. Uzbekistan became the USSR's largest cotton producer and the third largest in the world.

[12] In the 1940s, the population increased, Kurds and Mesketian Turks, who had been deported from the Caucasus, came to Uzbekistan, and all was stable until the 1980s. When Leonid Brezhnev became head of the Communist Party of the USSR, in 1983, he changed the Uzbek authorities. The new first secretary of the local CP, Inamjon Ousmankhodjaev, revealed that official cotton production statistics had been systematically falsified by his predecessor.

[13] The Uzbek corruption case was the biggest scandal in the Soviet Union at the time, and triggered large amounts of arrests, lawsuits against 4,000 civil servants and expulsions from the party. However, this turned out to be just another political purge, as it did not bring about any substantial changes.

[14] From 1985 onward, the changes initiated by President Mikhail Gorbachev, the deterioration of the economic situation and the weakening of the Soviet Communist Party's centralized authority gave rise to increased ethnic and religious conflicts in Uzbekistan. The Uzbeks, predominantly Sunni Mus-

ENVIRONMENT

Uzbekistan is bounded in the north and northwest by Kazakhstan, in the southeast by Tajikistan, in the northeast by Kyrgyzstan and in the south by Afghanistan. Part of its terrain is flat, while the rest is mountainous. There are two main rivers and more than 600 streams, some of which are diverted for irrigation, while others are utilized for hydroelectric projects. The plains are located in the northwest and the center (the Ustyurt Plateau, the Amu Darya Valley and the Kyzylkum desert), while the mountains are in the southeast (the Tien-Shan and Gissar and Alay ranges). It has a hot, dry climate on the plains, and is more humid in the mountains. There are large deposits of natural gas, oil and coal. Among the most pressing environmental problems are the salinization of the soil as a result of monoculture, desertification, drinking water and air contamination.

SOCIETY

Peoples: Uzbeks, 71.4%; Russians, 11%; Tajiks, 4.7%; Kazakhs, 4.1% (1991). **Religions:** Muslim (Sunni). **Languages:** Uzbek (official), Russian, Tajik. **Political Parties:** People's Democratic Party; National Progress Party; Birlik Democratic Party; Islamic Renaissance Party.

THE STATE

Official Name: Ozbekistan Jumhuriyäti. **Capital:** Tashkent, 2,120,000 inhab. (1992). **Other cities:** Samarkand, 372,000 inhab.; Namangan, 333,000 inhab.; Andijon, 302,000 inhab.; Bukhara, 235,000 inhab. **Government:** Islam Karimov, President since March 24, 1990. Legislature: Supreme Assembly, (single-chamber), with 150 members elected by direct popular vote every five years (replaced the Supreme Soviet). **National Holiday:** September 1t, Independence (1991). **Armed Forces:** 25,000 (1995). **Paramilitaries:** (National Guard) 700.

STATISTICS

DEMOGRAPHY

Urban: 41% (1995). **Annual growth:** 2.4% (1991-99). **Estimate for year 2000:** 26,000,000. **Children** per **woman:** 4.1 (1992).

EDUCATION

School enrolment: Primary (1993): 79% fem., 79% male. Secondary (1993): 92% fem., 96% male. University: 33% (1993).

COMMUNICATIONS

6.6 **telephones** per 100 inhabitants (1993).

ECONOMY

Per **capita GNP:** $960 (1994). **Annual growth:** -2.30% (1985-94). **Annual inflation:** 109.10% (1984-94). **Currency:** 1.200 som-kupon = 1$ (1993). **Cereal imports:** 4,151,000 metric tons (1993). **Fertilizer use:** 1,566 kgs per ha. (1992-93). **Imports:** $3,243 million (1994). **Exports:** $3,543 million (1994). **External debt:** $1,156 million (1994), $52 per capita (1994). **Debt service:** 3.2% of exports (1994).

ENERGY

Consumption: 1,886 kgs of Oil Equivalent per capita yearly (1994), 3% imported (1994).

---

lims, resisted the Party's anti-clerical campaigns.

[15] The USSR's intervention in Afghanistan against fellow-Sunni Muslims only worsened the situation, and all of these factors combined to create a hostile climate toward Moscow and the Russian minority residing in Uzbekistan. The most serious consequence was the conflict in the region of Fergana in June 1989, and conflicts in Namangan, in December 1990.

[16] In August 1991, the Uzbek Soviet (Parliament) approved the Law of Independence. On December 12, in Alma Ata, the Uzbek delegation signed the agreement creating the Commonwealth of Independent States (CIS), which marked the end of the USSR.

[17] On the 29th of the same month, presidential elections were held with a referendum supporting independence. Islam Karimov, the former first secretary of the Uzbek Communist Party, was elected president.

[18] In January 1992, students demonstrating in Tashkent against the shortage of bread and the high price of staple goods were fired upon by police; six people were killed and many were injured. The Uzbekistan government has gradually started to liberalize the economy by deregulating some prices, following the example of other former Soviet republics. In an attempt to reduce the tension, authorities announced an increase in student grants.

[19] At the same time, a leadership struggle in the upper echelons of the religious (Islamic) hierarchy, reinforced the secular power of the president.

[20] Karimov imposed a harsh, authoritarian style in an attempt to maintain the country's political stability. When the deregulation of prices resulted in student protests, the government responded with severe repressive measures. One such demonstration resulted in two students being killed and several injured.

[21] In March, arrests were made in the city of Namangan, a bastion of Muslim opposition. The Islamic center was attacked; with several leaders of the Birlik Democratic Party and other religious leaders being arrested.

[22] In March 1992, the presidents of the parliaments of seven member republics of the CIS – Armenia, Belarus, Kazakhstan, Kyrgyzstan, the Russian Federation, Tajikistan and Uzbekistan – agreed to create an inter-parliamentary assembly, for consultation and coordination. That same month Uzbekistan was admitted to the UN as a new member.

[23] In June, Karimov defined a new economic course for Uzbekistan, based upon the Southeast Asian development model: eliminating dissidence and modernizing the economy by adopting free-market policies. The opposition officially went underground.

[24] The deregulation of prices, privatization and financial and fiscal reforms which have been implemented since 1992, have been far less drastic in Uzbekistan than in other former Soviet republics.

[25] During 1993, Uzbekistan tightened political links with other former soviet republics in Central Asia, establishing diplomatic relations with Hungary and signing bilateral agreements with Turkey, Afghanistan and Pakistan.

[26] Criticized for the supposed slowness of his economic reforms, Karimov decided to launch a privatization plan in January 1994, announcing price increases of basic goods and energy of around 300 per cent.

[27] Several countries of Central Asia tried to design a program to defend the environment in the region of the Aral Sea. However, one of the conditions needed to prevent the drying of the lake, according to Western experts, would mean reducing the irrigation of cotton crops in Uzbekistan.

[28] On a diplomatic front, the country aimed to contain the Russian and Iranian influences in the region, particularly from 1995 onwards. Thus, Karimov actively supported the US embargo on Iran calling for the creation of a "common Turkmenistan" between the nations of the region, making it understood that it was threatened by the "imperialist" tendencies of Russia.

Vanuatu

Population: 165,000
(1994)
Area: 12,190 SQ KM
Capital: Port Vila

Currency: Vatu
Language: Bislama English and French

Vanuatu

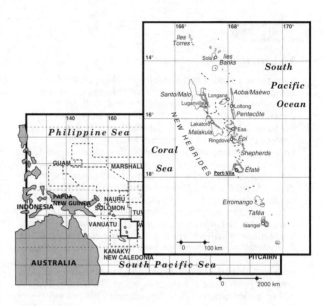

The first colonization of Polynesia is still a mystery to anthropologists and historians. Thor Heyerdahl's highly publicized expedition in his "Kon Tiki" tried to demonstrate that people could have reached the South Pacific islands from America. But linguistic, cultural and agricultural similarities relate Melanesians to Indonesians.

[2] Sailing westwards, Polynesians reached Vanuatu around 1400 BC. These navigators crossed and populated the entire Pacific Ocean from Antarctica, south of New Zealand, to Hawaii, far north of the Equator and as far as Easter Island on the eastern edge of the Pacific Ocean. Their culture was highly developed, Polynesians domesticated animals and developed some subsistence crops; they manufactured ceramics and textiles, organized their societies with a caste system and, in some cases, possessed a historical knowledge which had been orally transmitted down the generations for centuries.

[3] On April 29 1605 the Portuguese-Spanish navigator, Pedro Fernandez de Quiros, was the first European to sight mountains which he believed to be part of the Great Southern Continent for which he was searching; he named the place, Tierra del Espíritu Santo ("Land of the Holy Spirit"). A century and a half later, Frenchman Louis Antoine de Bougainville, sailed around the region and demonstrated that it was not part of Australia but rather a series of islands. In 1774, James Cook drew the first map of the archipelago, calling it the New Hebrides after the Scottish Hebrides Islands.

[4] Shortly thereafter, traders arrived and felled all the aromatic sandalwood forests. Lacking other attractive natural resources, the islands became the source of a semi-

enslaved labor force. The workers were either taken by force or purchased from local leaders in exchange for tobacco, mirrors and firearms. In the 19th century, the Civil War in the United States created an increase of international cotton prices and Indochina's plantations became so lucrative that the slave trade continued long after slavery had been officially abolished.

[5] Popular resentment of this plundering caused the murders of more European missionaries in the New Hebrides than in any other area of the Pacific.

[6] During almost all of the 19th century, the archipelago was on the dividing line between the French (in New Caledonia) and British (in the Solomon islands) zones of influence, and these nations finally decided to share the islands instead of fighting over them. In 1887, a Joint Naval Commission was established, and in 1906, the Condominium was formalized. The native islanders soon refered to it as the "pandemonium".

Most of the neighbouring archipelagos achieved independence in the 1970s.

[7] The formula of shared domination envisaged the joint upkeep of some basic services; the post, radio, customs, public works, but left each power free to develop other services. Consequently, there were two police forces, two monetary systems, two independent health services and two school systems ruled by two representatives on the islands.

[8] Melanesians were relegated to being "stateless" in their own country, and they were not considered citizens until a legislative assembly was established in the territory in 1974. Until then only British or French people were entitled to citizenship or land ownership.

[9] Most of the neighboring archipelagos achieved independence in the 1970s. This encouraged the foundation of the National Party of the New Hebrides (today the Vanuatu Party) in 1971; the Party organized grass-roots groups on all of the islands on the basis of

the Protestant parochial structure. The Party became the country's leading political force and demanded total independence, in opposition to various "moderate" pro-French parties that preferred to maintain the colonial situation. When the Party won two-thirds of the vote in 1979, this convinced the British that independence could no longer be postponed. They also carried out a strong anti-nuclear policy and turned Vanuatu into the only nation among the Pacific islands to become a member of the Non-Aligned Movement.

[10] During that period French agents and US business encouraged the growth of the Na Griamel separatist group in Espíritu Santo. This group began as a popular protest movement against the sale of property to North American hotels. The group's leader, Jimmy Stevens, received $250,000, arms and a radio from the Phoenix Foundation, an ultra-right US organization. In return for concessions to install a casino and, presumably, cover for illicit activities from Stevens' "Republic of Vemarana".

[11] Stevens visited France to obtain the support of President Giscard d'Estaing, and the French police did nothing to prevent the rebellion and expulsion of Vanuatu Party followers from the island. Meanwhile francophone support of the other islands encouraged small "moderate" pro-French parties to oppose the government of Walter Lini, who was accused of being authoritarian,

too centralist and too close to Britain and Australia.

[12] Finally, Lini, supported by forces from Papua New Guinea, managed to disarm the separatists, deporting Stevens and his followers, who were granted asylum in Kanaky/New Caledonia.

[13] Vanuatan independence was declared on July 30 1980. Measures were immediately taken to return the land held by foreigners to the Melanesians; the school system was unified and a national army was created.

[14] In February 1981, the newly-arrived itinerant ambassador of Vanuatu, Barak Sope, was refused entry at the Noumea airport by the French government, to prevent his participation in the Melanesian Independence Front Congress in New Caledonia. Retaliation was immediate; Lini's government declared the French ambassador 'persona non grata' and requested that the French diplomatic mission be reduced to five members, and this incident delayed the cooperation program between the governments.

[15] Government revenues depend heavily on foreign aid and the country's exports cover only half of the cost of its imports. Since the country is a tax haven with more than 60 banks established in Port Vila, the capital, and over 1,000 absentee firms in the territory, there is no solid strategy to ensure provision of finances for the Public Treasury. One proposed project calls for opening the country to international maritime registration under Vanuatu's flag.

WORKERS

1994

MEN 62%   WOMEN 38%

AGRICULTURE 68%   INDUSTRY 8%   SERVICES 24%

ENVIRONMENT

Vanuatu is a Melanesian archipelago, of volcanic origin, comprising more than 70 islands and islets many of them uninhabited. It stretches for 800 km, in a north-south direction in the South Pacific about 1,200 km east of Australia. Major islands are: Espiritu Santo, Malekula, Epi, Pentecost, Aoba, Maewa, Paama, Ambrym, Efate, Erromango, Tanna and Aneityum. Active volcanoes are found in Tanna, Ambrym and Lopevi and the area is subject to earthquakes. The land is mountainous and covered with dense tropical forests. The climate is tropical with heavy rainfall, moderated by the influence of the ocean. The subsoil of Efate is rich in manganese and the soil is suitable for farming. Fishing is a traditional economic activity. The land tenure system has contributed to general soil depletion (deforestation, erosion). Rising sea levels - a 20 cm increase in Vanuatu's tides has been estimated for the next four decades - will affect both inhabited and uninhabited coastal areas. There is also a risk that salt water (from rising tides) will seep into ground water, threatening water supplies

SOCIETY

Peoples: The people are basically Melanesian, with 3% European (British and French) and smaller groups from Vietnam, China and other Pacific islands. **Religions:** Mainly Christian. **Languages:** Bislama, English and French are official. Other Melanesian languages are also spoken. **Political Parties:** Union of Moderate Parties; United National Party; People's Democratic Party; Unity Front; Tan Union, Frei Melanesio and Na Griamel. **Social Organisations:** Vanuatu Trade Union Congress (VTUC).

THE STATE

Official Name: Ripablik blong Vanuatu. **Capital:** Port Vila, on Efate Island, 30,000 inhab. (1992). **Other cities:** Luganville (Santo), 6,900 inhab.; Port Olry, 884 inhab.; Isangel, 752 inhab.(1989). **Administrative Divisions:** 11 Government Regions, 2 Municipalities. **Government:** Jean-Marie Leye, President; Maxime Carlot Korman, Prime Minister. The unicameral Representative Assembly is made up of 46 representatives elected every 4 years. **National Holiday:** July 30, Independence Day (1980).

DEMOGRAPHY

Annual growth: 1.1% (1992-2000). **Children** per **woman:** 5.3 (1992).

EDUCATION

Primary school teachers: one for every 31 students (1992).

COMMUNICATIONS

2.5 **telephones** per 100 inhabitants (1993).

ECONOMY

Per capita GNP: $1,150 (1994).
Annual growth: -0.30% (1985-94).
Consumer price index: 100 in 1990; 117.4 in 1991.
Currency: 112 vatu = 1$ (1994).

[16] In February 1984, President Ati George Sokomanu resigned. He justified his action by claiming that the constitution did not adequately protect the head of state or his ministers, or guarantee "coherent management". However, after this governmental crisis, the resigning president agreed to participate in the March 1984 elections, and was re-elected.

[17] The second Vanuatu Development Plan (1987-1991) was effected to achieve the following goals: to ensure balanced regional and rural development, to take better advantage of the country's natural resources, to provide for the faster development of human resources and to promote development of the private sector. The plan suffered a serious set-back in February 1987, with hurricane Ulna, which caused an estimated $36 million damage and affected about 34 per cent of the population.

[18] In the December 1987 elections, Lini won over Barak Sope, who attempted an unsuccessful challenge for the party leadership. Barak Sope accepted the cabinet post of minister of tourism and immigration, but in May 1988 he was relieved of his responsibilities by Lini, due to violent demonstrations by his followers. Finally, Sope split from the ruling party and founded his own political group, the Melanesian Progressive Party.

[19] In 1988, former president Sokomau, together with Barak Sope, tried unsuccessfully to oust Prime Minister Walter Lini. They were both arrested and imprisoned. Fred Timakata was appointed the new president.

Vanuatan independence was declared on July 30 1980. Measures were immediately taken to return the land held by foreigners to the Melanesians; the school system was unified and a national army was created.

[20] On August 7 1991, the ruling party of Vanuatu voted in favor of deposing Walter Lini as prime minister. The minister of foreign affairs, Donald Kalpokas, was appointed to this post.

[21] In October, Walter Lini founded the United National Party. The Union of Moderate Parties, led by Maxime Carlot Korman, won the elections held on December 2.

[22] In January 1992, Carlot Korman formed a coalition with the UNP. Lini's sister Hilda was named minister of health, occupying the cabinet position to which the UNP - as a member of the coalition - was entitled. At this time, French was re-established as an official language. Minister of Finance Willy Jimmy's economic plan established a series of priorities, including exports, foreign investment, agrarian reform and free primary education.

[23] Carlot Korman launched a program to diversify agriculture in 1993, a year after hurricane Betsy had destroyed 30 per cent of the harvest. In 1994 the government decided to eliminate 200 of the 4,800 posts in the existing public service. One of the arguments for this was the disproportionate increase of state employees since 1985, when there had been 3,300.

[24] On the international front, Vanuatu was the only country in the South Pacific Forum which did not join in this organization's protest against the French nuclear tests in Polynesia in September. Following the elections in November, the governing coalition took 29 of the 50 seats at stake, Carlot Korman was replaced by Serge Vohor. However, Vohor resigned two months later and, in February 1996, Carlot Korman returned to his post as head of government.

Venezuela

Population: 21,177,000 (1994)
Area: 912,050 SQ KM
Capital: Caracas

Currency: Bolívar
Language: Spanish

Venezuela

Cumanagotos, Tamaques, Maquiritares, Arecunas, and other Carib groups inhabited the northern tip of South America when Christopher Columbus arrived on these shores in 1498. Local buildings, constructed on poles, reminded the Spaniards of the appearance of Venice, so they named the country Venezuela.

[2] During the colonial period, Venezuela was organized as a Captaincy General of the Viceroyalty of New Granada , and in the 18th century it became the most important farming colony, producing mainly cocoa. A local aristocracy of "mantuanos" developed, a landowning class of European extraction, who used African slave labor on their plantations. The "pardos" ("blacks") who constitute the overwhelming majority of the population are descended from the slaves.

[3] Two of the greatest leaders of the Latin American independence movement were born in the Captaincy: Francisco de Miranda and Simón Bolívar. On April 19 1810, the wars of independence broke out after the constitution of a Cabildo or Assembly. Miranda became the commander of the army, envisioning a vast American Confederacy to be known as Colombia, which would crown an Inca emperor. His dreams did not however come to fruition as he was captured by the Spaniards in 1811, and later died in prison.

[4] Bolívar took up Miranda's project for American liberation and was backed, in principle, by the Mantuan oligarchy. In a swift campaign (1812-13) he took over the country, and was able to install a government in Caracas. His plans for independence did not contemplate changes in social structure, and he was not supported by the masses who lived in the plains (*llaneros*), most of whom were mulattos living under the severe repression of white creole masters who they hated. Their liberation movement was led by the Spanish loyalist General José Tomás Boves, who defeated Bolívar in 1814. Boves abolished slavery and redistributed the land among the people.

[5] Having lost the first republic, Bolívar went into exile, working closely with the Haitian president, Alexandre Sabés Petión, who helped him to add a social dimension to his American revolutionary project. Bolívar returned to Venezuela, and, championing popular demands, he won mass support.

[6] Accompanied by other important military leaders like Antonio José de Sucre, Mariño, José A. Páez and Arismendi, he carried out successful military campaigns in the northern half of the continent. He subsequently founded Bolivia in what had hitherto been known as "Upperu".

[7] In 1819, the Angostura Congress created the new republic of "Gran Colombia", uniting Colombia, Ecuador, Panama and Venezuela. In 1830, after Bolívar's death, General José Antonio Páez declared Venezuela's secession from Gran Colombia, establishing it as an independent nation.

[8] For decades, Venezuelan politics revolved around the *caudillo*, or leader, Páez. His political successor, Antonio Guzmán Blanco, was determined to modernize Venezuela. He carried out the aim to a certain extent, introducing new technology, new means of communication, and reforming the country's legal code.

[9] Juan Vicente Gómez took power in 1908 ruling for 17 years, and his government made an attempt to eliminate the *caudillo* system. He gave free access to the multinational oil companies, which set up oil extracting operations primarily on the Lake Maracaibo oil fields.

[10] In 1935 his former aide, General Eleazar López Contreras, took office. He was succeeded, in 1941, by General Isaías Medina Angarita, who laid the foundations for freer political activity by legalizing the Democratic Action Party (AD).

[11] It was a long time before majority demands were met, however, and this situation was aggravated by the unrest within the army. A civilian-military movement took power, led by Rómulo Betancourt, the leader of AD, and general Marcos Pérez Jiménez, and they held the country's first free elections in 1947. The writer Rómulo Gallegos, of the AD, was elected president only to be overthrown in 1948 by yet another military coup, installing the harsh dictatorship of Marcos Pérez Jiménez.

[12] Ten years later, in January 1958, Pérez Jiménez was overthrown by a popular revolt and he fled the country. Venezuela then entered a stable democratic period under a coalition government formed by Democratic Action, COPEI (the Christian Democrats) and the Republican Democratic Union (URD). This stability was largely achieved because of the influx of massive oil revenues, improved relations with the US, and expanded political rights. Unfortunately, the ensuing economic growth brought little change to the lives of the poor majority, with popular discontent resulting in guerrilla warfare led by the Communist Party, the Movement of the Revolutionary Left (which split from AD), and other left-wing groups.

[13] In 1960, Venezuela sponsored the formation of the Organization of Petroleum Exporting Countries (OPEC). Sixteen years later, during the presidency of social democrat Carlos Andrés Pérez, Venezuelan reserves of oil and iron were nationalized. Pérez also supported the creation of the Latin American Economic System (SELA) and argued in favor of a New International Economic Order.

[14] The Christian Democrats were returned to power in the December 1978 elections, but the administration of President Luis Herrera Campins brought no substantial change to reformist policy.

[15] Venezuela was the world's

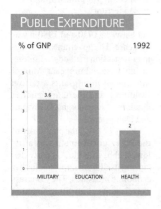

PUBLIC EXPENDITURE

% of GNP — 1992

- MILITARY 3.6
- EDUCATION 4.1
- HEALTH 2

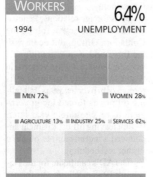

WORKERS

1994 — **6.4%** UNEMPLOYMENT

- MEN 72%
- WOMEN 28%

- AGRICULTURE 13%
- INDUSTRY 25%
- SERVICES 62%

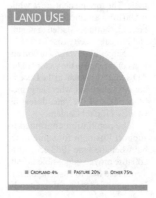

LAND USE

- CROPLAND 4%
- PASTURE 20%
- OTHER 75%

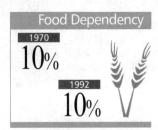

Food Dependency

1970
10%

1992
10%

Foreign Trade

MILLIONS $ 1994

IMPORTS
7,710
EXPORTS
15,480

Illiteracy

1995

9%

PROFILE

ENVIRONMENT

The country comprises three main regions. In the north, the Andes extend for about 1,000 km along the Caribbean coast, with other mountain chains in the East, and high mountains to the south. The central Orinoco Plains are a livestock farming area. In the southeast, highlands of ancient rock and sandstone extend to the borders with Brazil and Guyana, forming Venezuelan Guyana. It is a sparsely inhabited area with thick forests, savannas, rivers, and some peculiar features: the two highest waterfalls in the world, the "tepuyes" or plateau mountains, and the rare Sarisarinama depths. Most of the population lives in the hilly north. The oil-rich Maracaibo lowlands and Gulf of Paria are on the coast. The country produces oil, iron ore, manganese, bauxite, tungsten and chrome. Among the country's most important environmental problems are deforestation and the degradation of the soil. In addition, the lack of sewage treatment facilities in the main urban and industrial centers has increased pollution in the country's rivers and the Caribbean Sea.

SOCIETY

Peoples: Venezuelans come from the integration of indigenous peoples, Afro-Caribbeans and European settlers. Today, indigenous peoples and Afro-Caribbeans each account for less than 7% of the population. In recent times, Venezuela has received more immigrants than any other

South American country. **Religions:** Mainly Catholic. **Languages:** Spanish, official and predominant; 25 native languages are spoken. **Political Parties:** Democratic Action (AD), member of the Socialist International; the Social Christian Party (COPEI), a member of the Christian Democrat International; the Movement To Socialism (MAS), founded in 1970; Radical Cause; Convergence, created in 1993 by Christian Democratic splinter groups; the People's Electoral Movement (MEP); the Venezuelan Communist Party (PCV); and the New Democratic Generation, linked to the Liberal International. **Social Organizations:** The Venezuelan Confederation of Workers (CTV) is the main trade union, and is controlled by AD. There are other trade unions which are clearly linked to political parties.

THE STATE

Official Name: República de Venezuela.
Administrative divisions: 21 states with partial autonomy (including the Federal District), 2 Federal territories.
Capital: Caracas, 4,000,000 inhab. (1990).
Other cities: Maracaibo 1,249,670 inhab.; Valencia 903,621 inhab. (1990).
Government: Rafael Caldera, president since February 2, 1994. Presidential system, congress with Chamber of Deputies and Senate (199 and 53 members respectively). **National Holiday:** July 5, Independence Day (1811). **Armed Forces:** 79,000, including 18,000 conscripts (1994). **Paramilitaries:** Cooperation Army, 23,000.

third largest oil exporter and it received its highest prices during the governments of Pérez and Herrera. At this time, Venezuela had the highest per capita income in Latin America, but the government was unable to manage the enormous amounts of money coming into the country. Huge state-run companies were created for the manufacture of iron ore, aluminum, cement, and for hydroelectric power, while most private companies were being subsidized. Today, the Venezuelan state still generates 95 per cent of the country's exports and is by far its largest business owner.

[16] Long-standing border disputes between Venezuela and its neighbors flared up again in 1981 and 1982. Disputes with Colombia over the Guajira peninsula, and with Guyana over the Esequibo region, continued. Meanwhile, after hold-

ing many conferences to discuss the issue of national waters, Venezuela established its territorial rights over huge areas of the Caribbean sea.

[17] In 1982, a sharp decrease in oil revenues, foreign debt, and a massive flight of private capital abroad, forced the government to take control of exchange rates and foreign trade. The foreign debt grew, while inflation, unemployment, and housing shortages started to mount, and critical poverty rose.

[18] Charges of government corruption contributed to the victory of the AD's Jaime Lusinchi in the December 1983 elections, where he gained 56 per cent of the vote, defeating COPEI's Rafael Caldera.

[19] The population continued to support Lusinchi for some time, but when the June 1983 economic collapse proved impossible to halt, Venezuela had to suspend $5 billion in service payments on its $37

billion debt, and the ensuing difficulties resulted in a drop in his popularity.

[20] In reaction to the economic problems, Lusinchi's policy comprised an austerity plan which gave meagre results, and a flawed "social pact" between employers and unions and, towards the end of his term, increasing state control over the economy.

[21] In its foreign relations, Venezuela resumed its active role in the Contadora Group, but dealt with the IMF and US in a "traditional" way. In January 1986, Venezuela rescheduled nearly two thirds of its foreign debt over ten years, but falling oil prices forced the government to renegotiate a year later. Before leaving office, Lusinchi announced a moratorium to obtain a further rescheduling.

[22] The AD's candidate for the December 1988 elections was former

president Carlos Andrés Pérez. Backed by the Confederation of Venezuelan Workers (CTV), he took 54.5 per cent of the vote, and Eduardo Fernández, the COPEI candidate, took 41.7 per cent.

[23] Increasing social tension could not be halted and 25 days later the most underprivileged sectors of the population responded with a wave of riots and looting. Police repression left more than 1,000 dead (246 according to the government), and 2,000 wounded or jailed.

[24] The populist tone was dropped, and IMF-backed economic restructuring measures were initiated. In December 1989, these policies were blamed for abstention rates of nearly 70 percent and broad gains by the Christian Democrats and left-wing parties, when Venezuelans elected their 20 state governors and 369 mayors for the first time.

[25] Ten AD governors were elected, and COPEI had four with the AD losing seats in many of its traditional strongholds. Most remarkable, however, were the left-wing victories in three states: in Bolívar, the trade union leader Andrés Velázquez of the Popular Cause party was elected; in Aragua, the Movement To Socialism (MAS) won an easy victory; and Anzoátegui elected a leader of the People's Electoral Movement (MEP).

[26] In the early 1990s, Venezuela's indigenous population totalled around 200,000, approximately one per cent of the population, but their future survival is threatened. Despite the fact that the government has clearly recognized Yanomami land rights, there has not been adequate protection of the indigenous population. They have continued to suffer persecution from landowners, farmers and government officials, while Brazilian miners continue to invade their land prospecting for gold within Venezuelan territory.

[27] The immigration rate has fallen, but between 1950 and 1980 it was immense. The government is bringing in selection methods, admitting mainly Eastern Europeans. Most of the present immigrants are from Colombia and there are smaller numbers of Ecuadorans, Peruvians, and Dominicans.

[28] Early in 1992, the popularity of the Pérez government was at its lowest ebb, with the AD party distancing itself from imposing further economic measures. The Congress exercised its controlling function

Maternal Mortality
1989-95

PER 100,000
LIVE BIRTHS
200

External Debt
1994

PER CAPITA
$1,740

by investigating crimes, but due to the corruption and inefficiency of the Judiciary these investigations very rarely led to prosecutions.

[29] On February 4 1992, a group of lieutenant-colonels staged a coup against President Carlos Andrés Pérez. The group was led by Francisco Arias and while the attempt failed, it re-emphasised the fact that administrative corruption and the economic crisis were the causes of instability. Arias and parachute commander Hugo Chavez, both belonging to the Bolívar-200 Military Movement, were arrested.

[30] This coup attempt had caused several dozen civilian and military casualties. The defence minister Fernando Ochoa officially stated that 51 military personnel had been injured, 19 had died, and an indeterminate number of soldiers had been arrested.

[31] On the day of the attempted coup, the Venezuelan president suspended all constitutional rights. This emergency measure, ratified by Congress, banned meetings, limited the freedom of press and movement and made it possible to arrest citizens without a warrant. An agreement was reached with the teachers trade unions, putting an end to a fortnight of strikes and police repression of students and teachers.

[32] On February 12, a military judge started to investigate the charges against the 13 officers and nearly a thousand soldiers who had staged the coup. The lieutenant-colonels who had led the insurrection were detained in the San Carlos barracks, awaiting sentence. Two days later Pérez lifted the emergency measures.

[33] Political and military tension increased again in early March with more trouble from the army. The president was asked to resign, it was reported that weapons had been stolen and military units were confined to barracks.

[34] On March 5 1992, in the light of the impending crisis, Pérez announced a popular referendum to reform the constitution, the creation of a national unity cabinet, a reform of the judiciary, and a "change of direction" in the economic policy instigating a further economic austerity program.

[35] In early April, a "National Civic Strike" was organized by students, labor unions and other organizations, to put pressure on the president to resign. The strike's or-ganizers also demanded the release of the military officers who had been arrested for participating in the February 4 coup attempt.

[36] While protest demonstrations against Carlos Andrés Pérez continued to increase throughout the country, the party in power (Democratic Action) and COPEI reached an agreement in late May, deciding to form an alliance which they believed would reinforce the stability of Venezuela's democratic system.

[37] Another coup attempt took place on November 27 1992. The Air Force played a key role in controlling the rebels. The ruling party only won in 7 of the 22 states where elections for governors, mayors, regional legislative assemblies and municipal councils were held.

[38] President Carlos Andres Pérez was suspended from office on May 21 1993, charged with the misappropriation of funds. He was provisionally replaced by Ramón Velázquez.

[39] Rafael Caldera, former president between 1969 and 1974 from the Social Christian Party, won the general elections on December 5 1993. He was supported by a wide range of sectors and political parties, with the backing of Convergencia and MAS being the most important factors in his success. Abstention reached 40 per cent as few over five million voters actually exercised their right to vote.

[40] According to official statistics, the annual rate of inflation reached nearly 40 per cent in 1993 and almost half of the population was living below the poverty line.

[41] Another source of social unrest was the repeated violation of human rights, both of ethnic minorities and of inmates in several prisons. On August 21 the murder of 16 Yanomami close to the Brazilian border at the hands of the "ga-rimpeiros" - Brazilian gold prospectors - was reported. Several members of the Yupca indigenous tribe were murdered by landowners in the state of Zulia.

[42] The repression of a riot in a Maracaibo jail by the security forces caused 122 deaths on January 3 1994. Built to accomodate 1,500, the prison housed 2,500 inmates.

[43] Due to his age – over 75 – former president Carlos Andrés Pérez was sentenced to house arrest. In his opulent residence overlooking Caracas, Pérez concluded his sentence in September 1996 and returned to the political scene, af-

firming he would be president once more. During his "confinement", Pérez was accused of having travelled abroad.

[44] At the end of the 1980s, Caracas turned into one of Latin America's most violent cities. A feeling of insecurity prevailed as the number of murders rose. In 1995 it was estimated that 10 per cent of Caracas residents carried a weapon.

[45] The economic crisis worsened after 1994. The banks' collapse began in February 1993 with the fall of the Banco Latino, the country's second commercial bank. In August 1995, 18 out of the 41 private banks had been investigated and 70 per cent of the deposits were being managed by the government.

[46] President Caldera suspended constitutional guarantees regarding real estate, private property and business. He also restricted trips abroad, meetings and the immunity against arbitrary arrests. Despite the Congress's vote to re-instate these rights, the president restricted them once more to prevent speculation and capital flight.

[47] Efforts to attract foreign investors failed in 1995 and the government decided to offer equal opportunities both to nationals and foreigners for the prospecting and exploitation of oil for the first time since the oil industry's nationalization in 1976.

[48] The price of fuel rose 100 per cent in September 1995 and March 1996 provoked a significant increase in the prices of other basic consumer goods.

Vietnam

Viêt-nam

Population: 72,039,000
(1994)
Area: 329,560 SQ KM
Capital: Hanoi

Currency: Dong
Language: Vietnamese

The Vietnamese nation was established after centuries of struggle against more powerful peoples. In the ninth century, they defeated the Chinese Han dynasty, ending nearly a thousand years of subjugation. In the 11th and 12th centuries, the Vietnamese succeeded in fighting off the Chams. They repulsed Genghis Khan, his Mongol hordes, and later his grandson Kublai Khan, when these were conquering the Asian world in the 18th century. In the 15th and 18th centuries the Vietnamese were victorious over the Ming and Ching Chinese dynasties, and in the 18th century they drove back the Khmers.

[2] In 1860, the French began to occupy Indochina, meeting with spontaneous, disorganized and badly armed resistance which, nonetheless, took them three decades to overcome. By 1900 the French had consolidated their domination over the region. Vietnam was divided into three parts: Tonkin in the north, Annam in the center and Cochin-China in the south. To eliminate Chinese cultural influence, the French taught Quoc Ngu, a Romanized writing script of the Vietnamese language. The French did not intend to help Vietnam's nationalists in the renewal and dissemination of national culture, but this system of notation facilitated reading and lowered printing costs when compared with the formerly used Chinese ideograms.

[3] In the 1920s, the first nationalist organizations were created. Marxist-Leninist parties appeared in different parts of Vietnam in 1929, achieving unity as the Indochina Communist Party the following year. This was founded by a patriot called Nguyen Ai Quoc, who adopted the name Ho Chi Minh. The party later split into three different organizations for Cambodia, Laos and Vietnam respectively. The latter was known as the Workers' Party until its 4th Congress in 1976, when it became the Communist Party of Vietnam once again.

[4] During World War II, the country was occupied by Japan. Vietnam's communists organized the resistance from the start, cooperating with the allies but staying true to their intent to defeat colonialism. In 1941, Ho Chi Minh founded the Viet Minh or League for the Independence of Viet Nam, conceived as a broad unity front to include workers, peasants, the petty-bourgeoisie and the nationalist bourgeoisie. By the time of the Japanese surrender, the Viet Minh army had become a powerful force with widespread popular support. On August 18 1945, Ho Chi Minh launched a general insurrection. Within two weeks the revolutionary forces took control of the entire country, and an independent republic was declared. Emperor Bao-Dai abdicated and offered to cooperate in advising the new regime.

[5] The newborn Vietnamese Democratic Republic was forced into warfare to prevent the return of the French colonialists. After nine years of fighting, the Vietnamese were finally victorious when they captured Dien Bien Phu on May 7 1954.

[6] The 1954 Geneva Agreements stipulated the withdrawal of the French and directed the Vietnamese to hold general elections in 1956. Viet Minh troops were to withdraw north of the 17th parallel. The United States set up the Ngo Dinh Diem regime in Saigon (present-day Ho Chi Minh), dividing the country and violating the Geneva agreements, as elections were prevented. In 1960, the organizations hostile to Diem's regime (democratic, socialist, nationalist and Marxist forces), became the National Liberation Front, led by lawyer Nguyen Huu Tho. The "second resistance" was then launched, this time against the successive military governments based in Saigon and, above all, against the United States, the true supporter, arms-supplier and policy-maker. US aid was initially limited to a handful of advisors; then troops were sent, a process that escalated until, by 1969, 580,000 US troops were in Vietnam. A larger tonnage of bombs was dropped in Vietnam than during the whole of World War II, and terrible chemical and biological weapons were used.

[7] The US spent $150 billion in Vietnam, destroyed 70 percent of the northern villages and left 10 million hectares of productive land barren. Despite their efforts, Saigon was finally liberated by the Viet Cong on April 30 1975. On July 2 1976, Vietnam was reunified under the name of the Socialist Republic of Vietnam.

[8] Since then Vietnam has not enjoyed stable peace. In January 1979, it went to war against Pol Pot's regime in Cambodia; Pol Pot had claimed vast portions of Vietnamese territory. No sooner had the Cambodian regime been overthrown, than China invaded Vietnam on its northern border (see Cambodia) as a "punitive" measure. China, in the meantime, had been a staunch ally of the ousted Cambodian regime. Subsequent clashes along the border in 1980 underscored the military superiority of the Vietnamese, but the border between the two countries remained unchanged.

[9] Vietnam considered Cambodia's union with the other Indo-chinese states to be essential for its se-

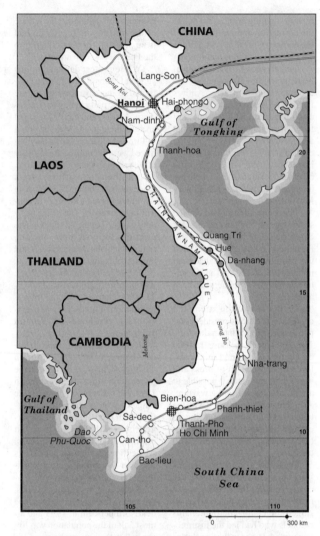

WORKERS

1994

■ MEN 53%   ■ WOMEN 47%

■ AGRICULTURE 67% ■ INDUSTRY 12% ■ SERVICES 21%

LAND USE

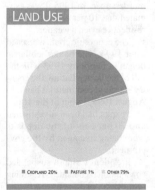

■ CROPLAND 20%   ■ PASTURE 1%   ■ OTHER 79%

PER CAPITA
$349

IMPORTS
4,440
EXPORTS
3,770

curity. Vietnam's support of the Heng Samrin government against the Khmer Rouge, Prince Sihanouk, and the National Liberation Front entailed costly military efforts and a drain on Vietnam's resources, particularly in terms of food, which was already scarce. From 1978 to 1981, Vietnam suffered floods, typhoons and droughts.

[10] After 1981, Vietnam faced active hostility from the Reagan administration, which stopped grain shipments provided by an American charitable association and blocked a Vietnamese aid program organized by the United Nations Development Program (UNDP).

[11] The 6th Congress confirmed that self-sufficiency was the aim in some key sectors. The agricultural sector planned to produce 20 million tons of cereal, and energy self-sufficiency was to be achieved with the construction of steam power plants and a new contract for oil prospecting with the USSR.

[12] In April 1985, several political prisoners were freed, and diplomatic ties with ASEAN countries and the United States were intensified in an effort to reach negotiated settlements for Indochina. Arrangements were made to cooperate with the US to search for the corpses of missing soldiers as well as to allow people loyal to the former government to leave the country, and to permit the exit of children of US servicemen.

[13] The legendary Le Duan, Secretary-General of the Communist Party of Vietnam since 1969 and a close collaborator of Ho Chi Minh, died on July 11 1986. The Congress designated 71 year-old Nguyen Van Linh, a former Viet Cong strategist, as the new Secretary-General.

[14] Van Linh proposed a "complete and radical renovation" of the political and economic system, the government was renewed and restructured, and a series of economic reforms was announced, loosening restrictions governing the commercializing of agricultural products and the establishment of private businesses. Laws were also passed to attract foreign investment and guarantee that foreign enterprises would not be nationalized.

[15] Toward the end of 1988, the government adopted austerity measures to combat inflation, including a marked devaluation of the dong (national currency), in relation to the dollar. In the process, many

PROFILE

ENVIRONMENT

A long and narrow country covering the eastern portion of Indochina along the Gulf of Tonkin and the South China Sea. The monsoon influenced climate is hot and rainy. Rainforests predominate and there is a well-supplied river system. The northern region is comparatively higher. There are two river deltas: the Song Koi in the north, and the Mekong in the south. Most of the population are farmers, and rice is the main crop. The north is rich in anthracite, lignite, coal, iron ore, manganese, bauxite and titanium. Textile manufacture, food products and mining are the major economic activities. The felling of trees for domestic use (firewood) and construction have contributed to deforestation. However, the most significant losses - particularly in the northern part of the country - are a result of the Vietnam War, specifically from the use of such chemical defoliants like "Agent Orange".

SOCIETY

Peoples: Most are native Vietnamese. Mountain groups of different ethnic origins - Tho, Hoa, Tai, Khmer, Muong, Nung - and descendants of the Chinese form the rest of the population. **Religions:** Mainly Buddhist; traditional religions. There are 2 million Catholics and 3 million followers of the Hoa-Hao and Cao-Dai sects.
Languages: Vietnamese (official) and languages of the ethnic minorities. **Political**

Parties: The Vietnamese Communist Party (Dan Cong San Vietnam), according to the constitution, "the leading organization of society", founded by Ho Chi Minh in 1931; the Socialist Party and the Democratic Party of Vietnam, founded respectively in 1946 and 1944, to organize intellectuals and the national bourgeoisie for the anti-colonialist struggle.
Social Organizations: The Federation of Unions of Vietnam (Tong Cong Doan Vietnam), founded in 1946, is the only union confederation and a WFTU member; the Vietnamese Women's Union, founded in 1930.

THE STATE

Official Name: Nu'ó'c Công Hòa Xâ Hôi Chu' Nghî'a Viêt Nam. **Administrative Divisions:** 39 Provinces, including the Urban Areas of Hanoi, Haiphong and Ho Chi Minh. **Capital:** Hanoi, 2,154,900 inhab. (1993). **Other cities:** Ho Chi Minh (formerly Saigon), 4,322,300 inhab.; Haiphong, 1,573,000 inhab.; Da-Nhang, 381,410 inhab. (1993). **Government:** Le Duc Anh, President of the Council of State. Vo Van Kiet, Prime Minister and head of government, elected in August 1991. The National Assembly is made up of 395 members elected by universal suffrage; it elects the members of the Council of State. The Executive Power is exercised by a Council of Ministers. **National Holiday:** September 2, Independence Day (1945). **Armed Forces:** 572,000 troops (1995). **Paramilitaries:** 5,000,000 Urban Defence Units, Rural Defence Units.

businesses were forced to close down, and the already high unemployment rate reached almost 30 per cent, while lack of food led to popular protests.

[16] However, a good rice harvest in 1988-89 resulted in a slight improvement in the living conditions of the population, as well as Vietnam's admission to the group of rice-exporting countries. By the end of 1989, they had exported around 1 millon tons of rice, valued at $250 million. Most of the rice was exported to western Africa, the Philippines, India, Sri Lanka and China.

[17] The political changes in Eastern Europe at the end of the 1980s were not well-received by Vietnamese communists since these had local repercussions, specifically growing demands for a multiparty political system. In the absence of such a system, dissenting voices sought expression within the Vietnamese

Communist Party (VCP) itself. Some sectors of the party demanded change toward a multiparty, parliamentary political system.

[18] Another group, calling itself the "Resistance Fighters' Group"; made up of southern revolutionaries who had fought against the French and the Americans for the country's independence, voiced its disagreement, in particular with the government's economic policy. This group, which described itself as a pressure group within the Party, urged its leaders to speed up the pace of reforms and the granting of political freedoms. The appearance of this group shed light upon the country's economic and social reality, characterized by the most dynamic economic center being in the surroundings of Ho Chi Minh City.

[19] In August, the Central Committee of the VCP accused the West of trying to destroy the socialist world.

Nguyen Van Linh urged his fellow-citizens to reject bourgeois liberalization, multiparty politics and opposition parties, whose aims were to make them renounce socialism. Even so, the participation of non-communist candidates was authorized in the 1989 National Assembly elections, and a year later, the government authorized the formation of the Vietnam Veterans Association. However, this liberalization was accompanied by a reaffirmation of the dominant role of the VCP, with the arrest of a number of campaigners for multiparty politics, and the dismissal of 18,000 public employees accused of corruption.

[20] During the 1980s, "illegal emigration" – the flight of Vietnamese toward neighboring countries – continued. The reluctance of Western countries to receive them has indefinitely prolonged their stays in

Illiteracy
1995

6%

Maternal Mortality
1989-95

PER 100,000
LIVE BIRTHS

105

STATISTICS

DEMOGRAPHY

Urban: 20% (1995). **Annual growth:** 1.2% (1992-2000). **Children** per **woman:** 3.7 (1992).

HEALTH

One **physician** for every 247 inhab. (1988-91). **Under-five mortality:** 46 per 1,000 (1995). **Calorie consumption:** 92% of required intake (1995). **Safe water:** 36% of the population has access (1990-95).

EDUCATION

Illiteracy: 6% (1995). **School enrolment:** University: 2% (1993). **Primary school teachers:** one for every 35 students (1990).

COMMUNICATIONS

88 **newspapers** (1995), 89 **TV sets** (1995) and 86 **radio sets** per 1,000 households (1995). 0.4 **telephones** per 100 inhabitants (1993). **Books:** 90 new titles per 1,000,000 inhabitants in 1995.

ECONOMY

Per **capita GNP:** $200 (1994). **Annual inflation:** 102.60% (1984-94). **Currency:** 8.125 dong = 1$ (1990). **Cereal imports:** 289,000 metric tons (1993). **Fertilizer use:** 1,347 kgs per ha. (1992-93). **Imports:** $4,440 million (1994). **Exports:** $3,770 million (1994). **External debt:** $25,115 million (1994), $349 per capita (1994). **Debt service:** 6.1% of exports (1994). **Development aid received:** $319 million (1993); $5 per capita; 3% of GNP.

ENERGY

Consumption: 105 kgs of Oil Equivalent per capita yearly (1994), -55% imported (1994).

---

overpopulated refugee centers. This has been especially true in Hong Kong; the British government began deporting them from there in December 1989, though this operation was later suspended because of international pressure.

[21] The 8th VCP Congress held in June 1991 agreed to continue the process of "building" socialism, a path considered to be "the only correct choice". Van Linh was replaced as the party's general secretary by Du Muoi, who had been the architect of the renewal process which had been initiated in 1986 (called Doi Moi). Tron My Moa, the first woman to be a part of the country's new leadership, became a member of the 9-member Secretariat of the Central Committee.

[22] The VCP confirmed the continued existence of a single-party system. The first step toward renewal and reform was a more flexible economic policy, specifically, a change to a mixed economy. This implied allowing private enterprise and foreign investment within pre-established parameters.

[23] In October 1991, the initiation of negotiations between the four Cambodian factions opened the way for the normalization of Sino-Vietnamese relations.

[24] In November 1991, the British government with the support of the US, began the forcible repatriation of 64,000 Vietnamese exiles from refugee camps within Hong Kong, a British dependency.

[25] The new constitution approved in April 1992 authorized independent candidates to run for election, but in the July legislative elections, 90 per cent of the candidates belonged to the VCP. In September, the new Assembly elected General Le Duc Anh president, known ally of Prime Minister Vo Van Kiet. In November, a commission of US senators visited the country to discuss the issue of the missing remains of over 2,000 US military killed in action.

[26] At the beginning of the 1990s Vietnam appeared to be ready for takeoff, following the example of its neighbors, the so-called "Asian tigers". Privatization and the liberalization of foreign investment led to an 8.3 per cent increase in the GDP in 1992. For 1993 and 1994, 8 and 9 per cent increases were expected respectively. In the meantime, the volume of exports has risen 20 per cent per year since 1991.

[27] As for foreign trade, Vietnam depends primarily on two products: rice and, to a lesser extent, oil. Since 1989, Vietnam has become the world's third largest exporter of rice, after Thailand and the United States. However, this dependence upon a single product – until oil production increases to the point where it can be considered an important export product – makes the country vulnerable to the vagaries of the international market. This problem is compounded by falling rice prices, and the loss of part of the rice crop each year because of a lack of storage facilities.

[28] According to Vietnamese standards, a "poor" family is one that can not afford 13 kg of rice person each month. A "hungry" family is one that can not afford 8 kg of rice person each month.

[29] The south has been the main beneficiary of the rice production and export boom in the Mekong delta. This has underscored the structural inequality between the two parts of the country. With more than five million inhabitants - almost double the size of Hanoi, the capital city - Ho Chi Minh City is the country's economic capital, adapting itself far better to the new era than the north.

[30] This expansion of trade is taking place in what is primarily an agricultural country. 57 million of the 72 million Vietnamese depend upon agriculture, the backbone of the economy and of Vietnamese society. As a result, the government has granted a number of benefits to farmers, such as loans and long-term land leases, tax cuts and the right to inherit up to 3 hectares of land. However, private land ownership has not yet been authorized.

[31] The effects of several decades of war are still evident, especially in the economic structure which was tailored to the demands of war. The devastation remains visible in a large part of the country, as a result of the use of lethal weapons like napalm or "Agent Orange", a pow-

erful defoliant. The country's infrastructure, in general – telecommunications, highways and energy production – slowly began to recover as a result of projects financed by the World Bank and Asian Development Bank loans.

[32] In 1993, the per capita GDP was among the lowest in Asia. Between $230 and $350 per year, it was roughly equivalent to one-sixth of Thailand's GDP, and one-twentieth of South Korea's. Unemployment and underemployment, taken together, continue to stand at 30 per cent, while the underground economy and contraband are on the increase.

[33] The US embargo of Vietnam began to be lifted in January 1993, when a US Senate commission concluded that there was no evidence of US prisoners of war in Vietnam. In February 1994, US President Clinton announced the end of the trade embargo, which had lasted 19 years.

[34] By adapting to international competition, it is likely that ecological damage will increase. Even today, the intensification of agriculture, particularly the intensive cultivation of rice, and the felling of trees for construction or timber exports have become serious environmental problems.

[35] The Communist Party continued to oppose political and economic liberalization, fearing a massive return of emigrants and the ensuing Western influence. However, Prime Minister Vo Van Kiet forced his government officials to learn foreign languages, especially English.

[36] Vietnam and the United States established formal diplomatic relations in August 1995. President Clinton stated his concern for the 2,200 US soldiers still considered to be lost in action in Southeast Asia.

[37] The rapprochement between the governments of Hanoi and Washington did not lead to a massive arrival of US firms, although many multinational corporations showed their interest in the Vietnamese economy, in anticipation of the consolidation of economic liberalization.

Virgin Is. (US)

Population: 100,000 (1994)
Area: 340 SQ KM
Capital: Charlotte Amalie
Currency: US Dollar
Language: English

Virgin Islands

As is the case in other Caribbean islands, the Virgin Islands were originally inhabited by Carib and Arawak Indians. The only gold on the islands was seen in the ornaments worn by the inhabitants and was of little economic interest. From the time of Columbus' arrival in 1493 onwards, the local population was persecuted and massacred until they were all destroyed by the second half of the 16th century.

[2] The western part of the archipelago was claimed by other countries after the departure of the Spaniards. The Netherlands gained control in the 18th century and organized crop cultivation; first of sugar cane and then of cotton. With this in mind, they imported African slaves, as in many other parts of Latin America and the Caribbean. At the height of Dutch colonization, there were 40,000 African slaves on the islands.

[3] During Abraham Lincoln's second presidential term, the United States failed to acquire the islands. But in 1917, after World War 1 when the US wanted to consolidate its presence in the area, $25 million was paid to the Danes for the islands and the 26,000 African former slaves who inhabited them.

[4] From that time on, the US initiated a series of changes in the territorial administration, ranging from the continuation of the legislation established by the Danes, to the new law approved in 1969. This instituted the election of a governor and vice-governor, the election of a non-voting Congressional delegate and the right of its inhabitants to vote in American elections. In a 1981 plebiscite, a proposed Constitution was voted against by 50 percent of the registered voters. In December 1984, the UN General Assembly reiterated the right of the islands to self-determination, and urged the local population to exercise their right to independence.

[5] The local economy is totally dependent upon the United States. It revolves, to a great extent, around an oil refinery owned by US-Ameranda Hess Corporation on the island of St Croix. Today the refinery is the largest of its kind in the world producing over 700,000 barrels of oil per day. The company's influence on local politics is very important, as it has 1,300 permanent and 1,700 temporary employees. The corporation also supplies all the islands' energy needs. In spite of this, tourism is still the island's main source of income.

[6] A referendum scheduled for November 1989 was to decide the future status of the islands. However, it could not held due to the fact that in September 1989 the islands were severely damaged by hurricane Hugo. According to official estimates, 80 per cent of the buildings were destroyed. The disaster caused an outburst of looting and unrest. Armed bands wandered the streets; some members of the police and National Guard also took part in the pillaging. The US sent more than 1,000 troops to the area; however, the disturbances had virtually ceased by the time they arrived and in the words of the locals, there was nothing left to steal.

[7] The US government declared the islands a disaster area and granted $500 million in humanitarian aid. However, damages were estimated at $1 billion.

[8] An aluminum oxide plant, which had been closed in 1985, was purchased in 1989 by an international group trading in raw materials. The plant has a capacity for an annual production of 700 thousand metric tons. Shipments to the US and Europe began in 1990.

[9] In 1995, Roy L Schneider became governor in place of Alexander A Farrelly (1987-1995).

PROFILE

ENVIRONMENT

It comprises the western area of the Virgin Islands, east of Puerto Rico. The territory includes three main islands - St Thomas, St John and St Croix - and approximately 50 uninhabited small islands. The mountainous terrain is of volcanic origin. The climate is tropical, but irregular rainfall makes farming difficult. A small amount of fruit and vegetables are grown in St Croix and St Thomas. Tourism is an important economic activity. There is a large oil refinery in St Croix, supplying the US market.

SOCIETY

Peoples: Most of the population is of African origin with a small Spanish-speaking Puerto Rican minority. 35 to 40% of the inhabitants are from other Caribbean islands; 10% from the US. **Religions:** Protestant and Catholic.
Languages: English, official. **Political Parties:** There are local representatives of the US Republican and Democrat parties and an Independent Citizens Movement (ICM).

THE STATE

Official Name: Virgin Islands of the United States.
Capital: Charlotte Amalie, 12,331 inhab. (1990).
Government: Governor Roy L Schneider was elected in 1995. The unicameral legislature is made up of 15 representatives (7 from St Thomas, 7 from St Croix and 1 from St John). **Armed Forces:** The US is responsible for the defence of the islands.

COMMUNICATIONS

59.3 **telephones** per 100 inhabitants (1993).

ECONOMY

Currency: US dollars.

Virgin Is. (British)

Population: 99,000 (1994)
Area: 150 SQ.KM
Capital: Road Town

Currency: US Dollar
Language: English

British Virgin Islands

The Virgin Islands, were named by Christopher Columbus in 1493 in honor of St Ursula. At that time the islands were inhabited by Caribs and Arawaks, but by the mid-16th century they had all been exterminated by the Europeans.

2 In the 18th century, the English gained definitive control of the easternmost islands of the archipelago, and, using African slave labor, they began cultivating sugar cane, indigo and cotton. By the mid-18th century, the islands' slave population had reached 7,000, outnumbering the European colonists 6 to 1. Slavery was eventually abolished in the islands in the 1830s.

3 From 1872 the islands joined the British colony of the Leeward Islands, which were administered under a federal system. This federation was dissolved in July 1956, but the governor of the Leeward Islands continued to be responsible for the administration of the Virgin Islands until 1960. In that year direct control over the islands passed into the hands of an administrator designated by the British Crown. Unlike the Leeward Islands, the British Virgin Islands did not belong to the East Indies Federation, which existed between 1958 and 1962, preferring to develop links with the American Virgin Islands, under US control.

4 Constitutional reform required elections to fill the seats of the Legislative Council. They were held in November 1979, and independent candidates won five of the nine disputed seats and the Virgin Islands Party (VIP) won the four remaining seats. Five years later, the independent candidates only retained one seat; however, by forming an alliance with the United Party (UP), they managed to have one of their members appointed prime minister.

5 The colony has been governed under several constitutions this century, and the ministerial form of government began in 1967 with the first open elections. In 1977 the Constitution was modified, granting greater self-government and carrying out changes in the electoral system. The responsibility for defence, internal security and foreign affairs remains in the hands of the British-appointed governor.

6 Tourism accounts for 45 per cent of the national income, while fishing is the traditional economic activity. Gravel and sand are important exports, and a "tax-haven" banking facility has been established. The mineral potential of the islands is being examined in the hope of finding deposits that can be exploited.

7 In August 1986, the British government dissolved the Legislative Council, calling a new election several days after the opposition presented a vote of "no-confidence" in the prime minister. In September the VIP won 5 of the 9 seats and the UP won 2. Lavity Stott, the VIP leader, was named chief minister.

8 In March 1988, Prime Minister Omar Hodge resigned under accusations of corruption, being replaced by Ralph O'Neal. Hodge denied the charges against him, and in March 1989 he created a new party, the Independent People's Movement and in early 1990 was cleared of the charges against him.

9 In the November elections, the VIP increased its majority to 6

> **Tourism accounts for 45 per cent of the national income, while fishing is the traditional economic activity.**

seats, while the IPM obtained 1 seat, and the independents 2. Stott maintained his post as head minister, and O'Neal was re-elected prime minister.

10 Due to the increase in drug trafficking, and following recommendations by the British government, legislation was introduced to control operations of international financial companies. These would need a licence to operate on the islands and their activities would be subject to periodic inspections, as the authorities believed most of the investments were coming from the drug trade.

11 In October 1991, Peter Alfred Penfold was named Governor.

Yemen

Yaman

Population: 14,785,000 (1994)
Area: 527,970 SQ.KM
Capital: Political Capital: San'a

Currency: Rials
Language: Arabic

When the Cretan civilization reached its highest point, another trade culture was flourishing on the Arabian peninsula which rapidly led to the appearance of numerous cities. Ma'in, Marib, Timna, and Najzan were the principle cities of the interior, along the caravan routes that brought aromatic fragrances from Dhufar (presently part of Oman) and Punt (Somalia). These extensive trade routes continued along the coast of the Red Sea as far as the Mediterranean markets, and from Taima on towards Mesopotamia.

[2] These cities were united and ruled by whichever trade center held greatest power at the time, first Mina and later the better-known Saba, cited in the Bible. The latter's ties with the African coast dated from the founding of the Ethiopian state of Axum (see Ethiopia).

[3] In spite of the rudimentary state of their fleet, saban merchants kept close links with Africa for centuries, and as a result Christianity was propagated among the Yemenis through the teachings of Ethiopian preachers during the fourth century. Shortly thereafter, the Himyarites (a Hebrew sect) took control in Southern Arabia, establishing Judaism as the official religion and provoking a series of conflicts between the groups.

[4] Ethiopians conquered the country in 525 and were driven out by the Persians in 570. At this time the Persians established their first contact with southern African trade, which was renewed centuries later when the Persians settled in Africa (see Tanzania: "The Zandj Culture in East Africa").

[5] By the time of Muhammad and the ensuing Arab unification, the region had suffered almost three centuries of conflict and invasions, resulting in the loss of much of its splendor. Even the Marib Dam, a monumental construction forming the center of the agricultural irrigation system, had collapsed from lack of maintenance. The downfall of this civilization triggered great population migrations towards Africa and the eastern part of the peninsula.

[6] By the end of the eighth century the borders of the Arabian Empire reached from north Africa and Spain in the west, to Pakistan and Afghanistan in the east. Damascus in Syria became the capital of the empire, where the foundations of a new culture were laid. Greco-Roman, Persian and Indian components blended to form the new dominant culture, with the Arabs reaching high levels of scholarship and philosophy. The Arabs formed the social elite, the ruling class, though little changed in the lives of the Yemenis and other subject peoples.

[7] Then, in the 16th century, the Ottoman expansion started. The Turks occupied a few of the coastal areas of the Red Sea, leaving inland areas and the southern coast independent, governed by an iman.

[8] In 1618, the British arrived in the area and established the East India Company in the port of Mukha (Mocha: origin of the name for a type of coffee).

[9] In the 19th century they expanded their presence. As a consequence of Muhammad Ali's conquest of the country, the British occupied the entire extreme south-western region (see Egypt) and took Aden, the best harbor in the region, to discreetly monitor Turkish activities. Meanwhile the Turks consolidated their inland dominion, which was finally achieved in 1872. Concessions were made to allow the iman to retain his position and also make the post hereditary rather than elective.

[10] Towards 1870, with the opening of the Suez Canal and the consolidation of Turkish domination over the northern part of Yemen, the Aden settlement acquired new importance in British global strategy; it was a key port on the Red Sea and ultimately gave them access to the new canal. Friendship and protectorate treaties were gradually signed with local leaders, but it was a slow process, only completed in 1934 when the British gained control of the entire territory, as far as the border with Oman.

[11] In 1911, Iman Yahya Ad-Din led a nationalist rebellion; and two years later the Turks recognized the iman's full authority over the territory in exchange for the formal acceptance of Turkish sovereignty. After World War I Yahya proclaimed himself independent sovereign of the whole of Yemen. Whereafter, conflicts arose with the Saudi Emir of Najd and with the British, in Aden.

[12] After uniting Hidjaz with his nation, Ibn Saud launched a rapid military offensive against Yahya. In the 1934 peace treaty, Yahya accepted the loss of territory in the northern part of the country in return for the cessation of aggression. The British stopped before the actual declaration of war, but internal rivalries were fuelled and groups actively opposed to the iman were encouraged; the iman was assassinated in 1948.

[13] In 1958, his successor, Ahmad, joined the United Arab Republic formed by Egypt and Syria, but he later withdrew in 1961. He was succeeded in 1962 by his son Muhammad Al Badr, who was later deposed by the Nasserist military, in September of that year. The former iman received both Saudi and British support, initiating a long civil war against the republican government, which lasted until 1970.

[14] The conflict was settled when a coup, staged by a group of the republican faction, put the moderate Al Iryani in power. In 1967, in South Yemen, the National Liberation Front, formed in 1963, took the port of Aden and declared independence and the completion of a "socialist" revolution.

[15] The new People's Democratic Republic of Yemen closed all the British bases in 1969, nationalized the banking system, took charge of foreign trade and the ship-building industry and initiated agrarian reform. In efforts to isolate the "Cuba of the Middle East", Saudi Arabia concluded that Al Iryani was the lesser of several possible evils.

[16] In October, 1972, despite ideological and political differences between North and South Yemen,

> In December 1978, the first free elections since independence were held.

WORKERS

1994

MEN 88% WOMEN 12%

AGRICULTURE 63% INDUSTRY 11% SERVICES 26%

LAND USE

CROPLAND 3% PASTURE 30% OTHER 67%

PROFILE

ENVIRONMENT

The Republic of Yemen is formed by the union of Democratic and Arab Yemen. North Yemen has the most fertile lands of the Arabian Peninsula. For that reason the country, together with the Hadhramaut Valley, used to be called "Happy Arabia". Beyond a semi-desert coastal strip along the Red Sea, lies a more humid mountainous region where the agricultural lands are found (sorghum is grown for internal consumption and cotton for export). The traditional coffee crop was replaced by qat, a narcotic herb. The climate is tropical with high temperatures especially in Tihmah - where rainfall is heavy - and in the eastern region. The southern territory is dry, mountainous and lacks permanent rivers. Two-thirds of its land area is either desert or semi-desert. Agriculture is concentrated in the valleys and oases (1.2% of the territory). Fishing is an important commercial activity. The country's boundaries include the island of Socotra, which due to its location at the entrance to the Gulf of Aden, has important strategic value. This island, which became part of South Yemen in 1967, has 17,000 inhabitants spread over 3,626 sq.km. The use of ground water beyond its capacity to regenerate has caused a decline in its levels.

SOCIETY

Peoples: Nearly all Arab. A small Persian minority lives along the sea coast. **Religions:** Muslims (Shiite and Sunni). **Languages:** Arabic, official and virtually the only language. **Political Parties:** Principal parties established as of 1989, before unification: in the north, the General Congress of the People, of President Saleh; and in the south, the Yemenite Socialist Party. The political expression of Islamic fundamentalists, who are stronger in the north, is the Islah Party.

THE STATE

Official Name: Al-Jumhouuriya al-Jamaniya. **Administrative Divisions:** 16 Provinces. **Political Capital:** San'a, 427,150 inhab. **Other cities:** Aden, 318,000 inhab, Al-Hudaydah (Hodeida), 155,000 inhab.; Taiz, 178,000 inhabj. **Government:** Since unification of the Yemen Arab Republic (North) and the People's Democratic Republic of Yemen (South) in May 1990, Presidential Council made up of 5 members (3 from the North, 2 from the South), led by Ali Abdullah Saleh. Single-chamber Parliament, with 301 members (159 northern deputies, 111 from the south, 31 non-elected members). **National Holiday:** May 22, unification (1990). **Armed Forces:** 39,500 (1995)

took power vowing to continue their predecessor's policies. This action cost al-Gashmi his life as he died in a bomb attack in June 1978.

[19] In October 1978, with the help of a Congress that was well supported by the people, the NLF founded the Yemeni Socialist Party. In December, the first free elections since independence were held to choose the 111 members of the People's Revolutionary Council from the 175 candidates. Abdel Fattah Ismail, secretary general of the Party, was named head of state, but resigned in April 1980 and was replaced by Prime Minister Ali Nasser Mohammed, one of the founders of the National Liberation Front.

[20] When South Yemen claimed territorial rights over the areas where the Algerian state-owned oil enterprise had discovered large deposits, Saudi Arabia intensified hostilities and US military presence

> As a consequence of the Gulf War, a million Yemenis were expelled from Saudi Arabia and other countries of the region in retaliation for Yemen's pro-Iraqi stance during the war.

in Saudi Arabia added to tensions in the area.

[21] Major Ali Abdullah Saleh was appointed president in 1978 but was unable to prevent internal dissensions from leading to armed conflict in January 1979. The National Democratic Front, which included the nation's progressive sectors, was on the brink of taking power, so with Saudi provocation the conflict was diverted into a war with Democratic Yemen. Syria, Iraq and Jordan intervened and their mediation led to a cease-fire and negotiations for the unification of the two Yemeni states, which had been suspended since 1972.

[22] In February 1985, the People's

Supreme Council made President Ali Nasser Muhammad step down from the office of prime minister, appointing Haider Abu Bakr Al-Atlas – linked to Fattah Ismail – as his successor. Nasser Muhammad resented the loss of power and conspired to recover his position. On January 13, a counter-coup unleashed a brief but intense armed confrontation leaving 10,000 dead, including Fattah Ismail. Nasser Muhammad was ousted and Haider Abu Bakr Al-Atlas was appointed president. The new head of state pledged to maintain the alliance with Ethiopia and Syria.

[23] In May 1981, both countries agreed to draw up a more accurate inventory of their mineral wealth. In 1985, the discovery of significant oil deposits on both sides of the border helped to establish close links between the governments. These deposits have turned Yemen into one of the main oil producers in the Arab world.

[24] President Saleh's successful efforts to balance internal and external pressures resulted in more favourable conditions for rapid re-unification.

[25] Finally, on May 22 1990, the Republic of Yemen was proclaimed, with the political capital in San'a (former capital of the Arab Republic of Yemen) and the economic capital in Aden (former capital of the Democratic Republic of Yemen).

[26] In a joint session of the Legislative Assemblies of the two states, held in Aden, a Presidential Council was elected, made up of General Ali Abdullah Saleh (ex-president of North Yemen), Kadi Abdul Karim al-Arshi, Salem Saleh Mohammed and Abdul Aziz Abdel Ghani. The members of the Council elected Ali Abdullah Saleh president of the united republic. Ali Al Beid was vice-president and General Haidar Abu Bakr Al-Atlas, ex-president of South Yemen, was also given a post in the new government.

[27] In May 1991, the Constitution was ratified in a national referendum: an overwhelming majority voted for freedom of expression and political pluralism. Islamic fundamentalist groups opposed to unification called for a boycott, finding the absence of the principles of sharia (law based on the Koran) unacceptable, including, among other things, the introduction of voting rights for women.

[28] The new constitution caused

Al Iryani signed a treaty with the revolutionary government of the People's Democratic Republic of Yemen (South Yemen) for a future merger of the two States.

[17] This ran counter to Saudi strategy and, in June 1974, Colonel Ibrahim al-Hamadi forced Al Iryani to resign and installed himself as ruler in San'a. Although initially accepted by Saudi's King Faisal, the young officer made a powerful

enemy when he challenged the landlords of the north in an effort to centralize power. He survived three assassination attempts, but was finally killed on October 11 1977, together with his brother, Lt-Colonel Abdallah Muhammad al-Hamdi.

[18] A junta led by Lt-Colonel Ahmed al-Gashmi, with Prime Minister Abdelaziz Abdul Ghani and Major Abdul al-Abdel Aalim,

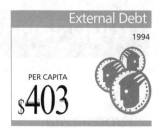

immediate changes. Some 53 political parties and 85 daily newspapers and periodicals have appeared throughout the country, political prisoners have been released and the freedom of expression is being fully taken advantage of.

[29] The unification of Yemen was an important step in the development of the region. In spite of the political turmoil experienced between British colonization and reunification, the Yemenis have always shared the common denominator of the Islamic religion.

[30] As a consequence of the Gulf War, a million Yemenis were expelled from Saudi Arabia and other countries of the region in retaliation for Yemen's pro-Iraqi stance during the Gulf War. However, Yemen made proposals to Saudi Arabia to settle border disputes, which had existed since 1930 when Saudi

> In response to the measures adopted by Saudi Arabia, the Yemeni government withdrew the contracts of thousands of foreign employees, including approximately ten thousand Arabian citizens employed as teachers.

Arabia had annexed three Yemeni provinces.

[31] At the same time, another million Yemenis returned to the country from Africa, mainly from Somalia. The sudden arrival of such a massive number of people from abroad had a negative effect on the economic situation within Yemen. Unemployment currently stands at over two million in a total population of ten and a half million. The returning workers, especially from Saudi Arabia, also deprived Yemen of an important source of foreign currency which came from the

money sent home by these migrant workers.

[32] In response to the measures adopted by Saudi Arabia, the Yemeni government withdrew the contracts of thousands of foreign employees, including approximately ten thousand Arabian citizens employed as teachers. This economic crisis has resulted in spiralling inflation and the rapid growth of the black market.

[33] After unification, in the middle of the crisis triggered by the Gulf War, the San'a government has not been able to begin the project to convert the old capital of South Yemen into a center of economic development. Strikes and demonstrations are symptomatic of increasing social tension which the government has tried to control under the conviction that stability would bring about an immedate increase in foreign investment.

[34] The war also resulted in the suspension of petrol refining in the Yemeni industrial processing plants in Aden, as the crude oil they normally worked with came from Iraq and Kuwait. US economic aid dropped from $22 million to $2.9 million. Saudi Arabia was even harsher on the country, simply suspending their grant of $70 million to the Yemeni government.

[35] In late 1991, Yemen was exporting around 300,000 barrels of crude oil per day and hoped to increase this figure. However, in 1992, oil production fell to 200,000 barrels, although it rose back once again the following year. Activity by local rebel groups and Saudi Arabia's territorial claims prevented exploitation of new oil deposits along border areas.

[36] Thirty-six per cent unemployment and price rises of basic consumer goods, led to a series of protest demonstrations in late 1992 and early 1993. At the same time, Islamic fundamentalism, strongly supported by the poorest sectors of the population, was blamed for a series of attacks against politicians from South Yemen.

[37] In March 1993, Saleh's General Congress of the People won the elections and obtained 122 seats, followed by the fundamentalist Reform (Islah) Party, with 62. The Socialist Party of then vice-president Salem El Baidh came in third place, with 56 seats. In order to weaken unified Yemen, a hypothetical "bad example" to the region's

monarchies, Saudi Arabia supported the fight for secession led by El Baidh.

[38] In May 1994, once again secessionists proclaimed the creation of a southern Yemen democratic republic and requested diplomatic support from Saudi Arabia and the Gulf states. However, they were defeated by forces loyal to the government. In July, the council of ministers adopted a plan of general amnesty in order to protect political pluralism. Furthermore, the government named Aden as the country's economic capital, which was regarded as a gesture towards southern Yemenites.

[39] In September, Socialist Party members were forced to leave the government, while Islah obtained six new places in the cabinet. The Constitution was modified, stating that sharia law would be the source of all Yemenite legislation.

[40] In February 1995, eleven parties formed a new alliance, the Op-

posing Democratic Coalition, seeking to obtain the power. The government signed an agreement draft with Saudi Arabia in which both states express their will to set permanent common borders and promote bilateral relations.

[41] In December, the landing of Eritrean troops on Hanish islands in the Red Sea, was considered an "armed aggression" by the Yemeni government, which led to an armed conflict. In May 1996, Yemen and Eritrea accepted an international arbitration to settle the quarrel.

Yugoslavia Fed. Rep.

Population: 10,520,000
(1994)
Area: 69,780 SQ KM
Capital: Belgrade

Currency: New Dinar
Language: Serb

Yugoslavija

In the 4th century BC, the Balkan Peninsula and the Adriatic coast were inhabited by Illyrian, Thracian and Panonian tribes, and they were also the site of Greek colonies. In the mid-2nd century, Rome defeated the alliance of the Illyrian peoples and began colonizing the new province of Illyria. Important Roman cities developed, such as Emona (present-day Ljubljana), Mursa (present-day Osijek) and Singidunum (present-day Belgrade). When the Roman Empire split into Eastern and Western regions, the border between the two ran straight through what is now Yugoslavian territory. Towards the end of Roman domination, Christianity was established in the region.

[2] In the 5th and 6th centuries AD, these territories were invaded by a number of nomadic tribes: Visigoths, Huns, Ostrogoths, Avars, Bulgars and Slavs. They imposed their own religious beliefs upon the people, but Christianity gradually took hold again, between the 9th and 11th centuries. From the 7th to the 13th centuries there were several feudal states. The Serbs were separate from these states, although they were unable to resist external pressure. Bosnia was subdued by Hungary, and the rest of the territory, as far as the state of Ducla, by Byzantium. Macedonia was divided up between Byzantium and Bulgaria (see Bosnia and Herzegovina, and Croatia).

[3] In the mid-11th century, under the reign of Stephen Nemanja (1168-1196), Serbia freed itself from Byzantine domination. The Serbian rulers of the Nemanja dynasty fought against the non-Christian religions which had been spread by the Bulgars. They received a royal title from the Pope in 1217, but the hoped-for propagation of the Catholic faith did not follow. In 1219, the Serbian Orthodox Church was founded, and mass began to be celebrated in Serbian. Under the reign of Stefan Dusan (1331-1355) the medieval Serbian State reached its apogee when it occupied Albania and Macedonia.

[4] The Ottoman Empire began its conquest of the Balkans in the mid-14th century, after the Battle of Kosovo in 1389. In the 14th and 15th centuries, the first of a series of migrations began, from Serbia and Bosnia to neighboring Slav regions and ultimately to Russia. In 1395, all of Macedonia came under the Ottoman Empire, and Bosnia, which had been a part of the Hungarian kingdom since the 12th century, was conquered in 1463. The Slav population of Bosnia became Muslim within a relatively short period of time. In 1465, the Turks occupied Herzegovina. By this time, Venetia had annexed the territories of Neretva and Zetina, along the coast. The city-state of Dubrovnik came under Hungarian control, and then, after 1526, became a part of the Ottoman Empire for 489 years.

[5] Between the 16th and 18th centuries, all of Yugoslavia's territories had been divided up: Serbia, Bosnia, Herzegovina, Montenegro and Macedonia belonged to the Ottoman Empire; Croatia, Slovenia, Slavonia, part of Dalmatia and Vaivodina belonged to the Hapsburgs; and Istria and Dalmatia belonged to the Venetian Republic. After the 1690 revolution was put down in Old Serbia, some 70,000 people took refuge in the Hapsburg Empire. The Ottoman Empire transferred Albanian Muslims to the abandoned territories of Kosovo and Motojia.

[6] After the Russo-Turkish war of 1768-74, Russia obtained the right to sponsor the Orthodox population of the Ottoman Empire, through the Treaty of Kuchuk-Kainardzhi. Austria seized the Balkans in 1797, as a result of the Napoleonic Wars, The Balkan *pashalik* (the north of Serbia), which belonged to the Ottoman Empire after the first Serbian uprising (1804-13), the Russo-Turkish war (1806-12) and the second Serbian uprising (1815), was granted internal autonomy. The political and military leaders Gueorgui Cherny (Karadjordje) and Milos Obrenovic founded Serbia's ruling dynasties. In 1829, Serbia became an independent principality within the Ottoman Empire, with Milos Obrenovic as its prince.

[7] In the 1878 Congress of Berlin, the Great Powers recognized the full independence of Serbia and Montenegro, which became kingdoms in 1882 and 1905, respectively. In the first Balkan war in 1912, Serbia, Montenegro, Greece, Romania and Bulgaria all formed an alliance, and in the second, in 1913, they warred against each other over the Ottoman Empire's domains. The end result was Macedonia's partition between Serbia, Greece and Bulgaria, while Serbia and Montenegro expanded their territories.

[8] Serbian resistance to the Austro-Hungarian Empire led to the assassination of the Austrian archduke Franz Ferdinand in 1914, in Sarajevo, the event that marked the beginning of World War I. After the war, which put an end to Austria-Hungary's empire, a kingdom of Serbs, Croats and Slovenes was founded, including Serbia, Montenegro and the territories of Slovenia, Croatia, Slavonia, Bosnia and Herzegovina.

[9] In 1929, the nation began to be called Yugoslavia: the land of the southern Slavs. The government remained in the hands of Serbians, and under the reign of Alexandr Karagueorgevich, it became an absolute monarchy. The regime's exclusivist policies gave rise to a strong anti-Serbian movement among Croats and other ethnic minorities, which led to the king's assassination in Marseilles, in 1934.

[10] At the beginning of World War II, Yugoslavia was neutral. In 1941, when Hitler attacked Yugoslavia, the country was so divided internally that it was easily subdued within a few days. The king and the members of the government fled the country, and the German Command carried out a policy of extermination against the Serbian and Muslim population.

[11] Two groups which were hostile towards each other launched the resistance movement: the nationalists loyal to the king – called "chetniks" – led by Draza Mihajlovic, and the partisans, under the leadership of Josip Brozxz, a Croat better known by his nom de guerre, Tito. This group was made up of communists in favor of a unified Yugoslavia, and anti-Nazi forces from all the republics with the exception of Serbia; it later became the Yugoslavian League of Communists (YLC).

[12] After bloody fighting against occupation troops and the Croatian "ustasha" (fascist) movement allied to the Germans, Tito emerged victorious. After the liberation of the country in May 1945, a Provisional Government was formed, led by Tito and supported by the Soviet Union and Britain.

[13] 2,000,000 Yugoslavians were killed during the war, and 3,500,000 were left homeless. When the war finally ended, the country was in ruins.

[14] On November 29 1945, a Constituent Assembly abolished the monarchy, proclaiming a federation of six republics: Slovenia and Croatia in the North, Serbia in the East, Bosnia-Herzegovina and Montenegro at the center, and Macedonia in the South. It also founded two autonomous provinces: Voyvodina and Kosovo, to the northeast and southeast of Serbia, respectively.

[15] The YLC joined the Comin-

ENVIRONMENT

Most of the country is taken up by mountain ranges, leaving one important plain north of the Sava River, a tributary of the Danube, where agricultural activities are concentrated. The climate is continental in this area, and Mediterranean along the coast. Rich deposits of lignite, lead, copper and gold in the mountains have permitted substantial industrial development. Tourism along the Adriatic is another major source of foreign currency.

SOCIETY

Peoples: Thirteen distinct ethnic groups are represented in the Yugoslav population. The major ones are: Serbs and Montenegrans. In Kosovo there is a large community of Albanians. **Languages:** Serb (official), Albanian, Montenegran, Hungarian. **Political Parties:** Socialist Party of Serbia (SPS), nationalist, led by Serbian president S Milosevic; Serbian Radical Party; Reformist Party, neoliberal, led by former prime minister A Markovic; Democratic Movement of Serbia; Socialist Party of Voivodina; Hungarian Democratic League of Voivodina; Green Party, among others. **Social Organizations:** Yugoslavia's five union federations form the League of Unions.

THE STATE

Government: Federal parliamentary republic, made up of the republics of Serbia and Montenegro. President, Zoran Lillic, since June 1993; Prime Minister, Radoje Kontic, since February 1993. Parliament, bicameral: citizens' assembly, with 138 members (Serbia: 108, Montenegro: 30), and assembly of the republics, with 40 members (20 from each republic). President of Serbia, Slobodan Milosevic; President of Montenegro, Momir Bulatovic. Kosovo and Voivodina are provinces within Serbia that were autonomous until 1990 and 1989, respectively. The government and parliament of Kosovo have been operating since 1991 and 1992, respectively, but are not recognized by Serbia. **National Holiday:** November 29, Proclamation of the Republic (1945).

---

form (Communist and Workers' Parties' Information Bureau) in 1947, but withdrew in the Spring of 1948 over disagreements with the Soviet CP leadership. The USSR initiated an embargo against Yugoslavia, which led to the strengthening of its ties with the West and the Third World.

[16] After a phase of economic centralization including the forced collectivization of agriculture, in 1950, Tito introduced the concept of "self-management". Its goals were to ensure workers direct, democratic participation in all decision-making processes concerning their living and working conditions, and to protect social democracy against the distortions and abuses of "Statism", bureaucracy and technocracy. The system was based on social (not state) ownership of the means of production and natural resources, managed directly by workers in their own and the community's interest. Similar forms of self-management were developed for trade and services.

[17] Tito was one of the founders of the Non-Aligned Movement. The Yugoslav leader defined non-alignment as the process whereby countries which were not linked to political or military blocs could take part in international issues, without being satellites of the major powers.

[18] The Yugoslavian leadership was charged with "revisionism" and isolated from the international Communist movement because of its neutral foreign policy and heterodox model of social and economic organization. Relations with the USSR slowly returned to normal after Stalin's death in 1953.

[19] That same year, an Agrarian Reform Law was approved authorizing private farming, and 80 per cent of the land returned to private hands. The combination of private economic activity and the self-management system led to average annual GNP growth rates of 8.1 per cent, between 1953 and 1965. In 1968, industrial production was twelve times greater than in 1950, the year in which the self-management system was first introduced. Likewise, industrial production was three times as important as agricultural production, in its impact upon the national economy. As a result, Yugoslavia's social and economic model was of great interest to the European left.

[20] The country's growth rate declined slightly toward the end of the 1960s. Even so, until the end of the 1970s, the annual growth rate exceeded 5 per cent.

[21] President Tito was aware of the inter-ethnic tensions in the Yugoslavia of his day, as well as the sharp contrasts between the socioeconomic situation of the industrialized north, and that of the underdeveloped south. In 1970, President Tito announced that after he stepped down, the country's leadership should be exercised by a body made up of the federated republics and the autonomous provinces.

[22] In 1971 and 1972, ethnic conflicts worsened, especially between Serbs and Croats, and Croatia presented a formal complaint against the confederated system. After 1974, there was an increase in separatist activity by Kosovo's Albanian majority.

[23] After Tito's death in April 1980, the executive power was vested in a collective presidential body, made up of a representative of each republic and autonomous province, and the president of the YLC with a rotating annual presidency. The new regime ratified Tito's policies of self-management socialism and non-alignment.

[24] Despite a strong economic position during the 1960s, both Serbia and Croatia became exporters of labor. Yugoslavia continued to export workers, and cash remittances from Yugoslavian workers in Western Europe, together with revenues from tourism, constituted an essential source of foreign currency to improve the balance of payments. This became more important in 1980 when the economic situation worsened.

[25] In March and April 1981, in the autonomous province of Kosovo, (bordering on Albania), there were riots, which recurred in 1988 and 1990. In Kosovo, 90 per cent of the population (1.9 million) is of Albanian origin. It is the poorest region of Yugoslavia, with unemployment reaching 50 per cent in 1990, with a per capita GNP of $730 while Serbia's was $2,200.

[26] According to the federal government, Kosovo housed nationalist forces and separatist extremists inspired and instigated from abroad, whose final objective was secession from Serbia and Yugoslavia. Many Serbs and Montenegrans left the area. Repression of the uprisings in Kosovo, that left a number of people dead and injured, led to mutual diplomatic recrimination between Belgrade and Tirana and the resignation of Kosovo's governor, Jusuf Zejnullahu, in March 1990. There was also tension in other republics due to the growth of militant Muslim and Catholic groups.

[27] Ethnic conflicts - coupled with inflation, which reached 90 per cent in 1986 and four figures in 1989 - were considered by the central committee of the Communist League to be rooted in deeper contradictions. During these years, several lawsuits dealing with government corruption exposed the fact that the system was crumbling. The Communist parties of Slovenia and Croatia announced their withdrawal from the YLC. In its January 1990 congress, the Yugoslavian League of Communists renounced its constitutional single-party role, and called on the parliament to draft a new constitution, eliminating the leading role assigned to the League in all spheres of Yugoslav life.

[28] In April of the same year, in the first multi-party elections to be held in Yugoslavia since World War II, nationalist groups demanding either secession or a confederated structure won in all of the republics except Serbia and Montenegro.

[29] In 1989, Yugoslavia's last socialist prime minister, Ante Markovic – a Croat and a neo-liberal –

Gypsies within borders

Since their arrival in Central and Western Europe at the beginning of the 15th century, the Sinti, Rom, Calé and Manush peoples have suffered persecution. However, they have managed to survive as a minority, in the midst of European societies.

[2] The term "gypsy" was probably coined by the non-gypsy majority from the word "Egyptian", based on the popular notion that they had originally come from Egypt. As, on their way to Europe, the Gypsies had passed through ancient Persia, Turkey and Greece, their own oral tradition gives their place of origin as "Little Egypt".

[3] In fact they seem to have originated in the Indian subcontinent. Linguistic studies have linked the Rom gypsies ("Rom" meaning "free person") with the "dom" Indian caste, and have shown similarities between Romany - their current language - and Sanskrit, as well as modern languages like Hindi, and to a lesser extent, Bengali, Punjabi, Gujarati and Rajasthani.

[4] According to recent studies, Gypsies left the Indian region of Punjab around the year 900, arrived in Persia by the 11th century and migrated into various European countries in small groups, between 1405 and 1430. They are first documented in present-day Germany in 1407; Switzerland in 1418; and France in 1419. They arrived in the Spanish colony which is now the Netherlands and Belgium, in 1420; in Italy in 1422; in Spain in 1425; and in Poland in 1428. They entered northern Europe and Russia in the early 16th century, but did not reach Siberia until the 18th century.

[5] Their constant migration did not result in assimilation and they retained some fundamental social laws which guaranteed the survival of their people, such as the sacredness of a promise, the solidarity of the members of the clan toward one another and the respect for the elderly, who represent the living memory and transmission of the experience of a culture that is basically oral.

[6] Right from their first arrival in Europe, the gypsy people and their itinerant, communal lifestyle were not accepted by the civil authorities and the Church, who at first tried to bar or expel them, and later tried to get them to settle down and assimilate into the dominant cultures, the most extreme example of this being the Nazi regime's extermination program.

[7] By 1500, in the central Europe of the Holy Roman Empire, blanket permission was issued for gypsies to be killed. In 1721, this vague "permission" became official policy when Emperor Charles VI ordered their extermination, though groups of armed gypsies resisted the order. In the mid-18th century, during the reigns of Maria Theresa and Joseph II (1740-1790), an assimilation policy replaced the Hapsburg persecution. These attempts at assimilation - which were eventually accepted as futile - coincided with the strategy of "settling" employed in Spain by Charles III (1759-1788). When the latter failed, the king ordered that all male gypsy children be deported to the American colonies, and female children to colonies in the East.

[8] Such policies as persecution, forced uprooting or assimilation, were echoed and amplified by modern states, after 1870. These processes were facilitated by the recording of censuses; in the case of Germany, aiding Nazi leaders - years later - to enforce genocidal measures against both the Jewish population, and hundreds of thousands of European gypsies.

[9] In post-war Europe, gypsies are finding the trend toward integration to be nearly as threatening to their survival as the explicit persecution of the past. For example, in modern-day Spain, new attempts are being made to get the gypsy population to find "regular" jobs and provide their children with formal education. The price to be paid for such "progress" could well be the loss of their traditional values, oral traditions, the bonds of the clan structure, and the role of the elderly.

[10] At present, the number of gypsies in Europe is estimated at over 15 million, mostly concentrated in the eastern and southeastern parts of the continent. Some experts have delineated three main gypsy groups: the Kalderash (the most numerous group, mostly in the Balkans and central Europe); the Gitanos (French Gitans, mostly in the Iberian Peninsula, North Africa, and south France, strong in the arts of entertainment); the Manush (French Manouches, also known as Sinti, mostly in France and Germany, often travelling show and circus people). Each of these main divisions were further divided into two or more subgroups distinguished by occupational specialization or territorial origin, or both.

launched a series of structural reforms in order to put an end to the crisis. The country eliminated customs duties on 90 per cent of its imports. As a result, imports increased by 11.9 per cent, while exports increased at a slower rate. There was minimal growth of the country's GDP and of its industrial production. Agricultural production increased by 5.2 per cent, and real income grew an average 26.6 per cent. Retail prices rose by percentages of 1,255 in 1989, and 586.6 in 1990.

[30] Yugoslavia had hoped that all the new attention on the part of Western Europe toward the countries of Eastern Europe would lead to a liberalization of its economy. However, in all of 1989, Yugoslavia did not receive a single dollar in foreign aid, and paid, during the same period between $3.7 and $3.8 billion to its creditors.

[31] As of 1990, the situation began to deteriorate even further, with the exception of foreign trade which began registering negative figures a year later. Industrial production fell by 18-20 per cent in Serbia and 13 in Montenegro - while the gross foreign debt of the entire Federation reached $16.295 million. Of this, more than $5.5 million corresponded to Serbia and Montenegro.

[32] With mounting social pressures because of the economic situation and the disintegrating state, Yugoslavian politics were polarized by two fundamentally opposed concepts. Kucan in Slovenia argued for decentralization to relieve the wealthier regions of the obligation to subsidize the more backward areas, while Slobodan Milosevic, charismatic president of the Serbian Communist League, proposed greater centralization and solidarity in the federation.

[33] In December 1990, the Croatian parliament adopted a new constitution, which established the right to withdraw from the federation. At the same time, a plebiscite in Slovenia endorsed independence. In the following months, disagreement over the reform of the federal system and the designation of the presidency gave rise to an insoluble crisis within Yugoslavia's collective leadership.

[34] On September 8 1991, Croatia and Slovenia declared their independence from the federation, and the Serbian population of Croatia declared its intention of separating from this country. The federal army, whose officials answered primarily to Serbia, intervened in Slovenia

and Croatia, stating that separation was a threat to Yugoslavia's integrity. War broke out, with many victims on both sides (see Bosnia and Herzegovina).

[35] War led to the destruction by federal troops of entire cities - like Osijek, Vukovar and Karlovac - as well as the occupation of nearly a fourth of Croatia's total land area. Areas affected included the territories of West and East Slavonia, as well as the area known as "Krajina", which toward the end of 1991 proclaimed itself the "Republic of Serbian Krajina".

[36] In December, both the president of the governing council, Stjepan Mesic, and Prime Minister Markovic resigned, the last representatives of a unified government. Markovic - like Mesic, a Croat - was unwilling to present the preliminary legislation for the 1992 budget, believing it to be biased against Croatia (81 per cent of the projected budget was earmarked for the federal army). By the end of 1991, there was a total of 550,000 refugees, 300,000 of them from occupied areas of Croatia.

[37] In 1992, within the territory which still defined itself as Yugoslavia, Serbian president Milosevic shored up his position by retiring 70 generals and admirals of the federal armed forces who were not definitely loyal to him. He also won Montenegran support for his plan to establish a unified Yugoslavia with its capital in Belgrade, through a plebiscite in that republic. The March 1 plebiscite was boycotted by the opposition.

[38] On January 15 1992, the European Community recognized Croatia and Slovenia as sovereign states. On April 27, the parliament of Serbian and Montenegran deputies announced the foundation of the new Federal Republic of Yugoslavia, a federation between Serbia and Montenegro with a parliamentary system of government.

[39] In the meantime, fighting continued in the regions characterized by inter-ethnic strife. Since April 1992, the heart of the fighting had been the republic of Bosnia-Herzegovina, which was recognized by the EC on April 7. In order to divest itself of responsibility for the aggression in Croatia and Bosnia-Herzegovina, in early May Belgrade announced that it was no longer in control of troops from what had formerly been the federal army, now fighting in the independ-

dent republics. Far from having the desired effect, Serbia's position led the EC to declare a trade embargo against Yugoslavia on May 28.

[40] However, the general elections held in the new federation on May 31 served to further strengthen Slobodan Milosevic's position. His Socialist Party won 70.6 per cent of the vote in Serbia, while the Socialist Democratic Party of Montenegro won 76.6 per cent. The democratic opposition, as well as the Albanian minorities and the Muslims of Sandchak, boycotted the elections; only 56 per cent of the electorate participated.

[41] With the election of Dobrica Cosic, a writer, as federal president on June 15, and Milan Panic, a businessman, as prime minister on July 14 Serbia's leaders projected an image of a greater willingness to negotiate. Unlike Milosevic, Panic agreed with the London peace proposals of August 26 and 27. Nevertheless, this new development did not prevent the intensification of Serbian aggression in Bosnia.

[42] In the presidential elections held in Serbia on December 20, Panic was defeated by Milosevic, and Radoje Kontic succeeded Panic as prime minister on February 9. A series of purges began within Serbian government institutions, in which more than a thousand state radio and television employees - as well as dissident intellectuals and professors - lost their jobs. Finally, on June 25 federal president Cosic was replaced by Zoran Lilic, a member of Milosevic's inner circle.

[43] The intransigence of the Serbian government was evident in Kosovo, where any attempt at independence was met with repression, within the framework of a policy of cultural annihilation. Serbia has refused to recognize Kosovo's parliament and its provincial government, which have been operating since May 24 1992. In 1993, Serbian police violently broke up a meeting in memory of Albanians who had been killed; in addition, they arrested several political party leaders and closed the Kosovo Academy of Sciences.

[44] As Milosevic and the ruling SPS party grew more authoritarian, international pressure increased and the social and economic crisis deepened. In 1993, 80 per cent of the federal budget was earmarked for the armed forces, and 20 per cent of the country's GNP went to supporting the Serbs in Bosnia and

Croatia. During the first nine months of the year, industrial-activity dropped 39 per cent in relation to 1992, and the GNP decreased from $25 billion in 1991, to $10 billion, 2 years later.

[45] Consequently, by the end of 1993, unemployment had reached 50 per cent. Given the fact that 80 per cent of the country's public spending was being financed by uncontrolled printing of currency, inflation shot up from 100 per cent in January to 20,190 per cent in November. The monthly minimum wage was approximately $150, enough to feed a family of four for 3 days.

[46] In the autumn, the government introduced the rationing of basic foodstuffs, sugar, flour and oil, and in December the first electricity cuts occurred. In early 1996, the income of 78 per cent of the families in the federation was no higher than $235 per month.

[47] The power of the Socialist Party of Serbia, led by Milosevic, grew from 1994 onwards. After the parliamentary elections of December 1993 they held 123 of the 250 seats in the Assembly and had formalized alliances with eight opposition deputies, gaining the majority in the chamber. Milosevic imprisoned his most important political rivals and tried to take a less central role in the war. This line of action included selective collaboration with the war crimes trials in The Hague.

[48] The main aim of Milosevic was that the UN should suspend its sanctions on Yugoslavia from May 1992. On September 24 1994, the international organisation decided to partially lift the measures for 100 days, allowing international flights, and cultural and sporting exchanges. This diplomatic success of Milosevic - previously accused of being the main instigator of the war in the Balkans - contributed to his

increasing popularity in the Federation.

[49] The Yugoslav president continued to play an important role in the peace process in Bosnia-Herzegovina during 1995. The political distancing between Yugoslavia and the Bosnian Serb leaders Radovan Karadzic and Ratko Mladic was at odds with the attitude adopted on the military front. The bloody conquest of Srebrenica and Zepa by the Bosnian Serbs in July contributed to the marginalization of the opposition practised by Milosevic.

[50] Belgrade continued to supply arms and troops to the "Serbian Republic of Krajina" in Croatia, during the first half of the year. However, Yugoslavia did not intervene when the Croats invaded the territory of Krajina in August. Some Serbian refugees were authorized to enter Yugoslavia and were housed in the province of Kosovo, amongst a majority Albanian population, along with groups of Hungarians and Croats expelled from Vojvodina.

[51] The popularity of Milosevic jumped up again following the signing of the peace agreement in Dayton, Ohio, though his major triumph came on December 14 1995, when the United States suspended the sanctions on Yugoslavia as a result of the signing of the Paris agreement. During the embargo, per capita income had fallen to half the previous amount and more than one and a half million people were unemployed.

[52] In August 1996, a month before the elections in Bosnia, 220,640 Bosnian refugees had registered to vote in the territories of Serbia and Montenegro.

STATISTICS

DEMOGRAPHY
Annual growth: 0.8% (1991-99).

HEALTH
Under-five mortality: 23 per 1,000 (1995). **Calorie consumption:** 115% of required intake (1995).

EDUCATION
Primary school teachers: one for every 22 students (1991).

COMMUNICATIONS
109 **newspapers** (1995), 102 **TV sets** (1995) and 94 **radio sets** per 1,000 households (1995). 18.0 **telephones** per 100 inhabitants (1993). **Books:** 103 new titles per 1,000,000 inhabitants in 1995.

ECONOMY
Consumer price index: 100 in 1990; 218.0 in 1991. **Currency:** 20 new dinars = 1$ (1990).

Zaire

Zaïre

Population: 42,540,000 (1994)
Area: 2,344,885 SQ KM
Capital: Kinshasa

Currency: Zaire
Language: French

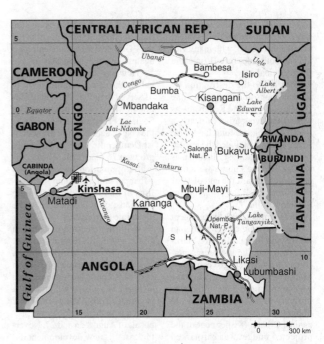

The first known state to emerge in what is now Zaire was the Luba kingdom, located in the Katanga (Shaba) region. The Luba kingdom was created in the 16th century when a warrior named Kongolo subdued the small chiefdoms in the area and established a highly centralized state. To the northwest was the Kuba, a federation of numerous chiefdoms that reached its peak in the 18th century. Dr Livingstone brought Zaire to the notice of the western world through the journalistic coverage of his explorations in Africa, between 1840 and 1870. In 1876, King Leopold II of Belgium founded the International African Association (later, the International Association of Congo), a private organization that financed Henry Stanley's expeditions. Stanley, a journalist and adventurer, succeeded in signing more than 400 trade and/or protectorate agreements with local leaders along the Congo River. These treaties, and the Belgian trading posts established at the mouth of the river, were used to devise a system for the economic exploitation of the Congo. The Berlin Conference, 1884 to 1885, decided that the "Free State of Congo" was the Belgian king's personal property. Consequently, Leopold's "Compagnie du Katanga" stopped British colonialist Cecil Rhodes' northward expansion.

[2] The Congolese population was subjected to extremely harsh working conditions, which did not change when they formally became a Belgian colony in 1908. Military force was systematically employed to suppress anticolonial opposition and to protect the flourishing copper mining industry in Katanga (now Shaba).

[3] In 1957, liberalizing measures permitted the formation of African political parties. This led countless tribal-based movements to enter the political arena, all trying to benefit from the general discontent. Only the National Congolese Movement, led by Patrice Lumumba, had a national outlook, opposing secessionist tendencies and supporting independence claims.

[4] In 1959, the police suppressed a peaceful political rally triggering a series of bloody confrontations. King Baudouin of Belgium tried to appease the demonstrators by promising independence in the near future, but European residents of the Congo reacted with more oppressive measures. Independence was finally achieved in 1960, with Joseph Kasavubu as president and Lumumba as prime minister. A few days later, Moses Tchombé, then prime minister of the Province of Katanga, initiated a secessionist movement.

[5] Belgium sent in paratroopers

The Congolese population was subjected to extremely harsh working conditions, which did not change when they formally became a Belgian colony in 1908.

and the United Nations, acting under US influence, intervened with a "peacekeeping force". Kasavubu staged a coup and arrested Lumumba, delivering him to Belgian mercenaries in Katanga who killed him. The civil war continued until 1963.

Secessionist activity ceased when Tchombé, who represented the neo-colonial interests, was appointed prime minister. With the help of mercenaries, Belgian troops and US logistical support, the Tchombé regime was able to stifle the nationalist opposition movement. In 1965, he was forced to resign by Kasavubu who was in turn overthrown by army commander Joseph Desiré Mobutu. According to representatives of transnationals, Mobutu was the only man in a position to restore the conditions required for them to continue operating there.

[6] Under the doctrine of "African authenticity", Mobutu changed the name of the country to Zaire and his own to Mobutu Sese Seko. However, his nationalism went little further than this, and the "Zairization" of copper, which he declared in 1975, only benefitted an already wealthy economic elite and the state bureaucracy.

[7] These measures caused some discomfort among US diplomats as Mobutu had offered Washington his services in the region.

[8] Zaire sheltered and actively supported the so-called National Front for the Liberation of Angola (FNLA). Mobutu encouraged secessionist groups in the oil-rich Angolan province of Cabinda, and Zaire's troops effectively cooperated with the South African racist forces in their war against the Angolan nationalists.

[9] Meanwhile, in Zaire guerrillas continued the struggle in the interior. In 1978 and 1979, the major offensive launched by the Congolese Liberation Front was checked with the aid of French and Belgian paratroopers and Moroccan and Egyptian troops, again with US logistical support.

[10] At the end of 1977, international pressure led to parliamentry elections being held for an institution which had been given limited legislative functions. This helped to divert international attention from human rights violations against students and intellectuals in the cities, the establishment of concentration camps for Mobutu's opponents and the brutal reception given to refugees who returned under an "amnesty" decreed in 1979.

[11] Zaire was the world's largest cobalt exporter, the fourth diamond exporter and ranks among the top ten world producers of uranium, copper, manganese and tin. 90 per cent of the cobalt used in the US aerospace industry comes from Zaire. But corruption was rampant throughout the country's administration, worsening the already unstable economic situation, and leading to soaring rates of unemployment.

[12] During 1980 and 1981, the major Western powers decided to seize direct control of the strategic mineral reserves in the country. The International Monetary Fund took

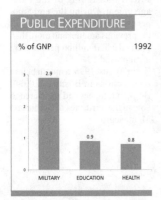

PUBLIC EXPENDITURE

% of GNP — 1992

- MILITARY: 2.9
- EDUCATION: 0.9
- HEALTH: 0.8

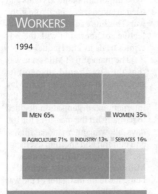

WORKERS

1994

MEN 65% ■ WOMEN 35%

■ AGRICULTURE 71% ■ INDUSTRY 13% ■ SERVICES 16%

special interest in Zaire's economy, facilitating the renegotiation of its foreign debt, while at the same time imposing drastic measures against corruption. The economy of Zaire fell under direct IMF control, and the Fund's representatives in Kinshasa began to supervise the country's accounts personally.

[13] At the end of 1982, positive results were visible to international creditors and European investors. Internally, however, the situation had deteriorated even further because of austere measures imposed by the IMF.

[14] In April 1981, Prime Minister Nguza Karl I Bond sought political asylum in Belgium; he presented himself to the western powers as a "decent alternative" in view of the official corruption in Zaire.

[15] During the elections of June 1984, Mobutu obtained 99.16 per cent of the vote.

[16] In February 1985, Zaire signed a security pact with Angola to improve relations between the two countries. These had deteriorated by the end of the previous decade because of Zaire's support of the FNLA and the Congolese Liberation Front, which operated from Angolan territory.

[17] In July 1984, Mobutu attempted to create a "League of Black African States" clearly destined to compete with the Arab League. The idea was emphatically rejected by all progressive African governments.

[18] In September 1986, it was discovered that the Reagan administration in the US had channelled $15 million of covert aid for Angolan mercenaries through Zaire, virtually transforming the country into an arms repository.

[19] In June 1989, Mobutu visited the United States, and Zaire obtained a loan of $20 million from the World Bank. A few days earlier, Mobutu had hosted a historic meeting in his native city of Gbadolite, where the president of Angola, Jose Eduardo dos Santos, and the leader of the counter-revolutionary UNITA, Jonas Savimbi, had accepted a cease-fire to negotiate a peaceful solution to the country's conflict.

[20] In April 1990, anticipating the process of democratization, which he considered imminent, Mobutu decided to take an even bolder step. He ended the one-party system, opened up the labor movement and promised to hold free elections

within a year. A rapid process of political organization began. Hundreds of associations and political groups of all kinds demanded legal recognition from the government. The extent of the popular reaction frightened the authorities, and on May 3, Mobutu issued a statement saying that no party had yet been legalized and that it would be necessary to modify the constitution before holding elections, because the head of state wished to "pre-

serve his authority without exposing himself to criticism".

[21] Students initiated demonstrations throughout the country, especially at the university in Lubumbashi, capital of the province of Shaba. The students began demanding the resignation of Mobutu, who reacted by sending in his presidential guard to stifle the protests.

[22] The troops stormed the university campus at dawn on May 11.

More than 100 students were assassinated and the terrified survivors fled to other provinces and Zambia, from where they condemned the massacre.

[23] President Mobutu managed to muffle the repercussions of the killings, but the European Community demanded an international investigation, and Belgium cut off all economic aid. The plan for opening up the country was shelved, at least for the time being.

[24] The massacre at Lubumbashi University generated a wave of repudiation which led to a series of strikes, like that at Gecamina, the country's most important mining company, a state-owned enterprise. In the United States there were repeated calls for the cessation of aid to Mobutu.

[25] In October 1990, under growing internal and external pressure, Mobutu decided to carry out a new political "democratization" process and he authorized the unrestricted creation of new political parties. In December, the opposition, grouped together under the Holy Union, a front made up of nine parties (including the four largest) demanded Mobutu's resignation, calling for a national conference to decide on the political future of Zaire without presidential intervention.

[26] In September 1991, Mobutu faced major popular uprisings throughout the country, triggered by a general price increase and the failure of a conference in August which had aimed to introduce democratic reforms. The uprisings caused several deaths and the intervention of hundreds of French and Belgian soldiers, allegedly to "protect foreign citizens residing in Zaire".

[27] In November 1991, the Holy Union formed a "shadow government", and appealed to the armed forces to depose Mobutu. The same month, the President appointed Nguza Karl I Bond as his new prime minister (his fifth in 1991). Nguza, a former opposition leader who had already been Mobutu's head of government ten years earlier, took office amidst a worsening economic crisis and growing international pressures, especially from the US.

[28] Early in 1992, the National Conference was set up. The opposition had long been awaiting this opportunity to press for constitutional reform and transition to democracy. In February of the same

STATISTICS

DEMOGRAPHY

Annual growth: 1% (1992-2000). **Children** per woman: 6.2 (1992).

HEALTH

One **physician** for every 14,286 inhab. (1988-91). **Under-five mortality:** 186 per 1,000 (1995). **Calorie consumption:** 89% of required intake (1995). **Safe water:** 27% of the population has access (1990-95).

EDUCATION

Illiteracy: 33% (1995).

COMMUNICATIONS

82 **newspapers** (1995), 82 **TV sets** (1995) and 88 **radio sets** per 1,000 households (1995). 0.1 **telephones** per 100 inhabitants (1993). **Books:** 82 new titles per 1,000,000 inhabitants in 1995.

ECONOMY

Annual growth: -1.00% (1985-94). **Currency:** 3.250 zaires = 1$ (1994). **Cereal imports:** 336,000 metric tons (1990).

---

year, Prime Minister Nguza Karl-I Bond suspended the Conference, causing a faction of the army to rebel, taking over a state-run radio station and demanding President Mobutu's resignation. Some hours later the rebels were defeated by troops loyal to the government. Thousands of demonstrators demanding the president's resignation

In November 1991, the Holy Union formed a "shadow government", and appealed to the armed forces to depose Mobutu.

and the reopening of the Conference were harshly repressed by the army, resulting in several deaths and many injuries.

[29] The European Community suspended financial aid to Zaire immediately, until the reinstatement of the National Conference. Meanwhile, representatives from the US, France, and Belgium agreed to increase the pressure on the Mobutu government to speed up political change.

[30] In March 1992, after meetings with Archbishop Monsegwo Pasinya, the Conference president, President Mobutu appeared on radio and television to announce the reopening of the National Conference. Etienne Tshisekedi, leader of the Holy Union, was made prime minister, to replace Nguza Karl-I Bond. The delegates representing the government and almost 160 groups resolved to return to the country's former name of "the Congo Republic", which had been changed by Mobutu.

[31] In a patent display of US displeasure with the Mobutu regime, Monsegwo Pasinya was received in Washington by Secretary of State James Baker, and by the Senate Foreign Relations Committee.

[32] Inter-ethnic strife erupted again in 1992. In Shaba, in the southwest of the country, there were outbreaks of violence after Karl-I Bond's dismissal from government. Lunda people, Karl-I Bond's tribal group, attacked members of the Luba community, Tshisekedi's people. Some 2,000 people were killed, and thousands of Luba left Shaba as their homes had been destroyed. Security forces eventually intervened several weeks after the fighting broke out.

[33] The twelve commercial banks operating in Zaire closed indefinitely in 1992, due to a lack of funds.

Inflation reached 16,500 per cent. According to a report by the Washington-based Population Crisis Committee, in 1992 Zaire was among the ten poorest countries in the world, in 88th place on a scale of one to a hundred.

[34] The government confiscated oil company funds, claiming that this was necessary in order to prevent a collapse of the national economy.

[35] In December 1992, galloping inflation caused the prime minister to declare that the 'zaire' would no longer be legal tender, placing a new currency in circulation. Regardless of this Mobutu ordered that troops receive their back pay in the old currency.

[36] In early 1993, battles broke out between the soldiers, furious over having been paid in worthless bills, and Mobutu's personal guard. This confrontation caused over 1,000 deaths in Kinshasa, and the capital suffered looting, arson and attacks by irate soldiers.

[37] The French ambassador, Philippe Bernard, was killed when the embassy was hit by tank fire. The French minister of defence insisted that the event was a deliberate hostile act.

[38] On February 24, Mobutu's soldiers and tanks surrounded the building housing the High Council of the Republic, a transitional body formed by the National Conference, demanding that the 800 legislators approve the old currency which Mobutu had returned to circulation.

[39] With the worsening of the situation, the US, Belgium and France sent a letter to Mobutu demanding that he resign in favor of a provisional government headed by Tshisekedi. Mobutu responded by dismissing Prime Minister Tshisekedi, in March 1993. Faustin Birindwa replaced him.

[40] Though Tshisekedi had not legally been appointed prime minister by the president, it was a moot point whether Mobutu was entitled to dismiss him. However the prime minister had never had any real power anyway, controlling neither the army nor any of the state agencies. Mobutu maintained control of the country from Gbadolite, near the border with the Central African Republic. He did not return to Kinshasa for several months, as it was reported that he feared for his personal safety.

[41] The United States suggested Belgium and France embargo

Mobutu's property - his personal fortune was estimated at around $4 billion. Meanwhile, amidst the monetary disorder and suspended foreign debt repayments, the informal economy took over in the urban areas.

[42] The economic and political uncertainty continued throughout 1994. The genocide in Rwanda and the arrival of masses of refugees – amongst whom were found thousands of members of the militias responsible for the massacre – created great tension in eastern Zaire.

Inter-ethnic strife erupted again in 1992. In Shaba, in the southwest of the country, there were outbreaks of violence after Karl-I Bond's dismissal from government.

[43] When the FPR guerrillas came to power in Rwanda, several Western nations, like France, began to reduce the pressure on Mobutu, newly considered as a potential ally following the victory of "the English-speaking Tutsis" in the neighboring country. This reinforced the president's power, facilitating the nomination of Léon Kengo Wa Dondo as prime minister, though this was immediately questioned by the opposition.

[44] The announcement in mid-1995, that the transition government would remain in power for two more years provoked a new wave of protests. In late July, the confrontations between police and demonstrators calling for Tshisekedi to be nominated prime minister left ten people dead.

[45] The tension between Zaire and Rwanda increased in 1996, after the Rwandan Hutu militias, supported by members of the Zairian armed forces began "ethinc cleansing" in the Masisi region, ousting and killing Tutsis who had lived in this zone of eastern Zaire for generations.

Zambia

Zambia

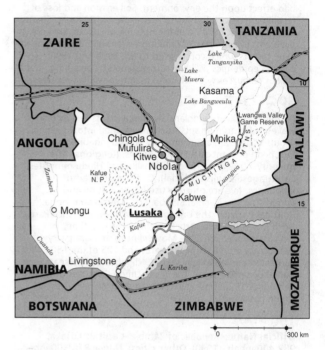

Population: 9,203,000 (1994)
Area: 174,000 SQ KM
Capital: Lusaka

Currency: Kwacha
Language: English

Archaeological evidence suggests that early humans roamed present-day Zambia between 2,000,000 and 1,000,000 years ago. Stone Age sites and artifacts are found in many areas. Early Iron Age peoples settled in the region with their agriculture and domesticated animals about 2,000 years ago. Ancestors of the modern Tonga tribe reached the region early in the 2nd millennium AD, but other modern peoples reached the country only in the 17th and 18th centuries from Zaire and Angola. Portuguese trading missions were established early in the 18th century at the confluence of the Zambezi and Luangwa rivers. The mighty state of Lunda began to lose power towards the end of the 18th century. The slave trade declined in the 19th century, weakening the authority of the Mwata Yambo (traditional Lunda leaders). This, coupled with the growing autonomy of Kazembes (provincial governors), gave rise to small local autocratic states. In 1835 a group of Bantu-speaking Ngoni settled in the Lake Nyasa-Luangwa watershed. The Suto people, the Kololo (Makololo), crossed the upper Zambezi and made themselves masters of Barotseland

2 The Portuguese crossed the country several times between 1798 and 1811, trying to establish a land route between Mozambique and Angola, but the turmoil stirred up by the Zulu campaigns frustrated their plans. Some nations, like the Sotos, driven out of their southern lands, managed to set up new states in southwestern Zaire around 1835.

3 In 1851, David Livingstone, a British missionary, sailed up the Zambezi river as far as Victoria Falls. He was later followed by merchants and explorers in the service of British-born millionaire Cecil Rhodes, who owned a vast fortune in South Africa and wanted to expand northwards. In 1889 the British Crown granted Rhodes, and his British-South Africa Company exclusive rights to establish a mining and trade monopoly in the Katanga area in present day Zaire. A year later Cecil Rhodes signed a treaty with Sotho ruler Lewanika and this protectorate was soon followed by colonial domination which created Northern Rhodesia.

4 The British-South Africa Company kept control of this territory in order to prevent the Portuguese from achieving their plan of joining Angola and Mozambique. In 1909, a railway line was built to the coast of the Indian Ocean and in 1924, Britain assumed direct control of the region, developing arable and livestock farms along the railway. South African and US mining companies increased their investment in copper, and 13 years later nearly 40,000 Africans worked in the mines which made massive profits as the labor came cheap. The miserable conditions of the miners led to protest campaigns and stimulated the formation of workers unions; the North Rhodesia African National Congress (NRANC), linked to South Africa's African National Congress (ANC), was born in this era. In 1952, Kenneth Kaunda, a primary school teacher, became NRANC secretary-general with Harry Nkumbula as president. Kaunda was to be Zambia's main leader in the pro-independence struggle.

5 In 1953, the British engineered a federation comprising Northern Rhodesia (Zambia), Southern Rhodesia (Zimbabwe) and Nyasaland (Malawi), assigning Zambia mining production and Zimbabwe, agricultural. The ANC had just launched its struggle for independence and against racial discrimination. When Nkumbula hesitated before a constitutional project designed to institutionalize European domination, Kaunda chose to leave the NRANC, founding the African National Council of Zambia (ANCZ) and boycotting the elections.

6 The ANCZ was promptly outlawed and Kaunda was arrested in 1959. Undaunted, his followers created the United National Independence Party (UNIP), which Kaunda chaired when he was released in 1960.

7 The UNIP was outlawed a few months later, when the British recognized the existence of massive popular support for the party's nationalist platform. In 1961, repression and marginalization of the majority led to an outbreak of violence in Zambia. By October 1964, the federation was dissolved, UNIP candidates had won the general election and proclaimed independence.

8 Zambia actively supported liberation movements in Angola and Mozambique. It nationalized its copper reserves, acted as a founding member of OCEC, sometimes referred to as the "copper OPEC", and served as host for the 3rd Summit Meeting of the Non-Aligned Countries Movement in 1970.

9 In 1974, with the inauguration of the Tan-Zam Railway, Zambia gained access to an ocean outlet dominated by colonial forces. In the same year, Angolan and Mozambican independence brought about a complete turnaround in the regional balance of power.

10 Zimbabwean independence in 1980 was welcomed by Zambia. However, forces from the South African army continued their incursions into Zambia, attacking Namibian refugee camps. In October, a coup against Kaunda, supported by the South African regime, failed.

11 The situation forced the government to decree a state of emergency and curfew.

12 President Kaunda tried to surround himself with former comrades-at-arms, generating friction with the younger sectors, who felt that after Zimbabwe's indepen-

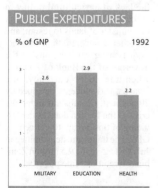

PUBLIC EXPENDITURES

% of GNP — 1992

MILITARY 2.6
EDUCATION 2.9
HEALTH 2.2

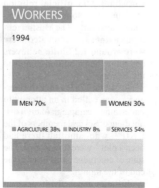

WORKERS

1994

MEN 70% — WOMEN 30%

AGRICULTURE 38% — INDUSTRY 8% — SERVICES 54%

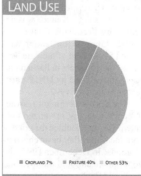

LAND USE

CROPLAND 7% — PASTURE 40% — OTHER 53%

ENVIRONMENT

A high plateau extends from Malawi on the east to the swamp region along the border with Angola in the west. The Zambezi River flows from north to south and provides hydroelectric power at the Kariba Dam. The climate is tropical, tempered by altitude. Mining is the main economic activity, as there are large copper deposits. Mining has had a bad effect upon the environment. Soil erosion and loss of fertility are associated with the overuse of fertilizers. Shanty towns account for 45% of the housing in Lusaka, linked with problems like the lack of drinking water and adequate health care, factors which contributed to the 1990 and 1991 cholera epidemics. Wildlife is threatened by poaching, and a lack of resources with which to maintain wildlife reserves.

SOCIETY

Peoples: 98% of Zambians are descendants of successive waves of Bantu migrants, currently divided into 73 ethnic groups. There are approximately 70,000 inhabitants of European descent and 15,000 Asians. **Religions:** The majority, 80 %, are Christian; Muslim and Hindu minorites. **Languages:** English (official). Of the many Bantu languages, five are officially used in education and administration; Nyanja, Bemba, Lozi, Luvale and Tonga. **Political Parties:** The United National Independence Party (UNIP), was the sole party until 1991. At that time, in anticipated elections, the Movement for a Multi-party Democracy won 81% of the votes and 125 of the 150 contested parliamentary seats; UNIP obtained 25 seats. **Social organizations:** The Trade Union Congress of Zambia comprises 16 unions. The newly-created Workers Trade Union has 2 million members.

THE STATE

Official Name: Republic of Zambia. **Capital:** Lusaka, 982,400 inhab. (1990). **Other cities:** Ndola, 376,300 inhab.; Kitwe, 348,600 inhab.; Mufulira,175,000 inhab. (1990). **Government:** Frederick Chiluba, President since the October 1991 elections. Legislative power: National Assembly, single-chamber, with 150 members, of which 10 are appointed directly by the president. His term is for 5 years. **National Holiday:** October 24, Independence Day (1964). **Armed Forces:** 21,600 (1995). **Parmilitaries:** Police Mobile Unit (PMU): 700; Police Paramilitary Unit (PPMU): 700.

22%

PER 100,000 LIVE BIRTHS

229

STATISTICS

DEMOGRAPHY

Urban: 42% (1995). **Annual growth:** 0.6% (1992-2000). **Estimate for year 2000:** 10,000,000. **Children** per **woman:** 6.5 (1992).

HEALTH

One **physician** for every 11,111 inhab. (1988-91). **Under-five mortality:** 203 per 1,000 (1995). **Calorie consumption:** 89% of required intake (1995). **Safe water:** 50% of the population has access (1990-95).

EDUCATION

Illiteracy: 22% (1995). **Primary school teachers:** one for every 44 students (1988).

COMMUNICATIONS

86 **newspapers** (1995), 87 **TV sets** (1995) and 83 **radio sets** per 1,000 households (1995). 0.9 **telephones** per 100 inhabitants (1993). **Books:** 100 new titles per 1,000,000 inhabitants in 1995.

ECONOMY

Per **capita GNP:** $350 (1994). **Annual growth:** -1.40% (1985-94). **Annual inflation:** 92.00% (1984-94). **Consumer price index:** 100 in 1990; 1,655.4 in 1993. **Currency:** 500 kwacha = 1$ (1993). **Cereal imports:** 353,000 metric tons (1993). **Fertilizer use:** 160 kgs per ha. (1992-93). **External debt:** $6,573 million (1994), $714 per capita (1994). **Debt service:** 31.5% of exports (1994). **Development aid received:** $870 million (1993); $97 per capita; 24% of GNP.

ENERGY

Consumption: 140 kgs of Oil Equivalent per capita yearly (1994), 29% imported (1994).

---

dence, they no longer needed to live in a climate of war. They even questioned the one-party system. This inter-generational conflict diminished after the October 1983 election, where President Kenneth Kaunda won 93 per cent of the votes, 10 per cent more than in 1978.

[13] In 1984, Zambia faced drought, the worsening of the economic crisis, and trade union demands for wage-increases. Severe food shortages affected 300,000 people especially in the south, east and western areas. The constant deterioration of world copper prices led the government to declare a price rise of up to 70 per cent on basic foodstuffs in July 1984.

[14] The rising cost of living has led Zambian trade unions to protest against IMF impositions. Various strikes broke out in early 1985 and were suppressed by security forces.

[15] Toward the end of that year, IMF conditions for a $200 million credit for the 1986-87 period, led to a decision to increase the price of corn meal. This caused violent protests in the mining region, in the northern part of the country. Three days of protests and looting left 15 people dead and $90 million of damage done. In May 1987, Kaunda changed his policy towards the IMF, and limited debt servicing to 10 per cent of the country's foreign exchange.

[16] With 45 per cent abstention, Kaunda was re-elected in 1988. The following elections were moved forward 2 years to 1991, because of the critical economic situation. This time, Kaunda was defeated by Frederick Chiluba, a former labor union leader who gained 81 per cent of the vote, and 125 of the 150 seats in Parliament. Kaunda resigned as UNIP party leader in January 1992.

[17] The new government restructured the economy, privatized state enterprises and doubled the price of corn meal and other basic consumer goods. In December 1992, the government adopted the IMF and the World Bank recommendations, devaluing the country's currency by 29 per cent, freeing the exchange market and deregulating foreign trade. In early 1992, southern Africa experienced the worst drought of the century, resulting in widespread food shortages.

[18] A meeting of Western countries called by the World Bank granted $400 million of food aid to Zambia, and the Paris Club agreed to restructure its debt.

[19] President Chiluba declared Christianity the official religion and banned the formation of a fundamentalist party. The Muslim religious authorities estimated there were around 1.2 Shiite believers and more than a million Sunnis in the country.

[20] In March 1993, a state of emergency was called in an attempt to undermine a campaign of civil disobedience. Several members of the UNIP – who admitted there had been a plot – were imprisoned. The state of emergency was lifted at the end of the month.

[21] Inflation reached 140 per cent in 1993. The government had to reduce public spending, promote the privatization of companies and fight drug trafficking in order to get the aid asked of the Paris Club in 1994.

[22] The University of Zambia was closed in April 1994, when 300 teachers and researchers were dismissed for striking over pay claims. The reduction of spending on public schools led to constant teachers' strikes.

[23] Accusations of corruption and of the lack of agricultural policy in 1995 led President Chiluba to ask his minister of lands to resign, and the others to declare their income. Shortly afterwards he ousted the governor of the Bank of Zambia, when the local currency, the kwacha, was suddenly devalued by 20 per cent. Chiluba attributed the crisis to the foreign debt, as the servicing of this ate up 40 per cent of GDP. The trade minister admitted in July that 5.5 of the 9.5 million Zambians live in extreme poverty.

Zimbabwe

Population:	10,778,000 (1994)
Area:	390,580 SQ KM
Capital:	Harare
Currency:	Zimbabwe Dollar
Language:	English

Zimbabwe

Between the Zambezi and Limpopo rivers "...one can find thousands of abandoned mines; cultivated terraces that covered entire mountains, irrigation canals, paths, and and wells twelve meters deep wells excavated in rock..." (Pierre Bertaux, French historian).

2 Among the ruins scattered over an extensive region including nearly 300 archeological sites, the most important cities were Khami, Naletali, DhloDhlo, Mapungubwe and the better known Zimbabwe with its high-walled enclosure. The Karangas, members of the present-day Shona people of the Sotho Bantu ethnic group, built these walls in the 10th century.

3 The Bantu ironworkers who settled in the region in the 5th century also discovered gold, copper and tin deposits. In a few centuries they developed sophisticated techniques for working these metals. In the eighth century, the rise of Arab-influenced trade centers on the coast provided a market for their goods, and the growth of trade resulted in a great expansion of this culture. When the Shirazis founded Sofala (in present-day Mozambique) in the 10th century, the Karanga state acquired an export market for its mining production. The "Monomotapa", the Karanga leader, imposed a tributary relation on the neighboring Muslim nation as he had done with other minor cultures of the area. Thus, Karangan supremacy was established over a region including parts of present-day Malawi.

4 This civilization established important trading connections with Asia and continued to develop until the mid-15th century when the Rotsi, a southern people belonging to the Shona, the same ethnic group as the Karanga, forced the Mono-motapa to withdraw towards the north and the coast. The Zimbabwe citadel and palace were taken over by the Rotsi, whose Changamira (king) extended his control over the mining area. He did not however succeed in controlling an area as vast as the ancient Karanga had done.

5 Portuguese presence brought about the end of the prosperous trade with the east and a consequent economic decline. Also, in reaction to the Europeans' greed for gold, the Shona miners filled in their mines, keeping only their ironworks functioning.

6 In 1834, the Zulus devastated the region. The Rotsis emigrated westwards; cities and farmlands, palaces and irrigation canals were abandoned and grass began to grow over the ancient walls of Zimbabwe. In the first half of the 19th century, the territory was divided between the Shona peoples in the northeast, and the Zulu kingdom of Matabele in the southeast. In 1889, Lobenguela, the Ndebele ruler, received a visit from Charles Rudd, envoy of the wealthy English adventurer-businessman Cecil Rhodes. In exchange for arms, a life pension and a steamship, Lobenguela granted exclusive rights for the exploitation of the country's mineral resources to Rhodes' British-South Africa Company (BSA).

7 The British government gave the BSA control over trade, immigration, communications and the police in the Ndebele territory. Since "Matabeleland", as the British called it, was an independent state, the concession of these privileges granted by a foreign government, needed "formal and free" approval from ruler Lobenguela who lived in Bulawayo. After Rhodes took control, 200 British settlers were established in the Shona territory. They were protected by 700 police officers and were authorized to live there, founding a fortified camp which they called Salisbury. The settlers had no legal title deeds to the land other than the backing of the British government who ignored the town boundaries.

8 Lobenguela witheld his approval for the BSA's activities. So Rhodes's agents created conflictual situations with him to provide a "reason" to depose him. In 1895, under the pretext of arresting some Ndebele who had stolen cattle from the Shona, Rhodes's "police" attacked Bulawayo, the capital of Lobenguela. The ruler was driven into the jungle and his nation and the Shona territory fell under BSA domination, under the name of Southern Rhodesia.

9 In 1960, settlers of European origin accounted for hardly 5 percent of the population but owned more than 70 per cent of the arable land. When the decolonization process began in Africa, Zambia and Malawi, formerly part of a federation with Southern Rhodesia, resisted further European domination and gained independence in 1964. In Rhodesia, the African National Congress (ANC) also intensified the struggle for independence. The colonial government of Prime Minister Ian Smith reacted declaring a "state of emergency" in November 1965. The British government urged Southern Rhodesia to gradually transfer power to the African native majority. Smith flatly refused and proclaimed independence on November 11 1965 in order to remain in power with his segregationist Rhodesian Front.

10 The rebel regime faced an embargo declared by the United Nations, though the blockade was systematically violated by the western powers. Guerrilla warfare organized by ZAPU (Zimbabwe African People's Union) and ZANU (Zimbabwe African National Union), was launched. The armed struggle grew increasingly intense and the Smith regime bombarded Zambia and Mozambique. These countries together with Angola, Botswana and Tanzania, formed the group of Front Line Countries to fight racism. In Zimbabwe, ZAPU and ZANU joined to form the Patriotic Front under the joint leadership of Joshua Nkomo and Robert Mugabe.

PUBLIC EXPENDITURES

% of GNP — 1992

MILITARY 4.3 — EDUCATION 10.6 — HEALTH 3.2

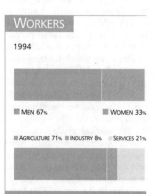

WORKERS

1994

MEN 67% — WOMEN 33%

AGRICULTURE 71% — INDUSTRY 8% — SERVICES 21%

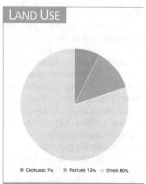

LAND USE

CROPLAND 7% — PASTURE 13% — OTHER 80%

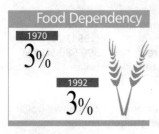

Food Dependency

1970
3%

1992
3%

Illiteracy

1995
15%

PROFILE

ENVIRONMENT

The country consists mainly of a high rolling plateau. Most of the urban population live in the High Veld, an area of fertile land, with moderate rainfall and mineral wealth. The climate is tropical, tempered by altitude. Soil depletion is very severe, above all on communal farms, where subsistence agriculture is practised.

SOCIETY

Peoples: The majority of Zimbabweans, 98%, are of Bantu origin from the Shona (founders of the first nation in the region) and the Ndebele group (a Zulu people arrived in the 19th century). **Religions:** Most of the population are Christian (45%), predominantly Anglican. Also African traditional beliefs are followed. **Languages:** English, official. Most of the people speak their own Bantu languages. **Political Parties:** Zimbabwe African National Union-Patriotic Front (ZANU-PF), ruling party, emerged from the fusion of the Zimbabwe African National Union (ZANU), led by Mugabe, and the Zimbabwe People's Union (ZAPU), led by Joshua Nkomo, in December 1987; Zimbabwe Unity Movement (ZUM), of Edgar Tekere; Regional ZANU-Ndonga party, of right-wing nationalist Ndabaningi Sithole; Emmanuel Magoche's Democratic Party split from the ZUM in 1991. **Social Organizations:** Organization of Rural Associations for Progress (ORAP).

THE STATE

Official Name: Republic of Zimbabwe. **Administrative Divisions:** 8 Provinces. **Capital:** Harare 1,184,200 inhab. (1992). **Other cities:** Bulawayo, 620,900 inhab.; Chitungwiza, 274,000inhab.; Mutare, 131,800 inhab.; Gweru, 124,700 inhab. (1992). **Government:** Robert Mugabe, President. There is a bicameral parliament with a 40-member Senate and a House of Assembly with 100 seats elected through universal suffrage. **National Holiday:** April 18, Independence Day (1980). **Armed Forces:** 45,000 (1995). **Paramilitaries:** 15,000 Republican Police Force, 2,000 Police Support Unit, 4,000 National Militia.

[11] In 1978 Smith and some African leaders opposed to the Patriotic Front signed an "internal agreement", which legalized their own political parties. In 1979, after fraudulent elections, Bishop Abel Muzorewa became premier and changed the name of the country to ZimbabweRhodesia. With the majority of parliamentary seats, the racist minority had the power to control the socio-economic and political system.

[12] However, guerrilla pressure was mounting, and finally the European government and its African allies were forced to negotiate. The British government agreed to supervise free elections arranged for February 1980, when Robert Mugabe's ZANU party won a landslide victory. By the Lancaster House agreements, signed on April 18 1980, Britain held power temporarily before transferring to the ZANU Party. Although the Europeans maintained some economic and political privileges, they lost their veto over possible constitutional changes.

[13] Meanwhile, the country's cattle herd had shrunk to a third of its previous size; thousands of kilometers of roads had been rendered useless, and more than two thirds of schools had remained closed for seven years. The medical and sanitary systems were also in serious disrepair, and various diseases, such as malaria, were increasingly in evidence among the population.

[14] Prime Minister Robert Mugabe offered generous cabinet participation to the ZAPU leadership, also calling segregationist politicians to form the government team. These steps were aimed at preventing old rivalries from interfering with national reconstruction, especially the ambitious National Development Plan. The gross national product grew by 7 per cent, farm production broke all records, and consumption reached higher levels than expected. However, Robert Mugabe had to face two major difficulties: South Africa's blockade on Zimbabwe's agricultural exports and political dissent between ZANU and ZAPU. In 1981 this led the prime minister to remove Nkomo from the home affairs ministry, though maintaining his ministerial rank.

[15] Political differences intensified. In 1982, Nkomo's followers created an armed movement called "Super-ZAPU". Growing political tension coincided with the start of the great drought which was responsible for a drop in agricultural production from 2 million tons in 1981 to 620,000 in 1983.

[16] Pressure from native farmers, who had hoped for a true agrarian reform after independence, clashed with the limitations imposed by the Lancaster House agreement. This treaty impeded the expropriation of European land-holdings and the British and Americans evaded giving Zimbabwe the resources promised for the purchase and distribution of land. In Matabeleland, Nkomo's followers explained that the difficulties stemming from the drought and the lack of funds were anti-Ndebele schemes of Mugabe, aimed against the country's second largest ethnic group after the Shona, and that they were all elements designed to create inter-tribal animosity.

[17] Towards the end of June 1985, Mugabe's Zimbabwe African National Union obtained a comfortable victory in parliamentary elections throughout the country, except in Matabeleland, while the majority of whites voted for the Rhodesian Front, created by Ian Smith, a fact which led the president to remind them that the privileges granted to former colonists under the Lancaster House agreement should not be considered unalterable.

[18] The political privileges enjoyed by whites were subsequently eliminated, though there were extensive privileges in other areas. These were evident by figures in a technical report on Zimbabwe's land tenure system: 4,500 farmers (most of them white) owned 50 per cent of the country's productive land while the 4.5 million peasants lived in communally-owned rural areas - known as "tribal lands", where the black population were moved to during the colonial era.

[19] The Commercial Farmers' Union of white farmers blocked many initiatives for rural relocation. They controlled 90 per cent of all agricultural production, paid a third of the country's salaries and exported 40 per cent of the country's goods. This situation existed despite the fact that a third of the properties belonging to its members were not exploitable, and that productivity could be increased, and that the great efficiency of the farmers was due to state support rather than the managerial talent of the farmers themselves.

[20] Two constitutional reforms were carried out in September 1987. Previously, 20 seats of the Assembly and 10 in the Senate had been reserved for whites, but this practice was abolished. In addition, an executive authority was assigned to the president, elected by parliament for a six-year period.

[21] Mugabe played an important role in the summit of the Movement of Non-Aligned Countries held in September in Harare, where he enthusiastically promoted the widespread adoption of sanctions against South Africa. In addition, Zimbabwe supported the government of Mozambique against the counter-revolutionaries of the National Resistance of Mozambique (RENAMO). In May 1987, some 12,000 Zimbabwean troops were stationed in Mozambique, while RENAMO led incursions into Zimbabwean territory during 1987 and 1988.

[22] In December 1987, Mugabe and Nkomo reached reconciliation, and the two major political parties agree to unification. This agreement was ratified in April 1988 creating the Patriotic Front of the Zimbabwe African National Union (ZANU-PF). Subsequently, with the help of the alliance of several white deputies within ZANU, the country has oriented itself, against the general trend in Africa, toward a one-party system.

[23] In the March 1990 elections, ZANU-PF obtained 116 of the 119 parliamentary seats. President Mugabe interpreted election results as a "popular mandate" in favor of

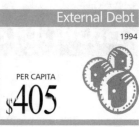

his idea of a one-party system. Nevertheless, the newly organized Zimbabwe Unity Movement (ZUM) obtained 15 per cent of the votes, and only 54 per cent of the people went to the polls, a significant abstention rate, compared to percentages of participation exceeding 90 per cent in the 1980 and 1985 elections. The opposition obtained almost 30 per cent of the votes in Harare and other urban centers, while ZAPU held its ground in rural areas.

24   In an economy with an average annual growth of 4 per cent, the share of agriculture in the GNP was 14 per cent in 1980, and rose to 20 per cent in 1990. That same year, Parliament approved land reform authorizing the government to expropriate land held by Europeans, at a price fixed by the State, and to redistribute it among the poor. The majority of the African population supported the law, deeming it an act of racial and economic justice. The white farmers, on the other hand, supported by legal experts and by the Catholic hierarchy, criticized it as a violation of the civil and human rights established under the Constitution.

25   The Organization of Rural Associations for Progress (ORAP) was inaugurated after Zimbabwean independence aimed at preserving indigenous culture and social organization. In 1991, it had half a million members in 16 villages, covering 3.5 of the country's 5 provinces. The basic structure of the ORAP is the "Amalema", a neighborhood association of 10 families, attempting to recreate the communal organization and production methods existing before colonization. ORAP operate under the hypothesis that the traditional lifestyles of the local population may do more to solve the country's problems than the paternalistic models imported by European settlers.

26   In June 1991, the ruling ZANU-PF decided to relinquish its Marxist-Leninist doctrine, eliminating all references to this and scientific socialism from its statute. ZANU-PF is still maintaining a socialist perspective for as long as it feels is appropriate for the Zimbabwean situation. Mugabe appealed for the doctrine of pure socialism to be abandoned, in favor of social democracy and a mixed economy.

27   In October 1991, the Zimbabwean Unity Movement split, creating the Democratic Party, led by Emmanuel Magoche, the former president of the ZUM.

28   Although Zimbabwe has a multiparty system, it operates similarly to a one-party regime. The absence of effective opposition to the ZANU-PF has meant that the ruling party has gradually come to control all state mechanisms, the police force, and the state administration. Although the ZANU-PF leadership reached a consensus decision to maintain the regime, the emergence of movements in favor of a multiparty system in South Africa, Angola, Namibia, Mozambique, and Zambia pressurized Zimbabwe into allowing the formal creation of several parties.

29   Late in 1991, President Mugabe reported an increase of 3.5 per cent in Zimbabwe's GNP from the previous year (1.4 points above the rate for the previous year). Yet the country faced a sudden increase in inflation, which jumped from 13.3 per cent in 1990 to 25 per cent in 1991. In January 1992, the government announced that a five-year reform plan would be implemented. Its aim would be to liberalize the economy, bring the budget deficit down to 5 per cent of the GNP, and to substantially increase job opportunities.

30   One of the first measures adopted in January 1992 was the elimination of 32,000 public sector jobs. Under pressure from the IMF, Mugabe began reducing the number of ministries.

31   Within the framework of public spending cuts, in May, the government reduced the education budget and announced that tertiary education would no longer be free. As a result, massive student protests broke out, culminating in the expulsion of 10,000 students from the University of Zimbabwe in June. Demonstrations repudiating this measure were harshly repressed by the police. The Labor Union Congress declared a general strike, protesting the government's economic policies.

32   A demonstration by students, women and children demanding better life conditions in the the gold mines region, east of Harare, culminated with three deaths under police fire. US elite troops, acting together with Zimbabwe troops, undertook military maneuvres. According to the government, these maneuvres were intended to prepare the troops that would operate in Somalia under the UN, but according to the opposition the purpose was to train them for domestic repression.

33   By the middle of 1993, the Union of Commercial Farmers,

supported by minority parties of European origin, accused the government of expropriating unproductive land. Mugabe threatened them with deportation in case they did not accept the agrarian reform. Nevertheless, he did not ask them to assure minimal life conditions for the rural workers. In 1994, a rural worker earned $30 per month in the bigger properties.

34   The countries that gave Zimbabwe economic assistance gathered in Paris in March 1995, a month later than planned, because Mugabe's government did not have their finance books in good order. Despite some local fear of reprisals for not having carried on the financial reforms demanded by the IMF "in due time", the country received a bigger amount than the figure initially announced. The foreign debt of Zimbabwe was equal to the PBI, so that 23 per cent of the national budget was earmarked for servicing the debt.

35   In October 1995, Ndabaningi Sithole, leader of the ZANU-Ndonga, and one of the two members of the Assembly not belonging to the ZANU-PF, was arrested for allegedly planning the assassination of the president. Mugabe's opponents attributed the arrest to Sithole's announcement of his candidacy for the presidency.

36   With the abstention of 68 per cent of the voters, the April 1996 elections resulted in Mugabe's victory with 93 per cent of the votes. Several opposition parties boycotted the elections since they questioned the electoral procedures. At the moment of the election the government held heavy debts with most public employees.

Bibliography

ABDEL-MALEK, A, ed - Contemporary Arab Political Thought. London, Zed, 1983.

AGENDA roja 1978.

AGUILAR DERPICH, J - Guyana: Otra vía al socialismo. Caracas, El ojo del camello, 1973.

ALGERIE: Guide economique et social. Dely, ANEP, 1987.

ALIANZAS políticas y procesos revolucionarios. Mexico, SEPLA, 1979.

ALMANAQUE BRASIL 1995/1996. Editora Terceiro Mundo, Rio de Janeiro, 1995.

AMARASINGAM, SP - The Industrialized Nations of the West and the Third World. Colombo, Tribune, 1982.

AMIN, S, et ali - Nuevo Orden Internacional. Mexico, Nueva Política, 1977.

AMNESTY INTERNATIONAL - Contra la pena de muerte. Madrid, 1992.

AMNESTY INTERNATIONAL - Guía de la Carta Africana de los derechos humanos y de los pueblos. Madrid, 1991.

AMNESTY INTERNATIONAL - Informes 1984-1996. Madrid, 1996.

AMNESTY INTERNATIONAL - Reino Unido:desigualdad ante la ley. Madrid, 1991.

ANUARIO del Centro de Investigaciones para la Paz 1988/1989: Paz, militarización y conflictos. Madrid, IEPALA, 1989.

ANNUAIRE Geopolitique Mondial. Paris, Machette, 1989. (Suppl. de Politique International.)

ARNOLD, M - The Testimony of Steve Biko. Suffolk, Granada Publ., 1979.

ARRUDA, M , et al - Transnational Corporations: A Challenge for Churches and Christians. Geneva, World Council of Churches, 1982.

ASIA 1990 Yearbook. Hong Kong, Dai Nippon, 1990.

ATLAS Barsa. Rio de Janeiro, Encyclopaedia Britannica, 1980.

ATLAS de L'humanité. Paris, Solar, 1983.

AZIZ, T - Reflexions sur les relations Arabo-Iraniennes. Baghdad, 1980.

BARDINI, R - Belice: historia de una nación en movimiento. Tegucigalpa, Universitaria, 1978.

BARDINI, R - El Frente Polisario y la lucha del pueblo saharaui. Tegucigalpa, CIPAAL, 1979.

BARRACLOUGH, G - Introducción a la Historia Contemporánea. Barcelona, Anagrama, 1975.

BASSOLS BATALLA, A - Geografía, subdesarrollo y regionalización. Mexico, Nuestro Tiempo, 1976.

BASSOLS BATALLA, A - Los recursos naturales. Mexico, Nuestro Tiempo, 1974.

BASTIDE, R - Las Américas negras. Madrid, Alianza, 1969.

BECKFORD, G - Persistent Poverty, Underdevelopment in Plantation of the Third World. 2a.ed. London, 1983.

BEDJAQUI, M - Towards a new International Economic Order. Paris, UNESCO, 1979.

BELFRACE, C & ARONSON, J - Something to Guard. New York, Columbia Univ. Press, 1978.

BELLER, WS., ed - Environmental and Economic Growth in the Smaller Caribbean Islands. Washington, Department of State, 1979.

BENOT, I - Ideologías de las independencias africanas. Barcelona, 1973.

BENZ, W - El siglo XX: Problemas mundiales entre los dos bloques de poder. Madrid, Siglo XXI, 1984.

BERGER, J. & MOHR, J - A seventh man. Middlesex, 1975.

BIANCO, L - Asia contemporánea. México, Siglo XXI, 1980.

BILAN Economique et Social 1984-1989-1991. Le Monde; Dossiers et Documents; Paris, 1992.

BOUSTANI, R & FARGUES, P - Atlas du Monde Arabe. Géopolitique et Societé. Paris, Bordas, 1990.

BRANDT, W.et alii - Das Ueberleben sichern. Kiepenheur y Witsch, Koln, 1980.

BRITTAIN, V., & SIMMONS, M - Third World Review. London, Guardian, 1987.

BROWN, L - Food or Fuel: New Competition for the World's Cropland. Washington, Worldwatch Institute, 1980.

BRUBAKER, S - Para vivir en la tierra. Mexico, Pax, 1973.

BRUHAT, J - Historia de Indonesia. Buenos Aires, EUDEBA, 1964.

BUETTNER, T, et al - Afrika: Geschichte von den Anfaengen bis zur Gegenwart. Koln, Pahl-Rugenstein, 1979.

BURCHETT, WG - Otra vez Corea. Mexico, ERA, 1968.

BURGER, J - The Gaia Atlas of first peoples. A future for the indigenous world. London, Gaia, 1990.

CALCAGNO, AE & JAKOBOWICZ, JM - El monólogo Norte-Sur y la explotación de los países subdesarrollados. México, Siglo XXI, 1981.

CALCHI NOVATI, G - La revolución argelina. Barcelona, Bruguera, 1970.

CALDWELL, M - The Wealth of Some Nations. London, Zed, 1977.

CARATINI, R - Dictionnaire des nationalités et des minorités en URSS. Paris, Larousse, 1990.

CARRERAS, J - Historia de Jamaica. Havana, Ed.Ciencias Sociales, 1984.

CARTA africana de Derechos Humanos y de los pueblos. Nairobi, 1981.

Adoptada por la Decimoctava Conferencia de Jefes de Estado y de Gobierno de la Organización de la Unidad Africana.

CASTRO, F - La crisis económica y social del mundo. Havana, OPCE, 1983.

CAVALLA, A - Geopolítica y Seguridad Nacional en América. Mexico, UNAM, 1979.

CEPAL - Anuario estadístico de América Latina. Santiago de Chile, 1981.

CEPAL - Estudio económico de América Latina y el Caribe. Santiago de Chile, 1984.

CEPAL - Notas sobre la economía y el desarrollo. Santiago de Chile, 1985.

CEPAL - La evolución de la economía de América Latina en 1983. Santiago de Chile, 1984.

CINCO conferencias Cumbres de los Paises No Alineados. Havana, 1979.

CIPOLLA, C - La explosión demográfica (entrevista). Barcelona, Salvat, 1973.

CLAIRMONTE, FF & CAVANAGH, H - Transnational Corporations and Services: The final frontier. Geneva, UNCTAD, 1984.

CLARK, C - Crecimiento demográfico y utilización del suelo. Madrid, 1968.

COLCHESTER, M & LOHMANN, L - The tropical forestry action plan; what progress?. 1990.

COMANDANTE de los pobres: testimonios sobre Omar Torrijos. Madrid, Centro de Estudios Torrijistas, 1984.

COMISION INTERNACIONAL SOBRE PROBLEMAS DE LA COMUNICACION - Un solo mundo, voces múltiples. Mexico, Fondo de Cultura, 1980.

CONNELL-SMITH, G - Los Estados Unidos y la América Latina. Mexico, 1977.

CONSTANTINO, R - The Philippines:A past revisited. Manila, 1975.

CONTRERAS, M & SOSA, I - Latinoamérica en el siglo XX. Mexico, UNAM, 1973.

CORDOVA, A., et alii - El Imperialismo. Mexico, Nuestro Tiempo, 1979.

CORM, G - Le Proche-Orient éclaté. Paris, La Découverte, 1983.

Le COURRIER des Pays de l'est. Paris, 1986.

CUEVA, A - El proceso de dominación política en el Ecuador. Quito, Critica, 1973.

CHALIAND, G - Revolution in the Third World. Middlesex, Penguin, 1979.

CHEE, Y - How big powers dominate the Third World. Penang, Third World Network, 1987.

CHOMSKY, N; STEELE, J & GITTINGS, J - Super Powers in Collision. Penguin, 1984.

CHRISTENSEN, Ch - The Right to Food: how to guarantee. New York, 1978.

DAVIDSON, B - The Liberation of Guiné. Middlesex, Penguin, 1971.

DAWISHA, K - Eastern Europe, Gorbachev and reform. the great challenge. 2a.ed. New York, Cambridge Univ. Press, 1990.

DEBRAY, R -La crítica de las armas. Madrid, Siglo XXI, 1975

DECRAENE, P - El Panafricanismo. Bs.As., EUDEBA, 1962.

DEGENHARDT, H, comp - Political dissention international guide to dissent, extraparliamentary, guerrillas and illegal political movements. London, Longman, 1983.

DE LA COURT, T; PICK, D & NORDQUIST, D - The nuclear fix. A guide to nuclear activities in the Third World. Amsterdam, Wise, 1982.

DENVERS, A - Points choc: l'environnement dans tous ses états. Editions 1, 1990.

DE SOUZA, H - El capital mundial. Montevideo, CUI, 1987. v.1.

DER FISCHER WELT ALMANACH 1993. Fischer Taschenbuch Verlag. Bonn, 1992.

DIFRIERI, H., et al - Geografía Universal. Bs.As., ANESA, 1971.

DORE, F - Los regímenes políticos en Asia. Mexico, Siglo XXI, 1976.

DRECHSLER, H -Africa del Sudoeste bajo la dominación colonial alemana. Berlin, Verlag, 1986.

DREIFUSS, R -A Internacional capitalista. Rio le Janeiro, Espaço e Tempo, 1986.

ECE - Economic Bulletin for Europe. New York, 1991.

ECE - Economic Survey of Europe in 1988-1989. New York, 1989.

ECE - Economic Survey of Europe in 1989-1990. New York, 1990.

ECE - Economic Survey of Europe in 1990-1991. New York, 1991.

ECKHOLM, E - The dippossessed of the earth: Land reform and Sustainable Development. Washington, Worldwatch Institute, 1979.

L'ECONOMIE de la drogue. Le Monde; Dossiers et Documments; Paris, feb 1990.

L'ECONOMIE de l'Espagne. Le Monde; Dossiers et Documents; Paris, Jan 1990.

ELLIOT, F - A Dictionary of Politics. Middlesex, Penguin, 1974.

EMMANUEL, A., AMIN, S, et al - Imperialismo y comercio Internacional.(El intercambio desigual). Madrid, Siglo XXI, 1977.

El EMPUJE del Islam. Madrid, IEPALA, 1989. (Africa Internacional, 7)

El SALVADOR. Alianzas políticas y proceso revolucionario. México, SEPLA, 1979.

ENTRALGO, A., ed - Africa. Havana, Ed.Ciencias Sociales, 1979.

ENZENSBERG, H - Zur Kritik der politischen Okologie. Berlin, Kursbuch, 1973.

L'ETAT du Monde 1989-1990.Annuaire économique et géopolitique mondial. Paris, La Découverte, 1989.

L'ETAT des religions dans le Monde. Paris, La Découvert, 1987.

L'ETAT du Monde. Paris, La Découverte, 1996.

L'ETAT du Tiers Monde. Paris, La Découverte, 1989.

The EUROPA Yearbook. London, Europa Pub., 1984.

FABER, G - The Third World and the EEG. Netherlands, 1982.

FAHRNI, D - Historia de Suiza. Ojeada a la evolución de un pequeño país desde sus orígenes hasta nuestros días. 2a.ed. Zurich, Pro Helvetia, 1984.

FALK, R & WAHL, P - Befreiungsbewegungen in Afrika. Pahl- Rugenstein, Koln, 1987.

FANON, F - Os condenados da Terra. Rio de Janeiro, Civilizaçao Brasileira, 1979.

FAO - Estado mundial de la agricultura y la alimentación. Rome, 1983.

FAO - Production Yearbook. Rome, 1993.

FERNANDEZ, W - El gran culpable.La responsabilidad de los Estados Unidos en el proceso militar uruguayo. Montevideo, Atenea, 1986.

FIELDHOUSE, DK - The Colonial Empires.A comparative survey from the eighteenth century. 2a.ed. London, MacMillan1982.

FIGUEROA ALCACES, E, ed - Antología de geografía histórica moderna y contemporánea. Mexico, UNAM, 1974.

FISHLOW, A, et al - Rich and Poor Nations in the World Economy. New York, McGraw Hill, 1978.

FORDHAM, P - The geography of African Affairs. Middlesex, Penguin.

FOREST Resources crisis in the Third World, Malasya, Sep.1986. Proceedings. Malaysia, 1987.

FRENTE POLISARIO -VII Anniversaire du Declenchement de la Lutte de Liberation nationale. Dep.Informations, 1980.

FRETLIN Conquers the Right to Dialogue. London, 1978.

GALEANO, E - As Véias Abertas de América Latina. Rio de Janeiro, Vozes, 1979.

GALTUNG, J. & O'BRIEN, P., et al - Self-reliance.A Strategy for development. London, IDS, 1980.

GANDHI, MK - An autobiography. Boston, Beacon Press, 1957.

GATT International trade 1981-1982. New York, 1982.

GAZOL SANTAGE, A - Los países pobres. México, FCE, 1987.

GENOUD, R - Sobre las revoluciones parciales del Tercer Mundo. Barcelona, Anagrama, 1974.

GEORGE, P - Geografía económica. Barcelona, Ariel, 1976.

GEORGE, P - Panorama del mundo actual. Barcelona, Ariel, 1970.

GEORGE, P - Geografía y medio ambiente, población y economía. Mexico, UNAM, 1979.

GEORGE, S - How the Other Half Dies. Middlesex, Penguin, 1980.

GEZE, F, et al - L'Etat du Monde-1982. Paris, Maspero, 1982.

GOUROU, P - L'Afrique. Paris, Hachette, 1970.

GRANT, JP - Situaçao Mundial da Infancia. New York, UNICEF, 1984.

GREENBERG, S, ed - Guiness book of Olympic Records. Bantam, 1988.

GREINER, B - Amerikanische Aussenpolitik von Truman bis heute. Koln, Pahl-Rugenstein, 1980.

GROUSSET, R - Historia de Asia. Buenos Aires, EUDEBA, 1962.

GUIA CIUDADANA sobre el Banco Mundial y el Banco Interamericano de Desarrollo, Red Bancos, Instituto del Tercer Mundo, Montevideo, 1996.

GRUNEBAM, GE, ed - El Islam. Desde la caída de Constantinopla hasta neustros días. Mexico, Siglo XXI, 1975.

GUIM, JB - Compendio de geografía Universal. Mexico, Bouret, 1875.

HALLIDAY, F - Arabia without Sultans. Middlesex, Penguin, 1974.

HALPERIN DONGHI, T - Historia contemporánea de América Latina. Madrid, Alianza, 1983.

HAMELINK, C - The Corporate Village. Rome, IDOC, 1977.

HANDWÖRTERBUCH INTERNATIONALE POLITIK. Wichard Woyke (Hrsg.). Leske Budrich. 1993.

HAYES, MD - Dimensiones de seguridad de los intereses de Estados Unidos en América Latina. Mexico, CIDE, 1981.

HERRERA, L & VAYRRYNEN, R, ed - Pace, development and New International Economic Order. Tampere, IPRA, 1979.

HUIZER, G - El potencial revolucionario del campesino en América Latina. Mexico, Siglo XXI, 1974.

HUMAN Rights Newsletter. 1991.

HUMANN, K. & BRODERSEN, I - Welt Aktuell'86. Hamburg, Roro, 1985.

IBGE - Tabulaçes Avançadas do Censo Demográfico. Rio de Janeiro, 1982.

IDB - External Debt and Economic Development in Latin America. Washington, 1984.

IDB - Progreso económico y social en América Latina. Washington, 1984.

ILO - Yearbook of Labour Statistics/ Anuario de Estadísticas del Trabajo. Geneva, 1995.

IIED - World Resources 1986. New York, 1986.

INDEX on Censorship. London. 1991-1992.

INSTITUTE OF RACE RELATIONS - Patterns of Racism. London, 1982.

INSTITUTO GEOGRAFICO DE AGOSTINI - Atlas Universal Geoeconómico. Barcelona, Teide, 1977.

INTERNATIONAL INSTITUTE FOR STRATEGIC STUDIES - The Military Balance 1993-94. London, Brassey's, 1993.

INTERNATIONAL INSTITUTE FOR STRATEGIC STUDIES - Strategic Survey 1982-1983. London, 1983.

INTERNATIONAL COMMISION OF JURISTS AND CONSUMERS' ASSOCIATION OF PENANG - Rural development and Human Rights in South East Asia. Penang, 1982.

INTERNATIONAL MONETARY FUND - Directory of Regional Economic Organizations and Intergovernmental Commodity Organizations. Washington, 1979.

INTERNATIONAL MONETARY FUND - Estadísticas Financieras Internacionales. Washington, 1985.

JALEE, P - El Tercer Mundo en la economía Mundial. Mexico, Siglo XXI, 1980.

JAULIN, R - La des-civilización. Política y práctica del etnocidio. Mexico, Nueva Imagen, 1979.

JONAS, S.& TOBIS, D -Guatemala. New York, NACLA, 1974.

JOYAUX, F - L'année internationale 1990. Paris, Hachette, 1989. Annuaire géopolitique de la Revue Politique Internationale.

KAPLAN, L - Revoluciones. Mexico, Extemporáneos, 1973.

KÄKÖNEN, J - The Mechanics of Neo-Colonialism. Mänttä, Finnish Peace Research Association, 1974.

KENT, G - Food Trade: The Poor Feed the Rich in the Ecologist Ecosystem. United Kingdom, 1985.

KETTANI, M - Ali, the Muslim Minorities. Leicester, Islamic Foundation, 1979.

KHADAFI, M - O livro verde. Tripoli, EPOEPD.

KHOR, K.P - Recession and the Malaysian Economy. Penang, Masyarakat, 1983.

KIDROM, M & SEGAL, R - The State of the World Atlas. London, Pan Books, 1981.

KINDER, H. & HILGEMANN, W - The Penguin Atlas of World History. New York, 1978.

KI-ZERBO, J - Historia del Africa Negra. Madrid, Alianza, 1980.

KLAINER, R, LOPEZ, D & PIERA, V - Aprender con los chicos: propuesta para una tarea docente fundada en los Derechos Humanos. Buenos Aires, MEDH, 1988.

KLARE, M, et al - Supplying Repression. Washington, IPS, 1981.

KUPER, L - Genocide: Its political use in the Twentieth Century. Middlesex, Penguin, 1981.

LACOSTE, Y -Geografía del subdesarrollo. Barcelona, Ariel, 1978.

LACOSTE, Y -Los países subdesarrollados. Buenos Aires, EUDEBA, 1965.

LAINO, D - Paraguai: Fronteiras e penetraçao brasileira. Sao Paulo, Global, 1979.

LEBEDEV, N - La URSS en la política mundial. Moscow, Progreso, 1983.

LEGER SIVARAD, R - World Military and Social Expenditures 1986. Washington, 1986.

LEGER SIVARAD, R - World Military and Social Expenditures 1989. Washington, 1989.

LEWIS, DE - Reform and revolution in Grenada:1950-1981. Havana, Casa de las Américas, 1984.

LEWYCKY, D.& WHITE, S - An African abstract;Council for International Cooperation. Manitoba, 1979.

LICHTHEIM, G - Imperialism. Middlesex, Penguin, 1974.

LINHARES, MY - A luta contra a metrópole (Asia e Africa). Sao Paulo, Brasilense, 1981.

LINIGER GOUMAZ, M - Guinée Equatoriale: de la dictadure des Colons á la Dictadure des Colonels. Geneva, Ed.Temps, 1982.

LINIGER GOUMAZ, M - De la Guinée Equatoriale Nguemiste. Geneva, Ed.Temps, 1983.

LINTER, B - OUTRAGE: Burma's struggle for democracy. Hong Kong, Review Publ., 1989.

LIPSCHUTZ, A -El problema racial en la conquista de América. Mexico, Siglo XXI, 1963.

LUNA, J -Granada, la nueva joya del Caribe. Havana, Ciencias Sociales, 1982.

McCOY, A.& JESUS, EDC - Philippine Social History. Manila, ASSA, 1982.

McEVEDY, C.& JONES, R -Atlas of world Population History. Kerala, 1986. Middlesex, Penguin, 1978

MAGDOFF, H - La empresa multinacional en una perspectiva histórica. Barcelona, 1980.

MAHATHIR BIN MOHAMED - The Malay dilema. Selangor, 1982.

MANORAMA Yearbook 1986. Kerala, 1986.

MASUREL, E - L'année 1989 dans le Monde:les principaux evénements en France et a la l'etranger. Paris, Gallinard, 1990.

MAX-NEEF, M - La economía descalza. Lima, 1985.

MAYOBRE MACHADO, J - Información dependencia y desarrollo;la prensa y el nuevo orden económico internacional. Caracas, Monte Avila, 1978.

MEILE, P - Historia de la India. Buenos Aires, EUDEBA, 1962.

MEJIA RICART, T - Breve historia dominicana. Santa Dominica, 1982.

MENENDEZ DEL VALLE, E - Angola, imperialismo y guerra civil. Barcelona, Akal, 1976.

MICHELINI, Z - Uruguay vencerá. Barcelona, Laja, 1978.

MOITA, L. Os Congressos da Frelimo do PAIGC e do MPLA, CIDAC. Lisbon, 1979.

MOORE LAPPE, F & COLLINS, J -El hambre en el mundo. Diez mitos. Mexico, COPIDER/FONAPAS, 1980.

MOREIRA, N - Modelo peruano. Rio de Janeiro, Paz e Terra, 1975.

MOREIRA, N - El Nasserrismo y la revolución del Tercer Mundo. Montevideo, EBO, 1970.

MPLA - Historia de Angola. Porto, Afrontamento, 1975.

MYERS, N., ed - The Gaia Atlas of Planet Management. London, Pan Books, 1985.

MYERS, N - The Gaia Atlas of future worlds challenge and Opportunity in Age of change. London, Gaia, 1990.

MYLLYMAKI, E.& DILLINGER, B - Dependency and Latin American Development. Finnish Peace Research Association. Tampere, 1977.

El NACIMIENTO de una nación: la lucha por la liberación de Nambia. London, Zed, 1985.

NKRUMAH, N - I Speak of Freedom. New York, Praeger, 1961.

NKRUMAH, N - Africa debe unirse. Buenos Aires, EUDEBA, 1965

NEARING, S & FREEMAN, J - La diplomacia del dólar. Mexico, SELFA, 1926.

NEDJATIGIL, ZM - The Crprus Conflict. Nicosia, A-Z Publ., 1981.

NYERERE, JK - El reto del Sur; informe de la Comisión Sur.

NYERERE, JK - Freedom and Socialism. London, Oxford Univ.Press, 1968.

NEWLAND, K - The Sisterhood of Man. New York, WW Norton, 1979.

OECD - Geographical Distribution of Financial Flows to Developing Countries. Paris, 1982.

OMAR TORRIJOS imágenes y voz. Panama, CET, 1985.

O MUNDO HOJE/93. Anuario Economico e Geopolitico Mundial. Ensaio, Sao Paulo, 1993.

ORTIZ MENA, A -América Latina en desarrollo. Washington, IDB, 1980.

ORTIZ QUESADA, F - Salud en la pobreza. Mexico, CEESTEM-Nueva Imagen, 1982.

OSBORNE, M - Region of Revolt. Focus of Southeast Asia. Middlesex, Penguin, 1970.

OSMANCZYK, EJ -Enciclopedia Mundial de relaciones internacionales y Naciones Unidas. Madrid, Fondo de Cultura, 1976.

PAIGC - História da Guiné e Ilhas de Cabo Verde. Porto, Afrontamento, 1974.

PAISES NO ALINEADOS, Los. Mexico, Diogenes, 1976.

PAISES NO ALINEADOS, Los. Prague, Prensa Latina-Orbis, 1979.

PAQUE, R., ed -Afrika antwortet Europa. Berlin, Ullstein, 1976.

PEASE GARCIA, H., ed -América Latina 80: Democracia y movimiento popular. Lima, DESCO, 1981.

PERROT, D.& PREISWEK, R - Etnocentrismo e Historia. Mexico, Nueva Imagen, 1979.

Le PETROLE, les matières de base et le développment. Algeria, Sonatrach, 1974.

PIACENTINI, P - O Mundo do petróleo. Lisbon, Tricontinental, 1984.

PIERRE-CHARLES, G - El caribe contemporáneo. Mexico, Siglo XXI, 1981.

PUTZGER, FW - Historicher Weltatlas. Berlin, Velhagen, 1961.

Quality of Life, from a common people's point of view, by PapyRossa Verlags GmbH & Co. KG, Köln & World Data Research Center, Ernst Fidel Fürntratt-Kloep, 1995, Hackás, Sweden.

QUIROGA SANTA CRUZ, M - El saqueo de Bolivia. Buenos Aires, Crisis, 1973.

RAGHAVAN, Ch - Un GATT sin cascabel. La Ronda Uruguay, una sigilosa reconquista del tercer Mundo. Montevideo, Red del Tercer Mundo, 1990

RAGHAVAN, Ch - Recolonization. GATT, the Uruguay round and the Third World. Penang, Third World Network, 1990

RAHMAN, MA - Grass-Roots Participation and self-reliance: experiences in South and South East. Asia, ILO, 1984.

RAMA, C - Historia de América Latina. Barcelona, Bruguera, 1978.

REFORMS in foreigns economic relations of Eastern Europe and the Soviet Union. Proceedings. New York, ECE, 1991. (Economic Studies, 2)

RETURN to the good earth. Third World Network Dossier, Penang, 1983.

REYES MATTA, F, ed - La noticia internacional. Mexico, ILET, 1977.

RIBEIRO, D - Las Américas y la civilización. Buenos Aires, Centro Editor, 1973.

RIBEIRO, S - Sobre a unidade no pensamento de Amilcar Cabral. Lisbon, Tricontinental, 1983.

RODNEY, W - How Europe Underdeveloped Africa. Washington, Howard Univ., 1974.

RODRIGUEZ, M - Haití, un pueblo rebelado. Mexico, 1982.

ROGER, M -Timor: hier la colonisation portugaise, aujourd'hui la résistance. Paris, L'Harmattan, 1977.

SAID, E - Orientalism. Middlesex, Penguin, 1985.

SALAS, RM - Estado de la población mundial. Barcelona, FNUA, 1983.

SARKAR, S, - Modern India 1885-1947, Madras, McMillan India, 1985.

SCHLESINGER, R -La Internacional Comunista y el Problema Colonial. Mexico, Pasado y Presente, 1974.

SCHMIEDER, O - Geografía de América Latina. Mexico, FCE, 1975.

SCHUON, F - Understanding Islam. London, Mandala Books, 1979.

SEAGER, J & OLSON, A -Women in the world. An International Atlas. London, Pan Books, 1986.

SECRETARIA DE AGRICULTURA (MEXICO) - El desarrollo Agroindustrial y la Economía Internacional. Mexico.

SEGAL, G - The world affairs companion. New York, Touchstone, 1991.

SELSER, G - Apuntes sobre Nicaragua. Mexico, CESTEM, 1981.

SERRYN, P - Le monde d'aujourd'hui. Paris, Bordas, 1981.

SHINNIE, M - Ancient African Kingdoms. New York, Mentor, 1980.

SHIVA, V - The violence of the green revolution, ecological degradation and political conflict in Punjab. India, 1989.

SITUACION en el mundo 1991, La. Buenos Aires, Sudamericana, 1991.

SIVARD, RL - World Military and Social Expeditures. Virginia, World Priorities, 1979.

SOCIAL WATCH - The Starting Point. Instituto del Tercer Mundo, Montevideo, 1996.

SOMALIA.MINISTRY OF INFORMATION - The Arab World. Magdishn, 1975.

SOY un soldado de América Latina. Panama, CET, 1981.

STANLEY, D - Eastern Europe on a shoestring. Australia, Lonely Planet, 1989.

STANLEY, D - South Pacific Handbook. Hong Kong, Moon Publ., 1982.

STATE OF THE WORLD 1991-1996. Worldwatch Institute, New York, Norton & Company, 1996.

STAVRIANOS, LS - Global Rift.The Third World Comes of Age. New York, William Morow, 1981.

SUAREZ, L - Los Países No Alineados. Mexico, FCE, 1975.

SUB-SHARAN AFRICA. From crisis to sustainable growth, a Long- Term perspective study. Washington, World Bank, 1989.

SUTER, K - West Irian, east Timor and Indonesia. London, 1979.

SWEEZY, P -Teoria do Desenvolvimento Capitalista. Sao Paulo, Abril, 1983.

TERRE DES FEMMES. Panorama de la situation des femmes dans le monde. Paris, La Découverte, 1982.

THE STATE of the World's Refugees 1995, United Nations High Commissioner for Refugees. UNHCR, New York, 1995.

THE UNIVERSAL ALMANAC 1994. Wright, John W. USA, 1993.

THIRD WORLD FOUNDATION -Third World Affairs 1987. London, 1987.

THOMAS, EJ., ed - Les travailleurs immigrés en Europe:quel status? Paris, UNESCO, 1981.

TIMERMAN, J - Israel: La Guerra más larga. Madrid, Mochnik, 1983.

TITO, JB - La misión histórica del Movimiento de No Alineación. Beograd, CAS, 1979.

TOMLINSON, A.& WHANNEL, G, ed - Five-ring circus: Money, power and politics at the Olympic Games. London, Pluto Press, 1984.

TORRIELLO GARRIDO, G - Tras la Cortina de Banano. Havana, Ciencias Sociales, 1979.

TORRIJOS, O - La quinta frontera. Costa Rica, Ed.Univ.Centroamericana, 1981.

TORRIJOS: figura, tiempo, faena. Panama, Lotería Nacional, 1981.

TOWARDS Socialist Planning. Tanzania, UCHEMI, 1980.

TRIBUNAL PERMANENTE DOS POVOS. Sessao sobre Timor-Leste. Lisbon, 1981.

TUNBULL, M - A Short History of Malaysia, Singapore and Brunei. Singapore, Graham Brash, 1980.

UL HAQ, M - La cortina de la pobreza. Mexico, FCE, 1978.

UN - ABC das Nacoes Unidas. New York, 1982.

UN - Concise Report on the World Population situation. New York, 1979.

UN - Demographic Yearbook. New York, 1984.

UN - Efectos de la empresas multinacionales en el desarrollo. New York, 1974.

UN - Final documents of the Eighth Conference of Heads of States or Government of Non-Aligned Countries. New York, 1986.

UN - A guide to the New Low of the Sea. New York, 1979.

UN - Informe del Comité Especial sobre descolonización. New York, 1975.

UN - Monthly Bulletin of Statistics. New York.

UN - Resolutions and Decisions. General Assembly. New York, 1995.

UN - Resolutions on the Palestine question. Beirut

UN - Statistical Yearbook, Fortieth issue. New York, 1995.

UN - Statistical Yearbook for Asia and the Pacific. New York, 1980.

UN - Terminology Bulletin. New York, 1979.

UN - A trust betrayed: Nambia. New York, 1974.

UN - UNCTAD VII. Actas. Geneva 1987.

UNCTAD - Handbook of International Trade and Development Statistics 1993. New York, 1994.

UNESCO - Geografía de América Latina. Barcelona, 1975.

UNESCO - Historia General de Africa. UNESCO-Tecnos, 1987.

UNESCO - Recherches en matière du relations raciales, Paris, 1965.

UNESCO - Résumé statistique. Paris, 1982.

UNESCO - Yearbook 1984. Paris, 1985.

UNDP - Human Development Reports 1980-1996. New York, 1996.

URSS Anuario 1991. Moscow, Novosti.

US ARMS CONTROL AND DISARMAMENT AGENCY - World Military Expediture and Arms Transfer 1972-1982. Washington, 1982.

VALDES VIVO, R - Etiopía:la Revolución desconocida. Havana, Ciencias Sociales, 1977.

VARELA BARRAZA, H - Africa: Crisis del poder político. Mexico, CEESTEM, 1981.

VARGAS, JA - Terminología sobre Derecho del Mar. Mexico, CEESTEM, 1979.

VENTURA, J - El Poder Popular en El Salvador. Mexico, SALPRESS, 1983.

VERDIEU, E & BWATSHIA, K - Les Eglises face au Nouvel Ordre Economique National e International. Quebec, CECI, 1980.

VERDIEU, E & BWATSHIA, K - Cooperation Technique des Pays en Development;est-ce Possible? Quebec, CECI, 1980.

VERDIEU, E & BWATSHIA, K - Liberation et Autonomie Collectives. Queébec, CECI, 1980.

VERLAG DIE WIRTSHAFT - Länder der Erde. Koln, Pahl- Rugenstein, 1982.

VIEL, B - La explosión demográfica. Mexico, PAX, 1974.

VILLALOBOS, J - Por qué lucha el FMLN. Radio Venceremos, 1983.

VILLEGAS QUIROGA, C & AGUIRRE BADANI, A - Estudio de la crisis y la nueva política económica en Bolivia. Bolivia, Centro de Estudios para el Desarrollo Laboral y Agrario.

VISION de Belice. Havana, Casa de las América, 1982.

VIVO ESCOTO, JA -Geografía Humana y Económica. Mexico, Patria, 1975.

WEISSMAN, S.et al- The Trojan Horse. A radical Look at Foreign Aid. San Francisco, Ramparts, 1974.

WETTSTEIN, G -Subdesarrollo y geografía. Mérida, Universidad de los Andes, 1978.

WHANNEL, G - Blowing the whistle: The politics of sport. London, Pluto, 1983.

WIENER, D - Shalom, Israels Friedensbewegung. Hamburg, Rowohlt, 1984.

WILLIAMS, N - Chronology of the modern World 1763-1965. Middlesex, Penguin, 1975.

WOLFE, A.et al - La cuestión de la Democracia. Mexico, UILA, 1980.

WORLD DEBT TABLES 1996. External debt of developing countries. Washington, 1996.

WORLD BANK - Atlas 1996. New York, 1996.

WORLD BANK - Energy in the Developing Countries. Washington, 1980.

WORLD BANK - World Development Report 1996. Washington, 1996.

WORLD FACTS and MAPS. Rand Mc Nally. USA, 1994.

WORLD Military Expenditure and Arms Transfer 1972-1982. Washington.1984.

WORLD RESOURCES INSTITUTE - The Environmental Almanac 1993. Houghton Mifflin Company, Boston & New York, 1992.

WORLD RESOURCES INSTITUTE - World Resources 1990-91. A guide to the global environment. New York, Oxford Univ., 1991.

WORLD RESOURCES INSTITUTE - World Resources 1986. New York, Basic Books, 1986.

WORSLEY, P - El tercer Mundo: una nueva fuerza en los asuntos internacionales. Mexico, Siglo XXI, 1978.

ZERAOUI, Z - El mundo árabe: imperialismo y nacionalismo. Mexico, CEESTEM-Nueva Imagen, 1981.

ZERAOUI, Z - Irán-Iraq: guerra política y sociedad. Mexico, Nueva Imagen, 1982.

ZIEGLER, J - Main Basse sur l'Afrique. Paris, Du Seuil, 1978.

PERIODICALS

Africa News, Durham, NC, USA
Afrique Mass-Media, Budapest, Hungary
Afrique Nouvelle, Dakar, Senegal
Afrique-Asie, Paris, France
ALAI, Montreal, Canada
ALDHU, Quito, Ecuador
Altercom, Mexico,
AMPO Japan-Asia Quarterly Review, Tokyo,
Análisis, Santiago de Chile,
APSI, Santiago de Chile, Chile
AQUI, La Paz, Bolivia
Barricada Internacional, Managua, Nicaragua
Bohemia, La Habana, Cuba
Boletín de Namibia, UNITED NATIONS, New York
Bulletin of Concerned African Scholars, Charlemont, MA, USA
Caribbean Monthly, Río Piedras, Puerto Rico
CEAL, Brussels, Belgium
Central America Update, Toronto, Canada
CERES, Rome, Italy
CILA, Mexico
Comercio Exterior, Mexico,
Contextos, Mexico,
Counterspy, Washington, DC, USA
CovertAction, Washington, USA
CRIE, Mexico,
Cuadernos del Tercer Mundo, Brasil,
Descolonización, UNITED NATIONS, New York
Diálogo Social, Panama,
Documentos FIPAD, Ginebra, Switzerland
Economie et Politique, Paris, France
El Caribe Contemporáneo, Mexico,
Eritrea in Struggle, New York, USA
Facts & Reports, Amsterdam, Netherlands
Far Eastern Economic Review, Asia 1990, Yearbook. Japan
Fortune International, Los Angeles, CA, USA
Freedomways, New York, USA
Gombay, Belize
IFDA, Nyon, Switzerland
Informe "R"- CEDOIN , Bolivia.
Indian and Foreign Review, New Delhi, India
Internews, Berkeley, USA
Isis, Rome, Italy
Ko-Eyú, Caracas, Venezuela
L'Economiste du Tiers Monde, Paris, France
Lateinamerika Nachrichten, Berlin
Latin America Weekly Report, London, United Kingdom
Le Monde Diplomatique, Paris, France
Monthly Review, New York, USA
Mujer-Fempress, Santiago, Chile.
Multinational Monitor, Washington, USA
NACLA Report on the Americas, New York, USA
New Outlook, Dar-es-Salaam, Tanzania
Newsfront International, Oakland, USA
Noticias Aliadas, Lima, Peru
Novembro, Luanda, Angola
O Correio da UNESCO, Paris-Rio de Janeiro,
Onze Wereld, Den Haag, Netherlands
Palestine, Beirut, Lebanon
Pensamiento Propio, Managua, Nicaragua
Philippine Liberation Courier, Oakland, USA
Politica Internazionale, Rome, Italy
Política Internacional, Belgrade, Yugoslavia
Resister (Bulletin of the Committee on South African War Resistance), London, United Kingdom
Revista del Centro de Estudios del Tercer Mundo, Mexico.
Revista del Sur, ITeM, Montevideo, Uruguay.
Sharing, Geneva, Switzerland
SIAL, Italy
Soberanía, Managua, Nicaragua
Southern Africa, New York, USA
Statesman's Yearbook, 1988-1989.
Tempo, Maputo, Mozambique
Tercer Mundo Económico, ITeM, Montevideo, Uruguay
The Black Scholar, Sausalito, CA, USA
The Ecologist, United Kingdom.
The CTC Reporter, UNITED NATIONS, New York.
Third World Quarterly, London, United Kingdom.
Third World Resurgence, Penang, Malaysia.
Tiempos Nuevos, Moscow
Tigris, Madrid
Tribune, Sri Lanka,
Tricontinental, Havana
Two Thirds, A journal of underdevelopment studies, Toronto, Canada
WISE (World Information Service on Energy), Netherlands
YEKATIT Quarterly, Adis Ababa, Ethiopia

NEWS AGENCIES

Interpress-IPS, INA, WAFA, SALPRESS, ANN, Prensa Latina, Angop, AIM, Shihata.
ALAI (Agencia Latinoamericana de Información).
SEM (Servicio Especial de la Mujer).

ELECTRONIC SOURCES

Roberto Ortiz de Zarate's Datasets: http://lgdx01.lg.ehu.es/~ziaorarr/
Rulers: http://www.geocities.com/Athens/1058/rulers.html
Le Monde Diplomatique online: http://www.ina.fr/CP/MondeDiplo/Thesaurus/thesaurus.fr.html
Folha de São Paulo: http://www.uol.com.br/fsp/
OneWorld News Service: http://www.oneworld.org/news/index.html
Fourth World Documentation Project: http://www.halcyon.com/FWDP/fwdp.html
Banco Interamericano de Desarrollo: http://www.iadb.org/
OMRI Daily Digest: http://www.omri.cz/Publications/Digests/Digest/Index.html
Il Manifesto in Rete: http://www.mir.it/mani/index.html
Electric Library: http://www.elibrary.com/id/2525/
United Nations: http://www.un.org/
The Washington Report's Resources Page: http://www.washington-report.org/links.html
Foundation for Middle East Peace: http://www2.ari.net/fmep/

Alphabetical Index

Sufanuvong, Tiao 349
sugar 92, 96, 108, 111, 126, 132, 133, 134,
140, 148, 149, 186, 192, 202, 209, 211,
215, 222, 224, 225, 232, 233, 234, 237,
249, 250, 275, 276, 277, 280, 287, 289,
290, 303, 307, 322, 323, 360, 368, 369,
382, 383, 386, 387, 388, 395, 399, 420,
448, 451, 455, 457, 464, 465, 468, 473,
482, 483, 487, 492, 504, 509, 523, 524,
525, 527, 538, 546, 547, 574, 585, 586,
593, 595
Suh Eui Hyun 342
Suharto 227, 304, 305
Suhrawardi, Hussain Shadeed 435
Sukarno, Ahmed 153, 269, 301, 303, 304,
305
Sukarnoputri, Megawati 305
Sukhumi 264
Suleyman the Magnificent 550
Sumatra 303, 304, 305
Sumgait 117, 118
Sun Yat Sen 183, 372
Sunia, Fofo 486
Sunni 84, 122, 234, 308, 310, 330, 344, 348,
353, 354, 355, 362, 436, 490, 506, 519,
533, 536, 549, 551, 553, 562, 563, 576,
588
Surabaya 304
Suriname 77, 79, 416, 418, 522, 523
Sushkevich, Stanislav 129
Sussu 284, 497, 498
Suu Kyi, Aung San 408
Suur Muna Magi 243
Suva 249, 250
Suvadiva, Republic of 375
Suvana Fuma 349
Suyab 333
Svacic, Petar 206
Svoboda, Jiri 215
Svoboda, Ludwik 215
Swahili 161, 196, 197, 336, 405, 537, 538,
539, 558, 595
Swains Island 486
Swan, John 136
Swane, Ambrose 524
SWAPO 92, 409, 410, 411
Swaziland 79, 80, 356, 405, 524, 525
Sweden 35, 49, 66, 78, 79, 83, 115, 217,
218, 242, 251, 297, 302, 307, 351, 363,
430, 431, 432, 458, 473, 526, 527, 528,
560
Switzerland 35, 49, 66, 73, 75, 76, 78, 79,
80, 81, 82, 137, 201, 247, 253, 307, 308,
529, 530, 531, 573, 592
Sykes-Picot agreement 532
Symonette, Roland 119
Syr Darya 347, 536, 575
Syria 77, 234, 235, 255, 306, 309, 310, 311,
315, 316, 317, 318, 329, 353, 354, 355,
361, 434, 439, 440, 467, 490, 532, 533,
534, 539, 551, 552, 587, 588
Syrian 108, 234, 306, 315, 316, 353, 354,
355, 440, 490, 497, 532, 533, 534, 552,
563

T

Tabatabai, Sayyid 306
Tadjoura 220, 221
Tajikistan 77, 85, 536
Taegu 340
Taha, Mahomoud 520
Tahirid, kingdom 535, 553
Tahiris 309
Tahiti 258, 457
Tahuantinsuyu 139, 229
Taima 587
Taino 209, 224, 464
Taipei 192, 438
Taiping 182
Taira 324
Tais, Tupuola 485
Taiwan 77, 182, 183, 184, 186, 188, 190,
191, 192, 271, 326, 339, 381, 438, 516
Tajidine Ben Said Massomde 196
Takashi, Hara 325
Takemura, Masayoshi 328
Takeshita 327, 328
Taki, Mohammed 197
Talal, Hussein Ibm 330
Talas 348
Taliban 85
Tallinn 242, 243

Tamara, Queen 263
Tamasese 485
Tambo, Oliver 508, 511
Tamils 302, 372, 373, 387, 408, 500, 516,
517, 518
Tampere 82, 252
Tamuz 310
Tan-Zam Railway 597
Tanaka, Kakuei 326
Tang 192
Tangiers 400
Tanna Island 578
Tanumafili, Malietoa 485
Tanzania 34, 45, 46, 54, 75, 79, 80, 160,
184, 196, 244, 302, 335, 336, 374, 404,
405, 433, 476, 477, 490, 509, 537, 538,
539, 557, 558, 559, 587, 600
Taraba 427
Tarabi 535
Tarahumara 390
Tarbagatay 334
Tarmiuah 311
Tarquinius Priscus 319
Tartu, Treaty of 243
Tashfin, Ysuf Ibn 402
Tashkent 347, 575, 576
Tasman, Abel 98
Tasmania Island 111, 112
Tassili 424
Tatar 183, 394, 469, 472, 473, 536
Tau Island 486
Taveuni Island 250
Taya, Ahmed Sidi Ould 383, 385
Taylor, Charles 359, 360, 498, 555
Taylor, Henry 110
Tbilisi 263, 264
Tchombé, Moses 594
tea 124, 182, 186, 187, 304, 336, 370, 383,
448, 450, 473, 495, 516, 517, 518, 558
Teannaki, Teatao 338
Tedzhen 553
Teheran 307, 308, 554
Tekrour 402
Tel Aviv 245, 316, 318, 439, 517, 524
Television Marti 210
Temaru, Oscar 258
Temasek 499
Temne 497, 498
Temuco 179
Temujin 396
Tenochtitlan 389
Tenorio, Pedro 429
Teotihuacan 389
Ter-Petrosian, Levon 108
Terai 414
Terreblanche, Eugene 512
Tete 404
Tetovo 367
Teutonic Order 351, 363, 458
Texaco 229, 491
Texas 65, 392, 568, 569, 570
Thailand 76, 77, 113, 154, 162, 326, 342,
349, 350, 407, 417, 539, 540, 541, 542,
584
Thakombau 249
Thatcher, Margaret 180, 567
Thessaloniki 273
Thessaly 272, 273, 366
Thiam, Habib 494
Thibaw 407
Thimphu 137
Third World Network 75, 79
Thracia 155, 156
Thuwaini 433
Tiahuanaco 139
Tiananmen 185
Tiber 248, 319
Tibesti 175, 176
Tibet 137, 183, 185, 186, 396, 413
Tibu 176, 424, 425
Tiflis 263
Tigranes 107
Tigre 241, 245
Tigrinya 241
Tigris 74, 309, 310, 311
Tihmah 588
Tilemsi 402
Timakata, Fred 578
timber 92, 98, 132, 239, 259, 260, 271, 305,
326, 350, 374, 426, 430, 448, 450, 470,
501, 504, 514, 523, 540, 559
Timbuktu 283, 378, 400
Timisoara 470, 471
Timna 587

tin 139, 140, 266, 303, 304, 372, 397, 425,
426, 473, 594, 595, 599
Tinduf 479
Tinian 429
Tinsulanonda, Prem 541, 542
Tiradentes 148
Tirana 86, 87, 591
Tiraspol 395
Tisza 295
Titicaca 139, 140
Tito (Josip Broz) 338
Tiv 427
Tjibaou, Jean Marie 331
Tlatelolco Massacre 391
Tlatelolco Treaty 484
Toba 103
Tobago Island 351
Togo 38, 54, 75, 77, 78, 79, 266, 270, 359,
426, 543, 544
Tojar 535
Tokara 325
Tokugawas 325
Tokyo 325, 326, 327, 397, 438, 545
Tolbert, William 359
Toloa, Letuli 486
Tomb Islands 318
Tombalbaye, Ngarta 175
Tonelli, Salvatore 248
Tonga 100, 249, 250, 371, 525, 545, 556,
597, 598
Tongatapu Islands 545
Tonkin 582, 583
Tonton-Macoutes 275, 290, 291
Tordesillas, Treaty 147, 148
Toro 557
Toronto 167
Torres, Camilo 193
Torres, Juan José 140
Torres, Vaez de 110
Torrijos, Omar 443, 444, 445
Torrijos-Carter treaty 440
torture 45, 89, 104, 143, 208, 221, 281, 336,
346, 357, 371, 408, 456, 491, 518, 549,
552, 573
Tosco, Agustin 104
Totonaca 390
Toungoo 407
Tournai 130
Toxandria 415
toxic waste 240, 243, 381, 422
Transcaucasia 107, 117, 118, 263, 264
Transjordan 315, 440
Transkei 356, 509
transnationals 64, 65, 76, 119, 126, 148,
176, 222, 224, 259, 279, 322, 323, 335,
361, 369, 409, 417, 447, 499, 522, 527,
558, 594
Transvaal 356, 507, 508, 509
Transylvania 114, 294, 295, 469, 470, 471,
474
Traore, Moussa 379
Trarza 383
Trejos, José Joaquin 201
Trelleborg 527
Trianon Treaty 294, 295
Tribhuvan Bir Bikram Shah Deva 413
Trincomalee 516
Trinidad and Tobago 77, 546, 547
Triple Alliance 102, 148, 266, 319, 320,
448, 450, 572
Tripoli 175, 354, 361, 362
Tripolitza 273
Trobriand Island 447
Tromelin Island 388
Troodos 213
Trotsky, Leon 474
Trovoada, Muguel 487, 488
Troy 272
Trucial states 562
Trudeau, Pierre 166
Trujillo, Rafael Leonidas 224
Truman, Harry 315, 339, 570
Tsachila 230, 231
Tsankov, Aleksandur 156
Tshisekedi, Etienne 596
Tsiranana, Philbert 368
Tsonga 405
Tsugaru 324
Tsumeb 409
Tsushima 324
Tswana 145, 146, 510
Tu 186, 545
Tuanku Ja'afar ibni al-Marhum Tuanku
Abdul Rahman 373, 374

Tuareg 176, 283, 378, 379, 423, 424, 425,
4797, 480
Tuat 402
Tubu 176, 425
Tudjman, Franjo 143, 144, 207, 208
Tuimalealiifano 485
Tujia 186
Tumart, Muhammad Ibn 402
Tumbuka 371
Tumin 347
Tungu 340
Tunis 176, 235, 380, 441, 548, 549
Tunisia 77, 88, 255, 282, 309, 361, 380,
400, 402, 548, 549
Tupac Amaru 451
Tupac Katari 139, 140, 141
Tupamaros 573
Tupé Guarani 147
Tupou IV, Taufa'ahau 545
Turajlic, Hakija 143
Turbay Ayala, Julio Cesar 193
Turkey 55, 78, 79, 81, 85, 107, 108, 117,
118, 142, 155, 156, 157, 212, 213, 221,
263, 264, 266, 267, 273, 274, 309, 310,
311, 348, 361, 363, 366, 367, 394, 460,
470, 532, 550, 551, 552, 554, 566, 570,
576, 592
Turkmenistan 347, 396, 535, 536, 553, 575
Turku 252
Turov 128
Tushpa 107
Tutsi 160, 161, 476, 477, 478
Tutu, Desmond 428, 510
Tutuila Island 486
Tuvalu 72, 77, 101, 255, 338, 412, 556
Tuzla 143
Tver 472
Tvrtko 142
Twa 161, 476, 477, 478
Twagiramungu, Faustin 478
Tyler, Wat 564
Tyre 353, 354
Tyrins 272
Tyrol 114, 115

U

UN 40, 42, 47, 58, 59, 62, 65, 66, 67, 75,
76, 78, 80, 81, 82, 84, 85, 97, 101, 104,
108, 114, 118, 124, 129, 137, 143, 144,
150, 153, 154, 157, 162, 163, 167, 180,
191, 196, 197, 202, 203, 204, 207, 208,
211, 213, 220, 228, 234, 237, 238, 240,
241, 246, 278, 280, 281, 285, 291, 310,
311, 315, 316, 317, 318, 326, 328, 330,
334, 338, 339, 346, 352, 354, 355, 356,
360, 361, 362, 364, 367, 376, 380, 381,
388, 393, 395, 397, 403, 406, 408, 409,
410, 421, 427, 429, 438, 440, 442, 445,
453, 456, 462, 464, 468, 475, 478, 479,
480, 485, 486, 490, 506, 509, 512, 515,
518, 520, 523, 527, 531, 534, 545, 554,
560, 576, 585, 593, 601
Uali, Mustafa Seyid al 479, 480
Ubico, Jorge 279
Uganda 38, 45, 46, 54, 79, 160, 335, 336,
337, 476, 477, 478, 519, 520, 521, 537,
538, 557, 558, 559
Uighur 186, 187, 333
UK 66, 70, 180, 188, 271, 312, 413, 457,
566, 567
Ukraine 66, 72, 77, 128, 129, 206, 364, 395,
432, 458, 459, 472, 473, 474, 475, 560,
561
Ulan Bator 397, 398
Ulate, Otilio 201
Ulivau, Na 249
Ulmanis, Karlis 351, 352
Ulna 578
Ummaia 532
UNDP 49, 76, 97, 165, 411, 583
UNESCO 31, 32, 76, 445
Ungo, Guillermo 237
UNHCR 45, 46, 76
UNICEF 35, 49, 50, 54, 76, 340, 411, 538
UNIFIL 355
UNITA 91, 92, 93, 184, 405, 410, 595
United Arab Emirates 467
United Brands 279
United Provinces of Central America 201
United States 568, 569, 570, 571
UNPROFOR 76, 143
Upolu Island 485

INSTITUTO DEL TERCER MUNDO

The Whole World
in Figures

Countries	Area (thousands square km)	Population (thousands) 1994	Demographic growth (%) 2000	Population estimate (millions) for year 2000	Children per woman 1992	Under-five mortality rate 1994	Calorie consump. % of required intake 1995	Primary school teachers/ student 1992
Afghanistan	652,090	22,789	1.6		6.9	257	89	
Albania	28,750	3,202	1.8	4	2.9	41	102	19
Algeria	2,381,740	27,422	1.4	31	4.3	65	104	27
Andorra	0,450	65	4.6		..			
Angola	1,246,700	10,442	2.3		6.6	292	85	32
Anguilla	0,960	8						
Antigua	0,440	67	0.3		1.7	23		
Aotearoa/NZ	270,990	3,493	0.8	4	2.1	9	113	16
Argentina	2,791,810	34,194	0.3	36	2.8	27	109	16
Armenia	29,800	3,748	1.4	4	2.8	32		
Australia	7,713,360	17,843	1.5	19	1.9	8	111	17
Austria	83,850	8,028	0.7	8	1.6	7	114	11
Azerbaijan	86,600	7,459	1.4	8	2.7	51		
Bahamas	13,880	272	0.6		2.1	28	103	21
Bahrain	0,680	557	0.5		3.7	20		18
Bangladesh	144,000	117,941	3.0	132	4.0	117	87	63
Barbados	0,430	260	1.2		1.8	10	110	17
Belarus	207,600	10,356	0.4	10	1.9	21		
Belgium	30,520	10,116	0.3	10	1.6	10	120	10
Belize	22,960	211	-0.1		4.5	41		25
Benin	112,620	5,325	1.5	6	6.2	142	93	40
Bermuda	0,050	63	1.3		..			17
Bhutan	47,000	675	3.8	2	5.9	193	98	31
Bolivia	1,098,580	7,237	1.5	9	4.7	110	88	25
Bosnia and Herzegovina	51,130	4,383	0.1		1.6	17		
Botswana	581,730	1,443	3.5	2	4.7	54	93	29
Brazil	8,511,970	159,128	0.8	172	2.9	61	102	23
Brunei	5,770	280	0.3		3.1	10	104	15
Bulgaria	110,910	8,435	-0.8	8	1.5	19	117	14
Burkina Faso	274,000	10,118	6.9	12	6.9	169	91	60
Burundi	27,830	6,183	3.5	7	6.8	176	88	63
Cambodia	181,040	9,951	3.1		4.5	177	90	
Cameroon	475,440	12,986	1.9	16	5.8	109	91	51
Canada	9,976,140	29,248	1.3	30	1.9	8	113	17
Cape Verde	4,030	372	3.1		4.3	73	102	33
Central African Republic	622,980	3,234	1.0	4	5.8	175	86	90
Chad	1,284,000	6,288	1.1	7	5.9	202	81	64
Chile	756,950	13,994	0.2	15	2.7	15	99	25
China	9,561,000	1,190,918	2.7	1,255	2.0	43	102	22
China-Hong Kong	1,040	6,061	0.2	6	1.4	6	105	27
Colombia	1,138,910	36,330	0.7	37	2.7	19	99	28
Comoros	2,230	485	1.9		6.7	126	83	36
Comoros-Mayotte	0,370	89	..		..			
Congo	342,000	2,577	1.6	3	6.6	109	98	66
Costa Rica	51,100	3,304	1.1	4	3.1	16	103	32
Côte d'Ivoire	322,460	13,841	1.4	17	6.6	150	99	37
Croatia	56,540	4,778	0.4		1.7	14		
Cuba	110,860	10,978	0.6	11	114	10		12
Cyprus	9,250	726	1.0		2.4	10	107	18
Czech Republic	79,000	10,333	0.0	11	1.9	10	115	18
Denmark	43,090	5,205	0.2	5	1.8	7	116	11
Djibouti	23,200	603	0.5		5.8	158		43
Dominica	0,750	72			2.5	21		29
Dominican Republic	48,730	7,622	1.2	8	3.0	45	95	47
East Timor	14,870	821						

Countries	Newspapers	TV sets	Radio sets	Books/ new titles per million inhab.	GNP per capita (US dollars)	Average annual growth (%)	Average annual inflation (%)	Exports (million $)	Imports (million $)	Energy consump. per capita (kg)	Energy imports as % of consump.
	(per 1,000 households)										
	1995	1995	1995	1995	1993	1985-94	1984-94	1994	1994	1994	1994
Afghanistan	86	84	89	101	280	..	..	..	..		..
Albania	97	99	92	107	340	..	33	116	596	422	28
Algeria	101	100	102	87	1,780	-3	22	8,594	8,000	1,030	-273
Andorra											
Angola	85	83	81	81	700	-7					
Anguilla											
Antigua						3					
Aotearoa	106	443	108	107	12,600	1	5	12,200	11,900	4,352	5
Argentina	101	103	108	101	7,220	2	.317	15,839	21,527	1,399	-21
Armenia					660	-13	139	209	401	667	87
Australia	111	107	114	104	17,500	1	4	47,538	53,400	5,173	-91
Austria	106	105	102	108	23,510	2	3	45,200	55,300	3,276	65
Azerbaijan					730	-12	123	682	791	1,414	-41
Bahamas	102	103	104	..		-1					
Bahrain	103	114	112	100		-1					
Bangladesh	84	83	80	83	220	2	7	2,661	4,701	65	31
Barbados	104	103	111	102		-0					
Belarus					2,870	-2	137	3,134	3,857	2,692	89
Belgium	103	105	104	107	21,650	2	3	137,394	125,762	5,091	77
Belize						5					
Benin	82	83	85	81	430	-1	3	..	..	18	-239
Bermuda						-1					
Bhutan	83	82	77	—	170	4					
Bolivia	95	100	107	95	760	2	20	1,032	1,209	307	-90
Bosnia and Herzegovina											
Botswana	92	85	90	101	2,790	7	12	1,845	1,638	380	55
Brazil	99	103	102	101	2,930	-0	900	43,600	36,000	691	38
Brunei	104	106	101	98		..					
Bulgaria	111	102	101	103	1,140	-3	42	4,165	4,160	2,786	63
Burkina Faso	81	83	78	78	300	-0	2	..	..	16	100
Burundi	82	82	81	—	180	-1	5	106	224	23	97
Cambodia	82	84	88	—	200	..					
Cameroon	86	87	90	81	820	-7	1	..	1,100	83	-525
Canada	108	109	110	103	19,970	0	3	166,000	155,072	7,795	-46
Cape Verde	84	86	92	87		2					
Central African Republic	81	82	81	—	400	-3	3	..	..	29	76
Chad	81	82	94	—	210	1	2	..	..	16	100
Chile	103	103	102	101	3,170	7	18	11,539	11,800	943	66
China	96	89	92	96	490	8	8	121,047	115,681	647	-1
China-Hong Kong	117	103	105	108	18,060	5	9	151,395	162,000	2,280	100
Colombia	100	101	94	93	1,400	2	26	8,399	11,883	613	-103
Comoros	81	82	90	—		-1					
Comoros-Mayotte											
Congo	87	83	86	82	950	-3	-0	..	..	147	-2,492
Costa Rica	102	102	101	98	2,150	3	18	2,215	3,025	558	41
Côte d'Ivoire	85	93	88	82	630	-5	0	..	2,000	170	82
Croatia			—			..	..	4,259	5,231	1,057	28
Cuba	102	103	101	101	1,170	..					
Cyprus	102	101	97	107		5					
Czech Republic	106	105	103	104	2,710	-2	12	14,252	15,636	3,902	13
Denmark	104	103	103	118	26,730	1	3	41,417	34,800	3,996	27
Djibouti	87	95	85	..		..					
Dominica						4					
Dominican Republic	97	100	93	101	1,230	2	29	633	2,630	340	89
East Timor											

Countries	Area (thousands square km) 1994	Population (thousands) 1994	Demographic growth (%) 2000	Population estimate (millions) for year 2000	Children per woman 1992	Under-five mortality rate 1994	Calorie consump. % of required intake 1995	Primary school teachers/ student 1992
Ecuador	283,560	11,227	1.2	13	3.5	57	94	29
Egypt	1,001,450	56,767	0.6	63	3.8	52	112	26
El Salvador	21,040	5,635	0.6	6	3.8	56	93	44
Equatorial Guinea	28,050	386	2.8		5.5	177		
Eritrea	125,000	3,482	..		..	200		
Estonia	45,100	1,499	0.1	2	1.8	23		25
Ethiopia	1,097,000	54,890	1.9	67	7.5	200	81	27
Fiji	18,270	767	0.8		3.0	27	103	31
Finland	338,130	5,089	0.4	5	1.9	5	108	14
France	551,500	57,928	0.6	59	1.8	9	114	12
French Guiana	90,000	141	..			..		
French Polynesia	4,000	219	2.7		3.2			14
Gabon	267,670	1,301	1.5	2	5.9	151	98	44
Gambia	11,300	1,079	2.4	1	6.5	213		30
Georgia	69,700	5,418	0.4	5	2.2	27		
Germany	356,910	81,516	0.6	81	1.3	7		
Ghana	238,540	16,639	1.4	20	6.1	131	88	29
Greece	131,990	10,426	0.5	11	1.4	10	118	19
Grenada	0,340	92			2.9	34		27
Guadeloupe	1,710	421	1.7		2.2			19
Guam	0,550	146	2.0		2.9			20
Guatemala	108,890	10,322	1.1	12	5.1	70	93	34
Guinea	245,860	6,425	2.5	8	6.5	223	88	49
Guinea-Bissau	36,120	1,044	2.4	1	6.0	231	95	25
Guyana	214,970	826	1.6	2.6	99	61		
Haiti	27,750	7,008	1.9	4.7	94	127		29
Honduras	112,090	5,750	1.5	7	4.9	54	91	38
Hungary	93,030	10,261	-0.5	10	1.8	14	115	12
Iceland	103,000	266	1.2		2.2	5	115	
India	3,287,590	913,600	1.1	1,016	3.7	119	92	63
Indonesia	1,904,570	190,389	2.7	206	2.9	111	101	23
Iran	1,648,000	62,550	1.0	75	5.5	51	110	32
Iraq	438,320	20,356	0.7		5.7	71	106	22
Ireland	70,280	3,571	0.0	4	2.0	7	119	25
Israel	21,060	5,383	2.9	6	2.7	9	109	16
Italy	301,270	57,120	0.2	58	1.3	8	115	12
Jamaica	10,990	2,497	0.9	3	2.7	13	100	38
Japan	377,800	124,961	0.4	127	1.5	6	105	20
Jordan	89,210	4,035	0.7	5	5.2	25	104	22
Kanaky/New Caledonia	19,080	178				48		20
Kazakhstan	2,717,300	16,811	1.0	18	2.7	90		
Kenya	580,370	26,017	2.8	31	5.1	97	89	31
Kiribati	0,730	78	2.0	3.8	276	78		
Korea Dem. Rep.	120,540	23,448	0.5	26	2.4	31	106	33
Korea Rep.	99,020	44,453	1.4	47	1.8	9	104	33
Kuwait	17,820	1,620	0.4		3.7	14	108	16
Kyrgyzstan	198,500	4,473	1.6	5	3.7	56		
Laos	236,800	4,748	2.9	6	6.7	138	97	29
Latvia	64,500	2,547	-0.1	3	1.8	26		
Lebanon	10,400	3,930	0.4		3.1	40	109	
Lesotho	30,350	1,942	3.2	2	4.8	156	92	51
Liberia	97,750	2,719	1.3		6.2	217	94	
Libya	1,759,540	5,218	0.6	6	6.4	95	114	12
Liechtenstein	0,160	31						
Lithuania	65,200	3,721	0.7	4	2.0			

622 THE WHOLE WORLD IN FIGURES

Countries	Newspapers (per 1,000 households) 1995	TV sets 1995	Radio sets 1995	Books/ new titles per million inhab. 1995	GNP per capita (US dollars) 1993	Average annual growth (%) 1985-94	Average annual inflation (%) 1984-94	Exports (million $) 1994	Imports (million $) 1994	Energy consump. per capita (kg) 1994	Energy imports as % of consump. 1994
Ecuador	102	100	102	96	1,200	1	48	3,820	3,690	517	223
Egypt	100	100	102	88	660	1	16	3,463	10,185	608	-67
El Salvador	101	100	105	81	1,320	2	15	844	2,250	219	58
Equatorial Guinea	82	83	102	92		2					
Eritrea					100	..					
Estonia					3,080	-6	77	1,329	1,690	3,552	42
Ethiopia	81	82	93	82	100	..	..	372	1,033	21	86
Fiji	103	87	112	91		2					
Finland	110	105	107	117	18,850	-0	4	29,700	23,200	5,954	62
France	102	104	106	105	22,490	2	3	235,905	230,203	3,839	47
French Guyana						..					
French Polynesia						..					
Gabon	86	88	87	—	4,960	-4	3	..	..	520	-2,268
Gambia	83	81	92	93	350	0	10	35	209	56	100
Georgia					580	..	228	381	744	572	81
Germany					23,560	..	..	427,219	381,890	4,097	58
Ghana	96	85	101	88	430	1	29	..	..	91	64
Greece	101	101	101	103	7,390	1	15	9,384	21,466	2,235	63
Grenada						..					
Guadeloupe						..					
Guam						..					
Guatemala	89	93	83	92	1,100	1	19	1,522	2,604	186	70
Guinea	81	83	79	—	500	1	19	..	..	65	87
Guinea-Bissau	83	81	79	—	240	2	66	32	63	37	100
Guyana	102	89	105	91		0					
Haiti	83	82	80	91	370	-5	13	73	292	47	70
Honduras	99	99	106	106	600	0	13	843	1,056	169	71
Hungary	107	105	103	103	3,350	-1	19	10,733	14,438	2,455	44
Iceland	111	103	105	119		0					
India	94	89	85	84	300	3	10	25,000	26,846	243	20
Indonesia	91	93	90	83	740	6	9	40,054	31,985	393	-101
Iran	90	96	100	100	2,200	..	23	13,900	20,000	1,565	-127
Iraq	99	100	101	82	1,036	..					
Ireland	108	104	106	105	13,000	5	2	34,370	25,508	3,136	70
Israel	106	103	103	104	13,920	2	18	16,881	25,237	2,815	96
Italy	101	106	107	103	19,840	2	6	189,805	167,685	2,710	81
Jamaica	100	101	103	88	1,440	4	28	1,192	2,164	1,112	100
Japan	119	119	119	102	31,490	3	1	397,000	275,000	3,825	82
Jordan	101	101	102	—	1,190	-6	9	1,424	3382	997	97
Kanaky/New Caledonia											
Kazakhstan					1,560	-6	150	3,285	4,205	3,710	-16
Kenya	88	84	90	87	270	0	12	1,609	2,156	107	82
Kiribati					850	..					
Korea Dem. Rep.	106	85	88	—		..					
Korea Rep.	107	102	114	107	970	8	7	96,000	102,348	3,000	85
Kuwait	112	110	107	103	19,360	1	..	11,614	21,716	7,615	-711
Kyrgyzstan					850	-5	101	340	459	715	76
Laos	83	83	89	88	280	..	24	300	564	38	-19
Latvia					2,010	-6	70	967	1,367	1,755	88
Lebanon	103	108	116	—	2,150	..					
Lesotho	89	83	83	—	650	1	14	..	..	..	..
Liberia	86	85	98	—	450	..					
Libya	86	101	100	90	5,310	..					
Liechtenstein						..					
Lithuania					1,320	-8	102	1,892	2,210	2,194	80

Countries	Area (thousands square km) 1994	Population (thousands) 1994	Demographic growth (%) 2000	Population estimate (millions) for year 2000	Children per woman 1992	Under-five mortality rate 1994	Calorie consump. % of required intake 1995	Primary school teachers/ student 1992
Luxembourg	3,000	404	1.0		1.7	9	120	13
Macau	0,020	444	3.2		2.1			
Macedonia	25,710	2,100	1.0		2.2	32		20
Madagascar	587,040	13,100	2.5	16	6.1	164	93	38
Malawi	118,480	9,532	2.9	11	6.7	221	90	68
Malaysia	329,750	19,669	1.4	22	3.5	15	102	20
Maldives	0,300	246	0.9		6.0	78	93	
Mali	1,240,190	9,524	2.4	12	7.1	214	91	47
Malta	0,320	368	0.7		2.1	12	109	21
Marshall Is.	0,200	54	3.9		..	92		
Martinique	1,100	383	1.0		2.0			
Mauritania	1,025,520	2,215	2.1	3	6.8	199	96	51
Mauritius	2,040	1,115	0.3	1	2.0	23	104	21
Mexico	1,958,200	88,543	0.7	99	3.2	32	108	30
Micronesia	0,700	104	2.4	4.8		29		
Moldova	33,700	4,350	0.5	4	2.3	36		
Monaco	0,001	33			..			20
Mongolia	1,566,500	2,363	1.0	3	4.6	76	98	28
Montserrat	0,100	11				14		
Morocco	446,550	26,367	1.0	30	3.8	56	105	28
Mozambique	801,590	15,463	3.9	20	6.5	277	80	53
Myanmar/Burma	676,580	45,581	1.4	52	4.2	109	98	35
Namibia	824,290	1,508	2.8	2	5.4	78	85	32
Nauru	0,021	10			..			20
Nepal	140,800	20,885	4.1	24	5.5	118	89	39
Netherlands	37,330	15,381	0.7	16	1.6	8	110	17
Netherlands Antilles	0,800	198	0.9		2.1			
Nicaragua	130,000	4,156	0.9	5	4.4	68	95	37
Niger	1,267,000	8,730	2.3	11	7.4	320	94	38
Nigeria	923,770	108,014	2.0	128	5.9	191	91	39
Northern Marianas	0,480	47	..		..			
Norway	323,900	4,337	0.5	4	1.9	8	111	6
Oman	212,460	2,098	3.4	2	7.2	27		27
Pakistan	796,100	126,284	1.7	148	5.6	137	92	41
Palau	0,494	16				35		
Palestine	0,380							
Panama	75,520	2,612	0.7	3	2.9	20	97	23
Papua New Guinea	462,840	4,197	1.6	5	4.9	95	94	31
Paraguay	406,750	4,788	1.4	6	4.6	34	103	23
Peru	1,285,220	23,238	0.6	26	3.3	58	90	28
Philippines	300,000	67,038	1.8	77	4.1	57	94	36
Pitcairn	0,005	..						
Poland	312,680	38,544	0.4	39	1.9	16	112	17
Portugal	92,390	9,902	-0.6	10	1.5	11	111	14
Puerto Rico	8,900	3,651	0.8	4	2.1		96	
Qatar	11,000	610	0.4		4.0	25		11
Reunion	2,510	640	1.5		2.3			
Romania	237,500	22,731	0.0	23	1.5	29	110	21
Russia	17,075,400	148,350	0.4	150	1.7	31		
Rwanda	26,340	7,755	1.9	9	6.2	139	85	58
Sahara (Western)	266,000	1,000						
St Helena	0,122	8						
St Kitts	0,360	41	1.4		2.6	41		21
St Lucia	0,620	160	1.0		3.2	22		29
St Vincent	0,390	110	2.3		2.5	23		20

Countries	Newspapers (per 1,000 households) 1995	TV sets 1995	Radio sets 1995	Books/ new titles per million inhab. 1995	GNP per capita (US dollars) 1993	Average annual growth (%) 1985-94	Average annual inflation (%) 1984-94	Exports (million $) 1994	Imports (million $) 1994	Energy consump. per capita (kg) 1994	Energy imports as % of consump. 1994
Luxembourg	105	102	103	111		1					
Macau											
Macedonia					820	..	..	1,120	1,260	..	..
Madagascar	84	85	94	84	220	-2	16	277	434	37	83
Malawi	83	82	95	84	200	-1	19	325	491	39	59
Malaysia	106	102	105	101	3,140	6	3	58,756	59,581	1,711	-66
Maldives	88	88	90	85		8					
Mali	83	82	80	87	270	1	3	..	..	22	80
Malta	103	113	104	110		5					
Marshall						..					
Martinique						..					
Mauritania	81	89	90	84	500	0	7	..	..	103	100
Mauritius	101	104	103	98	3,030	6	9	1,347	1,926	387	92
Mexico	103	102	100	91	3,610	1	40	61,964	80,100	1,577	-55
Micronesia						..					
Moldova					1,060	..	..	618	672	962	99
Monaco						..					
Mongolia	106	90	88	103	390	-3	46	324	223	1,079	15
Montserrat											
Morocco	86	99	100	—	1,040	1	5	4,013	7,188	307	95
Mozambique	83	82	79	82	90	4	53	..	1,000	40	74
Myanmar/Burma	85	82	85	85	220	..	27	771	886	..	..
Namibia	98	85	89	99	1,820	3	11	1,321	1,196	..	..
Nauru											
Nepal	84	82	80	82	190	2	12	363	1,176	23	84
Netherlands	104	105	106	108	20,950	2	2	155,554	139,795	4,558	9
Netherlands Antilles						..					
Nicaragua	101	100	103	83	340	-6	1311	352	824	241	84
Niger	82	83	83	101	270	-2	0	..	..	37	83
Nigeria	85	88	92	84	300	1	30	9,378	6,511	162	-484
Northern Mariana						..					
Norway	106	103	102	109	25,970	1	3	34,700	27,300	5,326	-636
Oman	94	116	107	—	4,850	0	0	5,418	3,915	2,347	-801
Pakistan	96	86	88	85	430	1	9	7,370	8,890	255	38
Palau											
Palestine											
Panama	100	102	96	95	2,600	-1	2	584	2,404	566	83
Papua New Guinea	87	82	82	—	1,130	2	4	2,640	1,521	236	-150
Paraguay	93	94	83	—	1,510	1	26	817	2,370	261	-141
Peru	101	100	100	86	1,490	-2	492	4,555	6,794	351	1
Philippines	100	94	92	85	850	2	10	13,304	22,546	364	70
Pitcairn											
Poland	103	104	102	102	2,260	1	98	17,000	21,400	2,563	5
Portugal	105	102	93	105	9,130	4	12	17,540	26,680	1,828	90
Puerto Rico	103	104	108		7,000	2					
Qatar	104	114	108	110		-2					
Reunion						..					
Romania	102	101	90	100	1,140	-4	62	6,151	7,109	1,750	27
Russia					2,340	-4	124	53,000	41,000	4,038	-52
Rwanda	84	81	82	88	210	-7	5	..	..	27	78
Sahara (Western)											
St Helena											
St Kitts											
St Lucia											
St Vincent											

Countries	Area (thousands square km) 1994	Population (thousands) 1994	Demographic growth (%) 2000	Population estimate (millions) for year 2000	Children per woman 1992	Under-five mortality rate 1994	Calorie consump. % of required intake 1995	Primary school teachers/ student 1992
Samoa	2,840	164	0.5		4.5	55		27
Samoa-American	0,200	55	..		..			15
São Tome & Principe	0,960	125	1.7		5.0	82	112	35
Saudi Arabia	2,149,690	17,985	0.5	22	6.4	36	106	14
Senegal	196,720	8,263	1.2	10	5.9	115	94	59
Seychelles	0,450	72	1.7		2.7	20		19
Sierra Leone	71,740	4,399	2.1	5	6.5	284	84	34
Singapore	0,620	2,930		3	1.8	6	107	26
Slovakia	49,030	5,347	0.4	6	2.0	15		22
Slovenia	20,250	1,989	0.6	2	1.5	8		18
Solomon Is.	28,900	365	3.0		5.8	32	92	25
Somalia	637,660	8,775	1.4	10	6.8	211	86	
South Africa	1,221,040	40,539	0.8	47	4.1	68	107	
Spain	504,780	39,143	0.2	39	1.2	9	114	21
Spain-Gibraltar	0,006	30	..					
Sri Lanka	65,610	17,865	1.3	19	2.5	19	94	29
Sudan	2,505,810	27,364	1.9	33	6.1	122	88	34
Suriname	163,270	407	1.3		2.8	33	101	23
Swaziland	17,360	906	3.0		6.6	107	100	32
Sweden	449,960	8,781	0.6	9	2.1	5	106	10
Switzerland	41,290	6,994	1.0	7	1.7	7	114	
Syria	185,180	13,844	0.8	17	6.1	38	109	25
Taiwan	35,981	21,268				9	105	
Tajikistan	143,100	5,751	2.8	7	5.1	81		21
Tanzania	945,090	28,817	2.9	33	6.3	159	92	36
Thailand	513,120	58,024	1.6	65	2.2	32	94	17
Togo	56,790	4,007	1.6	5	6.5	132	91	59
Tokelau	0,010	2						
Tonga	0,750	101	-0.3		3.6	24		24
Trinidad and Tobago	5,130	1,295	0.7	1	2.8	20	106	26
Tunisia	163,610	8,815	0.9	10	3.8	34	107	26
Turkey	779,450	60,840	1.9	68	3.4	55	110	29
Turkmenistan	488,100	4,406	2.5	5	4.2	87		
Turks and Caicos	0,430	14				31		
Tuvalu	0,002	13				56		21
Uganda	235,880	18,592	2.4	22	7.1	185	91	35
Ukraine	603,700	51,921	0.3	52	1.8	25		17
United Arab Emirates	83,600	1,877	0.6	2	4.5	20	114	17
United Kingdom	244,880	58,395	0.3	59	1.8	7	110	20
United States	9,363,500	260,650	0.9	276	2.1	10	117	
Uruguay	177,410	3,163	0.3	3	2.3	21	101	21
Uzbekistan	447,400	22,378	2.4	26	4.1	64		
Vanuatu	12,190	165	1.1		5.3	259		31
Venezuela	912,050	21,177	0.4	24	3.6	24	98	23
Vietnam	331,690	72,039	1.2	83	3.7	46	92	35
Virgin Is. (Us)	0,340	100	-1.1		2.6			
Virgin Is. (Br)	0,150	99				29		
Yemen	527,970	14,785	2.6	17	7.6	112	93	29
Yugoslavia Fed. Rep.	102,170	10,520	0.8		2.1	23	115	22
Zaire	2,344,860	42,540	1.0		6.2	186	89	
Zambia	752,610	9,203	0.6	10	6.5	203	89	44
Zimbabwe	390,760	10,778	2.3	12	4.6	81	92	38

Sources: See Contents (pages 12/13).

Countries	Newspapers (per 1,000 households) 1995	TV sets 1995	Radio sets 1995	Books/ new titles per million inhab. 1995	GNP per capita (US dollars) 1993	Average annual growth (%) 1985-94	Average annual inflation (%) 1984-94	Exports (million $) 1994	Imports (million $) 1994	Energy consump. per capita (kg) 1994	Energy imports as % of consump. 1994
Samoa											
Samoa-American											
São Tome & Principe						-2					
Saudi Arabia	101	108	104	85	7,510	-2	3	38,600	22,796	4,744	-435
Senegal	84	89	87	82	750	-1	3	..	..	102	100
Seychelles						5					
Sierra Leone	83	84	97	82	150	-0	67	115	150	73	100
Singapore	107	107	104	106	19,850	6	4	96,800	103,000	6,556	100
Slovakia					1,950	-3	10	6,587	6,823	..	..
Slovenia					6,490	..	..	6,828	7,304	1,506	19
Solomon	88	81	90	—		2					
Somalia	81	84	80	—	120	-2					
South Africa	97	100	101	101	2,980	-1	14	25,000	23,400	2,253	-33
Spain	101	106	97	108	13,590	3	7	73,300	92,500	2,414	69
Spain-Gibraltar											
Sri Lanka	100	90	96	100	600	3	11	3,210	4,780	111	83
Sudan	89	95	100	—	480	-0					
Suriname	101	101	107	—		2					
Swaziland	90	86	91	—		-1					
Sweden	106	104	104	111	24,740	-0	6	61,292	51,800	5,603	36
Switzerland	114	104	105	116	35,760	1	4	66,200	64,100	3,603	59
Syria	91	97	102	83	1,160	-2					
Taiwan	105	106	109	104							
Tajikistan					470	-11	104	531	619	642	55
Tanzania	85	82	78	83	90	1	33	519	1,505	34	83
Thailand	102	101	95	101	2,110	9	5	45,262	54,459	770	59
Togo	82	83	98	—	340	-3	3	..	..	46	100
Tokelau											
Tonga						0					
Trinidad and Tobago	103	105	104	100	3,830	-2	7	1,867	1,131	4,549	-89
Tunisia	97	100	97	90	1,720	2	6	4,660	6,580	590	-7
Turkey	101	103	92	100	2,970	1	66	18,106	23,270	955	56
Turkmenistan					1,230	..	59	2,176	1,690	3,198	-116
Turks and Caicos											
Tuvalu											
Uganda	82	83	85	—	180	2	75	421	870	23	58
Ukraine					2,210	-8	297	11,818	14,177	3,292	43
United Arab Emirates	107	102	105	100	21,430	0	..	19,700	21,100	12,795	-470
United Kingdom	108	105	110	109	18,060	1	5	205,000	227,000	3,754	-9
United States	104	111	123	101	24,740	1	3	513,000	690,000	7,905	19
Uruguay	104	102	104	102	3,830	3	74	1,913	2,770	623	68
Uzbekistan					970	-2	109	3,543	3,243	1,886	3
Vanuatu						0					
Venezuela	103	103	106	98	2,840	1	36	15,480	7,710	2,331	-245
Vietnam	88	89	86	90	170	..	103	3,770	4,440	105	-55
Virgin Is. (Us)						..					
Virgin Is. (Br)											
Yemen	86	88	80	—	520	..	..	..	..	214	-406
Yugoslavia Fed. Rep.	109	102	94	103		..					
Zaire	82	82	88	82	220	-1					
Zambia	86	87	83	100	380	-1	92	..	..	140	29
Zimbabwe	95	89	86	89	520	-0	20	..	..	432	26

National Distributors

AUSTRALIA
NI Australia,
7 Hutt Street,
Adelaide 5000, SA.
Tel: 08 8232 1563
Fax: 08 8232 1887
E-mail: sandyl@caa.org.au

For book shop distribution:
Bush Books
PO Box 1370,
Gosford South,
NSW 2250.
Tel: 04 323 3274
Fax: 02 9212 2468

BELGIUM
NCOS
Vlasfabriekstraat 11,
1060 Brussel.
Tel: 322 539 2620
Fax: 322 539 1343

CANADA
NI Canada,
1011 Bloor Street W.,
Ste 300, Toronto,
Ontario M6H 1M1
Tel: 416 588 6478
Fax: 416 537 6435
E-mail: nican@web.net

For book shop distribution:
Garamond Press
77 Mowat Avenue,
Suite 403, Toronto,
Ontario M6K 3E3
Tel: 416 516 2709
Fax: 416 516 0571
E-mail: garamond@web.net

DENMARK
Mellemfolkeligt Samvirke
Verdenshjørnet
Borgergade 14, 1300
Copenhagen.
Tel: 3332 6244
Fax: 3315 6243

GERMANY
Lamuv Verlag
Postfach 2605 D-37016,
Gottingen.
Tel: 551 44024
Fax: 551 41392

NETHERLANDS
Novib Publications
PO Box 30919, 2500 GX
The Hague.
Tel: 70 3421 777
Fax: 70 3614 461
E-mail:
admin@novib.antenna.nl

**NEW ZEALAND/
AOTEAROA**
NI Aotearoa
P O Box 4499,
Christchurch.
Tel: 03 365 6153
Fax: 03 365 6153

NORWAY
Arning Publications
Møllergata 12, 0179
Oslo.
Tel: 22 42 70 77
Fax: 22 42 03 51

SWEDEN
Hillco Media Group
Storgatan 7B, S-753 31,
Uppsala.
Tel: 18 13 36 33
Fax: 18 12 36 63
E-mail:
jacob.lindquist@hillco.se

UNITED KINGDOM
New Internationalist
Publications
55 Rectory Road, Oxford
OX4 1BW Oxon.
Tel: 01865 728181
Fax: 01865 793152
E-mail: newint@gn.apc.org

Oxfam Publications
274 Banbury Road,
Oxford OX2 7DZ Oxon.
Tel: 01865 313168
Fax: 01865 313925
E-mail:
publish@oxfam.org.uk

For book shop distribution:
Central Books
99 Wallis Road, Hackney,
London E9 5LN.
Tel: 0181 986 4854
Fax: 0181 533 5821
E-mail:
mark@centbooks.demon.co.uk

UNITED STATES
Humanities Press
International Inc.
165 First Avenue,
Atlantic Highlands,
NJ 07716
Tel: 908 8721441
Fax: 908 8720717

URUGUAY
Instituto del Tercer Mundo
Juan Jackson, 1136,
CP 11200, Montevideo.
Tel: 8 249 6192
Fax: 8 241 9222
E-mail:
item@chasque.apc.org